2018 International Power Electronics Conference (IPEC-Niigata 2018 –ECCE Asia-)

Niigata, Japan
20-24 May 2018

Pages 2153-2847

IEEE Catalog Number: CFP1854I-POD
ISBN: 978-1-5386-4190-3

Copyright © 2018, IEEJ Industry Applications Society
All Rights Reserved

*** This is a print representation of what appears in the IEEE Digital Library. Some format issues inherent in the e-media version may also appear in this print version.

IEEE Catalog Number: CFP1854I-POD
ISBN (Print-On-Demand):   978-1-5386-4190-3
ISBN (Online):   978-4-88686-405-5

Additional Copies of This Publication Are Available From:

Curran Associates, Inc
57 Morehouse Lane
Red Hook, NY 12571 USA
Phone: (845) 758-0400
Fax: (845) 758-2633
E-mail: curran@proceedings.com
Web: www.proceedings.com

TABLE OF CONTENTS

THREE-PHASE INDUCTIVE POWER TRANSFER SYSTEM WITH 12 COILS FOR RADIATION NOISE REDUCTION 69

Keisuke Kusaka ; Jun-Ichi Itoh

SECONDARY-SIDE-ONLY CONTROL FOR SMOOTH VOLTAGE STABILIZATION IN WIRELESS POWER TRANSFER SYSTEMS WITH CONSTANT POWER LOAD 77

Giorgio Lovison ; Takehiro Imura ; Hiroshi Fujimoto ; Yoichi Hori

CONSTANT CURRENT CHARGING AND THE MAXIMUM SYSTEM EFFICIENCY TRACKING FOR WIRELESS CHARGING SYSTEMS EMPLOYING DUAL-SIDE CONTROL 84

Zhenjie Li ; Xiaoliang Huang ; Kai Song ; Jinhai Jiang ; Chunbo Zhu ; Zhijiang Du

ELECTRIC FIELD COUPLING TYPE HIGH POWER WIRELESS POWER TRANSFER WITH LEAKAGE ELECTRIC FIELD STRUCURE 88

Mitsuru Masuda

TRANSFER POWER ANALYSIS OF CAPACITIVELY ISOLATED OUTLET AND PLUG (CAPISOP) USING SERIES RESONANCE 94

Hirohito Funato ; Koki Amano ; Takuya Hatsumi ; Junnosuke Haruna

WIDE VOLTAGE GAIN RANGE LLC DC/DC TOPOLOGIES: STATE-OF-THE-ART 100

Qi Cao ; Zhiqing Li ; Haoyu Wang

DUAL HALF-BRIDGE LLC RESONANT CONVERTER WITH HYBRID-SECONDARY-RECTIFIER (HSR) FOR WIDE-OUPUT-VOLTAGE APPLICATIONS 108

Jae-Il Baek ; Chong-Eun Kim ; Keon-Woo Kim ; Min-Su Lee ; Gun-Woo Moon

A STUDY ON THE ANALYSIS AND CONTROL OF NO-LOAD CHARACTERISTICS OF LLC RESONANT CONVERTER FOR PLASMA PROCESS 114

Min-Jun Kwon ; Woo-Cheol Lee

MECHANISM OF CURRENT IMBALANCE IN LLC RESONANT CONVERTER WITH CENTER TAPPED TRANSFORMER 118

Mitsuru Sato ; Shingo Nagaoka ; Takeshi Uematsu ; Toshiyuki Zaitsu

PERFORMANCE STUDY OF HIGH-POWER HALF-BRIDGE INTERLEAVED LLC CONVERTER 123

Hung-I Hsieh ; Hui-Lung Chiu ; Guan-Chyun Hsieh

MULTI-CHIP SIC MOSFET POWER MODULES FOR STANDARD MANUFACTURING, MOUNTING AND COOLING 130

Alberto Castellazzi ; Asad Fayyaz ; Emre Gurpinar ; Abdallah Hussein ; Jianfeng Li ; Bassem Mouawad

AN ALTERNATIVE METHOD TO ACCURATELY DETERMINE THE THERMAL RESISTANCE OF SIC MOSFET STRUCTURES WITH DISCRETE DIODES 137

Andras Vass-Varnai ; Young Joon Cho ; Gabor Farkas ; Marta Rencz

HEAT-RESISTANT PACKAGING TECHNOLOGY FOR WIDE BANDGAP POWER DEVICES AND THERMAL RELIABILITY TESTING 142

K. Suganuma ; H. Zhang ; S. Nagao ; C. Chen ; T. Sugahara ; A. Shimoyama ; A. Suetake

VERIFICATION OF IDENTIFICATION ACCURACY OF LOSS CALCULATED BY INVERSE THERMAL ANALYSIS 148

Yuki Ikari ; Kazushige Nakao

PACKAGING ARCHITECTURES FOR SILICON CARBIDE POWER ELECTRONIC MODULES 153

H. Alan Mantooth ; Simon S. Ang

DEVELOPMENT OF A HOMO-POLAR BEARINGLESS MOTOR WITH CONCENTRATED WINDING FOR HIGH SPEED APPLICATIONS 157

Dai Suzuki ; Takaaki Oiwa

HIGH-SPEED SLOTLESS PERMANENT MAGNET MACHINES: MODELLING AND DESIGN FRAMEWORKS 161

S. Jumayev ; K.O. Boynov ; E.A. Lomonova ; J. Pyrhonen

DEVELOPMENT AND PERFORMANCE OF HIGH-SPEED SPM SYNCHRONOUS MACHINE 169

Kota Kawanishi ; Keisuke Matsuo ; Takayuki Mizuno ; Koji Yamada ; Takashi Okitsu ; Kouki Matsuse

1.2KW 100,000RPM HIGH SPEED MOTOR FOR AIRCRAFT 177

Takehiro Jikumaru ; Gen Kuwata

COMPARATIVE EVALUATION OF Y-INVERTER AGAINST THREE-PHASE TWO-STAGE BUCK-BOOST DC-AC CONVERTER SYSTEMS 181

Michael Antivachis ; Dominik Bortis ; David Menzi ; Johann W. Kolar

DC-POWERED OFFICE BUILDINGS AND DATA CENTRES : THE FIRST 380 VDC MICRO GRID IN A COMMERCIAL BUILDING IN GERMANY 190
Tilo Pueschel

RECENT TREND IN POWER ELECTRONICS FOR ICT SYSTEMS 196
Hiroshi Nakao ; Yu Yonezawa ; Yoshiyasu Nakashima

GREEN BASE STATION USING ROBUST SOLAR SYSTEM AND HIGH PERFORMANCE LITHIUM ION BATTERY FOR NEXT GENERATION WIRELESS NETWORK (5G) AND AGAINST MEGA DISASTER 201
M. Nakamura ; K. Takeno

OPTIMIZATION OF MAINTENANCE BY FAILURE PREDICTION CONSIDERING INSTANTANEOUS AND CUMULATIVE EFFECTS OF EXTERNAL ENVIRONMENTS 207
Kaisei Kanetani ; Masahiro Yamazaki ; Tadatoshi Babasaki ; Hideaki Kim ; Tatsushi Matsubayashi

HYBRID CONVERTERS WITH REDUCED INDUCTOR LOSS FOR INTEGRATABLE POWER CONVERSION 213
Gab-Su Seo ; Hanh-Phuc Le

ENERGY SAVING SYSTEM TREND FOR HARBOR CRANE WITH LITHIUM ION BATTERY 219
Hidemasa Yoshihara

INVERTER DRIVE OF DYNAMOMETERS FORAUTOMOTIVE EVALUATION SYSTEM 227
Shizunori Hamada ; Toshimichi Takahashi ; Nobutaka Kezuka ; Masaju Kouketsu ; Shingo Ishigaki

EXPERIMENTAL INVESTIGATION OF PROTOTYPE ALL-SIC CONVERTER FOR ULTRA-HIGH-SPEED ELEVATOR 233
Kazuhisa Mori ; Kaoru Katoh ; Yohei Matsumoto ; Tatsushi Yabuuchi ; Naoto Ohnuma

HIGH-VOLTAGE, LARGE-CAPACITY CONVERTER TECHNOLOGIES AND THEIR APPLICATIONS 238
Daisuke Yoshizawa ; Paul Bixel ; Masahiko Tsukakoshi

HIGHER RADIAL SUSPENSION FORCE OF MAGNETIC BEARING ON CENTRIFUGAL COMPRESSOR FOR HVAC 244
Yuji Nakazawa ; Yusuke Irino ; Atsushi Sakawaki ; Kazunobu Ohyama

NOVEL SWITCHING CONTROL METHOD FOR FULL-BRIDGE DC-DC CONVERTERS FOR IMPROVING LIGHT-LOAD EFFICIENCY USING REVERSE RECOVERY CURRENT 250
Fumihiro Sato ; Takae Shimada ; Takayuki Ouchi

A 800V/14V SOFT-SWITCHED CONVERTER WITH LOW-VOLTAGE RATING OF SWITCH FOR XEV APPLICATIONS 256
Byeongwoo Kim ; Kangsan Kim ; Sewan Choi

HIGH SPEED CONTROL METHOD FOR SUPERPOSING HIGH-FREQUENCY-HIGH-SINUSOIDAL-CURRENT WITH DC CURRENT TO ANALYZE BATTERY AC IMPEDANCE 261
Jin Xu ; Toshihiko Kishimoto ; Noboru Shimosato

EV BMS WITH TIME-SHARED ISOLATED CONVERTERS FOR ACTIVE BALANCING AND AUXILIARY BUS REGULATION 267
Z. Gong ; B.A.C. Van De Ven ; Y. Lu ; Y. Luo ; K. Gupta ; C. Da Silva ; H.J. Bergveld ; O. Trescases

A DRIVING CIRCUIT WITH PARTIAL POWER REGULATION FOR RGB LED LAMPS 275
You-Chun Huang ; Yu-Jen Chen ; Yong-Jyun Li ; Chin-Sien Moo

FPGA-BASED DYNAMIC DUTY CYCLE AND FREQUENCY CONTROLLER FOR A CLASS-E2DC-DC CONVERTER 282
Sanghyeon Park ; Juan Rivas-Davila

DESIGN METHODOLOGY OF 3 KW INDUCTION HEATING SYSTEM FOR BOTH LOW RESISTANCE AND HIGH RESISTANCE CONTAINERS IN A SINGLE BURNER 289
Si-Hoon Jeong ; Hwa-Pyeong Park ; Jee-Hoon Jung

MULTI-RESONANT INVERTER REALIZING DOWNSIZING AND LOSS REDUCTION FOR ALL-METALLIC IH COOKTOP 296
Takayuki Hirokawa ; Makoto Imai ; Atsushi Fujita

TEMPERATURE ESTIMATION OF ALUMINUM ELECTROLYTIC CAPACITOR UNDER ACTUAL CIRCUIT OPERATION 302
Kazuki Urata ; Toshihisa Shimizu

DESIGN AND EVALUATION OF CURRENT DISTRIBUTION IN POWER MODULE 309
Takaaki Ibuchi ; Eisuke Masuda ; Tsuyoshi Funaki

DEVELOPMENT OF IMPEDANCE-SOURCE INVERTER USING SIC-MOSFET 313
Ryuji IIjima ; Thilak Senanayake ; Takanori Isobe ; Hiroshi Tadano

CONTROL METHODOLOGY FOR REALIZATION OF 100KW HEECS CHOPPER WITH 99.5% EFFICIENCY 318
Yukinori Tsuruta ; Atsuo Kawamura

IRON LOSS REDUCTION IN THE CORES OF INDUCTION HEATING COILS FOR SMALL-FOREIGN-METAL PARTICLE DETECTOR WITH A 400-KHZ SIC-MOSFETS HIGH-FREQUENCY INVERTER 324

Takuya Shijo ; Yuki Uchino ; Yujiro Noda ; Hiroaki Yamada ; Toshihiko Tanaka

FREQUENCY TRACKING BURST-MODE PDM-CONTROLLED CLASS-D ZERO VOLTAGE SOFT-SWITCHING RESONANT CONVERTER FOR INDUCTIVE POWER TRANSFER APPLICATIONS 329

Yoichiro Tabata ; Tomokazu Mishima ; Tatsuya Kido

REDUCED-ORDER DYNAMICAL MODELS OF TUNED WIRELESS POWER TRANSFER SYSTEMS 337

Hongchang Li ; Jingyang Fang ; Yi Tang

DYNAMIC MODELLING AND CLOSED LOOP CONTROL OF TRANSMITTER PARALLEL AND RECEIVER SERIES COMPENSATED IPT TOPOLOGY FOR EV APPLICATIONS 342

Suvendu Samanta ; Akshay Kumar Rathore

DEVELOPMENT OF INDUCTIVE POWER TRANSFER SYSTEM FOR EXCAVATOR UNDER LARGE LOAD FLUCTUATION : CONSIDERATION OF RELATIONSHIP BETWEEN LOAD VOLTAGE AND RESONANCE PARAMETER 348

Jun-Ichi Itoh ; Kent Inoue ; Keisuke Kusaka

WIRELESS POWER TRANSFER SYSTEM USING THREE-PHASE TO SINGLE-PHASE MATRIX CONVERTER 356

Yuji Hayashi ; Hiromasa Motoyama ; Takaharu Takeshita

DESIGN OF A REDUCED-ORDER OBSERVER FOR SENSORLESS CONTROL OF DUAL-ACTIVE-BRIDGE CONVERTER 363

Nguyen Duy Dinh ; Goro Fujita

IMPROVED LOAD TRANSIENT RESPONSE OF A DUAL-ACTIVE-BRIDGE CONVERTER 370

Sheng-Zhi Zhou ; Chuan Sun ; Song Hu ; Guo Chen ; Xiaodong Li

MODULATION AND ACTIVE MIDPOINT CONTROL OF A THREE-LEVEL THREE-PHASE DUAL-ACTIVE BRIDGE DC-DC CONVERTER UNDER NON-SYMMETRICAL LOAD 375

Philipp Joebges ; Anton Gorodnichev ; Rik W. De Doncker

A NOVEL SWITCHING ALGORITHM TO IMPROVE EFFICIENCY AT LIGHT LOAD CONDITIONS FOR THREE-PHASE DAB CONVERTER IN LVDC APPLICATION 383

Hyun-Jun Choi ; Si-Hoon Jung ; Jee-Hoon Jung

DESIGN OF A HIGH-FREQUENCY DUAL-ACTIVE BRIDGE CONVERTER WITH GAN DEVICES FOR AN OUTPUT POWER OF 3.7 KW 388

Philipp Schülting ; Christian Winter ; Rik W. De Doncker

EXPLORATION OF THE DESIGN AND PERFORMANCE SPACE OF A HIGH FREQUENCY 166 KW/10 KV SIC SOLID-STATE AIR-CORE TRANSFORMER 396

Piotr Czyz ; Thomas Guillod ; Florian Krismer ; Johann W. Kolar

NOVEL CALCULATION METHOD OF IRON LOSS OF GAPPED INDUCTORS USING LOSS MAP 404

Yoshihiro Miwa ; Toshihisa Shimizu

VERIFICATION OF THE REDUCTION OF THE COPPER LOSS BY THE THIN COIL STRUCTURE FOR INDUCTION COOKERS 410

Morimasa Hataya ; Koki Kamaeguchi ; Eiji Hiraki ; Kazuhiro Umetani ; Takayuki Hirokawa ; Makoto Imai ; Hideki Sadakata

CONDITION MONITORING OF ELECTROLYTIC CAPACITOR BASED ON ESR ESTIMATION AND THERMAL IMPEDANCE MODEL USING IMPROVED POWER LOSS COMPUTATION 416

Sundararajan Prasanth ; Mohamed Halick ; Mohamed Sathik ; Firman Sasongko ; Tan Chuan Seng ; Peng Yaxin ; Rejeki Simanjorang

TEST SETUP FOR CHARACTERISATION OF BIASED MAGNETIC HYSTERESIS LOOPS IN POWER ELECTRONIC APPLICATIONS 422

Min Luo ; Drazen Dujic ; Jost Allmeling

A FAST OPEN-CIRCUIT FAULT DIAGNOSIS SCHEME FOR MODULAR MULTILEVEL CONVERTERS WITH MODEL PREDICTIVE CONTROL 428

Dehong Zhou ; Shunfeng Yang ; Yi Tang

AN ONLINE OPEN-CIRCUIT FAULT DIAGNOSIS AND FAULT TOLERANT SCHEME FOR THREE-PHASE AC-DC CONVERTERS WITH MODEL PREDICTIVE CONTROL 434

Dehong Zhou ; Yi Tang

THE LIFETIME ASSESSMENT OF A MICRO-INVERTER FOR PV APPLICATIONS 439

Tohihiro Shimao ; Koji Kato ; Youichi Ito ; Akio Iwabuchi ; Yongheng Yang ; Frede Blaabjerg

ONLINE HEALTH MONITORING OF MULTIPLE MOSFETS IN A GRID-TIED PV INVERTER USING SPREAD SPECTRUM TIME DOMAIN REFLECTOMETRY (SSTDR) 446
Sourov Roy ; Faisal Khan

AN IMPROVED EQUIVALENT MODEL FOR A LONG PV STRING UNDER PARTIAL SHADING CONDITIONS 453
Xiaoyang Wang ; Huiqing Wen ; Xingshuo Li

OPTIMIZED FLUX-WEAKENING CONTROL OF INDUCTION MOTOR FOR TORQUE ENHANCEMENT IN VOLTAGE EXTENSION REGION 459
Zhen Dong ; Yong Yu ; Bo Wang ; Qinghua Dong ; Dianguo Xu

IMPROVED PERFORMANCE OF CFTC-BASED DIRECT TORQUE CONTROL OF INDUCTION MACHINES BY INCREASING TORQUE LOOP BANDWIDTH 466
Ibrahim Mohd Alsofyani ; June-Hee Lee ; Byung-Moon Han ; Kyo-Beum Lee

μ-ANALYSIS EVALUATION OF A NOVEL COMBINED CURRENT-AND-SPEED CONTROL FOR INDUCTION MOTORS VIA ILQ DESIGN METHOD 471
Shuto Omori ; Hiroshi Takami ; Masashi Nakamura

LOSS MINIMIZATION CONTROL OF SENSORLESS SCALAR-CONTROLLED INDUCTION MOTOR DRIVES CONSIDERING IRON LOSS 478
Nguyen Anh Tan ; Dong-Choon Lee

TUNING OF INDUCTION MOTOR DRIVE WITH TORQUE SENSOR 483
Hajime Kubo ; Yugo Tadano

QUASI-TWO-LEVEL CONVERTER FOR OVERVOLTAGE MITIGATION IN MEDIUM VOLTAGE DRIVES 488
F. Bertoldi ; M. Pathmanathan ; R. S. Kanchan ; K. Spiliotis ; J. Driesen

A MEDIUM-VOLTAGE THREE-PHASE AC-DC CONVERTER CONSISTING OF CASCADED THREE-LEVEL BOOST-TYPE RECTIFIERS AND AN OPEN-END WINDING TRANSFORMER 495
Ryoji Tsuruta ; Hiromitsu Suzuki ; Ritaka Nakamura

A FAULT TOLERANT CONTROL STRATEGY FOR THE DELTA-CONNECTED CASCADED CONVERTER 503
Ping-Heng Wu ; Po-Tai Cheng

COOLING PERFORMANCE IMPROVEMENT OF HEAT SINK BY OSCILLATING HEAT PIPE ADDITION AND DESIGN FOR ENVIRONMENT OF OSCILLATING HEAT PIPE REFRIGERANT 511
Kuan-Chung Tey ; Kenichiro Suzuki

COMPACT LARGE CAPACITY GAS TURBINE STATIC STARTER 517
Hironori Kawaguchi ; Shigeyuki Nakabayashi ; Akinobu Ando ; Hiroshi Ogino ; Yasuaki Matsumoto ; Ikuto Udagawa ; Takahiro Ohta

VOLTAGE REFERENCE MODIFICATION SCHEME FOR RESONANCE SUPPRESSION IN LCL-FILTERED INVERTERS WITH DISCONTINUOUS PWM METHOD 521
Hyeon-Sik Kim ; Seung-Ki Sul

PARAMETRIC ROBUSTNESS ANALYSIS FOR PARALLEL FEEDFORWARD COMPENSATION BASED ACTIVE DAMPING OF LCL GRID CONNECTED INVERTER 528
Muhammad Talib Faiz ; Muhammad Mansoor Khan ; Xu Jianming ; Muhammad Ali ; Houjun Tang

OPEN-LOOP-BASED ISLAND-MODE VOLTAGE CONTROL METHOD FOR SINGLE-PHASE GRID-TIED INVERTER WITH MINIMIZED LC FILTER 534
Satoshi Nagai ; Jun-Ichi Itoh

EXPERIMENTAL VALIDATION OF ADAPTIVE CURRENT INJECTING METHOD FOR GRID-SYNCHRONIZATION IMPROVEMENT OF GRID-TIED REGS DURING SHORT-CIRCUIT FAULT 542
Shaokang Ma ; Hua Geng ; Geng Yang ; Bo Liu

ADAPTIVE CONTROL OF GRID-VOLTAGE FEEDFORWARD FOR GRID-CONNECTED INVERTERS BASED ON REAL-TIME IDENTIFICATION OF GRID IMPEDANCE 547
Roni Luhtala ; Tuomas Messo ; Tomi Roinila

MODEL BASED TUNING OF PROPORTIONAL RESONANT CONTROLLERS FOR VOLTAGE SOURCE INVERTERS 555
Stefan Almér ; Thomas Besselmann ; Mario Schweizer

AN SOC-BASED PLATFORM FOR INTEGRATED MULTI-AXIS MOTION CONTROL AND MOTOR DRIVE 560
Yongping Sun ; Ming Yang ; Yangyang Chen ; Wangpin He ; Dianguo Xu

VARIABLE SWITCHING FREQUENCY STRATEGY FOR ENHANCED SETTLING PERFORMANCE OF POSITION CONTROL WITHIN INVERTER LOSS LIMIT 565
Choongin Lee ; Jung-Ik Ha

TWO-WHEEL CANE FOR WALKING ASSISTANCE..571
Phi Van Lam ; Yasutaka Fujimoto

FALL PREVENTION AND VIBRATION SUPPRESSION OF WHEELCHAIR USING RIDER MOTION STATE..575
Isseki Takahashi ; Toshiyuki Murakami

STABILIZATION METHOD FOR RESIDENTIAL DC SYSTEM BASED ON PASSIVITY CRITERION..583
Hiroaki Kakigano

A NOVEL CONTROL APPROACH TO MULTI-TERMINAL POWER FLOW CONTROLLER FOR NEXT-GENERATION DC POWER NETWORK..588
Kenji Natori ; Yuta Nakao ; Yukihiko Sato

DC MICROGRID FOR TELECOMMUNICATIONS SERVICE AND RELATED APPLICATION..593
Keiichi Hirose

MVDC DISTRIBUTION GRIDS FOR ELECTRIC VEHICLE FAST-CHARGING INFRASTRUCTURE..598
Marco Stieneker ; Benedict J. Mortimer ; Arne Hinz ; Adolf Müller-Hellmann ; Rik W. De Doncker

REVIEW OF RESONANT GATE DRIVER IN POWER CONVERSION..607
Bainan Sun ; Zhe Zhang ; Michael A.E. Andersen

A LOW PROFILE HIGH FREQUENCY LED DRIVING SYSTEM BASED ON AIRCORE PLANAR INDUCTOR..614
Yueshi Guan ; Xihong Hu ; Shu Zhang ; Yijie Wang ; Dianguo Xu ; Wei Wang

ANALYSIS AND COMPENSATION OF DEAD-TIME EFFECT IN SIC-DEVICE-BASED HIGH-SWITCHING-FREQUENCY INVERTERS..619
Qingzeng Yan ; Xibo Yuan ; Xiaojie Wu ; Yiwen Geng

CONTROL AND PERFORMANCE OF NEW ASYMMETRICAL OPERATION FOR SWITCHED-CAPACITOR-BASED RESONANT CONVERTERS..626
Hadi Setiadi ; Hideaki Fujita

HIGH-FREQUENCY RESONANT CONVERTER WITH SYNCHRONOUS RECTIFICATION FOR HIGH CONVERSION RATIO AND VARIABLE LOAD OPERATION..632
Lei Gu ; Kawin Surakitbovorn ; Juan Rivas-Davila

SMART PV INVERTERS FOR SMART GRID APPLICATIONS..639
Cheng-Jhen Yang ; Terng-Wei Tsai ; Yi-Chan Li ; Cheng-Yu Tang ; Yaow-Ming Chen ; Yung-Ruei Chang

HIGH-VOLTAGE BI-DIRECTIONAL HALF-BRIDGE THREE-LEVEL SERIES RESONANT CONVERTER WITH FREQUENCY MODULATION CONTROL..645
Lee Sih-Yi ; Jhang Jynu-Jhe ; Lin Jing-Yuan ; Hsieh Yao-Ching ; Chiu Haung-Jen

A CONTROL STRATEGY FOR FLYING-START OF SHAFT SENSORLESS PERMANENT MAGNET SYNCHRONOUS MACHINE DRIVE..651
Zih-Cing You ; Sheng-Ming Yang

CONTACTLESS EV POWER TRACK SYSTEM WITH SEGMENT-EXCITED INDUCTIVELY COUPLED STRUCTURE..657
Jia-You Lee ; Yu-Chi Wang ; Chih-Yi Liao

DRIVING TEST EVALUATION OF SENSORLESS VEHICLE DETECTION METHOD FOR IN-MOTION WIRELESS POWER TRANSFER..663
Katsuhiro Hata ; Kensuke Hanajiri ; Takehiro Imura ; Hiroshi Fujimoto ; Yoichi Hori ; Motoki Sato ; Daisuke Gunji

A SYSTEM DESIGN METHOD OF HIGH-FREQUENCY CLASS-D INVERTER FOR WIDEBAND CURRENT CONTROL..669
Hiroki Kurumatani ; Seiichiro Katsura

ANALYSIS OF INTERIOR PERMANENT MAGNET TWO DEGREES OF FREEDOM MOTOR BASED ON CROSS-COUPLED STRUCTURE..675
Yoshiyuki Hatta ; Tomoyuki Shimono

STUDY COMPARISON BETWEEN FIREFLY ALGORITHM AND PARTICLE SWARM OPTIMIZATION FOR SLAM PROBLEMS..681
Mounia Janah ; Yasutaka Fujimoto

BANDWIDTH LIMITATIONS IN FORCE CONTROL OF A SERIES ELASTIC ACTUATOR WITH BACKLASH AND QUANTIZATION..688
Hanul Jung ; Chan Lee ; Sehoon Oh

ROTOR SHAPE OPTIMIZATION OF INTERIOR PERMANENT MAGNET SYNCHRONOUS MOTORS WITH CONCENTRATED WINDINGS BY CONSIDERING END-LEAKAGE FLUX..693
Katsumi Yamazaki ; Hiroki Narushima

LOSS ANALYSIS OF PERMANENT-MAGNET SYNCHRONOUS MACHINES CONSIDERING IN-PLANE EDDY CURRENT IN ELECTRICAL STEEL SHEETS...699

Hideki Ohguchi ; Satoshi Imamori ; Katsumi Yamazaki ; Haiyan Yui ; Masao Shuto

STUDY ON INFLUENCE OF DIFFERENCE IN STRUCTURE OF CONCENTRATED WINDING IPMSMS OBTAINED BY AUTOMATIC DESIGN 704

A. Ura ; M. Sanada ; S. Morimoto ; Y. Inoue

CARRIER HARMONIC LOSS REDUCTION TECHNIQUE ON DUAL THREE-PHASE PERMANENT-MAGNET SYNCHRONOUS MOTORS WITH PHASE-SHIFT PWM...711

Yoshihiro Miyama ; Haruyuki Kometani ; Kan Akatsu

FLUX INTENSIFYING PM-MOTOR WITH VARIABLE LEAKAGE MAGNETIC FLUX TECHNIQUE...718

Masahiro Aoyama ; Toshihiko Noguchi

CONTINUOUS OPERATION CONTROL OF PMSM IN THE CASE OF DC POWER SUPPLY LOSS...726

Jongwon Heo ; Keiichiro Kondo

MODEL PREDICTIVE CONTROL FOR MULTIPHASE MOTOR DRIVES – A TECHNOLOGY STATUS REVIEW 732

A. Tenconi ; S. Rubino ; R. Bojoi

INFLUENCE OF FAST SWITCHING SEMICONDUCTORS ON THE WINDING INSULATION SYSTEM OF ELECTRICAL MACHINES 740

Kay Hameyer ; Andreas Ruf ; Florian Pauli

CENTRALIZED CONTROL OF MODULAR MULTI RECTIFIER FOR MOTOR DRIVE APPLICATIONS UNDER UNBALANCED GRID...746

Yipeng Song ; Pooya Davari ; Frede Blaabjerg

VECTOR CONTROL OF MAGNETICALLY MODULATED MOTOR FOR POWER SPLITTING OF HEV APPLICATION 753

Toshihiko Noguchi ; Sawanth Krishna Machavolu ; Masahiro Aoyama ; Yuto Motohashi

IMPEDANCE-BASED STABILITY EVALUATION OF VIRTUAL SYNCHRONOUS MACHINE IMPLEMENTATIONS IN CONVERTER CONTROLLERS 759

Eneko Unamuno ; Atle Rygg ; Mohammad Amin ; Marta Molinas ; Jon Andoni Barrena

STABLE POWER SUPPLY METHOD FOR HOUSEHOLD APPLIANCES VIA VIRTUAL SYNCHRONOUS GENERATOR IN SINGLE-PHASE THREE-WIRE MICROGRID 767

Yuko Hirase ; Hidehiko Nakagawa ; Eiji Yoshimura ; Shogo Katsura ; Kensho Abe ; Osamu Noro ; Kazushige Sugimoto ; Kenichi Sakimoto

A NOVEL OSCILLATION DAMPING METHOD OF VIRTUAL SYNCHRONOUS GENERATOR CONTROL WITHOUT PLL USING POLE PLACEMENT 775

Jia Liu ; Yushi Miura ; Toshifumi Ise

OPERATION OF A MODULAR MULTILEVEL CONVERTER CONTROLLED AS A VIRTUAL SYNCHRONOUS MACHINE...782

Salvatore D'arco ; Giuseppe Guidi ; Jon Are Suul

ASSESSMENT OF VIRTUAL SYNCHRONOUS MACHINE BASED CONTROL IN GRID-TIED POWER CONVERTERS...790

Chi Li ; Igor Cvetkovic ; Rolando Burgos ; Dushan Boroyevich

RESEARCH ON THE BLOCKCHAIN-BASED INTEGRATED DEMAND RESPONSE RESOURCES TRANSACTION SCHEME 795

Shengnan Zhao ; Yang Li ; Beibei Wang ; Huiling Su

INDIRECT CURRENT CONTROL FOR SEAMLESS TRANSFER OF UTILITY INTERACTIVE INVERTER...803

Kyungbae Lim ; Injong Song ; Jaeho Choi

STUDY OF AC POWER INTERCHANGE AND DC POWER INTERCHANGE FOR MICRO GRID SYSTEMS 809

Kazuto Yukita ; Daiki Owaki ; Shunsuke Horie ; Toshiro Matsumura ; Yasuyuki Goto

STABILITY ENHANCEMENT STRATEGY FOR ISLANDING MICROGRID WITH MULTI-TYPE INVERTERS BASED ON HYBRID IMPEDANCE MODELLING 815

Meiqin Mao ; Yong Ding ; Yatao Shen ; Liuchen Chang

DC POWERED DATA CENTER WITH 200 KW PV PANELS 822

Keiichi Hirose

INFLUENCES OF DETERIORATION IN CAPACITOR AND INDUCTOR ON CURRENT SENSORLESS STATIC MODEL DC-DC CONVERTER...826

Fujio Kurokawa ; Masashi Taguchi ; Jizhe Wang ; Hidenori Maruta ; Nobumasa Matsui

CAPACITIVE DIVIDER BASED PASSIVE START-UP METHODS FOR FLYING CAPACITOR STEP-DOWN DC-DC CONVERTER TOPOLOGIES...831
Michael Halamicek ; Tom Moiannou ; Nenad Vukadinovic ; Aleksandar Prodic

HIGH VOLTAGE GAIN INTERLEAVED ACTIVE-CLAMP FORWARD (IACF) CONVERTER HAVING REDUCED PRIMARY CONDUCTION LOSS...838
Yeonho Jeong ; Mu-Hyun Park ; Gun-Woo Kim ; Byoung-Hee Lee ; Gun-Woo Moon

CONTROL OF SWITCHING-CAPACITOR BASED BUCK-BOOST CONVERTER...845
M. Veerachary ; Vasudha Khubchandani

IMPROVEMENT OF UPLOAD TRANSIENT RESPONSES FOR ULTRA HIGH STEP-DOWN CONVERTER...851
Y.T. Yan ; K.I. Hwu

POWER ELECTRONICS AND CONTROL TECHNOLOGIES FOR HOUSEHOLD WASHER...856
Toru Niki

DEVELOPMENT OF ROOM AIR CONDITIONER WITH TWIN-PROPELLER FANS...860
Takamasa Uemura ; Tomoya Fukui ; Kenichi Sakoda

ELECTROLYTIC CAPACITOR-LESS SINGLE-PHASE TO THREE-PHASE INVERTER WITH HARMONICS SUPPRESSION CONTROL FOR AIR CONDITIONER...866
Nobuo Hayashi ; Takuro Ogawa ; Tomoisa Taniguchi ; Morimitsu Sekimoto

LATEST DEVELOPMENT OF SIC POWER MODULE-BASED SINGLE-STAGE AC-AC RESONANT CONVERTER FOR HIGH-FREQUENCY INDUCTION HEATING APPLICATIONS...872
Tomokazu Mishima

AN OPTIMIZED CONTROL STRATEGY TO IMPROVE THE CURRENT ZERO-CROSSING DISTORTION IN BIDIRECTIONAL AC/DC CONVERTER BASED ON V2G CONCEPT...878
Lei Jing ; Xiaoqing Wang ; Bodong Li ; Maohang Qiu ; Bo Liu ; Min Chen

PER-PHASE CONTROL STRATEGY OF THE THREE-PHASE FOUR-WIRE INVERTER...883
Yi-Chan Li ; Terng-Wei Tsai ; Cheng-Jhen Yang ; Yaow-Ming Chen ; Yung-Ruei Chang

OPPORTUNITIES FOR PERFORMANCE IMPROVEMENT OF SINGLE-PHASE POWER CONVERTERS THROUGH ENHANCED AUTOMATIC-POWER-DECOUPLING CONTROL...889
Huawei Yuan ; Sinan Li ; Wenlong Qi ; Siew-Chong Tan ; S. Y. Ron Hui

ZERO VOLTAGE SWITCHING SCHEME FOR FLYBACK CONVERTER TO ENSURE COMPATIBILITY WITH ACTIVE POWER DECOUPLING CAPABILITY...896
Hiroki Watanabe ; Jun-Ichi Itoh

MODEL PREDICTIVE FAULT TOLERANT CONTROL OF BIDIRECTIONAL AC/DC CONVERTER WITH VOLTAGE BALANCE OF SPLIT CAPACITOR...904
Nan Jin ; Chongyan Zhao ; Leilei Guo

PWM STRATEGY FOR PARALLEL OPERATION OF THREE PHASE CONVERTERS TIED TO GRID...911
Hyun-Sam Jung ; Seung-Ki Sul

PRACTICAL ISSUES AND IMPLEMENTATION CIRCUITS OF THE DIGITAL-ANALOG HYBRID FULL FEED-FORWARD METHOD WITH UNIPOLAR AND BIPOLAR MODULATIONS...917
Xin Zhang ; Henry S. H. Chung ; Zhixun Ma

AN AC-DC POWER CONVERTER FOR ELECTROLYTIC CAPACITOR-LESS LED DRIVER WITH HIGH LUMINOUS EFFICACY...922
Kwon-Sik Park ; Byuong-Jun Seo ; Kyoung-Suk Kang ; Eui-Cheol Nho

AN IMPROVED CASCADED DUAL-BUCK INVERTER...927
Usman Ali Khan ; Honnyong Cha ; Ashraf Ali Khan ; Heung-Geun Kim ; Wilson Eberle ; Liwei Wang

A SINGLE-SWITCH INTEGRATED-STAGE LED DRIVER BASED ON CUK AND CLASS-E CONVERTER...934
Shu Zhang ; Yijie Wang ; Xiaosheng Liu ; Yan Zhou ; Dianguo Xu

A FAULT-TOLERANT PARALLEL INVERTER APPLIED TO MICRO-GRID...939
Xiangyue Shi ; Jinjie Peng ; Zhifeng Qiu ; Wei Xiong

STABILITY ANALYSIS OF GRID-CONNECTED CONVERTERS WITH ADD-ON VOLTAGE SUPPORT FUNCTIONALITY USING REPETITIVE CONTROL...946
Y. Zhang ; M. G. L. Roes ; M. A. M. Hendrix ; J. L. Duarte

ADAPTIVE SERIES STABILIZER MODULE FOR THE GRID CONNECTED INVERTER UNDER VARIABLE GRID CONDITIONS...953
Xin Zhang

AN IMPROVED DROOP CONTROL BASED SMOOTH TRANSFER CONTROL STRATEGY...957
Xin Meng ; Jinjun Liu ; Zeng Liu ; Ronghui An

FREQUENCY RESPONSE ANALYSIS OF LOAD EFFECT ON DYNAMICS OF GRID-FORMING INVERTER 963

Matias Berg ; Tuomas Messo ; Teuvo Suntio

A NEW CONTROL METHOD FOR TRIPLE-ACTIVE BRIDGE CONVERTER WITH FEED FORWARD CONTROL 971

Takanobu Ohno ; Nobukazu Hoshi

ANALYSIS OF PFM OPERATION MODEL FOR CAPACITOR CHARGER RESONANT TOPOLOGY WITH ENERGY DOSAGE 977

Pengyu Jia ; Yiqin Yuan ; Shengwen Fan ; Zhenyu Shan

AN ACTIVE-CLAMPED CURRENT-FED HALF-BRIDGE DC-DC CONVERTER WITH THREE SWITCHES 982

Truong-Duy Duong ; Minh-Khai Nguyen ; Young-Cheol Lim ; Joon-Ho Choi

A HIGH GAIN QUASI SINGLE STAGE LLC RESONANT DC/DC CONVERTER WITH COUPLED INDUCTOR AND PARTIAL ACTIVE CLAMP 987

Chongcan Huo ; Xiaogao Xie ; Shuai Jiang ; Hanjing Dong

SUPPRESSION OF RIPPLE CURRENT IN HIGH STEP-UP DC-DC CONVERTER UTILIZING COCKCROFT-WALTON CIRCUIT WITH INDUCTOR 992

Takumi Yasuda ; Masataka Minami ; Shin-Ichi Motegi ; Masakazu Michihira

AN OPTIMAL DESIGN METHOD CONSIDERING TRANSFORMER PARASITIC CAPACITANCE OF LLC RESONANT CONVERTERS 998

Naizeng Wang ; Xu Yang ; Mofan Tian ; Haiyang Jia ; Guangzhao Xu ; Zhenwei Li

COMPARISON OF HARMONIC LINEARIZATION AND HARMONIC STATE SPACE METHODS FOR IMPEDANCE MODELING OF MODULAR MULTILEVEL CONVERTER 1004

Jing Lyu ; Xin Zhang ; Jingjing Huang ; Jianwen Zhang ; Xu Cai

AN IMPROVED PHASE-SHIFTED PWM FOR A FIVE-LEVEL HYBRID-CLAMPED CONVERTER 1010

Kui Wang ; Nianzhou Liu ; Zedong Zheng ; Yongdong Li

INTEGRATED CONTROL METHODS FOR ASYMMETRICAL CASCADED H-BRIDGE RECTIFIER 1015

Wenjing Dai ; Jie Chen ; Xin Chen ; Chunying Gong

TRANSIENT VOLTAGE STRESS MODELING FOR SUBMODULES OF MODULAR MULTILEVEL CONVERTERS UNDER GRID VOLTAGE SAGS 1021

Zhijian Yin ; Yongheng Yang ; Huai Wang

SVPWM STRATEGY BASED ON MULTILEVEL 3LNPC-CR 1027

Xiaoqiong He ; Pengcheng Han ; Xiaolan Lin ; Yi Wang ; Xu Peng

THE MULTIPLE DEGREE OF FREEDOM BASED NEUTRAL POINT POTENTIAL CONTROL OF THREE LEVEL NEUTRAL POINT CLAMPED CONVERTERS 1032

Bo Guan ; Shinji Doki

A MODIFIED PHASE-SHIFTED PWM TECHNIQUE FOR THE GRID-CONNECTED HYBRID CASCADED CONVERTER 1038

Yu-Chen Su ; Po-Tai Cheng

NOVEL T-TYPE DUAL-BUCK INVERTER WITH MINIMUM NUMBER OF INDUCTORS 1046

Tien-The Nguyen ; Honnyong Cha ; Bang Le-Huy Nguyen ; Heung-Geun Kim

CONTROL OF DIRECT AC/AC MODULAR MULTILEVEL CONVERTER IN RAILWAY POWER SUPPLY SYSTEM 1051

Shuguang Song ; Jinjun Liu ; Shaodi Ouyang ; Xingxing Chen ; Baojin Liu

WIRELESS POWER TRANSFER: CRITICAL REVIEW OF RELATED STANDARDS 1062

Mohamad Abou Houran ; Xu Yang ; Wenjie Chen ; Mehdi Samizadeh

COMPARATIVE STUDY OF SINGLE-PHASE FUNDAMENTAL COMPONENT FREQUENCY ESTIMATION SCHEMES UNDER TIME-VARYING HARMONIC DISTORTION OPERATION 1067

E. B. Kapisch ; J. L. Duarte ; C. A. Duque

A COMPREHENSIVE DEAD-TIME COMPENSATION METHOD FOR A THREE-PHASE DUAL-ACTIVE BRIDGE CONVERTER WITH HYBRID MODULATION SCHEMES 1073

Jingxin Hu ; Zhiqing Yang ; Rik W. De Doncker

EVALUATION OF A HIGH-FREQUENCY REACTOR WITH A NEW WIRE GUIDE FOR A TOROIDAL CORE 1080

Hideki Ayano ; Akira Fujimura ; Yoshihiro Matsui

CORE LOSS EVALUATION IN POWDER CORES: A COMPARATIVE COMPARISON BETWEEN ELECTRICAL AND CALORIMETRIC METHODS 1087

Yuki Ishikura ; Jun Imaoka ; Mostafa Noah ; Masayoshi Yamamoto

MODELING, MAGNETIC DESIGN, AND SIMULATION METHODS CONSIDERING DC SUPERIMPOSITION CHARACTERISTIC OF POWDER CORES USED IN POWER CONVERTERS 1095

Jun Imaoka ; Kenkichiro Okamoto ; Masahito Shoyama ; Yuki Ishikura ; Mostafa Noah ; Masayoshi Yamamoto

MODELLING AND DESIGN OF A MEDIUM FREQUENCY TRANSFORMER FOR HIGH POWER DC-DC CONVERTERS 1103

Miloš Stojadinovic ; Jürgen Biela

EVALUATION OF INDUCTOR LOSSES ON Z-SOURCE INVERTER CONSIDERING AC AND DC COMPONENTS 1111

Ryuji IIjima ; Naoki Kamoshida ; Rene Alexander Barrera Cardenas ; Takanori Isobe ; Hiroshi Tadano

AN INTEGRATING STRUCTURE OF OUTPUT FILTER FOR GRID CONNECTED INVERTER BASED ON FMLF TECHNIQUE 1118

Jie Ma ; Yenan Chen ; Pingping Chen ; Wenxing Zhong ; Dehong Xu

NEW SCREENING METHOD FOR IMPROVING TRANSIENT CURRENT SHARING OF PARALLELED SIC MOSFETS 1125

Junji Ke ; Zhibin Zhao ; Peng Sun ; Huazhen Huang ; James Abuogo ; Xiang Cui

PSPICE MODELING AND APPLICATION FOR SIC POWER MOSFET TO EVALUATE THE POWER LOSS IN FULL-BRIDGE CONVERTER 1131

Juan Wei ; Fei Lin ; Zhongping Yang ; Xianjin Huang ; Chanjuan Xiao ; Hao Zhang ; Wencai Liang

ALL-SIC MODULE PACKAGING TECHNOLOGY 1137

Kento Shirata ; Norihiro Nashida ; Hideyo Nakamura ; Yoshitaka Nishimura

A NEW SMALLEST 1200V INTELLIGENT POWER MODULE FOR THREE PHASE MOTOR DRIVES 1141

Minsub Lee ; Miran Baek ; Junbae Lee ; Daewoong Chung

DESIGN AND ENHANCEMENT OF ESD RELIABILITY IN CIRCULAR UHV 300-V NLDMOS POWER COMPONENTS 1145

Shen-Li Chen ; Yi-Hao Chao ; Chih-Ying Yen ; Jen-Hao Lo ; Chun-Ting Kuo ; Yu-Lin Lin ; Yi-Hao Chiu ; Pei-Lin Wu ; Yu-Lin Jhou

A TECHNOLOGY ANALYSIS OF VOLTAGE SHARING IN SERIES CONNECTED POWER DEVICES 1149

Z Davletzhanova ; O Alatise ; R Bonyadi ; J Ortiz-Gonzalez ; T Dai ; M Jennings ; L Ran ; P Mawby

FAILURE MECHANISM ANALYSIS AND PHYSICS-OF-FAILURE LIFETIME PREDICTION METHOD FOR PRESS-PACK THYRISTOR OF CONVERTER VALVE 1157

Ning Liang ; Zhigang Zhang ; Yating Gou ; Cuicui Liu ; Zebin Yang ; Jiangnan Chen ; Fang Zhuo ; Feng Wang

SURGE VOLTAGE ABSORPTION BY A SILICON CARBIDE AVALANCHE-DIODE WITH P-N STRUCTURE 1162

K. Koseki ; Y. Tanaka

CALCULATION OF THYRISTOR RELIABILITY PARAMETER OF UHVDC CONVERTER VALVE IN HEMP ENVIRONMENT 1167

Zhigang Zhang ; Yating Gou ; Cuicui Liu ; Zebin Yang ; Xiaotong Du ; Jiangnan Chen ; Fang Zhuo ; Feng Wang ; Yuanliang Lan ; Caiwang Sheng

GENERALIZED STACKELBERG GAME-THEORETIC APPROACH FOR JOINTED ENERGY AND RESERVE COORDINATION OF ELECTRIC VEHICLES 1172

Tianyang Zhao ; Xuewei Pan ; Lei Li ; Fei Zhao ; Can Wang

IMPEDANCE INFLUENCE ANALYSIS OF PHASE-LOCKED LOOPS ON THREE-PHASE GRID-CONNECTED INVERTERS 1177

Yuncheng Wang ; Xin Chen ; Yang Zhang ; Jie Chen ; Chunying Gong

PULSE-INJECTION-BASED SENSORLESS CONTROL METHOD WITH IMPROVED DYNAMIC CURRENT RESPONSE FOR PMSM 1183

Hechao Wang ; Kaiyuan Lu ; Dong Wang ; Frede Blaabjerg

INFLUENCE OF PARAMETER VARIATIONS ON OPERATING CHARACTERISTICS OF MTPF CONTROL FOR DTC-BASED PMSM DRIVE SYSTEM 1189

Keisuke Fujii ; Yukinori Inoue ; Shigeo Morimoto ; Masayuki Sanada

A QUIET POSITION SENSORLESS CONTROL FOR AN IPMSM BASED ON EXTENDED EMF AND VOLTAGE INJECTION SYNCHRONIZED WITH PWM CARRIER 1196

Yuki Ishii ; Hiroki Yamashita ; Hisao Kubota

STUDY OF TORQUE RIPPLE REDUCTION AND TORQUE BOOST BY MODIFIED TRAPEZOIDAL MODULATION 1202

Satoshi Joryo ; Kazuto Tatsumi ; Toshimitsu Morizane ; Katsunori Taniguchi ; Noriyuki Kimura ; Hideki Omori

FAULT DIAGNOSIS METHOD OF CURRENT SENSOR FOR PERMANENT MAGNET SYNCHRONOUS MOTOR DRIVES 1206

Guoqiang Zhang ; Guoxin Wang ; Gaolin Wang ; Junya Huo ; Lianghong Zhu ; Dianguo Xu

SENSORLESS SPEED CONTROL OF DIESEL-GENERATOR SYSTEMS BASED ON MULTIPLE SOGI-FLLS 1212
Ngoc Dat Dao ; Dong-Choon Lee ; Dae-Sik Lim

ROBUSTNESS OF SIMPLIFIED SPEED-SENSORLESS VECTOR CONTROL FOR INDUCTION MOTOR 1217
Naoki Akao ; Mineo Tsuji ; Shin-Ichi Hamasaki

MAXIMUM TORQUE CONTROL REFERENCE FRAME BASED ON A TORQUE MAP FOR IPMSMS WITH LARGE INDUCTANCE VARIATION 1223
Kazuki Ohta ; Takumi Ohnuma ; Shinji Doki

PMSM MODEL DISCRETIZATION IN CONSIDERATION OF PARK TRANSFORMATION FOR CURRENT CONTROL SYSTEM 1228
Masamichi Inoue ; Shinji Doki

PSEUDO-RANDOM HIGH-FREQUENCY SINUSOIDAL VOLTAGE INJECTION BASED SENSORLESS CONTROL FOR IPMSM DRIVES 1234
Guoqiang Zhang ; Huiying Wang ; Gaolin Wang ; Junya Huo ; Lianghong Zhu ; Dianguo Xu

AT-NPC 3-LEVEL INVERTER-FED INDUCTION MOTOR VECTOR CONTROL WITH NEUTRAL POINT VOLTAGE CONTROL 1240
K. Sudo ; M. Tsuji ; S. Hamasaki ; T. Fukuoka ; H. Ichinose

INVESTIGATION OF VARIOUS POSITION ESTIMATION ACCURACY ISSUES IN PULSE-INJECTION-BASED SENSORLESS DRIVES 1246
Hechao Wang ; Kaiyuan Lu ; Dong Wang ; Frede Blaabjerg

POSITION SENSORLESS CONTROL OF SWITCHED RELUCTANCE MOTOR USING ESTIMATED PWM PHASE VOLTAGE 1253
Y. Nakazawa ; K. Ohyama ; H. Fujii ; H. Uehara ; Y. Hyakutake

EXPERIMENTAL CONFIRMATION OF THRUST AND ATTRACTIVE FORCE CONTROL OF LINEAR INDUCTION MOTOR BY TWO DIFFERENT FREQUENCY COMPONENTS 1259
Kenta Sannomiya ; Toshimitsu Morizane ; Noriyuki Kimura ; Hideki Omori

GA BASED OPTIMIZED TRAJECTORIES OF ROTATING SPEED AND D-Q AXIS CURRENTS FOR AN IPMSM 1264
Shuta Kumagai ; Kaoru Inoue ; Toshiji Kato

2-DEGREE-OF-FREEDOM DEADBEAT CONTROL WITH DISTURBANCE COMPENSATION FOR PMSM DRIVE SYSTEM USING FPGA 1270
Arata Takahashi ; Shotaro Takakura ; Tomoki Yokoyama

EXTENDED EMF-BASED SIMPLE IPMSM SENSORLESS VECTOR CONTROL USING COMPENSATED CURRENT CONTROLLER 1276
Takatoshi Inoue ; Yasumasa Hamabe ; Mineo Tsuji ; Shin-Ichi Hamasaki

FULL-BAND OUTPUT IMPEDANCE MODEL OF VIRTUAL SYNCHRONOUS GENERATOR IN DQ FRAMEWORK 1282
Li Wenbing ; Wang Jianhua ; Song Jingyu ; Luo Fangfang ; Gao Shang ; Wu Zaijun

AN MTPA CONTROL METHOD OF A PMSM AND A SYNRM BASED ON A DTC IN THE STATOR FLUX LINKAGE SYNCHRONOUS FRAME 1289
Gimpei Itoh ; Yukinori Inoue ; Shigeo Morimoto ; Masayuki Sanada

EEMFS EXCITED BY SIGNAL INJECTION FOR POSITION SENSORLESS CONTROL OF PMSMS AND THEIR PERFORMANCE COMPARISON BY USING IMAGINARY ELECTROMOTIVE FORCE 1295
Takumi Nimura ; Shota Kondo ; Shinji Doki ; Mutuwo Tomita

HARMONIC CURRENT CANCELLATION METHOD FOR PMSM DRIVE SYSTEM USING RESONANT CONTROLLERS 1301
Dongsheng Li ; Yoshitaka Iwaji ; Yasuo Notohara ; Ken Kishita

ESTIMATION ERROR ANALYSIS OF STATOR FLUX OBSERVER FOR DTC-BASED PMSM DRIVES 1308
Atsushi Shinohara ; Kichiro Yamamoto

APPLICATION OF FICTITIOUS REFERENCE ITERATIVE TUNING TO CONTROLLER DESIGN FOR VARIOUS MACHINES 1315
Hidehiro Ikeda ; Kazuya Goto ; Feili Zhang ; Kazuya Kayashima ; Tsuyoshi Hanamoto

HIGH EFFICIENCY CONTROL FOR PERMANENT MAGNET MOTOR DRIVE SYSTEM WITH FUEL CELLS CONNECTED IN SERIES WITH ELECTRIC DOUBLE-LAYER CAPACITORS 1322
Kichiro Yamamoto ; Fumiya Ohdera ; Atsushi Shinohara

COMPARATIVE STUDY OF SPEED RIPPLE REDUCTION BY VARIOUS CONTROL METHODS IN PMSM DRIVE SYSTEMS WITH PULSATING LOAD 1329
Yuma Komaru ; Yukinori Inoue ; Shigeo Morimoto ; Masayuki Sanada

ESTIMATION OF THE PARAMETERS OF THE SERVO DRIVE SYSTEM USING PARTICLE SWARM OPTIMIZATION ALGORITHM 1336

Helin Zhu ; Jae Hyuk Choi ; Sang Uk Park ; Jusuk Lee ; Hyong Gun Lee ; Hyung Soo Mok

A PROGRAMMABLE BATTERY TEST SYSTEM WITH ENERGY RECYCLING FEATURE BASED ON SINUSOIDAL LOADING TECHNIQUE 1341

Chang-Hua Lin ; Guan-Jung Chen ; Hwa-Dong Liu ; Kun-Feng Chen

DEVELOPMENT OF LARGE-CAPACITY CONVERTER FOR BATTERY ENERGY STORAGE SYSTEMS 1346

Hiroyoshi Komatsu ; Tatsuji Katayama ; Noriko Kawakami

ANALYSIS AND COMPARISON OF DC/DC TOPOLOGIES IN PARTIAL POWER PROCESSING CONFIGURATION FOR ENERGY STORAGE SYSTEMS 1351

Maria C. Mira ; Zhe Zhang ; A. E. Michael Andersen

TWO-STAGE PROTECTION FOR MULTI-CHANNEL POWER ELECTRONIC CONVERTERS FED LARGE ASYNCHRONOUS HYDRO-GENERATING UNIT 1358

R. R. Semwal ; Anto Joseph

CURRENT SHARING CONTROL FOR SERIES-PARALLEL CHANGEOVER USING BATTERY AND ELECTRIC DOUBLE-LAYER CAPACITOR BANK 1364

Taisei Nishino ; Keisaku Isozaki ; Naoki Kogai ; Kyungmin Sung

CONTROL METHOD OF ENERGY STORAGE SYSTEM TO IMPROVE OUTPUT POWER OF PCS 1370

Mikiya Ishibashi ; Hitoshi Haga ; Kenji Arimatsu ; Koji Kato

A CONTROL STRATEGY OF MMC BATTERY ENERGY STORAGE SYSTEM BASED ON ARM CURRENT CONTROL 1376

Liu Danqing ; Wang Guangzhu ; Ou Zhujian ; Liu Jiaxing

EQUIVALENT RESISTANCE CONTROL FOR MAXIMUM POWER TRANSFER METHOD OF PIEZOELECTRIC ELEMENT IN VIBRATION POWER GENERATION 1381

Kenya Takamura ; Hiroaki Yamada ; Toshihiko Tanaka ; Tomoharu Yada ; Hajime Fujiwara

DC BUS VOLTAGE STABILIZATION FOR CASCADED POWER CONVERTER BY INTEGRATING AN EXTRA PORT INTO LOAD SIDE PSFB 1386

Jiang You ; Weiyan Fan ; Mengyan Liao

COMMON MODE CURRENT REDUCTION OF THREE-PHASE CASCADED MULTILEVEL TRANSFORMERLESS INVERTER FOR PV SYSTEM 1391

Wenjie Wang ; Ke Chen ; Lijun Hang ; Anping Tong ; Yiliang Gan

CURRENT SHARING/VOLTAGE SHARING CONTROL STRATEGY FOR CASCADED DC/DC CONVERTER IN PHOTOVOLTAIC DC COLLECTION SYSTEM 1397

Bo Chen ; Yi Wang ; Yanjun Tian ; Shilei Wei

PCC VOLTAGE COMPENSATION OF PV INVERTER WITH ACTIVE POWER DECOUPLING CIRCUIT 1403

Duck-Hwan Hwang ; Jung-Yong Lee ; Younghoon Cho

A NOVEL PARTIAL SHADING DETECTION ALGORITHM UTILIZING POWER LEVEL MONITORING OF PHOTOVOLTAIC PANELS 1409

Thusitha Randima Wellawatta ; Sung-Jin Choi

BOOST INTEGRATED THREE-PHASE SOLAR INVERTER USING CURRENT UNFOLDING AND ACTIVE DAMPING METHODS 1414

N. Ha Pham ; Tomoyuki Mannen ; Keiji Wada

LINEAR ACTIVE DISTURBANCE REJECTION CONTROL FOR ISOLATED THREE-PORT CONVERTER 1421

Jiang You ; Mengyan Liao ; Weiyan Fan

STABILITY CONSTRAINED GAIN OPTIMIZATION OF DROOP CONTROLLED CONVERTERS IN DC NANOGRIDS 1426

Soumya Bandyopadhyay ; Laura Ramirez-Elizondo ; Pavol Bauer

SIC BASED SSPC FOR HIGH VOLTAGE SPACE APPLICATIONS 1435

D. Marroquí ; A. Garrigós ; José M. Blanes ; R. Gutiérrez

AN IMPROVED VOLTAGE-TYPE GRID-CONNECTED CONTROL STRATEGY FOR COMPENSATING UNBALANCED VOLTAGE 1442

Liu Hongpeng ; Zhou Jiajie ; Wang Wei

DUAL TWO-STAGE ISOLATED BIDIRECTIONAL DC-DC CONVERTER FOR DC GRID STORAGE 1447

Gabriel Tibola ; Jorge L. Duarte

MODULAR MULTILEVEL CONVERTER WITH CAPACITOR VOLTAGE SELF-BALANCING USING REDUCED NUMBER OF VOLTAGE SENSORS 1455

Taiyuan Yin ; Yue Wang ; Xiaolei Wang ; Shiyuan Yin ; Shumin Sun ; Guanglei Li

PLUG AND OUTLET IN HOUSEHOLD DC LOW VOLTAGE MICRO-GRID POWER DISTRIBUTION..1460

Worapong Pairindra ; Surin Khomfoi

PERFORMANCE PROGRAMMING TECHNIQUE FOR MULTI-STAGE DC POWER DISTRIBUTION SYSTEMS 1465

Syam Kumar Pidaparthy ; Hansang Kim ; Yeonjung Kim ; Byungcho Choi

COORDINATION CONTROL FOR PARALLELED INVERTERS BASED ON VSG FOR PV/BATTERY MICROGRID 1472

Meiqin Mao ; Cheng Qian ; Liuchen Chang ; Yan Du

ADAPTIVE VOLTAGE CONTROL SCHEME FOR DAB BASED MODULAR CASCADED SST IN PV APPLICATION 1478

Tao Liu ; Yang Xuan ; Xu Yang ; Peng Xu ; Yang Li ; Lang Huang ; Xiang Hao

SIX-STEP MMC-BASED HIGH POWER DC-DC CONVERTER..1484

Stefan Milovanovic ; Dražen Dujic

COMBINED DC POWER FLOW CONTROLLER FOR DC GRID 1491

Yongning Chi ; Xizhou Du ; Siqi Liu ; Xu Cai

AN APPROACH FOR THE EMULATION OF DC GRID ADMITTANCES: IMPLEMENTATION ON A BUCK CONVERTER 1498

Enrique Rodriguez-Diaz ; Fracisco D. Freijedo ; Drazen Dujic ; Juan C. Vasquez ; Josep M. Guerrero

A COMPOUND CONTROLLER FOR POWER FLOW AND SHORT-CIRCUIT FAULT IN DC GRID..1504

Han Ye ; Wu Chen ; Pengpeng Pan ; Xiaokun He

DESIGN PROCEDURE AND CONTROL OF A HYBRID CIRCUIT BREAKER WITH ADAPTABLE PULSE CURRENT INJECTION 1509

Andreas Jehle ; Jürgen Biela

A PRAGMATIC SOH AND SOC CO-ESTIMATOR FOR LITHIUM-ION BATTERIES IN SMART GRID APPLICATIONS..1517

Kaiyuan Li ; King Jet Tseng ; Feng Wei ; Boon-Hee Soong

MODELING AND STABILITY ANALYSIS OF PARALLEL DROOP-CONTROLLED AND CURRENT-CONTROLLED INVERTERS..1524

Shike Wang ; Zeng Liu ; Jinjun Liu ; Ronghui An

DIRECT WIRELESS BATTERY CHARGING SYSTEM 1530

Woo-Seok Lee ; Jin-Hak Kim ; Shin-Young Cho ; Il-Oun Lee

AN IMPROVED PWM SCHEME TO ACHIEVE ZERO-VOLTAGE SWITCHING FOR ALL DEVICES IN THREE-PHASE ISOLATED MATRIX RECTIFIER 1537

Xuerui Lin ; Yunwei Ryan Li ; Jahangir Afsharian ; Dewei David Xu

FIXED-FREQUENCY HF GATE DRIVER BY A PUSH-PULL SELF-EXCITATION LC OSCILLATOR HAVING A CAPACITANCE TRANSISTOR..1543

Naoyuki Ishibashi ; Takuya Mizushima ; Masahiko Hirokawa ; Akihiko Katsuki

A FLEXIBLE REDUCED CAPACITOR VOLTAGES STRATEGY FOR VARIABLE-SPEED DRIVES WITH MODULAR MULTILEVEL CONVERTER 1549

Fangzhou Zhao ; Guochun Xiao ; Daoshu Yang ; Zhiqian Wu ; Xin Meng

A LEAKAGE FLUX CANCELLATION TECHNIQUE FOR SERIES-PARALLEL COMBINED RESONANT CIRCUITS WITH ASYMMETRIC ROTARY TRANSFORMERS USED FOR ULTRASONIC SPINDLE DRIVE..1554

Jun Imaoka ; Masahito Shoyama

A NOVEL STRUCTURAL HEALTH MONITORING SYSTEM WITH WIRELESS POWER AND BI-DIRECTIONAL DATA TRANSFER..1562

Yujin Jangs ; Keon-Woo Kim ; Moo-Hyun Park ; Nayoung Lee ; Gun-Woo Moon

CONTROL STRATEGY FOR STARTER GENERATOR IN UAV WITH MICRO JET ENGINE 1567

Jun-Ichi Itoh ; Kazuki Kawamura ; Hiroyuki Koshikizawa ; Kazuyuki Abe

STUDY ON THE INFLUENCE OF VOLTAGE VARIATIONS FOR NON-INTRUSIVE LOAD IDENTIFICATIONS..1575

Yu-Hsiu Lin ; Shun-Kang Hung ; Men-Shen Tsai

BASIC EXPERIMENT OF A MAGLEV SYSTEM FOR A FLEXIBLE STEEL PLATE WITH CURVATURE: FUNDAMENTAL CONSIDERATION ON LEVITATION STABILITY UNDER DISTURBANCE 1580

Makoto Tada ; Kazuki Ogawa ; Takayoshi Narita ; Hideaki Kato ; Hiroyuki Moriyama

PERFORMANCE OF HYBRID MAGNETIC LEVITATION CONTROL SYSTEM FOR THIN STEEL PLATE BY EMS AND PMS: EXPERIMENTAL EVALUATION OF APPLYING OPTIMAL GAP AND ARRANGEMENT OF PMS...1586

Yasuaki Ito ; Yoshiho Oda ; Kengo Okuno ; Toshiki Suzuki ; Masahiro Kida ; Takayoshi Narita ; Hideaki Kato ; Hiroyuki Moriyama

A PRACTICAL LITHIUM-ION BATTERY MODEL BASED ON THE BUTLER-VOLMER EQUATION 1592

Kaiyuan Li ; King Jet Tseng ; Feng Wei ; Boon-Hee Soong

BONDING TECHNOLOGY USING COLD-ROLLED AG SHEET IN DIE-ATTACHMENT APPLICATIONS 1598

Seungjun Noh ; Chanyang Choe ; Chuantong Chen ; Hao Zhang ; Katsuaki Suganuma

HIGH-FREQUENCY SELF-DRIVEN SYNCHRONOUS RECTIFIER CONTROLLER FOR WPT SYSTEMS 1602

Akihiro Konishi ; Kazuhiro Umetani ; Eiji Hiraki

AUTOMATIC RESONANCE FREQUENCY TUNING METHOD FOR REPEATER IN RESONANT INDUCTIVE COUPLING WIRELESS POWER TRANSFER SYSTEMS 1610

Masataka Ishihara ; Kazuhiro Umetani ; Eiji Hiraki

INDUCTIVE POWER TRANSFER FOR T5 FLUORESCENT LAMP LIGHTING SYSTEM 1617

Chung-Chuan Hou ; Tang-Jung Chen ; Ching-Chen Chen ; Chen-Wei Chang ; Po-Wei Wang

AN IMPLEMENT 1.5 MHZ OF INDUCTION HEATING FOR ALUMINUM BASED ON VACUUM TUBE OSCILLATOR CIRCUIT..1622

A. Bilsalam ; P. Chanmontree ; S. Supanyapong ; V. Chunkag

SINGLE-INDUCTOR MULTIPLE-OUTPUTS DIMMABLE LED DRIVER WITH BUCK CONVERTER..1626

Ta-Wei Huang ; Wei-Jing Tseng ; Jun-Xian Huang

A SOFT-SWITCHED THREE-LEVEL T-TYPE INVERTER WITH AUXILIARY COMMUTATED POLES 1634

Apollo Charalambous ; Xibo Yuan

CARRIER-BASED REALIZATION OF ARBITRARY SPACE-VECTOR PWM METHODS FOR THREE-LEVEL INVERTERS 1642

Somboon Sangwongwanich ; Supakorn Paiboon

MULTI-LEVEL TOPOLOGY BASED LINEAR AMPLIFIER FAMILY FOR REALIZATION OF NOISE-LESS INVERTERS..1649

Hidemine Obara ; Tatsuki Ohno ; Atsuo Kawamura

A NEW ZERO-VOLTAGE SWITCHING THREE-LEVEL CONVERTER WITH REDUCED RECTIFIER VOLTAGE STRESS 1655

Keon-Woo Kim ; Cheon-Yong Lim ; Dong-Kwan Kim ; Yu-Jin Jang ; Gun-Woo Moon

MODEL PREDICTIVE CONTROL OF A THREE-LEVEL NPC RECTIFIER WITH A SLIDING MANIFOLD TERM 1661

Xiaonan Gao ; Wei Tian ; Xicai Liu ; Zhenbin Zhang ; Ralph Kennel

H∞ CONTROL-BASED VIBRATION SUPPRESSION IN ROBOT ARM WITH STRAIN WAVE GEARING 1666

Tran Vu Trung ; Makoto Iwasaki

FINE FORCE SENSORLESS FORCE CONTROL BASED ON FRICTION-FREE DISTURBANCE OBSERVER 1673

Ohishi Kiyoshi ; Naoki Kamiya ; Toshimasa Miyazaki ; Yuki Yokokura

KINEMATICS AND TRACKING CONTROL OF A FOUR AXIS ANTENNA FOR SATCOM ON THE MOVE 1680

Oguz Kaan Hancioglu ; Mustafa Celik ; Ugur Tumerdem

POSITION SENSORLESS POSITION CONTROL FOR DUAL SOLENOID ACTUATOR 1687

Sakahisa Nagai ; Atsuo Kawamura

CAE TECHNOLOGY APPLICATION TREND FOR LARGE-CAPACITY POWER ELECTRONICS DEVELOPMENT 1692

Teruo Yoshino ; Kuniaki Nagasaka ; Shigeaki Nakabayashi ; Ikuto Udagawa ; Isamu Tominaga ; Junya Konno

XILINX SYSTEM GENERATOR BASED MODELLING OF FINITE STATE MPC........................1698

Vijay Kumar Singh ; Ravi Nath Tripathi ; Tsuyoshi Hanamoto

POWER HARDWARE-IN-THE-LOOP SETUP FOR STABILITY STUDIES OF GRID-CONNECTED POWER CONVERTERS 1704

Tommi Reinikka ; Henrik Alenius ; Tomi Roinila ; Tuomas Messo

PASSIVITY-BASED LCL FILTER DESIGN OF GRID-CONNECTED VSCS WITH CONVERTER SIDE CURRENT FEEDBACK 1711

Shih-Feng Chou ; Xiongfei Wang ; Frede Blaabjerg

ADAPTIVE CONTROL OF DC POWER DISTRIBUTION SYSTEMS: APPLYING PSEUDO-RANDOM SEQUENCES AND FOURIER TECHNIQUES...1719
Tomi Roinila ; Hessamaldin Abdollahi ; Silvia Arrua ; Enrico Santi

AN IMPROVED FINITE-SET MODEL PREDICTIVE TORQUE CONTROL FOR INTERIOR PERMANENT MAGNET SYNCHRONOUS MOTOR DRIVES..1724
Xinan Zhang ; Gilbert Foo ; Tung Ngo

PREDICTIVE TORQUE CONTROL FOR FIVE PHASE INDUCTION MOTOR DRIVE WITH COMMON MODE VOLTAGE REDUCTION 1730
Apekshit Bhowate ; Mohan Aware ; Sohit Sharma ; Yogesh Tatte

INDIRECT MATRIX CONVERTER FOR PERMANENT-MAGNET-SYNCHRONOUS-MOTOR DRIVES BY IMPROVED TORQUE PREDICTIVE CONTROL..1736
Yun Jang ; Yeongsu Bak ; Kyo-Beum Lee

PREDICTIVE DC-LINK CURRENT CONTROL BASED ON IPMSM DISCRETE STATE EQUATION FOR INVERTER WITHOUT INDUCTOR OR ELECTROLYTIC CAPACITOR.............1741
Yousuke Akama ; Kodai Abe ; Kiyoshi Ohishi ; Yuki Yokokura ; Koji Kobayashi ; Tatsuki Kashihara

NEW SEARCH ALGORITHM OF MODEL PREDICTIVE CONTROL TO REDUCING CALCULATION AMOUNT FOR IMPROVING STEADY CURRENT CONTROL PERFORMANCE..1747
Masahiro Shimaoka ; Shinji Doki

DISTRIBUTED POWER SHARING STRATEGY FOR ISLANDED MICROGRIDS WITHOUT FREQUENCY AND VOLTAGE DEVIATIONS...1752
Tuan V. Hoang ; Hong-Hee Lee

LIFETIME-ORIENTED DROOP CONTROL STRATEGY FOR AC ISLANDED MICROGRIDS...................1758
Yanbo Wang ; Dong Liu ; Fujin Deng ; Dao Zhou ; Zhe Chen

EXPERIMENT ON HIERARCHICAL CONTROL BASED POWER QUALITY ENHANCEMENT FOR STANDALONE MICROGRID...1764
Darith Leng ; Sompob Polmai ; Kittichot Soontorntaweesub

A DISTRIBUTED PREDICTIVE CONTROL STRATEGY BASED ON STATE ESTIMATOR FOR ISLANDED MICROGRID..1771
Mi Dong ; Li Li ; Xiaoyu Tian

MAXIMUM POWER POINT TRACKING METHOD FOR PV MODULE UNDER WIDE RANGE VARYING IRRADIANCE LEVELS...1777
Hwa-Dong Liu ; Chang-Hua Lin

DUAL MPPT CONTROL AND FIELD TESTING FOR SWITCHED CAPACITOR-BASED CELL-LEVEL POWER BALANCING UTILIZING DIFFUSION CAPACITANCE OF PHOTOVOLTAIC CELLS..1782
Masatoshi Uno ; Yota Saito ; Masaya Yamamoto ; Shinichi Urabe

SERIES RESONANT DC-DC CONVERTER WITH DUAL-MODE RECTIFIER FOR PV MICROINVERTERS...1788
Yanfeng Shen ; Huai Wang ; Zhan Shen ; Yongheng Yang ; Frede Blaabjerg

VOLTAGE-REFERENCE ACTIVE POWER DECOUPLING BASED ON BOOST CONVERTER FOR SINGLE-PHASE BRIDGE INVERTER..1793
Shuang Xu ; Meiqin Mao ; Riming Shao ; Liuchen Chang

A SINGLE-PHASE COMMON GROUND BOOST INVERTER FOR PHOTOVOLTAIC APPLICATIONS..1799
Tan-Tai Tran ; Minh-Khai Nguyen ; Young-Cheol Lim ; Joon-Ho Choi

STUDY FOR FURTHER INTRODUCTION OF THE ELECTRONIC FREQUENCY CONVERTERS TO THE TOKAIDO SHINKANSEN 1803
Toshimasa Shimizu ; Ken Kunomura ; Masahiko Kai ; Hiroki Miyajima ; Teruhisa Matsui

COUNTERMEASURE FOR PARTIAL TURN-OFF OF THYRISTOR CHANGEOVER SWITCH INTRODUCED TO TOHOKU SHINKANSEN SHIN-YONO SECTIONING POST 1810
Yuki Mizumoto ; Nobuhito Kurosawa

HARDWARE–IN–THE–LOOP REAL–TIME SIMULATION EXPERIMENT PLATFORM FOR TRACTION POWER SUPPLY SYSTEM BASED ON DSPACE-XSIM....................................1816
Runze Zhang ; Fei Lin ; Zhongping Yang ; Hu Cao ; Yuping Liu

EVALUATING THE NON-SINUSOIDAL AND NON-SYMMETRIC REGIMES FROM A RAILWAY SUPPLYING SUBSTATION 1822
Ileana-Diana Nicolae ; Petre-Marian Nicolae ; Radu-Florin Marinescu

A FUNDAMENTAL TRAIN RUNNING EXPERIMENT FOR A BASIC PERFORMANCE VERIFICATION OF A TRAIN POWER DEMAND CONTROL SYSTEM BY DECENTRALIZED CONTROL ALGORITHM 1828
Yusuke Oki ; Tomoyuki Ogawa ; Yoko Takeuchi ; Tatsuhito Saito ; Jun'ichiro Kawaguchi

VERIFICATION OF SIC BASED MODULAR MULTILEVEL CASCADE CONVERTER (MMCC) FOR HVDC TRANSMISSION SYSTEMS 1834

Y. Ishii ; T. Jimichi

CONTROL OF A 6.6-KV TRANSFORMERLESS STATCOM BASED ON THE MMCC-SDBC USING SIC MOSFETS 1840

Laxman Maharjan ; Toshihisa Tajyuta ; Hiroshi Shinohara ; Akio Suzuki ; Akio Toba

ISOLATED THREE–PHASE AC/DC CONVERTER USING A SOFT–SWITCHING TECHNIQUE FOR BATTERY CHARGER 1847

Yuto Matsui ; Kazuma Suzuki ; Takaharu Takeshita ; Wataru Kitagawa

IMPLEMENTATION OF A MINIATURIZED SIC INVERTER 1854

Hideaki Fujita ; Cristian Andres Garces Guajardo

DESIGN CONSIDERATION OF FLYING CAPACITOR MULTILEVEL INVERTERS USING SIC MOSFETS 1860

Yukihiko Sato ; Kenji Natori

A CONTROL METHOD OF OVERVOLTAGE SUPPRESSION ACROSS THE DC CAPACITOR IN A GRID-CONNECTION CONVERTER USING LEG SHORT-CIRCUIT OF POWER MOSFETS DURING THE INITIAL CHARGE 1866

Tomoyuki Mannen ; Keiji Wada

THE ESSENTIAL RELATIONSHIP BETWEEN DEADBEAT PREDICTIVE CONTROL AND CONTINUOUS-CONTROL-SET MODEL PREDICTIVE CONTROL FOR PWM CONVERTERS 1872

Bi Liu ; Tao Chen ; Wensheng Song

DEADBEAT CONTROL FOR MULTI-LEVEL INVERTER USING 1MHZ MULTISAMPLING METHOD FOR UTILITY INTERACTIVE SYSTEM 1877

Ryosuke Kikuchi ; Ryunosuke Araumi ; Tomoki Yokoyama

1MHZ MULTISAMPLING DEADBEAT CONTROL WITH DISTURBANCE COMPENSATION METHOD FOR THREE PHASE PWM INVERTER 1883

Hiroaki Ueta ; Tomoki Yokoyama

MODULAR MULTILEVEL CONVERTER REPLACED ONE MODULE WITH HIGH VOLTAGE IGBT 1890

Kazunobu Oi ; Kenta Takasho ; Yugo Tadano

INCREASED EFFICIENCY AND REDUCED REALIZATION EFFORT OF DSBC AND DSCC MODULAR MULTILEVEL CONVERTERS (MMCS) 1896

A. Hillers ; J. Biela

COMMON-MODE VOLTAGE INJECTION TECHNIQUES FOR QUASI TWO-LEVEL PWM-OPERATED MODULAR MULTILEVEL CONVERTERS 1904

Jakub Kucka ; Axel Mertens

CURRENT TRACKING AND CELL-VOLTAGE LIMITATIONS OF MODULAR MULTILEVEL CONVERTERS WITH DIRECT DIGITAL CONTROL 1912

T.-F. Wu ; T.-C. Chou ; K.-E. Lin ; T.-Y. Li

SWITCHING LOSS ANALYSIS OF SIC-MOSFET BASED ON STRAY INDUCTANCE SCALING 1919

Keiji Wada ; Masato Ando

MODELING AND OPTIMIZATION OF DISPLACEMENT WINDINGS FOR TRANSFORMERS IN DUAL ACTIVE BRIDGE CONVERTERS 1925

Zhan Shen ; Yanfeng Shen ; Zian Qin ; Huai Wang

OPTIMIZED SELECTION AND UTILIZATION OF DC-LINK CAPACITOR IN A SINGLE-PHASE PV GRID INVERTER SYSTEM 1931

Caspar Collins ; Li Ran

AN EVALUATION CIRCUIT FOR DC-LINK CAPACITORS USED IN A HIGH-POWER THREE-PHASE INVERTER WITH CONDITION MONITORING 1938

Kazunori Hasegawa ; Ichiro Omura ; Shin-Ichi Nishizawa

RECENT MARKET AND TECHNICAL TRENDS IN COPPER ROTORS FOR HIGH-EFFICIENCY INDUCTION MOTORS 1943

Daniel Liang ; Victor Zhou

OVERVIEW OF THE LATEST RESEARCH AND DEVELOPMENT FOR COPPER DIE-CAST SQUIRREL-CAGE ROTORS 1949

Shu Yamamoto

A NOVEL HEAT-RESISTANT INSULATION-PROCESSING AGENT APPLICABLE TO COPPER DIE-CAST SQUIRREL-CAGE ROTORS 1955

Junichi Uchida ; Yuki Sueuchi ; Naosumi Kamiyama

INSULATION-PROCESSING OF COPPER DIE-CAST SQUIRREL-CAGE ROTOR ON MOTOR EFFICIENCY IN HIGH-SPEED OPERATION OVER 10,000 R/MIN 1960

Hideaki Hirahara ; Akira Tanaka ; Shu Yamamoto

HIGH-PRECISION ROTOR POSITION ESTIMATION FOR HIGH-SPEED SPMSM DRIVE BASED ON STATE OBSERVER AND HARMONIC ELIMINATION 1966
Peng Yang ; Xi Xiao ; Meng Zhang ; Shkodyrev Vyacheslav

HARMONIC LOSS REDUCTION IN HIGH SPEED MOTOR DRIVE SYSTEMS BY FLYING CAPACITOR MULTILEVEL INVERTER 1972
Anudari Tumurbaatar ; Sae Mochidate ; Koji Yamaguchi ; Tomohiro Matsuda ; Yukihiko Sato

CURRENT SOURCE TYPE PMSG WIND TURBINE SYSTEM WITH THREE-PHASE THREE-SWITCH BUCK-TYPE RECTIFIER FOR MACHINE-SIDE CONVERTER 1977
Beomseok Chae ; Tahyun Kang ; Yongsug Suh

A STUDY OF 10MW LOAD COMMUTATED INVERTER FOR GAS-TURBINE START-UP 1985
An Hyunsung ; Cha Hanju

PROTOTYPING OF 500 KVA MEDIUM FREQUENCY TRANSFORMER FOR OFFSHORE DIRECT-CURRENT COLLECTION GRID 1991
Tomoyuki Hatakeyama ; Naoyuki Kurita ; Mamoru Kimura

PSCAD/EMTDC AND RTDS SIMULATION ANALYSIS OF MULTIVENDOR MULTI-TERMINAL HVDC SYSTEM CONNECTED TO OFFSHORE WINDFARMS 1997
Hiroshi Suwa ; Takuro Arai ; Takahiro Ishiguro ; Tohru Yoshihara ; Mamoru Kimura ; Tsuneshisa Wachi ; Takahiro Horikoshi ; Tatsuhito Nakajima

INTEROPERABILITY OF MODULAR MULTILEVEL CONVERTERS AND 2-LEVEL VOLTAGE SOURCE CONVERTERS IN A LABORATORY-SCALE MULTI-TERMINAL DC GRID 2003
Salvatore D'arco ; Atsede G. Endegnanew ; Giuseppe Guidi ; Jon Are Suul

PRINCIPLE EXPERIMENT OF CURRENT COMMUTATED HYBRID DCCB FOR HVDC TRANSMISSION SYSTEMS 2011
Ryuta Hasegawa ; Kazuhisa Kanaya ; Yushi Koyama ; Toshiaki Matsumoto ; Takahiro Ishiguro

A THREE-INPUT CENTRAL CAPACITOR DC/DC CONVERTER 2016
Jiaxin Liu ; Feng Gao

SERIES/PARALLEL SWITCHING CIRCUITS USING POWER MOSFETS FOR PHOTOVOLTAIC MODULES 2022
Masamichi Tanemo ; Koki Matsudate ; Shinichi Nomura

MODULARIZED EQUALIZATION ARCHITECTURE BASED ON SWITCHED CAPACITOR CONVERTER TO VIRTUALLY UNIFY MISMATCHED PHOTOVOLTAIC PANEL CHARACTERISTICS 2030
Masatoshi Uno ; Masaya Yamamoto

BUCK-BOOST TYPE MPPT CIRCUIT SUITABLE FOR PHOTOVOLTAIC GENERATION OF VEHICLE INSTALLATION 2036
Fumihisa Kano ; Yuji Kasai ; Hideki Kimura ; Kouhei Sagawa ; Junnosuke Haruna ; Hirohito Funato

VERIFICATION TEST OF ENERGY-EFFICIENT OPERATIONS AND SCHEDULING UTILIZING AUTOMATIC TRAIN OPERATION SYSTEM 2042
Shoichiro Watanabe ; Yasuhiro Sato ; Takafumi Koseki ; Eisuke Isobe ; Jun Kawashita

THE DIRECT BENEFIT OF SIC POWER SEMICONDUCTOR DEVICES FOR RAILWAY VEHICLE TRACTION INVERTERS 2047
Shingo Makishima ; Kazuki Fujimoto ; Keiichiro Kondo

THE LOSS CHARACTERISTICS OF PSFB ZVS DC-DC CONVERTER APPLIED TO THE AUXILIARY POWER SYSTEM 2051
Xianjin Huang ; Juan Zhao ; Fei Lin

SURVEY ON ELECTROMAGNETIC INTERFERENCE ANALYSIS FOR TRACTION CONVERTERS IN RAILWAY VEHICLES 2058
Zhichang Yang ; Hong Li ; Chao Feng ; Yanfeng Jiang ; Fei Lin ; Zhongping Yang

DEVELOPMENT OF TRACTION MOTOR FOR NEW ZERO - EMISSION VEHICLE 2066
Akinobu Iwai ; Satoshi Honjo ; Hirofumi Suzumori ; Toshio Okazawa

EMC DESIGN AND DEVELOPMENT METHODOLOGY FOR TRACTION POWER INVERTERS OF ELECTRIC VEHICLES 2073
Isao Hoda ; Jia Li ; Hiroki Funato

SIMULATION-DRIVEN DESIGN OPTIMIZATION OF A MULTILAYER EMC INPUT FILTER 2078
Fatou Diouf ; Nadim Sakr ; Anna Gheonjian

EV TRACTION INVERTER EMPLOYING DOUBLE-SIDED DIRECT-COOLING TECHNOLOGY WITH SIC POWER DEVICE 2082
Takashi Hirao ; Masami Onishi ; Yusuke Yasuda ; Akihiro Namba ; Kinya Nakatsu

AN OVERVIEW OF STABILITY IMPROVEMENT METHODS FOR WIDE-OPERATION-RANGE FLYBACK CONVERTER WITH VARIABLE FREQUENCY PEAK-CURRENT-MODE CONTROL...2086

Ching-Hsiang Cheng ; Ching-Jan Chen ; Shinn-Shyong Wang

DESIGN AND IMPLEMENTATION OF A HIGH POWER DENSITY ACTIVE-CLAMPED FLYBACK CONVERTER 2092

Yu-Chen Liu ; Bing-Siang Huang ; Cheng-Hung Lin ; Katherine A. Kim ; Huang-Jen Chiu

OPTIMIZED VARIABLE ON-TIME CONTROL FOR LED LIGHTING DRIVER.................2097

Jizhe Wang ; Haruhi Eto ; Fujio Kurokawa

DESIGN OF MULTIMODE BATTERY CHARGER WITH DYNAMIC VOLTAGE TRACKING CONTROL...2102

Pang-Jung Liu ; Lin-Hao Chien ; Song-Kai Lee ; Ang-Tung Chen

DUAL-SLOT POWER-PICKUP STRUCTURE FOR CONTACTLESS STRIP INDUCTIVE POWER TRACK SYSTEM 2107

Jia-You Lee ; I-Lin Chen ; Chien-Tzu Ko

DISCONTINUOUS SVM TECHNIQUE FOR THREE-LEG VSI FED BALANCED/UNBALANCED TWO-PHASE LOADS 2113

Supanut Charoensuksirikul ; Yuttana Kumsuwan

REDUCTION OF POWER LOSSES BASED ON GENERALIZED TWO-LEVEL PWM ALGORITHM FOR A NINE-SWITCH VSI 2121

Neerakorn Jarutus ; Yuttana Kumsuwan

SIC-BASED THREE-PHASE QUASI-Z-SOURCE INVERTER VERSUS THE TWO-STAGE TOPOLOGY - A COMPARISON 2129

Kornel Wolski ; Mariusz Zdanowski ; Jacek Rabkowski

DC-SIDE CIRCUIT IMPLEMENTATION OF A THREE-PHASE INVERTER FOR BALANCING PHASE-LEG CAPACITOR CURRENTS 2137

Takashi Hirao ; Keiji Wada ; Toshihisa Shimizu

A THREE-PHASE HYBRID SWITCHED-BOOST INVERTER 2145

Minh-Khai Nguyen ; Tan-Tai Tran ; Hoan-Tien Luong ; Kyoung-Won Lee ; Youn-Ok Choi ; Geum-Bae Cho

THE EFFECT OF BUILT-IN CR SNUBBER CAPACITOR INTO THE POWER MODULE 2149

Ryotaro Hata ; Shigeki Nishiyama

EVALUATION OF NOVEL HYBRID PROTECTION BASED ON PYROSWITCH AND FUSE TECHNOLOGIES...2153

Tomokazu Sakuraba ; Rémy Ouaida ; Song Chen ; Thibaut Chailloux

OPTIMAL DESIGN OF A MAGNETICALLY COUPLED FILTER FOR HIGH EFFICIENCY, LOW COST AND LOW VOLUME DC-DC BATTERY STORAGE CONVERTER 2158

Timothé Delaforge ; Robert Pasterczyk ; Mickaël Robert ; Hervé Chazal ; Jean-Luc Schanen ; Sébastien Mariethoz

HIGH POWER/CURRENT INDUCTOR LOSS MEASUREMENT WITH SHUNT RESISTOR CURRENT-SENSING METHOD..2165

Pin Yu Huang ; Toshihisa Shimizu

SENSITIVITY ANALYSIS OF MEDIUM FREQUENCY TRANSFORMER DESIGN.................2170

Marko Mogorovic ; Drazen Dujic

STANDARD MODELS FOR POWER ELECTRONIC SYSTEM SIMULATION......................2176

Koichi Shigematsu ; Hiroki Ishikawa ; Taku Noda ; Kentarou Fukushima ; Yoichi Sekiba ; Yusuke Kouno ; Takashi Abe ; Takayuki Sekisue ; Shinji Katoh

MODELING AND MODEL PARAMETER EXTRACTION OF WIDE BANDGAP POWER SEMICONDUCTOR DEVICE, PACKAGE, AND CIRCUIT FOR SIMULATING FAST SWITCHING BEHAVIOR...2181

Tsuyoshi Funaki

STABILITY ANALYSIS METHODS OF A GRID-CONNECTED INVERTER IN TIME AND FREQUENCY DOMAINS...2186

Toshiji Kato ; Kaoru Inoue ; Taiki Sakiyama

FINITE ELEMENT METHODS FOR MULTI-OBJECTIVE OPTIMIZATION OF A HIGH STEP-UP INTERLEAVED BOOST CONVERTER...2193

Wilmar Martinez ; Camilo Cortes ; Ahmad Bilal ; Jorma Kyyra

HIGH FIDELITY REAL-TIME SIMULATION OF MULTI-LEVEL CONVERTERS 2199

Jost Allmeling ; Niklaus Felderer ; Min Luo

AN ENHANCED HIGH FREQUENCY PULSATING VOLTAGE INJECTION METHOD BASED ON IMMUNE ALGORITHM FOR SENSORLESS IPMSM DRIVES...............................2204

Yanping Zhang ; Zhonggang Yin ; Chao Du ; Youyun Wang ; Xiangdong Sun

POSITION ESTIMATION ACCURACY IMPROVEMENT FOR MAGNETIC SALIENCY BASED SENSORLESS CONTROL INCLUDING CROSS-COUPLING FACTOR 2210

Keita Shimamoto ; Shinya Morimoto

SENSORLESS DRIVE IN THE LOW SPEED REGION AND AUTO-TUNING METHOD FOR PERMANENT MAGNET SYNCHRONOUS MOTORS 2216

Naofumi Nomura ; Shinichi Higuchi

HIGH STABILITY V/F CONTROL OF PMSM USING STATE FEEDBACK CONTROL BASED ON N-T COORDINATE SYSTEM 2224

Yosuke Matsuki ; Shinji Doki

STABILIZATION METHOD USING EQUIVALENT RESISTANCE GAIN BASED ON V/F CONTROL FOR IPMSM WITH LONG ELECTRICAL TIME CONSTANT 2229

Jun-Ichi Itoh ; Takato Toi ; Koroku Nishizawa

SINGLE-PHASE SOLID-STATE TRANSFORMER USING MULTI-CELL WITH AUTOMATIC CAPACITOR VOLTAGE BALANCE CAPABILITY 2237

Jun-Ichi Itoh ; Kazuki Aoyagi ; Keisuke Kusaka ; Masakazu Adachi

A DEVELOPED DUAL MMC ISOLATED DC SOLID STATE TRANSFORMER AND ITS MODULATION STRATEGY 2245

Yan Li ; Chao Liu ; Xu Cai

DC FAULT RIDE-THROUGH OF A THREE-PHASE DUAL-ACTIVE BRIDGE CONVERTER FOR DC GRIDS 2250

Jingxin Hu ; Shenghui Cui ; Rik W. De Doncker

A COMPOUND 10KV DVR SYSTEM BASED ON SOLID STATE TRANSFORMER STRUCTURE 2262

Yaqian Zhang ; Jianzhong Zhang ; Xing Hu ; Zakiud Din

A DUAL-ENERGY-SOURCE UNINTERRUPTIBLE POWER SUPPLY (UPS) 2270

Hao Wang ; Dehong Xu ; Binci Xu ; Haijin Li ; Ye Zhu

INFLUENCE OF WIND POWER FORECASTS ON EQUITABLE DISTRIBUTION METHOD OF WIND POWER CURTAILMENT 2278

Daisuke IIoka ; Hiroumi Saitoh

COMPARISON OF OPTIMIZED DEMAND OF EGS FOR MINIMIZING FUEL CONSUMPTION AND EGS MODEL WITH POWER GRID FREQUENCY USING A HPSPITAL LOAD WITH PV 2283

Yuji Mizuno ; Teppei Baba ; Fujio Kurokawa ; Nobumasa Matsui

COORDINATED DFIG WIND TURBINES AND SOLAR PV GENERATORS FOR INTER-AREA OSCILLATION DAMPING 2287

Tossaporn Surinkaew ; Issarachai Ngamroo

ENERGY MANAGEMENT USING A QUICK CHARGER WITH STORAGE BATTERIES FOR ELECTRIC VEHICLES 2292

Taku Ishibashi ; Toyonari Shimakage ; Norikazu Takeuchi ; Takaaki Kikuchi ; Midori Nonogaki

A METHOD FOR JUNCTION TEMPERATURE ESTIMATION UTILIZING TURN-ON SATURATION CURRENT FOR SIC MOSFET 2296

Hui-Chen Yang ; Rejeki Simanjorang ; Kye Yak See

FIELD BUS FOR DATA EXCHANGE AND CONTROL OF MODULAR POWER ELECTRONIC SYSTEMS WITH HIGH SYNCHRONISATION ACCURACY 2301

Stefan Rietmann ; Simon Fuchs ; André Hillers ; Jürgen Biela

ANALYTICAL INVESTIGATION ON ASYMMETRIC LCC COMPENSATION CIRCUIT FOR TRADE-OFF BETWEEN HIGH EFFICIENCY AND POWER 2309

Kodai Takeda ; Takafumi Koseki

PROBABILISTIC PCA-SUPPORT VECTOR MACHINE BASED FAULT DIAGNOSIS OF SINGLE PHASE 5-LEVEL CASCADED H-BRIDGE MLI 2317

Nagendra Vara Prasad Kuraku ; Yigang He ; Murad Ali

A STUDY ON EDGE SUPPORTED ELECTROMAGNETIC LEVITATION SYSTEM: FUNDAMENTAL CONSIDERATION ON LEVITATION PERFORMANCE OF THIN STEEL PLATE 2324

Yoshiho Oda ; Yasuaki Ito ; Kengo Okuno ; Masahiro Kida ; Toshiki Suzuki ; Takayoshi Narita ; Hideaki Kato ; Hiroyuki Moriyama

APPLICATION OF FACTS DEVICES FOR A DYNAMIC POWER SYSTEM WITHIN THE USA 2329

Jan Paramalingam ; Fuminori Nakamura ; Akihiro Matsuda ; Daisuke Yamanaka ; Taichiro Tsuchiya

CAPACITOR VOLTAGE BALANCING IN SEMI-FULL-BRIDGE SUBMODULE WITH DIFFERENTIAL-MODE CHOKE : (INVITEDPAPER) 2335

Kalle Ilves ; Yuhei Okazaki ; Nan Chen ; Muhammad Nawaz ; Antonios Antonopoulos

RESEARCH ON KEY TECHNOLOGY AND EQUIPMENT FOR ZHANGBEI 500KV DC GRID 2343

Hui Pang ; Xiaoguang Wei

WHAT LED TO SUCCESS IN ACADEMIC RESEARCH ON THE FAMILY OF MODULAR MULTILEVEL CASCADE CONVERTERS? 2352
Hirofumi Akagi

OPERATING PRINCIPLE OF CURRENT RESONANT CONVERTER USING AIR CORE TRANSFORMER FOR ISOLATED POWER SUPPLY ON CHIP 2360
Seiya Abe ; Hikaru Kaishakuji ; Satoshi Matsumoto

ANALYSIS FOR HIGH-FREQUENCY LLC RESONANT CONVERTER WITH PLANAR TRANSFORMER AT LIGHT-LOAD CONDITION 2365
Keon-Woo Kim ; Jae-Il Baek ; Yeonho Jeong ; Ki-Mok Kim ; Gun-Woo Moon

A NOVEL FULL DIGITAL CONTROL H-BRIDGE DC-DC CONVERTER FOR POWER SUPPLY ON CHIP APPLICATIONS 2370
Shigeki Nakano ; Toshiomi Oka ; Seiya Abe ; Satoshi Matsumoto

A HIGH-EFFICIENCY POWER SUPPLY FROM MAGNETIC ENERGY HARVESTERS 2376
Cheon-Yong Lim ; Yeonho Jeong ; Keon-Woo Kim ; Feel-Soon Kang ; Gun-Woo Moon

OPPORTUNITIES FOR LEVERAGING LOW-VOLTAGE GAN DEVICES IN MODULAR MULTI-LEVEL CONVERTERS FOR ELECTRIC-VEHICLE CHARGING APPLICATIONS 2380
Mojtaba Ashourloo ; Mohammad Shawkat Zaman ; Miad Nasr ; Olivier Trescases

A NEW CONTROL STRATEGY FOR MODULAR MULTILEVEL CONVERTER OPERATING IN QUASI TWO-LEVEL PWM MODE 2386
Chao Wang ; Kui Wang ; Zedong Zheng ; Yongdong Li

A CURRENT-SOURCE TYPE MMC WITH DELTA-CONNECTED ARMS FOR SMES 2393
Yushi Miura ; Toshifumi Ise

NEW MODULE WITH ISOLATED HALF BRIDGE OR ISOLATED FULL BRIDGE FOR MODULAR MEDIUM VOLTAGE CONVERTER 2400
Yunpeng Si ; Yifu Liu ; Qin Lei

DEVELOPMENT OF A 700-V-CLASS REVERSE-BLOCKING IGBT FOR ADVANCED T-TYPE NEUTRAL POINT-CLAMPED POWER CONVERSION SYSTEM 2404
Hiroki Wakimoto ; Haruo Nakazawa ; David H. Lu ; Takashi Matsumoto ; Yoichi Nabetani

CERAMIC EMBEDDING AS PACKAGING SOLUTION FOR FUTURE POWER ELECTRONIC APPLICATIONS 2410
Hoang Linh Bach ; Tobias Maximilian Endres ; Daniel Dirksen ; Sigrid Zischler ; Christoph Friedrich Bayer ; Andreas Schletz ; Martin März

MICROELECTROMECHANICAL SYSTEM (MEMS) RESONATOR: A NEW ELEMENT IN POWER CONVERTER CIRCUITS FEATURING REDUCED EMI 2416
A N M Wasekul Azad ; Sourov Roy ; Abu Saleh Imtiaz ; Faisal Khan

A LUMPED THERMAL MODEL INCLUDING THERMAL COUPLING EFFECTS AND BOUNDARY CONDITIONS FOR CAPACITOR BANKS 2421
Qiusheng Wang

HYSTERESIS MODELING OF MAGNETIC DEVICES BASED ON RELUCTANCE NETWORK ANALYSIS 2426
Yoshiki Hane ; Kenji Nakamura

OPTIMAL SIZING AND PLACEMENT OF SOLAR POWERED CHARGING STATION UNDER EV LOADS PENETRATION USING ARTIFICIAL BEE COLONY TECHNIQUE 2430
Yuttana Kongjeen ; Kulsomsup Yenchamchalit ; Krischonme Bhumkittipich

A COMPARISON OF AVERAGE MODEL, SAMPLED-DATA MODEL AND MULTI-FREQUENCY MODEL BASED ON DC/DC CONVERTERS 2435
Xiangpeng Cheng ; Jinjun Liu ; Zeng Liu ; Yiming Tu ; Danhong Xue

SMALL-SIGNAL DISCRETE-TIME MODELING AND DIGITAL CONTROL OF THE BI-DIRECTIONAL DC/DC CONVERTERS 2441
Jia Yaoqin ; Xu Yingchun ; Hou Yijie

ENERGY MANAGEMENT OF HYDROGEN-STORAGE PHOTOVOLTAIC GENERATION SYSTEM WITH A FUNCTION OF SUPPRESSING SHORT-PERIOD COMPONENTS 2449
Yuuki Machida ; Akihisa Goto ; Akiko Takahashi ; Shigeyuki Funabiki

A DYNAMIC BATTERY CHARGING APPROACH FOR ENERGY TRADING IN THE SMART GRID 2456
Avinash Sharma ; Akshay Kumar Rathore ; Rajesh Kumar

A FORCED COMMUTATION METHOD OF THE SOLID-STATE TRANSFER SWITCH IN THE UNINTERRUPTED POWER SUPPLY APPLICATIONS 2462
Meng-Jiang Tsai ; Jiuyang Zhou ; Po-Tai Cheng

ONLINE INTERNAL IMPEDANCE MEASUREMENTS OF LI-ION BATTERY USING PRBS BROADBAND EXCITATION AND FOURIER TECHNIQUES: METHODS AND INJECTION DESIGN................2470

Jussi Sihvo ; Tuomas Messo ; Tomi Roinila ; Roni Luhtala

A DC CURRENT FLOW CONTROLLER FOR MESHED HVDC GRIDS 2476

Viktor Hofmann ; Mark-M. Bakran

AN ISOLATED SOFT-SWITCHING HYBRID-SOURCE DC-DC CONVERTER FOR DC OFFSHORE WIND FARMS................2484

Shenghui Cui ; Jingxin Hu ; Marco Stieneker ; Rik W. De Doncker

A TRANSFORMERLESS MULTI-CELL SOLID-STATE FAULT CURRENT LIMITER FOR MEDIUM VOLTAGE POWER SYSTEM................2490

Pantarote Techama ; Sompob Polmai ; Chanin Bunlaksananusorn

A NOVEL DC POWER FLOW CONTROLLER FOR HVDC GRIDS WITH DIFFERENT VOLTAGE LEVELS................2496

Ya'nan Wu ; Han Ye ; Wu Chen ; Xiaokun He

DESIGN AND CONTROL OF SINGLE-PHASE GRID-CONNECTED PHOTOVOLTAIC MICROINVERTER WITH REACTIVE POWER SUPPORT CAPABILITY 2500

Geon-Hong Min ; Kyung-Hwan Lee ; Jung-Ik Ha ; Myong Hwan Kim

OPTIMAL SIZE AND MULTI-OBJECTIVE CONTROL OF BATTERY ENERGY STORAGES IN DISTRIBUTION SYSTEM WITH HIGH PENETRATION OF DISTRIBUTED PV GENERATORS 2505

Meiqin Mao ; Lei Zhou ; Yangyang Wang ; Liuchen Chang

MISSION PROFILE-ORIENTED CONTROL FOR RELIABILITY AND LIFETIME OF PHOTOVOLTAIC INVERTERS 2512

Ariya Sangwongwanich ; Yongheng Yang ; Dezso Sera ; Frede Blaabjerg

DISCONTINUOUS CURRENT MODE CONTROL FOR MINIMIZATION OF THREE-PHASE GRID-TIED INVERTER IN PHOTOVOLTAIC SYSTEM 2519

Hoai Nam Le ; Jun-Ichi Itoh

A THEORETICAL ANALYSIS ON STATIC CHARACTERISTICS OF VOLTAGE BASED CONTROL METHOD AND CURRENT BASED CONTROL METHOD FOR THE WAYSIDE ENERGY STORAGE SYSTEM IN DC-ELECTRIFIED RAILWAY 2527

Hiroyasu Kobayashi ; Keiichiro Kondo ; Diego Iannuzzi

IMPROVEMENT OF A DC ELECTRICAL RAILWAY SIMULATOR USING ARTIFICIAL INTELLIGENCE 2534

Alvaro J. Lopez-Lopez ; Ramon R. Pecharroman ; Antonio Fernandez-Cardador ; Asuncion P. Cucala

FEEDING-LOSS REDUCTION BY HIGHER-VOLTAGE DC RAILWAY FEEDING SYSTEM WITH DC-TO-DC CONVERTER................2540

Hidenori Shigeeda ; Hiroaki Morimoto ; Kazuhiko Ito ; Toshiyuki Fujii ; Naoki Morishima

MODELING AND SIMULATION OF NOVEL RAILWAY POWER SUPPLY SYSTEM BASED ON POWER CONVERSION TECHNOLOGY 2547

Minwu Chen ; Ruofei Liu ; Shaofeng Xie ; Xiaofang Zhang ; Yimin Zhou

COMPARATIVE STUDY ON FRONT-END PARAMETER IDENTIFICATION METHODS FOR WIRELESS POWER TRANSFER WITHOUT WIRELESS COMMUNICATION SYSTEMS.................2552

Sinan Li ; S. Y. Ron Hui

A NEW TYPE OF WIRELESS V2X SYSTEM WITH A DUAL-ACTIVE BIDIRECTIONAL SINGLE-ENDED CONVERTER AND OPTIMIZED SIC-MOSFET 2558

Hideki Omori ; Aoto Yamamoto ; Naoki Mukaiyama ; Masahito Tsuno ; Kenji Fukuda ; Hisato Michikoshi ; Noriyuki Kimura ; Toshimitsu Morizane

METAL OBJECT DETECTION SYSTEM WITH PARALLEL-MISTUNED RESONANT CIRCUITS AND NULLIFYING INDUCED VOLTAGE FOR WIRELESS EV CHARGERS.................2564

Seog Y. Jeong ; Van X. Thai ; Jun H. Park ; Chun T. Rim

WIRELESS EV CHARGING SYSTEM WITHOUT AIR-GAP AND MISALIGNMENT 2569

Wenxing Zhong ; Dehong Xu

FIXED SLOPE CARRIER PWM FOR INDIRECT MATRIX CONVERTER 2576

Tzung-Lin Lee ; Chun-Yao Hung ; Yen-Wen Chen ; Wen-Mei Huang

CARRIER-BASED OVERMODULATION STRATEGY FOR MATRIX CONVERTERS.................2581

Paiboon Kiatsookkanatorn ; Somboon Sangwongwanich

THREE-PHASE TO HIGH-FREQUENCY SINGLE-PHASE MATRIX CONVERTER : A FREQUENCY CONTROL SUITABLE FOR SOFT SWITCHING.................2589

Wataru Kodaka ; Satoshi Ogasawara ; Koji Orikawa ; Masatsugu Takemoto ; Takashi Hyodo ; Hiroyuki Tokusaki

TWO-STEP COMMUTATION FOR ISOLATED DC-AC CONVERTER WITH MATRIX CONVERTER.................2596

Shunsuke Takuma ; Jun-Ichi Itoh

A DC-LINK CAPACITOR VOLTAGE OSCILLATION REDUCTION METHOD FOR A MODULAR MULTILEVEL CASCADE CONVERTER WITH SINGLE DELTA BRIDGE CELLS (MMCC-SDBC) 2604

Takaaki Tanaka ; Huai Wang ; Frede Blaabjerg

OPTIMIZED DECOUPLING CONTROL OF FLYING CAPACITOR IN ANPC FIVE-LEVEL INVERTER 2611

Fusheng Wang ; Deyou Zheng ; Jianing Wang ; Fei Li ; Fang Liu ; Shuying Yang ; Zhen Xie

CASCADED DUAL-BUCK AC-AC CONVERTER USING COUPLED INDUCTORS 2619

Sanghun Kim ; Duekjin Jang ; Heung-Geun Kim ; Honnyong Cha

INSTANTANEOUS POWER LOSS CALCULATION FOR MMC BASED ON VIRTUAL ARM MATHEMATICAL MODEL 2625

Yin Shiyuan ; Wang Yue ; Yin Taiyuan ; Nie Cheng ; Duan Guozhao ; Wang Zhang

COMPARISON OF CURRENT CONTROL STRATEGIES IN MODULAR MULTILEVEL CONVERTER 2630

Jianzhao Wei ; Anirudh Budnar Acharya ; Lars Norum ; Pavol Bauer

MODEL PREDICTIVE CONTROL OF A MODULAR MULTILEVEL CONVERTER WITH AN IMPROVED CAPACITOR BALANCING METHOD 2638

Shichong Zhang ; Baodong Bai ; Dezhi Chen

HIGH STEP-UP DC-DC CONVERTER BASED ON MULTI-CELL COUPLED INDUCTOR DIODE-CAPACITOR NETWORK 2646

Xinying Li ; Yan Zhang ; Jinjun Liu ; Pengxiang Zeng

NOVEL ACTIVE CLAMPING STEP-DOWN DC-DC CONVERTER WITH LOWER VOLTAGE STRESS 2653

Chi-Hsuan Hsu ; Jun-Min Jian ; Jiann-Fuh Chen ; Hsuan Liao

DESIGN AND EVALUATION OF A MAGNETICALLY-LOOSELY-COUPLED INDUCTOR FOR A FOUR-PHASE INTERLEAVED BOOST CHOPPER 2660

Hiroki Kowatari ; Toshinori Kitamura ; Nobukazu Hoshi

A SYNCHRONOUS-REFERENCE-FRAME I-V DROOP CONTROL METHOD FOR PARALLEL-CONNECTED INVERTERS 2668

Mingshen Li ; Yonghao Gui ; Zheming Jin ; Yajuan Guan ; Josep M. Guerrero

TRANSIENT STABILITY IMPACT OF THE PHASE-LOCKED LOOP ON GRID-CONNECTED VOLTAGE SOURCE CONVERTERS 2673

Heng Wu ; Xiongfei Wang

COMPREHENSIVE ANALYSIS OF VIRTUAL IMPEDANCE-BASED ACTIVE DAMPING FOR LCL RESONANCE IN GRID-CONNECTED INVERTERS 2681

Teng Liu ; Zeng Liu ; Jinjun Liu ; Yiming Tu ; Zipeng Liu

A COMPARATIVE STUDY OF THE TRADITIONAL FS-MPC AND THE PROPOSED CSF-PCC FOR THE THREE-PHASE GRID-CONNECTED INVERTERS 2688

Zhixun Ma ; Xin Zhang ; Jingjing Huang

CONSTANT SWITCHING-FREQUENCY PREDICTIVE- CURRENT-CONTROL METHOD WITH A DICHOTOMY SOLUTION FOR THE GRID-TIED INVERTERS 2692

Zhixun Ma ; Xin Zhang ; Jingjing Huang ; Zhao Bin ; Lyu Jing

OBSERVER-BASED ACTIVE DAMPING FOR GRID-CONNECTED CONVERTERS WITH LCL FILTER 2697

Y. Zhang ; M. G. L. Roes ; M. A. M. Hendrix ; J. L. Duarte

CONDUCTION LOSS ANALYSIS AND OPTIMIZATION DESIGN OF FULL BRIDGE LLC RESONANT CONVERTER 2703

Yugang Yang ; Lifei Zhang ; Tianshu Ma

FULL-BRIDGE T-TYPE ISOLATED DC/DC CONVERTER WITH WIDE INPUT VOLTAGE RANGE 2708

Dong Liu ; Yanbo Wang ; Fujin Deng ; Zhe Chen

RESEARCH ON HIGH EFFICIENCY LLC DC-DC CONVERTER BASED ON SIC MOSFET 2714

Pengcheng Han ; Xiaoqiong He ; Haijun Ren ; Zhiqing Zhao ; Xu Peng

AN IMPROVED DUAL PHASE SHIFT CONTROL STRATEGY FOR DUAL ACTIVE BRIDGE DC-DC CONVERTER WITH SOFT SWITCHING 2718

Miao Hong ; Gao Xuanjie ; Zeng Chengbi ; Duan Shujiang

DEVELOPMENT OF AN SIC HIGH-FREQUENCY PWM INVERTER USING A THICK MULTILAYER PCB TO MINIMIZE STRAY INDUCTANCE 2725

Kohsuke Ishikawa ; Satoshi Ogasawara ; Masatsugu Takemoto ; Koji Orikawa

FAST SWITCHING PLANAR POWER MODULE WITH SIC MOSFETS AND ULTRA-LOW PARASITIC INDUCTANCE 2732

Arash Edvin Risseh ; Hans-Peter Nee ; Konstantin Kostov

EXPERIMENTAL EVALUATION OF INVERTER SYSTEM CONSISTING OF 4-PARALLEL GAN DEVICES UNIT 2738

Yoshiya Ohnuma ; Satoshi Miyawaki ; Fumiya Hattori ; Masayoshi Yamamoto

IMPACT OF THE THERMAL-INTERFACE-MATERIAL THICKNESS ON IGBT MODULE RELIABILITY IN THE MODULAR MULTILEVEL CONVERTER 2743

Yi Zhang ; Huai Wang ; Zhongxu Wang ; Yongheng Yang ; Frede Blaabjerg

NANOSCALE INVESTIGATION OF THE POWER MOSFET BY THE AFM/KFM/SCFM 2750

Mizuki Nakajima ; Yuuki Uchida ; Nobuo Satoh ; Hidekazu Yamamoto

SIMULATION ANALYSIS OF OPTIMUM GATE DRIVING CONDITIONS OF IGBTS 2756

Satoshi Sugahara ; Masaki Kawakami ; Kousuke Kamakura

IMPROVEMENT OF THE I2T CAPABILITY FOR XEV ACTIVE SHORT CIRCUIT PROTECTION BY COMBINATION OF RC-IGBT AND LEADFRAME TECHNOLOGIES 2764

Keiichi Higuchi ; Hayato Nakano ; Akihiro Osawa ; Akio Kitamura ; Shunji Takenoiri ; Daisuke Inoue ; Souichi Yoshida ; Hiromichi Gohara

INVESTIGATION OF SWITCHING BEHAVIOR OF AN IGBT UNDER SOFT TURN-OFF IN APPLICATION FOR DUAL-ACTIVE BRIDGE CONVERTERS 2768

Eri Ogawa ; Yuichi Onozawa ; Rik W. De Doncker

600 V HIGH VOLTAGE GATE DRIVER IC (HVIC) WITH 1.0 MHZ HIGH FREQUENCY OPERATION FOR LLC CURRENT RESONANT POWER SUPPLY 2774

Masaharu Yamaji ; Masashi Akahane ; Takahide Tanaka ; Akihiro Jonishi ; Hidetomo Ohashi ; Masahiro Sasaki ; Hitoshi Sumida

AN INTEGRATED VOLTAGE AND CURRENT BALANCING STRATEGY OF SERIES-PARALLEL CONNECTED IGBTS 2780

Xiaotong Du ; Fang Zhuo ; Haotian Sun ; Hao Yi ; Yanlin Zhu

THERMAL DESIGN AND ANALYSIS OF A CABLE CHARGER USED FOR PORTABLE ELECTRONICS 2785

Mofan Tian ; Xu Yang ; Naizeng Wang ; Yang Chen ; Laili Wang

PARASITIC INDUCTANCE DESIGN CONSIDERATIONS TO SUPPRESS GATE VOLTAGE OSCILLATION OF FAST SWITCHING POWER SEMICONDUCTOR DEVICES 2789

Yusuke Sugihara ; Kimihiro Nanamori ; Masayoshi Yamamoto ; Yasuki Kanazawa

THE EXAMINATION OF INCREASING OPERATION SPEED OF CONSEQUENT POLE TYPE AXIAL GAP MOTOR FOR HIGHER OUTPUT POWER DENSITY 2796

Toru Ogawa ; Tomohira Takahashi ; Masatsugu Takemoto ; Satoshi Ogasawara ; Hideaki Arita ; Akihiro Daikoku

BASIC STUDY OF PMASYNRM WITH BONDED MAGNETS FOR TRACTION APPLICATIONS 2802

Marika Kobayashi ; Shigeo Morimoto ; Masayuki Sanada ; Yukinori Inoue

STUDY ON ROTOR STRUCTURE SUITABLE FOR IMPROVING POWER DENSITY AND EFFICIENCY IN IPMSMS FOR AUTOMOTIVE APPLICATIONS 2808

R. Imoto ; M. Sanada ; S. Morimoto ; Y. Inoue

EXAMINATION OF THE DEMAGNETIZATION SUPPRESSION EFFECT OF PLACING FLUX BARRIERS IN AN IPMSM USING RARE-EARTH BONDED MAGNETS 2814

Takashi Umeda ; Masayuki Sanada ; Shigeo Morimoto ; Yukinori Inoue

A NOVEL POLE-CHANGING METHOD WITH A MULTIPLE THREE-PHASE INVERTER 2820

Yuki Hidaka ; Taiga Komatsu ; Hideaki Arita

STARTING CHARACTERISTICS OF AN ULTRA-LIGHTWEIGHT MOTOR USING MAGNETIC RESONANCE COUPLING 2826

Kenta Takishima ; Kazuto Sakai

DESIGN AND BASIC CHARACTERISTICS ANALYSIS OF TOROIDAL WINDING AXIAL GAP INDUCTION MOTOR 2832

Ryosuke Sakai ; Yukihiro Yoshida ; Katsubumi Tajima

MAGNET ARRANGEMENT SUITABLE FOR LARGE AIR GAP LENGTH IN LINEAR PM VERNIER MOTOR 2836

Tatsuya Ninomiya ; Abdulaziz Gasim ; Shoji Shimomura

MICRO ELECTROMAGNETIC VIBRATION ENERGY HARVESTER WITH MECHANICAL SPRING AND IRON FRAME FOR LOW FREQUENCY OPERATION 2842

Yecheng Shen ; Kaiyuan Lu ; Yongming Xia

MEASUREMENT OF TWO-LEVEL INVERTER INDUCED CURRENT SLOPES AT HIGH SWITCHING FREQUENCIES FOR CONTROL AND IDENTIFICATION ALGORITHMS OF ELECTRICAL MACHINES 2848

Simon Decker ; Andreas Liske ; Daniel Schweiker ; Johannes Kolb ; Michael Braun

A NEW TOPOLOGY OF SWITCHED-CAPACITOR MULTILEVEL INVERTER FOR SINGLE-PHASE GRID-CONNECTED WITH ELIMINATING LEAKAGE CURRENT 2854

Mehdi Samizadeh ; Xu Yang ; Bagher Karami ; Wenjie Chen ; Mohamad Abou Houran ; Adib Abrishamifar ; Abdolreza Rahmati

AN INTERLEAVED BUCK-CASCADED BUCK-BOOST INVERTER FOR PV GRID-CONNECTION APPLICATIONS 2860

Chien-Hsuan Chang ; Chun-An Cheng ; Hung-Liang Cheng

A NOVEL PV ARRAY CONNECTION STRATEGY WITH PV-BUCK MODULE TO IMPROVE SYSTEM EFFICIENCY 2866

Chi Shao ; Wenjie Wang ; Lijun Hang ; Anping Tong ; Shitao Wang

A COMMON-MODE VOLTAGE REDUCTION FOR TWO-STAGE THREE-PHASE TRANSFORMERLESS PV INVERTERS 2871

Adisak Promyoo ; Surapong Suwankawin

A GRID-CONNECTED PV-ENERGY STORAGE SYSTEM WITH SYNCHRONOUS GENERATOR CHARACTERISTICS 2877

Huadian Xu ; Jianhui Su ; Ning Liu ; Yong Shi ; Yan Du

A TRANSFORMERLESS BIDIRECTIONAL DC-DC CONVERTER BASED ON POWER UNITS WITH UNIPOLAR AND BIPOLAR STRUCTURE FOR MVDC INTERCONNECTION 2882

Lejia Sun ; Fang Zhuo ; Feng Wang ; Hao Yi ; Baohui Ma

NEW MODULATION CONTROL OF CONVERTER SYSTEM APPLIED FOR OFFSHORE WIND FARMS 2887

Naoki Kawabata ; Noriyuki Kimura ; Toshimitsu Morizane ; Hideki Omori

SPHERE DECODING BASED LONG-HORIZON PREDICTIVE CONTROL OF THREE-LEVEL NPC BACK-TO-BACK PMSG WIND TURBINE SYSTEMS 2895

Ferdinand Grimm ; Zhenbin Zhang ; Ralph Kennel

BASED ON PCHD AND HPSO SLIDING MODE CONTROL OF D-PMSG WIND POWER SYSTEM 2901

Lijun Hou ; Xuemei Zheng ; Chao Wang ; Yangman Li ; Haoyu Li

ESTABLISHMENT AND DYNAMIC CONTROL OF WIND INDUCTION GENERATOR 2907

M. Z. Lu ; V. K. Ganisetti ; C. M. Liaw

MIDDLE FREQUENCY SOLID STATE TRANSFORMER FOR HVDC TRANSMISSION FROM OFFSHORE WINDFARM 2914

Noriyuki Kimura ; Toshimitsu Morizane ; Isao Iyoda ; Kazushige Nakao ; Tomoki Yokoyama

SIMULATION OF WIND POWER GENERATION SYSTEM USING SWITCHED RELUCTANCE GENERATOR AND CAPACITOR-LESS AC-AC CONVERTER 2921

Guyuan Ji ; Kazuhiro Ohyama

VARIABLE FREQUENCY CONTROL AND FILTER DESIGN FOR OPTIMUM ENERGY EXTRACTION FROM A SIC WIND INVERTER 2932

Abdallah Hussein ; Alberto Castellazzi

EXPERIMENTAL VERIFICATIONS OF UPFC USING DEADBEAT CONTROL WITH 3-PHASE UNBALANCED COMPENSATION 2938

Shin-Ichi Hamasaki ; Hiroto Fukuda ; Syohei Tokumaru ; Mineo Tsuji

A CONTROL METHOD FOR TWO TYPES OF THREE-PHASE TRANSFORMERLESS UNIFIED POWER QUALITY CONDITIONER 2944

Fujian Li ; Guochun Xiao ; Fangzhou Zhao ; Shuai Zhang ; Baojin Liu

DESIGN OF CUSTOMER-END CONVERTER SYSTEMS FOR LOW VOLTAGE DC DISTRIBUTION FROM A LIFE CYCLE COST PERSPECTIVE 2948

A. Mattsson ; P. Nuutinen ; T. Kaipia ; P. Peltoniemi ; J. Karppanen ; V. Tikka ; A. Lana ; P. Pinomaa ; P. Silventoinen ; J. Partanen

A CONTROL METHOD OF DC CAPACITOR VOLTAGE IN MMC FOR HVDC SYSTEM USING NEGATIVE SEQUENCE CURRENT 2956

Hanis Afiqah Binti Jaffar ; Ahmad Arif Bin Abd Rahman ; Hiroaki Kakigano

A COORDINATE AND DISTRIBUTED CONTROL SCHEME FOR MULTILEVEL AND MULTI-STAGE MEDIUM VOLTAGE SOLID STATE TRANSFORMER 2963

Jintong Nie ; Liqiang Yuan ; Qing Gu ; Jianning Sun ; Zhengming Zhao

AN IMPROVED HARMONIC POWER SHARING SCHEME OF PARALLELED INVERTER SYSTEM 2969

Liu Hongpeng ; Liu Xiaoxi ; Zhang Wei ; Wang Wei

THE GRID IMPEDANCE ADAPTATION DUAL MODE CONTROL STRATEGY IN WEAK GRID 2973

Ming Li ; Xing Zhang ; Ying Yang ; Pengpeng Cao

TRANSMISSION POWER ANALYSIS AND CONTROL OF THE DC TRANSFORMER IN HYBRID AC/DC MICROGRID 2980
Jingjin Huang ; Xin Zhang ; Tengfei Zhang

A NOVEL FLEXIBLE INTERCONNECTION SCHEME FOR MICROGRID TO OPTIMIZE THE CAPACITY OF ENERGY STORAGE SYSTEM (ESS) 2986
Zhou Jianqiao ; Zhang Jianwen ; Cai Xu ; Li Zhuyong ; Wang Jiacheng ; Zang Jiajie

VSC CONTROL AND PARAMETERS DESIGN BASED ON VIRTUAL SYNCHRONOUS GENERATOR 2992
Fang Liu ; Meng Wang ; Zhen Xie ; Fusheng Wang ; Jinxin Deng ; Xing Zhang

MULTI-TARGET VIRTUAL RESISTANCE CONTROL STRATEGY IN A 400 HZ LOW VOLTAGE MICROGRID 2997
Yuze Li ; Xuejun Pei ; Zhi Chen ; Hanyu Wang ; Yong Kang

AN ADAPTIVE POWER COMPENSATION STRATEGY FOR THE VOLTAGE STABILIZATION OF LCL-VSC BASED MICROGRIDS 3002
Sheng Xu ; Wu Cao ; Dongchen Fan ; Jianfeng Zhao ; Shunyu Wang

RESONANCE DETECTION STRATEGY FOR MULTIPLE GRID-CONNECTED INVERTERS-BASED SYSTEM USING CASCADED SECOND-ORDER GENERALIZED INTEGRATOR 3010
Wu Cao ; Dongchen Fan ; Kangli Liu ; Jianfeng Zhao ; Liheng Ruan ; Xiaojun Wu

HARMONIC STABILITY ASSESSMENT BASED ON GLOBAL ADMITTANCE FOR MULTI-PARALLELED GRID-CONNECTED VSIS USING MODIFIED NYQUIST CRITERION 3015
Wu Cao ; Dongchen Fan ; Kangli Liu ; Jianfeng Zhao ; Liheng Ruan ; Xiaojun Wu

THE AC TRACTION POWER SUPPLY SYSTEM FOR URBAN RAIL TRANSIT BASED ON NEGATIVE SEQUENCE CURRENT COMPENSATOR 3020
Tianshu Zhao ; Xu Peng

GRID CONNECTED POWER GENERATION CONTROL METHOD FOR Z-SOURCE INTEGRATED BIDIRECTIONAL CHARGING SYSTEM 3025
Xu Jia ; Guoming Chuai ; Haonan Niu ; Qianfan Zhang

AN ISOLATED PFC CONVERTER WITH HARMONIC MODULATION TECHNIQUE FOR EV CHARGERS 3030
Byung-Kwon Lee ; Jun-Young Lee ; Dong-Hun Kang

HIGHLY DYNAMIC SWITCHING FREQUENCY-BASED CALCULATION OF POWER QUANTITIES, FUNDAMENTAL WAVEFORMS, AND RMS VALUES OF INVERTER-FED ELECTRICAL MACHINES 3034
Alexander Stock ; Johannes Teigelkötter ; Johannes Büdel

DESIGN AND ANALYSIS OF HIGH VOLTAGE POWER SUPPLY FOR INDUSTRIAL ELECTROSTATIC PRECIPITATORS 3040
Shengwen Fan ; Yiqin Yuan ; Pengyu Jia ; Zhigang Chen ; Haisi Li

LOAD SHARING OPERATION IN N+1 UPS SYSTEM BY USING HARMONIC SHARING CONTROL METHOD 3046
Prashant Patel ; Sagar Naina ; Utsav Patel ; Premal Patwa

RESEARCH ON CAPACITY OPTIMIZATION OF PV-WIND-DIESEL-BATTERY HYBRID GENERATION SYSTEM 3052
Cailing Zhu ; Furong Liu ; Sheng Hu ; Shu Liu

A NUMERICAL ANALYSIS AND IMPROVEMENT OF OUTPUT CHARACTERISTICS IN DIFFERENT PASSIVE RECTIFIERS BASED ON VIBRATION GENERATORS 3058
Tomoki Sakabe ; Masataka Minami ; Shin-Ichi Motegi ; Masakazu Michihira

CIRCUIT MODELING APPROACH FOR ANALYZING TRIBOELECTRIC NANOGENERATORS FOR ENERGY HARVESTING 3063
Bo-Kyung Yoon ; Jeong Min Baik ; Katherine A. Kim

GENERAL POWER ELECTRIC CONVERTER MODEL 3069
Jingwen Xie

A MODULAR CONVERTER- AND SIGNAL-PROCESSING-PLATFORM FOR ACADEMIC RESEARCH IN THE FIELD OF POWER ELECTRONICS 3074
Rüdiger Schwendemann ; Simon Decker ; Marc Hiller ; Michael Braun

CONTROL IC FOR BOOST-FLYBACK CONVERTER FOR ENERGY HARVESTING APPLICATIONS 3081
Jhih-Sian Li ; Kai-Hui Chen ; Jui-Hung Lai ; Jun-Xian Huang

NEW CONCEPT OF THE DC-DC CONVERTER CIRCUIT APPLIED FOR THE SMALL CAPACITY UNINTERRUPTIBLE POWER SUPPLY 3086
Dang Minh Huynh ; Yoichi Ito ; Shinji Aso ; Koji Kato ; Kenji Teraoka

COMPARATIVE STUDY ON THE PERFORMANCE OF DUAL-PHASE TAPPED-INDUCTOR BOOST CONVERTER AND INTERLEAVED BOOST PARALLEL-INPUT SERIES-OUTPUT CONVERTER IN 40 TO 400V APPLICATIONS 3092

Niño Christopher Ramos ; Tsuyoshi Funaki

A NEW STANDBY STRUCTURE INTEGRATED WITH BOOST PFC CONVERTER FOR SERVER POWER SUPPLY 3100

Jae-Il Baek ; Jae-Kuk Kim ; Jae-Bum Lee ; Moo-Hyun Park ; Gun-Woo Moon

NONISOLATED TWO-CHANNEL LED DRIVER WITH SIMPLE SNUBBER 3107

Jong-Woo Kim ; Jung-Kyu Han ; Jih-Sheng Lai

DESIGN AND IMPLEMENTATION OF SINGLE-PHASE ASYMMETRIC MULTILEVEL STATCOM 3112

Hao Chen ; Yang Han ; Ping Yang ; Congling Wang ; Josep M. Guerrero

SUBMODULE VOLTAGE BALANCING AND LOSS EQUALISATION IN ALTERNATE ARM CONVERTERS BASED ON VIRTUAL VOLTAGES 3117

Georgios Konstantinou ; Harith R. Wickramasinghe ; Salvador Ceballos ; Josep Pou

BALANCED CONDUCTION LOSS DISTRIBUTION AMONG SMS IN MODULAR MULTILEVEL CONVERTERS 3123

Zhongxu Wang ; Huai Wang ; Yi Zhang ; Frede Blaabjerg

SIMPLIFICATION OF MODEL PREDICTIVE CONTROL FOR MODULAR MULTILEVEL CONVERTER THROUGH DIRECT VOLTAGE LEVEL SELECTION 3129

Xingxing Chen ; Jinjun Liu ; Shaodi Ouyang ; Shuguang Song ; Rui Luo

FAMILY OF INTEGRATED MULTI-INPUT MULTI-OUTPUT DC-DC POWER CONVERTERS 3134

Bang Le-Huy Nguyen ; Honnyong Cha ; Tien-The Nguyen ; Heung-Geun Kim

LOW-COMPLEXITY STATE-SPACE BASED SYSTEM IDENTIFICATION AND CONTROLLER AUTO-TUNING METHOD FOR MULTI-PHASE DC-DC CONVERTERS 3140

Marc Kanzian ; Harald Gietler ; Christoph Unterrieder ; Matteo Agostinelli ; Michael Lunglmayr ; Mario Huemer

A PHASE-SHIFT DOUBLE FULL-BRIDGE (PSDB) CONVERTER WITH THREE SHARED LEADING-LEGS 3145

Junjie Zhu ; Qinsong Qian ; Shengli Lu ; Weifeng Sun ; Le Zhang

DUAL ACTIVE BRIDGE SYNCHRONOUS RECTIFIED STEP-DOWN CONVERTER 3151

Chien-Chun Huang ; Chang-Lin Tsai ; Tsung-Lin Tsai ; Yao-Ching Hsieh ; Huang-Jen Chiu ; Jing-Yuan Lin

ACCURATE IMPEDANCE MODEL OF GRID-CONNECTED INVERTER FOR SMALL-SIGNAL STABILITY ASSESSMENT IN HIGH-IMPEDANCE GRIDS 3156

Tuomas Messo ; Roni Luhtala ; Aapo Aapro ; Tomi Roinila

MODELING OF UNBALANCED THREE-PHASE GRID-CONNECTED CONVERTERS WITH DECOUPLED TRANSFER FUNCTIONS 3164

Wei Liu ; Xiongfei Wang ; Frede Blaabjerg

PREDICTING VOLTAGE CHARACTERISTIC OF CHARGING MODEL FOR LI-ION BATTERY WITH ANN FOR REAL TIME DIAGNOSIS 3170

Minella Bezha ; Naoto Nagaoka

IMPEDANCE MODELING AND STABILITY ANALYSIS OF THE CASCADED THREE-PHASE SYMMETRIC SYSTEMS USING COMPLEX TRANSFER FUNCTIONS 3176

Teng Liu ; Zeng Liu ; Jinjun Liu ; Yiming Tu ; Zipeng Liu

ACOUSTIC NOISE REDUCTION OF 12/8 POLES SRM WITHOUT EFFICIENCY DROP USING SIMPLE CURRENT WAVEFORMS 3182

Kyohei Kiyota ; Kenji Amei ; Takahisa Ohji ; Jun Jisaki ; Masanobu Nakai

STUDY OF SWITCHED RELUCTANCE MOTOR DIRECTLY DRIVEN BY COMMERCIAL THREE-PHASE POWER SUPPLY 3186

Masaki Takahashi ; Kohei Aiso ; Kan Akatsu

DOUBLE STATOR AXIAL-FLUX SWITCHED RELUCTANCE MOTOR FOR ELECTRIC CITY COMMUTERS 3192

Hiroki Goto

TORQUE RIPPLE REDUCTION USING ASYMMETRIC FLUX BARRIERS IN SYNCHRONOUS RELUCTANCE MOTOR 3197

Yuuto Yamamoto ; Shigeo Morimoto ; Masayuki Sanada ; Yukinori Inoue

ON-BOARD SINGLE-PHASE ELECTRIC VEHICLE CHARGER WITH ACTIVE FRONT END 3203

Theodore Soong ; Peter W. Lehn

A BIDIRECTIONAL BUFFERED CHARGING UNIT FOR EV'S (BBCU) 3209

Gabriel Fernandez

RECONFIGURABLE CONVERTER WITH MULTIPLE-VOLTAGE MULTIPLE-POWER FOR E-MOBILITY CHARGING 3215

Mohamed S A Dahidah ; He Liu ; Vassilios G. Agelidis

DEVELOPMENT OF A SERIES HYBRID ELECTRIC VEHICLE LABORATORY TEST BENCH WITH HARDWARE-IN-THE-LOOP CAPABILITIES................3223
Poria Fajri ; Nima Lotfi ; Mehdi Ferdowsi

NEW THREE-PHASE STATIC TRANSFER SWITCH USING AC SSCB................3229
Seung-Min Song ; Jin-Young Kim ; In-Dong Kim

HARMONICS COMPENSATION IN HIGH FREQUENCY RANGE OF ACTIVE POWER FILTER WITH SIC-MOSFET INVERTER IN DIGITAL CONTROL SYSTEM................3237
Shin-Ichi Hamasaki ; Kengo Nakahara ; Mineo Tuji

CONTROL OF BUCK-BOOST DIRECT MATRIX CONVERTER WITH LOW VOLTAGE RIDE-THROUGH CAPABILITY................3243
Nico Remus ; Martin Leubner ; Wilfried Hofmann

AN IMPROVED PLL BASED SEAMLESS TRANSFER CONTROL STRATEGY................3251
Xin Meng ; Jinjun Liu ; Zeng Liu ; Ronghui An

EFFICIENT URBAN RAILWAY DESIGN INTEGRATING TRAIN SCHEDULING, ONBOARD ENERGY STORAGE, AND TRACTION POWER MANAGEMENT................3257
Warayut Kampeerawar ; Takafumi Koseki ; Fulin Zhou

OPTIMAL CONTROL METHOD OF AN ENERGY STORAGE SYSTEM FOR ENERGY SAVING................3265
Yoko Takeuchi ; Tomoyuki Ogawa ; Keisuke Sato ; Hiroaki Morimoto ; Tatsuhito Saito

START-UP AND TRANSIENT OPERATION OF A BIDIRECTIONAL CHOPPER WITH AN AUXILIARY CONVERTER................3273
Hamzeh J. Ahmad ; Haruna Ohnishi ; Makoto Hagiwara

EXPERIMENTAL RESULTS OF QUASI-OPTIMAL CHARGING CURRENT PATTERNS TO REDUCE THE INTERNAL HEAT GENERATION OF THE LITHIUM-ION BATTERY................3280
Yoshiaki Taguchi ; Gaku Yoshikawa

DEVELOPMENT OF TEST METHODS AND EVALUATION RESULTS FOR 500KV HVDC CONVERTER................3286
Keisuke Hattori ; Asuka Ohtake ; Takayoshi Kamejima ; Haruhisa Wada

DISSIPATION LOOP FOR SHOOT-THROUGH FAULTS IN HVDC CONVERTER CELLS................3292
Keijo Jacobs ; Staffan Norrga ; Hans-Peter Nee

A SUPPRESSION METHOD OF HARMONIC INSTABILITY IN LINE-COMMUTATED CONVERTERS APPLYING ACTIVE HARMONIC FILTERS................3299
Kenichiro Sano ; Toshiaki Kikuma ; Tatsuhito Nakajima ; Junya Kanno

EXPERIMENT OF SEMICONDUCTOR BREAKER USING SERIES-CONNECTED IEGTS FOR HYBRID DCCB................3304
Kazuyasu Takimoto ; Hiroshi Takenaka ; Toshiaki Matsumoto ; Takahiro Ishiguro

STUDY OF EMI CAUSED BY BUCK CONVERTER ON CONTROLLER AREA NETWORK................3309
Ryo Shirai ; Toshihisa Shimizu

A STUDY ON REDUCTION TECHNIQUES OF A WIDEBAND COMMON-MODE VOLTAGE PRODUCED BY A PWM INVERTER................3315
Shotaro Takahashi ; Satoshi Ogasawara ; Masatsugu Takemoto ; Koji Orikawa ; Michio Tamate

A MODIFIED DISCONTINUOUS PWM FOR COMMON-MODE VOLTAGE ELIMINATION IN 3-LEVEL 4-LEG PWM CONVERTER SYSTEM................3323
Seon-Ik Hwang ; Jun-Hyung Jung ; In-Ho Cho ; Jang-Mok Kim ; Yung-Deug Son

EMI ANALYSIS OF FULL-SIC INTEGRATED POWER MODULE................3329
Xiliang Chen ; Wenjie Chen ; Yu Ren ; Liang Qiao ; Yilin Sha ; Xu Yang

EXPERIMENTAL VERIFICATION OF COUPLING EFFECT AND POWER TRANSFER CAPABILITY OF DYNAMIC WIRELESS POWER TRANSFER................3332
Chan Anyapo ; Nithiphat Teerakawanich ; Chowarit Mitsantisuk ; Kiyoshi Ohishi

NEIGHBORING EFFECTS ON THE DEACTIVATED INVERTER IN A SEGMENTED DYNAMIC WIRELESS EV CHARGING SYSTEM................3338
Qingwei Zhu ; Yanjie Guo ; Lifang Wang ; Shufan Li ; Chenglin Liao

MULTIPLE EXCITING VOLTAGE CONTROL FOR MAXIMIZATION OF MULTI-HOP WIRELESS POWER TRANSFER EFFICIENCY................3344
Masato Sasaki ; Masayoshi Yamamoto

GENERAL ANALYTICAL MODEL FOR INDUCTIVE POWER TRANSFER SYSTEM WITH EMF CANCELING COILS................3349
Keita Furukawa ; Keisuke Kusaka ; Jun-Ichi Itoh

STABILITY INFLUENCE OF FILTER COMPONENTS PARASITIC RESISTANCE ON LCL-FILTERED GRID CONVERTERS................3357
Hiroaki Matsumori ; Toshihisa Shimizu ; Frede Blaabjerg ; Xiongfei Wang ; Dongsheng Yang

REAL-TIME ESTIMATION CONTROL OF INDUCTANCE PARAMETERS USING DUST CORE MATERIALS FOR PWM INVERTER 3363
Kazu Imai ; Takuma Yoshino ; Ohasi Shunsuke ; Tomoki Yokoyama

CONTROL DESIGN OF OUTPUT-STAGE FILTERLESS SINUSOIDAL-WAVE INVERTER 3369
Shinichi Hiroshige ; Kenji Yamanaka ; Masahide Hojo

SERIES REACTIVE POWER COMPENSATOR WITH REDUCED CAPACITANCE FOR HYBRID TRANSFORMER 3375
Yuki Takahashi ; Takanori Isobe ; Hiroshi Tadano

AN INSIGHT INTO THE VOLTAGE RISING BEHAVIOR DURING TURN-OFF PROCESS OF SERIES CONNECTED SIC MOSFETS ON CIRCUIT LEVEL 3383
Panrui Wang ; Feng Gao ; Yang Jing ; Yufeng Chen ; Lei Zhang

PARALLELING SIX 320A 1200V ALL-SIC HALF-BRIDGE MODULES FOR A LARGE CAPACITY POWER STACK 3390
David Hongfei Lu ; Hiromu Takubo ; Sho Takano ; Yuhei Suzuki

3.3KV ALL-SIC MODULE FOR ELECTRIC DISTRIBUTION EQUIPMENT 3396
Ryohei Takayanagi ; Katsumi Taniguchi ; Satoshi Kaneko ; Naoyuki Kanai ; Keishirou Kumada ; Motohito Hori ; Yoshinari Ikeda ; Kouji Maruyama ; Itsuo Kawamura

PRESENT STATUS OF SIC BASED POWER CONVERTERS AND GATE DRIVERS – A REVIEW 3401
Abhijit Choudhury

METHOD OF APPLYING FORCE DISTRIBUTION FUNCTION FOR LINEAR SWITCHED RELUCTANCE MOTOR DRIVEN BY CURRENT SOURCE INVERTER 3406
Tadashi Hirayama ; Shuma Kawabata

A NOVEL DRIVE CIRCUIT FOR SWITCHED RELUCTANCE MOTORS WITH BIPOLAR CURRENT DRIVE 3412
Hiroki Ishikawa ; Yuma Uesugi ; Seiya Sakurai

TORQUE RIPPLE MINIMIZATION CONTROL OF SRM BASED ON NOVEL MOTOR MODEL CONSIDERING MUTUAL COUPLING EFFECT 3418
Sungyong Shin ; Naruse Hikaru ; Takashi Kosaka ; Nobuyuki Matsui

COMPARISON OF HIGH FREQUENCY VOLTAGE INJECTION METHODS FOR SHAFT SENSORLESS CONTROL OF WOUND-FIELD FLUX SWITCHING MACHINE 3426
Hong-Quan Nguyen ; Sheng-Ming Yang

DESIGN AND EXPERIMENTAL VERIFICATION OF A DAB MEDIUM FREQUENCY TRANSFORMER FOR A 6.6KV/200V SOLID STATE TRANSFORMER 3431
Rene Barrera-Cardenas ; Takanori Isobe ; Terazono Katsushi ; Tadano Hiroshi

RESEARCH ON THE UNBALANCED COMPENSATION RANGE OF DELTA-CONNECTED CASCADED H-BRIDGE MULTILEVEL SVG 3439
Rui Luo ; Yingjie He ; Yiming Tu ; Xingxing Chen ; Jinjun Liu

STATIC SYNCHRONOUS COMPENSATOR TO STABILIZE GRID VOLTAGE FOR WIND AND PHOTOVOLTAIC POWER PLANT 3450
Ryota Okuyama ; Naoki Morishima ; Yusuke Ashizaki ; Yohei Itaya

LARGE EQUALIZATION CURRENT CONTROL STRATEGY FOR SERIES CONNECTED BATTERY PACKS BASED ON BUCK-BOOST CONVERTER 3455
Xinbo Liu ; Zhuo Gao ; Xuehao Huang ; Yaohan Zou

A MULTI-PORT BIDIRECTIONAL POWER CONVERSION SYSTEM FOR REVERSIBLE SOLID OXIDE FUEL CELL APPLICATIONS 3460
Xiang Lin ; Kai Sun ; Jin Lin ; Zhe Zhang ; Wei Kong

SELF-PREHEATING METHOD FOR LI-ION BATTERY USING BATTERY IMPEDANCE ESTIMATOR 3466
Dong-Kwan Kim ; Young-Dal Lee ; Sang-Hyun Ha ; Yu-Jin Jang ; Gun-Woo Moon

ACTIVE ANTI-ISLANDING TECHNIQUE WITH REDUCED NON-DETECTION ZONE FOR CENTRALIZED INVERTERS 3471
Prashant Jain ; Vivek Agarwal ; Bishnu Prasad Muni ; Eswar Rao ; Deepak Gehlot ; S. Gautam Kumar

DEVELOPMENT OF SIC APPLIED TRACTION SYSTEM FOR SHINKANSEN HIGH-SPEED TRAIN 3478
Kenji Sato ; Hirokazu Kato ; Takafumi Fukushima

DEVELOPMENT OF A HIGH POWER DENSITY AUXILIARY CONVERTER BASED ON 1700V 225A SIC MOSFET FOR TRAMS 3484
Liu Hao ; Fei Lin ; Zhongping Yang ; Hu Cao ; Meng Xia

EXPERIMENTAL TESTS RESULTS OF DAMPING CONTROL WITH OVER VOLTAGE RESISTOR FOR REGENERATIVE BRAKE CONTROL OF RAILWAY VEHICLE 3490
Natsuki Kawagoe ; Febry Pandu Wijaya ; Hiroyasu Kobayashi ; Keiichiro Kondo ; Tetsuya Iwasaki ; Akihiko Tsumura ; Takumi Nagashima ; Yoshinori Yamashita ; Ryota Gondo

COILS LAYOUT OPTIMIZATION OF DYNAMIC WIRELESS POWER TRANSFER SYSTEM TO REALIZE OUTPUT VOLTAGE STABLE..3495
Yi Wang ; Fei Lin ; Zhongping Yang ; Panpan Cai ; Zhiyuan Liu

QUICK CHARGER FOR A BATTERY USING MODULAR MATRIX CONVERTER (MMXC).....................3501
Kazuma Suzuki ; Takaharu Takeshita

VARIABLE OUTPUT VOLTAGE CONTROL OF AN ISOLATED BI-DIRECTIONAL AC/DC CONVERTER WITH A SOFT-SWITCHING TECHNIQUE..3507
Takumi Hamaguchi ; Kazuma Suzuki ; Wataru Kitagawa ; Takaharu Takeshita

A NEW MODULATION METHOD APPLYING OPTIMAL DUTY CYCLE AND PHASE SHIFT FOR BIDIRECTIONAL ISOLATED THREE-PHASE AC/DC CONVERTER BASED ON MATRIX CONVERTER..3514
Koji Shigeuchi ; Jin Xu ; Noboru Shimosato ; Yukihiko Sato

DECOUPLING CONTROL METHOD FOR ELIMINATING DC BIAS FLUX OF HIGH FREQUENCY TRANSFORMER IN A BIDIRECTIONAL ISOLATED AC/DC CONVERTER..........................3522
Kensuke Sakuma ; Koji Shigeuchi ; Jin Xu ; Noboru Shimosato ; Yukihiko Sato

INTERLEAVED VOLTAGE-DOUBLER BOOST CONVERTER FOR POWER FACTOR CORRECTION..3528
Bo-Jia Huang

ZVS INTERLEAVED TOTEM-POLE BRIDGELESS PFC CONVERTER WITH PHASE-SHIFTING CONTROL..3533
Moo-Hyun Park ; Jae-Il Baek ; Jung-Kyu Han ; Cheon-Yong Lim ; Gun-Woo Moon

A ZERO-VOLTAGE-SWITCHING TOTEM-POLE BRIDGELESS BOOST POWER FACTOR CORRECTION RECTIFIER HAVING MINIMIZED CONDUCTION LOSSES..........................3538
Young-Dal Lee ; Chong-Eun Kim ; Jae-Il Baek ; Dong-Kwan Kim ; Gun-Woo Moon

POWER-FACTOR-CORRECTION WITH POWER DECOUPLING FOR AC-TO-DC CONVERTER..3544
Wan-Jung Chen ; Tsung-Hsi Wu ; Yao-Ching Hsieh ; Chin-Sien Moo ; Po-Hsiang Wen

DESIGN AND ANALYSIS OF THE DISTRIBUTED CONTROLLER FOR THE MODULAR MULTILEVEL CASCADED CONVERTER..3549
Ping-Heng Wu ; Yu-Chen Su ; Po-Tai Cheng

ASYMMETRIC MIXED MODULAR MULTILEVEL CONVERTER TOPOLOGY IN HYBRID BIPOLAR HVDC TRANSMISSION SYSTEMS..3557
Joon-Hee Lee ; Jae-Jung Jung ; Seung-Ki Sul

HIGH POWER MEDIUM VOLTAGE 10 KV SIC MOSFET BASED BIDIRECTIONAL ISOLATED MODULAR DC–DC CONVERTER..3564
Sayan Acharya ; Ritwik Chattopadhyay ; Anup Anurag ; Satish Rengarajan ; Yos Prabowo ; Subhashish Bhattacharya

MULTI-LEVEL POWER CONVERTER USING SERIES-CONNECTED SOLID-STATE TRANSFORMERS..3572
Yuichi Mabuchi ; Yuki Kawaguchi ; Kimihisa Furukawa ; Mitsuhiro Kadota ; Mizuki Nakahara ; Akihiko Kanoda

CAPACITOR VOLTAGE CONTROL OF MMC-STATCOM DURING UNBALANCED AC SYSTEM FAULT..3578
Kaho Nada ; Takeshi Kikuchi ; Tsuguhiro Takuno ; Toshiyuki Fujii ; Ryosuke Uda ; Takashi Sugiyama

SIC BASED POWER SEMICONDUCTOR IN APPLICATIONS - ASPECTS AND PROSPECTS.....................3584
Peter Friedrichs

ELECTROMAGNETIC MODELING APPROACHES TOWARDS VIRTUAL PROTOTYPING OF WBG POWER ELECTRONICS..3588
Ivana Kovacevic-Badstübner ; Daniele Romano ; Giulio Antonini ; Jonas Ekman ; Ulrike Grossner

SILICON BASED DEVICES FOR DEMANDING HIGH POWER APPLICATIONS..3596
A. Kopta ; J. Vobecky ; M. Rahimo ; T. Wikström ; U. Vemulapati ; C. Papadopoulos ; C. Corvasce ; M. Andenna ; F. Dugal ; F. Fischer ; S. Hartmann

RECENT PROGRESS IN HIGH TO ULTRA-HIGH-VOLTAGE SIC POWER DEVICES: DEVELOPMENT AND APPLICATION..3603
Y. Yonezawa

DYNAMIC DRIFT EFFECTS IN GAN POWER TRANSISTORS: CORRELATION TO DEVICE TECHNOLOGY AND MISSION PROFILE..3607
Joachim Würfl ; Eldad Bahat-Treidel ; Oliver Hilt ; Maria Troppenz ; Mihaela Wolf ; Jan Böcker ; Carsten Kuring ; Sibylle Dieckerhoff

COMPENSATION METHOD OF RADIAL UNBALANCE FORCE AT FAILURE OF A MOTOR SECTION IN A D-Q AXIS CURRENT CONTROL BEARINGLESS MOTOR..........................3613
Masahide Ooshima

A BEARINGLESS SYNCHRONOUS RELUCTANCE SLICE MOTOR WITH ROTOR FLUX BARRIERS 3619

Thomas Holenstein ; Thomas Nussbaumer ; Johann W. Kolar

PARAMETER IDENTIFICATIONS OF CURRENT-FORCE FACTOR AND TORQUE CONSTANT IN SINGLE-DRIVE BEARINGLESS MOTORS 3627

Hiroya Sugimoto ; Akira Chiba

DAMPENING OF AXIAL VIBRATIONS IN A BEARINGLESS FLUX-SWITCHING SLICE MOTOR BY FIELD CURRENT REGULATION 3632

Bianca Klammer ; Karlo Radman ; Wolfgang Gruber

ANALYSIS AND DESIGN OF A BEARINGLESS AXIAL-FORCE/TORQUE MOTOR WITH FLEX-PCB WINDINGS 3640

Nobuyuki Kurita ; Walter Bauer ; Gerald Jungmayr ; Wolfgang Gruber ; Wolfgang Amrhein

A PLOTTER-BASED AUTOMATIC MEASUREMENT AND STATISTICAL CHARACTERIZATION OF MULTIPLE DISCRETE POWER DEVICES 3644

Michihiro Shintani ; Benjamin Dauphin ; Kazuki Oishi ; Masayuki Hiromoto ; Takashi Sato

A NOVEL HIGH-SPEED SIC MOSFET DRIVER WITH A LOW SWITCH-VOLTAGE STRESS 3650

Xiuqin Wei ; Yuchong Sun ; Hiroo Sekiya

ENHANCEMENT OF DRIVING CAPABILITY OF GATE DRIVER USING GAN HEMTS FOR HIGH-SPEED HARD SWITCHING OF SIC POWER MOSFETS 3654

Takafumi Okuda ; Takashi Hikihara

DESIGN AND EXPERIMENTAL VERIFICATION OF ROBOT ARM OPERATION FOR POWER PACKET DISPATCHING SYSTEM 3658

Tomoki Yokoyama ; Ryunosuke Araumi ; Kazunori Asada ; Takashi Ando

A RESOURCE SHARING MODEL IN A POWER PACKET DISTRIBUTION NETWORK 3665

H. Ando ; R. Takahashi ; S. Azuma ; M. Hasegawa ; T. Yokoyama ; T. Hikihara

DECOUPLED DSOGI-PLL FOR IMPROVED THREE PHASE GRID SYNCHRONISATION 3670

A. A. Nazib ; D. G. Holmes ; B. P. Mcgrath

A DEVIATION ELIMINATION CONTROL BASED ON AUTONOMOUS CURRENT-SHARING CONTROLLER FOR THE PARALLEL-CONNECTED INVERTERS IN AC MICROGRIDS 3678

Yajuan Guan ; Wei Feng ; Baoze Wei ; Wenzhao Liu ; Mingshen Li ; C. Juan Vasquez ; M. Josep Guerrero

SISO TRANSFER FUNCTIONS FOR STABILITY ANALYSIS OF GRID-CONNECTED VOLTAGE-SOURCE CONVERTERS 3684

Hongyang Zhang ; Lennart Harnefors ; Xiongfei Wang ; Jean-Philippe Hasler ; Hans-Peter Nee

A COMMUNICATION-INDEPENDENT REACTIVE POWER SHARING SCHEME WITH ADAPTIVE VIRTUAL IMPEDANCE FOR PARALLEL CONNECTED INVERTERS 3692

Ronghui An ; Zeng Liu ; Jinjun Liu ; Shike Wang

DESIGN AND INTEGRATION OF THE BI-DIRECTIONAL ELECTRIC VEHICLE CHARGER INTO THE MICROGRID AS EMERGENCY POWER SUPPLY 3698

Yang Song ; Pengcheng Li ; Yuanliang Zhao ; Shuai Lu

STABILITY IMPACT OF PV INVERTER GENERATION ON MEDIUM VOLTAGE DISTRIBUTION SYSTEMS 3705

Ye Tang ; Rolando Burgos ; Chi Li ; Dushan Boroyevich

1MW POWER CONDITIONING SYSTEM WITH MULTIPLE DC INPUTS FOR PVS AND BATTERIES 3711

Yasuaki Furusho ; Yasuyuki Noto ; Kansuke Fujii

A ROBUST AND FLEXIBLE DC-LINKED 3-PHASE ENERGY MANAGEMENT SYSTEM WITH ADAPTIVE DROOP CONTROL STRATEGY 3717

Yue Ma ; Yuki Ishikura ; Hitoshi Tsuji ; Kazuaki Mino

MAXIMUM POWER POINT TRACKING CONTROL FOR SMALL HYDROELECTRIC GENERATION 3723

Kazuya Azegami ; Masashi Takiguchi ; Junya Yano ; Hirohiko Tsutsumi ; Toshitake Masuko

DESIGN AND EXPERIMENTAL VERIFICATION OF A THREE-PHASE DUAL-ACTIVE BRIDGE CONVERTER FOR OFFSHORE WIND TURBINES 3729

Takushi Jimichi ; Murat Kaymak ; Rik W. De Doncker

OPTIMIZED BIDIRECTIONAL PFC RECTIFIERS & INVERTERS - SI VS. SIC VS. GAN IN 2L AND 3L TOPOLOGIES - 3734

Jonas Wyss ; Jürgen Biela

A STANDARD BLOCK OF "SERIES CONNECTED SIC MOSFET" FOR MEDIUM/HIGH VOLTAGE CONVERTER 3742

Qin Lei ; Chunhui Liu ; Yunpeng Si ; Yifu Liu

DESIGN AND TESTING OF 1 KV H-BRIDGE POWER ELECTRONICS BUILDING BLOCK BASED ON 1.7 KV SIC MOSFET MODULE 3749
Jun Wang ; Rolando Burgos ; Dushan Boroyevich ; Zeng Liu

A FLYBACK CONVERTER WITH SIC POWER MOSFET OPERATING AT 10 MHZ: REDUCING LEAKAGE INDUCTANCE FOR IMPROVEMENT OF SWITCHING BEHAVIORS 3757
Kazuki Hashimoto ; Takafumi Okuda ; Takashi Hikihara

A STUDY ON LOAD FLUCTUATION OF ISOLATED DC-DC CONVERTER WITH CLASS PHI-2 INVERTER USING GAN-HFET 3762
Yuta Yanagisawa ; Yushi Miura ; Hiroyuki Handa ; Tetsuzo Ueda ; Toshifumi Ise

SINGLE-INDUCTOR MULTIPLE-OUTPUT CURRENT-SOURCE CONVERTER WITH IMPROVED CROSS REGULATION AND SIMPLE CONTROL STRATEGY 3768
Zheng Dong ; Xiaolu Lucia Li ; Chi K. Tse

LIMIT OPERATING FREQUENCY OF PEAK CURRENT-MODE CONTROL DC-DC CONVERTER CONSIDERING TURN-OFF DELAY TIME 3773
Ryo Ute ; Kazuya Fujiwara ; Jun Imaoka ; Masahito Shoyama

A NOVEL SINGLE SWITCH HIGH FREQUENCY DC/DC CONVERTER AND ITS MATHEMATIC MODEL 3780
Yueshi Guan ; Xihong Hu ; Shu Zhang ; Yijie Wang ; Dianguo Xu ; Wei Wang

ANALYSIS OF CLOSED LOOP OPERATION OF AN ISOLATED BIDIRECTIONAL DAB DC-DC CONVERTER WITH LC COUPLING 3785
Bruno Yukio Enomoto ; Kelly C. M. Carvalho ; Lourenço Matakas Junior ; Wilson Komatsu

ISOLATED AC/DC CONVERTER USING SIMPLE PWM STRATEGY 3791
Naoki Hirose ; Yuto Matsui ; Takaharu Takeshita

ANALYSIS OF ONE PHASE LOSS OPERATION OF THREE-PHASE ISOLATED BUCK MATRIX-TYPE RECTIFIER WITH EIGHT-SEGMENT PWM SCHEME 3797
Jahangir Afsharian ; Dewei David Xu ; Bin Wu ; Bing Gong ; Zhihua Yang ; Jun-Ichi Itoh

NOVEL ISOLATED BIDIRECTIONAL INTEGRATED DUAL THREE-PHASE ACTIVE BRIDGE (D3AB) PFC RECTIFIER 3805
F. Krismer ; E. Hatipoglu ; J. W. Kolar

LOAD VOLTAGE REGULATION METHOD FOR AN ISOLATED AC-DC CONVERTER WITH POWER DECOUPLING OPERATION 3813
Shohei Komeda ; Hideaki Fujita

OPTIMAL DESIGN OF A LOW COST 20KW 99.1% EFFICIENCY ACTIVE ZCS ISOLATED DC-DC CONVERTER 3820
Timothé Delaforge ; Sébastien Mariéthoz

SOFT-SWITCHING ANALYSIS AND PFM CONTROL METHOD OF BIDIRECTIONAL DC/DC CONVERTER TOPOLOGY 3825
Yijie Wang ; Haoyu Wang ; Hongyu Song ; Dianguo Xu

A FULLY SOFT-SWITCHED PWM DC-DC CONVERTER USING AN ACTIVE-SNUBBER-CELL 3833
Hai N. Tran ; Adhistira M. Naradhipa ; Sunju Kim ; Ali Tausif

FLYING CAPACITOR RESONANT POLE INVERTER WITH DIRECT INDUCTOR CURRENT FEEDBACK 3840
Sjef J. Settels ; Jorge L. Duarte ; Jeroen Van Duivenbode

DESIGN OF A GAN-BASED WIRELESS POWER TRANSFER SYSTEM AT 13.56 MHZ TO REPLACE CONVENTIONAL WIRED CONNECTION IN A VEHICLE 3848
Kawin Surakitbovorn ; Juan Rivas-Davila

EFFICIENCY MAXIMIZATION OF INDUCTIVE POWER TRANSFER SYSTEM BY IMPEDANCE AND SWITCHING FREQUENCY CONTROL IN SECONDARY-SIDE CONVERTER 3855
Ryosuke Ota ; Dannisworo S. Nugroho ; Nobukazu Hoshi

ANALYSIS OF OPTIMAL OPERATION FREQUENCY RANGE FOR BATTERY CHARGING IN WPT SYSTEM 3863
Yongbin Jiang ; Min Wu ; Junwen Liu ; Yue Wang ; Laili Wang ; Hailong Zhang

INITIAL CURRENT INJECTION METHOD OF A DIRECT THREE-PHASE TO SINGLE-PHASE AC/AC CONVERTER FOR INDUCTIVE CHARGER 3870
Ferdi Perdana Kusumah ; Jorma Kyyrä

MISSION PROFILE EMULATOR FOR PERMANENT MAGNET SYNCHRONOUS MACHINE BASED ON THREE-PHASE POWER ELECTRONIC CONVERTER 3877
Yubo Song ; Ran Cheng ; Ke Ma

A VARIABLE DC BUS VOLTAGE BASED POWER HARDWARE-IN-THE-LOOP EMULATION OF ELECTRIC MOTORS WITH WIDE VARIATION IN INTERFACE FILTER INDUCTANCE 3884
Tsai-Fu Wu ; Mitradatta Misra ; Ying-Yi Jhang ; Chang-Jun Yang ; Yin-Chi Xu

COPPER LOSS MINIMIZATION CONTROL AT ZERO OUTPUT VOLTAGE FOR ELECTROLYTIC CAPACITOR-LESS INVERTER 3890
Kodai Abe ; Haruya Kada ; Kiyoshi Ohishi ; Hitoshi Haga ; Yuki Yokokura

ARMATURE TEMPERATURE ESTIMATION INSENSITIVE TO ROTOR FLUX VARIATION FOR SPMSM 3896
Toshiki Sano ; Kiyoshi Ohishi ; Yuki Yokokura ; Hiroki Iwata ; Yuji Ide ; Daigo Kuraishi ; Akihiko Takahashi

VIRTUAL SYNCHRONOUS GENERATOR CONTROL WITH RELIABLE FAULT RIDE-THROUGH CAPABILITY BY ADOPTING MODEL PREDICTIVE CONTROL 3902
Jonggrist Jongudomkarn ; Jia Liu ; Toshifumi Ise

RESHAPING QUADRATURE-AXIS IMPEDANCE OF THREE-PHASE GRID-CONNECTED CONVERTERS FOR LOW-FREQUENCY STABILITY IMPROVEMENT 3910
Yi Tang ; Jingyang Fang ; Xiaoqiang Li ; Hongchang Li

COMPARISON BETWEEN TRADITIONAL DROOP AND A NEW AUTONOMOUS CONTROL SCHEME FOR PARALLEL INVERTERS 3916
Mohammad Bani Shamseh ; Teruo Yoshino ; Atsuo Kawamura

A NOVEL MICROGRID POWER SHARING SCHEME ENHANCED BY A NON-INTRUSIVE FEEDER IMPEDANCE ESTIMATION METHOD 3924
Baojin Liu ; Zeng Liu ; Jinjun Liu ; Ronghui An ; Shuguang Song

DEVELOPMENT OF A 3.2MW PHOTOVOLTAIC INVERTER FOR LARGE-SCALE PV POWER PLANTS 3929
Naoya Shibata ; Tsuguhiro Tanaka ; Masahiro Kinoshita

IMPEDANCE-BASED STABILITY ANALYSIS OF LARGE-SCALE PV STATION UNDER WEAK GRID CONDITION CONSIDERING SOLAR RADIATION FLUCTUATION 3934
Yiming Tu ; Jinjun Liu ; Teng Liu ; Xiangpeng Cheng

EXPERIMENTAL VERIFICATION OF GRID-CONNECTION OF A PV CONVERTER USING A SYMMETRICALLY CONNECTED BOOST CONVERTER FOR A HIGH-LEG DELTA TRANSFORMER 3940
Daiki Yamaguchi ; Hideaki Fujita

A NOVEL SINGLE- STAGE HIGH-FREQUENCY BOOST INVERTER CASCADED BY RECTIFIER-INVERTER SYSTEM FOR PV GRID-TIE APPLICATIONS 3945
Hamdy Radwan ; Mahmoud A. Sayed ; Takaharu Takeshita ; Adel A. Elbaset ; G. Shabib

NINE SWITCHES MATRIX CONVERTER USING BI-DIRECTIONAL GAN DEVICE 3952
Takashi Hirota ; Kentaro Inomata ; Daisuke Yoshimi ; Masato Higuchi

A MODEL PREDICTIVE DUAL CURRENT CONTROL METHOD FOR INDIRECT MATRIX CONVERTER FED INDUCTION MOTOR DRIVES 3958
Mei Yang ; Chen Lisha ; Liang Wang ; Yunwei Li

FAULT TOLERANT PREDICTIVE CONTROL OF THREE-LEVEL NEUTRAL-POINT-CLAMPED BACK-TO-BACK POWER CONVERTERS 3965
Zhenbin Zhang ; Xicai Liu ; Kejun Cai ; Feng Gao ; Ralph Kennel

TWO-STAGE OPTIMIZATION BASED PREDICTIVE TORQUE CONTROL WITH REDUCED COMPLEXITY FOR A THREE-LEVEL INVERTER DRIVEN INDUCTION MOTOR 3971
Ilham Osman ; Dan Xiao ; Faz Rahman

DESIGN CHALLENGES OF SIC DEVICES FOR LOW- AND MEDIUM-VOLTAGE DC-DC CONVERTERS 3979
Georges Engelmann ; Alexander Sewergin ; Markus Neubert ; Rik W. De Doncker

DESIGN AND TESTING OF 6 KV H-BRIDGE POWER ELECTRONICS BUILDING BLOCK BASED ON 10 KV SIC MOSFET MODULE 3985
Jun Wang ; Slavko Mocevic ; Jiewen Hu ; Yue Xu ; Christina Dimarino ; Igor Cvetkovic ; Rolando Burgos ; Dushan Boroyevich

HIGH POWER MEDIUM VOLTAGE CONVERTERS ENABLED BY HIGH VOLTAGE SIC POWER DEVICES 3993
Sanket Parashar ; Ashish Kumar ; Subhashish Bhattacharya

SOFT-SWITCHING – THE KEY TO HIGH POWER WBG CONVERTERS 4001
Deepak Divan ; Zheng An ; Prasad Kandula

SIC: TECHNOLOGY ENABLER FOR MV DC/DC GALVANICALLY INSULATED MODULAR CONVERTERS 4009
S. Alvarez ; M. Bellini ; U. Vemulapati ; F. Canales ; M. Rahimo

A BEARINGLESS SLICE MOTOR WITH A SOLID IRON ROTOR FOR DISPOSABLE CENTRIFUGAL BLOOD PUMP 4016
Tadahiko Shinshi ; Ryo Yamamoto ; Yoshiki Nagira ; Junichi Asama

REDUCED HARDWARE PARALLEL DRIVE FOR NO VOLTAGE BEARINGLESS MOTORS 4020
Eric L. Severson

DUAL FIELD-ORIENTED CONTROL OF BEARINGLESS MOTORS WITH COMBINED WINDING SYSTEM 4028
Wolfgang Gruber ; Siegfried Silber

OPEN-CIRCUIT FAULT TOLERANT STUDY OF BEARINGLESS MULTI-SECTOR PERMANENT MAGNET MACHINES 4034
G. Valente ; L. Papini ; A. Formentini ; C. Gerada ; P. Zanchetta

BALANCE CONTROL OF SPLIT CAPACITOR POTENTIAL FOR MAGNETICALLY LEVITATED MOTOR SYSTEM USING ZERO-PHASE CURRENT 4042
Takaaki Oiwa

ASYMMETRICAL HALF-BRIDGE CONVERTER WITH ZERO DC-OFFSET CURRENT IN TRANSFORMER USING NEW RECTIFIER STRUCTURE 4049
Jung-Kyu Han ; Jong-Woo Kim ; Seung-Hyun Choi ; Jih-Sheng Lai ; Gun-Woo Moon

CIRCULATING CURRENT-LESS PHASE-SHIFTED FULL-BRIDGE CONVERTER WITH NEW RECTIFIER STRUCTURE 4054
Jung-Kyu Han ; Gun-Woo Moon

A BI-DIRECTIONAL CURRENT DETECTION USING CURRENT TRANSFORMERS FOR BI-DIRECTIONAL DC-DC CONVERTER 4059
Seiji Iyasu ; Yuji Hahashi ; Yuuichi Handa ; Kimikazu Nakamura ; Keiji Wada

A 10 MHZ GANFET BASED ISOLATED HIGH STEP-DOWN DC-DC CONVERTER 4066
Prasanth Thummala ; Dorai Babu Yelaverthi ; Regan Zane ; Ziwei Ouyang ; Michael A. E. Andersen

ANALYSIS AND DESIGN OF A PARALLEL RESONANT CONVERTER FOR CONSTANT CURRENT INPUT TO CONSTANT VOLTAGE OUTPUT DC-DC CONVERTER OVER WIDE LOAD RANGE 4074
Tarak Saha ; Hongjie Wang ; Baljit Riar ; Regan Zane

NOVEL SINUSOIDAL INPUT CURRENT SINGLE-TO-THREE-PHASE Z-SOURCE BUCK+BOOST AC/AC CONVERTER 4080
M. Haider ; D. Bortis ; J. W. Kolar ; Y. Ono

SIMPLE PWM STRATEGY OF A MATRIX CONVERTER FOR MINIMIZING OUTPUT VOLTAGE HARMONICS 4088
Takuya Oshima ; Takaharu Takeshita

NOVEL THREE-LEVEL BACK-TO-BACK CONVERTERS: STRUCTURE, MODULATION METHOD, AND EXPERIMENT 4096
S. Sangwongwanich ; K. Niyomsatian ; S. Samermurn ; S. Nuchnoi ; S. Suwankawin

MODEL PREDICTIVE CONTROL USING SUBDIVIDED VOLTAGE VECTORS FOR CURRENT RIPPLE REDUCTION IN AN INDIRECT MATRIX CONVERTER 4104
Keon Young Kim ; Yeongsu Bak ; Jin-Hyuk Park ; Kyo-Beum Lee

DC-LINK RIPPLE CURRENT REDUCTION IN BACK-TO-BACK CONVERTERS WITH DPWM 4109
Anatolii Tcai ; Kyo-Beum Lee

AN ANALYSIS OF CLASS DE VOLTAGE-SOURCE PARALLEL RESONANT INVERTER 4114
Takeshi Kondo ; Tsuyoshi Inaba ; Yoshikazu Sakai ; Hirotaka Koizumi

AN IMPROVEMENT ON EXTENDED IMPEDANCE METHOD TOWARDS EFFICIENT STEADY-STATE ANALYSIS OF HIGH-FREQUENCY CLASS-E RESONANT INVERTERS 4122
Junrui Liang

OUTPUT POWER CAPABILITY COMPARISONS OF CLASS-E POWER AMPLIFIERS WITH HARMONIC RESONANCE 4127
Hiroo Sekiya ; Xiuqin Wei ; Yuchong Sun

A CLASS Φ2 RESONANT BUCK CONVERTER WITH RIPPLE INJECTION BURST CONTROL METHOD 4133
Min Lin ; Masahiko Hirokawa

PRACTICAL DESIGN TECHNIQUE FOR HIGH POWER DENSITY LLC RESONANT CONVERTER 4139
Shingo Nagaoka ; Hiroyuki Onishi ; Koji Takatori ; Toshiyuki Zaitsu ; Takeshi Uematsu

OPERATIONAL STUDY AND PROTECTION OF A SERIES RESONANT CONVERTER WITH DC CURRENT INPUT APPLIED IN DC CURRENT DISTRIBUTION SYSTEMS 4145
Hongjie Wang ; Tarak Saha ; Baljit Riar ; Regan Zane

A STUDY ON IMPROVEMENT OF POWER UTILIZATION RATE OF ENERGY SYSTEMS WITH PVS AND BATTERIES 4151
Hiroaki Endo ; Masakatsu Kurisaka ; Tsutomu Ueno ; Yusuke Yoshioka ; Kaoru Inoue ; Toshiji Kato

A NOVEL DC DISTRIBUTION NETWORK WITH MULTI-LEVEL BUS VOLTAGES AND ITS ENERGY MANAGEMENT SYSTEM DESIGN 4157
Jingjin Huang ; Xin Zhang ; Zhixun Ma ; Jianfang Xiao

A NOVEL DC-SIDE-PORT IMPEDANCE MODELING OF MODULAR MULTILEVEL CONVERTERS BASED ON HARMONIC STATE SPACE METHOD 4162
Jing Lyu ; Xin Zhang ; Zhixun Ma ; Xu Cai

AN IMPROVED MASTER-SLAVE CONTROL FOR THREE-PORT CONVERTER BASED DISTRIBUTED DC GRID-CONNECTED PV SYSTEM 4168
Siyue Jiang ; Kai Sun ; Hongfei Wu ; Haixu Shi ; Xiaofeng Dong ; Syed Muhammad Raza Kazmi

SENSORLESS POSITION ESTIMATION, PARAMETER IDENTIFICATION AND CONTROL INTEGRATION FOR PERMANENT MAGNET SYNCHRONOUS MACHINES USING CURRENT DERIVATIVE MEASUREMENTS 4174
M.X. Bui

DYNAMIC PERFORMANCE IMPROVEMENT OF BIDIRECTIONAL SWITCHED-CAPACITOR DC/DC CONVERTER BY RIGHT-HALF-PLANE ZERO ELIMINATION 4181
Ding Kaicheng ; Zhang Yan ; Liu Jinjun ; Zeng Pengxiang ; Zhang Jinshui

A MATRIX BASED ISOLATED BIDIRECTIONAL AC-DC CONVERTER WITH LCL TYPE INPUT FILTER FOR ENERGY STORAGE APPLICATION 4186
Prathamesh Pravin Deshpande ; Amit Kumar Singh ; Sanjib Kumar Panda

ON A STUDY OF VOLTAGE DIVIDING CLASS Φ AMPLIFIER 4193
Katsutoshi Hirayama ; Tadashi Suetsugu ; Yudai Furukawa ; Fujio Kurokawa

A DPWM BASED CONTROL STRATEGY TO INTEGRATE PHOTOVOLTAIC SYSTEM AND BATTERY STORAGE USING GRID CONNECTED THREE-LEVEL T-TYPE INVERTER 4198
Mohammad M. Hashempour ; Yue-Ting Tsai ; T. L. Lee

IMPEDANCE MEASUREMENT OF MEGAWATT-LEVEL RENEWABLE ENERGY INVERTERS USING GRID-FORMING AND GRID-PARALLEL CONVERTERS 4205
Matias Berg ; Tuomas Messo ; Tomi Roinila ; Henrik Alenius

IMPROVED VIRTUAL INDUCTANCE BASED CONTROL STRATEGY OF DFIG UNDER WEAK GRID CONDITION 4213
Ran Fang ; Wenjia Chen ; Xueguang Zhang ; Dianguo Xu

CONTROL OF VSC-HVDC FOR WIND FARM INTEGRATION WITH REAL-TIME FREQUENCY MIRRORING AND SELF-SYNCHRONIZING CAPABILITY 4220
Renxin Yang ; Chen Zhang ; Xu Cai ; Gang Shi ; Jing Lyu

A STUDY ON STEADY-STATE CHARACTERISTICS OF SERIES-CONNECTED WIND FARM USING AN EXPERIMENTAL SET OF LABORATORY SIZE 4227
Fujio Tatsuta ; Shoji Nishikata

A NOVEL ISLANDING DETECTION METHOD WITH TWO-PHASE MAGNIFICATION INSPECTION 4233
Jian-Tang Liao ; Shun-Hao Yeh ; Hong-Tzer Yang

Author Index

Evaluation of Novel Hybrid Protection Based on Pyroswitch and Fuse Technologies

Tomokazu Sakuraba[1*], Rémy Ouaida[2], Song Chen[3], Thibaut Chailloux[2]
[1] MERSEN Japan K. K., Tokyo, Japan, [2] MERSEN France SB, France
[3] MERSEN Shanghai, Shanghai, China.

*E-mail: tomokazu.sakuraba@mersen.com

Abstract— In battery applications, a DC contactor is usually used in series with a current-limiting fuse to clear respectively low and high fault currents. However, classic protection devices have operation limits due to maximum and minimum breaking capacities of each device. In this paper, the self-triggered pyrofuse (Xp-ST) based on conventional fuses and pyroswitch is introduced. This protection solution can be installed as a standalone device. The Xp-ST has an advantage of selectivity. The selectivity is more easily tunable by using a low voltage fuse for the trigger fuse. As an experimental result under the DC condition of 1000 V with 15.6 kA fault current, the proposed self-triggered pyrofuse cleared the fault in 1.5 ms and has limited the current to 7 kA. Furthermore, reliability of the Xp-ST is validated by simulations and aging tests for cycling performances. The cycling performances of the Xp-ST can satisfy the requirements from the EV application with up to 1000 V.

Keywords— EV application, Hybrid over current protection, Fuses, Pyro technics

I. CONTEXT

The recent collective awareness on air pollution and global warming have led to a rapid growth of the electric vehicle (EV) market [1-4]. Not only is the number of electric vehicles increasing, but the trend of evolution also seems to be an increasingly high power and operating voltage [5-6]. This increase poses new problems in the protection of the battery, the system and the user.

Usually, a DC contactor and a current-limiting fuse are connected in series with the battery (Fig. 1) [7-8]. Table 1 gathers an example of possible requirements for the next generation of EV. While satisfying these requirements, the protection devices must protect the battery from any faults. Moreover, the load profile in EV applications is complex due to the battery charging, engine acceleration, regenerating braking and so on [10]. Thus, finding the appropriate protection becomes more complicated and may require an innovative solution.

II. PROTECTION SOLUTIONS FOR EV APPLICATION

In this section, the technical requirement from EV application in term of battery protection is clarified. Firstly, the limitations of current-limiting fuses and DC contactor or DC power relay are introduced. Second,

from the specifications of the batteries, the requirements for the protection of EV are clarified. Finally, through the comparison between the technical requirements from EV application and the limitation of the conventional protection device, the demands for the protection devices are discussed.

A. DC contactor and power relay

Whereas the power contactors and the power relays are widely used for the AC applications, the demands from the DC applications like PV, EV, EES are increasing as well. For instance, as shown in Fig. 1, the power contactors are positioned at a DC line for the battery protection. DC contactors are usually designed to open repeatedly under load and DC power relay are designed to open repeatedly under no load and only a few times under load. Both offer an important safety feature: a galvanic isolation between the load and the battery. However, when increasing the DC voltage, mechanical switches suffer from contact erosion and arcing chamber fatigue. Thus, DC contactors and power relays have operation limitations, especially to clear high fault current.

They can operate from zero up to a maximum current, which is in general far from the maximum short-circuit current. So a DC contactor is generally not sufficient to fully protect this kind of system, and usually they are used in series with a fuse.

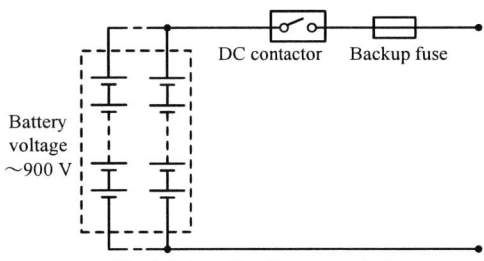

Fig. 1. Schematic of battery protection.

Table 1. Requirements from EV application to battery protection based on [2], [5], [9]

DC voltage [9]	900 V
Current rating [9]	500 A
Operation temperature [2], [5]	-40 to +90 °C

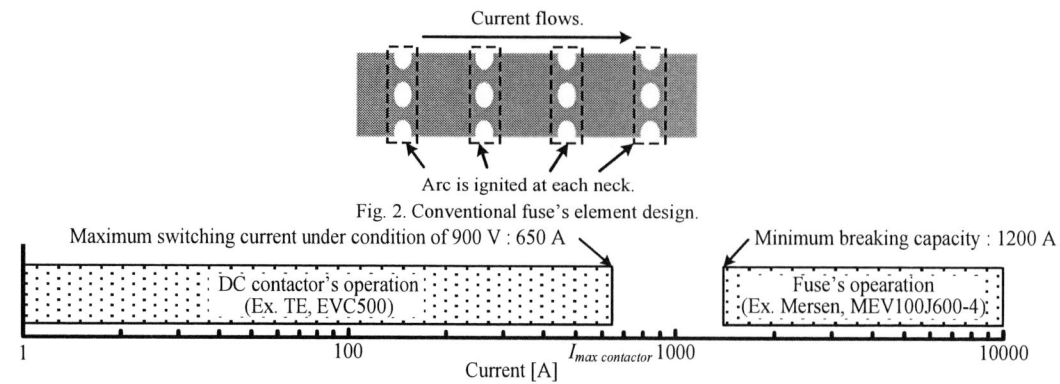

Current flows.

Arc is ignited at each neck.

Fig. 2. Conventional fuse's element design.

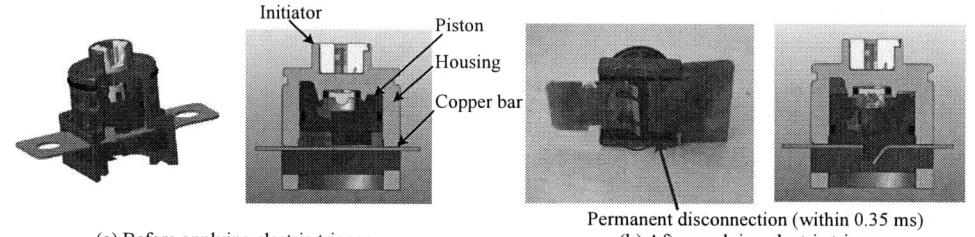

Maximum switching current under condition of 900 V : 650 A

Minimum breaking capacity : 1200 A

DC contactor's operation
(Ex. TE, EVC500)

Fuse's opearation
(Ex. Mersen, MEV100J600-4)

1 100 $I_{max\ contactor}$ 1000 10000

Current [A]

Fig. 3. Operation region of DC contactor of EVC500 and current-limiting fuse of MEV100J600-4.

Initiator Piston
Housing
Copper bar

Permanent disconnection (within 0.35 ms)

(a) Before applying electric trigger   (b) After applying electric trigger

Fig. 4. Pyroswitch operations. (a) Before applying electric trigger, current flows through copper bar with low resistance.
(b) After detection of over current, pressure pulse accelerates insulated piston. Insulted piston cut copper bar within 0.35 ms.

B. Current-limiting fuse

The current-limiting fuses are the most reliable current limiting devices. The fuses are also used for the battery protection of the EV application. The fuses are good at clearing high fault current compared to their nominal current by igniting the element described in Fig. 2. However, they have operation limitations, especially to clear low fault current. Some fuse manufacturers give a characteristic parameter called Minimum Breaking Capacity. For example, Mersen MEV100J600-4 fuse is rated for 600A and has a MBC of 1200A. It means that a fault current between 600A and 1200A will not be enough to melt the fuse, so the system is not protected by the fuse for this range of current. Moreover, at 1200A the fuse will melt but it will take some time, meaning the system has to endure the fault current for a significant duration..

C. Issues with protection requirements

As briefly explained in sections II.A and II.B, each of the classic protections has its own limitation: difficulties to open low fault current for fuses and high fault current for DC contactors.

To grant the safety of the system and the user, the protection devices must operate for any current, from zero up to the maximum short circuit current. Choosing the appropriate protection consists in matching the contactor and the fuse (Fig.3). The reason of the difficulty relies in the fact that a fuse will need a long time to melt at his MBC and in the fact that the relay is not able to withstand that much current for too long.

For example, to meet the requirements in Table 1, we selected the EVC500 power relay (TE connectivity) and

for the fuse, we can select Mersen MEV100J600-4 or Bussmann FWJ-600A. Fig. 3 shows the protection limitation of Mersen fuse and the power contactor. Under the DC condition of 900 V, EVC500 has the maximum limitation for the switching current I_{max} of 650 A. On the other hand, the MEV100J600-4 has the MBC of 1200 A. Thus, the classic protection devices cannot protect the battery in region of 650 A-1200 A in low fault.

Moreover, EVC500 has a short-time current limit of 1 kA-60 s, while the fuse has the maximum time of 200 s to clear 1200 A. This means that the contactor suffer damages on the contact part or be broken in worst case.

III. PROPOSED HYBRID PROTECTION

In previous works [11-12], Mersen has tested a solution reducing the fuse's MBC and the clearing time by using a hybrization of a fuse and a pyroswitch described in Fig. 4. This hybrid protection (called pyrofuse or Xp-S) is shown in Fig. 5 . This pyrofuse has displayed evidences of its good performances compared to regular fuses: capacity to clear very low fault current, shorter clearing time at low fault current compact and light. Moreover, the power dissipation is very low. For example, Mersen's 400A pyrofuse is 81% less resisitve than the 400A fuse from the MEV series (MEV100J400-4) in Fig. 6. However, this pyrofuse needs an additional electronic circuit and a power supply. In some applications, because the current control command is not straightforward, the standalone operation is required for the simple electrical installation to ensure the simple electrical installation.

1) Self-triggered operation principle

Fig. 7 shows the structure of the proposed self-triggered pyrofuse (Xp-ST). The Xp-ST consists of a trigger fuse in series with the pyrofuse shown in Fig. 5 [13]. In this solution, the trigger fuse is used both as a current sensor and as an energy harvester to ignite the pyrofuse. In nominal conditions, the circuit current flows through the trigger fuse and the pyroswitch. When a fault current occurs, an electrical arcing arises inside the trigger fuse and a voltage drop is created across the fuse terminals. At this moment, a part of the arcing energy is transferred to the initiator of the pyroswitch through a resistance. The later helps limiting the current flow and controls the opening speed of the pyroswitch. Once the pyroswitch is ignited, the pyrofuse operates and clear the fault. It should be noted that the trigger fuse only starts arcing but it does not clear the fault.

Consequently, the proposed self-triggered pyrofuse clears a fault current as a standalone device.

2) Advantages of the proposed pyroswitch

This design keeps several advantages of the original Xp-S design.

Since most of the current flows in through the pyroswitch and the trigger fuse in the normal condition, the parallel fuse current rating can be less than the nominal current. Thus, when subjected to any fault current, the parallel fuse will always open quickly.

On the other hand, the trigger fuse is not designed to clear the fault current, only to trigger the rest of the device. It means it is designed for a lower voltage than the nominal one, leading to a less long and less resistive fuse.

Furthermore, by selecting the right trigger fuse and the right parallel fuse, the self-triggered pyrofuse has a tunable selectivity. This feature can be more easily achieved thanks to the low voltage trigger fuse.

IV. EVALUATION OF PROPOSED PROTECTION

In this section, at first, we introduce the limitation of the classic protections. After that, the experimental validations of the proposed self-trigger pyrofuse under the condition of 1000 V are presented. Finally, we discuss the cycling performance and aging test for the self-triggered pyrofuse for the reliability.

A. Breaking capacity test

In order to evaluate the self-triggered pyrofuse, the experimental verification are shown in this section.

Fig. 8 shows the experimental waveforms of the self-triggered pyrofuse under 1000 V and a 15.6 kA prospective fault current. The current and the voltage are depicted in blue and red, respectively. In order to understand the operation principle of the self-triggered pyrofuse, the waveforms can be divided in 4 steps in Fig. 8; first, the beginning of the short circuit with the increase of the current through the trigger fuse. When the current reaches a threshold (energy-based), a voltage drop arises and the current is limited during this step, a portion of the arc energy is transferred to the pyroswitch initiator. The pyroswitch opens and the current flows to the parallel fuse. The parallel fuse is crossed by a current over his rating and starts to operate, the circuit energy is

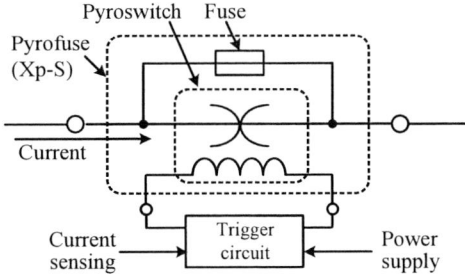

Fig. 5. Hybrid protection device (Xp-S) [13]. Xp-S is constructed from pyroswitch and fuse. This device needs the current control command to operate the pyroswitch.

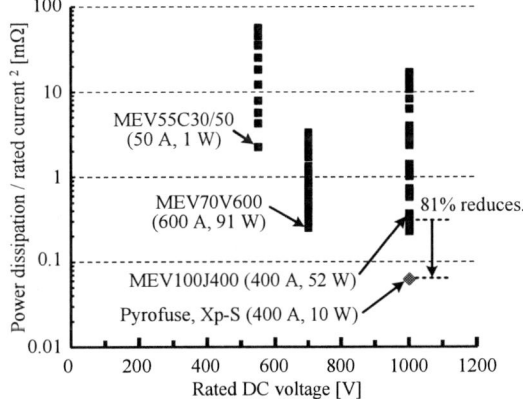

Fig. 6. Relation between rated DC voltage and power dissipation of MEV series fuses and Xp-S.

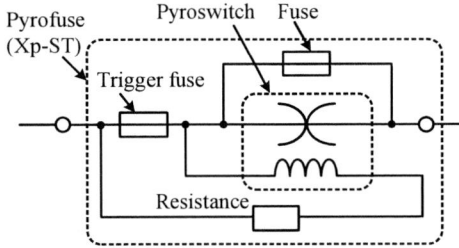

Fig. 7. Hybrid self-triggered pyrofuse (Xp-ST) [13].

absorbed, which causes an overvoltage and a current drop. As a result, the proposed self-triggered pyrofuse cleared the fault in 1.5 ms and has limited the current to 7 kA.

B. Cycling performance simulation

In order to validate the reliability of the proposed self-trigger pyrofuse, the cycling performance is estimated thanks to a simulation software developed by Mersen. As described in session II-C, in steady state, most of the current flows through the trigger fuse and the pyroswitch. Because the pyroswitch consists mainly in a simple copper bar in the current path, a long lifetime can be expected. Thus, the cycling performance of Xp-ST is mostly limited by the trigger fuse.

The cycling performance of the fuses depends on the installation environment (temperature, humidity, altitude, connection condition, cooling performance), especially the fuse's load profile. In the EV application, the load profile for the battery is complex for acceleration: braking (regeneration), battery charging and so on [10].

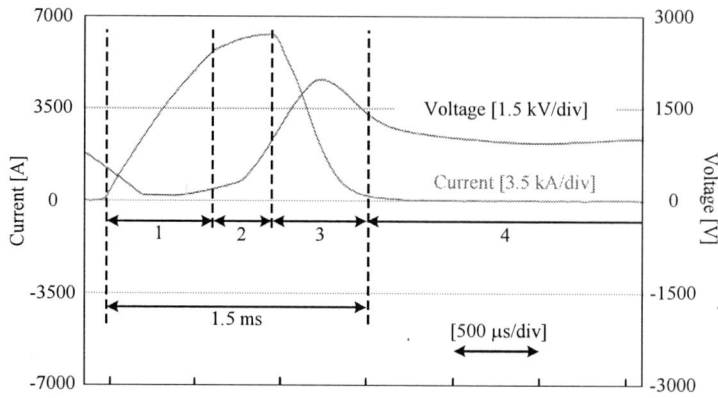

Fig. 8. Experimental result of Xp-ST in short-circuit test under condition of 1000 VDC with 15.6-kA fault current. The peak current and the clearing time are 6.3 kA and 1.5 ms, respectively.

Table 2. Fuse's load profiles and estimated results for cycling performance.

No.	Mode	Fuse average load profiles	Estimated cycling number	Frequency of load
#1	Battery charging [14]	120 A – 1000 s 50 A – 2200 s	1.4×10^{13}	1 charge / day
#2	Battery charging [15]	144 A – 900 s 16 A -1200 s	9.4×10^{11}	1 charge / day
#3	Acceleration load 1 [15]	847 A – 60 s 0 A – 120 s	2,900	1 load / race
#4	Acceleration load 2 [15], [16]	847 A – 30 s 0 A - 120 s	9,300	1 load / race
#5	Acceleration load 3 [15], [17]	700 A – 10 s 250 A – 20 s 0 A– 30 s	44,000	1 load / race

Table 3. Aging test conditions and experimental results

Test No.	Temperature	Test load	Test result
1st test	25 °C	847 A – 30 s 0 A – 120 s	11,500 (Still alive)
2nd test	25 °C	700 A – 10 s 250 A – 20 s 0 A– 30 s	45,000 (Still alive)

Table 2 and 3 show the load profiles estimated and simulated in this paper [14-17] and the results of the cycling performance. In these simulations, the case #3 for the acceleration is the most severe situation for the fuse. Even in this case, the fuse have the estimated cycling number of 2,900. From these results, it was confirmed that the Xp-ST has excellent cycling performance to satisfy the requirements for the battery protection from EV applications.

C. Reliability test by aging test

The validation of the simulation for the cycling performance of the proposed self-trigger pyrofuse was confirmed by the experimental verification. Table 3 summarizes the result of the aging test under the condition of the load profiles of #4 and #5 with 25 °C [14]. As a result, the cycling number is beyond 2,000 required from the Li-ion battery and it still arrive [1]. To summarize, the proposed self-trigger fuse has the enough potential in terms of cycling performance and breaking capacity for EV applications with up to 1000 V.

V. CONCLUSION & FUTURE PROSPECTS

In battery applications, usually, a DC contactor is used in series with a current-limiting fuse to clear both low and high fault currents. However, with increasing operating voltage, classic protection devices have limitations due to maximum and minimum breaking capacities of each device.

In this paper, the self-triggered pyrofuse (Xp-ST) based on conventional fuses and pyroswitch, is introduced. This protection solution can be installed as a standalone device. The Xp-ST has better selectivity. The selectivity is more easily tunable by using a low voltage fuse for the trigger fuse. As an experimental result under the DC condition of 1000 V with 15.6 kA fault current, a peak current was limited less than 7 kA and a clearing time is 1.5 ms. Furthermore, the reliability of the Xp-ST was validated by simulations and aging tests for cycling performances. From these tests, the cycling performances of the Xp-ST can satisfy the requirements from the EV application with up to 1000 V.

In practical application, the safe disconnection of the battery can be required not only because of an overcurrent but also in case of an independent event like a car crash or a starting fire. Thus we would like to develop a device able to self-operate when subjected to an overcurrent, but also to operate remotely when externally triggered.

Another possible future work could also be to improve the maximum breaking current of the DC contactor.

REFERENCES

[1] J. Y. Yong, V. K. Ramachandaramurthy, K. M. Tan, N. Mithulananthan : "A review on the state-of-the-art technologies of electric vehicle its impacts and prospects", Renewable and Sustainable Energy Reviews, Vol. 49, pp. 365-385 (2015)

[2] J. Taggart : "Ambient Temperature Impacts on Real-World Electric Vehicle Efficiency & Range", IEEE Transportation Electrification Conference and Expo (ITEC), No. 1168, (2017)

[3] K. Kusaka, J. Itoh : "Input Impedance Matched AC-DC Converter in Wireless Power Transfer for EV Charger", 15th International Conference on Electrical Machines and Systems (ICEMS2012), No. LS2A-2, pp. (2012)

[4] M. Vasiladiotis, B. Bahrani, N. Burger, A. Rufer : "Modular Converter Architecture for Medium Voltage Ultra Fast EV Charging Stations: Dual Half-Bridge-based Isolation Stage", 2014 International Power Electronics Conference (IPEC-Hiroshima 2014 - ECCE ASIA), pp. 1386-1393 (2014)

[5] Y. Murakami, Y. Tajima, S. Tanimoto : "Air-Cooled Full-SiC High Power Density Inverter Unit", EVS27 Invertenational Battery, Hybrid and Fuel Cell Electric Vehicle Symposium, (2013)

[6] T. S. Bryden, A. J. Cruden, G. Hilton, B. H. Dimitrov, C. Ponce de Leon, A. Mortimer : "Off-vehicle Energy Store Selection for High Rate EV Charging Station", 6th Hybrid and Electric Vehicles Conference (HEVC 2016), pp. 166-174 (2016)

[7] Y. C. Hsieh, J. L. Wu, Q. Y. Kuo : "A Li-ion Battery String Protection System", International Conference on Applied Electronics (AE), pp. 169-172 (2011)

[8] Rajasekhar MV, Parandhamaiah Gorre : "High Voltage Battery Pack Design for Hybrid Electric Vehicles", IEEE International Transportation Electrification Conference (ITEC), pp. 95-101 (2015)

[9] National Organisation Hydrogen and Fuel cell technology : "[online]https://www.now-gmbh.de/en/news/press/fastcharge-project-examines-recharging-at-up-to-450-kw" (2017)

[10] N. Zhao, N. Schofield, R, Yang, R. Gu : "Investigation of DC-Link Voltage and Temperature Variations on EV Traction System Design", IEEE Trans. On Industry Applications, Vol. 53, No. 4, pp. 3707-3718 (2017)

[11] R. Ouaida, J. F. de Palma, G. Gonthier : "New over current protection technology addressing DC Transportation", 2016 IEEE Transportation Electrification Conference and Expo (ITEC), No. TS13-4 (2016)

[12] R. Ouaida, M. Berthou, D. Tournier, J. F. de Palma : "State of Art of Current and Future Technologies in Current Limiting Devices", IEEE First International Conference on DC Microgrids (ICDCM), pp. 175-180 (2015)

[13] R. Ouaida, J. F. de Palma : "DC GRIDS New Over Current Protection", 18th European Conference on Power Electronics and Applications (EPE'16), No. DS3i-123 (2016)

[14] M. Shirk, J. Wishart : "Effects of Electric Vehicle Fast Charging on Battery Life and Vehicle Performance", SAE 2015 World Congress & Exhibition, Vol. , No. 2015-01-1190, pp. (2015)

[15] D. W. Cooke : "Design of a Lithium Ion Battery pack for 400 MPH Electric Landspeed Racing", Presented in Partial Fulfillment of the Requirements for Graduation with Distinction in the Ohio State University, pp. 51, 54, 67, 84 (2012)

[16] J. Lee : "Vehicle Inertia Impact on Fuel Consumption of Conventional and Hybrid Electric Vehicles Using Acceleration and Coast Driving Strategy", Dissertation submitted to the faculty of the Virginia Polytechnic Institute and State University in partial fulfillment of the requirements of the degree of Doctor of Philloophy, pp. 77 (2009)

[17] A. Affanni, A. Bellini, G. Franceschini, P. Guglielmi, C. Tassoni : "Battery Choice and Management for New-Generation Electric Vehicles", IEEE Trans. On Industrial Electronics, Vol. 52, No. 5, pp. 1343-1349 (2005)

Optimal design of a magnetically coupled filter for high efficiency, low cost and low volume dc-dc battery storage converter

Timothé Delaforge*, Robert Pasterczyk[†], Mickaël Robert[†], Hervé Chazal[‡], Jean-Luc Schanen[‡], Sébastien Mariethoz*

*Bern University of Applied Sciences, Power Electronics Laboratory, Bienne, Switzerland
Email: timothe.delaforge@bfh.ch
[†]Schneider Electric ITB, Grenoble, France
[‡]Grenoble Electrical Engineering Laboratory G2ELab, Grenoble, France
Email: herve.chazal@g2elab.grenoble-inp.fr

Abstract—The paper presents the optimal design of a magnetically coupled filter for a 500 kVA dc/dc converter. The converter is used for batteries storage applications in UPS systems. The high efficiency and low cost ;14% of converter total price saving; are obtained by using a magnetically coupled filter rather than classical chokes. The design of the filter is achieved using a multi-objective optimization algorithm. Several solutions are compared, including interleaved chokes, coupled monolithic interleg-transformer and coupled separated interleg-transformers. Two prototypes were built. Tests and validations were performed in an industrial environment.

Index Terms—analytical component modeling, dc-dc battery storage, magnetic coupled filter, NSGA-II optimization.

I. INTRODUCTION

In battery storage applications, the energy flows several times from the grid to the storage through the dc-dc converter. For the system to be efficient and competitive, the converter and its passive components must guarantee high efficiency. In these UPS applications no galvanic isolation is required so classical hard-switching legs with inductor filter are well suited. The power stage design is done independently of this paper and results in two independent 250 kVA power blocks with each three interleaved dual 450 A-1.2 kV IGBT legs operated at 4 kHz.

A. Paper contributions: optimal design of a passive filter for a 500 kVA dc-dc battery storage converter

The paper presents the optimal design and prototype validation of a passive filter for two times three legs 250 kVA dc-dc converter. The main contributions are:

1) the application of sophisticated models to the formulation of the design of the filter as an optimization problem; the optimization algorithm is derived from algorithms available in literature [1]–[3];
2) the introduction of separated chokes and interleg transformers for inductive filter with low price, low volume and high efficiency;

3) the comparison between several technological solutions for magnetic filter. Interleaved chokes, monolithic and separated interleg transformers;
4) a fault tolerant coupling strategy;
5) the realization of prototypes and their validation at full converter power;
6) complete set of tests, such as control of the power, the coupling electrical behavior, noise, EMC, temperature;
7) a coupled filter that enables a reduction of 62.5% of the filter weight and a reduction of 55.5% of the filter price.

B. Magnetic filter and coupled solutions

Converter performances enhancement is usually achieved through active devices and command improvement. Yet passives still contribute in a large part of cost and losses in converters Fig. 1. Interleaved topologies are proposed to reduce capacitor and coupled topologies to reduce magnetics in [4]–[12]. These studies are however not applied on industrial converter, as they do not take into account standards and mechanical integration for examples. The solutions they propose are mainly intercell monolithic transformers, which have important drawbacks in comparison with the proposed solution as will be presented in last section of this paper.

C. Advantages of the proposed filter solution

The proposed optimal design of the filter has the following advantages over existing solutions:

1) the filter is designed using optimization algorithm and libraries of available components; this dramatically reduce the costs of prototyping and the time to market;
2) the filter is based on separated interleg transformer. The inductance and coupling function are split into two components. This results in a better design flexibility, a reduction of the total weight, an improved modularity and a better tolerance to disturbance;
3) the filter is fault tolerant and auto-balanced;

The 2018 International Power Electronics Conference

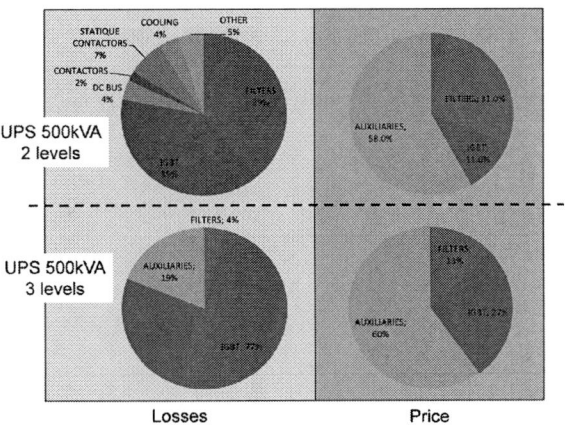

Fig. 1: Breakdown of losses and prices of two topologies 500 kVA UPS. Internal sources.

4) the filter is designed addressing industrial requirements: i.e. feasibility, EMC, competitive price, mechanical integration, noise, etc;

II. MAGNETIC COUPLED FILTER OUTLOOK

A. Objectives of coupled solution

The dc-dc converter is a three interleaved IGBT legs as shown in Fig. 2. An inductive filter must be implemented to reduce the current ripple on the batteries for both charge and discharge modes. The classical interleaving of the legs is a phase shift of the modulator carriers. For three legs a 120° phase shift is applied to the carriers. This solution reduces the current ripple on the batteries but does not reduce the ripple in the chokes and in IGBT legs.

To reduce the current ripple on the legs and to reduce the filtering need and the IGBT losses, solutions based on coupled inductors were proposed [4]–[13].

B. Interleg transformer

1) Topology: To significantly reduce the filter size, interleg transformers are used. Several topologies have been proposed [14]. For the optimization, we focus on the most relevant that are the cyclic cascade configuration Fig. 3, the cyclic parallel configuration Fig. 4 and the secondary loop configuration Fig. 5. The combinatorial configurations will not be considered because they require too many transformers for higher number of legs and they are then not economically viable.

2) Technological choice: The authors choose to split inductance and coupling functions in two separated components. The inductance part of classical interleg transformer and the transformer part. The reasons are:

1) Experience shows better capability to achieve an optimal design of the whole filter. The main reason is that different magnetic materials being better for each function;
2) No leakage inductance is required in transformer for the coupling part, limiting EMI in the air and additional winding loss;

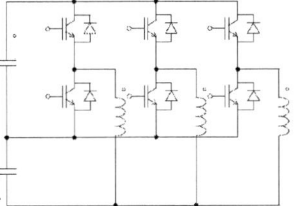

Fig. 2: dc-dc three interleaved legs solution with classical inductor filter.

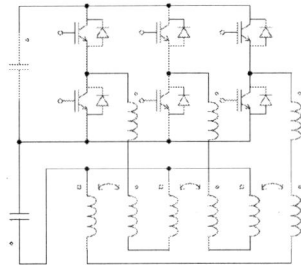

Fig. 3: dc-dc three interleaved legs solution with cyclic series coupled filter.

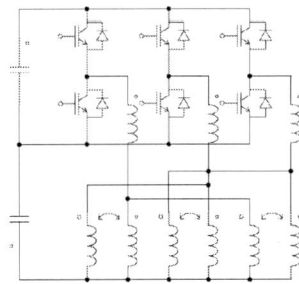

Fig. 4: dc-dc three interleaved legs solution with cyclic parallel coupled filter.

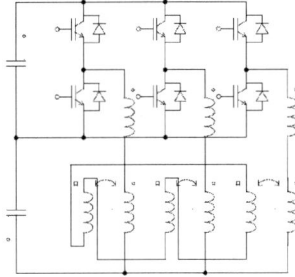

Fig. 5: dc-dc three interleaved legs solution with secondary loop coupled filter.

3) This assembling allows floating potentials giving capability to auto-balance the legs;
4) Even with saturation of the transformer, the interleave

operation of the filter is still possible.

3) analytical study: The previous interleg topologies, the cyclic cascade, the cyclic parallel and the secondary loop configuration are retained. The aim of the coupled filter is to reduce the current ripple in order to reduce both the losses and the required inductance. The formulas for the current ripple are calculated with separated inductance and mutual parts.

The current ripple in a leg for the interleaved solution is given by:

$$\Delta I = \frac{1}{L} \cdot V \cdot \Delta t \quad (1)$$

With V the choke voltage and L the inductance value.

For the cyclic cascade configuration the ripple on leg i is given by:

$$\Delta I = \frac{L \cdot V_i + M \cdot \sum_k V_k}{L \cdot (L + 3M)} \cdot \Delta t \quad (2)$$

With V_i the leg i voltage, L the inductance value and M the coupling value.

For the cyclic parallel configuration the ripple on leg i is given by:

$$\Delta I = \frac{L \cdot (L \cdot V_i + \frac{M}{2} \cdot (2V_i + V_{i+1})) + (\frac{M}{2})^2 \cdot \sum_k V_k}{(L + \frac{M}{2})^3 - (\frac{M}{2})^3} \cdot \Delta t \quad (3)$$

For the secondary loop the ripple on leg i is given by:

$$\Delta I = \frac{3L \cdot V_i + M \cdot \sum_k V_k}{L \cdot (3L + 3M)} \cdot \Delta t \quad (4)$$

The maximums of current ripple are compared on Fig. 6 for the worth case of our converter, i.e. end of discharge. The interleaved configuration requires three time the inductance value for same ripple as we have a three coupled legs configuration for the other coupled topologies. The secondary loop and cyclic parallel are almost equivalent in terms of ripple. The cyclic series has a lower ripple than other for same inductance and coupling values.

Note that for a correct coupling and thus minimum ripple the mutual inductance from transformer must be bigger than inductance.

III. MULTI-OBJECTIVE OPTIMIZATION MODEL

The objective of the optimization is to find the Pareto frontier, that features the best achievable trade-off between filter price, volume and the converter efficiency (filter + IGBT). To this aim four optimizations are launched, one for each filter configuration from II-B.

A. Optimization definition

In order for the resulting design of the optimization to be close to final product and limits expensive prototyping iterations, libraries of available components and materials with their properties are implemented. Each solution is computed from a set of existing components with market price and efficiency very close to reality. The libraries contain

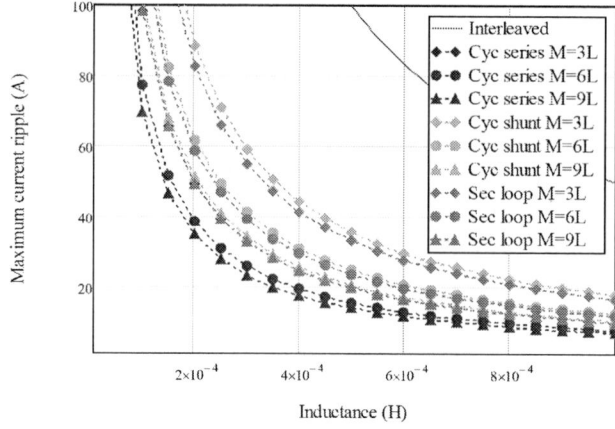

Fig. 6: dc-dc three legs, comparison of leg maximum current ripple for several solutions of filter.

coefficients for each component that are used as inputs to analytical models. To deal with discrete possibilities of design an optimization algorithm derived from NSGA-II [1], [2] is used.

The objectives are:

1) minimization of the price
2) minimization of the volume
3) maximization of the efficiency including filter + IGBT

The constraints are

1) thermal limitations of components
2) the filter still operating with 5% of current imbalance between legs
3) maximum available volume in converter cabinet

The optimization inputs are:

1) the library of magnetic materials
2) the size of toroidal choke and transformer core
3) the winding properties (number of turns and parallel wires, wires diameter and material)

B. Optimization models

1) IGBT: The IGBT have been selected previous to this work. The parameters to compute conduction and switching losses are fitted from manufacturer datasheets.

2) Core loss model: Both inductor and interleg transformer cores are modeled using the LossMap model from [15]. This model takes into account the current bias effect on the magnetic losses. The parameters for several materials, powder / amorphous / ferrite and nanocrystallin are implemented in a library with magnetic hysteresis, price of cores, density and other properties.

3) Winding model: The windings frequency resistance and then the losses are computed from [16]. The winding can be either aluminum or copper.

The 2018 International Power Electronics Conference

IV. OPTIMIZATION RESULTS

A. Catalog sizes

The results of the optimization are presented in Fig. 7 and Fig. 8. Fig. 7 presents the best trade-off between the filter losses and the filter price for 6 legs, with 2times 250 kVA blocks of 3 coupled legs. It is shown that the best solutions are obtained with the cyclic cascade configuration as found in Fig. 6. The interleaved configuration has the same price and loss Pareto frontier than the other coupled solutions. The reason for that is that only six chokes are required and no transformer. Yet Fig. 8 shows that with only six chokes, the interleaved solutions are heavier than coupled solutions that have six chokes and six transformers.

The Paretos present non-smooth curves. The breaks correspond to powder materials selected for the chokes. The material for the inlerleg transformers is always the same, medium permeability amorphous.

Each solution on the paretos is made from components available in catalogs and can be directly manufactured. No iteration is required.

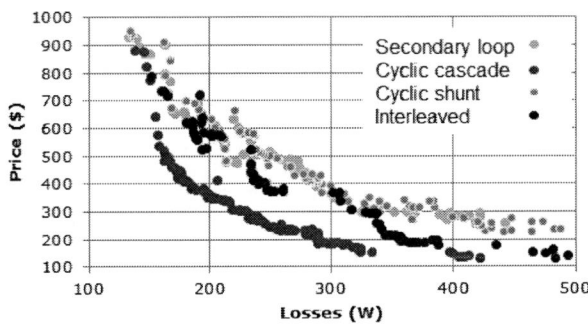

Fig. 7: Paretos between filter loss and filter price from optimizations on different filter topologies.

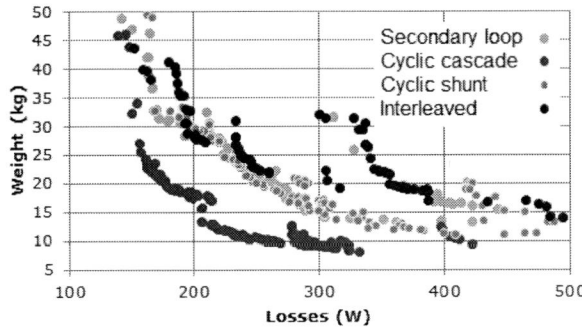

Fig. 8: Paretos between filter loss and filter volume from optimizations on different filter topologies.

B. Optimal size

The optimization is based on available size of toroid core from suppliers. Customized sizes can however be realized with a one time cost to build a customized matrix. For large series this cost can be recouped with achievable gain. Fig. 9 shows the gain achieved for the cyclic cascade configuration with free size of toroid core for both chokes and transformers. For same performance an average of 15% of the filter cost can be saved.

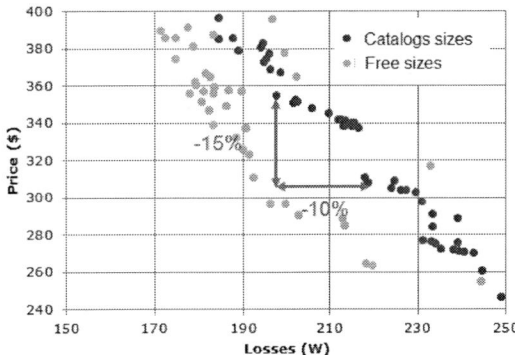

Fig. 9: Two paretos from optimization of cyclic cascade configuration. One is done with catalog size for magnetic cores. The other is done with free size of cores.

C. Feedback

As presented previously, in order to obtain good coupling properties and reduce the current ripple, the mutual inductance of the transformer must be bigger than the inductance value. Meanwhile, for the whole filter to tolerate differential currents, classical transformer material with high permeability cannot be employed while they guarantee a high mutual they saturate with small differential currents. That is why middle permeability amorphous materials are well suited for coupled filter. They assure a good mutual while supporting high differential current.

The optimal sizes for inductance and transformer are reached with value of inductance and mutual that result in an equal repartition of the leg voltage on inductance and transformers.

V. PROTOTYPE EXPERIMENTAL VALIDATION

A. Nominal electrical behavior of the filter

The prototype is build from one solution of the cyclic cascade configuration pareto. The chokes are realized with low permeability powder core and the interleg transformer with medium permeability amorphous. Classical solid round conductors are used. One block of the three coupled legs is tested Fig. 10.

The nominal operation of the filter is presented in upper screenshot in Fig. 11 without any control. The filter has a 40 A ripple current with a 220 A mean current. Even with 1.3% current imbalance between leg 1 and 2 the filter works

2161

perfectly. The cyclic configuration allows interleg transformer with floating voltage guarantying auto-balance of the coupling. A start of the UPS from idle mode to full charge is shown on lower screenshot Fig. 11. The filter is coupled and balanced during the whole starting phase.

Fig. 10: 250 kVA prototype. One block of three magnetically coupled leg with 450 A-1.2 kV IGBT and separated choke+transformer per leg configuration.

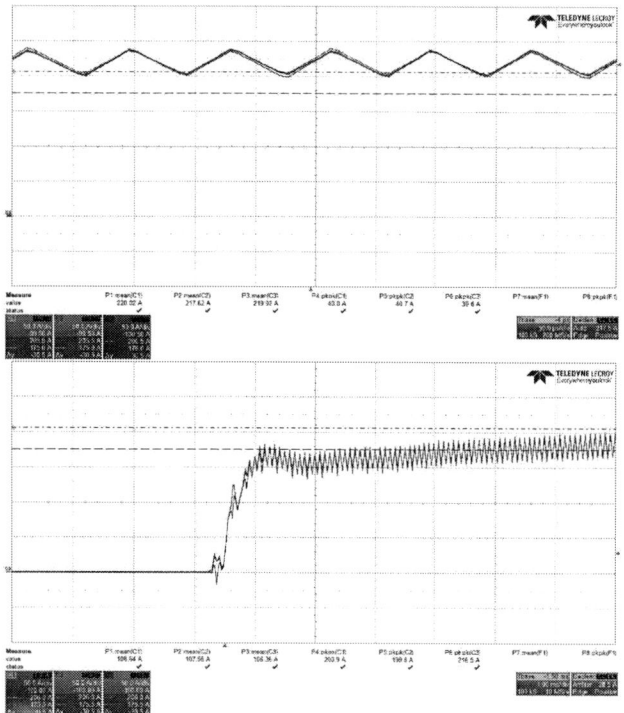

Fig. 11: 250 kVA prototype. Upper screenshot, nominal filter operation with no control, 220 A per leg with 40 A ripple. Lower screenshot, start of the dc-dc from idle to full load. The filter is balanced during whole phase.

B. Fault mode operation

One of the main advantage of separated interleg transformer configuration is the capability to operate in fault mode. With several interleg transformers, a smart assembling can result with floating potential for one or more transformers.
The gain is that even with unpaired value of mutual and inductance on each leg, the voltages will be automatically balanced on the floating transformers and result in balanced leg currents Fig. 12.

The proposed configuration choke + interleg transformer with cyclic cascade configuration is exposed to an artificial over differential current situation as illustrated in Fig. 13.

On top, a two coupled legs configuration is tested. It is shown that when the differential current is too high the transformers saturate and their voltages drop. When the differential current is lower the transformers are working and the coupling is restored. On the rest of the period the legs operate as classical interleaved legs.

On bottom, the three legs topology is submitted to a huge differential current. Once again one of the transformers is saturated. In this three legs configuration, only one leg works with interleave equivalent ripple while the two other legs still operate as two coupled legs.

The voltage drop on the saturated transformer is distributed on the other transformers and inductances.

Finally, to demonstrate the importance of floating transformer, the same case is applied on a two coupled leg configuration with no floating transformer Fig. 14. This case is similar to a monolithic interleg transformer. The current ripple on the unbalanced leg becomes huge with the saturation of the transformer.

This could result in the explosion of the converter semiconductors.

This operating mode lead to an emergency stop of the converter while the previous mode can preserve the operation of the converter or lead to a derated operation but still functional.

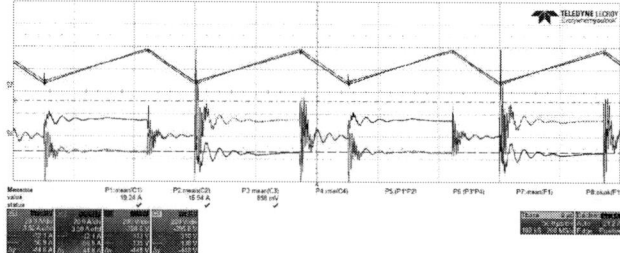

Fig. 12: 500 kVA prototype. Two legs with two interleg coupling transformers. Currents and transformers voltages. Difference between inductance values $6.67\mu H$ 2.5%, difference between mutual values $313\mu H$ 10%.

The 2018 International Power Electronics Conference

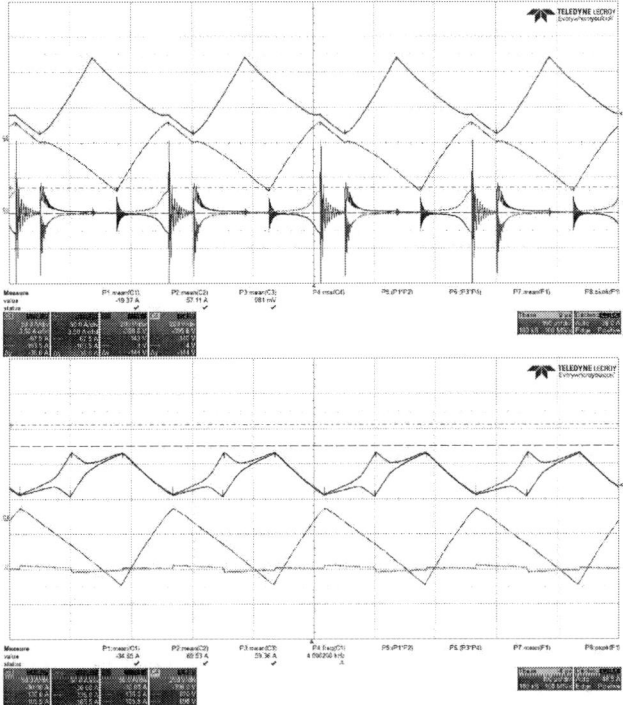

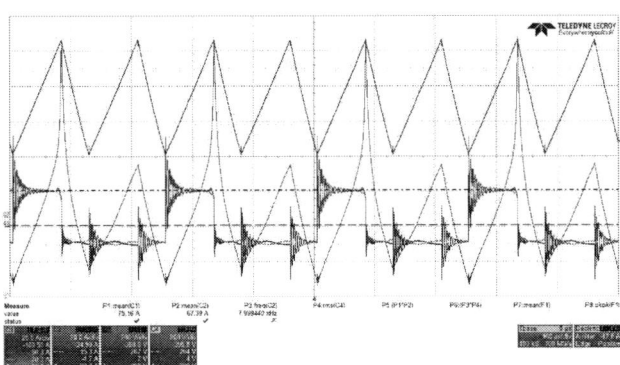

Fig. 13: 500 kVA prototype. Unbalanced leg current artificial operation to show filter operation. The unbalanced leg has a saturation of its coupling transformer but keeps choke capability. Other legs are lightly impacted. Top two coupled legs, bottom three coupled legs.

Fig. 14: 500 kVA prototype. Two legs with two interleg coupling transformers. Currents and transformers voltages. No floating transformer case, saturation of the transformer with differential current.

VI. SEPARATED INTERLEG TRANSFORMER VS MONOLITHIC TRANSFORMER

The classical monolithic transformer solution is also build from an optimization with the three legs converter topology Fig. 15. The same constraints are applied. The monolithic solution is selected with the constraint to fit in the same box that the separated component solution. The conclusion of the comparison gives:

1) The price and the weight of the monolithic solution are the same than the separated solution;
2) The monolithic requires a complicated control to prevent saturation of the core with current unbalanced situation;
3) It can not operate in fault mode because of the saturation of the core;
4) The use of leakage flux for coupling leads to high magnetic field in the air with consequences: LEM measure perturbation, vibration of converter metallic parts, higher EMI.
5) The use of air-gap even with silicon creates higher noise and winding losses.
6) The total weight is equal to separated solution. But it is one piece of 33 kg difficult for mechanical assembling.
7) The monolithic can only be use for one configuration. The separated solution can be use in another power rated converter with same current per leg.

Fig. 15: 500 kVA monolithic interleg transformer prototype.

VII. CONCLUSION

The paper has presented the optimal design and the prototype validation of a magnetically coupled filter for a multi-leg dc-dc converter. The design has been done with optimization models and a multi-objective algorithm. It is based on available libraries of components to limit time of prototyping phase.

A technological solution is proposed with one choke and one interleg coupling transformer per leg. Two prototypes are build from proposed solution and classical monolithic solution and compared. Tests and validation to full power are presented.

The capability of fault mode operation of separated solution with floating transformers is proved. This particular capability

2163

enables the converter derated operation and prevent high peak currents causing semiconductor damage.

The coupled filter lead to a reduction of 62.5% of the filter weight and a reduction of 55.5% of the filter price that is 14% of the total converter price. The efficiency of the filter is 98.8%.

REFERENCES

[1] K. Deb, S. Agrawal, A. Pratap, and T. Meyarivan, "A Fast Elitist Nondominated Sorting Genetic Algorithm for Multi-objective Optimisation: NSGA-II," in *Proceedings of the 6th International Conference on Parallel Problem Solving from Nature*, ser. PPSN VI. London, UK, UK: Springer-Verlag, 2000, pp. 849–858. [Online]. Available: http://dl.acm.org/citation.cfm?id=645825.668937

[2] J. Knowles and D. Corne, "The pareto archived evolution strategy: a new baseline algorithm for pareto multiobjective optimisation," in *Proceedings of the 1999 Congress on Evolutionary Computation-CEC99 (Cat. No. 99TH8406)*, vol. 1, 1999, p. 105 Vol. 1.

[3] E. Zitzler, K. Giannakoglou, D. Tsahalis, J. Periaux, K. Papailiou, T. F. (eds, E. Z. Ler, M. Laumanns, and L. Thiele, *SPEA2: Improving the Strength Pareto Evolutionary Algorithm For Multiobjective Optimization*, 2002.

[4] P.-L. Wong, P. Xu, P. Yang, and F. C. Lee, "Performance improvements of interleaving vrms with coupling inductors," *IEEE Transactions on Power Electronics*, vol. 16, no. 4, pp. 499–507, Jul 2001.

[5] J. Li, C. R. Sullivan, and A. Schultz, "Coupled-inductor design optimization for fast-response low-voltage dc-dc converters," in *APEC. Seventeenth Annual IEEE Applied Power Electronics Conference and Exposition (Cat. No.02CH37335)*, vol. 2, 2002, pp. 817–823 vol.2.

[6] W. Wu, N.-C. Lee, and G. Schuellein, "Multi-phase buck converter design with two-phase coupled inductors," in *Twenty-First Annual IEEE Applied Power Electronics Conference and Exposition, 2006. APEC '06.*, March 2006, pp. 6 pp.–.

[7] E. Laboure, A. Cuniere, T. A. Meynard, F. Forest, and E. Sarraute, "A theoretical approach to intercell transformers, application to interleaved converters," *IEEE Transactions on Power Electronics*, vol. 23, no. 1, pp. 464–474, Jan 2008.

[8] F. Forest, E. Labouré, T. A. Meynard, and V. Smet, "Design and comparison of inductors and intercell transformers for filtering of pwm inverter output," *IEEE Transactions on Power Electronics*, vol. 24, no. 3, pp. 812–821, March 2009.

[9] T. Meynard, B. C. Laplace, F. Forest, and E. Labouré, "Parallel multicell converters for high current: Design of intercell transformers," in *2010 IEEE International Conference on Industrial Technology*, March 2010, pp. 1359–1364.

[10] B. Cougo, T. Friedli, D. O. Boillat, and J. W. Kolar, "Comparative evaluation of individual and coupled inductor arrangements for input filters of pv inverter systems," in *2012 7th International Conference on Integrated Power Electronics Systems (CIPS)*, March 2012, pp. 1–8.

[11] D. O. Boillat and J. W. Kolar, "Modeling and experimental analysis of a coupling inductor employed in a high performance ac power source," in *2012 International Conference on Renewable Energy Research and Applications (ICRERA)*, Nov 2012, pp. 1–18.

[12] C. Nan and R. Ayyanar, "A 1 mhz bi-directional soft-switching dc-dc converter with planar coupled inductor for dual voltage automotive systems," in *2016 IEEE Applied Power Electronics Conference and Exposition (APEC)*, March 2016, pp. 432–439.

[13] S. Sanchez, D. Risaletto, F. Richardeau, T. Meynard, and E. Sarraute, "Pre-design methodology and results of a robust monolithic inter cell transformer (ict) for parallel multicell converter," in *IECON 2013 - 39th Annual Conference of the IEEE Industrial Electronics Society*, Nov 2013, pp. 8198–8203.

[14] I. G. Park and S. I. Kim, "Modeling and analysis of multi-interphase transformers for connecting power converters in parallel," in *PESC97 Record 28th Annual IEEE Power Electronics Specialists Conference. Formerly Power Conditioning Specialists Conference 1970-71. Power Processing and Electronic Specialists Conference 1972*, vol. 2, Jun 1997, pp. 1164–1170 vol.2.

[15] T. Delaforge, "Optimal sizing of passive components in power converters using discrete methods," in *PhD*, Mars 2016.

[16] T. Delaforge, H. Chazal, J. L. Schanen, and R. J. Pasterczyk, "Copper losses evaluation in multi-strands conductors formal solution based on the magnetic potential," in *2015 IEEE Energy Conversion Congress and Exposition (ECCE)*, Sep. 2015, pp. 3057–3063.

High Power/Current Inductor Loss Measurement with Shunt Resistor Current-sensing Method

Pin Yu Huang* and Toshihisa Shimizu
Tokyo Metropolitan University, Tokyo, Japan
*E-mail: eins0620@gmail.com

Abstract— With the high-power-density converters requirement increasing, the power loss analysis of inductive components requested further accurate. Some of the inductors are operated at high frequency and current conditions. However, the high precisely measurement instruments, B–H analyzer, only can provide maximum $\pm$ 6A AC ripple current condition at present. In this paper, a low-cost measurement method by using the shunt resistor as a high current sensor is proposed. To abbreviate the measurement deviation cause by the parasitic inductance of shunt resistor, the Runge-Kutta calculation method is applied to figure out the inductor current. Furthermore, a shunt resistor connecting configuration is proposed to reduce the parasitic inductance influence. To demonstrate the feasibility, an inductor is measured under 160 A (RMS) current, 10 kHz switching frequency, and 50 % duty cycle square-wave-voltage measurement conditions as an example.

Keywords— *Inductor loss estimation; High current detection; B–H Analyzer; High-power density; High current shunt resistor.*

I. INTRODUCTION

Thanks to advances in technology, more and more power electronics applications are developed to meet the needs of human life. How to enhance the power density of converters become an important issue in power converter design. With the significant advancement of the semiconductor devices, e.g. the SiC and GaN materials, the switching frequency of power converters have enable to be increased. The volume of inductor can be effectively reduced because of the improvement of the operating frequency. More attention is paid to the estimation of inductor loss. Several iron loss calculation methods are proposed to realizing the high-power-density converter [1]-[9]. In addition, the accurate power losses experimental measurement of inductive components can aid researchers and designers in inductor investigation and loss calculation verification.

At present, the inductor is operated at over 100A current condition in more and more high-power applications. The B–H analyzer has high accuracy to measure the inductor characteristics. However, it has only maximum $\pm$ 30 A DC bias current and $\pm$ 6A AC ripple current testing conditions. It cannot deal with the high current measurement condition.

In [7], the calorimetric determination method was proposed, which has high accuracy and immune against EMI phase errors features. However, the copper and iron losses of inductor cannot be distinguished, and a high cost temperature chamber is required. In [8], the inductor loss measurement was established as measuring the secondary winding voltage and primary winding current of inductive components. It is appropriate for analyzing the inductor losses. A shunt resistor with very small associated stray inductance and capacitance is required. However, the parasitic components of current-sensing resistor will be increased corresponding to the power rating of resistor while measuring higher power inductor.

To detect the high inductor current conditions, the current probe is one of the candidates. However, in [8], the author noted that the phase shift errors of current probe strongly influence the measurement accuracy. Therefore, in [9], the phase error correction method for current probe was proposed to solve this problem. The measurement accuracy by using current probe can be greatly improving. However, the measurement results is very sensitive to the corrected phase difference, as a result, the high precisely phase difference measurement instrument, B–H analyzer, is required for current probe compensation. Shunt resistor is one of the low-cost current-sensing candidates. However, the volume of resistor is increased corresponding to the measurement power rating. Using more paralleled resistors make the current unbalance distribution of each resistors and the increasing of parasitic components lead to the current measurement deviation.

In this paper, a shunt resistor paralleled-connected configuration is proposed for measuring inductor loss under high current amplitude condition. Moreover, the Runge-Kutta calculation of differential equation solution is applied to figure out the inductor current that the result excluded parasitic inductance influence from the sensing voltage of shunt resistor

II. RUNGE-KUTTA METHOD FOR INDUCTOR CURRENT CALCULATION

Fig. 1 is the generally used equivalent model of shunt resistor which include the series-connected parasitic inductor and parallel-connected parasitic capacitance. To simplify the analysis, the parasitic capacitance is not considered in this paper. Further, the influence of parasitic capacitance is small enough to be ignored.

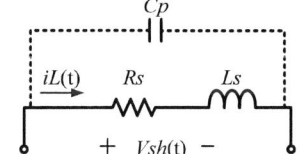

Fig. 1. Equivalent circuit of shunt resistor.

While using the shunt resistor as the current detection of inductor loss measurement, the parasitic inductance generate additional voltage resulted in the measurement deviation, as shown in Fig. 2.

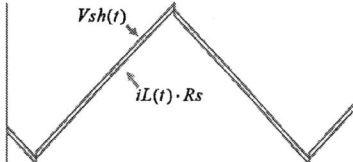

Fig. 2. Voltage waveform of shunt resistor.

The practical current of inductor through shunt resistor could be found as a first order differential equation as expressed in (1).

$$\frac{d}{dt} iL(t) = \frac{Vsh(t)}{Ls} - \frac{iL(t) \cdot Rs}{Ls} \qquad (1)$$

The Runge-Kutta method is a good candidate for numerical solutions of differential equations. In this paper, the current of inductor is calculated by using the classic Runge-Kutta method, RK4. The explicit Runge-Kutta mathematically calculation will not describe in this paper. Several software, e.g. Matlab, could help in executing the calculation.

III. PROPOSED SHUNT RESISTOR

Fig. 3(a) is one of the generally used DC shunt resistors for ammeter applications. However, it is more suitable for the low frequency application. In order to improve the accuracy of Runge-Kutta method calculation, the resistance and parasitic inductance of shunt resistor, Rs and Ls, have to be considered as constant real number in (1). Therefore, a shunt resistor which has good frequency and temperature characteristics is required to obtaining precisely solution of the Runge-Kutta method, as shown in Fig. 3(b).

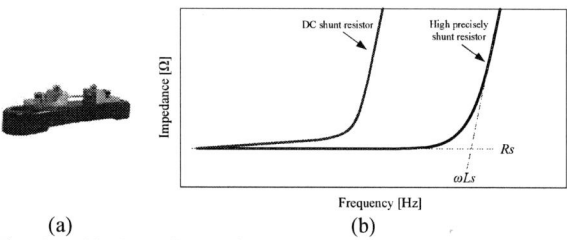

(a)   (b)

Fig. 3. (a) Generally used DC shunt resistor and (b) Frequency characteristics of recommended shunt resistor.

In addition, the parallel-connected shunt resistor configuration is generally used when the power rating of inductor measurement is increased, as shown in Fig. 4. However, the connecting wires make additional parasitic inductor and increased follow the number of parallel unit. In Fig. 5, each of current loops contains different parasitic inductances that the unbalanced current wave shape resulted in the calculation and measurement difficulties increased.

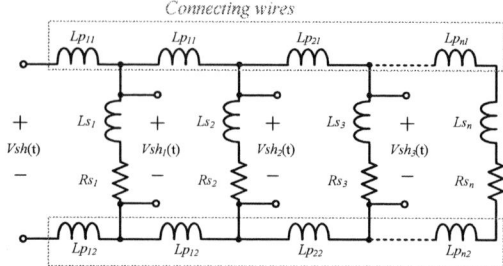

Fig. 4. Equivalent circuit of parallel-connected shunt resistor.

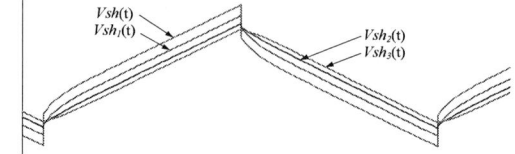

Fig. 5. Current waveforms of each shunt resistor in parallel-connected configuration.

Consequently, the equivalent circuit of proposed connecting concept is illustrated in Fig. 6. To ensure that the value of parasitic inductance on each paralleled component are as close as possible, the symmetrical construction is required.

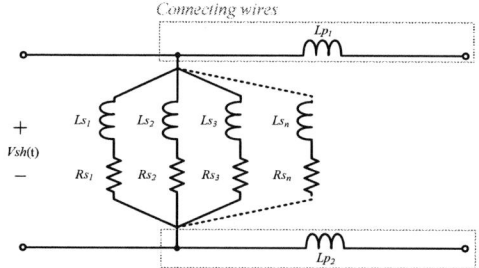

Fig. 6. Equivalent circuit of proposed shunt resistor connecting method.

Fig. 7 is the practical view and equivalent circuit of the proposed parallel-connected shunt resistors bank which used in this paper. It consists of five shunt resistors units. The issue of parasitic inductance unbalanced can be improved by means of the symmetrical construction. In addition, more paralleled units are possible to use when measurement power rating is increased.

In addition, the position of current signal measurement point is become a critical issue to improve the detection accuracy because the volume of shunt resistor will be greatly increased while measuring the high power rating inductor loss. For example, the wide distance of two measurement terminals resulted in long grounding wires of probe and easily coupling the magnetic flux cause by the high measurement current.

In Fig. 7 (b), the red circle indicated the current signal measurement point of the proposed configuration. Two terminals are set as close as possible, therefore, the BNC adaptor can be used to connect the oscilloscope probe tip directly. The measurement accuracy can be improved by minimizing the grounding wire of probe.

2166

The 2018 International Power Electronics Conference

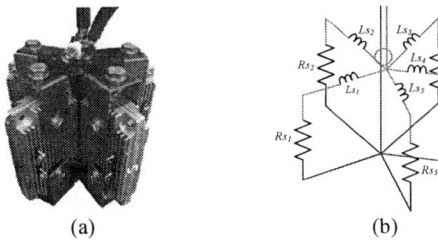

Fig. 7. (a) Practical view and (b) equivalent circuit of shunt resistors which proposed in this paper.

Each shunt resistor unit consists of two paralleled-connected ultra-precision shunt resistors, PSBX1R000B, which has very good frequency and temperature responses characteristics. Fig. 8 shows the single component, the vertical, and the lateral views of one shunt resistors unit, respectively. In Fig. 8(b) and Fig. 8(c), the resistors are paralleled and connected back to back that the heat sink of resistors has the biggest contact area. This configuration can help the resistors have well heat distribution in high power rating inductor loss measurement.

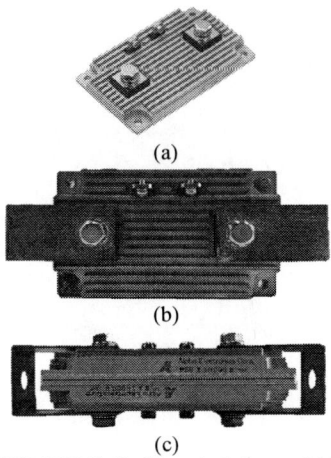

(a)

(b)

(c)

Fig. 8. (a) PSBX1R000B (1 Ω), (b) vertical view, and (c) lateral view of one shunt resistors unit.

To realize the Runge-Kutta method, the parasitic components of proposed shunt resistors bank are measured and calculated by Fig. 9, (2), and (3), respectively. The calculation results of the proposed shunt resistors bank are listed in TABLE I.

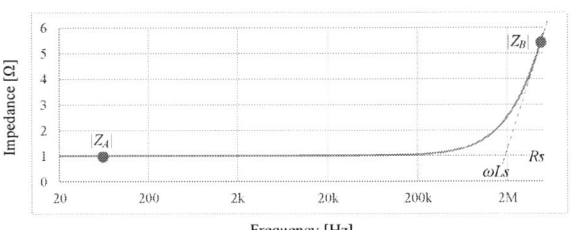

Fig. 9. Frequency characteristic of proposed shunt resistors bank and the calculation of parasitic components.

$$Ls = \sqrt{\frac{\left|Z_B\right|^2 - \left|Z_A\right|^2}{\omega_B^2 - \omega_A^2}} \qquad (2)$$

$$Rs = \sqrt{\left|Z_A\right|^2 - \left(\omega Ls\right)^2} \qquad (3)$$

TABLE I
SPECIFICATION OF PROPOSED SHUNT RESISTORS BANK

	Value
Used Ultra-precision Resistor	PSBX1R000B (1 Ω)
Number of Paralleled Resistors	10
Total Resistance, Rs	0.1009 Ω
Parasitic inductance, Ls	67.59 nH

IV. MEASUREMENT DEVIATION CORRECTION

In order to realize accurate measurement, two important calibration steps of oscilloscope are introduced as follow.

A. Deskew Correction for Phase Error

The phase displacement between the two voltage probes should be corrected [8]. The relative error in percentage, k, can be expressed as (4), where θ denotes the actual phase angle between voltage and current and $\Delta\varphi$ the phase error between the voltage and current sensor.

$$k = \frac{\cos\left(\theta + \Delta\varphi(f)\right) - \cos\left(\theta\right)}{\cos\left(\theta\right)} \times 100 \ [\%] \qquad (4)$$

One of the simple solutions is to utilize the deskew function of the oscilloscope to adjust the phase difference between two probes.

B. DC Offset Correction of Measurement Probes

The DC offset errors are inherent in the oscilloscopes and probes. The value of DC offset errors will be change with the different attenuations and scales of channel. The inductor loss measurement deviation of DC offset error can be calculated in (5) to (8), where the phase displacement is not considered.

$$iL(t) = iL(t)_{original} + iL_{dc\_offset} \qquad (5)$$

$$VL(t) = VL(t)_{original} + VL_{dc\_offset} \qquad (6)$$

$$P_{loss} = \frac{1}{T}\int\left[VL(t)\cdot iL(t)\right]dt \qquad (7)$$

$$P_{loss\_error} = VL_{dc\_offset}\cdot\left(iL_{average} + iL_{dc\_offset}\right) \qquad (8)$$

Fig. 10 shows calculation results of the inductor power loss deviation when the probes have different DC offset errors. The calculation results of inductor power loss deviation are made under 100 A (RMS) current, ± 100 V voltage, 10 kHz switching frequency, and 50 % duty cycle square-wave-voltage conditions. The DC offset errors are represented in a percentage of average current and peak voltage of inductor, respectively. As shown in Fig. 10, the DC offset error of voltage probe is greatly affect to the inductor power loss measurement. Only 0.5 V (50 V×0.01=0.5 V) DC offset error of inductor voltage leads the inductor loss measurement to 50 W deviation. To abbreviate these problems, the DC offset and phase calibrations are strongly recommended before beginning the inductor power loss measurement.

2167

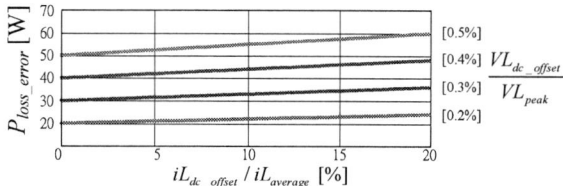

Fig. 10. Inductor power loss deviation corresponding to different DC offset errors of probes under 50 % duty cycle, $\pm$ 100 V square-wave-voltage, 10 kHz switching frequency, and 100 A (RMS) current conditions.

V. MEASUREMENT SETUP

Fig. 11 shows the inductor measurement system which used in this study. It is an energy circulating measurement system. The device under test (DUT) of measured inductor is set to the output inductor of buck converter. Thus, the inductor is operated and measured under square-wave-voltage condition. In addition, the voltage and current of measured inductor can be controlled to target value through system feedback control. The output capacitor of buck converter, C_2, is considered as an input capacitor of the boost converter and then transfer the energy back to the input capacitor of buck converter, C_1. Therefore, the power supply only need to provide the energy for inductor power loss and system energy dissipation. The proposed shunt resistors bank is series connected to the DUT for inductor current detection.

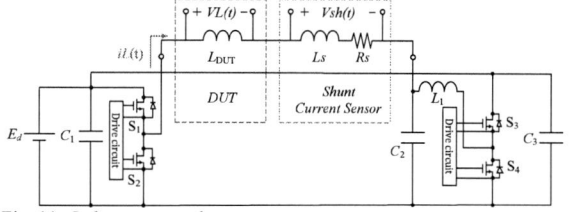

Fig. 11. Inductor power loss measurement system.

A low attenuation probe is recommend to maximize the oscilloscope's vertical sensitivity. In addition, the sufficient bandwidth and high input impedance of measurement instrument is preferred. Therefore, the 10X passive probes with IWATSU isolation amplifier (SE - 6000/SE - 6010) are applied to the measurement system.

In addition, the proposed shunt resistors bank is put in the water cooling tank to enhance the measurement power rating. Meanwhile, the shunt resistor can operate under well temperature distribution condition.

VI. EXPERIMENTAL RESULTS

To demonstrate the measurement method feasibility, the total loss of inductor is measured as an example. The measurement conditions are listed in TABLE II.

TABLE II
INDUCTOR MEASUREMENT CONDITIONS

	Value
Inductance	209 μH
Switching frequency	10 kHz
Voltage of inductor	$\pm$ 100 V
Duty cycle	50 %
Inductor current	10 A (RMS) ~ 160 A (RMS)

Fig. 12. is the measurement results of proposed method and high precisely power analyzer, PPA5530. The maximum measured current range of common used power analyzer, PPA5530, is limited to 30 Ampere. As shown, the results of two different methodologies are basically matched under 30A current conditions. It could be found that two methods have same accuracy on inductor loss measurment. Furthermore, the proposed method provide a potential solution to measure the inductor loss at high current conditions.

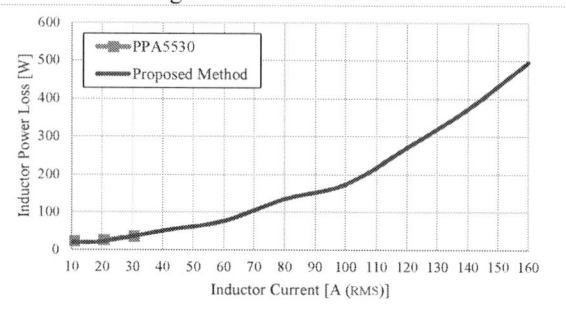

Fig. 12. Total loss comparison between power analyzer, PPA5530, and proposed method.

The Runge-Kutta method is applied to calculate the practical inductor current. Without considering the parasitic inductance effect to shunt resistor detecting measurement method, the significant error was occurred as shown in Fig. 13.

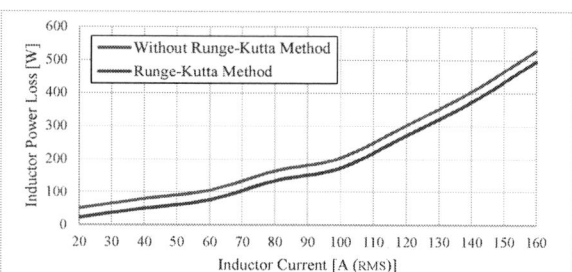

Fig. 13. Measurement results with corresponding Runge-Kutta calculation method.

DC offset errors of oscillation scopes will happen when using different probes. Moreover, the DC offset errors will be changed with different display scales while the measurement conditions are increased. Fig. 14 shows the measurement results without DC offset correction. As shown, the deviation increasing significantly when the high current measurement conditions because large scale of channel on oscillation scope lead to decrease the minimum unit accuracy of measuring waveform.

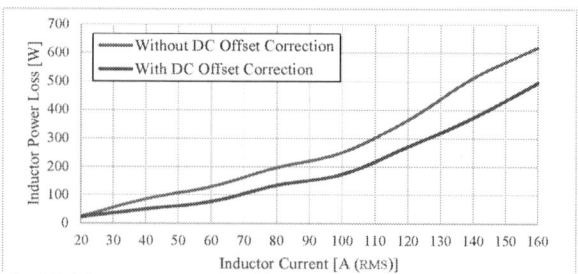

Fig. 14. Measurement results with corresponding DC offset correction.

Fig. 15 shows the temperature distribution of proposed shunt resistors bank under 100 A (RMS) inductor current condition. The temperature of shunt resistor is uniform distributed by means of the water cooling tan and designed configuration.

 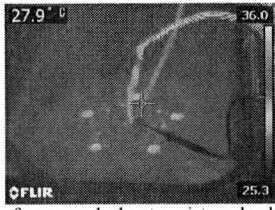

Fig. 15. Temperature distribution of proposed shunt resistors bank under 100 A (RMS) inductor current condition.

VII. CONCLUSIONS

A measurement method of high power rating inductor loss measurement is achieved by means of shunt resistor current detection. The proposed connecting configuration of shunt resistors make identical parasitic inductance on each of paralleled current loops. In addition, by using the Runge-Kutta calculation method, the inductor current can be figured out to minimum the parasitic inductor effects of current detection. The inductor is measured under 10 ~ 160 A (RMS) current, 10 kHz switching frequency, and 50 % duty cycle square-wave-voltage measurement conditions. The measurement results with proposed method are compared with power analyzer, PPA5530, under 10 ~ 30A (RMS) current conditions. The inductor loss measurement under high current conditions is extremely sensitive with measurement instruments and detecting components, such as temperature, DC offset errors, phase errors, practice components, and so on.

Power analyzer and B–H analyzer provide high precisely loss measurement for inductor loss but limit in low current conditions. A potential high current inductor loss measurement method with a low-cost shunt resistor current detection is proposed.

ACKNOWLEDGMENT

The authors gratefully acknowledge the contribution of Yoshio Bizen and Hiroaki Matsumori for their technical support. This work was supported by Council for Science, Technology and Innovation (CSTI), Cross-ministerial Strategic Innovation Promotion Program (SIP), "Next-generation power electronics." (Funding agency: NEDO)

REFERENCES

[1] M. Sippola and R. E. Sepponen, "Accurate prediction of high-frequency power-transformer losses and temperature rise," *IEEE Trans. Power Electron.*, vol. 17, no. 5, pp. 835–847, Sep. 2002.

[2] K. Venkatachalam, C. R. Sullivan, T. Abdallah, and H. Tacca, "Accurate prediction of ferrite core loss with nonsinusoidal waveforms using only Steinmetz parameters," in *Proc. IEEE Workshop Comput. Power Electron.*, 2002, pp. 36–41.

[3] S. Iyasu, T. Shimizu, and K. Ishii, "A novel iron loss calculation method on power converters based on dynamic minor loop," in *Proc. Eur. Conf. Power Electron. Appl.*, 2005, pp. 2016–2022.

[4] T. Shimizu and S. Iyasu, "A practical iron loss calculation for AC filter inductors in PWM inverters," *IEEE Trans. Ind. Electron.*, vol. 56, no. 7, pp. 2600–2009, Jul. 2009.

[5] J. Muhlethaler, J. Biela, J. W. Kolar, and A. Ecklebe, "Improved Core-Loss Calculation for Magnetic Components Employed in Power Electronic Systems," *IEEE Transactions on Power Electronics*, vol. 27, no. 2, pp. 964-973, Feb. 2012.

[6] N. Kurita, K. Onda, K. Nakanoue and K. Inagaki, "Loss Estimation Method for Three-Phase AC Reactors of Two Types of Structures Using Amorphous Wound Cores in 400-kVA UPS," *IEEE Transactions on Power Electronics*, vol. 29, no. 7, pp. 3657-3668, July 2014.

[7] D. Christen, U. Badstuebner, J. Biela, and J. W. Kolar, "Calorimetric Power Loss Measurement for Highly Efficient Converters," *The 2010 International Power Electronics Conference*, Sapporo, 2010, pp. 1438-1445.

[8] V. J. Thottuvelil, T. G. Wilson, and H. A. Owen, "High-frequency measurement techniques for magnetic cores," *IEEE Transactions on Power Electronics*, vol. 5, no. 1, pp. 41-53, Jan. 1990.

[9] H. Matsumori, T. Shimizu, K. Takano and H. Ishii, "Evaluation of Iron Loss of AC Filter Inductor Used in Three-Phase PWM Inverters Based on an Iron Loss Analyzer," *IEEE Transactions on Power Electronics*, vol. 31, no. 4, pp. 3080-3095, April 2016.

The 2018 International Power Electronics Conference

Sensitivity Analysis of Medium Frequency Transformer Design

Marko Mogorovic and Drazen Dujic
Power Electronics Laboratory - PEL
École Polytechnique Fédérale de Lausanne - EPFL
Station 11, CH-1015 Lausanne
marko.mogorovic@epfl.ch, drazen.dujic@epfl.ch

Abstract—This paper discusses the technical challenges and trade-offs tied to design of medium frequency transformers (MFTs) for medium-voltage (MV) high-power (HP) power electronic applications, namely emerging solid state transformers (SSTs). A detailed analysis of the factors influencing the MFT operation and limiting the design range is performed. A dedicated MFT design optimization algorithm is used to generate the set of all feasible transformer designs for the given electric requirements, taking into account different MFT geometry ratios, materials and operating frequencies. Design sets are generated for various combinations of design criteria thus exposing the general trends and impact of different design requirements on the feasible design space.

NOMENCLATURE

A_p	-	MFT area product
P_n	-	MFT nominal power
K_f	-	Excitation waveform coefficient
K_u	-	Window utilization coefficient
B_m	-	Peak flux density
J	-	Current density
f	-	Switching frequency
Δ	-	Penetration ratio

I. INTRODUCTION

Novel high-power medium-voltage DC-DC converter technologies are needed to support the development of the emerging MVDC grids as well as various traction applications [1]. The most popular solutions recurring in the literature are based on multiple stages of dual active bridge (DAB) or series resonant converter (SRC) topologies, as displayed in Fig. 1, connected in series at the MV side and in parallel at the low-voltage side. As can be seen in Fig. 1, the central component of any such switched-mode DC-DC power supply topology is the medium frequency transformer (MFT), providing both the galvanic insulation and input-output voltage matching.

The research interest in the area has intensified recently, both in academic and industrial domain, dealing with modeling and optimization [2]–[5] and specific design challenges such as insulation coordination [6] or multi-winding design [7].

In contrast to traditional line frequency transformers (LFTs), normally operating at low grid frequency with sinusoidal voltage and current excitation, MFTs operate on higher

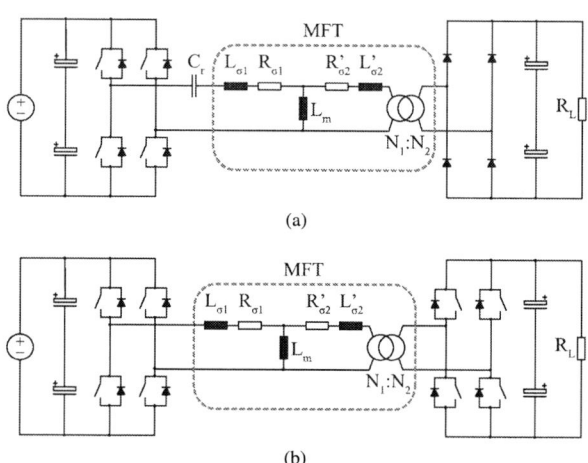

Fig. 1. DC-DC converter topologies commonly used within SSTs: (a) SRC converter (b) DAB converter

switching frequencies with square voltage and, in general, non-sinusoidal current waveforms characteristic for the given power electronic converter topology. This has implications on MFT losses and dielectric withstand requirements. Moreover, correct design of electric parameters is essential for proper operation of these converters and therefore imposes strict requirements on the accuracy of the corresponding models [8].

The main motivation for operating a transformer at high frequency is the potential for decrease in size, according to approximate relation (1).

$$A_p \approx \frac{P_n}{K_f K_u B_m J f} \tag{1}$$

This has many positive implications: easier integration, less material utilization, lower investment cost and environmental footprint etc. However, size decrease implies decreased cooling surfaces resulting in higher temperature gradients unless additional cooling effort is introduced [9].

These technologies have already been deployed in low-voltage low-power applications with great success, achieving the expected power densities. However, the increased dielectric withstand and power processing requirements, characteristic for aimed MV applications, affect the maximum achievable power density. Depending on the application requirements,

2170

The 2018 International Power Electronics Conference

TABLE I.   MFT Scaling Laws For Constant Material Utilisation - Magnetic Flux And Current Density

Variable	Formula	Proportion	Simplified Generic Shell MFT Structure
Cooling Surface	$S_c = C_1 l^2$	k^2	
Volume and Mass	$M = \gamma V = C_2 l^3$	k^3	
Current	$I = J S_{Cu}$	k^2	
Induced Voltage	$U = C_3 f B_m S_{Fe}$	$f k^2$	
Apparent Power	$P = UI$	$f k^4$	
DC Resistance	$R_{DC} = N \rho l / S_{Cu}$	$1/k$	
Copper Losses	$P_{Cu} = F(f) R_{DC} I^2$	$F(f) k^3$	
Core Losses	$P_{Fe} = K f^a B_m^b V$	$f^a k^3$	
Temperature Rise	$\Delta\theta = (P_{Cu} + P_{Fe})/(\alpha S_c)$	$k(F(f) + f^a)$	
Relative Losses	$P_r = (P_{Cu} + P_{Fe})/P$	$(F(f) + f^a)/(kf)$	
Relative Cost	$\epsilon = M/P$	$1/(kf)$	

Where: l, k - spatial dimensions, C_i - proportionality constants, γ - MFT density, $F(f)$ - skin and proximity effect correction factor

different design choices have direct impact on MFT characteristics.

An MFT is a complex system with coupled multi-physics which makes understanding the effects of certain design changes rather abstract and difficult to grasp. This paper discusses in detail the trends and different design outcomes in both qualitative and quantitative sense. Impact of different parameters is analyzed in a structured and intuitive manner thus exposing different design trade-offs, potential gains and expectation limitations.

II.   Qualitative Analysis of MFT Design Trade-Offs

The approximate relation (1) provides a very simplified estimation of the transformer size, in function of electric requirements and selected design alternatives, suitable for fast design. However, it does not show the effects on the MFT internal characteristics such as temperature gradients and relative cost. A more comprehensive step-by-step qualitative analysis of the scaling laws of each variable characterizing the MFT in mechanical and electric sense, under the assumption of equal material utilization in terms of current and flux densities, is described in Table I.

Starting from basic relations such as calculation of arbitrary MFT surfaces, volume and weight, it is possible to derive the scaling laws for more complex characteristics which are not so intuitively obvious, such as temperature rise, relative losses and relative cost. As can be seen, the relative cost is reverse proportional to both size and frequency. Therefore, from the material quantity and cost point of view, HP MFTs appear more attractive than their smaller counterparts.

However, it can be seen that the temperature rise is proportional to the linear spatial dimension and the sum of the additional winding ($F(f)$) and core (f^a) loss correction factors associated to high frequency effects. Consequently, the temperature gradients increase with the increase of transformer size (processing power), as well as the frequency. Depending on the type of insulation, additional insulation reinforcement may as well substantially increase the thermal resistances towards the ambient (e.g. solid type insulation). Therefore, the frequency, power processing and voltage domain where the described scaling can be preserved without additional cooling effort is limited.

On the other hand, relative losses decrease reversely proportional to size increase, indicating that higher power rated transformers should yield better efficiency. The frequency dependency is a function of winding ($F(f)$) and core (f^a) loss frequency correction factors. The frequency exponent of core materials, $a > 1$, and therefore the relative core losses increase with the frequency increase in the amount depending on the material properties. Skin and proximity correction factor ($F(f)$) is negligible for low frequencies ($\Delta < 1$), but increases exponentially at higher frequencies ($\Delta > 1$) [2]. Therefore, relative winding losses have a minimum at some frequency where penetration ratio is around one. Consequently, depending on the core material properties, and loss distribution between the core and the windings, the total relative losses are a convex function of frequency either having a minimum point or an increasing trend in the entire frequency range, depending on the initial gradient.

There are many coupled effects that influence the scaling of the MFT making it a rather difficult task to map the domain of possible designs or fairly assess the design improvement potential of each design choice on its own.

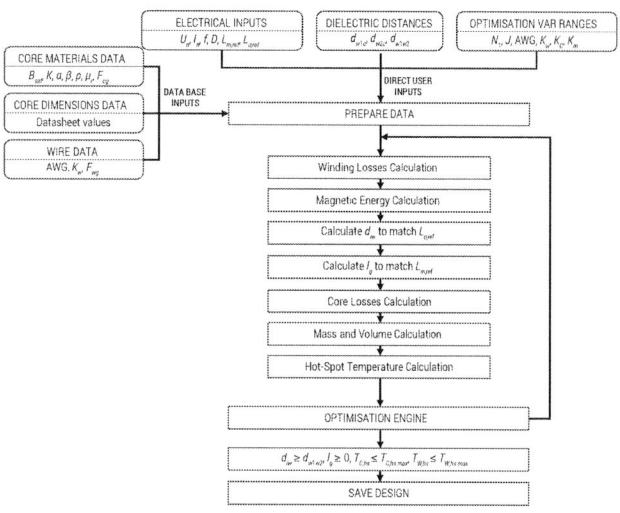

Fig. 2. A brute-force model-based MFT design optimization algorithm [10]

III. QUANTITATIVE ANALYSIS OF MFT DESIGN TRADE-OFFS

This section provides a method of analyzing the MFT design trade-offs in an integral manner using a sophisticated MFT design tool. Without loss of generality, the design is fixed to an air-insulated, air-cooled, N87 Si-ferrite core, AWG 32 litz-wire winding MFT with electrical specifications, as listed in Table II.

TABLE II. MFT PROTOTYPE ELECTRIC SPECIFICATIONS

P_n	V_1	V_2	$L_{\sigma1}, L'_{\sigma2}$	L_m
100 kW	750 V	750 V	4.2 μH	750 μH

A brute force MFT design optimization algorithm [10], capable of generating sets of all feasible MFT designs for given electric and dielectric specifications, as illustrated in Fig. 2, is used to map various design choice influences and trade-offs. All geometry ratios and relative field densities are considered as optimization variables whose each variation fully defines one MFT design. The maximum allowed hot-spot temperatures of the windings and the core are selected as 150 °C and 100 °C, respectively, corresponding to the selected material properties.

The feasible MFT design sets are generated for a family of frequencies in range from 1 kHz to 100 kHz in logarithmic scale and displayed as efficiency versus volumetric and weight power density plots, as given in Fig. 3. These plots reveal the trend related to the influence of frequency increase on the disposition of the feasible MFT designs. The upper boundary of the feasibility sets represents the Pareto front corresponding to the trade-off between the efficiency and power density. On the other hand, the lower boundary is set by the thermal limitations, showing how, at the set boundary, a smaller design must have a higher efficiency in order to compensate for the decrease of the cooling surfaces and remain within the defined

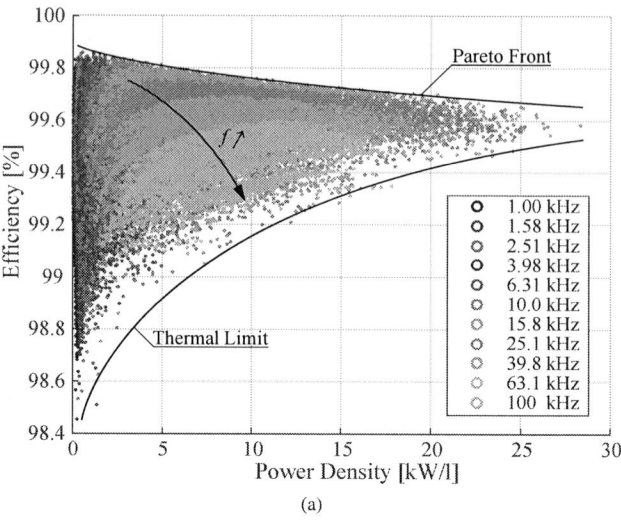

(a)

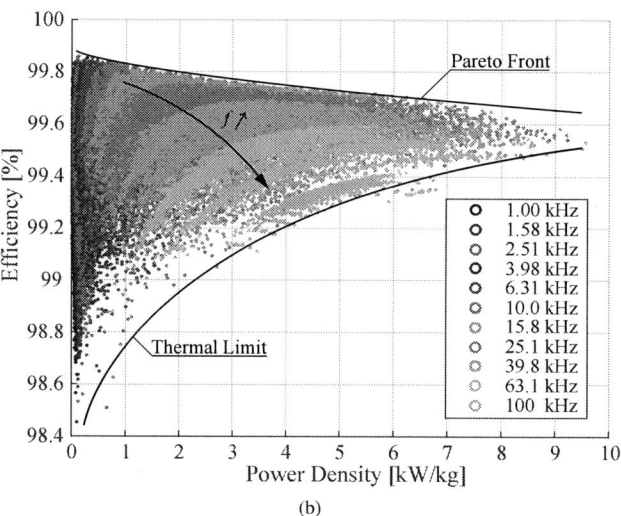

(b)

Fig. 3. Efficiency versus volumetric (a) and weight (b) power density plots of all mathematically feasible MFT designs generated with design optimization algorithm (around 12 million designs, uniformly down-sampled to 66000 for faster rendering). Designs are organized in color groups based on operating frequency

temperature limitations.

It can as well be seen how the feasible design sets at different frequencies reach the maximum achievable efficiency point at different power densities. The loss density within the core and the windings is higher at higher frequencies and therefore the increase of efficiency cannot be achieved by mere increase of cross-sections. Therefore, with the increase of frequency, where additional loss density associated to frequency effects is higher, maximum efficiency points are shifted towards higher power densities. Analyzing only the efficiency versus power density trade-off for any frequency, all of the designs which have lower power density than this critical value are sub-optimal as the increase of power density would allow for an increase in efficiency as well.

The 2018 International Power Electronics Conference

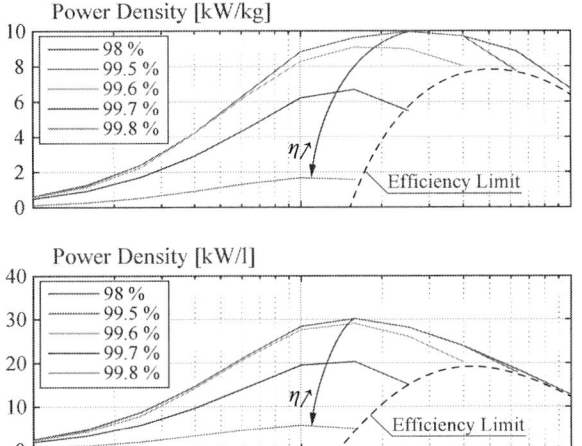

Fig. 4. Maximum achievable weight (top) and volumetric (bottom) power density versus operating frequency plots for different minimum efficiency constraints

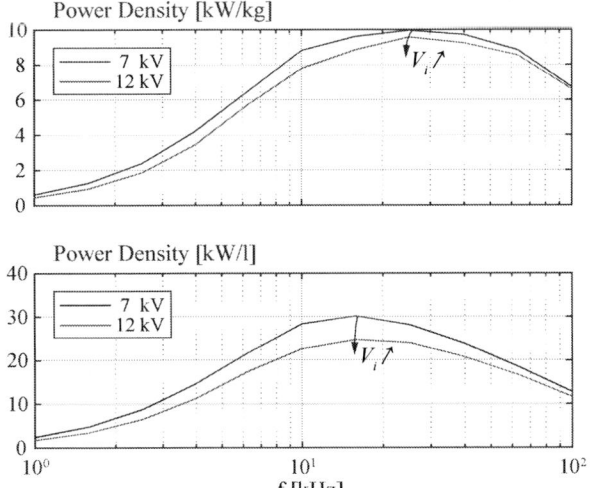

Fig. 5. Maximum achievable weight (top) and volumetric (bottom) power density versus operating frequency plots for different insulation requirements for the secondary winding

Beside these direct conclusions, with some simple post-processing of these feasible MFT design sets, it is possible to expose all sorts of complex design trends featuring compound constraints.

It is especially interesting to analyze the potential increase in power density that can be achieved with the increase of operating frequency under specified minimum efficiency constraint. This trend can easily be identified by filtering of the maximum MFT design feasibility sets from Fig. 3 by minimum allowed efficiency and selection of designs with maximum power density for each frequency.

The family of curves, representing maximum achievable weight and volumetric power densities versus operating frequency, for different minimum efficiency constraints, are displayed in Fig. 4. It can be seen, that it is not possible to achieve the scaling such as estimated with (1) for higher frequencies. For each minimum efficiency constraint, there exists a Pareto optimal frequency at which the maximum achievable MFT power density is the highest. For higher frequencies, the additional frequency dependent losses start to dominate and it is not possible to maintain the scaling. Furthermore, it can be seen that, as the minimum efficiency constraint is tightened, the maximum achievable power densities decrease, as well as the feasible design set frequency range. It is also interesting to notice that Pareto optimal frequencies are lower for MFT designs with higher efficiency requirement.

On the other hand, MV MFT insulation coordination is not a straight-forward task and it depends on many different details which are not easy to take into account in detail at the optimization stage. However, for design comparison purposes, without the loss of generality, it can be claimed that depending on the chosen insulation material, higher blocking voltage requirements (V_i) will yield larger dielectric distances (d_i), proportional to the corresponding dielectric strength (V_b) according to

$$d_i \approx k_s k_{pd} \frac{V_i}{V_b} \qquad (2)$$

where k_{pd} is a PD test standard voltage front multiplier from IEC 60664-1 international standard and $(k_s > 1)$ is a safety margin factor that takes into account the partial discharges and depends on the application and the material.

Therefore, it is possible to perform a simplified analysis of the influence of the required insulation level, by comparing two feasible MFT design sets with two different minimum dielectric distance constraints, corresponding to two different insulation voltage requirements, as displayed in Fig. 5. As can be seen, the maximum achievable power density is lower in case of the MFT design with higher insulation voltage. This is an expected result, as the higher insulation level requires larger dielectric distances and therefore more volume is occupied for this function. Design optimization algorithm optimizes the active part of the transformer, windings and the core, whereas the minimum insulation distances must be respected for a chosen dielectric.

Consequently, dielectric properties of the insulation material determine the sensitivity of the volume and weight of the MFT design to the increase of insulation requirements. Air insulated designs, as featured design choice in this paper, will require large volume to support MV applications, due to relatively poor dielectric properties of air compared to oil or solid, but relatively little added weight, only due to larger mean lengths of the core and the windings, as can be seen in Fig. 5.

Provided that a reliable data base of various insulating materials is available, insulation coordination could as well be integrated into design optimization, thus allowing the selection of the most optimal alternative.

2173

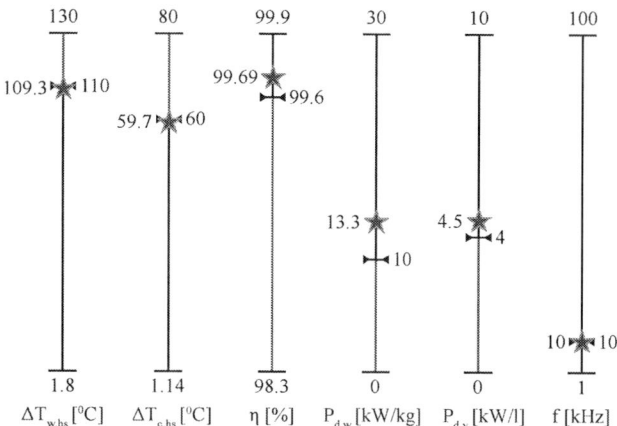

Fig. 6. Position of the optimal 10 kHz MFT design within a multidimensional design space. Scales show the full range of possible MFT characteristics corresponding to the maximum feasibility set at 10 kHz operating frequency. Performance filters, narrowing down the range of each MFT characteristic based on the desired specifications, are specified with black boundary markers and quantified with the value on the right side of each scale. Specifications of the optimal MFT design are depicted with the purple star marker and quantified with the value on the left side of each scale.

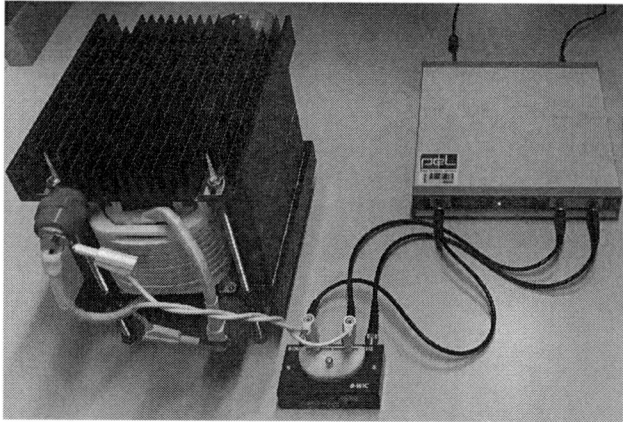

Fig. 7. Measurement of the optimal MFT prototype electric parameters using Bode 100 vector network analyzer

IV. MFT PROTOTYPE

In order to verify the accuracy of the used models and the overall design optimization algorithm, an optimal MFT prototype, for a resonant converter operating at 10 kHz, was realized with the aforementioned materials, design choices and electric specifications, as provided in Table II.

The optimal design was selected from the feasible MFT design set for 10 kHz using sophisticated performance filters that allow to arbitrarily narrow down the design ranges based on the desired specifications. The position of the optimal design within a multidimensional design space of interest is displayed in Fig. 6, together with the corresponding filter boundaries. While the discussion about various Pareto optimal fronts is important, as it gives a good insight into the limitations of what is the maximum theoretically achievable performance and provides a good platform for comparison of different combinations of design choices and materials, when it comes to prototyping, it is important to take into account safety margins, especially for phenomena with high stochasticity such as natural convection cooling.

It can be seen that the optimal design is not placed on the boundary of any of the parameter ranges. In order to achieve a thermally robust design, the maximum allowed core and winding hot-spot temperature rise constraint (filter) is tightened by a safety margin of 20 °C, thus generating a new Pareto front between efficiency and power density. The most optimal design is selected on this new Pareto front by adjusting the efficiency and power density filters in a desired manner.

In this way, it is possible to take into account different uncertainties for each MFT characteristic, caused either from limited modeling accuracy or manufacturing and assembly imperfections. The result is a robust Pareto optimal solution that properly takes into account the uncertainty of different models and processes relative to their importance for the given application.

The realized prototype and the electric parameter measurement setup consisting of Bode 100 vector network analyzer are displayed in Fig. 7. The measured electric parameters correlate very well to the reference values, whereas the hot spot temperatures remain within the set safety margins, as described in more detail in [8]–[10].

V. CONCLUSION

One of the main drivers for SST operation at medium frequency is the potential to substantially decrease the size of magnetic components. In order to fully benefit from this concept, an MFT design optimization is required to properly take into account all of the effects associated to medium-frequency high-power operation at MV.

Design optimization of an MFT is a complex task featuring coupled multi-physics and a multitude of different application specific constraints - e.g. electrical parameters, efficiency, temperature rise, weight, volume, height, width etc. A large number of various design choices and materials makes proper classification or comparison of different designs very difficult.

This paper provides a detailed analysis of the key MFT design trends and influences of various factors in a systematic way. A maximum set of feasible MFT designs has been generated for a fixed variation of MFT design choices and a characteristic electrical requirements for resonant converter operation, exposing all the limitations and trade-offs of such design.

By post processing of these maximum feasibility design sets, families of curves, representing various Pareto optimal fronts, are exposed one by one in a systematic manner, providing direct insight into the influence of each optimization variable on the design parameters of interest. This analysis provides a great platform for design comparison, exposing the maximum potential gains that can be achieved by manipulation of each optimization variable in relation to others.

However, when it comes to absolute design, modeling and manufacturing uncertainties have to be taken into account properly in relation to the sensitivity of the application to the corresponding MFT parameters. It is shown that, by means of feasible design set filtering, it is possible to set appropriate safety margins for each MFT design property of interest thus offsetting the initial Pareto fronts and generating an optimal robust design.

Finally, even though the analysis has been performed for one fixed combination of design choices and materials, the conclusions can intuitively be extrapolated to other variations in a qualitative sense by appropriately taking into account their relative characteristics.

REFERENCES

[1] M. Claessens, D. Dujic, F. Canales, J. K. Steinke, P. Stefanutti, and C. Vetterli, "Traction Transformation: A Power-Electronic Traction Transformer (PETT)," *ABB Review, No: 1/12*, pp. 11–17, 2012.

[2] I. Villar, "Multiphysical Characterization of Medium-Frequency Power Electronic Transformers," PhD thesis, EPFL Lausanne, Switzerland, 2010.

[3] G. Ortiz, "High-Power DC-DC Converter Technologies for Smart Grid and Traction Applications," PhD thesis, ETH Zurich, Switzerland, 2014.

[4] M. Bahmani, "Design and Optimization Considerations of Medium-Frequency Power Transformers in High-Power DC-DC Applications," PhD thesis, Chalmers University of Technology Gothenburg, Sweden, 2016.

[5] U. Drofenik, "A 150kW Medium Frequency Transformer Optimized for Maximum Power Density," in *2012 7th International Conference on Integrated Power Electronics Systems (CIPS)*, Mar. 2012, pp. 1–6.

[6] T. Gradinger, U. Drofenik, and S. Alvarez, "Novel insulation concept for an mv dry-cast medium-frequency transformer," in *19th European Conference on Power Electronics and Applications (EPE'17 ECCE Europe)*, Warsaw, Poland, 2017., pp. 1–10.

[7] S. Isler, T. Chaudhuri, D. Aguglia, and A. Bonnin, "Development of a 100 kW, 12.5 kV, 22 kHz and 30 kV insulated medium frequency transformer for compact and reliable medium voltage power conversion," in *Proceedings of the 19th European Conference on Power Electronics and Applications (EPE 2017 - ECCE Europe), Warsaw, Poland*, 2017.

[8] M. Mogorovic and D. Dujic, "Medium frequency transformer leakage inductance modeling and experimental verification," in *IEEE Energy Conversion Congress and Exposition (ECCE) 2017*, Cincinnatti, OH, USA, 2017., pp. 419–424.

[9] ——, "Thermal modeling and experimental verification of an air cooled medium frequency transformer," in *19th European Conference on Power Electronics and Applications (EPE'17 ECCE Europe)*, Warsaw, Poland, 2017., pp. 1–10.

[10] ——, "Medium frequency transformer design and optimization," in *Power Conversion and Intelligent Motion - (PCIM) 2017*, Nuremberg, Germany, 2017., pp. 423–430.

Standard models for
Power Electronic System Simulation

Koichi Shigematsu[*], Hiroki Ishikawa[†], Taku Noda[‡], Kentarou Fukushima[‡], Yoichi Sekiba[§],
Yusuke Kouno[¶], Takashi Abe[‖], Takayuki Sekisue[**] and Shinji Katoh[††]

[*]CYBERNETSYSTEMS Co. Ltd., [†]Gifu University,
[‡]Central Research Institute of Electric Power Industry, [§]Denryoku Computing Center, [¶]Toshiba Co.,Ltd.,
[‖]Nagasaki University, [**]ANSYS Japan, [††]Kobe City College of Technology

Abstract—**Cooperative Study Group on Practical Modeling and Simulation Techniques for Power Electronics Simulations, created under the Industry Applications Society of the IEEJ (Institute of Electrical Engineers of Japan), is active to develop standard models for simulations related to power electronics systems. The purpose of developing the standard models is to use in transient simulations related to power electronics for system simulation. Under the study group, four working groups (WGs) such are Smart Grid WG, the Motor Drive WG, the Automotive WG and the Power Conversion WG have been created for different fields of simulation. This paper provides the simulation models and examples such are voltage regulation equipment, pv and battery model for Smart Grid, DC modor model and its characteristics research for Motor Drive, engine, alternator, gear and ECU model for Automotive and some electric power conversion models.**

I. INTRODUCTION

Recently, effective use of energy has become an important social objective in order to reduce carbon dioxide emissions for the purpose of suppressing global warming. For this reason, power electronics technology capable of freely converting electric energy into a highly convenient form has increased its presence compared with the conventional technology, and at the same time, the importance of the simulation technology is increasing. On the other hand, from the technical point of view, in recent years simulation has entered a stage where it is frequently used for actual equipments development/design and problem solving. Under such social and technological backgrounds, the society of Industry Applications of the IEEJ Semiconductor Power Conversion Technology Committee, in order to develop the standard model required for simulation in the field of power electronics, "Cooperative Study Group on Practical Modeling and Simulation Techniques for Power Electronics Simulations" established. This group has a smart grid working group (WG), a motor drive WG, an automobile WG and a power conversion WG are installed and standard components of various constituent elements required for simulation in each field we have been developing several models [1], [2], [3], [4], [5].

Here, the purpose and significance of the standard model will be described. For example, in the case of the Smart Grid WG, standard models of the components of the distribution system, such as distribution substations, high voltage / low

voltage distribution lines, various distribution equipment, residential photovoltaic power generation systems, and consumer devices connected are developed. Engineers and researchers will be able to perform various analysis related to smart grid by combining these models. In addition, designers of power electronics equipments such as power conditioners are unfamiliar about power distribution technology, and on the other hand, technicians in the field of power distribution are often unfamiliar with power electronics technology, and standard models are not detailed parts. The standard model proposed in this paper consists of the equivalent circuits and control method which are considered to be the most fundamental for the device, and then the circuit constant and the control constant has been selected. Therefore, it seems that it can be utilized to a large extent for the purpose of education and research which put emphasis on practical aspect. For realistic simulation, it is usual to spend a great deal of time investigating what kind of model should be used for constituent elements to be analyzed, and what kind of parameter values should be set. In that points, the proposed standard model can be thought of as one solution to solve this problem. This paper excerpts and outlines the standard model described in the reference [6] summarized by the Cooperative Study Group.

II. STANDARD MODEL

A. Standard Model for Smart Grid

For the purpose of utilizing it for instantaneous value analysis related to smart grid, we developed a standard model for the components of the distribution substation for distribution and the consumer's equipment connected to power grid systems. The reason for modeling from the distribution substation to the customer's equipment is that the residential solar power generation is assumed as the main renewable energy in the Japanese smart grid, which is assumed to be [7], [8]. This is because it will be introduced to general homes in the distribution system. In other words, based on the assumption that Japanese type smart grid will be realized mainly in distribution system and general housing, we developed a standard model of its individual components. Fig. 1 illustrates the distribution system and the image of the general house connected. Table I shows a list of standard models that developed development or modeling guidelines. Here, the status of Web publishing is the situation on May 12 in 2017. (The same applies to the

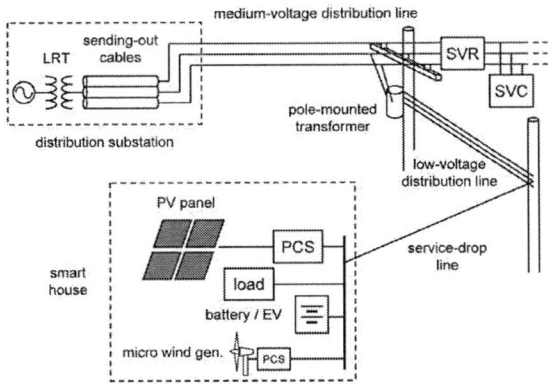

Fig. 1. The image of Smart grid system and models

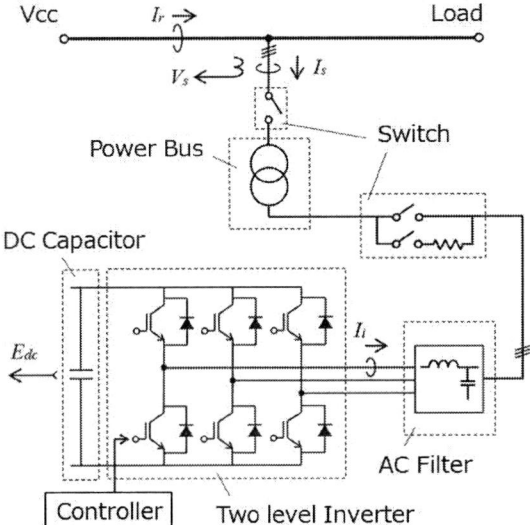

Fig. 2. STATCOM model

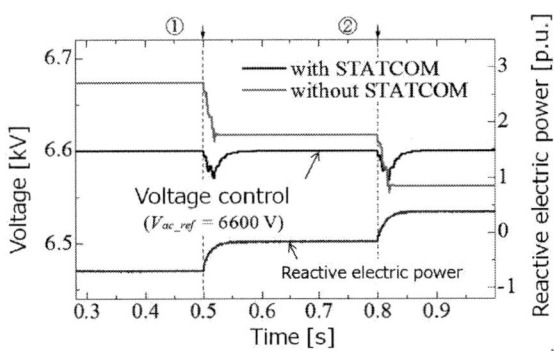

Fig. 3. Simulation results of STATCOM

following table.) Also, these models are shown in the Fig 1 in the distribution system and the image of the general residence connected there. In this paper, the outline of STATCOM, PV panel, lithium ion battery model will be described.

1) STATCOM: In the electric power system, a reactive power compensator may be installed for the purpose of adjusting the voltage and the power factor. A reactive power compensator is a kind of FACTS (Flexible AC Transmission System) [9] appliances to which power electronics technology is applied, and is a device that adjusts reactive power using semiconductor switching elements. Here, the self-excited reactive power compensator installed in the distribution system is called STATCOM (STATic synchronous COMpensator) for distribution and modeled. Fig. 2 shows the STATCOM main circuits model. This model consists of DC - AC converter, DC capacitor, AC filter, interconnection transformer, switchgear, voltage / current measurement system and control system. Fig. 3 shows the analysis results when the target power factor is 0.98 (delay from the system) in the power factor constant control operation mode. As a result, when the power distribution STATCOM is not in operation, the power factor has changed from the lead side to the lag side, but after the operation, the power factor of the power receiving point is controlled to maintain the target value.

2) PV Panel: PV panel is a main device that can convert solar energy directly into electric energy by utilizing the photovoltaic effect. Since the output of the PV panel depends on the solar radiation intensity, there is a fear that the voltage control of the electric power system and the frequency control may be

TABLE I.   SMART GRID STANDARD MODELS

Model name	Publication method	Simulator	Web publication
Distribution line	Equiv. circuits	XTAP	–
SVR	Equiv. circuits	XTAP	–
SVC, STATCOM	Equiv. circuits	XTAP	–
Transformer	Equiv. circuits	–	–
PV Panel	Equation	Maple	✓
PCS	Equiv. circuits	PSIM	–
Battery	Equiv. circuits	Simplorer	✓
Micro Wind gen	Equiv. circuits	XTAP	–

adversely affected by the sudden change of the generated electric power with the change of the solar radiation intensity due to the weather. Therefore, in order to quantitatively calculate the influence on the electric power system due to the sudden change in PV panel output, an instantaneous value analysis model that can simulate PV panel voltage / current change accompanying change in solar radiation intensity is needed. Here we developed a standard model for PV panels based on the document [10]. Figure 4 shows the equivalent circuit of the standard model of the PV panel. This model consists of four elements, a diode connected in parallel with a current source simulating photovoltaic current and a shunt resistance R_p, and a resistor R_s connected in series, and is called a single-diode model [10]. The shunt resistance R_p represents the effect of the leakage current at the pn junction and the series resistance R_p is the contact resistance between pn junction and metal terminal. Also, among the parameters of the equivalent circuit shown in Fig. 4, R_p and R_s are not listed in the products sheet, and it is necessary to calculate the fundamental simultaneous equations. We actually publish a sheet for which parameters are to be obtained on the Web to be described.

$$I_{mp} = I_{pv} - I_0 \left[\exp\left(\frac{V_{mp} + R_s I_{mp}}{aV_t} \right) - 1 \right] - \frac{V_{mp} + R_s I_{mp}}{R_p} \tag{1}$$

$$\left. \frac{dP}{dV} \right|_{mp} = I_{mp} - V_{mp} \frac{\frac{I_0}{aV_t} \exp\left(\frac{V_{mp} + R_s I_{mp}}{aV_t} \right) + \frac{1}{R_p}}{1 + \frac{R_s I_0}{aV_t} \exp\left(\frac{V_{mp} + R_s I_{mp}}{aV_t} \right) + \frac{R_s}{R_p}} = 0 \tag{2}$$

Fig. 5 shows the calculation results of the output characteristics of the created PV panel model. The calculation of output characteristics was made by connecting a variable resistor to the model terminal and changing the resistance value from 0 to ∞. Calculation results well reproduce the measured values of the panel output characteristics described in the document [12], indicating the validity of this model.

B. Motor Drive

Motor drive systems are widely applied to home electronics and industrial applications. Conventionally, the range of application of the motor drive system has been expanding with the demand for energy saving, noise / vibration countermeasures, and high-performance control also in equipment that realized basic performance only with a motor due to improvement of motor characteristics and control strategies. Fig. 6 shows the general configuration of the motor drive system. The purpose of constructing the motor drive system is to control the mechanical quantity of the rotating system which is the motor output, the torque, the rotational speed, and the position, and the means is the input power control of the motor by electric power conversion using the drive circuit. Therefore, in the motor drive system, a motor model that shows the relationship between the quantity of electricity input to the motor and the amount of machine in the rotating system as

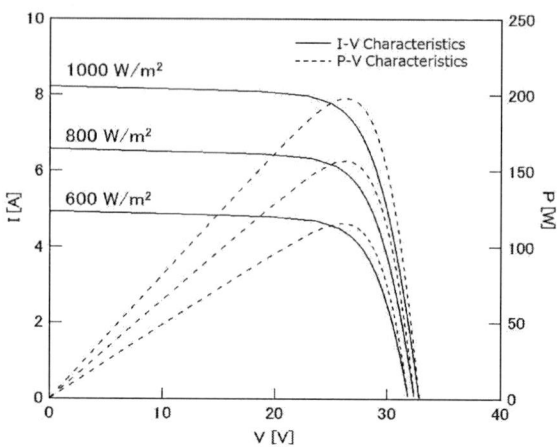

Fig. 5. Output characteristic calculation result of PV panel model

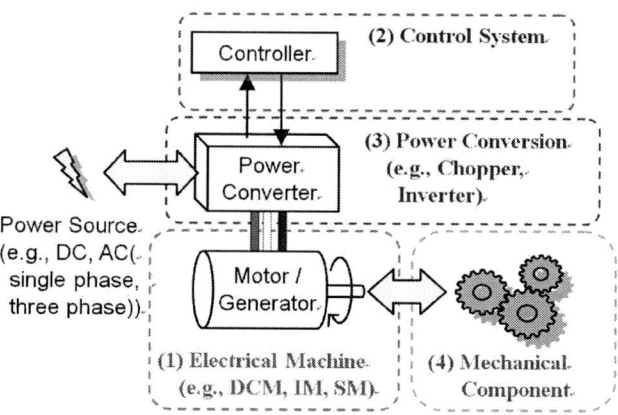

Fig. 6. General configuration of the motor drive system

output is important. Table II shows a list of standard models that developed development or modeling guidelines.

The electrical equivalent circuit in the armature circuit of the DC motor is shown in Fig. 7. Motor drive WG investigated the motor constants and summarized the trend. In this WG, IEEE Trans. on IA (2009-2011), IEEE Trans. on PELS (2008-2010), IEEE Trans. on IE (2009-2010), IPEC-Sapporo 2010, The Institute of Electrical Engineers of Semiconductor Power Conversion Study (2009-2011), The Journal of the Institute of Electrical Engineers of Japan (2008-2011), The 2009 Annual Meeting Record IEEJ, The Journal of the Power Electronics Society (2009), published in the Internet, product catalogs,

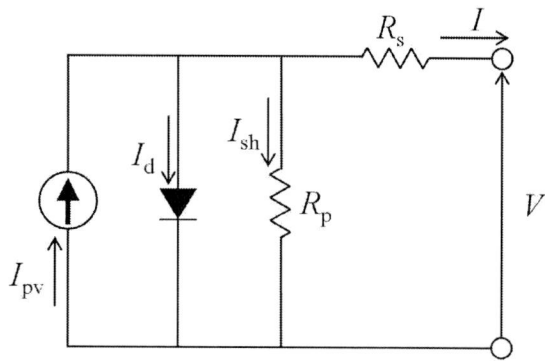

Fig. 4. Equivalent circuit model for PV panel

TABLE II.   STANDARD MOTOR MODELS

Model name	Publication method	Simulator	Web publication
DC Motor	Equation	Maple	✓
Induction Motor	Equiv. circ. model	Maple	✓
Controller	Block diagram	maple	✓
Driv circ.	Block diagram	—	—
Load	Block diagram	—	—

2178

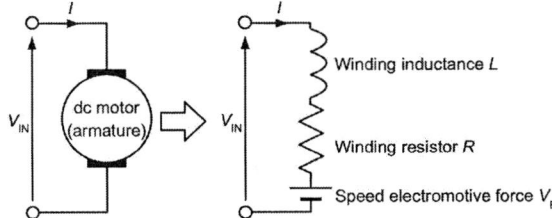

Fig. 7.   Equivalent circuit for DC Motor

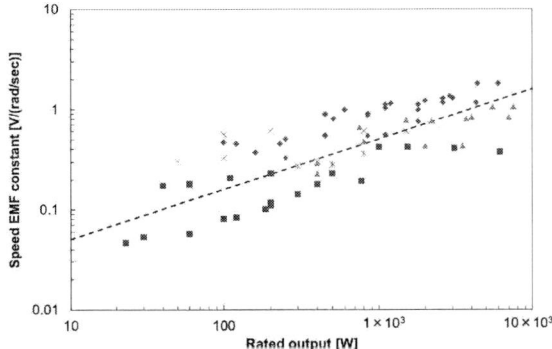

Fig. 8.   Rated speed and rated power factor characteristics

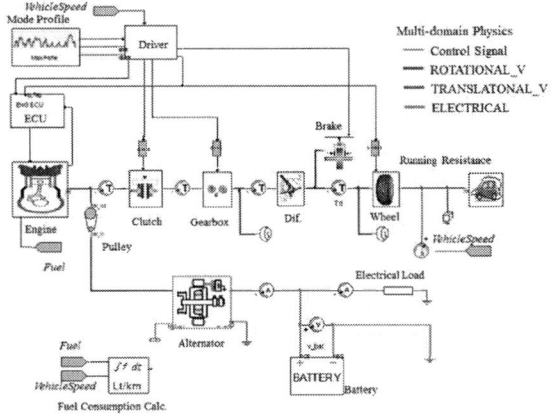

Fig. 9.   Entire simulation model for fuel efficiency calculation

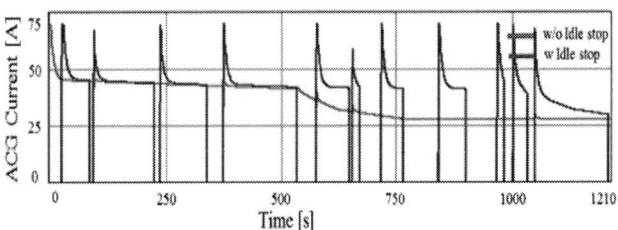

Fig. 10.   Effect of idling stop

etc., now or in the past. We investigated the motor constant of the 85 DC motors examined, 77 models (rated output 23 W to 7.5 kW) for which the motor constants necessary for constructing the standard model were released for each rated revolution number. The typical output/speed electromotive force coefficient characteristic is shown in Fig. 8 as its typical characteristic.

C. Automotive

Recently, a reduction in CO_2 emission and an improvement of fuel consumption are an important subject in development of the vehicle systems. The new technologies has been applying to vehicle systems, for example, the no idling control, the energy regeneration of HEV and PHEV, the weight saving and high efficiency by electrical power-train, and thermal recycle technology[13]. Therefore, vehicle systems have continued to increase in complexity, the simulation technology is essential on the planning and development stage[14]. A number of papers have reported about the fuel consumption and optimum design using vehicle system simulation[15], [16]. Table III shows a list of standard models that developed development or modeling guidelines.

Fig. 9 shows the entire simulation model of fuel efficiency calculation. In this model, the accelerator and brake are adjusted so that the driver model follows the target vehicle speed, so the performance of the driver model greatly affects fuel economy. Vehicles in the model simulate small passenger cars with MT (Manual Transmission), displacement of 1300 cc, vehicle weight 1 t class. We estimated the fuel economy

when traveling in compact/light vehicles in JC 08 mode using the simulation model shown in Fig. 9. The simulation result is shown in Fig. 10. It was confirmed that improvement of fuel efficiency is 20.3 km/L in the normal driving mode and 22.1 km/L in the idling stop running mode and 9% improvement. In addition, when the regulation voltage of the alternator was increased by 0.5 V at all times, it was seen that the fuel economy efficiency improved by 2 to 5%, and the influence of electrical behavior on fuel consumption was confirmed.

The technical report[6] also described application examples to HEV. In HEV system simulation, the dynamics of the vehicle is very important from the viewpoint of energy consumption. However, the vehicle also considered what was described as a multi-body composed of complex components of each component being linked via links and force elements. Here, we modeled dynamics in the direction of travel of the car. The one degree of freedom was only running direction, and rolls, pitches and yaw were neglected like the movement in the horizontal and vertical directions. Internally, considering

TABLE III.   AUTOMOTIVE MODELS(EXCERPT)

Model name	Publication method	Simulator	Web publication
driver	language model	VHDL-AMS	✓
ECU	language model	VHDL-AMS	✓
engine	language model	VHDL-AMS	✓
crutch	language model	VHDL-AMS	✓
transmission	language model	VHDL-AMS	✓
alternator	language model	VHDL-AMS	✓

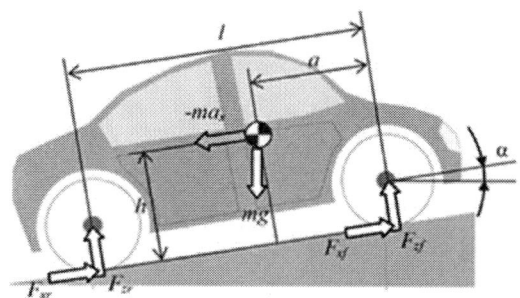

Fig. 11. Vehicle model

the two tire parts, the connection between the axle of the front wheels and the axles of the rear wheels was to output the normal force and the transmission force between the vehicle and the axle. The vehicle model shown in Fig.11.

D. Power Conversion

In the simulation of the power conversion system, it is easier to understand an example than to provide a individual standard model. Therefore, practical examples were created rather than standard models. Therefore, details of the models are not described in the report[6], but some example models are shown in the table IV and will release on the website sequentially.

TABLE IV.   POWER CONVERSION MODELS(EXCERPT)

Model name	Publication method	Simulator	Web publication
Up converter	Equiv. circuits	MapleSim	✓
Down converter	Equiv. circuits	MapleSim	✓
Single PWM inverter	Equiv. circuits	Simplorer	✓
DC/DC and Single inverter	Equiv. circuits	Simplorer	✓
Three phase inverter	Equiv. circuits	Simplorer	✓

III.   IMPLEMENTATION AND DISCLOSURE

The standard models introduced in this paper and the models posted in the report had been published on the web page. Details will be reported at the presentation.

IV.   CONCLUSION

It is necessary to develop the models for power electronics simulation. In that sense, the standard models proposed in this technical report is each developed using a specific simulator. However, this does not mean that each model was developed exclusively for a specific simulator, but it was developed in consideration of not using unique parts so as to operate with as many simulators as possible. An another standard interface called Functional Mockup Interface (FMI) is being developed and applied worldwide. This is to combine models and simulators via the state space signal flow interface, and expectations are gathering for future model exchange and commonality. Furthermore, from now on, simulator vendors and others are also expecting to create a standard model of this technical report for their own program and to make it public regardless of charge or free.

REFERENCES

[1] T.Kato, N.Noda, K.Fukushima and H.Ishikawa,"Toward Development of Standard Models for Simulation of Power Electronics and Automotive Systems -General Remarks-", Trans. The 2012 Annual Meeting Record IEEJ, Vol.4, No.S13-1, pp.1-2 (2012). (in Japaneses)

[2] T.Noda, Y.Kabasawa, K.Fukushima, H.Tokuda, H,Ishikawa, J.Ichihara, S.Kato, H.Kakigano, T.Sekisue, T.Kato, N.Kumura, Y,Kuroe, R.saito, J.Shinomura and M.Matsui, "Developmet of Standard Models for Transient Simulations of Smartgrids:Present State and its Plan", The 2012 Annual Meeting Record IEEJ, Vol.4, No.S13-1, pp.1-2 (2012). (in Japanese)

[3] H.Ishikawa, T.Kato, N.Umeda, S.Kato, K.Tago, T.Sekisue, T.Abe, S.Ohashi, S.Ogasawara, Y.Kuroe, H.Kubota, K.Shigematsu, J.Shimomura, T.Horiguchi, T.Morizane and K.Yasui, "Investigation for Sstandard Parameters of Motor Drive Systems", The 2012 Annual Meeting Record IEEJ, Vol.4, No.S13-3, pp.7-10 (2012). (in Japanese)

[4] T. Noda, Y. Kabasawa, K. Fukushima, H. Tokuda, H. Ishikawa, J. Ichihara, S.Kato, H. Kakigano, T.Sekisue, T. Kato, N. Kimura, Y. Kuroe, R. Saito, J.Shimomura and M. Matsui, "Present State and Future Plan of Standard Model Development for Smart Grid Simulations," IEEE COMPEL 2012, Kyoto, Japan, 2012.

[5] H. Ishikawa, T. Kato, N. Umeda, S. Kato, K. Tago, T. Sekisue, T. Abe, S.Ohashi, S. Ogasawara, Y. Kuroe, Y. Kubota, K. Shigematsu, J. Shimomura, T.Horiguchi, T. Morizane and K. Yasui, "Investigation into Standard Parameters of Motor Drive Systems," IEEE COMPEL 2012, Kyoto, Japan, 2012.(in Japanese)

[6] Cooperative Study Group on Generic Model Development for Power Electronics System Simulations, 「Generic Models for the Simulation of Power Electronics Systems:Smart-Grid, Motor-Drive and Automotive Applications」, IEEJ Technical Report, No1382, 2016. (in Japanese)

[7] Akihiko Yokoyama,"Development of Smarter Grid (I)", IEEJ,130(2), pp. 94-97, 2010. (in Japanese)

[8] Akihiko Yokoyama,"Development of Smarter Grid (II)", IEEJ, 130(3), 163-167, 2010. (in Japanese)

[9] N. G. Hingorani and L. Gyugyi, "Understanding FACTS: Concepts and technology of flexible AC transmission systems," IEEE Press, 2000.

[10] Marcelo Gradella Villalva, Jonas Rafael Gazoli, and Ernesto Ruppert Filho : "Comprehensive Approach to Modeling and Simulation of Photovoltaic Arrays", IEEE Trans. Power Electron., vol. 24, no. 5, May. 2009.

[11] W. Xiao, W. G. Dunford, and A. Capel : "A novel modeling method for photovoltaic cells," in Proc. IEEE 35th Annu. Power Electron. Spec. Conf.(PESC), 2004, vol. 3, pp. 19501956.

[12] Sharp Electronics ND198U1F : (in Japanese)

[13] Investigating R&D Committee on Technologies for Automotive ElectricPower Management : "Technologies forAutomotive Electric Power Management", IEEJ Technical Report No.1268 (2012). (in Japanese)

[14] Investigating R&D Committee on power electronics for automobile : "Power Electronics for Automobiles", IEEJ Technical Report No.1182 (2010). (in Japanese)

[15] Cooperation study group of modeling and simulation techniques of electronic systems : "Modeling and simulation techniques of electronic systems", IEEJ Technical Report No.1114 (2008). (in Japanese)

[16] K.Shigematsu, T.Sekisue, K.Tsuji : "Automotive System Simulation", Proc. of the 2006 JIASC, Vol.I, No.1, pp.51-56 (2006). (in Japanese)

Modeling and model parameter extraction of wide bandgap power semiconductor device, package, and circuit for simulating fast switching behavior

Tsuyoshi Funaki

Div. Electrical, Electronic and Information Eng., Graduate School of Eng., Osaka University, Osaka, Japan
E-mail:funaki@eei.eng.osaka-u.ac.jp

Abstract- Wide band gap power semiconductor devices have superiority in fast switching operation, when compared to Si IGBT and PiN diodes for voltage range from several hundred to several kilo volts. The fast switching operation gives high di/dt and dv/dt, which results in inducing surge voltage and EMI noise for poorly designed circuit. The precise dynamical model of power device and peripheral component is necessary in estimating these transient phenomenon. This paper presents the dynamical model of wide bandgap power semiconductor devices, package of power device, and circuit wiring. The fixture configurations to extract model parameter is also presented. The static blocking and conducting condition for power device is characterized with curve tracer. The terminal capacitance of power device dominates switching dynamic characteristics. Then, voltage dependency of terminal capacitances are characterized with the developed fixture, which can cope with normally on transistor characterization. The electro magnetic analysis of package and circuit to identify parasitic component is also addressed in the paper. The switching behaviors of power device and the phenomenon in the circuit are discussed based on the model and extracted model parameters.

I. Introduction

A power semiconductor device is a key component in power electronics system. The behavior of power semiconductor devices gives non-linear characteristics of circuit operation. Therefore, it is important to know the characteristics of power semiconductor device in understanding phenomenon occurring in power electronics system. The recent development of wide band gap semiconductor device enables fast switching and high temperature operation for high voltage and high power electronics system. This makes the design of power conversion circuit difficult in EMC and heat management. Then, the concept of system integration by front loading design is expected to be utilized. The front loading design is co-ordination of heterogeneous physical phenomenon in the same system, which is assessed through numerical simulation. Therefore, the accurate device model for simulation is indispensable for system integration. This paper discusses the electrical modeling of packaged MOSFET and Schottky barrier diode (SBD) power device. The characterization system for extracting model parameters for high voltage power device is also addressed.

II. MODELING AND PARAMETER EXTRACTION OF POWER DEVICE

A power semiconductor device functions as a switch in power electronics circuit. The on and off binary state model is enough for system level simulation. However, precise device simulation which takes into account of dynamic behavior of potential, electron and hole density distribution in the device with mesh dividing in 2 or 3 dimension, is necessary for analyzing designed device behavior [1]. But, it is too finite and not suitable for circuit simulation. Power device model used for circuit simulation should have intermediate preciseness between ideal switch and device simulation model. This section discusses the modeling and parameter extraction of power semiconductor device for high voltage power electronics simulation. The power device model for circuit simulation can be categorized into 3 types as physics, semi-physics and empirical model [2, 3]. The following discussions are based on physics based model.

A. SiC Schottky barrier diode (SBD)

The static current and voltage (I-V) characteristics of PN and Schottky junction in diode is expressed as eq. (1).

$$I_{ak} = I_0 \left(e^{\frac{qv_d}{nkT}} - 1 \right) \qquad (1)$$

Where, I_{ak}: terminal current, v_d: junction voltage, n: emission factor, k: Boltzman constant, q: unit charge, T:junction temperature, I_0: saturation current.

The voltage drop in drift region is not negligible for power device in large current conduction. Then, the terminal voltage V_{ak} is given as eq. (2).

$$V_{ak} = v_d + Ri_d \qquad (2)$$

Where, R: on resistance.

The measured I_{ak}-V_{ak} characteristics of SiC SBD at room temperature with curve tracer in Fig. 1(a) is shown in Fig. 1(b). The parameters in eqs. (1) and (2) are extracted from the measured results. The I_{ak} - V_{ak} characteristics of the model is also shown in Fig. 1(b).

The 2018 International Power Electronics Conference

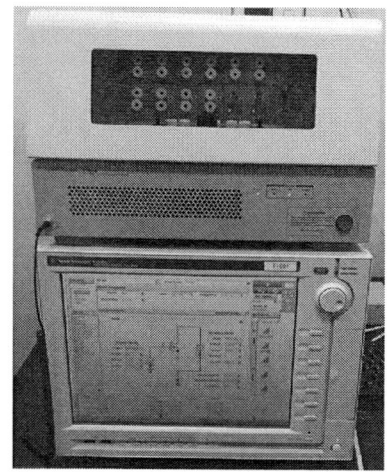

(a) Curve tracer (Agilent 1505B)

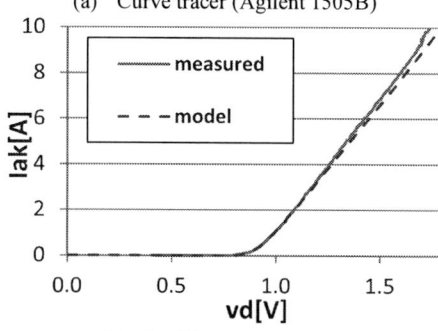

(b) I_{ak}-V_{ak} characteristics

Fig. 1. I_{ak}-V_{ak} characteristic of diode.

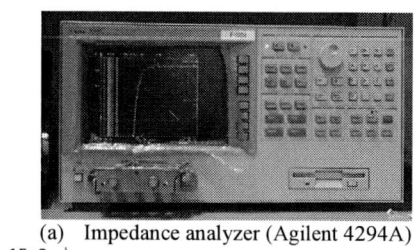

(a) Impedance analyzer (Agilent 4294A)

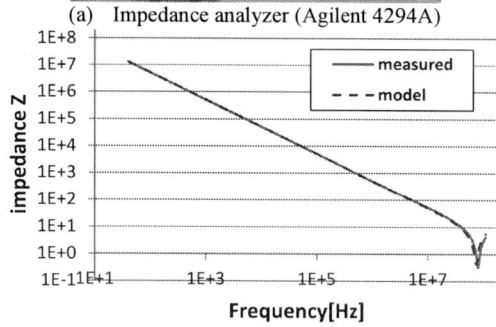

(b) Z-f characteristics

Fig. 2. Z-f characteristic of diode.

A power device in blocking condition can be modeled as a capacitor, which resulting from depletion of semiconductor and insulation layer. The wiring in the package and lead terminal has parasitic inductance. Then, the frequency characteristics of impedance for power device in blocking condition can be approximated as a series connected RLC equivalent circuit model. The

frequency characteristic of impedance for SiC SBD in zero bias voltage condition, which is measured with impedance analyzer in Fig. 2(a), is shown as Fig. 2(b). The characteristic of equivalent series connected RLC circuit model with identified parameters is also shown in Fig. 2(b).

(a) Parameter analyzer with high voltage bias T (Keithley 4200-SCS)

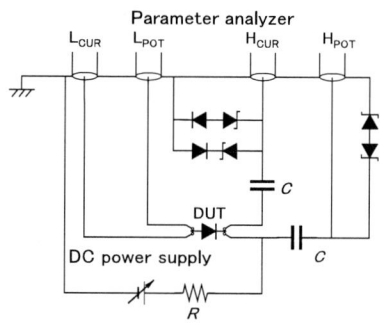

(b) Circuit diagram of bias T

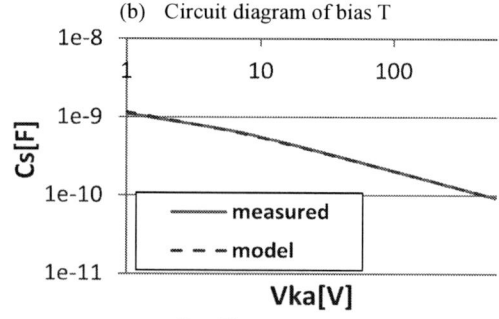

(c) C_{ak}-V_{ak} characteristic

Fig. 3. C_{ak}-V_{ak} characteristic of diode.

The terminal capacitance of power device changes with applied voltage, which depends on the impurity concentration in semiconductor to be depleted. The conventional impedance analyzer cannot apply high bias voltage. Then, the voltage dependency of terminal capacitance is measured with the parameter analyzer and high voltage bias T shown in Fig. 3(a). Bias T, whose circuit diagram is shown in Fig. 3 (b), is a fixture to impose bias dc voltage and measurement ac signal on DUT (device under test). But, it blocks to apply bias dc voltage on the

2182

measurement instrument. The measured and modeled capacitance – voltage (C-V) characteristic is shown in Fig. 3(c).

B. SiC MOSFET

The operation region of MOSFET can be divided into three regions by blocking, saturation, and linear. The I_{ds}-V_{ds} characteristic is expressed as eq. (3) [4].

$$
I_{ds} =
\begin{bmatrix}
0 & V_{gs} \leq V_{th} \\
\frac{k}{2}\left(V_{gs} - V_{th}\right)^2\left(1 + \lambda V_{gs}\right) & V_{th} < V_{gs} \leq V_{ds} + V_{th} \\
k\left(V_{gs} - V_{th} - \frac{V_{ds}}{2}\right)V_{ds}\left(1 + \lambda V_{gs}\right) & V_{ds} + V_{th} < V_{gs}
\end{bmatrix}
\tag{3}
$$

Where, I_{ds}: drain current, V_{ds}: drain voltage, V_{gs}: gate voltage, V_{th}: threshold gate voltage , k: trans conductance, λ: channel-length modulation parameter.

The measured I_{ds}-V_{ds} characteristic of SiC MOSFET is shown in Fig. 4(a). SCT2080KE is a SiC MOSFET, and the knee voltage of body diode is high stemming from wide band gap of SiC. The conduction loss of body diode is not negligible. Then, SiC SBD is connected in anti-parallel with SiC MOSFET for SCH2080KE to reduce conduction loss in reverse conduction operation through diode. SiC SBD lowers the knee voltage in reverse conduction. It also avoids PN junction degradation by body diode conduction.

The threshold gate voltage and trans-conductance are identified from the I_{ds}-V_{gs} characteristic shown in Fig. 4(b).

The parasitic inductances in the package of MOSFET are also identified from the frequency characteristic of impedance. The measurement of voltage dependency of capacitance for three terminal device of MOSFET requires additional blocking capacitor or inductor to the bias T in fixture of C-V characterization system for MOSFET shown in Fig. 4(c) [5]. The measured input Ciss, output Coss and reverse transfer capacitance Crss are shown in Fig. 4(d)(e)(f). The dynamic behavior of MOSFET is modeled with I-V and C-V characteristic as SBD.

III. EXPERIMENT AND SIMULATION

The dynamic behavior of power device is evaluated with experiment and simulation.

Figure 5(a) shows the circuit diagram of experimented power conversion circuit with inductive load. The numerical simulation is conducted with the equivalent circuit model shown in Fig. 5(b), which features the parasitic component in power device, passive component and wiring in circuit and package. The mutual inductances among lead wires in power device also can be modeled in Fig. 5(b). The experimented and numerically simulated transient behavior of voltage and current in turn on and off operation of power device is shown in Fig. 6, 7, 8.

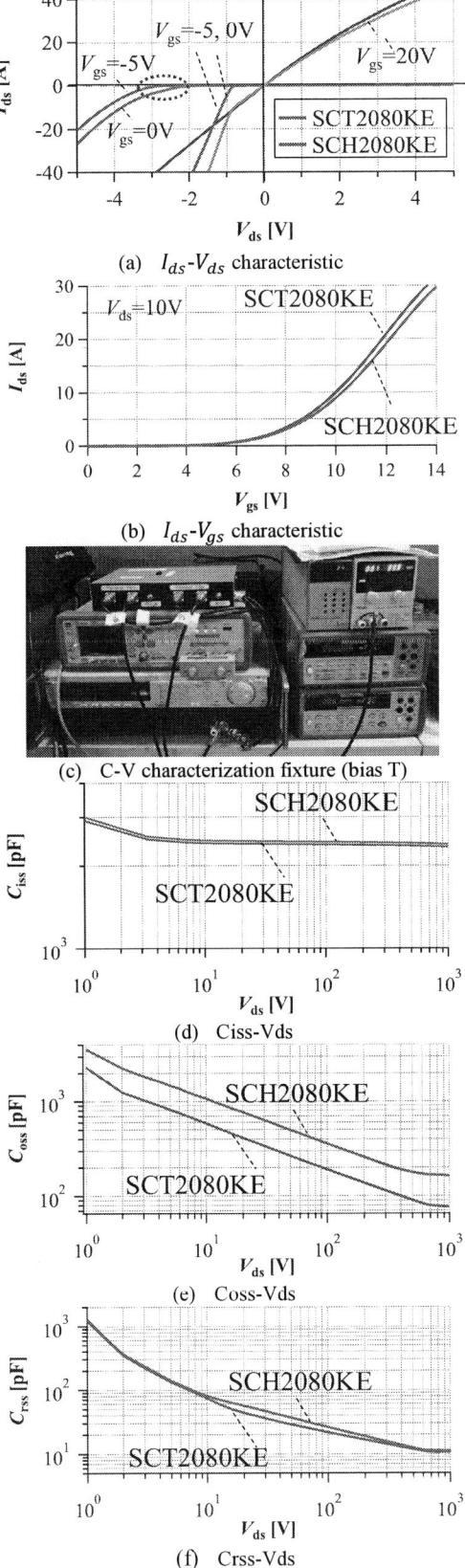

(a) I_{ds}-V_{ds} characteristic

(b) I_{ds}-V_{gs} characteristic

(c) C-V characterization fixture (bias T)

(d) Ciss-Vds

(e) Coss-Vds

(f) Crss-Vds

Fig. 4. Static characteristic of MOSFETs.

The behaviors in turn off operation of diode is shown in Fig. 6. The diode current decreases in accordance with the turn on operation of corresponding MOSFET. The simulated peak reverse current coincides with the experimental result. Slight oscillation in experimental result is not found clearly in the simulation result. The timing of anode voltage rise up in the simulation slightly advances to the experimental result.

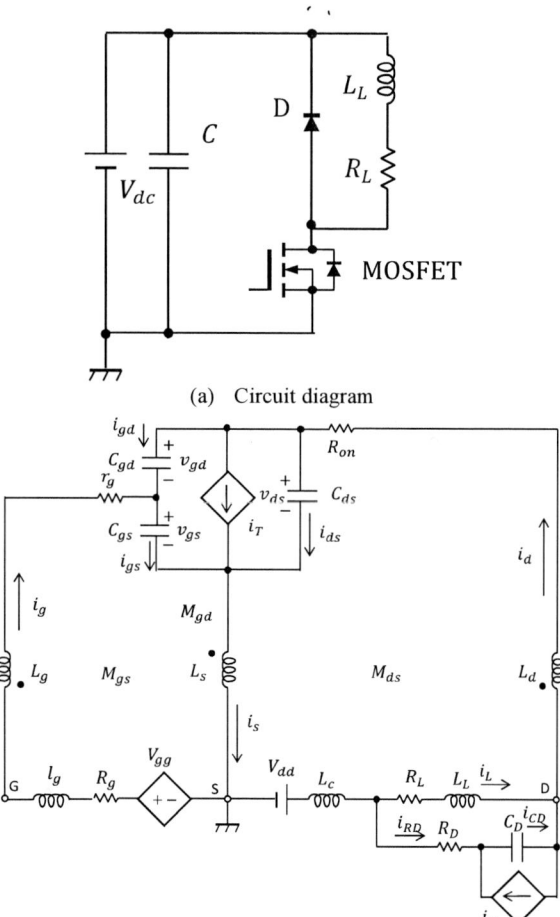

(a) Circuit diagram

(b) Equivalent circuit for simulation
Fig. 5. Evaluation circuit.

The voltage and current response in MOSFET for turn on operation is shown in Fig. 7. The gate voltage is pulled down -15V in off steady state. The gate driver charges input capacitance of MOSFET at the onset of turn on operation and gate voltage rises up. The drain current remains 0A and holds blocking condition until gate voltage reaches up to threshold voltage. The calculated gate voltage response almost coincide with the experimental result for Vgs<0V. But, it begins to mismatch around Vgs=0V, which may attributed to model error in gate voltage dependency of input capacitance and internal parasitic gate resistance.

The MOSFET transits saturation region with lowering drain voltage, when gate voltage exceeds threshold voltage. The drain current builds up during Miller effect

period, where gate voltage hardly rises up and shows plateau behavior. The gate driver injects current into large effective reverse transfer capacitance Crss during Miller effect period. The gate voltage begins to rise up again, when creep out saturation region and ingresses linear region. The simulated model response in gate voltage Vgs and drain voltage Vds, and drain current almost coincide with the experimented result.

MOSFET turns off with lowering gate voltage. The gate driver draws current from input capacitance Ciss of MOSFET. The drain voltage Vds and current Ids remain unchanged for gate charge extraction while gate voltage is in linear region of MOSFET. The response of simulated gate voltage slightly differ with the experimental result. The input capacitance model for simulation is identified in blocking condition of MOSFET. Because, the measurement of input capacitance in saturation region of MOSFET with flowing drain current and applying drain voltage is not feasible, because of significant self-heat up of MOSFET in saturation region operation. The input capacitance of MOSFET at the beginning of turn off is expected to be different from model value. This difference in input capacitance displaces the timing to build up drain voltage and close off drain current.

The drain voltage builds up in saturation region with flowing drain current for inductive load. Then, drain current commutates to the other arm. The plateau of gate voltage is not clearly observed in the experiment. The negative spike voltage, which is observed in the measured gate voltage around t=400us, is resulting from high di/dt of turning off drain current. This high di/dt of drain current induces spike voltage in drain voltage with interacting parasitic inductance of wiring in the arm. The peak voltage and damping of ringing oscillation of simulation is hasher than experiment. Then, more precise modeling, which can simulate fast transient phenomenon, is required.

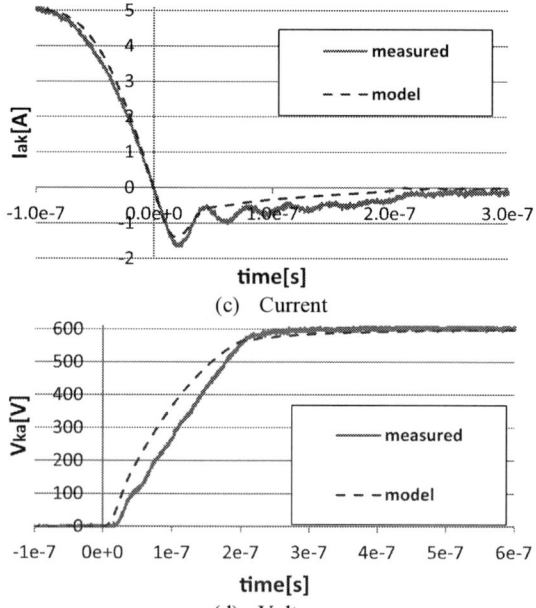

(c) Current

(d) Voltage
Fig. 6. Switching behavior of diode.

2184

The 2018 International Power Electronics Conference

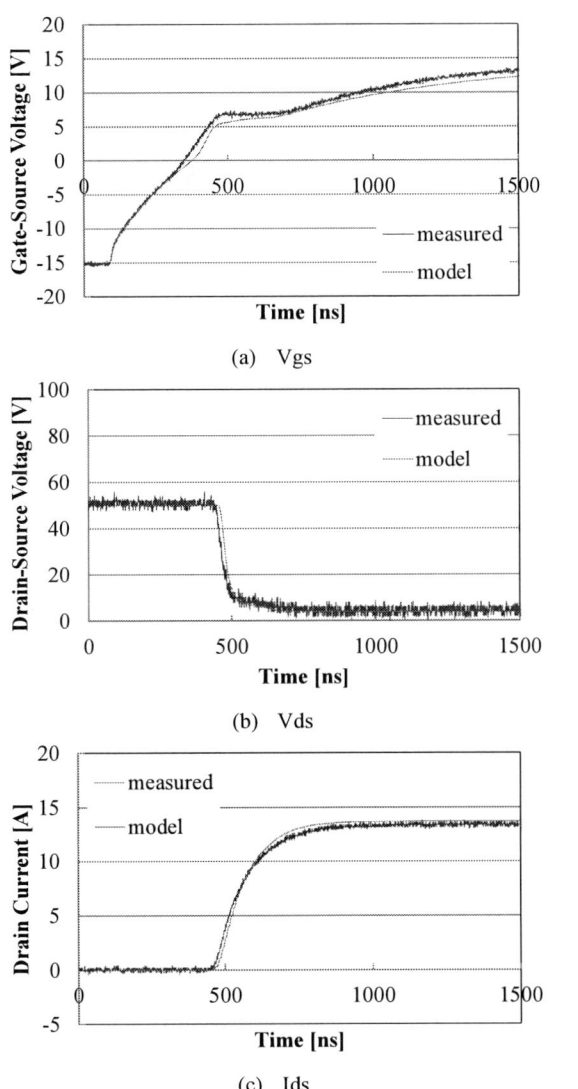

(a) Vgs

(b) Vds

(c) Ids

Fig. 7. Switching behavior of MOSFET in turn on operation.

IV. CONCLUSION

This paper discussed the modeling of dynamic characteristics and parasitic component in unipolar power device and circuit wiring to utilize in designing and analyzing power conversion circuit with wide band gap power semiconductor device. The static I-V characteristics, parasitic terminal capacitance and wiring inductance is modeled for simulating transient behaviors of circuit voltage and current in switching operation. The parameter extraction system and fixtures for device modeling is addressed in this paper. The behaviors of physics based model used in the analysis does not completely correspond with the actual device behavior, which stems from non-ideality of device characteristics. The further study for semi-empirical device model, which consist of physics based and experimental based equations,

is required for accurate modeling of actual device and circuit.

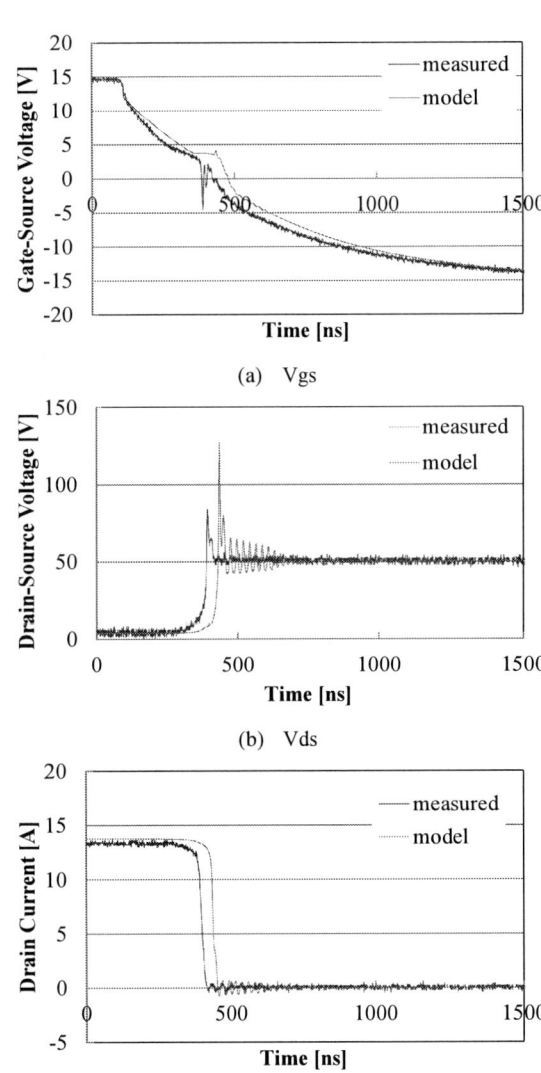

(a) Vgs

(b) Vds

(c) Ids

Fig. 8. Switching behavior of MOSFET in turn on operation

REFERENCES

[1] A. S. Wong, A. R. Neureuther, "The intertool profile interchange format: a technology CAD environment approach," IEEE Trans. CADICS, Vol.10, No.9, pp. 1157--1162, 1991.

[2] H. Mantooth, et al., "Modeling of Wide Bandgap Power Semiconductor Devices-Part I, " IEEE trans. PELS, vol.62, no.2, pp.423-433, 2015.

[3] E. Santi, et al. "Modeling of Wide-Bandgap Power Semiconductor Devices-Part II," IEEE trans. PELS, vol.62, no.2, pp.434-442, 2015.

[4] G. Massobrio, P. Antognetti, Semiconductor device modeling with SPICE, pp.170-171, 1993.

[5] T. Funaki, et al., "Measuring Terminal Capacitance and Its Voltage Dependency for High-Voltage Power Devices", IEEE trans. on Power Electronics, Vol. 24, No. 6, pp.1486-1493, (2009) .

Stability Analysis Methods of a Grid-Connected Inverter in Time and Frequency Domains

Toshiji Kato, Kaoru Inoue, Taiki Sakiyama
Department of Electrical Engineering, Doshisha University
Kyotanabe, Kyoto, 610-0321, JAPAN
E-mail: tkato@mail.doshisha.ac.jp

Abstract—A grid-connected inverter must be designed to assure stable operation. Stability analysis of the system is one of the most important design processes and it is calculated mainly by four types of analysis methods both in time and frequency domains. This paper investigates and compares these stability analysis methods. In frequency domain, according to the impedance method, frequency characteristics of an output impedance or admittance are calculated and investigated. Three frequency analysis methods, the analytical derivation method, the numerical frequency analysis method that is based on an automatic and numerical formulation and solution algorithm, and the simulator-based frequency scan method, are often used and useful. According to the impedance method, it is sufficient when the real part of the admittance of the inverter is positive and passive to ensure stability. Three types of passivity-based design examples are shown. In time domain, a unique analysis method which is based on the eigenvalue analysis of the Poincaré map or the sensitivity matrix of state-variables for perturbations along an operating trajectory, has been developed and it is also useful. These four methods are investigated for their features. Application example cases for grid-connected inverters are investigated and compared between the methods by simulation and experiment.

I. INTRODUCTION

A grid-connected inverter is becoming more important for photovoltaic power generation and smart grid systems, and it must be designed to assure stable operation. Stability analysis of the system is one of the most important design process and it is calculated mainly by four types of the analysis methods both in time and frequency domains whose features are seldom discussed.

This paper investigates and compares these four analysis methods. In frequency domain, the impedance method is mainly used because its principle is simple and useful[1]-[8]. First frequency characteristics of the output impedance or admittance of an object system are calculated. Then they are checked if they satisfy, for example, the Nyquist criterion or not. In this process, there are mainly three methods to calculate the frequency characteristics. One is the analytical derivation method for a comparatively simple case. Another is the numerical frequency analysis method[9] that is based on general-purpose simulation algorithms[10]-[13] and is suitable for a more complex linear case, for example, including digitally controlled subsystem. The other is the simulator-based frequency scan method which calculate numerical responses for a given excitation signal by FFT[14]. According to the impedance method, it is sufficient when the real part of the admittance of the inverter is positive and passive to ensure

stability. Three types of passivity-based design examples are shown[15]-[32].

It is possible to analyze the system transient behavior for stability also in time domain. An alternative analysis method, which is based on the eigenvalue analysis of the Poincaré map or *the sensitivity* matrix for perturbations along an operating trajectory, is also useful because it can analyze even a nonlinear system[33]-[39] mainly by the shooting method[40]-[45].

Application example cases for grid-connected inverters are investigated and compared between the methods for their features by simulation and experiment[46].

II. STABILITY ANALYSIS OF A GRID-CONNECTED INVERTER

A. Stability analysis in Time- and Frequency-domain

Stability of a grid-connected inverter is discussed. A single-phase inverter with an LCL filter in Fig.1(a) is adopted as an example and it is controlled with a digital scheme. System stability can be analyzed both in frequency and time domains. In frequency domain, first frequency characteristics of the output admittance of the object system are calculated. Then the characteristics are investigated mainly according to the impedance method because its principle is simple, useful, and widely applicable. However, it is applicable only to a linear system and the averaging method is necessarily applied to eliminate switching operations as in Fig.1(b).

In time domain, the system is analyzed to get a sort of transition matrix or a sensitivity matrix between state variables for one periodic cycle. The matrix includes complete information about interactions between state variables. This time domain method does not need application of the averaging method and the system can be nonlinear. According to the eigenvalue analysis, it is possible to investigate stability which is related with nonlinear conditions including harmonics.

B. Impedance Method in Frequency-Domain

The basic principle of the impedance method[2] is briefly reviewed to analyze stability of the grid-connected single-phase inverter with an LCL filter in Fig.1. The inverter is expressed with the Norton's circuit of the output admittance $Y_o(s)$ and the inverter control current $I_c(s)$. The grid system is expressed with the Thevenin's circuit of the equivalent system voltage $V_s(s)$ and the line impedance $Z_s(s)$. The inverter can be expressed with an equivalent model in Fig.2. The output terminal voltage to the grid system $\hat{V}_s(s)$ becomes (1).

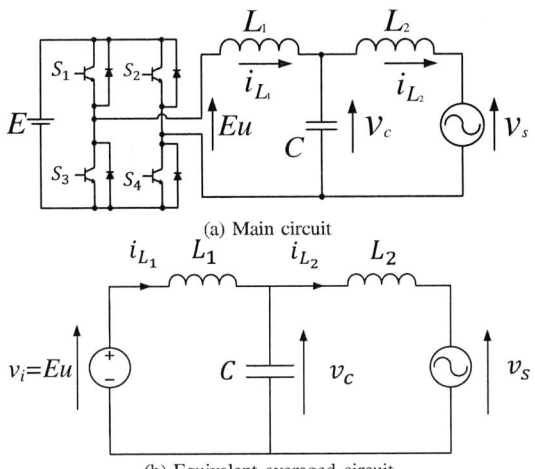

(a) Main circuit

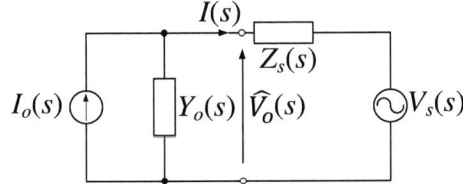

(b) Equivalent averaged circuit

Fig. 1.   Grid-connected inverter with an LCL filter.

Fig. 2.   Frequency dependent equivalent model for stability analysis.

$$\hat{V}_s(s) = \frac{V_s(s) + Z_s(s)I_c(s)}{1 + Z_s(s)Y_o(s)} \qquad (1)$$

Applying the Nyquist criterion to the denominator $1 + Z_s(s)Y_o(s)$, the system is stable when the phase of $Y_o(s)Z_s(s)$ is less than 180° or $|Y_o(s)Z_s(s)| < 1$ at the cross frequency of 180°. It is equivalent that the phase difference of $Y_o(s)$ and $Y_s(s)$ is less than 180° or $|Y_o| < |Y_s|$ at the cross frequency.

C. Eigenvalue Analysis using Poincaré Map in Time Domain

Nonlinear dynamics of a grid system is described as follows where x is a state-space variable vector and t is time.

$$\frac{dx}{dt} = f(t, x) \qquad (2)$$

The value x_T at $t = T$, which is integrated for one period T from an initial value x_0 at $t = 0$, is as follows.

$$x_T = \int_0^T f(\tau, x(\tau))d\tau + x_0 \qquad (3)$$

The dynamic behavior is described in a Poincaré map where x_0 starts from the Poincaré section at $t = 0$ and comes back to x_T in the section at $t = T$. The periodic steady-state (PSS) condition is that the two boundary values at $t = 0$ and T are equal and it is described as follows.

$$x_T = P(x_0) = x_0 \qquad (4)$$

where $P(\ )$ is mapping relation from x_0 to x_T, and it is generally a nonlinear function with respect to x_0[40]-[45].

It is possible to analyze stability of the system by this mapping relation. When all absolute eigenvalues of the sensitivity matrix $\frac{\partial x_T}{\partial x_0}$ at a PSS condition are less than one, the system is stable because variations of the initial values are not magnified.

III.   ANALYTICAL ANALYSIS METHOD BASED ON THE IMPEDANCE METHOD IN FREQUENCY-DOMAIN: METHOD I

A. State-Equations and Control Principle of Grid-Connected Inverter

A digital control scheme is described for the grid-connected voltage-source inverter connected to a utility source of voltage v_s as in Fig.1. Frequency characteristics of the output admittance of the system can be derived analytically. An LCL filter of L_1, C, L_2 is inserted at the inverter output to convert the voltage input v_i to the current output i_{L2} and to reduce its harmonic components sufficiently. The main circuit includes switching operations which are nonlinear and it is averaged and its approximate linearized circuit is reduced. State variables of L_1, C, L_2 are denoted as i_{L_1}, i_{L_2}, v_C and their vector form sampled at $t = iT$ is as $x[i]$ where T is a sampling time interval. Its input voltage $v_i[i] = Eu[i]$ is generated according to its input $u[i]$ where E is DC voltage of the inverter source. Discrete state equations for the output current of L_2, $y[i] = i_{L2}[i]$, are as follows.

$$\left. \begin{array}{rcl} x[i+1] & = & Ax[i] + bu[i] + hv_s[i] \\ y[i] & = & cx[i] \end{array} \right\} \qquad (5)$$

B. Analytical Derivation of Output Admittance

The feedback(FB) control loop in Fig.3 is designed to track the output current to a desired sinusoidal waveform according to the optimal control principle. The state equations of the system in frequency domain are as follows.

$$\left. \begin{array}{rcl} sX(s) & = & A_cX(s) + b_cU(s) + h_cV_s(s) \\ U(s) & = & -fS_mX(s) + G(i_r^* - cS_mX(s)) \end{array} \right\} \qquad (6)$$

From the above two equations, the following equation is derived.

$$(sI - A_c + b_cfS_m + b_cGcS_m)^{-1}X(s) \qquad (7)$$

The analytical output admittance Y_o is derived as follows.

$$Y_o = -\frac{I_{L2}}{V_s} = -c(sI - A_c + b_cfS_m + b_cGcS_m)^{-1}h_c \quad (8)$$

The 2018 International Power Electronics Conference

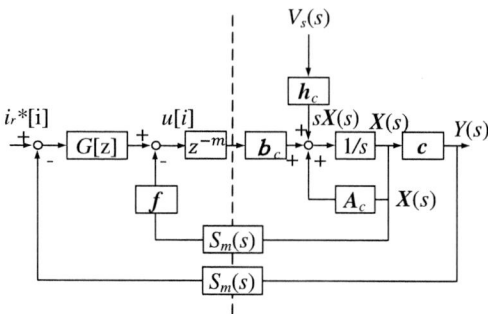

Fig. 3. Feed-back control block diagram of grid-connected inverter.

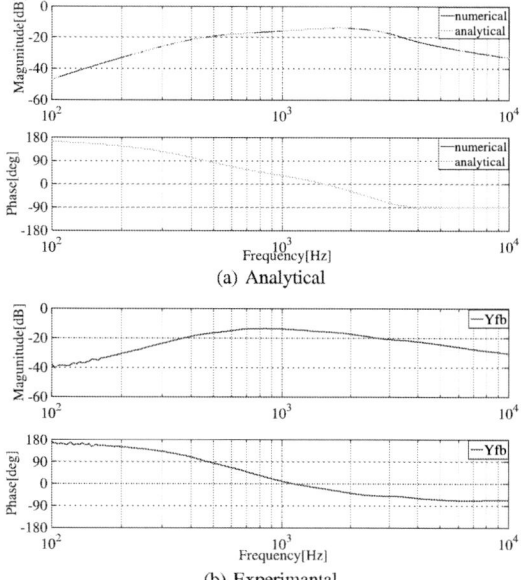

(a) Analytical

(b) Experimantal

Fig. 4. Calculated and measured frequency characteristics of the output admittance.

C. Calculated Output Frequency Characteristics

The above described method is applied to the single-phase grid-connected inverter to investigate frequency characteristics of the output admittance. The input DC voltage is 200V, the system voltage 100V(rms), $L_1, L_2 = 0.76$mH, and $C = 9.3\mu$F. The calculated characteristics by the analytical derivation is shown in Fig.4(a).The characteristics are compared with experimental characteristics in Fig.4(b) where the system is controlled with a DSP-based digital control system at a sampling time of $T = 100\mu$sec. The analytically calculated and measured characteristics are close.

IV. THE IMPEDANCE METHOD USING FREQUENCY-ANALYSIS SIMULATOR: METHOD II

The analytical method described in the above is simple and useful for a comparatively small system. A system becomes more and more complex, the analytical method is more difficult to apply not only because the system equation becomes too

complex to handle manually but also because selection of state variables becomes complex due to inductor cutset and capacitor loop restrictions.

The general-purpose frequency analysis method of a grid inverter for frequency characteristics has no such restrictions because it is simple and general. It consists of three processes as follows: (1) the decomposition process of an object system into components of circuit and control elements; (2) a modeling process of each component for its frequency characteristics; (3) computation steps of formulation and solution of the system equations from the component characteristics, and desired characteristics are computed.

The third computation process consists of the following three steps: input, formulation, and solution steps. Before this process, all components are modeled for their frequency characteristics and their formulation stamps must be prepared in advance. First circuit information of an object system is inputted for element connection, type, and parameter data. Circuit equations for the system are then formulated according to the element stamps at a sampled frequency. Finally they are solved for desired characteristics. This formulation and solution steps are repeated at all sampled frequencies.

V. THE IMPEDANCE FREQUENCY-SCAN METHOD USING GENERAL-PURPOSE TIME-DOMAIN SIMULATOR: METHOD III

Output admittances of an inverter system can be computed also by a time-domain simulator. A perturbation small-signal ΔV_p is inserted directly at the output of the inverter and its response current ΔI_p is computed as in Fig.5. The computation principle is as follows.

1) The original circuit under test is analyzed for its steady-state operation I_p.
2) The perturbed circuit with a small excitation signal ΔV_p of a frequency f_i is analyzed for its steady-state operation $I_p + \Delta i_p$ where its period is $T_i = 1/f_i$.
3) An admittance $Y_o(f_i)$ for the frequency f_i is calculated from division of the difference signal $\Delta I_p(f_i)$ by the perturbation voltage $\Delta V(f_i) = \Delta V_p$ using FFT.

$$Y_o(f_i) = \frac{\Delta I_p(f_i)}{\Delta V(f_i)} \qquad (9)$$

4) The above steps are repeated for necessary f_i samples.

The above steps must be iterated for necessary f_i samples because the system is nonlinear and its operating condition is slightly changed.

VI. COMPARISON OF THE COMPUTED RESULTS BETWEEN THE THREE FREQUENCY ANALYSIS METHODS I-III

Computed frequency characteristics by the analytical, numerical, and simulated frequency scan methods are shown in Fig.6. The results by the analytical and the numerical methods coincide because they are identical in principles and they are only different from the ways. Two types of the results by the simulated frequency scan method are also shown in the figure.

2188

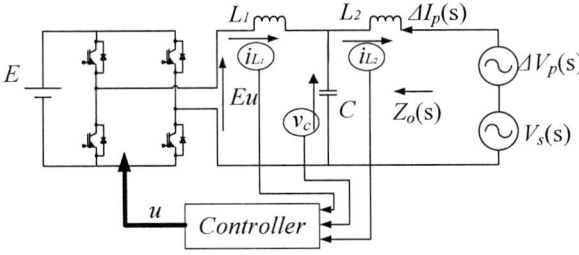

Fig. 5. Simulator-based frequency scan method.

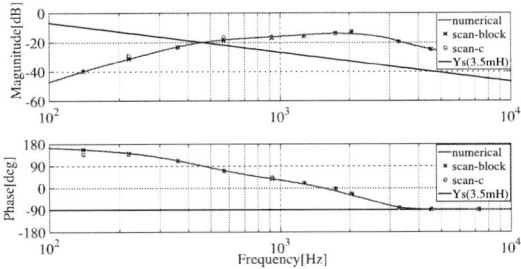

Fig. 6. Comparison of computed frequency characteristics by analytical, numerical, and simulated frequency scan methods.

The result by the control scheme realized by simulator control blocks is shown by cross marks and it is very close to the analytical results. The result by the control scheme realized by a C program is shown by circle marks and it is also close to the analytical results. The results by the three frequency analysis methods coincide well except small averaging and numerical errors.

VII. EQUIVALENCE BETWEEN PHASE AND DQ DOMAINS

In a three-phase system, the equation (1) is expressed with its matrix equation form. Applying the Nyquist criterion to the denominator, its stability is analyzed similarly. The matrix equation can be diagonalized based on eigenvectors of the admittance matrix and it is decoupled into three independent components, which correspond to the symmetrical coordinate components when relation between phases is cyclic.

$$\begin{bmatrix} V_0 \\ V_p \\ V_n \end{bmatrix} = \begin{bmatrix} 1 & 1 & 1 \\ 1 & a & a^2 \\ 1 & a^2 & a \end{bmatrix} \begin{bmatrix} V_u \\ V_v \\ V_w \end{bmatrix} \quad (10)$$

where V_j ($j = u, v, w$) is a phasor in the phase domain, V_j ($j = 0, p, n$) is a phasor in the sequence domain, and $a = \varepsilon^{j2\pi/3}$.

Furthermore, there are following relations in the sequence and the dq domains[8].

$$\begin{bmatrix} V_d(\omega_{dq}) \\ V_q(\omega_{dq}) \end{bmatrix} = \sqrt{\frac{3}{2}} \begin{bmatrix} 1 & 1 \\ -j & j \end{bmatrix} \begin{bmatrix} V_p(\omega_p) \\ V_n(\omega_n) \end{bmatrix} \quad (11)$$

$$\begin{bmatrix} V_p(\omega_p) \\ V_n(\omega_n) \end{bmatrix} = \frac{1}{\sqrt{6}} \begin{bmatrix} 1 & j \\ 1 & -j \end{bmatrix} \begin{bmatrix} V_d(\omega_{dq}) \\ V_q(\omega_{dq}) \end{bmatrix} \quad (12)$$

where

$$\left. \begin{array}{l} \omega_p = \omega_{dq} + \omega_1 \\ \omega_n = \omega_{dq} - \omega_1 \end{array} \right\} \quad (13)$$

The eigenvalues in the two domains are the same and the two stability analyses are proved to be equivalent.

VIII. PASSIVITY-BASED STABILITY ANALYSIS

A. Passivity-Based Stability Crteria

The Nyquist criterion in the impedance method essentially checks if the cross point of eigenvalues of $Z_s Y_o$ at the negative real axis exceeds -1 or not. The line impedance Z_s is passive and its phase stays between $\pm 90°$. If the phase of the output admittance Y_o of the positive sequence (and the negative sequence) can be kept between $\pm 90°$, there is no cross point at the negative real axis and the system is always stable even if the admittance becomes large. It is sufficient when the real part of the admittance of the inverter is positive and passive to ensure stability.

$$Y_{re}(s) > 0 \quad (14)$$

B. Passivity-Based Design Methods

1) Energy-Based Method: In this subsection, three types of passivity-based design examples are shown. First, the energy-function-based methods like the Lyapunov function method are such control examples[15]-[19]. For this example, output impedances of cases operated by a single inverter are described [19]. Measured frequency characteristics of the Lyapunov control are shown in Fig.7(a). There is a cross point of the two amplitudes of $Z_o = Z_L$ and $Z_s = j\omega L_s$ ($L_s = 4$mH) around at 800Hz. However, the phase varies only around between $0°$ and $90°$ and this inverter is passive, and the phase difference is less than $180°$. This means the inverter is always stable evenfor large L_s values.

2) Passivity-Based Design: Development of a control method which has good tracking and stability characteristics has been desired. However, there is a certain limit for a control only with FB loops. To solve this limit, addition of a FF control loop, as a new passive design freedom, has been proposed[24]-[29]. Furthermore, a FB and FF stabilization method, which has a good tracking characteristic by the FB optimal control and has a passive characteristic for stability by the FF control, has been proposed[30]. The method consists of two steps. In the first step, the state FB control gains are designed to have a good sinusoidal tracking characteristic by the optimal control. The designed characteristic of the output impedance may not be passive at certain frequency bands. In the second step, conductive characteristics are added to the bands by the FF control to make the total characteristic passive and to ensure stability.

2189

The 2018 International Power Electronics Conference

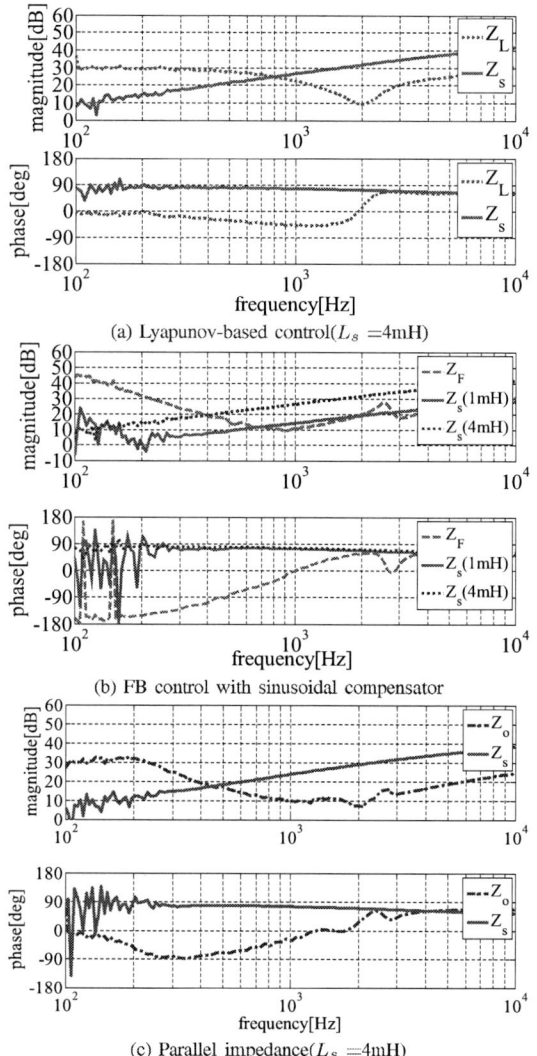

(a) Lyapunov-based control(L_s =4mH)

(b) FB control with sinusoidal compensator

(c) Parallel impedance(L_s =4mH)

Fig. 7. Measured frequency characteristics of inverter output and line inductor impedances.

3) Stabilization by Parallel Connection with Passive Inverters: It is possible to make a non-passive inverter totally passive with addition of a passive inverter in parallel[30]-[32]. In the previous example of Fig.7, the figure (b) shows frequency characteristics of a FB control case with the sinusoidal compensator. This case is not passive and the phase of $Z_s Y_o$, or the phase difference between Z_s and Z_o, becomes 180° around at 450Hz for L_s = 4mH. There is a cross point of the two amplitudes around at 400Hz and $|Z_s|$ becomes larger than $|Z_o|$ at 450Hz. This means the inverter is unstable for L_s = 4mH. Frequency characteristics of the Lyapunov control case are shown in Fig.7(b) as described in the above.

When these two inverters are connected in parallel, the total system has the impedance in Fig.7(c) where the phase varies within ±90°. The total system becomes passive and it is stabilized by addition of the passive Lyapunov inverter.

The above results by the proposed stability analysis are

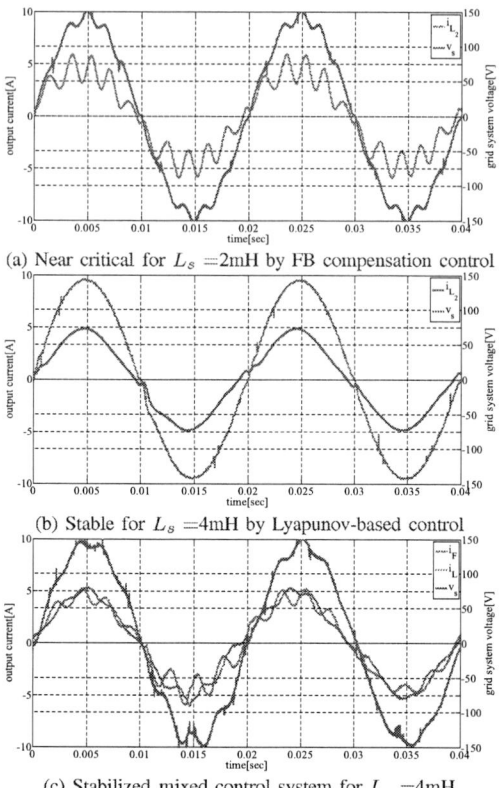

(a) Near critical for L_s =2mH by FB compensation control

(b) Stable for L_s =4mH by Lyapunov-based control

(c) Stabilized mixed control system for L_s =4mH

Fig. 8. Measured output voltage waveforms by the FB compensation and the Lyapunov-based controls.

investigated validated from measured output currents. The output current and the line voltage waveforms for the FB compensated case are shown in Fig.8(a) for L_s = 2mH. The inverter is stable. However, the waveforms have large oscillations of 450Hz which corresponds to the frequency at the phase difference of 180° and the inverter is not operated at all for L_s = 4mH because it becomes unstable. The output current and the line voltage waveforms for the Lyapunov case are shown in Fig.8(b) for L_s = 4mH. The inverter is stable without any oscillations. Results of a combined case with the FB compensated and the Lyapunov-controlled inverters for L_s = 4mH are shown in Fig.8(c). The output current waveforms have a little oscillations of 450Hz. However, the combined system is stable and the FB inverter is stabilized the passive Lyapunov inverter.

IX. STABILITY ANALYSIS USING POINCARÉ MAP IN TIME-DOMAIN: METHOD IV

A. Generalized Sensitivity Matrix Computation for Digital Control

As described in Section II C, the stability analysis method using Poincaré map finally reduced to eigen value analysis of the sensitivity matrix for one periodic cycle. The matrix for the digital control case is expressed as follows[39].

Time lengths of the switch modes are determined at each digital sample clock time T or at the beginning of each switch

2190

TABLE I.   EIGENVALUES OF THE SENSITIVITY MATRIX.

eigenvalues	abs. val.
-0.0912 + j 0.3468	0.3586
-0.0912 - j 0.3468	0.3586
0.0000 + j 0.0000	0.0000
0.0000 + j 0.0000	0.0000
0.0000 + j 0.0000	0.0000

cycle. For example, when one period consists of N_T switch cycles and each switch cycle consists of m_i circuit modes, the sensitivity matrix is as follows where ϕ_{ji} stands for the transition matrix of the j-th circuit mode at the i-th switch cycle.

$$\frac{\partial \boldsymbol{x}_T}{\partial \boldsymbol{x}_0} = \prod_{i=1}^{N_T} \phi_{m_i\ i} \left(\phi_{m_i-1\ i} \left(\cdots \left(\phi_{1i} + \dot{\boldsymbol{x}}_{1i} \frac{\partial t_{1i}}{\partial \boldsymbol{x}_{0i}} \right) \right. \right.$$
$$\left. \left. \cdots \right) + \dot{\boldsymbol{x}}_{m_i-1\ i} \frac{\partial t_{m_i-1\ i}}{\partial \boldsymbol{x}_{0i}} \right) \qquad (15)$$

When all absolute eigenvalues of the sensitivity matrix are less than one, the system is stable because variations of the initial values are not magnified.

B. Eigenvalue Analysis Result of the Sensitivity Matrices

The system line inductor value is gradually increased and the system is analyzed for its stability by the Poincaré Map method. The inductor value $L = 3.402\text{mH}$ was critical. The reason why $L = 3.402\text{mH}$ is critical is that the harmonic component affects the inverter output saturation. Due to the harmonics of i_{L2}, the control input $u[i]$ becomes to be unit which is the limit of the output voltage. This means that the critical condition is determined by the harmonic components and the inverter output saturation. Eigenvalues of the sensitivity matrix of the two boundary values are shown in Table I and their maximum absolute value is 0.3586 and less than one. The stability analysis is practically possible by the average circuit and it works. However, the stability analysis without the averaging technique by this Poincaré Map is useful to derive a precise critical condition by simulation, considering harmonic and inverter saturation effects.

X.   CONCLUSIONS

This paper investigated four types of stability analysis methods for a grid inverter system. In conclusion, they are distinguished depending on what case they are utilized.

Generally the impedance method is useful for stability analysis of a grid system. It is based on calculation of frequency characteristics of the output admittance because the characteristics include rich system behavior information and they are utilized to tune design parameters with enough stability margins. The frequency characteristics can be calculated for most basic cases by the analytical method. The objective system becomes more and more complex, the analytical operations becomes too difficult to apply. The frequency analysis simulator method is useful to analyze a general linear time-invariant system because it formulates and solves the system equations automatically and numerically according to a general-purpose simulator algorithm. These two methods need the averaging approximation to linearize the switching operations and such effects should be considered. The simulator frequency scan method does not need the averaging approximation and it computes a frequency response by FFT for a perturbed signal generally and numerically. However, this method needs such repeated manual operations and it needs a lot of computing efforts by the user.

According to the impedance method, it is sufficient when the real part of the admittance of the inverter is positive and passive to ensure stability. Three types of passivity-based design examples are shown.

The Poincaré or the sensitivity matrix analysis method is general ans analyzed in time-domain, and it does not include any approximation. It calculates a sensitivity matrix for state variables along a PSS trajectory which includes complete transient behavior information and checks all eigenvalues of the matrix to be less than one. This method is generally applicable even to a nonlinear system and it can find a critical stability condition considering nonlinear conditions, for example, saturation of inverter outputs and harmonics. However, it needs considerations of complex variations of switch timings along the operating trajectory and it has difficulties in general computation of the sensitivity matrix at present. Development of this general computation algorithm is a future topic.

Recently instability due to a negative admittance of a PLL circuit has been researched in many papers. As a future subject, this topic is a good example to investigate various features of the stability analysis methods.

ACKNOWLEDGMENT

This work was supported by (MEXT/JSPS) KAKENHI Grant Number JP17K06324.

REFERENCES

[1] R.D. Middlebrook: "Input filter considerations in design and application of switching regulators" , *IEEE Trans. on Power Electronics*, Vol.22, pp.1402-1409, (2007).

[2] J. Sun: "AC power electronic systems: stability and power quality", *Control and Modeling for Power Electronics(COMPEL)*, pp.1-10, (2008).

[3] Y.A. Familiant, J. Huang, K.A. Corzine, and M. Belkhayat: "New techniques for measuring impedance characteristics of three-phase ac power systems," IEEE Transactions on Power Electronics, Vol.24, No.7, pp.1802 - 1810, 2009.

[4] J. Sun: "Impedance-based stability criterion for grid-connected inverters", *IEEE Transactions on Power Electronics*, Vol.26, No.11, pp.3075-3078, (2011).

[5] A. Riccobono and E. Santi: "Comprehensive review of stability criteria for DC power distribution systems," IEEE Transactions on Industry Applications, Vol.50, No.5, pp.3525 - 3535, 2014.

[6] M. Cespedes and J. Sun: "Impedance modeling and analysis of grid-connected voltage-source converters", *IEEE Transactions on Power Electronics*, Vol.29, No.3, pp.1254-1261, (2014).

[7] T. Kato, K. Inoue, Y. Akiyama, K. Ohashi : "Stability analysis for grid-connected three-phase inverter with LCL filters," IEEE 16th Workshop on Control and Modeling for Power Electronics (COMPEL), pp.1 - 7, 2015.

[8] A. Rygg, M. Molinas, C. Zhang, and X. Cai: "A modified sequence-domain impedance definition and its equivalence to the dq-domain impedance definition for the stability analysis of ac power electronic systems," IEEE Journal of Emerging and Selected Topics in Power Electronics, Vol.4, No.4, pp.1383 - 1396, 2016.

[9] T. Kato, K. Inoue, and Y. Takami: "General-purpose computation method of a power converter for frequency characteristics: application to stability analysis of a grid inverter," IEEE Journal of Emerging and Selected Topics in Power Electronics, Vol.5, No.4, pp.1466 - 1473, (2017).

[10] G.D. Hachtel, R. Brayton, and F.G. Gustavson : "The sparse tableau approach to network analysis and design," IEEE Trans. Circuit Theory, Vol.CT-18, pp.101-113, (1971)

[11] C. H. Ho, A. E. Ruehli, and P. A. Brennan : "The modified nodal approach to network analysis," IEEE Trans. Circuit and Systems, CAS-22, pp.504-509, (1975)

[12] L. O. Chua and P. M. Lin; Computer Aided Analysis of Electronic Circuits : Algorithms and Computational Techniques, Prentice-Hall, (1975)

[13] J. Vlach and K. Singhal ; Computer Method for Circuit Analysis and Design Van Nostrand Reinhold, (1983)

[14] T. Kato, K. Inoue, and H. Kawabata: "Stability analysis of grid-connected inverter system", Control and Modeling for Power Electronics(COMPEL), pp.1-5, (2012).

[15] S.R. Sanders and G.C. Verghese: "Lyapunov-based control for switched power converters," IEEE Transactions on Power Electronics, Vol. 7 , No. 1, pp.17-24, 1992.

[16] A.M. Stankovic, G. Escobar, R. Ortega, and S.R. Sanders: "Energy-based control in power electronics," from Chapter 8.3 of Nonlinear Phenomena in Power Electronics by S. Banerjee and G.C. Verghese, IEEE Press.

[17] R. Ortega, A. Loría, P.J. Nicklasson, and H. Sira-Ramírez: Passivity-based control of Euler-Lagrange Systems, Springer Verlag, 2010.

[18] T. Kato, K. Inoue, and M. Ueda: "Lyapunov-based digital control of grid-connected inverter with an LCL filter," IEEE J. Emerg. Sel. Topics Power Electron., Vol. 2, No.4, pp.942-948, 2014.

[19] T. Kato, K. Inoue, and M. Ishida: "Investigation of stabilities of Lyapunov-based digital control for grid-connected inverter," IEEE Energy Conversion Congress and Exposition, pp.2394-2399, 2015.

[20] A. Riccobono and E. Santi: "Positive feed-forward control of three phase voltage source inverter for DC input bus stabilization," IEEE Applied Power Electronics Conference and Exposition (APEC), pp.741-748, 2011.

[21] A. Riccobono and E. Santi: "Positive feed-forward control of three-phase voltage source inverter for DC input bus stabilization with experimental validation," IEEE Transactions on Industry Applications, pp.169-177, 2013.

[22] A. Riccobono, J. Siegers, and E. Santi: "Stabilizing positive feed-forward control design for a DC power distribution system using a passivity-based stability criterion and system bus impedance identification," IEEE Applied Power Electronics Conference and Exposition - APEC 2014, pp.1139 - 1146, 2014.

[23] J. Siegers, S. Arrura, and E. Santi: "Stabilizing controller design for multi-bus MVDC distribution systems using a passivity based stability criterion and positive feed-forward control," IEEE Energy Conversion Congress and Exposition, pp.5080-5187, 2015.

[24] L. Harnefors, L. Zhang, and M. Bongiorno: "Frequency-domain passivity-based current controller design," IET Power Electron., Vol.1, No.4, pp.455-465, 2008.

[25] L. Harnefors, A.G. Yepes, A. Vidal, and J. Douval-Gandoy: "Passivity-based stabilization of resonant current controllers with consideration of Time Delay," IEEE Trans. on Power Electron., Vol.29, No.12, pp.6260-6263, 2014.

[26] L. Harnefors, A. G. Yepes, A. Vidal, and J. Doval-Gandoy: "Passivity-based controller design of grid-connected VSCs for prevention of electrical resonance instability," IEEE Ind. Electron., Vol. 62, No. 2, pp.702-710, 2015.

[27] L. Harnefors, X. Wang, A.G. Yepes, and F. Blaabjerg: "Passivity-based stability assessment of grid-connected VSCs - an overview - ," IEEE J. Emerg. Sel. Topics Power Electron., Vol.4, No.1, pp.116 - 125, 2016.

[28] L. Harnefors, R. Finger, X. Wang, H. Bai, and F. Blaabjerg: "VSC input-admittance modeling and analysis above the Nyquist frequency for passivity-based stability assessment," IEEE Transactions on Industrial Electronics, 2017.

[29] X. Wang, F. Blaabjerg, and P.C. Loh: "Passivity-based stability analysis and damping injection for multi-paralleled voltage-source converters with LCL filters," IEEE Transactions on Power Electronics, 2017.

[30] T. Kato, K. Inoue, and Y. Nakajima: "Stabilization of grid-connected inverter system with feed-forward control," IEEE Energy Conversion Congress and Exposition (ECCE), pp.3375 - 3382, 2017.

[31] T. Kato, K. Inoue, and M. Ishida: "Stabilization of grid-connected inverter system with mixed control methods," IEEJ Static Power Converter Workshop, SPC-16-025, January, 2016.(In Japanese)

[32] H. Bai, X. Wang, and F. Blaabjerg: "Passivity enhancement in renewable energy source based power plant with paralleled grid-connected VSIs," IEEE Transactions on Industry Applications, Vol.53, No.4, pp.3793 - 3802, 2017.

[33] Y. Kuroe: "Computer methods to analyze stability and bifurcation phenomena," in Sec. 4.5 of Nonlinear Phenomena in Power Electronics by IEEE Press, 2001

[34] O. Dranga, B. Buti, and I. Nagy: "Stability analysis of a feedback-controlled resonant DC-DC converter," IEEE Transactions on Industrial Electronics, Vol.50, No.1, pp.141 - 152, 2003.

[35] S.K. Mazumder, A. H. Nayfeh, and D. Boroyevich: "An investigation into the fast- and slow-scale instabilities of a single phase bidirectional boost converter," IEEE Transactions on Power Electronics, Vol.18, No.4, pp.1063 - 1069, 2003.

[36] S. Eren, M. Pahlevaninezhad, A. Bakhshai, and P.K. Jain: "Composite nonlinear feedback control and stability analysis of a grid-connected voltage source inverter With LCL Filter," IEEE Transactions on Industrial Electronics, Vol.60, No.11, pp.5059 - 5074, 2013.

[37] S. Eren, A.Bakhshai, and P. Jain: "Geometric analysis of grid-connected VSI with LCL-filter using Poincaré map," 2015 IEEE Energy Conversion Congress and Exposition (ECCE), pp.1948 - 195, 2015.

[38] A.E. Aroudi, D. Giaouris, H.H.C. Iu, and I.A. Hiskens: "A review on stability analysis methods for switching mode power converters," IEEE Journal on Emerging and Selected Topics in Circuits and Systems, Vol.5, No..3, 2015.

[39] T. Kato, K. Inoue, and Y. Takami: "Stability analysis using Poincaré map in the time-domain for grid-connected inverter," 2017 IEEE 18th Workshop on Control and Modeling for Power Electronics (COMPEL), pp.1-7, 2017.

[40] T.J. Aprlle and T. N. Trick : "Steady-state analysis of nonlinear circuits with periodic inputs," Proc. IEEE, Vol. 60, pp. 108-114, Jan. 1972.

[41] L. O. Chua and P. M. Lin : Computer Aided Analysis of Electronic Circuits: Algorithms and Computational Techniques, Prentice-Hall, 1975.

[42] R. C. Wong : "Accelerated convergence to the steady-state solution of closed-loop regulated switching-mode systems as obtained through simulation," in IEEE PESC' 87 Rec., pp. 682-692.

[43] Y. Kuroe, T. Maruhashi, and N. Kanayama: "Computation of sensitivities with respect to conduction time of power semiconductors and quick determination of steady state for closed-loop power electronic systems," PESC Record, pp. 756-764, 1988.

[44] D. G. Bedrosian and J. Vlach : "An accelerated steady-state method for networks with internally controlled switches," IEEE Trans. Circuits Syst.I, vol. 39, no. 7, pp. 520-530, 1992.

[45] T. Kato and W. Tachibana : "Periodic steady-state analysis of an autonomous power electronic system by a modified shooting method," IEEE Transactions on Power Electronics, Vol.13, No.3, pp.522 - 527, 1998.

[46] T. Kato, K. Inoue, and Y. Donomoto: "Fast current-tracking control for grid-connected inverter with an LCL filter by sinusoidal compensation," IEEE Energy Conversion Congress and Exposition, pp.2543 - 2548, 2011.

Finite Element Methods for Multi-objective optimization of a High Step-up Interleaved Boost Converter

Wilmar Martinez[1,2*], Camilo Cortes[2], Ahmad Bilal[3], and Jorma Kyyra[3]

[1] KU Leuven, Campus Diepenbeek, Belgium
[2] Universidad Nacional de Colombia, Bogota, Colombia
[3] Aalto University, Espoo, Finland
*E-mail: wilmar.martinez@kuleuven.be

Abstract— **High step-up converters have been widely used in renewable energy systems and, recently, in automotive applications, due to their high voltage gain capability. Moreover, in these applications, efficiency and high-power density are usually required, although these characteristics are commonly opposite objectives. Therefore, a multi-objective optimization is quite useful in order to comply with both requirements of high efficiency and small size. In that sense, Finite Element Methods can be effective to complement optimization methods. This paper presents a procedure to optimize the efficiency and the volume of a high step-up converter that utilizes a coupled inductor with three windings installed in only one core. This optimization procedure is carried out using 3D and 2D Finite Element Method simulations. In this procedure, a complete modeling of power losses, size, and flux density is evaluated by comparing different materials and dimensions. The results of this modeling stage are introduced into a multi-objective optimization algorithm to obtain a Pareto front. Finally, the optimization methodology is validated by experimental tests.**

Keywords— *Finite Element Methods, High Step-Up Converter, Efficiency, Power Density.*

I. INTRODUCTION

High Step-Up (HSU) converters are able to boost low voltages to obtain much higher voltages required by the application load. Usually, HSU converters are used in systems where the power source has low voltage storage cells that need to power industrial applications such as Uninterruptible Power Systems-UPS, communication systems, renewable energy appliances, grid-connected systems, electric mobility, among others [1]-[7].

These converters become useful due to their advantages of high voltage gain, which is quite higher in comparison with the conventional and the interleaved boost topologies. Conventional topologies require an extremely high duty cycle to achieve the required gain, and high duty cycles generate large losses. Moreover, parasitic components play an important role in HSU converters because they hamper the conventional voltage-gain at high duty cycles. When a large duty cycle

is in operation, the voltage-gain tends to decrease because of the parasitic components [7]-[9].

Nevertheless, most of the HSU converters evidence limitations of efficiency and power density because of the complex circuitry with abundant components that increase its size. These additional components may be bulky and heavy, increase the volume and mass of the circuit and compromise the converter power density and, in many cases, the efficiency as well [10]-[12].

Consequently, multi-objective optimization of efficiency and power density of these converters has become needed to obtain suitable designs for the applications mentioned above. These optimization procedures are more required in applications where power density and efficiency are critical, like electric mobility or renewable energies [13]-[19].

This study analyses an HSU converter with a novel coupled-inductor capable of increasing the power density and the voltage gain. This converter acquires an outstanding performance thanks to the combination of the techniques of magnetic coupling and interleaving phases, which are effective for downsizing inductive and capacitive components in DC-DC converters [20]-[22]. Also, from the physics point of view, a reduction in the magnetic core size could represent a direct reduction on the iron losses [23]. Consequently, this HSU converter offers an outstanding high voltage gain without the addition of many semiconductors or bulky inductors and capacitors.

In addition, an optimization procedure is implemented to design an HSU converter suitable for applications where high-power density and efficiency are required. In this procedure, 3D and 2D Finite Element Method (FEM) simulations are conducted in COMSOL to complement the analytical models, and estimate the power losses in the magnetic core and shielding materials, the magnetic stray fields, and the temperature rise in the materials.

This paper is organized in four sections. First, the review of the operating principle of the analyzed converter is presented. The ideal and non-ideal voltage gain is reviewed as well. Second, the optimization procedure is presented; the power loss analysis and the FEM simulations are conducted. Then, the Pareto front is obtained from the optimization procedure and from the 3D FEM simulations. Finally, a 1kW prototype, built for the validation of the theoretical analysis, is experimentally tested.

II. REVIEW OF THE HIGH STEP-UP CONVERTER

A. Circuit Configuration

The analyzed HSU converter, shown in Fig. 1, is a two-phase interleaved boost converter composed of a magnetic coupled-inductor with three windings that are installed in only one magnetic core. This core can be constructed with different three-leg shapes (usually EE, EI, EER, EC cores). Two windings, L_1 and L_2, are connected to the power source and a central winding L_c (usually installed in the center of the three-leg core) is located between the anodes of D_3 and D_4. For convenience, the positive terminal of Lc is defined as the node where the cathode of D_1 and the anode of D_3 are connected. In addition, this HSU converter has two power switches, S_1 and S_2, which are alternatively commuted with a phase difference of 180-degrees between them. Finally, the converter is composed of four diodes D_1-D_4, and one output capacitor C_o, as well.

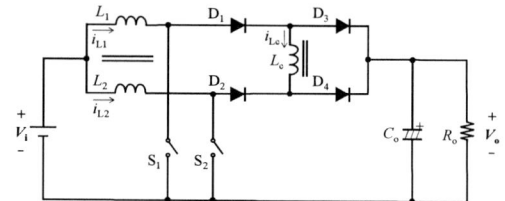

Fig. 1. High step-up converter with coupled inductor.

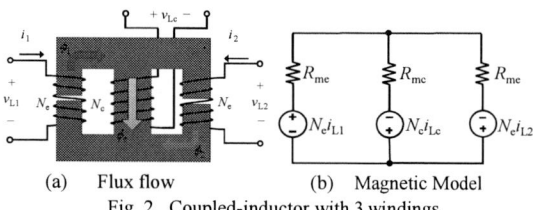

(a)　Flux flow　　　(b)　Magnetic Model

Fig. 2. Coupled-inductor with 3 windings.

Fig. 2 depicts the representation of the coupled-inductor shown in Fig. 1 using an EE core. The magnetic component of this converter has three windings that share the same core. The outer windings are directly coupled. An air-gap is placed in each outer leg to suppress DC flux induction. Three different magnetic fluxes are circulating through the core (Fig. 2(a)) and three magnetic reluctances can be defined (Fig. 2(b)). ϕ_1, ϕ_2 and ϕ_c are the external and central magnetic fluxes in the EE core; Ne and Nc are the number of turns of the external windings and the central one, respectively; R_{me} and R_{mc} are the magnetic reluctances of the external and central legs, respectively; and i_{L1}, i_{L2}, and i_{LC} are the currents flowing through each winding.

B. Operating Principle

Fig. 3 shows the operating waveforms of the converter under an ideal and a continuous conduction operation. Fig. 3 shows the cases of duty cycle d when it is lower and higher than 50%. As it is well known, when a converter has an interleaved two-phase operation, it has at least two duty cycle cases with different performances.

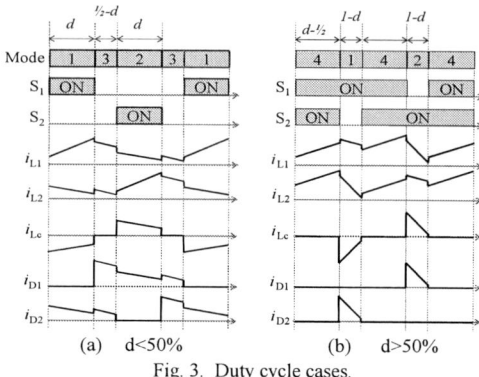

(a)　d<50%　　　(b)　d>50%

Fig. 3. Duty cycle cases.

As many other two-phase interleaved converters, the analyzed converter present four operating modes. Each operating mode is presented in Fig. 4. Modes 1, 2 and 3 correspond to the case of duty cycles d<0.5, and Modes 1, 2 and 4 are presented in the case d>0.5. The complete description of each operating mode and the steady-state analysis are presented in [22].

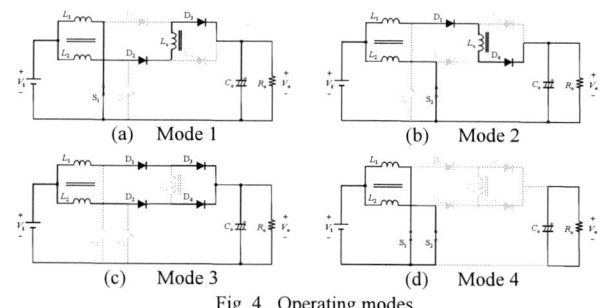

(a)　Mode 1　　　(b)　Mode 2

(c)　Mode 3　　　(d)　Mode 4

Fig. 4. Operating modes.

To put in context the effectiveness of this converter, its voltage gain is reviewed. Considering the analysis of [22] and the definition of N as the ratio between the number of turns of the central leg and the external legs ($N=Nc/Ne$), the voltage gain M of both duty cycle cases is derived as follows:

$$M_{d<0.5} = \frac{V_o}{V_i} = \frac{1+N}{(1+N)-d(1+2N)} \tag{1}$$

$$M_{d>0.5} = \frac{V_o}{V_i} = \frac{1+N}{1-d} \tag{2}$$

where V_o and V_i are the output and input voltages.

Fig. 5 shows the ideal voltage gain with different turn ratios, based on (1) and (2).

2194

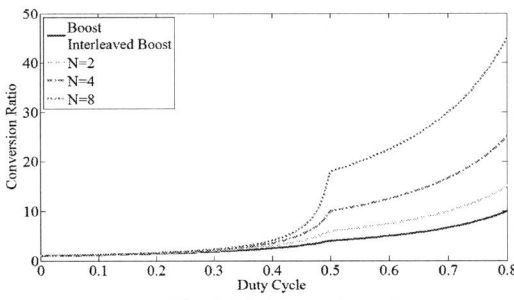

Fig. 5. Ideal conversion ratio.

Nevertheless, as it was mentioned above, parasitic components are important in HSU converters. The non-ideal voltage gain can be calculated as shown in Fig. 6. Fig. 6 considers different cases of ratios between the parasitic resistances R_L and the load R_o and proves the importance of parasitic components in HSU converters. It is possible to conclude that when the ratio between the parasitic resistances in the windings and the resistance of the load increases, the voltage gain is increasingly affected [20].

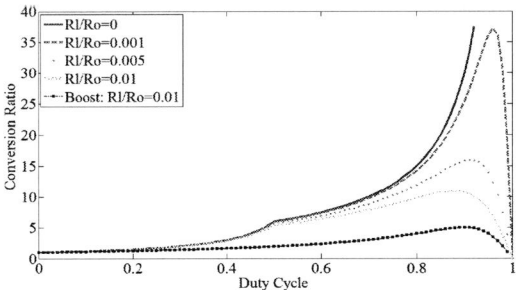

Fig. 6. Non-Ideal conversion ratio vs. Duty cycle.

III. OPTIMIZATION PROCEDURE

The first analysis of the optimization procedure is the power loss analysis. Then, several constrains and objective functions are defined. In the case of the converter efficiency, it is optimal when the following objective function is minimized:

$$
\min \begin{bmatrix} P_{Loss} = P_{Lcore}(f_{sw},k_{core}) + P_{Lcopper}(f_{sw},k_w) + \\ P_C(f_{sw},k_C) + P_{TR}(f_{sw},k_{TR}) + P_D(f_{sw},k_D) \end{bmatrix} \quad (3)
$$

where each k corresponds to the particular factors of each component needed to calculate their individual power losses. In addition, f_{sw} is the switching frequency and it is one of the main variables that affect the optimization procedure. As it is well-known, increasing the switching frequency (until certain values) can be beneficial for reducing the converter size but harmful for the power conversion efficiency.

Moreover, in order to calculate and minimize the power losses and the converter size, an analytical algorithm is conducted. This analytical calculation is carried out following the steps of the design procedure presented in Fig. 7.

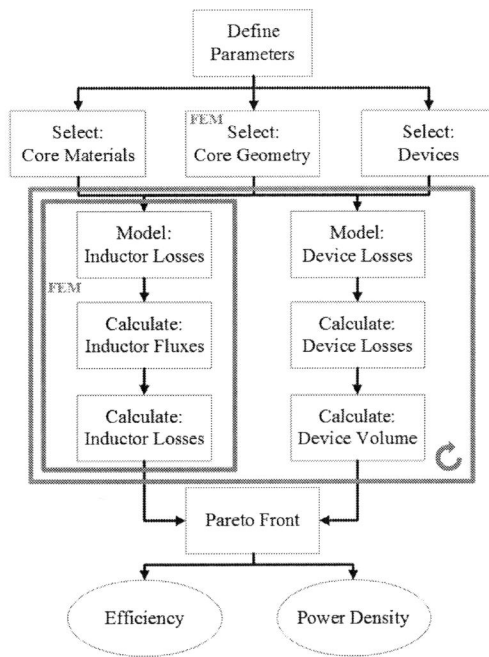

Fig. 7. Design procedure.

In the design procedure of Fig. 7, it is highlighted that part of it is conducted through FEM simulation. The coupled-inductor is simulated with the evaluation of magnetic and electrical fields through time dependent and frequency domain studies.

Taking as a case study of this particular procedure, the parameters presented in Table I were chosen to validate the effectiveness of the proposed methodology.

TABLE I
CIRCUIT PARAMETERS

Input Voltage V_i [V]	80
Output Voltage V_o [V]	200
Turn Ratio N	2
Power P [kW]	1
Switching Frequency f_{sw} [kHz]	15-200

In this procedure, several soft-magnetic materials were considered. In addition, silicon, silicon carbide, and gallium nitride semiconductors were selected for this optimization study, as well as film, electrolytic, and multi-layer ceramic capacitors. All these components are evaluated in the procedure of Fig. 7, considering different switching frequencies, number of turns, possible core geometries, and coupling factors. Table II shows a review of the characteristics of the magnetic materials selected for this case study. Table III and Table IV show the characteristics of the Mosfets and Diodes selected for this study, respectively. In total, 70 Mosfets, 70 Diodes, and 30 magnetic materials were considered in this case study. Diodes and Mosfets were selected at ratings of 650V and average 20A.

TABLE II
MAGNETIC MATERIAL CHARACTERISTICS

Material	Relative Permeability	Losses [mW/cm³]*	Saturation Flux [T]**
Micrometals Powdered Iron	35~75	600~1300	~0.5
Magnetics Kool M	60~125	~200	~1
Magnetics Molypermalloy	60~550	87~890	~0.8
Magnetics High Flux	14~160	290~1280	~1.5
Magnetics Ferrite	2000~3000	5~38	0.45~0.47
TDK Ferrite	2000~2500	14~21	~0.35
FairRite	~1500	~85	~0.5

*Loss at 100kHz 50 mT   **Remanent flux excluded

TABLE III
MOSFET CHARACTERISTICS

Manufacturer	Ciss [nF]	Coss [pF]	Ron [mΩ]
STM	0.78~9.6	~300	45~299
Fairchild	1~1.415	~35	165~250
Infineon	0.95~1.5	~750	26~280
ROHM SiC	~0.852	~55	60~196
Panasonic GaN	~0.405	~71	70~290
On Semi GaN	~0.760	~26	340~350
GaN Systems	~0.520	~130	25~560
Transphorm GaN	0.72~1.13	~56	72~180

TABLE IV
DIODE CHARACTERISTICS

Manufacturer	trr [ns]	Qr [nF]	Vd [V]
Fairchild	50~80	0.29~65	0.7~1.5
ROHM SiC	0~60	0.21~0.73	1.05~1.55
CREE SiC	~	0.29~1.1	1.45~1.8
Infineon SiC	0 ~13.7	0.19~1.14	1.35~1.7

IV.   FEM PROCEDURE

A. Optimization Solvers using FEM

As Fig. 7 shows, FEM procedures were conducted for two main purposes. The first one was the core geometry optimization using a 2D FEM simulation and applying a topology optimization procedure for the case of an EE core.

The second FEM procedure was used for the calculation and modeling of magnetic fluxes and core losses. In this case a 3D FEM simulation was conducted based in the results of the 2D procedure.

As it is well-known, different optimization procedures can be conducted in FEM software. As an example, COMSOL offers different optimization modules capable of solving different problems with both Gradient-Free and Gradient-Based methods. In these modules, several methods can be implemented, *inter alia,* Random, 1st and 2nd order approximate gradient, Linear, quadratic [24].

B. 2D FEM Simulation

As it was mentioned above, a 2D FEM simulation was conducted I order to implement a topology optimization algorithm. In this context, the main dimensions of a three- leg EE core (see Fig. 2) were evaluated in this algorithm using the parameters of the case study presented in Table I.

Fig. 8 shows the case of the core EC70 PC40 manufactured by TDK driven at a frequency of 200kHz. Fig. 8a shows that the evaluated core is oversized because it has an average magnetic flux of nearly 150mT when its saturation flux is 250mT (having in mind the remanent flux). In addition, Fig. 8b shows the arrow surface magnetic flux, where it is possible to see the effect of the direct coupling between the external windings.

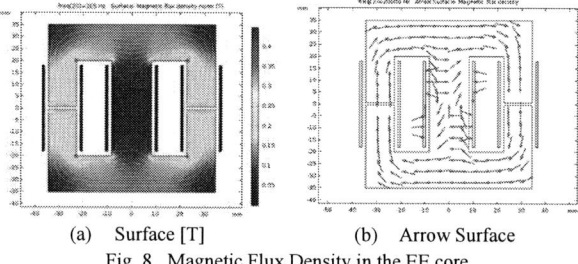

(a)   Surface [T]   (b)   Arrow Surface

Fig. 8. Magnetic Flux Density in the EE core.

As a result of this topology optimization procedure, the core EE50 PC40 was selected for the case study described above. This core is much smaller than the one that was originally selected and that have been tested in previous studies [20][22].

C. 3D FEM Simulation

A 3D FEM simulation was conducted in order to calculate the magnetic fluxes and the power losses in the evaluated cores. At the same time, these simulations served as a validation of the coupled-inductor operation. This validation was possible via the verification of the lack of saturation in the core.

Fig. 9 shows the 3D FEM component of the originally selected core (EC70). This component was modeled with a mesh of 300 thousand element with sizes between 15 and 3 mm.

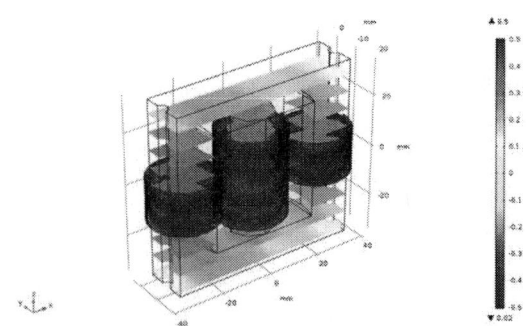

Fig. 9. FEM Model of the HSU Coupled-Inductor.

After the 3D FEM simulations, the loss evaluation and power density of transistors, diodes, and capacitors of the HSU converter are considered in order to find the suitable operating conditions that offer the lowest power losses and the highest power density.

V. CONVERTER TEST RESULTS

Taking into account the components of Table II-IV, the parameters of Table I, and the optimization method using FEM described above, the full design procedure of Fig. 7 was conducted.

As a result, the Pareto-Front of Fig. 10 was obtained. Fig. 10 shows many points that correspond to all the possible designs of the coupled inductor at different conditions in the selected HSU converter. In this figure, efficiencies higher than 90% and their possible power densities are presented. As a matter of fact, it is important to notice that the efficiency and power density values correspond to the case of the inductor only, without considering cooling and casing systems.

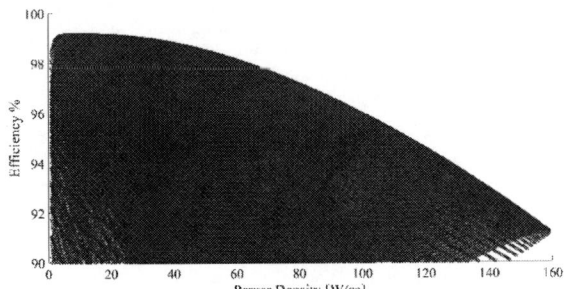

Fig. 10. Efficiency – Power Density Pareto representation of the coupled inductor.

To validate the optimization procedure presented in the previous sections, an experimental validation was conducted. Therefore, two 1kW-prototypes were constructed with a core EE50, SiC Diodes, and one Multilayer Ceramic Capacitor. One prototype was constructed with SiC Mosfets and the other with GaNFETs. These circuits were tested using the parameters of Table I. The efficiency tests were measured with a Precision Power Analyzer. Fig. 11 shows the prototype of the HSU converter with GaNFETs.

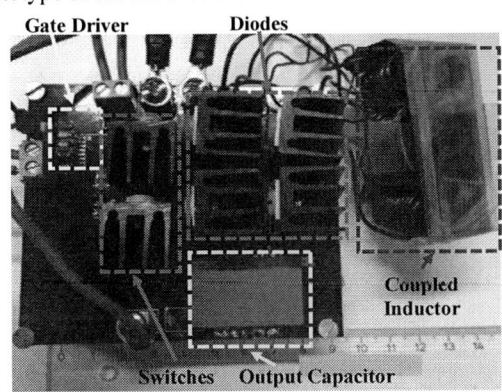

Fig. 11. Prototype of the HSU converter with GaNFETs.

In addition, efficiency and volume measurements were conducted. The 1kW prototypes were tested at the conditions that the Pareto front provided. In these measurements, additional parasitic components such as the ESR of the output capacitor and the ESR of the terminals were neglected in the optimization procedure. Therefore, it is considered to take them into account in future optimization procedures.

In that sense, four switching frequencies were tested for the case of both converters. Fig. 12-15 show the efficiency and the voltage gain measured during the experimental tests.

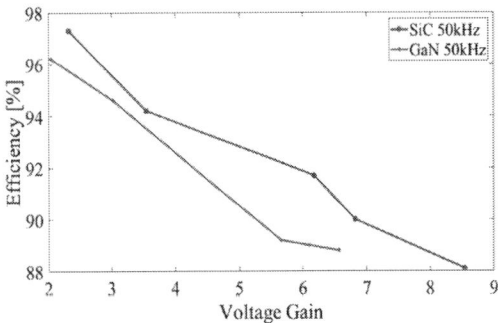

Fig. 12. Efficiency vs. Voltage Gain at 50kHz.

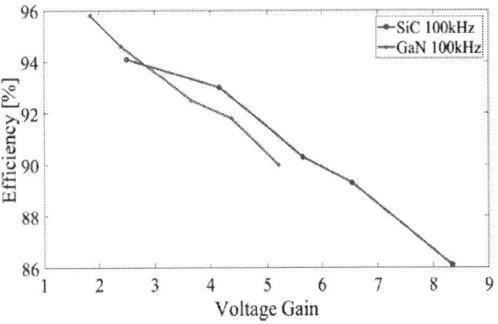

Fig. 13. Efficiency vs. Voltage Gain at 100kHz.

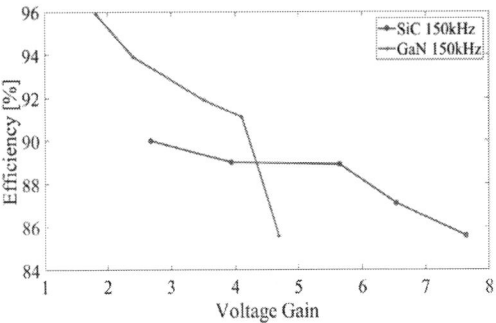

Fig. 14. Efficiency vs. Voltage Gain at 150kHz.

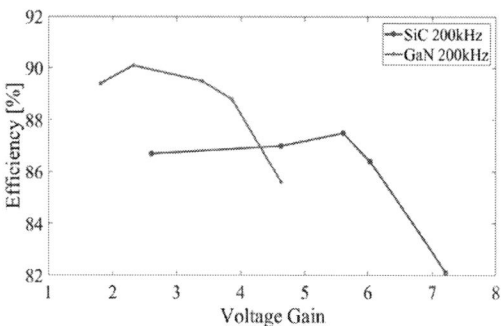

Fig. 15.Efficiency vs. Voltage Gain at 200kHz.

It is possible to notice that the voltage gain that the GaNFETs presented was lower than the one of the SiC Mosfets. This behavior was observed due to the parasitic conditions of the GaN switches. On the other hand, at frequencies higher than 100kHz, GaN switches presented higher efficiency than the prototypes with SiC switches.

It is important to mention that using multi-objective optimization procedures always leads to different solutions that must be considered with the use of constrains and limitations. In the case of applications where high voltage gain is required, the use of SiC would be suitable considering the selected case study.

Finally, the constructed prototype showed a power density around 2.5W/cc, without casing and cooling systems (despite the heatsinks).

VI. CONCLUSIONS

An optimization methodology that combines analytical calculation of power losses and power density with 2D and 3D FEM simulations of the coupled inductor in a novel HSU converter is presented and evaluated in this paper. This optimization procedure takes into account emerging technologies of semiconductors, novel magnetic materials, different core geometries and sizes, as well as several design conditions like number of turns, magnetic coupling, and switching frequency. This methodology was conducted with an analytical calculation of an optimization algorithm where FEM simulations were integrated. As a result, the efficiency – power density Pareto front was obtained. It was found that this procedure is effective to select the suitable core size for a converter with specific parameters.

REFERENCES

[1] L. Schmitz et al, "Design optimization of a high step-up DC-DC converter for photovoltaic microinverters," *2017 IEEE International Telecommunications Energy Conference (INTELEC)*, pp. 432-437, 2017.

[2] T. S. Lakshmi and P. R. Rao, "High voltage gain boost converter for micro grid application," *2012 Annual IEEE India Conference (INDICON)*, pp. 323-328, 2012.

[3] G. Cao et al, "A high step-up modular DC/DC converter for photovoltaic generation integrated into DC grids," *43rd Annual Conference of the IEEE Industrial Electronics Society*, pp. 4421-4426, 2017.

[4] W. Martinez, C. Cortes, M. Yamamoto, J. Imaoka and K. Umetani, "Total Volume Evaluation of High Power Density Non-Isolated DC-DC Converters with Integrated Magnetics for Electric Vehicles," *IET Power Electronics*, vol.10, no.14, pp. 1755-4535, Nov. 2017.

[5] G. Wu, X. Ruan and Z. Ye, "High Step-Up DC-DC Converter Based on Switched Capacitor and Coupled Inductor," *IEEE Transactions on Industrial Electronics*, vol. 65, no.7, pp. 5572-5579, 2018.

[6] H. C. Liu and F. Li, "Novel High Step-Up DC–DC Converter with an Active Coupled-Inductor Network for a Sustainable Energy System," *IEEE Transactions on Power Electronics*, vol. 30, no.12, pp. 6476-6482, 2015.

[7] C.-T. Pan, C.-F. Chuang, and C.-C. Chu, "A Novel Transformer-less Adaptable Voltage Quadrupler DC Converter with Low Switch Voltage Stress," *IEEE Trans. Power Electron.*, vol. 29, no. 9, pp. 4787–4796, Sep. 2014.

[8] Z. Zhang, and L. Zhou, "Analysis and design of isolated flyback voltage-multiplier converter for low-voltage input and high-voltage output applications," *IET Power Electron.*, vol. 6, no. 6, pp. 1100–1110, 2013.

[9] K.-C. Tseng, C.-C. Huang, and C.-A. Cheng, "A Single-Switch Converter with High Step-Up Gain and Low Diode Voltage Stress Suitable for Green Power-Source Conversion," *IEEE J. Emerg. Sel. Top. Power Electron.*, vol. 4, no. 2, pp. 363–372, Jun. 2016.

[10] W. Martinez, J. Imaoka, Y, Itoh, M. Yamamoto and K. Umetani "A Novel High Step-Down Interleaved Converter with Coupled Inductor," *IEEE International Telecommunications Energy Conference - INTELEC*, pp. 1-6, Oct, 2015.

[11] M. Pavlovsky, G. Guidi, and A. Kawamura, "Assessment of Coupled and Independent Phase Designs of Interleaved Multiphase Buck/Boost DC-DC Converter for EV Power Train," *IEEE Trans. Power Electron.*, vol. 29, no. 6, pp. 2693–2704, 2014.

[12] K. Park, G. Moon, and M. Youn, "Non-isolated high step-up stacked converter based on boost-integrated isolated converter," *IEEE Trans. Power Electron.*, vol. 26, no. 2, pp. 577–587, 2011.

[13] J. M. Myrzik and M. Calais, "String and module integrated inverters for single-phase grid connected photovoltaic systems-a review," *Power Tech Conference*, pp. 1-8, 2003.

[14] D. Pal, H. Koniki and P. Bajpai, "Central and micro inverters for solar photovoltaic integration in AC grid," *National Power Systems Conference (NPSC)*, pp. 1-6, 2016.

[15] S. B. Kjaer, J. K. Pedersen and F. Blaabjerg, "Power inverter topologies for photovoltaic modules-a review," *37th IAS Annual Meeting. Conference*, pp. 782-788, 2002.

[16] S. B. Kjaer, J. K. Pedersen and F. Blaabjerg, "A review of single-phase grid-connected inverters for photovoltaic modules," *IEEE Trans. Ind. Appl.*, vol. 41, no.5, pp. 1292-1306, 2005.

[17] B. Xiao et al, "Modular cascaded H-bridge multilevel PV inverter with distributed MPPT for grid-connected applications," *IEEE Trans. Ind. Appl.*, vol. 51, no.2, pp. 1722-1731, 2015.

[18] M. Hirakawa et al., "High power density interleaved DC/DC converter using a 3-phase integrated close-coupled inductor set aimed for electric vehicles," in 2010 *IEEE Energy Conversion Congress and Exposition*, pp. 2451–2457, 2010.

[19] B. Ahmad, W. Martinez, and J. Kyyra, "Efficiency Optimization of Interleaved High Step-Up Converter," *9th International Conference on Power Electronics, Machines and Drives - PEMD*, pp. 1-6, Apr, 2018.

[20] W. Martinez, C. Cortes, J. Imaoka and M. Yamamoto, "Parasitic Resistance Effect on the Voltage-Gain of High Step-Up DC-DC Converters for Electric Vehicle Applications," *IET Power Electronics*, vol.PP, no.99, (11 Pages), Jan. 2018

[21] W. Josias de Paula, D. de S. Oliveira Júnior, D. de C. Pereira, and F. L. Tofoli, "Survey on non-isolated high-voltage step-up dc–dc topologies based on the boost converter," IET Power Electron., vol. 8, no. 10, pp. 2044–2057, Oct. 2015.

[22] W. Martinez, J. Imaoka, M. Yamamoto, and K. Umetani, "High Step-Up Interleaved Converter for Renewable Energy and Automotive Applications," in 4th International Conference on Renewable Energy Research and Applications, pp. 1–6. 2015.

[23] W. Martinez, S. Odawara, and K. Fujisaki, "Iron Loss Characteristics Evaluation Using a High Frequency GaN Inverter Excitation," *IEEE Transactions on Magnetics*, vol. 53, no. 11, pp. 1-7, Jun. 2017.

[24] W. Frei, "Optimization with COMSOL Multiphysics," COMSOL Tokyo Conference 2014, pp. 1-55, 2014.

The 2018 International Power Electronics Conference

High Fidelity Real-Time Simulation of Multi-Level Converters

Jost Allmeling, allmeling@plexim.com – Niklaus Felderer, felderer@plexim.com – Min Luo, luo@plexim.com

Plexim GmbH, Technoparkstrasse 1, 8005 Zürich

Abstract—Control schemes of multi-level converters are comprehensive and challenging to test on real hardware targets under various operation conditions. Hardware-In-the-Loop (HIL) simulations allow the replacement of the actual plant with an appropriate dynamic model, allowing testing of the control algorithm in a safe environment. The large number of switching elements in multi-level topologies increases the complexity of the corresponding state space model and hence its execution time on a real-time platform. This work presents a method that allows the modeling of converter topologies with an arbitrary number of output voltage levels with only two diodes and controlled current and voltage sources. The models offer high fidelity as they work under various conditions such as discontinuous conduction mode, finite dead time and rectifier mode.

I. INTRODUCTION

Voltage source converter (VSC) topologies with multiple output voltage levels are employed in different areas of power electronic application such as drives and traction, high voltage DC transmission (HVDC) or static synchronous compensator (STATCOM) for flexible AC transmission [1]–[3]. The increased number of voltage levels allows, for example, decreasing the current ripple in electrical drives and increasing the total voltage rating of a converter by using individual semiconductor switches of lower voltage rating [4]. Due to the large number of electrical switches in these systems the control and modulation schemes are comprehensive [5]. Hardware-In-the-Loop (HIL) simulation can be used to emulate the circuit topology and represent a virtual target for the controller to drive. In this manner different operating conditions such as fault cases, system start-up and transient load changing can be analyzed, avoiding potential damage to the power circuit in the real system.

A challenge of real-time simulation of multi-level topologies is the high number of switching devices. If the ideal switch model proposed by [6] is employed to model the switching transitions, the simulator must apply a different state space description of the system for each switching combination. If there are n independent switches in a converter, the number of state space descriptions grows exponentially with 2^n [6]. As every real-time simulation platform offers limited memory to store these computed state space matrices, the number of independent switching elements n imposes a hard limit on how many levels can be simulated.

This limitation can be overcome by using a concept called sub-cycle averaging which takes advantage of hybrid converter models instead of individual ideal switches [7]. The following sections introduce this concept to cascaded and neutral point clamped (NPC) inverter topologies. The accuracy and efficiency of the developed hybrid models are investigated with the example of a three-phase five-level active neutral point clamped (5L-ANPC) converter.

II. SUB-CYCLE AVERAGING REVISITED

This section gives a brief overview on the topic of sub-cycle averaging, a more comprehensive explanation of the fundamental idea of modeling can be found in [7]. Sub-cycle averaging consists of two parts:

- An average switching signal $\bar{s}(t)$ which is obtained by averaging the input switching signal $s(t)$ over an interval equal to the model discretization step-size T_{disc}. The average value is calculated with the time average method (TAM) [8].
- A hybrid converter model consisting of controlled current and voltage sources and a diode pair.

The controlled sources reflect the behavior of the forced commutated switches as a function of the switching signal where the diode pair is used to distinguish between a positive and negative phase current. These diodes capture the natural commutation of current in a switching network. Their state depends on the applied switching signals $s(t)$ together with the externally connected circuit. This means the rectifier mode and discontinuous conduction mode of a half-bridge can also be analyzed in simulation.

Besides the advantage that hybrid models are able to process information about the duty cycle of $s(t)$ with much higher resolution than its switched counterpart, multi-level VSCs require only one diode pair per phase [7]. This brings the benefit that the dynamic behavior of p series-connected half-bridges can be modeled with only two diodes per arm instead of $2p$ electrical switches. This reduces the model complexity significantly.

An example of a hybrid model for a half-bridge (Fig. 1a) is given in Fig. 1b. The analysis of the model is carried out in two steps, one for a positive phase current i_{ph}^+ where the lower diode conducts and another for a negative phase i_{ph}^- current where the upper diode conducts. The diodes never conduct at the same time and are turned-on exclusively. Both diodes, however, can be blocking at the same time, for example in discontinuous conduction mode [9]. The functions controlling the voltage and current sources are found by investigating the basic case when the average switching signals $\bar{s}^+$ and $\bar{s}^-$ are either 0 or 1. For these two cases the output voltages are given

2199

The 2018 International Power Electronics Conference

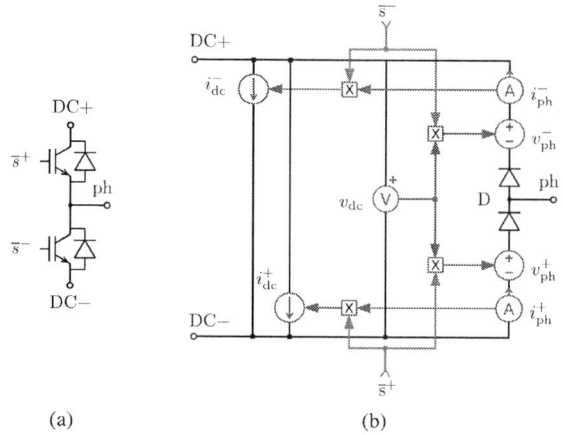

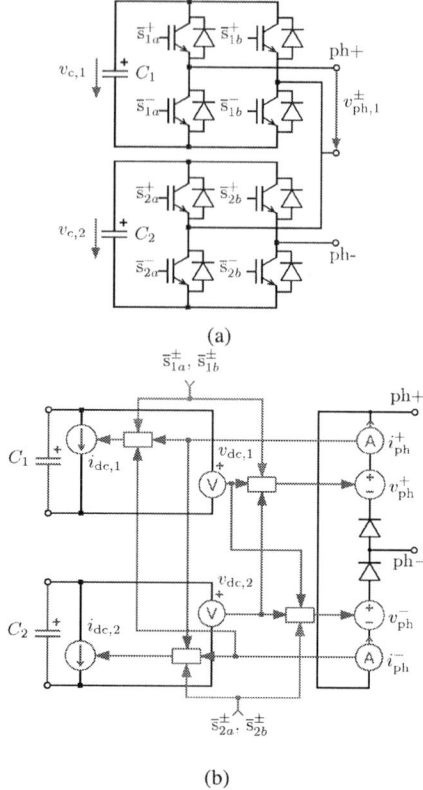

(a)   (b)

Fig. 1: Switched (a) and hybrid model (b) of a half-bridge.

by

$$
\boxed{\overline{s}^+ = 1, \overline{s}^- = 0} \qquad v_{ph}^+ = 1 \cdot v_{dc}, \qquad v_{ph}^- = 0 \cdot v_{dc}
$$

$$
\boxed{\overline{s}^+ = 0, \overline{s}^- = 1} \qquad v_{ph}^+ = 0 \cdot v_{dc}, \qquad v_{ph}^- = 1 \cdot v_{dc}.
$$

For values of $\overline{s}^\pm$ between 0 and 1 these voltages scale linearly. Therefore, the inputs of the controlled voltage sources are a multiplication of the switching signal $\overline{s}^\pm$ with the DC voltage, see Fig. 1b. Similarly the relations for the DC side currents i_{dc}^+ and i_{dc}^- can be derived.

III. HYBRID MODELS OF MULTI-LEVEL TOPOLOGIES FOR SUB-CYCLE AVERAGING

This section shows the hybrid models for different multi-level topologies. A hybrid model of series-connected half-bridges is already introduced in [7] which is similar to the model presented in [10]. Hybrid models for other cascaded converter topologies are simple to derive once the hybrid model of a single cell is known. It requires only the summation of all individual cell voltages on the AC output, while the DC currents are kept the same.

A. Series-connected full-bridges

To derive the hybrid model of series-connected full-bridges in Fig. 2a, a single full-bridge is analyzed first. The final expression for the phase output voltage is given in Eq. 1.

$$
v_{ph,1}^+ = \underbrace{\underbrace{-v_{dc,1}}_{A} \underbrace{+\overline{s}_{1a}^+ \cdot v_{dc,1}}_{B} +\overline{s}_{1b}^- \cdot v_{dc,1}}_{C} \tag{1}
$$

The explanation of this equation is as follows: if all active switches are in blocking state (all gate inputs are zero) and a positive phase current flows, the phase output voltage is $v_{ph,1}^+ = -v_{dc,1}$ (term A in Eq. 1). If only the gate signal $\overline{s}_{1,a}^+$ is 1, the phase output voltage $v_{ph,1}^+$ is zero (term B in Eq. 1). If both gate signals $\overline{s}_{1,a}^+$ and $\overline{s}_{1,b}^-$ are 1 the phase output

voltage is given by $v_{ph,1}^+ = +v_{dc,1}$ (term C in Eq. 1). The total phase output voltage of the cascaded full-bridges is calculated by adding the individual phase voltages. The hybrid model equations for the switching network in Fig. 2a follow the form

$$
v_{ph}^+ = \sum_{\ell=1}^{n} v_{dc,\ell} \cdot (\overline{s}_{\ell a}^+ + \overline{s}_{\ell b}^- - 1) \tag{2}
$$

$$
v_{ph}^- = \sum_{\ell=1}^{n} v_{dc,\ell} \cdot (\overline{s}_{\ell a}^- + \overline{s}_{\ell b}^+ - 1) \tag{3}
$$

$$
i_{dc,\ell} = (\overline{s}_{\ell a}^+ + \overline{s}_{\ell b}^- - 1) \cdot i_{ph}^+ + (\overline{s}_{\ell a}^- + \overline{s}_{\ell b}^+ - 1) \cdot i_{ph}^- \quad, \tag{4}
$$

where $n = 2$ denotes the number of full-bridges and ℓ the index of one individual full-bridge. The complete hybrid model is demonstrated in Fig. 2b. Please note that the underlying structure is the same as that for the series-connected half-bridges in [7] and only the functions in calculating the reference values of the individual voltage and current sources differ.

B. Three-level NPC half-bridge

Hybrid models of cascaded multi-level topologies allow the reduction in the number of independent switches, facilitating a real-time simulation. This is also the case for multi-level

2200

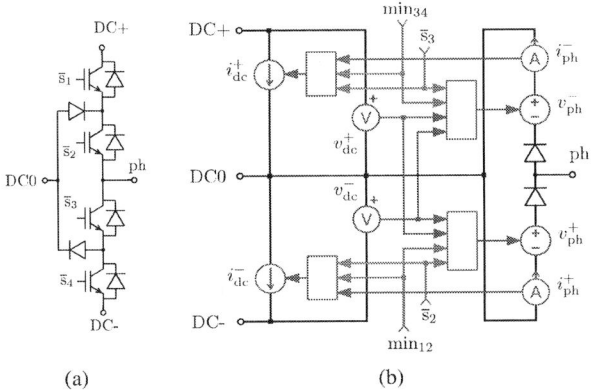

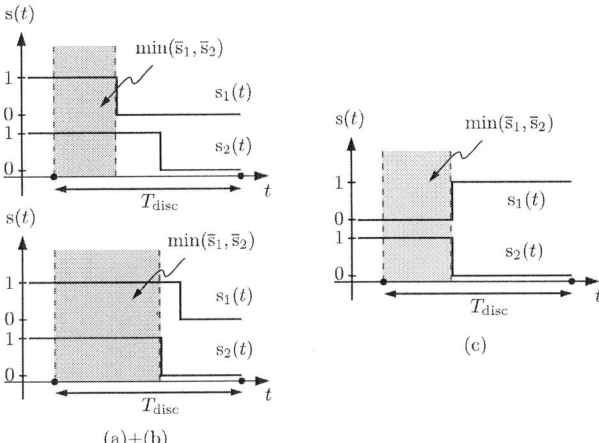

Fig. 3: Switched (a) and hybrid model (b) of a three-level NPC half-bridge

Fig. 4: Modulation of two high-side switches. Figs. (a)+(b) show modulation schemes that the model can handle. Fig. (c) shows a situation where sub-cycle averaging is not able to predict the correct phase output voltage.

topologies without series-stacking such as the three-level neutral point clamped (NPC) half-bridge. The underlying structure of the hybrid model for the NPC half-bridge shown in Fig. 3b is also similar to that of the half-bridge in Fig. 1. However, the neutral point (DC0 terminal) must be added, in turn requiring two voltage sensors and two current sources on the DC side. This allows the capture of the DC capacitor voltages imbalance and, in addition, testing of balancing control algorithms [11]. The functions of the controlled voltage and current sources take the form

$$v_{\mathrm{ph}}^+ = v_{\mathrm{dc}}^+ \cdot \min_{12} - v_{\mathrm{dc}}^- \cdot (1 - \overline{s}_2) \tag{5}$$

$$v_{\mathrm{ph}}^- = v_{\mathrm{dc}}^- \cdot \min_{34} - v_{\mathrm{dc}}^+ \cdot (1 - \overline{s}_3) \tag{6}$$

$$i_{\mathrm{dc}}^+ = i_{\mathrm{ph}}^+ \cdot \min_{12} - i_{\mathrm{ph}}^- \cdot (1 - \overline{s}_3) \tag{7}$$

$$i_{\mathrm{dc}}^- = i_{\mathrm{ph}}^- \cdot \min_{34} - i_{\mathrm{ph}}^+ \cdot (1 - \overline{s}_2) \quad . \tag{8}$$

The term $\min_{jk}$ is defined as

$$\min_{jk} := \min(\overline{s}_j, \overline{s}_k) . \tag{9}$$

As can be seen in these expressions, the positive DC voltage (v_{dc}^+) can only appear at the phase terminal for a positive phase output current (i_{ph}^+) if the two upper switches ($\overline{s}_1, \overline{s}_2$) are conducting. This situation is illustrated in Fig. 4a+b, where the positive DC voltage v_{dc}^+ only appears at the phase output terminal during the interval indicated in grey. Case 4b is not a typical modulation scheme and puts unnecessary voltage stress on switch 2. Nevertheless, the hybrid model predicts the correct output voltage. The limitation of sub-cycle averaging in this regard is illustrated in Fig. 4c. The averaging method TAM can only extract the average value of two input signals ($s_1(t)$ and $s_2(t)$) rather than the phase relation and the number of switching transitions within a discretization time-step T_{disc}. Therefore, the calculated output voltage is incorrect if the high-state periods of the switching signals s_1 and s_2 have a certain shift relative to each other. In case of Fig. 4c and Eq. 5 this means the first term delivers $v_{\mathrm{dc}}^+ \cdot \overline{s}_2$ instead of zero.

C. Five-level ANPC half-bridge

The hybrid model of a five-level ANPC half-bridge is derived analogously to the three-level case in the previous section. One important difference is that the five-level half-bridge has two additional terminals Vc+ and Vc-, compared to only four (DC+, DC-, DC0, ph) of the three-level half-bridge. The hybrid model in Fig. 3b is extended with the current source i_c and volt-meter v_c at the bottom of Fig. 5b. The expressions for the controlled sources are derived analogously to the three-level half-bridge case and take the following form

$$v_{\mathrm{ph}}^+ = v_{\mathrm{dc}}^+ \cdot \min_{15} + v_c \cdot (\overline{s}_6 - \overline{s}_5) - v_{\mathrm{dc}}^- \cdot (1 - \max_{35}) \tag{10}$$

$$v_{\mathrm{ph}}^- = -v_{\mathrm{dc}}^- \cdot \min_{48} + v_c \cdot (\overline{s}_8 - \overline{s}_7) + v_{\mathrm{dc}}^+ \cdot (1 - \max_{28}) \tag{11}$$

$$i_{\mathrm{dc}}^+ = i_{\mathrm{ph}}^+ \cdot \min_{15} - i_{\mathrm{ph}}^- \cdot (1 - \max_{28}) \tag{12}$$

$$i_{\mathrm{dc}}^- = i_{\mathrm{ph}}^- \cdot \min_{48} - i_{\mathrm{ph}}^+ \cdot (1 - \max_{35}) \tag{13}$$

$$i_c = i_{\mathrm{ph}}^+ \cdot (\overline{s}_6 - \overline{s}_5) - i_{\mathrm{ph}}^- \cdot (\overline{s}_8 - \overline{s}_7) . \tag{14}$$

The function $\max_{jk}$ is defined analogously to the previously defined $\min_{jk}$ function in Eq. 9. The additional maximum function $\max_{jk}$ is introduced to cover the following mechanism: if a positive phase output current flows and none of the switches conduct, the phase output voltage is $-v_{\mathrm{dc}}^-$. However, if one of the switches 3 or 5 conducts, the output voltage becomes to be larger or equal to zero. The minimum function (like the first term in Eq. 10) is required because only if both switches $s_1(t)$ and $s_5(t)$ conduct the positive DC voltage v_{dc}^+ can appear at the phase output terminal.

As for three-levels, the short circuit state of the five-level half-bridge can not be analyzed, because only one diode may conduct current at a time.

IV. RESULTS

This section analyzes the efficiency and accuracy of the hybrid model of a five-level ANPC half-bridge in a real-time

The 2018 International Power Electronics Conference

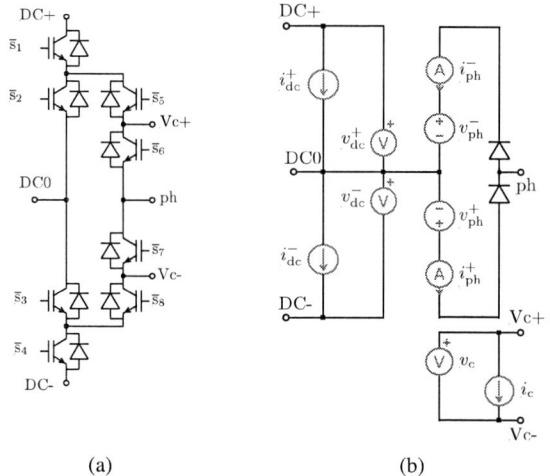

(a)   (b)

Fig. 5: Switched (a) and hybrid model (b) of a five-level NPC half-bridge. For the sake of simplicity, the signal flow in calculating the controlled current and voltage sources are omitted.

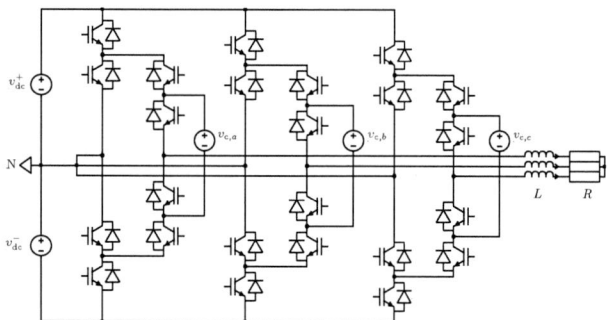

Fig. 6: Test circuit schematic with three five-level half-bridges modelled by either individual switches or hybrid models and symmetric RL-load. The model parameters are: $v_{dc} = 12\,\text{kV}$, $L = 1\,\text{mH}$, $R = 10\,\Omega$ and $f_{sw} = 5\,\text{kHz}$.

simulation. To focus on the performance of the hybrid model, the simplified application example of a three-phase inverter with constant RL-load in Fig. 6 is investigated. All DC capacitors are replaced with perfectly balanced ideal voltage sources (i.e. $v_{dc}^+ = v_{dc}^- = v_{dc}/2$ and $v_{c,a} = v_{c,b} = v_{c,c} = v_{dc}/4$).

The modulation scheme described in [12] is used to drive the individual switches of the ANPC half-bridges together with a sinusoidal modulation index with phase-shift 0°, -120° and 120° for the phases a, b and c. The total number of switches in Fig. 6 is 24, resulting in more than 10^7 possible switching combinations. With the hybrid model given in Fig. 5b, this number can be drastically reduced to only $3^3 = 27$, as each half-bridge contributes only two individual diodes that never conduct at the same time. This reduces the total number of switching combinations by a factor of over $6 \cdot 10^5$.

To verify the fidelity of the proposed approach, a reference

model in Fig. 6 is first established in PLECS using 24 separate IGBTs. These are modelled as ideal switches following the method proposed in [6]. The 24 ideal switches allow all possible switching combinations. A variable step solver ensures that physical states (inductor currents) are accurately integrated and each ideal switching transition is hit exactly. This avoids sampling issues that could possibly occur for fixed time-step solvers in conjunction with modulators. The simulation results for the line-to-neutral phase voltages and phase currents are shown in Fig. 7. For clarity, the simulation result of only one phase (a) is displayed. A symmetric sinusoidal phase current appears at the RL-load, with some ripple due to the switching nature of the converter.

In a next step the five-level half-bridges are replaced by the proposed module (Fig. 5b), then discretized with fixed time-step and generated as C-code for real-time simulation. In the first simulation scheme, the discretization step-size T_{disc} is $4\,\mu\text{s}$, and in the second simulation scheme T_{disc} is chosen to be equal to the switching period ($200\,\mu\text{s}$). The code is later executed on the PLECS RT Box real-time target. The 24 digital gate signals are sampled at the inputs with a resolution of $7.5\,\text{ns}$ and averaged by a high speed FPGA using TAM [8]. The result for the $4\,\mu\text{s}$ discretization in Fig. 7 shows good agreement with the reference solution (with 24 separate IGBTs) in both voltage and current. The resolution of the second simulation scheme is limited to the large discretization step-size and coincides with the classical converter average models that average over one complete switching cycle using state-space or circuit averaging [13], [14]. To further verify the hybrid converter model, the modulation scheme in [12] is adapted, introducing different redundant switching states. In both cases, the simulation results are the same as in Fig. 7.

V. CONCLUSION

The concept of sub-cycle averaging is successfully applied to multi-level topologies such as cascaded full-bridges and three- and five-level (active) NPC converters. Sub-cycle averaging takes advantage of the high resolution of the averaged switching signals. Additionally, hybrid models of multi-level converters allow the reduction of the high number of individual switches to only two diodes and controlled voltage and current sources. This drastically lowers the number of switching combinations and therefore also the size of the generated code and the maximum execution time on a real-time simulation platform [7]. In case of a three-phase five-level ANPC converter the number of switching combinations is reduced by five orders of magnitude. The high fidelity of switched converter models is preserved in their hybrid counterparts, so that dead times, rectifier mode and discontinuous conduction mode of inverter half-bridges are accounted for. The hybrid model of a three-phase five-level ANPC converter shows good agreement in a real-time simulation with a reference solution obtained with a variable time-step solver and purely switched converter model.

2202

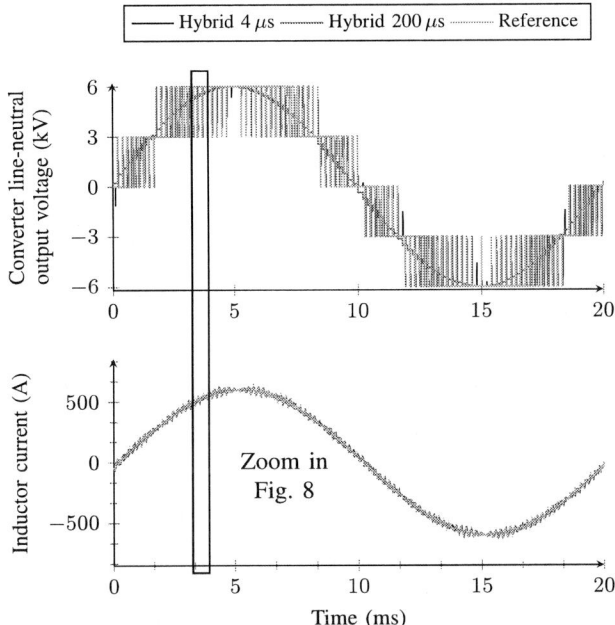

Fig. 7: Simulation results for phase a.

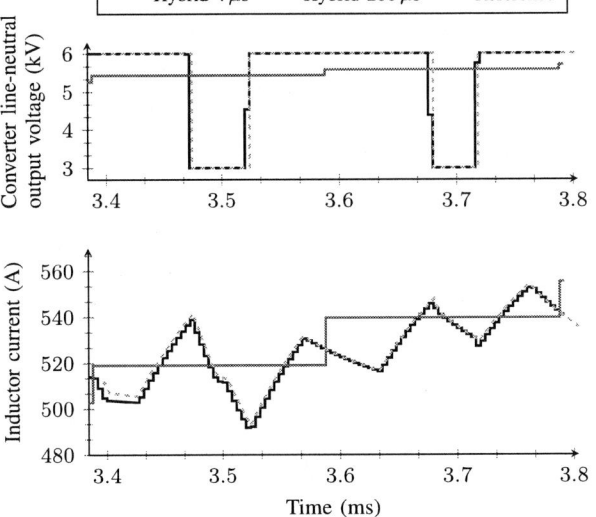

Fig. 8: Close-up of rectangular area in Fig. 7.

REFERENCES

[1] A. Nabae, I. Takahashi, and H. Akagi, "A new neutral-point-clamped pwm inverter," *IEEE Transactions on Industry Applications*, vol. IA-17, no. 5, pp. 518–523, Sept 1981.

[2] S. Allebrod, R. Hamerski, and R. Marquardt, "New transformerless, scalable modular multilevel converters for hvdc-transmission," in *2008 IEEE Power Electronics Specialists Conference*, June 2008, pp. 174–179.

[3] W. Song and A. Q. Huang, "Fault-tolerant design and control strategy for cascaded h-bridge multilevel converter-based statcom," *IEEE Trans-*

actions on Industrial Electronics, vol. 57, no. 8, pp. 2700–2708, Aug 2010.

[4] J. Rodriguez, J.-S. Lai, and F. Z. Peng, "Multilevel inverters: a survey of topologies, controls, and applications," *IEEE Transactions on Industrial Electronics*, vol. 49, no. 4, pp. 724–738, Aug 2002.

[5] S. Debnath, J. Qin, B. Bahrani, M. Saeedifard, and P. Barbosa, "Operation, control, and applications of the modular multilevel converter: A review," *IEEE Transactions on Power Electronics*, vol. 30, no. 1, pp. 37–53, Jan 2015.

[6] J. H. Allmeling and W. P. Hammer, "PLECS – piece-wise linear electrical circuit simulation for simulink," in *Power Electronics and Drive Systems, 1999. PEDS '99. Proceedings of the IEEE 1999 International Conference on*, vol. 1, 1999, pp. 355–360 vol.1.

[7] J. H. Allmeling and N. C. Felderer, "Sub-cycle average models with integrated diodes for real-time simulation of power converters," in *2017 IEEE Southern Power Electronics Conference*, vol. 3, Dec 2017, pp. 705–710.

[8] K. L. Lian and P. W. Lehn, "Real-time simulation of voltage source converters based on time average method," *IEEE Transactions on Power Systems*, vol. 20, no. 1, pp. 110–118, Feb 2005.

[9] R. W. Erickson and D. Maksimovic, *Fundamentals of power electronics*. Springer Science & Business Media, 2007.

[10] N. Ahmed, L. Ängquist, S. Mahmood, A. Antonopoulos, L. Harnefors, S. Norrga, and H. P. Nee, "Efficient modeling of an mmc-based multiterminal dc system employing hybrid hvdc breakers," *IEEE Transactions on Power Delivery*, vol. 30, no. 4, pp. 1792–1801, Aug 2015.

[11] H. du Toit Mouton, "Natural balancing of three-level neutral-point-clamped pwm inverters," *IEEE Transactions on Industrial Electronics*, vol. 49, no. 5, pp. 1017–1025, Oct 2002.

[12] K. Wang, L. Xu, Z. Zheng, and Y. Li, "Capacitor voltage balancing of a five-level anpc converter using phase-shifted pwm," *IEEE Transactions on Power Electronics*, vol. 30, no. 3, pp. 1147–1156, March 2015.

[13] G. W. Wester and R. D. Middlebrook, "Low-frequency characterization of switched dc-dc converters," *IEEE Transactions on Aerospace and Electronic Systems*, vol. AES-9, no. 3, pp. 376–385, May 1973.

[14] R. D. Middlebrook and S. Cuk, "A general unified approach to modelling switching-converter power stages," in *1976 IEEE Power Electronics Specialists Conference*, June 1976, pp. 18–34.

An Enhanced High Frequency Pulsating Voltage Injection Method Based on Immune Algorithm for Sensorless IPMSM Drives

Yanping Zhang[1*], Zhonggang Yin[1], Chao Du[1], Youyun Wang[2] and Xiangdong Sun[1]

1 Department of Elctrical Engineering, Xi'an University of Technology, Xi'an, China
2 Tianshui Electric Drive Research Institute, Tianshui, China
*E-mail: zhangyanpingmao@126.com

Abstract— The fixed bandwidth filter is used in traditional high frequency pulsating voltage injection (HFPVI) based on a sensorless vector control scheme for permanent magnet synchronous motors (PMSM) drives, which restricts the working speed region and dynamic performance of HFPVI. In order to solve this proplem, a novel method of HFPVI by appling the immune algorithm (IA) is proposed in this paper. The IA is applied into HFPVI, the bandwidth of filter can be adjusted on-line in the demodulation process. Thus, the working speed region of HFPVI is enlarged, and the dynamic performance of sensorless PMSM drive is improved by using the IA. The correctness and the effectiveness of the proposed method are verified by the various experimental results.

Keywords— *High frequency pulsating voltage injection, immune algorithm, permanent magnet synchronous motors, sensorless.*

I. INTRODUCTION

The accurate rotor position is required in permanent magnet synchronous motors (PMSM) vector control system [1]. However, the high-resolution sensors lead to an increase in cost and size along with a decrease in reliability [2]. Therefore, various sensorless methods have been investigated, which can be divided into two main categories based on back-EMF [3]-[4] and the machine saliency.

The machine saliency-based sensorless drive approach, known as the high frequency (HF) voltage injection method, it shows satisfactory performance in the low speed region or at zero speed. The HF voltage injection method consists of the HF rotating voltage injection (HFRVI) [5]-[7] and the HF pulsating voltage injection (HFPVI) [8]-[9].

The HFRVI method injects HF voltage into the stationary reference frame. The HFcurrents caused by the injected HF voltages contain rotor position information. To obtain the rotor position from the HF currents by using synchronous reference frame filter (SRFF) and two low pass filters (LFP). However, these filters will deteriorate the dynamic performance and restrict the working speed region of HFRVI. For the HFPVI technique, the HF voltage is injected into estimated the *d*-axis (or *q*-axis). It shows satisfactory estimation performance for interior PMSMs (IPMSM) and surface-mounted PMSM at zero or low speed [10]. However, LPFs are necessary to obtain the position information in the demodulation process of HFPVI. Since the injected signal frequency is fixed, it is theoretically possible to use

a fixed-bandwidth LPF. However, it is not easy to select the optimal bandwidth of the LPF. On the one hand, it is best to have a high bandwidth when the PMSM is running under dynamic operating conditions to improve the filter's convergence speed. On the other hand, a narrow bandwidth will allow the LPF to cut off any unwanted frequency components under steady-state operating conditions to enhance the accuracy of position estimation. This selection is usually a compromise between these two opposite requirements. To solve this issue, a variable band neural network filter is proposed to replace the fixed bandwidth filter in [11]. The experimental results show that the method improves the working speed region of the HFPVI, but the implementation of the algorithm is relatively complex.

Recently, immune algorithm (IA) is widely used to solve complex parameter optimization problems have been reported in the literature [12]–[14]. In [12], identification of the optimal switching laws of inverters by IA was presented. In [13], IA is applied to optimal dead-time elimination control sequences to improve the current waveform quality. In [14], IA is introduced into internal model controller to improve dynamic performance.

In this paper, a novel method of the HFPVI by using the IA (IA-HFPVI) is proposed to improve the performance of the HFPVI sensorless drive. The IA is introduced to HFPVI, the bandwidth of filter can be adjusted on-line in the demodulation process. As a consequence, the working speed region of HFPVI is enlarged, and the dynamic performance of sensorless IPMSM drives is improved by using the IA. A 2.0 kW IPMSM drive platform is used to verify the conclusion.

II. CONVENTIONAL SENSORLESS ALGORITHM USING HFPVI FOR IPMSM

A. HFPVI Method and its Current Response

The stator voltage equation of an IPMSM in the *d-q* axis can be expressed as

$$
\begin{bmatrix} u_d \\ u_q \end{bmatrix} = R_s \begin{bmatrix} i_d \\ i_q \end{bmatrix} + \begin{bmatrix} L_d & 0 \\ 0 & L_q \end{bmatrix} \frac{d}{dt} \begin{bmatrix} i_d \\ i_q \end{bmatrix}
$$
$$
+ \omega_r \begin{bmatrix} 0 & -L_q \\ L_d & 0 \end{bmatrix} \begin{bmatrix} i_d \\ i_q \end{bmatrix} + \begin{bmatrix} 0 \\ \omega_r \psi_f \end{bmatrix} \quad (1)
$$

where u_d, u_q and i_d, i_q are the stator voltages and currents in the d-q axis, L_d and L_q are the d-q axis inductances, R_s is the stator resistance, ω_r is the rotor electrical speed, and Ψ_f is the permanent magnet flux linkage. It is assumed that the frequency of the injected HF is much higher than the motor running frequency, (1) can be simplified as

$$\begin{bmatrix} u_{dh} \\ u_{qh} \end{bmatrix} = \begin{bmatrix} L_d & 0 \\ 0 & L_q \end{bmatrix} \frac{d}{dt} \begin{bmatrix} i_{dh} \\ i_{qh} \end{bmatrix} \qquad (2)$$

where the subscript "h" means the HF component. The transformation matrix $T(\Delta\theta_r)$ from actual d-q axis to estimated d-q axis is

$$T(\Delta\theta_r) = \begin{bmatrix} \cos(\Delta\theta_r) & \sin(\Delta\theta_r) \\ -\sin(\Delta\theta_r) & \cos(\Delta\theta_r) \end{bmatrix} \qquad (3)$$

where $\Delta\theta_r = \theta_r - \theta_e$, θ_r is the actual positon, θ_e is the estimated position.

From (2) and (3), the HF voltage equation in the estimated d-q axis can be obtained as follow

$$\begin{bmatrix} u_{dh}^e \\ u_{qh}^e \end{bmatrix} = T^{-1}(\Delta\theta_r) \begin{bmatrix} L_d & 0 \\ 0 & L_q \end{bmatrix} T(\Delta\theta_r) \frac{d}{dt} \begin{bmatrix} i_{dh}^e \\ i_{qh}^e \end{bmatrix} \qquad (4)$$

where the superscript "e" means the estimated d-q axis.

Injected HF voltage can be expressed as

$$\begin{bmatrix} u_{dh}^e \\ u_{qh}^e \end{bmatrix} = V_c \begin{bmatrix} \cos\omega_h t \\ 0 \end{bmatrix} \qquad (5)$$

where u_{dh}^e and u_{qh}^e are injected d-axis and q-axis HF voltage in the estimated d-q axis, V_c and ω_h are the amplitude and angular speed of the injected voltage, respectively.

According to (4) and (5), the HF current can be obtained as follow

$$\begin{bmatrix} i_{dh}^e \\ i_{qh}^e \end{bmatrix} = \frac{V_c \sin\omega_h t}{\omega_h(L^2 - \Delta L^2)} \begin{bmatrix} L - \Delta L\cos(2\Delta\theta_r) \\ -\Delta L\sin(2\Delta\theta_r) \end{bmatrix} \qquad (6)$$

where $\Delta L = (L_d - L_q)/2$ and $L = (L_d + L_q)/2$.

B. Current Signal Demodulation

From (6), $-i_{qh}^e \sin\omega_h t$ can be expressed as

$$i_{qhs}^e = -i_{qh}^e \sin\omega_h t = \frac{V_h}{2}\left[\frac{\Delta L\sin 2\Delta\theta_r}{\omega_h(L^2 - \Delta L^2)} - \frac{\Delta L\sin 2\Delta\theta_r \cos 2\omega_h t}{\omega_h(L^2 - \Delta L^2)} \right] \qquad (7)$$

As can be seen in (7), The low frequency part of i_{qhs}^e is proportional to the position error $\Delta\theta_r$, it can be obtained by applying an LPF. (8) shows the result after ideal LPF

$$i_{qhl}^e = \mathrm{LPF}(i_{qhs}^e) = \frac{V_h \Delta L}{2\omega_h L_d L_q}\sin(2\Delta\theta_r). \qquad (8)$$

LPF is ususlly designed as a first order LPF, it is described as

$$f(s) = \frac{1}{\tau s + 1} \qquad (9)$$

where τ is the filter time constant. Since the injected signal frequency is fixed, it is theoretically possible to use a fixed-bandwidth LPF. However, it is not easy to select the optimal bandwidth of the LPF. On the one hand, it is best to have a high bandwidth when the PMSM is running under dynamic operating conditions to improve the filter's convergence speed. On the other hand, a narrow bandwidth will allow the LPF to cut off any unwanted frequency components under steady-state operating conditions to enhance the accuracy of position estimation. This selection is usually a compromise between these two opposite requirements. To solve this proplem, the IA is introduced to HFPVI, the bandwidth of filter can be adjusted on-line in the demodulation process.

III. HFPVI BASED ON IA FOR IPMSM

A. Immune System and Algorithm

The IA usually uses the immune system's characteristics of learning and memory to solve the parameter optimal problems. Fig. 1 shows feedback regulation of the immune response system.

In Fig. 1, the mechanism of immune response is: when the antigen invades the human body and is digested by the surrounding cells, the information is transmitted to the Helper T-cells (Th), and the Th stimulate B cells, Killer T-cells (Tk) and Suppressor T-cells (Ts), and B cells to produce antibodies to eliminate the invading antigens. When antigens are more, Th are secreting accelerated, meanwhile, Ts are decrease that will produce more B cells to eliminate the invading antigens. Then antigens are decrease, and Ts are more that will suppress the production of Th, which will reduce the number of B cells. The immune response system tends to balance after some time.

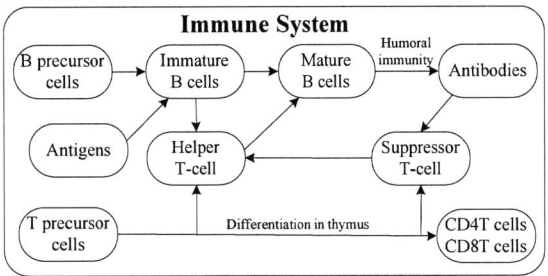

Fig. 1. Feedback regulation of the immune response system.

B. Proposed Solution Algorithm

In this paper, IA is planted in HFPVI to optimize the time constant of LPF. The aim of this paper is to enlarge working speed region of HFPVI, and improve dynamic performance of sensorless drive system. In order to better understand the IA optimization process, the corresponding relationship between the immune system and the IA is shown in Table 1.

The input of the IA is the deviation of the current e_{qh} at time instant k, which can be considered as the antigens

$$e_{qh}(k) = i_{qhl}^e(k) - i_{qhl}^e(k-1). \qquad (10)$$

The process of activating Th cells to eliminate antigens by IA can be expressed as

$$\mathrm{Th}(k) = k_1 e_{qh}(k) \qquad (11)$$

where k_1 is the stimulating factor of the Th. Considering the effect of feedback regulation in Fig.1, it is assumed that the effect of Ts cells on B cells is Ts (k), and Ts (k) can be expressed as

$$Ts(k) = k_2 g(\Delta i(k)) \cdot e_{qh}(k) \qquad (12)$$

where $g(\Delta i(k))$ is a nonlinear function that is associated with antigens density. The B cells activated by Th and Ts can be expressed as

$$B(k) = Th(k) - Ts(k)$$
$$= K[1 - \mu g(\Delta i(k))] \cdot e_{qh}(k) = k_s \cdot e_{qh}(k) \qquad (13)$$

where $K = k_1$, and $\mu = k_2/k_1$ is used to control the stability effect parameter.

TABLE I
ANALOGY BETWEEN IMMUNE SYSTEM AND IMMUNE ALGORITHM

Immune System	Immune Algorithm
Antigen	$i_{qhl}^e(k) - i_{qhl}^e(k-1)$
Antibody	k_s
Recognition of antigen	Identification of the optimization
Lymphocyte differentiation	Maintenance of good solutions
The kth generation cell reproduced by (antigen, antibodies etc)	The kth sampling time in the discrete system
The kth generation antigen density	The error $e_{qh}(k)$ between the given value and the output value at kth sampling time
The kth generation B-cells density	The expected output of the IA

In this paper, $g(\Delta i(k))$ can be defined by

$$g(\Delta i(k)) = 1 - \exp(-\Delta i(k)^2 / b) \qquad (14)$$

where b is an adjustable parameter. Here, Fig. 2 indicates the function $g(\Delta i(k))$ curves when b varied.

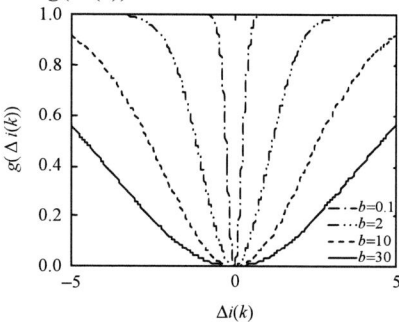

Fig. 2. $g(\Delta i(k))$ function curves when parameter b varied.

C. Design of the Proposed Demodulation Process

The LPF is used in the HF current demodulation process in (8). However, it is not easy to select the optimal bandwidth of the LPF. The choice is normally a tradeoff in transient and steady state conditions. In order to enlarge working speed region of the HFPVI and improve dynamic performance of sensorless drive system, IA is introduced into the HFPVI.

Here, a LPF can be described as

$$f(s) = \frac{1}{(\tau s + 1)^n} \qquad (15)$$

where τ is the time constant of the LPF. The LPF is designed as a first order LPF. Therefore, The IA LPF (IALPF) can be expressed as

$$f(s) = \frac{1}{k_s \tau s + 1} \qquad (16)$$

where k_s is an adjustment factor obtained by IA.

Therefore, (8) can be expressed as

$$i_{qhl}^e = IALPF(i_{qhs}^e) \approx \frac{V_h \Delta L}{2\omega_h L_d L_q} \sin 2\Delta \theta_r . \qquad (17)$$

To obtain the rotor position from (17), a Luenberger rotor position observer is designed. The mechanical equation of IPMSM can be expressed as

$$T_e = \frac{J}{n_p} \frac{d\omega_r}{dt} + T_L + \frac{R_\Omega \omega_r}{n_p} \qquad (18)$$

where T_e is the electromagnetic torque, J is the moment of inertia, n_p is the pole pair number, R_Ω is the viscous coefficient, T_L is the load torque. When ω_r is used as estimated speed ω_e, (18) can be rewritten as follow

$$\frac{d\omega_e}{dt} = \frac{n_p}{J} T_e - \frac{n_p}{J} T_L - \frac{R_\Omega}{J} \omega_e . \qquad (19)$$

As can be seen in (19), when $T_L = T_e$, the estimated speed can track the actual speed. Therefore, the accuracy of the eatimated speed is directly related to $-(n_p/J)T_L$, and the error compensator instead of $-(n_p/J)T_L$ can be expressed as

$$-\frac{n_p}{J}\hat{T}_L = (K_p' + \frac{K_i'}{s} + K_d' s)(\theta_r - \theta_e) . \qquad (20)$$

(20) can be rewritten as follow

$$-\hat{T}_L = (K_p + \frac{K_i}{s} + K_d s)(\theta_r - \theta_e) \qquad (21)$$

where $K_p = K_p' J / n_p$, $K_i = K_i' J / n_p$ and $K_d = K_d' J / n_p$.

Through the above analysis, if the damping torque is ignored, the observer and signal demodulation block diagram is shown in figure 3.

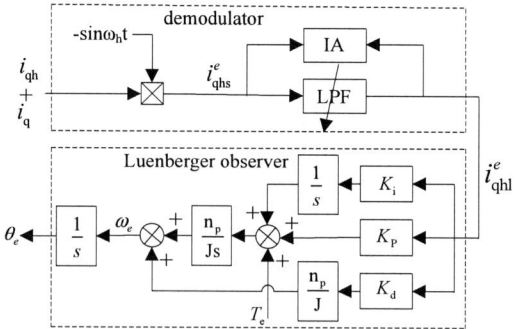

Fig. 3. Current demodulator and Luenberger observer.

D. IA-HFPVI Sensorless Control System Based Vector Control

The speed estimation method based on IA-HFPVI is formed by HFPVI and IA. Moreover, in order to enlarge the working speed region of HFPVI, and improve dynamic performance of sensorless drive system, the filter time constant can be adjusted on-line in the demodulation process by using IA. Sensorless drive scheme block for IPMSMs based on the IA-HFPVI is shown in Fig. 4.

The HF voltage is injected at the estimated d-axis.

2206

Fundamental currents and HF currents are included in the stator currents. The fundamental currents are extracted by the LPF as the feedback of the current regulator. The HF currents are extracted by the demodulator. i_{qh1}^e is used as the input of Luenberger observer, the rotor position and speed can be obtained.

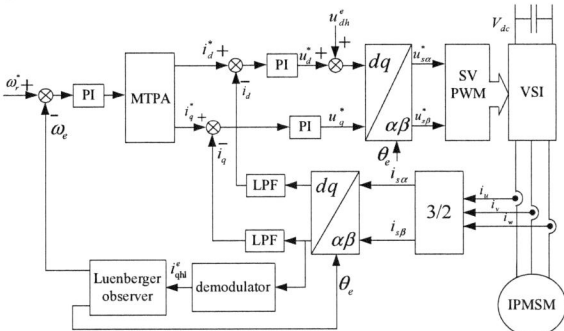

Fig. 4. Sensorless drive scheme block for IPMSMs based on the IA-HFPVI.

IV. EXPERIMENTAL RESULTS

In order to verify the effectiveness of IA-HFPVI, experiments were carried out on a 2.0 kW IPMSM vector control platform. The parameters of the motor used in the experiment are shown in Table II.

TABLE II
IPMSM PARAMETERS

Parameter	Value	Unit
Power	2.0	kW
Voltage rating	380	V
Current rating	5.5	A
Frequency rating	50	Hz
Rated speed	1500	r/min
Pole pairs	2	-
Stator Resistance	2.73	Ω
d-axis Inductance	12.5	mH
d-axis Inductance	31.18	mH
Rated Torque	12.73	N·m
Moment Inertia	0.011	kg·m²

When IPMSM is set at different speed, the experimental result based on IA-HFPVI is shown in Fig.5. In Fig.5, the motor runs at 6 stages, including 300, 600, 750, 450, 150 and 15 r/min, respectively, which contains the medium and low speed range, it shows that IA-HFPVI has good tracking performance at medium and low speed.

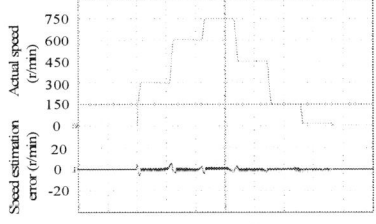

Fig. 5. Speed tracking performance based on at medium and low speed IA-HFPVI.

Fig. 6 shows the position response and the speed response based on IA-HFPVI when the set speed changes

from +150 r/min to -150 r/min. The speed and position waveforms are shown in Fig. 6(a) and in Fig. 6(b), respectively. The estimated speed error and estimated position error are within ±12 r/min and within ±0.13 rad, respectively. The vector control drive based on IA-HFPVI can operate both in motoring and regenerating mode.

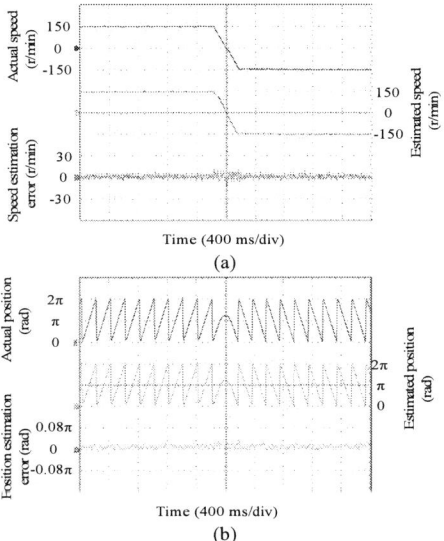

Fig. 6. Estimated speed and position based on IA-HFPVI at switching point of reversal. (a) Speed waveforms. (b) Position waveforms.

Fig. 5 and Fig. 6 prove the correctness of the sensorless method based on IA-HFPVI.

Fig. 7, 8 and 9 show the waveforms of estimated rotor position based on HFPVI and IA-HFPVI in steady state with no load. Fig. 7 shows estimated position waveforms at 15 r/min comparison with HFPVI and IA-HFPVI. The position estimation errors are within ±0.09 rad and ±0.07 rad based on HFPVI and IA-HFPVI, respectively. In Fig. 8, the estimated position waveforms at 300 r/min comparison with HFPVI and IA-HFPVI are shown. The position estimation errors are within ±0.15 rad and ±0.1 rad based on HFPVI and IA-HFPVI, respectively. Fig. 9 shows estimated position waveforms at 600 r/min comparison with HFPVI and IA-HFPVI. The position estimation errors are within ±0.22 rad and ±0.13 rad based on HFPVI and IA-HFPVI, respectively. It can be seen from Fig. 7, 8 and 9 that position estimation errors increase with increased of the running speed based on HFPVI and IA-HFPVI, but position estimation error based on IA-HFPVI the smaller than HFPVI. Therefore, the working speed region of IA-HFPVI is enlarged under ensuring the accuracy of estimated position.

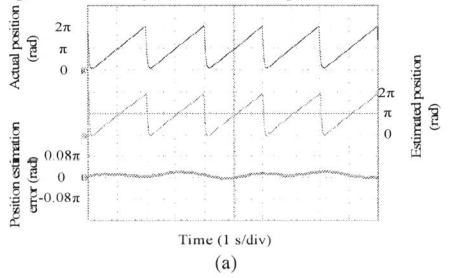

(a)

The 2018 International Power Electronics Conference

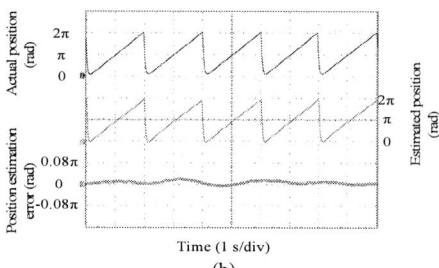

(b)

Fig. 7. Position estimation waveforms at 15 r/min. (a) HFPVI. (b) IA-HFPVI.

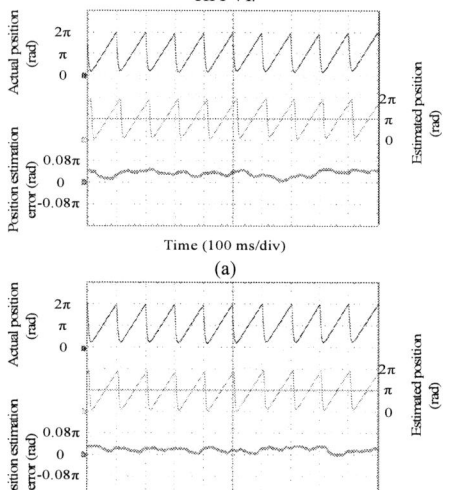

(a)

(b)

Fig. 8. Position estimation waveforms at 300 r/min. (a) HFPVI. (b) IA-HFPVI.

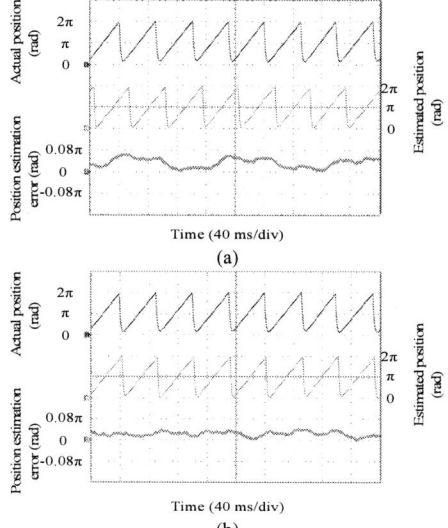

(a)

(b)

Fig. 9. Position estimation waveforms at 600 r/min. (a) HFPVI. (b) IA-HFPVI.

Fig. 10 and 11 show the fast acceleration experimental waveforms from 0 r/min to 300 r/min based on HFPVI and IA-HFPVI, respectively. Fig. 10(a) and Fig. 11(a) show speed waveforms based on HFPVI and IA-HFPVI, respectively. Fig. 10(b) and Fig. 11(b) show position waveforms based on HFPVI and IA-HFPVI, respectively. In the whole speed acceleration, the experimental results

show that IA-HFPVI estimated speed errors and position errors are significantly reduced compared with HFPVI in the entire acceleration process.

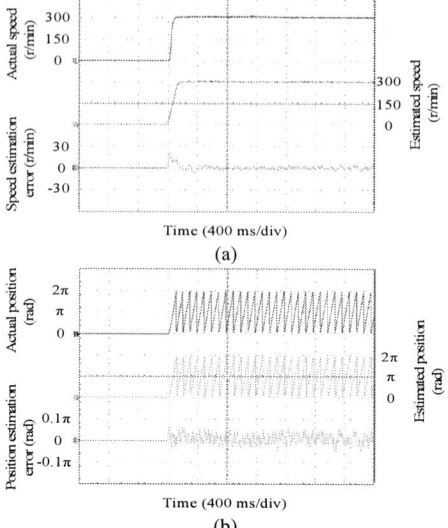

(a)

(b)

Fig. 10. The fast acceleration experimental waveform from 0 r/min to 300 r/min based on HFPVI. (a) Speed waveforms. (b) Position waveforms.

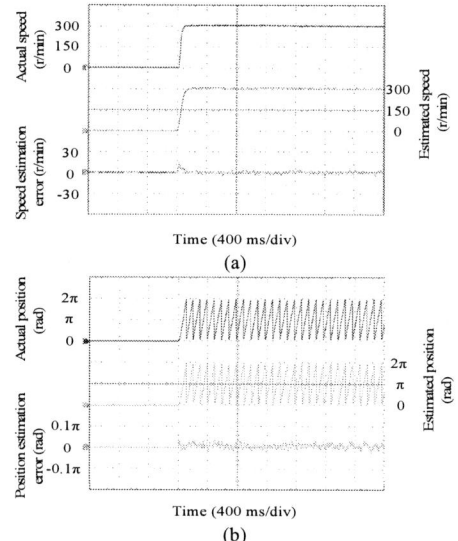

(a)

(b)

Fig. 11. The fast acceleration experimental waveform from 0 r/min to 300 r/min based on IA-HFPVI. (a) Speed waveforms. (b) Position waveforms.

Fig.12 shows sensorless operation result with step rated load disturbance at 150 r/min. Fig. 12(a) and Fig. 12(b) show the experimental results of HFPVI and IA-HFPVI, respectively. It can be seen that the maximum speed estimation error reduces to 13 r/min from 32 r/min during with step rated load disturbance.

2208

The 2018 International Power Electronics Conference

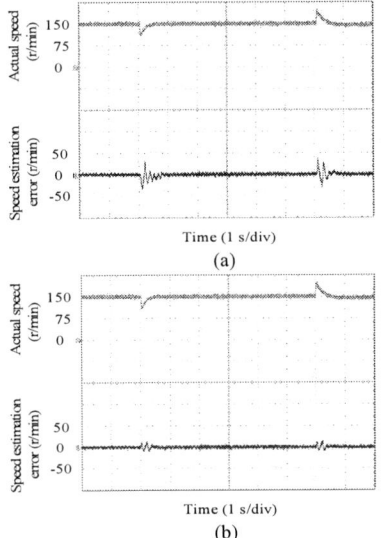

Time (1 s/div)

(a)

Time (1 s/div)

(b)

Fig. 12. Sensorless operation result with step rated load disturbance at 150 r/min. (a) HFPVI. (b) IA-HFPVI.

V. CONCLUSION

This paper has proposed a sensorless method for IPMSM based on IA-HFPVI. The IA is introduced to HFPVI, the bandwidth of filter can be adjusted on-line in the demodulation process. The correction and the effectiveness of IA-HFPVI have been verified at a 2.0 kW IPMSM vector control platform. As a consequence, the working speed region of HFPVI is enlarged, and the dynamic performance of sensorless IPMSM drives is enhanced by using the IA.

REFERENCES

[1] G. L. Wang, L. Yang, and G. Q. Zhang, "Comparative investigation of pseudorandom high-frequency signal injection schemes for sensorless IPMSM drives," *IEEE Trans. on Power Electronics*, vol. 32, no. 3, pp. 2123-3132, 2017.

[2] D. L. Liang, J. Li, and R. H. Qu, "Sensorless control of permanent magnet synchronous machine based on second-order sliding-mode observer with online resistance estimation," *IEEE Trans. on Industrial Applications*, vol. 53, no. 4, pp. 3672-3682, 2017.

[3] Y. B. Zbede, S. M. Gadoue, and D. J. Atkinson, "Model predictive MRAS estimator for sensorless induction motor drives," *IEEE Trans. on Industrial Electronics*, vol. 33, no. 6, pp. 3511-3520, 2016.

[4] Z. G. Yin, G. Y. Li, C. Du, J. Liu, and X. D. Sun, "An adaptive speed estimation method based on strong tracking extended Kalman filter with least-square for induction motors," *IEEE Trans. on Power Electronics*, vol. 17, no. 1, pp. 149-160, 2017.

[5] X. Luo, Q. P. Tang, and A. W. Shen, "PMSM sensorless control by injecting HF pulsating carrier signal into estimated fixed-frequency rotating reference frame," *IEEE Trans. on Industrial Electronics*, vol. 63, no. 4, pp. 2294-2303, 2016.

[6] S. Kim, J. H. Im, and E. Y. Song, "A new rotor position estimation method of IPMSM using all-pass filter on high-frequency rotating voltage signal injection," *IEEE Trans. on Industrial Electronics*, vol. 63, no. 10, pp. 6499-6509, 2016.

[7] S. Medjmadj, D. Diallo, and M. Mostefai, "PMSM drive position estimation: contribution to the high-frequency injection voltage selection issue," *IEEE Trans. on Energy Conversion*, vol. 30, no. 1, pp. 349-358, 2015.

[8] Q. Q. Tang, A. W. Shen, and X. Luo, "PMSM sensorless control by injecting HF pulsating carrier signal into ABC frame," *IEEE Trans. on Power Electronics*, vol. 32, no. 5, pp. 3767-3776, 2017.

[9] G. L. Wang, R. F. Yang, and D. G. Xu, "DSP-based control of sensorless IPMSM drives for wide-speed-range operation," *IEEE Trans. on Industrial Electronics*, vol. 60, no. 2, pp. 720-727, 2013.

[10] S. C. Yang and R. D. Lorenz, "Comparison of resistance-based and inductance-based self-sensing controls for surface permanent-magnet machines using high-frequency signal injective," *IEEE Trans. on Industry Applicitions*, vol. 48, no. 5, pp. 977-986, 2012.

[11] A. Accetta, M. Cirrincione, and M. Pucci, "Sensorless control of PMSM fractional horsepower drives by signal injection and neural adaptive-band filtering," *IEEE Trans. on Industrial Electronics*, vol. 59, no. 3, pp. 1355-1366, 2012.

[12] J. X. Yuan, J. B. Pan, and W. L. Fei, "An immune-algorithm-based space-vector PWM control strategy in a three-phase inverter," *IEEE Trans. on Industrial Electronics*, vol. 60, no. 5, pp. 2084-2093, 2013.

[13] J. X. Yuan, Z. Zhao, and B. C. Chen, "An immune-algorithm-based dead-time elimination PWM control strategy in a single-phase inverter," *IEEE Trans. on Power Electronics*, vol. 30, no. 7, pp. 3964-3875, 2015.

[14] Q. Zhu, Z. G. Yin, and Y. Q. Zhang, "Research on two-degree-of-freedom internal model control strategy for induction motor based on immune algorithm," *IEEE Trans. on Industrial Electronics*, vol. 63, no. 3, pp. 1981-1992, 2016.

Position Estimation Accuracy Improvement for Magnetic Saliency Based Sensorless Control Including Cross-Coupling Factor

Keita Shimamoto[1*] and Shinya Morimoto[2]
1 Tsukuba Research Laboratory, YASKAWA Electric Corporation, Ibaraki, Japan
2 Corporate Research & Development Center, YASKAWA Electric Corporation, Fukuoka, Japan
*E-mail: Keita.Shimamoto@yaskawa.co.jp

Abstract— This paper describes improvement method of position estimation accuracy for signal-injection based sensorless control. Conventional signal-injection based sensorless position estimation methods are based on voltage equation without incorporating mutual inductances between d-axis and q-axis. Therefore as a result, position estimation error is caused by the neglected mutual inductances. In addition, the mutual inductances depend on current, and position. It is difficult to know the actual values of mutual inductances. Therefore, in this paper, improvement method of position estimation accuracy without knowing the cross coupling factor is proposed. The experimental results show the validity of the proposed method.

Keywords— *Position Estimation; Estimation Accuracy; High Frecuency Signal Injection; Cross Coupling Factor*

I. INTRODUCTION

Interior Permanent Magnet Synchronous Machines (IPMSM) have become popular in industrial applications. In order to utilize Field Oriented Control (FOC) for the IPMSM drive, the accurate rotor position information is essential. Thus, position sensors such as optical encoders or resolvers are usually employed. Removing position sensors have some advantages in sensor cable saving, higher torque density, and higher robustness to machine vibration. Therefore, position estimation method without position sensors has attracted wide attention [1].

Various position estimation methods have been developed. Suitable methods for machine startup and low-speed operation are based on the magnetic saliency. The rotor-position is estimated from the current variation caused by the applied voltage. The typical methods are indirect flux detection by on-line reactance measurement method, high frequency voltage signal injection methods, and modified Pulse Width Modulation (PWM) methods [2]-[4]. As the algorithms are different, the frequency and waveform of applied voltage are different, respectively. In order to improve the sensorless control performance, position estimation algorithms have following two problems to be solved.
1: A time delay of estimation
2: Estimation errors

One of the solutions to the first problem is that the higher frequency voltage signal is used since estimation period can be shorter. In high frequency injection methods, the frequency of injected voltage signal needs to be close to the carrier frequency [5]. Modified PWM methods use Space Vector PWM (SVPWM). The vector pattern is modified to amplify the current variation in a carrier period [6]. As a result, estimation period of both high frequency injection methods and modified PWM methods can be close to the carrier period. One of the solutions to the second problem is accounting for cross-coupling factor of IPMSM. Generally, the estimation algorithms are based on voltage equation without incorporating mutual inductances in dq-axis. According to conventional method, position estimation error is caused by neglecting the cross-coupling factor in dq-axis. The cross-coupling factor depends on machine design, load current, and position. Therefore, it is difficult to know the cross-coupling factor from mutual inductances in dq-axis. One of the methods to determine the factor is finite-element analysis. Estimation algorithms accounting for cross-coupling factor determined by finite-element analysis show improvement in the position estimation accuracy [7][8]. However, this method cannot be applied to general purpose inverters which are not given the specified motor whose cross-coupling factor is determined by the analysis. In addition, data collection before the running of equipment is not practical in many industrial applications. Therefore, position estimation method accounting for cross-coupling factor without analytical values or data collection.

In this paper, a position estimation accuracy improvement method is proposed. In this method, mutual inductances are used in the voltage equation to accounting for cross-coupling factor. In the proposed algorithm, the cross-coupling factor from mutual inductances is not required to know. They are canceled out by mathematical algorithm. In order to compare the estimation accuracy, the proposed method is applied to a modified PWM method described in [6]. The validity of the proposed method is shown by experimental results.

The 2018 International Power Electronics Conference

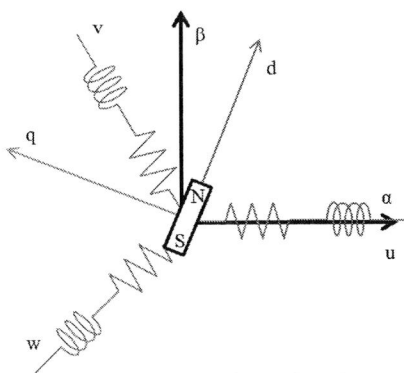

Fig. 1 Coordinate of IPMSM.

TABLE I
PARAMETERS IN THIS PAPER

V	Voltage	α,β	Fixed System Coordinates
i	Current	$d.q$	Rotated Coordinates
R	Resistance	x	Mutual Inductance Component
L,l	Inductance	est	Estimation Component
θ	Position	prop	Proposed Method
ω	Angular Velocity	conv	Conventional Method
t	Time	sin	Sine wave Component
a	Sensitivity parameter	cos	Cosine wave Component
ψ	Flux Linkage	res	Response
ref	Reference	V_a,V_b	Efficient Voltage Vectors
r	Rotation	A,B,C	Arbitrary coefficient

II. PROPOSED METHOD

In this section, the estimation algorithm accounting for dq-axis mutual inductances is shown. The Fig. 1 shows the coordinates of IPMSM. The α-β frame is fixed to the stator winding. The α-axis coincides with the u-phase direction. The d-q frame expresses the rotating reference frame and the d-axis coincides with the N-pole. Parameters in this paper are shown in Table I. The general voltage equation of IPMSM is expressed as equation (1)

$$\begin{bmatrix} V_\alpha \\ V_\beta \end{bmatrix} = R_a \begin{bmatrix} i_\alpha \\ i_\beta \end{bmatrix} + \begin{bmatrix} L+l\cos 2\theta & l\sin 2\theta \\ l\sin 2\theta & L-l\cos 2\theta \end{bmatrix} \frac{d}{dt}\begin{bmatrix} i_\alpha \\ i_\beta \end{bmatrix}$$
$$+\psi\begin{bmatrix} -\omega\sin\theta \\ \omega\cos\theta \end{bmatrix} \qquad (1)$$
$$L = \frac{L_d+L_q}{2}, \quad l = \frac{L_d-L_q}{2}$$

On the other hand, the voltage equation of IPMSM accounting for dq-axis mutual inductances is expressed as equation (2).

$$\begin{bmatrix} V_\alpha \\ V_\beta \end{bmatrix} = R\begin{bmatrix} i_\alpha \\ i_\beta \end{bmatrix} + \begin{bmatrix} L(1,1) & L(1,2) \\ L(2,1) & L(2,2) \end{bmatrix}\frac{d}{dt}\begin{bmatrix} i_\alpha \\ i_\beta \end{bmatrix} + \psi\begin{bmatrix} -\omega\sin\theta \\ \omega\cos\theta \end{bmatrix}$$
$$L(1,1) = L + l\cos 2\theta - L_x \sin 2\theta$$
$$L(2,1) = l\sin 2\theta + l_x + L_x \cos 2\theta \qquad (2)$$
$$L(1,2) = l\sin 2\theta - l_x + L_x \cos 2\theta$$
$$L(2,2) = L - l\cos 2\theta - L_x \sin 2\theta$$
$$L = \frac{L_d+L_q}{2}, \quad l = \frac{L_d-L_q}{2},$$
$$L_x = \frac{L_{dq}+L_{qd}}{2}, \quad l_x = \frac{L_{qd}-L_{dq}}{2}$$

The conventional estimation algorithm is shown in equation (3).

$$2\theta_{conv}^{est} = \tan^{-1}\left(\frac{f_{\sin}\left(\frac{di_\alpha}{dt},\frac{di_\beta}{dt}\right)}{f_{\cos}\left(\frac{di_\alpha}{dt},\frac{di_\beta}{dt}\right)}\right) = \tan^{-1}\left(\frac{\sin 2\theta}{\cos 2\theta}\right) \qquad (3)$$

The numerator and denominator of the equation (3) show the sine and cosine current variation that utilizes the magnetic saliency in stationary coordinates. For example, $f_{\sin}$ and $f_{\cos}$ can be written as equation (4) and (5).

$$f_{\sin}\left(\frac{di_\alpha}{dt},\frac{di_\beta}{dt}\right) = A\sin 2\theta \qquad (4)$$

$$f_{\cos}\left(\frac{di_\alpha}{dt},\frac{di_\beta}{dt}\right) = A\cos 2\theta \qquad (5)$$

The coefficient A is related to the injected voltage and the inductances. If dq mutual inductances are not neglected, the sine and cosine functions can be represented as equation (6) and equation (7).

$$f_{\sin}\left(\frac{di_\alpha}{dt},\frac{di_\beta}{dt}\right) \propto \left(l\sin 2\theta + L_x \cos 2\theta\right) \qquad (6)$$

$$f_{\cos}\left(\frac{di_\alpha}{dt},\frac{di_\beta}{dt}\right) \propto \left(l\cos 2\theta - L_x \sin 2\theta\right) \qquad (7)$$

Multiplying equation (6) and equation (7) by sin2θ and cos2θ and following set of equations are obtained.

$$f_{\sin}\left(\frac{di_\alpha}{dt},\frac{di_\beta}{dt}\right)\sin 2\theta = l\sin^2 2\theta + L_x \sin 2\theta \cos 2\theta \qquad (8)$$

$$f_{\sin}\left(\frac{di_\alpha}{dt},\frac{di_\beta}{dt}\right)\cos 2\theta = l\sin 2\theta \cos 2\theta + L_x \cos^2 2\theta \qquad (9)$$

$$f_{\cos}\left(\frac{di_\alpha}{dt},\frac{di_\beta}{dt}\right)\sin 2\theta = l\sin 2\theta \cos 2\theta - L_x \sin^2 2\theta \qquad (10)$$

$$f_{\cos}\left(\frac{di_\alpha}{dt},\frac{di_\beta}{dt}\right)\cos 2\theta = l\cos^2 2\theta - L_x \sin 2\theta \cos 2\theta \qquad (11)$$

From equation (8) to (11), equation (12) can be derived.

$$\frac{f_{\sin}\left(\frac{di_\alpha}{dt},\frac{di_\beta}{dt}\right)\cos 2\theta - f_{\cos}\left(\frac{di_\alpha}{dt},\frac{di_\beta}{dt}\right)\sin 2\theta}{f_{\sin}\left(\frac{di_\alpha}{dt},\frac{di_\beta}{dt}\right)\sin 2\theta + f_{\cos}\left(\frac{di_\alpha}{dt},\frac{di_\beta}{dt}\right)\cos 2\theta} = \frac{L_x}{l} \qquad (12)$$

Furthermore, equation (12) can be rewritten as equation (13).

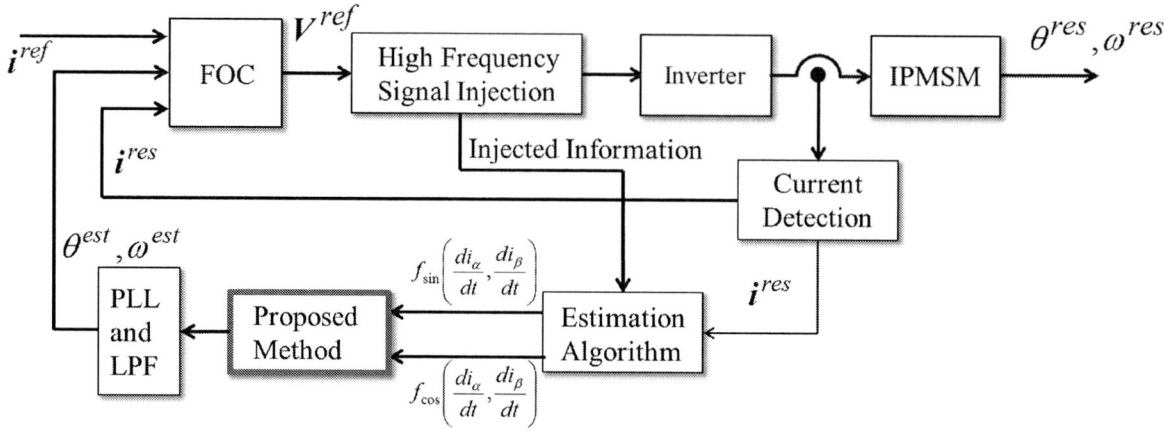

Fig.2. Block Diagram of Control System with Proposed Method.

$$\begin{bmatrix} f_{\sin}\left(\dfrac{di_\alpha}{dt},\dfrac{di_\beta}{dt}\right) \\ f_{\cos}\left(\dfrac{di_\alpha}{dt},\dfrac{di_\beta}{dt}\right) \end{bmatrix} = l\begin{bmatrix} 1 & \dfrac{L_x}{l} \\ -\dfrac{L_x}{l} & 1 \end{bmatrix}\begin{bmatrix} \sin 2\theta \\ \cos 2\theta \end{bmatrix} \quad (13)$$

From equation (13), equation (14) is obtained.

$$\begin{bmatrix} \sin 2\theta \\ \cos 2\theta \end{bmatrix} = \dfrac{1}{l\left(1+\left(\dfrac{L_x}{l}\right)^2\right)}\begin{bmatrix} 1 & -\dfrac{L_x}{l} \\ \dfrac{L_x}{l} & 1 \end{bmatrix}\begin{bmatrix} f_{\sin}\left(\dfrac{di_\alpha}{dt},\dfrac{di_\beta}{dt}\right) \\ f_{\cos}\left(\dfrac{di_\alpha}{dt},\dfrac{di_\beta}{dt}\right) \end{bmatrix} \quad (14)$$

Therefore, estimated position from proposed method is expressed as equation (15).Since L_x/l is obtained by equation (12), position can be estimated.

$$2\theta_{prop}^{est} = \tan^{-1}\left(\dfrac{\sin 2\theta}{\cos 2\theta}\right)$$

$$= \tan^{-1}\left(\dfrac{f_{\sin}\left(\dfrac{di_\alpha}{dt},\dfrac{di_\beta}{dt}\right) - \dfrac{L_x}{l}f_{\cos}\left(\dfrac{di_\alpha}{dt},\dfrac{di_\beta}{dt}\right)}{\dfrac{L_x}{l}f_{\sin}\left(\dfrac{di_\alpha}{dt},\dfrac{di_\beta}{dt}\right) + f_{\cos}\left(\dfrac{di_\alpha}{dt},\dfrac{di_\beta}{dt}\right)}\right) \quad (15)$$

III. PROPOSED METHOD IN DISCRETE TIME

This section gives the proposed method in discrete time [k]. In actual operation (8) to (11) cannot be obtained, since actual position is unknown. Therefore, last estimated position is utilized for the multiplication. It is assumed that the difference between the actual position and the last estimated position is negligibly small, since the sampling time is very small. The assumption is expressed by equation (16).

$$\theta[k] \approx \theta^{est}[k-1] \quad (16)$$

Therefore, equations from (8) to (11) can be rewritten as equations from (17) to (20).

$$f_{\sin}[k]\sin 2\theta_{prop}^{est}[k-1]$$
$$\approx l\sin^2 2\theta[k] + L_x\sin 2\theta[k]\cos 2\theta[k] \quad (17)$$

$$f_{\sin}[k]\cos 2\theta_{prop}^{est}[k-1]$$
$$\approx l\sin 2\theta[k]\cos 2\theta[k] + L_x\cos^2 2\theta[k] \quad (18)$$

$$f_{\cos}[k]\sin 2\theta_{prop}^{est}[k-1]$$
$$\approx l\sin 2\theta[k]\cos 2\theta[k] - L_x\sin^2 2\theta[k] \quad (19)$$

$$f_{\cos}[k]\cos 2\theta_{prop}^{est}[k-1]$$
$$\approx l\cos^2 2\theta[k] - L_x\sin 2\theta[k]\cos 2\theta[k] \quad (20)$$

As a result, L_x/l and estimated position can be obtained from the same algorithm expressed by equations from (8) to (11). Moreover, as the estimated position is calculated from current information, the estimation accuracy depends on the accuracy of the current information. That is, the estimated position becomes noisy in the case of much current information noise. In that case, the assumption might have bad effect on the position estimation. Therefore, the sensitivity parameter "a" is defined. By using the sensitivity parameter, estimated position can be rewritten as equation (21).

$$2\theta_{prop}^{est}[k]$$
$$= \tan^{-1}\left(\dfrac{f_{\sin}[k] - \dfrac{aL_x}{l}f_{\cos}[k]}{\dfrac{aL_x}{l}f_{\sin}[k] + f_{\cos}[k]}\right) \quad (21)$$

Block diagram of sensorless control system with proposed method is shown in Fig.2. As Fig.2 shows that no additional information is required for the proposed method. $f_{\sin}$ and $f_{\cos}$ are output of conventional position estimation methods. Therefore, the proposed method is easily attached to the conventional position estimation methods based on voltage equation expressed by equation (2).

IV. Applied To A Modified PWM Method

This section shows an estimation algorithm of characteristic PWM method with accounting for dq-axis inductance as an example. The applied method is described in [6]. This method uses modified SVPWM to amplify the current variation. High frequency component of equation (2) is expressed in equation (22) by subscript 'h', with the determinant B of the inductance matrix.

$$\frac{d}{dt}\begin{bmatrix} i_{\alpha h} \\ i_{\beta h} \end{bmatrix} = \frac{1}{|B|}\begin{bmatrix} L(2,2) & -L(1,2) \\ -L(2,1) & L(1,1) \end{bmatrix}\begin{bmatrix} v_{\alpha h} \\ v_{\beta h} \end{bmatrix} \qquad (22)$$

Fig. 3 shows the conceptual diagram of SVPWM. V_a and V_b are vector notations of efficient voltage vectors. V_a represents V_1, V_2, and V_4, whereas V_b represents V_3, V_5, and V_6. Equation (22) with V_a is expressed as (23), and (22) with V_b is expressed as (24).

$$\frac{d}{dt}\begin{bmatrix} i_{\alpha hVa} \\ i_{\beta kVa} \end{bmatrix} = \frac{1}{|B|}\begin{bmatrix} L(2,2) & -L(1,2) \\ -L(2,1) & L(1,1) \end{bmatrix}\begin{bmatrix} v_{\alpha hVa} \\ v_{\beta hVa} \end{bmatrix} \qquad (23)$$

$$\frac{d}{dt}\begin{bmatrix} i_{\alpha hVb} \\ i_{\beta kVb} \end{bmatrix} = \frac{1}{|B|}\begin{bmatrix} L(2,2) & -L(1,2) \\ -L(2,1) & L(1,1) \end{bmatrix}\begin{bmatrix} v_{\alpha hVb} \\ v_{\beta hVb} \end{bmatrix} \qquad (24)$$

In order to correspond to the direction of the direction of output voltage vector with α-axis, rotational coordinate transformation is used. The transformation is rotation by the angle of the output voltage vector as shown in Fig.4. As a result, the voltage vector component of β-axis direction is zero. Therefore, equations (25) and (26) can be obtained.

$$\frac{d}{dt}\begin{bmatrix} i_{\alpha hrVa} \\ i_{\beta hrVa} \end{bmatrix} = \frac{v_{\alpha hrVa}}{|B|}\begin{bmatrix} L_{rVa}(1,1) \\ L_{rVa}(2,1) \end{bmatrix}$$

$$L_{rVa}(1,1) = L - l\cos(2\theta - \theta_{Va}) + L_x\sin(2\theta - \theta_{Va}) \qquad (25)$$

$$L_{rVa}(2,1) = -l\sin(2\theta - \theta_{Va}) - l_x - L_x\cos(2\theta - \theta_{Va})$$

$$\frac{d}{dt}\begin{bmatrix} i_{\alpha hrVb} \\ i_{\beta hrVb} \end{bmatrix} = \frac{v_{\alpha hrVb}}{|B|}\begin{bmatrix} L_{rVb}(1,1) \\ L_{rVb}(2,1) \end{bmatrix}$$

$$L_{rVb}(1,1) = L - l\cos(2\theta - \theta_{Vb}) + L_x\sin(2\theta - \theta_{Vb}) \qquad (26)$$

$$L_{rVb}(2,1) = -l\sin(2\theta - \theta_{Vb}) - l_x - L_x\cos(2\theta - \theta_{Vb})$$

Assuming the magnitude V_a and V_b are same. Therefore, equation (27) is obtained from equations (25) and (26).

$$\begin{bmatrix} \dfrac{di_{\alpha hrVa}}{dt} - \dfrac{di_{\alpha hrVb}}{dt} \\ -\left(\dfrac{di_{\beta hrVa}}{dt} - \dfrac{di_{\beta hrVb}}{dt}\right) \end{bmatrix}$$

$$= C\begin{bmatrix} l\sin(2\theta - \theta_{Va} - \theta_{Vb}) + L_x\cos(2\theta - \theta_{Va} - \theta_{Vb}) \\ l\cos(2\theta - \theta_{Va} - \theta_{Vb}) - L_x\sin(2\theta - \theta_{Va} - \theta_{Vb}) \end{bmatrix} \qquad (27)$$

Here, $f_{\sin}$ and $f_{\cos}$ can be defined as equation (28)

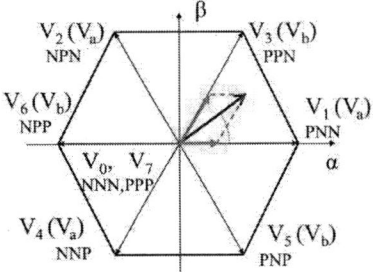

Fig. 3 Space Vector PWM.

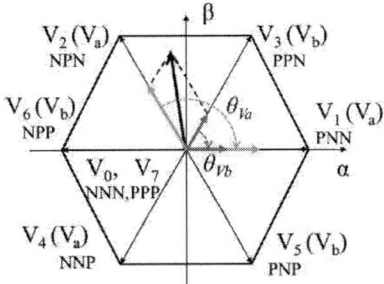

Fig. 4 Rotational coordinate transformation.

$$\begin{bmatrix} f_{\sin}\left(\dfrac{di_\alpha}{dt}, \dfrac{di_\beta}{dt}\right) \\ f_{\cos}\left(\dfrac{di_\alpha}{dt}, \dfrac{di_\beta}{dt}\right) \end{bmatrix}$$

$$= \begin{bmatrix} \cos(\theta_{Va} + \theta_{Vb}) & -\sin(\theta_{Va} + \theta_{Vb}) \\ \sin(\theta_{Va} + \theta_{Vb}) & \cos(\theta_{Va} + \theta_{Vb}) \end{bmatrix}\begin{bmatrix} \dfrac{di_{\alpha hrVa}}{dt} - \dfrac{di_{\alpha hrVb}}{dt} \\ -\left(\dfrac{di_{\beta hrVa}}{dt} - \dfrac{di_{\beta hrVb}}{dt}\right) \end{bmatrix}$$

$$= C\begin{bmatrix} l\sin 2\theta + L_x\cos 2\theta \\ l\cos 2\theta - L_x\sin 2\theta \end{bmatrix} \qquad (28)$$

Equation (28) is equal to equations (6) and (7). The procedure of calculating from (22) to (28) is same as the calculation without accounting for mutual inductance in dq-axis. Therefore, the proposed method can be applied to many kinds of position estimation methods with equation (6) and (7) in the procedure of calculating.

V. Experiments

In this section, experimental results are shown. The experimental setup, the accuracy of position estimation, and the velocity control performance are explained.

A. Experimental Setup

The experimental setup is shown in Fig.5. The characteristics of the test motor are shown in Table II. Current sensors are connected to A/D converter integrated in Field Programmable Gate Array (FPGA).

2213

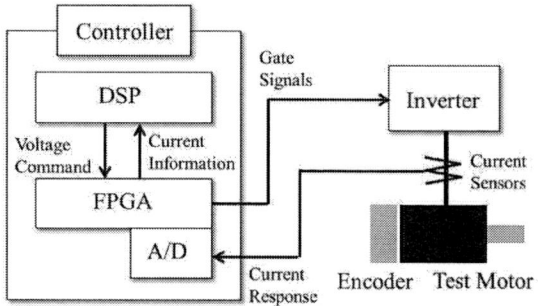

Fig.5. Experimental Setup.

TABLE II
CHARACTERS OF TEST MOTOR

Rated Power	400 [W]
Rated Velocity	87.5 [Hz]
Rated Current	1.8 [Arms]
Stator Resistance	2.247 [Ω]
Nominal Inductance	L_d:22.47 [mH], L_q:32.5 [mH]
PM Flux Linkage	0.2242 [V·s/rad]

TABLE III
CONTROL PARAMETERS

Carrier period	4000 [Hz]
Control period	8000 [Hz]
Current Control Bandwidth	120 [Hz]
Velocity Control Bandwidth	5 [Hz]
Attenuation Constant in Velocity Control	0.7
Position Control Bandwidth	0.83 [Hz]
PLL Bandwidth	60 [Hz]
LPF Bandwidth	40 [Hz]
Sensitivity Parameter a	0.8

The current information was sent to Digital Signal Processor (DSP) and position was estimated. By using the estimated position, position or velocity control based on FOC was achieved. On the other hand, actual position information is obtained by the rotary encoder attached to the test motor. The actual position is only used for the comparison with estimated values. The proposed method is applied to a sensorless control method, which is based on magnetic saliency [6]. Therefore, the output PWM from the inverter is characteristic PWM based on Space Vector PWM. The motor angular velocity is expressed as the ratio of the test motor rated velocity. The control parameters are shown in Table III.

B. Accuracy of Position Estimation

The accuracy is evaluated by sensorless position control mode. The error between the estimated position and actual position detected by encoder measures the accuracy of position estimation. A position controller and a velocity controller are added to the control system shown in Fig.2. The position controller is a proportional

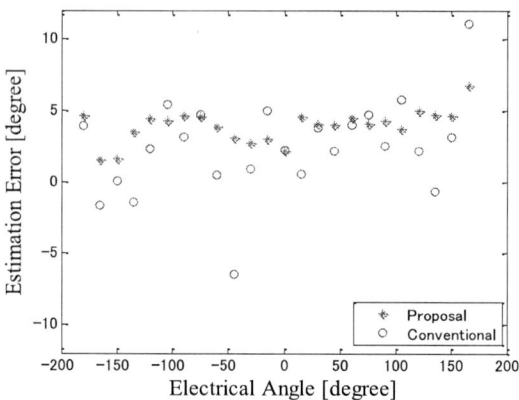

Fig.6. Estimation error at each electrical angle

TABLE IV
QUANTITATIVE COMPARISON OF ESTIMATION ERROR

	Proposal [degree]	Conventional [degree]
Average error	3.9	2.4
Max error	5.7	11.0
Minimum error	1.5	-6.4
Unbiased sample variance	1.3	11.0

controller. The velocity controller is a proportional-integral controller. The errors between estimated position and actual position are shown in Fig. 6. The drawn data are average values of the position error sampled in 100ms with the interval of 15 degree electrical angle. Red asterisk dots show the estimated values with proposed method and blue dots show the estimated values without proposed method. According to Fig.4, the data spread of estimation error is decreased. In addition, the max value of the estimation error is also decreased by using proposed method. The quantitative comparison of the result is shown in Table IV. Table IV shows that the unbiased sample variance of the estimation error with proposed method is smaller than that of estimation error without proposed method. The smaller unbiased sample variance of the estimation error provides smooth rotation or lower machine noise.

C. Performance of Velocity Control

Similarly performance of velocity is evaluated by sensorless speed control. A velocity controller is added to the control system shown in Fig.2. The velocity controller is a proportional-integral controller. Fig.7 and Fig.8 show the steady state response of velocity control. The red line shows the estimated angular velocity, and the blue broken line shows the actual velocity response. The vibration amplitude of red line in Fig.7 is smaller than that of red line in Fig.8. Therefore, the vibration of estimated angular velocity can be small by using proposed method. Furthermore, Fig.9 shows the sound pressure level of the audible noise. The sound pressure level was measured during the steady-state response of

The 2018 International Power Electronics Conference

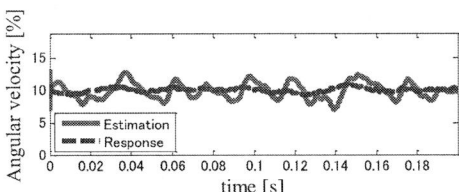

Fig.7. Estimated velocity and actual velocity
(with proposed method)

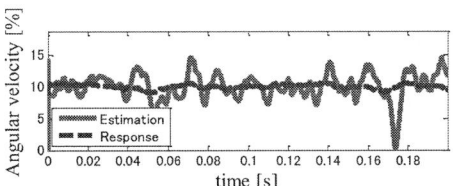

Fig.8. Estimated velocity and actual velocity
(without proposed method)

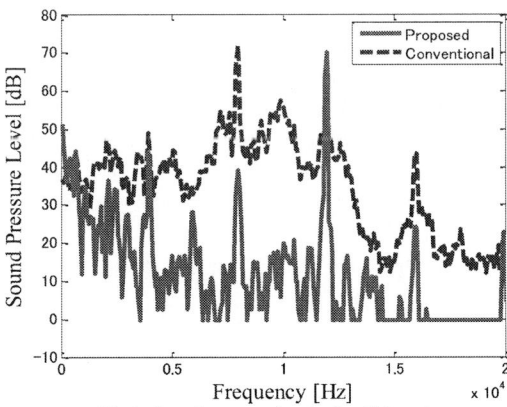

Fig.9. Sound pressure level of audible noise

the angular velocity control. The velocity command was 10 %. The red line shows the sound pressure level with proposed method, and the blue broken line shows that without proposed method. By using proposed method, the sound pressure level was decreased from 5 kHz to 10 kHz. This frequency band is in the audible range.

VI.   CONCLUSION

In this paper, an improved method of position estimation accuracy was proposed. The method utilizes voltage equation accounting for mutual inductances in dq-axis. In the proposed method, cross-coupling effect of mutual inductances is not required to be known since it is canceled out. Experimental results show that the proposed method not only improves the estimation accuracy, but also improves the control performance of velocity control, and decreases the sound pressure level of the audible noise.

REFERENCES

[1]   S. -K. Sul and S. Kim "Sensorless Control of IPMSM: Past , Present, and Future," *IEEJ Journal IA*, Vol. 1, No.1, pp. 15-23, 2012.

[2]   M. Schrodl and M. Lambeck, "Statistic Properties of the INFORM-Method in Highly Dynamic Sensorless PM Motor Control Applications Down to Standstill," *European Power Electronics and Drives*, vol.13, no.3,   pp. 22-29, Mar. 2003.

[3]   J. –I. Ha, K. Ide, T. Sawa, and S. –K. Sul, "Sensorless Position Control and Initial Position Estimation," *Conference Record of the 2001 IEEE Industry Applications Conference*, 2001, pp. 2607-2613, Sept. 30-Oct. 4 2001.

[4]   S. Ogasawara, and H. Akagi, "Implementation and position control performance of a position-sensorless IPM motor drive system based on magnetic saliency," *IEEE Trans. Ind. Appl.*, Vol.34, pp. 806-812, Jil./Aug. 1998.

[5]   S. Kim, J. –I. Ha, and S. –K. Sul, "PWM Switching Frequency Signal Injection Sensorless Method in IPMSM," *IEEE Trans. Ind. Appl.*, Vol. 48, Issue 5,   pp. 1576- 1587, 2012.

[6]   K. Shimamoto, S. Morimoto, and S. Fukumaru, "Sensorless control based on position estimation by switching operation of modified PWM," *The 42nd Annual Conference of the IEEE Industrial Electronics Society,* pp.2898-2903, 2016

[7]   Y. Li, Z. Q. Zhu, D. Howe, C. M. Bingham, D. A. Stone, "Improved Rotor-Position Estimation by Signal Injection in Brushless AC Motors, Accounting for Cross-Coupling Magnetic Saturation," *IEEE Trans. on Industry Applications*, vol. 45, no. 5, pp. 1843-1850, 2009.

[8]   J. Liu, M. Ma, and L. Li, "A Position Error Compensation way for Sensorless Linear Motor drive Using High Frequency Injection," *The 16h International Symposium on Electromagnetic Launch Technology*, 2012

Sensorless Drive in the Low Speed Region and Auto-tuning Method for Permanent Magnet Synchronous Motors

Naofumi Nomura[1*], and Shinichi Higuchi[1]
1 Fuji Electric Co.,Ltd., Suzuka-City, Mie, Japan
*E-mail: nomura-naofumi@fujielectric.com

Abstract— This paper presents a sensorless drive in the low speed region and auto-tuning method for permanent magnet synchronous motors (PMSMs). In the proposed sensorless drive, it is possible to drive various PMSMs stably. And, in the proposed auto-tuning method, it is possible to detect saliency of rotor and magnetization characteristics without rotor position information, which suits sensorless drive. Experimental results of the proposed method are also obtained. According to the proposed method, the sensorless drive is stable in various kinds of PMSMs. And motor parameters measured by auto-tuning have high accuracy for sensorless drive.

Keywords— *Permanent magnet synchronous motor, sensorless drive, auto-tuning, adaptive identifier.*

I. INTRODUCTION

The permanent magnet synchronous motor (PMSM) is a suitable alternative to the induction motor to save energy and reduce the size in various applications such as air conditioners, industrial drives, etc. For lower cost and higher reliability of the PMSM-based drive, the elimination of the position-sensor (sensorless drive) has been developed [1][2][4][5][6].

One of important problems in sensorless drives is improvement of the starting characteristics. A PMSM is classified into an interior permanent magnet synchronous motor (IPMSM) and a surface permanent magnet synchronous motor (SPMSM), and these electrical characteristics are different. Therefore, the electrical characteristics of PMSMs are considered and the drive method suitable for each is chosen.

As the technique which can be used for the period of low speed and standstill, for IPMSMs, several rotor position estimation methods based on saliency of the rotor are reported so far [1][6]. For SPMSMs, the rotor position estimation method based on the nonlinear magnetization characteristics caused by the magnet of the rotor is reported [2]. And also, by using the estimated rotor position, the sensorless drive in the low speed region which includes zero speed was realized.

To implement sensorless drive technique, motor parameters are necessary. An auto-tuning method of motor parameters for PMSMs is proposed [3]. However, the conventional auto-tuning method needs information of the rotor positon, and in the case of sensorless drive, the value of motor parameters that are auto-tuning targets and unknown values are needed for position detection. Therefore development of an auto-tuning method which suits sensorless drives is a problem.

This paper presents a sensorless drive in the low speed region and an auto-tuning method for PMSMs. In the proposed sensorless drive, it is possible to drive stably various PMSMs. And, in the proposed auto-tuning method, it is possible to measure saliency of rotor and magnetization characteristics without rotor position information, which suits sensorless drives. Experimental results of the proposed method are also obtained. According to the proposed method, sensorless drive works stably. And motor parameters measured by auto-tuning have high accuracy.

II. OUTLINE OF SENSORLESS DRIVE FOR PMSM

In the high speed region, the induced voltage of the motor is used for rotor position estimation, and sensorless drive can be achieved using estimated rotor position. However, in the low speed region, because the induced voltage is low, it is necessary to drive using another method.

As a drive method suitable for the low speed region, "Open loop drive" is often used [4]. Open loop drive achieves speed control by controlling frequency of the current vector in frequency reference of the rotor. This method can achieve sensorless drive in the low speed region easily, and it can be applied to various PMSMs.

When the difference between direction of the PMSMs magnetic pole and direction of the current vector is too large, stability of open loop control is degraded. As a result, at the time of a drive start, a rotor reverses and the maximum torque decreased. Therefore, improvement of the starting characteristics is needed.

The first method to solve this problem is to keep the frequency of the current vector in a small constant value and wait until a rotor position becomes parallel with the current vector before starting drive. This method can be applied to most types of PMSMs, and is easy. However, this method makes the start-up time longer.

The second method is to detect an initial rotor position using the inductance change dependent on the rotor position before starting drive, and start to drive based on this position. By using this method, start-up time can be made shorter and stability is improved.

Table I shows the comparison of initial positioning methods for PMSMs. In Table I, Method#1 is the first method already indicated, and it doesn't detect the rotor position before starting drive. In the case of Method#2, the initial rotor position is detected based on saliency of the rotor and magnetic saturation. This method can be applied to IPMSMs which can detect magnetic saturation characteristics. In the case of Method#3, the initial rotor position is detected based on magnetic saturation. This method can be applied to SPMSMs which can detect magnetic saturation characteristics.

TABLE I
THE COMPARISON OF INITIAL POSITIONING METHODS FOR PMSM

No.	Initial positioning method	Applicable PMSM	
		Rotor structure	Magnetic saturation
#1	Open loop drive by a low frequency	IPMSM SPMSM	Unnecessary
#2	Detect saliency and magnetic saturation	IPMSM	Can detect
#3	Detect magnetic saturation	SPMSM	Can detect

III. INITIAL ROTOR POSITION ESTIMATION FOR IPMSM (METHOD#2)

A. Initial Rotor Position Estimation Based on Saliency

The estimation of the rotor position for IPMSMs is based on saliency of the rotor.

Figure 1 shows the definition of the gamma-delta axis. The gamma-delta axis is defined as the estimated axis of PMSM's d-q axis.

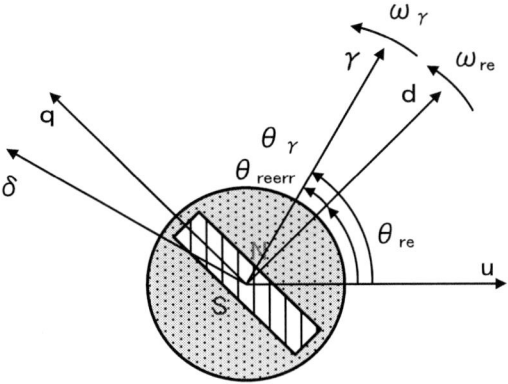

Fig. 1. The definition of gamma-delta axis.

When a high frequency voltage is injected on the gamma-axis as expressed in equation (1), high frequency current flows to the gamma-axis and the delta-axis as expressed in equation (2) and (3).

$$v_\gamma = V_{\gamma h} \cos \theta_h \tag{1}$$
$$i_\gamma = I_{\gamma h} \sin \theta_h \tag{2}$$
$$i_\delta = I_{\delta h} \sin \theta_h \tag{3}$$

where,

$$\theta_h = \omega_h t \tag{4}$$

v_γ : gamma-axis voltage
i_γ, i_δ: gamma-axis and delta-axis current
$V_{\gamma h}$: gamma-axis high frequency voltage
$I_{\gamma h}$, $I_{\delta h}$: gamma-axis and delta-axis high frequency current
θ_h : angle of high frequency voltage
ω_h : angular frequency of high frequency voltage
t: time

The gamma-axis and delta-axis high frequency current $I_{\gamma h}$, $I_{\delta h}$ are defined in Fourier coefficients as expressed in equation (5) and (6).

$$I_{\gamma h} = \frac{1}{\pi} \int_0^{2\pi} i_\gamma \sin \theta_h \, d\theta_h \tag{5}$$
$$I_{\delta h} = \frac{1}{\pi} \int_0^{2\pi} i_\delta \sin \theta_h \, d\theta_h \tag{6}$$

When the phase of the current is later than the phase of the voltage, a sign of $I_{\gamma h}$ and $I_{\delta h}$ are defined as "positive", when it isn't so, they are defined as "negative".

$I_{\gamma h}$ and $I_{\delta h}$ will be the function of the position estimate error θ_{reerr} as shown in figure 2. $I_{\gamma h}$ and $I_{\delta h}$ are expressed in the following equation.

$$I_{\gamma h} = \frac{1}{2}\left(I_{dh} + I_{qh}\right) + \frac{1}{2}\left(I_{dh} - I_{qh}\right) \cos 2\theta_{reerr} \tag{7}$$
$$I_{\delta h} = -\frac{1}{2}\left(I_{dh} - I_{qh}\right) \sin 2\theta_{reerr} \tag{8}$$

where,

$$\theta_{reerr} = \theta_\gamma - \theta_{re} \tag{9}$$
$$I_{dh} = \frac{V_{\gamma h}}{\omega_h L_d} \tag{10}$$
$$I_{qh} = \frac{V_{\gamma h}}{\omega_h L_q} \tag{11}$$

θ_{re} : rotor position, θ_γ : estimated rotor position
I_{dh}, I_{qh} : d-axis and q-axis high frequency current
L_d, L_q : d-axis and q-axis inductance

According to the equation (8), when θ_{reerr} is small, θ_{reerr} can be calculated approximately by the following equation.

$$\theta_{reerr} \cong -\frac{1}{I_{dh} - I_{qh}} I_{\delta h} \tag{12}$$

The rotor position and rotor speed can be calculated from the position estimated error θ_{reerr} calculated by the equation (12) [1][6].

To calculate θ_{reerr} from the equation (12), values of I_{dh} and I_{qh} are necessary. And also, to improve the calculation precision of θ_{reerr}, it is necessary to adjust the gamma-axis high frequency voltage $V_{\gamma h}$ so that I_{dh} and I_{qh} may become large enough, and they have to be below the maximum output current of the invertor supplied to a PMSM.

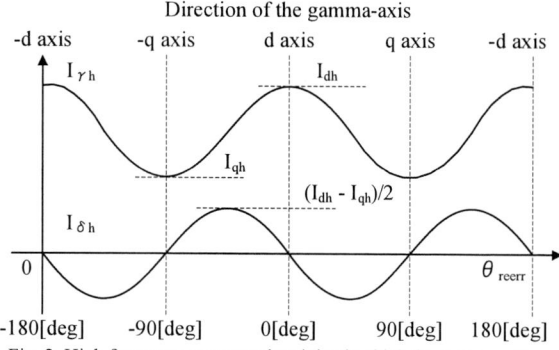

Fig. 2. High frequency current when injecting high frequency voltage on gamma-axis of IPMSM.

B. Auto-Tuning Method

Figure 3 shows the block diagram of the proposed auto-tuning method. This method injects a high frequency voltage to a PMSM, and relations between the position estimate error θ_{reerr} and the high frequency current $I_{\gamma h}$, $I_{\delta h}$, which are indicated in figure 2, are measured.

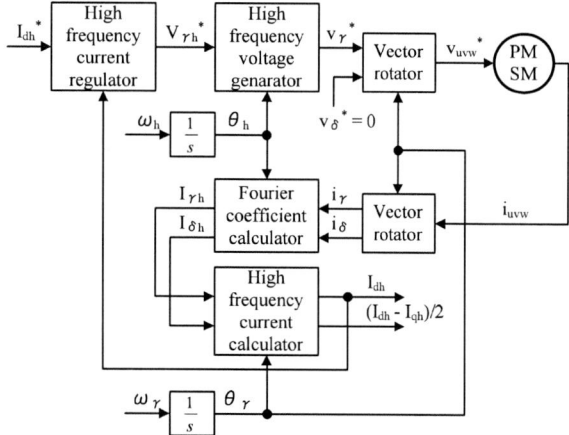

Fig. 3. The block diagram of the auto-tuning of initial rotor position estimate for IPMSM (Method#2).

The block diagram on figure 3 is explained below.

The gamma-axis voltage reference v_γ^* is controlled to a high frequency whose angular frequency is ω_h. The angular frequency of gamma-axis ω_γ is controlled to be constant.

Figure 4 shows the block diagram of the high frequency current regulator [3].

From equation (10), I_{dh} is expressed in the following equation.

$$I_{dh} = -B_{dh}V_{\gamma h} \tag{13}$$

where,

$$-B_{dh} = \frac{1}{\omega_h L_d} \tag{14}$$

B_{dh} : d-axis high frequency susceptance

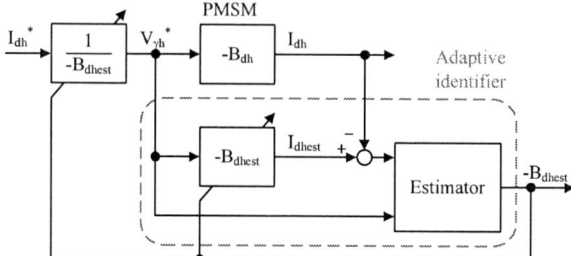

Fig. 4. The block diagram of the high frequency current regulator for IPMSM (Method#2).

The amplitude of the gamma-axis high frequency voltage reference $V_{\gamma h}^*$ is regulated so that the d-axis high frequency current I_{dh} will be the current reference I_{dh}^* by feed forward control based on the equation (13).

$$V_{\gamma h}^* = -\frac{1}{B_{dhest}}I_{dh}^* \tag{15}$$

where,

$V_{\gamma h}^*$: gamma-axis high frequency voltage reference
I_{dh}^* : d-axis high frequency current reference
B_{dhest} : the estimated value of B_{dh}

The estimated value B_{dhest} reaches to the true value by the estimator. Detail of the estimator will be described in the next section.

As a result, the gamma-axis high frequency voltage reference $V_{\gamma h}^*$ is operated accurately, and the d-axis high frequency current reaches to the d-axis high frequency current reference I_{dh}^*. This current control is stable and highly accurate up to the high frequency area, even though motor parameters are unknown before parameter tuning is executed.

Equation (13) is a linear model, where the input signal is $V_{\gamma h}$ and the output signal is I_{dh}. Therefore, it is possible to estimate the parameter B_{dh} by the adaptive identifier, and the parameter reaches to the true value.

Here, the estimated value of I_{dh} is denoted as I_{dhest}, and the estimated value of B_{dh} is denoted as B_{dhest}. The estimator adjusts B_{dhest} so that the deflection between I_{dh} and I_{dhest} may become zero based on the adaptive identifier as expressed in the following equation.

$$-B_{dhest} = \int \{-\Gamma V_{\gamma hN}I_{dherrN}\}dt \tag{16}$$

where,

$$V_{\gamma hN} = \frac{V_{\gamma h}{}^*}{N} \tag{17}$$

$$I_{dherrN} = \frac{I_{dhest} - I_{dh}}{N} \tag{18}$$

$$N = \sqrt{\rho + V_{\gamma h}{}^{*2}} \tag{19}$$

I_{dhest} : the estimated value of I_{dh}

Γ: gain, ρ: constant (positive)

$I_{\gamma h}$ and $I_{\delta h}$, which are the θ_h components of i_γ and i_δ, are calculated from Fourier coefficients of the gamma-delta axis current i_γ and i_δ as expressed in equation (5) and (6).

$I_{\gamma ha0}$, which is the DC component of $I_{\gamma h}$ and $I_{\gamma hc2}$, which is the $2\theta_\gamma$ component of $I_{\gamma h}$, and $I_{\delta hc2}$, which is the $2\theta_\gamma$ component of $I_{\delta h}$, are calculated from Fourier coefficients of $I_{\gamma h}$ and $I_{\delta h}$ as expressed in equation (20),(21), and (22).

$$I_{\gamma ha0} = \frac{1}{\pi}\int_0^\pi I_{\gamma h}\, d\theta_\gamma \tag{20}$$

$$I_{\gamma hc2} = \sqrt{I_{\gamma ha2}{}^2 + I_{\gamma hb2}{}^2} \tag{21}$$

$$I_{\delta hc2} = \sqrt{I_{\delta ha2}{}^2 + I_{\delta hb2}{}^2} \tag{22}$$

where,

$$I_{\gamma ha2} = \frac{2}{\pi}\int_0^\pi I_{\gamma h}\cos 2\theta_\gamma\, d\theta_\gamma \tag{23}$$

$$I_{\gamma hb2} = \frac{2}{\pi}\int_0^\pi I_{\gamma h}\sin 2\theta_\gamma\, d\theta_\gamma \tag{24}$$

$$I_{\delta ha2} = \frac{2}{\pi}\int_0^\pi I_{\delta h}\cos 2\theta_\gamma\, d\theta_\gamma \tag{25}$$

$$I_{\delta hb2} = \frac{2}{\pi}\int_0^\pi I_{\delta h}\sin 2\theta_\gamma\, d\theta_\gamma \tag{26}$$

I_{dh} and $(I_{dh} - I_{qh})/2$ are calculated by the following equations.

$$I_{dh} = I_{\gamma ha0} + I_{\gamma hc2} \tag{27}$$

$$\frac{1}{2}\left(I_{dh} - I_{qh}\right) = I_{\delta hc2} \tag{28}$$

IV. INITIAL ROTOR POSITION ESTIMATION FOR SPMSM (METHOD#3)

A. Initial Rotor Position Estimation Based on Magnetic Saturation

The estimation of the rotor position for SPMSMs is based on the nonlinear magnetization characteristics of the stator core.

When a high frequency voltage is injected on the delta-axis as expressed in equation (29), high frequency current flows to the delta-axis as expressed in equation (30a) and (30b).

$$v_\delta = V_{\delta h}\cos\theta_h \tag{29}$$

i) $0 \le \theta_h < \pi$

$$i_\delta = I_{\delta hP}\sin\theta_h \tag{30a}$$

ii) $\pi \le \theta_h < 2\pi$

$$i_\delta = I_{\delta hN}\sin\theta_h \tag{30b}$$

where,

v_δ : delta-axis voltage
$V_{\delta h}$: delta-axis high frequency voltage
$I_{\delta hP}$, $I_{\delta hN}$: delta-axis high frequency current on the positive side and the negative side

$I_{\delta hP}$ is defined by the amplitude of the positive side of the delta-axis high frequency current. $I_{\delta hN}$ is defined by the amplitude of the negative side of the delta-axis high frequency current. $I_{\delta hP}$ and $I_{\delta hN}$ are defined in Fourier coefficients as expressed in equation (31) and (32).

$$I_{\delta hP} = \frac{2}{\pi}\int_0^\pi i_\delta \sin\theta_h\, d\theta_h \tag{31}$$

$$I_{\delta hN} = \frac{2}{\pi}\int_\pi^{2\pi} i_\delta \sin\theta_h\, d\theta_h \tag{32}$$

$I_{\delta hP}$ and $I_{\delta hN}$ will be the function of the position estimate error θ_{reerr} as shown in figure 3.

When θ_{reerr} is zero, the delta-axis in which a high frequency voltage is injected, is parallel with the q-axis, and high frequency current flows to the right angle direction with the magnetic flux excited by a permanent magnet of a rotor. In this case, $I_{\delta hP}$ is equal to $I_{\delta hN}$, because magnetization characteristics are equal.

On the other hand, when θ_{reerr} is -90[deg] of an electrical angle, the delta-axis is parallel with the d-axis, and high frequency current flows to the parallel direction with the rotor flux. In this case, $I_{\delta hP}$ is larger than $I_{\delta hN}$, because the magnetic flux which is generated by the stator current and the magnetic flux excited by a permanent magnet of a rotor are added, and inductance decreases because of magnetic saturation.

$I_{\delta hP}$ and $I_{\delta hN}$ are expressed in the following equation.

$$I_{\delta hP} = \frac{1}{2}(I_{dhP} + I_{dhN}) - \frac{1}{2}(I_{dhP} - I_{dhN})\sin\theta_{reerr} \tag{33}$$

$$I_{\delta hN} = \frac{1}{2}(I_{dhP} + I_{dhN}) + \frac{1}{2}(I_{dhP} - I_{dhN})\sin\theta_{reerr} \tag{34}$$

where,

$$I_{dhP} = \frac{(L_q + L_1)V_{\delta h}}{\omega_h L_d L_q} \tag{35}$$

$$I_{dhN} = \frac{(L_q - L_1)V_{\delta h}}{\omega_h L_d L_q} \tag{36}$$

I_{dhP}, I_{dhN} : d-axis high frequency current on the positive side and the negative side

2219

L_1 : fluctuation of d-axis inductance caused by magnetic saturation

From the equation (33) and (34), the following equation can be derived.

$$\frac{1}{2}(I_{\delta hP} - I_{\delta hN}) = -\frac{1}{2}(I_{dhP} - I_{dhN})\sin\theta_{reerr} \quad (37)$$

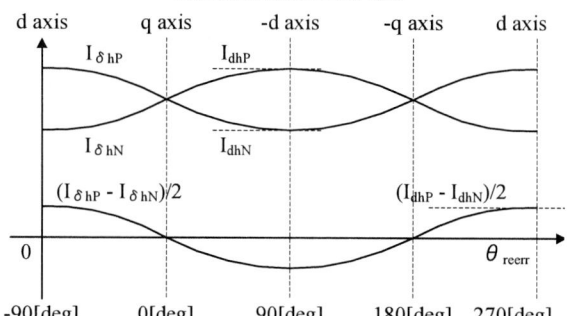

Fig. 5. High frequency current when injecting high frequency voltage on delta-axis of SPMSM.

According to the equation (37), when θ_{reerr} is small, θ_{reerr} can be calculated approximately by the following equation.

$$\theta_{reerr} \cong -\frac{1}{I_{dhP} - I_{dhN}}(I_{\delta hP} - I_{\delta hN}) \quad (38)$$

The rotor position and rotor speed can be calculated from the position estimated error θ_{reerr} calculated by the equation (38).

To calculate θ_{reerr} from the equation (38), values of I_{dhP} and I_{dhN} are necessary. And also, to improve the calculation precision of θ_{reerr}, it is necessary to adjust a delta-axis high frequency voltage $V_{\delta h}$ so that I_{dhP} and I_{dhN} may become large enough, and they have to be below the maximum output current of the invertor supplied in a PMSM.

B. Auto-Tuning Method

Figure 6 shows the block diagram of the proposed auto-tuning method. This method injects a high frequency voltage on a PMSM, and relations between the position estimate error θ_{reerr} and the high frequency current $I_{\delta hP}$, $I_{\delta hN}$, which are indicated on figure 5, are measured.

The block diagram on figure 6 is explained below.

The delta-axis voltage reference v_δ^* is controlled to high frequency whose angular frequency is ω_h. The angular frequency of gamma-axis ω_γ is controlled to be constant.

Figure 7 shows the block diagram of the high frequency current regulator. This block diagram is similar to figure 4.

From equation (35), I_{dhP} is expressed in the following equation.

$$I_{dhP} = -B_{dhP}V_{\delta h} \quad (39)$$

where,

$$-B_{dhP} = \frac{L_q + L_1}{\omega_h L_d L_q} \quad (40)$$

B_{dhP} : d-axis high frequency susceptance on the positive side

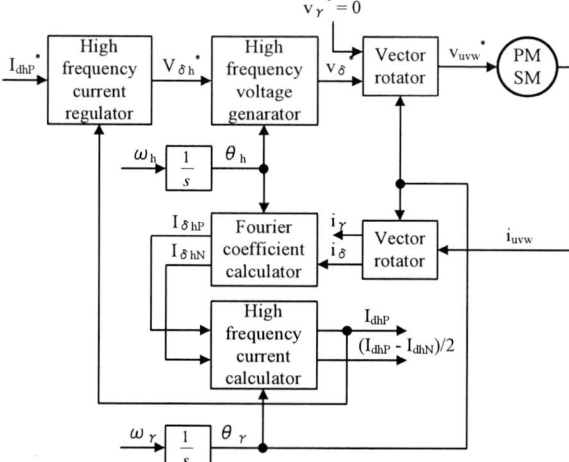

Fig. 6. The block diagram of the auto-tuning of initial rotor position estimate for SPMSM (Method#3).

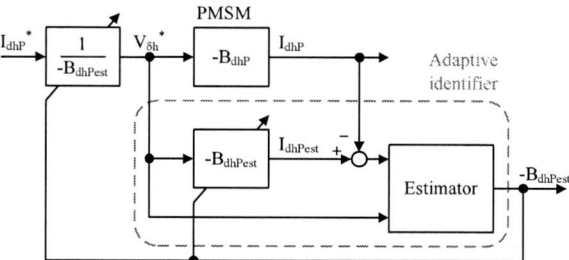

Fig. 7. The block diagram of the high frequency current regulator for SPMSM (Method#3).

The amplitude of the delta-axis high frequency voltage reference $V_{\delta h}^*$ is regulated so that the d-axis high frequency current I_{dhP} will be the same as the current reference I_{dhP}^* by feed forward control based on the equation (39).

$$V_{\delta h}^* = -\frac{1}{B_{dhPest}}I_{dhP}^* \quad (41)$$

where,

$V_{\delta h}^*$: delta-axis high frequency voltage reference
I_{dhP}^* : d-axis high frequency current reference on the positive side
B_{dhPest} : the estimated value of B_{dhP}

The estimated value of I_{dhP} is denoted as I_{dhPest}, and the estimated value of B_{dhP} is denoted as B_{dhPest}. The estimator adjusts B_{dhPest} so that the deflection between I_{dhP} and I_{dhPest} may become zero based on the adaptive identifier as expressed in the following equation.

$$-B_{dhPest} = \int \{-\Gamma V_{\delta hN} I_{dhPerrN}\} dt \qquad (42)$$

where,

$$V_{\delta hN} = \frac{V_{\delta h}{}^*}{N} \qquad (43)$$

$$I_{dhPerrN} = \frac{I_{dhPest} - I_{dhP}}{N} \qquad (44)$$

$$N = \sqrt{\rho + V_{\delta h}{}^{*2}} \qquad (45)$$

I_{dhPest} : the estimated value of I_{dhP}
Γ: gain, ρ: constant (positive)

$I_{\delta hP}$ and $I_{\delta hN}$, which are the θ_h components of i_δ, are calculated from Fourier coefficients of the delta-axis current i_δ as expressed in equation (31) and (32).

$I_{\delta hPa0}$, which is the DC component of $I_{\delta hP}$, $I_{\delta hPc1}$, which is the θ_γ component of $I_{\delta hP}$, and $I_{\delta hPc2}$, which is the $2\theta_\gamma$ component of $I_{\delta hP}$, are calculated from Fourier coefficients of $I_{\delta hP}$ and $I_{\delta hN}$ as expressed in equation (46), (47), and (48).

$$I_{\delta hPa0} = \frac{1}{2\pi} \int_0^\pi \{I_{\delta hP} + I_{\delta hN}\} d\theta_\gamma \qquad (46)$$

$$I_{\delta hPc1} = \sqrt{I_{\delta hPa1}{}^2 + I_{\delta hPb1}{}^2} \qquad (47)$$

$$I_{\delta hPc2} = \sqrt{I_{\delta hPa2}{}^2 + I_{\delta hPb2}{}^2} \qquad (48)$$

where,

$$I_{\delta hPa1} = \frac{1}{\pi} \int_0^\pi \{I_{\delta hP} \cos\theta_\gamma - I_{\delta hN} \cos\theta_\gamma\} d\theta_\gamma \quad (49)$$

$$I_{\delta hPb1} = \frac{1}{\pi} \int_0^\pi \{I_{\delta hP} \sin\theta_\gamma - I_{\delta hN} \sin\theta_\gamma\} d\theta_\gamma \quad (50)$$

$$I_{\delta hPa2} = \frac{1}{\pi} \int_0^\pi \{I_{\delta hP} \cos 2\theta_\gamma + I_{\delta hN} \cos 2\theta_\gamma\} d\theta_\gamma \quad (51)$$

$$I_{\delta hPb2} = \frac{1}{\pi} \int_0^\pi \{I_{\delta hP} \sin 2\theta_\gamma + I_{\delta hN} \sin 2\theta_\gamma\} d\theta_\gamma \quad (52)$$

I_{dhP} and $(I_{dhP} - I_{dhN})/2$ are calculated by the following equations.

$$I_{dhP} = I_{\delta hPa0} + I_{\delta hPc1} + I_{\delta hPc2} \qquad (53)$$

$$\frac{1}{2}(I_{dhP} - I_{dhN}) = I_{\delta hPc1} \qquad (54)$$

V. EXPERIMENTAL RESULTS

Experiments have been made with two types of PMSMs. Table II shows specifications of test motors. Table III shows specifications of the inverter, which controls voltage of tested PMSMs. Table IV and V show conditions of the initial rotor position estimation and auto-tuning for IPMSMs (Method#2). Table VI and VII show conditions of the initial rotor position estimation and auto-tuning for SPMSMs (Method#3).

TABLE II
SPECIFICATIONS OF TEST MOTOR

Motor Name	PMSM#1	PMSM#2
Rated power	7.5kW	4kW
Base speed	10000r/min	8000r/min
Pole	4pole	6pole
Base frequency	333.3Hz	400Hz
Rated current	29A	15.5A
Rated voltage	175V	175V
d-axis inductance	0.48mH	0.67mH
q-axis inductance	0.57mH	0.80mH

TABLE III
SPECIFICATIONS OF THE INVERTER

Input power-supply voltage	AC200V
PWM carrier frequency	8kHz
Control cycle	250µs

TABLE IV
CONDITIONS OF INITIAL ROTOR POSITION ESTIMATION FOR IPMSM (METHOD#2)

Gamma-axis high frequency voltage: $V_{\gamma h}$	0.5pu(0-p)
Angular frequency of high frequency voltage: ω_h	2kHz

TABLE V
CONDITIONS OF AUTO-TUNING FOR IPMSM (METHOD#2)

d-axis high frequency current reference: $I_{dh}{}^*$	0.8pu(0-p)
Angular frequency of high frequency voltage: ω_h	2kHz
Angular frequency of gamma-axis: ω_γ	20Hz
Gain: Γ	10.0
Constant: ρ	0.01pu

TABLE VI
CONDITIONS OF INITIAL ROTOR POSITION ESTIMATION FOR SPMSM (METHOD#3)

Delta-axis high frequency voltage: $V_{\delta h}$	0.65pu(0-p)
Angular frequency of high frequency voltage: ω_h	1kHz

TABLE VII
CONDITIONS OF AUTO-TUNING FOR SPMSM (METHOD#3)

d-axis high frequency current reference: $I_{dhP}{}^*$	1.6pu(0-p)
Angular frequency of high frequency voltage: ω_h	1kHz
Angular frequency of gamma-axis: ω_γ	20Hz
Gain: Γ	10.0
Constant: ρ	0.01pu

Figure 8 shows acceleration characteristics of initial positioning Method#1, for PMSM#1.

Figure 9 shows acceleration characteristics of initial positioning Method#2, for PMSM#2. From of the results of figure 8 and 9, by using Method#2, the initial positioning time can be reduced to 0.2[sec] from 2[sec], and the torque ripple when starting can be reduced.

Figure 10 shows the estimated speed and rotor position of initial positioning Method#2, for PMSM#2. From of the results of figure 10, the mean of the estimated speed is equal to the true value, and estimated rotor position follows the true position.

Figure 11 shows the current and the voltage on auto-tuning Method#2, for PMSM#2. From of the result of figure 11, it is clear that the parameter estimate and current control are stable. The currents become under the steady state in 1.0[sec] from the start of auto-tuning. The relation between the high frequency voltage and high frequency currents obtained by auto-tuning corresponds with the result of measurement. By applying auto-tuning, the high frequency voltage and the high frequency current can be controlled to the optimal value, and the initial rotor position estimation can be achieved stably.

Figure 12 shows acceleration characteristics of initial positioning Method#3 for PMSM#2. From of the result of figure 12 and 8, by using Method#3, the initial positioning time can be reduced to 0.1[sec] from 2[sec], and the torque ripple when starting can be reduced.

Figure 13 shows the estimated speed and rotor position of initial positioning Method#3, for PMSM#2. From of the results of figure 13, the mean of the estimated speed is equal to the true value, and the estimated rotor position follows the true position.

Figure 14 shows the current and the voltage on auto-tuning Method#3, for PMSM#2. From of the result of figure 14, it is clear that parameter estimate and current control are stable. Currents become under the steady state in 1.0[sec] from the start of auto-tuning. The relation between the high frequency voltage and high frequency currents obtained by auto-tuning corresponds with the result of measurement. By applying auto-tuning, the high frequency voltage and the high frequency current can be controlled to the optimal value, and initial rotor position estimation can be achieved stably.

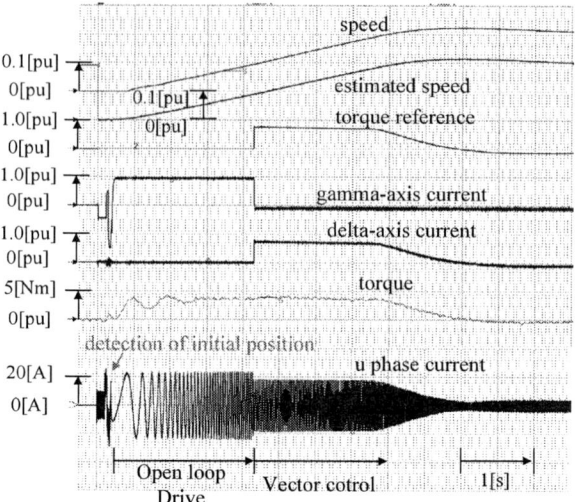

Fig. 9. Acceleration characteristics (Method#2, for PMSM#2).

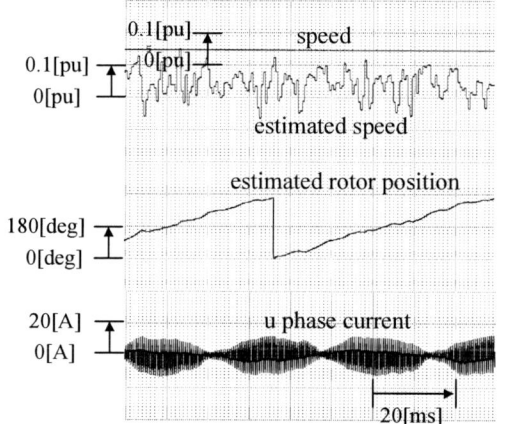

Fig. 10. Estimated speed and rotor position (Method#2, for PMSM#2).

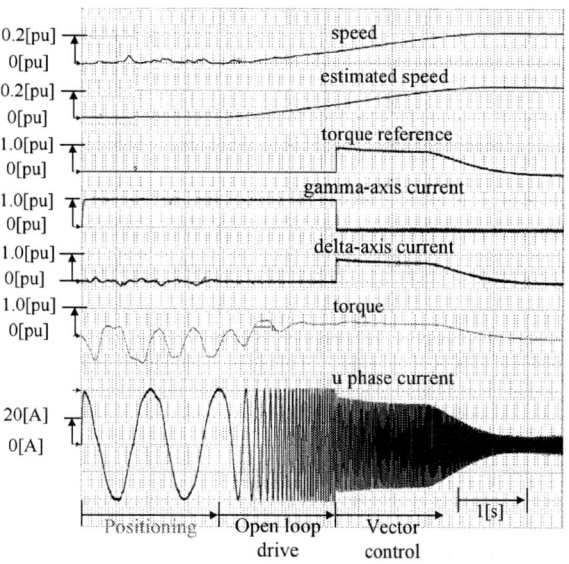

Fig. 8. Acceleration characteristics (Method#1, for PMSM#1).

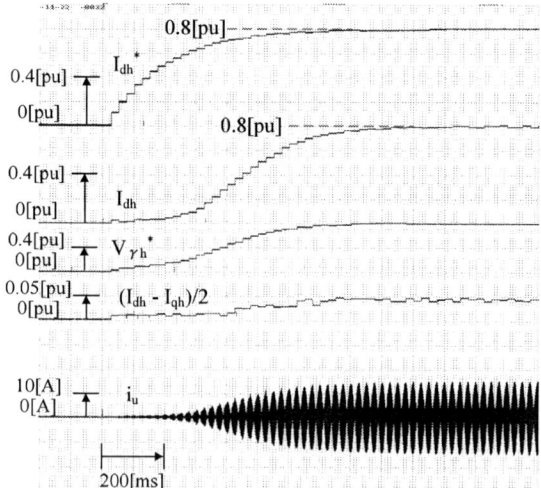

Fig. 11. Auto-tuning characteristics (Method#2, for PMSM#2).

The 2018 International Power Electronics Conference

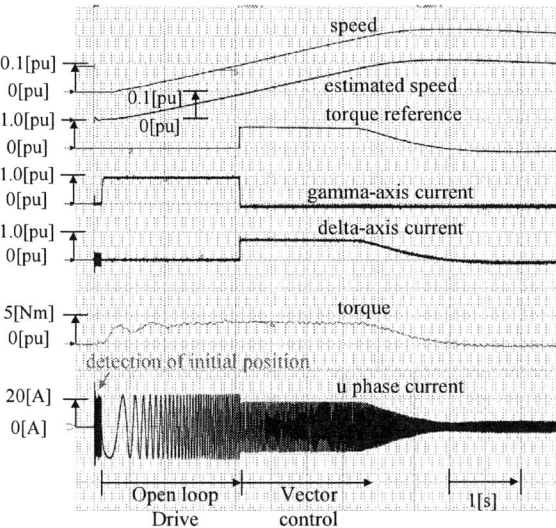

Fig. 12. Acceleration characteristics (Method#3, for PMSM#2).

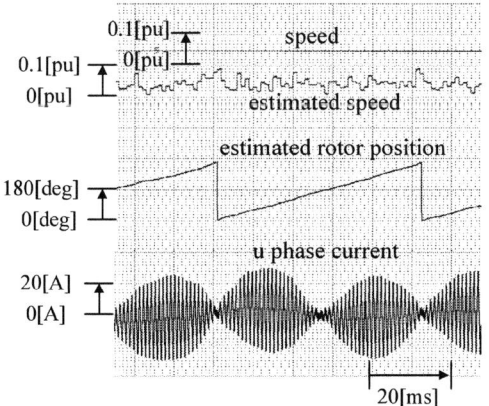

Fig. 13. Estimated speed and rotor position (Method#3, for PMSM#2).

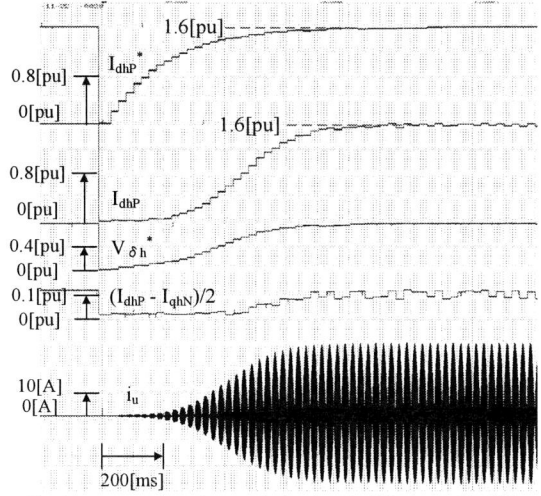

Fig. 14. Auto-tuning characteristics (Method#3, for PMSM#2).

VI. CONCLUSIONS

Sensorless drive in the low speed region and the auto-tuning method for PMSMs has been presented and its validity has been confirmed experimentally. According to the proposed sensorless drive method, acceleration characteristics are stable by using an initial positioning method which is chosen based on the electric characteristics of PMSMs. And also, according to the proposed auto-tuning method which is suitable for sensorless drive, parameter estimate and current control are stable.

REFERENCES

[1] T. Aihara, A. Toba, T. Yanase, A. Mashimo, and K. Endo, "Sensorless Torque Control of Salient-Pole Synchronous Motor at Zero-Speed Operation", IEEE Trans. Power Electron., Vol. 14, No. 1, January 1999, p.202-208

[2] S. Nakashima, Y. Inagaki, and I. Miki, "Sensorless initial rotor position estimation of surface permanent-magnet synchronous motor", IEEE Trans. Ind. Appl., Vol.36, No.6, pp.1598-1603 (2000-11/12)

[3] N. Nomura, and S. Higuchi, "Auto-tuning Method of Inductances for Permanent Magnet Synchronous Motors", IPEC-Hiroshima 2014, pp.1522-1528

[4] S. Shinnaka, "A New Simple Vector control Method for Starting Up Sensorlsee Drive of Permanent-Magnet Synchronous Motors", IEEJ Trans. IA, Vol.126, No.3, 2006, pp.225-236 (in Japanese)

[5] N.Nomura, A. Toba, T. Yamasaki, S. Ozaki, and H. Ohsawa, "Position-Sensorless Drive of the Interior Permanent Magnet Synchronous Motor for Wide Speed Range", EPE2001-Graz, 2001

[6] Yi Li, Z. Q Zhu, David Howe, Chris M. Bingham, and Dave A. Stone, "Improved Rotor-Position Estimation by Signal Injection in Brushless AC Motors, Accounting for Cross-Coupling Magnetic Saturation", IEEE Trans. Ind. Appl., Vol.45, No.5, pp.1843-1850 (2009-9/10)

High Stability V/f Control of PMSM Using State Feedback Control Based on n-t Coordinate System

Yosuke Matsuki[1*] and Shinji Doki[2]

1 DENSO CORPRATION, Kariya, Aichi, Japan

2 NAGOYA UNIVERSITY, Nagoya Aichi, Japan

*E-mail: YOUSUKE_MATSUKI@denso.co.jp

Abstract— **This paper proposes a new V/f control method for permanent magnetic synchronous motors (PMSM) without position sensor. The proposed method uses state feedback control based on a n-t coordinate system. The t-axis is a tangent line of a constant voltage ellipse, and the n-axis is a normal line of the ellipse. The t-axis current is utilized to keep the system poles in the stable plane in low speed of motor and to achieve high stability.**

Keywords— *PMSM, V/f control, Sensorless control*

I. INTRODUCTION

PMSMs (Permanent Magnet Synchronous Motors) have been employed in home appliances, industrial equipment and automobiles owing to high efficiency, high power density and maintenance free. Recently, PMSMs are applied to also accessory motors of automobiles; Cooling fans and pumps, as well as main motors. This is caused by the tightening of CO_2 emission regulations and fuel economy regulations [1].

In general, current vector control is used to control PMSMs and can achieve high response and high efficiency. However, this method takes a high cost for current sensors and a position sensor [2]. Therefore, this method is not used to control accessory motors. Sensorless controls [2]-[5] are widely used for accessory motors.

V/f control [6]-[7] is a kind of sensorless controls. This method can control rotational speed of a motor without a position sensor. However, open-loop V/f control is inherently unstable. Thus, V/f control with a stabilizing loop is proposed in previous researches. In [6] and [7], the stabilizing loop is the state feedback control based on γ-δ coordinate system. In γ-δ coordinate system, the γ-axis is a vertical line of a voltage vector and the δ-axis is a parallel line of the vector. This method is widely used for home appliances and industrial equipment. Nevertheless, there is a problem with this method for accessory motors of automobiles. The problem is that the disturbance suppression of this method is low in low speed of a motor. There are various disturbances of accessary motors; wind, obstacles and so on. Accessary motors are required to keep driving under load variation caused by these disturbances regardless of speed of a motor. We suppose the problem of this method is caused

by the system poles movement to the unstable plane in low speed of motor.

In this paper, we propose a new V/f control method. The proposed method uses state feedback control based on a n-t coordinate system [8]-[9] to solve the problem. The n-axis is a normal line of a constant voltage ellipse, and the t-axis is a tangent line of the ellipse. The current vector moves on the t-axis by change in load regardless of speed of a motor. Therefore, the system poles and the disturbance suppression in low speed can be improved via state feedback control using t-axis current.

In section II, a conventional motor model and the proposed motor model of V/f control are defined. These models demonstrate the mechanism of the system poles and the disturbance suppression improvement in low speed. The simulation and experimental results in low speed, medium speed and high speed are given in section III, and conclusions are given in section IV.

II. MOTOR MODELS OF V/F CONTROL

Fig. 1 shows a fundamental block diagram of V/f control of PMSM. The voltage equation of PMSM based on d-q coordinate system and based on γ-δ coordinate system are represented by

$$\begin{bmatrix} v_d \\ v_q \end{bmatrix} = \begin{bmatrix} R+pL & -\omega_e L \\ \omega_e L & R+pL \end{bmatrix} \begin{bmatrix} i_d \\ i_q \end{bmatrix} + \omega_e K_e \begin{bmatrix} 0 \\ 1 \end{bmatrix} \quad (1)$$

$$\begin{bmatrix} v_\gamma \\ v_\delta \end{bmatrix} = \begin{bmatrix} R+pL & -\omega_e * L \\ \omega_e * L & R+pL \end{bmatrix} \begin{bmatrix} i_\gamma \\ i_\delta \end{bmatrix} + \omega_e K_e \begin{bmatrix} \sin\delta \\ \cos\delta \end{bmatrix} \quad (2)$$

where R is the stator resistance, L is the inductance. In this paper, the saliency is neglected to simplify an expression. K_e is the back EMF constant and ω_e, ω_e* are the actual and reference electric angular velocity. v_d, v_q, i_d and i_q are d-/q- axis voltages and currents. v_γ, v_δ, i_γ and i_δ are γ-/δ- axis voltages and currents. Fig. 2 illustrates d-q coordinate system and γ-δ coordinate system. δ is the phase angle and obtained by

$$p\delta = \omega_e * -\omega_e \quad (3)$$

θ_e, θ_e* is the electric angle and the angle of α-axis and γ-axis. θ_e* is described as

$$\theta_e^* = \int P_n \omega_m^* \, dt = \int \omega_e^* \, dt \qquad (4)$$

where ω_m, ω_m^* is the actual and reference mechanical angular velocity and P_n is the number of pole pairs. ω_m can be written as

$$p\omega_m = (T - T_L)/J \qquad (5)$$

where T, T_L is the motor torque and load torque, J is the moment of inertia of the motor. T is obtained by

$$T = P_n K_e (i_\gamma \sin \delta + i_\delta \cos \delta) \qquad (6)$$

From (1) to (6), The linearized equation of state of V/f control can be formed as follows [6].

$$p\mathbf{x} = \mathbf{A}\mathbf{x} + \mathbf{B}\mathbf{u} \qquad (7)$$

$$\mathbf{x} =^t \begin{bmatrix} \Delta i_\gamma & \Delta i_\delta & \Delta \omega_e & \Delta \delta \end{bmatrix}$$

$$\mathbf{u} =^t \begin{bmatrix} \Delta v_\gamma & \Delta v_\delta & \Delta \omega_e^* & \Delta T_L \end{bmatrix}$$

$$\mathbf{A} = \begin{bmatrix} -\dfrac{R}{L} & \omega_{eo}^* & -\dfrac{K_e}{L}\sin\delta_o & -\dfrac{K_e}{L}\omega_o\cos\delta_o \\[2mm] -\omega_{eo}^* & -\dfrac{R}{L} & -\dfrac{K_e}{L}\cos\delta_o & \dfrac{K_e}{L}\omega_o\sin\delta_o \\[2mm] \dfrac{P_n^2 K_e}{J}\sin\delta_o & \dfrac{P_n^2 K_e}{J}\cos\delta_o & 0 & \dfrac{P_n^2 K_e}{J}I_{do} \\[2mm] 0 & 0 & -1 & 0 \end{bmatrix}$$

$$\mathbf{B} = \begin{bmatrix} 1/L & 0 & I_{\delta o} & 0 \\ 0 & 1/L & -I_{\gamma o} & 0 \\ 0 & 0 & 0 & -P_n/J \\ 0 & 0 & 1 & 0 \end{bmatrix}$$

where Δ means small change in the value. On γ-δ coordinate system, Δv_γ is described as

$$\Delta v_\gamma = 0 \qquad (8)$$

Furthermore, approximation (9) can be obtained under the condition that electrical time constant of the motor is much smaller than mechanical time constant.

$$p\begin{bmatrix} \Delta i_\gamma & \Delta i_\delta \end{bmatrix} = \mathbf{0} \qquad (9)$$

A. Conventional method

A conventional method [7] uses state feedback control based on γ-δ coordinate system as shown in Fig. 3. The δ-axis current i_δ is used to compensate the reference electric angular velocity ω_e^* and is used for stabilization. The block of MTPA Control is used to achieve high efficiency drive and is not used for stabilization.

The equation of state of the conventional method is described as

$$p\mathbf{x} = \mathbf{A}\mathbf{x} + \mathbf{B}(\mathbf{u} - \mathbf{F}\mathbf{x})$$
$$= (\mathbf{A} - \mathbf{B}\mathbf{F})\mathbf{x} + \mathbf{b}\mathbf{u} = \hat{\mathbf{A}}\mathbf{x} + \mathbf{B}\mathbf{u} \qquad (10)$$

where $\mathbf{F}$ is the feedback gain matrix as (11).

$$\mathbf{F} = \begin{bmatrix} 0 & 0 & 0 & 0 \\ 0 & 0 & 0 & 0 \\ 0 & 0 & 0 & 0 \\ 0 & k_\omega & 0 & 0 \end{bmatrix} \qquad (11)$$

(10) can be transformed to (12) using (8) and (9).

$$p\mathbf{x}_2 = \hat{\mathbf{A}}_2 \mathbf{x}_2 + \mathbf{B}_2 \mathbf{u}_2 \qquad (12)$$

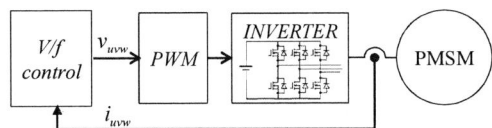

Fig. 1. Fundamental block diagram of V/f control of PMSM

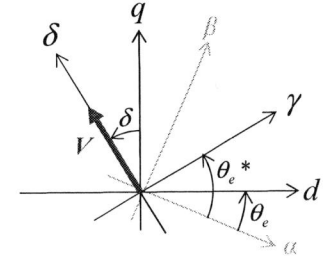

Fig. 2. Definitions of d-q and γ-δ coordinate system.

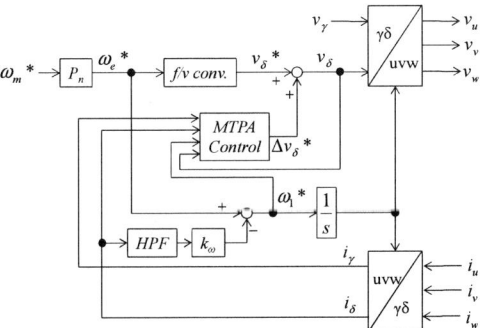

Fig. 3. Block diagram of the conventional method.

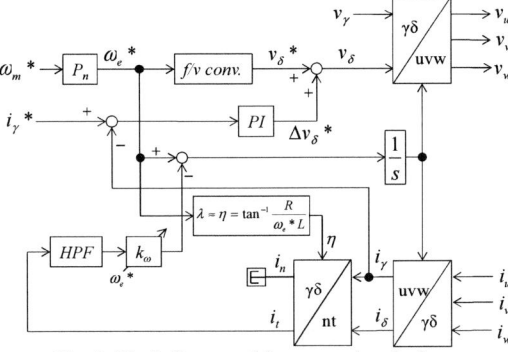

Fig. 4. Block diagram of the proposed method.

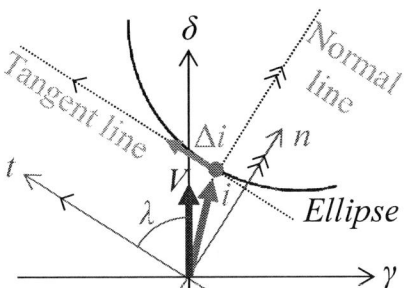

Fig. 5. Definitions of n-t coordinate system.

where matrix $\hat{\mathbf{A}}_2$ is represented by

$$\hat{\mathbf{A}}_2 = \begin{bmatrix} \hat{a}_{11} & \hat{a}_{12} \\ \hat{a}_{21} & \hat{a}_{22} \end{bmatrix} \tag{13}$$

$$\hat{a}_{11} = a_{11} + \frac{P_n^2 K_e^2}{JLE'} k_\omega I_{qo} \cos \delta_o \tag{14}$$

$$\hat{a}_{12} = a_{12} - k_\omega I_{do} \cos \delta_o \frac{P_n^2 K_e^2}{JLE'} \omega_{eo} \tag{15}$$

$$\hat{a}_{21} = a_{21} - k_\omega \frac{K_e}{LE'} \left(-\frac{R}{L} \cos \delta_o + \omega_{eo} * \sin \delta_o \right) \tag{16}$$

$$\hat{a}_{22} = -k_\omega \frac{K_e \omega_{eo}}{LE'} X \tag{17}$$

$$X = R \sin \delta_o / L + \omega_{eo} * \cos \delta_o \tag{18}$$

$$E' = E - k_\omega \left(RI_{yo} / L + k_\omega \omega_{eo} * I_{\delta o} \right)$$

System poles are obtained by

$$s = \frac{\hat{a}_{11} + \hat{a}_{22}}{2} \pm j \sqrt{-\hat{a}_{12}\hat{a}_{21} - \left(\frac{\hat{a}_{11} + \hat{a}_{22}}{2} \right)^2} \tag{19}$$

Stability of the system depends on a real part of the poles. From (19), the real part of the poles of V/f control is mainly decided by $\hat{a}_{22}$ because $\hat{a}_{11}$ is very small. X in $\hat{a}_{22}$ contains reference electric angular velocity $\omega_{eo}*$. We suppose this characteristic causes system poles movement to the unstable plane and detarioration of the disturbance suppression in low speed.

B. Proposed method

On the other hand, the proposed method uses state feedback control based on n-t coordinate system [8]-[9] as shown in Fig. 4. The t-axis current i_t is used to compensate the reference electric angular velocity ω_e*. n-t coordinate system is illustrated as Fig. 5. The n-axis is a normal line of a constant voltage ellipse, and the t-axis is a tangent line of the ellipse. Δi is the small current vector by change in small load. Δi is on the t-axis for following two reasons: First, change in load cause speed error of ω_e and ω_e*. As a result, δ is changed by the speed error as shown in (3). Second, the ellipse is a locus due to the change of δ.

λ is the angle of δ–axis and t–axis and described as (20). Here, approximation (21) is used under the condition of $K_e \gg Li_q$ and $i_d=0$.

$$\lambda = \eta - \delta_o \approx \eta \tag{20}$$

$$\eta = \tan^{-1} \frac{R}{\omega_{eo} * L}$$

$$\delta_o = \tan^{-1} \frac{-RI_{do} + \omega_{eo} Li_{qo}}{Ri_{qo} + \omega_{eo}(Li_{do} + K_e)} \approx \tan^{-1} \frac{Li_{qo}}{K_e} \approx 0 \tag{21}$$

Fig. 6 illustrates η by change in rotational speed of a test motor, which parameters are displayed in Table I. In high speed, η is close to 15° and Δi is close to the δ-axis. However, in low speed, η is close to 90° and Δi is not close to δ-axis. The problem of the conventional method in low speed is attributed to this characteristic of Δi, and the same characteristic also cause a performance improvement in the proposed method.

TABLE I
FUNDAMENTAL PHYSICAL CONSTANTS

Symbol	Meaning	Value
P_n	Pole pairs	5
R	Stator resistance	50 mΩ
L	Inductance	60μH
K_e	Back EMF constant	3.2mV/(rad/s)
J	Moment of inertia	2.5e^{-4}kgm^2
T_s	Sampling period	50μs
F_s	Inverter carrier frequency	20kHz

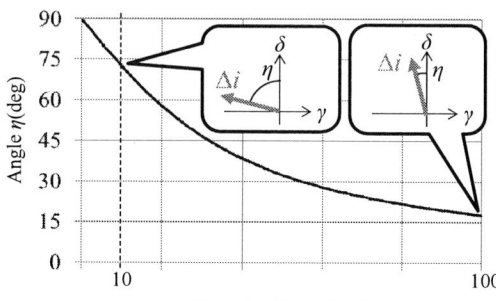

Fig. 6. Relation of the rotational speed and angle η.

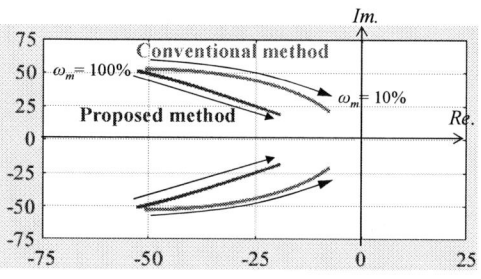

Fig. 7. Root locus. (k_ω=2)

The equation of state of V/f control based on n-t coordinate system can be obtained using (7) and rotation matrix $\mathbf{R}$, represented as (22).

$$\mathbf{R} = \begin{bmatrix} \cos\eta & \sin\eta & 0 & 0 \\ -\sin\eta & \cos\eta & 0 & 0 \\ 0 & 0 & 1 & 0 \\ 0 & 0 & 0 & 1 \end{bmatrix} \tag{22}$$

The equation of state of the proposed method is obtained as same as (10). Besides, the equation can be transformed as same as (12). The term $\hat{a}_{22}$ in the proposed method is described as

$$\hat{a}_{22} = -k_\omega \frac{K_e \omega_{eo}}{LE''} X' \tag{23}$$

$$X' = \sqrt{(R/L)^2 + \omega_{eo} *^2} \cos \delta_o \tag{24}$$

$$E'' = E - k_\omega \left(RI_{no} / L - \omega_{eo} * I_{to} \right)$$

As mentioned earlier, the real part of the poles are mainly decided by $\hat{a}_{22}$. X' can be transformed to (25) using (21) and (24).

$$X' = \sqrt{(R/L)^2 + \omega_{eo} *^2} \tag{25}$$

Then, X' gets closer to R/L with a decrease in rotational speed of the motor. Meanwhile, X in the conventional

The 2018 International Power Electronics Conference

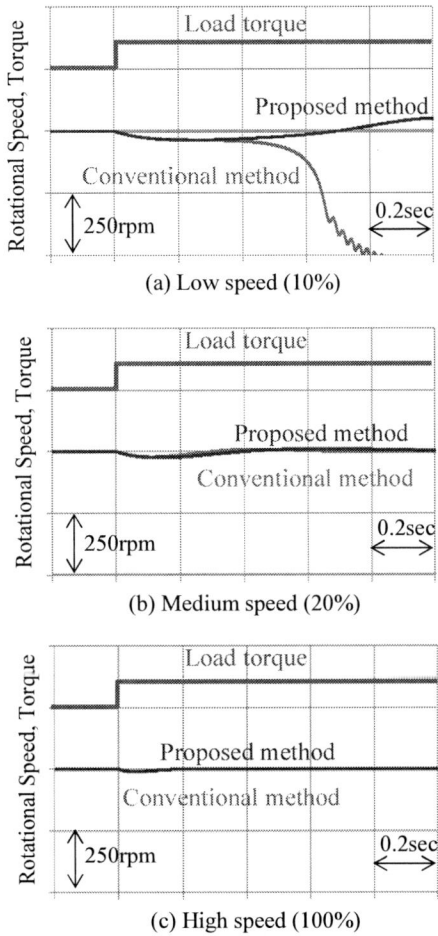

(a) Low speed (10%)

(b) Medium speed (20%)

(c) High speed (100%)

Fig. 8. Simulation results.

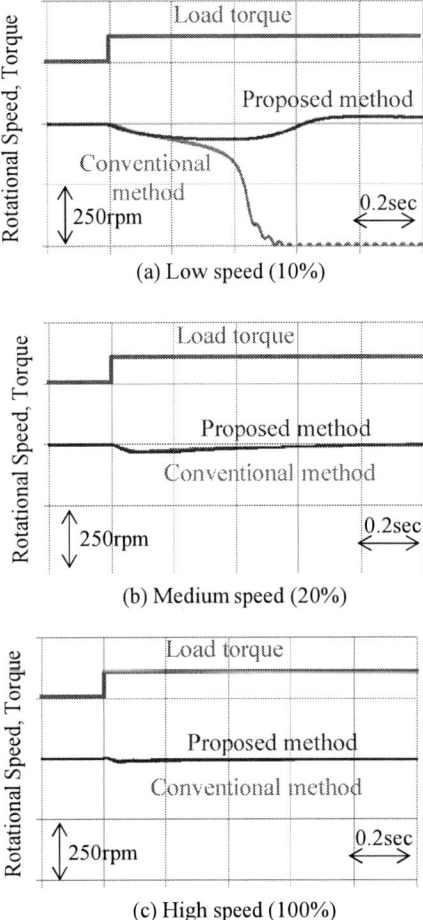

(a) Low speed (10%)

(b) Medium speed (20%)

(c) High speed (100%)

Fig. 9. Experimental results.

method gets closer to 0 with a decrease in rotational speed because X can be transformed to (26) using (18) and (21).

$$X = \omega_{eo}* \tag{26}$$

Fig. 7 shows the root locus of the methods. The poles move to the right a with decrease in rotational speed. However, the poles of the proposed method are always to the left of the poles of the conventional method. This result means that the proposed method is stable compared with the conventional method especially in low speed. Therefore, the proposed method can solve the problem in the conventional method; low disturbance suppression in low speed. This result is caused by the difference between X' and X.

III. SIMULATION AND EXPERIMENTAL RESULTS

Simulation results are illustrated as Fig. 8. Fig.8(a) is the results in low speed, and the conventional method steps out under load variation. In contrast, the proposed method can keep driving in low speed. Fig. 8(b) and (c) are the results in medium speed and high speed and load variation are the same as Fig. 8(a). Both methods can keep driving under load variation in medium speed and

high speed. Moreover, transient speed error under load variation increase with decrease in rotational speed. This trend occurs in both of the conventional method and the proposed method. We assume that this is caused by the poles movement in Fig. 7 and remaining ω_e in $\hat{a}_{22}$, as shown in (23). The proposed method can improve the performance compared with the proposed method, however there is still room for improvement in low speed.

Experimental results are illustrated as Fig. 9. The results are approximately the same as the simulation results.

IV. CONCLUSIONS

In this paper, we propose a new V/f control method for PMSM. The proposed method using state feedback control based on a n-t coordinate system. The proposed method improves the stability and the disturbance suppression, and can keep driving under load variation in low speed. The simulation and experimental results verified the effectiveness of the proposed method.

REFERENCES

[1] M. Ball and M. Wietschel, "The Hydrogen Economy: Opportunities and Challenges," *Cambridge University Press,* 2010.

[2] S. Sumita, K. Tobari, S, Aoyagi and D. maeda, "A simplified Sensorless Vector Control Based pm Average DC Bus Current for Fan Motor," *IEEJ Trans. on Industry Applications*, vol. 130, no. 11, pp. 1233-1240, 2010.

[3] M.J. Corley and R.D. Lorenz, "Rotor Position and Velocity Estimation for a Salient-Pole Permanent Magnet Synchronous Machine at Standstill and High Speeds," *IEEE Trans. on Industry Applications*, vol. 34, no. 4, pp. 784-789, 1998.

[4] S. Ichikawa, Z. Chen, M. Tomita, S. Doki and S. Okuma, "Sensorless Controls of Salient-Pole Permanent Magnet Synchronous Motors Using Extended Electromotive Force Models," *IEEJ Trans. on Industry Applications*, vol. 122, no. 12, pp. 1088-1096, 2002.

[5] K. Sakamoto, Y. Iwaji and T. Endo, "A Simplified Vector Control of Position Sensorless Permanent Magnet Synchronous Motor for Electrical Household Appliances," *IEEJ Trans. on Industry Applications*, vol. 124, no. 11, pp. 1133-1140, 2004.

[6] J. Ito, J. Toyosaki and H. Ohsawa, "High Performance V/f Control Method for PM Motor," *IEEJ Trans. on Industry Applications*, vol. 122, no. 3, pp. 253-259, 2002.

[7] H.M. Tuan, M. Kato, T. Toi and J. Ito, "A Comparison on V/f Control and Sensorless Field Oriented Control based on Measuring Power Consumption of Air Conditioner," *2017 Annual Meeting of The Institute of Electrical Engineers of japan*, 4-152, pp. 262-263, 2017.

[8] Y. Matsuki and H. Kajiura, "High Response Torque Control of IPMSM Based on a New Coordinate System Suitable for Voltage Amplitude and Phase Control," *IEEJ Trans. on Industry Applications*, vol. 135, no. 11, pp. 1123-1129, 2015.

[9] Y. Matsuki and S. Doki, "High response voltage phase torque control of IPMSM for both linear and over-modulation range of inverter," *2017 19th European Conference on Power Electronics and Applications (EPE'17 ECCE Europe)*, pp. 1-7, 2017.

The 2018 International Power Electronics Conference

Stabilization Method Using Equivalent Resistance Gain Based on V/f Control

for IPMSM with Long Electrical Time Constant

Jun-ichi Itoh, Takato Toi, and Koroku Nishizawa
Department of Electrical, Electronics and Information Engineering
Nagaoka University of Technology
Nagaoka, Niigata, Japan
itoh@vos.nagaokaut.ac.jp, t_toi@ stn.nagaokaut.ac.jp, koroku_nishizawa@stn.nagaokaut.ac.jp

Abstract— This paper proposes a novel feedback control loop based on a damping control in a V/f control in order to stabilize interior permanent magnet synchronous motors (IPMSMs) with a long electrical time constant. A problem of the conventional damping control is that ignored roots move to the unstable region due to the conventional damping gain K_1. In addition, the ignored roots are apt to become unstable because of its long electrical time constant. Therefore, a novel method is proposed in order to solve this instability problem. In this paper, first, a boundary condition of stable region is derived based on state equation. Then, a novel current feedback loop of the current is added to an output voltage command. As experimental results, the motor becomes unstable with the conventional damping control under a rated speed of 0.9p.u. and a rated torque of 0.7p.u. Under the common operation condition, the motor is stabilized by employing the novel feedback control loop.

Keywords— *Damping control; IPMSM; Root locus; V/f control*

I. INTRODUCTION

Recently, interior permanent magnet synchronous motors (IPMSMs) are widely utilized due to their high efficiency and high power density [1–6]. V/f control is generally employed for applications such as pumps and fans, where high dynamic performance is not demanded, because of its effectiveness and simplicity instead of a sensorless field-oriented-control [7–8].

In case that an open-loop V/f control without any feedback loop is applied for an IPMSM, persistent oscillation occurs in motor speed. Thus, the control system becomes unstable due to this oscillation. Therefore, the employment of the feedback loop of an active current is proposed as one of stabilizing methods [9]. By suppressing the oscillation with this damping control, stable operation is achieved. In this method, the parameters of the damping control are decided by equations which are acquired from second-order state equation. In addition, the parameters of the damping control are calculated by an auto-tuning method. With this auto-tuning method, the motor parameters are not

necessary in advance. However, due to the influence of ignored roots, this design method still cannot stabilize some IPMSMs because these roots move to the right in the s-plane in accordance with the increase of the conventional feedback loop gain K_1. In addition, the ignored roots are apt to become unstable in the long electrical time constant. In particular, the electrical time constant of a high-speed motor becomes long because of its small winding resistance. Therefore, a high-speed motor tends to become unstable based on V/f control with the conventional damping control using active current feedback.

In this paper, a novel feedback loop is proposed in order to solve the instability problem. The originality of this paper is an additional gain K_2 which is the gain in the proposed feedback loop in order to stabilize the high speed IPMSMs by increasing the winding resistance equivalently. The contribution of this paper is that the application of the V/f control is enlarged to the long electrical time constant IPMSM by adding a feedback loop.

This paper is organized as follows; first, the conventional damping control is introduced. Next, an auto-tuning method for the conventional damping control is explained. In addition, the stability of the V/f control is analyzed in order to clarify the boundary condition of the unstable region with the conventional damping control method. Next, the stability analysis is conducted with proposed feedback loop. From the experimental results, the equivalent resistance gain K_2 stabilizes the unstable V/f control system.

II. V/F CONTROL FOR IPMSM

A. Damping Control Based on V/f Control

Fig. 1 shows the relation between $\gamma\delta$-frame and dq-frame. The d-axis is defined as the direction of the flux vector of the permanent magnet. The q-axis is defined as the electro motive force vector. On the other hand, the V/f control is implemented on the $\gamma\delta$-frame. The δ-axis is aligned with the direction of the inverter output voltage, whereas the γ-axis is defined as the δ-axis delayed by 90 degrees.

Fig. 2 shows the V/f control block diagram with the conventional damping control. The V/f control is based on the γδ-frame. The constant oscillation occurs in the high-speed region due to the resonance between the inertia of the motor and the inductance when the IPMSMs are driven by the open loop V/f control. Thus, the δ-axis current, which represents the active component, is utilized in order to stabilize the oscillation.

B. Conventional Design Method for Damping Control Based on V/f Control

First, in order to analyze the stability of the V/f control, the IPMSM model at steady state is linearized. The linearized fifth-order state equation is expressed by (1) below.

Here, i_γ is the γ-axis current, i_δ is the δ-axis current, ω is the motor speed, ω^* is the command value of motor speed, δ is the load angle, x is the output of integrator from high pass filter (HPF), p is the differential operator, R is the winding resistance, ψ_m is the field flux linkage, V_f is the f/v conversion ratio, K_1 is the damping gain, τ is the time constant of HPF, P_f is the number of pole pairs, and J is the inertia of the motor. In addition, L_0 and L_1 are defined as

$$L_0 = \frac{L_d + L_q}{2} \quad \text{.. (2),}$$

$$L_1 = \frac{L_q - L_d}{2} \quad \text{.. (3).}$$

The state variables are i_γ, i_δ, ω, δ, and x. Root locus is obtained from fifth-order state equation.

Table I shows the motor parameters of two IPMSMs. In this paper, the stability of the two motors is analyzed. Note that the motor with a rated power of 3.7 kW is defined as motor A, and the other is motor B [10]. In addition, electrical time constants %X_{Ld} / %R and %X_{Lq} / %R of motor B are longer than that of motor A by 7.6 times and 3.4 times.

Fig. 3 shows the eigenvalue plot of the motor A when

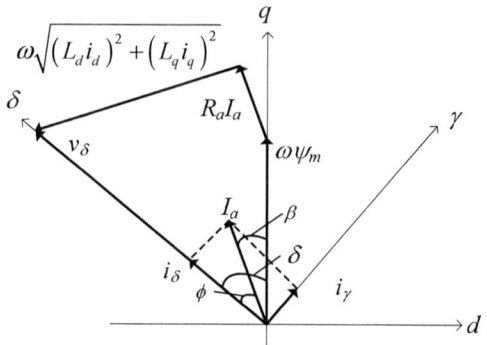

Fig. 1. Relation between γδ-frame and dq-frame.

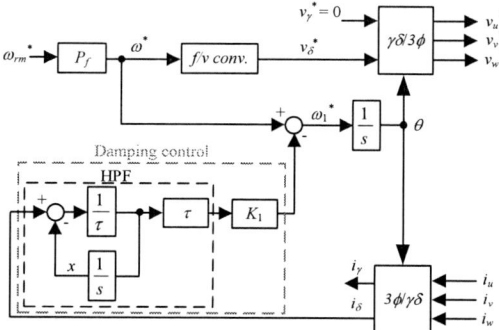

Fig. 2. Block diagram of V/f control based on γδ-frame with conventional damping control.

the damping gain K_1 is varied at rated speed and no load. Note that No. 1-5 represent eigenvalues of fifth order state equation. In the conventional design method, the damping gain K_1 is decided to become multiple roots of No. 2 and No. 3 in order to suppress the overshoot of the motor speed. As shown in Fig. 3, the value of 0.15p.u. is the optimum value for the damping gain K_1. In addition, all roots are located in the left half of the s-plane; therefore, the system is stable.

From fifth-order state equation, the approximated second-order state equation is expressed by (4).

$$(1)$$

$$p\begin{bmatrix} \Delta\omega \\ \Delta\delta \end{bmatrix} = \begin{bmatrix} 0 & \dfrac{3}{2}\dfrac{P_f^2\psi_m^2}{JL_q} \\ -1 & -K_1\dfrac{\psi_m}{L_q} \end{bmatrix}\begin{bmatrix} \Delta\omega \\ \Delta\delta \end{bmatrix} + \begin{bmatrix} 0 \\ 1 \end{bmatrix}\Delta\omega^* \quad\ldots\ldots\ldots(4)$$

It is noted that second-order state equation is derived under the condition of high-speed region and no load in order to simplify the analysis of the stability because the influence of load is relatively smaller than the influence of motor speed [11–12].

From (4), characteristic equation is expressed

$$s^2 + K_1\frac{\psi_m}{L_q}s + \frac{3}{2}\frac{P_f^2\psi_m^2}{JL_q} = 0 \quad\ldots\ldots\ldots(5).$$

where s is the Laplace operator.

Then, the damping coefficient ζ and the natural angular frequency ω_n are expressed by (6) and (7).

$$\zeta = \frac{\psi_m K_1}{2\omega_n L_q} \quad\ldots\ldots\ldots(6)$$

$$\omega_n = \sqrt{\frac{3}{2}}\frac{P_f\psi_m}{\sqrt{JL_q}} \quad\ldots\ldots\ldots(7)$$

As mentioned before, the damping gain K_1 is set in order to become multiple roots of No. 2 and No. 3.

On the other hand, the cutoff frequency ω_c of the HPF needs to be lower than the natural angular frequency ω_n in order to suppress the oscillation of the motor speed.

Fig. 4 shows the eigenvalue plot of the Motor A when the damping gain K_1 is varied under different cutoff frequency ω_c at rated speed and no load. In the figure, the root of No. 5 is on the real axis regardless of the damping gain K_1 when the cutoff frequency is set as 1/20 of the natural angular frequency ω_n.

From above considerations, the damping gain K_1 and the cutoff frequency ω_c are expressed by (8) and (9).

$$K_1 = \frac{2\omega_n L_q}{\psi_m} \quad\ldots\ldots\ldots(8)$$

$$\omega_c = \frac{\omega_n}{20} = \frac{1}{20}\sqrt{\frac{3}{2}}\frac{P_f\psi_m}{\sqrt{JL_q}} \quad\ldots\ldots\ldots(9)$$

It is noted that the damping coefficient ζ is set as 1 in (8).

C. Auto-tuning Method for Parameters of Conventional Damping Control

In this section, an auto-tuning method for the conventional damping control is explained. In the auto-tuning, the parameters are identified in order to calculate the damping gain K_1 and cutoff frequency ω_c based on the equations (8) and (9).

The voltage command v_δ is expressed by

$$v_\delta^2 = \left(\omega\psi_m + RI_a\right)^2 + \left(\omega L_q I_a\right)^2 \quad\ldots\ldots\ldots(10).$$

It is noted that the equation (10) is derived when $i_d = 0$ control is achieved.

Table I.
Motor parameters of two IPMSMs.

	Motor A	Motor B
Rated mechanical power P_m	3.7 kW	3 kW
Rated speed ω_n	1800 r/min	12000 r/min
Rated current I_n	14 A	17.3 A
Number of pole pairs P_f	3	2
d-axis inductance L_d	6.2 mH	2.04 mH
q-axis inductance L_q	15.3mH	2.24 mH
Winding resistance R	0.69 Ω	0.133 Ω
Field flux linkage ψ_m	0.27 Vs/rad	0.107 Vs/rad
Inertia moment J	0.037 kgm²	0.0013 kgm²
Electrical time constant $\%X_{Ld}/\%R$	5.05	38.5
Electrical time constant $\%X_{Lq}/\%R$	12.5	42.3

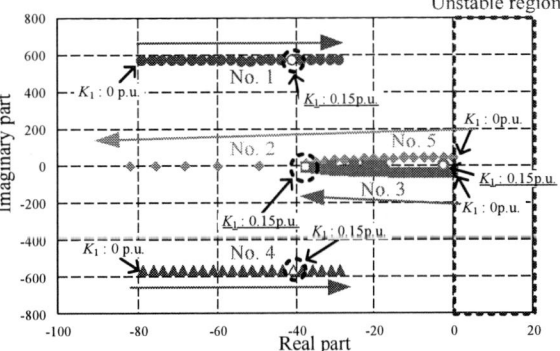

Fig. 3 Loot locus of fifth-order state equation when damping gain K_1 is increased at no-load and rated speed with motor A.

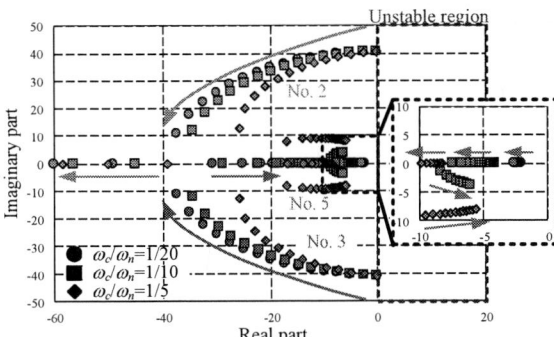

Fig. 4. Loot locus of fifth-order state equation under different cutoff frequency ω_c at no-load and rated speed.

From (10), the field flux linkage ψ_m is identified at the steady state by using

$$\hat{\psi}_m = \frac{\sqrt{v_\delta^2 - \left(\omega^*\hat{L}_q I_a\right)^2} - \hat{R}I_a}{\omega^*} \quad\ldots\ldots\ldots(11).$$

Here, v_δ is the voltage command, $\hat{L}_q$ is the identified value of q-axis inductance, $\hat{R}$ is the identified value of winding resistance, I_a is the output current.

Next, the winding resistance R is identified in DC test at a standstill as (12).

2231

$$\hat{R} = \frac{V_{DC}\left(D - f_{sw}T_d\right)}{1.5I_u} \quad\text{............................(12)}$$

Here, V_{DC} is the DC-link voltage, D is the duty ratio, f_{sw} is the switching frequency, T_d is the dead time, and I_u is the U-phase current at the steady state.

Next, the q-axis inductance L_q is derived from the relation of the reactive power on the dq-axis and γδ-axis.

The reactive power on the dq-axis is expressed by (13) under $i_d = 0$ control.

$$Q_{dq} = \omega L_q I_a^2 \quad\text{..............................(13)}$$

On the other hand, the reactive power on the γδ-axis is expressed by (14)

$$Q_{\gamma\delta} = v_\delta i_\gamma \quad\text{................................(14).}$$

The reactive power on each axis is corresponded. Therefore, the identified q-axis inductance is expressed as (15) from (13) and (14).

$$\hat{L}_q = \frac{v_\delta i_\gamma}{\omega^* I_a^2} \quad\text{..............................(15)}$$

The natural angular frequency is identified based on the hill-climbing method by injecting the sinusoidal wave into the speed command. The output current becomes the maximum value at the natural angular frequency. Therefore, the frequency of the injected sinusoidal wave is varied in order to search the maximum value of the output current.

Fig. 5 shows the flowchart of the auto-tuning. First, the winding resistance R is identified in DC test at standstill. Then, the motor is accelerated up to 0.5p.u. because it is impossible to identify the field flux linkage ψ_m and the q-axis inductance L_q at standstill as shown in (11) and (15). Next, the q-axis inductance L_q and the field flux linkage ψ_m are identified after applying maximum torque per ampere (MTPA) control based on the hill-climbing method [13]. The motor parameters are not necessary in the MTPA control because the operation point is searched by the relation of the output current. In addition, the natural angular frequency is identified by the relation between the frequency of the injected sinusoidal wave and the magnitude of the output current based on the hill-climbing method.

Fig. 6 shows the block diagram of the V/f control during the auto-tuning. From the figure, the motor is accelerated without the HPF of the damping control.

With this auto-tuning for the conventional damping control, the motor A, which is stable with the conventional damping control, is stabilized.

D. Unstable Condition with Conventional Damping Control

Fig. 7 shows the eigenvalue plot of the motor B when the damping gain K_1 is varied at the same condition of rated speed and no load. As shown in Fig. 7, the roots of No. 2 and No. 3 are multiple roots when the damping gain K_1 is 0.05p.u. However, the roots of No.1 and No.4

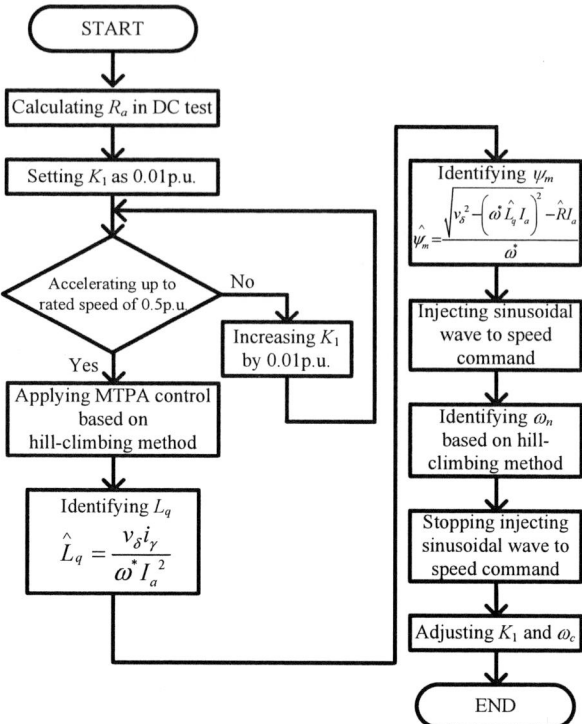

Fig. 5 Flowchart of auto-tuning method for conventional damping control based on V/f control.

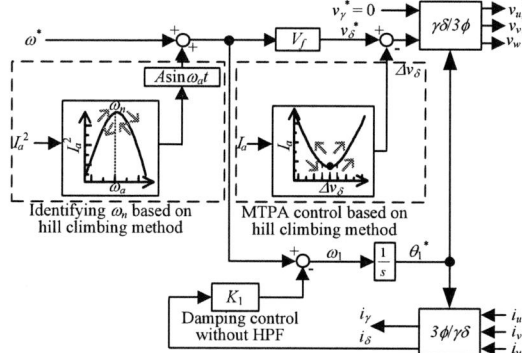

Fig. 6 Block diagram of V/f control when tuning parameters.

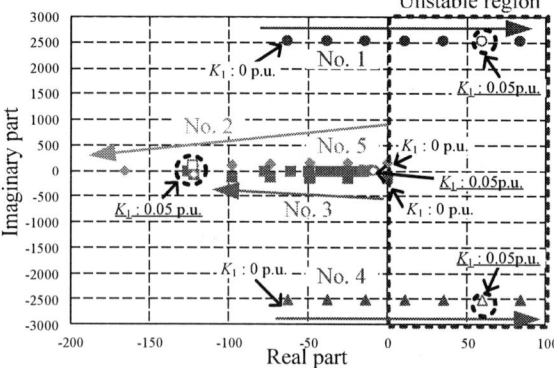

Fig. 7 Loot locus of fifth-order state equation when damping gain K_1 is increased at no-load and rated speed with motor B.

move into the right half of the s-plane. Therefore, these roots make the system unstable.

It is concluded from the above considerations that the motor B becomes unstable due to the roots of No. 1 and No. 4 when the damping gain K_1 is decided by the conventional design method. It is noted that these roots are ignored when the fifth order state equation is linearized to second-order state equation.

E. Boundary Condition of Conventional Damping Control

In this section, the unstable condition with only the damping gain K_1 is derived. As mentioned before, the analysis is conducted at no load condition in order to simplify the analysis of the stability.

First, the fifth-order state equation is linearized into fourth-order state equation when the cutoff frequency is much smaller than the natural angular frequency of the second-order state equation. Then, the characteristic equation is expressed by (16).

$$s^4 + a_3 s^3 + a_2 s^2 + a_1 s + a_0 = 0 \quad \text{...............................(16)}$$

where

$$a_3 = \frac{R}{L_d} + \frac{R}{L_q} \quad \text{..(17)},$$

$$a_2 = \omega_0^2 + \frac{3 P_f^2 \psi_m^2}{2 J L_q} + \frac{R^2}{L_d L_q} \quad \text{.............(18)},$$

$$a_1 = K_1 \frac{\omega_0^2 \psi_m}{L_q} + \frac{3 P_f^2 \psi_m^2 R}{2 J L_d L_q} \quad \text{.................(19)},$$

$$a_0 = \frac{3 P_f^2 \omega_0^2 \psi_m^2}{2 J L_q} \quad \text{...................................(20)}.$$

Here, ω_0 is the steady state value of the motor speed.

From the equations (16)–(20) that the coefficients of the fourth-order characteristic equation are positive regardless of the value of the motor parameters. Therefore, the necessary condition of the Routh-Hurwitz stability criterion is satisfied. Thus, the unstable condition is derived from the Routh table.

Table II shows the Routh table which is acquired from the equations (16)–(20). The coefficients of the left end column are focused in order to evaluate the stability of the system. The unstable condition, where the coefficients b_1 and c_1 are negative, is expressed as (21) and (22).

$$\frac{L_d}{R} + \frac{L_d}{R} \frac{\omega_n^2}{\omega_0^2} + \frac{R}{L_q} \frac{1}{\omega_0^2} - \frac{\psi_m}{L_d + L_q} \left(\frac{L_d}{R} \right)^2 K_1$$
$$- \frac{3}{2} \frac{P_f^2 \psi_m^2}{J(L_d + L_q)} \frac{L_d}{R} \frac{1}{\omega_0^2} < 0$$
$$\text{..(21)}$$

$$b_1' \left\{ \frac{\psi_m}{R} K_1 + \frac{L_q}{L_d} \left(\frac{\omega_n}{\omega_0} \right)^2 \right\} - \frac{L_d + L_q}{R} \left(\frac{\omega_n}{\omega_0} \right)^2 < 0 \text{ ....(22)}$$

Here, b_1' is the left-side term of the inequality equation (21).

Table II.
Routh table of fourth order state equation of V/f control with only damping gain K_1.

s^4	$a_4 = 1$	$a_2 = \omega_0^2 + \dfrac{3}{2} \dfrac{P_f^2 \psi_m^2}{JL_q} + \dfrac{R^2}{L_d L_q}$	$a_0 = \dfrac{3}{2} \dfrac{\omega_0^2 P_f^2 \psi_m^2}{JL_q}$
s^3	$a_3 = \dfrac{R}{L_d} + \dfrac{R}{L_q}$	$a_1 = K_1 \dfrac{\omega_0^2 \psi_m}{L_q} + \dfrac{3}{2} \dfrac{P_f^2 \psi_m^2}{JL_q} \dfrac{R}{L_d}$	0
s^2	$b_1 = \dfrac{a_3 a_2 - a_4 a_1}{a_3}$	a_0	0
s^1	$c_1 = \dfrac{b_1 a_1 - a_3 b_2}{b_1}$	0	0
s^0	a_0	0	0

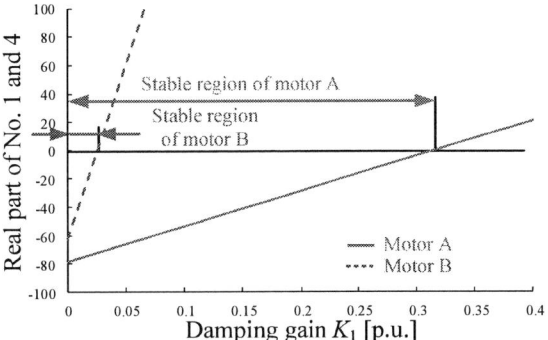

Fig. 8. Real part of No.1 and 4 when damping gain K_1 is increased under rated speed and no load.

It is concluded from the inequality equations (21)(22) and Tables I–II, the left-side term are positive when the damping gain K_1 of 0.15p.u. is designed in the motor A. On the other hand, b_1', which is the left-side term of (21), is negative when the damping gain K_1 of 0.05p.u. is employed in the motor B. Therefore, the system is unstable. These consequences are corresponded to the eigenvalue plots as shown in Fig. 3 and 6.

Fig. 8 shows the real parts of the roots of No. 1 and 4 when the damping gain K_1 is varied. The real parts $\alpha_{2,3}$ of No. 2 and No. 3 are expressed

$$\alpha_{2,3} = -\frac{K_1 \psi_m}{2 L_q} \quad \text{...(23)}.$$

Then, the real parts $\alpha_{1,4}$ of No. 1 and No. 4 are expressed

$$\alpha_{1,4} = \frac{K_1 \psi_m}{2 L_q} - \frac{R}{2} \left(\frac{1}{L_d} + \frac{1}{L_q} \right) \quad \text{..............(24)}.$$

The real parts of No. 2 and No. 3 decrease proportionally to the damping gain K_1. On the other hand, the real parts of No. 1 and No. 4 increase proportionally to the damping gain K_1. Therefore, the stable condition is limited depending on the damping gain K_1. In addition, the stable region of the motor B is smaller than that of the motor A. In particular, the high-speed motor with small winding resistance is more apt to become unstable as expressed in (24). It is concluded from above considerations that there is unstable condition depending on the roots of No. 1 and No. 4. Therefore, a novel method is necessary in order to make the roots of No. 1 and No. 4 move into left-half of s-plane.

2233

III. PROPOSED ADDITIONAL FEEDBACK LOOP

A. Principle of Stabilizing Method with Additional Feedback Loop

Fig. 9 shows the control diagram of the V/f control with the novel feedback loop. As shown in Eq. (21), the third term in the left-side term of the inequality equation becomes larger when the winding resistance increases. Furthermore, the negative term, which is the fourth term, becomes smaller when the winding resistance increases. In other words, it is possible to stabilize the system when the term of the winding resistance is increased equivalently. Therefore, a novel feedback loop is added to the inverter voltage command v_δ^*. The feedback loop consists of an additional gain K_2 and the δ-axis current which is filtered by the HPF.

B. Stability Analysis with Additional Feedback Loop

The stability analysis is conducted under the same procedure in chapter II, section C. The unstable condition is derived from the Routh table. It is expressed as

$$\frac{L_d}{R} + \frac{L_d}{R}\frac{\omega_n^2}{\omega_0^2} + \frac{R+K_2}{L_q}\frac{1}{\omega_0^2} - \frac{\psi_m}{(R+K_2)L_d + RL_q}\frac{L_d^2}{R}K_1$$

$$- \frac{3}{2}\frac{P_f^2\psi_m^2}{J}\frac{R}{(R+K_2)L_d + RL_q}\frac{L_d}{R}\frac{1}{\omega_0^2} < 0$$

.... (25),

$$b_1^{\cdot}\left\{\frac{\psi_m}{R}K_1 + \frac{L_q}{L_d}\left(\frac{\omega_n}{\omega_0}\right)^2\right\} - \frac{(R+K_2)L_d + RL_q}{R^2}\left(\frac{\omega_n}{\omega_0}\right)^2 < 0$$

.... (26).

where $b_1^{"}$ is the left-side term of (25).

It is shown in (25) that the additional gain K_2 is added to the numerator of the third term in left-side by adding new feedback loop. In addition, the denominators of the fourth and fifth terms are also increased by the additional gain K_2. Therefore, the winding resistance is equivalently increased by the additional gain K_2 in order to stabilize the system.

Fig. 10 shows the eigenvalue plot when K_2 is increased at the rated speed and no load. It is noted that the damping gain K_1 is decided as multiple roots of No. 2 and No. 3. The roots of No. 1 and No. 4 move into the left side of s-plane in accordance with the increase of the additional gain K_2. This stabilization is not achievable with only damping gain K_1.

IV. EXPERIMENTAL RESULTS

A. Auto-tuning for Conventional Damping Control

In this section, the effectiveness of the auto-tuning method for the conventional damping control is confirmed in the experiment with the motor A. It is noted that the DC test is conducted at standstill. In addition, the other parameters are identified under the condition of a rated speed of 0.5p.u. and a rated torque of 0.27p.u.

Fig. 11 shows the waveform of the DC test. The

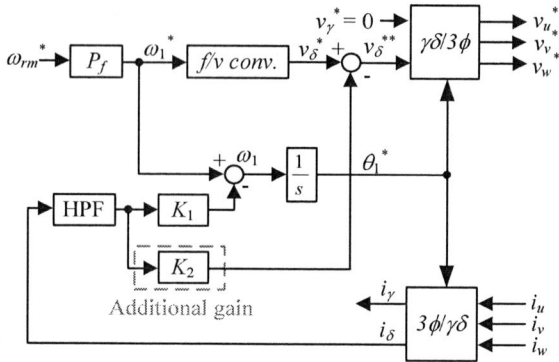

Fig. 9. Block diagram of V/f control based on $\gamma\delta$-frame with proposed feedback loop to output voltage command.

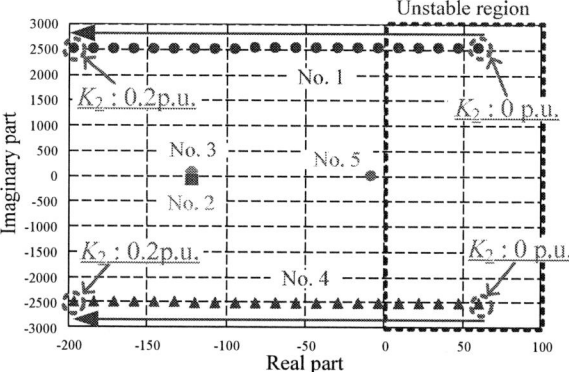

Fig. 10. Root locus when damping gain K_2 is increased under rated speed and no load with motor B.

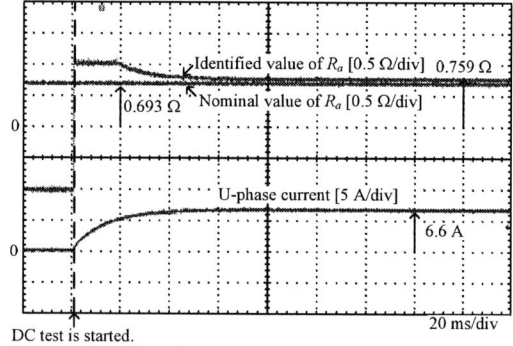

Fig. 11 Waveforms of DC test.

identified value of the winding resistance converges to the nominal value in the steady state. In general, winding resistance of a general-purpose motor is relatively small. Therefore, the duty ratio is set as 5% in order to prevent overcurrent in the test. From the figure, the winding resistance is identified with an error of 9.6%.

Fig. 12 shows each identified parameter and calculated damping gain K_1. From the figure, the identification and the calculation are conducted in 44 s. The frequency of the injected sinusoidal wave is varied each 1 Hz in the identification of the natural angular frequency. In addition, one period of the sinusoidal wave needs to be injected at least. Thus, the injecting time of the sinusoidal wave on each frequency is set as 1 s. Therefore, the tuning time of the natural angular frequency is relatively long compared with the other parameters.

Fig. 13 shows the step response before and after the auto-tuning. In Fig. 13 (a), the damping gain is set as 0.01p.u. In addition, the HPF of the damping control is not utilized. On the other hand, the damping gain K_1 is set as 0.13p.u. and the cutoff frequency is 1/20 of the identified natural angular frequency. In the figure, the motor speed oscillates when the step speed command is employed before applying the auto-tuning. In addition, the steady-state error remains in the motor speed because the HPF is not implemented. On the other hand, the motor speed follows the command without an overshoot.

Fig. 14 shows the step responses with the optimal damping gain and the tuned damping gain. From the figure, the overshoot with the tuned damping gain is almost same as the optimal damping gain.

Through the experimental results, the effectiveness of the auto-tuning method is confirmed.

B. Stabilization with Additional Feedback Loop

In order to confirm the validity of the additional feedback loop, the experiment is conducted. It is noted that Motor A which is shown in Table I is used in the experiment. In addition, inductors of 10 mH are connected in series in order to evaluate the motor with the long time constant.

Fig. 15 shows the waveforms of γ, δ-axis current and U-phase current at 0.9p.u. of rated speed and 0.7p.u. of rated torque when the damping gain K_1 is varied from 0.1p.u. to 0.2p.u. Note that the additional gain K_2 is set as 0 at this condition. The diverging oscillation occurs in γ, δ-axis current after the damping gain K_1 is varied to 0.2p.u. Then, the motor is stopped due to the overcurrent detection. Each frequency of the oscillation is 81 Hz and 78 Hz, whereas, the frequency which is obtained from the eigenvalue plot is 81 Hz. The experimental value agrees with the theoretical value with error of 5.5%. The system becomes unstable due to the influence of the ignored roots depending on the damping gain K_1.

Fig. 16 shows the waveforms of γ, δ-axis current and U-phase current at 0.9p.u. of rated speed and 0.7p.u. of rated torque when the additional gain K_2 is varied. The system is stabilized by the additional gain K_2 even under the unstable condition with only the damping gain K_1. In addition, the current ripple is also suppressed by the additional gain K_2.

From the experimental results, the effectiveness of the additional feedback loop is confirmed.

V. CONCLUSION

In this paper, the designing method for the conventional damping control is explained. In addition, the auto-tuning method is introduced for the motor which is stable with the conventional damping control. Then,

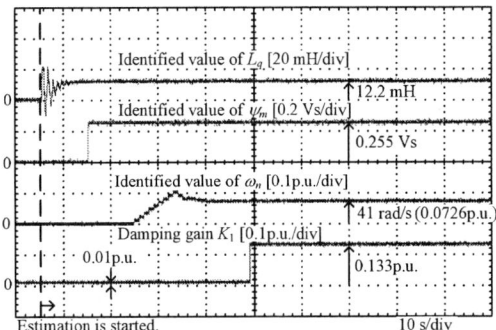

Fig. 12 Waveforms of auto-tuning each parameter.

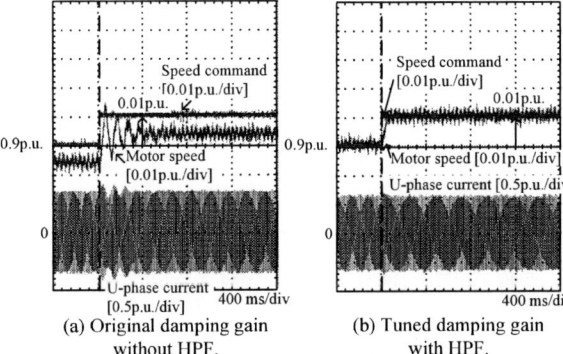

(a) Original damping gain without HPF.　(b) Tuned damping gain with HPF.

Fig. 13 Waveforms of step response before and after auto-tuning under 0.9p.u. of rated speed and 0.8p.u. of rated torque.

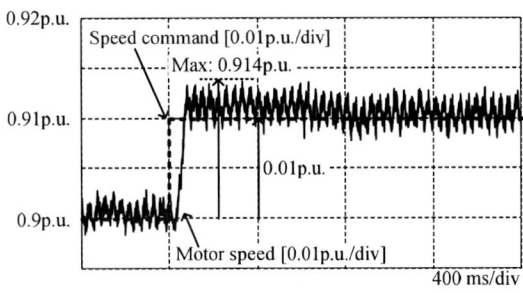

(a) With optimal damping gain ($K_1 = 0.16$p.u.).

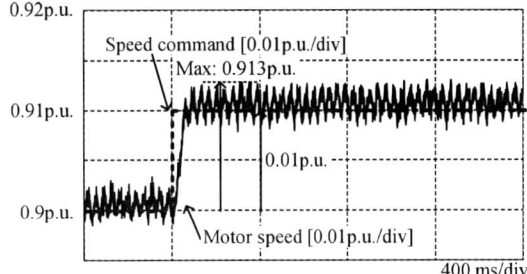

(b) With tuned damping gain ($K_1 = 0.13$p.u.).

Fig. 14 Waveforms of step response with different damping gain K_1 under 0.9p.u. of rated speed and 0.8p.u. of rated torque.

the unstable condition due to the damping gain K_1 is derived based on the state equation regarding the V/f control for IPMSMs. From the unstable condition, it is clarified that the high-speed motor, which the electrical time constant is long, is apt to be unstable. The validity of the analysis is confirmed by the eigenvalue plot and the experiment. In order to solve this instability problem, the novel feedback loop is added to the inverter voltage command in order to increase the winding resistance. As a result, the stable condition is achieved by the additional equivalent resistance gain K_2 even under the unstable condition with only damping gain K_1. In the experiment, first, the effectiveness of the auto-tuning for the conventional damping control is confirmed. Next, the effectiveness of K_2 is confirmed under 0.9p.u. of rated speed and 0.7p.u. of rated torque.

REFERENCES

[1] K. Kondo, S. Doki : "Position Estimation System for PMSM Position Sensorless Control in Inverter Overmodulation Drive", IEEJ J. Industry Applications, Vol.6, No.3, pp.165-172, 2017

[2] R. Takahashi, K. Ohishi, Y. Yokokura, H. Haga, T. Hiwatari : "Stationary Reference Frame Position Sensorless Control Based on Stator Flux Linkage and Sinusoidal Current Tracking Controller for IPMSM", IEEJ J. Industry Applications, Vol.6, No.3, pp.181-191, 2017

[3] A. Shinohara, Y. Inoue, S. Morimoto, M. Sanada : "Correction Method of Reference Flux for Maximum Torque per Ampere Control in Direct-Torque-Controlled IPMSM Drives", IEEJ J. Industry Applications, Vol.6, No.1, pp.12-18, 2017

[4] R. Tanabe, K. Akatsu : "Advanced Torque and Current Control Techniques for PMSMs with a Real-time Simulator Installed Behavior Motor Model", IEEJ J. Industry Applications, Vol.5, No.2, pp.167-173, 2016

[5] X. Ji, T. Noguchi : "Online q-axis Inductance Identification of IPM Synchronous Motor Based on Relationship between Its Parameter Mismatch and Current", IEEJ J. Industry Applications, Vol.4, No.6, pp.730-731, 2015

[6] M. Miyamasu, K. Akatsu : "Efficiency Comparison between Brushless DC Motor and Brushless AC Motor Considering Driving Method and Machine Design", IEEJ J. Industry Applications, Vol. 2, No.1, pp.79-86, 2013

[7] P. D. Chandana Perera, F. Blaabjerg, John K. Pederson, P. Thøgersen : "A sensorless, Stable V/f Control Method for Permanent-Magnet Synchronous Motor Drives", IEEE Trans. on Industry Applications, Vol. 39, No. 3, pp. 783-791, 2003

[8] G. L. Kruger, A. J. Grobler, S. R. Holm : "Improved non-ideality compensation for the V/f controlled permanent magnet synchronous motor", Industrial Technology, pp. 331-336, 2013

[9] J. Itoh, N. Nomura, H. Ohsawa: "A comparison between V/f control and position-sensorless vector control for the permanent magnet synchronous motor", in Proc. Power Conversion Conf., Vol. 3, pp. 1310-1315, 2002

[10] D. Sato, J. Itoh: " Open-loop control for permanent magnet synchronous motor driven by square-wave voltage and stabilization control", Energy Conversion Congress and Exposition, EC-0319, 2016

[11] Motion Control, INTECH, pp. 439-458, 2010

[12] Z. Tang, X. Li, S. Dusmez, B. Akin : "A New V/f Sensorless MTPA Control for IPMSM Drives", IEEE Trans. on Power Electronics, Vol. 31, No. 6, pp. 4400-4415, 2016

[13] J. Itoh, T. Toi, M. Kato : "Maximum Torque per Ampere Control Using Hill Climbing Method Without Motor Parameters Based on V/f Control", European Conference on Power Electronics and Applications, DS3d-Topic40283, 2016

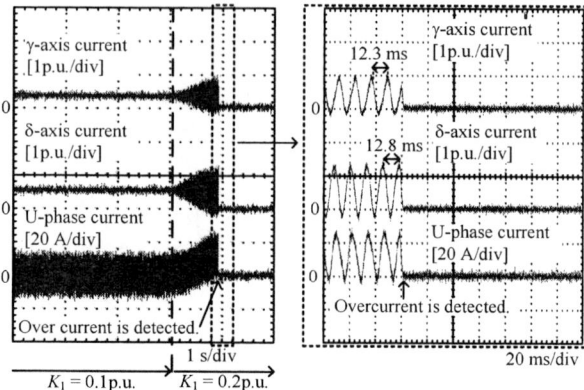

Fig. 15. Operation waveform without feedback loop K_2 when damping gain K_1 is varied from 0.1p.u. to 0.2p.u. under 0.9p.u. of rated speed and 0.7p.u. of rated torque.

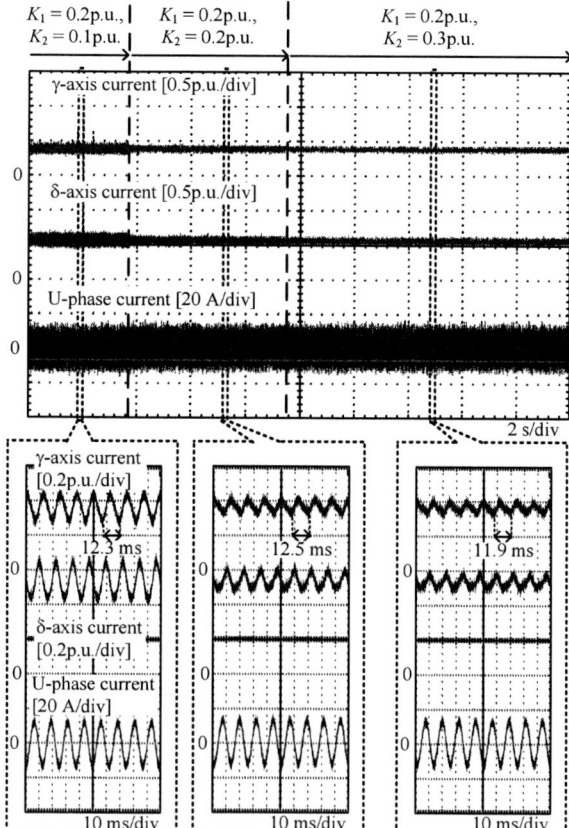

Fig. 16. Waveforms when additional gain K_2 is varied under 0.9p.u. of rated speed and 0.7p.u. of rated torque.

Single-phase Solid-State Transformer using Multi-cell with Automatic Capacitor Voltage Balance Capability

Jun-ichi Itoh, Kazuki Aoyagi, Keisuke Kusaka and Masakazu Adachi
Department of Electrical, Electronics and Information Engineering
Nagaoka University of Technology
Nagaoka, Niigata, Japan
itoh@vos.nagaokaut.ac.jp, aoyagi@stn.nagaokaut.ac.jp, kusaka@vos.nagaokaut.ac.jp,
m_adachi@stn.nagaokaut.ac.jp

Abstract- In this paper, a single-phase solid-state transformer (SST) system based on a multilevel topology using multi-cell is proposed. The proposed SST has an automatic capacitor voltage balancing capability on a primary side due to use of a resonant DC-DC converter. The main contribution of this paper is revealing the fundamental loss design of the proposed topology connected with a 6.6-kV grid. It is predicted that the maximum efficiency of full model SST reaches 99%. The miniature model SST is tested to confirm the fundamental operation with an input voltage of 1320 V, which is 1/5 of the full model. As a result, the sinusoidal input current is obtained with a total harmonic distortion of 4.3%. Besides, the bidirectional operation is verified. Then, it is confirmed that the primary side capacitor voltage of each cell is kept constant and balanced without a voltage balance control. Moreover, the loss analysis is derived in each part and compared with that of the experimental result. The error of the loss between the experimental result and the calculation is less than 5%.

Keywords— Solid-State Transformer; Power factor correction converter; Resonant DC-DC converter; High-frequency transformer;

I. INTRODUCTION

Recently, a DC distribution system has been actively researched in order to achieve the energy-saving of the data-center or the large building [1] [2]. The DC distribution system is possible to achieve down-sizing and high efficiency compared with an AC distribution system [3]. Moreover, a 6.6-kV AC power grid is employed as one of the power sources.

In general, a transformer is applied between the AC power grid and the DC distribution system in order to achieve the step-down from 6.6 kV to the distribution voltage of several hundred volts and obtain galvanic isolation between the AC power grid and the DC distribution system [4] [5]. However, the conventional transformers are bulky and heavy because the transformers operate at a grid frequency, e.g., 50 Hz or 60 Hz.

As one of the solutions, a transformer-less converter is proposed [6]. In the five-level diode-clamped converter, the high voltage IGBTs with the voltage rating more than 4.5 kV are required. In addition, the balancing circuit is required as the auxiliary circuit in order to correct the unbalanced voltage among four capacitors in the DC link. Besides, it is operated by low switching frequency because the loss of high voltage rating devices is large compared to low voltage rating devices. Consequently, large passive components are required due to low switching frequency in order to suppress harmonic distortion of input current and voltage ripple of output DC voltage.

As another solution, solid-state transformers (SST) have been attracting attention [7]–[9]. SST is possible to significantly reduce the system volume compared to the conventional transformer by introducing high-frequency switching with the spread use of silicon carbide (SiC) and gallium nitride (GaN) devices [10]. SST simultaneously achieves the insulation and the step-down function by using a high-frequency transformer. To sum up, SST has following advantages [11]:

· Reduced size and weight of the system
· Harmonic suppression
· Active/ Reactive power control
· AC voltage adjustment
· DC voltage output

Besides, it is possible to compose the cell converters based on multilevel topologies, which enable to reduce a required voltage rating of the switching devices. The cell converters allow using switching devices with low on-state resistance and high-speed switching. In addition, the switching frequency is equivalently increased. Thus, the inductor volume can also be reduced [12]. Consequently, SST achieves high voltage rating with high switching frequency by multilevel topologies.

However, the number of the switches still increases greatly due to the multistage cell. Moreover, the control system, which drives the main circuit, is complicated because the number of the gate signal is increased with increased number of the switch [13]. Moreover, a balance control for each capacitor on the cells is required in the multilevel system, e.g., modular multilevel converter (MMC) [14]. The balance control may cause an unstable if a feedback control has a delay due to isolation or signal transmission [15]. Furthermore, the DC link capacitor with a large capacitance is required in order to maintain the constant capacitor voltage in each cell [16].

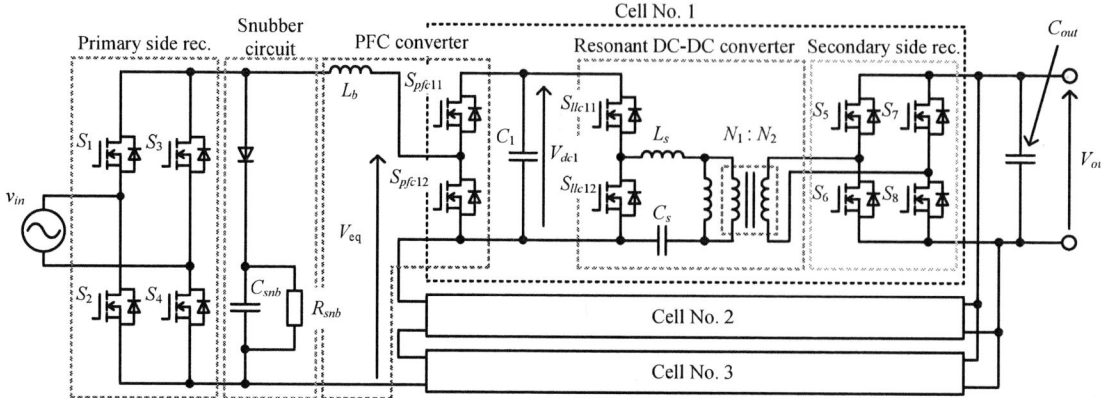

Fig. 1. Circuit configuration of proposed bidirectional single-phase SST.

In this paper, a simple circuit configuration of the single-phase SST is proposed. The proposed SST, which achieves capacitor voltage balance among cells without a complex voltage balance control with small primary side capacitor. Additionally, the proposed circuit reduces the number of the switching devices compared to the conventional circuit. The originalities of in this paper are proposing a new SST topology and the automatic voltage balancing method using a resonant DC-DC converter, which is connected in parallel in the secondary side. The contributions of this paper are that the volume of system is reduced and the control is simple. The proposed SST system is experimentally tested under a derating voltage of 1320 V which is 1/5 of the full model. In addition, the bidirectional operation is verified with the input voltage of 200 V to validate a loss distribution. Then, the proposed SST is designed for a 6.6-kV grid as the input voltage.

II. SYSTEM CONFIGURATION OF PROPOSED SST

A. Circuit configuration

Figure 1 shows the circuit configuration of the proposed SST. At the primary side, the output of the full bridge rectifier is connected to all cells in series. The devices with high voltage rating are required on the primary side. However, it is possible to use the devices with the slower switching speed because these switches are operated at the grid frequency. This system is characterized by a multi-cell input stage based on the full bridge rectifier. Each cell consists of a boost-type power-factor-correction (PFC) converter and a resonant DC-DC converter. PFC converter controls the input current to sinusoidal waveform with the unity power factor. Moreover, the rectified voltage is equally divided among each PFC converter because each cell at the primary side is connected in series. Thus, the voltage per cell is reduced. Consequently, at the primary side, it is possible to apply the switching device with low voltage rating and low on-state resistance. In the resonant DC-DC converter, the volume of the transformer is reduced because the transformer operates at high frequency.

In addition, a large capacitor is used in the output of a general PFC converter. In this system, small capacitors is

used because the proposed circuit decouples a power pulsation at twice the grid frequency in the secondary side.

Table I shows the comparison of the switch number between the proposed SST and the conventional SST which includes a PWM rectifier and a dual active bridge converter [17]. Note that the number of cell is calculated by the rated voltage of the switch. As shown in table 1, the number of devices is reduced by 30% compared to conventional SST. The reason is because the proposed SST uses only one rectifier for each cell. Consequently, the proposed circuit increases the utilization rate of circuit compared the MMC.

B. Control system

Figure 2 shows the control block diagram of the proposed circuit. The proposed control includes an automatic current control (ACR) for the boost inductor current. In the ACR, the boost inductor current is controlled into full wave rectified waveform in order to correct the power factor of the grid side. Hence, the inductor current command value I_L^* is given by

$$I_L^* = I_{amp} \left| \sin(\omega t) \right| \quad\quad (1),$$

where I_{amp} is the amplitude command value of the boost inductor current. Inductor current command I_L^* is generated by the multiplication of I_{amp} and the full-wave rectified waveform with same phase as the input voltage.

In the triangular wave comparator, the gate signal for PFC is generated by phase shifted carriers. Thus, the input voltage is equally divided because the switching timing is different. In addition, it is possible to use the switching device with low voltage rating. Note that the ripple current is reduced because the inductor voltage is reduced by the

TABLE I. COMPARISON OF SWITCHING DEVICE BETWEEN CONVENTIONAL SST AND PROPOSED SST.

Rated Voltage	Number of cell	Number of Switching devices	
		Conventional SST (PWM rec. + DAB)	Proposed SST
3.3 kV	6	72	52
1.7 kV	11	132	92
1.2 kV	16	192	132

The 2018 International Power Electronics Conference

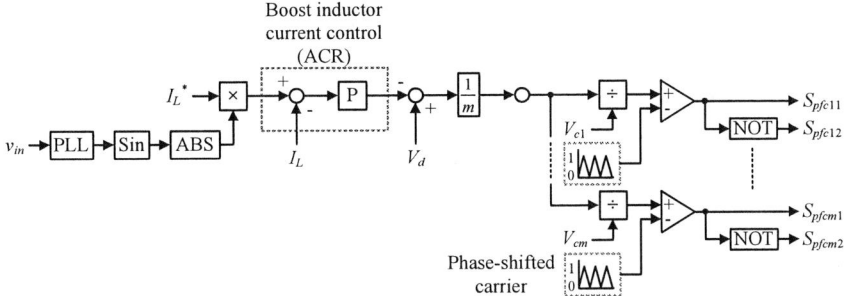

Fig. 3. Control block of PFC converter in proposed circuit.

series connection in the PFC converter. Then, the phase shift angle θ is given by (2).

$$\theta = \frac{2k}{m}\pi \ (k = 0, 1, \cdots, m-1) \ \ \dots\dots\dots\dots (2)$$

Meanwhile, the balance control of the primary side capacitor voltage V_{dc1} is not required in this system. The voltage of the primary side capacitor may be imbalanced due to the variation of capacitances or a difference of transient response in general multilevel topology. However, due to the parallel connection of the resonant DC-DC converters, which is operated with constant duties, at the secondary side, the output voltage of each cell is automatically adjusted. Consequently, the primary side capacitor voltage is naturally clumped by the voltage which is decided by the turn ratio and the secondary voltage because the cell with low voltage is fed more power from the cell with high voltage. Thus, the voltage management on the high-voltage side is not required.

Figure 3 shows the switching pulse generation of the primary side rectifier. The switching pulse is generated by comparing the input voltage v_{in} and the thresholds voltage of positive/negative (V_{thp}/V_{thn}). The switching states are following:

- $v_{in} > V_{thp}$
  Turn on S_1 and S_4
  Turn off S_2 and S_3
- $v_{in} < V_{thn}$
  Turn on S_2 and S_3
  Turn off S_1 and S_4
- $V_{thn} < v_{in} < V_{thp}$
  Turn off S_1, S_2, S_3, and S_4

In the power running operation, the current flows the body diode of MOSFET. In the case of regeneration operation, the current flows the snubber circuit.

Table II shows the switching state of the primary side rectifier, the resonant DC-DC converter, and the secondary side rectifier. In the primary side rectifier, the switching frequency is set to the grid frequency in order to achieve the polarity inversion. In the resonant DC-DC converter, the switching frequency is set to the resonant frequency in order to achieve ZCS, and the switches are modulated with duty ratio of 50%. Hence, the closed-loop control for the resonant DC-DC converter is unnecessary. Consequently, the control is simple in this system because the current

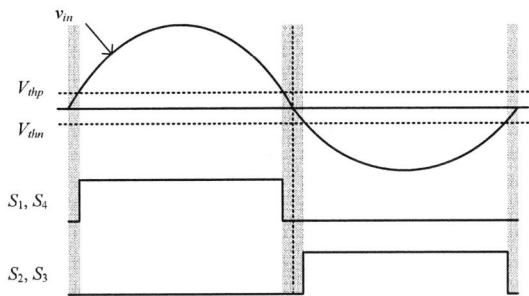

Fig. 2. Pulse generation for primary side rectifier.

TABLE II. SWITCHING MODE OF RESONANT DC-DC CONVERTER AND RECTIFIER.

	$S_1 \sim S_4$	$S_{llc11} \sim S_{llc12}$	$S_5 \sim S_8$
Switching frequency	50 Hz	50 kHz ($= f_o$)	
Duty ratio	50%		

control is achieved by only the switches of PFC (S_{pfc}). In the secondary side rectifier uses the switching pulse synchronized with the resonant DC-DC converter.

III. DESIGN OF PROPOSED SST

A. Snubber circuit

In the proposed SST, a snubber circuit is used in order to achieve the bidirectional operation. At regeneration operation, the continuous current flow is secured because the boost inductor current flows into the snubber circuit during the dead-time. Moreover, the snubber circuit also has to absorb all energy which is generated in the boost inductor when all gates are off with over current detection. Thus, the capacitor of the snubber circuit is given by

$$C_{snb} \geq \frac{L_b I_{max}^2}{\Delta V^2} \ \ \dots\dots\dots\dots\dots\dots\dots\dots(3),$$

where I_{max} is current at over current detection, ΔV is voltage rise value of capacitor. The resistor of the snubber circuit is given by

$$R_{snb} \leq \frac{2V_{clamp}{}^2}{L_b I_{max}^2 f_{sw\_rec}} \ \ \dots\dots\dots\dots\dots\dots(4),$$

2239

where f_{se_rec} is the switching frequency of the primary side rectifier, and V_{clamp} is the clamp voltage. The clamp voltage is designed to have a margin with respect to the rated voltage of device. In the miniature model SST, the margin is 20%.

B. Power-factor-correction (PFC) converter

The boost converter corrects the power factor of the grid side because the boost inductor current is controlled into full-wave rectified waveform same as general PFC circuit [18] [19]. The boost inductor L_b in the PFC circuit is given by

$$L_b = \frac{\sqrt{2}V_{in}}{4f_{eq}\Delta I_{Lb}} \quad\text{.. (5),}$$

where ΔI_{Lb} is the ripple current of the inductor current, and f_{eq} is the equivalent switching frequency of the output voltage V_{eq}. Then, the equivalent switching frequency f_{eq} is given by

$$f_{eq} = m \times f_{sw} \quad\text{.. (6),}$$

where m is the number of cells, f_{sw} is the switching frequency of the PFC circuit. Each cell is operated by phase-shifted carrier. Consequently, the switching frequency component in V_{eq} is increased in proportional to the number of cells. Thus, the size of the boost inductor is reduced because the inductance is inversely proportional to frequency.

C. Resonant DC-DC converter

The resonant DC-DC converter generates a high-frequency voltage for the isolated transformer [20]. The high-frequency operation leads to the minimization of the isolation transformer. In addition, the zero current switching (ZCS) is achieved by the series resonance between the inductor L_s and the capacitor C_s. ZCS greatly reduce the switching loss of the proposed SST system.

Furthermore, the leakage inductance is designed to be negligibly smaller than the excitation inductance. Then, the switching frequency f_o of the resonant DC-DC converter is given by (7). From resonance frequency, and the duty ratio of the switch is set to 50%.

$$f_o = \frac{1}{2\pi\sqrt{L_s C_s}} \quad\text{.. (7)}$$

In the proposed SST, the operation mode is always the boost operation with respect of the primary side voltage. Thus, the turn ratio of the transformer is designed by

$$N = \frac{N_1}{N_2} \geq \frac{\sqrt{2}V_{in}}{2m\lambda V_{out}} \quad\text{.. (8),}$$

where N_1 and N_2 are the number of turns for primary/secondary side of the high-frequency transformer,

TABLE III. CIRCUIT PARAMETER OF PROPOSED SST FOR THE MINIATURE MODEL

Parameter	Symbol	Value
Input voltage	v_{in}	1320 V_{rms}
Rated output power	P_{out}	2 kW
Rated output voltage	V_{out}	320 V
Snubber capacitor	C_{snb}	0.2 μF
Sunbber resistance	R_{snb}	2.5 MΩ
Boost inductor	L_b	24 mH (%Z = 0.87%)
Primary side capacitor	C_1	48 μF
Resonant capacitor	C_s	204 nF
Leakage inductor	L_s	50 μH
Secondary side capacitor	C_{out}	8200 μF
Switching frequency of rec.	f_{sw_rec}	50 Hz
Switching frequency of PFC	f_{sw_pfc}	10 kHz
Resonant frequency	f_o	50 kHz
Number of cells	m	3
Trans turns ratio	N_1/N_2	1.0

V_{in} is the input voltage, V_{out} is the output voltage, and λ is the modulation index of the boost converter.

IV. EXPERIMENTAL RESULTS

A. Power running operation

Table III shows the specifications and the circuit parameters. In this experiment, the fundamental operation is verified with the input voltage of 1320 V which is 1/5 of the full model. The prototype has three cells.

Figure 4 shows the waveforms of the input voltage, the input current, and the output voltage. The operation of the miniature model without the any large distortion is confirmed. At the input side, it is confirmed that the unity power factor between the input voltage and the input current is achieved. The input current THD is 4.3% at the rated load. At the output side, the step-down operation is achieved because the output voltage is regulated to 320 V.

Figure 5 shows the primary side capacitor voltage of each cell when the output power is changed from 0.8p.u. to 1.0p.u. (2 kW). It is observed that the primary side capacitor voltage is balanced even during transient response. Moreover, the maximum value of the primary side capacitor voltage also shows same value for all cells. Thus, it is confirmed that the primary side capacitor voltage is balanced among all cells without the balance control even when the output power suddenly changes.

Figure 6 shows the output voltage of all cells. From Fig. 6(a), the input voltage is equally divided to each cell because the output voltage of all cells forms balanced multilevel waveform. In Fig. 6(b), it is also confirmed that the equivalent switching frequency f_{eq} is 30 kHz. The equivalent switching frequency is determined by the switching frequency in PFC and the number of the cells.

Figure 7 shows the relationship between efficiency and input power factor. The maximum efficiency is 89.5% at the rated load. The reason is that the percentage of loss in devices against the power becomes low. The input power

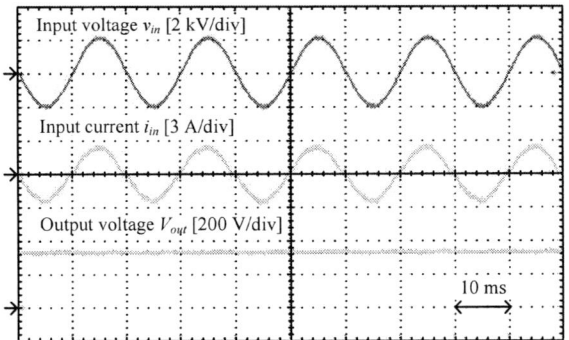

Fig. 4. Operation waveform at power running.

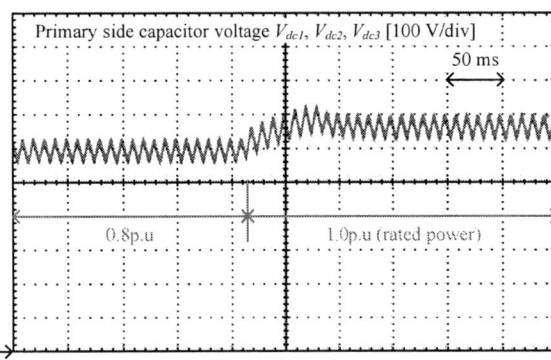

Fig. 5. Primary side capacitor voltage in each cell.

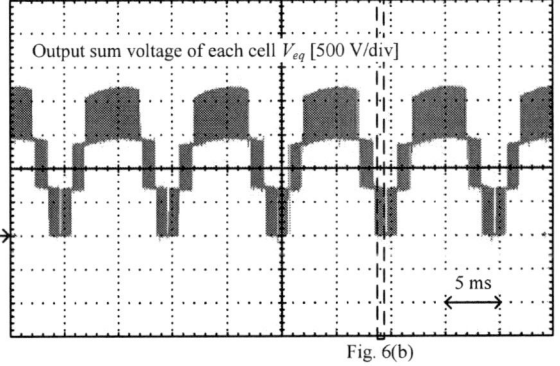

(a) Whole figure

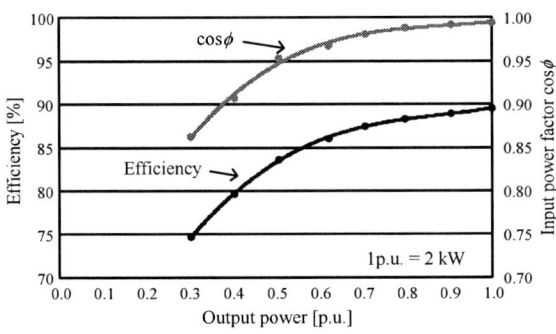

(b) Enlarged figure

Fig. 6. Output voltage of all cells at power running.

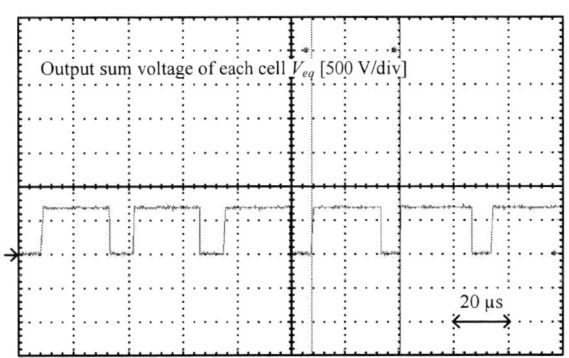

Fig. 7. Characteristic of efficiency and power factor.

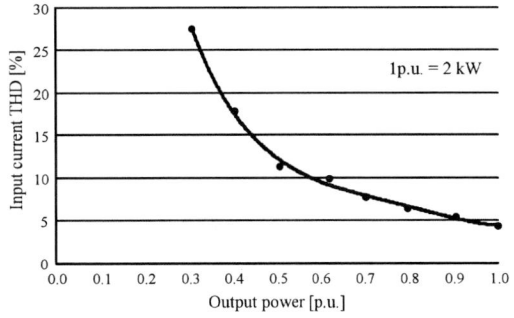

Fig. 8. Characteristic of input current THD

factor is over 0.95 with the output power from 0.5p.u. to 1.0p.u.

Figure 8 shows the relationship of the input current THD and the output power of the SST. It is confirmed that the input current THD is large when the output power is low. The reason is that the rate of the low-order harmonics component appears remarkably with respect to the fundamental component because the input current is low when the output power is low.

B. Bidirectional operation

In this experiment, the miniature model SST is tested to confirm the fundamental operation with the input voltage of 200 V due to the limitation of the experimental facilities. Note that the regeneration power supply is connected to

the output side in order to achieve the regeneration operation.

Figure 9 shows the bidirectional operation of SST when the switching from power running to regeneration is tested. In the power running operation, it is confirmed that the unity power factor between the input voltage and the input current is achieved. On the other hand, it is confirmed that the input current is reversed against the input voltage in the regeneration operation. The input current THD of 4.2% is also confirmed. In the output sum voltage of each cell, it is confirmed that the waveform is four-level staircase voltage. Furthermore, an equivalent switching frequency f_{eq} of 30 kHz is also confirmed. Thus, the stable operation of the miniature model without the any large distortion is achieved even when the operation abruptly changes.

2241

Figure 10 shows the primary side capacitor voltage of each cell in the bidirectional operation. It is observed from the waveform that the primary side capacitor voltage is balanced among all cells without the balance control even when the operation changes abruptly.

V. LOSS ANALYSIS AND ESTIMATION FOR FULL MODEL

From Fig. 1, the loss of SST is separated following components:
- (i) Primary side diode bridge
- (ii) Switching devices of PFC converter
- (iii) Switching devices of the resonant DC-DC converter
- (iv) Secondary side rectifier

Table IV shows the selected devices in each part. In the proposed SST, the rated voltage of 3.3 kV is used in the primary side rectifier.

The current which flows into the electrolytic capacitor includes not only the power ripple component but also the switching frequency component from the inverter. Thus, it is very difficult to derive analytically. Hence, the capacitor ripple current is derived by simulation [21]. The capacitor ripple current is the function of the output power factor angle φ and the modulation index λ, which is a nonlinear value. Then, the effective value of the capacitor ripple current is given by

$$I_{rms\_cap} = K_{cap}(\varphi, \lambda) I_{out} \quad \text{............................ (9),}$$

where I_{out} is the average value of the output current, and K_{cap} is the coefficient which is obtained by the simulation.

Figure 11 shows the simulation result of K_{cap}. The modulation index , which expresses the ratio of the voltage per cell and the dc-link voltage, is 0.94 in the miniature model SST. Therefore, from Fig. 11, K_{cap} (1.0, 0.94) is 0.83.

A. Primary side diode bridge

The loss of switches, which is calculated by the on-voltage of the switch and the current through the switch, is given by

$$P_{con} = \frac{1}{2\pi} \int_0^{\pi} v_{on} i_{sw} d\omega t \quad \text{.................................... (10),}$$

where v_{on} is the on-voltage of the switch, i_{sw} is the current through the switch. In this case, v_{on} and i_{sw} are given by.

$$v_{on} = r_{on} \sqrt{2} \frac{P}{V_{in}} \sin(\omega t) + v_0 \quad \text{.............................. (11),}$$

$$i_{sw} = \sqrt{2} \frac{P}{V_{in}} \sin(\omega t) \quad \text{... (12),}$$

where r_{on} is the on-resistance of the switch, P is the rated power of SST. In (11), v_0 is defined as zero because the MOSFETs are used in the prototype. Moreover, the phase

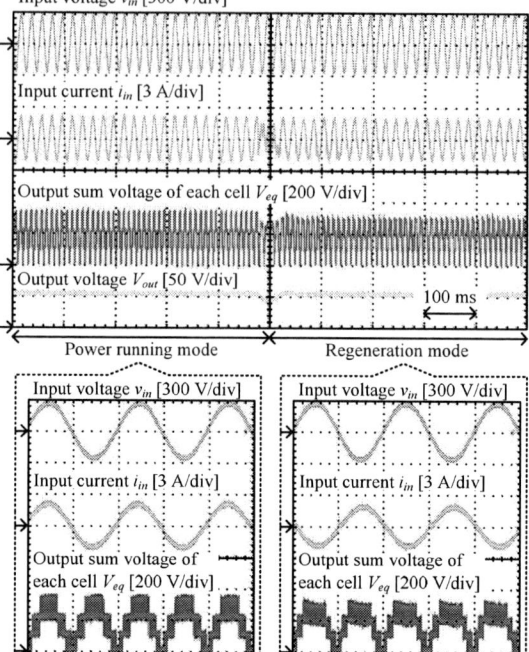

Fig. 9. Bidirectional operation of proposed circuit.

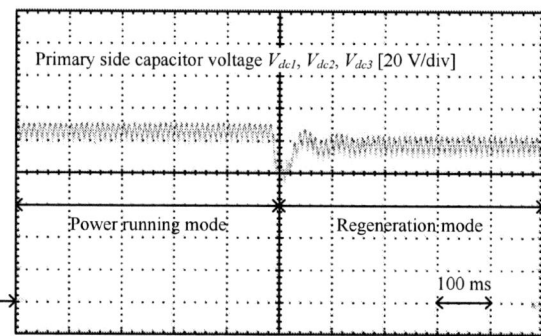

Fig. 10. Primary side capacitor voltage of each cell at bidirectional operation.

difference between the input voltage and the input current is not considered because the power factor is 1. The loss of the switches in the primary side rectifier is given by (13).

$$P_{con\_pri\_rec} = \frac{1}{2} r_{on} \left(\frac{P}{V_{in}}\right)^2 \quad \text{...................................(13)}$$

B. PFC converter

The conduction loss of the switches in PFC is given by

$$P_{con\_PFC} = \frac{1}{2} r_{on} I_L^2 \quad \text{...(14),}$$

2242

where I_L is the current effective value through the boost inductor.

On the other hand, the switching loss of the switches, which is assumed that it is directly proportional to the voltage and the current of the switch, is given by

$$P_{sw\_PFC} = \frac{1}{\pi} \frac{e_{on} + e_{off}}{E_{nom} I_{nom}} \frac{V_{dc}}{V_{cell}} \frac{P}{m} f_{sw} \quad \dots\dots\dots\dots (15),$$

where V_{dc} is the voltage of the primary side capacitor, P is the rated power, m is the number of cell, f_{sw} is the carrier frequency, e_{on} and e_{off} are the turn-on and the turn-off energy per switching from datasheet, E_{nom} and I_{nom} are the voltage and the current under the measurement condition of the switching loss from the datasheet, and V_{cell} is the input voltage of each cell.

C. Resonant DC-DC converter

The loss of switches in the resonant DC-DC converter is only the conduction loss because ZCS assume achieving over all operation regions.

Therefore, the conduction loss is given by

$$P_{con\_LLC} = \frac{1}{2} R_{on} \left(\frac{N_2}{N_1}\right)^2 \frac{I_{out}^2 - I_{rms\_cap}^2}{m^2} \quad \dots\dots\dots (16).$$

At the secondary side, the conduction loss is given by

$$P_{con\_sec\_rec} = \frac{1}{2} R_{on} \frac{I_{out}^2 - I_{rms\_cap}^2}{m^2} \quad \dots\dots\dots\dots (17).$$

D. High-frequency transformer

Iron loss, which occurs in the high-frequency transformer, is calculated by the magnetic flux density and the characteristic of the core. The AC magnetic flux density B_{ac} is given by

$$B_{ac} = \frac{V_{out}}{4 f_o A_e N} \quad \dots\dots\dots\dots\dots\dots\dots\dots\dots\dots\dots (18),$$

where A_e is the effective cross-section of the core, and N is the turns ratio of the transformer. The core loss value is given by the characteristic graph between the core loss value vs. the magnetic flux density, which is obtained from the core material, and the magnetic flux density which is calculated from (19). Therefore, the iron loss is given by

$$P_{iron\_loss} = P_{cv} V_e \quad \dots\dots\dots\dots\dots\dots\dots\dots\dots\dots\dots (19),$$

where V_e is the effective volume of core.

The high frequency transformer of full model SST is designed by Gecko MAGNETICS which uses improved-improved Generalized Steinmetz Equation (i²GSE) in order to calculate the iron loss of the transformer [22]. Consequently, it is possible to select the optimum core shape, core material and winding shape. From the analysis

TABLE IV. SELECTED DEVICES OF PROTOTYPE FOR BIDIRECTIONAL OPERATION.

Circuit topology	Part	Type	Maximum ration
Single-phase rectifier	$S_1 \sim S_4$	-	3300 V
PFC converter	$S_{pfc11} \sim S_{pfc12}$	SCT2080KE	1200 V 40 A
Resonant DC-DC converter	$S_{llc11} \sim S_{llc12}$		
Secondary side rectifier	$S_5 \sim S_8$		

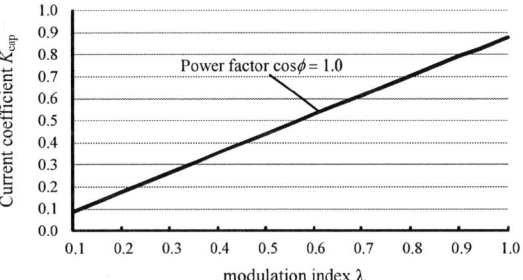

Fig. 11. Current coefficient of output capacitor.

of Gecko MAGNETICS, it is confirmed that the loss is minimum by using EPCOS N95 as the core in the full model SST.

E. Loss distribution

Figure 12 shows the loss distribution obtained from the experiment and the calculation of the bidirectional operation. Note that the loss is normalized with the experimental loss as 100%. The error of the loss between the experiment and the calculation is less than 5%.

Figure 13 shows the loss distribution of the full model SST. The loss distribution is calculated with assuming 6.6-kV input voltage and a 10-kVA rated power. Then, the number of cells is 15 because 1.2-kV switching devices can be used. From this consideration, a 99% efficiency at the rated power is expected. Moreover, it is possible to reduce the loss by applying synchronous rectification in the secondary rectifier.

VI. CONCLUSION

This paper has proposed a miniature model SST, which has a capacitor voltage balance capability without a control. The fundamental operation of SST was confirmed with the input voltage of 1320 V which is 1/5 of the full model from the experimental results. As a result, the sinusoidal waveform of the input current was obtained without any large distortion at the primary side. In addition, the bidirectional operation is confirmed in the proposed SST with the input voltage of 200 V. Furthermore, the average voltage of the primary side capacitor are stable and balanced among all cells without a voltage balance control.

Finally, the loss equation was derived at each part of the system and compared with experimental result. As a result, the error of the loss between the experimental result and the calculation is less than 5%. It is predicted that the maximum efficiency of full model SST is 99%.

REFERENCES

[1] T. Tanaka, Y. Takahashi, K. Natori, and Y. Sato "High-Efficiency Floating Bidirectional Power Flow Controller for Next-Generation DC Power Network," IEEJ J. Industry Applications, vol. 7, no. 1, pp. 29-34, (2018)

[2] R. Chattopadhyay, S. Bhattacharya, N. C. Foureaux, A. M. Silva, B. Cardoso F., H. de Paula, I. A. Pires, P. C. Cortizio, L. Moraes, and J. A. de S. Brito: "Low-Voltage PV Power Integration into Medium Voltage Grid Using High-Voltage SiC Devices," IPEC 2014, pp.3225-3232, (2014)

[3] T. Nakanishi, K. Orikawa, J. Itoh: "Modular Multilevel Converter for Wind Power Generation System Connected to Micro-Grid," ICRERA2014, No. 219, (2014)

[4] L. Wang, D. Zhang, Y. Wang, B. Wu, and H. S. Athab: "Power and Voltage Balance Control of Novel Three-Phase Solid-State Transformer Using Multilevel Cascaded H-Bridge Inverters for Microgrid Application," IEEE Trans. On Power Electronics, Vol. 31, No. 4, pp.3289-3301, (2016)

[5] T. Nakanishi, and J. Itoh, "Control Strategy for Modular Multilevel Converter based on Single-phase Power Factor Correction Converter," IEEJ J. Industry Applications, vol.6, no.1, pp.46-57, (2017).

[6] N. Hatti, Y. Kondo, and H. Akagi, "Five-Level Diode-Clamped PWM Converters Connected Back-to-Back for Motor Drives," IEEE Trans. On Industry Applications, Vol.44, No.4, pp.1268-1276, (2008)

[7] M. Nakahara, and K. Wada, "Loss Analysis of Magnetic Components for a Solid-State-Transformer," IEEJ Journal of Industry Applications, Vol.4, No.7, pp.387-394, (2015)

[8] D. Ronanki, and S. S. Williamson: "Evolution of Power Converter Topologies and Technical Considerations of Power Electronic Transformer based Rolling Stock Architectures," IEEE Trans. On Transportation Electrification, (2017)

[9] X. Yu, X, She, X. Zhou and A. Q. Huang: "Power Management for DC Microgrid Enabled by Solid-State Transformer," IEEE Trans., Vol.5, No.2, pp.954-965 (2014)

[10] A. Q. Huang, Q. Zhu, L. Wang, and L. Zhang, "15 kV SiC MOSFET: An Enabling Technology for Medium Voltage Solid State Transformers," CPSS Trans., Vol.2, No.2, pp.118-130 (2017)

[11] J. W. Kolar and G. Ortiz: "Solid-State-Transformers: Key Components of Future Traction and Smart Grid Systems", IPEC 2014, pp.22-35 (2014)

[12] T. Nakanishi, and J. Itoh, "Design Guidelines of Circuit Parameters for Modular Multilevel Converter with H-bridge Cell," IEEJ J. Industry Applications, vol.6, no.3, pp.231-244, (2017)

[13] H. Hwang, X. Liu, J. Kim and H. Li: "Distributed Digital Control of Modular-Based Solid-State Transformer Using DSP+FPGA," IEEE Trans. On Industrial Electronics, Vol.60, No.2, pp.670-680, (2013)

[14] T. Nakanishi, and J. Itoh, "Capacitor Volume Evaluation based on Ripple Current in Modular Multilevel Converter," 9th International Conference on Power Electronics, No. WeA1-5, (2015)

[15] J. Shi, W. Gou, H. Yuan, T. Zhao and A. Q. Huang: "Research on Voltage and Power Balance Control for Cascaded Modular Solid-State Transformer," IEEE Trans. On Power Electronics, Vol. 26, No. 4, pp.1154-1166, (2011)

[16] T. Isobe, H. Tadano, Z. He, and Y. Zou: "Control of Solid-State-Transformer for Minimized Energy Storage Capacitors," IEEE ECCE, pp.3809-3815, (2017)

[17] J. E. Huber, and J. W. Kolar: "Solid-State Transformer: On the Origins and Evolution of Key Concepts," IEEE Industrial Electronics Magazine, Vol. 10, pp.19-28, (2016)

[18] T. Nussbaumer, K. Raggl, and J. W. Kolar: "Design Guidelines for Interleaved Single-Phase Boost PFC Circuits," IEEE Trans. On Industrial Electronics, Vol. 56, No. 7, pp.2559-2573, (2009)

[19] Y. Hayashi, Y. Matsugaki, and T. Ninomiya, "Capacitively Isolated Multicell Dc-Dc Transformer for Future Dc Distribution System," IEEJ J. Industry Applications, vol. 6, no. 4, pp. 268–277, (2017)

[20] M. Sato, R. Takiguchi, J. Imaoka, and M. Shoyama: "A Novel Secondary PWM Controlled interleaved LLC Resonant Converter for Load Current Sharing," IPEMC 2016, pp.2276-2280, (2016)

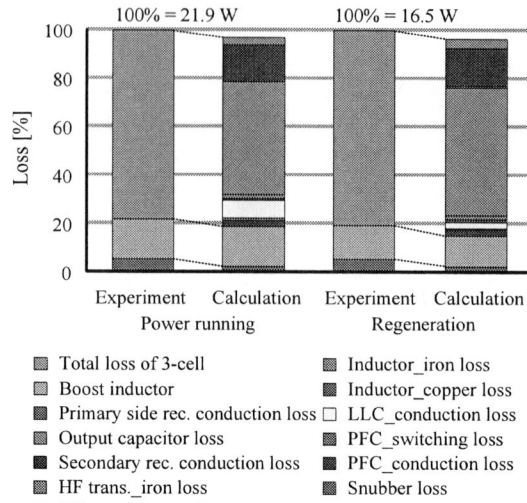

Fig. 12. Loss distribution result by experiment and calculation at bidirectional operation. (Explanatory note corresponds to each color of graph.)

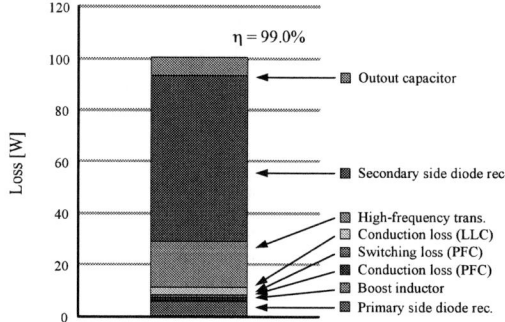

Fig. 13. Loss distribution of 6.6 kV/ 10 kW full model SST by calculation.

[21] J. Itoh, T. Sakuraba, H. N. Le, K. Kusaka: "Requirements for Circuit Components of Single-Phase Inverter Applied with Power Decoupling Capability toward High Power Density," 18th European Conference on Power Electronics and Applications (EPE'16), DS2a 0291, (2016)

[22] J. Muhlethaler, J. Biela, J. W. Kolar, and A. Ecklebe: "Improved Core-Loss Calculation for Magnetic Components Employed in Power Electronic Systems," IEEE Trans. On Power Electronics, Vol. 27, No. 2, pp.964-973, (2012)

A Developed Dual MMC Isolated DC Solid State Transformer and Its Modulation Strategy

Yang CHEN[1], *Student Member, IEEE*, Yan LI[2], Miao ZHU[*1], *Senior Member, IEEE*, Chao LIU[2] and Xu CAI[1]

1 School of Electronic Information and Electrical Engineering, Shanghai Jiao Tong University, China, 200240
2 China Electric Power Research Institute, China, 100192
*E-mail: miaozhu@sjtu.edu.cn

Abstract— DC solid state transformer will be a key device in DC power grid in the future. This paper proposes a novel topology of MMC based DC solid state transformer. Unlike the common back-to-back converters, the AC side of the two MMC converters are connected via a middle-frequency AC transformer. A modulation method and corresponding submodule balancing strategy has been presented in the paper. Compared to those sine-type modulation technologies, the proposed Quasi-2-Level (Q2L) modulation has a greater power transmission capacity. Additionally, Q2L modulation can significantly reduce dv/dt stress on the isolated AC transformer, which protects converters from electromagnetic interference. A corresponding submodule balancing strategy based on Q2L modulation is also proposed in this paper, without any effect on AC waveform or requirement on arm current measurement. The effectiveness of the modulation and balancing strategy has been validated by simulation results.

Keywords— *Solid state transformer, MMC, Quasi-2-level modulation.*

I. INTRODUCTION

With the development of power electronics technology, DC transmission has become an alternative option for modern power system [1]. DC grid has higher reliability, lower transmission loss and simpler control system compared to AC grid [2-3]. Several demonstration projects have been constructed in China and other countries. A ± 200kV flexible DC transmission system with five terminals has been put to use in Zhoushan, Zhejiang Province. The system had the highest voltage level, largest power capacity and most terminals in the world in 2014. A new flexible DC power grid is under construction in Zhangbei, Hebei Province. The ±500kV 9000MW system is a milestone of DC grid development in the world. Wind, solar and batteries will be connected together via the DC grid.

With the development of DC grid, there will be different voltage levels in the grid. DC solid state transformer (DC-SST) has the ability to connect and exchange energy between different voltage levels. As a key device in DC grid, DC-SST is a popular research topic nowadays. Isolated DC-SST topologies are suitable for high-voltage DC (HVDC) and medium-voltage DC

(MVDC) applications. There have been three main topologies for isolated DC-SST so far. They are series-connected IGBTs [4-5], input series output series (ISOS) structure and modular multilevel converter (MMC). The concept of series-connected IGBTs is simple and clear, but the voltage imbalance among IGBTs in series is difficult to overcome. ABB Company has done several HVDC projects with its series connected IGBT valves so far. However, the technology has not been applied for ultra-high-voltage DC (UHV-DC) projects such as ± 500kV. Input-series output-series DC-SST often consists of cascaded DABs [6]. It possesses some distinctive features such as standardized modular manufacturing and flexibility of power extension. However, the power sharing strategy among modules is complex. As an advanced flexible solution, the modular multi-level converter (MMC) has been widely applied in HVDC scenarios [7]. It is a promising method for high-voltage and high-power application. Currently, researches on MMC mainly focus on HVDC applications with fundamental frequency modulation. MMCs applied in SSTs are often combined with DABs in secondary side and always have high frequency isolated transformers, and the control strategy is complex. New topology and modulation strategy should be developed to take advantages of MMC for DC grid.

A novel DC-SST topology is proposed in this paper. Dual MMCs connected back-to-back via a medium frequency isolated transformer is adopted in the topology. The adoption of medium frequency transformer will help increase power density and reduce the volume of the whole device. Thus, the modulation method for medium frequency is important. Compared with PWM modulation and nearest level modulation, fundamental frequency is more suitable for proposed DC-SST since AC waveform quality is not the main consideration. A quasi-2-level modulation is proposed for the topology. The modulation method has greater power transmission ability. A new capacitor voltage balancing control strategy is provided in the paper. The proposed capacitor voltage balancing strategy does not need to measure arm current at all. Detailed simulation results are provided and verifies the theoretical analysis.

II. TOPOLOGY AND OPERATION PRINCIPLES OF PROPOSED DC-SST

Traditional SST often consists of MMC and DABs. MMC often serve at the higher voltage side for power supply [8-9]. The topology at the lower voltage side is series or paralleled DABs to achieve high voltage or large current. Those topologies are complex, and it is difficult to design the corresponding control strategy. A new topology is presented in this paper adopting MMCs in both primary and secondary side. MMC structure at different side will be different to match the voltage level of the transformer. A typical topology of the proposed DC-SST is shown in Fig. 1.

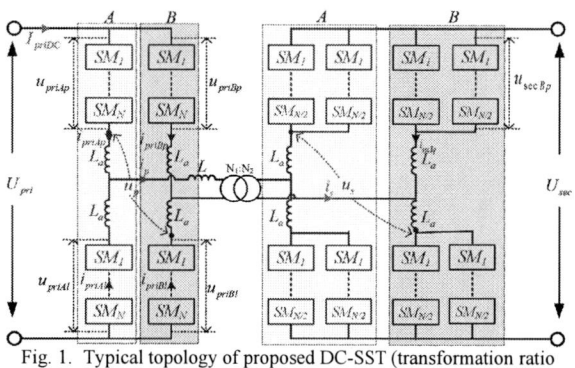

Fig. 1. Typical topology of proposed DC-SST (transformation ratio 1:2).

Here in Fig. 1, the voltage transforming ratio is 2:1. The topology can be regarded as two equivalent full bridge converters connected via a medium frequency transformer, and each equivalent bridge arm is composed of MMC submodules. The DC-SST input at the primary side consists of two phases, and each phase has one branch with N submodules. The arms at the secondary side have two branches with N/2 submodules. In this way, switches of both sides share the same voltage and current in this topology. Fig. 2 shows detailed structure of submodules in this topology. The submodules in Fig. 2 has two main working states as shown in Table 1. S means the state of the submodule. S=1 means the module is put into the circuit and the output voltage is u_C, while S=0 means the submodule is shorted out and the output voltage is 0. By changing the switching state of switches, input voltage can be controlled. It is not permitted to open or close the two switches simultaneously.

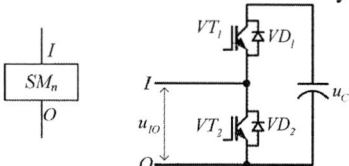

Fig. 2. Detailed structure of submodule.

TABLE I
OPERATION STATUS OF SUBMODULES

Status	U_{IO}	S
On	U_C	1
Off	0	0

Due to the working status from Table I, modulation waveforms of each submodule in one single switching period is shown in Fig. 3(a). Length of the switching cycle is 2π. The two-level modulation method determines the duty ratio to be 50%. As shown in Fig. 3, red part means on state of each submodule and the length is π (50%).

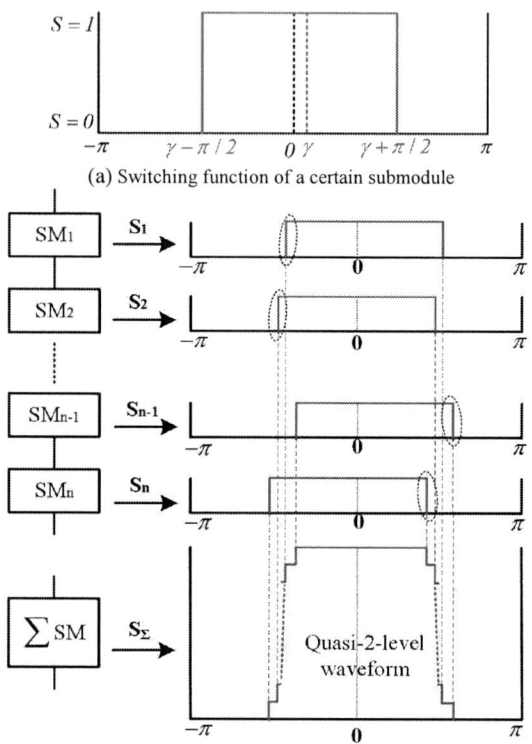

(a) Switching function of a certain submodule

(b) Switching function of each arm

Fig. 3. Switching function of MMC.

The problem of common 2-level modulation is that it often brings large electromagnetic interference (EMI). Here in the proposed strategy, the modulation wave has an active phase shift to avoid large EMI. As shown in Fig. 3(b), the total switching function is a quasi-2-level waveform. Each submodule will have a phase shift δ_i. By setting different phase shift of each submodule in the same arm, dv/dt can be reduced and the capacity voltage balance may be maintained as well. There are three basic requirements for the modulation:

1) Modulation waves of upper arm and lower arm are symmetric waveforms about the cycle center.

2) Each arm is an independent modulation unit. Units share the same normal phase shift. Phase difference between upper and lower arm are fixed.

3) Total amount of modules of upper and lower arms in the same phase is a constant value (N).

To satisfy requirements above, a typical phase shift array is provided. If odd submodules are chosen, phase shift array may be $\{-(N-1)/2, -(N-1)/2+1, \ldots 0 \ldots, (N-1)/2-1, (N-1)/2\}$. If even submodules are chosen, phase shift array may be $\{-N/2, -N/2+1, \ldots, N/2-1, N/2\}$. Every element in the array will be assigned to a certain submodule in the same arm. The theoretical output waveforms are shown in Fig.4.

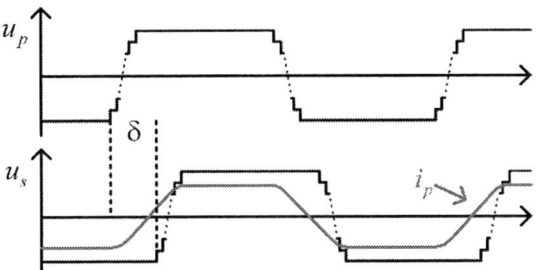

Fig. 4 Voltage and current waveforms of primary and secondary sides.

The equivalent circuit of the DC-SST in Fig. 1 is shown in Fig. 5. The topology is equivalent to two AC voltage sources u_p and u_s connected with an inductor Leq. Power transmission can be controlled by controlling phase angle δ. The quansi rectangular waveform in Fig.5 certainly has the ability to transmit more power than sinusolid waveforms.

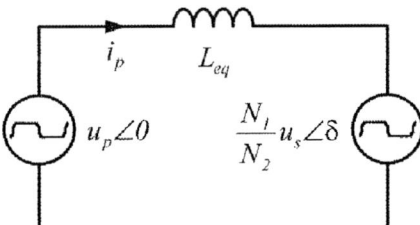

Fig. 5. Equivalent circuit of proposed DC-SST

III. SUBMODULE CAPACITOR VOLTAGE BALANCING STRATEGY

The balancing of capacitor voltage is a dynamic process [10-11]. Submodules will be in a balanced state if they do not gain or lose any energy in one circle. Here in the proposed modulation method, a single switching cycle is a complete energy absorbing or releasing process. Take a submodule as an example. The energy that a single submodule absorbs in one cycle can be calculated by the expression below.

$$E_{absorb} = \frac{U_{pri\_DC}}{N} \int_0^{T_s} S_k i_{pLA} dt \qquad (1)$$

Where: S_k is the switching function of $\mathrm{k^{th}}$ submodule, and its Fourier expansion is as below.

$$S_k = \frac{1}{2} + \frac{2}{\pi} \sum_{n=1}^{\infty} \left\{ \frac{(-1)^{2n-1}}{2n-1} \cos[(2n-1)\omega t - \gamma_{k,2n-1}] \right\} \qquad (2)$$

Where: $\gamma_{k,2n-1}=(2n-1)\gamma_k$.

As common operating principle of MMC indicates, upper lower arms share the output current i_p, and there is another circuit current i_c. Thus lower arm current can be calculated below.

$$i_{pLA} = -i_c + \frac{1}{2} i_p \qquad (3)$$

As (1) shows, the energy in a cycle is related to both circuit current i_c. and AC current i_p. The circuit current i_c consists 2 parts. They are DC current and even circuit current. The equation of ic is shown as below.

$$i_c = \frac{1}{2} I_{priDC} + \sum_{m=1}^{\infty} I_{c,2m} \cos(2m\omega t + a_{2m}) \qquad (4)$$

The first part is DC component and the rest is even circulating current. The switch function in (2) consist of only odd componments. The integration of the product of switching function S_k and circuit current i_c is 0 in a cycle. Here in Fig. 6, Q means the electrical charge in a cycle. By setting differen phase shift for a particular sub module, the voltage balance can be maintained. Since duty ratio of each switch in a submodule is 0.5, the DC component of primary side is the same among different submoules. From the analysis above, common-mode current has the same energy effect among submodules in the same arm. AC components will give different energy to submodules with different phase shift.

Take S_i and S_j as an examle. Q_i and Q_j are the integration of the product of switch function and AC component of arm current as Fig. 6 shows. The physical meaning of Q is the quantity of electricity absorbed or released in a single period. Different electricity will be accumulated with different phase shift. In this way, submodule voltage may be balanced by setting different phase shift.

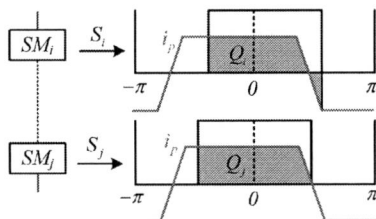

Fig. 6. Electric charging process of submodules in one period.

Take primary side as an example. The primary voltage u_p is as follow.

$$u_p = \frac{4U_{Pri\_DC}}{N\pi} \times$$
$$\sum_{k=1}^{n} \sum_{n=1}^{\infty} \left\{ \frac{(-1)^{n+1}}{2n-1} [\cos(2n-1)\omega t \cos\gamma_{k,2n-1}] \right\} \qquad (5)$$

The harmonic components of u_p is

$$u_{p,2n-1} = (-1)^{n+1} U_{p,2n-1} \cos(2n-1)\omega t \qquad (6)$$

The harmonic components of i_p is

$$i_{p,2n-1} = (-1)^{n+1} I_{p,2n-1} \cos[(2n-1)\omega t + \phi_{p,2n-1}] \qquad (7)$$

Where:$\phi_{p,2n-1}$ is power angle of $U_{p,2n-1}$ and $I_{p,2n-1}$. After putting (6) and (7) into (1), the absorbed energy E_{absorb} can be calculated as bellow.

$$E_{absorb} = E_{const} + \sum_{n=1}^{\infty} E_{2n-1} \qquad (8)$$

E_{const} is introduced by common-mode current. It is the same in all submodules in the same arm. E_{2n-1} is introduced by AC current components. It's different in different submodules. It can be calculated by (9).

$$E_{2n-1} = \frac{T_s U_{\text{Pri\_DC}} I_{p,2n-1}}{(2n-1)N\pi} \cos(\gamma_{k,2n-1} + \phi_{p,2n-1}) \quad (9)$$

It can be proved that energy of the kth submodule accumulated in a cycle is only related to the phase shift γ_k, while the AC current doesn't have any influence according to (9). Thus voltage balancing strategy will be closely related to phase shift. Right phase shifts for different submodules can be achieved by tracking present phase shifts and voltage fluctuations of the submodules.

IV. SIMULATION RESULTS

In order to verify the proposed topology and modulation strategy, simulation based on Fig. 1 has been carried out with Matlab/Simulink. The simulation circuit is shown in Fig. 7.

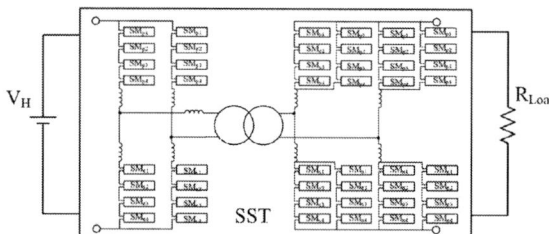

Fig. 7. Simulation circuit.

Simulation parameters are shown in Table II. Nominal power of both simulation and experimental systems are 2 kVA. To verify the power transmission ability, there is a resistive load in the secondary side.

TABLE II
SIMULATION PARAMETERS

Symbol	Meaning	Value
u_p	Normal Primary Voltage	800V
u_s	Normal Secondary Voltage	400V
$N_1{:}N_2$	SST Transforming Ratio	2:1
f	Switching Frequency	1000Hz
N	Submodule number	8/4
C	Capacitor Value	1.8mF
La	Arm inductor	25μH
γ	Normal phase shift	0.01π

Output voltage waveforms are shown in Fig. 8. Fig. 8(a) is the output voltage at primary side. The peak voltage is 800V and has sharp rising and falling edges. Since there are 8 submodules in each arm in primary side, the output voltage has 9 levels as shown in Fig.8(b). The voltage transforming ratio is 2:1 as expected.

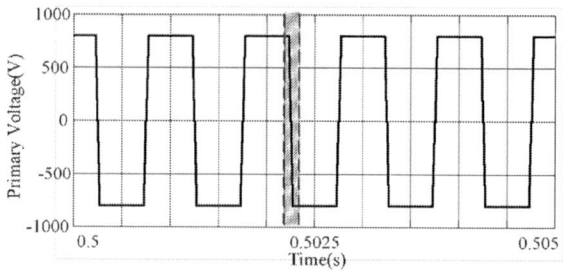

(a)   No-load voltage at primary side

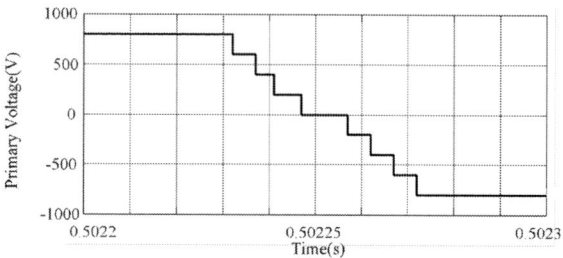

(b)   Detailed no-load voltage waveform

Fig. 8 Output voltage and current waveforms in no-load condition.

In order to verify the power transmission ability, a 10 Ω register is connected to the seconary side as a load. Fig. 9 shows the output voltage and current waveforms. (a) shows the voltage of primary side and secondary side. The terminal voltage ratio is 800:400 as expected. There isn't any DC current through transformer as shown in (b). Output load current is almost constant as shown in (c).

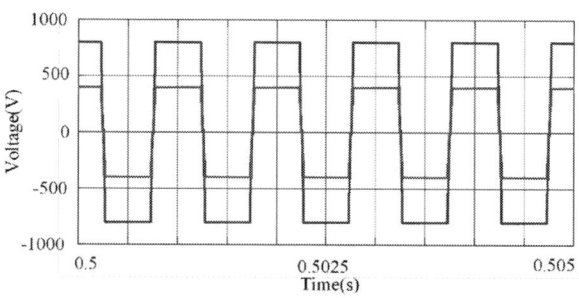

(a)   Voltage waveforms at both primary and seconary side

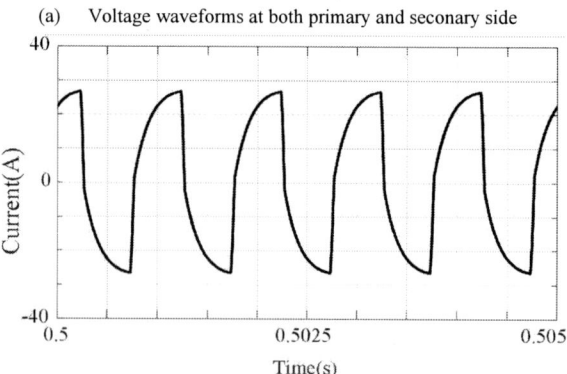

(b) Current waveform at primary side

2248

The 2018 International Power Electronics Conference

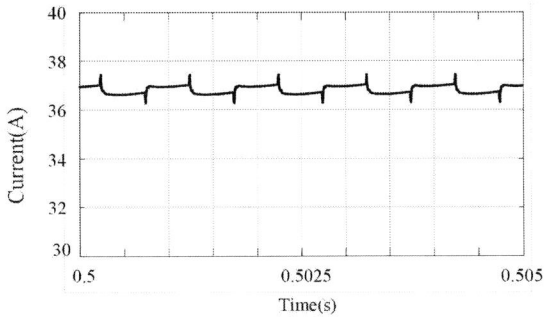

(c) Load Current waveform
Fig. 9 Output voltage and current waveforms

Fig.10 shows the performance of the proposed voltage balancing strategy. Since there are 8 submodules in each arm at the primary side, capacitor voltage of each submodule at the primary side should be 800/8=100V. Capacitor voltage at the secondary side should be 100V as well. The capacitor voltage had been set to 100V in advance. As shown in Fig. 10(a), voltages begin to diverge without any capacitor voltage balancing strategy applied. Then the voltage balancing strategy is put into use at about t=0.06s. Capacitor voltages begin to converge right away indicating that the proposed voltage balancing strategy work well. Fig. 10(b) shows that capacitor voltages in secondary side maintain balance as well.

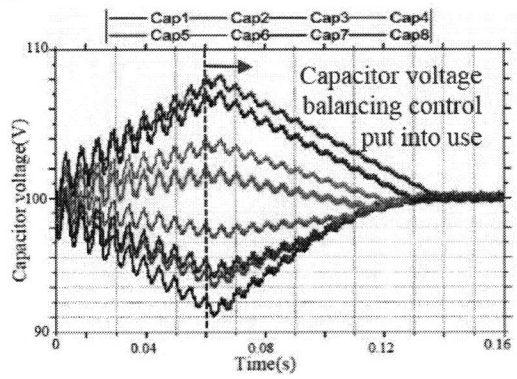

(a) Capacitor voltage waveform in primary side

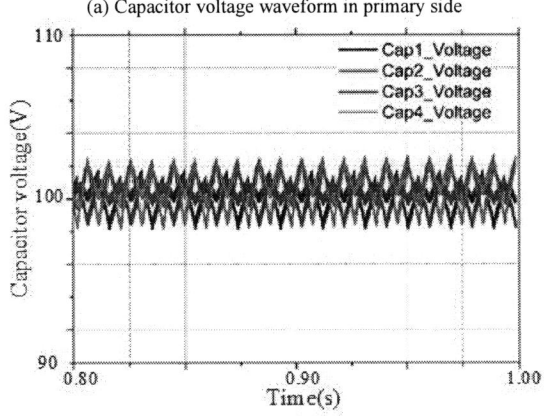

(b) Capacitor voltage waveform in secondary side
Fig. 10 Voltage balancing performance

The simulation results above prove that the

fundamental functions of proposed topology and modulation strategy is working well.

V. CONCLUSIONS

A developed dual MMC isolated DC solid state transformer is presented in this paper. The corresponding quasi-2-level modulation has been given as well as its capacitor voltage balancing strategy. The proposed topology is straightforward, and EMI of the proposed modulation is much smaller than that of classical 2-level modulation. The capacitor voltage balancing strategy is effective and does not need to measure arm current additionally. Power can be transmitted precisely via proposed DC-SST.

ACKNOWLEDGEMENT

The authors' work is financially supported by Research Program of State Grid Corporation of China (Research on operation control and protection technologies of systems with large-scale renewable energy connected to the weakly-synchronized sending-end DC power grid).

REFERENCES

[1] N. Barnes, D. Hertem, S. P. Teeuwsen, et al. "HVDC Systems in Smart Grids." *Proceedings of the IEEE*, vol. 105, no. 11, pp. 2082-2098, 2017.

[2] Jianjun. Ma, Yan Li, Miao Zhu and X. Cai, " Parallel Operation of Distributed Voltage Balancers for Bipolar DC System with Improved Reliability and Efficiency," *IECON 2017 - 43th Annual Conference of the IEEE Industrial Electronics Society*, Beijing, 2017.

[3] Jianjun Ma, M. Zhu, Qiong Li and X. Cai, "From 'Voltage Balancer' to 'Interlinking Converter'—A Shift of Operation Concept for Bipolar DC Distribution System," *IECON 2017 - 43th Annual Conference of the IEEE Industrial Electronics Society*, Beijing, 2017.

[4] G. Christian. "Fast high-power/high-voltage switch using series-connected IGBTs with active gate-controlled voltage-balancing." *Conference Proceedings of Applied Power Electronics Conference and Exposition*, 1994.

[5] D. Sha, Z. Guo, T. Luo, et al. "A general control strategy for input-series–output-series modular dc–dc converters." *IEEE Trans. on Power Electronics*, vol. 29, no. 7, pp. 3766-3775, 2014.

[6] B. Zhao, Q. Song, J. Li, et al. "Comparative Analysis of Multilevel-High-Frequency-Link and Multilevel-DC-Link DC-DC Transformers Based on MMC and Dual-Active-Bridge for MVDC Application." *IEEE Trans. on Power Electronics*, advance published, 2017.

[7] S. Debnath, J. Qin, B. Bahrani, et al. "Operation, control, and applications of the modular multilevel converter: A review." *IEEE Trans. on Power Electronics*, vol. 30, no. 1, pp. 37-53, 2015.

[8] H. ZHU, M. ZHU, J.W. ZHANG and X. CAI, "Topology and Operation Mechanism of Monopolar-to-Bipolar DC-DC Converter Interface for DC Grid," in *Proc. 9th Inter. Power Electronics Conf. -ECCE ASIA*, China, May. 2016

[9] S. Kenzelmann, A. Rufer, D. Dujic, et al. "Isolated DC/DC Structure Based on Modular Multilevel Converter." *IEEE Trans. on Power Electronics*, vol. 30, no. 1, pp. 89-98, 2015.

[10] J.J. MA, M. ZHU, X. CAI and Y.W. LI, "Configuration and Operation of DC Microgrid Cluster Linked Through DC-DC Converter," in *Proc. 11th IEEE Conf. on Industrial Electronics and Applications, ICIEA*, China, Jun. 2016.

DC Fault Ride-Through of a Three-Phase Dual-Active Bridge Converter for DC Grids

Jingxin Hu, Shenghui Cui and Rik W. De Doncker
Institute for Power Generation and Storage Systems
E.ON Energy Research Center, FEN Research Campus
RWTH Aachen University, Aachen, Germany
Email: post_pgs@eonerc.rwth-aachen.de

Abstract—DC fault ride-through capability is a highly desired feature of three-phase dual-active bridge (DAB3) converters in a dc distribution grid. However, the state-of-the-art modulations schemes for DAB3 converters such as single phase-shift and asymmetrical duty-cycle control have a limited dc fault ride-through capability due to either the resulting high peak currents or a limited power transfer capability and the lack of $0\,V$ ride-through capability. In this paper, a space-vector-based asymmetrical duty-cycle control method is proposed allowing the DAB3 converter to ride through a dc fault down to $0\,V$ with a controllable peak current of the devices and feature a significantly higher power transfer capability at low voltage-ratios. Moreover, soft switching turn-on is still realized for all semiconductor devices during the dc fault ride-through operation. The proposed method can be combined with the asymmetrical duty-cycle control and single phase-shift modulation to achieve a fast-dynamic control of the DAB3 converter even during dc faults. Validity of the proposed method is verified by simulation of a $300\,kW$ **DAB3 converter.**

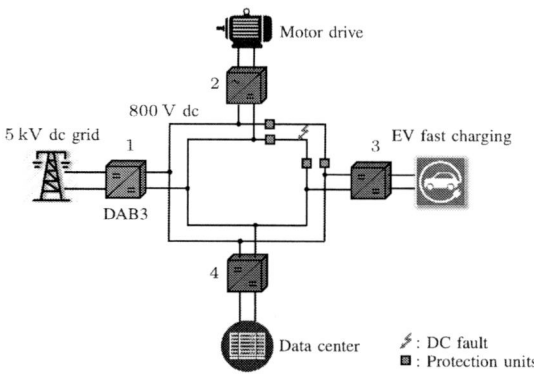

Fig. 1: Layout of a dc distribution system

I. INTRODUCTION

In compliance with the targets of reducing $CO2$ emission and saving energy, a vast amount of inherently dc loads e.g. electric vehicles (EV), data centers, variable-speed drives, distributed renewable energy generation and battery energy storage systems are connected to the distribution grid. Consequently, it makes dc distribution grids more advantageous than its conventional ac counterpart in terms of efficiency, footprint, control complexity and flexibility [1].

As the key component in the dc distribution grid, the isolated dc-dc converter connecting medium-voltage and low-voltage dc distribution grids is usually assigned with multiple functionalities e.g. stepping up/down dc voltages, providing galvanic isolation, transferring power efficiently, regulating dc grid voltage and riding through dc faults. Among numerous isolated dc-dc converter topologies, the three-phase dual-active bridge (DAB3) converter, firstly proposed in 1991 for aerospace applications [2], is an attractive candidate mainly due to its inherent soft-switching capability, wide operation range and reduced filter size [3]–[6]. It is also widely used in renewable energy generation, energy storage systems and railway applications [7]–[11].

In a dc distribution system as shown in Figure 1, the DAB3 converter interconnects a $5\,kV$ medium-voltage dc grid and $800\,V$ low-voltage dc distribution grid that incorporates loads of a motor drive, a data center and a EV fast charging station. It is assumed that the grid voltage is regulated by the DAB3 converter. In case of a dc fault in the low-voltage dc distribution grid, the fault is inherently isolated from the

medium-voltage side by the DAB3 converter, and by blocking all the gate signals the DAB3 converter could shut down safely. However, in this manner a blackout of the complete low-voltage dc distribution grid occurs. If the DAB3 converter can't ride through dc faults, selective protection measures cannot be realized. Consequently, the blackout region of the dc grid cannot be minimized. It is obvious that this leads to a longer power outage time for the dc grid, which is not desired economically.

To increase the reliability and availability of the dc distribution grid, the dc fault ride-through (FRT) capability of the DAB3 converter is required with the following objectives: a) ride through the fault even at $0\,V$, b) transfer the highest possible power to the connected loads and support the dc voltage of the faulty grid, c) limit the peak current of the semiconductor devices in the DAB3 converter, d) limit the thermal stress of the DAB3 converter, f) a fast-dynamic control. During the dc FRT operation, the fault location is detected, and subsequently the adjacent protection units disconnect the faulty part from the dc grid as shown in Figure 1.

However, most of the research has focused on the internal fault analysis of the single-phase dual-active bridge converter (DAB1) [12]–[15], and very few literatures have discussed about the dc FRT of the dual-active bridge converter [16], [17]. In this paper, the dc FRT capability of the state-of-the-art modulation schemes for the DAB3 converters i.e. single phase-shift (SPS) modulation [2] and asymmetrical duty-cycle control (ADCC) [18] [19] is thoroughly analyzed. It is found that the SPS modulation is not suitable for dc FRT due to high peak currents and the hard-switching operation in the fault condition, although it exhibits a good power transfer capability. Thanks to the reduced duty cycles in the ADCC, the current stress is limited, and soft switching can be realized in fault conditions. However, the ADCC shows very limited power transfer capability for low dc voltages, and more importantly it cannot be operated at 0 V fault voltage. This makes ADCC insufficient to support the dc FRT operation of the DAB3 converter. To tackle these limitations, a space-vector-based asymmetrical duty-cycle control (SV-ADCC) method is proposed for the dc FRT operation. The proposed method can ride through a dc fault down to 0 V with a controllable peak current of the devices and a significantly higher power transfer capability in low voltage conditions. Moreover, soft switching can be realized for all devices. The proposed method can be combined with ADCC and SPS to achieve a fast-dynamic control.

II. DC FAULT ANALYSIS OF STATE-OF-THE-ART MODULATION SCHEMES

The schematic of the DAB3 converter is depicted in Figure 2, which employs 10 kV SiC MOSFETs for the input bridge of 5 kV dc voltage, and 1.7 kV Si IGBTs for the output bridge of 800 V dc voltage. A dc fault on the output bridge is investigated in the remaining part of the paper. Detailed parameters of the simulated DAB3 converter are presented in Table I.

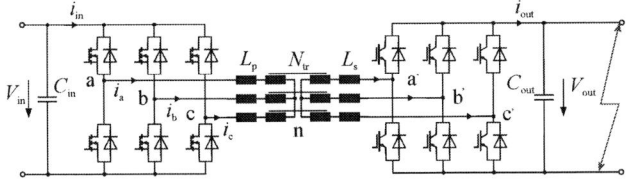

Fig. 2: Schematic of a DAB3 converter

TABLE I: Parameters of the 300 kW DAB3 converter

Parameters	Value
Input dc voltage V_{in}	5 kV
Output dc voltage range V_{out}	$640 - 960$ V
Rated output dc voltage $V_{\mathrm{out,rated}}$	800 V
Switching frequency f	20 kHz
Transformer turns ratio N_{tr}	25/4
Total leakage inductance $L_\sigma = L_{\mathrm{p}} + N_{\mathrm{tr}}^2 L_{\mathrm{s}}$	208 µH
Input dc-link capacitor C_{in}	100 µF
Output dc-link capacitor C_{out}	100 µF
Rated power P_{rated}	300 kW

A. Single Phase-Shift Modulation

The conventional and simplest control for the DAB3 converter is the SPS modulation. As shown in Figure 3, in this modulation scheme, the three-phase converters operate in block mode with fixed duty-cycles of 50%, and the transferred power can be regulated by only one control variable i.e. load angle φ. The transferred power can be characterized by (1), where $d = N_{\mathrm{tr}} V_{\mathrm{out}}/V_{\mathrm{in}}$. Detailed analysis of the ZVS boundaries can be found in [2]. The operation boundary of the SPS modulation

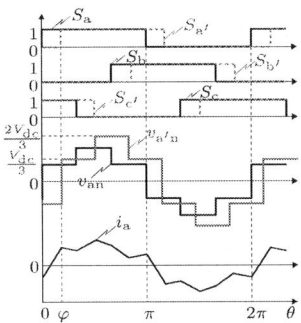

Fig. 3: Switching pattern of the SPS modulation $(0 < \varphi < \frac{\pi}{3})$

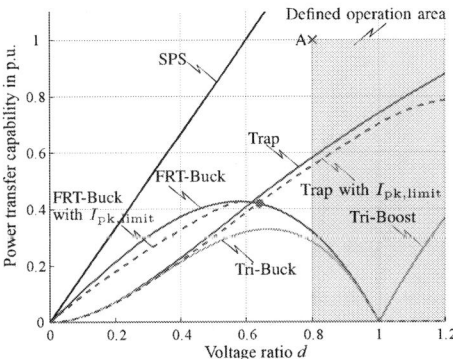

Fig. 4: Operation range of SPS, ADCC and the proposed FRT control

of the DAB3 converter is shown in Figure 4 with a per unit power of $P_{\mathrm{pu}} = 300$ kW. The voltage range of the DAB3 converter at the normal operation is considered to be $\pm 20\,\%$, while the transferred power at the normal operation ranges from 0 to 1 p.u.. The defined operation area of the DAB3 converter is also shown in Figure 4. It should be mentioned that only the case of an unidirectional power flow is considered in this paper, since the proposed method and the corresponding analysis can be extended to the reversed power flow direction similarly.

$$P_{\mathrm{SPS}} = \frac{d V_{\mathrm{in}}^2}{2\pi f L_\sigma}\left(\frac{2\varphi}{3} - \frac{\varphi^2}{2\pi}\right), \quad \text{for } 0 \le \varphi \le \frac{\pi}{3} \quad (1)$$

Although in the SPS modulation the DAB3 converter can still theoretically transfer a certain amount of power in low voltage-ratio conditions as shown in Figure 4, the peak current of the devices increases dramatically with decreasing output voltage. Given that the DAB3 converter in this application operates only within the grey area, the highest peak current of 86.3 A is found at the operation point A ($d = 0.8$, $P = 1.0$ p.u.). Simulation results of the operation point A are presented in Figure 5. However, in the 0 V fault condition (worst-case scenario), the device peak current reaches 134.9 A as shown in Figure 6 although there is no power transfer. This is because the transformer current is produced by the voltage difference between the primary side and secondary side of the transformer applied on the equivalent leakage inductor. When

2251

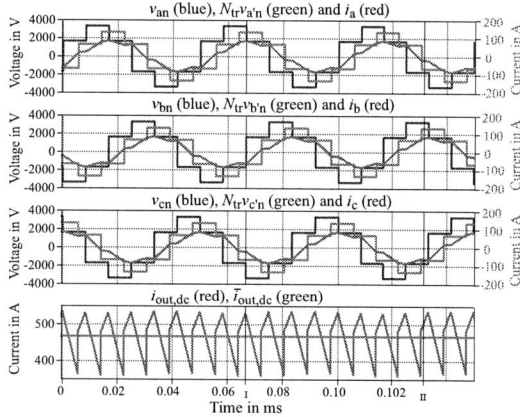

Fig. 5: Simulation waveforms of SPS modulation at point A
$P = 300\,\text{kW}$, $d = 0.8$, $I_{\text{a,pk}} = 86.3\,\text{A}$

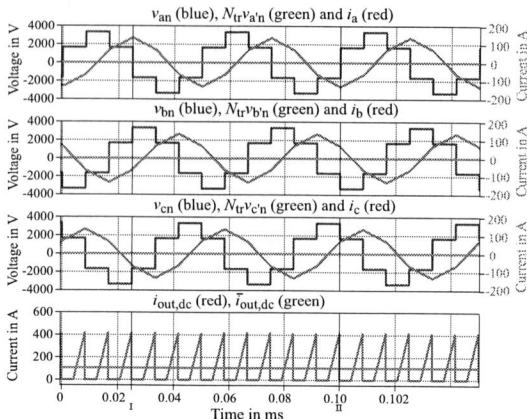

Fig. 6: Simulation waveforms of SPS modulation at the worst case scenario $P = 0\,\text{kW}$, $d = 0$, $I_{\text{a,pk}} = 134.9\,\text{A}$

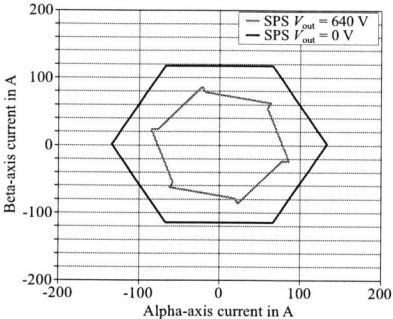

Fig. 7: Three-phase transformer currents in $\alpha\beta$ domain for SPS mode

the output voltage reduces to $0\,\text{V}$, the secondary voltage of the transformer decreases to $0\,\text{V}$ as well, and the voltage applied on the leakage inductor increases dramatically. Consequently, high currents flow through the transformer. At the beginning of the fault transient, the peak current is even higher than $134.9\,\text{A}$ due to the overshoot in the transformer current. This indicates that the current magnitude is more than $75\,\%$ than the current at the operation point A. From the thermal perspective, power converters are normally not designed to ride through this condition. When the three-phase transformer currents are transformed from abc to $\alpha\beta$ coordinates, they can be expressed by a hexagon trajectory as shown in Figure 7. Taking the current trajectory for $V_{\text{out}} = 640\,\text{V}$ as the benchmark, it is clearly seen that the current trajectory for $V_{\text{out}} = 0\,\text{V}$ is significantly larger. Moreover, it should be noted that the output bridge of the DAB3 converter are hard switched in the low voltage-ratio conditions. The corresponding thermal stress could result in damage of the semiconductor devices. Therefore, the SPS modulation is not suitable for the dc FRT operation.

It should be mentioned that the device peak current is also inversely proportional to the switching frequency. That means with a variable frequency control in the SPS mode i.e. increasing the switching frequency by a factor of 1.56, the device peak current can be maintained as $86.3\,\text{A}$ in the given fault condition. One of the drawbacks is that the switching losses will increase not only due to increased switching frequency but also due to the hard-switched secondary bridge at the low voltage ratio. In addition, the transformer design gets complicated for a variable frequency in high power applications. Therefore, only the constant frequency operation of the DAB3 converter is considered in this paper.

B. Asymmetrical Duty-Cycle Control

The ADCC method has been proposed in [18] to realize the soft-switching operation of the DAB3 converter in a wide load and a wide voltage-ratio condition, which could be used for dc fault conditions. By introducing the duty cycles of the input and output bridge as two additional control variables, the DAB3 converter can operate in three soft-switching oper-

ation modes i.e. triangular current buck operation mode (Tri-Buck), triangular current boost operation mode (Tri-Boost) and trapezoidal current mode (Trap). For the dc fault condition of the output bridge, the converter operates at a low voltage ratio ($d < 1$). Since Tri-Boost is not valid for $d < 1$, only Tri-Buck and Trap modes are considered as shown in Figure 8. The operation boundaries of the three modes can be calculated using (2) as found in [18]. The operation range of the three ADCC modes can be seamless combined as shown in Figure 4. Applying the maximum peak current of $I_{\text{pk,limit}} = I_{\text{pk,A}} = 86.3\,\text{A}$ as an additional constraint, the maximum power of Trap mode is slightly reduced, while the Tri-Buck mode is not influenced.

$$P_{\text{ADCC}} = \begin{cases} \dfrac{d^2 V_{\text{in}}^2 \varphi^2}{\pi^2 f L_\sigma (1-d)} & \text{Tri} - \text{Buck} \\[3mm] \dfrac{d V_{\text{in}}^2 \varphi^2}{\pi^2 f L_\sigma (d-1)} & \text{Tri} - \text{Boost} \\[3mm] \dfrac{d V_{\text{in}}^2}{9\pi^2 f L_\sigma (1+d)^2} \cdot (6\pi\varphi(1+d^2) - \\ 9\varphi^2(1+d+d^2) - \pi^2(1-d)^2) & \text{Trap} \end{cases}$$

(2)

As presented in Figure 4, at a very low voltage ratio e.g. $d = 0.1$, the Tri-Buck mode would be selected other than the Trap mode. It can be seen in Figure 9 that the peak

The 2018 International Power Electronics Conference

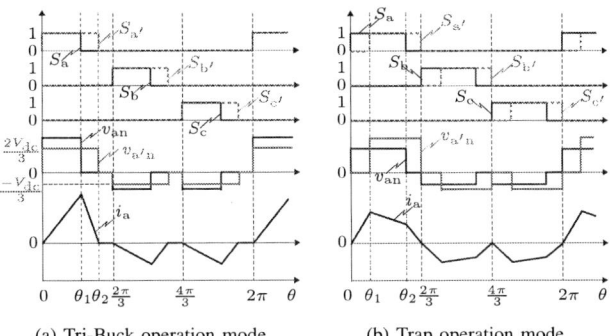

(a) Tri-Buck operation mode   (b) Trap operation mode

Fig. 8: Switching patterns of the ADCC modulation

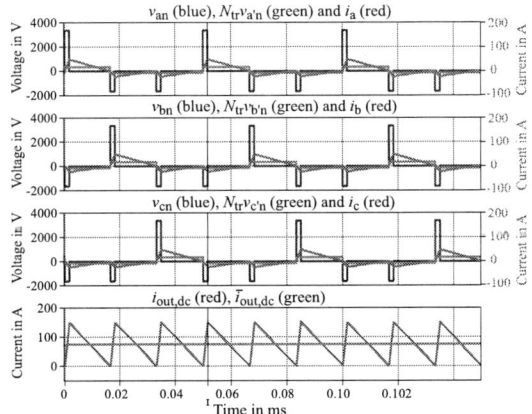

Fig. 9: Simulation waveforms of Tri-Buck modulation at $P = 6\,\text{kW}$, $d = 0.1$, $I_{\text{a,pk}} = 24\,\text{A}$

current in the Tri-Buck mode is effectively limited to 24 A, because the duty cycle of the upper devices in the input bridge D_1 is limited. This indicates that the DAB3 converter works in the safe operating area, because the current is much lower than the maximum peak current at operation point A. It should be mentioned that in the Tri-Buck mode, the condition $D_1 = d \cdot D_2$ is required, where D_2 is the duty cycle of the upper devices in the output bridge. Although D_2 is maximized to $1/3$ to transfer the maximum power in the low voltage ratio condition, D_1 is ultimately constrained by the voltage ratio d. The current trajectory in $\alpha\beta$ coordinates is shown in Figure 10. In the Tri-Buck mode, the current trajectory becomes an "Y" shape which has a significantly smaller magnitude than the hexagon trajectory of the SPS mode at the operation point A. Moreover, when $d = 0$ in the worst case, $D_1 = 0$ which indicates that the Tri-Buck is not valid in this case. In summary, the Tri-Buck and Trap modes of the ADCC method could be used for the dc FRT operation in low voltage ratio conditions. However, the limitations are the lack of 0 V FRT capability and the limited power transfer capability in the low-voltage fault conditions.

III. PROPOSED SPACE-VECTOR-BASED ASYMMETRICAL DUTY-CYCLE CONTROL (SV-ADCC)

To enhance the dc FRT capability of the ADCC method, it is clear from Section II that the duty cycle D_1 needs to

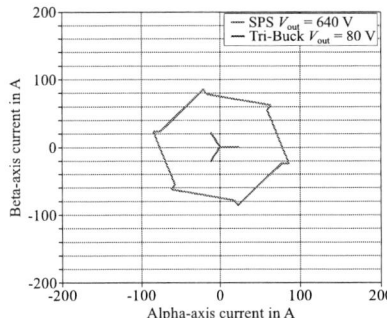

Fig. 10: Three-phase transformer currents in $\alpha\beta$ domain for SPS mode and Tri-Buck mode

be further increased. Note that in the Tri-Buck mode only three voltage vectors i.e. $\vec{V}_{100}$, $\vec{V}_{010}$ and $\vec{V}_{001}$ are applied for each bridge as shown in Figure 11(a). $\vec{V}_{\text{p}}$ refers to the voltage vector of the primary bridge, and $\vec{V}_{\text{s}}$ refers to the voltage vector of the secondary bridge. In each one third of the switching period, the current starts from the origin to the peak, and then returns to the origin following a "Y" shape trajectory in $\alpha\beta$ coordinate.

In order to increase D_1 while maintaining soft switching, the other three voltage vectors are utilized. According to the operation principle of the Tri-Buck mode [18], D_2 has to be maximized to $1/3$ to transfer more power, since the output voltage is low. Therefore, the proposed SV-ADCC method introduces the three remaining voltage vectors i.e. $\vec{V}_{011}$, $\vec{V}_{101}$ and $\vec{V}_{110}$ of the input bridge to the Tri-Buck mode to form a new modulation scheme i.e. FRT-Buck as shown in Figure 12. For example, in the first one third of the switching period, not only $\vec{V}_{\text{p},100}$ but also the opposite voltage vector $\vec{V}_{\text{p},011}$ is applied to the input bridge. To realize the zero-current switching (ZCS), condition (3) needs to be fulfilled.

$$V_{\text{in}} \cdot D_{\text{p},100} - V_{\text{in}} \cdot D_{\text{p},011} = N_{\text{tr}} V_{\text{out}} \cdot D_{\text{s},100}, \qquad (3)$$

where $D_{\text{p},100}$, $D_{\text{p},011}$ and $D_{\text{s},100}$ correspond to the relative applied time of the voltage vectors $\vec{V}_{\text{p},100}$, $\vec{V}_{\text{p},011}$ and $\vec{V}_{\text{s},100}$ respectively. Referred to Figure 12, $D_{\text{p},100} = \frac{\theta_1}{2\pi}$, $D_{\text{p},011} = \frac{1}{3} - \frac{\theta_2}{2\pi}$ and $D_{\text{s},100} = \frac{1}{3}$. In other words, by using the opposite voltage vectors of the input bridge, the volt-seconds of the input bridge can be compensated or even self balanced. Therefore, the peak current can be controlled by $D_{\text{p},100}$ and $D_{\text{p},011}$. More importantly, even at 0 V fault condition, the proposed method is still applicable with the condition of $D_{\text{p},100} = D_{\text{p},011}$. The current trajectory of the proposed FRT-Buck mode in the $\alpha\beta$ coordinate is presented in Figure 11(b).

The piecewise linear transformer current i_{a} can be calculated in the following steps. Substituting θ_1, θ_2 and $D_{\text{s},100} = \frac{1}{3}$ into (3), we can obtain

$$\theta_1 + \theta_2 = \frac{2\pi N_{\text{tr}} V_{\text{out}}}{3 V_{\text{in}}} + \frac{2\pi}{3}. \qquad (4)$$

For $0 \leq \theta < \theta_1$,

$$i_{\text{a}} = \frac{2(V_{\text{in}} - N_{\text{tr}} V_{\text{out}})}{3\omega L_\sigma} \cdot \theta, \qquad (5)$$

2253

The 2018 International Power Electronics Conference

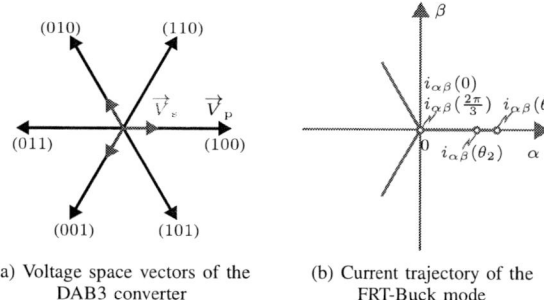

(a) Voltage space vectors of the DAB3 converter

(b) Current trajectory of the FRT-Buck mode

Fig. 11: Voltage space vectors and current trajectory in $\alpha\beta$ coordinate

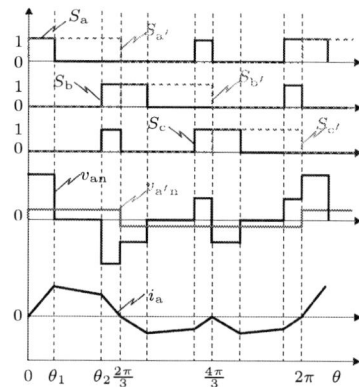

Fig. 12: Switching pattern of the proposed FRT-Buck operation mode

where $\omega = 2\pi f$.

For $\theta_1 \le \theta < \theta_2$,

$$i_a = i_a(\theta_1) - \frac{2N_{\mathrm{tr}}V_{\mathrm{out}}}{3\omega L_\sigma} \cdot (\theta - \theta_1). \quad (6)$$

For $\theta_2 \le \theta < \frac{2\pi}{3}$,

$$i_a = i_a(\theta_2) - \frac{2(V_{\mathrm{in}} + N_{\mathrm{tr}}V_{\mathrm{out}})}{3\omega L_\sigma} \cdot (\theta - \theta_2). \quad (7)$$

Applying (4) into (5) - (7), it is obtained that $i_a(\frac{2\pi}{3}) = 0$, which verifies the ZCS condition given in (3). The positive peak currents are obtained from (5) - (7), where $i_a(\theta_1) > i_a(\theta_2)$.

$$\begin{cases} i_a(\theta_1) = \dfrac{2(V_{\mathrm{in}} - N_{\mathrm{tr}}V_{\mathrm{out}})\theta_1}{3\omega L_\sigma} \\ i_a(\theta_2) = \dfrac{2(V_{\mathrm{in}} + N_{\mathrm{tr}}V_{\mathrm{out}})\theta_1}{3\omega L_\sigma} - \dfrac{2N_{\mathrm{tr}}V_{\mathrm{out}}(V_{\mathrm{in}} + N_{\mathrm{tr}}V_{\mathrm{out}})}{9\omega V_{\mathrm{in}}L_\sigma} \end{cases} \quad (8)$$

The power transfer capability of the FRT-Buck mode can be characterized by (9), which is also shown in Figure 4.

$$P_{\mathrm{FRT-Buck}} = \frac{2dV_{\mathrm{in}}^2}{fL_\sigma}\left(-\left(\frac{\theta_1}{2\pi} - \frac{1+d}{6}\right)^2 + \frac{1-d^2}{36}\right) \quad (9)$$

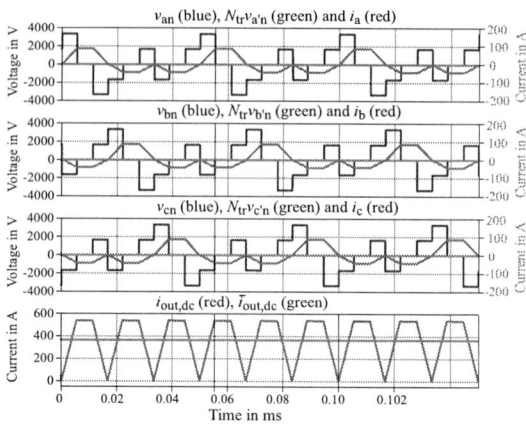

Fig. 13: Simulation waveforms of the proposed FRT-Buck modulation at $P = 0\,\mathrm{kW}$, $d = 0$, $I_{a,\mathrm{pk}} = 86.3\,\mathrm{A}$

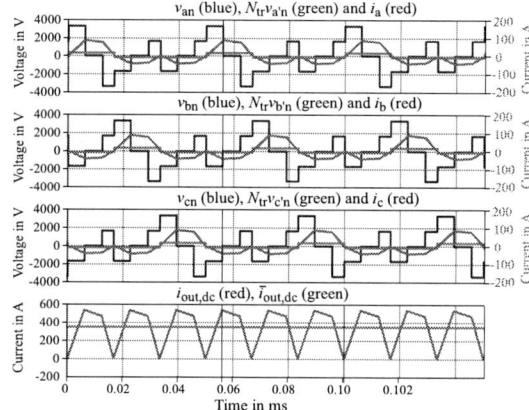

Fig. 14: Simulation waveforms of the proposed FRT-Buck modulation at $P = 28.1\,\mathrm{kW}$, $d = 0.1$, $I_{a,\mathrm{pk}} = 86.3\,\mathrm{A}$

To realize a controllable peak current in dc fault conditions, $i_a(\theta_1)$ has an upper limit of $I_{\mathrm{pk,limit}}$. Consequently, this gives a constraint for θ_1,

$$\theta_1 \le \frac{3\omega L_\sigma I_{\mathrm{pk,limit}}}{2(V_{\mathrm{in}} - N_{\mathrm{tr}}V_{\mathrm{out}})}. \quad (10)$$

Substituting (10) into (9), the maximum transferred power of the FRT-Buck mode with a limited peak current of $I_{\mathrm{pk,limit}} = 86.3\,\mathrm{A}$ is obtained, which is shown in Figure 4. It can be seen that the maximum transferred power is slightly reduced compared to the theoretical power transfer capability without current constraints. Nevertheless, it still exhibits a significantly higher power transfer capability than Tri-Buck especially at very low voltage-ratios. Simulation results in Figure 13 present the $0\,\mathrm{V}$ FRT operation with the proposed method. Although no power can be transferred to the load side in this condition, a relatively large output dc current will recharge the dc-link capacitor after the fault clearance. The current trajectory of FRT-Buck is shown in Figure 15 which coincides well with the theoretical analysis in Figure 11(b). Figure 14 presents the simulation results of the FRT-Buck mode at $d = 0.1$. Compared to the Tri-Buck mode in Figure 9, the FRT-Buck mode operates at the maximum peak current of

2254

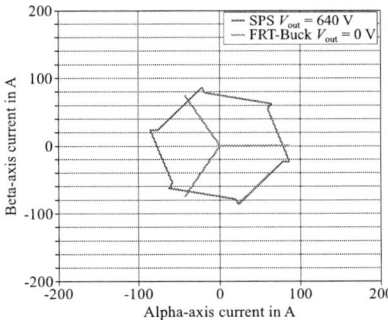

Fig. 15: Three-phase transformer currents in $\alpha\beta$ domain for SPS mode and FRT-Buck mode

TABLE II: Summary of simulation results for the 300 kW DAB3 converter

Mode	d	I_{pk}	$\bar{i}_{\mathrm{out,dc}}$	P	Comment
SPS	0.8	86.3 A	468.8 A	300 kW	Point A
SPS	0	134.9 A	468.3 A	0	—
Tri-Buck	0.1	24.0 A	75 A	6 kW	—
Tri-Buck	0	0	0	0	No gate signals
FRT-Buck	0.1	86.3 A	363.8 A	28.1 kW	—
FRT-Buck	0	86.3 A	351.2 A	0	—

$I_{\mathrm{pk}} = 86.3$ A, and transfers 28.1 kW power instead of only 6 kW for the Tri-Buck mode. With the proposed FRT-Buck operation mode, the DAB3 converter can not only ride through dc faults down to 0 V with a controllable peak current but also transfer more power in low voltage ratio conditions. In the end, the simulation results of SPS, ADCC and FRT-Buck are summarized in Table II.

It is also noticed that with the proposed method each primary switch is turned on and off once more than the Tri-Buck mode. However, as shown in Figure 16, for upper switches in the primary bridge, ZVS is realized for the additional turn-on operation, and ZCS is realized for the additional turn-off operation. Correspondingly, as for lower switches in the primary bridge, ZCS is realized for the additional turn-on operation, and the additional turn-off is in hard-switching mode. Since the additional turn-off of the lower device occurs with a current that equals only to half of the peak current, the resulted loss is not critical. As shown in Figure 17, the secondary bridge realizes ZCS turn-on and turn-off for all devices. Therefore, no hard-switching turn-on occurs in the FRT-Buck mode. The additional conduction loss takes only a small part of the total loss, since the conduction time is short and the rms currents are low during the conduction time. The soft-switching feature of the FRT-Buck mode makes it also possible to be applied as one of the normal operation modes for the low voltage ratio. As shown in Figure 4, the soft-switching range of the DAB3 converter in low voltage ratio conditions is further extended by the proposed FRT-Buck mode. Since the transformer current starts and ends at zero current in every switching cycle, a smooth mode transition can be achieved between the FRT-Buck mode and the other three ADCC modes without any overshoot current. This allows for a fast dynamic

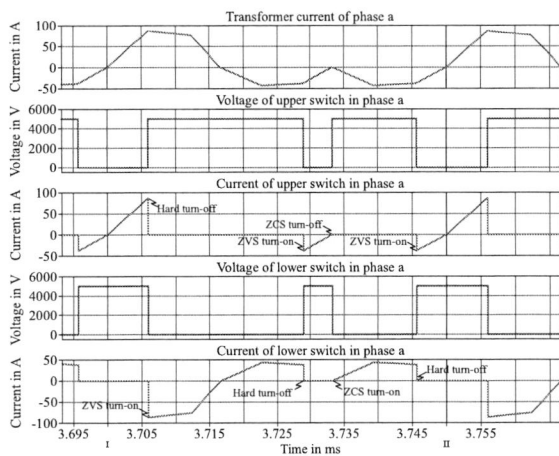

Fig. 16: Simulation waveforms of the secondary bridge device in FRT-Buck mode at $P = 28.1$ kW, $d = 0.1$, $I_{\mathrm{a,pk}} = 86.3$ A

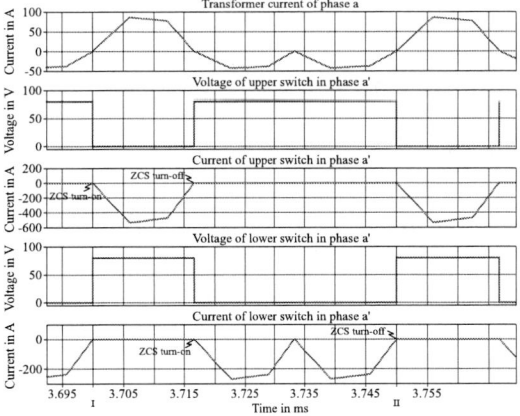

Fig. 17: Simulation waveforms of the primary bridge device in FRT-Buck mode at $P = 28.1$ kW, $d = 0.1$, $I_{\mathrm{a,pk}} = 86.3$ A

control of the DAB3 converter with hybrid modulation mode operation (SPS+ADCC+FRT-Buck), which will be discussed in future publications.

IV. CONCLUSION

In this paper, a comprehensive dc fault analysis of the state-of-the-art modulation schemes has been performed for the DAB3 converter. The SPS modulation is not suitable for the dc FRT operation due to the resulted high peak currents, while the dc FRT capability of the ADCC modulation is limited due to the lack of 0 V FRT capability and the limited power transfer capability at low voltage-ratios. A SV-ADCC method has been proposed by introducing the opposite voltage space vectors of the input bridge to realize a soft-switching dc FRT operation down to 0 V with a controllable peak current. With the proposed method, the power transfer capability of the DAB3 converter is significantly increased in low voltage ratio conditions. Due to the soft-switching feature of the proposed FRT-Buck mode, the soft-switching range of the DAB3 converter is further extended.

Acknowledgement

Funded by the Federal Ministry of Education and Science (BMBF, FKZ03SF0490), Flexible Electrical Networks (FEN) Research Campus.

References

[1] R. W. D. Doncker, "Power electronic technologies for flexible dc distribution grids," in *2014 International Power Electronics Conference (IPEC-Hiroshima 2014 - ECCE ASIA)*, May 2014, pp. 736–743.

[2] R. W. De Doncker, D. M. Divan, and M. H. Kheraluwala, "A three-phase soft-switched high-power-density dc/dc converter for high-power applications," *IEEE transactions on industry applications*, vol. 27, no. 1, pp. 63–73, 1991.

[3] M. H. Kheraluwala, R. W. Gasgoigne, D. M. Divan, and E. Bauman, "Performance characterization of a high power dual active bridge dc/dc converter," in *Conference Record of the 1990 IEEE Industry Applications Society Annual Meeting*, Oct 1990, pp. 1267–1273 vol.2.

[4] S. P. Engel, M. Stieneker, N. Soltau, S. Rabiee, H. Stagge, and R. W. D. Doncker, "Comparison of the modular multilevel dc converter and the dual-active bridge converter for power conversion in hvdc and mvdc grids," *IEEE Transactions on Power Electronics*, vol. 30, no. 1, pp. 124–137, Jan 2015.

[5] N. Soltau, H. Stagge, R. W. D. Doncker, and O. Apeldoorn, "Development and demonstration of a medium-voltage high-power dc-dc converter for dc distribution systems," in *2014 IEEE 5th International Symposium on Power Electronics for Distributed Generation Systems (PEDG)*. IEEE, 2014, pp. 1–8.

[6] N. H. Baars, J. Everts, C. G. E. Wijnands, and E. A. Lomonova, "Performance evaluation of a three-phase dual active bridge dc-dc converter with different transformer winding configurations," *IEEE Transactions on Power Electronics*, vol. 31, no. 10, pp. 6814–6823, Oct 2016.

[7] P. Joebges, J. Hu, and R. W. D. Doncker, "Design method and efficiency analysis of a dab converter for pv integration in dc grids," in *2016 IEEE 2nd Annual Southern Power Electronics Conference (SPEC)*, Dec 2016, pp. 1–6.

[8] J. Hu, P. Joebges, and R. W. De Doncker, "Maximum power point tracking control of a high power dc-dc converter for pv integration in mvdc distribution grids," in *Applied Power Electronics Conference and Exposition (APEC), 2017 IEEE*. IEEE, 2017, pp. 1259–1266.

[9] M. Stieneker, B. J. Mortimer, N. R. Averous, H. Stagge, and R. W. De Doncker, "Optimum design of medium-voltage dc collector grids depending on the offshore-wind-park power," in *Power Electronics and Machines for Wind and Water Applications (PEMWA), 2014 IEEE Symposium*. IEEE, 2014, pp. 1–8.

[10] S. Inoue and H. Akagi, "A bidirectional dc-dc converter for an energy storage system with galvanic isolation," *IEEE Transactions on Power Electronics*, vol. 22, no. 6, pp. 2299–2306, Nov 2007.

[11] N. H. Baars, H. Huisman, J. L. Duarte, and J. Verschoor, "A 80 kw isolated dc-dc converter for railway applications," in *2014 16th European Conference on Power Electronics and Applications*, Aug 2014, pp. 1–10.

[12] A. M. Airabella, G. G. Oggier, L. E. Piris-Botalla, C. A. Falco, and G. O. Garca, "Open transistors and diodes fault diagnosis strategy for dual active bridge dc-dc converter," in *2012 10th IEEE/IAS International Conference on Industry Applications*, Nov 2012, pp. 1–6.

[13] A. M. Airabella, G. G. Oggier, and G. O. Garca, "Semi-conductors faults analysis in dual active bridge dc-dc converter," *IET Power Electronics*, vol. 9, no. 6, pp. 1103–1110, 2016.

[14] E. Ribeiro, A. J. M. Cardoso, and C. Boccaletti, "Fault analysis of dual active bridge converters," in *IECON 2012 - 38th Annual Conference on IEEE Industrial Electronics Society*, Oct 2012, pp. 398–403.

[15] A. Virdag, T. Hager, J. Hu, and R. W. D. Doncker, "Short circuit behavior of dual active bridge dcdc converter with low resistance dc side fault," in *2017 IEEE 8th International Symposium on Power Electronics for Distributed Generation Systems (PEDG)*, April 2017, pp. 1–6.

[16] B. Zhao, Q. Song, J. Li, Q. Sun, and W. Liu, "Full-process operation, control, and experiments of modular high-frequency-link dc transformer based on dual active bridge for flexible mvdc distribution: A practical tutorial," *IEEE Transactions on Power Electronics*, vol. 32, no. 9, pp. 6751–6766, Sept 2017.

[17] Y. A. Harrye, K. H. Ahmed, and A. A. Aboushady, "Dc fault isolation study of bidirectional dual active bridge dc/dc converter for dc transmission grid application," in *IECON 2015 - 41st Annual Conference of the IEEE Industrial Electronics Society*, Nov 2015, pp. 003 193–003 198.

[18] J. Hu, N. Soltau, and R. W. De Doncker, "Asymmetrical duty-cycle control of three-phase dual-active bridge converter for soft-switching range extension," in *Energy Conversion Congress and Exposition (ECCE), 2016 IEEE*. IEEE, 2016, pp. 1–8.

[19] J. Hu, Z. Yang, N. Soltau, and R. W. De Doncker, "A duty-cycle control method to ensure soft-switching operation of a high-power three-phase dual-active bridge converter," in *Future Energy Electronics Conference and ECCE Asia (IFEEC 2017-ECCE Asia), 2017 IEEE 3rd International*. IEEE, 2017, pp. 866–871.

Gap in pagination due to withheld paper.

Pages 2257-2261

The 2018 International Power Electronics Conference

A Compound 10kV DVR System based on Solid State Transformer Structure

Yaqian Zhang, Jianzhong Zhang*, Xing Hu, Zakiud Din
School of Electrical Engineering, Southeast University, Nanjing, China
*E-mail: jiz@seu.edu.cn

Abstract—**A new simplified dynamic voltage restorer (DVR) topology is proposed to compensate for voltage abnormality in this paper. Inspired by the general three-stage solid state transformer (SST) structure, the proposed DVR structure can not only realize voltage compensation for medium voltage grid, but also possess a smaller volume compared to the conventional DVR system by replacing all the line frequency transformers with high frequency transformers. Compensation criteria and control strategies on system level and stage level are given and analyzed. Finally, simulations are carried out and verify the feasibility and validity of the proposed topology and control method.**

Keywords— *DVR, power quality, high frequency transformer, SST.*

I. INTRODUCTION

Renewable energy has been exploited in very large scale in recent decades to provide cleaner electricity for the worldwide society. With the massive distributed power generations and power electronics devices introduced into power grid, the occurrence of various abnormal grid conditions, such as voltage sags, swells, unbalance and harmonics, are common and likely to threaten the performance of certain loads. The power losses caused by long-distance transmission will also decrease voltage quality and voltage amplitude. In this case, dynamic voltage restorer (DVR) is installed in distribution network interface to compensate the load-side voltage for abnormal power grid conditions.

Fig. 1 shows a conventional DVR system, where a three-phase 10kV/380V line frequency transformer (LFT), a three-phase coupling 1:1 transformer, a shunt connected three-phase converter, a series connected three-phase inverter and a capacitor bulk are included. It is agreed that the existence of LFT, also known as iron-steel based step-down transformer, which is put on AC side to isolate medium voltage (MV) from low voltage (LV) side, makes the system expensive and bulky. The application of high frequency transformer (HFT) may decrease system volume greatly since higher frequency offers more efficient power transmission and smaller iron size. Reference [1] proposed a HFT topology to realize power transmission in high power density. However, the introduction of uncontrolled diode rectifiers could not realize bidirectional power flow, which cannot meet the requirement of DVR system.

With the rapid development of power electronics technologies in power system, solid state transformer

(SST) has been proposed to reform the traditional structure of power grid, which performs more intelligently and more flexibly compared with traditional transformer [2]. There has been many researches on SST in various aspects including topology [3]- [6], control strategy [7]- [9], applications [10]- [13], modeling [14] [15] and prototype [16]- [19].

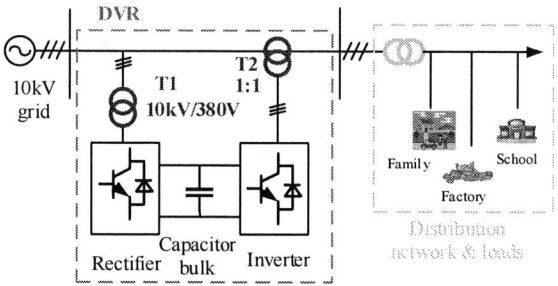

Fig.1 Conventional DVR configuration

It is suggested that SST is able to act as both power transformer and power quality controller at the same time [2]. However, this will lead to great challenges for the capacity, control and reliability of the grid-forming SST system, which has been kept as a worldwide hot issue in recent years. The adoption of cascaded SST topology in this paper demands little on power capacity and power transmission but asks for strict voltage control and accurate power flow control. The compound system is the combination of SST topology and DVR control.

The paper is organized as follows. The topology is put forward in Section II, system power flow and corresponding modular control are also analyzed. The control on system level and accurate phase detection of grid voltage are introduced in Section III. Section IV is devoted to the simulations and results analyses, after which conclusion is made in section V.

II. TOPOLOGY

Compared with the conventional "back-to-back" structure, shown in Fig. 1, a new DVR structure based on SST is proposed in Fig. 2. It is shown that two LFT, 10kV/380V T1 and coupling transformer T2, are deleted. Instead, the input stage is directly connected with 10kV grid due to the addition of DC/DC converter with HFT. In SST section of Fig .2, there includes three stages which all have detailed inner topology shown in Fig .3.

A medium voltage multiple modular converter (MMC) is adopted in input stage. The AC side of MMC is

2262

directly connected with grid and DC side provides a common medium voltage bus for the system. MMC is also one of the interfaces to obtain power from grid or feed power to grid according to different work modes of DVR.

The transmission stage, structured as dual active bridge (DAB), is the most important part to complete system power flow control since the power flow of the three-stage DVR is bidirectional, and there contains HFT to realize the isolation between input stage and output stage. Since the HFT works under very high frequency, it helps to realize much more efficient power transmission and attain much higher power density for DVR compared to LFT. Inside the transmission stage, DAB units are connected in input series and output parallel (ISOP) structure, with input connected with MVDC bus and output parallel connected into LVDC voltage.

The output stage is series connected with the loads and provides demanded compensation voltage according to the DVR control signal. Each phase is equipped with its own single-phase inverter to lower down the control complexity of output. Also, the third stage is the other power interface of DVR which should keep equivalent absolute value and corresponding direction of DVR inner power flow with the input interface when ignoring inner power losses of DVR system.

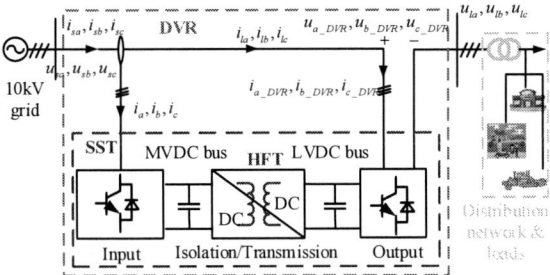

Fig. 2 Proposed DVR configuration (the simplified version)

A. System Power Flow

To explain how DVR performs its voltage correction function for the distribution network clearly, system power flow of DVR is analyzed in this part.

Equations (1-3) are given according to Fig .2, where u_{sa}, u_{sb}, u_{sc} and i_{sa}, i_{sb}, i_{sc} are three phase voltage and current on grid side, u_{la}, u_{lb}, u_{lc} and i_{la}, i_{lb}, i_{lc} are three phase voltage and current on load side, similarly, $u_{a\_DVR}, u_{b\_DVR}, u_{c\_DVR}$ and $i_{a\_DVR}, i_{b\_DVR}, i_{c\_DVR}$ represent the three phase output voltage and current of DVR, i_a, i_b, i_c are three phase input current of DVR.

$$
\begin{aligned}
i_{sa} &= i_a + i_{la} \\
i_{sb} &= i_b + i_{lb} \\
i_{sc} &= i_c + i_{lc}
\end{aligned}
\tag{1}
$$

$$
\begin{aligned}
i_{a\_DVR} &= i_{la} \\
i_{b\_DVR} &= i_{lb} \\
i_{c\_DVR} &= i_{lc}
\end{aligned}
\tag{2}
$$

$$
\begin{aligned}
u_{a\_DVR} &= u_{sa} - u_{la} \\
u_{b\_DVR} &= u_{sb} - u_{lb} \\
u_{c\_DVR} &= u_{sc} - u_{lc}
\end{aligned}
\tag{3}
$$

With detailed diagram of power flow in different modes shown in Fig. 4, where all the capital U and I represent root mean square value of their corresponding voltage and current (u and i) with the identical subscript, P_s, P_l, P_{in_DVR} and P_{out_DVR} are the grid power, load power, DVR input power on MMC AC side and DVR output power on inverter AC side. P_{trans} possesses no physical meaning but a power value only for calculation. All the power is calculated as follows.

$$
\begin{aligned}
P_s &= u_{sa}*i_{sa} + u_{sb}*i_{sb} + u_{sc}*i_{sc} \\
P_l &= u_{la}*i_{la} + u_{lb}*i_{lb} + u_{lc}*i_{lc}
\end{aligned}
\tag{4}
$$

$$
P_{trans} = u_{sa}*i_{la} + u_{sb}*i_{lb} + u_{sc}*i_{lc}
\tag{5}
$$

$$
\begin{aligned}
P_{in\_DVR} &= u_{sa}*i_a + u_{sb}*i_b + u_{sc}*i_c \\
P_{out\_DVR} &= -(u_{a\_DVR}*i_{la} + u_{b\_DVR}*i_{lb} + u_{c\_DVR}*i_{lc})
\end{aligned}
\tag{6}
$$

Neglecting the power losses of DVR system, equation (7) can be given.

$$
\begin{aligned}
P_l &= P_s \\
P_s &= P_{trans} + P_{in\_DVR} \\
P_l &= P_{trans} - P_{out\_DVR}
\end{aligned}
\tag{7}
$$

In Fig. 4, the values of U_{la}, U_{lb}, U_{lc} are supposed to be constant to maintain the normal operation of load side, as well as I_{la}, I_{lb}, I_{lc}, P_s and P_l. All the discussed power flow only exits within the DVR without influencing the power transmission from grid to load.

On the condition that grid voltage swells occur, according to (3), with U_{sa}, U_{sb}, U_{sc} rising, u_{a_DVR}, u_{b_DVR}, u_{c_DVR} will output compensation voltage at the same polarity as marked in Fig. 2, which causes the positive power output of DVR system. Meanwhile, since P_s offers the equivalent power as the load side demands and maintains constant, I_{sa}, I_{sb}, I_{sc} will decline with the swells of grid voltage. According to equation (1), since I_{la}, I_{lb}, I_{lc} should be kept unchanged, the value of I_a, I_b, I_c will be kept below zero, which means i_a, i_b, i_c will flow in opposite direction from that is marked in Fig .3.

In the above case, the part of output stage that is series inserted in grid acts as an equivalent load, with absorbed power flowing in feedback direction to input stage. For the occasion of voltage sags, the analysis process shown in Fig .4 is similar.

As illustrated in Fig. 4, the series connected part of DVR system acts as a controlled load with power flowing in feedback direction in case that DVR compensates for voltage swells. Under voltage sagging condition, the series connected part offers power to load side as a controlled voltage source with power flowing in feedforward direction inside DVR structure.

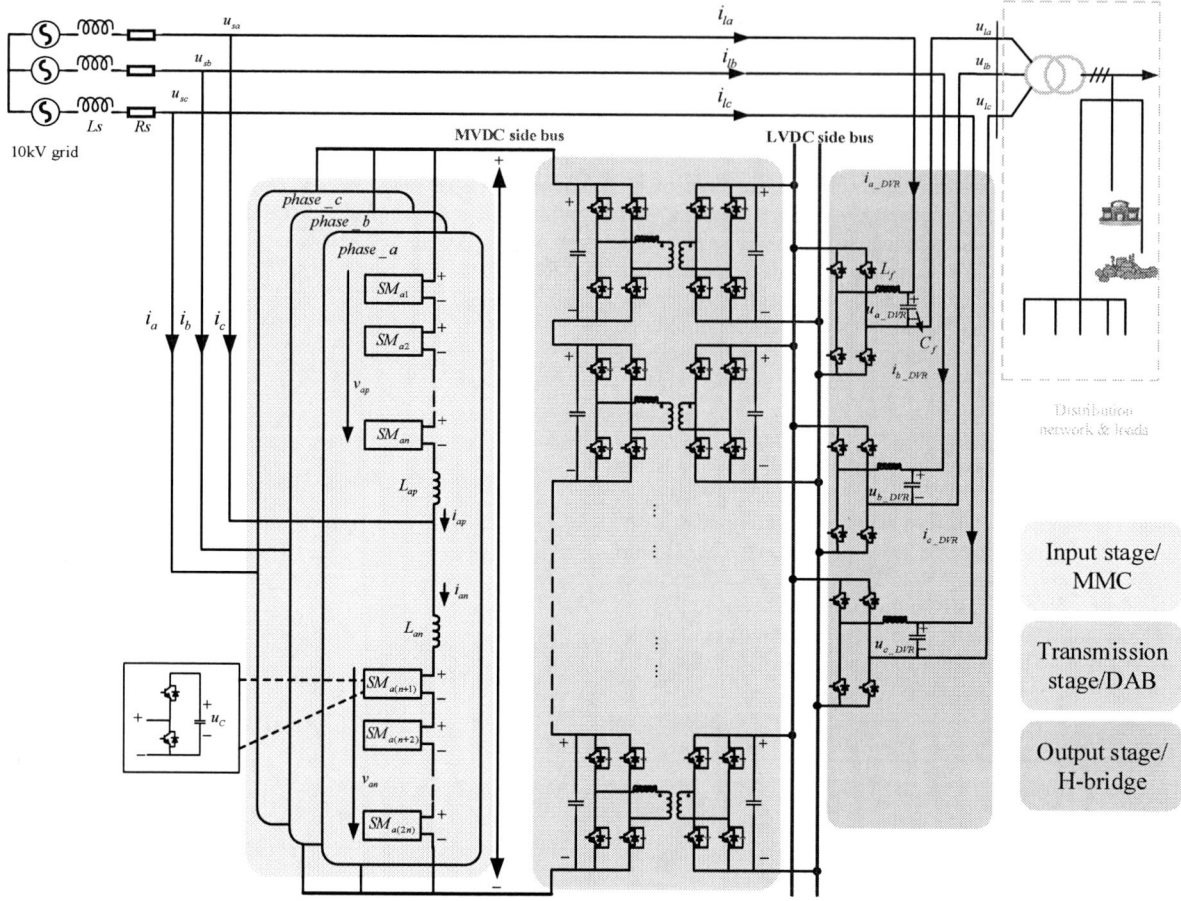

Fig. 3 Proposed DVR configuration with DAB (the detailed version)

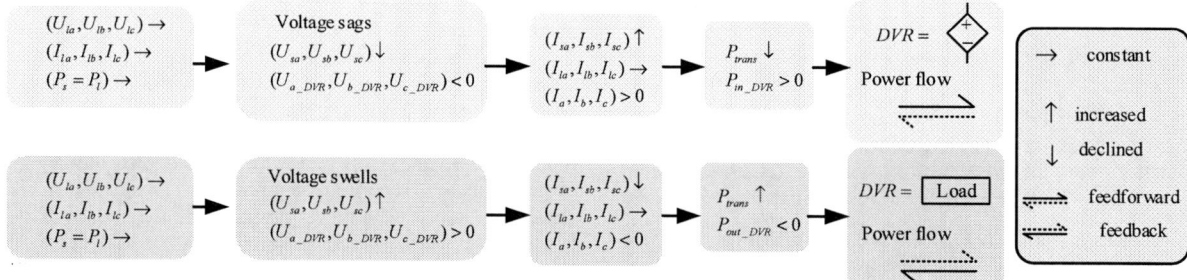

Fig. 4 Power flow in different operation modes of DVR

B. MMC based Input stage

The three-stage cascaded structure is the most commonly used SST topology due to its total decoupling of each stage. Cascaded H-bridge and MMC structure are most adopted to realize the voltage conversion from MVAC to LVDC or MVDC. In comparison, MMC can provide more flexible DAB number selection which have to be identical with the number of H-bridge submodules in cascaded H-bridge structure. In addition, the submodule in cascaded H-bridge structure usually keeps low capacitor DC voltage and this will lead to the limitation of voltage level and power level of DAB units. Therefore, in this paper, MMC is selected as the input stage for the proposed SST based DVR topology. And the output

MVDC voltage will be shared by the series connected input of several identical DAB units.

Fig. 5 shows the DC output control of MMC, where u_{a_ref}, u_{b_ref}, u_{c_ref} refer to fundamental modulation waveforms for three phases of MMC. The control strategy is designed according to different operation modes of DVR system. The reference current amplitude of inner loop is attained from the PI controller with input of difference between MVDC u_{H_DC} and its reference voltage.

Fig. 5(a) shows the control under voltage sagging condition, in which case the input stage works as a rectifier and the power flows in feedforward direction, Fig. 5(b) shows the control under voltage swelling condition, in which case the input stage works in inverter mode and the power flows in feedback direction.

2264

The 2018 International Power Electronics Conference

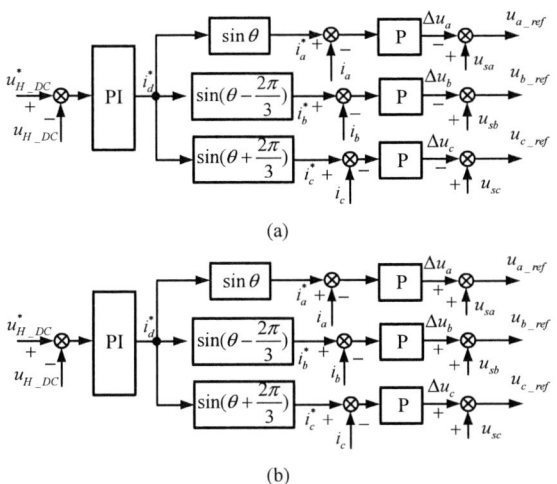

(a)

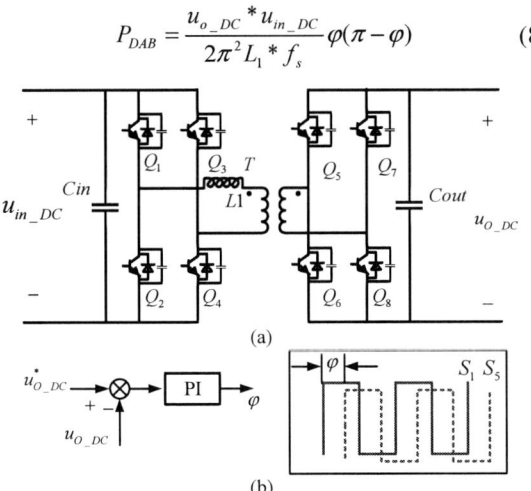

(b)

Fig. 5 Voltage control of MVDC on stage level under (a) feedforward and (b) feedback power flow condition

C. DAB based transmission stage

DAB structure is used as transmission stage in the proposed SST based DVR. In Fig .6(a), inside the DAB unit, HFT is included to isolate MV side from LV side. The volume of HFT is much smaller compared with that of LFT. As mentioned in Section II, this stage is the key one to decide the direction of power flow inside system.

The direction of DAB power flow is controlled by phase difference φ between the two H-bridges in different sides. Power flows in feedforward direction with $0 < \varphi < \pi$ and in feedback direction with $-\pi < \varphi < 0$. In equation (8), P_{DAB} is the transmission power a single DAB can offer, L_1 is the leakage inductance of HFT, f_s is the switching frequency, u_{o_DC} and u_{in_DC} are the output voltage and input voltage of DAB unit.

$$P_{DAB} = \frac{u_{o\_DC} * u_{in\_DC}}{2\pi^2 L_1 * f_s} \varphi(\pi - \varphi) \qquad (8)$$

(a)

(b)

Fig .6 (a) Topology and (b) control of DAB unit

In Fig .6(b), the phase difference φ is obtained by PI regulator with deviation input of referred DC value and measured output DC voltage. Fig .7 shows the modulation pulses S_1-S_8 of each switch Q_1-Q_8, u_{Bri1} and u_{Bri2} are the

voltage of HFT interfaces shown in Fig .6(a) and they co-decide the slope of inductance current i_L. Since the phase difference only exits between bridges not in H-bridge arms, the modulation shown in Fig .6(b) and Fig .7 is called single phase shift (SPS) modulation.

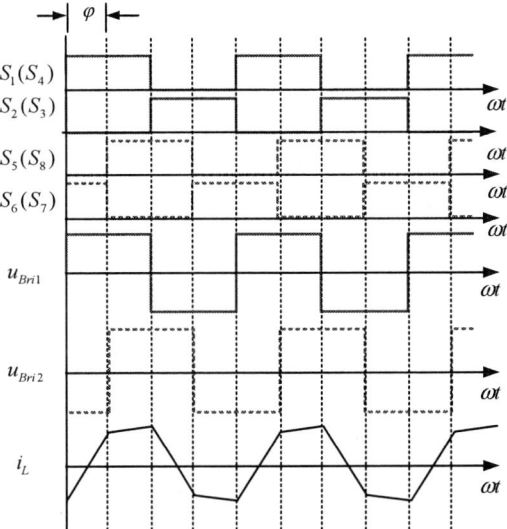

Fig.7 SPS modulation of DAB

D. H-bridge based output stage

The inverters are all single-phase H-bridges and driven by the same low voltage DC bus, their outputs are connected in series with three phase distribution network. In Fig. 8, the control strategy of inverter stage is also designed according to different operation modes. Fig. 8(a) shows the control diagram with DVR working in feedforward power flow mode to compensate for voltage sags, in which case each H-bridge performs as an inverter. Fig. 8(b) shows the control process in case of voltage swells compensation.

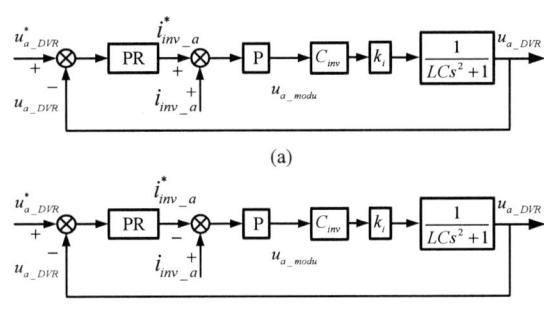

(a)

(b)

Fig. 8 Voltage control on output stage under (a) feedforward and (b) feedback power flow condition

Outer voltage loop introduces PR controller to realize the static error free tracking control of AC voltage, inner current loop adopts proportion regulator to enhance the response speed of the system, u_{DVR}^* is obtained from detection system which will be discussed in next section.

III. SYSTEM CONTROL

In this section, the appropriate compensation criteria and voltage control approach will be discussed and

2265

theoretical analysis will be given. In-phase voltage compensation criterion is adopted with an improved phase lock loop (PLL) structure based on the second-order generalized integrator-quadrature signal generator (SOGI-QSG). In terms of the voltage control on system level, combination of feedback and feedforward control are introduced considering both the stability and rapidity of system.

A. Compensation Criteria

Voltage compensation criteria of the DVR are divided into the following four kinds according to how to determine the reference voltage in different application cases: in-phase voltage compensation, minimum power compensation, reactive power compensation and full voltage compensation. In-phase voltage compensation (also called minimum voltage compensation) is adopted in this paper due to its simplification of algorithm without complex power calculation compared with the other three approaches.

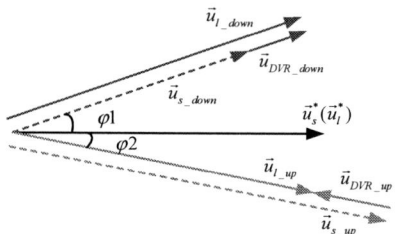

Fig. 9 In-phase voltage compensation criterion

In-phase voltage compensation criterion is shown in Fig. 9, where $\vec{u}_s^*(\vec{u}_l^*)$ is the grid voltage vector in normal condition and ideal reference load voltage vector, $\vec{u}_{l\_down}$ and $\vec{u}_{l\_up}$ are the actually adopted reference load voltage vector, $\vec{u}_{s\_down}$ and $\vec{u}_{s\_up}$ are grid voltage vector, φ_1 and φ_2 are the phase of $\vec{u}_{s\_down}$ and $\vec{u}_{s\_up}$, $\vec{u}_{DVR\_down}$ and $\vec{u}_{DVR\_up}$ are the compensation voltage vector output by DVR system.

$$\left|\vec{u}_{l\_up}\right| = \left|\vec{u}_{s\_up}\right| - \left|\vec{u}_{DVR\_up}\right|$$
$$\left|\vec{u}_{l\_down}\right| = \left|\vec{u}_{s\_down}\right| + \left|\vec{u}_{DVR\_down}\right| \qquad (9)$$
$$\left|\vec{u}_{l\_up}\right| = \left|\vec{u}_{l\_down}\right| = \left|\vec{u}_s^*\right| = \left|\vec{u}_l^*\right|$$

In in-phase voltage compensation criterion, $\vec{u}_{l\_down}$ and $\vec{u}_{l\_up}$ are determined according to the amplitude of $\vec{u}_s^*(\vec{u}_l^*)$ and the phase of real-time grid voltage phase φ_1 and φ_2. It should be noted that the recommended compensation criterion demands strict detection of grid phase.

B. SOGI-QSG based PLL

The accuracy of phase detection could be influenced by many factors including grid voltage unbalance and harmonic components. Reference [20] proposed an improved phase detection method to eliminate the disturbances. Fig. 10(a) is the diagram of overall PLL,

with SOGI-QSG shown in Fig. 10(b). In Fig. 10(a), u_a, u_b, u_c are the tested real-time grid voltage, SOGI-QSG is introduced to abstract the positive-sequence voltage component if voltage asymmetry occurs. u_α^+, u_β^+ are the positive-sequence voltage in two-phase stationary coordinate, ω' is the real-time angle frequency and θ is the tested phase. Usually, u_q^* is set as 0.

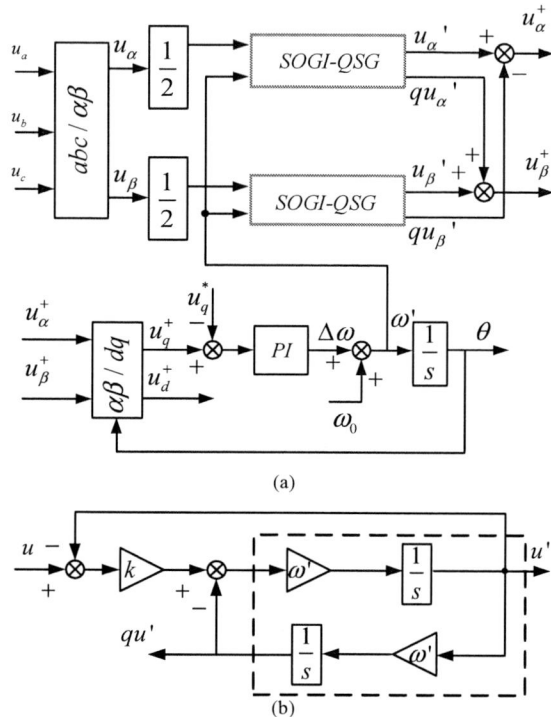

(a)

(b)

Fig. 10 PLL configuration based on SOGI-QSG. (a) structure of PLL; (b) inner structure of SOGI-QSG

All the calculation process in Fig. 10 is given as follows. To simplify the expressions, $u_{abc}, u_{\alpha\beta}, u_{\alpha\beta}^+$ are introduced to represent the voltage matrix vector.

$$u_{abc} = \begin{bmatrix} u_a \\ u_b \\ u_c \end{bmatrix}, u_{\alpha\beta} = \begin{bmatrix} u_\alpha \\ u_\beta \end{bmatrix}, u_{\alpha\beta}^+ = \begin{bmatrix} u_\alpha^+ \\ u_\beta^+ \end{bmatrix} \qquad (10)$$

Equation (11-16) shows the calculation process to abstract positive-sequence voltage component from grid voltage $u_{\alpha\beta}^+$ and the phase of $u_{\alpha\beta}^+$ will be adopted.

$$u_{\alpha\beta} = [T_{\alpha\beta}] * u_{abc}, [T_{\alpha\beta}] = \frac{2}{3} \begin{bmatrix} 1 & -\dfrac{1}{2} & -\dfrac{1}{2} \\ 0 & \dfrac{\sqrt{3}}{2} & -\dfrac{\sqrt{3}}{2} \end{bmatrix} \qquad (11)$$

$$u_{abc}^+ = [T_+] * u_{abc} \qquad (12)$$

$$[T_+] = \begin{bmatrix} 1 & a & a^2 \\ a^2 & 1 & a \\ a & a^2 & 1 \end{bmatrix}, a = e^{j\frac{2\pi}{3}} \qquad (13)$$

2266

$$u_{abc}^{+} = [T_{+}] * u_{abc} = [T_{+}] \cdot [T_{\alpha\beta}]^{-1} * u_{\alpha\beta} \quad (14)$$

$$u_{\alpha\beta}^{+} = [T_{\alpha\beta}] * u_{abc}^{+} = [T_{\alpha\beta^{+}}] * u_{\alpha\beta} \quad (15)$$

$$[T_{\alpha\beta^{+}}] = [T_{\alpha\beta}] \cdot [T_{+}] \cdot [T_{\alpha\beta}]^{-1} = \begin{bmatrix} 1 & -q \\ q & 1 \end{bmatrix}, q = e^{j-\frac{\pi}{2}} \quad (16)$$

C. Voltage Control

For voltage control strategy on system level, this paper adopts the combination of feedback control and feedforward control.

Fig. 11 shows the voltage control strategy, where u_l is the load voltage, u_s is the grid voltage, u_s^* is the standard grid voltage and the referred load voltage. C_{inv} is the equivalent inverter transfer function of output stage. The solid line leads the feedforward part of the reference compensation voltage u_{DVR}^* and is important to maintain the rapidity of the system, while the dash line leads the feedback part to enhance the accuracy of voltage control and stability of system. Hence, u_{DVR}^* can be given as follows and will be used as reference voltage for DVR output stage.

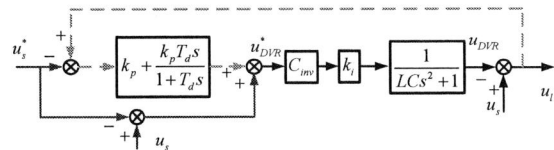

Fig. 11 Voltage control on system level

$$u_{DVR}^* = G_{PD}(-u_l^* + u_l) + (-u_s^* + u_s)$$
$$u_l^* = u_s^* \quad (17)$$
$$G_{PD} = k_p + \frac{k_p T_d s}{1 + T_d s}$$

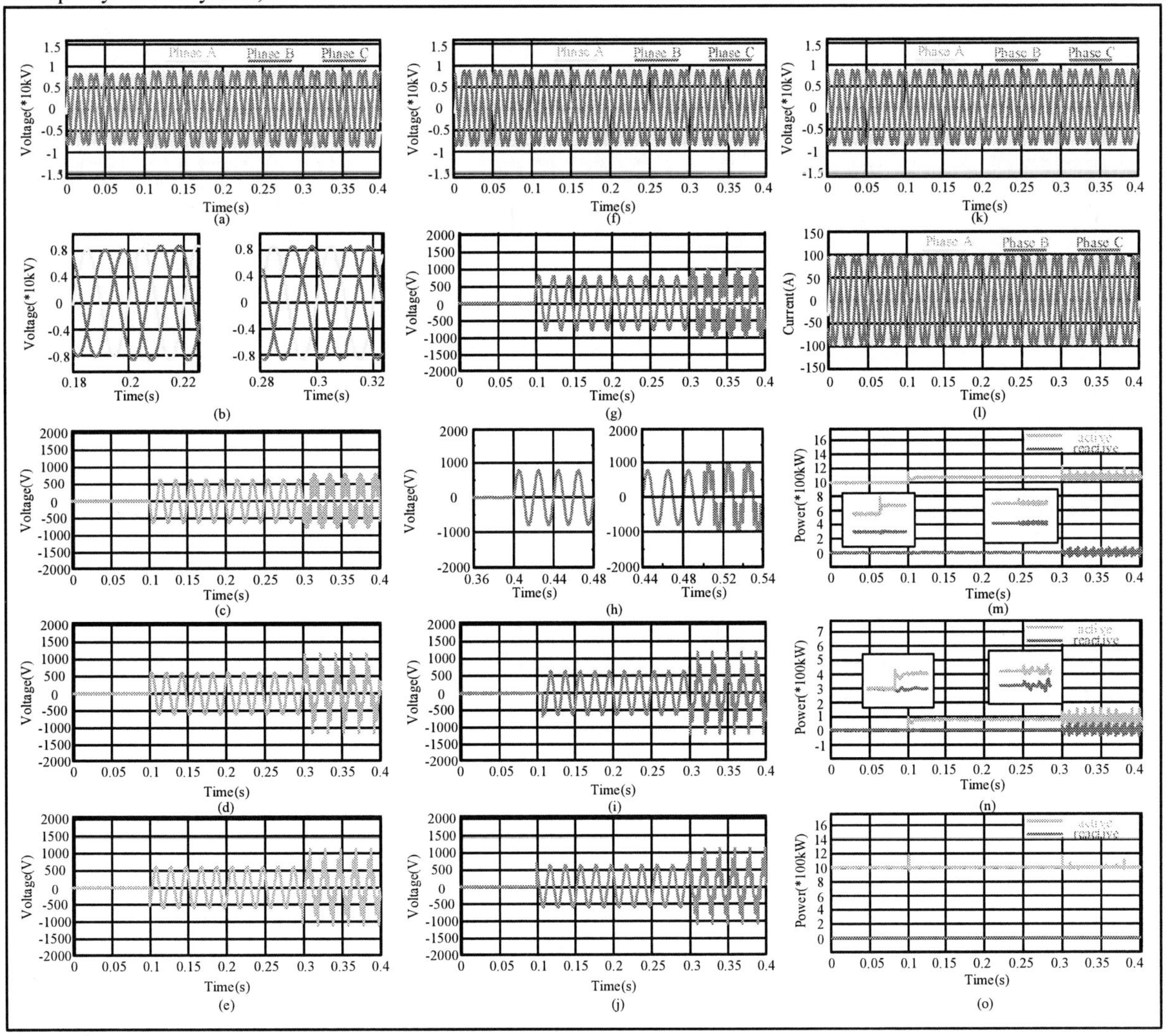

Fig .12 (a) Grid voltage, (b) voltage fluctuation at 0.2s and 0.3s, (c) reference compensation voltage for phase A, (d) reference compensation voltage for phase B, (e) reference compensation voltage for phase C, (f)reference load voltage, (g) DVR output voltage of phase A, (h) compensation voltage at 0.2s and 0.3s, (i) DVR output voltage of phase B, (j) DVR output voltage of phase C, (k) load voltage after compensation, (l) load current after compensation, (m) fluctuation of P_{trans}, (n) power output by DVR, (o) load power.

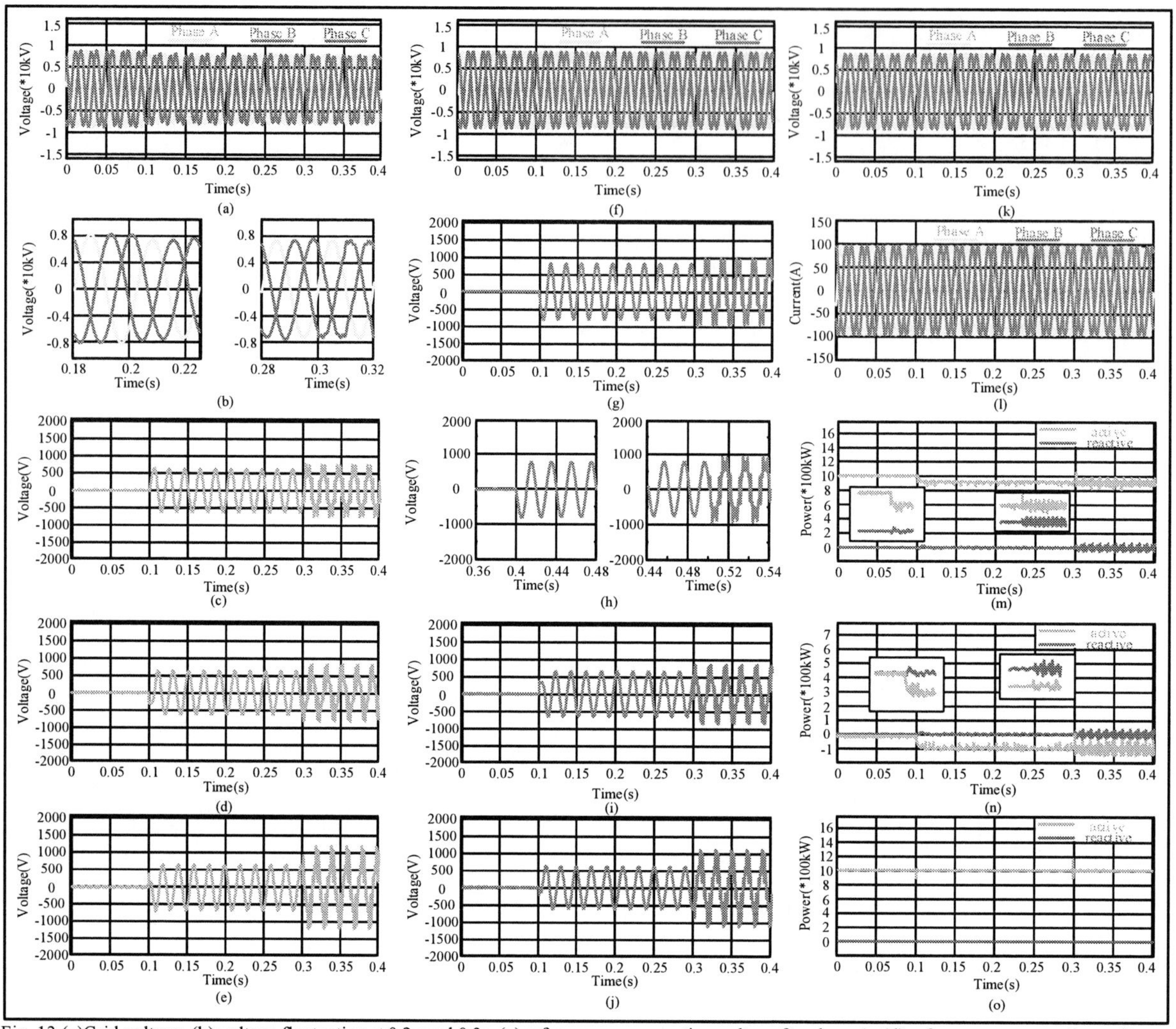

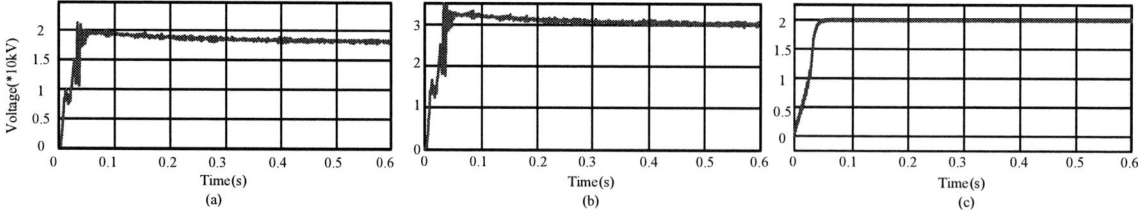

Fig .13 (a)Grid voltage, (b) voltage fluctuation at 0.2s and 0.3s, (c) reference compensation voltage for phase A, (d) reference compensation voltage for phase B, (e) reference compensation voltage for phase C, (f)reference load voltage, (g) DVR output voltage of phase A, (h) compensation voltage at 0.2s and 0.3s, (i) DVR output voltage of phase B, (j) DVR output voltage of phase C, (k) load voltage after compensation, (l) load current after compensation, (m) fluctuation of P_{trans}, (n) power output by DVR, (o) load power.

Fig .14 (a) MVDC voltage output by MMC, (b) input voltage of each DAB unit, (c) output voltage of transmission stage.

IV. SIMULATIONS

Table I shows the simulation parameters of each stage. MMC in input stage converts 10kV AC voltage into 18kV DC voltage. There are six DAB units series connected to share the 18kV DC voltage and for each unit 3kV DC voltage will step down to 2kV. The output of DAB units is in parallel connection to support the 2kV LVDC bus which is the input of inverter stage. Three inverters will output voltage to compensate for grid abnormality on load side.

In Fig .12, the voltage swell in grid is given and all the relative current, voltage and power are shown. In Fig .12(a), from 0-0.2s, the grid operates in normal condition, voltage swell occurs at 0.2s and harmonic voltage components are introduced at 0.3s. Fig .12(f) gives the reference load voltage, according to which giving the three-phase voltage to be compensated in Fig .12(c)-(e). Fig.12 (g)-(j) is the three-phase DVR output voltage, which has opposite phase with the reference output voltage. The DVR system starts working from 0.2s. After compensation, Fig .12(k) and Fig .12(l) show the load voltage and current. From Fig .12(m) to

Fig .12(o), P_{trans}, P_{out_DVR} and P_{load} are shown, at 0.2s, P_{trans} and P_{out_DVR} both experience a 100kW stepup while the reactive power gains fluctuation from 0.3s after harmonics are introduced.

Table I simulation parameters

MMC/ Input stage 10kV AC<-->18kV DC	Bridge inductance	2mH
	Switching frequency	500Hz
	submodule capacitor	2mF
DAB/ Transmission stage 3kV DC<-->2kV DC	Switching frequency	20kHz
	Frequency for MFT	20kHz
	Ratio for MFT	3000:2000
	Resonant inductance	10μH
	Output capacitor	2.2mF
Inverters/ Output stage 2kV DC<-->	Filter inductance	2mH
	Filter capacitor	20μF
	Switching frequency	10kHz

In Fig .13, the voltage sag in grid is given, and the voltage, current and power are all given in the similar sequence as Fig .12. In addition, there exits an asymmetry from 0.2s to 0.3s, the voltage of phase C has a smaller voltage sag of amplitude than the other two phases. In that case, phase C has a lower reference compensation voltage and voltage output. Also, the detection of grid phase keeps accurate without being influenced by the voltage asymmetry, in Fig .13(c)-(e) and Fig.13 (g)-(j), all the voltage stays the same phase with grid.

Fig .14 shows the MVDC and LVDC of system in case that voltage swells occur from 0.2-0.4s and voltage sags occur from 0.4-0.6s. As introduced in Section II, the input of transmission stage is in series connection which will weaken the voltage stabilization ability of capacitor, therefore, the voltage on MV side has some turbulence. However, the output of DAB unit is in parallel connection and this will multiply the capacity of energy and keep the LVDC voltage more smooth.

V. CONCLUSIONS

A new simplified SST based DVR system with bi-directional power flow is proposed in this paper. Two traditional line frequency transformers are substituted by HFT to realize more efficient power transmission, which has improved the power density and reduced the system volume greatly.

The operation modes and power flows of the DVR are discussed in details. The control strategy on system level and stage level are both given and verified in this paper. validity of proposed topology and control is verified by simulations.

REFERENCES

[1] T. Jimichi, H. Fujita and H. Akagi, "A Dynamic Voltage Restorer Equipped With a High-Frequency Isolated DC–DC Converter," *IEEE Transactions on Industry Applications*, vol. 47, no. 1, pp. 169-175, 2011.

[2] J. E. Huber and J. W. Kolar, "Solid-State Transformers: On the Origins and Evolution of Key Concepts," *IEEE Industrial Electronics Magazine*, vol. 10, no. 3, pp. 19-28, 2016.

[3] Z. Sun, Y. Li, Z. Li, P. Wang and H. Zhu, "An accelerated imulation model for the isolation stage of the smart energy router

system," *2015 18th International Conference on Electrical Machines and Systems (ICEMS)*, Pattaya, 2015, pp. 1537-1540

[4] L. He, J. Zhang, C. Cheng and T. Li, "A Bidirectional Bridge Modular Switched-Capacitor-Based Power Electronics Transformer," *IEEE Transactions on Industrial Electronics*, vol. 65, no. 1, pp. 718-726, 2018.

[5] M. Vasiladiotis and A. Rufer, "A Modular Multiport Power Electronic Transformer With Integrated Split Battery Energy Storage for Versatile Ultrafast EV Charging Stations," *IEEE Transactions on Industrial Electronics*, vol. 62, no. 5, pp. 3213-3222, 2015.

[6] H. Fan and H. Li, "High-Frequency Transformer Isolated Bidirectional DC–DC Converter Modules With High Efficiency Over Wide Load Range for 20 kVA Solid-State Transformer," *IEEE Transactions on Power Electronics*, vol. 26, no. 12, pp. 3599-3608, 2011.

[7] J. Tian, C. Mao, D. Wang, J. Lu, X. Liang and Y. Liu, "Analysis and control of electronic power transformer with star-configuration under unbalanced conditions," *IET Electric Power Applications*, vol. 9, no. 5, pp. 358-369, 2015.

[8] X. Wang, J. Liu, S. Ouyang, T. Xu, F. Meng and S. Song, "Control and Experiment of an H-Bridge-Based Three-Phase Three-Stage Modular Power Electronic Transformer," *IEEE Transactions on Power Electronics*, vol. 31, no. 3, pp. 2002-2011, 2016.

[9] M. Zhang, Z. Du, X. Lin and J. Chen, "Control Strategy Design and Parameter Selection for Suppressing Circulating Current Among SSTs in Parallel," *IEEE Transactions on Smart Grid*, vol. 6, no. 4, pp. 1602-1609, 2015.

[10] N. Parseh and M. Mohammadi, "Solid State Transformer (SST) interfaced Doubly Fed Induction Generator (DFIG) wind turbine," *2017 Iranian Conference on Electrical Engineering (ICEE)*, Tehran, 2017, pp. 1084-1089.

[11] Q. Chen, N. Liu, C. Hu, L. Wang and J. Zhang, "Autonomous Energy Management Strategy for Solid-State Transformer to Integrate PV-Assisted EV Charging Station Participating in Ancillary Service," *IEEE Transactions on Industrial Informatics*, vol. 13, no. 1, pp. 258-269, 2017

[12] J. Feng, W. Q. Chu, Z. Zhang and Z. Q. Zhu, "Power Electronic Transformer-Based Railway Traction Systems: Challenges and Opportunities," *IEEE Journal of Emerging and Selected Topics in Power Electronics*, vol. 5, no. 3, pp. 1237-1253, 2017.

[13] I. Syed and V. Khadkikar, "Replacing the Grid Interface Transformer in Wind Energy Conversion System With Solid-State Transformer," *IEEE Transactions on Power Systems*, vol. 32, no. 3, pp. 2152-2160, 2017.

[14] T. Zhao, J. Zeng, S. Bhattacharya, M. E. Baran and A. Q. Huang, "An average model of solid state transformer for dynamic system simulation," *2009 IEEE Power & Energy Society General Meeting*, Calgary, AB, 2009, pp. 1-8.

[15] H. Krishnamurthy and R. Ayyanar, "Stability analysis of cascaded converters for bidirectional power flow applications," *INTELEC 2008 - 2008 IEEE 30th International Telecommunications Energy Conference*, San Diego, CA, 2008, pp. 1-8.

[16] M. Leibl, G. Ortiz and J. W. Kolar, "Design and Experimental Analysis of a Medium-Frequency Transformer for Solid-State Transformer Applications," *IEEE Journal of Emerging and Selected Topics in Power Electronics*, vol. 5, no. 1, pp. 110-123, 2017.

[17] D. Wang *et al.*, "A 10-kV/400-V 500-kVA Electronic Power Transformer," in *IEEE Transactions on Industrial Electronics*, vol. 63, no. 11, pp. 6653-6663, Nov. 2016.

[18] A. Q. Huang, Q. Zhu, L. Wang and L. Zhang, "15 kV SiC MOSFET: An enabling technology for medium voltage solid state transformers," in *CPSS Transactions on Power Electronics and Applications*, vol. 2, no. 2, pp. 118-130, 2017.

[19] K. Mainali *et al.*, "A Transformerless Intelligent Power Substation: A three-phase SST enabled by a 15-kV SiC IGBT," *IEEE Power Electronics Magazine*, vol. 2, no. 3, pp. 31-43, 2015.

[20] M. Ciobotaru, V. Agelidis and R. Teodorescu, "Accurate and less-disturbing active anti-islanding method based on PLL for grid-connected PV Inverters," *2008 IEEE Power Electronics Specialists Conference*, Rhodes, 2008, pp. 4569-4576

A Dual-Energy-Source Uninterruptible Power Supply (UPS)

Hao Wang, Dehong Xu, Binci Xu, Haijin Li, Ye Zhu

Institute of Power Electronics, Zhejiang University

Zheda Road 38, Hangzhou, China

Email: xdh@cee.zju.edu.cn

Abstract-In this paper, a dual-energy-source uninterruptible power supply system (DES UPS) is investigated. There are two energy sources with independent infrastructures in the system, grid and natural gas pipeline. It boosts the reliability of traditional UPS. Grid is the primary energy source, while natural gas generator (NG-Gen) acts as the secondary energy source for long-time backup. Firstly, reliability and source cost of the system are evaluated. Then, cold-start characteristic of NG-Gen is analyzed and the seamless transfer strategy of DES UPS is proposed based on the cold-start characteristic of NG-Gen. It achieves the uninterruptible load power supply when system operation mode changes at the grid fault. Then the control diagram is designed accordingly and state of the system during working mode transfer is analyzed. Finally, the concept and control strategy of DES UPS are verified in a 10kW prototype.

Keywords: dual energy sources, uninterruptible power supply, seamless transfer

I. INTRODUCTION

In order to avoid huge losses caused by power outages, uninterruptible power supplies are widely used to ensure continuous power supply for critical loads, such as datacenters, semiconductor manufacturing, hospitals, etc. Nowadays, growing requirements for reliability and power rating bring severe challenges to traditional UPS, due to short expectancy and limited energy storage capacity of lead-acid battery.

Traditionally, the diesel generator (D-Gen) is used as a long-time backup power source. However it occupies a large space for oil tank and causes some air pollution. Besides, both operation and maintenance cost of D-Gen are higher. Many researches have been done about integrating alternative energy into power supply system to prolong its backup time and reliability. Fuel cell (FC) is introduced into UPS in [1] and [2], due to its zero emission and long backup time. But FC has relatively short lifetime and high price, which limits its present applications. In [3], natural gas generator (NG-Gen) is connected to the grid and acts as a schedulable power source. The energy storage unit is adopted to cooperate with NG-Gen to provide the dynamic power. In [4], NG-Gen is added to smooth output power fluctuation of the photovoltaic (PV). In [5], a DC microgrid with gas generator, PV and secondary battery is proposed. The author focuses on the coordinate control strategy of different power converters. In [6], a control scheme to increase fuel efficiency is designed for an isolated power

system. The generator operation is optimized according to the load level.

NG–Gen has been used in many different systems to improve their performance, but the design of NG-Gen for UPS is not well documented..

In this paper, a dual-energy-source uninterrupted power supply system (DES UPS) is investigated. Grid works as the primary energy source while NG-Gen acts as the secondary energy source for long-time backup. Firstly, reliability and cost advantages of the system are evaluated. Then, dynamic cold-start characteristic of NG-Gen is analyzed and the method to compensate the power gap caused by the cold-start of NG-Gen is proposed. It achieves the uninterruptible load power supply when system operation mode changes. After that, the control diagram which can realize seamless transfer between different operation modes at grid failure is designed accordingly. State of the system during working mode transfer is analyzed in detail. Finally, the concept and control strategy of DES UPS are verified in a 10kW prototype.

II. COMPARISON ABOUT RELIABILITY AND COST

As the power supply of critical load, reliability is the most important factor to be considered when building a UPS system. With reliability data of different sources shown in TABLE I [7], MTBF of UPS with different backup sources is calculated using reliability model given in [8].

TABLE I
RELIABILITY PARAMETERS OF DIFFERENT SOURCES

	Grid	Lead-acid	Li	NG-Gen
Operating failure rate(h^{-1})	0.005	16.6×10^{-5}	8.3×10^{-5}	3.3×10^{-4}
Backup failure rate(h^{-1})	-	1.1×10^{-5}	1.1×10^{-5}	3.8×10^{-6}
Repair rate(h^{-1})	0.167	-	-	-

In order to simplify the analysis, we only consider source failures here. Calculation results are shown in Fig.1.

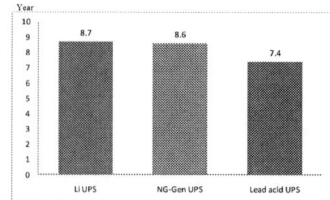

Fig. 1. MTBF of UPS with different standby source

From Fig.1, we can see that UPS based on Li-battery and NG-Gen has the longer MTBF than lead-acid based UPS. Lead-acid based UPS acts worse due to shorter life of the lead-acid battery.

Cost data of different sources is given in TABLE II [9], [10]. Source cost based on 10kW load according to different backup time is shown in Fig.2.

TABLE II
COST RELATED DATA OF LEAD-ACID BATTERY /NG-GEN/D-GEN

	Lead-acid	NG-Gen	D-Gen
Initial cost	200$/kWh	330$/kW	117$/kW
Fuel cost	0.08$/kWh	0.3$/kWh	0.78$/kWh

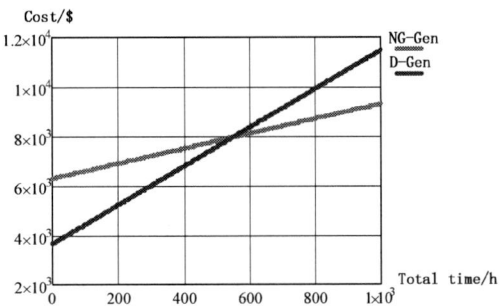

Fig. 2. Source cost according to different backup time

Source cost of D-Gen and NG-Gen based UPS is higher than lead-acid battery based UPS in short backup time situation. But cost of battery rises quickly with the increasing of backup time. D-Gen and NG-Gen are more economical for long-time backup applications.

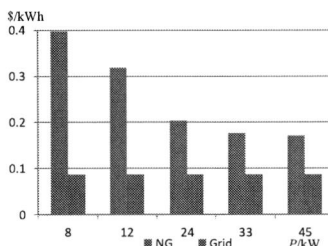

Fig.3. Cost comparison of NG-Gen and D-Gen

Fig.3 shows cost of NG-Gen and D-Gen according to total operating time. Initial cost of NG-Gen is higher than D-Gen, but total cost is less than D-Gen with the increasing of the operating time due to lower fuel cost.

According to dynamic and pollution data of D-Gen and NG-Gen in Table III [9], it can be seen that NG-Gen and D-Gen have similar dynamic characteristics, but NG-Gen has much lower pollutant emission than that of D-Gen. Besides, NG-Gen can easily supplement natural gas from the public infrastructure, which can realize long-time backup. Above all, NG-Gen is a reasonable choice to be used as backup source in DES UPS, due to its relatively high reliability, economic advantage and lower pollution in long backup time situation.

TABLE III
OPERATION DATA COMPARISON OF D-GEN AND NG-GEN

	NG-Gen	D-Gen
Transient frequency regulation rate	-4%~4%	-5.2%~5.9%
Frequency stabilization time	≤4s	≤7s
Frequency fluctuation rate	≤0.5%	≤1%
Transient voltage regulation rate	-12%~12%	-13.6%~10.4%
Voltage stabilization time	≤1.5s	≤2.4s
Co emission	≤10ppmv	≥300ppmv
Nox emission	≤100ppmv	≥1000ppmv

III. CHARACTERISTIC OF NATURAL GAS GENERATOR

Generating cost is tested based on a 50 kW NG-Gen. Gas consumption data and cost comparison between power generating of NG-Gen and grid is given in Fig.4 and Fig.5 respectively.

From Fig.4 it can be seen that the fuel consumption of NG-Gen per kWh under light load is significantly higher than that at rated power due to incomplete fuel combustion. It declines with the increasing of load and reaches a steady state after 30kW.

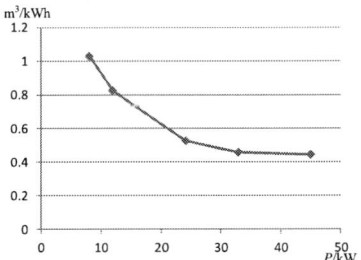

Fig. 4. Gas consumption at different load level

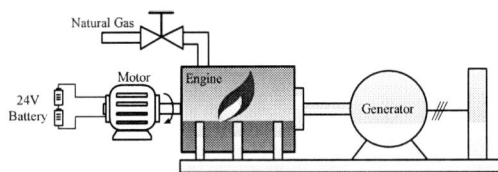

Fig. 5. Power generating cost comparison between NG-Gen and grid

As shown in Fig.5, power generating cost of the NG-Gen is higher than grid at all load level. Even at 45kW, cost of NG-Gen is twice of the grid. So, NG-Gen should stay in cold backup mode when grid is normal.

Fig. 6. Structure of NG-Gen

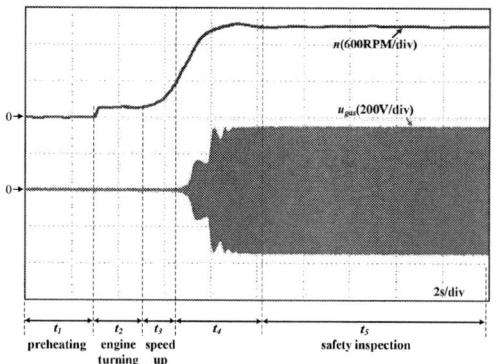

Fig.7. Cold start waveform of NG-Gen

Typical structure of NG-Gen is shown in Fig.6. A period of time is needed for a cold standby NG-Gen to provide a stable output power. To reduce the inrush current, NG-Gen is required to start without load. The cold start process is presented in Fig.7. After receiving the start command, NG-Gen preheats for about 3 seconds firstly. Then the motor is turned by the starting battery at about 200 RPM. Soon the fuel is ignited and the generator is driven by the engine. After a short regulation, start process finishes when the speed and voltage are stable. After that, there is a safety inspection to ensure the normal operation of NG-Gen. About 10 seconds is needed for NG-Gen to finish its cold start process.

Since NG-Gen has 10 seconds delay before providing load power, it will cause power interruption of load when grid fails. Therefore an energy buffer is needed to support load during the transferring period between the grid and the NG-Gen. In this paper, super capacitor (SC) is used to fill the energy gap.

IV. DESCRIPTION OF DES UPS SYSTEM

Topology of DES UPS proposed in this paper, is shown in Fig.8. Grid and NG-Gen are connected to a common DC bus through AC/DC converter #1/#2 respectively. Load gets power from DC bus through a DC/AC converter. SC is connected to the DC bus through a bidirectional DC/DC converter. Bypass is used to connect grid and load when DC/AC fails.

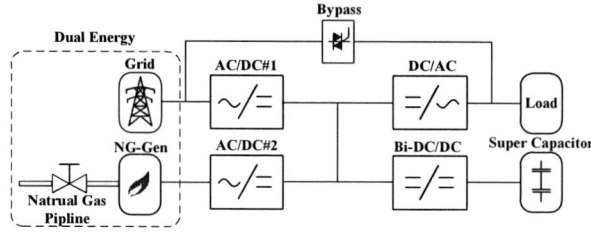

Fig. 8. Topology of DES UPS

There are three working modes in the system: grid mode, NG-Gen mode and bypass mode. Power flow of different working modes is shown as Fig.9.

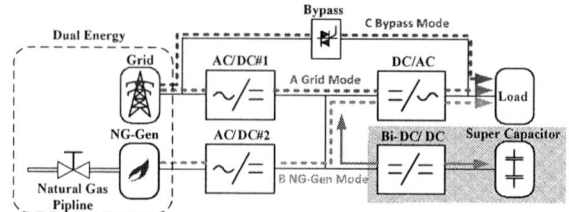

Fig .9. Power flow of different working modes

A. Grid Mode: In this mode, load is powered by grid through AC/DC#1 and DC/AC.

B. NG-Gen Mode: In this mode, NG-Gen feeds load through AC/DC#2 and DC/AC.

C. Bypass Mode: In this mode, load is feed by grid through bypass switches.

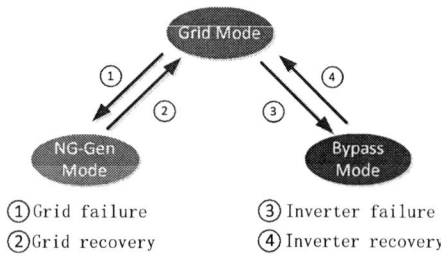

Fig. 10. Transferring logic between different working modes

Transferring logic between different working modes is given in Fig.10. DES UPS works in grid mode when grid is normal. If grid fails, working mode should transfer from grid mode to NG-Gen mode. Super capacitor here maintains the DC bus voltage through a bi-directional DC/DC (built using series buck/boost topology) during working mode transfer from grid mode to NG-Gen mode to ensure uninterrupted power supply of load. When grid recovers, NG-Gen mode should be switched back to grid mode to reduce fuel consumption.

If DC/AC fails, working mode should transfer from grid mode to bypass mode. When DC/AC recovers, bypass mode should be switched back to grid mode to guarantee quality of power supply.

Realization of seamless transfer from grid mode to NG-Gen mode is analyzed in detail in next section.

V. SEAMLESS TRANSFER SCHEME FROM GRID NORMAL MODE TO NG-GEN MODE

Fig.11 presents control scheme and communication architecture of the DES UPS system. All of AC/DC#1, AC/DC#2 and bi-DC/DC have ability to control bus voltage and maintain voltage balance of positive/negative bus. Four-wire structure is used to handle unbalance AC load. Besides voltage control loop, bi-DC/DC has an SC voltage loop to maintain a certain energy level of SC. Working mode of the system is supervised by a host computer.

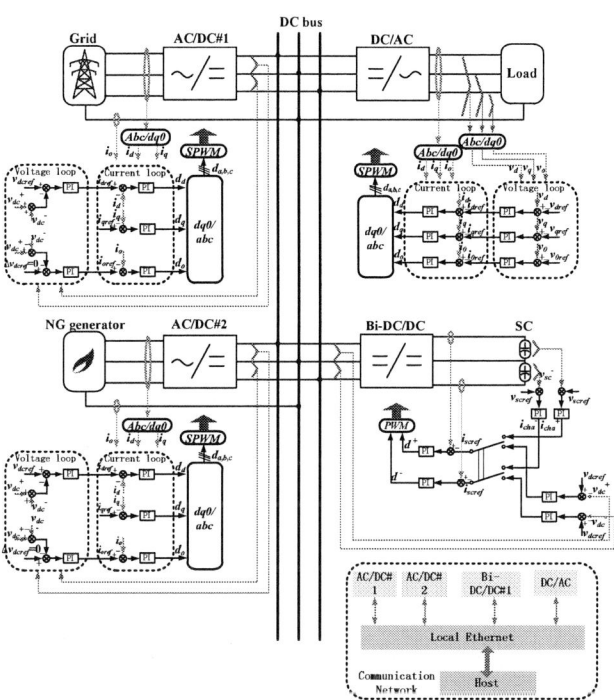

Fig. 11. Control and communication architecture of DES UPS

When the grid fails, DES UPS system will experience four stages transition from the grid mode to NG-G mode as shown in Fig.12 and Fig.13 respectively.

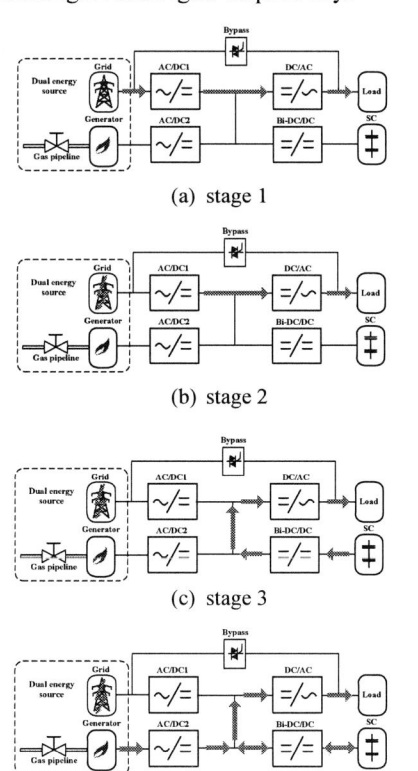

(a) stage 1

(b) stage 2

(c) stage 3

(d) stage 4

Fig. 12. State transfer from grid mode to NG-Gen mode

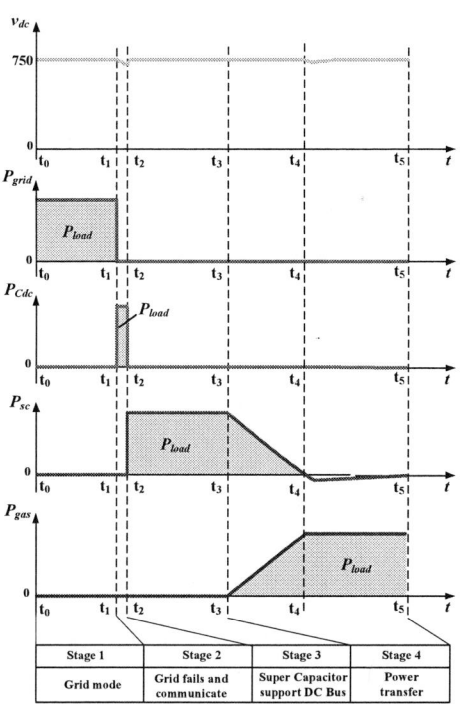

Fig. 13. Power transfer when gird fails

Stage1: *Grid mode (0-t1)*

In this stage, grid is normal and NG generator stays in cold backup condition due to the considerations of generating cost and lifespan. DES UPS works in grid mode and the DC bus voltage is sustained at 750V by AC/DC converter #1.The bi-DC/DC converter works at floating charge mode. The grid supplies total power of load.

Stage2: *Grid fails and bus capacitor holds up the DC bus voltage (t1-t2)*

At the moment of t_2, grid fails. Then the AC/DC converter #1 detects the failure and sends message to the host. At the same, it blocks the IGBT drive and disconnects the contactor. Then, host sends start order to NG generator and DC voltage control order to bi-DC/DC. During the process of detection and signal transmission, load power is supplied by DC-bus capacitors and DC bus voltage decreases. In order to ensure the quality of invert voltage, the DC bus voltage sag should be checked as follows:

$$\Delta E = \frac{1}{2} C \cdot V_{dc}^2 - \frac{1}{2} C \cdot \left(V_{dc} - \Delta V_{dc} \right)^2 = P_{load} \cdot t_{delay} \quad (1)$$

In this equation, C=8.25mF, P_{load}=10kW, t_{delay}=5ms. We can obtain the DC bus voltage sag.

$$\Delta V_{dc} = V_{dc} - \sqrt{V_{dc}^2 - \frac{2P_{load} \cdot t_{delay}}{C}} = 9.1V$$

So the DC bus voltage is about 740.9V after detection delay at 10kW load. It is still enough for DC/AC converter to keep invert voltage constant.

Stage3: *Super Capacitor supports DC bus voltage (t2-t3)*

In order to avoid the excessive voltage drop caused by abnormal condition, bi-DC/DC will switch from charge mode to bus voltage control mode when receiving order from host or the bus voltage drops to 720V.

According to the calculating result in stage2, in the normal condition, bus voltage will drop to about 740V before host order reaches bi-DC/DC. At the moment of t_3, bi-DC/DC converter receives the order and maintains the DC bus voltage at 750V immediately. At the same time, NG-Gen starts with no load. In this stage, SC provides all the energy required by the load. According to the characteristic of NG-Gen, SC should have enough energy to support the load for at least 10s until cold start finishes. Working state of DES UPS is shown as Fig.14.

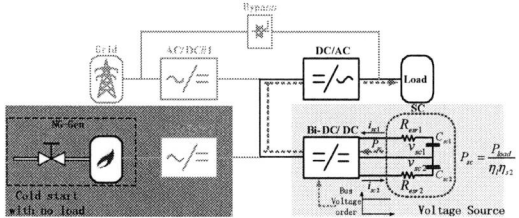

Fig. 14. Working state of DES UPS in stage3

In order to calculate the energy requirement for the super capacitor at this stage, a series resistance equivalent circuit model of SC shown in Fig.15(a) is used. Output characteristic curve of SC under constant power load is shown in Fig.15(b).

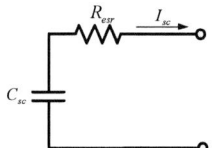

(a) Series resistance equivalent circuit model of SC

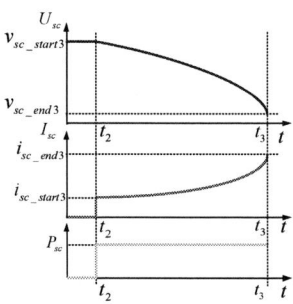

(b) Output characteristic curve of SC

Fig.15 Equivalent circuit model and output characteristic of SC

In Fig.14, C_{sc1} and C_{sc2} is the capacitance of SC1 and SC2 respectively. R_{esr1} and R_{esr2} is the equivalent series resistance of SC1 and SC2 respectively. i_{sc1} and i_{sc1} is the current of SC1 And SC2 respectively. v_{sc1} and v_{sc2} is the voltage of C_{sc1} and C_{sc2} respectively. In the ideal condition, $C_{sc1} = C_{sc2} = C_{sc}$, $R_{esr1} = R_{esr2} = R_{esr}$, $i_{sc1} = i_{sc2} = i_{sc}$, $v_{sc1} = v_{sc2} = v_{sc}$. The efficiency of DC/AC and bi-DC/DC is η_l and η_{s2} respectively. Following equation can be derived from Fig.14.

$$\begin{cases} 2(v_{sc}i_{sc} - i_{sc}^2 R_{esr}) = \dfrac{P_{load}}{\eta_l \eta_{s2}} \\ i_{sc} = -C_{sc}\dfrac{dv_{sc}}{dt} \end{cases} \quad (2)$$

The energy consumed from SC in this stage is shown as

$$E_{sc1} = \int_{t_2}^{t_3} i_{sc}^2(t)R_{esr}dt + \frac{P_{load}}{2\eta_l \eta_{s2}} \cdot t_{cold} \quad (3)$$

$$= \frac{1}{2}C_{sc}v_{sc\_start3}^2 - \frac{1}{2}C_{sc}v_{sc\_end3}^2$$

Where, v_{sc_start3} and v_{sc_end3} represent start and finish voltage of SC in stage3. t_{cold} is cold start time of NG-Gen. The values of parameters are shown in TABLE IV.

TABLE IV
VALUES OF PARAMETERS

Parameter	Value	Parameter	Value
C_{sc}	5F	P_{load}	10kW
R_{esr}	280mΩ	η_l	0.98
t_{cold}	10s	η_{s2}	0.98

Supposing that $v_{sc\_start3} = 275V$. With values given in TABLE IV, at the end of stage3, SC voltage can be calculated using (1) and (2) as $v_{sc\_end3} = 227V$.

In addition, the SC design should ensure that the SC can provide enough power for the load even at final time of the cold start stage as shown in the following:

$$(v_{sc\_end3} - I_{limit}R_{esr})I_{limit} \geq \frac{P_{load}}{2\eta_l \eta_{s2}} \quad (4)$$

I_{limit} is the current limit of SC. Supposing that $I_{limit} = 120A$, we can obtain $v_{sc\_end3} > 77V$ using (4). So that energy stored in SC is enough for cold start period when set $v_{sc\_start3} = 275V$. According to (3), total energy consumption in stage3 is 120.46kJ.

Stage4: *Power transfer between Super Capacitor and NG-Gen (t3-)*

At the moment of t_3, the cold start of NG-Gen finishes. Central controller gets the start finish signal of NG-Gen and sends the transfer signal to bi-DC/DC controller and AC/DC converter #2. AC/DC converter #2 starts to hold the DC bus voltage. Bi-DC/DC converter transfers from voltage source to current source at the same time. To decrease the dash current during power transition, the output power of super capacitor decreases to zero slowly. State of DES UPS and power transfer is shown as Fig.16(a). At the moment of t_4, power output of SC falls to zero and load is supported by NG-Gen. Bi-DC/DC switches to charge mode and energy storage of super capacitor recovers. State of DES UPS is shown in Fig.16(b).

The 2018 International Power Electronics Conference

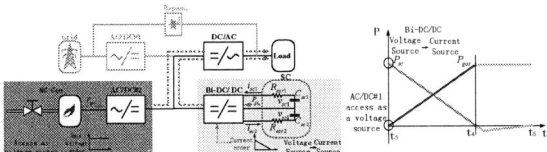

(a) Power transfer between Ng-Gen and SC

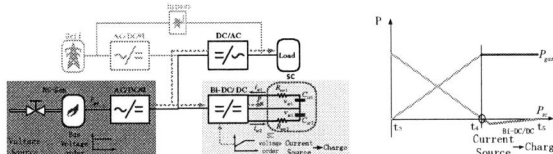

(b) Energy recovery of SC

Fig.16 Working state of DES UPS in stage4

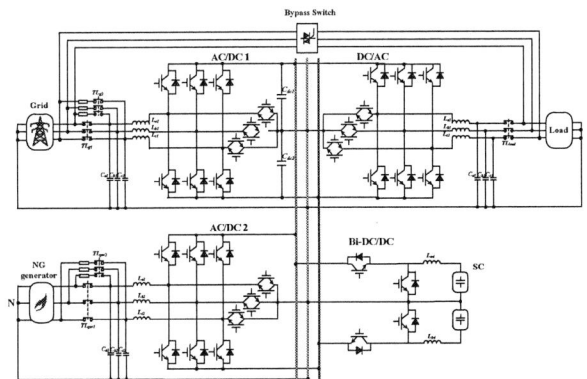

Fig. 17. Topology of proposed DES UPS

In the process of power transmission in Figure 14 (a), SC is still in the state of energy release. The decline rate of bi-DC/DC output power k_{p_sc} and the rise rate of AC/DC#2 output power k_{p_gas} can be expressed as

$$2k_{p\_sc} = k_{p\_gas} = v_{dc} \cdot k_{i\_sc} \qquad (5)$$

Where k_{i_sc} is the current decline rate set by controller. The time required for the whole process can be estimated as

$$t_{trans} = t_4 - t_3 = \frac{P_{load}}{\eta_l \cdot v_{dc} \cdot k_{i\_sc}} \qquad (6)$$

Supposing that $k_{i\_sc} = -8A/s$, $t_{trans} = 2.8s$ according to (6).

Following equation can be derived from Fig.16(a)

$$\begin{cases} v_{sc}i_{sc} = \dfrac{P_{load}}{\eta_l\eta_{s2}} \cdot \dfrac{t_{trans}-t}{t_{trans}} + i_{sc}^2 R_{esr} \\ i_{sc} = -C_{sc}\dfrac{dv_{sc}}{dt} \end{cases} \qquad (7)$$

Energy consumption in this stage can be calculated as

$$E_{sc2} = \int_{t_3}^{t_4} i_{sc}^2(t)R_{esr}dt + \frac{P_{load}}{2\eta_l\eta_{s2}} \cdot t_{trans} \qquad (8)$$

$$= \frac{1}{2}C_{sc}v_{sc\_start4}^2 - \frac{1}{2}C_{sc}v_{sc\_end4}^2$$

Supposing that $P_{load} = 10kW$, efficiency of DC/AC $\eta_l = 0.98$, efficiency of bi-DC/DC $\eta_{s2} = 0.98$, $t_{cold} = 10s$, $C_{sc} = 5F$, $R_{esr} = 280m\Omega$, $v_{sc\_start4} = v_{sc\_end3} = 227V$. At the end of stage3, SC voltage can be calculated using (7) and (8) as $v_{sc\_end4} = 213V$. According to (8), total energy consumption in stage4 is 32.92kJ.

VI. EXPERIMENTAL PLATFORM AND RESULT

A 10 kW DES UPS platform is built to verify the seamless transfer control scheme. The topology of DES UPS circuits is shown in Fig.12. All source and storage components are connected to DC bus through power converters respectively. Both AC/DC and DC/AC converters use topology of T-type three-level. Bi-DC/DC converter uses the topology of series buck/boost.

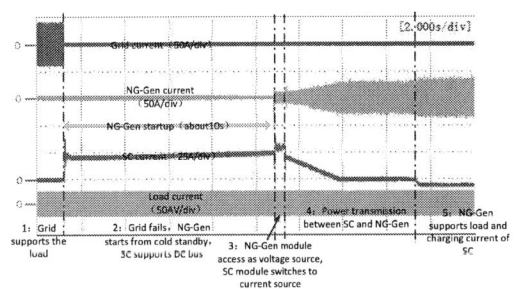

(a) Overall transfer waveform

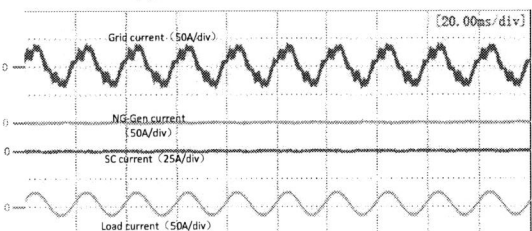

(b) Zoom-in waveform of grid normal mode

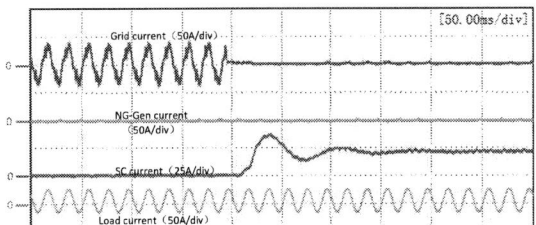

(c) Zoom-in waveform of grid failure

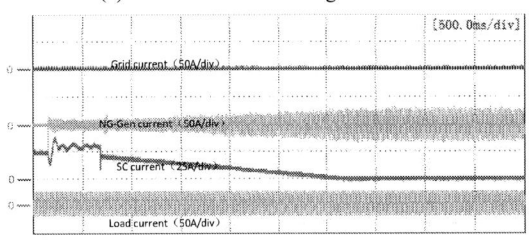

(d) Zoom-in waveform of NG-Gen access and power transfer

2275

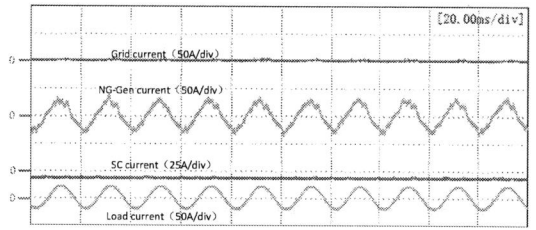

(e) Zoom-in waveform of NG-Gen mode

Fig. 18. Waveform of work mode transfer at grid fails

Experiment results of working mode transfer from grid mode to NG-Gen mode when grid fails at 10kW load is shown in Fig.18. Overall waveform of working mode transfer is shown in Fig.18(a). Zoom-in waveform of steady state of grid mode is shown in Fig.18(b). At first, load is supported by grid, NG-Gen stays in cold standby mode and super capacitor stays in floating charge mode. Zoom-in waveform of grid failure is shown in Fig.18(c). At the moment of grid fails, super capacitor starts to discharge to maintain bus voltage and NG-Gen begins to start with no load. Power supply of load is uninterrupted. When NG-Gen finishes its start, there is a power transfer between super capacitor and NG-Gen. Zoom-in waveform is shown in Fig.18(d). It can be indicated that the whole transfer is seamless and power supply of load is uninterruptible. After the transfer, energy of super capacitor is recovered and load is supported by NG-Gen like shown in Fig.18(e). From the experimental results, the control strategy guarantees NG-Gen cold start with no load during the transfer. Power supply of load stays normal during the whole process. Seamless transfer is realized by using the proposed control strategy.

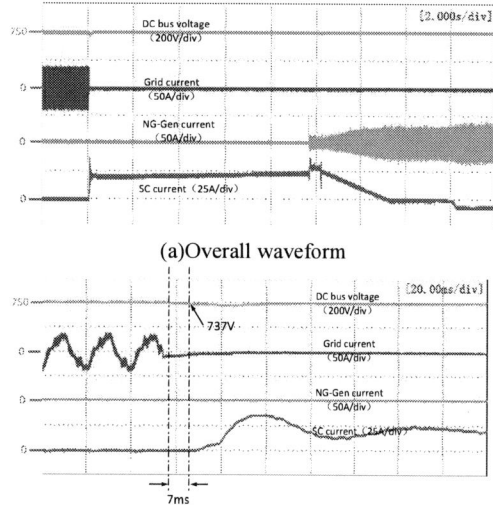

(a) Overall waveform

(b) Zoom-in waveform of grid failure moment

Fig. 19 Waveform of DC-Bus voltage during mode transfer

The transient of the DC-bus voltage directly affects the quality of the load voltage, the condition of the auxiliary power supply, etc. Therefore, the dc-bus voltage is important indicator for the DES UPS system during the whole transfer. Fig. 19 shows the DC bus voltage during the transfer from grid mode to NG-Gen mode. At first, DC bus voltage is controlled to be 750V by grid connected AC/DC#1. When grid fails, during the time of detect and delay, load is powered by DC bus capacitor and bus voltage falls to about 737V in about 7ms before bi-DC/DC receives bus voltage control order from host like shown in Fig.19(b). Then, although the bus voltage is still higher than 720V, host order of control bus voltage reaches bi-DC/DC, so it begins to control bus voltage. Bus voltage is controlled back to 750V by bi-DC/DC during cold start time of NG-Gen and remains steady during power transfer between NG-Gen and SC. It can be seen that the DC-bus voltage is maintained smoothly and the system works well.

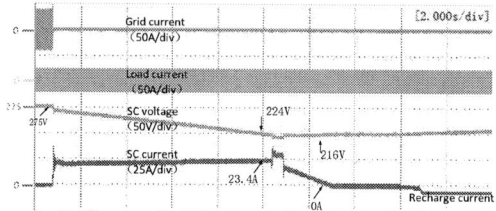

Fig.20 Voltage and current of SC during work mode transmission

Fig.20 gives the current and voltage transition of SC during working mode transfer. Floating charge voltage is set at 275 V under grid mode. The voltage drops to 224 V and current increases to 23.4 A after supporting DC bus during cold start period of NG-Gen. When power transfer is finished, the voltage falls to 216V and current falls to 0A. Then bi-DC/DC transfers to charge mode to restore energy of super capacitors. The experimental results are matched to the theoretical analysis.

VII. CONCLUSION

This paper proposes a dual-energy-source UPS with two independent energy sources. NG-Gen is introduced into DES UPS to achieve redundancy of power source. Compared with traditional UPS, there are advantages in reliability, economics and environmental friendliness for the DES UPS which uses NG-Gen as a secondary energy source. The three operation modes of DES UPS are introduced. According to the cold-start characteristics of NG-Gen, seamless transfer strategy is proposed. When grid fails, super capacitor is used to support the load during cold start time of NG-Gen. It ensures the uninterruptible load power. Then, the control diagram of seamless transfer strategy is provided. Besides, state of the system during working mode transfer is analyzed and the energy capacity requirement of super capacitors is calculated. Finally, the concept and seamless control strategy of DES UPS are verified in the 10kW prototype.

ACKNOWLEDGMENT

This work was supported by the National Natural Science Foundation of China (51337009).

REFERENCES

[1] Wenping Zhang, Dehong Xu, Xiao Li, Ren Xie, Haijin Li, DezhiDong, Chao Sun, Min Chen, "Seamless Transfer Control Strategy for Fuel Cell Uninterruptible Power

Supply System", *IEEE Transactions on Power Electronics*, vol.28, no.2, pp.717,729, Feb. 2013

[2] Xiao Li, Wenping Zhang, Haijin Li, Ren Xie, Min Chen, Guoqiao Shen, Dehong Xu, "Power Management Unit With Its Control for a Three-Phase Fuel Cell Power System Without Large Electrolytic Capacitors", *IEEE Transactions on Power Electronics*, vol.26, no.12,pp.3766,3777, Dec. 2011

[3] Katiraei, Farid, et al. "Microgrids management." *Power and Energy Magazine*, IEEE 6.3 (2008): 54-65.

[4] Johnson, Jay, et al. "PV Output Smoothing using a Battery and Natural Gas Engine-Generator." Sandia National Laboratories (SNL-NM), Albuquerque, NM (United States), SAND2013-1603 (2013).

[5] Kakigano, Hiroaki, et al. "DC Micro-grid for Super High Quality Distribution—System Configuration and Control of Distributed Generations and Energy Storage Devices", *Power Electronics Specialists Conference*, 2006. PESC'06. 37th IEEE. IEEE, 2006.

[6] Lee, Joon-Hwan, Seung-Hwan Lee, and Seung-Ki Sul. "Variable-speed engine generator with supercapacitor: Isolated power generation system and fuel efficiency." Industry Applications, IEEE Transactions on 45.6 (2009): 2130-2135.

[7] "Research and development of power electronic device for energy storage system based on redundancy multiple energy", National High Technology Research and Development Program of China (863 Program).

[8] Haijin Li, Wenping Zhang, Dehong Xu. High-Reliability Long-Backup-time Super UPS with Multiple Energy Sources.

[9] "Gas Turbine Packaged Sets and Diesel Generating Sets in the Very Large Room IDC : Application and Comparison", *Telecom Power Technology*, Mar.25,2010,V01.27 No.2

[10] "Long-Term vs. Short-Term Energy Storage Technologies Analysis", Sandia National Laboratories.

The 2018 International Power Electronics Conference

Influence of Wind Power Forecasts on Equitable Distribution Method of Wind Power Curtailment

Daisuke Iioka[1*], and Hiroumi Saitoh[1]
1 Dept. of Electrical Engineering, Tohoku University, Sendai, Japan
*E-mail: daisuke.iioka.b5@tohoku.ac.jp

Abstract— The aim of this work is the investigation of influence of wind farm area on the equitable distribution of minimized curtailment obtained by the optimization problem which calculates the curtailment plan for the wind farms interconnected to the power grid. 3-area wind power profiles has been applied to the optimization problem, so that the annual simulation from April 2014 to March 2015 could be carried out. The equitability of curtailment for wind farms has been assessed by the variation of energy loss of wind farms by the curtailment. It was found that the variation of energy loss of wind farms obtained by the curtailment plan for the next day and the current day doesn't depend on the area of wind farms and the capacity of wind farms. The simulation results demonstrated the versatility of the proposed algorithm for the curtailment in spite of the difference of area of wind farms.

Keywords— curtailment, optimization problem, over supply of electricity, wind power

I. INTRODUCTION

The penetration of renewable energy power such as photovoltaic power system (PV) and wind power farm (WF) may result in the over supply of electricity in the power grid. Although the transmission system operator (TSO) controls the thermal generating unit to prevent the power grid from the imbalance, curtailment of renewable energy power will be required owing to the weather condition and power demand [1] [2].

Several investigations have been conducted on the method of curtailment for the wind power farms. The influence of wind power curtailment on a unit commitment and economic dispatch model was presented in a previous paper [3]. From the viewpoint of minimization of curtailment for wind power farm, the application of optimization problem [4][5], state estimation [6], demand response [7], energy storage [8][9], reactive power management [10], virtual power player, active network management [11] to the wind power curtailment have been reported. There have been several attempts of influence of wind power curtailment on the operation of the transmission system operator (TSO)[12] [13]. From the view point of economic benefits for owners of wind power farm, TSO should distribute the curtailment of wind energy power equitably among the owners of wind power

farm. However, a comprehensive understanding of the method of equitable distribution of wind power curtailment is still lacking.

We have proposed a basic algorithm of equitable distribution of minimized curtailment for wind farms interconnected to the power grid [5]. The algorithm consists of three major phases: annual curtailment plan, plan for the next day and plan for the current day, which are based on wind power forecasts and optimization problems. Since the output of wind farm are explained by the characteristics of area such as weather condition, the accuracy of curtailment equitability obtained by the proposed optimization problem depends on the wind power forecast. The objective of the present work has been evaluation of accuracy of curtailment equitability by simulations. Three-area wind power profiles have been applied to the simulations. The influence of area of wind farm on the curtailment equitability has been investigated.

II. POWER SYSTEM MODEL

Fig. 1 shows a configuration of power system model. This system consists of thermal generation units, pumped storage hydro plants, PV units, 3 – wind farm group and load demand. The output power from the thermal generating units, PV units, wind farm group WF_1, WF_2, and WF_3 are expressed by P_G, P_{PV}, P_1, P_2, and P_3, respectively. The power consumption of load and hydro plants are given by P_D and P_P, respectively. The capacity of PV units, wind farm group WF_1, WF_2 and WF_3 are 1000 MW, S_1=1000[MW], S_2=1500[MW] and S_3 [MW], respectively.

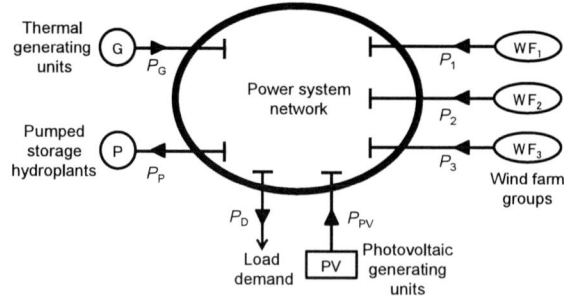

Fig. 1. Power system model.

2278

The following inequality shows the power grid imbalance that the supply power is larger than demand power in the power system model.

$$P_G + P_{PV} + P_1 + P_2 + P_3 > P_D + P_P \qquad (1)$$

It is assumed that the TSO requires the curtailment for the wind power farm WF_1, WF_2 and WF_3 if the power system balance is described by (1). The magnitude of curtailment for each wind farms is calculated by the optimization problem described in chapter III.

III. Proposed Method of Wind Power Curtailment

A. Outline of Wind Power Curtailment Plan

Fig. 2 shows the upper limit of power from wind farm group. The horizontal axis and vertical axis correspond to time and output of wind farm group, respectively. The area in gray which is over the upper limit of output shown in Fig. 2 expresses the energy loss of wind power by the curtailment. If the integrated areas in gray for wind farm groups WF_1, WF_2 and WF_3 in one year are equal to each other, we conclude that the curtailment of wind power are distributed equitably.

The proposed method [5] is composed of three-level curtailment plan: the annual plan, the plan for the next day, and the plan for the current day, which are obtained by the optimization problem.

B. Annual Plan

The annual energy loss C_j for wind farm WF_j is defined by the expression below.

$$C_j = \sum_{d=1}^{350} \sum_{i=1}^{24} \left(1 - \frac{\sum_{k=0}^{100} x(d,i,j,k) \cdot k}{100} \right) \qquad (2)$$

where $x(d, i, j, k)$ is decision variable expressed by zero-one variable, d, i, and k are date ($1 \leq d \leq 350$), hour ($1 \leq i \leq 24$) and upper limit of output ($0 \leq k \leq 100$), respectively. The parameter d, i, j and k are defined as integer. If $x(d, i, j, k) = 1$, the upper limit of wind farm WF_j is equal to k [%] on the date d at the time i.

For the purpose of equitable distribution of minimized curtailment for the annual plan, the objective function for 0-1 integer programming is defined by the expression below.

$$\text{Minimize} \quad \max\{C_1, C_2, C_3\} \qquad (3)$$

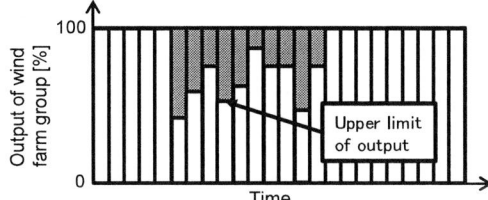

Fig. 2. Upper limit of wind farm groups.

The power balance in the grid expressed by (4) is one of the constraints for this problem, where S_j is the capacity of wind farm WF_j.

$$\sum_{j=1}^{3} \sum_{k=0}^{100} x(d,i,j,k) \cdot \frac{k}{100} \cdot S_j \leq P_D + P_P - P_G - P_{PV} \qquad (4)$$

C. Plan for the Next Day and the Current Day

The curtailment plans for the next day and the current day are also developed by 0-1 integer programming. In the case that the date of the next day is equal to D, the objective function of the curtailment plan for the next day is expressed by (5).

$$\text{Minimize} \quad \max \left\{ \begin{array}{l} \left| C_1^{\text{ref}}(D) - C_1^{\text{cum}} - C_1^{\text{tmr}}(D) \right|, \\ \left| C_2^{\text{ref}}(D) - C_2^{\text{cum}} - C_2^{\text{tmr}}(D) \right|, \\ \left| C_3^{\text{ref}}(D) - C_3^{\text{cum}} - C_3^{\text{tmr}}(D) \right| \end{array} \right\} \qquad (5)$$

where $C_j^{\text{ref}}(D)$ is obtained by the annual plan, which represents the integrated energy loss for the wind farm WF_j until the date D at the time $i = 24$. C_j^{cum} represents the result of curtailment which are obtained by the plan for the next day and the current day until the date D-1 at the time $i = 24$. $C_j^{\text{tmr}}(D)$ shown in (6) expresses the energy loss during the next day, which is given by the decision variable $x(d, i, j, k)$.

$$C_j^{\text{tmr}} = \sum_{i=1}^{24} \left(1 - \frac{\sum_{k=0}^{100} x(D,i,j,k) \cdot k}{100} \right) \qquad (6)$$

The objective function depends on the energy loss obtained by the annual plan.

The objective function of the plan for the current day is given in the similar terms. One of the terms is the energy loss in one hour which is expressed by the decision variable $x(d, i, j, k)$. The others correspond to the energy loss obtained by the annual plan and the result of curtailment.

IV. Influence of Forecast Data on Wind Power Curtailment

A. Specifications in Input Data

Annual plan, plan for the next day and plan for the current day from April 2014 to March 2015 have been obtained by the optimization problem described in chapter III. The input data of thermal generating units, pumped storage hydro plants, PV units and load demand are measured in an electric utility. Wind power outputs are based on the wind velocity data obtained by the meteorological stations at Fukushima, Iwate and Akita in Japan. 3-area wind power output are applied to WF_1, WF_2 and WF_3 as described in TABLE I. The influence of area of wind power farm on the equitable distribution of curtailment has been investigated.

Fig. 3 shows the example of time variation in supply and demand in the power grid. This figure is obtained by the input data for the plan for the current day on 4 May 2014. Data case shown in TABLE I is (A), the capacity of WF_3 is 600 MW. When the total supply power expressed

2279

by red line is larger than the total demand, the power balance in the grid is in the state of oversupply. Total time and energy of oversupply in Fig. 3 are 11 hours and 5300MWh, respectively.

The influence of data case shown in TABLE I and capacity of WF_3 on the total hour and energy of oversupply has been investigated. Fig. 4 (a) and (b) show the total hour and energy loss of oversupply from April 2014 to March 2015 against the data case, respectively. Fig. 4 indicates that the total hour and energy of oversupply depends on the data case. The total hour and energy for case (D) is the largest among the six cases. On the contrary, the total hour and energy for case (C) is the smallest. The difference between case (C) and (D) shown in Fig. 4 is probably due to the output of wind power during periods of low demand. As shown in TABLE I, the difference of input data between case (C) and (D) is found in wind farm data for Akita and Fukushima.

Fig. 5 shows the wind power profile for the plan for the current day in Akita and Fukushima from April 2014 to March 2015. From Fig. 5, Akita and Fukushima are nearly equal in the integrated hours which the wind power is larger than 80 % of rated capacity. On the contrary, during the period of demand under 9000MW, the integrated hours in Akita is 77 hours while that in Fukushima is 128 hours. Since the capacity of WF_2 is larger than that of WF_3, the case (D) which Fukushima data is applied to WF_2 shows a tendency to oversupply.

B. Effect of Curtailment Plan on Power Curve and Energy Loss

Fig. 6 shows the integrated energy loss obtained by the annual curtailment plan for case (A). This figure is obtained by the optimization problem mentioned in chapter III. The capacity of WF_3 is 600 MW. The horizontal and vertical axis show time and energy loss by curtailment, respectively. Since the energy loss C_1, C_2 and C_3 in one year are equal to each other, it was found that the annual plan distributes the curtailment power to wind power farms equitably.

TABLE I
AREA LIST OF WIND FARM

Case	Wind farm 1	Wind farm 2	Wind farm 3
(A)	Fukushima	Akita	Iwate
(B)	Fukushima	Iwate	Akita
(C)	Iwate	Akita	Fukushima
(D)	Iwate	Fukushima	Akita
(E)	Akita	Iwate	Fukushima
(F)	Akita	Fukushima	Iwate

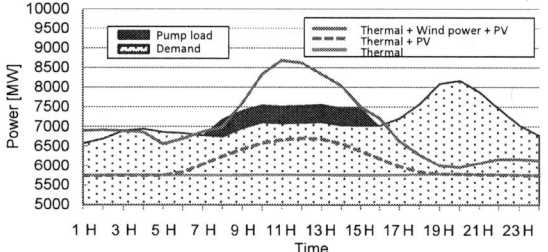

Fig. 3. Time variation in power for the current day without curtailment.

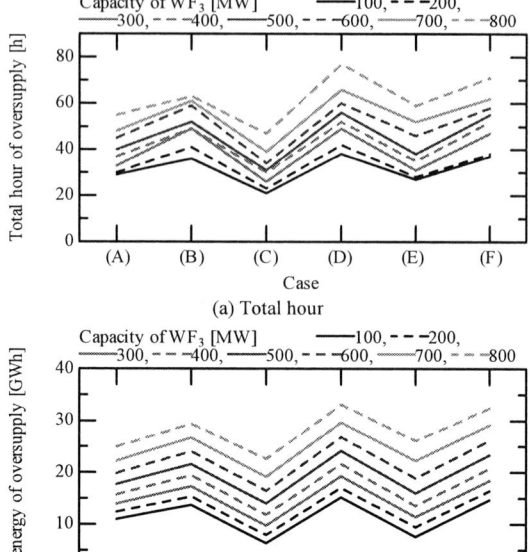

(a) Total hour

(b) Total energy

Fig. 4. Total hour and energy of oversupply against data case shown in TABLE I.

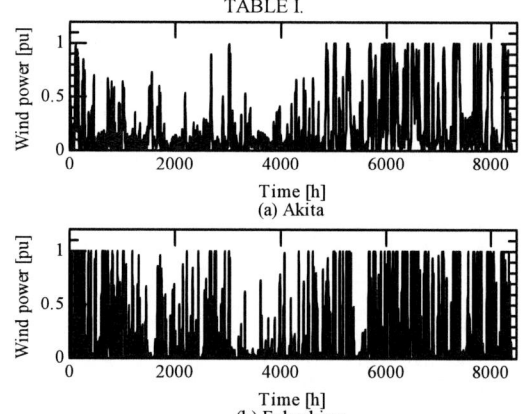

(a) Akita

(b) Fukushima

Fig. 5. Wind power profiles in Akita and Fukushima for the plan for the current day.

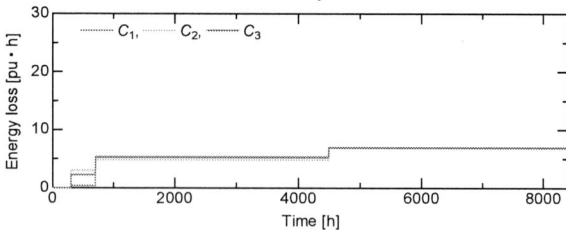

Fig. 6. Time variation in energy loss obtained by annual curtailment plan for case (A) data.

Fig. 7 shows an example of effect of curtailment obtained by the plan for the next day and the current day on the power curve with oversupply shown in Fig. 3. It was found that there is no oversupply in Fig. 7.

Fig. 8 shows the integrated energy loss obtained by the plan for the next day and the current day based on the annual curtailment plan shown in Fig. 6. The energy loss C_1 in one year is slightly larger than C_2 and C_3. This

2280

implies that the curtailment plan for the next day and the current day has a little imbalance compared with the annual curtailment plan from the view point of equitable distribution of curtailment.

C. Influence of Area of Wind Power on Energy Loss by Curtailment

Fig. 9 shows the relationship between the capacity of WF_3 and energy loss C_1, C_2 and C_3 in one year. The energy loss is calculated by the plan for the current day in the case of (A). The energy loss by the curtailment increases with the capacity of WF_3. Similarly to the Fig. 8, there is a certain amount of variation in the energy loss C_1, C_2 and C_3 in one year.

The variation in the energy loss are evaluated by (7), where $\overline{C}$ expresses the average of C_1, C_2 and C_3. The equitable distribution of curtailment requires a small variation C_{stdev}.

$$C_{\text{stdev}} = \sqrt{\frac{1}{3} \sum_{j=1}^{3} \left(C_j - \overline{C} \right)^2} \qquad (7)$$

Fig. 10 shows the variation C_{stdev} against the data case. Although the total hours and energy of oversupply shown in Fig. 4 depends on the data case and capacity of WF_3, the dependence of the variation C_{stdev} on the data case and capacity of WF_3 is unclear shown in Fig. 10.

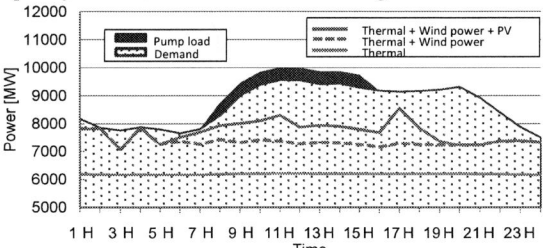

Fig. 7. Time variation in power for the current day with curtailment.

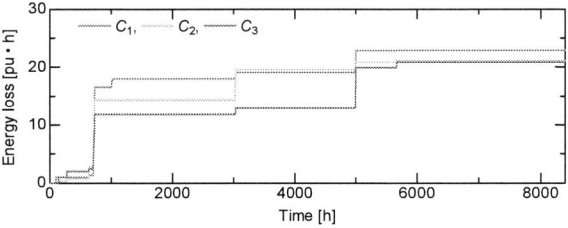

Fig. 8. Time variation in energy loss obtained by plan for the next day and the current day for case (A) data.

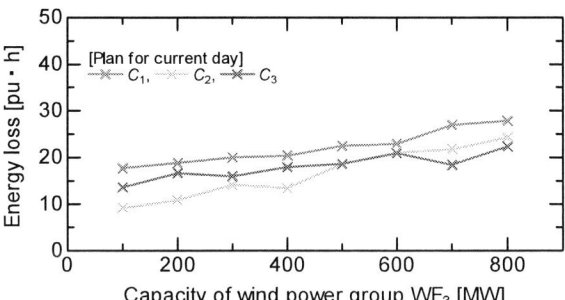

Fig. 9. Energy loss as a function of capacity of wind power group for case (A) data.

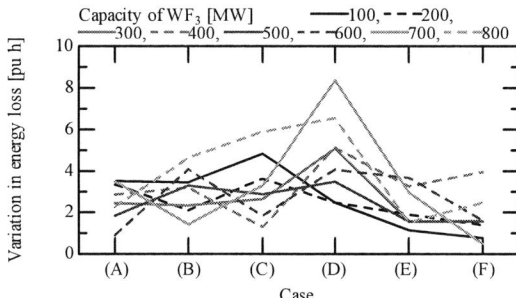

Fig. 10. Variation in energy loss by curtailment against data case shown in TABLE I.

For the purpose of reduction of variations in the energy loss shown in Fig. 10, the validity of curtailment power has been investigated by (8).

$$P_{\text{EX}} = \sum_{d=1}^{350} \sum_{i=1}^{24} \left((P_D + P_P) - (P_G + P_{PV} + P_{WF}) \right)$$

$$P_{\text{WF}} = \sum_{j=1}^{3} \left(\frac{\sum_{k=0}^{100} x(d,i,j,k) \cdot k}{100} S_j \right) \qquad (8)$$

The parameter P_{EX} given by (8) expresses excessive energy loss obtained by the optimization problem. Since the equitable distribution of minimized curtailment is defined as the objective function for the optimization problem, the energy loss by the curtailment is sometimes more than the minimum necessary amount of the energy loss.

Fig. 11 shows the excessive energy loss against the data case. Compared with Fig. 4, it was found that the excessive energy loss increases with the total hour and energy of oversupply.

Fig. 12 expresses the relationship between the variation C_{stdev} and the excessive energy loss P_{EX} shown in Figs. 10 and 11. Although there is a slightly correlation between C_{stdev} and P_{EX} by appearances of Fig. 12, the correlation coefficient is 0.084. We reached the tentative conclusion that the effect of reduction of excessive energy loss on the equitable distribution or curtailment is unclear until the confirmatory analyses by using other area data.

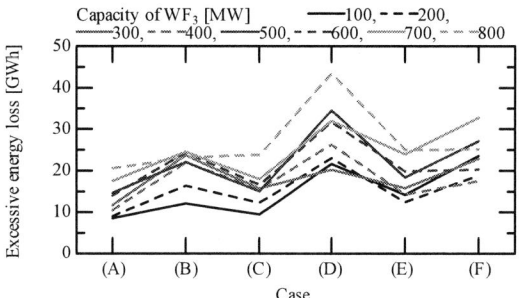

Fig. 11. Variation in energy loss by curtailment against data case shown in TABLE I.

The 2018 International Power Electronics Conference

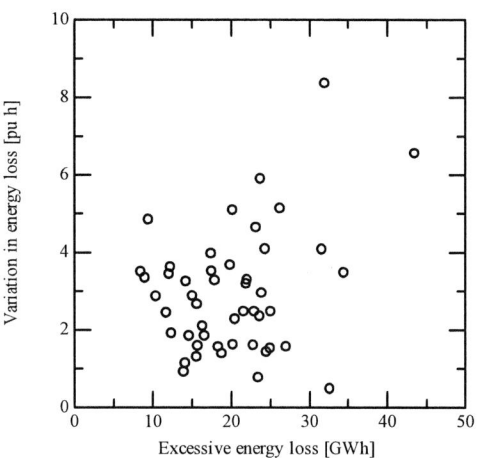

Fig. 12. Relationship between variation C_{stdev} and excessive energy loss P_{EX}.

V. CONCLUSION

Three-area wind power profile has been applied to the basic algorithm for equitable distribution of minimized curtailment so that the accuracy of equitability could be evaluated. The main results obtained are shown below.

1. From the simulation without curtailment, the total hour and energy of oversupply in the power grid depends on the area of wind farm. .

2. It is not clear whether there is any cause-and-effect relation between the variation of energy loss of wind farms obtained by the curtailment plan for the next day and the current day and the area of wind farms. This indicates that the proposed algorithm doesn't depend on the area of wind farms.

3. The excessive energy loss by the curtailment is probably due to the total hour and energy of oversupply in the power grid without the curtailment.

ACKNOWLEDGMENT

This presentation is based on results obtained from a project commissioned by the New Energy and Industrial Technology Development Organization (NEDO).

REFERENCES

[1] EirGrid, SONI, "Annual Renewable Energy Constraint and Curtailment Report 2016," 2017

[2] S. M. Martínez, E. G. Lázaro, A. H. Escribano, M. C. Carretón and A. Molina-Garcia, "Wind Power Curtailment Analysis under generation flexibility requirements: The Spanish case study," *2015 IEEE Power & Energy Society General Meeting*, Denver, CO, 2015

[3] M. Martin Almenta, D. J. Morrow, R. J. Best, B. Fox and A. M. Foley, "An Analysis of Wind Curtailment and Constraint at a Nodal Level," in *IEEE Transactions on Sustainable Energy*, vol. 8, no. 2, pp. 488-495, April 2017

[4] M. Shimamura, D. Iioka and H. Saitoh, "Basic Study on Decision Method of Equitable Distribution of Wind Power Curtailments Based on Accumulation of Past Error," in The International Conference on Electrical Engineering (ICEE) 2016, Okinawa, Japan, 2016

[5] D. Iioka, and H. Saitoh, "Proposal of Equitable Distribution Method of Wind Power Curtailment during Excess-Supply of Electricity in Power System", *IEEJ Trans. PE*, Vol.138, No.2, pp. 82-89, 2018 (in Japanese)

[6] R. Liu, A. K. Srivastava, D. E. Bakken, A. Askerman and P. Panciatici, "Decentralized State Estimation and Remedial Control Action for Minimum Wind Curtailment Using Distributed Computing Platform," in *IEEE Transactions on Industry Applications*, vol. 53, no. 6, pp. 5915-5926, Nov.-Dec. 2017

[7] E. McKenna, P. Grünewald, M. Thomson, "Going with the wind: temporal characteristics of potential wind curtailment in Ireland in 2020 and opportunities for demand response," *IET, Renewable Power Generation*, vol. 9, pp. 66-77, 2015

[8] X. Dui, G. Zhu and L. Yao, "Two-stage optimization of battery energy storage capacity to decrease wind power curtailment in grid-connected wind farms," in *IEEE Transactions on Power Systems*, vol. PP, no. 99, pp. 1-1

[9] B. Cleary, A. Duffy, A. OConnor, M. Conlon and V. Fthenakis, "Assessing the Economic Benefits of Compressed Air Energy Storage for Mitigating Wind Curtailment," in *IEEE Transactions on Sustainable Energy*, vol. 6, no. 3, pp. 1021-1028, July 2015

[10] T. Niu, Q. Guo, H. Jin, H. Sun, B. Zhang and H. Liu, "Dynamic reactive power optimal allocation to decrease wind power curtailment in a large-scale wind power integration area," in *IET Renewable Power Generation*, vol. 11, no. 13, pp. 1667-1678, 2017

[11] D. Boldt, P. Faria and Z. Vale, "Study and analysis of wind curtailment situations and developing an appropriated methodology for its management," *2015 IEEE Eindhoven PowerTech*, Eindhoven, 2015, pp. 1-6

[12] Y. Gu and L. Xie, "Fast Sensitivity Analysis Approach to Assessing Congestion Induced Wind Curtailment," in *IEEE Transactions on Power Systems*, vol. 29, no. 1, pp. 101-110, Jan. 2014

[13] S. Martín-Martínez, E. Gómez-Lazaro, A. Molina-Garcia and A. Honrubia-Escribano, "Impact of wind power curtailments on the Spanish Power System operation," *2014 IEEE PES General Meeting | Conference & Exposition*, National Harbor, MD, 2014, pp. 1-5

Comparison of Optimized Demand of EGs for Minimizing Fuel Consumption and EGs Model with Power Grid Frequency Using a Hpspital Load with PV

Yuji Mizuno[1*], Teppei Baba[2], Fujio Kurokawa[1] and Nobumasa Matsui[1]

1 Institute for Innovative Science and Technology, Nagasaki Institute of Applied Science, Nagasaki, Japan
2 Graduate School of Engineering Master's Program, Nagasaki Institute of Applied Science, Nagasaki, Japan
*E-mail: MIZUNO_Yuji@NiAS.as.jp

Abstract— **The purpose of this paper is to compare of optimized demand of emergency generators (EGs) for minimizing fuel consumption and EGs model with power grid frequency using a hospital load with a photovoltaic power generation system (PV). When a power failure occurs caused by disasters or utility grid trouble, EGs are installed in a large hospital, that start up within a few tens of seconds after the power failure and supply the power to the whole hospital. In previous work, we have proposed an optimum method of energy scheduling for a combination of the EGs with PV using linear programing due to minimization of fuel consumption for driving EGs as long as possible. However, especially in cloudy sky, the output fluctuation of the PV becomes large, and the influence on PV power balance in the hospital cannot be ignored. This paper presents a comparison of the optimized demand of EGs for minimizing fuel consumption using linear programming and our developed EGs model considering grid frequency with MATLAB/ Simulink using the actual hospital load for one week.**

Keywords— *The authors shall provide up to 4 keywords or phrases (in alphabetical order and separated by commas) to help identify the major topics of the paper.*

I. INTRODUCTION

Since the Great East Japan Earthquake and the Fukushima Nuclear Power Plant accident on March 11th, 2011, the needs for independent, decentralized, renewable energy supply systems have become more urgent as ever. To take a simple example, installations of backup power systems, such as the PV, gas cogeneration systems and a sodium-sulfur battery, are becoming popular measures against the power failure and an energy-saving in the hospital. Furthermore, on April 14th, 2016, the Kumamoto Earthquake in Japan was hit by a magnitude 7.3 quake with many aftershocks in the days; accordingly, a large-scale blackout occurred. Many hospitals could not keep medical services due to power failure occurs for seven days [1]. Based on disasters with high frequency of occurrence, it follows that the disaster management plan is formulated and implemented [2]. Although a power failure occurs

caused by disasters or utility grid trouble, the EGs installed in a large hospital, that start up within a few tens of seconds after the power failure and supply the power to the whole hospital, so medical services can keep. But, at the same time, it is clear that the fuel is reduced in volume by power generation, an operation term of the EGs are limited. Therefore, replenishing the fuel is to be need, but the severing of distribution such as roads and railways is made worse in large-scale disaster. To supply the power to the whole hospital as long as possible using the EGs, it is necessary to minimize the fuel consumption by optimum energy scheduling of EGs.

In our previous works, first, we have proposed an optimum method of energy scheduling using linear programing due to minimization of fuel consumption of the EGs when the power failure occurred using actual hospital load data of the hospital that installed EGs with rated output 1,000kVA [3]. Then, it was shown that when assuming an emergency power system of a combination of the EGs with the PV, it is appropriate to installation of the PV with 20% capacity against the needing hospital load. Next, as mentioned above, we have optimized assuming the combination of the EGs with the PV with 20% capacity, it was confirmed that combinations a plurality of small capacity generators is needed because of the power supply is unstable with EGs with rated output 1,000 kVA, consequently, an optimum energy scheduling has been able to evaluate in combinations of EGs of three units with rated output 200kW and one unit with rated output 400kW by four in total for minimization of fuel consumption of the EGs [4][5]. However, especially in cloudy sky, the output fluctuation of the PV becomes large, and the influence on PV output power balance in the hospital cannot be ignored. A combination of a diesel generator with the PV and a battery for island mode in a micro grid that consists of the generator and the battery with 100% capacity and PV with 50% for a power load have been discussed, while keeping efficient operation conditions of each micro grid component in the recent literature [6]. The hospital is planning to install the PV and the battery due to reduce a cost of the power. According to battery makers, the battery capacity is recommended approximately 50% of a renewable energy system. So, we have proposed an estimation method of the optimum capacity of the battery based on the measured PV output data to realize a stable power supply in the hospital by combined use of the PV and the EGs for an island mode. Proposed method is that the charged/discharged power can be calculated between the measured PV output data and the moving average its value every 30 minutes. It was clarified that the constraint condition of the optimum battery capacity for island operation mode is estimated in rainy three months [7].

The power balance of grid is not only the output fluctuation of the PV but also the influence of EG on the governor control with droop cannot be ignored. It is concerned that governor control of EG increases fuel consumption due to overshoot and undershoot with large grid frequency fluctuations.

This paper presents a comparison of optimized demand of EGs for minimizing fuel consumption using linear programming and our developed EGs considering power grid frequency model using the hospital load with the photovoltaic power generation system (PV) in MATLAB/ Simulink.

II. EMERGENCY POWER SYSTEM

Figure 1 shows an emergency power system combined of a diesel generator with the PV and a battery, which based on installed system in a large hospital. It consists of the PV, the battery connected to a DC bus line via a bidirectional DC/DC converter while the hospital load, the EGs and utility grid are connected to an AC bus line. The AC bus is equivalent to domestic electricity, which is connected to the DC bus via a bidirectional DC/AC inverter. When the power failure occurs, several the EGs supply the power to the whole hospital and the battery controls the voltage of the DC bus to a constant value by the directional DC/DC converter. With the higher

penetration of renewable energy sources into the system, the hospitals are actively being installed even in Japan.

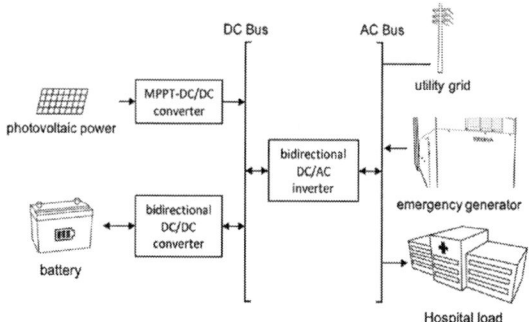

Fig. 1. Overview of the emergency power system in a large hospital.

III. EGs AND CONTROL MODEL

Figure 2 shows the diagram of EGs with power grid frequency model using a hospital load with PV. Here, the hospital load and PV output are used measured data. But, the measured of PV output data is scale up from experimental facility. Smoothing of PV output is assumed by a battery that uses a moving average of the measured PV output per second. An optimized demand is made to use linear programing for the EGs for minimizing fuel consumption [3][4][5]. An input of linear programing is difference of an actual hospital load and a moving average of the measured PV output.

An EG model consists of fuel model, gas turbine model and generator model [8]. A power grid frequency model uses both all EG output and the actual hospital load. Controller model has two role of load control and frequency control via droop.

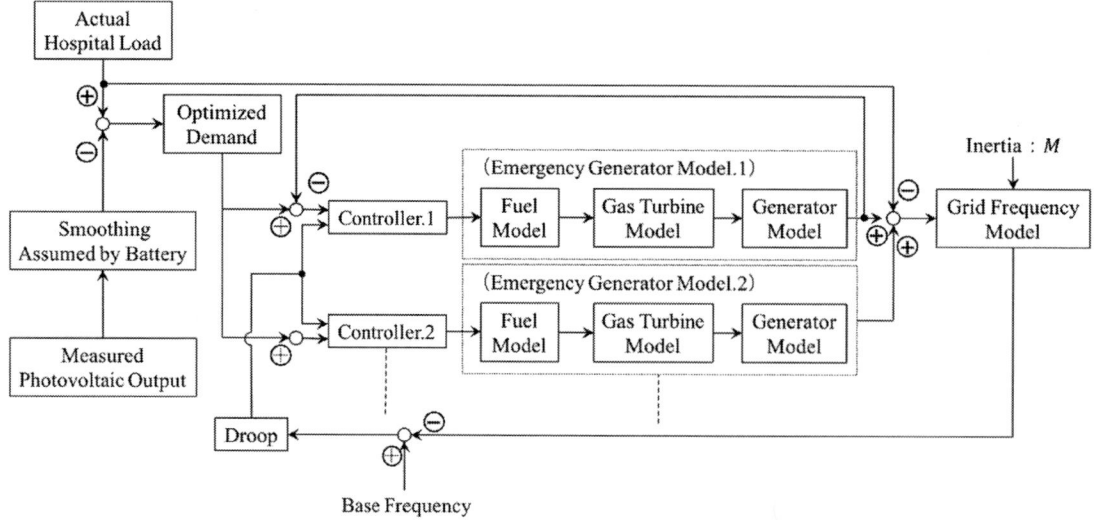

Fig. 2. Diagram of EGs with power grid frequency model using a large hospital.

IV. CASE STUDY

Calculation of a comparison of the optimized demand of EGs for minimizing fuel consumption using linear programming and our developed EGs with power grid frequency model using the actual hospital load for a week in MATLAB/Simulink are described in this section.

A. Simulation condition

The actual hospital load for one day is shown in Fig. 3. Prior to 6:00 is all the almost constant of low load, the load is increased because of preparation of for breakfast after 6:00. After 8:00, the load of the daytime during outpatient services time indicates broad peak. In this figure, both demand power of August and February are displayed that the blue color shows August and February is the red color. The maximum load is 910kW in August, and February is denoted 734kW. The differences of the both month data cause of the use an air conditioner of summer season.

In this study, the input data of linear programing is the actual hospital load for one week in August in the large hospital.

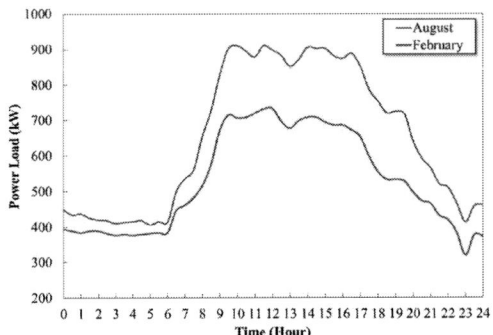

Fig. 3. Actual hospital load of one day in the large hospital.

B. Simulation Results

Results of calculation using the actual power load for one week in the large hospital are shown in Fig. 4 and Fig. 5. The top of figure indicates the frequency deviation is the output from the power grid frequency model. Second from the top shows output of EG_1, third from the top denotes output of EG_2, bottom is output of EG_3. Blue lines are output from model respectively, red lines show the optimized demand. Fig. 4 indicates the result that calculation schedule of the optimized demand shows every 60 seconds. On the other hand, it is every 1800 seconds, as shown in Fig. 5.

Table I summarizes the results of fuel consumption by changing the calculation schedule of the optimized demand in 60s, 300s, 600s and 1800s. It is clarified that the calculation schedule of demand should be short-term. However, as the difference of fuel consumption is a few, the calculation schedule can be expected to adjust long-term for a controller with prediction method that needs high performance calculation. The long-term calculation schedule has a demerit because the deviation of power grid frequency is high fluctuation.

TABLE I
THE RESULT OF SIMULATION CASE

Calculation Schedule	Fuel Consumption
60s	35501.3kg
300s	35514.4kg
600s	35529.7kg
1800s	35609.3kg

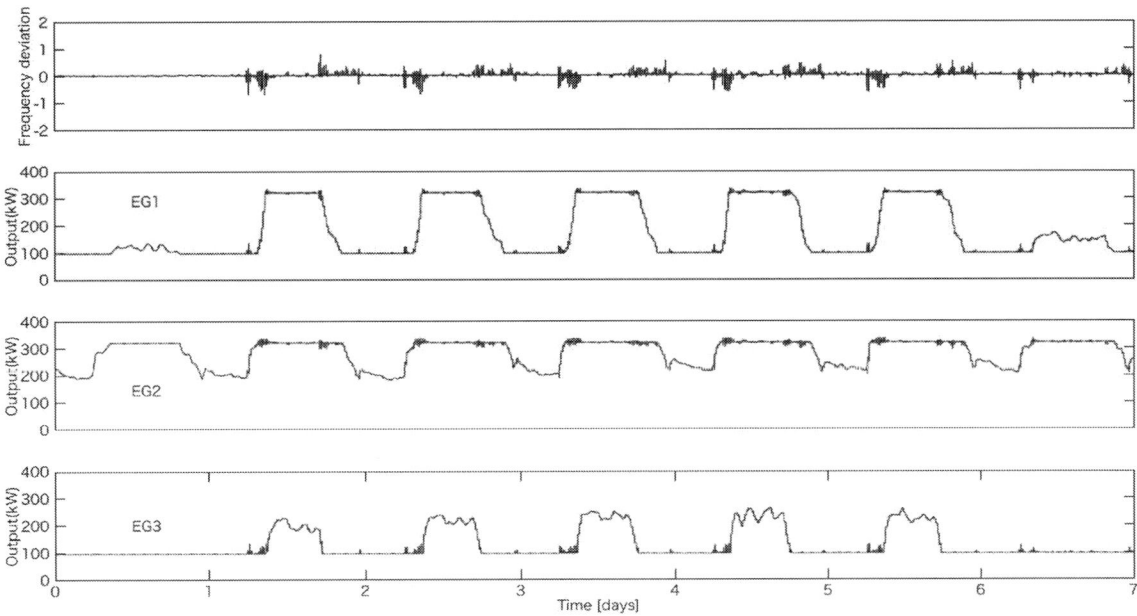

Fig. 4. Simulation case: the calculation schedule of the optimized demand in 60s.

2285

The 2018 International Power Electronics Conference

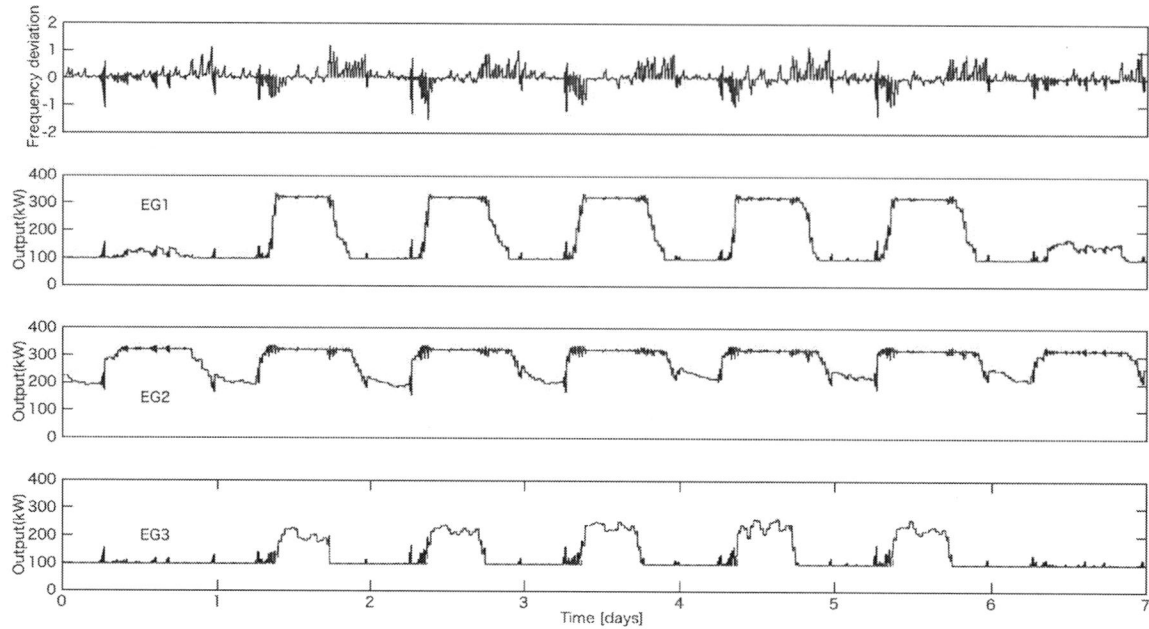

Fig. 5. Simulation case: the calculation schedule of the optimized demand in 1800.

V. CASE STUDY

This paper presents a comparison of optimized demand of EGs for minimizing fuel consumption using linear programming and our developed EGs considering power grid frequency model using the hospital load for a week in summer season with PV in MATLAB/Simulink. The results are summarized as follows:

(1) EGs considering power grid frequency model using the hospital load with PV is developed in MATLAB/ Simulink.

(2) The optimized demand of EGs is compared its with the fuel consumption using EGs and control model is calculated.

(3) It is clarified that the calculation schedule of demand should be short-term. However, as the difference of fuel consumption is a few, the calculation schedule can be expected to adjust long-term for a controller with prediction method that needs high performance calculation.

REFERENCES

[1] "About the damage situation to affect an earthquke to assume Kumamoto," Cabinet Office of Japan, 6th July 2016.

[2] "The Disaster Mnanagement Plan," Ministry of Health, Labor and Welfare, July 2017.

[3] Yuji Mizuno, Yoshito Tanaka, Fujio Kurokawa and Nobumasa Matsui, "A New Approach of Optimum Energy Scheduling of Emergency Generators Using Linear Programing in a Large Hospital," in Proc. of International Conference on Renewable Energy Research and Application (ICRERA), pp. 832-836, Nov. 2016.

[4] Teppei Baba, Yuji Mizuno, Yasuma Uchida and Nobumasa Matsui, "An Operation Method of Emergency Generators with Photovoltaics in a Large Hospital," Energy Engineering in Electronics and Communications, IEICE technical report, Vol. 116, No. 429, pp. 55-58, 2017.

[5] Teppei Baba, Yuji Mizuno, Yoshito Tanaka, Masaharu Tanaka, Fujio Kurokawa, Ilhami Colak and Nobumasa Matsui, "Comparison of Optimum Energy Scheduling of Emergency Generators of a Large Hospital with Renewable Energy System Using Mathematical Programing Method," in Proc. of International Conference on Renewable Energy Research and Application (ICRERA), pp. 519-523, Nov. 2017.

[6] Swaminathan Ganesan, Ramesh V and Umashankar S, "Hybrid Control of Microgrid with PV, Diesel Generator and BESS," International Journal of Renewable Energy Reserch, Vol. 7, No 3, pp. 1317-1323, 2017.

[7] Yuji Mizuno, Teppei Baba, Yoshito Tanaka, Masaharu Tanaka, Fujio Kurokawa, Ilhami Colak and Nobumasa Matsui, "Estimation of Optimum Capacity of Battery by Combined Use of a Renewable Energy System and Distributed Emergency Generators in a Large Hospital," in Proc. of International Conference on Renewable Energy Research and Application (ICRERA), pp. 515-518, Nov. 2017.

[8] Katsutoshi Hiromasa, Yoshiki Takabayashi, Koji Shiota, Kenichi Tanomura, Satoshi Osaki, Yukihiro Onoue and Kimihiko Shimomura, "A Study of Generation Control System in Consideration of Renewable Energy Source," The transactions of the Institute of Electrical Engineers of Japan. B, A publication of Power and Energy Society, Vol. 137, No 2, pp. 93-101, 2007.

Coordinated DFIG Wind Turbines and Solar PV Generators for Inter-area Oscillation Damping

Tossaporn Surinkaew and Issarachai Ngamroo
Department of Electrical Engineering, Faculty of Engineering,
King Mongkut's Institute of Technology Ladkrabang,
Bangkok 10520, Thailand
Issarachai.ng@kmitl.ac.th

Abstract- **Intermittent power injection from renewable energy sources such as wind and solar farms may cause low damping of critical inter-area oscillation modes. However, such renewable sources may be located in some areas with higher controllability of inter-area oscillations than the conventional synchronous generators. By controlling the reactive power output of such renewable sources, the superior damping effect of inter-area oscillations can be anticipated. This paper proposes the robust control design of power oscillation dampers of wind turbines with doubly-fed induction generator and solar photovoltaic generators to damp inter-area oscillations in large-scale power systems. The coordinated robust controllers are designed to achieve the desired damping and robustly operate against system uncertainties such as noises, external disturbance, and intermittent power. Study results indicate that the damping effect of coordinated wind and solar farms is higher than that of conventional power system stabilizers under various operating conditions.**

I. INTRODUCTION

In large-scale power systems, the most vital stability problem is the inter-area oscillation with poor damping in a range of frequency 0.2-0.8 Hz [1]. Moreover, with immense penetration of wind and solar farms to power grids, intermittent power injections may not only deteriorate power quality but also adversely affect power system stability. Therefore, the most challenging task is controlling such uncertain sources to improve system stability [2].

To stabilize power oscillation without degradation of active power, the stabilizing signal is generally applied to the reactive power control loop of wind and solar generators. In previous works, several applications have been accomplished by using wind or solar farms to stabilize power oscillation and maintain system stability. For an application of wind farms, the power oscillation damper (POD) in the reactive power control loop of the wind turbine with doubly-fed induction generator (DFIG) is used to stabilize the inter-area power oscillation [3]-[5]. On the other hand, applying solar photovoltaic (SPV) farms to stabilize inter-area oscillation by equipping POD in the reactive control loop has been done in [6]. In these works, DFIG and SPV equipped with PODs have been successfully utilized in maintaining power system stability. Regarding the locations of DFIG and SPV, they might contribute higher controllability of inter-area oscillation than those of the synchronous generators

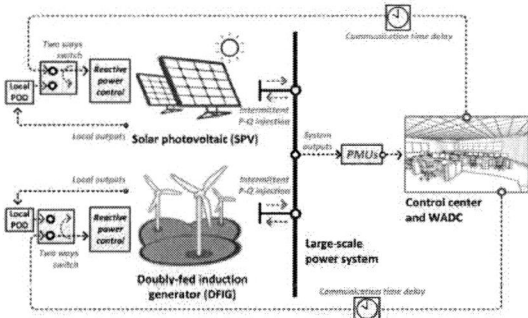

Fig. 1. Coordinated control of DFIG and SPV.

equipped with power system stabilizers (PSS). Consequently, such DFIG and SPV are expected to contribute more damping effect than PSS.

This paper presents a robust control design of coordinated DFIG and SPV to stabilize multiple inter-area modes. The suitable stabilizing DFIG and SPV with high controllability and observability of inter-area oscillations are selected by the geometric measure. The robust control technique is applied to design PODs of DFIG and SPV. Simulation study is conducted to confirm the stabilizing effect of the coordinated DFIG and SPV in comparison with the conventional PSS.

II. SYSTEM MODELING

Fig. 1 illustrates a coordinated control of DFIG and SPV in large-scale power systems. The coordinated control consists of the two-level controller. The two ways switch is used to choose the control level. For the first level, the wide-area damping controller (WADC) is primarily used to damp out the power oscillation. The input signal of WADC with high observability of the corresponding inter-area mode is obtained by phasor measurement units (PMU). For the second level, the local POD will be consequently activated when communication failure occurs. In the local level, the power flows from nearby transmission lines are utilized. Normally, the WADC is used to damp out inter-area oscillation. Communication time delay is also taken into account in WADC. When communication failure occurs, it causes a malfunction in the WADC. To tackle this problem, the local POD is activated and used to damp out inter-area oscillation with local power flow nearby

The 2018 International Power Electronics Conference

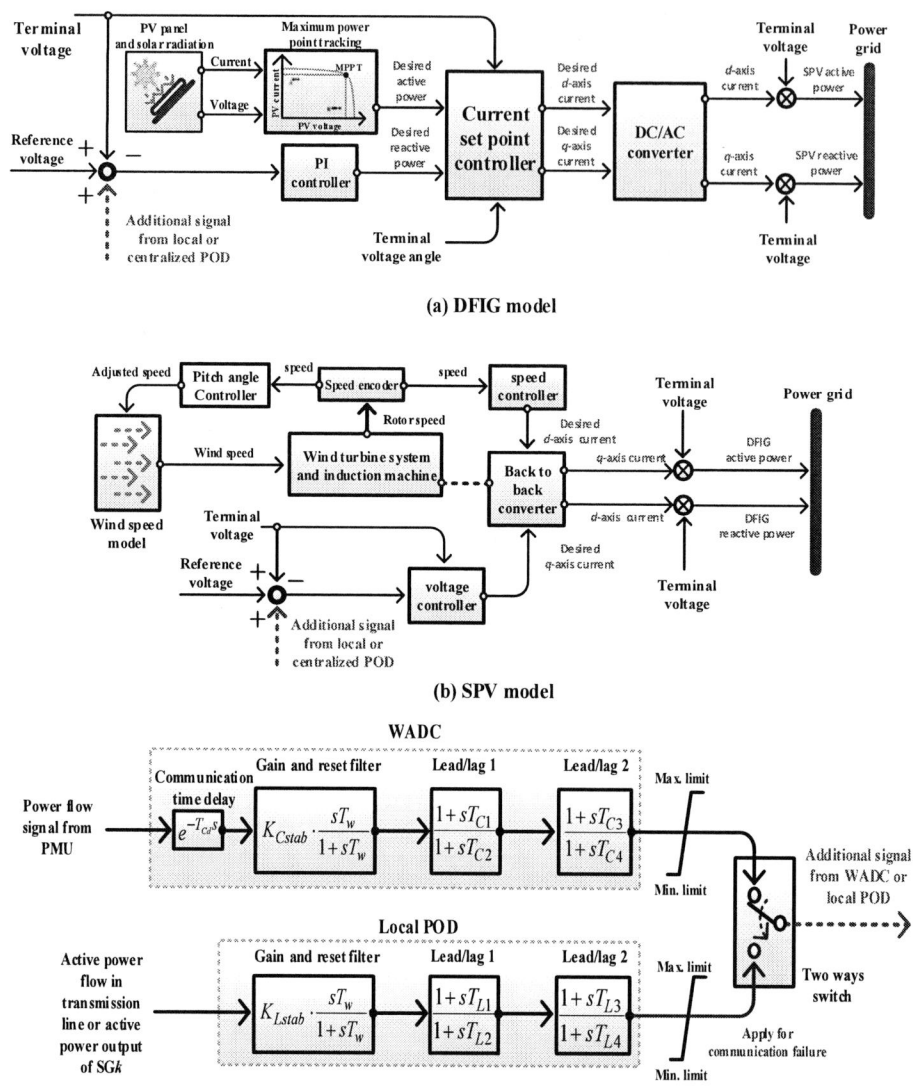

(a) DFIG model

(b) SPV model

(c) two-level POD

Fig. 2. DFIG, SPV and two-level POD models.

transmission lines. To avoid active power degradations, the stabilizing signals from both levels are added to reactive power control loops of DFIG and SPV. The DFIG and SPV models are depicted in Fig. 2(a) and (b), respectively. The structure of POD in both levels which is a 2nd-order lead/lag compensator is shown in Fig. 2(c).

The POD parameters are optimized by the proposed design. The power system in Fig. 1 can be represented by the linearized state equation as

$$\Delta \dot{\mathbf{X}} = A\Delta \mathbf{X} + B\Delta \mathbf{U}$$
$$\Delta \mathbf{Y} = C\Delta \mathbf{X} + D\Delta \mathbf{U} \tag{1}$$

where Δ is a small deviation, A, B, C and D are state, input, output and feed-forward matrixes, respectively, $\mathbf{X}$, $\mathbf{Y}$ and $\mathbf{U}$ are state, output and input vectors. The suitable

stabilizing devices and input signals can be determined by geometric measure of controllability and observability as [7]

$$g_{cg}(m) = \cos(B_g, \psi_m) = \left| B_g^T \times \psi_m \right| / \left\| \psi_m \right\| \left\| B_g^T \right\| \tag{2}$$

$$g_{oh}(m) = \cos(C_h, \phi_m) = \left| C_h \times \phi_m \right| / \left\| \phi_m \right\| \left\| C_h \right\| \tag{3}$$

where g_{cg} and g_{oh} are geometric measure of controllability and observability, respectively, B_g is the gth column of B, C_h is the hth row of C, ψ_m is left eigenvectors of the mth oscillation mode, ϕ_m is right eigenvectors of the mth oscillation mode, the superscript T represents the transpose of matrix, and $|\ |$ and $\|\ \|$ are the Modulus and Euclidean norms of matrix, respectively.

The 2018 International Power Electronics Conference

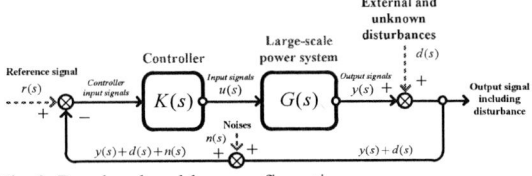

Fig. 3. Regular closed-loop configuration.

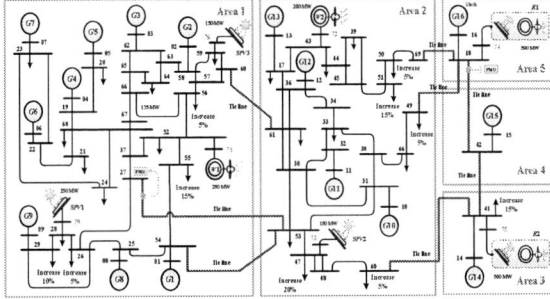

Fig.4 Study system (Base 100 MVA, 60Hz).

III. PROPOSED ROBUST COORDINATED DESIGN

To design POD parameters, damping performance and robustness are taken into account in the design process. For the damping performance, the multiple pole placement approach is applied by the following function.

Maximize
$$J_{damp} = \sum_{OS=1}^{nOS} \left(\delta_{\zeta,OS} \times \zeta_{a,OS} \right) \quad (4)$$

Subject to
$$\zeta_{a,OS} \geq \zeta_{spec,OS}$$

where OS is the inter-are mode index and nOS is total inter-area modes, when $OS=1,...,nOS$, $\zeta_{a,OS}$ and $\zeta_{spec,OS}$ are actual and specified damping ratios corresponding to the OS^{th} inter-area mode, respectively, J_{damp} is the damping performance index, $\delta_{\zeta,OS}$ is the weighting factors corresponding to each OS^{th} mode. When J_{damp} is maximized, the desired damping ratio can be obtained.

For robustness against system uncertainties, a general closed-loop with unknown disturbances is demonstrated in Fig. 3, where $y(s)$ and $u(s)$ are outputs of system and controller, respectively, $d(s)$, $r(s)$ and $n(s)$ are external disturbances, reference signal and noises, respectively [8]. The following relationship can be written by

$$y(s) = \left\langle (I+G(s)K(s))^{-1}G(s)K(s)r \right\rangle + \left\langle (I+G(s)K(s))^{-1}d \right\rangle$$
$$- \left\langle (I+G(s)K(s))^{-1}G(s)K(s)n \right\rangle$$
$$u(s) = \left\langle K(s)(I+G(s)K(s))^{-1}r \right\rangle - \left\langle K(s)(I+G(s)K(s))^{-1}d \right\rangle$$
$$- \left\langle K(s)(I+G(s)K(s))^{-1}n \right\rangle$$
$$(5)$$

Let $H_1= (I+G(s)K(s))^{-1}$ and $H_2=(I+G(s)K(s))^{-1}G(s)K(s)$. As a result, the robust performance index J_∞ can be achieved by minimizing the ∞-norm as

Minimize

$$J_\infty = \delta_{\infty,1} \left\| \begin{matrix} H_2 \\ K(s)H_1 \end{matrix} \right\|_\infty + \delta_{\infty,2} \left\| \begin{matrix} H_1 \\ -K(s)H_1 \end{matrix} \right\|_\infty + \delta_{\infty,3} \left\| \begin{matrix} -H_2 \\ -K(s)H_1 \end{matrix} \right\|_\infty$$
$$(6)$$

TABLE I SUITABLE STABILIZING DEVICES AND INPUT SIGNALS

Inter-area mode	Conventional PSS		Coordinated DFIG and SPV	
	Stabilizing device	Input signals	Stabilizing device	Input signals
1st	G8 (g_{cg}=0.42), G5 (g_{cg}=0.51)	P_{27-53} (g_{oh}=0.74)	R2 (g_{cg}=0.81), W1 (g_{cg}=0.69)	P_{27-53} (g_{oh}=0.74)
2nd	G11 (g_{cg}=0.47)	P_{27-53} (g_{oh}=0.61)	SPV3 (g_{cg}=0.75)	P_{27-53} (g_{oh}=0.61)
3rd	G15 (g_{cg}=0.59)	P_{18-42} (g_{oh}=0.85)	SPV3 (g_{cg}=0.72)	P_{18-42} (g_{oh}=0.85)

TABLE II DAMPING RATIOS OF OSCILLATION MODES

Inter-area modes	Without controller	Conventional PSS	Coordinated DFIG and SPV
1st	-1.03 %	4.27 %	5.00 %
2nd	1.23 %	4.21 %	4.47 %
3rd	-0.11 %	3.80 %	4.23 %

It can be analyzed in (6) that the first term hints the tracking reference, the second term implies the disturbance attenuation, and the third term insinuates the noises rejection, where $\delta_{\infty,1}$, $\delta_{\infty,2}$ and $\delta_{\infty,3}$ are weighting factors corresponding to first, second, and third terms of (6), respectively. When J_∞ is minimized, the controller robustness can be obtained.

The optimization function achieving both damping performance and robustness can be formulated by

Minimize
$$\left(1/J_{damp} \right) + J_\infty \quad (7)$$

Subject to
$$0 \leq \left(\delta_{\zeta,os}, \delta_{\infty,1}, \delta_{\infty,2}, \delta_{\infty,3} \right) \leq 1,$$
$$0.01 \leq K_{stab} \leq 5, \quad 0.01 \leq T_{lead/lag} \leq 1$$
$$\zeta_{a,OS} \geq \zeta_{spec,OS}$$

where K_{stab} is stabilizing gain and $T_{lead/lag}$ is lead/lag time constants. $\zeta_{spec,OS}$ is set at 2 %. Note that, since WADC and local POD operate at different time, they are optimized separately by (7).

IV. SIMULATION RESULTS

Fig. 4 shows a modified IEEE New England and New York system which consists of 16 synchronous generators (SGs) with 5 areas. DFIGs and SPVs are located in areas 1, 2, 3 and 5. The proposed coordinated DFIG and SPV referred to as "*Coordinated DFIG and SPV*" is compared with SG equipped conventional PSS referred to as "Conventional PSS" Note that both *Coordinated DFIG and SPV* and *Conventional PSS* are designed by (7).

By using (3), suitable stabilizing devices and input signals of *Coordinated DFIG and SPV* and *Conventional PSS* are given in Table I. The PMUs can be suitably located at the buses 18 and 27 to measure the input signals. The values of g_{cg} of inter-area modes of DFIG and SPV are much higher than that of synchronous generator. As a result, the stabilizing signal from the coordinated DFIG and SPV is more effective than that of synchronous generators with PSSs.

2289

The 2018 International Power Electronics Conference

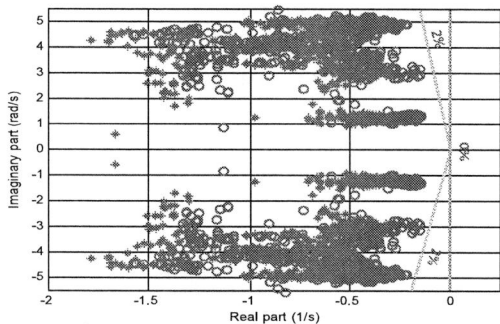

Fig. 5. Eigenvalues plot of 10,000 events by MCS; grey dot: *Conventional PSS*, red star: *Coordinated DFIG and SPV*.

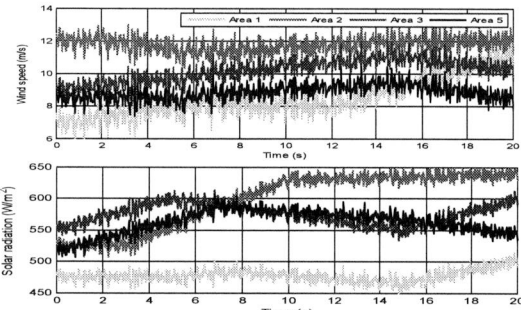

Fig. 6. Intermittent wind speeds and solar radiations.

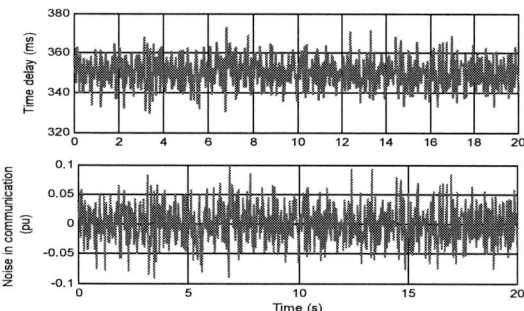

Fig. 7. Variable time delay and noise.

After optimization by (7), Table II demonstrates damping ratios of inter-area modes. Without controller, the damping ratios of the second mode are lower than 2% while those of the first and third modes are negative, the system is unstable. *Coordinated DFIG and SPV* and *Conventional PSS* are able to improve the damping ratios of all three modes. Although using the same objective function in the optimization, *Coordinated DFIG and SPV* can achieve greater damping ratio due to higher g_{cg}.

Fig. 5 illustrates the plot of eigenvalues corresponding to oscillation modes of 10,000 events by Monte Carlo Simulation (MCS) by randomly varying power outputs of all synchronous generators in all areas between -25% to 25%, DFIGs and SPVs power outputs between -5% to 5%, and time delay between 200ms to 700ms. Obviously, the damping ratios in case of *Coordinated DFIG and SPV* are higher than those of *Conventional PSS*.

Next, time simulation is conducted by the Power System Analysis toolbox vol. 2.1.8 (PSAT) [9]. In time simulation, the different patterns of wind speed and solar radiation in each area are applied as illustrated in Fig. 6. The total transmitted and received time delays of WADC are set at 350ms. Besides, the uncertainty due to variable time delay and noise in the measurements as depicted in Fig. 7 are applied to time simulation.

When the system operating point is changed by increasing power outputs of all synchronous generators in area 1 by 25%, and DFIGs and SPVs in system by 2.5%, Fig. 8 shows time simulation results of four case studies under different faults and without communication failure. Clearly, the damping effect of wide-area *Coordinated DFIG and SPV* is higher than that of wide-area *Conventional PSS*.

With communication failure, Fig. 9 depicts time simulation results of four case studies. The local *Conventional PSS* fails to suppress power oscillations in cases 2 and 3. On the contrary, the local *Coordinated DFIG and SPV* robustly suppress the power oscillations in all cases.

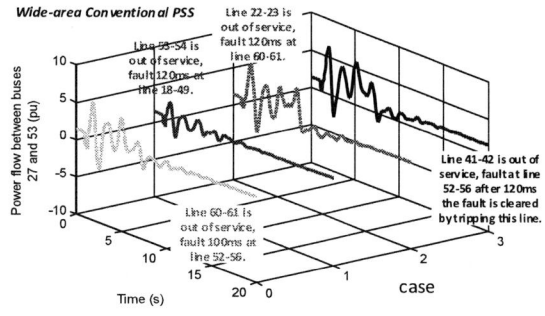

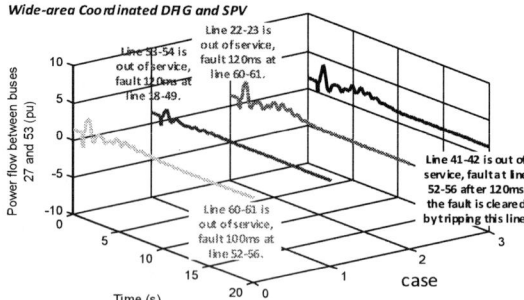

Fig. 8. Time simulation results (without communication failure).

2290

The 2018 International Power Electronics Conference

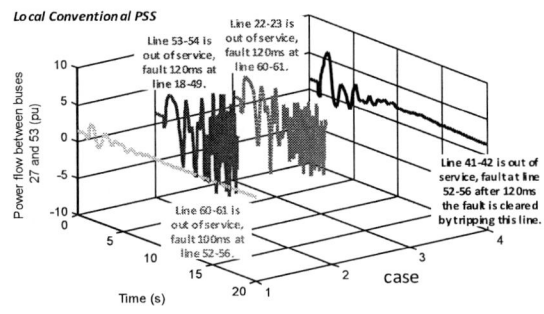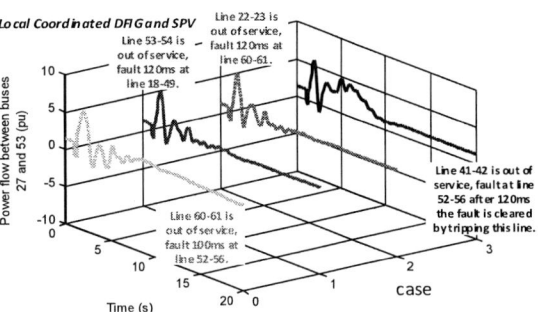

Fig. 9. Time simulation results (with communication failure).

V. CONCLUSION

The coordinated DFIG wind turbines and SPV generators for inter-area power oscillation damping has been proposed in this paper. Study results can be summarized as follows.

(1) When the locations of DFIG wind turbines and SPV generators yield higher controllability of oscillation modes than those of PSSs, the better stabilizing effect by DFIG and SPV can be obtained.

(2) As confirmed by small signal stability and transient stability study results, both wide-area and local levels of coordinated DFIG and SPV provides superior damping effect and robustness to PSSs under intermittent power, noises in communication, variable time delays, and faults.

REFERENCES

[1] G. Roger, "*Power system oscillations*", Springer Inc., US, 2000.

[2] I. Ngamroo, "Review of DFIG wind turbine impact on power system dynamic performances," *IEEJ Trans. Electrical and Electronics Engineering*, vol. 12, no. 3, pp. 301-311, 2017.

[3] D. Rimorov, G. Joós, and I. Kamwa, "Design and implementation of combined frequency/oscillation damping controller for type 4 wind turbines", *Proc. Power Systems Computation Conference 2016*, pp. 1-7, 2016.

[4] G. Tsourakis, B. M. Nomikos, and C. D. Vournas, "Contribution of doubly fed wind generators to oscillation damping", *IEEE Trans. Energy Convers.*, vol. 24, no. 3, pp. 783-791, 2009.

[5] L. L. Fan, H. P. Yin, and Z. X. Miao, "On active/reactive power modulation of DFIG-based wind generation for interarea oscillation damping", *IEEE Trans. Energy Convers.*, vol. 26, no. 2, pp. 513-521, 2011.

[6] R. Shah, N. Mithulananthan, and K. Y. Lee, "Large-scale PV plant with a robust controller considering power oscillation damping", *IEEE Trans. Energy Convs.*, vol. 28, no.1, pp. 106-116, 2012.

[7] A. Heniche, and I. Kamwa, 'Control loops selection to damp inter-area oscillations of electric networks,' *IEEE Trans. Power Syst.*, vol. 17, no.2, pp.378-384, 2002

[8] S. Skoqestad, and I Postlethwaite, "*Multiple feedback control: analysis and design*", John Wiley & Sons Inc., 2005.

[9] F. Milano, "*Power system analysis toolbox version 2.1.8*", University College Dublin, 2013.

2291

Energy management using a quick charger with storage batteries for electric vehicles

Taku Ishibashi*, Toyonari Shimakage, Norikazu Takeuchi, Takaaki Kikuchi and Midori Nonogaki
NTT FACILITIES, INC., Tokyo, Japan
*E-mail: ishiba32@ntt-f.co.jp

Abstract— We developed an Electric Vehicle Quick Charger (EVQC) incorporating storage batteries and conducted a verification field test. This EVQC is equipped with demand control features which reduce economic burdens by suppressing the peak power load for EV charging, while also saving charging time by reducing the input power from the utility through the use of power stored in internal batteries. These batteries are charged by using surplus power from the PV generation system on the rooftop or from night-time power. We also verified the method for monitoring the deterioration state of these daily-use storage batteries.

In this paper, the field test results of the EVQC incorporating storage battery systems are reported.

Keywords— *Demand control, Electric vehicle quick charger, Storage batteries*

I. INTRODUCTION

In recent years, EVs have been attracting increased attention due to their low environmental load.

To promote EV use, it is necessary to increase the number of highly convenient charging stations which provide quick-charge services. However, there are a number of problems with building more stations.

An EV could be charged in a short period of time with quick chargers, but it may not be practical to increase the contracted amount of power (per 30 min) by pushing up the peak electricity demand or to use a number of quick chargers at the same time. Either of these solutions could result in too much stress on the electric grid [1-3]. Simply suppressing the charging power could be a solution but this makes the charging time longer and ends up ruining the convenience.

On the other hand, those PV generators which are increasingly being installed on the rooftop of apartment complexes and ordinary residences generate electric power in the daytime and their peak power demand is morning and night. This leads to an imbalance in power demand, even causing harmful effects on the grid if more power is generated than is used by the buildings.

To resolve these problems, we developed an EVQC equipped with an internal storage battery system and conducted a verification field test. The storage batteries are charged using surplus daytime PV power or night-time power which has lower demand. The power stored in the batteries suppresses the power reversal caused by PV generation as well as demand peak in the buildings by charging EVs.

Furthermore, we verified the method for monitoring the deterioration state of these daily-use storage batteries by estimating their capacities.

II. CONFIGURATION

The EVQC was installed in a typical apartment complex for testing. The specifications of the EVQC are shown in Table I, and the system configuration and its overview is shown in Figure 1 and 2 respectively. The EVQC incorporating storage batteries consists of four 10kW quick charger units, a 10kW battery discharger unit, an information transmission unit, and a controller, which is equipped with lithium ion batteries as well. The information transmission unit collects information on the storage batteries including their demand and State of Charge (SOC). The controller regulates each converter unit operation based on the information collected by the information transmission unit and demands for charging EVs. The information cooperation method is in accordance with the CHAdeMO protocol stipulated by the CHAdeMO convention. Also, items measured for speculating the storage battery capacity are shown in Table II.

TABLE I Specification

Systems	Items	Units	Specifications
EVQC (AC input)	Phase and wire	-	3-phase, 3-wire
	Rated voltage	V	200
	Power factor	-	Over 0.95
EVQC (DC output)	Range of output voltage	V	50-500
	Range of output current	A	0-125
	Range of output power	kW	0-50
	Efficiency	%	Over 90
Battery	Rated capacity	Ah	85
	Range of operational voltage	V	237.6-369.6
	Range of discharge current	A	0-300

The 2018 International Power Electronics Conference

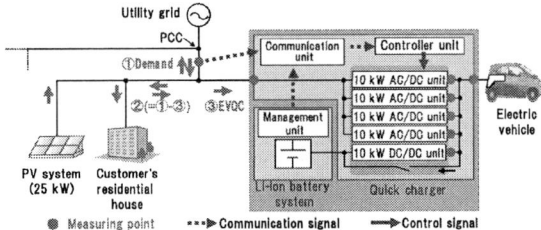

Fig. 1. Configuration of the system

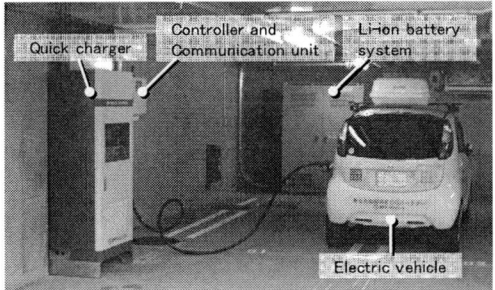

Fig. 2. Picture of battery-equipped EVQC

TABLE II SPECIFICATIONS

Items	Units	Acquisition interval
Cell voltage	V	Per minute
Battery total voltage	V	Per second
Charge and discharge current	A	Per second

III. CONTROL METHOD

A. EV charging method by demand controlling

The power supplied to EVs from the utility can exceed the service-contract maximum, but only for a few minutes, and EVs are charged without having to exceed the limit over a 30 minute average.

First, when the EVQC receives an EV charging request, a virtual contracted power (P_{Cv}) shown in equation (1) is calculated, and the difference between P_{Cv} and the building power consumption is supplied from the utility to use for quick charging. Applying P_{Cv} makes it possible to secure more power for charging EVs without exceeding the service-contract maximum over a 30minute average, versus simply setting the upper limit to the target value.

$$P_{Cv} = P_C + \frac{Pc \cdot t_s - \int_0^{t_s} P_D(t)dt}{30 - t_s} \quad (1)$$

P_C [kW] stands for the contracted power amount, $P_D(t)$ [kW], receiving power including EVQC power consumption and building power consumption, and t_s [min], the time when EV charging starts($0 \leqq t_s < 30$).

Second, whether or not to use the storage batteries is determined by their SOC. When the SOC is sufficient, they will be discharged for EV charging. When the SOC of the batteries reaches the operational lower limit during EV charging, they will be stopped.

B. Battery charging

The batteries are charged in two different ways while an EV is not being charged.

The first way is to charge them by using power from the utility, which is implemented within the range of contracted power amount, carefully observing the receiving power amount. Also, the SOC for starting and stopping charging can be set to specific hours, which makes it possible to set the system such that daytime charging can be minimized and more charging operation can be carried out during the night.

The second way is to charge them by using surplus power from PV generation. The difference between PV generating power and the whole building power consumption is used as the surplus power. However, as PV power fluctuates a lot, surplus power multiplied by the moving average is applied as the charging power amount in order to suppress mechanical burdens caused by too frequently switching between charging and not, which might occur around the point when the surplus power value is 0.

These two methods enable batteries to be charged by using power from the utility during the night when there is sufficient power available, and also enable surplus daytime PV power to be effectively utilized.

C. Battery-capacity estimation method

In this verification, the estimated capacity was calculated by putting the collected data applicable to specific conditions into equation (2).

$$C_e = I_{c/d} (SOC_b − SOC_a) \quad (2)$$

C_e [Ah] is estimated capacity, $I_{c/d}$ [Ah] is charging/discharging is integrated current, SOC_b [%] is SOC before charging/discharging, SOC_a [%] is SOC after charging/discharging.

The SOC differs even with the same stable voltage level between after-charge and after-discharge conditions depending on battery types (Fig.3). A feature of this battery-capacity estimation method is that after-charge and after-discharge curves are used separately when the SOC is calculated based on the measured voltage data.

Specific conditions for the battery-capacity estimation are shown in Fig.4. Condition 1 examines whether charging or discharging had continued for more than a certain minimum time and amount to minimize measurement errors. Condition 2 and 3 examine whether the voltage had been stable without any charge or discharge performance before and after the charge or discharge performance that meets condition 1. The data that meet all these three conditions are put into equation (2) to calculate the estimated capacity.

2293

Example data for capacity estimation is shown in Fig.5. In this figure, T3 shows more than a certain amount and duration of discharge performance, and no charge or discharge had been performed in T2 and T4. All of this meets the conditions for implementing capacity estimation. To put a data value into equation (2), the discharge amount in T3 is calculated. After that, the SOC level with voltage A and voltage B are calculated using the after-discharging curve and the after-charging curve respectively, both of which are based on the OCV (Open Circuit Voltage) – SOC characteristics.

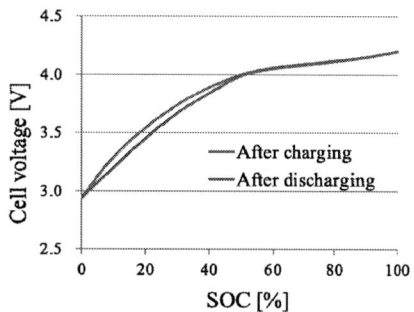

Fig. 3. OCV-SOC characteristics

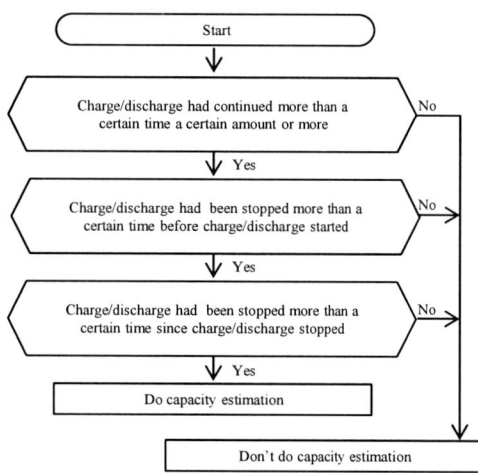

Fig. 4. Flowchart

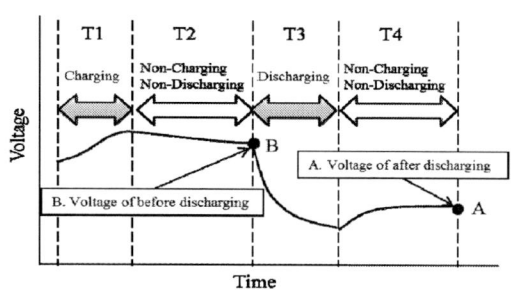

Fig. 5. Example of capacity estimation

IV. RESULTS

A. Storage battery discharge and demand control

The demand control state in which the upper limit of receiving power in a company apartment complex is set to be 30kW is graphically depicted in Fig. 6.

The demand control method allows receiving power in the company apartment to exceed the contracted amount for a few minutes, but the quick charger unit is controlled so the 30 minute average will not exceed the contracted amount. Also, the power stored in the batteries is used to charge the EV. The combination of these two procedures enables it to effectively utilize power from the utility while the average contracted power never exceeds 30 kW, preventing the longer period of charging time by using the discharging power from the storage batteries.

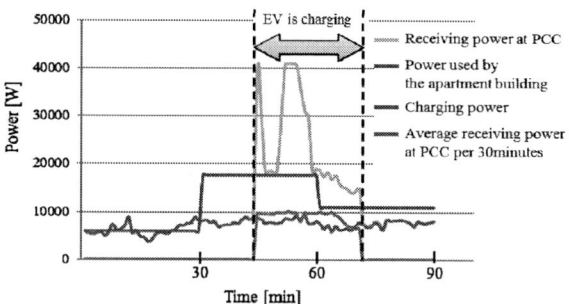

Fig. 6. Demand control and storage battery discharge

B. Battery charge

The batteries are charged by using night-time power which has less demand or surplus power from PV generation. Charging operation by using surplus power is shown in Fig.7.

When surplus power is available (when the apartment power consumption including the PV power is below zero), storage batteries are charged. This reduces the reverse current where the power at the receiving point is below zero.

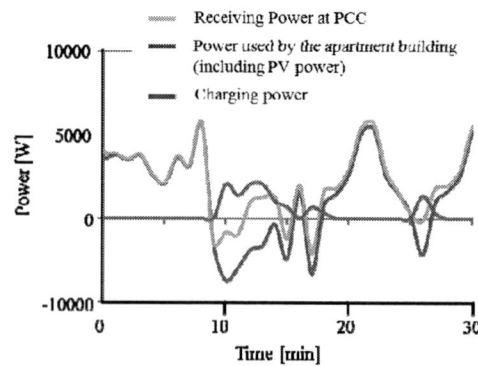

Fig. 7. Charging operation by using surplus power

C. Battery-capacity estimation method

This time, the conditions for implementing capacity estimation are defined as charge or discharge performance for 10min or more as well as 10A or more.

The capacity estimation results based on the method introduced in this paper, and the capacity data gained by closely testing in the factory are shown in Fig.8. The average error rate was 3.62% when charging data was used and 3.17% when discharging data was used. Both of these figures show good results.

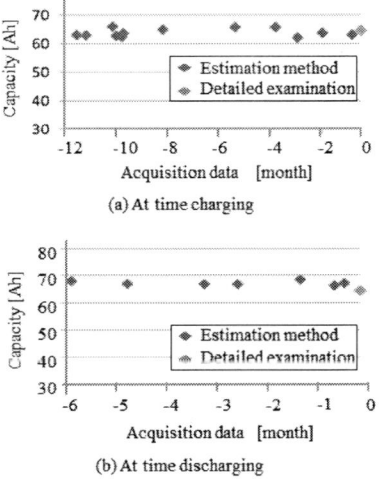

(a) At time charging

(b) At time discharging

Fig. 8. Results of capacity estimation

V. CONCLUSION

The operational data showed that the demand control procedure successfully charged an EV by using the power stored in the batteries. The data also verified that the reverse current could be suppressed by changing the amount of charging power to the batteries depending on the amount of surplus power.

We confirmed that the battery-capacity estimation method examined here was effective. This method enables the deterioration state to be determined by daily charge and discharge data.

This study and field test was performed with the support of the New Energy Promotion Council (NEPC).

REFERENCES

[1] Mustapha Aachiq and Iwafune Yumiko, "Evaluation of the effect of reducing the surplus PV power through the usage of EV's battery "EEJ National Convention 2013, No.6-122, pp. 220-221

[2] Daisuke Satoya, Yasuhiko Miyashita, Daiki Yamashita and Ryuichi Yokoyama, "Improvement of PV utilization by optimal operation of V2H" IEEJ National Convention 2014, No.6-200, pp. 373-374

[3] Toyonari Shimakage, Taku Ishibashi, Masashi Baba, Norikazu Takeuchi and Takaaki Kikuchi, "Operation results of a battery-equipped quick charger for electric vehicles having a demand side management function" IEICE SOCIETY Conference 2013, BS-5-4, S27-28

The 2018 International Power Electronics Conference

A Method for Junction Temperature Estimation Utilizing Turn-on Saturation Current for SiC MOSFET

Hui-Chen Yang[1*], Rejeki Simanjorang[2] and Kye Yak See[1]

[1] School of Electrical and Electronic Engineering, Nanyang Technological University, Singapore

[2] Applied Technology Group, Rolls-Royce Singapore Pte. Ltd., Singapore

*E-mail: HYANG011@e.ntu.edu.sg

Abstract— Keeping the operating junction temperature of a SiC MOSFET within safe and tolerable range is vital. Besides safety reasons, it also extends the device's life span as thermal transient is identified as one of the major stressor for device aging. Conventional methods use thermal sensors for temperature monitoring but these sensors require extra space and are unable to track transient temperature variation due to relatively slow response time. This paper proposes a new temperature sensor-less method to estimate junction temperature by using the device's square root of saturation current ($\sqrt{I_D}$) and threshold voltage (V_{th}) for temperature monitoring. Both parameters can be extracted from device's saturation current and the experimental results have demonstrated the feasibility of this temperature estimation method. The proposed method looks promising as an alternative for future electronic health monitoring (EHM) system.

Keywords— *Junction temperature estimation, saturation current, SiC MOSFET, threshold voltage.*

I. INTRODUCTION

Owing to the excellent material property for high temperature operation, silicon carbide (SiC) MOSFET has gained its popularity for high power density converter (HPDC) design to accommodate the harsh operating environment of mission-critical systems, such as aerospace application. Early detection of abnormalities of these power devices is vital before they develop into catastrophic failures. Hence, the necessity the development of electronic health monitoring (EHM) technology. Monitoring the junction temperature during the operation of SiC MOSFETs is essential since thermal stress is closely correlated to device wear-out. Direct measurement of device temperature requires sensors integrated on the chip or being part of it. However, such sensors occupy the chip's active area and compromise the device's current carrying capability. Today, many device manufacturers have incorporated temperature sensor into the package to estimate device temperature with physical-based thermal models. However, the validity of the estimation relies on the accurate device package thermal model and tedious analytical analysis [1]. Besides, the slow response of the sensor affects the capability of transient temperature measurement in real-time.

Temperature-dependent device parameters such as on-state voltage, on-state resistance, short-circuit current, as well as dynamic behaviors during on-off transition provided straightforward alternatives [2]. Nonetheless, the wide-bandgap nature of SiC MOSFET makes direct measurement of temperature-dependent parameters challenging. For example, the measurement of on-state resistance, as SiC MOSFET is less temperature sensitive compared to silicon (Si) devices [3]. Typically, the temperature dependency of the on-state resistance of a commercially available 1.2 kV/36 A SiC MOSFET from Cree Inc. is around 0.5 mΩ/°C [4]. Such low sensitivity implies that any slight measurement error can lead to misleading information. Similarly, the threshold voltage (V_{th}) exhibits sensitivity below 10 mV/°C. Past studies had put in effort on making V_{th} measurement on-line [5-7] by capturing gate-to-source voltage (V_{GS}) when device current is zero during turn-on transition. This method, however, requires high-speed measurement circuits. Otherwise, the switching time has to be extended with large external gate resistor. Another method is the use of dynamic behaviors during switching transition, such as drain current switching rate and turn-on delay, as temperature-dependent parameters [8-9]. The required sensing resolution typically ranges from nanosecond to picosecond, and is more stringent for SiC MOSFET, which requires current sensor with very large bandwidth.

To address above-mentioned concerns on temperature estimation, this paper proposes a new sensor-less solution to mitigate the aforementioned measurement error and the measurement circuits' bandwidth requirement. Furthermore, the proposed method serves as an aid for EHM system's development. The rest of the paper is organized as follows. Section II provides the theoretical background of the proposed temperature-dependent parameters. Section III describes the proposed method in details. The experiment results are presented in Section IV and finally, Section V concludes this paper.

II. TEMPERATURE-DEPENDENT PARAMETERS

During the switching of SiC MOSFETs, it always experiences saturation region before it exhibits the resistive behavior when it fully turned on, as illustrated in Fig. 1.

It illustrates the four phases during turn-on process (phase A to D) with gate-to-source voltage ($V_{GS}(t)$), device current ($I_D(t)$) and drain-to-source voltage ($V_{DS}(t)$) shown. The saturation region incorporates phase B and parts of

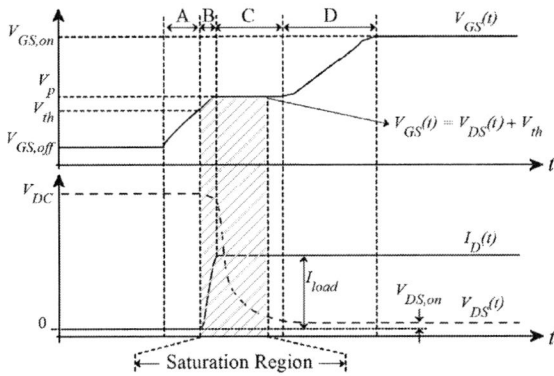

Fig. 1. Turn-on timing of SiC MOSFET. As $V_{GS}(t)$ increases from the nominal turn-off voltage ($V_{GS,off}$) to the turn-on voltage ($V_{GS,on}$), $I_D(t)$ changes from zero to the full load current (I_{load}). At the same time, $V_{DS}(t)$ drops from V_{DC} to the on-state voltage ($V_{DS,on}$) of the SiC MOSFET.

phase C (Miller plateau period) as defined by the shaded region marked in Fig. 1, where $V_{th} < V_{GS}(t) < (V_{DS}(t) + V_{th})$. Since the SiC MOSFET is operating at full load current during phase C, the proposed method will utilize the device current during phase B to avoid excessive switching loss.

The device current during saturation region is temperature-dependent and it can be expressed by equation (1) [10]. In (1), constants μ, Z, C_{ox}, L, ΔL, and T are the carrier mobility, the length of active area in vertical dimension defined for device with vertical structure (Z), unit gate oxide capacitance (C_{ox}), channel length (L), the channel length reduction due to depletion, which is subject to increasing drain-to-source voltage (ΔL), and temperature (T), respectively.

$$I_D(T) = \frac{\mu(T)C_{ox}Z}{2L}\left[V_{GS} - V_{th}(T)\right]^2\left(\frac{L}{L-\Delta L}\right) \quad (1)$$

The first temperature estimator available is $\sqrt{I_D}$. Its temperature dependency is given in (2).

$$\frac{d\sqrt{I_D(T)}}{dT} = \frac{k}{\sqrt{\mu(T)}} \cdot \left\{\frac{1}{2}\left[V_{GS} - V_{th}(T)\right] - \mu(T)\frac{dV_{th}(T)}{dT}\right\}$$

$$(2)$$

where $k = \sqrt{\dfrac{C_{ox}Z}{2L}\left(\dfrac{L}{L-\Delta L}\right)}$.

Since the device is not fully turn-on, the temperature effect on μ can be neglected, as compared to the temperature effect on V_{th} [8]. Thus, both $k/\sqrt{\mu(T)}$ and $\mu(T)(dV_{th}(T)/dT)$ terms in (2) can be treated as constants. As V_{th} has a negative temperature coefficient [11], $\sqrt{I_D}$ is expected to have linear increments as temperature rises.

V_{th} is another temperature estimator, which can be obtained from two sets of device saturation current I_{D1} and

I_{D2} under known gate-to-source voltage V_{GS1} and V_{GS2}, respectively. Under the same temperature condition, V_{th} can be expressed by (3), where $a = \sqrt{I_{D2}/I_{D1}}$.

$$V_{th} = \frac{aV_{GS1} - V_{GS2}}{a - 1} \quad (3)$$

III. THE PROPOSED METHOD

Device saturation current can be obtained by a multi-level gate control method as illustrated in Fig. 2. When a test mode enable/disable signal $test\_en$ is applied, two test voltages (V_{GS1}, V_{GS2}) will be generated during the saturation region. The corresponding saturation currents can be obtained (I_{D1}, I_{D2}).

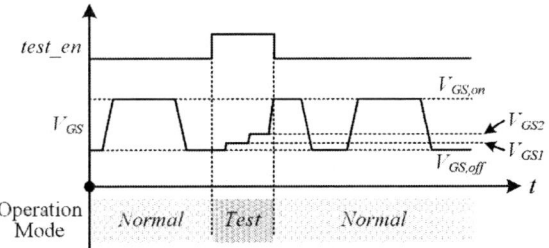

Fig. 2. Multi-level gate control method for obtaining saturation current information. For the ease of illustration, the Miller plateau period is not drawn here.

Multi-level gate control is achieved by adding a multi-level output (MLO) generation block to the conventional isolated gate driver as shown in the inductive load configuration in Fig. 3. The MLO generation block is inserted between node A and G by simply cut the turn-on gate resistor path so that the minimum change is made on the traditional driving scheme. The three switches ($S1$ to $S3$) in the MLO generation block are controlled by $test\_ctrl$, $Vtest\_ctrl$ and $Vadj\_ctrl$, respectively. Test mode is enabled/disabled by switch $S1$ on/off. When test mode is enabled and $S2$ is switched on, two sets of test voltages can be generated by turning $S3$ on and off as given by equation (4).

$$\begin{cases} V_{GS1} = V_{ref} - V_s & when\ S3\ off. \\ V_{GS2} = \left(1 + \dfrac{R1}{R2}\right)V_{ref} - V_s & when\ S3\ on \end{cases} \quad (4)$$

Detailed operation is illustrated in Fig. 4, where *PWM* is a double pulse for gate control in the inductive load testing. The SiC MOSFET is switching normally between nominal $V_{GS,on}$ and $V_{GS,off}$ when test mode is disabled ($test\_ctrl$ is low). When test mode is enabled ($test\_ctrl$ is high) while $S2$ is not turned on yet ($Vtest\_ctrl$ is low), V_{GS} is slowly pulled to $AVSS$ through resistor R_{GS}. Then, V_{GS} is set to V_{GS1} by turning $S2$ on while $S3$ remains off ($Vadj\_ctrl$ is low), and the corresponding device current I_{D1} is recorded. Finally, V_{GS} is set to V_{GS2} by turning $S3$ on

2297

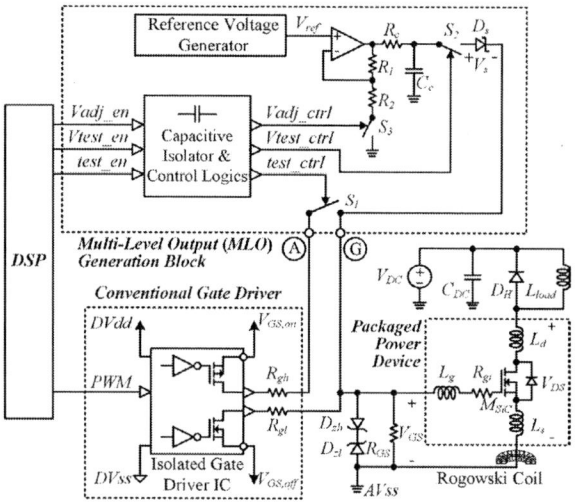

Fig. 3. Proposed multi-level gate control circuit.

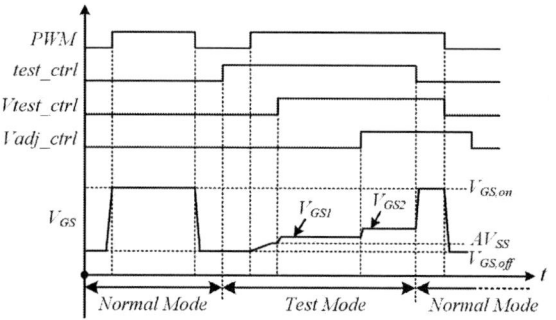

Fig. 4. Timing for multi-level gate control. Switches *S1* to *S3* in Fig. 3 turned on when the corresponding control signal is high.

(*Vadj_ctrl* is high) to capture another device current information I_{D2}. The normal operation can be resumed once the settled current information is captured. All the switches in the circuit should not switching at the same time to avoid current shoot-through.

With the measured saturation current information, the V_{th} at a preset temperature can be obtained by (3) with the two sets of test voltages shown in (4).

IV. EXPERIMENTAL RESULTS

A clamped inductive load test circuit is designed and built with double-pulse input to emulate the switching operation of a SiC MOSFET. Fig. 5 shows the measurement setup and the double pulse test waveforms at room temperature based on the circuit shown in Fig. 3 with V_{GS1} and V_{GS2} designed to be 3.86 and 4.68 V, respectively. The timing periods of V_{GS1} and V_{GS2} are set to be relatively longer than the settling time they required for the ease of observation. It can be observed that the actual current settling time is around 3 μs. In practical application, device temperature can be estimated as long as the current settling time is well covered by clock duty time. For example, the current settling time can be sufficiently covered with switching frequency up to 50 kHz with 50%

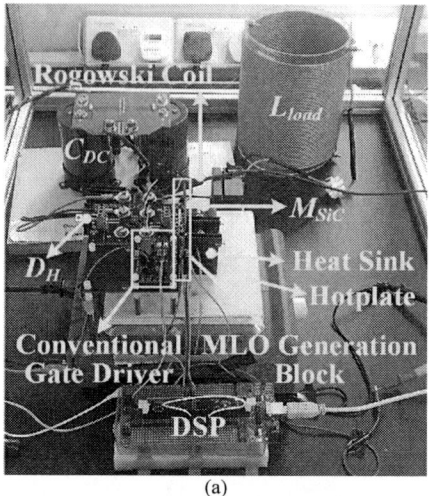

(a)

(b)

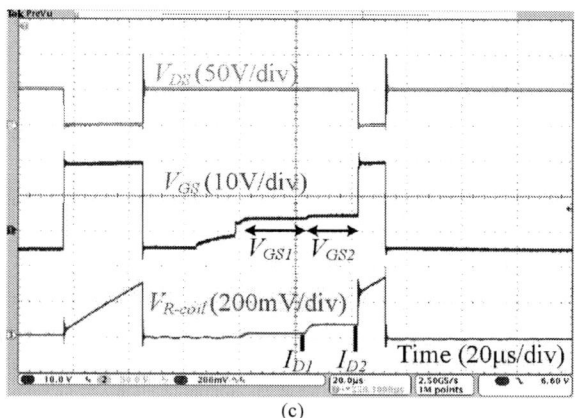

(c)

Fig. 5. Measurement setup. (a) Overall view. (b) Side view. (c) Double-pulse test waveform at room temperature.

duty cycle. A commercial SiC MOSFET M_{SiC} (C2M0080120D, *Cree Inc.*) is fixed on top of a heatsink and the temperature is controlled by placing it on a hotplate (UC150, *Stuart*). The preset temperature is read by an analog temperature sensor (LMT86, *Texas Instruments*), which is placed adjacent to M_{SiC}. A 246 μH inductor is chosen as a load (L_{load}), and a SiC Schottky diode (C4D20120A, *Cree Inc.*) is used as a freewheeling diode

2298

(D_H). The MLO generation block and conventional gate driver are designed on separated printed circuit boards (PCBs) for flexibility. The device current is measured by a Rogowski coil (CWTMini HF06B, *PEM*) with sensitivity of 50 mV/A. It can be either observed by an oscilloscope or processed by data acquisition system directly. Double-pulse test is performed at least 30 minutes after the temperature is set to allow M_{SiC} to reach thermal equilibrium where junction temperature can be assumed to be reached that of the hotplate. Soft turn-on is designed to avoid drastic current change for fast settling.

In this demonstration, current data is obtained by averaging 5000 data points captured by oscilloscope (MDO4104B-3, *Tektronix*) right before switching as indicated by the black bars in Fig. 5-(c). Five measurements are taken to average out the random error contributed by oscilloscope. The calculated V_{th} are shown in Fig. 6-(a).

The mean threshold voltages are plotted in Fig. 6-(b) together with the V_{th} extracted from datasheet [11]. Both curves show negative temperature dependency. The temperature sensitivity of the measured data shows close agreement with the datasheet.

The temperature dependency of the measured square root of device current is plotted in Fig. 7. All five measurements of $\sqrt{I_{D1}}$ and $\sqrt{I_{D2}}$ have shown excellent linearity. The mean data are summarized in Fig. 8 and the good linearity suggests the promising applicability for effective temperature monitoring.

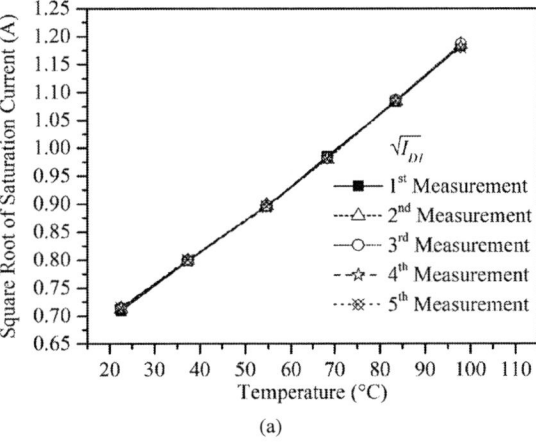

(a)

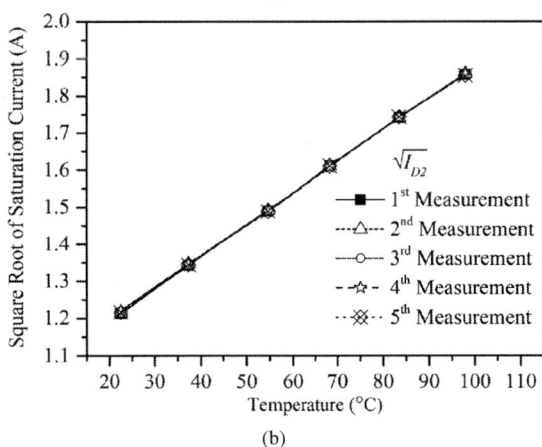

(b)

Fig. 7.   (a) Measured $\sqrt{I_{D1}}$ (at V_{GS} = 3.86V) and (b) $\sqrt{I_{D2}}$ (at V_{GS} = 4.68V).

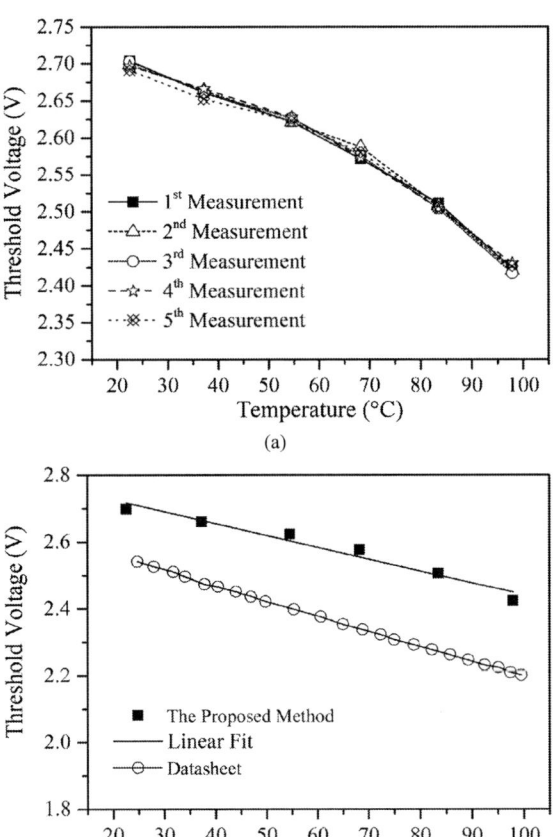

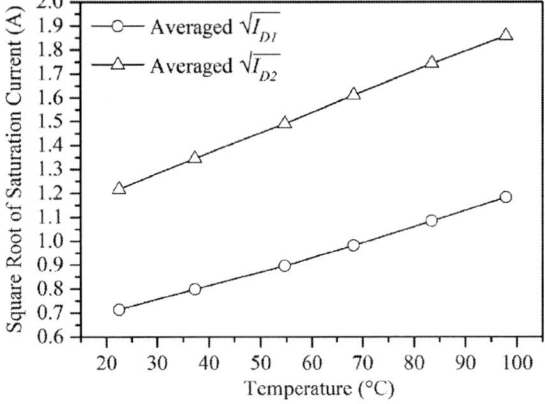

Fig. 8.   Averaged square root of saturation currents of the five measurements of $\sqrt{I_{D1}}$ and $\sqrt{I_{D2}}$.

Fig. 6.   (a) V_{th} obtained from the proposed multi-level gate control method. (b) Mean V_{th} of the five measurements, linear fit of the measurement data and V_{th} extracted from datasheet.

V. CONCLUSIONS

This paper proposes a new method for SiC MOSFETs' junction temperature estimation. This method leverages on the square root of device saturation current ($\sqrt{I_D}$) and threshold voltage (V_{th}) as temperature estimators. The experiments have successfully extracted these parameters from device saturation current, and the results have demonstrated the feasibility of the proposed method for junction temperature estimation. For timing-critical applications, the proposed two-step gate control can be reduced to one-step gate control, the square root of saturation current is now adopted as device temperature estimator in lieu of threshold voltage owing to its highly linear relationship with respect to temperature change.

ACKNOWLEDGMENT

This work was conducted within the Rolls-Royce@NTU Corporate Lab with the support from the National Research Foundation (NRF) Singapore under the Corp Lab@University Scheme.

REFERENCES

[1] L. Wei, J. McGuire and R. A. Lukaszewski, "Analysis of PWM Frequency Control to Improve the Lifetime of PWM Inverter," *IEEE Trans. Ind. Appl.*, vol. 47, no. 2, pp. 922–929, Mar./Apr. 2011.

[2] N. Baker, M. Liserre, L. Dupont and Y. Avenas, "Improved reliability of power modules: A review of online junction temperature measurement methods," *IEEE Ind. Electron. Mag.*, vol. 8, no. 3, pp. 17–27, Sep. 2014.

[3] L. Zhang, P. Liu, S. Guo and A. Q. Huang, " Comparative study of temperature sensitive electrical parameters (TSEP) of Si, SiC and GaN power devices," in *Proc. IEEE 4th Workshop Wide Bandgap Power Devices and Applications*, Fayetteville, AR, 2016, pp. 302–307.

[4] Y. Zhang and Y. C. Liang, "A simple approach on junction temperature estimation for SiC MOSFET dynamic operation within safe operating area," in *Proc. Energy Conversion Congr. And Expo.*, Montreal, Canada, 2015, pp. 5704–5707.

[5] H. Chen, B. Ji, V. Pickert and W. Cao, "Real-time temperature estimation for power MOSFETs considering thermal aging effects," *IEEE Trans. Device Mater. Rel.*, vol. 14, no. 1, pp. 220–228, Nov. 2013.

[6] H. Chen, V. Pickert, D.J. Atkinson and L.S. Pritchard, "On-line monitoring of the MOSFET device junction temperature by computation of the threshold voltage," in *Proc. 3rd IET Int. Conf. Power Electronics, Machines and Drives*, Dublin, Ireland, 2006, pp. 440–444.

[7] B. Strauss and A. Lindemann, "Measuring the junction temperature of an IGBT using its threshold voltage as a temperature sensitive electrical parameter (TSEP)," in *Proc. 13th Int. Multi-Conf. Systems, Signals & Devices*, Leipzig, Germany, 2016, pp. 459–467.

[8] J. O. Gonzalez, O. Alatise, J. Hu, L. Ran and P. Mawby, "Temperature sensitive electrical parameters for condition monitoring in SiC power MOSFETs," in *Proc. 8th IET Int. Conf. Power Electronics, Machines and Drives*, Glasgow, UK, 2016, pp. 1–6.

[9] D. Barlini, M. Ciappa, M. Mermet-Guyennet and W. Fichtner, "Measurement of the transient junction temperature in MOSFET devices under operating conditions," *Microelectronics Rel.* vol. 47, no. 9–11, pp. 1707–1712, Aug. 2007.

[10] R. Fu, A. Grekov, J. Hudgins, A. Mantooth and E. Santi, " Power SiC DMOSFET model accounting for nonuniform current distribution in JFET region," *IEEE Trans. Ind. Appl.*, vol. 48, no. 1, pp. 181–190, Jan/Feb. 2012.

[11] Cree Inc.," Silicon carbide power MOSFET C2M™ MOSFET technology N-channel enhancement mode," C2M0080120D datasheet, Oct. 2015.

Field Bus for Data Exchange and Control of Modular Power Electronic Systems with High Synchronisation Accuracy

Stefan Rietmann, Simon Fuchs, André Hillers and Jürgen Biela
Laboratory for High Power Electronic Systems (HPE)
ETH Zürich, Switzerland
Email: rietmann@hpe.ee.ethz.ch

Abstract—In this paper, a new field bus protocol based on the IEEE 802.3 Ethernet standard is proposed. It enables fast data exchange between and synchronised control in the modules of modular power electronic systems. With increasing switching frequency, a highly accurate synchronisation of the different modules is a necessary requirement for a control bus. The proposed field bus protocol provides a stable and efficient scheme including a novel data frame structure and allows a synchronisation accuracy of $\pm 4\,$ns on the 1 GBit Ethernet standard. For validation of the new protocol a prototype system has been built and measurement results are provided. Eventually, the implementation has been found to meet the specification during testing.

I. INTRODUCTION

In power electronic systems, more and more modular converter structures are applied due to their increased availability (redundancy), simple scaling and improved output quality (e.g. less current ripple by interleaving [1]). The most prominent example for modular converters is the modular multilevel converter (MMC). For the MMC various possibilities for distributed control systems can be found in literature. For example, distributed control methods [2], distributed protection control [3] or modulation methods that allow to distribute the switching signal generation on the modules [4]. As for these distributed control methods, computational power is required on each module/cell (e.g. an FPGA), also more complex communication systems for data exchange and/or communication can be implemented in the modules. Fig. 1 shows an exemplary hardware implementation of such distributed control systems. On each of the modules in the shown MMC stack, there is an FPGA control board connected to an optical communication bus. As can be seen in Fig. 1, the size of the module hardware is increased only very little by the FPGA control board. With a field bus communication system, the number of data connections can be drastically reduced compared to a point-to-point implementation of a master controller to each module/cell. For a daisy chained setup, maximum two lines per slave are required (as e.g. with EtherCAT). The communication system for such modular converter systems typically has the following properties:

1) As all modules are of the same type and require the same communicated data, the individual data frames received and transmitted by the modules are also of the same size.

2) The amount of exchanged data per module is low. It is typically in the range of 5 to 10 Bytes per communication and/or switching cycle.

3) The data exchange on the bus system is bidirectional with respect to all slaves and the master. Every bus member can read from and write to the bus.

4) All modules need to run on the same clock.

5) All modules have to refer to a common point in time, such that they can perform simultaneous or synchronised switching actions (e.g. interleaving). Especially for high switching frequencies, the accuracy of this synchronisation is required to be high (low nanosecond range).

6) Some communication error detection has to be present to protect the system from e.g. implementing wrong reference values.

7) The hardware requirements should be rather low, such that not to much FPGA area is occupied by the communication logic itself rather than the control logic.

In [5], the CAN bus is used to control an MMC. The synchronization accuracy was found to be $\pm 20\,\mu$s. In [6]–[8] the commercial EtherCAT field bus system has been used

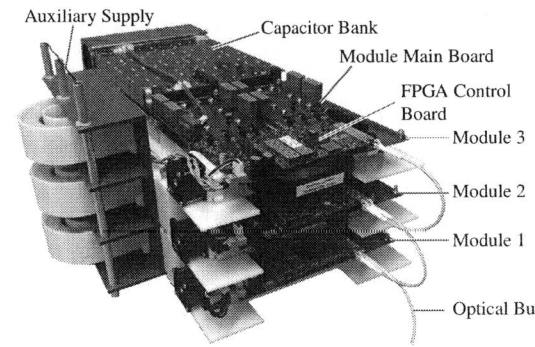

Fig. 1. (a) Photo of a MMC stack (3 modules) using the proposed SyCCo-Bus. It can be seen, that each module is equipped with the same FPGA control hardware. (b) Photo of one MMC modules showing the FPGA control board used for this paper.

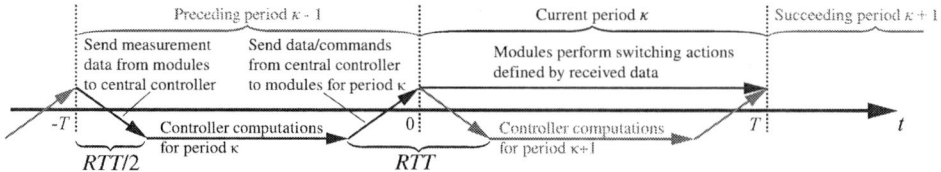

Fig. 2. Typical control timing for modular converters with a central control unit: The central controller receives measurement data from the modules and performs the controller computations based on this data. After that it sends the resulting commands to the modules which perform the switching actions accordingly. Note, that the communication delay, the round-trip time RTT, of the communication system shortens the time available for the control computations to $T - T_\mathrm{d}$, where $T = 1/f_\mathrm{sw}$. The computations and communication actions relevant for the current period κ in which the switching actions are performed are drawn black.

to control an MMC topology. EtherCAT features a synchronization option, that is claimed to result in a jitter of less than $1\,\mu s$ [9]. However, in literature the jitter of the synchronisation signals ranges between $\pm20\,\mathrm{ns}$ [8] and $15\,\mu s$ [6]. For achieving a more precise synchronisation accuracy of $\pm5\,\mathrm{ns}$, a basic concept of a daisy-chained field bus protocol has is proposed in [10]. It is based on $100\,\mathrm{Mbit}$ Ethernet and dedicated physical layer ICs in combination with an FPGA. The master/slaves are interconnected with bidirectional fiber optic lines such that one interconnection per slave results. The protocol tolerates different frame sizes for the individual slaves connected to the bus system and does not require to have knowledge on the number of slaves prior to the operation. This results in a rather complicated start-up procedure and occupies a rather large amount of logic elements/registers in the FPGAs. Therefore, in this paper a simplified frame structure is proposed in order to reduce the implementation effort and significantly reduce the latency of the protocol from [10]. With the new frame structure the amount of received/transmitted data (the frame size) per slave is equal for all slaves and the number of slaves is known a priori. These assumptions lead to a much simpler, more robust and less logic consuming implementation compared to [10]. The delay introduced by the data passing through each slave is also drastically reduced, because all operations on the frame can be performed within one clock cycle and no buffering of multiple frame parts is necessary. The implemented system including the proposed protocol is referenced as **S**ynchronous-**C**onverter-**C**ontrol-Bus (SyCCo-Bus). In addition, the hardware setup was changed to Altera Cyclone V GX FPGAs enabling an all integrated transceiver setup using Altera IP cores. This features the Gigabit Ethernet standard increasing the data throughput.

The paper is organized as follows. Section II describes the various requirements necessary for a bus system in modular converter systems. In section III the applied field bus concept, the operating principles and the novel frame structure are introduced. The individual module time synchronization scheme and the expected synchronization accuracy are explained. Section IV refers to the actual hardware implementation including an FPGA based prototype system. The VHDL implementation of the protocol is described in section V. In section VI the concept is tested and particular timing measurement results are shown. The paper concludes in section VII.

II. Data Exchange Requirements in Modular Converter Systems

For explaining the requirements for the data exchange in modular converter systems, a medium voltage MMC setup with $N = 90$ modules and a switching frequency of $f_\mathrm{sw} = 10\,\mathrm{kHz}$ based on the research work of [11] serves as an example in this section. In an MMC, all modules (slaves) have to submit their capacitor voltages ($12\,\mathrm{bit}$) to the central control unit (master) once every switching period (necessary for PWM modulation [4]). Note, that the PWM modulation are done on the modules itself instead of modulations calculated on a central control unit. For protection reasons it can be reasonable to transmit also current measurements ($12\,\mathrm{bit}$) on the modules (cf. [3]). Furthermore, some information on the semiconductor/module-state ($8\,\mathrm{bit}$) as well as temperature information ($8\,\mathrm{bit}$) could be transmitted from the slaves to the master. All this data sums up to a worst case scenario of $5\,\mathrm{Byte}$ per slave. The central control unit has to distribute data to the modules. The duty cycle for the upcoming switching period ($10 - 12\,\mathrm{bit}$) as well as the max. and min. values for current and module voltage ($4 \times 12\,\mathrm{bit}$). This sums up to $8\,\mathrm{Byte}$ of data per slave. For the presented protocol, both directions have the same frame structure, such that one needs to use $8\,\mathrm{Byte}$ data/module for both directions plus a CRC byte for data integrity check.

The frame has to be communicated every switching period, such that in the considered example a minimum data rate of $N \cdot f_\mathrm{sw} \cdot (8 + 1)\,\mathrm{Byte} = 65.8\,\mathrm{Mbit/s}$ is required for each direction plus the protocol overhead.

This value does, however, not contain information about the delay introduced by the communication system. The larger the delay, the less time there is for computations on the central control unit (cf. Fig. 2). Thus, the communication delay should be lower than half a switching period for a complete communication cycle. This results in a maximum cycle time of $0.5/f_\mathrm{sw} = 50\,\mu s$ which is equal to $550\,\mathrm{ns}$ per slave or a delay of $69.5\,\mathrm{ns}$ per slave and byte of data. If one calculates the bus data rate necessary for the time the communication is actively transmitting and receiving a frame, a minimum of $90 \cdot (8 + 1)\,\mathrm{Byte}/50\,\mathrm{us} = 129.6\,\mathrm{Mbit/s}$ results.

The accuracy of the synchronisation should be higher than the possible implementation accuracy of the PWM generated on the modules. For a $10\,\mathrm{bit}$ PWM, this results in a minimum synchronisation accuracy of $1/f_\mathrm{sw}/(2^{10} - 1) = 97.8\,\mathrm{ns}$.

The 2018 International Power Electronics Conference

1..7	1	1	m	1	...	m	1
Preamble	SFD	SCB	Data$_1$	CRC$_1$	...	Data$_N$	CRC$_N$

Fig. 3. Frame schematic: The frame is divided into two distinct sections, the header and the data section. The header section mainly contains the preamble and the SFD bytes. Furthermore, the data section consists out of multiple slave specific sections of which each contains a slave data and a CRC section. Note that the number of possible slaves is currently limited to 255 by the current frame structure since the SCB contains 8 bit. This limitation can be solved by increasing the number of SCB bytes.

One can state quite high requirements regarding the data rate for a realistic MMC system. Thanks to the rather low switching frequency in MMC applications, the requirements concerning the synchronisation accuracy and the communication delay are less demanding. Nevertheless, for systems with higher switching frequencies, the synchronisation accuracy must be much higher. In [12], 100 kHz with the same duty cycle precision (10 bit) would result in a required accuracy of 9.5 ns.

III. OPERATING PRINCIPLES

A. Field Bus Concept

The proposed field bus protocol presumes a daisy-chained network topology. Considering the data flow direction each module is linked to its successor over the forward path and hence the predecessors are attached on the backward path. Compared to a star-like topology this field bus concept does only allow broadcast message transfers, meaning that the complete set of information is always sent to all participants. To initialize a transmission process the master module is issuing the data frame. During the transmission on the forward path each slave module has the chance to read data from the bus and write information back to the bus. A slave being at the end of the forward path chain detects its state as last slave, exchanges information with the bus frame and returns the frame on the backward path to its predecessor.

B. Frame Structure

The frame structure as shown in Fig. 3 and especially the header partition is mainly based on the Gigabit Ethernet frame defined in IEEE 802.3 [13]. In contrast to this standard only the key elements such as the preamble and the start frame delimiter (SFD) octets have been adopted. The destination and address bytes have been omitted since the master is always addressing all slaves. As a replacement for the addressing scheme a frame internal counter byte, the Slave Counter Byte (SCB), is introduced. The complete system is not limited to any number of slaves. Nevertheless, it is worth to note that a slave count exceeding 255 slaves would require a larger SCB section. Due to the delay introduced by each slave, one has to consider the increasing round trip time (RTT) and therefore a decreasing data rate per slave for an increasing number of slaves. The static payload region contains a distinct data region and an additional CRC byte per slave. A complete frame consist of:

- Preamble, 7 octet
- Start-of-Frame Delimiter (SFD), 1 octet
- Slave Counter Byte (SCB), 1 octet
- Data, m byte per slave
- CRC, 1 byte per slave

C. Module Data Exchange

The data exchange between the bus and the slaves is a time critical process since its duration directly contributes to the overall system latency. This latency is denoted in Fig. 4 as $T_{\text{Forward},i}$ and $T_{\text{Backward},i}$, respectively. Achieving a low latency data exchange will result in a high bus throughput. The static structure of the frame allows an efficient and simple data exchange with low latency on the slaves. A finite state machine (FSM) representation of the internal data exchange logic is given in Fig. 5. The complete frame is looped through each slave module. The internal state machine is triggered by the arrival of the preamble and prepares to detect the SFD byte. Note that a slave does not count the number of preamble bytes rather than it checks for the specific SFD pattern. Next, the SCB byte is read and incremented on the fly. From the value previously read SCB value the slaves can estimate their position in the system. Furthermore, the number of bytes per slave are defined in advance and hence the position of the slave specific data in the frame can be calculated after the appearance of the SFD. The module has to remain in the WaitForSlot state until the first byte of its own data arrives. At this point the state changes to Exchange Data. The incoming frame is stored in internal registers and the data which has been prepared for the master module in advance is replacing the read data on the bus. Since all data is looped through the slaves a simple multiplexer logic as illustrated in Fig. 7 is sufficient to change the output between the incoming data frame and the stored, slave specific data. As a final step, the CRC byte which has been calculated during the Exchange Data state is written onto the bus. Eventually, the slave changes back to the Idle state and the remaining incoming data is passed on. The complete frame is received and further transmitted by every module but only the module specific part of the frame is modified.

D. Synchronization

The different modules do not share a common time base nor the same base clock. Furthermore, different boot durations, clock shifts or transmission delays add a non negligible amount of clock jitter to the complete system. As mentioned in the introduction, it is essential for power electronic systems to run synchronous. Therefore, a scheme which provides a common time base and a common base clock is necessary. The proposed protocol aims at a synchronizing accuracy of $\frac{T}{2}$ where T denotes the period.

By synchronizing the modules the internal logic has to rely on the same. Therefore a reference clock has to be shared with every single module. For this purpose Ethernet uses 8b/10b en- and decoding to ensure enough alternating 1 and 0 bits

2303

The 2018 International Power Electronics Conference

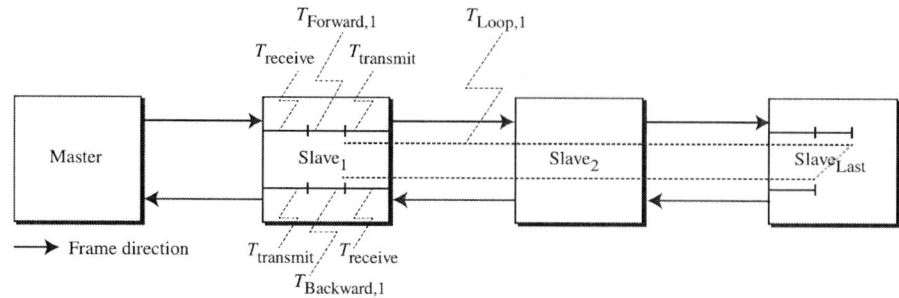

Fig. 4. Synchronisation Schematic: The illustration describes the round trip time measurement of a single slave. Note that the forward path and the backward path are equally long in terms of communication time.

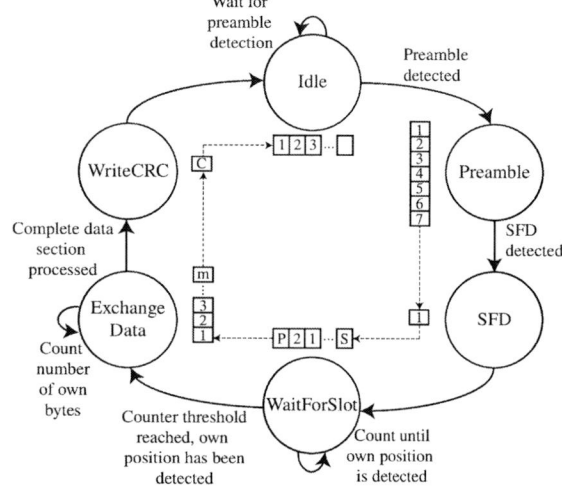

Fig. 5. The Data Exchange Scheme describes the processing of the novel static data frame. The inner illustration shows the number of bytes processed through every step in the complete processing of the data frame.

[13] (cf. Return-to-Zero). This then allows to perform a clock data recovery (CDR) on each module.

With the proposed frame structure, the SFD byte has been introduced as common time token. It is received twice by each slave in each communication round, once on the forward path and once on the backward path. The time measured between the arrival of the SFD on the forward path and on the backward path is referenced as $T_{\text{loop},i}$ as depicted in Fig. 4. The static frame structure as well as the requirement for homogeneous modules allows the conclusion that the transmission time on the forward path is equal to the transmission time on the backward path. Note that the physical wire over which the information is transmitted is the same for the forward and the backward path. In contrast to [10] this allows a simple and accurate determination of a single point in time on which all modules can equally rely.

For this protocol the common time token is used to calculate a waiting time $T_{\text{valid},i}$ for each individual slave. It describes a module specific point in time when every slave has received and processed the slave specific data section of the frame.

This means that after detecting the SFD byte, slave i has to wait for $T_{\text{valid},i}$ until the last slave has completely processed the current frame. Hence, it can be used as a starting point for e.g. switching activities in the converter modules. The measured time $T_{\text{loop},i}$ includes the slave internal latency $T_{\text{Forward},i}$ and $T_{\text{Backward},i}$, the transceiver delay T_{receive} and $T_{\text{transmit},i}$ and the physical transmission delay as indicated in Fig. 4.

First of all, each bus module is determining the point in time where the data has been received by each single participant. The modules are then counting the time between the event of sending the SFD byte to the forward path and receiving the SFD on the backward path (cf Fig. 4). Due to the requirement of homogeneous modules and the static frame construction the data processing delay $T_{\text{Forward},i}$ and $T_{\text{Backward},i}$ on each slave can be safely assumed to be equal for coarse synchronization. By introducing the same delay on the backward path as on the forward path the time measured can be divided by two to determine the point in time where the last slave module receives the frame. Again, thanks to the static frame structure it is known by each slave how much time has to be taken into account until the last slave has completely received its data and CRC bytes. This event is denoted by the slave data valid signal (SlaveDValid_S) which is then set high on all modules. Hence, for the T_{valid} we can state:

$$T_{\text{valid}} = \left(\frac{\text{counter value}}{2} + 1 + n \cdot (m+1) \right) \cdot f_{\text{slave}}^{-1} \quad (1)$$

The measured $T_{\text{valid},i}$ values are used during the next communication cycle. This has two important consequences. First the complete process needs a starting frame which does not transmit vital information rather than a start-up sequence. After the starting frame the synchronisation time is re-evaluated in each frame cycle again. This allows to compensate for possible temperature and environmental effects. Furthermore, the measurements do not depend on each other, therefore no subsequent errors are possible.

IV. HARDWARE

In order to demonstrate the capabilities of the proposed field bus implementation, a prototype has been developed. The prototype is part of the custom-made high-speed communication and computing platform shown in Fig. 6 (a) which has been designed at HPE, ETH Zurich. The platform is based on an

The 2018 International Power Electronics Conference

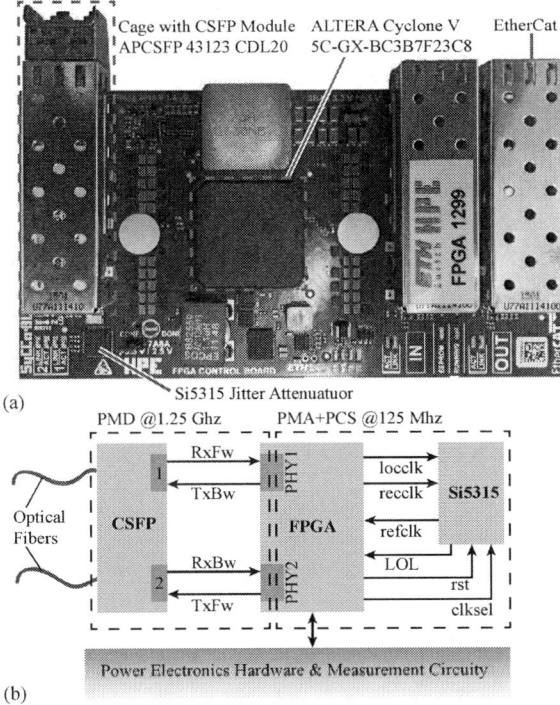

(a)

(b)

Fig. 6. (a) The developed custom FPGA board including the actual Altera Cyclone V GX FPGA, the transceiver interfaces (CSFP) and a clock jitter attenuator (Si5315). (b) Simplified schematic of the hardware and their interaction signals.

Altera Cyclone V GX FPGA which both runs the firmware of the bus system as well as all specific user-related computation and control tasks involved with the operation of the power electronic systems, for which the platform can be used. In the following, it is briefly explained how the physical layer hardware of the communication system has been designed with an emphasis on a compact realization and how the synchronization of the individual clock domains is achieved.

A. Small Footprint Physical Medium Attachment

In order to keep the board space occupied by the communication hardware to a minimum, the internal multi-purpose high-speed serial transceiver PHY IP cores (cf Fig. 6) of the Cyclone V FPGA are used in conjunction with a common *compact small form factor pluggable* (CSFP) fiber-optic transceivers. The CSFP module presents the physical dependent sublayer (PMD) and attaches directly to the respective inputs / outputs of PHY IP cores of the Cyclone V FPGA using logic-level high-speed differential-mode signaling. The driving of the optics for sending and receiving on the physical medium is handled by the CSFP module.

As opposed to the more widespread *small form factor pluggable* (SFP) transceivers (which work exactly the same way as the CSFP modules), CSFP modules incorporate two independent transceivers in the same form factor. This is achieved by sending and receiving data on different wavelengths of

light on a single fiber. The two sockets for fibers of the CSFP module shown in Fig. 6 (a) in fact belong to two independent data channels.

A block-diagram of the physical layer is shown in Fig. 6 (b). The physical medium attachment (PMA) and the physical coding sublayer (PCS) are provided by the PHY IP cores of the Cyclone V FPGA. This way, the space occupied by the physical layer hardware is minimized. Compared to the implementation presented in [10], external physical-layer ICs are thus no longer necessary.

B. Synchronization Hardware

As explained in section III, the presented bus system uses the recovered clock information of the incoming bit-stream as a reference clock for each node, similar to the implementation proposed in [10]. Because the recovered clock is not immediately available at startup, the system boots up with a free running clock and switches hitless over to the recovered clock as soon as the recovered clock is available and stable. The base clock frequency of the transceivers is 125 MHz which is internally multiplied to the 1.25 GHz with which the data is transferred on the medium.

The switchover is handled by the Si5315 jitter-attenuator PLL. In the beginning, the Si5315 starts up with a free-running clock on each board (locclk). This clock is fed back to the FPGA as the reference clock (refclk) for the PMA+PCS as well as the user logic to initiate a link with the previous slave. As soon as the PMA+PCS of the PHY IP core of the Cyclone V FPGA have synchronized to the incoming bitstream of the previous module, the Si5315 is commanded (clksel) to switchover to the recovered receive clock (recclk).

The switchover is performed hitless in incremental steps over consecutive clock periods to prevent a loss-of-lock. The use of an external IC like the Si5315 is neccessary, because the internal general purpose PLLs of the Cyclone V are not specified for transparent clock switchover and do not meet the jitter requirements for clocking the PHY IP cores in this type of application.

V. FPGA IMPLEMENTATION

The proposed SyCCo-Bus protocol has been implemented on the FPGA using the hardware/components described in section IV. The necessary steps to provide a running protocol implementation are described in subsection V-A and V-B.

A. Start-Up

In steady state all slaves run on the master clock that is distributed via the bus. The start-up of the slaves nevertheless has to be executed with a local clock because the clock data recovery (CDR) does not work here yet. To suppress potential oscillations from the CDR during start-up, all slaves do not send any data (and therefore also no clock) to the next slave via their FW path. The master continuously sends a filler frame ('Comma' [13]). At the beginning of the start-up, the first slave synchronizes on the incoming clock recovered from the data received from the master and switches its Si5315 output

2305

to this recovered clock before taking its FW path out of the reset. Now, the second slave is supplied with commas it can synchronize on to recover the first slave's clock which is equal to the master clock. By this mechanism, one slave after the other can synchronize on the master clock before the protocol and therefore the data transmission is started.

B. Protocol Implementation

The complete protocol implementation is differentiated into two distinct modules, the master module and the slave module respectively. As depicted in Fig. 4 the master module only contains a single interface which is connecting it with the subsequent slave module chain.

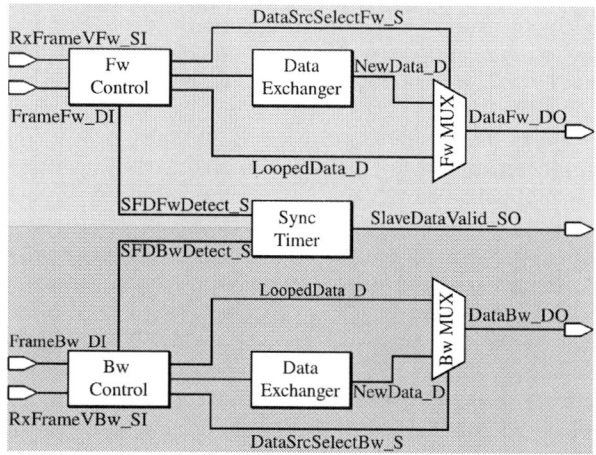

Fig. 7. The slave module is illustrated in a schematic way to show the logical separation between the forward path (blue) and the backward path (red).

In the following the basic structure of the slave modules are explained. Note that the master modules structure is similar to the slave modules structures but includes some more control logic since it initializes the communication rounds.
The modules are differentiated into two separate paths, the forward path and the backward path as illustrated in Fig. 7. Since the transceiver on the FPGA module itself will detect an incoming frame due to the preamble structure the actual slave/master modules can rely on a frame valid signal (RxFrameVFw_SI / RxFrameVBw_SI) issued by the transceiver. Each module then contains a forward and a backward control instance which reacts on this particular valid signal. The control modules are responsible of reading the current byte received from the frame. For the different input bytes such as explained in section 3 the status and control signals of the controller modules output changes. It is especially worth to mention the SFD detect signal which triggers the synchronization process by starting/stopping the counter mechanism. The data exchange sub-module is responsible for the read and write processes on the bus. It is combined with an output multiplexer which is associated with the control module to decide whether the input data frame or the new data gathered from the module has to be sent. The decision is based on state of the current input frame since the a slave module is only allowed to write

to its own space in the frame. The data exchange module in the backward is optional as the data does not necessarily have to be exchanged again in the same communication round. Therefore a simplified version of the data exchanger can be used. As a central unit to both paths the synchronization timer triggered by the two distinct SFD detect signals issues a valid signal. The data valid signal is representing a common time token on every module as described in section III-D.

VI. MEASUREMENT RESULTS

The measurements have been done using different bus configurations with one master and up to eight slave modules as depicted in Fig. 8. In particular, the synchronisation accuracy has been evaluated to show that the synchronisation method is working as proposed.
The measured round-trip time (RTT) describes the time used by the master to write a complete frame on the bus, send it to all participants and read it back from the backward path. This time leads to the data rate and the data rate per slave which is compared to a theoretical value, the maximum possible data rate.

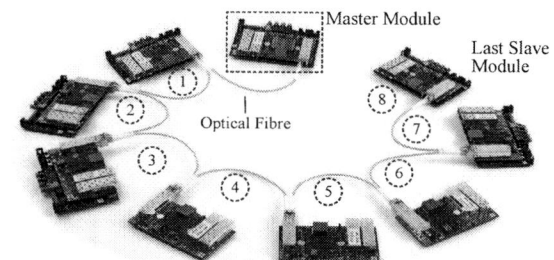

Fig. 8. Exemplary configuration of the SyCCo-Bus with eight slaves and one master module.

A. Synchronisation Accuracy

The synchronisation accuracy has been measured by connecting the internal generated SlaveDValid_S signal to an output pin to observe it with an oscilloscope. Since this signal serves as the common time token for each module over the whole bus, the falling/rising edges observed at the output pins of the different participants are expected to occur at least in the range of one clock cycle (± 4 ns). Fig. 9 shows the synchronisation accuracy measured on a bus configuration with eight slaves and one master. The results show that the current implementation of the bus protocol allows an accuracy of ± 4 ns between the different modules. The measurement has been repeated several times to confirm the shown results, while the configuration has been altered to eliminate potential dependencies between the modules. Also, a long term test with 14 hours run time and more than $4.3 \cdot 10^6$ measurement points has been conducted. The test showed that the signals, and therefore the internal clock signals, are drifting in the range of less than 100 ps. It is important to mention that the synchronisation accuracy inside the stated range is depended on the start up process of the FPGA board and on the

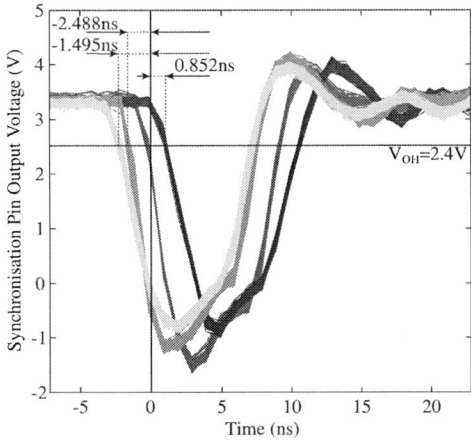

Fig. 9. The measured synchronisation accuracy evaluated over a data set of $2 \cdot 10^3$ trigger points. The signals have been triggered for $V_{OH} = 2.4\,\text{V}$. Note that this capture is only exemplary. The distribution of the accuracy in the proposed $\pm 4\,\text{ns}$ depends on the power-on process of the FPGA device and the PHY itself.

TABLE I
CYCLE TIME MEASUREMENTS

Bytes/Slave	2 Slaves	4 Slaves	8 Slaves
8	$0.984\,\mu s$	$1.904\,\mu s$	$3.172\,\mu s$
16	$1.112\,\mu s$	$2.152\,\mu s$	$4.224\,\mu s$
32	$1.368\,\mu s$	$2.664\,\mu s$	$5.248\,\mu s$
64	$1.880\,\mu s$	$3.688\,\mu s$	$7.296\,\mu s$
128	$2.896\,\mu s$	$5.728\,\mu s$	$11.384\,\mu s$

individual start up progress of the internal PHY. This leads to a phase difference between the different synchronisation signals which has shown to be non-deterministic. Note that this phase differences denotes the phase differences between the internal $125\,\text{MHz}$ Therefore, a simple constant phase offset would not be enough to achieve higher accuracy. Whereas, the long term test showed that almost zero additional phase shift has to be expected and therefore further corrections are only necessary in larger time intervals.

B. Communication Delay

The communication delay has been evaluated in terms of the systems RTT. The RTT is taken as a bare measure to evaluate the delays on the various different participants. The measured RTT for different bus configurations are listed in table I. For a further investigation of the individual delays occurring on the bus, the contributors influencing the RTT are included in (2):

$$
\begin{aligned}
\text{RTT} =& (2 \cdot T_{\text{Trans}} + T_{\text{Prot}}) \cdot N_{\text{Slave}} \\
&+ T_{\text{Clk}} \cdot (N_{\text{Byte}} \cdot N_{\text{Slave}} + N_{\text{Header}})
\end{aligned}
\tag{2}
$$

T_{Prot} and T_{Trans} describe the protocol internal and physical delay, respectively. Since the size of the complete frame matters N_{Slave}, the total number of slaves and N_{Byte}, the number of bytes per slave, have to be taken into account. Additionally,

N_{Header} is the protocol overhead due to the preamble, the SFD and the SCB octets (cf. section III-B).

All values except for the T_{Trans} are known based on the implementation. In addition those values are adjustable. Whereas, T_{Trans} can only be measured since it includes the physical transmission time via the fibre optical cable and the processing time inside the provided PHY IP and the CSFP modules. The transmission time has been found to be in the range of $390\,\text{ns}$ In table I one can see that RTT is almost but not completely doubling, while leaving the number of bytes per slave constant but increasing the number of participants in the system by a factor of two. This is because of the previously mentioned header which acts as a constant offset to the RTT. On the other hand, leaving the number of slaves in the system constant but doubling the number of bytes per slave shows that exactly the added bytes per slave have to be processed.

C. Data Rate

The maximum possible data rate is determined by the underlying link speed but limited by the protocol header. By defining the header and the number of interframe gap bytes (N_{IFG}) one can calculate the data rate achievable on the physical link itself. This can be done by:

$$
\text{Data Rate} = \frac{N_{\text{Slave}} \cdot N_{\text{Byte}} \cdot \text{link speed}}{N_{\text{Header}} + N_{\text{Slave}} \cdot N_{\text{Byte}} + N_{\text{IFG}}}
\tag{3}
$$

To calculate the maximum data rate N_{IFG} has to be equal zero, meaning the the transmission medium is always occupied. Based on the frame structure proposed in Fig. III-B, the data

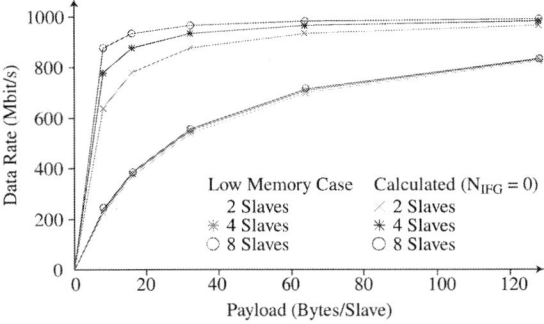

Fig. 10. A Comparison between the theoretical and the measured data rate: The theoretical data rate (blue, green and purple lines) assumes that no gap between two distinct frames is necessary. In fact the protocol will need a certain gap between each frame to reset the every FSM, clean the memory and to make sure that data will not be overwritten. The red, orange and yellow line show the impact of the IFG in the current implementation. Aiming towards a small time gap between two subsequent frames will lead to an substantial increase in necessary memory since every incoming data transmission needs to be stored until its individual SlaveDValid occurs.

rate has been calculated and is depicted in Fig. 10. In fact, the maximum possible data rate is a theoretical metric since it does not include necessary time gaps between subsequent frames and time delays on the individual modules. Therefore, a comparison with the data rate calculated from the measured RTT has been done as well in Fig. 10. Those values have been evaluated by introducing a minimal interframe gap (IFG) to satisfy the requirements of the current implementation. Since the data received from the previous frame can not be set valid on any module until the SlaveDValid_S signal occurs, the data needs to be latched. If the IFG should be smaller than the period of the SlaveDValid_S signal, additional memory or registers are necessary to latch more than one single data set.

The comparison between the maximum and the measured values shows, that the influence of the modules individual time delay and the IFG is not negligible. Furthermore, one can see that the data rate increases with increasing frame size and converges to the physical link speed. The more bytes per frame are sent over the channel, the more negligible is the header and a potential IFG. As indicated in section III-B, an

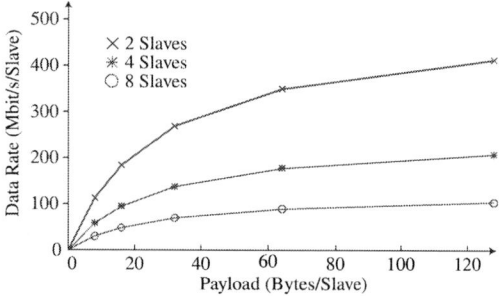

Fig. 11. Data rate per slave: An increasing number of slaves leads to a decreasing data rate per slave, since more participants on the bus will lead to a larger frame. This larger frame eventually results in a longer RTT which means that the individual slaves need to wait longer until they can issue the SlaveDValid_S resulting in a larger IFG.

increasing amount of slaves results in a longer RTT. In Fig. 10 no obvious difference between the different configurations is noticeable since the graph depicts the throughput of the complete bus.

A further interesting and important metric is the data rate per slave, since this is eventually determining how many communication rounds are necessary to complete a message. Fig. 11 shows, that an increasing amount of participants will result in a data rate reduction of each individual slave. With the presented hardware it is possible to run two SyCCo-Bus systems in parallel, as the master still has one free CSFP port. This can significantly reduce the round-trip time (RTT), because the transmission delay introduced per slave (390 ns) is much larger as the additionally generated overhead (80 ns). It can thus be assumed that the parallel operation saves approximative half of the RTT.

Of course, one would like to have both bus systems to be synchronous as well. This can easily be achieved by two SyCCo-Bus masters that are implemented on one FPGA on the Master FPGA Board exchanging their measured RTTs (counter value in (1)) such that the difference in their cycle time is known as Δc. The one having the lower counter value has to wait for $\Delta c/2$ cycles before starting to send the next frame, such that the SlaveDataValid signal (cf. Fig. 7) is synchronous on all slaves in both bus systems.

VII. CONCLUSION

In this paper, the bus protocol for the Synchronous-Converter-Control-Bus-System has been presented. The proposed protocol with the novel static frame structure has been shown to work synchronous in the range of ± 4 ns. Furthermore, the implementation of the core of the protocol has been shown and the necessary hardware has been evaluated and built. Eventually, the round-trip time and data rate of different bus configurations have been measured and evaluated. A comparison of the implemented prototype system with a time optimal system has been shown. Furthermore, the prototype system has been found to work for a real world application since the synchronisation accuracy of ± 4 ns and the data rate of more than 700 MBit/s are both exceeding the necessary requirements of [11]. Since the data rate per slave starts to drop the more slaves are introduced into the system a possible countermeasure has been presented based on the already existing hardware.

REFERENCES

[1] G. Tsolaridis and J. Biela, "Interleaved Hybrid Control Concept for Multiphase DC-DC Converters," in *IEEE Energy Conversion Congress and Exposition (ECCE)*, Oct. 2017.

[2] M. Hagiwara and H. Akagi, "Control and experiment of pulsewidth-modulated modular multilevel converters," *IEEE Transactions on Power Electronics*, vol. 24, no. 7, pp. 1737–1746, July 2009.

[3] A. Hillers, H. Tu, and J. Biela, "Central control and distributed protection of the DSBC and DSCC modular multilevel converters," in *IEEE Energy Conversion Congress and Exposition (ECCE)*, Sept 2016.

[4] S. Fuchs, S. Beck, and J. Biela, "High output voltage precision PWM for modular multilevel converters," in *19th European Conf. on Power Electronics and Applications (EPE)*, Sept 2017.

[5] M. A. Parker, L. Ran, and S. J. Finney, "Distributed control of a fault-tolerant modular multilevel inverter for direct-drive wind turbine grid interfacing," *IEEE Transactions on Industrial Electronics*, vol. 60, no. 2, pp. 509–522, Feb 2013.

[6] P. D. Burlacu, L. Mathe, and R. Teodorescu, "Synchronization of the distributed pwm carrier waves for modular multilevel converters," in *2014 International Conference on Optimization of Electrical and Electronic Equipment (OPTIM)*, May 2014.

[7] C. L. Toh and L. E. Norum, "Implementation of high speed control network with fail-safe control and communication cable redundancy in modular multilevel converter," in *2013 15th European Conference on Power Electronics and Applications (EPE)*, Sept 2013.

[8] ——, "A high speed control network synchronization jitter evaluation for embedded monitoring and control in modular multilevel converter," in *2013 IEEE Grenoble Conference*, June 2013.

[9] "EtherCAT Technology Group," https://www.ethercat.org, Accessed: 02-26-2018.

[10] C. Carstensen, R. Christen, H. Vollenweider, R. Stark, and J. Biela, "A Converter Control Field Bus Protocol for Power Electronic Systems with a Synchronization Accuracy of ± 5ns," in *2015 17th European Conference on Power Electronics and Applications (EPE'15 ECCE-Europe)*, Sept 2015.

[11] A. Hillers and J. Biela, "Increased efficiency and reduced realization effort of DSBC and DSCC modular multilevel converters (MMCs)," in *International Power Electronics Conference (IPEC)*, May 2018.

[12] G. Tsolaridis and J. Biela, "Modular, highly dynamic and ultra-low ripple arbitrary current source for plasma research," in *2017 IEEE 21st International Conference on Pulsed Power (PPC)*, June 2017, pp. 1–4.

[13] IEEE, "IEEE standard for Ethernet," *IEEE Std 802.3-2015 (Revision of IEEE Std 802.3-2012)*, pp. 1–4017, March 2016.

Analytical Investigation on Asymmetric LCC Compensation Circuit for Trade-off between High Efficiency and Power

Kodai Takeda* and Takafumi Koseki*

*Department of Electrical Engineering and Information Systems, The University of Tokyo, Tokyo, Japan
E-mail: k_takeda, koseki@koseki.t.u-tokyo.ac.jp

Abstract—In a wireless power transfer system, a circuit configuration, where the main coil is connected to capacitors in series and parallel respectively and also connected to an inductor in series, is called LCC compensation. There was a systematic methodology founded for deciding appropriate amount of series compensation in symmetric parameter designed LCC compensation circuit. Although it has multiple resonances in itself, previous researchers did not explain explicitly a mechanism of them and theoretical analysis of efficiency are insufficient. In this paper, the working of resonances is investigated, and a trade-off relation between high efficiency and high output power is analyzed. Based on the analysis, asymmetric parameter design for LCC is proposed. Experimentally, the effect of primary and secondary tuning parameters is revealed, and validity of the proposed design is verified.

Keywords—*Circuit design, Dynamic wireless transmission, Electric vehicle, Inductive power transfer,*

I. INTRODUCTION

Electric vehicles (EVs) are getting more pervasive because of their high kinetic and energy performance[1]. A driving range of EVs is too short as inter-city transportation, and we have to spend much time to recharge them. Dynamic wireless power transfer system (DWPTS) has been investigated extensively to mitigate this problem. DWPTS allows long operating range by introducing wireless power transfer (WPT) technology in a road. Fig. 1 shows a basic outline of WPT. According to the combination of compensation circuits, the circuits are given names. In DWPTS, power is transferred from coils installed in a road to ones attached to a car. To charge a car sufficiently, facilities on road side should be long and powerful enough. Considering these characteristics of DWPTS, there are two main targets. One is to reduce costs per unit length of DWPTS to install long section, and the other is to improve transfer power.

Several structures of DWPT have been proposed[2], [3]. The structures are divided into two groups according to the length of coils at the road side. Length of the coils is from several to hundreds of meters long in the running direction. Though the group has a simple structure and control, the significant difference of size of coils between the road and car side lowers magnetic coupling between the coils, which causes a drop of power transmission efficiency and makes electromagnetic emission larger. The other group with coil equal in size to car side and up

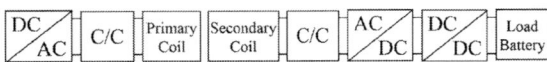

Fig. 1. A basic outline of the WPT system. C/C represents compensation circuit of road side (primary) and car side (secondary) respectively. Actually, DC-DC converter is often inserted to control output voltage, but in this paper, DC-DC converter are not used.

to a few meters long are advantageous from the viewpoint of efficiency and electromagnetic emission. The structure, however, needs more frequent car position detection than the other group to know when to turn on a source, which makes control and structures more complex. Authors in [4][5] have proposed sensorless car position detection methods, which use coils as sensors. Though a group with short coils and parallel connection to a single source is more simple than a group with short coils and one by one connection, additional switches could make an unintentional surge and could be broken easily compared with other components. Moreover, in a no-load condition, a circuit impedance of S/S circuit, which contains capacitors in series to primary and secondary coils and has been widely studied as a compensation circuit for WPT, could be nearly equal to 0 and large current would flow into a source and could be dangerous. To realize DWPT without switches, LCC compensation circuit, which increases an impedance of the circuit in a no-load condition, has been proposed. A circuit which has LCC as primary side compensation and S as secondary compensation is called LCC/S and LCC/S should be compared with LCC/LCC called Double-LCC which has LCC both primary and secondary side. There are comparison studies between Double-LCC and S/S. They demonstrate that S/S can achieve the higher efficiency because of fewer components than LCC/LCC [6]. On the other hand, Double-LCC emits less magnetic field than S/S [7]. Moreover, [8] showed that a source current becomes automatically small in a no-load condition. LCC compensation circuit has multiple resonance circuits in itself as shown in Fig. 2. Previous researchers revealed its characteristics by using superposition theory or introducing several new variables, which provide design methods to decide the parameters [9], [10], [11], [12]. On the other hand, in many studies, a symmetric circuit is supposed, and they do not explicitly explain the mechanism of each resonance circuit of LCC. As a result, there is no established systematic asymmetric design of Double-LCC circuit. Furthermore, comparison

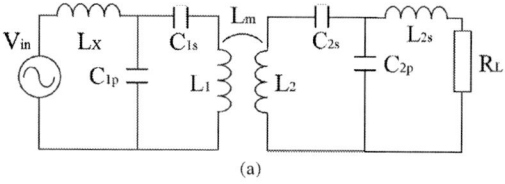

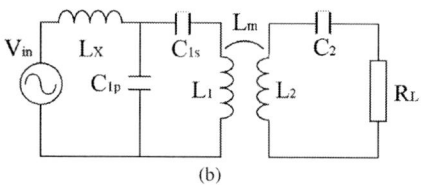

Fig. 2. Two compensation circuits are shown. (a) This circuit is called Double-LCC. In this paper, however, to prevent confusing, we call it LCC/LCC. (b) LCC/S

of LCC/S and LCC/LCC is not done and how to choose compensation circuit is unclear.

In this paper, LCC compensation circuit is investigated theoretically by focusing on two parameters. Based on the analysis, a systematic asymmetric design of Double-LCC circuit and LCC/S is proposed. After the proposal, an effect of the two parameters on output power and efficiency is discussed and verified by calculation. Finally, validation of the analysis and the asymmetric circuit design is confirmed by experiments.

Fundamental analysis for power design is done following Section II, theoretical investigation of each resonance circuit in LCC circuit by comparing LCC/LCC and LCC/S compensation circuits from the viewpoint of efficiency and power is done. Two parameters are also explained. In Section III, effect of the two parameters are explained, and asymmetric circuit design is proposed. Numerically, 6 cases are designed based on the previous analysis. In Section IV, experimental verification is done in 10 W prototype. A conclusion is stated in Section V.

II. FUNDAMENTAL THEORETICAL ANALYSIS

A. Analysis Outline

Fig. 2 shows LCC/LCC and LCC/S, which have LCC as a primary circuit and LCC or S as a secondary compensation circuit respectively. To simplify the analysis, several assumptions are supposed to be introduced. First, in a real system, a square wave is applied as an input voltage and contains high order harmonics. However, compensation circuits work as a bandpass filter. As a result, it is assumed that an applied input voltage includes a fundamental sinusoidal wave of the square wave. Second, coils used for WPT have relatively small resistance compared to their reactance. Therefore their resistance is ignored in the analysis of output power. Finally, values of components are constant even if the coupling of coils changes.

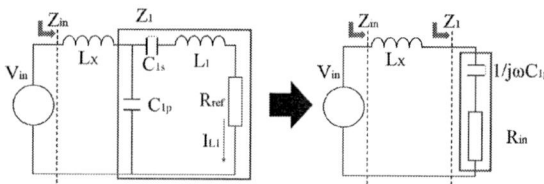

Fig. 3. An equivalent circuit of primary compensation are shown on the left side. R_{ref} is reflected resistance from secondary side and I_{L1} represents current flowing into primary coil. On the right side, the transformed equivalent circuit with combined impedance of components on the left side encircled with a blue frame is shown.

B. Primary Circuit Analysis

First of all, primary LCC is analyzed. The equivalent circuit, where C_{1s} and C_{1p} represent capacitors connected to the primary coil in series and parallel respectively, is shown Fig. 3. A reflected impedance from the secondary side is complicated. The impedance, however, makes it more challenging to keep unity power supply, especially in DWPTS. To mitigate this problem, usually, secondary compensation circuits which can keep unity power consumption are chosen. Therefore, R_{ref} is assumed as a pure resistance. Transferred power to the secondary side can be calculated as

$$P_{\mathrm{trans}} = R_{\mathrm{ref}}|I_{L1}|^2. \tag{1}$$

ω is source angular frequency. If L_X is ignored, I_{L1} can be derived as

$$I_{L1} = \frac{V_{\mathrm{in}}}{j\left(\omega L_1 - \frac{1}{\omega C_{1s}}\right) + R_{\mathrm{ref}}}. \tag{2}$$

V_{in} represents source voltage and a variable a refers to (3).

$$a = \frac{\omega L_1 - 1/\omega C_{1s}}{\omega L_1} \tag{3}$$

a describes how much compensate primary inductance by C_{1s} and is called uncompensated inductance coefficient in this paper. Substituting (3) into (2), I_{L1} becomes

$$I_{L1} = \frac{V_{\mathrm{in}}}{ja\omega L_1 + R_{\mathrm{ref}}}. \tag{4}$$

It is seen that I_{L1} is regulated by L_1. As a decreases, I_{L1} becomes larger, as a result, P_{trans} gets larger from (1). A combined impedance Z_1 shown in Fig. 3 can be calculated as (6). When resonance condition $C_{1p} = 1/\omega^2 a L_1$ is achieved, an imaginary part of Z_1 is constant and independent of magnetic coupling coefficient as (7), and a real part of it becomes (8). If an imaginary part of (6) is made into 0, the imaginary part varies according to the coupling.

$$
\begin{aligned}
Z_1 &= (ja\omega L_1 + R_{\mathrm{ref}})//\frac{1}{j\omega C_{1p}} & (5)\\[2mm]
&= \frac{R_{\mathrm{ref}}}{(\omega C_{1p}R_{\mathrm{ref}})^2 + (1 - \omega^2 C_{1p}aL_1)^2} \\
&\quad + j\frac{\omega L_1(1 - \omega^2 C_{1p}aL_1) - \omega C_{1p}R_{\mathrm{ref}}^2}{(\omega C_{1p}R_{\mathrm{ref}})^2 + (1 - \omega^2 C_{1p}aL_1)^2} & (6)\\[2mm]
&= \frac{(\omega aL_1)^2}{R_{\mathrm{ref}}} - j\frac{1}{\omega C_{1p}} & (7)
\end{aligned}
$$

2310

$$R_{\text{in}} = \frac{(a\omega L_1)^2}{R_{\text{ref}}} \tag{8}$$

When a secondary coil has gone, R_{ref} gets nearly 0 and Z_1 drastically increases. Consequently, little current gets to flow into a source.

It is obvious that unity power factor can be achieved by inserting L_X, whose value is (9), in series as shown in Fig. 3. The source current I_{in} is given by (10). A voltage applied to C_{1p} changes from source voltage because of inserted L_X. Gv_1 is defined as a ratio of those voltages and defined as (11).

$$L_X = \frac{1}{\omega^2 C_{1p}} \tag{9}$$

$$I_{\text{in}} = \frac{V_{\text{in}}}{R_{\text{in}}} \tag{10}$$

$$Gv_1 = \frac{|V_{Z_1}|}{|V_{\text{in}}|} = \frac{|Z_1|}{|Z_{\text{in}}|} = \sqrt{1 + \left(\frac{R_{\text{ref}}}{a\omega L_1}\right)^2} \tag{11}$$

$|I_{L1}|$ with L_X becomes Gv_1 times larger than without L_X.

$$|I_{L1}| = \left|\frac{Gv_1 V_{\text{in}}}{ja\omega L_1 + R_{\text{ref}}}\right| \tag{12}$$

$$|I_{L1}| = \frac{V_{\text{in}}}{a\omega L_1} \tag{13}$$

Substituting (11) in (12), $|I_{L1}|$ can be derived as (13). It can be easily understood that the constant current is realized by inserted L_X. Moreover, source current can be controlled by a value of a because I_{L1} is in inverse portion to a.

C. Secondary Circuit Analysis

V_{ind} is a voltage induced at secondary coil by primary coil current. It can be derived as (14) where L_m is a mutual inductance between primary and secondary coils and defined as (15).

$$V_{\text{ind}} = j\omega L_m I_{L1} \tag{14}$$

$$L_m = k\sqrt{L_1 L_2} \tag{15}$$

V_{out} is equal to V_{ind} under S compensation circuit because a combined impedance of the secondary circuit is equal to load resistance at resonance condition. On the other hand, to calculate V_{out} of LCC compensation as shown in Fig. 4, secondary LCC circuit is analyzed as being done in primary one. L_{2s}, R_L and C_{2p} can be regarded as L_1, R_{ref} and C_{1p} of the primary LCC respectively. Z_{Leq} is equivalent load impedance seen from dotted line shown in Fig. 4. Gv_2 is a voltage gain between V_{ind} and $V_{Z_{Leq}}$. $V_{Z_{Leq}}$ is an applied voltage of C_{2p}. Under the perfect resonance condition given by (16) and (17),

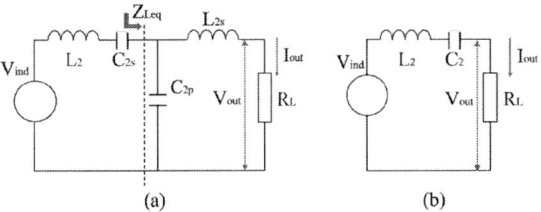

Fig. 4. The equivalent circuit of secondary compensation circuit.

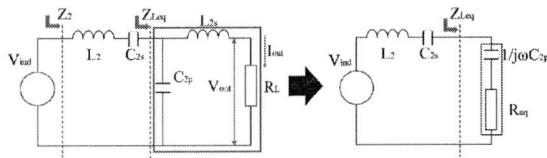

Fig. 5. An equivalent circuit of secondary compensation are shown on the left side. R_L is a load resistance and I_{out} represents current flowing into the load. On the other side, the transformed equivalent circuit with combined impedance of components on the left side encircled with a blue frame is shown.

they can be derived as

$$C_{2p} = \frac{1}{\omega^2 L_{2s}} \tag{16}$$

$$\omega L_2 - \frac{1}{\omega C_{2s}} = \frac{1}{\omega C_{2p}} \tag{17}$$

$$Z_{L_{eq}} = (j\omega L_{2s} + R_L) // \frac{1}{j\omega C_{2p}}$$

$$= \frac{(\omega L_{2s})^2}{R_L} + \frac{1}{j\omega C_{2p}} \tag{18}$$

$$Gv_2 = \sqrt{1 + \left(\frac{R_L}{\omega L_{2s}}\right)^2}. \tag{19}$$

Because V_{ind} is a proportional to I_{L1}, V_{ind} can be regarded as a constant voltage source under a constant coupling coefficient. The value of output voltage of the LCC circuit is (20).

$$|V_{\text{out}}| = |Gv_2 V_{\text{ind}}| \left|\frac{R_L}{j\omega L_{2s} + R_L}\right| \tag{20}$$

Substituting (13), (14), (15) and (19) in (20), the equations of output voltage of both circuits are as following.

$$V_{\text{out}}(S) = \frac{k}{a}\sqrt{\frac{L_2}{L_1}} V_{\text{in}} \tag{21}$$

$$V_{\text{out}}(LCC) = \frac{k}{a}\sqrt{\frac{L_2}{L_1}}\frac{R_L}{\omega L_{2s}} V_{\text{in}} \tag{22}$$

D. Output Power

Output power can be derived by (21) and (22) as following.

$$P_{\text{out}}(S) = \frac{L_2}{L_1}\frac{k^2}{a^2}\frac{V_{\text{in}}^2}{R_L} \tag{23}$$

$$P_{\text{out}}(LCC) = \frac{L_2}{L_1}\frac{k^2}{a^2}\frac{V_{\text{in}}^2}{R_{\text{eq}}} \tag{24}$$

TABLE I.   SPECIFICATION OF THE CIRCUIT USED FOR CALCULATION

Symbol	Value	Symbol	Value	Symbol	Value
f	85 kHz	L_{2s}	60 μH	R_{2s}	90 mΩ
V_{in}	180 V	L_X	40 μH	R_X	60 mΩ
		L_1	156 μH	R_1	300 mΩ
		L_2	163 μH	R_2	320 mΩ

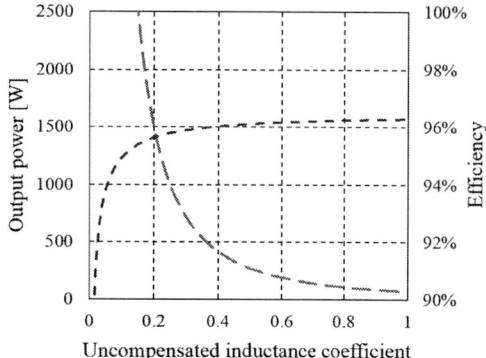

Fig. 6.   Calculation result of output power at maximal efficiency and the efficiency according to each a is shown. Red and blue lines represent output power and efficiency respectively. The output power at maximal efficiency has significantly changed according to a, but maximal efficiency decreases when a is small.

Compared with LCC/S and LCC/LCC, a secondary LCC circuit converts load resistance into R_{eq} as shown in Fig. 5. R_{eq} is given by (25).

$$R_{\text{eq}} = \frac{(\omega L_{2s})^2}{R_L} \qquad (25)$$

Output power is strongly dependent on load resistance. In LCC/S circuit, an output voltage is constant, and power is in inverse portion to R_L. In LCC/LCC circuit, an output current is constant and power is proportional to R_L.

E. Efficiency

In the real system, coils have their resistance whose values are represented by R_i, and i is identical to a subscript of its inductor. For example, R_1 is winding resistance of L_1. The system efficiency η of LCC/LCC can be derived as

$$\eta = \frac{R_{\text{in}}}{R_x + R_{\text{in}}} \frac{R_{\text{ref}}}{R_1 + R_{\text{ref}}} \frac{R_{\text{eq}}}{R_2 + R_{\text{eq}}} \frac{R_L}{R_{2s} + R_L}, \quad (26)$$

where

$$R_{\text{eq}} = \frac{(\omega L_{2s})^2}{R_{2s} + R_L} \qquad (27)$$

$$R_{\text{ref}} = \frac{(\omega L_m)^2}{R_2 + R_{\text{eq}}} \qquad (28)$$

$$R_{\text{in}} = \frac{(\omega a L_1)^2}{R_{\text{ref}} + R_1}. \qquad (29)$$

In LCC/S, efficiency can be derived as

$$\eta = \frac{R_{\text{in}}}{R_X + R_{\text{in}}} \frac{R_{\text{ref}}}{R_1 + R_{\text{ref}}} \frac{R_L}{R_2 + R_L}, \qquad (30)$$

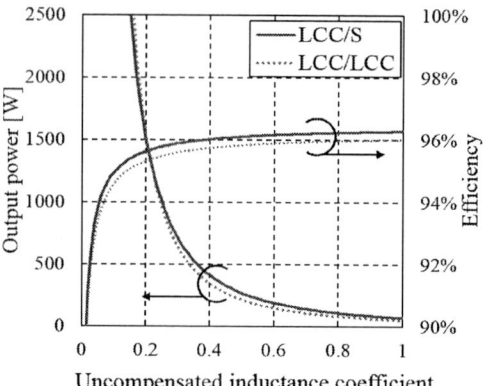

Fig. 7.   Calculation result of output power at maximal efficiency and maximal efficiency of LCC/S and LCC/LCC. Output power of both circuits are almost same, and efficiency of LCC/LCC is slightly lower than that of LCC/S.

where

$$R_{\text{ref}} = \frac{(\omega L_m)^2}{R_2 + R_L} \qquad (31)$$

$$R_{\text{in}} = \frac{(\omega a L_1)^2}{R_{\text{ref}} + R_1}. \qquad (32)$$

Load resistance which maximizes η is called optimal load resistance. The optimal load of LCC/S can be calculated as (33) [13] with k and Q factor, where Q can be defined by dividing reactance of coil by its resistance.

$$R_{Lopt} = R_2 \sqrt{\frac{(1+F)(1+aQ_1Q_X+F)}{1+aQ_1Q_X}} \qquad (33)$$

$$\text{where } F = k^2 Q_1 Q_2 \qquad (34)$$

On the other hand, an optimal load of LCC/LCC hasn't been studied precisely. Therefore, the optimal load is derived as shown in the appendix and the result is (35).

$$R_{Lopt} = \sqrt{\frac{[R_S + FR_{2s}][FR_{2s} + R_S(1+aQ_1Q_X)]}{(1+F)(1+F+aQ_1Q_X)}} \qquad (35)$$

$$\text{where } R_S = \frac{(\omega L_{2s})^2}{R_2} + R_{2s} \qquad (36)$$

III.   DESIGN OF ASYMMETRIC LCC COMPENSATION CIRCUIT

In this section, LCC/LCC and LCC/S are compared from the viewpoint of output power and efficiency, and asymmetric design of LCC circuit is described. The values used for calculation are shown in TABLE I. Other parameters such as capacitance can be calculated from (9), (16) and (17).

A. Effects of Uncompensated Inductance Coefficient

In LCC circuit, output power and efficiency has a trade-off relation and to transfer higher output power. It is necessary to agree with a drop of efficiency to achieve

The 2018 International Power Electronics Conference

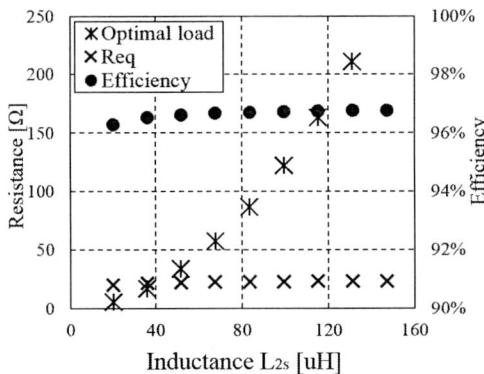

Fig. 8. Calculation result of maximal efficiency, an optimal load and R_{eq} versus L_{2s}. Efficiency and R_{eq} are practically constant. On the other hand, an optimal load increases drastically.

TABLE II. SPECIFICATION OF THE SYSTEM

Symbol	Value	Symbol	Value	Symbol	Value
f	85 kHz	k	0.2	V_{DC}	20 V
gap	70 mm				
size of L_1	200×300 mm	Turns	20 T		
size of L_2	200×200 mm	Turns	30 T		

higher output power. A calculation result of output power at maximal efficiency and maximal efficiency of LCC/S is shown in Fig. 6 when a changes from 0 to 1. The output power at a maximal efficiency significantly increases as a decreases. On the other hand, maximal efficiency decreases in small a sharply.

a is an essential factor for determined transferred power because it controls current of the primary coil. According to (21) and (22), P_{out} depends on the value of a once primary and secondary coils and a load condition are determined. The decrease in efficiency is caused by R_{in} shown in (26) and (30). R_{in} is getting small as a becomes smaller. It means that the first fraction of the equations gets smaller and efficiency also gets smaller. From the viewpoint of loss, smaller a increases primary current and loss drastically.

B. Effects of L_{2s}

From (25) and Fig. 5, R_{eq} of LCC/LCC can be assumed as R_L of LCC/S under $R_{2s} \ll R_L$ condition. Under the condition, the consumption power of R_{eq} becomes almost identical to that of R_L of LCC/LCC. Therefore, output power at maximal efficiency of LCC/S and LCC/LCC are almost same. A calculation result of output power at maximal efficiency under conditions of TABLE I is shown in Fig. 7. In the calculation, Q_X is assumed constant and 100. Moreover, efficiency of LCC/LCC slightly drops from that of LCC/S.

Fig. 8 shows calculation results of R_{Lopt} and R_{eq} of LCC/LCC. The results prove that R_{eq} is almost constant in any L_{2s} under the optimal condition. Moreover, maximal efficiency is constant and almost irrelevant to L_{2s}, namely, L_{2s} converts load resistance and does not strongly affect output power and efficiency. As a result, LCC/LCC can achieve maximal efficiency at no matter what load resistance by adjusting L_{2s}.

C. Decision of Circuit Structure and Parameters

In previous research, it is said that LCC compensation is suitable for bidirectional WPT because of its symmetric

circuit configuration. However, in DWPTS, it is better to feed back regenerative power to a battery pack which is rarely fully charged than to the road because sending back regenerative power through WPT lowers utilization of it. It is also challenging to use regenerated power fed back to the road effectively. Moreover, primary and secondary compensation coils have different effects as already studied in the previous section.

Because LCC at the secondary side works as an impedance converter, LCC/LCC is useful if there is a big difference between an optimal load and a real load such as batteries or motors. On the other hand, if the difference is small or active control is adopted at the secondary side such as a DC-DC converter, LCC/LCC causes an increase in complexity of a secondary system and decrease in efficiency compared with LCC/S. Thus, in the cases, LCC/S might be better than LCC/LCC.

In proposed asymmetric design, except for a and L_{2s}, all parameters such as load resistance and source voltage are given. The proposed design starts with thinking of LCC/S. To find a which satisfies desired output power and maximal efficiency at the same time, (23) and (33) must be used. Because output power at maximal efficiency of the two is not so different, a of LCC/S is adopted to LCC/LCC. Then, an approximated optimal load of LCC/LCC can be derived by solving $R_{eq} = R_{Lopt}$, where R_{Lopt} is an optimal load of LCC/S and R_L is reference load resistance of LCC/LCC. When R_{Lopt} is given by (33). The optimal load can be achieved by adequate adjustment of L_{2s} given by (37) in LCC/LCC.

$$\omega L_{2s} \approx \sqrt{R_L R_{Lopt}} \qquad (37)$$

IV. EXPERIMENTAL VALIDATION

A. Experimental Setup

Fig. 9 is an equivalent circuit diagram for an experiment, and experimental setups are shown in Fig. 10. In this circuit, SCT2080KE SiC power MOSFETs (ROHM) are adopted to a full-bridge inverter as a high-frequency power source. Coils are made from 160-strand AWG 42 Litz-wire and shown in Fig. 11. As a rectifier, FMX-4202S (SANKEN ELECTRIC) are used for bridge diodes, and 1000 μF capacitor is used for a smoothing capacitor. A load is pure resistance. Values of experimental components are shown in TABLE I.

B. Experimental Test Cases

Here, a and L_{2s} are determined numerically. Specifications for circuit design are same as shown in TABLE I

2313

The 2018 International Power Electronics Conference

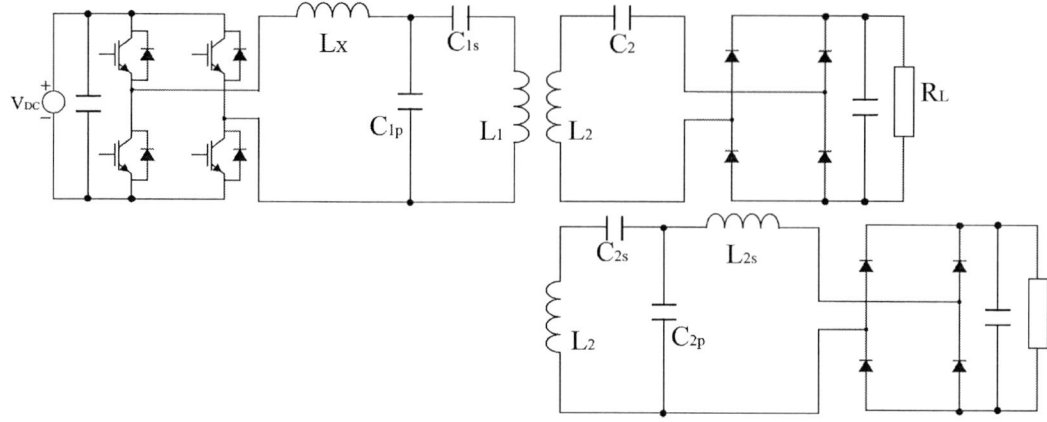

Fig. 9. Experimental equivalent circuit to verify validation of proposed design method. LCC/S is a upper circuit and LCC/LCC is a lower circuit.

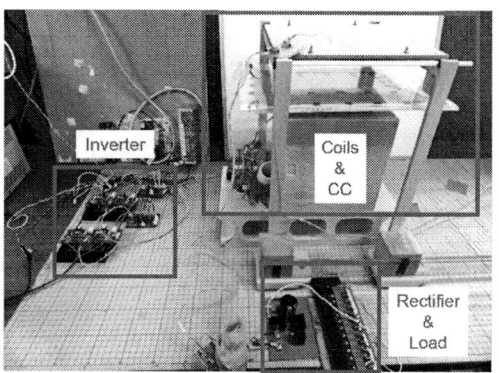

Fig. 10. Picture of experimental setups used for verification of validation.

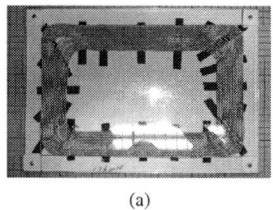 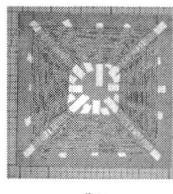

(a)   (b)

Fig. 11. Main coils for WPT. (a) Primary coil. (b) Secondary coil.

and TABLE II. Desired output power is 10 W and three circuit structures are prepared shown in TABLE III. Each structure also has two cases. One is 20 Ω, the other is 50 Ω. In LCC/S, the values are substituted in (23) and a is calculated. Case 1 is $a = 0.26$, Case 2 is $a = 0.17$. In LCC/LCC, if a symmetric design is adapted, Case 3 is $L_X = L_{2s} = 40\ \mu H$, Case 4 is $L_X = L_{2s} = 50\ \mu H$. In proposed asymmetric LCC design, optimal load of LCC/S is about 20 Ω thus, L_{2s} can be calculated by substituting 20 in (37). Case 5 is $L_{2s} = 37\ \mu H$, Case 6 is $L_{2s} = 60$ μH. The summary of cases is shown in TABLE III.

C. Results and Discussion

Experimental results are shown in TABLE IV. In this results, a forward voltage drop caused by a rectifier

TABLE III. EXPERIMENTAL TEST CASES

Structure	Reference	L_X	L_{2s}	R_X	R_{2s}
LCC/S	20 Ω	39 μH	–	60 mΩ	–
	50 Ω	26 μH	–	50 mΩ	–
LCC/LCC sym.	20 Ω	39 μH	39 μH	60 mΩ	60 mΩ
	50 Ω	50 μH	50 μH	80 mΩ	80 mΩ
LCC/LCC asym.	20 Ω	39 μH	37 μH	60 mΩ	60 mΩ
	50 Ω	39 μH	60 μH	60 mΩ	90 mΩ

TABLE IV. EXPERIMENTAL RESULTS IN EACH TEST CASE

Structure	Case	Reference load	P_{out}	Efficiency
LCC/S	case 1	20 Ω	9.1 W	88.9 %
	case 2	50 Ω	9.2 W	84.8 %
LCC/LCC sym.	case 3	20 Ω	10.4 W	84.3 %
	case 4	50 Ω	7.9 W	84.4 %
LCC/LCC asym.	case 5	20 Ω	10.4 W	84.3 %
	case 6	50 Ω	10.6 W	86.4 %

diode is considered and compensated by calculation. An increase of output power was observed in Fig. 12 when a becomes smaller. Moreover, since load resistance of case 2 is different from the optimal load, efficiency of case 2 becomes smaller than that of case 1. On the other hand, maximal efficiency only achieves at a specific load and efficiency drops at load resistance which is different from the optimal load. Maximal efficiency of LCC/S and LCC/LCC are compared in Fig. 13. As theoretical analysis revealed, maximal efficiency of LCC/S is higher than LCC/LCC.

case 3 and case 5 are identical by chance. In LCC/LCC, designed output power is achieved at both 20, and 50 Ω and efficiency stays high. The proposed asymmetric design achieves designed output power and also guarantees maximal efficiency. On the other hand, symmetric design can achieve designed output power but cannot ensure maximal efficiency.

V. CONCLUSION

We analyzed the characteristics of LCC compensation circuit and verified that the capacitor connected to the

2314

The 2018 International Power Electronics Conference

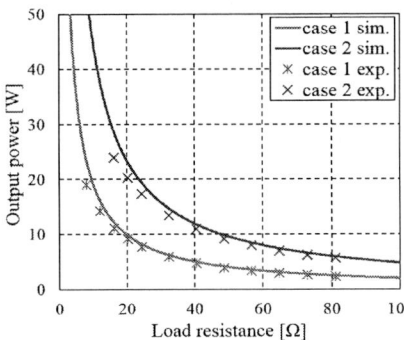

Fig. 12. Load resistance versus output power in case 1 and 2. Red and blue solid lines represent calculation result of case 1 and case 2 respectively. Marks represent experimental results.

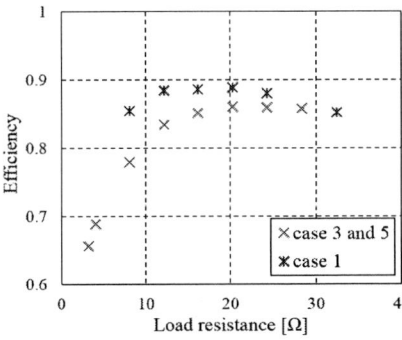

Fig. 13. Experimental load resistance versus efficiency of LCC/S and LCC/LCC. Maximal efficiency of LCC/S is higher than that of LCC/LCC.

primary coil in series and the value of secondary compensation inductor are important to design output power and efficiency respectively. Moreover, the theoretical analysis reveals previous symmetric circuit design cannot always satisfy maximal efficiency. This paper established a theoretical investigation of a systematic methodology for asymmetric LCC compensation circuit to decide appropriate uncompensated compensation coefficient in the primary circuit as well as secondary compensation inductance or change of circuit configuration. Besides, experimental verification shows the validity of proposed asymmetric circuit design.

APPENDIX

Substituting (27) – (29) in (26), η can be (A.1). With k and Q factor, (A.1) can be transformed into (A.2).

$$\eta = \frac{(a\omega L_1)^2}{[R_X R_1 + (a\omega L_1)^2] + \dfrac{R_X(\omega L_m)^2}{R_2 + \frac{(\omega L_{2s})^2}{R_{2s}+R_L}}}$$
$$\cdot \frac{(\omega L_m)^2}{R_1 R_2 + (\omega L_m)^2 + R_1 \frac{(\omega L_{2s})^2}{R_{2s}+R_L}}$$
$$\cdot \frac{(\omega L_{2s})^2}{R_2(R_{2s}+R_L) + (\omega L_{2s})^2} \frac{R_L}{R_{2s}+R_L} \quad \text{(A.1)}$$

$$\eta = \frac{C_0 R_L}{C_1 R_L^2 + C_2 R_L + C_3} \quad \text{(A.2)}$$

where

$$R_0 = \frac{(\omega L_{2s})^2}{R_2} \quad \text{(A.3)}$$

$$C_0 = a k^4 Q_1^2 Q_X Q_2^2 R_0 \quad \text{(A.4)}$$

$$C_1 = (1+F)(1+aQ_1 Q_X + F) \quad \text{(A.5)}$$

$$C_2 = (1+F)[R_0(1+aQ_1 Q_X) + R_{2s}(1+aQ_1 Q_X + F)]$$
$$+ (1+aQ_1 Q_X + F)[R_0 + R_{2s}(1+F)] \quad \text{(A.6)}$$

$$C_3 = [R_0 + R_{2s}(1+F)]$$
$$[R_0(1+aQ_1 Q_X) + R_{2s}(1+aQ_1 Q_X + F)]. \quad \text{(A.7)}$$

(A.2) is differentiated with respect to R_L and when the value of derivative equals to 0, R_L makes efficiency maximal and the optimal load is (A.8).

$$R_L = \sqrt{\frac{C_3}{C_1}} \quad \text{(A.8)}$$

REFERENCES

[1] Y. Hori, "Future vehicle driven by electricity and Control-research on four-wheel-motored "UOT electric march II"," *IEEE Transactions on Industrial Electronics*, vol. 51, no. 5, pp. 954-962, Oct. 2004.

[2] L. Chen, G. R. Nagendra. J. T. Boys, and G. A. Covic, "Double-coupled systems for IPT roadway applications," *IEEE Journal of Emerging and Selected Topics in Power Electronics*, vol. 3, no. 1, pp. 37-49, 2015.

[3] S. Li and C. Mi, "Wireless Power Transfer for Electric Vehicle Applications," *IEEE Journal of Emerging and Selected Topics in Power Electronics*, vol. 3, no. 1, p. 4-17, 2014.

[4] G. R. Nagendra, L. Chen, G. A. Covic, and J. T. Boys, "Detection of EVs on IPT highways," *Conf. Proc. - IEEE Appl. Power Electron. Conf. Expo. - APEC*, vol. 2, no. 3, pp. 1604-1611, 2014.

[5] D. Kobayashi, K. Hata, T. Imura, H. Fujimoto, and Y. Hori, "Sensorless Vehicle Detection Using Voltage Pulses in Dynamic Wireless Power Transfer System," *EVS29 Symposium*, pp. 1-10, 2016.

[6] W. Li, H. Zhao, J. Deng, S. Li, and C. C. Mi, "Comparison Study on SS and double-sided LCC compensation topologies for EV/PHEV Wireless Chargers," *IEEE Trans. Veh. Technol.*, vol. 65, no. 6, pp. 4429-4439, 2016.

[7] T. Campi, S. Cruciani, F. Maradei, and S. Member, "Near-Field Reduction in a Wireless Power Transfer System Using LCC Compensation," *IEEE Trans. Electromagnetic compatibility*, vol. 59, no. 2, pp. 686-694, 2017.

[8] F. Lu, S. Member, H. Zhang, and S. Member, "A Dynamic Charging System With Reduced Output Power Pulsation for Electric Vehicles," *IEEE Trans. Ind. Electron.*, vol. 63, no. 10, pp. 6580-6590, 2016.

[9] S. Li, W. Li, J. Deng, T. D. Nguyen, and C. C. Mi, "A Double-Sided LCC Compensation Network and Its Tuning Method for Wireless Power Transfer," *IEEE Trans. Veh. Technol.*, vol. 64, no. 6, pp. 2261-2273, 2015.

[10] S. Zhou and C. Chris Mi, "Multi-Paralleled LCC Reactive Power Compensation Networks and Their Tuning Method for Electric Vehicle Dynamic Wireless Charging," *IEEE Trans. Ind. Electron.*, vol. 63, no. 10, pp. 6546-6556, 2016.

[11] Q. Zhu, L. Wang, Y. Guo, C. Liao, and F. Li, "Applying LCC Compensation Network to Dynamic Wireless EV Charging System," *IEEE Trans. Ind. Electron.*, vol. 63, no. 10, pp. 6557-6567, 2016.

[12] W. Li, C. C. Mi. S. Li, J. Deng, T. Kan, and H. Zhao, "Integrated LCC Compensation Topology for Wireless Charger in Electric and Plug-in Electric Vehicles," *IEEE Trans. Ind. Electron.*, vol. 62, no. 7, pp. 4215-4225, 2015.

[13] Y. Geng, B. Li, Z. Yang, F. Lin, and H. Sun, "A High Efficiency Charging Strategy for a Supercapacitor Using a Wireless Power Transfer System Based on Inductor/Capacitor/Capacitor(LCC) Compensation Topology," Energies 2017, 10, 135.

Probabilistic PCA-Support Vector Machine Based Fault Diagnosis of Single Phase 5-Level Cascaded H-Bridge MLI

Nagendra Vara Prasad.Kuraku, Yigang He * and Murad Ali

School of Electrical Engineering and Automation, Hefei University of Technology, Hefei, P.R.China

*E-mail: 18655136887@163.com

Abstract— In present era, Multilevel Inverters (MLIs) are very popular in many industrial and renewable energy applications. The fast and accurate fault diagnosis is very important for improving the reliability. The present study proposes a novel fault diagnosis method based on the Probabilistic Principle Component Analysis (PPCA) and Support Vector Machine (SVM) for controlled switches in single phase Cascaded H-Bridge Multilevel Inverter (CHMLI). The output voltage signals under different fault conditions of the CHMLI are taken as fault features by using Phase Shift PWM technique. PPCA is used to optimize the data and reduce dimension of fault features. Finally, SVM classifier is used to diagnose the different fault modes. An experimental setup of CHMLI has been designed to validate the proposed fault diagnosis method. The simulation and experimental results show that by using PPCA-SVM, we can improve the accuracy of the fault location and reduce the time taken to diagnosis the fault in CHMLI.

Keywords— *Fault Diagnosis, CHMLI, PPCA, SVM*

I. INTRODUCTION

In recent years the MLI plays an important role in all power electronic applications, because of their advantages such as low electromagnetic interference (EMI) caused by high dv/dt, high voltage capability, draws input current with low distortion and high efficiency with low switching frequency control methods [1]. CHMLI consists of several number of H-Bridge converters to increase the output voltage levels. More number of voltage levels is an attractive feature that will reduce the harmonics in output waveform [1], by increasing voltage levels the number of switches also will increase, it will cause the circuit complexity to increase, and the reliability to reduce.

Reliability for CHMLI has been an important issue to get continuous supply to the load and to improve the power conversion efficiency, for example in Hybrid Electric Vehicle the motor needs continues power supply from the inverter, and a sudden failure in the power semiconductor devices may cause interruption in the power supply [2]. The most frequent faults occur in variable-speed AC (VSAC) drives are due to failures of power semiconductor devices, which are about 38% of the total faults occur in the systems [3]. Hence, the identification of possible faults and its diagnosis is paramount importance.

The Open Circuit (OC) fault of the switch is detected with Short Circuit (SC) fault detection time in [4], but they used voltage sensors at desired locations, it will lead to increase the total cost of the system. In [5] by using histogram of the trajectory and Fourier series in stator current pattern, an Switch is open to evaluate fault detection in VSI is identified, in this method the time taken to detect the fault location is more, which is about 10ms. By inverter output PWM voltage signal, an online Fault isolation & detection method is presented in [6]. A novel and fast fault identifier designed to identify the SC faults and OC faults by analyzing the performance of MLI drive using multi-resolution analysis based on wavelet is presented in [7, 8], drawback of these methods are not suitable for the variable load drive systems, because the voltage and current signals are used for the wavelet analysis.

In this diagnosis technique based on PPCA-SVM algorithm has been implemented in this paper to evaluate the fault location and reduce the time taken to diagnosis the fault in CHMLI. First the data taken from the output voltage signals of sensitive load is given to PPCA, and then the Probability PCA is used to reduce dimensions and data is optimized also. The characteristics signals have been taken as the fault features when the power devices (MOSFET's) are in OC state. Finally, the SVM is identify and classify the fault location. An experimental setup of CHMLI has been built it to be validate the proposed PPCA-SVM.

II. ANALYSIS OF FAULT FEATURES IN CHMLI

There are many reasons to get faults in the MLI, but due to high thermal and electrical stresses in switching devices, there could be a high chance to get faults in the CHMLI. The present study mainly focused on open circuit faults, SC faults are implementing by fuses, so that SC faults is to be converted into OC faults in the inverter [9]. The OC faults are divided into two categories, first one is simple faults, defined as one switch fault at a time, and second one is complicated faults, defined as any two switches fault at a time. In this paper, simple faults and complicated faults are considered to diagnosis the faults.

To test the proposed topology, a single phase CHMLI is taken for OC faults. CHMLI and the output voltage signals with the load resistance R = 50Ω. The simulation model of CHMLI is controlled by PSPWM [10].

Phase shifted PWM (PS-PWM) technique is mainly used to reduce the Total Harmonic Distortion (THD) in the MLI [11], in each module, the reference signal is sine wave and carrier waveform is triangular, carrier signals are depends on the output voltage levels, like 5 level output voltage using (5-1=4) 4 carrier waveforms, but these four carrier waveforms are different phase shifted to each other, because output voltage is depends on varying of gate pulse of switches. The phase shift between each carrier signal is θ (in degrees).

$$\theta = \frac{360^{\circ}}{n-1} \qquad (1)$$

From equation (1), θ is 90^0 for each carrier signal, and the pulse signals given to the switches are produced by comparing the carrier signals and reference signal. Here t_d = 2.4µs is applied based on driver circuit used in the 5-level CHMLI experiment.

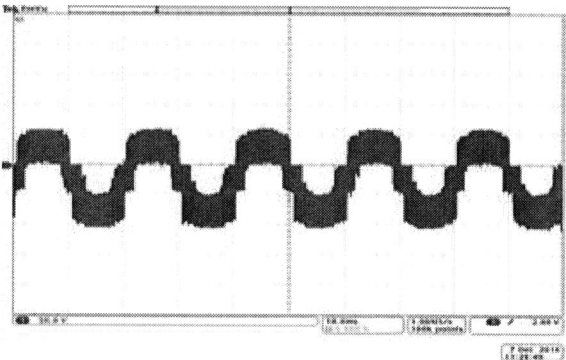

(b)

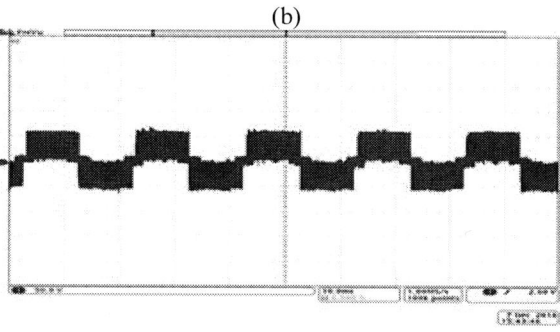

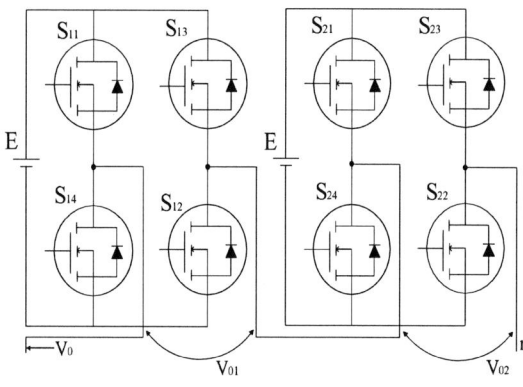

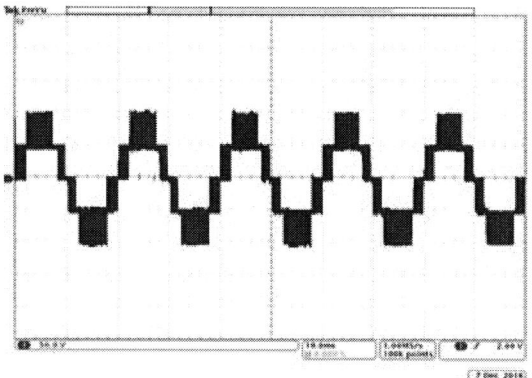

(a)

(c)

Fig. 1. Output Voltages of the CHMLI under Different Fault Conditions
(a) No-Fault Condition
(b) Single Fault Condition (at S_{11})
(b) Double Fault Condition (at S_{12} & S_{12}).

In CHMLI we can get output voltage & output current signals, here only the output voltage signals are taken as characteristic i/p signals to the classifier, current signals are dependent only load currents does not depends on fault conditions, so that voltage signals are considered has a feature extraction input signals. Under different fault conditions are shown in Fig. 1.

The proposed PPCA-SVM topology for fault diagnosis of CHMLI is shown in Fig. 2. The data taken from the output voltage signals of the load is very huge, so it is very difficult for the feature classifier (SVM) to classify the different faults. Many feature extractions are available to optimize and dimensional reduction of the data, in this paper PPCA is used to dimensional reduction of the output voltage signal data. PPCA data is classified into different classifications by using feature classifier as shown in TABLE I Here total 37 faults are divided into six categories, the first category is fault free, second one is simple faults and the remaining categories are complicated faults, which means the faults occur because of failure of two switches at a time in different combinations. The reduced data is given to SVM classifier, which will identify the faults in different switching positions.

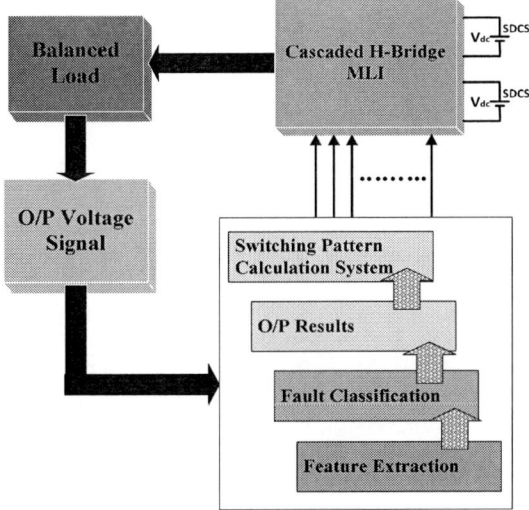

Fig. 2. Fault Diagnosis System Structure

The SVM classifies the total faults into different classifications. The classification results are shown TABLE I; and the results are converted decimal codes into binary codes. Classification model output's are [1 1 1 1 1 1 1 1 0]^T when the OC fault occurs at S_{24}, the output voltage signal is shown in Fig. 1. The classification model outputs for different faults are presented in TABLE I; in this '0' represents the faulty condition and '1' represents the healthy condition. To control the multilevel inverter, we should give the switching sequences given by the SVM.

TABLE I
FAULT LABELS & CLASS

Open Circuit Fault Modes	Labels	Fault Class
Normal	[1 1 1 1 1 1 1 1 1]^T	1
S_{11}	[1 0 1 1 1 1 1 1 1]^T	2
S_{12}	[1 1 0 1 1 1 1 1 1]^T	3
S_{13}	[1 1 1 0 1 1 1 1 1]^T	4
S_{14}	[1 1 1 1 0 1 1 1 1]^T	5
S_{21}	[1 1 1 1 1 0 1 1 1]^T	6
S_{22}	[1 1 1 1 1 1 0 1 1]^T	7
S_{23}	[1 1 1 1 1 1 1 0 1]^T	8
S_{24}	[1 1 1 1 1 1 1 1 0]^T	9

III. FAULT DIAGNOSIS METHOD

In the proposed fault diagnosis Probability PCA-Support Vector Machine method is explained as follows: firstly, initialized the parameter setting of PPCA & SVM, and then the output voltage signals from inverter are converted into sample signals, and the sample signals are given to PPCA. Here PPCA is used to optimize the data, and the optimized data is given to the SVM. Then the SVM classifies the different faults occur in the inverter. This process is going on until the expected goal reached.

A. Feature Extraction (Probabilistic PCA)

Many authors proposed Principal Component Analysis (PCA) to analyze the data and extract variables with similar properties. But the main disadvantage of the PCA is absence of a generative model or model of probability density is [12]. To overcome this problem, the PPCA is used in this paper. A latent variable model seeks to relate a d- dimensional observed data vector v:

$$v = y(x;U) + \varepsilon \qquad (2)$$

Where U is the parameters, ε is an x dimensional independent noise, and x is q- dimensional vector of latent variable, Equation (2) induces a corresponding data space in distribution and the model parameters is determined by Maximum-Likelihood techniques. Perhaps the latent variable model is varied by various statistical factor analysis, in which the linear function [13] of x:

$$v = Ux + \xi + \varepsilon \qquad (3)$$

Latent variables are independent as well as Gaussian variance, so $x \sim \mathfrak{N}(0, I)$. The noise model also Gaussian isotropic such that $\varepsilon \sim \mathfrak{N}(\sigma^2 I, 0)$, $\sigma^2 I$ diagonal matrix and size is $(d * q)$, U matrix contains the loading factor, the observation vectors is $v \sim \mathfrak{N}(\xi, \mathbb{C})$ are normally distributed, the covariance model is $\mathbb{C} = \sigma^2 I + UU^T$. The corresponding [14] likelihood ML estimator for ξ is depends on mean value of original data, to be estimates for σ^2 an U may be considered as [15] by iterative procedure to maximization of θ. 'x' is the latent variable of conditional distribution of given observed v, calculated based on Bayes' rule with Gaussian:

$$\mathbb{P}(x|v) \sim \mathfrak{N}(M^{-1}U^T(v - \xi), \sigma^2 M^{-1}) \qquad (4)$$

Where we have defined $M = U^T U + \sigma^2 I$. Note that M is of size $q * q$ while $\mathbb{C}$ is $d * d$.

$$U_{ML} = U_q(A_q - \sigma^2 I_q)^{1/2} \aleph \qquad (5)$$

Where U_q is principal Eigenvectors and corresponding eigenvalues are $\lambda_1, \lambda_2, \ldots, \lambda_q$ as well as diagonal matrix is Λ_q, and $\aleph$ is an arbitrary $q * q$ rotational orthogonal matrix Equation (5).

$$\xi = \frac{1}{R}\sum_{n=1}^{R} v_i \qquad (6)$$

$$\sigma_{ML}^2 = \frac{1}{d-q}\sum_{i=q+1}^{d}\lambda_i \tag{7}$$

Note that in PCA feature extraction assumed as $U = U_q$, but in PPCA feature extraction model is using based on Maximum likelihood is taken as optimal sense of matrix of $\left(\Lambda_q - \sigma^2 I_q\right)^{1/2}$, this matrix depends on the U_q of individual column vector, the probability distribution is v given as x.

$$p(v/x) = (2\pi\sigma^2)^{-1/2} exp\left(-\frac{1}{2\sigma^2}\|v - U - \bar{v}\|\right) \tag{8}$$

Where $U = U_q\left(\Lambda_q - \sigma^2 I_q\right)^{1/2}x$.

Hence, the reconstructed data point ML is

$$\theta = U_q\left(\Lambda_q - \sigma^2 I_q\right)^{1/2}x + \bar{v} \tag{9}$$

In that case, the reduction data map is

$$\ell = \left(\Lambda_q - \sigma^2 I_q\right)^{1/2}U_q^T(v - \bar{v}) \tag{10}$$

Total average reconstruction error after minimization of least square,

$$\varepsilon = \frac{1}{n}\sum_{i=1}^{n}\|v_i - U_q U_q^T(v_i - \bar{v}) - \bar{v}\|^2 \tag{11}$$

These reconstruction & reduction maps are adopted in [16].

B. Feature Classification (SVM)

SVM classifier is based on supervised learning algorithm. SVM is used for both regression analysis and classification. The training set is considered has either new training set or non-probabilistic binary linear classifier. In the new training, sets are mapped into the same space & category, and classes are divided based on different set of object comprises [17].

SVM is used to find the distance between any two classes [18] of hyper plane. In that hyper plane has to be largest distance b/w the nearest training data point of any classes. Input data is generalized by means of the kernel technique that is introduced in the following section. If the set of training data points in SVM are linear separable then it is called Linear SVM, and the set of training data points in SVM are nonlinear separable then it is called nonlinear SVM [33]. Given training set Z, including m points,

$$Z = \sum_{k=1}^{m}(P_k, Q_k), \begin{cases} P_k \in \Re^n \\ Q_k \in \{-1,1\} \end{cases} \tag{12}$$

Every P_k has a n dimensional real vector.

$$c.P - r = 0 \tag{13}$$

c represent the normal vector and P is the set of plane, we can select two hyper plane equations [19] [section 7.2] like,

$$c.P - r = \pm 1 \tag{14}$$

Any two hyper planes distance is $\frac{2}{\|c\|}$, i.e. to minimize the $\|c\|$. And also prevent the data points from falling into the margin, so to add the constraint for each $'K'$ is either.

$$\mathcal{E}_k(c.P_k - r) \geq 1 \ for \begin{cases} P_k \ eclass - 1 \\ k \in \{1,...,m\} \end{cases} \tag{15}$$

The SVM can be formulated and problem of to minimize $(mc,r)\|c\|$.

IV. ANALYSIS OF SIMULATION & EXPERIMENTAL RESULTS

In this section, simulation & experimental results of the proposed fault diagnosis method for 1-Ø, 5-level CHMLI is discussed. For simulation data, we have taken output voltage waveforms of the inverter under different fault conditions. To be evaluate the effectiveness of proposed PPCA-SVM method, we considered the simple faults as well as complicated faults. The parameters taken to test the proposed diagnosis method shows in TABLE II the switching frequency f_c is 20kHz, the m_a is varied from 0.5 to 0.95 to take data samples, the measuring time and the sampling time in each simulation are 0.2s and 40µs respectively, so that sampling number is 501.

TABLE II. SIMULATION & EXPERIMENTAL PARAMETERS

Simulation Parameters		
	Description	Values
V_{dc}	DC-link Voltage	100 V
f_r	Fundamental Frequency	50Hz
R_{load}	Resistive Load	50Ω
Experimental Parameters		
V_{dc}	DC-link Voltage	50V
R_{load}	Load	50Ω
f_r	Switching Frequency	25kHz

To verify the fault diagnosis performance of the PPCA-SVM, (1515*501) simulation samples are taken from the output voltage signals of the CHMLI. The value 501 is the number of samples taken by varying the ma from 0.5 to 1, and the value 1515 is depends on number of faults considered in the inverter, here 15 faults are taken to test more accurate performance of the proposed technique. To test the effectiveness of the PPCA-SVM fault diagnosis, Gaussian noise 10% is add to the simulation samples.

FFT analysis is used for signal preprocessing of the voltage signals getting from the inverter. The first PC of PCA contains 90% of the validation data is contained by the first PC of the PCA, and 92.5% of the validation data contained by the first PC of the PPCA when q=10 PC's. Compared to first PC of PCA, the first Principle Components (PC) of PPCA is more.

2320

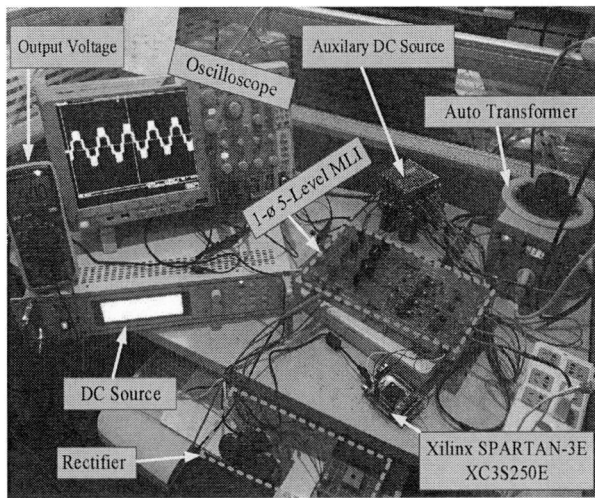

Fig. 3. Experimental System Setup

To validate the proposed PPCA-SVM algorithm, the experimental setup of the CHMLI has been built based on offline FPGA XC3S250E, the experimental system setup is shown in Fig. 3. And the simulation and experimental parameters in TABLE II, and the driver circuit consists is integrated modules is IR21844. MOSFET 17N80C3 combination of power switch transistor and diode is used in this system. In this experiment, PSPWM technique is used to control the

To further more evaluate, the proposed PPCA-SVM method is compared with FFT-PCA-SVM for different kernel functions of the SVM classifier, those are

Quadratic Kernel function and Gaussian Kernel Function. Each method is running for 88 times and the experimental results of the proposed and conventional fault diagnosis techniques are illustrated in TABLE III. The total data given to the classifier is divided into Train & Test Data. Based on different groups of Train Data and Test Data, the total data is categorized into three categories. In each category, the total data is same, but it contains different groups of Train Data and Test Data.

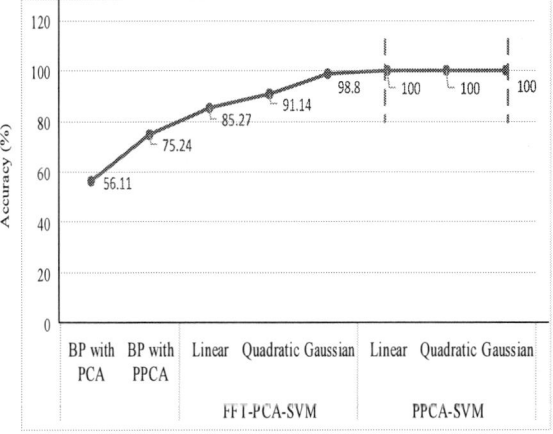

Fig. 4. Fault Diagnosis Accuracy for different methods

TABLE III

COMPARISON OF THE SIMULATION & EXPERIMENTAL RESULTS OF THE DIFFERENT FAULT DIAGNOSIS TECHNIQUES

Categories	Train Data (Tr) = % of Feature Extraction Data	Simulation Results								Exp. Results	
		Back Propagation with PCA		Parameters		FFT-PCA-SVM		PPCA-SVM		PPCA-SVM	
		Accuracy (%)	Time (ms)	Kernel Functions	Kernel Scale	Accuracy (%)	Time (ms)	Accuracy (%)	Time (ms)	Accuracy (%)	Time (s)
I	90%	56.59	32.51	Quadratic	K=1	98.5	637	**100**	3.8	98.4	0.0323
				Gaussian	Fine K=0.79	99.07	172.15	**100**	9.617	95.19	0.0514
					Medium K=3.2	82.1	751.2	**100**	3.8	96.97	0.0424
II	85%	56.11	48.16	Quadratic	K=1	91.14	589.7	**100**	0.044	100	0.0493
				Gaussian	Fine K=0.79	98.8	145.9	**100**	10.527	99.14	0.0642
					Medium K=3.2	89.61	684.7	**100**	0.983	99.57	0.0287
III	75%	55.28	51.56	Quadratic	K=1	96.7	154.5	**100**	0.381	100	0.0269
				Gaussian	Fine K=0.79	99.02	157.6	**100**	11.841	98.05	0.0873
					Medium K=3.2	80	171.2	**100**	0.362	98.52	0.0265

Here Quadratic and Gaussian kernel functions are used in SVM, and the range of used kernel parameters is (0.06 to 1.5), fault diagnosis accuracy is 100%, with the increase of kernel parameters. Therefore, for all used SVM kernel functions, the fault diagnosis kernel parameter is initialized to 0.7. In the first category, the total data is divided into 90% of Train Data and 10% of Test data. This data is given to the BP & SVM classifiers. For PCA-BP classifier, the average diagnostic validation accuracy is 56.59% and the time taken to diagnosis the process is 32.51ms. In FFT-PCA-SVM, the fault diagnosis accuracy & the diagnosis time for quadratic kernel function and Gaussian kernel function are 98.5% & 637ms and 82.1% & 751.2ms respectively. In PPCA-SVM, the fault diagnosis accuracy is maximum (that is 100%) for all kernel functions compared to the existing methods as shown in Fig.4, by using PPCA-SVM, the fault diagnosis time is also reduced compared to other existing methods.

In TABLE III, it is clearly shows that in every category compared to PCA-BP & Fast Fourier Transform-PCA-Support Vector Machine, the proposed PPCA-SVM method improved the diagnosis accuracy of 100% and reduced the diagnosis process time for different kernel functions of SVM.

IV. CONCLUSIONS

In this paper, a PPCA-SVM algorithm-based fault diagnosis technique is proposed for CHMLI. To evaluate and the effectiveness of the proposed method, PPCA-SVM is compared with PCA-BP and Fast Fourier Transform-PCA-Support Vector Machine, also compared with different kernel functions of the SVM classifier. Dimension reduction of PPCA data, PPCA feature extraction is introduced. The proposed technique gives an accurate & fast diagnosis for not only single fault occur at a time, also for more than two faults occur a time. Thus, the operation of the inverter is continuous and improve the reliability of the CHMLI. The PPCA feature extraction with SVM given better accurate fault diagnosis and reduce the diagnosis time compared to PCA feature extraction with SVM classifier. Experiment is also conducted to validate the proposed method. The experimental & simulation results are presented in Table III are conclude the Proposed PPCA-SVM fault diagnosis gives maximum accurate fault diagnosis & reduced the fault diagnosis process time compared to all the conventional methods.

ACKNOWLEDGMENT

This work was supported by the State Key Program of National Natural Science Foundation of China under Grant No. 51637004, The National Natural Science Foundation of China under Grant No. 51577046,

Equipment Research Project in Advance Grant No. 41402040301, The National Key Research and Development Plan "Important Scientific Instruments and Equipment Development" Grant No. 2016YFF0102200.

REFERENCES

[1] S. Mariethoz, "Systematic Design of High-Performance Hybrid Cascaded Multilevel Inverters With Active Voltage Balance and Minimum Switching Losses," *IEEE Transactions on Power Electronics*, vol. 28, pp. 3100-3113, 2013.

[2] U. R. Prasanna and A. K. Rathore, "Dual Three-Pulse Modulation-Based High-Frequency Pulsating DC Link Two-Stage Three-Phase Inverter for Electric/Hybrid/Fuel Cell Vehicles Applications," *IEEE Journal of Emerging and Selected Topics in Power Electronics*, vol. 2, pp. 477-486, 2014.

[3] F. W. Fuchs, "Some Diagnosis Methods for Voltage Source Inverters in Variable Speed Drives with Induction Machines - A Survey," in *Industrial Electronics Society, 2003. IECON '03. The 29th Annual Conference of the IEEE*, vol. 2, pp. 1378-1385, 2003.

[4] T. Wang, J. Qi, H. Xu, Y. Wang, L. Liu, and D. Gao, "Fault Diagnosis Method Based on FFT-RPCA-SVM for Cascaded-Multilevel Inverter," *ISA Transactions*, vol. 60, pp. 156-163, 2016.

[5] R. L. d. A. Ribeiro, C. B. Jacobina, E. R. C. d. Silva, and A. M. N. Lima, "Fault Detection of Open-Switch Damage in Voltage-Fed PWM Motor Drive Systems," *IEEE Transactions on Power Electronics*, vol. 18, pp. 587-593, 2003.

[6] D. U. Campos-Delgado, J. A. Pecina-Sanchez, D. R. Espinoza-Trejo, and E. R. Arce-Santana, "Diagnosis of Open-Switch Faults in Variable Speed Drives by Stator Current Analysis and Pattern Recognition," *IET Electric Power Applications*, vol. 7, pp. 509-522, 2013.

[7] M. Alavi, D. Wang, and M. Luo, "Short-Circuit Fault Diagnosis for Three-Phase Inverters Based on Voltage-Space Patterns," *IEEE Transactions on Industrial Electronics*, vol. 61, pp. 5558-5569, 2014.

[8] R. A. Keswani, H. M. Suryawanshi, and M. S. Ballal, "Multi-Resolution Analysis for Converter Switch Faults Identification," *IET Power Electronics*, vol. 8, pp. 783-792, 2015.

[9] M. Aminian and F. Aminian, "A Modular Fault-Diagnostic System for Analog Electronic Circuits Using Neural Networks With Wavelet Transform as a Preprocessor," *IEEE Transactions on Instrumentation and Measurement*, vol. 56, pp. 1546-1554, 2007.

[10] J. Chavarria, D. Biel, F. Guinjoan, C. Meza, and J. J. Negroni, "Energy-Balance Control of PV Cascaded Multilevel Grid-Connected Inverters Under Level-Shifted and Phase-Shifted PWMs," *IEEE Transactions on Industrial Electronics*, vol. 60, pp. 98-111, 2013.

[11] B. Li, R. Yang, D. Xu, G. Wang, W. Wang, and D. Xu, "Analysis of the Phase-Shifted Carrier Modulation for Modular Multilevel Converters," *IEEE Transactions on Power Electronics*, vol. 30, pp. 297-310, 2015.

[12] M. E. T. a. C. M. Bishop. (2006, Mixtures of Probabilistic Principle Component Analysers. MIT Press, 443-482.

[13] M. E. T. C. M. Bishop.: Probabilistic Principal Component Analysis. Journal of the Royal Statistical Society, pp. 13, sept 1999.

[14] S. Benameur, M. Mignotte, F. Destrempes, and J. A. D. Guise. "Three-Dimensional Biplanar Reconstruction of Scoliotic Rib Cage Using The Estimation of A Mixture of Probabilistic Prior Models". *IEEE Transactions on Biomedical Engineering*, vol. 52, pp. 1713-1728, 2005.

[15] E. Jon, K. Dong Kook, and K. Nam Soo.: Robust Correlation Estimation for EMAP-Based Speaker Adaptation. IEEE Signal Processing Letters, vol. 8, pp. 184-186, 2001.

[16] K. Dong Kook and K. Nam Soo.: Rapid Speaker Adaptation Using Probabilistic Principal Component Analysis. IEEE Signal Processing Letters, vol. 8, pp. 180-183, 2001.

[17] S.-W. Lin, K.-C. Ying, S.-C. Chen, and Z.-J. Lee, "Particle Swarm Optimization for Parameter Determination and Feature Selection of Support Vector Machines," *Expert Systems with Applications,* vol. 35, pp. 1817-1824, 11// 2008.

[18] C. Delpha, H. Chen, and D. Diallo, "SVM Based Diagnosis of Inverter Fed Induction Machine Drive: A New Challenge," in *IECON 2012 - 38th Annual Conference on IEEE Industrial Electronics Society*, pp. 3931-3936,2012.

[19] S. Yin, X. Zhu, and C. Jing, "Fault Detection Based On A Robust One Class Support Vector Machine," *Neurocomputing,* vol. 145, pp. 263-268, 12/5/ 2014.

The 2018 International Power Electronics Conference

A Study on Edge Supported Electromagnetic Levitation System: Fundamental Consideration on Levitation Performance of Thin Steel Plate

Yoshiho Oda[1], Yasuaki Ito[1*], Kengo Okuno[1], Masahiro Kida[1], Toshiki Suzuki[1],
Takayoshi Narita[2*], Hideaki Kato[2] and Hiroyuki Moriyama[2]
1 Course of Mech. Eng., Tokai University, 4-1-1 Kitakaname, Hiratsuka-shi, 259-1292, Japan
2 Dep. of Prime Mover Eng., Tokai University, 4-1-1 Kitakaname, Hiratsuka-shi, 259-1292, Japan
*narita@tsc.u-tokai.ac.jp

Abstract— Recent years, high quality thin steel plate is used for many industrial products. A production line of thin steel plate includes a transport system utilizing the friction force induced by contact. However, the deteriorates of surface quality from contact with rollers is a problem. As a solution to this problem, non-contact conveyance systems applying electromagnetic levitation technology have been proposed. From the previous study, control horizontal displacement of thin steel plate using electromagnet. Using horizontal displacement control, we confirmed improvement performance of magnetic levitation system. By using an electromagnet installed horizontally direction, a vertical force to support the steel plate and a horizontal force to suppress elastic vibration act on the steel plate. Focusing on these forces, we have proposed a magnetic levitation system for a steel plate with simultaneous horizontal positioning control and vertical non-contact support using a horizontal electromagnet. In this study, we carried out levitation experiment using magnetic levitation system. We verify levitating characteristic of thin steel plate.

Keywords— *Electromagnet, Thin steel plate, edge supported, magnetic lavetation.*

I. INTRODUCTION

A production line of industrial products which includes many rollers transports utilizing the friction force induced by contact. However, the deteriorates of surface quality from contact is a problem. Thin steel plate is one of the industrial products. As a solution of this problem, in the thin steel plate production line, non-contact conveyance systems applying electromagnetic levitation technology have been proposed [1-3]. Author installed an electromagnet not only vertical direction but also horizontal direction. Using horizontal displacement control, we confirmed improvement performance of magnetic levitation system [4,5]. By using an electromagnet installed horizontally direction a vertical force to support the steel plate and a horizontal force to suppress elastic vibration act on the steel plate. Focusing on these forces, we have proposed a magnetic levitation system for a steel plate with simultaneous horizontal positioning control and vertical non-contact support using

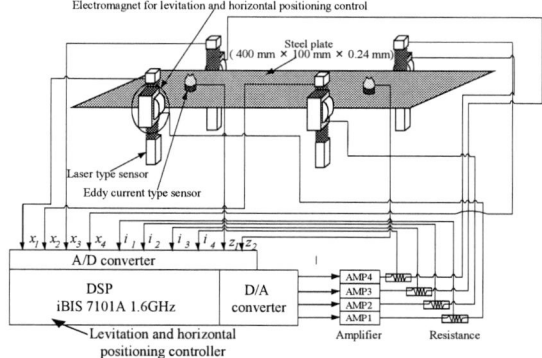

Fig. 1. Electromagnetic levitation system for steel plate using only electromagnets installed in horizontal direction.

a horizontal electromagnet. Author carried out electromagnetic field analyze and experiment about static deflection. We confirmed that the proposed system generated an enough suction force to levitate a steel plate [6]. Author made trial experimental device and have successfully levitating thin steel plate with a thickness of 0.24 mm [7]. In the previous study, analysis shown proposed magnetic levitation system generated a suspension force is effectively generated as the thickness of the steel plate is thin. However, we have not experimental considered about thinner flexible steel plate. In this study, we considered thin steel plate with a thickness of 0.19 mm which is thinner and more flexible than previous study. We carried out levitation experiment based on support characteristic obtained from electromagnetic field analysis using finite element method(FEM). We considered levitation characteristic and verify the effectiveness of proposed system.

II. EDGE SUPPORTED LEVITATION SYSTEM

Figure 1 shows the edge supported electromagnetic levitation system and Fig. 2 shows a photograph of the electromagnetic levitation system during the levitation of a steel plate. Figure 3 shows photograph of electromagnetic levitation system from the top view. This

2324

The 2018 International Power Electronics Conference

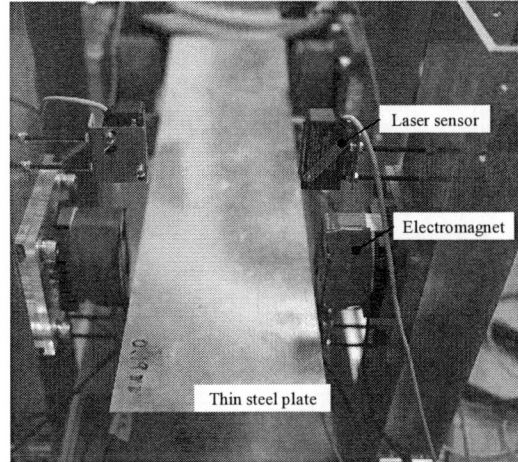

Fig. 2. Photograph of electromagnetic levitation system during the levitation.

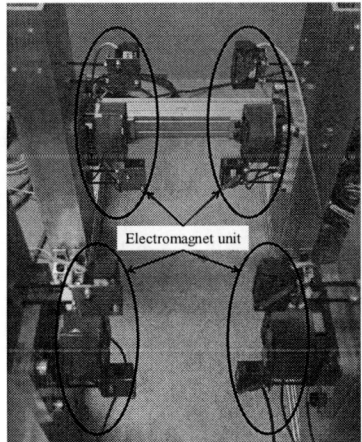

Fig. 3. Photograpf of electromagnetic levitation system from the top view.

Fig. 4. Unit of electromagnet.

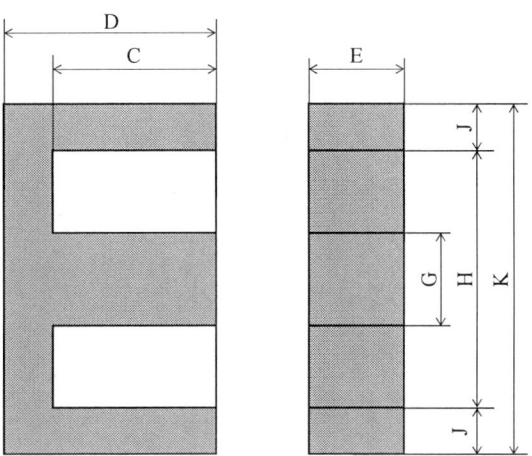

Fig. 5. Schematic illustration of E type core.

TABLE I
SPECIFICATION OF ELECTROMAGNET.

Symbol	Value
C	28×10^{-3} m
D	36×10^{-3} m
E	16×10^{-3} m
G	16×10^{-3} m
H	44×10^{-3} m
J	8×10^{-3} m
K	60×10^{-3} m
Resistance	$10\ \Omega$

device consists of four electromagnetic unit. Figure 4 shows one of the electromagnetic unit. One electromagnetic unit is installed to one pillar. Electromagnetic unit are consisting of one laser sensor and one electromagnet. The levitating object is a rectangular galvanized steel plate (material SS400) whose length is 400 mm, width is 100 mm, and thickness is 0.19 mm. Two electromagnets facing each other are installed in the longitudinal direction near the edge of the thin steel plate. Four laser sensors (which measure the displacement by the cutoff amount of a belt-like laser beam) are used to measure the horizontal displacement of the edge of the steel plate. Thereby, the steel plate is controlled by non-contact positioning by the electromagnets at a distance of 5 mm from the edge of the steel plate. Furthermore, the control law is calculated by detecting the current flowing in each electromagnet from the measured external resistance and inputting a total of eight measurement values into a digital signal processor from an A/D converter. One of the e-type core installed in this system is schematically shown in Fig.5. Table 1 shows each parameter of electromagnet. The E-type core material is ferrite with an enamel wire of 0.5 mm diameter wound around it 1005 times. To measure the displacement of the steel plate to evaluate it's levitation state, eddy current type non-contact displacement sensors are installed as shown in Fig. 6.

2325

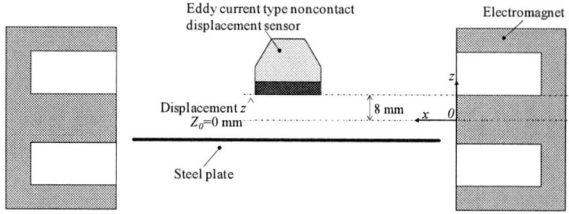

Fig. 6. Position of eddy current type noncontact displacement sensor.

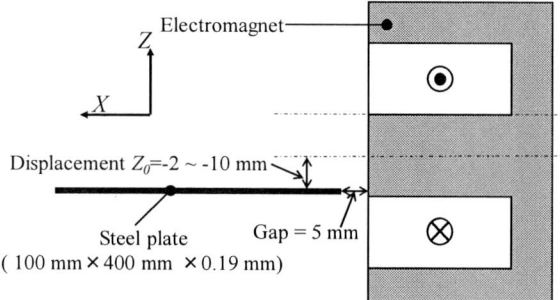

Fig. 7. Electromagnetic field analysis model.

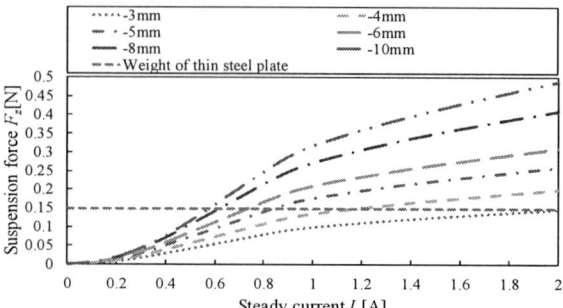

Fig. 8. Relationship between vertical attractive force for each displacement.

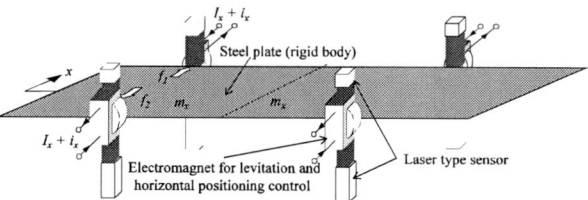

Fig. 9. Experimental model of electromagnetic suspension force.

III. SUSPENSION FORCE ANALYSIS OF ELECTROMAGNET USING FINITE ELEMENT METHOD

A. FE model and analysis condition

We performed suspension force analysis using FEM. We considered support characteristic of thin steel plate with a thickness of 0.19 mm. Analysis model shows Fig. 7. Electromagnet used same one as the electromagnetic levitation system in previous study [5]. The levitating object is a rectangular galvanized steel plate (material SS400) whose length is 400 mm, width is 100 mm, and thickness is 0.19 mm. Analysis condition decided as follows. Range of steady current flowing in electromagnet is 0.1A to 2.0A. The distance between electromagnet core surface and edge of steel plate is Gap. Gap is 5 mm. Steel plate displacement of vertical direction is Z_0. Range of Z_0 is -3mm to -10mm.

B. Analitical result by FEM

Fig. 6 shows relationship between steady current I_x and suspension force F_z of vertical direction. Dashed line （0.146 N） in the figure shows steel plate weight divided to 4. It is possible to levitate the steel plate at a steady current value corresponding steel plate weight. From analytical result in Fig. 7, we clarified steel plate is possible to levitate at analysis condition within this chapter.

IV. CONTROL MODEL

Although the flexible thin steel plate exhibits elastic vibration in the vertical direction, it can be regarded as a rigid body in the horizontal direction. The proposed system virtually divides the steel plate into two parts as shown in Fig. 7. We model the motion of the steel plate in the horizontal direction using a 1-DOF model that actively controls each part. The same static attractive force is applied by the two electromagnets installed to sandwich the steel plate, and the equilibrium position of the steel plate is at the same distance from each electromagnet. The displacement of the steel plate from equations and circuit equations are given by Eqs. (1) - (4). Also, the attractive force of the electromagnets at the equilibrium point was linearized.

$$m_x \ddot{x} = f_1 - f_2 = f_x \tag{1}$$

$$f_x = \frac{4F_x}{X_0}x + \frac{4F_x}{I_x}i_x \tag{2}$$

$$\frac{d}{dt}i_x = -\frac{L_{xeff}}{L_x}\cdot\frac{I_x}{X_0^2}\dot{x} - \frac{R_x}{2L_x}i_x + \frac{1}{2L_x}v_x \tag{3}$$

$$L_x = \frac{L_{xeff}}{X_0} + L_{xlea} \tag{4}$$

Using the state vector, Eqs. (1)-(4) are written as the following state equations:

$$\dot{x} = A_x\,x + B_x\,v_x \tag{5}$$

$$x = \begin{bmatrix} x & \dot{x} & i_x \end{bmatrix}^{\mathrm{T}},$$

$$A_x = \begin{bmatrix} 0 & 1 & 0 \\ \dfrac{4F_x}{m_x X_0} & 0 & \dfrac{4F_x}{m_x I_x} \\ 0 & -\dfrac{L_{xeff}}{L_x}\cdot\dfrac{I_x}{X_0^2} & -\dfrac{R_x}{2L_x} \end{bmatrix},$$

$$B_x = \begin{bmatrix} 0 & 0 & \dfrac{1}{2L_x} \end{bmatrix}^{\mathrm{T}}$$

TABLE □
FEEDBACK GAIN OF F_x

f_x	f_v	f_i
1.04×10^3	3.66×10^2	4.2×10^1

TABLE □
COEFFICIENT OF CONTROL MODE

Symbol	Value
X_0	5×10^{-3} m
L_{xeff}	1.25×10^{-5} H
L_{xlea}	1.89×10^{-1} H
L_x	1.92×10^{-1} H
R_x	10 Ω
M_x	2.96×10^{-2} kg

where F_x: magnetic force of the electromagnets in the equilibrium state [N], X_0: gap between steel plate and electromagnet in the equilibrium state [m], I_x: current of the electromagnets in the equilibrium state [A], i_x: dynamic current of the electromagnets [A], L_x: inductance of one electromagnet coil in the equilibrium state [H], R_x: resistance of the electromagnets coils [Ω], v_x: dynamic voltage of the electromagnets [V], L_{xeff}/X_0: effective inductance of the one electromagnet coil [H], and L_{xlea}: leakage inductance of the one magnet coil [H].

V_x is obtained from feedback on state variable x as following equation.

$$v_x = -F_x x$$
$$F_x = [f_x \quad f_v \quad f_i] \tag{6}$$

Feedback gain F_x at equation (6) was obtained by trial and error. Feedback gain F_x shows at table 2. Each parameter shows at table 3.

V. LEVITATION EXPERIMENT

We carried out levitation experiment using thin steel plate with a thickness of 0.19 mm. When the steady current was 1.0 A, we could levitate a steel plate in the non-contact mode. We measured the vertical and horizontal displacements when the steel plate was levitating, and the standard deviation of the displacement was calculated in both directions. Figure 8 shows the time history of the displacement of the steel plate in the horizontal direction. Figure 9 shows the time history of the displacement of the steel plate in the vertical direction. The standard deviation in the horizontal direction was 0.0551 mm, which was suppressed to 0.1 mm or less. We confirmed from the standard deviation of the displacement and the time history of the displacement in the horizontal direction that the position of the steel plate in the horizontal direction can be controlled. It was

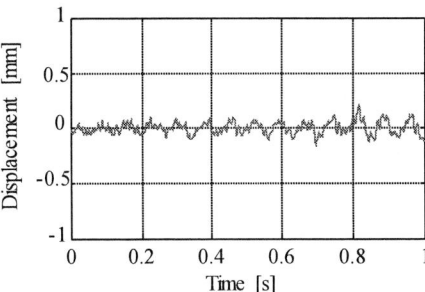

Fig. 10. Time history of displacement of the steel plate in horizontal direction.

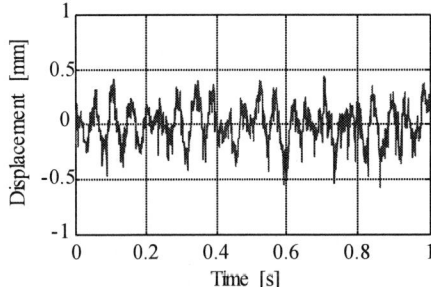

Fig. 11. Time history of displacement of the steel plate in vertical direction.

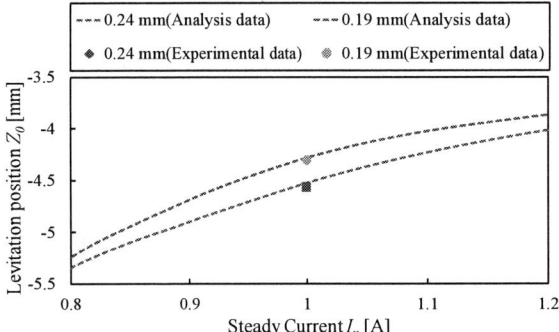

Fig. 12. Relationship between steady current Ix and levitation position Z_0

confirmed that sustained levitation can be achieved from the time history of the displacement in the vertical direction. We confirmed that it is possible to levitate thin and flexible thin steel plate with a thickness of 0.19 mm by using proposed equipment.

VI. MEASUREMENT EXPERIMENTAL OF STEEL PLATE LEVITATING POSITION IN VERTICAL DIRECTION

Steady current is set to 1.0A, and the other experimental conditions are same at chapter 4. We measured steel plate levitating position of Z direction. The levitation position was measured five time for each condition. Those average of result is experimental value. Figure 12 shows measured the levitation position about thin steel plate with a thickness of 0.19 mm. Furthermore, result of the levitation position about thin steel plate with a thickness of 0.24 mm is also shown in Fig. 12.

Furthermore, Fig. 12 dashed line shows relationship between displacement Z_0 and steady current I_x of each steel plate obtained from Fig. 7. Levitation position is -4.57 mm at thin steel plate with a thickness of 0.24 mm. Levitation position is -4.31 mm at thin steel plate with a thickness of 0.19 mm. Levitation position of thin steel plate with a thickness of 0.19 mm was compared to the other steel plate levitation position. 板 As a result, levitation position about thin steel plate with a thickness of 0.19 mm increased by 0.26mm.

VII. CONCLUSION

In this study, we carried out electromagnetic field analyze thin steel plate with a thickness of 0.19 mm and levitation experiment based on support characteristic obtained from electromagnetic field analysis. We confirmed that it is possible to levitate thin and flexible thin steel plate with a thickness of 0.19 mm by using proposed equipment. Furthermore, we performed measurement experiment about steel plate stationary levitating position. The stationary levitating position of the steel plate with a thickness of 0.19 mm is upper than 0.24 mm in the case of same steady current. We compared with electromagnetic field analysis. Steel plate levitation position obtained from analysis agreed with experimental result. From this result, proposed magnetic levitation system generated a suspension force is effectively generated as the thickness of the steel plate is thin. These shows effectiveness from experimentally. From the above, we confirmed effectiveness about proposed electromagnetic levitation system.

REFERENCES

[1] Y. Takada, T. Kimura and T. Nakagawa, Influence of Inductance Properties on a Magnetic Levitation for Thin-Steel Plates, *IEEE Trans. on Magnetics*, vol.52, no.11.

[2] K. Matsushima, F. Kato, T. Ohji, K. Amei, M. Sakui, Characteristics of Cogging Force to a Non-Magnetic Thin Plate Created by an AC Ampere Type Linear Maglev Conveyance System, *Journal of JSAEM*, vol. 21, no. 2, pp.296-301, 2013.

[3] F. Sun, K. Oka, Magnetic Suspension Using Variable Flux Path Control Mechanism with Permanent Magnet (Simultaneous Suspension Experiment of Two Iron Balls), *Transactions of the JSME Series C*, vol. 78, no. 792, pp.2771-2780, 2012, (in Japanese).

[4] T. Narita, T. Kurihara and H. Kato, A Study on the stability improvement by electromagnetic force applied to edge of the non-contact gripping thin steel plate, *Mechanical Engineering Journal*, vol. 3, no. 6, pp.15-00376, 2016.

[5] M. Kida, T. Suzuki, Y. Oda, T. Narita, H. Kato and H. Moriyama, Effects on Levitation Characteristics by Plate Thickness of Thin Steel Plate on Magnetic Levitation Transport System Using Horizontal Positioning Control *Transaction of the Magnetics Society of Japan (Special Issues)*, vol.1, no.1, pp. 76-81, 2017 (in Japanese).

[6] T. Narita, M. Kida, T. Suzuki and H. Kato, Study on Electromagnetic Levitation System for Ultrathin Flexible Steel Plate Using Magnetic Field from Horizontal Direction, *Journal of the Magnetic Society of Japan*, Vol. 41, pp. 14-19, 2017.

[7] Y. Oda, M. Kida, T. Suzuki, Y. Ito, A. Endo, L. Xiaojun, T. Narita, H. Kato and H. Moriyama, Edge Supported Electromagnetic Levitation System for Flexible Steel Plate -Experimental Consideration on Levitation Performance-, *Proceeding of 26th MAGDA conference in Kanazawa*, pp. 205-210.

The 2018 International Power Electronics Conference

Application of FACTS Devices for a Dynamic Power System within the USA

Dan Sullivan, Senior Member[1*], Bryan Buterbaugh, Member[1] and Jan Paramalingam
Fuminori Nakamura[2], Akihiro Matsuda[2] and Daisuke Yamanaka[2]
Taichiro Tsuchiya[3]
1 FACTS & HVDC, Mitsubishi Electric Power Products Inc., Warrendale, PA, USA
2 Power Systems Engineering Project Group, Mitsubishi Electric Corporation, Kobe, Japan
3 Power Electronics Department, Toshiba Mitsubishi-Electric Industrial Systems Corporation, Kobe, Japan
*E-mail: dsullivan@ieee.org

Abstract- **This paper presents the application, technology and design of recent Flexible AC Transmission System (FACTS) projects in the USA, including a brief overview of three case study examples that describe recent FACTS applications. Renewable integration, generation retirement, and applied system reliability standards to the U.S. transmission system are currently the primary drivers for most applications.**

Keywords— Flexible AC Transmission Systems (FACTS) Fault Induced Delayed Voltage Recovery (FIDVR), Statci Synchronous Compensator (STATCOM), Static Var Compensator (SVC)

I. STATUS OF THE U.S. POWER SYSTEM

Since the beginning of deregulation and the separation of generation and transmission systems in the electric power industry in the mid 1990s, voltage stability and reactive power-related system restrictions have been a concern for electric utilities. With deregulation came a federal "open access" rule to accommodate competition that requires utilities to accept generation and load sources at any location in the existing transmission system. This open access structure has challenged transmission owners to continually maintain system security and reliability, while at the same time trying to minimize costly power flow congestion in transmission corridors. More recently, there has been significant focus on reducing carbon emissions by decommissioning thermal coal generation plants (Figure 1 indicates over 65 GW actual or planned retirements from 2008 to 2020) and integrating more renewable energy resources into the transmission and distribution power systems.

With increasing levels of renewable penetration and decommissioning of existing thermal plants, the U.S. transmission system is facing increasing challenges with voltage control and stability. Fluctuating power output from renewables combined with weaker network conditions from thermal generation retirements has resulted in difficulties to maintain a reliable and robust network.

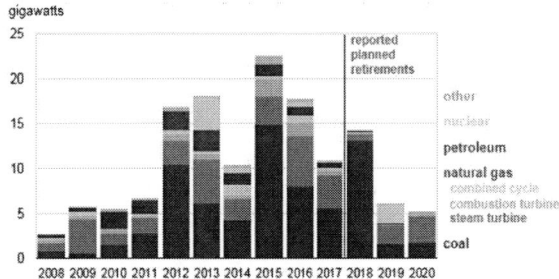

Fig. 1. US utility-scale electric generating capacity retirements (2008-2020) [1]

Reactive power plays an important role in transmission of power over long distances. A transmission system that is able to support its reactive requirements locally is well positioned to handle varying system conditions, system disturbances, changes in generation portfolio, and varying flow patterns.

Electric utilities continue to look to advanced power electronic technologies to provide solutions for the reliable, secure and efficient utilization of power systems.

When voltage instability or congestion problems are observed during the planning study process, cost effective solutions must be considered for such problems. Traditional solutions to congestion and voltage instability problems were to install new costly transmission lines that are often faced with public resistance, or mechanically-switched capacitor banks that have limited benefits for dynamic performance due to switching time and frequency.

One approach to solving this problem is the application of Flexible AC Transmission System (FACTS) technologies, such as the Static Var Compensator (SVC) and Static Synchronous Compensator (STATCOM). FACTS technologies are founded on the rapid control response of power electronic-based reactive power controls.

2329

Power electronics based systems such as FACTS and HVDC technologies provide proven technical solutions to allow for improved power system operation with minimal infrastructure investment, in response to generation retirement, renewable interconnection and enforcement of system reliability.

Furthermore, there is a recent trend in the US power system condition with larger portions of the transmission grid becoming more susceptible to very weak system strengths. This creates unique problems for protection systems and voltage control in this system condition. One region of the US that exhibits this condition is in the western portion of Texas. This region has recently shown signs of rapid load growth for oil and gas processing activities, while nearby regions of Texas have large penetration of wind generation. STATCOM technology can operate better than SVC in weak systems, even up to short ratios of one (this is the ratio of system MVA to FACTS device Mvar capacity).

II. FACTS TECHNOLOGIES

Over the last several years, there were numerous installations of FACTS in the United States [2-4]. FACTS have proven to be cost-effective solutions to a wide range of the power system needs, and have given utilities the option to delay new transmission line construction by increasing capacity on existing lines and/or providing dynamic control and compensation of the system voltages. FACTS systems are available in different forms such as SVC, thyristor controlled series capacitors (TCSCs), STATCOM, and unified power flow controllers (UPFCs).

SVC and STATCOM are categorized as shunt dynamic reactive compensation that rapidly inject or absorb volts-ampere-reactive (vars) to support power system voltage during and immediately following system disturbances.

SVCs employ thyristor-based switching devices that are line-commutated and require the thyristor gate to turn on every half cycle (or as required by the control system). STATCOMs employ insulated gate bipolar transistors (IGBT) or gate commutated thyristors (GCT) that are self-commutated, thereby allowing the switching device to be turned on and off, independent of source voltage. This added control of the STATCOM switching device allows for faster control response when compared to SVC, with potential for faster response time and improved performance in weak power system networks.

While both technologies provide similar function in the power system to regulate the voltage and both perform well for most application, the STATCOM technology has a few advantages that are preferred by transmission owners and operators. For example, STATCOM is more resilient to changes in power system conditions such as variation in harmonic levels, while the SVC may have harmonic filters that need to be evaluated as the power system undergoes any significant changes.

The STATCOM also has a smaller footprint and less outdoor equipment than SVC. In any case, both technologies have performed well for their prospective applications.

Recent transmission STATCOMs have been applied with typical capacities between +/-100 to +/-200 Mvar, with some large units at +/-250 Mvar. SVCs have been applied at capacities up to +500 to +600 Mvar. Connection voltages for transmission SVCs and STATCOMs range from 115 kV to 765 kV in USA.

The case studies discussed in sections III, IV, and V of this paper describe SVC and STATCOM projects executed in the USA by Mitsubishi Electric Power Products Inc. (MEPPI) and their Japan parent company Mitsubishi Electric Corporation. These applications apply power electronic equipment and controls manufactured by Toshiba Mitsubishi Electric Industrial Company (TMEIC).

III. CASE STUDY 1: SVC APPLICATION EXAMPLE

Entergy recently commissioned a +200/-100 Mvar SVC at the Alden 138kV substation in Conroe, Texas (the Western Region of their territory) as shown in Figure 2. Voltage stability studies were performed to examine fault induced delayed voltage recovery (FIDVR) characteristics and address the requirements of North American Electric Reliability Council (NERC) Transmission Planning (TPL) standards.

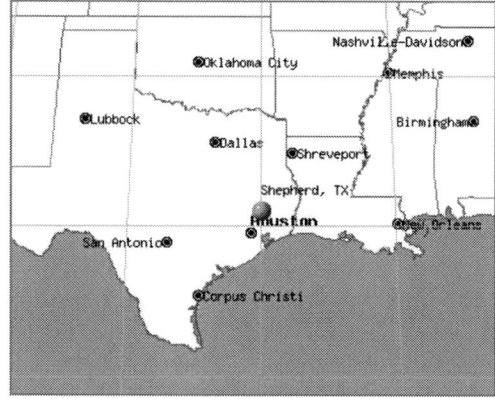

Fig. 2. Conroe Texas; Alden SVC location

Planning studies for years 2016 through 2024 indicated that select system disturbances could lead to a FIDVR event. In the past, FIDVR was alleviated with an undervoltage load shedding scheme; however, the new NERC planning standards no longer allow load shedding for select contingencies. To alleviate the FIDVR issue, a +200/-100 Mvar SVC was recommended at the Alden 138 kV station. Installing an SVC at the Alden station addressed the following objectives:

1) Compliance with NERC criteria
2) Mitigation of FIDVR issues
3) Elimination of existing UVLS scheme
4) Redundancy for existing nearby SVC

2330

The SVC's +200/-100 Mvar continuous rating is derived from one thyristor-controlled reactor (TCR) branch with continuous var absorption capacity of 0 to -165 Mvar, one 135 Mvar thyristor-switched capacitor (TSC) branch, and 65 Mvar of harmonic filters. A one-line diagram of the Alden SVC is shown in Figure 3.

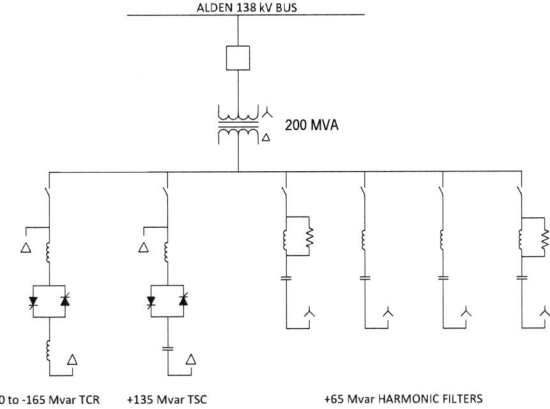

Fig. 3. Alden SVC simplified one-line diagram

A major part of the SVC design studies is the harmonic performance analysis, which addresses the design of the ac harmonic filters required for the SVC. The filter configuration and capacity are designed to limit harmonic voltage distortion at the high voltage point of connection within the required values. The harmonic filters are also designed so that no resonance is created between the power system and the SVC. This analysis is carried out in two stages. The first stage is the calculation of the harmonic sectors (system harmonic impedance envelope) based on modeling of the transmission system around the Alden 138 kV bus. The second stage applies the result of stage 1 with the operation of the SVC to ensure harmonic filters effectively meet the harmonic performance requirements. The harmonic distortion levels at the point of interconnection were calculated at a maximum of 2.45% total harmonic distortion (THD), which is within the specified requirements.

Stability analyses were performed to re-examine key cases that Entergy identified during initial planning studies to verify the SVC controls the power system's dynamic performance. The transient analyses were performed to determine and validate key control parameters and evaluate the impact of system disturbances on the SVC electrical equipment. A step response test analysis was performed in accordance with IEEE Std. 1031-2011 to ensure proper control stability, verity the SVC's control parameter settings, and confirm the response time is less than 50 ms. The response time must be achieved for both minimum and maximum system strengths (1284 MVA to 3046 MVA). In general, the SVC response time decreases and overshoot/settling time increases with decreasing system short-circuit power.

The Alden SVC is designed with the following main control functions:

• *Dynamic Voltage Control*: This control is implemented by the AVR control based on three-phase average power frequency voltage. In this control function, the SVC output closely follows the demand relative to power system voltage.

• *Steady-State Voltage Control:* The steady-state output of the SVC is controlled based on a floating voltage reference level (VREF) with upper and lower voltage limits. This operates in coordination with super-imposed reactive output power control. In steady-state voltage control, the SVC reactive power output slowly returns to a preset steady-state value (QREF) so that a significant amount of its reactive capacity is held in reserve.

• *Coordinated Control of 138 kV MSCs:* This control coordinates the operation of local (on the 138kV Alden Bus) and remote MSCs based on the steady-state voltage control function.

IV. Case Study 2: SVC Application Example

The Indianapolis Power & Light Company (IPL) commissioned a +300/-100 Mvar SVC in May of 2016 at the 138 kV Southwest substation In Indianapolis, Indiana as shown in Figure 4. The primary need for the SVC is driven by local generation retirement and resulting symptom of FIDVR of the 138 kV transmission system.

Fig. 4. Indianapolis Indiana; Southwest SVC location

This project was driven by Environmental Protection Agency (EPA) requirements that forced most of the coal generation on IPL's 138 kV transmission system to be retired or converted to natural gas by 2016. This resulted in a transitional year between 2016 and 2017 during which the coal units would be retired and a new combined cycle natural gas generation unit would become operational in 2017. System studies showed that the dynamic voltage response on the 138 kV transmission system would not be within design criteria and that transmission system disturbances could result in a significant loss of load.

The design criteria for the SVC capacity was selected to replace the reactive power capacity of the coal based generation scheduled for retirement. While the SVC capacity was not a one for one replacement, some system disturbances may not be able to completely mitigate the FIDVR issues, resulting in potential load shedding risk during the Gap Year assuming the largest 138 kV generating unit is out-of-service. Following the gap year, additional new generation is scheduled to be installed to help meet the design criteria.

To mitigate the delayed voltage recovery issues for all faults, additional reactive support was considered for the IPL area. The first option was to apply a 300 Mvar fixed capacitor (switched shunts), but found to be ineffective in solving FIDVR issue. The second option applied a 300 Mvar SVC at the Southwest 138 kV bus.

The speed of the SVC's response was found to be faster than that of the generators' excitation systems, thus more effectively preventing motor stalling and improving overall system performance.

Both SVC and STATCOM technologies were considered for the solution. The required ratings for each technology to meet the system contingency criteria was determined. For SVC at 138 kV, a nominal continuous rating of -100 Mvar inductive to +300 Mvar capacitive was required, while STATCOM nominal continuous rating of -100 Mvar inductive to +250 Mvar capacitive (with 125% short-term overload) at 138 kV was considered.

The SVC includes a 300 MVA main coupling transformer, power electronics and passive harmonic filtering devices connected on the secondary bus of the transformer. The power electronics and harmonic filtering devices consist of one -200 Mvar TCR, one +200 Mvar TSC, one set of +100 Mvar harmonic filters consisting of a 5th and 7th harmonic filter. A simplified one-line diagram of the SVC design is shown in Figure 5. Additional detail on this SVC application can be found in [3].

V. Case Study 3: MMC-based STATCOM Application Example

The Outer Banks (OBX) is a 200-mile-long (320 km) string of barrier islands on the east coast of the United States that stretches from southeastern Virginia downward the entire coast of North Carolina as illustrated in Figure 6. The islands separate interior waters from the Atlantic Ocean and have become a major international tourist destination known for its subtropical climate and wide expanse of open beachfront [6]. Population of towns making up the Outer Banks is around 58,000, however during summer months (June – August) the population can increase 3 to 5 times.

The wide variation of load levels (i.e., population) that can exist in the Outer Banks, coupled with the large usage of air-conditioning loads in summer months, results in significantly different power system conditions depending on the time of year. These fluctuating load conditions require careful consideration of power system operation and voltage control, including reinforcement of the power system to manage voltage during and after system disturbances.

In 1997, Dominion Energy installed an SVC at the Colington substation with the purpose of regulating voltage and managing FIDVR caused by the air-conditioning loads. [2, 5] This original SVC developed issues with limited spares availability, reduced reliability and had legacy issues with multiple single points of failure. Additionally, the severe coastal weather and contamination had compromised the integrity of the outdoor equipment.

Dominion Energy evaluated their options and chose to replace the SVC at Colington with STATCOM technology. The primary application and requirement for the STATCOM is dynamic restoration of the 115kV system voltage recovery to 70% of nominal in 2.5 seconds following an N-1-1 system contingency, per Potomac-Jersey-Maryland (PJM) guidelines.

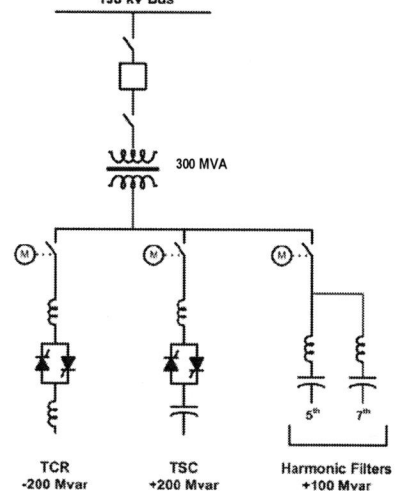

Fig. 5. Southwest SVC simplified one-line diagram

Fig. 6. Outer Banks of North Carolina, USA (Google Earth)

The main purpose of the Colington STATCOM is to:

- Continuously regulate and support the 115kV voltage at the Colington substation in the Outer Banks area of North Carolina under normal and transient conditions of the power system.

- Provide a dynamic reactive power response following an N-1-1 system contingency (meeting NERC and PJM requirements) due to the loss of two 230kV transmission lines, including restoration of the 115kV voltage at the Colington substation to 70% of nominal in 2.5 seconds. This dynamic reactive support is needed so that the transmission system remains stable in the Outer Banks area of North Carolina for FIDVR problems.

A. STATCOM Design

The main purpose of the STATCOM is to continuously regulate and support the 115kV voltage at Colington substation in the Outer Banks area of North Carolina under normal and transient conditions of the power system. Dynamic reactive support is needed so that the transmission system remains stable in the Outer Banks area of North Carolina for FIDVR.

The STATCOM system is connected to the 115 kV transmission system through a 125 MVA, 3-phase transformer.

A simplified one-line diagram is shown in Figure 7 below.

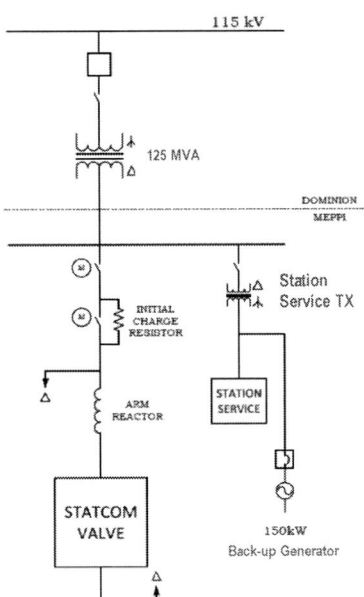

Fig. 7. Colington STATCOM simplified one-line diagram

The Colington STATCOM features the *SVC-Diamond*™ technology that employs Voltage Sourced Converter (VSC), Modular Multilevel Converter (MMC) topology as illustrated in Figure 8 with a continuous rating of +/-125 Mvar.

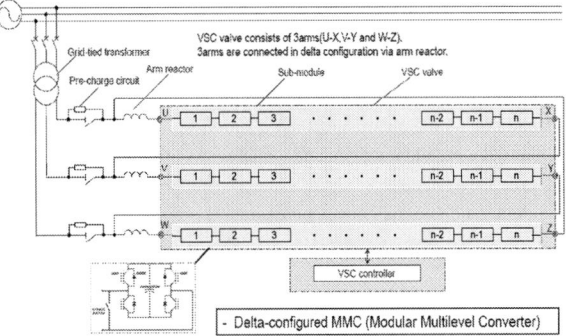

Fig. 8. Colington STATCOM converter configuration

The fundamental control objective of the STATCOM is to maintain a desired voltage at the high-voltage bus (i.e., regulate transmission bus). This is achieved by raising or lowering the inverter bus voltage to either inject or absorb vars with the power system. Capacitive reactive power always flows from higher voltage magnitude to lower voltage magnitude, so the adjustment of the inverter voltage is achieved by sequentially switching sub-modules to create a synthesized voltage waveform that contains low harmonic levels. The control system for the MMC shown in Figure 9 provides gate signals to the submodules whose coordinated operation provides the appropriate inverter voltage, thus resulting in the desired reactive power output.

The VSC controller for this application also has applied supplementary controls for unbalance voltage control, susceptance control, no-load stand-by control to reduce losses near zero output, and cyclical gain adjustment.

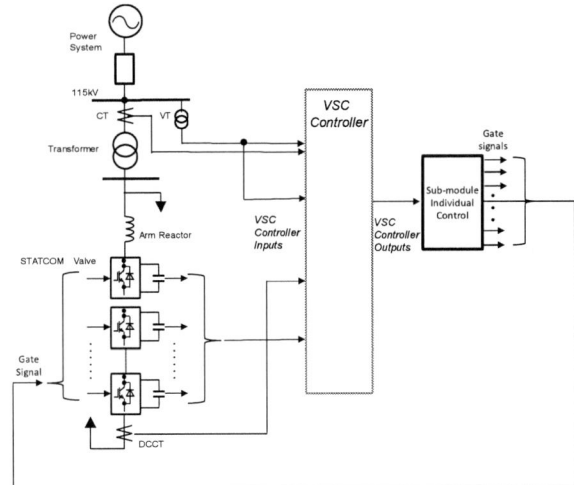

Fig. 9. Simplified SVC-Diamond converter control

B. STATCOM Site and Arrangement

The new STATCOM is be located within the footprint of the SVC equipment as shown in Figure 10 below.

The general scope of replacement project removed all existing SVC equipment (except for control building, which is reused for storage), and install new STATCOM. Dominion Energy also chose to connect the new STATCOM at a new position within the substation (connect to transmission line as opposed to substation bus voltage).

Fig. 10. Colington substation with old SVC

The Colington STATCOM site is exposed to severe weather conditions, resulting in flooding or storm surge and heavy salt contamination. To address flooding, elevated structures were used to keep all main equipment and the control building three feet above grade. Stainless steel was used for all outdoor control cabinets, enclosures, and heat exchangers.

A key feature of the STATCOM is the reduced amount of outdoor substation equipment such as capacitors and reactors when compared to the SVC one-line diagrams in Section III and IV of this paper. This is one of the inherent benefits associated with the STATCOM. The new STATCOM installation is shown in Figure 11, and was successfully placed into commercial operation on June 15, 2017.

Fig. 11. Colington substation with STATCOM

VI. Conclusion

This paper has described the current status of the application of FACTS devices such as SVC and STATCOM in United States. Renewable integration, generation retirement, and applied system reliability standards are currently the primary drivers for applications of SVCs and STATCOMs. This paper presents three case study examples to demonstrate recent FACTS applications. While the SVC is a mature solution, STATCOM technology has recently come to the forefront with its inherent benefits. In any case, both technologies have performed well for their prospective applications.

VII. References

[1] US Energy Information Administration (eia) website, https://www.eia.gov/todayinenergy/detail.php?id=34452

[2] D. Sullivan, R. Pape, J. Birsa, et al "Managing Fault-Induced Delayed Voltage Recovery in Metro Atlanta with the Barrow County SVC," FACTS Panel Session, IEEE PES Power Systems Conference and Exposition, Seattle, WA, March 2009.

[3] Grainger, B., Reed, G.F., Kempker, M., Sullivan, D.J., et al, "Technical Requirements and Design of the Indianapolis Power & Light 138 kV Southwest Static Var Compensator," IEEE PES T&D Conference and Exposition, Dallas, TX, May 2016.

[4] G. Reed, J. O'Connor, S. Varadan, "Power System Planning Analysis and Functional Requirements of the Progress Energy Carolinas Jacksonville Static VAR Compensator," FACTS/Power Electronics Panel Session, IEEE PES

[5] L. Taylor, S. Hsu, "Transmission Voltage Recovery Following a Fault Event in Metro Atlanta Area," IEEE PES Summer Mtg., Seattle Washington, July 2000.

[6] Wikipedia, https://en.wikipedia.org/wiki/Outer_Banks

Capacitor Voltage Balancing in Semi-Full-Bridge Submodule with Differential-Mode Choke

(Invited Paper)

Kalle Ilves, Yuhei Okazaki, Nan Chen, Muhammad Nawaz, Antonios Antonopoulos

ABB Coporate Research, ABB AB, Västerås, Sweden

*E-mail: kalle.ilves@se.abb.com

Abstract—This paper presents a semi-full-bridge submodule implementation with an integrated differential-mode choke for modular multilevel converters. The semi-full-bridge submodule comprises two capacitors that are connected in either series or parallel to generate a positive voltage, while the capacitors are connected only in parallel to generate a negative voltage. A capacitor voltage difference appearing between two capacitors during the series connection period is inevitable due to the capacitance mismatch. This results in a large current spike during the parallel connection period. The paper evaluates the effect of the differential-mode choke, increased modulation index, and second-order harmonic injection. Numerical analysis indicate that the differential-mode choke can provide significant reduction in the peak current during parallel connection of the capacitors. The numerical results are validated by simulation results from a single semi-full-bridge module as well as a full converter. It is concluded that with proper control and dimensioning of the submodule capacitors and differential-mode choke, the current-spike can be reduced to levels in the same range as the peak arm current.

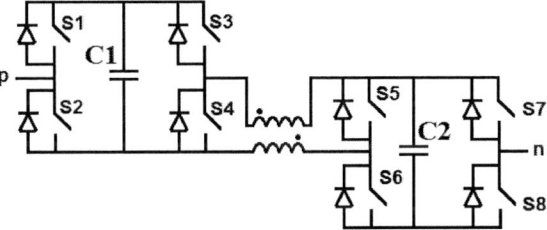

Fig. 1. Semi-full-bridge with magnetically coupled inductors forming a differential-mode choke.

(SFB) submodule can offer reasonable compromise in increased power rating of power devices without loosing the above mentioned two functionalities offered by an all full-bridge implementation [9].

I. INTRODUCTION

The modular multilevel converter (MMC) have been paid attention from industry and academia for various high-voltage and high-power applications. The MMC consists of multiple submodules connected in series, each submodule of which is typically either half- or full-bridge converter cell. The full-bridge submodule offers (1) dc-fault current blocking and (2) higher alternating- to direct-voltage ratio, leading to elimination of dc breaker [1] and reduced submodule capacitance [2], [3]. The total power rating of power devices in the full-bridge is, however, doubled compared to the half-bridge submodule, resulting in increased conduction losses.

Several submodule configurations have been investigated in the literature to find the optimum submodule configuration having similar advantages to full-bridge submodule but reduced total power rating of devices. For example, double-clamp submodules [1], [4] allow dc-fault blocking with fewer devices compared to the full-bridge solution. Other possible solutions include various types of asymmetric full-bridge cells [5], [6] or mixed cell configurations combining half- and full-bridge submodules [7], [8]. Such solutions can provide similar advantages to full-bridge MMC with reduced number of power devices. However, the mixed cell configuration is limited in its operating range due to the energy balancing between half-bridge and full-bridge submodules. The semi-full-bridge

One of the challenges of the SFB is that it has to deal with parallel connection of two capacitors. The main challenge here is that even a relatively small voltage difference between the two capacitors can result in a large current spike. In [10] a numerical and experimental evaluation of the balancing current at parallel connection is presented. It is found that at certain conditions the resulting current-spike may exceed the rated value of the repetitive peak collector current of the devices. Therefore, this paper proposes a differential-mode choke to be used, as shown in Fig. 1. The aim of the choke is to reduce the peak current when the capacitors are connected in parallel. To further reduce the peak-current during the parallel connection, the effect of the differential-mode choke is evaluated in combination with second-order harmonic injection, which can reduce the voltage ripple and thereby the voltage imbalance between the capacitors in the SFB.

The outline of this paper is as follows. In Section II the semi-full-bridge with differential-mode choke and its principles of operation are described. In Section III the voltage difference which can be expected between the capacitors is evaluated and the impact of harmonic injection is discussed. This is followed by a brief discussion on the resulting peak-current when the capacitors are connected in parallel in Section IV. The theoretical results are then compared with simulation results in V.

TABLE I. SWITCHING STATES OF PROPOSED SFB

Voltage	S1	S2	S3	S4	S5	S6	S7	S8
$-v_c$	0	1	1	0	0	1	1	0
v_c	1	0	1	0	0	1	0	1
0	0	1	0	1	1	0	1	0
$2v_c$	1	0	0	1	1	0	0	1

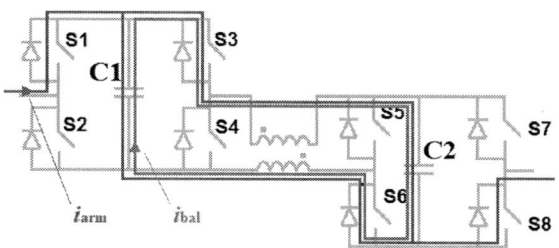

Fig. 2. Arm current (red) and balancing current (blue) in semi-full-bridge with magnetically coupled inductors.

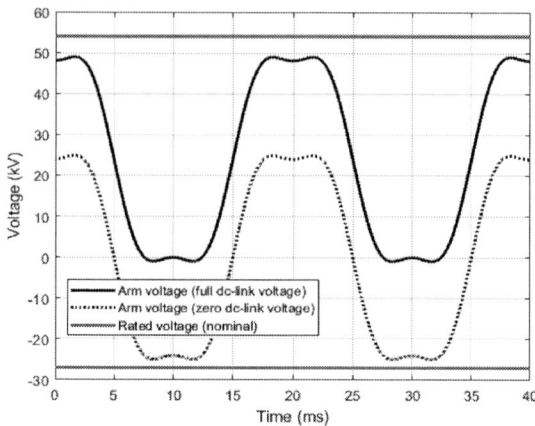

Fig. 3. Arm voltage at normal operation with full dc-link voltage (black, solid) and during fault conditions with zero dc-link voltage (black, dotted) including available positive and negative voltage in converter arms (red, solid).

II. SFB WITH DIFFERENTIAL-MODE CHOKE

Fig. 1 shows a semi-full-bridge configuration utilizing a differential-mode choke. It is observed that for this configuration, the middle device which is shared between the two full-bridges need to be divided into two switches (S_4 and S_5), thus forming a double-connection of FB modules, or two-port cells, as discussed in [11]–[13]. Similar to S_3 and S_6, the switches S_4 and S_5 only conduct half of the arm current. Furthermore, since the devices S_4 and S_5 are connected in parallel through the differential-mode choke, both the conduction losses and installed silicon in the proposed semi-full-bridge implementation are similar to that in [9].

Table I lists the combinations of switching states used to insert the available voltage levels. Similar to the implementation in [9], the proposed SFB can produce $\pm v_c$, 0, and $+2v_c$. By inspecting Table I and the circuit illustrated in Fig. 1, it can be concluded that each of the two magnetically coupled inductors in Fig. 1 always carry half of the arm current. Accordingly, the differential mode choke is not part of a commutation loop for the switching states listed in Table I and it will therefore not affect the switching losses.

When the capacitors are connected in series to generate $2v_c$, the capacitor voltages will diverge if the capacitance of C_1 is not identical to C_2. In the following study, a spread of $\pm5\%$ in the capacitance values is assumed. Any voltage difference between the capacitors will generate a large current spike when the capacitors are connected in parallel. Hereafter, the voltage difference between the two capacitors will be referred to as the voltage imbalance ΔV, and the circulating current between the capacitors at parallel connection, as illustrated in Fig. 2, will be referred to as balancing current.

A. Modulation Index and Negative Voltage-Levels

As the SFB can provide negative voltages it is possible to let the peak-to-peak alternating voltage exceed the dc-link voltage. An example of this is shown in Fig. 3 with

modulation index 1.2. Here, the modulation index m is defined as

$$m = \frac{2\hat{V}_{\text{ac}}}{V_{\text{dc}}} \tag{1}$$

where $\hat{V}_{\text{ac}}$ is the alternating voltage amplitude generated by the converter arms and V_{dc} is the dc-link voltage. This means that if the modulation index according to (1) equals 1.2, each arm must be able to generate a positive voltage corresponding to 102% of the dc-link voltage, if third-order harmonic injection is used. Since the sum of the upper and lower arm voltages must equal the dc-link voltage, this will also require negative voltages to be inserted, corresponding to 2% of the dc-link voltage.

However, in case of a zero dc-link voltage, for example due to a dc-link fault, as much as 52% of the dc-link voltage must be inserted with negative polarity. It should be noted that here the negative voltage which is inserted at zero dc-link voltage corresponds to slightly more than half of the positive voltage which is inserted at full dc-link voltage conditions. Depending on the required modulation margins, nominal operation at modulation index 1.2 can still be possible. That is, if it can be accepted that the modulation margin reduced at zero dc-link voltage conditions. This can be seen in Fig. 3 where the red lines indicate the available positive and negative voltages. The requested arm voltage is within these limits both for the nominal and zero dc-link voltage case. However, the available margin is clearly reduced at zero dc-link voltage.

III. CAPACITOR VOLTAGE IMBALANCE

The resulting peak-current after the parallel connection of the two capacitors is closely related to the voltage difference between the capacitors. This voltage difference is in turn affected by the energy variation in the arm, the size of the submodule capacitors, and the switching frequency. The energy variation can, however, be reduced by increasing the modulation index and injecting a second-order harmonic, as shown in Fig. 4. The second-order harmonic injection which is considered here is defined as

$$i_2 = k_2 m \hat{I}_{ac} \cos(2\omega t - \varphi). \quad (2)$$

If k_2 equals 0.25, the second-order harmonic in the arm energies can be eliminated [14]. However, to reduce the amount of injected current the case with k_2 equal to 0.20 is also considered in the present study.

A. Submodule Capacitance and Energy Storage

The size of the capacitors relates to the nominal energy stored in the converter. Since there are two capacitors per submodule, the energy stored in one SFB module is given by

$$E_{SFB} = C_0 V_0^2 \quad (3)$$

where V_0 is the nominal capacitor voltage. If third-order harmonic injection is used, the alternating voltage generated by each SFB is

$$\hat{V}_{1,SFB} = \frac{2mV_0}{(1 + m\frac{\sqrt{3}}{2})(1 + \delta)} \quad (4)$$

where δ is the modulation margin. Here δ is defined such that its relation to the modulation reference $s_{ref} \in [-0.5, 1.0]$ is given by

$$s_{ref} = \frac{1 + m[\cos(\omega t) - \frac{1}{6}\cos(3\omega t)]}{(1 + \frac{\sqrt{3}}{2}m)(1 + \delta)}. \quad (5)$$

That is, the peak value of s_{ref} multiplied with $(1 + \delta)$ equals unity. It should be noted that here direct modulation is assumed, that is, the modulation reference s_{ref} is not adjusted to compensate for the capacitor voltage ripple or injection of second-order harmonic. This simplification is applied here to limit the number of variables affecting s_{ref}. Accordingly, s_{ref} is only a function of time, modulation index, and modulation margin, which simplifies the analysis.

The apparent power transfer per semi-full-bridge is obtained as half of the product of (4) and the amplitude of the alternating current per arm. That is,

$$S_{SFB} = \frac{1}{2}\frac{mV_0\hat{I}_{ac}}{(1 + m\frac{\sqrt{3}}{2})(1 + \delta)} \quad (6)$$

where $\hat{I}_{ac}$ is the amplitude of the alternating phase current. The nominal energy storage per power transfer can then be expressed as the ratio of (3) and (6). That is,

$$\tau = \frac{C_0 V_0 2(1 + m\frac{\sqrt{3}}{2})(1 + \delta)}{m\hat{I}_{ac}}. \quad (7)$$

The nominal energy storage τ is closely related to the capacitor voltage ripple in the converter cells. As the SFB modules insert the requested voltage in the arm, the product of the inserted voltage and arm current corresponds to the power transfer to or from the submodule capacitors, causing a ripple in the stored energy in the capacitors. This energy ripple, normalized with respect to the power transferred by each submodule, is shown in Fig. 4. It can be observed that both increased modulation index and second-order harmonic injection can allow a

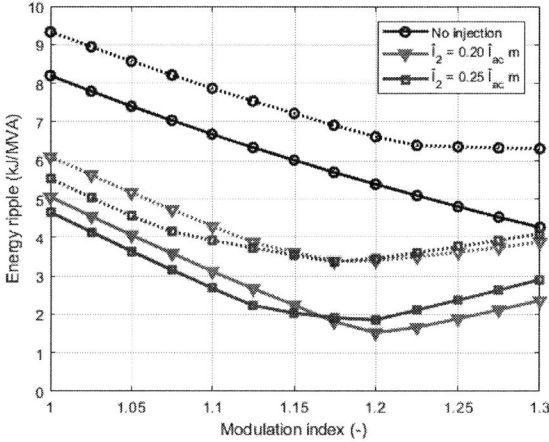

Fig. 4. Normalized peak-to-peak energy ripple in converter arms as function of modulation index with $\phi = 10$ deg. (solid) and $\phi = 30$ deg. (dotted) for different amplitudes of injected second-order harmonic.

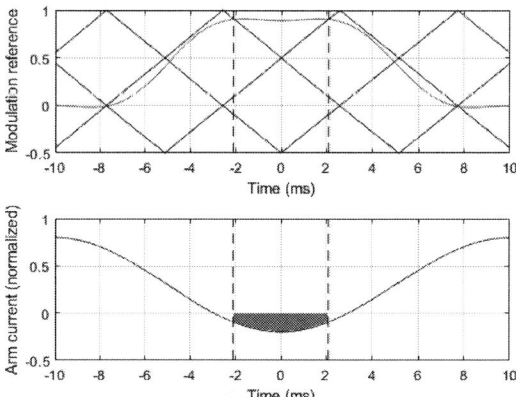

Fig. 5. Example of reference and carrier waveforms (upper) and charge transferred to the capcitors when connected in series (blue area, lower graph).

significant reduction in the energy ripple in the submodule capacitors. Furthermore, it can be concluded that the reduced amplitude, i.e. $k_2 = 0.20$, of the second-order harmonic performs equally well or even better compared to the case when $k_2 = 0.25$

B. Charge Transfer to Capacitors

If the submodule capacitance is changed, the voltage variations and the resulting peak-current during the parallel connection are both affected. However, the charge which is transferred to the capacitors during the series-connection stays approximately the same. A small variation is expected since the modulation reference is adjusted in order to compensate the capacitor voltage ripple. As a simplification, the variations in the modulation reference for compensating the capacitor voltage ripple are neglected. In this way, a more general relation between the energy storage and the capacitor voltage imbalance in the SFB can be derived.

The charge transferred to the capacitors during the series-connected switching state is denoted Q_s. The voltage imbalance between the two capacitors can then be expressed as

$$\Delta V_c = \frac{Q_s}{C_1} - \frac{Q_s}{C_2} \tag{8}$$

where C_1 and C_2 are the capacitances of the two capacitors in one SFB as shown in Fig. 1. The charge Q_s is illustrated in Fig. 5 as blue color in the normalized arm current waveform. Obviously, the value of Q_s depend on the angular offset of the carrier waveforms. To determine the worst-case scenario, Q_s will hereafter refer to the maximum charge which can be obtained for any angular offset in the carrier waveforms. It should be noted that for each combination of reference waveform, carrier frequency, and arm current, a different value of Q_s is obtained.

Assuming the worst-case scenario, the capacitance values are given by

$$C_1 = C_0(1 - k) \tag{9a}$$
$$C_2 = C_0(1 + k) \tag{9b}$$

where k indicates the maximum possible spread among the capacitance values (e.g. $k = 0.05$ indicate that the difference between the capacitance values is $\pm 5\%$ from the nominal value). Substituting (9) in (8) yields

$$\Delta V_c = \frac{Q_s}{C_0}\left(\frac{2k}{1 - k^2}\right). \tag{10}$$

Solving (7) for C_0 and substituting in (10) gives

$$\Delta V_c = \frac{Q_s 2 V_0 (1 + m\frac{\sqrt{3}}{2})(1 + \delta)}{m\tau \hat{I}_{ac}}\left(\frac{2k}{1 - k^2}\right). \tag{11}$$

For a given power angle φ, carrier frequency f_c, and modulation reference s_{ref}, the charge Q_s is proportional to the alternating current amplitude. Accordingly, the charge can be expressed as coulombs per ampere alternating current, or seconds. This charge can be pre-calculated numerically to obtain the function

$$T_Q(m, f_c, \varphi) = \frac{Q_s(m, f_c, \varphi)}{\hat{I}_s}. \tag{12}$$

The values of Q_s in (12) are calculated by first assuming a carrier frequency, modulation reference, and power angle to determine the arm current waveform. The value of Q_s must be calculated numerically for each operating point by performing a sweep in the angular offset of the carrier waveforms.

Substituting (12) in (11) gives

$$\frac{\Delta V_c}{V_0} = \frac{T_Q 2(1 + m\frac{\sqrt{3}}{2})(1 + \delta)}{m\tau}\left(\frac{2k}{1 - k^2}\right). \tag{13}$$

It is observed that the voltage difference ΔV is proportional to the nominal submodule voltage V_0 and inversely proportional to the normalized energy storage τ. Regarding the relation between ΔV_c and the modulation index m it is important to remember that T_Q is a function of

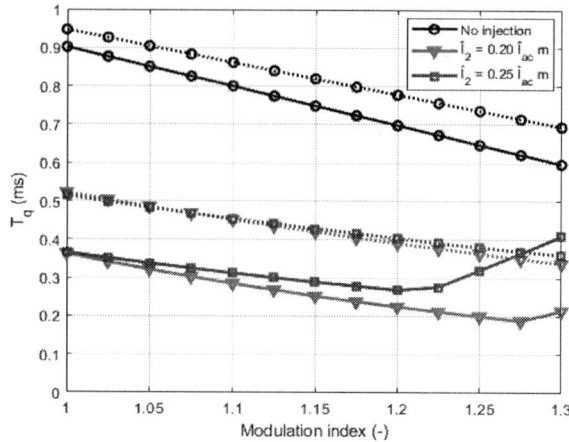

Fig. 6. Functions $T_Q(m, f_c = 95.83\mathrm{Hz}, \varphi = 10\mathrm{deg.})$ (solid) and $T_Q(m, f_c = 95.83\mathrm{Hz}, \varphi = 10\mathrm{deg.})$ (dotted) with $k = 0.05$, $\tau = 25$ kJ/MVA, and $\delta = 0.1$, for different amplitudes of injected second-order harmonic.

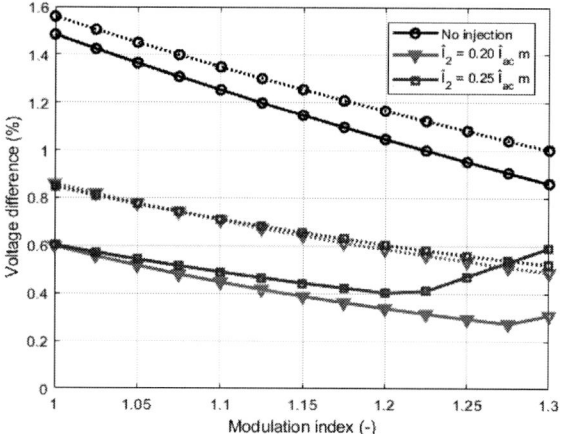

Fig. 7. Voltage difference between the two capacitors with 95.83 Hz carrier frequency, $k = 0.05$, $\tau = 25$ kJ/MVA, $\delta = 0.1$, $\phi = 10$ deg. (solid) and $\phi = 30$ deg. (dotted) for different amplitudes of injected second-order harmonic.

the modulation index as well. The function T_Q is shown in Fig. 6 for selected values of f_c and φ.

With the calculated values of T_Q, the voltage difference ΔV can be evaluated. The resulting voltage difference is shown in Fig. 7 for different levels of second-order harmonic injection. It is observed that increasing the modulation index can help to reduce the capacitor voltage imbalance. For example, with a 10-degree power angle, no harmonic injection, modulation index 1.0, and 25 kJ/MVA, the voltage difference is 1.5% of the nominal capacitor voltage. For a 3.0 kV submodule this corresponds to 45 V. However, if the modulation index is increased to 1.20 and a second-order harmonic with an amplitude corresponding to 20% of $\hat{I}_{ac}m$ is injected, the voltage difference is reduced to 0.34%. For a 3.0 kV submodule, this corresponds to 10 V.

The 2018 International Power Electronics Conference

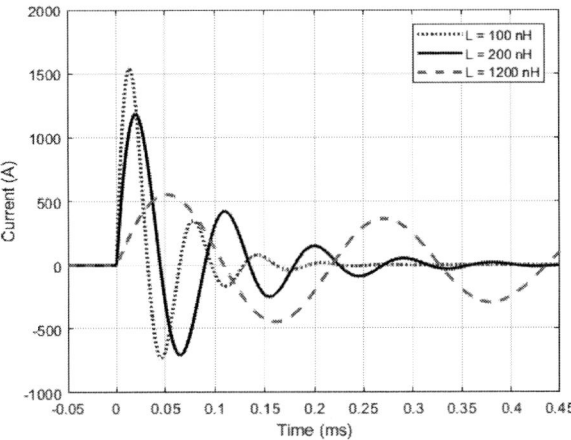

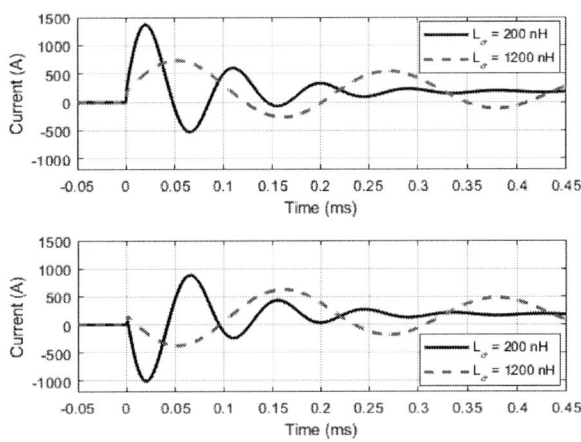

Fig. 8. Balancing current for different loop inductance with 1.95 V forward voltage drop per device, 4.6 mΩ total resistance, and 25 V difference in capacitor voltages.

Fig. 9. Current in devices S_3 (upper) and S_6 (lower) at parallel connection with 360 A arm current, 25 V voltage imbalance, 200 nH loop inductance (solid, black), and 1200 nH loop inductance (dashed, red).

C. Peak-Current at Parallel-Connection

When the parallel-connected state is used, any voltage imbalance ΔV between the capacitors will lead to a balancing current circulating in the loop formed by the two capacitors and devices S_3 and S_6, as shown in Fig. 2. As discussed in [10], this balancing current can be expressed as

$$i_{rd}(t) = \frac{\Delta V_c - 2V_{\mathrm{pn}}}{\sqrt{2\frac{L_\sigma}{C_0} - \frac{R_{eq}^2}{4}}} e^{-\frac{R_{eq}t}{2L_\sigma}} \sin(\omega_c t) \qquad (14)$$

where V_{pn} is the average built-in potential in the devices in the conducting loop, R_{eq} is the total resistance of the busbars and devices in the loop, ω_c is given by

$$\omega_c = \sqrt{\frac{2}{L_\sigma C_0} - \frac{R_{eq}^2}{4L_\sigma^2}}, \qquad (15)$$

and L_σ is the total inductance in the loop. It is observed that the amplitude of the balancing current is proportional to the voltage difference after subtracting the built-in potentials of the semiconductor devices.

As an example, an average forward voltage drop of 1.95 V per device, 4.6 mΩ resistance, and 2.0 mF capacitors with 25 V voltage difference is considered. The resulting balancing current for different loop inductances are shown in Fig. 8. Here the 100 and 200 nH cases represents possible cases without a differential-mode choke and the 1200 nH case is included to show the effect of a significantly higher loop inductance, which can be achieved using a differential-mode choke.

The peak balancing current in Fig. 8 is in the range 300 to 1700 A. To put these numbers in context, consider a case with 3.0 kV nominal capacitor voltage, modulation index 1.20, modulation margin $\delta = 0.1$, and nominal energy storage of 25 kJ/MVA. In such case, the 2.0 mF which was used in this calculation example implies an alternating current amplitude of 900 A. At the modulation

index 1.20 and unity power factor this gives a peak arm current of 720 A. However, it should be noted that the devices S_3–S_6 conduct only half of the arm current, meaning that the peak value of the nominal current through these devices is only 360 A.

The results in Fig. 8 indicate that the peak balancing current is 1.18 kA for the case with 200 nH loop inductance. The actual current through the devices S_3 and S_6 will be the sum and difference the balancing current and half of the arm current, as shown in Fig. 2. The current through the devices S_3 and S_6 is shown in Fig. 9 under the assumption that the arm current equals 360 A at the moment when the capacitors are connected in parallel. It is observed that the peak current through device S_3 is 1.36 kA. Allowing such high surge current would lead to unacceptable overrating of the devices S_3 and S_6. For the case with 1200 nH loop inductance, which can be achieved using a differential-mode choke, the peak current is 0.73 kA, which is is within a more feasible range for the considered example.

IV. SIMULATION RESULTS

To validate the calculated ΔV and resulting balancing current, the MMC specified in Table II is simulated in PSCAD. The simulated system has a stiff dc-link voltage and the MMC is connected to the ac-grid through a transformer with 0.1 p.u leakage reactance. To compensate for the capacitor voltage ripple, the converter is using open-loop energy estimation as described in [15]. The alternating current is controlled using a P-type controller in combination with a feed-forward term in the dq-reference frame. The transformation angle for the dq-reference frame is obtained from a PLL tracking the voltage at the PCC.

2339

TABLE II.	SPECIFICATION OF CONSIDERED SYSTEM
Dc-link voltage	48.0 kV
Arm alternating voltage amplitude	28.8 kV (35.3 kV L-L,rms)
Nominal modulation index	1.20
Alternating current amplitude	900 A
Submodule power angle φ	10.0 deg.
Output power	38.9 MVA
Arm inductance	15 mH (15%)
Submodule capacitance	2.0 mF (25 kJ/MVA)
Nominal capacitor voltage	3.0 kV
Number of SFB modules per arm	9
Carrier frequency	97.2 Hz
Modulation margin δ	0.1

TABLE III.	SIMULATED ΔV AND ARM CURRENT.		
Case		Δ**V**	**Arm current**
Steady-state without harmonic injection		31.7 V	328 A
Steady-state with harmonic injection		10.5 V	125 A
Step-down transient without harmonic injection		19.3 V	189 A
Step-down transient with harmonic injection		19.3 V	269 A
Step-up transient without harmonic injection		47.9 V	253 A
Step-up transient with harmonic injection		26.8 V	179 A

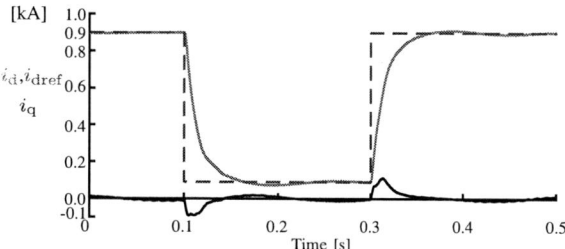

Fig. 10. Simulation Waveforms d and q axes currents during active power transient from 38.9 MVA to 3.98 MVA.

A. Voltage Imbalance

To compare the resulting voltage imbalance in steady state and during transients, two step changes in the alternating current reference is simulated. That is, a step-down from the rated current to 10% of the rated value is simulated at the time t=0.1 seconds. This is followed by a step-up response at the time t=0.3 seconds. The simulated d- and q-axis currents are shown in Fig. 10

To evaluate the impact of harmonic injection in the circulating current, the same operating conditions are simulated both with and without second-order harmonic injection in Fig. 11(a) and (b), respectively. It is observed that the second-order harmonic injection increases the peak-value of the arm current from 720 A to 940 A. On the other hand, the capacitor voltage ripple is significantly reduced when second-order harmonic injection is used. In fact, the peak-to-peak voltage ripple in the sum capacitor voltages is reduced to 1.84 kV from 5.44 kV. This corresponds to a 66% reduction.

To compare the results with the calculated energy ripple in Section III-A, the simulated voltages can be converted to stored energy in the capacitors. It is found that the simulated voltages correspond to a energy ripple of 1.7 kJ/MVA and 5.4 kJ/MVA for the cases with and without second-order harmonic injection, respectively. It is concluded that the simulated capacitor voltages are in agreement the previously calculated values in Fig. 4.

To evaluate the resulting ΔV between the capacitor voltages, one SFB in each arm is adjusted such that the capacitance values are 1.9 mF and 2.1 mF, respectively. A sample and hold circuit is implemented to monitor the voltage difference and arm current at the moment when the capacitors are connected in parallel. The sample and hold circuit is also used to monitor the arm current at the

moment when the capacitors are connected in parallel. The results from all six arms are compared and the worst case is noted both for steady-state operation and the two transient responses. The observations are summarized in Table III.

According to the calculated values, the expected ΔV is 10.1 V for the case when second-order harmonic is used and 31.4 V for the case when second-order harmonic is not used. It is concluded that the observed values in the simulation are fairly close to the expected values. It should be noted that in the theoretical derivation a few simplifications were made. For example, the impact of on the modulation reference by the capacitor voltage ripple compensation and second-order harmonic injection was neglected. Accordingly, some minor deviations from the theoretical results can be expected.

The results in Table III reveal that the resulting voltage difference can be significantly higher during transients compared to steady state operation. When second-order harmonic injection is used, a significant reduction in the voltage difference can be observed both for the steady-state operation and the transient responses. It should be noted that the voltage imbalance is closely related to the voltage ripple in the capacitors. In the considered simulation case, both the voltage ripple and the voltage imbalance are reduced by approximately 66–67% in steady state operation when second-order harmonic injection is used.

The simulated cases provide a few examples of voltage imbalances that could occur in the SFB under different operating conditions. It should be pointed out that the simulated cases are merely examples and the results may vary depending on type of transient, control implementation, etc. The study on how different modulation and balancing techniques affect the capacitor voltage imbalance is, however, outside the scope of this paper.

B. Balancing Current at Parallel Connection

According to the results in Sections III-A and IV-A, the voltage imbalance between the two capacitors is in the range of 10–50 V for the considered cases. To evaluate the resulting current in the semiconductor devices and the impact of the differential-mode choke, a model of the SFB module is implemented in Simulink. The devices are modeled with a constant voltage drop and on-state resistance. The constant voltage drop in the simulated devices is 2.0 V for the diodes and 1.9 V for the IGBTs. The on-state resistance is 1.9 mΩ in the diodes and 2.7 mΩ in the IGBTs.

2340

The 2018 International Power Electronics Conference

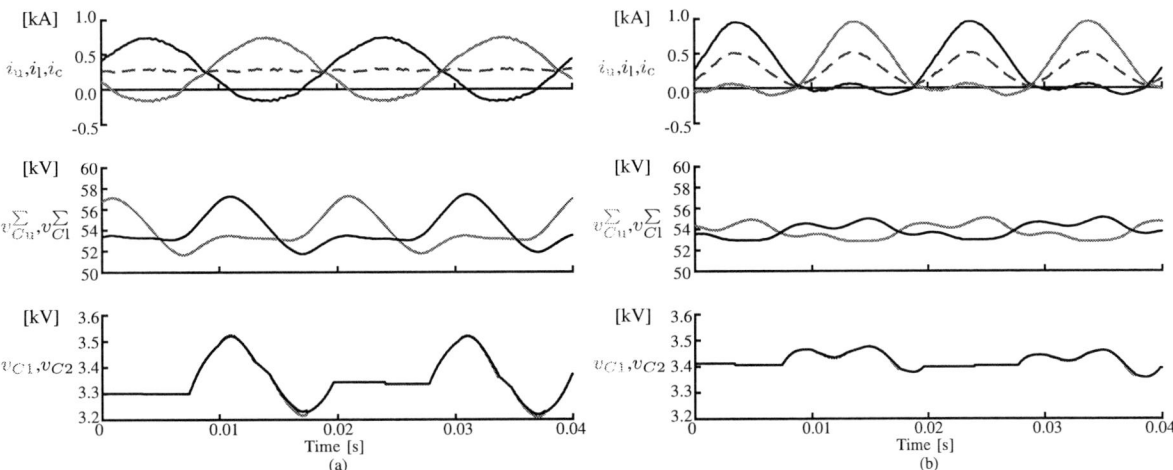

Fig. 11. Simulation waveforms of arm current (red and black), circulating current (blue), total available capacitor voltages, and individual capacitor voltages at $m = 1.2$. (a) Without second-order harmonic injection. (b) Second-order harmonics injection.

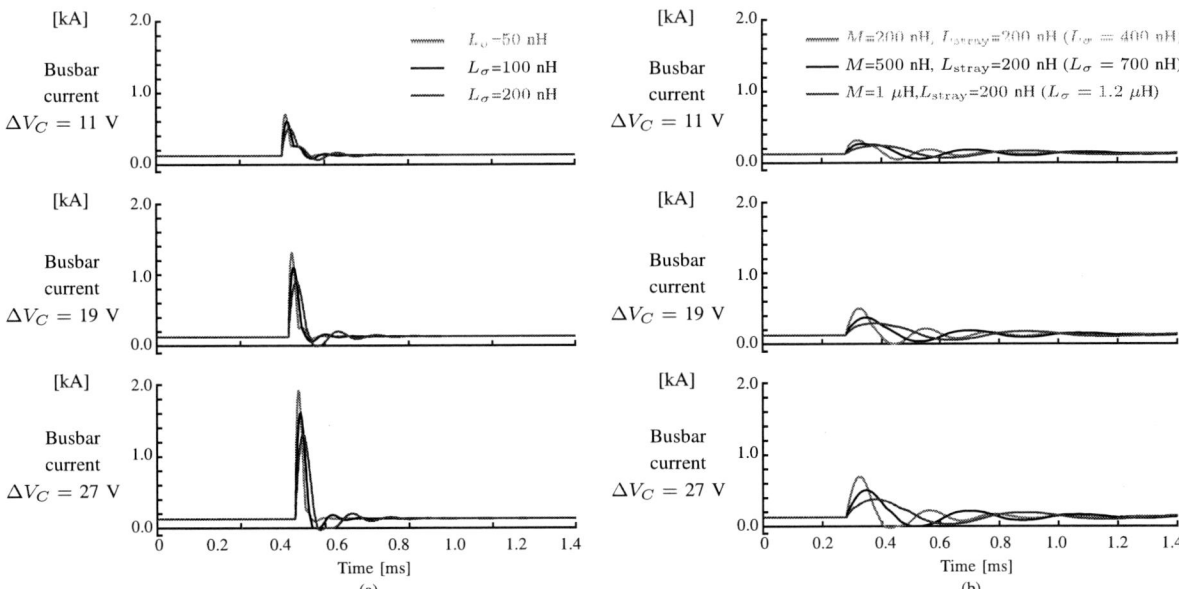

Fig. 12. Simulation waveforms of busbar current with different ΔV_C and inductance values under the arm current of 250 A. (a) Loop inductance L_σ. (b) Differential-mode choke with the loop inductance of $L_\sigma = M + L_{\mathrm{stray}}$.

The impact of different stray inductances and differential-mode chokes on the balancing current are shown in Fig. 12(a) and (b), respectively, for the voltage difference of 11 V, 19 V, and 27 V between the two capacitors. The differential-mode choke is implemented in the simulation model as two coupled inductors as shown in Fig. 1. Three different values of the mutual inductance, i.e., 0.2 μH, 0.5 μH and 1.0μH are considered with the same loop inductance of 200 nH. The coupling coefficient of the differential-mode inductors is 0.99. Accordingly, leakage inductances for these cases are 4 nH, 10 nH, and 20 nH, respectively. It is observed that although the balancing current is affected by the loop inductance, there

is still a significant overcurrent at a voltage difference of 27 V when no differential-mode choke is used.

If a 1.0 μH differential-mode choke is used, the peak balancing current can be reduced to less than 0.4 kA, even for the case with 27 V voltage difference between the capacitors. The actual current through the devices S_3 and S_6 would, however, be the sum and difference of the arm current and the balancing current in Fig. 12. According to the simulation results in Section IV-A, the arm current is in the range of 100–300 A when the capacitors are connected in parallel. The simulated current in the devices S3 and S6 are shown in Fig. 13 for the simulated voltage

2341

The 2018 International Power Electronics Conference

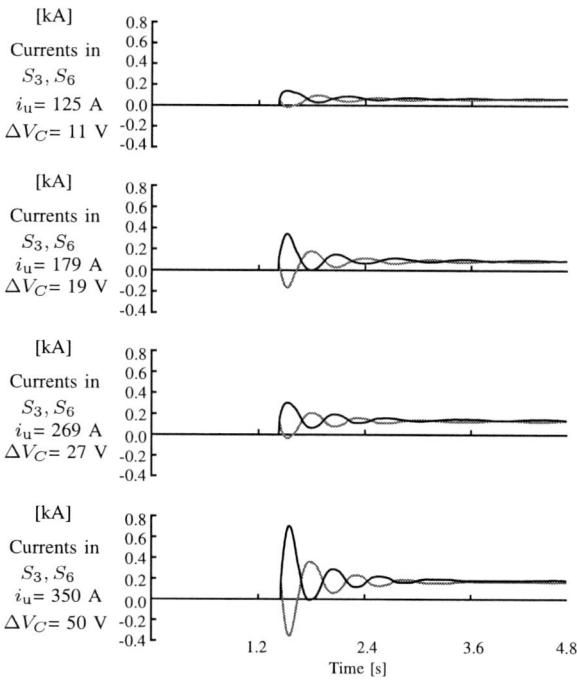

Fig. 13. Simulation waveforms of power device currents in various arm currents and voltage differences.

differences and arm currents. It is observed that in the worst case, a 50 V difference with 350 A arm current, the peak current is still limited to 0.7 kA, which is the same as the peak-value of the arm current for the case without harmonic injection.

V. CONCLUSIONS

The semi-full-bridge is an attractive alternative to the full-bridge submodule as well as mixed cell solutions for modular multilevel converters. With proper ratings it can provide boosting of alternating voltage amplitude, operation with reduced dc-link voltage, and dc-fault blocking capability. A challenge with the semi-full-bridge is, however, the current-spike appearing when the two capacitors are connected in parallel. It is found that if the relative size of the capacitors is constant in terms of kJ per MVA, the voltage imbalance is roughly proportional to the nominal capacitor voltage. In some cases, this voltage imbalance can lead to surge currents exceeding the allowed repetitive peak current of the semiconductor devices. The peak current can be significantly reduced by coupled inductors forming a differential-mode choke. The coupling of the inductors is such that the flux generated by the arm current is canceled out and only the differential-mode current resulting from a capacitor voltage imbalance between the capacitors in the semi-full-bridge is affected. Simulation results indicate that the peak current in the devices that are affected by the parallel connection of the capacitors can be reduced to values in the same range or even lower than the peak arm current. In fact, for steady

state conditions, the simulated peak current was well below the peak arm current indicating that a differential-mode choke could be one possible solution for handling the parallel connection of the capacitors in the semi-full-bridge.

REFERENCES

[1] R. Marquardt, "Modular multilevel converter: An universal concept for HVDC-networks and extended DC-bus-applications," *Proc. ECCE Asia IPEC*, Jun. 2010, pp. 502–507.

[2] L. Baruschka and A. Mertens, "Comparisons of cascaded h-bridge and modular multilvel converters for bess application," *Proc. ECCE*, Sept. 2011, pp. 909–916.

[3] K. Ilves, S. Norrga and H. P. Nee "On energy variations in modular multilevel converters with full-bridge submodules for Ac-Dc and Ac-Ac applications," in *Proc. European Conference on Power Electronics and Applications (EPE)* , Lille, 2013, pp. 1-10.

[4] R. Marquardt, "Modular Multilevel Converter topologies with DC-Short circuit current limitation," 8th International Conference on Power Electronics - ECCE Asia, Jeju, 2011, pp. 1425-1431.

[5] X. Li, W. Liu, Q. Song, H. Rao and S. Xu, "An enhanced MMC topology with DC fault ride-through capability," IECON 2013 - 39th Annual Conference of the IEEE Industrial Electronics Society, Vienna, 2013, pp. 6182-6188.

[6] J. Qin, M. Saeedifard, A. Rockhill and R. Zhou, "Hybrid Design of Modular Multilevel Converters for HVDC Systems Based on Various Submodule Circuits," in IEEE Transactions on Power Delivery, vol. 30, no. 1, pp. 385-394, Feb. 2015.

[7] G. P. Adam, K. H. Ahmed, and B. W. Williams, "Mixed cells modular multilevel converter," in IEEE-ISIE, Jun. 2014, pp. 1390–1395.

[8] A. Nami, J. Liang, F. Dijkhuizen and P. Lundberg, "Analysis of modular multilevel converters with DC short circuit fault blocking capability in bipolar HVDC transmission systems," 2015 17th European Conference on Power Electronics and Applications (EPE'15 ECCE-Europe), Geneva, 2015, pp. 1-10.

[9] K. Ilves, L. Bessegato, L. Harnefors, S. Norrga, and H. P. Nee, "Semi-full-bridge submodule for modular multilevel converters," *Proc. ECCE Asia ICPE*, Jun. 2015, pp. 1067–1074.

[10] S. Heinig, K. Jacobs, K. Ilves, S. Norrga, and H. P. Nee, "Implications of capacitor voltage imbalance on the operation of the semi-full-bridge submodule," *Proc. European Conference on Power Electronics and Applications (EPE)* , Sept. 2017.

[11] A. Nami, J. Liang, F. Dijkhuizen and G. D. Demetriades, "Modular Multilevel Converters for HVDC Applications: Review on Converter Cells and Functionalities," in IEEE Transactions on Power Electronics, vol. 30, no. 1, pp. 18-36, Jan. 2015.

[12] C. Dahmen, F. Kapaun and R. Marquardt, "Analytical investigation of efficiency and operating range of different Modular Multilevel Converters," 2017 IEEE 12th International Conference on Power Electronics and Drive Systems (PEDS), Honolulu, HI, USA, 2017, pp. 336-342.

[13] D. Gunasekaran, S. Yang and F. Z. Peng, "A cascaded two-port bridge multilevel converter with automatic voltage balancing capability," 2015 IEEE Energy Conversion Congress and Exposition (ECCE), Montreal, QC, 2015, pp. 3564-3569.

[14] M. Winkelnkemper, A. Korn and P. Steimer, "A modular direct converter for transformerless rail interties," 2010 IEEE International Symposium on Industrial Electronics, Bari, 2010, pp. 562-567.

[15] L. Angquist, A. Antonopoulos, D. Siemaszko, K. Ilves, M. Vasiladiotis and H. P. Nee, "Open-Loop Control of Modular Multilevel Converters Using Estimation of Stored Energy," in IEEE Transactions on Industry Applications, vol. 47, no. 6, pp. 2516-2524, Nov.-Dec. 2011.

Research on Key Technology and Equipment for Zhangbei 500kV DC Grid

Guangfu Tang, *Member, IEEE*, Hui Pang, Zhiyuan He, *Member, IEEE, Xiaoguang Wei*
Global Energy Interconnection Research Institute, Beijing, China
gftang@geiri.sgcc.com.cn

Abstract — **With the capacity of new energy resources such as wind power and photovoltaic in Zhangbei region increasing rapidly, the reliable transmission and consumption capabilities of large-scale and high-proportion new energy resources are urgently improved. Considering the advantages of the VSC-HVDC transmission technology, Zhangbei project is designed as symmetrical bipolar system which adopts half-bridge modular multi-level converter (MMC) topology and overhead lines for power transmission. In this paper, a system scheme is presented, and some results are provided. Meanwhile, a novel modular cascaded hybrid DC breaker concept is proposed, and 500kV VSC valve and DC breaker prototype are developed to meet the requirement of the Zhangbei four-terminal dc grid.**

Index Terms—**DC grid, DC breaker, overhead line, Voltage Soured Converter, VSC-HVDC.**

I. INTRODUCTION

Due to the increasing consumption of primary energy and the degradation of the environment, a side effect of global economic development, plans have been made by countries around the world to encourage the evolution of energy sources from fossil energy towards clean energy. As a result, this trend accelerates the development and utilization of large-scale renewable and clean energy. The current and future development of electric power industry in China will focus on following points: facilitating the efficient exploitation and utilization of renewable clean energy, promoting the optimized allocation and transformation of power resources in a large scale, securing the reliable power supply, and improving the economical and flexible operation of power system.

Based on a new generation voltage source converter, high voltage direct current (VSC-HVDC) transmission technology has captured an increasing number of research. Comparing this approach with Current Source Converter (CSC) HVDC, the VSC-HVDC is superior in circuit performance, this includes independent control of both active and reactive power flows, no requirement for filters or reactive power compensation equipment, and fixed voltage polarity for reversed power flows. Thus, the VSC-HVDC offers a safe and efficient solution for the integration and consumption of intermittent renewable sources. So far, most of VSC-HVDC projects running in the commercial world adopt the point to point transmission mode. Comparing with the multiple point to point transmission modes, the flexible DC grids are more reliable, economical and flexible in high voltage and large capacity. With the exploitation of renewable energy like wind power and photovoltaic,

integration of large-scale clean energy demands the development and establishment of DC grids

The Zhangbei region, due to its geographical advantage, has been installed a lot of renewable and clean energy based power stations and this trend has no stop sign until 2020. However, due to the asymmetric distribution of renewable energy resources and energy consumption, and characteristics of intermittency and fluctuation of renewable energy, the demand of the integration and consumption of large-scale renewable energy resources becomes difficult to realize. Fortunately, the VSC based DC grid technology, as described in this paper, offers a good solution to the aforementioned problems.

Until now industrial VSC-HVDC projects under operation has grown dramatically, also including Zhoushan five-terminal $\pm$ 200kV/1000MW(total capacity) and Xiamen $\pm$ 320kV/1000MW VSC-HVDC projects in China. Therefore, in order to improve the connection and consumption capability of large-scale renewable energy resources in Zhangbei region, the $\pm$500kV/3000MW four-terminal DC Grid project is proposed by State Grid Corporation of china (SGCC).

This article is organized as follows. In Section II, the design of Zhangbei 500kV DC grid is described in terms of the profile of Zhangbei DC grid, the system configuration scheme and technical parameter requirements for key equipment. In Section III and IV, the key equipment, including the VSC valve converter and DC circuit breaker, are introduced respectively. Section V presents the research and experimental results of the prototype of VSC valve converter and DC circuit breaker, which serves as the key equipment of the DC grid system. Section VI concludes this paper.

II. DESIGN OF ZHANGBEI 4-TERMINAL HVDC GRID

A. Profile of Zhangbei DC Grid

With the rapidly increasing capacity of new energy resources in Zhangjiakou district, China, the reliable transmission and consumption capabilities of large-scale and high-proportion of new energy resources are urgently improved. However, the AC power grids in Zhangjiakou are relatively weak, and the transmission capability improvement of the 500kV AC system is still limited even by increasing the dynamic reactive power support. Meanwhile, the power sources of Bashang grid in Zhangjiakou are almost all renewable energy resources. It has no voltage support provided by the traditional synchronous generators. All of these lead to the difficult transmission and consumption of renewable energy resources.

Zhangjiakou district is adjacent to the Beijing - Tianjin - Tangshan load center. Considering the advantages of the VSC-HVDC transmission technology, it would be a positive demonstration to apply VSC-HVDC technology in Zhangbei district to solve the problem of renewable energy resources transmission and consumption. Meanwhile, it will also be on operation for "Green Winter Olympics" held by Zhangjiakou and Beijing. SGCC is planning to construct the world's first VSC-HVDC grid demonstration project based on overhead line in Zhangbei district. Figure 1 shows the geographical location of the Zhangbei VSC-HVDC grid demonstration project.

Figure 1. Geographical location of Zhangbei four-terminal DC grid

Zhangbei VSC-HVDC grid demonstration project adopts four-terminal ring, and the operating configuration is bipolar topology with metallic return line as shown in Figure 2. The main parameters of the system are shown in table I. In Zhangbei four-terminal VSC-HVDC grid, the Zhangbei and Kangbao converter stations are the sending terminals collecting local wind power. The Fengning converter station is a regulating terminal connected to a local pumped-storage hydroplant which can suppress the fluctuation of wind power. The Beijing converter station is the receiving terminal used to supply stable and clean power for Beijing.

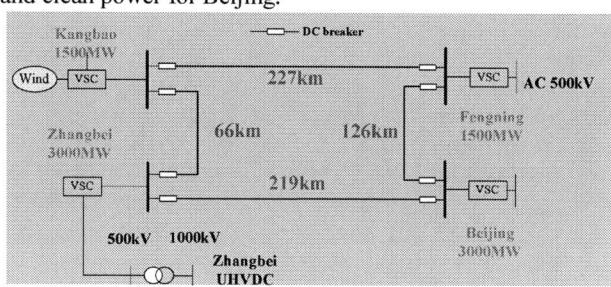

Figure 2. Design schematic of Zhangbei four-terminal DC grid

TABLE I.
MAIN PARAMETERS OF ZHANGBEI DC GRID

Parameter	Zhangbei	Beijing	Kangbao	Fengning
AC Voltage/kV	220	220	500	500
DC Voltage/kV	±500	±500	±500	±500
Capacity of Converter Station/MW	3000	3000	1500	1500

When the system is in normal operation condition, the Zhangbei and Kangbao converter stations simultaneously inject power into the DC grid, and serve to maintain voltage stability and power balance of the whole system. The Fengning and Beijing converter stations receive electrical power from the DC side according to their load demand. Figure 3 shows the power flow in four-terminal DC grid.

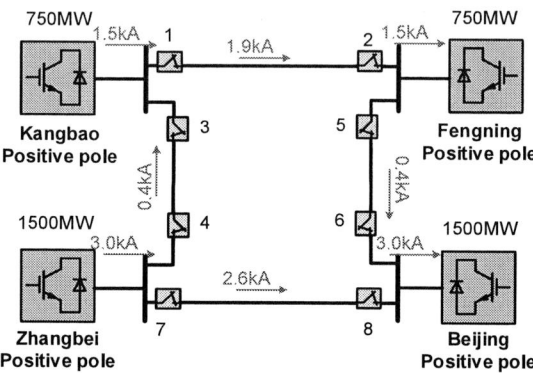

Figure 3. Power flow distribution of Zhangbei four-terminal DC grid

B. System configuration scheme

Considering the cost due to transmission distance and voltage level increment, the overhead lines are adopted for Zhangbei VSC-HVDC grid. The use of overhead lines may increase the fault probability of dc lines. The dc fault current spreads very fast due to very low damping in VSC-HVDC transmission system [1] and it is difficult to be interrupted because of no zero crossing point. The converters in VSC-HVDC grid consists of numerous power electronic semiconductor devices and the capability for withstanding over current is very limited. Therefore the fast fault clearance and isolation is required in VSC-HVDC grid. If it is unable to isolate faulty areas quickly, multiple converters will be blocked and result in power transmission interruption of the HVDC grid [2]. In order to decrease the DC fault impact, the DC grid should have the capability of fast fault clearance and isolation.

Presently, the modular multi-level converter is widely used in the practical VSC-HVDC projects as shown in Figure 5(a). The sub-module (SM) topology mainly includes the half-bridge SM as shown in Figure 5(b) and full-brideg SM as shown in Figure 5(c). The half-bridge SM based MMC can not block the dc fault current after blocking, but the full-bridge SM based MMC has the capability to block the dc fault current. There are two kinds of main technical schemes proposed to achieve this aim according to the fault characteristic of the different MMCs. One is based on the HVDC circuit breaker which is used to break the faulty DC line, and the half-bridge SM based MMCs are also used. The other one is based on the converter with dc fault clearance capability which uses the SMs as similar as full-bridge, and fast mechanical disconnectors are required to isolate the faulty areas [2]. The two schemes will be compared and one of them is selected for Zhangbei DC grid.

(a) Scheme I-Half-bridge MMC with HVDC Breakers

The first technical scheme is shown in Figure 6, the converter adopts half bridge MMC submodule [3], and the HVDC breakers will be placed at both ends of each DC overhead line [4]. If this technical scheme is adopted, the converter is no need to have the fault-blocking capability and the failure recovery of DC grids requires auto-reclose function for the use of the overhead lines. In order to decrease the DC fault impact on the interconnected AC and DC system, it allows to block the nearby converter station, and the power transmission of the DC grid is uninterrupted.

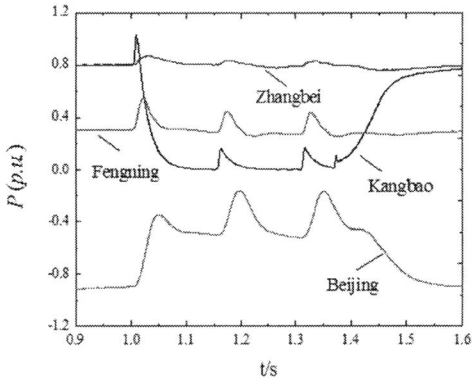

Figure 8. The active power of converter station

It can be seen that the DC voltage is maintained by the converter station at the remote end. Once the fault is cleared, the blocked converter station can be deblocked at short time and resumes operation.

(b) Scheme II-Full-bridge MMC with fast mechanical disconnectors

The second technical scheme is shown in Figure 9, the converter is based on MMC with fault current blocking capability [5-6]. The DC mechanical disconnectors are placed at both ends of each overhead DC line. If this technical scheme is adopted, the DC mechanical disconnector is required to isolate the fault. It needs to block all converters in order to realize fault clearance when the faults occur at DC side, and isolate the converter and DC faulty line by openning the DC mechanical disconnectors. This technical scheme will cause the power transmission interruption which has the great impact on interconnected AC system.

Figure 6. The first technical route

The simulation results of the permanent pole to pole fault on the dc line between Kangbao and Fengning are shown in Figure 7 and Figure 8.

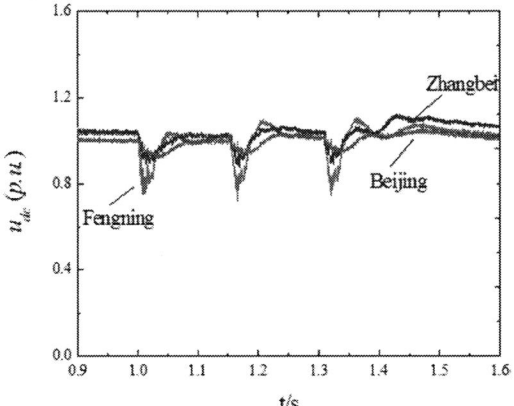

Figure 7. The DC voltage of converter station

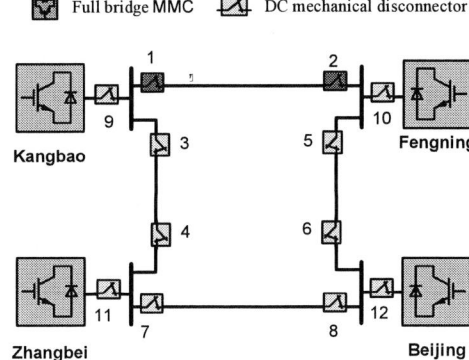

Figure 9. The second technical route

The simulation results of the permanent pole to pole fault on the dc line between Kangbao and Fengning are shown in Figure 10 and Figure 11.

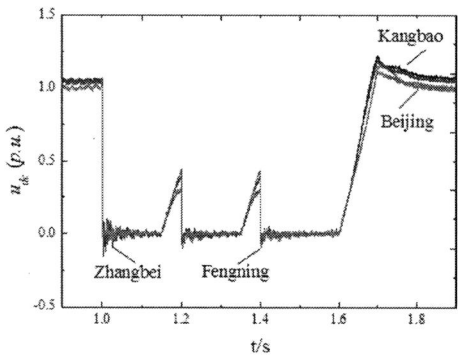

Figure 10. The DC voltage of converter station

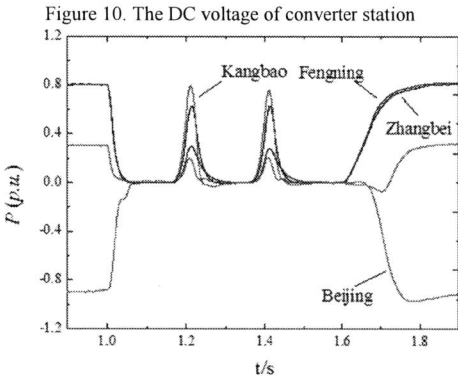

Figure 11. The active power of converter station

It can be seen that the interruption time of the power transmission is long caused by the DC fault, and the impact on the AC system is relatively large, which may cause frequency and voltage fluctuation of the weak power system such as wind farms connection.

(c) Comparison of the two schemes

Comparing the two technical schemes, it can be seen that the first technical scheme will not cause the global block in the failure period, and the failure recovery speed is faster relatively. However, it requires high speed fault location and DC circuit breaker interruption. The second technical scheme requires relatively lower speed for fault location, but it will cause power transfer interruption for a long time which has great impact on the AC system. Table II summarizes the advantages and disadvantages of each scheme.

TABLE III.
SUMMARY OF ADVANTAGES AND DISADVANTAGES OF THE TWO SCHEMES

Items	Scheme I	Scheme II
Advantages	1. Uninterrupted power transmission 2. Fast fault recovery	1. Lower speed requirement for fault location
Disadvantages	1. Fast fault location and breaking time requirements 2. Higher overall investment	1. Power transmission interruption 2. Higher investment for converters and power loss
Technical Difficulties	1. Development of 500kV HVDC breaker	1. Development of 500kV fast mechanical

2.Ultra high speed protection	disconnectors

According to the comparision of the two technical schemes, scheme I is more suitable for dc fault clearance and isolation in DC grid. Scheme II will lead to the outage of entire DC grid in case of partial failure, and therefore it is suitable for point-to-point VSC-HVDC transimission applications. In Zhangbei DC grid project, the scheme I is selected.

C. Technical parameter requirements for key equipment

As described above, the application of overhead lines will certainly cause a great increase in the probability of temporary faults. If a pole-to-pole short-circuit fault occurs on the DC side, the most serious transient current will impose on the whole system. Therefore, to ensure the safe, reliable and continuous operation of the whole system to the maximum extent, the protection principle for Zhangbei VSC-HVDC grid is that none or one at most converter station is blocked due to overcurrent before fault isolation. To meet the above requirements, on the one hand, the MMC should be able to withstand fault current for a certain time to avoid blocking as far as possible. On the other hand, the protection system should be able to detect and locate DC faults quickly and the HVDC breaker should reduce the transient current before it reaches the threshold for blocking MMC. Therefore, it is required to make clear quantitative calculation of transient current levels of DC system, based on which the requirements of technical parameters of key equipment is to be put forward.

The time for protection system to detect and locate DC fault is 3ms, which is the fastest speed the protection device can achieve currently and is also the design parameter of protection system for Zhangbei VSC-HVDC grid. Based on the HVDC breaker which has been put into operation in Zhoushan multi-terminal HVDC system and the technic development, the designed time for HVDC breakers in Zhangbei project to break the fault current is 3 ms, which is the same as in Zhoushan project, but with a capability of breaking a larger fault current than in Zhoushan. So It's 6ms (3ms + 3ms) in total from the time the fault occurs until the fault current begins to decay, that means, it is judged if the converters need to be blocked by comparing the fault current at 6ms with their threshold.

When a pole to pole fault of DC side occurs, all the converters will inject short circuit current into the faulted area. The major short circuit current components include the contributions from the discharge of converter capacitors and AC grid. In order to limit the transient current during fault clearance, DC reactors are needed to be installed. Two DC reactor configuration schemes, namely centralized and distributed configuration, are proposed as shown in Figure 12(a) and 12(b) respectively. According to the further analysis, it is obtained that the DC loop reactance of distributed configuration is larger than that of centralized configuration. Therefore, the spread speed of fault current is slower in distributed configuration. Thus the distributed configuration of

DC reactor is recommended for Zhangbei project.

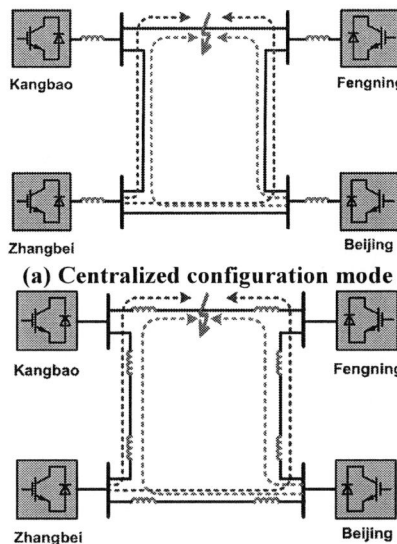

(a) Centralized configuration mode

(b) Distributed configuration mode

Figure 12. The influence on fault current under different configuration mode of DC reactor

The simulation result for the maximum line transient current at different time without blocking any MMC after a pole-to-pole short-circuit fault occurs at different locations on DC side in distributed configuration is shown in Table III. It can be seen from Table III, increasing the DC reactance can effectively reduce the transient fault current. However, the increase of DC reactance will make the DC system dynamic performance worse. Thus DC reactance cannot be too large.

TABLE III.
THE MOST SERIOUS LINE TRANSIENT CURRENT AT DIFFERENT TIME

DC react or	1ms/ kA	2ms/ kA	3ms/ kA	4ms/ kA	5ms/ kA	6ms/ kA	7ms/ kA	10ms/ kA
100 mH	6.6	10.5	13.6	14.7	18.0	19.5	19.7	22.4
150 mH	5.5	8.2	10.9	12.6	12.9	15.3	17.5	19.3
200 mH	4.8	7.0	9.1	11.1	12.9	13.0	13.7	18.0

The overcurrent threshold for blocking MMC of Kangbao and Fengning stations are 4 kA and that of Zhangbei and Beijing stations are 6 kA. Taking the DC reactance as 200mH, there will be 2 converters blocked at 6ms after a fault occurs between Kangbao and Zhangbei. This is because Kangbao - Zhangbei line is very short and the line impedance is small. To meet the requirement that the maximum number of blocked converters does not exceed one, a further optimization is made by increasing the DC reactance between Kangbao and Zhangbei to 300mH. **Error! Reference source not found.**13 shows the maximum transient current to be cut off by DC breaker after failure, and the value at 6ms is 12.984kA. Taking into account a certain margin and the situation of expansion to the 7 terminals in the future, the breaking ability of DC breaker should not be less than 20kA/3ms.

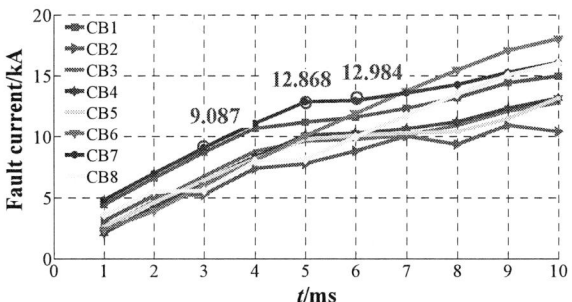

·Figure 13. The maximum transient current to be cut off by DC breaker at 6ms after failure

III. R&D OF VSC CONVETER

The converter valve used in the Zhangbei DC transmission system is of a bipolar half-bridge modular multilevel topology, and the sub-module is its fundamental functional unit, as shown in Figure 14. It can be seen that each converter is made up of three phase units, and one phase unit contains two valve bridge arms, which are cascaded through the same number of sub-modules. Two fully-controlled IGBTs and one capacitor can be found in the topology of each power sub-module unit. As for the working principle of converter, the power sub-module can be switched in or out by controlling the fully-controlled switches. Therefore, the bridge arm voltage amounts to sum of the voltage contributed from all switched-in power sub-module units. The DC voltage of the converter is superposed by the voltage of the top bridge arm and the bottom bridge arm together. Thereafter, by changing the number of power sub-modules to be switched in, the required AC voltage can be formed [7].

Since IGBT is the core semiconductor component of the converter valve. Its rated operating voltage depends on the voltage ripple of the sub-module DC capacitor, failure rate of components, manufacturing level of the metal film of the DC capacitor, and the sub-module protection. Considering the above factors, the status quo of system application, operation reliability and cost-effectiveness of the converter valve, the rated operating voltage of each sub-module is set to 2.2 kV.

Based on the layout of converter valve, each converter arm contains two double-row 4-layer valve towers. Each layer consists of 6 valve modules and each valve module contains 6 power sub-modules. One valve module consists of multiple converter valve sub-modules. According to the principles of modularization and standardization, the valve module is designed to be a standardized unit. Depending on the voltage class, the number of cascaded valve modules is selected to satisfy the system needs and to increase its general use. Taking into account the rated operating voltage of each sub-module, the reliability and operation conditions, 264 power sub-modules including 20 redundant sub-modules are applied in each converter arm in Zhangbei DC Grid Project.

The valve tower consists of valve modules, supporting structure parts, conductive bus-bar, cooling water pipes, voltage-sharing structure parts, optical fibers, and auxiliary supporting parts. The modularized multilevel converter valve is of supporting tower structure featured with air insulation and water cooling equipment (as shown in Figure 16). The hybrid structure of converter valve is reflected into the following two aspects. Firstly, the self-supporting structure is adopted in the converter valve tower, and meanwhile, the suspended structure is taken in the power sub-module unit. The prototype of ± 500kV VSC converter valve tower can be seen in Figure 15.

Figure 15. Prototype of ±500kV VSC Converter Valve

The main parameters for ±500kV VSC converter valve are presented as follows:
- Rated voltage: 535kV
- Rated Capacity: 3000MW
- Switching Impulse Withstand Voltage: 1175kV
- Lightning Impulse Withstand Voltage: 1425kV
- 100ms Transient Withstand Current: 32kA
- Anti-seismic Level (Seismic Fortification): Level 9

Since the converter valve is capable of withstanding the transient current up to 32kA within 5 cycles of 100ms, the above mentioned breaking ability on DC breaker being 20kA/3ms will not cause any damage or faults on converter valve [8].

IV. R&D OF DC CIRCUIT BREAKER

Zhangbei project is designed as symmetrical bipolar system which adopts half-bridge modular multi-level converter (MMC) topology and overhead lines for power transmission. 535kV dc breaker must be configured at both ends of the transmission lines to allow for rapid clearance of the dc faults such as lighting stroke, and realize fast reclosing after fault clearance. Figure 16 illustrates the proposed diode based H-bridge hybrid HVDC circuit breaker. The topology contains three paralleled branches: the main branch, the transfer branch and the absorption branch. The main branch, carrying nominal current, contains an ultra-fast mechanical disconnector in series connection with several IGBT-based H-bridge module units. The transfer branch, undertaking the function of fault current breaking, comprises a large quantity of cascaded diode-based H-bridge module units, of which the advantage is that it can reduce the number of IGBTs by half without sacrificing the breaking capability. The absorption branch is made of MOVs to limit the transient interruption voltage (TIV) and to dissipate the magnetic energy stored in the DC system.

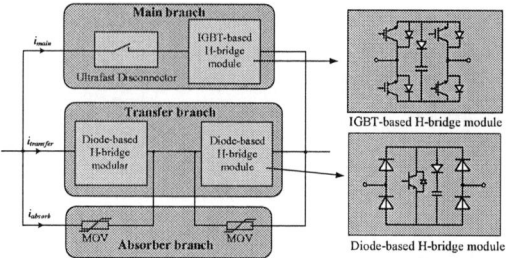

Figure 16. H-bridge modular cascaded hybrid DC breaker and two types of semiconductor module

The operation principle of IGBT based H-bridge module has been introduced in reference [9]. The H-bridge module appearing in the transfer branch adopts the diode full-bridge rectifier structure that consists of 4 fast recovery DIODEs and 2 press pack IGBTs as well as a snubber circuit. This topology provides a bidirectional current path as UFD has thereby allow this hybrid circuit breaker operating bidirectional transmission as well.

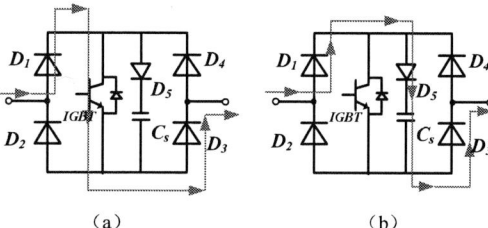

Figure 17: The operating state of diode based H-bridge. (a) Conducting state. (b) blocking state.

The diode based H-bridge module has two operating states named the conducting state and blocking state, as shown in figure 17. During the conducting state, the current loop includes D_1-IGBT-D_3, it is worth noting that no matter current direction that from D_1-D_3 or D_4-D_2 the current direction flowing through the IGBTs keep constant. As the trip signal is received that the blocking state occurs, IGBTs are turned-off and the current starts to flow through the snubber capacitor. This means the voltage across snubber capacitor could be charged to certain value until MOV clamping.

The key operation waveforms of the proposed hybrid DC breaker around the instant of current interrupting are shown in Figure 18. The main branch, with low on-state impedance, conducts load current during normal operation. When a fault occurs, the breaker receives the breaking demand sent from VSC-HVDC system at t1, the semiconductor modules of main branch is switched off to force the fault current commutated to the transfer branch. Until the main branch current decreases to zero at t2, the UFD is opened. After the UFD reaches enough distance and is capable of withstanding the TIV, the semiconductor modules of the transfer branch are switched off at t3 and the MOVs are activated. As a result, the fault current transfers to the absorption branch during the period of t3~t4. The interruption process of the breaker ends at t5 when the fault current decrease to zero.

The reclosing order is received at t6 after successful arc

extinguishing and the deionization of a DC fault, and then the semiconductor modules of the transfer branch turn on. If the DC fault still exists, the modules block immediately at t7 when the current exceeds the protection threshold (ITH) to reduce the inrush fault current. Otherwise, the main branch will be closed and the breaker is put into operation.

Figure 18. Schematic voltage and current waveforms during current breaking and reclosing

The features of the proposed hybrid HVDC breaker are as follows:
- Excellent fault current breaking capability
- Better voltage distribution
- UFD arc-less separation
- Fast reclosing capability
- Free scalability assured by modular design

In order to reduce the footprint of the breaker, components including the UFDs, the semiconductor switches and the MOVs, are all integrated and installed at a 500kV high potential platform as shown in Figure 19. Thanks to the frame structure design of valve support, the seismic performance can reach up to 9.0 on the Richter scale. With the modular design, this breaker can be easily scaled for higher voltage levels.

Figure 19. The 500kV hybrid HVDC breaker prototype

To echo to the need of Zhangbei DC grid, the parameter requirements for the breaker are:
- Rated voltage: 535kV
- Rated current: 3kA
- Breaking time（not including fault clear time）:≤3ms
- Interrupting current: ≥25kA
- Dielectric withstand in open state: ≥800kV
- Reclose time: ≤300ms

V. TEST OF CONVERTER VALVE AND DC BREAKER

As the key equipment configured in DC grid, the converter and the circuit breaker have captured an increasing concern on their operation safety and reliability. For this purpose, it is essential to carry out a series of tests on the converter and the

circuit breaker to check whether their designs meet all the specified engineering requirements.

A. Test for Converter Valve

IEC 62501 has defined all the electrical type tests for converter valves, mainly including operational test, short-circuit test, IGBT overcurrent turn-off test, EMC test and dielectric test.

Among the aforementioned test items, the operational test is to check the adequacy of converter valves under the worst repetitive conditions and to demonstrate correct interaction between valve electronics and power circuits. The operational test mainly includes maximum continuous operating duty test, maximum temporary over-load operating duty test, minimum DC voltage test. The maximum continuous operating duty test is to reproduce the maximum continuous IGBT/diode junction temperatures based on the worst operating conditions of the converter in service. The maximum temporary over-load operating duty test is for the valve specified for temporary over-load conditions. The minimum DC voltage test is to verify the correct performance of the valve designs in which energy for the valve electronic circuits is extracted from the voltage appearing between the valve terminals. As required by IEC 62501, the test conditions shall represent the worst scenarios of a valve in service, such as the operating voltage and current, coolant flow rate and inlet temperature, switching frequency and modulating pattern, etc. All these test conditions are to ensure that the maximum IGBT junction temperature could be achieved during the test. A safety factor, 1.05, needs to be considered when deriving the testing voltage and current for the maximum continuous operating duty test and similarly, a safety factor, 0.95, needs to be considered for the testing voltage of the minimum DC voltage test. The minimum duration required for both tests are 30 minutes and 10 minutes, respectively.

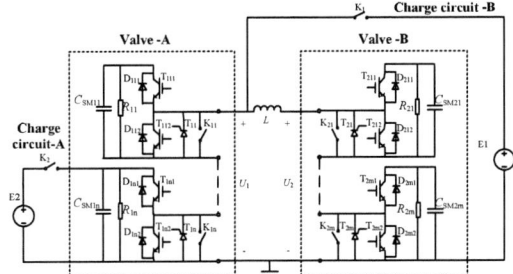

Figure 20. Operational test circuit of converter valves.

Due to the large number of sub-modules configured in MMC and its high voltage and large capacity, it is unrealistic to carry out operational test on the whole converter in the laboratory. Consequently, a scaled-down equivalent test circuit, as shown in Figure 20, was proposed to carry out operational test. By exchanging power between two valves, AC and DC current is generated in test circuit so as to be equivalent to the current stress under actual operating conditions. The typical waveform of valve current and voltage during operational test are shown in Figure 21.

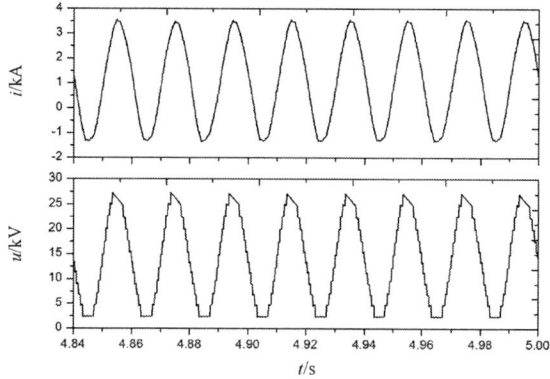

Figure 21. Typical waveform of valve current and voltage during operational test.

B. Test for HVDC circuit breaker

As a new kind of high-end electric equipment in power electronic field, hybrid DC circuit breaker has such operating principles and conditions that are different from those of conventional AC circuit breakers or DC circuit breakers in medium-low voltage application fields [10]. So far, there has been no international or national standard that can be referred to for its electrical test [11]. The complete design of type test proposed in this paper is based on operation test and insulation test. Breaking test is the core of operation test.

LC power supply creates high current to simulate the fault current by discharging the pre-charged capacitor to the inductor, mainly utilizing a quarter of cycle period at LC oscillation current rise stage. The advantages of LC power supply include large impulse power, high energy density, easy realization of charge and discharge control, as well as wider range of power supply parameter regulation. Under failure transient state, LC oscillation power supply is the test power supply closest to real VSC-HVDC grid system, di/dt of test current is determined by initial test voltage and test reactance, and the reduction rate of test voltage is determined jointly by capacitor's capacity and reactance value [12].

In order to verify the core functionality of 500kV DC circuit breaker, overall breaking test on whole machine has been implemented. Figure 22 shows the breaking test circuit with LC oscillation power supply. Once the capacitors C are fully charged, the power source is isolated by turning-off mechanical switch K. After breakover of DC circuit breaker, the thyristor valve T is triggered to start test and the DC circuit breaker will finally complete breaking. With multiple-trigger of the thyristor valve, reclosing test will be achieved. The test inductors L are used for regulating the test current rising rate and the final test breaking current amplitude.

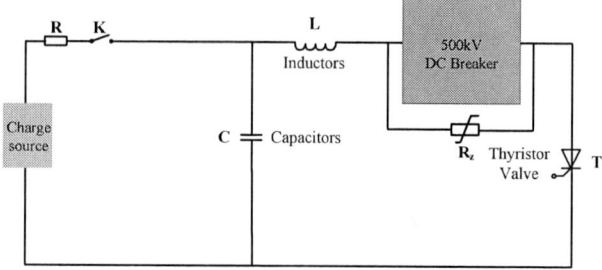

Figure 22. Breaking test circuit for 500kV breaker prototype.

In July 2017, the developed DC breaker prototype passed the type test. Figure 24 shows the waveform of the reclosing test of the DC breaker. Reclosing test includes two short-circuit current breaking processes. In first breaking operation, the breaker starts to operate 1.6ms after the thyristor valve has been triggered and takes 2.6ms to turn off 26kA current, and the TIV in this process exceeds 810kV. The second breaking operation will dial with 9kA current some 300 milliseconds later.

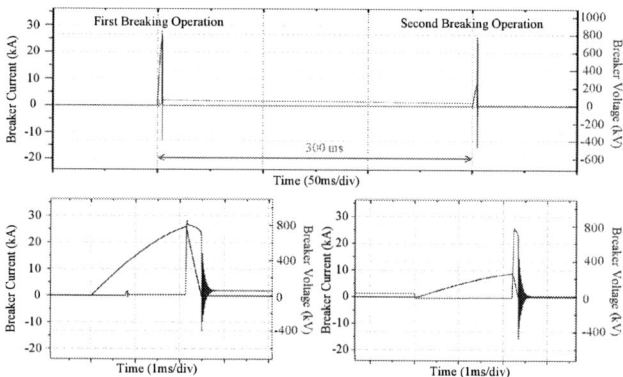

Figure 23. Reclosing test waveforms of 500 kV hybrid HVDC breaker.

Table IV lists the insulation test results for the 500kV DC breaker prototype.

TABLE IV
INSULATION TEST

Test Item	Test Result
Terminal-to-ground DC voltage withstand test	856 kV/1 min, 588 kV/3 h
Terminal-to-ground switching impulse voltage withstand test	1175 kV
Inter-terminal DC voltage withstand test	856kV/5 s, 588 kV/10 min

VI. CONCLUSION

Zhangbei project adopts half-bridge modular multi converter (MMC) and uses overhead lines, so that 535kV DC breaker must be configured at the both end of line to clearthe potential dc fault such as the lighting strike. In this paper, a system scheme is presented, and some results are provided. Meanwhile, a novel modular cascaded hybrid DC breaker concept is proposed, and 500kV VSC valve and DC breaker prototype are developed to meet the requirement of the Zhangbei four-terminal dc grid. The project will be put into operation by 2019.

The technological advantages that are available in large-scale renewable energy integration have made DC power grid technology viable, attracting global attention. In particular, this technology is in great demand in Europe, North Africa, and northwestern China, where renewable energy generates a relatively large proportion of electricity power. In the coming decades, the DC power grid technology will embrace rapid development. The experience gained from designing, developing, and constructing the Zhangbei DC power grid will serve as a great reference for development and application of this technology worldwide.

REFERENCES

[1] Chengyu Li, Chengyong Zhao, Jianzhong Xu, et al. "A Pole-to-Pole Short-Circuit Fault Current Calculation Method for DC Grids", IEEE Transactions on Power Systems, 2017, 11pages.

[2] Leterme W, Hertem D V. "Classification of Fault Clearing Strategies for HVDC Grids", Cigre Lund Symposium 2015, Sweden.

[3] Li, Rui, et al. "Continuous operation of radial multi-terminal HVDC systems under DC fault", IEEE Transactions on Power Delivery, 2016, 11pages.

[4] Sano, Kenichiro, and Masahiro Takasaki. "A surge-less solid-state DC circuit breaker for voltage-source-converter-based HVDC systems", IEEE Transactions on Industry Applications, 2014, 10.

[5] Cui S, Sul S K. "A Comprehensive DC Short-Circuit Fault Ride Through Strategy of Hybrid Modular Multilevel Converters (MMCs) for Overhead Line Transmission", IEEE Transactions on Power Electronics, 2016, 17pages.

[6] Oliveira R, Yazdani A. "A Modular Multilevel Converter with DC Fault Handling Capability and Enhanced Efficiency for HVDC System Applications", IEEE Transactions on Power Electronics, 2016, 12pages.

[7] ADAM G P, ANAY A O, BURT G. Multi-terminal DC transmission system based on modular multilevel converter[C]//Proceedings of the 44th International Universities Power Engineering Conference, Glasgow, Scotland, UK, 2009:1-5.

[8] NAMI A, LIANG J, DIJKHUIZEN F, et al. Analysis of modular multilevel converters with DC short circuit fault blocking capability in bipolar HVDC transmission systems[C]//17th Power Electronics and Applications of Conference, Geneva, 2015, 1-10.

[9] G.F. TANG, X. G. WEI, W.D. ZHOU, "Research and Development of a Full-bridge Cascaded Hybrid HVDC Breaker for VSC-HVDC Applications," CIGRE 2016 Session, Paris, paper A3-117, Sep.2016.

[10] X. Ding, G. F. Tang, and M. X. Han, "Current-breaking test method based on LC source for hybrid HVDC circuit breaker," Electric Power Construction, vol.38, no. 08, pp. 2-9, 2017.

[11] Ding, G. F. Tang, and M. X. Han, "Characteristic parameters extraction and application of the hybrid DC circuit breaker in MMC-HVDC," Proceedings of the CSEE, vol.38, no. 01, pp. 32-39, 2018.

[12] X. Ding, G. F. Tang, and M. X. Han, "Design and equivalence evaluation of type test for hybrid DC circuit breaker," Power System Technology, vol.42, no. 01, pp. 72-78, 2018.

The 2018 International Power Electronics Conference

What Led to Success in Academic Research on the Family of Modular Multilevel Cascade Converters?

Hirofumi Akagi
Department of Electrical and Electronic Engineering
Tokyo Institute of Technology, Tokyo, Japan
E-mail: akagi@ee.titech.ac.jp

Abstract—**This paper provides a comprehensive discussion on the topology and terminology of the family of modular multilevel cascade converters, including their chronological review. It is followed by answering the following question: "What motivated the author to apply phase-shifted-carrier PWM to the multilevel converters?" The PWM is achievable from integrating inter-cluster balancing or inter-arm balancing control into the middle layer of a hierarchical control system consisting of three layers. This integration makes it easy to expand the phase-shifted-carrier PWM to any bridge-cell or chopper-cell count per cluster or arm.**

Keywords—Circulating currents, grid-tied applications, hierarchical control, modular multilevel cascade converters, and pulsewidth modulation (PWM).

I. INTRODUCTION

A. Chronological Review

In 1971, McMurry patented a basic circuit configuration based on multiple single-phase full-bridge (H-bridge) cells connected in cascade [1]. In 1981, Alesina and Venturini published a theoretical paper on generalized circuit configurations based on multiple full-bridge or half-bridge cells connected in cascade [2]. However, they were unable to verify their circuit configurations experimentally in the 1970s and 1980s. Remarkable technology developments over the last four decades have enabled to put their circuit configurations on the market.

Around 1994, Robicon Corporation, presently a part of Siemens AG, commercialized medium-voltage high-power motor drives combining three-phase "cascaded H-bridge" multilevel inverters with three-phase phase-shifted multi-winding transformers [3]. The emergence of the products on the market surprised and impressed research scientists and engineers who were engaged in research and development on medium-voltage motor drives. The circuit configuration per phase is the cascade connection of the ac output terminals of multiple H-bridge cells. Since 1994, the cascaded H-bridge inverters have been recognized as the origin of modern multilevel inverters. However, their manufacturers have suffered from the phase-shifted multiwinding transformers that are complicated, heavy, costly, and prone to failure.

In 1996, Lai and Peng presented a high-voltage static synchronous compensator (STATCOM) for reactive-power control in power transmission systems [4], [5]. This circuit configuration is referred to as a "cascade multilevel converter," which is the same as the cascaded H-bridge inverter except for having no phase-shifted multiwinding transformer. The STATCOM

is characterized by adopting staircase modulation (SCM), in which the actual switching frequency is equal to the line frequency.

In 2006, Akagi and his coauthors presented a practical paper on a three-phase STATCOM based on the cascade multilevel converter characterized by phase-shifted-carrier PWM [6], [7]. It consists of three bridge cells per cluster, using a total of 36 IGBTs. Experimental waveforms obtained from a downscaled STATCOM rated at 200 V and 10 kVA were included to verify the validity and effectiveness of voltage-balancing control of all the floating or flying dc capacitors. As discussed in section IV, hierarchical control was introduced to the voltage-balancing control, in which inter-cluster balancing control was integrated into the middle layer. This integration resulted in providing excellent voltage-balancing performance even in transient states, making sinusoidal the three-phase currents at the ac side of the STATCOM with neither harmonic nor switching-ripple filter.

From 2003 to 2005, Marquardt and his coauthors presented a series of innovative papers on high-power conversion systems intended for applications to HVDC (high-voltage direct current) transmission systems [8]–[11]. They named the power conversion systems as "modular multilevel converter (MMC)," and made a lucid description of the circuit configurations and its basic principles of operation. However, they disclosed neither experimental result nor waveform, and made no detailed description of how to realize not only capacitor-voltage balancing but also pulsewidth modulation (PWM).

In 2008, Akagi and his coauthor presented a seminal paper on a single-leg MMC or DSCC (double-star chopper-cell) inverter characterized by phase-shifted-carrier PWM at the last IEEE PESC [12], followed by its Transactions paper [13]. These two papers provided experimental verification of the effectiveness and validity of capacitor-voltage balancing. One main reason for succeeding in their experiment was to define the circulating current that plays an essential role in power conversion, as discussed in section V. Another was to integrate a circulating-current control loop into the hierarchical control system stemming from that of the three-phase STATCOM presented in 2006.

In 2009, a few research papers on modular multilevel converters or DSCC converters were presented [14], [15]. However, these papers included no concept of any circulating current, made no description of capacitor-voltage balancing, and disclosed neither experimental result nor waveform.

Since 2010, the two seminal papers [12], [13] have spurred

2352

many scientists and engineers in academia and industry to do further research on MMCs or DSCC converters including the concept of the circulating current in the single-leg DSCC inverter or slightly-modified circulating currents in three-leg (three-phase) DSCC inverters.

Nowadays, DSCC converters have been applied to long-distance HVDC transmission systems, BTB (back-to-back) systems with the same or different line frequencies, and medium-voltage high-power motor drives.

B. Terminology Issue and its Solution

Since the technical term "modular multilevel converter" has a broad meaning, the name/designation makes it difficult for the beginners of power electronics to distinguish one circuit configuration from the others. In other words, the term "modular" does not have enough information to identify circuit configurations. Moreover, it is reasonable to refer to it as a "cascade multilevel converter" because the modular multilevel converter is based on the "cascade" connection of multiple bidirectional choppers.

On the other hand, the multilevel inverter commercialized by Robicon Corporation is now referred to as a "cascade multilevel inverter" [5], "cascaded H-bridge inverter" [16], or "chain-link multilevel inverter" [17]. This means that different manufactures or companies use different names, like trade names. However, the cascade multilevel inverter can be considered also as a "modular multilevel inverter" because it is based on the cascade connection of "modular" H-bridge cells. This may lead to the following confusion: When a power electronics engineer uses either "modular multilevel converter" or "cascade multilevel converter" in his/her technical paper/article or presentation, the other engineers cannot identify the circuit configuration or may have a misunderstanding about it in the worst case [18], [19].

Avoiding the above confusion may allow the use of given/first names and a family/last name for the modern multilevel converters including six different circuit configurations. The family/last name of "modular multilevel cascade converters" merges together the two terms, "cascade multilevel converters" and "modular multilevel converters." On the other hand, the following given/first names are assigned to the six family members [18], [19]:

- SSBC (single-star bridge-cell) or a "cascade multilevel converter with star configuration" in [5].

- DSBC (double-star bridge-cell) or an "MMC with full-bridge submodules" in [20].

- TSBC (triple-star bridge-cell) or a "modular multilevel matrix converter" in [21]

- SDBC (single-delta bridge-cell) or a "cascade multilevel converter with delta configuration" in [5], [22].

- DDBC (double-delta bridge-cell) or a "hexagonal converter" in [23].

- DSCC (double-star chopper-cell) or an "MMC with half-bridge submodules" in [20].

Unless these given/first names cause any confusion or misunderstanding, it is possible to eliminate the family/last name.

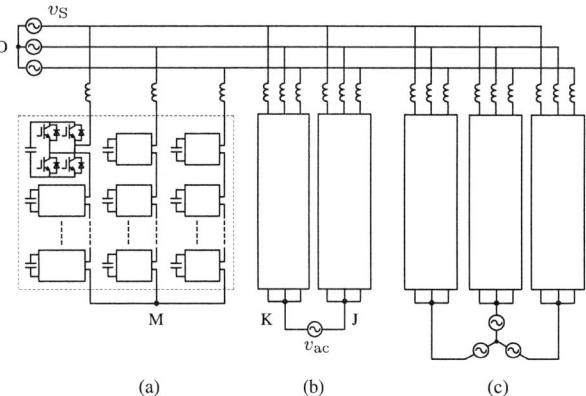

(a)   (b)   (c)

Fig. 1. Modular multilevel xSBC converters, where x = S, D, and T. Each of the five white boxes is the same as the SSBC surrounded by four straight dashed lines in the left. (a) The SSBC (single-star bridge-cell) converter. (b) The DSBC (double-star bridge-cell) converter. (c) The TSBC (triple-star bridge-cell) converter [19].

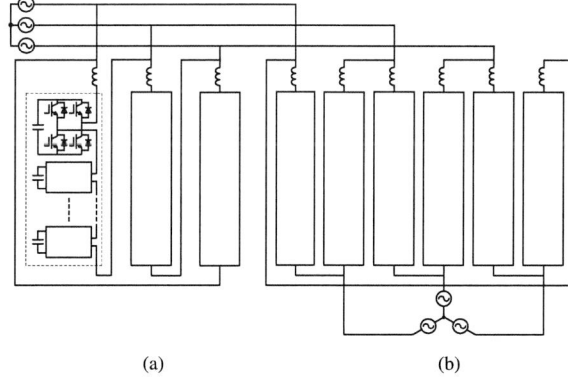

(a)   (b)

Fig. 2. Modular multilevel xDBC converter, where x = S and D. Each of the eight white boxes is the same as the cluster surrounded by four straight dashed lines in the left. (a) The SDBC (single-delta bridge-cell) converter. (b) The DDBC (double-delta bridge-cell) converter [19].

II. Modular Multilevel Cascade Converters

A. Circuit Configurations of Six Family Members

Figs. 1 to 3 summarize six circuit configurations that belong to the family of modular multilevel cascade converters. The reader may ask a question, "What is the concept or base common to the family?" The author can answer it as follows; "modular" structure, "multilevel" voltage, and "cascade" connection. This enables both reader and author to use the term "modular multilevel cascade converters" as their family name. However, only the family name cannot identify the individual family members.

Fig. 1 shows the following multilevel converters; the single-star bridge-cell (SSBC) converter for (a), the double-star bridge-cell (DSBC) converter for (b), and the triple-star bridge-cell (TSBC) converter for (c). One reason for their naming is that the three converters are based on one (single), two (double), and three (triple) set(s) of three star-configured clusters, respectively. The other reason is that each cluster consists of multiple "bridge cells" connected in cascade. This

The 2018 International Power Electronics Conference

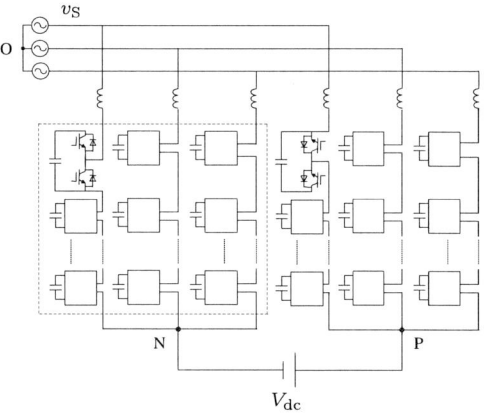

Fig. 3. Modular multilevel DSCC (double-star chopper-cell) converter [19].

paper refers to a string of the multiple cascaded bridge cells as a "cluster," to distinguish it from the already-existing terms "arm" and "leg" that are used for traditional three-phase two-level voltage-source PWM converters or inverters consisting of six power switching devices.

Fig. 1(a) has no current path or branch between the cluster mid-point M and the supply-voltage neutral point O. Hence, no current would flow between points M and O even if any voltage was injected as v_{MO} between points M and O. Appropriate adjustment of v_{MO} allows the SSBC converter to achieve inter-cluster balancing control [24]–[32]. Note that the intentionally-injected voltage v_{MO} does not come across any line-to-line voltage at the ac side of the SSBC converter.

Fig. 1(b) has a current path between two mid-points J and K, unlike Fig. 1(a). The voltage difference between v_{JO} and v_{KO} is given by

$$v_{JK} = v_{JO} - v_{KO} = v_{ac}. \tag{1}$$

Appropriately adjusting of both v_{JO} and v_{KO} allow the DSBC converter to achieve bidirectional power conversion between the three-phase supply voltage sources v_S and the single-phase ac voltage source v_{ac}.

Fig. 2 shows the following multilevel converters; the single-delta bridge-cell (SDBC) converter for (a) and the double-delta bridge-cell (DDBC) converter for (b). One reason for their naming is that the two converters are based on one (single) and two (double) set(s) of three delta-connected clusters, respectively. The other reason is that each cluster consists of multiple "bridge cells" connected in cascade.

Fig. 3 shows the circuit configuration of the double-star chopper-cell (DSCC) converter. The naming is based on two (double) sets of three star-connected clusters. Each cluster consists of multiple bidirectional "chopper cells" connected in cascade. The DSCC converter can be considered as two identical single-star chopper-cell (SSCC) converters. However, three clusters in one SSCC converter are opposite in voltage polarity to those in the other.

B. Topological Discussion on the DSBC and DSCC Converters

The DSBC converter shown in Fig. 2(b) can connect a single-phase ac voltage source v_{ac} between two mid-points J and K. On the other hand, the DSCC converter shown in Fig. 3 can connect a dc voltage source V_{dc} between two mid-points P and N. The dc voltage source is given by

$$V_{dc} = v_{PO} - v_{NO}. \tag{2}$$

The dc voltage in (2) can be considered as a special case of the ac voltage source v_{ac} in (1). This leads to imposing the following constraint on V_{dc}.

$$V_{dc} \geq 2\sqrt{2}V_S/\sqrt{3}, \tag{3}$$

where V_S is the three-phase supply line-to-line rms voltage. Whenever the above constraint is effective in Fig. 2(b), each bridge cell always produces a positive voltage at the arm side. This means that the constraint imposed by (3) enables each bridge cell to be replaced with each chopper cell. Hence, it is concluded that the DSCC converter is a special case of the DSBC converter in terms of circuit configuration because it can be derived from the DSBC converter, as mentioned above.

III. A DOWNSCALED BTB (BACK-TO-BACK) SYSTEM USING TWO DSCC CONVERTERS

This section presents a three-phase BTB system without dc-link capacitor between the two identical 17-level DSCC converters [33]. Each DSCC converter phase-shifted-carrier PWM consists of eight chopper cells per arm.

A. Circuit Configuration

Fig. 4 shows the overview of the three-phase 200-Vac, 400-Vdc, 10-kW, 50-Hz BTB system designed, built, and tested in the author's laboratory. It includes a practical starting circuit consisting of a magnetic contactor and a current-limiting resistor at each ac side. Note that the two inductors per leg in Fig. 3 are replaced with a single center-tapped inductor L_Z per leg. This replacement leads to significant reductions in volume and weight [13]. The ac terminals of each DSCC converters are connected to the three-phase 200-V ac mains through a per-phase ac-link inductor L_{ac}, a starting circuit and a line-frequency transformer for galvanic isolation. The positive directions of i_{dc} and p are defined as the direction from DSCC-A to DSCC-B (left to right).

Fig. 5 presents the photograph of the 400-V_{dc}, 10-kW BTB system, including the digital controller and data acquisition systems. This BTB system was used in the following experiments.

Table I summarizes the circuit parameters of Fig. 4, which were used in the following experiment. Since the dc-link voltage reference was set to $v_{dc}^* = 400$ V, the dc reference voltage of each dc floating or flying capacitor was set to $v_C^* = 50$ V (= 400 V/8). Each triangular-carrier frequency is set at 450 Hz. Note that the 16 triangular carriers were phase-shifted each other by 22.5° (= 360°/16). Therefore, each DSCC converter produced a 17-level (line-to-neutral) waveform with a voltage step as low as 25 V (= 50 V/2) at the ac terminals, with the help of the center-tapped inductor.

2354

The 2018 International Power Electronics Conference

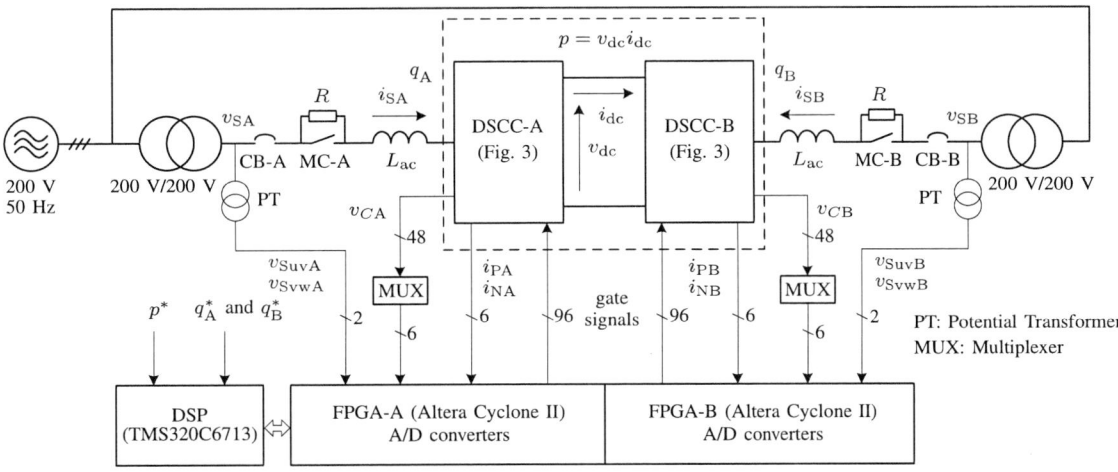

Fig. 4.   Overview of the three-phase 200-V, 10-kW, 50-Hz downscaled BTB system with no dc-link capacitor between DSCC-A and DSCC-B [33].

TABLE I.   CIRCUIT PARAMETERS OF FIG. 4 [33].

Rated power	P	10 kW
Nominal line-to-line rms voltage	V_S	200 V
Nominal line frequency	f_S	50 Hz
Chopper-cell count per arm	n	8
AC-link inductor	L_{ac}	2 mH (16%)
Center-tapped inductor	L_Z	3 mH (24%)
Starting resistor	R	20 Ω
DC-link reference voltage	V_{dc}^*	400 V
DC-capacitor reference voltage	V_C^*	50 V
DC capacitor	C	6.6 mF
Unit capacitance constant [34]	H_C	40 ms at 50 V
Triangular-carrier frequency	f_C	450 Hz
Equivalent carrier frequency	$2nf_C$	7.2 kHz
Dead or blanking time		8 μs

The value in ( ) is on a three-phase 200-V, 10-kW, 50-Hz base

Fig. 5.   Photograph of the experimental BTB system of Fig. 4 [33].

B. Hierarchical Control and Operating Performance

The voltage control of all the floating dc capacitors is characterized by the following hierarchical control system consisting of three layers:

1) Overall voltage control.
2) Inter-arm balancing control.
3) Intra-arm balancing control.

Subsection IV-B makes a detailed description of the hierarchical control system.

Fig. 6 show experimental waveforms in DSCC-A under steady and transient states, where the active-power reference p^* was changed from 10 kW to -10 kW with a ramp function in 20 ms. Note that q_A^* and q_B^* were set to zero. This means that DSCC-A changed its operation from the rated rectification to the rated inversion whereas DSCC-B did it from the rated inversion to the rated rectification. Such an extremely fast response can enhance transient system stability and frequency regulation in "contingency situations," for example, sudden disconnections of large-capacity synchronous generators from either power system. However, such a fast response may not be required under normal operating conditions.

According to [33], the experimental waveforms obtained from the BTB system agreed well with simulated waveforms obtained from the software package, "PSCAD/EMTDC" under the same operating conditions, circuit parameters, and control gains. This means that both experiment and simulation are reliable enough to investigate a more practical system [35] and fault-ride-through (FRT) performance [36].

IV.   HIERARCHICAL CONTROL FOR AN SSBC-BASED STATCOM WITH PHASE-SHIFTED-CARRIER PWM

A. Background and Motivation

In the 1990s, flexible ac transmission system (FACTS) devices such as unified power flow controllers (UPFCs) and

2355

The 2018 International Power Electronics Conference

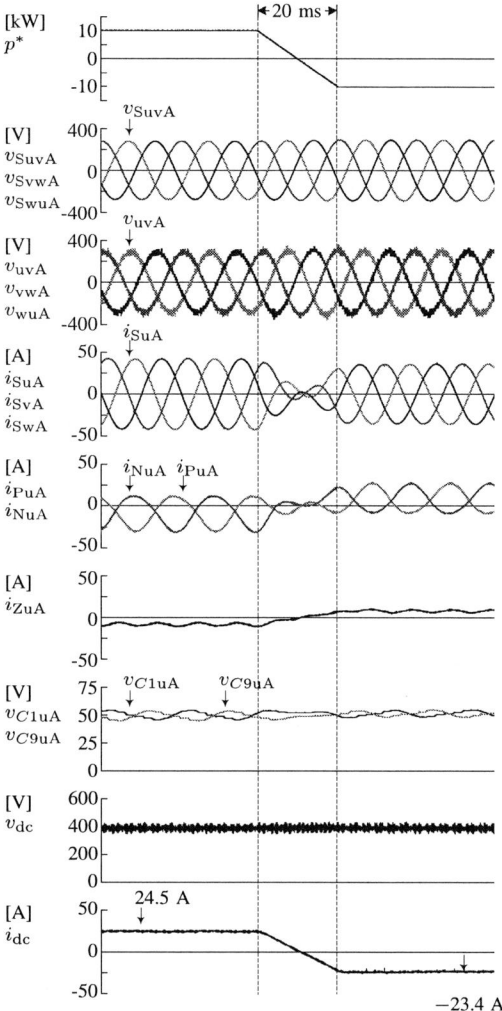

Fig. 6. Experimental waveforms to a ramp change in p^* from 10-kW (rated) rectification to 10-kW inversion where $q_A^* = q_B^* = 0$ [33].

STATCOMs using gate-turn-off (GTO) thyristors were installed on power transmission systems. At present, actually-operating FACTS devices in a broad sense are a few in Japan, and are not so many even in the world. The GTO thyristors had been replaced gradually with insulated-gate bipolar transistors (IGBTs). Since 1999, IGBTs have been applied to the three-level neutral-point-clamped (NPC) PWM inverter [37] for driving main traction motors in Japanese bullet trains or "Shinkansen." This epoch-making event made the author interested in a transformerless SSBC-based STATCOM using IGBTs for direct installation on the 6.6-kV industrial and utility distribution systems in Japan.

In 2002, Akagi and his graduate students commenced doing comprehensive research on a medium-voltage transformerless STATCOM based on the SSBC converter shown in Fig. 1(a). His long experience on power electronics had helped to recognize the superiority of phase-shifted-carrier PWM to staircase modulation (SCM) for medium-voltage SSBC converters. The

reason was that IGBTs have better switching performance as well as much more compact gate-drive and auxiliary circuits than GTO thyristors. The better switching performance leads to higher switching frequencies, thus resulting in bringing excellent current-control performance to the SSBC converters. In fact, phase-shifted-carrier PWM has the following practical advantage: The actual switching frequency of each IGBT is exactly equal to the triangular-carrier frequency, independent of operating conditions including transient states. This is welcomed by design engineers of the SSBC converters. However, one of the most crucial issues at that time was how to realize voltage balancing of all the floating or flying capacitors in the SSBC converters characterized by phase-shifted-carrier PWM.

B. Hierarchical Control

In 2002, Akagi inquired of himself the following question: "To what should the top priority be given in academic research on capacitor-voltage balancing for an SSBC-based STATCOM characterized by phase-shifted-carrier PWM?" His answer was that the top priority is capable of easy expansion to any SSBC-based STATCOM, irrespective of a bridge-cell count per cluster. This priority initiated him to introduce the so-called "hierarchical control system" to the STATCOM. This system consists of the following three layers:

- The top layer takes part of overall voltage control.

- The middle layer is responsible for inter-cluster balancing control [31] or cluster balancing control [6], [7].

- The bottom layer is responsible for intra-cluster balancing control [31] or individual balancing control [6], [7].

Fig. 7 depicts the concept of the overall voltage control. Here, the modular multilevel SSBC converter shown in Fig. 1(a) would act as if it were a traditional three-phase two-level PWM converter equipped with a single floating or flying dc capacitor. The capacitor voltage is given as an arithmetic average of all the dc capacitors in the SSBC converter.

Fig. 8 shows the concept of the inter-cluster balancing control. Here, the SSBC converter would act as if it were three single-phase two-level PWM converters, each of which is equipped with a single floating flying dc capacitor. The capacitor voltage in each cluster is given as an arithmetic average of all the dc capacitors in the corresponding cluster. This control layer plays an important role in achieving voltage balancing among the three clusters. It relies on injecting an appropriate amount of zero-sequence voltage v_{MO} between points M and O in Fig. 8, or superimposing a small amount of three-phase negative-sequence currents on the three-phase supply currents drawn from the ac mains.

Fig. 9 shows the concept of the intra-cluster balancing control. It is so straightforward that it contributes to voltage balancing of all the capacitors inside each cluster. Note that the three clusters are independent of each other.

In summary, the integration of the inter-cluster balancing control into the hierarchical control system has led to succeeding in academic research on the SSBC converters for STATCOMs [24]–[26] and battery energy storage systems

The 2018 International Power Electronics Conference

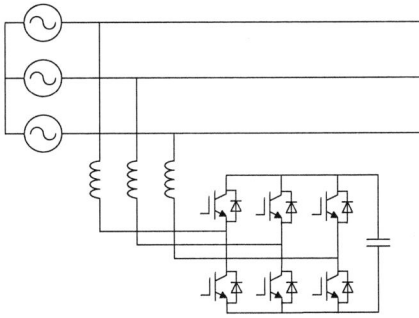

Fig. 7. Concept of the overall voltage control in the top layer.

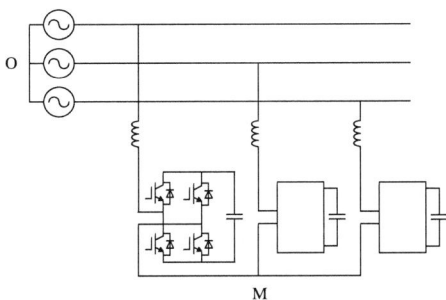

Fig. 8. Concept of the inter-cluster balancing control in the middle layer.

[27]–[30], and on the SDBC converters for STATCOMs [22] and utility-scale photovoltaic systems [31], [32].

V. CIRCULATING CURRENTS

A. Background and Motivation

Around 2005, Akagi had a chance of reading a series of Marquardt's papers on two slightly-different modular multi-level converters [8]–[11], or more strictly, DSCC and DSBC converters. At a glance, Akagi had an intuition that the two modular multilevel converters show considerable promise as high-voltage ac-to-dc and ac-to-ac bidirectional power converters in terms of circuit configuration and operating principle.

In 2005, Akagi's research group succeeded in experimental verification of the three-phase SSBC-based STATCOM with phase-shifted-carrier PWM, as described in the previous section [6], [7]. This experience made him convinced of conducting academic research on the DSCC converter with phase-shifted-carrier PWM, paying attention to capacitor-voltage balancing.

B. Circulating Current in a Cycloconverter

The DSCC converter shown in Fig. 3 is more complicated in capacitor-voltage balancing than the SSBC converter shown in Fig. 1(a). What led to success in academic research on the DSCC converter can be summarized as follows: Akagi gained experience of research on naturally-commutated cyclo-converters operating with either circulating-current mode or non-circulating-current mode in the 1970s. In other words, his research experience of the cycloconverters made him recognize

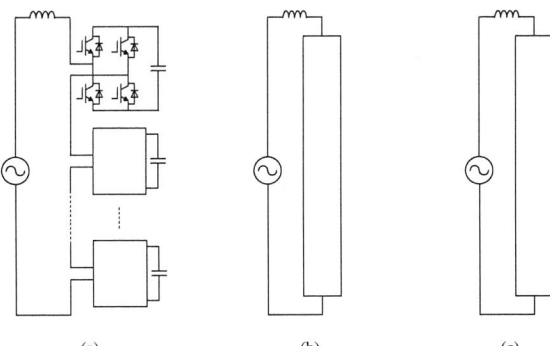

(a)   (b)   (c)

Fig. 9. Concept of the intra-cluster balancing control in the bottom layer. (a) The a-phase cluster. (b) The b-phase cluster. (c) The c-phase cluster.

well what the circulating current of the cycloconverters operating with circulating-current mode was, and how it contributed to power conversion.

Fig. 10 shows a naturally-commutated cycloconverter operating with circulating-current mode. It consists of a three-phase twelve-pulse transformer, two three-phase full-bridge positive and negative thyristor converters, and a center-tapped inductor. This cycloconverter feeds electric power from the three-phase ac mains to a single-phase R-L load at a much lower frequency than the line frequency. The cycloconverter has both positive and negative thyristor converters operated always, unlike the other cycloconverter with non-circulating-current mode. Since both thyristor converters continue carrying positive dc currents, the circulating current i_z should be controlled so as to satisfy the following relations, taking into account the positive directions of all the three loop currents in Fig. 10.

For the positive thyristor converter, $i_Z + i_L/2 > 0.$ (4)

For the negative thyristor converter, $i_Z - i_L/2 > 0.$ (5)

C. Circulating Current in a DSCC Inverter

Fig. 11 shows a single-phase single-leg DSCC inverter with an R-L load. The following equation regarding three branch currents i_P, i_N, and i_L exists at nodes M and N;

$$i_L = i_P - i_N. \quad (6)$$

The above equation means that two independent branch currents among the three should be taken into account.

Fig. 11(a) shows two independent loop currents when a pair of i_P and i_N is selected. This selection is so straightforward that almost all beginners of power electronics do so. However, it would be an unfavorable selection from the following reasons: The waveforms of i_P and i_N include dc and ac components even in ideal operating conditions, where the ac component is related to the output frequency. On the other hand, when i_L is selected as one independent loop current, either i_P or i_N should be selected as the other independent loop current. However, this selection would be unfavorable because it leads to "asymmetry" in terms of control.

Fig. 11(b) shows two independent loop currents when a pair of i_L and i_Z is selected. This selection is so favorable

2357

as to fully solve the above issues caused by the selection of a pair of i_P and i_N or a pair of i_L and either i_P or i_N. The waveform of i_Z includes only a dc component under ideal operating conditions. This fact makes the current control of i_Z simple, easy, and accurate.

The above interesting consideration can be discussed from a different point of view. The following relation exists among the four loop currents in Fig. 11(a) and (b):

$$\begin{bmatrix} i_L \\ i_Z \end{bmatrix} = \begin{bmatrix} 1 & -1 \\ 0.5 & 0.5 \end{bmatrix} \begin{bmatrix} i_P \\ i_N \end{bmatrix}. \tag{7}$$

Since the determinant of the two-dimensional matrix in (7) is unity, the inverse matrix exists as follows:

$$\begin{aligned} \begin{bmatrix} i_P \\ i_N \end{bmatrix} &= \begin{bmatrix} 1 & -1 \\ 0.5 & 0.5 \end{bmatrix}^{-1} \begin{bmatrix} i_L \\ i_Z \end{bmatrix} \\ &= \begin{bmatrix} 0.5 & 1 \\ -0.5 & 1 \end{bmatrix} \begin{bmatrix} i_L \\ i_Z \end{bmatrix}. \end{aligned} \tag{8}$$

Equations (7) and (8) mean that the two-dimensional matrix achieves a reversible linear transformation of current from a mathematical point of view [38]. This transformation is somewhat similar to the well-known "three-phase to two-phase transformation," that is often used in modeling, analysis, and control of three-phase circuits and motors without zero-sequence voltage or current.

The single-leg DSCC inverter in Fig. 11(b) has the following constraint on the circulating current from active-power balance between the dc and ac sides:

$$2EI_Z = V_L I_L \cos\theta, \tag{9}$$

where I_Z is the dc component of i_Z, V_L and I_L are the load voltage and current in rms, and $\cos\theta$ is the load power factor.

D. Similarity and Difference in Circulating Current

Fig. 10 looks somewhat similar to Fig. 11 in circuit configuration. This enabled the author to define the circulating current i_Z for the DSCC inverter, as discussed in the previous subsection. However, the following differences exist in circulating current between the cycloconverter and the DSCC inverter.

- Equations (4) and (5) mean strict constraints on the circulating current in the cycloconverter because each thyristor converter has the nature of unidirectional current flow. However, power balance between the three-phase ac input and the single-phase ac output has no relation to the circulating current. This means that the cycloconverter should regulate the circulating current to a constant value that should be larger than the peak load current, independent of the electric power converted by the cycloconverter.

- Equation (9) results from power balance between the dc input and the ac output. It is a strict constraint on the circulating current in the DSCC inverter. However, the peak load current itself imposes no constraint on the circulating current.

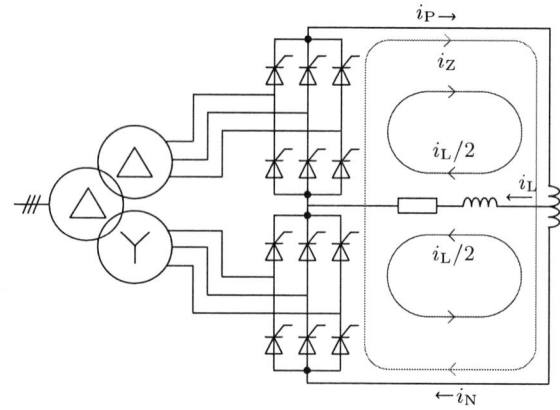

Fig. 10. A naturally-commutated cycloconverter with a single-phase R-L load, operating with circulating-current mode.

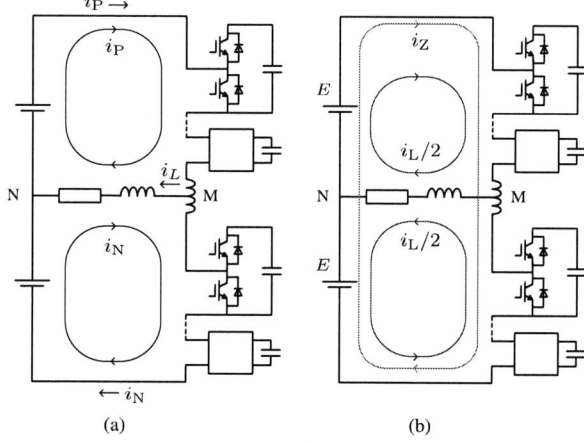

(a) (b)

Fig. 11. Two independent loop currents in a single-leg DSCC inverter. (a) When a pair of i_P and i_N is selected. (b) When a pair of i_Z and i_L is selected.

The single-phase DSCC inverter shown in Fig. 11 consists of a single leg and a center-tapped inductor. Therefore, it allows the single circulating current to circulate through the two dc voltage sources and the inductor, without flowing into, or out of, the R-L load. On the other hand, the three-phase DSCC inverter in Fig. 3 consists of three legs. Thus, it has two additional independent loop currents circulate through two legs among the three, which do not flow through the dc voltage sources. As a result, ac components included in the individual circulating currents have a freedom to select their frequencies and amplitudes from a practical point of view. This freedom makes use of inter-arm balancing among the six DSCC arms [38].

VI. Conclusion

This paper has made an intensive description of modular multilevel converters or modular multilevel cascade converters characterized by phase-shifted-carrier PWM, with focus on hierarchical control and circulating current. It would be interesting to the reader that unless the author had experienced any research on naturally-commutated cycloconverters operating

The 2018 International Power Electronics Conference

with non-circulating-current or circulating-current mode, it would have taken much more time to notice the existence of the circulating current, and then, to define it. In the worst case, he would have failed in achieving voltage balancing of all the floating or flying capacitors in a single-leg DSCC inverter, followed by thee-leg DSCC converters and inverters for grid-tied applications and motor drives. Finally, this paper concludes that the author owes his success in research to phase-shifted-carrier PWM, hierarchical control, and the circulating current resulting from similarity or analogy between the cycloconverter in Fig. 10 and the DSCC inverter in Fig. 11.

REFERENCES

[1] W. McMurry, "Fast responce stepped-wave switching power converter circuit," U. S. Patent, 3 581 212, May 24, 1971.

[2] A. Alesina and M. G. B. Venturini, "Solid-state power conversion: A Fourier analysis approach to generalized transformer synthesis," *IEEE Trans. Circuits Syst.*, vol. 28, no. 4, pp. 319–330, Jul./Aug. 1981.

[3] P. W. Hammond, "A new approach to enhance power quality for medium voltage ac drives," *IEEE Trans. Ind. Appl.*, vol. 33, no. 1, pp. 202–208, Jan./Feb. 1997.

[4] J. S. Lai, and F. Z. Peng, "Multilevel converters—A new breed of power converters," *IEEE Trans. Ind. Appl.*, vol. 32, no. 3, pp. 509–517, 1996.

[5] F. Z. Peng and J. S. Lai, "Dynamic performance and control of a static var generator using cascade multilevel inverters," *IEEE Trans. Ind. Appl.*, vol. 33, no. 3, pp. 748–755, Mar./Jun. 1997.

[6] T. Yoshii, S. Inoue, and H. Akagi, "Control and performance of a medium-voltage transformerless cascade PWM STATCOM with star topology,' in *Proc. IEEE IAS Annual Meeting*, Oct. 2006, pp. 1716–1723.

[7] H. Akagi, S. Inoue, and T. Yoshii, "Control and performance of a transformerless cascade PWM STATCOM with star configuration," *IEEE Trans. Ind. Appl.*, vol. 43, no. 4, pp. 1041-1049, Jul./Aug. 2007.

[8] R. Marquardt and A. Lesnicar, "A new modular voltage source inverter topology," in *Proc. EPE*, Sep. 2003.

[9] A. Lesnicar and R. Marquardt, "An innovative modular multilevel converter topology suitable for a wide power range," in *Proc. IEEE Bologna Power Tech*, 2003, vol. 3, pp. 23–26.

[10] M. Glinka and R. Marquardt, "A new ac/ac-multilevel converter family applied to a single-phase converter," in *Proc. IEEE-PEDS*, 2003, vol. 1, pp. 16–23.

[11] M. Glinka and R. Marquardt, "A new ac/ac multilevel converter family," *IEEE Trans. Ind. Electron.*, vol. 52, no. 3, pp. 662–669, 2005.

[12] M. Hagiwara and H. Akagi, "PWM control and experiment of modular multilevel converters," *Proc. IEEE PESC*, Jun. 2008, pp. 154–161.

[13] M. Hagiwara and H. Akagi, "Control and experiment of pulsewidth-modulated modular multilevel converters," *IEEE Trans. Power Electron.*, vol. 24, no. 7, pp. 1737–1746, Jul. 2009.

[14] A. Antonopoulos, L. Angquist, and P. Nee, "On dynamics and voltage control of the modular multilevel converter," in *Proc. EPE*, Sep. 2009.

[15] S. Rohnwe, S. Bernet, M. Hiller, and R. Sommer, "Pulse width modulation scheme for the modular multilevel converter," in *Proc. EPE*, Sep. 2009.

[16] Y. Fukuta and G. Venkataramanan, "DC bus ripple minimization in cascaded H-bridge multilevel converters under staircase modulation," in *Proc. IEEE IAS Annual Meeting*, Oct. 2002, pp. 1988–199.

[17] C. Oates, "A methodology for developing 'chainlink' converters," in *Proc. EPE*, Sep. 2009.

[18] H. Akagi, "Classification, terminology, and application of the modular multilevel cascade converter (MMCC)," *IEEE Trans. Power Electron.*, vol. 26, no. 11, pp. 3119–3130, Nov. 2011.

[19] H. Akagi, "Multilevel converters: Fundamental circuits and systems," *Proceedings of the IEEE*, vol. 105, no. 11, pp. 2048–2065, Nov. 2017.

[20] J. Dorn, H. Gambach, J. Strauss, T. Westerweller, and J. Alligan, "HVDC and power electronic systems for overhead line and insulated cable applications," in *Proc. CIGRE San Francisco Colloquium*, 2012, no. B4-8.

[21] C. Oates and G. Mondal, "DC circulating current for capacitor voltage balancing in modular multilevel matrix converter," in *Proc. EPE*, Aug. 2011.

[22] M. Hagiwara, R. Maeda, and H. Akagi, "Negative-sequence reactive-power control by a PWM STATCOM based on a modular multilevel cascade converter (MMCC-SDBC)," *IEEE Trans. Ind. Appl.*, vol. 48, no. 2, pp. 720–729, Mar./Apr. 2012.

[23] L. Baruschka and A. Mertens, "A new three-phase ac/ac modular multilevel converter with six branches in hexagonal configuration," *IEEE Trans. Ind. Appl.*, vol. 49, no. 3, pp. 1400–1410, May/Jun. 2013.

[24] C. Lee, B. Wang, S. Chen, S. Chou, J. Huang, P. Cheng, H. Akagi, and P. Barbosa, "Active power balancing control pf a STATCOM based on the cascaded H-Bridge PWM converter with star configuration," *IEEE Trans. Ind. Appl.*, vol. 50, no. 6, pp. 3893–3901, Nov./Dec. 2014.

[25] J. I. Y. Ota, Y. Shibano, N. Niimura, and H. Akagi, "A phase-shifted PWM D-STATCOM using a modular multilevel cascade converter (SSBC)–Part I: Modeling, analysis, and design of current control," *IEEE Trans. Ind. Appl.*, vol. 51, no. 1, pp. 279–288, Jan. 2015.

[26] J. I. Y. Ota, Y. Shibano, and H. Akagi, "A phase-shifted PWM D-STATCOM using a modular multilevel cascade converter (SSBC)–Part II: Zero-voltage-ride-through capability," *IEEE Trans. Ind. Appl.*, vol. 51, no. 1, pp. 289–296, Jan. 2015.

[27] L. Maharjan, S. Inoue, H. Akagi, and J. Asakura, "State-of-charge (SOC)-balancing control of a battery energy storage system based on a cascade PWM converter," *IEEE Trans. Power Electron.*, vol. 24, no. 6, pp. 1628–1636, Jun. 2009.

[28] L. Maharjan, T. Yamagishi, and H. Akagi, "Active-power control of individual converter cells for a battery energy storage system based on a multilevel cascade PWM converter," *IEEE Trans. Power Electron.*, vol. 27, no. 3, pp. 1099–1107, Mar. 2012.

[29] N. Kawakami, S. Ota, H. Kon, H. Akagi, H. Kobayashi, and N. Okada, "Development of a 500-kW modular multilevel cascade converter for battery energy storage systems," *IEEE Trans. Ind. Appl.*, vol. 50, no. 6, pp. 3902–3910, Nov./Dec. 2014.

[30] J. I. Y. Ota, T. Sato, and H. Akagi, "Enhancement of performance, availability, and flexibility of a battery energy storage system based on a modular multilevel cascade converter (MMCC-SSBC) " *IEEE Trans. Power Electron.*, vol. 31, no. 4, pp. 2791–2799, Jan. 2016.

[31] P. Sochor and H. Akagi, "Theoretical comparison in energy-balancing capability between star- and delta-configured modular multilevel cascade inverters for utility-scale photovoltaic systems," *IEEE Trans. Power Electron.*, vol. 31, no. 3, pp. 1980–1992, Mar. 2016.

[32] P. Sochor and H. Akagi, "Theoretical and experimental comparison between phase-shifted PWM and level-shifted PWM in a modular multilevel cascade SDBC inverters for utility-scale photovoltaic applications," *IEEE Trans. Ind. Appl.*, vol. 53, no. 5, pp. 4696–4707, Sep./Oct. 2017.

[33] K. Sekiguchi, P. Khamphakdi, M. Hagiwara, and H. Akagi, "A grid-level high-power BTB (Back-To-Back) system using modular multilevel cascade converters without common DC-link capacitor," *IEEE Trans. Ind. Appl.*, vol. 50, no. 4, pp. 2648–2659, Jul./Aug. 2014.

[34] H. Fujita, S. Tominaga, and H. Akagi, "Analysis and design of a dc voltage-controlled static var compensator using quad-series voltage-source inverters," *IEEE Trans. Ind. Appl.*, vol. 32, no. 4, pp. 970–977, Jul./Aug. 1996.

[35] F. Sasongko, K. Sekiguchi, K. Oguma, M. Hagiwara, and H. Akagi, "Theory and experiment on an optimal carrier frequency of a modular multilevel cascade converter with phase-shifted PWM," *IEEE Trans. Power Electron.*, vol. 32, no. 7, pp. 5058–5069, Jul. 2017.

[36] K. Oguma and H. Akagi, "Low-voltage-ride-through (LVRT) control of an HVDC transmission system using two modular multilevel DSCC converters," *IEEE Trans. Power Electron.*, vol. 32, no. 8, pp. 5931–5942, Aug. 2017.

[37] A. Nabae, I. Takahashi, and H. Akagi, "A new neutral-point-clamped PWM inverter," *IEEE Trans. Ind. Appl.*, vol. 17, no. 5, pp. 518–523, Sep./Oct. 1981.

[38] N. Niimura and H. Akagi, "Decoupled control of a three-phase modular multilevel cascade converter based on double-star chopper-cells," (in Japanese) *IEE Japan*, vol. 132-D, no. 11, pp. 1055–1064, Nov. 2012.

The 2018 International Power Electronics Conference

Operating Principle of Current Resonant Converter Using Air Core Transformer for Isolated Power Supply on Chip

Seiya Abe, Hikaru Kaishakuji and Satoshi Matsumoto
Kyushu Institute of Technology
*E-mail: abe@ele.kyutech.ac.jp

Abstract— **A power-SoC can realize compact and thin power supply although the power capacity is very small (low voltage input and low voltage / current output). The isolated power supplies are required to achieve the hybrid connection of series and parallel in order to adapt the high input voltage and high current output applications. On the other hand, the switching frequency tends to be very high due to small passive components. However, the loss of the magnetic material becomes severe problem. In this paper, the air core transformer is adapted and the current resonant converter is evaluated as an example of isolated power-SoC. Moreover, the isolated power supply is implemented on PCB in order to confirm the operating principle of the isolated power-SoC.**

Keywords— *Isolated power-SoC, Air core transformer, Current resonant converter.*

I. INTRODUCTION

Recently, a compact and thin power supply is strongly demanded. This is because the mobile information communication devices such as smart phone, tablet and so on are becoming light and thin. The most effective way to realize the compact and thin power supply is to reduce the size of the inductors and capacitors. A power supply on chip (power-SoC) which integrates MCU, switching devices, control units and passive components on the same chip is shown in Fig. 1, has been paid attentions [1-4]. Because of this, the power-SoC can realize ultimate compact and thin power supply.

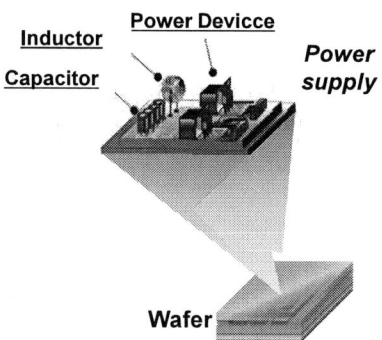

Fig. 1. Power Supply on Chip (Power-SoC).

From the view of the power supply configuration so far, the battery voltage is input the DC-DC converter then low dropout (LDO) regulator makes desired voltage for LSI as shown in Fig. 2 (a) [5]. In the future, the power supply configuration will move to only DC-DC converter configuration without LDO regulator as shown in Fig. 2 (b). In order to realize this system, the high efficiency DC-DC converter is necessary. The advantage of the power-SoC is high efficiency at light load condition, so the power-SoC is suitable for mentioned above system.

On the other hand, the switching frequency tends to be reaches from several tens to hundreds MHz due to small passive components [6-8]. In this case, the loss of the magnetic material becomes severe problem. Thin film magnetic materials have been developed for high frequency applications which is enough performance for magnetic shield. However, they have some issue for the power applications.

In addition, a power-SoC can achieve the compact and thin power supply although the power capacity is very small (low voltage input and low voltage / current output).

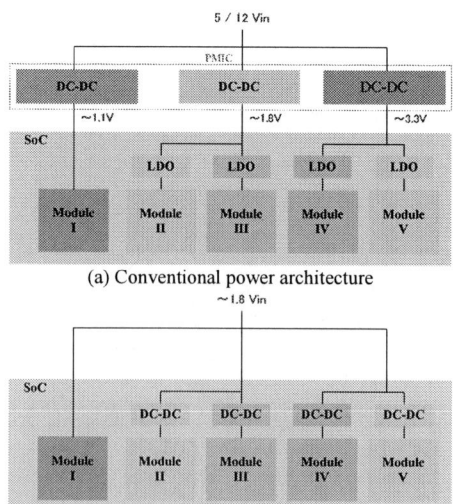

(a) Conventional power architecture

(b) Future architecture with power-SoC
Fig. 2. Power architecture for mobile equipment.

2360

In order to adapt the high input voltage and high current output applications, the isolated power supplies are required to achieve the hybrid connection of series and parallel.

In this paper, the air core transformer is adapted and the current resonant converter is evaluated as an example of isolated power-SoC. Moreover, the isolated power supply is implemented on PCB in order to confirm the operating principle of the isolated power-SoC.

II. CONFIGURATION OF ISOLATED POWER SUPPLY ON CHIP

The isolated power supply is realized by using two spiral inductors on the basis of principal of transformer. The proposed isolated power supply with air core transformer is shown in Fig.2. The isolated power-SoC is manufactured by 3D stacking technology through LSI and MEMS process, which embedded passive components. Hence, the length of interconnection can be minimized. In this case, the loss and voltage variation due to the line resistance and impedance.

Moreover, the coupling factor of the transformer depends on the thin of the isolated layer such as SiO_2, and the circuit topology depends on the coupling factor. When the coupling factor nearly equal 1, the PWM type converter such as Flyback converter can adapt. On the other hand, resonant type converter is suitable for the low coupling factor case.

In this paper, the resonant type converter especially current resonant converter is evaluated for isolated power-SoC. Moreover, the isolated power supply is implemented on PCB in order to confirm the operating principle of the isolated power-SoC.

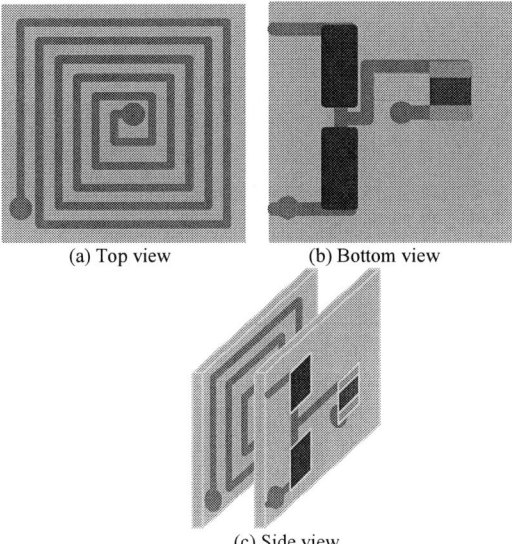

(a) Top view   (b) Bottom view

(c) Side view

Fig. 2. Proposed isolated power supply on chip.

III. EVALUATION OF AIR CORE COIL AND TRANSFORMER

In the isolated power-SoC, the performances of the air core coil and transformer are very important. Especially, the detail discussion of the internal resistance which affects to power conversion efficiency is necessary.

The frequency characteristics of the inductance and the internal resistance are shown in Fig. 3. In this case, the thickness and width of the coil pattern set to be 100um and 0.6mm, respectively. The frequency dependency of the inductance is very small as shown in Fig. 3 (a), and the inductance is increased with increasing the number of turns. On the other hand, the frequency dependency of the internal resistance is very large as shown in Fig. 3 (b), and the internal resistance is increased with increasing the frequency. This resistance increase is caused by skin effect and proximity effect. In this case, since the pattern width is not so wide and inter-pattern distance comparatively wide, the influence of the proximity effect is considered to be dominant.

The pattern width dependency of the inductance and the internal resistance are shown in Fig. 4. In this case, the thickness of the coil pattern and the frequency set to be 100um and 1MHz, respectively. The pattern width dependency of the inductance is low although the inductance is slightly decreased with increasing the line width as shown in Fig. 4 (a). On the other hand, the pattern width dependency of the internal resistance is very large as shown in Fig. 4 (b). When the pattern width is increased, the cross sectional area becomes large, so the resistance decreases.

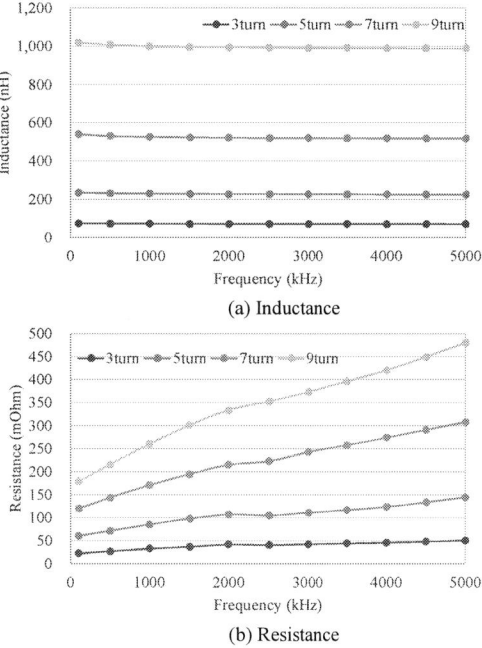

(a) Inductance

(b) Resistance

Fig. 3. Dependency of number of turns.

The 2018 International Power Electronics Conference

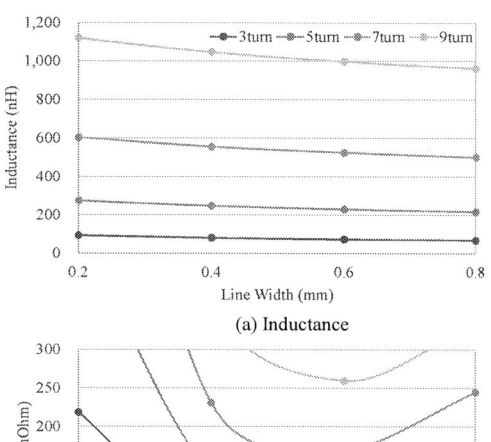

(a) Inductance

(b) Resistance

Fig. 4. Dependency of line width

However, when the pattern width exceeds a certain pattern width, the resistance becomes large. This is because the pattern width becomes wider so that the inter-pattern distance becomes narrower and the influence of the proximity effect becomes greater. In this case, the optimal line width is 0.6mm as shown in Fig. 4 (b).

IV. CHARACTERISTICS OF CURRENT RESONANT CONVERTER

The circuit configuration of the current resonant converter for the isolated power-SoC is shown in Fig. 5. The current resonant converter has four sections which are switch network of primary side, resonant network with air core coil and resonant capacitor, isolation and switch network of secondary side.

A. Analytical Discussions

The fundamental harmonics approximation (FHA) technique is usually used for the analysis of current resonant converter [9-15]. In this way, the AC equivalent model showing in Fig. 6 is used with the f-matrix calculation. Moreover, its operating characteristics can be evaluated by AC analysis. The characteristics equations of the current resonant converter showing in Fig. 5 derive as following equations;

$$
\begin{bmatrix} V_1 \\ I_1 \end{bmatrix} = \begin{bmatrix} 1 & r_1 + s(L_1 - M) + \frac{1}{sC_{r1}} \\ 0 & 1 \end{bmatrix} \begin{bmatrix} 1 & 0 \\ \frac{1}{sM} & 1 \end{bmatrix} \begin{bmatrix} 1 & r_2 + s(L_2 - M) + \frac{1}{sC_{r2}} \\ 0 & 1 \end{bmatrix} \begin{bmatrix} V_2 \\ -I_2 \end{bmatrix}
$$

$$
= \begin{bmatrix} A & B \\ C & D \end{bmatrix} \begin{bmatrix} V_2 \\ I_2 \end{bmatrix} \tag{1}
$$

where, $L_1 = L_2 = L$, $C_{r1} = C_{r2} = C_r$, $r_1 = r_2 = r$

$$
\begin{cases}
A = 1 + \frac{1}{sM} \left\{ r + s(L - M) + \frac{1}{sC_r} \right\} \\
B = \left[2 + \frac{1}{sM} \left\{ r + s(L - M) + \frac{1}{sC_r} \right\} \right] \left\{ r + s(L - M) + \frac{1}{sC_r} \right\} \\
C = \frac{1}{sM} \\
D = 1 + \frac{1}{sM} \left\{ r + s(L - M) + \frac{1}{sC_r} \right\}
\end{cases} \tag{2}
$$

Input impedance

$$
Z_{in} = \frac{AR_{ac} + B}{CR_{ac} + D} \tag{3}
$$

Output impedance

$$
Z_o = \frac{B}{A} \tag{4}
$$

Output voltage

$$
Vo = \frac{V_{in}}{2} \frac{R_{ac}}{AR_{ac} + B} \tag{5}
$$

where, the AC equivalent resistance is given by following equation.

$$
R_{ac} = \frac{2}{\pi^2} R_L \tag{6}
$$

The input and output impedance characteristics of light and heavy load condition are shown in Fig. 7, respectively. In output impedance characteristic, there are three peaks of a resonant peak and two anti-resonant peaks. The lowest frequency peak is series resonant peak fsr, the middle range peak is parallel resonant peak fpr and the highest frequency peak is fpr'.

The output impedance does not depend on the load condition, and it is constant. However, the input impedance depends heavily on the load condition as shown in Fig. 7. In heavy load condition, three peaks appear on the input impedance as shown in Fig. 7 (a), and only one peak appears in light load condition as shown in Fig. 7 (b).

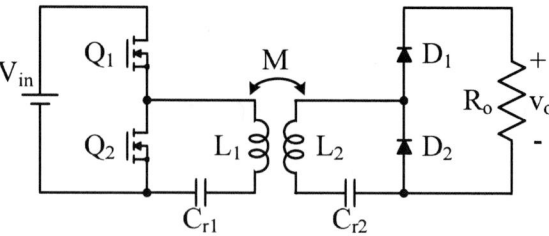

Fig. 5. Current resonant converter.

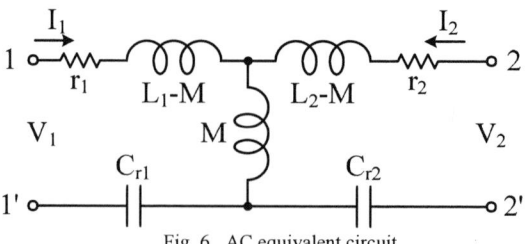

Fig. 6. AC equivalent circuit.

2362

The input impedance which is intimately involved with zero voltage switching (ZVS) operation is very important. For ZVS operation, the input impedance needs to be inductive. There are two ZVS operation region, fpr'<fs<fpr and fs>fpr in heavy load case, and the ZVS operation region is only one, fs>fzin (fzin : peak frequency of Zin) in light load condition.

The frequency characteristics of the voltage conversion ratio are shown in Fig. 8. This characteristics are similar to input impedance characteristics, and three peaks appear on the input impedance in heavy load condition, and only one peak appears in light load condition as shown in Fig. 8.

The resonant peak fpr' is moved along the blue line of gain curve depending on load condition, and when the resonant peak fpr' overlaps with fpr, the high frequency side resonant peak fsr becomes completely invisible.

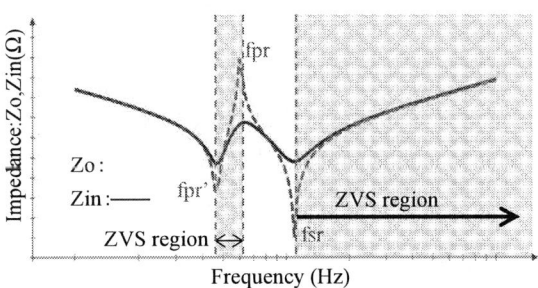

(a) Heavy load condition

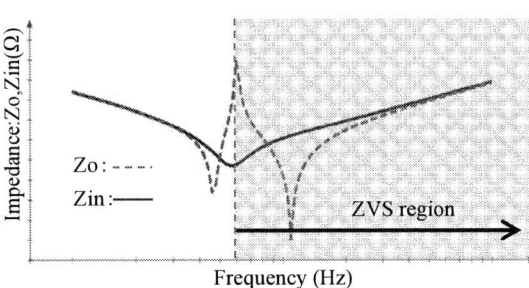

(b) Light load condition
Fig. 7. Impedance characteristics.

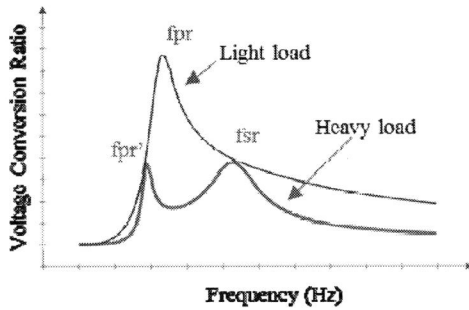

Fig. 8. Output voltage characteristics.

Figure 9 shows the analytical result of the output voltage characteristic. The circuit parameters are adapted in Table. 1. The Q-factor of this circuit set to be small, so the output voltage characteristics are broad. In 100 ohm and 50 ohm case, there is little variation in the output voltage. On the other hand, the output voltage is gradually decreased in 10 ohm case.

B. Experimental Verifications

In order to evaluate the performance of the current resonant converter with air core transformer, the prototype circuit is implemented, and the circuit parameters are also used in Table. 1. In this case, the series resonant frequency is around 340kHz, so ZVS operation can be achieved for almost operating range. Moreover, the diodes are used for secondary side rectification.

Figure 10 shows the output voltage and efficiency characteristics. These results are similar to analytical results. In 100 ohm and 50 ohm case, there is little variation in the output voltage at low frequency range. However, the output voltage is gradually decreased at higher frequency range in these cases. The output voltage of 10 ohm case is gradually decreased for all frequency range.

Furthermore, the peak efficiency is around 65% in 100 ohm and 50 ohm case, and 54% in 10 ohm case.

TABLE I
CIRCUIT PARAMETERS

Symbol	Description	Value
Vin	Input voltage	5 V
L	Self-inductance	1 uH
r	Internal resistance	0.3 ohm
Cr	Resonant capacitance	2.2 uF
k	Coupling factor	0.9

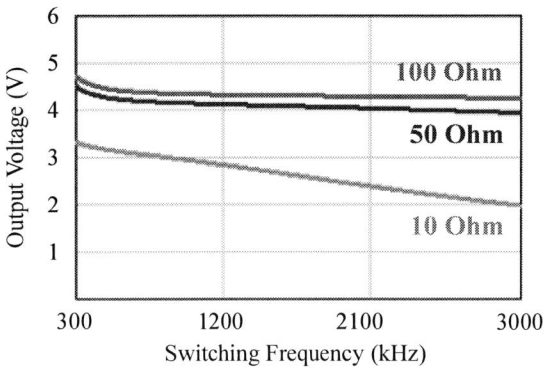

Fig. 9. Analytical result of output voltage characteristic.

The 2018 International Power Electronics Conference

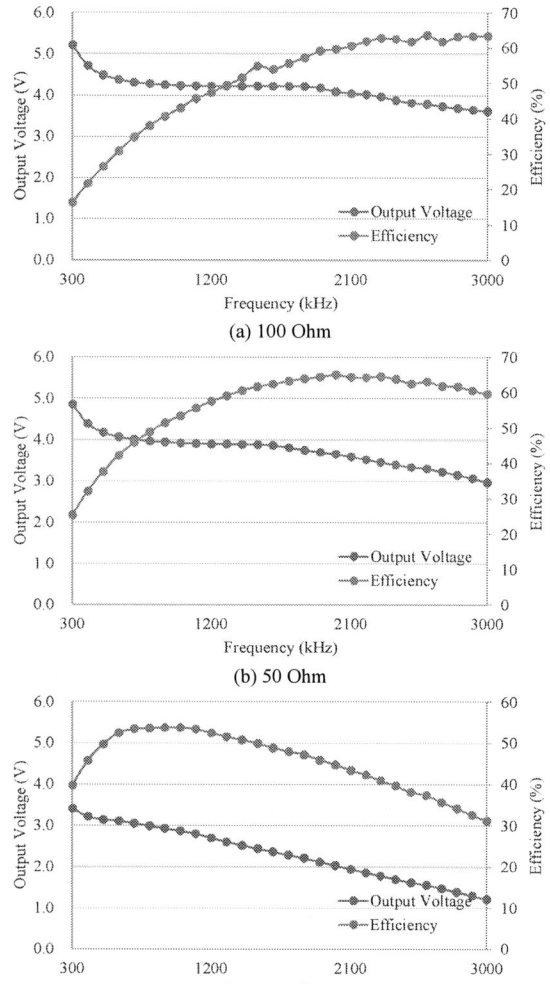

(a) 100 Ohm

(b) 50 Ohm

(c) 10 Ohm

Fig. 10. Experimental results of output voltage and efficiency characteristics.

V. CONCLUSIONS

This paper was investigated the operating principle of the isolated power-SoC. The current resonant converter was adapted as an isolated power-SoC. Here, the prototype evaluation board was implemented by PCB, and the operating characteristics ware analytically and experimentally confirmed. As a result, the shape of the voltage characteristic curve was similar between the analytical and experimental results. Moreover, the power conversion efficiency peak was around 65 % in 100 ohm and 50 ohm case.

ACKNOWLEDGMENT

This work was supported by JSPS KAKENHI Grant Number JP16K18060.

REFERENCES

[1] S. Matsumoto, M. Mino, and T. Yachi: "Integration of a power supply part for a system on silicon", IEICE Trans. Communication and Computer Science, vol.80-A, No.2, pp.276-282, 1997

[2] International Workshop on Power Supply On Chip 2010(PwrSoc' 10) http://www.powersoc.org/index.php

[3] A. Prodic: "High-Performance Mixed-Signal Controllers for On-Chip Integrated SMPS" International Workshop on Power Supply On Chip 2010(PwrSoc' 10).

[4] H. Meyvaert and E. Micas: "The importance of fully-integrated CMOS: Cost Effective Integrated DC-DC Converters", International Workshop on Power Supply On Chip 2010(PwrSoc' 10).

[5] I. Ranmuthu, "Challenges of Integration of Power Supplies on Chip", Power SoC 2016

[6] S. Roy, "Challenges in magnetics for PwrSoC – Development in high-frequency magnetics, materials and integration", International Workshop on Power Supply On Chip 2010 (PwrSoc'10).

[7] E. Friedman, "Small Area Power Converter for Application to Distributer On-Chip Power Delivery", International Workshop on Power Supply On Chip 2010 (PwrSoc'10).

[8] B. Chan, "Air Core and Magnetic Core Transformers for Isolated Power Converter", International Workshop on Power Supply On Chip 2016 (PwrSoc'16).

[9] V. Vorperian, S. Cuk, "A Complete DC Analysis of The Series Resonant Converter, " PESC'82, pp. 85-100, 1982

[10] X. Fang, H. Hu, L. Chen, A. Amirahmadi, J. Shen, I. Batarseh," Operation Analysis and Numerical Approximation for the LLC DC-DC Converter," APEC'12, pp. 870-876, 2012

[11] J. F. Lazar and R. Martinelli, "Steady-State Analysis of the LLC Series Resonant Converter," APEC '01, pp. 728-735, 2001.

[12] Ashoka K. S. Bhat,"A Generalized Steady-State Analysis of Resonant Converters Using Two-Port Model and Fourier-Series Approach," IEEE Trans. on P. E., vol. 13, No. 1, pp.142-151, 1998

[13] H. Huang,"FHA-Based Voltage Gain Function with Harmonic Compensation for LLC Resonant Converter," APEC'10, pp. 1770-1777, 2012

[14] V. Vorperian, "High-Q Approximation in The Small-Sgnal Analysis of Resonant Converters," PESC'85, pp. 707-715, 1985

[15] R. L. Steigerwald, "A Comparison of Half-Bridge Resonant Converter Topology," IEEE Trans. On PE, Vol. 3, No. 2, pp. 174-182, 1988.

Analysis for High-frequency LLC Resonant Converter with Planar Transformer at Light-load Condition

Keon-Woo Kim[1]*, Jae-Il Baek[1], Yeonho Jeong[1], Ki-Mok Kim[1,2] and Gun-Woo Moon[1]
1 School of Electrical Engineering, KAIST, Daejeon, Republic of Korea
2 Gumi Campus of Korea Polytechnic College, Gumi, Republic of Korea
*E-mail: rainbowdot@kaist.ac.kr

Abstract- **A high-frequency LLC converter has been widely employed to achieve high power density. However, it suffers from output voltage regulation problem under light-load conditions, which is light-load regulation problem. In this paper, a new synchronous rectifier (SR) control method is presented to improve the light-load regulation capability of high-frequency LLC converter. The proposed control method applies the extended SR turn-on time and the output voltage is regulated by the negative current flowing through SRs. Based on the equivalent circuit modeling of the LLC converter, the voltage gain of LLC converter with proposed control method can be calculated. Therefore, by reducing the voltage gain by negative current, the output voltage can be regulated with fixed switching frequency. The validity of a proposed control method is confirmed by the prototype with 330-380 VDC input and 750W (12 V/62.5 A) output.**

I. INTRODUCTION

Recently, high power density of power systems has become a very important issue in many fields, such as LED drivers, Flat TVs, server powers, and electric vehicle battery chargers [1]-[5]. To achieve high power density by reducing the size of passive components such as transformers and inductors, the switching frequency should be increased. However, the main drawback of high switching frequency converter is high switching losses. To minimize the switching losses, the soft switching techniques like zero-voltage-switching (ZVS) and zero-current-switching (ZCS) have applied. Among soft-switched converters, a LLC resonant converter could be a great candidate because of many advantages such as wide ZVS range, low electromagnetic interference, and wide gain range over narrow frequency variation [6]-[9]. In a high switching frequency LLC resonant converter, slim profile transformer is necessary for high power density because maximum height is determined by transformer. At the same time, heat dissipation capability becomes important, because the generated heat is difficult to be removed in high power density converters. In order to satisfy the abovementioned conditions, the planar transformers (PTs) are widely used due to low profile and high heat dissipation capability [10]-[12].

In spite of advantages of PT, it has large parasitic capacitance which leads to serious problem for LLC resonant converter [13]-[15]. In order to design high

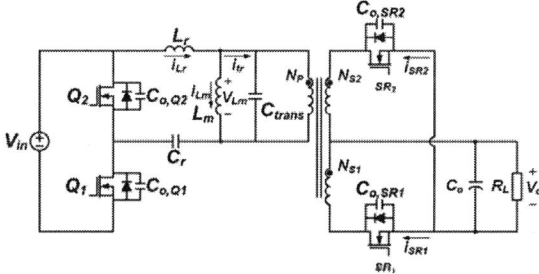

Fig. 1. Circuit diagram of the LLC converter considering the parasitic capacitors.

power density transformer, Printed circuit board (PCB) should be used as the windings of the planar magnetics [16]. Overlapping PCB tracks and high voltage difference between tracks result in significant large parasitic capacitance which can be modeled by one capacitor, C_{trans}, as shown in Fig. 1 [14]. Due to high parasitic capacitance, the voltage gain increases as switching frequency increases under the light load conditions which means the output voltage is not regulated [17]-[18]. Moreover, output capacitance, C_{oss}, of SR also accelerates regulation problem similar to parasitic capacitance of planar transformer. Low output voltage applications such as telecommunication applications, server powers utilize a synchronous rectifier (SR) because of the conduction loss of the rectifier. In order to decrease the drain-source on state resistance, $R_{ds,on}$, SRs are connected in parallel of the SR, and it causes the large C_{oss} of the SR.

During last few years, many papers have been studied to solve the regulation problem [18]-[24]. The work in [19] decreases the gain curve by connecting dummy load under the light load condition. Due to simple structure, dummy load is normally used in many applications such as server power and adapter applications. However, the dummy load consumes constant power loss which decreases the power efficiency. Research in [20]-[23] suggested changing control method to regulate output voltage. The work in [20]-[21] solved the voltage regulation problem through phase-shifted gate signals between primary switch legs. However, this method is only available with full bridge inverter structure. The work in [22]-[23] developed burst mode strategy in order to improve output voltage regulation capability. However, high parasitic capacitance of planar transformer and C_{oss} of SR, result in a large output voltage ripple because of enlarged burst mode operation region. Therefore, output capacitor size increases which reduces the power density. The work in [18] and [24] developed the PCB winding layout to minimize PT parasitic capacitance. But the C_{oss} of SR still leads to poor light load regulation characteristic.

In this paper, mathematical analysis of LLC resonant converter according to SR control is explained. Based on analysis, the appropriate SR control method is proposed to solve the light load regulation problem, while it maintains the switching frequency. The operation and performance of the proposed converter are verified by a prototype with 330-380VDC input and 750 W (12 V/62.5 A) output.

II. MODE ANALYSIS

Key waveforms of the LLC resonant converter considering the parasitic capacitors with proposed control method are described in Fig. 2. For the simple analysis, the following two assumptions have been made:

1) All active power devices are ideal switches except for their parallel body diodes and parasitic capacitors C_{oss}.

2) Output capacitor is large enough that the output voltage can maintain constant value, V_o.

Operations are divided into ten modes. Since modes 6-10 [t_6 - t_{10}] have similar operations with modes 1-5 [t_0 - t_5], only modes 1-5 [t_0 - t_5] will be explained.

Mode 1 [t_0 - t_1] : Mode 1 starts when drain-source voltage of SR_1 becomes zero and resonant inductor current, i_{Lr}, starts to flow through the SR. In this mode, resonant capacitor, C_r, and resonant inductor, L_r, resonate each other. Because primary switch, Q_1, is turned off, primary current flows through the body diode of Q_1. Unlike general LLC resonant converter, initial point of the current flow through the SR, i_{SR}, has positive value. Large parasitic capacitors require large ZVS energy, thus i_{Lr} drops. Difference between magnetizing inductor current, i_{Lm} and i_{Lr} is reflected to secondary side, and i_{SR} has positive initial value.

Mode 2 [t_1 - t_2]: Mode 2 starts when Q_1 and secondary side rectifier, SR_1, is turned on. Operation in this mode is almost same with mode 1. Difference is that primary current flows through the Q_1 and secondary current flows through the SR, not body diode.

Mode 3 [t_2 - t_3]: Mode 3 starts when i_{Lr} and i_{Lm} become equal. In this mode, negative current flows through the SR. Negative current transfer the power from secondary side to primary side. Therefore, the voltage gain of LLC resonant converter is reduced. The duration of mode 3 is enlarged under the light load condition.

Mode 4 [t_3 - t_4]: Mode 4 starts when SR_1 is turned off. In this mode, magnetizing inductor, L_m, and parasitic capacitors join the resonance.

Mode 5 [t_4 - t_5] : Mode 5 starts when switch Q_1 is turned off. In this mode, the ZVS operation of the primary side switches is performed. Due to large parasitic capacitors, large ZVS energy is required i_{Lr} drops a lot.

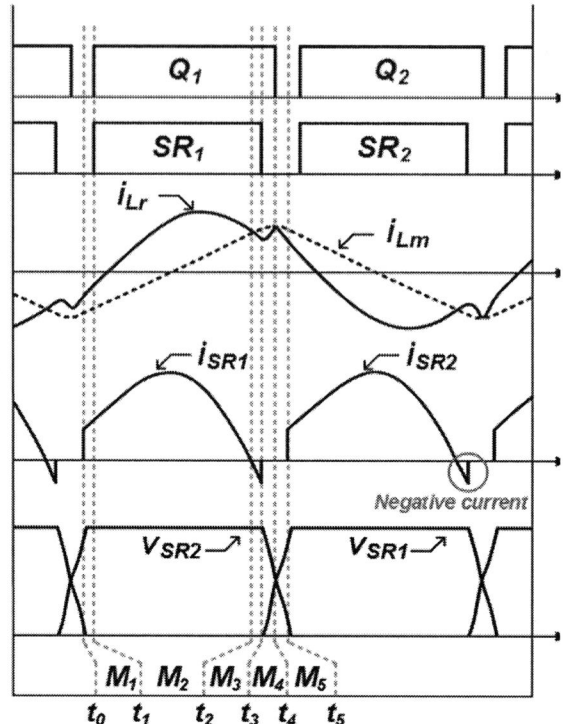

Fig. 2. Key waveforms of the LLC converter with the proposed control method.

III. EQUIVALENT CIRCUIT MODELING AND VOLTAGE GAIN

As mentioned in Section I, the parasitic capacitance of planar transformer and C_{oss} of SR cause the light load regulation problem. If C_{oss} of SR is reflected to the primary side, these two parasitic capacitors can be simplified to one capacitor, C_{pa}. Simplified capacitor can be calculated as follow:

$$C_{pa} = C_{trans} + \frac{2n_{SR}C_{oss,SR}}{n^2}, \tag{1}$$

where n_{SR} is the number of SRs in parallel and n is a transformer turns ratio.

Due to C_{pa}, equivalent circuit of the LLC resonant converter is changed. In order to consider the interaction between secondary side of LLC resonant converter and the C_{pa}, new equivalent circuit model has been previously researched [25]-[27]. They have used Rectifier compensated first harmonic approximation (RCFHA) technique to derive equivalent circuit model of LLC resonant converter. In this section, RCHFA technique is used to analyze equivalent circuit model considering the negative current flows through SR.

The AC equivalent circuit of the LLC converter using the proposed control method is shown in Fig. 4. Using RCFHA technique, the values of the equivalent resistor, R_{eq}, and equivalent capacitance, C_{eq}, can be calculated as follow:

2366

$$R_{eq} = \frac{\sin^2 \phi_{np} + \sin^2 \phi_{neg} + 2\cos \phi_{neg} - 2}{2\pi^2 f_s C_{pa}}, \quad (2)$$

$$C_{eq} = \frac{\pi C_{pa}}{(\phi_{np} - \phi_{neg}) - \sin \phi_{np} \cos \phi_{np} - \sin \phi_{neg} \cos \phi_{neg} + 2\sin \phi_{neg}} \quad (3)$$

where Φ_{neg} is the negative conduction angle, Φ_{np} is the non-positive conduction angle as shown in Fig. 3.

Based on the equivalent circuit, voltage gain of the LLC converter using the proposed control method can be calculated. Firstly, an input impedance, $Z_{in}(j\omega_s)$, and the fundamental components of the input resonant current, $i_{Lr(1)}$, can be obtained as follows:

$$Z_{in}(j\omega_s) = j\omega_s L_r + \frac{1}{j\omega_s C_r} + \frac{j\omega_s L_m \left(R_{eq} + \frac{1}{j\omega_s C_{eq}} \right)}{R_{eq} + j\omega_s L_m + \frac{1}{j\omega_s C_{eq}}}, \quad (4)$$

$$i_{Lr(1)} = \frac{2V_{in}}{\pi |Z_{in}(j\omega_s)|}, \quad (5)$$

where V_{in} is the input voltage, and ω_s is the angular switching frequency.

Relationship between $i_{Lr(1)}$ and $I_{Ctr,peak}$ is determined by the impedance ratio between L_m and Z_{eq}, and the peak value of i_{Ctr}, $I_{Ctr,peak}$, can be solved as follows:

$$I_{Ctr,peak} = \left| \frac{j\omega_s L_m}{R_{eq} + 1/(j\omega_s C_{eq}) + j\omega_s L_m} \right| i_{Lr(1)}. \quad (6)$$

Under the steady-state conditions, the average output current, I_o, is same with the amount of current transfer to the secondary side of the transformer. Therefore, I_o can be calculated as follows:

$$I_o = \frac{nI_{Ctr,peak}(\cos \phi_{neg} + \cos \phi_{np})}{\pi}. \quad (7)$$

Using (4)-(7), the voltage gain of the LLC converter using the proposed control method, M_{LLC}, is obtained as follows:

$$M_{LLC} = \frac{4n^2 R_L}{\pi^2} \frac{(\cos \phi_{neg} + \cos \phi_{np})}{|Z_{in}(j\omega_s)|} \left| \frac{j\omega_s L_m}{R_{eq} + 1/(j\omega_s C_{eq}) + j\omega_s L_m} \right|, \quad (8)$$

where R_L is the load resistance.

Using (8), the voltage gain of the LLC converter with the proposed control method according to the maximum negative conduction time, $t_{neg,max}$, is represented in Fig. 5. According to the increase of $t_{neg,max}$, the voltage gain is decreased, and it enables the light-load regulation with low switching frequency, f_s.

IV. PROPOSED CONTROL METHOD

Fig. 6 shows the conceptual control diagram of proposed control method. In the conventional method, output voltage is regulated by changing f_s according to change of the load conditions. In the proposed control method, the output voltage is regulated by pulse frequency modulation (PFM) under the heavy load

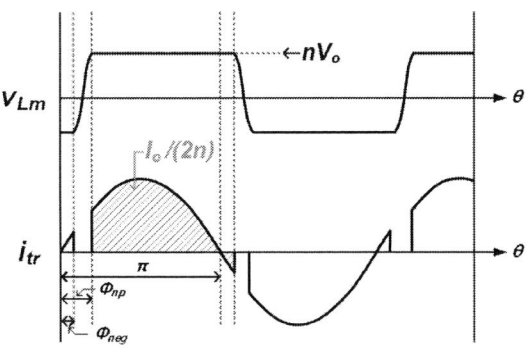

Fig. 3. Simplified waveforms of LLC converter with proposed converter.

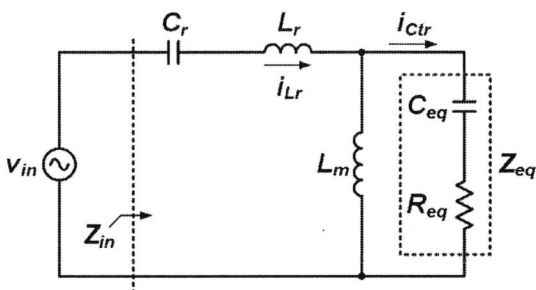

Fig. 4. Equivalent circuit of the LLC converter using the proposed control method.

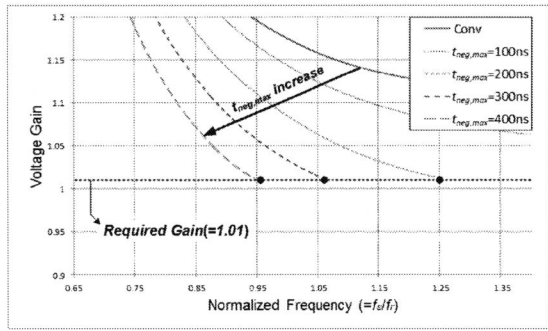

Fig. 5. Voltage gain of the LLC resonant converter according to $t_{neg,max}$ at no-load conditions.

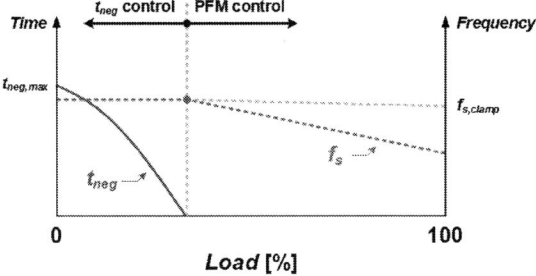

Fig. 6. Conceptual control diagram for the proposed control method.

condition which means f_s increases as load current decreases. After the f_s reaches the frequency limit, $f_{s,clamp}$, the output voltage is regulated only by negative current flowing through SR. According to increase of negative

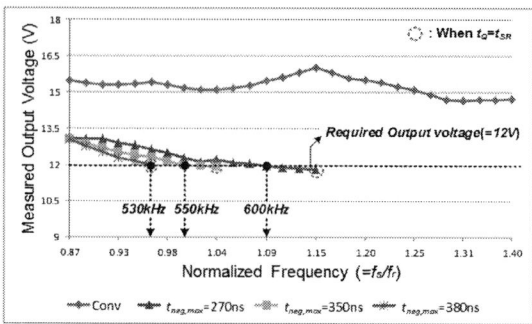

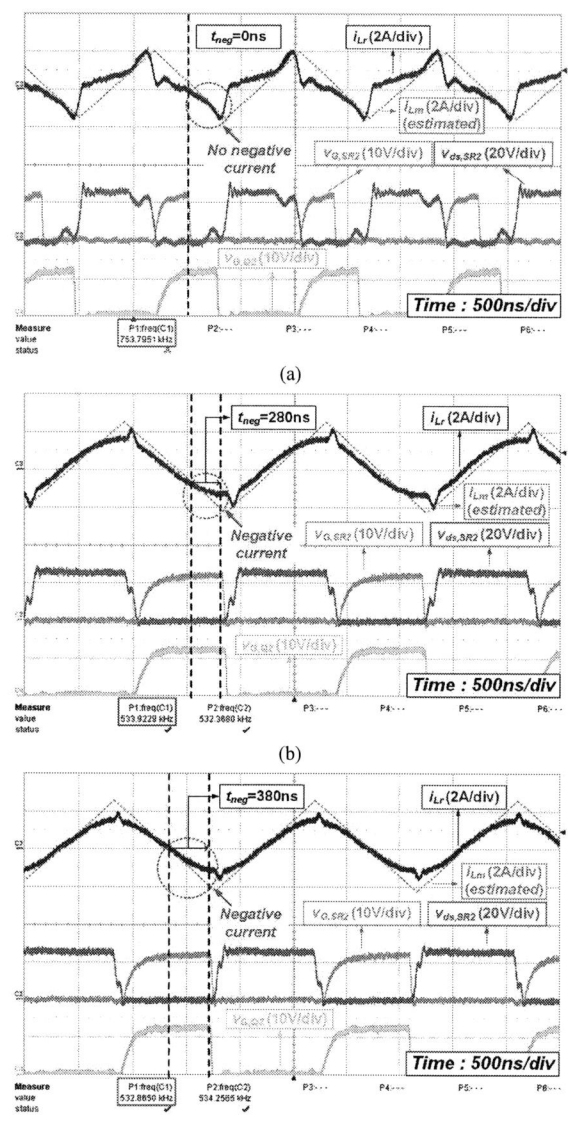

Fig. 8. Measured output voltage of the LLC resonant converter according to $t_{neg,max}$ at no-load condition.

voltage=12 V, rated output power=750 W, resonant frequency=550 kHz, and power density=152.68 W/in³.

Fig. 8 shows the experimental waveforms of the LLC converter under light-load conditions. Assume the estimated waveform as dotted orange color is the current of L_m. As shown in Fig. 7(a), when the proposed control method is not adapted, the negative current does not flow through SR. Because the output voltage is regulated by PFM, the operating switching frequency is high. When the proposed control method is adapted, V_o can be regulated with the constant f_s as shown in Fig. 7(b) and (c). According to decrease of load conditions, t_{neg} is enlarged.

Fig. 8 shows the measured output voltage of the LLC resonant converter according to $t_{neg,max}$ at no-load conditions. According to the increase of $t_{neg,max}$, V_o and $f_{s,clamp}$ is decreased. It is clear that the LLC resonant converter with the negative current flowing through SRs can improve the light-load regulation capability.

VI. CONCLUSION

This paper proposes the SR control method to solve the light-load regulation problem of the LLC converter. The modified AC equivalent circuit model and the voltage gain of the LLC converter are calculated. By extending t_{SR}, the voltage gain of the LLC converter is reduced. Compare with the conventional control method, the proposed control method can regulate the output voltage with low switching frequency under light load conditions. Moreover, there are no additional components to utilize the proposed control method, thus it can achieve high power density. Consequently, the proposed control method is suitable for high power density applications such as server powers and telecommunication applications, and so on.

ACKNOWLEDGMENT

This work was supported by the National Research Foundation of Korea (NRF) grant funded by the Korea government (MSIP)(No. 2016R1A2B2010328).

REFERENCES

[1] Q. Luo, S. Zhi, C. Zou, W. Lu, and L. Zhou, "An LED driver with dynamic high-frequency sinusoidal bus voltage regulation for

Fig. 7. Experimental waveforms of the LLC resonant converter. (a) Using the conventional control method at 10% load conditions. (b) Using the proposed control method at 10% load conditions. (c) Using the proposed control method at no-load conditions.

conduction time, t_{neg}, the amount of the negative current increases and the voltage gain decreases. Thus, the output voltage can be regulated by increasing t_{neg} when load current decreases. When SR gate signal is the same with the primary switch gate signal, t_{neg} becomes $t_{neg,max}$. In order to ensure the output voltage regulation over entire load conditions, $f_{s,clamp}$ should be limited.

V. EXPERIMENTAL RESULTS

The proposed control method is implemented with 750 W LLC converter with a Texas Instruments digital power controller UCD3138064, and the design specification is as following: input voltage=330-380 VDC, output multistring applications," IEEE Trans. Power Electron., vol. 29, no. 1, pp. 491–500, Jan. 2014.

[2] B. Erkmen and I. Demirel, "A very low profile dual output LLC resonant converter for LCD/LED TV applications," IEEE Trans. Power Electron., vol. 29, no. 7, pp. 3514–3524, Jul. 2014.

[3] W. Zhang, F. Wang, D. J. Costinett, L. M. Tolbert, and B. J. Blalock, "Investigation of Gallium Nitride Devices in High Frequency LLC Resonant Converter," IEEE Trans. Power Electron, vol. 32, no. 1, pp.571-583, Jan. 2017.

[4] D. Huang, S. Ji, and F. C. Lee, "LLC resonant converter with matrix transformer," IEEE Trans. Power Electron., vol. 29, no. 8, pp. 4339–4347, Aug. 2014.

[5] B. Whitaker et al., "A high-density, high-efficiency, isolated on-board vehicle battery charger utilizing silicon carbide power devices," IEEE Trans. Power Electron., vol. 29, no. 5, pp. 2606–2617, May 2014.

[6] U. Kundu, K. Yenduri, P. Sensarma, "Accurate ZVS Analysis for Magnetic Design and Efficiency Improvement of Full-Bridge LLC Resonant Converter," IEEE Trans. Power Electron, vol. 32, no. 3, pp.1703-1706, Jan. 2017.

[7] B. Yang, F. C. Lee, A. J. Zhang, and G. Huang "LLC resonant converter for front end DC/DC conversion," in Proc. Appl. Power Electron. Conf. Expo, 2002, pp. 1108-1112.

[8] B. C. Kim, K. B. Park, C. E. Kim, B. H. Lee, and G. W. Moon, "LLC resonant converter with adaptive link-voltage variation for a high-power density adapter," IEEE Trans. Power Electron., vol. 25, no. 9, pp. 2248–2252, Sep. 2010.

[9] S. Y. Chen, Z. R. Li, and C. L. Chen, "Analysis and design of single stage ac/dc LLC resonant converter," IEEE Trans. Ind. Electron., vol. 59, no. 3, pp. 1538–1544, Mar. 2012.

[10] J. Sun and V. Mehrotra, "Orthogonal winding structures and design for planar integrated magnetics," IEEE Trans. Ind. Electron., vol. 55, no. 3, pp. 1463–1469, Mar. 2008.

[11] J. Lu and F. Dawson, "Characterizations of high frequency planar transformer with a novel comb-shaped shield," IEEE Trans. Magn., vol. 47, no. 10, pp. 4493–4496, Oct. 2011.

[12] Y. Guan, Y. Wang, D. Xu, and W. Wang, "A 1MHz Half-Bridge Resonant DC/DC Converter Based on GaN FETs and Planar Magnetics," IEEE Trans. Power Electron., vol. 32, no. 4, pp. 2876–2891, Apr. 2017.

[13] F. Blache, J. P. Keradec, and B. Cogitore, "Stray capacitance of two winding transformer: equivalent circuit, measurments, calculation and Lowering," IEEE Ind. application Society Annual Meeting, Vol. 2, 1994, pp 1211-1217.

[14] H. Y. Lu, J. G. Zhu, and S. Y. R. Hui, "Experimental determination of stray capacitances in high frequency transformers," IEEE Trans. Power Electron, vol. 18, no. 5, pp. 1105-1112, Sep. 2003.

[15] M. Pahlevaninezhad, D. Hamza, and P. K. Jain, "An improved layout strategy for common-mode EMI suppression applicable to high frequency planar transformers in high-power DC/DC converters used for electric vehicles," IEEE Trans. Power. Electron., vol. 29, no. 3, pp. 12111228, Mar. 2014.

[16] E. C. W. de Jong, B. J. A. Ferreira, and P. Bauer, "Toward the next level of PCB usage in power electronic converters," IEEE Trans. Power Electron., vol. 23, no. 6, pp. 3151–3163, Nov. 2008.

[17] A. Hariya, H. Yanagi, Y. Ishizuka, K. Matsuura, S. Tomioka, and T. Ninomiya, "Influence of parasitic components on MHz-level frequency LLC resonant DC-DC converter," in Proc. 41st Annu. Conf. IEEE Ind. Electron. Soc. (IECON), Yokohama, Japan, Nov. 2015, pp. 4842-4847

[18] M. Saket, N. Shafiei, and M. Ordonez, "LLC Converters with Planar Transformers: Issues and Mitigation," IEEE Trans. Power Electron., vol. 32, pp. 4524-4542, June. 2017.

[19] B. H. Lee, M. Y. Kim, C. E. Kim, K. B. Park and G. W. Moon, "Analysis of LLC Resonant Converter Considering 26 Effects of Parasitic Components," in Proc. Telecommunications Energy Conference, INTELEC'09, pp. 1-6, Oct. 2009.

[20] J. -H. Kim, C. -E. Kim, J. -K. Kim, J. -B. Lee, and G.-W. Moon, "Analysis on load-adaptive phase-shift control for high efficiency full-bridge LLC resonant converter under light-load

conditions," IEEE Trans. Power Electron., vol. 31, no. 7, pp. 4942–4955, Jul. 2016.

[21] Yu-Kang Lo, Chung-Yi Lin, Min-Tsong Hsieh, and Chien-Yu Lin, "Phase-Shifted Full-Bridge Series-Resonant DC-DC Converters for Wide Load Variations," IEEE Trans. Ind. Electron., vol. 58, no. 6, pp. 2572-2575, Jun. 2011.

[22] N. Shafiei, M. Ordonez, M. Craciun, C. Botting, and M. Edington, "Burst mode elimination in high-power LLC resonant battery charger for electric vehicles," IEEE Trans. Power Electron., vol. 31, no. 2, pp. 1173–1188, Feb. 2016

[23] [23] W. Feng, F. C. Lee, and P. Mattavelli, "Optimal Trajectory Control of Burst Mode for LLC Resonant Converter," IEEE Trans. Power Electron., vol. 28, no. 1, pp. 457–466, Jan. 2013.

[24] M. Pahlevaninezhad, D. Hamza, and P. K. Jain, "An improved layout strategy for common-mode EMI suppression applicable to high frequency planar transformers in high-power DC/DC converters used for electric vehicles," IEEE Trans. Power. Electron., vol. 29, no. 3, pp. 12111228, Mar. 2014

[25] N. Shafiei, M. Ordonez, S. R. Cove, M. Craciun, and C. Botting, "Accurate modeling and design of LLC resonant converter with planar transformers," in Proc. IEEE 2015 Energy Conversion Congress and Exposition (ECCE), 2015, pp. 5468-5473.

[26] Y. A. Ang, C. M. Bingham, M. P. Foster, D. A. Stone, and D. Howe, "Design oriented analysis of fourth-order LCLC converters with capacitive output filter," IEE Proc. Electron. Power Appl., vol. 152, no. 2, pp. 310–322, Mar. 2005.

[27] N. Shafiei, M. Pahlevaninezhad, H. Farzanehfard, A. Bakhshai, and P. Jain, "Analysis of a fifth-order resonant converter for high-voltage dc power supplies," IEEE Trans. Power Electron., vol. 28, no. 1, pp. 85–100, Jan. 2013.

A Novel Full Digital control H-bridge DC-DC converter for power Supply on Chip applications

Shigeki Nakano, Toshiomi Oka, Seiya Abe, and Satoshi Matsumoto
Graduate School of Kyushu Institute of Technology, Kitakyushu, Japan
smatsu@ele.kyutech.ac.jp

Abstract- **In this paper, we propose a new control algorithm for a novel full digital control H-bridge DC-DC converter which can output the power in both boost and buck mode only rewriting three digital code such as input voltage, output voltage, and maximum load current. In addition, the proposed algorithm improves the transient response.**

I. INTRODUCTION

Point of load (POL) becomes key parts as increasing the clock frequency of LSIs and reducing the power consumption of LSIs. In order to suppress the voltage fluctuation at the transient states due to the voltage drop caused by the wiring resistance or the parasitic inductance, it is necessary to put POLs in the immediate vicinity of LSI. In order to achieve the ultimate miniaturization of power supply, power supply on chip (power-SoC), which can implement power semiconductor devices, passive components, and control circuits on silicon wafers has attracted attention and studied in recent years (Fig.1)[1-6]. Power-SoC has some challenges. First, PWM control is hard to use high frequency (>10MHz) DC-DC converters. In order to realize power-SoC, high frequency switching of several tens MHz is required. Therefore, a control technology other than the PWM control must be developed. Second, the power capacity per one POL is small. When we use power-SoC, the volume is very small and the power density is high because the passive components of them are small. Therefore, it operates with high efficiency at light load, however the efficiency drops considerably at heavy load. In order to solve these two problems, we proposed a control technique based on parallel connected POLs shown in Fig. 2[7-10]. The proposed control technique can realize high frequency switching and high efficiency operation over the wide load range. Figure. 3 shows the previously proposed control algorithm [8, 10]. First, the output voltage is read by FPGA through AD converter. Next, output voltage ($V_{out(0)}$) is compared with target voltage ($V_{set}, V_{set + V_N}$) and number of working POLs are changed one by one to regulate the output voltage. However, in this control algorithm, it takes much time to change to the appropriate number of POLs to switch the number of POLs to be operated one by one. In addition, the power-SoC is promising for using mobile equipment because it can save the space. H bridge DC-DC converter (Fig. 4) is attractive for mobile equipment because they are usually powered by battery. For the equipment powered by battery, H-bridge DC-DC converter is convenient because it is possible to output buck-boost mode.

Fig. 1. Power Soc.

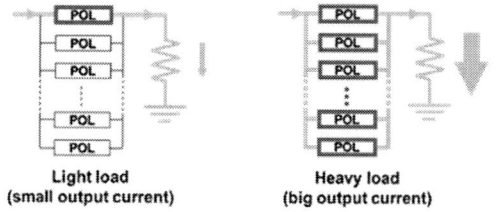

Light load
(small output current)

Heavy load
(big output current)

Fig. 2. The control technology of POLs[7-10].

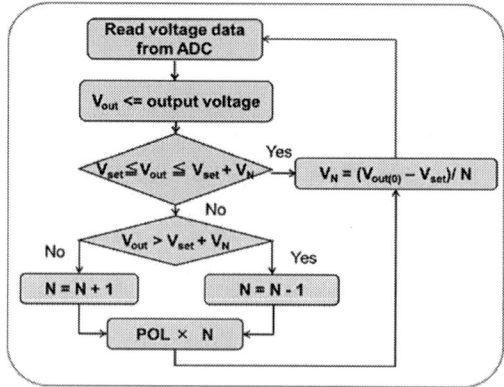

Fig. 3 Our previously proposed control algorithm[8,10]

The 2018 International Power Electronics Conference

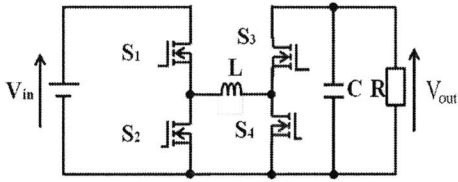

Fig. 4. H bridge DC-DC converter.

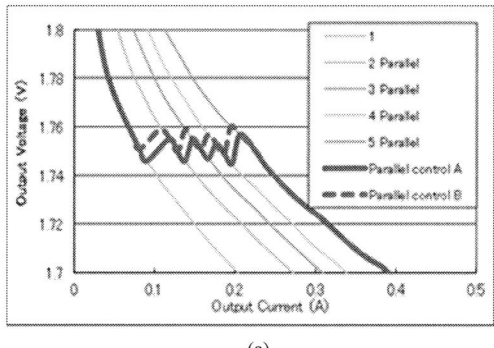

(a)

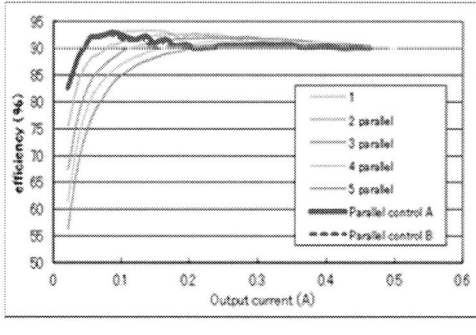

(b)

Figs. 5 (a) Load and (b) Efficiency characteristics of the buck mode.

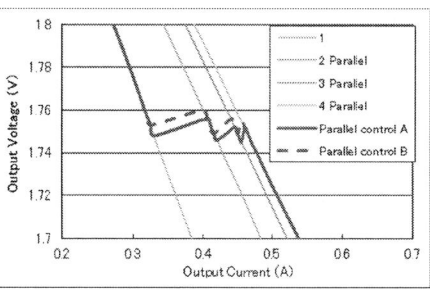

(a)

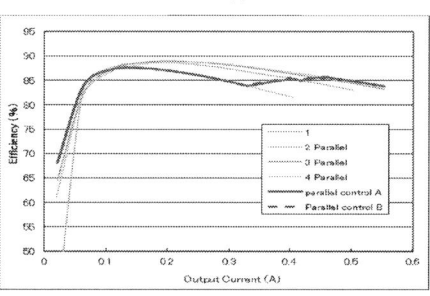

(b)

Figs. 6 (a) Load and (b) Efficiency characteristics of the boost mode.

$$V_{out} = DV_{in} - \frac{r}{N}I \qquad (1)$$

$$V_{out} = \frac{1}{D}V_{in} - \frac{r}{N}I \qquad (2)$$

$$V_{out} = \frac{D}{D'}V_{in} - \frac{r}{N}I \qquad (3)$$

D is the duty ratio of "on time", D' is the duty ratio of "off time", V_{in} is the input voltage, r is the internal resistance, N is the number of POLs connected in parallel connected in parallel, and I is the output current. In this technique, D and r are constant. Thus, output voltage changes according to I from the equations (1) to (3). Therefore, the output voltage is regulated by adjusting the operating number (N) of POLs according to the load current. Figures. 5 and 6 show an experimental results of the load characteristics and the efficiency characteristics of the proposed control technique using equation (1) (2)[7]. Input voltage of buck mode is 4.0 V, and boost mode is 1.2V. From results of Figs. 5(a) and 6(a), the slope of the voltage drop becomes gradual as the number of POLs increases. As indicated by the red line in these figures, the output voltage can be regulated by switching the number of POLs according to the load current. Similarly, it is possible to maintain high efficiency over the wide load range as shown in Figs. 5(b) and 6(b).

In this paper, we propose a new control algorithm for a novel full digital control H-bridge DC-DC converter which can output the power in both boost and buck mode only rewriting three digital code such as input voltage, output voltage, and maximum load current.

II. DESCRIPTION OF THE NEW CONTROL ALGORITHM FOR H-BRIDGE DC-DC CONVERTER

Our previously proposed control technique can regulate the output voltage by changing the number of working POL according to load current without feedback loop [7-9]. For example, a small number of POLs are operated at light load, and a large number of POLs are operated at heavy load. Expressions (1) to (3) show output voltage conversion equations of POL in back mode, boost mode, buck-boost mode, respectively.

The 2018 International Power Electronics Conference

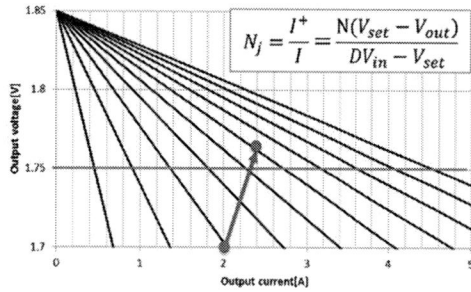

Fig. 7. Behavior of the control algorithm.

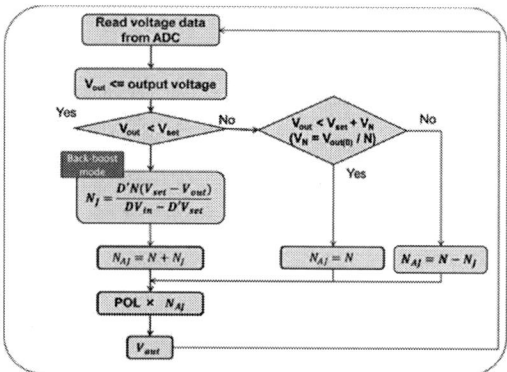

(a)

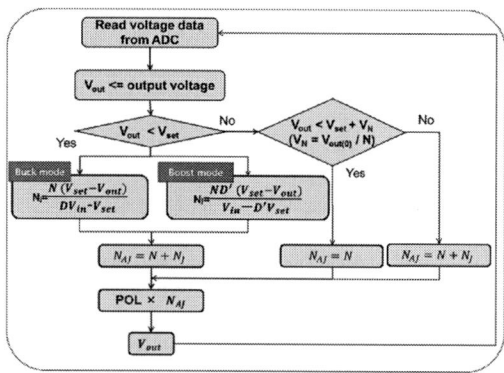

(b)

Fig. 8. New control algorithm.
(a) Control algorism of back-boost mode for (i)
(b) Control of buck and boost mode for (iv)

Figure 7 illustrates dependence of output voltage on output current at each number of working POLs. The slope becomes   gradual with increasing number of working POLs. The slope is decided by number of working POLs. The voltage drop between target voltage and present output voltage is also defined by number of working POLs. From results of Fig. 7, the flowchart of the proposed control algorism for Back-boost mode are shown in Figs. 8 (a) and (b).

The formula used in the proposed algorithm is explained in the range of (i) to (iv) below.

(i) $V_{out} < V_{set}$ (Back-boost mode)
(ii) $V_{out} < V_{set}$ (Buck mode)
(iii) $V_{out} < V_{set}$ (Boost mode)
(iv) $V_{out} > V_{set}$ (Back-boost mode)
(v) $V_{out} > V_{set}$ (Buck mode)
(vi) $V_{out} > V_{set}$ (Boost mode)

When the output voltage is lower than the target voltage as shown in (i), the number of operating POLs increase. When the output current is increased by sudden load change, the slope $\triangle N$ of the V-I characteristic is as follows when the number of converted converters is N.

$$\triangle N = -\frac{DV_{in} - D'V_{set}}{D'NI} \tag{4}$$

From the equation (4), the current increment I^+ is as follows when the output voltage drops to xV.

$$I^+ = \frac{D'NI(V_{set} - V_{out})}{DV_{in} - D'V_{set}} \tag{5}$$

From the above, the number N_J of converters to be increased is as follows.

$$N_J = \frac{I^+}{I} = \frac{D'N(V_{set} - V_{out})}{DV_{in} - D'V_{set}} \tag{6}$$

Likewise, the number of converters N_J that increases when calculating in (ii) and (iii) is as follows when they are calculated in these times.

$$N_J = \frac{I^+}{I} = \frac{N(V_{set} - V_{out})}{DV_{in} - V_{set}} \tag{7}$$

$$N_J = \frac{I^+}{I} = \frac{D'N(V_{set} - V_{out})}{V_{in} - D'V_{set}} \tag{8}$$

From equations (6) to (8), the number of POLs (N_{AJ}) to be operated can be determined as follows.

$$N_{AJ} = N + N_J \tag{9}$$

Also, as shown in (iv), when the output voltage is higher than the target voltage, the output voltage is set to the target voltage by decreasing the number. When reducing the number, hysteresis V_N is used to avoid oscillation of the output voltage. The conditional expression for reducing one POL is as follows.

$$V_{set} + V_N \cdot 2 > V_{out} \geq V_{set} + V_N \cdot 1 \tag{10}$$

The number of N_J to be reduced is given as follows by equation (10).

$$N_J = \frac{V_{out} - V_{set}}{V_N} \tag{11}$$

Accordingly, the number of POLs running (N_{AJ}) is obtained as follows.

2372

$$N_{AJ} = N - N_J \qquad (12)$$

Also in (v) and (vi), the number of units can be controlled in the same way as in (iv).

Figure. 9 illustrates digital control system used in in this study [8-10]. FPGA controls parallel connected POLs by fedding the controll signal from FPGA to the switches (S1-S4) of H bridge DC-DC converter. We can set up any output voltages and only input three digital codes such as input voltage, output voltage, and maximum current to FPGA. So, we can get an advantage that it is not necessary to replace external parts and adjasting the parameters.

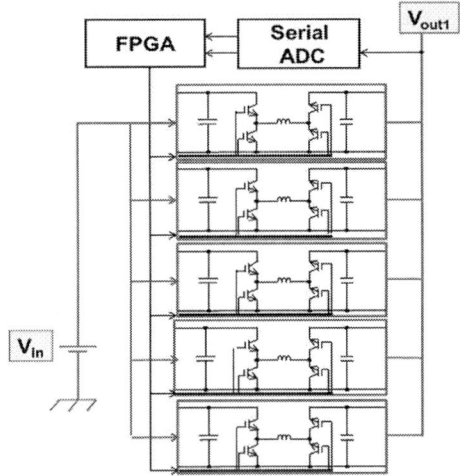

Fig. 10. Block diagram of experimental system.

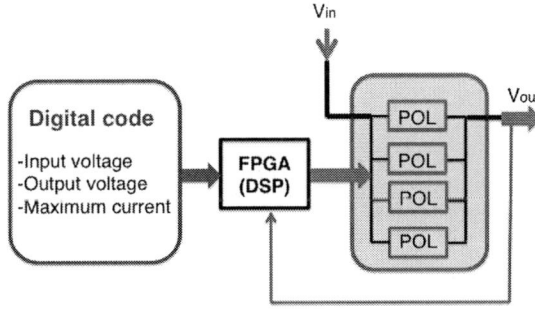

Fig. 9. The system controlling by FPGA[8-10].

III. EXPERIMENTAL RESULTS AND DISCUSSIONS

Figure 10 illustrates the block diagram of experimental system for evaluations. H-bridge DC-DC converter is used.

Figure 11 shows the picture of experimental circuits.

Table 1 shows the circuit parameters and system specifications for evaluations. In this system, five DC-DC converters connected in parallel were used. The output voltage is inputted to the FPGA after it was digitized by a serial type AD converter, and a gate drive signal is outputted converters based on a control algorithm. The target voltage (V_{set}) was set to 1.75 V. In this experiment, we use buck mode of H bridge DC-DC converter.

Transient response of the (a) conventional and (b) newly proposed control algorithm are shown in Figs. 12 (a) and (b), respectively. The buck mode is used. The load current (I_{out}) is changed from 70 mA to 300 mA. In Fig. 12 (a), the output voltage stabilized around 1.73 V at 460 µs. after sudden load change. In Fig 12 (b), after sudden load change, the output voltage stabilized around 1.73 V at 420 µs..

Fig. 11. The circuit used in experiment.

TABLE I
CIRCUIT PARAMETERS AND SYSTEM SPECIFICATIONS.

Symbol	Description	Value
V_{in}	Input voltage	4 V
V_{out}	Output voltage	1.75 V
L_o	Inductance	250 µH
r_{DC}	Internal resistance	100 mOhm
C_o	Output Capacitance	4.7 µF
f_s	Switching frequency	300k Hz
D	Duty ratio	0.5, 1
N	Number of active POLs	1-5

Transient response of the (a) conventional and (b) newly proposed control algorithm are shown in Figs. 13 (a) and (b), respectively. The buck mode is used. The load current (I_{out}) is changed from 315 mA to 70 mA. In Fig. 13 (a), the output voltage stabilized around 1.752 V at 400 µs. after sudden load change. In Fig 13 (b), after sudden load change, the output voltage stabilized around 1.752 V at 300 µs.. These results shows that the newly control algorithm shows faster response time than the conventional one.

The 2018 International Power Electronics Conference

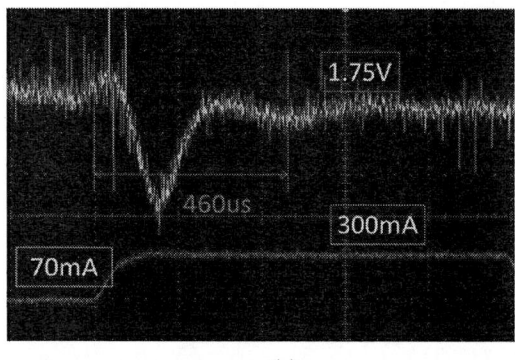

(a)

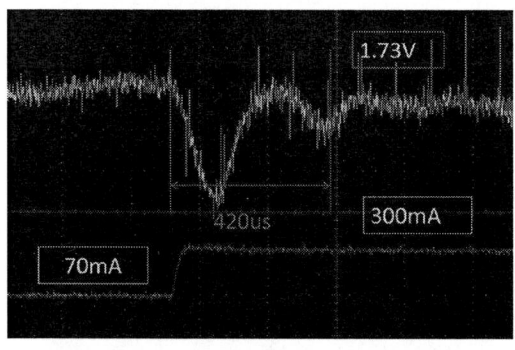

(b)

Figs. 12. Output voltage waveform when output current is changed from 70 mA to 300 mA (buck mode).

(a) Conventional[10] (b) Proposed

We obtained output voltage waveform for transient response using simulations (MATLAB/ Simulink, Figs. 14 (a), (b) and (c)). The output voltage waveforms are shown in Figs. 15 (a) and (b). The circuit parameters and specifications are liasted in Table II. The transient response improved by increasing switching frequency.

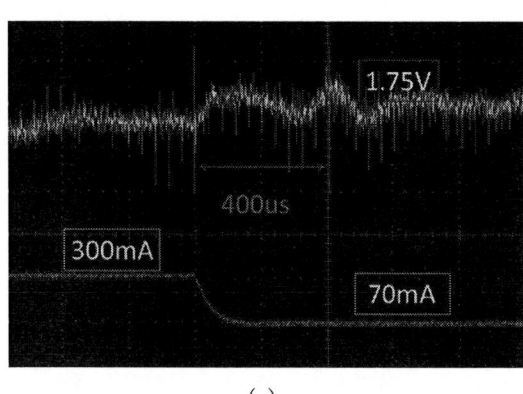

(a)

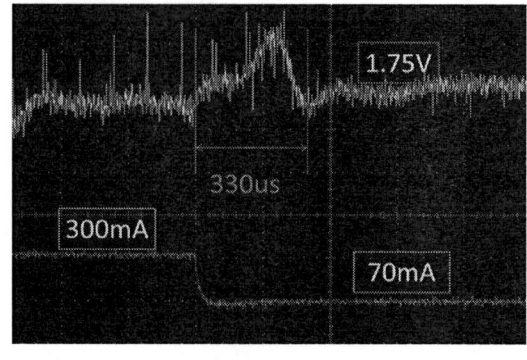

(b)

Fig. 13. Output voltage waveform when output current is changed from 315 mA to 70 mA (buck mode).

(a) Conventional[10] (b) Proposed

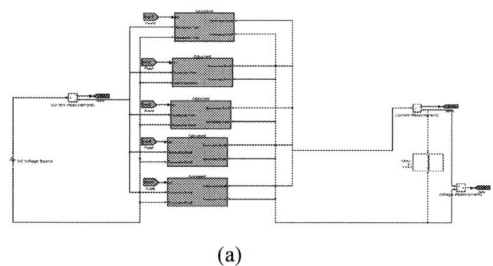

(a)

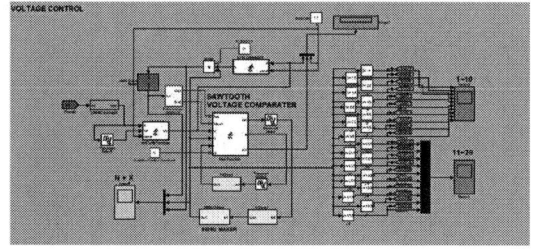

(b)

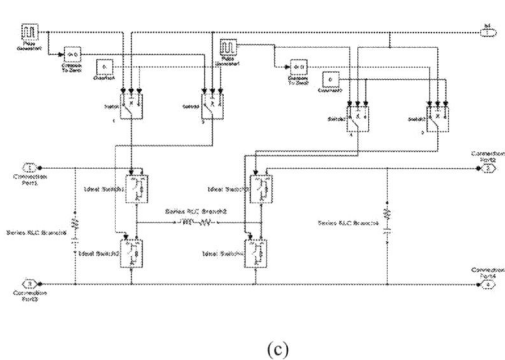

(c)

Figs. 14. Simulated H bridge converter.
(a) Structural diagram of the entire configuration circuit
(b) H bridge DC-DC convertor
(c) Structure drawing of control unit

2374

TABLE II
CIRCUIT PARAMETERS AND SYSTEM SPECIFICATIONS

Symbol	Meaning	Value	
V_{in}	Input voltage	4 V	4 V
V_{out}	Output voltage	1.75 V	1.75 V
L_o	Inductance	19 nH	25 µH
r_{DC}	Internal resistance	170 mOhm	170 mOhm
C_o	Output Conductance	47 µF	470 µF
f_s	Switching frequency	30 MHz	300 kHz
D	Duty ratio	0.5, 1	0.5, 1
N	Number of active POLs	1-5	1-5

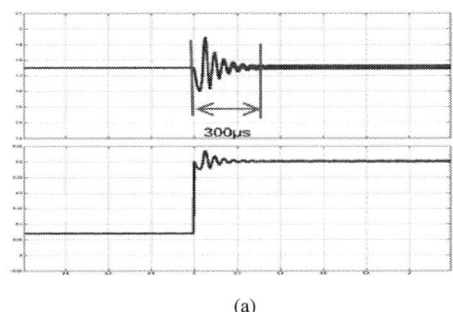

(a)

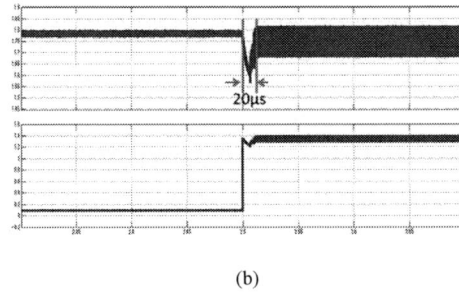

(b)

Fig. 15 Output voltage waveform when output current is changed from 70 mA to 300 mA (buck mode).
(a) f_s=300kHz (b) f_s=30MHz

IV. CONCLUSIONS

We proposed a novel full digital control H-bridge DC-DC converter and its control algorithm. The proposed control algorithm can operate boost / buck the voltage simply by inputting input voltage, output voltage and maxim load current without changing external parts. The proposed digital control H-bridge DC-DC converter can regulate output with high efficiency by only changing three digital code. In addition, the newly proposed control algorism can realize faster response.

REFERENCES

[1] S. Matsumoto, M. Mino, and T. Yachi: "Integration of a power supply part for a system on silicon", IEICE Trans. Communication and Computer Science, vol.80-A, No.2, pp.276-282, 1997.

[2] S. Matsumoto "Future Power Electronics for Realizing Sustaining Society", International Workshop on Power Supply On Chip 2010 (PwrSoc'10), Session 6.6,2010.

[3] T.P.Chow, "GaAs p-HEMT based power ICs for high frequency switching converter applications", International Workshop on Power Supply On Chip 2010(PwrSoc'10), Session 2-3, 2010.

[4] D. Anderson, "Applications of power SIP/pwrSoC products,", International Workshop on Power Supply On Chip 2012(PwrSoc'12), Session 1-6, 2012.

[5] F. Carobolante, " Power Supply on Chip: from R&D to commercial products", International Workshop on Power Supply On Chip 2014(PwrSoc'14), Plenary Session 3, 2014.

[6] S. Sanders, "The Road to Integrated Power Conversion via the Switched Capacitor Approach", International Workshop on Power Supply On Chip 2014(PwrSoc'14), Plenary Session 2, 2014.

[7] T. Yamamoto, J. Rikitake, S. Matsumoto, S. Abe, "A New Control Strategy for Power Supply on Chip Using Parallel Connected DC-DC Converter", IEEE 10th International Conference on Power Electronics and Drive Systems (PEDS) 2013, pp. 109-112, 2013.

[8] T. Yamamoto, S.Abe, and S.Matsumoto, "A Novel Concept of Digitally Controlled Multiple Output POL for Power Supply on Chip", International Telecommunication Energy Conference 2014(Intelec2014), PO-33, 2014 .

[9] M. Higashida, T. Yamamoto, S. Abe, and S. Matsumoto, "A Concept of Field Programmable Power Supply Array Utilizing Power Supply on Chip-- Fully digital controlled multiple input and output voltages POL –" ,17th European Conference on Power Electronics and Applications (EPE 2015, ECCE Europe)   LS1e.4, 2015.

[10] T. Hashiguchi, S. Abe, S. Matsumoto, "A fully digitally controlled multiple input and output voltage buck-boost POL for power supply on chip", International Telecommunication Energy Conference 2015(Intelec2015), pp.1081-1096, 2015.

[11] T Ninomiya, R Shibahara, S Abe, " Control Characteristics of a Matrix-POL Power Supply System "   IEICE Tech.vol.111, No.400, EE2011-34, pp.19-24, Jan.2012.

The 2018 International Power Electronics Conference

A High-Efficiency Power Supply from Magnetic Energy Harvesters

Cheon-Yong Lim[1*], Yeonho Jeong[1], Keon-Woo Kim[1], Feel-Soon Kang[2], and Gun-Woo Moon[1]
1 School of Electrical Engineering Korea Advanced Institute of Science and Technology, Daejeon, Korea
2 Electronics & Control Engineering Hanbat National University, Daejeon, Korea
* E-mail: yong0491@kaist.ac.kr

Abstract— Magnetic energy harvesting is a promising technology for a self-powered sensor, because it rarely depends on the weather condition. In order to increase the power density, maximizing the harvested power is important. In this paper, the analysis for power harvesting according to varying primary current is prevailed, and a new design guideline of primary voltage for maximizing harvested power is suggested. This analysis is distinct in that the effect of magnetizing inductance was taken into the consideration. To confirm the validity of this paper, the experiments were prevailed with a prototype of $4 \sim 7$ Arms primary current.

Keywords— *High-efficiency, magnetic energy harvester, self-powered sensor.*

I. INTRODUCTION

Sensors are required to be self-powered, where battery maintenance or power wiring are challenging. Energy harvesting technology offers a great opportunity to make sensors self-powered. Among several ambient energy sources (wind, temperature, solar, magnetic field, etc.), magnetic energy harvesting has been a promising technology, because it rarely depends the weather condition and it can achieve a relatively high power density [1-4]. Magnetic energy harvester can extract power from a current-carrying conductor. In this scheme, a core is mounted on a conductor, as shown in Fig. 1. When primary current i_{pri} flows through the conductor, the core acts like a transformer and electric power is transferred from the conductor to the secondary side. Since the harvested output power P_o is directly relevant to the core size, to maximize P_o is essential for increasing the power density. In order to maximize P_o, a precise analysis is required. The analysis in [4] has revealed that P_o can be maximized when the core is in the vicinity of saturating point. In [4], in order to obtain P_o in a restricted circumstance, nano-crystalline core, of which permeability is over 150,000, was inevitably used. With this ultra-high permeability, the magnetizing inductance L_m can be assumed to be very large and the effect of L_m was ignored in the analysis. However, nano-crystalline core is too expensive to be commercially adopted yet. In this reason, ferrite cores, of which permeability is several thousands, are widely used in magnetic energy harvesters [1-3]. In this case, the effect of L_m must be considered to

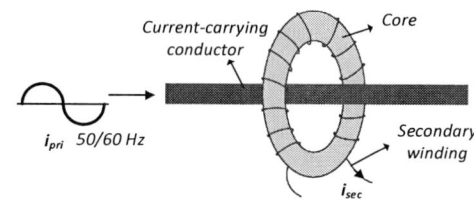

Fig. 1. Energy harvester mounted on a power line.

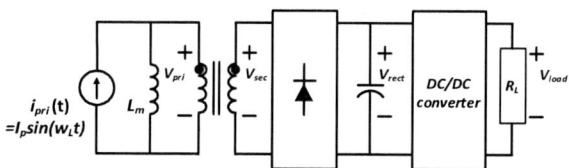

Fig. 2. Schematic for power harvesting system.

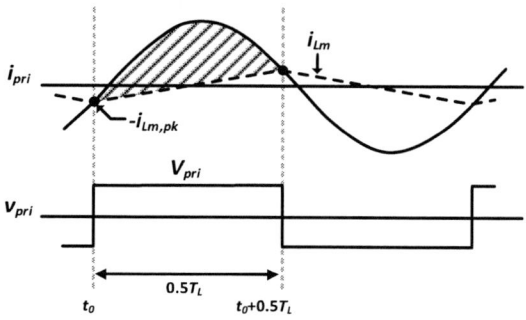

Fig. 3. Periodic i_{pri} and i_{Lm} waveforms.

obtain an accurate analysis for maximizing P_o.

In this paper, more accurate analysis is performed by taking the L_m effect into the consideration. Based on the analysis, a new optimal operating point to maximize P_o according to i_{pri} is presented.

II. POWER HARVESTING ANALYSIS

A. Structure of the Power Harvesting System

Fig. 2 shows the schematic for magnetic energy harvesting system. A core, which is hanging on the conducting wire, can be presented as a transformer. For simplicity, the parasitic components such as leakage inductance and parasitic resistances are omitted. The load is a supercapacitor through a rectifier and dc/dc converter, which can be considered as a constant voltage source. When the rectifiers are conducting, the secondary winding voltage V_{sec} is equal to the rectified voltage V_{rect} with alternative polarity. The primary winding voltage V_{pri} is determined as V_{sec} divided by turn ratio N.

B. Analysis for Power Harvesting

Fig. 3 shows i_{pri} and magnetizing current i_{Lm} waveforms. The averaged harvested output power P_o can be determined as follows:

$$P_o = \frac{1}{0.5T_L} \int_{t_s}^{t_s+0.5T_L} V_{rect} \frac{\left(i_{pri}(t) - i_{Lm}(t)\right)}{N} dt, \quad (1)$$

where T_L is line period and t_s is the start time of transferring current.

Since i_{pri} meets i_{Lm} at t_s, following equations can be established as follows:

$$i_{pri}(t_s) = i_{Lm}(t_s),$$
$$I_p \sin(w_L t_s) = -i_{Lm}{}^{pk}, \quad (2)$$

where w_L is line angular frequency and $i_{Lm}{}^{pk}$ is the peak value for i_{Lm}, as shown in Fig. 3.

Since the offset of i_{Lm} is zero, $i_{Lm}{}^{pk}$ is determined as the half of the ripple of i_{Lm}. Since the ripple of i_{Lm} can be obtained as follows:

$$\Delta i_{Lm} = \frac{\int_{t_s}^{t_s+0.5T_L} V_{pri} dt}{L_m} = \frac{V_{pri} \cdot 0.5T_L}{L_m}, \quad (3)$$

$i_{Lm}{}^{pk}$ can be expressed as follows:

$$i_{Lm}{}^{pk} = 0.5\Delta i_{Lm} = \frac{1}{4f_L L_m} V_{pri}, \quad (4)$$

where f_L is the line frequency.

By substituting (4) for $i_{Lm}{}^{pk}$, (2) can be expressed as follows:

$$I_p \sin(w_L t_s) = -\frac{1}{4f_L L_m} V_{pri}. \quad (5)$$

From (5), t_s can be obtained as follows:

$$t_s = -\frac{1}{w_L} \sin\left(\frac{1}{4\sqrt{2}f_L i_{pri,rms} L_m} V_{pri}\right), \quad (6)$$

where $i_{pri,rms}$ is root-mean-square value for i_{pri}.

Since the offset of i_{Lm} is zero, the magnetizing charge over $0.5 T_L$ is zero and by substituting (6) for t_s, (1) can be reduced as follows:

$$P_o = \frac{1}{0.5T_L} \int_{t_s}^{t_s+0.5T_L} V_{rect} \frac{I_p \sin(w_L t)}{N} dt$$
$$= \frac{2\sqrt{2}i_{pri,rms}}{\pi} V_{pri} \cos(w_L t_s). \quad (7)$$

From (6), it can be noted that as $i_{pri,rms}$ is decreased, t_s has the larger negative value. Thus, from (7) the cosine

term for P_o is decreased. On the other hand, in the previous research, since L_m was assumed to be very large, t_s was assumed to be zero and cosine term for P_o was assumed as 1.

C. Maximizing Harvested Power

From (7), it can be noted that P_o is relevant to $i_{pri,rms}$ and V_{pri}. The extremum P_o point can be easily found for the case of V_{pri} by differentiating P_o with respect to V_{pri} and solving for extrema as follows:

$$0 = 1 - \left(V_{pri,extm} \frac{1}{4f_L L_m i_{pri}}\right)^2, \quad (8)$$

where $V_{pri,extm}$ is the primary voltage where the extremum P_o can be obtained.

From (8), $V_{pri,extm}$ can be expressed as follows:

$$V_{pri,extm} = 4f_L L_m i_{pri,rms}, \quad (9)$$

and by substituting (9) for V_{pri}, (7) can be expressed as follows:

$$P_{o,extm} = \frac{2V_{pri}}{\pi} i_{pri,rms}. \quad (10)$$

(10) might fit well, with an ideal core without the magnetic saturation. However, in practice, the saturation condition must be considered for obtaining the optimal point of V_{pri}. Since little energy can be harvested once the magnetic flux density is saturated in the core, the operating point for V_{pri} is restricted under the saturation primary voltage V_{sat}, where V_{sat} can be obtained as follows:

$$V_{sat} = \frac{2A_e B_{sat}}{0.5T_L} = 4A_e B_{sat} f_L, \quad (11)$$

where A_e is the effective cross-sectional area of the core and B_{sat} is the saturation flux density.

Consequently, the optimal primary voltage $V_{pri,opt}$, where the maximized power can be harvested, can be classified into 3 cases: $V_{pri,extm} > V_{sat}$, $V_{pri,extm} = V_{sat}$, or $V_{pri,extm} < V_{sat}$. In the first case, where $V_{pri,extm} > V_{sat}$, $V_{pri,opt}$ is at V_{sat}, because V_{pri} is restricted under V_{sat}. P_o can be easily obtained by substituting V_{sat} for V_{pri} in (7). In the second case, where $V_{pri,extm} = V_{sat}$, it is obvious that $V_{pri,opt}$ is at V_{sat}. P_o can be easily obtained by substituting V_{sat} for V_{pri} in (10). In the third case, where $V_{pri,extm} < V_{sat}$, the maximized P_o can be obtained when $V_{pri,opt}$ is at $V_{pri,extm}$ rather than at V_{sat}. P_o can be easily obtained by substituting $V_{pri,extm}$ for V_{pri} in (10). In each case for the proposed analysis, the maximum P_o and operating point are summarized in Table I.

From (9) and (11), the boundary condition, where the optimal V_{pri} is deviated from the V_{sat}, can be expressed as follows:

$$V_{pri,extm} \le V_{sat}$$
$$4f_L L_m i_{pri,rms} \le 4A_e B_{sat} f_L$$
$$i_{pri,rms} \le \frac{l_c}{\mu_0 \mu_r} B_{sat} = \alpha, \quad (12)$$

2377

The 2018 International Power Electronics Conference

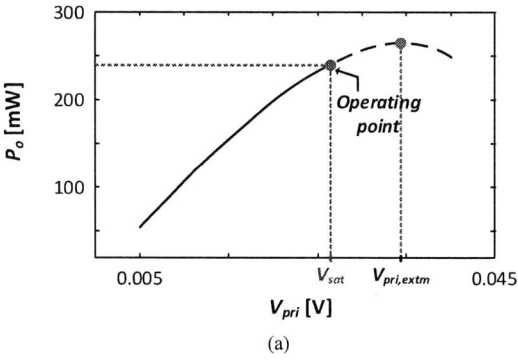

(a)

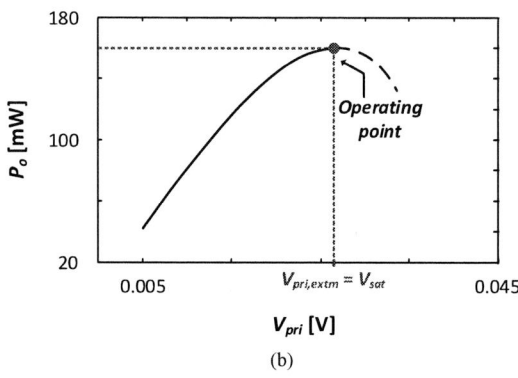

(b)

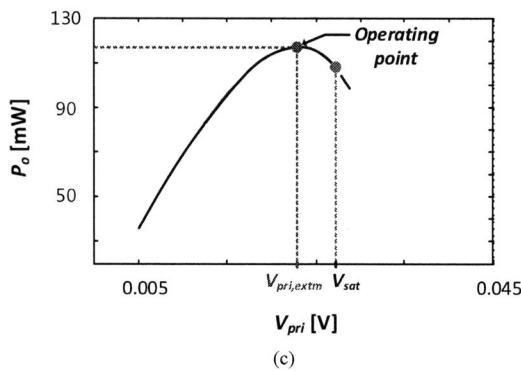

(c)

Fig. 4. P_o according to V_{pri}. (a) $V_{pri,max} > V_{sat}$, where $i_{pri} = 12$ A$_{rms}$, (b) $V_{pri,max} = V_{sat}$, where $i_{pri} = 9.4$ A$_{rms}$, and (c) $V_{pri,max} < V_{sat}$, where $i_{pri} = 8$ A$_{rms}$.

where μ_0 and μ_r are the permeability of the air and the relative permeability for the core, l_c is the length of the magnetic path, and α is the value for $i_{pri,rms}$ at the boundary condition.

From (12), it can be pointed out that when $i_{pri,rms}$ becomes smaller than α, the optimum V_{pri} is deviated from the V_{sat}. Fig. 4 shows P_o according to V_{pri} with different $i_{pri,rms}$. The parameters for the core are listed in Table II. Here, α was obtained as 9.4 A$_{rms}$. Fig. 4(a) is the case where $V_{pri,extm} > V_{sat}$, where $i_{pri,rms} = 12$ A$_{rms}$. Since the operating point for V_{pri} is restricted under V_{sat}, the operating point to maximize P_o is at V_{sat}. Fig. 4(b) is the case where $V_{pri,extm} = V_{sat}$, where $i_{pri,rms} = 9.4$ A$_{rms}$. Still, the operating point is at V_{sat}. Fig. 4(c) is the case where

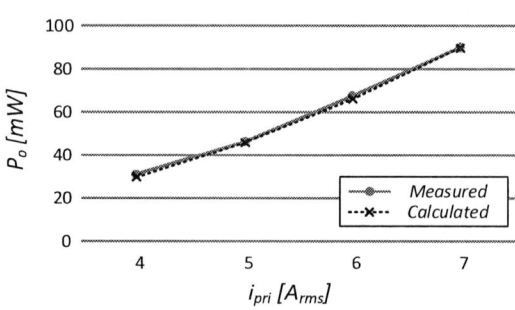

Fig. 5. Experimentally measured P_o and the calculated P_o at the adjusted V_{pri} according to i_{pri}.

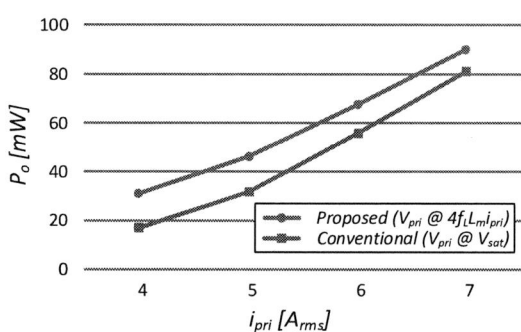

Fig. 6. Measured P_o according to i_{pri}.

TABLE I.
$V_{PRI,OPT}$ AND P_O IN EACH CASES

Case	$V_{pri,opt}$	P_o
(a) $V_{pri,extm} > V_{sat}$	V_{sat}	$\dfrac{2\sqrt{2}V_{sat}}{\pi} i_{pri,rms} \cos(w_L t_s)$
(b) $V_{pri,extm} = V_{sat}$	V_{sat}	$\dfrac{2V_{sat}}{\pi} i_{pri,rms}$
(c) $V_{pri,extm} < V_{sat}$	$V_{pri,extm}$ $(=4f_L L_m i_{pri})$	$\dfrac{8 f_L L_m}{\pi} i_{pri,rms}^2$

TABLE II.
CORE PARAMETERS

Description	Specification
Saturation flux density B_{sat}	0.35 T
Effective cross-sectional area A_e	36×10^{-5} m^2
Length of magnetic path l_c	0.19 m
Relative permeability μ_r	5670
Magnetizing inductance L_m	12 μH

$V_{pri,extm} < V_{sat}$, where $i_{pri,rms} = 8$ A$_{rms}$. In this third case, the operating point should be adjusted according to i_{pri}.

III. EXPERIMENTAL RESULTS

In order to prove the effectiveness of the proposed analysis, the experiments were prevailed with a prototype of $4 \sim 7$ A$_{rms}$ $i_{pri,rms}$, where the distinction of the proposed analysis can be figured out. The parameters for core are listed in Table II. Fig. 5 shows the experimentally measured P_o and the calculated P_o at the adjusted V_{pri} according to i_{pri}. It can be noted that the excellent agreement was obtained.

Fig. 6 shows the measured P_o of the conventional method and the proposed method according to i_{pri}. For conventional method, it always operates at V_{sat} regardless of i_{pri}. On the other hand, for proposed method, the optimal V_{pri} is adjusted according to i_{pri}, resulting in a higher P_o.

IV.   CONCLUSION

In this paper, the analysis for power harvesting according to varying i_{pri} is prevailed, and a new design guideline of V_{load} for maximizing harvested power is suggested. It was shown that V_{load} should be adjusted according to i_{pri} to maximize P_o. The feasibility of the proposed analysis is validated with an $4 \sim 7$ A$_{rms}$ $i_{pri,rms}$ prototype.

V.   ACKNOWLEDGMENT

This work was supported by the National Research Foundation of Korea (NRF) grant funded by the Korea government (MSIP) (No. 2016R1A2B2010328)

REFERENCES

[1] S. Yuan, Y. Huang, J. Zhou, C. Song, Q. Xu and G. Yuan, "A High Efficiency Helical Core for Magnetic Field Energy Harvesting," accepted by in *IEEE Transactions on Power Electronics*, DOI: 10.1109/TPEL.2016.2610323.

[2] WeiWang, Xueliang Huang, Linlin Tan, Jinpeng Guo and Han Liu, "Optimization Design of an Inductive Energy Harvesting Device forWireless Power Supply System Overhead High-Voltage Power Lines", *Energies*, v9, n4, April 1, 2016

[3] P. Li, Y. Wen, Z. Zhang, and S. Pan, "A high-efficiency management circuit using multi winding up conversion current transformer for powerline energy harvesting," *IEEE Trans. Ind. Electron.*, vol. 62, no. 10, pp. 6327–6335, Oct. 2015.

[4] J.Moon, S. Leeb, "Analysis model for magnetic energy harvesters," *IEEE Trans. Power Electron.*, vol. 30, no. 8, Aug. 2015.

Opportunities for Leveraging Low-Voltage GaN Devices in Modular Multi-level Converters for Electric-Vehicle Charging Applications

Mojtaba Ashourloo, Mohammad Shawkat Zaman*, Miad Nasr, Olivier Trescases
Electrical & Computer Engineering, University of Toronto, Canada
*E-mail: Shawkat.Zaman@utoronto.ca

Abstract—**Modular multi-level converters (MMCs), already well-established in high-voltage, high-power AC-DC conversion, can potentially bring advantages in lower-power applications, such as on-board chargers in electric vehicles (EVs). The availability of mature, high-quality GaN devices with low voltage ratings have made it worthwhile to consider the MMCs for these applications, due to its limited voltage gradients and higher AC-side power quality. To investigate these possibilities, a simulated 6-level MMC is compared against an experimentally-validated two-level EV charger. Both converters are designed for a maximum power level of 6.6 kW and compatible with 240 V and 400 V AC-side and DC-link voltages, respectively. The study reveals that the MMC offers great promise in terms of power-quality improvement and AC-side filtering requirements, and the need for large sub-module capacitances to maintain the module voltages is counterbalanced by the reduced requirements for EMI filtering and DC-link decoupling.**

Keywords—*Electric-vehicle charging, EMI performance, Low-voltage GaN, Single-phase Modular Multi-level Converter (MMC).*

I. INTRODUCTION

The Modular Multi-level Converter (MMC) is considered as the most promising and state-of-the-art developed topology for high-voltage (a few hundred kVs), high-power (hundreds of kWs) applications [1]–[5]. Compared to the conventional two-level voltage-source converters (VSCs), the MMC offers many advantages such as modularity, scalability, lower losses, limited voltage gradients, and higher AC voltage quality [6], [7]. The modularity of the MMC also lends itself to higher fault tolerance, which is attractive in safety-critical applications. Furthermore, the series connection of sub-modules (SMs) enables the use of semiconductor switching devices with rated voltages much lower than the maximum application voltage. As such, applying this concept to applications with existing two-level solutions can enable the use of lower-voltage switches with notably better Figures of Merit (FoMs). The MMC topology has typically been considered too costly and complex for power levels below hundreds of kWs due to the large number of components and connections. However, with increased levels of integration and co-packaging options, it has become worthwhile to consider MMCs for lower-power applications, such as on-board Electric Vehicles (EV) chargers. In particular, high-quality GaN devices with rated voltages below 100 V are becoming sufficiently mature [8] to be considered for automotive converters. This work investigates the feasibility of leveraging these devices using the MMC topology for the non-isolated single-phase stage of an on-board EV charger

delivering only a few kWs of power, a range where the benefits of MMCs have not yet been closely examined, especially with regards to the impact on Electromagnetic Interference (EMI) filtering and passive volume.

This paper is organized as follows. Section II outlines the basic concepts and terminology related to MMCs. Section III describes some of the component-level requirements for effective utilization of the MMC architecture. Section IV compares the performance parameters of MMC against a two-level converter prototype, specifically AC-side waveform quality and losses, as well as the sizes of passive and active devices. Finally, Section V summarizes the findings of this work.

II. MMC OPERATING PRINCIPLE

The basic configuration of an N-level single-phase MMC is shown in Fig. 1(a). As compared to the conventional two-level full-bridge (FB) architecture, shown in Fig. 1(b), an MMC is composed of two legs, each consisting of an upper arm and a lower arm. The arms are identical, each being made up of N series-connected half-bridge (HB) SMs and an arm inductor, L_{arm}. Each HB SM, in turn, contains two power transistors as the switching devices and a capacitor as the energy storage element, with a nominal average voltage of $V_{cap} = V_{DC}/N$, where V_{DC} is the DC-link voltage. Given such an N-level MMC, the generated AC-side voltage can have either $N+1$ or $2N+1$ levels, depending on the modulation scheme used. Though the latter option can bring additional advantages (and challenges), this work employs the Level-Shifted Sinusoidal Pulse-Width Modulation (LS-SPWM) scheme with $N+1$ levels as a reasonable compromise between complexity and flexibility.

Detailed descriptions of the operational principles and control of MMCs can be found in the literature [9]–[17] and are not covered in this paper.

III. COMPONENT CONSIDERATIONS

Though the superiority of the MMC topology over two-level architectures has been demonstrated for high-voltage, high-power applications [12], [17], whether MMC can be effective at replacing conventional two-level converters in low-power applications, such as on-board EV chargers, is critically dependent on the availability of high-quality lower-voltage switching devices. As shown in Fig. 1, an N-level MMC has $2N$ times more switching devices than the two-level converter.

The 2018 International Power Electronics Conference

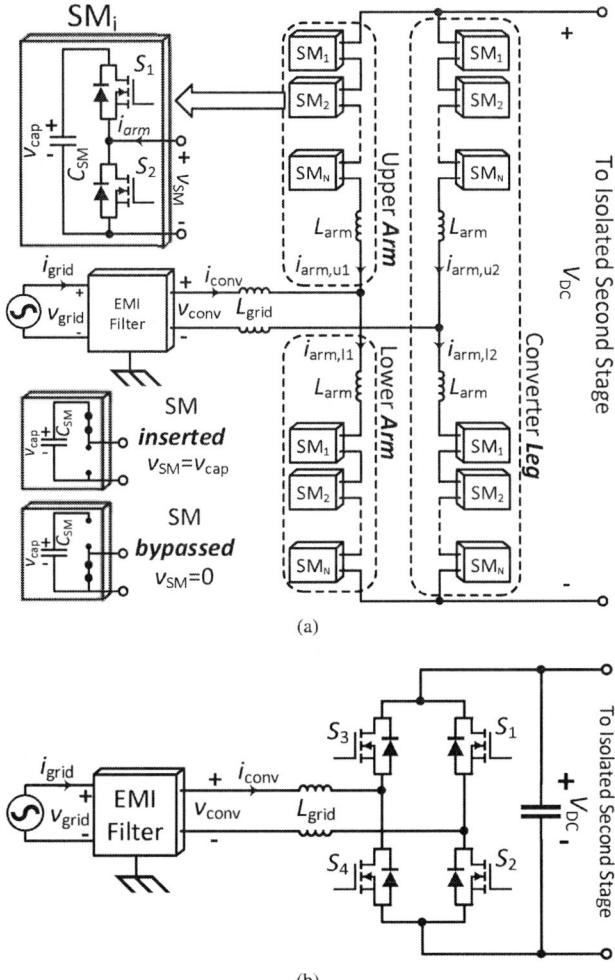

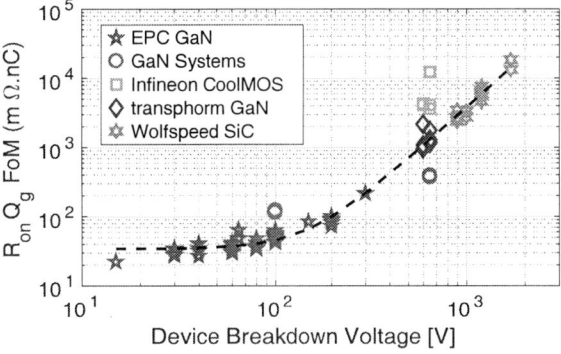

Fig. 2. $R_{on}Q_g$ FoM for selected commercial devices at $T_j = 25°C$.

On the other hand, the AC-side power quality of MMCs and the ability to use lower-voltage power switches hinge upon maintaining the SM capacitor voltage within an acceptable range. In steady state, as the SMs are inserted into the conduction path, their capacitors are charged or discharged, depending on the arm current direction. Consequently, the capacitors must be sized to limit the voltage ripple at the maximum power level. The SM capacitor size is governed by [19]

$$C_{SM} \geq \frac{\Delta E}{2NV_{cap}^2 K_v},\qquad(1)$$

where ΔE is the maximum energy deviation of the capacitor over one cycle and K_v is the ratio of maximum voltage deviation to nominal average capacitor voltage.

Given unity power factor operation and a $\pm 10\%$ variation in the grid voltage, the capacitor size requirement can be alternatively expressed as [19]

$$C_{SM} \geq \frac{NP_{out,max}}{6\pi f_{grid}V_{DC}^2 K_v},\qquad(2)$$

where $P_{out,max}$ is the maximum active power at the output, and f_{grid} is the AC grid frequency.

Fig. 1.   (a) Typical single-phase MMC architecture. (b) Competing two-level converter.

Consequently, to maintain comparable switching and conduction losses despite the higher number of devices, the MMC requires devices with FoM improvements in the order of N. To illustrate this point, the $R_{on}Q_g$ FoM of various commercially-available devices with respect to their breakdown voltages are shown in Fig. 2. From 600-1000 V (typical voltage range in two-level EV chargers) to sub-100 V, device FoM improves in a steeper-than-linear manner, which demonstrates the potential for MMCs. However, the FoM improvements diminish below 100 V, pointing to a practical limit to increasing N in an MMC.

Though MMCs have an increased number of components and connections compared to their two-level competitors, the reduced voltage rating lends itself to miniaturization, decreasing the volume and complexity of an MMC solution. As an example, gate drivers and other auxiliary circuits may be monolithically integrated, and/or co-packaged with the SM power switches. Switch and driver integration, such as in the LMG5200 module (an 80-V GaN HB power stage) from Texas Instruments [18], can be particularly beneficial for taking full advantage of GaN power stages.

IV. PERFORMANCE COMPARISON BETWEEN MMC AND TWO-LEVEL CONVERTERS

The effectiveness of the MMC against two-level converters in terms of various performance metrics is evaluated in this section. A two-level EV charger prototype, shown in Fig. 3(a) and described in [20], is chosen as a representative example of two-level converters. The two-level converter operates in hysteretic Continuous Conduction Mode (CCM) at the peaks and troughs of the AC-side current, and in Boundary Conduction Mode (BCM) through the zero crossings. This mixed operation is referred to as the dual-mode control scheme, as described in [21]. The measured converter waveforms and efficiency are shown in Figs. 3(b) and 3(c), respectively. Table I summarizes the relevant specifications of the two-level prototype and the MMC.

The two-level prototype employs the 900-V SiC MOSFETs to take advantage of its higher thermal conductivity and robustness, even though these devices do not have the best FoM at this voltage rating, as shown in Fig. 2. In contrast, the EPC2029 80-V GaN device is selected as the MMC power

2381

switch, since it has one of the best FoMs among commercially-available low-voltage power devices. Given the DC-link voltage of 400 V and the voltage rating of the MMC power device, 6 SMs per arm are required. Using (2), SM capacitors of 2.2 mF are used to obtain voltage ripples below ±10%, which satisfies the power device rating while preserving acceptable AC voltage quality.

Though these capacitors are significant in size, the increased volume is counterbalanced by two factors. First, these capacitors only carry $V_{cap} = V_{DC}/N$, which allows the use of higher-density components such as film capacitors to alleviate the increase in volume. Second, the SM capacitors also serve as the distributed decoupling capacitors for the DC link. To have a constant DC-side voltage, the number of inserted SMs in each leg needs to be constant and equal to N. Therefore, during normal operation, each converter leg effectively presents N series-connected module capacitors to the DC link. Consequently, the total effective capacitance seen by the DC link, for an MMC topology, is given by $C_{link,eff} = 2C_{SM}/N$. This essentially eliminates the need for additional link capacitors, which are a significant part of the overall volume for the two-level converter, as shown in Section IV-C. Note that the decoupling (active or otherwise) required to mitigate the double-grid-frequency power ripple at the DC link is common to both converters and does not impact the following discussions. It is also instructive to note that both converters have approximately the same effective link capacitance: 840μF and 900μF for the two-level converter and the MMC, respectively.

An accurate `Cadence` simulation model is built for the 6-level MMC system under investigation. To accurately capture the impact of the switching devices, Spectre models provided by the device manufacturer (EPC) are used, and the switches are driven by realistic gating signals with 10 ns deadtime. The AC-side current of the MMC is controlled by a linear Proportional-Resonant (PR) controller.

As shown in Fig. 3(a), the EMI filter inductors of the two-level prototype clearly occupy a dominant portion of the overall converter volume. This is a key motivator behind employing the MMC topology, which can contribute to reducing the size of these components, as highlighted in Section IV-A. Furthermore, the reduction in EMI filter size due to the higher power quality of the MMC can compensate for the increased volume due to SM capacitors.

A. MMC Power Quality and EMI Filter Sizing

The simulated AC-side current waveforms and spectra for both the two-level converter and the MMC, at the maximum power level and without an EMI filter, are shown in Figs. 4 and 5, respectively. In particular, a mixed-signal simulation is used to cover all quantization effects in the two-level converter due to the complex modulation scheme. Furthermore, the spectra are obtained using a standard Line Impedance Stabilization Network (LISN), LI-150. As shown in Fig. 4(b), the MMC current clearly stands out as being nearly sinusoidal even without an EMI filter. This implies that smaller filter components would be needed to meet emission standards. Furthermore, the MMC spectrum has a notably lower noise floor, though the peak levels of the two spectra are comparable.

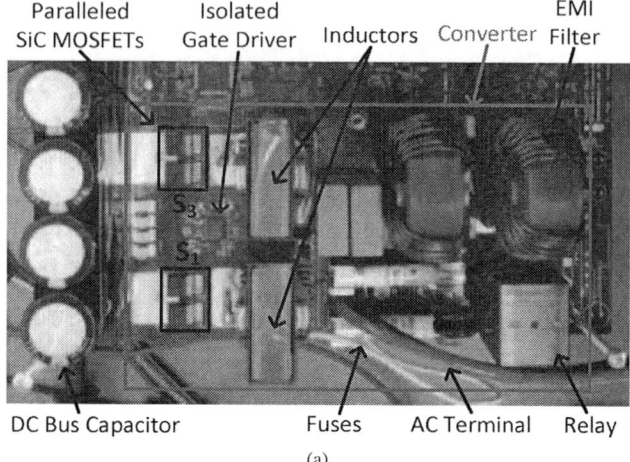

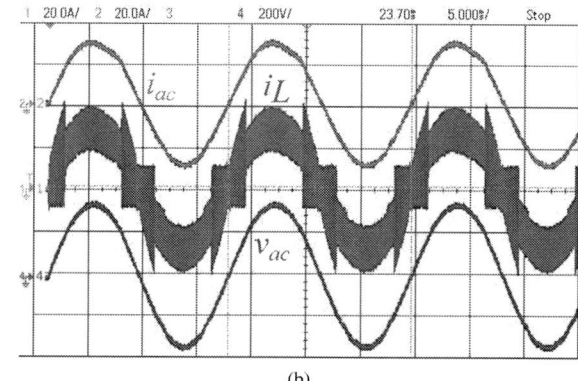

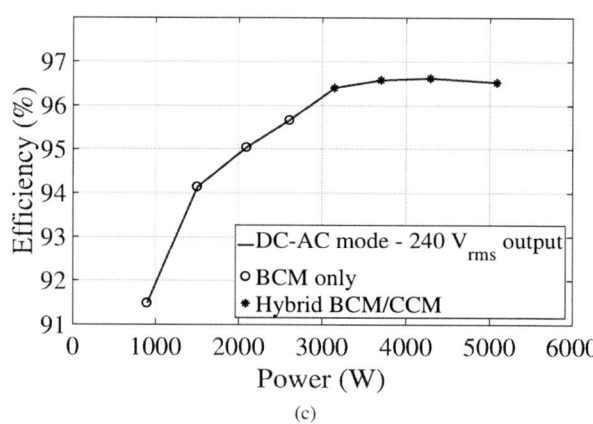

Fig. 3. (a) The on-board two-level EV charger used in the comparison, and its measured (b) waveforms and (c) efficiency.

However, unlike the MMC, the two-level spectrum lacks any sharp frequency peaks due to its variable-frequency operation. In addition, the magnetic core losses in the various inductors are expected to be lower in the MMC since its current contains less distortion.

The same EMI filter structure is used for both converters, which is shown in Fig. 6. Table II lists the component sizes required for meeting CISPR-22 Class-B conducted emission standards. The MMC EMI filter has more than 5× smaller

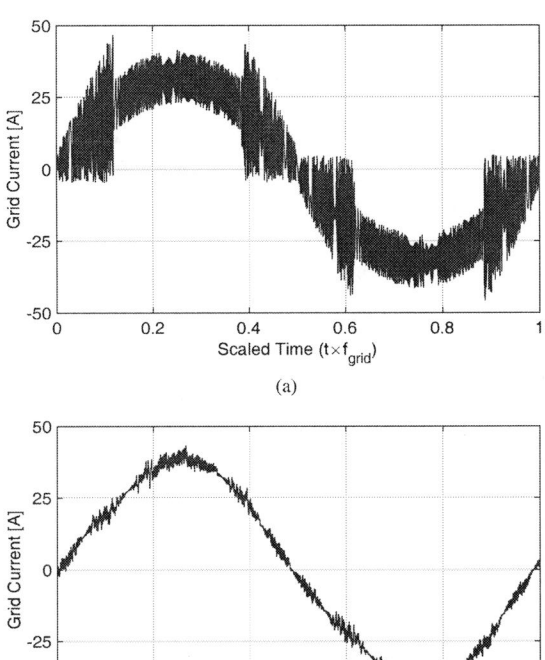

(a)

(b)

Fig. 4. Simulated AC-side current waveforms without an EMI filter at maximum power level, for (a) the two-level converter and (b) the MMC.

TABLE I. CONVERTER SPECIFICATIONS

Parameter	SiC Two-level Converter	6-level GaN MMC
Maximum Power Level, P_{out}	6.6 kW	
Nominal Power Level, P_{nom}	5.0 kW	
RMS Grid Voltage, $V_{grid,RMS}$	240 V	
Grid Frequency, f_{grid}	60 Hz	
DC-Link Voltage, V_{DC}	400 V	
Grid Inductance, L_{grid}	25 µH×2	
Switching Frequency, f_{sw}	250 kHz (maximum)	125 kHz (fixed)
Switching Device	Two parallel SiC (C3M0065090, 900 V)	GaN (EPC2029, 80 V)
Control Method	Dual-mode BCM-CCM	LS-SPWM
Arm Inductance, L_{arm}	–	1 µH
DC-Link Capacitance, C_{link}	840 mF ±20% 600 V	–
SM Capacitance, C_{SM}	–	2.7 mF ±20%, 80 V

inductors than those of the two-level prototype, which alone represents a considerable amount of volume and cost savings. Moreover, thanks to the inherent higher power quality of the MMC, magnetic cores with lower saturation ratings can be used, providing additional cost savings.

The simulated AC-side current spectra for both the MMC and the two-level converter, with the EMI filter from Fig. 6, are shown in Fig. 7. Both converters meet the CISPR-22 standard for Class-B equipment.

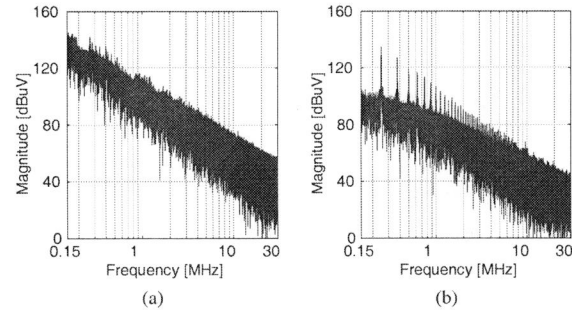

(a)   (b)

Fig. 5. Simulated AC-side current spectra using LI-150 LISN without an EMI filter at maximum power level, for (a) the two-level converter and (b) the MMC.

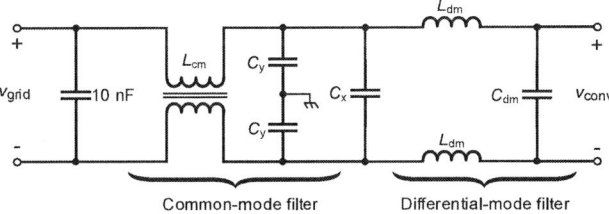

Fig. 6. The EMI filter structure used in this work.

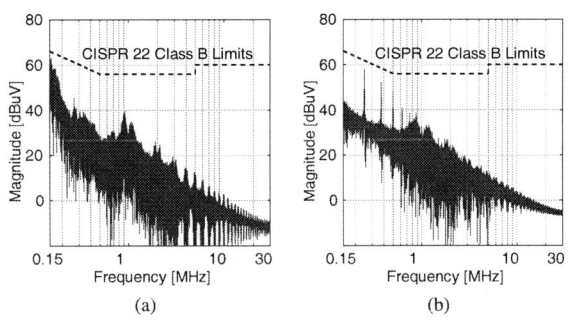

(a)   (b)

Fig. 7. Simulated AC-side current spectra using LI-150 LISN with EMI filter at maximum power level, for (a) the two-level converter and (b) the MMC.

TABLE II. EMI FILTER COMPONENTS

Parameter	SiC Two-level Converter	6-level GaN MMC
L_{cm}	1500 µH	275 µH
L_{dm}	140 µH	25 µH
C_{dm}	14 µF	14 µF
C_x	33 nF	33 nF
C_y	25 nF	25 nF

B. Loss Breakdown at Nominal Power

Fig. 8 shows the various loss components of both converters at the nominal power level. The EMI filter losses for the MMC are assumed to be the same as those of the two-level converter. Gate-drive losses are calculated to be well below 1 W for both converters and therefore are not included.

Despite the increased number of power switches, the MMC topology manages to achieve only half as much losses as the two-level converter. The primary reason is that the MMC

2383

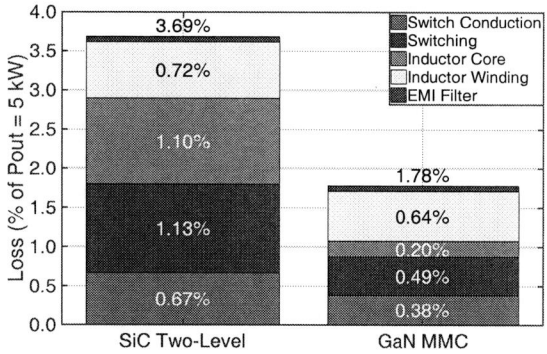

Fig. 8. Loss breakdown for the two-level converter (measured) and the MMC (simulated) at nominal power level.

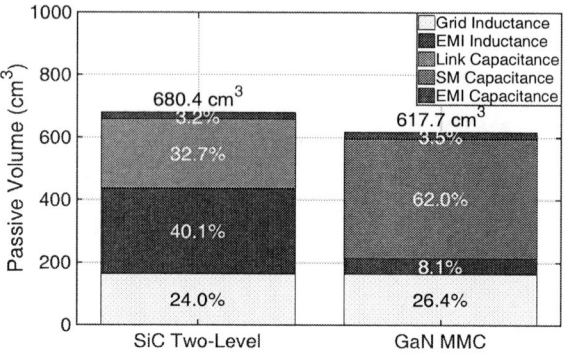

Fig. 9. Passive volume breakdown for the two-level converter (implemented) and the MMC (estimated).

configuration results in significant reduction of magnetic core losses, due to its smoother AC-side current waveform. The reduced conduction and switching losses of the GaN devices, due to better device FoM and reduced module voltages, also contribute to lowering the total loss. Furthermore, the reduced losses and distributed heat sources in the MMC can reduce the required heat sink volume and simplify thermal management.

C. Volume Comparison of the Converters

Though the final converter size depends on many different factors, the volume of the passive components (i.e., inductors and capacitors) is the major contributor for this application area. Fig. 9 and Table III compare the volume of the largest passive components in the two-level converter to their counterparts in the MMC topology. For the two-level converter, the implemented volumes of the passives are reported; the inductors are built from the ETD-44 core utilizing N97 ferrite core material. In the case of the MMC, capacitor volumes are obtained from commercially-available parts with the required specifications (with sufficient margin for variations), whereas inductor volumes are extrapolated from the two-level components.

It is clear from Table III that the increase in SM capacitor volume in the MMC is completely negated by the reduction in EMI filter magnetics and link capacitance. As a result, the passive volume in the MMC is actually 9% lower than the two-level converter. The SM capacitor volume can be further reduced by increasing the allowed voltage ripple on the SMs to fully utilize the voltage rating of the power devices.

Another aspect to consider with regards to the two converters is the area consumed by the power devices. Since the power switches are typically of much lower height than the passives, it is sufficient to consider their footprint on the printed circuit board, rather than their volume. Table IV compares the area of the power devices.

Despite the massive increase in the number of power devices in the MMC, the lower voltage rating and compact chip-scale packaging of the GaN devices lead to less than half the active area compared to the two-level converter. This indicates that the HB power stage within the MMC SMs can easily be upgraded to more complex topologies (such as a full

TABLE III.   VOLUME COMPARISON OF PASSIVE COMPONENTS

Component	SiC Two-level Converter (Implemented)	6-level GaN MMC (Estimated)
Grid Inductance, L_{grid}	163.2 (25 µH ×2)	163.2 (25 µH ×2)
Arm Inductance, L_{arm}	–	Negligible
EMI Filter Inductance, $L_{cm} + L_{dm}$	273.0 (1500 µH + 2×140 µH)	48.8 (275 µH + 2×25 µH)
Link Capacitance, C_{link}	222.7 (LGN2X121MELB45, 0.12 mF, 600 V ×7)	–
SM Capacitance, C_{SM}	–	383.2 (80MXG2700MEFC22X40, 2.7 mF, 80 V ×24)
EMI Filter Capacitance, $C_{dm} + C_x + C_y$	21.5	21.5
Total Volume	680.4	617.7

TABLE IV.   AREA COMPARISON OF ACTIVE DEVICES

Parameter	SiC Two-level Converter	6-level GaN MMC
Device	SiC, C3M0065090	GaN, EPC2029
Footprint Area	1.62 cm×1.02 cm	0.46 cm×0.26 cm
Number of Devices	4 × 2	6 × 4 × 2
Total Area	13.2 cm²	5.7 cm²

bridge) in order to take advantage of their benefits without incurring a significant area penalty.

V. CONCLUSIONS

The simulated results of the single-phase 6-level MMC system show clear promises when compared to a two-level EV charger. The need for relatively large sub-module capacitance for effective implementation at sub-100 kW power levels is negated by the reduction in EMI filter components and DC-link capacitance. The reduction in magnetic core losses, increased fault tolerance, and easier thermal management, bolstered by the availability of high-quality lower-voltage GaN power devices, clearly make the MMC topology an attractive contender for EV charging applications.

REFERENCES

[1] R. Marquardt, "Modular Multilevel Converter: An universal concept for HVDC-Networks and extended DC-Bus-applications," in *Power Electronics Conference (IPEC), 2010 International*, June 2010, pp. 502–507.

[2] T. Hammons, V. Lescale, K. Uecker, M. Haeusler, D. Retzmann, K. Staschus, and S. Lepy, "State of the Art in Ultrahigh-Voltage Transmission," *Proceedings of the IEEE*, vol. 100, no. 2, pp. 360–390, Feb 2012.

[3] J. Mei, B. Xiao, K. Shen, L. Tolbert, and J. Y. Zheng, "Modular Multilevel Inverter with New Modulation Method and Its Application to Photovoltaic Grid-Connected Generator," *Power Electronics, IEEE Transactions on*, vol. 28, no. 11, pp. 5063–5073, Nov 2013.

[4] L. Guan, X. Fan, Y. Liu, and Q. Wu, "Dual-mode Control of AC/VSC-HVDC Hybrid Transmission Systems with Wind Power Integrated," *Power Delivery, IEEE Transactions on*, vol. PP, no. 99, pp. 1–1, 2014.

[5] X. Chen, H. Sun, J. Wen, W.-J. Lee, X. Yuan, N. Li, and L. Yao, "Integrating Wind Farm to the Grid Using Hybrid Multiterminal HVDC Technology," *Industry Applications, IEEE Transactions on*, vol. 47, no. 2, pp. 965–972, March 2011.

[6] B. Gemmell, J. Dorn, D. Retzmann, and D. Soerangr, "Prospects of multilevel VSC technologies for power transmission," in *Transmission and Distribution Conference and Exposition, 2008. T 00026;D. IEEE/PES*, April 2008, pp. 1–16.

[7] Y. Zhang, G. Adam, T. Lim, S. Finney, and B. Williams, "Voltage source converter in high voltage applications: Multilevel versus two-level converters," in *AC and DC Power Transmission, 2010. ACDC. 9th IET International Conference on*, Oct 2010, pp. 1 5.

[8] [Online]. Available: https://epc-co.com/epc/DesignSupport/eGaNFETReliability.aspx

[9] A. Yazdani and R. Iravani, *Voltage-Sourced Converters in Power Systems:Modeling, Control, and Applications*, 1st ed. Wiley-IEEE Press, 2010.

[10] Q. Tu, Z. Xu, and L. Xu, "Reduced Switching-Frequency Modulation and Circulating Current Suppression for Modular Multilevel Converters," *Power Delivery, IEEE Transactions on*, vol. 26, no. 3, pp. 2009–2017, July 2011.

[11] J.-W. Moon, C.-S. Kim, J.-W. Park, D.-W. Kang, and J.-M. Kim, "Circulating Current Control in MMC Under the Unbalanced Voltage," *Power Delivery, IEEE Transactions on*, vol. 28, no. 3, pp. 1952–1959, July 2013.

[12] M. Perez, S. Bernet, J. Rodriguez, S. Kouro, and R. Lizana, "Circuit Topologies, Modeling, Control Schemes, and Applications of Modular Multilevel Converters," *Power Electronics, IEEE Transactions on*, vol. 30, no. 1, pp. 4–17, Jan 2015.

[13] A. Shojaei and G. Joos, "An improved modulation scheme for harmonic distortion reduction in modular multilevel converter," in *Power and Energy Society General Meeting, 2012 IEEE*, July 2012, pp. 1–7.

[14] M. Saeedifard and R. Iravani, "Dynamic Performance of a Modular Multilevel Back-to-Back HVDC System," *Power Delivery, IEEE Transactions on*, vol. 25, no. 4, pp. 2903–2912, Oct 2010.

[15] G. Konstantinou, M. Ciobotaru, and V. Agelidis, "Selective harmonic elimination pulse-width modulation of modular multilevel converters," *Power Electronics, IET*, vol. 6, no. 1, pp. 96–107, Jan 2013.

[16] P. Meshram and V. Borghate, "A Simplified Nearest Level Control (NLC) Voltage Balancing Method for Modular Multilevel Converter (MMC)," *Power Electronics, IEEE Transactions on*, vol. 30, no. 1, pp. 450–462, Jan 2015.

[17] "Guide for the development of models for hvdc converters in a hvdc grid," CIGRE WG B4-57, July 2013.

[18] T. Instruments, *LMG5200 80-V, 10-A GaN Half-Bridge Power Stage datasheet (Rev. D)*, Apr. 2016, revised Mar. 2017. [Online]. Available: http://www.ti.com/product/LMG5200

[19] M. M. C. Merlin and T. C. Green, "Cell capacitor sizing in multilevel converters: cases of the modular multilevel converter and alternate arm converter," *Institution of Engineering and Technology: Power Electronics*, vol. 8, no. 3, pp. 350–360, 2015.

[20] M. Nasr, K. Gupta, C. da Silva, C. H. Amon, and O. Trescases, "SiC Based On-Board EV Power-Hub with High-Efficiency DC Transfer Mode through AC Port for Vehicle-to-Vehicle Charging," in *Proc. IEEE Applied Power Electronics Conference and Exposition*, March 2018.

[21] S. Chung, M. Nasr, D. Guirguis, M. Otsuka, S. Poshtkouhi, D. K. W. Li, V. Palaniappan, D. Romero, C. Amon, R. Orr, and O. Trescases, "Thermal and electrical co-design of a modular high-density single-phase inverter using wide-bandgap devices," in *Proc. IEEE Applied Power Electronics Conference and Exposition*, March 2016, pp. 1350–1357.

The 2018 International Power Electronics Conference

A New Control Strategy for Modular Multilevel Converter Operating in Quasi Two-Level PWM Mode

Chao Wang[1*], Kui Wang[1], Zedong Zheng[1] and Yongdong Li[1]
1 Department of Electrical Engineering, Tsinghua University, Beijing, China
*E-mail: wangchao16@mails.tsinghua.edu.cn

Abstract—**Modular multilevel converter (MMC) is difficult to operate when driving a motor with large constant load torque at low speed, due to the large voltage fluctuation of submodule (SM) capacitors on this condition. Quasi two-level PWM is an alternative control method to solve this problem. This paper proposes a new control strategy to implement quasi two-level PWM for a three-phase MMC inverter. Arm current commutation is realized by control delay time of each PWM signal, and voltage of SM capacitors are kept balanced by change the connecting relationship between PWM signals and SMs, without requirements of a high-frequency PWM control. The control system is verified by simulation results and compared with high frequency circulating current injection method under same load condition. Although the total harmonic distortion (THD) is relatively large due to the sacrifice of multilevel operation, the voltage fluctuation of proposed method is only 28.5% of conventional circulating current injection method, and what is more, the arm current stress is also reduced by 24%.**

Keywords—*Arm current control, capacitor voltage fluctuation, modular multilevel converter (MMC), quasi two-level PWM.*

I. INTRODUCTION

Since direct series connection technology of power device is especially hard to be mastered, multilevel converter is a better solution for medium- and high-voltage applications. Compared with most commercially used multilevel topologies, such as neutral point clamped (NPC), flying capacitor (FC), and cascaded H-bridge (CHB) converter, MMC have some unique advantages: modular structure, easy to scale voltage, convenient to operate in four-quadrant applications and low expense to design fault redundancy [1,2]. In spite of these prominent advantages, the most drawback of MMC is the inherent voltage fluctuation of SM capacitors. According to [3], the voltage fluctuation is direct proportion to the amplitude of load current and inversely proportional to the output frequency, which results in large voltage fluctuation when MMC inverter drives a motor with large constant load torque at low speed. This problem renders it difficult to popularize the adoption of MMC in variable-speed drives application.

Special control method needs to be applied to overcome this problem. The first category of method is to inject high-frequency common mode voltage in MMC output voltage and circulating current in arm current, and by utilizing the low-frequency power component generated by them, most of the inherent low-frequency power of every arm can be counteracted. Thus, the arm power only contains little low-frequency component and some high-frequency component, resulting in reduction of voltage fluctuation. References [4-7] traversed this kind of strategy, using sinusoidal-wave, square-wave or trapezoidal-wave as the injected high-frequency component. According to [4-7], the magnitude of injected high frequency circulating current is in direct proportion to the load current. As a consequence, in spite of the reduction of SM capacitors voltage fluctuation, arm current stress increases largely when load current is near to its rated value, leading to higher power loss and total cost of the whole system. Moreover, the high frequency and large amplitude common-mode voltage may threaten lifespan of the motor bearings [4]. Another kind of method is proposed in [8], that is decreasing the average voltage of submodule capacitors when MMC operates at low frequency. In this way, although the voltage fluctuation is still relatively large, the peak voltage of capacitors do not exceed the maximum value limited by safety operation area (SOA) of MMC on account of the decrease of its average value. However, the low speed range is limited, which is not lower than one third of the base speed.

Recently, a new approach is proposed in [9,10], which altered the PWM mode of MMC from multilevel PWM to quasi two-level PWM. Different from the two methods mentioned above, the operation mode of MMC is changed in this approach to eliminate longtime charge and discharge of SM capacitors actively. As a result, the energy variation of MMC arms reduces significantly, thus voltage fluctuation is limited to a diminutive value. This method sacrifices multilevel output voltage but maintains other basic function of MMC, which means MMC is still capable to be applied in medium- and high-voltage application without requirements of device serial connection. The control strategy proposed in [9,10] consists of a carrier frequency PWM which controls the output current and a high frequency PWM which is higher than 25kHz to implement the circulating current control and balance of SM capacitors' voltage, causing that it may be hard to realize in a practical high voltage and high

This work was supported by the National Natural Science Foundation of China (Grant No. 51777110).

2386

power system based on IGBT devices. In addition, the value of arm inductances is much lower than regular designed MMC inverter, and this may cause that the PWM mode of MMC inverter is hard to be switched to multilevel PWM mode at high frequency.

This paper proposes a new control strategy for MMC operating in quasi two-level PWM mode. To implement this operation mode, arm current control and capacitor voltage balance control are discussed in detail. Compared with the control system proposed in [9,10], high-frequency PWM is not necessary anymore, therefore the whole control system is simplified. Moreover, based on this control method, the value of arm inductances of MMC discussed in this paper is nearer to regular value, thus it is more suitable for practical MMC system.

This paper is organized as follows. First of all, the basic theory of quasi two-level PWM mode is introduced in Section II to illustrate the main problem needs to be solved when quasi two-level PWM is applied to MMC. Then, Section III elaborates the detail of the control strategy. To verify the control strategy, Section IV presents the simulation results of a 2-MVA MMC-based variable-speed motor drive system. The results suggest that the voltage fluctuation of proposed method is only 28.5% of conventional circulating current injection method and arm current stress is also reduced by 24%. Finally, Section V gives the conclusion.

II. BASIC THEORY OF QUASI TWO-LEVEL PWM MODE

The general topology of a three-phase MMC is shown in Fig.1. Every phase leg consists of two arms, and each arm is cascaded by N identical half-bridge SMs and an arm inductor. In conventional multilevel PWM mode, several SMs of both upper arm and lower arm are connected to the DC bus to generate a multilevel voltage waveform at the output terminal, which means each arm needs to supply load current at any time. When a submodule capacitor is connected to the DC bus, arm current flows through it, generating voltage fluctuation. This means the large voltage fluctuation is caused by a long time charge and discharge of large arm currents to capacitors. From this point of view, the large voltage fluctuation would decrease if the longtime charge and discharge are eliminated, which needs to avoid high voltage and large current appearing in an arm at the same time.

The equivalent circuit model of one phase leg (here is phase leg A) of MMC is shown as Fig.2, the cascaded half-bridge SMs of each arm are replaced by controlled voltage sources, which is quite similar to an ordinary half-bridge circuit when regarding controlled voltage sources as power devices. With regard to a half-bridge circuit, neither top nor bottom power device conducts current when its junction capacitor supports DC bus and vice versa. Except for the switching process, high voltage and large current do not appear in one device at the same time. If supplants power devices by MMC arms, the goal aforementioned is achieved. Quasi two-level PWM mode of MMC inverter derives from this phenomenon, and the ideal operation

process of phase A is shown is Fig.3. This mode requires to control upper arm and lower arm conduct load current alternately. At steady state, when the load current flows through either arm, all SMs of that arm are bypassed and those of the other one are connected to the DC bus. As a consequence, there is no long time charge or discharge to submodule capacitors at steady state under ideal conditions, so that the large voltage fluctuation is suppressed. The voltage fluctuation is mainly caused by energy variation during the current commutation between two arms.

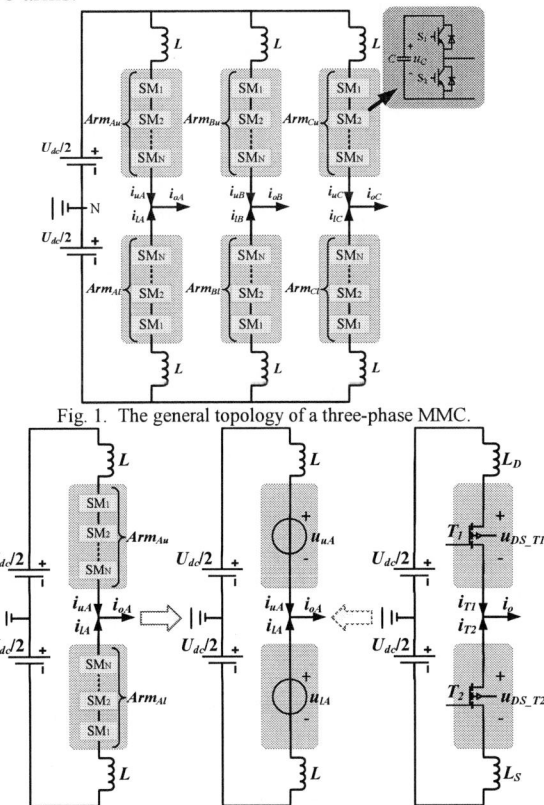

Fig. 1. The general topology of a three-phase MMC.

Fig. 2. The equivalent circuit model of one phase leg of MMC and comparison with half-bridge circuit.

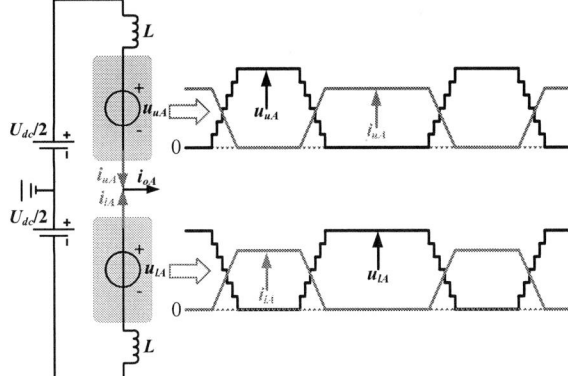

Fig. 3. The ideal quasi two-level PWM operation process of phase A.

To achieve this control objective, three critical points need to be considered. First of all, extra high dv/dt results in the maturing of insulation of motor windings and bearings, thus the output voltage of MMC cannot be an ideal square waveform as that of an ideal two-level inverter. This indicates that the SMs cannot switch at the same time during the arm current commutation. On the contrary, every SM needs to switch one by one with some time interval to realize a step voltage waveform to decrease dv/dt of MMC output voltage as shown in Fig.3. Consequently, there should be a delay time among PWM signals of SMs. Secondly, since all SMs of an arm needs to be connected or bypassed at steady state, redundant switch states of multilevel PWM mode are no longer exist. Therefore, the generally used method such as voltage sorting and carrier phase shifting PWM with capacitor voltage closed-loop are not available anymore, which means a different control method is required to keep the voltage of SM capacitors balanced. Finally, the most important problem is arm current control. The current commutation of two-level converter is between two devices, which is controlled by gate driver signals, and the stray inductance is quite small so the current commutation finishes instantly. As for MMC, the current commutation is between two arms with large arm inductors, which impedes the current commutation. As a result, proper voltage needs to be applied on arm inductor to force the load current transfer from one arm to the other in a short time. On account of the existence of charge and discharge for SM capacitors in arm commutation interval, the commutation time needs to be as short as possible.

III.   PROPOSED CONTROL SYSTEM

A. Delay Control for SMs of One Arm

The delay control scheme is shown in Fig.4. The original PWM signal is generated the same as a two-level inverter, after delayed by delay algorithm, a new PWM signal is produced. Then, the new PWM signal is utilized as the input signal to produce the next PWM signal. Generally, the delay time between two adjacent PWM signals is configured to several microseconds to ensure a relatively short time current commutation, which is able to adjust as required.

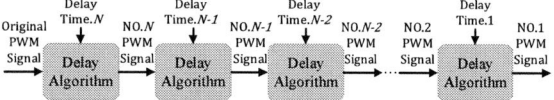

Fig. 4. The delay control scheme.

B. Arm Current Control

To elucidate the arm current control strategy proposed by this paper, the current commutation situation of an MMC with five SMs in each arm in which i_{uA} decreases and i_{lA} increases when the load current i_{oA} is positive is considered as an example. This indicates the whole voltage of phase A leg U_{phaseA} needs to be larger than DC bus voltage U_{dc} to induce a negative circulating current to help finish current commutation. Neglecting the voltage of arm inductor, U_{phaseA} is defined as

$$U_{phaseA} = u_{uA} + u_{lA} \qquad (1)$$

Where u_{uA} and u_{lA} are the voltage of cascaded half-bridge SMs of upper and lower arm respectively.

Assuming that the voltage of all capacitors is maintained to their rated value U_C. To achieve the increment and decrease of arm current, one control freedom degree of quasi two-level PWM is the delay time of each PWM signal. By increasing the delay time of several SMs in lower arm, as shown in Fig.5, U_{phaseA} is larger than U_{dc} at each voltage step, satisfying the requirement aforementioned. At steady state, i_{uA} or i_{lA} supplies all load current i_{oA}, thus the steady amplitude of i_{uA} and i_{lA} need to be controlled accurately. To achieve this goal, the delay time of each SM needs calculation.

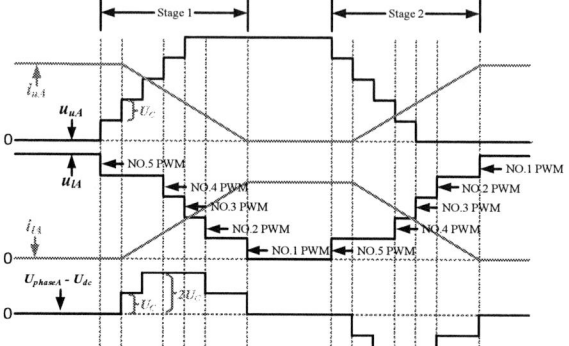

Fig.5. Quasi two-level PWM operation process when the load current i_{oA} is positive.

At first, every PWM signal is numbered according to the order of its edge, as shown in Fig.5. Assuming the minimum delay time between two adjacent PWM signals is T_d, which is called unit delay, and all PWM signals of upper arm are delayed by T_d to limit dv/dt. Ignoring the voltage fluctuation, the voltage difference between U_{phaseA} and U_{dc} is always an integral multiple of submodule rated voltage U_C which is defined as unit voltage difference. The unit variation of lower arm current Δi_{l_unit} caused by U_C lasting for T_d is

$$\Delta i_{l\_unit} = T_d \frac{U_C}{2L} \qquad (2)$$

Defining P as the number of unit delay, it can be analyzed from Fig.5 that PWM signals with different numeration delayed by same P resulting in different Δi_{lA}. For instance, if NO.4 PWM signal is delayed by T_d and the others remain unchanged, Δi_{lA} equals to $4 \cdot \Delta i_{l\_unit}$. It should be noted that the delay time discussed here does not contain the basic delay time T_d of two adjacent PWM signals, which is an additional delay to implement current commutation. Moreover, if NO.4 PWM signal is delayed by $2 \cdot T_d$, Δi_{lA} equals to $8 \cdot \Delta i_{l\_unit}$ (the others remain unchanged). Analyzing other signals, it can be concluded that if NO.i PWM signal is delayed by $P \cdot T_d$, Δi_{lA} caused by this delay control equals to $i \cdot P \cdot \Delta i_{l\_unit}$. What is more, it also can be proved that the Δi_{lA} caused by delay of any PWM signal does not by influenced by delay of others. That is no matter how delay time of other signals change, as long as i and P are determinate, Δi_{lA} caused by delay control of NO.K PWM signal always equals to $i \cdot P \cdot \Delta i_{l\_unit}$. For phase B and phase C, the results are unaltered. This rule can be generalized to an MMC inverter with N SMs in each arm, which is summarized in Table I. What calls for special attention is that any P_i, $i \in \{1,2,3,\cdots,N\}$ should be a natural number, and Δi_{lA_i} represents the increment of arm current caused by the delay time of NO.i PWM signal.

TABLE I
THE RELATIONSHIP BETWEEN DELAY TIME OF PWM SIGNAL AND INCREMENT OF ARM CURRENT

Numeration of PWM Signal of Each Arm	Delay Time	Increment of Arm Current
1	$P_1 \cdot T_d$	$\Delta i_{lA\_1} = P_1 \cdot \Delta i_{l\_unit}$
2	$P_2 \cdot T_d$	$\Delta i_{lA\_2} = 2P_2 \cdot \Delta i_{l\_unit}$
3	$P_3 \cdot T_d$	$\Delta i_{lA\_3} = 3P_3 \cdot \Delta i_{l\_unit}$
...	...	...
N-2	$P_{N-2} \cdot T_d$	$\Delta i_{lA\_N-2} = (N-2)P_{N-2} \cdot \Delta i_{l\_unit}$
N-1	$P_{N-1} \cdot T_d$	$\Delta i_{lA\_N-1} = (N-1)P_{N-1} \cdot \Delta i_{l\_unit}$
N	$P_N \cdot T_d$	$\Delta i_{lA\_N} = N \cdot P_N \cdot \Delta i_{l\_unit}$

Then, supposing i_{lA} varies from zero to i_{oA} in this situation, that is the total current increment Δi_{lA_total} equals to i_{oA}. Δi_{lA_total} can be expressed by a rational number D multiplies Δi_{l_unit}, that is

$$\Delta i_{lA\_total} = \Delta i_{oA} = D\Delta i_{l\_unit} \quad (3)$$

To implement current commutation, the sum of Δi_{lA_i} (i is the numeration of PWM signal) should be equals to Δi_{lA_total}, but noticing Δi_{lA_i} must be an integral multiple of Δi_{l_unit}, the equation can be derived as follow

$$\sum_{i=1}^{N} \Delta i_{lAi} = \lfloor D \rfloor \Delta i_{l\_unit} \quad (4)$$

$\lfloor D \rfloor$ represents rounding D up to an integer to ensure the steady value of i_{lA} can reach to i_{oA}. In accordance with the rule shown in TABLE I, (4) is equivalent to

$$\sum_{i=1}^{N} i \times P_i = \lfloor D \rfloor \quad (5)$$

Finally, after calculating $\lfloor D \rfloor$, this problem is transformed to allocate delay time of each PWM signal to make up (5). It is obviously that if all PWM signals are delayed by one unit delay, the PWM signal with larger numeration results in larger i_{lA}. Thus, to accelerate the variation speed of i_{lA}, the delay time of the PWM signal with larger numeration should be given priority. According to this rule, the allocation method of delay time is shown in Fig.6. Calculating $\lfloor D \rfloor$ divides N to determine P_i of NO.N PWM signal to generate $N \cdot P_N \cdot \Delta i_{l\_unit}$, after this, there are no more than $(N-1) \cdot \Delta i_{l\_unit}$ remained to be produced, which can be produced by one of PWM signals from 1 to $(N-1)$ with one unit delay T_d.

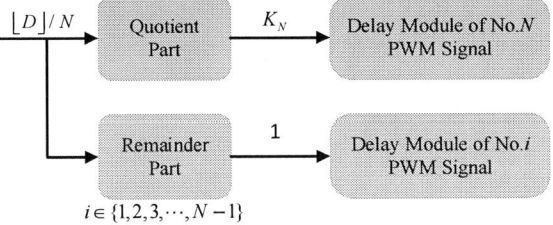

Fig.6. The allocation method of delay time.

When the current commutation situation turns into i_{uA} increases and i_{lA} decreases, the current commutation can be implemented automatically due to U_{phaseA} is lower than U_{dc} as shown in Fig.5. If i_{oA} changes to negative, the delay time of upper arm should be increased while that of lower arm remains unchanged. The discussion above can be applied to any phase of MMC and other current commutation situations.

Having done all aforementioned, the final task is to allocate PWM signals to SMs by using the allocation strategy illustrated in voltage balancing control.

C. Capacitor Voltage Balance Control

Due to there is no redundant switch state in quasi two-level PWM mode, a new control algorithm is required to keep voltage balanced. Since a group of delayed PWM signals has been generated, the order of connection between this PWM signal group and the controlled SMs of one arm is a control freedom degree. In order to illustrate the voltage balance control method proposed by this paper, the current commutation situation shown in Fig.5 is still utilized as an example.

As for Stage 1, the current commutation process is that i_{uA} decreases from i_{oA} to 0 while i_{lA} increases from 0 to i_{oA} and SMs of upper arm are connected to DC bus successively while SMs of lower arm are bypassed successively. For SMs of upper arm, it is obviously that capacitors of them are charged in this situation, considering that i_{oA} is positive. As a consequence, the capacitor that is connected to DC bus later than other capacitors is charged by i_{uA} for a longer time, leading to larger voltage fluctuation. Therefore, to equilibrate capacitors' voltage, SMs with lower voltage need to be connected to DC bus earlier. For SMs of lower arm, capacitors of them are discharged in this situation, thus the capacitor that is bypassed later than other SMs is discharged by i_{lA} for a longer time, resulting in lager voltage fluctuation. For this reason, SMs with lower voltage need to be bypassed earlier.

While as for Stage 2, the current commutation process is that i_{uA} increases from 0 to i_{oA} while i_{lA} decreases from i_{oA} to 0 and SMs of upper arm are bypassed successively while SMs of lower arm are connected to DC bus successively. The same as Stage 1, capacitors of upper arm is charged and that of lower arm are discharged. Thus, capacitors with higher voltage need to be bypassed earlier for capacitors of upper arm while capacitors with higher voltage need to be connected to DC bus earlier.

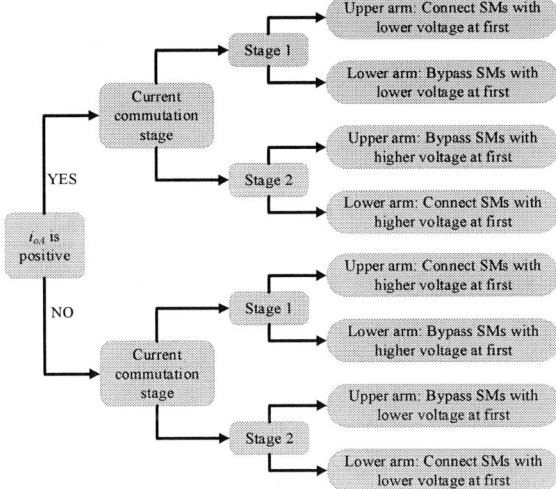

Fig.7. The whole capacitor voltage balance control method.

When i_{oA} is negative, the charge and discharge state in Stage 1 and Stage 2 of upper arm and lower arm is altered, thus the control direction aforementioned should be reversed. As a conclusion, the whole capacitor voltage balance control method is shown in Fig.7.

IV. SIMULATION RESULT

To verify the control method proposed by this paper, a 2-MVA MMC inverter with ten SMs per arm driving an induction motor (IM) is simulated in by MATLAB/Simulink 2015a. The detailed simulation parameters are shown in TABLE II. The arm inductance is relatively lower than regular value to reduce the commutation time.

TABLE II
SIMULATION PARAMETERS

MMC parameters	
DC bus voltage	$U_{dc} = 12000$ V
Number of SMs per arm	$N = 12$
Nominal SM capacitor voltage	$U_C = 1000$ V
SM capacitance	$C = 3$ mF
Arm inductance	$L = 0.5$ mH
Switching frequency	$f_{sw} = 2$ kHz
Unit delay	$T_d = 2 \mu s$
IM parameters	
Rated capacity	$S = 2$ MVA
Rated rotor speed	$n_N = 1380$ r/min
Rated electromagnetic torque	$T_N = 12000$ N·m
Rated RMS line-to-line voltage	$U_{LL} = 6$ kV
Rated phase current magnitude	$\hat{I}_O = 272$ A

To compare the control performance, MMC is controlled by the control method described in Section III and the high-frequency component injection method proposed in [4], respectively. The IM is controlled by vector control, with load torque equals to T_N, which can ensure the output current of MMC reaches its rated value. The target rotor speed is set to 100 r/min to verify the effect of voltage fluctuation suppression. During the whole star-up process, the max electromagnetic torque of IM is limited to $1.25T_N$. It should be noted when high-frequency component injection method is applied, the arm inductance is set to a regular value 2mH.

Fig.8. shows the simulated operating waveform using high-frequency component injection method. After a pre-excitation time last for 0.625s, the load torque of IM alters from 0 to T_N instantaneously, and the output current of MMC rise rapidly to produce a large electromagnetic torque by vector control. As presented in Fig.8, the steady peak to peak value of voltage fluctuation at steady state reaches 280V, which indicates the voltage fluctuation rate exceeds 10% (a regular limitation of MMC). The voltage fluctuation rate can decrease only if enlarge the value of SM capacitors. In addition, due to the injection of high frequency circulating current, the peak value of arm currents reaches to 420A, 120A in excess of the magnitude of output current.

Fig.9. shows the simulated operating waveform using

method proposed by this paper. The start-up process is the same as aforementioned. It can be observed from Fig.9 that the steady peak to peak value of voltage fluctuation is only 80V, which reduces by 71.5% in contrast to Fig.8. Moreover, although each arm needs to supply the output current alternately, the peak value of arm currents is 320A, also reduced by 24% compared with Fig.8. Therefore, the voltage fluctuation and the arm current stress are both reduced by quasi two-level PWM method.

Fig.10 presents the comparison of motor current between high-frequency component injection method and quasi two-level PWM method, and it is undeniable that the total harmonic distortion (THD) of motor current is relatively larger when using quasi two-level PWM method because of sacrifice of the multilevel output waveform, which also results in relatively larger ripple of electromagnetic torque of IM. Notwithstanding this drawback, because there is no need to inject high-frequency common mode voltage, this isolation stress of motor bearings is much lighter. Consequently, this method is more practical to operate for longtime for medium- and high-voltage motors driven by MMCs.

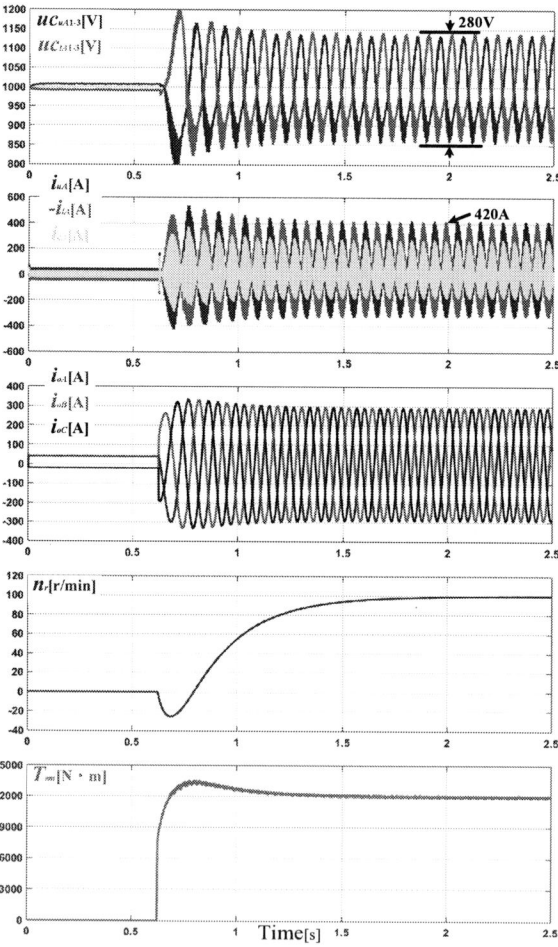

Fig.8. Simulated operating waveform using high-frequency component injection method.

The 2018 International Power Electronics Conference

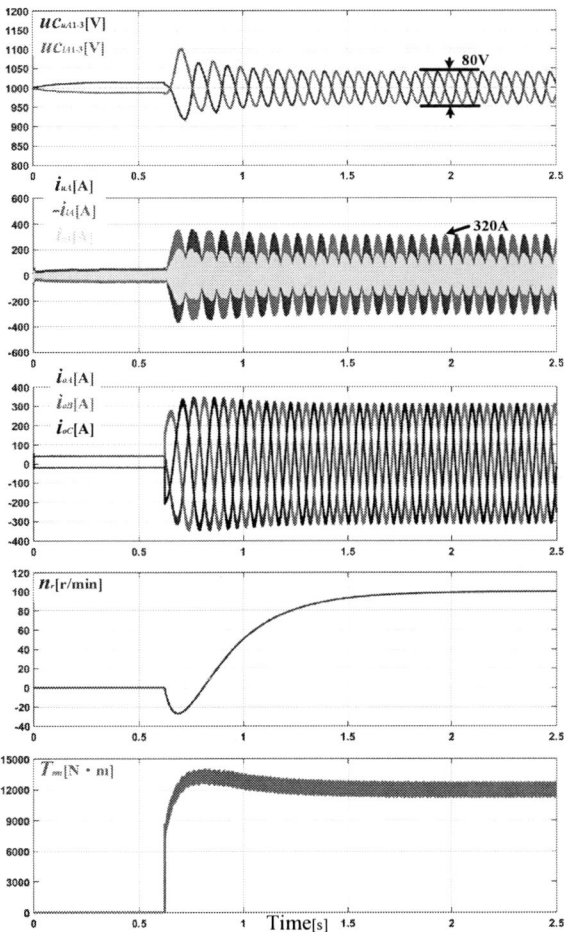

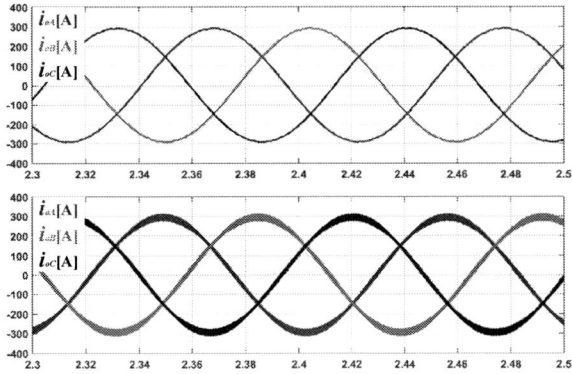

Fig.9. Simulated operating waveform using quasi two-level PWM mode proposed by this paper.

Fig.10. Comparison of motor current between high-frequency component injection method and quasi two-level PWM method proposed by this paper. (a) High-frequency component injection method. (b) Quasi two-level PWM method.

To verify the implementation of quasi two-level PWM mode, Fig.11 presents the detailed waveform of arm voltage and arm current. It can be aware that upper arm and lower arm supply the output current alternately. When all SMs of either of the two arms are connected to the DC bus, the current of that arm is close to zero, and vice versa. It also can be noticed that PWM control lower arm SMs is delayed to implement current commutation, which is

coincident with the current commutation situation shown Fig.5. What is more, from the steady waveform of u_{uA} and u_{lA}, high-frequency PWM used in [9,10] is not applied, leading to less power loss.

The validity verification of capacitor voltage balance control method is presented in Fig.12, the operating condition is the same as Fig.9, excepting that balance control algorithm is canceled at 1.5s and reemployed at 1.7s. It is evident that capacitor voltage begins to diverge as soon as the control algorithm is canceled, but it can be balanced after control algorithm is reemployed.

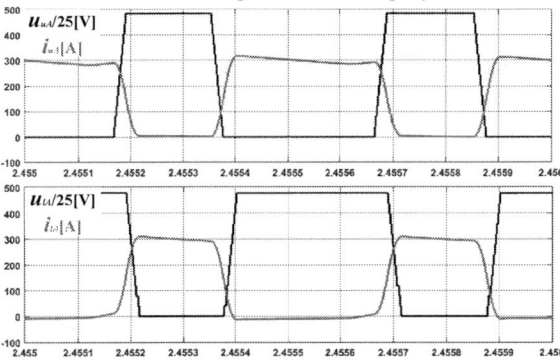

Fig.11. Detailed waveform of arm voltage and arm current of phase A using quasi two-level PWM method.

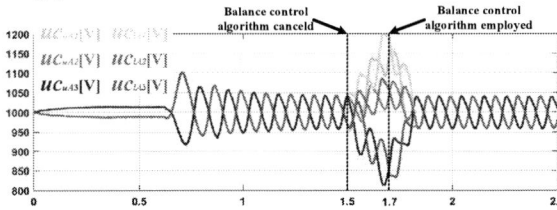

Fig.12. The validity verification of capacitor voltage balance control method.

V. Conclusion

This paper proposed a new control method to implement the quasi two-level PWM mode of MMC. The arm current control is mainly discussed in this paper, and calculation method of delay time is concluded. The control method is verified by a 2-MVA MMC-based variable-speed motor drive system. Compared with the high-frequency injection method, the voltage fluctuation and the arm current stress are both reduced, especially the voltage fluctuation, which is only 28.5% high-frequency injection method. What is more, there is no need to employ a high-frequency PWM, leading to simplification of control system and decrease of power loss. Notwithstanding it limitation of relatively larger current ripple, quasi two-level mode is quite suitable for MMC-based motor drive system operating in low speed and large load torque.

Future research will focus on the implementation of an experimental prototype. And to realize a smooth transition between low speed and high speed (multilevel PWM mode should be used to reduce the harmonics of output current), the switching method between quasi two-level PWM and multilevel PWM will also be studied.

2391

REFERENCES

[1] A. Lesnicar and R. Marquardt, "An innovative modular multilevel converter topology suitable for a wide power range," in *Proc. IEEE Power Tech. Conf.,* Bologna, Italy, Jun. 23–26, 2003, vol. 3, pp. 1–6.

[2] M. A. Perez, S. Bernet, J. Rodriguez, S. Kouro, and R. Lizana, "Circuit topologies, modeling, control schemes, and applications of modular multilevel converters," *IEEE Trans. Power Electron.,* vol. 30, no. 1, pp. 4–14, Jan. 2015.

[3] M. Hagiwara, K. Nishimura, and H. Akagi, "A medium-voltage motor drive with a modular multilevel PWM inverter," *IEEE Trans. Power Electron.,*vol. 25, no. 7, pp. 1786–1799, Jul. 2010.

[4] B. Li, S. Zhou, D. Xu, R. Yang, D. Xu, C. Buccella, C. Cecati, "An Improved Circulating Current Injection Method for Modular Multilevel Converters in Variable-Speed Drives," in *IEEE Transactions on Industrial Electronics,* vol. 63, no. 11, pp. 7215-7225, Nov. 2016.

[5] J. Jung, H. Lee, and S.-K. Sul, "Control strategy for improved dynamic performance of variable-speed drives with modular multilevel converter," *IEEE J. Emerging Sel. Topics Power Electron.,* vol. 3, no. 2, pp. 371–380, Jun. 2015.

[6] K.Wang, Y. Li, Z. Zheng, and L. Xu, "Voltage balancing and fluctuation suppression method of floating capacitors in a new modular multilevel converter," *IEEE Trans. Ind. Electron.,* vol. 60, no. 5, pp. 1943–1954, May 2013.

[7] M. Hagiwara, I. Hasegawa, and H. Akagi, "Start-up and low-speed operation of an electric motor driven by a modular multilevel cascade inverter," *IEEE Trans. Ind. Appl.,* vol. 49, no. 4, pp. 1556–1565, Jul. /Aug. 2013.

[8] A. Antonopoulos, L. A¨ngquist, L. Harnefors, and H.-P. Nee, "Optimal selection of the average capacitor voltage for variable-speed drives with modular multilevel converters," *IEEE Trans. Power Electron.,* vol. 30, no. 1, pp. 227–234, Jan. 2015.

[9] J. Kucka and A. Mertens, "Control for Quasi Two-Level PWM Operation of Modular Multilevel Converter," *2016 IEEE 25th International Symposium on Industrial Electronics (ISIE),* Santa Clara, CA, 2016, pp. 448-453.

[10] A. Mertens and J. Kucka, "Quasi Two-Level PWM Operation of an MMC Phase Leg With Reduced Module Capacitance," in *IEEE Transactions on Power Electronics,* vol. 31, no. 10, pp. 6765-6769, Oct. 2016.

A Current-source Type MMC with Delta-connected Arms for SMES

Yushi Miura[1*] and Toshifumi Ise[1]
1 Osaka University, Osaka, Japan
*E-mail: miura@eei.eng.osaka-u.ac.jp

Abstract— A current-source type modular multilevel converter (CMMC) is proposed for a power conditioning system (PCS) of a superconducting magnetic energy storage system (SMES). The CMMC has delta-connected arms, and each arm is composed of parallel-connected H-bridge cells. This circuit configuration can eliminate a high voltage large dc capacitor that is necessary for a conventional voltage-source inverter and reduce the capacity of the PCS. The delta connection of arms makes it possible to use circulating current to balance the superconducting coil current among arms. In addition, the parallel connection of cells enables the SMES to achieve the plug-in/out operations of cells, which is required in case of the quench of a superconducting coil. In this paper, the circuit configuration and its control scheme are proposed. Operations of charge and discharge of the SMES with the CMMC are demonstrated as well as plug-in/out operations of a cell through numerical simulation.

Keywords— *current source type, delta connection, MMC, SMES*

I. INTRODUCTION

The use of renewable energy sources such as solar power generation and wind power generation is growing year by year in the world. Since the output power of these renewable energy sources varies with the weather conditions, a large amount of installation of the renewable energy sources may bring about negative impact on the power system and therefore the compensation of the power fluctuation of renewable energy sources is one of the main issues.

With this background, the importance of large-capacity energy storage systems (ESSs) has increased. The storage battery such as lead-acid battery and lithium-ion battery is most commonly used for the ESS. The battery stores the energy as chemical energy and therefore input/output of electrical power involves chemical reaction. It causes the performance deterioration with the count of charge and discharge.

For high repetition application, a flywheel system is a suitable ESS. The flywheel system stores the energy as the kinetic energy of a rotational wheel/disk and the performance deterioration with the count of charge and discharge does not occur. However, in general, the standby power losses of the flywheel system is large and therefore the flywheel system is unsuitable for long-period power compensation.

A superconducting magnetic energy storage system (SMES) is one of promising ESSs to compensate power fluctuations due to the renewable energy sources. The SMES stores the energy as the magnetic energy in superconducting (SC) coils. The SMES has advantages over other ESSs: high repeating count of charge and discharge, a fast response, easy measurement of stored energy and so on. In the past decades, the development of high-temperature superconductors such as Bismuth Strontium Calcium Copper Oxide (BISCO) and Yttrium Barium Copper Oxide (YBCO) promoted the scale-up in capacity and high performance of SMES. However, it is still difficult to construct a large-scale SMES. High magnetic field and large current generate huge electromagnetic forces and the supporting structure of superconducting coils also becomes huge and complex. This leads to the increase in the weight of structures that must be cryogenically cooled. Additionally, the large-scale SMES requires the large-capacity power conditioning system (PCS) shown in Fig. 1(a), which is commonly used for the SMES and composed of a chopper and an inverter. As the power treated by the SMES becomes larger, the PCS also becomes larger. A conventional current-source type inverter (CSI) can be also employed for the SMES as shown in Fig. 1(b) [1]. However, the output current has generally pulse waveforms with the large step change that has the same

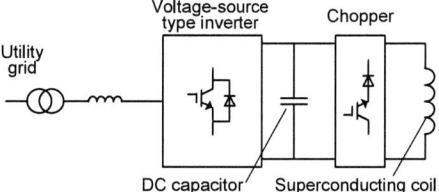

(a) PCS with a voltage-source type inverter and a chopper.

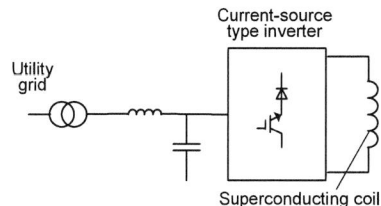

(b) PCS with a conventional current-source type inverter.
Fig. 1. A configuration of the PCS for SMES.

amplitude of the large SC coil current, and therefore an ac filter would be large.

To solve this problem, the concept of modularization of the SMES has been proposed [2], i.e. a large-capacity SMES is composed of multiple small-capacity SMESs. The modularity gives the SMES the feasibility of construction of a large-capacity ESS, and the flexibility of the operation such as the control of the number of the operating SC coils. However, each SC coil requires its own PCS shown Fig. 1(a) and the total capacity of the PCSs becomes large.

In the past decade, for high power application, many types of modular multilevel converters (MMCs) have been vigorously investigated [3], and almost of them are voltage-source type converters. Each of their arms has series-connected H-bridge cells/chopper cells and each cell has a capacitor that functions as a voltage source. Therefore they are suitable for high voltage applications. For the ESS, the MMC with batteries was also proposed [4].

On the other hand, current-source type converters are suitable for large current applications such as the SMES because the PCS has to treat large current. In this paper, we proposes a current-source type modular multilevel

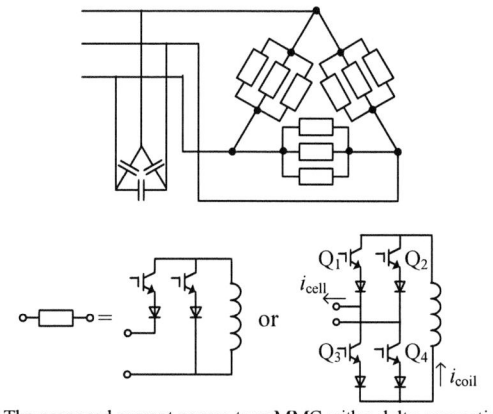

(a) The proposed current-source type MMC with a delta connection of arms, a chopper and a H-bridge cell.

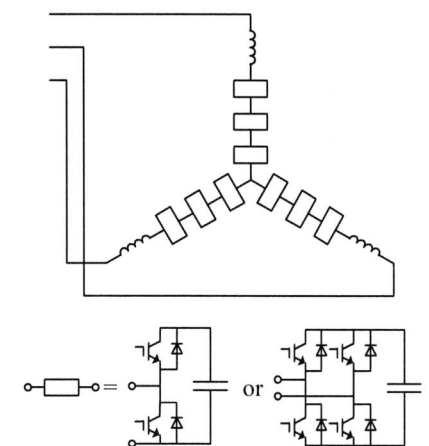

(b) The conventional voltage-source type MMC with a star connection of arms, a chopper and a H-bridge cell.
Fig. 2. Dual circuits of MMC and cells.

converter (CMMC) applied to the SMES, which is the dual circuit of the voltage-source type MMC. The proposed CMMC has three arms that are connected in delta and each arm is composed of H-bridge cells/chopper cells that are connected in parallel. In contrast to the voltage-source type MMCs (VMMCs), each cell of the CMMC has an inductor that functions as a current source, instead of a capacitor in the VMMC. Therefore, the CMMC can achieve multilevel current waveform unlike the conventional CSI in Fig. 1(b). Additionally, the CMMC does not require the inverter and large dc capacitor in the conventional PCS shown in Fig. 1(a) and therefore the capacity of the converter can be greatly reduced.

Moreover, the parallel connection of cells enables the CMMC to achieve plug-in/out operations of each cell to change the number of operating cells depending of stored energy. From the viewpoint of quench protection of a SC coil, the plug-in/out operations are also indispensable. The quench of the SC coil is the phenomenon where the coil suddenly loses superconductivity due to thermal and magnetic disturbance. In case of the quench, the SC coil must be disconnected from the main circuit and connected to the discharge resistor. The stored energy of the quenched SC coil must be consumed in the discharge resistor rapidly, otherwise, the SC coil may be damaged by overheating. The CMMC can disconnect the quenched SC coil without a shut-down of the whole SMES and continue the operation.

In the following sections, the CMMC with delta-connected arms and its control scheme are presented. The operations of the SMES based on the CMMC including plug-in/out operations are demonstrated through numerical simulation.

II. CURRENT-SOURCE TYPE MMC

The proposed CMMC is composed of three arms that are connected in delta configuration. The circuit configuration is shown in Fig. 2 together with a VMMC composed of star connected arms. The CMMC is the duel circuit of the VMMC, and therefore each cell has an inductor instead of a capacitor. Either chopper cell or H-bridge cell can be used for the CMMC, and in this paper, the H-bridge cell is employed. Since the chopper cell cannot output negative direction current, in order for the CMMC to output ac current to the utility grid, circulating current has to flow continuously through the delta-connected arms. On the other hand, the H-bridge cell is able to output bidirectional current. When switches Q_1 and Q_4 in Fig. 2(a) are turned on, the cell output positive current ($i_{cell} = i_{coil}$). Conversely, when Q_2 and Q_3 are turned on, negative current is output ($i_{cell} = -i_{coil}$). When (Q_1, Q_3) or (Q_2, Q_4) are in the on-state, the coil current circulates in the cell, and the output current becomes zero ($i_{cell} = 0$). Therefore, when the H-bridge cell is employed, the circulating current is unnecessary under a steady-state balanced condition. Additionally, the CMMC requires filter capacitors in shunt on the ac side instead of interconnection reactors of the VMMC.

2394

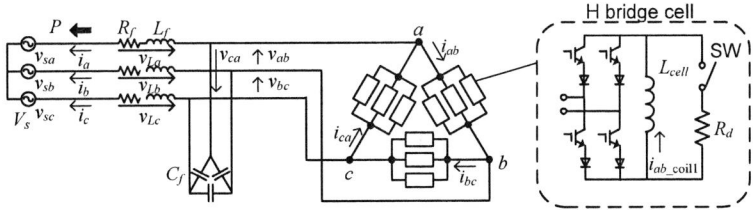

Fig. 3. A circuit configuration of the CMMC for simulation.

This configuration can reduce the capacity of the PCS compared with the conventional PCS shown in Fig. 1(a) because the inverter is unnecessary. Moreover, the large dc capacitor can be eliminated. The CMMC also achieves the multilevel waveforms of output current and it contributes to downsizing of the ac filter compared with the CSI in Fig. 2(b). In addition, the parallel connection of cells makes it possible to achieve plug-in/out operations of cells. Thus, the number of operating cells can be controlled depending on the stored energy of the SMES. In case of a quench of the SC coil, the quenched SC coil can be demagnetized separately while the other cells operate continuously.

III. CONTROL SCHEME

This section presents the control scheme of the CMMC for the SMES. Fig. 3 shows the circuit configuration of the CMMC investigated in this paper. Each arm of the CMMC has three H-bridge cells and each arm has a SC coil. An LC filter is installed on the ac side.

A. Power Control

The SMES inputs/outputs active power P and generate reactive power Q following their power references of P^* and Q^*, respectively. Since the CMMC is a current-source type converter, a desired current can be directly output. Fig. 4 shows the block diagram of the power control of the CMMC. The current control is conducted on a dq coordinate. From Proportional-Integral (PI) controllers of the power control, the current command i_d^* and i_q^* are obtained. In contrast to the capacitor voltage in the VMMC is kept at constant, the SC coil current in the CMMC varies depending on the stored energy. Therefore, the output current value of the CMMC is limited by the SC coil currents. Therefore, the limiter for the output current is installed. Since the deviation in current of the SC coils exists, the average value of all the SC coil current $i_{\text{coil\_ave}}$ is used for the limiter, which is calculated by the following equation,

$$i_{\text{coil\_ave}} = \frac{1}{3N} \sum_{j=1}^{3N} i_{\text{coil\_}j} \tag{1}$$

where $i_{\text{coil\_}j}$ is the SC coil current of the j-th cell, and N is the number of cells per arm, i.e. $N = 3$ in this paper. Using the inverse dq transformation, line current references i_a^*, i_b^* and i_c^* are obtained from i_d^* and i_q^*. The references of output current component for each arm, $i_{ab\_ac}^*$, $i_{bc\_ac}^*$ and $i_{ca\_ac}^*$ are calculated as

$$\begin{bmatrix} i_{ab\_ac}^* \\ i_{bc\_ac}^* \\ i_{ca\_ac}^* \end{bmatrix} = \frac{1}{3} \begin{bmatrix} -1 & 1 & 0 \\ 0 & -1 & 1 \\ 1 & 0 & -1 \end{bmatrix} \begin{bmatrix} i_a^* \\ i_b^* \\ i_c^* \end{bmatrix} \tag{2}$$

Since the CMMC has the LC filter on the ac side, the oscillation due to the LC resonance occurs. To suppress the oscillation, the voltage across the filter reactor, v_{Lfa}, v_{Lfb} and v_{Lfc}, are fed back with a gain h.

B. Coil Current Balancing Control with Circulating Current

Under the balanced condition of the current of the SC coils, circulating current is unnecessary. However, when unbalance occurs, especially in case of plug-in/out operations of a cell, in order to balance the SC coil current among arms, the circulating current is used.

The circulating current i_{cir} is defined as the current that flows through the delta circuit and does not flow out to the utility grid. It is expressed as

$$i_{\text{cir}} = \sqrt{2} I_{\text{cir}} \sin(\omega t + \theta_l + \theta_{\text{cir}}) \tag{3}$$

where I_{cir} is the RMS value of the circulating current; ω is a fundamental angular frequency; θ_l is the phase difference between line-to-line voltage v_{ab} and utility voltage v_{sa}; θ_{cir} is the phase difference between v_{ab} and i_{cir}. Giving the line-to-line voltage as

$$v_{ab} = \sqrt{2} V_{\text{line}} \sin(\omega t + \theta_l) \tag{4}$$

where V_{line} is the RMS value of v_{ab}, average power that are input into each arm P_{ab}, P_{bc}, P_{ca} are calculated as

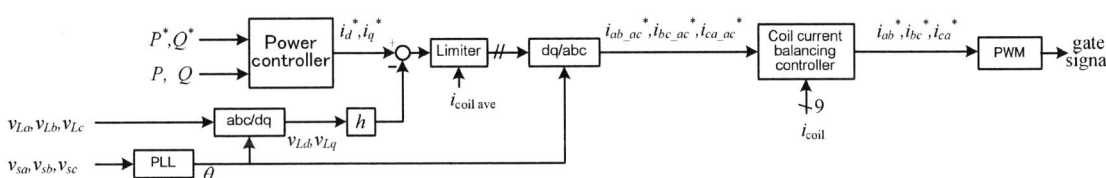

Fig. 4. Control block diagram of the CMMC.

$$\begin{bmatrix} P_{ab} \\ P_{bc} \\ P_{ca} \end{bmatrix} = V_{\text{line}} I_{\text{cir}} \begin{bmatrix} \cos\theta_{\text{cir}} \\ \cos\left(\theta_{\text{cir}} - 2\pi/3\right) \\ \cos\left(\theta_{\text{cir}} + 2\pi/3\right) \end{bmatrix} \tag{5}$$

Therefore, we can control the power that is input into each arm by changing the RMS value I_{cir} and phase angle θ_{cir}. It should be noted the sum of power in (5) is equal to zero and it indicates that balancing of the arm currents with the circulating current requires no active power from the utility grid.

The circulating current should flow to balance the SC coil current among the arms. The reference of the circulating current $i_{\text{cir}}{}^{*}$ is given as

$$i_{\text{cir}}^{*} = \sqrt{2} I_{\text{cir}}^{*} \sin(\omega t + \theta_l + \theta_{\text{cir}}^{*}) \tag{6}$$

The average of the SC coil current in each arm $i_{xy\_\text{coilave}\_}$ $(xy = ab, bc, ca)$ is calculated as

$$i_{xy\_\text{coilave}} = \frac{1}{N_{op}} \sum_{j=1}^{N_{op}} i_{xy\_\text{coil}\_j} \tag{7}$$

where $i_{xy\_\text{coil}\_j}$ is the SC coil current of the jth cell and N_{op} is the number of the cells that operate in arm xy. If one cell is disconnected due to a quench of the SC coil, N_{op} is equal to N-1. The deviation ε_{xy} of the operating cell coil current from the average value of all the SC coil $i_{\text{coil\_ave}}$ is calculated as

$$\varepsilon_{xy} = i_{xy\_\text{coilave}} - i_{\text{coil\_ave}} \tag{8}$$

The reference value $I_{\text{cir}}{}^{*}$ in (6) is calculated using the maximum deviation $\varepsilon_{\max} = \max\,(\varepsilon_{ab}, \varepsilon_{bc}, \varepsilon_{ca})$ and a gain K as

$$I_{\text{cir}}^{*} = K \varepsilon_{\max} \tag{9}$$

The phase angle $\theta_{\text{cir}}{}^{*}$ is determined by the arm that has the maximum deviation $\varepsilon_{\max}$,

$$\theta_{\text{cir}}^{*} = \begin{cases} 0 & \text{if } \varepsilon_{ab} = \varepsilon_{\max}, \\ -\dfrac{2\pi}{3} & \text{if } \varepsilon_{bc} = \varepsilon_{\max}, \\ \dfrac{2\pi}{3} & \text{if } \varepsilon_{ca} = \varepsilon_{\max}. \end{cases} \tag{10}$$

Finally, the current references $i_{xy}{}^{*}$ for arm xy is given as the sum of the ac current component reference $i_{xy\_ac}{}^{*}$ in (2) and circulating current reference $i_{\text{cir}}{}^{*}$,

$$i_{xy}^{*} = i_{xy\_ac}^{*} + i_{\text{cir}}^{*} \tag{11}$$

Applying this control, the current balance among the arms can be achieved quickly even after the plug-in/out

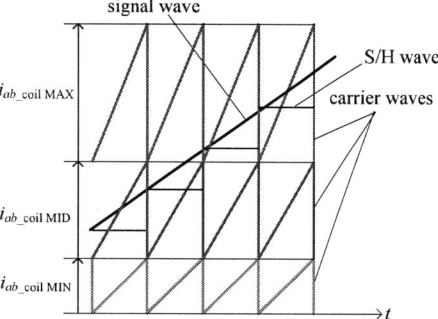

Fig. 5. Level shifted PWM technique: in the case of charging SC coils in arm ab; $i_{ab\_\text{coilMAX}} > i_{ab\_\text{coilMID}} > i_{ab\_\text{coilMIN}}$; $N = 3$.

of cells.

C. Level Shifted PWM technique

A level shifted PWM technique is employed to make pulse pattern for the CMMC. When the number of the cells per arm is N, N sawtooth carrier waves are provided. Fig. 5 shows a signal wave and three carrier waves for arm ab. Each carrier wave has the same amplitude of each SC coil current. A switching signal for each cell is obtained by comparison of the sampled and held (S/H) signal wave and each carrier wave. In order to balance current of the SC coils in the same arm, the carrier waves are allocated in order of the magnitude of SC coil current. When the line-to-line voltage is applied to charge the SC coils, the carrier waves are arranged from the bottom in ascending order by the coil current magnitude as shown in Fig.5. Thus, the line-to-line voltage is applied to the SC coil that has the smallest coil current for the longest period among the SC coils in the same arm. Conversely, when the line-to-line voltage is applied in the direction to discharge the SC coils, the carrier waves are arranged in descending order. Thus, in every sampling period, sorting of the SC coil current is required.

Using this technique, the balance of the current of the SC coils in the same arm is achieved.

IV. SIMULATION

To demonstrate operations of the SMES based on the CMMC, the numerical simulation was conducted using software PSCAD. The circuit configuration for simulation is shown in Fig. 3 and the major parameters are summarized in TABLE I. The sampling frequency was set to 3 kHz. The inductance of the SC coil L_{cell} was 0.2 H. Equivalent resistance R_{eq} of both the IGBT and diode was set to 0.01 Ω and therefore when the SC coil current circulates in the cell through four switching devices, the current decreases. In this simulation, the direction of power flow to discharge the SMES was set to be positive.

A. Start Up Operation

The simulation of start-up of the CMMC was conducted first. The CMMC requires no dc power supplies for initial charging, and the SC coils are charged by the power from the ac side.

TABLE I
Major Parameters of the CMMC

Parameters	Value
Inductance of superconducting coil, L_{cell}	0.2 H
The total number of cells	9
The number of cells per arm, N	3
Cell type	H-bridge
Equivalent resistance of IGBT, R_{eq}	0.01 Ω
Equivalent resistance of diode, R_{eq}	0.01 Ω
Filter capacitor, C_f	40 μF(Δ)
Filter inductor, L_f	2 mH
Resistance of the filter inductor, R_f	0.01 Ω
Discharge resistor, R_d	20 Ω
Sampling frequency	3 kHz
Utility voltage	1500 V
Utility frequency	60 Hz

2396

The 2018 International Power Electronics Conference

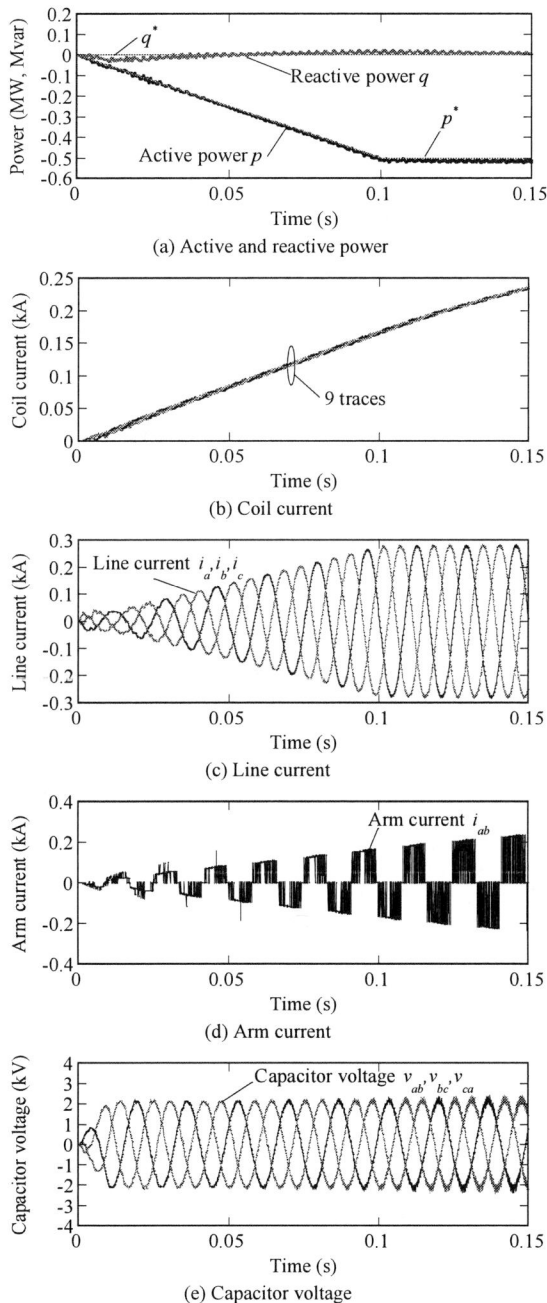

(a) Active and reactive power

(b) Coil current

(c) Line current

(d) Arm current

(e) Capacitor voltage

Fig. 6. Simulation of start-up operation.

Fig. 6 shows the simulation result of the start-up of the CMMC. The active power reference p^* increased at a rate of -5 MW/s from 0 to 0.1 s and then kept at -0.5 MW, meanwhile the reactive power q^* was set to zero during the period as shown in Fig. 6(a). The current of the SC coils increased almost linearly as shown in Fig. 6(b), and following it, the amplitudes of arm current and ac current increased as shown in Figs. 6(c) and (d), respectively. Meanwhile, Fig. 6(e) shows that the capacitor voltage was hardly influenced by the increase in current. These results indicate that the CMMC can be easily started up without other power supplies and complex control.

B. Power Control

Fig. 7 shows the simulation result of power control of the CMMC. The references of active power p^* and reactive power q^* were given with the change rates of the references of +/-0.5 MW/ms and +/-0.5 Mvar/ms, respectively, as shown in Fig. 7(a). Fig. 7(a) shows that both the instantaneous active and reactive power p and q well followed the references and were controlled independently.

The waveform of the arm current i_{ab} and its time-expanded waveform are shown with its reference i_{ab}^* in Fig. 7(b) and Fig. 8, respectively. From these results, it is confirmed the number of levels in the current waveform changed as the output power changed. The number of levels depends on both the output power of the CMMC and the current of the SC coils.

The waveforms of line current i_a, i_b, i_c are shown in Fig. 7(c). The amplitude of line current also changed as the output power changed. Fig. 7(d) shows the capacitor voltage v_{ab}, v_{bc}, v_{ca}. The transient oscillation of the line current and capacitor voltage due to resonance of the LC filter was effectively suppressed thanks to the resonance suppression control. The maximum voltage reached around 4 kV during the period when the CMMC output the power of 3 MW. The maximum output power may be limited from the viewpoint of the protection of switching devices.

Figs. 7(e) and 7(f) show the current waveforms of 9 SC coils and the whole stored energy of the CMMC, respectively. The whole stored energy E_{SMES} is calculated as the sum of stored energy of all SC coils.

$$E_{SMES} = \frac{1}{2} L_{cell} \sum_{xy=ab,bc,ca} \sum_{j=1}^{N} i_{xy\_coil\_j}^2 \quad (12)$$

At the beginning, each SC coil was charged with a current of around 0.5 kA. When active power p was negative, the SC coils were charged, as a result, the coil current and the stored energy increased. Conversely, when p was positive, the SC coils were discharged, and then the coil current decreased. However, even during the period when p was zero, the coil currents decreased. This is because the stored energy consumed in the equivalent resistance R_{eq} of the switching devices. On the other hand, the change of q did not affect the coil currents. Despite the changes of p and q, the balance of 9 SC coil currents was achieved successfully.

The result of simulation verifies that the CMMC can achieve the control of active/reactive power and the balance of SC coil current sufficiently using the proposed control scheme.

C. Plug-in/out Operations

The simulation of the plug-in/out operations of a cell was also conducted. The case where a quench of one of the SC coils in arm ab occurred during the period the CMMC output active power of 1 MW was assumed as shown in Fig. 9(a). After detection of the quench, the quenched coil was disconnected at $t = 0.65$ (s) by gate-block of the switches of H-bridge cell and connected to a discharge resistor R_d by closing the switch SW in Fig. 3. The resistance of R_d was set to 20 Ω, which was

2397

considerably large to shorten the computational time of the simulation. Therefore, after connection to the resistor,

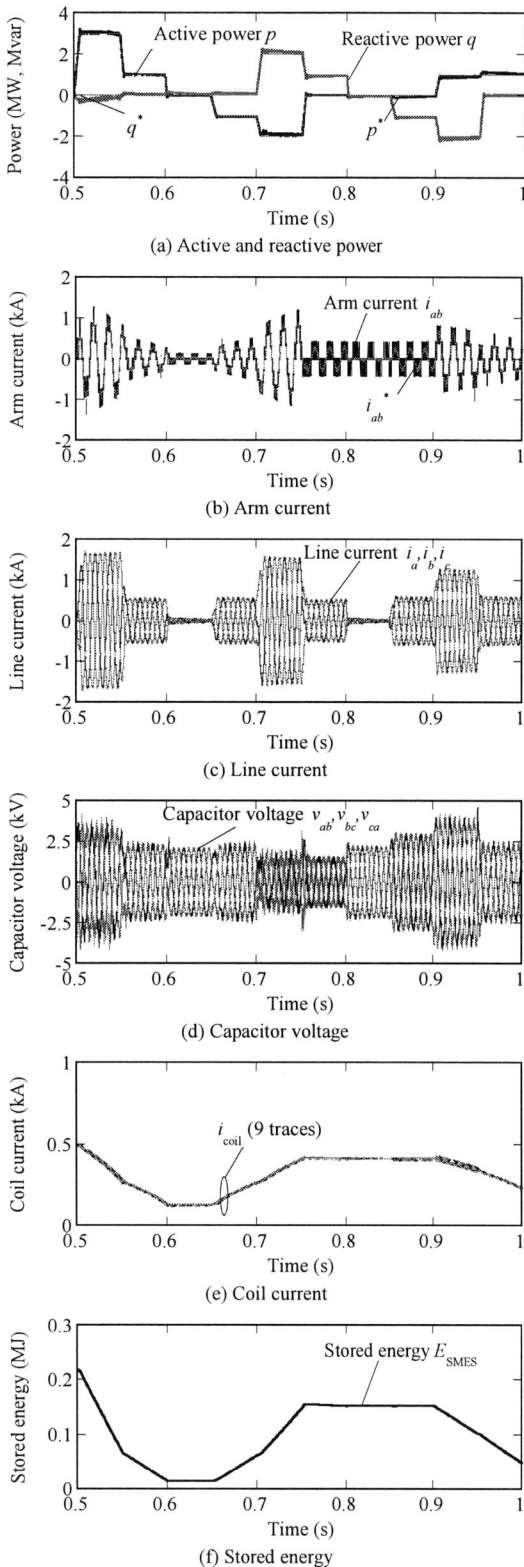

(a) Active and reactive power

(b) Arm current

(c) Line current

(d) Capacitor voltage

(e) Coil current

(f) Stored energy

Fig. 7. Simulation of power control.

Fig. 8. Arm current: time-expanded waveform of Fig. 6(b).

the coil current decreased rapidly to zero as shown in Fig. 9(b). At $t = 0.70$ (s), the CMMC changed output power from 1 to 0 MW. At $t = 0.85$ (s), the SC coil was reconnected to the main circuit. From $t = 0.90$ (s), in order to charge the reconnected SC coil quickly, the CMMC was charged with a power of -2 MW.

Figs. 9(c) and 9(d) show waveforms of the line current and capacitor voltage, respectively. The waveform of arm current of arm ab that had the disconnected SC coil is shown in Fig. 9(e). Figs. 9(a)-(e) indicate the following features of the CMMC: the first is that the CMMC was able to continue to output the power even after disconnecting the quenched SC coil. The amplitudes of the line current and capacitor voltage nearly unchanged and the active power was kept at 1 MW from 0.65 to 0.70 s. The second is that the CMMC was able to charge the reconnected SC coil even under the condition that input power from the utility grid was almost zero. The current of reconnected SC coil increased after reconnection from 0.85 to 0.9 s, meanwhile the SC coil current of the other cells decreased. The third is that after starting the charge of the SMES from 0.9 s, the reconnected SC coil was charged rapidly, and the balance of current of all the SC coils was achieved in a short time thanks to the coil current balancing control with circulating current.

The plug-in/out operations of a cell that were indispensable to the SMES were successfully demonstrated by this simulation.

V. DISCUSSION

The capacitor voltage in the LC filter increases substantially when the CMMC outputs large current. It depends on the impedance of the LC filter as well as the impedance of distribution line that connects the CMMC to the utility grid. Therefore the LC filter should be designed carefully considering the rise of voltage as well as the reduction of harmonics. Additionally, the transformer for interconnection of the CMMC to the medium voltage/ high voltage utility grid should be designed adequately considering its impedance. Otherwise, the output power should be limited for the viewpoint of suppression of the rise of voltage.

The number of levels of the current waveform changes depending on the output current of the CMMC and the current of the SC coils. It would cause the variation of the amount of harmonics contained in the output current and affect the LC filter design. The further analysis on the harmonics is required for design of the LC filter.

The 2018 International Power Electronics Conference

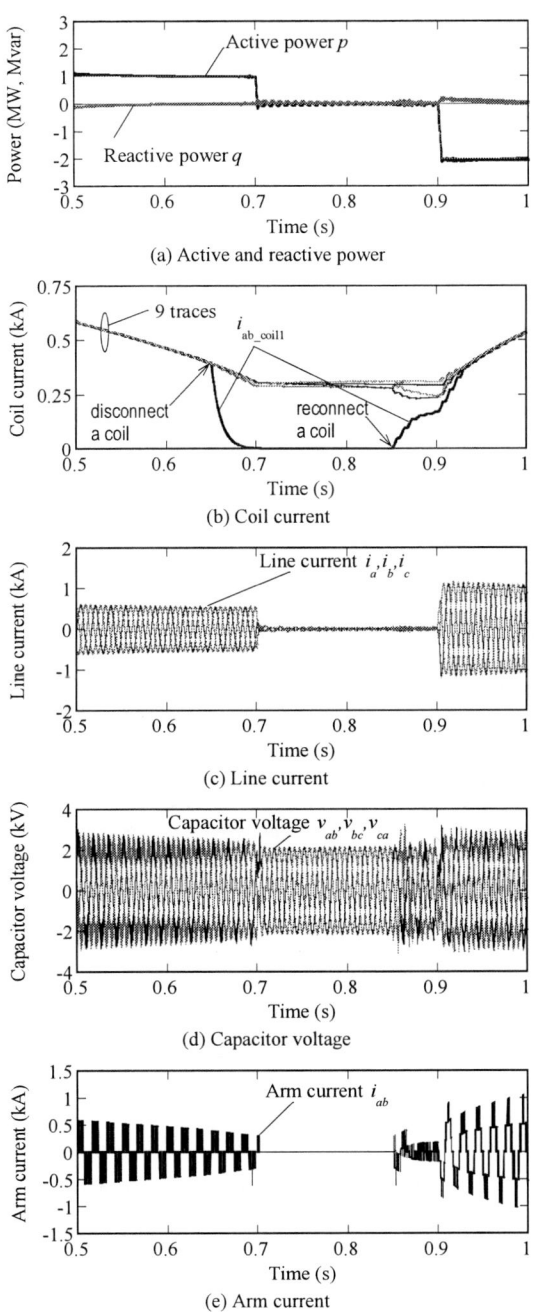

(a) Active and reactive power

(b) Coil current

(c) Line current

(d) Capacitor voltage

(e) Arm current

Fig. 9. Simulation of plug-in/out operations.

demonstrated through numerical simulation. The plug-in/out operations of a cell with a sufficient coil current balance were also demonstrated. These results indicate that the proposed CMMC is one of the promising PCSs for the SMES.

References

[1] Y. Miura et al, "Development of an IGBT converter for a magnet power supply," Proc. of the 20th Symp. on Fusion Tech, pp. 743-746, 1998.

[2] S. Nomura et al. : "Feasibility Study on Large Scale SMES for Daily Load Leveling Using Force-Balanced Helical Coils", IEEE Trans. on Appl. Conductivity, Vol.23, No.3, pp.1-4, 2013.

[3] H. Akagi, "Classification, Terminology, and Application of the Modular Multilevel Cascade Converter (MMCC)," IEEE Trans. on Power Electronics, vol. 26, no. 11, pp. 3119-3130, 2011.

[4] J. Inacio et al. "Enhancement of Performance, Availability, and Flexibility of a Battery Energy Storage System Based on a Modular Multilevel Cascaded Converter (MMCC-SSBC)," IEEE Trans. on Power Electronics, vol. 31, no. 4, pp.2791-2799, 2016.

VI.   Conclusions

The current-source type modular multilevel converter (CMMC) with delta-connected arms was proposed as the PCS for the SMES. This circuit configuration can reduce the capacity of the PCS compared to the conventional PCS composed of an inverter and a chopper, and make it possible to achieve the plug-in/out operations of a cell, which is indispensable to the SMES. The control scheme of the CMMC including the circulating current control to balance the current of SC coils was also proposed. The charge and discharge operations of the SMES were

New Module with Isolated Half Bridge or Isolated Full Bridge for Modular Medium voltage converter

Yunpeng Si, Yifu Liu, Qin Lei

Electrical, Computer and Energy Engineering, Arizona State University, Tempe, USA

yunpengs@asu.edu, yliu457@asu.edu, Qin.Lei@asu.edu

Abstract-This paper proposes a standard module of "Isolated Half Bridge (IHB)" or "Isolated Full Bridge (IFB)" and its application in MVDC/MVAC system by using modular structure. The fundamental and 2nd harmonic current ripple can be eliminated through connecting the DC links of three phase on the same level. Energy density and system efficiency can be largely improved since the capacitor size is reduced. In addition, energy storage devices can be integrated in the proposed topology and thus make it feasible for smart grid and marine applications. Detailed simulation results of comparison between proposed topology and traditional MMC have been presented. Moreover, medium voltage drive starting from zero frequency is also simulated. Modular DC/DC converters with energy storage are discussed as well.

Keywords—Isolated Half Bridge (IHB), MVDC/MVAC system, energy storage, medium voltage drive, modular DC/DC converters

I. INTRODUCTION

Compared to non-modular structured topologies, modular structured topologies such as Modular Multilevel Converter (MMC), Cascaded H-Bridge (CHB) and Solid State Transformer (SST) are the best candidates for MVDC/MVAC system due to following reasons. Firstly, the modular topologies when used with wide-band-gap devices have higher bandwidth due to their greater number of voltage levels, lower dv/dt and higher possible carrier frequencies; Secondly, the modular topologies have higher reliability; Thirdly, the modular topologies can integrate low voltage energy storage devices such as battery and supercapacitor directly by putting them on the dc link of each module. In addition, the modular topologies are able to achieve higher efficiency and require a much smaller filter.

However, the traditional modular topologies have the following drawbacks: (1) If distributed energy storage devices are used, the fundamental and 2nd harmonic current in the dc link will cause significant loss in the submodule battery/supercapacitor due to its large ESR. One solution is to use another capacitor with boost DC/DC converter to absorb the ripple [1]. One of the drawbacks of this method is the large size film capacitor which needs to take the full rating low frequency current. Besides, the current control is hard to achieve and the switching loss of this additional boost converter is comparable to the half bridge. (2) The size of the additional film capacitor which absorbs the full rating fundamental and 2nd harmonic current is forced to be big. One existing solution for improving the power density of MMC is GE's MEMC [2]. This topology is a 3-level flying capacitor structure with the MMC module in the position of the flying capacitor. This method can reduce the capacitor size by 50%. However, the major drawback is that the thyristor used in this topology needs to be forced commutation, which has a high failure rate. This in turn increases the control complexity; (3) When the modular topologies are used for variable frequency such as drives, it is very hard to reverse or even start from zero frequency under full torque. High amplitude high frequency circulating current injection has been adopted by ABB [3]. But the disadvantage of this method is high switching and conduction loss at low frequency operation. Another alternative, variable DC link method, has been proposed by Siemens SINAMICS SM120 in order to keep energy ripple constant at different frequencies [4]. However, it introduces large number of devices and capacitors since several DC-DC converters with diode rectifiers need to be implemented.

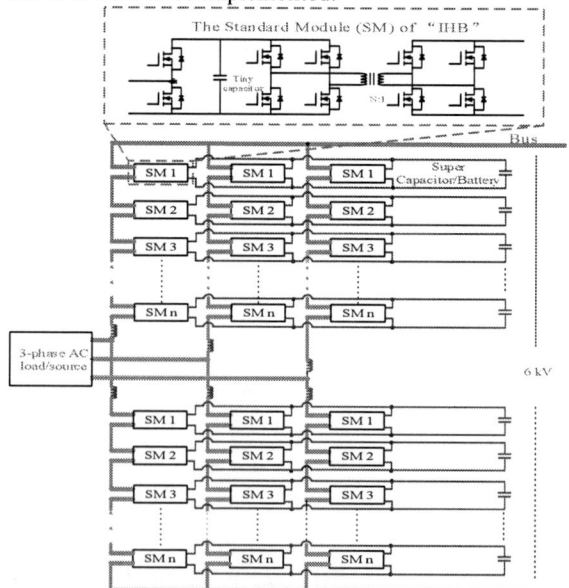

Fig. 1. Proposed topology of a "Modular Isolated Multilevel Converter (MIMC)" with supercapacitor/battery using the proposed standard module of "Isolated Half Bridge (IHB)"

The proposed standard module of "Isolated Half Bridge (IHB)" and "Isolated Full Bridge (IFB)", when used in DC-AC inverters, can eliminate the fundamental and 2nd harmonic current in the module capacitor completely by combining the three-phase arm currents on the same level together.

II. PROPOSED APPROACH AND INTELLECTUAL MERIT

The circuit schematic of the proposed standard module used alongside MMC structure together with the MMC that integrate the supercapacitor/battery inside are shown in Fig. 1. In this topology, the secondary sides of the modules at different levels can also be connected in series or parallel. The combined three phase energy ripple is

Table I. Quantitative metrics of proposed MIMC compared with traditional MMC

Metric	MMC	MIMC	MIMC + SC
Power (W)	150KW	150KW	150KW
Voltage (V):	6-kV dc and 3.3 kV ac	6-kV dc and 3.3 kV ac	6-kV dc and 3.3 kV ac
Module Delink Voltage	1KV	1KV	1KV
Efficiency (%)	98	97	96.5
Power density (W/inch³)	115	634	87
Switching f.	20kHz	20kHz	20kHz
Dev. type & count	C2M0025170D*96	C2M0025170D*192	C2M0025170D*192
Dev. Spec.	1.7kV, 72A	1.7kV, 72A	1.7kV, 72A
Film Cap/module	107uF	2.7uF*2	2.7uF*2
Film Capacitor type and count	1300V/14.5A/12μF, B32776G1126, Quantity: 9*48	1300V/5A/2.7μF, B32776T1275, Quantity: 2*48	1300V/5A/2.7μF, B32776T1275, Quantity: 2*48
Supercapacitance type and count			VHC 2R3 127 QG, 300F/ 2.3V/ 3A Quantity: 50*48
Total film Cap.Vol.	1302 inch3	86.4 inch 3	86.4 inch 3
Total Supercap size	0	0	1488 inch3
Isolated DCDC conv power density	0	1KW/ inch3	1KW/ inch3
Total Vol. of DC Conv.	0	150 inch3	150 inch3
Total Vol. of the module	27 inch3	4.8 inch3	36 inch3
Total Volume	1302 inch3	236 inch3	1724 inch3

zero, thus the capacitor voltage is constant. The proposed Modular Isolated Multilevel Converter (MIMC) has advantages over the existing solutions in the following aspects. (1) Supercapacitor/battery can be inserted since the low frequency current ripple in the module's capacitor has been eliminated; (2) Lower voltage supercapacitor/battery can be selected as the isolated dc-dc converter can buck the secondary voltage; (3) As long as the three phase modules on the same level are always connected, different numbers of modules can be connected in parallel on the secondary side, to create a flexible current rating for the supercapacitor/battery pack; (4) All the secondary sides can be connected in parallel to interface with the supercapacitor/battery pack. This supercapacitor/battery pack can be grounded. Therefore, the common mode voltage due to floating the supercapacitor/battery is avoided, and the common mode current caused by high dv/dt is eliminated; (5) The additional capacitor which has to be used with the supercapacitor/battery to absorb the fundamental and 2nd

harmonic current can be replaced by a tiny capacitor which only needs to take the switching ripple current. The size of the arm inductor is also reduced due to the elimination of the second order harmonic current in the phase leg. For the isolated dc-dc converter which is used in the proposed topologies, its size can be reduced by making the converter soft-switching at hundreds of kilo-Hertz; (6) In MIMC, the phase energy is automatically balanced by controlling the current of the dc-dc converter. The dc link current in each half bridge still needs to have the same fundamental and 2nd harmonic ripple as usual, since the energy ripple in each phase still has the same form; (7) In the dc short circuit fault, the isolated dc-dc converter can prevent the discharge of the supercapacitor/battery through the diodes by stop switching the DC-DC converters.

Table I shows the quantitative metrics that compares the proposed MIMC with the traditional MMC, A theoretical 150kW system with 1.7V/72A SiC MOSFET has been chosen. From this table, the following conclusions can be made based on the data: (1) The proposed MIMC has significant power density improvement compared to traditional MMC. The MIMC with supercapacitors has lower power density because of the large size of the integrated supercapacitors; (2) The efficiency of MIMC will be a slightly lower than MMC due to the additional losses of the dc-dc converter; (3) The MIMC has a larger number of devices than MMC, which could result in a slightly lower reliability and higher cost. However, this could be mitigated by simply increasing the number of redundancy; (4) The required arm inductor in the MIMC is smaller than that in the MMC.

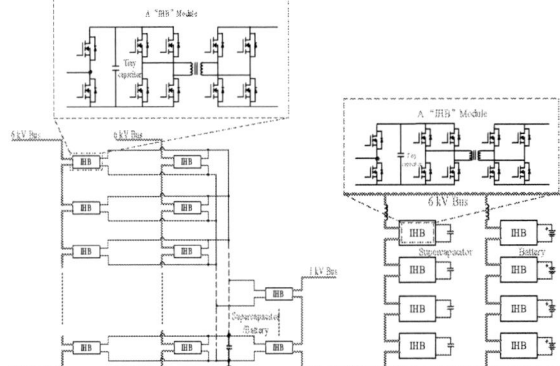

Fig. 2. (a). Proposed topology of "Single-phase Isolated Multilevel Converter (SIMC)" with the proposed standard module of "Isolated Half Bridge (IHB)" (b) Proposed topology of mixed branches of supercapacitor SIMC and battery SIMC

The proposed standard module of IHB and IFB, when used as DC/DC converters, can transfer the energy between buses through the supercapacitor/battery link and can also behaves as a dc-breaker during fault conditions. The proposed dc-dc converter topologies are shown in Fig. 2 (a) and (b). To provide both high power density and high energy density, the branch with the supercapacitors and the batteries can be used together, as shown in Fig. 2 (b). In details, the proposed topologies have the following advantages over the existing solutions.

2401

(1) The same set of supercapacitor/battery can provide energy to the pulse load on both dc buses. In the traditional topologies, separated DC/DC converters need to be used to connect the supercapacitor/battery to dc busses, as well as to connect the two buses. This limits the bandwidth and the efficiency at the pulse load; (2) It is capable of using a single high current supercapacitor/battery pack for a full leg, or for several, and ground it; (3) Similar to MIMC, these DC/DC converters possess the freedom to connect the DC/DC transformer secondary in series or in parallel; (4) The standard module could also use high frequency transformers with different turn ratios, so the lower voltage supercapacitor/battery can be used; (5) The modularity allows a significant reduction in the number of different spare parts. The same module could be used for several of the proposed topologies

III. PRELIMINARY RESULTS

A. Simulation results of MIMC at steady state

For comparison purpose, the simulation results of traditional MMC, MIMC with small capacitor, and MIMC with supercapacitor at the rated point are presented here. The schematic and the parameters are shown in Fig. 3. The dc current rating is 25A and the output current is 40A. The difference between MMC, MIMC with small capacitor and MIMC with supercapacitor is: the module capacitor used in MMC is 600uF, in MIMC with small capacitor is 10uF, and in MIMC with supercapacitor is 10F per phase

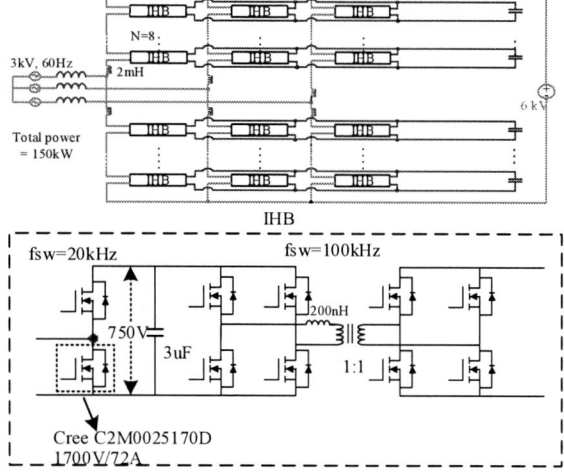

Fig. 3. Schematic of the MIMC in the simulation

The following conclusions can be reached:
1. The ac terminal voltage and current of the half bridge are very similar for three cases. The only difference is that MIMC has lower voltage harmonics compared to MMC since the ω and 2ω voltage ripple on the capacitor were eliminated. For the same harmonics results, MIMC requires a significantly smaller arm inductor.

2. The biggest difference is the capacitor voltage. It is flat in MIMC but with ω and 2ω ripple for MMC, although the capacitance in MIMC with small capacitor was set 100x smaller. It needs to be noted that figure b has a little bigger capacitor voltage ripple than figure c only because the supercapacitor in figure c has much higher capacitance than the film capacitor in figure b.

3. The phase energy and arm energy ripples are the same for MIMC with small capacitor, MIMC with supercapacitor and MMC, since the energy ripple is only related to the terminal ac voltage and current. However, MIMC with small capacitor has very small dc energy due to its small capacitance even becoming negative for sections of the waveforms. Negative means that ripple is bigger than the average.

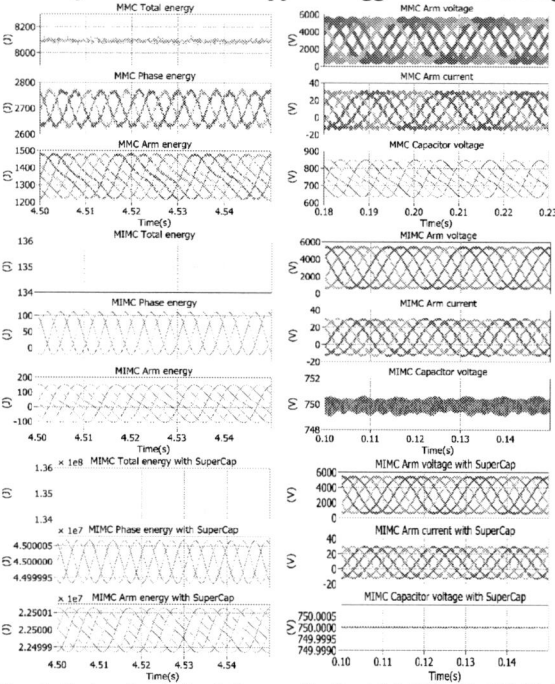

Fig. 4. System level simulation results for (a) MMC (b) MIMC (c) MIMC with supercapacitor. (The waveforms from the left column show the total energy, phase energy, arm energy and the right column show arm voltage, arm current, capacitor voltage)

B. Simulation results of the detailed behaviors of the isolated dc-dc converter in "MIMC with supercapacitor"

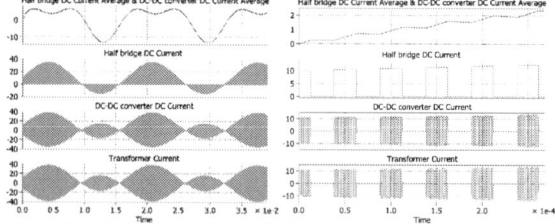

Fig. 5. Simulation results for the isolated dc-dc converter in MIMC with supercapacitor (a) Zoomed out (b) Zoomed in (The waveforms from the top to bottom are: half bridge DC current Average and DC-DC converter DC current average, half bridge dc link current,

The following conclusions can be made according to the simulation waveforms of the dc-dc converter:

1. The half bridge dc link current is a PWM AC current with a floating average of a combination of the fundamental and 2nd harmonic components and a frequency of the half bridge switching.
2. The dc-dc converter primary side dc link current is a PWM AC current with the same floating average as the half bridge dc, but with the frequency of the dc-dc converter switching.
3. The dc link capacitor only absorbs the high frequency current.
4. The transformer current is an AC current with an envelope of the dc link current.

C. Operation of MIMC at variable speed drive

One important advantage of the proposed topologies is being able to start from zero frequency with a small capacitance. This could be of advantage if the converter is connected to a generator and is asked to operate as a turbine starter. The following conditions have been simulated: the converter AC side output voltage is ramping with frequency with a constant V/f ratio of 3kV/60Hz, while the current is kept at a constant value; the frequency range is from 0 to 60Hz. The schematic of the simulation is exactly the same as that in Fig. 3 except that the ac voltage is ramping up with a constant V/f ratio from zero to 60Hz. The simulation results are shown in Fig 6.

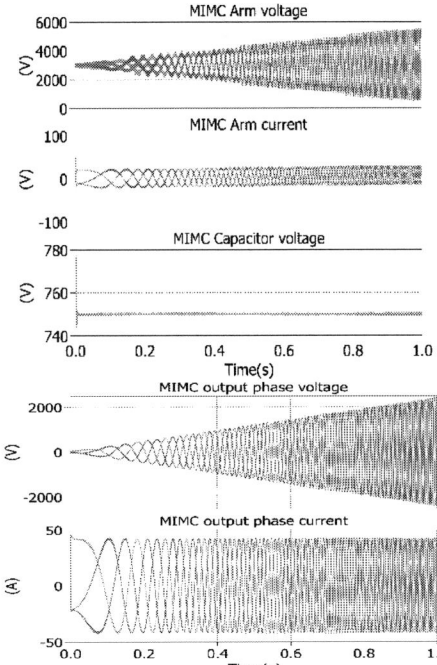

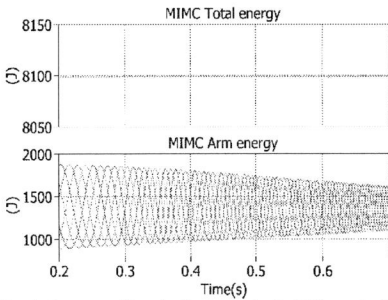

Fig. 6. Simulation results at the constant V/f controlled variable frequency motor drive for "MIMC with supercapacitor": (a) from top to bottom: arm voltage, arm current, half bridge capacitor voltage; (b) from top to bottom: phase voltage, phase current; (c) from top to bottom: total energy, arm energy.

The following conclusions can be made according to this variable frequency simulation waveforms:
1. The same as the conventional MMC, as the frequency increases, the arm energy ripple amplitude decreases.
2. Different as with the MMC, the voltage of the capacitor in the half bridge does not contain the low frequency components of the output frequency. It remains constant during the full transient.

IV. CONCLUSIONS AND FUTURE WORKS

This paper proposes a standard module of "Isolated Half Bridge (IHB)" or "Isolated Full Bridge (IFB)". An MIMC topology is introduced involving these standard modules. In addition, two modular DC/DC converters are also proposed. According to the simulation results, the fundamental and 2nd harmonic current ripple can be eliminated through connecting the DC link capacitors of three phase on the same level together. Therefore, the capacitor size can be significantly reduced which lead to huge improvement of energy density and system efficiency. Moreover, energy storage devices can be integrated in the proposed topology and thus make it feasible for smart grid and marine applications. A medium voltage drive starting from zero frequency is also simulated. Detailed experiment results will be presented in full paper.

REFERENCES

[1] Ciccarelli, F., G. Clemente, and D. Iannuzzi. "Energy storage management control based on supercapacitors using a modular multilevel inverter topology for electrical vehicles." Clean Electrical Power (ICCEP), 2013 International Conference on. IEEE, 2013.

[2] Zhang, D., Datta, R., Rockhill, A., Lei, Q., & Garces, L., "The modular embedded multilevel converter: A voltage source converter with IGBTs and thyristors." Energy Conversion Congress and Exposition (ECCE), 2016 IEEE. IEEE, 2016.

[3] A. Antonopoulos, L. Ängquist, S. Norrga, K. Ilves, L. Harnefors and H. P. Nee, "Modular Multilevel Converter AC Motor Drives With Constant Torque From Zero to Nominal Speed," in IEEE Transactions on Industry Applications, vol. 50, no. 3, pp. 1982-1993, May-June 2014.

[4] http://www.industry.siemens.com/drives/global/en/converter/mvdr ives/pages/sinamics-sm120-cm.aspx

The 2018 International Power Electronics Conference

Development of a 700-V-class Reverse-Blocking IGBT for Advanced T-type Neutral Point-Clamped Power Conversion System

Hiroki Wakimoto[1*], Haruo Nakazawa[1], David H. Lu[2], Takashi Matsumoto[3] and Yoichi Nabetani[3]

[1] Fuji Electric Co., Ltd. 4-18-1, Tsukama, Matsumoto, Nagano, Japan
[2] Fuji Electric Co., Ltd. 1, Fuji, Hino, Tokyo, Japan
[3] University of Yamanashi, 4-3-11, Takeda, Kofu, Yamanashi 400-8511, Japan
*wakimoto-hiroki@fujielectric.com

Abstract— **In this study, we developed a 700-V reverse-blocking insulated gate bipolar transistor (RB-IGBT), which is suitable for a large-capacity advanced T-type neutral-point-clamped (AT-NPC) three-level power conversion system. As this device can tolerate a larger turn-off surge voltage than the existing 600-V device, the turn-off speed can be higher and the turn-off loss can be reduced by 30%. Improving the p+ collector activation method reduces the reverse leakage current and increases the device operation temperature by 25 °C. When the 700-V devices are applied to the bidirectional switches of the rectifier stage of the AT-NPC system, the total loss of the bidirectional switch can be reduced by 10% compared with the application to 600-V devices. We have developed a power semiconductor module product mounting the 700-V RB-IGBTs for the bidirectional switches. The modules are installed in the large-capacity uninterrupted power supply machine "7300WX-T3U," achieving a high conversion efficiency of 97.5%.**

Keywords— *bidirectional switch, leakage current, reverse-blocking capability, three-level power converter*

I. INTRODUCTION

The development of efficient power conversion systems is an important issue in the field of power electronics. Multi-level power converters are one of the most effective approaches to improving the power conversion efficiency of power converters. Neutral point-clamped (NPC) three-level power converters, which have diodes that clamp the output to a neutral point, have been proposed [1]. It is widely used to reduce the power loss and realize a smaller filter size. Figure 1(a) illustrates a conventional NPC three-level power converter topology. However, this system requires too many power semiconductor devices.

To address this problem, a T-type NPC three-level power converter system, with bidirectional switches connected between the output and neutral point, has been proposed. For the bidirectional switch part, both forward and reverse-blocking (RB) capabilities are required [2, 3].

The bidirectional switches are generally configured by serially connecting a semiconductor switching device such as an insulated gate bipolar transistor (IGBT) or MOSFET and a diode such as a PiN diode or Schottky barrier diode, as shown Fig. 1(b). Because a conventional

IGBT or MOSFET has no RB capability, a serially connected diode is required. At a current conducting mode, current flows through both the switching device and diode and the conduction loss is the sum of those two devices. Thus, the conduction loss becomes large.

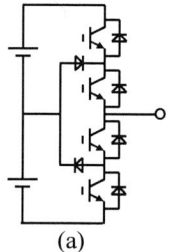

(a)

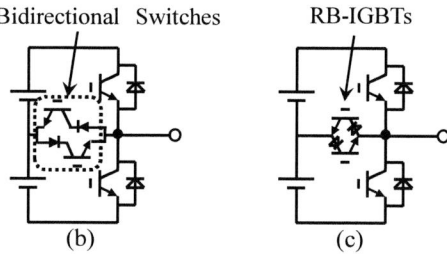

Bidirectional Switches RB-IGBTs

(b)   (c)

Fig. 1. NPC three-level power converter topology. (a) Conventional NPC, (b) T-type NPC with IGBTs and diodes, and (c) Advanced T-type NPC with RB-IGBTs.

For the bidirectional switches in the advanced T-type NPC (AT-NPC) configuration shown in Fig. 1(c), RB-IGBTs have been developed in the voltage range from 600 to 1700 V [4–8]. By applying the RB-IGBTs, no additional diodes for the RB are required and the conduction loss can thus be significantly reduced [2, 3].

The AT-NPC three-level topology with the RB-IGBTs are applied to power conversion equipment such as an uninterrupted power supply (UPS) and power conditioning system to improve the conversion efficiency. The AT-NPC three-level power conversion system with RB-IGBTs is illustrated in Fig. 2.

2404

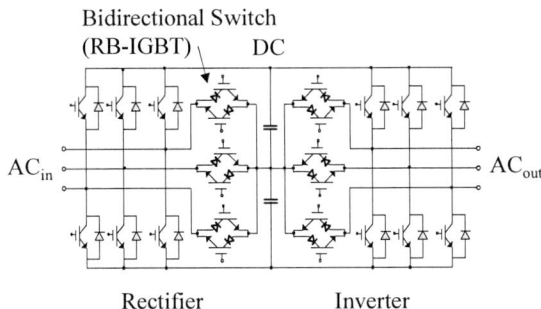

Fig. 2. AT-NPC three-level power conversion circuit applying RB-IGBTs.

The existing 600-V RB-IGBT was optimized for small- or medium-capacity equipment. Using large-capacity power conversion equipment with a large parasitic inductance, an excessive surge voltage is applied when the RB-IGBT is turned off and easily exceeds the rated voltage of the RB-IGBT. Therefore, in the case of applying the 600-V RB-IGBTs to large-capacity power conversion equipment, these switching devices should be driven slowly, compared to the application to small- or medium-capacity equipment. Consequently, it causes larger turn-off switching losses.

The turn-off waveforms of the 600- and 700-V rated-voltage RB-IGBT are illustrated in Fig. 3. Increasing the rated voltage by 100 V improves the tolerance for the surge voltage in the turn-off period, enables it to be turned off quicker by using a smaller gate resistance, and it is possible to achieve a lower turn-off loss for a 700-V RB-IGBT.

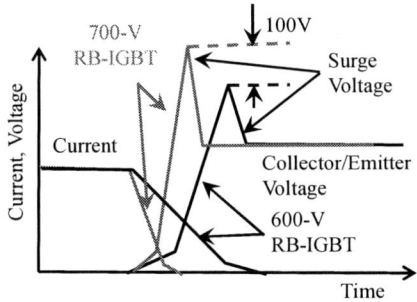

Figure 3. Illustration of turn-off waveforms of 700- and 600-V rated RB-IGBTs.

II. DEVICE STRUTURE AND FABRICATION PROCESS

A schematic of the cross-sectional structure of the RB-IGBT is illustrated in Fig. 4. The difference between an RB-IGBT and a normal IGBT is the deep p^{++} isolation region, which connects the front-side edge termination structure to the back-side p$^+$ collector layer at the die edge. The edge termination region relaxes the electric field in the voltage blocking mode and ensures a desired breakdown voltage (BV). The p^{++} isolation region is formed by extreme temperatures and longtime dopant diffusion.

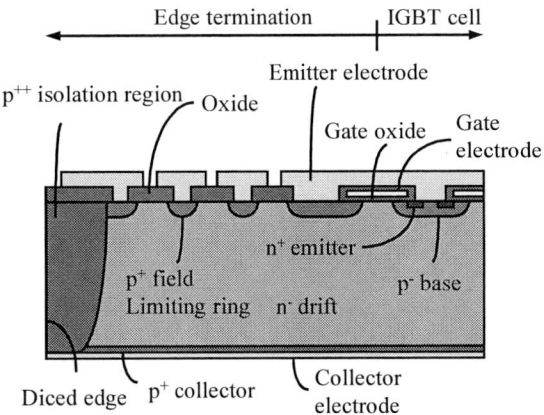

Fig. 4. Schematic of cross-sectional structure of RB-IGBT.

In the forward-blocking mode for normal IGBT operation, a positive voltage is applied to the collector for the emitter. The gate voltage with respect to the emitter (V_{GE}) should be 0 V or less to keep the MOS channel off. In this mode, a depletion layer expands from the surface side with the MOS structure, as indicated by the yellow line in Fig. 5.

The blue line in Fig. 5 represents the depletion boundary in the RB mode when a negative voltage is applied to the collector. The depletion region extends from both the p$^+$ collector / n$^-$ drift region and the p^{++} isolation region / n$^-$ drift region, thus accomplishing the RB capability.

In a simplified model, the BV can be described by Eq. (1).

$$BV = \frac{qN_D}{2K_S\varepsilon_0}W^2$$

(1)

where N_D is the donor density, K_S is the semiconductor relative permittivity, ε_0 is the permittivity in vacuum, and W is the depletion region thickness. It indicates that the thicker the n$^-$ drift region where the depletion region extends, the greater the increase in the BV. However, increasing the n$^-$ drift region thickness inevitably causes an increase in the drop in the ON-state forward conduction voltage ($V_{CE(sat)}$).

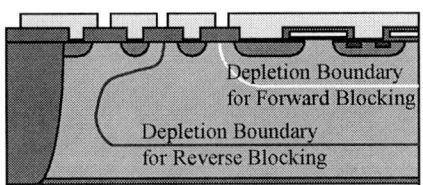

Fig. 5. Schematic of depletion boundary in forward-blocking and RB modes.

For a normal IGBT without the p^{++} isolation region, electrical carriers are generated at the diced edge in the RB mode, causing a large leakage current. As a result, the RB capability cannot be maintained.

The 700-V RB-IGBT cell structure is designed to be identical to the existing 600-V RB-IGBT. The n^- drift resistivity, wafer thickness, and edge termination structure are redesigned.

Next, the device manufacturing process is described. In the initial step of the device fabrication process, the p^{++} isolation region is formed. In order to acquire higher BVs, not only must the n^- drift region thickness increase, but the p^{++} isolation region must also expand while suppressing the crystal defect generation because of the additional thermal budget. The IGBT cells and edge termination structure on the front side are manufactured using the same fabrication process as the normal IGBT. Then, the thickness of the device wafer is reduced to a value according to the required BV. After that, boron ions are implanted and activated by thermal annealing for the formation of the p^+ collector. Finally, the collector electrode is formed and divided into each die.

III. ELECTRICAL PROPERTY AND ANALYSIS

A. Voltage-Blocking Capability

Figure 6 shows a comparison of the bidirectional blocking characteristics with the 600- and 700-V RB-IGBT. The horizontal axis (V_{CES}) represents the collector voltage with respect to the emitter under the condition where the gate is shorted to the emitter. The vertical axis represents the current conducting through the collector. The junction temperature (T_j) is 25 °C. The BVs for the 700-V RB-IGBT are over 800 V in both directions and are over 100-V higher than those of the 600-V device because of the thicker n^- drift region.

Figure 7 shows the RB leakage current of the 700-V RB-IGBT at $V_{GE} = 0$ or $+15$ V. V_{EC} and I_{EC} indicates the emitter voltage with respect to the collector and the current flowing at that time. As a feature of the RB-IGBT, the reverse leakage current greatly differs when the MOS channel is open or closed. When the MOS channel is open, for example, under $V_{GE} = +15$ V, the leakage current of the p–i–n diode composed of the p^+ collector / n^- drift / n^+ emitter is measured. In contrast, when the MOS channel is closed, which is $V_{GE} = 0$ V or less, a larger reverse leakage current flows because of the amplification effect of the parasitic transistor. By changing V_{GE} to 0 or $+15$ V, the reverse leakage current at $V_{EC} = 800$ V is reduced by a factor of ten, as can be seen in Fig. 7. As the reverse leakage current affects the tolerance of the device operating temperature, it is desirable to be as small as possible. Therefore, applying a gate voltage of $V_{GE} = +15$ V is recommended in the RB mode of the RB-IGBT.

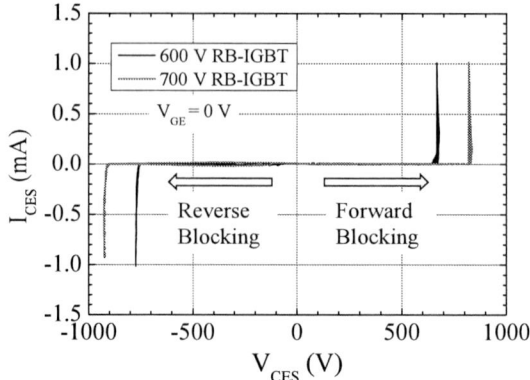

Fig. 6. BV comparison of 600- and 700-V RB-IGBTs at $T_j = 25$ °C.

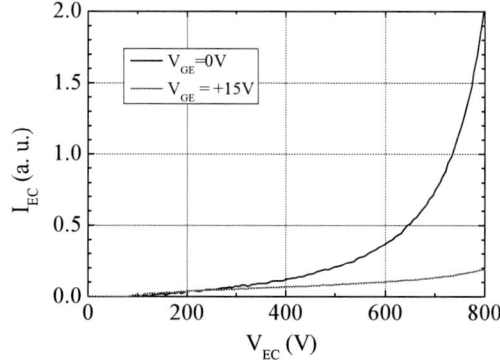

Fig. 7. Reverse leakage current at $V_{GE} = 0$ or $+15$ V and $T_j = 25$ °C.

The reverse leakage current at $V_{GE} = +15$ V and the high T_j are indicated in Fig. 8. At $T_j = 125$ °C, the reverse leakage current represented with a red solid line for the 700-V RB-IGBT is reduced to one tenth or less, lower than that of the black solid line of the 600-V RB-IGBT. In the RB mode, crystal defects near the p^+ collector / n^- drift region greatly influence the reverse leakage current. In the 700-V RB-IGBT, the method of activating the p^+ collector layer improves and the reverse leakage current is greatly reduced. Less leakage current (red dotted line) compared to the 600-V device enables 150 °C operation for the 700-V RB-IGBT.

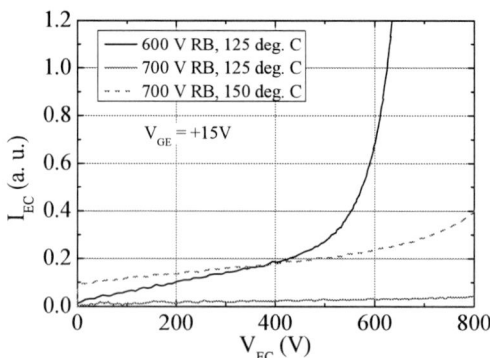

Fig. 8. Reverse leakage current of 600 and 700-V RB-IGBT at $V_{GE} = +15$ V and high T_j.

The deep-level transient spectroscopy (DLTS) signals when the reverse bias is applied to the junction of the p+ collector / n⁻ drift are presented in Fig. 9. With regard to the black line for the 600-V RB-IGBT, three peaks toward the negative direction of the y-axis appear, indicating that there exist three kinds of crystal defects with deep levels in the semiconductor bandgap. Such deep levels cause an increase in leakage current. The peak height is proportional to the deep level density. On the red line of the 700-V RB-IGBT, each peak height is lower. The improvement in the p+ collector activation method reduces the crystal defects near the p–n junction and improves the quality of the junction [9].

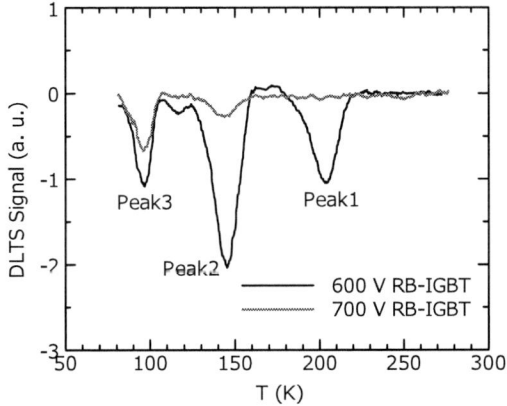

Fig. 9. DLTS signals when reverse bias is applied to the junction of the p+ collector / n⁻ drift.

B. Forward-Conducting I–V Characteristics

The forward-conducting I–V curves measured at V_{GE} = +15 V (i.e., open MOS channel) and T_j = 150 °C are presented in Fig. 10 and, whatever the allowable operation, T_j is 125 °C for the 600-V RB-IGBT in the power modules. The I–V curve of the normal 600-V IGBT and diode pair (our 6th generation products) serially connected to realize a bidirectional-blocking capability is plotted as a blue line in Fig. 10. The 700-V RB-IGBT (red line) exhibits a 0.1-V greater drop in forward-conducting voltage at the rated current of I_{CE} = 100 A than the 600-V device (black line). This is because the n⁻ drift region is expanded to improve the voltage-blocking capability. Although $V_{CE(sat)}$ is larger than the 600-V RB-IGBT, it is 0.5 V lower than that of the normal IGBT and diode pair at I_{CE} = 100 A, maintaining the great advantage in terms of the current conduction loss.

C. Turn-off Characteristics

Figure 11 presents the turn-off waveforms for the test modules of the rated current of 300 A at T_j = 150 °C, V_{DC} = 400 V, and L_S = 120 nH, where V_{DC} is the DC voltage and L_S is the stray inductance of the measurement circuit. Although the reverse leakage current of the 600-V RB-IGBT is large, the normal turn-off operation involving the forward-blocking mode is possible even at T_j = 150 °C.

The gate resistances are adjusted so the turn-off surge voltage due to the relatively large L_S remains within the rated voltage of 600 or 700 V, respectively. Because the 700-V RB-IGBT has a larger tolerance for the turn-off surge voltage, a smaller gate resistance is selected, and the turn-off speed can be higher. Therefore, the turn-off loss of the 700-V RB-IGBT can be reduced lower than that of the 600-V device.

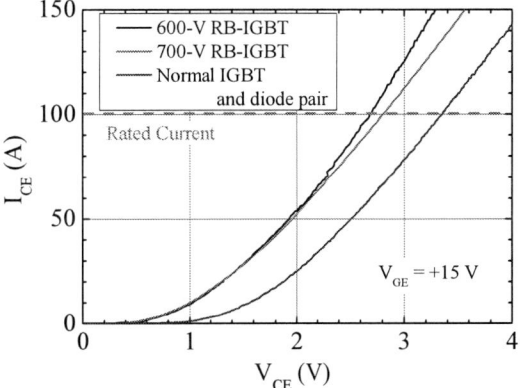

Fig. 10. Forward-conducting I–V characteristics comparison at T_j = 150 °C.

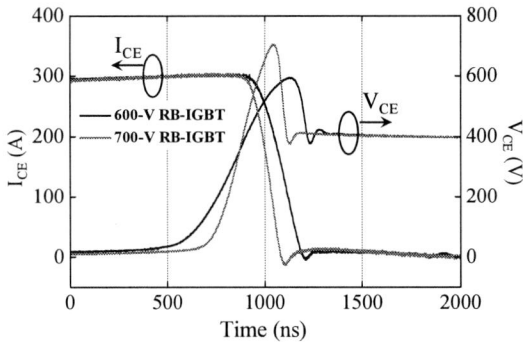

Fig. 11. Turn-off waveform comparison of 300-A rated test module at V_{DC} = 400 V, L_S = 120 nH, and T_j = 150 °C.

Figure 12 presents a plot of the correlation between $V_{CE(sat)}$ and the turn-off loss (E_{off}) Tj = 150 °C. For the 700-V RB-IGBT, three kinds of devices with different electrical properties by the change in the carrier lifetime are measured. The carrier lifetime is controlled by the electron beam irradiation method, which is generally used in power semiconductor devices. Comparing with the same $V_{CE(sat)}$ of the 600-V RB-IGBT, the E_{off} of the 700-V device is approximately 30% lower because of the higher turn-off speed, achieving an excellent trade-off correlation.

Generally, when the switching speed increases, the time rate of the voltage change (dV/dt) and current change (di/dt) become large. With such a fast time rate of change, it would be possible to increase the electromagnetic interference (EMI) noise. As the 700-V RB-IGBT is designed for increasing switching speeds, it is necessary to examine the EMI noise level in applications where it will be used.

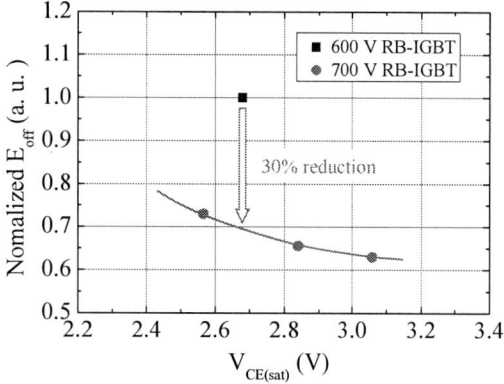

Fig. 12. Correlation between forward-conducting voltage drop $V_{CE(sat)}$ and turn-off loss E_{off} at $T_j = 150\ °C$

D. Semiconductor Loss for Power Conversion System

The estimated power loss comparison for the bidirectional switch parts in the rectifier stage of the AT-NPC three-level power conversion system shown in Fig. 2 at a carrier frequency of 15 kHz are presented in Fig. 13.

At this carrier frequency, the turn-off loss P_{off} has a larger contribution to the total loss than the conduction loss P_{sat}. The reverse recovery loss P_{rr} is negligibly small. Although the $V_{CE(sat)}$ of the 700-V device is slightly larger than the 600-V device, the total loss of the 700-V RB-IGBT can be reduced by 10% as P_{off} can be reduced. Compared to the configuration of the normal IGBT and diode pair, the power loss is reduced by 16%. As the carrier frequency increases, the superiority of the 700-V device further increases.

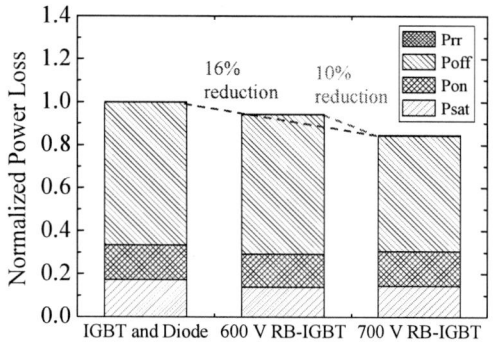

Fig. 13. RB-IGBT power loss comparison for the NPC switches in the rectifier stage at a carrier frequency of 15 kHz.

In the inverter stage, the power loss of the RB-IGBT is greatly affected by the conduction loss. The power loss reduction is barely obtained when the 700-V RB-IGBT with the slightly larger $V_{CE(sat)}$ than the 600-V device is applied to the inverter stage. However, it is possible to

adjust the $V_{CE(sat)}$–E_{off} characteristics suitable for the inverter stage by the carrier lifetime control. When the carrier lifetime is lengthened, the characteristic shifts to the low $V_{CE(sat)}$ side on the red line in Fig. 12, and E_{off} increases. For applications where the conduction loss dominates the total loss, even if E_{off} increases somewhat, applying devices with a low $V_{CE(sat)}$ may potentially reduce the total loss. If applying the 700-V RB-IGBT to the inverter stage, there is an advantage that the operating junction temperature can be raised from 125 to 150 °C because of the lower leakage current in the RB mode.

Now, we are developing a newly designed 700-V RB-IGBT with a lower conduction loss. It is possible to reduce $V_{CE(sat)}$ by improving the carrier distribution in the ON-state by changing the gate structure of the IGBT cells. Figure 14 shows a comparison of the forward conduction I–V characteristics of the existing 700-V RB-IGBT model and those of the improved device calculated by device simulation software. A $V_{CE(sat)}$ reduction of 0.4 V would be expected to contribute to a reduction in power loss even in the inverter stage.

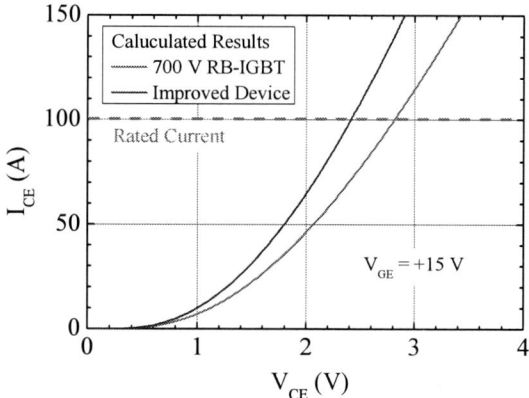

Fig. 14. Calculated I–V characteristics of existing 700-V RB-IGBT and improved device at $T_j = 150\ °C$.

IV. POWER SEMICONDUCTOR MODULE AND APPLICATION TO POWER ELECTRONICS PRODUCTS

We have constructed the power semiconductor module "2MBI480WH-070-53," mounting a plurality of the 700-V RB-IGBT dies for the bidirectional switches of the AT-NPC three-level power conversion system in one package. The maximum rating collector-emitter voltage and collector current are 700 V and 480 A, respectively. An equivalent circuit diagram is illustrated in Fig. 15.

Figure 16 shows an exterior photo of our large-capacity UPS product "7300WX-T3U" with the three-phase four-wire system. The above power semiconductor module mounting 700-V RB-IGBT dies are applied to the rectifier stage of this product. Furthermore, SiC Schottky barrier diodes are adopted, achieving a high conversion efficiency of 97.5%.

2408

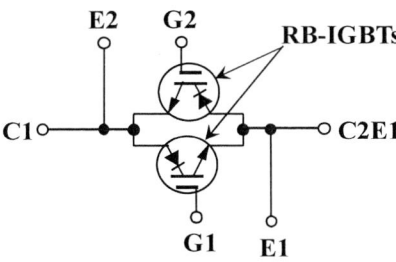

Fig. 15. Equivalent circuit diagram of power semiconductor module "2MBI480WH-070-53."

Fig. 16. Exterior photo of large-capacity UPS "7300WX-T3U."

V. CONCLUSIONS

In this study, we developed a 700-V RB-IGBT suitable for a large-capacity AT-NPC three-level power conversion system. The RB-IGBT is capable of reverse-bias blocking by forming a p^{++} isolation region at the diced edge.

By optimizing the device design parameters and device structure, such as the n^- drift resistivity, wafer thickness, and edge termination structure, the BV was raised by 100 V compared to the existing 600-V device and reaches 800 V or more. Improving the activation method, the p^+ collector in the 700-V device leads to a reduction in the reverse leakage current, and the device operating temperature can be increased from 125 to 150 °C.

The 700-V RB-IGBT has a lower $V_{CE(sat)}$ of 0.5 V at the rated current than the IGBT and diode configuration, which is advantageous. As it is designed with a thicker n^- drift region, $V_{CE(sat)}$ inevitably increases by 0.1 V more than the 600-V device.

Because of the greater improvement in the RB capability than the 600-V device, the tolerance for the turn-off surge voltage increases, enabling higher turn-off speeds to be achieved and E_{off} to be reduced by 30%.

When the 700-V devices are applied to the bidirectional switches of the rectifier stage of the AT-NPC power conversion system, the total loss of the bidirectional switch part can be reduced by 16% compared to the IGBT and diode configuration and by 10% compared to the 600-V devices at a carrier frequency of 15 kHz. When applied to

the inverter stage, as the conduction loss is larger than that of the 600-V device, there is little benefit to loss reduction. We have been developing the RB-IGBT with a reduced $V_{CE(sat)}$ by improving the gate structure.

We have constructed the power semiconductor module "2MBI480WH-070-53," mounting a plurality of the 700-V RB-IGBT dies for the bidirectional switches of the AT-NPC three-level power conversion system. The modules are installed in our large-capacity UPS product "7300WX-T3U," achieving a high conversion efficiency of 97.5%.

ACKNOWLEDGMENT

The authors would like to thank Mr. T. Muramatsu for his contribution to the fabrication and testing of the RB-IGBTs.

REFERENCES

[1] A. Nabae, I. Takahashi, and H. Akagi, "A new neutral-point-clamped PWM inverter," *IEEE Trans. Ind. Appl.*, vol. IA-17, no. 5, pp. 518–523, 1981.

[2] M. Yatsu, K. Fujii, S. Takizawa, Y. Yamakata, Y. Okuma, K. Komatsu, and H. Nakazawa, "A study of high efficiency UPS using advanced three-level topology," *in Proc. PCIM Europe 2010*, pp. 550–555.

[3] K. Komatsu, S. Okita, H. Nakazawa, S. Igarashi, and T. Fujihira, "Advanced neutral point-clamped (A-NPC) IGBT module for industrial application," *in Proc. PCIM China 2010*, pp. 170–174.

[4] H. Wakimoto, M. Ogino, D. H. Lu, S. Takizawa, H. Nakazawa, M. Yatsu, and Y. Takahashi, "600 V reverse blocking IGBTs with low on-state voltage," *in Proc. PCIM Europe 2011*, pp. 317–322.

[5] N. Tokuda, M. Kaneda, and T. Minato, "An ultra-small isolation area for 600V class reverse blocking IGBT with deep trench isolation process (TI-RB-IGBT)," *in Proc. 16th ISPSD 2004*, pp. 129–132.

[6] M. Takei, T. Naito, and K. Ueno, "The reverse blocking IGBT for matrix converter with ultra-thin wafer technology," *Proc. IEE – Circuits, Devices and Systems*, vol. 151, no. 3, pp. 156–159, 2004.

[7] T. Naito, M. Takei, M. Nemoto, T. Hayashi, and K. Ueno, "1200V reverse blocking IGBT with low loss for matrix converter," *in Proc. 16th ISPSI 2004*, pp. 125–128.

[8] D. H. Lu, M. Ogino, T. Shirakawa, H. Nakazawa, and Y. Takahasi, "1700V reverse-blocking IGBTs with V-groove isolation layer for multi-level power converters*in Proc. PCIM Europe 2012*, pp. 815–821.

[9] H. Wakimoto, H. Nakazawa, T. Matsumoto, and Y. Nabetani, "DLTS analysis of p/n junction implanted with boron to n type Si substrate," *in Proc. ICDS 2017*, p. 34.

Ceramic Embedding as Packaging Solution for Future Power Electronic Applications

Hoang Linh Bach[1*], Tobias Maximilian Endres[1], Daniel Dirksen[1], Sigrid Zischler[1], Christoph Friedrich Bayer[1],
Andreas Schletz[1], Martin März[1, 2]

1 Devices and Reliability, Fraunhofer Institute for Integrated Systems and Device Technology IISB, Erlangen, Germany
2 Chair of Electric Power Engineering, Friedrich-Alexander University Erlangen-Nürnberg, Nuremberg, Germany
*E-mail: linh.bach@iisb.fraunhofer.de

This paper proposes a novel packaging concept for power electronic applications on basis of embedding power devices in ceramic circuit carriers, such as direct bonded copper (DBC) substrates. The semiconductor devices are assembled into laser structured DBC substrates and then sealed with a copper cover afterwards. This proposed method is an alternative solution to printed circuit board (PCB) embedding and low-temperature co-fired ceramic (LTCC) based multilayer technologies, which are insufficient for high power applications due to the limited temperature resistance and current carrying capacity. The feasibility study confirmed that the DBC embedding approach was successfully implemented by using laser technology combined with conductive gluing, solder, and silver sintering processes.

Keywords— Ceramic chip embedding, cost-effective power electronics packaging solution, multilayer DBC substrates, WBG packaging for industrial production

I. INTRODUCTION

For future power electronic applications, high performance and high reliability are the core goals aimed by the application. Trends as miniaturization and three-dimensional integration result in challenges concerning realizable high temperature over 250 °C, high active and passive temperature cycling capability as well as low parasitic inductance. High temperature packages could take out the full advantage of wide bandgap (WBG) devices [1]. Hutzler et al. [2] have presented that the lifetime of the appropriate die attach materials can be improved by increasing the coolant temperature T_{min} in power cycling tests. For the long-term protection of Micro-Electro-Mechanical Systems (MEMS) and Radio Frequency (RF) electronic applications, several hermetic glass-to-metal, ceramics-to-metal and glass encapsulation technologies have been introduced [3-5]. Organic materials are still problematic in terms of thermal degradation [6]. Novel organic insulated materials such as Isola B and Benzo, which are capable of high temperature up to 220 °C, have been invented during the course of the project HELP [7]. For both materials, shrinking and discoloration still address an issue. In terms of inorganic material selection, various chip embedding concepts for advanced power modules packages based on short vertical vias in LTCC substrates have been

developed [8] [10]. However, high current carrying capability is essential for high power applications. Therefore, a promising concept is the embedding of semiconductor and sensor devices in DBC substrates with thick copper metallizations. Embedding chips in DBC substrates enables the design of hermetic, high temperature packages for power electronic devices. Combined with electrical vias, whose simple and cost-effective manufacturing processes have been presented in the previous work [9], low parasitic inductive power modules based on multilayer DBC stacks can be achieved.

As a first stage of the development of such a novel power module approach, a DBC embedding technology for WBG semiconductors was investigated. In this study, single processing steps of this concept are evaluated and presented regarding technological feasibility and industrial capability. Furthermore, the exploitation of the full potential of WBG devices with such an embedding technology is discussed.

II. CONCEPT OF CERAMIC EMBEDDING

Recent works on power devices embedded in ceramics have been introduced, especially by using LTCC technology. In [10] a wire bondless half-bridge 3-D stacked power module was investigated by attaching silicon carbide (SiC) MOSFETs and schottky barrier diodes (SBD) into a LTCC substrate, in order to reduce the parasitic inductance (**Fig. 1**).

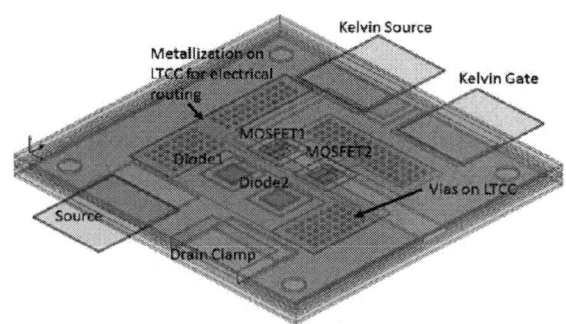

Figure 1. Stand-alone power module consisting of a switching position according to Dutta et al. [10]

Zhang et al. [8] fabricated an embedded power module, which is based on LTCC substrate including conductive vias. First, a SiC MOSFET and a SiC diode were embedded. In the next step, two additional DBC substrates were attached at the top and bottom side (**Fig. 2**). This wire-bondless module design enables double-sided cooling, which can be advantageous for the thermal management.

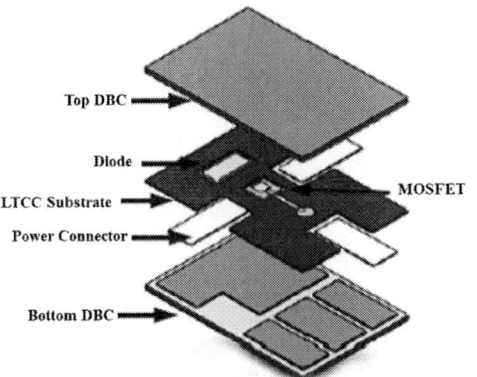

Figure 2. Exploded view of the LTCC based double-sided cooling module according to Zhang et al. [8]

However, for high power applications complete LTCC packaging solutions are insufficient, due to their low thermal conductivity and limited metallization thickness. Whereas DBC substrates based on Al_2O_3, AlN, or Si_3N_4, can provide significant higher heat dissipation compared to LTCC (five times or higher). Furthermore, the DBC substrate manufacturing process enables copper metallization pads up to 0.8 mm thick (dependent on ceramic material and thickness). Therefore, extremely high current density of state of the art WBG devices can be covered.

A. Overview of Processing Steps of Embedding Semiconductor Devices in DBC Substrates

As a first step, an embedding concept for DBC substrates has been created in the scope of this study (**Fig. 3**). The semiconductor devices used were IGBT[3] bare dies (650 V, 200 A) from Infineon. The IGBT bare dies were selected for this work in order to investigate the practicability of the DBC embedding process in the first step. After that, the new findings can be used for dealing with SiC bare dies which are more difficult to handle in terms of chip size and chip contacting. The DBC substrate consisted of a 0.38 mm thick Al_2O_3 layer and 0.1 mm thick copper layers.

Before embedding the IGBT into the DBC substrate, a fitting cavity layout has to be produced. This was realized through a laser structuring process (**Fig. 3, a**). In the next step, contact surfaces (gate and emitter) are printed with conductive adhesive, solder, or silver sinter materials. Afterwards, the IGBT is flipped and positioned in the structured DBC substrate, followed by a chip bonding process (**Fig. 3, b**). In the final step, the DBC package is completely sealed with a copper cover (**Fig. 3, c**).

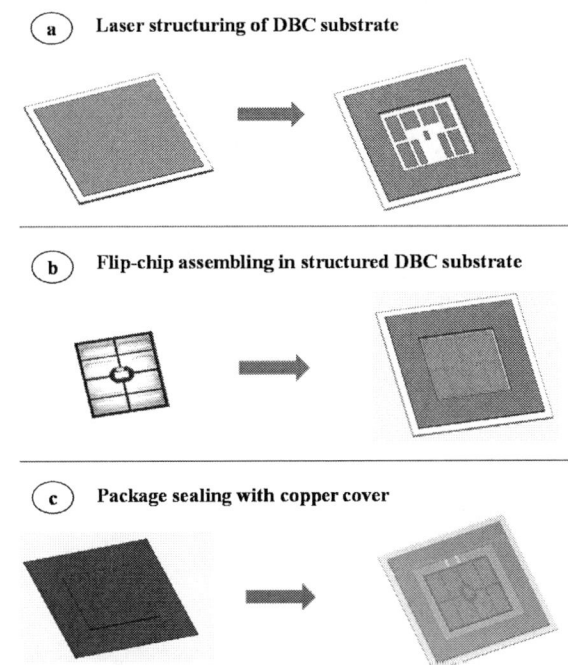

Figure 3. Approach of embedding chip in DBC substrate, chip size 9.73 mm x 10.23 mm, DBC substrate size 20 mm x 20 mm, copper cover size 18 mm x 18 mm

B. Laser Structuring of DBC Substrates

Compared to other alternative subtractive technologies such as chemical etching, the laser processing of ceramic materials can be in the long-term view a more cost-effective and efficient solution. Normally, additional masks are needed for etching, in order to achieve the desired structures. Furthermore, new masks have to be produced if the substrate layout is modified. This issue also considers acid compositions for etching metal and ceramic materials. Especially for etching ceramics such as Al_2O_3, which has a very high corrosion resistance, most of the acids are ineffective except hydrofluoric acid [11]. By using laser technology, an efficient material removal of metal and ceramic is given. Further advantages are the flexible design possibilities for chip and electrical via structures. The layout of the DBC substrate does not have to be defined before its manufacturing.

Different laser systems have been tested in the scope of this work, in order to achieve the required structures of Al_2O_3 and Cu. The challenge within this process step was the laser treatment of the Al_2O_3 ceramic. The structures have to be in the micrometer range. For an optimal chip assembly, the remaining inner ceramic area has to be smooth. Especially for the sintering, a layer thickness of 30 μm has to be achieved, which is adapted to the thickness of the silver bond line after the drying process (**Fig. 4**). When soldering the IGBT, the ceramic thickness has to be 130 μm or lower for matching the solder preform thickness.

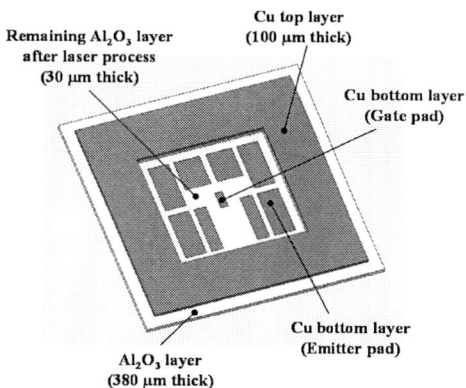

Figure 4. Required DBC structure for chip embedding concept, structured Cu and Al_2O_3 layer for gate and emitter pads of IGBT chip, top side view

If the thickness of the inner Al_2O_3 layer does not match the die attach thickness, a fully covered connection of the chip pads cannot be guaranteed. Especially for the pressure related silver sinter process, fractures in the chip can occur, resulting from the unevenly applied pressure. On the backside of the DBC substrate, the bottom Cu layer is structured in two separated pads, for the gate and emitter contacting of the final package (**Fig. 5**).

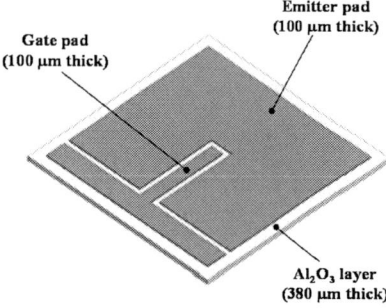

Figure 5. Required DBC structure for chip embedding concept, layout of Cu bottom layer, separated pads for gate und emitter contacts, bottom side view

The first experiments were carried out with a nanosecond ultraviolet (UV) laser (355 nm wavelength). An example of a structured DBC substrate is depicted in **Fig. 6**. The area of the DBC Cu top layer (10.13 mm x 10.63 mm) was completely removed. In the next step, the Al_2O_3 ceramic was partially structured and openings have been created for the gate and emitter contacts (Cu bottom layer).

The energy dispersive X-ray (EDX) analysis showed remained Cu_2O and Al_2O_3 content on the surface of the Cu bottom layer after the laser process. Visible dark long shape patterns were detected, resulting from the interactions between the laser and the Cu surface (**Fig. 6-7**). One reason for this appearance is the ablation of the cavities, which is performed sequentially through a laser beam. Depending on the sequence set up of the laser beam (line movement), these long shape patterns can occur.

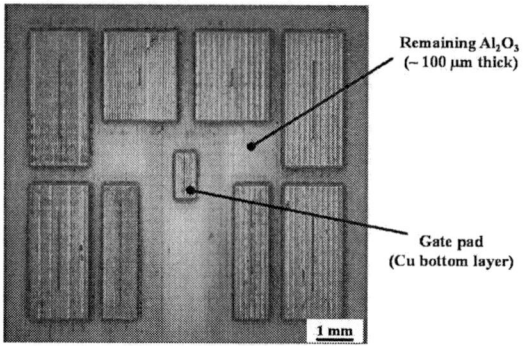

Figure 6. Structured DBC substrate after the laser process, nanosecond UV laser, 355 nm wavelength, structured gate pad area 1.45 mm x 0.68 mm

The Al_2O_3 removal has been carried out effectively. The required area for the gate and emitter pads fitted to the IGBT chip size. However, the remaining Al_2O_3 layer thickness fluctuated strongly between 50 μm to 160 μm. A constant and precise Al_2O_3 layer thickness could not be achieved with the UV laser, due to the limited laser pulse length (nanosecond range). Furthermore, small cracks were detected on the Cu surface of several specimens. Therefore, further experiments have been carried out with a femtosecond laser system, in order to prevent deformation through heat influx, and to improve the precision of the material removal.

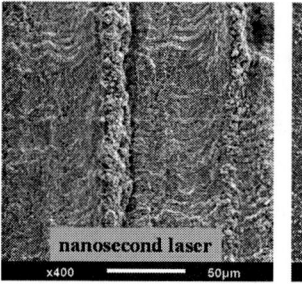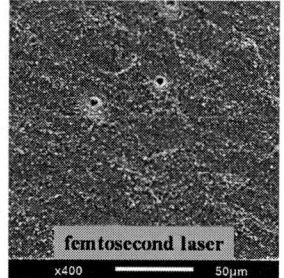

Figure 7. Scanning electron microscope (SEM) measurement of Cu surface after the laser treatment

According to the measured results, the femtosecond laser system (1030 nm, 800 fs) significantly showed a higher precision regarding the processing of Cu and Al_2O_3. The Al_2O_3 layer thickness was structured to 25 μm - 40 μm. This result was reproducible over a number of tests (10 specimens). No traces of Cu_2O and discoloration were detected on the Cu and Al_2O_3 surfaces (**Fig. 8**). The remaining Al_2O_3 was darker after the laser processing, which has the appearance of a strong oxidation. More detailed analysis showed that the color of the Al_2O_3 was not an oxidation. This optical appearance is dependent on the layer thickness. The Al_2O_3 layer is more transparent by decreasing the thickness, which results in a visible show-through of the DBC bottom Cu layer. Therefore, no critical damage or oxidation was detected by using the femtosecond laser system.

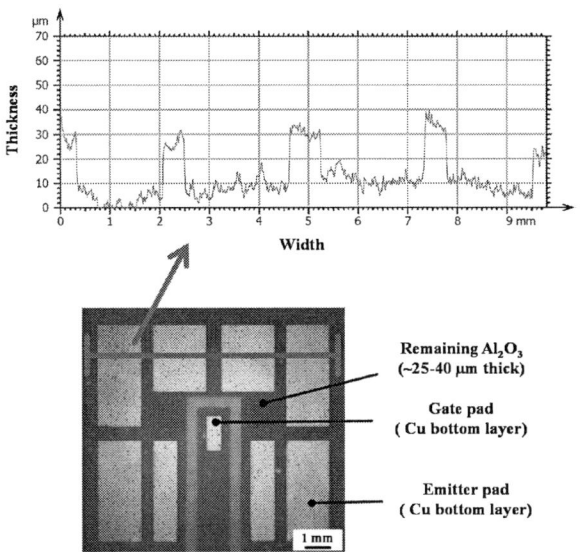

Figure 8. Structured DBC substrate after the laser process, femtosecond laser (800 fs), 1030 nm wavelength

The femtosecond laser system proved to be an efficient method for structuring Cu and Al_2O_3 material. Depending on the desired chip and circuit layout, the DBC substrate can be flexibly customized. Therefore, the preconditions for further ceramic embedding concepts are given for upcoming investigations, for example the layout structuring of half-bridge or full-bridge topologies.

C. Embedding by Conductive Gluing

After the laser processing and analyzing of the samples, different methods of chip assembling have been investigated in the next step. The DBC substrates structured by both, nanosecond and femtosecond laser systems, have been used to evaluate the influence of the laser on the die attach process. First, a conductive adhesive was tested (**Fig. 9**). This pretest was done in order to secure the electrical functionality of the packaging concept.

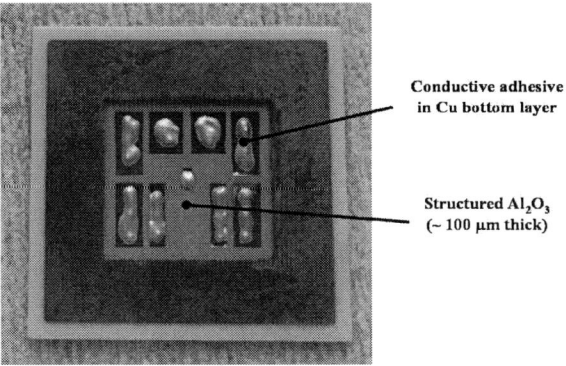

Figure 9. Conductive adhesive (Panacol Ecolit 3655) in gate and emitter Cu pads of DBC substrate, structured with nanosecond laser

In the next step, a flipped IGBT chip was positioned into the DBC substrate, followed by an oven curing process. The curing was performed at 150 °C for 30 minutes. After that, the setup was sealed with a copper based cover. The size of the copper cover was adjusted to the structured DBC substrate. The same conductive adhesive was used for the sealing process. In order to test the electrical functionality of the embedded package, a demonstrator was produced by assembling the DBC package on a vapor chamber (**Fig. 10**).

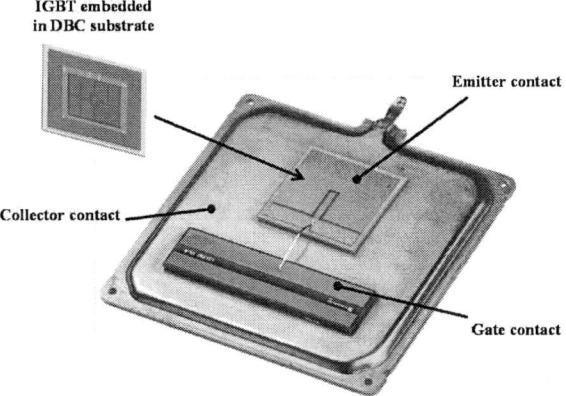

Figure 10. Demonstrator for electrical measurement, DBC copper cover side attached to the surface of the vapor chamber with conductive adhesive

The measurement of the embedded IGBT package showed that the current and blocking characteristics correspond to the values from the data sheet. The measured gate leakage current is far below the specified value of 600 nA (measured value < 300 pA). Thus, the electrical functionality of the designed DBC package was validated.

D. Embedding by Soldering

In the following experiments a further die attach process has been investigated. Here, the focus was on evaluating the feasibility of the embedding process by using solder technology. Both DBC samples structured by the nanosecond and femtosecond laser have been tested. Solder preforms ($T_g = 220$ °C) from Pfarr Stanztechnik GmbH were sectioned and positioned into the DBC cavities before the oven process. In the next step, the DBC substrates were soldered at 240 °C under nitrogen atmosphere. During the solder process, formic acid was added for three minutes to remove oxide from the copper surface. The analysis afterwards showed significant differences in terms of the joining quality (**Fig. 11**). On the Cu surface structured with the nanosecond laser, the wetting behavior of the solder was characterized. The solder compressed itself and could not cover the defined Cu pads completely due to the insufficient surface quality of the structured Cu. In order to ensure an optimal wetting of the solder, the Cu surface has to be smooth, and free of oxide.

DBC substrate structured by nanosecond laser	DBC substrate structured by femtosecond laser

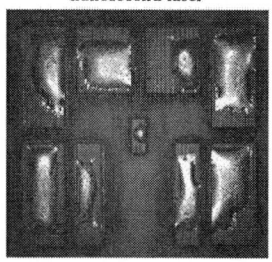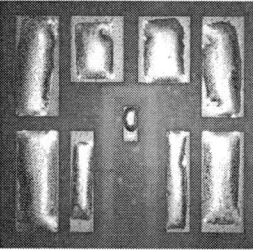

Figure 11. Influences of different laser systems on the joining quality of solder preforms, solder preforms in structured DBC cavities after the oven process, heat-up rate 50 K/min, peak temperature 240 °C, holding time 3 minutes

The solder quality on the DBC sample structured with the femtosecond laser was comparatively better in terms of wetting behavior. The solder material spread throughout the Cu pads, which is a sufficient condition for joining the IGBT in the next step. The bond line between Cu, solder and chip is depicted in **Fig. 12.**

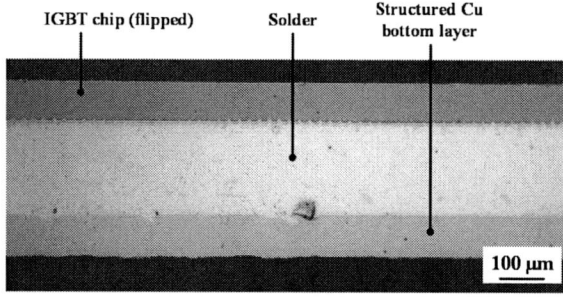

Figure 12. Cross section view of a flipped IGBT soldered into a DBC substrate, DBC substrate structured by a femtosecond laser

The cross section analysis showed an optimal wetting behavior of the soldered sample. The solder layer was completely free of voids. Especially for the connection between solder and Cu layer, a high joining quality has been achieved. Thus, it can be mentioned that a femtosecond laser treatment of DBC substrates has no negative effects on the soldering process afterwards.

E. Embedding by Silver Sintering

A further die attach method tested was sintering technology. The challenge here was to carry out the embedding process by using silver based material. Silver sintering preforms are generally more brittle than solder preforms such as SAC-solder (SnAgCu), which make it difficult to cut the preforms into pieces and to position it into the DBC cavities without any cracks or damage. Thus, the experiments were carried out with silver sinter paste. A silver sinter paste from Heraeus (LTS 295-26P2) has been tested for the embedding process. First, the sinter paste was dispensed on the structured Cu pads. A flipped IGBT was placed into the DBC substrate afterwards, followed by a pressure-less sinter process at 250 °C for 30 minutes under oxygen atmosphere. The analysis after the experiment showed that the chip could

not be successfully assembled into the DBC substrate. The reason for the failed joining includes the possible insufficient drying of the sinter paste solvent. The solvent needs to be dried out during the heating process. In this embedding concept, the sinter paste is placed into DBC cavities, covered by the IGBT. This impedes the leaking of the solvent during the drying process. Furthermore, it is not guaranteed that the air atmosphere in the oven can drift into the cavities effectively. Therefore, the focus was on a pressure-related sintering process in the next experiments.

The same sinter paste was used for the pressure-related sintering. By using the transfer-printing method, defined sinter paste layers were printed on the gate and emitter pad of the IGBT. In the next step the IGBT was flipped and sintered into the DBC cavity at 260 °C for 3 minutes. The set pressure was 5 MPa. Only DBC substrates structured with the nanosecond laser were tested at first. The substrates were pre-cleaned with formic acid before sintering. The evaluation showed that 50 % of the IGBT chips were damaged after the sinter process (**Fig. 13**). Due to the inaccurate structure of the remained Al_2O_3 layer, the chip broke at the critical locations. Furthermore, the sinter paste could not fully connect to the Cu layer and the IGBT pads. There were partial voids between the sinter and Cu layer, resulting from the insufficient Cu surface quality after the laser treatment.

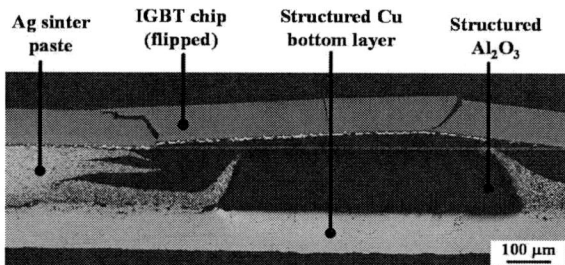

Figure 13. Cross section view of a damaged IGBT chip after pressure-related sintering, DBC structured by a nanosecond laser

However, undamaged samples were also produced in the scope of this experiment. Half of the assembled specimens showed a sufficient joining quality between chip, sinter paste, and structured Cu layer. In order to ensure high sinter process reliability, a high-precision ablation of the Al_2O_3 layer is essential. Therefore, it is to be expected that the sintering into DBC samples structured with a femtosecond laser will show better results. Further tests will be carried out in upcoming studies. There is a high potential to improve the sinter quality by optimizing the laser treatment and the sinter paste assembly. Furthermore, a pretreatment of the DBC substrate, such as silver coating of the structured Cu layer, can enhance the joining quality.

III. CONCLUSIONS

A novel packaging concept based on the embedding of semiconductor devices in a ceramic circuit carrier such as DBC substrates was presented. Different laser systems and die attach methods have been investigated in terms of feasibility. Two laser systems with a pulse length in nanosecond and femtosecond range have been tested to prove the suitability for the ablation of Cu and Al_2O_3. A higher process accuracy and stability have been achieved with the femtosecond laser. A first electrical functional demonstrator was manufactured by assembling and sealing a semiconductor device. The embedding process was realized by conductive gluing, soldering, and silver sintering. A significant influence of the laser systems on the joining quality of the die attach material was validated. Therefore, laser systems with a pulse length in the femtosecond range turned out to be more sufficient for the embedding process, due to their high-precision material removal. In general, the laser process can be rated as a potential application for ceramic embedding approaches. Especially in terms of design flexibility, the ceramic substrate layout can be structured as desired.

The work is ongoing to further improve the embedding process reliability, with a strong focus on fast processing. Experiments will be carried out in future studies to investigate aspects concerning high temperature operation, high active and passive cycling capability of the package. Furthermore, application-oriented layouts, such as WBG half-bridge and full-bridge topologies, embedded in multilayer DBC stacks, will be presented. The final packages can be used for specific applications such as hybrid electric vehicles, solar inverters, down-hole oil drilling, and geothermal instrumentation.

REFERENCES

List only one reference per reference number according to the following samples:

[1] F. P. McCluskey, T. Podlesak and R. Grzybowski, "High temperature electronics," *CRC Press*, 1989.

[2] A. Hutzler, A. Tokarski, and A. Schletz, "Extending the lifetime of power electronic assemblies by increased cooling temperatures," *Microelectronics Reliability 53.9*, pp. 1774-1777, 2013.

[3] F. Aguirre and D. Schatzel, "High Density Packaging Technologies for EF Electronics in Small Spacecraft," *Aerospace Conference*, 2017.

[4] J. H. Chang, Y. Liu, and Y. Tai, "Long term glass-encapsulated packaging for implant electronics," *Micro Electro Mechanical Systems*, 2014.

[5] H. Xu, M. Broas, H. Dong, V. Vuorinen, T. Suni, S. Vähänen, P. Monnoyer, and M. Paulasto-Kröckel, "Reliability of wafer-level SLID bonds for MEMS encapsulation," *Microelectronics Packaging Conference*, 2013.

[6] P. Evangelopoulos, E. Kantarelis, and W. Yang, "Investigation of the thermal decomposition of printed circuit boards (PCBs) via thermogravimetric analysis (TGA) and analytical pyrolysis (Py-GC/MS)," *Journal of Analytical and Applied Pyrolysis 115*, pp. 337-343, 2015.

[7] K. Tröger, R. K. Darka, T. Neumeyer, V. Altstädt, J. Keller, and A. Fathi, "Tailored Benzoxazines as Novel Resin Systems for Printed Circuit Boards in High Temperature E-mobility Applications," *AIP Conference Proceedings*, pp. 678-682, 2014.

[8] H. Zhang, S. S. Ang, H. A. Mantooth, and S. Krishnamurthy, "A High Temperature, Double-sided Cooling SiC Power Electronics Module," *Energy Conversion Congress and Exposition*, pp. 2877-2883, 2011.

[9] L. Bach, Z. Yu, S. Letz, C. F. Bayer, U. Waltrich, A. Schletz, and M. März, "Vias in DBC Substrates for Embedded Power Modules," unpublished.

[10] A. Dutta and S. S. Ang, "A 3-D stacked wire bondless silicon carbide power module," *Wide Bandgap Power Devices and Applications (WiPDA)*, 2016.

[11] Verband der Keramischen Industrie e. V., "Brevier Technische Keramik," *Fahner Verlag*, 2013.

Microelectromechanical System (MEMS) Resonator: A New Element in Power Converter Circuits Featuring Reduced EMI

A N M Wasekul Azad[1], Sourov Roy[1], Abu Saleh Imtiaz[2] and Faisal Khan[1]

1 Computer Science and Electrical Engineering, University of Missouri-Kansas City, Kansas City, MO, USA

2 Globalfoundries, San Francisco, USA

Email: aax9c@mail.umkc.edu

Abstract- **This paper presents a new passive device, which could be successfully used in switching power converters to replace inductors, and capacitors. Microelectromechanical System (MEMS) resonators are traditionally used in phase locked loop (PLL) based digital clock circuits, but could be effectively used for power conversion. The major advantage of MEMs resonators is the non-magnetic energy storage with inductor-like properties. Moreover, they could be fabricated in today's conventional silicon process without requiring any major modifications. This integration makes it a very attractive choice for implantable circuits where inductor-free design is a major criterion. MEMs resonators can be used as inductors or capacitors by changing the switching frequency, and they are not impacted by strong magnetic or electric fields, therefore making them safe, and compact for MRI friendly implantable circuits. In addition, this resonator based circuits produce substantially less electro-magnetic interference (EMI) because inductors are not used in the circuit. This paper describes the various features of this new device, applications in resonant power converters, and future uses especially biomedical applications of MEMs resonators.**

Keywords- *MEMS Resonator, inductor on chip, high Q device, inductor less circuit, low EMI, small footprint.*

I. INTRODUCTION

Today's implantable devices need to be safe, and non-invasive to the surrounding body tissues to minimize rejection. Most of these devices require power conditioning, which is in most cases accomplished by AC-DC and DC-DC converter circuits [1]. In most power conditioning circuits, voltage conversion involves magnetic components, which introduces EMI as a byproduct of high di/dt and slew rates [2]. EMI is harmful for health, and continued research effort is in place to minimize EMI caused by power converters. In this paper, a new power conditioning component, Microelectromechanical System (MEMS) based resonator, proposed in a half bridge series resonant converter topology, demonstrated the ability to control output voltage by changing the switching frequency of switching MOSFETs. This resonator focuses on replacing the traditional inductor in the switching converter circuit, which can provide superior EMI performance in a smaller footprint [3]. Moreover, due to the compact size and high inductance density of the resonator [3], it has the potential to be monolithically integrated. *Piezoelectric transformers showing similar features like the MEMS resonators have* *been historically used in power converter circuits [4] [5], but they suffer from larger footprint, and cannot be implemented in existing silicon process.* Moreover, the input capacitance of piezo electric transformer adds up with the MOSFET output capacitance in a converter which complicates ZVS (zero voltage switching) operation of converter [6]. In addition, those piezoelectric devices have been used to realize transformers, and not to replace inductors or capacitors in a resonant converter circuit. Our proposed device is 100% compatible with existing Si process and can be used to replace inductors or capacitors.

II. APPLICATIONS OF MEMS RESONATORS AND THEIR POTENTIAL IN IMPLANTABLE ELECTRONICS

MEMS devices are miniature structures which are fabricated on silicon substrates using available semiconductor processing techniques. The process is called micro machining, and the devices usually range from a few hundred microns to a few millimeters in length [7]. Traditionally, MEMS devices are used as high Q on-chip elements for RF and IF band pass filtering, accurate and stable clock generation, improved phase noise performance actuator in oscillator circuit in GSM arena. These devices eradicate the disadvantages of a passive off-chip high Q element in terms of size, cost, power consumption, and they are immune to thermal variations, and aging [8]. Increased use of MEMS technology can also be found in large and small scale optical switches, tunable lasers, and variable optical attenuators [7]. MEMS technology can be classified into several categories depending on the material used (e.g., silicon, silicon carbide, glass, GaAs, Polysilicon, AlN etc.), micromachining process (e.g., surface, bulk, 3-D growth), and applications (e.g., optical MEMS, bio MEMS etc.) [8]. However, the use of the MEMS resonator as an element in power conditioning circuit replacing the passive components (i.e., inductors, capacitors) is unprecedented till date. Hence, introducing a DC-DC converter using MEMS resonator is a novel approach with a lot of potential.

The number of people being subjected to treatment through implantable devices is increasing rapidly worldwide. Many of these implantable devices need regulated power to remain functional, and operate properly. Several energy harvesting and transfer techniques are available which can power implantable devices. However, every energy transfer mechanism requires voltage regulation to cater for different input voltages, and power requirements of implantable device loads. At present, power conditioning in implantable devices is mainly performed by low dropout regulators (LDO) [9, 10], switched capacitor converters [11], and in some cases output of a full-bridge rectifier is directly used without conditioning. The downside of using LDO is the heat dissipation inside human tissues which poses serious health hazards, and lack of control over voltage regulation. Switched capacitor circuits have advantages of some degree of voltage regulation, and can be monolithically integrated. However, limited voltage regulation capability, and requirement for significant number of switches render this option less efficient. The proposed MEMS resonator may provide a wide range of output voltages by changing the switching frequency of the switching devices which can cater for varying load demands. The proposed MEMS resonator may replace the bulky inductor of the DC-DC converter, which consumes a significant amount of space in the embedded system, therefore potentially ushering a new era of on-chip DC-DC converter fabrication. Another potential advantage of replacing inductor is the improved EMI characteristic which is beneficial in circuit level as well as inside the body atmosphere around implantable devices.

III. MEMS RESONATORS: THEIR DESIGN, FABRICATION AND CHARACTERISTICS

One of the co-authors fabricated a ring-shaped Aluminium Nitride (AlN) piezoelectric microresonator in a CMOS compatible process. The final fabricated prototype achieved a resonant frequency as low as 87.28 MHz and a motional resistance of 36.728 Ω. The prototype also achieved significantly small footprint having surface area of 0.11 mm^2, and height of about 2.5 µm. Therefore, the achieved inductance density of the prototype was significantly higher than the existing piezoelectric transformer (PT) based converters, which eliminates the need for thick piezoelectric or metal films [3]. The equivalent circuit of a MEMS resonator can be derived using the Van-Dyke Butterworth equivalent circuit used in [12], and is presented in Fig. 1.

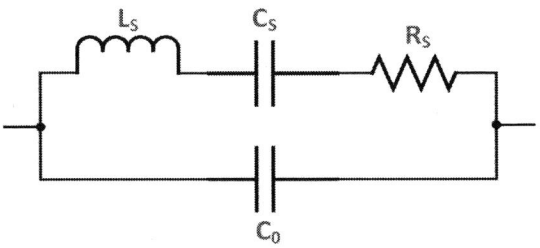

Fig. 1: Van-Dyke Butterworth equivalent Circuit.

The electrical model of the resonator is fundamentally

an LC tank. Series connection of motional inductance L_s, motional resistance R_s, motional capacitance C_s along with shunt capacitance C_o in parallel form the LC tank. A small capacitor is added at the output to filter out the high frequency components. The series LC tank circuit leads to a series resonant frequency (f_s) while the shunt capacitance contributes to the anti-resonant frequency (f_p). *The resonator behaves like an inductor between a resonant and anti-resonant frequency, and within this frequency range, the phase angle is around 90 degrees.* However, it behaves like a capacitor outside this range, and the phase angle sharply drops to -90 degrees. The equivalent impedance becomes lowest at resonant frequency, and it attains maximum value at anti-resonant frequency. In Fig. 2, the impedance, and phase response of the resonator are illustrated.

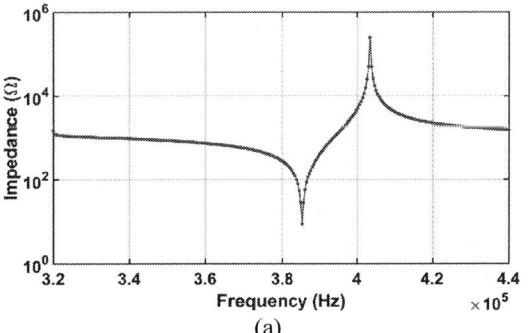

(a)

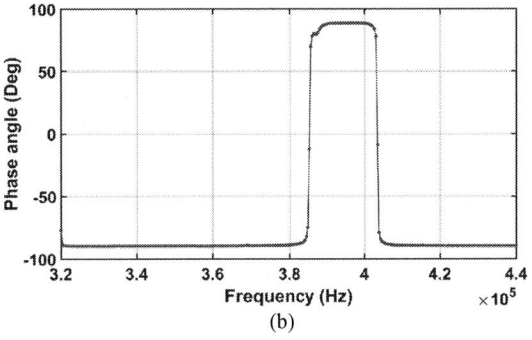

(b)

Fig. 2: Frequency response of a MEMS resonator, (a) impedance characteristics, (b) phase characteristics.

Neglecting the motional resistance, the resonant frequency of the resonator can be calculated from the following expression.

$$f_s = \frac{1}{2\pi\sqrt{L_s C_s}}$$

Once the resonant frequency is known, the anti-resonant frequency can be calculated from here:

$$f_p = f_s\sqrt{1 + \frac{C_s}{C_p}}$$

Between f_s and f_p, the resonator predominantly behaves like an inductor along with the lossy component R_s.

Motional parameters L_s and C_s are calculated from the well-established equations as follows [13].

$$C_s = \left(\frac{f_p}{f_s} - 1\right) * 2 * C_o \ldots\ldots\ldots\ldots\ldots\ldots(1)$$

$$\text{And } L_s = \frac{1}{4*\pi^2*f_s^2*C_s} \ldots\ldots\ldots\ldots\ldots\ldots(2)$$

The shunt static capacitance C_o can be measured by an LCR meter at a frequency which is far away from the resonant frequency as described in [14]. At resonant frequency, motional reactance of the series arm of the resonator becomes zero, and its equivalent impedance drops to the minimum value. Therefore, maximum output can be obtained at this point. As the operating point shifts away from this point, the equivalent impedance increases, and subsequently the output gradually diminishes. This unique behavior provides output voltage controllability by precisely controlling the switching frequency of the active switching devices incorporated in the half bridge converter block. Therefore, MEMs resonator based converter offers output voltage regulation capability without the integration of magnetic components like inductors.

IV. MEMS RESONATORS IN A SERIES RESONANT CONVERTER

Many of the existing implantable devices need regulated power to remain active and functional. Wireless power transfer through inductive power link [9], photovoltaic infrared energy harvesting [15], ultrasound energy harvesting [16], thermal energy harvesting through temperature gradient, piezoelectric material based vibrational energy harvesting [17] are some of the methods for energy harvesting inside human body. A series resonant topology, which is basically a DC-AC inverter was primarily chosen to demonstrate the various advantageous features of the proposed MEMS resonator to achieve power conversion. Higher switching frequency used in resonant converters enables smaller passive component size, which leads to smaller size of the overall converter, facilitating monolithic integration in biomedical applications. The proposed MEMS based resonant converter operates at 400 kHz. In our experiment, a commercial ceramic resonator (mimicking a MEMS resonator), ECS Inc. ZTB400P was used to emulate the characteristic of any future MEMS resonator. The series resonant frequency (f_s), and parallel resonant frequency (f_p) were measured as 384,500 Hz and 403,400 Hz from the frequency response obtained from the LCR meter (Hioki IM 3536). The authors are presently working on designing a low-frequency MEMS resonator.

V. SIMULATION AND EXPERIMENTAL RESULTS

The obtained values of the L_s, C_s and C_o were 4.88 mH, 35 pf, and 384 pf respectively. An EPC 9032 GaN development board was used in this experiment to build the resonant converter. It is incorporated with gate drivers, and two EPC 2024 GaN MOSFETs, which are arranged in

half-bridge configuration. The DC input voltage (V_{in} in Fig. 3a) was 5 V, while two different load impedances of 1 kΩ and 2.4 kΩ were used in the experimental setup separately. The schematic in Fig. 3a was simulated in PSIM, and the circuit board shown in Fig. 3b was used to generate the experimental results.

Frequency sweep function from Keysight 33500B function generator was used to vary the switching frequency of the MOSFETs resulting in variable output voltages using the resonator. Fig. 4(a) shows the output voltage as a function of switching frequency for two different loads: 1 kΩ and 2.4 kΩ. Like any series resonant converter with a discrete LC tank, the output voltage can be controllable within a range by changing the switching frequency of this resonator based converter, and the resonator behaves like a true inductor in the circuit. Fig. 4(b) shows the comparison between the simulation and experimental results for 1 kΩ load, and these results closely match. Fig. 4(c) shows the same comparison for a 2.4 kΩ load, and the results also match closely. Therefore, the commercial resonator closely emulates the characteristics exhibited by discrete LC tank based resonant converter in PSIM simulation.

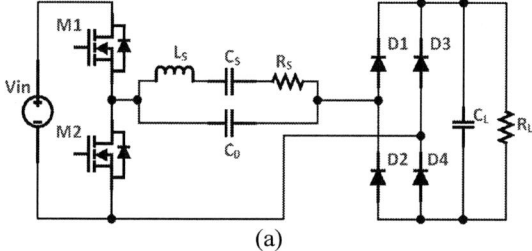

(a)

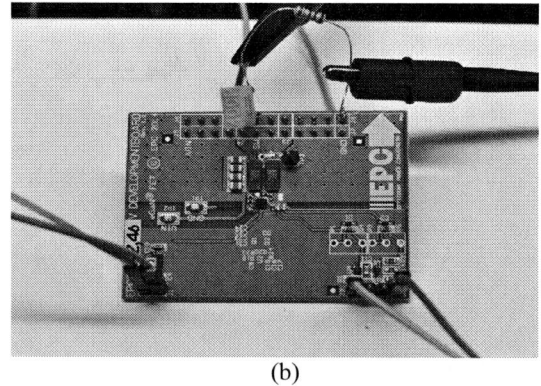

(b)

Fig. 3: (a) PSIM simulation circuit diagram, (b) experimental setup using EPC board.

Fig. 4(d) shows the simulated rectified output voltage as functions of frequency and load impedance. As expected, the output voltage is a strong function of the load impedance, and a control loop is needed to stabilize the output voltage at different loading conditions.

This converter might achieve better EMI performance compared to the discrete L-C based converter [11]. One of the co-authors performed conducted EMI measurement for

2418

The 2018 International Power Electronics Conference

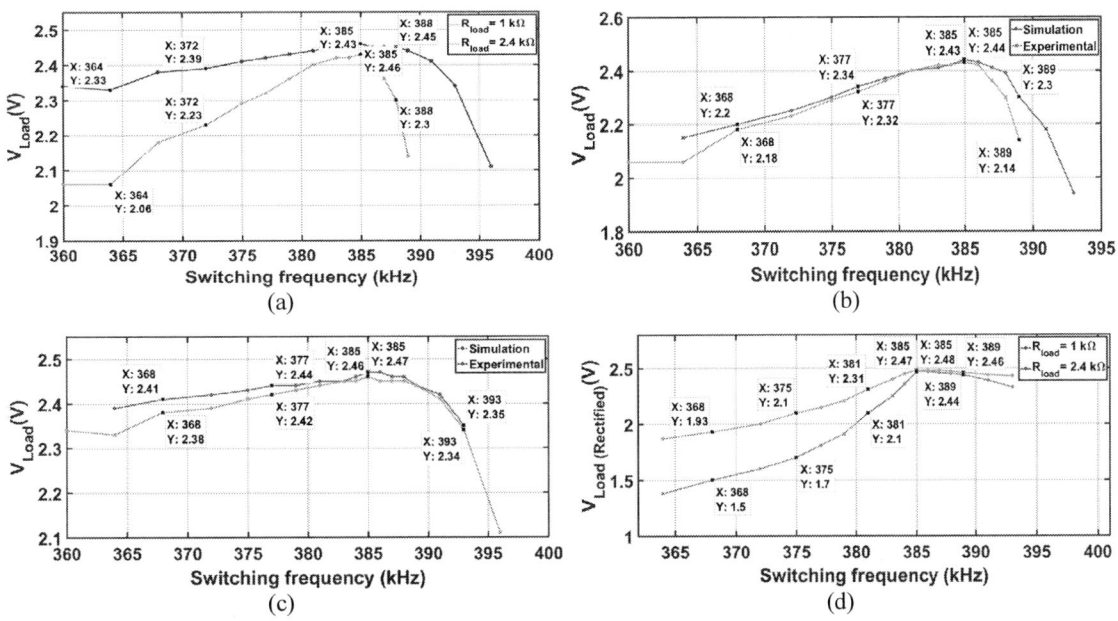

Fig. 4. (a) Experimental output voltage with different loads, (b) comparison of simulation and experimental output voltage at 1 kΩ, (c) comparison of simulation and experimental output voltage at 2.4 kΩ, (d) comparison of simulated rectified output voltages at different loading conditions and frequency.

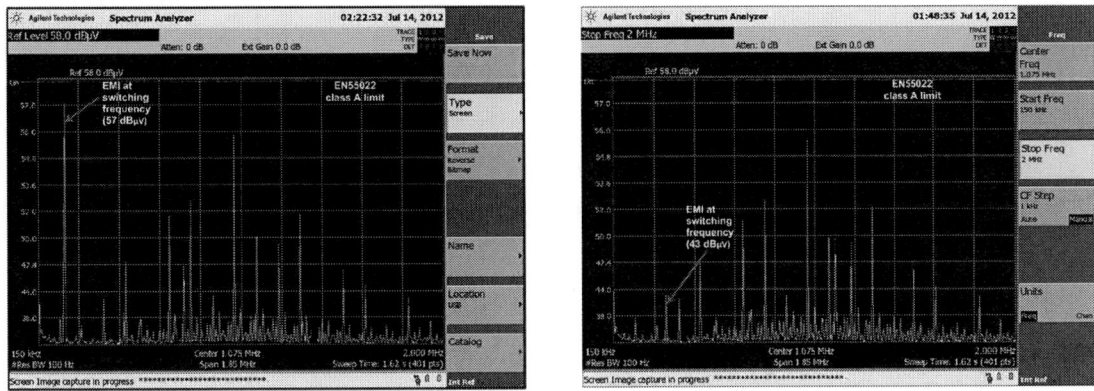

Fig. 5: (a) Measured EMI: discrete L-C series resonant inverter LED driver (b) Measured EMI: Commercial resonator based series resonant inverter LED driver [18].

both discrete L-C based and commercial resonator based series resonant converter driven LED driver [18]. The resonator based setup demonstrated marked improvement in EMI performance compared to the discrete L-C based setup when measured with a spectrum analyzer. Fig. 5(a) and Fig. 5(b) illustrate comparison of EMI generation (57 dB as opposed to 43 dB) at resonant switching frequency of the resonator for both of the setups. As the energy is stored in the form of high frequency vibration, the EMI produced from the resonator is almost non-existent, and the overall EMI produced from the circuit is caused by the PCB traces and other lumped elements. With careful PCB design and components selection, the EMI level could be reduced substantially, and this feature is highly attractive

for implantable electronics.

VI. CONCLUSION AND FUTURE WORK

Historically, mechanical resonators have never been used as inductors, although piezoelectric transformer based lamp drivers exist. It is possible to fabricate resonators using conventional silicon processing steps (CMOS), and it has been shown in [3] with fabrication details. This work is the continuation of that research, and this paper shows that it is possible to use mechanical resonators to implement power (voltage) conversion. The resonator design shown in [3] has very high switching frequency, makes it difficult to design a power converter around it. The present work demonstrated a power converter using a commercially available mechanical resonator (emulating a Si MEMS resonator) with similar characteristics but offering much smaller resonant frequency. The authors are presently working on the design and fabrication of low frequency (below 5 MHz) resonator on Si so that the entire power converter could be implemented on a single wafer, making it highly attractive for implantable electronics. Moreover, feasibility of fabricating a new prototype MEMS microresonator having different geometrical shapes, and dimensions is being analyzed. The objective of tuning the shape, interconnection and dimension is to achieve lower switching frequency, higher current rating and reduced motional resistance. In this paper, ceramic resonator based half bridge series resonant converter has been presented. Both the experimental and simulation results provide strong evidence of the possibility that MEMS resonators could be used instead of inductors in resonant converter circuits. The authors believe that this resonator based circuits are likely to open a new horizon in implantable electronics because of high Q design, low EMI and compact footprint.

REFERENCES

[1] A. Yakovlev, S. Kim and A. Poon, "Implantable biomedical devices: Wireless powering and communication," in IEEE Communications Magazine, vol. 50, no. 4, pp. 152-159, April 2012.

[2] Wei Zhang, M. T. Zhang, F. C. Lee, J. Roudet and E. Clavel, "Conducted EMI analysis of a boost PFC circuit," Proceedings of APEC 97 - Applied Power Electronics Conference, Atlanta, GA, USA, 1997, pp. 223-229 vol.1.

[3] A. M. Imtiaz, F. H. Khan and J. S. Walling, "Contour-Mode Ring-Shaped AlN Microresonator on Si and Feasibility of Its Application in Series-Resonant Converter," in IEEE Transactions on Power Electronics, vol. 30, no. 8, pp. 4437-4454, Aug. 2015.

[4] R.-L. Lin, "Piezoelectric transformer characterization and application of electronic ballast," Ph.D. dissertation, Virginia Tech, Blacksburg, VA, USA, 2001William D. George, Myron C. Selby, and Reuben Scolnik, "Electrical Characteristics of Quartz-Crystal Units and Their Measurement," Research Paper RP1774, Volume 38, March 1947

[5] E. M. Baker, W. Huang, D. Y. Chen, and F. C. Lee, "Radial mode piezoelectric transformer design for fluorescent lamp ballast applications," IEEE Trans. Power Electron., vol. 20, no. 5, pp. 1213–1220, Sep. 2005.

[6] E. L. Horsley, A. V. Carazo, N. Nguyen-Quang, M. P. Foster and D. A. Stone, "Analysis of Inductorless Zero-Voltage-Switching Piezoelectric Transformer-Based Converters," in IEEE Transactions on Power Electronics, vol. 27, no. 5, pp. 2471-2483, May 2012.

[7] A. Neukermans and R. Ramaswami, "MEMS technology for optical networking applications," in IEEE Communications Magazine, vol. 39, no. 1, pp. 62-69, Jan 2001.

[8] C. T. c. Nguyen, "MEMS technology for timing and frequency control," in IEEE Transactions on Ultrasonics, Ferroelectrics, and Frequency Control, vol. 54, no. 2, pp. 251-270, February 2007.

[9] C. Y. Wu, X. H. Qian, M. S. Cheng, Y. A. Liang and W. M. Chen, "A 13.56 MHz 40 mW CMOS High-Efficiency Inductive Link Power Supply Utilizing On-Chip Delay-Compensated Voltage Doubler Rectifier and Multiple LDOs for Implantable Medical Devices," in IEEE Journal of Solid-State Circuits, vol. 49, no. 11, pp. 2397-2407, Nov. 2014.

[10] Y. P. Lin and K. T. Tang, "An Inductive Power and Data Telemetry Subsystem With Fast Transient Low Dropout Regulator for Biomedical Implants," in IEEE Transactions on Biomedical Circuits and Systems, vol. 10, no. 2, pp. 435-444, April 2016.

[11] O. Al-Terkawi Hasib, M. Sawan and Y. Savaria, "A Low-Power Asynchronous Step-Down DC–DC Converter for Implantable Devices," in IEEE Transactions on Biomedical Circuits and Systems, vol. 5, no. 3, pp. 292-301, June 2011.

[12] M. D. Bellar, T. S. Wu, A. Tchamdjou, J. Mahdavi and M. Ehsani, "A review of soft-switched DC-AC converters," in IEEE Transactions on Industry Applications, vol. 34, no. 4, pp. 847-860, July-Aug. 1998.

[13] William D. George, Myron C. Selby, and Reuben Scolnik, "Electrical Characteristics of Quartz-Crystal Units and Their Measurement," Research Paper RP1774, Volume 38, March 1947.

[14] A. Arnau, T. Sogorb and Y. Jimenez, "A new method for continuous monitoring of series resonance frequency and simple determination of motional impedance parameters for loaded quartz-crystal resonators," in IEEE Transactions on Ultrasonics, Ferroelectrics, and Frequency Control, vol. 48, no. 2, pp. 617-623, March 2001.

[15] E. Moon, D. Blaauw and J. D. Phillips, "Subcutaneous Photovoltaic Infrared Energy Harvesting for Bio-implantable Devices," in IEEE Transactions on Electron Devices, vol. 64, no. 5, pp. 2432-2437, May 2017.

[16] M. Donohoe, S. Balasubramaniam, B. Jennings and J. M. Jornet, "Powering In-Body Nanosensors With Ultrasounds," in IEEE Transactions on Nanotechnology, vol. 15, no. 2, pp. 151-154, March 2016.

[17] M. Wahbah, M. Alhawari, B. Mohammad, H. Saleh and M. Ismail, "Characterization of Human Body-Based Thermal and Vibration Energy Harvesting for Wearable Devices," in IEEE Journal on Emerging and Selected Topics in Circuits and Systems, vol. 4, no. 3, pp. 354-363, Sept. 2014.

[18] A. M. Imtiaz and F. H. Khan, "Film bulk acoustic resonator (FBAR) based power converters: A new trend featuring EMI reduction and high power density," 2013 Twenty-Eighth Annual IEEE Applied Power Electronics Conference and Exposition (APEC), Long Beach, CA, USA, 2013, pp.1485-1491.

A Lumped Thermal Model Including Thermal Coupling Effects and Boundary Conditions for Capacitor Banks

Haoran Wang[1], *IEEE Student Member*, Qiusheng Wang[2] and Huai Wang[1], *IEEE Senior Member*

1. Center of Reliable Power Electronics (CORPE), Department of Energy Technology, Aalborg University
Pontoppidanstraede 101, Aalborg 9220, Denmark
2. Anyang Vibrator Co., Ltd (Group), Anyang, China
hao@et.aau.dk and hwa@et.aau.dk

Abstract—Capacitors are widely used in power electronic converters to buffer the pulsation power, filter the harmonics and support voltage for stable operation. For these applications where single capacitor can not fulfill the voltage rating or capacitance requirements, capacitor bank is always used as the energy buffer by connecting several capacitors in parallel for larger capacitance, or in series for higher voltage capability. The existing design considerations for the capacitor bank are in terms of voltage ripple, electromagnetic-interference, power loss, weight and volume. With more stringent constrains on volume for high power density applications, reliability as well as temperature of capacitor banks should be considered into the design phase. In order to estimate the temperature of capacitor bank with sufficient accuracy and computational efficiency, this paper proposes a lumped thermal model for a capacitor bank by taking into account the boundary conditions (e.g., ambient temperature, power loss for each cell) and the thermal coupling effect among the capacitor cells. Considering the variable boundary with different loading conditions in FEM simulation, fast thermal resistance extraction method is investigated to obtain the self-heating and coupling thermal impedance. Critical uneven temperature distribution among capacitors in a bank may be observed. A case study of an electrolytic capacitor bank is presented to verify the accuracy of the proposed method.

I. Introduction

Capacitors are widely used in power electronic converters to buffer the pulsation power, filter the harmonics and support voltage for stable operation [1]. In some applications, they are also used to provide sufficient energy during the hold-up time [2]. For these applications where single capacitor can not fulfill the voltage rating or capacitance requirements, capacitor bank is always used as the energy buffer by connecting several capacitors in parallel for larger capacitance, or in series for higher voltage rating [3].

The design considerations for a capacitor bank is dependent on the system requirements and user constraints. A discussion on how to make the choice of capacitor cell for a buck converter is given in [3]. Various constraints are considered: voltage ripple, differential-mode electromagnetic-interference level, power loss, weight and volume. With more stringent reliability constrains brought by automotive, aerospace and energy industries, the design of a capacitor bank encounters the following challenges [4]:

a) capacitor cells are one kind of the stand-out components in terms of failure rate in field operation of power electronic systems;

b) more stringent constrains on volume and thermal dissipation of capacitors as the trends for high power density and reliability capacitor banks;

c) cost reduction pressure from global competition dictates minimum design margin of capacitors without undue risk.

The efforts to overcoming the above challenges can be divided into three categories: a) analyze temperature and lifetime distribution in the design phase [5]; b) Optimize the capacitor bank design to achieve proper robustness margin and cost-effectiveness [6]; c) adopt novel longer lifetime capacitor technologies [7]. From the above three categories, it can be noted that thermal modeling of a capacitor bank is essential to the understanding and optimization of its reliability aspect performance. The methods to obtain the temperature distribution are experimental characterizations or FEM simulations, however, which are time consuming and and can not be used for the optimization procedure in the design phase. Recent years, analytical thermal models for single capacitor have been explored, which offer a faster way to estimate the hot-spot temperature of capacitors [5, 6, 8]. These models can be programmed as a script function, therefore, can be used for system-level parameter optimizations in the design phase. However, these models are not sufficient for a capacitor bank. Two further issues need to be overcome:

a) Thermal modeling based on individual capacitors may become invalid due to the possible uneven temperature distribution within a capacitor bank causing by thermal coupling and finite boundary conditions [9].

b) Thermal impedances, both self-heating one and thermal couple one, are the key components of the thermal model, are essential to capacitor bank thermal stress analysis. They are usually obtained by experimental characterization or simulation. In FEM simulation, the boundary can not be adaptively changed with the loading conditions, so thermal impedance measured from different loading conditions could be different, which introduce non-negligible deviation. How to extract the thermal impedance time efficiently is worth to investigation. Although infinite environmental boundary can eliminate the

The 2018 International Power Electronics Conference

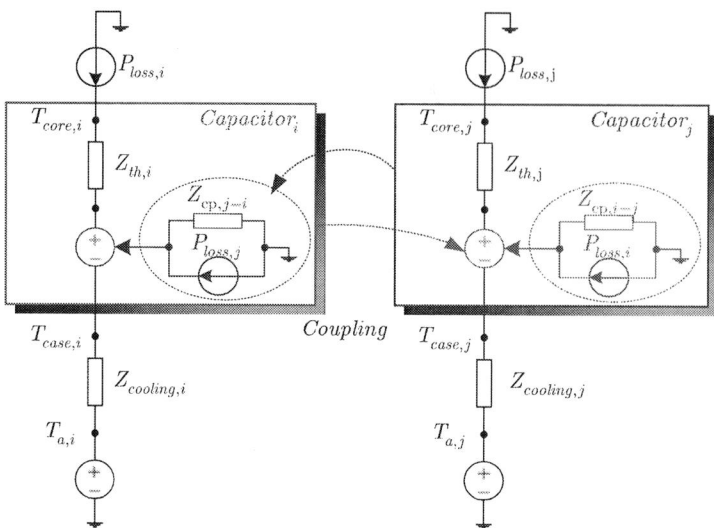

Fig. 1. The lumped thermal model of a capacitor bank considering the thermal coupling effect.

deviation, it implies heavy computational burden.

This paper aims to propose a lumped thermal model for a capacitor bank by taking into account the thermal coupling effects and boundary conditions. It can be applied to different field applications of power electronic converters to estimate the temperature distribution. Further more, a FEM based thermal impedance extraction method considering loading condition is studied to acquire the key parameters of the thermal model. Thermal impedance values due to both self-heating and thermal-coupling are characterized from the FEM simulations, while critical thermal distribution is obtained from circuit simulations. The rest of this paper is as follows: Section II introduces the proposed lumped thermal model; Section III studies a time-efficient thermal impedance extraction method from FEM simulations; Section IV demonstrates two case studies of a capacitor bank to verify the accuracy of the proposed thermal model and thermal impedance extraction method.

II. A LUMPED THERMAL NETWORK MODEL

A mathematical thermal model is built in this section to represent the temperature distribution of a capacitor bank, which estimates the self-heating as well as thermal coupling effects from other capacitor cells in the bank. Indeed, this model should make a superposition of the self-heating and thermal coupling effects. In the conventional capacitor thermal modeling, the self-heating effect is modeled as the conventional RC-lumped network for each individual capacitors in the bank. RC network starts from the hot-spot of each capacitor, end in the ambient, including the thermal impedance from hot-spot to case and from case to ambient as shown in Fig. 1. Beside the self-heating thermal impedance, thermal coupling effect defined as a temperature gradient is also considered in this lumped thermal model. To represent this effect, two subcircuits with blue and red dashed line, respectively, are added as shown in Fig. 1 is incorporated, which include a power loss

source from the neighboring capacitor and coupling thermal impedance. Based on this, a lumped thermal network is obtained as shown in Fig. 1 with two capacitors as an example. In Fig. 1, $P_{\mathrm{loss,i}}$ and $P_{\mathrm{loss,j}}$ are the power loss of the capacitor i and j, respectively. $T_{\mathrm{core,i}}$, $T_{\mathrm{case,i}}$, $T_{\mathrm{a,i}}$ are the hot-spot, case and ambient temperature of capacitor i, while $T_{\mathrm{core,j}}$, $T_{\mathrm{case,j}}$, $T_{\mathrm{a,j}}$ are for the capacitor j. The thermal branches are extracted for the critical monitoring points from hot-spot to case $Z_{\mathrm{th,i}}$ and $Z_{\mathrm{th,j}}$, case to ambient temperature $Z_{\mathrm{cooling,i}}$ and $Z_{\mathrm{cooling,j}}$. $Z_{\mathrm{cp,j-i}}$ and $Z_{\mathrm{cp,i-j}}$ are the two coupling thermal impedance between capacitor j to i.

Roughly speaking, the temperature raise across the capacitor is proportional to the power dissipation in steady state. The temperature response in steady state can be calculated applying thermal resistance between target point and the reference point. In order to study the thermal coupling effects among the capacitors and to find the temperature raise in critical points, more detailed thermal impedances are needed. This information is essential for the optimized design of capacitor banks. To consider the self-heating and thermal coupling effects, the lumped thermal model of the capacitor bank can be written in the matrix shown by (2). T_i with $i \in 1, ...m$ is the monitoring point temperature, $P_{\mathrm{loss,j}}$ with $j \in 1, ...n$ is the power losses on each capacitor, $T_{\mathrm{a,i}}$ with $i \in 1, ...m$ is the reference ambient temperature at the monitoring points, and Z_{mn} ($m \neq n$) is the coupling thermal impedance between the monitoring point and the reference point. In particular, Z_{mn} ($m = n$) is the self-heating thermal impedance.

III. THERMAL IMPEDANCE EXTRACTION FROM FEM SIMULATIONS

Thermal impedance in the thermal model of the capacitor bank in (2) can be extracted from the experiment or FEM simulation. Compared with experiment, the thermal impedance extraction from FEM simulation is faster and easier to be implemented. However, different from the ambient in experimen-

$$
\begin{bmatrix} T_1 \\ T_2 \\ \dots \\ T_m \end{bmatrix} = \begin{bmatrix} Z_{11} & Z_{12} & \dots & Z_{1n} \\ Z_{21} & Z_{22} & \dots & Z_{2n} \\ \dots & \dots & \dots & \dots \\ Z_{m1} & Z_{m2} & \dots & Z_{mn} \end{bmatrix} \begin{bmatrix} P_{\mathrm{loss},1} \\ P_{\mathrm{loss},2} \\ \dots \\ P_{\mathrm{loss},n} \end{bmatrix} + \begin{bmatrix} Z_{\mathrm{cooling},1} & 0 & \dots & 0 \\ 0 & Z_{\mathrm{cooling},2} & \dots & 0 \\ \dots & \dots & \dots & \dots \\ 0 & 0 & \dots & Z_{\mathrm{cooling},n} \end{bmatrix} \begin{bmatrix} P_{\mathrm{loss},1} \\ P_{\mathrm{loss},2} \\ \dots \\ P_{\mathrm{loss},n} \end{bmatrix} + \begin{bmatrix} T_{\mathrm{a},1} \\ T_{\mathrm{a},2} \\ \dots \\ T_{\mathrm{a},m} \end{bmatrix} \tag{2}
$$

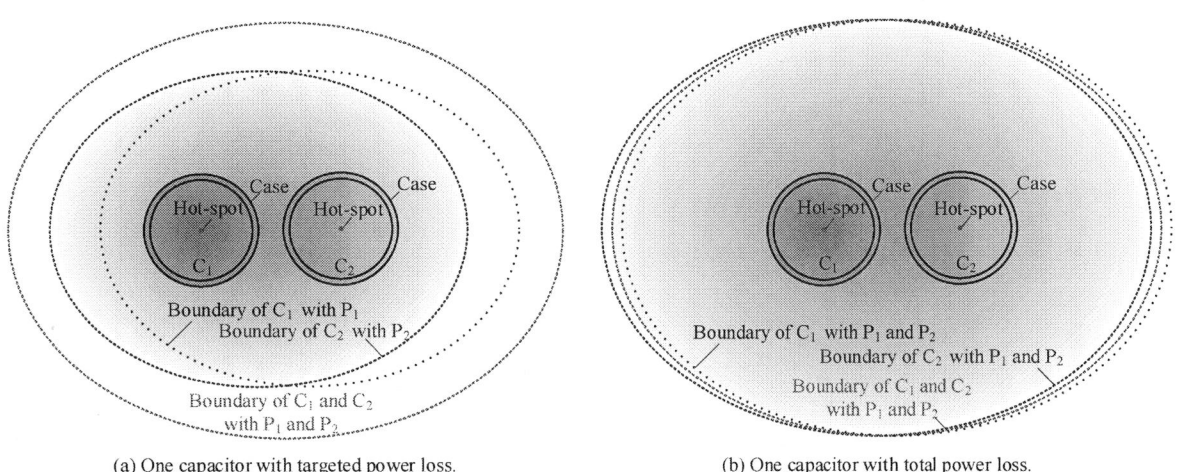

(a) One capacitor with targeted power loss.　　　　(b) One capacitor with total power loss.

Fig. 2. The boundary comparisons of two capacitors with different loading conditions.

tal measurement, the boundary range in simulation can not be infinite. For different loading conditions, the boundary ranges as shown in Fig. 2 will impact the accuracy of the thermal impedance extraction. Meanwhile, the thermal behavior of the air is changing with the temperature and loading condition. It indicates that for accurate thermal impedance extraction in a simulation environment, the thermal impedance should be extracted with targeted loading condition to minimize the boundary discrepancy.

Fig. 2 shows the diagram of the difference between boundary in impedance measurement process and in rated operating condition. From Fig. 2 (a), it can be seen that if C_1 and C_2 inject P_1 and P_2 individually, the boundary is the blue and black dashed line, while the boundary of the capacitor bank in rated operation condition with total power loss $P_1 + P_2$ is the red dashed line. The difference of the two boundaries will introduce thermal impedance discrepancy, further impact the temperature estimation. Compared with Fig. 2 (a), Fig. 2 (b) shows the negligible difference between the boundaries. Because the total power loss $P_1 + P_2$ is applied for C_1 and C_2 individually, which is the same with the normal operation, the boundary in the impedance extraction has small difference with that of the rated operation.

The method used in this paper to extract the thermal network is based on the extraction of thermal impedances from the FEM analysis by using a step response analysis. The RC networks can then be used in circuit simulators, such as PSpice and PLECS, to calculate temperatures. It accelerates the simulation time with acceptable accuracy compared with FEM analysis.

To extract the responses, a step power loss input is applied to each capacitor cell individually, and the temperature responses from the intended monitoring points are extracted. If only the self-heating of the monitoring point is intended to be calculated, step response analysis can be applied to extract the equivalent thermal RC network. If the thermal coupling effect is also considered, the thermal impedance for one node can be extracted based on superposition principle as the summation of a self-heating thermal impedance and the coupling thermal impedances from the other capacitor. The methodology to find the thermal response of the capacitor bank is detailed in the following:

a) The geometry and material information of the capacitor bank are imported to or drawn in a FEM simulation tool;

b) A step response analysis is performed for all capacitors, by applying the step power loss to an individual capacitor of interest, respectively,;

c) Identify the self-heating response of the individual capacitor and its thermal coupling effect to other capacitors;

d) The extracted temperature response is divided by the power loss of individual capacitor.

The thermal resistance and the thermal impedance are given by

$$
R_{\mathrm{th,a-b}} = \frac{T_{\mathrm{a}} - T_{\mathrm{b}}}{P_{\mathrm{loss,self}}} \tag{1}
$$

$$
Z_{\mathrm{th,a-b}}(t) = \frac{T_{\mathrm{a}}(t) - T_{\mathrm{b}}(t)}{P_{\mathrm{loss,self}}} \tag{2}
$$

where T_{a} and T_{b} are the temperatures in two adjacent points and $P_{\mathrm{loss,self}}$ is the power loss of the individual capacitor. Moreover, the coupling thermal impedance can be written as

$$
Z_{\mathrm{cp,a-b}}(t) = \frac{T_{\mathrm{a}}(t) - T_{\mathrm{b}}(t)}{P_{\mathrm{loss,cp}}} \tag{3}
$$

The 2018 International Power Electronics Conference

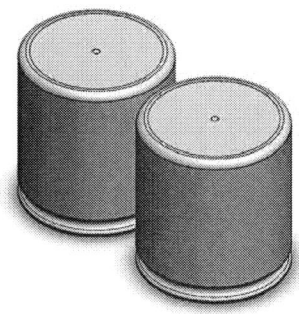

Fig. 3.   Geometry of a capacitor bank with two cells.

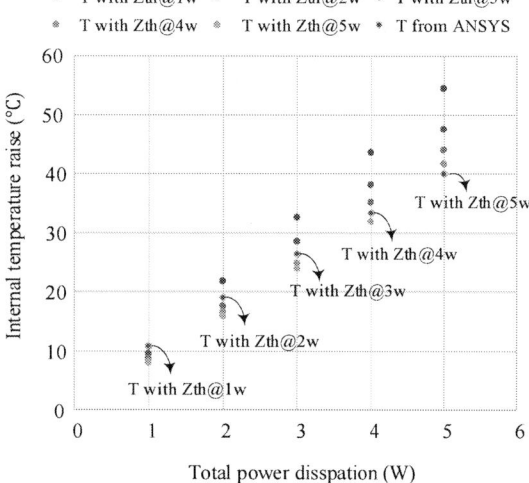

Fig. 4.   Comparison between the temperature raise estimated results with the thermal resistance measured from different power disspation and the temperature raise FEM simulation results.

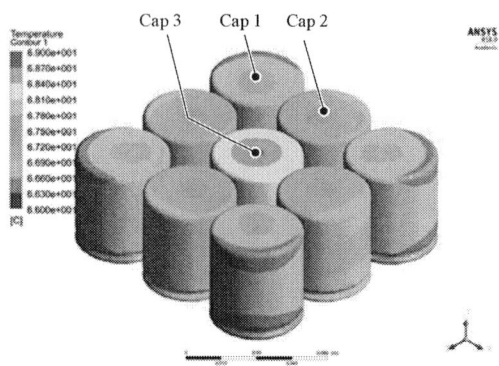

Fig. 5.   Temperature distribution of a capacitor bank with nine cells by FEM simulation.

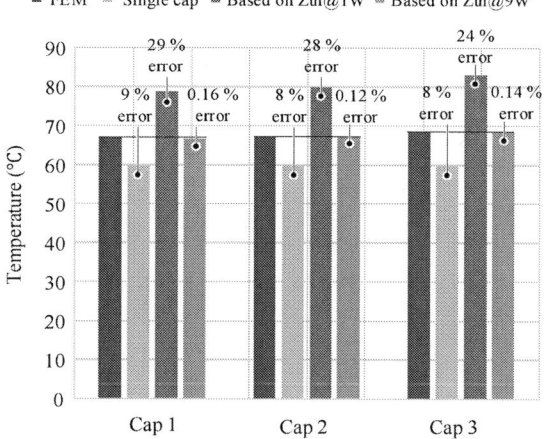

Fig. 6.   Comparison of temperature in FEM simulation, temperature estimated from single capacitor, temperature estimated from thermal model with thermal impedance values extracted at different power loss levels.

where $P_{\mathrm{loss,cp}}$ is the power loss of the neighbor capacitor of interest.

IV. CASE STUDY

In this section, two case studies are presented to verify the accuracy of the thermal model and the thermal resistance extraction method. First case is two capacitor cells in a bank. The voltage rating is 450 V. Capacitance of this bank is 1120 μF (560 μF*2) and the total power loss is 2 W. For each cell, the diameter is 40 mm and height is 45 mm. The distance between the capacitor cells is 2 mm. As shown in Fig. 3, the geometry of the capacitor is drawn ifor FEM simulation in Icepack

In FEM simulation, set the power loss as 1 W to 5 W for the capacitor cell i and measure the internal temperature raise of capacitor cell i and j. Then the thermal resistance can be calculated based on (1). Based on the thermal resistance calculated from different power loss, temperature raise can be acquired which are shown in Fig. 4. It can be seen that the simulation results are compatible with the estimated results

which are based on the thermal resistance extracted from the total power loss injection to one cell. As an example where the total power loss is 2 W, the temperature estimation should be based on the thermal resistance extracted from 2 W injection to one cell. From this case study, it can be seen that the temperature estimation of the capacitor bank for different loading conditions should be calculated carefully, where the thermal resistance matrix should be obtained from a similar loading condition.

The second case study is based a capacitor bank with nine electrolytic capacitors connected in parallel. The voltage rating is 450 V. Capacitance of this bank is 5040 μF (560 μF*9). The ambient temperature in simulation is 45 $^\circ C$. For each cell, the diameter is 40 mm and height is 45 mm. The distance between the capacitor cells is 2 mm. The current is shared among nine capacitors, and the power disspation for each cell is 1 W.

The simulation results are shown in Fig. 5. For single capacitor, the temperature raise with 1 W power loss is 15 $^\circ C$. However, from the simulation results, it can be seen that the thermal coupling is significant, which leads to uneven thermal

2424

distribution within the capacitor bank.

The layout of the capacitor bank leads to three different hot-spot temperature of Cap 1, Cap 2 and Cap 3 as shown in Fig. 5 and Fig. 6, because different levels of different thermal coupling. The first bars of each capacitor temperature in Fig. 6 is the FEM thermal simulation results. The second bar shows the estimated temperature without consideration of the thermal coupling effect, which are around 10 % lower the simulation results. The estimated hot-spot temperature based on different thermal impedance extraction method is also shown in Fig. 6. The results based on the thermal impedance extracted with total power disspation 9 W presents smaller estimated error from 0.14 % to 0.16 %, while the results based on 1 W power injection present a larger error from 24 % to 29 %. The temperature difference between the highest and the lowest is 2.5 $°C$, which could result in shorter lifetime. From this case study, it can be seen that the thermal resistance should be extracted at specified loading condition to guarantee the accuracy of the temperature estimation.

V. Conclusions

This paper proposes a lumped thermal model for a capacitor bank, considering the thermal coupling effects and boundary conditions. It enables a time-efficient estimation of the temperature distribution within the capacitor bank with insignificant errors benchmarked with FEM simulation results. Moreover, the extracted thermal impedance model for the capacitor bank can be programmed as a function for system optimization design. The case study outcomes suggest that thermal impedances extracted at the targeted loading conditions could minimize the boundary discrepancy as well as the error of the thermal stress analysis.

References

[1] R. W. Erickson and D. Maksimovic, "Fundamentals of power electronics," MA, Norwell: Kluwer, 2001.

[2] A. Lazaro, A. Barrado, J. Pleite, R. Vazquez, J. Vazquez, and E. Olias, "Size and cost reduction of the storage capacitor in ac/dc converters under hold-up time requirements," in *Proc. IEEE PESC*, vol. 4, Jun. 2003, pp. 1959–1964 vol.4.

[3] P. Pelletier, J. M. Guichon, J. L. Schanen, and D. Frey, "Optimization of a dc capacitor tank," *IEEE Trans. Ind. App.*, vol. 45, no. 2, pp. 880–886, Mar. 2009.

[4] H. Wang and F. Blaabjerg, "Reliability of capacitors for dc-link applications in power electronic converters- an overview," *IEEE Trans. Ind. Appl.*, vol. 50, no. 5, pp. 3569–3578, Sep. 2014.

[5] Y. Yang, K. Ma, H. Wang, and F. Blaabjerg, "Instantaneous thermal modeling of the dc-link capacitor in photovoltaic systems," in *Proc. IEEE APEC*, Mar. 2015, pp. 2733–2739.

[6] M. L. Gasperi and N. Gollhardt, "Heat transfer model for capacitor banks," in *Proc. IEEE Industry Applications Conference*, vol. 2, Oct. 1998, pp. 1199–1204.

[7] Nippon-Chemi-con, "Aluminum capacitors group chart," Available: https://www.chemi-con.co.jp/e/catalog/pdf/al-e/al-sepa-e/001-guide/al-groupchart-e-171001.pdf, 2017.

[8] J. Rajmond and P. Dan, "Thermal modeling of through hole capacitors," in *Proc. IEEE SIITME*, Oct. 2012, pp. 227–232.

[9] Z. Wang, F. Yan, M. Xu, Z. Wang, X. Wang, and Z. Xu, "Influence of external factors on self-healing capacitor temperature field distribution and its validation," *IEEE Trans. Plasma Sci.*, vol. 45, no. 7, pp. 1680–1688, Jul. 2017.

The 2018 International Power Electronics Conference

Hysteresis Modeling of Magnetic Devices based on Reluctance Network Analysis

Yoshiki Hane[1*] and Kenji Nakamura[1]
1 Tohoku University, Graduate School of Engineering, Sendai, Japan
*E-mail: yoshiki.hane.t5@dc.tohoku.ac.jp

Abstract— **In research and development of electrical machines, establishment of a method for quantitatively calculating iron loss including magnetic hysteresis behavior is required. In a previous paper, a novel magnetic circuit model incorporating a play model, which is one of the phenomenological models of magnetic hysteresis, was proposed. It was clear that the proposed model can calculate the hysteresis loop of the magnetic reactor with high speed and accuracy. This paper describes that the play model is applied to reluctance network analysis (RNA), in order to estimate the iron loss of electric machines with more complex shape such as electric motors.**

Keywords— *Landau-Lifshitz-Gilbert (LLG) equation, magnetic circuit model, play model, reluctance network analysis (RNA)*

I. INTRODUCTION

In recent years, development of high-efficiency electrical equipment is required from the viewpoint of global environmental issues and energy saving. It is necessary to establish a method for quantitatively calculating iron loss including magnetic hysteresis behavior, in order to further improve the efficiency of the electrical equipment. The magnetic hysteresis modeling is roughly divided into a physical model and a phenomenological model. A micro magnetic simulation using the Landau-Lifshitz-Gilbert (LLG) equation, which is one of the physical model, can simulate the hysteresis behavior and the magnetization distribution inside magnetic material. Hence, it can express micro magnetic phenomena in detail, including domain wall motion, magnetic anisotropy, and interaction between magnetizations. However, since the analytical model is large and complicated, it is difficult to apply it to the analysis of electrical equipment. To overcome the above problem, the reference [1] presented a method for expressing the magnetic hysteresis of a silicon steel sheet by providing several assumptions in the micro magnetic simulation such as not directly considering domain wall motion, *etc*.

In a previous paper, a novel magnetic circuit model incorporating the above method [2], [3] was proposed. In the proposed model, dc hysteresis is expressed by the LLG equation in the reference [1], and eddy current loss and anomalous eddy current loss are represented by the elements of the magnetic circuit. It was clear that the proposed model can calculate iron loss including minor

loops of several kinds of magnetic materials with high accuracy [4]. However, the calculation time tends to be longer because the LLG equation requires repeat convergence calculation.

To solve the above problem, we have focused on a play model [5], which is one of the phenomenological models. In general, to derive the play model, a large number of dc hysteresis loops with different maximum magnetic flux densities are measured, which was difficulty in practical use, but we have proposed a method for calculating dc hysteresis loops by the above-mentioned LLG equation. As a result, only one or two measured dc hysteresis loops, which are used to determine the parameters of the LLG equation, are required to derive the play model. In addition, we combined the above play model with the magnetic circuit model, and calculated iron loss including minor loops with high accuracy and speed [6]. However, this method has been applied only for the objects with simple shapes such as a ring core.

The authors proposed a reluctance network analysis (RNA), which expresses an analysis object by one reluctance network. All the reluctances can be determined by *B-H* curve of the material and dimensions [7]. The RNA has some advantages such as simple model, fast calculation, and easy coupling with external electric circuits and motion equation. The RNA has been applied to the calculation of characteristics of various electric machines including transformers and motors. However, a method for expressing magnetic hysteresis is not established for RNA.

This paper describes that the play model is applied to reluctance network analysis (RNA), in order to estimate the iron loss of electric machines with more complex shape such as electric motors.

II. PLAY MODEL

In the play model, as shown in Fig. 1, an arbitrary hysteresis loop can be expressed by multiplying play hysterons with different widths by shape functions. Although a large number of measured dc hysteresis with different maximum flux densities is required to derive the play model in general, the proposed method requires only one or two measured dc hysteresis because the LLG equation is used to calculate the dc hysteresis loops.

Fig. 2 shows the measured dc hysteresis of a non-oriented silicon steel with a thickness of 0.2 mm and the calculation results obtained from the LLG equation. In the experiment, the hysteresis loop at frequency $f = 5$ Hz was assumed as a dc hysteresis loop. It is clear that the calculated values by the LLG equation and the measured values are in good agreement in various maximum flux densities.

Fig. 3 shows the hysteresis loops calculated by the LLG equation. As shown in this figure, a large number of hysteresis loops which is used to derive a play model can be obtained without experiments.

Fig. 4 shows the previously proposed magnetic circuit model incorporating the play model [6]. In the model, dc hysteresis is represented by the play model, and eddy current loss and anomalous eddy current loss are represented by the elements of the magnetic circuit.

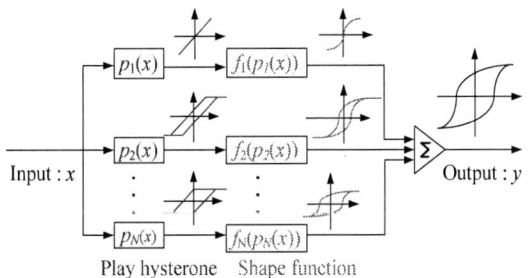

Play hysterone   Shape function

Fig. 1 Schematic diagram of a play model.

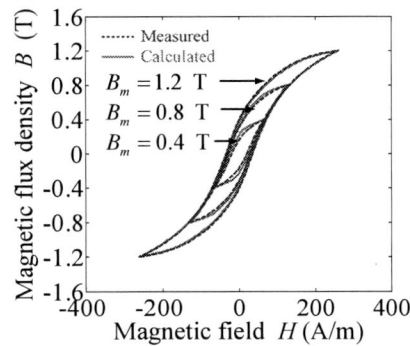

Fig. 2 Measured and calculated dc hysteresis of non-oriented silicon with a thickness of 0.2 mm.

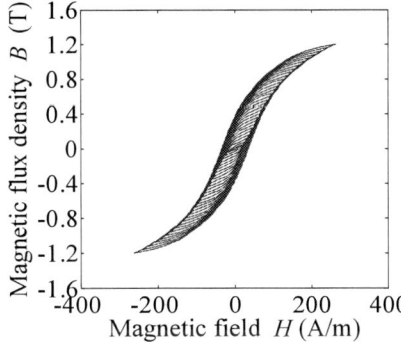

Fig. 3 Calculated dc hysteresis loops of the non-oriented silicon with a thickness of 0.2 mm from Bm = 0.4 T to 1.2 T at intervals of 0.04 T.

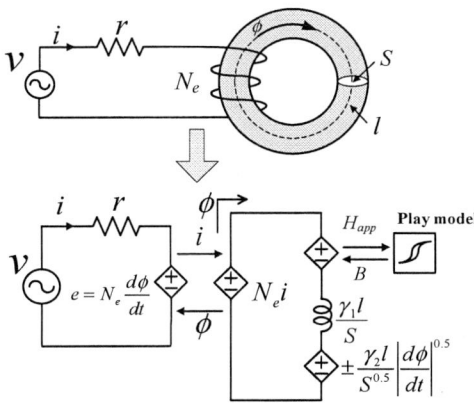

Fig. 4 Magnetic circuit model incorporating play model.

III. RNA MODEL INCORPORATING PLAY MODEL

In this chapter, first, a method for driving the conventional two-dimensional RNA model [7].

As shown in Fig. 5, the magnetic core is divided into multiple elements. In addition, an air space surrounding the magnetic core is included in the analytical region and divided so that leakage flux from the magnetic core can be taken into consideration. Each divided element is replaced with four reluctances as shown in the figure. Among them, reluctances in the rolling direction are needed to be determined in consideration of the nonlinear magnetic characteristics. In this paper, the magnetic nonlinearity is given by

$$H = \alpha_1 B + \alpha_m B^m \tag{1}$$

where α_1 and α_m are coefficients. The order m is determined by the strength of the nonlinearity of the B-H curve.

From the equation (1), the relationship between the magnetomotive force (MMF) f and the magnetic flux ϕ in each reluctance can be expressed by the following equation using the average cross sectional area S and the average magnetic path length l of each element.

$$f = Hl$$
$$= \frac{\alpha_1 l}{S}\phi + \frac{\alpha_m l}{S^m}\phi^m$$
$$= \left(\frac{\alpha_1 l}{S}\phi + \frac{\alpha_m l}{S^m}\phi^{m-1} \right) \tag{2}$$

The parentheses in the equation (2) represents nonlinear reluctance.

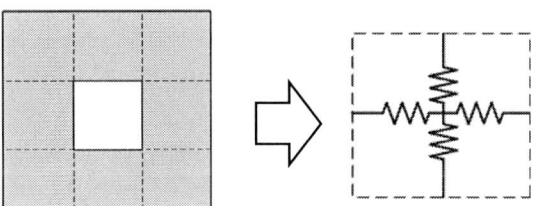

Fig. 5 Unit magnetic circuit.

On the other hand, reluctances perpendicular to the rolling direction are needed to be determined in consideration of the flux passing through the nonmagnetic layer between steel sheets. In general, the reluctance R_m is given by the following equation using the magnetic permeability μ.

The laminated steel sheet is constructed by Si steel sheets with magnetic permeability μ_s and nonmagnetic layer with vacuum permeability μ_0 at a ratio of $d_f : (1-d_f)$, where the space factor of the magnetic core is d_f. Hence, the effective magnetic permeability μ' is given by the following equation.

$$\frac{1}{\mu'} = \frac{d_f}{\mu_s} + \frac{1-d_f}{\mu_0} \tag{4}$$

In the equation (4), the flux flowing perpendicular to the steel sheet is small and magnetic saturation does not occur, so the permeability μ_s is sufficiently larger than the vacuum permeability μ_0. Thus, the equation (4) can be approximated as follows.

$$\frac{1}{\mu'} \cong \frac{1-d_f}{\mu_0} \tag{5}$$

Therefore, the reluctance perpendicular to the steel sheet is given by

$$R_{ml} \cong \frac{(1-d_f)l}{\mu_0 S} \tag{6}$$

Reluctances in an air space surrounding the magnetic core are simply given by the following equation using the dimensions of the divided elements and the vacuum permeability μ_0.

$$R_{ma} \cong \frac{l}{\mu_0 S} \tag{7}$$

Fig. 6 shows an example of a two-dimensional RNA model of cut core. As shown in the figure, MMF due to the winding current is placed in the portion where the winding is applied. As shown in the figure, the conventional RNA model consists of only reluctances and MMF. Hence, the magnetic hysteresis behavior is not taken into consideration. Therefore, in this paper, the play model described in Chapter 2 is applied to the RNA model.

Fig. 7 shows the shape and dimensions of the cut core used for measurement and simulation. The excitation frequency in the experiment is 100 Hz and the maximum flux density is 1.2 T. In addition, the parameters γ_1 and γ_2 of the circuit elements of the RNA model can be determined by approximating the core loss curves of the material, where $\gamma_1 = 0.0072$ and $\gamma_2 = 0.2830$, respectively.

Fig. 8 shows the proposed RNA model incorporating the play model. In the conventional RNA model, the nonlinear reluctance was derived based on the equation (2), but in the proposed model, it is given as the play model and the circuit elements representing eddy current loss and anomalous eddy current loss. On the other hand, the reluctances of lamination direction of the cut core and the surrounding space are given by the equations (6) and (7), respectively.

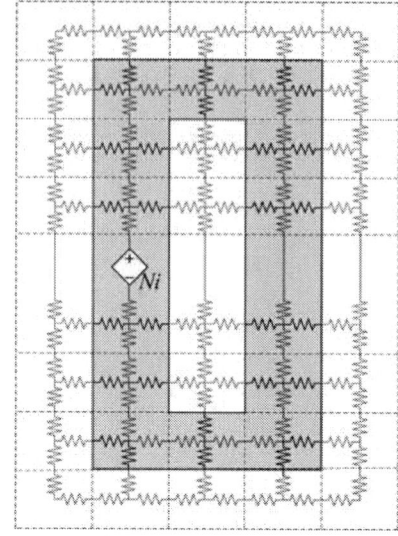

Fig. 6 An example of two-dimensional RNA model.

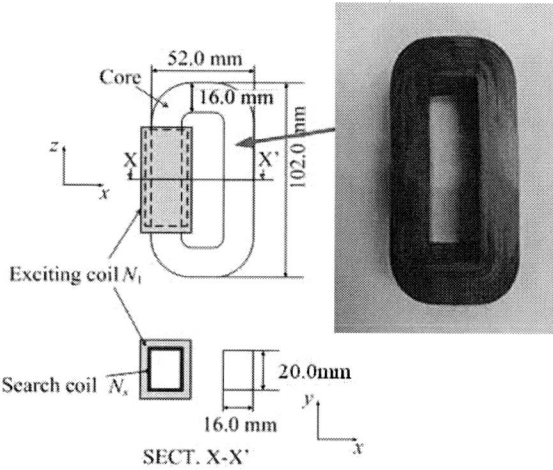

Fig. 7 Shape and dimensions of the cut-core used in the experiment.

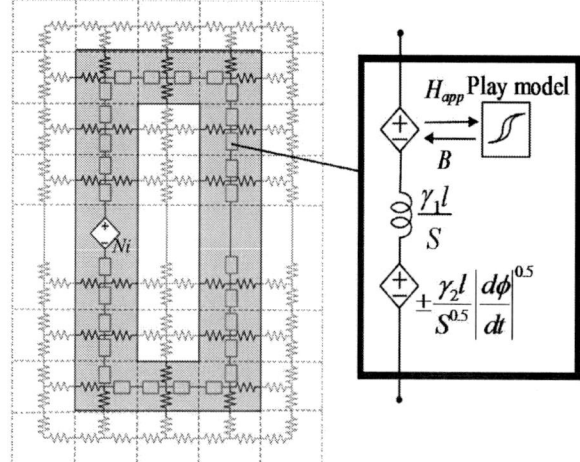

Fig. 8 RNA model incorporating the play model.

Using the proposed RNA model, exciting current and hysteresis loop are calculated when the cut core is excited by PWM voltage shown in Fig. 9. Fig. 10 shows calculated and measured waveforms of exciting current. From the figure, it reveals that calculated and measured current are in good agreement even if the PWM voltage is applied to the cut core.

Fig. 11 indicates calculated hysteresis loop in a certain divided element of the RNA model. As shown in the figure, using the proposed model, it is understood that magnetic hysteresis inside the iron core, which is difficult to measure, can be calculated including minor loops.

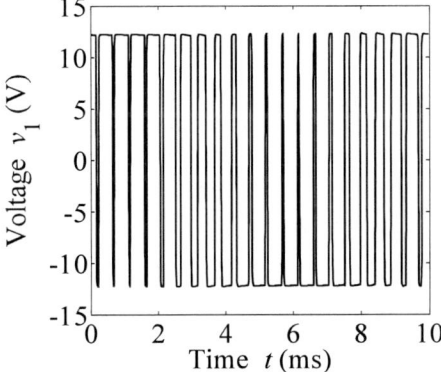

Fig. 9 Exciting voltage waveform.

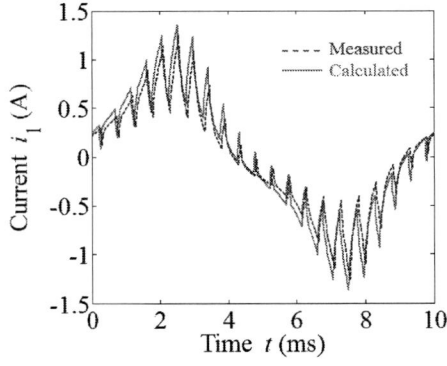

Fig. 10 Measured and calculated waveforms of the exciting current.

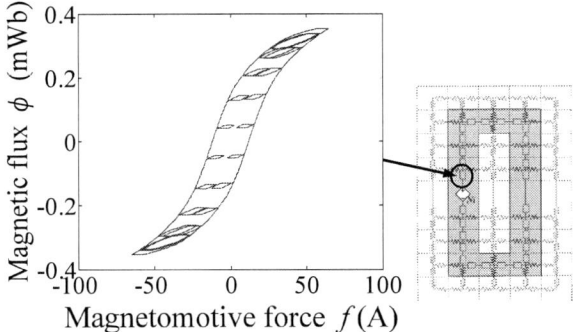

Fig. 11 Calculated hysteresis loop in a certain divided element of the RNA model.

IV. CONCLUSION

This paper presented a novel RNA model incorporating the play model, in order to estimate the iron loss of electrical machines including hysteresis behavior with more complex shape such as electric motors. First, we conducted a fundamental examination using a cut core of non-oriented silicon steel. Next, the proposed method is applied to the analysis of surface permanent magnet (SPM) motor.

As a result, it was clear that the proposed RNA model can calculate the magnetic hysteresis and the iron loss with high accuracy. In addition, the magnetic hysteresis inside the iron core, which is difficult to measure, can be drawn.

Next, we plan to verify the validity of the calculated iron loss and local hysteresis loop, and apply the proposed method to electrical machines other than the PM motor, such as the switched reluctance motor.

REFERENCES

[1] A. Furuya, J. Fujisaki, Y. Uehara, K. Shimizu, H. Oshima, Y. Murakami, and N. Takahashi, "Iron Loss Analysis of the Electrical Steel Sheet under the High Frequency Excitation", *The Papers of Joint Technical Meeting on "Magnetics", IEEJ*, SA-13-6, RM-13-6 (2013).

[2] H. Tanaka, K. Nakamura, and O. Ichinokura, *IEEJ Trans.FM*, Vol. 134, No. 4, pp. 243-249 (2014).

[3] H. Tanaka, K. Nakamura, and O. Ichinokura, "Calculation of Iron Loss in Soft Ferromagnetic Materials using Magnetic Circuit Model Taking Magnetic Hysteresis into Consideration", *Journal of the Magnetics Society of Japan*, Vol. 39, No. 2, pp. 65-70 (2015).

[4] H. Tanaka, K. Nakamura, and O. Ichinokura, "Dynamic Analysis of Amorphous Transformer in Switching Power Converters Based on Magnetic Circuit Method with LLG Equation", *MMM-Intermag 2016*, FJ-08 (2016).

[5] S. Bobbio, G. Miano, C. Serpico, and C. Visone, "Models of Magnetic Hysteresis Based on Play and Stop Hysteresis", *IEEE Trans. Magn.*, Vol. 33, No. 6, pp. 4417-4426 (1997).

[6] H. Tanaka, K. Nakamura, and O. Ichinokura, *The Papers of Joint Technical Meeting on "Magnetics", IEEJ*, MAG-16-141 (2016).

[7] K. Nakamura and O. Ichinokura, "Reluctance Network Based Dynamic Analysis in Power Magnetics", *IEEJ Trans.FM*, Vol.128, No.8, pp. 506-510 (2008).

Optimal Sizing and Placement of Solar Powered Charging Station under EV loads Penetration using Artificial Bee Colony Technique

Yuttana Kongjeen[1], Kulsomsup Yenchamchalit[2] and Krischonme Bhumkittipich[1*]

[1]Department of Electrical Engineering, Rajamangala University of Technology Thanyaburi, Pathumthani, Thailand
[2]Faculty of Industrial Technology,Thepsatri Rajabhat University, Lopburi, Thailand
*E-mail: krischonme.b@en.rmutt.ac.th

Abstract–This paper has proposed the optimization technique for finding the optimal sizing and placement of solar powered electric vehicles (EVs) charging station under EVs load using artificial bee colony (ABC), when the photovoltaic (PV) system was installed on roof of each charging station. The artificial bee colony algorithm was selected to solve the optimal condition under the proposed EV modellings on voltage dependent characteristics and PV placement modelling. On the hand, the IEEE 33 bus test system was used to express the load voltages deviation (LVD), under the minimum power loss condition as an objective function. The simulation method was divided into two conditions. With the first condition, optimal values of EV, LVD, and real power loss without PV installation had to be found. The second condition considers the first condition including optimal PV installation. The simulation results showed that, the optimal sizing of EVs, LVD, and total real power loss under the first condition were 50 kW, 0.155, and 241.547 kW, respectively. On the second condition, it was found that the optimal sizing and position of PV which was installed, measured at 250.758 kW on Bus No.15. The proposed system can reduce the LVD, and total real power loss of about 0.021 and 218.142 kW, respectively. Therefore, the study of EVs penetration with optimal PV placement on the power distribution system could be improved with voltage stability in term of voltage profiles.

Index-Terms–artificial bee colony, electric vehicle loads, load voltage deviation, photovoltaic, voltage stability

I. Introduction

Nowadays, the energy consumption of the electrical component has been increasing in the electrical power system and emerging load, because of the development of nanotechnology which consist materials that use high-powered technology of electronics. From this outcome of improved technology, it had caused an increase in the need for high quality components as a major requirement and number of energy sources [1]. The most significant load in the near future is called the Electric Vehicle loads (EVs); which will be connected into the power grids and will transfer power energy more efficiently from the grids to vehicles (G2V) by storing energy in battery packs. Meanwhile, the large scale of EVs penetration had been affected in the peak demand and high power loss; in regards to the stability of the electrical power system. To counter this issue, the energy management on power supply and demand side will be needed to maintain to optimal condition.

The impact of EVs in the power system, showed in terms of voltage stability, and loading margin. When EVs penetration had been increased, it could be described in [1]. The modelling of EVs was presented on the fast charging station; which had shown an extremely high impact on the power system. When oscillatory stability was compared to other load types the following factors must be taken into account: power constant (P), current constant load (I), and impedance constant (Z) [2].

Photovoltaic (PV) system is one type of distribution generator that forms renewable energy source which is directly connected to the distribution system of the power network. However, inappropriate placement may cause high system loss, and operation capital cost. Therefore, optimal PV placement can improve the power network performance; in terms of voltage profiles, low system loss, and an increase to power system stability. In order for there to be a more stable system put in place, the verity of the optimization technique is to solve the optimal DGs placement by using Artificial Bee Colony algorithm (ABC) [3-4]. The aim of the optimal EVs placement and PV placement was to improve the sizing and location under minimization of low voltage deviation and low power loss. This study was solved under the assumption of the following factors: EVs has a voltage dependent load, and PV has a voltage setting value of 1 p.u.

The remainder of the study was written as follows: Section II introduces the ABC algorithm and brief in importance methodology. The electric vehicle loads for power flow analysis showed the following: the static load models, forward-backward sweep, load voltage deviation, and total real power loss on electrical power system were presented in Section III. Section IV presents the methodology. The simulation results were presented in Section V. Finally, the conclusion and discussion were given in Section VI.

II. Artificial Bee Colony optimization Technique

The ABC method was one of the best optimization tools, which proposed in [4]; based on the intelligent behavior of the honey bee swarm. The ABC algorithm had provided a population based search procedure; which used individuals called foods position that had been modified by the artificial bees, in respect to time and discovery of food source location. In ABC algorithm there were three groups of bees: worker bees, onlookers and scouts. Consequently, when each natural/artificial bee was separated, the number of food sources around the hive became equal to the number of employed bees in the colony. The employed bees and the onlookers flied around in a multidimensional search space area; by using knowledge that have learned and had shared the information of these food sources, and adjust their position with other bees. Therefore, the nectar amount from the food sources of new position was of high quality (fitness) than that of the previous one in their memory. Thus, the ABC process had combined the local search method and the global search method that had been carried out by employed and onlooker bees and scouts, respectively [4].

Therefore, food source was selected based on the probability value P_i associated with that of an onlooker bee as (1) following .

$$P_i = \frac{fit_i}{\sum_{i=1}^{SN} fit_i} \quad (1)$$

where fit_i is the fitness value of solution, i and SN is the number of food sources that had been defined from the number of employed bees or onlooker bees, respectively.

Consequently, to select a coordinate food position from produce as $\mathbf{V}_i = \left[v_{i,1}, v_{i,1} ..., v_{i,D} \right]$ and from old one $\mathbf{X}_i = \left[x_{i,1}, x_{i,1} ..., x_{i,D} \right]$ in memory is listed in the following formula below:

$$v_{i,j} = x_{i,j} + \phi_{i,j} \left(x_{i,j} - x_{k,j} \right) \quad (2)$$

where $k \in \{1, 2, ..., SN\}$ and $j \in \{1, 2, ..., D\}$ are randomly chosen indexes; k had to be different from i; and D was the number of variables (problem dimensions); $\phi_{i,j}$ is a random number in the range limits. From each candidate source, position in production and evaluation by the artificial bee was compared with the old to select a new food source; that was objective to equal or better quality than the old source and replace the old location with the new. In order for food position to not be improved from the limit value on food sources; it's assumed that it had been abandoned and the abandoned source was x_i and $j \in \{1, 2, 3 ..., D\}$. From the formula above, the scout had discovered a new food source to be replaced with as x_i following.

$$x_{i,j} = x_{\min, j} + \text{rand}(0,1)\left(x_{\max, j} - x_{\min, j} \right) \quad (3)$$

III. ELECTRIC VEHICLE LOADS (EVs) FOR POWER FLOW ANALYSIS

A. Electric Vehicle Loads Modelling

Recently, the EVs modelling were proposed in [2] by representing voltage dependent characteristics of the fast charging station that consist of three parts: AC to DC converter, Buck converter and battery, respectively. Therefore, EVs could be represented as (4) and (5) following.

$$P_{EV} = P_0 \left(b + a \left(\frac{V}{V_0} \right)^{\alpha} \right) \quad (4)$$

$$Q_{EV} = P \times \tan(\theta) \quad (5)$$

Where b is the power constant equal to 0.93, a voltage dependent is equal to 0.07, α is an exponential indices of load which is equal to -3.107, θ is the power factor of the connected load equal to 0.97, P_0 that referred to the real power consumption of load at voltage V_0. Therefore, the real and reactive powers of EV load were given by P and Q, respectively.

B. A backward-forward sweep [5]

The radial distribution system was used to analyze the power flow analysis, under the rearrangement of each branch of feed to each bus by deviding in layer from root node or source node to the last node. Using the backward - forward sweep methodology(BFS) from proposed to solve the problem, had returned the minimized value of fitness function in the selection process of ABC algorithms. Therefore, the main topic of a backward – forward sweep could be described as the following.

- A forward sweep is a voltage drop calculation with branch current. The purpose of forward propagation is to calculate the voltages at each node by starting from the feeder source node to the last node. The feeder substation voltage was set at its rated value. During the forward sweep compute process, the power in each branch was held at a constant value which had been obtained in backward propagation.
- A backward sweep is a current or power flow solution with possible voltage updates. The power flow started from the branches in the last layer and moved towards the branches to the first layer that were connected to the root node. The updated effective power flowed in each branch and had been obtained in the backward propagating computation. Therefore, the node voltage of the previous iteration were considered. In summary, this means voltage values were obtained in the forward propagation and were held at a constant during the backward propagation. Therefore, the power flows were constantly updated in each branch. The current or power flows were estimated along the feeder using the backward path.

C. Load voltage deviation (LVD)

The LVD used to solve the bus voltage deviation, was affected from the load increase into the electrical power system. The LVD needed to minimize the value of the load voltage bus and could be described in (6) [1].

$$LVD = \sum_{k}^{n} \left(\frac{V_k^{ref} - V_k}{V_k^{ref}} \right)^2 \qquad (6)$$

Generally, the voltage references V_k^{ref} were considerate to a set amount of 1 p.u., and effort to manage the load and had improved the electrical power system nearly or close to zero.Meanwhile, the voltage each bus V_k were delivered from compute of the porposed method.

D. Total active power loss

The total power loss(P_{Loss}) in an electrical power system was affected from the electrical network configuration and had energized power to each load. Consequencely, the problem needed to analyze and be prepared in the optimal condition for maximizing power transmission to the customer or the consumer load in a suitable condition. Therefore, the active power loss in the electrical power system was calculated by using (7) as following [6].

$$P_{Loss} = \sum_{i=1}^{N_L} g_{i,j} \left\{ V_i^2 + V_j^2 - 2V_i V_j Cos \left(\delta_i - \delta_j \right) \right\} \qquad (7)$$

Where i and j are indices from bus to bus, g was conductance of transmission, NL is total number of the feeder, V is the voltage magnitude of bus i or bus j, and δ is the angle of bus i or bus j, respectively.

IV. METHODOLOGY

In order to perform the study, the IEEE33 bus radial distribution systemwas considered; to solve the optimal site and sizing of EVs and PV. The methodology had modified the backward – forward sweep algorithm and had applied the ABC algorithm in the conditions which were set and proposed. Fitness function used from (8) had subjected to minimize the LVD of the power system.Meanwhile, the boundary of constraints were defined on voltage in the limits as (9) and single possible PV position to place in the electrical power system as (10) could be described as written in the following formula below.

$$\text{Min } f = \sum_{i=1}^{N} LVD_i \quad i \in N \qquad (8)$$

Constrained

$$V_{min} \le V_i \le V_{max} , i \in N \qquad (9)$$

$$2 \le PV_i \le N, i \in N \qquad (10)$$

where i and N are indices of bus number, and the total amount of buses, respectively. Meanwhile, V_i , V_{max} and V_{max} are voltage buses from i indices, and had voltage limits on the boundary max limit equal to 1.05 and the minimum limit equal to 0.95, respectively.

The radial IEEE33 bus distribution network was used for simulation studies and had been selected to analyze the effects of EVs penetration and PV placement. This system was a 12.66kV, which had 33 buses and 3 laterals

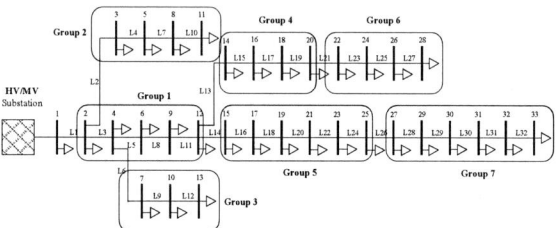

Fig. 1. IEEE33 bus test system and EVs group.

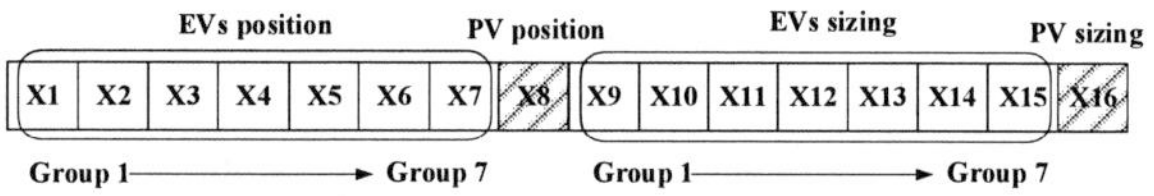

EVs position							PV position		EVs sizing						PV sizing
X1	X2	X3	X4	X5	X6	X7	X8	X9	X10	X11	X12	X13	X14	X15	X16

Group 1 ———————→ Group 7 Group 1 ———————→ Group 7

Fig. 2. The variable of EVs and PV to find optimal site and sizing into the electrical power system.

TABLE I
POSSIBLE POSITION AND SIZING OF EV LOADS (1-7) AND PV

Item/Group	Possible Position (Bus No.)	Sizing limit(kW)	
		Min	Max
1	2,4,6,9,12	50	500
2	3,5,8,11	50	500
3	7,10,13	50	500
4	14,16,18,20	50	500
5	22,24,26,28	50	500
6	15,17,19,21,23,25	50	500
7	27,29,30,31,32,33	50	500
PV	2-33	100	2000

as shown in Fig.1. The IEEE33 bus radial distribution system consists of 33 load points with a total of 3.72 MW and 2.3 MV loads. Additionally, the active power loss and reactive power loss had a total of 211.1180 kW and 143.205 kV which measured at 1 p.u. of the voltage reference, respectively [7]. In order for everything to work efficiently, it was assumed that the location or site of EVs placement on seven groups; only had one possible station to be in. Meanwhile, the PV location could be installed from Bus No.2 to Bus No.33 and had a rate limit of 100 kW to 2,000 kW, Voltage setting equal 1.0 p.u. with the sizing of PV, respectively.

Therefore, the variable to decode of the possible position and sizing each relervant could be showed as Fig.2, that consists of 16 bit or variables(Xn) from randomize of ABC for solving the optimization problem.

Algorithm: the ABC to solve the optimal condition(Basic)
1:    # **Step1**:Initailization
2:    Read data from the power system(lines,bus,load)
3:    Define the boundary of EV and PV from TABLE I
4:    Define the parameter of ABC.
5:    # **Step2**: ABC algorithim
While(i<=iteration&torlerance) {
6:    $x_{i,j} = x_{\min,j} + \mathrm{rand}(0,1)(x_{\max,j} - x_{\min,j})$
7:    $\mathrm{Min}\ f = \sum_{i=1}^{N} LVD_i\ \ i \in N$ (Applied BFS)
8:    $P_i = \dfrac{fit_i}{\sum\limits_{i=1}^{SN} fit_i}$
9:    $v_{i,j} = x_{i,j} + \phi_{i,j}(x_{i,j} - x_{k,j})$
i=i+1
}
10:   #**Step 3**: calculate the total power loss
11:   $P_{Loss} = \sum_{i=1}^{NL} g_{i,j}\left\{ V_i^2 + V_j^2 - 2V_iV_j Cos(\delta_i - \delta_j) \right\}$
12:   #**Step 4**: Showed the results(total power loss,voltage profiles)
13:   #**Step 5**: terminate

Table I shown group 1 to group 7 of EVs and PV had consisted of possible positions on bus and sizing; which had a minimum limit and maximum value of EVs and PV, respectively. When supposing EVs sizing was ranged between 50 – 500 kW, the assumption made on a EVs load was able to use the EVs charging station and was divided on the distribution of the electrical power system and had randomized positions. The ABC algorithm was applied for finding the optimal condition which was defined from the objective function.

As Fig.2, had shown that sixteen variables of EVs and PV, were used in the ABC algorithm which followed the EVs position, PV position, EVs sizing and PV sizing, respectively. Therefore, the number of colony size (employed bees and onlooker bees) had an equal amount of 200; with a number of food sources equal to half the size of the colony, a food source which couldn't be improved through limit trials had been abandoned by its employed bees with an equal amount of 100, and a number of variables that were equal to 16 with a defined max cycle equal to 1000, respectively.

V. SIMULATION AND RESULTS

In order to analyze the optimal EVs and PV placement were divided into three cases: Case1, original load of the power system, Case2 include EVs into original load of the power system by using ABC algorithm and Case3 include original load, EVs and PV by using ABC algorithm, respectively.

Table II showed Case 1 to Case 3 from proposed to fine the optimal site and sizing of EVs and PV. In order to find the optimal EVs site and sizing on Case 2 showed the optimal site and optimal sizing are 2, 3, 7, 14, 15, 24, 27 and 50kW of EVs sizing, respectively. Meanwhile, total active power loss and LVD was 241.547kW and

0.155. Obviously, the LVD after EVs placement was reduced from Case1 (Based case). So that, in condition from Case3 will be installed PV in optimal position of the electrical power system which is optimal site equal Bus No.15 and sizing of the PV equal 250.785kW, respectively. From Case 3 were found active power loss and LVD reduced at 218.142 kW and 0.021 that showed the PV could be an improved system with reduction of +3.3 % ,and a load voltage deviation reduction of -86.71% from the based case of the electrical power system.

TABLE II SIMULATION RESULTS

Case	Opt. Location (Bus)	Opt. Sizing (kW)	Before		After	
			Loss (kW)	LVD	Loss (kW)	LVD
1	-	-	211.118 (0%)	0.158 (0%)	-	-
2	EVs= 2,3,7,14, 15,24,27	50,50,50, 50,50,50 (+9.41%)	241.547 (+14.41%)	0.155 (-1.90%)	-	-
3	EVs= 2,3,7,14, 15,24,27 PV=15	50,50,50, 50,50,50,50 (+9.41%) 250.785	241.547 (+14.41%)	0.155 (-1.90%)	218.142 (+3.3%)	0.021 (-86.71%)

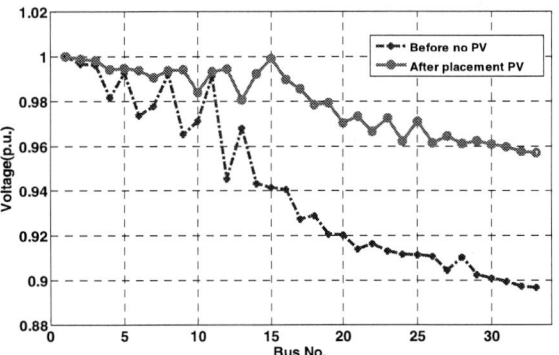

Fig. 3. Comparison of EVs penetration between, before, and after PV placement

Accordingly, the optimal EVs placement and the PV placement both had used the ABC algorithm. From the proposed simulation results, Fig.3 showed that the voltage profiles for each bus with the electrical power system, and the voltage profiles before squared on a dash line and after placement PV could be circled with a bold line, respectively. When the voltage profiles were in placement, the PV showed that PV was installed on Bus no.15 and had a size of 250.785 kW. Therefore, the PV could improve the voltage profile for the electrical power system. However, in a large scale condition of EVs into the electrical power system, it was confirmed that the effect to power system needed to be managed; in order to be kept in optimal condition and had incline/improvement in the stability of the power system in terms of voltage and system loss.

VI. CONCLUSION

The ABC algorithm was presented in this paper for the application; to solve the optimal problems presented in the electrical power system. The proposed methods that were applied to solve the optimal placement were that of the EVs and PV on the electrical power system. The voltage dependent of EVs and PV were also integrated the power constants of the conventional load by using ABC which has been proposed. The objective function was to use LVD for the location of an optimal site and sizing of EVs and PV. The simulation results showed LVD had a reduction of -1.19 % and power loss increase of +14.41 %, when EVs were installed and reduced after PV was installed. The LVD had a reduction of -86.71 % and had a power loss increase of +3.3 % in the electrical power system. Therefore, the optimal EVs in terms of location and size should be considered and needed; in order to find other energy sources for reducing energy consumption with EVs penetration on the electrical power system. Furthermore, in order to analyze the impact of EVs on the electrical power system it should be considered that in a multi objective; it's important to find a suitable case and have planning for the electrical power system in near the future.

REFERENCES

[1] Y.Kongjeen and K. Bhumkittipich, "Modeling of electric vehicle loads for power flow analysis based on PSAT," *13th International Conference on Electrical Engineering/ Electronics, Computer, Telecommunications and Information Technology (ECTI-CON)*, pp.1-6, Y2016.

[2] C.H. Dharmakeerthi, N. Mithulananthan, T.K. Saha, "Impact of Electric Vehicle Load on Power System Oscillatory Stability," *Australasian Universities Power Engineering Conference (AUPEC 2013)*, pp.1-6, Y2013.

[3] SD.Meera Shareef and T.Vinod Kumar, "A Review on Models and Methods for Optimal Placement of Distributed Generation in Power Distribution System," *IJEAR*, vol. 4, Issue Spl-1, Jan - June 2014; ISSN: 2348-0033 (Online).

[4] D. Karaboga and B. Basturk, "A powerful and Efficient Algorithm for Numerical Function Optimization: Artificial Bee Colony (ABC) Algorithm," *Journal of Global Optimization*, vol. 39, Issue: 3, pp: 459-171, November 2007.

[5] G. Meerimatha, G. Kesavarao and N. Sreenivasula, "A novel Distribution System Power Flow Algorithm using Forward Backward Matrix Method," *IOSR Journal of Electrical and Electronics Engineering(IOSR-JEEE)*, vol.10, pp.46-51, 2015.

[6] P.Sindhu Priya and N. Chaitanya Kumar Ready, "Optimal Placement of the DG in Radial Distribution System to Improve the Voltage Profiles," *International Journal of Science and Research (IJSR)*, vol.4, 2015.

[7] K. Bhumkittipich, "Control Techniques of Voltage Stability," Triple ED: Thailand, pp.101-102, 2015.

The 2018 International Power Electronics Conference

A Comparison of Average Model, Sampled-data Model and Multi-frequency Model Based on DC/DC Converters

Xiangpeng Cheng*, Jinjun Liu, Zeng Liu, Yiming Tu and Danhong Xue

State Key Laboratory of Electrical Insulation and Power Equipment, Xi'an Jiaotong University

Xi'an, China

*E-mail:alexcheng1994@163.com

Abstract— The widely-used state-space average model based on DC/DC converters loses its accuracy in high frequency domain, which makes it ineffective to predict the stability of some specific systems. In order to solve this problem, several modified models were provided. This paper mainly focuses on two modified models, i.e. the sampled-data model and the multi-frequency model. By comparing them with traditional average model, a unified model structure is proposed to show the key differences of these models. Through this structure, the basic reason for the frequency limitation of average model can be revealed clearly. And the accuracy and possible application sceneries of these models are discussed. Simulation results are presented as verifications.

Keywords— DC/DC converters, comparison, sampling effect, frequency limitation.

I. INTRODUCTION

The analysis tools developed in linear and time-invariant systems could not be directly applied to switching power converters because of the effects caused by switches and regulators. In order to solve this problem, the state-space average model was proposed by Cuk [1] in 1976 to transform the power electronic converters into a linear and time-invariant system, which made the design of the compensator become accurate and easy.

But in some specific application sceneries, for example, the current-mode control DC/DC converters and high-bandwidth voltage regulators, the average model loses its accuracy in high frequency domain, which makes it ineffective to predict the stabilities of these systems. So based on the average model, several modified models were proposed to remedy the shortcomings. Among them, the sampled-data model [2]-[4] and the multi-frequency model [5]-[7] can be both used to detect the high-frequency characteristic of DC/DC converters, although their modeling processes are totally different. The sampled-data model was proposed by Arthur R.Brown and Middlebrook in 1981 to successfully predict the instability of DC/DC converters under current-mode control. And the multiple-frequency model was proposed by Yang Qiu in 2005 to handle the high-frequency characteristics of high-bandwidth buck voltage regulators. But in this paper, both

This work was supported by the National Natural Science Foundation of China under Grant 51437007.

models will be developed to analyze general DC/DC converters under linear analog control.

The main work of this paper is to make an intrinsic comparison of the average model, the sampled-data model and the multi-frequency model. Based on the analysis, a unified model structure is proposed to describe all of these models. Through this structure, the key differences of these models can be viewed directly. And it can reveal the basic reason for the frequency limitation of average model and give a clear answer of the effective frequency range. Based on that, the possible application sceneries of these three are discussed.

In order to combine the three models into a unified model structure, the small-signal linearization of state-space equation is adopted as the first step of the modeling process, which is different with the traditional method used in average model and multi-frequency model. Then based on this linear but still time-variable equation, different simplifications are required for different models, but actually these simplifications are all strongly related with the sampling effect of the switches, that is why these models can be described with a unified model structure. Finally, simulations about buck and boost converters by Saber are presented as verifications.

II. THE SMALL-SIGNAL LINEARIZATION OF STATE-SPACE EQUATION

For a DC/DC converter operating in the continuous conduction mode under constant frequency PWM control, if $x(t)$ is defined as the state vector, $v_g(t)$ as the source voltage, T_s as the switching period, the state-space equation for it can be easily obtained:

$$\dot{x} = A_1 x + b_1 v_g \quad nT_s < t < (n + d_n)T_s$$
$$\dot{x} = A_2 x + b_2 v_g \quad (n+d_n)T_s < t < (n + 1)T_s \quad (1)$$
$$n = 0,1,2,3, \ldots \ldots$$

Here A_1 and A_2 are the matrices determined by the converter topologies in different switching states; b_1 and b_2 represent the effects of source voltage; d_n is the duty ratio in the *nth* switching-cycle.

If the switching function is defined as below:

$$d(t) = \begin{cases} 1 & nT_s < t < (n+d_n)T_s \\ 0 & (n+d_n)T_s < t < (n+1)T_s \end{cases} \quad (2)$$

$$d'(t) = 1 - d(t) \quad (3)$$

Then the unified state-space equation can be derived:

$$\dot{x} = [d(t)A_1 + d'(t)A_2]x + [d(t)b_1 + d'(t)b_2]v_g \quad (4)$$

Considering that frequency-domain analysis is only applied to the linear system, the small-signal linearization adapted from Packard [8] is a necessary step for all three models.

Assume that each variable in Eq. (4) contains a steady-state value and a small-signal perturbation when the system is working on the steady-state as below:

$$\begin{cases} v_g = V_g + \widetilde{v_g} \\ x = X + \tilde{x} \\ d(t) = D(t) + \tilde{d}(t) \end{cases} \quad (5)$$

These expressions are then substituted into Eq. (4) and eliminates the steady-state components in both sides:

$$\begin{aligned} \dot{\tilde{x}} = &[D(t)A_1 + D'(t)A_2]\tilde{x} + [D(t)b_1 + D'(t)b_2]\widetilde{v_g} \\ &+ [(A_1 - A_2)X + (b_1 - b_2)V_g]\tilde{d}(t) \\ &+ [(A_1 - A_2)\tilde{x} + (b_1 - b_2)\widetilde{v_g}]\tilde{d}(t) \end{aligned} \quad (6)$$

In Eq. (6), the final term is the product of two small-signal perturbations which can be neglected. After the small-signal linearization, a linear but still time-variant equation can be derived:

$$\begin{aligned} \dot{\tilde{x}} = &[D(t)A_1 + D'(t)A_2]\tilde{x} + [D(t)b_1 + D'(t)b_2]\widetilde{v_g} \\ &+ [(A_1 - A_2)X + (b_1 - b_2)V_g]\tilde{d}(t) \end{aligned} \quad (7)$$

Finally, the perturbation of switching function should be detected carefully. Fig. 1 shows the detailed waveforms of $d(t)$, $D(t)$ and $\tilde{d}(t)$.

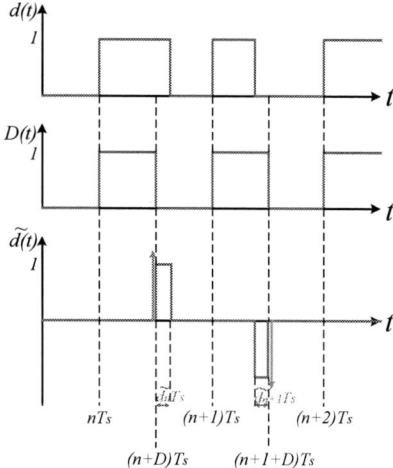

Fig. 1. Waveforms of $d(t)$, $D(t)$ and $\tilde{d}(t)$.

When the perturbation of d_n is small enough, the $\tilde{d}(t)$ will be a series of narrow pulses, which can be replaced by the ideal pulses (the blue lines with arrows in Fig. 1). So a delta function $\tilde{p}(t)$ can be defined to replace switching function perturbation $\tilde{d}(t)$:

$$\tilde{d}(t) \approx \tilde{p}(t) = \sum_{n=-\infty}^{+\infty} \widetilde{d_n} T_s \delta(t - (n+D)T_s) \quad (8)$$

Finally, the result of small-signal linearization is:

$$\begin{aligned} \dot{\tilde{x}} = &[D(t)A_1 + D'(t)A_2]\tilde{x} + [D(t)b_1 + D'(t)b_2]\widetilde{v_g} \\ &+ [(A_1 - A_2)X + (b_1 - b_2)V_g]\tilde{p}(t) \end{aligned} \quad (9)$$

III. SIMPLIFICATIONS FOR DIFFERENT MODELS

Based on the small-signal signal linearization, different simplifications are used for different models.

A. The state-space average model

The average model is a rigorous linear and time-invariant state-space equation, which is driven by a time-continuous duty ratio function. But in Eq. (9), there exist two offensive components: one is the time-variant coefficients, and the other one is the time-discontinuous switching function [2].

First, if the switching period is small enough, which means the ripples of state values can be neglected in steady state, the time-variant coefficients can be replaces by their moving average values:

$$\frac{1}{T_s}\int_{t-T_s}^{t}[D(\tau)A_1 + D'(\tau)A_2]\,d\tau = DA_1 + D'A_2 \quad (10)$$

$$\frac{1}{T_s}\int_{t-T_s}^{t}[D(\tau)b_1 + D'(\tau)b_2]\,d\tau = Db_1 + D'b_2 \quad (11)$$

$$D = 1 - D' \quad (12)$$

Where the D represents the steady-state value of duty ratio. Actually, a very persuasive mathematical theorem called *straight-line approximation* [2] can be used to support this simplification. The only requirement for this step is the guarantee of the relatively small switching period comparing with the system's time constant, which is usually satisfied by the converters with PWM control.

Second, moving average can also be utilized to remedy the time-discontinuous switching function. Assume that $\tilde{u}(t)$ is a time-continuous function and has a specific relationship with $\widetilde{d_n}$:

$$\tilde{u}((n+D)T_s) = \widetilde{d_n} \quad (13)$$

Then $\tilde{p}(t)$ can be represented as below:

$$\tilde{p}(t) = \left[\sum_{n=-\infty}^{+\infty} T_s \delta(t - (n+D)T_s)\right]\tilde{u}(t) \quad (14)$$

Apply moving average on the delta function part of Eq. (14):

$$\frac{1}{T_s}\left[\sum_{n=-\infty}^{+\infty} T_s \delta(\tau - (n+D)T_s)\right]d\tau = 1 \quad (15)$$

So finally $\tilde{u}(t)$ can be used to replace $\tilde{p}(t)$, and the state-space average model is obtained:

$$\begin{aligned} \dot{\tilde{x}} = &[DA_1 + D'A_2]\tilde{x} + [Db_1 + D'b_2]\widetilde{v_g} + \\ &[(A_1 - A_2)X + (b_1 - b_2)V_g]\tilde{u}(t) \end{aligned} \quad (16)$$

The Laplace transform should be applied on Eq. (16) in order to detect its frequency domain characteristics:

$$s\tilde{x}(s) = A\tilde{x}(s) + b\widetilde{v_g}(s) + K\tilde{u}(s) \quad (17)$$

$$A = DA_1 + D'A_2$$
$$b = Db_1 + D'b_2$$
$$K = (A_1 - A_2)X + (b_1 - b_2)V_g$$

Although a linear and time-invariant average model is derived, but the second approximation lacks reasonable mathematical support. This simplification allows $\tilde{u}(t)$ to affect the system at all time, rather than at a specific instant in each switching cycle, which introduces considerable errors into the model

B. The sampled-data model

The *straight-line approximation* is still remained to derive the sampled-data model because it's a reasonable approximation to handle the time-variant coefficients. But the sampled-data model will not utilize the moving average method to deal with the time-discontinuous switching function, which is the main difference with the average model [2].

If $\tilde{u}^*(t)$ is defined as:

$$\tilde{u}^*(t) = \tilde{p}(t) = [\textstyle\sum_{n=-\infty}^{+\infty} T_s\, \delta(t - (n + D)T_s)]\tilde{u}(t) \quad (18)$$

Then the sampled-data model can be easily derived as below:

$$\dot{\tilde{x}} = [D(t)A_1 + D'(t)A_2]\tilde{x} + [D(t)b_1 + D'(t)b_2]\widetilde{v_g} + $$
$$[(A_1 - A_2)X + (b_1 - b_2)V_g]\tilde{u}^*(s) \quad (19)$$

The Laplace transform of Eq. (19) is:

$$s\tilde{x}(s) = A\tilde{x}(s) + b\widetilde{v_g}(s) + K\tilde{u}^*(s) \quad (20)$$

Where $\tilde{u}^*(s)$ is the Laplace transform of $\tilde{u}^*(t)$. By using the complex multiplication theorem for the Laplace transform, a very clear and useful relationship between $\tilde{u}^*(s)$ and $\tilde{u}(s)$ can be derived [9]:

$$\tilde{u}^*(s) = \sum_{n=-\infty}^{+\infty} \tilde{u}(s + jn\omega_s) \quad (21)$$

C. The multi-frequency model

In frequency domain, if the feedback control signal $\tilde{u}(s)$ only contains single frequency component ω_p ($0 < \omega_p < \omega_s$), then abundant frequency components will be generated in the sampled input control signal $\tilde{u}^*(s)$, as shown in Fig.2.

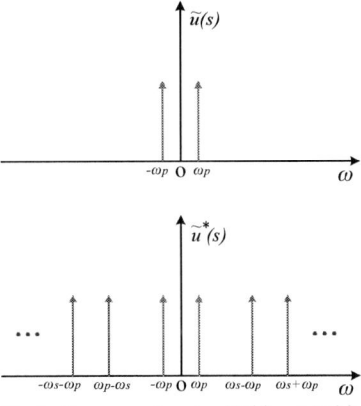

Fig. 2. Frequency domain of $\tilde{u}(s)$ and $\tilde{u}^*(s)$.

Actually, the feedback control loop of DC/DC converter will be designed as a low-pass filter ($f_{bandwidth} < f_s$), which means that high-frequency components will be attenuated to a negligible amount and there is no need to take all the frequency components of $\tilde{u}^*(s)$, i.e. $\tilde{u}(\omega_p)$, $\tilde{u}(\omega_s \pm \omega_p)$, $\tilde{u}(2\omega_s \pm \omega_p)$, $\tilde{u}(3\omega_s \pm \omega_p)\cdots$ into consideration.

In the multi-frequency model, only two frequency components lower than switching frequency ω_s, i.e. $\tilde{u}(\omega_p)$, $\tilde{u}(\omega_s - \omega_p)$ are maintained to make a balance between accuracy and simplification. Of course, it also utilizes the *straight-line approximation* to remedy the time-variant coefficients. So the Laplace transform of multi-frequency model is shown as below:

$$s\tilde{x}(s) = A\tilde{x}(s) + b\widetilde{v_g}(s) + K\tilde{u}'(s) \quad (22)$$
$$\tilde{u}'(s) = \tilde{u}(\sigma \pm j\omega_p) + \tilde{u}(\sigma \pm j(\omega_s - \omega_p))$$
$$\sigma = 0, \text{when in frequency domain analysis.}$$

D. The unified modelling structure

For DC/DC converters, the feedback signal from state values goes through the comparator and compensator to be the final feedback control signal $\tilde{u}(s)$, so actually it's a linear path for small signal to flow. The Laplace transform of the control path can be presented as:

$$\tilde{u}(s) = -H_e\tilde{x}(s) \quad (23)$$

Where H_e is a matrix determined by the control design.

Now, combining both the power plant and feedback control models, the modelling structures for different modelling methods are shown as below:

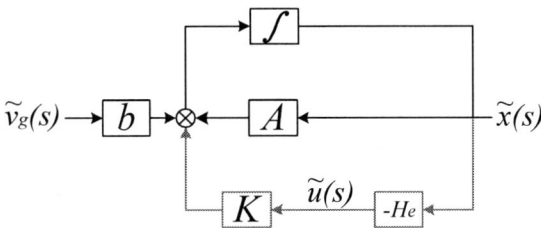

Fig. 3. Modelling structure for the average model

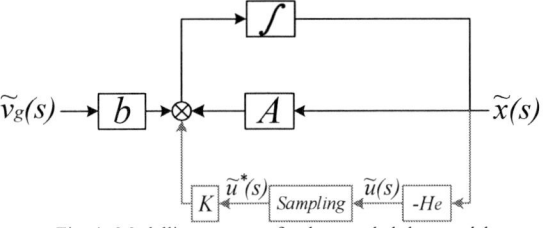

Fig. 4. Modelling structure for the sampled-data model

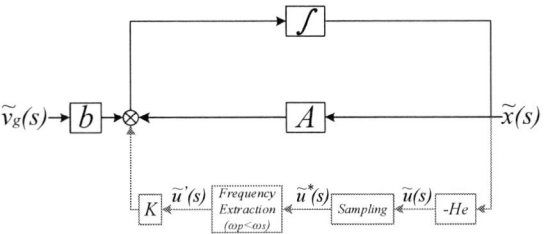

Fig. 5. Modelling structure for the multi-frequency model

By comparing these structures, a unified modelling structure can be prospected to describe all three models:

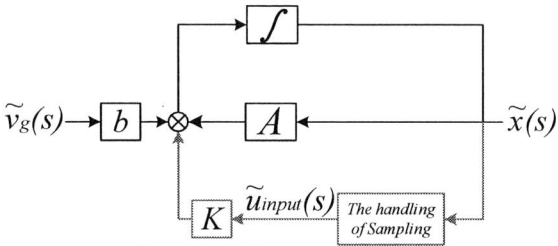

Fig. 6. The unified modelling structure

Through it, it's obvious that the different handling of sampling effects caused by the switches are the key characteristics for different models. When the bandwidth of the system is small enough ($f_{bandwidth} \ll f_s$), there is no need to consider the sampling effects of the switch, which means the average model can maintain accuracy from 0 Hz to f_s and is good enough for the control design.

Now, the loop gain should be derived to guide the design of the real system control design.

For the average model, the relationship between $\widetilde{v_g}(s)$ and $\tilde{u}(s)$ can be obtained from Fig. 3:

$$\tilde{u}(s) = \frac{-H_e (sI - A)^{-1} b}{1 + H_e (sI - A)^{-1} K} \widetilde{v_g}(s) \qquad (24)$$

So the loop gain is:

$$T(s) = H_e (sI - A)^{-1} K \qquad (25)$$

For the sampled-data model, because all the frequency components are taken into consideration, it's not convenient to derive an analytic form of loop gain. But the relationship with the loop gain of average model can be given [9]:

$$T_s^*(s) = \left[T(s)\right]^* = \sum_{n=-\infty}^{+\infty} T(s + jn\omega_s) \qquad (26)$$

For the multi-frequency model, in order to obtain the analytic result of loop gain, the Fig. 5 should be developed to describe the details of signal flow paths as Fig. 7:

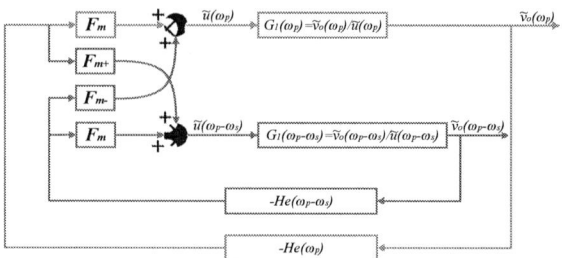

Fig. 7. Signal flow paths for multi-frequency model

Here Fm, $Fm+$ and $Fm-$ represent the mathematical relationships between two frequency components which are determined by the specific control strategy. Usually the describing function should be utilized to calculate them. Here the loop gain of DC/DC converter under voltage control with trailing-edge natural sampling strategy is given as an example [7]. The results of Fm, $Fm+$ and $Fm-$ are given directly without detailed process due to the derivation complexity [7]:

$$F_m = 1/V_R$$
$$F_{m+} = e^{jD2\pi}/V_R \qquad (27)$$
$$F_{m-} = e^{-jD2\pi}/V_R$$

Then the loop gain of the multi-frequency model can be derived as below:

$$T_s^{'}(s) = \frac{T(s)}{1 + T(s - j\omega_s)} \qquad (28)$$

Here $T(s)$ represents the loop gain of the average model.

From Eq. (26) and Eq. (28), the relationships between these models can also be viewed directly. If the bandwidth of $T(s)$ is narrow enough, $T_s^*(s)$ and $T_s^{'}(s)$ will matches well with it at the frequency domain from 0 Hz to near switching frequency f_s, which means that the average model is good enough for the design. But in some using sceneries, large bandwidth of loop gain is necessary in order to guarantee the rapid reactions to perturbations, then the average model can't be utilized to design the control because the failing in high frequency domain. Although the sampled-data model should be more accurate than the multi-frequency model due to the considerations of all the frequency components of the input control signal, but it's difficult to obtain the analytic form of loop gain for some

TABLE I: A COMPARISON SUMMARY OF THESE THREE MODELS

	Average model	Multi-frequency model	Sampled-data model
Loop gain	$T(s)$	$T_s^{'}(s) = \dfrac{T(s)}{1 + T(s - j\omega_s)}$	$T_s^*(s) = \sum\limits_{n=-\infty}^{+\infty} T(s + jn\omega_s)$
Frequency limitation	If the bandwidth of loop gain is small enough, the effective frequency range can reach near switching frequency f_s.	Maintain accuracy between $0 \sim fs$.	Maintain accuracy in all frequency domain.
Application scenery	Suitable for all DC/DC converters under linear control with narrow bandwidth of the loop gain.	Especially suitable for DC/DC converters under voltage-mode control with high bandwidth, e.g. CPU voltage regulators.	A good description model for all DC/DC converter under linear control. Especially suitable for DC/DC converters under current-mode control.

2438

specific control methods. Consequently, the multi-frequency model is a good choice for design when the loop gain bandwidth is relatively high, for example, near $\frac{1}{2}f_s$.

In Table I, the detailed comparisons about the frequency limitations and application sceneries are shown to guide the use of these models.

And a simulation comparison between the average model and the multi-frequency model will be shown in the next part to verify these conclusions.

IV. SIMULATION VERIFICATIONS

For the buck converter, two different average model loop gain designs, i.e. 5k Hz and 20k Hz are applied when the switching frequency is 50k Hz. And the theoretical analyses by the average model and the multi-frequency model are compared with the Saber simulation result as shown in Fig. 8 and Fig. 9:

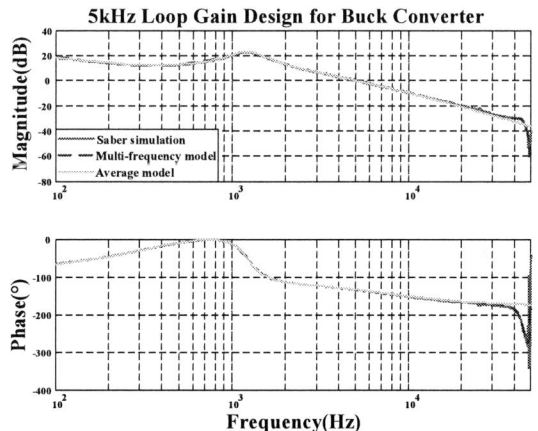

Fig. 8. Loop gain comparisons for Buck converter with 5k Hz design

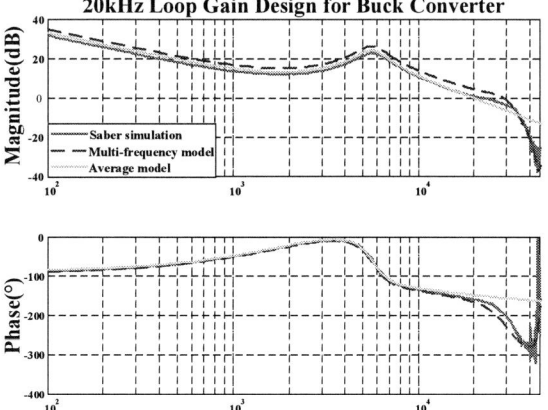

Fig. 9. Loop gain comparisons for Buck converter with 20k Hz design

When the loop gain is relatively high comparing with the switching frequency, the average model loses its accuracy in high-frequency domain, but multi-frequency model can successfully predict the magnitude attenuation and phase delay.

For the boost converter, the average model loop gain

designs are changed to 1.2k Hz and 10k Hz when the switching frequency is still 50k Hz. The comparison results are shown in Fig. 10 and Fig. 11：

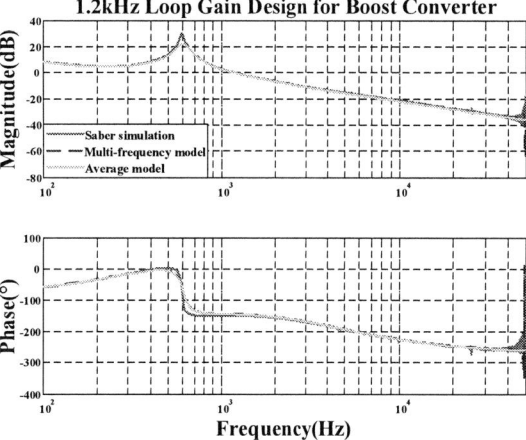

Fig. 10. Loop gain comparisons for Boost converter with 1.2k Hz design

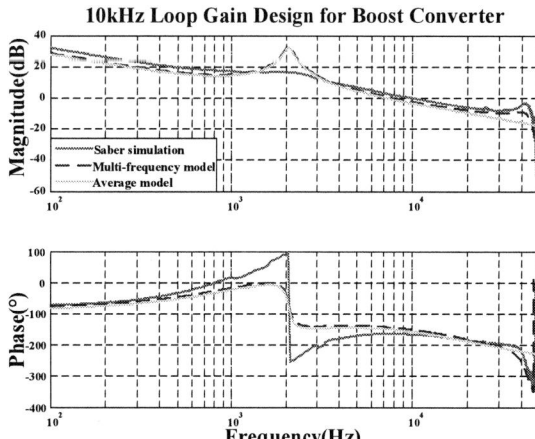

Fig. 11. Loop gain comparisons for Boost converter with 10k Hz design

When the loop gain bandwidth goes higher, the boost converter can't satisfy the small-signal linearization approximation as well as the buck converter, which will heavily effect the accuracy of loop gain prediction around the resonant frequency. But in high-frequency domain, the multi-frequency model is still a better choice than average model.

V. CONCLUSIONS

This paper proposes a unified modelling structure to analyze the average model, the sampled-date model and the multi-frequency model. And the key difference of these models is how to handle the sampling effects caused by the switches. When the bandwidth of system loop gain is much smaller than the switching frequency, the average model is good enough for the control design. When the bandwidth is relatively high, the sampled-data model and the multi-frequency model should be used to make sure the effective frequency range of the control design. And the multi-frequency model is more suitable for practical implementation because it just takes necessary frequency components into consideration.

REFERENCES

[1] Slobodan Cuk, "Modelling, Analysis, and Design of Switching Regulators", *PhD thesis, California Institute of Technology*, November 1976; also, NASA Report CR-135174.

[2] A. R. Brown and R. D. Middlebrook, "Sampled-data modeling of switching regulators," *in Proc. Power Electronics Specialists Conf.*, June 29-July 3, 1981, pp. 349-369.

[3] G. Verghese, C. Bruzos, and K. Mahabir, "Averaged and sampled-data models for current-mode control: A reexamination," *in IEEE Power Electron. Specialists Conf.* Rec., 1989, pp. 484–491.

[4] Y.-W. Lo and R. J. King, "Sampled-data modeling of the average-input current-mode-controlled buck converter," *IEEE Trans. Power Electron.*, vol. 14, no. 5, pp. 918–927, Sep. 1999.

[5] Y. Qiu, M. Xu, K. Yao, J. Sun, and F. C. Lee, "The multi-frequency small-signal model for buck and multiphase interleaving buck converters," *in Proc. IEEE Appl. Power Electron. Conf.*, 2005, pp. 392–398.

[6] Y. Qiu, "High-frequency modeling and analyses for buck and multiphase buck converters," *Ph.D. dissertation*, Virginia Polytechnic Institute and State University, Blacksburg, 2005.

[7] Y. Qiu, M. Xu, J. Sun, and F. C. Lee, "A generic high-frequency model for the nonlinearities in buck converters," *IEEE Trans. Power Electron.*, vol. 22, no. 5, pp. 1970–1977, Sep. 2007.

[8] Dennis J. Packard, "Discrete Modeling and Analysis of Switching Regulators", *PhD thesis*, California Institute of Technology, May 1976; also, Report No. M76-43, Hughes Aircraft Co., Aerospace Groups, Culver City, Calif.

[9] Eiiahu I. Jury, Sampled-Data Control Systems, John Wiley 8c Sons, 1958.

Small-Signal Discrete-time Modeling and Digital Control of the Bi-directional DC/DC Converters

Jia Yaoqin, Xu Yingchun and Hou Yijie
School of Electrical Engineering, Xi'an Jiaotong University, Xi'an, China
E-mail: yaotsin@mail.xjtu.edu.cn

Abstract— Bi-directional DC/DC converters have been widely used in many applications because of its energy bi-directional flow. This paper proposes the small-signal discrete-time model of bi-directional buck/boost converter at a selected operating point and the accuracy of the model is verified in comparison to the traditional continuous-time average model. Also, the digital PI controller is designed for the converter based on the discrete-time model. In addition, a unified digital current controller is utilized and suitable for two working modes. A set of effective digital PI voltage controller parameters are also designed. Simulation results of the bi-directional buck/boost converter under start-up condition and load transient condition in MATLAB are presented. Finally, two hardware testing platforms are employed and the experimental results are presented to validate the proposed methods.

Keywords— *Bidirectional DC/DC converter, Discrete-time modeling, Digital controller design*

I. INTRODUCTION

Traditional DC-DC converters usually operate in only one direction, but if replace the unidirectional device with the bi-directional controlled switch or change the topology structure by series and parallel connections, then numerous bi-directional DC/DC converters can be obtained [1-3]. In general, the voltage polarity of the bi-directional DC-DC converter keeps constant and its current direction can be adjusted according to the demand so as to realize bi-directional energy flow. The bi-directional converters have been widely used in electric vehicles, distributed generation, energy storage systems, power quality control, renewable energy generation and superconducting energy storage and other fields [4-7]. Due to its broad application, it has attracted the attention of a large number of researchers. The main research directions are topology, parameter design, modeling and controller design [8-13].

With the development of the semiconductor industry, the price of the digital control chips is decreasing constantly. It is possible to introduce digital control into products. In addition, the digital control has many excellent merits [15-17]. It can realize novel complex control algorithm and has high flexibility rather than analog control. What's more, it has higher anti-jamming capacity hence becoming more and more popular.

Many scholars have done a lot of research on bi-directional converters. The main emphasis in [1-4] is on topology. [14] shows how to design optimal parameters. [8,10] focus on different control methods. [9,12] has displayed how to build the model of bi-directional

converter. However, the final result is a continuous-time model and it adopts analog control system. [13] builds a bi-directional model and the converter is controlled by digital controller, whereas the model is still in continuous-time domain. Despite using digital control, the controller is designed in s-domain and then is discretized by some discretization methods. The final result is not in conformity with the initial design requirements, so it is still not sufficiently appropriate enough.

In this paper, main research object is bi-directional buck/boost converter, which is applied to the battery charging/discharging, and the research is based on the previous research [13] and makes use of the discrete-time modeling method in [16]. The influence of sampling and modulation is taken into consideration in modeling of the converter. In the end, a small-signal discrete-time model of the buck/boost bi-directional DC-DC converter is built. Further, a unified digital current controller is designed and it can be available for two modes and adapt to a wider range of operating point. Two digital voltage controllers are also designed for two working modes at the same time.

The structure of the paper is as follows. The proposed system specifications and the modeling procedure are described in Section II. In Section III, the design of the digital PI controller is discussed. The simulation results of the converter are also presented. The validity of modeling and design method are verified by the experiments performed in Section IV. The conclusion is given in Section V.

II. DISCRETE-TIME MODELING

Fig. 1 shows the topology of the bi-directional buck/boost converters. Where, r_{dc} and r_b are the internal resistance of two power sources respectively and are usually ignored in order to simplify the analysis. r_{C1}, r_{C2} and r_L are parasitic resistance of capacitors and inductor. S_1 and S_0 are IGBTs in parallel with diodes. This topology is evolved from conventional buck or boost converter by replacing the unidirectional freewheel diode with MOSFET or IGBT in parallel with a diode. Our main research is its application in battery charging and discharging, so both sides of the converter are voltage source.

Interleaved multi-phase technology is used to reduce ripple and energy storage element in many engineering applications at present [18-20]. Although the topology is more complex, the analysis can be simplified to the analysis of half-bridge buck/boost converter [20]. Thus the research achievement of this simple topology can be extended to other complex situations.

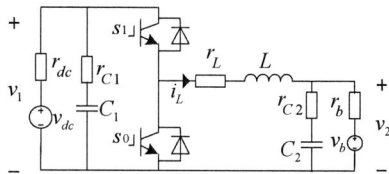

Fig. 1. Bi-directional buck/boost converter.

Design specifications and power stage parameters are summarized in TABLE I.

TABLE I
SYSTEM PARAMETERS

Parameter	Value
High voltage side V_{dc}	380V
Low voltage side V_c	255-325V
Power rating P_0	25kW
Switching frequency f_s	5kHz
Filter inductance L	0.4mH
Inductor series resistance r_L	0.1Ω
Filter capacitance C_2	420μF
Capacitance C_1	5500μF
Capacitor C_1 ESR r_{C1}	0.2mΩ
Capacitor C_2 ESR r_{C2}	1.6mΩ

Fig. 2 shows the modulation schematic of the symmetric triangular carrier. u_c is carrier wave, N_r is the maximum of carrier wave, u is the input of PWM modulation module, U is steady state value of u, $c(t)$ is PWM signal, $c_s(t)$ is PWM signal corresponding to the U, t_{ct} is sampling and calculation time, D is duty ratio, T_s is switching period and k is an integer.

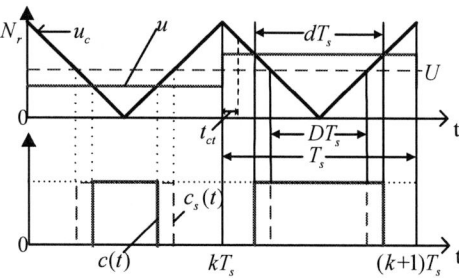

Fig. 2. Symmetric triangular modulation.

Complementary drive signals with certain dead time protection are employed to control the high-side and low-side switch and the converter has three types of working states as in Fig. 3. The two topside waveforms are drive signals and the other three are the inductor current waveforms and it is always continuous in three working modes. The direction of inductor current is always positive in the buck (charging) mode and negative in the boost (discharging) mode and the third current waveform is alternating between positive and negative in the critical mode. The direction of average current in the critical

mode depends on the value of i_{Lmax} and i_{Lmin}. The average current of the third waveform is too small to charge battery in high power with small current ripple, so the converter is needed to work in the first two modes for charging and discharging battery, and the parameters of circuit are designed to enable the converter to work in the first two modes.

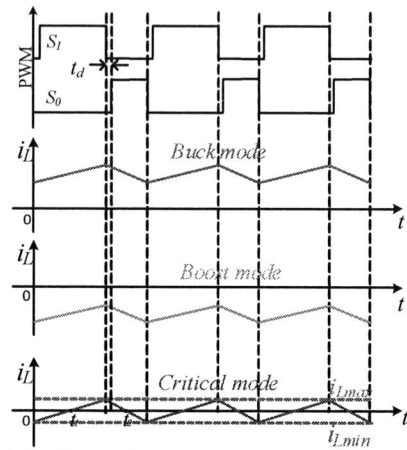

Fig. 3. The working states of buck/boost converter.

The influence of sampling and modulation are considered in modeling, because these two factors will have effect on the operation of the converter [21-22]. As in Fig. 4, sample and transform the regulated quantity and then update the duty cycle command at peak of triangular carrier waveform. When the circuit reaches steady state, the input of PWM module will be constant. The intersection of modulating wave and carrier wave are corresponding to the peak and valley point of inductor current. Due to the adoption of symmetric triangular modulation, this sampling instant is the midpoint and it is average value of the regulated quantity. Therefore, the mean values of the circuit are sampled and can be used to realize the control of average value in the converter.

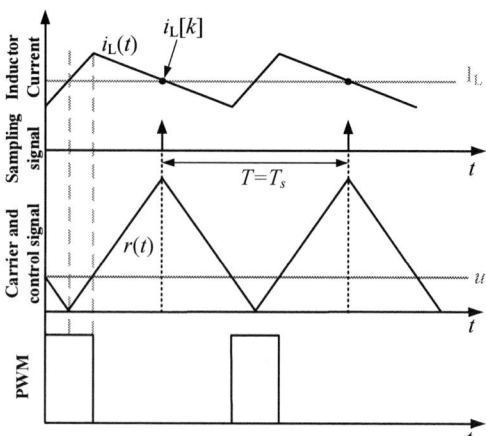

Fig. 4. The relationship among current, sampling signal and PWM.

Discrete-time modeling method is then used for

modeling the bi-directional buck/boost converter and the process is as follows:

1) Similar to the state-space average modeling method, the state equations in each sub-topology can be described as

$$\begin{cases} \dot{x} = A_n x + B_n v \\ y = C_n x \end{cases} \tag{1}$$

Where state vector x is $[i_L \ v_c]^T$, i_L is inductor current and v_c is capacitor voltage . Input vector v is input voltage v_g and output state y is output voltage v_o, while $n \in \{0,1\}$ denotes two different topological states corresponding to S_0 is on and S_1 is on respectively.

Then describe the system with the nonlinear equation

$$\dot{x} = (dA_1 + d'A_0)x + (dB_1 + d'B_0)v \tag{2}$$

Where d represents the duty ratio, and $d'=1-d$.

Solve the equation (2) by (3) in one switch period

$$x(t) = e^{A_n(t-t_0)}x(t_0) + \int_{t_0}^{t} e^{A_n(t-\tau)}B_n V d\tau \tag{3}$$

Where t_0 is initial time of one switching period and it can be set any value. V is considered as a constant value of v because we just want to focus on the control to state transfer function.

Eventually, get the nonlinear state equation

$$x[k+1] = f(x[k], v[k], d[k]) \tag{4}$$

Where f is a nonlinear vector function.

2) Let $x[k+1] = x[k] = X$, $v[k] = V$, $d[k] = D$ to find the operating point of the converter

$$X = (I - e^{A_0\frac{T_s}{2}(1-D)}e^{ADT_s}e^{A_0\frac{T_s}{2}(1-D)})^{-1} * [-e^{A_0\frac{T_s}{2}(1-D)}e^{ADT_s}A_0^{-1}B_0(I - e^{A_0\frac{T_s}{2}(1-D)}) -$$
$$e^{A_0\frac{T_s}{2}(1-D)}A_1^{-1}B_1(I - e^{ADT_s}) - A_0^{-1}B_0(I - e^{A_0\frac{T_s}{2}(1-D)})]V \tag{5}$$

3) Deduce the small-signal discrete-time state-space description of the system by successively perturbing and linearizing the nonlinear function in the neighborhood of operating point

$$\hat{x}[k+1] = \Phi\hat{x}[k] + \Upsilon\hat{u}[k]$$
$$\hat{y}[k] = \delta\hat{x}[k] \tag{6}$$

Where $^\wedge$ is the small-signal components sign relative to their dc components. Matrices Φ and γ represent state matrix and control-to-state matrix in small-signal, respectively, as in Eq. (7)

$$\begin{cases} \Phi = e^{A_0\frac{T_s}{2}(1-D)}e^{ADT_s}e^{A_0\frac{T_s}{2}(1-D)} \\ \Upsilon = \frac{T_s}{2N_r}e^{A_0\frac{T_s}{2}(1-D)}(I + e^{ADT_s})P \end{cases} \tag{7}$$

Where $P \triangleq (A_1 - A_0)X + (B_1 - B_0)V$

Matrices δ, represents the converter output matrix pertaining to the sub-topology in which sampling occurs

$$\delta = \begin{cases} C_0 \ , & \text{Sampling occurs during subtopology } S_0 \\ C_1 \ , & \text{Sampling occurs during subtopology } S_1 \end{cases} \tag{8}$$

Apply the z transform to the above equation (6)

$$z\hat{x}[z] = \Phi\hat{x}[z] + \Upsilon\hat{u}[z]$$
$$\hat{y}[z] = \delta\hat{x}[z] \tag{9}$$

Assume output vector $y = [i_L \ v_{out}]^T$, where v_{out} can be different in two modes, v_1 in boost mode and v_2 in buck mode. The small-signal discrete-time transfer function becomes

$$G[z] \triangleq \frac{\hat{y}[z]}{\hat{u}[z]} = \begin{bmatrix} G_{iu}[z] \triangleq \frac{\hat{i}_L[z]}{\hat{u}[z]} \\ G_{vu}[z] \triangleq \frac{\hat{v}_{out}[z]}{\hat{u}[z]} \end{bmatrix} = \delta(zI - \Phi)^{-1}\Upsilon \tag{10}$$

Eventually, from (10) the control to output discrete-time transfer function $G_i(z)$ and $G_u(z)$ can be derived.

When the input and output are all voltage sources, the model of the system can be simplified to first-order system [13]. So, just the most serve situation that the output of the converter is a resistive load is required to consider when modeling the converter.

For the buck mode, the low-side switch is regarded as freewheel diode. The converter can be simplified, as in Fig. 5, and its operating principle is the same as buck circuit, so it's not repeated here. State vector $x=[i_L \ v_c]^T$, input vector $v=v_{dc}$, output vector $y=[i_L \ v_b]^T$, and follow the above steps and have

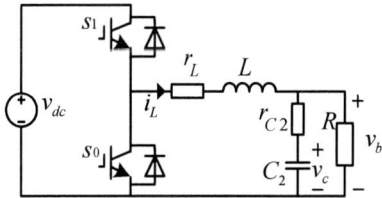

Fig. 5. The buck/boost converter in buck mode.

$$A_1 = A_0 = \begin{bmatrix} -\dfrac{r_L + R \| r_{C2}}{L} & -\dfrac{R}{(R+r_{C2})L} \\ \dfrac{R}{(R+r_{C2})C_2} & -\dfrac{1}{(R+r_{C2})C_2} \end{bmatrix},$$

$$B_1 = \begin{bmatrix} \dfrac{1}{L} & 0 \end{bmatrix}^T, \quad B_0 = 0,$$

$$C_1 = C_0 = \begin{bmatrix} 1 & 0 \\ R \| r_{C2} & \dfrac{R}{R+r_{C2}} \end{bmatrix} \tag{11}$$

The output voltage is in the range of 255-325V, so the steady state operating point is not unique. In order to obtain the discrete model, a special point is selected as a working point, that is output voltage $V_b=325$V and input voltage $V=V_{dc}=380$V. Therefore, the duty ratio $D = 325/380 = 0.855$ and consider the full load condition to find the equivalent load resistance R=4.225Ω. The state vector in steady state is $X=[75 \ 318]^T$.

The discrete transfer function near the steady-state operating point is obtained

$$G_{iu}(z) = \frac{0.4639z - 0.4126}{z^2 - 1.629z + 0.8492}$$
$$G_{vu}(z) = \frac{0.1082z + 0.1044}{z^2 - 1.629z + 0.8492} \tag{12}$$

Conventional averaged small-signal model of the converter is [15]

$$G_{iu}(s) = \frac{C_2(r_{C2}+R)s+1}{N(s)}$$

$$G_{vu}(s) = \frac{r_{C2}RC_2s+R}{N(s)} \tag{13}$$

Where

$N(s)=LC_2(r_{C2}+R)s^2+(L+C_2r_{C2}R+C_2r_LR)s+r_L+R.$

And Eq. (13) is discretized directly [17]

$$G_{iuz} = Z[G_{iu}(s)]$$

$$G_{vuz} = Z[G_{uu}(s)] \tag{14}$$

A valid correction to Eq. (13) with Delay t_d and t_d $=0.5T_s$ [21,22] is

$$G_{iudelay}(s) = G_{iu}(s)e^{-st_d}$$

$$G_{vudelay}(s) = G_{uu}(s)e^{-st_d} \tag{15}$$

In summary, the magnitude and phase bode plots of Eq. (12) to Eq. (15) are shown in Fig. 6.

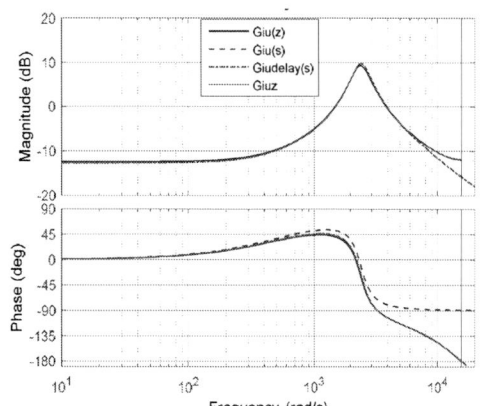

(a) Bode plot of the control-to-output current transfer functions.

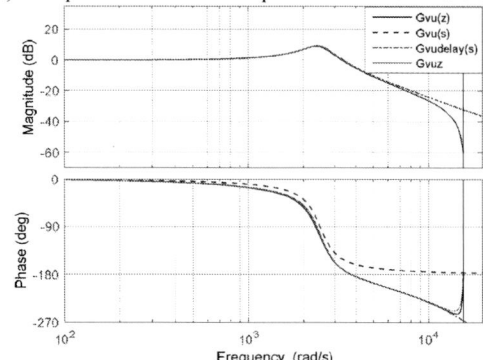

(b)Bode plot of the control-to-output voltage transfer functions.
Fig. 6 Bode plot of the control-to-output current and voltage transfer functions.

The magnitude responses are indeed quite similar, with a small departure visible in the high frequency section. On the other hand, the comparison between the phase responses of the four models reveals the additional phase lag caused by the loop delay, correctly modeled by the discrete-time transfer function but absent in the s-domain averaged model. A small departure between the z-domain transfer function and the $G_{udelays}$ is visible close to the Nyquist rate, due to the fact that aliasing effects are not entirely absent.

Similarly, for the boost mode, converter is simplified as shown in Fig. 7, and its operating principle is the same as boost circuit, so it's not repeated here. The coefficient matrix is (16), state vector $x=[i_L \quad v_c]^T$, input vector $v=v_b$, and output vector $y=[i_L \quad v_{dc}]^T$.

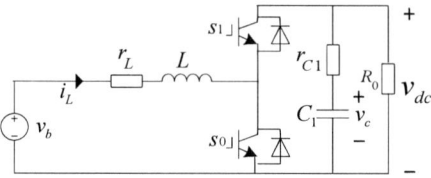

Fig. 7 The buck/boost converter in boost mode.

$$A_1 = \begin{bmatrix} -\dfrac{r_L}{L} & 0 \\ 0 & -\dfrac{1}{(R_0+r_{C_1})C_1} \end{bmatrix}$$

$$A_0 = \begin{bmatrix} -\dfrac{r_L+R_0r_{C_1}}{(R_0+r_{C_1})L} & -\dfrac{R_0}{(R_0+r_{C_1})L} \\ \dfrac{R_0}{(R_0+r_{C_1})C_1} & -\dfrac{1}{(R_0+r_{C_1})C_1} \end{bmatrix} \tag{16}$$

$$B_1 = B_0 = \begin{bmatrix} \dfrac{1}{L} & 0 \end{bmatrix}^T$$

$$C_1 = \begin{bmatrix} 1 & 0 \\ 0 & \dfrac{R_0}{R_0+r_{C_1}} \end{bmatrix}, \quad C_0 = \begin{bmatrix} 1 & 0 \\ R_0 \parallel r_{C_1} & \dfrac{R_0}{R_0+r_{C_1}} \end{bmatrix}$$

Selecting input voltage $V=V_b=255$V, output voltage $V_{dc}=380$V, considering the full load condition to find the equivalent load resistance $R_0=5.776\Omega$, then $X=[94 \quad 366]^T$. The discrete-time transfer function near the steady-state operating point is obtained as

$$G_{iu}(z) = \frac{0.4708z-0.4647}{z^2-1.937z+0.9452}$$

$$G_{vu}(z) = \frac{-0.0031z+0.0140}{z^2-1.937z+0.9452} \tag{17}$$

III. DIGITAL CONTROLER DESIGN

Just one of the switches needed to control is regarded as active switch and the other one use complementary drive signal in the converter. Use the method of state space averaging technique to get the relational expression [13], as in (18), between average inductor current and steady state duty ratio. The internal resistance of power source is ignored in order to simplify the analysis.

$$I_L = \frac{DV_{dc}-V_b}{D^2r_{dc}+r_b+r_L} \tag{18}$$

Where r_{dc} and r_b are often ignored in the analysis, so (18) can be simplified as a linear equation, and follow the diagram in [13] the relationship between current and duty ratio is shown in Fig. 8. It shows how to control the inductor current by adjusting the duty ratio. When D is larger than the critical duty ratio $D_{critical}$, the converter is working in charging area; when D is smaller than $D_{critical}$,

2444

the converter is working in discharging area. D_{min} and D_{max} are set to limit saturation and dramatic dynamic response. Besides, the inductor current is responsive to the change in duty ratio. This is the basis of using unified current controller.

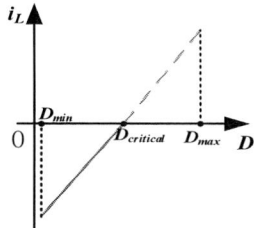

Fig. 8. The relationship between current and duty ratio.

[13] develops the unified model of the bi-directional buck/boost converter and considers many different situations. Neglecting the internal resistance in both voltage sources, the model can be simplified to first - order system, and the stability of this situation can easily meet when design the controller. Therefore, consider the worst scenario in designing the controller where the output is a resistive load and use equivalent resistance to represent different work conditions and design the digital PI controller later.

The converter is supposed to realize the function of constant voltage and constant current output. Only one current loop control in constant current mode and double loop in constant voltage mode, that is inner current loop and outer voltage loop, are required. The voltage and current loop PI controller make the converter stable, fast and accurate in buck and boost mode. Fig. 9 shows the boost modes control diagram. It is similar to the buck mode and the only difference is the voltage loop.

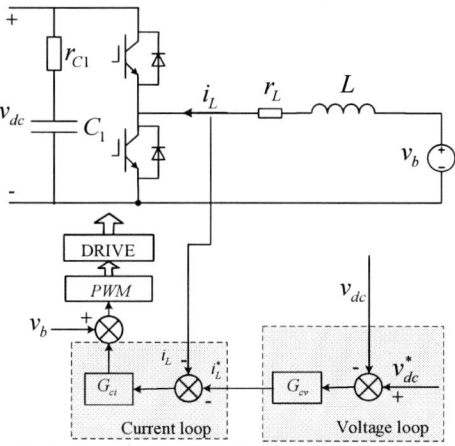

Fig. 9 The overall control diagram of boost converter.

The structure of PI controller is shown in Fig. 10 and the digital PI transfer function is

$$G_c(z) = K_p + \frac{K_i}{1 - z^{-1}} \qquad (19)$$

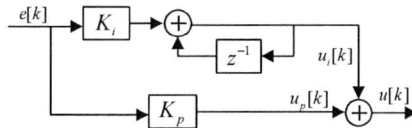

Fig. 10 Block diagram of a digital PI compensator.

From section II, the control to inductor current transfer function of the converter in both buck and boost mode are obtained, and then design the unified current controller in the worst case. The bode plot of the two modes is shown in Fig. 11. The converter has less margin in boost mode in high frequency, thus, design the unified current controller in boost mode.

After that, use the bilinear transform to map z-domain transfer function to p-domain and then directly design the controller in p-domain. P-domain is similar to the s-domain, so that the conventional continuous controller design method can be directly used. Finally, the controller parameters can be obtained in z-domain by mapping p-domain to z-domain again.

In order to adapt to the wide battery voltage variation, feedforward of battery voltage is also added to the control loop and the dynamic response will be improved.

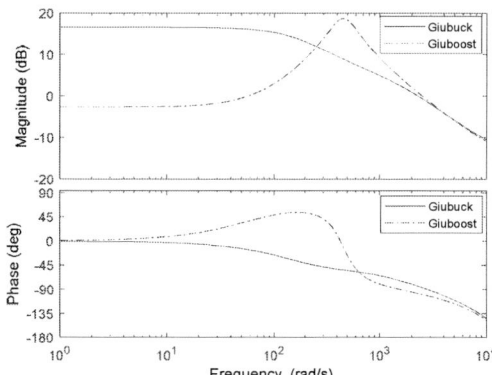

Fig. 11 Bode plot of control to inductor current transfer function in two modes.

Set the target crossover frequency and phase angle margin of the current loop to $f_{iu}=f_s/6$、 $\varphi_{im}=60°$ [16,17], respectively, and use the bilinear transform design method

$$G_{ci}(z) = 2.0174 + \frac{0.0905}{1 - z^{-1}} \qquad (20)$$

Then use the same method and get the buck and boost mode voltage controller. According to the limiting factor of the crossover frequency of the voltage loop, the crossover frequency and the phase angle margin of the selection voltage in buck mode are $f_{uu} = f_s / 30$ and $\varphi_{um} = 76°$, respectively. The PI voltage controller of buck mode is

$$G_{cv}(z) = 0.390949 + \frac{0.050735}{1 - z^{-1}} \qquad (21)$$

The target crossover frequency and phase angle margin of the voltage loop in boost mode are $f_{uu}=f_s/40$,

φ_{um}=70° respectively. The PI voltage controller of boost mode is

$$G_{cv}(z) = 6.6868 + \frac{0.2028}{1-z^{-1}} \qquad (22)$$

Simulation of the designed unified current controller in both modes is as in Fig. 12.

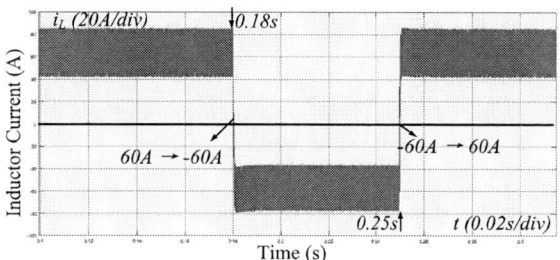

Fig. 12 Simulation results of inductor current.

From Fig. 12, the inductor current can be controlled well when the reference value is altered. Before 0.18s, the set value of inductor current is 60A and change the reference to -60A at 0.18s, then the current decreases fast and reaches stable state in a very short period of time. The similar result is also shown at 0.25s.

The effectiveness of the voltage controller is verified in two modes. The transient response of the voltage loop in buck mode is shown in Fig. 13. Before 0.1s, the outer voltage loop is not included and it has only the current loop controller. Then change the current mode to voltage mode at 0.1s and set the voltage value is 325V. After 0.004s, the voltage gets stable and is equal to the reference value.

In boost mode, the input voltage is set as the constant voltage source 325V and the output is a controllable current source instead of the load to simulate the variation of load. The reference voltage value is set to 380V. The closed-loop transient performance of the designed controller is illustrated in Fig. 14. It shows the response to the change of load of the boost converter 0A to 65A (full power charge) ,65A to -65A (full power discharge) and -65A to 65A at 0.1s, 0.2s and 0.3s, respectively. As can be seen from Fig. 14, the maximum overshoot is about 6.58%(25V). In spite of the wide range of the variation of load, the closed-loop transient performance still meets the requirements.

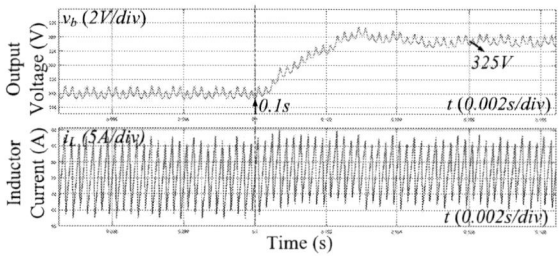

Fig. 13 Simulation results of output voltage and inductor current in buck mode.

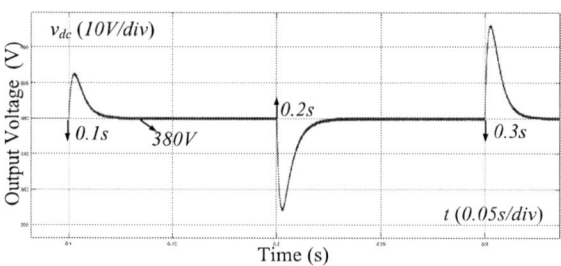

Fig. 14 Simulation result of output voltage in boost mode .

IV. EXPERIMENTAL RESULTS

Firstly, verification of the unified digital current controller using PE-Expert3 platform which is the production of Myway company is done. Regenerative power supply is used as input source and the output source is Panasonic LC-P12100ST 12V battery. The value of inductance is 1mH and its series resistance is 94mΩ. Besides the filter capacitor is not used to simplify the experiment. In this system, use the same controller design method which is shown above and obtain the PI current controller as

$$G_{ci}(z) = 0.0488 + \frac{0.0301}{1-z^{-1}} \qquad (23)$$

The Fig. 15 shows the dynamic process of constant current charge and discharge. Fig. 15(a) shows the inductor current dynamic response to the reference value variation from -2A to 2A and Fig. 15(b) shows the inductor current dynamic response to the reference value variation from 5A to -5A. This result proves that the unified digital current controller can work well in two modes and the modeling and design method are all effective.

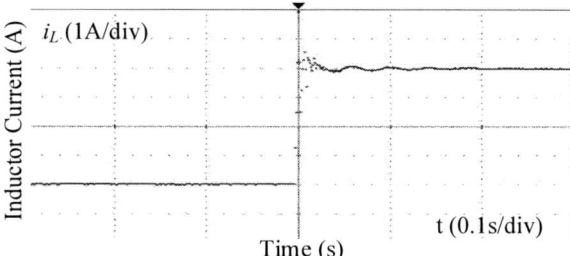

(a) The current reference value is changed from -2A to 2A.

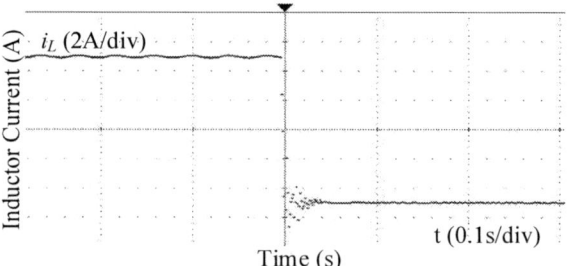

(b) The current reference value is changed from 5A to -5A.
Fig. 15 Experiment results of inductor current

2446

The 2018 International Power Electronics Conference

After that do experiments in another platform and its parameters has been shown in TABLE I. Experimental results for the system are as follows. As the DC side of the inductor current can't be measured in this platform, the measured current is the output of LC filter.

Fig. 16(a) shows the output current in the constant current charge mode from zero to the command value. The settling time of output current is about 10ms and the regulating process has almost no overshoot. Fig. 16(b) is the output current ripple in the constant current charge mode and the ripple is about 4A.

Fig. 17 shows the dynamic waveform of output voltage when DC/DC converter works in the Boost mode. Figure. 17(a) shows the dynamic response of step load from no-load to half-power discharge. Fig.17(b) shows the response of step load from half-power discharge to no-load. As can be seen from Fig.17, the output voltage overshoot is about 1.6% and the settling time is about 20ms, meeting the performance requirements.

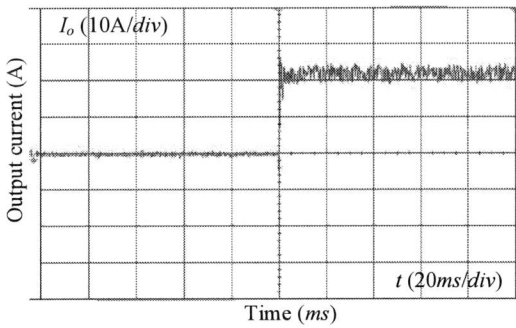

(a) Output current in constant current charge mode.

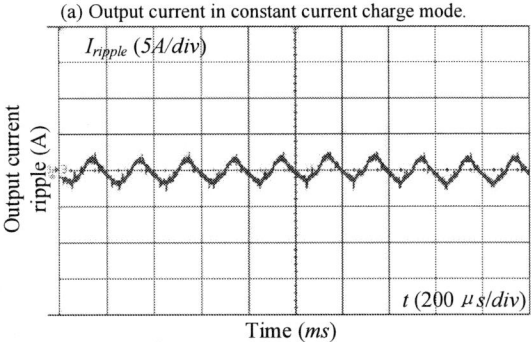

(b) Ripple of output current in constant current charge mode.

Fig. 16 Experiment results of output current and steady current ripple.

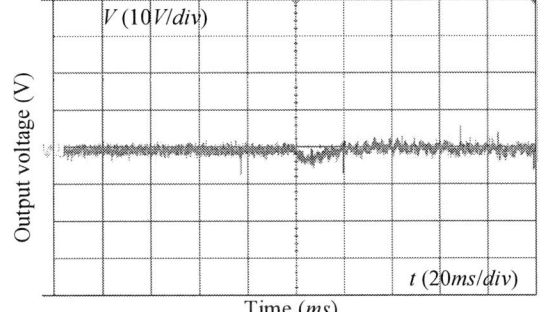

(a) Experiment results of output voltage.
(no-load to half-power discharge)

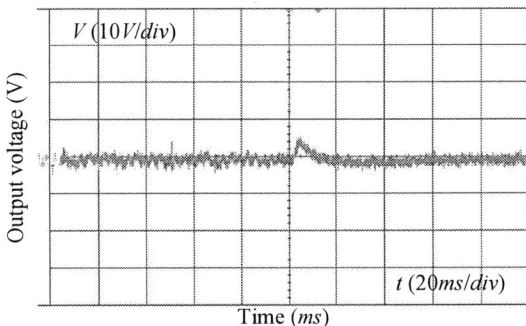

(b) Experiment results of output voltage.
(half-power discharge to no-load)

Fig. 17 Experiment results of output voltage to step load.

V. CONCLUSIONS

In this paper, the small-signal discrete-time model for the bidirectional buck/boost converter in buck and boost mode is established. And on this basis, the digital controller design flow based on bilinear transform is used to design the digital PI controller. The unified digital PI current controller is designed and the parameters of each voltage controller in Buck and Boost mode are also obtained. The simulation in MATLAB verifies the correctness of the model and digital controller design method. Finally, two hardware testing platform are employed and the experimental results further verify the model, control strategy and controller parameters design are correct and effective.

REFERENCES

[1] R. Zgheib R, Kamwa I, Al-Haddad K. Comparison between isolated and non-isolated DC/DC converters for bidirectional EV chargers[C]. Industrial Technology(ICIT), IEEE International Conference on. IEEE, 2017: 515-520.

[2] Schupbach R M, Balda J C. Comparing DC-DC converters for power management in hybrid electric vehicles[C]. Electric Machines and Drives Conference, 2003. IEMDC'03. IEEE International. IEEE, 2003, 3: 1369-1374.

[3] Zhiguo K, Chunbo Z, Shiyan Y, et al. Study of bidirectional DC-DC converter for power management in electric bus with supercapacitors[C]. Vehicle Power and Propulsion Conference, 2006. VPPC'06. IEEE. IEEE, 2006: 1-5.

[4] Jiang J, Bao Y, Wang L Y. Topology of a bidirectional converter for energy interaction between electric vehicles and the grid[J]. Energies, 2014, 7(8): 4858-4894.

[5] Erb D C, Onar O C, Khaligh A. Bi-directional charging topologies for plug-in hybrid electric vehicles[C]. Applied Power Electronics Conference and Exposition (APEC), 2010 Twenty-Fifth Annual IEEE. IEEE, 2010: 2066-2072.

[6] C Yoo C G, Lee W C, Lee K C, et al. Transient current suppression scheme for bi-directional DC-DC converters in 42 V automotive power systems[C]. Applied Power Electronics Conference and Exposition, 2005. APEC 2005. Twentieth Annual IEEE. IEEE, 2005, 3: 1600-1604.

[7] Ni L, Patterson D J, Hudgins J L. High power current sensorless bidirectional 16-phase interleaved DC-DC converter for hybrid vehicle application[J]. IEEE Transactions on Power electronics, 2012, 27(3): 1141-1151.

[8] Su N, Xu D, Chen M, et al. Study of bi-directional buck-boost converter with different control methods[C]. Vehicle Power and Propulsion Conference, 2008. VPPC'08. IEEE. IEEE, 2008: 1-5.

[9] Yang S, Goto K, Imamura Y, et al. Dynamic characteristics model of bi-directional DC-DC converter using state-space averaging

2447

method[C]. Telecommunications Energy Conference (INTELEC), 2012 IEEE 34th International. IEEE, 2012: 1-5.

[10] Pfaelzer A, Weiner M, Parker A. Bi-directional automotive 42/14 Volt bus DC/DC converter[R]. SAE Technical Paper, 2000.

[11] Rehman M M U, Zhang F, Zane R, et al. Control of bidirectional DC/DC converters in reconfigurable, modular battery systems[C]. Applied Power Electronics Conference and Exposition (APEC), 2017 IEEE. IEEE, 2017: 1277-1283.

[12] Mane J A, Jain A M. Design, modelling and control of bidirectional DC-DC converter (for EV)[C]. Emerging Research in Electronics, Computer Science and Technology (ICERECT), 2015 International Conference on. IEEE, 2015: 294-297.

[13] Zhang J, Lai J S, Yu W. Bidirectional DC-DC converter modeling and unified controller with digital implementation[C]. Applied Power Electronics Conference and Exposition, 2008. APEC 2008. Twenty-Third Annual IEEE. IEEE, 2008: 1747-1753.

[14] Ni L, Patterson D J, Hudgins J L. High power current sensorless bidirectional 16-phase interleaved DC-DC converter for hybrid vehicle application[J]. IEEE Transactions on Power electronics, 2012, 27(3): 1141-1151.

[15] Erickson R W, Maksimovic D. Fundamentals of power electronics[M]. Springer Science & Business Media, 2007.

[16] Buso S, Mattavelli P. Digital control in power electronics[J]. Lectures on power electronics, 2006, 1(1): 1-158.

[17] Franklin G F, Powell J D, Workman M L. Digital control of dynamic systems[M]. Menlo Park, CA: Addison-wesley, 1998.

[18] Chang C. Current ripple bounds in interleaved DC-DC power converters[C]. Power Electronics and Drive Systems, 1995., Proceedings of 1995 International Conference on. IEEE, 1995: 738-743.

[19] Neacsu D O, Bonnice W, Holmansky E. On the small-signal modeling of parallel/interleaved buck/boost converters[C]. Industrial Electronics (ISIE), 2010 IEEE International Symposium on. IEEE, 2010: 2708-2713.

[20] Jung M, Lempidis G, Hölsch D, et al. Optimization considerations for interleaved DC-DC converters for EV battery charging applications, in terms of partial load efficiency and power density[C]. Power Electronics and Applications (EPE'15 ECCE-Europe), 2015 17th European Conference on. IEEE, 2015: 1-9.

[21] VandeSype D M, DeGusseme K, DeBelie F M L L, et al. Small-signal z-domain analysis of digitally controlled converters[J]. IEEE Transactions on Power Electronics, 2006, 21(2): 470-478.

[22] Van de Sype D M, De Gusseme K, Van den Bossche A P, et al. Small-signal Laplace-domain analysis of uniformly-sampled pulse-width modulators[C]. Power Electronics Specialists Conference, 2004. PESC 04. 2004 IEEE 35th Annual. IEEE, 2004, 6: 4292-4298.

The 2018 International Power Electronics Conference

Energy Management of Hydrogen-Storage Photovoltaic Generation System with a Function of Suppressing Short-Period Components

Yuuki Machida[1*], Akihisa Goto[1], Akiko Takahashi[1] and Shigeyuki Funabiki[1]

1 Graduate School of Natural Science and Technology, Okayama University, Okayama, Japan

*E-mail: yuuki.machida@s.okayama-u.ac.jp

Abstract— The output power of photovoltaic generation (PV) systems changes with the variation in the solar irradiance and temperature of the PV panel surface. The change in output power influences the power quality of the power system. Therefore, a PV system with an electrolyzer (ELY) is proposed to prevent the degradation of the power quality. The proposed system converts the fluctuating components in the PV power fluctuations into hydrogen, which is supplied to fuel cell vehicles. To realize a stable hydrogen supply chain, it is necessary to manage the hydrogen produced by the ELY. This paper proposes a novel method for energy management of a hydrogen-storage PV system with a function of suppressing the short-period components in the PV power. It is verified that the proposed energy management method is effective for reducing the fluctuations of the PV power and the establishment of a hydrogen supply chain.

Keywords— *Photovoltaic generation, Electrolyzer, Hydrogen storage, Energy management*

I. INTRODUCTION

In recent years, a large number of photovoltaic generation (PV) systems have been installed in numerous countries as a measure to counter global warming. Particularly, the installation of large-capacity PV systems is being promoted globally. The large-capacity PV systems connected to the grid produce power fluctuations [1]. The power fluctuations in the range of load frequency control (LFC) and economic load dispatching control (EDC) degrades the power quality of the grid.

Power smoothing control methods using an energy storage system (ESS) have been proposed and discussed in terms of its effect on power smoothing. A battery and a system combining a fuel cell and an electrolyzer have been proposed as an ESS [2]–[7]. The authors have also proposed a smoothing control method for the ESS using ELY. This control method smooths the fluctuations of the PV power by converting the fluctuating power components into hydrogen [8]. The hydrogen produced may be used as fuel for fuel cell vehicles (FCVs). The amount of hydrogen produced varies across days because the amount of solar radiation changes with the weather

during the day. Therefore, it is necessary to develop a system for managing the amount of hydrogen produced.

This paper proposes a novel control method for managing the amount of hydrogen produced for the hydrogen-storage PV system with ELY. The proposed method adjusts the reverse power flowing to the grid based on the amount of hydrogen stored in the hydrogen storage tank. If the amount of stored hydrogen is less than the target value, the power to the ELY is supplied from the grid during nighttime to replenish the shortage of hydrogen.

II. PV SYSTEM WITH ELECTROLYZER

A. System Configuration

Fig. 1 shows the hydrogen-storage PV system with the ELY. This system consists of the PV system and the ESS incorporating an ELY and a hydrogen storage tank. The component of short-period fluctuation in the PV power that degrades the power quality of the grid is converted to hydrogen by the ELY. The produced hydrogen is stored in the hydrogen storage tank. The hydrogen stored in the hydrogen storage tank is used as fuel for FCVs. Meanwhile, the component of long-period fluctuation in the PV power flows to the high-voltage distribution system. To regulate the amount of hydrogen supplied to the FCVs, the proposed method adjusts the reverse power flowing to the grid. When the amount of hydrogen stored in the hydrogen storage tank is less than the target value, the power to the ELY is supplied from the grid in order to produce hydrogen during nighttime.

B. Smoothing of PV power

A command value of the power supplied to the ESS for producing hydrogen, P^*_{ES}, is expressed by the following equation.

$$P^*_{ES} = P_{PV} - P^*_{S} \qquad (1)$$

where P_{PV} is the PV power and P^*_{S} is the command value

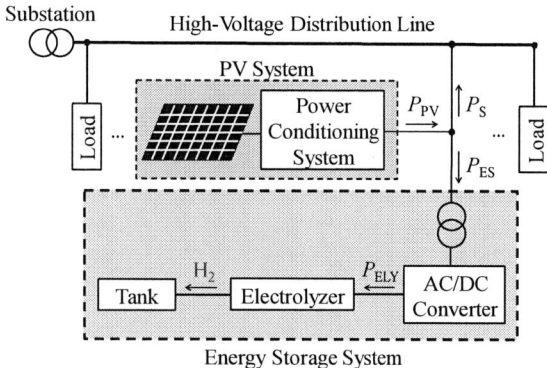

Fig. 1. System Configuration of hydrogen-storage PV System.

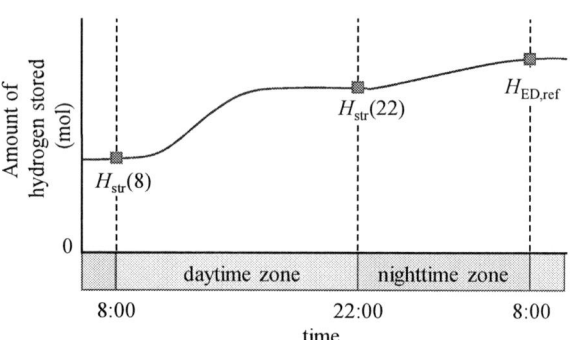

Fig. 2. Time zone of hydrogen management in nth control day.

of the reverse power flowing to the grid. P^*_S is expressed by the following equation.

$$P^*_S(k) = \beta(k)P_{EMA}(k) \qquad (2)$$

where $\beta(k)$ is a coefficient indicating the local minimum value of P_{PV}/P_{exp} in the section with the time window T_W (h). P_{EMA} is the power after smoothing the PV power, and k is the control period. Because the ESS cannot output power, P^*_S is smaller than P_{PV}. Therefore, P_{EMA} is multiplied by β ($0 \leqq \beta \leqq 1$) to determine P^*_S. $\beta(k)$ is selected based on the smallest value β_1 and the second smallest value β_2 among the local minimum values [8]. $P_{EMA}(k)$ is expressed by

$$P_{EMA}(k) = (1-\alpha)P_{EMA}(k-1) + P_{PV}(k) . \qquad (3)$$

where α is the smoothing constant. Therefore, the component of long-period fluctuation in the PV power flows to the grid. Meanwhile, the component of short-period fluctuation in the PV power flows to the ELY.

III. MANAGEMENT OF AMOUNT OF HYDROGEN

A. Method of Managing Amount of Hydrogen

We propose a method to manage the hydrogen production and the amount of hydrogen stored in the hydrogen storage tank. The number of days for managing hydrogen and the target amount of hydrogen produced during the period of managing hydrogen are defined in the proposed method. The target amount of hydrogen produced in a day is determined based on the amount of hydrogen stored in the hydrogen storage tank on the start time for managing hydrogen. Furthermore, the proposed method adjusts the power flowing to the grid such that the amount of hydrogen stored in the hydrogen storage tank approaches the target value. Fig. 2 shows the time zone of hydrogen management. In the time-of-day rate plan, the daytime zone is from 8:00 to 22:00 and nighttime zone is from 22:00 to 8:00 of the following day [9]. Therefore, we propose a method for managing hydrogen in the two time-zones. The start time for managing hydrogen on a day is 8:00, and the end time is

8:00 on the following day. The control method for each time zone is described below.

B. Daytime Zone

The target amount of hydrogen produced in the daytime zone $H_{DT,ref}$ (mol) is determined through the following procedure. First, the amount of hydrogen $H_{str}(8)$ stored at 8:00 is detected. Next, the target amount of hydrogen stored at the end time $H_{ED,ref}$ on the nth control day is calculated using the following equation.

$$H_{ED,ref}[n] = n \times \frac{H_{D,ref}}{N_D} \qquad (4)$$

where N_D (day) is the number of days for managing hydrogen, and $H_{D,ref}$ (mol) is the target amount of hydrogen produced during the period of managing hydrogen. Further, n (day) ($1 \leq n \leq N_D$) indicates the nth control day. Using the $H_{ED,ref}$ in (4) and the detected $H_{str}(8)$, the target amount of hydrogen produced on the nth control day, $H_{DT,ref}$, is determined using the following equation.

$$H_{DT,ref}[n] = H_{ED,ref}[n] - H_{str}(8)[n] \qquad (5)$$

It is necessary to bring the amount of hydrogen produced on the nth control day closer to $H_{DT,ref}$. Therefore, the command value of the reverse power to the grid, $P^*_{S,mod}$, is obtained by multiplying P^*_S by ε ($0 \leq \varepsilon \leq 1$).

$$P^*_{S,mod} = P^*_S \times \varepsilon \qquad (6)$$

ε is a coefficient required when determining the the command value of the reverse power to the grid. ε in (6) is determined based on $H_{DT,ref}$ calculated in (5). ε corresponding to $H_{DT,ref}$ is determined using past data of the PV power from April 2010 to March 2011. The period from January to June is defined as the first half of the year (first half), and that from July to December is defined as the second half of the year (second half). The value of ε for the first half and the second half are determined using the PV power of each period. ε is derived through the following steps.

Step 1: P^*_S is calculated using (2). The command value of the reverse power flowing to the grid, $P^*_{S,mod}$, is determined at each 0.1 step of ε in (6).

2450

Step 2: P^*_S in (1) is replaced with $P^*_{S,mod}$ to obtain P^*_{ES}.

Step 3: The amount of hydrogen produced in one day is calculated for ε by integrating the amount of hydrogen produced by the ELY.

Step 4: Classification of the day according to the range of the amount of hydrogen is executed.

Step 5: Count the number of days classified according to the range of the amount of hydrogen is executed. The ε for which the number of days is the largest is selected from the same range of amount of hydrogen.

Step 6: Replace the amount of hydrogen in Step 5 with $H_{DT,ref}$, and determine ε in (6) based on $H_{DT,ref}$.

C. Nighttime Zone

When the amount of hydrogen produced during daytime does not attain $H_{DT,ref}$, the ELY must obtain power from the grid in order to produce hydrogen. Therefore, the proposed method decides at 22:00 whether the ELY receives power from the grid. The amount of hydrogen stored in the hydrogen storage tank at 22:00, $H_{str}(22)$, is detected. Then, the amount of hydrogen produced from the following day to the end of the period of managing hydrogen on the nth control day, H_{total}, is estimated using the following equation.

$$H_{total}[n] = (N_D - n)(H_{PV} + H_{NT,max}) \qquad (7)$$

where H_{PV} (mol/day) is the estimated amount of hydrogen produced using the PV power for each day. $H_{NT,max}$ (mol/day) is the maximum amount of hydrogen produced using power from the grid in the nighttime zone for each day. H_{PV} is a constant arbitrarily determined at the start of managing hydrogen.

When the sum of H_{total} and $H_{str}(22)$ is smaller than $H_{D,ref}$, the ELY is supplied power from the grid. Meanwhile, when the sum of H_{total} and $H_{str}(22)$ is larger than $H_{D,ref}$, it is not necessary to supply power to the ELY from the grid. When the power to the ELY is supplied from the grid, the command value of hydrogen produced using power from the grid in the nighttime zone on the nth control day, H^*_{NT} (mol), is obtained using the following equation.

$$H^*_{NT}[n] = H_{D,ref} - (H_{total}[n] + H_{str}(22)[n]) \qquad (8)$$

The energy from the grid that is necessary to produce H^*_{NT} on the nth control day, W_{NT} (Ws), is calculated by

$$W_{NT}[n] = \frac{H^*_{NT}[n]}{V_{HELY}} \qquad (9)$$

where V_{HELY} (mol/Ws) is the amount of hydrogen produced at 1 Ws. The command value of power from the grid to the ELY on the nth control day, P^*_{NT} (W), is expressed by

$$P^*_{NT}[n] = \frac{W_{NT}[n]}{3600 \times \eta_{trans} \times \eta_{conv}} \qquad (10)$$

where η_{trans} and η_{conv} are the efficiencies of the transformer and the AC/DC converter, respectively. P^*_{NT}

is assumed to be constant in the nighttime zone of 10 hours and, equal to $P_{NT,max}$ when P^*_{NT} is larger than $P_{NT,max}$.

IV. EVALUATION METHOD

The proposed method is evaluated based on the electricity charge incurred in the nighttime zone and the reduction in LFC and EDC flowing to the grid. The electricity rate at nighttime is 9.68 JPY/kWh [9]. The power component in the LFC flowing to the grid, P_{LFC} (W), is calculated by

$$P_{LFC} = \sqrt{\sum_{m=f_2}^{f_1} P_S(m)^2} \qquad (11)$$

where $P_S(m)$ is the reverse power flowing to the grid, and f_1 and f_2 are the boundary frequencies of the LFC. $f_1 = 1/120$ and $f_2 = 1/1200$ because the period of the components in the LFC is from 2 minutes to 20 minutes. Similarly, the component in the EDC, P_{EDC} (W), is calculated by substituting $f_1 = 1/120$ and $f_2 = 0$ in (11) because the period of EDC is 20 minutes or higher. The reduction in LFC flowing to the grid, G_{LFC} (dB), is expressed by the following equation.

$$G_{LFC} = 20 \times \log_{10} \frac{P_{LFC(total)}}{P_{PV,LFC(total)}} \qquad (12)$$

where $P_{LFC(total)}$ (W) is the total value of P_{LFC} for a year, and $P_{PV,LFC(total)}$ (W) is the total value of the component of LFC included in the PV power for a year. Similarly, The reduction in EDC flowing to the grid, G_{EDC} (dB), is calculated by replacing $P_{LFC(total)}$ in (12) with $P_{EDC(total)}$ and $P_{PV,LFC(total)}$ in (12) with $P_{PV,EDC(total)}$. Here, $P_{EDC(total)}$ (W) is the total value of P_{EDC} for a year, and $P_{PV,EDC(total)}$ (W) is the total value of the component of EDC included in the PV power for a year.

V. VERIFICATION OF PROPOSED METHOD

A. Conditions of Managing Amount of Hydrogen

The rating of the PV system is 800 kW. The conversion efficiencies of the transformer and the AC/DC converter are both 98%. The efficiency of the ELY is 80% [10], [11] and the operation efficiency of the hydrogen storage tank is 100%. The constant of smoothing control α is 5.10×10^{-2} when the gain of (2) is -10 dB at the frequency 1/1200 Hz. Furthermore, the control cycle is 30 sec, T_w is 2 h, and N_D is 28 days.

B. Determination of Parameters

The determination method of ε is described following. The number of days in the first half classified in the range of the amount of hydrogen (H_{amount}) counted in the step 5 described above is shown in Table I(a), and the number of days in the second half is shown in Table I(b). The value of ε for which the number of days is the largest is selected in the same the range of amount of hydrogen in

2451

TABLE I
NUMBER OF DAYS FOR ε AND AMOUNT OF HYDROGEN
(a) First half

		ε										
		0	0.1	0.2	0.3	0.4	0.5	0.6	0.7	0.8	0.9	1.0
	$0 \leq H_{amount} < 5$	4	5	6	6	7	9	11	15	17	30	52
	$5 \leq H_{amount} < 10$	12	12	12	18	19	22	30	37	47	62	58
	$10 \leq H_{amount} < 15$	15	20	28	26	30	34	32	49	56	47	41
Amount of	$15 \leq H_{amount} < 20$	25	24	26	33	35	41	46	42	41	28	22
hydrogen	$20 \leq H_{amount} < 25$	27	30	26	28	36	37	43	25	13	11	5
H_{amount}	$25 \leq H_{amount} < 30$	25	25	30	30	25	27	13	10	4	0	0
(kmol)	$30 \leq H_{amount} < 35$	21	25	19	22	23	8	3	0	0	0	0
	$35 \leq H_{amount} < 40$	18	14	25	15	3	0	0	0	0	0	0
	$40 \leq H_{amount}$	31	23	6	0	0	0	0	0	0	0	0

(b) Second half

		ε										
		0	0.1	0.2	0.3	0.4	0.5	0.6	0.7	0.8	0.9	1.0
	$0 \leq H_{amount} < 5$	10	11	12	12	12	12	14	17	22	32	50
	$5 \leq H_{amount} < 10$	7	8	11	11	15	18	21	42	49	57	62
	$10 \leq H_{amount} < 15$	15	18	18	24	35	44	47	41	52	42	32
Amount of	$15 \leq H_{amount} < 20$	23	30	34	35	32	33	42	38	22	16	4
hydrogen	$20 \leq H_{amount} < 25$	31	24	26	27	29	32	20	9	3	1	0
H_{amount}	$25 \leq H_{amount} < 30$	20	22	22	27	21	8	4	1	0	0	0
(kmol)	$30 \leq H_{amount} < 35$	17	22	18	10	4	1	0	0	0	0	0
	$35 \leq H_{amount} < 40$	16	7	7	2	0	0	0	0	0	0	0
	$40 \leq H_{amount}$	9	6	0	0	0	0	0	0	0	0	0

each of the first half and the second half. For example, in the case of $0 \leq H_{amount} < 5$ shown in Table I(a), because the number of days of $\varepsilon = 1.0$ is the largest, $\varepsilon = 1.0$ for $0 \leq H_{amount} < 5$ is selected. However, if ε for which the largest number of days is plurality, the larger ε is selected. For example, in the case of $25 \leq H_{amount} < 30$ shown in Table I (a), ε with the largest number of days is 0.2 and 0.3. In this case, where ε for which the largest number of days is a plurality, the larger ε is selected; therefore, $\varepsilon = 0.3$ is selected for $25 \leq H_{amount} < 30$. Table II shows ε of each of the range of amount of hydrogen.

ε in (6) is determined according to $H_{DT,ref}$ calculated by (5). Replace H_{amount} in Table II with $H_{DT,ref}$, and determine ε in (6) based on $H_{DT,ref}$ in (5). Because $H_{DT,ref}$ is different on each day, ε is different on each day.

The optimum values of H_{PV} and $P_{NT,max}$ are decided by using the data of the PV power from April 2010 to March 2011, where the data of 28 continuous days is defined as one period data. H_{PV} and $P_{NT,max}$ for the first half and the second half are determined using each of ε shown in Table II.

The amount of hydrogen stored in the hydrogen storage tank during the period of managing hydrogen is calculated for H_{PV} from 10 kmol/day to 35 kmol/day and $P_{NT,max}$ from 10 kW to 100 kW. The average error of hydrogen production for a year is calculated for the pair of H_{PV} and $P_{NT,max}$; then, the pair is selected when the error is the smallest. This is the absolute error between the amount of hydrogen produced and the target value in the period of managing hydrogen.

TABLE II
ε FOR THE HIGHEST INCIDENCE OF PRODUCING AMOUNT OF HYDROGEN
(a) First half

Amount of hydrogen H_{amount} (kmol)	ε
$0 \leq H_{amount} < 5$	1.0
$5 \leq H_{amount} < 10$	0.9
$10 \leq H_{amount} < 15$	0.8
$15 \leq H_{amount} < 20$	0.6
$20 \leq H_{amount} < 25$	0.6
$25 \leq H_{amount} < 30$	0.3
$30 \leq H_{amount} < 35$	0.1
$35 \leq H_{amount} < 40$	0.2
$40 \leq H_{amount}$	0

(b) Second half

Amount of hydrogen H_{amount} (kmol)	ε
$0 \leq H_{amount} < 5$	1.0
$5 \leq H_{amount} < 10$	1.0
$10 \leq H_{amount} < 15$	0.8
$15 \leq H_{amount} < 20$	0.6
$20 \leq H_{amount} < 25$	0.5
$25 \leq H_{amount} < 30$	0.3
$30 \leq H_{amount} < 35$	0.1
$35 \leq H_{amount} < 40$	0
$40 \leq H_{amount}$	0

Next, the amount of hydrogen stored in the hydrogen storage tank during the period of managing hydrogen is calculated using the selected H_{PV} by changing $P_{NT, max}$ from 10 kW to 400 kW. $P_{NT,max}$ is selected when the variation of the average error of hydrogen production and the rate of hydrogen production in the nighttime zone for a year is –0.1% or less with increasing $P_{NT,max}$. The rate of hydrogen production in the nighttime zone is the ratio of the amount of hydrogen produced using the power of the grid and the amount of hydrogen produced for a day.

Table III shows the optimum values of H_{PV} and $P_{NT,max}$ for $H_{D,ref}$ for the first half and the second half selected in the above procedure. H_{PV} for all $H_{D,ref}$ of each of the first half and the second half is 10 kmol/day. However, $P_{NT,max}$ of each the first half and the second half are different. Therefore, ε and $P_{NT,max}$ are defined for a year. H_{PV} is fixed to 10 kmol/day, and the management of the amount of hydrogen for all the combinations of ε shown in Table II and $P_{NT,max}$ shown in Table III for each of $H_{D,ref}$ in each period of the first half and the second half is carried out. Therefore, 16 types of the management of the amount of hydrogen ($H_{D,ref}$: four types, H_{PV} : one type, $P_{NT,max}$: two types, ε : two types) are carried out in each period of the first half and the second half. Based on the result, the combination of ε, H_{PV}, and $P_{NT,max}$ with the smallest average error of hydrogen production is designated as the set of optimum values of each of the three. However, when the average errors of hydrogen production are identical, the combination with the smallest rate of hydrogen production in the nighttime zone is designated as the set of optimum values. The average error of hydrogen production and the rate of hydrogen production in the nighttime zone for each combination of $H_{D,ref}$, H_{PV}, $P_{NT,max}$, and ε in each period of the first half and the second half is shown in Table IV. The first half of the ε item shown in Table IV indicates that the amount of hydrogen is managed by using the ε shown in Table II(a). Meanwhile, the second half of the ε item shown in Table IV implies that the amount of hydrogen is managed using the ε shown in Table II(b). A combination of H_{PV}, $P_{NT,max}$ and ε for each of $H_{D,ref}$ is selected in the above procedure.

Table V(a) shows the optimal values of each $H_{D,ref}$ for the first half determined from Table IV(a), and Table V(b) shows the optimal values of each $H_{D,ref}$ for the second half determined from Table IV(b). Table V(c) show the optimum values of each $H_{D,ref}$ for a year determined from Table V(a) and Table V(b). The first half of the ε item shown in Table V indicates that the amount of hydrogen is managed by using the ε shown in Table II(a). Meanwhile, the second half of the ε item shown in Table V implies that the amount of hydrogen is managed using the ε shown in Table II(b). Comparing Table V(a) with Table V(b), only the optimum values of $H_{D,ref}$ = 420 kmol are different. Furthermore, focusing on ε in the first half and the second half, only ε for $H_{D,ref}$ = 420 kmol in the first half is different. Because it is desirable to set ε = second half for a year for all $H_{D,ref}$, we examined whether it is feasible to set ε = second half for a year. When $H_{D,ref}$ = 420 kmol, H_{PV} = 10 kmol/day, $P_{NT,max}$ = 110 kW, and ε

TABLE III
OPTIMUM VALUE OF H_{PV} AND $P_{NT,max}$

(a) First half

$H_{D,ref}$ (kmol)	280	420	560	700
H_{PV} (kmol/day)	10	10	10	10
$P_{NT,max}$ (kW)	20	120	210	300

(b) Second half

$H_{D,ref}$ (kmol)	280	420	560	700
H_{PV} (kmol/day)	10	10	10	10
$P_{NT,max}$ (kW)	110	110	170	210

= first half, the average error of hydrogen production in the first half is 0.47%. Meanwhile, when $H_{D,ref}$ = 420 kmol, H_{PV} = 10 kmol/day, $P_{NT,max}$ = 120 kW, and ε = second half, the average error of hydrogen production in the first half is 0.59%. The average error of hydrogen production difference between the above two conditions is 0.12%, which is negligible. Therefore, replacing the optimum values of $H_{D,ref}$ = 420 kmol in the first half with the optimal values of $H_{D,ref}$ = 420 kmol in the second half dose not affected the average error of hydrogen production. Based on the above, the optimal values in the second half are adopted for the optimum values of each $H_{D,ref}$ for a year.

C. Result of Hydrogen Management

Fig. 3 shows the simulation results of managing the amount of hydrogen for $H_{D,ref}$ = 420 kmol, $P_{NT,max}$ = 120 kW and H_{PV} = 10 kmol/day from January 2 to 30, 2011. The proposed hydrogen management method calculates $H_{DT,ref}$ and determines ε based on $H_{DT,ref}$ each day. Therefore, ε is changed each day to adjust the amount of hydrogen produced. If the target amount of hydrogen is not achieved with only the amount of hydrogen produced by using the PV power in the daytime zone, power from the grid is used to adjust the amount of hydrogen in the nighttime zone. Therefore, at the end of the period of managing hydrogen, the amount of hydrogen stored in the hydrogen storage tank is adjusted to be the target amount of hydrogen produced during the period of managing hydrogen ($H_{D,ref}$) by using power from the grid in the nighttime zone. Consequently, the amount of hydrogen stored in the hydrogen storage tank coincides with 450 kmol on the final day of the period of managing hydrogen. Therefore, the simulation result is consistent with the operation of the proposed hydrogen management method.

Table VI shows the average error of hydrogen production of each $H_{D,ref}$ for a year. The average errors of hydrogen production for a year for $H_{D,ref}$ = 280 kmol is approximately 3%. Meanwhile, The average errors of hydrogen production for a year for $H_{D,ref}$ = 420 kmol, 560 kmol, and 700 kmol is less than 1%. Because the average errors of hydrogen production for a year of all $H_{D,ref}$ is small, the values shown in Table V(c) are effective.

TANBLE IV
Average Error of Hydrogen Production for Each Combination of ε, H_{PV}, and $P_{NT,max}$
(a) First half

$H_{D,ref}$ (kmol)	H_{PV} (kmol/day)	P_{NT_max} (kW)	ε	Average error of hydrogen production (%)	Rate of hydrogen production in the nighttime zone (%)
280	10	20	first half	5.23	0.36
280	10	110	first half	5.15	0.19
280	10	20	second half	4.86	0.46
280	10	110	second half	4.74	0.10
420	10	110	first half	0.47	1.52
420	10	120	first half	0.52	1.57
420	10	110	second half	0.63	1.60
420	10	120	second half	0.59	1.45
560	10	170	first half	0.32	2.96
560	10	210	first half	0.32	2.44
560	10	170	second half	0.32	2.39
560	10	210	second half	0.32	2.11
700	10	210	first half	0.25	9.94
700	10	300	first half	0.25	8.97
700	10	210	second half	0.25	9.19
700	10	300	second half	0.25	8.20

(b) Second half

$H_{D,ref}$ (kmol)	H_{PV} (kmol/day)	P_{NT_max} (kW)	ε	Average error of hydrogen production (%)	Rate of hydrogen production in the nighttime zone (%)
280	10	20	first half	2.72	2.14
280	10	110	first half	1.48	2.69
280	10	20	second half	2.22	2.04
280	10	110	second half	0.94	2.53
420	10	110	first half	0.54	4.58
420	10	120	first half	0.57	4.49
420	10	110	second half	0.53	3.76
420	10	120	second half	0.51	3.76
560	10	170	first half	0.37	7.99
560	10	210	first half	0.37	7.81
560	10	170	second half	0.37	7.36
560	10	210	second half	0.37	7.26
700	10	210	first half	0.29	15.4
700	10	300	first half	0.29	15.0
700	10	210	second half	0.29	15.2
700	10	300	second half	0.27	14.9

TABLE V
Optimum Values of ε, H_{PV}, and $P_{NT,max}$
(a) First half

$H_{D,ref}$ (kmol)	280	420	560	700
ε	second half	first half	second half	second half
H_{PV} (kmol/day)	10	10	10	10
$P_{NT,max}$ (kW)	110	110	210	300

(b) Second half

$H_{D,ref}$ (kmol)	280	420	560	700
ε	second half	second half	second half	second half
H_{PV} (kmol/day)	10	10	10	10
$P_{NT,max}$ (kW)	110	120	210	300

(c) For a year

$H_{D,ref}$ (kmol)	280	420	560	700
ε	second half	second half	second half	second half
H_{PV} (kmol/day)	10	10	10	10
$P_{NT,max}$ (kW)	110	120	210	300

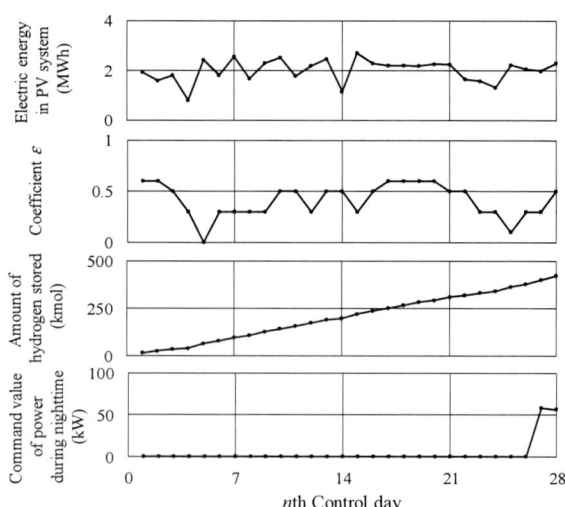

Fig. 3. Simulation results from January 2 to 30, 2011.

TABLE VI
AVERAGE ERROR OF HYDROGEN PRODUCTION FOR EACH $H_{D,ref}$ FOR
A YEAR

$H_{D,ref}$ (kmol)	280	420	560	700
Average error of hydrogen production for a year (%)	2.8	0.55	0.34	0.26

D. Evaluation of the proposed method

Fig. 4 shows the reduction in LFC and EDC. The reduction in LFC becomes lower than −10 dB because the filter gain α decided is less than −10 dB in the LFC range. Meanwhile, the reduction in EDC is −10 dB or less in some cases; however, its value increases with increase in $H_{D,ref}$. In the case of $H_{D,ref}$ = 280 kmol, the electric energy supplied from the grid to the ELY in the nighttime zone is 3.8 MWh for a year, so that the electricity charge is approximately 40 000 JPY. Then, the electric charges are approximately 120 000 JPY and 270 000 JPY for $H_{D,ref}$ = 420 kmol and 560 kmol, respectively. Meanwhile, the charge for $H_{D,ref}$ = 700 kmol exceeds 800 000 JPY and is expensive. To reduce the electricity charges during nighttime, it is necessary to reduce the amount of hydrogen produced by using power from the grid. Therefore, it is necessary to consider a method to reduce the amount of hydrogen produced using power from the grid.

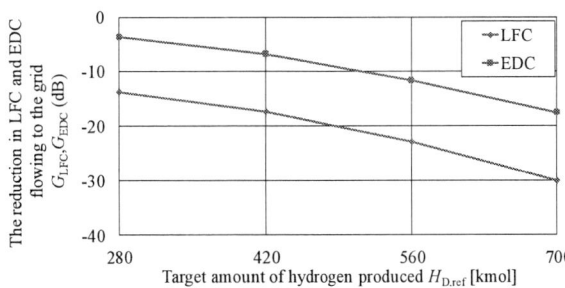

Fig. 4. Reduction in LFC and EDC.

amount of hydrogen produced by using power from the grid. Therefore, it is necessary to consider a method to reduce the amount of hydrogen produced using power from the grid.

VI. CONCLUSIONS

We proposed a novel method to manage the amount of hydrogen by regulating the power flowing to the grid by adjusting the coefficient ε in the command of reverse power. A simulation was executed to demonstrate the effectiveness of the proposed method. It was clarified that changing ε each day according to the deviation between the amount of hydrogen stored in the hydrogen storage tank and its target value is effective for managing the hydrogen production.

The proposed method is evaluated based on the electricity charge incurred in the nighttime zone and the reduction in LFC and EDC flowing to the grid. The reduction in LFC becomes lower than −10 dB because the filter gain α decided is less than −10 dB in the LFC range. Meanwhile, the reduction in EDC is −10 dB or less in some cases; however, its value increases with increase in $H_{D,ref}$. It is clarified that the proposed management of the amount of hydrogen is effective. The electricity charges incurred in the nighttime zone were approximately 40 000 JPY, 120 000 JPY, and 270 000 JPY for $H_{D,ref}$ = 280 kmol, 420 kmol, and 560 kmol for a year, respectively. However, it became more than 800 000 JPY for $H_{D,ref}$ = 700 kmol. To reduce the electricity charges during nighttime, it is necessary to reduce the

REFERENCES

[1] A. Sangwongwinich, Y. Yang, F. Blaabjerg, and D. Sera, "Delta Power Control Strategy for Multistring Grid - Connected PV Inverters," *IEEE Trans. Ind. Appl.*, Vol.53, No.4, PP.3862-3870, 2017.

[2] X. Li, D. Hui, and X. Lai, "Battery Energy Storage Station (BESS) – Based Smoothing Control of Photovoltaic (PV) and Wind Power Generation Fluctuation," *IEEE Trans. Sustain. Energy*, Vol.4, No.2, pp.464-473, 2013.

[3] G. Wang, M. Ciobotaru, and V. G. Agelidis, "Power Smoothing of Large Soral PV Plant Using Hybrid Energy Storage," *IEEE Trans. Sustain. Energy*, Vol.5, No.3, pp.834-842, 2014.

[4] S. Kim, S. Bae, Y. C. Kang, and J. Park, "Energy Management Based on the Photovoltaic HPCS with an Energy Storage Device," *IEEE Trans. Ind. Electron.*, Vol.62, No.7, pp.4608-4617, 2015.

[5] S. G. Tesfahunegn, Ø. Ulleberg, P. J. S. Vie, and T. M. Undeland, "PV Fluctuation Balancing Using Hydrogen Storage - a Smoothing Method for Integration of PV Generation into the Utility Grid," *Energy Procedia*, pp.1015-1022, 2011.

[6] B. Escobar, J. Hernández, R. Barbosa, and Y. Verde-Gómez, "Analytical Model as a Tool for the Sizing of a Hydrogen Production System Based on Renewable Energy: The Mexican Caribbean as a Case of Study," *Int. J. Hydrogen Energy*, pp.12562-12569, 2013.

[7] P. Poggi, C. Darras, M. Muselli, and G. Piglet, "The PV - Hydrogen Platform - PV Output Power Fluctuations Smoothing," *Energy Procedia*, pp.607-616, 2014.

[8] N. Okada, A. Takahashi, J. Imai, and S. Funabiki, "Discussion on Time Window for Power Smoothing Control in Photovoltaic System using a water electrolyzer," *2017 Kansai Joint Convention Institutes Elec. Eng.* (in Japanese)

[9] The Chugoku Electric Power Company, Incorporated, "Table for Electric Charge Unit Price (for Businesswork)," http://www.energia.co.jp/elec/b_menu/h_volt3/pricelist_gyoumu.html (2017.8.6).

[10] M. Fisher, "Review of Hydrogen Production with Photovoltaic Electrolysis System," *Hydrogen Energy.*, Vol.11, No.8, pp.495-501, 1986.

[11] T. Maeda, N. Endo, S. Suzuki, and K. Goshome, "Performance of 5Nm3/h class Unitized Reversible Fuel Cell System with Photovoltaic Power Generation," *Int. Conf. Elec. Eng. 2016 (ICEE2016)*.

A Dynamic Battery Charging Approach for Energy Trading in the Smart Grid

Avinash Sharma and Akshay Kumar Rathore
Department of Electrical & Computer Engineering,
Concordia University, Montreal, Canada
Email: avinashmnit30@gmail.com, akshay.k.rathore@ieee.org

Rajesh Kumar
Department of Electrical Engineering
MNIT Jaipur, India
Email:rkumar@ieee.org

Abstract—Battery systems are going to play a major role in the emergence of the smart grid system. In this paper, we examined the effect of the battery system dynamics in the energy trading from the end-user perspective. In particular, a novel energy trading framework called Dynamic Battery Charging is developed to use the battery storage for energy trading strategically. Here, the market is considered to be partially decentralized with a two-way trade framework. The end-user is free to decide the amount of energy to bilaterally trade with the central grid. The proposal is tested and validated through a case study of three different load profiles (different in scale) in three energy markets. The simulation results clearly show the profitability of the proposed strategy in all test benchmarks. The proposed strategy resulted in $10 - 30\%$ profit in Ontario, California and New-York market. Further, the prospect of reduced battery prices makes the proposal even more attractive.

I. INTRODUCTION

Last few decades have witnessed a rapid increase in the energy demand around the world. With an ever-increasing human population, growing popularity of energy-dependent lifestyle, industrial energy demand and a shift towards Electric Vehicles (EV), the overall energy demand is certain to continue rise at this rate. If the WEC estimates [1] are to be believed then by the year 2050 the energy consumption is estimated to increase by at least 30%. Along with this, the conventional energy sources are running out of popularity due to their role in increased air pollution and greenhouse emissions. Further, the recent studies [2], [3] have shown the long-term benefits of the renewable energy sources on economic growth of a country. All this together has resulted in an increasing demand for the clean sources. The front-runner in this shift towards renewable sources is the solar power which allows for a large scale decentralization of the energy generation and distribution process.

This change towards renewable sources, although inevitable, requires a large scale and efficient integration with the present electricity systems [4]. However, this process is not without problems. Renewable sources like solar and wind power tend to be highly unpredictable and unstable. They tend to vary in supply with no correlation to the changes in demand. Further, the conventional power grids are not flexible enough, and due to their highly centralized and homogeneous nature, increased integration of renewable energy can lead to stability

and reliability issues in the system [5], [6]. The future penetration of renewable sources is rather unavoidable. This calls for a shift towards more sophisticated smart grid based approaches [7]. In general, a smart grid is a system that allows the small-scale producers to generate and sell electricity. In contrast to the conventional systems, these systems are heterogeneous and decentralized in nature and incorporate a variety of resources like smart meters, intelligent power distribution devices, two-way communications, advanced sensors, and energy storage systems. These systems together would facilitate the increased penetration of renewable sources, electric vehicles and micro-grids into the present electricity infrastructure.

However, this shift towards smart grids requires an increased participation by the consumers in the process of electricity generation and distribution. To facilitate this transition, the energy markets would have to be a lot more open and must allow the customers to sell the surplus electricity back to the grid. Demand side management and response mechanisms would be an integral part of this system [8]. In this respect, energy storage units like battery are going to play a significant role in incentivizing the consumer for adopting locally available renewable sources like solar power [9]. These units allow the participating user to store energy as a reserve or for enabling smart energy trade. However, deployment of storage units in the system comes with its problem, both for the overall grid system and for the consumer [10], [11]. The problems can range from economic viability to system stability. However, the rapidly decreasing cost of battery systems [12], [13] in the recent years has made them a lot more attractive and economically feasible to be deployed for use in small-scale PV systems. Companies like Tesla are investing heavily in improving the battery technology and further reducing their overall cost. Recent events like negative electricity pricing in California due to massive solar energy influx, further confirms the prospect of energy trading to be attractive. To incentivize the battery usage among end-user, we need to come up with proper trading strategies to enable the user to participate and gain profit in an open energy market. Researchers are working on different such strategies in various market conditions. In [14], a non-cooperative game between storage units is proposed and is solved using a game theory based approach. In [15], authors formulated a stochastic programming based approach to select optimal energy and reserve bids for the storage units while working in a scenario where a group of independently-operated investor-owned storage units seeks to offer both energy and reserve. In [16], a cooperative energy trading approach is proposed in coordinated multipoint (CoMP) systems powered by smart grids. In [17] a profit maximization strategy is proposed for micro-generation unit working in a mostly centralized market with incentives for energy trading. In [18], authors proposed a heuristic approach called Hybrid Immune Algorithm, for auction based distributed energy resource management in Smart Grid. In [19], BM reinforcement scheme [20] is used to attain Nash equilibrium in a constrained energy trading game between players with incomplete information. Various other studies used reinforcement learning based approaches [21], [22] in various types of energy trading scenarios. While other studies [23], [24] used game theory based approaches for the same. In [25], used a Continuous Double Auction (CDA) based mechanism for congestion management with a microgrid based energy trading scenario. However, little work has been done to incorporate the battery dynamics for energy trading in a relatively centralized energy market where a little decentralization has occurred due to the emergence of local microgrids, but the central grid still has most important role market dynamics and electricity distribution.

In this paper, a detailed study about the use of battery systems for energy trading was conducted from the end-user perspective by incorporating battery dynamics in the form of constraints in the overall energy trading framework. This was formulated in terms of an energy trading model. Further, a strategy called Dynamic Battery Charging Algorithm (DBC) was developed to find the optimal battery charging states during the overall trading process. For validating the performance of the proposed strategy, the algorithm was tested on nine different scenarios. For this, a case study of three different load profiles (a small scale residential user, a Walmart Supercenter as a medium scale commercial user and an automotive assembly unit as a large-scale industrial user) was done. Further, all three profiles were tested in three electricity spot markets in different states (Ontario, California and New York). Using the real-time traces, three estimators were trained for predicting one hour-ahead price, solar power, and load demand. This was used to test and validate the proposed strategy in real time scenarios.

The remainder of this paper is organized as follows. Section II formulates the energy trading model from the end user perspective. Section III introduces the proposed energy trading scheme called Dynamic Battery Charging Algorithm (DBC) and presents various strategies tested later on. Section IV presents the overall experimental setup to test and validate the proposed strategy. Section V shows the simulation results and Section VI draws the conclusions.

II. Energy Transfer Model Formulation

The proposed model considers the trading operation from the end user perspective. The end user is directly connected to a central smart grid system that allows a two-way electricity trade. The model formulates the loss incurred by the end user

while buying/selling electrical power over a time-period. All the variables are considered to be discrete with each time-slot taken as 1 hour and denoted by $'t'$. The system consists of a load, a PV system for electricity generation, a battery to store energy and a central utility to buy energy deficit or sell energy surplus. Here d_t denotes the load energy requirement in the time slot t. s_t denotes the solar energy harvested using the PV system (excluding the energy loss during the harvesting process) in time slot t.

b_t indicates the initial charge of the battery (at the start of time slot t), and Δb_t denotes the change in charge of the battery in time slot t. Positive Δb_t denotes charging of the battery while negative Δb_t denotes discharging of the battery. This change can be formulated as Eq.(1). $b^{capacity}$ represents the maximum charge accumulation capacity of the battery and b^{min} represents the reserved battery charge. Thus in any time-slot t the battery charge cannot go beyond this range. Eq.(2) accounts for this constraint. As represented by the Eq.(3), Δb_t is limited by maximum charging rate (db^{crate}) and maximum discharging rate (db^{drate}). Eq.(4), limits the battery discharge ($-\Delta b_t$) in slot t to remain within the present battery charge. u_t denotes the amount of electricity traded from the utility in time slot t. Positive u_t indicates that energy is brought from the utility while its negative value denotes that the energy is sold to the utility. The total energy supply and energy demand in a system (Fig. 1) are always balanced. This can be formulated as Eq.(5).

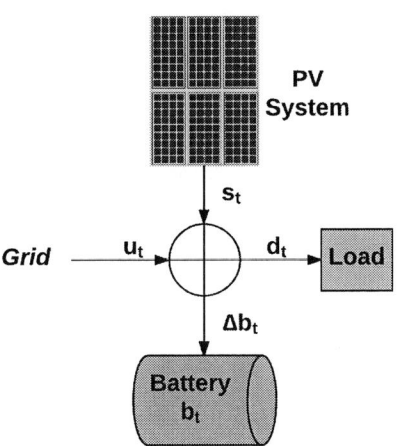

Fig. 1. System Energy Transfer Model

$$b_{t+1} = b_t + \Delta b_t \tag{1}$$

$$b^{min} \leq b_t \leq b^{capacity} \tag{2}$$

$$db^{drate} \leq \Delta b_t \leq db^{crate} \tag{3}$$

$$-\Delta b_t \leq b_t - b^{min} \tag{4}$$

$$u_t + s_t = \Delta b_t + d_t \tag{5}$$

$$\ell_t = \begin{cases} p_t * u_t & \text{if } u_t \geq 0 \\ \beta_1 * \beta_2 * p_t * u_t & \text{if } u_t < 0 \end{cases} \tag{6}$$

$$\beta = \begin{cases} 1 & \text{if } u_t \geq 0 \\ \beta_1 * \beta_2 & \text{if } u_t < 0 \end{cases} \tag{7}$$

$$\ell_t = \beta * p_t * u_t = \beta * p_t * [\Delta b_t + d_t - s_t] \tag{8}$$

$$\ell_{total} = \sum_t \beta * p_t * [\Delta b_t + d_t - s_t] \tag{9}$$

$$\ell_{av} = \lim_{T \to \infty} \frac{1}{T} \sum_t \beta * p_t * [\Delta b_t + d_t - s_t] \tag{10}$$

$$\ell_{av}(\tau, T) = \frac{1}{T} \sum_{t=\tau}^{\tau+T-1} \beta * p_t * [\Delta b_t + d_t - s_t] \tag{11}$$

$$p_t(\tau) = E_p[p_t; \tau] \tag{12}$$

$$d_t(\tau) = E_d[d_t; \tau] \tag{13}$$

$$s_t(\tau) = E_s[s_t; \tau] \tag{14}$$

$$L(\Delta b(\tau); \tau, T) \approx \ell_{av}(\tau, T)$$
$$\approx \frac{1}{T} \sum_{t=\tau}^{\tau+T-1} \beta * p_t(\tau) * [\Delta b_t(\tau) + d_t(\tau) - s_t(\tau)] \tag{15}$$

The variable p_t denotes the price at which the consumer can buy the electricity from the grid. Whenever the energy is bought from the grid ($u_t \geq 0$), the consumer incurs a net loss represented by $p_t * u_t$. In contrast to this, the selling price is assumed to be less than p_t. This is to factor in the cost incurred by the central grid for running the energy market. This can be formulated by using a constant parameter called price coefficient (β_1) with a value between 0 and 1. So, in this case, the selling price is taken as $\beta * p_t$. Further, there exists an energy loss (AC/DC conversion loss and transmission loss) when selling the electricity to the grid. This loss is approximated by using another constant parameter called loss coefficient (β_2). So, the net energy sold in this case can be approximated as $\beta_2 * u_t$. While selling ($u_t \leq 0$), the consumer gains a profit which can be represented in terms of negative loss as $\beta_1 * \beta_2 * p_t * u_t$. Overall, the monetary loss (ℓ_t) incurred in time slot t is formulated as given by Eq.(6). The two trading situations are combined by taking another parameter called trading coefficient (β) as represented in Eq.(7). This can be used to combine the two conditions in Eq.(6) to form loss equation as shown in Eq.(8).

The Eq.(9) represents the total loss ℓ_{total} incurred by the consumer. The objective of every consumer based strategy is to minimize this loss or rather gain a significant profit. This can also be achieved by minimizing the average loss ℓ_{av} as represented by Eq.(10). The Eq.(11) represents the

average loss incurred by the consumer during T consecutive time slots starting from time slot (τ). For calculating $\ell_{av}(\tau, T)$ (as given in Eq.(11)), the accurate values of p_t, Δb_t, d_t and s_t are required to be known for time slots τ, $\tau + 1$.... and $\tau + T - 1$ at time slot τ. This can be approximated by estimating the price (p), load demand (d) and harvested solar energy (s) for all the future time slots. $p_t(\tau)$, $d_t(\tau)$ and $s_t(\tau)$ in Eq.(12), 13 and 14 represent the forecasted estimations of p_t, d_t and s_t at τ time slot. By using above estimations, the Eq.(11) can be approximated as a function of the change in battery charge. This leads to the loss function given in Eq.(15). Here, $\Delta b(\tau) = [\Delta b_\tau(\tau), \Delta b_{\tau+1}(\tau), ..., \Delta b_{\tau+T-1}(\tau)]$. By minimizing the loss function (Eq.(15)) for different time slot pairs (τ, T), we can achieve the end user task of reducing the overall monetary loss of consumer. This would give the optimal values of battery charge for different time slots.

III. DYNAMIC BATTERY CHARGING ALGORITHM (DBC)

$$f_\tau : \min_{\Delta b(\tau)} \frac{1}{T} \sum_{t=\tau}^{\tau+T-1} \beta * p_t(\tau) * [\Delta b_t(\tau) + d_t(\tau) \\ - s_t(\tau)] \quad s.t.((2)), ((3)), ((4)) \tag{16}$$

$$\Delta b^*(\tau) = \operatorname*{argmin}_{\Delta b(\tau)} \frac{1}{T} \sum_{t=\tau}^{\tau+T-1} \beta * p_t(\tau) * [\Delta b_t(\tau) \\ + d_t(\tau) - s_t(\tau)] \quad s.t.((2)), ((3)), ((4)) \tag{17}$$

$$b_\tau^* = b_{\tau-1}^* + \Delta b_\tau^*(\tau) \tag{18}$$

$$F_\tau(\Delta b) = \frac{1}{T} \sum_{t=\tau}^{\tau+T-1} [\beta * p_t(\tau) * [\Delta b_t(\tau) + d_t(\tau) - s_t(\tau)] \\ + c_1 * (b(t) - b^{min})^2 + c_2 * (b^{capacity} - b(t))^2 \\ + c_3 * (\Delta b(t) - db^{drate})^2 + c_4 * (db^{crate} - \Delta b(t))^2 \\ + c_5 * (b(t) + \Delta b(t) - b_{min})^2] \tag{19}$$

Let t_f be the total number of time slots for which we have to predict the battery charge state. The aim of the DBC algorithm is to accurately estimate these values. Prior to estimating the optimal battery charge values, three separate time series estimators, E_p, E_d and E_s, are trained for price, load and solar power prediction. For this a machine learning technique called Support Vector Machine [26] was used to train the three time series regression models [27]–[29]. At any time slot τ, these time series estimators can be used to predict the values of p_t, d_t and s_t for future time slots ($t > \tau$).

The Eq.(16) represent the minimization problem at a particular time-slot "τ". So, in total there are t_f different minimization problems from time slots $\tau = 1$ to $\tau = t_f$. Solving the minimization problem (f_τ) at particular time slot τ yields the optimal battery charge change (Eq.(17)) from $t = \tau$ to $t = \tau + T - 1$ w.r.t. to the present time slot τ. In Eq.(17), $\Delta b^*(\tau) = [\Delta b_\tau^*(\tau), \Delta b_{\tau+1}^*(\tau), ..., \Delta b_{\tau+T-1}^*(\tau)]$. Of these, the battery charge values for future time slots ($t > \tau$) are affected by inaccuracies in predictions of price, load and solar

2458

Algorithm 1 DBC
1: Train Estimators: E_p, E_d and E_d.
2: **for** $\tau = 0 : t_f$ **do**
3:   **for** $t = \tau + 1 : \tau + T$ **do**
4:     Estimate $p_t(\tau)$, $d_t(\tau)$ and $s_t(\tau)$
5:   Solve f_τ using optimizer
6:   $b_\tau^* \leftarrow b_{\tau-1}^* + \Delta b_\tau^*(\tau)$

---

power. So, these later values are discarded and the change in battery charge at τ ($\Delta b_\tau^*(\tau)$) is used to calculate the optimal battery charge at that time slot (Eq.(18)). By solving the minimization problem for all time slots, the required optimal battery charge pattern can be obtained. Eq.(19) represents the overall minimization objective function (F_τ) incorporating all the battery constraints (Eq.(2),(3),(4)). This is done by adding a penalty to the objective function whenever any of the constraint is violated. c_1, c_2, c_3, c_4 and c_5 are penalty coefficients for these constraints. They are given a positive value if the respective constraint is violated else they are 0.

Any iterative optimization algorithm can be used to solve f_τ in step 7 of the Algorithm 1. All the iterative optimization algorithms can be classified into two main types: gradient based and non-gradient based. In this paper, both of these optimization algorithms are tested with momentum based gradient descent (MGD) representing the gradient-based method [30] and differential evolution (DE) [31] representing non-gradient method. Apart from these, three other strategies (Without Solar Harvesting, Greedy Approach and Reinforcement Learning) are also tested against DBC algorithm in the paper

IV. EXPERIMENTAL SETUP

In the following simulations, we tested the proposed algorithms using the hourly traces of electricity prices (p_t), harvested energy (s_t) and electricity demand (d_t). This was done using three different load profiles (Residential, Commercial, and Industrial) in three different electricity markets (Ontario [32], California [33], and New York [34]). The data was collected for a one year period from 1/1/2012-12/31/2012. The solar data for these areas were collected from National Solar Radiation Database (NSRDB) [35].

For all the scenarios, parameter β is set to be 0.8. As each time slot is of $1hr$, the value t_f is set to 8760 hours. Further, the value of T was taken as 8. Value of T was decided by testing the algorithm on a sample dataset of 1 month for the California Commercial scenario. Larger values of T induces errors in the prediction due to error in predicting the values of p_t, d_t and s_t. We considered the batteries to be deep cycle batteries. b^{min} was considered to be 20% of the battery capacity ($b^{capacity}$). Further, maximum charging rate (db^{crate}) and maximum discharging rate (db^{drate}) were also considered to be 50% of $b^{capacity}$. These limits were selected considering the specification of the battery. The residential user was considered to be equipped with solar panels with a total capacity of $4kW$. The annual electricity consumption of the user was around $10MWh/year$. Further, the user

was equipped with a battery bank of $5kWh$ total capacity ($b^{capacity}$). For commercial user case, we considered load profile of an average Walmart Supercenter [36] with a mean annual electricity consumption of $5000MWh/year$. The unit was considered to be equipped with solar panels of $1MW$ capacity with a total battery capacity of $2.5MWh$. Further, for the industrial load, we considered an automotive assembly unit with an average annual electricity consumption of $75000MWh/year$. The unit was considered to be equipped with solar panels of $15MW$ capacity with a total battery capacity of $37.5MWh$.

V. SIMULATION RESULTS AND ANALYSIS

For simulating the proposed system, we considered deep cycle battery systems. In particular, we used "Thundersky Winston LiFePO4 Battery" (Model No: WB-LYP1000AHC(A)). These batteries are known to last $5000 - 7000$ cycles before losing 30% of its capacity. For this study, the cycle life for the battery system is taken to be 6000 cycles. The price of each battery is $1000\$$. So, cost per kWh of battery system is $350\$/kWh$. Table I shows the net monetary loss accumulated by end-user in the year 2012 in different market scenarios. The algorithm with minimum loss performs the best. In general, the order of monetary loss from higher to lower is: DBC:DE > RL > DBC:GD > Greedy. DBC:DE easily outperformed all other approaches in all the tested scenarios. Fig.2 shows the loss accumulated by all the tested algorithms up to a certain time slot in the year 2012. Different colored lines are used to represent all the algorithms. Clearly, in both cases (Ontario residential and Ontario commercial) the cumulative loss up to each time slots is minimum in the case of DBC:DE algorithm. This is evident from the fact that green line representing DBC:DE algorithm is always below all the other lines.

TABLE I.   NET MONETARY LOSS(IN $\$$) USING DIFFERENT
APPROACHES(R-RESIDENTIAL, C-COMMERCIAL, I-INDUSTRIAL)

		Without Energy Harvesting	Greedy	RL	DBC:GD	DBC:DE
Ont.	**R**	347	157	112	109	**7**
	C	161787	110771	48733	79928	**37239**
	I	24306056	1664863	894665	1198309	**561225**
Cal.	**R**	308	51	13	42	**-45**
	C	150490	81010	56138	65171	**34930**
	I	2257833	1215076	838270	995203	**524405**
N.Y.	**R**	357	119	66	89	**16**
	C	173884	109695	71229	87803	**60180**
	I	2611878	1648641	1223453	1223616	**905617**

Fig.3 shows the proposed algorithm's decision-making pattern for a particular day. The plot shows the change pattern in energy traded (u) and change in battery charge (Δb) w.r.t. changes in market price (p), solar power generation (s) and load demand (d) in a set of 24 time-slots. The changes in energy traded (u) and battery state (b) represent the decision-making process of the tested algorithm. Clearly, the time slots with decreased electricity prices are followed by a positive u value representing that the energy was bought during those time slots. Similarly, for time slots with high electricity prices the energy was sold to the grid represented by its negative values in those time slots. Further, during time slots with

The 2018 International Power Electronics Conference

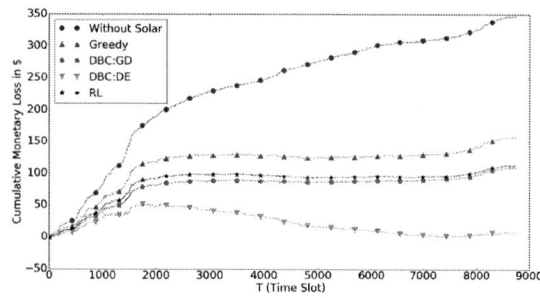

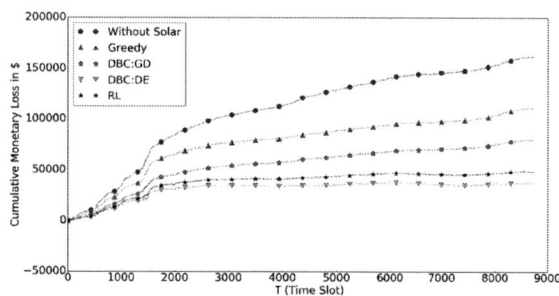

Fig. 2. Cumulative loss curve a) Ontario residential and b) a) Ontario commercial

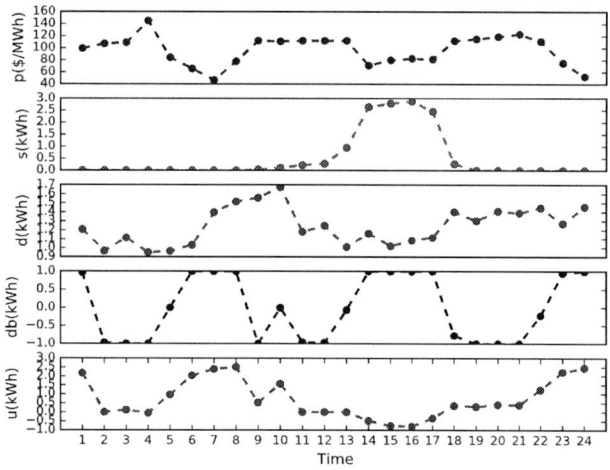

Fig. 3. DBC:DE decision pattern in Ontario Residential Case

high solar power generation the battery is getting charged (represented by positive Δb) and extra energy is sold back (represented by negative u) to the grid.

Table II shows the profit generated by the use of battery units in a solar system. It lists the values of number of battery cycles (n_c), Profit/cycle (P_b/n_c), Estimated total profit (EP_b) and profit/kWh (NEP_b) of the battery. P_b is simply the difference between loss accumulated by Greedy and DBC:DE approach i.e. profit due to the deployment of the battery system. Considering that the price/kWh of the battery system

TABLE II. ESTIMATED PROFIT GENERATION DUE TO USE OF BATTERY

		Greedy	DBC:DE	n_c	P_b/n_c	$EP_b(\$)$	NEP_b
Ont.	**R**	157	7	421.6	0.36	2135	426.9
	C	110771	37239	402	182.92	1097493	439.0
	I	1664863	561225	401.6	2748.10	16488616	439.7
Cal.	**R**	51	-45	325.1	0.30	1772	354.4
	C	81010	34930	291.8	157.92	947498	379.0
	I	1215076	524405	291.2	2371.81	14230859	379.5
N.Y.	**R**	119	16	315.2	0.33	1961	392.1
	C	109695	60180	292	169.57	1017432	407.0
	I	1648641	905617	291.4	2549.84	15299053	408.0

would be around ($350\$/kWh$), the use of the battery in Ontario, California and New York markets gave a much higher estimated profit per kWh for all the tested scenarios. This easily justifies the use of the battery systems for energy trading in these states. Further, due to expected reduction in battery prices in the near future, energy trading will soon be even more profitable.

VI. CONCLUSION

The paper has presented an energy trading strategy called DBC to empower the end-user to strategically trade electricity with the central grid. The central grid was assumed to have a 2-way energy communication as its integral part. The proposal was tested in three different markets (Ontario, California, and New-York) using three load profiles (Residential, Commercial, and Industrial). Due to the different market and environmental conditions, the use of the battery systems for energy trading may not be a smart choice in all the states. Still the results prove that the proposed strategy was profitable in all tested states. It easily accumulated a profit of $10-30\%$ profit all test scenarios. With ever decreasing battery prices, the proposed strategy would be even more profitable in the near future.

REFERENCES

[1] C. Frei, R. Whitney, H.-W. Schiffer, K. Rose, D. A. Rieser, A. Al-Qahtani, P. Thomas, H. Turton, M. Densing, E. Panos, *et al.*, "World energy scenarios: Composing energy futures to 2050," tech. rep., Conseil Francais de l'energie, 2013.

[2] U. Lehr, C. Lutz, and D. Edler, "Green jobs? economic impacts of renewable energy in germany," *Energy Policy*, vol. 47, pp. 358–364, 2012.

[3] M. Bhattacharya, S. R. Paramati, I. Ozturk, and S. Bhattacharya, "The effect of renewable energy consumption on economic growth: Evidence from top 38 countries," *Applied Energy*, vol. 162, pp. 733–741, 2016.

[4] T. Adefarati and R. Bansal, "Integration of renewable distributed generators into the distribution system: a review," *IET Renewable Power Generation*, vol. 10, no. 7, pp. 873–884, 2016.

[5] P. Denholm and M. Hand, "Grid flexibility and storage required to achieve very high penetration of variable renewable electricity," *Energy Policy*, vol. 39, no. 3, pp. 1817–1830, 2011.

[6] Y. Wang, V. Silva, and M. Lopez-Botet-Zulueta, "Impact of high penetration of variable renewable generation on frequency dynamics in the continental europe interconnected system," *IET Renewable Power Generation*, vol. 10, no. 1, pp. 10–16, 2016.

[7] V. C. Gungor, D. Sahin, T. Kocak, S. Ergut, C. Buccella, C. Cecati, and G. P. Hancke, "A survey on smart grid potential applications and communication requirements," *IEEE Transactions on Industrial Informatics*, vol. 9, no. 1, pp. 28–42, 2013.

2460

[8] P. Palensky and D. Dietrich, "Demand side management: Demand response, intelligent energy systems, and smart loads," *IEEE transactions on industrial informatics*, vol. 7, no. 3, pp. 381–388, 2011.

[9] B. Dunn, H. Kamath, and J.-M. Tarascon, "Electrical energy storage for the grid: a battery of choices," *Science*, vol. 334, no. 6058, pp. 928–935, 2011.

[10] D. Lindley, "Smart grids: The energy storage problem," *Nature News*, vol. 463, no. 7277, pp. 18–20, 2010.

[11] J. Lassila, J. Haakana, V. Tikka, and J. Partanen, "Methodology to analyze the economic effects of electric cars as energy storages," *IEEE Transactions on smart grid*, vol. 3, no. 1, pp. 506–516, 2012.

[12] B. Nykvist and M. Nilsson, "Rapidly falling costs of battery packs for electric vehicles," *Nature Climate Change*, vol. 5, no. 4, pp. 329–332, 2015.

[13] A. A. Asif and R. Singh, "Further cost reduction of battery manufacturing," *Batteries*, vol. 3, no. 2, p. 17, 2017.

[14] Y. Wang, W. Saad, Z. Han, H. V. Poor, and T. Başar, "A game-theoretic approach to energy trading in the smart grid," *IEEE Transactions on Smart Grid*, vol. 5, no. 3, pp. 1439–1450, 2014.

[15] H. Akhavan-Hejazi and H. Mohsenian-Rad, "A stochastic programming framework for optimal storage bidding in energy and reserve markets," in *Innovative Smart Grid Technologies (ISGT), 2013 IEEE PES*, pp. 1–6, IEEE, 2013.

[16] J. Xu and R. Zhang, "Cooperative energy trading in comp systems powered by smart grids," *IEEE Transactions on Vehicular Technology*, vol. 65, no. 4, pp. 2142–2153, 2016.

[17] S. Chen, N. B. Shroff, and P. Sinha, "Energy trading in the smart grid: From end-user's perspective," in *Signals, Systems and Computers, 2013 Asilomar Conference on*, pp. 327–331, IEEE, 2013.

[18] B. Ramachandran, S. K. Srivastava, C. S. Edrington, and D. A. Cartes, "An intelligent auction scheme for smart grid market using a hybrid immune algorithm," *IEEE Transactions on Industrial Electronics*, vol. 58, no. 10, pp. 4603–4612, 2011.

[19] H. Wang, T. Huang, X. Liao, H. Abu-Rub, and G. Chen, "Reinforcement learning for constrained energy trading games with incomplete information," *IEEE Transactions on Cybernetics*, 2016.

[20] A. S. Poznyak and K. Najim, *Learning automata and stochastic optimization.* 1997.

[21] D. Urieli and P. Stone, "Tactex'13: a champion adaptive power trading agent," in *Proceedings of the 2014 international conference on Autonomous agents and multi-agent systems*, pp. 1447–1448, International Foundation for Autonomous Agents and Multiagent Systems, 2014.

[22] H. Wang, T. Huang, X. Liao, H. Abu-Rub, and G. Chen, "Reinforcement learning in energy trading game among smart microgrids," *IEEE Transactions on Industrial Electronics*, vol. 63, no. 8, pp. 5109–5119, 2016.

[23] W. Tushar, J. A. Zhang, D. B. Smith, H. V. Poor, and S. Thiébaux, "Prioritizing consumers in smart grid: A game theoretic approach," *IEEE Transactions on Smart Grid*, vol. 5, no. 3, pp. 1429–1438, 2014.

[24] J. Lee, J. Guo, J. K. Choi, and M. Zukerman, "Distributed energy trading in microgrids: A game-theoretic model and its equilibrium analysis," *IEEE Transactions on Industrial Electronics*, vol. 62, no. 6, pp. 3524–3533, 2015.

[25] P. Vytelingum, S. D. Ramchurn, T. D. Voice, A. Rogers, and N. R. Jennings, "Trading agents for the smart electricity grid," in *Proceedings of the 9th International Conference on Autonomous Agents and Multiagent Systems: volume 1-Volume 1*, pp. 897–904, International Foundation for Autonomous Agents and Multiagent Systems, 2010.

[26] N. I. Sapankevych and R. Sankar, "Time series prediction using support vector machines: a survey," *IEEE Computational Intelligence Magazine*, vol. 4, no. 2, 2009.

[27] J. Che and J. Wang, "Short-term electricity prices forecasting based on support vector regression and auto-regressive integrated moving average modeling," *Energy Conversion and Management*, vol. 51, no. 10, pp. 1911–1917, 2010.

[28] E. Ceperic, V. Ceperic, and A. Baric, "A strategy for short-term load forecasting by support vector regression machines," *IEEE Transactions on Power Systems*, vol. 28, no. 4, pp. 4356–4364, 2013.

[29] J. Zeng and W. Qiao, "Short-term solar power prediction using a support vector machine," *Renewable Energy*, vol. 52, pp. 118–127, 2013.

[30] G. Goh, "Why momentum really works," *Distill*, vol. 2, no. 4, p. e6, 2017.

[31] K. Fleetwood, "An introduction to differential evolution," in *Proceedings of Mathematics and Statistics of Complex Systems (MASCOS) One Day Symposium, 26th November, Brisbane, Australia*, 2004.

[32] "Independent electricity system operator," *http://www.ieso.ca*.

[33] "California iso," *http://oasis.caiso.com*.

[34] "New york independent system operator," *http://www.nyiso.com*.

[35] "National solar radiation database," *https://nsrdb.nrel.gov/*.

[36] C. B. E. Efficiency, "Walmartsaving energy, saving money through comprehensive retrofits.,"

A Forced Commutation Method of the Solid-state Transfer Switch in the Uninterrupted Power Supply Applications

Meng-jiang Tsai, Jiuyang Zhou, and Po-tai Cheng

Abstract—This paper investigates the effects of forced commutation mechanism on the solid-state transfer switches of the uninterrupted power supplies (UPSs). The economical mode provides a solution to effectively reduce the loss induced by the switching devices. However, the SCR based bypass switches have a worse dynamics in turn-off characteristic, so the critical load may suffer from severe grid faults for a long time. In this paper, a control mechanism is presented to force thyristor-based switches to be commutated. Besides, this study also shows how different grid faults affect the performance of forced commutation. Finally, the simulation result and laboratory test results are provided to validate its effectiveness.

Index Terms—Forced commutation, fault detection, islanding mode, natural commutation, solid-state switch, three-phase three-wire, thyristor, uninterrupted power supply.

I. INTRODUCTION

Uninterrupted power supply (UPS) has become a popular solution in the industry, like enterprise IT, commercial telecom, data center, cloud computing area, and so on. Meanwhile, there is a strong demand for the UPS system to provide more reliable and secure electrical power supply for the critical devises [1] [2].

The conventional UPS systems can be classified as the online, and off-line UPSs. Fig. 1 shows the diagram of UPS systems, comprising of ac/dc rectifier, dc/ac inverter, and the solid-state transfer switches (STS). Such circuit can provide the customers with the flexible choice between the on-line mode and the economical mode (off-line mode). In on-line mode, the load is powered by the inverter, so it provides excellent capability in being immune to the grid frequency variation. However, this also leads to the increased switching loss of the power devices. Instead, the economical mode is a very cost-effective and convenient approach to enhance the efficiency of UPS since the loads are powered through the STS system. Nevertheless, the commutation issue of thyristor-based STS system during grid faults must be adequately addressed in order to take full advantage of this mechanism.

Thyristor-based STS system provides many advantages of low cost and low conduction loss compared with the gate turn off thyristor or integrated gate-commutated thyristor-based systems, but the natural commutation mechanism of thyristor must take the long time to turn it off. Besides, it is significantly affected by the load and the conditions of grid fault, an other power issues, and this even prolongs the transfer process from the economical mode to on-line mode. As disturbances occur in the utility, the typical off-line UPS should take over the

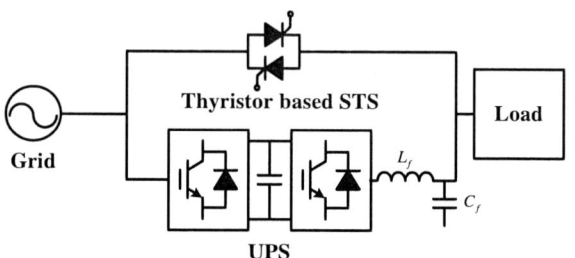

Fig. 1. Simplified system diagram of uninterrupted power supply.

load within 1.0 to 5.0ms to prevent any interruptions [3]–[5], so the natural commutation may not meet these requirements, and degrades the effectiveness of the economical mode. In order to address this issue, Chen *et. al* presented an impulse-commutated STS (ICSTS) which utilizes the LC resonance to force commutate the thyristors [6]. However, the extra design of circuit increase the control complexity and cost.

This paper presents a control mechanism to speed the commutation of thyristor-based STS system. The analysis indicates the commutation of thyristor-based STS system is closely related to the conditions of the grid faults and the loads, so the critical power quality issues, like voltage sag and non-unity power factor, are taken into consideration. Besides, this study also presents the whole transfer process, including fault detection, forced commutation mechanism, and islanding mode. In the end, the simulation result and the laboratory test results are presented to validate the aforementioned analysis.

II. OPERATION PRINCIPLE

As shown in Fig. 2, all control stages can be classified as the grid-connection, fault detection, the forced commutation, and islanding stage. In economical mode, the inverter are standby during the grid-connection period until the grid fault is detected. Besides, such controller can confirm fault-detection within 5ms, and the forced commutation mechanism can accelerate the overall transfer process.

A. Fault Detection

Fig. 3 shows the control diagram to detect the grid voltage fault. The grids voltages ($v_{g,abc}$) are sensed, and transformed to derive their synchronous-reference frame components ($v_{g,q}$, $v_{g,d}$), and these synchronous-reference-frame transformation is defined in (2). Note that ω represents for the line frequency of

The 2018 International Power Electronics Conference

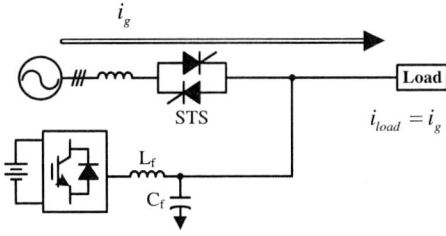

Stage 1. Grid connection

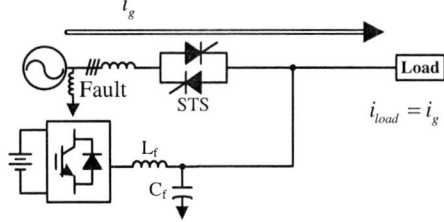

Stage 2. Fault detection period

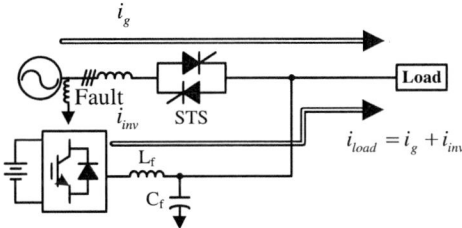

Stage 3. Forced commutation

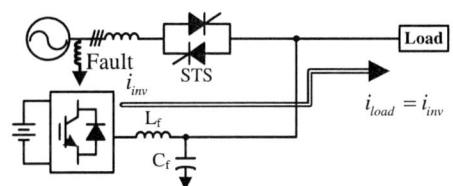

Stage 4. Islanding mode

Fig. 2. Commutation stage of UPS's economical mode during grid faults.

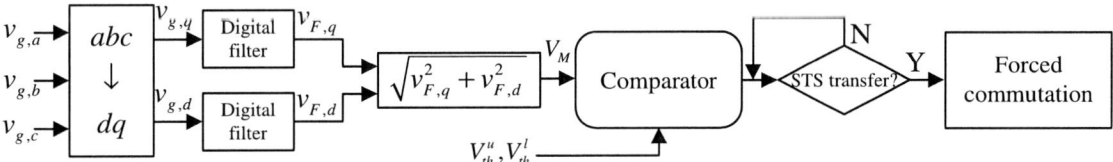

Fig. 3. Grid fault detection for voltage sag or swell.

grid. The digital filter is used to attenuate the transient voltage spike and the 120Hz voltage-ripple caused by the unbalanced sequence-component. Subsequently, V_M is compared with the threshold voltages (V_{th}^u, V_{th}^l) to detect whether the grid fault occurs or not, which the normal grid operation is confirmed as V_{th} between V_{th}^u and V_{th}^d. $v_{pre,q}$ and $v_{pre,d}$ are the prefault voltages of the grid, and the threshold voltages are derived in

$$\begin{cases} V_{th}^u = 1.1 V_{pre} \\ V_{th}^l = 0.9 V_{pre} \end{cases} \text{, where } V_{pre} = \sqrt{v_{pre,q}^2 + v_{pre,d}^2}. \quad (1)$$

B. Force commutation

Fig. 4 shows the waveform during the forced commutation. At the stage 1 and stage 2, the load currents are mainly supplied by the gird, and $i_{load} = i_g$. The forced commutation mechanism is activated once the grid fault is detected, and the gate signals of STS (S_{STS}) is set to zero. Due to the characteristic of thyristor, the STS system cannot turns off until the current through STS system (i_g) is reduced to zero. In this transfer process (stage 3), the inverter reduces the grid current to speed turning thyristor off, and $i_{load} = i_g + i_{inv}$.

Fig. 5 shows the controller diagram of forced commutation. Based on (2), the inverter voltages ($v_{load,a}$, $v_{load,b}$, and $v_{load,c}$), the inverter currents ($i_{inv,a}$, $i_{inv,b}$, and $i_{inv,c}$), and the

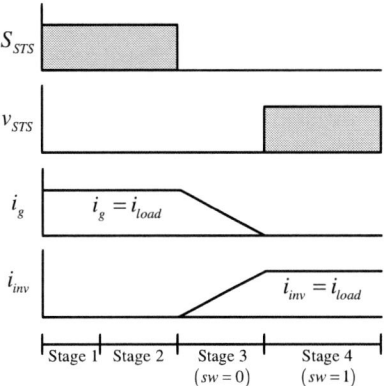

Fig. 4. Waveform during the forced commutation.

load current ($v_{load,a}$, $v_{load,b}$, and $v_{load,c}$) are transformed into the synchronous reference frame components ($v_{load,q}$, $v_{load,d}$, $i_{inv,q}$, $i_{inv,d}$, $i_{load,q}$, and $i_{load,d}$). At $sw = 0$, the inverter's currents track the load currents with the predictive current control, and the inverter's voltages are taken as the feedfowrd term. When the current errors are attenuated below threshold current (I_{th}), it is transferred to the next stage ($sw = 1$), and

2463

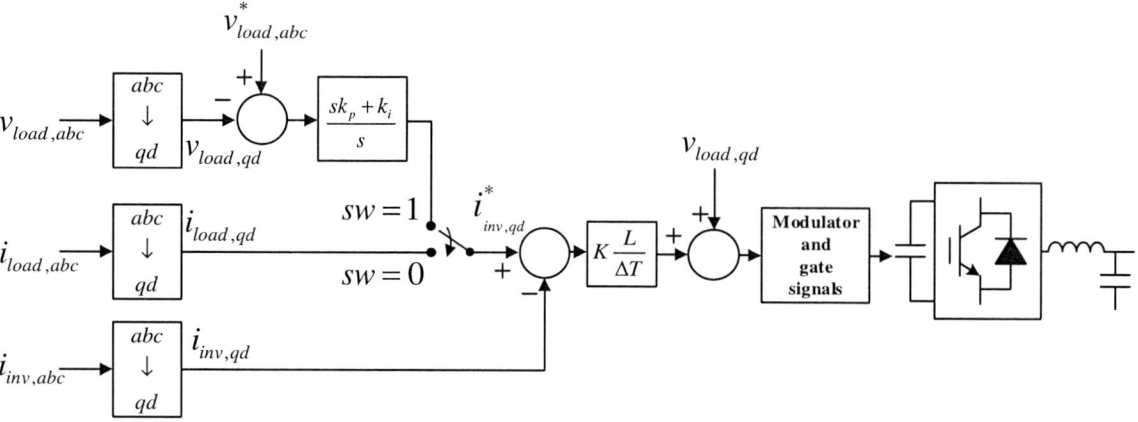

Fig. 5. Control diagram of the forced commutation mechanism.

$$
\begin{bmatrix} v_{x,a} \\ v_{x,b} \\ v_{x,c} \end{bmatrix} = \begin{bmatrix} 1 & 0 \\ -\frac{1}{2} & -\frac{\sqrt{3}}{2} \\ \frac{1}{2} & \frac{\sqrt{3}}{2} \end{bmatrix} \begin{bmatrix} v_{x,\alpha} \\ v_{x,\beta} \end{bmatrix} = \begin{bmatrix} 1 & 0 \\ -\frac{1}{2} & -\frac{\sqrt{3}}{2} \\ \frac{1}{2} & \frac{\sqrt{3}}{2} \end{bmatrix} \left(\begin{bmatrix} \cos \omega t & \sin \omega t \\ -\sin \omega t & \cos \omega t \end{bmatrix} \begin{bmatrix} v_{x,q} \\ v_{x,d} \end{bmatrix} \right)
$$

$$
\begin{bmatrix} i_{x,a} \\ i_{x,b} \\ i_{x,c} \end{bmatrix} = \begin{bmatrix} 1 & 0 \\ -\frac{1}{2} & -\frac{\sqrt{3}}{2} \\ \frac{1}{2} & \frac{\sqrt{3}}{2} \end{bmatrix} \begin{bmatrix} i_{x,\alpha} \\ i_{x,\beta} \end{bmatrix} = \begin{bmatrix} 1 & 0 \\ -\frac{1}{2} & -\frac{\sqrt{3}}{2} \\ \frac{1}{2} & \frac{\sqrt{3}}{2} \end{bmatrix} \left(\begin{bmatrix} \cos \omega t & \sin \omega t \\ -\sin \omega t & \cos \omega t \end{bmatrix} \begin{bmatrix} i_{x,q} \\ i_{x,d} \end{bmatrix} \right)
$$

(2)

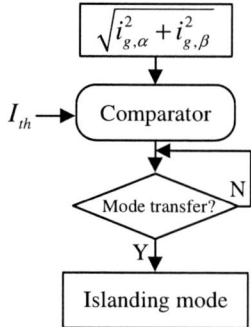

Fig. 6. Current detection for the mode transfer of islanding mode.

these processes are shown in Fig. 6. Note that sw is the transfer signal from the forced commutation mode to the islanding mode.

The performance of the forced commutation is closely related to the instantaneous current output capacity of inverter. Based on KVL equation, the instantaneous current error of inverter (Δi) can be expressed as

$$
\Delta i = \frac{\Delta T}{L}(v_{inv} - v_{load}),
$$

(3)

where v_{inv} is the voltage output of inverter, and ΔT is denoted . Assuming that the sampling delay and switching error are

neglected, v_{inv} can be shown as

$$
v_{inv} = v_{load} + K\frac{L}{\Delta T}(i_{load} - i_{inv}).
$$

(4)

Then, the compensation dynamics can be rewritten as

$$
\Delta i = K(i_{load} - i_{inv}),
$$
$$
\text{where } i_{load} = f(v_{load}).
$$

(5)

Note that the load currents can be expressed by the function of the load voltages. Equation (5) shows the forced commutation capability is significantly affected by the voltage and current waveforms of the loads. Among three phases, there are the different instantaneous value of current and voltage, so the different turn-off sequences are generated. Based on this relationship, the forced commutation can further be divided into two stages, as shown in Fig. 7. Note that the red dash-line represents for the current path between the load and grid.

1) Turn-on of STS of Three Phases: The early stage of the forced commutation requires the superior dynamics of current compensation to attenuate the current through STS. The inverter must bear the heavy burden of modulation, and it even trigs the margin of linear modulation. Equation (6) shows the Kirchhoff current laws (KCL) equations in this stage.

$$
\begin{cases} i_{inv,a} = i_{g,a} + i_{load,a} \\ i_{inv,b} = i_{load,b} + i_{load,b} \\ i_{inv,c} = i_{g,c} + i_{load,c} \end{cases}
$$

(6)

2464

The 2018 International Power Electronics Conference

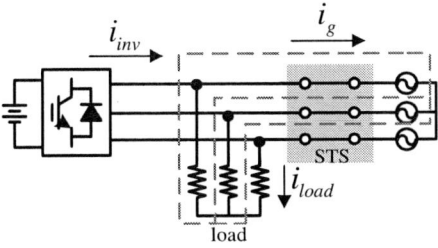

(a) Stage 1. turn-on of STS of three phases

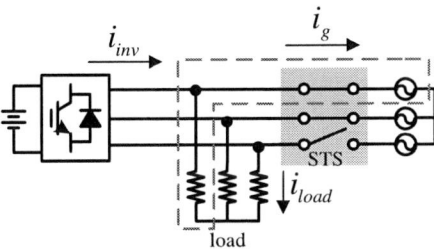

(b) Stage 2. turn-off of single phase STS

Fig. 7. Two stages during the forced commutation period (a)turn-off of STS of three phases (b)turn-off of STS of single phase.

In addition, the period of the first stage (T_1) is dominated by the phase first turning STS off.

2) Turn-off of STS of Single Phase: In this stage, the inverter is operated with the single-phase current compensation, and the KCL equations can be shown as

$$i_{inv,a} = i_{g,a} + i_{load,a}, \text{ where } \begin{cases} i_{g,a} = -i_{g,b} \\ i_{inv,a} = -i_{g,b} \\ i_{load,a} = -i_{load,b} \end{cases} \quad (7)$$

According to (7), the time of the second stage (T_2) is simultaneously affected by the two of three phases. Then, the overall commutation time can be derived as

$$t_{com} = T_1 + T_2; \quad (8)$$

C. Islanding mode

In this mode, the critical loads are isolated from the grid fault, so the inverter must be taken as an alternate voltage source, and regulate the load voltage at rated voltage magnitude and frequency. Fig. 5 shows the control diagram at islanding mode ($sw = 1$). In synchronous reference frame ($v_{inv,q}$, $v_{inv,d}$), a conventional integral and proportional controller is used to regulate the voltage error between the rated commands ($v^*_{inv,q}$, $v^*_{inv,d}$) and the feedback signals ($v_{inv,q}$, $v_{inv,d}$). Subsequently, the generated current commands are fed into the predictive current controller to derive the modulation voltages and their output gate signals. The islanding mode must continue to maintain until the grid faults is cleared.

The typical UPS system includes the more than one modules to share the burden of large power-demands and enhance the reliability of overall system. Therefore, the conventional

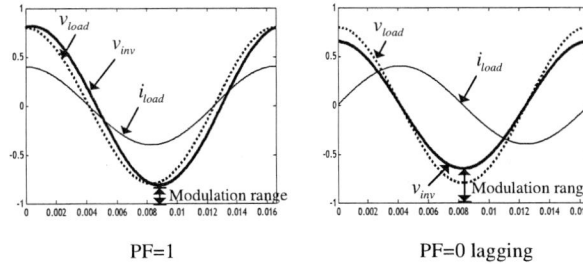

Fig. 8. Waveforms of voltages and currents with the different power factors.

methods, like the droop control and the master-slave control, are presented to provide the excellent performance to address these issues [7]–[11]. These issues about power sharing issues are beyond the scope of this paper, and they are not be discussed in this study.

III. RELATIONSHIP OF POWER ISSUES AND FORCED COMMUTATION

A. Power Factor of Load

Based on (5), the performance of forced commutation is closely associated with the load current. Fig. 8 shows the waveforms of voltages and currents with the different power factors (PFs). When the forced commutation starts at the trough of v_{inv}, the inverter with unity PF has narrow linear modulation range to compensate the corresponding peak of load currents, so the insufficient current dynamics may leads to a long commutation period. Notice the narrow modulation range easily trigs the margin of linear modulation. On the other hand, the load with PF=0 lagging has larger modulation range to compensate the peak of load current. Therefore, the inverter has the better performance of the forced commutation when the load current at lower PF.

B. Voltage Sag

Fig. 9 shows the current and voltage waveforms of the normal grid and 50% voltage sag. When the 50% voltage sags, there is the reduced results in the load currents and inverter voltages, respectively, so the inverter can provide more modulation range than the normal grid operation. Besides, the required compensation current is also decreased due to the reduced load current. As a result, the voltage sag can provide a great help in speeding the forced commutation.

C. Voltage Swell

Fig. 10 shows the current and voltage waveforms of the normal grid and 20% voltage swell. When the 20% voltage swell, both of the load currents and inverter voltage are forced to be increased, so the modulation range is reduced compared with the normal grid operation. Besides, the required compensation current is increased due to the increased load current, so the voltage swell degrades the effectiveness of the forced commutation mechanism.

2465

The 2018 International Power Electronics Conference

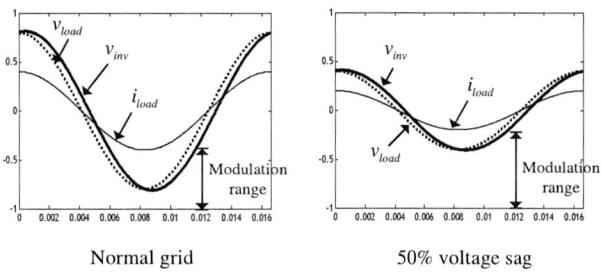

Normal grid   50% voltage sag

Fig. 9. Current and voltage waveforms of the normal grid and 50% voltage sag.

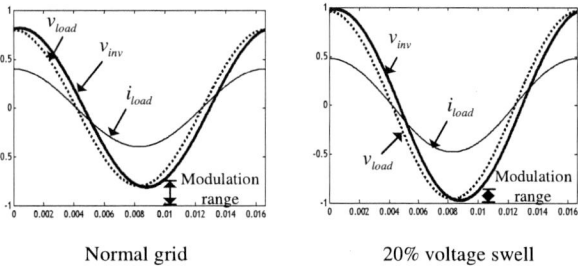

Normal grid   20% voltage swell

Fig. 10. Current and voltage waveforms of the normal grid and 20% voltage sag.

IV. IMPACT OF LARGE FILTER CAPACITOR ON FORCED COMMUTATION

In the large power-scale UPS system, the low switching frequency must be designed to meet the demands of the switching loss, so the passive filters must be selected for the low corner frequency. The conventional voltage-source-inverter designs the filter inductor as 5% per unit to strike the great balance between the modulation utilization and the quality of current output. In order to achieve the aforementioned requirements, a large filter capacitor must be used in such UPS system.

As Fig. 1 shows, such large filter capacitors are always parallel-connected with load, and the large filter capacitor current is generated. In order to address this issue, an extra capacitor current command is calculated as follows;

$$i_C^* = k\frac{\Delta v_{load}}{\Delta T}, \tag{9}$$

where is the k is a constant value depended on the filter capacitor. Therefore, the overall command of the inverter is

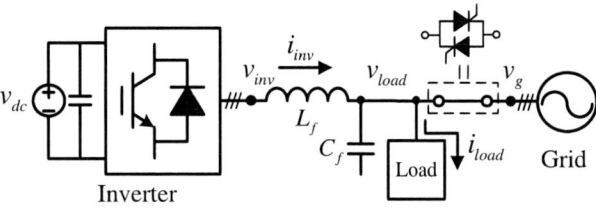

Fig. 11. Simplified circuit diagram and laboratory test bench.

TABLE I
PARAMETER OF CIRCUIT

Parameter	Value
Filter capacitor (C_f)	600 μF
Filter inductors (L_f)	0.162 mH
Line to line voltage	480V
DC bus voltage	780V
Fundamental frequency	60 Hz
Switching frequency	5k Hz

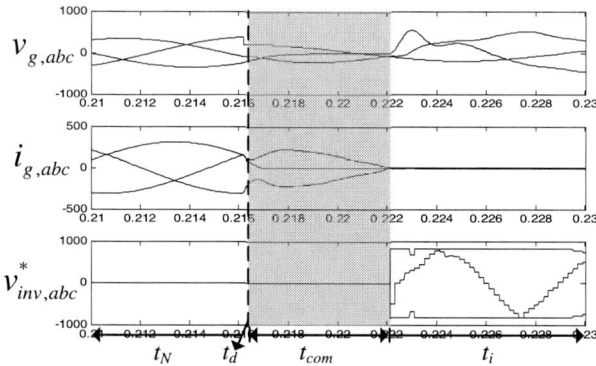

Fig. 12. Simulation waveforms as the natural commutation and the 50% voltage sag. Scales: v_g: V/s, i_g: A/s, and v_{inv}^*: V/s.

derived as

$$i_{inv}^* = i_C^* + i_{load}. \tag{10}$$

V. SIMULATION RESULT

The simulation circuit is shown in Fig. 11, and the circuit parameters are based on a 240kVA UPS system, as shown in the TABLE I. The simulation waveforms include the four stages, like grid-connection, fault detection, commutation of STS, and islanding mode, and their period are denoted with t_N, t_d, t_com, and t_i. Fig. 12 and Fig. 13 respectively show the simulation waveforms operated natural commutation and forced commutation as the 50% voltage sag in grid. The natural commutation takes the long time (t_cm=5.6ms) to reduce the current through thyristor-based STS to zero compared with the forced commutation, so the load under natural commutation must bear the long-time distortion from grid faults. In the forced commutation, the inverter trigs the linear modulation range in the early stage in order to provide the great current dynamics, but it only needs 1.2ms to turn thyristor-based STS off. Fig. 14 shows simulation waveforms of grid voltages, currents through thyristor-based STS, and modulation voltages as the forced commutation and the 20% voltage swell. As figure shows, there is a longer commutation process (t_cm=2.1 ms) in 20% voltage swell compared with the 50% voltage sag since the voltage swell easily leads to the insufficient linear modulation range, but it still has a better result than the natural commutation.

2466

The 2018 International Power Electronics Conference

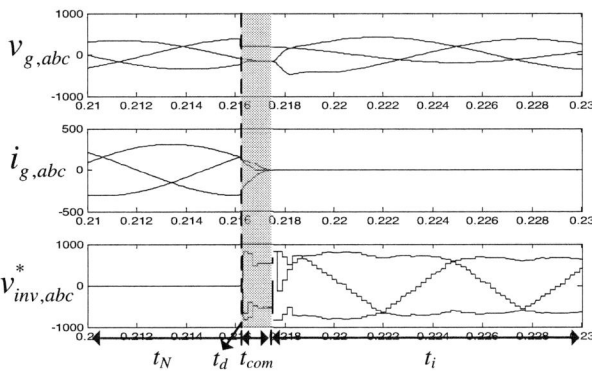

Fig. 13. Simulation waveforms as the forced commutation and the 50% voltage sag. Scales: v_g: V/s, i_g: A/s, and v^*_{inv}: V/s.

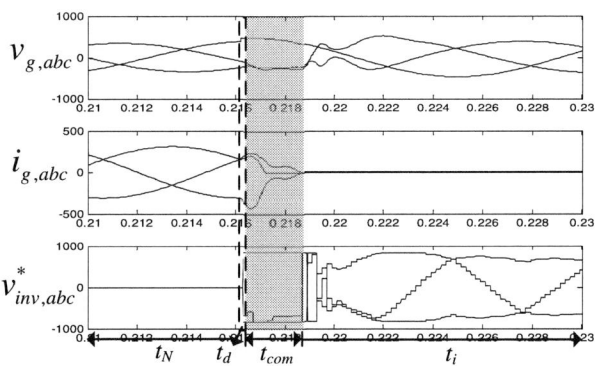

Fig. 14. Simulation waveforms as the forced commutation and the 20% voltage swell. Scales: v_g: V/s, i_g: A/s, and v^*_{inv}: V/s.

VI. LABORATORY TEST RESULTS

The circuit configuration is shown in Fig. 11, and their parameters are shown in TABLE II. The test conditions include the different filter capacitor, $C_f = 10\mu$ and $C_f = 14.7\mu$. The measured signals include the three phase current through STS (i_{ga}, i_{gb}, i_{gc}) and the three-phase voltages of load ($v_{loada}, v_{loadb}, v_{loadc}$).

A. $C_f = 10\mu F$

Fig. 15 and Fig. 16 respectively show the current waveform of thyristor based STS at the natural commutation and the forced commutation. As figures show, the natural commutation takes 6.35ms to turn the thyristor-based STS off, but the forced commutation just takes 0.63ms to turn it off. The commutation time can be apparently decreased by the proposed control mechanism. Fig. 17 and Fig. 18 respectively show the voltage waveform of the loads at the natural commutation and the forced commutation. The figures indicate the forced commutation leads to the less distortion in the load voltage since the commutation period is much reduced.

TABLE II
PARAMETER OF CIRCUIT

Parameter	Value
Filter capacitor(C_f)	10 μF
Filter inductors (L_f)	2 mH
Line to line voltage	220V
DC bus voltage	400V
Fundamental frequency	60 Hz
Switching frequency	10k Hz

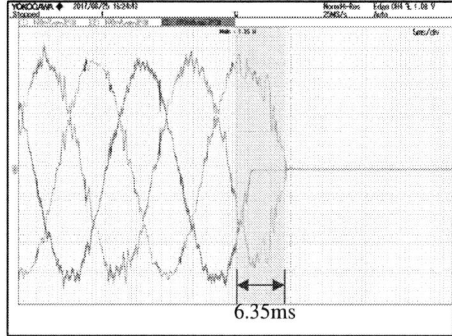

Fig. 15. Current waveform of thyristor based STS at the natural commutation.

B. $C_f = 14.7\mu F$

This case is used to validate the impact of larger filter capacitor. Fig. 19 shows the forced commutation without consideration of capacitor current. As figure shows, the commutation time is 1.9ms, which is longer than the case with $C_f=10\mu$F. Fig. 20 shows the forced commutation with consideration of capacitor current. Under consideration of the capacitor currents, the forced commutation process can be effectively reduce to 0.8ms. In addition, Fig. 17 and Fig. 18 respectively show the voltage waveforms of the forced commutation with and without consideration of capacitor current. As figures show, the consideration of filter capacitor current provides load voltages with a short time in distortion. Therefore, the impact of capacitor on forced commutation can be validated.

VII. SUMMARY

In this paper, a forced commutation mechanism is presented to improve the commutation of UPSs economical mode during grid faults. The analysis indicates the performance of forced commutation is closely related to the compensation dynamics of the predictive current control, so the different load conditions have a significant impact on the commutation effectiveness. The grid faults also affect this mechanism a lot, which voltage sag and voltage swell respectively lead to the beneficial and bad effect on the commutation mechanism. In addition, the use of large filter capacitor also leads to a significant effect on the force commutation mechanism. Finally, the simulation results and laboratory test results are

2467

The 2018 International Power Electronics Conference

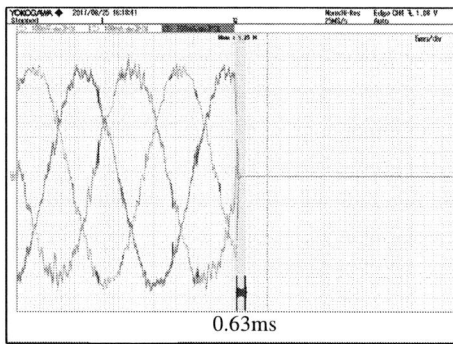

Fig. 16. Current waveform of thyristor based STS at the forced commutation.

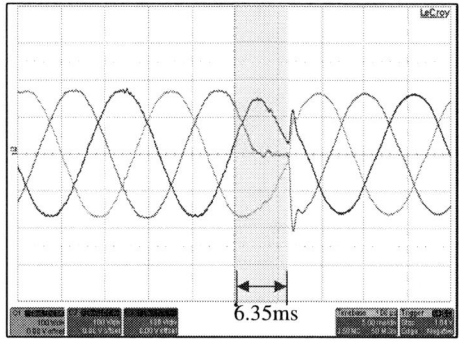

Fig. 17. Voltage waveform of the loads at the natural commutation.

supported to verify the effectiveness of the proposed forced commutation.

REFERENCES

[1] R. GN, "Emerging trends in uninterrupted power supplies: Patents view," in *2016 Biennial International Conference on Power and Energy Systems: Towards Sustainable Energy (PESTSE)*, Jan 2016, pp. 1–5.

[2] M. S. Racine, J. D. Parham, and M. H. Rashid, "An overview of uninterruptible power supplies," in *Proceedings of the 37th Annual North American Power Symposium, 2005.*, Oct 2005, pp. 159–164.

[3] M. J. Ryan and R. D. Lorenz, "A high performance sine wave inverter controller with capacitor current feedback and ldquo;back-emf rdquo;

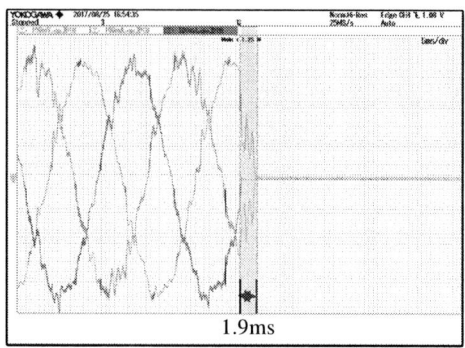

Fig. 19. Current waveform of thyristor based STS without consideration of capacitor current.

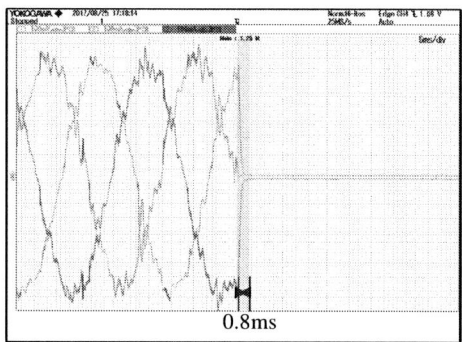

Fig. 20. Current waveform of thyristor based STS with consideration of capacitor current.

decoupling," in *Power Electronics Specialists Conference, 1995. PESC '95 Record., 26th Annual IEEE*, vol. 1, Jun 1995, pp. 507–513 vol.1.

[4] M. J. Ryan, W. E. Brumsickle, and R. D. Lorenz, "Control topology options for single-phase ups inverters," *IEEE Transactions on Industry Applications*, vol. 33, no. 2, pp. 493–501, Mar 1997.

[5] ——, "Control topology options for single-phase ups inverters," *IEEE Transactions on Industry Applications*, vol. 33, no. 2, pp. 493–501, Mar 1997.

[6] P. T. Cheng and Y. H. Chen, "Design of an impulse commutation bridge for the solid-state transfer switch," *IEEE Transactions on Industry Applications*, vol. 44, no. 4, pp. 1249–1258, July 2008.

[7] M. C. Chandorkar, D. M. Divan, and R. Adapa, "Control of parallel

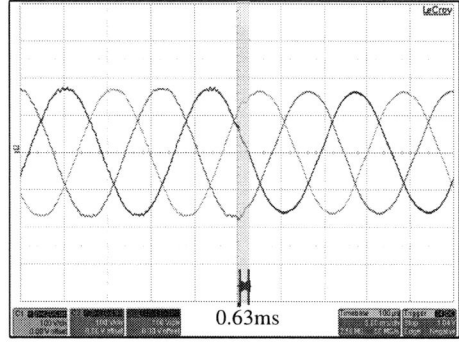

Fig. 18. Voltage waveform of the loads at the forced commutation.

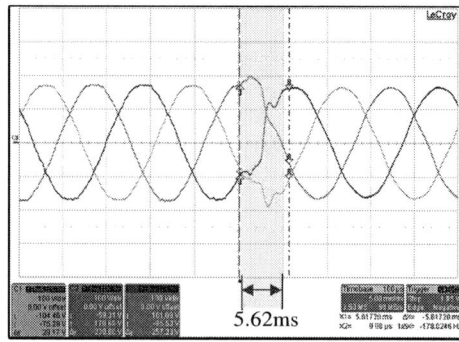

Fig. 21. Voltage waveform of loads without consideration of capacitor current.

2468

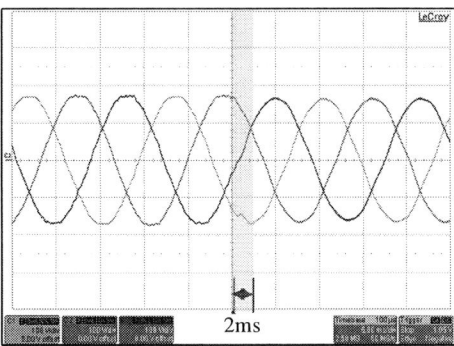

Fig. 22. Voltage waveform of loads with consideration of capacitor current.

connected inverters in standalone ac supply systems," *IEEE Transactions on Industry Applications*, vol. 29, no. 1, pp. 136–143, Jan 1993.

[8] M. C. Chandrokar, D. M. Divan, and B. Banerjee, "Control of distributed ups systems," in *Power Electronics Specialists Conference, PESC '94 Record., 25th Annual IEEE*, Jun 1994, pp. 197–204 vol.1.

[9] K. D. Brabandere, B. Bolsens, J. V. den Keybus, A. Woyte, J. Driesen, and R. Belmans, "A voltage and frequency droop control method for parallel inverters," *IEEE Transactions on Power Electronics*, vol. 22, no. 4, pp. 1107–1115, July 2007.

[10] P. Piagi and R. H. Lasseter, "Autonomous control of microgrids," in *2006 IEEE Power Engineering Society General Meeting*, 2006, pp. 8 pp.–.

[11] C. K. Sao and P. W. Lehn, "Autonomous load sharing of voltage source converters," *IEEE Transactions on Power Delivery*, vol. 20, no. 2, pp. 1009–1016, April 2005.

Online Internal Impedance Measurements of Li-ion Battery Using PRBS Broadband Excitation and Fourier Techniques: Methods and Injection Design

Jussi Sihvo[1], Tuomas Messo[1], Tomi Roinila[2], Roni Luhtala[2]

1 Laboratory of Electrical Energy Engineering, Tampere University of Technology, Tampere, Finland
2 Laboratory of Hydraulics and Automation, Tampere University of Technology, Tampere, Finland

Abstract—The internal impedance of a battery has been shown to vary as a function of state-of-charge (SOC) and state-of-health (SOH) which are important parameters defining the battery conditions in Li-Ion batteries. Therefore, the online monitoring of the internal impedance can be used as a useful SOC- and SOH-estimation parameter in battery management systems. Recent studies have shown methods based on broadband injections which make it possible to perform rapid Li-Ion battery impedance measurements. The studies have not, however, fully considered the properties of the applied injections nor the applied computation methods. This work extends the studies, and presents systematic steps for designing an excitation signal for battery internal impedance measurements. A pseudo-random binary sequence (PRBS) is applied and the internal impedance is obtained through Fourier techniques. The presented methods can be used for rapid and accurate impedance measurements, and thus, as a valuable online tool in SOC- and SOH-estimation algorithms. Practical measurements are shown from a commercial nickel-manganese-cobalt (NMC) Li-Ion battery.

I. INTRODUCTION

The number of applications involving lithium-ion batteries has greatly increased in the recent years due to the global ambitions in moving towards renewable energy production. Due to the relatively high power and energy density, lithium-ion batteries are suitable for energy storage and transportation applications such as electric vehicles and grid-tied energy storages [1]–[3]. Such applications often have more strict requirements for the battery state-of-charge (SOC) and state-of-health (SOH) estimation in battery-management-system (BMS) [3] than small-scale applications. The SOC is a parameter that states the available charge in the battery and is thus one of the most important estimates of the battery condition. The SOH is a parameter that defines the capacity of the battery with respect to a new battery capacity. These parameters are affected by non-linear behavior of the battery quantities such as terminal voltage, temperature, charge/discharge current and internal resistance [1], [3]–[5].

It has been shown that the battery internal impedance changes as a function of SOC and SOH. Therefore, online monitoring of the internal impedance can provide useful data for the battery state estimation in the BMS. The battery internal impedance is conventionally measured by using electrochemical impedance spectroscopy (EIS) [3], [5]–[7]. In the method, a sine wave perturbation is used as a current reference, and the

battery terminal voltage and current are measured to obtain the internal impedance. However, injecting single sine waves is a slow method, and therefore, not ideal for online measurements. Long measurement time causes the battery steady-state to change during the measurement which leads to corrupted results [7].

A highly potential alternative to the conventional impedance-measurement technique is to apply a broadband injection such as the pseudo-random binary sequence (PRBS) and Fourier techniques [8]. In the method, the PRBS is used as a current reference to discharge the battery, and the battery current and voltage are measured. Discrete-fourier-tranform (DFT) is then applied to the measurements to obtain the frequency response of the impedance. Applying such technique makes it possible to measure the battery impedance in a fraction of time compared to conventional techniques. Moreover, the method is an attractive solution for online applications as it can be implemented within the control system of a DC-DC charger. The measurement method based on broadband injection has already been applied in [7], [9]–[11]. However, more comprehensive optimization of the method is considered in this paper.

This work presents systematic steps for designing a PRBS injection for optimized internal impedance measurements of a lithium-ion battery pack. The rest of the paper is organized as follows. The SOC and SOH and their dependency to internal impedance of the battery is reviewed in Section II. System identification and PRBS injection design are presented in Sections III and IV. Practical measurements are shown from a commercial Nickel-Manganese-Cobalt (NMC) Li-Ion battery in Section V, and conclusion are drawn in Section VI.

II. SOC AND SOH DEPENDENCY ON INTERNAL IMPEDANCE

The energy available in a battery is defined by the SOC. The SOC indicates the remaining charge of the battery with respect to rated full charge. A conventional algorithm for the SOC estimation is *Coulomb counting* in which the battery current is integrated over time and compared against the nominal capacity as

$$\text{SOC}(t) = \text{SOC}(t_0) - \left(\frac{1}{Q_0}\int_{t_0}^{t} i_{\text{bat}}dt\right)100\%, \quad (1)$$

where Q_0 is the nominal capacity of the battery, i_{bat} is the battery current, and t_0 and t are adjacent time instant at which the SOC is calculated. In reality, *Coulomb counting* method is inaccurate since Q_0 varies as a function of temperature, charge/discharge current and life cycles. More sophisticated algorithms presented in [3], [4] are used in real applications.

The aging of the battery is analyzed by the SOH that indicates the actual capacity in batteries with respect to nominal capacity. The SOH also indicates when the battery is incapable of meeting the requirements of the application the battery is designed for and should be replaced. The SOH is typically analyzed through the resistive component of the internal impedance but there are also other definitions for SOH depending on the application. [3], [5]

It is widely observed that the battery internal impedance has a SOC dependency at very low frequency components, around and below 1 Hz [6], [7], [12]. The resistive component of the impedance spectrum for the SOH estimation is present at relatively high frequencies, around 1 kHz [5]. In reality, the impedance can change quite radically as observed in [5] and [6] as terminal voltage, temperature, discharge/charge current and cycle life change. Therefore it would be beneficial to monitor the internal impedance of the battery in real-time. However, small magnitude of the internal impedance and relatively long time constants of the battery make the impedance real-time monitoring quite challenging and the measurement methods must be carefully designed.

III. SYSTEM IDENTIFICATION USING EXTERNAL INJECTION

By simply measuring the current and voltage during battery charging and discharging, no information of the internal impedance is obtained. This is due to the parasitic properties of the internal impedance. Instead, by perturbing the battery with an excitation signal and measuring the changes in current and voltage, more relevant information can be obtained. However, different excitation signals have different properties and effect on the results [9]. To extract the desired information of the system, the excitation signal needs to be carefully designed.

Fig.1 shows a typical setup where the device under test, presented by an impulse-response function $g(t)$, is to be identified. The system is perturbed by the excitation $x(t)$, which yields the corresponding output response $y(t)$. The measured signals are corrupted with noise, as presented by $e(t)$ and $r(t)$. The measured excitation and output response can now be denoted by $x_e(t)$ and $y_r(t)$. The noises are assumed to resemble white noise and are uncorrelated with $x(t)$ and $y(t)$. All of the signals are assumed to be zero mean sequences. The injection point of the excitation depends on the application. For battery impedance measurements, the transfer function of the impedance can be given as follows

$$Z(j\omega) = \frac{V(j\omega)}{I(j\omega)}, \qquad (2)$$

where $V(j\omega)$ and $I(j\omega)$ are the measured and Fourier transformed voltage and current. The impedance $Z(j\omega)$ can there-

fore be regarded as a system impulse response (g(t) in Fig. 1) in the frequency domain. Considering the fact that the battery is realized as a voltage source, the battery current is the quantity that should be perturbed and the injection point is located at the input side of the system.

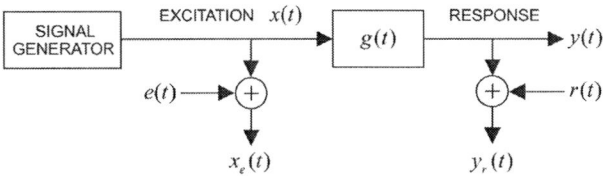

Fig. 1: Typical measurement set up.

IV. PSEUDO-RANDOM BINARY SEQUENCE (PRBS)

Pseudo-random binary sequence (PRBS) is a broadband injection which contains energy at several frequencies. It is a binary signal which has only two levels. Thus, it can be generated with lighter processors compared to sine wave generation which has theoretically infinite number of levels. One of the most useful and simplest forms of the PRBS is a maximum-length-binary-sequence (MLBS). The spectral energy content of the MLBS is largely controllable by design, and therefore, the signal is well suited for many practical applications in system identification. To avoid confusion, term PRBS is used in the further sections of this paper, representing the MLBS. [8]

The PRBS has limited number of different possible signal lengths N restricted to

$$N = 2^n - 1, \qquad (3)$$

where n is a positive integer. Due to the broadband characteristics of the PRBS, the sequence contains energy at several frequencies up to the generating frequency f_{gen}. The frequency components in the PRBS are discrete and linearly distributed according to the frequency resolution given by

$$f_{res} = \frac{f_{gen}}{N}, \qquad (4)$$

which is also the lowest frequency that can be measured. Since every unity addition to n doubles the N in (3), the frequency resolution can be more freely designed by adjusting the f_{gen} in (4). The power spectrum of the PRBS is given as

$$\Phi_{PRBS} = a^2 \frac{(N+1)}{N} \frac{\sin^2(\pi i/N)}{(\pi i/N)^2}, \quad i = \pm 1, \pm 2, \dots \qquad (5)$$

where a is the amplitude and i is the sequence harmonic number. Fig. 2 shows a power spectrum of the PRBS generated at 100 Hz. The figure shows that the power drops to zero at the signal generation frequency. Therefore, reliable information cannot be acquired at frequencies beyond and close to f_{gen}. [8]

In a typical frequency-response identification, the applied excitation signal should have an approximately equal amount

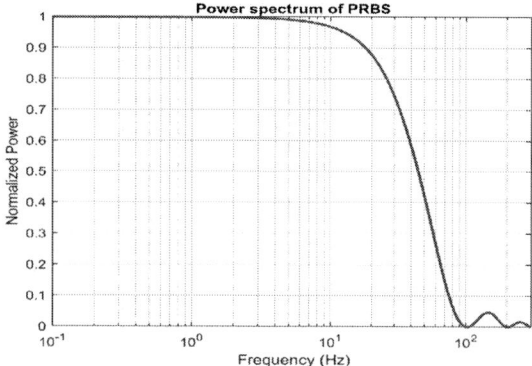

Fig. 2: Power spectrum of the PRBS.

A. Frequency-response computation

There are several techniques for computing the frequency-response function from input-output data generated by a broadband injection [8]. In the paper, an arithmetic averaging procedure is proposed to obtain the impedance $Z(j\omega)$ as

$$Z(j\omega) = \frac{1}{P} \sum_{k=1}^{P} \frac{V_k(j\omega)}{I_k(j\omega)}, \qquad (6)$$

where P denotes the number of injected periods, and V and I are the measured voltages and currents. In this method, the measurements from both input and output sides are segmented and Fourier transformed according to the length of the injection, after which (6) is applied.

V. EXPERIMENTS

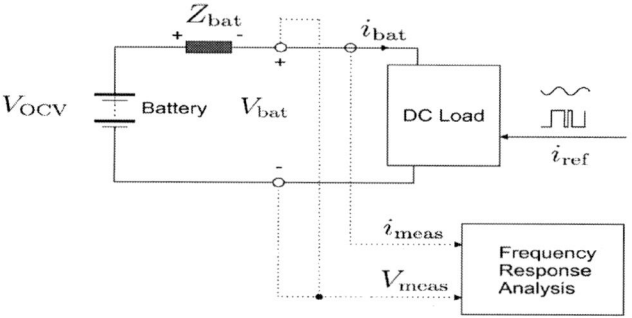

Fig. 3: Practical setup for battery impedance measurement.

Fig. 3 shows the practical measurement setup applied in the experiments. In the figure, V_{OCV} is the battery open-circuit voltage, Z_{bat} is the internal impedance of the battery and V_{bat} is the voltage between battery terminals. V_{bat} is therefore different from V_{OCV} when the battery draws any current. During the measurements, the SOC is monitored by (1). The PRBS is injected into the circuit as a current reference i_{ref} to the DC-load device. The battery is then discharged by the current i_{bat}. The current and terminal voltage of the battery are measured (i_{meas} and V_{meas}) and the frequency response is then calculated. The practical setup used for the experiments is shown in Fig. 4 where the applied devices are as follows.

1) NMC Lithium-Ion battery pack (36V, 2.6Ah)
2) DC-load used for discharge of the battery and as a link to the excitation signals
3) Venable frequency response analyzer used for creating the excitation signal and for data acquisition during sine sweep measurements
4) NI USB-6363 DAQ device used for creating the excitation signal and for data acquisition during the PRBS measurements
5) Voltage probe
6) Current probe
7) Oscilloscope for visualizing the measurements and excitation signals

of energy at the frequencies where the system is identified. According to Fig. 2, the energy in the PRBS is clearly spread over the harmonics in a non-uniform manner. However, approximately uniform energy can be achieved by generating the sequence with a sufficiently high frequency. Typically, the spectrum is considered to be flat until the power has dropped -3 dB. This part of the excitation signal is known as an effective frequency band limited to roughly $0.45 f_{gen}$. Thus, the generation frequency of the signal has to be selected such that the effective frequency band covers the frequency band where the system is identified. Other design variables of the PRBS include the length of one PRBS period (N), number of injection periods (P) and the amplitude of the injection (a).

Considering battery internal impedance measurements, the impedance should be measured at frequencies below 1 Hz where the SOC dependency is present. On the other hand, the resistive component of the impedance is used for the SOH analysis which is typically present at around 1 kHz. The need for a very small frequency resolution is challenging with battery impedance measurements. It can be seen from (4) that small f_{res} and relatively high f_{gen} equals to large N which should be kept as small as possible. Large N increases the memory requirements, and buffer sizes for the signal generator and the data acquisition.

The number of excitation periods P should be defined by the desired accuracy (variance) of the measurements, memory limitations of data acquisition and the time used for the measurements. For battery impedance measurements, the most important restriction for P is the measurement time which should be kept as short as possible.

The injection amplitude a should be large enough to have a sufficient impact on the system. On the other hand, a should be low enough to avoid too large disturbance of the identified system, and to keep the system at steady-state. For battery impedance measurements, the amplitude must be relatively high because the magnitude of the internal impedance is very small.

The 2018 International Power Electronics Conference

Fig. 4: Laboratory setup of the battery impedance measurements

Due to the characteristics of the DC-load used for measurements, the perturbations are generated with an offset that will ensure the discharge operation when given to the DC-load as a current reference. That is, the PRBS current should not go to zero since it was observed to corrupt the results. The measurements were carried out at a temperature of $25^{\circ}C$.

A. Injection design

In order to keep the injection length (N) relatively low, the measurements were carried out by using two separate 1023-bit-long PRBS with different f_{gen} and f_{res}. The first sequence was used to measure the low-frequency part of the impedance spectrum with dense frequency resolution. The second sequence with more sparse resolution was used for the higher frequencies. Similar *double-injection* method is applied for fuel cell measurements in [13].

For the low-frequency PRBS, the starting point of the excitation design is to determine the lowest frequency, that is desired. However, too dense resolution will prolong the measurements to be insufficient for online measurements [8]. For the experiments, the frequency resolution for the low-frequency PRBS was selected as 100 mHz. Such resolution can be achieved by selecting a 1023-bit-length PRBS and a generation frequency of 102.3 Hz. For the high-frequency PRBS, it is important to select the generation frequency so that it covers sufficiently the highest frequency of interest. The highest frequency to be measured was selected as 5 kHz in order to cover the resistance region in impedance spectrum at around 1 kHz. Due to the effective frequency band restricted to $0.45 f_{gen}$, a generation frequency of 12 kHz was considered appropriate to cover the frequencies up to 5 kHz. The sampling frequency was set to four times higher than generating frequency for both sequences to avoid aliasing in the measurements.

The internal impedance of the battery is usually very small, and thus, a relatively large current is required to produce a measurable change in the terminal voltage [6]. This means that a relatively large injection amplitude must be used. The amplitude was decided to be kept well below the battery C-rate to ensure that the battery conditions are not changing too much during the measurements. A peak-to-peak amplitude of 1 A was chosen for both PRBS injections. As discussed, an offset is required for the injection. This was chosen as 1 A. The peak-current for the injection is therefore 1.5 A which is well below the C-rate of the measured battery pack (2.6 A). The parameters used for the measurements are shown in Table I. The measurements were also carried out by using a sine sweep-based network analyzer (reference measurements).

	Sine sweep	low-freq. PRBS	high-freq. PRBS
Offset	1A	1A	1A
Amplitude a_{p-p}	1A	1A	1A
Averaged periods P	10	10	50
frequency band	100mHz - 5kHz	100mHz - 102.3Hz	11.7Hz - 5kHz
generation freq. f_{gen}	-	102.3Hz	12kHz
sampling freq. f_s	-	$4f_{gen}$	$4f_{gen}$
integer n	-	10	10

TABLE I: Injection parameters for measurements

B. Measurement results and observations

Figs. 5 and 6 show the impedance frequency response as a function of SOC, measured by the sine sweeps and the PRBS respectively. The frequency responses at five different SOCs are shown in Figs. 7 - 11. The figures show that the results obtained by the PRBS match relatively well to the reference at a wide frequency band. However, Figs. 7 - 11 show some amount of distortion in the measurement results. The impedance at 20% of SOC is significantly distorted below 10 Hz. Small deviation and offset is also introduced below 1 Hz to impedances at SOC values of 25% and 100%.

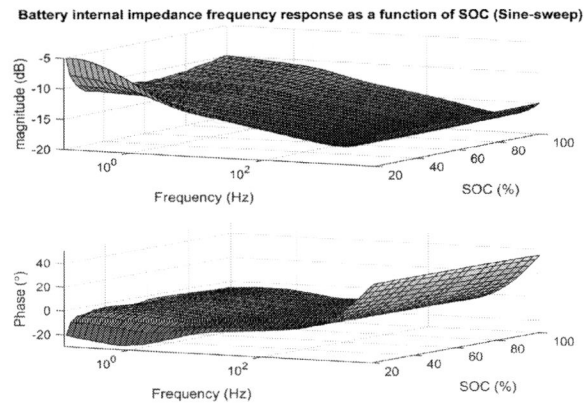

Fig. 5: Sine sweep measurements of battery impedance as a function of frequency and SOC.

2473

The 2018 International Power Electronics Conference

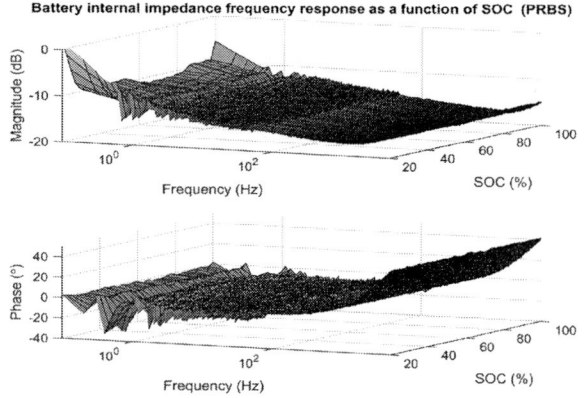

Fig. 6: PRBS measurements of battery impedance as a function of frequency and SOC.

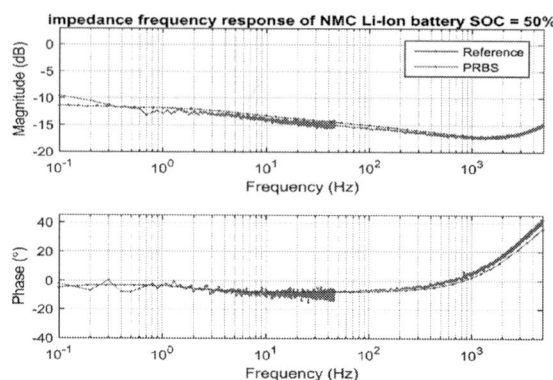

Fig. 9: Internal Impedance of NMC-Li-ion battery measured with different excitations at SOC = 50%

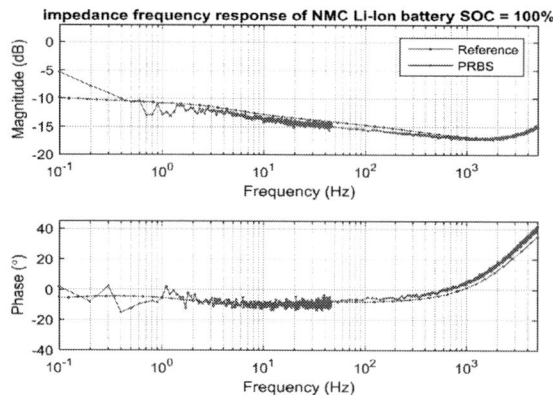

Fig. 7: Internal Impedance of NMC-Li-ion battery measured with different excitations at SOC = 100%

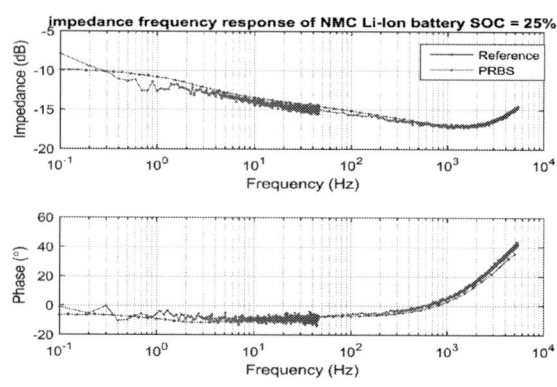

Fig. 10: Internal Impedance of NMC-Li-ion battery measured with different excitations at SOC = 25%

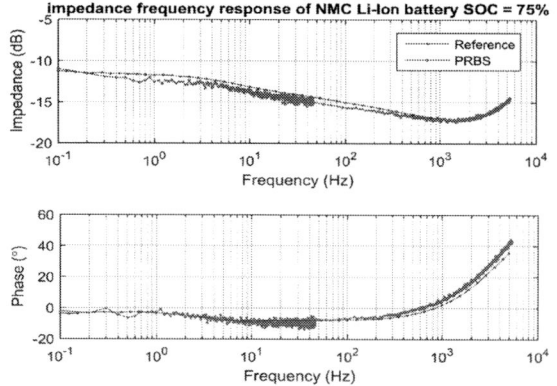

Fig. 8: Internal Impedance of NMC-Li-ion battery measured with different excitations at SOC = 75%

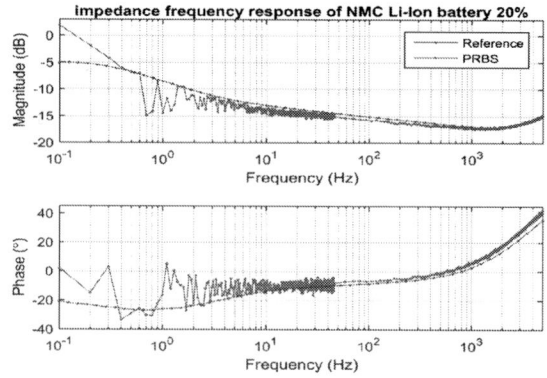

Fig. 11: Internal Impedance of NMC-Li-ion battery measured with different excitations at SOC = 20%

2474

Distortions in the results are mostly caused by the varying SOC during a measurement. It was observed that the amount of distortion and bias in the impedance spectrum is somewhat proportional to the derivative of the V_{OCV} as illustrated in Fig. 12. For example, the voltage derivative is lowest at SOC value of 75% where the impedance is obtained most accurately when comparing Figs. 7 - 11. A change in the voltage indicates that the steady-state conditions of the battery have been changed and the impedance is not the same anymore, as expected.

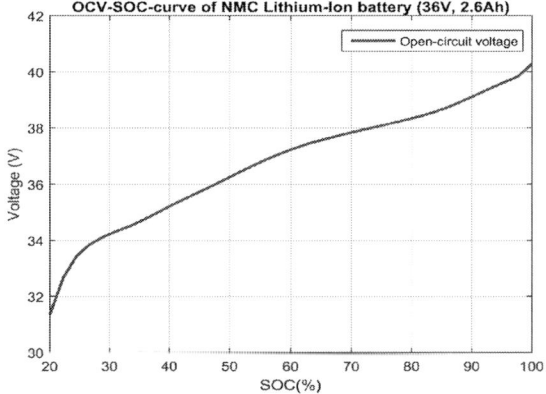

Fig. 12: Measured open-circuit voltage of NMC-Li-ion battery as a function of SOC

Similar observation about corrupted impedance measurements was done also in [9], which concluded that this is due to non-linear behavior of the battery pack. Errors in the PRBS measurements can be avoided by making the measurements significantly faster. However, this would yield to more sparse frequency resolution or reduced averaging, and therefore, less comprehensive characterization of the internal impedance. Considering the SOH analysis, the method already gives reliable results at around 1 kHz.

VI. CONCLUSIONS

This paper has presented online methods for obtaining the internal impedance from a battery. In the methods, a pseudo-random binary sequence (PRBS) is injected to the battery as a current reference, the battery current and voltage responses are measured, and Fourier techniques are applied to obtain the internal impedance of the battery. Practical measurements were shown from a commercial NMC Lithium-Ion battery pack. The experiments show that the presented method accurately produces the internal impedance of the battery. The impedance is obtained in a fraction of a time compared to conventional techniques based on sine sweep injections. Only the frequencies around 1 Hz and below are slightly distorted due to the non-linear behavior of the battery. The presented methods can be efficiently used in online applications for defining the SOC and SOH in battery management systems.

REFERENCES

[1] J. Yang, B. Xia, W. Huang, and C. Mi, "On-board State-of-health Estimation Based on Charging Current Analysis for LiFePO 4 batteries," pp. 5229–5233, 2017.

[2] T. Mesbahi, N. Rizoug, P. Bartholomeus, R. Sadoun, fouad Khenfri, and P. LE MOIGNE, "Dynamic model of Li-Ion Batteries Incorporating Electrothermal and Ageing Aspects For Electric Vehicle Applications," *IEEE Transactions on Industrial Electronics*, vol. 65, no. 2, pp. 1–1, 2017.

[3] P. Weicker, *A Systems Approach to Lithium-Ion Battery Management*. Norwood, UNITED STATES: Artech House, 2013.

[4] X. Zhang, J. Wu, and G. Kang, "SOC estimation of Lithium battery by UKF algorithm based on dynamic parameter model," *2016 13th International Conference on Ubiquitous Robots and Ambient Intelligence, URAI 2016*, pp. 945–950, 2016.

[5] D. I. Stroe, M. Swierczynski, a. I. Stan, V. Knap, R. Teodorescu, and S. J. Andreasen, "Diagnosis of lithium-ion batteries state-of-health based on electrochemical impedance spectroscopy technique," *Energy Conversion Congress and Exposition (ECCE), 2014 IEEE*, pp. 4576–4582, 2014.

[6] A. Zenati, P. Desprez, and H. Razik, "Estimation of the SOC and the SOH of Li-ion batteries, by combining impedance measurements with the fuzzy logic inference," *IECON Proceedings (Industrial Electronics Conference)*, pp. 1773–1778, 2010.

[7] R. Al Nazer, V. Cattin, P. Granjon, M. Montaru, and M. Ranieri, "Broadband identification of battery electrical impedance for HEVs," pp. 2896–2905, 2013.

[8] K. Godfrey, *Perturbation Signals for System Identification*. Prentice Hall, 1994.

[9] R. Al Nazer, V. Cattin, P. Granjon, M. Montaru, M. Ranieri, and V. Heiries, "Classical EIS and square pattern signals comparison based on a well-known reference impedance," *World Electric Vehicle Journal*, vol. 6, no. 3, pp. 800–806, 2013.

[10] A. J. Fairweather, M. P. Foster, and D. A. Stone, "Battery parameter identification with Pseudo Random Binary Sequence excitation (PRBS)," *Journal of Power Sources*, vol. 196, no. 22, pp. 9398–9406, 2011.

[11] S. Nejad, D. T. Gladwin, and D. A. Stone, "A hybrid battery parameter identification concept for lithium-ion energy storage applications," *IECON Proceedings (Industrial Electronics Conference)*, pp. 1980–1985, 2016.

[12] J. Sihvo, "Internal impedance measurement techniques and charger dynamics for lithium-ion batteries," MSc, Tampere University of Technology, 2017.

[13] P. Manganiello, G. Petrone, M. Giannattasio, E. Monmasson, and G. Spagnuolo, "FPGA implementation of the EIS technique for the on-line diagnosis of fuel-cell systems," *IEEE International Symposium on Industrial Electronics*, 2017.

The 2018 International Power Electronics Conference

A DC Current Flow Controller for Meshed HVDC Grids

Viktor Hofmann* and Mark-M. Bakran

Department of Mechatronics, University of Bayreuth, Center of Energy Technology, Bayreuth, Germany
*E-mail: Viktor.Hofmann@Uni-Bayreuth.de

Abstract— **This paper proposes a modular and scalable current flow controller (CFC) for meshed High Voltage Direct Current (HVDC) grids. The exchange of active power is performed directly between two DC branches and the circuit can be installed directly at a DC node without any additional AC grid. The CFC is especially suitable for medium CFC voltage ratings and neither an insulation to ground nor a transformer is needed. The basic design as well as the functional principle is presented and a control and balancing method is described, which allows a cell capacitor balancing and switching frequency control. Finally, full-scale simulation results validate the functionality of the proposed CFC.**

Keywords— *Control, Current Flow Controller, HVDC, Modulation.*

I. INTRODUCTION

There is a constant increase of renewable energy generation all over the world. Mostly, the renewable energy production is located in remote areas and has to be transported efficiently to the centers of consumption. To meet the requirements of its high efficient utilization, it is necessary to develop reasonable transmission technologies [1]. The DC power transmission technology has advantages over AC ones with regard to unit cost, transmission efficiency, transmission capacity, footprint and transmission distance [2]-[5]. Consequently, the high voltage direct current (HVDC) transmission technology is recognized as an advantageous approach for the long-distance bulk power transmission [6], [7] and is gaining more and more attraction for the large scale integration of renewable energy sources [8]. There are several applications in China [9], India [10] and Europe [11], [12], where the HVDC technology is already in use.

However, the realization of all the mentioned benefits is not without many technical challenges. A major concern is the power flow control within a grid. In an AC grid, the power flow control can be achieved by manipulating the reactive power. However, there is no reactive power in a DC grid. The current flow depends on the resistances of the transmission lines and is achieved by adjusting the dc voltage at each terminal. Consequently, there is no direct control on the dc currents that flow on each dc branch. This can result in transmission overload or under-utilization of some dc lines. This makes a current flow control indispensable.

In a DC grid, the current flow control has to be done by injecting or extracting active power. Many innovative circuits have been introduced in literature to handle this task. They can roughly be categorized into three types: variable series-connected resistors, series-connected voltage sources with an external power exchange and series-connected voltage sources with an interline power exchange. These possible current flow controllers (CFC) are illustrated in Fig. 1.

A very simple way to achieve the power flow control is adding additional resistances into a dc branch. In [13] and [14] two methods of inserting variable resistances were proposed. These methods have advantages of simple structure and control, but have too high power losses to be acceptable and can only realize a unidirectional power flow adjustment. Another way to handle this task is inserting a controllable voltage source in series to a DC line. In [15] a thyristor-based converter topology was presented which requires an AC connection. Another voltage source-based current flow controller is described in [16] where IGBTs are used instead of thyristors. In both circuits, an auxiliary transformer establishes the AC connection. These transformers are expensive and require a large footprint due to a specialized manufacturing and design as well as a high transformation ratio. In [17] a current flow controller is proposed which eliminates the need to connect an AC system and enables a direct power

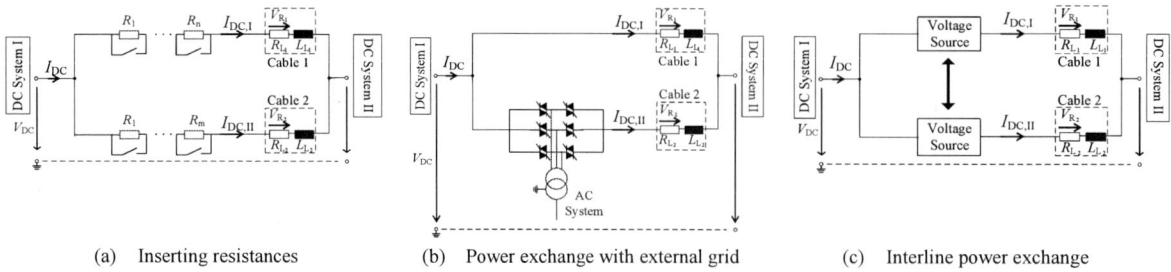

(a) Inserting resistances   (b) Power exchange with external grid   (c) Interline power exchange

Fig. 1. Possible solutions to perform a current flow control in a DC grid.

2476

exchange between two DC lines through an internal AC link.

A possible topology which completely eliminates the constraint of using a transformer, is described in [18], [19] and provides a compact and cheap solution. The disadvantage is, that this circuit can only use a single voltage level and is not scalable to higher voltage ratings. A modular and scalable HVDC current flow controller is described in [20], [21]. This circuit is easily scalable and can apply any desired amount of voltage levels. Consequently, it offers a suitable solution for higher CFC voltage ratings. The disadvantage of this circuit is the high semiconductor effort for lower voltage ratings.

This paper presents a current flow controller which fills the gap of medium CFC voltage ratings. Besides the presentation of the basic design and the functional principle, this paper will present a control and balancing method, which allows a cell capacitor balancing as well as a switching frequency control. The functionality of the proposed circuit will be validated by full-scale simulation results.

II. HVDC CURRENT FLOW CONTROLLER

A. System Configuration

In this paper, a monopole meshed DC grid is considered. A schematic diagram of a possible implementation of the CFC within a DC grid is illustrated in Fig. 2. The CFC can be installed directly at a DC node and is able to control the current flow of the connected cables.

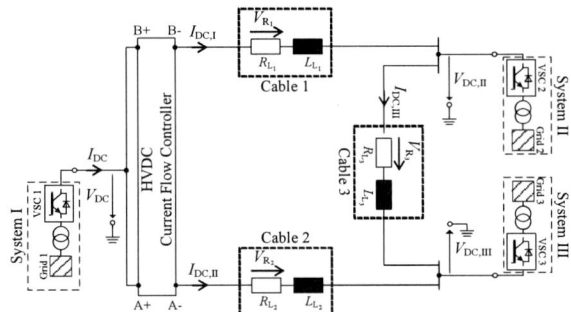

Fig. 2. Meshed grid configuration with the HVDC current flow controller.

B. Basic Design

The basic design of the proposed CFC is illustrated in Fig. 3. It shows the basic cell of the CFC which consists of two full bridge switching units (FBS), four bidirectional switches (BS) and a cell capacitor. With this cell, it is possible to modulate three different voltage states between the terminals A+ / A- (V_A) and B+ / B- (V_B):

- a positive cell capacitor voltage ($+V_C$),
- a negative cell capacitor voltage ($-V_C$)
- and a zero voltage level.

The terminals B+/B- provide a zero voltage level, if the cell capacitor is clamped between the terminals A+/A- and vice versa. To reach higher voltage ratings, several (n)

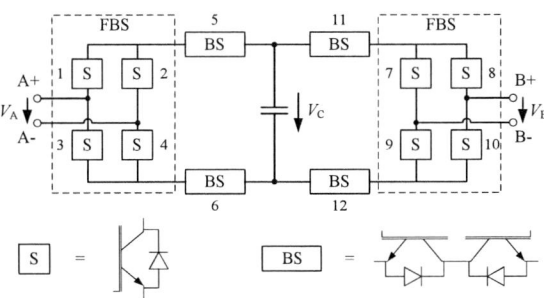

Fig. 3. Basic design of the proposed HVDC current flow controller.

basic cells can be connected in series. This is illustrated in Fig. 4. For the sake of clarity, the illustration only shows the point-to-point connection of two DC systems by two parallel lines which are represented by a line resistance R_L and a line inductance L_L. The bidirectional switches in a

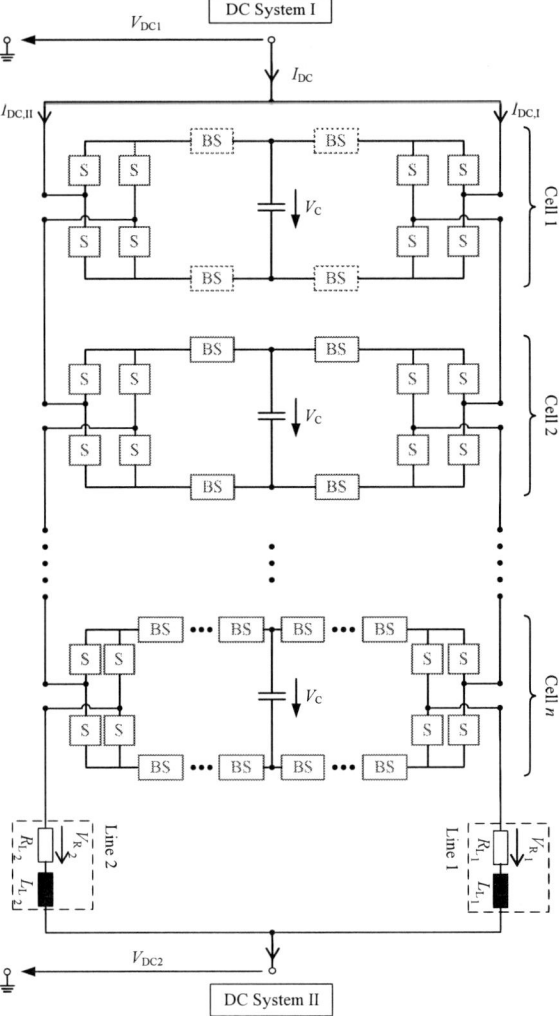

Fig. 4. Implementation of the proposed CFC as a series connection of n basic cells to achieve higher voltage ratings. (Illustrated in a point-to-point connection of two DC systems)

2477

cell stage n are used to block the sum of the cell voltages of the previous stages. Consequently, the needed voltage rating of a BS V_{BS} increases in every cell stage and is given by:

$$V_{BS,n} = \sum_{i=1}^{n-1} V_{C,i} \tag{1}$$

This is necessary to keep the DC current of each branch separated. The use of the BS in the first cell stage is not necessary and would not change the functionality. Only the current load of some semiconductors vary in asymmetric operation points.

C. Functional Principle

The CFC consists of n basic cells as it is illustrated in Fig. 4, where each cell stage has the same cell capacitor voltage rating V_C. The voltage modulation is performed by switching the capacitors with a positive, negative or zero voltage state into the different cables. Consequently, the CFC is able to modulate a desired voltage level V_1 and V_2 in each DC line and can be seen as an adjustable voltage source. A detailed description of the modulation and control scheme will be presented in chapter III.

The functional principle of the proposed CFC will be explained on the basis of the equivalent circuit diagram, shown in Fig. 5. For the sake of clarity, only a point-to-point connection of two DC systems by two parallel lines will be discussed. Nevertheless, the CFC can also be applied in a multi-terminal grid. The DC current I_{DC} as well as the line resistances R_{L1} and R_{L2} are assumed to be given. The line inductances can be neglected for a steady state operation. Furthermore, two auxiliary variables are introduced:

$$a := \frac{I_{DC,I}}{I_{DC}} \tag{2}$$

$$b := \frac{R_{L2}}{R_{L1}} \tag{3}$$

a is the current distribution factor and describes the relative amount of the DC current flowing in line 1. In principle, it is possible to realize any desired current distribution. The value b describes the resistance ratio between the line resistances 2 and 1. Thus, the currents $I_{DC,I}$ and $I_{DC,II}$ in the DC branches 1 and 2 can be specified as:

$$I_{DC,I} = a \cdot I_{DC} \tag{4}$$

$$I_{DC,II} = (1 - a) \cdot I_{DC} \tag{5}$$

The line voltage levels V_1 and V_2 which have to be modulated to achieve a desired operation point, can be described by these parameters. For a steady state operation the equation of mesh M_1 is given by:

$$V_1 + V_{R1} = V_2 + V_{R2} \tag{6}$$

V_{R1} and V_{R2} are the voltage drops due to the line resistances R_{L1} and R_{L2}:

$$V_{R1} = R_{L1} \cdot a \cdot I_{DC} \tag{7}$$

$$V_{R2} = R_{L1} \cdot b \cdot (1 - a) \cdot I_{DC} \tag{8}$$

To ensure a safe operation, the power balance of the CFC has to be fulfilled. Otherwise the cell capacitors would be charged or discharged over time:

$$P_{V1} = -P_{V2}$$
$$V_1 \cdot I_{DC,I} = -V_2 \cdot I_{DC,II} \tag{9}$$

By using the indicated equations above, the voltage levels V_1 and V_2 are calculated:

$$V_1 = R_{L1} \cdot I_{DC} \cdot (1 - a) \cdot [b \cdot (1 - a) - a] \tag{10}$$

$$V_2 = -R_{L1} \cdot I_{DC} \cdot a \cdot [b \cdot (1 - a) - a] \tag{11}$$

The minimum amount of cell stages n_{min} is given by the sum of the absolute voltage ratings of V_1 and V_2:

$$n_{min} = \frac{|V_1| + |V_2|}{V_C}$$
$$= \begin{cases} \frac{|R_{L1} \cdot I_{DC} \cdot [b \cdot (1-a) - a]|}{V_C} & \text{if } 0 < a < 1 \\ \frac{|R_{L1} \cdot I_{DC} \cdot (1-2a) \cdot [b \cdot (1-a) - a]|}{V_C} & \text{else} \end{cases} \tag{12}$$

III. OPERATION AND CONTROL SYSTEM OF THE CURRENT FLOW CONTROLLER

This chapter presents the operation and control system of the CFC. A schematic overview of the system is shown in Fig. 6. The reference line voltages V_1^* and V_2^* are generated by a high-level CFC control, which includes a branch current control and a total energy control. The reference voltages are transmitted to the modulator which performs a cell selection and finally generates the control pulses for each cell. The following sections present a description of the different system parts.

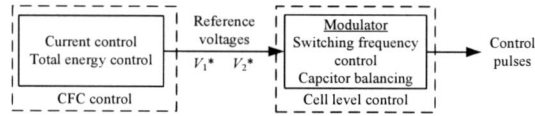

Fig. 6. Schematic system overview.

A. Switching States

The CFC is assembled with n basic cells which perform the desired voltage modulation. The possible switching states of one cell are listed in Table 1 and illustrated with the appropriate current path in Table 2. Each voltage state

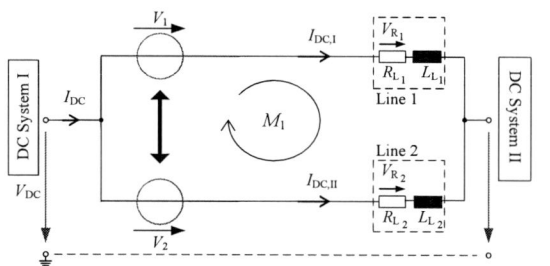

Fig. 5. Equivalent circuit diagramm of the CFC for a steady state operation in a point-to-point connection of two DC systems.

TABLE I
SWITCHING STATES OF A BASIC CELL OF THE CFC

Nr.	S1	S2	S3	S4	BS 5	BS 6	S7	S8	S9	S10	BS 11	BS 12	V_A	V_B
1.1	1	0	0	1	1	1	1	1	0	0	0	0	+	0
1.2	1	0	0	1	1	1	0	0	1	1	0	0	+	0
2.1	1	1	0	0	0	0	1	0	0	1	1	1	0	-
2.2	0	0	1	1	0	0	1	0	0	1	1	1	0	-
3.1	0	1	1	0	1	1	1	1	0	0	0	0	-	0
3.2	0	1	1	0	1	1	0	0	1	1	0	0	-	0
4.1	1	1	0	0	0	0	0	1	1	0	1	1	0	+
4.2	0	0	1	1	0	0	0	1	1	0	1	1	0	+

1: IGBT turned on; 0: IGBT turned off

+: $+V_C$; -: $-V_C$

offers two different freewheeling paths for the current in a DC branch and each cell state has a bidirectional current path. To reduce the semiconductor load, the two freewheeling paths are used alternately during the active voltage modulation.

B. Modulation Scheme

A possible way to perform the voltage modulation is to split the total reference voltage equally on each cell. Consequently, each cell would permanently modulate a voltage level $V_{mod,1/2}$ in line 1 and line 2 which is given by:

$$V_{mod,1/2} = \frac{V_{1/2}^*}{n} \tag{13}$$

The voltage modulation scheme and sequence of switching actions is predefined and could be performed by switching the cell capacitor of each cell in a charging state (and a subsequent zero voltage state) in one line and afterwards in a discharging state (and a subsequent zero voltage state) in the other line ([22], [23]). The energy balance of the cell capacitors can be handled by controlling the duty cycle $D_{1/2}$, which is given by the time $t_{\pm V_{1/2}}$, where the cell capacitor is switched into a line and the time of the appropriate zero voltage state t_{V_0}:

$$D_{1/2} = \frac{t_{\pm V_{1/2}}}{t_{\pm V_{1/2}} + t_{V_0}} \tag{14}$$

Furthermore, a pulse-width modulation (PWM) can be used to increase the voltage accuracy. Because of the several cell stages, this modulation strategy is not an ideal solution for the presented CFC. It would be necessary to perform a separated energy control on each cell and the switching frequency of the semiconductors as well as the energy fluctuation of the cell capacitors can only be influenced by the modulation frequency.

Therefore, this paper presents a different approach which eliminates the mentioned drawbacks. A modulator performs the cell selection and determines the switching moment of every cell. The operating principle of the proposed modulation algorithm will be explained by means of Fig. 7. There are two kinds of switching events: one is performed due to the tolerance band condition and the other one due to the voltage-time-area.

In the first step of the modulation algorithm it is checked whether each cell capacitor voltage $V_{C,i}$ fulfills the desired tolerance band condition, which is defined by the relation between the maximum admissible cell capacitor voltage $V_{C,max}$ and the nominal cell capacitor voltage $V_{C,0}$. This relation is described by the parameter k:

$$k = \frac{V_{C,max}}{V_{C,0}} - 1 \tag{15}$$

TABLE II
ILLUSTRATION OF THE CURRENT PATHS OF THE DIFFERENT SWITCHING STATES

Nr.	Cell state	V_A / V_B	Nr.	Cell state	V_A / V_B
1.1		$+V_C / 0$	1.2		$+V_C / 0$
2.1		$0 / -V_C$	2.2		$0 / -V_C$
3.1		$-V_C / 0$	3.2		$-V_C / 0$
4.1		$0 / +V_C$	4.2		$0 / +V_C$

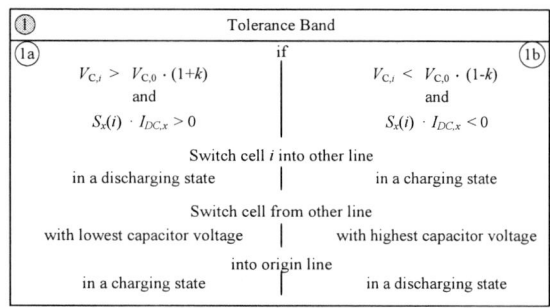

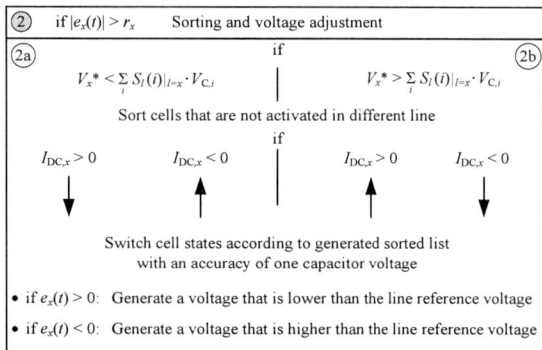

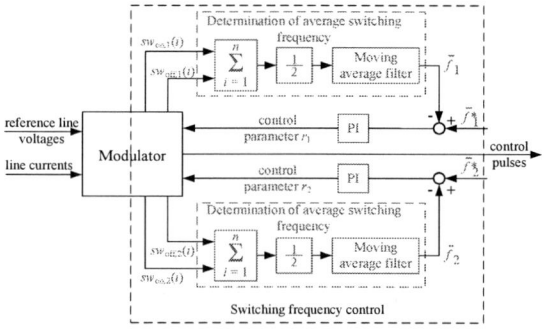

Fig. 7. Operating principle of the proposed modulation algorithm.

If the cell capacitor voltage is higher (lower) than the maximum (minimum) admissible voltage and the capacitor is charged (discharged) in the current switching state, the cell will be switched in the other line in a discharging (charging) state. At the same time, the cell with the lowest (highest) cell capacitor voltage is switched from the other line into the first one, in a charging (discharging) state. Note that the switching operation itself is not performed in this step and only the changed cell state will be listed on a virtual list. The switching state of a single cell i is described by S:

$$S_l(i) = \begin{cases} 1_1 & \text{cell } i \text{ is inserted positive in line 1} \\ -1_1 & \text{cell } i \text{ is inserted negative in line 1} \\ 0_0 & \text{cell } i \text{ is bypassed} \\ 1_2 & \text{cell } i \text{ is inserted positive in line 2} \\ -1_2 & \text{cell } i \text{ is inserted negative in line 2} \end{cases} \quad (16)$$

The second kind of switching event is performed, if the voltage error $e_x(t)$ reaches a certain value r_x:

$$|e_x(t)| > r_x \quad (17)$$

The voltage error is calculated by integrating the difference of the reference line voltage V_x^* and the modulated line voltage V_x in line x:

$$e_x(t) = \int_{t'=0}^{t} \left(V_x(t') - V_x^*(t') \right) \mathrm{d}t' \quad (18)$$

The second step creates a sorted list of cells. Therefor the sum of all cell capacitor voltages, which are set to ON state in line x is calculated at first $\left(\sum_i S_l(i) \big|_{l=x} \cdot V_{C,i} \right)$ and compared with the line reference voltage. If the sum of the activated cell capacitor voltages is greater than the line reference voltage, it is necessary either to set activated cells to OFF state or to set bypassed cells to a negative ON state. Therefor a sorted list is generated which contains a ranking of the cells that have to be used first for this procedure. In this case, only the cells are used which are currently not activated in the other line. The cells are sorted by cell capacitor voltages and the sorting is performed in an ascending or descending order in dependence of the line current $I_{DC,x}$. The appropriate sorting specifications of all cases are indicated in Fig. 7. Afterwards, the algorithm chooses the cells, which will be turned on or turned off according to the generated list. The voltage modulation is always performed with an accuracy of one cell capacitor voltage level. If the voltage error is positive (negative), the modulated output voltage will be less (greater) than the line reference voltage.

The last step of the modulation algorithm generates the control pulses for each cell and the cells are finally switched.

This modulation method enables a switching frequency control of different groups of semiconductors. As it can be seen in Table II, the semiconductor switches 1-6 (according to Fig. 3) are only stressed by the current in line 2 $(I_{DC,II})$ and the semiconductor switches 7-12 are only stressed by the current in line 1 $(I_{DC,I})$. Consequently, there are two groups of semiconductors which can have a completely different current load. The switches 1-6 of each cell represent the first group (highlighted in green in Fig. 4) and the switches 7-12 of each cell represent the second semiconductor group (highlighted in blue in Fig. 4). By controlling the switching frequency of these groups, it is possible to influence the semiconductor losses as well as the energy fluctuation of the cell capacitors.

The minimum switching frequency is always given by the line current $I_{DC,x}$ and the desired tolerance band factor k (if the cell capacitor is not discussed). A higher current rating as well as a lower tolerance band factor lead to a higher switching frequency. Above this limit, the switching frequency can be adjusted by controlling the parameter r_x, which influences the trigger time of a

Fig. 8. Schematic overview of the switching frequency control system.

2480

switching event. The average switching frequency of a group $\bar{f}_x$ is defined as the sum of all turn on / off events $sw_{\mathrm{on/off},x}$ in a period T_{avg}:

$$\bar{f}_x = \frac{1}{T_{\mathrm{avg}}} \cdot \sum_{i=1}^{n} \frac{sw_{\mathrm{on},x}(i) + sw_{\mathrm{off},x}(i)}{2} \tag{19}$$

The average switching frequency control is performed by a PI-controller and a schematic overview of the control system is illustrated in Fig. 8. Due to the fact, that this control system works on cell level and the control commands are executed by the modulator, this control system does not influence the high-level CFC control in any way. The last chapter will present full-scale simulation results which validate the functionality of the control system.

C. Control Strategy

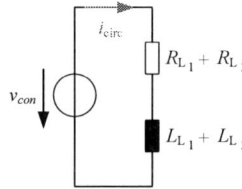

Fig. 9. Equivalent electric circuit for the high-level control.

This section presents the high-level control strategy of the CFC. It includes the control of the DC branch current and the energy control of the cell capacitors. The line current control is performed by modulating a voltage in one or both lines which generates a circulating current in the DC branches. Consequently, the resulting current in the DC branches is given by:

$$I_{\mathrm{DC,I}} = I_{\mathrm{DC,I,0}} + i_{\mathrm{circ}} \tag{20}$$

$$I_{\mathrm{DC,II}} = I_{\mathrm{DC,II,0}} - i_{\mathrm{circ}} \tag{21}$$

$I_{\mathrm{DC,I/II,0}}$ represent the initial branch current. For the DC branch current control it is not important, which voltage source performs the voltage modulation. Only the modulated voltage difference is essential. For that reason, the DC branch control voltage $v_{\mathrm{con,DC}}$ is splitted equally on both voltage sources.

Due to the used modulation scheme, the energy control of the cell capacitors can be reduced to a total energy control. Consequently, it suffices to monitor and control the total energy of all cells $w(t)$, which is given by:

$$w(t) = \sum_{i=1}^{n} \frac{1}{2} \cdot C \cdot V_{\mathrm{C},i}^2(t)$$
with C: cell capacitance $\tag{22}$

The reference energy W^* is:

$$W^* = n \cdot \frac{1}{2} \cdot C \cdot V_{\mathrm{C},0}^2 \tag{23}$$

The energy control is also performed by modulating a voltage $v_{\mathrm{con,W}}$. If both voltage sources modulate the same voltage level, the circulating current remains unchanged. However, the circulating current has a different polarity in both voltage sources. Consequently, it is possible to increase the charging power of one voltage source and decrease the discharging power of the other one at the same time and vice versa. This enables a control of the total energy.

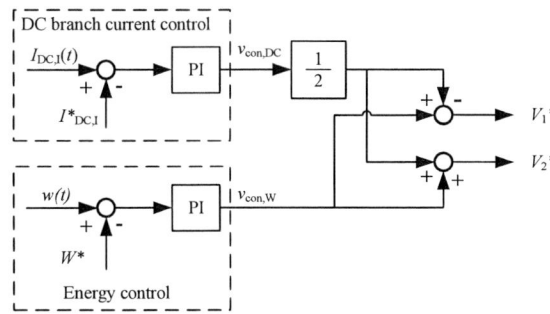

Fig. 10. Schematic overview of the high-level control system.

That means, that the DC branch current control is realized by modulating a voltage difference between the voltage sources and the energy control is performed by modulating an identical voltage offset. The control is performed by a PI-controller and the control system is illustrated in Fig. 10. The reference line voltages are finally:

$$V_1^* = v_{\mathrm{con,W}} - \frac{1}{2} v_{\mathrm{con,DC}} \tag{24}$$

$$V_2^* = v_{\mathrm{con,W}} + \frac{1}{2} v_{\mathrm{con,DC}} \tag{25}$$

IV. SIMULATION RESULTS

To verify the functionality of the proposed CFC, simulation studies are performed in MATLAB/Simulink. The investigated system is a point-to-point connection of two DC systems according to Fig. 5. The used simulation parameters are listed in Table III. The DC current I_{DC} is assumed to be impressed. In section A transient load changes are performed, which show the functionality of the CFC as well as the dynamic performance. Section B focuses on the verification of the presented switching frequency control. The simulation will only be performed for a chosen system configuration and a fixed amount of cell stages. Other system configurations enable a lower amount of cells or require a higher amount. The CFC and also the modulation algorithm are also working with different system configurations and deliver the same results.

TABLE III
SIMULATION PARAMETERS

Number of cell stages n	12
Cell capacitor C	4.5 mF
Nominal cell capacitor voltage $V_{\mathrm{C},0}$	2.2 kV
Tolerance band factor k	0.05
DC current I_{DC}	2 kA
Length line 1	500 km
Length line 2	1000 km
Line resistance R_{L}	1.21 Ω / 100 km
Line inductance L_{L}	84 mH / 100 km

The 2018 International Power Electronics Conference

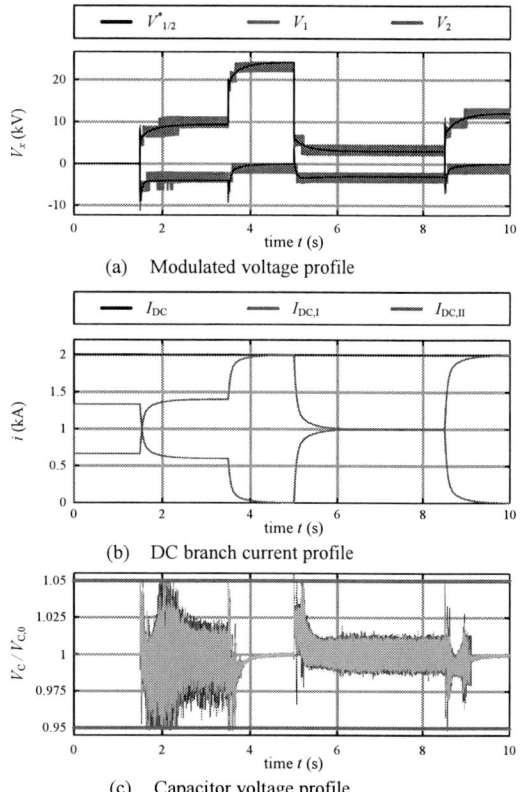

(a) Modulated voltage profile

(b) DC branch current profile

(c) Capacitor voltage profile

Fig. 11. Simulation results.

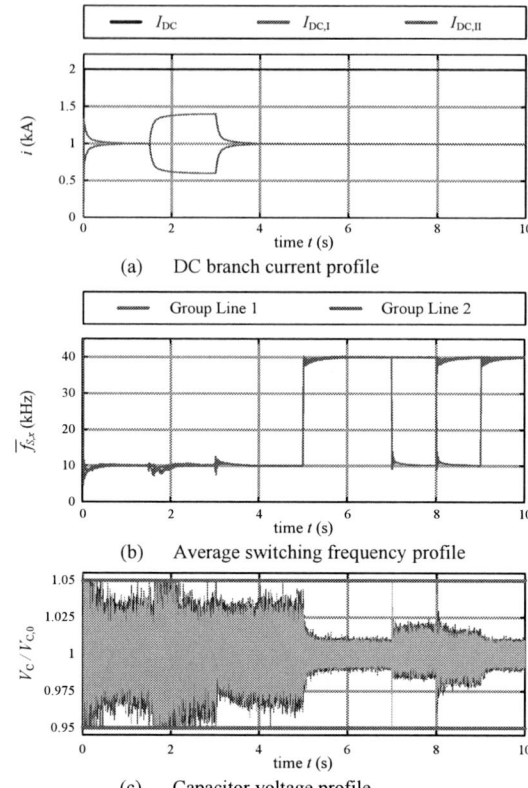

(a) DC branch current profile

(b) Average switching frequency profile

(c) Capacitor voltage profile

Fig. 12. Simulation results.

A. Dynamic Performance

The simulation results are illustrated in Fig. 11. The simulation starts with an idle mode of the CFC. In this case, the power flow is only determined by the line resistances and 66% of the DC current is carried by line 1 (Fig. 11b). In this time, no voltage modulation is performed by the CFC (Fig 11a). At $t_1 = 1.5$s, the CFC is activated and the current distribution factor is set to $a = 0.3$. The desired current rating is reached after an appropriate time and the cell capacitors stay well balanced (Fig. 11c). The next power step is performed at $t_2 = 3.5$s, where the current distribution factor is set to $a = 0$ (total DC current in line 2). In this case, the capacitor voltage ripple is minimal, because the voltage modulation is performed only in line 1, where the line current is $I_{DC,I} = 0$. A symmetric current distribution is modulated at $t_3 = 5$s and finally, a total current flow through line 1 is performed at $t_4 = 8.5$s.

The simulation approves the functionality of the CFC and the presented modulation algorithm. The reference values are reached in an appropriate time after each load step and the energy is well balanced between the single cells. The desired tolerance band of $\pm5\%$ of the nominal cell capacitor voltage is not exceeded (red line in Fig. 11c) at any time.

B. Switching Frequency Control

The simulation results of the second case study are

presented in Fig. 12. The CFC is activated at $t_1 = 0$s and the current distribution factor is set to $a = 0.5$ (symmetric current distribution). The average switching frequency of both semiconductor groups is set to $\bar{f}_{1/2} = 10$kHz. First of all, the dynamic behavior of the switching frequency controller is investigated. The reference switching frequency remains unchanged and the current distribution factor is set to $a = 0.3$ at $t_2 = 1.5$s and back to $a = 0.5$ at $t_3 = 3$s. As it can be seen in Fig. 12b, the switching frequency control performs a very constant operation, when the system reaches a steady state. The influence of transient load steps is acceptable.

The current distribution factor remains unchanged for further investigations and the system is held at a symmetric operation point. At $t_4 = 5$s, the switching frequency of both groups is increased to $\bar{f}_{1/2} = 40$kHz. The reference value is reached after a short time and the system continues the stable operation. Due to the higher switching frequency, the energy fluctuation of the cell capacitors is reduced clearly (Fig. 12c).

Finally, the asymmetric operation is examined. Only the switching frequency of group 1 is set to $\bar{f}_1 = 10$kHz at $t_5 = 7$s while the switching frequency of group 2 remains at $\bar{f}_2 = 40$kHz. The switching frequencies are exchanged at $t_6 = 8$s and at $t_7 = 9$s both groups continue the operation at $\bar{f}_{1/2} = 40$kHz. The decrease of the switching frequency of only one group leads also to a higher energy fluctuation.

2482

These investigations approve the functionality of the switching frequency control. Different operation points were handled and the steady state operation is reached after a short time. Even dynamic load steps have a low effect on the switching frequency control.

V. Conclusion

In this paper, a modular and scalable DC current flow controller was described. With this circuit, it is possible to control the load flow in a meshed DC grid or even to generate a zero DC current in one DC branch. It offers a possibility to perform an interline power exchange without the need of a transformer or an insulation to ground and is especially suitable for medium CFC voltage ratings.

In addition to the description of the basic design and the functional principle, a control and balancing method was presented which allows a cell capacitor balancing with a reduced control effort and a separated switching frequency control of the relevant semiconductor groups.

The functionality of the proposed CFC and the appropriate modulation algorithm was validated by full-scale simulation results.

References

[1] N. Flourentzou, V. G. Agelidis, and G. D. Demetriades, "VSC-Based HVDC Power Transmission Systems: An Overview," *IEEE Transactions on Power Electronics*, vol. 24, no. 3, pp. 592-602, 2009.

[2] J. Guo, G. Yao, Z. Xu, and Y. Yin., "The investigation on Chinese nationwide interconnected grid in future," *Power Engineering Society Winter Meeting*, pp. 6-11, 2000.

[3] T. An, G. Tang, and W. Wang, "Research and application on multi-terminal and DC grids based on VSC-HVDC technology in China," *IET High Voltage*, vol. 2, no. 1, pp. 1-10, 2017.

[4] L. Yao, H. Cui, J. Zhuang, G. Li, B. Yang, et al., "A DC power flow controller and its control strategy in the DC grid," *IPEMC-ECCE Asia*, pp. 2609-2614, 2016.

[5] W. Chen, and X. Cui, "Foreword for the Special Section on AC and DC Ultra High Voltage Technologies," *CSEE Journal of Power and Energy Systems*, vol. 1, no. 3, pp.1-2, 2015.

[6] M. P. Bahrman, and B. K. Johnson, "The ABCs of HVDC transmission technologies," *IEEE Power and Energy Magazine*, vol. 5, no. 2, pp. 32-44, 2007.

[7] M. K. Bucher, R. Wiget, G. Andersson, and C. M. Franck, "Multiterminal HVDC Networks – What is the Preferred Topology?," *IEEE Transactions on Power Delivery*, vol. 29, no. 1, pp. 406-413, 2014.

[8] R. Majumder, C. Bartzsch, P. Kohnstam, E. Fullerton, A. Finn, et al., "Magic bus: High-Voltage DC on the New Power Transmission Highway," *IEEE Power Energy Magazine*, vol. 10, no. 6, pp. 39-49, 2012.

[9] X. Guo, S. Zhao, Y. Wang, G. Bu, and Q. Guo, "Discussion on cascade-connected multiterminal UHVDC system and its application," *IEEE Power and Energy Society General Meeting*, pp. 1-5, 2012.

[10] X. Li, Z. Yuan, J. Fu, Y. Wang, T. Liu, et al., "Nanao multi-terminal VSC-HVDC project for integrating large-scale wind generation," *IEEE PES General Meeting*, pp. 1-5, 2014.

[11] T. M. Haileselassie, and K. Uhlen, "Power system security in a meshed North Sea HVDC grid," *Proceedings of the IEEE*, vol. 101, no. 4, pp. 978-990, 2013.

[12] A. L'Abbate, F. Careri, R. Calisti, and S. Rossi, "The Impact of HVDC in the development of the pan-European system: Focus on Italy-South East Europe ties," *IEEE PowerTech 2017*, pp. 1-6, 2017.

[13] D. Jovcic, M. Hajian, H. Zhang, and G. Asplund, "Power flow control in dc transmission grids using mechanical and semiconductor based DC/DC devices," *IET ACDC 2012*, pp. 1–6, 2012.

[14] Q. Mu, J. Liang, Y. Li, and X. Zhou, "Power flow control devices in DC grids," *IEEE Power Energy Society General Meeting*, pp. 1–7, 2012.

[15] E. Veilleux, and B.-T. Ooi, "Multiterminal HVDC With Thyristor Power-Flow Controller," *IEEE Transactions on Power Delivery*, vol. 27, no. 3, pp. 1205-1212, 2012.

[16] S. Balasubramaniam, J.Liang, and C. E. Ugalde-Loo, "An IGBT Based Series Power Flow Controller for Multi-Terminal HVDC Transmission," *IEEE UPEC*, pp. 1-6, 2014.

[17] M. Ranjram, and P. W. Lehn, "A Three-Port Power Flow Controller for HVDC Grids," *ICPE-ECCE Asia*, pp. 1815-1822, 2015.

[18] C. Barker, and R. Whitehouse, "A Current Flow Controller for Use in HVDC Grids," *IET ACDC 2012*, pp. 1-5, 2012.

[19] F. Hassan, R. King, R. Whitehouse, and C. Barker, "Double modulation control (DMC) for dual full bridge current flow controller (2FB-CFC)," *EPE'15 ECCE-Europe*, pp. 1-9, 2015.

[20] V. Hofmann, and M.-M. Bakran, "A modular and scalable HVDC Current Flow Controller," *EPE'15 ECCE-Europe*, pp. 1-9, 2015.

[21] V. Hofmann, and M.-M. Bakran, "Design Optimization of an MMC Based HVDC Current Flow Controller," *EPE'17 ECCE-Europe*, pp. 1-10, 2017.

[22] N. Deng et al., "A DC Current Flow Controller for Meshed Modular Multiterminal HVDC Grids," *CSEE Journal of Power and Energy Systems*, vol. 1, no. 1, pp. 43-51, 2015.

[23] J. Sau-Bassols, E. Prieto-Araujo, and O. Gomis-Bellmunt, "Modelling and Control of an Interline Current Flow Controller for Meshed HVDC Grids," *IEEE Transactions on Power Delivery*, vol. 32, no. 1, 2017.

The 2018 International Power Electronics Conference

An Isolated Soft-Switching Hybrid-Source DC-DC Converter for DC Offshore Wind Farms

Shenghui Cui[*], Jingxin Hu, Marco Stieneker and Rik W. De Doncker

Institute for Power Generation and Storage Systems, RWTH Aachen University, Aachen, Germany

*scui@eonerc.rwth-aachen.de

Abstract- **In this paper, an isolated dc-dc converter for dc offshore wind farms which combines a voltage-source converter and a current-source converter is proposed. The proposed converter can achieve soft switching inherently on both input and output sides over full power range. Based on the proposed dc-dc converter, a single-stage offshore wind farm configuration which consists of distributed dc platforms can be built up. Compared to the existing offshore wind farm configurations, the proposed concept can achieve lower power-conversion losses, higher power density, and lower investment costs. The validity of the proposed concept is verified by a full-scale simulation of a 640 MW offshore wind farm.**

I. INTRODUCTION

Nowadays, wind energy is one of the most important renewable energy sources, and the offshore wind energy is getting more and more crucial due to its large capacity and relatively stable power profile compared to the onshore wind energy.

For remote offshore wind farms, it is becoming a standard measure to transmit the collected wind energy to the shore via HVDC subsea cables [1-3] since the transmission distance of the HVAC subsea cable is limited due to its considerable capacitive reactive currents.

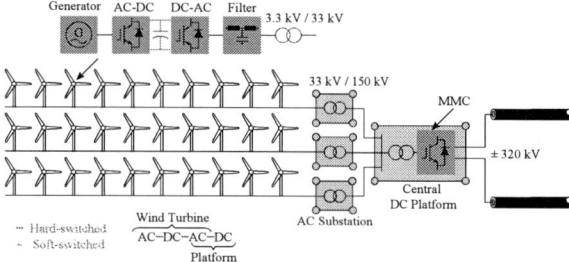

Fig. 1. Typical configuration of today's remote offshore wind farms.

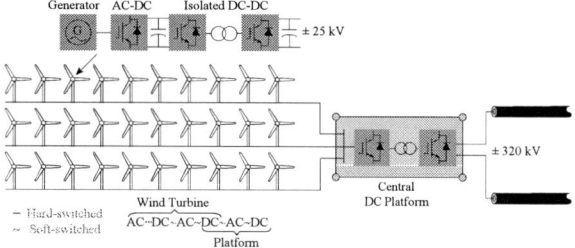

Fig. 2. Configuration of offshore wind farms with dc collectors and a central dc platform.

In Fig. 1, the typical configuration of today's remote offshore wind farms is presented [1-2]. In each wind turbine, the wind energy is converted to electricity via the generator and then rectified to a dc form by a pulse-width modulated (PWM) rectifier. Afterwards, the energy is inverted from dc to ac via a hard-switched inverter and filters, and then the ac voltage is stepped up to medium voltage via a distribution transformer. The wind energy generated by wind turbines is collected in the ac collectors and the voltage is stepped up via transformers in ac substations. Finally, all electricity is collected to the central dc platform and is rectified to dc by a modular multilevel converter (MMC) [4] for HVDC subsea transmission.

The configuration shown in Fig. 1 has the following drawbacks: a) in each wind turbine and ac substation there is a bulky line-frequency transformer, however, power density is a crucial factor for offshore applications, b) in each wind turbine there are a hard-switched inverter and the ac filter leading to high losses, and the distribution of the inverters in individual wind turbines increases the effort for the maintenance and repair which are a significant issue in offshore applications, c) in ac collector grids, reactive power control [5] and harmonic current suppression [6-7] must be addressed by significant efforts.

As an alternative to address the aforementioned issues, an offshore wind farm configuration with dc collectors is proposed and discussed by numerous publications [8-10]. In such a configuration, there is an isolated dc-dc converter with the dual-active bridge (DAB) [11] topology inside each wind turbine to boost the rectifier dc-link voltage to a medium voltage. The electricity is collected via medium-voltage dc collectors, and the voltage is stepped up to high voltage in the central dc platform. Since the DAB converter is soft switched and the transformer operates at a medium frequency in the kilohertz range, power density and efficiency are significantly improved compared to the conventional approach in Fig. 1. However, this leads to significantly higher total device ratings (TDR) of power-semiconductor devices and consequently results in considerable investment cost.

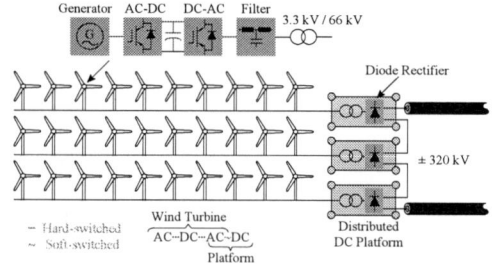

Fig. 3. Configuration of DRU-based offshore wind farms.

Another cost-effective approach for onshore grid access via HVDC subsea cables is the modular diode rectifier unit (DRU) based configuration proposed by Siemens in 2015 as shown in Fig. 3 [12-13]. The electric energies collected in the ac collectors are rectified to dc via DRUs in distributed dc platforms, and the dc-links of the DRUs are connected in series to build up a high voltage for the HVDC transmission. This approach presents lower investment cost compared to the ones shown in Fig. 1 and Fig. 2, however, bulky transformers still remain in wind turbines and platforms, and the inverters are still distributed individually in wind turbines. Moreover, significant efforts should be made to suppress the low-order harmonic currents introduced by the diode rectifiers and to control the power flow (especially the reactive power) in ac collectors.

In this paper, a new single-stage compact and efficient configuration of offshore wind farms is proposed as a promising alternative. In the proposed configuration, the hard-switched inverter, the ac filter, and the bulky transformer in each individual wind turbine are omitted. The electricity is collected by the medium-voltage dc collector and the voltage is stepped up by the proposed hybrid-source dc-dc converters in distributed dc platforms. Since the proposed hybrid-source dc-dc converter is fully soft-switched and the employed transformer operates at a medium frequency of hundreds Hertz range, both the efficiency and the power density can be significantly improved in a system point of view.

II. PROPOSED CONFIGURATION OF OFFSHORE WIND FARMS

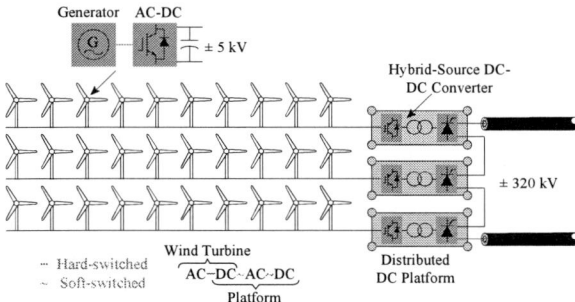

Fig. 4. Proposed single-stage configuration of offshore wind farms.

The proposed configuration of offshore wind farms is shown in Fig. 4. In each turbine the hard-switched inverter, the ac filter, and the bulky line-frequency transformer are omitted, and the dc-link within the turbines is directly connected to the dc collector grid which is in connection with the input side of the corresponding distributed dc platform. Consequently, the power density of the wind turbine can be significantly improved and the mechanical structure is simplified. Moreover, the number of distributed power-electronic converters can be reduced remarkably to lower the effort for maintenance and repair.

In each distributed dc platform, the proposed hybrid-source dc-dc converter is employed to provide the galvanic isolation and to step up the voltage. On the input side of the proposed dc-dc converter is a self-commutated voltage-source converter (VSC) so that the dc-link voltage can be regulated at a constant value, namely the rated dc-link voltage of the PWM rectifier in the wind turbine. On the output side of the proposed dc-dc converter is the thyristor-based current-source converter (CSC), and the

output sides of the proposed dc-dc converters in distributed dc platforms are connected in series and share the same dc-link current.

From the dc-link within the wind turbine to the dc-link of the HVDC transmission line there are two power conversion stages i.e. the dc-ac conversion and the ac-dc conversion in both the proposed configuration and the DRU-based one as depicted in Fig. 3 and Fig. 4. However, there are significant differences in various aspects as follows: a) From efficiency point of view, in the DRU-based configuration the dc-ac power conversion in the wind turbine is realized in a hard-switching manner, and the ac filters for the switching-frequency harmonics suppression cause losses as well. While in the proposed configuration, the dc-ac power conversion located in the platform operates in a soft-switching manner and no ac filters are required. b) From power density point of view, in the DRU-based configuration a bulky line-frequency transformer is employed for each wind turbine as well as in each platform. On the contrary, in the proposed concept the bulky transformer in the wind turbine is omitted. In addition, the operation frequency of the transformer in the platform can be elevated to medium frequency in the range of hundred hertz whereby the size and weight can be significantly reduced. c) From system control point of view, since the collector is in a form of dc instead of ac in the proposed concept, the reactive power control and harmonic suppression issues are absent and the system control can be simplified. Moreover, since the output side of the proposed hybrid-source dc-dc converter is the current-source type based on thyristors instead of diodes, the output-side dc-link voltages of distributed platforms can vary in a wide range by controlling the firing angle to maintain the system stability in case of power unbalance of wind turbine clusters connected to different platforms. d) From maintenance point of view, the dc-ac inverters are centralized in the proposed concept instead of distributed individually in wind turbines, and the effort for the maintenance can be significantly reduced.

The typical voltage of a medium-voltage wind turbine generator is 3.3 kV and a practical maximum voltage of 6.6 kV is realistic taking into account trends in wind generator technology [14-15]. This results in a dc-link voltage of 10 kV in maximum, which is still much lower than the ac collector voltage (66 kV) in the DRU-based wind farm concept. Thus, one might argue that the proposed configuration leads to higher conduction losses in the collectors. However, since the losses caused by the hard-switched inverter, filters and the transformer in the wind turbines are saved, the proposed configuration would still result in advantageous overall efficiency compared to the ac collector [16].

This paper focuses on the topology and the operation principle of the proposed hybrid-source dc-dc converter. The system-level control of the proposed offshore single-stage dc wind farm configuration and a communication-free coordinated control among the distributed dc platforms and the onshore station will be discussed in details in future dedicated publications.

III. PROPOSED HYBRID-SOURCE DC-DC CONVERTER

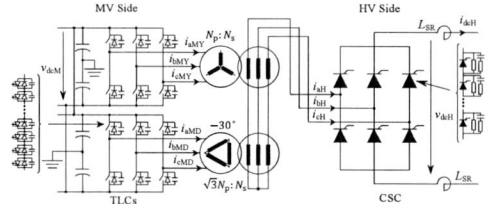

Fig. 5. Circuit diagram of the proposed hybrid-source dc-dc converter.

The proposed hybrid-source dc-dc converter is shown in Fig. 5. There are two two-level converters (TLCs) connected in parallel on the MV side and a thyristor-based CSC on the HV side. The MV side and the HV side are connected via two cascade-connected open-winding transformers in yd/Y configuration. Both TLCs operate in the six-step mode, and the TLC connected to the d-winding transformer lags 30° with respect to the other one in connection with the y-winding transformer. On the MV side self-commutated press-pack devices e.g. IGCTs are connected in series to block the dc-link voltage, and on the HV side thyristors are connected in series. Each thyristor is connected with a static balancing resistor and a dynamic balancing RC snubber.

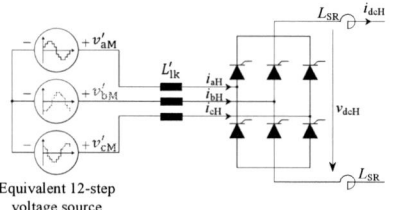

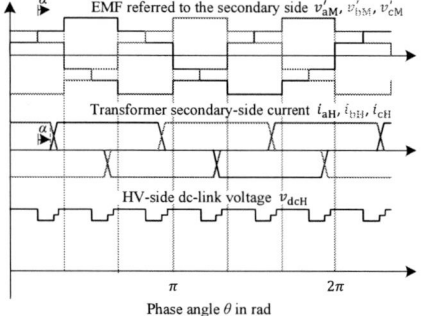

Fig. 6. Equivalent circuit of the proposed converter referred to the HV side.

Fig. 7. Operation principle of the proposed converter.

The equivalent circuit of the proposed hybrid-source dc-dc converter referred to the HV side is shown in Fig. 6. Since the TLCs on the MV side are 30° phase shifted and the transformer's secondary sides are connected in series, the TLCs present a three-phase 12-step equivalent voltage source behind the transformer leakage inductor.

The operation principle of the proposed converter is depicted in Fig. 7. When the firing angle α is larger than 0°, the instantaneous phase voltage of 'a' phase is larger than that of 'b' phase and the thyristors of 'a' phase are forward biased. Therefore, the thyristors of 'a' phase can be triggered on to commutate the dc-link current as long

as α is larger than 0° as in a line-commutated converters (LCCs) for HVDC transmissions. By increasing the firing angle it is possible to reduce the mean dc-link voltage on the HV side, and the mean dc-link voltage is approaching zero when α approaches 90°. The mean dc-link voltage on the HV side is calculated as (1)-(3) in accordance with the firing angle α in different ranges where L'_{lk} represents the transformer leakage inductance referred to the secondary side.

$$\bar{v}_{dcH} = \left(1 + \frac{2}{3}\sqrt{3} - \frac{\sqrt{3}}{\pi}\alpha\right)\frac{N_s}{N_p}v_{dcM} - \frac{3}{\pi}\omega L'_{lk}i_{dcH}, \, 0° < \alpha < 30° \quad (1)$$

$$\bar{v}_{dcH} = \left(\frac{3}{2} + \frac{2}{3}\sqrt{3} - \frac{3+\sqrt{3}}{\pi}\alpha\right)\frac{N_s}{N_p}v_{dcM} - \frac{3}{\pi}\omega L'_{lk}i_{dcH}, \, 30° < \alpha < 60° \quad (2)$$

$$\bar{v}_{dcH} = \left(\frac{3}{2} + \sqrt{3} - \frac{3+2\sqrt{3}}{\pi}\alpha\right)\frac{N_s}{N_p}v_{dcM} - \frac{3}{\pi}\omega L'_{lk}i_{dcH}, \, 60° < \alpha < 90° \quad (3)$$

Similar to the case of the LCC, the mean dc-link voltage on the HV side is determined by not only the firing angle but also the transformer leakage inductor and the HV-side dc-link current. The remarkable difference compared to the LLC is that the HV-side mean dc-link voltage is proportional to the firing angle instead of its sine value, and this feature makes an arcsine operator not necessary anymore in the converter controller.

During turn-on process of the thyristor, its di/dt rate is inherently limited by the transformer leakage inductor and the risk of hot spot caused by high di/dt can be avoided. Consequently, the transformer leakage inductor results in the commutation angle as the same as in the LCC as shown in Fig. 7. Since all the thyristors are turned on and turned off at zero current, they are inherently soft switched in a zero-current switching (ZCS) manner. In the practical loss calculation, the switching loss introduced by the thyristor reverse recovery and the energy damp of the dynamic balancing RC snubber during the turn-on transient of the thyristor should be taken into account.

One of the important characteristics of the thyristor-based CSC is that it always draws inductive reactive current from the ac side regardless of the magnitude or direction of the active power [17]. On the other hand, the TLCs can inherently operate in zero-voltage switching (ZVS) if it feeds inductive reactive current into the ac side as in the DAB converter [18-19]. Due to the presence of the inductive reactive current the snubber capacitors connected to the self-commutated devices in the TLCs are fully charged and discharged during the voltage commutation, and the self-commutated devices can be turned on at zero voltage due to the full discharge of the snubber capacitors connected in parallel [18]. Thus, the inductive reactive current drawn by the CSC on the HV side charges and discharges the snubber capacitors on the MV side during the voltage commutation, and all self-commutated switches on the MV side are soft switched in a ZVS manner.

In difference to LCC-HVDC applications, the operation frequency of the transformer in the proposed converter can be elevated from the line frequency to the medium frequency in a range of hundred hertz. Thus, the size and the weight of the transformer can be significantly reduced compared to the one in the platform of the DRU-based wind farm. One of the main limiting factors of the

2486

transformer frequency is the hold-off time of the thyristor due to the reverse recovery. It should be noted that, since the CSC in the proposed dc-dc converter operates in the rectifier mode and the firing angle is far away from 180°, the risk of the commutation failure is relatively small [17].

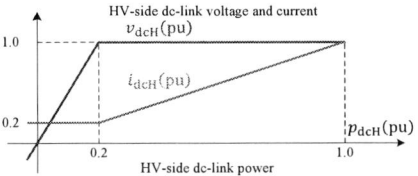

Fig. 8. Systematic operation principle of the proposed single-stage offshore wind farm.

Similar to the DRU-based offshore wind farm concept, the onshore converter is the full-bridge sub-module based modular multilevel converter (MMC) in the proposed offshore wind farm concept so that the HV-side dc-link voltage can vary in a wide range.

In the proposed offshore wind farm concept, the hybrid-source dc-dc converter plays a role of regulating the MV-side dc-link voltage by controlling the HV-side power. The systematic operation principle of the wind farm is shown in Fig. 8. In strong wind situation, the onshore MMC regulates the HV-side dc-link voltage at its rated voltage, and the hybrid-source dc-dc converter regulates the HV-side dc-link current to manipulate the HV-side power. When the wind energy decreases, then the HV-side dc-link current decreases correspondingly. To keep the continuous conduction mode (CCM) operation of the thyristor-based CSC in the hybrid-source dc-dc converter, the onshore MMC transits from the voltage control mode to the current control mode in calm wind situation. Then the hybrid-source dc-dc converter manipulates the HV-side dc-link voltage to control the HV-side power in order to regulate the MV-side dc-link voltage.

In the DRU-based offshore wind farm concept, the diode rectifier is a unidirectional power converter and it cannot deliver electricity from the onshore side to the offshore ac collector grid. However, the auxiliary loads e.g. yaw and pitch control motors and the control system still require power even in no wind situation. Thus, the so-called umbilical cable should be employed to transmit electricity from onshore to offshore which results in extra costs. On the contrary, the hybrid-source dc-dc converter is a bidirectional power converter and no extra hardware

is required for the reverse power transmission.

The transformer connection between the TLCs and the CSC is of interest. One might argue that a simple y/Y transformer connection would also fit the circuit operation. The waveforms of the proposed converter with a y/Y transformer connection instead of a yd/Y (in Fig.5) connection are shown in Fig. 9. When a y/Y transformer connection is employed, the TLCs on the MV side present a three-phase 6-step equivalent voltage source instead of a 12-step one. It is clearly shown in Fig. 9, that when the firing angle is smaller than 30° the phase voltages of 'a' phase and 'b' phase are identical. It means that the minimum firing angle (corresponding to the maximum mean dc-link voltage) should be designed larger than 30° for normal operation of the circuit. Consequently, it introduces a significant amount of reactive current and results in increased device conduction loss and transformer winding losses.

Another benefit that can be gained by employing a yd/Y transformer connection is that, the dc-link capacitor size on the MV side can be reduced to half since the dominant 6th-order dc-link current ripple is reduced to half due to the phase-shift operation of two TLCs. In addition, the size of the smoothing reactor L_{SR} can also be reduced to half since the HV-side dc-link voltage ripple is reduced as shown in Fig. 7 and Fig. 9.

IV. SIMULATION RESULTS

To verify the validity of the proposed concept a computer simulation is performed in PLECS software. A ±320 kV, 640 MW dc offshore wind farm is simulated. The offshore wind farm includes three distributed dc platforms and the output voltage of each platform is 213.3 kV. The wind farm is connected to the onshore station via a 100 km HVDC cable, and the onshore station is emulated by a variable dc voltage source. In each platform there are two proposed hybrid-source dc-dc converters. The MV sides of two converters are connected in parallel, and the HV sides of them are connected in series. The operations of two converters are 30° phase shifted so that the 6th-order dc-link harmonic currents on both MV and HV sides are eliminated. Consequently, 12th-order harmonic currents become dominant and both the MV-side dc-link capacitor and the HV-side dc-link smoothing reactor can

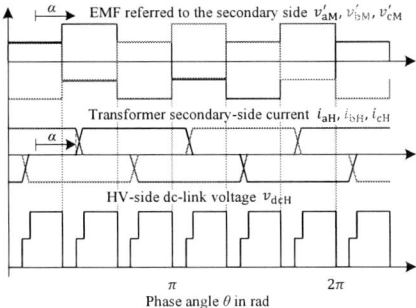

Fig. 9. Waveforms of the proposed converter with y/Y transformer connection.

Table I. Parameters of the simulated proposed converter

MV-side TLCs	
Converter quantity	2
Rated dc voltage	10 kV
Rated power	53.3 MW
Switch	4.5 kV device ×4 per valve
Snubber capacitor	1 μF (per device)
Transformers	
Transformer quantity	2
Turns ratio (N_{pY}:N_{pD}:N_s)	2 : 2$\sqrt{3}$: 13
Leakage inductance (L'_{lk})	10 mH
Fundamental frequency	200 Hz
HV-side CSC	
Rated dc voltage	106.7 kV
Rated dc current	1 kA
Rated power	106.7 MW
Smoothing reactor	10 mH

be significantly reduced. Detailed parameters of the simulated converter are presented in Table I.

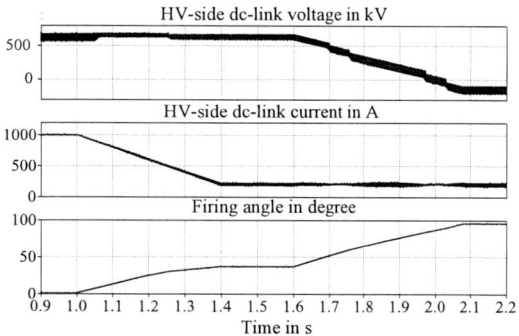

Fig. 10. Waveforms of wind farm HVDC-link voltage and current, and the firing angle of the hybrid-source dc-dc converter.

In the simulated scenario, it is assumed that the active powers of the three distributed platforms are identical. When the power of the wind farm is larger than 0.2 p.u., then the HVDC-link voltage is fixed at 1.0 p.u. for minimum transmission losses and the HVDC-link current is manipulated to regulate the power flow. When the wind farm power is getting smaller than 0.2 p.u., then the HVDC-link current is controlled constant at 0.2 p.u. to maintain the continuous-conduction operation of the CSC of the proposed dc-dc converter, and the HVDC-link voltage is manipulated to regulate the power flow.

From t=1.0 s to t=1.4 s, the wind farm power decreases from 1.0 p.u. to 0.2 p.u.. And from t=1.6 s to t=2.1 s, the wind farm power decreases from 0.2 p.u. to -0.04 p.u.. Thus, the proposed system is able to transmit power from onshore to the wind farm to provide power for the auxiliary loads e.g. pitch and yaw control motors in case of a no wind situation. When the wind farm power is 1.0 p.u., the firing angle is 0.5° and the wind farm HVDC-link voltage is 1.0 p.u (640 kV) as shown in Fig. 10. When the wind farm power decreases, the firing angle increases to maintain the HVDC-link voltage as the same in the LCC-HVDC transmission. When the power decreases to -0.04 p.u., the firing angle is 95° which is still much lower than 180°.

Voltage across the switch valves of the TLC on the MV side and their corresponding signals are shown in Fig. 10 in different power situations. As stated in Section III, the TLC converter can realize ZVS soft switching in the full power range since the inductive reactive current drawn by the CSC helps commutating the snubber capacitor voltages. Thus, the proposed converter presents very low power-semiconductor losses compared to the DRU-based wind farm configuration.

V. CONCLUSION AND FUTURE WORK

In this paper, a new single-stage dc offshore wind farm configuration is proposed. The proposed configuration is based on distributed dc platforms where the proposed hybrid-source dc-dc converter is employed. The proposed converter can realize soft switching over a wide power range. Moreover, the transformer frequency of the converter can be elevated to the medium frequency range.

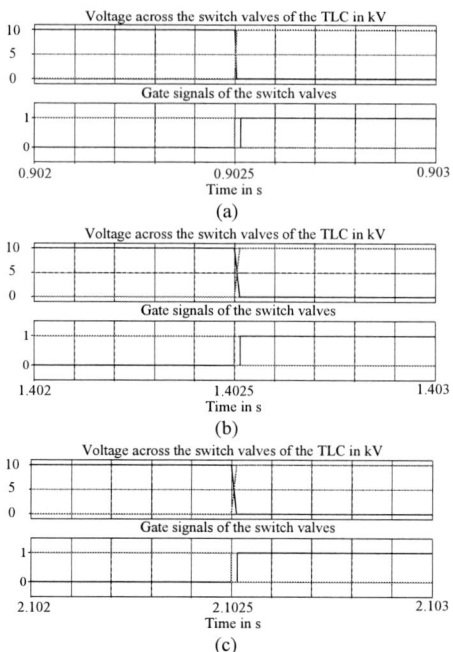

Fig. 11. Voltage across and gate signals of the switch valves of the TLC in different wind farm power situations: a) 1.0 p.u., b) 0.2 p.u., c) -0.04 p.u..

Compared to the existing offshore wind farm configurations, the proposed one presents higher power density, lower investment cost and lower power conversion losses. Validity of this work is verified by time-domain simulations.

REFERENCES

[1] P. Bresesti, W. L. Kling, R. Hendriks, and R. Vailati, "HVDC connection of offshore wind farms to the transmission system," *in IEEE Trans. on Energy Conversion*, vol. 22, pp. 37-43, 2007.

[2] A. Abdalraham, E. Isabegovic, "DolWin1 – Challenges of connecting offshore wind farms," *in IEEE International Energy Conference*, pp. 1-10, 2016.

[3] E. M. Callavik, P. Lundberg, M. P. Bahrman, and R. P. Rosenqvist, "HVDC technologies for the future onshore and offshore grid," *in CIGRE Symposium Grid of the Future*, Kansas, USA, 2012.

[4] A. Lesnicar, R. Marquardt, "An innovative modular multilevel converter topology suitable for a wind power range," *in IEEE Bologna Power Tech Conference*, vol. 3, 2003.

[5] V. S. Pappala, M. Wilch, S. N. Singh, I. Erlich, "Reactive powe management in offshore wind farms by adaptive PSO," *in International Conference on Intelligent Systems Applications to Power Systems*, pp. 1-8, 2007.

[6] L. Kocewiak, J. Hjerrild, C. L. Bak, "Harmonic analysis of offshore wind farms with full converter wind turbines," *in International Conference on Large-Scale Integration of Wind Power Into Power Systems*, 2009.

[7] H. Liu, J. Sun, "Voltage stability and control of offshore wind farms with ac collection and HVDC transmission," *in IEEE Journal of Emerging and Selected Topics in Power Electronics*, vol. 2, pp. 1181-1189, 2014.

[8] C. Meyer, M. Hönig, A. Peterson, and R. W. De Doncker, "Control and design of dc grids for offshore wind farms," *in IEEE Trans. on Industry Applications*, vol. 43, pp. 1475-1482, 2007.

[9] T. Jimichi, M. Kaymak, and R. W. De Doncker, "Comparison of single-phase and three-phase dual-active bridge dc-dc converters with various semiconductor devices for offshore wind turbines," *in IEEE International Future Energy Electronics Conference and ECCE Asia*, pp. 591-596, 2017.

[10] G. Shi, X. Cai, C. Sun, Y. Chang, and R. Yang, "All-dc offshore wind farm with parallel connection: An overview," *in IET International Conference on AC and DC Power Transmission*, pp. 1-6, 2016.

[11] R. W. De Doncker, D. M. Divan, M. H. Kheraluwala, "A three-phase soft-switched high-power-density dc/dc covnerter for high-power applications, " *in IEEE Trans. on Industry Applications*, vol. 27, pp. 63-73, 1991.

[12] T. Hammer, *et al.*, "Diode-rectifier HVDC link to onshore power systems: Dynamic performance of wind turbine generators and reliability of liquid immersed HVDC diode rectifier units," *in CIGRE Conference Technical Committee B4*, 2016.

[13] S. Seman, N. T. Trinh, R. Zurowski, S. Kreplin, "Modeling of the diode-rectifier based HVDC transmission solution for large offshore wind power plants grid access," *in International Workshop on large-scale integration of wind power into power systems*, 2016.

[14] H. Polinder, *et al.*, "Trends in wind turbine generator systems," *in IEEE Journal of Emerging and Selected Topics in Power Electronics*, vol. 1, pp. 174-185, 2013.

[15] V. Yaramasu, B. Wu, P. C. Sen, S. Kouro, M. Narimani, "High-power wind energy conversion systems: state-of-the-art and emerging technologies," *in Proceedings of the IEEE*, vol. 103, pp. 740-788, 2015.

[16] M. Stieneker, "Analysis of medium-voltage direct-current collector grids in offshore wind parks," *PhD Dissertation*, E.ON Energy Research Center, RWTH Aachen University, 2017.

[17] C.-K. Kim, *et al.*, "HVDC Transmission: Power Conversion Applications in Power Systems," *Wiley Press*, 2009.

[18] S. Cui, N. Soltau, R. W. De Doncker, "A high step-up ratio soft-switching dc-dc converter for interconnection of MVDC and HVDC grids, " *in IEEE Trans. on Power Electronics*, 2017. (Early Access)

[19] R. U. Lenke, "A contribution to the design of isolated dc-dc converters for utility applications," *PhD Dissertation*, E. ON Energy Research Center, RWTH Aachen University, 2012.

A Transformerless Multi-Cell Solid-State Fault Current Limiter for Medium Voltage Power System

Pantarote Techama[1], Sompob Polmai[2], Chanin Bunlaksananusorn[3]

Faculty of Engineering, King Mongkut's Institute of Technology Ladkrabang, Bangkok, Thailand

[1]E-mail: pantarote.techama@gmail.com, [2]E-mail: sompob.po@kmitl.ac.th, [3]E-mail: chanin.bu@kmitl.ac.th

Abstract—**This paper proposes a transformerless solid-state fault current limiter composed of multiple cells of single-phase bridge controlled rectifier with DC reactor and an AC reactor connected in parallel with cascaded rectifier cells performing current limiting in steady state. The proposed fault current limiter can be used in medium-voltage power system without power transformer so the system is more compact. With modular design of each rectifier cell, the construction and maintenance of the system could also be simplified. In this paper, the circuit designs, the computer simulation and the construction of the solid-state fault current limiter prototype are presented. The simulation and experimental results show that the proposed fault current limiter can operate properly.**

Keywords—*solid-state fault current limiter , fault , distribution system , bridge-type FCL*

I. INTRODUCTION

At the present, the demand of using electricity is increased and the government promote electrical generation from renewable energy, e.g., PV, wind turbine, and biomass. The expansion of the electrical grid by new installations or interconnections and the proliferation of renewable energy resources increase the fault current level in the power system. Changing new protection devices to cope with the increased fault current level is expensive and complicates the setting for protection coordination. Fault current limiter (FCL) can help solve this problem because it is transparent (zero impedance) during normal operation of the power system. During fault the FCL's impedance increases, the fault current is limited to the safety operating range of the installed protection devices.

Several types of FCL have been proposed such as resonance-type, series switch-type and bridge-type [1-5]. The solid-state switch FCL has been applied to medium and high voltage power system because the solid-state switch has higher voltage and current rating, is simple to control, has high reliability and doesnt need thermal management that reduce cost to thermal management system [1]. At high voltage level, transformer is needed. However, at medium voltage level, transformer-less FCL is considered possible used in the distribution feeders.

The solid-state bridge FCL with DC and AC reactors has been prosed and studied [1-4]. When fault occurred, both reactors will automatically limit fault current at the first peak without delay. After a short period, the DC

reactor is bypassed and only the AC reactor limits the fault current. Having the AC reactor helps reduce the heat and size of the DC reactor.

When applying the FCL in medium or high voltage system. FCL with transformer has been used [4]. The disadvantages of using transformer are high volume, weight, and cost [4]. In this paper, a transformerless multi-cell solid-state bridge FCL (MC-SSB FCL) is proposed and developed. Each bridge cell is modular or has the same design and specifications. The proposed transformerless MC-SSB FCL would be more compact, simple to build, easy to maintenance and high reliability.

In this paper, the circuit designs, the computer simulation and the construction of the solid-state fault current limiter prototype are presented. The simulation and experimental results will be discussed.

II. CONSIDERATION AND PRINCIPLE OF OPERATION

A. FCL configuration for medium voltage system

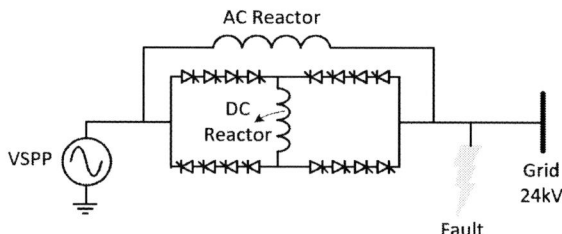

Fig. 1. Single-phase solid-state bridge fault current limiter

Fig. 1 shows the construction of a single-phase SSB FCL. The proposed SSB FCL is expected to be used in very small power producer (VSPP) feeder that has 24kV voltage rating and 10 MVA power capacity. The step-down transformer is not used to reduce the cost and installation space. the SSB thyristors must withstand the system voltage up to 19.59kV, this result in using four 6.5kV-rating thyristors connecting in series or sixteen thyristors per phase in total. In this case only one large DC reactor is used. An alternative approach is to use four bridge cells connecting in series. Each cell has four 6.5kV-rating thyristors and a quarter size DC reactor. This is called multi-cell solid-state bridge fault current limiter (MC-SSB FCL) in this paper. These cells are identical,

2490

and the MC-SSB FCL shares the concept of modular design for construction, control and maintenance. The proposed MC-SSB FCL is shown in Fig. 2, to increase the reliability, another cell is added to the system, and each cell also have one by-pass switch. If one cell fails, it can be bypassed and the redundant cell can be activated.

B. Operation principles

The FCL have three modes of operation; charging mode, normal operation mode, and fault condition mode. *Charging mode*:For simplicity, in this study, the DC reactor will be naturally charged by the system voltage to the peak value of the line current within a several cycles. In practical, an auxiliary power supply can be used to pre-charge the dc reactor before inserting the FCL into the feeder. The auxiliary power supply also compensates for conduction loss in thyristor and DC reactor to avoid line voltage distortion. *Normal operation mode*: When the DC reactor current reaches steady state, the reactance of the DC reactor disappears, almost the line current passes through the DC reactor. The forward voltage of the thyristors and resistance of the DC reactor cause a small voltage drop across the AC reactor resulting in a small line current pass through the AC reactor. *Fault condition mode*: When fault occurs, the fault current rises above the DC reactor current, the fault current will pass through both DC and AC reactors. The DC reactor current reaches its maximum designed value within the first half cycle and all of bridge cells will be disconnected from the system by turning off one arm of the bridge thyristors. The DC reactor current freewheels through another arm of the bridge thyristors. All of the fault current flows through the AC reactor and is limited only by AC reactor impedance. To suppress temperature rising in the DC reactor due to high current, the bridge cells are reconnected to the system again by entering the inversion control. The DC reactor current is discharged into the system. When the DC current level was a little above the normal peak line current, the bridge cells enter the freewheeling again for maintaining the current level.

After fault is cleared, the bridge cells enter the charging and normal operation, respectively. Fig. 3 shows the state diagram of the operation of the FCL.

The value of AC reactor is determined by the fault current level that allow pass through the system I_{sc} as described by (1), where ω is angular frequency of the system and U_n is the system voltage.

$$L_a = \frac{U_n}{\omega I_{sc}} \quad (1)$$

The optimal value of each DC reactor is determined in (2) [2].

$$L_d = \frac{1}{4} \frac{2U_{lm}}{\omega I_m} \quad (2)$$

where Im is the DC reactor current in steady state. U_{lm} is the peak voltage of DC reactor. The DC reactor current within a half cycle of fault condition does not exceed two times the normal DC reactor current.

C. Compensation for voltage distortion during normal operation

In practical, internal resistance of DC reactor rdc, and voltage drop across thyristor VT represent power losses and lower power quality in the system during normal operation. So, charging circuits are used to solve this problem. In normal operation mode, the DC reactor current is discharged via internal resistance of DC reactor and thyristors that cause the DC reactor current level lower than the line peak current. So, the voltage distortion of the load side is noticeable. The role of charging circuit, which is a controllable DC voltage source, is to compensate for voltage drop across the thyristors and DC reactor and also to maintain the DC reactor current level above the peak line current to avoid voltage distortion during rapid line current raise.

III. SIMULATION RESULTS

A. Simulation results without charging circuit in 24kV

In this section, the simulation of MC-SSB FCL is discussed. The important parameters for the simulation are shown in Table I for 24kV system. The system configuration is shown in Fig. 4.

The waveforms of fault current with and without FCL are compared in Fig. 5. The fault current is limited to

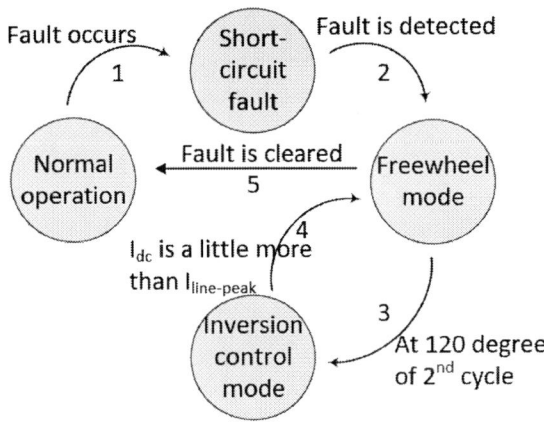

Fig. 3. State diagram of the operation of the FCL

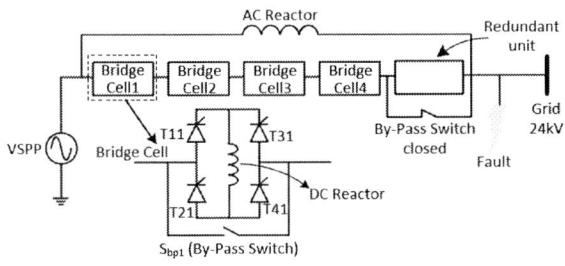

Fig. 2. Multi-cell solid-state bridge fault current limiter

2491

the designed value in steady-state. The DC reactor current waveform is shown in Fig. 6. The DC reactor current reaches it maximum value about twice the normal value at the first half cycle. The current stops increasing due to the bridge cells enter the freewheeling mode. About one cycle later, the bridge cells enter inversion mode, the DC reactor current decreases rapidly. A little later, the bridge cells enter the freewheeling mode again the DC reactor current gradually decreases. Fig. 7 shows the voltage across the AC reactor. There is small voltage drop across the AC reactor at normal operation due to thyristor forward voltage and DC reactor resistance. During fault, nearly all the system voltage drops across the AC reactor and the fault current is limited by the AC reactor impedance.

Fig. 8 and Fig. 9 show the voltage drop across each thyristor in each bridge cell. An equal voltage sharing between the cells is obtained.

TABLE I.   SIMULATION PARAMETERS AT 24KV

Symbol	Meaning	Value
V_s	Source voltage (line-line) (rms)	24 kV
f	Power system frequency	50 Hz
r_s	Source resistance	0.1 Ω
r_L	Line resistance	0.1 Ω
V_T	Voltage drop across thyristor	1.1 V
V_D	Voltage drop across diode	0.8 V
L_{dc}	DC reactor inductance (per cell)	91 mH
r_{dc}	DC reactor resistance (per cell)	0.2 Ω
L_{ac}	AC reactor inductance	37 mH
r_{ac}	AC reactor resistance	0.2 Ω
r_{load}	Load resistance	58.3 Ω
L_{load}	Load inductance	139 mH
r_f	Fault resistance	5 Ω
V_{charge}	Charging voltage (rms)	220 V

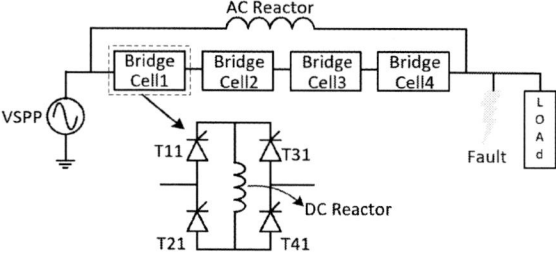

Fig. 4.   Solid-state bridge FCL configuration without charging

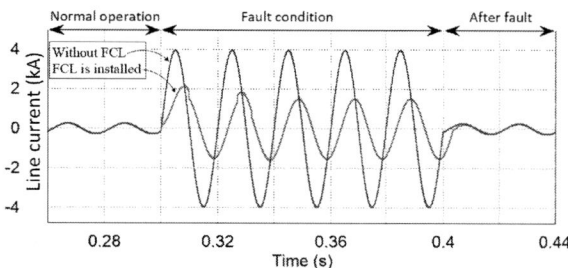

Fig. 5.   Comparing line current between FCL is installed and without FCL in normal operation, fault condition and after fault

B. Simulation results with charging circuit in 24kV

In this section, we add the charging circuits in each cell and bypass switch S_{bpAC}. The charging circuit parameters that consist of voltage drop across diode and V_{charge} are shown in Table I. The system configuration is shown in Fig. 10. In Fig. 11 shows the comparing DC reactor current with and without charging circuits. If the charging

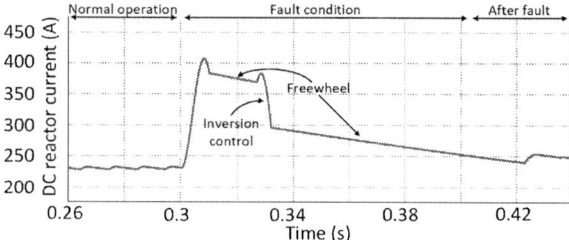

Fig. 6.   DC reactor current in first bridge cell in normal operation, fault condition and after fault

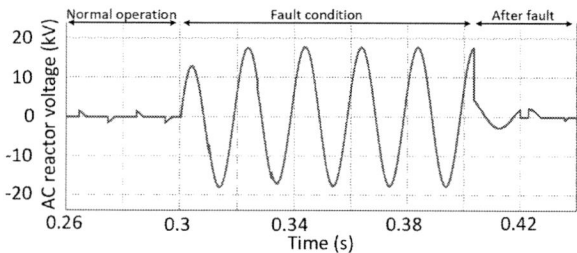

Fig. 7.   AC reactor voltage in normal operation, fault condition and after fault

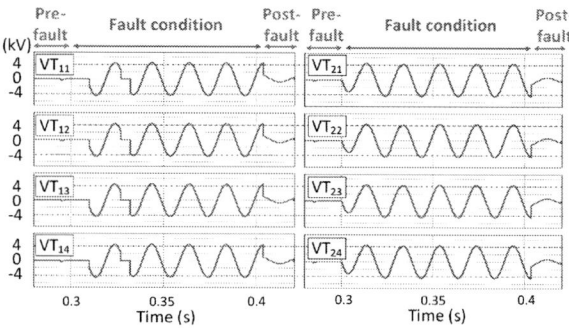

Fig. 8.   Voltage drop across thyristors T1 and T2 in every bridge cell

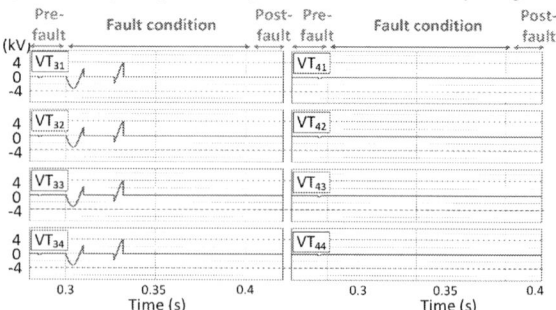

Fig. 9.   Voltage drop across thyristors T3 and T4 in every bridge cell

circuits are not installed, the DC reactor current level is below the peak line current, So the distortion at the peak load voltage is noticeable as shown in Fig. 12.

In the case of using charging circuits in the bridge cell, the by-pass switch $S_{bpAC}$is inserted in parallel with AC reactor during DC reactors charging state.

operation principle: Pre-charge state, the S_{bpAC} will be closed to conduct all of the line current. In charge state, T1, T2, TC1 and TC2 in every bridge cell is full conduction to charge the DC reactor current level. When the DC reactor current level is above the peak

line current, the TC1 and TC2 will be controlled with some firing angle to maintain the DC reactor current level. Subsequently, T3 and T4 will be in full conduction and the line current will pass through bridge cell. The next, S_{bpAC} will be opened to change the conduction path to the bridge cell and enter the normal operation. When the charging circuits are installed in every bridge cell and DC reactor current level is maintained above the peak line current, the voltage drop across all bridge cell is almost to zero and no current flow through AC reactor and there is no distortion on load voltage side.

IV. EXPERIMENTAL RESULTS

A. Experimantal results without charging circuit

Due to various limitations, a prototype of MC-SSB FCL has been built for a single-phase 220 V system to verify the operations. The picture of the system is shown in Fig. 13. The important parameters of the prototype are shown in Table II. The charging circuit is not used.

Fig. 14 shows AC reactor voltage and line current. In fault condition, the line current is limited to the designed level around 20Arms in steady-state. The voltage across the AC reactor during normal condition is easily observed owning to operation at low voltage system. Fig. 15 shows the DC reactor current and AC reactor current. Under freewheeling and inversion control the DC reactor current decreases faster compared to the simulation results due to lower time constant of the prototype system. The

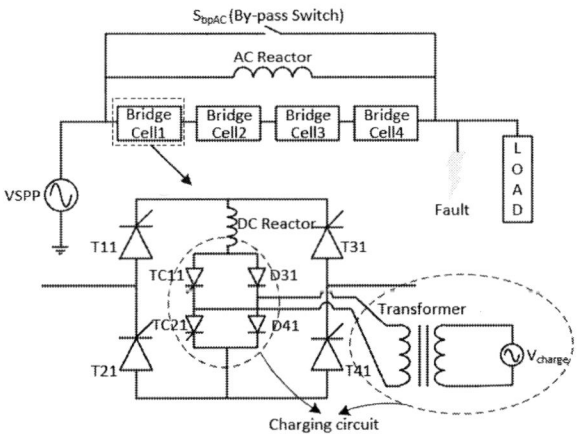

Fig. 10. Solid-state bridge FCL configuration with charging circuits in simulation

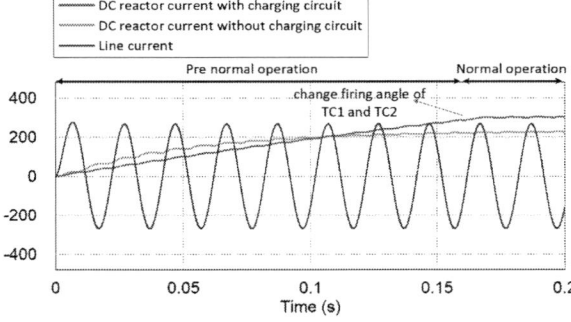

Fig. 11. Comparing between DC reactor current with and without charging circuits

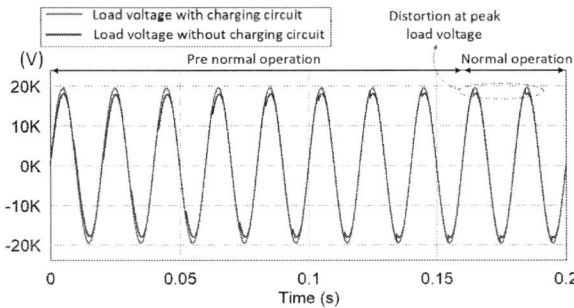

Fig. 12. Comparing load voltage with and without charging circuits

TABLE II. EXPERIMENTAL PARAMETERS AT 220V

Symbol	Meaning	Value
V_s	Source voltage (line-neutral) (rms)	220 V
f	Power system frequency	50 Hz
L_{dc}	DC reactor inductance (per cell)	70 mH
r_{dc}	DC reactor resistance (per cell)	0.2 Ω
L_{ac}	AC reactor inductance	35 mH
r_{ac}	AC reactor resistance	2.4 Ω
r_{load}	Load resistance	35.2 Ω
L_{load}	Load inductance	84 mH
r_f	Fault resistance	2.93 Ω
V_{charge}	Charging voltage (rms)	12 V

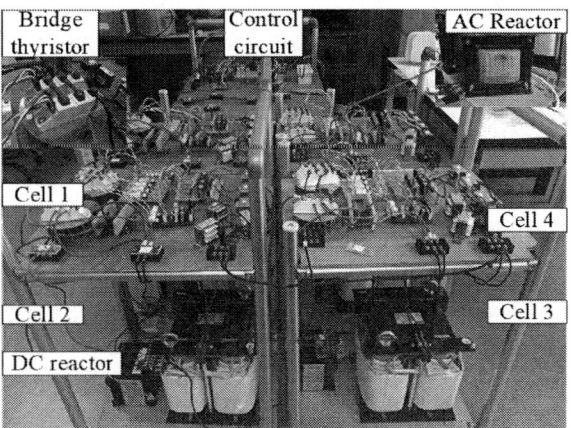

Fig. 13. FCL prototype in laboratory operating at 220V

operation of the bridge cells is in accordant with those of the simulations.

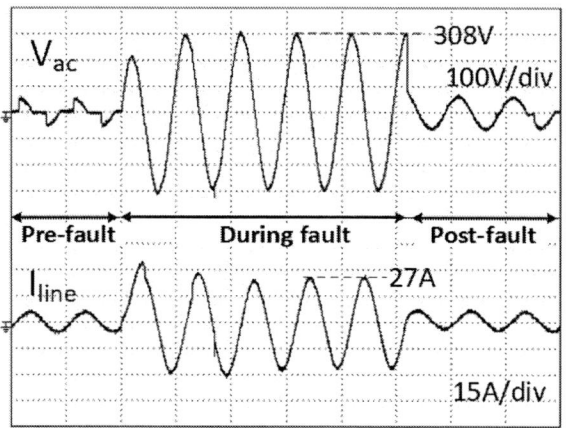

Fig. 14. AC reactor voltage and line current (voltage/division = 100V, current/division = 15A, time/division = 20ms)

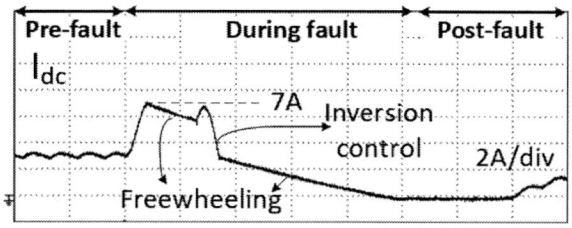

Fig. 15. DC reactor current in first bridge cell (current/division = 2A, time/division = 20ms)

Fig. 16 shows the voltage drop across each thyristor in each bridge cell. Almost equal voltage sharing between each cell is obtained. The waveforms are quite identical to those of the simulations on 24kV system. Fig. 17 shows the DC reactor current, line current and output voltage in normal operation. The distortion is noticeable when the DC reactor current is below the load peak current.

B. Experimantal results with charging circuit

In this experiment, the charging circuits, which includes 12V isolation transformer, two thyristors and two diodes, are added to each cell. Fig. 18 shows AC reactor voltage and line current. The waveforms in fault condition are similar to those of the experiments without charging circuits. However, the AC reactor voltage in normal operation is nearly zero so that no distortion occurs on the output voltage. Fig. 19 shows DC reactor current and AC reactor current in fault condition. After inversion control was applied, the DC reactor current dropped below the line peak current. The DC reactor was recharged by controlling TC1 and TC2 in full conduction, after reaching the required level they were controlled with some firing angle to maintain the DC reactor current level.

Fig. 20 shows the voltage drop across each thyristor in each cell. Equal voltage sharing between each cell is obtained.

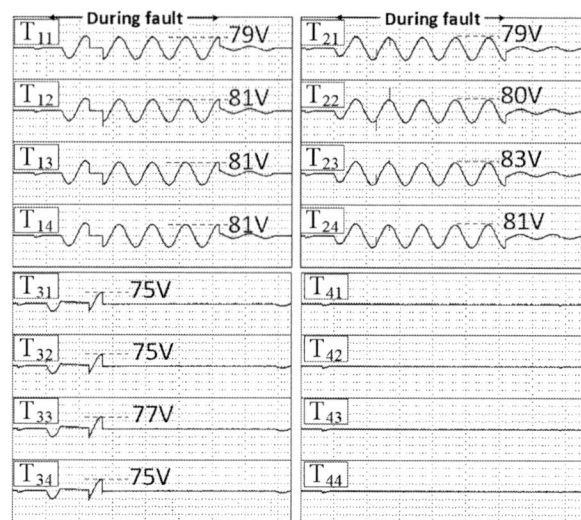

Fig. 16. Voltage drop across thyristors in every bridge cell (voltage/division = 50V, time/division = 20ms)

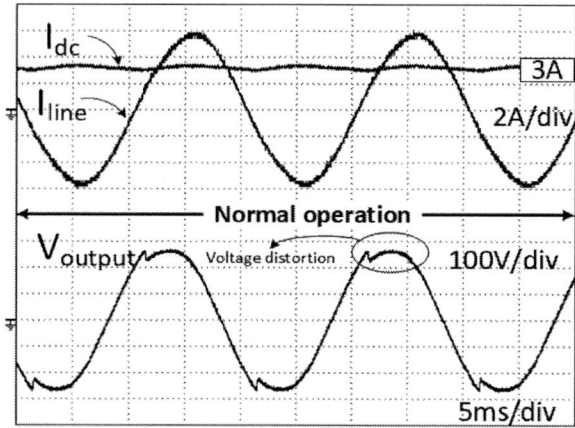

Fig. 17. DC reactor current, line peak current, output voltage (voltage/division = 50V, time/division = 5ms)

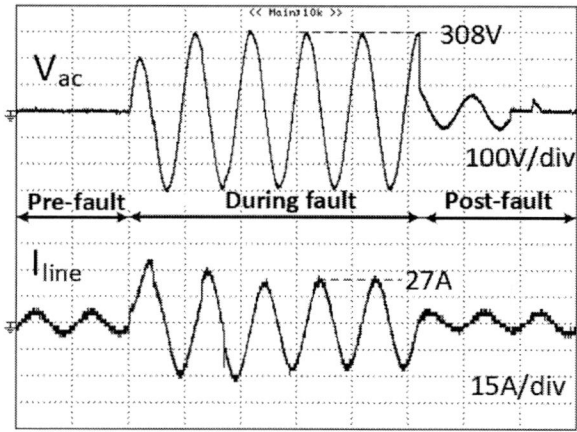

Fig. 18. AC reactor voltage, line current with charging circuit (voltage/division = 100V, current/division = 15A, time/division = 20ms)

2494

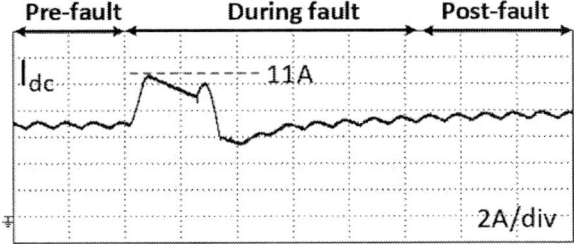

Fig. 19. DC reactor current with charging circuit (current/division = 2A, time/division = 20ms)

The output voltage distortion disappears when the DC reactor current level is maintained above the line peak current. This can be confirmed in Fig. 21.

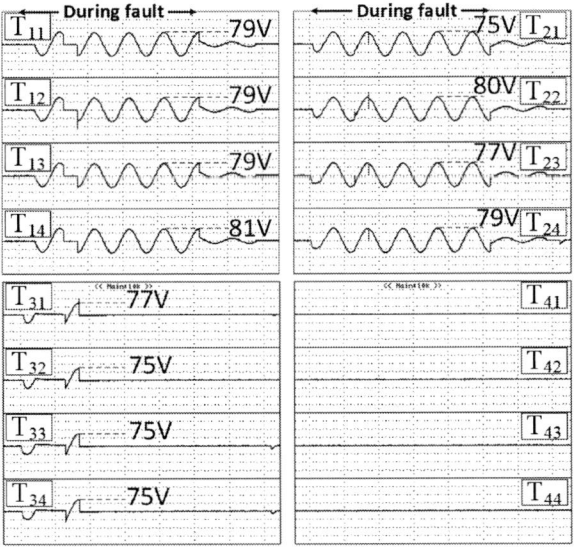

Fig. 20. Voltage drop across thyristors in every bridge cell with charging circuit (voltage/division = 50V, time/division = 20ms)

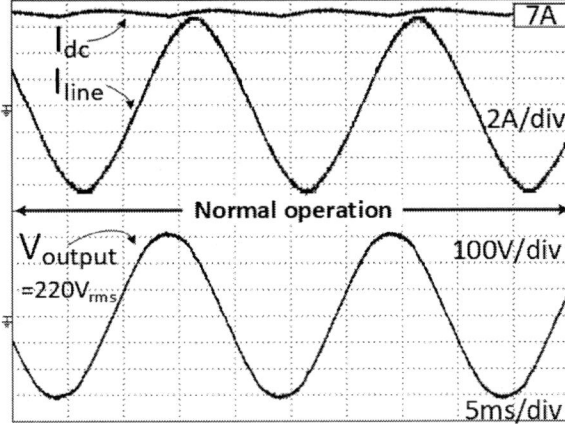

Fig. 21. DC reactor current, line peak current, output voltage with charging circuit (voltage/division = 50V, time/division = 5ms)

V. CONCLUSION

A tranformerless MC-SSB FCL has been proposed and the operation principle has been explained. The simulation results for the 24kV system has been carried out. The simulation results confirm the principle of operation. A 220V prototype single-phase FCL has been built to validate to validity of the proposed control method. The experimental results show that the prototype FCL operates as expected. The line current is limited to the designed value. The voltage sharing between each bridge cell is acceptable. The voltage across AC reactor during normal operation can be mitigated using auxiliary power supply for each bridge cell.

REFERENCES

[1] Alexander Abramovitz and Keuye Ma Smedley, Survey of Solid State Fault Current Limiters IEEE Transections on Power Electronics., Vol. 27. no.6, June 2012. pp. 2770-2782.

[2] Z. Lu, D. Jiang, and Z. Wu. A new topology of fault-current limiter and its parameters optimization Proc. IEEE Power Electronics Specialist Conf. 2003, pp. 462465.

[3] H. J. Boenig and D. A. Paice. Fault current limiter using a superconducting coil IEEE Trans. Magn., vol. 19, no.3, May 1983. pp. 1051-1053.

[4] GangChen, Daozhuo Jiang, Zhengyu Lu and Zhaolin Wu. A New Proposal for Solid State Fault Current Limiter and Its Control Strategies IEEE Power Engineering Society General Meeting. 2004, Vol. 2. June 2004. pp.1468-1473.

[5] H. Radmanesh, S.H. Fathi, and G.B. Gharehpetian, Novel high performance DC reactor type fault current limiter, Electric Power Systems Research, vol. 122, pp. 198-207, 2015.

A Novel DC Power Flow Controller for HVDC Grids with Different Voltage Levels

Ya'nan Wu[1], Han Ye[2], Wu Chen[2*], Xiaokun He[2]

1 State Key Laboratory of Advanced Power Transmission Technology (Global Energy Interconnection Research Institute), Beijing, China

2 Center for Advanced Power-Conversion Technology and Equipment, School of Electrical Engineering Southeast University, Nanjing, China

*E-mail: chenwu@seu.edu.cn

Abstract-As direct current transmission (HVDC) has been improving rapidly, the power flow control of which is increasingly popular and can be realized by the DC power flow controller. In this paper, a novel DC power flow controller (DCPFC) is proposed with the advantages of simpler strategy, little side effect on direct current system, no auxiliary power source and wide applications. A simulation model with two different DC grids was built in the software PLECS, where the performance of the proposed controller was tested. As a consequence, the simulation results show that the proposed DCPFC would have great performance and prospect.

Index Terms—HVDC; DC power; DC power flow controller (DCPFC).

I. INTRODUCTION

As an efficient approach for grid integration and power transmission over long distances, the voltage-source converter based on high voltage direct current (VSC-HVDC) [1] transmission can be applied for large-scale renewable energy plants. Compared with the ac implement, the dc one has advantages of simple facilities, high transmission efficiency and stability [2]. However, many challenges will be encountered and many aspects need to be taken into consideration, for example, the dc power flow control within multi-terminals [3]. The dc power flow controller can be adopted to regulate line current and power configuration, and it is favorable if the dissipation losses can be minimized and thermal risk could be limited within reasonable range. Different from the ac grids, no reactive power is existed in dc systems, as well as the reactance and phase angle. Hence, only the transmission line resistances and node voltages, in dc grids, can be adjusted for dc power flow regulation.

Kinds of structures have been reported to realize the dc power flow control and a series of developments are achieved [4]-[11]. Although the former approach of adjusting the dc resistance [4] has simple structure and easy control, the problems of accompanying dissipation losses from resistors and the bulky cooling system cannot be neglected, which may increase the weight and complexity of the controller, as well as the total cost. When it comes to the latter approach of dc node voltage adjustment, there are two ways to achieve the dc power flow control generally. The first way is to employ a dc

This work is supported by the State Key Laboratory of Advanced Power Transmission Technology Beijing China (Grant No. GEIRI-SKL-2017-010).

transformer [5], [6], which can role not only as the connections among different voltage level dc grids but also as transmission line current regulators in direct current grids with the similar voltage levels. However, the high-power (up to dozens megawatts or even higher) dc transformer is always required and considerably high costs and losses become the main restrict. The second one presents as an assistant voltage source inserted into the direct current transmission feeder [9], [10]. In such cases, to consume/supply power from/to one system, as least one external ac or dc source is needed. Moreover, a colossal low-frequency high-voltage transformer is sometimes required to isolate the dc grid and ac source, which leads to large volume and footprint.

In order to realize the current flow regulation, the concept of interline controller can also be adopted [11], [12]. For instance, a direct current power configuration device was proposed [11] based on the concept of interline controller ideas. An interline DC power flow controller (IDCPFC) was proposed in [12], which was validated by a HVDC grid of three-terminal ring-type. The power flow control can be realized through two capacitors and ovonic dc-to-dc device. However, the ovonic power regulation may not be realized for this IDCPFC. Topology in [13], [14] can realize bi-directional power flow control, but with two lines in one DC grid. In this paper, a modified DCPFC structure is presented, which cannot merely work in all quadrants, achieving wide applications, but in one grid or two disconnecting grids with different voltage levels.

II. CIRCUIT CONFIGURATION AND OPERATION PRINCIPLES

This proposed DCPFC is just exhibited in Fig. 1, where two capacitors C_1 and C_2, each with a bypass switch (S_1 and S_2), are inserted into Line1 and Line2 operating as variable voltage sources, respectively. Each capacitor is paralleled with a unit which is composed of two inductors, two switches with body diodes and two diodes. $L_1 \sim L_4$ are coupled inductors and made with one core, $Q_1 \sim Q_4$ are four IGBTs with anti-parallel diodes $D_1 \sim D_4$, $D_{11} \sim D_{44}$ are four reverse-blocking diodes.

It can be found that C_1 and C_2 are shorted once S_1 and S_2 are both turned on. As a result, this DCPFC loses the function of line current control. Hence, the current configuration can be regulated by the proper operation of

$Q_1 \sim Q_4$ only when S_1 and S_2 are both turned off. The main idea of interline current flow controller is to insert a positive resistance element into one transmission line and a negative one into another, where the positive one could help that the DC voltage difference between both ends of one line opposes and the current in the transmission route is then reduced. It can be equal to increasing the unitary line impedance through an auxiliary impedance. In the contrast, the negative one has opposite result. To make it more clear, two selected working modes out of nine of this presented DCPFC will be illustrated in the following and the standard directions of electrical quantities are plotted in Fig. 1.

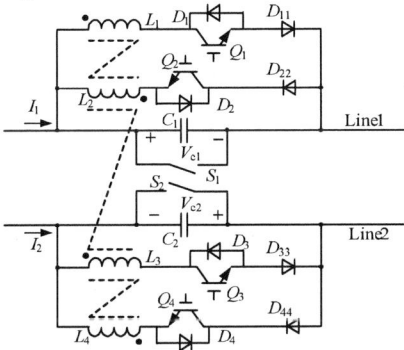

Fig. 1: Proposed IDCPFC.

Mode 1: I_1 and I_2 have the same direction

Without loss of generality, when the current direction of I_1 and I_2 is just as shown in Fig. 1 during this mode, the control aim is to decrease I_1 and increase I_2, in which an equivalent positive impedance and an equivalent negative one must be inserted into Line1 and Line2, respectively. Hence, the voltage polarities of V_{c1} and V_{c2} are consistent with Fig. 1, and energy should be transferred from C_1 to C_2 to make the system balanced and stable.

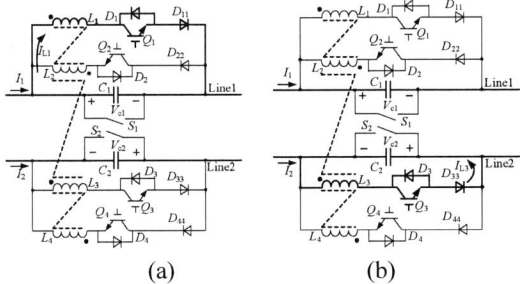

(a)   (b)

Fig. 2: Working states in Mode 1. (a) Only Q_1 is ON. (b) Only Q_3 is ON.

In this case, Q_1 should switch on during the first stage, based on the voltage direction and power delivery route, while other three switches should be turned off. As shown in Fig. 2(a), C_1, L_1, Q_1 and D_{11} form the circuit Loop1, where inductor current i_{L1} increases linearly assuming V_{c1} keeps unchanged during a switching period. During the second stage shown in Fig. 2(b), Q_1 is turned off and Q_3 on. Then the circuit Loop3 is formed by Q_3, D_{33}, C_2 and L_3, where i_{L3} decays linearly, indicating the energy is transferred from L_3 to C_2. When the second stage ends it will return back to the first stage repeatedly.

It can be found that, in an operation cycle, the energy is transferred from C_1 to C_2 via L_1 and L_3. According to the above analysis, Q_2 and Q_4 are always OFF, and Q_1 and Q_3 are switched out of phase in this case. And it is easy to infer the switch state when the I_1 and I_2 are both in negative direction, in which Q_1 and Q_3 are always OFF while Q_2 and Q_4 switched out of phase.

Mode 2: I_1 and I_2 have opposite directions

I_1 and I_2 have different flow directions, namely I_1 is positive while I_2 negative in this mode. Accordingly, the regulation requirement for power flow is to increase I_1 and reduce I_2, implying that the equivalent impedance of Line1 would be decreased, and Line2 would be added. Hence, V_{c2} has the similar direction as the one in Fig. 1, while V_{c1} has the opposite direction. To keep capacitors stable, power should be transferred from C_2 to C_1.

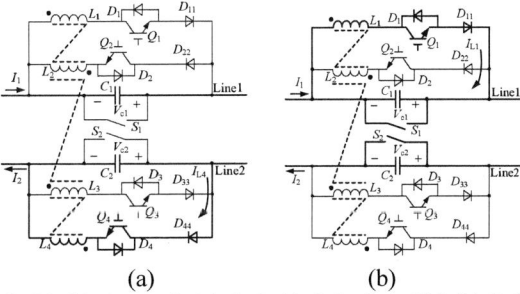

(a)   (b)

Fig. 3: Working states in Mode 2. (a) Only Q_4 is ON. (b) Only Q_1 is ON.

As discussed in Mode 1, Q_4 is turned on and other three switches are turned off during the first stage, as shown in Fig. 3(a). Partial energy of C_2 is delivered and stored into $L4$ in this state. During the second stage shown in Fig. 3(b), Q_1 is turned on and Q_4 is turned off, leading to energy of L_4 is transferred to C_1 via L_1 with the help of coupled inductor. It can be found that, in an operation cycle, the energy is transferred from C_2 to C_1 via the coupled inductors. In conclusion, $Q2$ and $Q3$ are always OFF and Q_1 and Q_4 are switched out of phase in this mode.

III. SIMULATION RESULTS

Two three-terminal HVDC grid simulation models at different voltage levels are built to verify the effectiveness of this presented DCPFC, as shown in Fig. 4. In the left model, terminal 3 (VSC3) works in a constant DC voltage mode and voltage value is 200 kV while the other two terminals (VSC1 and VSC2) are set in constant power modes with P_1=300 MW and P_2=120 MW, respectively. In the right model, terminal 6 is maintained at a constant voltage V_6=100 kV while VSC 4 and VSC5 operate in constant power, with P_4=150 MW and P_5=60 MW, respectively. Without loss of generality, the proposed DCPFC located between terminal 3 and terminal 6, in which C_1 is added into Line3 and C_2 into Line5, can realize power transmission between two lines from different grids. Meanwhile, the described DCPFC is equal to two ideal voltage sources which exchange energy with each other, highlighted in Fig.4. The switching frequency is 4 kHz, C_1= C_2=5 mF and

$L_1=L_2=400$ μH. The detail information of all transmission lines is listed in Table I.

TABLE I
TRANSMISSION LINES INFORMATION

DC Cable Parameters	Line 1	Line 2	Line 3	Line 4	Line 5	Line 6
Length/km	150	300	250	150	300	250
Resistance/Ω	1.5	3	2.5	1.5	3	2.5
Inductance/mH	60	120	100	60	120	100

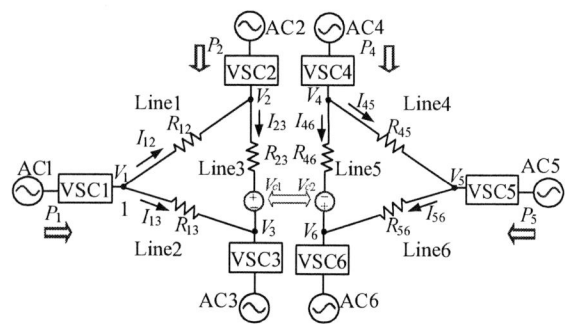

Fig. 4: Two three-terminal HVDC grids with DCPFC.

Based on the above transmission lines information, all unknown line currents and terminal voltages are $I_{12}=0.42$ kA, $I_{13}=1.06$ kA, $I_{23}=1.01$ kA, $V_1=203.17$ kV, $V_2=202.53$ kV, $I_{45}=0.41$ kA, $I_{46}=1.04$ kA, $I_{56}=1.00$ kA, $V_4=103.12$ kV and $V_5=102.50$ kV when the controller is not inserted in the system.

A. Steady-State Operation

Assuming that the target value of I_{23} is 0.25 kA and with the foregone values of V_3, P_1, P_2, I_{23} (0.25 kA), V_6, P_4, P_5 and Table I, It can be obtained that $I_{13}=-0.33$ kA, $I_{13}=1.80$ kA, $I_{23}=0.25$ kA, $V_1=205.88$ kV, $V_2=205.38$ kV, $I_{45}=0.26$ kA, $I_{46}=1.20$ kA, $I_{56}=0.85$ kA, $V_4=102.51$ kV, $V_5=102.12$ kV, $V_{c1}=5.25$ kV and $V_{c2}=1.10$ kV. The simulation results of line currents and voltages are exhibited in Fig. 5. The DCPFC does not work before 4 s and starts to regulate the current flow from the instant $t=4$ s. As shown in Fig. 5, the theoretical analysis is consistent with simulation waveforms and semiconductors Q_2 and Q_4 in the controller have a good performance.

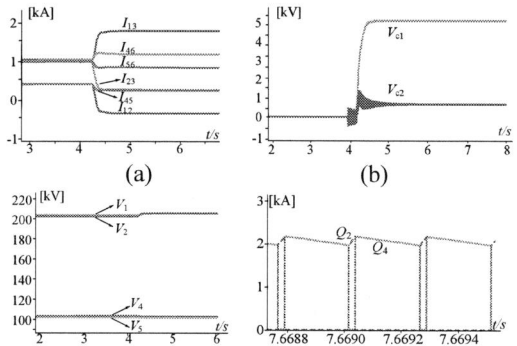

(c)   (d)

Fig. 5: Simulation waveforms when the DCPFC starts to work. (a) Transmission line currents. (b) V_{c1} and V_{c2}. (c) V_1, V_2, V_4 and V_5. (d) Currents flowing through Q_2 and Q_4.

B. Terminal Loss with Unchanged Line Current

The terminal loss mode is simulated by shutting down the VSC1 in the left grid while the right grid work in normal state. Before that, $P_1=300$ MW and $P_2=120$ MW are injected into VSC1 and VSC2, respectively, in which the target value of I_{23} is 0.25 kA. From $t=4$ s, no energy is injected into VSC1 which implies $P_1=0$. With given values of V_3, P_1 (0), P_2, I_{23} (0.25 kA), P_4, P_5, V_6 and Table I, $V_1=201.56$ kV, $V_2=201.04$ kV, $I_{12}=-0.35$ kA, $I_{13}=0.35$ kA, $V_4=103.00$ kV, $V_5=102.42$ kV, $I_{45}=0.38$ kA, $I_{46}=1.07$ kA, $I_{56}=0.97$ kA, $V_{c1}=0.93$ kV and $V_{c2}=0.22$ kV can be obtained after terminal 1 breaks down. Fig. 6 exhibits the current and voltage waveforms. At 4.35 s, the dc grids restore to normal working state. The control aim I_{23} is up to 0.25 kA after oscillation in a short time and other lines keep working in a new state when VSC1 is removed from the system.

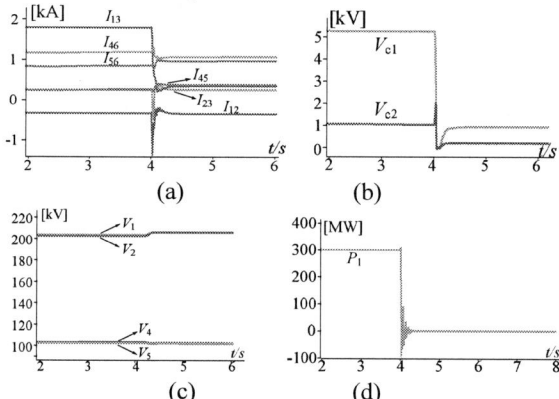

Fig. 6: Simulation waveforms when VSC1 breaks down while I_{23} keeps unchanged. (a) Transmission line currents. (b) V_{c1} and V_{c2}. (c) V_1, V_2, V_4 and V_5. (d) The injected power of VSC1.

C. Power Flow Reversal

The simulation of this subsection is conducted to demonstrate the ability of DCPFC to change the direction of power flow. Before the power flow reversal, the power of VSC 1~2 and VSC 4~5 has the same value as set in last subsection before $t=4$ s, as well as the target value of I_{23}. At instant $t=4$ s, the target value of I_{23} is set to -0.25 kA. With given V_3, P_1, P_2, I_{23} (-0.25 kA), P_4, P_5, V_6 and Table I, one can get $V_1=206.83$ kV, $V_2=208.07$ kV, $I_{12}=-0.83$ kA, $I_{13}=2.28$ kA, $V_4=103.31$ kV, $V_5=102.75$ kV, $I_{45}=0.82$ kA, $I_{46}=1.57$ kA, $I_{56}=1.10$ kV, $V_{c1}=8.70$ kV and $V_{c2}=-1.39$ kV after the direction of power in Line3 is changed, as shown in Fig. 7. From the waveforms, the direction of power in Line3 is changed and V_{c1} has the same polarity as the precondition, while V_{c2} has the opposite polarity. The simulation results show that the proposed DCPFC could operate well even for power reversal condition.

2498

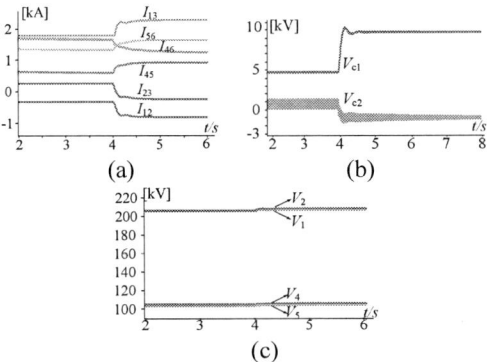

(a)

(b)

(c)

Fig. 7: Simulation waveforms when the direction of power in Line3 is changed. (a) Transmission line currents. (b) V_{c1} and V_{c2}. (c) V_1, V_2, V_4 and V_5.

IV. CONCLUSION

This paper proposed a novel DC current flow controller, which can be used in future grids with different voltage levels, owing the advantages such as low cost for active switches, flexible control strategy, bidirectional current flow control and no external power sources. There are as many as nine operation modes and two of them are labored. This DCPFC has been validated by simulations results based on two meshed three-terminal HVDC grids in PLECS.

REFERENCES

[1] Weixing Lu and Boon-Teck Ooi, "DC overvoltage control during loss of converter in multiterminal voltage-source converter-based HVDC (M-VSC-HVDC)," *IEEE Trans. Power Del.*, vol. 18, no. 3, pp. 915–920, Jul. 2003.

[2] D. Jovcic and N. Strachan. "Offshore wind farm with centralised power conversion and DC interconnection," *IET Generation, Transmission & Distribution*, Vol. 3, no. 6, pp. 586–595, Jun. 2009.

[3] B. R. Andersen, "HVDC grids: Overview of CIGRE activities and personal views," Available online: http://b4.cigre.org/Publications/ Other- Documents/SC-B4-presentations-and-papers, 2014.

[4] D. Jovcic, M. Hajian, H. Zhang, and G. Asplund, "Power flow control in dc transmission grids using mechanical and semiconductor based dc/dc devices," *IET International Conference on AC and DC Power Transmission (ACDC)*, 2012, pp. 1–6.

[5] D. Jovcic and B. Ooi, "Developing dc transmission networks using dc transformers," *IEEE Trans. Power Del.*, vol. 25, no. 4, pp. 2535–2543, 2010.

[6] K. Natori, H. Obara, K. Yoshikawa, B. Hiu, and Y. Sato, "Flexible power flow control for next-generation multi-terminal DC power network," *IEEE Energy Conversion Congress and Exposition (ECCE)*, 2014, pp. 778–784.

[7] S. Rodrigues, R. T. Pinto, P. Bauer and J. Pierik, "Optimal power flow control of VSC-based multiterminal DC network for offshore wind integration in the north sea," *IEEE Journal of Emerging and Selected Topics in Power Electronics*, vol. 1, no. 4, pp. 260–268, Dec. 2013.

[8] L. Xu, L. Yao, "DC voltage control and power dispatch of a multi-terminal HVDC system for integrating large offshore wind farms", *IET Renew. Power Gener.*, vol. 5, no. 3, pp. 223–233, 2011.

[9] E. Veilleux and B. Ooi, "Multi-terminal HVDC with thyristor power-flow controller," *IEEE Trans. Power Del.*, vol. 27, no. 3, pp. 1205–1212, 2012.

[10] S. Balasubramaniam, J. Liang, and C. Ugalde-Loo, "An IGBT based series power flow controller for multi-terminal HVDC transmission," *International Universities' Power Engineering Conference (UPEC)*, 2014, pp. 1–6.

[11] C. Barker and R. Whitehouse, "A current flow controller for use in HVDC grids," I*ET International Conference on AC and DC Power Transmission (ACDC)*, 2012, pp. 1–5.

[12] W. Chen, X. Zhu, L. Yao, X. Ruan, Z. Wang, and Y. Cao, "An interline DC power-flow controller(IDCPFC) for multiterminal HVDC system," *IEEE Trans. Power Del.*, vol. 30, no.4, pp. 2027–2036, 2015.

[13] Chen W, Zhu X, Yao L, et al. "A Novel Interline DC Power-Flow Controller (IDCPFC) for Meshed HVDC Grids," *IEEE Transactions on Power Delivery*, vol.31,no.4,pp. 1719-1727,2016.

[14] Ning G, Chen W, Zhu X. "A novel interline DC power flow controller for meshed HVDC grids," *IEEE Energy Conversion Congress and Exposition (ECCE)*, pp.1-7, 2016.

The 2018 International Power Electronics Conference

Design and Control of Single-Phase Grid-Connected Photovoltaic Microinverter with Reactive Power Support Capability

Geon-Hong Min[1], Kyung-Hwan Lee[1], Jung-Ik Ha[1]* and Myong Hwan Kim[2]
[1]Department of Electrical and Computer Engineering, Seoul National University, Seoul, Republic of Korea
[2]Solar System Development Team Solar R&D Lab, LG Electronics, Gumi, Republic of Korea
*E-mail: jungikha@snu.ac.kr

Abstract- **This paper proposes a single-phase photovoltaic (PV) microinverter capable of supplying reactive power to the grid. In the past, researchers mainly focused on obtaining high efficiency, low cost, and high boost ratio. However, as the solar power generation has increased in recent years, the grid regulations have been changed so that PV inverters provide reactive power for the grid stability. Therefore, the proposed PV microinverter is constructed by the bidirectional boost/buck dc-dc converter for reactive power support and achieving high step-up ratio by using coupled inductors. In addition to the design guideline, this paper presents the control strategy of the proposed microinverter. Unlike traditional two-stage microinverters, the dc-link voltage is shaped like the rectified grid voltage. The experimental results verify that the 320-W laboratory prototype can supply reactive power and reaches 96.5% California Energy Commission (CEC) efficiency at unity power factor operation.**

Keywords— Microinverter, Photovoltaic (PV), Reactive power, Single-stage inverter.

I. INTRODUCTION

The photovoltaic (PV) energy is being commonly used with the increasing interest in renewable energy. In some cases, PV modules are connected in series feeding a string inverter. However, due to mismatch caused by shading and panel orientation, amount of power generated by PV strings can be significantly reduced [1]–[3]. Therefore, instead of connecting PVs in series, single PV ac module system with the maximum power point tracker (MPPT) is used to resolve the mismatch problem. In this system, an inverter is attached to a single PV module to provide power to the grid. This inverter is called as the microinverter and various studies about topologies and control method of the microinverter are in progress.

Depending on the topology used, microinverter can be classified into single-stage [2]-[4] and two-stage [5], [6] configuration. In a single-stage configuration, the output voltage of the converter is shaped like a sinusoidal voltage and dc-dc converter controls ac current toward the grid while full-bridge acts merely as an unfolder to inject sinusoidal current to the grid. Two-stage configuration, on the contrary, has the constant dc link voltage provided by the converter followed by the inverter that controls ac current toward the grid with PWM switching. In this paper,

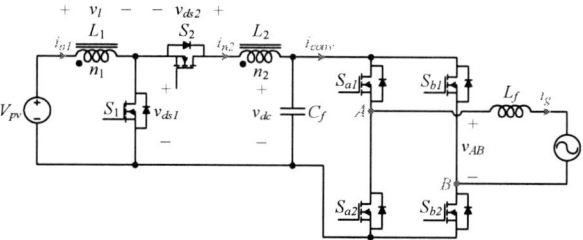

Fig. 1.   Circuit diagram of the PV microinverter.

single-stage topology is chosen due to its simple structure and efficiency since only dc-dc converter is switching at high frequency.

In the past, these microinverters were designed without reactive power capability. Therefore, conventional microinverters could only provide active power to the grid. However, regulations are being altered for PV system to have reactive power support in the case of grid fault [7]. Therefore, research is being made for a microinverter topology that has reactive power capability [8], [9]. However, most of the research is based on reactive power support in two-stage topology and reactive power capability on single-stage PV microinverter topology is often overlooked.

In this paper, design and control of single-stage PV microinverter capable of supporting reactive power to the grid are proposed. The proposed topology consists of a bidirectional boost/buck dc-dc converter with coupled inductors for high step-up ratio and an unfolder injecting ac current to the grid. In addition to the design of PV microinverter, the control strategy is also proposed. The system is able to inject desired reactive power to the grid with the single-stage topology using the proposed control method. The result from the experiment confirms the viability of the proposed system.

II. THE PV MICROINVERTER TOPOLOGY

In order to supply reactive power to the grid, the bidirectional dc-dc converter should be used. Fig. 1 shows the circuit diagram of the proposed PV microinverter topology. The circuit is constructed using a bidirectional boost/buck dc-dc converter with coupled inductors and a

2500

full-bridge inverter. The chosen converter is bidirectional thus capable of supporting reactive power to the grid and can achieve high step-up ratio by using coupled inductors with turn ratio n (n_2/n_1). The full-bridge inverter is operated either as an unfolder, injecting sinusoidal ac current to the grid or as an inverter depending on the coupled capacitor voltage v_{dc}. Both operations of the boost/buck converter and full-bridge inverter will be explained in this paper.

A. Steady-State Analysis of the Boost/Buck Converter

The operation of the dc-dc converter differs depending on the power transfer direction. When power is transferred from the input to the output, the converter operates as boost converter while the converter operates as buck converter for reverse power direction.

Fig. 2 shows current and voltage waveform of the dc-dc converter for boost direction. In this operation, switch S1 is first turned on increasing the input current i_{n1}. The peak current of the input current i_{pk1} can be calculated as

$$i_{pk1} = \frac{V_{pv}}{L_1} t_{on,s1}, \quad (1)$$

where $t_{on,s1}$ is a turn on period for switch S1. After switch S1 is turned off, switch S2 is turned on reducing the input current and transferring power to the output. The peak current of output current i_{pk2} is

$$i_{pk2} = \frac{i_{pk1}}{n+1}. \quad (2)$$

If the converter is operating in discontinuous conduction mode (DCM), the switch S2 is turned on until output current i_{n2} becomes zero. Let voltage across the inductor L_1 be V_l then the voltage across L_2 is nV_l where n is the turn ratio of the coupled inductors. Therefore, V_l, when

switch S2 is turned on, is

$$V_l = \frac{V_{pv} - v_{dc}}{n+1}. \quad (3)$$

Since the average voltage across the inductor should be zero, turn on time of switch S2 can be calculated as

$$t_{on,s2} = \frac{(n+1)L_1 i_{pk1}}{v_{dc} - V_{pv}}, \quad (4)$$

where $t_{on,s2}$ is a turn-on time of switch S2. In DCM operation, after output current i_{n2} reaches zero, both switches S1 and S2 are turned off and current starts to resonate. The resonant frequency is expressed as

$$f_{res} = \frac{1}{T_{res}} = \frac{1}{2\pi\sqrt{L_1\left\{C_{ds1} + (n+1)^2 C_{ds2}\right\}}}, \quad (5)$$

where f_{res} is a resonant frequency and T_{res} is a resonant period. In order to achieve valley switching to increase the efficiency, half of the resonant period is added to the switching period as shown as

$$T_{s,conv} = t_{on,S1} + t_{on,S2} + \frac{T_{res}}{2}, \quad (6)$$

where $T_{s,conv}$ is a switching period of the converter.

When the converter is operating in buck mode, switch S2 is turned on before switch S1 as shown in Fig. 3. In this mode, direction of the input and output current is different and therefore, equation (1) and (4) should be rewritten as

$$i_{pk1} = -\frac{V_{pv}}{L_1} t_{on,S1}, \quad (7)$$

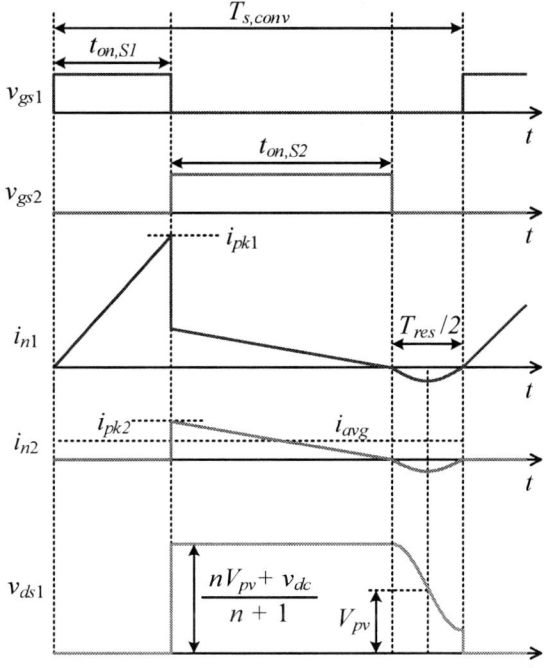

Fig. 2. Key waveforms in boost direction.

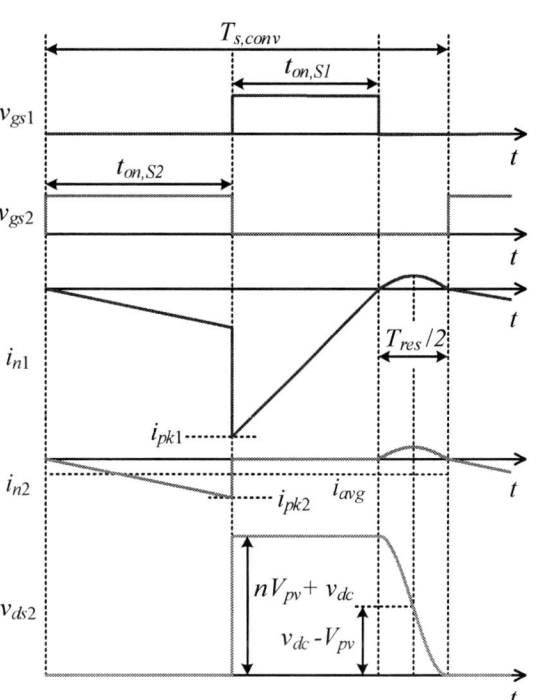

Fig. 3. Key waveforms in buck direction.

2501

$$t_{on,S2} = -\frac{(n+1)L_1 i_{pk1}}{v_{dc} - V_{pv}}. \tag{8}$$

B. Operation Modes of the PV Microinverter

In single-stage microinverter, full-bridge is operated as unfolder as expressed in Fig. 4 as mode I, where i_g is a grid current. In this operation mode, if the grid current is positive switch S_{a1} and S_{b2} is turned on while S_{b1} and S_{a2} are turned off. The reverse is true for the negative grid current. However, due to the limitation of the chosen dc-dc converter topology, the output voltage of the converter v_{dc} should be higher than the input voltage V_{pv}. Therefore, full-bridge can no longer be operated as an unfolder if the grid voltage is below V_{pv}. In this case, operation modes are switched and full-bridge acts as an inverter controlling the grid current. This operation is denoted as mode II in Fig. 4 and in this mode, switch S2 is turned on and S1 is turned off in the dc-dc converter to maintain the output voltage v_{dc} as V_{pv}.

III. THE CONTROL STRATEGY

Different control strategy should be implemented depending on the mode of operation. In mode II, the full-bridge is operating as an inverter and controls the grid current using a conventional current controller. The dc-dc converter does not operate in this mode and therefore no control strategy is needed.

In mode I, however, since full-bridge only operates as an unfolder, the converter needs to control the grid current. The grid current reference $i_g{}^*$ for the converter can be calculated as

$$i_g{}^* = I_{grid} \sin\left(\theta_g - \varphi_g\right) sign\left(\sin\theta_g\right), \tag{9}$$

where θ_g is the current angle of the grid voltage, φ_g is the phase of the grid current compared to the grid voltage and I_{grid} is the magnitude of the desired grid current. In order to provide this current, the average of the output current i_{n2} should be equal to the current reference. Using equation (2), (4) and (8) following equation can be derived

$$\frac{1}{2} i_{pk2} t_{on,s2} = \frac{L_i i_{pk1}{}^2}{2\left(v_{dc} - V_{pv}\right)} = \left| i_g{}^* \right| T_{s,conv}, \tag{10}$$

where $v_{dc} = V_{grid} \left| \sin\left(\theta_g\right) \right|$ in mode I. Assuming resonant frequency is much higher than switching frequency $T_{res}/2 \ll T_{s,conv}$, the switching period of the converter can be calculated

$$T_{s,conv} = T_{on,s1} + T_{on,s2} = L_1 \left| i_{pk1} \right| \frac{\left(v_{dc} + nV_{pv}\right)}{V_{pv}\left(v_{dc} - V_{pv}\right)}, \tag{11}$$

From (10) and (11) relationship between i_{pk1} and grid current reference can be calculated

$$i_{pk1} =$$
$$2\left| i_g{}^* \right| \left(n + \frac{V_{grid}}{V_{pv}} \left| \sin\theta_g \right| \right) sign\left\{ \sin\theta_g \sin\left(\theta_g - \varphi_g\right) \right\}. \tag{12}$$

Therefore, current envelop of i_{pk1} according to grid voltage angle θ_g can be made as shown in Fig. 5. Using this envelope, peak current control is used to control the grid

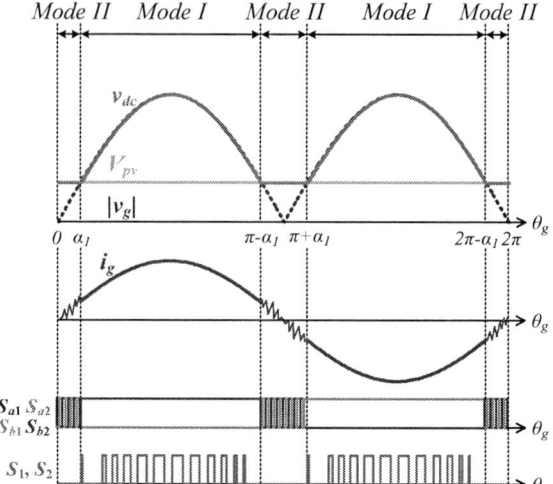

Fig. 4. Key waveforms at different operation mode.

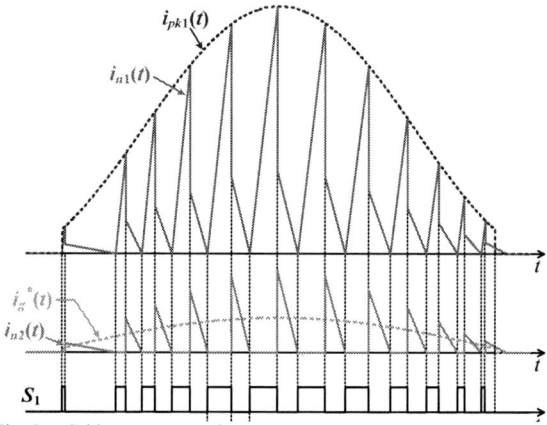

Fig. 5. Grid current control.

current.

The limitation of this control strategy is the fact that switching frequency can be drastically increased at some points. For instance, when power factor does not equal to one and grid current $i_g{}^*$ is close to zero, the switching period is reduced to almost zero. This drastically increases the switching frequency of the converter, which will increase the switching loss. Therefore, when switching frequency becomes too long, the switching period is extended by adding more than one resonant period changing equation (6) into

$$T_{s,conv} = T_{on,s1} + T_{on,s2} + N_{res}\left(\frac{T_{res}}{2}\right), \tag{13}$$

where N_{res} is the number of resonant periods added to the switching period. The peak current is then recalculated using equation (10) with new switching period value.

There are also things to consider when transitioning the operation mode. At the boundary between mode I and mode II, v_{dc} and V_{pv} are almost the same increasing the switching period substantially since $V_{pv}\text{-}v_{dc} \approx 0$ in (11). When the power factor is close to one, the value of i_{pk1} is minimal at the boundary thus switching period does not increase as much. However, when the power factor is below one, i_{pk1} does not equal to zero and switching period

2502

The 2018 International Power Electronics Conference

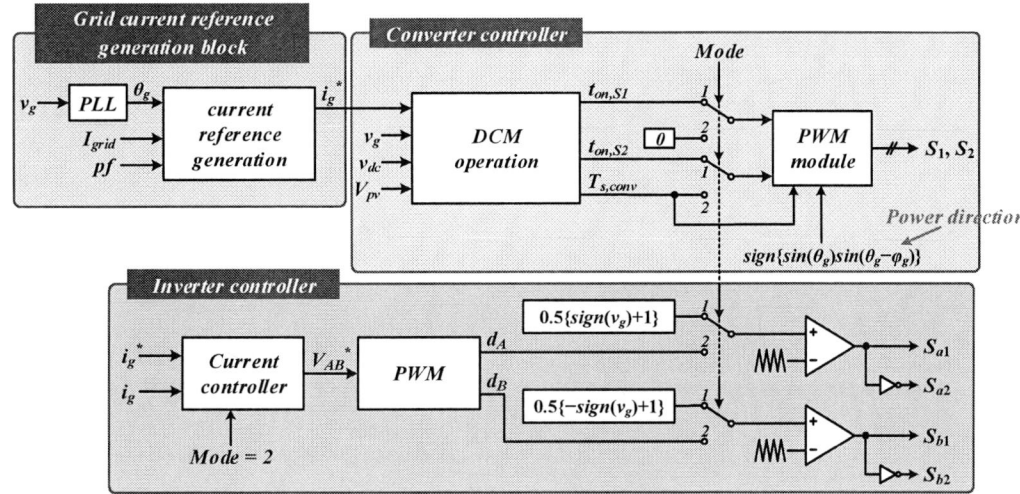

Fig. 6.   Control block diagram.

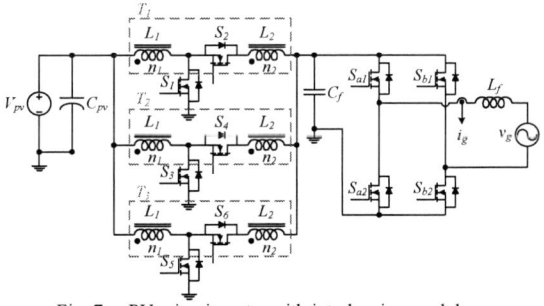

Fig. 7.   PV microinverter with interleaving modules.

becomes too long. Therefore, the maximum limit of switching period must be set in order to prevent converter from operation in continuous conduction mode (CCM). The overall control block diagram of the proposed system is shown in Fig. 6.

TABLE I
PARAMETERS FOR EXPERIMENTS

Parameters		Value
C_f		500 [nF]
Transformer parameters	L_1	20.5 [uH]
	L_2	510 [uH]
	n_1:n_2	1:5
L_f		L_{f1} = 516 [uH] L_{f2} = 520 [uH] Coupling coefficient k = 0.972

IV.   EXPERIMENTAL RESULT

In this experiment 3 interleaved converter modules, as shown in Fig. 7, are used to provide up to 320W power to the grid. Parameters for the experiment is shown in Table I. The voltage of PV is set to 32V and the grid voltage is 240Vrms. Fig. 8 shows the

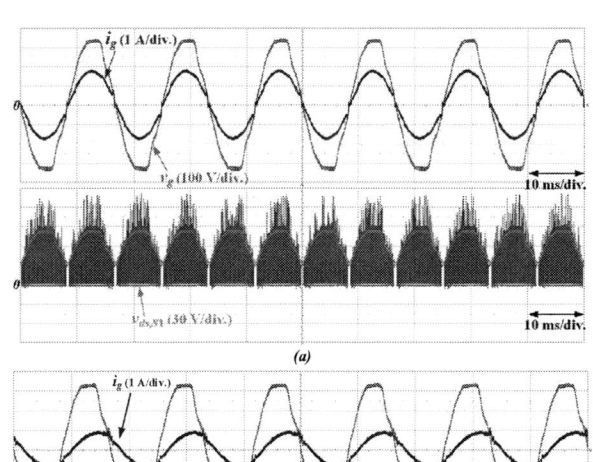

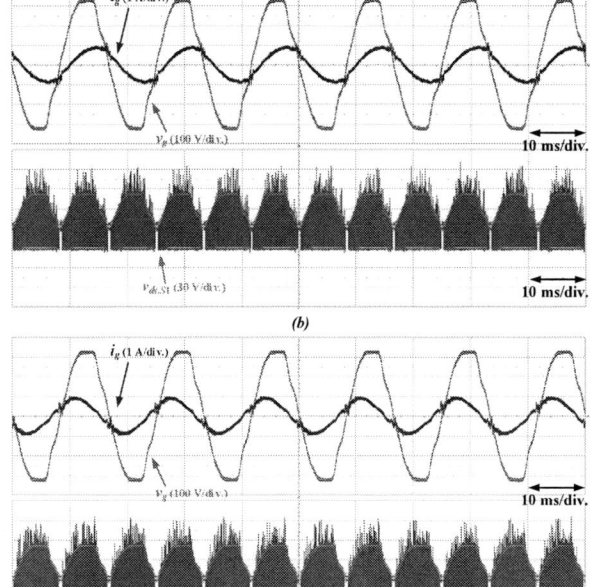

Fig. 8. The grid voltage and current waveforms (a) power factor = 1 (b) power factor = 0.8 lagging (c) power factor = 0.8 leading.

2503

The 2018 International Power Electronics Conference

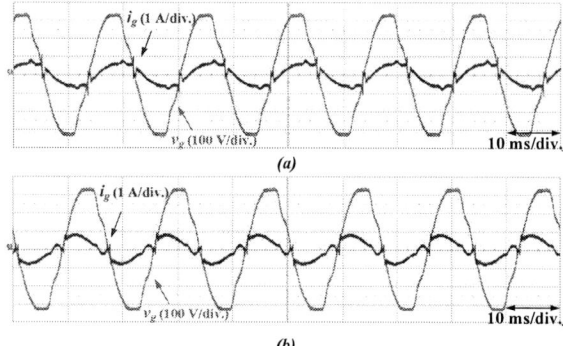

(a)

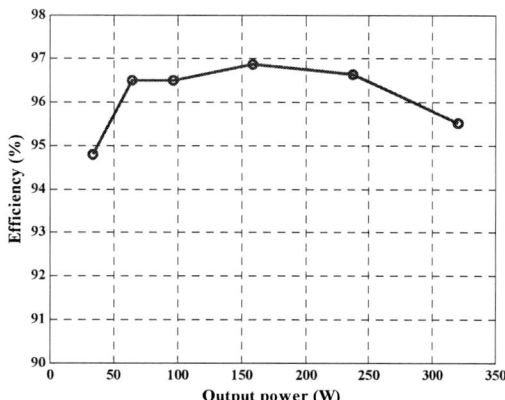

(b)

Fig. 9. The grid voltage and current waveforms without mode II (a) power factor = 0.8 lagging (b) power factor = 0.8 leading.

Fig. 10. Efficiency of microinverter at unity power factor.

experimental results of the proposed PV microinverter. Fig. 8(a) shows the voltage and current waveforms at full load when the power factor is one. The voltage and current waveforms denoted as v_g and i_g are the grid voltage and current, while $v_{ds,s1}$ is a voltage across the switch S_1. The experimental results when the phase of the current is lagging, Fig. 8(b), and leading, Fig. 8(c), with power factor 0.8 are also displayed. It can be seen that there is some current spike when grid voltage is almost zero. This is due to a PWM switching of the full-bridge inverter in mode II. If there is no PWM switching of the full-bridge inverter, the current distortion near zero grid voltage aggravates as shown in Fig. 9. The efficiency of the system at unity power is measured at different power load. The measured efficiency of the PV microinverter is shown in Fig. 10. The California Energy Commission (CEC) efficiency of the PV microinverter is 96.5%.

V. Conclusion

In this paper, design and control of PV microinverter with reactive power supply capability are presented. The steady-state operation of the buck/boost converter used in this microinverter is analyzed and operation mode depending on the grid voltage is explained. Using equations from the steady-state analysis, the grid current control using peak current envelop is described. Using this

control strategy, reactive power can be supplied using single-stage PV microinverter topology. The experimental result shows proposed PV microinverter is capable of providing reactive power with 96.5% efficiency.

References

[1] W. Xiao, N. Ozog and W. G. Dunford, "Topology Study of Photovoltaic Interface for Maximum Power Point Tracking," *IEEE Trans. on Industrial Electronics*, vol. 54, Issue. 3, pp. 1696-1704, April 2007.

[2] S. H. Lee, W. J. Cha, J. M. Kwon and B. H. Kwon, "Control Strategy of Flyback Microinverter With Hybrid Mode for PV AC Modules," *IEEE Trans. on Industrial Electronics*, vol. 63, issue. 2, pp. 995-1002, Sept. 2015.

[3] Y. Li and R. Oruganti, "A Low Cost Flyback CCM Inverter for AC Module Application," *IEEE Trans. on Power Electronics*, vol. 27, issue. 3, pp. 1295-1303, Aug. 2011.

[4] N. Sukesh, M. Pahlevaninezhad and P. K. Jain, "Analysis and Implementation of a Single-Stage Flyback PV Microinverter With Soft Switching," *IEEE Trans. on Industrial Electronics*, vol. 61, Issue. 4, pp. 1819-1833, May 2013.

[5] S. M. Chen, T. J. Liang, L. S. Yang and J. F. Chen, "A Boost Converter with Capacitor Multiplier and Coupled Inductor for AC Module Applications," *IEEE Trans. on Industrial Electronics*, vol. 60, issue. 4, pp. 1503-1511, Sept. 2011.

[6] S. A. Arshadi, B. Poorali, E. Adib and H. Farzanehfard, "High Step-Up DC-AC Inverter Suitable for AC Module Application," *IEEE Trans. on Industrial Electronics*, vol. 63, issue. 2, pp. 832-839, Sept. 2015.

[7] Y. Yang, P. Enjeti, F. Blaabjerg and H. Wang, "Wide-Scale Adoption of Photovoltaic Energy: Grid Code Modifications Are Explored in the Distribution Grid," *IEEE Industry Applications Magazine*, vol. 21, issue. 5, pp. 21-31, June. 2015.

[8] Y. Yang, F. Blaabjerg and Z. Zou, "Benchmarking of Grid Fault Modes in Single-Phase Grid-Connected Photovoltaic Systems," *IEEE Trans. on Industry Applications*, vol. 49, issue. 5, pp. 2167-2176, April. 2013.

[9] Y. K. Wu, J. H. Lin and H. J. Lin, "Standard and Guidelines for Grid-Connected Photovoltaic Generation Systems: A Review and Comparison," *IEEE Trans. On Industry Applications*, vol. 53, issue. 4, pp. 3205-3216, March. 2017.

Optimal Size and Multi-objective Control of Battery Energy Storages in Distribution System with High Penetration of Distributed PV Generators

Meiqin Mao[1]*, Lei Zhou[1], Yangyang Wang[1] and Liuchen Chang[2]

1.Electrical and Automation Engineering Department, Hefei University of Technology, Hefei, China
2.Electrical and Computer Engineering Department, university of new Brunswick, Fredericton, Canada
*E-mail: mmqmail@163.com

Abstract- **High penetration of PV into distribution system brings about side effects, such as local voltage rise because of light local loads and increase of power loss in the network. The application of Battery Energy Storages (BES) may be one of possible solutions to mitigate such side effects of PV systems. This paper presents a multi-objective optimal design and control method of BES used in distribution system with high penetration of distributed PV generators, such as smoothing the output power of PV generators, minimizing the loss in the network and peak shaving and valley filling by considering the capability of line connected with PVs, and an economic optimal model of BES sizing with maximum net income is proposed. Taking a real distribution system of 35kV with 60MW of PVs in total in China as an example, simulations are performed to test the proposed methods. The simulation results show that by the proposed optimal designed method and multi-objective control strategy has better economic performances than single objective control while the voltage of key nodes can be kept within the limits with minimized capacity of BES. The method proposed in this paper is beneficial to promote the high penetration of PV generators into power system.**

I. INTRODUCTION

Over the past decade, the development of photovoltaic generation systems (PVs) has been booming all over the world thanks to its pollution free advantage and persisting price drop of modules. For example, the installation of PVs reaches 130 GW by end of 2017 in China, increasing by 82.6 percent compared with that in 2016[1]. However, the high penetration of PVs into utility brings about severe challenges to the secure and reliable operation of power system, which has been attracting lots of attentions from academia and industry [2].

The current researches show that Battery Energy Storages (BES) has the potential to enable the high integration of PVs into the smart grid by providing various ancillary services, such as frequency regulating, power quality improvement, power fluctuation equilibrium and peak load shifting [3]. According to different locations for battery tank to be integrated into the power system, the application of BES can be grouped into two situations: source side integration and grid side integration.

For source side integration, the BES is installed and operated by PV plant operators. In this case, BES is used to reach two control objectives. One is to smooth fluctuation of PV output power [4-6], and the other is to track the predicted output power of PVs [7-8]. For the former objectives, the BES is controlled to be charged or discharged so that the output power of PVs ramps up or down at set rate, while for the later objectives, the BES is controlled to compensate the prediction errors of output power of PVs because the randomness and intermittence of the insolation.

For grid side integration, BES is installed and operated by gird operators. In this case, the main application scenarios of energy storage include peak shaving and valley filling [9], frequency modulation [10], voltage regulation [11] and minimizing network loss.

No matter for source side integration or grid side integration, the BES is usually designed and controlled with single operation objective in current literature. However, for single objective-oriented control such as peak shaving and valley filling, BES is in idle state for most time of a day until the peak or valley occurs, which will lead to the low utilization of energy storage systems. As a result, the economic performance of BES is poor. To address the issue above, a few of the multi-objective location optimization methods of BES are proposed such as a chaotic multi-objective genetic algorithm for optimal size of BES in a microgrid [12] and or multi-objective optimization of BES just used for source side integration[13].

In this paper, a BES operating with multi-objective for the distribution system with a high penetration of PVs is investigated. The optimal size and control strategies of the BES for both source side integration and grid side integration are proposed and discuss the economic model of BES, which has the objective function of maximum net income. The methods are validated by a real distribution system of 35kV with 60MW of PVs in total in China.

The rest of paper is arranged as follows. The next section proposes an optimal sizing method for BES with mutil-objective operation control, and discuss the economic performance of BES capacity. The Section III provides the simulation results of proposed BES control methods in a real distribution system followed by conclusion in Section IV.

II. OPTIMAL SIZING METHOD FOR BES WITH MULTI-OBJECTIVE OPERATION CONTROL

In different operation scenarios, different control objectives and control methods will be used for BES, and the configuration principle of BES is also different. On the one hand, the configuration of BES has its basic principle, such as realizing specific objective with minimizing the capacity, which is regardless of where it is located. On the other hand, different operation scenarios may have different effects on the configuration of BES. Therefore, it is necessary to summarize the factors influencing configuration results, which will be discussed in the following segments.

A. Methods of BES configuration in mutil-objective

In this paper, the objective of smoothing output power fluctuation of PVs is combined with that of peak shaving and valley filling.

a. Smoothing output power of PV

For smoothing the fluctuation of output power of PVs, the first order low-pass filter (FLF) method is used in this paper to obtain the fluctuant components balanced by the BES through optimizing the filter time constant in order that the maximum fluctuation of the output from FLF can be limited to expected values. The transfer function of FLF is expressed as (1).

$$H(s) = \frac{1}{1+Ts} \qquad (1)$$

Where, T is filter time constant. The larger T is, the smoother the output of FLF will be.

Fig.1 shows the flow chart of smoothing power fluctuation using FLF. The PV output power P_{pv} is filtered by the FLF，which provides the expected smooth output power $P_{rpv}(t)$ fed to the grid. The difference between $P_{rpv}(t)$ and $P_{pv}(t)$ is the reference power that BES should compensate. $P_0(t)$ is the total output power from BES and PV.

The relationship between $P_{pv}(s)$ and $P_{rpv}(s)$ can be expressed as (2).

$$P_{rpv}(s) = \frac{P_{pv}(s)}{1+Ts} \qquad (2)$$

After the discretization of (2), recurrence formulas for P_{ref} and P_b are obtained as (3).

$$P_{rpv}(t) = \frac{T}{T+\Delta t}P_{rpv}(t-\Delta t) + \frac{\Delta t}{T+\Delta t}P_{pv}(t) \qquad (3)$$

Where Δt is the time interval, which is set as 5 minutes in this paper, t is the current time while t-Δt is the last time.

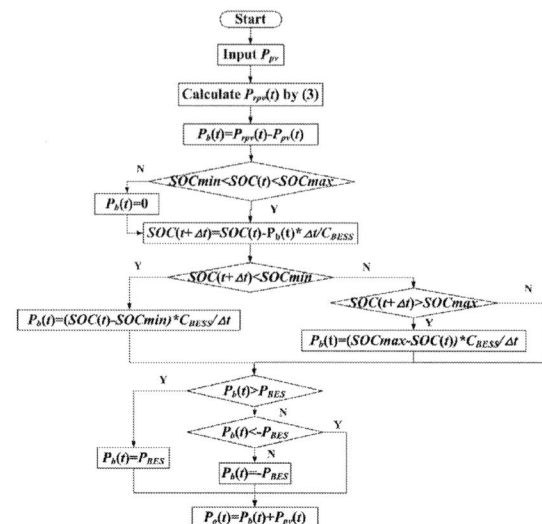

Fig. 1. Flow chart of smoothing power fluctuation

The power for BES can be derived by (4)

$$P_b(t) = P_{rpv}(t) - P_{pv}(t) \qquad (4)$$

To quantitively describe the objective for power fluctuation smoothing, we introduce the parameter of fluctuation rate δ(t) of PV output power which is defined in (5)

$$\delta(t) = (P_{pv}(t) - P_{pv}(t-\Delta t))/\Delta t \qquad (5)$$

Where, $P_{pv}(t)$ and $P_{pv}(t-\Delta t)$ are the output power of PVs at time slot t and t-Δt, respectively .

After FLF, the value of δ(t) can be improved. Therefore, we can set the expected value δ_{exp} after FLF so that the T can be optimized to maintain the maximum fluctuation of PV output power less than δ_{exp}.

So it is important to determine the value of δ_{exp}. When the value is large, the effect of power smoothing will be bad. In contrast, the smaller the δ_{exp} is, the larger the capacity of BES is. Thus, in order to choose the proper value of δ_{exp}, we use the probability method in this paper, , which is described as follows.

Firstly, we analyze the annual data of PV output power with time interval of 5 minutes and calculate the power fluctuation data by (5). As a result, the frequency distribution histogram of PV fluctuation rate can be obtained, which is shown in Fig.2.

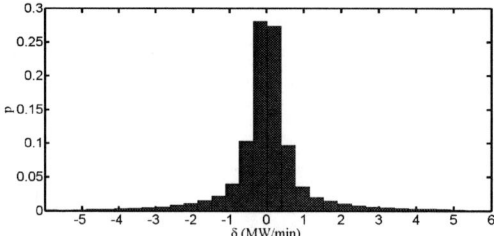

Fig. 2.Frequency distribution histogram of PV fluctuation rate

The confidence can be applied here to determine the value of δ_{exp}. For example, if we choose 85% as the confidence, we can get two values of δ: -0.706 (MW/min) and 0.615 (MW/min). Then we compare the absolute value of the two values and choose the 0.706 (MW/min) as the value of δ_{exp}. That is to say, a value of confidence corresponds to a value of δ_{exp}. It is reasonable for the confidence to be in the interval of [0.85, 0.95], so the proper interval of δ_{exp} is [0.706, 1.464] (MW/min).

For a practical PVs, its $\delta(t)$ can be controlled smaller than δ_{exp} by selecting proper T in (1) because T is the key parameter to calculate the reference power of BES. The value of T can be obtained when δ_{exp} is determined.

In order to get the value of T, we adopt the back tracking method of discrete variable, which is shown as follows:

Step 1: Choosing an initial value of T, such as 0, calculate $P_{rpv}(t)$ by (3).

Step 2: Calculating the maximum value of $\delta(t)$.

Step 3: Making a judgment. If (6) is satisfied, the value of T will be obtained. Otherwise, let $T=T+0.1$ and go back to the step1.

As a result, the feasible domain of T is [6.7,11.7].

$$\max[\delta(t)] = k \cdot \delta_{exp} \qquad 0 < k < 1 \qquad (6)$$

b. Peak shaving and valley filling

For peak shaving and valley filling, an upper and lower bound method used in this paper to decide the optimal size of BES.

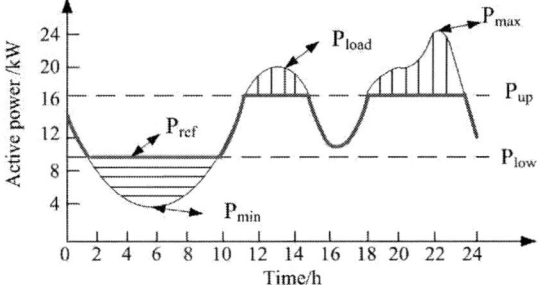

Fig. 3. Upper and lower bound method

The principle of peak shaving and valley filling is shown in Fig.3. When load value is larger than the upper boundary P_{up}, BES will be discharged meanwhile when load value is smaller than P_{low}, BES will be charged. P_{up} and P_{low} can be calculated by formula (7) and (8)

$$P_{up} = (1 + a\frac{\Delta P}{P_{av}})P_{av} \qquad (7)$$

$$P_{low} = (1 - a\frac{\Delta P}{P_{av}})P_{av}$$

$$\beta = \frac{P_{max} - P_{min}}{P_{max}} \times 100\% \qquad (8)$$

Where, P_{up} and P_{low} are the upper and lower bound of load curve. P_{max} and P_{min} are the maximum and minimum value of loads, P_{av} is the average value of the daily load. a is a coefficient set by users. β is the difference of peak and valley.

In order to get the value of a, we can also use the back tracking method of discrete variable.

Step 1: Choosing an initial value of a, such as 1, calculate β by (8).

Step 2: Make a judgment. If (9) is satisfied, the value of a will be obtained. Otherwise, let a=a-0.01 and go back to the first step.

$$\beta = l \cdot \beta_{exp} \qquad 0 < l < 1 \qquad (9)$$

Where β_{exp} is the expected value of β. Considering the effect of peak shaving and valley filling, the reasonable interval of β_{exp} is [30%,40%] in existing literatures. Therefore, we can get the feasible domain of a: [0.16~0.19].

As it is stated above, BES is used to compensate the shadow area in Fig.3. The output power of BES can be calculated in formula (10).

$$P_b(t) = P_{load}(t) - P_{ref}(t) \qquad (10)$$

Where $P_{load}(t)$, $P_{ref}(t)$ and $P_b(t)$ are the load value, the expected load value and the output power of BES at time t.

c. Multi-objectives scenario

For smoothing power fluctuation of PVs, BES will be in an idle state at night, which will be a waste of BES. Therefore, to make better use of BES, the objective of smoothing output power fluctuation of PVs is combined with that of peak shaving and valley filling. In order to accomplish the multi-objective control of BES, one day is divided into 3 periods by the proposed method with corresponding control objectives as follows:

Period 1 (0:00-5:00): BES is used for valley filling;

Period2 (5:00-19:00): BES is used for smoothing power fluctuation of PV and peak shaving and valley filling;

Period 3 (19:00-23:55): BES is used for peak shaving.

The flow chart for calculating BES capacity of multi-objective method is shown in Fig.4. The data of PV output power and load value with the time interval of 5 minutes need to be input.

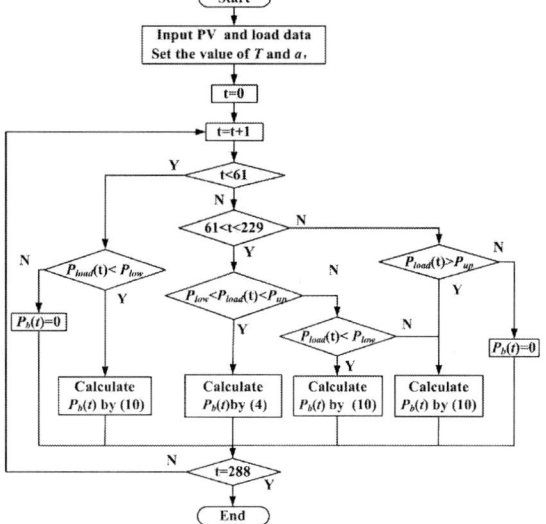

Fig. 4. Flow chart of BES power and capacity calculation

In Fig.4, the data of PVs and load for a day are divided into 288 data points with the time interval of 5minutes. $P_{Load}(t)$ and $P_b(t)$ are the load value and output power of BES at current time t, respectively.

2507

The capacity of BES can be calculated as follows:

$$P_{BES} = \max |P_b(t)| \qquad (11)$$

$$C_d = \sum_{t=1}^{288} P_b(t) \cdot \Delta t \qquad P_b(t) > 0 \qquad (12)$$

$$C_c = -\sum_{t=1}^{288} P_b(t) \cdot \Delta t \qquad P_b(t) < 0$$

$$C_{BES} = \frac{\max\{C_d, C_c\}}{SOC_{max} - SOC_{min}} \qquad (13)$$

Where P_{BES} is the power capacity of BES; C_{BES} is the energy capacity of BES; SOC_{max} is the maximum value of SOC; SOC_{min} is the minimum value of SOC.

B. Economic optimization of BES capacity

In section A, we discuss the control objectives and corresponding control strategies for mutil-objective, and feasible regions of T and a have been given.

In this section, we discuss the economic model of BES, which has the objective function of maximum net income, and calculate the optimal values of T and a.

When the control strategy of BES is determined, the location of the BES has little influence on the effect of peak shaving and valley filling. However, the network loss caused by the system operation is different when the location of BES changes. Therefore, the optimal location of BES needs to be considered.

The economic model of BES is expressed as (14):

$$\max f(T, a, N_i) = (C_s + C_{kw} + C_{td} + C_{loss} - \frac{C_{AC}}{365}) \qquad (14)$$

s.t.

$$V_{min} \le V_i \le V_{max} \qquad (i = 0, 1, 2, \dots, N) \qquad (15)$$

$$\begin{cases} P_{G,i} - P_{L,i} = \sum_{j=1}^{n} V_i V_j \left(G_{ij} \cos\theta_{ij} + B_{ij} \sin\theta_{ij} \right) \\ Q_{G,i} - Q_{L,i} = \sum_{j=1}^{n} V_i V_j \left(G_{ij} \cos\theta_{ij} - B_{ij} \sin\theta_{ij} \right) \end{cases} \qquad (16)$$

Where T, a and N_i are the optimization variables, N_i is the BES location Node i, $f(T,a,N_i)$ refers to the daily net income of BES, C_s, C_{kw}, C_{td} and C_{loss} are benefits of energy arbitrage, transmission support, reducing power cut loss and reducing network loss, C_{AC} is the annual cost of BES. V_{min} and V_{max} are the lower and upper limit of voltage value. V_i is the voltage of node i. $P_{G,i}$ and $Q_{G,i}$ are active and reactive power of power source in node i, $P_{L,i}$ and $Q_{L,i}$ are active and reactive power of load in node i. G_{ij} and B_{ij} are real part and imaginary part of admittance between node i and node j. θ_{ij} is the voltage phase difference between node i and node j.

The node voltage constraints are shown in (15), and power flow constraints are shown in (16).

Before establishing the economic model, we first analyze the costs and benefits of BES.

1) Costs of BES

The costs of BES include initial investment cost C_{cs}, operation and maintenance cost C_{om}, the costs is shown as (17).

$$C_{cs} = C_e E + C_P P \qquad (17)$$
$$C_{om} = C_o P + C_m P$$

Where C_e and C_p are capacity construction cost and power construction cost of BES. E and P are the energy capacity and power capacity of BES. C_o and C_m are operation cost and maintenance cost of BES.

The annual cost of BES can be expressed as (18)

$$C_{AC} = C_{cs} \frac{i(1+i)^n}{(1+i)^n - 1} + C_{om} \qquad (18)$$

Where i is the discount rate, we consider it 5%, while n is the life of BES.

2) Benefits of BES

a. Benefit of Energy Arbitrage C_s

BES monitors local electricity prices to store energy when the price is low to be utilized when electricity prices are high. This is commonly referred to as arbitrage. The benefit of energy arbitrage is shown in (19).

$$C_{dis} = \sum_{t=1}^{288} m(t) \cdot P_b(t) \cdot \Delta t \qquad P_b(t) > 0$$

$$C_{ch} = -\sum_{t=1}^{288} m(t) \cdot P_b(t) \cdot \Delta t \qquad P_b(t) < 0 \qquad (19)$$

$$C_s = C_{dis} - C_{ch}$$

Where $P_b(t)$ is the power of BES at time t. $m(t)$ is the electricity price at time t, Δt is 5 minutes.

b. Benefit of Transmission Support C_{kw}

BES may provide services to ease congestion on electricity transmission networks by storing energy during heavy transmission periods to be released during less congested periods. The use of this service can prolong the life of infrastructure and defers system upgrades. The benefit of prolong the life of infrastructure is shown in (20).

$$C_{kw} = W \cdot \theta \cdot C_k \qquad (20)$$

Where W is the peak load shifting value. θ is the unit capacity investment cost of infrastructure. C_k is uniform annual value coefficient of infrastructure.

c. Benefit of reducing power cut loss C_{td}

BES actively monitors the output power of PV to eliminate unexpected fluctuations, so that reduce the loss of abandoned PV power, the benefit of reducing power cut loss is shown in (21)

$$C_{td} = E \cdot A_s \cdot R_{IEA} \qquad (21)$$

Where A_s is the average blackout rate. R_{IEA} is evaluation rate of user's loss.

d. Benefit of reducing network loss C_{loss}

$$C_{loss} = m_t \sum_{t=1}^{288} \sum_{l=1}^{M} [P_{E,l}(t, N_i) - P_{B,l}(t, N_i)] \cdot \Delta t \qquad (22)$$

Where $P_{B,l}(t,N_i)$ and $P_{E,l}(t,N_i)$ are injected active power and output active power of line l in time t and BES located in node i, $P_{loss}(t)$ is the network loss at time t. m_t is the electricity price bought from power grid.

The 2018 International Power Electronics Conference

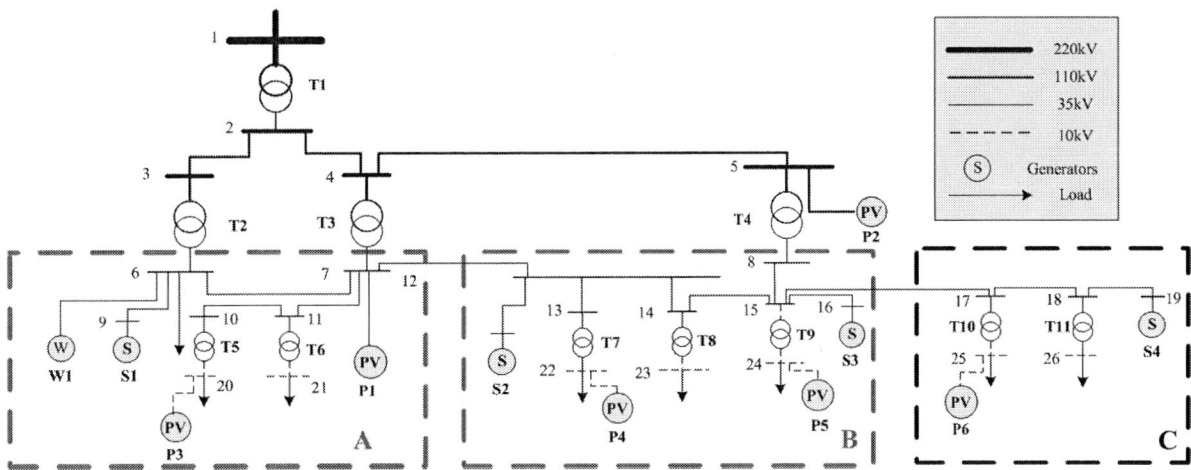

Fig.5. Real distribution system of 35kV with 190MW of PVs in total in China

3) c. SOC constraints of BES

$$SOC_{min} \leq SOC(t) \leq SOC_{max} \qquad (23)$$

Where SOC_{min} and SOC_{max} are the lower and upper limit of SOC value. $SOC(t)$ is the SOC value at time t.

d. Power constraints of BES

$$|P_b(t)| < P \qquad (24)$$

In this paper, the data fitting toolbox in matlab is used to solve the economic model of BES.

C. Factors influencing BES configuration

When the mutil-objective is determined, the main factor that may affect the output characteristics of the BES is the location of BES. The location of BES will significantly affect the grid loss. The variables are listed as follows.

1. Integrated voltage level of BES.
2. Electrical distance between BES and source nodes.
3. Electrical distance between BES and load nodes.

III. SIMULATION

In this paper, we use a real distribution system of 35kV with 60MW of PVs in total in China to test the proposed method by simulation, the power system is shown in Fig.5.

A. Simulation cases

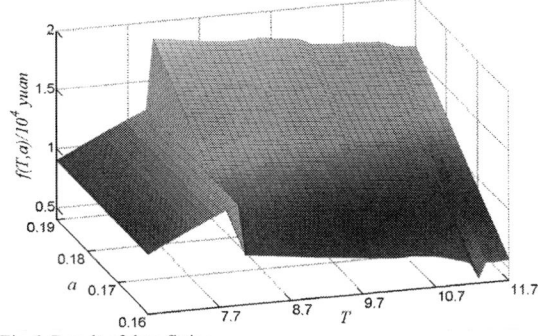

Fig.6. Result of data fitting

The economic model of BES is solved by data fitting toolbox in Matlab. The result of data fitting shown in Fig.6 illustrates the relationship between the objective function and the variables. The daily net income $f(T,a)$ will take the maximum value of 18540 *yuan/day* when the values of T and a refer to 8.054 and 0.19. According to equations (2)-(5), the fluctuation rate can be calculated, δ_{max} is 0.847 (MW/min), the confidence is 0.88.

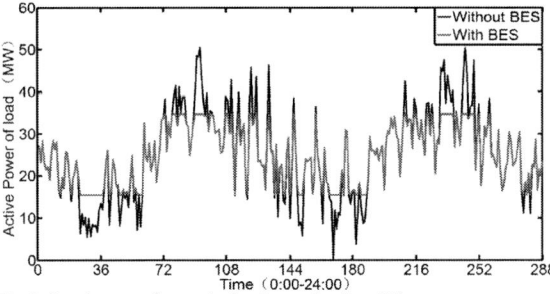

Fig. 7. Power of PV with BES and without BES

In Fig.7, each time interval is 5 minutes, black line refers to the output power curve of the PV while red line refers to the combined output power of PV and BES. From Fig.7 we can see that the participation of BES makes the curve much smoother than before. The maximal power fluctuation is reduced from 11.42MW to 4.49MW.

Fig. 8. Load curve after peak shaving and valley filling

In Fig.8, each time interval is 5 minutes, black line refers load curve without BES while red line refers to the load curve with BES. It can be seen that BES can also

2509

accomplish the function of peak shaving and valley filling. The peak valley difference has been reduced from 50.46MW to 19.17MW.

TABLE I
RESULTS OF BES CONFIGURATION

Objective	Capacity
1.smoothing power fluctuation of PV	7MW/13MWh
2.peak shaving and valley filling	16/29MWh
3.Multi-objective of 1 and 2	16/32MWh

The capacity of BES calculated is listed in Table I. We can find that the capacity of BES with multi-objective method is 32MWh, which is much smaller than the results of two single objectives (13+29=42Mwh).

What's more, the cost of BES configuration is listed in Table II .

TABLE II
COSTS OF BES CONFIGURATION

Objective	C_{CS}(MILLION YUAN)	C_{OM} (MILLION YUAN)	Total cost(MILLION YUAN)
1	11.30	1.952	13.26
2	25.39	4.271	29.66
3	26.96	4.271	31.23

We can find that the annual cost of BES with multi-objective method is 31.23 million yuan, which is much smaller than the results of two single objective (13.26+29.66=42.92 million yuan). The results show that the cost of BES will be much cheaper with the multi-objective method.

Fig.9, Fig.10 and Fig.11 show the SOC of BES for different objectives. We can see that the SOC of BES can be controlled in the range of [0.1, 0.95] set in this paper, which verifies the feasibility of the proposed BES control method.

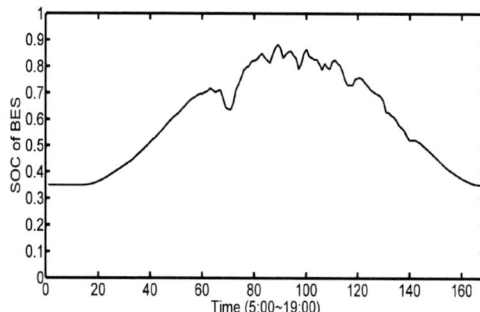

Fig.9. SOC of BES for smoothing power fluctuation of PV

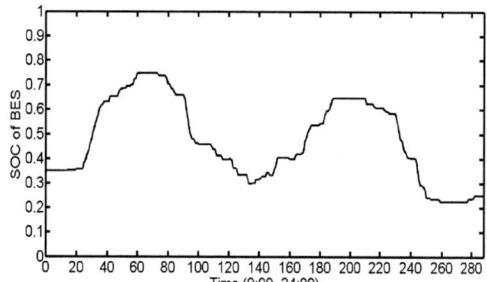

Fig.10. SOC of BES for peak shaving and valley filling

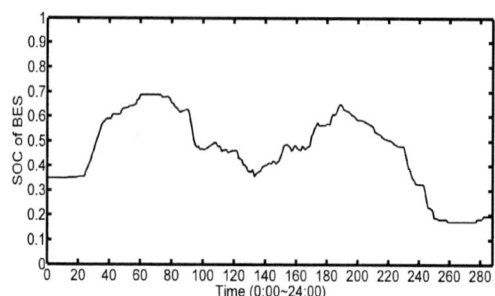

Fig.11. SOC of BES for multi-objective

The active power of BES for different objectives in daily operation is shown in Fig.12 to Fig14. It can be seen that the BES for multi-objective has the highest utilization rate.

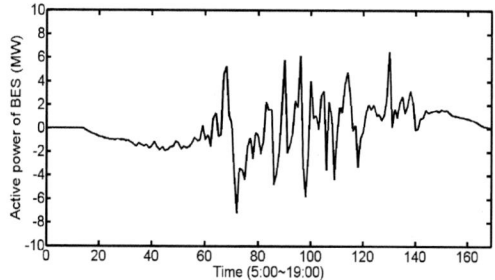

Fig.12. Active power of BES for smoothing power fluctuation of PV

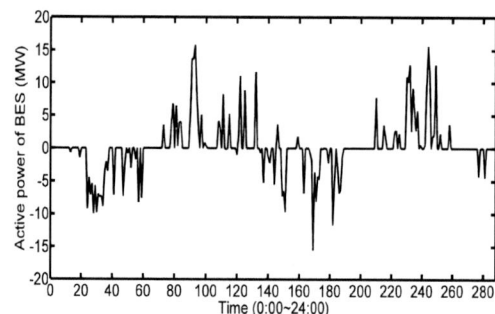

Fig.13. Active power of BES for peak shaving and valley filling

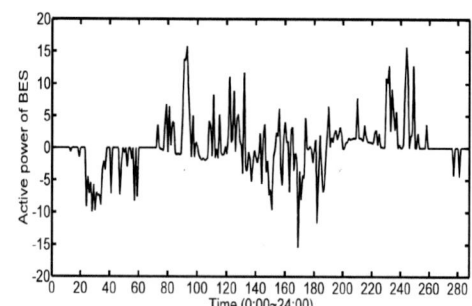

Fig.14. Active power of BES for multi-objective

We can calculate the utilization rate of BES in daily operation for each objective by means of the figures. The calculation results are shown in Table III.

TABLE III
UTILIZATION RATE OF BES

Objective	Action Times of BES	Utilization rate of BES
1	156/288	54.17%
2	105/288	36.46%
3	203/288	70.49%

As it can be seen in Fig.5, the network is divided into 3 areas (A, B and C). Of all the areas, the load value in Area B is the largest while that in Area C is the smallest.

To find the influence to network loss of the location of BES, we connect BES to different nodes in different areas. The results can be seen in Table IV.

TABLE IV
NETWORK LOSS OF DIFFERENT SITUATIONS

Upper value of peak	BES node location in A	BES node location in B	BES node location in C	Network loss (MWh)
	3	4	5	114.885
88% of load peak	6	12	17	115.854
	21	23	26	107.257

From the results of network loss, it can be seen that when the Upper value of peak is lower, the storage capacity is larger and the network loss is smaller. In several cases of discussion, the value of network loss is relatively small when the energy storage is connected to the load side and lower voltage level.

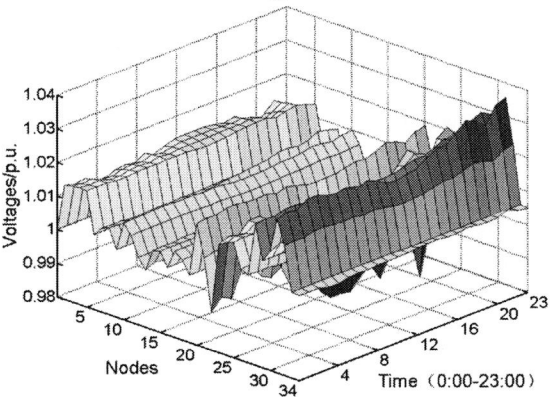

Fig. 15. Node voltage at different time

From Fig.15 we can see that the node voltage can be kept within the limits when BES is applied in mutil-objective.

IV. CONCLUSION

This paper presents a multi-objective optimal design and control method of BES used in distribution system with high penetration of distributed PV generators. The simulation results show that the annual cost of BES in multi-objectives is 31.23 million yuan, and the sum annual cost of two BESs in single objective is 42.92 million yuan, so the multi-objectives control reduces the annual cost by 11.69 million yuan, the utilization rate of BES increased from 54.17% and 36.46% to 70.49%. The network loss simulations show that the value of network

loss is relatively small when the energy storage is connected to the load side and lower voltage level, and the voltage of key nodes can be kept within the limits.

ACKNOWLEDGMENT

This work was supported in part by the National Natural Science Foundation of China under Grant 51577047, in part by the International Science & Technology Cooperation Project of Anhui Province under Grant 1604b0602015.

REFERENCES

[1] China National Renewable Energy central, "China renewable energy industry development report 2017," J. China economic publication house., in press.

[2] J. Zhao, C. Wan, Z. Xu and J. Li, "Impacts of Large-scale Photovoltaic Generation Penetration on Power System Spinning Reserve Allocation," *2016 IEEE Power and Energy Society General Meeting (PESGM)*, 2016, pp. 1-5.

[3] H. Xie, H. Chen, W. Yan, W. Gao, Q. Sun and J. Ma, "Application of Energy Storage in High Penetration Renewable Energy System," *2017 IEEE International Conference on Electro Information Technology (EIT)*, 2017, pp. 188-193.

[4] N. Yan, W. Li and Z. Xing, "Capacity Allocation Method in Active Distribution Network Based on Hybrid Energy Storage," *Transactions of China Electrotechnical Society*, vol. 32, no.19, pp. 180-186, 2017.

[5] Q. Jiang and H. Hong, "Wavelet-Based Capacity Configuration and Coordinated Control of Hybrid Energy Storage System for Smoothing Out Wind Power Fluctuations," *IEEE Trans. on Power Systems*, vol. 28, no. 2, pp. 1363-1372, 2013.

[6] Y. Sun, X. Tang and X. Sun, "Research on Energy Storage Capacity Allocation Method for Smoothing Wind Power Fluctuations," *Proceedings of the CSEE*, vol. 37, Supplement, pp. 88-97, 2017.

[7] C. Wang, Z. Liang, J. Liang, et al. "Modeling the temporal correlation of hourly day-ahead short-term wind power forecast error for optimal sizing energy storage system," *International Journal of Electrical Power and Energy Systems*, vol. 98, pp. 373-381, 2018.

[8] X. Li, T, Zhou and J. Huang, "The hybrid energy storage system capacity configuration in tracking wind power project output," *Acta Energiae Solaris Sinica*, vol. 37, no.9, pp. 2194-2200, 2016.

[9] X. Li, G. Geng, Y. J, et al. "Study on optimal allocation of battery energy storage in distribution network considering the actual operation," *Power System Protection and Control*, vol. 45, no.9, pp. 88-94, 2017.

[10] X. Li, J. Huang, Y. Chen, et al. "Review on large-scale involvement of energy storage in power grid fast frequency regulation," *Power System Protection and Control*, vol. 44, no.7, pp. 145-153, 2016.

[11] Y. Wang, K. Tan, X. Peng, et al. "Coordinated control of distributed energy-storage systems for voltage regulation in distribution networks," *IEEE Transactions on Power Delivery*, vol. 31, no.3, pp. 1132-1141, 2016.

[12] M. Liu, C. Wang, L. Guo, et al. "An optimal design method of mutil-objective based island microgrid," *Automation of Electric Power Systems*, vol. 36, no.17, pp. 34-39, 2012.

[13] X. Tan, H. Wang, L. Zhang, et al. "Mutil-objective optimization of Hybrid energy storage and assessment indices in microgrid," *Automation of Electric Power Systems*, vol. 38, no.8, pp. 7-14, 2014.

The 2018 International Power Electronics Conference

Mission Profile-Oriented Control for Reliability and Lifetime of Photovoltaic Inverters

Ariya Sangwongwanich, Yongheng Yang, Dezso Sera, and Frede Blaabjerg
Department of Energy Technology, Aalborg University, Aalborg DK-9220, Denmark
ars@et.aau.dk, yoy@et.aau.dk, des@et.aau.dk, fbl@et.aau.dk

Abstract—With the aim to increase the competitiveness of solar energy, the high reliability of Photovoltaic (PV) inverters is demanded. For PV applications, the inverter reliability and lifetime are strongly affected by the operating condition that is referred to as the mission profile (i.e., solar irradiance and ambient temperature). Since the mission profile of PV systems is location-dependent, the inverter reliability performance and lifetime expectation can vary accordingly. That is, from the reliability perspective, PV inverters with the same design metrics (e.g., component selection) may be over- or under-designed under different mission profiles. This will increase the overall system cost, e.g., initial cost for over-designed cases and maintenance cost for under-designed cases, which should be avoided. This paper thus explores the possibility to adapt the control strategies of PV inverters to the corresponding mission profiles. With this, similar reliability targets (e.g., component lifetime) can be achieved even under different mission profiles. Case studies have been carried out on PV systems installed in Denmark and Arizona, where the lifetime and the energy yield are evaluated. The results reveal that the inverter reliability can be improved by selecting a proper control strategy according to the mission profile.

Index Terms—PV inverters, lifetime, reliability, mission profile, control, power device, capacitor.

I. INTRODUCTION

There is a strong demand to further reduce the cost of PV energy, in order to increase its competitiveness and enable more renewable energy harvesting [1]. For instance, the U.S. Department of Energy has set a target to reduce the cost of PV energy from 0.18 USD/kWh (in 2016) to 0.05 USD/kWh by 2030 (for residential PV systems in the USA) [2]. The similar cost reduction tendency is also expected in other countries globally [3]–[5]. In order to achieve this target, PV systems should be improved in several aspects. Among those, enhancing the reliability and lifetime of PV inverters has high potential for a significant cost reduction [5]. The field experience has shown that the PV inverter failure contributes to a large portion of the unexpected operating and maintenance cost [6]–[8]. This gives a negative impact to the overall cost of energy in addition to the energy production loss during the inverter down-time periods. Thus, avoiding PV inverter replacements during the entire lifespan of PV power plants (e.g., 20 years) is one of the keys to the cost reduction [2].

Accordingly, the reliability engineering approach has recently been more involved in the design phase of PV inverters (in general, power electronic systems) [9]–[12]. This is normally referred to as a Design for Reliability (DfR) approach, as it is illustrated in Fig. 1. Following the DfR approach, the

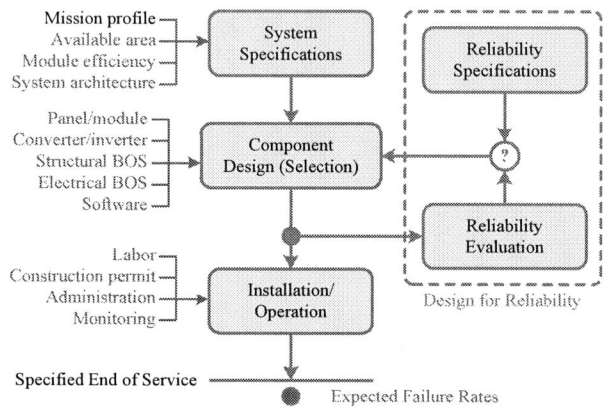

Fig. 1. Flow diagram of the Design for Reliability (DfR) approach applied to the design of power electronics in PV systems (BOS: Balance of System).

reliability specification (e.g., the lifetime target) is defined and it should be fulfilled during the design phase. In that respect, the lifetime prediction tool plays an important role in assessing the reliability of the designed inverter under given operating conditions (e.g., the mission profile of the installation site).

In the prior-art research, it is suggested that the reliability and lifetime of power electronic systems (e.g., PV inverters) are strongly affected by the operating conditions [13]–[18], referred to as mission profiles. Thus, the mission profile is usually required as an input of the DfR process. For PV applications, the solar irradiance and the ambient temperature are normally considered as the components of a mission profile, as they determine the PV power production (i.e., the PV inverter loading). Since the solar irradiance and ambient temperature are location-dependent (due to the climate condition of the installation site), the mission profile can vary significantly, and thus the reliability of PV inverters [15]–[18]. From the design perspective, this is a challenge for the DfR approach, where the concept of "one design fits all" is difficult to be achieved. For instance, if the PV inverter is designed to achieve the lifetime of 20 years under cold climate conditions (e.g., low average solar irradiance level), there is a high risk that the same inverter design (e.g., component selection and cooling system design) will not fulfill the reliability target when it is installed in a hot climate region (e.g., high average solar irradiance level).

2512

On the other hand, the PV inverter designed with respect to the hot climate condition with strong average solar irradiance and high ambient temperature will be considered as an over-designed case for other installation sites with cold climate conditions. This is not preferable in the DfR concept, as it will increase the overall system cost, e.g., initial cost for over-designed cases and maintenance cost for under-designed cases. Moreover, applying different inverter designs according to installation sites is impractical with respect to the cost.

Actually, the inverter control strategies can affect the reliability and lifetime performances in addition to the mission profile. For instance, PV power variations (reflecting mission profile characteristics) induce thermal fluctuations on the inverters. Hence, limiting the maximum feed-in power can smooth the temperature variations and lower the thermal loading to some extent [19]–[21]. This contributes to improved lifetime, which can also be seen in smart de-rating control strategies [22]. Furthermore, at the switching timescale, the thermal loading of the PV inverters can be regulated [23]. This opens a direction to enhance the reliability and lifetime of PV inverters through a proper control, where the mission profiles are considered.

In light of the above, a Power Limiting Control (PLC) scheme is employed in this paper to enhance the PV inverter reliability, where mission profiles are also considered. The proposed strategy is applied to 6-kW single-phase PV inverters. In § III, the lifetime evaluation of PV inverters is presented, where two mission profiles in Denmark and Arizona are used. The results in § IV demonstrate that the same reliability target (e.g., the lifetime target of 20 years) can be achieved under both mission profiles with the proposed control strategy. Finally, concluding remarks are provided in § V.

II. SINGLE-PHASE GRID-CONNECTED PV INVERTERS

A. System Description

The system configuration and control structure of a single-phase grid-connected PV systems are shown in Fig. 2 and its parameters are given in Table I. Here, a DC-DC converter is employed to step up the PV array voltage v_{pv} to match the minimum required DC-link voltage and also provide the control of PV power extraction [24]. This is normally achieved through the regulation of the PV voltage, whose reference (v_{pv}^*) is determined by a Maximum Power Point Tracking (MPPT) algorithm. However, the PLC strategy can also be implemented in the control of the DC-DC converter, instead of the MPPT algorithm, to limit the PV power extraction to a certain level (below the maximum available power) [25], [26]. The extracted power is then delivered to a full-bridge DC-AC inverter (PV inverter), which provides grid-integration control (i.e., current control, grid synchronization) [27].

Regarding the power components, IGBT devices from [28] are used. The cooling system (e.g., heat sink sizing) is designed to ensure that the power device maximum junction temperature is 100 °C at 120% of the rated power (i.e., 7.2 kW). The dc-link is realized by connecting two electrolytic capacitors (2200 μF/350 V) from [29] in series.

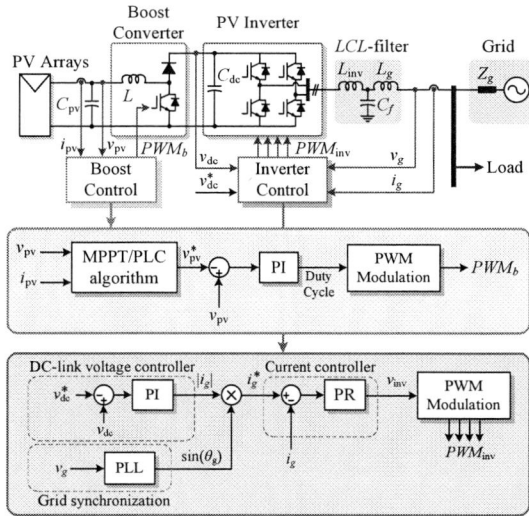

Fig. 2. System configuration and control structure of a two-stage single-phase grid-connected PV system (MPPT: Maximum Power Point Tracking, PLC: Power Limiting Control, PI: Proportional Integral, PR: Proportional Resonant, PLL: Phase-Locked Loop, PWM: Pulse Width Modulation).

TABLE I
PARAMETERS OF THE TWO-STAGE SINGLE-PHASE PV SYSTEM (FIG. 2).

PV inverter rated power	6 kW
Boost converter inductor	$L = 1.8$ mH
DC-link total capacitance	$C_{dc} = 1100$ μF
LCL-filter	$L_{inv} = 4.8$ mH, $L_g = 2$ mH, $C_f = 4.3$ μF
Switching frequencies	Boost converter: $f_b = 16$ kHz, Full-Bridge inverter: $f_{inv} = 8$ kHz
DC-link reference voltage	$v_{dc}^* = 450$ V
Grid nominal voltage (RMS)	$V_g = 230$ V
Grid nominal frequency	$\omega_0 = 2\pi \times 50$ rad/s

B. Power-Limiting Control (PLC) Strategy

Instead of always tracking the Maximum Power Point (MPP), the PV output power P_{pv} can be limited at a certain level P_{limit} below the available PV power P_{avai}. This operation can be achieved by regulating the operating PV voltage below the MPP, as it is demonstrated in Fig. 3. This is called power-limiting control in the literature, which is normally required when the available PV power becomes higher than the PV inverter rated power P_{rated} [25]. This situation usually occurs in the PV system with an over-sized PV array (i.e., the PV array is intentionally designed to have higher rated power than the inverter in order to gain more energy yield during the low solar irradiance condition) [30]. Another incident is due to the solar irradiance reflection from the cloud, resulting in the solar irradiance level higher than 1000 W/m². Conventionally, the power-limit level is selected as the inverter rated power to ensure the safety of the inverter [30]. However, it should be pointed out that the PLC strategy is capable of flexibly

The 2018 International Power Electronics Conference

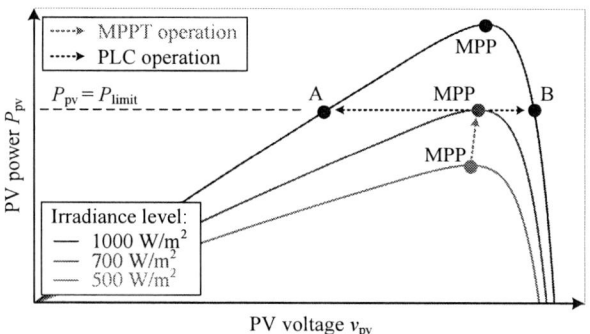

Fig. 3. Operational principle of the PV system with the Power-Limiting Control (PLC) strategy, e.g., operating point of the PV array is regulated at A or B in order to limit the extracted PV power at $P_{pv} = P_{limit}$.

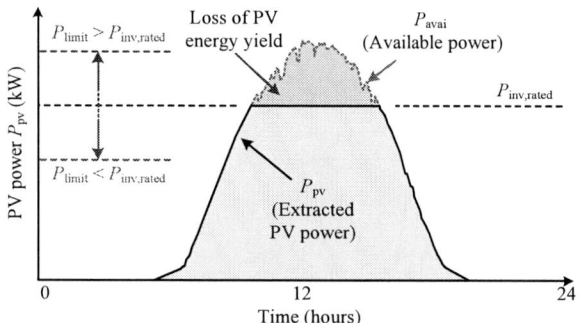

Fig. 4. PV power extraction with the Power-Limiting Control (PLC) strategy (P_{avai}: available PV power, P_{pv}: extracted PV power, P_{limit}: power-limit level, $P_{inv,rated}$: PV inverter rated power).

regulating the extracted PV power at any power level below the available power P_{avai} (i.e., $0 \leq P_{pv} < P_{avai}$), as it is illustrated in Fig. 4. This flexible power controllability is suitable to be employed in the mission profile-oriented control strategy, which will be analyzed in this paper. More details regarding the design and implementation of the PLC strategy have been discussed in [25].

III. RELIABILITY ASSESSMENT OF PV INVERTERS

The mission profile is important in the reliability assessment and lifetime prediction of PV inverters. Thus, it is usually considered during the reliability evaluation process as it is illustrated in Fig. 5. From the mission profile, the PV inverter loading (e.g., power loss of the component) is determined from the PV panel model and the control strategy. Then, the power losses are applied to the thermal models of the components (e.g., power device and capacitor) to obtain the thermal loading during the operation, which is required for the lifetime model. This procedure will be discussed in the following and the mission profiles in Denmark and Arizona will be applied. The lifetime of the components in the PV inverter (e.g., power devices and capacitor) will be evaluated, where the 20-year lifetime is selected as a reliability target.

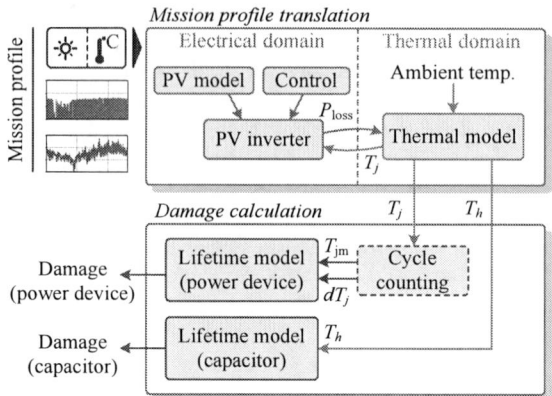

Fig. 5. Reliability assessment of PV inverters based on mission profiles [12].

A. Mission Profiles

The mission profiles recorded in Denmark and Arizona are used in this study, as they are shown in Fig. 6. It can be seen from Fig. 6 that the average solar irradiance level in Arizona is constantly high through the year, while the average solar irradiance level in Denmark is relatively low through November to February. The same trend is applied to the ambient temperature profile. The mission profiles in Denmark and Arizona represent the installation site in a cold and hot climate condition, respectively. It can be expected from the mission profile that the PV power production of the PV system in Arizona will be higher than that in Denmark.

When translating the mission profile into the inverter loading (following Fig. 5), it can be expected that the PV inverter installed in Arizona will experience higher loading during the operation. In that case, the reliability-critical components in the system (e.g., power devices and capacitor) will be subjected to higher thermal stresses than those installed in Denmark. Consequently, the reliability and lifetime of the PV inverter under the two installation sites can differ considerably, which will be demonstrated in the following.

B. Damage Calculation

For the reliability-critical components in the PV inverter such as power devices and capacitors, the main cause of component wear-out failures is related to the thermal stress. In the case of power devices (e.g., IGBT), the thermal cycling is one of the main stress factors that cause bond-wire lift-off and solder fatigue after a number of thermal cycles, which can be determined from the lifetime model as

$$
N_f = A \times (\Delta T_j)^{\alpha} \times (ar)^{\beta_1 \Delta T_j + \beta_0} \times \left[\frac{C + (t_{on})^{\gamma}}{C + 1} \right] \\
\times \exp\left(\frac{E_a}{k_b \times T_{jm}} \right) \times f_d
$$

(1)

where N_f is the number of cycles to failure [31]. In (1), the thermal cycle amplitude ΔT_j, the mean junction temperature T_{jm}, and cycle period t_{on} are the stress levels obtained from the cycle counting algorithm, while the lifetime model parameters are given in Table II.

2514

The 2018 International Power Electronics Conference

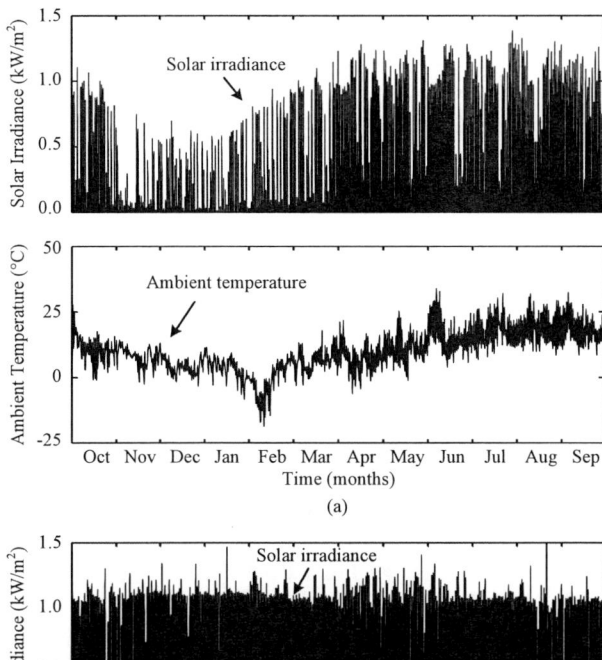

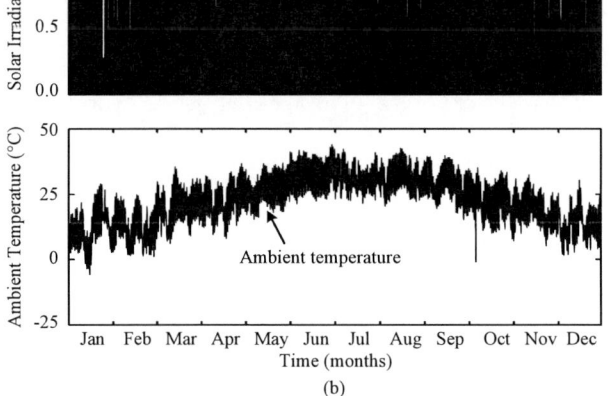

Fig. 6. Yearly mission profiles (i.e., irradiance and ambient temperature with a sampling rate of 5 mins per sample) in: (a) Denmark and (b) Arizona.

Normally, it is assumed that the contribution of each thermal cycle to the failure of power device is accumulated linearly and independently during operation following the Miner's rule as

$$AD = \sum_i \frac{n_i}{N_{fi}} \qquad (2)$$

where n_i is the number of cycles at a certain stress level (T_{jm}, ΔT_j, and t_{on}), and N_{fi} is the number of cycles to failure calculated from (1) at that stress condition. Here, AD is the accumulated damage of the power device during operation. When the damage is accumulated to unity (i.e., $AD = 1$), the power device is considered to reach its end-of-life.

The DC-link capacitor is another lifetime-limiting component in the PV inverter, where the hotspot temperature T_h is

TABLE II
PARAMETERS OF THE LIFETIME MODEL OF AN IGBT MODULE [31].

Parameter	Value	Experimental condition
A	3.4368×10^{14}	
α	-4.923	$64 \text{ K} \leq \Delta T_j \leq 113 \text{ K}$
β_1	-9.012×10^{-3}	
β_0	1.942	$0.19 \leq ar \leq 0.42$
C	1.434	
γ	-1.208	$0.07 \text{ s} \leq t_{on} \leq 63 \text{ s}$
f_d	0.6204	
E_a	0.06606 eV	$32.5\,°\text{C} \leq T_j \leq 122\,°\text{C}$
k_B	$8.6173324 \times 10^{-5} \text{ eV/K}$	

TABLE III
PARAMETERS OF THE LIFETIME MODEL OF A CAPACITOR [29].

Parameter	Symbol	Value
Rated lifetime (at V_{rated} and T_m)	L_m	3000 hours
Rated operating voltage	V_{rated}	350 V
Rated operating temperature	T_m	105°C

the main stress parameter. The lifetime model of the capacitor (e.g., aluminum electrolytic capacitor) is given as

$$L_f = L_m \times \left(4.3 - 3.3\frac{V_{\text{op}}}{V_{\text{rated}}}\right) \times 2^{\left(\frac{T_m - T_h}{10}\right)} \qquad (3)$$

in which L_f is the time-to-failure under the thermal stress level of T_h and the voltage stress level of V_{op} [32], and the other parameters are given in Table III [29].

Then, the Miner's rule can also be applied to the lifetime calculation of the capacitor as

$$AD = \sum_i \frac{l_i}{L_{fi}} \qquad (4)$$

where l_i is the operating time for a set of T_h and V_{op} (e.g., the mission profile time resolution) and L_{fi} is the time-to-failure calculated from (3) at that specific stress condition.

C. Case Study

Following the reliability assessment method in Fig. 5, the damage occurred in the power device and capacitor during the operation can be calculated and used as a reliability metric. For instance, the operation with high accumulated damage indicates low reliability and a high failure rate of the component. In this case study, the MPPT operation is applied to demonstrate the mission profile-dependency of the PV inverter reliability. Notably, for the installation site in Denmark, the rated installed power of the PV arrays is 8.4 kW, which is 1.4 times higher than the PV inverter rated power. In this case, the PV arrays are over-sized, which is practical for the installation site with relatively low solar irradiance conditions (e.g., Denmark) [30].

By applying the mission profiles in Fig. 6, the corresponding damage of the component in the PV inverter installed in Denmark and Arizona can be obtained, as shown in Fig. 7(a) and (b), respectively. For the mission profile in Denmark, it can be seen in Fig. 7(a) that only small damage occurs in the

2515

The 2018 International Power Electronics Conference

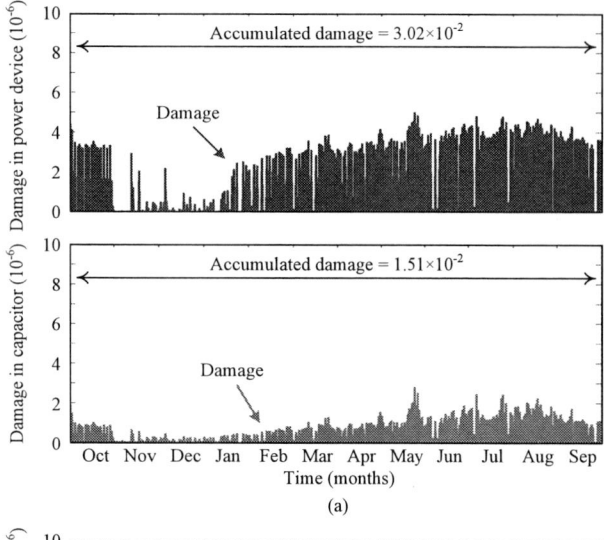

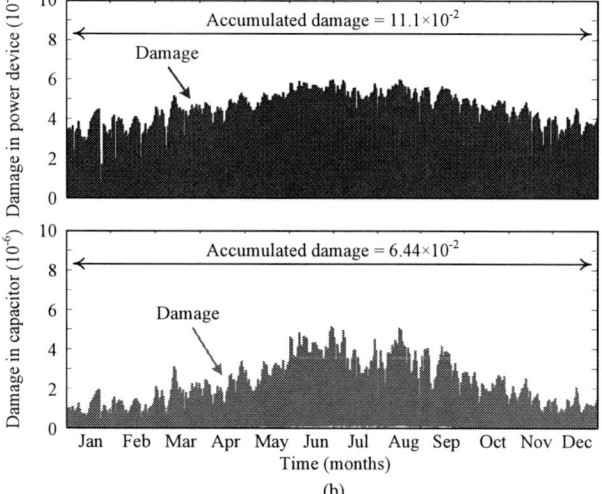

Fig. 7. Damage in the power device and capacitor of the PV inverter under one-year mission profile in: (a) Denmark and (b) Arizona.

power device and capacitor of the inverter during winter (e.g., November to February) due to low solar irradiance conditions. In fact, most of the damage occurs from April to August. The accumulated damage over one year of the power device and capacitor in the PV inverter is $AD = 3.02 \times 10^{-2}$ per year and $AD = 1.51 \times 10^{-2}$ per year, respectively. This corresponds to the component lifetime of 33 years for the power device and 66 years for the capacitor. Accordingly, the reliability target (i.e., the component lifetime of 20 years) is fulfilled with the designed inverter under the mission profile in Denmark.

For the PV inverter installed in Arizona, the damage in the power device and capacitor is relatively high through the entire year, as it is shown in Fig. 7(b), which reflects the mission profile characteristics. In that case, a one-year operation under the Arizona mission profile contributes to the accumulated

damage of $AD = 11.1 \times 10^{-2}$ per year for the power device and $AD = 6.44 \times 10^{-2}$ per year for the capacitor. Thus, the power device is expected to fail after 9 years, while it is 15 years for the capacitor. In this case, the reliability target (i.e., the component lifetime of 20 years) is not fulfilled for the given inverter design.

IV. MISSION PROFILE-ORIENTED CONTROL STRATEGY

As shown previously, the designed PV inverter cannot fulfill the reliability target in the Arizona case, while it is considered to be over-designed when installed in Denmark. In the following, the PLC strategy is applied to reshape the inverter reliability according to the mission profile.

A. Control for Reliability

As discussed in § II-B, the PLC strategy can be employed to flexibly regulate the extracted PV power (i.e., PV inverter loading) during the operation. However, there is always a trade-off between the PV inverter loading improvement and the PV energy yield, which needs to be considered when applying the PLC strategy. For instance, decreasing the power-limit level of the PLC strategy will reduce the peak-load of the PV inverter during the operation. This will certainly be beneficial to the PV inverter reliability, as it will reduce the thermal stress of the components. However, the energy yield will also be reduced due to the power curtailment. On the other hand, more PV energy yield can be gained by increasing the power-limit level, but the PV inverter loading will also increase, which decreases the PV inverter reliability.

B. Lifetime Evaluation

Following the above consideration, the power-limit level should be increased for the PV inverter installed in Denmark, since it is considered to be an over-designed case compared to the lifetime target of 20 years. In that case, more energy yield can be gained with a reduced margin in terms of reliability performance (e.g., lower component lifetime). Notably, the power-limit can be increased up to 120 % of the inverter rated power, following the design in § II in order to ensure that the components still operate within the safe operating area (according to [28] and [29]). The lifetime of the power device and capacitor of the PV inverter installed in Denmark under different power-limit levels are demonstrated in Fig. 8(a). From the result, it can be seen that the power-limit should not be increased to more than 108.5 % of the inverter rated power, which is the case when the lifetime target of 20 years is marginally fulfilled for the power device.

In contrast, the PV inverter in Arizona should operate with a reduced power-limit level to improve the reliability, since the pre-designed inverter cannot achieve the reliability target. The evaluation results in Fig. 8(b) show that the power device lifetime of 20 years can be achieved, if the power-limit level is kept at 87.5 % of the inverter rated power. By further decreasing the power-limit below 87.5 % of the inverter rated power, the component lifetime can be further increased but it will also result in more energy yield loss. This is not preferable from the cost-of-energy point of view.

2516

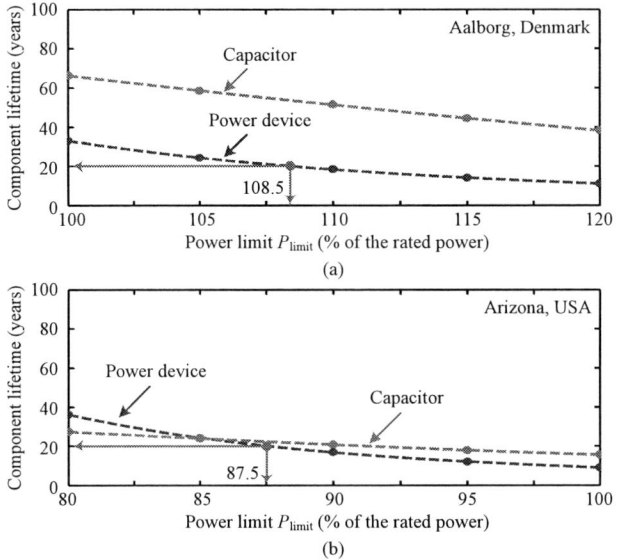

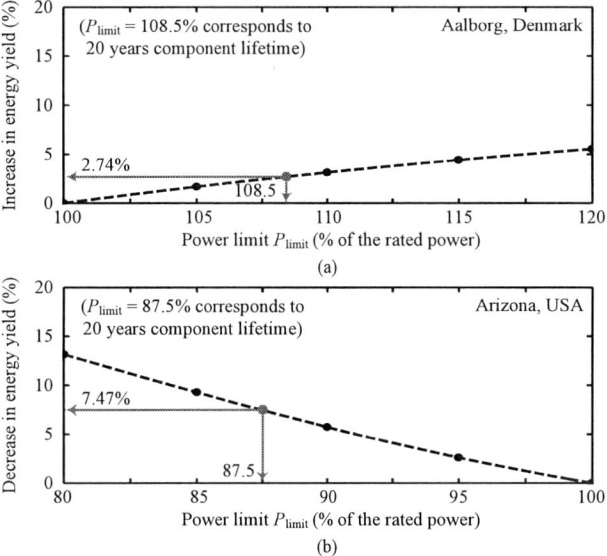

Fig. 8. Lifetime of power device and capacitor with different power-limit levels under one-year mission profile in: (a) Denmark and (b) Arizona.

Fig. 9. Impact of the energy yield of the PV inverter with different power-limit levels under one-year mission profile in: (a) Denmark and (b) Arizona.

C. PV Energy Yield

As a trade-off of the PLC strategy, the energy yield has to be considered together with the reliability improvement. The relative increase/decrease in the PV energy yield (compared to the case with the MPPT operation) with different power-limit levels is evaluated and shown in Fig. 9. For the mission profile in Denmark, more PV energy can be extracted by increasing the power-limit level above the inverter rated power. By increasing the power-limit level to 108.5 % of the inverter rated power (i.e., when the obtained lifetime is 20 years), the energy yield is increased by 2.74 %. For the case of PV inverter installed in Arizona, 7.47 % of the energy yield needs to be sacrificed to achieve a lifetime target of 20 years.

The above results (Figs. 8 and 9) suggest that the PLC strategy can be employed to minimize the overall cost of solar energy concerning the total energy yield together with the operation and maintenance cost (e.g., cost associated with the inverter failure). For instance, the multi-objective optimization problem to minimize the life-cycle cost of the overall PV system should be used for determining the optimal power-limit level for each mission profile.

V. CONCLUSION

In this paper, a mission profile-oriented control strategy for PV inverters has been presented. The proposed control strategy is based on the power-limiting control scheme, which has been adaptively applied according to the mission profile characteristic. A case study of the mission profiles in Denmark and Arizona has been carried out, where the reliability target is specified as the component lifetime of 20 years. For the Denmark case, where the inverter is over-designed, the energy yield can be increased up to 2.74 % by allowing the PV inverter to operate above the rated power. In contrast, the PV inverter installed in Arizona cannot fulfill the lifetime target with the conventional MPPT control, when the same inverter design of the Denmark case is adopted. However, by limiting the feed-in power at 87.5 % of the designed inverter rated power, the power device lifetime can be prolonged to 20 years with the compromise of 7.47 % reduction in the energy yield.

REFERENCES

[1] REN21, "Renewables 2017: Global Status Report (GSR)," 2017. [Online]. Available: http://www.ren21.net/.

[2] National Renewable Energy Laboratory, "On the path to sunshot: The role of advancements in solar photovoltaic efficiency, reliability, and costs," Tech. Rep. No. NREL/TP-6A20-65872, 2016.

[3] Fraunhofer ISE, "Current and future cost of photovoltaics. long-term scenarios for market development, system prices and LCOE of utility-scale PV systems," February, 2015. [Online]. Available: http://www.pv-fakten.de/.

[4] M. Taylor, P. Ralon, and A. Ilas, "The power to change: Solar and wind cost reduction potential to 2025," International Renewable Energy Agency (IRENA), Tech. Rep., Jun. 2016.

[5] KIC InnoEnergy, "Future renewable energy costs: solar photovoltaics," Tech. Rep., 2015.

[6] L. M. Moore and H. N. Post, "Five years of operating experience at a large, utility-scale photovoltaic generating plant," *Progress Photovoltaics: Res. Appl.*, vol. 16, no. 3, pp. 249–259, 2008.

[7] A. Golnas, "PV system reliability: An operator's perspective," *IEEE J. of Photovolt.*, vol. 3, no. 1, pp. 416–421, Jan. 2013.

[8] P. Hacke, S. Lokanath, P. Williams, A. Vasan, P. Sochor, G. TamizhMani, H. Shinohara, and S. Kurtz, "A status review of photovoltaic power conversion equipment reliability, safety, and quality assurance protocols," *Renew. Sustain. Energy Rev.*, vol. 82, pp. 1097–1112, 2018.

[9] H. Wang, M. Liserre, and F. Blaabjerg, "Toward reliable power electronics: Challenges, design tools, and opportunities," *IEEE Ind. Electron. Mag.*, vol. 7, no. 2, pp. 17–26, Jun. 2013.

[10] N. C. Sintamarean, F. Blaabjerg, H. Wang, F. Iannuzzo, and P. de Place Rimmen, "Reliability oriented design tool for the new generation of grid connected pv-inverters," *IEEE Trans. Power Electron.*, vol. 30, no. 5, pp. 2635–2644, May 2015.

[11] K. Ma, H. Wang, and F. Blaabjerg, "New approaches to reliability assessment: Using physics-of-failure for prediction and design in power electronics systems," *IEEE Power Electron. Mag.*, vol. 3, no. 4, pp. 28–41, Dec. 2016.

[12] Y. Yang, A. Sangwongwanich, and F. Blaabjerg, "Design for reliability of power electronics for grid-connected photovoltaic systems," *CPSS Trans. Power Electron. Appl.*, vol. 1, no. 1, pp. 92–103, 2016.

[13] H. Huang and P. A. Mawby, "A lifetime estimation technique for voltage source inverters," *IEEE Trans. Power Electron.*, vol. 28, no. 8, pp. 4113–4119, Aug. 2013.

[14] M. Musallam, C. Yin, C. Bailey, and M. Johnson, "Mission profile-based reliability design and real-time life consumption estimation in power electronics," *IEEE Trans. Power Electron.*, vol. 30, no. 5, pp. 2601–2613, May 2015.

[15] S. E. D. Leon-Aldaco, H. Calleja, F. Chan, and H. R. Jimenez-Grajales, "Effect of the mission profile on the reliability of a power converter aimed at photovoltaic applications - a case study," *IEEE Trans. Power Electron.*, vol. 28, no. 6, pp. 2998–3007, Jun. 2013.

[16] A. Anurag, Y. Yang, and F. Blaabjerg, "Reliability analysis of single-phase PV inverters with reactive power injection at night considering mission profiles," in *Proc. of ECCE*, pp. 2132–2139, Sep. 2015.

[17] C. Felgemacher, S. Araujo, C. Noeding, P. Zacharias, A. Ehrlich, and M. Schidleja, "Evaluation of cycling stress imposed on IGBT modules in PV central inverters in sunbelt regions," in *Proc. of CIPS*, pp. 1–6, Mar. 2016.

[18] A. Sangwongwanich, Y. Yang, D. Sera, and F. Blaabjerg, "Lifetime evaluation of grid-connected PV inverters considering panel degradation rates and installation sites," *IEEE Trans. Power Electron.*, vol. 33, no. 2, pp. 1225–1236, Feb. 2018.

[19] Y. Yang, H. Wang, F. Blaabjerg, and T. Kerekes, "A hybrid power control concept for PV inverters with reduced thermal loading," *IEEE Trans. Power Electron.*, vol. 29, no. 12, pp. 6271–6275, Dec. 2014.

[20] M. Andresen, G. Buticchi, and M. Liserre, "Thermal stress analysis and mppt optimization of photovoltaic systems," *IEEE Trans. Ind. Electron.*, vol. 63, no. 8, pp. 4889–4898, Aug. 2016.

[21] Y. Yang, E. Koutroulis, A. Sangwongwanich, and F. Blaabjerg, "Pursuing photovoltaic cost-effectiveness: Absolute active power control offers hope in single-phase PV systems," vol. 23, no. 5, pp. 40–49, Sep. 2017.

[22] I. Vernica, K. Ma, and F. Blaabjerg, "Optimal derating strategy of power electronics converter for maximum wind energy production with lifetime information of power devices," *IEEE J. Emerg. Sel. Topics Power Electron.*, vol. 6, no. 1, pp. 267–276, Mar. 2018.

[23] D. A. Murdock, J. E. R. Torres, J. J. Connors, and R. D. Lorenz, "Active thermal control of power electronic modules," *IEEE Trans. Ind. App.*, vol. 42, no. 2, pp. 552–558, Mar. 2006.

[24] S.B. Kjaer, J.K. Pedersen, and F. Blaabjerg, "A review of single-phase grid-connected inverters for photovoltaic modules," *IEEE Trans. Ind. Appl.*, vol. 41, no. 5, pp. 1292–1306, Sep. 2005.

[25] A. Sangwongwanich, Y. Yang, F. Blaabjerg, and H. Wang, "Benchmarking of constant power generation strategies for single-phase grid-connected photovoltaic systems," *IEEE Trans. Ind. App.*, vol. 54, no. 1, pp. 447–457, Jan.-Feb. 2018.

[26] A. Sangwongwanich, Y. Yang, and F. Blaabjerg, "Development of flexible active power control strategies for grid-connected photovoltaic inverters by modifying MPPT algorithms," in *Proc. IFEEC 2017 - ECCE Asia*, pp. 87–92, Jun. 2017.

[27] F. Blaabjerg, R. Teodorescu, M. Liserre, and A.V. Timbus, "Overview of control and grid synchronization for distributed power generation systems," *IEEE Trans. Ind. Electron.*, vol. 53, no. 5, pp. 1398–1409, Oct. 2006.

[28] *SGP30N60*, Infineon Technologies AG, 2007, rev. 2.3.

[29] *Type 381LX / 383LX 105 C High Ripple, Snap-In Aluminum*, Cornell Dubilier, Liberty, SC, USA. [Online]. Available: http://http://www.cde.com/resources/catalogs/381-383.pdf

[30] SolarEdge, "Oversizing of SolarEdge inverters, technical note," Tech. Rep., July 2016.

[31] U. Scheuermann, R. Schmidt, and P. Newman, "Power cycling testing with different load pulse durations," in *Proc. of PEMD 2014*, pp. 1–6, Apr. 2014.

[32] *Application guide, Aluminum Electrolytic Capacitors*, Cornell Dubilier, Liberty, SC, USA. [Online]. Available: http://www.cde.com/catalogs/AEappGUIDE.pdf

The 2018 International Power Electronics Conference

Discontinuous Current Mode Control for Minimization of Three-phase Grid-Tied Inverter in Photovoltaic System

Hoai Nam Le[1*] and Jun-ichi Itoh[2]

1 Department of Electrical, Electronics and Information Engineering, Nagaoka University of Technology, Nagaoka, Japan
2 Department of Science of Technology Innovation, Nagaoka University of Technology, Nagaoka, Japan

*E-mail: lehoainam@stn.nagaokaut.ac.jp

Abstract— **In this paper, a current control method of discontinuous current mode (DCM) is proposed for a three-phase grid-tied inverter in order to minimize inductors without worsening current total harmonic distortion (THD). In a conventional continuous current mode (CCM) control, current THD increases as an inductor value is reduced, because a zero-clamping phenomenon occurs due to dead-time. In the proposed DCM current control, a zero-current interval is intentionally controlled and a dead-time-induced error voltage is simply compensated with a conventional dead-time compensation. The validation of the control method is confirmed by simulation and a 700W-prototype. As simulation results, compared to the conventional CCM current control, the current THD is reduced by 97.6% with the proposed DCM current control, whereas the inductor volume is reduced by 70%. In the experiments, the current THD is maintained below 5% over load range from 0.3 p.u. to 1.0 p.u. even when the inductance impedance is reduced to 0.5% of the inverter total impedance.**

Keywords— three-phase grid-tied inverter, continuous current mode, discontinuous current mode

I. INTRODUCTION

In the last decade, researches on photovoltaic system (PV) have accelerated due to an increasing demand of renewable and sustainable energy sources [1]-[3]. In the PV system, H-bridge three-phase grid-tied inverters are generally employed as an interface between solar panels and three-phase grid. In such grid-tied inverters, a grid filter is required to connect between an output of the inverter and the grid in order to filter out the current harmonics and to meet grid current harmonic constraints as defined by standards such as IEEE-1547 [4]. Due to the observation that inductors in the grid filter occupy a major volume of the inverter, an inductor value of the grid filter is necessarily reduced in order to minimize the grid filter as well as the inverter. However, this reduction of the inductor value implies a design of a high switching current ripple due to a high dc-link voltage to inductance

ratio. This high current ripple results in a current distortion phenomenon called zero-current clamping, where a current distortion increases notably as the switching current ripple increases [5]-[9].

Due to the zero-current clamping effect, the dead-time-induced error voltage exhibits a strong nonlinear behavior around zero-current crossing points. Hence, conventional dead-time compensation methods such as, e.g. two-level approximation compensation method (ACM) [10]-[11], linear ACM [12] and three-level ACM [13]-[15], cannot compensate for this nonlinear behavior of the dead-time-induced error voltage. Several compensation methods for the nonlinearity of the dead-time-induced error voltage such as adaptive dead-time compensation method and turn-off transition compensation method have been proposed to deal with this nonlinearity behavior and to reduce the zero-crossing current distortion [16]-[17]. Nevertheless, both methods exhibit the requirements which restrict the employment over a wide range of application. Adjustment mechanism parameters for the adaptive dead-time compensation must be properly tuned for each individual system [16]. Meanwhile, accurate device parameters, e.g. parasitic capacitances, are required for the turn-off transition compensation method [17].

In this paper, a current control for the three-phase grid-tied inverter operated in discontinuous current mode (DCM) is proposed in order to minimize the grid filter without worsening the current distortion. In order to deal with the zero-current clamping effect, the inverter is intentionally operated in DCM instead of a conventional continuous current mode (CCM). In other words, the zero-current interval with the DCM operation is controlled, enabling a proper compensation for the nonlinear behavior. Consequently, the conventional dead-time compensation method, i.e. two-level ACM, can be employed simply in order to compensate the dead-time-induced error voltage and reduce the current

2519

distortion. The contribution of this paper is that the proposed DCM current control is implemented with the conventional dead-time compensation method to simply compensate the dead-time-induced error voltage. The effectiveness of the proposed current control is confirmed by simulations and experiments.

II. DISCONTINUOUS-CURRENT-MODE CURRENT CONTROL

A. Zero-crossing distortion

Figure 1 depicts the H-bridge three-phase grid-tied inverter with a *LCL*-based grid filter. The minimization of the *LCL* filter generates a current with a high ripple in the inductors *L*. The filter stage with L_f and C_f can suppress the high-order current harmonics in order to meet grid current harmonic constraints as defined by standards such as IEEE-1547 [4].

Figure 2 describes the zero-crossing current distortion phenomenon. As the current ripple increases with the minimized *LCL* filter, the current distortion increases notably around the zero-crossing points due to the zero-current clamping effect, making the dead-time-induced error voltage become nonlinear. Therefore, the employment of the conventional two-level ACM just further increases the current distortion [17].

B. Discontinuous-current-mode operation

Figure 3 indicates the phase clamping selection in six cycles of traditional discontinuous pulse width modulation (DPWM), and the inverter output current waveform in one switching period during 0°-60° region. In order to simplify the control of DCM, only two phase currents should be controlled, whereas the current of the third phase is the summation of the currents of the first two phases. Hence, DPWM is employed in order to satisfy this control condition. During each 60-degree time region in DPWM, one phase is clamped to P or N polarity of dc-link voltage as shown in Fig. 3(a), whereas the other two phases are modulated to control separately two inverter output currents as shown in Fig. 3(b). Note that D_1 and D_2 indicate the duty ratios of the first and the second intervals of the first controlled current, D_3 and D_4 indicate the duty ratios of the first and the second intervals of the second controlled current, whereas D_5 depicts the duty ratio of the zero-current intervals. The duty ratios in Figure 3 are calculated as follows.

The inductor voltage of the inverter-side inductor *L*

during a switching period T_{sw} is given by (1),

$$L \frac{di_{avg\_u}}{dt} = D_1 \left(V_{dc} - v_{uo} + v_{vo} \right) + D_2 \left(-v_{uo} + v_{vo} \right) \dots (1)$$

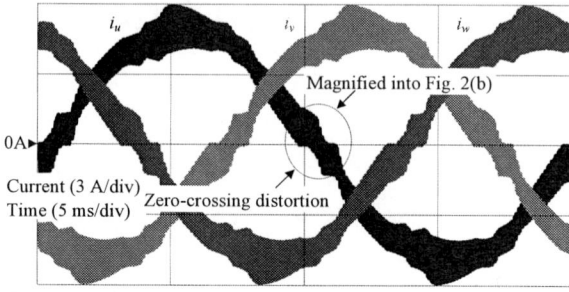

(a) Zero-crossing distortions in inverter output currents

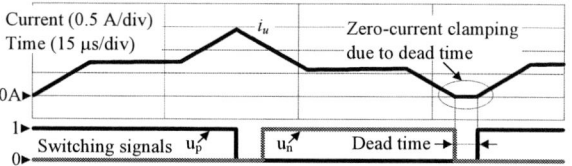

(b) Zero-current clamping effect due to dead-time

Fig. 2. Zero-current distortion phenomenon. The current distortion increases with the high current ripple due to the dead-time.

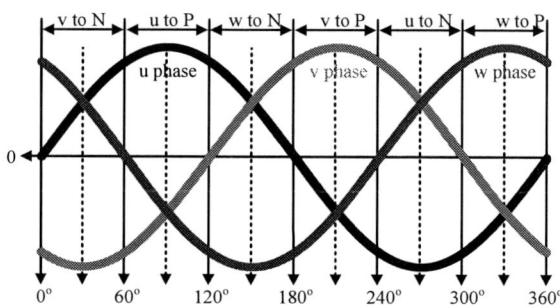

(a) Phase clamping selection in six cycles of DPWM

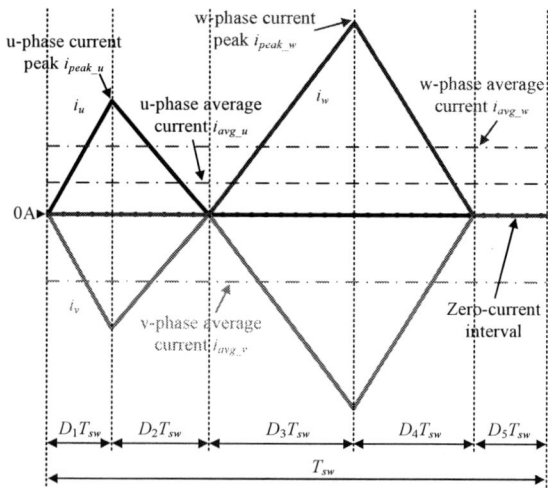

(b) Inverter output current waveform in one switching period during 0°-60° region

Fig. 3. Six 60-degree time regions of DPWM and inverter output current in DCM. In order to simplifying the DCM control, DPWM is employed, controlling only two phase currents at the same time.

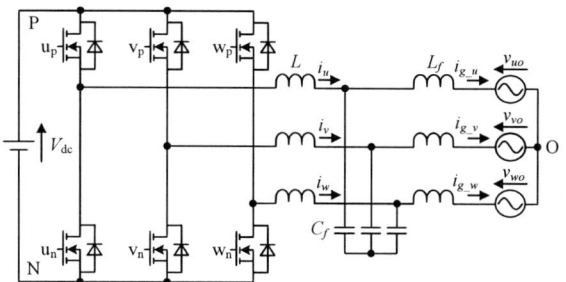

Fig. 1. H-bridge grid-tied three-phase inverter. This topology is employed due to its simple control and construction.

2520

where V_{dc} is the dc-link voltage, v_{uo} and v_{vo} are the grid phase voltages. The average current i_{avg_u} and the current peak i_{peak_u} shown in Figure 3 is expressed as,

$$i_{avg\_u} = \frac{i_{peak\_u}}{2}(D_1 + D_2) \quad\text{.......................(2)}$$

$$i_{peak\_u} = \frac{V_{dc} - (v_{uo} - v_{vo})}{2L}D_1 T_{sw} \quad\text{..............(3)}$$

Substituting (3) into (2), and solving the equation for the duty ratios D_2, then the duty ratio D_2 is expressed by (4),

$$D_2 = \frac{4Li_{avg\_u}}{D_1 T_{sw}(V_{dc} - v_{uo} + v_{vo})} - D_1 \quad\text{..............(4)}$$

Substituting (4) into (1) in order to remove the duty ratio D_2 and representing (1) as a function of only the duty ratio D_1, (5) is obtained [18].

$$L\frac{di_{avg\_u}}{dt} = D_1 V_{dc} - \frac{4Li_{avg\_u}(v_{uo} - v_{vo})}{D_1 T_{sw}(V_{dc} - v_{uo} + v_{vo})} \quad\text{..............(5)}$$

Substituting the differential of the inductor current di_{avg_u}/dt in (5) as zero and the duty ratio D_1 is expressed as in (6),

$$D_1 = 2\sqrt{\frac{i_{avg\_u} L f_{sw}(v_{uo} - v_{vo})}{V_{dc}(V_{dc} - v_{uo} + v_{vo})}} \quad\text{..................(6)}$$

where f_{sw} is the switching frequency. Then, substituting the differential of the inductor current di_{avg_u}/dt in (1) as zero and the duty ratio D_2 is expressed as in (7),

$$D_2 = \frac{D_1(V_{dc} - v_{uo} + v_{vo})}{v_{uo} - v_{vo}} \quad\text{..........................(7)}$$

Similarly, the duty ratios D_3 and D_4 shown in Figure 3 can be expressed as in (8)-(9),

$$D_3 = 2\sqrt{\frac{i_{avg\_w} L f_{sw}(v_{wo} - v_{vo})}{V_{dc}(V_{dc} - v_{wo} + v_{vo})}} \quad\text{..................(8)}$$

$$D_4 = \frac{D_3(V_{dc} - v_{wo} + v_{vo})}{v_{wo} - v_{vo}} \quad\text{..........................(9)}$$

Figure 4 shows the control system of the three-phase grid-tied inverter operating completely in DCM, whereas Table I depicts the values for the duty calculation and the switching signal output in each 60-degree time region. When the grid operates normally, the inverter only has to regulate the grid current following the sinusoidal waveform. First, the 60-degree time region is detected by detected values of the grid phase voltage v_{uo}, v_{vo}, and v_{wo}. Then, the phase current references and the phase voltages are distributed to input values of a duty calculation based on the detected 60-degree time region as shown in Table I. In the duty calculation step, the duty ratios D_1-D_5 are expressed as follows. Note that the calculation of the duty ratios D_1-D_5 is similar to that of the duty ratios D_1-D_4 shown in Figure 3.

$$D_1 = 2\sqrt{\left|\frac{i_{1\_ref} L f_{sw}(v_1 - v_2)}{V_{dc}(V_{dc} - v_1 + v_2)}\right|} \quad\text{..................(10)}$$

$$D_2 = \frac{D_1(V_{dc} - v_1 + v_2)}{v_1 - v_2} \quad\text{..........................(11)}$$

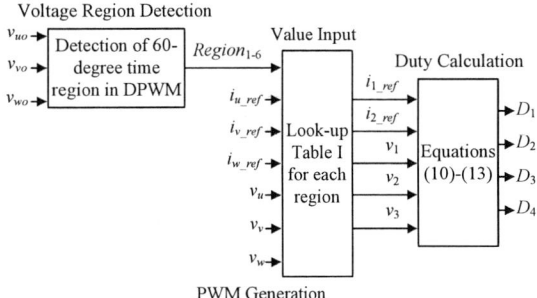

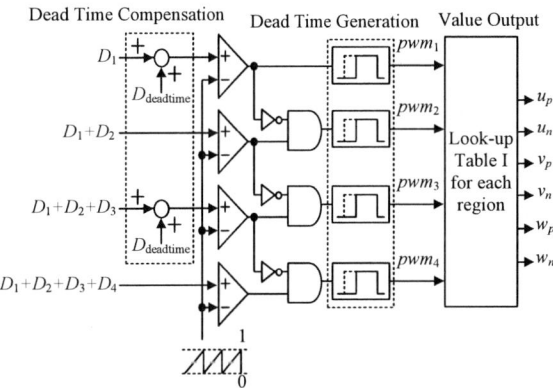

Fig. 4. Control system of the three-phase grid-tied inverter operating completely in DCM. The dead-time-induced error voltage is compensated simply when the inverter is intentionally operated in DCM because the zero-current interval is controlled.

TABLE I
LOOK-UP VALUES FOR DUTY CALCULATION AND PWN OUTPUT.

Region Variable	0°-60°	60°-120°	120°-180°	180°-240°	240°-300°	300°-360°
i_{1_ref}	i_{u_ref}	i_{w_ref}	i_{v_ref}	i_{u_ref}	i_{w_ref}	i_{v_ref}
i_{2_ref}	i_{w_ref}	i_{v_ref}	i_{u_ref}	i_{w_ref}	i_{v_ref}	i_{u_ref}
v_1	v_u	$-v_w$	v_v	$-v_u$	v_w	$-v_v$
v_2	v_v	$-v_u$	v_w	$-v_v$	v_u	$-v_w$
v_3	v_w	$-v_v$	v_u	$-v_w$	v_v	$-v_u$
u_p	pwm_1	1	pwm_3	pwm_2	0	pwm_4
u_n	pwm_2	0	pwm_4	pwm_1	1	pwm_3
v_p	0	pwm_4	pwm_1	1	pwm_3	pwm_2
v_n	1	pwm_3	pwm_2	0	pwm_4	pwm_1
w_p	pwm_3	pwm_2	0	pwm_4	pwm_1	1
w_n	pwm_4	pwm_1	1	pwm_3	pwm_2	0

$$D_3 = 2\sqrt{\left|\frac{i_{2\_ref} L f_{sw}(v_3 - v_2)}{V_{dc}(V_{dc} - v_3 + v_2)}\right|} \quad\text{................(12)}$$

$$D_4 = \frac{D_4(V_{dc} - v_3 + v_2)}{v_3 - v_2} \quad\text{....................(13)}$$

$$D_5 = 1 - D_1 - D_2 - D_3 - D_4 \quad\text{..................(14)}$$

where i_{1_ref} and i_{2_ref} are the first and second controlled currents in each 60-degree time region, and v_1, v_2 and v_3 are the voltages corresponding to the controlled currents.

For instance, during the 0°-60° time region, the controlled currents are i_u and i_w as shown in Fig. 3. Therefore, the input values to i_1, i_2, v_1, v_2, and v_3 are i_{u_ref}, i_{w_ref}, v_{uo}, v_{vo} and v_{wo}, respectively, as shown in Table I.

Next, the dead-time compensation is introduced at the first step of PWM generation. The duty ratio which compensates for the dead-time-induced error voltage, is expressed as follow,

$$D_{deadtime} = f_{sw}T_{deadtime} \quad \cdots\cdots\cdots\cdots\cdots\cdots\cdots\cdots (15)$$

where $T_{deadtime}$ is the dead-time. The dead-time-induced error voltage is simply compensated as shown in Fig. 4 because when the inverter is intentionally operated in DCM, the zero-current interval is under control. The compensated duty ratios are then compared with the sawtooth waveform to generate the PWM signals. In order to avoid the simultaneous turn-on of both switching devices in one leg, the typical dead-time generation is used to delay the turn on. Finally, the PWM signals are distributed to the switching devices corresponding to each 60-degree time region of DPWM based on Table I. Note that if the outputs pwm_2 and pwm_4 are utilized as shown in Table I, the inverter is operated under synchronous switching; otherwise, if the outputs pwm_2 and pwm_4 are set to zero, the inverter is operated under asynchronous switching.

III. SIMULATION RESULTS

Table II shows the circuit parameters to evaluate the operation of the inverters, whereas Figure 5 depicts the inductor volume against the inductor impedance. The inverter-side inductors L in Fig. 1 occupy a majority of the inverter volume. Therefore, the minimization of L is mainly focused in this paper. Generally, the inductor value is expressed as a grid filter impedance scaled to the inverter total impedance $\%Z_L$ [21]. In particular, three designs of the grid filter impedance are evaluated. As shown in Fig. 5, the inverter-side inductor L volume is minimized by 70% when the inductor impedance $\%Z_L$ is reduced from 2.5% to 0.075%.

Figure 6 shows the inverter output currents and the average currents of the conventional CCM current control and the proposed DCM current control at rated load with three inductor designs from Fig. 5. As the current ripple increases, i.e. the decrease in the inductor impedance, the current with the conventional CCM current control distorts notably around the zero-crossing points. Consequently, the current THD increases from 1.5% to 9.8% when the inductor impedance $\%Z_L$ is reduced from 2.5% to 0.075%. On the other hand, when the inverter is operated in DCM, the zero-current interval can be controlled and the dead-time-induced error voltage can be compensated simply as shown in Fig. 4. Therefore, even with the minimized inductor impedance of 0.075%, the low current THD of 0.3% is achieved with the proposed DCM current control.

Figure 7 depicts the load step response of the proposed DCM current control. As shown in Fig. 7(a), even under the sudden load step between the load of 0.1 p.u. and the load of 1.0 p.u., the stable inverter operation and the

TABLE II
SIMULATION PARAMETERS.

Circuit Parameter		
V_{DC}	DC link Voltage	500 V
v_g	Line-to-line Voltage	200 Vrms
P_n	Nominal Power	3 kW
f_g	Grid Frequency	50 Hz
Z_b	Total Impedance	13.3 Ω
f_{sw}	Switching Frequency	40 kHz
$T_{deadtime}$	Dead-time	500 ns
L_1	1st Inductor Value	1061 µH (2.5%)
L_2	2nd Inductor Value	254.6 µH (0.6%)
L_3	3rd Inductor Value	31.8 µH (0.075%)
Current Controller Parameter		
ζ	Damping Factor	0.7
f_c	Cutoff Frequency	1 kHz

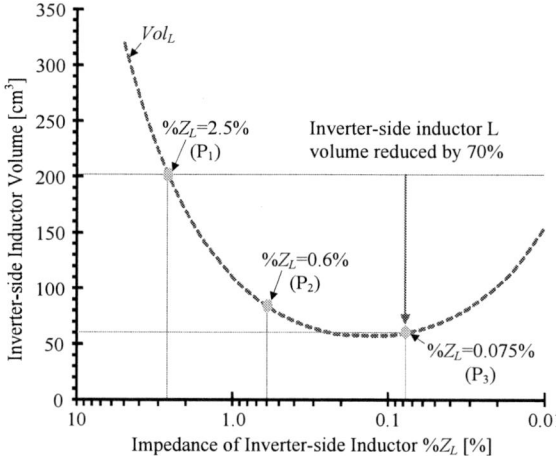

Fig. 5. Relationship between filter volume and inductor impedance at switching frequency of 40 kHz. The inductor volume can be minimized greatly when reducing the inductor impedance.

balanced three-phase currents are still achieved with the proposed control.

Figure 8 shows the current THD characteristics of the conventional CCM current control and the proposed DCM current control with three inductor designs from Fig. 5. At rated load with the inductor impedance of 0.075%, the current distortion of the proposed DCM current control is reduced by 97.6% compared to the conventional CCM current control. Note that the current THD of the conventional CCM current control with the inductor impedance of 0.075% or 0.6% has a tendency to decrease at light load. The reason is when the average current is significantly smaller than the current ripple, the current mode is no longer CCM but triangular current mode (TCM) [22]-[23]. In TCM, all the turn on of the switching devices is zero voltage switching. On other words, the dead-time-induced error voltage does not occur in TCM. Hence, the current distortion due to the zero-clamping phenomenon disappears at light load.

2522

The 2018 International Power Electronics Conference

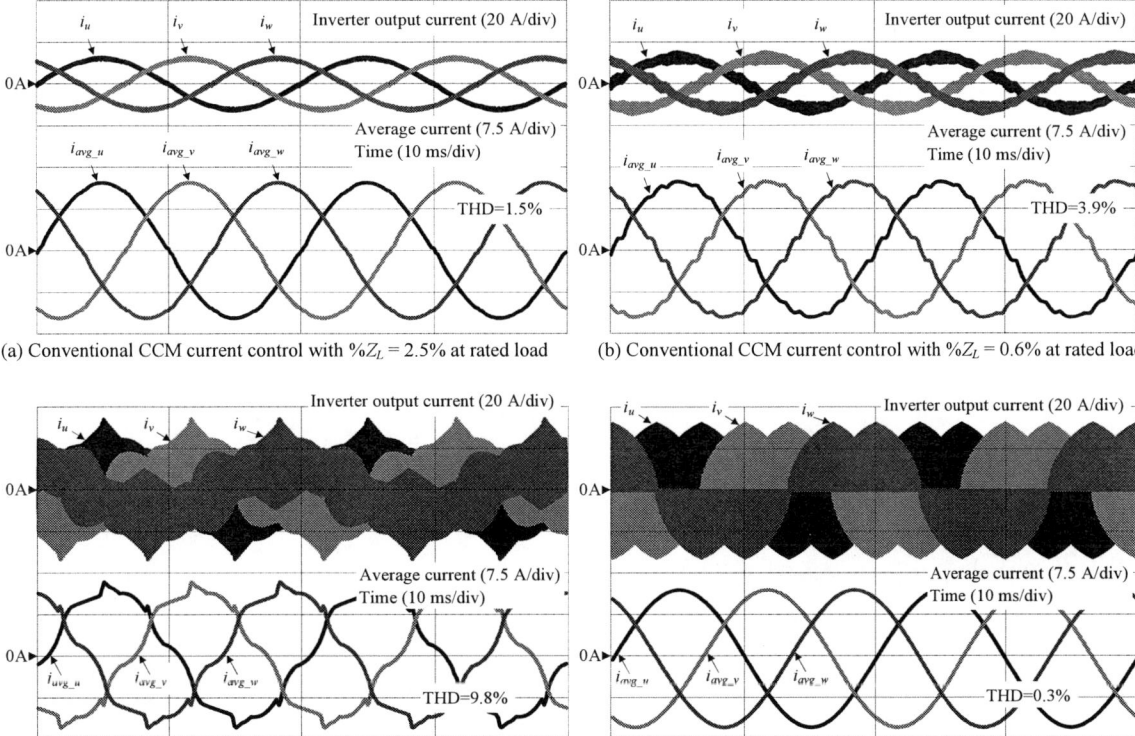

(a) Conventional CCM current control with %Z_L = 2.5% at rated load   (b) Conventional CCM current control with %Z_L = 0.6% at rated load

(c) Conventional CCM current control with %Z_L = 0.075% at rated load   (d) Proposed DCM current control with %Z_L = 0.075% at rated load

Fig. 6. Inverter output currents and average currents of conventional CCM current control and proposed DCM current control at rated load. The current THD of the conventional CCM current control increases with the reduction of the grid filter impedance, whereas the current THD of the proposed current control is still low.

IV. LABORATORY SETUP

Table III shows the experimental parameters, whereas figure 9 depicts the prototype of the miniature three-phase grid-tied inverter. In order to operate the inverter under DCM over entire load range with the switching frequency of 20 kHz, the inverter-side inductor value is designed at 80 µH, whose impedance is 0.5% of the total inverter impedance.

A. Discontinuous-current mode operation

Figure 10 depicts the three-phase grid-tied inverter DCM operation waveform at rated load. In Figure 10(a), the phase difference between the grid current of u phase and the grid u-phase voltage is almost zero, i.e. the unity-power-factor operation. Furthermore, even with the small inverter-side inductor impedance of 0.5%, the low current THD of 2.4% is still achieved. As shown in Figure 10(b)-(c), the three-phase inverter output currents are similar to those shown in Figure 6(c), i.e. the operation of the proposed DCM control is confirmed.

Figure 11 shows the grid phase voltages and the grid currents of u phase and w phase at the normal operation and at step-up load change. At the normal operation, the three-phase grid current is well balance and the low current THD of 2.4% is achieved for all three-phase grid current. At the step-up load change from 0.1 p.u. to 1.0 p.u., the stable current response is confirmed. Note that

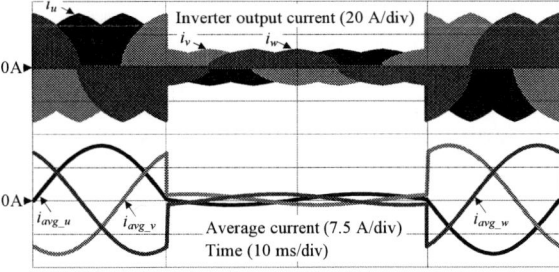

Fig. 7. Load step response between load of 0.1 p.u. and load of 1.0 p.u.. The stable inverter operation is confirmed even at load step change.

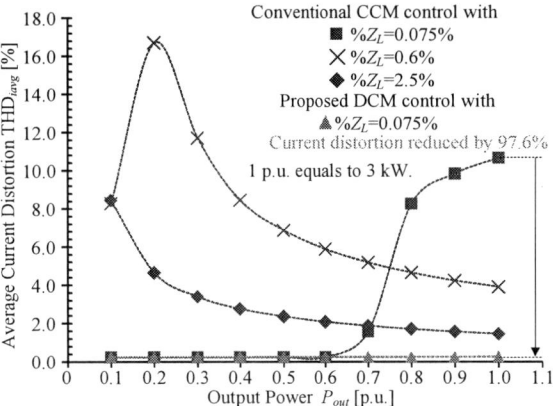

Fig. 8. Current THD characteristics of conventional CCM current control and proposed DCM current control with three different inductor designs. With the proposed DCM current control, the current THD is maintained below 5% over entire load range from 0.1 p.u. to 1.0 p.u..

2523

TABLE III
EXPERIMENTAL PARAMETERS.

Circuit Parameter		
V_{DC}	DC link Voltage	300 V
v_g	Line-to-line Voltage	100 Vrms
P_n	Nominal Power	700 W
f_g	Grid Frequency	50 Hz
Z_b	Total Impedance	4.8 Ω
f_{sw}	Switching Frequency	20 kHz
$T_{deadtime}$	Dead-time	500 ns
L	Inductor Value	80 μH (0.5%)

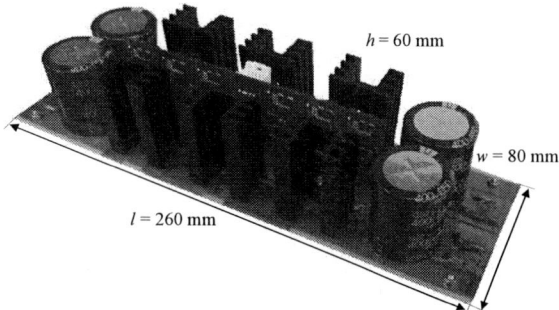

Fig. 9. Prototype of miniature three-phase grid-tied inverter.

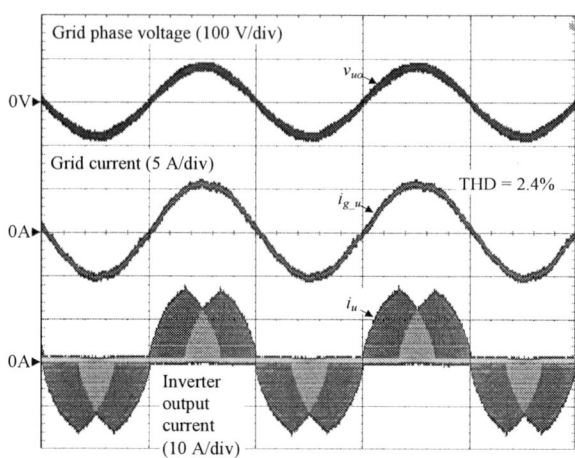

(a) Grid phase voltage, grid current, and inverter output current of u phase

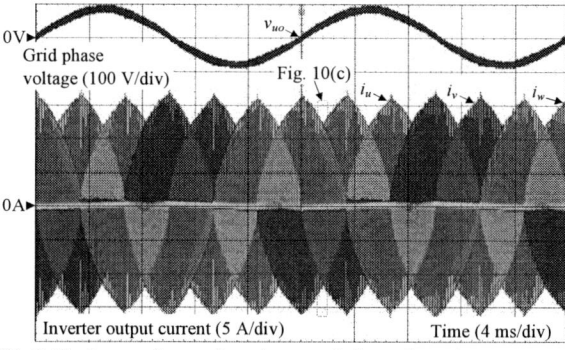

(b) Grid phase voltage of u phase, and three-phase inverter output currents

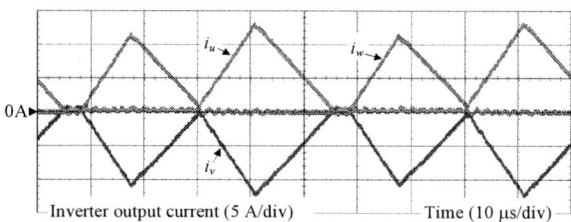

(c) Zoom-in three-phase grid current from Fig. 10(b)

Fig. 10. Three-phase grid-tied inverter DCM operation waveform at rated load. The three-phase inverter output currents shown in Fig. 10(b) are similar to those shown in Fig. 6(c). This confirms the operation of the proposed DCM control.

the three-phase grid currents are still balance both before and after the step-up load change.

Figure 12 depicts the current THD characteristics of the proposed DCM current control. The current THD is maintained below 5% over load range from 0.3 p.u. to 1.0 p.u., which satisfies the current THD constraint in IEEE-1547, even when the inductance impedance is reduced to 0.5% of the inverter impedance. The increase of the current THD at light load can be explained due to the high occupation of the reactive current flowing through the filter capacitor. Therefore, in order to reduce the current THD at light load, the DCM control should also consider the effect of the reactive current in the filter capacitor.

B. Efficiency comparison between asynchronous switching and synchronous switching in DCM

Figure 13 shows the asynchronous switching and synchronous switching in DCM. In the asynchronous switching, the corresponding switches are turned after the period D_1T_{sw} and D_3T_{sw} finish. Therefore, the current has to flow through the diode. In the next generation switching devices such as SiC or GaN, the forward voltage of the inverse diode in such devices is generally higher than that of the conventional MOS-FET devices. Consequently, the conduction loss with the asynchronous switching is higher that of the synchronous switching, where the current flows through the FET part. As shown in Figure 13, the synchronous switching can also be applied into DCM in the same manner as the conventional CCM. Consequently, the conduction loss of the switching device is reduced.

Figure 14 depicts the efficiency comparison between asynchronous switching and synchronous switching in DCM. The application of the DCM synchronous switching reduces the conversion loss by 33% compared to the DCM asynchronous switching at rated load. Furthermore, the maximum efficiency of 97.8% is achieved at rated load.

V. CONCLUSION

In this paper, the DCM current control was proposed to the grid-tied three-phase inverter in order to minimize the grid filter volume without worsening the current THD. When the inverter is operated under DCM, the zero-

The 2018 International Power Electronics Conference

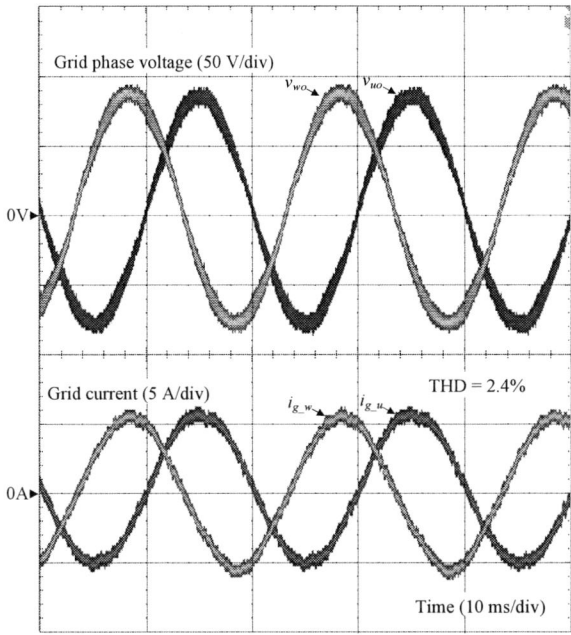

(a) Grid phase voltages and grid currents of u phase and w phase at normal operation

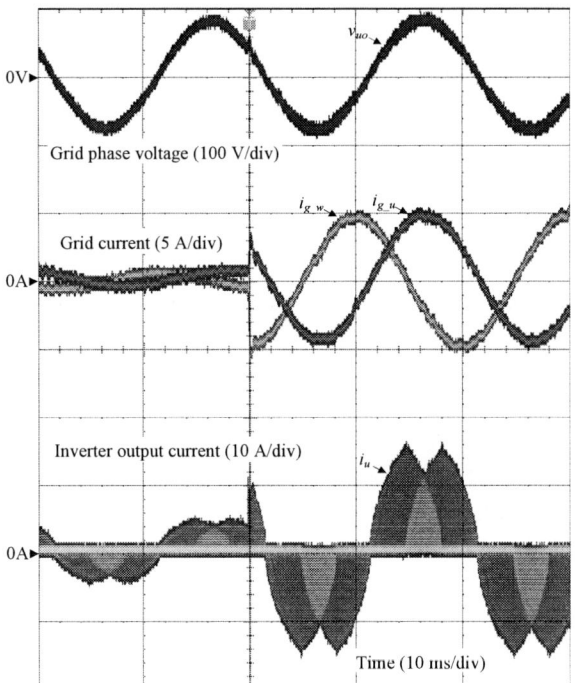

(b) Current response of step-up load change

Fig. 11. Grid phase voltages and grid currents of u phase and w phase at normal operation and at step-up load change. Three-phase currents are balanced at normal operation as well as at step-up load change.

current interval can be controlled, compensating simply the dead-time-induced error voltage. Consequently, when the inductor impedance is reduced to 0.075%, the current THD of the proposed DCM current control is reduced by 97.6% compared to the conventional CCM current control as simulation results. In the experiments, the

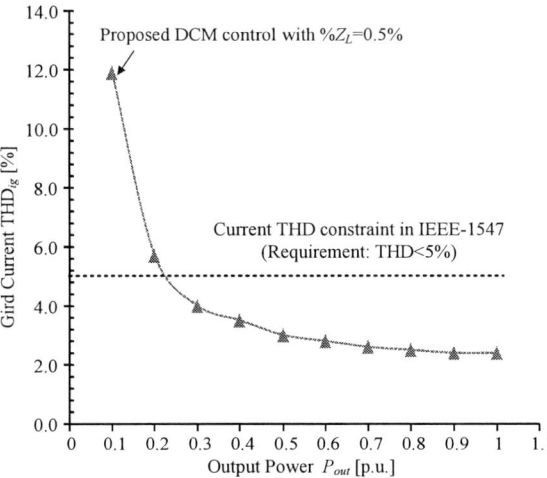

Fig. 12. Current THD characteristics of proposed DCM current control. The current THD is maintained below 5% over load range from 0.3 p.u. to 1.0 p.u., which satisfies the current THD constraint in IEEE-1547, even when the inductance impedance is reduced to 0.5% of the inverter impedance.

current THD was maintained below 5% over load range from 0.3 p.u. to 1.0 p.u. even when the inductance impedance is reduced to 0.5% of the inverter impedance.

In the future work, the DCM current feedback control will be considered in order to eliminate the circuit-parameter dependency.

REFERENCES

[1] M. Matsui, T. Sai, B. Yu, and X. D. Sun," A New Distributed MPPT Technique using Buck-only MICs Linked with Controlled String Current", IEEJ Journal of Industry Applications, vol.4, no.6, pp.674-680, 2015.

[2] R. Chattopadhyay, S. Bhattacharya, N. C. Foureaux, I. A. Pires, H. de Paula, L. Moraes, P. C. Cortizio, S. M. Silva, B. C. Filho, and J. A. de S. Brito," Low-Voltage PV Power Integration into Medium Voltage Grid Using High-Voltage SiC Devices", IEEJ Journal of Industry Applications, vol.4, no.6, pp.767-775, 2015.

[3] S. Yamaguchi, and T. Shimizu," Single-phase Power Conditioner with a Buck-boost-type Power Decoupling Circuit", IEEJ J. Industry Applications, vol.5, no.3, pp.191-198, 2016.

[4] *IEEE Application Guide for IEEE Std 1547, IEEE Standard for Interconnecting Distributed Resources with Electric Power Systems*, IEEE Standard 1547.2-2008, 2009.

[5] Y. Wang, Q. Gao and X. Cai, "Mixed PWM for Dead-Time Elimination and Compensation in a Grid-Tied Inverter," in *IEEE Transactions on Industrial Electronics*, vol. 58, no. 10, pp. 4797-4803, Oct. 2011.

[6] J. W. Choi and S. K. Sul, "A new compensation strategy reducing voltage/current distortion in PWM VSI systems operating with low output voltages," in *IEEE Transactions on Industry Applications*, vol. 31, no. 5, pp. 1001-1008, Sep/Oct 1995.

[7] S. Bolognani, L. Peretti and M. Zigliotto, "Repetitive-Control-Based Self-Commissioning Procedure for Inverter Nonidealities Compensation," in *IEEE Transactions on Industry Applications*, vol. 44, no. 5, pp. 1587-1596, Sept.-Oct. 2008.

[8] J. M. Schellekens, R. A. M. Bierbooms and J. L. Duarte, "Dead-time compensation for PWM amplifiers using simple feed-forward techniques," *The XIX International Conference on Electrical Machines - ICEM 2010*, Rome, 2010, pp. 1-6.

[9] Y. Wang, Q. Gao and X. Cai, "Mixed PWM for Dead-Time Elimination and Compensation in a Grid-Tied Inverter," in *IEEE Transactions on Industrial Electronics*, vol. 58, no. 10, pp. 4797-4803, Oct. 2011.

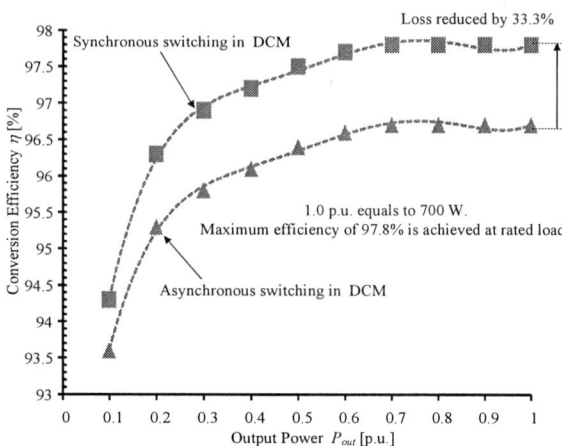

Fig. 14. Efficiency comparison between asynchronous switching and synchronous switching in DCM. The application of the DCM synchronous switching reduces the conversion loss by 33% compared to the DCM asynchronous switching at rated load.

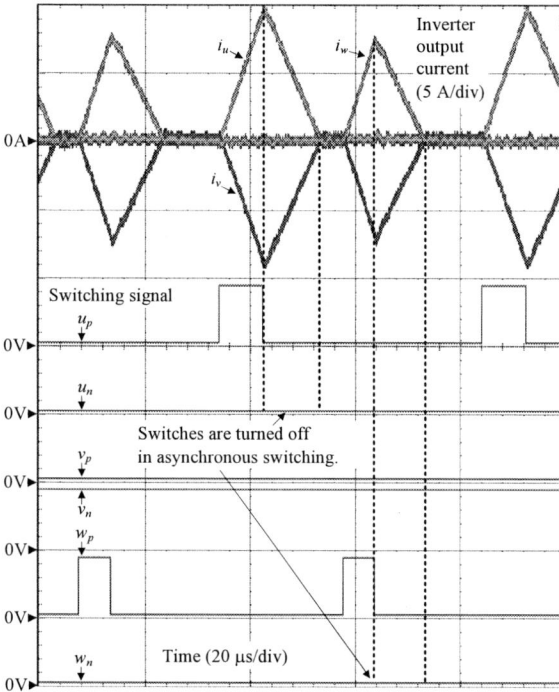

(a) Asynchronous switching in DCM

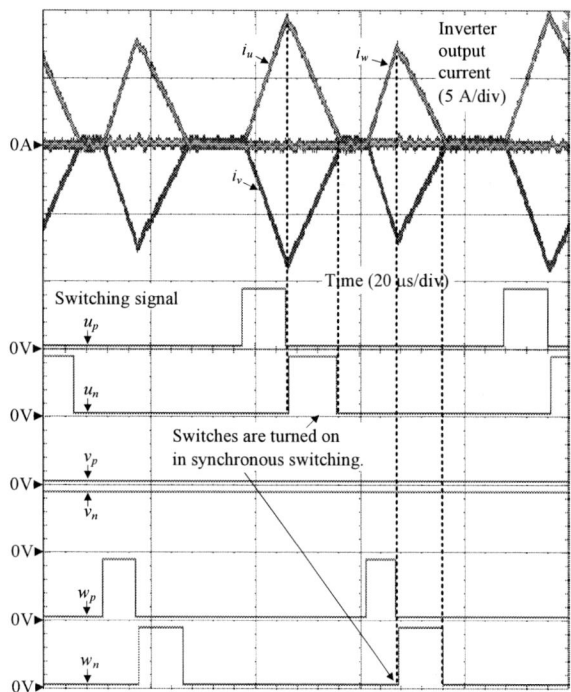

(b) Synchronous switching in DCM

Fig. 13. Asynchronous switching and synchronous switching in DCM. Similar to the conventional CCM, the synchronous switching can also be applied into DCM. This makes the current flow through the FET part instead of the diode. Consequently, the conduction loss of the switching device can be reduced.

[10] S. Jeong and M. Park, "The analysis and compensation of dead-time effects in PWM inverters," *IEEE Trans. Ind. Electron.*, vol. 38, no. 2, pp. 108–114, Apr. 1991.

[11] Seon-HwanHwang and Jang-MokKim, "Dead time compensation method for voltage-fed PWMinverter," *IEEE Trans. Energy Convers.*, vol. 25, no. 1, pp. 1–10, Mar. 2010.

[12] A. C. Oliveira, C. B. Jacobina, A. M. N. Lima, and E. R. C. da Silva, "Dead-time compensation in the zero-crossing current region," *in Proc. IEEE 34th Annu. Power Electron. Spec. Conf.*, 2003, vol. 4, pp. 1937–1942.

[13] Z. Guo and F. Kurokawa, "Control and PWM modulation scheme for dead-time compensation of CVCF inverters," *in Proc. IEEE 31st Int. Telecommun. Energy Conf.*, 2009, pp. 1–6.

[14] J. M. Schellekens, R. A. M. Bierbooms, and J. L. Duarte, "Dead-time compensation for PWM amplifiers using simple feed-forward techniques," *in Proc. IEEE Int. Conf. Elect. Mach.*, 2010, pp. 1–6.

[15] M. A. Herr´an, J. R. Fischer, S. A. Gonz´alez, M. G. Judewicz, and D. O. Carrica, "Adaptive dead-time compensation for grid-connected PWM inverters of single-stage PV systems," *IEEE Trans. Power Electron.*, vol. 28, no. 6, pp. 2816–2825, Jun. 2013.

[16] M. A. Herran, J. R. Fischer, S. A. Gonzalez, M. G. Judewicz and D. O. Carrica, "Adaptive Dead-Time Compensation for Grid-Connected PWM Inverters of Single-Stage PV Systems," in *IEEE Transactions on Power Electronics*, vol. 28, no. 6, pp. 2816-2825, June 2013.

[17] T. Mannen and H. Fujita, "Dead-Time Compensation Method Based on Current Ripple Estimation," in *IEEE Transactions on Power Electronics*, vol. 30, no. 7, pp. 4016-4024, July 2015.

[18] J. Sun, D. M. Mitchell, M. F. Greuel, P. T. Krein and R. M. Bass, "Averaged modeling of PWM converters operating in discontinuous conduction mode," in *IEEE Transactions on Power Electronics*, vol. 16, no. 4, pp. 482-492, Jul 2001.

[19] K. De Gusseme, D. M. Van de Sype, A. P. M. Van den Bossche and J. A. Melkebeek, "Digitally controlled boost power-factor-correction converters operating in both continuous and discontinuous conduction mode," in *IEEE Transactions on Industrial Electronics*, vol. 52, no. 1, pp. 88-97, Feb. 2005.

[20] H. N. Le, K. Orikawa and J. I. Itoh, "Circuit-Parameter-Independent Nonlinearity Compensation for Boost Converter Operated in Discontinuous Current Mode," in *IEEE Transactions on Industrial Electronics*, vol. 64, no. 2, pp. 1157-1166, Feb. 2017.

[21] M. Liserre, F. Blaabjerg, and S. Hansen, "Design and Control of an LCL-Filter-Based Three-Phase Active Rectifier," *IEEE Trans. Power Electron.*, vol. 41, no. 5, pp. 1281-1291, Nov. 2005.

[22] C. Marxgut, F. Krismer, D. Bortis and J. W. Kolar, "Ultraflat Interleaved Triangular Current Mode (TCM) Single-Phase PFC Rectifier," in *IEEE Transactions on Power Electronics*, vol. 29, no. 2, pp. 873-882, Feb. 2014.

[23] H. N. Le and J. I. Itoh, "Wide-Load-Range Efficiency Improvement for High-Frequency SiC-Based Boost Converter With Hybrid Discontinuous Current Mode," in *IEEE Transactions on Power Electronics*, vol. 33, no. 2, pp. 1843-1854, Feb. 2018.

2526

The 2018 International Power Electronics Conference

A Theoretical Analysis on Static Characteristics of Voltage Based Control Method and Current Based Control Method for the Wayside Energy Storage System in DC-electrified Railway

Hiroyasu Kobayashi[1*], Keiichiro Kondo[1] and Diego Iannuzzi[2]

1 Department of Electrical and Electronic Engineering, Chiba University, Chiba, Japan
2 Department of Electrical Engineering and Information Technologies, University Federico II of Naples, Naples, Italy
*E-mail: hkobayashi@chiba-u.jp

Abstract— As one of the solution for energy saving in DC-electrified railway system, wayside energy storage systems (WESSs) are recently studied. In the previous study, current based control, which regulates current of ESSs, and voltage based control, which regulates the filter capacitor (FC) voltage, have been proposed as a power control method for WESSs. However, the difference between these two power control methods has not discussed before. In this paper, a theoretical comparison between voltage based power control method and current based power control method is carried out focusing on each static characteristic. Furthermore, an influence of distance between a WESS and a rail vehicle on characteristics of each control is also considered. Though the study in this paper, theoretical comparison between both control methods is carried out. Results of this paper contribute to establish a method to design WESSs which realize maximum energy saving effect with minimum energy capacity of energy storage devices.

Keywords—DC-electrified railway system, Power control strategy, Static characteristics, Wayside energy storage system

I. Introduction

Railways are energy-saving transportation measure because of the utilization of the regenerative brake and the low running resistance. However, if there are no other powering trains, regenerating train cannot apply the regenerative brake in the DC-electrified railway system. This is because diode rectifiers are generally used in substations in DC-electrified railway. This deteriorates the energy efficiency of the railway transportation system. Considering this technical issue, a lot of researches on a method for energy saving in DC-electrified railway system are carried out. As one of the solutions to cope with this technical issue, it is effective to apply energy storage devices (ESDs). There are two type of introduction of ESDs, one is to introduce ESDs along wayside of DC-electrified railway [1]-[5], and the other is to introduce on board [6]-[10]. In the case of the on board ESSs, regenerative brake energy can be stored to the on board ESD completely. However, the energy saving effect of wayside energy storage systems (WESSs) is limited because of the feeder resistance between the WESS and the vehicle. However, WESSs generally have an advantage in enough space for ESDs because it is implemented on the ground. On the other hand, space for ESDs is limited in the case of on board ESSs. Considering these features and the cost for ESDs, it is necessary for the power control strategy of WESSs to realize maximum energy saving effect with minimum energy capacity.

"Current based control," which control charge or discharge current of ESDs are commonly applied as a power control method of the WESS [3] [11]. Against this conventional method, we proposed "voltage based control," which control voltage of feeder line side [12-14]. In railway vehicle traction, it is common to apply current control of the traction motor. From this viewpoint, the load for wayside ESD can be regarded as a current source. Thus, the power converter of the wayside ESD should be controlled as a voltage source so as to control the power flow to the load which behaves as a current source.

For the design of WESS for the purpose of energy saving, it is desirable to realize maximum energy saving effect with minimum energy capacity as mentioned before. According to reference [12] and [13], the proposed voltage based control has an advantage for this purpose. However, a theoretical comparison between the voltage based power control method and current based power control method has not been studied before. In this paper, an analysis on the static characteristics of both control method is carried out. Furthermore, the effect of the distance between the vehicle and the ESD on the static characteristics of both power control strategies is also discussed. This is because the voltage drop at the feeder resistance which varies according to the position of the vehicle has an influence on the power flow between the vehicle and the ESS.

2527

The 2018 International Power Electronics Conference

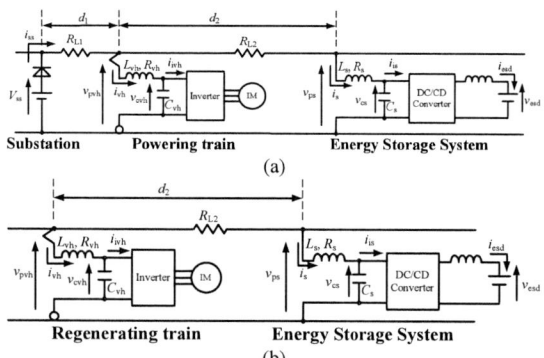

(a)

(b)

Fig. 1. Assumed circuit configuration: (a) vehicle powering, (b) vehicle regenerating

TABLE I
DEFINITION OF VARIABLES

Meaning	Symbol	Unit
Output voltage of the substation	v_{ss}	[V]
Output voltage of the vehicle	v_{pvh}	[V]
Output voltage of the ESS	v_{ps}	[V]
FC voltage of the vehicle	v_{cvh}	[V]
FC voltage of the ESS	v_{cs}	[V]
Terminal voltage of the ESD	v_{esd}	[V]
Output current of the substation	i_{ss}	[A]
Output current of the vehicle	i_{vh}	[A]
Output current of the ESS	i_s	[A]
Current of the traction inverter of the vehicle	i_{ivh}	[A]
Current of the DC/DC converter of the ESS	i_{is}	[A]
Current of the ESD	i_{esd}	[A]
Reference value of FC voltage of the ESS	v_{cs}^{*}	[V]
Reference value of current of the ESD	i_{esd}^{*}	[A]
Distance between the substation and the vehicle	d_1	[km]
Distance between the vehicle and the ESS	d_2	[km]
Duty of the DC/DC converter of the ESS	α	-
FL resistance of the vehicle	R_{vh}	[Ω]
FL resistance of the ESS	R_s	[Ω]
FL inductance of the vehicle	L_{vh}	[H]
FL inductance of the ESS	L_s	[H]
FC capacitance of the vehicle	C_{vh}	[F]
FC capacitance of the ESS	C_s	[F]
Feeder resistance between the substation and the vehicle	R_{L1}	[Ω]
Feeder resistance between the vehicle and the ESS	R_{L2}	[Ω]

II. Power control strategy for WESS

A. Assumed Circuit Configuration

Assumed circuit configuration is shown in Fig. 1. Fig. 1 (a) shows the circuit configuration when the vehicle powering, and Fig. 1 (b) shows the circuit configuration when the vehicle regenerating. Table I shows definition of each variable. In Fig. 1(b), the substation is assumed to be turned-off. Here, steady state equivalent circuits for each case are shown in Fig. 2. In DC-electrified railway system, it is common to apply current control for traction motor drive. Therefore, the inverter and traction motor are represented by current source model in Fig. 2. Also, in terms of the model of ESD, voltage source model and current source model are corresponding to voltage based

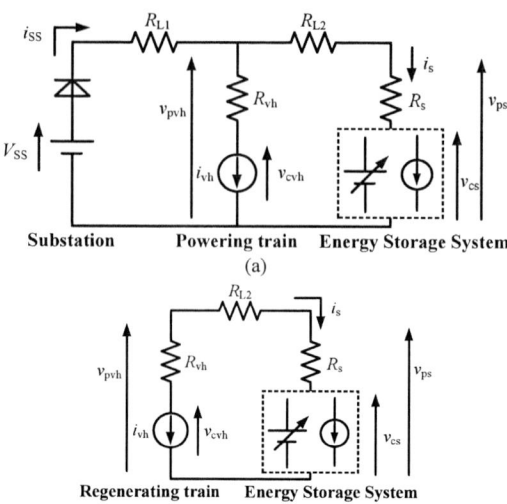

(a)

(b)

Fig. 2. Steady state equivalent circuit model: (a) vehicle powering, (b) vehicle regenerating

power control method and current based power control method, respectively.

From circuit equations in Fig. 2 (a), (1)–(5) are introduced.

$$V_{ss} - v_{pvh} = R_{L1} i_{ss} \tag{1}$$

$$v_{ps} - v_{pvh} = -R_{L2} i_s \tag{2}$$

$$i_{vh} = i_{ss} - i_s \tag{3}$$

$$v_{pvh} = v_{cvh} + R_{vh} i_{vh} \tag{4}$$

$$v_{ps} = v_{cs} + R_s i_s \tag{5}$$

In Fig. 2 (b), (2), (4) and (5) are also established. Furthermore, (6) is introduced from the circuit equation in Fig. 2(b).

$$i_{vh} = -i_s \tag{6}$$

B. Voltage Based Control Strategy

Fig. 3 shows a block diagram of voltage based power control system. In Fig. 3, v_{cs}^{*} and i_{esd}^{*} stands for the reference value of FC voltage v_{cs} and current of ESD i_{esd}, respectively. Also, k_{vp} stands for a proportional gain of Automatic Voltage Regulator (AVR), and α stands for duty of the DC/DC converter. In this paper, i_{esd} equals to i_{esd}^{*} because Automatic Current Regulator (ACR) is considered as ideal in this analysis. Also, v_{cs} equals to v_{cs}^{*} because the AVR is assumed as ideal. Concerning to v_{cs}^{*}, a method to determine v_{cs}^{*} is shown in Fig. 4. v_{cs}^{*} is determined according to (7).

$$v_{cs}^{*} = V_{mid}^{*} + k_{Vref}(v_{ps} - V_{mid}^{*}) \tag{7}$$

Where k_{Vref} stands for the slope of the pattern for

2528

The 2018 International Power Electronics Conference

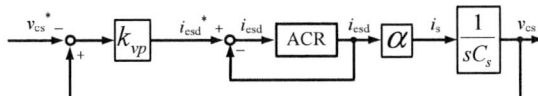

Fig. 3. Block diagram of voltage based control system.

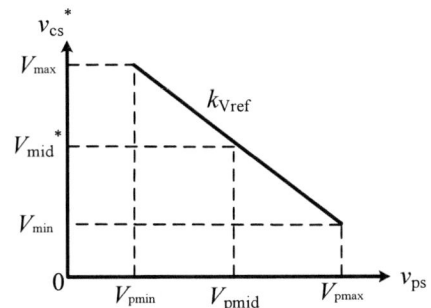

Fig. 4. Voltage based pattern for determination of v_{cs}^*.

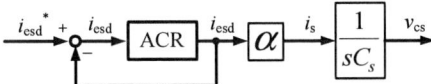

Fig. 5. Block diagram of current based control system.

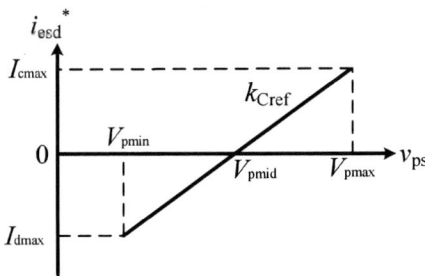

Fig. 6 Current based pattern for determination of i_{esd}^*.

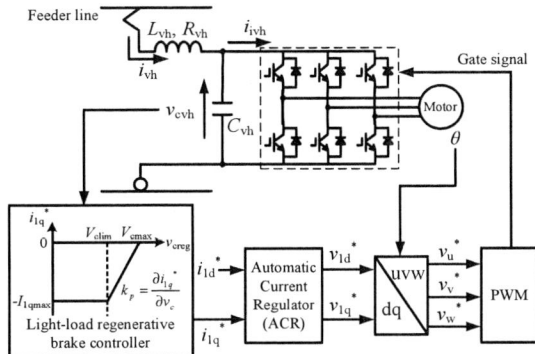

Fig. 7. Control pattern of torque current in "light-load regenerative brake control."

which is located far away from regenerating vehicle by increasing voltage of regenerating vehicle. On the other hand, in the regeneration, the motor torque control according to the input voltage of the traction inverter is commonly applied to railway vehicle for the sake of control of regenerative power corresponding to the load power. This control in the regeneration is called as "light-load regenerative brake control (Fig. 7)." When this control worked, the shortage of the braking force is compensated by the mechanical brake to keep the constant deceleration ratio.

Equation (10) is established because the power of AC side and DC side of the inverter should be equaled. Here, v_{1d} and i_{1d} is d-axis voltage and current, respectively. v_{1q} and i_{1q} is q-axis voltage and current, respectively.

$$v_{cvh} i_{vh} = v_{1d} i_{1d} + v_{1q} i_{1q} \tag{10}$$

In this analysis, the loss of inverter is not considered, and $v_{1d} i_{1d}$, which stands for the stator cupper loss by the stator d-axis current, is small enough compared with $v_{1q} i_{1q}$. The inverter is modeled in this paper as (10').

$$v_{cvh} i_{vh} \cong v_{1q} i_{1q} \tag{10'}$$

The light-load regenerative brake control which is shown in Fig. 7 is expressed by (11). In Fig. 7, I_{qmax} is the maximum q-axis current. V_{cmax} is fixed to avoid overvoltage of traction inverters. In terms of the control gain k_p, which is corresponding to V_{clim}, increasing k_p makes it possible to transmit regenerative power at high v_{cvh} during regeneration.

$$i_{1q}^* = \begin{cases} -I_{1qmax} & (v_{cvh} \leq V_{clim}) \\ -k_p(V_{cmax} - v_{cvh}) & (V_{clim} < v_{cvh} \leq V_{cmax}) \\ 0 & (V_{cmax} < v_{cvh}) \end{cases} \tag{11}$$

If q-axis current i_{1q} is completely controlled as its reference value i_{1q}^*, (12) is obtained by substituting (11) into (10').

determination in Fig. 4. In (7), V_{mid}^* is set to the same value as V_{pmid} in this analysis.

C. Current Based Control Strategy

Fig. 5 shows a block diagram of current based power control system. In Fig. 5, i_{esd} equals to i_{esd}^* because ACR is considered as ideal the same as Fig. 3. Then, (8) is introduced based on Fig. 5.

$$i_s = \alpha i_{esd}^* \tag{8}$$

Fig. 6 shows a method to determine. i_{esd}^* according to v_{ps}. i_{esd}^* is determined according to (9).

$$i_{esd}^* = k_{Cref}(v_{ps} - V_{pmax}) + I_{cmax} \tag{9}$$

In (9), k_{Cref} stands for the slope of the pattern in Fig. 6.

D. Static Characteristics of Vehicle Regenerating

In DC-electrified railway system, it is possible to transmit regenerative power to the other powering vehicle

2529

TABLE II
PARAMETERS FOR THE THEORETICAL COMPARISON

Meaning	Symbol	Value	Unit
Feeder circuit and regenerative brake control			
Feeder resistance	R_e	0.05	[Ω/km]
Distance between substation and vehicle	d_1	5.0	[km]
Distance between vehicle and ESD	d_2	5.0	[km]
No-load voltage of substation	V_{ss}	1620	[V]
FL resistance of vehicle	R_{vh}	0.1	[Ω]
FL resistance of ESD	R_s	0.1	[Ω]
FC capacitance of ESD	C_s	0.15	[F]
Voltage to start to apply light-load regenerative brake control	V_{clim}	1700	[V]
Voltage to stop applying light-load regenerative brake control	V_{cmax}	1800	[V]
Maximum q-axis current for 1 motor	I_{1qmax}	150	[A]
Number of motors of the assumed train	-	4	-
Parameters of voltage based control and current based control			
Maximum charge current	I_{cmax}	1500	[A]
Maximum discharge current	I_{dmax}	-1500	[A]
Maximum reference voltage of voltage based control	V_{max}	1800	[V]
Minimum reference voltage of voltage based control	V_{min}	1440	[V]
Maximum voltage of connecting point to feeder line	V_{pmax}	1800	[V]
Minimum voltage of connecting point to feeder line	V_{pmin}	1440	[V]
Medium voltage of connecting point to feeder line	V_{pmid}	1620	[V]
Proportional gain of AVR	k_{vp}	13.7	[A/V]
Duty of DC/DC converter	α	0.44	-
Time constant of AVR	T_{avr}	0.025	[sec]

TABLE III
SPECIFICATIONS OF THE ASSUMED ESD (LITHIUM-ION BATTERY)

Meaning	Value	Unit
Rated voltage of the LiB module	29.6	[V]
Storable energy of LiB module	1406	[Wh]
Maximum current of LiB module	125	[A]
Number of series LiB module	24	
Number of parallel LiB module	12	

$$i_{vh} = \begin{cases} -I_{1qmax}v_{1q}v_{cvh}^{-1} & (v_{cvh} \le V_{clim}) \\ -k_p v_{1q}(V_{cmax}v_{cvh}^{-1}-1) & (V_{clim} < v_{cvh} \le V_{cmax}) \\ 0 & (V_{cmax} < v_{cvh}) \end{cases} \tag{12}$$

III. Theoretical analysis on static characteristics

A. Conditions for Comparison between voltage based control and current based control

For the comparison of the static characteristics for each control method, conditions for the comparison should be considered. Table II shows established conditions for the comparison. Also, Table III shows the assumed specification of ESD in this analysis (Lithium-ion Battery). In terms of current based control strategy, the I_{cmax} and I_{dmax} are determined according to the rated current of the ESD. On the other hand, the maximum voltage V_{max} in voltage based control is set to V_{cmax} in the maximum voltage of light-load regenerative brake

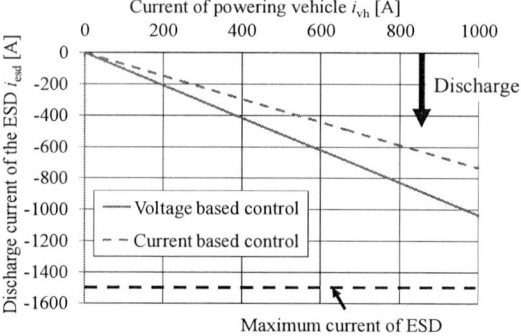

Fig. 8. Static characteristics of the ESD current i_{esd} when the vehicle powering.

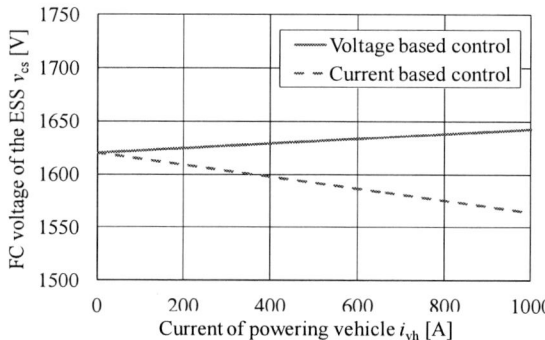

Fig. 9. Static characteristics of FC voltage of the ESS v_{cs} when the vehicle powering.

control (Fig. 7). V_{pmin}, V_{pmid} and V_{pmax} are set to the same values among the voltage based control and the current based control. k_{vp} is designed to realize AVR with time constant T_{avr} based on Fig. 3. Equation (13) shows the closed loop transfer function $G_{Vcl}(s)$. In (13), the ACR is assumed ideal.

$$G_{Vcl}(s) = \frac{1}{1+\dfrac{C_s}{\alpha k_{vp}}s} \tag{13}$$

Based on (13), k_{vp} which realizes the AVR with time constant T_{avr} can be designed by (14).

$$k_{vp} = \frac{C_s}{\alpha T_{avr}} \tag{14}$$

B. A static analysis during vehicle powering

The static characteristics of the ESS when the vehicle powering are shown in Fig. 8-10 using (1)–(5), (7)–(9) and TABLE II–III. Fig. 8 and Fig. 9 show the characteristics of current of the ESD i_{esd} and FC voltage of ESS v_{cs}, respectively. Fig. 8 shows that ESD current i_{esd} in the case of voltage based control is higher than that

The 2018 International Power Electronics Conference

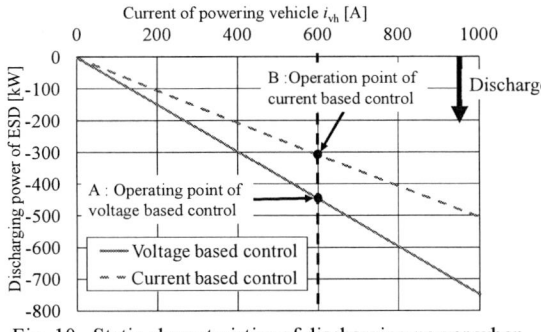

Fig. 10. Static characteristics of discharging power when the vehicle powering.

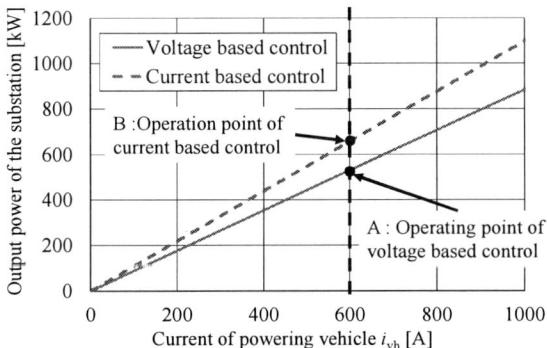

Fig. 11. Static characteristics of output power of the substation when the vehicle powering.

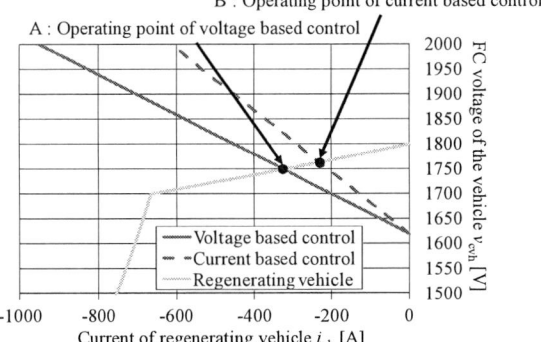

Fig. 12. Static characteristics when the vehicle regenerating.

of current based control. Fig. 8 also shows that i_{esd} doesn't exceed its maximum current in all the range of i_{vh} in both control strategies. Furthermore, i_{esd} in the case of current based control is less than that of voltage based control because i_{esd} is restricted by the pattern in Fig. 6. This restriction results in less discharge current of ESD. From Fig. 9, the voltage drop of v_{cs} is obliviously decreased in the case of voltage based control because of the effect of AVR in voltage based control strategy.

Fig. 10 and Fig. 11 shows the static characteristic of discharging power of the ESD and output power of the substation, respectively. As shown in Fig. 10, the output power in the case of voltage based control is higher than the case of current based control. For example, if the vehicle consumes 600 A, the values of output power in the case of voltage based control and current based control are 446 kW (operating point A) and 308 kW (operating point B), respectively. Fig. 11 shows that the output power of the substation in the case of voltage based control is lower than that of current based control. For example, if the vehicle consumes 600 A, the values of output power of the substation in the case of voltage based control and current based are 530 kW (operating point A) and 658 kW (operating point B). This result shows the contribution of voltage based control to reduction of output power of the substation.

C.   A static analysis during vehiecle regenerating

The static characteristics when the vehicle powering are shown in Fig. 12 using (2), (4)–(9) and TABLE II–III. In terms of the static characteristics of regenerating vehicle, (12) is utilized to describe. Fig. 12 shows that v_{cvh} in the case of voltage based control is decreased compared to that of current based control because of the effect of AVR. Considering the light-load regenerative brake control, regenerative brake power is increased by keeping lower v_{cvh}. In Fig. 12, point A and B stands for the operating point of voltage based control and current based control, respectively. From the coordinate of the operating point A, the regenerative power in the case of voltage based control is 568 kW. On the other hand, the regenerative power in the case of current based control is 407 kW from the operating point B. From this result, the effect on increase of regenerative brake power in voltage based control is verified.

D.   An analysis on the distance between the vehicle and the ESD

In railway system, feeder resistance between vehicles and other elements, such as substations, ESDs and so on, is always changed according to the position of vehicles. Therefore, R_{L1} and R_{L2} in Fig. 1 also vary according to the position of the vehicle. The feeder resistance occurs voltage drop, and that has an influence on the power flow in the DC-electrified railway system. In this section, an analysis on the effect of the distance between the vehicle and the ESD is investigated. d_1 and d_2 vary from 0 km to 10 km. The distance between the substation and the ESS is constant (10 km): the sum of d_1 and d_2 is set to 10km in this analysis. Also, i_{vh} when the vehicle powering is assumed to 600 A.

Fig. 13 shows the result of i_{esd} when the vehicle powering. As shown in Fig. 13, voltage based control strategy has the advantage form the viewpoint of assisting the powering vehicle in all the range of d_2. Fig. 12 also shows that the difference of i_{esd} between voltage based control and current based control becomes less when d_2 becomes longer. This is because the power supply from the substation becomes dominant in the case

2531

The 2018 International Power Electronics Conference

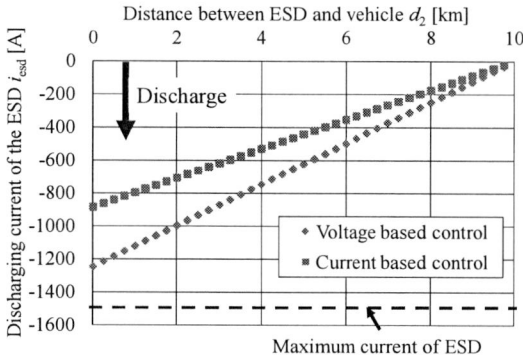

Fig. 13. Static characteristics of i_{esd} in powering when d_2 varies.

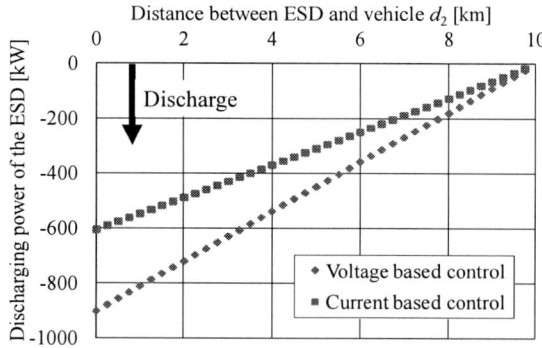

Fig. 14. Results of static analysis on the effect of d_2 when the vehicle powering.

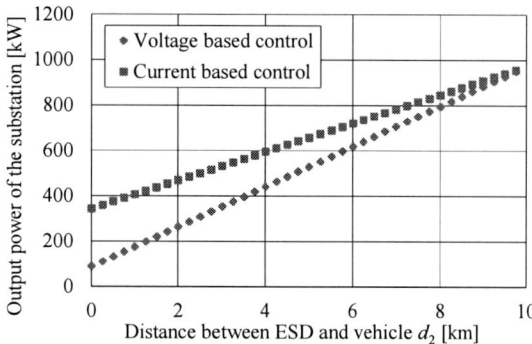

Fig. 15. Static characteristics of output power of the substation in powering when d_2 varies.

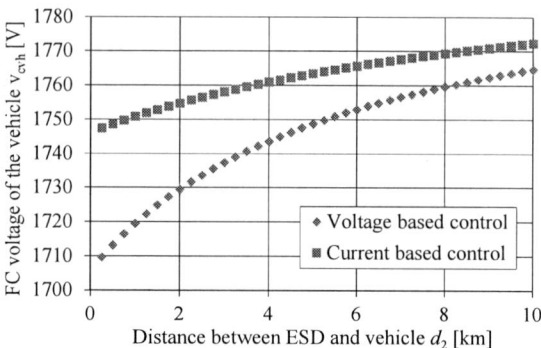

Fig. 16. Static characteristics of v_{cvh} in regenerating when d_2 varies.

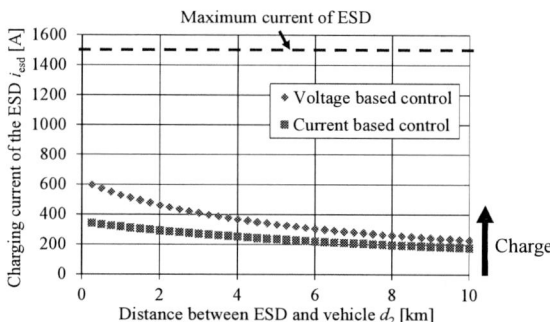

Fig. 17. Static characteristics of i_{esd} in regenerating when d_2 varies.

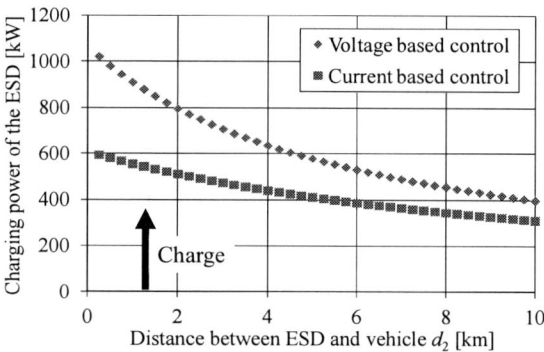

Fig. 18. Results of static analysis on the effect of d_2 when the vehicle regenerating.

of longer d_2. Fig. 13 also shows that i_{esd} isn't over the maximum current of the ESD in both control strategies. Fig. 14 and Fig. 15 show the results of output power of the ESD and the substation, respectively. Both figures show the advantage of voltage based control in all the range of d_2.

Fig. 16 and Fig. 17 show the result of FC voltage of the vehicle and charge current i_{esd} when the vehicle regenerating, respectively. Fig. 18 also shows the charge power of the ESD. As shown in Fig. 16, v_{cvh} in the case of voltage based control is lower compared to the case of current based control in all the range of d_2. As discussed in chapter III-C, lower v_{cvh} results in higher regenerative

brake power considering the light-load regenerative brake control. Therefore, charge current i_{esd} and power in the case of voltage based control become higher than that of current based control in all the range of d_2 [see Fig. 17 and Fig. 18]. Fig. 17 also shows that i_{esd} isn't over the maximum current of the ESD in both control strategies.

IV. CONCLUSIONS

In this paper, a theoretical analysis on static characteristics of voltage based control and current based control for WESSs is carried out. Also, the effect of the

2532

distance between vehicle and ESD on the static characteristics of both power control strategies is investigated.

According to the obtained static characteristics of both control strategies, current of the ESD is restricted by its referential pattern in the case of current based control. This results in decrease of charge and discharge power of the ESD. On the other hand, higher charge and discharge power can be obtained in the case of voltage based control. It can be mentioned that controlling FC voltage of the ESS has the advantage compared to the controlling output current of ESS. Results when the vehicle moves also shows the advantage of voltage based control strategy in the all the range of distance between the vehicle and the ESS (0 km to 10 km).

Results of this paper contribute to establish a method to design wayside ESSs which realize maximum energy saving effect with minimum energy capacity.

REFERENCES

[1] A. B. Turner, "A Study of Wayside Energy Storage Systems (WESS) for Railway Electrification," *IEEE Trans Ind. Appl.*, vol. 1A-20, no. 3, pp. 484-492, May. /Jun. 1984.

[2] F. Ciccarelli, D. Iannuzzi, K. Kondo, and L. Fratelli, "Line-Voltage Control Based on Wayside Energy Storage Systems for Tramway Networks," *IEEE Trans. Power Electron.*, Vol. 31, no. 1, pp. 884-899, Jan. 2016.

[3] Z. Li, S. Hoshina, N. Satake, and M. Nogi, "Development of DC/DC converter for Battery Energy Storage Supporting Railway DC Feeder Systems," *IEEE Trans. Ind. Appl.*, vol. 52, no. 5, pp. 4218-4224, Sep./Oct. 2016.

[4] S. de la Torre, A. J. Sanches-Racero, J. A. Aguado, M. Reyes, and O. Martinez, "Optimal Sizing of Energy Storage for Regenerative Braking in Electric Railway Systems," *IEEE Trans. Power Systems*, vol. 30, no. 3, pp. 1492-1500, May. 2015.

[5] T. Ratniyomchai, S. Hillmansen, and P. Tricoli, "Optimal Capacity and Positioning of Stationary Supercapacitors for Light Rail Vehicle Systems," in *Proc. IEEE SPEEDAM*, Ischia, Italy, 2014, pp.807-812.

[6] Masamichi Ogasa, "Application of Energy Storage Technologies for Electric Railway Vehicles—Examples with Hybrid Electric Railway Vehicles," *IEEJ Trans. Electrical and Electronic Engineering*, vol. 5, Issue 3, pp. 304-311, May. 2010.

[7] D. Iannuzzi, E. Pagano, P. Tricoli, "The use of energy storage systems for supporting the voltage needs of urban and suburban railway contact lines," Energies, 6 (4), pp. 1802-1820, Mar. 2013, doi: 10.3390/en6041802

[8] F. Ciccarelli, D. Iannuzzi, D. Lauria, "Supercapacitors-based energy storage for urban mass transit systems," Proceedings of the 14th European Conference on Power Electronics and Applications (EPE 2011), art. no. 6020597. ISBN: 978-161284167-0

[9] J. P. Torreglosa, P. Garcia, L. M. Fernandez, and F. Jurado, "Predictive Control for the Energy Management of a Fuel – Cell – Battery - Supercapacitor Tramway," *IEEE Trans. Ind. Informat.*, vol. 10, no. 1, pp.276-285, Feb. 2014.

[10] T. Saito and K. Kondo, "Implementation method of loss observer to power controller for overhead line and supercapacitor hybrid electric railway vehicle," *IEEJ Trans. Electrical and Electronic Engineering*, vol. 11, Issue S2, pp. S108-S115, Dec. 2016.

[11] F. Ciccarelli, D. Iannuzzi, P. Tricoli, "Speed-based supercapacitor state of charge tracker for light railway vehicles," Proceedings of the 14th European Conference on Power Electronics and Applications (EPE 2011), art. no. 6020254.ISBN: 978-161284167-0

[12] H. Kobayashi, J. Asano, T. Saito and K. Kondo: "A Power Control method to Save Energy for Wayside Energy Storage Systems in DC-electrified Railway System" , *IEEJ Trans. Ind. Appl.*, Vol.135, No.4, pp. 386-394, Apl. 2015. (in Japanese)

[13] H. Kobayashi, S. Akita, T. Saito, and K. Kondo, "A Voltage Basis Power Flow Control for Charging and Discharging Wayside Energy Storage Devices in the DC-electrified Railway System," in *Proc. 19th Int. Conf. Electr. Mech. Syst. (ICEMS)*, Chiba, Japan, 2016, pp. 1-6

[14] F. Ciccarelli, D. Iannuzzi, D. Lauria and P. Natale, "Optimal Control of Stationary Lithium-Ion Capacitor-Based Storage Device for Light Electrical Transportation Network," *IEEE Trans. Trans. Electrification*, vol. 3, no. 3, pp. 618-631, Sept. 2017, doi: 10.1109/TTE.2017.2739399

Improvement of a DC electrical railway simulator using artificial intelligence

Alvaro J. Lopez-Lopez[1*], Ramon R. Pecharroman[1], Antonio Fernandez-Cardador[1] and Asuncion P. Cucala[1]

1 Institute for Research in Technology, ICAI School of Engineering, Comillas Pontifical University, Madrid, Spain

*E-mail: alvaro.lopez@iit.comillas.edu

Abstract- **Electrical railway simulators play a critical role in mass rapid transit system (MRTS) studies. In most cases, MRTSs are DC-electrified systems which include elements that exhibit different electrical states, i.e. traction substations may be in ON or OFF modes and braking trains may be in power or voltage (rheostat) modes. This adds complexity to the electrical problem to be solved by the simulator.**

The simulator developed by the authors in previous works includes a module in charge of determining the electrical states of all the elements in the system. The block, based on heuristic rules, demands high computation times under certain circumstances.

This paper presents an upgrade of the heuristic block where artificial intelligence (AI) is used to obtain the electrical states of substations and trains. A neural network (NN) classification model is applied and compared with the previous approach by means of set of simulations. The results show that the NN approach outperforms the previous one.

Keywords— Electrical multi-train simulation, Machine Learning, Mass Rapid Transit Systems.

I. INTRODUCTION

Research on improving the electrical topology of mass rapid transit systems (MRTSs) focuses on several hot topics. Some examples are the way the system provides service to running trains in terms of power and voltage [1], the optimization of the regenerative energy use [2]-[6], etc.

Regardless their scope, in the majority of cases MRTSs electrical studies are carried out by means of electrical multi-train simulators. In general, these simulators must handle the time evolution of trains' positions and powers and the solution of the load flow problem for each time instant under analysis. There are two features of the load-flow problem in an MRTS that make it complex-to-solve: 1) loads (trains) exhibit a nonlinear behavior, so node voltages must be obtained with iterative methods; 2) electrical power substations and braking trains exhibit different electrical states depending on the rest of the elements in the system, which makes it necessary to verify iteration solutions not only from the mismatch perspective, but also checking out that element rules are accomplished.

Whereas the former condition is common to the classical load-flow problem in power systems and has been thoroughly tackled with methods like the Newton-Raphson's, the latter is a particular characteristic of MRTS which must be carefully handled to find the load-flow solution, and to make it in a reasonable time.

Therefore, one of the main concerns when designing the core of an electrical railway simulator is to deal with a module that makes it possible to find the states of the elements that feature different electrical modes [power

substations (PSs) and braking trains (BTs)] [7]-[11]. In the simulator developed by the authors' workgroup, this task is tackled by means of a heuristic module that changes the states of the multi-state devices as a function of the evolution of the partial load-flow solution results [12].

In this paper, we present a novel method to find the electrical states of substations and braking trains in MRTSs which is based on neural networks (NNs). The aggregation of information from previous simulations into a classification model allows including the electrical states of the system elements into the NR initial guess, reducing the number of iterations and thus boosting the computational efficiency of the simulator.

The paper is organized as follows:

- Section II. poses the main characteristics of the problem we tackle in this paper and presents the previous heuristic solution in the simulator developed by the authors.
- Then, Section III. presents the model to identify the element electrical states that we propose in this paper.
- Once the model is presented, Section IV. includes the main results both from the model training and simulator efficiency perspectives.

II. ELECTRICAL-STATE DETERMINATION: BACKGROUND

As stated in Section 0, one of the main tasks to be tackled for the solution of DC railway load flows is to determine the states of power substations and braking trains. It is important to note that this task consists in a combinatorial problem which possible solutions soar as the number of PSs and BTs in the system increases.

In general, load flow solvers must verify that partial solutions, i.e., results of the iterative solution method before convergence is attained; accomplish the physical rules of the system. E.g.: diode PSs do not handle negative currents and BT voltages may not be higher than the maximum voltage allowed.

The simulator developed by the authors' workgroup makes use of the NR algorithm. Therefore, for each iteration it calculates the system Jacobian matrix and updates the decision variables so that the mismatches are improved. Then, as described in Section 0, the solution algorithm must verify that the contour conditions of the multi-state elements in the system are not violated. To do that, the algorithm calls a checkout routine that is also in charge of finding a new guess for the system electrical states if necessary. The flowchart of this routine is presented in Fig. 1. The module checks whether the partial solution violates the system rules. If it does not, it is validated; if it does, the states of the elements that violate rules are changed and the solution is tagged as invalid. Note that once a type of element violates the system rules, the routine performs state changes and exits (it does not continue to check other elements). This strategy and the order of the state checkouts has been observed to lead to the best convergence results, measured as the number of iterations required to obtain the actual electrical configuration of the system.

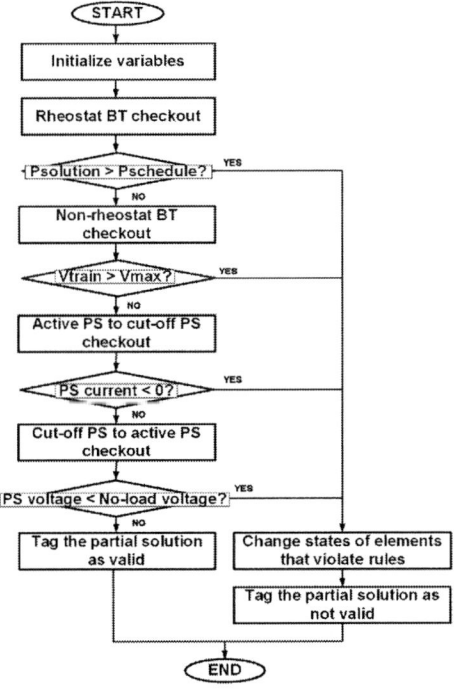

Fig. 1 Flowchart of the partial-solution validation module.

This module has been designed to avoid deadlocks, which may take place if the routine bounces between two stable electrical-state guesses. Nevertheless, in some difficult-to-solve configurations, the number of iterations devoted to find the system electrical topology affects the computational efficiency of the system. This may be improved by enhancing the initial electrical-state guess, which is currently set to a flat guess where all PSs are in ON mode and all BTs are in power mode (not rheostatic braking).

III. MODEL FOR THE DETERMINATION OF MRTS ELECTRICAL STATES

It is clear that the identification of the electrical states at each instant consists in a classification problem. Thus, the problem must be tackled by selecting a type of classification model and by identifying the variables that have a clear influence in the electrical states.

Regarding the former, for it is known to be highly performant if properly designed, we have selected a NN model both for the PS and BT electrical-state determination. Regarding this, it is important to note that the determination of the electrical states of PSs and BTs

must be performed before solving the load flow, which influences the available features for the model.

Fig. 2 represents a simplified situation with 2 PSs and 2 trains. For each element, before starting the solution process only positions for all elements and scheduled powers for trains are available. There is no *a priori* information about voltages, for these are determined by the load flow.

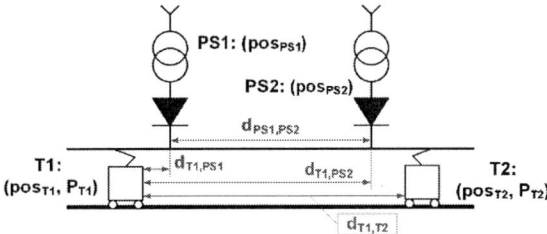

Fig. 2 Variables for the determination of the electrical states.

Nevertheless, the information available has a large influence on the element states. The studies in [13], [14] showed the influence of the relative locations between trains and PSs in the exchange of energy between the elements of the system. In this line, transforming the locations of elements to distance between elements may ease the classification.

Regarding these previous analyses, we propose two model structures for the two types of elements which electrical states must be determined.

The PS model is made up of a set of predictors computed from the locations of trains and PSs and the power consumed by trains. Equation (1) presents the vector of train powers for a certain system operated with a headway such that there are N trains running on the tracks.

$$\boldsymbol{X}_{PT} = [x_{PT1}, \dots, x_{PTN}] \tag{1}$$

$$x_{PTi} = \begin{cases} P_{Ti} & if \quad Train\ i\ is\ in\ the\ sytem \\ 0 & otherwise \end{cases} \tag{2}$$

Equation (2) presents the power assigned to each train on the tracks. It must be noted that in the load flow simulation approach the end-of-line operations are not included and therefore the number of running trains varies slightly. Since the number of features of the model must be fixed, we set the train power to 0 in these situations.

Equation (3) presents the vector of distance features included in the PS model for a system with M PSs.

$$\boldsymbol{X}_D \tag{3}$$
$$= [d_{T1,PS1}, \cdots, d_{T1,PSM}, \cdots, d_{TN,PS1}, \cdots, d_{TN,PSM}]$$

$$d_{PTi,PSj} = Fpos_{Ti} - pos_{PSj} \tag{4}$$

$$Fpos_{Ti} = \begin{cases} pos_{Ti} & if \quad Train\ i\ is\ in\ the\ sytem \\ P1 & otherwise \end{cases} \tag{5}$$

The signed distance between each train and PS is calculated as the difference between the train position and the PS one (Equation (4)). The train position is previously transformed: if a certain train is not present in the line for a given sample, its position is set to P1, which must be set to a large number (Equation (5)).

Equation (6) presents the set of features included in the PS model.

$$X = [X_{PT} \quad X_D \quad sign(X_D) \circ X_D^{\circ 2}] \tag{6}$$

The powers and distances presented in Equations (1) and (3) are complemented by the signed squared distances, which has been expressed in Equation (6) by the Hadamard product. The reason for the increase in the classification power derived from these features is that the important distance between elements is not the Euclidean but the electrical one.

The model includes M outputs that are the states of the M PSs in the system.

Regarding the BT model, it must be noted that the nature of this classification problem is rather different from the PS state one. First, the state prediction is restricted to braking trains, which leads to a number of predictions that varies from snapshot to snapshot. To tackle this situation, samples are generated by referencing all the PSs and the rest of trains to the BT to which the prediction applies. Then, the BT splits the line into two subsets: the stretch of line before the BT (including all the PSs and trains on this part of the line) and the stretch after the BT.

To have a fixed feature structure, the model reserves place for N-1 trains and M PSs before and after the BT which state is to be predicted. If there are less trains or PSs in the stretch before or after the BT, the spare predictors are filled with fictitious elements.

Equation (7) shows the structure of train powers before and after the BT, where spare train locations before and after are filled with zeros (Equation (8)).

$$X_{PTB} = [x_{PTB1}, \dots, x_{PTBN-1}] \tag{7}$$
$$X_{PTA} = [x_{PTA1}, \dots, x_{PTAN-1}]$$

$$x_{PTBi} = \begin{cases} P_{TBi} & if \ \ Train \ i \ is \ in \ the \ sytem \\ 0 & otherwise \end{cases} \tag{8}$$

Equation (9) presents the structure of N-1 distances before and after the BT. Since the line is already split in the before-after structure, the sign is no longer necessary for distances (Equation (10)). Spare trains are located in a far-away position (P1, Equation (11)).

$$X_{DTB} = [d_{T,TB1}, \cdots, d_{T,TBN-1}] \tag{9}$$
$$X_{DTA} = [d_{T,TA1}, \cdots, d_{T,TAN-1}]$$

$$d_{T,TBi} = pos_T - Fpos_{TBi} \tag{10}$$
$$d_{T,TAi} = Fpos_{TAi} - pos_T$$

$$Fpos_{TBi} = \begin{cases} pos_{TBi} & if \ TB_i \ is \ in \ the \ sytem \\ P1 & otherwise \end{cases} \tag{11}$$

In the BT model, the distances between the BT and PSs in both before and after stretches are included as shown in Equations (12) and (13). In this case, it is possible (it will be frequent, indeed) to have less than M PSs in the before or after stretch. For these features, a large enough position is assigned to this PS (P2, Equation (14))

$$X_{DPSB} = [d_{T,PSB1}, \cdots, d_{T,PSBM}] \tag{12}$$

$$X_{DPSA} = [d_{T,PSA1}, \cdots, d_{T,PSAM}]$$

$$d_{T,PSBj} = pos_T - Fpos_{PSBj} \tag{13}$$
$$d_{T,PSAj} = Fpos_{PSAj} - pos_T$$

$$Fpos_{PSBj} = \begin{cases} pos_{PSBj} & if \ PSB_j \ is \ in \ the \ sytem \\ P2 & otherwise \end{cases} \tag{14}$$

Finally, Equation (16) shows the features included in the BT model, which includes the net consumed power in the snapshot (see Equation (1)) and the power, distances to trains and distances to PSs in the before and after stretches. The squared distances are also included for the same reason that in the PS model (Equation (15)).

The model includes a single output that corresponds to the state of the BT under analysis.

$$X_B = [X_{PTB} \quad X_{DTB} \quad X_{DPSB} \quad X_{DTB}^{\circ 2} \quad X_{DPSB}^{\circ 2}] \tag{15}$$
$$X_A = [X_{PTA} \quad X_{DTA} \quad X_{DPSA} \quad X_{DTA}^{\circ 2} \quad X_{DPSA}^{\circ 2}]$$

$$X = \left[\sum_i x_{PTi} \quad X_B \quad X_A \right] \tag{16}$$

TABLE I summarizes the predictors included in each model.

TABLE I: MODELS FOR PS AND BT STATE CLASSIFICATION

Model		Inputs	Outputs
PS	1.	Train powers	ON / OFF MODE
	2.	Distances trains - PSs	
	3.	Squared distances trains - PSs	
BT	1.	Addition of train powers	POWER / VOLTAGE MODE
	2.	Train power B&A	
	3.	Distances to trains B&A	
	4.	Distances to PSs B&A	
	5.	Square distances to trains and PSs B&A	

IV. RESULTS

In this section we present the results obtained in this research from two perspectives: 1) the NN training, in order to verify that the models presented have good accuracy and generalization properties; and 2) the simulation of the system, to measure the improvement in the computational efficiency of the simulator derived from the application of the electrical-state determination models.

In both cases, the experiments consist in a set of electrical simulations performed in the case study-system presented in [13].

The following list contains the MRTS operation characteristics included in the experiments in this section:

- The headway has been set to 15 minutes.
- Dwell times and time shift between terminal stations are not fixed nor deterministic. Their values have been varied following the probability distributions of the model presented in [14].
- The traffic-disturbances included are not as large as to make it necessary to introduce a traffic regulation system.
- As explained in [14], a 15-minute headway traffic scenario consists of 900 snapshots that contain the

2536

evolution in time of trains' positions and powers.

A. Training

The NN models have been trained making use of the functions included in the MATLAB Neural Network Toolbox.

The classification error is presented in percent terms and expresses the number of instances where the model could not properly assess the state of a multi-state element. Then, for the selected model, the confusion matrix and Receiver Operating Characteristic (ROC) curve are presented to refine the classification accuracy.

In order to guarantee the adequate generalization capacity of the model, in all cases the set of samples used for the NN training has been split using 60 % of the samples for training, 20 % for validation (to decide when to stop training) and 20 % for test (to calculate the classification error).

The scaled conjugate gradient algorithm has been used to obtain the optimum NN coefficients. We have measured performance by means of the cross-entropy, and 6 validation checks have been used to prematurely stop the training process (2000 epochs otherwise).

PS Model

The first analysis performed consists in assessing the required number of samples in the training set. To do that, the NN layer structure has been set to two hidden layers of 250 and 50 neurons, respectively. Then, the number of samples has been varied from 1e3 to 300e3. Fig. 3 shows the NN learning curves obtained in this analysis. It may be observed that the model is not able to generalize until 50e3 samples are included in the input set. Then, a slight error decreasing trend is observed as the number of samples increases. The model selected to be used in the electrical multi-train simulator is the one obtained with 150e3 samples.

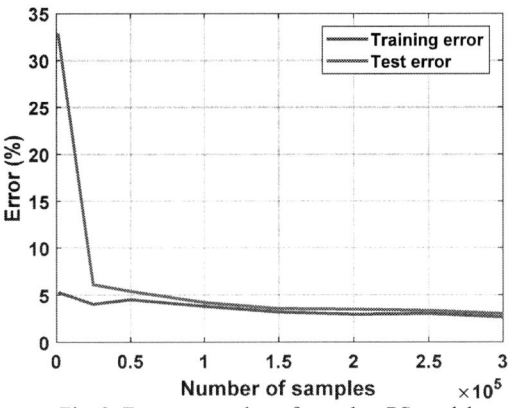

Fig. 3 Error vs. number of samples. PS model.

TABLE II presents the confusion matrix obtained with the selected model. The results are normalized by dividing by the total number of samples. It shows an accuracy in line with the expected one, and a slightly larger number of false negative predictions than false positive ones. Nevertheless, the sensitivity and specificity of the model meet a high standard.

TABLE II: Confusion matrix for the PS model

	PREDICTED **ON**	PREDICTED **OFF**
ACTUAL **ON**	0.51	0.02
ACTUAL **OFF**	0.01	0.46

Fig. 4 shows the ROC curve of the PS model computed in a validation set that is independent from the input model to the training stage, i.e. these samples were not even used for stopping the training stage. Only the results for PS 5 are shown. The rest of PSs follow a similar pattern. The area under curve (AUC) result is 0.98.

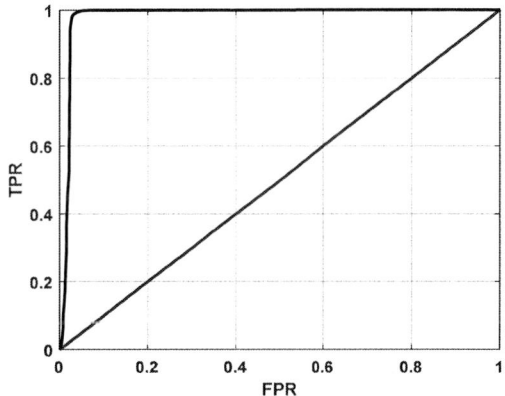

Fig. 4 Validation ROC curve for PS model (PS 5).

BT Model

The structure of the NN used for this model is the same than the one used for the PS model: a two-hidden-layer NN with 250 and 50 neurons, respectively. Fig. 5 presents the results obtained when the number of samples is varied from 1e3 to 300e3. The BT classification problem has revealed as a more complicated task than the BT one. Consequently, the BT model is less accurate than the PS one. The model accuracy is increased with the number of samples until around 200e3 samples. From that point on, it stabilizes. The accuracy for the validation set is in the range of 12.5 %, which is a poor value in comparison with the values obtained with the PS model. These results suggest that it could be interesting to look for new predictors (features) to increase the model accuracy. The model selected for the simulation exercise is the one with 200e3 samples.

The 2018 International Power Electronics Conference

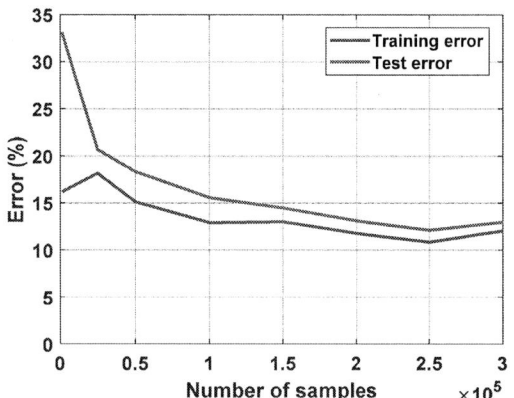

Fig. 5 Error vs. number of samples. BT model.

TABLE III presents the confusion matrix obtained with the selected BT model. The results are normalized by dividing by the total number of samples. It shows an accuracy in line with the expected one, and a slightly larger number of false negative predictions than false positive ones.

TABLE III: CONFUSION MATRIX FOR THE BT MODEL

	PREDICTED **ON**	PREDICTED **OFF**
ACTUAL **ON**	0.51	0.09
ACTUAL **OFF**	0.06	0.34

Fig. **6** shows the ROC curve of the BT model computed in a validation set that is independent from the input model to the training stage, i.e., like in the PS case, these samples were not even used for stopping the training stage. Now the AUC is 0.93. It is important to note that if the simulator shows to be more sensitive to false positives or vice-versa, the overall computation performance of the load flow may be increased without changing the model by simply varying the prediction threshold. The way the model would shift from sensitive to specific may precisely be observed in this ROC curve.

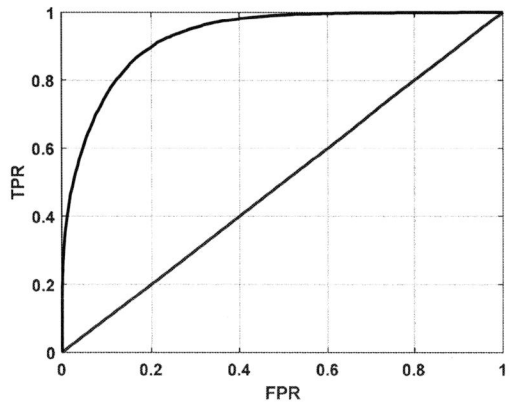

Fig. 6 Validation ROC curve for BT model.

B. Simulation

In this section we present the results obtained by including an initial state-guess step in the NR load flow solver.

We have performed the simulation of 10 15-min headway traffic scenarios, which account for 9000 snapshots.

Potential improvement

Prior to the presentation of the results obtained with the classification models in Subsection A. , Fig. 7 shows the number of iterations that takes NR algorithm to obtain the solutions of the load flows in the 9000 snapshots. The results are expressed in terms of the relative frequency observed for each number of iteration, and the figure includes the base case simulation (with the module presented in Section II.) and the case where the states of PSs and BTs are known *a priori* **without errors** (the so-called best guess case).

Whereas around 10 % of the snapshots are solved in 4 or less iterations in the base case, more than 50 % of them are solve with this number of iterations if the states are perfectly known before starting the NR process.

However, we observe a spike in 8 iterations for the perfect state prediction case (10 % of cases are solved with this number of iterations). In these cases, there are some state changes caused by the transient behavior of the NR method in its way to finding the load-flow solution. Of course, the checkout module cannot be disabled, for it guarantees that no wrong solution is validated, but it could be revisited to modulate the degree of confidence in electrical states. In this approach, the algorithm could skip some state changes if the initial states are likely to be true.

Regarding the computation-time results, the base case takes in average 2.7 seconds per traffic scenario. The best-guess case saves 26 % of the simulation time in average terms.

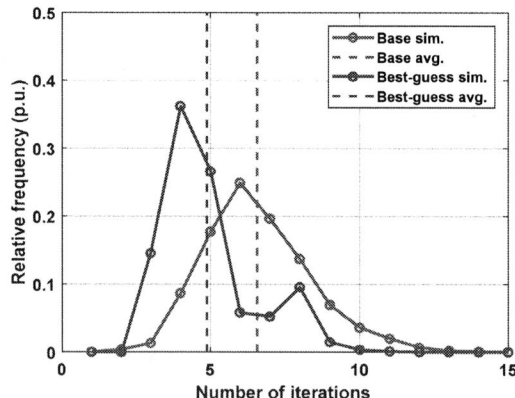

Fig. 7 Improvement in average number of NR iterations and simulation time when all states are known *a priori* without errors.

Actual improvement

Based on the results shown in Section IV. B. , it is clear that the models presented in Section III. will not be perfectly accurate in the electrical-state determination.

Fig. 8 presents the results obtained when the PS and BT models are used. The solution for the perfect model has been conserved to make it possible to assess the effect of the classification error in the simulation results. It can be observed that the inaccuracy of the classification models

2538

affects the simulation performance in terms of average number of NR iterations. Nevertheless, the initial state guess outperforms the flat-start approach, leading to 20 % savings in computation time.

To improve the results, in order to take full advantage of the potential improvement in the simulation performance from the electrical-state initial guess, both the BT model and the checkout module must be reviewed, searching for: 1) more accurate BT state classification and 2) better dynamic performance of the electrical-state changing of the checkout routine.

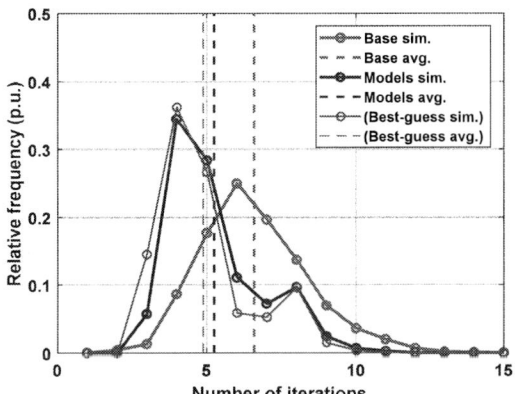

Fig. 8 Improvement in average number of NR iterations and simulation time with the PS and BT models.

V. CONCLUSIONS & FUTURE WORK

This paper presents a module based on NNs which improves the performance of an electrical multi-train simulator.

The module implements two classification models that outputs the electrical states of PSs and BTs, making it possible to start the load flow solution process with an improved guess of the electrical topology of the system.

As a result, 20 % computation time is obtained with respect to the previous model.

The analyses in the paper show that savings could be larger. To make it possible, further developments in the way the model is applied and in the models themselves (especially in the BT one) will be performed.

In the first line, it could be interesting to analyze the sensitivity of the solver to false positive and false negative predictions. This analysis is likely to lead to better computation results by simply biasing the model (changing the classification threshold and thus moving the models in the ROC curves).

In the second line, the results obtained in this paper suggest that there is room to improve the classification accuracy of the BT model by including new predictors or transforming the current ones.

Another interesting exercise would consist in comparing the computational performance of the multi-train electrical solver when the initial guess is not obtained from a classification model but by a regression one (which could implicitly give the electrical states as well). This will be tackled in future works in our research group.

Finally, the dynamic performance of the electrical-state

changing routine should be performed as well to adapt to the refined initial guess obtained by the new NN approach.

REFERENCES

[1] R. Takagi, "Energy Saving Techniques for the Power Feeding Network of Electric Railways," *IEEJ Transactions on Electrical and Electronic Engineering*, vol. 5, *(3)*, pp. 312-316, 05, 2010.
[2] C. H. Bae, "A simulation study of installation locations and capacity of regenerative absorption inverters in DC 1500 V electric railways system," *Simulation Modelling Practice and Theory*, vol. 17, *(5)*, pp. 829-838, 5, 2009.
[3] H. J. Chuang, "Optimisation of inverter placement for mass rapid transit systems by immune algorithm," *IEE Proceedings - Electric Power Applications*, vol. 152, *(1)*, pp. 61-71, 2005.
[4] S. de la Torre *et al*, "Optimal Sizing of Energy Storage for Regenerative Braking in Electric Railway Systems," *IEEE Transactions on Power Systems*, vol. 30, *(3)*, pp. 1492-1500, 2015.
[5] B. Wang *et al*, "An Improved Genetic Algorithm for Optimal Stationary Energy Storage System Locating and Sizing," *Energies*, vol. 7, *(10)*, pp. 6434, 2014.
[6] H. Xia *et al*, "Optimal Energy Management, Location and Size for Stationary Energy Storage System in a Metro Line Based on Genetic Algorithm," *Energies*, vol. 8, *(10)*, pp. 11618-11640, 2015.
[7] B. Mellitt, Z. Mouneimne and C. Goodman, "Simulation study of DC transit systems with inverting substations," *IEE Proceedings B - Electric Power Applications*, vol. 131, *(2)*, pp. 38-50, 1984.
[8] Y. Cai, M. R. Irving and S. H. Case, "Iterative techniques for the solution of complex DC-rail-traction systems including regenerative braking," *IEE Proceedings-Generation, Transmission and Distribution*, vol. 142, *(5)*, pp. 445-452, 09/01, 1995.
[9] M. Z. Chymera *et al*, "Modeling Electrified Transit Systems," *Vehicular Technology, IEEE Transactions On*, vol. 59, *(6)*, pp. 2748-2756, 2010.
[10] P. Arboleya, G. Diaz and M. Coto, "Unified AC/DC Power Flow for Traction Systems: A New Concept," *IEEE Transactions on Vehicular Technology*, vol. 61, *(6)*, pp. 2421-2430, 01/01, 2012.
[11] M. Coto, P. Arboleya and C. Gonzalez-Moran, "Optimization approach to unified AC/DC power flow applied to traction systems with catenary voltage constraints," *International Journal of Electrical Power & Energy Systems*, vol. 53, *(0)*, pp. 434-441, 12, 2013.
[12] Á J. López-López *et al*, "Assessment of energy-saving techniques in direct-current-electrified mass transit systems," *Transportation Research Part C: Emerging Technologies*, vol. 38, pp. 85-100, 1, 2014.
[13] A. J. López-López *et al*, "Smart traffic-scenario compressor for the efficient electrical simulation of mass transit systems," *International Journal of Electrical Power & Energy Systems*, vol. 88, pp. 150-163, 6, 2017.
[14] A. J. López-López *et al*, "Improving the traffic model to be used in the optimisation of mass transit system electrical infrastructure," *Energies*, vol. 10, *(8)*, pp. 1134, 2017.

The 2018 International Power Electronics Conference

Feeding-loss Reduction by Higher-voltage DC Railway Feeding System with DC-to-DC Converter

Hidenori SHIGEEDA[1], Hiroaki MORIMOTO[1], Kazuhiko ITO[1*], Toshiyuki FUJII[2]and Naoki MORISHIMA[3]

1 Power Supply Technology Division, Railway Technical Research Institute, Tokyo, Japan
2 Advanced Technology R&D Center, Mitsubishi Electric Corporation, Kobe, Japan
3 Power Electronics Department, Toshiba Mitsubishi-Electric Industrial Systems Corporation, Kobe, Japan
*E-mail: ito.kazuhiko@rtri.or.jp

Abstract— Reduction of the power loss is desired to reduce train operation energy in the DC railway feeding system, which is a lower-voltage and higher-current system in comparison with AC railway feeding system. It is difficult for existing DC lines to raise a feeding voltage due to costs of both dual-voltage vehicles and infrastructure although higher voltage is effective to solve feeding loss reduction. Therefore, the authors investigated renewed DC railway feeding system with a focus on feeding-loss reduction. This system consists of higher-voltage feeder and DC-to-DC converters in addition to the existing feeding circuit and enables to feed vehicles with conventional-voltage power. This paper reports the effect of feeding-loss reduction and other expected benefits brought by this system.

Keywords— *DC feeding system: High voltage: .Feeding loss: Power converter: conversion efficiency*

I. INTRODUCTION

The DC railway feeding system (hereinafter referred to as the DC feeding system) supplies relatively low voltage compared with the AC railway feeding system (hereinafter referred to as the AC feeding system). Since trains of this system do not need on-board transformers and rectifiers, their relevant equipment are easy to be insulated; they are less expensive than AC trains. Therefore, this system is considered to be suitable for short distance / high density lines such as commuting lines. On the other hand, since this system utilizes heavy current, there are some problems such as large feeding loss due to the resistance of the overhead contact line and rail, large voltage drop, etc. Therefore, to maintain proper voltage and current, this system requires more substations than AC feeding system dose. From the viewpoint of the energy saving of the DC feeding system, it is desirable to set the feeding voltage from 1500 V or less to a higher voltage , in order to reduce the feeding loss which accounts for 5% of feeding energy [1]. Therefore, studies are being made to boost the feeding voltage to 3000 V which has been generally adopted overseas so far [2]. However, the system of setting the feeding voltage to 3000 V has not been adopted in Japan so far due to problems such as cost increase of both trains and fixed installations, voltage switching, etc.

On the other hand, the AC feeding system can boost the transmission voltage of the substation, without changing the voltage between the overhead contact line and the rail by using auto-transformer (AT). This system is called AT feeding system already used in practical use. A similar circuit configuration system can be realized on a DC feeding system by using a DC-to-DC converter (hereinafter referred to as power converter) instead of the AT, and there are examples studied mainly as measures against voltage drop [3]. This DC feeding system with power converter (hereinafter referred to as the high voltage DC feeding system) can provide various advantages of high voltage without changing the supply voltage to the train.

This paper reports on the results of energy conservation research focusing on the high voltage DC feeding system.

II. FEEDING SYSTEM IN JAPANESE ELECTRIC RAILWAY

A. DC Feeding System

A substation of the DC feeding system receives three-phase power at a substation and supplies DC power converted to an appropriate voltage by a transformer to the train. Figure 1 shows general circuit configuration of the DC feeding system. The feeder line is connected in parallel with the adjacent substations, and the feeder line and the contact wire are connected at intervals of 250 m. A silicon rectifier is used as a general transformer of substation. The standard voltage in Japan is 1500 V, 750 V and 600 V. In the case of the 1500 V system, the substations interval is about 3 km to 5 km for high

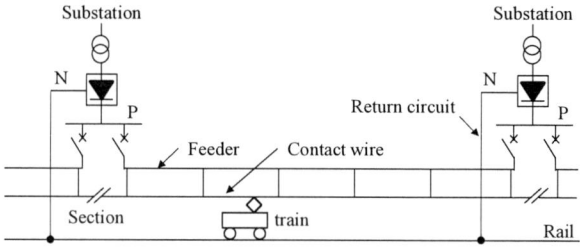

Fig.1. Circuit configuration of the DC feeding system.

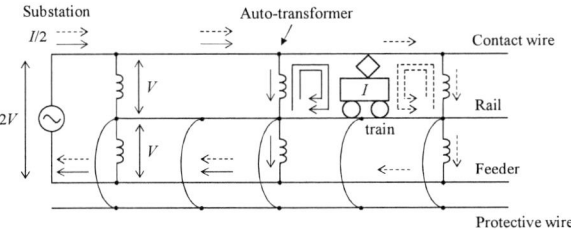

Fig.2. Circuit configuration of the AC feeding system [4].

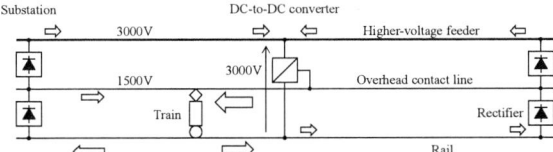

Fig.3. Circuit configuration of high voltage direct current system

density lines and about 10 km to 15km for intercity lines [4].

B. AC Feeding System

In the AC feeding system, three-phase power is provided at a substation and converted into two pairs of single-phase power with a phase difference of 90 degrees by a three-phase two-phase conversion transformer. There are two types of feeding system: BT feeding system using a boosting transformer (BT) and AT feeding system using an AT. Figure 2 shows the general circuit configuration of the AT feeding system. The feeding voltage of substation is stepped down by half by the AT arranged at about 10 km intervals and supplies to train. The feeding current is 1/2 of the overhead contact line current, since the feeding voltage is twice the overhead contact line voltage. The standard voltage of the overhead contact line voltage is 20 kV for the conventional line and 25 kV for the High-speed lines.

III. CIRCUIT CONFIGURATION OF HIGH VOLTAGE DC FEEDING SYSTEM

This paper describes a system in which the supply voltage to the train is the standard value of 1500 V, and the transmission voltage of the substation is increased to 3000 V or more, namely the value more than double of the transmission voltage. Then, we examined the high voltage DC feeding system that can be configured based on this premise. The proposed system has a configuration that a positive higher-voltage feeders is added to the conventional feeding system that is superior in terms of protection and cost and maintenance. In addition, the high voltage power supply is arranged to have a configuration consisting of only a conventional rectifier, as to be easy to control voltage. Fig.1 shows the circuit configuration and the current flow to the train.

On this system, DC-to-DC converters that boost the overhead contact line to a high voltage are arranged at the substation and at the middle point between the substations. The voltage of the higher-voltage feeders of this system is controlled by DC-to-DC converter at the substation in association with that at the middle point, although the overhead contact line voltage fluctuates according to the load current.

IV. FEEDING LOSS OF HIGH VOLTAGE DC FEEDING SYSTEM

A. Calculation Condition

The basic investigation of the feeding loss was carried out by use of the model consisting of two substations and one electric train as shown in Figure 3. The feeding loss was calculated by the steady state analysis simulation under the condition that the electric train moved at the intervals of 0.25km between the substations while consuming constant electric power. The simulation was carried out on the high voltage DC feeding system and also on the conventional feeding systems (two cases of

TABLE I
Calculation Condition of Feeding Loss

Substation rectifier		Constant-voltage source + equivalent internal resistance model Conventional rectifier : 1500 V, 3000 V (Conventional power scheme only) High voltage rectifier : 3000 V, 6000 V, 9000 V Equivalent internal resistance : Voltage fluctuation rate 8 %
Substation DC-to-DC converter		Low voltage side power = Ideal converter for high voltage side power (Do not consider conversion loss) Control of the voltage of the high voltage feed line at the constant level
High voltage feeder line voltage		3000 V, 6000 V, 9000 V
Substation interval		4 km, 8 km
Device of middle point DC-to-DC converter	Circuit model	The same ideal conversion device as substation equipment Control to minimize power loss
	Installation position	(a) $x/2$ [km] (1 point) (b) $x/4$, $x/2$, $3x/4$ [km] (3 points)
Train load		Constant-voltage source (4500 kW)
Line constant	overhead contact line	0.024 Ω/km
	Rail	0.017 Ω/km
	higher-voltage feeders	0.056 Ω/km

The 2018 International Power Electronics Conference

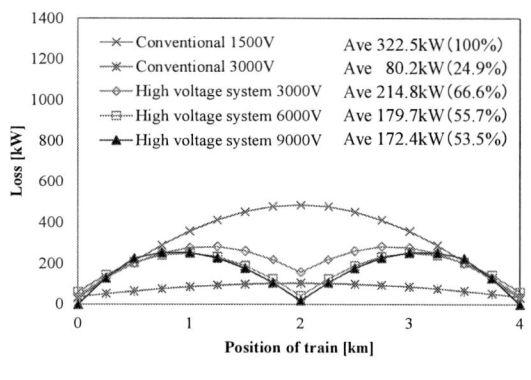

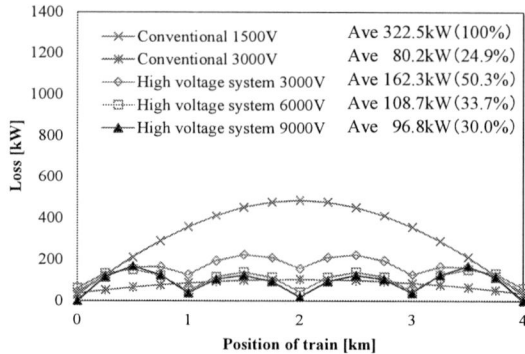

(a) Device at the middle point (1point)

(b) Device at the middle point (3 point)

Fig.4. Feeding loss (Substation interval 4 km).

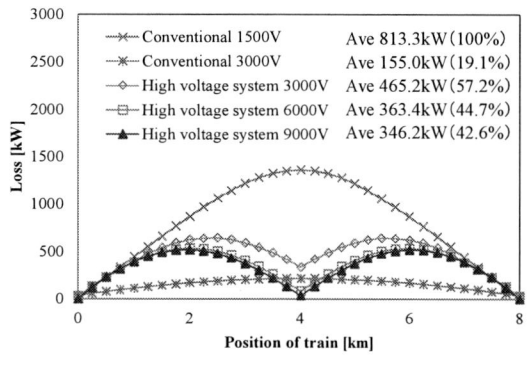

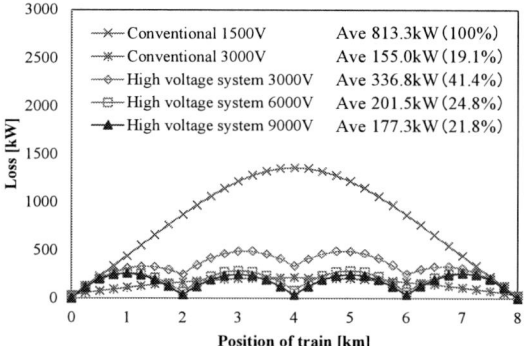

(a) Device at the middle point (1point)

(b) Device at the middle point (3 point)

Fig.5. Feeding loss (Substation interval 8 km).

1500 V and 3000 V), and the simulation results were compared. Table I shows the calculation conditions.

The power converter of the substation was assumed to control the voltage of the high voltage feeder so that it keeps the constant voltage. The output voltage of the intermediate power converter was set at such value as to minimize that the feeding loss each time the train moves.

B. Simulation Results of Feeding Loss

Figure 4 shows the calculation results of feeding loss in the case where the substation interval is 4 km, and Figure 5 shows the calculation results of feeding loss in the case where the substation interval is 8 km. In each figure, the results of the high voltage DC feeding system and those of the conventional feeding system are compared. In addition, the values of feeding loss averaged over the section between the substations for each case, and the ratio of the average values to that for the 1500 V conventional feeding system are also shown. The feeding loss represents only the Joule loss of the overhead contact line, rail and high voltage feeder, excluding the loss of the rectifier and the power converter.

In the conventional feeding system, the feeding loss is inversely proportional to the square of the voltage. Therefore, the feeding loss is reduced to 1/4 when the feeding voltage is boosted from 1500 V to 3000 V. In the

simulation, the train was modeled as the constant power source. However, the load current changes in accordance with the voltage at a pantograph position, therefore further feeding loss reduction can be expected in practical.

In the high voltage DC feeding system, the feeding loss is reduced because a part of the loading current flows through the high voltage feeder, and the feeding loss reduction effect is maximized in the case where a train is located at the position of the power converter. On the other hand, in the case where there is a train at the middle between the power converters, the feeding loss is maximized because a load current suitable for 1500 V electrical overhead contact line and rail flows. As a result, the feeding loss of 3000 V high voltage DC feeding system has a more reduction effect than that of the 1500 V conventional feeding system. But its reduction effect is lower than the feeding loss of 3000 V conventional feeding system. However, the loss reduction effect can be improved by increasing the number of intermediate power converter introduced.

The effect is also improved by boosting the voltage of the high voltage system to the value higher than 3000 V, but the effect in the case of boosting from 6000 V to 9000 V is smaller than that in the case of boosting from 3000 V to 6000 V. This is because the amount of the feeding loss generated in the high voltage feeder accounts for smaller percentage of the loss than that generated in

2542

The 2018 International Power Electronics Conference

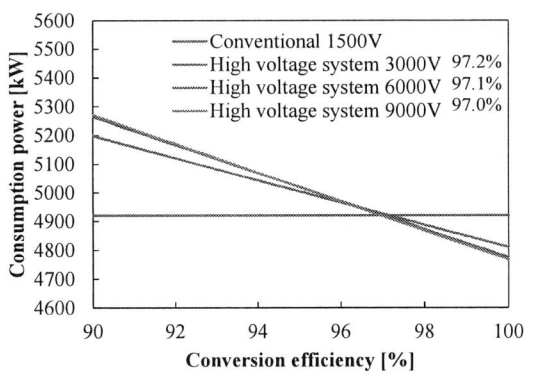
(a) Device of middle point (1 point)

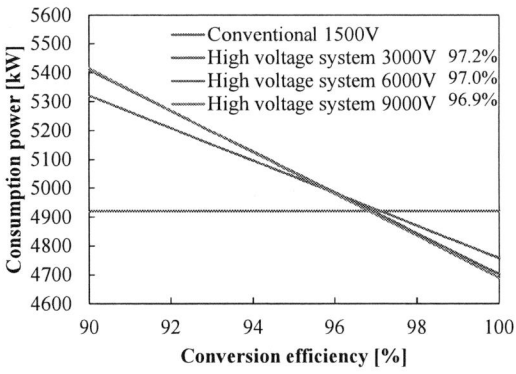
(b) Device of middle point (3 point)

Fig.6. Average consumption power of feeding system (Substation interval 4 km).

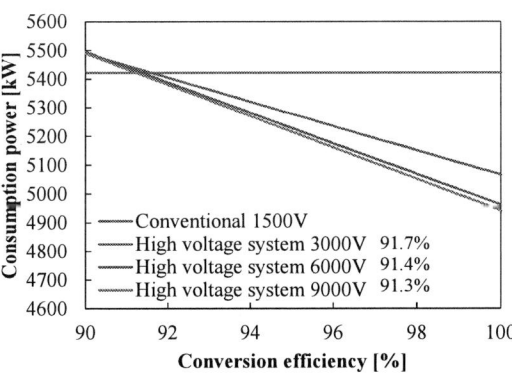
(a) Device of middle point (1 point)

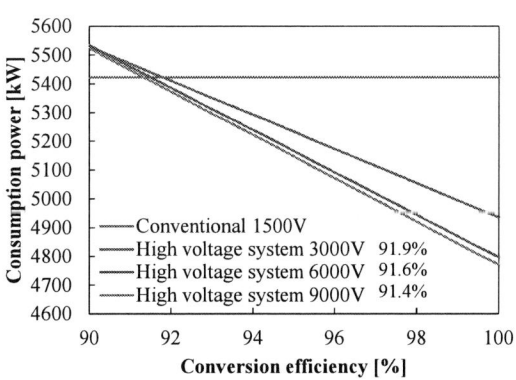
(b) Device of middle point (3 point)

Fig.7. Average consumption power of feeding system (Substation interval 8 km).

the overhead contact line and the rail, thus the feeding loss reduction effect due to boosting the voltage to the high level is relatively lowered. Therefore, in the high voltage DC feeding system, it is necessary to select suitable voltage for the high voltage system taking the capital investment for loss reduction and insulation strengthening into consideration.

In the case of three intermediate power converters introduced between the two substations with 8 kilometers interval, or in the case of a high voltage system voltage of 6000 V or more with one intermediate power converters, the average values of the feeding loss of the respective cases are equal to that of the case of the 1500 V conventional feeding system with the substation interval of 4 km. Therefore, in terms of feeding loss, it is possible to extend the substation interval without increasing the feeding loss by adopting the high voltage DC feeding system.

C. Simulation Results Considering Conversion Loss

In the preceding section, comparison of only the loss of the electric line among the relevant systems considered was discussed. Actually, it is necessary to consider the conversion loss in the rectifier and the power converter. In this section, we consider the conversion efficiency of the device necessary for achieving energy saving throughout the system.

For the respective cases in Figures 4 and 5, we calculated the average output power of the rectifier and the power converter in the cases of 1500 V conventional feeding system, and high voltage DC feeding system. Figures 6 and 7 show the result of calculation of the average power consumption of the whole feeding circuit with adoption of the conversion efficiency of the power converter as a parameter. The conversion efficiency of the rectifier was assumed to be 98%. The conversion efficiency of the power converter required to reduce the power consumption of high voltage DC feeding system more than that of the conventional feeding system of 1500 V was 98% or more in the case of 4km interval of the substation, and 92% or more in the case of 8 km interval. Namely the shorter the substation interval is, the more the efficient power converter is required. On the other hand, the influence of the voltage of the high voltage system and the number of intermediate power converters on the required conversion efficiency is small.

If the conversion efficiency the required level for energy saving, the total power consumption tends to decrease as the voltage of the high voltage system is higher, but the difference between the total power consumption in the case of 6000 V and that in the case of 9000 V is slight.

2543

V. Consideration For Putting The System To Practical Use

A. Effect of High Voltage

Generically, the following items can be cited as the effects of the high voltage of the DC feeding system on the ground equipment [5].

(1) Reduction of feeding loss
(2) Improvement in regeneration efficiency
(3) Improvement of safety
(4) Reduction of number of substations
(5) Reduction of contact wire wear

In this chapter, we will examine whether or not these effects exist in the case of using the high voltage DC feeding system, and organize other methods which should be compared with those of high voltage DC feeding system.

1) Feeding Loss Reduction

If it is possible to realize the required conversion efficiency of the power converter as shown in Chapter 3, the feeding loss can be reduced.

Other methods to be compared regarding reduction of feeding loss include the addition of substations, making the voltage of the conventional feeding system high, and adoption of the superconducting cable. Furthermore, if the route has more tracks than double, simultaneous up and down line feeding system and tie feeding system between up and down line are also candidates.

2) Improvement in Regeneration Efficiency

It is said that increasing the voltage of the conventional feeding system makes the regeneration efficiency improve, since the distance from the regenerative train to the powering train receiving the regenerative energy is increased in proportion to the square of the voltage ratio. In the high voltage DC feeding system, introducing a bidirectional power converter such as a step-up / step-down chopper, enable to supply regenerative energy to distant powering trains, thereby improving regeneration efficiency.

Other methods to be compared regarding improvement of regeneration efficiency include making the voltage of the conventional feeding system high, adoption of voltage controllable equipment such as thyristor rectifier, power storage device, and adoption of the superconducting cable. Furthermore, if the route has more tracks than double, simultaneous up and down line feeding system and tie feeding system between up and down line are also candidates.

3) Improvement of Safety

Boosting the voltage of the conventional feeding system makes it easier to protect the system in case of the fault, since the load current decreases and the detection sensitivity of the fault current is improved. In the high voltage feeding system, the train current has the same voltage as that of the conventional feeding system. Therefore, it can not be expected to improve safety in case of such low transportation density lines that only one train exists between substations. However, in the high transportation density route, there is a possibility that the sensitivity of the protective relay can be increased since the maximum current supplied from the substation decreases.

Other methods to be compared regarding safety improvement include the addition of substations, making the voltage of the conventional feeding system high, and adoption of the superconducting cable.

4) Reduction in Number of Substations

In general, the substation intervals of the DC feeding system are determined in consideration of voltage drop and electrolytic corrosion (leakage current from rail). In case of increasing the voltage of a conventional feeding system, in terms of voltage drop, the substation interval can be increased by square times of the voltage ratio at the maximum, in the case of low transport density lines. In terms of electrolytic corrosion, the substation interval can be expanded by the square root times of the voltage ratio at the maximum. Figure 8 shows an example of the results of calculating the pantograph voltage of the train in the study of Chapter 3 with respect to the high voltage DC feeding system. In addition, the results shown in Figure 8 was calculated by giving priority to minimizing the feeding loss. Even with the expanded substation interval, the pantograph voltage of the high voltage DC feeding system can be secured at the level equal to that of the conventional feeding system without expansion of the substation interval.

Figure 9 shows an example of the calculation results of the rail potential. The area of the positive rail potential is smaller in the high voltage feeding system than that in the conventional feeding system, indicating that the leakage current from the rail is decreased. Accordingly it can be said that it is possible to expand the substation interval from the viewpoint of electrolytic corrosion. In the high voltage feeding system, an intermediate power converter is required between substations. However, since the intermediate power converter does not require power receiving from the power companies, it can be installed in small space compared to the substation. Other methods to be compared in terms of reduction in number of substation include making the voltage of the conventional feeding system high, adoption of power storage device, and adoption of the superconducting cable.

5) Reduction of Contact Wire Wear

It is said that increasing the voltage of the conventional feeding system makes electric wear of the contact wire decreased, since the collected current of the train decreases. In the high voltage DC feeding system, the current of the train is the same as that of the conventional feeding system. Therefore, it can not be expected to

The 2018 International Power Electronics Conference

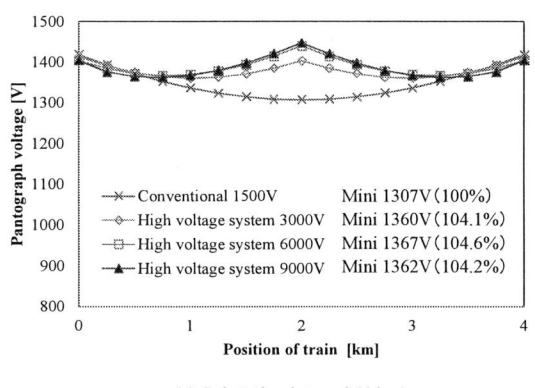
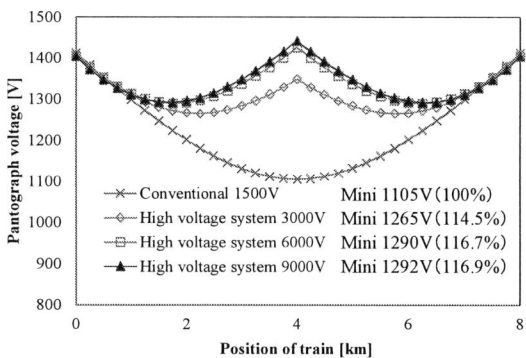

(a) Substation interval (4 km)   (b) Substation interval (8 km)

Fig.8. A calculation example of pantograph voltage (1 intermediate apparatus).

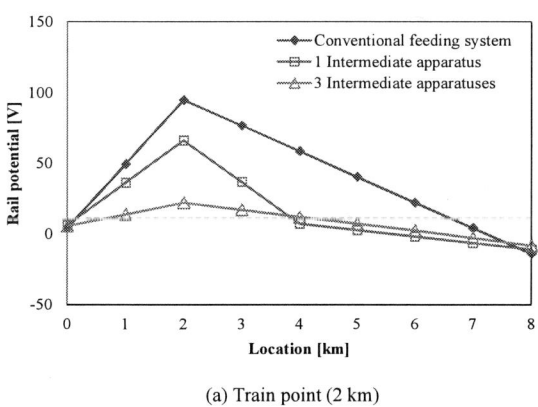
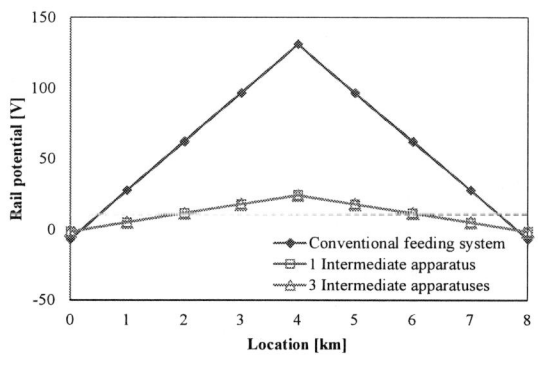

(a) Train point (2 km)   (b) Train point (4 km)

Fig.9. A calculation example of rail potential (Substation interval 8 km).

reduce the wear of the contact wire.

B. Technical Issues for Putting the Sysytem into Practical use

A summary of features of the high voltage direct current system is shown in the following.

- It is possible to obtain the effect of the high voltage such as the reduction of the feeding loss, the improvement of the regeneration efficiency, and the reduction of the number of substations without changing the feeding voltage to the train.
- Space saving compared to the substation, since the intermediate power converter does not require power receiving from the power companies.
- It is necessary to newly lay a high voltage feeder wire.
- In order to realize the reduction of the feeding loss, a high-efficiency power converter is required.

Toward putting the high voltage DC feeding system into practical use, we show the technological development tasks that are considered necessary for quantitatively evaluating these merits and demerits as given below.

- Development of an operation power simulator capable of verifying energy saving effect on actual train operation.

- Development of power converter control method by simulator.
- Development of protection (failure mode, fault detection method, protection method, etc.) of the power converter.
- Study on insulation coordination of high-voltage system (insulation method of electrical path, abnormal voltage protection, ground voltage boost at ground fault, etc.).
- Improvement of performance of power converter (improvement of conversion efficiency, cooperation with electric power storage device etc.).

This paper showed the contents of study intended for minimizing the feeding loss. In this case, it is necessary for the power converter to output most of the load on the train when the train is in immediate position of the power converter. Therefore, a large capacity power converter is required. However, there is a possibility of reduction of the feeding loss, even if the equipment capacity is set to about 1/2 of the train loading. In putting the system into practical use, it is necessary to optimize the system together with the examination of the above problems.

VI. Summary

This paper shows the results of verifying the basic characteristics of the proposed high voltage direct current feeding method by simulation, mainly from the viewpoint of energy saving of DC feeding method.

It was clarified that the power loss of the high voltage DC system can be lower than that of the conventional DC power system, and that the voltage of the high voltage system should be set at an appropriate level. In addition, as for the level of the conversion efficiency of the power conversion device necessary for low loss, it was found out that the shorter the substation interval becomes, the higher efficiency is required of the device.

From now on, we will develop a traction power simulation system. After that, we will promote more concrete research and development concerning the introduction effect of this method and the control method of the electric power conversion device etc.

References

[1] T. Ogawa, Y. Iino, H. Tanaka, "Analysis of energy consumption for potential reduction effect of energy-saving technologies of electric railway" *The papers of technical meeting on transportation and electric railway*, IEE Japan, TER-17-050, 2017(in Japanese)

[2] Research committee on high voltage DC feeding system, "Report on high voltage DC feeding system" *Technical report of IEEJ*, Vol.2, No.295, IEEJ, 1989 (in Japanese)

[3] Ladoux, P : Une nouvelle structure d'alimentation des catenaires 1500V: le systeme 2x1500V, Revue Generale des Chemins de Fer, pp.21-31, 2006

[4] Editorial committee on Electric Railway Handbook, "Electric Railway Handbook" Corona Publishing Co., LTD, pp.498, pp.531, 2007 (in Japanese)

[5] J. Ito, T. Ito, "Effect of high voltage DC electric railway" *The Papers of Technical Meeting on Transportation and Electric Railway*, IEE Japan, TER-93-9, 1993(in Japanese)

Modeling and Simulation of Novel Railway Power Supply System Based on Power Conversion Technology

Minwu Chen[1*], Ruofei Liu[1], Shaofeng Xie[1], Xiaofang Zhang[2], Yimin Zhou[2]
1 School of Electrical Engineering, Southwest Jiaotong University, Chengdu, China
2 WENZHOU MASS TRANSIT RAILWAY INVESTMENT GROUP CO., LTD, Wenzhou, China
*E-mail: chenminwu@swjtu.edu.cn

Abstract— **A novel railway power supply system is presented in this paper to improve power quality and eliminate neutral section of the conventional traction power supply system. The system adopts single-phase combined transformer and Power Flow Controller (PFC) based on Modular Multilevel Converter (MMC). Its equivalent mathematical model has been built according to the method of electrical transformation and the theory of balance compensation. To achieve the goal of negative sequence and reactive comprehensive compensation, the operation current of each port of PFC can be calculated. Moreover, the control approach of PFC is studied. Finally, the effectiveness and correctness of the proposed system are verified in MATLAB/Simulink surroundings.**

Keywords—**electrified railway, power flow controller, modular multilevel converter, comprehensive compensation, simulation**

I. INTRODUCTION

In China, the conventional single-phase 25kV AC electrified railway has been widely adopted, while it is faced with some outstanding power quality problems such as harmonics, unbalanced current and reactive power. These problems directly affect the industrial gird. In addition, due to the existence of neutral section insulators at exits of substations, the efficiency and safety of the electric locomotives are affected, especially in the surroundings of high speed or heavy haul railway [1].

In order to solve these problems, a variety of power supply schemes are proposed and a number of technologies are introduced in TPSS [2-4]. In [5], a railway power system that uses a three-phase step-down transformer working with three-phase to single-phase converter is presented, which could fundamentally solve the problems above by adopting the optimized topology. However, the system has a great influence on the fluctuation of traction load and the converter has large capacity as well as a relatively high cost. In [6], a traction power supply system with YNvd type transformer and Active Power Compensator (APC) is presented, which can effectively eliminate the neutral section, balance three-phase voltage and filter harmonics. However, its compensator capacity as well as cost is too high under the condition of full compensation algorithm. In China, a 10 MVA/27.5 kV co-phase system has been designed,

established and experimented in a practical operating railway. The practice shows that the scheme of the co-phase power supply system has a great advantage in improving power quality and strengthening traction power supply performance [7].

Meanwhile, for the sake of meeting the requirements of high voltage and large capacity engineering applications, the PFC, as the core device of the co-phase power supply system, is concerned with its research of topology, compensation strategy and control strategy currently. In China, the cascaded H bridge chain topology is widely used in PFC, which has advantages of high power grade and low output harmonic content. Unfortunately，with the increase in the voltage level, the design and manufacture of multi winding isolation matching transformer will become extremely complex. Recently, the modular multilevel converter (MMC), which has been developed as a novel multilevel voltage source converter topology, not only inherits the advantages of cascaded topology in the number of devices and modular structure, but also can be easily operated in back-to-back systems due to the presence of common DC buses.

In this paper, a novel railway power system with single-phase combined transformer and Power Flow Controller (PFC) based on MMC is presented. In traction side, single-phase feeder connection scheme is adopted, the continuous current can be supplied for electric locomotives without neutral sections. In three-phase power grid side, the PFC topology based on MMC is proposed and its port compensation currents are calculated according to the limit value of power quality standard in China. Finally, the correctness of the proposed system is tested in the MATLAB/Simulink surroundings.

II. MODELING OF COMBINED CO-PHASE POWER SUPPLY SYSTEM

A. Electrical Structure of Novel System

As illustrated in Fig. 1, the substation in the novel system is composed of a single-phase combined transformer and a single-phase back-to-back converter called PFC. In the traction side, the transformer T1 and

PFC provide continuous fundamental frequency current to electric locomotive with the same-phase voltage, thus the neutral section at exit of substation can be cancelled. In PFC, a single-phase MMC-based converter is adopted. Owning to easy expansion of its serial sub-module, it can directly connect to the traction side without the isolation matching transformer so as to reduce the cost. A step-down transformer (T2) is used to connect the PFC with three-phase 110/220kV grid in China, which can meet the requirement of voltage matching for power electronic devices.

B. Optimized Topology of PFC

A special topology of PFC and its sub-module is proposed and shown in Fig. 2. The PFC, a single-phase back-to-back converter, is composed of a MMC structure with four bridges. Compared with the two bridge structure, the common DC bus voltage can be reduced and the voltage stabilizing control link of the DC side capacitor can be eliminated. Compared with the three bridge structures, the unbalanced power allocation of the bridge arm and the problem of control complexity caused by positive and negative sequence decomposition can be avoided. In PFC, the power module of phase α performs active power balance and the phase β module performs comprehensive compensation. The Modular Multilevel Converter (MMC) has many advantages such as higher modularity, lower output THD, and higher voltage scalability compared with other multilevel circuits, so this topology is better for the occasion of high voltage and high capacity power conversion. [8-9].

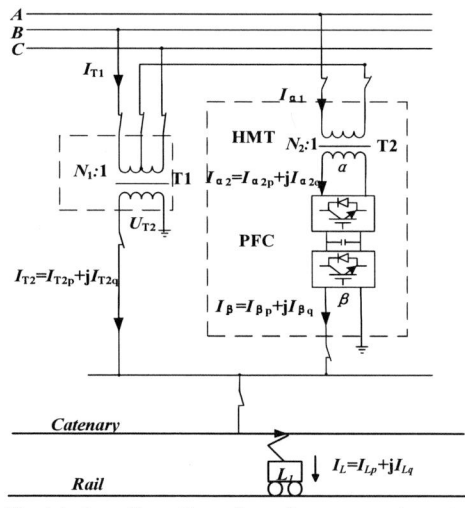

Fig.1. Electrical configuration of novel power supply system

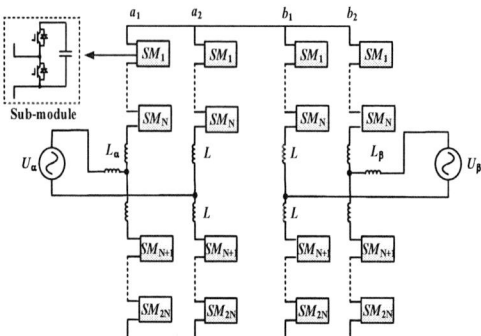

Fig.2. MMC topology of PFC

III. MATHEMATICAL MODEL OF NOVEL SYSTEM AND COMPENSATION CURRENT CALCULATION

A. Mathematical Model of Novel System

In Fig. 1, the three-phase voltages in the grid side are denoted as $U_A \angle A, U_B \angle B, U_C \angle C$. U_{T1} and $U_{\alpha 1}$ are the voltages in the grid side of the T1 and T2 separately. The port voltages of the converter in the grid side and in the traction side are expressed as U_α and U_β respectively. The three-phase grid-side power factor and three-phase voltage unbalance degree are presented as K and ε_U. $k_1 = |U_{T1}|/|U_{\alpha 1}|$ can be obtained.

According to the principle of power balance, the three-phase currents in grid side should satisfy (1).

$$\begin{cases} U_{T1}I_{T1p} + U_{\alpha 1}I_{\alpha 1p} = U_L I_{Lp} \\ K = \dfrac{U_L I_{Lp}}{\sqrt{(U_{T1}I_{T1q} + U_{\alpha 1}I_{\alpha 1q})^2 + (U_L I_{Lp})^2}} \end{cases} \quad (1)$$

In view of the principle of magnetic potential balance and current balance, the three-phase currents in grid side can be calculated by (2).

$$\begin{bmatrix} \dot{I}_A \\ \dot{I}_B \\ \dot{I}_C \end{bmatrix} = \frac{1}{\sqrt{3}K_1} \begin{bmatrix} -1 & \sqrt{3} \\ -1 & -\sqrt{3} \\ 2 & 0 \end{bmatrix} \begin{bmatrix} \dot{I}_{\alpha 2} \\ \dot{I}_{T2} \end{bmatrix} \quad (2)$$

By using the symmetric component method, the negative current in grid side is deduced by (2)

$$\dot{I}^- = \frac{1}{12\sqrt{3}}\left[-\dot{I}_{\alpha 2} + \sqrt{3}\dot{I}_{T2} + (-\dot{I}_{\alpha 2} - \sqrt{3}\dot{I}_{T2})e^{j240^\circ} + 2\dot{I}_{\alpha 2} e^{j120^\circ} \right] \quad (3)$$

Considering China's power quality standards, the three-phase voltage unbalance ratio ε_U can be evaluated by (4).

$$\varepsilon_U = \frac{\sqrt{3}|I^-|U_S}{S} \times 100\% \quad (4)$$

S, U_S are short-circuit capacity and line-to-line voltage respectively.

Due to the connection of the transformer HMT in Fig.1, the following formula (5) can be derived.

$$U_{\alpha 1} I_{\alpha 1} = U_{\alpha 2} I_{\alpha 2} \tag{5}$$

According to the topology structure of PFC, both sides of the compensation port meet the power balance shown in (6) below.

$$\begin{cases} U_{\alpha 2} I_{\alpha 2} = U_{\beta} I_{\beta p} \\ U_{\beta} I_{\beta p} = U_L I_{Lq} - U_{T2} I_{T2q} \end{cases} \tag{6}$$

B. Comprehensive Compensation Currents Calculation

Assuming the short circuit capacity of the power system is S (MVA), the traction load current is $I_L(t)$, and the power factor angle is ϕ_L. The three-phase grid-side power factor is K and three-phase voltage unbalance degree is ε_U. Based on the analysis above, the comprehensive compensation current of PFC can be calculated by the formula (3)-(6), respectively, as shown in (7)-(9) below:

$$\begin{cases} I_{\alpha 1q} = \dfrac{-2\sqrt{3}K_N I_{Lp}\sin(\psi_A + \frac{1}{2}\psi_B - \frac{3}{2}\psi_C + 90^o) + (K_C - K_N)I_{Lq}(2\sqrt{3}\cos(\psi_A + \frac{1}{2}\psi_B - \frac{3}{2}\psi_C + 90^o) - 4k_1)}{16k_1^2 + 12 - 16\sqrt{3}k_1\cos(\psi_A + \frac{1}{2}\psi_B - \frac{3}{2}\psi_C + 90^o)} \\[3mm] I_{\alpha 1p} = -\dfrac{K_N I_{Lp}(2\sqrt{3}\cos(\psi_A + \frac{1}{2}\psi_B - \frac{3}{2}\psi_C + 90^o) - 4k_1) + 2\sqrt{3}(K_C - K_N)I_{Lq}\sin(\psi_A + \frac{1}{2}\psi_B - \frac{3}{2}\psi_C + 90^o)}{16k_1^2 + 12 - 16\sqrt{3}k_1\cos(\psi_A + \frac{1}{2}\psi_B - \frac{3}{2}\psi_C + 90^o)} \end{cases} \tag{7}$$

$$\begin{cases} I_{\alpha 2q} = \dfrac{-12K_N I_{Lp}\sin(\psi_A + \frac{1}{2}\psi_B - \frac{3}{2}\psi_C + 90^o) + (K_C - K_N)I_{Lq}(12\cos(\psi_A + \frac{1}{2}\psi_B - \frac{3}{2}\psi_C + 90^o) - 8\sqrt{3}k_1)}{16k_1^2 + 12 - 16\sqrt{3}k_1\cos(\psi_A + \frac{1}{2}\psi_B - \frac{3}{2}\psi_C + 90^o)} \\[3mm] I_{\alpha 2p} = -\dfrac{K_N I_{Lp}(12\cos(\psi_A + \frac{1}{2}\psi_B - \frac{3}{2}\psi_C + 90^o) - 8\sqrt{3}k_1) + 12(K_C - K_N)I_{Lq}\sin(\psi_A + \frac{1}{2}\psi_B - \frac{3}{2}\psi_C + 90^o)}{16k_1^2 + 12 - 16\sqrt{3}k_1\cos(\psi_A + \frac{1}{2}\psi_B - \frac{3}{2}\psi_C + 90^o)} \end{cases} \tag{8}$$

$$\begin{cases} I_{\beta q} = \dfrac{-2\sqrt{3}k_1 K_N I_{Lp}\sin(\psi_A + \frac{1}{2}\psi_B - \frac{3}{2}\psi_C + 90^o) + I_{Lq}(K_C(3 - 2\sqrt{3}k_1\cos(\psi_A + \frac{1}{2}\psi_B - \frac{3}{2}\psi_C + 90^o)) - K_N(2\sqrt{3}k_1\cos(\psi_A + \frac{1}{2}\psi_B - \frac{3}{2}\psi_C + 90^o) - 4k_1^2))}{4k_1^2 + 3 - 4\sqrt{3}k_1\cos(\psi_A + \frac{1}{2}\psi_B - \frac{3}{2}\psi_C + 90^o)} \\[3mm] I_{\beta p} = -\dfrac{K_N I_{Lp}(12k_1\cos(\psi_A + \frac{1}{2}\psi_B - \frac{3}{2}\psi_C + 90^o) - 8\sqrt{3}k_1^2) + 12k_1(K_C - K_N)I_{Lq}\sin(\psi_A + \frac{1}{2}\psi_B - \frac{3}{2}\psi_C + 90^o)}{8\sqrt{3}k_1^2 + 6\sqrt{3} - 24k_1\cos(\psi_A + \frac{1}{2}\psi_B - \frac{3}{2}\psi_C + 90^o)} \end{cases} \tag{9}$$

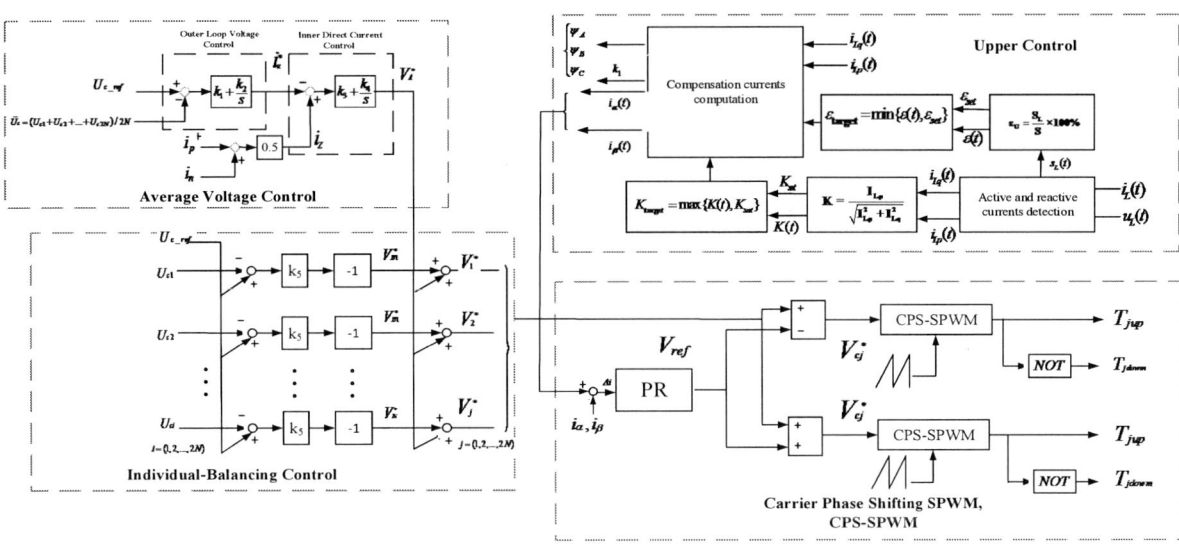

Fig.3. Control method of PFC

Where

$$\begin{cases} K_N = \dfrac{27.5 I_L - 10 S \varepsilon_U}{27.5 I_L} \\[3mm] K_C = 1 - \dfrac{\sqrt{1-K^2}}{K \tan \phi_L} \end{cases}$$

IV. CONTROL METHOD OF PFC

The control method block of PFC is shown in Fig.3. According to the method of electrical transformation and the theory of balance compensation, the compensation current is calculated in the upper control. Simultaneously, the lower control adopts direct current control to make the actual current follow the reference compensation current. In addition, the MMC-based PFC has a great deal of individual DC capacitors and its operation voltage need to be stable. Therefore, the DC capacitor voltage control algorithm needs to be added to the lower control, which includes average voltage control and individual-balancing control. The average voltage control, which is made up of outer loop voltage control and inner direct current control, can suppress circulation current. The individual-balancing control can make DC capacitor charge or discharge separately. Finally, carrier phase shifting SPWM generator module can command power modules to produce the accurate compensation currents [10].

V. SIMULATION AND ANALYSIS

The model of the proposed novel system under imbalanced voltage ($\psi_A = 0^0, \psi_B = 106^0, \psi_C = -106^0$) is performed in MATLAB/Simulink surroundings. The parameters setting of the system are illustrated in Table 1. In three cases, the fluctuation and nonlinearity of the traction load can be simulated and the effect of several compensation conditions can be analyzed. Fig. 3 shows the time-varying traction load current in traction side. Fig. 4 shows the three-phase currents in grid side. The compensation currents and voltages of PFC are indicated in Fig. 5~ Fig. 6. As a result, the simulation results validate the correctness and effectiveness of the design scheme of the proposed novel system and the topological structure of PFC.

TABLE I.

BASIC PARAMETERS OF POWER GRID AND TRACTION LOAD

Setting	Power Grid		Traction load		Compensation targets	
	S (MVA)	U_S (kV)	S_L (MVA)	$\cos\phi_L$	PF	ε_U (%)
Case 1			10	0.8	0.8	–
Case 2	1000	110	20	0.85	1	0
Case 3			30	0.9	0.9	2

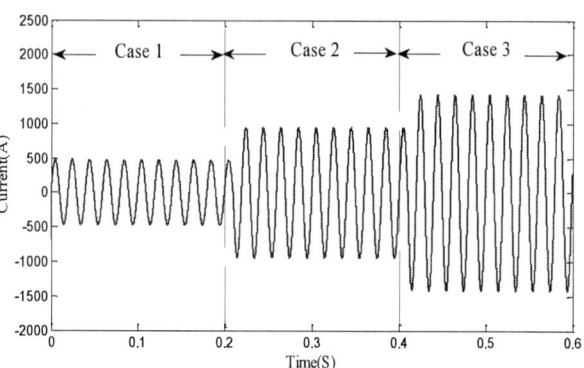

Fig.3. Traction load with time-varying characteristics

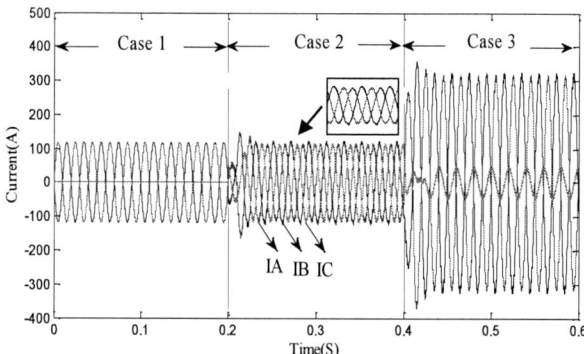

Fig.4. Three-phase currents in power grid side

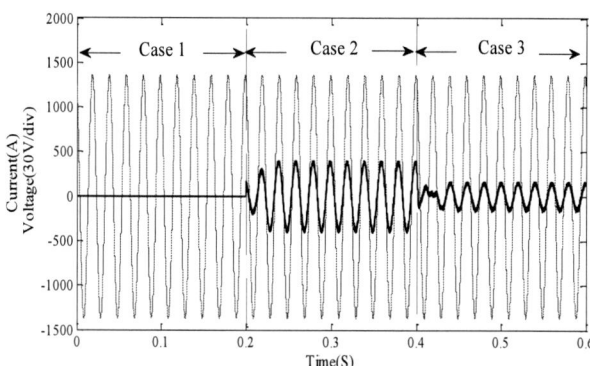

Fig.5. Voltage and current of α port

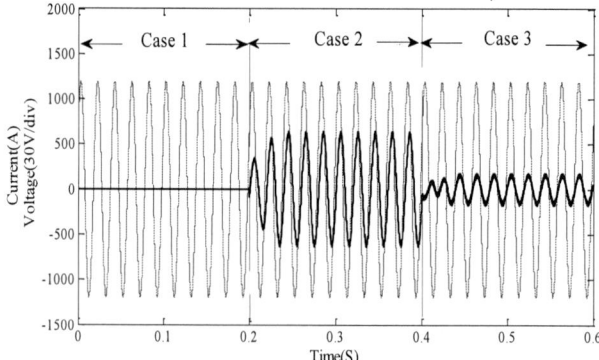

Fig.6. Voltage and current of β port

VI. CONCLUSION

A novel traction power system based on power conversion technology is discussed in this paper, which can cancel the neutral section at exits of substations and effectively improve the power quality problems caused by traction load. First of all, the equivalent mathematical model of the novel system has been established by the basis of electrical transformation relations. Meanwhile, the compensation strategy of negative sequence and reactive power has been implemented, which can be effectively applied to the PFC control scheme. Finally, the availability of the PFC's topology and the correctness of the PFC's control strategy have been validated by simulation. The novel traction power supply system is valuable to study more and the simulation results will be used to conduct the future experiments and work.

REFERENCES

[1] Qunzhan Li, "New generation traction power supply system and its key technologies for electrified railways," Journal of Modern Transportation, vol.23, no.1, pp.1-11, 2015.

[2] Xiaozhou Zhu, Minwu Chen, Shaofeng Xie and Jie Luo, "Research on new traction power system using power flow controller and Vx connection transformer," 2016 IEEE International Conference on Intelligent Rail Transportation (ICIRT), Birmingham, 2016, pp. 111-115.

[3] Deng Ming-li, Guang-ning Wu, Zhang Xueyuan, Fan Chun-lei, He Chang-hong, Ye Qiang, "The simulation analysis of harmonics and negative sequence with Scott wiring transformer," Condition Monitoring and Diagnosis, 2008. CMD 2008. International Conference on, vol., no., pp.513-516, Apr, 2008.

[4] Minwu Chen, Qunzhan Li, Clive Roberts, Stuart Hillmansen, Pietro Tricoli, Ning Zhao, Ivan Krastev, "Modelling and performance analysis of advanced combined co-phase traction power supply system in electrified railway," in IET Generation, Transmission & Distribution, vol. 10, no. 4, pp. 906-916, Mar, 2016.

[5] Xiaoqiong He, Zeliang Shu, Xu Peng, Qi Zhou, Yingying Zhou, Qijun Zhou, Shibin Gao, "Advanced Co-phase Traction Power Supply System Based on Three-Phase to Single-Phase Converter," Power Electronics, IEEE Transactions on, vol.29, no.10, pp. 5323 - 5333, Oct. 2014.

[6] Zeliang Shu, Shaofeng Xie, Qunzhan Li, "Single-Phase Back To-Back Converter for Active Power Balancing, Reactive Power Compensation, and Harmonic Filtering in Traction Power System," Power Electronics, IEEE Transactions on, vol.26, no.2, pp.334-343, Feb. 2011.

[7] Yanling Zhao, Liping Zhao, Qunzhan Li, "Some key problems on co-phase traction power supply device," Information Technology and Applications (IFITA), 2010 International Forum on, vol.3, no., pp. 444-449, July 2010.

[8] Yan Zhao, NingYi Dai and BaoAn, "Application of three-phase modular multilevel converter (MMC) in co-phase traction power supply system," 2014 IEEE Conference and Expo Transportation Electrification Asia-Pacific (ITEC Asia-Pacific), Beijing, 2014, pp. 1-6.

[9] Liu W, Zhang K, Chen X, et al. "Simplified model and sub-module capacitor voltage balancing of single-phase AC/AC modular multilevel converter for railway traction purpose," IET Power Electronics, vol. 9, no.5, pp. 951-959, 2016.

[10] P. M. Meshramand V. B. Borghate, "A simplified nearest level control (NLC) voltage balancing method for modular multilevel converter(MMC)," IEEE Trans. Power Electron, vol.30,no.1, pp. 450–462, Jan. 2015.

Comparative Study on Front-End Parameter Identification Methods for Wireless Power Transfer Without Wireless Communication Systems

Sinan Li[1] and S. Y. (Ron) Hui[1, 2*]

1 Department of Electrical and Electronic Engineering, The University of Hong Kong, Hong Kong, China
2 Department of Electrical and Electronic Engineering, Imperial College London, London, U. K.
*E-mail: ronhui@eee.hku.hk

Abstract— In this paper, five front-end parameter identification methods are compared for wireless power transfer applications by considering estimation accuracy, valid frequency range, the maximum number of identifiable parameters, execution time, scalability, and minimum sampling rate. The front-end parameter identification methods are shown to be time efficient, accurate and powerful and demonstrate huge potentials as compared to methods with wireless commutation systems. It is envisaged that front-end parameter identification techniques can be used as a powerful tool in future WPT applications.

Keywords— *Front-end, parameter identification, wireless power transfer, static and dynamic charging.*

I. INTRODUCTION

In typical Wireless Power Transfer (WPT) applications, the load power at the receiver-side of the system must be closely monitored and regulated for energy efficiency and safety considerations. The idea of wireless feeding back the load information from the receiver side to the transmitter side through a wireless communication system was previously reported in [1]–[5]. To date, the Wireless Power Consortium (WPC) - 1.2.2 extended power profile has specified a two-way communication (receiver-to-transmitter and transmitter-to-receiver) interface for power transfer, and has mandated that a power receiver must report to the power transmitter its received power in a Received Power Packet [6]. Most commercial wireless transmitter control ICs nowadays have also integrated into them such a communication interface, based on, for instance, amplitude-shift keying (ASK) and/or frequency-shifting keying (FSK) modulation [7]–[9]. While a WPT system with a wireless communication system enjoys (i) real-time load monitoring with improved efficiency and security, (ii) free-positioning operation, (iii) flexibility in power receiver design, and (iv) scalability to applications with multiple receivers, the following limitations have been observed:

(i) Wireless communication system requires specially designed electronic circuitries at both transmitter and receiver side of a WPT system. The need for extra circuitries increases the overall bill-of-materials and power losses in the WPT system;

(ii) With ASK and FSK modulation, the communication signals are modulated over the power transfer signals. The resultant large and high-frequency voltage/current ripples presented at the output of the WPT system lead to increased conduction losses (due to increased RMS current) and linear-regulator losses (e.g., an LDO is typically cascaded to the output of the power receiver in mobile charging applications);

(iii) Based on Maximum Energy Efficiency principle [10], the energy efficiency of a WPT system is also determined by factors, such as the coupling coefficient, the quality factor of coils, the topology of the compensation network, and the operating frequency. Although the load power can be regulated online through wireless communication, an optimal energy efficiency may not be retained adaptively in the event of (1) a misalignment of transmitter and receiver coils for static charging applications, (2) dynamic charging with a load in motion (where the coupling coefficient is time-varying), (3) tolerance and aging of the resonant components. In these scenarios, the operating condition deviates from the initial design that assumes perfect coil alignment and nominal component values.

In this paper, a range of front-end parameter identification methods without wireless communication systems is studied. Based on the electrical information from the transmitter side only, these methods can identify one or several key system parameters that can be used for simultaneous load monitoring and efficiency optimization. Front-end parameter identification methods can be particularly useful in applications where cost, size, and energy efficiency are of critical importance. Comparisons on the (i) estimation accuracy (ii) valid frequency range (iii) maximum number of identifiable parameters (iv) execution time (v) scalability (vi) sampling rate are included in this study. The underlying relationship between these methods and future research trend are also highlighted.

II. FRONT-END PARAMETER IDENTIFICATION METHODS

A. Secondary (Receiver-side) resonant mode method

This work is supported by the Hong Kong Research Grant Council under GRF Project 17203517.

This frequency-domain approach takes advantage of the "reflected impedance" concept that can be explained with a two-coil parallel-parallel (P-P) compensated WPT system in Fig. 1(a) [11]. Its equivalent coupling model is shown in Fig. 1(b) for the ease of analysis. In Fig. 1, the subscripts "1" and "2" are used to denote the transmitting- and the receiving-side parameters, respectively, V_s is the input voltage, M is the mutual inductance, L and C are the self-inductance and capacitances, respectively, and R_L is the load. An extra inductor L_{11} is employed to turn the voltage source V_s into a current source for driving the transmitter coil.

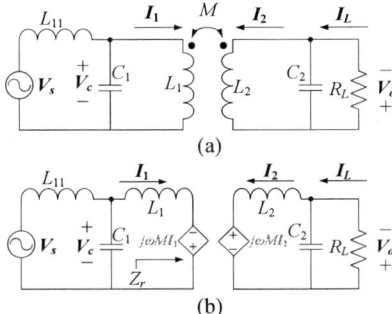

(a)

(b)

Fig. 1. (a) A two-coil WPT system and (b) its equivalent coupling model.

The reflected impedance is defined as the ratio of V_{r1} (the electromotive force induced in the transmitter coil, i.e., $V_{r1} = j\omega M I_1$, by the receiver-side current I_2) over I_1 as

$$Z_r = \frac{V_{r1}}{I_1} = \frac{\omega^2 M^2}{Z_2}, \quad (1)$$

where Z_2 is the total receiver-side impedance which can be calculated as

$$Z_2 = j\omega L_2 + \frac{1}{j\omega C_2 + 1/R_L}. \quad (2)$$

Substitution for Z_2 in (1) using (2) leads to

$$Z_r = \frac{\omega^2 M^2 R_L}{R_L^2 \left(\omega^2 C_2 L_2 - 1\right)^2 + \omega^2 L_2^2}$$
$$+ j\frac{-\omega^3 M^2 \left[C_2 R_L^2 \left(\omega^2 C_2 L_2 - 1\right) + L_2\right]}{R_L^2 \left(\omega^2 C_2 L_2 - 1\right)^2 + \omega^2 L_2^2}. \quad (3)$$

An important characteristic of Z_r is that it contains the information of the load (i.e., R_L) and the coupling (i.e., M). To simplify parameter identification, the system is deliberately operated at the receiver coil's resonant frequency $\omega = \omega_2$, where $\omega_2^2 C_2 L_2 = 1$, and thus (3) reduces to

$$Z_r = \frac{M^2 R_L}{L_2^2} - j\frac{\omega M^2}{L_2}, \quad (4)$$

where the load and the coupling information still retains. Now, with the assistance of Z_r, the transmitter-side electrical information which can be conveniently measured can be calculated as a function of the load and the coupling information. For instance, by applying Kirchhoff's Voltage Law (KVL) over the equivalent

circuit on the transmitter side while utilizing (4), one yields the capacitor voltage V_c of C_1 relating to R_L and M as [12]–[14]

$$V_c = \underbrace{\frac{M}{L_2} I_L R_L}_{V_o} + j\left(\omega L_1 - \frac{\omega M^2}{L_2}\right) I_1. \quad (5)$$

As a result, both M and the output voltage V_o information are readily available by analyzing the real and imaginary part of V_c.

As illustrated above, the secondary resonant mode method is very simple and time efficient. It has been successfully implemented for inductive power transfer (IPT) applications for closed-loop output voltage control in [12]–[14]. However, the main limitations of this approach are as follows:

(i) the method neglects the parasitic resistances in the transmitter and receiver coil in the equivalent coupling model of Fig. 1(b), which could lead to substantial inaccuracy in the estimated parameters;

(ii) the method is viable only at the receiver-side resonant frequency. Dynamic frequency variation for applications, e.g., energy efficiency optimization or frequency-modulation based output voltage control, is impossible;

(iii) the method is not general for any receiver-side compensation topologies. For instance, with a series compensated receiver, the WPT system has a Z_r without an imaginary part (as shown in (6)) when operating at the secondary resonant frequency. In this case, the coupling and the load information cannot be explicitly resolved as they are coupled in the real part of Z_r.

$$Z_r = \frac{\omega^2 M^2}{R_L}. \quad (6)$$

B. Energy equilibrium function based method

The second approach still follows the reflected impedance concept but with a different treatment of Z_r and having a freedom in the operating frequency [15]. Specifically, the method involves two phases of calculation: (i) a direct Z_r calculation phase in the time domain and (ii) an equation solving phase based on the relationship between Z_r, the coupling and the load in the frequency domain.

Firstly, instead of KVL, three energy functions, i.e., an energy storage function H, an input power function S, and a dissipation power function W, are defined for analyzing the transmitter-side equivalent circuit of the WPT system. Suppose Z_r is inductive-resistive at frequency ω satisfying

$$Z_r = R_r + j\omega L_r, \quad (7)$$

then, in the case of a series compensated transmitter, H, S, W are defined as

$$H = \frac{1}{2}\left(L_1 + L_r\right) i_1^2 + \frac{1}{2} C_1 v_{c1}^2, \quad (8)$$

$$S = v_s i_1, \quad (9)$$

$$W = i_1^2 \left(R_1 + R_r\right), \quad (10)$$

where C_1 is the transmitter-side compensating capacitor,

and v_{c1} is its instantaneous voltage.

By studying the relationships between H, S, and W, one is able to determine R_r and L_r, the real and the imaginary part of $\mathbf{Z_r}$. Fig. 2 depicts the steady-state voltage and current waveforms in the series-compensated transmitter coil. At steady state, the storage function H, being the sum of the inductive (i.e., L_1+L_r) and capacitive (i.e., C_1) energy is time-invariant in the transmitter. That is,

$$H = \frac{1}{2}C_1 v_{c1}^{\ 2}\left(t_1\right) = \frac{1}{2}\left(L_1 + L_r\right)i_1^{\ 2}\left(t_2\right), \qquad (11)$$

where the time instances t_1 and t_2 are the peaks of v_{c1} and i_1, respectively. Based on (11), L_r can be instantly resolved as

$$L_r = \frac{C_1 v_{c1}^{\ 2}\left(t_1\right)}{i_1^{\ 2}\left(t_2\right)} - L_1. \qquad (12)$$

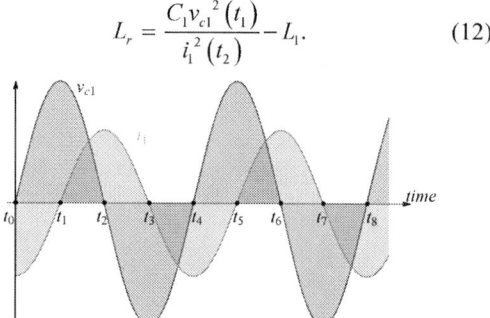

Fig. 2. Voltage and current waveforms on the transmitter side.

Noting that only the peaks of i_1 and v_{c1} need to be measured, a low sampling rate ADC is sufficient for determining L_r. Following similar procedures, R_r can be readily resolved with the peak v_{c1} measurement only.

As the real and the imaginary part of $\mathbf{Z_r}$ are now available, the coupling and the load can be further deducted. Take a parallel compensated receiver as an example. The two unknowns, M and R_L, can be derived by equating the real and the imaginary part of (3) and (7) [16], which formulate two equations. Evidently, the method has no constraints on the operating frequency provided that $\mathbf{Z_r}$ is inductive-resistive.

Compared to method A, the major advantages of method B are (i) its application to a wider operating frequency range and (ii) the low sampling rate requirement. Nonetheless, as method B relies on the exact measurement of the peak current/voltage information, the accuracy of parameter identification is highly sensitive to the phase delay in the zero current/voltage detection circuit and system noises. For example, [16] reported a maximum estimation error of 8% and 12.9% for M and R_L, respectively for a series-series (S-S) compensated WPT system, after taking an average of 1000 identification results.

C. Energy injection method

In [17], a novel energy injection method is proposed for load detection. Other than the full cycle or the peak voltage/current information, only the envelope of the transmitter-side current has to be determined. The method involves two modes of operation. The first one is the energy injection mode, with some initial energy injected into the WPT system at $t = \tau_0$ in order to excite the coils (see Fig. 3). At $t = \tau_1$, the energy injection mode ends and the system transits into the free resonant mode. The amplitude of the transmitter-side current i_1 will then gradually decay due to the damping from the parasitic losses and the real power consumed by the load. As a result, the decay rate of the amplitude of i_1, or, the envelopes of i_1, depends directly on the coupling and the loading profile. The current envelope can thus be utilized for determining the coupling and the loading condition.

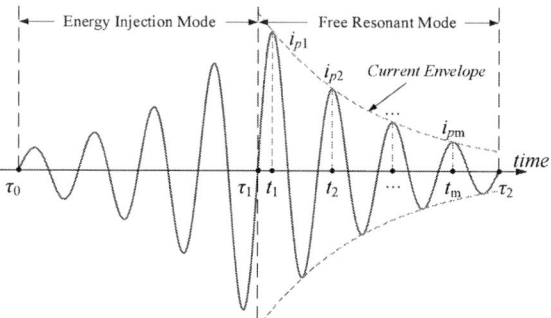

Fig. 3. Dynamic current waveforms in the transmitting coil.

The current envelope can be analytically derived by solving a differential equation governing the targeted WPT system. In [18], a 4th-order differential equation is established for a two-coil S-S compensated WPT system, as shown in Fig. 4. By solving the characteristic equation, R_L is mathematically solved and proved to be approximately inversely-proportional to the exponential decay rate of i_1's amplitude. Consequently, R_L can be determined upon detection of the envelope of i_1.

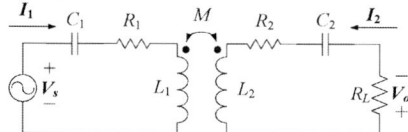

Fig. 4. A two-coil S-S compensated WPT system.

The energy injection method is more accurate for load identification than the previous two methods because (i) it is based on a dynamic system model (i.e., a differential equation) instead of a phasor model viable only at the steady state, and hence more robust against noise interference, (ii) it requires no zero current/voltage detection circuit and eliminates the phase delay issue. Despite these advantages, the energy injection method possesses several major constraints for practical applications:

(i) Due to the energy injection and the free resonant mode of operation, the energy is transferred to the load in a discrete manner which is unattractive for applications where continuous load regulation and efficient power transfer are needed. Therefore, this method can be used only for initial load detection prior to system startup. For instance, it can be used for detecting the loading condition (i.e. whether there is a load or not) for kitchen applications;

(ii) Unlike Z_r providing two degrees of system information (a real and an imaginary part), the amplitude decay rate, being a real number, offers only one degree of information. Therefore, the coupling and the load information cannot be solved simultaneously. For example, R_L can be determined while assuming M is pre-known, as reported in [17], [18];

(iii) Solving a high-order differential equation for a time-domain solution is generally nontrivial. In [18], several approximation steps are needed to simplify the derivation process, substantially affecting the accuracy of the identified parameters. Experimental results in [18] showed a maximum load estimation error of over 7% with this method.

D. Phasor-equation solving method

The phasor-equation solving method seeks to establish a formal and systematical treatment of the underlying relationships between various system parameters and front-end electrical quantities. Again, considering an S-S compensated two-coil WPT system in Fig. 4, the system can be described in a compact matrix form using phasors as

$$\begin{bmatrix} V_s \\ 0 \end{bmatrix} = \begin{bmatrix} Z_1 & j\omega M \\ j\omega M & Z_2 + R_L \end{bmatrix} \begin{bmatrix} I_1 \\ I_2 \end{bmatrix}, \qquad (13)$$

where

$$Z_1 = R_1 + j\left(\omega L_1 - 1/\omega C_1\right), \qquad (14)$$

$$Z_2 = R_2 + j\left(\omega L_2 - 1/\omega C_2\right), \qquad (15)$$

$$V_s = V_s \angle 0 = V_s, \qquad (16)$$

$$I_1 = I_1 \angle \theta_1 = I_1 \cos\theta_1 + jI_1 \sin\theta_1, \qquad (17)$$

$$I_2 = I_2 \angle \theta_2 = I_2 \cos\theta_2 + jI_2 \sin\theta_2, \qquad (18)$$

where θ_1 and θ_2 are the phase angles of I_1, I_2, respectively, and V_s, I_1, I_2 are expressed in the complex number form. The phasor equation set (13) actually formulates four equations. By separating the real and the imaginary part with the aid of (14)–(18), (13) is decoupled as:

$$\begin{cases} R_1 I_1 \cos\theta_1 - A I_1 \sin\theta_1 - \omega M I_2 \sin\theta_2 = V_s \\ R_1 I_1 \sin\theta_1 + A I_1 \cos\theta_1 + \omega M I_2 \cos\theta_2 = 0 \\ \omega M I_1 \sin\theta_1 + B I_2 \sin\theta_2 - (R_2 + R_L) I_2 \cos\theta_2 = 0 \\ \omega M I_1 \cos\theta_1 + B I_2 \cos\theta_2 + (R_2 + R_L) I_2 \cos\theta_2 = 0 \end{cases} \qquad (19)$$

where $A = \omega L_1 - 1/\omega C_1$, and $B = \omega L_1 - 1/\omega C_2$. The new equation set (19) indicates that a maximum of four unknowns can be solved. For instance, one could choose to solve an unknown set $\{M, R_L, I_2, \theta_2\}$ consisting of four parameters by providing the rest of the parameters $\{L_1, L_2, C_1, C_2, R_1, R_2\}$ and input states $\{V_s, \theta_1, I_1\}$. Once the unknown parameter set is determined, the output power can be further deducted for closed-loop control [19]–[21].

The above procedures for a two-coil system can be conveniently extended to a multi-coil system with different compensation topologies in each coil. One general conclusion that can be made here is at most $2n$ unknowns are identifiable for an n-coil system. An interesting example is reported in [19], demonstrating for the first time that high-fidelity and real-time load monitoring for an eight-coil series compensated WPT system is feasible. In this work, the unknown set is selected as $\{I_2$–I_8, θ_2–$\theta_8, R_L\}$ comprising 15 unknowns. Instead of resorting to a complex-number form equation set as (19), [19] shows that R_L can be solved systematically and efficiently using Cramer's rule for an n-coil system directly from the phasor form equation similar to that shown in (13). It is interesting to explore whether 16 unknowns can be identified based on a complex-number form equation set.

In many applications, identification of more system parameters, other than just M and R_L, is demanded. For instance, one might need to know C_1 and C_2 for a two-coil system as their values can deviate substantially from their nominal values, resulting in sub-optimal energy efficiency; in applications with multiple receivers, the load connected to each receiver needs to be identified for individual power monitoring and control; for Domino-Resonator WPT systems with multiple intermediate coils [22], the mutual inductances between each coil, being highly sensitive to coil dimensions, orientations, and relative positions, need to be determined because they are difficult to calculate or measure accurately. Mathematically, the only feasible means to identifying more system parameters based on method D is to have additional system equations. One intuitive method to achieve this is to perform frequency sweeping where data are collected at several discrete frequencies [23]. With measurements at two operating frequencies, $4n$ equations are then available for an n-coil system, meaning that a maximum of $4n$ unknowns may be solved. In [21], an approach without frequency sweeping is reported for a two-coil system. This method utilizes the harmonic information to formulate the additional equations needed. An alternative parameter identification method for an n-coil system is reported with the use of active decoupling switches [24]. As a result, an n-coil system is decoupled into several two-coil subsystems whose parameter identification then follows (13) or (19).

E. Curve-fitting method

The final method reported in the literature which seems more capable than other alternatives is the curve-fitting method. In particular, this method can be regarded as a special predication-error-minimization (PEM) approach (without modeling the system noise) that lies within the general system identification domain [25].

In [26], system identification is treated in the frequency domain. Given a general phasor equation set for a series compensated n-coil system (e.g. (13) where $n=2$), and assuming all system parameters $\{L_1...L_n, C_1...C_n, R_1... R_n, M_{12}...M_{(n-1)n}, R_L\}$ are known (where M_{ij} is the mutual inductance between the i-th and the j-th coil), then the input impedance $Z_{in}(\omega)$ of the WPT system at frequency ω as defined in (20) can be determined. Here, the $\hat{}$ symbol represents a predicted value given the full parameter set.

The 2018 International Power Electronics Conference

$$\hat{Z}_{in}(\omega) = \frac{V_s(\omega)}{I_1(\omega)} = f\begin{pmatrix} L_1, L_2, ... L_n, \\ C_1, C_2, ... C_n, \\ R_1, R_2, ... R_n, R_L, \\ M_{12}, M_{23}, ... M_{(n-1)n} \end{pmatrix}. \quad (20)$$

Meanwhile, the true $Z_{in}(\omega)$ can also be yielded experimentally through frequency sweeping. Therefore, by minimizing the (quadratic) prediction error given by

$$\varepsilon(\omega|\Theta) = \int_{-\infty}^{\infty} \left| \hat{Z}_{in}(\omega) - Z_{in}(\omega) \right|^2 d\omega \quad (21)$$

(where Θ is the unknown parameter set to be determined) through various of optimization algorithms, the true parameter set can be identified (note: prediction errors can be constructed in other forms). In [26], a genetic algorithm (GA) is employed to identify $\Theta = \{C_1, C_2, C_3, M_{12}, M_{13}, M_{23}\}$ for a 3-coil system with $Z_{in}(\omega)$ measured at 130 discrete frequencies around the system resonant frequency. Good agreement has been achieved with the measured values.

In [27], [28], system identification is achieved in the time-domain. The idea follows the classical PEM procedure by (i) establishing a general differential equation describing the system (ii) building a one-step-ahead predictor for the system output (iii) minimizing the prediction error between the predicted and the measured system output through optimization algorithms. In particular, v_s and i_1 are selected as system input and output for the purpose of front-end parameter identification. The time-domain curve-fitting method is especially useful for systems involving high nonlinearities when first harmonic approximation is inaccurate (e.g. with a diode bridge rectifier on the receiver side and with low-Q coils). It is demonstrated that simultaneous identification of parameter set $\{L_1, \omega_1, \omega_2, k, Q_L\}$ and output states $\{i_2(t), v_o(t)\}$ can be achieved with errors less than 5% for an S-S compensated two-coil WPT system loaded with a diode rectifier (Here ω_1 and ω_2 are the resonant frequency of the transmitter and the receiver, k is the coupling coefficient, and Q_L is the loaded quality factor). The number of unknowns identified is actually more than what is theoretically possible with the phasor-equation solving method.

Despite its effectiveness, a major limitation of the reported curve-fitting method is its long execution time, since both reported optimization engines are iterative algorithms. Even with a multi-core processor (operating at several GHz clock frequency), [26] reports a 3 minutes computation time to find the unknown parameters using GA, while [27] reports a 15 minutes time with a modified particle swarm optimization kernel. At present stage, it seems that the curve-fitting method is not yet ready for static and dynamic charging application where real-time control is demanded.

III. COMPARISONS OF FRONT-END PARAMETER IDENTIFICATION METHODS

Based on the discussion in Section II, the five front-end parameter identification methods are compared regarding several key figure-of-merits, e.g. estimation accuracy, scalability, operating frequency range, execution time, etc. The results are summarized in Table I.

The underlying relationships between each method are:

(i) Method D is a generalized method for method A and B, as the latter two are still based on frequency domain phasor analysis;

(ii) Method C can be obtained by applying an inverse Laplace transformation of the phasor equation set governing the overall system used in Method D. Therefore, Method C can be regarded as a time-domain interpretation of method D;

(iii) Both frequency- and time-domain system identification that have been utilized in Method E are also inherently correlated. Analyses in [25] have proved mathematically that minimizing quadratic prediction error in the time-domain for any linear time-invariant models can be interpreted as an alternative way of smoothing the Empirical Transfer-Function Estimate (ETFE) in the frequency domain. Here, ETFE, an analyzing tool for identifying system transfer functions based on power spectrum analysis, is actually a more general method for minimizing the prediction error of input impedance in (21).

TABLE I. COMPARISONS OF FRONT-END PARAMETER IDENTIFICATION METHODS

Method	A. Secondary resonant mode	B. Energy equilibrium function based	C. Energy injection	D. Phasor-equation solving	E. Curve-fitting
Level of prediction accuracy	Low		Moderate	High	
Maximum number of identifiable parameters for an n-coil system without frequency sweeping	2		1	$\leq 2n$	$\geq 2n$
Operating frequency	Fixed at secondary resonant frequency	Wider frequency range (an inductive reflected impedance must be ensured)		Full frequency range*	
Execution time	Fast			Relatively fast	Slow
Sampling rate	High	Low	Very Low	High	
Frequency sweeping required?	No			Depends on compensation topologies and number of parameters to identify	

* For some compensation topologies, such as an S-S compensated 2-coil system, parameter identification is unviable at the secondary resonant frequency with the phasor-equation solving method [20].

IV. CONCLUSIONS

This paper examines five front-end parameter identification methods for wireless power transfer (WPT) applications reported in the literature. These methods are also compared regarding key figure-of-merits, such as estimation accuracy, scalability, execution time, etc., which are important criteria for practical applications. It is revealed that parameter estimation based on phasor equations offers faster execution time, enabling real-time and closed-loop power regulation, while the disadvantages include a low prediction accuracy or a limited number of identifiable parameters. The curve-fitting methods based on system identification offer powerful tools for parameter identification, exhibiting huge potentials for future WPT applications. Advantages of curve-fitting methods include concurrent improved estimation accuracy and more identifiable parameters. One of their major limitations at current stage is the slow execution time. It is envisaged that, with more advanced and fast algorithms to be developed in the near future, the curve-fitting methods may eventually achieve an execution time that is suitable for real-time applications, e.g. static and dynamic charging applications.

REFERENCES

[1] Guoxing Wang *et al.*, "A closed loop transcutaneous power transfer system for implantable devices with enhanced stability," in *IEEE International Symposium on Circuits and Systems*, pp. 17–20.

[2] P. Si, A. P. Hu, S. Malpas, and D. Budgett, "A Frequency Control Method for Regulating Wireless Power to Implantable Devices," *IEEE Trans. Biomed. Circuits Syst.*, vol. 2, no. 1, pp. 22–29, Mar. 2008.

[3] J. T. Boys, C.-Y. Huang, and G. A. Covic, "Single-phase unity power-factor inductive power transfer system," 2008, pp. 3701–3706.

[4] H. L. Li, A. P. Hu, G. A. Covic, and ChunSen Tang, "A new primary power regulation method for contactless power transfer," in *International Conference on Industrial Technology (ICIT)*, 2009, pp. 1–5.

[5] N. Y. Kim, K. Y. Kim, J. Choi, and C.-W. Kim, "Adaptive frequency with power-level tracking system for efficient magnetic resonance wireless power transfer," *Electron. Lett.*, vol. 48, no. 8, p. 452, 2012.

[6] Wireless Power Consortium, "The Qi Wireless Power Transfer System Power Class 0 Specification, Version 1.2.2," 2016.

[7] IDT, "P9242-R Datasheet - Wireless Power Transmitter for 15W Applications," 2017.

[8] NXP Semiconductors, "WCT101XDS Datasheet," 2017.

[9] Texas Instruments, "bq500412 Datasheet - Low System Cost, Wireless Power Controller for WPC TX A6," 2013.

[10] S. Y. R. Hui, W. Zhong, and C. K. Lee, "A Critical Review of Recent Progress in Mid-Range Wireless Power Transfer," *IEEE Trans. Power Electron.*, vol. 29, no. 9, pp. 4500–4511, Sep. 2014.

[11] C.-S. Wang, O. H. Stielau, and G. A. Covic, "Design considerations for a contactless electric vehicle battery Charger," *IEEE Trans. Ind. Electron.*, vol. 52, no. 5, pp. 1308–1314, Oct. 2005.

[12] U. K. Madawala and D. J. Thrimawithana, "A single controller for inductive power transfer systems," in *Annual Conference of IEEE Industrial Electronics (IECON)*, 2009, pp. 109–113.

[13] U. K. Madawala and D. J. Thrimawithana, "New technique for inductive power transfer using a single controller," *IET Power Electron.*, vol. 5, no. 2, pp. 248–256, 2012.

[14] D. J. Thrimawithana and U. K. Madawala, "A primary side

controller for inductive power transfer systems," in *IEEE International Conference on Industrial Technology*, 2010, pp. 661–666.

[15] Zhi-hui Wang, Xiao Lv, Yue Sun, Xin Dai, and Yu-peng Li, "A simple approach for load identification in current-fed inductive power transfer system," in *IEEE International Conference on Power System Technology (POWERCON)*, 2012, pp. 1–5.

[16] X. Dai, Y. Sun, C. Tang, Z. Wang, Y. Su, and Y. Li, "Dynamic parameters identification method for inductively coupled power transfer system," in *IEEE International Conference on Sustainable Energy Technologies (ICSET)*, 2010, pp. 1–5.

[17] Z.-H. Wang, Y.-P. Li, Y. Sun, C.-S. Tang, and X. Lv, "Load detection model of voltage-fed inductive power transfer system," *IEEE Trans. Power Electron.*, vol. 28, no. 11, pp. 5233–5243, Nov. 2013.

[18] S. Hu, Z. Liang, Y. Wang, J. Zhou, and X. He, "Principle and application of the contactless load detection based on the Amplitude decay rate in a transient process," *IEEE Trans. Power Electron.*, vol. 32, no. 11, pp. 8936–8944, Nov. 2017.

[19] J. Yin, D. Lin, C.-K. Lee, and S. Y. R. Hui, "A systematic approach for load monitoring and power control in wirewess power transfer systems without any direct output measurement," *IEEE Trans. Power Electron.*, vol. 30, no. 3, pp. 1657–1667, Mar. 2015.

[20] J. Yin, D. Lin, T. Parisini, and S. Y. Hui, "Front-end monitoring of the mutual inductance and load resistance in a series-series compensated wireless power transfer system," *IEEE Trans. Power Electron.*, pp. 1–1, 2015.

[21] J. P. W. Chow and H. S. H. Chung, "Use of primary-side information to perform online estimation of the secondary-side information and mutual inductance in wireless inductive link," in *Applied Power Electronics Conference and Exposition (APEC)*, 2015, pp. 2648–2655.

[22] W. X. Zhong, Chi Kwan Lee, and S. Y. Hui, "Wireless power domino-resonator systems with noncoaxial axes and circular structures," *IEEE Trans. Power Electron.*, vol. 27, no. 11, pp. 4750–4762, Nov. 2012.

[23] Jian Yin, Deyan Lin, Chi Kwan Lee, T. Parisini, and S. Y. Hui, "Front-end monitoring of multiple loads in wireless power transfer systems without wireless communication systems," *IEEE Trans. Power Electron.*, vol. 31, no. 3, pp. 2510–2517, Mar. 2016.

[24] X. Dai, X. Li, and Y. Li, "Cross-coupling coefficient estimation between multi-receivers in WPT system," in *PELS Workshop on Emerging Technologies: Wireless Power Transfer (WoW)*, 2017, pp. 1–4.

[25] L. Ljung, *System identification : theory for the user.* Prentice Hall PTR, 1999.

[26] D. Lin, J. Yin, and S. Y. R. Hui, "Parameter identification of wireless power transfer systems using input voltage and current," in *Energy Conversion Congress and Exposition (ECCE)*, 2014, pp. 832–836.

[27] J. P.-W. Chow, H.-H. Chung, C.-S. Cheng, and W. Wang, "Use of transmitter-side electrical information to estimate system parameters of wireless inductive links," *IEEE Trans. Power Electron.*, vol. 32, no. 9, pp. 7169–7186, Sep. 2017.

[28] J. P.-W. Chow, H. S.-H. Chung, and C.-S. Cheng, "Use of transmitter-side electrical information to estimate mutual inductance and regulate receiver-side power in wireless inductive link," *IEEE Trans. Power Electron.*, vol. 31, no. 9, pp. 6079–6091, Sep. 2016.

A New Type of Wireless V2X System with a Dual-Active Bidirectional Single-Ended Converter and Optimized SiC-MOSFET

Hideki Omori[1][*], Aoto Yamamoto[1], Naoki Mukaiyama[1], Masahito Tsuno[2], Kenji Fukuda[3],
Hisato Michikoshi[3], Noriyuki Kimura[1] and Toshimitsu Morizane[1]

1 Osaka Institute of Technology, Osaka, Japan

2 Nichicon Corporation, Kyoto, Japan

3 Advanced Power Electronics Research Center National Institute
of Advanced Industrial Science and Technology (AIST), Ibaraki, Japan

*E-mail: hideki.omori@oit.ac.jp

Abstract- This paper deals with attractive bidirectional HF converter as transmitter and receiver of Wireless Power Transfer (WPT) for vehicle to home (V2H) / Building (V2B) / Community (V2C) smart energy management architecture. A new type of bidirectional WPT system with a dual-active seamless-controlled voltage-source single-ended resonant HF converter is proposed and tested for an energy storage system of a smart home / Building / Community. Moreover, described is a loss and reliability evaluation of a new SiC-MOSFET for the proposed system.

I. INTRODUCTION

The latest developments on a variety of wireless unidirectional resonant type power conversion transfer architectures; S/S, S/P, P/S and P/P resonant circuit topologies, which are based upon a principle of resonant IPT have attracted special interest for electric vehicle (EV)/ automated guided vehicle (AGV) battery charging system. [1][2][3]

In addition to hot-lighted topics of the HF power electronics technologies, bidirectional wireless power transfer architectures on the basis of dual resonant inverter-HF rectifier topology and vice versa can be investigated for smart house from an application point of view. In particular, WPT for EV/plug-in HEV can be effectively, conveniently economically utilized as an energy storage device. Accordingly, in smart house, bidirectional resonant WPT have been able to recognize as one of key technologies. [4][5]

On the other hand, the authors have developed so far for the generic circuit topologies of cost-effective voltage source single switch HF inverter family operating under the principle of zero voltage switching (ZVS) and zero current switching (ZCS) S-SW transition except for series/parallel resonant HF bridge inverter family and single-switch sub-resonant HF inverter circuit applied for WPT architectures has some inherent features as simple configuration, lower cost, and high efficiency except for high peak voltage across single power switch. [6]

It is noted that voltage source single-ended sub-resonant ZVS HF inverters-HF rectifiers topology

suitable for H2EV and EV2H modes and unidirectional single-ended sub-resonant ZVS HF inverter-diode rectifier scheme. [7]

This paper presents a novel type WPT architecture composed of bidirectional single-ended HF resonant inverter topology for smart home.

Furthermore, this paper introduces a special dual-active seamless control method in bidirectional single-ended HF resonant inverter.

In the first place, bidirectional WPT operating under H2EV and EV2H modes incorporating single-ended inverter and vice versa is proposed the smart house.

In the second place, a proposed bidirectional WPT based on newly-developed dual-active seamless controlled single-ended inverter is built and evaluated. Experimental result of cooperation between EV and smart house which connected by the newly developed wireless V2H system with a seamless method is carried out here in.

Finally, a proposed bidirectional WPT based on newly-developed SiC-MOSFET type single-ended inverter is built and tested. The power loss analysis of SiC-MOSFET inverter is evaluated as compared with RC-IGBT from an experimental point of view.

II. A WIRELESS BIDIRECTIONAL SINGLE-ENDED SUB-RESONANT INVERTER

Developed is a bidirectional IPT with a single-ended converter. Figure 1 shows the progress of bidirectional wireless power supply system. WPT is established to operate with a single-ended inverter. Resonant capacitor is added in wireless EV charging system pick-up side for improvement of output power and actual efficiency. HF rectifier in receiving coil side circuit is changed to a half-wave rectifier using an antiparallel diode of active switch. Then, the pick-up side circuit is changed to a single-ended inverter the same as the primary circuit. A bidirectional resonant IPT is established.

The 2018 International Power Electronics Conference

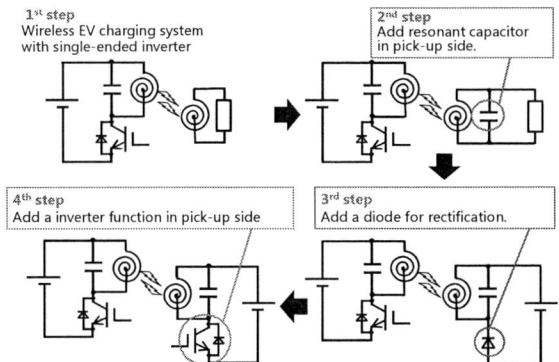

Fig. 1. The progress for bidirectional WPT system.

A cost effective configuration by a single-ended inverter with IPT principle for bidirectional power transfer is shown in Figure 2. In the vehicle to home mode, capacitance C_1 can operate as resonant capacitance for IPT and capacitance C_2 operates as resonant capacitance to achieve for ZVS. On the other hand, in the EV battery charging mode, the capacitance C_1 operates as resonant capacitance for ZVS and the capacitance C_2 operates as a resonant capacitance.

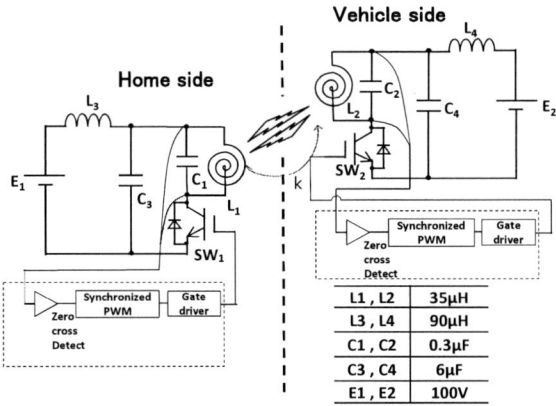

Fig. 2. Bidirectional WPT using single-ended converter.

Figure 3 shows measured operating waveforms of bidirectional WPT. The ZVS scheme can be achieved completely. Measured operating waveforms for the power flow from vehicle to home mode are also shown and the H2EV operation scheme can be completed.

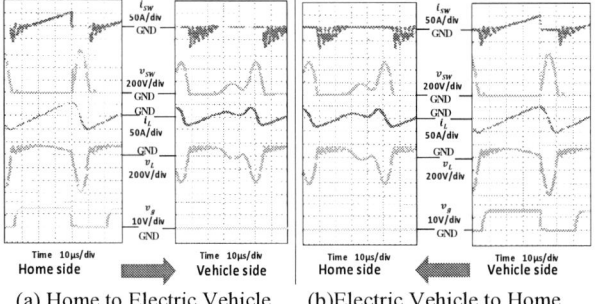

(a) Home to Electric Vehicle   (b) Electric Vehicle to Home

Fig. 3. Measured operating waveforms of propsed WPT system.

III. A SEAMLESS WIRELESS V2H SYSTEM WITH A BIDIRECTIONAL SINGLE-ENDED CONVERTER

A. Operating Characteristics of Dual-Active Seamless WPT

Proposed is a new bidirectional WPT system with a dual-active seamless controlled single-ended converter for wireless V2H. Figure 4 shows a construction of wireless V2H system using a dual-active seamless control method. Electric power is wireless-transferred automatically from higher voltage battery side to lower voltage battery side. Therefore, batteries in EV and smart house act as if they were parallel connected batteries. The proposed bidirectional WPT system can transfer electric power in appropriate direction with a simple constitution.

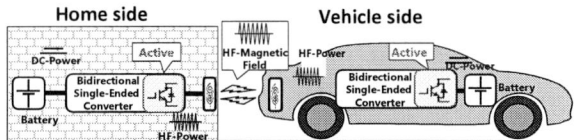

Fig. 4. A schematic wireless V2H system with a dual-active seamless control method.

Figure 5 shows simulated transfer current vs. voltage difference between primary battery voltage E_1 and secondary battery voltage E_2 in the dual-active seamless WPT. Direction of the transfer current is seamlessly changed corresponding to the voltage difference between primary and secondary battery. The transfer current changes continuously for voltage difference value. In addition, the transfer current value can be controlled by conduction time T_{ON} of the power devices. The primary side means smart-house side and the secondary side means EV side. When the home side battery energy decreases in comparison with the vehicle side battery, power is transferred from vehicle to home. On the other hand, when the vehicle side battery energy decreases in comparison with the home side battery, the power is transferred from home to vehicle. Therefore, required energy is transferred on required direction automatically.

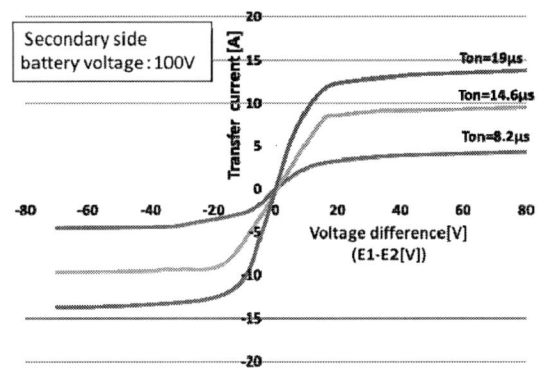

Fig. 5. Simulated transfer current characteristics of the newly developed wireless V2H.

Figure 6 shows measured operating waveforms of the proposed dual-active seamless WPT for wireless V2H

2559

system in 27 kHz operation. Both converters are synchronized and power flows from home to vehicle automatically.

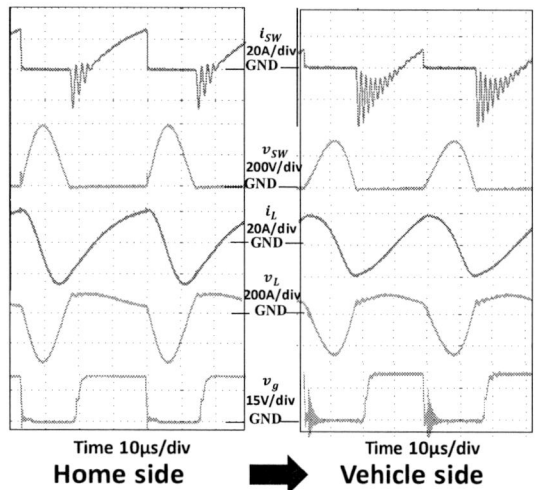

Fig. 6. Measured operating waveforms of the proposed dual-active seamless WPT. (E_1 = 95V, E_2 = 80V)

Figure 7 and Figure 8 show a circuit diagram of self-synchronized PWM and its operating waveforms respectively. The single-ended inverter is synchronized with its resonant coil zero-cross voltage point by this control block so as to keep zero-volt switching operation under various load conditions. Zero-cross point which is detected by coil voltage VL1 triggers a sawtooth wave circuit to turn over as indicated in Figure 8.

When the sawtooth wave crosses a threshold voltage, PWM output signal changes to high, then the power switch is turned on after the delay time Td which has been selected to keep zero voltage turn on.

Since coil voltages are influenced each other with mutual coupling of two coils, therefore, operating waveforms of vehicle and home unit are synchronized automatically.

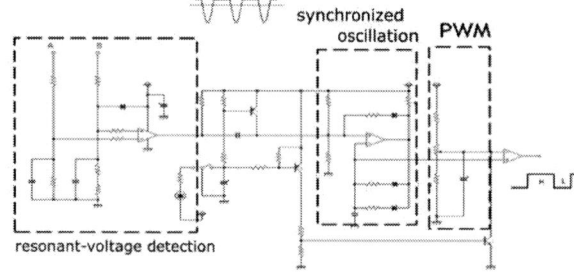

Fig. 7. a circuit diagram of synchronized PWM

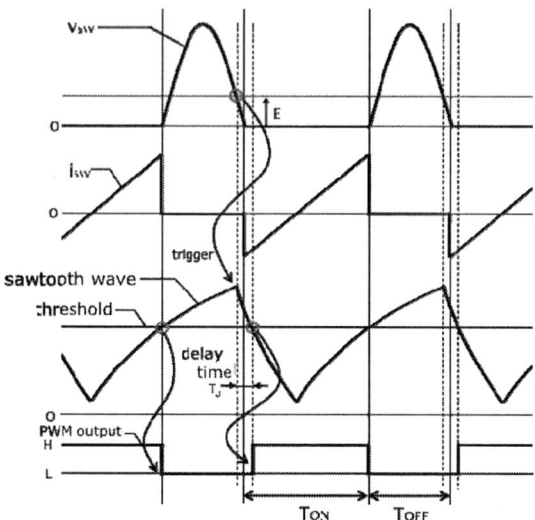

Fig. 8. operating waveforms of synchronized PWM

B. An Advanced Smart House in Connection with Dual-Active Seamless Wireless V2H

Dual-active seamless WPT is applied to the smart home as V2H as depicted in Figure 9. Smart house core block controls peak shift operation which charges an energy storage system with surplus electric power by PV in daytime and night electricity, and supplies required electric power from the energy storage system when the power loads in home are heavy. The advanced smart house can operate as peak cut system. Furthermore, this system can use electric power constantly by demand.

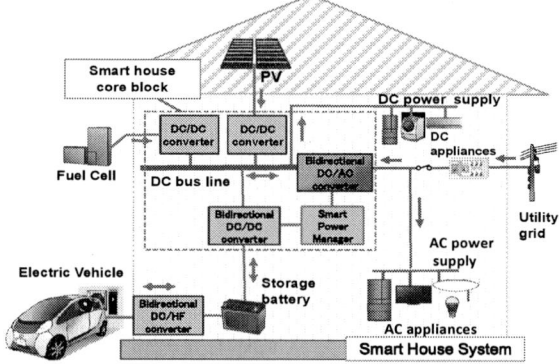

Fig. 9. A smart house DC micro grid architecture including seamless wireless bidirectional EV2H system.

Cooperation between EV and the advanced smart house connected by the wireless V2H system with a seamless operation is evaluated in experimental point of view. A configuration of the cooperation is depicted in Figure 10. A set value of the smart house core block is 400W electric power constant using. When the power consumption P_{Load} of the load is more than 400W, EV battery and smart house battery compensate power shortage of utility grid. On the other hand, when P_{Load} is less than 400W, the batteries are charged by surplus power.

Figure 11 illustrates an experimental result of constant utility power control in the proposed advanced smart house. P_{SH} means a power flow from the batteries to DC bus. P_{V2H} means a power flow from EV to the home battery. Utility power which is a difference of P_{Load} and P_{SH} is controlled about 400W constantly as shown in Figure 11 (a). In Figure 11 (b), P_{V2H} supplements the smart house battery in bidirectional way.

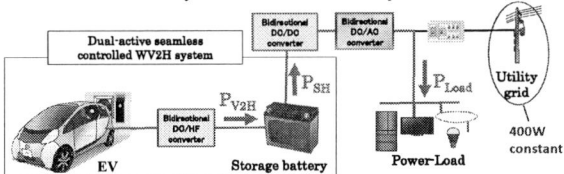

Fig. 10.   A smart house in connection with seamless WV2H.

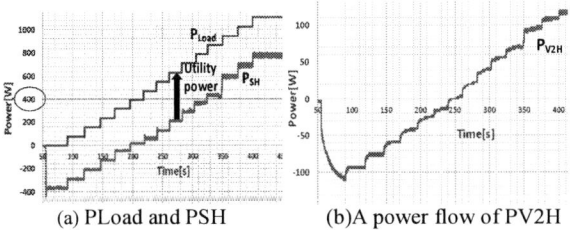

(a) PLoad and PSH   (b)A power flow of PV2H

Fig. 11.   An experimental result of constant utility power control   in the proposed advanced smart house.

IV. EVALUATION OF POWER DEVICES LOSS-TEMPERATURE CHARACTERISTICS

A. Power Loss Evaluation of Power Switching Devices

Figure 12 describes power losses in power semiconductor device of the power feeding converter.

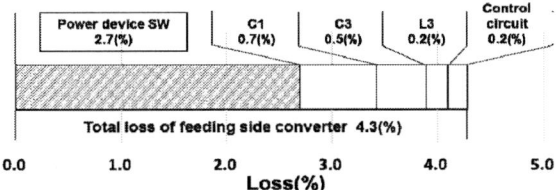

Fig. 12. Power loss analysis of power semiconductor switch in WPT transmitting side HF inverter.

It is noted that power switching device in the single-ended inverter occupies the largest power loss in the power converter. Figure 13 represents operating waveforms of SW in Figure 2. A RC-IGBT loss is calculated by eq. (1), and a MOSFET loss is calculated through eq. (2).

$$P_{loss} = f\int_{T_D}(v_F)(-i_C)dt + f\int_{T_{ON}}v_{CE(sat)}i_C dt + f\int_{T_{off}}v_{CE}i_C dt \ (1)$$

$$P_{loss} = f\int_{T_D}(v_F)(-i_D)dt + f\int_{T_{ON}}R_{ON}i_D{}^2 dt + f\int_{T_{off}}v_{DS}i_D dt \ (2)$$

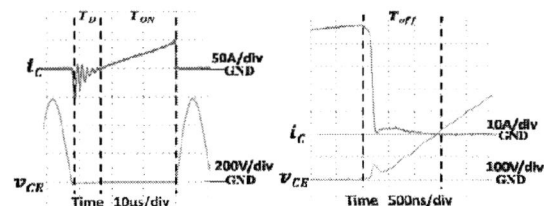

Fig. 13. Operating waveforms of power device SW.

Two types of 1200V Si-IGBT and a SiC-MOSFET are selected for comparative study. All devices have same size of 5mm square chip. High speed type of IKW40N120H3 has short tail current but high $V_{CE(sat)}$. On the other hand, low speed type of IHW40N120R3 has large tail current but low $V_{CE(sat)}$. A SiC-MOSFET of TPECMS12V78CA3 which is called IEMOSFET (Implantation & Epitaxial MOSFET) has been newly developed by The National Institute of Advanced Industrial Science and Technology (AIST). The SiC-MOSFET has tail-less turn-off and very low ON-resistance. Table 1 indicates the IGBTs and SiC-MOSFET specifications.

TABLE 1
SPECIFICATION OF THE POWER DEVICES

①Si-IGBT (Infineon) : 1200V/80A
IHW40N120R3 :Low speed

Chip size	5.0mm×5.0mm	tf	177[ns]
Breakdown voltage	1200[V]	VCE[sat]	1.6[V]@40A
Current capacity	80[A]	Vth[typ]	5.8[V]

②Si-IGBT (Infineon) : 1200V/80A
IKW40N120H3 :High speed

Chip size	5.0mm×5.0mm	tf	143[ns]
Breakdown voltage	1200[V]	VCE[sat]	2.1[V]@40A
Current capacity	80[A]	Vth[typ]	5.8[V]

③SiC-MOSFET(AIST) : 1200V/78A
IEMOSFET

Chip size	5.0mm×5.0mm	tf	55[ns]
Breakdown voltage	1200[V]	Ron	36.1[mΩ]@40A
Current capacity	78[A]	Vth	3.73[V]

Operating frequency was decided 25kHz to avoid audible noise range and large switching loss. However,85kHz band(81.38~90kHz) has been selected for the standard operating frequency of wireless EV charging system by the International Standard. Then, newly studied are operating characteristics of power devices under 85kHz wireless power transfer system.

TABLE 2. POWER DEVICE LOSSES AT 85kHz. (Pin = 1200W)

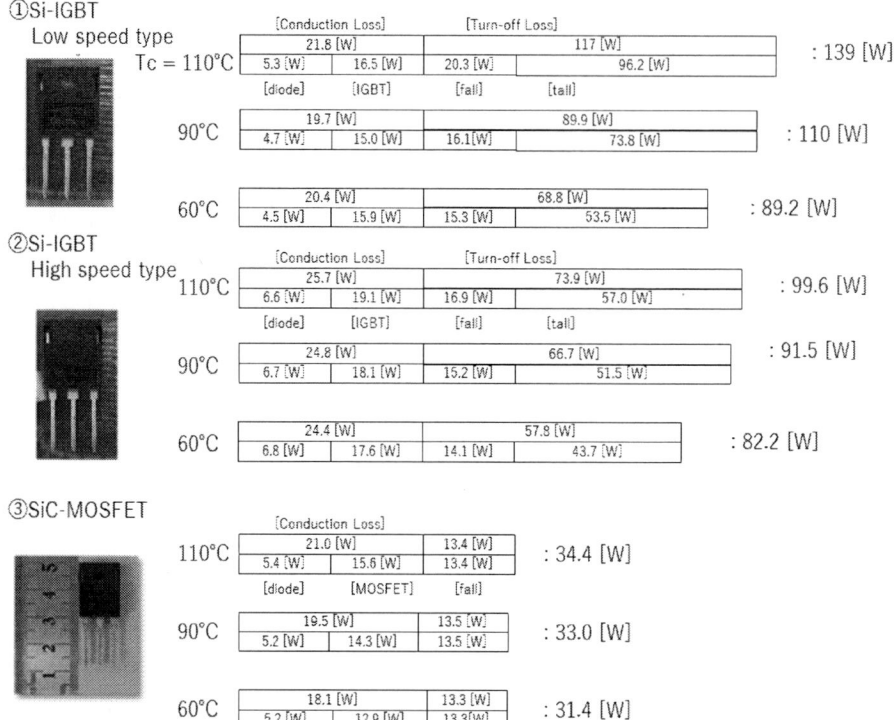

①Si-IGBT Low speed type

Tc = 110°C
[Conduction Loss] — 21.8 [W] (5.3 [W] [diode], 16.5 [W] [IGBT]) | [Turn-off Loss] — 117 [W] (20.3 [W] [fall], 96.2 [W] [tail]) : 139 [W]

90°C
19.7 [W] (4.7 [W], 15.0 [W]) | 89.9 [W] (16.1 [W], 73.8 [W]) : 110 [W]

60°C
20.4 [W] (4.5 [W], 15.9 [W]) | 68.8 [W] (15.3 [W], 53.5 [W]) : 89.2 [W]

②Si-IGBT High speed type

110°C
[Conduction Loss] — 25.7 [W] (6.6 [W] [diode], 19.1 [W] [IGBT]) | [Turn-off Loss] — 73.9 [W] (16.9 [W] [fall], 57.0 [W] [tail]) : 99.6 [W]

90°C
24.8 [W] (6.7 [W], 18.1 [W]) | 66.7 [W] (15.2 [W], 51.5 [W]) : 91.5 [W]

60°C
24.4 [W] (6.8 [W], 17.6 [W]) | 57.8 [W] (14.1 [W], 43.7 [W]) : 82.2 [W]

③SiC-MOSFET

110°C
[Conduction Loss] — 21.0 [W] (5.4 [W] [diode], 15.6 [W] [MOSFET]) | 13.4 [W] (13.4 [W] [fall]) : 34.4 [W]

90°C
19.5 [W] (5.2 [W], 14.3 [W]) | 13.5 [W] (13.5 [W]) : 33.0 [W]

60°C
18.1 [W] (5.2 [W], 12.9 [W]) | 13.3 [W] (13.3 [W]) : 31.4 [W]

B. Temperature Characteristics

In general, since power switching devices in converter generate heat by device loss, device temperature rises too high to keep stable operation under insufficient performance of cooling setup. Furthermore, thermal runaway which causes fatal damage to the device easily occurs when the device has poor loss-temperature characteristics. Then, loss-temperature characteristics of SiC-MOSFET and Si-IGBTs are comparatively studied herein. Table 2 describes experimental results of loss-temperature characteristics of them. Accumulated amount of losses during conduction time, fall period and tail period are shown in Figure 14.

The total loss of SiC-MOSFET is 33W at 90°C which is about 1/3~1/4 of Si-IGBT, because SiC-MOSFET which has very low ON-resistance and tail-less turn-off has the lowest conduction and switching loss compared with 2 types of Si- IGBT. The device loss of 3% is allowable level for implementation to actual application, generally. The SiC-MOSFET has the loss-temperature characteristics of 0.2%/°C while the Si-IGBTs have that of 0.4~1.3%/°C. Poor loss temperature characteristics of IGBTs are caused by the tail-loss which extremely increases by temperature rise, because principal component of IGBT switching loss is tail loss which is caused by tail current.

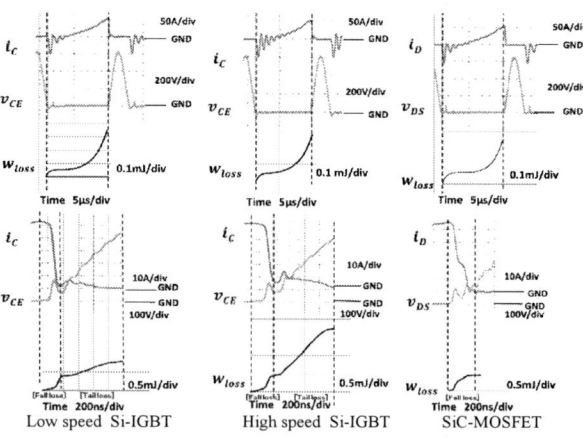

Low speed Si-IGBT   High speed Si-IGBT   SiC-MOSFET

Fig. 14.   Accumulated amount of power device loss at 85kHz(measured).

C. Avalanche Resistance Characteristics

Since the single-ended converter system uses resonance to achieve zero voltage switching operation, a power device must have high withstand voltage for operation under high resonant switch voltage. Avalanche resistance of the power device is effective to establish high reliability of the system. Then, avalanche resistance of SiC-MOSFET is compared with that of RC-IGBT.

Figure 15 shows switch voltage and current waveforms of SiC-MOSFET in avalanche-resistance evaluation. Table 3 describes experimental result of avalanche energy and current of SiC-MOSFET and Si-IGBT. The avalanche energy of the SiC-MOSFET is extremely superior to Si-IGBT.

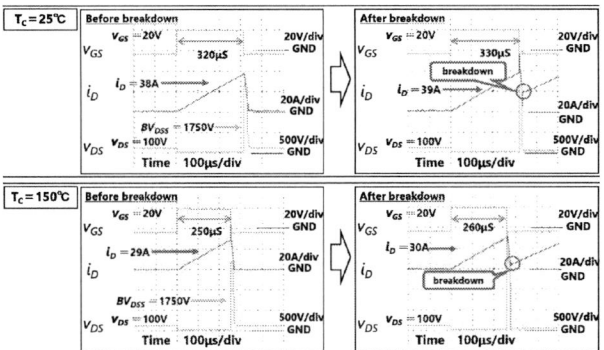

Fig. 15. Operating waveforms of avalanche resistance of SiC-MOSFET

TABLE 3
SPECIFICATION OF THE POWER DEVICES.

①Si-IGBT(Infineon): 1200V80A IHW40N120R3(Low speed type)

T_C [°C]	Avalanche Energy [mJ]	Avalanche current [A]
25	0.89	1.9
150	0.48	1.3

②SiC-MOSFET(AIST): 1200V78A IEMOSFET TPECMS12V78CA3

T_C [°C]	Avalanche Energy [mJ]	Avalanche current [A]
25	612	38
150	357	29

V. CONCLUSION

It should be noted that the simplest bidirectional single-ended WPT has been implemented for smart house.

Characteristics of IPT using the special dual-active seamless single-ended converter for wireless V2H system have been illustrated. The experimental result has confirmed that the advanced smart house with seamless wireless V2H system is well function for constant utility power operation and EV battery supplement.

The newly developed wireless V2H system with the bidirectional single-ended converter has high efficiency on 85kHz operation, high reliability against thermal runaway and high immunity for electro-magnetic disturbance by using the new SiC-MOSFET.

In the future, attractive single SiC-MOSFET voltage source type dual-active seamless controlled single-ended converter with SiC-MOSFET proposed here has to be investigated for an application of vehicle to community.

ACKNOWLEDGMENTS

A part of this work has been implemented under a joint research project of Tsukuba Power Electronics Constellations (TPEC).

This work was supported by JSPS Grants-in-Aid for Scientific Research Grant Number 17K06325.

REFERENCES

[1] Toshihiro Kai, Throngnumchai Kraisorn, Yuusuke Minagawa," A Study on Receiver Circuit Topology of Non-contact Charger for Electric Vehicle", Proc. Japan Industry Applications Society Conf. Okinawa, Japan, August 2011, CD-ROM, Paper No.2-16

[2] Jin Huh, Wooyoung Lee, Gyu-Hyeong Cho Byunghun Lee, Chun-Taek Rim, "Characterization of Novel Inductive Power Transfer Systems for On-Line Electric Vehicles", Proc. Annual IEEE Applied Power Electronics Conference and Exposition, APEC, Texas, USA, March 2011

[3] Shingo Machino, Masahiro Kozako, Katsuhiko Harada, Masayuki Hikita, Kazuyuki Hotta, Yoshinori Kataoka," Construction and Characteristics of Wireless Resonance Type Inductive Power Supply"(Japanese), *Proc. Japan Industry Applications Society Conf.* Okinawa, Japan, August 2011, CD-ROM, Paper No.2-13

[4] Duleepa J. Thrimawithana, Udaya K. Madawala, Yu Shi, "Design of a Bi-Directional Inverter for a Wireless V2G System", Proc. IEEE International Conf. on Sustainable Energy Technologies, ICSET, Kandy, Sri Lanka, December 2010

[5] Suvendu Samanta, Akshay Kumar Rathore, Duleepa J. Thrimawithana, "Bidirectional Current-Fed Half-Bridge (C)(LC)–(LC) Configuration for Inductive Wireless Power Transfer System ", Trans. on IEEE Vol.53, No.4, July/August.2017

[6] Hideki Omori , Mutsuo Nakaoka, "Generic Circuit Topologies and Their Performance Evaluations of Single-Ended Resonant High-Frequency Inverters for Induction-Heated Cooking Appliances", *Trans. on IEE Japan,* pp. 150-159, Vol. 117-D, No. 2, 1997,.

[7] Yuichi Iga, Yuhei Kubo, Hideki Omori, Toshiaki Morizane, Noriyuki Kimura, "Wireless EV Charging System with a Single-Ended Resonant Inverter", Proc. Japan Institute of Power Electronics Conf. p.11, Osaka, Japan, Dec. 2011

[8] Yuichi Iga, Hideki Omori, Hiroki Fukuoka Noriyuki Kimura, Toshimitsu Morizane , Kunio Nakagawa Yoshimichi Nakamura Mutuo Nakaoka, "A New Bidirectional Resonant IPT EV Charging System with Single-Ended Inverter for Wireless V2H" International Conference on Electrical Drives and Power Electronics 2013, Dubrovnik – Croatia, Oct. 2013

[9] Shinya Ohara, Hideki Omori, Kenji Fukuda, Hisato Michikoshi, Noriyuki Kimura, Toshimitsu Morizane, Mutsuo Nakaoka, "Comparative Study of IGBT and SiC-MOSFET in a Wireless V2H System with a New Bidirectional Single-Ended ZVS Converter", Internatinal Conference on Power Electronics and Motion Control, Jul.2016

[10] Biao Zhao, Qiang SongWenhua Liu, Member, Yandong Sun, "Overview of Dual-Active-Bridge Isolated Bidirectional DC–DC Converter for High-Frequency-Link Power-Conversion System" IEEE transaction on power electronics, Vol.29, No.8, 2014

[11] Ryota Kondo, Yusuke Higaki, Masaki Yamada, "Proposition and Experimental Verification of a Bi-Directional Isolated DC/DC Converter for Battery Charger-Discharger of Electric Vehicle", *Trans. on IEE Japan,* pp. 61-70, Vol. 136, No.1 , 2016

[12] M. Kitabatake, S. Kazama, C. Kudou, M. Imai, A. Fujita, S. Sumiyoshi and H. Omori; "4H-SiC-DIMOSFET Powerevice for Home Appliances", International Power Electronics Conference, Sapporo, Japan, June. 2010

[13] Shinya Ohara, Hideki Omori, Noriyuki Kimura, Toshimitsu Morizane,Mutuo Nakaoka,Yoshimichi Nakamura, "A New V2H System with Single-Ended Inverter Drive Bidirectional Wireless Resonant IPT" International Power Electronics and Application Conference and Exposition 2014, Shanghai-China , Nov.2014

Metal Object Detection System with Parallel-mistuned Resonant Circuits and Nullifying Induced Voltage for Wireless EV Chargers

Seog Y. Jeong, Van X. Thai, Jun H. Park, and Chun T. Rim

Graduate program of Energy Technology, GIST, Gwangju, Korea

*E-mail:{seogyong86, vanthaixuan, junhyeong816, ctrim}@gist.ac.kr

Abstract- **In this paper, a metal object detection (MOD) system, a kind of foreign object detection (FOD), which is based on mistuned resonant circuits and utilizes variation of self-inductance of a sensing pattern, is newly proposed for wireless electric vehicle (EV) chargers. The sensing pattern that consists of multiple loop coil sets is mounted on the transmitting (Tx) pad of an EV charger, where a loop coil set has two coils connected in series with opposite polarity to cancel out the induced voltage generated by the Tx coil. Variation of self-inductance of the loop coil set is detected by a parallel-resonant circuit, driven by a current source and operating at near 1 MHz. To increase the detection sensitivity of the proposed MOD system, instead of an exact resonant frequency, a mistuned operating frequency near the -3dB point is utilized for the parallel-resonant circuit. Through simulations and experiments, it is found that the proposed MOD system detects not only horizontal but also standing upright metal objects. A prototype MOD system, operating at 85 kHz to satisfy the standard J2954, was fabricated to verify its feasibility. The results showed that output voltage change of the proposed MOD system becomes 22.7 % for a piece of aluminum foil of 3 x 3 cm^2 and 40.9 % for 100 Korean Won coin, respectively.**

Keywords— Metal object detection (MOD), foreign object detection (FOD), wireless EV charger, mistuned resonance, blind zone

I. INTRODUCTION

As interest in and demand for electric vehicles (EVs) have increased in recent years, inductive power transfer systems (IPTSs) based on magnetic coupling between two coils has been extensively studied for convenient and safe charging of EVs [1]-[3]. During the transfer of power, transmitting (Tx) and receiving (Rx) coils generate a strong AC magnetic field. When foreign objects such as metal objects approach the wireless EV charger, an eddy current is generated inside the objects by the AC magnetic field. In particular, when a large amount of eddy current flows inside a material with high resistance such as metal objects, high temperature by ohmic loss will eventually lead to combustion. To avoid this problem, a metal object detection (MOD) system, employed for of foreign object detection (FOD), should transfer information on the metal presence to the controller of the IPTS and the user through the communication. While this is critical to prevent unwanted accidents, few studies to solve this problem have been reported [4]-[10].

A method of detecting the system parameters, which detects the quality factor, voltage and current of the Tx or Rx coils with and without metal objects, was proposed in [4]-[5]. However, this approach cannot clearly distinguish between the effect of metal objects and the effect of misalignment of the Rx pad since the parameters of the IPTS are changed by the position of the Rx pad as well as the load condition. In the meantime, MOD systems using an additional sensing pattern, where the loop coils are connected in the reverse direction to cancel out induced voltage, have been introduced by KAIST and WiTricity [6]-[8]. Although the method using the magnetic field generated by the Tx coil is effective, the sensing area is limited by the magnetic field generated by the Tx coil, which means there are some special areas called "blind zones" where metal objects cannot be detected or only can be partly detected [6]. In the case of multi-layer construction of the sensing pattern, the majority of blind zones at the intersection of loop coils are resolved, however, blind zones may remain at points where the magnetic field generated by the Tx coil is weak or only the magnetic field component in the horizontal direction is present.

In this paper, to resolve the problem of blind zones, the MOD system, that utilizes the self-inductance change of a sensing pattern and a mistuned parallel-resonant circuit is newly proposed for wireless EV charger applications. In order to increase the detection sensitivity of the proposed MOD system, a mistuned operating frequency near -3dB point is utilized instead of the exact resonant frequency of the parallel-resonant circuit.

II. PRINCIPLE OF PROPOSED MOD SYSTEM

The key idea of the proposed MOD system is to measure the impedance of the sensing pattern considering the reflected impedance by metal objects, as shown in Fig. 1.

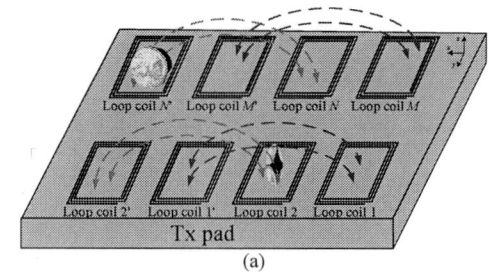

(a)

This work was supported by the Korea Evaluation Institute of Industrial Technology (KEIT) and the Ministry of Trade, Industry, and Energy (MOTIE) of the Republic of Korea under Grant 10052912.

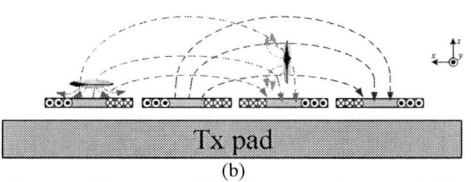

(b)

Fig. 1. Concept of the proposed MOD system, where metal objects are placed on the Tx pad. (a) Bird's eye view. (b) Front view, where loop coil n is connected in series with n'.

The equivalent circuit model under the influence of metal objects around the sensing pattern based on mutual coupling modeling is shown in Fig. 2.

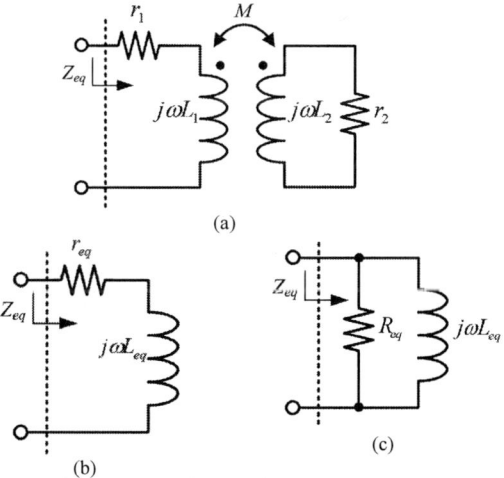

(a)

(b)

(c)

Fig. 2. (a) Equivalent circuit of the sensing pattern under the influence of metal objects. (b) Simplified equivalent circuit model expressed by an inductor with a series resistor (c) Simplified equivalent circuit expressed by an inductor with a parallel resistor.

When metal objects are close to the sensing pattern, the equivalent impedance should be changed. The equivalent circuit model expressed by an inductor with a series resistor or a parallel resistor, respectively, can be expressed as

$$Z_{eq,s} = j\omega\alpha_s L_1 + \beta_s r_1, \tag{1a}$$

$$Z_{eq,p} = j\omega\alpha_p L_1 + \beta_p R_1 \tag{1b}$$

where L_1 and R_1 are the self-inductance and equivalent internal resistance of the sensing pattern without metal objects. R_1 is an equivalent parallel resistor of the sensing pattern without any metal objects. α, β_s, and β_p are defined as the variation ratio of L_1, r_1, and R_1, as follows:

$$\alpha = \alpha_s \approx \alpha_p = 1 - \frac{L_2}{L_1} \cdot \frac{(\omega M)^2}{r_2^2 + \omega^2 L_2^2}, \tag{2a}$$

$$\beta_s = 1 + \frac{r_2}{r_1} \cdot \frac{(\omega M)^2}{r_2^2 + \omega^2 L_2^2}, \quad \because \beta_p = \frac{\alpha^2}{\beta_s} \tag{2b}$$

where L_2 and r_2 are the equivalent inductance and resistance of the metal objects and M is the mutual

inductance between the sensing pattern and the metal objects. Here, the subscripts "s", "p", "w", and "wo" denote "series", "parallel", "with metal objects", and "without metal objects", respectively.

The conventional sensing methods, such as use of an AC signal without any modulation, makes the system more complex and costly due to the noise and high operating frequency. Moreover, it is difficult to employ the MOD system directly in wireless EV charger applications because the impedance variation by a metal object may not be sufficient compared to the effects of noise. As a remedy for this problem, a resonance topology is applied to the MOD system to amplify the impedance variation. The proposed MOD system consists of the sensing pattern, a parallel-resonant circuit, its driving circuit excited by a sine wave current source with angular frequency ω_s, and filters. The parallel-resonant circuit consists of an inverting amplifier, unidirectional MOSFETs S_n, blocking capacitors C_n, a resonant capacitor C_p, and input side resistance R_{in}, as shown in Fig. 3.

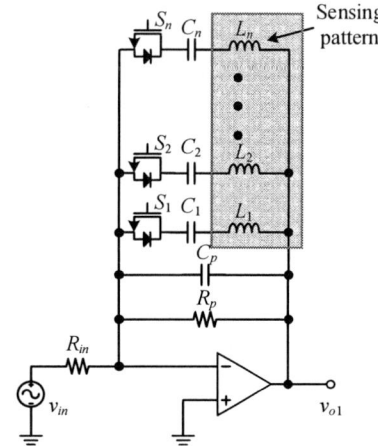

Fig. 3 Proposed parallel-resonant circuit with inverting amplifier.

Not only to increase the sensitivity of the proposed MOD system but also to simplify the circuit at the same time, unidirectional MOSFET switches control the parallel-resonant circuit time-divisionally. Common source terminals of all switches are connected to the virtual ground of the op-amp to simplify its driving circuit. A parallel feedback resistor R_p represents the equivalent resistance of the sensing pattern. If the equivalent series resistor of the sensing pattern changes too much with respect to temperature variation, the MOD system can be stabilized by inserting an additional external resistor. However, it should be noted that if an excessively small resistor is inserted, Q of the parallel-resonance circuit decreases, leading to decreased sensitivity. The series capacitor C_n with a sufficiently large capacitance is connected in series with the switch to block the DC component. The parallel capacitor C_p forms a feedback loop to adjust ω_r of the parallel-resonant circuit. For easy control, ω_r should be kept at the same value for all sensing patterns, as follows:

2565

The 2018 International Power Electronics Conference

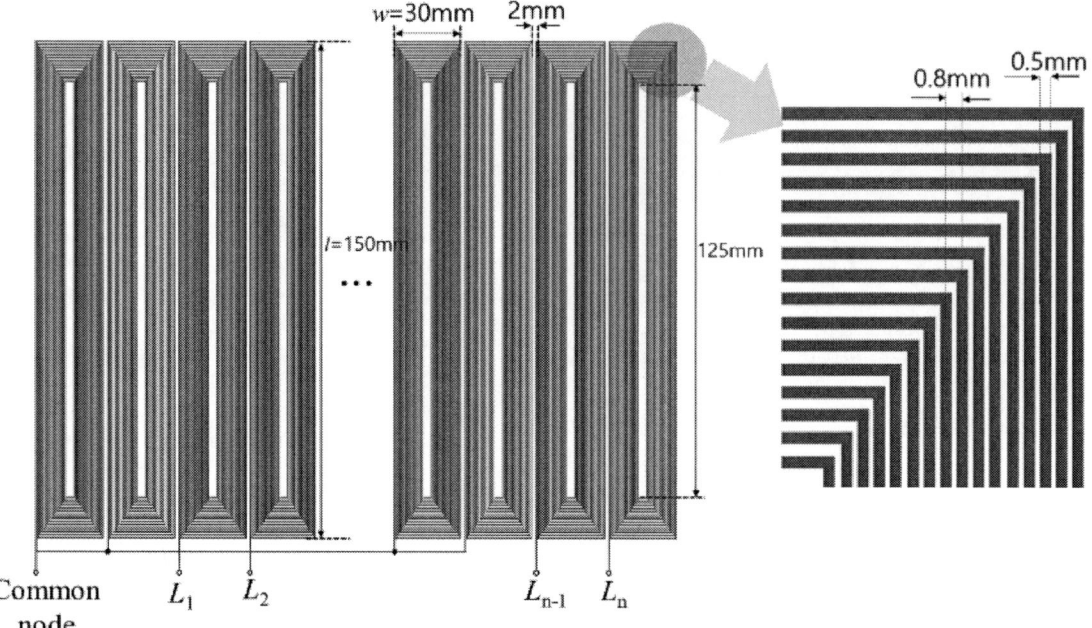

Fig. 5. Overall structure of the proposed sensing pattern.

$$\omega_r = \frac{1}{\sqrt{L_1(C_1 \, // \, C_p)}} = ... = \frac{1}{\sqrt{L_n(C_n \, // \, C_p)}} . \quad (3)$$

The transfer function of voltage gain is given as follows:

$$G(\omega) = \frac{V_{0,1}(\omega)}{V_{in}(\omega)} = \frac{1}{R_{in}} \cdot \frac{\beta_p R_1}{1 + j\beta_p R_1 \left(\omega C - \dfrac{1}{\omega \alpha L_1} \right)} . \quad (4)$$

The operating angular frequency of the proposed MOD system is mistuned to -3dB frequency ω_{3dB} from ω_r since it is not easy to accurately track ω_r under the condition with a high Q, as shown in Fig. 4.

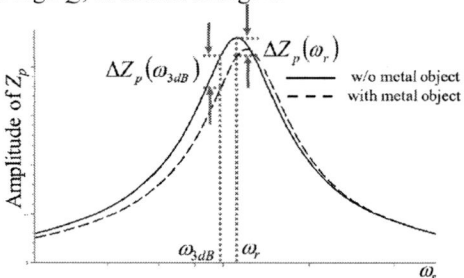

Fig. 4. Impedance characteristics for mistuned parallel-resonant at ω_{3dB}.

For example, if the operating frequency is incorrectly set as a higher value than ω_r, then when the metal objects are placed near the sensing pattern, there is a point at which the impedances are equal to each other. As a result, the MOD system cannot detect the metal objects due to the large amount of noise absorbed into the circuit. The second reason for employing the mistuned resonant condition is higher sensitivity. The variation of impedance of ω_{3dB}, $\Delta Z_p(\omega_{3dB})$ in the area below ω_r is given as follows:

$$\Delta Z_p(\omega_{3dB}) \approx R_1 \left(\frac{1}{\sqrt{2}} - \beta_p \bigg/ \sqrt{1 + \left\{ Q_p \beta_p \frac{(1-\alpha)}{\alpha} \right\}^2} \right). \quad (5)$$

The sensing pattern, which consists of multiple loop coil sets in order to increase the sensitivity, is mounted on the Tx pad, where each loop coil set has two coils connected in series with opposite polarity to cancel out the induced voltage generated by the Tx coil. Self-inductance change of the loop coil set is detected by the parallel-resonant circuit described in the previous section. Inductors of the sensing pattern, i.e., L_1, L_2, ... and L_n, are connected to common node as shown in Fig. 5.

There are two major points to consider when designing the sensing pattern. First, the inductance variation and the number of sensing patterns should be designed in a tradeoff relationship. The inductance variation by metal objects increases as the size of the sensing pattern becomes smaller; however, a number of sensing patterns is required to cover entire area of the Tx pad. Moreover, the signal processing can be complex, not only because of long calculation time but also because of the response time by system dynamics. Second, the induced voltage on the sensing pattern generated by the switching power system in the IPTS should be taken into account. Although there are filters to eliminate the effect of induced voltage in the proposed MOD system, the sensing pattern still should be carefully designed in order to cancel out the induced voltage. A simple method for this purpose is to adjust the number of turns of each loop coil sets so that the amount of linkage flux by the two loop coils is equal. To reduce the induced voltage, two loop coils can be connected in anti-series to cancel out the magnetic flux.

2566

III. SIMULATION AND EXPERIMENTAL RESULTS

To identify the behavior of the magnetic field of the sensing pattern by the metal objects, FEM MAXWELL simulations with only two loop coil sets are considered, as shown in Fig. 6. The simulation results for inductance variation depending on the height of the metal objects lying horizontally with the sensing pattern are shown in Fig. 7. The inductance varies by 2.8 % for an air-gap of 1 cm with a thin metal foil of 3 x 3 cm². The self-inductance always decreases for air-gap decrease, whereas there is no change in the self-inductance when the metal is placed vertically with sensing pattern. To cope with a blind zone by angle, another loop coil set lies alternately one by one. By measuring the self-inductance of the adjacent loop coil set, it can be determined that there is no blind zone on the Tx pad over entire area. Moreover, the effect on the Rx pad is also important.

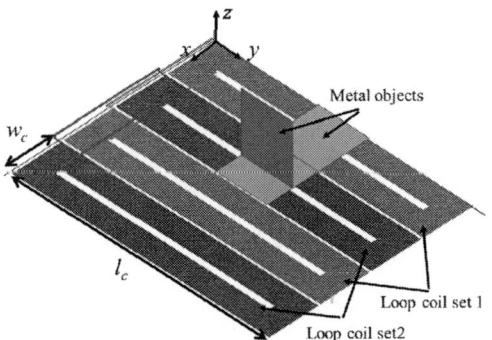

Fig. 6. Overall configuration of the FEM 3D simulation for $N = 16$, w_c = 3 cm, and l_c = 15 cm.

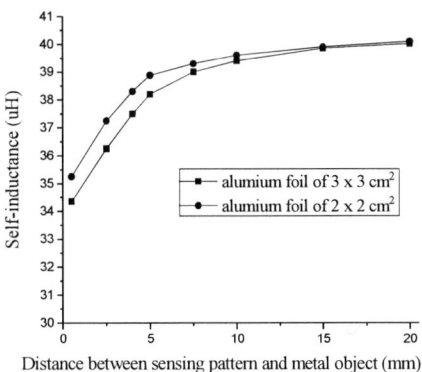

Fig. 7. Simulation results of self-inductance variation with respect to distance between sensing pattern and metal object.

To verify the operation of the proposed MOD system, only two sensing coil are simulated by the PSIM 9.1.1 circuit simulation tool where all circuit parameters including filters and peak detector are designed based on the results in previous chapter. When the self-inductance is changed from 40.05 uH to 38.20 uH (about variation of 5 %) the output voltage decreases from 4.7 V to 3.2 V, falling to 68 % compared to the case without the metal objects, with a dynamic response of 10 msec, as shown in Fig. 7.

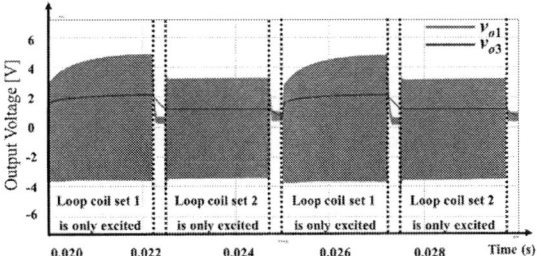

Fig. 8. Simulation results for operation verification of the proposed MOD system.

The experiment was set up as shown in Fig. 9. The Tx pad for the experiment was 724 x 628 mm² in size and fabricated sensing pattern were put on the Tx pad where the size of the sensing pattern was 362 x 314 mm² and it occupied one quarter of the pad. The size of the loop coil was determined to 3 x 15 cm² considering detection of metal objects with a size of 3 x 3 cm² and the number of loop coil sets being 50 or less.

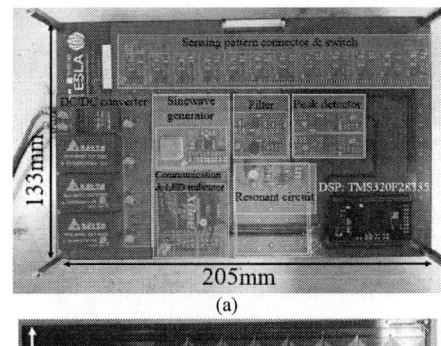

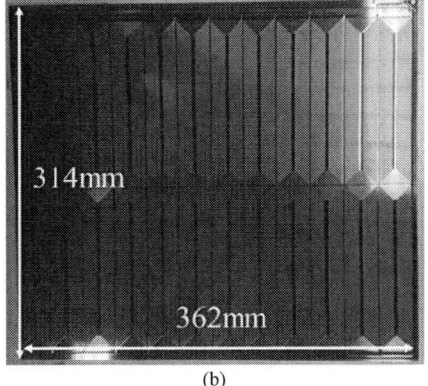

Fig. 9. Experiment condition. (a) the sensing circuit (b) the sensing pattern (c) various metal objects.

The open voltage of the sensing pattern was measured when Tx current of 20 A_{rms}, satisfying 6.6 kW power transfer, was applied to monitor how much noise is coming in. As shown in Fig. 10, 0.7 V_{rms} with fundament frequency of 85 kHz and 6 V_{pp} high frequency switching noise were induced. When a 3 x 3 cm² piece of aluminum foil is put at the center of the loop coil set 1, the output voltage decreased from 2.2 V to 0.5 V, as shown in Fig. 10 where the 85 kHz noise only remains by less than 40mV, in terms of signal processing, it can be rendered negligible by using various methods such as a filtering algorithm.

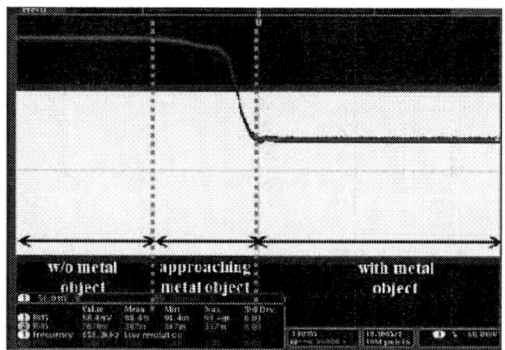

Fig. 10. Envelop of output voltage $v_{o1}(t)$ when 3 x 3 cm² piece of aluminum foil is put at the center of the loop coil set 1.

Experiments results for various metal objects and position are shown in Table. I. The voltage of the proposed MOD system decreased by more than 30 % for four metal specimens at aip-gap of 5 mm. The possibility of the proposed MOD system for wireless EV charger applications thus has been verified.

TABLE I

OUTPUT VOLTAGE VARIATION FOR VARIOUS SPECIMENS.

Metal types	Relative permeability	Conductivity	Output voltage variation
100 KRW coin (Diameter: 24mm)	75.0	4.82 x 10⁷	0.9 V
500 KRW coin (Diameter: 26.5mm)	75.0	4.82 x 10⁷	1.2 V
Aluminum foil of 2 x 2 cm²	1.0	3.54 x 10⁷	0.7 V
Aluminum foil of 3 x 3 cm²	1.0	3.54 x 10⁷	1.7 V

IV. CONCLUSION

In this paper, a new MOD system that can detect entire area on the Tx pad of wireless EV charger applications without any blind zone has been proposed. A parallel-resonant circuit with mistuned operating frequency near the -3dB point has been used to increase the detection sensitivity of metal objects. In this way, it can detect very small metal objects regardless of their position and orientation on the Tx pad. It has also been confirmed that multiple loop coil sets can be operated by only one signal processing circuit with electronic switches. A prototype of the MOD system on the Tx pad, operating at 85 kHz to satisfy the standard J2954, was fabricated, and it showed that the output voltage was reduced by 22.7 % for a piece

of aluminum foil of 3 x 3 cm² and 40.9 % for 100 Korean Won coin, respectively.

REFERENCES

[1] J. Huh, S. W. Lee, W. Y. Lee, G. H. Cho, and Chun T. Rim, "Narrow-width inductive power transfer system for online electrical vehicles," *IEEE Trans. on Power Electron.*, vol. 26, no. 12, pp. 3666-3679, Dec. 2011.

[2] W. Y. Lee, J. Huh, S. Y. Choi, X. V. Thai, J. H. Kim, E. A. Al-Ammar, M. A. El-Kady, and Chun T. Rim, "Finite-width magnetic mirror models of mono and dual coils for wireless electric vehicles," *IEEE Trans. on Power Electron.*, vol. 28, no. 3, pp. 1413-1428, Mar. 2013.

[3] Suyong Choi, J. Huh, S. Lee, and Chun T. Rim, "New cross-segmented power supply rails for roadway powered electric vehicles," *IEEE Trans. on Power Electron.*, vol. 28, no. 12, pp. 5832-5841, Dec. 2013.

[4] S. Fukuda, H. Nakano, Y. Murayama, T. Murakami, O. Kozakai and K. Fujimaki, "A novel metal detector using the quality factor of the secondary coil for wireless power transfer systems," *IEEE International Microwave Workshop Series on Innovative Wireless Power Transmission: Technologies, Systems, and Applications (IMWS)*, May 2012, pp. 241-244.

[5] Z. N. Low, J. J. Casanova, P. H. Maier, J. A. Taylor, R. A. Chinga, and J. Lin, "Method of load/fault detection for loosely coupled planar wireless power transfer system with power delivery tracking," *IEEE Trans. on Industrial Electron.*, vol. 57, no. 10, pp. 1478-1486, Apr. 2010.

[6] Seog Y. Jeong, Hyung G. Kwak, Gi C. Jang, Su Y. Choi, and Chun T. Rim, "Dual-purpose non-overlapping coil sets as metal object and vehicle position detections for wireless stationary EV chargers," *IEEE Trans. on Power Electron.*, Accepted.

[7] Gi C. Jang, Seog Y. Jeong, Hyeong G. Kwak, Chun T. Rim, "Metal object detection circuit with non-overlapped coils for wireless EV chargers," *IEEE SPEC 2016*, Dec. 2016, pp 1-6.

[8] Simon Verghese, Morris P. Kesler, Katherine L. Hall, and Herbert Toby Lou, "Foreign object detection in wireless energy transfer systems," Patent US20130069441 A1, (Witricity Corporation), filed on Sep. 2011.

[9] Ji-Won Jeong, Seung-Hee Ryu, Byoung-Kuk Lee, and Hee-Jun Kim "Tech tree study on foreign object detection technology in wireless charging system for electric vehicles," *2015 IEEE INTELEC*, Oct. 2015, pp. 1-4.

[10] Xian Zhang, Yao Jin, Qingxin Yang, Zhaoyang Yuan, Hao Meng, and Zhaohui Wang "Detection of metal obstacles in wireless charging system of electric vehicle," *2017 IEEE PELS Workshop on Emerging Technologies: Wireless Power Transfer (WoW)*, May 2017, pp. 89-92.

The 2018 International Power Electronics Conference

Wireless EV Charging System without Air-Gap and Misalignment

Wenxing Zhong[1] and Dehong Xu[1*]

1 College of Electrical Engineering, Zhejiang University, Hangzhou, China

*E-mail: xdh@zju.edu.cn

Abstract—Large air-gap and misalignment in static wireless charging of electric vehicles (EVs) cause two critical safety issues: human exposure to electromagnetic field and metal object overheating. Moreover, a large air-gap and a misalignment will degrade efficiency performance and complicate the design of a wireless charging system. This paper proposes a static wireless EV charging system with a mechanical structure to ensure zero air-gap and zero misalignment. The mechanical structure is purely passive and does not contain any motors or drives. Based on this structure, a pair of couplers which have a coupling coefficient of 0.94 are designed and analyzed. The performance of the charging system is investigated and compared with an ordinary charging system which has a large air gap.

Keywords— Wireless power transfer, wireless charging, inductive coupling, resonant coupling, inductive power transfer.

I. INTRODUCTION

As more and more countries have announced the plans to stop selling fossil fuel cars, electric vehicle (EV) will definitely have a fast development in the coming future. Compared with traditional cable charging of EV, wireless charging is much more convenient and elegant. Recently, a lot of research efforts have be devoted to wireless EV charging, including static wireless charging [1] [2] and dynamic charging [3]. The major challenge for dynamic charging will take more time to be practically implemented before the static wireless charging are widely used. Static wireless charging for home users seem more promising and the commercialization process has been initialed. However, there are still a few critical limitations for static wireless charging:

1. Safety concern of human exposure to magnetic field created by the wireless charging system. Now most of the existing approaches are based on two loosely coupled coils with an air-gap typically in the range of 100 mm to 200 mm. The large air-gap will result in a strong leakage flux in the air outside the charging area. Some scientific committees of international organizations, such as the Institute of Electrical and Electronics Engineers (IEEE), has developed standards or guidelines to define safe exposure levels [4]. In order to show compliance with these guidelines, it is a common practice to employ shielding layers with ferrite plates and conducting plates.

2. Ignition risk due to conducting material near or in the charging zone. If conducting material such as a small piece of aluminum plate gets closer to the charging system, the alternating magnetic flux will generate large eddy current in the material. Hence, the material will end up with a high temperature which might ignite nearby objects [5].

3. Misalignment, which not only cause an efficiency drop but also increase the leakage flux because of a poorer coupling. Moreover, the design of a wireless charging system will become more difficult in order to fulfill all the system requirements in a large coupling range.

In order to eliminate the air-gap and misalignment in a wireless EV charging system, the method using a robot arm to move the receiver coil to the perfectly coupled location is proposed in [6]. However, this kind of motor-based solutions significantly increase the construction cost and the complexity of the system.

In this paper, a purely passive structure without using any power supply or motor is proposed to get rid of the air-gap and misalignment between the transmitter and receiver coils in a wireless charging system. Based on this structure, a 3.7 kW wireless charging system is designed. Simulations show that the volume of the couplers for such a system could be less than 1/10 of that for a large-air-gap system. Meanwhile, the system efficiency can be increased. Measurement results are provided to verify the analysis.

II. MECHANICAL STRUCTURE

In general, the mechanical structure of the proposed charging system utilizes the physical contact and force between two parts in the transmitter and receiver sides to move and guide the primary pad so that the primary coil will eventually move to a location which is perfectly aligned with the secondary coil and the air-gap between two coils can be basically zero as well.

Fig. 1 shows the transmitter part of the proposed system. The primary coil along with its shielding layers is installed in the primary pad. The pad can be turned anticlockwise around Pivot-1 which is connecting the pad and Bar-1. In the position of Pivot-1, a spring is installed to make sure that the pad will move back to the initial position when external force disappears. Bar-1 and Bar-2

2569

are connected at Pivot-2. Bar-1 is able to rotate towards the ground around Pivot-2. Similarly, a spring is installed to generate an upwards force whenever Bar-1 is pushed down and shifts from its initial position. There are another pivot and spring connecting Bar-2 and the pole fixed to the ground. Thereby, Bar-2 is able to rotate along the ground around this joint and Spring-3 makes Bar-3 bake to its initial position after external force is removed.

Fig. 2 is the receiver part of the proposed system. It consists of a secondary pad in which a receiver coil and its shielding layers are embedded. The secondary pad will be installed right beneath the chassis of the car. Under the secondary pad, there is a smooth plate and along the edge of the plate, there is a belt. Another component in the receiver part is the roller. When a car moves towards the transmitter, the roller will hit the primary pad first and avoid a hard collision. Therefore the roller will be installed in front of the car and perpendicular to the moving direction of the car (see Fig. 3).

Fig. 3 illustrates how the primary part motions when the car starts to push it and keeps moving forward. When a car moves into a parking slot, the roller in front of the car will firstly hit the primary pad as shown in Fig. 3 (a). Then as the car keeps moving in, Bar-1 in the transmitter side will be pushed down as shown in Fig. 3 (b), because the pad can only rotate anticlockwise. After the roller moves over the center of the pad as shown in Fig. 3 (c), the pad will be fully attached to the plate due to the upwards force generated by Spring-2. Thus, there will be virtually zero air-gap between the primary pad and the secondary pad.

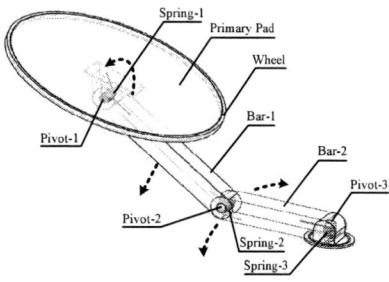

Fig. 1. Transmitter part of the proposed system.

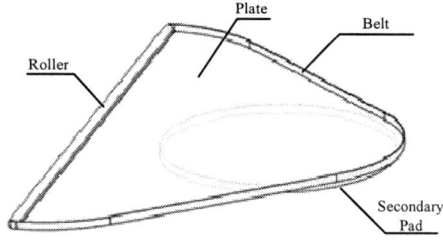

Fig. 2. Receiver part of the proposed system.

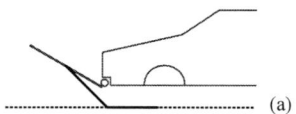

(a)

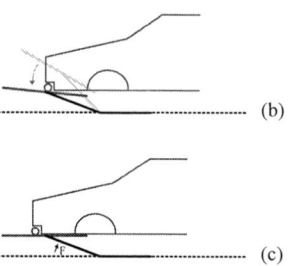

(b)

(c)

Fig. 3. Side views of the car and the primary part at different positions.

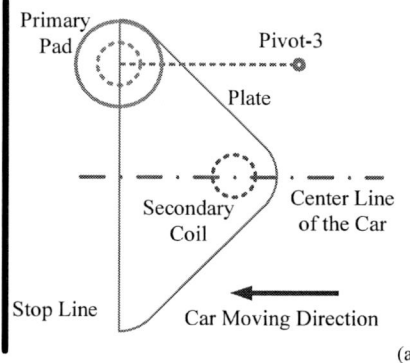

(a)

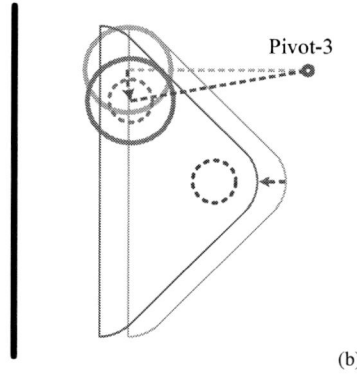

(b)

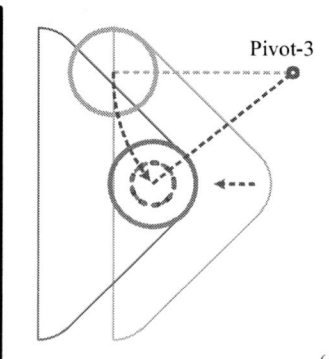

(c)

2570

The 2018 International Power Electronics Conference

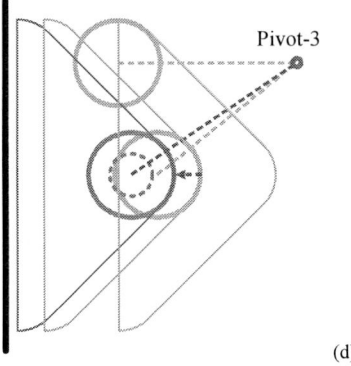

(d)

Fig. 4. Top views of the proposed system showing the position changes of the primary pad and the secondary plate.

Fig. 4 demonstrates how the pad will be moved to the perfect location which provides a zero-misalignment between the primary and secondary coils. When the primary pad has been fully attached to the plate in the receiver side as shown in Fig. 3 (c), the primary pad should be embraced by the ">" shape plate. Fig. 4 (a) shows the worst position of the primary pad where a largest allowable shift from the center line of the car appears. Two dotted circles stand for the primary and secondary coils. Then the car keeps moving forward and the belt around the plate will exert a force to the primary pad which makes Bar-2 to rotate around Pivot-3, as shown in Fig. 4 (b). Eventually, the rotation will stop after the primary pad goes to the set position where the primary coil is perfectly aligned with the secondary coil as shown in Fig. 4 (c). So the car should stop moving at this moment. However, it is difficult to precisely control the movement of a car, so the bars in the transmitter part can be designed to be extendable which will then provide some tolerance to the finale position of the car as shown in Fig. 4 (d).

III. COUPLERS DESIGN

TABLE I
SYSTEM SPECIFICATIONS

Symbol	Meaning	Value
P_{out}	Rated Output Power	3.7 kW
U_{out}	Output Voltage	400 V
f	Operating Frequency	85 kHz

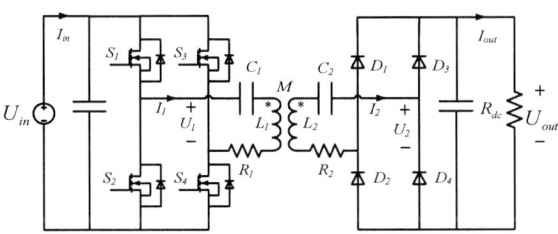

Fig. 5. Circuit diagram of the wireless charging system.

Based on the assumption of a perfect coupling between the primary and secondary coils, the couplers are designed. The basic specifications of the charging system are listed in TABLE I and the circuit diagram of the system is shown in Fig. 5. SS compensation is adopted. The design objectives include:

1) To achieve a high coils' efficiency of 99%;
2) To minimize the magnetic flux to which human or foreign objects are exposed;
3) To minimize the volume of the system.

ANSYS Maxwell is used to analyze and calculate the inductances and resistances of the coils. Then the optimum secondary compensation capacitance (C_2) is found with Matlab Optimization Tool for a given rated load resistance.

Finally, a pot-shape ferrite is chosen (see Fig. 6) and the overall structure of the pads is shown in Fig. 7. The designed parameters are shown in Fig. 8 and TABLE II. The total thickness of the pads including the cases is 20 mm. The cover layer above the coil in Fig. 7 is 2 mm in thickness which implies a 4 mm gap between the primary and secondary coils. The coupling coefficient is as high as 0.94. According to simulations, further decreasing the gap between the coils will not increase the efficiency of the system due to an increasing proximity effect.

Fig. 6. Coil and pot-shape ferrite.

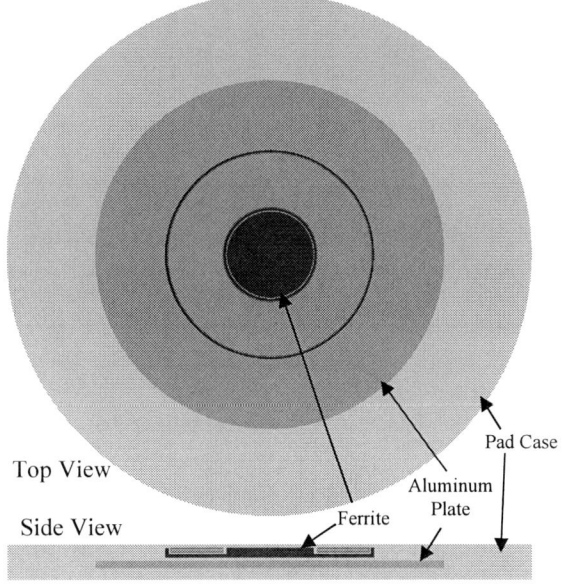

Fig. 7. Overall structure of the primay and secondary pads.

The 2018 International Power Electronics Conference

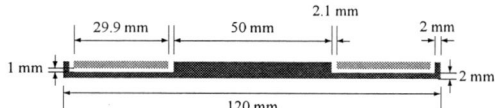

Fig. 8 Geometry parameters of the couplers.

TABLE II
DESIGNED PARAMETERS OF THE PROPOSED SYSTEM

Parameters	Value
Strand diameter of the litz wire	0.1 mm
No. of strands	300
No. of turns	13
Thickness of the pads	20 mm
Diameter of the pads	300 mm
Diameter of the aluminum plates	200 mm
Thickness of the aluminum plates	4 mm
Gap between the coils	4 mm
Gap between the ferrite and aluminum	2 mm

An existing ordinary wireless charging system using the same litz wire is presented here for comparison purpose. The structure of the couplers used in this system is shown in Fig. 9 and the geometric parameters of the coils are listed in TABLE III. Ferrite bars and aluminum plates are used for coupling enhancement and shielding. The practical setup of the system is shown in Fig. 10.

Fig. 9. Coupler of the ordinary system.

Fig. 1. TABLE III GEOMETRIC PARAMETERS OF THE ORDINARY SYSTEM

Parameter	Value
Edge length of coil	350 mm
Coil structure	33 turns and 1 layer
Length of ferrite bars	370 mm
Width of ferrite bars	20 mm
Thickness of ferrite bars	6 mm
No. of ferrite bars for a coupler	11
Edge length of aluminum plate	450 mm
Thickness of aluminum plate	4 mm
Gap between ferrite and aluminum	8 mm
Gap between coil and ferrite	3.5 mm
Air gap	200 mm

Fig. 10. Practical setup of the ordinary system.

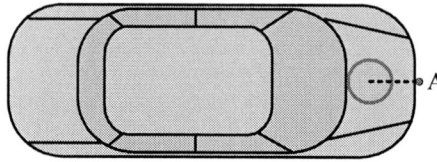

Fig. 11. Testing point of magnetic flux density.

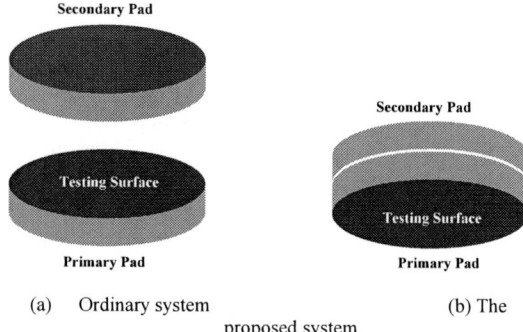

(a)   Ordinary system   (b) The proposed system

Fig. 12. Testing surfaces of magnetic flux density on the charging pad cases.

To evaluate the magnetic flux to which human or foreign objects are exposed, the maximum magnetic flux density at two locations are recorded in simulations under full load condition. One location, as shown in Fig. 11, is Point A which is the nearest possible spot where human will be exposed to EM field. Another location is the case surface of the pads where foreign objects might be exposed to the magnetic flux. For an ordinary system, the highest magnetic flux density is generated on the top surface of the primary pad as shown in Fig. 12 (a). In the proposed system, since two pads will overlap, the strongest magnetic flux density appears at the bottom of the primary pad as shown in Fig. 12 (b).

TABLE IV
PERFORMANCE COMPARISON BETWEEN PROPOSED SYSTEM AND ORDINARY SYSTEM

	Proposed System without Aluminum Plates	Proposed System with Aluminum Plates	Ordinary System without Aluminum Plates	Ordinary System with Aluminum Plates
η (coils)	99.02%	98.86%	97.67%	94.09%

2572

B_{max} at A	1.41 uT	**1.3 pT**	24.97 uT	**19.76 uT**
B_{max} at surface	1.08 mT	**7.11 uT**	8.49 mT	**15.65 mT**

Based on finite-element simulations and theoretical calculations, the comparison is done on the efficiency and the above defined maximum magnetic flux density between the proposed system and the ordinary system. Results are listed in TABLE IV. The proposed system can reach a near 99% coils' efficiency when generating a much lower magnetic flux density at the interested spots.

TABLE V gives the volume comparison between the proposed system and the ordinary system. For copper and ferrite, 93-95% reduction can be achieved. For aluminum, 84% reduction can be achieved. In general, the major construction material required in the proposed system is roughly only 1/10 of that for an ordinary system.

TABLE V
VOLUME COMPARISON BETWEEN PROPOSED SYSTEM AND ORDINARY SYSTEM

	Proposed System	Ordinary System	**Percentage Reduction**
Wire length	3.435 m	67.98 m	**94.95%**
Ferrite	6.309×10^{-5} m^3	9.768×10^{-4} m^3	**93.54%**
Aluminum	2.513×10^{-4} m^3	1.62×10^{-3} m^3	**84.49%**

IV. EFFECT ON HARMONICS OF THE PROPOSED SYSTEM

The above efficiency is calculated based on pure sinusoidal voltage and current. However in practice, the voltage applied to the primary resonator will be typically a square wave generated by a full-bridge inverter. The square wave contains high-order harmonics. In order to estimate the system efficiency more precisely, the performance of the designed system is studied with a circuit simulator LTspice.

It is found that the proposed system has a much lower input impedance for high-order harmonics compared with that of the ordinary system. Therefore, the currents in the primary and secondary windings will contain large high-order harmonics. As a result, system efficiency will be lower than the value calculated with only fundamental frequency, because the system operates at non-resonant frequencies for high-order harmonics.

Paper [7] provides an analysis on the harmonics of the square wave generated by a full-bridge inverter. The mathematical expression is repeated here.

$$V_{pk} = \left(\frac{4V_{dc}}{k\pi} \right) \sin\left(\frac{k\pi D}{2} \right) \sin\left(\frac{k\pi}{2} \right)$$

(1)

where V_{pk} is the peak voltage of the kth harmonic of the the square wave; V_{dc} is the DC input voltage of the inverter; D is the duty ratio of the inverter which can be regulated by using phase-shift control. Fig. 13 shows normalized voltage harmonics (normalized by a factor $4V_{dc}/\pi$) redrawn from [7]. From (1) and Fig. 13, the harmonics in the square wave can be varied by regulating the duty ratio of the inverter. For example, when $D = 1$, the normalized fundamental, third and fifth harmonics are 1, 0.33 and 0.20, respectively; however, if $D = 0.7$, the normalized fundamental, third and fifth harmonics are 0.89, 0.052 and 0.14, respectively. With $D = 0.7$, third harmonic can be largely eliminated.

The calculated efficiencies with only fundamental frequency are 98.98% and 95.04% for the proposed system and the ordinary system, respectively. TABLE VI shows the simulation results with square-waves. A 0.5% efficiency drop is observed in the proposed system if $D = 1$. While using $D = 0.7$, the efficiency drop reduces to 0.3%. For the ordinary system, the simulated efficiency is basically the same as the calculated one. Note that core loss is not considered in the efficiencies here. Therefore, the efficiencies are higher that the corresponding ones in TABLE IV.

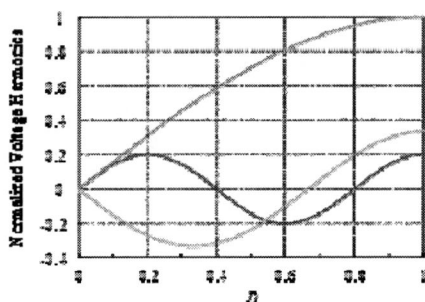

Fig. 13. Voltage harmonics vs duty ratio of a phase controlled inverter (redrawn from [1]).

TABLE VI
EFFICIENCY COMPARISON WITH SQUARE-WAVE

	Proposed System with $D = 1$	Proposed System with $D = 0.7$	Ordinary System with $D = 1$
Coils' Efficiency	98.51%	98.71%	95.03%
DC-to-DC Efficiency	95.77%	96.41%	93.38%

V. EXPERIMENTAL VERIFICATION

The mechanical part is still under construction but the design of the primary part has been completed as shown in Fig. 14. It should be noted that the optimization of the mechanical part is not a concern of this paper. Therefore, the focus is put on the electrical system consisting of the couplers, the inverter, the rectifier and the load, in the experiments.

A 3.7 kW prototype is built for experiments. Fig. 15 shows the structure of a coupler. An aluminum plate with a diameter of 200 mm is placed at the bottom. The designed core in Fig. 6 is not commercially available, so the final constructed core, as shown in Fig. 15, is not exactly the same as the designed one. As a result, the measured coupling coefficient is only about 0.91 compared with 0.94 obtained from FEA. However, it turns out that the system efficiency is not degraded much due to the decrease of the coupling.

Fig. 14. Mechanical design of the primary part.

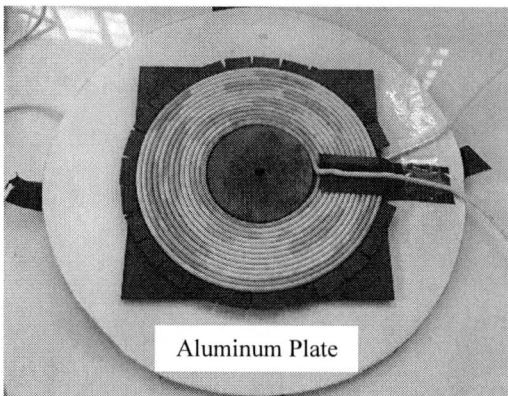

Fig. 15. Coupler of the prototype.

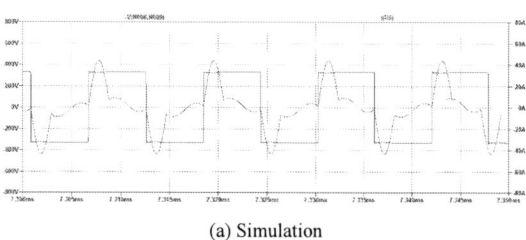

(a) Simulation

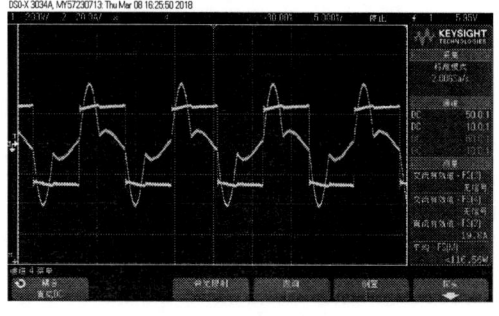

(b) Measurements

Fig. 16. Waveforms of the ouput voltage and current of the inverter.

Fig. 16 shows the measured output waveforms of the inverter when the duty ratio is 1, compared with the

simulation results. The measurements agree well with the simulation results.

(a) $D = 1$

(b) $D = 0.7$

Fig. 17. Measured DC-to-DC efficiencies at rated output power ($R_L = 30$ Ω) when (a) $D = 1$ and (b) $D = 0.7$.

The efficiencies of the proposed system are measured and recorded in Fig. 17. The measured DC-to-DC efficiency is 95.12%, while the simulation result is 95.77% as shown in TABLE VI. It should be noted that in simulations, core loss is neglected. When the duty ration is adjusted to 0.7, the system efficiency is improved to 95.36% which is also consistence with the above analysis.

VI. CONCLUSION

A novel wireless charging system for EVs is proposed. The mechanical structure of the proposed system ensures the coils will be aligned perfectly with negligible air gap. The coupling between the coils is maximized which brings significant reductions in the volume of the system and also push up the power transfer efficiency. Moreover, the dangerous magnetic flux can be reduced to a negligible level and thereby, human exposure and ignition risks vanish. Experimental results are provided to verify the proposed system.

ACKNOWLEDGMENT

This project is supported by the by the Fundamental Research Funds for the Central Universities of China.

REFERENCES

[1] G. A. Covic and J. T. Boys, "Inductive Power Transfer," *Proceedings of the IEEE*, vol. 101, no. 6, pp. 1276-1289, June 2013.

[2] R. Bosshard, J. W. Kolar, J. Mühlethaler, I. Stevanović, B. Wunsch and F. Canales, "Modeling and η-α-Pareto Optimization of Inductive Power Transfer Coils for Electric Vehicles," *IEEE Journal of Emerging and Selected Topics in Power Electronics*, vol. 3, no. 1, pp. 50-64, March 2015.

[3] C. C. Mi, G. Buja, S. Y. Choi and C. T. Rim, "Modern Advances in Wireless Power Transfer Systems for Roadway Powered Electric Vehicles," *IEEE Transactions on Industrial Electronics*, vol. 63, no. 10, pp. 6533-6545, Oct. 2016.

[4] *IEEE Standard for Safety Levels with Respect to Human Exposure to Radio Frequency Electromagnetic Fields, 3 kHz to 300 GHz. International Committee on Electromagnetic Safety, The Institute of Electrical and Electronics Engineers, Inc. 3 Park Avenue*, IEEE C95.1, IEEE Standards Dept., New York, NY, USA, 2005.

[5] M. Young, "The PWM strategy on DC-DC converter," *IEEJ Journal of Industry Applications*, vol. 28, no. 15, pp. 123-129, 1989.

[6] G. Eason, B. Noble, and I. N. Sneddon, "On certain integrals of Lipschitz-Hankel type involving products of Bessel functions," *IEEE Trans. on Power Electronics*, vol. 247, no. 8, pp. 529-551, 1995.

[7] Z. Pantic, K. Lee, S. M. and Lukic, "Multifrequency inductive power transfer," *IEEE Trans. Power Electron.*, vol. 29, no. 11, pp. 5995–6005, Nov. 2014.

The 2018 International Power Electronics Conference

Fixed Slope Carrier PWM for Indirect Matrix Converter

Tzung-Lin Lee, Chun-Yao Hung, Yen-Wen Chen and Wen-Mei Huang

Department of Electrical Engineering, National Sun Yat-sen University, Kaohsiung, TAIWAN

Abstract— Indirect matrix converter (IMC) typically requires zero current commutation (ZCC) at its rectifier stage. This paper proposes a fixed slope carrier modulation for IMC. According to homothetic triangle, the modulating functions in the inverter stage are scaled by the rectifier stage to accomplish ZCC. Compared with space vector modulation (SVM) or variable slope carrier modulation, the fixed slope carrier PWM benefits in implementation and reliability. The feasibility of the proposed method is verified by experimental results.

Keywords—Indirect matrix converter (IMC), zero current commutation (ZCC), fixed slope carrier modulation

I. INTRODUCTION

Matrix converters (MCs) have received much attention in AC/AC conversion, such as motor driving, wind turbine and microgrid. MC circuit is typically classified into two types: direct matrix converter (DMC) in Fig. 1 and indirect matrix converter (IMC) in Fig. 2. As shown, three-phase DMC consists of nine fully controlled four-quadrant switches. Three-phase IMC contains both rectifier and inverter stages in cascaded connection. DMC and IMC are able to generate variable-amplitude variable-frequency output from AC input directly with no bulky DC link capacitor. Due to lacking of decoupling capacitor, switching commutation of MC needs careful consideration to obey continuity of inductor current. IMC possesses less complexity in terms of control and commutation, which is the significant advantage compared with DMC [1-3].

Generally, the rectifier stage of IMC is modulated by using current space vector, while the inverter stage is controlled by voltage space vector. IMC can simply apply zero current commutation (ZCC) to coordinate switching action of two stages to accomplish commutation process. In this way, the rectifier stage is switched during a freewheeling interval of the inverter stage [4]. Basically, ZCC can be realized by using space vector modulation (SVM) [8]. SVM benefits in flexible arrangement of switching pattern, but complex vector operation increases computational load. In contrast, carrier-based modulation provides reliable solution to implementing ZCC for IMC, but variable slope carrier is required for the inverter stage [5]. In [6-7], a carrier-based PWM was proposed for IMC by using SVM. In this method, a fixed slope carrier can be used to modulate the inverter stage. But formulation and related analysis are complex and not straightforward.

This work was supported by Ministry of Science and Technology of TAIWAN under grant 105-2221-E-110-066-MY2.

This paper proposes a carrier-based pulse width modulation (PWM) for IMC. The modulating functions of the inverter stage are scaled by the rectifier stage according to principle of homothetic triangle so that the fixed slope carrier can be used to implement the same ZCC as variable slope carrier modulation. Formulation of the modulating functions in the fixed slope carrier is addressed with straightforward explanation. Experimental results verify the proposed method.

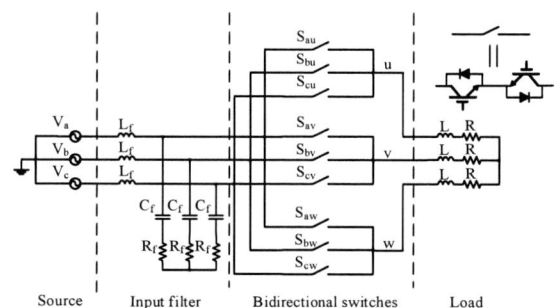

Fig. 1. Circuit topology of DMC.

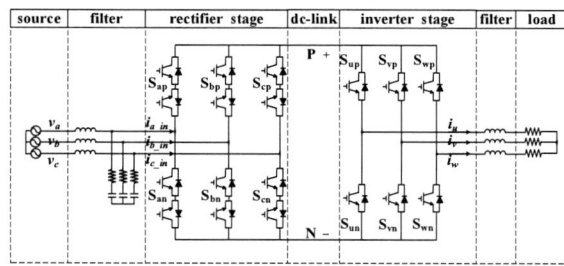

Fig. 2. Circuit topology of IMC.

II. VARIABLE SLOPE CARRIER MODULATION

According to input current commands i_{a_ref}, i_{b_ref}, i_{c_ref}, modulation of the rectifier stage can be divided into six sectors as shown in Fig. 3[5]. Trapezoidal modulation is applied to generate control signals of the rectifier stage as shown in Fig. 4. As given in Table I, the phase with maximum amplitude is clamped, while the other phases in other arms are switched according to current commands. Note that m_{xp}, m_{xn} (x=a,b,c) mean the modulating functions of the upper switches and the lower switches, respectively. Their corresponding switching signals are denoted by h_{xp} and h_{xn} (x=a,b,c). As shown in Table I, each sector contains two modulating functions. They generate non-complementary switching states, which cannot be used in the rectifier stage. In order to

2576

produce complementary signals, one of them is selected to be the active modulating function in each sector.

The inverter stage is modulated by sinusoidal PWM considering variable dc-link voltage induced by the rectifier stage. In order to accomplish ZCC, modulation of the inverter stage must be arranged based on switching timing of the rectifier stage. Fig. 5 shows an example of modulation arrangement in sector 1. Note that m_{xp} (x=u,v,w) means the modulating functions of the upper switch in the inverter stage. Switching functions of the inverter stage are denoted by h_{xp} and h_{xn} (x=u,v,w). As can be seen, the switching timing of the rectifier stage determines both rising and falling intervals of the carrier in the inverter stage. Thus variable slope carrier is required for the inverter stage to accomplish ZCC.

TABLE I
MODULATING FUNCTIONS OF RECTIFIER STAGE

	Sector 1	Sector 2	Sector 3	Sector 4	Sector 5	Sector 6
m_{ap}	1	$-\dfrac{i_{a\_ref}}{i_{c\_ref}}$	0	0	0	$-\dfrac{i_{a\_ref}}{i_{b\_ref}}$
m_{bp}	0	$-\dfrac{i_{b\_ref}}{i_{c\_ref}}$	1	$-\dfrac{i_{b\_ref}}{i_{a\_ref}}$	0	0
m_{cp}	0	0	0	$-\dfrac{i_{c\_ref}}{i_{a\_ref}}$	1	$-\dfrac{i_{c\_ref}}{i_{b\_ref}}$
m_{an}	0	0	$-\dfrac{i_{a\_ref}}{i_{b\_ref}}$	1	$-\dfrac{i_{a\_ref}}{i_{c\_ref}}$	0
m_{bn}	$-\dfrac{i_{b\_ref}}{i_{a\_ref}}$	0	0	0	$-\dfrac{i_{b\_ref}}{i_{c\_ref}}$	1
m_{cn}	$-\dfrac{i_{c\_ref}}{i_{a\_ref}}$	1	$-\dfrac{i_{c\_ref}}{i_{b\_ref}}$	0	0	0

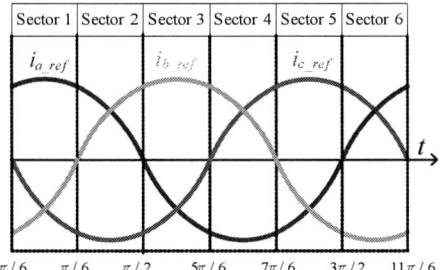

Fig. 3. Definition of sectors.

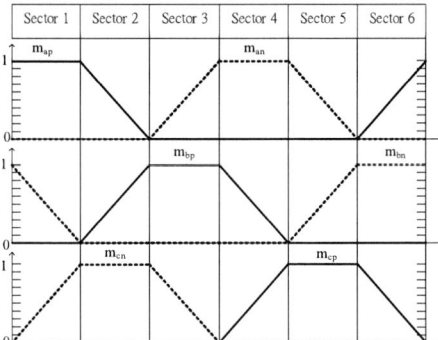

Fig. 4. Trapezoidal modulation of the rectifier stage.

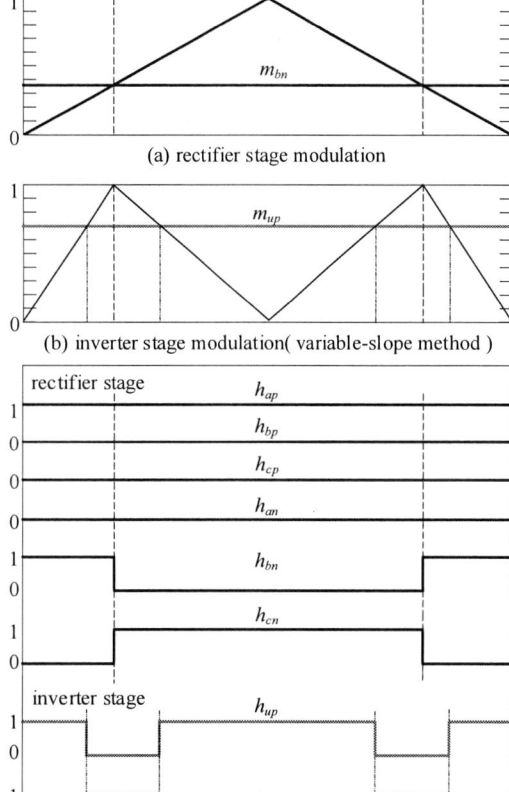

(a) rectifier stage modulation

(b) inverter stage modulation(variable-slope method)

(c) The switching states of the rectifier and inverter stages in sector 1

Fig. 5. Variable slope carrier modulation.

III. PROPOSED MODULATION

A. Rectifier stage

Since switching timing of the rectifier stage affects switching arrangement of the inverter stage, a continuous modulating function of the rectifier stage might be better. Based on TABEL I, the modulation function in the rectifier stage can be simplified to one signal m_{rec} as shown in TABLE II and Fig. 6. TABLE III shows the modulating signal h_{rec}, which is obtained by comparing m_{rec} with the triangular carrier. Waveforms of switching signals for all switches are given in Fig. 7.

TABLE II
PROPOSED MODULATION FUNCTIONS OF RECTIFIER STAGE

	Sector 1	Sector 2	Sector 3	Sector 4	Sector 5	Sector 6
m_{rec}	$-\dfrac{i_{b\_ref}}{i_{a\_ref}}$	$-\dfrac{i_{b\_ref}}{i_{c\_ref}}$	$-\dfrac{i_{c\_ref}}{i_{b\_ref}}$	$-\dfrac{i_{c\_ref}}{i_{a\_ref}}$	$-\dfrac{i_{a\_ref}}{i_{c\_ref}}$	$-\dfrac{i_{a\_ref}}{i_{b\_ref}}$

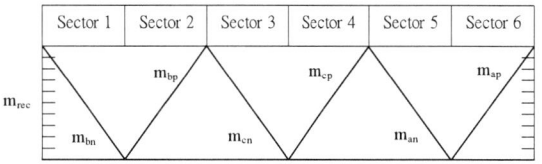

Fig. 6. The proposed modulation function in the rectifier stage.

2577

TABLE III
SWITCHING SIGNALS OF THE RECTIFIER STAGE

	Sector 1	Sector 2	Sector 3	Sector 4	Sector 5	Sector 6
h_{ap}	1	$\overline{h_{rec}}$	0	0	0	h_{rec}
h_{bp}	0	h_{rec}	1	$\overline{h_{rec}}$	0	0
h_{cp}	0	0	0	h_{rec}	1	$\overline{h_{rec}}$
h_{an}	0	0	$\overline{h_{rec}}$	1	h_{rec}	0
h_{bn}	h_{rec}	0	0	0	$\overline{h_{rec}}$	1
h_{cn}	$\overline{h_{rec}}$	1	h_{rec}	0	0	0

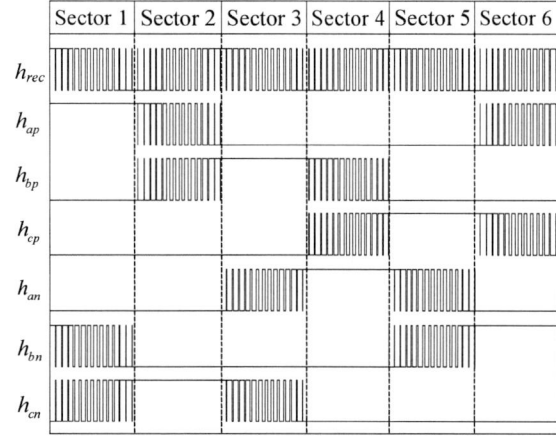

Fig. 7. Switching waveforms of the rectifier stage for IMC.

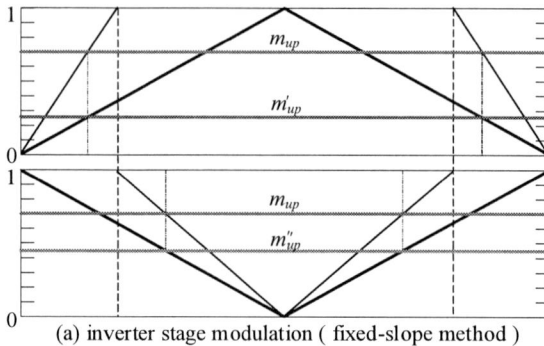

(a) inverter stage modulation (fixed-slope method)

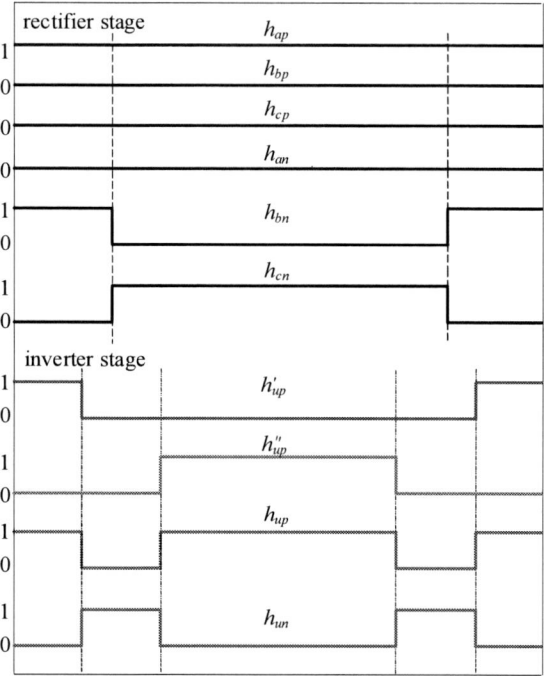

(b) The switching states of the rectifier and inverter stages in sector 1

Fig. 8.   Fixed slope carrier modulation.

B. Inverter stage

In this paper, we propose fixed slope carriers to replace a variable slope carrier but the same switching pattern is retained. As shown in Fig. 8, the inverter stage uses two fixed slope carriers. One of them is the same carrier as that of the rectifier stage. The other one is out of phase with respect to that of the rectifier stage. In order to accomplish ZCC, the modulating functions m_{xp} (x=u,v,w) of the inverter stage need to be scaled.

Fig.9 shows relationship of variable slope carrier and fixed slope carrier during the left-half period. The peak of variable slope carrier locates at time $m_{rec}*T_s$, at which switching action happens in the rectifier stage. On the other hand, the peak of the fixed slope carrier is located at time Ts. By principle of homothetic triangle, we can get T_{x1} and T_{y1} expressed as (3.1) and (3.2), respectively.

$$T_{x1} = \left(1 - m_{up} \cdot m_{rec}\right) \cdot T_s \qquad (3.1)$$
$$T_{y1} = T_{x1} - \left(1 - m_{rec}\right) \cdot T_s = \left(1 - m_{up}\right)m_{rec} \cdot T_s \qquad (3.2)$$
$$m'_{up} = m_{rec} \cdot m_{up} \qquad (3.3)$$

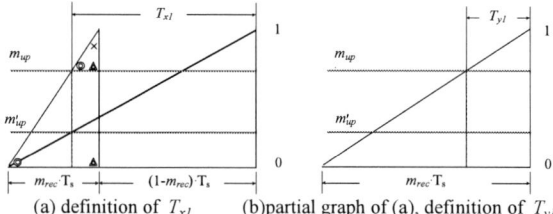

(a) definition of T_{x1} (b)partial graph of (a), definition of T_{y1}

Fig. 9. Relationship between variable slope carrier modulation and fixed slope carrier modulation during the left-half period.

After simplification, the modulation signal m'_{up} is scaled to $m_{rec}*m_{up}$. Thus, m_{rec} is the scaling factor to transfer from variable slope modulation to fixed slope one during left-half period. The other half part of one period is shown in Fig.10. As can be seen, the carrier is out-of-phase with respect to the carrier applied in the left-half period. Similarly, T_{x2} and T_{y2} can be obtained as (3.4) and (3.5) by using principle of homothetic triangle.

The scaled modulation signal m''_{up} is equal to $(1-m_{rec})*m_{up}$.

$$T_{x2} = m''_{inv} \cdot T_s \quad (3.4)$$
$$T_{y2} = (1 - m_{rec}) \cdot T_s - T_{x2} = (1 - m_{rec}) \cdot T_s - m''_{inv} \cdot T_s \quad (3.5)$$
$$m''_{up} = (1 - m_{rec}) \cdot m_{up} \quad (3.6)$$

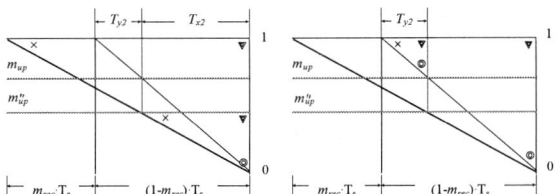

(a) and (b) are triangles similar used in the derivation

Fig.10. Relationship between variable slope carrier modulation and fixed slope carrier modulation during the right-half period.

Thus two modulating signals, m'_{up} and m''_{up} with corresponding carriers are able to generate the same switching pattern as that of variable slope carrier modulation. Note that those analyses also work for the other phases. After obtaining the switching functions, all switching signals of the inverter stage can be reconstructed by using OR logic operation as shown in Fig. 11.

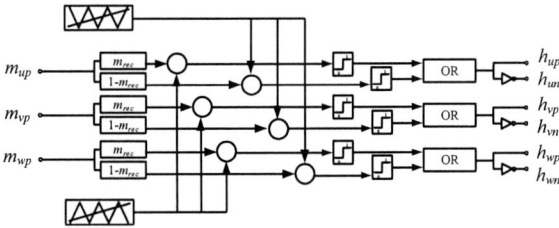

Fig. 11 Block diagram of generating switching signals.

TABLE IV
CIRCUIT PARAMETERS

Symbol	Meaning	Value
L_F	Input filter inductor	0.4 mH
C_F	Input filter capacitor	20 μF
R_F	Input filter resistance	2 Ω
L	Output filter inductor	10 mH
R	Output three-phase load	25 Ω

IV. EXPERIMENTAL RESULTS

Circuit parameters are given in TABLE IV. In the experiments, the switching frequencies of the rectifier stage and the inverter stage are set at 5 kHz and 10 kHz respectively. Fig. 12 shows that the switching action in the rectifier stage happens only when the inverter stage is during a freewheeling state. Thus, ZCC is verified.

Fig. 13 and Fig. 14 show the input and output waveforms at the same frequency 60Hz. Fig. 15 and Fig. 16 show output voltage at 475Hz and 1Hz, respectively. As can be seen, the inverter is able to generate balanced three-phase voltages and currents.

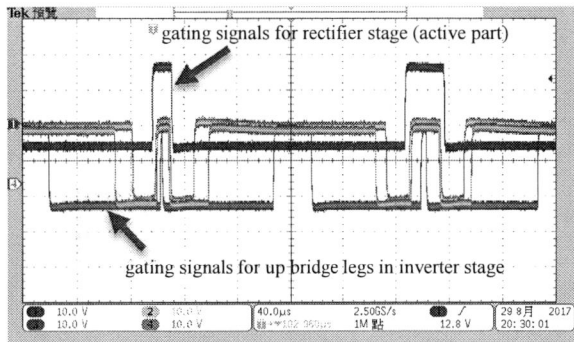

Fig. 12. Waveforms of ZCC.

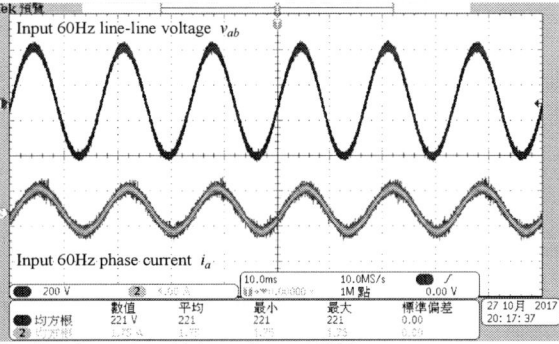

Fig. 13. Input sinusoidal waveforms at 60 Hz.

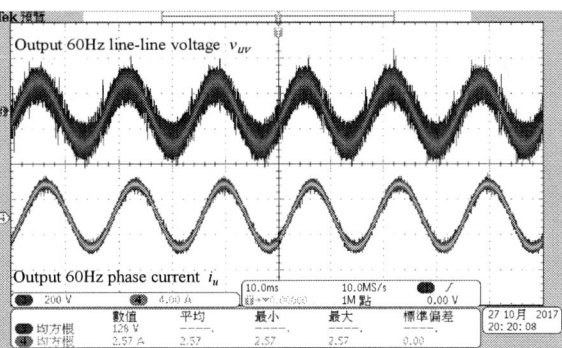

Fig. 14. Output sinusoidal waveforms at 60 Hz.

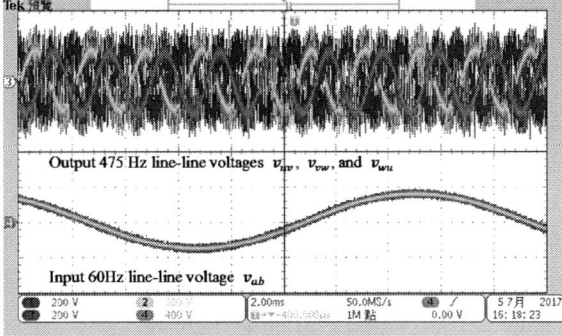

Fig. 15. Output sinusoidal waveforms at 475 Hz.

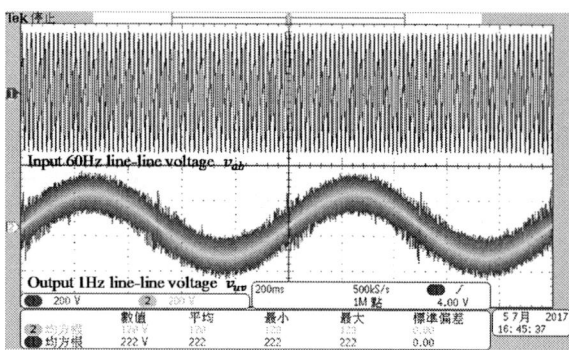

Fig. 16.   Output sinusoidal waveformsat 1 Hz.

V. CONCLUSION

A fixed slope carrier modulation for IMCs is presented in this paper. Based on the so-called homothetic triangle, the modulating functions in the inverter stage are scaled by the rectifier stage to accomplish ZCC. This method is able to generate the same switching pattern arranged by SVM. Theoretical analysis is given and experimental verification is provided. The results show that the inverter can generate sinusoidal voltage at frequency 60Hz, 1Hz and 475Hz, respectively.

REFERENCES

[1] J. W. Kolar, M. Baumann, F. Schafmeister and H. Ertl, "Novel Three-Phase AC-DC-AC Sparse Matrix Converter," *APEC. Seventeenth Annual IEEE Applied Power Electronics Conference and Exposition*, vol.2, pp. 777-791, 2002.

[2] A. Ecklebe, A. Lindemann and S. Schulz, "Bidirectional Switch Commutation for a Matrix Converter Supplying a Series Resonant Load," *IEEE Transactions on Power Electronics*, vol. 24, no. 5, pp. 1173-1181, 2009.

[3] M. Hamouda, H. F. Blanchette and K. Al-Haddad, "Indirect Matrix Converters' Enhanced Commutation Method," *IEEE Transactions on Industrial Electronics*, vol. 62, no. 2, pp. 671-679, 2015.

[4] V. K. Khanna, *The Insulated Gate Bipolar Transistor (IGBT): Theory and Design*, NJ, Piscataway: IEEE Press, Aug. 2003. ISBN 0-471-23845-7.

[5] B. Wang and G. Venkataramanan, "A Carrier Based PWM Algorithm for Indirect Matrix Converters," *2006 37th IEEE Power Electronics Specialists Conference*, pp. 1-8, 2006.

[6] Dinh-Tuyen Nguyen, Hong-Hee Lee, and Tae-Won Chun, "A Carrier-Based Pulse Width Modulation Method for Indirect Matrix Converters," Journal of Power Electronics (JPE), Vol. 12, No. 3, pp. 448-457, May 2012

[7] T. D. Nguyen and H. H. Lee, "Generalized carrier-based PWM method for indirect matrix converters," *2012 IEEE Third International Conference on Sustainable Energy Technologies (ICSET)*, Kathmandu, 2012, pp. 223-228.

[8] H. M. Nguyen, H. H. Lee and T. W. Chun, "Input Power Factor Compensation Algorithms Using a New Direct-SVM Method for Matrix Converter," in *IEEE Transactions on Industrial Electronics*, vol. 58, no. 1, pp. 232-243, Jan. 2011.

The 2018 International Power Electronics Conference

Carrier-Based Overmodulation Strategy for Matrix Converters

Paiboon Kiatsookkanatorn[1*] and Somboon Sangwongwanich[2]
1 Department of Electrical Engineering, Rajamangala University of Technology Suvarnabhumi, Suphanburi, Thailand.
2 Department of Electrical Engineering, Faculty of Engineering, Chulalongkorn University, Bangkok, Thailand.
*e-mail: paiboon.k@RMUTSB.ac.th, e-mail: somboona@chula.ac.th

Abstract—**In this paper, carrier-based overmodulation for matrix converters is proposed. From the viewpoint of carrier-based modulation, there are two cases of overmodulation which correspond to (1) over output-voltage range and (2) over input-current range. Three simple strategies to limit the modulating functions under overmodulation are then studied. It is also clarified how the switching patterns and space vector utilization change in the overmodulation range. And the effects of the switching patterns and space-vector utilization on the current and voltage distortions are compared. Finally, the simulation results reveal that the error of fundamental amplitude between the output voltage and the desired output voltage is small with the so-called <2u1d>PWM, while using the max-input-phase as the reference in the carrier-based modulation gives low output voltage distortion.**

Keywords— Matrix converters, carrier-based modulation, overmodulation.

I. INTRODUCTION

A matrix converter in Fig. 1 is a power electronic device that converts a fixed frequency and amplitude input AC voltage to a variable frequency and amplitude output AC voltage without the need for an energy storage like a back-to-back converter does [1]. However, one of the limitations of matrix converters when compared to the back-to-back converter is that the maximum modulation index is 0.866. Therefore, there have been many research works that proposed approaches to increase the modulation index. For examples, the overmodulation using carrier-based control has been proposed [2] and [3]. In [2], the overmodulation compensation concept of the voltage-source inverter has been adopted to matrix converters. However, the input harmonic currents are rather high, and it does not show how the modulating functions or the reference signals of the carrier-based modulation are limited during overmodulation. The indirect modulation method has been modified complicatedly in [3] to obtain overmodulation of the input-side and output-side duty cycles. A control strategy to reduce low-frequency current ripples in the overmodulation range is presented in [4] using the beatless control. Though the output harmonic currents can be reduced but are still high. In [5], comparison between the square and trapezoidal wave overmodulation strategies for indirect matrix converters has been discussed, and a remedy for commutation of the rectifier side is also proposed. Overmodulation technique for direct space-vector modulation is studied in [6] following the

overmodulation idea used for the two-level inverter. Recently, the research in [7] propose a technique for indirect space-vector overmodulation that results in low input and output harmonic currents.

From the previous review, it can be said that most of the overmodulation researches are done for the indirect space-vector modulation and focus mainly on the voltage transfer ratio rather than the distortion of the input and output quantities of the matrix converters. Also, many of the overmodulation algorithms do not deal directly with the instantaneous voltage or current commands, and may restrict its practical application.

The objective of this paper is therefore to present simple carrier-based overmodulation technique for direct matrix converters based on the instantaneous input and output voltages. Comparison among several overmodulation strategies will be given from the viewpoint of fundamental voltage error and harmonic contents of both the output voltages and input currents. It will also be revealed how the switching patterns and space-vector utilization are changed in the overmodulation range, and how it affects the voltage and current distortions and the output voltage error. All the analysis will be justified by simulation.

II. STRUCTURE AND BASIC EQUATIONS OF MATRIX CONVERTERS

Fig. 1 shows the simplified structure of a matrix converter. The relationships between the voltage and the current of input and those of output can be described by (1)-(3).

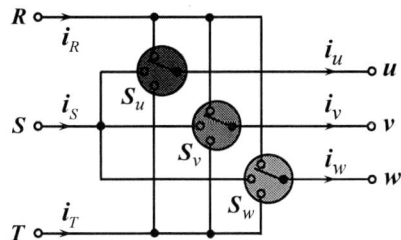

Fig. 1. Simplified structure of a matrix converter.

$$\begin{bmatrix} u \\ v \\ w \end{bmatrix} = \underbrace{\begin{bmatrix} m_{11} & m_{12} & m_{13} \\ m_{21} & m_{22} & m_{23} \\ m_{31} & m_{32} & m_{33} \end{bmatrix}}_{\mathbf{M}} \begin{bmatrix} R \\ S \\ T \end{bmatrix} = \begin{bmatrix} u^{*}+v_{Z} \\ v^{*}+v_{Z} \\ w^{*}+v_{Z} \end{bmatrix} \quad (1)$$

2581

$$\begin{bmatrix} i_R \\ i_S \\ i_T \end{bmatrix} = \underbrace{\begin{bmatrix} m_{11} & m_{12} & m_{13} \\ m_{21} & m_{22} & m_{23} \\ m_{31} & m_{32} & m_{33} \end{bmatrix}^T}_{\mathbf{M}^T} \begin{bmatrix} i_u \\ i_v \\ i_w \end{bmatrix} \qquad (2)$$

$$0 \le m_{ij} \le 1, \quad \sum_{j=1}^{3} m_{ij} = 1, \quad i = \{1,2,3\}, j = \{1,2,3\} \qquad (3)$$

The modulation matrix $\mathbf{M}$ described in (1) which gives the required output voltages and unity input power factor can be written in the form of the zero-voltage matrix $\mathbf{M_0}$ and the modulation matrix $\mathbf{M}'$ as shown in (4).

$$\mathbf{M} = \mathbf{M}' + \mathbf{M_0} \qquad (4)$$

where

$$\mathbf{M}' = \begin{bmatrix} m'_{11} & m'_{12} & m'_{13} \\ m'_{21} & m'_{22} & m'_{23} \\ m'_{31} & m'_{32} & m'_{33} \end{bmatrix} = \frac{1}{R^2 + S^2 + T^2} \begin{bmatrix} u^* \\ v^* \\ w^* \end{bmatrix} \begin{bmatrix} R \\ S \\ T \end{bmatrix}^T$$

and

$$\mathbf{M_0}' = \begin{bmatrix} X & Y & Z \\ X & Y & Z \\ X & Y & Z \end{bmatrix}$$

Here R, S, T and u^*, v^*, w^* are the input and desired output voltages, respectively. v_z is the zero voltage. i_R, i_S, i_T and i_u, i_v, i_w are the input current and the output current, respectively.

III. SWITCHING PATTERNS OF MATRIX CONVERTERS USING CARRIER-BASED DIPOLAR MODULATION

A method to generate the signals to drive the switches in carrier-based dipolar modulation has been proposed in [8]. The technique compares the reference signals u_P, u_N with two carrier waves as shown in Fig. 2.

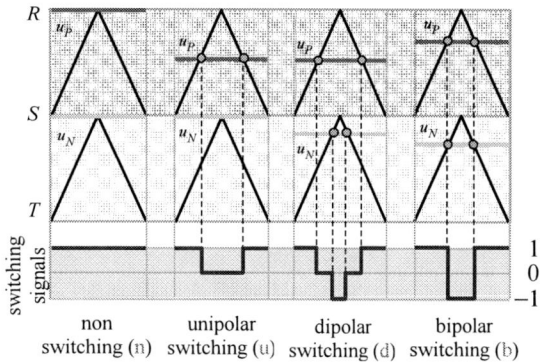

Fig. 2. Carrier-based Dipolar Modulation.

A. Choosing Input Phase as the Reference to Determine the Switching Sequences

Changing the input-phase reference in the carrier-based dipolar modulation will result in three different switching sequences as shown in Fig. 4. Assume that $|R| = \max(|R|, |S|, |T|)$, $|S| = \min(|R|, |S|, |T|)$ and $|T| = \mathrm{mid}(|R|, |S|, |T|)$, respectively.

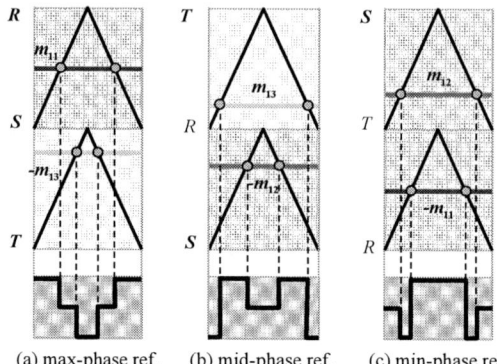

(a) max-phase ref.   (b) mid-phase ref.   (c) min-phase ref.

Fig. 3. Switching sequence when choosing different input phase as the reference(for instance, when R=max, S=mid and T=min)

1) Mid-Phase Reference: From (1) the output voltage referred to the mid-phase voltage 'S' can be rewritten as (7).

$$\begin{bmatrix} u - S \\ v - S \\ w - S \end{bmatrix} \triangleq \underbrace{\begin{bmatrix} u_P \\ v_P \\ w_P \end{bmatrix}}_{u_P} + \underbrace{\begin{bmatrix} u_N \\ v_N \\ w_N \end{bmatrix}}_{u_N} \triangleq \begin{bmatrix} m_{11} \\ m_{21} \\ m_{31} \end{bmatrix} [R-S] - \begin{bmatrix} m_{13} \\ m_{23} \\ m_{33} \end{bmatrix} [S-T] \quad (7)$$

The PWM switching signals can be generated using a dipolar modulation. The upper and lower reference signals $[U_P], [U_N]$ in the double-carrier waves are obtained by normalizing (8) with the respective bus voltages as:

$$[U_P] = \begin{bmatrix} m_{11} \\ m_{21} \\ m_{31} \end{bmatrix} = \begin{bmatrix} m'_{11} + X \\ m'_{21} + X \\ m'_{31} + X \end{bmatrix} \ge 0, \ [U_N] = -\begin{bmatrix} m_{13} \\ m_{23} \\ m_{33} \end{bmatrix} = -\begin{bmatrix} m'_{13} + Z \\ m'_{23} + Z \\ m'_{33} + Z \end{bmatrix} \le 0 \quad (8)$$

Based on (7)-(8), Fig. 4 illustrates for the mid-phase reference case how the reference signals U_P and U_N which are used to generate the PWM switching signals can be obtained from the commanded output voltages v_o^* (u, v, w). Other phase reference cases can also be done in the same way.

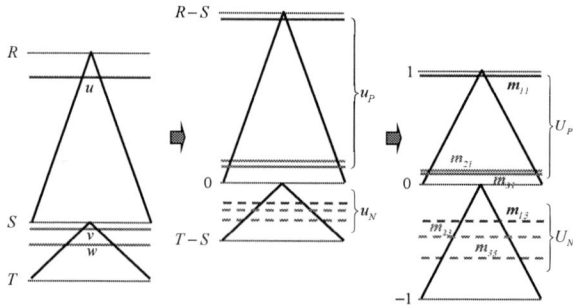

Fig. 4. Process of carrier-based dipolar modulation for generating reference signals with mid-phase reference.

2) Max-Phase Reference: If the max-phase voltage 'R' is chosen as a reference, the reference signals U_P, U_N can be determined by (9)-(10).

$$\begin{bmatrix} u-R \\ v-R \\ w-R \end{bmatrix} \triangleq \underbrace{\begin{bmatrix} u_P \\ v_P \\ w_P \end{bmatrix}}_{u_P} + \underbrace{\begin{bmatrix} u_N \\ v_N \\ w_N \end{bmatrix}}_{u_N} \triangleq \begin{bmatrix} m_{13} \\ m_{23} \\ m_{33} \end{bmatrix} [T-R] - \begin{bmatrix} m_{12} \\ m_{22} \\ m_{32} \end{bmatrix} [R-S] \quad (9)$$

$$[U_P] = [m_{i3}] = \begin{bmatrix} m_{13} \\ m_{23} \\ m_{33} \end{bmatrix} \geq 0, \quad [U_N] = -[m_{i2}] = -\begin{bmatrix} m_{12} \\ m_{22} \\ m_{32} \end{bmatrix} \leq 0. \quad (10)$$

3) Min-Phase Reference: The reference signals when choosing the min-phase voltage 'T' as a reference can also be determined by (11)-(12).

$$\begin{bmatrix} u-T \\ v-T \\ w-T \end{bmatrix} \triangleq \underbrace{\begin{bmatrix} u_P \\ v_P \\ w_P \end{bmatrix}}_{u_P} + \underbrace{\begin{bmatrix} u_N \\ v_N \\ w_N \end{bmatrix}}_{u_N} \triangleq \begin{bmatrix} m_{12} \\ m_{22} \\ m_{32} \end{bmatrix} [S-T] - \begin{bmatrix} m_{11} \\ m_{21} \\ m_{31} \end{bmatrix} [T-R] \quad (11)$$

$$[U_P] = [m_{i2}] = \begin{bmatrix} m_{12} \\ m_{22} \\ m_{32} \end{bmatrix} \geq 0, \quad [U_N] = -[m_{i1}] = -\begin{bmatrix} m_{11} \\ m_{21} \\ m_{31} \end{bmatrix} \leq 0 \quad (12)$$

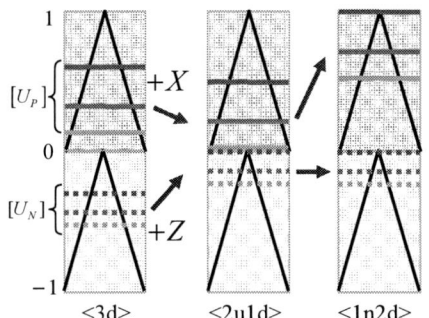

Fig. 5. PWM modes with different zero voltage matrix

B. Zero-Voltage Matrix Addition to Determine PWM Modes

According to (8), the values of X, Y and Z will determine which kind of modes the three output phases will be. Two PWM modes shown in Fig.5 are explained in the following.

1) <2u1d> PWM: This PWM mode means that the two output phases are in unipolar mode, and one output phase is in dipolar mode.

$$\begin{rcases} X = -\min(m'_{11}, m'_{21}, m'_{31}) \\ Y = \ 1-X-Z \\ Z = -\min(m'_{13}, m'_{23}, m'_{33}) \end{rcases} \quad (13)$$

2) <1n2d> PWM: This PWM mode means that one output phase is clamped to a particular input phase, and two other output phases are in dipolar mode.

$$\begin{rcases} X = 1 - \max(m'_{11}, m'_{21}, m'_{31}) \\ Y = 1 - X - Y \\ Z = -\min(m'_{13}, m'_{23}, m'_{33}) \end{rcases} \text{ or } \begin{rcases} X = -\min(m'_{11}, m'_{21}, m'_{31}) \\ Y = 1 - Y - Z \\ Z = 1 - \max(m'_{13}, m'_{23}, m'_{33}) \end{rcases} \quad (14)$$

$$\underbrace{\qquad\qquad}_{if \ |R| \geq |T|} \qquad \underbrace{\qquad\qquad}_{if \ |R| < |T|}$$

After selecting the switching sequence and the PWM modes as mention previously, the resultant switching patterns and space-vector utilization are shown in Table I for the case that the commanded output voltages are $u* > v* > w*$.

TABLE I
SWITCHING PATTERNS AND SPACE-VECTOR UTILIZATION
WHEN $0.5 < Q \leq 0.866$

PWM mode	<2u1d>			<1n2d>		
Phase Ref.	R	S	T	R	S	T
Space vectors	RTT	RRS	SSS	RTT	RRR	RSS
	RRT	RSS	RSS	RRT	RRS	RTS
	RRS	RST	RTS	RRR	RRT	RTT
	RSS	SST	RTT	RRS	RST	RRT
	SSS	STT	RRT	RSS	RTT	RRR

The relationship between the well-known space-vector modulations and a carrier-based modulation was shown in [8]. As from Table I, <2u1d> PWM with R phase reference will be equivalent to the indirect space-vector modulation in [9]-[11] and <2u1d> PWM with S phase reference will be equivalent to the direct space-vector modulation in [12[. Moreover, <1n2d> PWM with T phase reference will be equivalent to the indirect space-vector modulation in [9].

IV. PROPOSED CARRIER-BASED OVERMODULATION

In overmodulation range, the matrix converters have two main problems:

- a large error between fundamental amplitudes of the output voltages and their commands, and
- a high level of low-order harmonics of the output voltage and the input current.

A. Overmodulation under Carrier-Based Modulation

The main tasks of modulation for the matrix converters are to (1) generate the desired output voltages, and (2) to control the input currents. Therefore, there are two overmodulation cases for matrix converters under which the modulating functions violate the constraint (3). To simplify the explanation, we will use the mid-phase reference case in Figs. 6-8 in the following discussion. In this case $U_P = m_{11}$, and $U_N = -m_{33}$. From Figs. 6-8, there are two cases of overmodulation, i.e.,

(I) over available-output-voltage range when the modulating function $m_{11} > 1$ or $m_{33} < 0$. This can happen if the modulation index q > 0.866 as shown in Figs. 6-7 and

(II) over controllable-input-current range when the modulating function $m_{21} + m_{23} > 1$ while $0 < m_{23} < 1$, $0 < m_{21} < 1$. Under this situation, $m_{22} < 0$ which does not satisfy (3). An example of this overmodulation case is shown in Fig. 8.

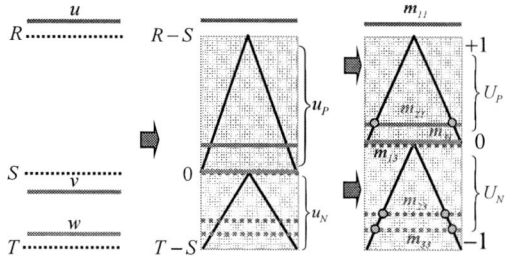

(a) command voltages (b) reference voltages (c) reference signals

Fig. 6 Overmodulation when $u > R$

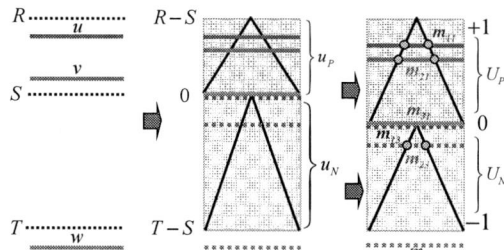

(a) command voltages (b) reference voltages (c) reference signals

Fig. 7 Overmodulation when $|w| > |T|$

Figs. 6(c)-7(c) are examples of the overmodulation case (I), wherein m_{11}, m_{21}, m_{31} are the reference signals of positive buses and m_{13}, m_{23}, m_{33} are the reference signals of negative buses, respectively. Fig. 6(c) illustrates that the matrix converter works in the overmodulation range with the reference signal U_P not complied with the given condition in (3). In this case, the u-phase output voltage will be distorted, and the input current will also be distorted as well. In the same way, the Fig. 7(c) illustrates that the matrix converter works in the overmodulation range with the reference signal U_N not complied with the condition (3). Again, in this case the w-phase output voltage together with the input currents will be distorted.

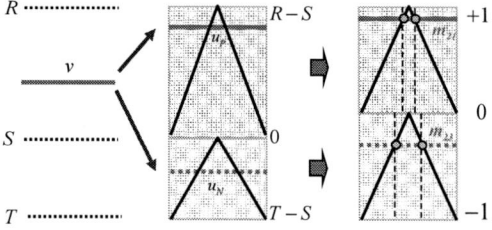

(a) command voltages (b) reference voltages (c) reference signals

Fig. 8 Overmodulation when $(U_P - U_N) > 1$

Fig. 8 is the example for overmodulation case (II). In this case, the v-phase output voltage is still within the range of the input voltages (R, S, T). However, to control the input current as required (PF=1), one has to split the v-phase command into the positive and negative references, and when this kind overmodulation occurs, the modulation of v-phase will change from the dipolar mode to the bipolar mode and beyond. Since bipolar

mode is the limit of the modulating function, the input current will not follow the command and will be distorted in general. For the v-phase output voltage, the distortion may or may not happen depending on how we handle this overmodulation, i.e., how we limit the modulating function.

B. Limitation of Reference Signals under Overmodulation

From Figs.6-8, there are two simple steps to limit the switching signals in the overmodulation range as follows.

Step 1: Check

$$\text{if } U_P > 1 \quad \text{then} \quad U_P = 1 \quad (17)$$
$$\text{if } U_N < 0 \quad \text{then} \quad U_N = 0 \quad (18)$$

Step 2: Check if $(U_P - U_N) > 1$ is true, then there are three strategies to control the reference signals of the relevant phase.

1) Strategy 1: Non switching

$$\text{If } |U_P| \geq |U_N| \quad \text{then} \quad U_P = 1, U_N = 0 \quad (19\text{-A})$$
$$\text{If } |U_P| < |U_N| \quad \text{then} \quad U_P = 0, U_N = -1 \quad (19\text{-B})$$

2) Strategy 2: Fixed bipolar mode

$$U_P(k) = U_P(k-1) \quad \text{and } U_N(k) = U_N(k-1) \quad (20)$$

where k denotes the kth sampling time.

3) Strategy 3: Proportional bipolar mode

$$U_P(k) \Leftarrow \frac{U_P(k)}{U_P(k) - U_N(k)} \quad \text{and } U_N(k) \Leftarrow \frac{U_N(k)}{U_P(k) - U_N(k)} \quad (21)$$

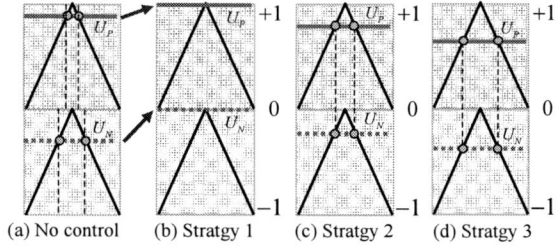

(a) No control   (b) Stratgy 1   (c) Stratgy 2   (d) Stratgy 3

Fig. 9. Control strategies of the reference signals when $(U_P - U_N) > 1$

Fig. 9 illustrates how the reference signals is limited when $(U_P - U_N) > 1$ according to the three strategies. For the strategy 1 the reference signal is adjusted to give non-switching mode shown in Fig. 9(b). For the strategy 2, the reference signals are fixed to the values when they enter the bipolar mode as shown in Fig. 9(c). Lastly, by the strategy 3, the reference signals are also kept in the bipolar mode but the ratios between U_P, U_N is adjusted according to (21) as illustrated in Fig. 9(d)

C. Switching Patterns in Overmodulation Range

According to Table I, when the output voltage is increased to the overmodulation range (q > 0.866), the new switching patterns and the new space-vector utilization with the R, S, and T phase reference are shown in Figs. 10-12, respectively.

2584

The 2018 International Power Electronics Conference

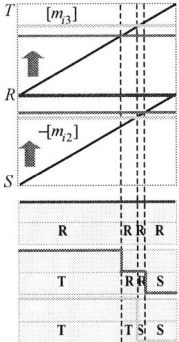 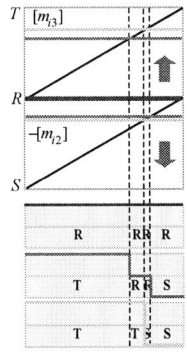

Fig. 10. <2u1d> PWM (Left) and <1n2d> PWM (Right) with R- phase reference.

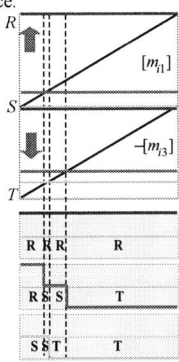 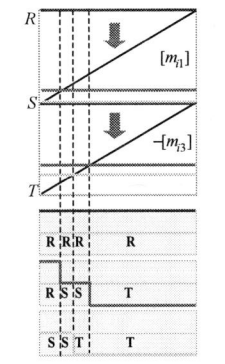

Fig. 11. <2u1d> PWM (Left) and <1n2d> PWM (Right) with S-phase reference

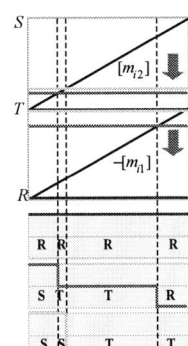 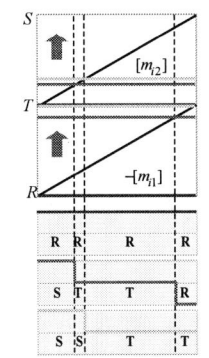

Fig. 12. <2u1d> PWM (Left) and <1n2d> PWM (Right) with T-phase reference.

Figs. 10-12 show that when the output voltage is increased to the overmodulation range, the space-vector utilizations for generating the voltage and current are reduced to four vectors and the zero vector is removed. Furthermore, there are also changes in the space-vector utilization and switching patterns as shown in Table II.

TABLE II
SWITCHING PATTERN AND SPACE-VECTOR UTILIZATION
IN THE OVERMODULATION RANGE WHEN Q>0.866

PWM	<2u1d>			<1n2d>		
Ref.	R	S	T	R	S	T
Space vectors	RTT	RRS	RSS	RTT	RRS	RSS
	RRT	RSS	RTS	RRT	RSS	RTS
	RRS	RST	RTT	RRS	RST	RTT
	RSS	RTT	RRT	RSS	RTT	RRT
	<1n1d1u>	<1n1d1u>	<1n1d1u>	<1n1d1u>	<1n1d1u>	<1n1d1u>

According to Table II, when performing in the overmodulation range, both <2u1d> and <1n2d> PWM converge to the same <1n1u1d> PWM.

V. SIMULATION RESULTS

Simulation is carried out to compare and analyze the current and voltage distortions for matrix converters in the overmodulation range. Simulation conditions are as shown in Table III.

TABLE III
SIMULATION CONDITIONS

input voltage: 380V,50Hz	load: $R = 40\Omega$, $L = 8mH$
output voltage: q*380V, 25Hz	modulation index q= 0.88, 0.92 and 1.0
input power factor: unity	switching frequency: 5 kHz

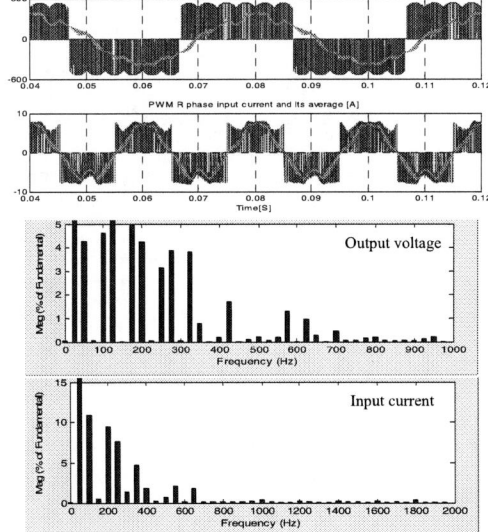

(a) No controlling

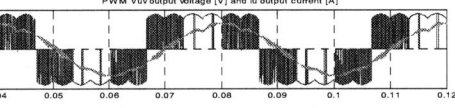

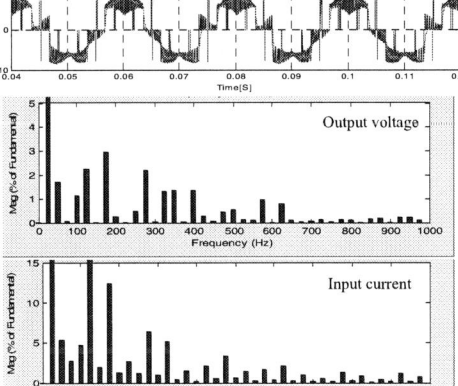

(b) strategy 1

2585

The 2018 International Power Electronics Conference

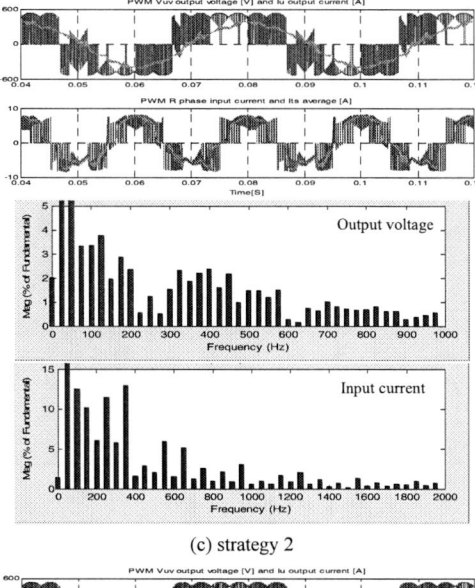

(c) strategy 2

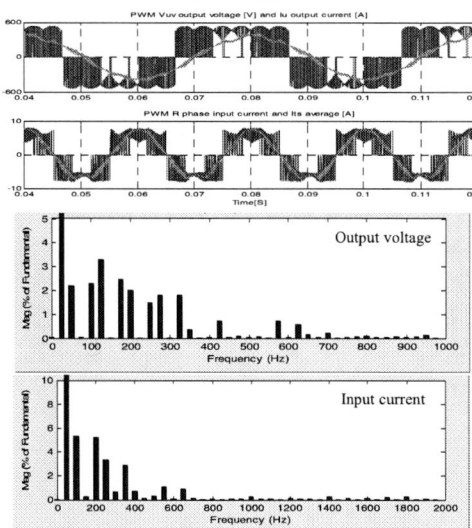

(d) strategy 3

Fig. 13. <1n2d> PWM with max-phase reference q=1

From simulation results of <1n2d> PWM with max-phase reference and q = 1 as shown in Fig. 13, the parameters of each strategy can be compared in Table IV

TABLE IV
PARAMETERS USED WHEN MODULATION INDEX Q=1

Switching pattern	Controlling Strategies	V_{o1m} (V)	V_{err} (V)	u_{uv} %THD	i_u %THD	i_R %THD
<1n2d> with Max Ref.	Non	489.6	49.29	53.64	13.59	18.53
	1	535.6	3.29	44.91	7.03	24.20
	2	515.9	22.99	48.15	11.18	24.22
	3	512.8	26.09	47.65	7.77	11.00

From Table IV, if the reference signals are not adjusted, the fundamental amplitude error and the distortions of the output voltage and the input current will be high. On the other hand, when the reference signals are limited according to the proposed strategies 1-3, the fundamental amplitude error and distortion problems are less severe. The three strategies give similar contents of

harmonic voltages. However, fundamental amplitude error in the output voltage is minimal by the strategy 1, and the input harmonic current becomes lowest for the strategy 3. Therefore, adjusting the ratio between U_P, U_N of the bipolar mode as in the strategy 3 improves the input harmonic current as compared to the fixed bipolar mode of the strategy 2.

Figs. 14-18 demonstrate some simulation results for q=0.92, 1.0 in the overmodulation range by the strategy 3. From the simulation, the results can be summarized as shown in Tables V-VII.

TABLE V
PARAMETERS USED WHEN MODULATION INDEX Q=0.88

PWM mode	phase Ref.	V_{o1m} (V)	V_{err} (V)	u_{uv} %THD	i_u %THD	i_R %THD
<2u1d>	Max	472.9	1.32	57.75	5.51	7.50
	Mid	473.1	1.12	49.58	5.92	9.26
	Min	473.1	1.12	60.09	7.11	7.50
<1n2d>	Max	472.7	1.52	57.77	5.22	7.68
	Mid	473.1	1.12	55.89	6.70	7.48
	Min	473.1	1.12	60.13	6.69	6.93

TABLE VI
PARAMETERS USED WHEN MODULATION INDEX Q=0.92

PWM mode	phase Ref.	V_{o1m} (V)	V_{err} (V)	u_{uv} %THD	i_u %THD	i_R %THD
<2u1d>	Max	490.1	5 67	53.62	5.55	7.83
	Mid	490.1	5 67	48 22	6.28	8.63
	Min	489.9	5 87	56 20	7.07	7.76
<1n2d>	Max	489.8	5.97	53.55	5.39	8.03
	Mid	490.0	5.77	51.76	6.64	7.71
	Min	489.3	6.47	56.24	6.91	7.71

TABLE VII
PARAMETERS USED WHEN MODULATION INDEX Q=1.0

PWM mode	phase Ref.	V_{o1m} (V)	V_{err} (V)	u_{uv} %THD	i_u %THD	i_R %THD
<2u1d>	Max	516.5	22.39	47.59	7.60	10.95
	Mid	516.6	22.29	46.27	8.14	10.79
	Min	516.6	22.29	50.43	8.46	10.92
<1n2d>	Max	512.8	26.09	47.65	7.77	11.00
	Mid	516.6	22.29	45.55	8.11	10.81
	Min	511.3	27.59	50.55	9.10	11.63

Tables V-VII reveal that the <2u1d> PWM gives less fundamental amplitude error than does the <1n2d> PWM. Moreover, it can be shown that the max-phase reference gives low distortion in the output voltage, while the mid-phase reference has less fundamental amplitude error.

VI. CONCLUSION

Simple carrier-based overmodulation technique for direct matrix converters which uses the instantaneous input and output voltages to calculate the relevant modulating functions or duty cycles of the switches, is proposed. Two kinds of overmodulation, i.e. the over output-voltage range and the over input-current range, are discussed from the viewpoint of modulating functions. Simple steps to generate and limit the modulating functions in the overmodulation range is then introduced together with three overmodulation strategies. Changes in the switching patterns and space-vector utilization under overmodulation are investigated. Simulation is carried out to compare the three overmodulation strategies in

2586

terms of the fundamental output voltage error and the input-current and output-voltage distortions. The simulation results reveal that the so-called <2u1d> PWM results in the smallest error of the fundamental output voltage, while choosing the maximum input phase as the reference in the double-carrier-based modulation helps to reduce the voltage distortion.

REFERENCES

[1] P. W. Wheeler, J. Rodriguez, J. C. Clare, L. Empringham, and A.Weinstein, "Matrix converters: A technology review," *IEEE Trans. Ind. Electron.*, vol. 49, no. 2, pp. 276–288, Apr. 2002.

[2] Y. D. Yoon *et al.*, "Carrier-based modulation technique for matrix converter," *IEEE Trans. Power Electronics*, vol. 21, No. 6, pp. 1691-1703, 2006.

[3] S. Thuta *et al.*, "Matrix converter overmodulation using carrier-based control: Maximizing the voltage transfer ratio," in *Proc. Power Electron. Specialists Conf.*, Rhodes, Greece, pp. 1727–1733, Jun.15–19, 2008.

[4] T. Yasuhiro *et al.*, "A novel control strategy for matrix converter in the over-modulation range," in *Proc. of PCC-Nagoya*, pp.1049-1055, 2007.

[5] G. T Chiang and J. Itoh, "Comparison of two overmodulation strategies in an indirect matrix converter," *IEEE Trans. Ind. Electron.*, vol. 60, no. 1, pp. 43–53, Jan. 2013.

[6] M. B. Amir, M. Mahammad, and R. M. Habib, "Two simple overmodulation algorithms for space modulated three-phase to three phase matrix converter," *IET Power Electron*, vol 7, no 7, pp. 1915–1924, 2014.

[7] Y. Xia *et al.*, "Research on a new indirect space-vector overmodulation strategy in matrix converter," *IEEE Trans. Ind. Electron.*, vol. 63, no. 2, pp. 1130-1141, 2016.

[8] P. Kiatsookkanatorn and S. Sangwongwanich, "A unified PWM method for matrix converters and its carrier-based realization using dipolar modulation technique," *IEEE Trans. Ind. Electron.*, vol. 59, no. 1, pp. 80-92, 2012.

[9] L. Huber and D. Borojevic, "Space vector modulated three-phase to three-phase matrix converter with input power factor correction," *IEEE Trans. Ind. Appl.*, vol. 31, no. 6, pp. 1234-1246, 1995.

[10] P. Nielsen et al, "Space vector modulated matrix convener with minimized number of switchings and a feed forward compensation of input voltage unbalance," in *Proc. PEDES*, pp. 833-839, 1996.

[11] H. J. Cha *et al.*, "An approach to reduce common-mode voltage in matrix converter," *IEEE Trans. Ind. Appl.*, vol. 39, no. 4, pp. 1151-1159, 2003.

[12] Y. Tadano *et al*, "Direct space vector PWM strategies for three-phase to three-phase matrix converter," in *Proc. of PCC-Nagoya*, pp.1064-1071, 2007.

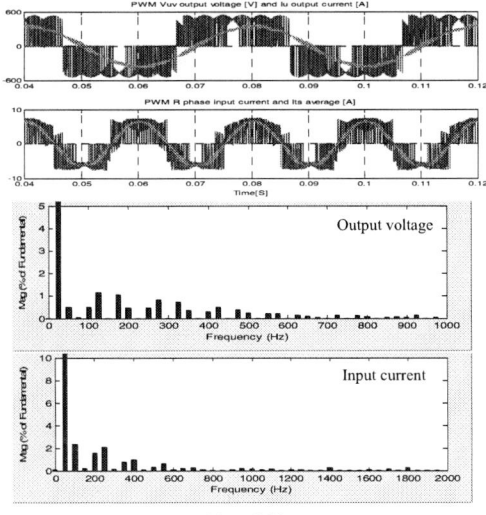

(a) q=0.92

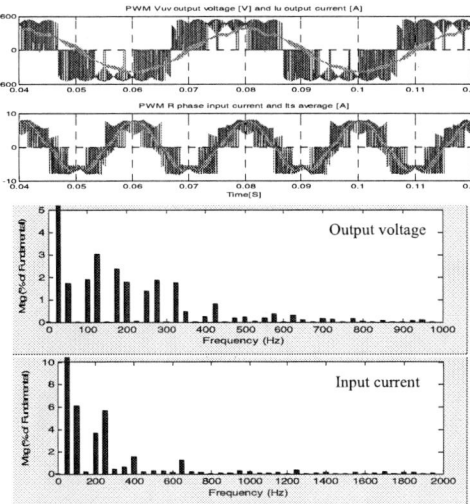

(b) q=1.0

Fig. 14. <1n2d> PWM with mid-phase reference (strategy 3)

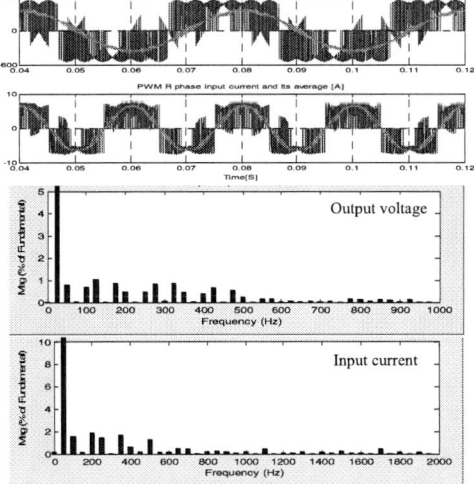

(a) q=0.92

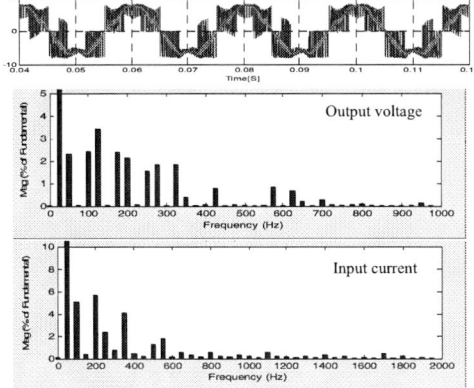

(b) q=1.0

Fig. 15. <1n2d> PWM with min-phase reference (strategy 3)

The 2018 International Power Electronics Conference

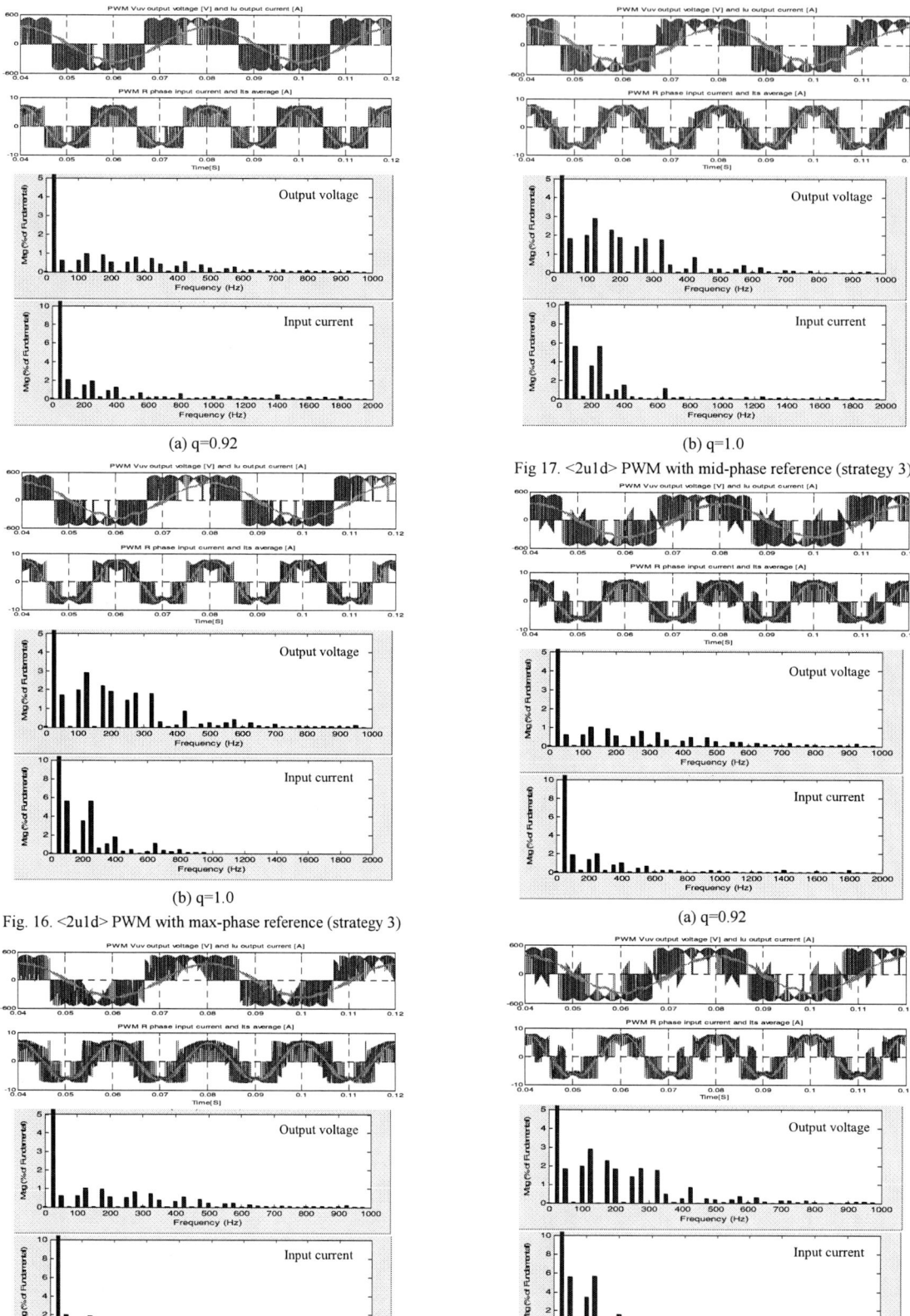

(a) q=0.92

(b) q=1.0

Fig 17. <2u1d> PWM with mid-phase reference (strategy 3)

(b) q=1.0

Fig. 16. <2u1d> PWM with max-phase reference (strategy 3)

(a) q=0.92

(a) q=0.92

(b) q=1.0

Fig 18. <2u1d> PWM with min-phase reference (strategy 3)

2588

Three-Phase to High-Frequency Single-Phase Matrix Converter
-A Frequency Control Suitable for Soft Switching-

Wataru Kodaka[1]*, Satoshi Ogasawara[1], Koji Orikawa[1], Masatsugu Takemoto[1],
Takashi Hyodo[2], Hiroyuki Tokusaki[2]

1 Graduate School of Information Science and Technology,
Hokkaido University, Sapporo, Hokkaido, Japan
2 Technology and Intellectual Property HQ, OMRON Corporation, Kyoto, Japan
*kodaka@ist.hokudai.ac.jp

Abstract—**Insulated AC/DC converters using matrix converters (MCs) have higher efficiency, smaller size, and longer lifespan compared with conventional ones. For downsizing of high-frequency (HF) transformers in these converters, developments of HF output MC control methods are active. It is expected to improve efficiency by applying soft switching techniques for these converters. However, to achieve soft switching, phase difference between an output voltage and a current is required. As a result, output power factor decreases. In this paper, a relationship between soft switching conditions and output power factor is revealed. Considering this relationship, the authors propose a new MC control method to achieve soft switching and maximum output power factor. Additionally, this method enables to suppress the distortion of input currents and reduce ripple of the output current. Also, the MC can control the output current and instantaneous reactive power. The authors confirm that experimental results show validity of the method.**

Keywords—*Matrix Converter, Insulated AC/DC converter, Soft Switching, Frequency Control*

I. INTRODUCTION

In industrial application, it is required to convert utility AC power to desirable DC power and to ensure electrical insulation between the DC and the utility AC in terms of safety. Power electronics technology make this conversion more flexible and higher efficiency. Thus, it is active to apply power electronics technology for AC/DC converters.

Generally, an insulated AC/DC converter consists of the following three conversion stages [1]–[3]; (1) a former rectifier converting utility AC to DC, (2) an inverter converting the DC to HF AC to ensure electrical isolation using an HF transformer, (3) a latter rectifier converting the HF AC to output DC. Owing to the many conversion stages, the efficiency of this AC/DC converter is relatively low, though a control method of this converter is generally easy. Furthermore, the conventional converter needs a large-size electrolytic capacitor for DC link, and it leads to increasing volume of the system and decreasing lifespan of the converter.

In recent years, to overcome these problems, developments and researches of MCs become actively [4]–[9].

The MC is able to realize direct conversion of utility power to arbitrary AC power without DC link. Therefore, an insulated AC/DC converter with the MC has only two conversion stages; (1) the MC converting utility AC to HF AC, (2) a latter rectifier converting the HF AC to output DC. As a result, the AC/DC converter becomes higher efficient. Also this converter realizes smaller size, and longer lifespan, because of the lack of an electrolytic capacitor.

In insulated AC/DC converters, transformers are dominant components in volume of the systems. Generally, downsizing of magnetic components is realized by increasing an operational frequency. However, for typical MC control techniques, output frequencies are very low, compared with its switching frequencies [2], [10]–[12]. This motivates to develop a novel MC control method with high-frequency. On the other hand, the higher switching frequency generates more switching losses [13].

The authors proposed an MC control method which an output frequency corresponds to its switching frequency [14]. However, this method has hard switching and be unable to adjust power factor of the MC output power. Hard switching causes large switching losses and makes efficiency less. Also, there are MC control methods for insulated AC/DC converters proposed in [12], [15], which achieve soft switching and suppress the source current harmonics. On the other hand, to achieve soft switching, an MC current must have a delayed phase behind a voltage. Therefore output power factor becomes low. Furthermore, required phase differences vary following variation of input voltage phases. To improve power factor, the MC needs to change the phase difference depending on the input voltage phases.

This paper introduces a new control method to achieve not only a high-frequency output, which equals to the switching frequency, but also soft switching, suppressing distortion of a input current, and adjusting power factor by a frequency control. Conditions for achieving soft switching and maximum output power factor are revealed, and the MC controls the output frequency to satisfy with these conditions responding to the input voltage phases.

The 2018 International Power Electronics Conference

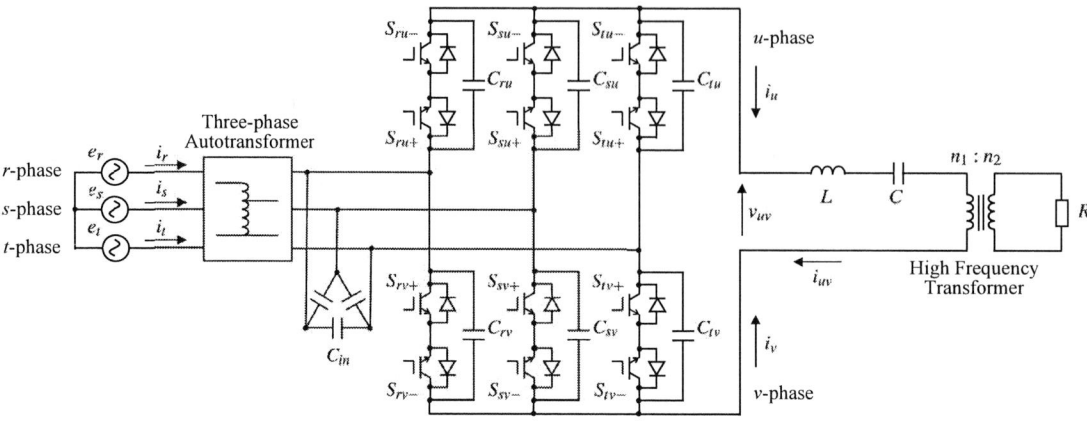

Fig. 1. A circuit of a three-phase to high-frequency single-phase matrix converter.

Furthermore, this method enables the MC to operate under utility input power factor and to output constant current amplitude. Experimental results show the validity and effectiveness of the method.

II. MATRIX CONVERTER TOPOLOGY

Fig. 1 shows an MC circuit. The MC input and output terminals are connected to a three-phase voltage source e_r, e_s, e_t and a single-phase load of output line voltage v_{uv}, respectively. The MC consists of six bidirectional switches $S_{ij}(i \in \{r, s, t\}, j \in \{u, v\})$. For soft switching operation, a snubber capacitor is connected to each bidirectional switch in parallel, and an LC resonant circuit is put between the MC and a HF transformer. the HF transformer has a primary winding of turns n_1 and a secondary winding of turns n_2. A resistance R imitates for a latter rectifier and an output DC circuit, thus the output of this experimental circuit is AC. It is desired that a bidirectional switch consists of two reverse-blocking IGBTs in order to become more efficiency. However in this MC, each bidirectional switch consists of two reverse-conduction IGBTs with free-wheeling diodes in antiparallel. Input capacitors C_{in} must be placed to absorb switching frequency components of input currents.

III. HIGH-FREQUENCY OUTPUT

From Fig. 1, we can consider the MC input side as a voltage source and the MC output side as a current source. Therefore, it is necessary to prevent the input side short circuit and the output side open circuit. If the input side becomes short circuit, an over current will flow through the input circuit and destroy the switches. Similarly, if the output side becomes open circuit, a voltage between the switch will be very large and destroy them. These constraints expressed as

$$\begin{cases} \zeta_{ru} + \zeta_{su} + \zeta_{tu} = 1 \\ \zeta_{rv} + \zeta_{sv} + \zeta_{tv} = 1 \end{cases} \quad (1)$$

where ζ_{ij} is S_{ij} ON time ratio to control period T_c.

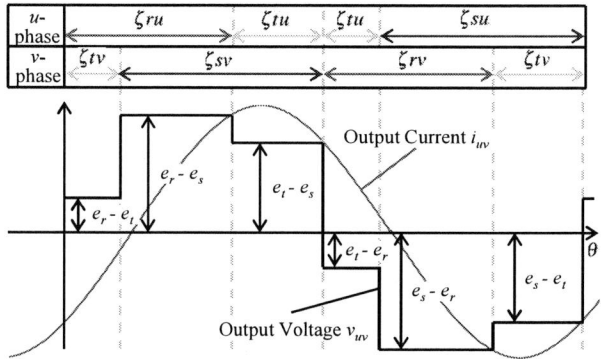

Fig. 2. Waveforms of the output voltage v_{uv} and current i_{uv} ($e_r > e_t > 0 > e_s$).

The authors have proposed the method in [14] to realize high-frequency output. Based on this method, the switching patterns are given as follows.

- When the medium input voltage is positive, the MC switches in the order of the maximum, the medium, and the minimum input voltage.

- When the medium input voltage is negative, the MC switches in the order of the minimum, the medium, and the maximum input voltage.

Additionally, in this method, the MC do not output zero voltage, because low voltage output makes it difficult to achieve soft switching. In other words, the MC considers the following constraints.

- A switch which connects an input phase of the maximum voltage to an output phase of the minimum voltage is OFF

- A switch which connects an input phase of the minimum voltage to an output phase of the maximum voltage is OFF

In the case of $e_r > e_t > 0 > e_s$, the above conditions

2590

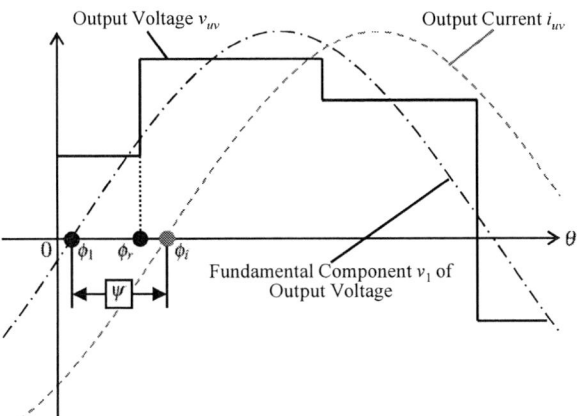

Fig. 3. A phase relationship to achieve soft switching.

can be written as

$$\begin{cases} \zeta_{su}, \zeta_{rv} = 0 & (v_{uv} > 0) \\ \zeta_{ru}, \zeta_{sv} = 0 & (v_{uv} < 0) \end{cases} \quad (2)$$

Therefore, an output voltage v_{uv} and an output current i_{uv} are illustrated as Fig. 2 in the case of $e_r > e_t > 0 > e_s$.

IV. SOFT SWITCHING

In [15], soft switching conditions are discussed in detail and given as follows:

- If v_{uv} increases, then $i_{uv} < 0$.

- If v_{uv} decreases, then $i_{uv} > 0$.

- A sign of i_{uv} must not change during charge or discharge of snubber capacitors C_s.

v_{uv}, i_{uv} in Fig. 3 are satisfied with these constraints. A phase ϕ_r is at a last rise point of v_{uv}. Fig. 3 shows that the output current phase ϕ_i lags behind the phase ϕ_r by charge or discharge time of the snubber capacitors in order to achieve soft switching. The phase difference between v_{uv} and i_{uv} affects power factor. Thus, C_s charge or discharge time is very important in terms of power factor.

Considering u-phase, Fig. 4 shows a commutation from s-phase to r-phase in the case of $e_r > e_s$. In Fig. 4, assuming that variation of the input voltages are slow enough than commutation time, they are regarded as DC. Mode 1-4 are intervals, which are called deadtime to prevent the input circuit from short. The capacitors C_{ru}, C_{su}, C_{tu} are snubber capacitors of C_s. Also, i_{uv} must be negative to achieve soft switching because of the conditions mentioned above. The switching states are shown in Table I.

In Mode 2, C_{ru}, C_{su}, C_{tu} cause charge or discharge and the current flows through all the capacitors of r-phase. The charge ΔQ during commutation is expressed as

$$\begin{aligned} \Delta Q &= (C_{ru} + C_{su} + C_{tu})(e_r - e_s) \\ &= 3C_s \Delta V \end{aligned} \quad (3)$$

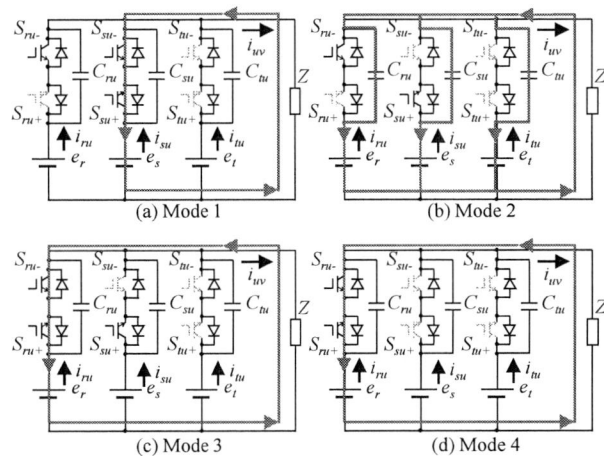

Fig. 4. A commutation from s-phase to r-phase in the u-phase ($e_r > e_s$).

TABLE I. THE SWITCHING STATES IN FIG. 4

	Mode 1	Mode 2	Mode 3	Mode 4
S_{ru+}	OFF	OFF	ON	ON
S_{ru-}	ON	ON	ON	ON
S_{su+}	ON	ON	ON	OFF
S_{su-}	ON	OFF	OFF	OFF
S_{tu+}	OFF	OFF	OFF	OFF
S_{tu-}	OFF	OFF	OFF	OFF

where ΔV is amount of a v_{uv} increment at the phase ϕ_r. On the other hand, while $i_{uv} < 0$, currents through the snubber capacitors can supply charge Q_s as follows.

$$Q_s = -\int_{\phi_r}^{\phi_i} (i_{ru} + i_{su} + i_{tu})\, dt = -\int_{\phi_r}^{\phi_i} i_{uv}\, dt \quad (4)$$

Q_s expresses the maximum charge to be supplied to C_{ru}, C_{su}, C_{tu} until the sign of i_{uv} changes. Thus for soft switching, $\Delta Q \le Q_s$ must be satisfied, because of the third soft switching condition mentioned above. This yields the following from Eqs. (3) and (4).

$$3C_s \Delta V \le -\int_{\phi_r}^{\phi_i} i_{uv}\, dt \quad (5)$$

For maximum output power factor, the MC needs to control the phase of i_{uv} to be minimum. Solving Eq. (5), a minimum i_{uv} phase to achieve soft switching is obtained.

There is upper of C_s charge or discharge time [15]. This condition is expressed as

$$T_{dis} \le T_{dead} \quad (6)$$

where T_{dis} is C_s charge or discharge time and T_{dead} is the time interval of Mode 2. If Eq. (6) is not satisfied, the input side will be short circuit. Thus, C_s charge or discharge must complete during Mode 2. In this experiment, C_s has small capacity and the charge or discharge time are much shorter than T_{dead}. Therefore, this constraint can be ignored.

V. Control Method

To control the MC, every variable ζ_{ij} has to be determined. Thus, we make control equations considering the MC conditions and obtain six ζ_{ij} by solving these equations for each control period T_c which is half period of the output period. This section introduces how to make and solve the control equations. The three-phase input voltages e_r, e_s, e_t are obtained from sensors. Then, we assume the following:

1) The output frequency f_{out} (a few kHz) is much higher than the input (generally 50 or 60 Hz). Therefore, the input voltages e_r, e_s, e_t are considered as DC during T_c.
2) Since the deadtime is enough shorter than T_c, the effect of deadtime is ignored.
3) The HF transformer has a enough small leakage inductance and a large magnetizing inductance. Thus, we regard the HF transformer as a ideal transformer.
4) Although the MC changes f_{out}, variation of f_{out} is enough slow and i_{uv} can be regarded as steady state.

A. Output Current

Let a fundamental component of v_{uv} to be v_1. v_1 is obtained by Fourier series developing v_{uv}.

$$v_1 = V_1 \sin(\theta - \phi_1) \qquad (\theta = \omega t) \qquad (7)$$

where

$$V_1 = \sqrt{a_1^2 + b_1^2} \qquad (8)$$

$$\phi_1 = -\tan^{-1}\left(\frac{a_1}{b_1}\right) \qquad (9)$$

$$a_1 = \frac{1}{\pi}\int_0^{2\pi} v_{uv}\cos\theta d\theta \qquad (10)$$

$$b_1 = \frac{1}{\pi}\int_0^{2\pi} v_{uv}\sin\theta d\theta \qquad (11)$$

and ω is an output angular frequency.

It is expected that i_{uv} has a sinusoidal waveform depending on v_1, because of the LC resonant circuit. Although i_{uv} maybe have a few of harmonics components, we ignore these components and express i_{uv} as

$$i_{uv} = I_{uv}\sin(\omega t - \phi_i) \qquad (12)$$

Assuming that power transfer from the MC to the load is almost associated to a fundamental component, we consider only the fundamental component (FHA method used by [16]). Let phasors of v_1 and i_{uv} to be $\dot{V}_1 = V_1 e^{-j\phi_1}$ and $\dot{I}_{uv} = I_{uv}e^{-j\phi_i}$ respectively. From the third assumption mentioned above, the MC output circuit is considered as an RLC series circuit. We express a phasor of it's impedance as $\dot{Z} = Ze^{j\psi}$. Then, relationships between $\dot{V}_1$ and $\dot{I}_{uv}$ are shown in Fig. 3 and written as

$$\dot{V}_1 = \dot{Z}\dot{I}_{uv} \qquad (13)$$

$$I_{uv} = \frac{V_1}{Z} = \frac{V_1}{\sqrt{R_{eq}^2 + (\omega L - 1/\omega C)^2}} \qquad (14)$$

$$\phi_i = \phi_1 + \psi = \phi_1 + \tan^{-1}\left(\frac{\omega^2 LC - 1}{\omega R_{eq}C}\right) \qquad (15)$$

where $R_{eq} = (n_1/n_2)^2 R$. ψ expresses the phase difference between v_1 and i_{uv}.

B. Instantaneous Reactive Power

The MC is able to control input current averages $\bar{i}_r, \bar{i}_s, \bar{i}_t$ arbitrarily in T_c by switching. At first, the MC input currents i_r, i_s, i_t are expressed the following with i_{uv} and switching function s_{ij}

$$\begin{bmatrix} i_r \\ i_s \\ i_t \end{bmatrix} = \begin{bmatrix} s_{ru} & s_{rv} \\ s_{su} & s_{sv} \\ s_{tu} & s_{tv} \end{bmatrix}\begin{bmatrix} i_u \\ i_v \end{bmatrix} = \begin{bmatrix} s_{ru} & s_{rv} \\ s_{su} & s_{sv} \\ s_{tu} & s_{tv} \end{bmatrix}\begin{bmatrix} i_{uv} \\ -i_{uv} \end{bmatrix} \qquad (16)$$

where

$$s_{ij} = \begin{cases} 1 & (S_{ij}\ \text{ON}) \\ 0 & (S_{ij}\ \text{OFF}) \end{cases}$$

From Fig. 2 and Eq. (16), in the case of $e_r > e_t > 0 > e_s$, $\bar{i}_r, \bar{i}_s, \bar{i}_t$ are calculated as follows with i_{uv} and ζ_{ij}.

$$\bar{i}_r = \frac{1}{\pi}\int_0^{\pi} i_r d\theta = \frac{1}{\pi}\int_0^{\pi - \zeta_{tu}\pi} i_{uv}d\theta$$

$$\bar{i}_s = \frac{1}{\pi}\int_0^{\pi} i_s d\theta = -\frac{1}{\pi}\int_{\zeta_{tv}\pi}^{\pi} i_{uv}d\theta \qquad (17)$$

$$\bar{i}_t = \frac{1}{\pi}\int_0^{\pi} i_t d\theta = \frac{1}{\pi}\int_{\pi - \zeta_{tu}\pi}^{\pi} i_{uv}d\theta - \frac{1}{\pi}\int_0^{\zeta_{tv}\pi} i_{uv}d\theta$$

The MC controls instantaneous reactive power q to adjust input power factor. A concept of the instantaneous reactive power is introduced by [17]. From [17], q is calculated as follows.

$$\begin{bmatrix} e_\alpha \\ e_\beta \end{bmatrix} = \sqrt{\frac{2}{3}}\begin{bmatrix} 1 & -\frac{1}{2} & -\frac{1}{2} \\ 0 & \frac{\sqrt{3}}{2} & -\frac{\sqrt{3}}{2} \end{bmatrix}\begin{bmatrix} e_r \\ e_s \\ e_t \end{bmatrix} \qquad (18)$$

$$\begin{bmatrix} i_\alpha \\ i_\beta \end{bmatrix} = \sqrt{\frac{2}{3}}\begin{bmatrix} 1 & -\frac{1}{2} & -\frac{1}{2} \\ 0 & \frac{\sqrt{3}}{2} & -\frac{\sqrt{3}}{2} \end{bmatrix}\begin{bmatrix} \bar{i}_r \\ \bar{i}_s \\ \bar{i}_t \end{bmatrix} \qquad (19)$$

$$q = \begin{bmatrix} e_\alpha & e_\beta \end{bmatrix}\begin{bmatrix} 0 & -1 \\ 1 & 0 \end{bmatrix}\begin{bmatrix} i_\alpha \\ i_\beta \end{bmatrix}$$
$$= \frac{1}{\sqrt{3}}\{(e_s - e_t)\bar{i}_r + (e_t - e_r)\bar{i}_s + (e_r - e_s)\bar{i}_t\} \qquad (20)$$

In this study, q is set to 0 in order to keep the input power factor in 1.

C. Soft Switching constraints

Eq. (5) expresses the soft switching constraints. By substituting Eq. (12) into Eq. (5), we obtain

$$\phi_i \geq \phi_r + \cos^{-1}\left(1 - 3\omega C_s \Delta V / I_{uv}\right) \qquad (21)$$

Then, in the case of $e_r > e_t > 0 > e_s$, Eq. (21) is calculated as follows.

$$\phi_i = \zeta_{tv}\pi + \cos^{-1}\left(1 - 3\omega C_s\left(e_t - e_s\right)/I_{uv}\right) + \phi_M \quad (22)$$

where ϕ_M expresses a margin phase to achieve soft switching certainly. It is desired to control the output power factor to be maximum. Considering only the fundamental component, the output power factor is

$$\cos(\phi_i - \phi_1) \qquad (23)$$

From Eq. (23), the power factor is maximum when ϕ_i is minimum, and Eq. (22) indicates lower limit of ϕ_i to achieve soft switching. Thus for the maximum power factor, it is required that $\phi_M = 0$.

D. Control Equations

Regarding I_{uv} and q as constant, it is enable to solve Eqs. (14), (20), and (22). I_{uv}, q are given as references. As an example, we consider the case of $e_r > e_t > 0 > e_s$. From Eqs. (1) and (2),

$$\begin{cases} \zeta_{ru} = 1 - \zeta_{tu}, & \zeta_{su} = 0 \\ \zeta_{rv} = 0, & \zeta_{sv} = 1 - \zeta_{tv} \end{cases} \qquad (24)$$

Considering Fig. 2 and Eq. (24), a_1 and b_1 are calculated as the function of ζ_{tu}, ζ_{tv} from Eqs. (10) and (11). Therefore, Eqs. (8) and (9) show that V_1 and ϕ_1 depend on ζ_{tu}, ζ_{tv}. By substituting Eqs. (12), (14), (15), and (17) into Eq. (20), it is obvious that q is function of ζ_{tu}, ζ_{tv}, and ω. Similarly substituting Eq. (15) into Eq. (22), Eq. (22) is confirmed to be function of ζ_{tu}, ζ_{tv}, and ω. Thus, Eqs. (14), (20), and (22) are simultaneous equations with three unknowns. Then, it is enable to solve these control equations for ζ_{tu}, ζ_{tv} and ω, and obtain all the time ratio ζ_{ij} from Eq. (24).

However, the control equations are very complex and it is difficult to solve mathematically. Thus, the control equations are solved by a numerical solving method. Also, the input phases in neighborhood zero cross cannot have solutions of the control equations. In such case, we insert of zero into one ON time ratio and neglect Eq. (20).

VI. EXPERIMENTAL SYSTEMS

Fig. 5 shows the MC experimental system. The MC unit is MWINV-2R020-MAT made by Myway Plus Corporation and IGBTs are 1MBK50D-060S. The control system consists of A/D converters, DSP and FPGA. This control system generates twelve gate signals based on the references I_{uv}, q and the input voltages.

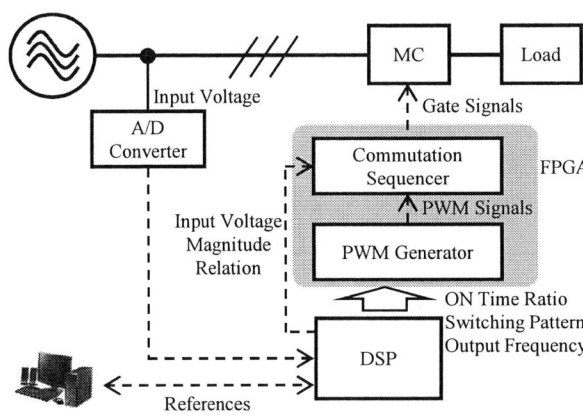

Fig. 5. An experimental system.

A. Calculations in DSP

DSP receives the input voltages from A/D converters and references I_{uv}, q from a terminal. From these values, DSP have to decide the switching pattern and selects the ON time ratio and the output frequency. However, it is difficult to perform the control mentioned above in real time processing. Assuming balanced three-phase input voltage, it is enable to determine ζ_{ij}, f_{out} uniquely when I_{uv}, q, and the input voltages are given. Therefore we solve the control equations in advance and set the solutions ζ_{ij}, f_{out} in a Look-up table. DSP obtains ζ_{ij}, f_{out} by referring to the table.

B. Processings in FPGA

FPGA includes PWM generators and commutation sequencers. Based on information send from the DSP, the PWM generators can generate PWM signals using triangle wave comparison method. The commutation sequencers generate the gate signals to drive the IGBTs from the PWM signals. This system uses four step commutation method [9] shown in Fig. 4. This method is a safe current commutation technique based on the input voltage magnitude. We set the dead time after turn on to 1.0 μs and after turn off to 3.0 μs. Thus, this MC spends 8.0 μs on one commutation.

C. Influence of Magnetizing Inductance

The HF transformer keeps electrical isolation between utility and the load. In this experiment, the transformer has a magnetizing inductance L_m and a leakage inductance L_r whose parameters are shown in Table II. The parameters of the experimental circuit is indicated in Table III. Although L_r is small enough compared with L, it is doubtful that L_m is so large to ignore this effect. The LC resonant circuit has a resonant angular frequency ω_r as

$$\omega_r = \frac{1}{\sqrt{LC}} = 46315 \text{ rad/s} \qquad (25)$$

Considering that the operational frequency is in the neighborhood of the resonant frequency, the impedance of the

TABLE II. THE PARAMETERS OF THE HF TRANSFORMER

Meaning	Symbol	Value
The magnetizing inductance of the transformer	L_m	5.01 mH
The total leakage inductance of the transformer	L_l	22.6 μH
The turn ratio of the transformer	$n_1 : n_2$	41:60

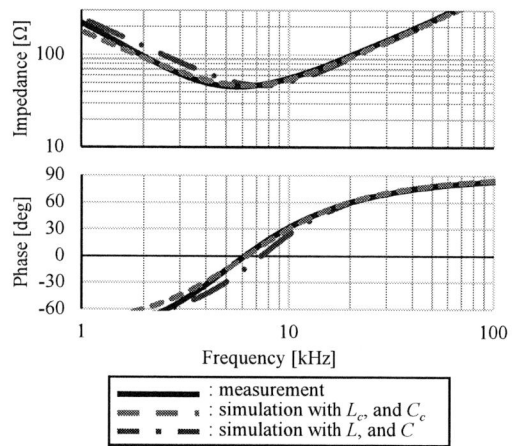

Fig. 6. Frequency characteristics of the output circuit impedance measured by a impedance analyzer, simulated with L, C, and simulated with L_c, C_c.

magnetizing inductance and the load while operation are the following.

$$\omega_r L_m = 232.04 \ \Omega \tag{26}$$

$$R_{eq} = R \left(\frac{n_1}{n_2} \right)^2 = 46.69 \ \Omega \tag{27}$$

From these results, the ignore of L_m is not acceptable. Thus, parameters of the resonant inductance and capacitance in the control system are decided to take effect of L_m into consideration. Since $R_{eq}^2 \ll \omega_r^2 L_m^2$, in the range of the operational frequency, the impedance of the MC output circuit is expressed as

$$R_{eq} + j \left\{ \omega L - \frac{1}{\omega \left(\frac{L_m C}{L_m - R_{eq}^2 C} \right)} \right\} \tag{28}$$

Therefore, the resonant inductance L_c and capacitance C_c in the control system are adjusted as follows.

$$L_c = L = 0.725 \text{ mH} \tag{29}$$

$$C_c = \frac{L_m C}{L_m - R_{eq}^2 C} = 893 \text{ nF} \tag{30}$$

Fig. 6 shows frequency characteristics of the output circuit impedance measured by a impedance analyzer, simulated with L, C, and simulated with L_c, C_c, respectively. The simulation result with L_c, C_c is more fit for the real circuit than the simulation result with L, C. Thus to apply the proposed method for the real circuit, Eqs. (29) and (30) are effective approximation to remove L_m influence.

TABLE III. THE PARAMETERS OF THE CIRCUIT AND THE VALUES OF THE REFERENCES

Meaning	Symbol	Value
The input line to line voltages	e_{rs}, e_{st}, e_{tr}	200 V_{rms}
The resistance	R	100 Ω
The resonant inductance	L	0.725 mH
The resonant capacitance	C	643 nF
The capacitance of snubber circuit	C_s	1 nF
The capacitance at MC input	C_{in}	10 μF
The margin phase	ϕ_M	5 deg
The reference of I_{uv}	I_{uv}	6.5 A
The reference of q	q	0

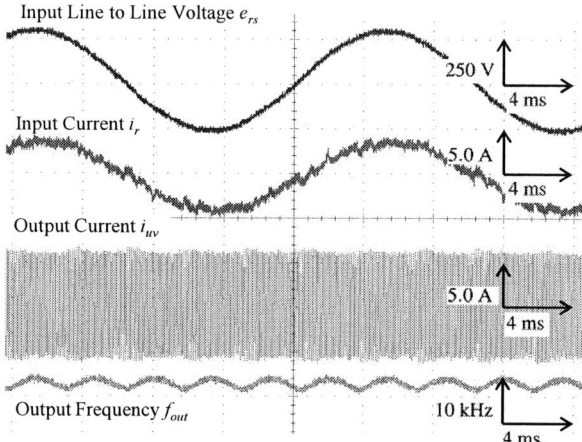

Fig. 7. Waveforms of an input line to line voltage e_{rs}, an input current i_r, an output current i_{uv}, and an output frequency f_{out}.

VII. EXPERIMENTAL RESULTS

Table III shows the parameters of the experimental circuit and the values of the references. The parameters L_c, C_c are decided by Eqs. (29) and (30). Considering an error in computation, the margin is given as $\phi_M = 5$ deg. The experimental results are shown in Fig. 7 and Fig. 8.

The waveforms of an input line to line voltage e_{rs}, an r-phase input current i_r, an output current i_{uv}, and an output frequency f_{out} are shown in Fig. 7. i_r has phase lead due to influence of the input capacitors. Considering this influence, i_r is in-phase with the input phase voltage and this result shows that utility power factor is obtained. i_{uv} has much higher frequency which is equal to switching frequency than i_r frequency. Therefore, it is confirmed that the MC achieves the high-frequency output. f_{out} waveform shows that the MC selects f_{out} suitable for soft switching according to the input voltage phases. Furthermore, although the MC changes f_{out}, a variation of i_{uv} amplitude is much small. This is because variation of f_{out} is much slow compared with i_{uv} frequency, and influence of i_{uv} transient is small.

Fig. 8 shows the enlarged waveforms of i_{uv} and an output voltage v_{uv}. It is clear that the zero cross of i_{uv} lags behind the phase ϕ_r of v_{uv}. Thus, v_{uv} and i_{uv} are satisfied with the soft switching constraints. Additionally, Fig. 8 shows a phase difference between v_{uv} and i_{uv} is very small while achieving soft switching. Therefore, we

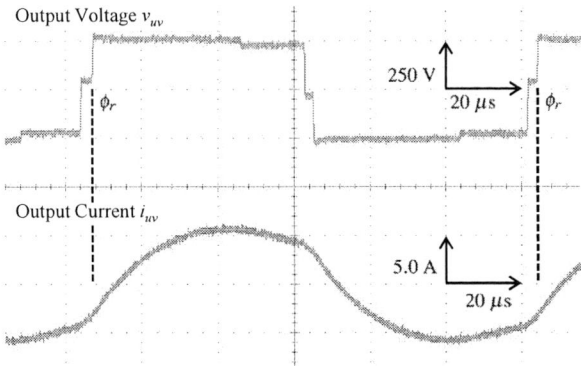

Fig. 8. Enlarged waveforms of an output voltage v_{uv} and the output current i_{uv}.

can ensure that the MC controls the output power factor to be maximum.

From Fig. 7, it is obvious that i_r has a sinusoidal waveform and the amplitude of i_{uv} is nearly constant. This is because assumption of sinusoidal i_{uv} is effectiveness. In fact, Fig. 8 shows that i_{uv} includes a few harmonics, but is almost sinusoidal by the reason of influence of the LC resonant circuit. THD of i_r is equal to 8.16 % and the ripple width of i_{uv} amplitude is 0.8 A.

VIII. CONCLUSION

This paper have described a new control method of a three-phase to high-frequency single-phase MC with an output resonant circuit for insulated AC/DC converters. The output frequency, which is equal to the switching frequency of the MC, is controlled to achieve soft switching on each switching device of the MC, and to maximize power factor of the MC output power for improving efficiency of the MC. An experimental system is composed and tested to confirm validity of the proposed control method. The experimental results show that the three-phase input currents are sinusoidal and in-phase with input phase voltages, and that amplitude of the high-frequency output current is kept constant. Furthermore, it is also indicated that soft switching is achieved and the output power factor is maximized from phase difference between the output voltage and current.

REFERENCES

[1] J. Deng, S. Li, S. Hu, C. C. Mi, and R. Ma, "Design methodology of LLC resonant converters for electric vehicle battery chargers," *IEEE Trans. Veh. Technol.*, vol. 63, no. 4, pp. 1581–1592, may 2014.

[2] S. Inoue and H. Akagi, "A Bi-Directional DC/DC Converter for an Energy Storage System," in *APEC 07 - Twenty-Second Annu. IEEE Appl. Power Electron. Conf. Expo.* IEEE, feb 2007, pp. 761–767.

[3] M. G. Egan, D. L. O'Sullivan, J. G. Hayes, M. J. Willers, and C. P. Henze, "Power-Factor-Corrected Single-Stage Inductive Charger for Electric Vehicle Batteries," *IEEE Trans. Ind. Electron.*, vol. 54, no. 2, pp. 1217–1226, 2007.

[4] P. Wheeler, J. Rodriguez, J. Clare, L. Empringham, and A. Weinstein, "Matrix converters: a technology review," *IEEE Trans. Ind. Electron.*, vol. 49, no. 2, pp. 276–288, apr 2002.

[5] T. Friedli and J. W. Kolar, "Comprehensive comparison of three-phase AC-AC Matrix Converter and Voltage DC-Link Back-to-Back Converter systems," in *2010 Int. Power Electron. Conf. - ECCE ASIA -.* IEEE, jun 2010, pp. 2789–2798.

[6] A. Ishiguro, T. Furuhashi, and S. Okuma, "A novel control method for forced commutated cycloconverters using instantaneous values of input line-to-line voltages," *IEEE Trans. Ind. Electron.*, vol. 38, no. 3, pp. 166–172, jun 1991.

[7] L. Wei and T. A. Lipo, "A novel matrix converter topology with simple commutation," *Conf. Rec. - IAS Annu. Meet. (IEEE Ind. Appl. Soc.*, vol. 3, pp. 1749–1754, 2001.

[8] S. Ratanapanachote, Han Ju Cha, and P. Enjeti, "A digitally controlled switch mode power supply based on matrix converter," *IEEE Trans. Power Electron.*, vol. 21, no. 1, pp. 124–130, jan 2006.

[9] K. Kato and J. I. Itoh, "Improvement of input current waveforms for a matrix converter using a novel hybrid commutation method," in *Fourth Power Convers. Conf. PCC-NAGOYA 2007 - Conf. Proc.*, 2007, pp. 763–768.

[10] D. Casadei, G. Serra, A. Tani, and L. Zarri, "Matrix converter modulation strategies: a new general approach based on space-vector representation of the switch state," *IEEE Trans. Ind. Electron.*, vol. 49, no. 2, pp. 370–381, apr 2002.

[11] N. Nguyen-Quang, D. Stone, C. Bingham, and M. Foster, "Single phase matrix converter for radio frequency induction heating," in *Int. Symp. Power Electron. Electr. Drives, Autom. Motion, 2006. SPEEDAM 2006.* IEEE, pp. 614–618.

[12] J. J. Sandoval, S. Essakiappan, and P. Enjeti, "A bidirectional series resonant matrix converter topology for electric vehicle DC fast charging," in *2015 IEEE Appl. Power Electron. Conf. Expo.* IEEE, mar 2015, pp. 3109–3116.

[13] W. Tabisz, P. Gradzki, and F. Lee, "Zero-voltage-switched quasi-resonant buck and flyback converters-experimental results at 10 MHz," *IEEE Trans. Power Electron.*, vol. 4, no. 2, pp. 194–204, apr 1989.

[14] A. Hadinata, S. Ogasawara, M. Takemoto, and T. Hyodo, "A control method for three-phase to high-frequency single-phase matrix converter based on instantaneous input voltages," in *2015 IEEE 2nd Int. Futur. Energy Electron. Conf.*, 2015, pp. 1–6.

[15] K. Suzuki, W. Kitagawa, and T. Takeshita, "Suppression control of source current harmonics of bi-directional isolated AC/DC converter using soft switching technique," in *2017 IEEE 3rd Int. Futur. Energy Electron. Conf. ECCE Asia (IFEEC 2017 - ECCE Asia).* IEEE, jun 2017, pp. 57–61.

[16] S. D. Simone, C. Adragna, C. Spini, and G. Gattavari, "Design-oriented steady-state analysis of LLC resonant converters based on FHA," in *Int. Symp. Power Electron. Electr. Drives, Autom. Motion, SPEEDAM 2006.*, 2006, pp. 200–207.

[17] H. Akagi, Y. Kanazawa, and A. Nabae, "Instantaneous Reactive Power Compensators Comprising Switching Devices without Energy Storage Components," *IEEE Trans. Ind. Appl.*, vol. IA-20, no. 3, pp. 625–630, 1984.

Two-step commutation
for Isolated DC-AC Converter with Matrix Converter

Shunsuke Takuma*, and Jun-ichi Itoh

Department of Electrical, Electronics and Information Engineering, Nagaoka University of Technology, Nagaoka, Japan
*E-mail: takuma_s@stn.nagaokaut.ac.jp

Abstract— This paper proposes a two-step commutation method for a three-phase-to-single-phase matrix converter. Conventional two-step commutation cannot be applied at all operation regions because a commutation failure due to the detection error of grid voltages still occurs. In the proposed two-step commutation, by modulating only one of two devices in a bi-directional switch and utilizing a zero vector to let the switches naturally turn-off, the commutation failures are avoided completely regardless of the voltage detection error. From experimental results, it is confirmed that the proposed two-step commutation has always safety operation to avoid commutation failure. The input current THD at 10 kW with the proposed two-step commutation is improved by 37% in comparison with the conventional four-step commutation.

Keywords— *Two-step commutation, Current direction estimation, Commutation failure.*

I.   INTRODUCTION

A lot of studies on Electric vehicles (EVs) and Plug-in hybrid vehicles (PHEVs) have been accelerated over the past decade. Compared to the gasoline vehicle, EVs or PHEVs still faces one of the main challenges, i.e. the long battery charging time. In order to solve this problem, high-power low-profile battery chargers are required [1-4]. In [5-9], isolated AC-DC converters using a matrix converter as a medium frequency AC-AC converter connected with the transformer at the primary side have been proposed. The matrix converter volume is expected to be greatly reduced compared to other topologies which employ a buffer capacitor because the buffer capacitor in the high-power application such as the rapid battery charger usually has to withstand a high current, which increases the capacitor volume.

Generally, the matrix converters are required a commutation sequence at the switching timing of the power devices to prevent short-circuit at a voltage source and open-circuit at inductive components. The conventional commutation method is separated into two types, the voltage commutation method based on the input voltage polarity and the current commutation method based on the output current direction [10-13]. The voltage commutation works reliably if the relationship of the input voltages is accurately obtained.

These commutation sequences are the four-step commutation which is divided into the four steps to avoid the open-circuit and the short-circuit. Each step is turned-on or turned-off a switch into the two bi-directional switches depended on the voltage polarity or the current direction. The commutation time which is longer than the switching speed of the switching devices is inserted among the first-step, the second-step, the third-step and the fourth-step. Therefore, the four-step commutation which is the voltage commutation or the current commutation is a complex commutation algorithm which greatly restricts the applicable control hardware. A life time and a reliability of the switching devices are decreased by the commutation failure which is based on the detection error

In order to simplify the control hardware for the matrix converter, two-step commutation methods have been proposed [5-11]. Commutation time in the two-step commutation is half of that in the four-step commutation. This results in a short commutation time with a simpler commutation algorithm. In particular, the two-step commutation in [11] is achieved by zero vectors which is the switching pattern of the additional circuit outputting the zero voltage at the input terminal. Another two-step commutation in [6-10] uses both the voltage polarity and the current direction for the commutation. However, the main problem of these conventional two-step commutations is that either the additional circuit is required or the commutation failures still occurs at the critical area due to a detection error of the input voltage. A solution for this problem is that the commutation method is switched between the conventional two-step commutation and the four-step commutation. However, this solution increases the number of switching step and requires the implement of the two commutation methods.

If the switching pulses are shorter than the dead-time, next commutation sequence starts before present commutation period has been completed and it causes input current distortions. In the low modulation index region, the input current is distorted by the output voltage error due to the conventional several-step commutation. It is necessary to reduce the number of the commutation step in order to reduce the output voltage error.

In this paper, the two-step commutation is proposed for a three-phase to single-phase matrix converter in order to improve the input current total harmonics distortion (THD) in the low modulation index region. The output voltage error is decreased by the proposed two-step commutation due to less commutation. In addition, the commutation failure does not occur regardless of the voltage detection error. The original idea of this paper is to turning-on only one of two devices in the bi-directional switches based on the current direction and to use a voltage vector which let the switches naturally turn-off. Therefore, the proposed two-step commutation is unnecessary to switch between the conventional two-step commutation and another commutation method to prevent the short-circuit at the critical area. The effectiveness of the proposed two-step commutation is evaluated with a 10-kW prototype thought experimental results.

II. CIRCUIT CONFIGURATION AND CONTROL METHOD

Figure 1 shows the isolated AC-DC converter with the three-phase to single-phase matrix converter. The proposed circuit consists of a LC filter to eliminate switching ripple component of the input current, a three-phase to single-phase matrix converter with bi-directional switches, a medium frequency transformer, a diode rectifier, and a smoothing inductor in the output DC side. In particular, the three-phase grid voltage is directly converted to medium frequency single-phase voltage by the matrix converter. Consequently, the volume of the transformer is significantly minimized because this transformer is operated with this medium frequency single-phase voltage.

III. TWO-STEP COMMUTATION METHOD

A. Problem of conventional commutation method

Figure 2 shows the equivalent circuit at each arm of the matrix converter. The equivalent circuit consists of the three voltage sources (maximum-phase-voltage V_{max}, middle phase-voltage V_{mid}, minimum phase-voltage V_{min}), current source which represents as the current at the transformer, and three bi-directional switches. Figure 2(b) and 2(c) show the transition situation from V_{max} to V_{mid} with the four-step voltage commutation and the conventional two-step commutation. If the actual voltage polarity does not agree with the detection voltage polarity, the short-circuit via the grid occurs in the commutation state of the four-step voltage commutation or in the steady state of the conventional two-step commutation after turning-on S_{mr}

B. Principle of two-step commutation

Figure 3 depicts the principle of the proposed two-step commutation, which is divided into two modes. Figure 3(a) shows the transition from an input phase to another input phase, whereas Figure 3(b) shows the transition from zero-vector state to an input phase. As shown in Fig. 3 (a), when the matrix converter starts to transit from

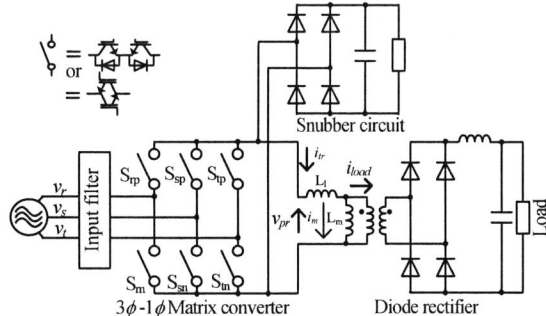

Fig. 1. Isolated AC-DC converter with three-phase-to-single-phase matrix converter.

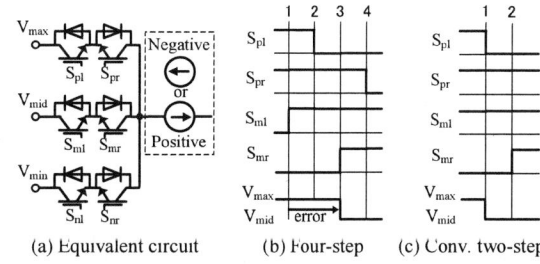

(a) Equivalent circuit   (b) Four-step   (c) Conv. two-step

Fig. 2. Conventional commutation step; If the acutal voltage does not agree with the detection voltage, the short-circuit occur by the commutation failure

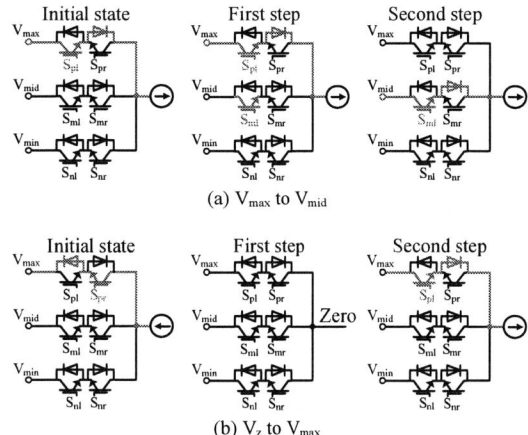

(a) V_{max} to V_{mid}

(b) V_z to V_{max}

Fig. 3. Proposed commutation sequence. The commutation step is always two-step by proposed sequence.

V_{max}-phase to V_{mid}-phase in case of the positive output current, only the switch S_{pr} is on at V_{max}-phase. At first step, only the switch S_{mr} at V_{mid}-phase turns-on. Then, at second step, the switch S_{pr} turns-off and the current commutates from V_{max}-phase to V_{mid}-phase. In this commutation mode, the short-circuit and the open-circuit are avoided regardless of the voltage detection error by the modulation of only one of two devices in the bi-directional switches. Therefore, the transition from an input phase to another input phase without any commutation failures is achieved. The initial state is on state of S_{pl} and then similar switching sequence are applied when the output current direction is negative. The initial state of the equivalent commutation model for each

2597

arm is express by (1)

$$\begin{cases} S_{xl}=1 & S_{xr}=0 & i_{load}>0 \\ S_{xl}=0 & S_{xr}=1 & i_{load}<0 \end{cases} \quad x=p,m,n \quad (1)$$

where subscript x indicates p (V_{max}-phase) or m (V_{mid}-phase) or n (V_{min}-phase) depend on the output line to line voltage V_{pr}. If the output voltage of the matrix converter is V_{max}-V_{min}, x of the upper side arm is p. In addition, x of the lower side arm is n.

Figure 4 shows the output voltage error of the commutation operation with the proposed two-step commutation. The output voltage is delayed by one-step time at the commutation from the high voltage phase to the low voltage phase because the current flow does not change until the second step. In contrast, the output voltage error does not occur when the commutation from the low voltage phase to the high voltage phase because the current flow changes after the first step. In last case, the output voltage is delayed by one-step time when the commutation from the zero vector to any vector. The reason is because the output voltage is zero when all switches is turned-off at first step after the zero vector.

Figure 5 shows the proposed two-step commutation sequence. The output voltage error due to the dead time is considered. The delay of one-step time occurs from the first step to the second step as shown in Fig. 2 (a) when the matrix converter transits from an input phase to another input phase in the positive output current. In the transition from the zero-vector state to another input phase, all switches have to be turned-off at the first step. Similarly, the delay of one-step time occurs from the first step to the second step as shown in Fig. 2 (b). Consequently, the compensation for the voltage error due to the delay of one-step time of the proposed two-step commutation is similar to that in the back-to-back converter, which is significantly simpler than the output voltage error compensation in the four-step commutation.

Figure 6 shows the half-cycle operation of one switching period including commutation in sector I. The relationship of phase-voltage is $v_r > v_s > v_t$.

(i) Initial state: V_1

S_{rp} and S_{nt} are on, whereas the output current direction is positive. Therefore, S_{pr} and S_{tn} can only conduct the current through the diode connected anti-parallel with the switching devices.

(ii) Commutation state V_1 to V_2

At first step, S_{sp} is turned-on. The current flow does not change because the R-phase voltage is higher than the S-phase voltage.

At second step, S_{rp} is turned-off. The output voltage is changed from V_{max}-phase to V_{mid}-phase. This matrix converter successfully outputs vector V_2

(iii) Commutation state V_2 to V_{zero}

At first step, S_{tp} and S_{nr} is turned-on, whereas the output current direction is still positive. At second step, S_{sp} and S_{nt} is turned-off. The output voltage polarity is changed to be opposite of the output current. Thus, the output current quickly decreases to zero. When the output

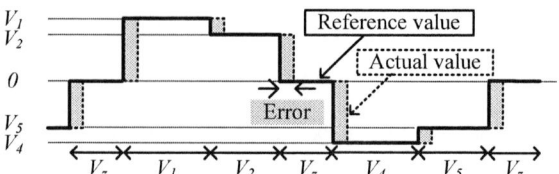

Fig. 4.   Output voltage error with proposed two-step commutation.

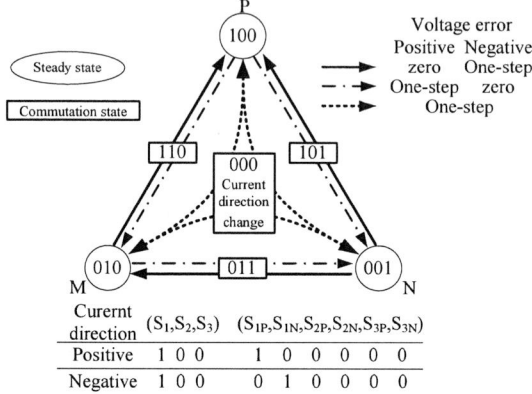

Fig. 5.   Proposed two-step commutation sequence.

Curernt direction	(S_1,S_2,S_3)	($S_{1P},S_{1N},S_{2P},S_{2N},S_{3P},S_{3N}$)
Positive	1 0 0	1 0 0 0 0 0
Negative	1 0 0	0 1 0 0 0 0

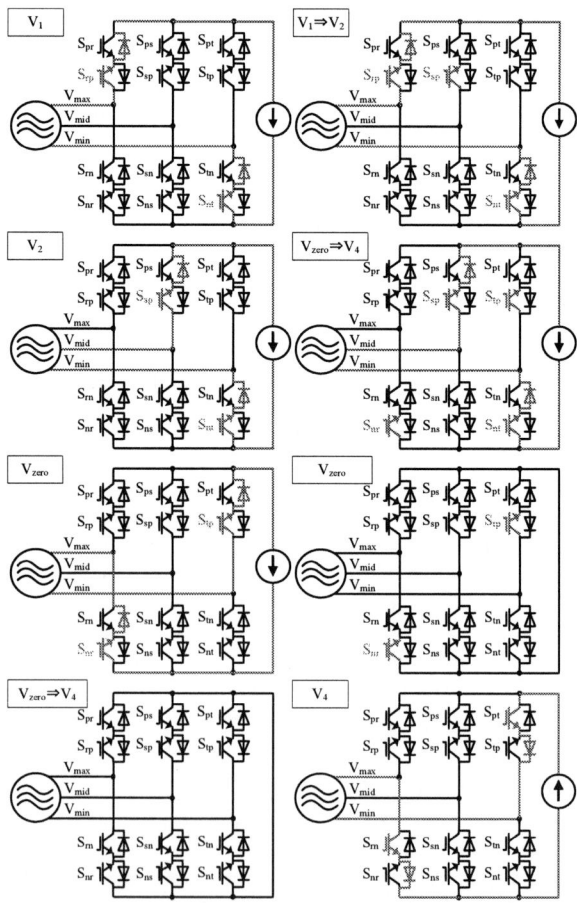

Fig. 6.   Transition situation during half cycle one switching period.

2598

current reaches zero, the output voltage also becomes zero, i.e. zero-vector state.

(iv) Commutation state V_{zero} to V_4

At first step, S_{tp} and S_{nr} can safely turned-off because the output current has become zero. At second step, $\underline{S_{rn}}$ and S_{pt} is turned-on and the matrix converter outputs the negative voltage.

In consequence, the proposed two-step commutation operates the matrix converter without any commutation failures regardless of the detection voltage error at the input voltage.

IV. INPLEMENT OF TWO-STEP COMMUTATION

A. Input current control for matrix converter

Figure 7 shows the space vector modulation (SVM) applied to the three-phase to single-phase matrix converter. The operation mode of SVM is divided by every 60 deg. (Sector I, II, III, IV, V and VI) of the input voltages. Output vectors which are close to the input voltage vector are selected. In sector I, V_1 and V_2 are used during the first half of the control period as the positive voltage, whereas V_4 and V_5 are used during the second half of the control period as the negative voltage. Note that the zero vector V_z which outputs the zero voltage, is decided by the sector. These duty reference T_1, T_2, and T_z are calculated by

$$T_1 = \frac{1}{|A|} \begin{vmatrix} v_\alpha & V_{2\alpha} \\ v_\beta & V_{2\beta} \end{vmatrix} \tag{2}$$

$$T_2 = \frac{1}{|A|} \begin{vmatrix} V_{1\alpha} & v_\alpha \\ V_{1\beta} & v_\beta \end{vmatrix} \tag{3}$$

$$T_Z = 1 - (T_1 + T_2) \left(\because |A| = \begin{vmatrix} V_{1\alpha} & V_{2\alpha} \\ V_{1\beta} & V_{2\beta} \end{vmatrix} \right) \tag{4}$$

B. Current direction estimation

Figure 8 shows the current estimation method to achieve the proposed two-step commutation. The current direction estimation is required for reducing the current sensor which has the high current ratio and the wide bandwidth. The output current is synchronized with the switching carrier by SVM. The switching signal for the zero vector is selected to reduce the output current up to zero. In order to achieve the two-step commutation, the zero current timing at the output side is necessary until

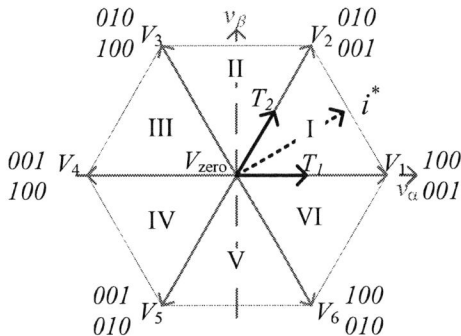

※ The number of "1" means switch is Turn on

Example $\begin{cases} 100 & Srp' = ON, Ssp' = OFF, Stp' = OFF \\ 001 & Srn' = OFF, Ssn' = OFF, Stn' = ON \end{cases}$

Fig. 7. Space vector modulation.

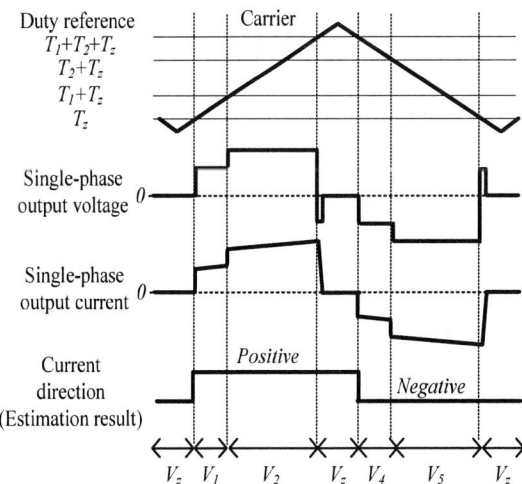

Fig. 8. Estimation principle of current direction. The output current is synchronized the switching carrier by SVM. Therefore, the output current direction is also same relationship.

end of the zero vector. The output voltage is clamped at the grid voltage during the zero vector when the output current thought via grid voltage. After that, the output current and voltage are also zero. The current direction estimation is achieved to keep the zero current at the output side until end of the zero vector.

C. Inplementation of proposed two-step commutation

Table I shows the switching table for the proposed two-step commutation. The switching table depends on the sector and the selected output vector. The switching pulse of the same vector is changed by the sector. According to the principle of the proposed two-step commutation as shown in Fig. 3, the gate signals are

TABLE I. SWITCHING TABLE FOR TWO-STEP COMMUTATION.

Sector	I			II			III			IV			V			VI		
Vector	V_1	V_2	V_z	V_2	V_3	V_z	V_3	V_4	V_z	V_4	V_5	V_z	V_5	V_6	V_z	V_6	V_1	V_z
Switching signal	100	010	001	010	010	001	010	001	100	001	001	100	001	100	010	100	100	010
	001	001	100	001	100	010	100	100	010	100	010	001	010	010	001	010	001	100

2599

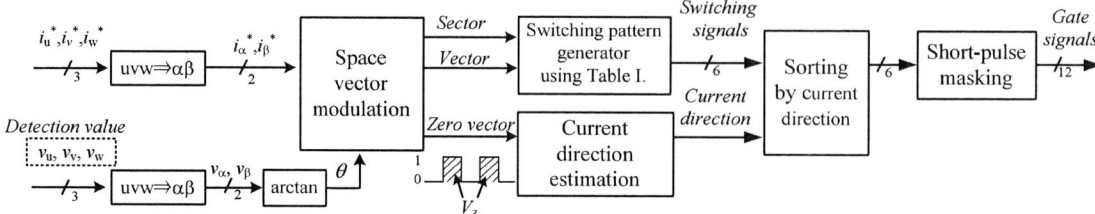

Fig. 9. Signal generation for proposed two-step commutation.

decided for the six bidirectional switches. For example, the output vector V_1 is selected in the sector I, S_{rp}, S_{tp} and S_{tn} are turned-on. S_{sp}, S_{rn} and S_{sn} are turned-off.

Figure 9 shows the gate signal generation for the proposed two-step commutation which requires the complex commutation algorism as the four-step commutation because the proposed two-step commutation only uses a switching table and the current direction estimation. The one of two devices in the bi-directional switches is operated by the gate signals depending on the current direction. Another one of that is turned-off during half cycle of the switching frequency. Finally, the short-pulse in the gate signals is masked because the short-pulse wide more than the sum of the raise time and the turn-on delay time required between the switching and next switching on the three-phase to single-phase matrix converter.

V. EXPERIMENTAL RESULTS

Table II shows the experimental conditions for 10 kW. The switching devices at the three-phase to single-phase matrix converter uses IGBT (MITSUBISHI ELECTRIC: CM400C1Y-24S). The one-step time t_d is decided by the switching characteristics of IGBT. The sum of the raise time and the turn-on delay time of this IGBT is shorter than 1.0 μs. Therefore, the one-step time sets to 1.0 μs.

Figure 10 (a), (b) shows the experimental waveforms of the three-phase to single-phase matrix converter at 10 kW with the four-step commutation method based on the grid voltage polarity (voltage commutation) and with the proposed two-step commutation method respectively. The commutation failure applying the voltage commutation occurs in the regions where the relationship of the grid voltages changes. As a result, the surge current is approximately 220 A. Consequently, the life time and the reliability of the switching devices is decreased. In addition, the input current is distorted by the commutation failure. The proposed method based on estimation of the output current direction achieves to avoid short-circuit at the grid voltage. Consequently, the input current distortion is low value by 2.9%. In addition, surge current is always suppressed by the proposed two-step commutation.

Figure 11(a)-(b) show the experimental waveforms with the four-step voltage commutation, and the extended waveforms at high modulation index, respectively. The input current has low THD by 4.9% and high-power

TABLE II. EXPERIMENTAL CONDITION.

Element	Symbol	Value
Three-phase AC voltage	v_{ac}	200 V
Input frequency	f	50 Hz
Rated output power	P_{out}	10 kW
Carrier frequency	f_c	20 kHz
Leakage inductance	L_l	0.4 μH
Turn ratio of transformer	$N_1:N_2$	1:2.4
Input filter	$L_f(\%Z)$	350 μH(2.3%)
	$C_f(\%Y)$	11 μF(4.7%)
Output filter	L	1.3 mH
	C	30 μF
Commutation time	t_d	1.0 μs

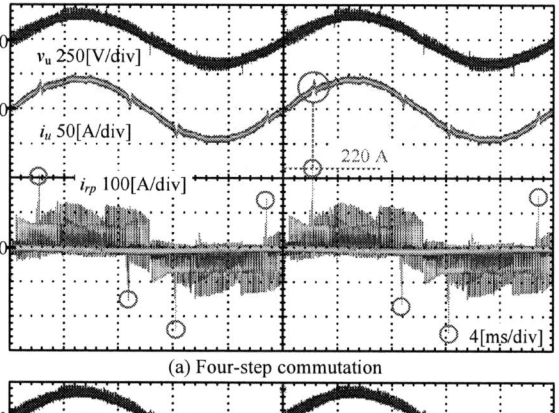

(a) Four-step commutation

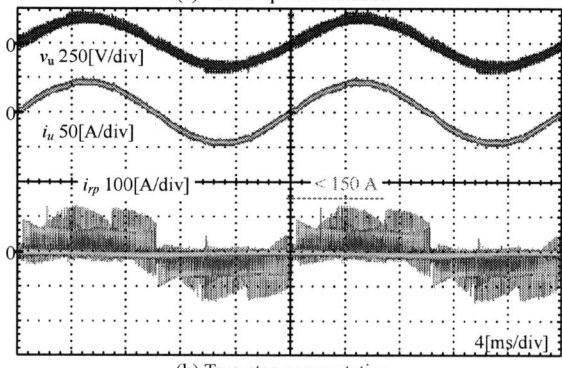

(b) Two-step commutation

Fig. 10. Conparison of device current at matrix converter at rated 10 kW.

factor as 0.99. However, the output current does not zero at zero vector.

Figure 12(a)-(b) show the experimental waveforms with the proposed two-step commutation, and the extended waveforms at high modulation index,

2600

The 2018 International Power Electronics Conference

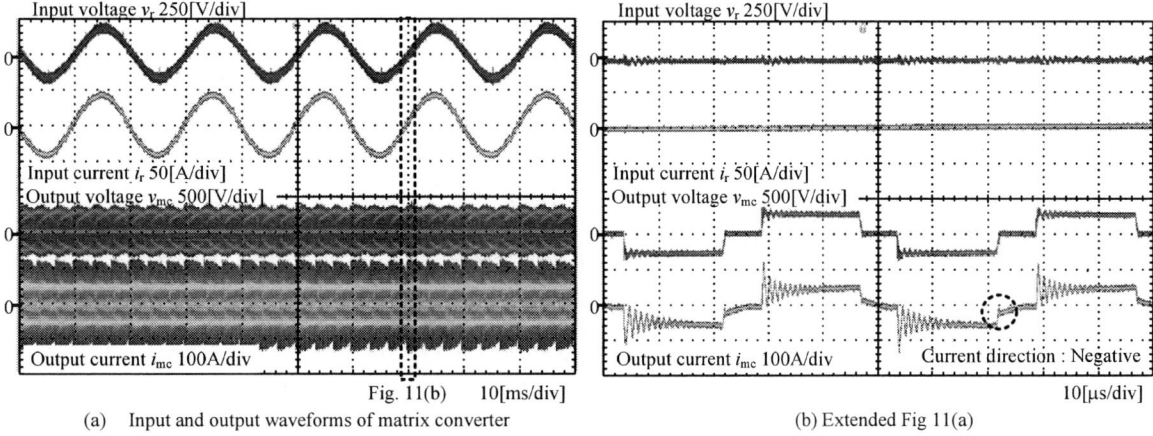

(a) Input and output waveforms of matrix converter (b) Extended Fig 11(a)

Fig. 12. Four-step voltage commutation with high modulation index (MI = 0.85) at rated power 10 kW.

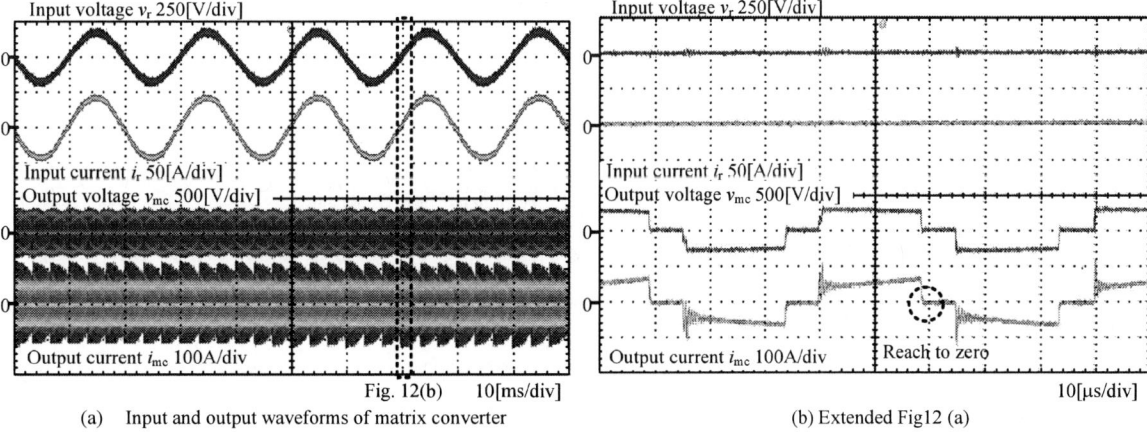

(a) Input and output waveforms of matrix converter (b) Extended Fig12 (a)

Fig. 13. Proposed two-step voltage commutation with high modulation index (MI = 0.85) at rated power 10 kW.

respectively. It is clear that the output current at the zero vector state quickly decreases to zero. Therefore, the surge voltage due to the output current does not occur when all switches turn-off. It confirms from this result that the proposed two-step commutation method operates the matrix converter without any commutation failures regardless of the input voltage detection error.

Figure 13 shows the characteristics of the input current THD with each commutation method at the high modulation index. The input current THD of the proposed two-step commutation is similar to one of the conventional four-step commutation. The commutation step is decreased by the proposed two-step commutation to keep the performance of input current control.

Figure 14(a)-(b) show the experimental waveforms with the conventional four-step voltage commutation, and the extended waveforms at low modulation index, respectively. The input current is distorted by the masking of the short-pulse.

Figure 15(a)-(b) show the experimental waveforms with the proposed commutation, and the extended waveforms at low modulation index, respectively. It confirms from this result that the proposed two-step commutation method operates the matrix converter

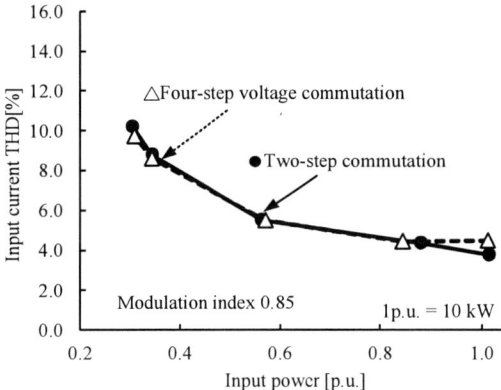

Fig. 11. Comparison of grid current THD at high modulation index. The input current THDs of the proposed two-step commutation and the conventional four-step commutation are also same.

regardless of the modulation index.

Figure 16 shows the distortion characteristics of each commutation methods at the low modulation index. The input current THD of the proposed two-step commutation is 7.7% at 10 kW. In the low modulation index, the proposed two-step commutation has the high performance in comparison with the conventional four-

The 2018 International Power Electronics Conference

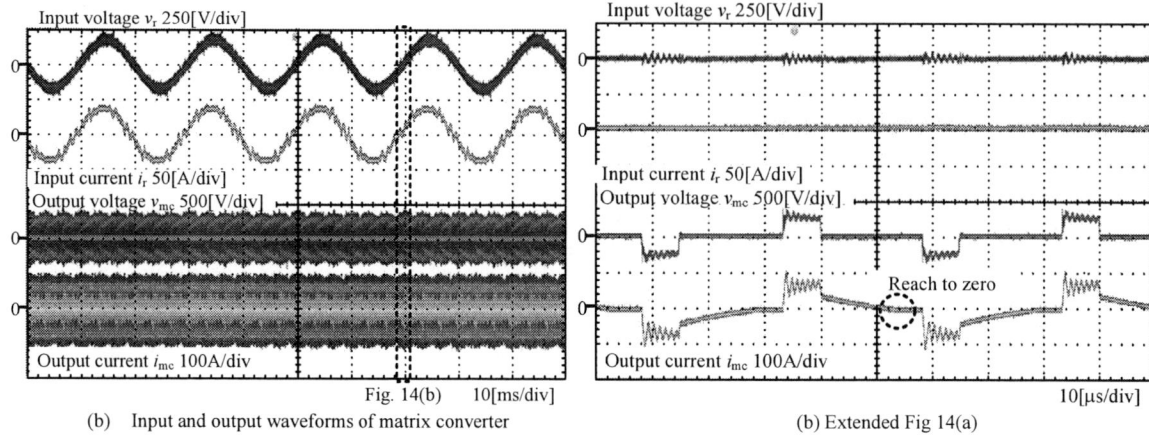

(b)   Input and output waveforms of matrix converter   (b) Extended Fig 14(a)

Fig. 15. Four-step voltage commutation with high modulation index (MI = 0.30) at rated power 10 kW.

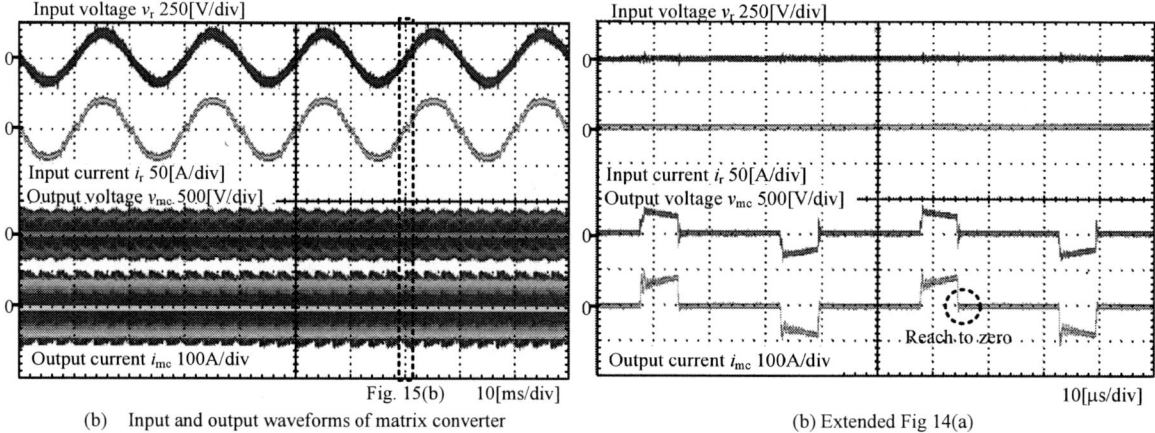

(b)   Input and output waveforms of matrix converter   (b) Extended Fig 14(a)

Fig. 16. Proposed two-step voltage commutation with high modulation index (MI = 0.30) at rated power 10 kW.

step commutation at entire loads.

Figure 17 shows the output voltage error of each commutation method. the output voltage error is less than 0.4% regardless of the modulation index. It is clear that the proposed two-step commutation method is unnecessary to compensate the output voltage error.

Figure 18 shows the distortion characteristics of each commutation methods against the modulation index at rated 10 kW. The input current THD of the proposed two-step commutation is 7.7% at 10 kW. The output voltage range is extended by 36% in case of same input current THD.

VI.   CONCLUSIONS

In this paper, the two-step commutation was proposed in the three-phase to single-phase matrix converter. Compared to the conventional two-step commutation, the proposed two-step commutation is always the safety operation regardless of the voltage detection error. In addition, the commutation algorithm of the proposed two-step commutation is much simpler than the conventional commutation. On the other words, the simple control hardware is employed for the matrix converter with the proposed two-step commutation. The

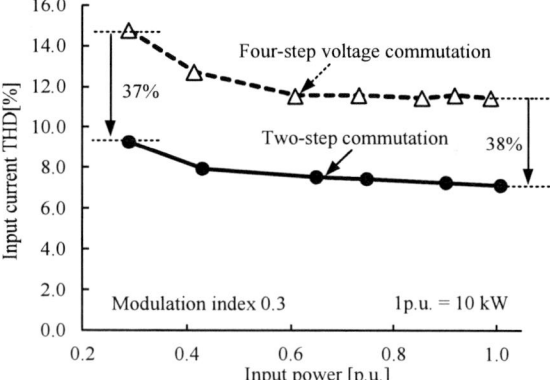

Fig. 14. Comparison of grid current THD with each commutation method at low modulation index. The input current THD is improved by 37 % with the proposed two-step commutation in comparison with the conventional four-step commutation.

input current THD with the proposed two-step commutation is improved by 38% at the low modulation index.

REFERENCES

[1] M. Pahlevaninezhad, P. Das, J. Drobnik, P. K. Jain and A. Bakhshai, "A New Control Approach Based on the Differential Flatness Theory for an AC/DC Converter Used in Electric Vehicles," in IEEE Transactions on Power Electronics, vol. 27, no. 4, pp. 2085-2103, April 2012.

[2] Kohei Aoyama, Naoki Motoi, Yukinori Tsuruta, and Atsuo Kawamura,"High Efficiency Energy Conversion System for Decreaces in Electric Vehicle Battery Terminal Voltage",IEEJ Journal of Industry Applications, vol.5, no.1, pp.12-19, 2016.

[3] S. Kim and F. S. Kang, "Multifunctional Onboard Battery Charger for Plug-in Electric Vehicles," in IEEE Transactions on Industrial Electronics, vol. 62, no. 6, pp. 3460-3472, June 2015.

[4] Fuka Ikeda, Toshihiko Tanaka, Hiroaki Yamada, and Masayuki Okamoto,"Constant DC-Capacitor Voltage-Control-Based Harmonics Compensation Algorithm of Smart Charger for Electric Vehicles in Single-Phase Three-Wire Distribution Feeders",IEEJ J. Industry Applications, vol.5, no.5, pp.405-406, 2016.

[5] R. Huang and S. K. Mazumder, "A Soft-Switching Scheme for an Isolated DC/DC Converter With Pulsating DC Output for a Three-Phase High-Frequency-Link PWM Converter," in IEEE Transactions on Power Electronics, vol. 24, no. 10, pp. 2276-2288, Oct. 2009.

[6] Mahmoud A. Sayed, Kazuma Suzuki, Takaharu Takeshita, Wataru Kitagawa : "PWM Switching Technique for Three-phase Bidirectional Grid-Tie DC-AC-AC Converter with High-Frequency", IEEE Transactions on Power Electronics, 2016

[7] R. Garcia-Gil, J. M. Espi, E. J. Dede and E. Sanchis-Kilders, "A bidirectional and isolated three-phase rectifier with soft-switching operation," in IEEE Transactions on Industrial Electronics, vol. 52, no. 3, pp. 765-773, June 2005.

[8] M. A. Sayed; K. Suzuki; T. Takeshita; W. Kitagawa, "PWM Switching Technique for Three-phase Bidirectional Grid-Tie DC-AC-AC Converter with High-Frequency Isolation," in IEEE Transactions on Power Electronics, vol.PP, no.99, pp.1-1

[9] A. K. Singh; E. Jeyasankar; P. Das; S. Panda, "A Single-Stage Matrix Based Isolated Three Phase AC-DC Converter with Novel Current Commutation," in IEEE Transactions on Transportation Electrification, vol.PP, no.99, pp.1-1

[10] A. Tajfar and S. K. Mazumder, "Sequence-Based Control of an Isolated DC/AC Matrix Converter," in IEEE Transactions on Power Electronics, vol. 31, no. 2, pp. 1757-1773, Feb. 2016.

[11] Lixiang Wei, T. A. Lipo and Ho Chan, "Robust voltage commutation of the conventional matrix converter," Power Electronics Specialist Conference, 2003. PESC '03. 2003 IEEE 34th Annual, 2003, pp. 717-722 vol.2.

[12] T. Schulte and G. Schröder, "Power loss comparison of different matrix converter commutation strategies," 2012 15th International Power Electronics and Motion Control Conference (EPE/PEMC), Novi Sad, 2012, pp. DS2c.9-1-DS2c.9-6.

[13] S. Tammaruckwattana, Chenxin Yue, Y. Ikeda and K. Ohyama, "Comparison of switching losses of matrix converters for commutation methods," 2014 16th European Conference on Power Electronics and Applications, Lappeenranta, 2014, pp. 1-10.

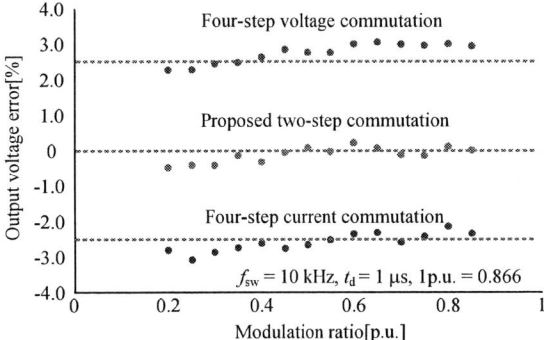

Fig. 17. Comparison of output voltage error.

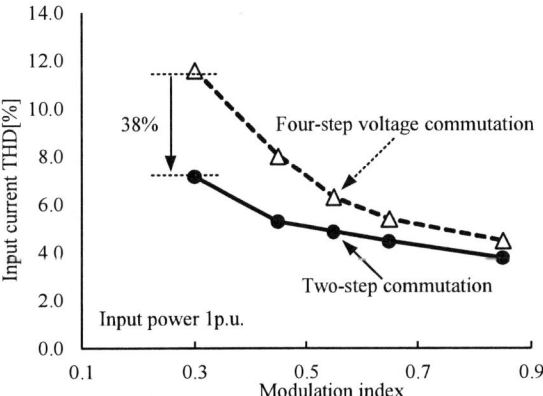

Fig. 18. Comparison of grid current THD with each commutation method at entire modulation index.

The 2018 International Power Electronics Conference

A DC-link Capacitor Voltage Oscillation Reduction Method for a Modular Multilevel Cascade Converter with Single Delta Bridge Cells (MMCC-SDBC)

Takaaki Tanaka[1]*, Huai Wang [2] and Frede Blaabjerg [2]
1 R&D Headquarters, Fuji Electric Co., Ltd, Tokyo, Japan
2 Department of Energy Technology, Aalborg University, Aalborg East, Denmark
*E-mail: tanaka-takaaki@fujielectric.com

Abstract— **This paper proposes a capacitor voltage oscillation reduction method by using third harmonic zero-sequence current for Modular Multilevel Cascade Converter (MMCC) with Single Delta Bridge Cells (SDBC). A practical case study on an 80 MVar/ 33 kV MMCC-SDBC based STATCOM is used to demonstrate the method. The impact of the third harmonic zero-sequence current level of the capacitor oscillation reduction and the electro-thermal stresses on IGBT modules is investigated. An optimal parameter of the current level is obtained by compromising the above two performance factors. The capacitor bank volume is reduced by 23 % by applying the proposed method.**

Keywords— Modular Mutilevel Cacade Converter, STATCOM, Capacitor voltage oscillation reducetion, Zero-sequence current.

I. INTRODUCTION

The Modular Multilevel Cascade Converter (MMCC) family with various circuit topologies has been widely used as a high-voltage and high-power power electronics converter used in HVDC transmission systems, STATCOMs, and motor drive applications. They have significant advantages compared to the conventional topology (e.g. two-level voltage source converter): such as easily scaled-up power and voltage ratings, lower harmonics distortion with low switching frequency for each switch device, transformer-less configuration and also modular/redundancy design [1].

A common disadvantage of each MMCC topology compared with the conventional topology is the dc-link capacitor voltage oscillation with second-order at grid frequency in each cell converter. To suppress the capacitor voltage oscillation below a certain value, the capacitor size becomes larger. In addition, when the carrier frequency of each cell converter for a PWM is designed at lower than a few hundred Hz, the capacitor voltage oscillation significantly increases more by the impact of the harmonic components of the carrier frequency, which is the practical case [2], [3].

Several capacitor voltage oscillation reduction methods for an MMCC with Double Star Chopper Cells

(DSCC) have been presented [4-10]. These methods reduce the voltage oscillation by injecting circulating current with second-order at grid frequency. It is noted that these papers are not considering the impact of the carrier frequency to the capacitor voltage oscillation, as stated previously and this influence cannot ignore where the carrier frequency is lower than a few hundred Hz.

On the other hand, any voltage oscillation reduction methods for other MMCC solutions with Single Star Bridge Cells (SSBC) and Single Delta Bridge Cells (SDBC) have not been proposed. Here, the SSBC and SDBC could inject zero-sequence voltage and zero-sequence current, respectively, instead of the circulating current, so there is a possibility to suppress the capacitor voltage oscillation using these zero-sequence components in both SSBC and SDBC.

In this paper, the capacitor voltage oscillation reduction method for one type of MMCC as a Single Delta Bridge Cells (SDBC) is proposed. Firstly, a practical 80 MVar/ 33 kV STATCOM based MMCC-SDBC for a case study is presented [11]. Secondly, the capacitor voltage oscillation reduction effect by using the proposed third harmonic zero-sequence current injection method is analyzed considering the impact of the low carrier frequency. Thirdly, the electrical losses and junction temperatures of each power semiconductor switch regarding the amplitude of the injected third harmonic zero-sequence current are simulated based on the case study model. Fourthly, optimum third-harmonic zero-sequence current to reduce the capacitor voltage oscillation is presented. Finally, the capacitor bank of each case is designed considering the required capacitance and reliability. The capacitor bank volume reduction effect by applying the proposed method is also discussed.

II. THE THEORETICAL BEHAVIOR OF THE DC-LINK CAPACITOR VOLTAGE

Fig. 1 shows the circuit configuration of the MMCC-SDBC having 80 MVar / 33 kV output as a case study. Table I shows the detail specification of the MMCC-SDBC. Phase-shift PWM modulation is chosen because

The 2018 International Power Electronics Conference

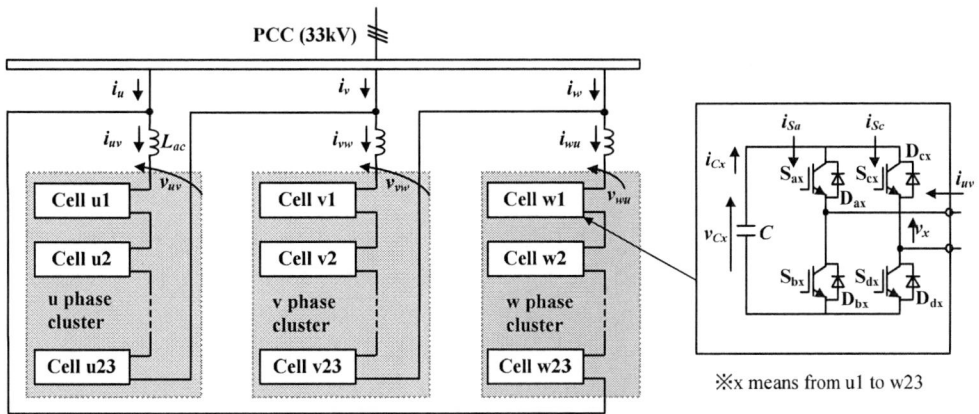

Fig. 1. The circuit configuration of the MMCC-SDBC.

of the advantage that the principal electro-thermal stresses of the IGBT modules and capacitors are equally distributed among cells in the same cluster

Here, the dc-link capacitor voltage of each cell-converter in the MMCC-SDBC with the conventional condition and applied the proposed voltage oscillation reduction method is analyzed theoretically.

A. The conventional operation principle

Fig. 2 shows with a solid line the conventional operation waveforms of the u1-cell converter as an example. When the MMCC-SDBC is in steady state and each dc-link capacitor voltage is balanced, the output voltage reference of each cell in the u-phase cluster e_{m_u} is given by

$$e_{m\_u} = M_a \sin \theta_m + e_{m\_z}$$
$$M_a \equiv \frac{\sqrt{2}\left(V_s \pm \omega_s L_{ac} I / \sqrt{3}\right)}{\left(N_{cell} / 3\right) V_{Cref}} \quad (1)$$

with the modulation factor: M_a, the phase angle of the u-phase cluster voltage: θ_m, each cell output voltage of the zero-sequence component: e_{m_z}, the nominal grid voltage: V_s, the grid frequency: ω_s, the inductance of each phase-cluster AC inductor: L_{ac}, the phase current at the PCC: I, the total cell number: N_{cell}, the dc-link voltage reference of each cell converter: V_{Cref}. The e_{m_z} is zero value in this condition. It is noted that the polarity of the voltage drop of the AC inductor depends on leading/lagging of the STATCOM output current.

The e_{c_u1} is the carrier signal of the u1-cell converter. The ϕ_{c_u1} is the phase delay of the e_{c_u1} from the e_{m_u} which has a zero value in Fig. 2. Each phase delay ϕ_{c_un} of the other cell converters in the u-phase cluster is given by

$$\phi_{c\_un}\Big|_{n=1,2\cdots,N_{cell}/3} = 2\pi \frac{n-1}{N_{cell}/3} + \phi_{c\_u\_ini} \quad (2)$$

with the number of the optional cell in the u-phase cluster: n, the initial phase of the u1-cell: $\phi_{c\_u\_ini}$. As an example, in case of the u1-cell, the phase delay ϕ_{c_u1} was set at 0 with $n = 1$ and $\phi_{c\_u\_ini} = 0$.

TABLE I
THE DETAILED SPECIFICATION OF THE MMCC-SDBC.

Rated power Q_r	± 80 MVA
Nominal grid voltage V_s	33 kVrms
Nominal cell DC-side voltage V_{Cref}	2600 Vdc
Nominal cell AC-side voltage v_x	1450 Vrms
Equivalent switching frequency f_{eq_sw}	10.35 kHz
Rated line current l I_r	808 Arms
Carrier frequency f_c	225 Hz
Grid frequency f_g	50 Hz
Interconnection inductor L_{ac}	7.8 mH (%L=6%)
dc-link capacitance each cell C_x	7.0 mF
IGBT module (rated 4500V / 900A)	MBN900D45A

The switching functions SW_{a_u1}, SW_{b_u1}, SW_{c_u1} and SW_{d_u1} show the on/off states of the power semiconductor switches S_{a_u1}, S_{b_u1}, S_{c_u1} and S_{d_u1}, respectively, which is determined by the PS-PWM of the e_{m_u1} and the e_{c_u1} by neglecting the dead time duration.

The u1-cell output current, as well as u-phase cluster current i_{uv}, is defined as

$$i_{uv} = \sqrt{\frac{2}{3}} I \sin\left(\theta_m + \phi_{pf}\right) + i_{zero} \quad (3)$$

where ϕ_{pf} is the power factor of the MMCC-SDBC, i_{zero} is the zero-sequence current. It is noted that the zero-sequence current i_{zero} is controlled to zero value where the all capacitor voltages are balanced.

The dc-link capacitor current i_{Cu1} of the u1-cell is given by using the switching functions as follows:

$$i_{Cu1} = -i_{uv}\left(SW_{a\_u1} - SW_{c\_u1}\right) \quad (4)$$

By integrating (4), the dc-link capacitor voltage v_{cu1} of the u1-cell is expressed by (5) with the initial value of the v_{cu1}: V_{Cu1_ini}.

Fig. 3(a) plots the v_{Cu1} based on the analytic model (5) and the simulation, respectively, under with $\phi_{c\_u1} = -3.11$ rad, $V_{Cu1\_ini} = 2480$ V as an example. The simulation condition is shown in section III. The analytic model corresponds reasonably well with the simulation result.

2605

The 2018 International Power Electronics Conference

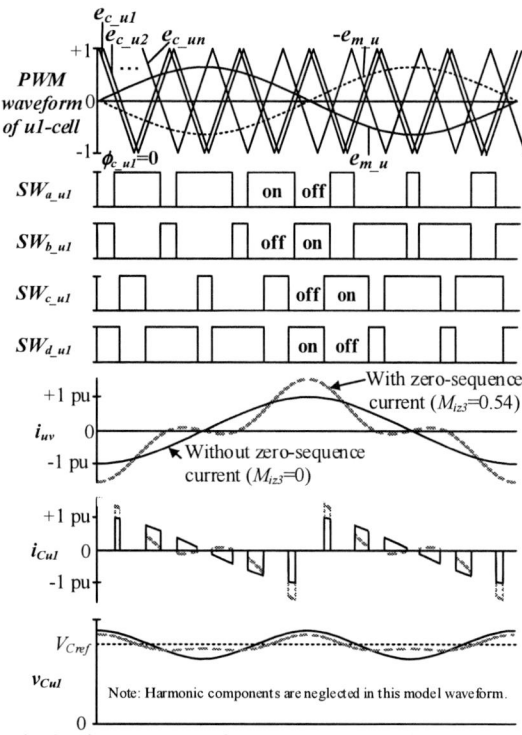

Fig. 2. The PWM waveforms and current in the u1-cell.

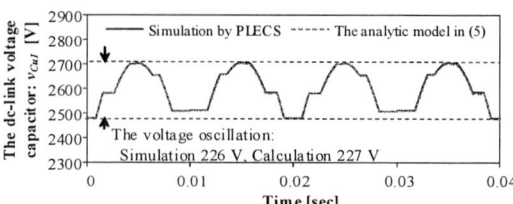

Note: $M_a = 0.827$ p.u., $M_{a3} = 0$ p.u. $\phi_{c\_u1} = -3.11$ rad, $V_{c\_ini} = 2480$ V
(a) Conventional operation as the $M_{iz3} = 0$ p.u.

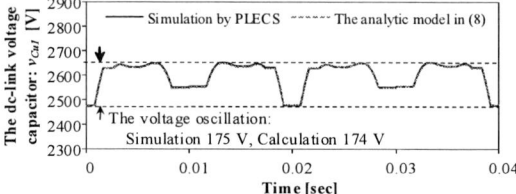

Note: $M_a = 0.827$, $M_{a3} = 0.0702$ p.u. $\phi_{c\_u1} = -3.11$ rad, $V_{c\_ini} = 2477$ V
(b) Proposed operation with $M_{iz3} = 0.5$ p.u.
Fig. 3. The u1-cell capacitor voltage waveforms.

where M_{iz3} is the amplitude ratio of the izero at the I, ϕ_{iz3} is the phase difference from θ_m.

In order to inject the i_{zero}, each cell converter requires additional output voltage. The additional output voltage reference e_{m_z} is given by

$$e_{m\_z} = M_{a3} \sin\left(3\theta_m + \frac{\pi}{2} + \phi_{iz3}\right) \tag{7}$$

$$M_{a3} = \sqrt{6}\omega_m L_{ac} I M_{iz3}$$

With regard to (1), (3), (4), (6) and (7), the v_{Cu1} injected the i_{zero} can be written as (8).

The M_{iz3} and ϕ_{iz3} to reduce the dc-link capacitor voltage are considered by using (8). For the sake of simplicity, only the first term in (8) is focused here. It is noted that the third and fourth term in (8) is the harmonic component associated with the carrier frequency f_c. Because the M_{a3} is much smaller than M_a, the second term in (8) is much smaller than the first term. It is obvious that, when ϕ_{iz3} is ϕ_{pf}, the oscillation of the v_{Cu1} by the double frequency of the f_g can be suppressed in response to the M_{iz3}.

B. The proposed capacitor voltage oscillation reduction method

Fig. 2 shows with a solid line the conventional operation waveforms of the u1-cell converter as an example. The i_{zero} is normally controlled at zero value where the all capacitor voltages are balanced. However, it is possible to output an optional value. A third harmonic zero-sequence current which has same angle and amplitude among the three phase clusters to be able to reduce the dc-link capacitor voltage oscillation is considered below.

The i_{zero} is defined as

$$i_{zero} = \sqrt{\frac{2}{3}} I M_{iz3} \sin(3\theta_m + \phi_{iz3}) \tag{6}$$

$$v_{Cu1} = -\frac{1}{C}\int i_{Cu1} dt + V_{Cu1\_ini} = -\sqrt{\frac{2}{3}}\frac{IM_a}{4\omega_m C}\sin(2\omega_m t + \phi_{pf}) \tag{5}$$

$$+\frac{1}{C}\int \sqrt{\frac{2}{3}} I \sin(\omega_m t + \phi_{pf}) \sum_{n=1}^{\infty} \frac{4}{n\pi}\cos\left(\frac{n\pi}{2}\right)\sin\left[\frac{n\pi}{2}\left\{M_a \sin\omega_m t + M_{a3}\sin\left(3\omega_m t - \phi_{iz3} + \frac{\pi}{2}\right)\right\}\right]\cos(n\omega_c t - n\phi_c)dt + V_{Cu1\_ini}$$

$$v_{Cu1} = -\sqrt{\frac{2}{3}}\frac{IM_a}{4\omega_m C}\left\{\sin(2\omega_m t + \phi_{pf}) - M_{iz3}\sin(2\omega_m t + \phi_{iz3}) + \frac{M_{iz3}}{2}\sin(4\omega_m t + \phi_{iz3})\right\}$$

$$+\sqrt{\frac{2}{3}}\frac{IM_{a3}}{4\omega_m C}\left\{\sin\left(2\omega_m t - \phi_{iz3} + \frac{\pi}{2} - \phi_{pf}\right) - \frac{1}{2}\sin\left(4\omega_m t - \phi_{iz3} + \frac{\pi}{2} + \phi_{pf}\right) - \frac{M_{iz3}}{3}\cos\left(6\omega_m t + \frac{\pi}{2}\right)\right\}$$

$$+\sqrt{\frac{2}{3}}\frac{I}{C}\int\left\{\sin(\omega_m t + \phi_{pf}) + M_{iz3}\sin(3\omega_m t + \phi_{iz3})\right\}\sum_{n=1}^{\infty}\frac{4}{n\pi}\cos\left(\frac{n\pi}{2}\right)\sin\left[\frac{n\pi}{2}\left\{M_a \sin\omega_m t + M_{a3}\sin\left(3\omega_m t - \phi_{iz3} + \frac{\pi}{2}\right)\right\}\right]$$

$$\cos(n\omega_c t - n\phi_{c\_u1})dt + V_{Cu1\_ini} \tag{8}$$

2606

Fig. 3(b) plots the v_{Cu1} based on the analytic model (8) and the simulation, respectively, under with ϕ_{C_u1} = - 3.11 rad, V_{Cu1_ini} = 2477 V as an example. These waveforms correspond reasonably well. The error of the oscillation amplitude of the v_{Cu1} is 3%.

C. The dc-link capacitor voltage oscillation

The capacitor voltage oscillation Δv_{Cu1} of the u1-cell is calculated with numeric calculations using equation (9) substituted (5) or (8) which period is a least common multiple among f_g and f_c.

$$\Delta v_{Cu1} = \max(v_{Cu1}) - \min(v_{Cu1}) \qquad (9)$$

Fig. 4 shows with a dot line plot of the capacitor voltage oscillation regarding of the M_{iz3} based on the analytic model (9) with ϕ_{c_u1} = -3.11 rad. The other condition is the same as Fig. 3. The circle mark shows the Δv_{Cu1} based on the simulation with the same condition of the dot line. These results correspond reasonably well with the maximum error of the Δv_{Cu1} = 3.7% at M_{iz3} = 0.6. The capacitor voltage oscillation of the other cells can also be calculated by using the same equation (8) with the different ϕ_{c_un}. The maximum and minimum capacitor voltage oscillation Δv_{c_max} and Δv_{c_min} regarding the ϕ_{c_un} are plotted by the solid lines, respectively. As an example, the capacitor voltage oscillation is reduced 24 % by injecting the M_{iz3} of 0.5 p.u.

III. POWER SEMICONDUCTOR LOSSES AND JUNCTION TEMPERATURES

The electrical and thermal simulations of the IGBT modules in the MMCC-SDBC regarding of the M_{iz3} are implemented by using the simulation software PLECS. The specification of the MMCC - SDBC is shown in Fig. 1 and Table 1. The reactive current reference is set to be the rated 1 p.u. The output current and the capacitor voltage control method are applied [12]. The junction temperatures are also simulated by using PLECS. The electrical-loss parameters and thermal impedance between junction and heatsink are based on Foster RC network, which are selected from the datasheets for the power modules [18]. As the temperature of the heatsink is normally much lower and more stable compared with the junction temperature in a properly designed converter system, the heat sink temperature is considered as a constant value at 60℃ in this paper.

Fig. 5 shows the total semiconductor losses and electrical efficiency of the MMCC-SDBC with respect to the M_{iz3}. It is noted that the total semiconductor loss increases when M_{iz3} is above 0.3 p.u. because the r.m.s. value of each cluster current increases. However, the total semiconductor loss is slightly reduced when M_{iz3} is below 0.3 because of the V_{ce} - I_c characteristics of the used bipolar devices.

Fig. 6 shows the instant peak and average junction temperature of the IGBT and Diode having maximum thermal stress among the 276 modules with respect to M_{iz3}. It is noted that the average junction temperature exhibits similar trends as the total loss. On the other hand,

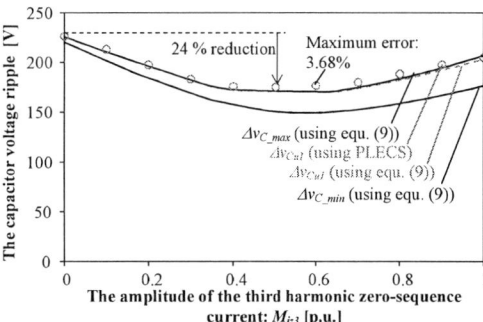

Fig. 4. The u1-cell capacitor voltage oscillation in respect to the amplitude ratio M_{iz3}.

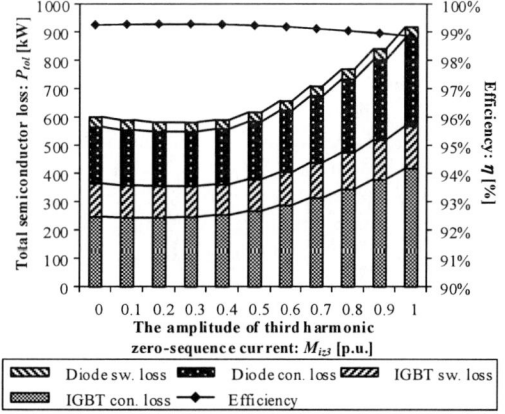

Fig. 5. The total semiconductor loss of the MMCC-SDBC.

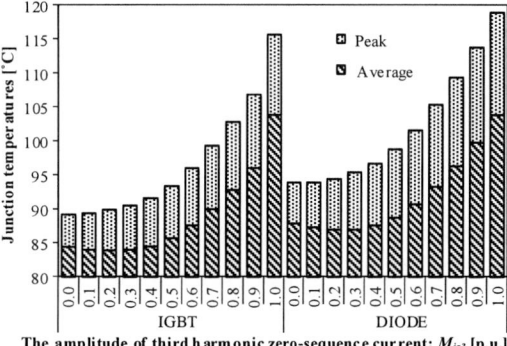

Fig. 6. The junction temperature of an IGBT and Diode switch chosen as the maximum in the all switches of the MMCC-SDBC.

the peak junction temperature increases as a function of the M_{iz3} because the peak value of the phase-cluster current increases regarding of the M_{iz3}.

Fig. 7 shows the total semiconductor losses and the capacitor voltage oscillation regarding of the M_{iz3} standardized at the zero value of the M_{iz3}. The M_{iz3} is selected to be 0.4 p.u. where the capacitor voltage oscillation could be reduced maximum without any total power semiconductor loss increase compared with the conventional method (M_{iz3} = 0).

2607

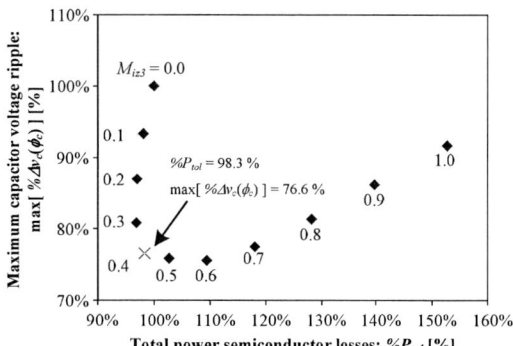

Fig. 7. The total semiconductor loss and the capacitor voltage oscillation as a function of the M_{iz3}.

Fig. 8 shows the electrical and thermal simulation waveforms with $M_{iz3} = 0$ (the conventional operation) and 0.4 (the proposed operation), respectively. The output phase currents between these cases are controlled adequately. The peak value of the cluster current for the proposed case increases 40 % because of injected the third harmonic zero-sequence current. In this result, the peak junction temperature of the power module in u1-cell increases 2 K. The other junction temperatures in the other cell converters have similar value because of selected Phase Shift PWM modulation. The capacitor voltage ripple for the proposed case decreases at 20 %. These waveforms show no evidence of abnormalities.

IV. CAPACITOR BANK DESIGN

The final purpose of the proposed capacitor voltage oscillation reduction method is the reduction of the entire capacitor bank size. Surely, the required capacitance is reduced by applying the proposed method, but the current through the capacitor bank is changed by the injected third harmonic zero-sequence current. The capacitor bank size depends not only on the required capacitance but also the current ripple through the capacitor bank [13]. In order to clarify the capacitor bank size reduction effect of the proposed method, the capacitor bank is designed in this section.

A. Capacitor bank structure

Table II shows the target specification of the capacitor bank for the MMCC-SDBC having the specification given by Table I. The type of the dc-link capacitor is selected to be metalized polypropylene film capacitor (MPPF-Cap), which is commonly chosen for the medium and high voltage class MMCC solutions because of high reliability [14], [15]. The capacitance of the dc-link capacitor bank with the conventional operation and the proposed method are set to 7 mF and 5.4 mF, respectively where the capacitor voltage ripples are suppressed to be less than 10%.

Fig. 9 shows the dc-link capacitor bank structure. The capacitor bank is constructed by series and parallel connection of a MPPF-Cap. Table III shows the capacitor bank specification of both the conventional and the proposed case. The capacitor bank is selected to use 970

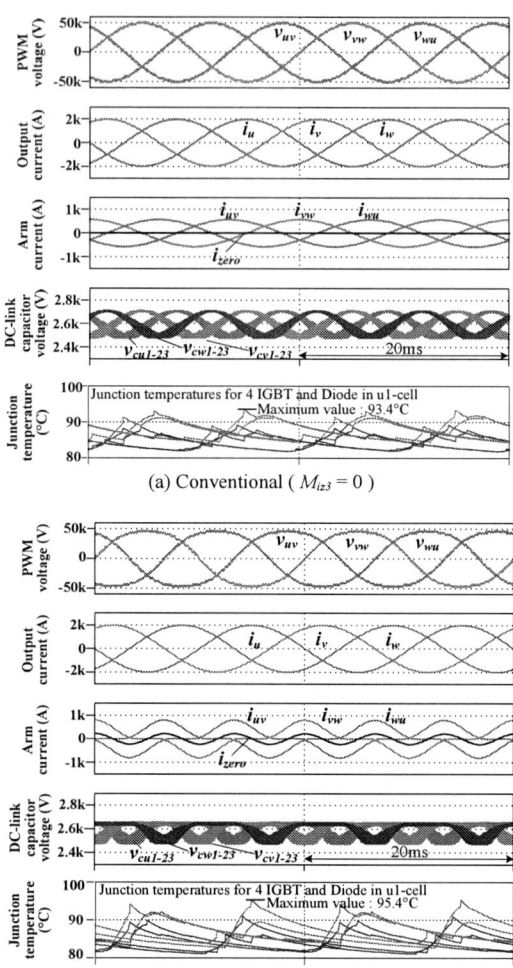

(a) Conventional ($M_{iz3} = 0$)

(b) Proposed with $M_{iz3} = 0.4$

Fig. 8. Key simulation waveforms of the MMCC-SDBC.

TABLE II
THE TARGET SPECIFICATION OF THE CAPACITOR BANK.

	Conventional method	Proposed method
Capacitor type	Metallized polypropylene film capacitor	
Rated dc voltage	2600 Vdc	
Capacitance	7 mF or more	5.4 mF or more
Expected life time	20 years or more	
Ambient temperature	45 °C	

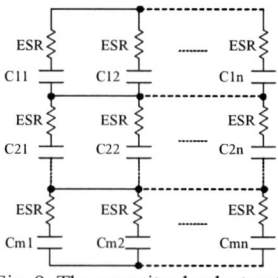

Fig. 9. The capacitor bank structure.

μF, 1000 Vdc, Type 947C MPPF-Cap from Cornell Dubilier [16]. In order to overcome the applied dc voltage to the capacitor bank, the series connection

TABLE III
THE CAPACITOR BANK SPECIFICATION.

	Conventional case	Proposed case
Used capacitor	970 μF, 1000 Vdc, Type 947C Polypropylene Film DC-Link Capacitors from Cornell Dubilier	
Series connection count: m	3	3
Parallel connection counts: n	22	17
Total count: $m \times n$	66	51
Total capacitance	7.11 mF	5.50 mF
Rated voltage	3000 Vdc	
Total capacitor volume	0.147 m³	0.113 m³

number is designed to be 3 for both cases. In order to provide the capacitor bank with the required capacitance, the parallel counts of the conventional and the proposed case are designed 22 and 17, respectively. It is noted that the capacitor bank volume is reduced 23% by applying the proposed method in this case study.

B. The failure rate of the individualf MPPF-Cap

Generally, MPPF-Caps have very long lifetime under the adequate operating condition including the applied dc-voltage, the core temperature of the capacitor and the humidity compared with Aluminum Electrolytic Capacitors, which have wear out failure modes. Such kind of components, it is important to estimate the reliability by using a failure rate. The failure rate means the mean number of failures per unit time [17]. Many manufacturers have investigated their own failure rate calculation method for their products. In this paper, the failure rate of the individual capacitor used in the entire capacitor bank under the conventional and proposed case is estimated by using the provided information from the capacitor manufacturer.

Fig. 10 and Fig. 11 show the current waveforms through the dc-link capacitor bank of the u-1 cell and an FFT analysis with or without the proposed method. The simulation conditions are the same as shown in Fig. 8. By applying the proposed method, the main current ripple component as the double fundamental frequency is reduced by 33% and distributes to higher frequency components. The r.m.s. values of the current through the capacitor bank in the conventional and proposed case are 473 Arms and 426 Arms, respectively.

In order to simplify the calculation, it is assumed that the current through each individual capacitors in the capacitor bank is balanced. The power loss of each capacitor is given by

$$P_{loss} = \sum_f ESR(f) \cdot I_C^2(f) \qquad (10)$$

where $ESR(f)$ is the Equivalent Series Resistance of the capacitor regarding of frequency f, I_C is the r.m.s. current through each capacitor in respect to f. The core temperature of the capacitor is given by

$$T_C = T_a + P_{Loss} \cdot R_{th} \qquad (11)$$

where T_a is the ambient temperature around the capacitor, R_{th} is the thermal resistance of the capacitor. The failure rate of the individual capacitor λ is calculated by using the provided method by the manufacturer, which is a function of the rated voltage of the capacitor V_r, the

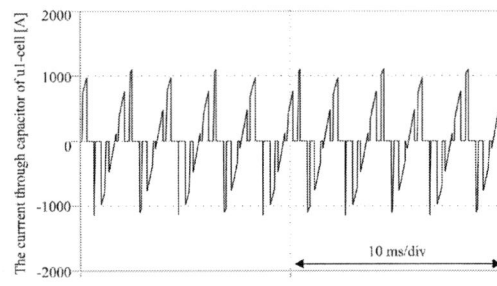

(a) Current waveform through the capacitor bank (473 Arms)

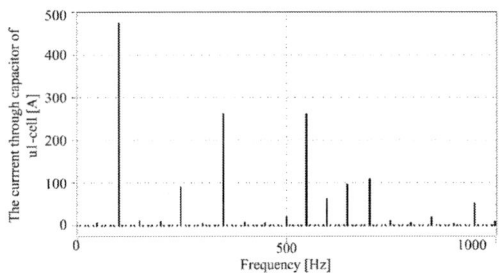

(b) FFT analysis

Fig. 10. The current through the capacitor bank under conventional condition.

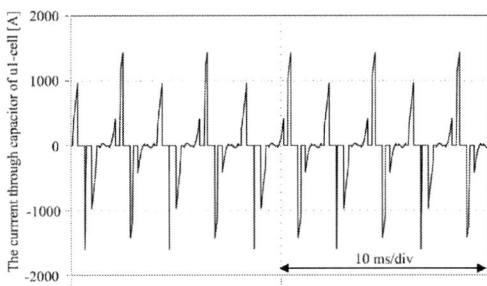

(a) Current waveform through the capacitor bank (426 Arms)

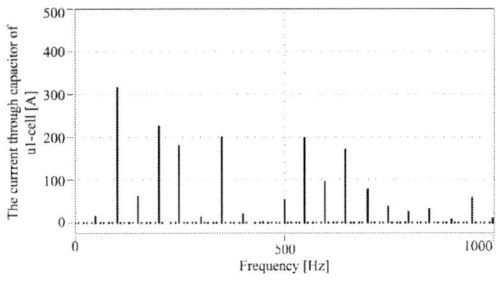

(b) FFT analysis

Fig. 11.The current through the capacitor bank using proposed method.

applied voltage of the capacitor V_a and the hotspot temperature of the capacitor, which is as follows:

$$\lambda = \lambda \left(\frac{V_a}{V_r}, T_C \right) \qquad (12)$$

Table IV shows the P_{loss}, T_c and failure rate of the individual capacitor between the conventional and proposed case. The power loss and hot spot temperature of the proposed capacitor case have slightly a higher value because of the smaller parallel capacitor count. In this result, the failure rate of the proposed case slightly gets worse by 5 FIT (17 %) compared to the conventional capacitor case. Here, the FIT means failures in time, and 1 FIT is 10^{-9} failure/hour.

C. The expected life time of the entire capacitor bank

The expected lifetime of the entire capacitor bank is estimated by using Mean Time to Failure (MTTF). The MTTF means expected time before a failure occurs. It gives the average time in which an item operates without failing. The MTTF of the entire capacitor bank $MTTF_{CB}$ is given by

$$MTTF_{CB} = \frac{1}{\lambda_{CB}} \qquad (12)$$

where λ_{CB} is the failure rate of the capacitor bank.

The λ_{CB} is calculated based on a Part-Count reliability model, which assumes any fault occurred to each of the components will cause the overall system to fail. The λ_{CB} is given by

$$\lambda_{CB} = \sum_{y=1}^{n} \left(\sum_{x=1}^{m} \lambda_{xy} \right) \qquad (13)$$

Table V shows the MTTF of the entire capacitor bank for each case. It is noted that the MTTF of the proposed case is better than the conventional case, because of the construction is using smaller total capacitor count. However, the MTTF of both cases is enough high value, and the satisfied target which is expected to be longer than 20 years.

V. CONCLUSIONS

In this paper, a capacitor voltage oscillation reduction method for an MMCC-SDBC by injecting the third harmonic zero-sequence current is proposed. The capacitor voltage oscillation is analyzed, considering the impact of the harmonics by the practical low carrier frequency. The electrical loss and junction temperature of each power semiconductor switch regarding of the amplitude of the zero-sequence current M_{iz3} are analyzed. In the case study, the capacitor voltage oscillation is reduced by 23 % without increasing the total power semiconductor losses at $M_{iz3} = 0.4$ p.u for the proposed method. The capacitor bank volume is also reduced 23 % by applying the proposed method.

REFERENCES

[1] H. Akagi, "Classification, Terminology, and Application of the Modular Multilevel Cascade Converter (MMCC), " *IEEE Trans. on Power Electron.*, vol. 26, no. 11, pp. 3119-3130, Nov. 2011.

[2] K. Ilves, L. Harnefors, S. Norrga, and H. Nee, "Analysis and Operation of Modular Multilevel Converters With Phase-Shifted Carrier PWM," *IEEE Trans. on Power Electron.*, vol. 30, no. 1, pp. 268-283, Jan. 2015.

[3] E. Behrouzian, M. Bongiorno, and R. Teodorescu, "Impact of Switching Harmonics on Capacitor Cells Balancing in Phase-Shifted PWM-Based Cascade H-Bridge STATCOM," *IEEE Trans. on Power Electron.*, vol. 32, no. 1, pp. 815-824, Jan. 2017.

[4] A. Marzoughi, R. Burgos, D. Boroyevich, and Y. Xue, " Analysis of Capacitor Voltage Ripple Minimization in Modular Multilevel Converter Based on Average Model," in *Proc. Control and Modeling for Power Electronics (COMPEL)*, 2015, pp. 1-7.

TABLE IV
FAILURE RATE CALCULATION OF EACH CAPACITOR.

	Conventional case	Proposed case
Power loss	1.09 W	1.45 W
Hot spot temperature	48.3 °C	49.4 °C
Failure rate	34.14 FIT	39.95 FIT

TABLE V
THE MTTF OF THE ENTIRE CAPACITOR BANKS.

	Conventional case	Proposed case
MTTF	50.66 Years	56.03 Years

[5] H. R. Parikh, R. S. M. Loeches, G. Tsolaridis, R. Teodorescu, L. Mathe, and S. Chaudhary, "Capacitor voltage ripple reduction and arm energy balancing in MMC-HVDC," in *Proc. International Conference on Environment and Electrical Engineering (EEEIC)*, 2016, pp. 1-6.

[6] H. Kim, S. Kim, Y. Chung, D. Yoo, C. Kim, and K. Hur, "Operating Region of Modular Multilevel Converter for HVDC With Controlled Second-Order Harmonic Circulating Current: Elaborating P-Q Capability," *IEEE Trans. on Power Delivery*, vol. 31, no. 2, pp. 493-502, Apr. 2016.

[7] K. Li, C. Li, F. C. Lee, M. Mu, and Z. Zhao, "Precise Control Law of MMC and Its Application in Reducing Capacitor Voltage Ripple by Injecting Circulating Current," in *Proc. International Conference on Electrical Machines and system (ICEMS)*, 2015, pp. 371-377.

[8] C.D. Townsend, R. Aguilera, P. Acuna, G. Konstantinou, J. Pou, G. Mirzaeva, and G.C. Goodwin, "Capacitance Minimization in Modular Multilevel Converters: Using Model Predictive Control to Inject Optimal Circulating Currents and Zero-Sequence Voltage," in *Proc. Annual Southern Power Electronics Conference (SPEC)*, 2016, pp. 1-6.

[9] M. A. Perez, S. Bernet, "Capacitor Voltage Ripple Minimization in Modular Multilevel Converters," in *Proc. International Conference on Industrial Technology (ICIT)*, 2015, pp. 3022-3027

[10] J. Pou, S. Ceballos, G. Konstantinou, V. G. Agelidis, R. Picas, and J. Zaragoza, "Circulating Current Injection Methods Based on Instantaneous Information for the Modular Multilevel Converter," *IEEE Trans. on Industrial Electronics*, vol. 62, no. 2, pp. 777-788, Feb. 2015

[11] T. Tanaka, H. Wang, K. Ma, and F. Blaabjerg, "Reactive Power Compensation Capability of a STATCOM based on Two Types of Modular Multilevel Cascade Converters for Offshore Wind Application," in *Proc. Internationa Future Energy Electronics Conference 2017 (ECCE-Asia 2017)*, 2017, pp. 1-6.

[12] M. Hagiwara, R. Maeda, and H. Akagi, "Negative-Sequence Reactive-Power Control by a PWM STATCOM Based on a Modular Multilevel Cascade Converter (MMCC-SDBC)," *IEEE Trans. Ind. Appl.*, vol. 48, no. 2, pp. 720-729, Mar./Apr. 2012.

[13] H. Wang, and F. Blaabjerg, " Reliability of Capacitors for DC-Link Applications in Power Electronic Converters — An Overview," *IEEE Trans. Ind. App.*, vol. 50, no. 5, pp. 3569-3578, Sep./Oct. 2014

[14] V. Najmi, J. Wang, R. Burgos, and D. Boroyevich, " High Reliability Capacitor Bank Design for Modular Multilevel Converter in MV Applications," in *Proc. Energy Conversion Congress and Exposition (ECCE)*, 2014, pp. 1051-1058.

[15] K. Sharifabadi, L. Harnefors, H. Nee, S. Norrga, and R. Teodorescu, Design, Control, and Application of Modular Multilevel Converters for HVDC Transmission Systems. Wiley-IEEE Press, 2016.

[16] Website of Cornell Dubilier Company, Type 947C Polypropylene, DC Link Capacitors. [Online]. Available: http://www.cde.com/resources/catalogs/947C.pdf

[17] Y. Song, and B. Wang, " Survey on Reliability of Power Electronics System," *IEEE Trans. on Power Electron.*, vol. 28, no. 1, pp. 591-604, Jan. 2013

[18] Website of Hitachi Power Semiconductor Device, Ltd., Product Lineup of 4500V-IGBT module. [Online]. Available: http://www.hitachi-power-semiconductor-device.co.jp/en/product/igbt/list/4500v.html

Optimized Decoupling Control of Flying Capacitor in ANPC Five-Level Inverter

Fusheng Wang[1], Deyou Zheng[1*], Jianing Wang[1],Fei Li[1],Fang Liu[1],Shuying Yang[1] and Zhen Xie[1]

1 Department of Electric Engineering and Automation, Hefei University of Technology,

P.R, China

*E-mail:1634326312@qq.com

Abstract—**Active neutral-point-clamped five-level (ANPC-5L) inverter has been gradually applied to the high-power photovoltaic grid-tied system. In order to solve the coupling problem between the flying capacitor voltages and neutral-point potential, a detailed mathematical derivation of the relationship between the flying capacitor voltages, output phase voltages and neutral-point potential is carried out based on the carrier phase-shifted modulation strategy. Then a decoupling control algorithm using modulation wave splitting is proposed. Finally, the above algorithm is applied to inverter startup, which can reduce the output current harmonic distortion rate during the establishment of flying capacitor voltages, and achieve the purpose of grid connection at startup. The experiment results prove the correctness and feasibility of the algorithm.**

Keywords—*Decoupling control, Modulation wave splitting, Fluctuation of neutral-point potential, Current distortion*

I. INTRODUCTION

In the case of medium voltage and high power, compared with the conventional three-level inverter, active neutral-point-clamped five-level (ANPC-5L) inverter has less switching losses and lower current harmonics[1-3]. The output level is redundant, which makes the control of flying capacitor voltage and neutral-point potential simpler and more reliable[4-6]. Therefore, ANPC-5L has attracted extensive attention of scholars worldwide[7].

The coupling between the flying capacitor voltages

This paper is supported by The National High Technology Research and Development of China 863 Program (2015AA050607).

and neutral-point potential has always been the key and difficult point in the research of ANPC-5L inverter. Due to the inherent characteristics of ANPC-5L topology, current will flow through the neutral-point of DC side during the control of flying capacitor voltage, which will affect the neutral-point potential, result in the output voltage waveform distortion and increase the harmonic content in output current. So, it does not meet the requirements of grid connection.

Due to differences in modulation strategies, the control methods for neutral-point potential and flying capacitor voltage are slightly different. In the existing literature, there are two main modulation strategies: carrier modulation and space vector modulation, where the carrier modulation is divided into carrier phase-shifted and carrier disposition modulation. In [8], a carrier phase-shifted modulation is proposed, which can control the flying capacitor charging and discharging time automatically equal in one carrier cycle. At the same time, the paper calculates the zero-sequence component added on the modulation wave by linear normalization method, which effectively controls the neutral-point potential. However, there is no reference to the coupling problem between the neutral-point potential and flying capacitor voltage. On the basis of carrier disposition, reference [9] is able to ensure the stability of flying capacitor voltage by controlling the state of charge and discharge for flying capacitor different during the adjacent carrier cycle. Under this condition, the expression of neutral-point current under the carrier disposition modulation strategy is derived, and the zero-sequence component is obtained by using the conservation of charge, and the relevant limiting principle is further proposed, which achieves better control effect of neutral-point potential. Although the coupling relationship between the two is mentioned in [9], a clear decoupling idea is not given. For space vector modulation, a simplified modulation method is presented in [10], and an universal method of voltage control for ANPC-5L flying capacitor is given by flow chart, that is, the suitable switching state is selected to control the flying capacitor voltage according to the phase current direction and the capacitor voltage. In the paper, the control of neutral-point potential is achieved by adjusting the function time ratio k of multiplex vectors. The coupling relationship between flying capacitor voltage and neutral-point potential is clearly put forward, but only a fuzzy solution to the problem is given in [10]. In summary, the independent control of flying capacitor voltage or neutral-point potential is relatively complete, but the analysis of coupling relationship between them is less, and no suitable decoupling scheme is put forward. So, the research in this aspect is particularly urgent.

In this paper, based on carrier phase-shifted modulation strategy, the relationship between the flying capacitor, output phase voltage and neutral-point potential is mathematically derived in detail to solve the above problems. Then, a decoupling control algorithm between the flying capacitor voltage and neutral-point potential is proposed. This algorithm can be applied to inverter startup in order to reduce the output current harmonic distortion rate during the establishment of flying capacitor voltage, and achieve the purpose of grid connection at startup. The experiment results prove the correctness and feasibility of the algorithm.

II. Topology and Operational Principle

Fig. 1 shows a single-phase bridge arm topology of ANPC five-level inverter. NP is the neutral-point of bus capacitor C_1 and C_2 in DC side. The switching function S_{k1}, S_{k2} and S_{k3} indicate the working states of the k phase switching devices, that is, 0 means off, 1 means on. The working states of $\overline{S_{k1}}$, $\overline{S_{k2}}$ and $\overline{S_{k3}}$ are opposite to that of S_{k1}, S_{k2} and S_{k3} respectively.

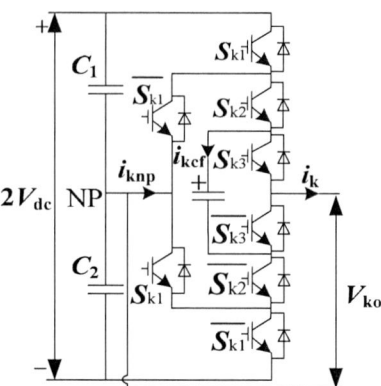

Fig.1. Structure of 5L-ANPC inverters

Taking NP as the zero potential reference point, each phase bridge arm can output five levels: $-V_{dc}$, $-0.5V_{dc}$, 0, $0.5V_{dc}$ and V_{dc}. Switching states and corresponding currents of ANPC-5L are given in Table I. It can be seen that the levels of $0.5V_{dc}$, $-0.5V_{dc}$ and 0 correspond to two redundant switch states respectively. The neutral-point current i_{knp} and the flying capacitor current i_{kcf} are both represented by the inverter output phase current i_k, and the current direction is shown in Fig. 1.

TABLE I

Switching States and Corresponding Currents of ANPC-5L

S	S_{k1}	S_{k2}	S_{k3}	V_{ko}/V	i_{kcf}/A	i_{knp}/A
V_1	1	1	1	V_{dc}	0	0
V_2	1	1	0	$0.5V_{dc}$	$+i_k$	0
V_3	1	0	1	$0.5V_{dc}$	$-i_k$	$+i_k$
V_4	1	0	0	0	0	$+i_k$
V_5	0	1	1	0	0	$+i_k$
V_6	0	1	0	$-0.5V_{dc}$	$+i_k$	$+i_k$
V_7	0	0	1	$-0.5V_{dc}$	$-i_k$	0
V_8	0	0	0	$-V_{dc}$	0	0

According to Table I, the formulas of V_{ko}, i_{kcf} and i_{knp} can be written as

$$
\begin{cases}
V_{ko} = \begin{cases} V_{dc}(S_{k2}+S_{k3})/2 & ,S_{k1}=1 \\ V_{dc}(S_{k2}+S_{k3}-2)/2 & ,S_{k1}=0 \end{cases} \\
i_{kcf} = (S_{k2}-S_{k3})i_k \\
i_{knp} = \begin{cases} i_k(1-S_{k2}) & ,S_{k1}=1 \\ i_k S_{k2} & ,S_{k1}=0 \end{cases}
\end{cases}
\tag{1}
$$

III. THEORETICAL ANALYSIS OF FLYING CAPACITIVE DECOUPLING CONTROL

A. Carrier Phase Shift Modulation Strategy

Defining the original modulation waves be m_{kref} (k=a,b,c), and then the modified modulation waves m_{k1} can be written as

$$m_{k1} = \begin{cases} m_{kref}, & 0 \le m_{kref} \le 1 \\ m_{kref}+1, & -1 \le m_{kref} < 0 \end{cases} \quad (2)$$

In the carrier phase shift modulation strategy, S_{k1} adopts low-frequency modulation mode as shown in (2), and the PWM signals of S_{k2} and S_{k3} are obtained by comparing the modulation wave m_{k1} with the carriers T_{r1} and T_{r2} respectively, as shown in Fig. 2. By the similarity triangle theorem, the duty cycle of S_{k2} and S_{k3} are the same as m_{k1}.

$$S_{k1} = \begin{cases} 1, & 0 \le m_{kref} \le 1 \\ 0, & -1 \le m_{kref} < 0 \end{cases} \quad (3)$$

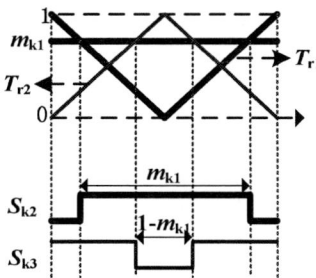

Fig.2. Work principle of S_{k2} and S_{k3}

Since the duty cycle of S_{k2} and S_{k3} are the same as m_{k1}, (3) is simplified as (4).

$$\begin{cases} V_{ko} = \begin{cases} V_{dc}m_{k1} & ,S_{k1}=1 \\ V_{dc}(m_{k1}-1) & ,S_{k1}=0 \end{cases} \\ i_{kcf} = (m_{k1}-m_{k1})i_k = 0 \\ i_{knp} = \begin{cases} i_k(1-m_{k1}) & ,S_{k1}=1 \\ i_k m_{k1} & ,S_{k1}=0 \end{cases} \end{cases} \quad (4)$$

According to (4), It can be found that carrier phase-shifted modulation strategy can control the flying capacitor voltage to be balanced under idea and steady-state condition. But, when the initial voltage of flying capacitor is not the expected target voltage, it can only be balanced at the undesired initial voltage. And the flexible control of flying capacitor voltage is unable to be achieved.

B. Mathematical Model of Split Modulation Wave

According to (1), it can be seen that the flying capacitor voltage control can be achieved by adjusting the duty cycle of S_{k2} and S_{k3}. Where duty cycle of S_{k2} and S_{k3} are only related to the modulation wave m_{k1}. In order to control the flying capacitor voltage effectively, this paper proposes an idea of modulation wave splitting based on carrier phase shift.

It can be seen that the coupling of flying capacitors and neutral-point potential control is inherent from Table I. Therefore, simply controlling the flying capacitors will inevitably causes an impact on the neutral-point potential, which is not conducive to the normal operation of the inverter.

In order to realize the decoupling control of neutral-point potential and flying capacitors, a general mathematical model of five-level inverter modulation wave splitting algorithm is established. S_{k1} still adopts the low frequency modulation mode, and two split modulation waves m_{k2} and m_{k3} are obtained by superimposing the components Δm_{k2} and $-\Delta m_{k3}$ on the basis of the modulation wave m_{k1}. The general expressions can be written as

$$\begin{cases} m_{k2} = m_{k1} + \Delta m_{k2} \\ m_{k3} = m_{k1} - \Delta m_{k3} \end{cases} \quad (5)$$

The splitting of the entire fundamental period is shown in Fig.3. Suppose actual value of the flying capacitor voltage is V_{kcf}, and then the output voltage V_{ko}, flying capacitor current i_{kcf} and neutral-point current i_{knp} are obtained from the principle of volt-second balance ,The general expressions can be written as

$$\begin{cases} V_{ko}^{'} = V_{ko} + \Delta m_{k2}V_{dc} - (\Delta m_{k2}+\Delta m_{k3})V_{kcf} \\ i_{kcf}^{'} = (\Delta m_{k2}+\Delta m_{k3})i_k \\ i_{knp}^{'} = i_{knp} - sign(V_{ko})\Delta m_{k2}i_k \end{cases} \quad (6)$$

In order to facilitate the discussion of the coupling between the flying capacitors and the neutral-point potential, (6) is simplified as (7).

$$\begin{cases} V_{opu} = \Delta m_{k2} - (\Delta m_{k2} + \Delta m_{k3})V_{kcfpu} \\ i'_{kcf} = (\Delta m_{k2} + \Delta m_{k3})i_k \\ \Delta i_{np} = \sum_{k=a,b,c} sign(V_{ko})\Delta m_{k2}i_k \end{cases} \quad (7)$$

Where V_{opu} is the per-unit voltage of the three-phase zero-sequence voltage, Δi_{np} is the fluctuation value of the neutral-point current caused by modulation wave splitting, V_{kcfpu} is the per-unit value of the three-phase flying capacitor voltage, and the expression of V_{kcfpu} can be written as

$$V_{kcfpu} = \frac{V_{kcf}}{V_{dc}} \quad (8)$$

1) Control mode A

The control mode A is operated under the condition of $\Delta m_{k2} = \Delta m_{k3} = \Delta m_k \neq 0$, and the (7) is simplified as the (9).

$$\begin{cases} V_{opu} = \Delta m_k(1 - 2V_{kcfpu}) \\ i'_{kcf} = 2\Delta m_k i_k \\ \sum_{k=a,b,c} sign(V_{ko})\Delta m_{k2}i_k = 0 \end{cases} \quad (9)$$

There are four variables in (9), which are Δm_a, Δm_b, Δm_c, and V_{opu}. And (9) is written as a vector equation with vector X as the unknown element as shown in (10).

$$AX = b \quad (10)$$

Where

$$\begin{cases} A = \begin{bmatrix} (1 - 2V_{acfpu}) & 0 & 0 & -1 \\ 0 & (1 - 2V_{bcfpu}) & 0 & -1 \\ 0 & 0 & (1 - 2V_{ccfpu}) & -1 \\ 2i_a & 0 & 0 & 0 \\ 0 & 2i_b & 0 & 0 \\ 0 & 0 & 2i_c & 0 \\ Sign(V_{ao})i_a & Sign(V_{bo})i_b & Sign(V_{co})i_c & 0 \end{bmatrix} \\ (11) \\ X = \begin{bmatrix} \Delta m_a & \Delta m_b & \Delta m_c & V_{opu} \end{bmatrix}^T \\ b = \begin{bmatrix} 0 & 0 & 0 & i'_{acf} & i'_{bcf} & i'_{ccf} & 0 \end{bmatrix}^T \end{cases}$$

According to (11), the rank of the coefficient matrix A is 4, and the rank of the augmented matrix B = (A, b) is 5, then R (A) <R (B). By mathematical criteria, (9) does not exist.

Therefore, the split mode has an influence on the phase current and the neutral-point potential based on the

control mode A, which can be unable to achieved decoupling control.

2) Control mode B

The control mode B is operated under the condition of $\Delta m_{k2} \neq \Delta m_{k3}$. and the (7) is simplified as the (12).

$$\begin{cases} V_{opu} = \Delta m_{k2} - (\Delta m_{k2} + \Delta m_{k3})V_{kcfpu} \\ i'_{kcf} = (\Delta m_{k2} + \Delta m_{k3})i_k \\ \sum_{k=a,b,c} sign(V_{ko})\Delta m_{k2}i_k = 0 \end{cases} \quad (12)$$

There are seven variables in (12), which are Δm_{a2}、Δm_{b2}、Δm_{c2}、Δm_{a3}、Δm_{b3}、Δm_{c3}、V_{opu}. Similarly，(12) is simplified into the form of (10). And the rank of the matrix A and the rank of the augmented matrix B = (A,b) are both calculated to be 7, then R (A) = R (B). By mathematical criteria, (12) exists.

Therefore, the split mode will not affect the phase current and the neutral-point potential based on the control mode B, which can realize purpose of decoupling control.

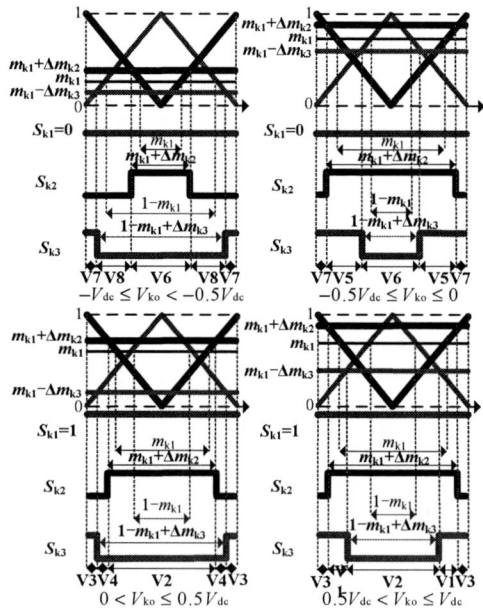

Fig. 3.   Modulation wave splitting schematics

IV. DECOUPLING CONTROL STRATEGY

In order to achieve the decoupling between flying capacitor voltage and neutral-point potential, and then eliminate the mutual influence of them in control, the conditions of $\Delta m_{k2} \neq \Delta m_{k3}$ must be satisfied.

After calculation, (12) can be written as

$$\sum_{k=a,b,c} sign(V_{ko})(V_{opu}i_k + V_{kcfpu}i_{kcf}^{'}) = 0 \qquad (13)$$

A pure proportional controller is used to achieve flying capacitor voltage tracking. The output of the regulator is the required charging and discharging current for flying capacitor. The control expression can be written as

$$i_{kcf}^{'} = K(0.5V_{dc} - V_{kcf}) \qquad (14)$$

Where $0.5V_{dc}$ is the flying capacitor voltage stability value, V_{kcf} is the actual value of the three-phase flying capacitor voltage, and K is the proportional coefficient of the pure proportional controller.

According to (13), expression of V_{opu} is shown in (15).

$$V_{opu} = -\frac{\sum\limits_{k=a,b,c} sign(V_{ko})V_{kcfpu}i_{kcf}^{'}}{\sum\limits_{k=a,b,c} sign(V_{ko})i_k} \qquad (15)$$

Where the range of V_{opu} can be written as

$$-1 - V_{min} \leq V_{opu} \leq 1 - V_{max} \qquad (16)$$

Where V_{max} and V_{min} are the minimal and maximal values of V_{ao}, V_{bo} and V_{co} respectively.

If V_{opu} exceeds this range of values, the boundary value is taken. But at this time, (13) is no longer equal to 0, which will destroy original balance of neutral-point potential. Therefore, it is necessary to recalculate the flying capacitor charging and discharging current according to the boundary value of V_{opu}. The expression can be written as

$$\begin{cases} i_{kcf}^{''} = i_{kcf}^{'}, & (-1 - V_{min} \leq V_{opu} \leq 1 - V_{max}) \\ i_{kcf}^{''} = i_{kcf}^{'}\dfrac{Q^{'}}{Q}, & (V_{opu} = 1 - V_{max} \text{或} -1 - V_{min}) \end{cases} \qquad (17)$$

Where the expression of Q and Q' as shown in (18).

$$\begin{cases} Q^{'} = -\sum\limits_{k=a,b,c} sign(V_{ko})(V_{opu}i_k) \\ Q = \sum\limits_{k=a,b,c} sign(V_{ko})(V_{kcfpu}i_{kcf}^{'}) \end{cases} \qquad (18)$$

Finally，V_{opu} and $i_{kcf}^{''}$ are introduced into (12)，and then the expression of Δm_{k2} and Δm_{k3} can be written as

$$\begin{cases} \Delta m_{k2} = V_{opu} + \dfrac{V_{kcfpu}i_{kcf}^{''}}{i_k} \\ \Delta m_{k3} = \dfrac{i_{kcf}^{''}}{i_k} - \Delta m_{k2} \end{cases} \qquad (19)$$

Where the range of Δm_{k2} and Δm_{k3} can be written as

$$\begin{cases} -m_{k1} \leq \Delta m_{k2} \leq 1 - m_{k1} \\ m_{k1} - 1 \leq \Delta m_{k2} \leq m_{k1} \end{cases} \qquad (20)$$

Due to the limiting conditions, Δm_{k2} and Δm_{k3} may be distorted, which will inevitably lead to three-phase unequal V_{opu}, and then cause phase current distortion. In order to avoid this problem, the algorithm uses a reduction in equal proportions to make Δm_{k2} and Δm_{k3} within the limits of the range.

Flow chart of decoupling algorithm is shown in Fig. 4. First of all, there are nine physical quantities of inverter need to be collected, that is, three-phase voltages (V_{ao}, V_{bo}, V_{co}), three-phase flying capacitor voltages(V_{acf}, V_{bcf}, V_{ccf}) and three-phase currents(i_a, i_b, i_c);Then the value of V_{opu} are obtained from the constraints and the value of $i_{kcf}^{'}$ is determined by the controller together with the V_{opu}. Finally, the zero-sequence voltages Δm_{k2} and Δm_{k3} are obtained, and then two groups of modulated waves m_{k2} and m_{k3} are obtained.

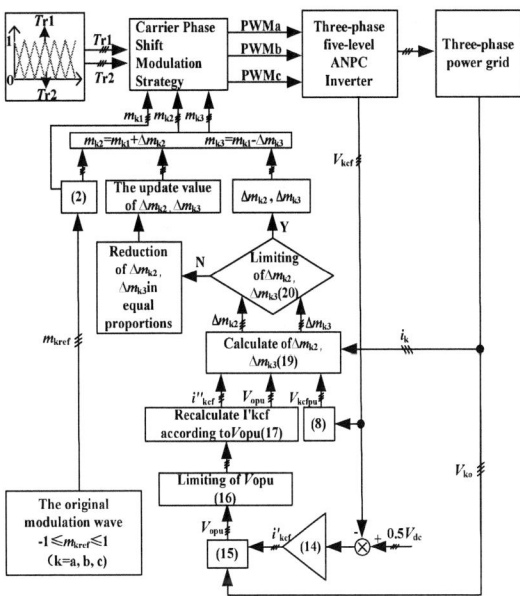

Fig.4. Flow chart of decoupling algorithm

The decoupling algorithm has the advantages of precise control, good stability, and independent control of flying capacitor and midpoint potential. At the same time, this algorithm does not affect the fundamental content of output currents during the control of the flying capacitor voltages. Therefore, It can be used in the process of establishment of flying capacitor voltages in ANPC-5L, which can achieve the purpose of grid connection at startup.

V. EXPERIMENTAL RESULTS

In order to further verify the feasibility and performance of the modulation strategy proposed in this paper, the prototype is designed and related experiments are carried out. IGBT is IRGP4063DPBF. The controller combines TMS320F28335 DSP with EPM1270T144I5NCPLD. The specific experimental parameters are shown in TABLE II.

TABLE II

EXPERIMENTAL PARAMETER

parameter	value
Carrier frequency	10kHz
DC side voltage	$2V_{dc}$=200V
Upper / lower bus capacitor	C_1=C_2=560μF
Three-phase flying capacitor	C_{fk}=470μF
R-L load	R=10Ω，L=1mH
Modulation degree	m=0.8

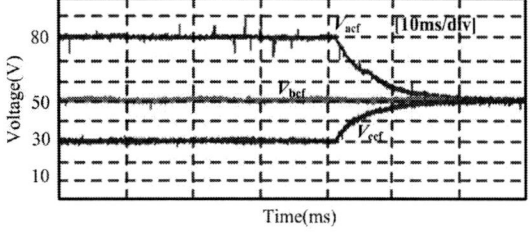

Fig.5. The flying capacitor voltage waveform in coupling state

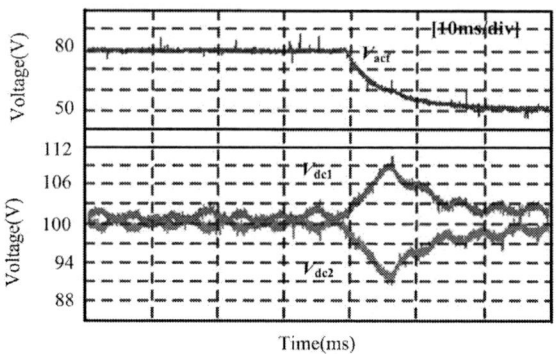

Fig.6. The fluctuation of neutral-point potential in coupling state

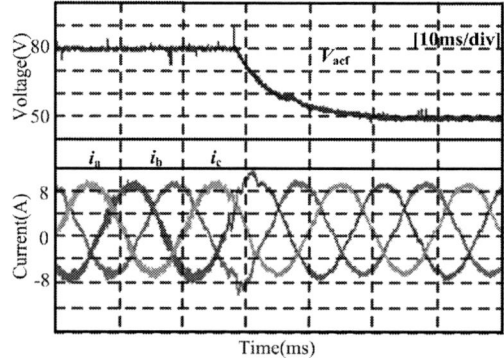

Fig.7. The waveform of current in coupling state

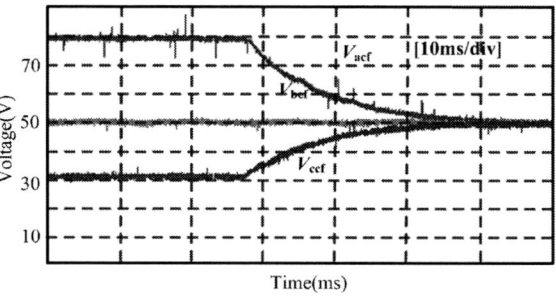

Fig.8. The flying capacitor voltage waveform in decoupling state

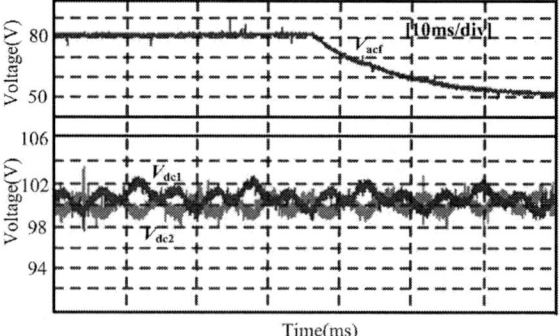

Fig.9. The fluctuation of neutral-point potential in decoupling state

The 2018 International Power Electronics Conference

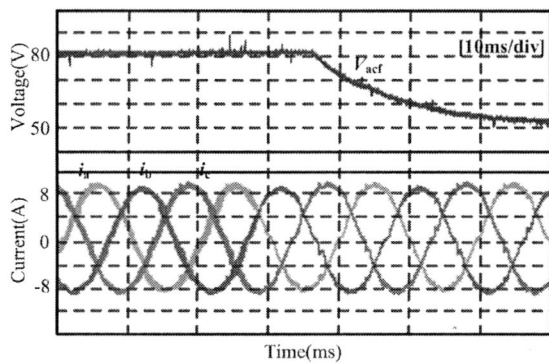

Fig.10. The waveform of current in decoupling state

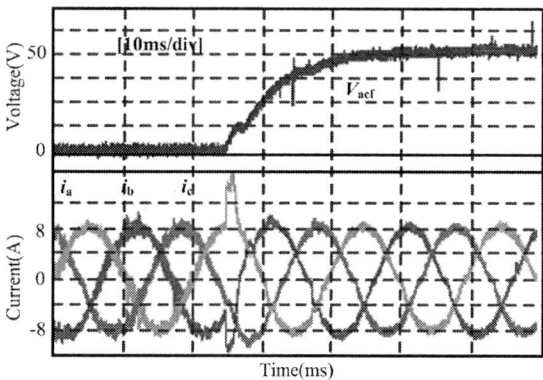

Fig.11. The waveform of start up in coupling state

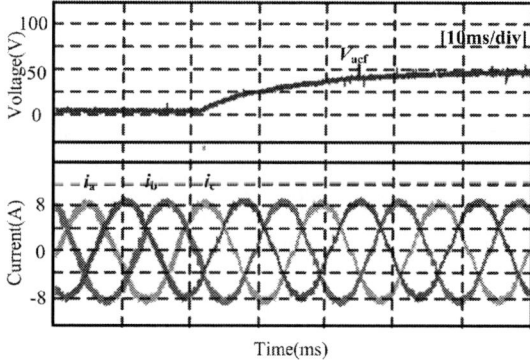

Fig.12. The waveform of start up in decoupling state

Coupling algorithm is based on the above control mode A. Fig. 5 is the flying capacitor voltage waveform in coupling state, the initial values of three-phase flying capacitors are 80V, 50V, 30V. After control algorithm, the voltages of the flying capacitor have been reached stable states. Fig. 6 is the fluctuation of neutral-point potential in coupling state, it can be seen the deviation of the neutral-point potential has occur and the offset could be as large as 16V when control algorithm take place. And it can be found that the three-phase output currents are obviously distorted from Fig. 7. It shows that the control of the flying capacitors in the coupled state will destroy the original midpoint balance and increase the output current distortion rate.

Under the condition of Table II, the waveform of decoupling algorithm is shown in Fig.8, Fig.9 and Fig.10. The initial values of three-phase flying capacitors are 80V, 50V and 30V. Compared with the waveform of the corresponding stages in Fig.5, Fig.6 and Fig.7, the fluctuation law of the midpoint potential is not affected, and the phase currents do not have obvious distortion. In summary, since the decoupling algorithm has no influence on the existing neutral-point potential balance during the control of flying capacitors, it can realize independent control of flying capacitors and midpoint potential. Meanwhile, the control of the algorithm does not cause large distortion to the phase currents.

Fig.11 is the waveform of start up in coupling state. It can be found that phase current distortion is more serious during the process of the voltage establishment, which does not achieve the purpose of grid connection smoothly. Fig.12 shows the waveform of start up in coupling state. It can be seen that there is no loss of the fundamental content of the phase currents during the process of the voltage establishment, and the distortion of the currents is small, which is favorable for the reliable grid connection of the inverter. Therefore, this algorithm can be used in the start-up process of ANPC-5L, the purpose of grid connection at startup can be achieved.

VI. CONCLUSION

Aiming at the coupling between the flying capacitor voltages and neutral-point potential, the following researches have been done.

1) The coupling between flying capacitor voltages and

neutral-point potential is inherent. The theoretical deduction and simulation results verify the deviation of neutral-point potential and the distortion of output current caused by this coupling.

2) Two new modulation waves are obtained by adding different zero-sequence components to original modulation wave. The inverter mathematical models obtained by adding different zero-sequence components are analyzed, and the decoupling condition is deduced.

3) The method solves the coupling problem between flying capacitor and neutral-point potential of ANPC five-level inverter, and then achieved an independent control of them. The algorithm is simple and the effect of control is accurate.

4) The algorithm greatly reduces the output current distortion rate during the transient control of flying capacitor.

5) The algorithm can achieve the purpose of that inverter connect to the gird successfully during the process of flying capacitor voltage establishment.

REFERENCES

[1] Guojun Tan, Zhan Liu, Zongbin Ye, et al. SVPWM Algorithm Based on Line Voltage Coordinate Transformation for ANPC-5L Inverter [J]. Proceedings of the CSEE, 2013, (30): 26-33.

[2] Yongdong Li, Xi Xiao, Yue Gao. Large capacity multilevel converter: principle, control, application[M]. Beijing: CSPM, 2005.

[3] Xiangning He, Alian Chen. Theory and application of multilevel converters[M]. Beijing: China Machine Press, 2006 (in Chinese).

[4] Kouro K, Malinowski M, Gopakumar K, et al. Recent advances and industrial applications of multilevel converters[J]. IEEE Transactions on Industrial Electronics, 2010, 57(8): 2553-2580.

[5] Pulikanti S R, Agelidis V G. Five-level active NPC converter topology: SHE-PWM control and operation principles[C]// Australasian Universities Power Engineering Conference. Perth, Australia: IEEE, 2007: 1-5.

[6] Peter Barbosa, Peter Steimer, J rgen Steinke, et al. Active neutral-point-clamped multilevel converters[C]//Proc. 36th IEEE PESC, 2005: 2296–2301.

[7] J. Meili, S. Ponnaluri, L. Serpa, et al. Optimized Pulse Patterns for the 5-Level ANPC Converter for High Speed High Power Applications[C]//Proc. 32nd IEEE Ind. Electron. (IECON), Nov. 2006: 2587-2592.

[8] Kui Wang, Xu Lie, Zedong Zheng, et al. Capacitor Voltage Balancing of a Five-Level ANPC Converter Using Phase-Shifted PWM[J]. IEEE Transactions on Power Electronics ,2015, 30(3): 1147-1156.

[9] Kui Wang, Zedong Zheng, Yongdong Li. Neutral-point Potential Balancing Problem of Five-level Active Neutral-point-clamped Inverter[J]. Proceedings of the CSEE, 2012, 32 (3): 30-35.

[10] Zhan Liu, Guojun Tan, Hao Li, et al. Research on Neutral-point Potential Balancing Problems of Active Neutral-point-clamped Five-level Inverter Based on Space Vector Pulse Width Modulation[J]. Proceedings of the CSEE, 2015, 35(24): 6499-6507.

Cascaded Dual-Buck AC-AC Converter Using Coupled Inductors

Sanghun Kim[1*], Duekjin Jang[1], Heung-Geun Kim[1] and Honnyong Cha[2]

1 Department of Electrical Engineering, Kyungpook National University, Daegu, Korea
2 Department of Energy Engineering, Kyungpook National University, Daegu, Korea
*E-mail: random61@naver.com

Abstract— A cascaded dual-buck ac-ac converter (CDBAC) possesses no shoot-through worries, no commutation problem, and can enhance efficiency by using power MOSFETs. The main drawback of a CDBAC is the use of more inductors than traditional cascaded multilevel ac-ac converter, which decreases the power density and increases the system cost. This paper presents a modified CDBAC using coupled inductors. In the proposed converter, the limiting inductors connected between units are integrated with one core. Because they share the current path where output inductor current is flowing. As a result, the total inductance, inductor footprints, and magnetic volume can all be reduced. To demonstrate the advantages of the proposed converter, the topology derivation, operation, and experimental results are presented.

Keywords— *coupled inductor, direct PWM ac-ac converter, dual-buck converter, switching cell.*

I. INTRODUCTION

For voltage regulation only, direct PWM ac–ac converters (DPACs) are widely used. Industrial applications of them include two-level converters [1-4], reactive power compensators [5-8], dynamic voltage restorers [9-11], and multilevel converters [12-13]. However, due to the serious commutation problem [1-2], the conventional DPACs must need either dedicated soft commutation strategy or lossy *RC* snubber circuits, so they have made very limited market penetration.

To overcome the commutation problem, a new type of dual-buck structured DPACs is firstly developed in [1]. Unlike conventional DPACs, its basic switching cell is implemented with dual buck structure [14-15] in which a switch leg consists of MOSFETs, external diodes, and coupled inductors. As this structure can resolve the shoot-through and dead-time problems of the traditional DPACs without dedicated soft commutation strategy, so this converter is highly reliable. Moreover, the body diodes of MOSFETs never conduct current anytime, which eliminates their reverse-recovery issues related switching loss. Therefore, the converter can boost the efficiency and increase the switching frequency.

The extension of this converter to cascaded multilevel ac-ac convert is introduced in [12]. It can obtain high ac output voltage with low-voltage-rating switching devices by cascading unit cells. However, this converter has somewhat bulky two coupled inductors for each unit. In

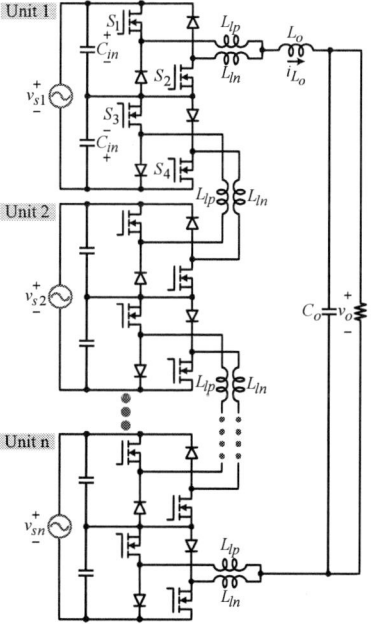

Fig. 1. Conventional cascaded dual-buck ac-ac converter [13].

addition, the phase shift PWM (PSP) control can reduce only the size of the output filter inductor except for the coupled inductors [13].

In order to improve this multilevel converter, a new cascaded dual-buck ac-ac converter is recently introduced in [13]. Fig. 1 shows this converter. It uses discrete inductors instead of employing coupled inductors. For the sake of clear description, the inductors split in each switch leg, i.e., L_{lp} and L_{ln} shown in Fig. 1, are called as a "limiting inductor (LI)". In this converter, all the LIs experience the increased effective frequency by PSP control, which can reduce the magnetic volume significantly. However, as the number of units increases, more LIs are needed and thus overall magnetic volume is increased. Besides, the converter has only 50% magnetics utilization, which is the main drawback of dual-buck structure. For example, the output current i_{Lo} is only flowing through L_{lp} at the positive half-cycle, and i_{Lo} is only flowing through L_{ln} at the negative half-cycle.

The 2018 International Power Electronics Conference

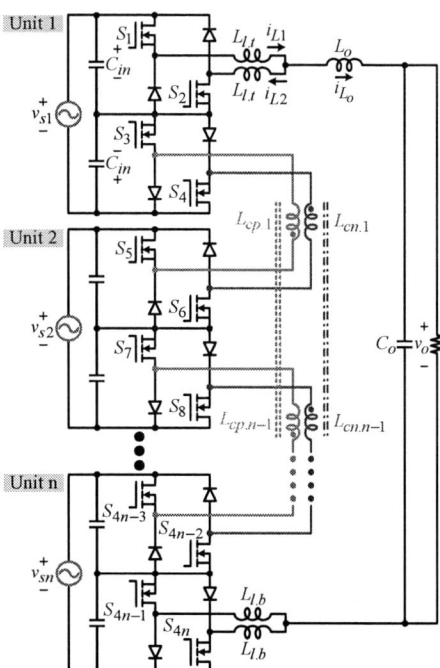

Fig. 2. Proposed cascaded dual-buck ac-ac converter using coupled inductors.

In this paper, a modified cascaded dual-buck ac-ac converter with coupled inductor is proposed to realize the minimization of the size of LIs. Note that the coupled inductor of the proposed scheme is different from that of [12]. According to the output current polarity, the two sets of LIs that share the current path of i_{Lo} are integrated with each one core by applying direct coupling method. Although the number of units increases, the total linkage flux of coupled inductor also increases, so high self-inductance value of each winding can be achieved. Consequently, the proposed scheme can reduce the overall self-inductance value, inductor footprints, and magnetic volume greatly.

II. PROPOSED CASCADED DUAL-BUCK AC-AC CONVERTER WITH COUPLED INDUCTORS

Fig. 2 shows the proposed converter. It looks similar to the conventional converter shown in Fig. 1, but it has different structure of LIs. Its LI structure is composed of two coupled inductors ($L_{cp.x}$, $L_{cn.x}$) and four discrete inductors ($L_{l.t}$, $L_{l.b}$) where $L_{cp.x}$ and $L_{cn.x}$ include the whole windings in each core. They work the same as the LIs that limit the shoot-through current, and also serve as filter inductors. As explain in Section I, the output inductor current (i_{Lo}) flows through the inductors ($L_{cp.x}$) for $i_{Lo} > 0$, and i_{Lo} flows through $L_{cn.x}$ for $i_{Lo} < 0$, Therefore, the two sets of inductors ($L_{cp.x}$, $L_{cn.x}$) can be integrated into each one core.

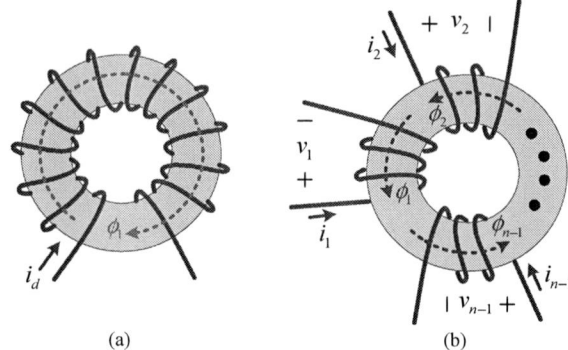

(a)   (b)

Fig. 3. Structure of discrete inductor and coupled inductor. (a) Discrete inductor. (b) Coupled inductor with (n-1)-windings.

As the number of units increases, more LIs are needed in the case of conventional converter. However, only two coupled inductors and four discrete inductors are needed in the proposed converter regardless of the number of units. In addition, the linkage flux of the coupled inductor increases with the number of units by using direct coupling method, so all the self-inductances of $L_{cp.x}$ and

$L_{cn.x}$ increase. As a result, when compared with the conventional converter under the same ripple condition of i_{Lo} , the overall value of self-inductance, inductor footprints, and magnetic volume can all be significantly reduced.

The input voltages having the same phase and magnitude are supplied by separate ac sources, which are provided by a multi-winding transformer. The proposed converter inherits all the advantages of the cascaded dual buck ac-ac converter, such as no commutation problem, no shoot-through worries, no PWM dead-time, and the use of power MOSFETs.

III. ANALYSIS OF THE COUPLED INDUCTOR

Fig. 3 shows the structure of discrete inductor and coupled inductor employed in the proposed converter. The LIs in Fig. 1 are also fabricated as Fig. 3(a). In Fig. 3(b), all the windings are wound on one core. All the winding currents must flow so that the magnetic fluxes generated from each winding are combined in the same direction, which is the general direct coupling method. The discrete inductor has the linkage flux generated from only one winding (ϕ_1), while the coupled inductor has the increased flux as follows:

$$\phi_{coupled} = \phi_1 + \phi_2 + \cdots + \phi_{n-1}. \tag{1}$$

From (1), it should be noted that each winding of the coupled inductor has the higher effective inductance than discrete inductor if all the winding currents in Fig. 3 are the same. Practically, the coupling coefficient (k) is an important factor of the coupled inductor. If a coupled inductor has 2-windings, k is defined as follows:

$$k_{12} = \frac{M_{12}}{\sqrt{L_1 L_2}} \tag{2}$$

2620

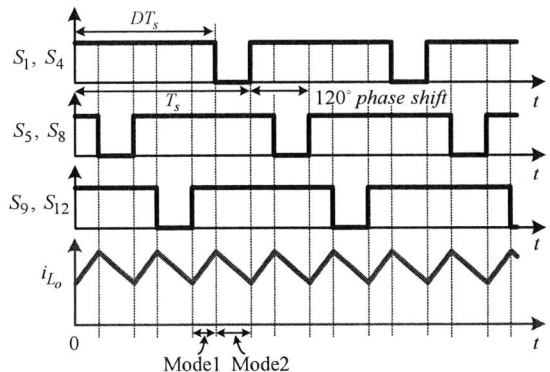

Fig. 4. Ideal gate signals and the waveforms of i_{Lo} in the proposed 3-unit cascaded converter.

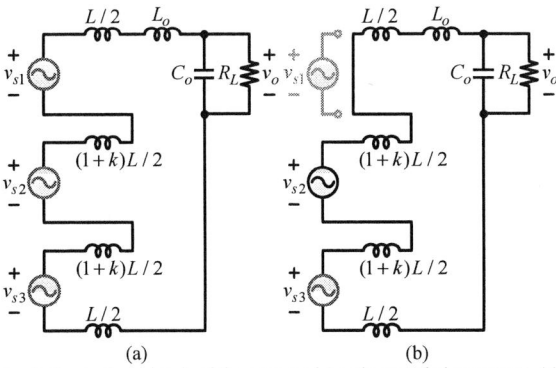

Fig. 5. Equivalent circuit of the proposed 3-unit cascaded converter. (a) Mode 1. (b) Mode 2.

where M_{12} is the mutual inductance. Considering k, the effective self-inductance of each winding of coupled inductor can be expressed through the following process. First, the general equation for the coupled inductor with (n-1)-winding can be expressed as

$$\begin{bmatrix} v_1 \\ v_2 \\ \vdots \\ v_{n-1} \end{bmatrix} = \begin{bmatrix} L_{11} & M_{12} & \cdots & M_{1n-1} \\ M_{21} & L_{22} & \cdots & M_{2n-1} \\ \vdots & \vdots & \ddots & \vdots \\ M_{n-11} & M_{n-12} & \cdots & L_{n-1n-1} \end{bmatrix} \cdot \frac{d}{dt} \begin{bmatrix} i_1 \\ i_2 \\ \vdots \\ i_{n-1} \end{bmatrix}. \quad (3)$$

In this paper, each winding of the coupled inductor has the same number of turns and the same winding current. In addition, it is arranged evenly in the core so that all the mutual inductances have the same value. Thus, it is possible to express the parameter as follows: $L_{11} = L_{22} = \cdots = L_{n-1n-1} = L$, $i_1 = i_2 = \cdots = i_{n-1} = i$, and all the mutual coupling coefficients (i.e., k_{12}, k_{13}, k_{14}, $\cdots$) have the same value, k. Therefore, the induced voltage, v_1 can be simply rewritten as

$$v_1 = (1+(n-2)k)L\frac{di}{dt}. \quad (4)$$

The effective self-inductance of one winding can be expressed as

$$L_{e.self} = (1+(n-2)k)L. \quad (5)$$

From (5), it is clear that the effective self-inductance increases with the number of units if $n > 2$.

IV. OPERATION OF THE PROPOSED CONVERTER AND ANALYSIS OF OUTPUT CURRENT RIPPLE

A. Operation of the Proposed Converter

Fig. 4 shows the gate signals of 3-unit cascaded converter as an example where D is defined as the time interval when S_1 and S_4 are turned on during one switching period. The modulation scheme of the proposed converter uses the PSP control. In each unit,

there are two switching pair: one is the top and bottom switches (e.g., S_1, S_4 in unit 1), the other is middle two switches (e.g., S_2, S_3 in unit 1), and they operate complementarily.

Fig. 5 shows the equivalent circuit diagrams of 3-unit cascaded converter regarding operation modes and equivalent inductance that will be addressed in next Section. There are only two operation modes in each unit. The input voltage is injected into the circuit when the top and bottom switch pair is turned on, while it is separated from the circuit when the middle switch pair is turned on, which is the freewheeling mode (see Fig. 5). When n-units are cascaded, the gate signals in each unit are $360°/n$ out-of-phase, and the equivalent output frequency increases to nf_{sw}. In Fig. 4, it can be seen that each unit is phase-shifted by 120°, 3-times the converter switching frequency is applied to the output inductor (L_o).

From [13], it was found that voltage gain of the proposed converter is identical to that of the conventional converter. The voltage gain is expressed as

$$\frac{v_o}{v_s} = nD \quad (6)$$

where $v_s = v_{s1} = v_{s2} = \cdots = v_{sn}$.

B. Analysis of Output Current Ripple

In this section, the comparison of the current ripple of the output inductor (i_{Lo}) between the proposed converter and the conventional one shown in Fig. 1 is investigated. Note that the only difference between two converters (i.e., the proposed and conventional converter) is LI structure. By considering total equivalent inductance of them, it is possible to compare the i_{Lo} ripples. If the LIs serving as filter inductor are large enough to maintain reasonable i_{Lo} ripple, L_o is not needed. In this analysis and experiment discussed in next Section, the inductance of L_o is zero ($L_o = 0$). For the sake of simplicity, it is assumed that the LIs of two converters have same inductance, i.e., $L_{lp} = L_{ln} = L_{l.t} = L_{l.b} = L_{cp.x} = L_{cn.x} = L$.

2621

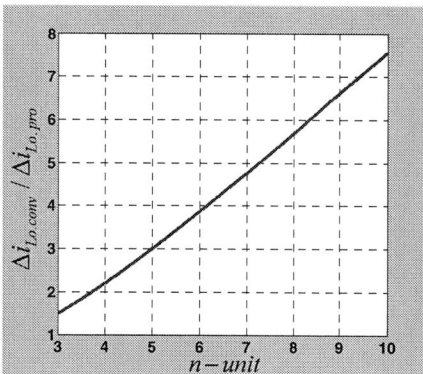

Fig. 6. Comparison of i_{Lo} ripples.

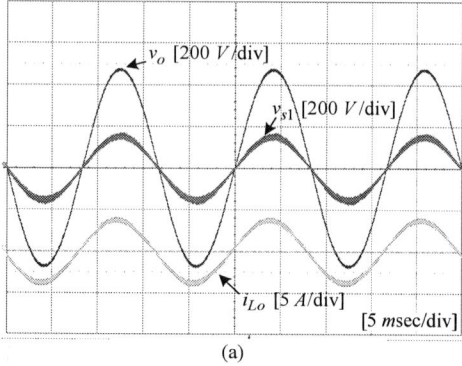

(a)

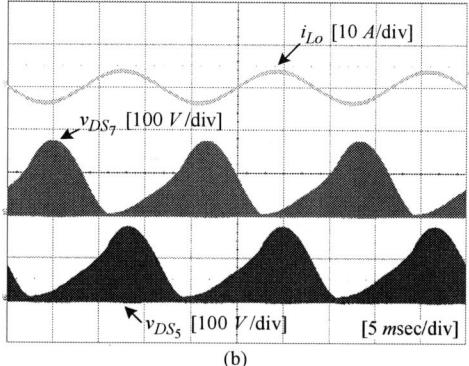

(b)

Fig. 7. Experimental results of the proposed 4-unit cascaded converter. (a) Waveforms of v_{s1}, v_o, and i_{Lo}. (b) Waveforms of i_{Lo}, v_{DS_5}, and v_{DS_7}.

During the normal operation of the converter in Fig. 1, considering circulating current element, currents are always flowing through every LI regardless of the operation modes. Hence, considering the current loop of i_{Lo}, each pair of L_{lp} and L_{ln} seems to be connected in parallel, so i_{Lo} sees its equivalent inductance, $L/2$. As a result, the total output equivalent inductance of the conventional n-unit cascaded converter can be expressed as

$$L_{t,conv} = \frac{n+1}{2}L. \tag{7}$$

Meanwhile, in Fig. 5, the one pair of the LIs between units (e.g., $L_{cp.1}$ and $L_{cn.1}$) has the increased equivalent inductance, $(1+k)L/2$ by direct coupling. From Fig. 5 and (5), the total output equivalent inductance of the proposed n-unit cascaded converter can be expressed as

$$L_{t,pro} = \frac{(n-1)(1+(n-2)k)}{2}L + L. \tag{8}$$

From (7) and (8), the following relation is obtained as

$$\frac{\Delta i_{Lo,conv}}{\Delta i_{Lo,pro}} = \frac{(n-1)(1+(n-2)k)+2}{n+1}. \tag{9}$$

The ratio in (9) is plotted versus n in Fig. 6 when k = 1. Fig. 6 shows that the i_{Lo} ripple of the proposed converter can decrease substantially as the number of units increases.

V. EXPERIMENTAL RESULTS

A 1-kW, 4-unit cascaded converter was designed and tested to prove the merits of the proposed scheme. The electrical specifications are as follows: $L_{lp} = L_{ln} = L_{l.t} = L_{l.b} = L_{cp.x} = L_{cn.x} = 50\ \mu H$, $C_{in} = 1.5\ \mu F$, $C_o = 3\ \mu F$, k = 0.64, D = 0.8, $f_{sw} = 50$ kHz, and $v_{s1} \sim v_{s4} = 110\ V_{rms}$. The four isolated input voltages ($v_{s1} \sim v_{s4}$) are generated by using input transformers (1:1) shown in Fig. 9(a). A resistive load is used for all the experimental results. Fig. 7(a) shows the waveforms

of the input voltage v_{s1}, the output voltage v_o, and i_{Lo}. Fig. 7(b) shows the waveforms of i_{Lo}, the drain-source voltages of switches S_5 and S_7 (v_{DS_5}, v_{DS_7}). It is verified that the proposed converter has high quality voltage and current waveforms without noticeable voltage overshoot in the switches, which indicates the converter has no commutation problem.

Fig. 8(a) shows the waveforms of the coupled inductor currents ($i_{cp.1} \sim i_{cp.3}$). As shown, coupled inductor currents have the same waveforms each other, and due to the direct coupling method, the linkage flux of the core is increased compared with separate cores used in conventional converter. Fig. 8(b) shows the waveforms of limiting inductor currents (i_{L1}, i_{L2}) at the top cell and i_{Lo}. It can be seen that i_{L1} and i_{L2} are always bigger than zero, so the body diodes of MOSFETs never conduct current anytime. This fact is the advantages of the dual-buck structure and it eliminates their reverse-recovery issues, which enables the converter to achieve high efficiency with fast switching frequency.

Fig. 9 shows the expanded i_{Lo} waveforms of the conventional and the proposed converter around the peak value of i_{Lo}. The conventional converter employs the same electrical specifications. From (9), when the converter has 4-unit cells and k=0.64, the ripple ratio is

2622

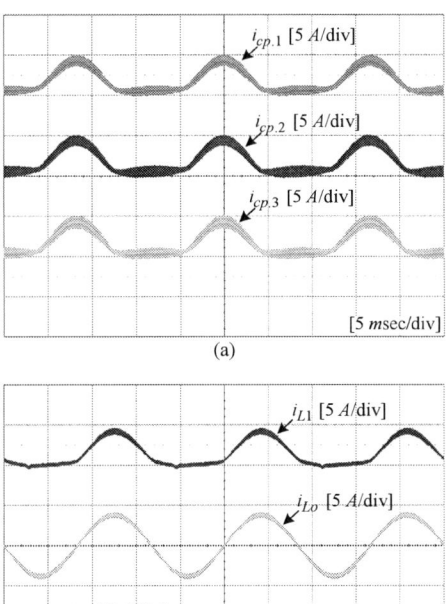

(a)

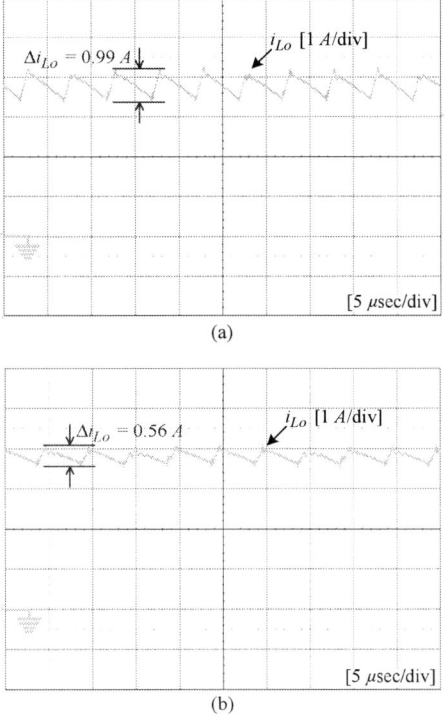

(b)

Fig. 8. Experimental results of the proposed 4-unit cascaded converter. (a) Waveforms of $i_{cp.1}$, $i_{cp.2}$, and $i_{cp.3}$. (b) Waveforms of i_{L1}, i_{L2}, and i_{Lo}.

(a)

(b)

Fig. 9. Expanded waveforms of i_{Lo} around the i_{Lo} peak. (a) Conventional. (b) Proposed.

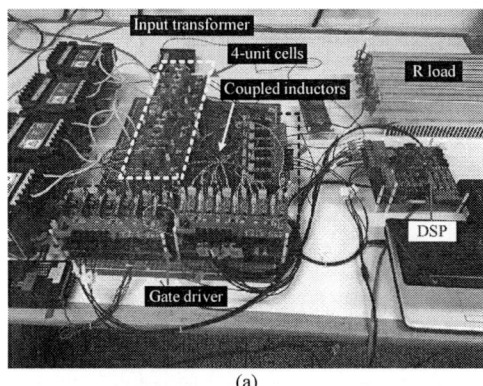

(a)

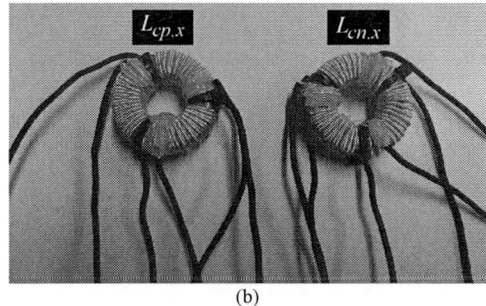

(b)

Fig. 10. Experimental setup. (a) A prototype converter system. (b) Two coupled inductors ($L_{cp.x}$ and $L_{cn.x}$).

about 1.9. As expected, the i_{Lo} ripple of the proposed converter shown in Fig. 9(b) is about half of that of Fig. 9(a).

Fig. 10(a) shows the 1-kW prototype converter system. The control and PWM switching process is realized by using a mixed environment digital signal processor (DSP 28335) and field programmable gate array (FPGA). Fig 10(b) shows the two coupled inductor employed in Fig. 10(a).

VI. CONCLUSIONS

In this paper, a new cascaded dual-buck ac-ac converter using coupled inductors was proposed. By employing coupled inductors with direct coupling method, the discrete limiting inductors between units can be integrated into one core, so the linkage flux of them is increased with the number of units. Consequently, when the conventional and proposed converter has the same limiting inductor value, the proposed one has greatly reduced current ripple of the output inductor. Although the number of units increases, the proposed converter needs only two coupled inductors and four discrete inductors. As a result, the total inductance, inductor footprints, and magnetic volume can be reduced. The experimental results for a 1-kW prototype verify the effectiveness of the proposed converter.

ACKNOWLEDGMENT

This work was supported by the Korea Institute of Energy Technology Evaluation and Planning(KETEP) and the Ministry of Trade, Industry & Energy(MOTIE) of the Republic of Korea (No. 20174030201490).

REFERENCES

[1] H. Shin, H. Cha, H.-G. Kim, and D.-W. Yoo, "Novel single-phase PWM ac–ac converters solving commutation problem using switching cell structure and coupled inductor," *IEEE Trans. Power Electron.*, vol. 30, no. 4, pp. 2137–2147, Apr. 2015.

[2] A. A. Khan, H. Cha, and H. F. Ahmed, "High-efficiency single-phase ac–ac converters without commutation problem," *IEEE Trans. Power Electron.*, vol. 31, no. 8, pp. 5655–5665, Aug. 2016.

[3] A. A. Khan, H. Cha, and H. F. Ahmed, "An improved single-phase direct PWM inverting buck–boost ac–ac converter," *IEEE Trans. Ind. Electron.*, vol. 63, no. 9, pp. 5384-5393, Sept. 2016.

[4] A. A. Khan, H. Cha, and H.-G. Kim, "Magnetic integration of discrete-coupled inductors in single-phase direct PWM ac–ac converters," *IEEE Trans. Power Electron.*, vol. 31, no. 3, pp. 2129-2138, Mar. 2016.

[5] S. Kim, H.-G. Kim, and H. Cha, "Reactive power compensation using switching cell structured direct PWM ac-ac converter," in *Proc. IEEE IPEMC-ECCE Asia*, 2016, pp. 1338–1344.

[6] A. Prasai, J. Sastry, and D. Divan, "Dynamic capacitor (D-CAP): An integrated approach to reactive and harmonic compensation," *IEEE Trans. Ind. Appl.*, vol. 46, no. 6, pp. 2518–2525, Nov./Dec. 2010.

[7] P. Ladoux, J. Fabre, and H. Caron, "Power-quality improvement in AC railway substations: the concept of chopper-controlled impedance," *IEEE Electrification Magazine*, vol.2, issue 3, pp. 6–15, Sept. 2014.

[8] G. Raimondo, P. Ladoux, A. Lowinsky, H. Caron, and P. Marino, "Reactive power compensation in railways based on AC boost choppers," *IET J. Electr. Syst. Transp.*, vol. 2, no. 4, pp. 169-177, Dec. 2012.

[9] S. Kim, H.-G. Kim, and H. Cha, "Dynamic voltage restorer using switching cell structured multilevel ac–ac converter," *IEEE Trans. Power Electron.*, vol. 32, no. 11, pp. 8406–8418, Nov. 2017.

[10] S. Kim, H.-G. Kim, and H. Cha, "Dynamic voltage restorer using PWM ac-ac converter with switching cell structure," in *Proc. IEEE ICPE-ECCE Asia, 2015*, pp. 946–951.

[11] J. Kaniewski, P. Szczesniak, M. Jarnut, and G. Benysek, "Hybrid voltage sag/swell compensators: a review of hybrid ac/ac converters," *IEEE Ind. Electron. Mag.*, vol. 9, no. 4, pp. 37–48, Dec. 2015.

[12] S. Kim, H.-G. Kim, and H. Cha, "A novel single-phase cascaded multilevel ac-ac converter without commutation problem," in *Proc. IEEE Energy Convers. Congr. Expo.*, 2014, pp. 556–562.

[13] A. A. Khan, H. Cha, J.-W. Baek, J. Kim, and J. Cho, "Cascaded dual-buck ac-ac converter with reduced number of inductors," *IEEE Trans. Power Electron.*, vol. 32, no. 10, pp. 7509–7520, Oct. 2017.

[14] P. W. Sun, C. Liu, J.-S. Lai, and C.-L. Chen, "Cascade dual buck inverter with phase-shift control," *IEEE Trans. Power Electron.*, vol.27, no. 4, Apr. 2012.

[15] F. Z. Peng, "Revisit power conversion circuit topologies - recent advances and applications," in *Proc. IEEE IPEMC, 2009*, pp. 188 - 192.

The 2018 International Power Electronics Conference

Instantaneous Power Loss Calculation for MMC Based on Virtual Arm Mathematical Model

Yin Shiyuan[*], Wang Yue , Yin Taiyuan, Nie Cheng, Duan Guozhao and Wang Zhang
State Key Lab of Electrical Insulation & Power Equipment, Xi'an Jiaotong University, Xi'an, China
*E-mail: yinshiyuan@stu.xjtu.edu.cn

Abstract— the present methods for MMC (Modular Multi-level Converter) valve loss estimation are either inaccurate or time consuming. This paper proposed a virtual arm mathematical model (VAMM). Based on this model, the valve loss of MMC could be calculated accurately as well as quickly. What's more, this calculation method is valid for various operating conditions, different modulation and voltage balancing strategies. Firstly, an introduction of VAMM was presented and the mathematical equations for evaluating valve loss were deduced. Then the comparison with 11-level detailed switching model was given to verify the accuracy of VAMM. Finally, the proposed method was carried out on the valve loss analysis of Zhangbei MMC-HVDC (High Voltage Direct Current) system, results showing a good agreement with traditional analytic calculation method. And the distribution characteristics of valve loss under various operating conditions were also analyzed.

Keywords— Modular Multi-level Converter (MMC), Power Loss Calculation, valve loss evaluation.

I. Introduction

The modular multi-level converter (MMC) was first introduced in 2001 [1] and has been widely used in high voltage and high power applications due to its excellent output waveforms, especially in high voltage direct current (HVDC) systems [2]. It's significant to evaluate the valve loss accurately in engineering project, which not only helps switch devices selection and heat dissipation systems design, but also benefits topology and control strategy optimization.

At present, two main methods are widely used to calculate the valve loss of MMC. The first one is to build detailed model in simulation software and collect the time-domain data to calculate power loss [3]. The results could be precise but it has high requirements on hardware and always costs too much time. Therefore it is not feasible for high voltage applications because of the large number of sub-modules and possible operation conditions. The second method is based on analytic calculation [4-6], using experience mathematical formulas, where the equivalent current is introduced, so a fast loss calculation could be achieved. However, it only obtains average

This work was supported by the National Key Research and Development Plan (2016YFB0900902).

power loss value in a period time and the results may be inaccurate since the switching frequency is hard to determine when using nearest level modulation (NLM). Besides, many calculation methods apply only to PWM scheme [7-9].

To realize a fast and accurate valve loss analysis, this paper presented an instantaneous power loss calculation method based on virtual arm mathematical model (VAMM). According to the main parameters and control strategies of the system, VAMM could produce instantaneous arm current, sub-module capacitor voltage fluctuation and switching actions, achieving precise valve power loss calculation. This method is especially suitable for fast MMC valve loss evaluation in high voltage applications.

II. Basic Theory of MMC and VAMM

A. Brief Introduction of MMC

The basic structure of MMC is shown in Fig. 1. The converter consists of three phase units, and each unit consists of one upper arm and one lower arm connected in series between two DC terminals. Each arm has N series-connected half-bridge submodules (SM) and one inductor L. Each SM comprises two IGBT plus antiparallel connected FWDs, one capacitor C_0. The nominal voltage of C_0 equals to U_{dc}/N.

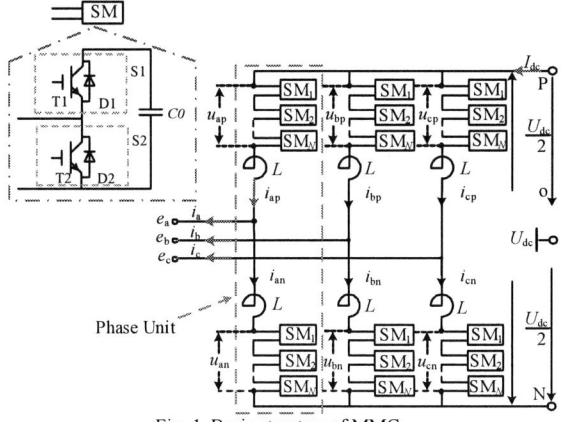

Fig. 1. Basic structure of MMC

In the steady state, the upper arm voltage and current of phase a are (neglecting the circulating current):

$$u_{ap} = \frac{U_{dc}}{2} - e_a = \frac{U_{dc}}{2}(1-m)\sin\omega t \qquad (1)$$

$$i_{ap} = \frac{I_{dc}}{3} + \frac{i_a}{2} = \frac{I_{dc}}{3} + \frac{1}{2}I_m\sin(\omega t - \varphi) \qquad (2)$$

where: m is modulation index, I_{m} is the peak value of the phase current, ω is the fundamental angular frequency, and φ is the phase of the phase current delay of the phase voltage.

Thus the instantaneous capacitor voltage of SM_{i} in upper arm of phase a is:

$$U_{Ci}(t) = U_{Ci}(t - \Delta t) + \frac{S_i(t)}{C_0}\int_{t-\Delta t}^{t} i_{ap}(\tau)\,\mathrm{d}\tau \qquad (3)$$

where Δt is the time interval between two control signals for SM_{i}; $S_i(t)$ is the switching function of SM_{i}, $S_i(t)=1$ means the SM_{i} being enabled, and $S_i(t)=0$ means the SM_{i} being bypassed.

B. Construction Theory of VAMM

The general method to simulate a MMC system is deriving the arm currents and voltages from the converter modulation process. By contrast, the VAMM firstly calculates the arm current and voltage from the main parameters (according to equation (1) and (2)). Then certain submodule selection algorithm would determine submodules' states. So the submodule capacitor voltages could be obtained from equation (3). Fig. 2 shows the structure of VAMM (taking the upper arm of phase a as example).

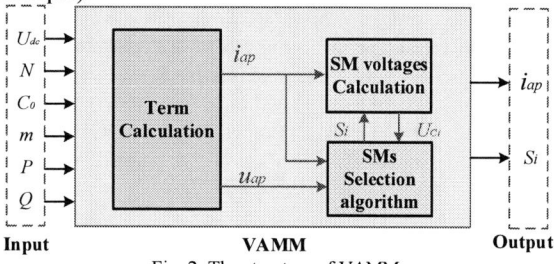

Input   **VAMM**   **Output**

Fig. 2. The structure of VAMM.

Once the real-time arm current and switching functions of every submodules are obtained, the valve power loss could be calculated according to section III.

III. VALVE LOSS CALCULATION OF MMC

According to IEC 62751-1, the total valve loss of MMC can be divided into five parts: the conduction loss of IGBT, the switching loss of IGBT, the conduction loss of FWD, the recovery loss of FWD, and their off-state blocking loss. But the off-state blocking loss could be neglected since it is far less than others.

The conduction power loss of IGBT and FWD is expressed as a function of current:

$$p_{Tcond}(i_C) = i_C \times (V_{T0} + R_{CE}i_C) \qquad (4)$$

$$p_{Dcond}(i_D) = i_D \times (V_{D0} + R_D i_D) \qquad (5)$$

where i_{C}, i_{D} is device conducting current; V_{T0}, V_{D0} is the voltage offset of the device, R_{CE}, R_{D} is the on-state resistance.

The switching energy per switching action of IGBT and the recovery energy of FWD are given in the datasheet, which can usually be approximated by second-order polynomials with high accuracy:

$$\begin{cases} E_{on}(i_C) = (a_{T1} + b_{T1}i_C + c_{T1}i_C^{\,2})k_1 \\ E_{off}(i_C) = (a_{T2} + b_{T2}i_C + c_{T2}i_C^{\,2})k_2 \\ E_{rec}(i_D) = (a_D + b_D i_D + c_D i_D^{\,2})k_3 \end{cases} \qquad (6)$$

where a_{i}, b_{i}, c_{i} are the coefficients of the second-order polynomials; k_i is the adjustment coefficient.

From section II, the switching actions of each device on submodules could be obtained, and the switching power loss during one control period could be expressed as:

$$P_{Tsw} = \frac{E_i}{T_s} \qquad (7)$$

$$P_{Dsw} = \frac{E_j}{T_s} \qquad (8)$$

where E_i is the switching energy caused by IGBT switching action, and $E_i=0$, E_{off} or E_{on} depending on the IGBT switching function; E_j is the switching energy caused by FWD switching actions, and $E_j=0$ or E_{rec} depending on the FWD switching functions. T_s is the control period.

IV. CASE STUDIES

A. VAMM Verification

To verify the accuracy of VAMM, an 11-level detailed switching model of MMC is established with time-domain simulation tool Matlab/Simulink. A VAMM having same parameters as 11-level MMC's upper arm of phase a is also established. Fig. 3 shows that there were no obvious differences in arm current between the two models. Fig. 4(a) shows that one submodule receives 17 pluses in one fundamental period in detailed switching model, where Fig. 4(b) shows one submodule receives 18 pulses in VAMM. Tiny and acceptable error exists because VAMM does not have a practical closed-loop control.

In particular, it takes 4 minutes to simulate 2 seconds using a detailed switching model, and the time required to simulate 2 seconds using a VAMM is only 7 seconds. So the VAMM is more efficient than detailed switching model when calculating the valve loss.

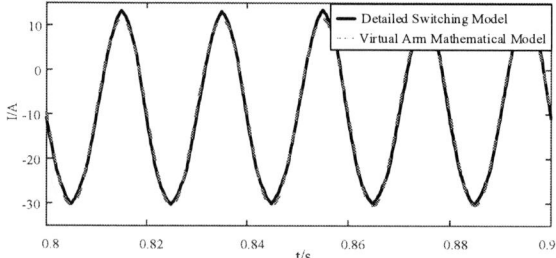

Fig. 3. Comparison of detailed switching model's arm current with VAMM result.

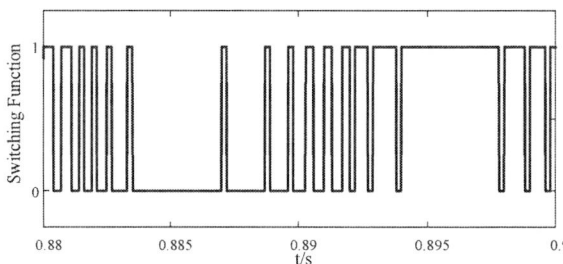

Fig. 4(a). Switching function of a submodule in detailed switching model.

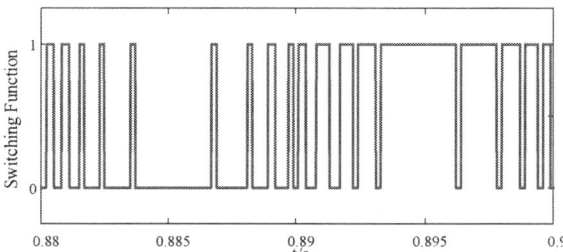

Fig. 4 (b). Switching function of a submodule in VAMM.
Fig. 4. Switching function of submodule.

B. Comparison with Traditional Analytic Calculation Methods

Kangbao station in Zhangbei HVDC system is used as an example to analyze the valve loss of MMC. The main parameters of this MMC system are listed in Table I. Assuming NLM and traditional capacitor voltage sorting strategy is applied. ABB IGBT modules (5SNA 3000K452300) are selected.

TABLE I
MAIN CIRCUIT PARAMETERS OF MMC

Items	values
Rated power/MW	750
DC bus voltage/kV	500
SM capacitance /mF	7
SM capacitor voltage /kV	2.155
Number of SMs per arm	232
modulation index	0.99

When MMC operates in rated power and rectifier mode, the calculated instantaneous power loss of an arm based on VAMM method is shown in Fig. 5. Traditional analytic calculation method [4] is also applied to compare with the results. But the traditional analytic calculation method can only get average loss, the average value of instantaneous power loss is also given. Fig. 5(a) shows the total conduction loss of the arm. And the average power losses evaluated from two methods have almost same value. Fig. 5(b) shows the total switching loss of the arm. Since the traditional analytic calculation method use equivalent current to evaluate loss where VAMM use real-time current, tiny errors do exist.

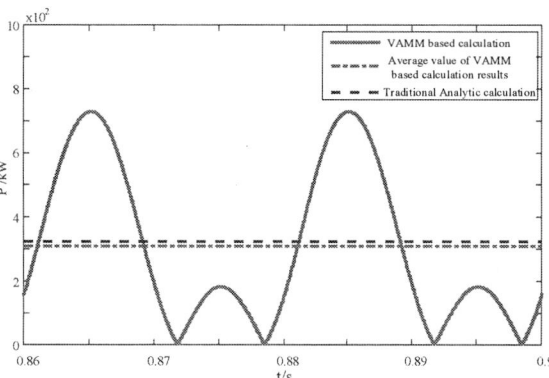

Fig. 5(a). Arm conduction power loss.

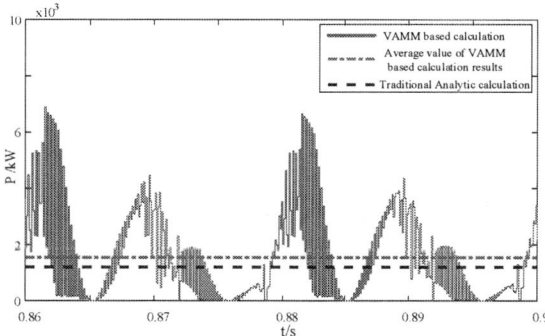

Fig. 5(b). Arm switching power loss.
Fig. 5. Arm power loss calculation results comparison.

C. Distribution Characteristics of Valve Loss

When MMC operates in different conditions, the conduction loss and switching loss would be distributed unequally in devices. In this paper, instantaneous power loss distributions under four typical operation conditions (φ=0, $\pi/2$, π, $3\pi/2$) are discussed using VAMM based loss calculation method.

Fig. 6 shows the instantaneous conduction and switching power loss distributions when φ=0. Note that the conduction power loss is proportional to the current and the direct current is negative according to Fig. 1. When the arm current reaches negative maximum value, the sub modules on the arm are almost bypassed, so the arm current would flow through D2, and the maximum instantaneous conduction power loss would be generated on the D2. For switching power loss: the arm current would flow through T1 and D2 more frequently due to the negative direct current and the switching actions of T1 are corresponding to D2 (When T1 turns on, D2 turns off and when T1 turns off, D2 turns on), but IGBT costs much higher switching energy than FWD, so T1 would generate maximum switching power loss.

In Fig. 7, when MMC operates as STATCOM (φ=$\pi/2$ or φ=$3\pi/2$), the conduction power loss generated on every device is similar because the direct current is zero. And switching power losses caused by IGBTs (T1, T2) are higher than FWD (D1, D2).

2627

The 2018 International Power Electronics Conference

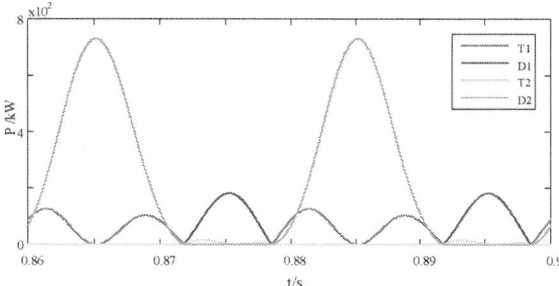

Fig. 6(a). Conduction power loss distribution on the arm

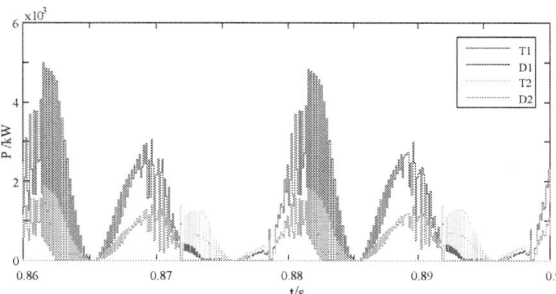

Fig. 6(b). Switching power loss distribution on the arm
Fig. 6. Power loss distribution when φ=0.

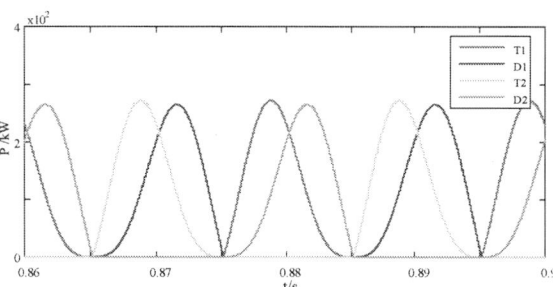

Fig. 7 (a). Conduction power loss distribution on the arm

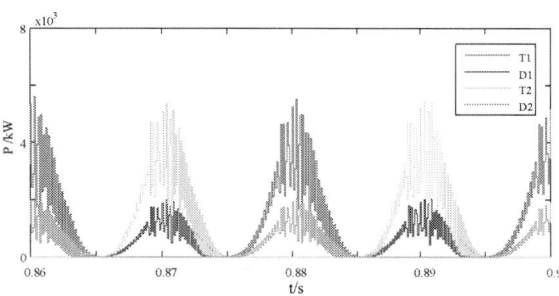

Fig. 7 (b). Switching power loss distribution on the arm
Fig. 7. Power loss distribution when φ=π/2 or φ=3π/2.

Fig. 8 shows when MMC operates as an inverter (φ=π), T2 would generate maximum conduction power loss because the direct current is positive. The switching actions of T2 are corresponding to D1 (When T2 turns on, D1 turns off and when T2 turns off, D1 turns on), but IGBT costs much higher switching energy than FWD, so T2 would generate maximum switching power loss.

Fig. 9 shows the average value of above calculation results. It's clearly that T2 would produce highest losses when MMC working as an inverter. So the heat

dissipation design for T2 should be carefully concerned. This conclusion is consistent with previous studies.

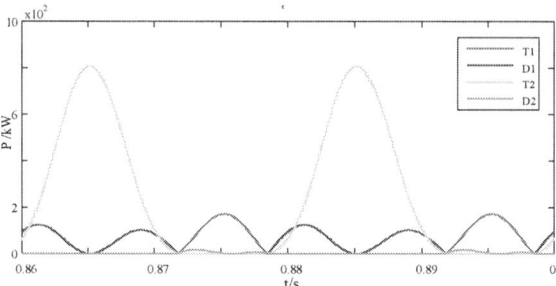

Fig. 8(a). Conduction power loss distribution on the arm

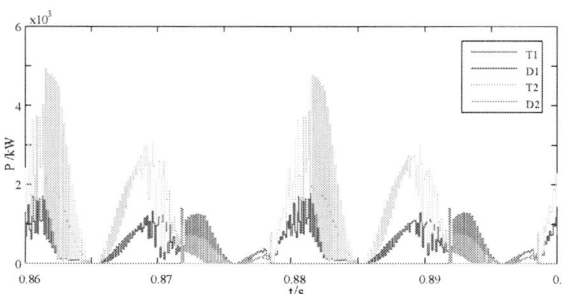

Fig. 8(b). Switching power loss distribution on the arm
Fig. 8. Power loss distribution when φ=π.

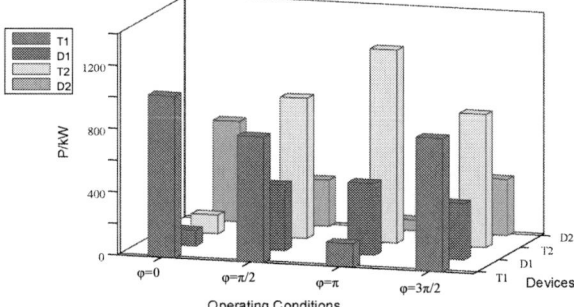

Fig. 9. Average power loss distribution under different operating conditions

In conclusion, switching loss obviously takes a larger part than conduction loss shown from Fig. 6 to Fig. 8, so it's necessary to adopt promoted voltage balancing strategies to reduce the switching frequency. And when φ takes different values, the loss distribution of the devices would vary greatly. When MMC operates as a rectifier or inverter, some certain device would generate relatively larger conduction or switching loss due to the existing of direct current.

V. CONCLUSION

An instantaneous power loss calculation method for MMC based on VAMM is presented in this paper. Both high efficiency and accuracy can be achieved. This method is convenient to evaluate the valve loss when MMC operates in various conditions and is valid for different modulation and voltage balancing strategies. So

2628

it is very suitable for fast loss evaluation in high voltage applications. And the calculation results in section IV show that it's necessary to reduce the switching frequency to decrease the power loss. And the valve loss distributions could be unequal, thus different selection criteria for devices would be considered.

REFERENCES

[1] R. Marquardt, "Stromrichterschaltungen mit Verteilten Energiespei-Chern," German Patent DE10103031A1, Jan. 24, 2001.

[2] S. Allebrod, R. Hamerski, and R. Marquardt, "New transformerless, scalable modular multilevel converters for HVDC-transmission," in Proc. IEEE Power Electron. Specialists Conf., 2008, pp. 174–179.

[3] A. D. Rajapakse, A. M. Gole, and P. L.Wilson, "Electromagnetic transients simulation models for accurate representation of switching losses and thermal performance in power electronic systems," IEEE Trans.Power Del., vol. 20, no. 1, pp. 319–327, Jan. 2005.

[4] H. Wang, G. Tang, Z. He, J. Cao, and X. Zhang, "Analytical approximate calculation of losses for modular multilevel converters," IET GENERATION TRANSMISSION & DISTRIBUTION, vol. 9, pp. 2455-2465, 2015.

[5] Z. Zhang, Z. Xu and Y. Xue, "Valve Losses Evaluation Based on Piecewise Analytical Method for MMC-HVDC Links," IEEE Transactions on Power Delivery, vol. 29, pp. 1354-1362, 2014.

[6] D. A. Montoya-Acevedo, A. Escobar-Mejia and M. Holguin-Londono, "Comparison of Two Analytical Methods to Estimate Power Losses in an HVDC Terminal Based on Modular Multilevel Converter," in International Conference on Power Electronics, 2016, pp. 58-63.

[7] S. Rohner, S. Bernet, M. Hiller, and R. Sommer, "Modulation, Losses, and Semiconductor Requirements of Modular Multilevel Converters," IEEE Transactions on Industrial Electronics, vol. 57, pp. 2633-2642, 2010.

[8] A. Christe and D. Dujic, "Virtual Submodule Concept for Fast Semi-Numerical Modular Multilevel Converter Loss Estimation," IEEE Transactions on Industrial Electronics, vol. 64, pp. 5286-5294, 2017.

[9] A. Babaie, B. Karami and A. Abrishamifar, "Improved Equations of Switching Loss and Conduction Loss in SPWM Multilevel Inverters," Power Electronics, Drive Systems & Technologies Conference, 2016.

[10] C. Oates and C. Davidson, "A comparison of two methods of estimating losses in the modular multi-level converter," in Proc. 14th Eur.Conf. Power Electron. Appl., 2011, pp. 1–10.

Comparison of Current Control Strategies in Modular Multilevel Converter

Jianzhao Wei[1], Anirudh Budnar Acharya[1*], Lars Norum[1] and Pavol Bauer[2]

1 Department of Electric Power Engineering, IME Faculty, NTNU, Trondhiem, Norway
2 Department of Electrical Sustainable Energy, EEMCS Faculty, TUD, Delft, Netherland
E-mail: anirudhb@ntnu.no

Abstract— There is a need to transmit the power generated from windfarm in remote locations to the residential and industrial load. One of the suitable and economic solution for such long distance transmission of power is the High Voltage DC (HVDC) transmission. The Modular Multilevel Converter (MMC) based HVDC transmission is a proven technology with very high power quality. A multi-objective control is used to control the MMC, one of the objectives is to deliver the power from windfarm to grid, the other objective of the control is to achieve stable operation of the MMC. In order to achieve a stable operation, the capacitors in the MMC should be balanced and the currents circulating within the phases of the MMC need to be controlled. Many control strategies have been proposed to suppress such Circulating Current (CC). However, there is no clear comparison between these control strategies.

In this paper control strategies for the MMC are reviewed specifically the circulating current suppressor. The cascaded control based on PI-controller and the Model Predictive Control (MPC) for outer current control are compared. Results show that traditional method has better steady state performance, while MPC based method has much faster dynamic response and has the advantage of involving less control loops in controlling systems with multiple control aspects. The control methods for capacitor voltage balancing and the circulating current suppression are simulated and compared.

Keywords—Modular Multilevel Converter (MMC), Circulating Current, Voltage Balancing Algorithm, Current Control

I. INTRODUCTION

With the increasing demand of large capacity and long distance energy transmission, such as that for offshore wind farm, the HVDC transmission is preferred and since over long distance the HVDC technology will be economical for future energy system[1]–[4]. In comparing with High Voltage Alternating Current (HVAC) system, the advantages of HVDC include: 1) the long distance transmission 2) low losses and voltage drop on transmission lines, 3) being able to connect two asynchronous system, 4) fast and accurate control of power flow and quick fault isolation, 5) cheaper cost for long distance transmission.

The MMC [5], [6] offers a modular solution with high power quality. It can be scaled to high voltage levels and produce low harmonics at the output. Such features are especially suitable in HVDC applications[7]–[11]. Intense research has been done on MMC during past few years. Each phase of the MMC has two arm (upper and lower), and each arm is a series connection of power electronics block referred as sub-module (SM). Each SM has a floating capacitor.

The floating capacitors in SM of the MMC needs to be balanced [12], [13]. As SM are inserted or bypassed at different parts of the power cycle, leading to asymmetric charging or discharging of individual capacitors, the capacitor voltage variation will occur, which make control system inaccurate and unstable[14]. Therefore, the SM capacitor voltage is required to be constant.

Due to the ripple in the inserted voltage, a CC flows between the phases of the MMC and does not appear in the output current [15], [16]. This increases the voltage ripple and power loss in the MMC [13]. So effective algorithms are needed for CC suppression.

For the MMC used in the HVDC system, the active and reactive power control are necessary for energy transmission purpose[17], [18]. This paper reviews the recent control strategies of the MMC associated with all the control aspects and compare different algorithms by simulation based on the application of HVDC system. In order to focus more on control algorithms, only one side MMC of HVDC system will be studied.

The rest of the paper is organized as follows. Section II introduces some background knowledge about MMC including MMC structure and mathematical model. In Section III, voltage balancing algorithms, CC suppression algorithms and current control algorithms are reviewed. After that, different strategies chosen to be simulated and compared will be introduced in Section IV. The simulation results are presented in Section V. Section VI includes conclusions and remarks.

II. STRUCTURE OF MODULAR MULTILEVEL CONVERTER

A. MMC structure

Fig.1 shows the structure of the MMC. Each phase is also referred as a leg. Each leg has two arms (upper and lower), both of which has a N series-connected identical SM (shown in dotted box). The topology of the SMs of the MMC can vary, the one shown in Fig.1 is half-bridge topology, which is the most popular one [13], [19] and will be used in this paper. The other variants of the SM topology are summarized in [13]. Each arm has a series inductor, which is used to limit the fault current through the arm of the MMC.

By controlling the switches in the SM, the output voltage of the SM V_{SM} can be equal to capacitor voltage or zero, which are called SM inserted state or bypassed state respectively. The desired output AC voltages are achieved by choosing correct state combinations of all SMs. More detailed operation principle can be found in [5], [6], [20].

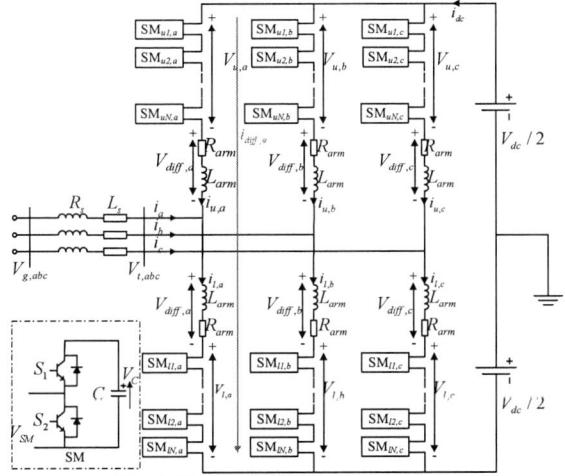

Fig. 1 The detailed diagram of MMC

B. Mathematical Model of MMC

As shown in Fig. 1, the inserted arm voltage is represented as $V_{k,j}$, the voltage $V_{diff,j}$ is the voltage appearing across the arm impedance. The circulating current (shown in red line for phase A) is $i_{diff,j}$. On the output side, the converter is connected to the grid. The grid voltage is represented as $V_{g,j}$ and the grid effective equivalent resistance is R_s and grid inductance is represented as L_s. The subscript 'k=u or l' based on upper or lower arm and subscript 'j=a, b or c' representing the phase. The DC link of the MMC is connected to DC voltage source with midpoint connected to ground. The directions of all quantities are shown in Fig. 1.

Applying KVL, the dynamic equations of the MMC in phase j can be expressed by:

$$\frac{V_{dc}}{2} - V_{u,j} - R_{arm}i_{u,j} - L_{arm}\frac{di_{u,j}}{dt} + R_s i_j + L_s \frac{di_j}{dt} - V_{g,j} = 0 \quad (1)$$

$$-\frac{V_{dc}}{2} + V_{l,j} + R_{arm}i_{l,j} + L_{arm}\frac{di_{l,j}}{dt} + R_s i_j + L_s \frac{di_j}{dt} - V_{g,j} = 0 \quad (2)$$

The output current is divided equally between the upper and lower arm. The DC current in each phase is dependent on the number of phases and, for the three phase MMC it is 1/3rd of the total DC current in each phase. Therefore, the arm current will be,

$$i_{u,j} = \frac{I_{dc}}{3} + i_{diff,j} + \frac{i_j}{2} \quad (3)$$

$$i_{l,j} = \frac{I_{dc}}{3} + i_{diff,j} - \frac{i_j}{2} \quad (4)$$

The circulating current $i_{diff,j}$ only represents the AC components as the DC current is expressed separately as $I_{DC}/3$.

Summing (1) and (2), and substituting i_j in (3) and (4), the outer dynamic equation for MMC is yielded,

$$(L_{arm} + 2L_s)\frac{di_j}{dt} = -(R_{arm} + 2R_s)i_j + V_{u,j} - V_{l,j} + 2V_{g,j} \quad (5)$$

Taking the difference between (2) and (1) and substituting $i_{diff,j}$ in (3) and (4), the inner dynamic equation for MMC will be,

$$V_{diff,j} = L_{arm}\frac{di_{diff,j}}{dt} + R_{arm}i_{diff,j} + R_{arm}\frac{I_{dc}}{3} = \frac{V_{dc}}{2} - \frac{V_{u,j}+V_{l,j}}{2} \quad (6)$$

As per (5), the output current is controlled by difference between the arm voltages $V_{u,j} - V_{l,j}$ and as per (6) the circulating current is controlled by the sum of arm voltages $V_{u,j} + V_{l,j}$ (or $V_{diff,j}$). Therefore, (5) and (6) is fundamental in controlling the MMC introduced.

The dynamics of the SMs is discussed in [13], the effective capacitance voltage is written as,

$$C\frac{dV_{cu,j}}{dt} = i_{u,j}\frac{n_{u,j}}{N} \quad (7)$$

$$C\frac{dV_{cl,j}}{dt} = i_{l,j}\frac{n_{l,j}}{N} \quad (8)$$

where $V_{ck,j}$ is the individual SM capacitor voltages, $n_{k,j}$ is the inserted number of SMs in upper and lower arms.

Equation (3) to (8) gives a generalized dynamic model of the MMC.

III. MMC CONTROL SCHEMES

In this section, the MMC control method for the SM capacitor voltage balancing, CC suppression and output AC current control are reviewed.

A. Voltage balancing Algorithms

The most common control used for balancing the SM capacitor voltages is sorting method. The sorting method measures the SM capacitor voltage, sorts them in ascending and descending order. If the arm current is positive, the SMs with lowest voltages are inserted. This will charge the capacitor and increase the voltage; otherwise, the SM with highest voltages are inserted, such that the capacitors will be discharged and voltages decrease [21]. This algorithm is simple and effective, however, unnecessary switching will occur and switching frequency increases. Many methods were proposed to solve the problem[12], [22]–[25]. In [26] a method based on carrier rotation is proposed. By rotating the carriers of PWM for each SM, the energy will be equally distributed to all SMs and voltage will be balanced. However, for more precise control, the capacitor voltages still need sorting. In [27] an averaging control and balancing

2631

control is proposed. Two closed-loop control are used to reduce the error between the individual SM capacitor voltage and the desired reference value. This method is complex compared to the sorting method but can achieve higher flexibility in order to control the SM voltages. In [28], MPC is used to balance the SM voltages. By measuring the arm current, the SM capacitor voltage can be predicted according to equation (7) and (8) and desired performance can be achieved by minimizing the cost function.

B. Circulating Current Suppression Control (CCSC)

In [14], [29], the control method based on energy control are proposed to suppress the CC by controlling the total energy and the difference of energy between upper and lower arm. In these methods, the CC includes both AC and DC components and only AC components need suppressing as the conclusion is made that the AC component of CC will break the energy balance between upper and lower arm, while DC component of the current is responsible for transfer of energy from AC side to DC side (and vice versa) of the MMC. Based on same idea of controlling energy, a similar control using mathematical optimization is proposed in [30] and it shows a better performance. Based on the idea of controlling the AC and DC component of circulating current separately, more methods are proposed [31], [32]. These methods suppress CC indirectly by controlling the energy or voltages of SMs. In addition, they can achieve single phase control, which is more flexible. Different from that, the CCSC proposed in [22] suppresses CC directly and treat three phases as a whole. It is concluded that the CC is a negative sequence component with double the grid frequency. The CC has not only double grid frequency component but also even harmonic components [16], [33], [34]. Therefore, in order to suppress these harmonics, the Proportional-Resonant (PR) controller and repetitive controller has been used. Hysteresis control is also used to suppress CC in defined band as discussed in [15]. In addition, MPC is used to eliminate circulating current by adding related constraints to cost function in [28].

C. AC side output current control

PI based control algorithm is the traditional method for AC side current control of voltage source converter. According to [17], [35], the control scheme for the MMC applied in HVDC system is same as that for a 2-level converters [3], [36]–[38]. It is divided into two control loops, the inner control loops has high bandwidth and the outer control loops has a relatively low bandwidth. Inner control loop is for the current controlled in d-q frame and outer loop can be achieved by active power (or DC voltage) control and reactive power control. It has the advantage of simplicity and robustness. Recently, MPC method is proposed as an alternative method for controller AC side current for 2-level VSC [39]–[43], as well as the MMC [8], [28], [44]. Different from PI based method, MPC highly depends on the mathematical model introduced in Section II B. All three control terms are

added into a single cost function, the desired performance is achieved by minimizing cost function. In addition, hysteresis current control method is proposed to control MMC [45], [46]. It is straightforward, but more effort is made on choosing voltage levels as the number of voltage level increases. Inherent nonlinear dynamics of the MMC makes the nonlinear modelling and control as a preferred choice of control to achieve better performance, as discussed in [47], [48]. But they involve high computational effort.

IV. COMPARISON OF VARIOUS CONTROL SCHEME

For each control aspects reviewed in last section, at least two methods with same control purpose are chosen and simulated in Matlab/Simulink environment for comparison. In addition, study of the MMC has been conducted for the HVDC transmission system and the system is shown in Fig. 2, in which the symbols have the same meaning as that in Fig. 1. The DC side is represented with the DC voltage sources. When the DC voltage control is applied, two resistors are connected as load on the DC side to represent the rest of the system.

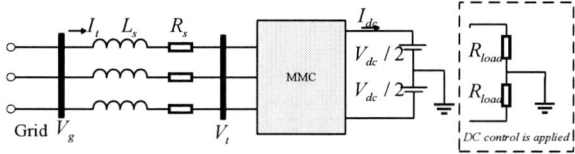

Fig. 2 The system considered for the simulation study.

In order to verify and compare different control algorithms, two cases are defined. Case 1 compares two CCSC methods and Case 2 aims to compare AC side current control algorithms and voltage balancing algorithms. Simulation results are shown in Section V.

A. Case 1: CCSC Comparison

In this case, two CCSC methods for comparison are defined (sorting method is used for both cases to balance voltages):

1) M1: Based on double line frequency negative sequence d-q frame [22].

The arm currents are measured and circulating current is calculated based on equation (3) and (4), which then are transferred to d-q frame but by setting the transfer angle double of fundamental one and input sequence to be negative. Two control loop are built for d and q axis respectively and the reference current is set to zero. The circulating currents before and after using the method are compared.

2) M2: Based on Energy control [14].

The voltages of SM capacitors on each phase are measured. The energy stored in the upper and the lower arm is calculated using the capacitance value. The control loops are developed to control the total energy and the energy difference. The steady state performance and dynamic response will be tested and the circulating current before and after using the methods are compared.

As mentioned in review section, *M1* is a direct method, while *M2* is indirect for CCSC, which makes the

2632

comparison more meaningful.

B. Case 2: Two overall control scheme comparison

Either of two overall control scheme includes all three control strategies. The traditional PI based method and relative modern MPC based method will be compared. The control diagrams are shown below.

1) M3: PI based method[3], [17]

In this method, AC side current control is achieved by PI controller, *M1* in Case 1 is used to suppress the CC and sorting method is used for SM capacitor voltage balancing. Here the DC voltage control is used. The overall control scheme is shown in Fig. 3.

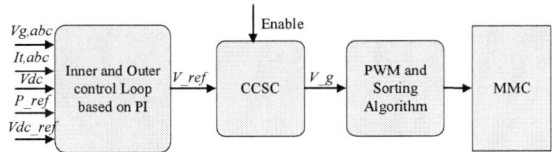

Fig. 3 Overall control Scheme of *M3*

2) M4: MPC based method[28]

In this method, all three control strategies are achieved by using MPC algorithm as shown in Fig. 4. The current reference is calculated by power equations according to desired active power and reactive power. In addition, the capacitor voltage balancing performance by using sorting algorithm and MPC is also compared.

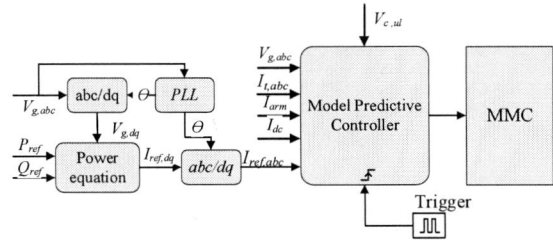

Fig. 4 Overall control Scheme of *M4*

C. Parameter Value

TABLE I
PARAMETERS FOR SIMULATION

Symbol	Meaning	Value
P	Active power	20 MW
Q	Reactive power	6.6 MVAr
V_g	Grid voltage (phase, peak)	14.14 kV
L_s	Grid side inductance	3.17 mH
R_s	Grid side resistance	0.062 Ω
$V_{dc}/2$	DC bus voltage	17.68 kV
N	Number of SMs per arm	6
C	SM capacitance	0.01 F
L_{arm}	Arm inductance	1.59 mH
R_{arm}	Arm resistance	0.1 Ω
V_c	SM capacitor voltage	5892 V
F_s	Carrier frequency	600 Hz
T_s	Sampling period (for MPC)	100 µs

The parameters of the model are summarized in TABLE I. The parameter values used in [22] is considered,

the modified parameters such as SM capacitor and inductance are chosen according to [49].

V. SIMULATING RESULTS

The cases defined in Section IV are simulated in Matlab/Simulink environment. The results are displayed in p.u. based on the value shown in TABLE I. In order to demonstrate the results better, the time periods from start up to steady state (0-0.4s) are ignored in the figures.

A. Case 1

1) M1

The simulating results are shown in Fig. 5. The CCSC algorithm is enabled at 0.5s as shown in Fig. 5(a). The circulating current is shown in Fig. 5(b) (the DC component is removed using a band-pass filter).

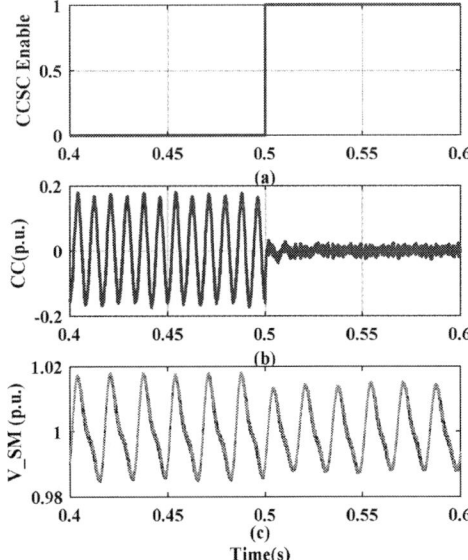

Fig. 5 (a) Enable signal for the CCSC; (b) Circulating current (c) Capacitor voltages of phase A

Before the CCSC was enabled, the peak magnitude of circulating current is 0.18 p.u. of the rated current at double grid frequency as expected. When the CCSC is enabled, the peak magnitude of circulating current is reduced to less than 0.03 p.u. Figure 5(c) shows the capacitor voltage of all SMs in phase A of the MMC, it is seen that the CCSC reduces the SM capacitor voltage ripple by 15%. The SM capacitor voltages are well balanced due to the usage of sorting method.

2) M2

The simulating results are shown in Fig. 6. The energy control is enabled from the beginning. At first, total energy and balance energy reference were at 1p.u. and 0 respectively until t=0.8s and t=1.3s, step changes were added to references one after another as shown in Fig. 6 (a) and (b). The total energy reference is upper limited by the SM capacitor voltage limitation, and lower limited by the maximum allowed modulation index otherwise it can be freely selected, while the balance energy reference should always be zero[14]. Here a step is added only for testing purpose.

2633

From Fig. 6(c) and (d), the SM capacitor voltages in both upper and lower arm follows the new references in less than 100ms with zero steady state error. In Fig. 6(e), the circulating current is suppressed to less than 0.1 p.u after enabling CSCC. The ripple of SM capacitor voltages in Fig. 6(c) or (d) are equal to 3% of the rated capacitor voltage.

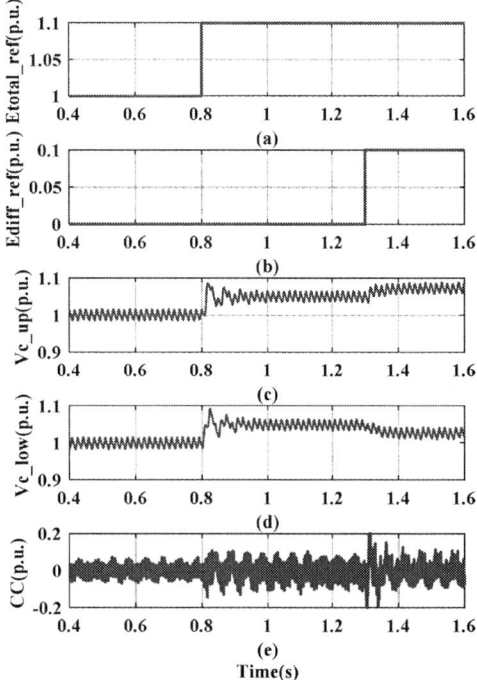

Fig. 6 (a) Reference for total energy controller (b) reference for balance controller (c) upper arm SM capacitor voltages of the MMC (d) lower arm SM capacitor voltages of the MMC (e) circulating current

3) Comparison between M1 and M2

The comparisons between two methods are summarized in TABLE II. As we can see, the method *M1* is better than *M2* in suppressing the CC and reducing the SM capacitor voltage ripple in the MMC. In addition, in case of *M1* it is relatively simple to tune the controller and synchronize it with other controllers. The main advantage of the energy control is that the DC energy can be controlled with much flexibility.

TABLE II
COMPARISON BETWEEN TWO CCSC METHODS (IN P.U. VALUE)

	No Controller	With M1	With M2
CC(peak)	0.18	<0.03	0.08
Voltage Ripple (p-p)	3.25%	2.75%	3%
Tuning		Easy	Hard
Other Function		No	Energy Control

B. Case 2

1) M3

The simulation model block diagram is based on Fig. 3. The results are shown in Fig. 7. At t=0.2s the CCSC is enabled as shown in Fig. 7(a) and at t=0.4 the reactive power has a step change from 0 p.u. to 0.33 p.u as shown in Fig. 7(b). It is seen that the reactive power tracks the reference in Fig. 8(b) with zero steady state error and the steady state is reached in 0.03s with no overshoot. The DC bus voltage, shown in Fig. 7(c), is not affected by the step change of reactive power and is kept at 1p.u. by DC control. From Fig. 7(d) and (e), the CCSC effectively reduced CC from about 0.15p.u. to almost zero both before and after the reactive power change. However, it brought about 3% noise on DC bus voltage as shown in Fig. 7(c). In addition, CCSC also injected oscillation on SM voltages with period about 2.5s, which is slow and the cause needs further study. Besides that the results verify the good performance of target control scheme. Also, it is proved that all three controllers can operate together well without unacceptable mutual effect.

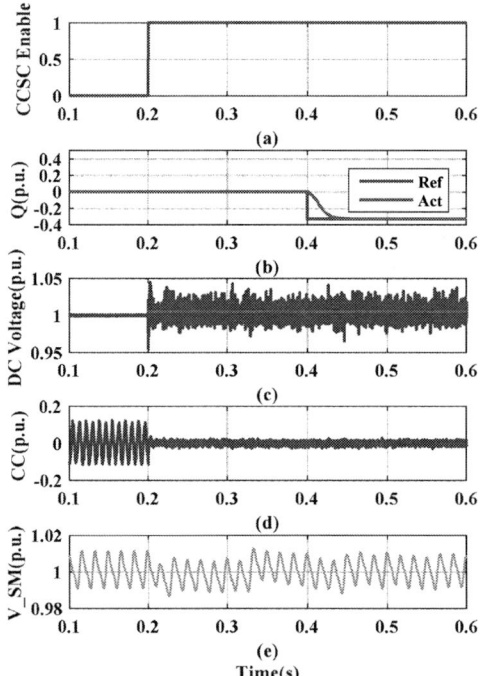

Fig. 7 (a) enable signal of CCSC (b) reactive power and reference value (c) DC bus voltage (d) circulating current (e) SM capacitor voltages in upper arm of phase A

2) M4

Based on Fig. 5, all three constraints are added to cost function. An empirical method is used to tune the weighting factors in [50] and value of these factors chosen to be $\lambda_v = 6$, $\lambda_{cir} = 1$ and $\lambda_{ac} = 1$ for voltage balancing control, CCSC and AC side current control respectively in this simulation.

The MPC function is triggered every 100 µs. In this way, the switching frequency is fixed and can be varied based on the trigger to MPC. Simulation results are shown in Fig. 8. As the time period from start up to steady state is longer than *M3*, the time starts from 1.5s.

The system is initially operated with zero active and reactive power reference as shown in Fig. 8(a) from t=1.5s to 1.6s and the capacitor voltage balancing and circulating current controller are enabled. At t= 1.6s, active power reference is changed as a step to 1p.u. in order to transfer active power from AC to DC side. At t= 1.8s, reactive power reference is varied as a step to transfer 0.33 p.u. power from DC to AC side.

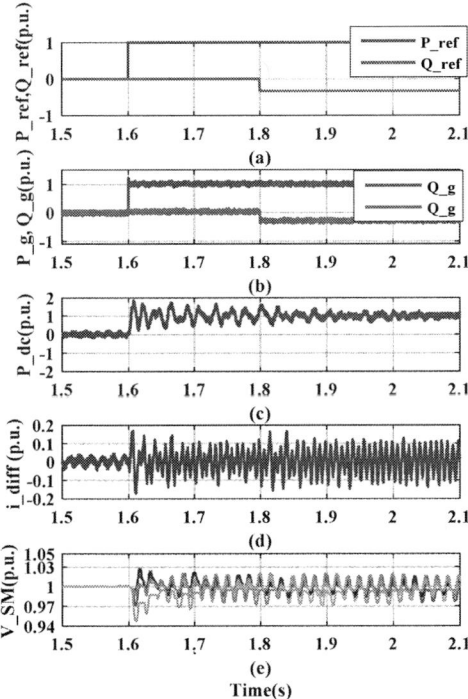

Fig. 8 (a) Active and reactive reference (b) the active and reactive power measured at grid bus (c) active power transferred to load (d) circulating current (e) SM capacitor voltages in the upper arm phase A of the MMC

Fig. 8(b) are the measured active and reactive power respectively. Comparing the simulation results, it is seen that both active and reactive power track the references with good dynamics responses, the rise time is 300µs. The steady state error is zero, but the ripples of active and reactive are about 5%, which is on the higher side than the desired value. Fig. 8(c) shows the power transferred to the DC side, which is changed from 0 to 1p.u after step change of active power reference as expected. Such variation causes power oscillations which is eventually damped in 500ms as seen from simulation, the overshoot is 100%, which increases the stress on the DC bus capacitor. Fig. 8(d) and (e) show the circulating current and SM capacitor voltages respectively. The CC was suppressed to 0.1p.u. and the ripple of SM capacitor voltage is kept under 5%. However, the SM capacitor voltages are not exactly balanced, the differences are up to 3% of rated value.

3) Comparison between M3 and M4

Two method are compared and summarized below:

1) **Performance evaluation of circulating current suppression and SM capacitor voltage balancing:**

Comparing Fig. 7(d), (e) and Fig. 8(d), (e) it is seen the both method controlled the circulating current and SM capacitor voltage in the desired range. Using method proposed in M3, the circulating current was suppressed to 0.03p.u. compared to 0.1p.u.(peak value) by MPC. The SM capacitor voltages are balanced in each arm and the voltage ripple is 2% and 5% for the MPC. In addition, 3% SM voltage difference is observed by using MPC. These all show better performance of PI based method over MPC.

2) **Reference tracking and response:** Comparing Fig. 8(b) and Fig. 7(b), for steady state performance, classical method (PI controller) tracks the reference with steady state error less than 0.01%, while MPC results in about 5% steady state error. However, MPC had considerably faster dynamic response, around 300µs to reach the new reference value after step change, while the reference is tracked in 30ms for PI based method.

3) **Complexity of the system.** For the MPC, the model of entire system is necessary, therefore, the system is sensitive to parameter variations. This increases computation efforts and reduces the robustness. With a single cost function, the three objectives of the control (output current, capacitor voltage balancing and circulating current suppression) are achieved. The weight factor can be selected to tune the response of the system. In case of the PI control, three loops are used to achieve the same objectives with different PI controllers. Proper tuning of the controllers are needed for synchronized control. However, in case of MPC no special care is needed for synchronized control.

VI. CONCLUSION

In this paper, control methods for the SM capacitor voltage balancing, circulating current suppression and output current control for the MMC are reviewed. Most promising control methods are simulated and compared in Simulink with windfarm connected HVDC system as case study. It is seen that the method based on double line-frequency d-q coordinate for CCSC has better performance over energy control method in suppressing the circulating current with in the MMC phases. While the advantage of energy control is that it can control the arm energy as desired with fast dynamics. Compared to MPC for the circulating current suppression, both double line-frequency d-q coordinate and energy control methods have better response. As for AC side output current control, classical PI control has better reference tracking compared to MPC which results is 5% steady state error. However, much faster dynamic response of the MPC is 10 times better than that of the classical PI control method.

REFERENCES

[1] O. Heyman, L. Weimers, and M.-L. Bohl, "HVDC-A key solution in future transmission systems," in *World Energy Congress-WEC*, 2010, pp. 12–16.

[2] W. Breuer, D. Povh, D. Retzmann, E. Teltsch, and X. Lei, "Role of HVDC and FACTS in future Power Systems," in *CIGER Symposium, Shang Hai*, 2004.

[3] A. Korompili, Q. Wu, and H. Zhao, "Review of VSC HVDC connection for offshore wind power integration," *Renew. Sustain. Energy Rev.*, vol. 59, pp. 1405–1414, Jun. 2016.

[4] M. P. Bahrman and B. K. Johnson, "The ABCs of HVDC transmission technologies," *IEEE Power Energy Mag.*, vol. 5, no. 2, pp. 32–44, 2007.

[5] A. Lesnicar and R. Marquardt, "An innovative modular multilevel converter topology suitable for a wide power range," in *Power Tech Conference Proceedings, 2003 IEEE Bologna*, 2003, vol. 3, p. 6-pp.

[6] M. Glinka and R. Marquardt, "A New AC/AC Multilevel Converter Family," *IEEE Trans. Ind. Electron.*, vol. 52, no. 3, pp. 662–669, Jun. 2005.

[7] I. C. Damian, M. Eremia, and L. Toma, "Advanced control of a modular multilevel high voltage direct current converter," in *2017 International Conference on ENERGY and ENVIRONMENT (CIEM)*, 2017, pp. 1–5.

[8] Z. Zhang, M. T. Larijani, W. Tian, X. Gao, J. Rodríguez, and R. Kennel, "Long-horizon predictive current control of modular-multilevel converter HVDC systems," in *IECON 2017 - 43rd Annual Conference of the IEEE Industrial Electronics Society*, 2017, pp. 4524–4530.

[9] K. Friedrich, "Modern HVDC PLUS application of VSC in modular multilevel converter topology," in *Industrial Electronics (ISIE), 2010 IEEE International Symposium on*, 2010, pp. 3807–3810.

[10] A. Nami, J. Liang, F. Dijkhuizen, and G. D. Demetriades, "Modular Multilevel Converters for HVDC Applications: Review on Converter Cells and Functionalities," *IEEE Trans. Power Electron.*, vol. 30, no. 1, pp. 18–36, Jan. 2015.

[11] N. Ahmed *et al.*, "HVDC SuperGrids with modular multilevel converters—The power transmission backbone of the future," in *Systems, Signals and Devices (SSD), 2012 9th International Multi-Conference on*, 2012, pp. 1–7.

[12] V. Hofmann and M. M. Bakran, "A capacitor voltage balancing algorithm for hybrid modular multilevel converters in HVDC applications," in *2017 IEEE 12th International Conference on Power Electronics and Drive Systems (PEDS)*, 2017, pp. 691–696.

[13] S. Debnath, J. Qin, B. Bahrani, M. Saeedifard, and P. Barbosa, "Operation, Control, and Applications of the Modular Multilevel Converter: A Review," *IEEE Trans. Power Electron.*, vol. 30, no. 1, pp. 37–53, Jan. 2015.

[14] A. Antonopoulos, L. Angquist, and H.-P. Nee, "On dynamics and voltage control of the modular multilevel converter," in *Power Electronics and Applications, 2009. EPE'09. 13th European Conference on*, 2009, pp. 1–10.

[15] X. Chen, J. Liu, S. Ouyang, S. Song, and H. Wu, "A modified circulating current suppressing strategy for nearest level control based modular multilevel converter," in *2017 IEEE Energy Conversion Congress and Exposition (ECCE)*, 2017, pp. 1817–1822.

[16] S. Yang, P. Wang, Y. Tang, M. Zagrodnik, X. Hu, and K. J. Tseng, "Circulating Current Suppression in Modular Multilevel Converters With Even-Harmonic Repetitive Control," *IEEE Trans. Ind. Appl.*, vol. 54, no. 1, pp. 298–309, Jan. 2018.

[17] E. N. Abildgaard and M. Molinas, "Modelling and Control of the Modular Multilevel Converter (MMC)," *Energy Procedia*, vol. 20, pp. 227–236, 2012.

[18] C. Damian, "A comprehensive approach to the modelling and control of a High Voltage Direct Current Modular Multilevel Converter," in *2017 10th International Symposium on Advanced Topics in Electrical Engineering (ATEE)*, 2017, pp. 649–654.

[19] E. Solas, G. Abad, J. A. Barrena, S. Aurtenetxea, A. Carcar, and L. Zajac, "Modular Multilevel Converter With Different Submodule Concepts—Part II: Experimental Validation and Comparison for HVDC Application," *IEEE Trans. Ind. Electron.*, vol. 60, no. 10, pp. 4536–4545, Oct. 2013.

[20] K. Sharifabadi, L. Harnefors, H. P. Nee, S. Norrga, and R. Teodorescu, *Design, control, and application of modular multilevel converters for HVDC transmission systems*. Chichester, West Sussex, United Kingdom: IEEE Press, Wiley, 2016.

[21] M. Saeedifard and R. Iravani, "Dynamic Performance of a Modular Multilevel Back-to-Back HVDC System," *IEEE Trans. Power Deliv.*, vol. 25, no. 4, pp. 2903–2912, Oct. 2010.

[22] Qingrui Tu, Zheng Xu, and Lie Xu, "Reduced Switching-Frequency Modulation and Circulating Current Suppression for Modular Multilevel Converters," *IEEE Trans. Power Deliv.*, vol. 26, no. 3, pp. 2009–2017, Jul. 2011.

[23] J. Qin and M. Saeedifard, "Reduced Switching-Frequency Voltage-Balancing Strategies for Modular Multilevel HVDC Converters," *IEEE Trans. Power Deliv.*, vol. 28, no. 4, pp. 2403–2410, Oct. 2013.

[24] M. Guan, Z. Xu, and H. Chen, "Control and modulation strategies for modular multilevel converter based HVDC system," in *IECON 2011-37th Annual Conference on IEEE Industrial Electronics Society*, 2011, pp. 849–854.

[25] K. Ilves, L. Harnefors, S. Norrga, and H.-P. Nee, "Predictive Sorting Algorithm for Modular Multilevel Converters Minimizing the Spread in the Submodule Capacitor Voltages," *IEEE Trans. Power Electron.*, vol. 30, no. 1, pp. 440–449, Jan. 2015.

[26] F. Deng and Z. Chen, "A Control Method for Voltage Balancing in Modular Multilevel

Converters," *IEEE Trans. Power Electron.*, vol. 29, no. 1, pp. 66–76, Jan. 2014.

[27] M. Hagiwara and H. Akagi, "Control and Experiment of Pulsewidth-Modulated Modular Multilevel Converters," *IEEE Trans. Power Electron.*, vol. 24, no. 7, pp. 1737–1746, Jul. 2009.

[28] Jiangchao Qin and M. Saeedifard, "Predictive Control of a Modular Multilevel Converter for a Back-to-Back HVDC System," *IEEE Trans. Power Deliv.*, vol. 27, no. 3, pp. 1538–1547, Jul. 2012.

[29] L. Ängquist, A. Antonopoulos, D. Siemaszko, K. Ilves, M. Vasiladiotis, and H.-P. Nee, "Inner control of modular multilevel converters-an approach using open-loop estimation of stored energy," in *Power Electronics Conference (IPEC), 2010 International*, 2010, pp. 1579–1585.

[30] G. Bergna *et al.*, "A Generalized Power Control Approach in ABC Frame for Modular Multilevel Converter HVDC Links Based on Mathematical Optimization," *IEEE Trans. Power Deliv.*, vol. 29, no. 1, pp. 386–394, Feb. 2014.

[31] J. Pou, S. Ceballos, G. Konstantinou, V. G. Agelidis, R. Picas, and J. Zaragoza, "Circulating Current Injection Methods Based on Instantaneous Information for the Modular Multilevel Converter," *IEEE Trans. Ind. Electron.*, vol. 62, no. 2, pp. 777–788, Feb. 2015.

[32] R. Darus, J. Pou, G. Konstantinou, S. Ceballos, and V. G. Agelidis, "Circulating current control and evaluation of carrier dispositions in modular multilevel converters," in *ECCE Asia Downunder (ECCE Asia), 2013 IEEE*, 2013, pp. 332–338.

[33] X. She, A. Huang, X. Ni, and R. Burgos, "AC circulating currents suppression in modular multilevel converter," in *IECON 2012-38th Annual Conference on IEEE Industrial Electronics Society*, 2012, pp. 191–196.

[34] Z. Li, P. Wang, Z. Chu, H. Zhu, Y. Luo, and Y. Li, "An Inner Current Suppressing Method for Modular Multilevel Converters," *IEEE Trans. Power Electron.*, vol. 28, no. 11, pp. 4873–4879, Nov. 2013.

[35] C. Xue and Z. Ma, "Capacitor voltage balancing and current control method of modular multilevel converter based on generalised averaging approach," *IET Power Electron.*, vol. 10, no. 15, pp. 2242–2247, Dec. 2017.

[36] G. Zhang and Z. Xu, "Steady-state model for VSC based HVDC and its controller design," in *Power Engineering Society Winter Meeting, 2001. IEEE*, 2001, vol. 3, pp. 1085–1090.

[37] C. Du, A. Sannino, and M. H. Bollen, "Analysis of the control algorithms of voltage-source converter HVDC," in *Power Tech, 2005 IEEE Russia*, 2005, pp. 1–7.

[38] S. Ruihua, Z. Chao, L. Ruomei, and Z. Xiaoxin, "VSCs based HVDC and its control strategy," in *Transmission and Distribution Conference and Exhibition: Asia and Pacific, 2005 IEEE/PES*, 2005, pp. 1–6.

[39] J. Han, Z. Ma, and D. Peng, "Analysis of Model Predictive Current Control for Voltage Source Inverter," *Res. J. Appl. Sci. Eng. Technol.*, vol. 6, no. 21, pp. 3986–3992, 2013.

[40] M. Preindl, E. Schaltz, and P. Thogersen, "Switching Frequency Reduction Using Model Predictive Direct Current Control for High-Power Voltage Source Inverters," *IEEE Trans. Ind. Electron.*, vol. 58, no. 7, pp. 2826–2835, Jul. 2011.

[41] S. Mariethoz and M. Morari, "Explicit Model-Predictive Control of a PWM Inverter With an LCL Filter," *IEEE Trans. Ind. Electron.*, vol. 56, no. 2, pp. 389–399, Feb. 2009.

[42] J. Rodrfguez, J. Pontt, P. Cortés, and R. Vargas, "Predictive control of a three-phase neutral point clamped inverter," in *Power Electronics Specialists Conference, 2005. PESC'05. IEEE 36th*, 2005, pp. 1364–1369.

[43] P. Cortes, M. P. Kazmierkowski, R. M. Kennel, D. E. Quevedo, and J. Rodriguez, "Predictive Control in Power Electronics and Drives," *IEEE Trans. Ind. Electron.*, vol. 55, no. 12, pp. 4312–4324, Dec. 2008.

[44] M. Vatani, B. Bahrani, M. Saeedifard, and M. Hovd, "Indirect Finite Control Set Model Predictive Control of Modular Multilevel Converters," *IEEE Trans. Smart Grid*, vol. 6, no. 3, pp. 1520–1529, May 2015.

[45] J. Mei, Y. Ji, X. Du, T. Ma, C. Huang, and Q. Hu, "Quasi-Fixed-Frequency Hysteresis Current Tracking Control Strategy for Modular Multilevel Converters," *J. Power Electron.*, vol. 14, no. 6, pp. 1147–1156, Nov. 2014.

[46] F. Martinez-Rodrigo, S. de Pablo, and L. C. Herrero-de Lucas, "Current control of a modular multilevel converter for HVDC applications," *Renew. Energy*, vol. 83, pp. 318–331, Nov. 2015.

[47] M. Vatani, M. Hovd, and M. Saeedifard, "Control of the Modular Multilevel Converter Based on a Discrete-Time Bilinear Model Using the Sum of Squares Decomposition Method," *IEEE Trans. Power Deliv.*, vol. 30, no. 5, pp. 2179–2188, Oct. 2015.

[48] P. Münch, D. Görges, M. Izák, and S. Liu, "Integrated current control, energy control and energy balancing of modular multilevel converters," in *IECON 2010-36th Annual Conference on IEEE Industrial Electronics Society*, 2010, pp. 150–155.

[49] M. Zygmanowski, B. Grzesik, and R. Nalepa, "Capacitance and inductance selection of the modular multilevel converter," in *Power Electronics and Applications (EPE), 2013 15th European Conference on*, 2013, pp. 1–10.

[50] P. Cortés *et al.*, "Guidelines for weighting factors design in model predictive control of power converters and drives," in *Industrial Technology, 2009. ICIT 2009. IEEE International Conference on*, 2009, pp. 1–7.

Model Predictive Control of a Modular Multilevel Converter with an improved capacitor balancing method

Shichong Zhang[1], Baodong Bai[1*] and Dezhi Chen[2]

School of Electrical Engineering, Shenyang University of Technology, Shenyang, Liaoning, China

*E-mail: baodongbai@163.com

Abstract— The modular multilevel converter (MMC) has a very good development prospect, however, still has a lot of problems and challenges, one of the essential problems is more intense computational burden with the voltage level of the MMC increasing. An improved model predictive control method that adopted establishing the cost functions and choosing the optimal switching state to realize sub-module capacitor voltage balancing control of the MMC is proposed in this paper. A relatively small number of sub-modules inserted or bypassed of each branch are selected by the cost function, and the number of calculating the cost function using this method is reduced to decrease computational burden. The proposed improved method is verified by simulation system based on 11-level MMC using MATLAB/Simulink.

Keywords—modular multilevel converter, MMC, model predictive control, capacitor voltage balancing control.

I. INTRODUCTION

The idea of MMC topology was first proposed in [1], and then the topology has been a hot spot for researchers in recent years [2]. In comparison with other multilevel converter topologies, the MMC has many attractive advantages, such as low switching frequency, low harmonic distortion and high efficiency. One important advantage of the MMC is the series connection of each sub-modules. The number of sub-modules with low voltages is changed to meet the different power and voltage levels. These advantages make it be used in HVDC [3] [4]. And the MMC also has wide applications in STATCOMs [5] [6] and medium voltage motor drives [7], and etc.

To date, most of the existing control methods for MMC are based on classical control method, linear control using pulse width modulation (PWM), which has been extensively studied over the last few decades [8–10]. Most controllers adopt proportional-integral (PI) control. In reference [11], PI control was adopted in circulating current suppression controller. For the sub-module capacitor voltage balancing, the conventional sorting algorithm has been widely used in [3][12][13], and the sub-module capacitor voltages of each branch are measured and sorted.

The model predictive control (MPC) provides a particular method of energy handling that uses the power converters as discrete and non-linear actuators. It is more flexible and optimizes many important parameters with a suitable cost function.

The MPC strategy has been widely used in industrial control. In recent years, the MPC has received extensive attention in the field of the power electronic converter [14]-[16]. There are also many literatures in regard to the MMC based on MPC strategy in [17]-[21].A control method using traditional MPC based on the MMC is proposed in [17]. In reference [18], a weighting factor is applied for the cost function to control the control system based on the MPC. The optimal combination of switching states is selected by calculating the cost function of the different combinations. The complex parameters tuning can be avoided and the control structure of the system can be simplified. However, the more intense computational burden was caused, because the number of switching states increases as the level increasing. The possible configurations of switching states of each phase leg are C_{2N}^N for per branch with N sub-modules using this method. The MPC strategy adopted the traditional method of rolling optimization is in [19]. In reference [20], FCS-MPC and cascaded PI control of the MMC was compared through the experiment, and the number of the combination of switching states is 2^{2N}. Thus, when the level of MMC is very high, the traditional MPC strategy is not suitable.

In order to decrease computational burden in the control system, reference [21] adopted an MPC strategy with maintaining capacitor voltage balancing by selecting inserted sub-module capacitor voltage according to the number of sub-modules inserted or bypassed of the upper (lower) branch, which is based on the predicted output capacitor voltage levels instead of the switching states.

This paper proposes an improved sub-module capacitor voltage balancing method that can decrease the number of calculating the cost function. In this way, only a relatively small number of sub-modules inserted or bypassed of each branch is selected and sorted by calculating the cost function, and the other sub-modules are not selected. Then, the sub-modules are inserted with lower voltage to recharge and higher voltage to discharge. The results using this method is the same as the conventional method. This method is evaluated based on simulation results in MATLAB /Simulink.

II. STRUCTURE OF THE MMC

A three-phase DC-AC structure of the MMC is shown in Fig. 1.

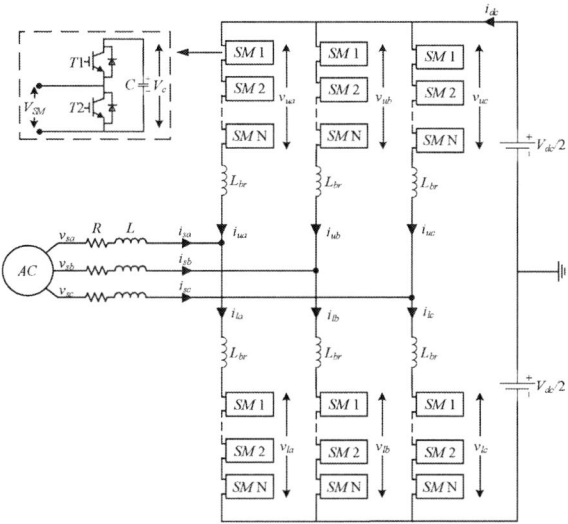

Fig. 1. Three-phase MMC topology

As shown in Fig. 1, each phase leg of the converter consists of an upper branch and a lower branch. Each branch consists of N series-connected sub-modules SMi, with $i \in (1,2,\cdots,N-1,N)$. An inductor L_{br} is added to each branch for suppressing the circulating current. The AC-side voltage of the MMC is represented by the grid voltage v_{sj} in series with the grid resistor R and inductor L. Where v_{uj} and v_{lj} is the upper branch and lower branch voltage of three-phase legs, V_{dc} denotes the DC side voltage.

Sub-module is comprised of two IGBTs and a capacitor C with voltage V_c that form a half-bridge, with switched on and switched off switching states. When the upper IGBT is turn on and the lower one is turn off, the sub-module can be switched on, the capacitor C is connected to the branch, and the terminal voltage of the module is equal to the capacitor voltage V_c. When the lower IGBT is turned on and the upper the one is turned off, the sub-module can be switched off, the terminal voltage of the module is zero.

III. MATHEMATICAL MODEL OF THE MMC

According to the circuit diagram in Fig. 1, the equations applying the *Kirchoff*'s voltage law for phase j are defined as

$$\frac{V_{dc}}{2} - v_{uj} - L_{br}\frac{di_{uj}}{dt} + L\frac{di_{sj}}{dt} + Ri_{sj} - v_{sj} = 0 \tag{1}$$

$$-\frac{V_{dc}}{2} + v_{lj} + L_{br}\frac{di_{lj}}{dt} + L\frac{di_{sj}}{dt} + Ri_{sj} - v_{sj} = 0 \tag{2}$$

The equations for the upper branch current i_{uj} and lower branch current i_{lj} of each phase leg are shown as

$$i_{uj} = -\frac{i_{sj}}{2} + \frac{i_{dc}}{3} + i_{cirj} \tag{3}$$

$$i_{lj} = \frac{i_{sj}}{2} + \frac{i_{dc}}{3} + i_{cirj} \tag{4}$$

where i_{cirj} represents circulating current of each phase leg and i_{dc} represents the DC side current. According to the equation (3) and (4), the circulating current and the phase current are expressed as

$$i_{cirj} = \frac{i_{uj} + i_{lj}}{2} - \frac{i_{dc}}{3} \tag{5}$$

$$i_{sj} = i_{uj} - i_{lj} \tag{6}$$

Thus, the external dynamic equation of the phase current is obtained as

$$v_{lj} - v_{uj} + L_{br}\frac{di_{sj}}{dt} + 2L\frac{di_{sj}}{dt} + 2Ri_{sj} - 2v_{sj} = 0 \tag{7}$$

And the internal dynamic equation of the branch circulating current is obtained as

$$\frac{V_{dc}}{2} - \frac{v_{uj} + v_{lj}}{2} - L_{br}\frac{di_{cirj}}{dt} = 0 \tag{8}$$

IV. THE MPC STRATEGY OF THE MMC

A. The AC-side Currents MPC Strategy

With the Euler approximation for the current derivative, the discrete AC-side current is deduced as follow

$$i_{sj}(t+T_s) = A[\frac{v_{lj}(t+T_s) - v_{uj}(t+T_s)}{2} \\ -v_{sj}(t+T_s) + \frac{A'}{T_s}i_{sj}(t)] \tag{9}$$

where $A = 1/(A'/Ts + R)$, $A' = L/2 + L$. Ts is the sampling period, $i_{sj}(t)$ is the measurement of phase j current at time t. And the equation of converter output voltage level as

$$v_{oj}(t+T_s) = \frac{v_{lj}(t+T_s) - v_{uj}(t+T_s)}{2} \tag{10}$$

Since the objective is to make the line currents to track reference currents, the cost function is expressed as

$$J_1 = |i^*_{sj}(t+T_s) - i_{sj}(t+T_s)| \tag{11}$$

The number of the presentable output voltage level is reduced to N+1[21], shown as

$$-\frac{N}{2}\frac{V_{dc}}{N} \le v^*_{oj} \le \frac{N}{2}\frac{V_{dc}}{N} \tag{12}$$

$$v^*_{oj} = \frac{V_{dc}}{N}[-\frac{N}{2}, -\frac{N-1}{2}, \cdots, \frac{N-1}{2}, \frac{N}{2}] \tag{13}$$

The reference voltages of the upper branch and lower branch of the MMC can be expressed as

$$v^*_{uj} = \frac{V_{dc}}{2} - v^*_{oj} \tag{14}$$

$$v^*_{lj} = \frac{V_{dc}}{2} + v^*_{oj} \tag{15}$$

$$v^*_{uj}, v^*_{lj} = [\frac{V_{dc}}{N}][0,1,2,\cdots,N-1,N] \tag{16}$$

The block diagram of the AC-side current MPC strategy is shown in Fig. 2. (Adapted from [21]).

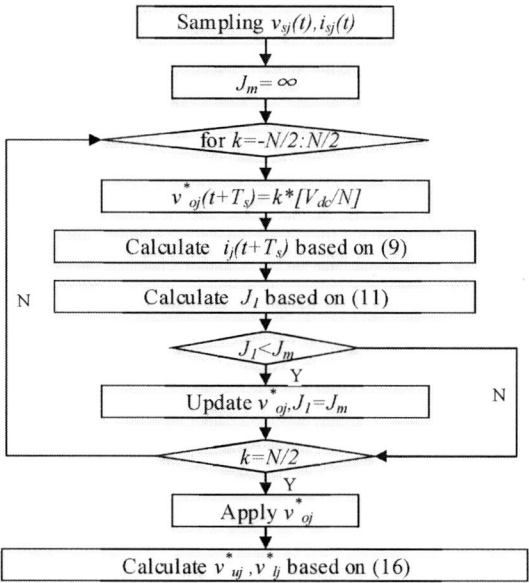

Fig. 2. The block diagram of AC-side current control.

B. The Circulating Currents MPC Strategy

With the Euler approximation, the discrete branch circulating current equation of each phase leg is defined as

$$i_{cirj}(t+T_s) = B[V_{dc} - v^*_{lj}(t+T_s) - v^*_{uj}(t+T_s)] \tag{17}$$

where $B = Ts/(2*L_{br})$. For the purpose of maintaining the inner current balancing, the voltage level v_{diffj} can be added in both branch output voltage references. According to the reference [21], the equations are defined as

$$\begin{aligned} i_{cirj}(t+T_s) = &B[V_{dc} - [v^*_{lj}(t+T_s) + v_{diffj} \\ &+ v^*_{uj}(t+T_s) + v_{diffj}]] + i_{cirj}(t) \end{aligned} \tag{18}$$

The external output characteristics of each phase leg of the MMC are determined by the difference between the voltage of the upper branch and lower branch. The internal current of each phase leg is affected by the sum of the DC side voltage and the voltage of the upper

branch and lower branch, having no relation with the external characteristics of the MMC. Therefore, under the condition that the DC-side voltage is stable, the circulating current can be controlled by changing the sum of the voltages of the upper branch and lower branch. Respectively, a compensation level v_{diff} is inserted to the upper branch and lower branch for the suppression of circulating current, the compensation level range is shown as

$$v_{diffj} = \frac{V_{dc}}{N}[-1,0,1] \tag{19}$$

Circulating current increases the loss of the converter and need to be suppressed, so the ideal value of circulating is zero. Hence, the cost function can be designed as

$$J_2 = |i_{cirj}(t+T_s)| \tag{20}$$

The voltage reference values of the upper branch and lower branch are defined as

$$v^*_{uj\_c} = v^*_{uj}(t+T_s) + v_{diffj} \tag{21}$$

$$v^*_{lj\_c} = v^*_{lj}(t+T_s) + v_{diffj} \tag{22}$$

The presentable voltage level is rewritten as

$$N_{uj} = \frac{v^*_{uj\_c}}{V_{dc}/N} = [0,1,2,\cdots,N-1,N] \tag{23}$$

$$N_{lj} = \frac{v^*_{lj\_c}}{V_{dc}/N} = [0,1,2,\cdots,N-1,N] \tag{24}$$

The block diagram of the circulating current control MPC strategy of the MMC is shown in Fig. 3.(Adapted from [21]).

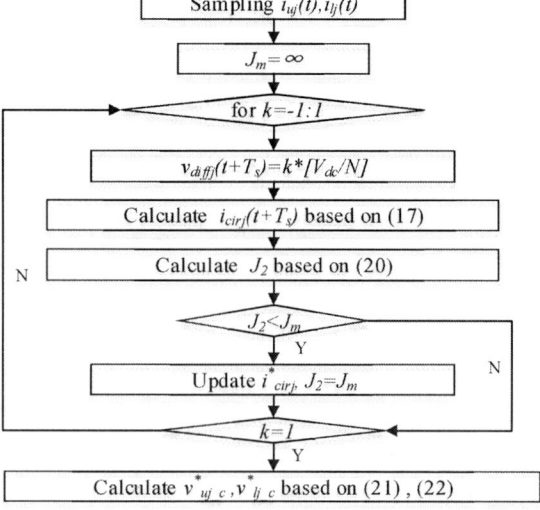

Fig. 3. The block diagram of circulating current control

C. Capacitor Voltages Balancing MPC Strategy

In reference [17], the value of the predictive sub-module capacitor voltage is defined as

$$V_{cij}(t+T_s) = V_{cij}(t) + \frac{i_m(t)}{C} T_s \quad \text{switched on SM} \quad (25a)$$

$$V_{cij}(t+T_s) = V_{cij}(t) \qquad \text{switched off SM} \quad (25b)$$

where $i_m(t) = i_{uj}(t)$ or $i_{lj}(t)$ for the sub-modules, the measured value of the branch current at the sub-module at time t.

The cost function is equation (26) in reference [21]. Here, the sub-module capacitor voltage is recharged, if $i_m(t)$ is positive, and discharged, if negative. Thus, the minimum value of the cost function decide the switched on status of the sub-modules.

$$J_3 = \frac{i_m T_s}{C} \left[\frac{V_{dc}}{N} - V_c(t+T_s) \right] \quad (26)$$

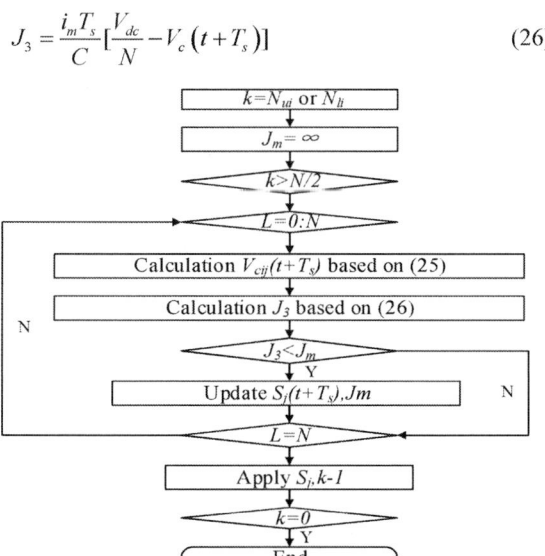

Fig. 4. The sub-module capacitor voltage balancing strategy
(Adopted from [21])

In this way, the number of calculating the cost function of the upper branch and lower branch are decreased to $N \times N$ for sub-module capacitor voltage balancing at $N+1$ level [21].

The block diagram in Fig. 4. is modified to Fig. 5.to reduce the number of calculating the cost function of the upper branch and lower branch.

According to reference [17], the total number of sub-modules inserted in each phase leg is N. The number of sub-modules inserted in the upper branch is assumed to be X, and in the lower branch is $N - X$. The number of calculating the cost function to select X sub-modules inserted in the upper branch is $N \times X$, and the number of calculating the cost function to select X sub-modules bypassed in the upper branch is also $N \times X$. The number of calculating the cost function to insert N sub-modules in each phase leg is $2N \times X$, and its range is $0 \sim N \times N$, if X is no more than $N/2$.

The same idea for X is more than $N/2$, and the $N - X$ is no more than $N/2$. The number of calculating the cost function to insert N sub-modules in each phase leg is $2N \times (N - X)$, and its range is also $0 \sim N \times N$.

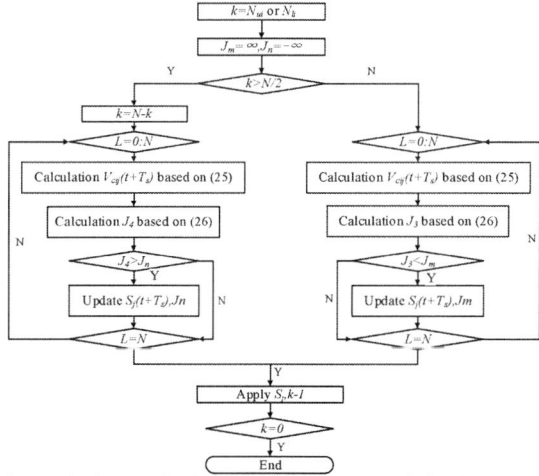

Fig. 5. The improved sub-module capacitor voltage balancing strategy

The sub-modules inserted are selected by the minimum value of the cost function. The sub-modules bypassed are selected by the maximum value of the cost function. The number of calculating the cost function under different MPC method is shown as Table I.

With the same idea, the cost function of the proposed another method is defined as

$$J_3 = \left| V_{cij}(t+T_s) \right| \quad (27)$$

The sub-module capacitor voltages of each branch are measured and calculated by (25). If the upper (lower) branch current is positive, $N_{uj}(N_{lj})$ sub-modules with the lowest voltages in the corresponding branch are selected by the cost function in (27), and then inserted. As a result, the corresponding sub-module capacitors are charged. If the upper (lower) branch current is negative, $N_{uj}(N_{lj})$ sub-modules with the highest voltages in the corresponding branch are selected by the cost function in (27), and then inserted. The block diagram of sub-module capacitor voltage balancing is shown in Fig. 6.

TABLE I
THE NUMBER OF CALCULATING THE COST FUNCTION UNDER DIFFERENT MPC STRATEGY

MPC strategies	Ac side current control	Circulating current control	Sub-module capacitor voltage balancing	Total
Inference [18]		C_{2N}^N		C_{2N}^N
Inference [21]		2^{2N}		2^{2N}
Inference [22]	N	3	$N \times N$	$3+N+N \times N$
This paper	N	3	$0 \sim N \times N$	$(3+N) \sim (3+N+N \times N)$

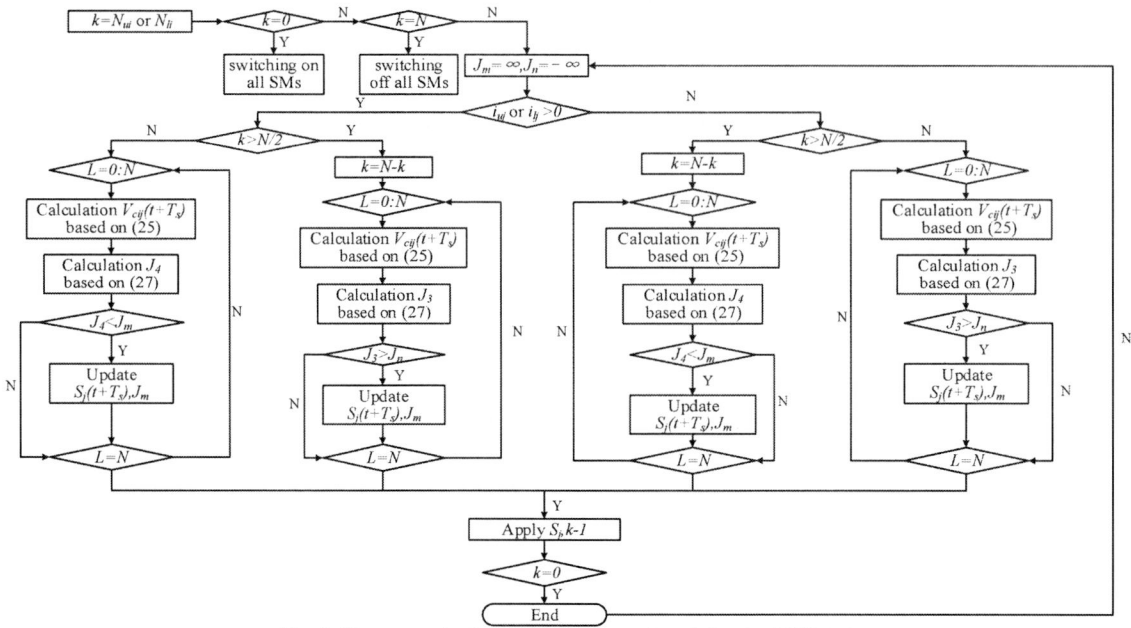

Fig. 6. The proposed sub-module capacitor voltage balancing MPC strategy

V. CONTROL OF THE GRID-CONNECTED MMC

The equation of transferring the *abc* models to *dq0* frame is expressed as

$$i_{abc} = M i_{dq} \tag{28}$$

Transformation matrix M is

$$M = \begin{bmatrix} \cos\theta & \cos(\theta - 2\pi/3) & \cos(\theta + 2\pi/3) \\ \sin\theta & \sin(\theta - 2\pi/3) & \sin(\theta + 2\pi/3) \\ 1/2 & 1/2 & 1/2 \end{bmatrix} \tag{29}$$

Active and reactive power of the equation of the system are obtained by

$$P = \frac{3}{2} v_d i_d \tag{30}$$

$$Q = -\frac{3}{2} v_d i_q \tag{31}$$

A diagram of the system control strategy is shown in Fig. 5.

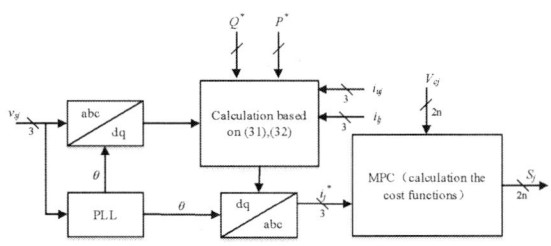

Fig. 7. The system control strategy.

VI. SIMULATION RESULTS

The simulation of an 11-level MMC is simulated in MATLAB/Simulink, and the MMC system parameters are provided in Table II.

TABLE II
MAIN CIRCUIT PARAMETERS

Items	Values
Active Power	8MW
Reactive Power	0
AC System Voltage	10kV
Nominal Frequency	50Hz
R	0.05ohm
L	15mH
L_{br}	15mH
Number of Sub-modules per branch	10
Sub-module Capacitor C	0.0075F
Sub-module Capacitor Voltage	2kV
DC Bus Voltage	20kV
Sampling Period	50us

The simulation results of the method in [21] and the improved method are shown in Fig. 5 using the same parameters. The active power reference value is set to 8 kW and reactive reference is set to zero at first, and then the active power reference value is changed to 5 kW at time 1.5 s while the reference value is maintained. Fig 10 show the AC-side voltages, active and reactive power, AC-side currents, the sum of three phases AC-side currents, DC-link current, sub-module capacitor voltages in phase A upper branch. The improved method is different from the original method in the number of calculating the cost function, but the effects are similar.

The 2018 International Power Electronics Conference

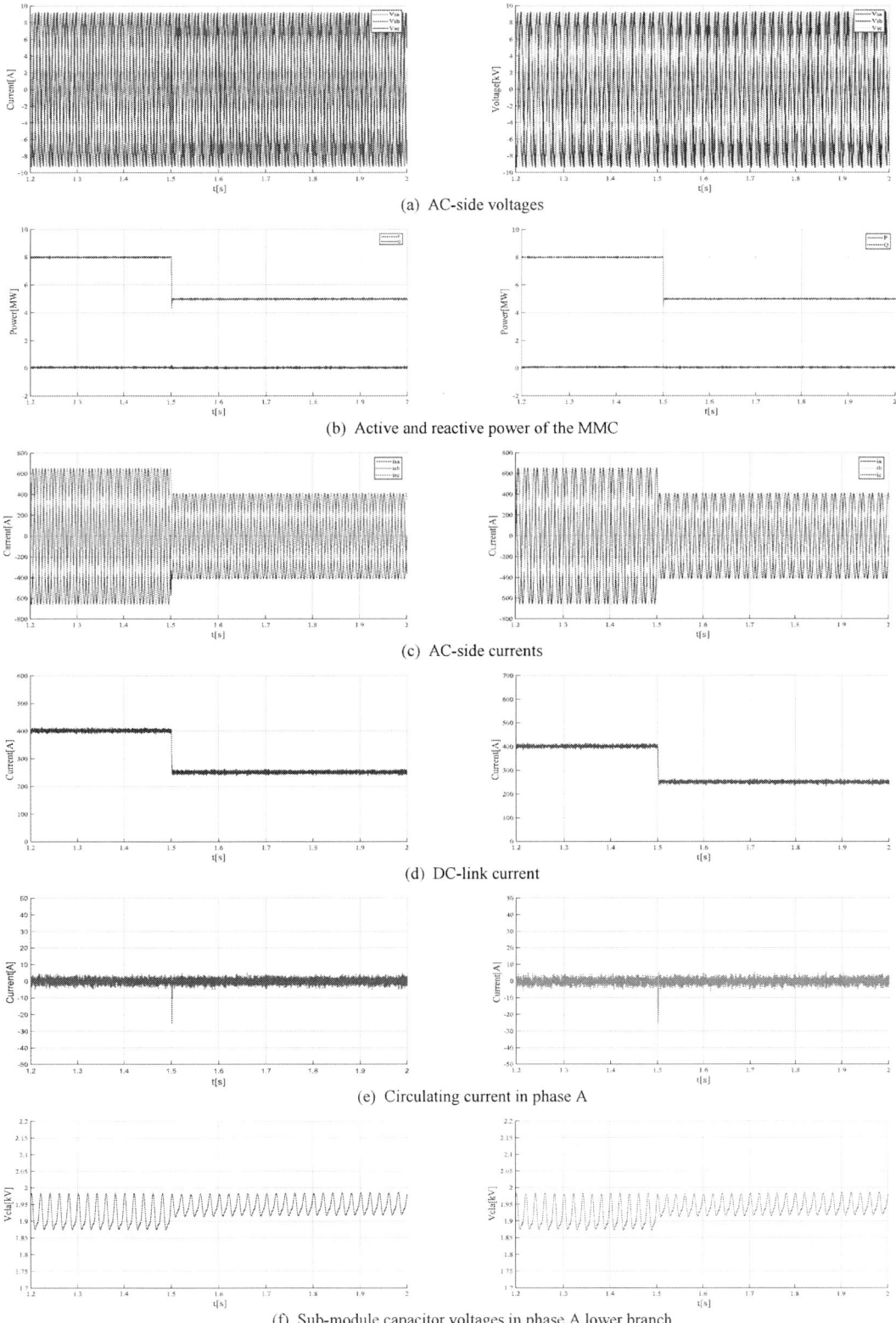

(a) AC-side voltages

(b) Active and reactive power of the MMC

(c) AC-side currents

(d) DC-link current

(e) Circulating current in phase A

(f) Sub-module capacitor voltages in phase A lower branch

Fig. 8. The simulation results of the original method and the improved method

2643

The 2018 International Power Electronics Conference

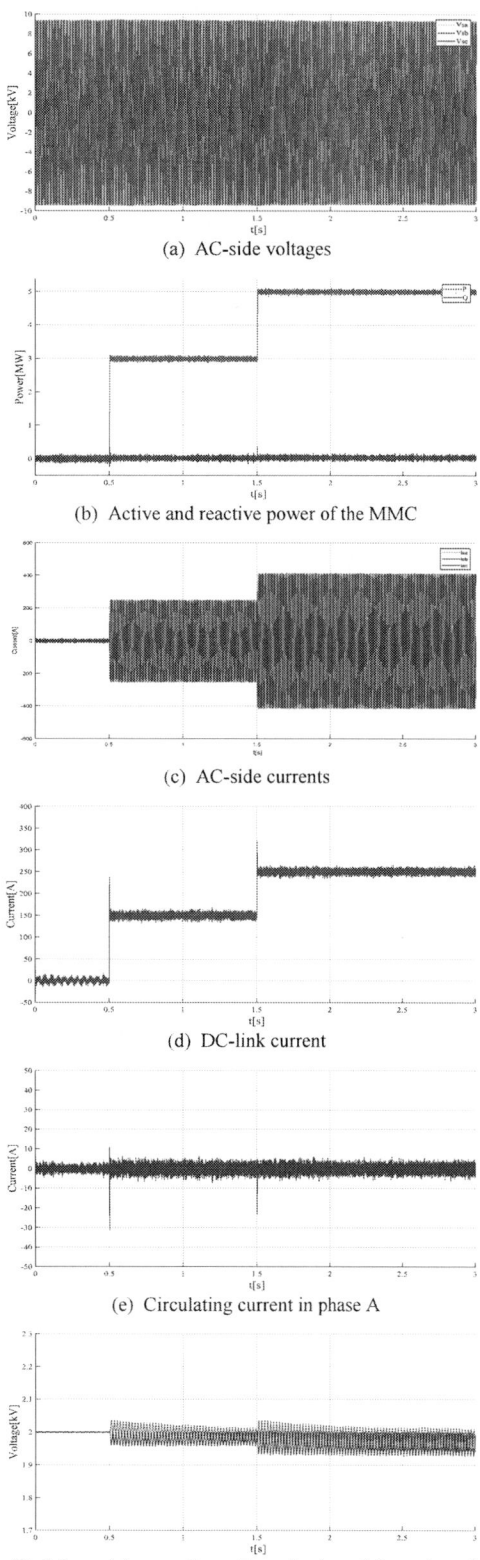

(a) AC-side voltages

(b) Active and reactive power of the MMC

(c) AC-side currents

(d) DC-link current

(e) Circulating current in phase A

(f) Sub-module capacitor voltages in phase A lower branch

Fig. 9. The simulation results of the proposed another method

The simulation results of the proposed another method is shown in Fig. 9. The active and reactive power of the system are controlled to zero at first, and then the reactive power is maintained to 0, the active power is set to 3 MW at 0.5s and 5 MW at 1.5s.

VII. CONCLUSIONS

An improved capacitor balancing method of MPC strategy to decrease the number of calculating the cost function is proposed based on MMC topology in this paper. The idea of proposed method is to selected a small number of sub-modules to inserted or bypassed in the each phase leg, and then to inserted sub-modules. Simultaneously, this idea can also be applied to some other capacitor balancing methods.

ACKNOWLEDGMENT

Thanks for the supporting of National Natural Science Foundation (Grant No. 51577122) and Department of education of Liaoning Province (Grant No. 20163409).

REFERENCES

[1] A. Lesnicar, R. Marquardt, "An innovative modular multilevel converter topology suitable for a wide power range," *IEEE Bologna Power Tech Conference Proceedings*, vol. 3, pp.23-26, 2003.

[2] S. Debnath, J. Qin, B. Bahrani, M. Saeedifard, and P. Barbosa, "Operation, control, and applications of the modular multilevel converter: A review," *IEEE Trans. on Power Electronics,* vol. 30, no. 1, pp. 37–53, 2015.

[3] M. Saeedifard, R. Iravani, "Dynamic performance of a modular multilevel back-to-back HVDC system," *IEEE Trans. Power Delivery*, vol. 25, no. 4, pp. 2903–2912, 2010.

[4] A. Nami, J. Liang, F. Di jkhuizen, and G. D. Demetriadis, "Modular multilevel converters for HVDC applications: Review on converter cells and functionalities," *IEEE Trans. on Power Electronics*, vol. 30, no. 1, pp. 18–36, 2015.

[5] H.M. Pirouz and M. T. Bina, "A transformerless medium-voltage STATCOM topology based on extended modular multilevel converters," *IEEE Trans. on Power Electronics*, vol. 26, no. 5, pp. 1534-1545, 2011.

[6] J. I. Y. Ota, Y. Shibano, N. Niimura, and H. Akagi, "A phase-shifted-PWM D-STATCOM using a modular multilevel cascade converter (SSBC)-Part I: Modeling, analysis, and design of current control," *IEEE Trans. on Industry Applications*, vol. 51, no. 1, pp. 279-288, 2015.

[7] M. Hagiwara, K. Nishimura, and H. Akagi, "A medium-voltage motor drive with a modular multilevel PWM inverter," *IEEE Trans. Power Electronics*, vol. 25, no. 7, pp. 1786–1799, 2010.

[8] G. S. Konstantinou, V. G. Agelidis, "Performance evaluation of half-bridge cascaded multilevel converters operated with multicarrier sinusoidal PWM techniques," *in Proc. IEEE Conf. Ind. Electron.* 2009.

[9] G. Konstantinou, M. Ciobotaru, and V. Agelidis, "Selective harmonic elimination pulse-width modulation of modular multilevel converters," *IET Power Electronics.*, vol. 6, no. 1, pp. 96–107, Jun. 2013.

[10] S. Fan, K. Zhang, J. Xiong, Y. Xue, "An improved control system for modular multilevel converters with new modulation strategy and voltage balancing control," *IEEE Trans. Power Electronics.* Vol. 30, no. 1, pp. 358–371, 2015.

[11] Q. Tu, Z. Xu, and L. Xu, "Reduced switching-frequency modulation and circulating current suppression for modular multilevel converters," *IEEE Trans. Power Delivery*, vol. 26, no. 3, pp. 2009–2017, 2011.

[12] S. Rohner, S. Bernet, M. Hiller, and R. Sommer, "Modelling, simulation and analysis of a modular multilevel converter for medium voltage applications," *IEEE International Conf. Industrial Technology (ICIT)*, 2010.

[13] S. Rohner, S. Bernet, M. Hiller, and R. Sommer, "Modulation, losses, and semiconductor requirements of modular multilevel converters," *IEEE Trans. on Industrial Electronics*, vol. 57, no. 8, pp. 2633-2642, 2010.

[14] P. Cortes, M. P. Kazmierkowski, R. M. Kennel, D. E. Quevedo, and J. Rodríguez, "Predictive control in power electronics and drives," *IEEE Trans. Ind. Electronics*, vol. 55, no. 12, pp. 4312–4324, Dec. 2008.

[15] J. Rodríguez, P. Cortes R. Kennel and M. P. Kazmierkowski, "Model predictive control-a simple and powerful method to control power converters," *IEEE 6th International Power Electronics and Motion Control Conference*, 2009.

[16] T. Geyer, "A comparison of control and modulation schemes for medium-voltage drives: Emerging predictive control concepts versus PWM-based schemes," *IEEE Trans. Ind. Applications.*, vol. 47, no. 3, pp. 1380–1389, 2011.

[17] J. Qin and M. Saeedifard, "Predictive control of a modular multilevel converter for a back-to-back HVDC system," *IEEE Trans. on Power Delivery*, vol. 27, no. 3, pp. 1538–1547, 2012

[18] P. Cortes, S. Kouro, B. La Rocca, R. Vargas, J. Rodriguez, J. I. Leon, S. Vazquez and L. G. Franquelo, "Guidelines for weighting factors design in model predictive control of power converters and drives," *IEEE International Conference on Industrial Technology*, 2009.

[19] A. Dekka, B. Wu, V. Yaramasu, N. Zargari, "Model predictive control with common-mode voltage injection for modular multilevel converter," *IEEE Trans. on Power Electronics*, vol. 32, no. 3, pp. 1767–1778, 2017.

[20] J. Bocker, B. Freudenberg, A. The, and S. Dieckerhoff, "Experimental comparison of model predictive control and cascaded control of the modular multilevel converter," *IEEE Trans. on Power Electronics*, vol. 30, no.1, pp. 422–430, 2014.

[21] J. Moon, J. Gwon, J. Park, D. Kang and J. Kim, "Model predictive control with a reduced number of considered states in a modular multilevel converter for HVDC system," *IEEE Trans. on Power Delivery*, vol. 30, no. 2, pp. 608-617, 2015.

The 2018 International Power Electronics Conference

High Step-Up DC-DC Converter Based on Multi-Cell Coupled Inductor Diode-Capacitor Network

Xinying Li[1], Yan Zhang[1*], Jinjun Liu[1], Pengxiang Zeng[1]

1 State Key Lab of Electrical Insulation and Power Equipment, Xi'an Jiaotong University, Xi'an, China

*E-mail: zhangyanjtu@163.com

Abstract— **Due to renewable energy sources output voltage is low and fluctuating,front boost DC-DC converter is indispensable in grid-tied system.High voltage conversion ratio can be realized through diode-capacitor network and cascade structure while inrush current is severe because of the directly charging through diode between the adjacent boost cells.In this article,a novel high step-up DC-DC converter based on multi-cell coupled inductor diode-capacitor network is proposed which conquers the inrush current problem and enhances the voltage boost capability remarkably.The proposed topology has the following advantages: 1)Increase the voltage gain and decrease the voltage stress of semiconductors. 2)realize the naturally switching off of all the diodes in the circuit and improve the efficiency. 3)The voltage stress of semiconductors is not related to voltage gain when cell number is even. 4)reduce turns ratio and volume of coupled inductor. 5)single switch working mode ease the control complexity and decrease the drive circuit cost.The simulation results and experiment outcome testify the validity of theoretical analysis.**

Keywords— *DC-DC converter;coupled inductor;diode capacitor;high conversion ratio;inrush current*

I. INTRODUCTION

Clean energy such as photovoltaic,wind turbine and fuel cell,the output voltage is pretty low and fluctuate within a wide range.Thus a front high boost DC-DC converter is necessary for grid-tied inverter to convert and deliver energy.High voltage conversion ratio,high efficiency,high reliability and high power density is the eternal goal for high boost DC-DC converter.

There already exist various methods to realize high voltage conversion ratio proposed in the literature[1]-[8].Compared with other methods,the switch-capacitor network boost technology has the advantage of high power density and high conversion ratio.While the efficiency is limited resulting from the inrush current in the circuit.

Fig.1 presents a kind of high voltage gain DC-DC converter based on diode-capacitor network [9].Through the operation of switch S,capacitor C_{i1} and C_{i2} are been charged in parallel while discharge in series to obtain the voltage boost capability($1 \leq i \leq N$).However when the switch S turn on,capacitor C_{11} and C_{12} connecting in series charge capacitor C_{21} through diode D_{22} and charge C_{22} through diode D_{21}.Inrush current will occur during this period and switching off period between the other

This work was supported in part by the State Key Laboratory of Electrical Insulation and Power Equipment under Grants EIPE16310 and the Power Electronics Science and Education Development Program of Delta Environmental and Educational Foundation underGrant DREG2016010.

adjacent boost cell as well.This will lead to converter high cost and low efficiency.As shown in Fig.2[10],this circuit is the single-cell diode-capacitor network high boost DC-DC converter.The only difference is the position of main switch.

Fig.3 shows the structure of proposed converter which combine these two topologies above with coupled inductor.The novel boost cell consist of secondary winding of coupled inductor and diode-capacitor network.With the increased cell number,the voltage stress of semiconductors decrease and voltage gain increase quickly.Because of the insertion of coupled inductor,this topology conquers the inrush current problem and enhances the voltage conversion ratio remarkably.

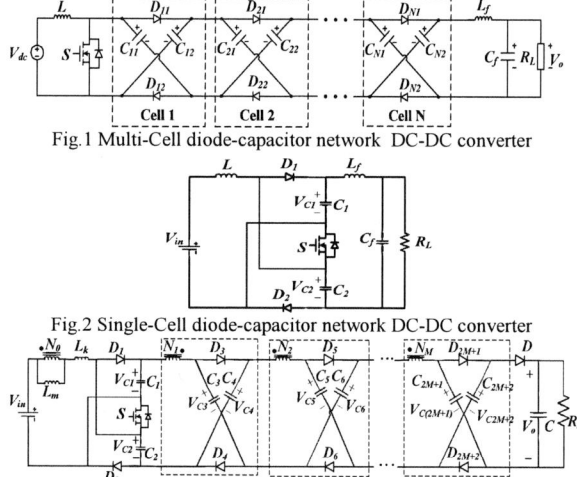

Fig.1 Multi-Cell diode-capacitor network DC-DC converter

Fig.2 Single-Cell diode-capacitor network DC-DC converter

Fig.3 Multi-Cell coupled inductor diode-capacitor network DC-DC converter

II. OPERATION PRINCIPLE

Supposing the cell number M is one,which is redrawn as Fig.4.The coupled inductor is modeled as a primary leakage inductor L_k, a magnetizing inductor L_m in parallel with an ideal transformer.k is the coupling coefficient which is expressed as $L_m/(L_m+L_k)$.Fig.5 gives the key waveforms during one switching cycle in detail.There are 5 modes during one period,as shown in Fig.6.Some assumptions are considered for simplify analysis of the proposed converter.

1)The switch and diodes are considered ideal and equivalent resistors and parasitic parameters are neglected.The forward voltage of diodes is zero.

2)All the capacitors are large enough.Thus the voltage across them are considered to be constant in one period of switching.

3)The value of magnetizing inductor L_m is pretty large,exciting current is a constant value.The coupling coefficient of coupled inductor k is equal to $L_m/(L_m+L_k)$,and turn ratio is equal to N_2/N_1.The equivalent series resistance of capacitors and parasitic resistance of the coupled inductor are neglected.

Mode 1($t_0 \sim t_1$):At $t=t_0$,switch S is turned on and its current increase from zero.The current flow path is shown in Fig.6(a).The voltage of magnetizing inductor still be negative due to the secondary winding.Thus the magnetizing current decrease and leakage inductor current increase rapidly.As a result,the current of diode D_5 decrease rapidly either.The energy in magnetizing inductor release to load in series with capacitor C_3 and C_4.This mode ends when leakage inductor current is equal to magnetizing current at t_1 moment.

Mode 2($t_1 \sim t_2$):Diode D_5 is turned off at t_1 moment.The secondary winding current flow through diode D_3 and D_4 instead.The current flow path is shown in Fig.6(b).Current i_{D3} and i_{D4} increase from zero and magnetizing current increase during this period.Switch S still be conducted.The voltage source charge the magnetizing inductor and the capacitor C_3,C_4 in parallel through the secondary winding in series with capacitor C_1 and C_2.Capacitor C_5 discharge to the load alone.This mode ends when switch S turn off.

Mode 3($t_2 \sim t_3$):At the t_3 moment,switch S turn off and diode D_1,D_2 turn on at the same time.The current flow path is shown in Fig.6(c).The voltage across the magnetizing inductor is positive,magnetizing current continue increase and leakage inductor current decrease rapidly.The secondary winding current and diode current i_{D3},i_{D4} decrease quickly as a result.Capacitor C_5 discharge to the load alone.This mode ends when leakage inductor current is equal to magnetizing current.

Mode 4($t_3 \sim t_4$):The current of diode D_3 and D_4 decrease to zero at t_3 moment.Therefore,diode D_3,D_4 is turned off and diode D_5 is turned on at this moment.The current flow path is shown in Fig.6(d).The voltage across the magnetizing inductor is negative so that magnetizing current start decreasing.The leakage inductor current continue decrease but ramp is bigger than magnetizing current.Magnetizing inductor transfer its energy to load in series with capacitor C_3 and C_4 through secondary winding.Besides,primary winding charge the capacitor C_1 and C_2 in parallel in series with voltage source together.This mode ends when leakage inductor current decrease to $i_{Lm}/(N+1)$.

Mode 5($t_4 \sim t_5$):At t_4 moment,the current of diode D_1 and D_2 decrease to zero so that diode D_1 and D_2 are turned off.The current flow path is shown in Fig.6(e).During this mode,magnetizing current continue

to decrease and leakage inductor current is equal to $i_{Lm}/(N+1)$.t_5 moment switch S is turned on and the next switching cycle begins.

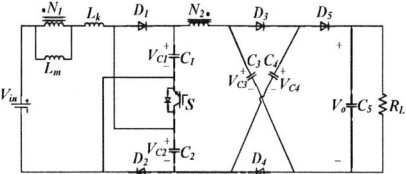

Fig.4 Proposed converter with single cell

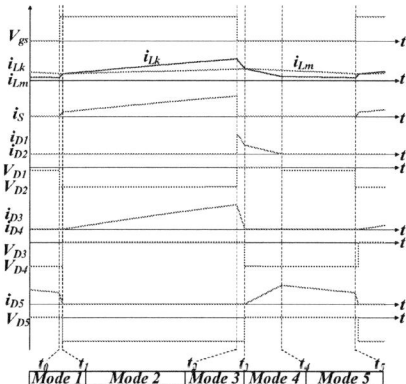

Fig.5 The key waveforms of proposed converter

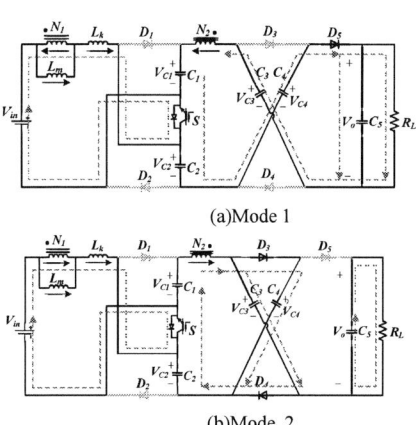

(a)Mode 1

(b)Mode 2

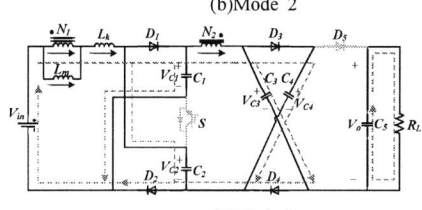

(c)Mode 3

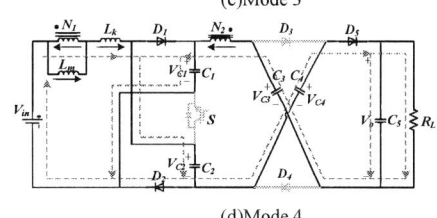

(d)Mode 4

2647

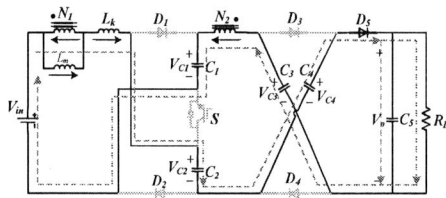

(e)Mode 5

Fig.6 Real circuit under different working mode

III. STEADY STATE ANALYSIS

Since the last time of mode 1 and mode 3 is pretty short comparing with the whole period,to simplify the analysis,these two modes are neglected.The simplified waveform is shown in Fig.7.The difference of leakage inductor current and magnetizing current divided by turn ratio N is the current of secondary winding.During mode 2,the integration of secondary winding current to time is the charge been transfered to capacitor C_3 and C_4.This is corresponding to the $1/N$ of shadow area which is above the magnetizing current part.While in the mode 4 and mode 5,capacitor C_3 and C_4 discharge in series so that the integration of secondary winding current to time is half of the charge discharged by capacitor C_3 and C_4.This is also corresponding to the $1/N$ of shadow area that is below the magnetizing current.Supposing the difference between magnetizing current and leakage inductor current is h at t_3 moment and magnetizing current is I_{Lm}.According to Fig.5(b),the peak current value of diode D_3 and D_4 is equal to $h/2N$ at t_3 moment.Also,the peak current value of diode D_1 and D_2 is $I_{Lm}/2$ at t_3 moment.In the mode 5,leakage inductor current is equal to secondary winding current and the turn ratio is N,so from KCL principle it is obtained that leakage inductor current is $I_{Lm}/(N+1)$.

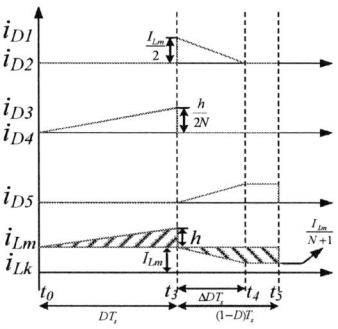

Fig.7 The simplified waveforms of proposed DC-DC converter

The average current value of diode D_1,D_2,D_3,D_4,D_5 is equal to output current flowing through resistor R_L because of charge balance principle.From this:

$$\Delta D \cdot \frac{I_{Lm}}{4} = D \cdot \frac{h}{4N} \tag{1}$$

From the analysis above,the surface of upper area is the double of below area,so there is equation (2).

$$\frac{1}{2} \cdot DT_s \cdot h = [2(1-D) - \Delta D]T_s \cdot \frac{N}{N+1} I_{Lm} \tag{2}$$

The ΔD can be obtained as following equation through equation (1) and (2).

$$\Delta D = \frac{4(1-D)}{N+3} \tag{3}$$

During mode 2,the voltage of magnetizing inductor and leakage inductor are denoted as V_{Lm}^2 and V_{Lk}^2 respectively, as shown in Fig.6(b).The following equations are obtained:

$$V_{Lm}^2 = kV_{in} \tag{4}$$

$$V_{Lk}^2 = (1-k)V_{in} \tag{5}$$

$$2V_{C1} + NkV_{in} = V_{C3} \tag{6}$$

The magnetizing voltage and leakage inductor voltage are expressed as V_{Lm}^4 and V_{Lk}^4 during mode 4,as shown in Fig.6(d).According to the KVL principle:

$$V_{in} = V_{Lm}^4 + V_{Lk}^4 + V_{C1} \tag{7}$$

$$V_{C1} + NV_{Lm}^4 = 2V_{C3} - V_{C5} \tag{8}$$

While in mode 5,the relationship between the magnetizing current and leakage inductor current is:

$$i_{Lk} = \frac{i_{Lm}}{N+1} \tag{9}$$

Due to the coupling coefficient k is equal to $L_m/(L_m+L_k)$,then there is:

$$V_{Lk}^5 = \frac{1-k}{k} \cdot \frac{1}{N+1} V_{Lm}^5 \tag{10}$$

At the same time for the whole circuit loop,it can be obtained as:

$$V_{in} - 2V_{C1} - NV_{Lm}^5 + 2V_{C3} - V_{C5} - V_{Lm}^5 - V_{Lk}^5 = 0 \tag{11}$$

Using the voltage balance on inductor L_m and L_k to yield:

$$\int_{t_0}^{t_3} V_{Lm}^2 dt + \int_{t_3}^{t_4} V_{Lm}^4 dt + \int_{t_4}^{t_5} V_{Lm}^5 = 0 \tag{12}$$

$$\int_{t_0}^{t_3} V_{Lk}^2 dt + \int_{t_3}^{t_4} V_{Lk}^4 dt + \int_{t_4}^{t_5} V_{Lk}^5 = 0 \tag{13}$$

From equation (3)-(13),the voltage gain expression is obtained as following:

$$G = \frac{(4-3D)N^3k^2 + (8-3D)N^2k^2 + (6-D)N^2k + (16-2D)Nk + DN^3k + 6}{2(1-D)(kN^2 + 2kN + 1)} \tag{14}$$

The relationship among voltage gain,duty cycle and coupling coefficient is shown in Fig.8.From this curve,it is observed that the fluctuation of coupling coefficient only has tiny influence to voltage gain.So when $k=1$,the ideal voltage gain expression is:

$$G = \frac{2N + 3 - ND}{1-D} \tag{15}$$

The voltage of capacitors C_1,C_2,C_3,C_4 are listed as following:

$$V_{C1} = V_{C2} = \frac{V_{in}}{1-D} \tag{16}$$

2648

$$V_{C3} = V_{C4} = \frac{N+2-ND}{1-D}V_{in} \qquad (17)$$

The voltage stress across switch S,diode D_1,D_2,D_3,D_4,D_5 are:

$$V_S = V_{D1} = V_{D2} = \frac{V_o}{2N+3-ND} \qquad (18)$$

$$V_{D3} = V_{D4} = V_{D5} = \frac{N+1}{2N+3-ND}V_o \qquad (19)$$

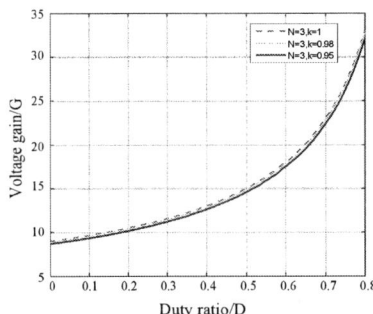

Fig.8 Voltage gain versus duty cycle under N=3 and various k

IV. COMPARISON WITH OTHER CONVERTERS

With increasing number of two-port coupled inductor diode-capacitor cells,the voltage conversion ratio of proposed converter can be further increased.The main circuit is shown as Fig.3 above.Supposing the turns ratio of coupled inductor is the same,that is $N_1:N_0=N_2:N_0=\cdots=N_M:N_0=N$,then the voltage gain and voltage stress of semiconductors can be derived as the following table.

TABLE 1
MULTI-CELL PROPOSED CONVERTER

Cell number	Voltage gain (V_o/V_{IN})	Voltage stress of main switch (V_S/V_o)	Voltage stress of diode D_1,D_2 $(V_{D1,D2}/V_o)$	Voltage stress of diode D (V_D/V_o)
M=1	$\frac{2N+3-ND}{1-D}$	$\frac{G-N}{G(N+3)}$	$\frac{G-N}{G(N+3)}$	$\frac{(G-N)(N+1)}{G(N+3)}$
M=2	$\frac{4(N+1)}{1-D}$	$\frac{1}{4(N+1)}$	$\frac{1}{4(N+1)}$	$\frac{2N+1}{4(N+1)}$
M=3	$\frac{8N+5-ND}{1-D}$	$\frac{G-N}{G(7N+5)}$	$\frac{G-N}{G(7N+5)}$	$\frac{(G-N)(3N+1)}{G(7N+5)}$
M=4	$\frac{6(2N+1)}{1-D}$	$\frac{1}{6(2N+1)}$	$\frac{1}{6(2N+1)}$	$\frac{4N+1}{6(2N+1)}$
...	...	...	...	...

It is easily observed from the table that when the cell number is even,the voltage stress of semiconductors is independent of voltage gain and only relate to output voltage and cell number.Fig.9 shows the relationship of voltage gain and boost duty cycle D, turns ratio of coupled inductor N, and the boost cell number M.The voltage gain increase quickly with larger cell number.

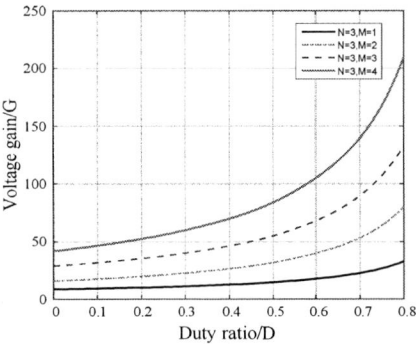

Fig.9 Voltage gain versus duty ratio under N=3

Table 2 shows the performance comparison that the proposed converter is compared with the traditional Boost converter,three single switch switch and coupled inductor DC-DC converters in [11],[12] and [13].When $k=1,N=3$,the comparison of voltage gain versus duty cycle,switch voltage stress versus voltage gain and output diode voltage stress versus voltage gain are shown in Fig.10.From Fig.10(a),when the duty cycle is lower than 6.5,the proposed converter has highest voltage gain.The proposed converter has the lowest switch voltage stress when the voltage gain range from 10 to 20 which is shown clearly in Fig.10(b).Also the proposed converter output diode voltage stress is better than converter in [13].

TABLE 2
COMPARISON WITH OTHER CONVERTERS

Topology	BOOST	CONVERTER IN[11]	CONVERTER IN[12]	CONVERTER IN[13]	PROPOSED CONVERTER
Numbers of active switches	1	1	2	1	1
Numbers of diodes	1	4	3	4	5
Voltage gain	$\frac{1}{1-D}$	$\frac{1+N+ND}{1-D}$	$\frac{(N+1)(2-D)}{1-D}$	$\frac{2+N+ND}{1-D}$	$\frac{2N+3-ND}{1-D}$
voltage stress of active switches	1	$\frac{G+N}{G(2N+1)}$	$\frac{G-N-1}{G(N+1)}$	$\frac{G+N}{2G(N+1)}$	$\frac{G-N}{G(N+3)}$
voltage stress of output diodes	1	$\frac{N(G+N)}{G(2N+1)}$	$\frac{(G-N-1)(N+1)}{G(N+1)}$	$\frac{G+N}{2G}$	$\frac{(G-N)(N+1)}{G(N+3)}$

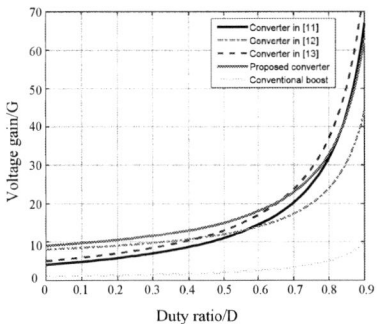

(a) Voltage gain comparison

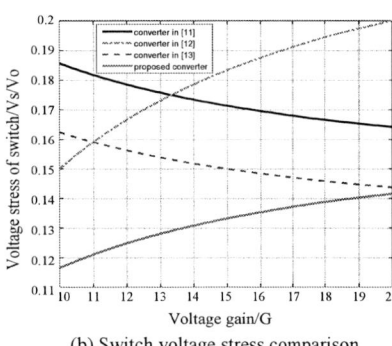

(b) Switch voltage stress comparison

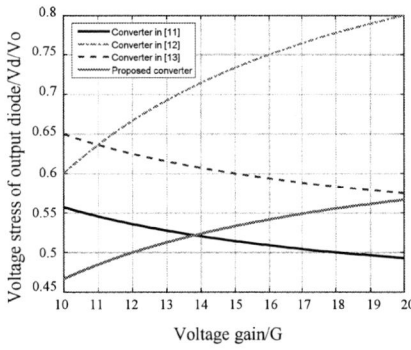

(c) Output diode voltage stress comparison

Fig.10 Comparison with other converters under N=3

V. DESIGN GUIDELINE

A. Coupled inductor design

Since the turns ratio of the coupled inductor determines the voltage stress of the switch and the operational duty-cycle of the converter, it is the key parameter in the circuit parameter design. Once the duty cycle and voltage gain is already known,the turns ratio can be obtained:

$$N = \frac{V_o(1-D) - 3V_{in}}{V_{in}(2-D)} \quad (20)$$

The magnetizing inductor can be designed by setting an acceptable current ripple on the magnetizing inductor, which is given by:

$$L_m = \frac{DV_{in}}{\Delta i_{Lm} f_s} \quad (21)$$

Where:Δi_{Lm} is current ripple of magnetizing inductor,f_s is the switch frequency.

B. Capacitors design

There are two parameters need to know when the capacitors are been chosen that is voltage stress and capacitance.The voltage stress has already been given in equation (15)-(17).The capacitance is determined by voltage ripple across it as shown below:

$$C_1 = C_2 = \frac{2V_o}{\Delta V_{C1} f_s R_L} = \frac{2V_o}{\Delta V_{C2} f_s R_L} \quad (22)$$

$$C_3 = C_4 = \frac{V_o}{\Delta V_{C3} f_s R_L} = \frac{V_o}{\Delta V_{C4} f_s R_L} \quad (23)$$

$$C_5 = \frac{(2D^2 + DN^2 + 2ND^2 + 5D + 2N + 2ND + 2)V_o}{\Delta V_{C5}(N+3)^2 f_s R_L} \quad (24)$$

Where:$\Delta V_{C1}, \Delta V_{C2}, \Delta V_{C3}, \Delta V_{C4}, \Delta V_{C5}$ is voltage ripple of capacitors.

C. Semiconductors

The key parameters for semiconductor is voltage stress and current stress.The voltage tress of semiconductors has been given above. The average current stress is given as following equations:

$$i_{D1,avg} = i_{D2,avg} = i_{D3,avg} = i_{D4,avg} = i_{D5,avg} = I_o \quad (25)$$

$$i_{S,avg} = \frac{4 + 4N - D - 3ND}{2(1-D)} I_o \quad (26)$$

The value of the maximum currents that flows through the semiconductors are given by:

$$i_{D1,peak} = i_{D2,peak} = \frac{(N+3)I_o}{2(1-D)} \quad (27)$$

$$i_{D3,peak} = i_{D4,peak} = \frac{2I_o}{D} \quad (28)$$

$$i_{S,peak} = \frac{4 + 4N - D - 3ND}{D(1-D)} I_o \quad (29)$$

According to (1),(2) and also referring to the Fig.7,the RMS value currents of the semiconductors can be derived by:

$$i_{D1,rms} = i_{D2,rms} = I_o \sqrt{\frac{N+3}{3(1-D)}} \quad (30)$$

$$i_{D3,rms} = i_{D4,rms} = \frac{2I_o}{\sqrt{3D}} \quad (31)$$

$$i_{D5,rms} = \frac{I_o}{N+1} \sqrt{\frac{3N^2 + 10N + 3}{3(1-D)}} \quad (32)$$

$$i_{S,rms} = I_o \sqrt{\frac{1}{3D} \cdot \frac{4 + 4N - D - 3ND}{1-D}} \quad (33)$$

VI. SIMULATION AND EXPERIMENT RESULTS

Simulation based on MATLAB/Simulink has been used to verify the working principle and theoretical analysis of proposed topology.The main circuit parameters are as following:V_{in}=30V,L_m=400uH,L_k=2uH,C_1=C_2=C_3=C_4=300uF,C_5=350uF,f_s=10kHz,N=3,D=0.6,R_L=300Ω.The simulation results presented at Fig.11 suit the theoretical value pretty well.Experiment parameters are the same with simulation except L_m=200uH,C_1=C_2=C_3=C_4=C_5=500uF,f_s=100kHz,N=2.Fig. 12 shows the experiment results that are the same with theoretical analysis.

2650

The 2018 International Power Electronics Conference

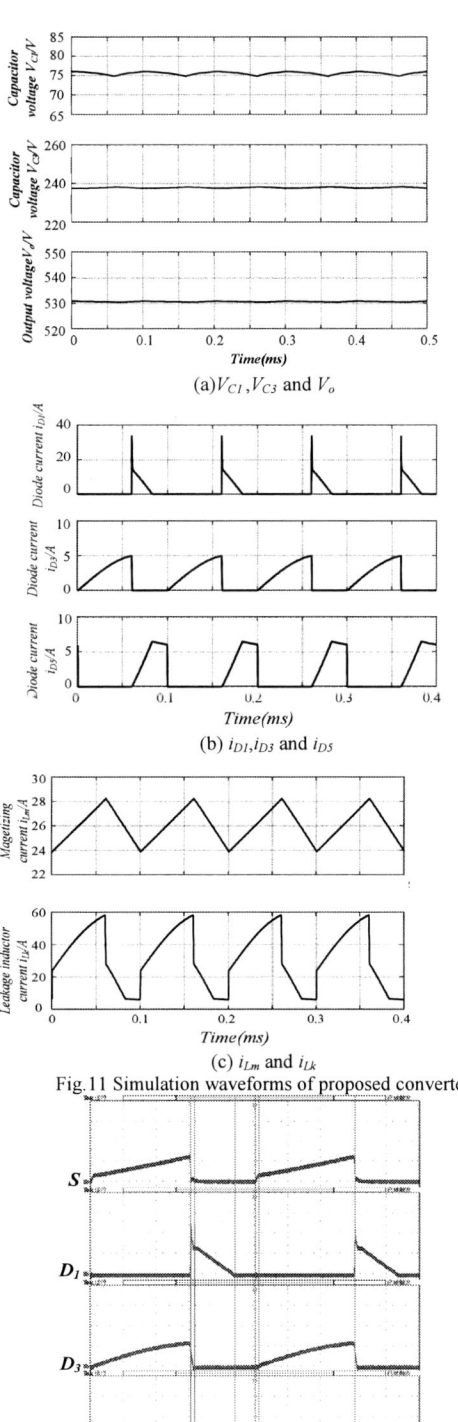

(a)V_{C1},V_{C3} and V_o

(b) i_{D1},i_{D3} and i_{D5}

(c) i_{Lm} and i_{Lk}

Fig.11 Simulation waveforms of proposed converter

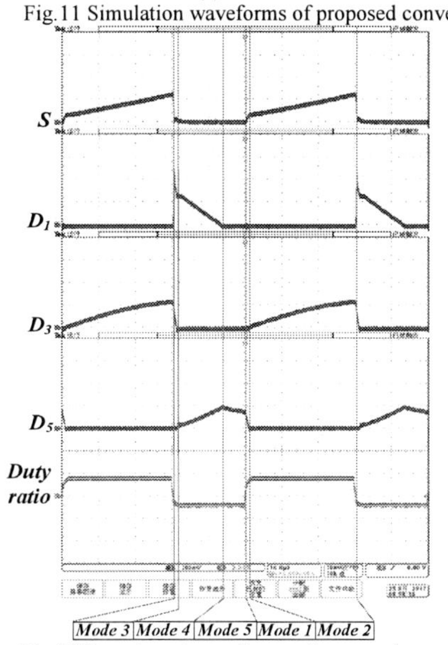

Fig.12 Experiment waveforms of proposed converter

VII. CONCLUSIONS

Compared with other converters,proposed converter has the following advantages:

Enhance the voltage conversion ratio remarkably and decrease the voltage stress on the semiconductor components;single switch working mode cuts the cost for drive circuit and simplify the control complexity;suppress the inrush current in switch-capacitor network and make all the diode turn off naturally;reduce turns ratio and volume of coupled inductor;the voltage stress of semiconductors is independent of voltage gain when the cell number is even.

The proposed converter is promising in the renewable generation system.

REFERENCES

[1] Y. Cao, V. Samavatian, K. Kaskani and H. Eshraghi, "A Novel Nonisolated Ultra-High-Voltage-Gain DC‐DC Converter With Low Voltage Stress," in IEEE Transactions on Industrial Electronics, vol. 64, no. 4, pp. 2809-2819, April 2017.

[2] Y. Jie, Y. Dongsheng, C. Hao, Z. Xiaoshu and C. He, "A single switch based integrated DC-DC converter with high step-up gain," 2016 14th International Conference on Control, Automation, Robotics and Vision (ICARCV), Phuket, 2016, pp. 1-5.

[3] Y. Ye, K. W. E. Cheng and S. Chen, "A High Step-up PWM DC-DC Converter With Coupled-Inductor and Resonant Switched-Capacitor," in IEEE Transactions on Power Electronics, vol. 32, no. 10, pp.

[4] K. C. Tseng, J. Z. Chen, J. T. Lin, C. C. Huang and T. H. Yen, "High Step-Up Interleaved Forward-Flyback Boost Converter With Three-Winding Coupled Inductors," in IEEE Transactions on Power Electronics, vol. 30, no. 9, pp. 4696-4703, Sept. 2015.

[5] T. Nouri, S. H. Hosseini, E. Babaei and J. Ebrahimi, "Interleaved high step-up DC‐DC converter based on three-winding high-frequency coupled inductor and voltage multiplier cell," in IET Power Electronics, vol. 8, no. 2, pp. 175-189, 2 2015.

[6] L. S. Yang, T. J. Liang, H. C. Lee and J. F. Chen, "Novel High Step-Up DC-DC Converter With Coupled-Inductor and Voltage-Doubler Circuits," in IEEE Transactions on Industrial Electronics, vol. 58, no. 9, pp. 4196-4206, Sept. 2011.

[7] M. S. Bhaskar, P. Sanjeevikumar, F. Blaabjerg, V. Fedák, M. Cernat and R. M. Kulkarni, "Non isolated and non-invertingCockcroft-Walton multiplier based hybrid 2Nx interleaved boost converterfor renewable energy applications," 2016 IEEE International Power Electronics and Motion Control Conference (PEMC), Varna, 2016, pp. 146-151.

[8] T. Nouri, N. Vosoughi, S. H. Hosseini and M. Sabahi, "A Novel Interleaved Nonisolated Ultrahigh-Step-Up DC‐DC Converter With ZVS Performance," in IEEE Transactions on Industrial Electronics, vol. 64, no. 5, pp. 3650-3661, May 2017.

[9] Yan Zhang, Zhuo Dong, Jinjun Liu, Xiaolong Ma, Xinying Li and Jiuqiang Han, "Modeling and controller design of high voltage gain DC-DC converter with multi-cell diode-capacitor network," 2016 IEEE 8th International Power Electronics and Motion Control Conference (IPEMC-ECCE Asia), Hefei, 2016, pp. 3066-3072.

[10] M. S. Bhaskar, P. Sanjeevikumar, F. Blaabjerg, V. Fedák, M. Cernat and R. M. Kulkarni, "Non isolated and non-inverting Cockcroft-Walton multiplier based hybrid 2Nx interleaved boost converter for renewable energy applications," 2016 IEEE International Power Electronics and Motion Control Conference (PEMC), Varna, 2016, pp. 146-151.

[11] B.Axelrod, Y.Beck and Y.Berkovich: "High step-up DC‐DC converter based on the switched-coupled-inductor boost converter and diode-capacitor multiplier: steady state and dynamics," IET Power Electron., vol. 8, pp. 1420‐1428, 2015.

[12] Hwu.K.I and Yau.Y.T: "High Step-Up Converter Based on Coupling Inductor and Bootstrap Capacitors With Active Clamping," IEEE Trans. Power Electron., 2014, 18, (1), pp. 2655 ‐2660.

2651

[13] J.Ai and M. Lin, "Ultra large Gain Step-Up Coupled-Inductor DC -DC Converter With an Asymmetric Voltage Multiplier Network for a Sustainable Energy System," in IEEE Transactions on Power Electronics, vol. 32, no. 9, pp. 6896-6903, Sept. 2017.

The 2018 International Power Electronics Conference

Novel Active Clamping Step-Down DC-DC Converter with Lower Voltage Stress

Chi-Hsuan Hsu[1], Jun-Min Jian[1*], Jiann-Fuh Chen[1] and Hsuan Liao[1]

1 Electrical Engineering, National Cheng Kung University, Tainan, Taiwan

*E-mail: yf89006@gmail.com

Abstract— **In this paper, a novel ZVS buck converter with lower voltage stress based on coupled-inductor and active clamp techniques has been proposed. This converter utilizes coupled-inductor to obtain higher step-down ratio than the conventional buck converter. Moreover, the proposed converter can achieve ZVS (Zero-voltage switching) on both switches from full load to extra-light load. Therefore, the efficiency at light load does not subject to the switching losses, which leads to higher overall efficiency.**

The ZVS conditions and the design considerations of proposed converter are described. The ZVS region can cover the full load range, if the value of the magnetic inductance and resonant inductance have been properly chosen. Finally, based on controller TMS320F28035, a converter with 156 V input voltage, 48 V output voltage, and 200 W output power is developed to prove the feasibility of the proposed converter. The highest efficiency is around 96.8% when operating at 125 W.

Keywords—Zero-voltage switching (ZVS), lower voltage stress, active clamping, coupled-inductor.

I. INTRODUCTION

Recently, switch-mode power supplies (SMPS) are widely used in residential, industrial and aerospace environment. Therefore, it is one ways for saving energy to raise the efficiency of the power converters. Then, the requirement of high efficiency power converters become more significant and aware.

Buck converter is one of the most important SMPS topology. It is widely used to convert a higher input voltage into a lower output voltage throughout the industry. There are many types of buck converters topology, such as, quasi-resonant buck converters, active-clamping buck converters, interleaved buck converters, quadratic buck converters, three-level buck converters, etc. In addition, coupled inductor or tapped inductor can be used to obtain larger conversion ratio in above-mentioned converters.

Quasi-resonant converters (QRCs) reduce the switching losses in PWM converters. However, the switches in QRCs withstand higher current stress, and these converters are controlled by variable switching frequency [1]. Active clamping buck converters can achieve zero-voltage switching easily, but the switches suffer from higher voltage stress than a conventional buck converter [2]-[3]. With the resonance between the

leakage inductance of coupled-inductor and the junction capacitors of switches, ZVS operation can easily achieve [4]-[5]. Otherwise, coupled-inductor is also used to obtain higher conversion ratio by adjusting the turns ratio [6]-[7]. To promote the voltage conversion ratio more, switched capacitors are applied in [8]. Besides coupled-inductor, tapped-inductor can also achieve high conversion ratios [9]. A modified form of interleaved buck converter use three additional switches to reduce the voltage stress to one fourth of the input voltage [10]. A double quadratic buck converter has lower voltage stress on the switches, though it has the same step-down conversion ratio as a conventional quadratic buck converter [11]. Three-level buck converters can also lower the voltage stress on switches effectively, but they increase the number of switches and cost [12]. What's more, interleaved buck converters, quadratic buck converters, and three-level buck converters are all hard switching control generally. Therefore, to develop a novel buck converter, which has all of the features of soft switching, low-voltage-stress, and high step-down ratio, is also the main motivation of this research. In this paper, a novel buck converter combined with these three advantages is proposed, and a 200 W prototype is implemented.

The circuit diagram of proposed converter is shown in Fig. 1. The proposed converter contains a DC input voltage V_I, two power switches S_1 and S_2, a coupled inductor L_m, a resonant inductor L_r, a resonant capacitor C_r, and a diode D_{sec}. Basically, the switches S_1 and S_2 are individually driven by two complementary PWM signals which are generated by digital controller. Moreover, the dead time is inserted between the gate signals for the switches S_1 and S_2 in practical applications to realize ZVS. With the resonant between L_r and C_r, ZVS of both switches can be realized respectively. Therefore, the switching losses are mitigated, which improves the efficiency of the converter, and the electromagnetic interference (EMI) will also be reduced.

The voltage stress of switches are much lower than that of conventional buck converter which makes it possible to use switches with lower voltage rating. As a result, both switching and conduction losses can be further reduced. In addition, because of lower voltage stress, lower energy is required to discharge from the parasitic capacitors of the switches S_1 and S_2. Therefore,

2653

the realization of ZVS of the switches becomes much easier.

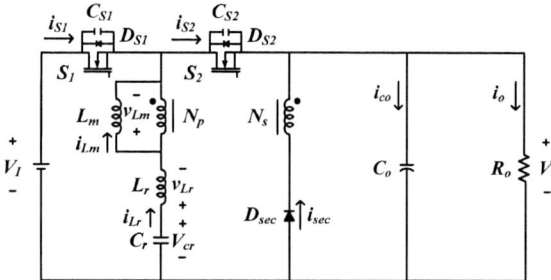

Fig. 1. Proposed converter.

II. OPERATING PRINCIPLE OF PROPOSED CONVERTER

In Fig. 1, the main parameters and reference directions of the proposed circuit are illustrated. Generally, the coupled inductor is modeled as the combination of the magnetizing inductor and the leakage inductor, which is used as the resonant inductor.

To facilitate the analysis operation, several assumptions are made as follows:

1) Switches S_1 and S_2 consist of the junction capacitances C_{S1} and C_{S2} and body diodes D_{S1} and D_{S2}; other parasitic components and voltage drop caused by the turn-on resistance are ignored. The diode D_{sec} is ideal.

2) Output capacitor C_o and resonant capacitor C_r are large enough that V_o and V_{cr} are regarded as constant values during a switching period.

3) The turns ratio of coupled inductor is $N_p : N_s = n : 1$.

Seven operating modes in a switching cycle will be presented. Several key waveforms are illustrated in Fig. 2. The current paths of each mode are depicted in Fig. 3.

Before t_0, switch S_1 is ON, and the energy is transferred to the load through the coupled inductor L_m. Diode D_{sec} is conducting. The voltage across L_m is equal to $-nV_o$, which is reflected from the secondary side. This mode ends when S_1 is OFF at $t = t_0$.

Mode I [t_0, t_1]:

This is the dead time during which both S_1 and S_2 are OFF. At $t = t_0$, switch S_1 is turned off. The resonant inductor current i_{Lr} starts to charge and discharge the junction capacitors C_{S1} and C_{S2} respectively. Once the drain–source voltage of S_2 reaches to zero, i_{Lr} flows through the body diode of S_2, D_{S2}. After the dead time, S_2 is turned on with ZVS.

Mode II [t_1, t_2]:

At $t = t_1$, S_2 is turned on with ZVS. i_{Lr} is diverted from D_{S2} to the $R_{ds(on)}$ of S_2, so that no significant energy is lost during the turn-on transient. The magnitude of i_{Lr}

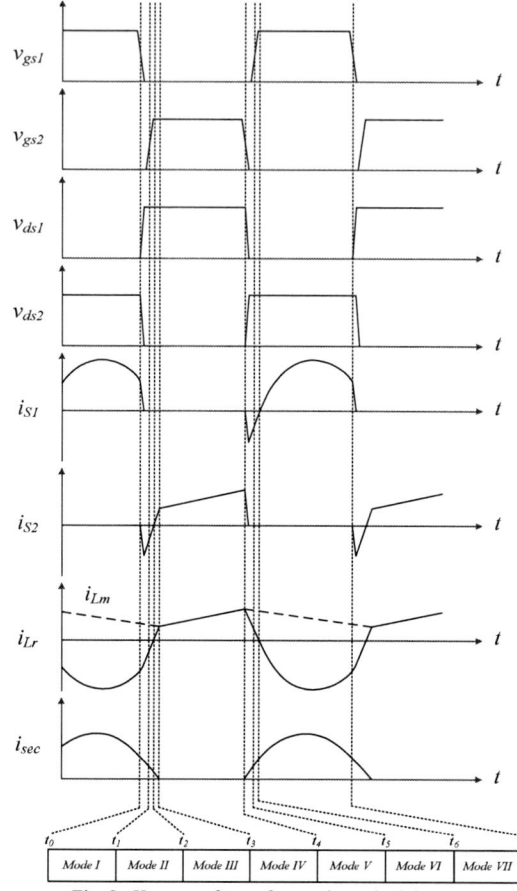

Fig. 2. Key waveform of operation principle.

decreases rapidly. And the current of D_{sec} is also decreasing quickly at the same time. This mode ends when $i_{Lr} = 0$ at $t = t_2$.

Mode III [t_2, t_3]:

The resonant inductor current i_{Lr}, which is zero at $t = t_2$, becomes a positive value. Due to the positive slope of L_r and the negative slope of L_m, the value of i_{Lr} reaches the value of i_{Lm} gradually. D_{sec} conducts until i_{Lr} equals to i_{Lm} at $t = t_3$, and then becomes reverse biased with a negative voltage equal to V_{Cr}.

Mode IV [t_3, t_4]:

At $t = t_3$, i_{sec} becomes zero. i_{Lr} is equal to i_{Lm} and the magnetic inductance L_m is no longer shunted by the load reflected to the primary side. L_m goes in series with L_r and participates to the resonance. This portion is similar to a straight line, as shown in Fig. 3. This mode ends when S_2 is turned off at $t = t_4$.

Mode V [t_4, t_5]:

This is the dead time during which both S_1 and S_2 are OFF. At $t = t_4$, switch S_2 is turned off. The resonant inductor current i_{Lr} starts to discharge and charge the

2654

junction capacitors C_{S1} and C_{S2}, respectively. Once the drain – source voltage of S_1 reaches to zero, i_{Lr} flows through the body diode of S_1, D_{S1}. D_{sec} becomes forward biased and starts conducting. After the dead time, S_1 is turned ON with ZVS at t = t_5.

Mode VI [t₅, t₆]:

At t = t_5, S_1 is turned on with ZVS. i_{Lr} is diverted from D_{S1} to the $R_{ds(on)}$ of S_1. The magnitude of i_{Lr} decreases rapidly. And the i_{sec} is increasing at the same time. The voltage across L_m is clamped at $-nV_o$, so that L_m does not participate in the resonance and C_r resonates with L_r only. This mode ends when $i_{Lr} = 0$ at t = t_6.

Mode VII [t₆, t₇]:

The inductor current i_{Lr}, which is zero at t = t_6, becomes a negative value. Since the diode D_{sec} is still conducting, the voltage across L_m is equal to $-nV_o$. Therefore, the magnetic inductor L_m is still not participating in the resonance, and C_r resonates with L_r only. This mode ends when S_2 is switched off at t = t_7.

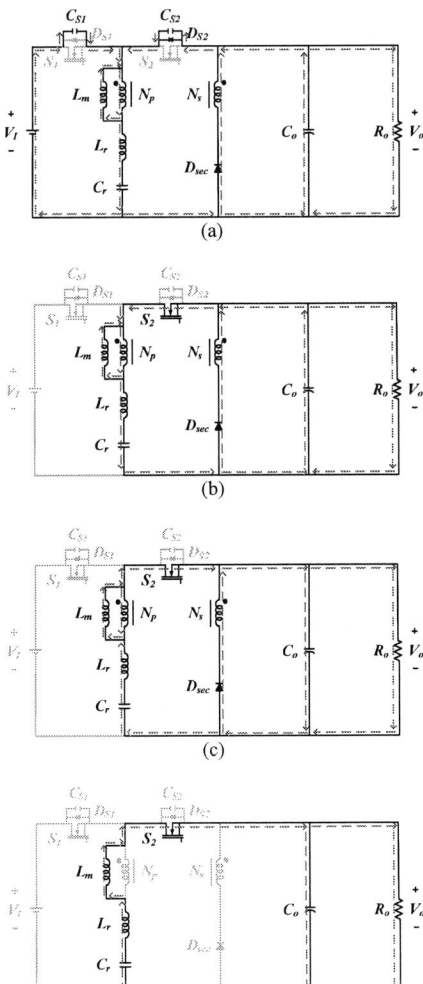

(a)

(b)

(c)

(d)

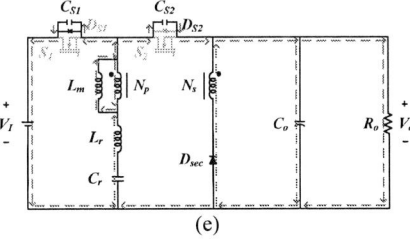

(e)

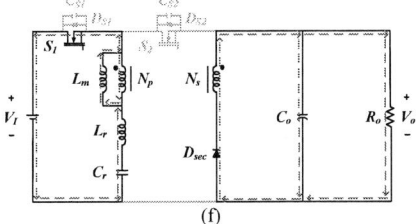

(f)

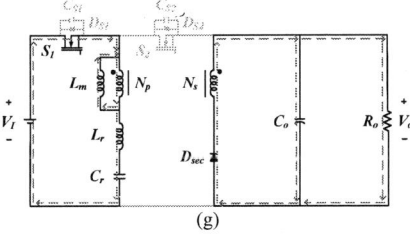

(g)

Fig. 3. The current paths of each mode.
(a) Mode I [t_0, t_1]. (b) Mode II [t_1, t_2].
(c) Mode III [t_2, t_3]. (d) Mode IV [t_3, t_4]. (e) Mode V [t_4, t_5].
(f) Mode VI [t_5, t_6]. (g) Mode VII [t_6, t_7].

III. STEADY-STATE ANALYSIS

In steady-state analysis, Mode IV and Mode VII are two fundamental operating modes for their long duration and the power flow in one switching period. The resonances occur in Mode VII as well. Consequently, these two modes play significant roles in steady-state analysis. Other modes last relatively short; therefore, they can be ignored.

During Mode IV, L_m is in series with L_r. Thus, the magnetizing current i_{Lm} is equal to i_{Lr}, and their expressions can be derived as follows:

$$i_{Lm}(t) = I_{Lm} - \Delta i_{Lm} + \frac{V_{Cr} - V_O}{L_m + L_r}(t - t_3) \, , \, t_3 \leq t \leq t_4 \quad (1)$$

$$i_{Lr}(t) = i_{Lm}(t) = I_{Lm} - \Delta i_{Lm} + \frac{V_{Cr} - V_O}{L_m + L_r}(t - t_3) \, , \, t_3 \leq t \leq t_4 \quad (2)$$

The ripple of magnetizing current i_{Lm} is given as (3).

$$\Delta i_{Lm} = \frac{V_{Cr} - V_O}{2(L_m + L_r)}(1 - D)T_{sw} \quad (3)$$

During Mode VII, The voltage across L_m is clamped at $-n \cdot V_o$ in this mode. Therefore, L_m isn't participating in the resonance. C_r is resonating with L_r only, and their expressions can be derived as follows:

$$i_{Lm}(t) = I_{Lm} + \Delta i_{Lm} - \frac{nV_O}{L_m}(t-t_4) \, , \, t_4 \le t \le t_7 \quad (4)$$

$$
\begin{aligned}
i_{Lr}(t) = \; & i_{Lr}(t_4)\cos\omega_r(t-t_4) \\
& - \frac{V_I - nV_O - v_{Cr}(t_4)}{Z_r}\sin\omega_r(t-t_4) \, , \, t_4 \le t \le t_7
\end{aligned} \quad (5)
$$

$$i_{Lr}(t_4) = I_{Lm} + \Delta i_{Lm} \quad (6)$$

In this mode, the resonant frequency ω_r and the characteristic impedance Z_r of the resonance between L_r and C_r can be written as:

$$\omega_r = \frac{1}{\sqrt{L_r \cdot C_r}} \quad (7)$$

$$Z_r = \sqrt{\frac{L_r}{C_r}} \quad (8)$$

Apply amp-sec balance law to the output capacitor C_O:

$$
\begin{aligned}
\int_{t_0}^{t_3} \left[i_{sec}(t) - i_O(t) \right] dt + \int_{t_3}^{t_4} \left[i_{Lm}(t) - i_O(t) \right] dt \\
+ \int_{t_4}^{t_7} \left[i_{sec}(t) - i_O(t) \right] dt = 0
\end{aligned} \quad (9)
$$

The expression of I_{Lm} can be founded by (9).

$$I_{Lm} = \frac{V_O}{R_O} - \frac{1}{T_{SW}}\int_{t_3}^{t_4}\left[i_{Lr}(t) \right] dt \quad (10)$$

The expression of $i_{Lr}(t_4)$ and $v_{Cr}(t_4)$ can be founded by equation (6) and (11), and the efficiency of proposed converter is assumed as η.

$$i_{Lr}(t_4) = \frac{P_O}{V_O} - \frac{P_O}{\eta V_I} - \frac{(V_O - V_{Cr})(1-D)T_{SW}}{2(L_m + L_r)} \quad (11)$$

$$v_{Cr}(t_4) = V_I - V_O - \frac{P_O}{2f_{SW}C_r \eta V_O} \quad (12)$$

By applying volt-sec balance law with these two modes on L_m (13) and L_r (14) respectively, then the expression of the output voltage V_o and the resonant capacitor voltage V_{Cr} can be derived as (15) and (16).

$$(V_{Cr} - V_O)kDT_{SW} - nV_O(1-D)T_{SW} = 0 \quad (13)$$

$$(V_{Cr} - V_O)(1-k)DT_{SW} - (V_I - nV_O - V_{Cr})(1-D)T_{SW} = 0 \quad (14)$$

$$V_O(V_I, D, n, k) = \frac{DkV_I}{n + Dk} \quad (15)$$

$$V_{Cr}(V_I, D, n, k) = \frac{n - Dn + Dk}{n + Dk}V_I \quad (16)$$

And the conversion ratio M can be derived from (15)

$$M(D, n, k) = \frac{V_O}{V_I} = \frac{Dk}{n + Dk} \quad (17)$$

$$k = \frac{L_m}{L_m + L_r} \quad (18)$$

The conversion ratios of this converter with different value of k versus the duty cycle is illustrated in Fig. 4.

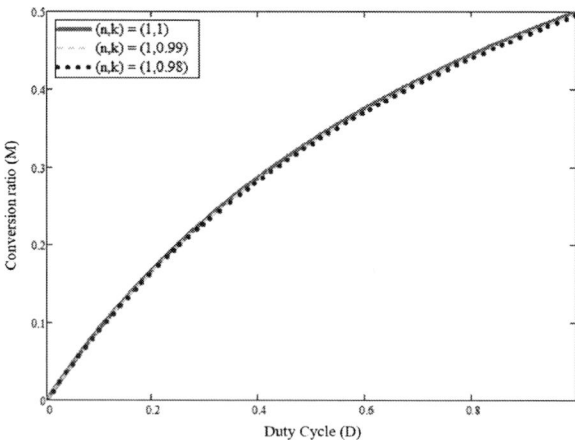

Fig. 4. Voltage conversion ratios of proposed converter with different value of k.

The conversion ratios of this converter with different value of n versus the duty cycle is illustrated in Fig. 5.

2656

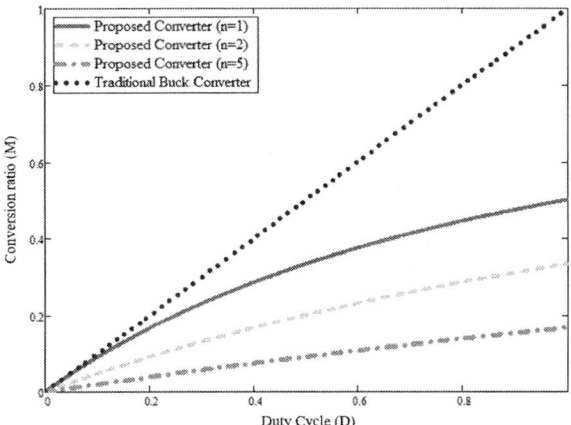

Fig. 5. Voltage conversion ratios of proposed converter with different value of n.

IV. DESIGN CONSIDERATIONS

Unlike other converters, proposed converter can achieve ZVS from full load to extra-light load by its own property. For switch S_1, both heavy load and light load can achieve ZVS in the same way. However, for switch S_2, the heavy load, the light load, and the extra-light load achieve ZVS in different ways. In this section, the ZVS condition of S_1 will be derived first. Secondary, the three conditions for heavy, light, and extremely light load will be introduced. Finally, the ZVS boundary map will be present.

The ZVS of S_1 can only occur if the resonant inductor L_r stores sufficient energy prior to the start of the resonance to transition both junction capacitances C_{S1} and C_{S2}. The ZVS constraint of switch S_1 is present as bellow:

$$\frac{1}{2}L_r\left[i_{Lr}(t_4)\right]^2 \geq \frac{1}{2}(C_{S1}+C_{S2})V_{ds}^2 \tag{19}$$

The expression of $i_{Lr}(t_4)$ has been derived in (11). In proposed converter, the drain source voltage V_{ds} is 108 V, which is equal to $V_I - V_O$. Assume the junction capacitance C_{S1} and C_{S2} of the switch are both 360 pF. Then, (19) can be combined with (11) to yield:

$$\frac{L_r}{L_m^2} \geq 107.14 \tag{20}$$

For switch S_2, the operation diagram of three different load conditions are shown in Fig. 6, Fig. 7 and Fig. 8.

In heavy load, the valley value of i_{Lm} is positive, and i_{Lr} must change its value from $i_{Lr\_heavy}(t_7)$ to $i_{Lm\_heavy}(t_b)$ after S_1 is turned off. If the value of $i_{Lr\_heavy}(t_7)$ is not large enough, S_2 cannot achieve ZVS. Therefore, the ZVS constraint under this condition can be written as (21).

$$P_o \geq \frac{1.323}{\sqrt{L_r}} - \frac{0.123}{L_m} \tag{21}$$

In light load, the valley value of i_{Lm} is negative, and the resonant inductor current will meet the value of $i_{Lm\_light}(t_c)$ before changing positive. The ZVS process of switch S_2 under this condition can be separated into two stages. In the first stage, L_r resonates with the junction capacitance alone, and its value changes from $i_{Lr\_light}(t_7)$ to $i_{Lm\_light}(t_c)$. While $t = t_c$, L_r is in series with L_m, and the stage II begins. L_m starts participating the resonance with the junction capacitance.

In extra-light load, the whole ZVS process has only one stage, which is similar to the effect of the stage II in light load condition. The ZVS constraint under light and extra-light load can be both written as (22).

$$P_O \leq \frac{0.015}{L_m} - \frac{0.207}{\sqrt{L_m}} \; or \; P_O \geq \frac{0.015}{L_m} + \frac{0.207}{\sqrt{L_m}} \tag{22}$$

To make sure the ZVS region is filled the whole load range, let the boundary of P_O of (21) greater than (22).

$$\frac{1.323}{\sqrt{L_r}} - \frac{0.123}{L_m} \leq \frac{0.015}{L_m} - \frac{0.207}{\sqrt{L_m}} \tag{23}$$

Then the relationship can be established as (24), and the proper area for full range ZVS of S2 is illustrated in Fig. 9.

$$L_r(L_m) \geq \frac{1}{\left(\dfrac{0.156}{\sqrt{L_m}} - \dfrac{0.104}{L_m}\right)^2} \tag{24}$$

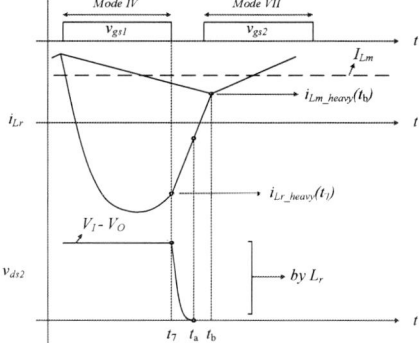

Fig. 6. Key waveforms of ZVS of S_2 (heavy load).

The 2018 International Power Electronics Conference

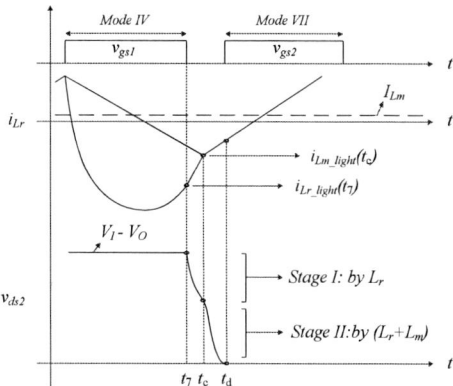

Fig. 7. Key waveforms of ZVS of S_2 (light load).

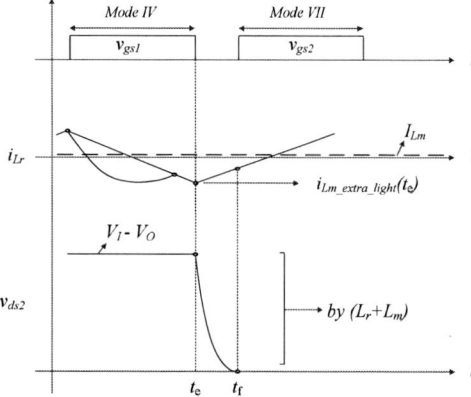

Fig. 8. Key waveforms of ZVS of S_2 (extra-light load).

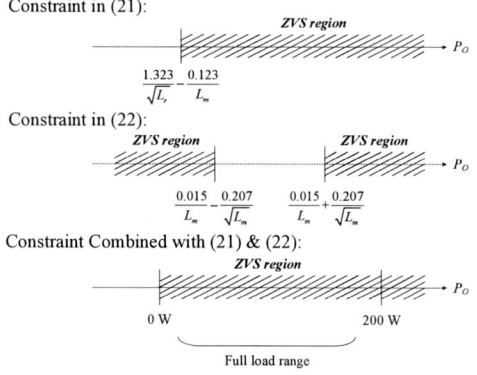

Fig. 9. The diagram of ZVS region of S_2.

With (20) and (24), if the magnetic inductance L_m = 150 μH, then the value of resonant inductance L_r must greater than 2.3 μH. Therefore, the value of L_r is chosen as 2.4 μH, and the value of C_r is obtained as 4.75 μF by (7).

V. EXPERIMENTAL RESULT

The major parameters and elements used in the prototype are shown in Table I.

TABLE I
SPECIFICATIONS OF THE PROTOTYPE

Symbol	Item	Part/Specification
V_{in}	Input DC voltage	156 V
V_o	Output DC voltage	48 V
P_o	Output power	200 W
f_{sw}	Switching frequency	50 kHz
S_1 and S_2	Switches	IRFB4227PbF
D	Diode	DPG60C300HB
N_p / N_s	Turns of winding	30 / 30 turns
n	Turns ratio	1
L_m / L_{lk}	Magnetic inductance / Leakage inductance	152.8 μH / 1.1μH
L_r	Resonant inductance	2.4 μH
C_r	Resonant capacitor	4.75 μF (ceramic capacitor)

To verity the ZVS property of proposed converter, the waveforms under different load conditions are present as bellow.

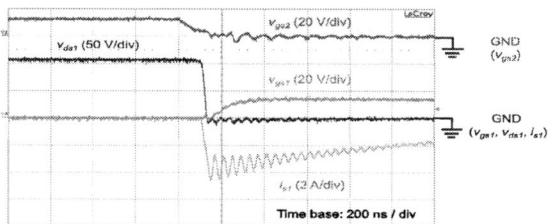

Fig. 10. ZVS of S_1 at full load.

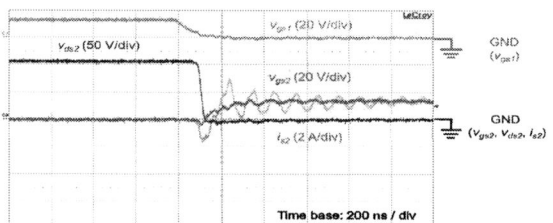

Fig. 11. ZVS of S_2 at full load.

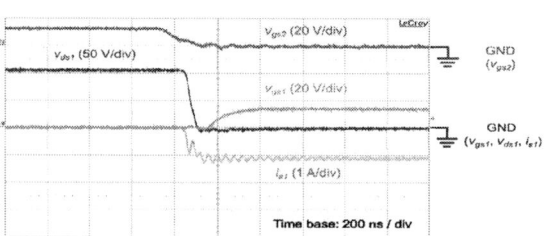

Fig. 12. ZVS of S_1 at 1% load.

2658

The 2018 International Power Electronics Conference

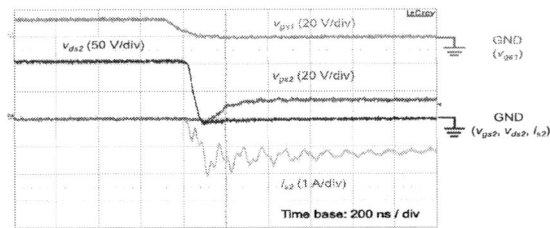

Fig. 13. ZVS of S_2 at 1% load.

VI. CONCLUSIONS

In this paper, the converter with 156 V input voltage, 48 V output voltage, and 200 W output power has been implemented to verify all the theoretical analysis. From the experiment waveforms, soft-switching performance of the proposed converter are proved. Both switches S_1 and S_2 can be turned on with ZVS from full load to extra-light load. Finally, the efficiency of the prototype reaches its peak value as 96.8%, when the output power is around 125 W.

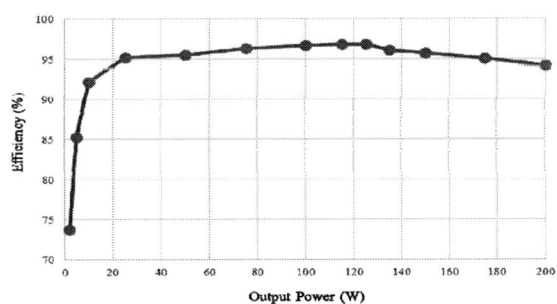

Fig. 14. Efficiency trend chart.

ACKNOWLEDGMENT

Authors would like to thank Ministry of Science and Technology of Taiwan under MOST 106-2218-E-006-024 for this paper achievement.

REFERENCES

List only one reference per reference number according to the following samples:

[1] K. H. Liu, R. O. Oruganti and F. C. Lee, "Resonant switches-topologies and characteristics", Proc. Power Electronic. Spec. Conf., pp. 62-67, 1985.

[2] C. M. C. Duarte and I. Barbi, "A family of ZVS-PWM active-clamping Dc-to-Dc converters: Synthesis analysis design and experimentation", IEEE Trans. Circuits Syst., vol. 44, pp. 698-704, Aug. 1997.

[3] Y. Ma, X. Wu, X. Xie, G. Chen and Z. Qian, "A new ZVS-PWM buck converter with an active clamping cell," IEEE IECON'07, pp. 1592-1597, 2007.

[4] S. S. Lee, "Step-down converter with efficient ZVS operation with load variation", IEEE Trans. Ind. Electronic, vol. 61, no. 1, pp. 591-597, Jan. 2014.

[5] K. I. Hwu, W. Z. Jiang and Y. T. Yau, "Ultrahigh step-down converter", IEEE Trans. Power Electronic, vol. 30, no. 6, pp. 3262-3274, Nov. 2015.

[6] K. I. Hwu, W. Z. Jiang and Y. T. Yau, "Nonisolated coupled-inductor-based high step-down converter with zero DC magnetizing inductance current and nonpulsating output current", IEEE Trans. Power Electronic, vol. 31, no. 6, pp. 4362-4377, Jun. 2016.

[7] G. Chen, Y. Deng, Y. Tao, X. He, Y. Wang and Y. Hu, "Topology derivation and generalized analysis of zero-voltage-switching synchronous dc–dc converters with coupled inductors", IEEE Trans. Ind. Electronic, vol. 63, no. 8, pp. 4805-4815, Aug. 2016.

[8] Y. P. Hsieh, J. F. Chen, L. S. Yang, C. Y. Wu and W. S. Liu, "High-conversion-ratio bidirectional DC–DC converter with coupled inductor", IEEE Trans. Ind. Electronic, vol. 61, no. 1, pp. 210-222, Jan. 2014.

[9] J. H. Park and B. H. Cho, "Nonisolation soft-switching buck converter with tapped-inductor for wide-input extreme step-down applications", IEEE Trans. Circuits Syst. I Reg. Papers, vol. 54, no. 8, pp. 1809-1818, Aug. 2007.

[10] P. C. Tsai, C. C. Feng and C. C. Chi, "A Novel Transformerless Interleaved High Step-Down Conversion Ratio DC-DC Converter With Low Switch Voltage Stress", Industrial Electronics IEEE Transactions on, vol. 61, pp. 5290-5299, Jan. 2014.

[11] F. L. de S´a, C. V. B. Eiterer, D. Ruiz-Caballero and S. A. Mussa, "Double Quadratic Buck Converter", in Proc. Brazilian Power Electronics Conference (COBEP), 2013.

[12] X. Ruan, B. Li, Q. Chen, S.C. Tan and C. K. Tse, "Fundamental Considerations of Three-Level DC-DC Converters: Topologies Analyses and Control", IEEE Transactions on Circuits and Systems, vol. 55, no. 11, pp. 3733-3743, Dec. 2008.

Design and Evaluation of A Magnetically-loosely-coupled Inductor for A Four-phase Interleaved Boost Chopper

Hiroki Kowatari[1]*, Toshinori Kitamura[1], Nobukazu Hoshi[1]
1 Department of electrical engineering, Tokyo University of Science, Chiba, Japan
*E-mail: 7314058@ed.tus.ac.jp

Abstract—An interleaved boost chopper is used between the battery module and the traction inverter in electric vehicles including fuel cell vehicles and hybrid electric vehicles as one of the applications. However, this circuit requires the increased number of components such as inductors and switching devices in comparison with a general chopper. This paper proposes a new type of magnetically-loosely-coupled inductor for a four-phase interleaved boost chopper and compares the characteristics between the proposed and non-coupled inductors used in the chopper. As a result in the experiment, it is shown that the magnitude of the current ripple in the proposed inductor is 74~83 % smaller than that of the non-coupled inductor. In addition, the power conversion efficiency of the chopper using the proposed inductor is higher than that using the non-coupled inductor in a case where the inductor current flows continuously.

Keywords—*Magnetically-coupled inductor, four-phase interleaved boost chopper, electric vehicle.*

I. Introduction

Recently, DC-DC converters are used between the battery modules and the traction inverter in electric vehicles (EVs) including fuel cell vehicles (FCVs) and hybrid electric vehicles (HEVs). In general, it is undesirable to make the battery voltage high in terms of safety and efficiency. On the other hand, higher voltage is required in order to get sufficient torque in high-speed range. Thus, DC-DC converter is used in some electric vehicles and regulates the dc-link voltage of the inverter according to the vehicle speed. The chopper must have capability of bidirectional power flow. As a chopper, interleaved boost chopper has been used. For example, a four-phase interleaved boost chopper is used in the *TOYOTA MIRAI* [1]. Although this chopper requires a lot of components such as inductors and switching devices compared with a general chopper, the magnitudes of the input and current ripple can be suppressed.

There are several reports [2]-[10] to decrease the number of components by using a magnetically-coupled inductor for interleaved circuits. Many types of magnetically-coupled inductor for two-phase interleaved boost choppers have been investigated [2]-[5]. The previous studies indicate that magnetically-loosely-coupled inductor is particularly useful for downsizing the inductor and improving the power conversion efficiency compared with other types of inductor [6], [7]. However, there are very few studies which report on magnetically-coupled

inductor for four-phase interleaved boost choppers. A magnetically-closed coupled inductor for this type of chopper is reported in [9], and a magnetically-loosely coupled inductor is proposed in [10]. The purpose of this study is to propose a new type of magnetically-loosely coupled inductor for a four-phase interleaved boost chopper and to verify the characteristics of this inductor. This proposed inductor has symmetrical structure in order to simplify analysis.

This paper consists of five sections. Section II describes the mathematical modeling of the proposed inductor. Section III shows that the comparison between the proposed and non-coupled inductors. In section IV, the experimental evaluation is described. Finally, the conclusion is illustrated in section V.

II. Mathematical Modeling of the Proposed Inductor

In this section, characteristics of the proposed inductor are described by using a mathematical modeling.

A four-phase interleaved boost chopper using the proposed inductor is shown in Fig. 1. Where V_i and V_o are the input- and output-voltage, respectively; i_{L1}, i_{L2}, i_{L3}, i_{L4} and i_S are the inductor current in each phase and the input current; S_1, S_2, S_3 and S_4 are the switching devices. These switching devices are controlled with PWM control. The phase of PWM carrier for each

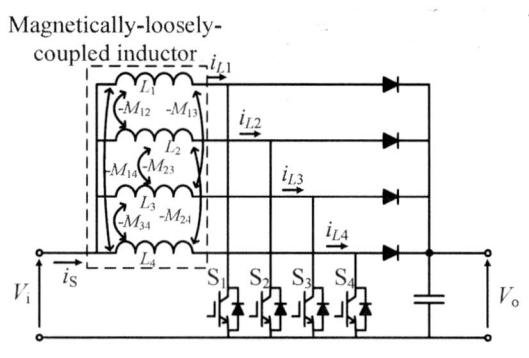

Fig. 1. Four-phase interleaved boost chopper using the proposed inductor.

switch is shifted with $\pi/4$ rad each other. The proposed inductor is the part surrounded with a dotted line in the figure. Where L_1, L_2, L_3 and L_4 are the self-inductance in each phase; M_{12}, M_{13}, M_{14}, M_{23}, M_{24} and M_{34} are the mutual inductance between each coil, respectively. For example, M_{12} means the mutual inductance between coils 1 and 2. Also, L_{lk1}, L_{lk2}, L_{lk3} and L_{lk4}, which are not shown in Fig. 1, are the leakage inductance. In this study, it is assumed that the self-inductance in all phases are same value L, the mutual inductances are same value M, and the leakage inductances are same value L_{lk} for simplification.

The appearance of the proposed inductor core is shown in Fig. 2. The winding wires are wound on each outer pole. The center pole has air gap. The direction of magnetic flux linkage generated by these coils are same in all outer poles.

In order to analyze this proposed inductor, the magnetic circuit analysis is conducted. The magnetic equivalent circuit of the proposed inductor is shown in Fig. 3. Where N_1, N_2, N_3 and N_4 are the number of turns on each outer pole, respectively; R_{mo1}, R_{mo2}, R_{mo3}, R_{mo4} and R_{mc} are the magnetic reluctance in each outer pole and the center pole; ϕ_{o1}, ϕ_{o2}, ϕ_{o3}, ϕ_{o4} and ϕ_c are the magnetic flux linkage in each outer pole and center pole, respectively. In this study, it is assumed that the number of turns on each outer pole is same value N; and the magnetic reluctance in each outer pole is same value R_{mo} for simplification.

The important parameters are the self-inductance L, the mutual inductance M, the leakage inductance L_{lk}, the magnitude of the input current ripple Δi_S, the maximum magnetic flux linkage in the outer pole ϕ_{omax} and center pole ϕ_{cmax}.

These inductances L, M and L_{lk} are derived from the magnetic circuit analysis and the values are expressed as the following equations.

$$L = N^2 \frac{R_{mo} + 3R_{mc}}{R_{mo}^2 + 4R_{mo}R_{mc}} \tag{1}$$

$$M = N^2 \frac{R_{mc}}{R_{mo}^2 + 4R_{mo}R_{mc}} \tag{2}$$

$$L_{lk} = L - 3M = N^2 \frac{1}{R_{mo} + 4R_{mc}} \tag{3}$$

To analyze the proposed inductor, the magnetic equivalent circuit analyses are conducted for DC component and AC component, separately.

A. DC component analysis

Here, Φ_o and Φ_c are the dc components in the magnetic flux linkage in the outer pole and the center pole, respectively; and I_L is the dc component in inductor current in each phase. The dc components in the magnetic flux linkage in the outer pole are calculated by

$$\Phi_{o1} = \Phi_{o2} = \Phi_{o3} = \Phi_{o4} = \Phi_o = \frac{NI_L}{R_{mo} + 4R_{mc}}. \tag{4}$$

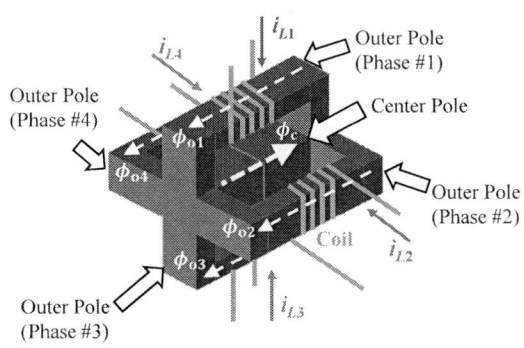

Fig. 2.   Appearance of the proposed inductor core.

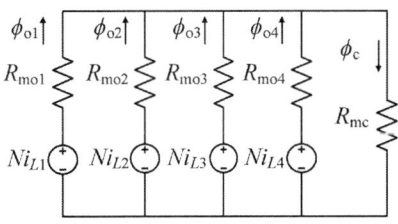

Fig. 3.   Magnetic equivalent circuit of the proposed inductor.

The dc component in the magnetic flux linkage in the center pole is calculated by

$$\Phi_c = 4\Phi_o = \frac{4NI_L}{R_{mo} + 4R_{mc}}. \tag{5}$$

B. AC component analysis

The current waveforms of the four-phase interleaved boost chopper depend on the switching timing. Therefore, the AC component analyses are conducted for the duty ratios in the cases $0 \leq d \leq 0.25$, $0.25 \leq d \leq 0.5$, $0.5 \leq d \leq 0.75$ and $0.75 \leq d \leq 1$, respectively.

Here, $\Delta\phi_{on}$ and $\Delta\phi_c$ are the ac components in the flux linkage in the outer pole and the center pole, respectively; the subscript n is phase number, i.e. n=1, 2, 3, 4; Δi_{Ln} is the magnitude of the inductor current ripple; Δi_S is the magnitude of the input current ripple; d is the duty ratio of the switches and T_S is the switching period. Then, the following equations are derived.

$$\Delta\phi_{on} = \frac{v_{Ln}}{N} dT_S \tag{6}$$

$$\Delta\phi_c = \Delta\phi_{o1} + \Delta\phi_{o2} + \Delta\phi_{o3} + \Delta\phi_{o4} \tag{7}$$

$$N\Delta i_{Ln} = R_{mo}\Delta\phi_{on} + R_{mc}\Delta\phi_c \tag{8}$$

$$\Delta i_{\mathrm{S}} = \Delta i_{L1} + \Delta i_{L2} + \Delta i_{L3} + \Delta i_{L4} \qquad (9)$$

C. Theoretical calculation

The values of ϕ_{omax}, ϕ_{cmax}, Δi_{S} and Δi_{L} are derived from the components analyses described in the above. The maximum magnetic flux linkage in the outer pole is given by

$$\begin{aligned}
\phi_{\mathrm{omax}} &= \Phi_{\mathrm{o}} + \frac{1}{2}\Delta\phi_{\mathrm{o}} \\
&= \frac{L_{\mathrm{lk}}I_{L}}{N} + \frac{V_{\mathrm{i}}dT_{\mathrm{S}}}{2N}.
\end{aligned} \qquad (10)$$

The maximum magnetic flux linkage in the center pole is given by

$$\phi_{\mathrm{cmax}} = \Phi_{\mathrm{c}} + \frac{1}{2}\Delta\phi_{\mathrm{c}}$$

$$= \begin{cases}
\frac{4L_{\mathrm{lk}}I_{L}}{N} + \frac{1-4d}{2N(1-d)}V_{\mathrm{i}}dT_{\mathrm{S}} \\
\qquad\qquad\qquad (0 \le d \le 0.25) \\
\frac{4L_{\mathrm{lk}}I_{L}}{N} + \frac{2-4d}{2N(1-d)}V_{\mathrm{i}}\left(d-\frac{1}{4}\right)T_{\mathrm{S}} \\
\qquad\qquad\qquad (0.25 \le d \le 0.5) \\
\frac{4L_{\mathrm{lk}}I_{L}}{N} + \frac{3-4d}{2N(1-d)}V_{\mathrm{i}}\left(d-\frac{1}{2}\right)T_{\mathrm{S}} \\
\qquad\qquad\qquad (0.5 \le d \le 0.75) \\
\frac{4L_{\mathrm{lk}}I_{L}}{N} + \frac{4-4d}{2N(1-d)}V_{\mathrm{i}}\left(d-\frac{3}{4}\right)T_{\mathrm{S}} \\
\qquad\qquad\qquad (0.75 \le d \le 1).
\end{cases} \qquad (11)$$

The magnitude of the input current ripple is given by

$$\Delta i_{\mathrm{S}} = \begin{cases}
\frac{1-4d}{L_{\mathrm{lk}}(1-d)}V_{\mathrm{i}}dT_{\mathrm{S}} & (0 \le d \le 0.25) \\
\frac{2-4d}{L_{\mathrm{lk}}(1-d)}V_{\mathrm{i}}\left(d-\frac{1}{4}\right)T_{\mathrm{S}} & (0.25 \le d \le 0.5) \\
\frac{3-4d}{L_{\mathrm{lk}}(1-d)}V_{\mathrm{i}}\left(d-\frac{1}{2}\right)T_{\mathrm{S}} & (0.5 \le d \le 0.75) \\
\frac{4-4d}{L_{\mathrm{lk}}(1-d)}V_{\mathrm{i}}\left(d-\frac{3}{4}\right)T_{\mathrm{S}} & (0.75 \le d \le 1).
\end{cases} \qquad (12)$$

The magnitude of the inductor current ripple is given by

$$\Delta i_{L} = \begin{cases}
\left(R_{\mathrm{mo}}\frac{V_{\mathrm{i}}}{N^2} + R_{\mathrm{mc}}\frac{4V_{\mathrm{i}}-3V_{\mathrm{o}}}{N^2}\right)dT_{\mathrm{S}} \\
\qquad\qquad\qquad (0 \le d \le 0.25) \\
\left(R_{\mathrm{mo}}\frac{2V_{\mathrm{i}}}{N^2} + R_{\mathrm{mc}}\frac{8V_{\mathrm{i}}-4V_{\mathrm{o}}}{N^2}\right)\left(d-\frac{1}{4}\right)T_{\mathrm{S}} \\
\quad + \left(R_{\mathrm{mo}}\frac{V_{\mathrm{i}}}{N^2} + R_{\mathrm{mc}}\frac{4V_{\mathrm{i}}-3V_{\mathrm{o}}}{N^2}\right)\left(\frac{1}{2}-d\right)T_{\mathrm{S}} \\
\qquad\qquad\qquad (0.25 \le d \le 0.5) \\
\left(R_{\mathrm{mo}}\frac{3V_{\mathrm{i}}}{N^2} + R_{\mathrm{mc}}\frac{12V_{\mathrm{i}}-3V_{\mathrm{o}}}{N^2}\right)\left(d-\frac{1}{2}\right)T_{\mathrm{S}} \\
\quad + \left(R_{\mathrm{mo}}\frac{2V_{\mathrm{i}}}{N^2} + R_{\mathrm{mc}}\frac{8V_{\mathrm{i}}-4V_{\mathrm{o}}}{N^2}\right)\left(\frac{3}{4}-d\right)T_{\mathrm{S}} \\
\qquad\qquad\qquad (0.5 \le d \le 0.75) \\
-\left(R_{\mathrm{mo}}\frac{V_{\mathrm{i}}-V_{\mathrm{o}}}{N^2} + R_{\mathrm{mc}}\frac{4V_{\mathrm{i}}-V_{\mathrm{o}}}{N^2}\right)(1-d)T_{\mathrm{S}} \\
\qquad\qquad\qquad (0.75 \le d \le 1).
\end{cases} \qquad (13)$$

The theoretical waveforms can be plotted based on the above equations with the circuit and inductor parameters. The example waveforms of the magnetic flux linkage $\phi_{\mathrm{o1}} \sim \phi_{\mathrm{o4}}$, ϕ_{c}, the input current i_{S}, and the inductor current $i_{L1} \sim i_{L4}$ in several duty ratios are shown in Fig. 4.

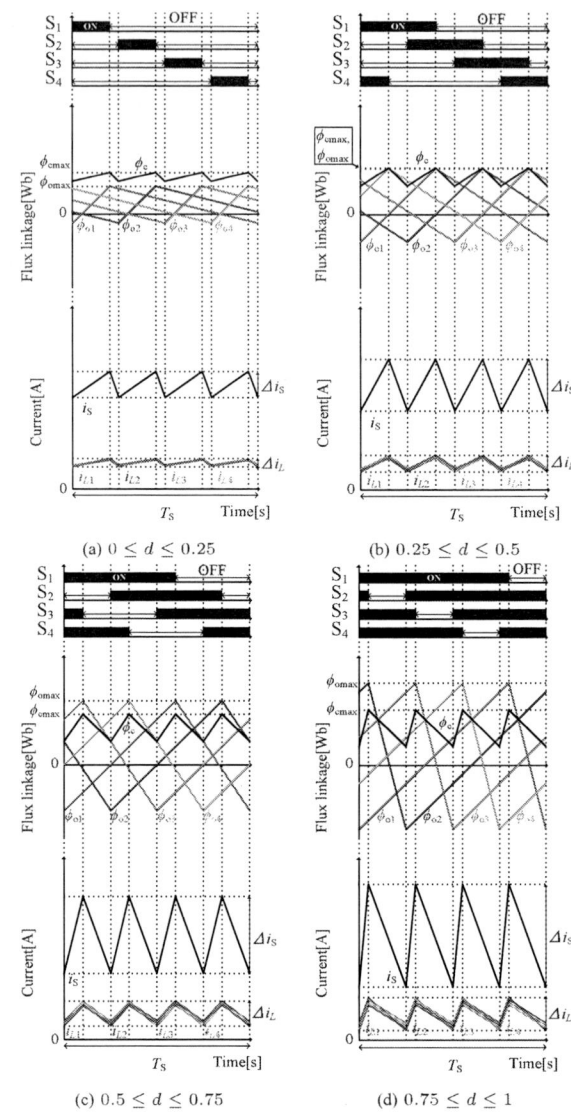

Fig. 4. Example waveforms of the magnetic flux linkage and the input- and inductor-current in several duty ratios.

D. Core assembly of the proposed inductor

Assembly drawing of the proposed inductor core is shown in Fig. 5. The proposed inductor core is made of the ferrite cores $TDK\ PC40\ EE80 \times 76 \times 20$. Firstly, four L-shape cores were made by cutting off two E-shape cores as shown in Fig. 5(a). Secondly, two E-shape cores with polished center pole were made by grinding the head of center pole of the core as shown in Fig. 5(b). Finally, these components are assembled as shown in Fig. 5(c). The outer poles of phases #1 and #3 are configured with two L-shape cores. In the upper- and under-parts of phases #2 and #4, outer poles and the center pole are not separated.

E. Constraint of duty ratio for obtaining target current ripple

Electromagnetic field analysis was conducted in order to estimate the inductor parameters. The inductor parameters are calculated by substituting the magnitudes of the input- and inductor-current ripple and the circuit parameters obtained by the analysis into (12) and (13). TABLE I shows the parameters of the proposed inductor obtained by magnetic analysis.

TABLE II shows the maximum ratings. Substituting these parameters into (10)~(13), the maximum magnetic flux density, the magnitudes of the input- and inductor-current ripples in the proposed inductor are calculated. The characteristics of the maximum magnetic flux density vs. the duty ratio are shown in Fig. 6. The characteristics of the magnitudes of the input- and inductor-current ripple vs. the duty ratio are shown in Fig. 7. The maximum magnetic flux density is below the saturation magnetic flux density (500 mT) over the whole range of the duty ratio as shown in Fig. 6. In order to obtain the target input current ripple ratio $\frac{\Delta i_S}{I_S}$, which is less than 20 %, the duty ratio must be less than 0.5 as shown in Fig. 7.

TABLE I. PARAMETERS OF THE PROPOSED INDUCTOR OBTAINED BY MAGNETIC ANALYSIS.

Magnetic reluctance in the outer pole	R_{mo}	110 kA/Wb
Magnetic reluctance in the center pole	R_{mc}	1354 A/Wb
Length of air gap	l_{g}	1 mm
Number of turns	N	15 turn
Self-inductance	L	1.5×10^3 μH
Mutual inductance	M	5.0×10^2 μH
Leakage inductance	L_{lk}	41 μH

TABLE II. MAXIMUM RATINGS.

Input voltage	V_{i}	40 V
Input current	I_{S}	40 Arms
Inductor current	I_L	10 Arms
Switching frequency	f_{sw}	20 kHz

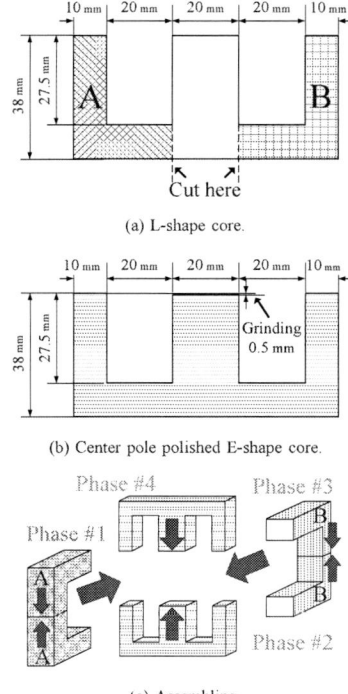

(a) L-shape core.

(b) Center pole polished E-shape core.

(c) Assembling.

Fig. 5. Assembly drawing of the proposed inductor core.

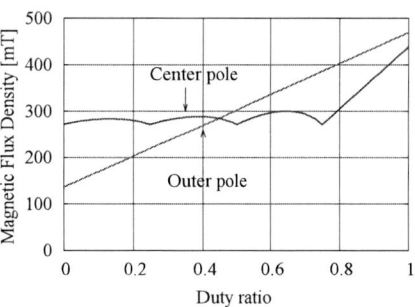

Fig. 6. Characteristics of the maximum magnetic flux density vs. the duty ratio.

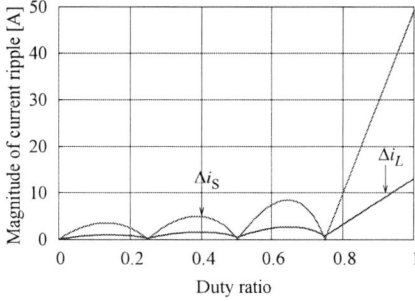

Fig. 7. Characteristics of the magnitudes of the input- and inductor-current ripple vs. the duty ratio.

III. The Comparison Between the Proposed and Non-Coupled Inductors

In this section, the comparison between the proposed and non-coupled inductors is described.

Each inductor is used in the four-phase interleaved boost chopper. Here, L_{non} is the self-inductance of the non-coupled inductor; ϕ_{nonmax} is the maximum magnetic flux linkage in the non-coupled inductor; $\Delta i_{S\text{non}}$ and $\Delta i_{L\text{non}}$ are the magnitudes of the input- and inductor-current ripples of the non-coupled inductor; respectively. The values of ϕ_{nonmax}, $\Delta i_{S\text{non}}$ and $\Delta i_{L\text{non}}$ are expressed as the following equations.

$$\phi_{\text{nonmax}} = \frac{L_{\text{non}}I_L}{N} + \frac{V_i dT_S}{2N} \tag{14}$$

$$\Delta i_{S\text{non}} = \begin{cases} \frac{1-4d}{L_{\text{non}}(1-d)}V_i dT_S & (0 \le d \le 0.25) \\ \frac{2-4d}{L_{\text{non}}(1-d)}V_i\left(d - \frac{1}{4}\right)T_S & (0.25 \le d \le 0.5) \\ \frac{3-4d}{L_{\text{non}}(1-d)}V_i\left(d - \frac{1}{2}\right)T_S & (0.5 \le d \le 0.75) \\ \frac{4-4d}{L_{\text{non}}(1-d)}V_i\left(d - \frac{3}{4}\right)T_S & (0.75 \le d \le 1) \end{cases} \tag{15}$$

$$\Delta i_{L\text{non}} = \frac{V_i}{L_{\text{non}}}dT_S \tag{16}$$

Assuming that L_{non} in the non-coupled inductor is equal to L_{lk} in the proposed inductor, from these equations, it is understood that the maximum magnetic flux linkage in outer poles and the magnitude of the input current ripple are same in the both inductors, respectively. Whereas, the magnitudes of the inductor current ripple are not same in the both inductors. TABLE III shows the theoretical circuit parameters. Fig. 8 shows theoretical waveforms of the input- and inductor-current in the both inductors. These waveforms are plotted by substituting the inductances ($L_{\text{lk}} = L_{\text{non}} = 41\ \mu\text{H}$) and the circuit

parameters shown in TABLE III into (12), (13), (15) and (16). It is understood that the frequency of the inductor current ripple in the proposed inductor is 4 times higher than the switching frequency as shown in Fig. 8. The inductor current waveforms in the non-coupled inductor are quite different from the waveforms in the proposed inductor.

Also, the characteristics of the magnitude of the inductor current ripple vs. the duty ratio in the both inductors are shown in Fig. 9. This characteristic is calculated by substituting the inductor parameters shown in TABLE I and the circuit parameters shown in TABLE III into (13) and (16). Since the magnitude of the inductor current ripple in the proposed inductor decreases, the copper loss reduction is expected.

IV. Experimental Evaluation

This section shows the evaluation of the validity of the derived mathematical model and the comparison between the characteristics of the both inductors in verification experiments.

A. The evaluation of the validity of the derived mathematical model

Appearance of the prototype of the proposed inductor is shown in Fig. 10. TABLE IV shows the circuit parameters used in the verification experiment.

The inductor current waveforms in each phase measured in the verification experiment are shown in Fig. 11. The measured inductor current waveforms are similar to the theoretical waveforms shown in Fig. 4.

Here, the theoretical magnitude of the inductor current ripple can be calculated by substituting the inductor parameters shown in TABLE I and the circuit parameters shown in TABLE IV. In this experiment, the theoretical value is 0.80 A. There are errors in the magnitudes of the inductor current ripple between the theory and the experiments as shown in Fig. 11. The magnitude of the inductor current ripple is affected by the circuit parameters and the inductor parameters. In this experiment, the circuit parameters are not much different from the designed

TABLE III. THEORETICAL CIRCUIT PARAMETERS.

Input voltage	V_i	40 V
Inductor current	I_L	2.5 Arms
Duty ratio	d	0.2
Switching frequency	f_{sw}	20 kHz

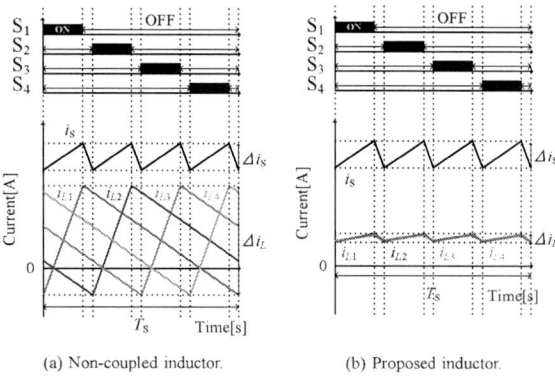

(a) Non-coupled inductor.　　　(b) Proposed inductor.

Fig. 8.   Theoretical waveforms of the input- and inductor-current in the both inductors.

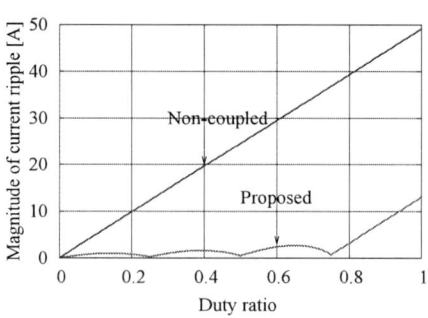

Fig. 9.   Characteristics of the magnitude of the inductor current ripple vs. the duty ratio.

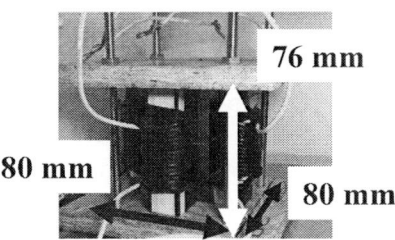

Fig. 10. Appearance of the prototype of the proposed inductor.

TABLE IV.   CIRCUIT PARAMETERS USED IN THE VERIFICATION
EXPERIMENT.

Input voltage	V_i	40 V
Input Power	P_{in}	400 W
Switching frequency	f_{sw}	20 kHz

values. Substituting the magnitudes of the input- and inductor-current ripple values and the circuit parameters into (12) and (13), the inductor parameters are obtained.

TABLE V shows the comparison of the inductor parameters between the designed and experimental values. Inductances are calculated with magnetic reluctances and the number of turns as shown in (1)-(3). Since the number of turns in each phase is same and 15, there is possibility that the magnetic reluctances in the prototype are different from the designed values.

From TABLE V, the cause of differences is expressed as follows. Firstly, the measured values of the magnetic reluctances in the outer pole are greater than the designed values of that. Secondly, the magnetic reluctances in Phase #1 and Phase #3 are greater than that of Phase #2 and Phase #4. The reason is that the prototype inductor has unexpected air gaps. Finally, the actual magnetic reluctance in the center pole is smaller than that of the designed values. In addition, fringing effect occurs. Since the cross-sectional area of the magnetic flux spreads in the air gap, the magnetic reluctance in the center pole decreased. To suppress these differences between the design and actual parameters, the proposed inductor has to be made by using metal mold or powder magnetic core with considering the fringing effect.

B. The evaluation of the prototype inductor

This subsection shows that the characteristics comparison between the proposed and non-coupled inductors in the verification experiment.

Firstly, the outer volumes of the both type cores are evaluated in the verification experiment. The prototypes of non-coupled inductors in the experiment were made by using TDK $PC40$ $EC120 \times 101 \times 30$. Here, the inductances of the non-coupled inductors were adjusted, and the magnitude of the input current ripple to be equal between the proposed and non-coupled inductors. The

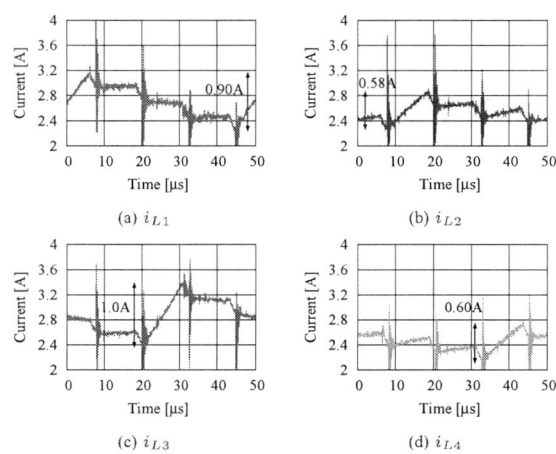

Fig. 11. The inductor current waveforms in each phase measured in the verification experiment.

TABLE V.   COMPARISON OF THE INDUCTOR PARAMETERS
BETWEEN THE DESIGNED AND EXPERIMENTAL VALUES.

Parameters	Designed value	Measured value
L_{lk1}		86 μH
L_{lk2}	41 μH	156 μH
L_{lk3}		75 μH
L_{lk4}		147 μH
L_1		4.7×10^2 μH
L_2	1.5×10^3 μH	7.3×10^2 μH
L_3		4.2×10^2 μH
L_4		7.0×10^2 μH
M_{12}		1.6×10^2 μH
M_{13}		0.8×10^2 μH
M_{14}	5.0×10^2 μH	1.5×10^2 μH
M_{23}		1.4×10^2 μH
M_{24}		2.7×10^2 μH
M_{34}		1.3×10^2 μH
R_{mo1}		404 kA/Wb
R_{mo2}	110 kA/Wb	221 kA/Wb
R_{mo3}		460 kA/Wb
R_{mo4}		235 kA/Wb
R_{mc}	1354 kA/Wb	410 kA/Wb

number of turns in the both inductors is 15. The air gap of the non-coupled inductors is 1.95 mm. The outer volume of the proposed inductor is 212,800 mm³ and that of the non-coupled inductors is 1,482,480 mm³. In conclusion, the outer volume of the proposed inductor is 85 % smaller than the non-coupled inductor in this experiment. The appearance of the prototypes of non-coupled and proposed inductors is shown in Fig. 12.

Secondly, the current waveforms are evaluated in the verification experiment. TABLE VI shows the circuit parameters in the verification experiment. The current waveforms measured in the verification experiment are shown in Fig. 13. Similar waveforms of the inductor current were observed in the theoretical analysis and the experiment as shown in Figs. 8 and 13.

Thirdly, the relationship between the magnitude of the inductor current ripple and the duty ratio is evaluated in the experiment. In the experiment, the input voltage and

2665

TABLE VI. CIRCUIT PARAMETERS IN THE VERIFICATION EXPERIMENT.

Input voltage	V_i	40 V
Output voltage	V_o	50 V
Input Power	P_{in}	500 W
Switching frequency	f_{sw}	20 kHz

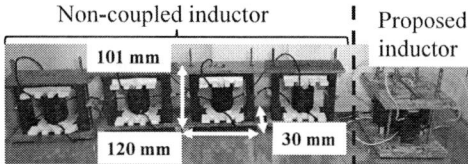

Fig. 12. Appearance of the prototypes of non-coupled and proposed inductors.

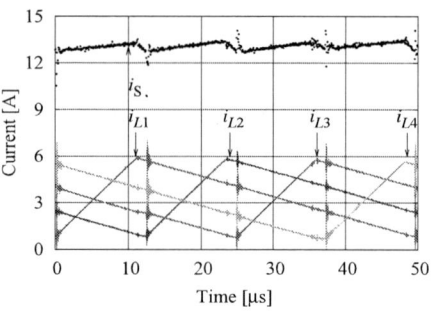

(a) Non-coupled inductor

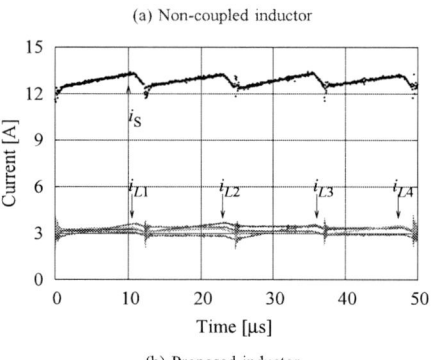

(b) Proposed inductor

Fig. 13. The current waveforms measured in the verification experiment.

the input power were fixed at the values shown in TABLE VI. In addition, the output voltage regulated by the duty ratio. The relationship between the magnitude of the inductor current ripple and the duty ratio was investigated in the verification experiment, and the result is shown in Fig. 14. The relationship between the magnitude of the inductor current ripple in the theoretical analysis and the experiment are not similar as shown in Figs. 9 and 14. The reason is that the parameters of the prototype the inductor used in the experiment are different from the designed values. However, it is found that the magnitude of the inductor current ripple can be reduced in the whole duty range in this experiment. In conclusion, the magnitude of the inductor current ripple of the proposed inductor is 74~83 % smaller than that of the non-coupled inductor in this experiment.

Finally, the power conversion efficiency comparison is conducted in the verification experiment. The circuit parameters in the experiment are same as the other experiments. Input power was changed from 100 to 900 W in this experiment. The power conversion efficiency measured in the experiment is shown in Fig. 15. The power conversion efficiency of the chopper using the proposed inductor is higher than that using the non-coupled inductor when the input power is over 400 W. The suppression of the magnitude of the inductor current ripple leads reduction of the copper loss and improvement of the power conversion efficiency.

The inductor current in the non-coupled inductor becomes discontinuous when the input power is under 400 W. Then, the switching devices can be turned-on with zero-current switching in that situation. This is the reason that the power conversion efficiency of the chopper using the non-coupled inductor is higher than that using the proposed inductor when the input power is under 400 W. In conclusion, the circuit power conversion efficiency of the chopper using the proposed inductor is higher than that using the non-coupled inductor in a case where the inductor current flows continuously.

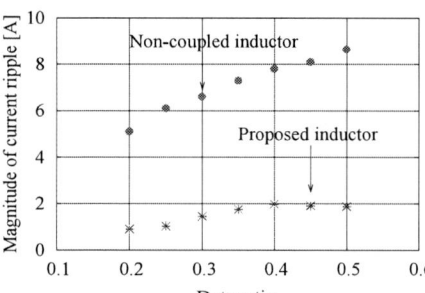

Fig. 14. Relationship between the magnitude of the inductor current ripple and the duty ratio in the experiment.

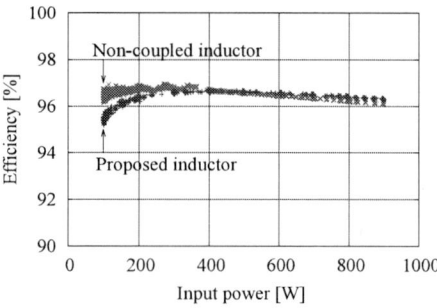

Fig. 15. Power conversion efficiency measured in the experiment.

V. CONCLUSION

This paper proposed the magnetically-loosely-coupled inductor for a four-phase interleaved boost chopper. In addition, theoretical analysis was conducted. As a result, it was found that the frequency of the inductor current ripple in the proposed inductor is 4 times higher than the switching frequency. Therefore, the magnitude of the inductor current ripple can be reduced than that of the non-coupled inductor when the magnitude of the input current ripple is equal. The both inductor current waveforms in experiments and theory are similar. However, the magnitudes of the ripples in the theory and experiment are not same because the inductor parameters of the prototype differ from the designed values. The reason is that the prototype of the proposed inductor has the unexpected air gaps on the bonded surfaces.

As a result in the experiment, the outer volume of the proposed inductor is 85 % smaller than that of the non-coupled inductor. In addition, the magnitude of the inductor current ripple of the proposed inductor is 74~83 % smaller than that of the non-coupled inductor. Moreover, the power conversion efficiency of the chopper using the proposed inductor is higher than that using the non-coupled inductor in a case where the inductor current flows continuously.

REFERENCES

[1] Y. Hasuke, H. Sekine, K. Katano and Y. Nonobe, "Development of Boost Converter for MIRAI," *SAE Technical Paper*, 6 pages, 2015.

[2] J. Imaoka and M. Yamamoto, "A novel integrated magnetic structure suitable for transformer linked interleaved boost chopper circuit," *IEEE Energy Conversion Congress and Exposition (ECCE)*, pp. 3279–3284, 2012.

[3] S. Kimura, J. Imaoka and M. Yamamoto, "Potential Power Analysis and Evaluation for Interleaved Boost Converter with Close-Coupled Inductor," *IEEE 10th International Conference on Power Electronics and Drive Systems (PEDS)*, pp. 26–31, 2013.

[4] K. Umetani, J. Imaoka, M. Yamamoto, S. Arimura and T. Hirano, "Evaluation of the Lagrangian Method for Deriving Equivalent Circuits of Integrated Magnetic Components: A Case Study Using the Integrated Winding Coupled Inductor," *IEEE Transactions on Industry Applications*, Volume. 51, pp. 547–555, 2015.

[5] J. Imaoka, M. Yamamoto, K. Umetani, S. Arimura and T. Hirano, "Characteristics Analysis and Performance Evaluation for Interleaved Boost Converter with Integrated Winding Coupled Inductor," *IEEE Energy Conversion Congress and Exposition (ECCE2013)*, pp. 3711–3718, 2013.

[6] W. Martinez, C. Cortes, M. Yamamoto, J. Imaoka and K. Umetani, "Total volume evaluation of high power density non isolated DC DC converters with integrated magnetics for electric vehicles," *IET Power Electronics*, Volume. 10, pp. 2010–2020, 2017.

[7] S. Kimura, J. Imaoka and M. Yamamoto, "Downsizing Effects of Integrated Magnetic Components in High Power Density DC-DC Converters for EV and HEV," *IEEE Energy Conversion Congress and Exposition (ECCE2014)*, pp. 5761–5768, 2014.

[8] M. Nakahama, M. Yamamoto, and Y. Satake, "Trans-linked Multiphase Boost Converter for Electric Vehicle," *IEEE Energy Conversion Congress and Exposition (ECCE2010)*, pp. 2458–2463, 2010.

[9] D. Ebisumoto, M. Ishihara, S. Kimura, W. Martinez, M. Noah and M. Yamamoto, "Design of a Four-Phase Interleaved Boost Circuit with Closed-Coupled Inductors," *IEEE Energy Conversion Congress and Exposition (ECCE2016)*, pp. 1–6, 2016.

[10] F. Velandia, W. Martinez, C.A.Cortes, M. Noah and M.Yamamoto, "Power loss analysis of multi-phase and modular interleaved boost DC-DC converters with coupled inductor for electric vehicles," *Power Electronics and Applications (EPE'16 ECCE Europe)*, pp. 1–6, 2016.

A Synchronous-Reference-Frame I-V Droop Control Method for Parallel-Connected Inverters

Mingshen Li[1*], Yonghao Gui[1], Zheming Jin[1], Yajuan Guan[1], and Josep M. Guerrero[1]

1 Department of Energy Technology, Aalborg University, Aalborg, Denmark

*E-mail: msh@et.aau.dk

Abstract-A simple and fast decentralized Current-Voltage (I-V) droop control method under the Synchronous-Reference-Frame (SRF) is proposed to share output current for parallel three-phase inverters with LC filter. The I-V droop characteristic is derived in accordance with the virtual impedance in SRF. Thus, each inverter enables to work in a voltage controlled mode where it controls the filter capacitor voltage. Moreover, a detailed state-space model of two parallel inverters is derived to analyze the control performance. Through the simulation and experiment validation, the proposed method provides accurate current sharing and faster transient response under inverter connection in comparison with the conventional droop control.

Abstract- Parallel inverters, I-V droop, SRF, Current sharing.

I. INTRODUCTION

As a significant operation mode, the islanded microgrid is typically equipped with converters and controlled to provide loads stable voltage and frequency [1], [2]. In order to satisfy the high power requirement, the multi-inverters are required to share load according to their power rating. The power droop control as an effective decentralized strategy has been commonly applied to the islanded mode, even the grid-tied mode. The voltage and frequency can be regulated to realize the active/reactive power sharing. However, the conventional droop technique has obvious drawbacks: active/reactive power coupling, sensitivity to line impedance, and slow dynamic response [3], [4].

To solve these problems, the virtual impedance control theme has been employed, which changes the inverter output characteristic in accordance to the line impedance.[5]-[7] However, these approaches are all based on the active power-voltage and reactive droop characteristics, and instantaneous active and reactive powers should be calculated, which means the low-pass(LPF) is necessary to average. The LPF will affect the transient response. Moreover, the power sharing performance is degraded when short lines with small impedance are used [8].

In order to improve the dynamic performance and stability, the current decoupling control based on droop has been investigated. In [9], a V-I characteristics droop control under SRF has been proposed to obtain the fast dynamics of parallel inverters system, and the SRF voltage and current are decoupled by using a new vector. However, the frequency fixed communication is employed. In [10], a novel and fast autonomous current sharing controller with virtual impedance loop under SRF

has been proposed, and it works well with inductive-resistive line impedance. However, the virtual impedance loop needs extra transformations which will burden the controller calculation.

To cope with the all the aforementioned, a simpler and faster controller is proposed in this paper. The I-V droop characteristic has been applied to generate the direct and quadrature current references. The droop coefficients are decided by the virtual impedance to realize the accurate sharing of currents. Moreover, the state-space equations of the two parallel inverters has been derived to model the system.

The paper is organized as follows. Section II introduces the proposed control structure and the control principle. In Section III, the state-space model of SRF I-V droop control is derived in details. In Section IV, the simulation and experiment results are given to validate proposed method under inverter connection. Finally, the conclusion and future works are highlighted in Section V.

II. PROPOSED SRF I-V DROOP CONTROL

The equivalent circuit of parallel multi-inverters system including output impedances, virtual impedances and line impedances are shown in Fig.1. and Fig.2. presents a two-terminal Thévenin equivalent branch in Laplace.

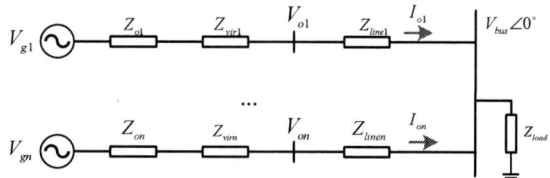

Fig. 1. Equivalent circuit of a parallel inverter system.

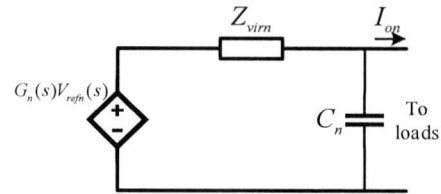

Fig. 2. Thévenin equivalent circuit of closed-loop inverter.

Considering that the $Z_{linen}(s)$ is very small in low-voltage microgrid, and $Z_{virn}(s)$ is the predominant component, according to Thévenin equivalent circuit, the

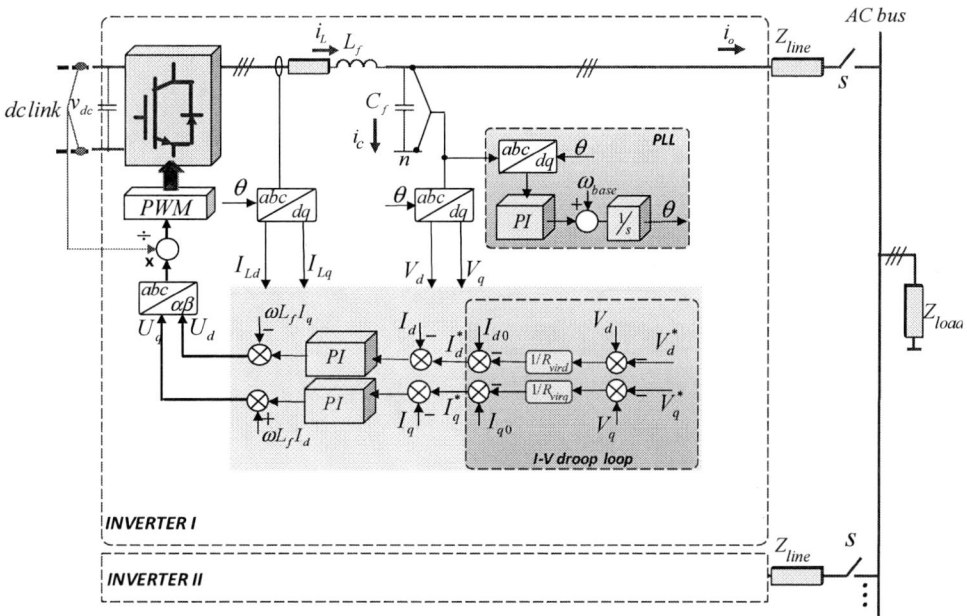

Fig. 3. The proposed control block for parallel inverters

voltage between the generated voltage $\vec{V}_{refn}$ and PCC voltage, $\vec{V}_{pcc}$, can be expressed in SRF:

$$\vec{V}_{refn} - \vec{V}_{pcc} = I_{dn}Z_{virdn} + jI_{qn}Z_{virqn} \quad (1)$$

Where I_{dn} and I_{qn} are the direct and quadrature components of output current respectively. Regardless of line impedance, the relationship of output currents in steady state among the parallel branches can be derived to:

$$I_{d1} : I_{d2} : \cdots : I_{dn} = \frac{1}{R_{vird1}} : \frac{1}{R_{vird2}} : \cdots : \frac{1}{R_{virdn}} \quad (2)$$

$$I_{q1} : I_{q2} : \cdots : I_{qn} = \frac{1}{R_{virq1}} : \frac{1}{R_{virq2}} : \cdots : \frac{1}{R_{virqn}} \quad (3)$$

Where R_{virdn} and R_{virqn} are the direct and quadrature virtual resistances of n-th inverter respectively.

Therefore, in order to share current flow, the d-axis and q-axis currents outputs of each inverter are able to regulated without communication by designing the reciprocals of virtual resistances. Consider that the voltage references of each inverter are equal, the active power P_n and reactive power Q_n of the n-th inverter also can be shared based on the virtual resistances. Consequently, (2) and (3) can be rewritten as:

$$P_1 : P_2 : \cdots : P_n = \frac{1}{R_{vird1}} : \frac{1}{R_{vird2}} : \cdots : \frac{1}{R_{virdn}} \quad (4)$$

$$Q_1 : Q_2 : \cdots : Q_n = \frac{1}{R_{virq1}} : \frac{1}{R_{virq2}} : \cdots : \frac{1}{R_{virqn}} \quad (5)$$

According to the above analysis, a new I-V droop can be obtained based on the direct and quadrature currents and voltage, which can be expressed as:

$$\begin{cases} I_d^* = I_{d0} - \dfrac{1}{R_{vird}}(V_d - V_d^*) \\[2mm] I_q^* = I_{q0} - \dfrac{1}{R_{virq}}(V_q - V_q^*) \end{cases} \quad (6)$$

Where V_d^*, V_q^* are the direct and quadrature current reference in SRF respectively, V_d, V_q are the instantaneous voltage components, I_d^*, I_q^* are the direct and quadrature current reference respectively, I_{d0}, I_{q0} are the nominal value of current components.

In case of active loads, the d-axis voltage can be regulated through the virtual resistance, which implies the droop characteristic to adapt the value of the d-axis current. Thus, the amplitude of output voltage can be regulated by the direct voltage reference. On the other hand, the quadrature current will affect the frequency of system. In SRF-PLL, the angular position is regulated through a feedback control loop which drives the V_q component to zero, and the PLL will compel the system to be stabilized at a certain point (50Hz) with zero phase delay if the quadrature voltage reference is zero.

Fig.3. shows the proposed control structure for parallel inverters, in which the system consists of the SRF-PLL, a I-V droop loop and PI current controllers. The direct and quadrature output voltage are independently controlled and rotated to generate the virtual current through the I-V droop relations (6). Then, the current controllers generate voltage references for the PWM signal to drive the IGBT inverter gates.

III. STAGE-SPACE MODEL FOR PROPOSED CONTROL

In order to analyze the output characteristics of I-V droop-controlled inverters, the state-space model of single inverter is established for a notional two parallel inverters system. To offer a fixed phase angle, the first inverter's phase angle can be the system reference [11], [12].

Regarding the I-V droop controller, the dq-axis current references can be generated by I-V droop relations. Therefore, we can obtain the input-output relationships from (6):

$$\begin{bmatrix} I_d^* \\ I_q^* \end{bmatrix} = \begin{bmatrix} I_{d0} \\ I_{q0} \end{bmatrix} - \begin{bmatrix} \dfrac{1}{R_{vird}} & 0 \\ 0 & \dfrac{1}{R_{virq}} \end{bmatrix} \begin{bmatrix} \varepsilon_d \\ \varepsilon_q \end{bmatrix} \quad (7)$$

Where $\varepsilon_d = V_d - V_d^*$, $\varepsilon_q = V_q - V_q^*$.

According to the current PI controllers, the state and output equations can be represented as:

$$\begin{bmatrix} U_d \\ U_q \end{bmatrix} = \begin{bmatrix} 0 & -\omega L_f \\ \omega L_f & 0 \end{bmatrix} \begin{bmatrix} I_d \\ I_q \end{bmatrix} + k_i \begin{bmatrix} \lambda_d \\ \lambda_q \end{bmatrix} + k_p \begin{bmatrix} \dot{\lambda}_d \\ \dot{\lambda}_q \end{bmatrix} \quad (8)$$

Where $\dot{\lambda}_d = I_d^* - I_d$, $\dot{\lambda}_q = I_q^* - I_q$.

According to Fig.3, the LC filter and resistive-inductive loads dynamics can be represented on a SRF as following:

$$\begin{cases} \dot{I}_{Ld} = \dfrac{1}{L_f}(-R_f I_{Ld} + V_{id} - V_d) + \omega_i I_{Lq} \\[2mm] \dot{I}_{Lq} = \dfrac{1}{L_f}(-R_f I_{Lq} + V_{iq} - V_q) + \omega_i I_{Lq} \\[2mm] \dot{V}_d = \dfrac{1}{C_f}(I_{Ld} - I_{od}) - \omega_i V_q \\[2mm] \dot{V}_q = \dfrac{1}{C_f}(I_{Lq} - I_{oq}) + \omega_i V_d \\[2mm] \dot{I}_{od} = \dfrac{-R_{Load}}{L_{Load}}I_{od} + \omega_i I_{od} \\[2mm] \dot{I}_{oq} = \dfrac{-R_{Load}}{L_{Load}}I_{oq} + \omega_i I_{oq} \end{cases} \quad (9)$$

Where R_f is the parasitic resistance of the inductor,

R_{load}, L_{load} is the loads resistor and inductor, and ω_i is the frequency generated by PLL.

Finally, combing the inverters, network and loads, the linearized model of the two parallel inverters system can be built into state-space model, which can be expressed as:

$$\dot{\mathbf{X}} = \mathbf{AX} + \mathbf{BU} \quad (10)$$

The states of the system under consideration are defined as:

$$\mathbf{X} = \begin{bmatrix} x_{inv1} & x_{inv2} & x_{load} \end{bmatrix}^T \quad (11)$$

Where the details of states matrix are:

$$\begin{cases} x_{inv1} = \begin{bmatrix} V_{dq1} & I_{Ldq1} & \lambda_{dq1} \end{bmatrix} \\ x_{inv2} = \begin{bmatrix} V_{dq2} & I_{Ldq2} & \lambda_{dq2} \end{bmatrix} \\ x_{load} = \begin{bmatrix} I_{odq1} & I_{odq2} \end{bmatrix} \end{cases} \quad (12)$$

The input vector of the system is defined as:

$$\mathbf{U} = \begin{bmatrix} V_{dq}^* & V_{loaddq} \end{bmatrix} \quad (13)$$

Therefore, the output admittance of inverters is depending on the inverters' dynamic, the filter and the common AC bus, and small-signal model can be built based on (10). The transfer matrix of state-space model can be derived by:

$$\mathbf{G} = \mathbf{C}(s\mathbf{I} - \mathbf{A})^{-1}\mathbf{B} \quad (14)$$

(10) and (14) can be solved by MATLAB symbolic math toolbox. due to limited contexts.

IV. SIMULATION AND EXPERIMENT RESULTS

The simulation has been conducted to compare the conventional droop control and the SRF I-V control. The system includes two parallel three-phase inverters with an LC filter. The electrical circuit and control system parameters are listed in Table I:

TABLE I
SIMULATION PARAMETERS

	Parameters	Value
Inverter and filter	DC voltage V_{dc}	650V
	Filter inductance L_f	1.8mH
	Filter capacitance C	25μF
	Local load	50+j2.83Ω
Droop control	k_{pP}, k_{iP}	5e-7,6e-6
	k_{pQ}, k_{iQ}	1e-5, 0
	Virtual resistance and inductance	2Ω, 8mH
Proposed I-V droop control	PLL proportional coefficient	1.4
	dq-axis inverter#1 virtual resistance R_{vird}, R_{virq}	2Ω,1Ω
	dq-axis voltage references V_d^*,V_q^*	100V,0V
	dq-axis normal current I_{d0}, I_{q0}	0A,0A
	PI controller k_p, k_i	0.0018,0.1

The Fig.4. shows the power-sharing performance of two control methods with resistive-inductive line impedance in the case that inverter2 is connected at 4.2s. From the transient response, the setting time of the conventional droop control is approximately 0.6s in Fig.4. (a) (b), and an offset of reactive current can be observed in Fig. 4. (b). However, the setting time of proposed method is approximately 0.08s, and the offset is well suppressed, which implies that the active and reactive powers are shared accurately, as shown in Fig.4. (b) and (c).

The 2018 International Power Electronics Conference

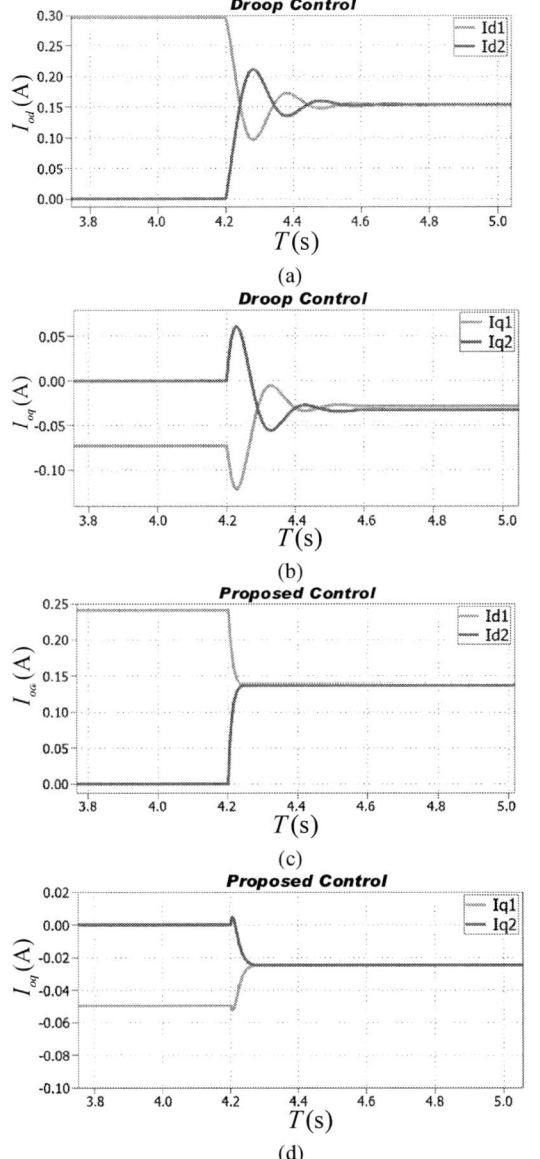

(a)

(b)

(c)

(d)

Fig.4. Comparison simulation results when the inverter connection under the conventional droop control ((a) and (b)) and the proposed method ((c) and (d)).

In order to validate the proposed control method, the system platform has been built. An islanded experimental Microgrid setup has been built, which includes two Danfoss 2.2kW inverters, a real-time dSPACE 1006 platform, LC filters, resistive load as shown in Fig.5. The experimental parameters are same as the simulation in addition to the d-axis voltage reference $V_d^* = 500V$ and the line impedance is $1\Omega+1.8$mH. The inverter connection case has been considered to test the performances of proposed controller with comparison to the conventional droop controller.

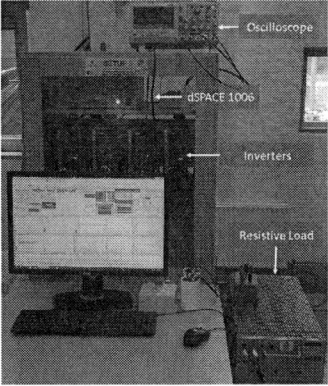

Fig.5. Configuration of the experimental platform

The experiment results of the SRF I-V control are presented in Fig. 6. When the inverter 2 is connected, the setting time of the direct and quadrature output currents is approximately 0.65s to compare with 1.8s setting time of conventional droop control, and the quadrature current is shared via the decoupling control with comparison the q-axis unshared current for droop control. Note that a small overshoot occurs due to inverters voltage error. In conclusion, the proposed method has increased the transient response greatly, and is able to achieve accurate current sharing.

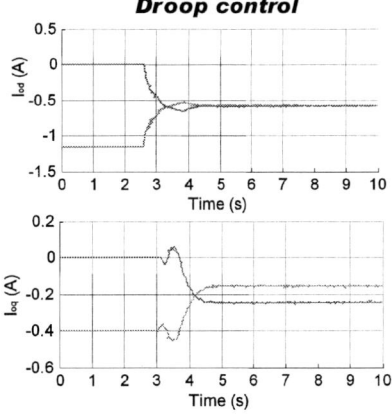

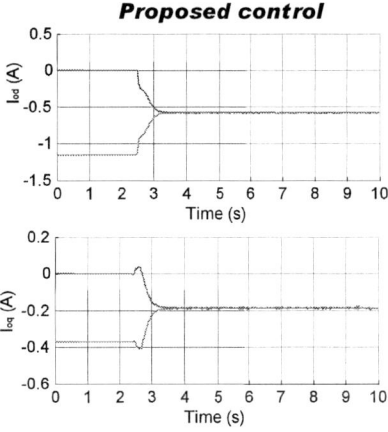

Fig.6 Experimental results of dq-axis current output when the inverter connection.

2671

V. CONCLUSION AND FUTURE WORK

A simplified SRF I-V droop control for parallel islanded inverters without communication is proposed, which is able to share the power accurately and get a fast transient response. The state-space model has been derived to analyze the controller performance. The simulation and experiment results verified the merits of the proposed method under the inverter connection situation. The full paper will discuss the bandwidth of control and show more experiment comparison results.

REFERENCES

[1] Shafiee, Qobad, Josep M. Guerrero, and Juan C. Vasquez. "Distributed secondary control for islanded microgrids—A novel approach." *IEEE Transactions on power electronics*, vol. 29, no. 2, pp. 1018-1031, Feb. 2014.

[2] Vasquez, Juan C., et al. "Modeling, analysis, and design of stationary-reference-frame droop-controlled parallel three-phase voltage source inverters." *IEEE Transactions on Industrial Electronics*, vol. 60, no. 4, pp. 1271-1280, Apr. 2013.

[3] Yao, Wei, et al. "Design and analysis of the droop control method for parallel inverters considering the impact of the complex impedance on the power sharing." *IEEE Transactions on Industrial Electronics,* vol. 58, no. 2, pp. 576-588, Feb. 2011.

[4] Kim, Jaehong, et al. "Mode adaptive droop control with virtual output impedances for an inverter-based flexible AC microgrid." *IEEE Transactions on Power Electronics* vol. 26, no. 3, pp. 689-701, Mar. 2011.

[5] Lu, Xiaonan, et al. "An improved droop control method for dc microgrids based on low bandwidth communication with dc bus voltage restoration and enhanced current sharing accuracy." *IEEE Transactions on Power Electronics* vol. 29, no. 4, pp. 1800-1812, Apr. 2014

[6] De Brabandere, Karel, et al. "A voltage and frequency droop control method for parallel inverters." *IEEE Transactions on power electronics*, vol. 22, no. 4, pp. 1107-1115, Jul. 2007.

[7] Li, Yun Wei, and Ching-Nan Kao. "An accurate power control strategy for power-electronics-interfaced distributed generation units operating in a low-voltage multibus microgrid." *IEEE Transactions on Power Electronics*, vol. 24, no. 12, pp. 2977-2988, Dec. 2009.

[8] Guerrero, Josep M., et al. "Decentralized control for parallel operation of distributed generation inverters using resistive output impedance." *IEEE Transactions on industrial electronic,* vol. 54, no. 2, pp.994-1004, Apr. 2007.

[9] Guan, Yajuan, et al. "A Dynamic Consensus Algorithm to Adjust Virtual Impedance Loops for Discharge Rate Balancing of AC Microgrid Energy Storage Units." *IEEE Transactions on Smart Grid*, Issue 99, Feb. 2017.

[10] Golsorkhi, Mohammad S., and Dylan DC Lu. "A control method for inverter-based islanded microgrids based on VI droop characteristics." *IEEE Transactions on Power Delivery,* vol. 30, no. 3, pp.1196-1204, Jun. 2015.

[11] Guan, Yajuan, et al. "A new way of controlling parallel-connected inverters by using synchronous-reference-frame virtual impedance loop—Part I: Control principle." *IEEE Transactions on Power Electronics* vol. 31, no. 6, pp. 4576-4593, Jun. 2016.

[12] Rasheduzzaman, Md, Jacob A. Mueller, and Jonathan W. Kimball. "An Accurate Small-Signal Model of Inverter- Dominated Islanded Microgrids Using dq Reference Frame." *IEEE Journal of Emerging and Selected Topics in Power Electronics* vol.2, no.4, pp. 1070-1080, Dec. 2014.

[13] Wang, Yanbo, et al. "Harmonic instability assessment using state-space modeling and participation analysis in inverter-fed power systems." *IEEE Transactions on Industrial Electronics* vol. 64, no.1, pp. 806-816, Jan. 2017.

The 2018 International Power Electronics Conference

Transient Stability Impact of the Phase-Locked Loop on Grid-Connected Voltage Source Converters

Heng Wu, Xiongfei Wang
Department of Energy Technology, Aalborg University, Aalborg, Denmark
E-mail: hew@et.aau.dk, xwa@et.aau.dk

Abstract—The phase-locked loop (PLL) is widely used in the grid-connected voltage source converter (VSC) for the purpose of grid synchronization. The impact of the PLL on the small-signal stability of VSC has recently been revealed, yet its influence on the transient stability of the VSC has seldom been addressed. This paper thus presents a comprehensive analysis on the transient stability effect of the PLL. It points out that the conventional equal-area criterion (EAC), which is used to analyze the transient stability of the synchronous generator (SG), cannot be extended to analyze the PLL effect, even though a dynamic analogy between the PLL and the swing equation of the SG can be found. The phase portrait is thus used in this work, and it shows that the long settling time and a high damping ratio of the PLL can enhance the transient stability of the VSC. Finally, the experimental tests are performed to verify the effectiveness of the theoretical analysis.

Keywords—*Transient stability, phase-locked loop, phase portrait, weak grid.*

I. INTRODUCTION

Nowadays, the stability of the power-electronic-based power system becomes a challenge due to the increasing penetration of voltage source converters (VSCs) in power grids. Basically, the stability problems can be categorized into two major groups: the small-signal stability problems and the large-signal stability problems [1]. For the small-signal stability analysis, VSCs are modeled by means of the small-signal linearization method, and thus the linear control theory can be used [2]. Numerous research works have been done on the small-signal stability of VSCs [3]-[4], revealing the dynamic impacts of different control loops, e.g. the active and the reactive power control loop [5], the direct voltage control loop [6], the vector current control loop [7], and the phase-locked loop (PLL) [8].

Compared to the small-signal stability study, less research attentions have been focused on large-signal stability problems of VSCs. In general, the large-signal stability issues can be categorized as the transient (angle) stability and the voltage stability [1]. A few research works can be found on the voltage stability study under different fault conditions [9]-[10]. However, the transient (angle) stability of VSCs, i.e. the ability to maintain synchronism with the power grid when subject to severe transient disturbances, e.g. the short-circuit fault on transmission lines, the loss of generators, or the step-

change of the large power load [1], has seldom been discussed. Further, the existing works related to the transient stability studies of VSCs are mainly focused on the dynamic influence of the active and reactive power control loops [11]-[13]. The effect of the current limitation on the transient stability of VSCs has recently been discussed in [14]-[16]. Yet, the influence of the PLL on the system transient stability is often overlooked [17]-[18]. Since the PLL acts as a key part in the VSC for the purpose of grid synchronization, the converter will lose synchronism with the power grid if its PLL is unstable. It is thus important to take the PLL effect into the transient stability analysis of the VSC. It has been pointed out that the PLL will be unstable if it does not have the equilibrium point after the transient disturbance [17]. In [18], the dynamic analogy between the PLL and the synchronous generator (SG) is revealed, and it points out that the transient instability of the PLL may also occur even if it has the equilibrium point after the disturbance. The equal-area criterion (EAC), which is used for the transient stability assessment of the SG [1], is adopted in [18] to analyze the transient stability of the PLL. Yet, this EAC-based analysis is only valid when the proportional gain (K_p) of the PLL is equal to zero, which is not a justified assumption in practice [19]. The EAC may lead to an inaccurate stability prediction for the PLL with a nonzero proportional gain.

To overcome the constraints of the EAC, the phase portrait is used in this paper to provide a systematic analysis on the transient stability impact of the PLL. It turns out that increasing the damping ratio and the settling time of the PLL can enhance the transient stability of the system.

The rest of this paper is organized as follows: Section II introduces a general control structure of the VSC, and formulates the nonlinear model of the PLL considering the grid impedance interaction. Section III then discusses the response of the PLL to the transient disturbance and elaborates that the PLL may still jeopardize the system transient stability even if it has the equilibrium operation point after the disturbance. Moreover, the constraints of the EAC for analyzing the transient stability impact of the PLL are also discussed. The phase portrait is then introduced for the transient stability analysis with different control parameters of the PLL. Two different

2673

grid impedance types, i.e. the resistive grid and the inductive grid, are considered in the analysis. Experimental tests are performed in Section IV for validating the theoretical analysis. Section V concludes this paper.

II. GRID-CONNECTED POWER CONVERTERS

Fig. 1 illustrates the simplified one-line diagram of a three-phase VSC using the typical vector current control, where L_f is the output filter of the converter and Z_g represents the grid impedance. The voltage at the Point of Common Coupling (PCC) is measured for synchronizing the VSC with the grid by means of the PLL. I_{dref} and I_{qref} are the references for the active current and the reactive current. The Proportional+ Integral (PI) controller is used for current regulation in the dq-frame to guarantee the zero steady-state tracking error [19].

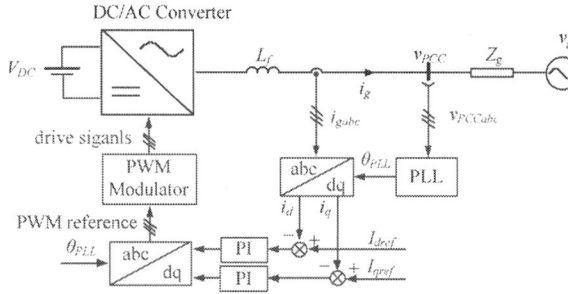

Fig. 1. Simplified one-line diagram of the three-phase VSC connecting to the weak ac network.

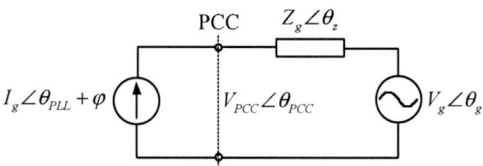

Fig. 2. Phasor representation of the simplified converter-grid system.

The timescale of the PLL is usually around 100 ms [20], which is well decoupled from the timescale of the current control loop (i.e. 1 ms - 10 ms) [20]. Hence, for the transient stability analysis of the PLL, the current loop can be regarded as a unity gain with the ideal reference tracking. Therefore, the converter shown in Fig. 1 can be simplified as an ideal current source in the transient stability analysis.

Fig. 2 shows the equivalent circuit of the VSC, where φ is the angle difference between the PCC voltage v_{PCC} and the grid current i_g, which is also called the power factor angle of the converter. I_g is the amplitude of the grid current. V_g, θ_g and V_{PCC}, θ_{PCC} are the amplitude and the phase angle of the grid voltage and the PCC voltage, respectively. θ_{PLL} is the phase angle detected by the PLL, and $\theta_{PLL}=\theta_{PCC}$ is expected in the steady state. θ_z represents grid impedance angle.

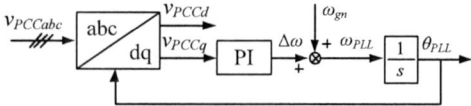

Fig. 3. Block diagram of the SRF-PLL.

Fig. 3 depicts the block diagram of the commonly used Synchronous Reference Frame PLL (SRF-PLL) [21]. The three phase voltages at the PCC are sampled and then transformed into the dq frame. The q-axis voltage is regulated by a PI controller for the phase tracking [21].

The PCC voltage in the dq frame can be expressed as [4]

$$v_{PCCd} = V_{PCC} \cos(\theta_{PCC} - \theta_{PLL})$$
$$v_{PCCq} = V_{PCC} \sin(\theta_{PCC} - \theta_{PLL}) \qquad (1)$$

Based on Fig. 3 and (1), the dynamic equation of the PLL can be expressed as

$$\theta_{PLL} = \int \left[\omega_{gn} + \left(K_p + K_i \int \right) v_{PCCq} \right] \qquad (2)$$

In the steady state, the integrator of the PI controller forces v_{PCCq} to become zero, and hence $\theta_{PLL}=\theta_{PCC}$, that is how the grid synchronization is realized by the PLL.

In the weak power grid where the grid impedance cannot be neglected, v_{PCC} is determined by the grid voltage and the current flowing through the grid impedance, as shown in Fig. 2. The relationship between v_{PCCd}, v_{PCCq}, v_g, i_g and Z_g can then be expressed as follows:

$$v_{PCCd} = V_g \cos(\theta_g - \theta_{PLL}) + I_g Z_g \cos(\theta_z + \varphi)$$
$$v_{PCCq} = V_g \sin(\theta_g - \theta_{PLL}) + I_g Z_g \sin(\theta_z + \varphi) \qquad (3)$$

Substituting (3) into (2), yielding

$$\theta_{PLL} =$$
$$\int \left\{ \omega_{gn} + \left(K_p + K_i \int \right) \left[V_g \sin(\theta_g - \theta_{PLL}) + I_g Z_g \sin(\theta_z + \varphi) \right] \right\} \qquad (4)$$

Define the angle difference between θ_{PLL} and θ_g as δ, i.e.

$$\delta = \theta_{PLL} - \theta_g \qquad (5)$$

Substituting (5) into (4), together with the simple relationship that $\theta_g=\int \omega_{gn} dt$, yielding

$$\delta = \int \left(K_p + K_i \int \right) \left[I_g Z_g \sin(\theta_z + \varphi) - V_g \sin \delta \right] \qquad (6)$$

which describes the dynamics of the PLL considering the grid impedance interaction. Based on (6), the equivalent control diagram of the PLL can also be plotted, as shown in Fig. 4.

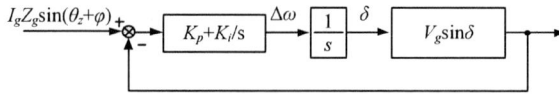

Fig. 4. Equivalent control diagram of the SRF-PLL under the weak grid condition.

III. TRANSIENT STABILITY ANALYSIS OF THE PLL CONSIDERING GRID IMPEDACNE INTERACTIONS

A. Transient Stability Mechnism of the PLL

The curve of $V_g \sin \delta$ is plotted as the solid line in Fig.

5. Obviously, if $I_g Z_g \sin(\theta_z + \varphi) > (V_g \sin\delta)_{max} = V_g$ or $I_g Z_g \sin(\theta_z + \varphi) < - (V_g \sin\delta)_{max} = -V_g$, the PLL does not have the equilibrium point and the system is certainly unstable, which has also been pointed out in [17]. Moreover, the PLL may still lose synchronism even if it has the equilibrium point after the large disturbance, which will be explained as follows.

Assuming that $I_g = I_{g1}$ before the disturbance, and the PLL operates at the point a where $I_{g1} Z_g \sin(\theta_z + \varphi) = V_g \sin\delta_0$, as shown in Fig. 5. The transient disturbance is considered as I_g is stepped up from I_{g1} to I_{g2}. As $I_{g2} Z_g \sin(\theta_z + \varphi) > V_g \sin\delta_0$ when the disturbance occurs, the output frequency of the PLL starts to increase, resulting in an increase in δ. The frequency continues to increase until it reaches the equilibrium point b, where $I_{g2} Z_g \sin(\theta_z + \varphi) = V_g \sin\delta_1$. However, as the output frequency of the PLL is still higher than the grid frequency ω_g at the point b, the δ continues to increase. Then the output frequency of the PLL begins to decrease after the point b due to the fact that $I_{g2} Z_g \sin(\theta_z + \varphi) < V_g \sin\delta$. Consequently, two possible operation scenarios can take place:

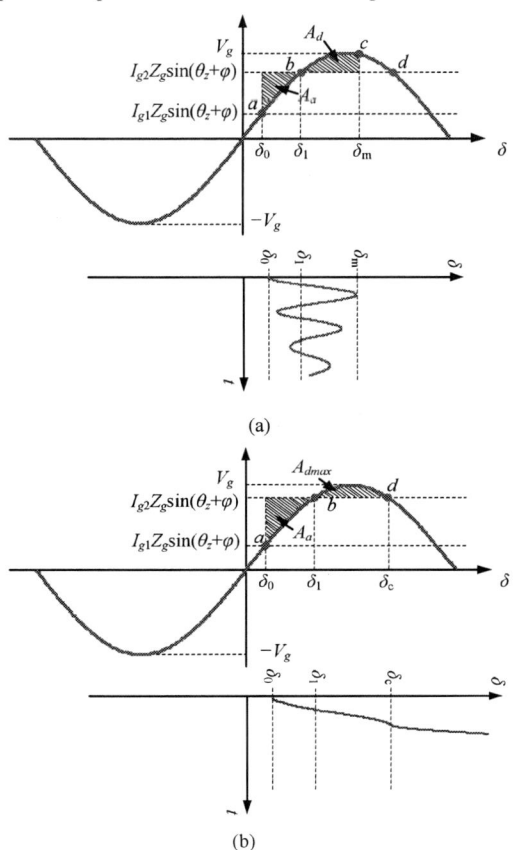

(a)

(b)

Fig. 5. Dynamic response of the PLL when I_g is stepped up from I_{g1} to I_{g2}. (a) Stable case (b) Unstable case

1) If the output frequency of the PLL recovers to the grid frequency before the unstable equilibrium point d, e.g. at the point c shown in Fig. 5(a). As $I_{g2} Z_g \sin(\theta_z + \varphi) < V_g \sin\delta_m$ still holds at the point c, the output frequency of the PLL further reduces and results

in a decrease in δ. Thus, the operating point retraces the $V_g \sin\delta$ curve from c to b, and then to a. The similar oscillation repeats several cycles and finally the PLL reaches to the equilibrium point b due to the damping effect, which means the system finally becomes stable.

2) If the output frequency of the PLL is still higher than the grid frequency even at the unstable equilibrium point d, as shown in Fig. 5(b). The output frequency of the PLL increases again after d as $I_{g2} Z_g \sin(\theta_z + \varphi) > V_g \sin\delta$ and δ also keeps increasing. As a result, the PLL loses synchronism with the power grid and the system becomes unstable.

B. *Principle and Constraints of the EAC*

Applying the differentiator on δ at both sides of (6) yields

$$\frac{d^2\delta}{dt^2} = K_i \left[I_g Z_g \sin(\theta_z + \varphi) - V_g \sin\delta \right]$$
$$+ K_p \left\{ \frac{d\left[I_g Z_g \sin(\theta_z + \varphi) \right]}{dt} - V_g \cos\delta \cdot \frac{d\delta}{dt} \right\} \quad (7)$$

The EAC can be used when K_p is equal to be zero, i.e.

$$\frac{d^2\delta}{dt^2} = K_i \left[I_g Z_g \sin(\theta_z + \varphi) - V_g \sin\delta \right] \quad (8)$$

For illustration, the basic principle of the EAC will be briefly discussed, and then the constraints of this method will be identified.

Multiplying both sides of (8) by $2d\delta/dt$ yields

$$2\frac{d\delta}{dt} \cdot K_i \left[I_g Z_g \sin(\theta_z + \varphi) - V_g \sin\delta \right]$$
$$= 2\frac{d\delta}{dt} \cdot \frac{d^2\delta}{dt^2} = \frac{d}{dt}\left(\frac{d\delta}{dt} \right)^2 \quad (9)$$

Then, integrating both sides of (9) leads to

$$\int_{\delta_0}^{\delta_m} 2K_i \left[I_g Z_g \sin(\theta_z + \varphi) - V_g \sin\delta \right] d\delta$$
$$= \left(\frac{d\delta}{dt} \right)^2 \quad (10)$$

The frequency deviation $d\delta/dt$ is initially zero, and it needs to become zero over certain time after the transient disturbance for the stable operation of the system, which consequently requires

$$2K_i \int_{\delta_0}^{\delta_m} \left[I_g Z_g \sin(\theta_z + \varphi) - V_g \sin\delta \right] d\delta = 0 \quad (11)$$

where δ_0 is the initial angle difference and δ_m is the maximum angle difference. As illustrated in Fig. 5(a), (11) is satisfied when area A_a is equal to area A_d. Similar with the definition in the SG-based systems, A_a can be defined as the acceleration area, where the output frequency of the PLL increases. A_d can be defined as the deceleration area, where the output frequency of the PLL decreases [1]. They are given by

$$A_a = \int_{\delta_0}^{\delta_1} \left[I_g Z_g \sin(\theta_z + \varphi) - V_g \sin\delta \right] d\delta \quad (12)$$

$$A_d = \int_{\delta_1}^{\delta_m} \left[V_g \sin\delta - I_g Z_g \sin(\theta_z + \varphi) \right] d\delta \quad (13)$$

Moreover, based on Fig. 5(b), the maximum deceleration area A_{dmax} can also be defined as

$$A_{d\max} = \int_{\delta_i}^{\delta_c} \left[V_g \sin\delta - I_g Z_g \sin(\theta_z + \varphi) \right] d\delta \qquad (14)$$

where δ_c is the corresponding angle difference at the unstable equilibrium point defined in Fig. 5(b). Hence, if A_a is smaller than $A_{d\max}$, the power angle δ_m which satisfies $A_a = A_d$ can always be found, and thus the system can keep stable after the transient disturbance.

Although the EAC is the straightforward method for the transient stability assessment, it is based on the simplified dynamic equation (8) by assuming $K_p = 0$, which resulting in a marginally stable system with zero phase margin [19]. Therefore, this assumption is less justifiable for the real applicable PLL, and the conclusion based on the EAC will be inaccurate when the effect of K_p is considered. Actually, the system can keep stable even if it does not satisfy the EAC, which will be shown in this work.

C. Basic Concept of the Phase Portrait

Due to the inaccurate stability prediction of the EAC, new tools are needed for the transient stability assessment. However, the dynamic equation of the PLL, i.e., equation (6), is inherent nonlinear, and it is difficult to solve nonlinear equations analytically [22]. To tackle this challenge, the concept of the phase portrait has been introduced in [22], providing a graphical solution to the first-order and second-order nonlinear systems. Therefore, this method can also be adopted here to assess the transient stability of the PLL with different controller parameters.

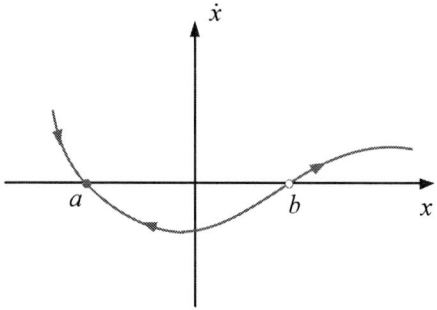

Fig. 6. An example of the phase portrait.

The basic concept of the phase portrait is introduced here based on a general nonlinear differential equation, which is given by

$$\dot{x} = f(x) \qquad (15)$$

where x is the state variable of concern. Instead of solving $x(t)$ analytically, the $\dot{x} - x$ curve can be easily plotted based on (15), as shown in Fig. 6, which is also called the phase portrait [22]. At each point, the changing trend of x is determined by its derivative $\dot{x}$, e.g., x will increase (moving rightwards in Fig. 6) if $\dot{x} > 0$, and will decrease (moving leftwards in Fig. 6) if $\dot{x} < 0$, which are indicated by the arrows in Fig. 6. Therefore, as long as the initial condition is determined, the system trajectory can be easily obtained based on the phase portrait.

D. Transient Stability Analysis of the PLL with Different Grid Impedance Types

The grid impedance is mainly resistive in the low-voltage distribution grid and inductive in the high-voltage transmission grid [1]. Based on the phase portrait, this part analyzes the transient stability impact of the PLL under these two grid conditions. Table I provides the main parameters used in this work. It is assumed that $Z_g = R_g$ in the resistive grid and $Z_g = \omega L_g$ in the inductive grid. The transient disturbance is considered as a step change of I_g.

Based on (12) and (14) and parameters listed in Table I, it can be calculated that $A_a = 53.1$ V·rad, $A_{d\max} = 26.5$ V·rad. As $A_a > A_{d\max}$, the conclusion of the EAC indicates that the PLL should become unstable after the transient disturbance. However, it will be proved in the following that the conclusion is too conservative and the transient stability of the PLL is highly dependent on its controller parameters.

TABLE I
MAIN CIRCUIT PARAMETERS

Symbol	Description	Value (p.u.)
V_{grms}	RMS value of grid voltage (phase to ground)	110 V (1)
f_g	Grid frequency	50 Hz (1)
I_m	Magnitude of grid current after disturbance	15.5 A (1)
L_f	Inductance of the output filter of the VSC	3 mH (0.096)
L_g	Grid impedance (inductive grid)	25mH (0.8)
R_g	Grid impedance (resistive grid)	7.8 Ω (0.8)

1) Resistive Grid : As a second-order dynamic system, the dynamics of the PLL can be characterized by its damping ratio (ζ) and setting time (t_s) [19], and their relationships with the controller parameters are listed as follows [19]:

$$\zeta = \frac{K_p}{2}\sqrt{\frac{V_g}{K_i}} \qquad (16)$$

$$t_s = \frac{9.2}{V_g K_p} \qquad (17)$$

For the resistive grid, the transient disturbance is considered as $I_{dref} = 0$, I_{qref} steps from 0 to I_m. Therefore, $\theta_z = 0°$, $\varphi = 90°$, $Z_g = R_g$. Substituting these into (7), yielding:

$$\frac{1}{K_i}\ddot{\delta} = I_m R_g - V_g \sin\delta - \frac{K_p V_g \cos\delta}{K_i} \cdot \dot{\delta} \qquad (18)$$

Based on (18), the phase portrait with different damping ratio and settling time can be plotted. For simplicity, only three typical trajectories with different controller parameters are given, as shown in Fig. 7, based on which, two important conclusions can be drawn:

1) In the resistive grid, the transient stability of the PLL is affected by its damping ratio, the conclusion can be easily understood from the control perspective. As mentioned in part A of this section, the transient instability only takes place when the overshoot is larger than δ_c, i.e. $\dot{\delta} > 0$ when $\delta = \delta_c$, as the dash dotted line shown in Fig. 7 (a). However, increasing the damping ratio can reduce the overshoot, which is helpful to

2676

enhance the transient stability of the PLL, the system can actually become stable even if $A_a > A_{dmax}$, as illustrated by the dash and solid lines shown in Fig. 7 (a).

2) In contrast, the settling time of the PLL has little effect on the system transient stability. As shown in Fig. 7 (b), due to the fact that the overshoot in the dynamic response of the PLL is only determined by the damping ratio, rather than the change of the settling time.

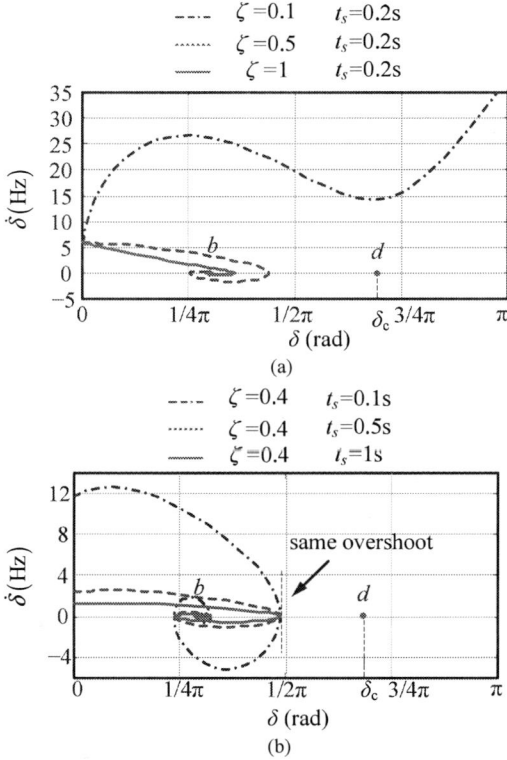

Fig. 7. Phase portrait of the PLL after I_{qref} is stepped up from 0 to I_m under the resistive grid condition. (a) t_s=0.2s, ζ=0.1,0.5,1. (b) ζ=0.4, t_s=0.1s, 0.5s, 1s.

2) Inductive Grid: The impedance of the grid inductance during the transient can be expressed as

$$Z_g = \omega L_g = \omega_n L_g + \dot\delta L_g \tag{19}$$

where ω_n is the nominal grid frequency. Based on the parameters in Table I, $R_g = \omega_n L_g$ is yielded, which means that if the impedance variation introduced by the frequency fluctuation, i.e. the last term in (19), is neglected, the conclusion of the transient stability of the PLL in the inductive grid will be the same with that in the resistive grid. However, it will be shown that this assumption is not valid, especially when the PLL has fast dynamics.

The transient disturbance is considered as I_{qref}=0, I_{dref} steps from 0 to I_m. Therefore, θ_z=90°, φ=0°. Substituting these into (7), the dynamic equation of the PLL in the inductive grid can be expressed as

$$\frac{(1 - K_p I_m L_g)}{K_i}\ddot\delta$$
$$= I_m \omega_n L_g + I_m \dot\delta L_g - V_g \sin\delta - \frac{K_p V_g \cos\delta}{K_i}\cdot\dot\delta \tag{20}$$

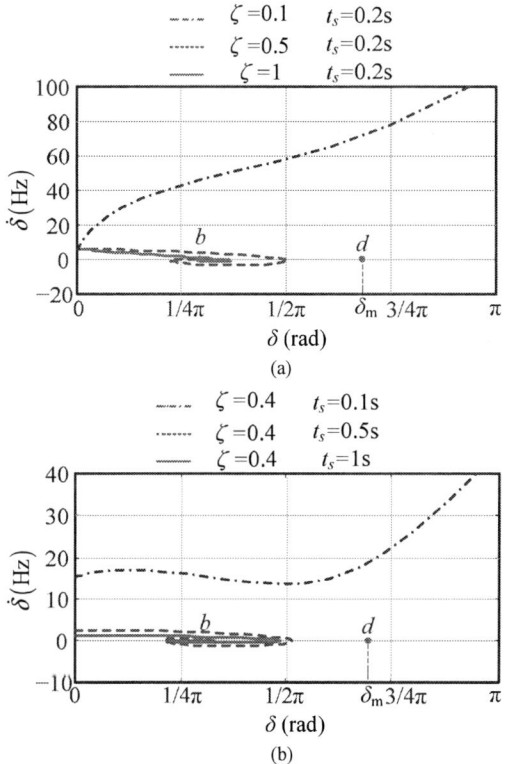

Fig. 8. Phase portrait of the PLL after I_{dref} is stepped up from 0 to I_m under the inductive grid condition. (a) t_s=0.2s, ζ=0.1,0.5,1. (b) ζ=0.4, t_s=0.1s, 0.5s, 1s.

Based on (20), the phase portrait can be plotted, as shown in Fig. 8. The controller parameters used here are exactly same as that used in the analysis under the resistive grid condition. Based on Fig. 8 (a), it can be concluded that the increase of the damping ratio helps to enhance the system transient stability. This conclusion is same as that in the resistive grid condition.

However, by comparing Fig. 7(b) with Fig. 8(b), it is found out that the influence of the settling time differs between the inductive grid and the resistive grid. The overshoot of the PLL is not affected by the settling time in the resistive grid. Yet, it increases with the reduced settling time in the inductive grid. Moreover, for the parameters ζ=0.4 and t_s=0.1s, the PLL is stable in the resistive grid but becomes unstable in the inductive grid after the disturbance, as illustrated by the dashed dotted line shown in Fig. 7(b) and Fig. 8(b). These phenomena can be explained as follows: The output frequency of the PLL begins to increase after the step up of I_g due to the fact that $I_g Z_g \sin(\theta_z+\varphi) > V_g \sin\delta$. The increased frequency, however, increases the value of grid impedance as $Z_g = \omega L_g$, and $I_g Z_g \sin(\theta_z+\varphi)$ is also increased subsequently. Then, the increased $I_g Z_g \sin(\theta_z+\varphi)$ further drives the output frequency of the PLL to increase, these iteration repeats and makes it much harder for the output frequency of the PLL to recover to the grid frequency, which finally degrades the transient stability performance of the PLL. The stability degradation will be more obvious with the large frequency derivation resulted from

The 2018 International Power Electronics Conference

the fast dynamics of the PLL, as shown in Fig. 8(b).

The analysis presented above reveals the necessity to take the frequency depended impedance into consideration in the transient stability analysis. Moreover, both slow dynamics and large damping are recommended for the transient stability enhancement of the PLL under the inductive grid.

Remark: The small-signal dynamics of the PLL can be optimized by choosing $\zeta=0.707$ [19]. However, based on the conclusion of this paper, ζ should be as large as possible for the transient stability enhancement. Therefore, it is of necessity to reinvestigate the parameters tuning guidelines to coordinate the requirement of the small-signal and large-signal stability of the PLL, which will be presented in the future work.

IV. EXPERIMENTAL RESULTS

To further verify the theoretical analysis, the experiments are carried out with a three-phase VSC. The control algorithm is implemented in the DS1007 dSPACE system, where the DS5101 digital waveform output board is used for generating the switching pulses, and the DS2004 high-speed A/D board is used for the voltage and current measurements. The active/reactive current references, active/reactive currents and the output frequency of the PLL are outputted through the DS2102 high-speed D/A board. A constant dc voltage supply is used at the dc-side, and Chroma grid simulator is used to generate the grid voltage. The transient disturbance is considered as a step up of I_g, which is exactly same as that described in Section III. The parameters used in the experiment are same with that used in the theoretical analysis, as shown in Table I

1) Resistive Grid: Fig. 9 shows experimental results of the reactive current reference, the output reactive current of the converter, and the output frequency variation of the PLL under the resistive grid condition. Three cases corresponding to the PLL with same setting time and different damping ratio are compared, and the output frequency of the PLL is limited from 0.5 pu to 1.5 pu in the experiment. Same to the theoretical analysis, the unstable response can be observed in Fig. 9(a) due to the

low damping ratio. The output frequency of the PLL is saturated at the upper limit and cannot recovers to the grid frequency after the disturbance, resulting in a 0.5 pu (25Hz) oscillation in the reactive current, which implies the loss of synchronism between the converter and the power grid. However, the system can be stable if the damping ratio is increased, as shown in Fig. 9 (b) and (c). Obviously, all the experimental results are in well accordance with the theoretical analysis.

Fig. 10 shows the experimental results of the system under the resistive grid condition. Three cases corresponding to the PLL with same damping ratio and different settling time are compared. Same to the theoretical analysis, the response time of the PLL varies with the different settling time, but the system is always stable in three cases, indicating that the settling time does not affect the transient stability of the system in the resistive grid. The experimental results further verify the theoretical analysis.

2) Inductive Grid: Fig. 11 shows experimental results of the active current reference, the output active current of the converter, and the output frequency variation of the PLL under the inductive grid condition. Three cases corresponding to the PLL with same setting time and different damping ratio are compared. It can be seen that the PLL may be unstable after the transient disturbance due to the low damping ratio, as shown in Fig. 11(a). Yet, its transient stability can be improved with the increased damping ratio, as shown in Fig. 11(b) and (c). Obviously, all the experimental results are in well accordance with theoretical analysis.

Fig. 12 shows the experimental results of the system under the inductive grid condition. Three cases corresponding to the PLL with same damping ratio and different settling time are compared. Unlike the case in the resistive grid, the settling time of the PLL not only affects its response time, but also affects its transient stability in the inductive grid. More specifically, the transient stability of the system is degraded with the reduced settling time ($\zeta=0.4$, $t_s=0.1$s), as the unstable case shown in Fig. 12(a). Obviously, all the experimental results are in well accordance with theoretical analysis.

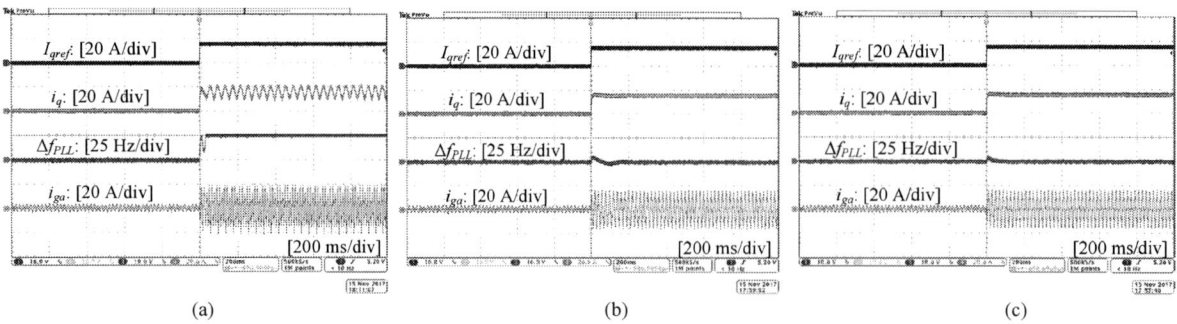

Fig. 9. Experimental results of the converter after I_{qref} is stepped up from 0 to I_m under the resistive grid condition. (a) t_s=0.2s, ζ=0.1. Unstable. (b) t_s=0.2s, ζ=0.5. Stable. (c) t_s=0.2s, ζ=1. Stable.

The 2018 International Power Electronics Conference

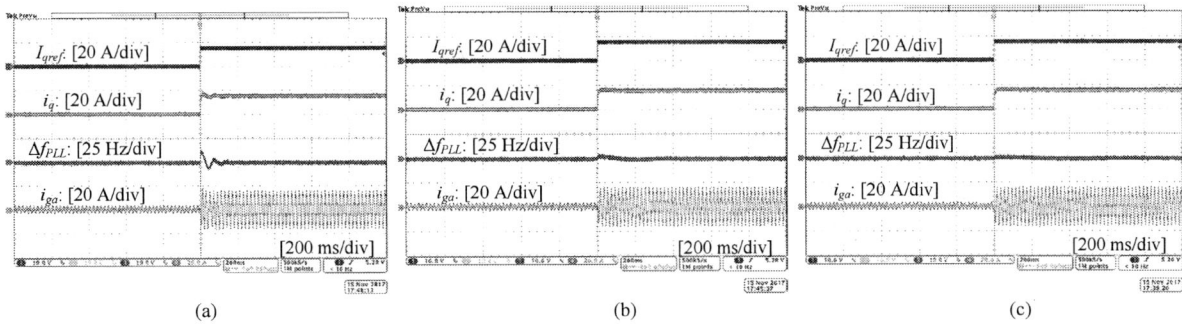

(a)   (b)   (c)

Fig. 10. Experimental results of the converter after I_{qref} is stepped up from 0 to I_m under the resistive grid condition. (a) t_s=0.1s, ζ=0.4. Stable. (b) t_s=0.5s, ζ=0.4. Stable. (c) t_s=1s, ζ=0.4. Stable.

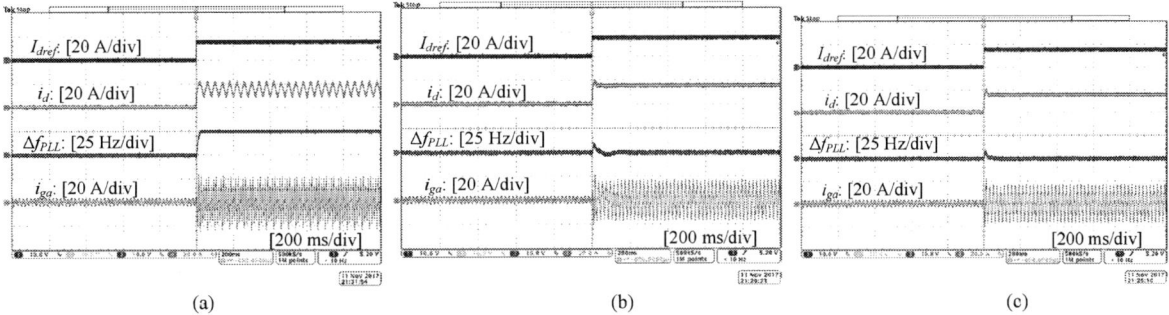

(a)   (b)   (c)

Fig. 11. Experimental results of the converter after I_{dref} is stepped up from 0 to I_m under the inductive grid condition. (a) t_s=0.2s, ζ=0.1. Unstable. (b) t_s=0.2s, ζ=0.5. Stable. (c) t_s=0.2s, ζ=1. Stable.

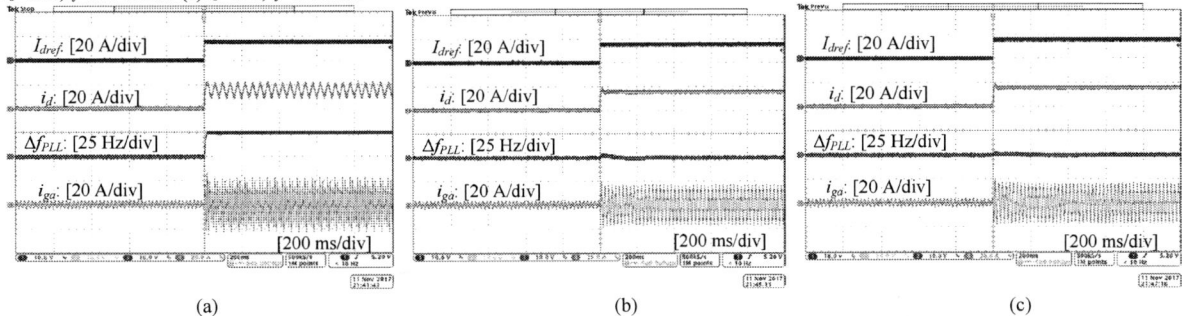

(a)   (b)   (c)

Fig. 12. Experimental results of the converter after I_{dref} is stepped up from 0 to I_m under the inductive grid condition. (a) t_s=0.1s, ζ=0.4. Unstable. (b) t_s=0.5s, ζ=0.4. Stable. (c) t_s=1s, ζ=0.4. Stable.

V. CONCLUSIONS

Transient instability of the PLL may occur even if it has the equilibrium point after the disturbance. However, the EAC cannot be simply extended to analyze the transient stability of the PLL as it is unable to access the impact of controller parameters, which has been proven to be crucial to the PLL's transient stability. To tackle this challenge, the phase portrait is used in this paper to perform a comprehensive study of the transient stability of the PLL. It is pointed out that the transient stability of the PLL is mainly affected by its damping ratio in the resistive grid but also affected by its settling time in the inductive grid. Generally, both slow dynamics and large damping are recommended for the transient stability enhancement of the PLL.

REFERENCES

[1] P. Kundur, *Power System Stability and Control.* New York, NY, USA: McGraw-Hill, 1994.

[2] R. W. Erickson and D. Maksimovic, *Fundamentals of Power Electronics*, 2nd ed. Norwell, MA, USA: Kluwer, 2001.

[3] L. Harnefors, M. Bongiorno, and S. Lundberg, "Input-admittance calculation and shaping for controlled voltage-source converters," *IEEE Trans. Ind. Electron.*, vol. 54, no. 6, pp. 3323-3334, Dec. 2007.

[4] X. Wang, F. Blaabjerg, and W. Wu, "Modeling and analysis of harmonic stability in ac power-electronics-based power system," *IEEE Trans. Power Electron.*, vol. 29, no. 12, pp. 6421-6432, Dec. 2014.

[5] H. Wu, X. Ruan, D. Yang, X. Chen, W. Zhao, Z. Lv, and Q. Zhong. "Small-signal modeling and parameters design for virtual synchronous generators". *IEEE Trans. Ind. Electron.*, vol. 63 no, 7, pp. 4292–4303, July. 2016.

[6] L. Harnefors, X. Wang, A. G. Yepes, and F. Blaabjerg, "Passivity-based stability assessment of grid-connected VSCs – an overview," *IEEE J.Emerg. Sel. Topics Power Electron.*, vol. 4, no. 1, pp. 116-125, Mar. 2016.

[7] D. Yang, X. Ruan, and H. Wu, "Impedance shaping of the grid-connected inverter with LCL filter to improve its adaptability to

2679

the weak grid condition," *IEEE Trans. Power Electron.*, vol. 29, no. 11, pp. 5795–5805, Nov., 2014.

[8] X. Wang, L. Harnefors, and F. Blaabjerg, "A unified impedance model of grid-connected voltage-source converters," *IEEE Trans. Power Electron., IEEE Trans. Power Electron.*, vol. 33, no. 2, pp. 1775-1787, Feb. 2018.

[9] M. J. Hossain, H. R. Pota, M. A. Mahmud, and R. A. Ramos, "Investigation of the impacts of large-scale wind power penetration on the angle and voltage stability of power systems," *IEEE Systems Journal*, vol. 6, no. 1, pp. 76–84, Mar. 2012

[10] M. Molinas, Jon. A. Suul, and T. Undeland, "Low voltage ride through of wind farms with cage generators: STATCOM versus SVC," *IEEE Trans. Power Electron.*, vol. 23, no. 3, pp. 1104-1117, May. 2008.

[11] M. V. A. Nunes, J. A. P. Lopes, H. H. Zurn, U. Bezerra, and R. Almeida, "Influence of the variable-speed wind generators in transient stability margin of the conventional generators integrated in electrical grids," *IEEE Trans. Energy Convers.*, vol. 19, no. 4, pp. 692–701, Dec. 2004.

[12] E. Vittal, M. O'Malley, and A. Keane, "Rotor angle stability with high penetrations of wind generation," *IEEE Trans. Power Syst.*, vol. 27, no. 1, pp. 353-362, Feb. 2012.

[13] M. Edrah, K. L. Lo, and O.A-Lara, "Impacts of high penetration of DFIG wind turbines on rotor angle stability of power systems," *IEEE Trans. Sustain. Energy.*, vol. 6, no. 3, pp. 759–766, July. 2015.

[14] H. Xin, L. Huang, L. Zhang, Z. Wang, and J. Hu, "Synchronous instability mechanism of P-f droop-controlled voltage source converter caused by current saturation," *IEEE Trans. Power Syst.*, vol. 31, pp. 5206-5207, 2016.

[15] L. Huang, L. Zhang, H. Xin, Z. Wang and D. Gan, "Current limiting leads to virtual power angle synchronous instability of droop-controlled converters," *in Proc. IEEE PES General Meeting*, 2016.

[16] L. Huang, H. Xin, Z. Wang, L. Zhang, K. Wu and J. Hu, "Transient stability analysis and control design of droop-controlled voltage source converters considering current limitation," *IEEE Trans. Smart Grid*, early access.

[17] D. Dong, B. Wen, D. Boroyevich, P. Mattavelli, and Y. Xue, "Analysis of phase-locked loop low-frequency stability in three-phase grid-connected power converters considering impedance interactions," *IEEE Trans. Ind. Electron.*, vol. 62, no. 1, pp. 310–321, Jan. 2015.

[18] C. Zhang, X. Cai, and Z. Li, "Transient stability analysis of wind turbines with full-scale voltage source converter," *Proceedings of the CSEE.*, vol. 37, no. 14, pp. 4018–4026, July. 2017. (in Chinese)

[19] R. Teodorescu, M. Liserre, and P. Rodriguez, *Grid Converters for Photovoltaic and Wind Power Systems*, Wiley Press, 2011

[20] Y. Hao, X. Yuan, and J. Hu, "Modeling of grid-connected VSCs for power system small-signal stability analysis in DC-Link voltage control timescale" *IEEE Trans. Power Syst.*, early access.

[21] S. K. Chung, "A phase tracking system for three phase utility interface inverters," *IEEE Trans. Power Electron.*, vol. 15, no. 3, pp. 431-438, May.2000.

[22] Steven H. Strogatz. *Nonlinear Dynamics and Chaos: With Applications to Physics, Biology, Chemistry, and Engineering.* Perseus Books, 1994

Comprehensive Analysis of Virtual Impedance-Based Active Damping for *LCL* Resonance in Grid-Connected Inverters

Teng Liu[*], Zeng Liu, Jinjun Liu, Yiming Tu, Zipeng Liu

State Key Laboratory of Electrical Insulation and Power Equipment, Xi'an Jiaotong University

Xi'an, China

*E-mail: teng.liu@stu.xjtu.edu.cn

Abstract—*LCL* filters have been widely applied as the interface between voltage source inverters (VSIs) and power grid. However, inherent *LCL* resonance complicates the design of current control or even threatens system stability. Extensive approaches have been proposed to deal with the resonance, among which the active damping (AD) methods realized by the feedback of filter state variables are proved to be effective and robust. In this paper, a comprehensive analysis of such AD by utilizing its virtual circuit property is presented. It is found that different filter state variables can be selected to achieve the effective damping functions by emulating proper virtual impedances (VIs). Meanwhile, the connection of the VI with the *LCL* network is not restricted by the selected state variable but mainly dependent on the applied AD controller, which, in result, enables a thorough investigation of such VI-based AD methods. Consequently, all possible forms of the AD controller to realize different connection types by utilizing different state variables are derived. The optimal selection of the AD controller together with the specific state variable is further analyzed, where the obtained results can be used to guide the design of such AD in practice. Finally, the correctness of the theoretical analyses are verified by the experimental results.

Keywords—active damping, grid-connected inverters, LCL resonance, virtual impedance.

I. INTRODUCTION

With fast development of distributed power generation systems (DPGSs), voltage source inverters (VSIs) have been widely used for integrating DPGSs into power grid [1]. To meet the grid codes, output filters should be equipped with VSIs. Compared with traditional *L* filters, *LCL* filters can provide better attenuation of switching ripples with smaller volumes and lower costs [2]-[3]. However, inherent *LCL* resonance characteristic imposes constraints on the current controller design, which may limit the current control loop bandwidth or even cause an unstable system [4].

To suppress the *LCL* resonance, extensive researches have been conducted [5]-[15]. One direct way is the so-called passive damping, which is realized by inserting passive resistors into the *LCL* filter network [5]. Even

This work was supported by the National Natural Science Foundation of China under Grant 51437007.

though it owns the merits of simple implementation and good robustness, it may not be preferred due to the inevitable extra power losses, especially in high-power applications [6].

An alternative way is the active damping (AD) to obtain a more efficient system. The existing AD methods can be divided into two main categories. One type is implemented by cascading a digital filter with the current regulator aiming at filtering out the *LCL* resonant peak [6]. However, system parameters should be well known to design the digital filter, which also makes such kind of AD methods quite sensitive to the variation of the system parameters. Another type is realized by feeding back the *LCL* filter state variables to form a dual-loop control [7]-[15]. It has been proved that different state variables, such as the inverter-side current [7], the filter capacitor current [8]-[12], the filter capacitor voltage [13], or the grid-side current [14]-[15], can all be selected to realize the effective damping performance. However, it should be noticed that different state variables need different AD controllers to achieve an effective damping performance. For example, when the capacitor current or the inverter-side current is fed back, the AD controller is usually implemented by a proportional gain [7]-[9]. However, when the feedback state variable is changed to the grid-side current, the AD controller should be modified to a first-order high-pass filter with negative sign accordingly [14]-[15]. Therefore, it is worthwhile to figure out all the possible forms of the AD controller when the feedback state variable to achieve the AD is determined.

An intuitive way to obtain appropriate forms of the AD controller is based on its equivalent virtual circuit. In [8], it has been proved that the proportional capacitor-current-feedback AD is equivalent to adding a virtual resistor in parallel with the filter capacitor when the time delay effect is neglected. The formed virtual resistor can thus dampen the *LCL* resonance effectively. When the feedback variable is changed to the inverter-side current, a proportional AD controller can also emulate a virtual resistor in series with the inverter-side filter inductor [7]. Recently, the equivalent virtual circuit property has also been revealed when the grid-side current is fed back by a first-order high-pass filter with negative sign, which is equivalent to adding a virtual impedance (VI) in parallel with the grid-side filter inductor. And this VI consists of a series RL damper in parallel with a negative inductance [15]. Therefore, the VI in different connections with LCL filter has specific relationships with the AD controller.

To thoroughly reveal all the possible AD controllers with different state variables for the effective damping, this paper proposes a mathematical way to derive the forms of the AD controller based on the virtual circuit property. It is obtained that an arbitrary state variable can be applied to emulate a VI in different connections with LCL filter network through different AD controllers, and the connection types will not be limited by the selected state variable. For instance, the feedback of the filter capacitor current can not only emulate a VI in series or parallel with the capacitor, but can also realize possible connections of VI with the inverter-side or grid-side filter inductor through proper AD controllers. As a result, all possible forms of the AD controller with different state variables are derived. Further, the optimal selection of the AD controller considering the availability of the sensors for measuring the filter state variable is analyzed, where the obtained conclusions can provide a guideline for the design of such AD in practice. Finally, experimental results validate the correctness of the theoretical analyses.

II. SYSTEM DESCRIPTION

Fig.1 shows the system structure of a three-phase grid-connected VSI system, where the inverter-side inductor L_1, the grid-side inductor L_2, and the filter capacitor C_f make up the LCL filters. L_g represents the grid impedance. The dc-link voltage V_{dc} is assumed to be constant for simplicity. v_{pcc} is the point of common coupling (PCC) voltage, whose phase angle θ is usually acquired by the synchronous reference frame phase-locked loop (SRF-PLL) for synchronizing the VSI with the grid. In practice, the bandwidth of the SRF-PLL is always designed much slower than that of the inner current control loop, which means the effect of the SRF-PLL on the high frequency resonance can be neglected [16].

For current regulation, either the inverter-side current i_1 or the grid-side current i_2 can be controlled to inject the demand power into the grid, where the control scheme can be implemented either in the stationary $\alpha\beta$ frame or in the rotating dq frame. In this paper, the current control in the $\alpha\beta$ frame is chosen to avoid nonlinearity introduced by the dq transformation. Herein, assuming a three-phase balanced grid-connected VSI system, the per-phase block diagrams of the current control with different AD, which are shown in Fig. 2, can thus be used for the following analyses. $G_c(s)$ is the current regulator implemented with a proportional-resonant (PR) controller expressed as

$$G_c(s) = k_p + \frac{k_r s}{s^2 + \omega_0^2} \qquad (1)$$

where ω_0 is the grid fundamental angular frequency. $G_d(s)$ is time delay effect existed in the digitally controlled VSI, which includes both the computational and PWM delays.

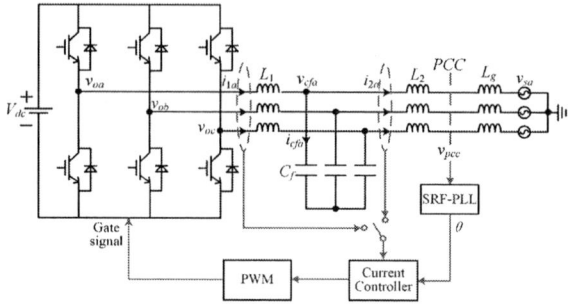

Fig. 1. System structure of a three-phase grid-connected VSI.

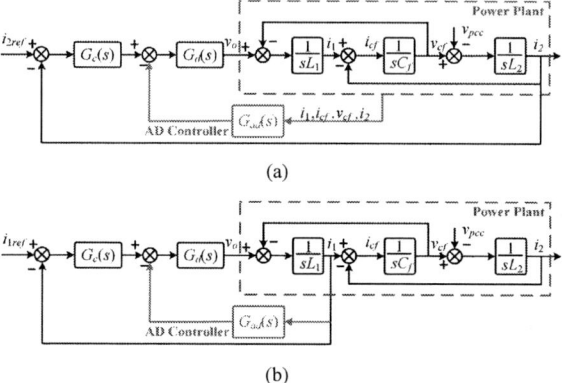

(a)

(b)

Fig. 2. Per-phase block diagrams of the current control with different AD. (a) grid-side current control. (b) inverter-side current control.

When the synchronous sampling scheme is applied, $G_d(s)$ can be expressed as

$$G_d(s) = e^{-1.5T_s s} \qquad (2)$$

where T_s is the sampling period [8].

It is worthwhile to notice that the selection of the state variable used for achieving the AD purpose is flexible when i_2 is fed back to control the injected power as shown in Fig. 2(a). In this case, i_1, i_2, filter capacitor current i_{cf}, or filter capacitor voltage v_{cf} can all be adopted to realize the effective damping performance. However, when the regulated current is changed to i_1, the same state variable i_1 may be the only choice to realize the AD simultaneously, as the current or voltage sensors to detect the other state variables may not be installed in most practical applications for saving the cost. The relevant current control diagram is shown in Fig. 2(b), where the AD controller $G_{ad}(s)$ is usually implemented by a simple proportional gain [7]. The proportional feedback of i_1 is equivalent to adding a virtual resistor in series with L_1 when the time delay effect is neglected, which thus can dampen the LCL resonance effectively.

According to the above analyses, this paper mainly focuses on the situation where i_2 is fed back to control the injected power. It is known that different state variables need different forms of $G_{ad}(s)$ to achieve the effective damping performance. However, most of the existing literatures presented $G_{ad}(s)$ directly without giving any explanation, which may not be helpful for a deeper insight of such AD methods. Hence, this paper derives all

2682

possible forms of $G_{ad}(s)$ with the different state variables based on the equivalent virtual circuit property, which will be demonstrated in the following part.

III. DERIVATION AND OPTIMAL SELECTION OF $G_{ad}(s)$ WITH DIFFERENT STATE VARIABLES

A. Derivation of $G_{ad}(s)$ With Different State Variables

To derive possible forms of $G_{ad}(s)$ with different state variables intuitively, Fig. 2(a) is equivalently transformed as shown in Fig. 3, where $G_p(s)$ is the transfer function from the output voltage of the VSI $v_o(s)$ to $i_2(s)$, and $G_t(s)$ represents the transfer function from $i_2(s)$ to the certain state variable fed back for realizing the AD. Herein, $G_p(s)$ can be derived based on Fig. 2(a), which is expressed as

$$G_p(s) = \frac{i_2(s)}{v_o(s)} = \frac{1}{s(L_1 L_2 C_f s^2 + L_1 + L_2)} \quad (3)$$

Fig. 3 can be further transformed to Fig. 4 to reveal the equivalent virtual circuit property introduced by the AD. Based on Fig. 4, it is clear that the components inside the dashed box can be regarded as the modified virtual power plant which is also represented by the transfer function from $v_o(s)$ to $i_2(s)$. Hence, the characteristic of the virtual power plant $G_{p\_m}(s)$, which is modified by the AD, can be derived as

$$G_{p\_m}(s) = \frac{i_2(s)}{v_o(s)} = \frac{G_p(s)}{1 + G_p(s) \cdot G_{ad}(s) \cdot G_d(s) \cdot G_t(s)} \quad (4)$$

Meanwhile, $G_{p\_m}(s)$ can also be obtained based on the introduced equivalent virtual circuit of the LCL filter network. In this paper, implementing a virtual impedance $Z_v(s)$ in parallel with C_f is taken as an example to demonstrate the derivation process of $G_{ad}(s)$. The relevant virtual circuit of the LCL filter network is shown in Fig. 5, based on which $G_{p\_m}(s)$ can be obtained as

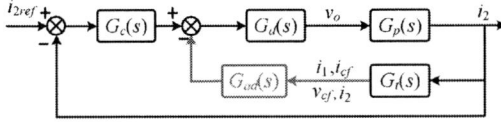

Fig. 3. Equivalent control block diagram of Fig. 2(a).

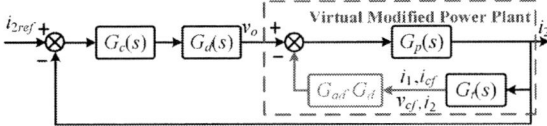

Fig. 4. Equivalent block diagram transformation to reveal the virtual circuit property of the AD.

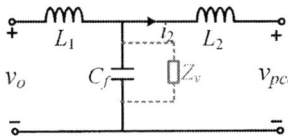

Fig. 5. Equivalent virtual circuit of the LCL filter network when $Z_v(s)$ is in parallel with C_f.

$$G_{p\_m}(s) = \frac{Z_v}{s\left[C_f L_1 L_2 Z_v s^2 + L_1 L_2 s + (L_1 + L_2) Z_v \right]} \quad (5)$$

Based on (3), (4) and (5), the form of $G_{ad}(s)$ to realize such $Z_v(s)$ in parallel with C_f can be derived as

$$G_{ad}(s) = \frac{1}{G_d(s) \cdot G_t(s)} \cdot \left(\frac{1}{G_{p\_m}(s)} - \frac{1}{G_p(s)} \right) \quad (6)$$

Further, if i_2 is fed back to realize the AD purpose, which makes $G_t(s) = 1$, the form of $G_{ad}(s)$ can finally be obtained as

$$G_{ad}(s) = \frac{L_1 L_2 s^2}{Z_v} \quad (7)$$

where $G_d(s) = 1$ is assumed. Even though the time delay will change the final form of the VI, and may also have negative effect on the damping performance due to the resulted negative virtual resistor in high frequency range [8], this assumption is still acceptable to derive the basic form of $G_{ad}(s)$, as the time delay effect can be partly compensated by different delay compensation techniques [8], [17]. In this paper, a lead-lag compensator proposed in [17] is directly cascaded with $G_{ad}(s)$ in the final control scheme to reduce the time delay effect.

From the above derivation process, it can be found that, even though the feedback state variable for the AD is the grid-side current, the formed virtual impedance can still be in parallel with C_f, which means the connection type of $Z_v(s)$ with LCL filter network will not be limited by the selected state variable.

Consequently, all the possible forms of $G_{ad}(s)$ with different state variables can be derived by following the similar process. The obtained results are listed in Table I. It is clear that the connection type of $Z_v(s)$ is mainly dependent on the specific form of $G_{ad}(s)$ rather than the selected state variable. With a proper $G_{ad}(s)$, an arbitrary state variable can be fed back to emulate $Z_v(s)$ in any connection with the LCL filter for achieving the effective damping performance.

B. Optimal Selection of $G_{ad}(s)$

After obtaining possible forms of $G_{ad}(s)$ with different state variables, the optimal selection of $G_{ad}(s)$ should be discussed. As suppressing the LCL resonance is the main topic of this paper, $Z_v(s)$ is realized as a virtual resistor, i.e., $Z_v(s) = R_v$. Other possible $Z_v(s)$, such as a virtual inductor or a more complicated VI for different purposes, will be investigated in the further work.

The first criterion for the optimal selection of $G_{ad}(s)$ is the complexity to implement $G_{ad}(s)$. The comparisons are made among different $G_{ad}(s)$ with the same state variable. For example, when i_2 is selected to realize the AD, three forms of $G_{ad}(s)$, which make the formed virtual resistor in series with L_1, in series with L_2 or in parallel with C_f, are much simpler than the other three forms of $G_{ad}(s)$. When the state variable is changed to i_{cf}, a proportional gain $L_1/(C_f R_v)$ is the simplest form of $G_{ad}(s)$. Consequently, all the simpler forms of $G_{ad}(s)$ are marked in red in Table I.

2683

TABLE I
DIFFERENT FORMS OF $G_{ad}(s)$ WITH DIFFERENT STATE VARIABLES USED FOR ACTIVE DAMPING

Virtual impedance connection type	Variable used for the active damping			
	Grid-side current i_2	Capacitor current i_{cf}	Inverter-side current i_1	Capacitor voltage v_{cf}
In series with L_1	$(1+L_2C_fs^2)\cdot Z_v$	$(1+\dfrac{1}{L_2C_fs^2})\cdot Z_v$	Z_v	$(C_fs+\dfrac{1}{L_2s})\cdot Z_v$
In parallel with L_1	$-\dfrac{(1+L_2C_fs^2)\cdot L_1^2s^2}{L_1s+Z_v}$	$-\dfrac{(1+L_2C_fs^2)\cdot L_1^2}{L_1L_2C_fs+L_2C_fZ_v}$	$-\dfrac{L_1^2s^2}{L_1s+Z_v}$	$-\dfrac{(1+L_2C_fs^2)\cdot sL_1^2}{L_1L_2s+L_2Z_v}$
In series with L_2	$(1+L_1C_fs^2)\cdot Z_v$	$\dfrac{(1+L_1C_fs^2)\cdot Z_v}{L_2C_fs^2}$	$\dfrac{(1+L_1C_fs^2)\cdot Z_v}{1+L_2C_fs^2}$	$(\dfrac{L_1C_f}{L_2}s+\dfrac{1}{L_2s})\cdot Z_v$
In parallel with L_2	$-\dfrac{(1+L_1C_fs^2)\cdot L_2^2s^2}{L_2s+Z_v}$	$-\dfrac{(1+L_1C_fs^2)\cdot L_2}{L_2C_fs+C_fZ_v}$	$-\dfrac{(1+L_1C_fs^2)\cdot L_2^2s^2}{(1+L_2C_fs^2)\cdot(L_2s+Z_v)}$	$-\dfrac{(1+L_1C_fs^2)\cdot sL_2}{L_2s+Z_v}$
In series with C_f	$-\dfrac{C_f^2L_1L_2Z_vs^4}{C_fZ_vs+1}$	$-\dfrac{C_fL_1Z_vs^2}{C_fZ_vs+1}$	$-\dfrac{C_f^2L_1L_2Z_vs^4}{(1+L_2C_fs^2)\cdot(C_fZ_vs+1)}$	$-\dfrac{C_f^2L_1Z_vs^3}{C_fZ_vs+1}$
In parallel with C_f	$\dfrac{L_1L_2s^2}{Z_v}$	$\dfrac{L_1}{C_fZ_v}$	$\dfrac{L_1L_2s^2}{(1+L_2C_fs^2)\cdot Z_v}$	$\dfrac{L_1s}{Z_v}$

It can be concluded that a simple proportional gain can be regarded as an optimal selection of $G_{ad}(s)$ when i_{cf} or i_1 is selected to realize the AD. However, if the state variable is changed to v_{cf} or i_2, there exist three possible selections of $G_{ad}(s)$ which can meet the requirement of simplicity.

To further compare these three $G_{ad}(s)$, the frequency responses of $G_{p\_m}(s)$ with different connections of the virtual resistor are plotted in Fig. 6 respectively. It is seen that $G_{ad}(s)$ to form a virtual resistor in parallel with C_f can easily achieve the satisfied damping performance. For the other two connections, even though the formed R_v can dampen the resonant peak, it will cause a flat and small magnitude gain in the low frequency range, which may lead to a relatively low current control bandwidth when the PR controller is adopted as the current regulator. Therefore, a virtual resistor connected in parallel with C_f is much preferred among all the connection types.

However, it should also be noticed that, when R_v is in

parallel with C_f, $G_{ad}(s)$ will contain the derivative term if the selected state variable is v_{cf} or i_2. And this derivative term is not practically feasible due to the possible noise amplification. Fortunately, this limitation can be solved by approximating the derivative term with a non-ideal generalized integrator [18].

To sum up, if the capacitor current sensor is available, $G_{ad}(s)$ realized only by a simple proportional gain can be regarded as the first choice to obtain a satisfied damping performance. However, in practice, the capacitor current sensor is usually unavailable for the purpose of saving the cost. At this time, the current sensor for i_2 or the voltage sensor for v_{cf} may be applied, where a virtual resistor in parallel with C_f is preferred to dampen the LCL resonant peak. If there is only the current sensor for i_1 available, the optimal selection of $G_{ad}(s)$ may be a proportional gain due to the simplest implementation.

IV. CASE STUDY OF THE GRID-SIDE CURRENT FEEDBACK ACTIVE DAMPING

As many existing literatures have already presented the AD realized by the feedback of i_{cf} or i_1, this paper only presents the grid-side current i_2 for realizing the AD to demonstrate the correctness of the theoretical analyses. It should be mentioned that choosing i_2 as the feedback variable can avoid the extra sensors, which thus can save the system cost.

A. Grid-side Current Feedback Active Damping

The per-phase block diagram of the grid-side current control with the grid-side current feedback AD is shown in Fig. 7. It is seen that the grid-side current i_2 is fed back for both injected power regulation and active damping purposes. Herein, the time delay $G_d(s)$ is considered to make the analysis more close to the real case.

Based on Table I, $G_{ad}(s)$ for obtaining a virtual resistor

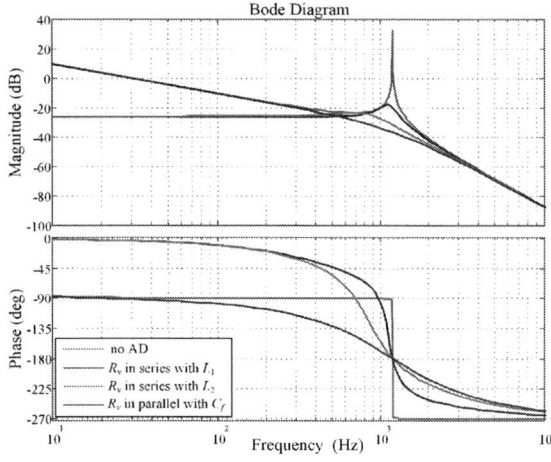

Fig. 6. Frequency responses of $G_{p\_m}(s)$ with different connections of the virtual resistor.

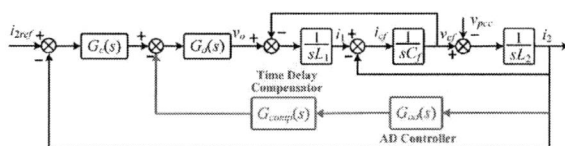

Fig. 7. Per-phase block diagram of the grid-side current control with the grid-side current feedback active damping.

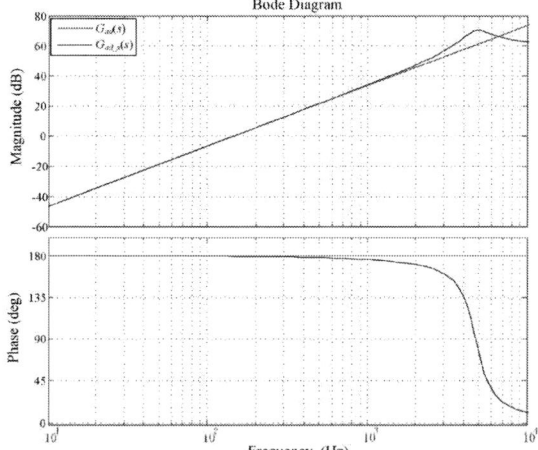

Fig. 8. Frequency responses of $G_{ad}(s)$ in (8) and $G_{ad\_x}(s)$ in (9).

in parallel with C_f can be expressed as

$$G_{ad}(s) = \frac{L_1 L_2 s^2}{R_v} \tag{8}$$

where R_v is the emulated virtual resistor. Its value can be determined by the desired damping coefficient.

As mentioned above, the derivative term s^2 in (8) can not be directly realized due to the severe high frequency noise amplification. To solve this issue, a second-order transfer function proposed in [19] is adopted to replace s^2 term. Consequently, the final form of $G_{ad\_x}(s)$ can be expressed as

$$G_{ad\_x}(s) = \frac{L_1 L_2}{R_v} \cdot \frac{\omega_c^2 \cdot s^2}{s^2 + \zeta s + \omega_c^2} \tag{9}$$

where ζ is used for damping to avoid the infinite gain at ω_c. The larger value of ω_c makes $G_{ad\_x}(s)$ more coincident with $G_{ad}(s)$. However, a larger magnitude gain at the high frequency range will be accompanied. Besides, the larger value of ζ ensures the flatter peak at ω_c. In this paper, $\omega_c = 3 \times 10^4$ rad/s and $\zeta = 10^4$ are chosen. The frequency responses of $G_{ad}(s)$ in (8) and $G_{ad\_x}(s)$ in (9) can thus be plotted in Fig. 8, respectively.

Besides, as the time delay $G_d(s)$ may have negative influence on the introduced virtual resistor, a second-order lead-lag compensator proposed in [17] should be added in cascaded with $G_{ad\_x}(s)$ to compensate the delay effect as shown in Fig. 7. Its expression is given as

$$G_{comp}(s) = K_c \cdot \frac{(1 + 0.75 T_s s)^2}{(1 + \lambda T_s s)^2} \quad 0 < \lambda < 0.75 \quad K_c \le 1 \tag{10}$$

where λ is defined as the compensation coefficient. A smaller λ means a better compensation capability. K_c is used to attenuate the high frequency noise amplification, whose value should be smaller than 1.

B. Damping Performance Evaluation

To illustrate the damping performance of this grid-side current feedback AD, the current control loop gain $T(s)$ can be derived based on Fig. 7, which is expressed as

$$T(s) = G_c(s) \cdot \frac{G_d(s) \cdot G_p(s)}{1 + G_{ad}(s) \cdot G_{comp}(s) \cdot G_d(s) \cdot G_p(s)} \tag{11}$$

From (11), it is clear that $T(s)$ without any damping can be obtained by setting $G_{ad}(s) = 0$. As a result, the frequency responses of $T(s)$ without and with the grid-side current feedback AD can be plotted in Fig. 9, where the main circuit parameters are listed in Table II.

Based on Fig. 9, it is seen that the frequency responses of $T(s)$ without any damping exist a resonant peak at the *LCL* resonance frequency, which will cause a negative $180°$ phase falling. According to the stability criterion, the system will be unstable. When the proposed grid-side current feedback AD is applied, this resonant peak can be suppressed effectively leading to a stable system. Besides, it can also be seen that the frequency responses of $T(s)$ with the AD is similar with those of $T(s)$ with an ideal virtual resistor in parallel with C_f, which thus proves the correctness of the theoretical analyses.

TABLE II
MAIN CIRCUIT PARAMETERS

Item	Symbol	Value
DC-link voltage	V_{dc}	400V
Grid voltage (*l-g*, rms)	v_s	120V
Switching frequency	f_{sw}	10kHz
Sampling frequency	f_s	10kHz
Inverter-side inductor	L_1	3.5mH
Grid-side inductor	L_2	1.75mH
Filter capacitor	C_f	15μF

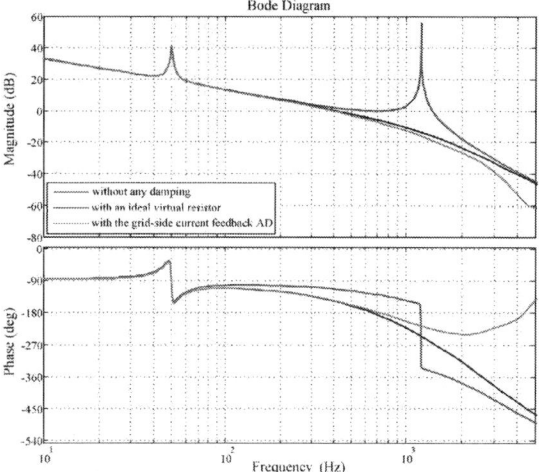

Fig. 9. Frequency responses of $T(s)$ without and with the proposed grid-side current feedback AD.

2685

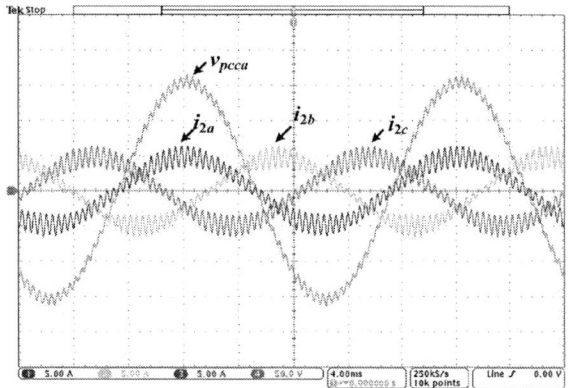

Fig. 10. The measured phase-*a* PCC voltage and three-phase grid-side currents when the VSI is operated without any damping.

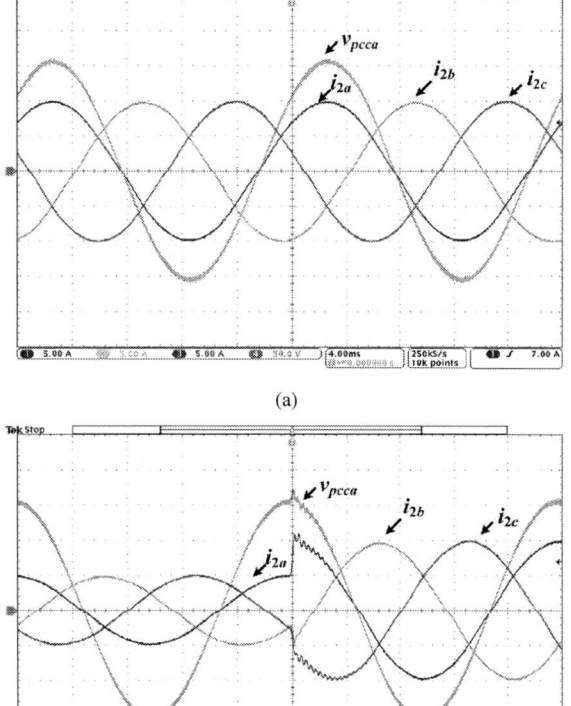

(a)

(b)

Fig. 11. The measured phase-*a* PCC voltage and three-phase grid-side currents when the VSI is operated with the proposed grid-side current feedback AD. (a) steady-state waveforms, (b) dynamic waveforms.

C. Experimental Results

To verify the effectiveness of such grid-side current feedback AD, a laboratory setup, where the main circuit parameters listed in Table II are applied, is built up. The structure is exactly the same with that shown in Fig. 1. The ac power grid is emulated by a *Chroma Regenerative Grid Simulator* 61860.

Firstly, the VSI is operated without any damping. The measured phase-*a* PCC voltage and three-phase grid-side currents are shown in Fig. 10. It is clear that the grid-side currents are severely oscillated indicating an unstable system, which is consistent with the analyses in Fig. 9.

Then, the proposed grid-side current feedback AD is applied. The measured steady-state waveforms are shown in Fig. 11(a), where the system can be well stabilized. Meanwhile, a satisfied dynamic performance is obtained by setting the current magnitude reference from 5A to 10A, whose measured waveforms are shown in Fig. 11(b). Consequently, the proposed grid-side current feedback active damping is proved to be effective to suppress the *LCL* resonance.

V. CONCLUSIONS

This paper presented a comprehensive analysis of the AD implemented by the feedback of the *LCL* filter state variables. It was found that different state variables can be adopted to achieve the effective damping function by emulating a proper VI. Besides, the connection type of VI with the *LCL* filter network is not constrained by the selected state variable but mainly dependent on the specific form of the AD controller. Therefore, all possible forms of the AD controller to realize the VI in different connections with the *LCL* filter network under different state variables are mathematically derived. Further, the optimal selection of the AD controller is discussed by considering the complexity of the controller's realization as well as the availability of the measuring sensors. The obtained conclusions may guide the design of such AD methods. Finally, a case study focusing on one particular AD, which is realized by the feedback of the grid-side current to emulate a virtual resistor in parallel with the filter capacitor, was presented. The damping performance evaluation in the frequency domain together with the experimental results have verified the correctness of the above theoretical analyses.

REFERENCES

[1] F. Blaabjerg, R. Teodorescu, M. Liserre, and A. V. Timbus, "Overview of control and grid synchronization for distributed power generation systems," *IEEE Trans. Ind. Electron.*, vol. 53, no. 5, pp. 1398–1409, Oct. 2006.

[2] M. Liserre, F. Blaabjerg, and S. Hansen, "Design and control of an *LCL*-filtered-based three-phase active rectifiers," *IEEE Trans. Ind. Appl.*, vol. 41, no. 5, pp. 1281-1291, Sept./Oct. 2005.

[3] Y. Tang, P. C. Loh, P. Wang, F. H. Choo, F. Gao, and F. Blaabjerg, "Generalized design of high performance shunt active power filter with output *LCL* filter," *IEEE Trans. Ind. Electron.*, vol. 59, no. 3, pp. 1443-1452, Mar. 2012.

[4] K. Jalili and S. Bernet, "Design of *LCL* filters of active-front-end two-level voltage-source converters," *IEEE Trans. Ind. Electron.*, vol. 56, no. 5, pp. 1674-1689, May 2009.

[5] R. N. Beres, X. Wang, F. Blaabjerg, C. L. Bak, and M. Liserre, "Comparative evaluation of passive damping topologies for parallel grid-connected converters with *LCL* filters," in *IEEE 2014 International Power Electron. Conf.*, 2014, pp. 3320-337.

[6] J. Dannehl, M. Liserre, and F. Fuchs, "Filter-based active damping of voltage source converters with *LCL* filter," *IEEE Trans. Ind. Electron.*, vol. 58, no. 8, pp. 3623-3633, Aug. 2011.

[7] J. Xu, S. Xie, C. Kan, and D. Ji, "An improved inverter-side current feedback control for grid-connected inverters with LCL filters," in *Proc. IEEE ICPE-ECCE Asia*, Jul. 2015, pp. 984-989.

[8] D. Pan, X. Ruan, C. Bao, W. Li, and X. Wang, "Capacitor-current-feedback active damping with reduced computation delay for improving robustness of *LCL*-type grid-connected inverter," *IEEE Trans. Power Electron.*, vol. 29, no. 7, pp. 3414-3427, Jul. 2014.

[9] D. Pan, X. Ruan, C. Bao, W. Li, and X. Wang, "Optimized controller design for *LCL*-type grid-connected inverter to achieve high robustness against grid-impedance variation," *IEEE Trans. Ind. Electron.*, vol. 62, no. 3, pp. 1537–1547, Mar. 2015.

[10] X. Li, X. W, Y. Geng, X. Yuan, C. Xia and X. Zhang, "Wide damping region for *LCL*-type grid-connected inverter with an improved capacitor-current-feedback method," *IEEE Trans. Power Electron.*, vol. 30, no. 9, pp. 5247-5259, Sep. 2015.

[11] X. Wang, F. Blaabjerg, and P. C. Loh, "Virtual RC damping of *LCL*-filtered voltage source converters with extended selective harmonic compensation," *IEEE Trans. Power Electron.*, vol. 62, no. 3, pp. 1537-1547, Mar. 2015.

[12] J. Dannehl, F. Fuchs, S. Hansen, and P. B. Thøgersen, "Investigation of active damping approaches for PI-based current control of grid-connected pulse width modulation converters with LCL filters," *IEEE Trans. Ind. Appl.*, vol. 46, no. 4, pp. 1509-1517, Jul. 2010.

[13] M. Malinowski and S. Bernet, "A simple voltage sensorless active damping scheme for three-phase PWM converters with an LCL filter," *IEEE Trans. Ind. Electron.*, vol. 55, no. 4, pp. 1876-1880, Apr. 2008.

[14] J. Xu, S. Xie, and T. Tang, "Active damping-based control for grid-connected LCL-filtered inverter with injected grid current feedback only," *IEEE Trans. Ind. Electron.*, vol. 61, no. 9, pp. 4746-4758, Sep. 2014.

[15] X. Wang, F. Blaabjerg, and P. C. Loh, "Grid-current-feedback active damping for *LCL* resonance in grid-connected voltage-source converters," *IEEE Trans. Power Electron.*, vol. 31, no. 1, pp. 213-223, Jan. 2016.

[16] S. G. Parker, B. P. McGrath, and D. G. Holmes, "Regions of active damping control for LCL filters," *IEEE Trans. Ind. Appl.*, vol. 50, no. 1, pp. 424-432, Jan./Feb. 2014.

[17] T. Liu, Z. Liu, J. J. Liu, and Y. Tu, "An improved capacitor-current-feedback active damping for LCL resonance in grid-connected inverters," in *Proc. IEEE IFEEC-ECCE Asia*, Jul. 2017, pp. 2111-2116.

[18] Z. Xin, X. Wang, P. C. Loh, and F. Blaabjerg, "Realization of digital differentiator using generalized integrator for power converters," *IEEE Trans. Power Electron.*, vol. 30, no. 12, pp. 6520-6523, Dec. 2015.

[19] T. Liu, Z. Liu, J. J. Liu, and Y. Tu, "Virtual impedance-based active damping for LCL resonance in grid-connected voltage source inverters with grid current feedback," in *Proc. 2016 IEEE ECCE*, Oct. 2016, pp. 1-8.

A Comparative Study of the Traditional FS-MPC and the Proposed CSF-PCC for the Three-Phase Grid-Connected Inverters

ZhiXun Ma[1,2], Xin Zhang[1*], Jingjing Huang[1,3]

1. School of Electrical and Electronic Engineering, Nanyang Technological University, Singapore
2. National Maglev Transportation Engineering Research Center, Tongji University, Shanghai
3. Electrical Engineering, Xi'an University of technology, China

E-mail: jackzhang@ntu.edu.sg*

Abstract- **Traditional finite set model predictive control (FS-MPC) and a proposed constant switching frequency predictive current control (CSF-PCC) for the three-phase grid-connected inverter are compared in this paper. Traditional FS-MPC is superior in terms of transient performance and flexibility. However, its variable switching frequency makes the total harmonics distortion (THD) of current spread in the whole switching range. With a voltage vector dichotomy solution approach, a constant frequency based CSF-PCC is proposed. According to the comparison with the traditional FS-MPC method, the proposed CSF-PCC not only can achieve the same good dynamic performance with the FS-MPC, but also has much smaller steady-state current ripple, as well as can reduce the current THD and locate the current harmonic in a certain range for easy filter design.**

I. INTRODUCTION

Grid-connected inverters play a key role in many industrial applications such as the integration of renewable energy sources and active-front-end converters for motor drives [1]. Therefore, the control of the grid-connected inverters is very important and has received many investigations [2].

Recent years, with the rapid development of computing, predictive control, which is the algorithm that predict the future behavior of a system to select the proper control action using a certain optimization criterion [3], becomes attractive in control of grid-connected inverters due to several advantages such as good dynamic response and straightforward inclusion of non-linearity and restrictions in the model and control.

Among the predictive control, the model predictive control (MPC) becomes more and more popular, thanks to it is easy to contain nonlinearities and constrains of the system in the control algorithm. According to the development trajectory of the MPC, two main types of the MPC have emerged successively: continuous MPC and finite set MPC (FS-MPC) [4, 5]. Firstly, the continuous MPC appears via being established on linear models and just extension to include constraints. In a way, the continuous MPC is belonging to the generalized predictive control (GPC) and degrades accuracy of the control for nonlinear systems such as power electronics systems. After the continuous MPC, the FS-MPC is proposed to consider only a finite set of possible switching states of the

inverter and solves the cost function for each of them and selects a switching state which minimizes the cost function. The FS-MPC is easy to be implemented when the prediction horizon is only one step and has very good dynamic performance for control of power converters. As a result, the FS-MPC is also often selected as a popular current control method for the existing three phase grid-connected inverters (GCIs) [6, 7].

Though the traditional FS-MPC method has a lot of advantages on the transient performance and flexibility of the three phase GCIs, it needs variable switching frequency which may make the total harmonic distortion (THD) of the current spread in the whole switching range and lead to large current ripple and difficult filter design. Therefore, a constant switching frequency predictive current control (CSF-PCC) for the three phase GCIs is proposed in this paper. The proposed CSF-PCC is actually a kind of constant frequency FS-MPC method, which can combine the basic FS-MPC with the PWM modulator together. It utilizes fast-proper dichotomy based solution algorithm to dynamically select & calculate the optimal voltage vector in the voltage space vector plane in each sampling interval. Compared to the traditional FS-MPC, the proposed CSF-PCC not only can keep the same good dynamic characteristics for the three phase GCI, but also can achieve a better steady state performance, as well as avoid the THD problem of the traditional DS-MPC to convenient GCI output filter design.

In this paper, both the traditional FS-MPC and the proposed CSF-PCC have been applied on the three phase GCI. A performance comparison between the traditional FS-MPC and the proposed CSF-PCC on the current control of the three phase GCI are presented as well. The comparison considers the steady state current ripple, dynamic current response and current THD.

II. MODELING OF THE EVALUATED THREE PHASE GCI

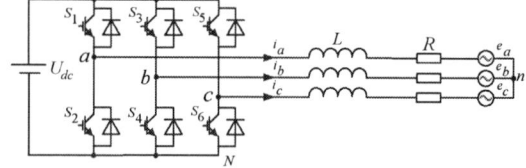

Fig. 1. The topology of the evaluated three-phase GCI.

The topology of a three-phase grid-connected inverter is shown in Fig. 1, which is ready for the analysis of the traditional FS-MPC based current control and CSF-PCC. This GCI is connected with the three-phase grid (e_a, e_b, e_c) through the filter inductances L and resistances R. Therefore, the current dynamic equations can be descripted as:

$$\begin{pmatrix} \frac{di_a}{dt} \\ \frac{di_b}{dt} \\ \frac{di_c}{dt} \end{pmatrix} = -\frac{R}{L}\begin{pmatrix} 1 & 0 & 0 \\ 0 & 1 & 0 \\ 0 & 0 & 1 \end{pmatrix}\begin{pmatrix} i_a \\ i_b \\ i_c \end{pmatrix} + \frac{1}{L}\begin{pmatrix} u_{aN}-e_a-u_{nN} \\ u_{bN}-e_b-u_{nN} \\ u_{cN}-e_c-u_{nN} \end{pmatrix} \quad (1)$$

where i_a, i_b, i_c are the output currents of the grid-connected inverter, u_{aN}, u_{bN}, u_{cN} are the output voltages of the inverter, u_{nN} is the voltage between the neutral point of the grid voltage and the negative point of dc-link, e_a, e_b, e_c are the three-phase grid voltages, respectively. Suppose that the three-phase voltage is balance, that is $e_a+e_b+e_c=0$, the current dynamics can be described in the stationary αβ frame by the vector equation

$$L\frac{di_s}{dt} = \boldsymbol{u}_s - R\boldsymbol{i}_s - \boldsymbol{e}_s \quad (2)$$

where $\boldsymbol{i}_s$ is the current vector, $\boldsymbol{u}_s$ is the voltage generated by the inverter; $\boldsymbol{e}_s$ is the supply line voltage.

The current vector is related to the phase currents by the equation

$$\boldsymbol{i}_s = \frac{2}{3}(i_a + \boldsymbol{a}i_b + \boldsymbol{a}^2 i_c) \quad (3)$$

where $\boldsymbol{a} = e^{j(2\pi/3)}$. The voltage vector is defined in a similar way

$$\boldsymbol{u}_s = \frac{2}{3}(u_{aN} + \boldsymbol{a}u_{bN} + \boldsymbol{a}^2 u_{cN}) \quad (4)$$

The voltage $\boldsymbol{u}_s$ is determined by the switching state of the inverter and the dc-link voltage, and can be also described by the equation

$$\boldsymbol{u}_s = \boldsymbol{S}U_{dc} \quad (5)$$

where U_{dc} is the dc-link voltage and $\boldsymbol{S}$ is the switching state vector of the inverter defined as

$$\boldsymbol{S} = \frac{2}{3}(S_a + \boldsymbol{a}S_b + \boldsymbol{a}^2 S_c) \quad (6)$$

where S_a, S_b, S_c are the switching states of each inverter leg, as shown in Fig. 1, and take the value of 0 if S_x is OFF, or 1 if S_x is ON ($x = a, b, c$).

The predicted current is calculated using the following discrete-time equation

$$\boldsymbol{i}_s(k+1) = \left(1 - \frac{RT_s}{L}\right)\boldsymbol{i}_s(k) + \frac{T_s}{L}[\boldsymbol{u}_s(k) - \boldsymbol{e}_s(k)] \quad (7)$$

which is obtained from the discretization of equation (2) with a sampling time of T_s. This discretization is made by approximating the derivative as the difference over one sampling interval

$$\frac{di_s}{dt} \approx \frac{i_s(k+1) - i_s(k)}{T_s}. \quad (8)$$

III. Review of the Traditional FS-MPC based Current Control

The traditional FS-MPC based current control scheme is shown in the block diagram in Fig. 2. As seen, the whole control strategy works in the stationary αβ frame. So, the measured grid side three phase voltage and current are firstly transformed to the stationary αβ frame using Clarke transformation. The block "FS-MPC" includes the

discrete-time model of equation (7). The value of the current in the next sampling interval $\boldsymbol{i}_s(k+1)$ for each of the different eight voltage vectors is calculated using this block. Therefore, eight current values are predicted one sampling step ahead. Then, on the block "Minimization of cost function", a cost function G, defined in (9), calculates the absolute error between the reference and predicted currents in the next sampling interval.

$$G = |i_\alpha^* - i_\alpha(k+1)| + |i_\beta^* - i_\beta(k+1)| \quad (9)$$

The voltage vector which minimizes the current error is selected and the respective switching state is applied to the GCI.

It is noted that, the traditional FS-MPC based current control strategy presents a variable switching frequency that depends on the sampling frequency and the operating conditions. The performance of this control strategy mainly depends on the sampling time. The higher the sampling frequency, the smaller the absolute current error. However, the switching frequency is increased as well. In industrial application, the switching frequency is limited by the power semiconductors. For this reason, the sampling frequency in the controller is usually selected considering the maximum switching frequency. Hence, the performance of the traditional FS-MPC based current control is usual not preferred compared with some traditional fixed switching frequency control schemes, especially on the control of the GCIs.

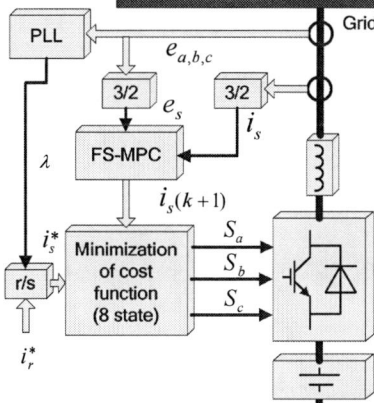

Fig. 2. Traditional FS-MPC based current control block diagram of the evaluated three-phase GCI.

IV. The Proposed Constant Switching Frequency Predictive Current Control (CSF-PCC)

The control scheme of the proposed CSF-PCC for grid-connected inverters is presented in Fig. 3. Firstly, the three-phase grid voltages and currents are measured and transformed to Cartesian coordinate by Clarke transformation. The predicted currents are obtained using the model predictive equations. Then, compared with the reference current, using dichotomy cost function solution algorithm, the optimal voltage vector is selected and sent to the PWM modulator. At last, the inverter receives the switching state and the currents are regulated as expected.

In the case of current control, the cost function is defined as the error between the reference current and the predicted current, and it is expressed as

$$G = |i_\alpha^* - i_\alpha(k+1)| + |i_\beta^* - i_\beta(k+1)| + g_I \quad (9)$$

$$g_I = \begin{cases} 0, & if \ i_s < I_{s.max} \\ \infty, & else \end{cases}$$

where g_I is added for the over current protection.

In the proposed CSF-PCC, the original idea of the cost function optimization algorithm is dichotomy. As shown in Fig. 4, the voltage vector search area is the circular plane with a radius of U_m–the maximum available voltage. In each switching interval, with $U_m/2$ amplitude and $\pi/4$ phase different with each other, eight voltage vectors are selected as the vector candidates at the initialization. At the first step of the iteration, one of the eight vectors are selected as the optimal vector based on the cost function. Next, based on the optimal vector selected in the last step, 18 new vectors, whose amplitude and phase angle are added or subtracted by half of last step optimal vector, are selected to solve the cost function. Hence, the search area decreases to half of the last step. After 14 steps iteration, the optimal vector is selected as the reference vector for PWM modulator.

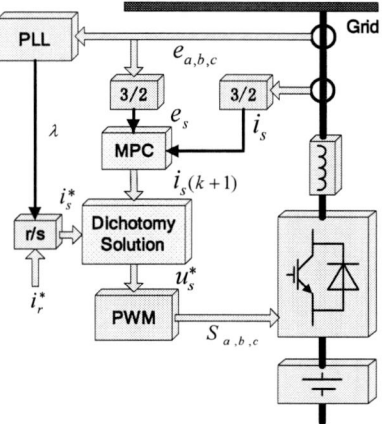

Fig. 3. Block diagram of the proposed CSF-PCC for evaluated three-phase GCI.

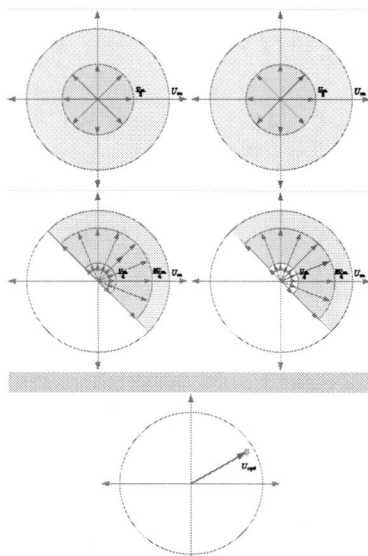

Fig. 4. Dichotomy based optimization algorithm of the cost function in the proposed CSF-PCC.

V. COMPARISON OF THE TRADITIONAL FS-MPC AND THE PROPOSED CSF-PCC WITH THEIR TEST RESULTS

The behaviors of the FS-MPC based current control and the proposed CSF-PCC scheme are evaluated and compared by a 2-kW grid-connected three-phase inverter. For practical issues, the sampling frequency of the traditional FS-MPC is set to 40 kHz to ensure the maximum switching frequency is 20 kHz, which is usual the highest switching frequency for IGBT in industrial applications. As a result, the average switching frequency of the traditional CSF-PCC is around 10 kHz. Hence, for fair comparison, the PWM switching frequency of the proposed CSF-PCC is also set to 10 kHz. The parameters of the evaluated three phase GCI are shown in Table 1.

TABLE I
PARAMETERS OF THE EVALUATED GCI

Parameter	Value
Rated power P_N	2 kW
DC-link voltage U_{dc}	582 V
Grid phase voltage peak e_{peak}	155 V (RMS 110V)
Filter inductance L	10 mH
Grid fundamental frequency	50 Hz
DC-link capacitance C	4700 μF
Line resistance R	0.01 Ω

A. Comparison of the steady state and dynamic performance with test results

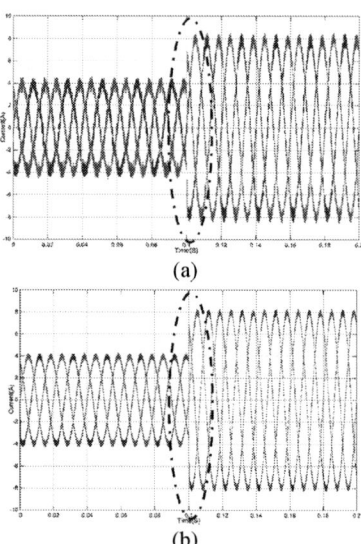

Fig. 5. Comparison of the steady state and dynamic performance: a) the traditional FS-MPC; b) the proposed CSF-PCC.

The steady state and dynamic performance of both control strategies is evaluated by an active current step test. As shown in Fig. 5, the active and reactive current values are set to 4 A and 0 A at the beginning, respectively. At the time of 0.1 s, the active current is given a step to 8 A, and the reactive current is still zero. It can be seen that the steady state current ripple of the proposed CSF-PCC is smaller than the traditional FS-MPC. For the transient state performance, the proposed CSF-PCC is as good as the traditional FS-MPC. For the clearer comparison, Fig. 6 show the detail dynamic process around the time of 0.1 s.

2690

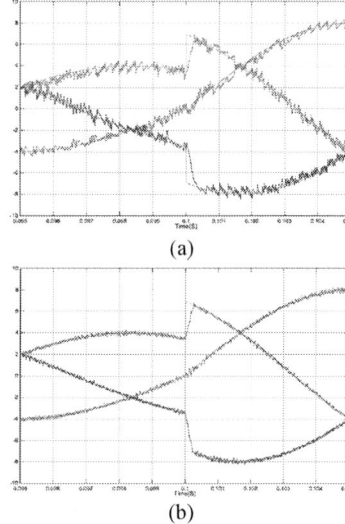

(a)

(b)

Fig. 6. Comparison of the detailed dynamic process of the active current step: a) the traditional FS-MPC; b) the proposed CSF-PCC.

B. Comparison of the GCI current THD with test results

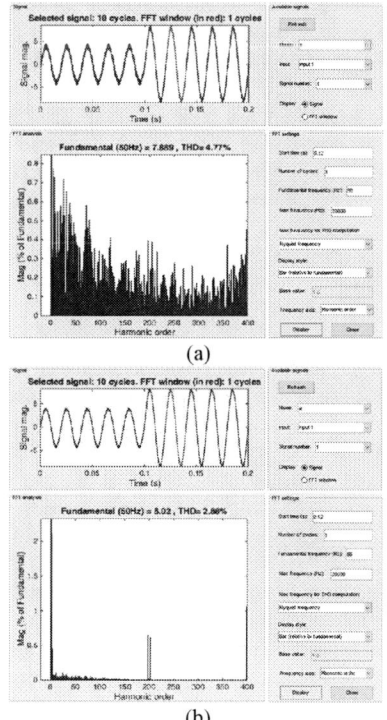

(a)

(b)

Fig. 7. Comparison of the GCI current THD: a) the traditional FS-MPC; b) the proposed CSF-PCC.

Fig. 7 presents the THD of the GCI current for the traditional FS-MPC and the proposed CSF-PCC respectively. THD of both control schemes are 4.77% for the traditional FS-MPC and 2.88% for the proposed CSF-PCC, respectively. In addition, the harmonics of the current controlled by the traditional FS-MPC spread in the whole range of switching frequency, but the proposed CSF-PCC make the harmonic regularly distributed according to the switching frequency. Thus, compared to

the traditional FS-MPC, the harmonic distribution of the current controlled by CSF-PCC is clearly located in a certain range which is easy for the filter design to remove the harmonic.

C. Summary of the comparison

The comparison of the traditional FS-MPC and the proposed CSF-PCC has been summarized in table II, which shows that the proposed CSF-PCC is more advantageous on steady-state current ripple, dynamic performance, current THD, filter design and modulation.

TABLE II
COMPARISON OF THE FS-MPC AND THE PROPOSED CSF-PCC

	Steady state	Dynamic response	Current THD	Filter design	Switching frequency
FS-MPC	*Large ripple*	*Quick*	*Large THD*	*Difficult*	*Variable*
Proposed CSF-PCC	*Small ripple*	*Quick*	*Small THD*	*Easy*	*Constant*

VI. CONCLUSION

The traditional FS-MPC has been reviewed and a constant switching frequency based CSF-PCC has been proposed. A fully comparison between the traditional FS-MPC and the proposed CSF-PCC on the three-phase GCI has also been carried out based on the steady state, dynamic performance and harmonics. According to the comparison, the proposed CSF-PCC not only keeps the same good dynamic performance with the traditional FS-MPC, but also enjoys the following advantages: a) current ripple of the GCI with the CSF-PCC is smaller than that with the traditional FS-MPC; b) current THD of the GCI with the CSF-PCC is smaller than that with the traditional FS-MPC; c) compared to the variable switching based FS-MPC, the proposed CSF-PCC is realized by constant frequency and hence, can locate the harmonic around the fundamental frequency & the switching frequency which is easy to design the filter to remove the harmonics. In addition, a 2-kW three phase GCI is also built whose test results verify the correctness of the above comparison.

REFERENCES

[1] P. Acuna, R. P. Aguilera, A. M. Y. M. Ghias et al., "Cascade-Free Model Predictive Control for Single-Phase Grid-Connected Power Converters," *IEEE Transactions on Industrial Electronics*, vol. 64, no. 1, pp. 285-294, 2017.

[2] T. Dragičević, "Model Predictive Control of Power Converters for Robust and Fast Operation of AC Microgrids," *IEEE Transactions on Power Electronics*, vol. PP, no. 99, pp. 1-1, 2017.

[3] S. Kouro, M. A. Perez, J. Rodriguez et al., "Model Predictive Control: MPC's Role in the Evolution of Power Electronics," *IEEE Industrial Electronics Magazine*, vol. 9, no. 4, pp. 8-21, 2015.

[4] J. Rodriguez, M. P. Kazmierkowski, J. R. Espinoza et al., "State of the Art of Finite Control Set Model Predictive Control in Power Electronics," *IEEE Transactions on Industrial Informatics*, vol. 9, no. 2, pp. 1003-1016, 2013.

[5] R. Kennel, A. Linder, and M. Linke, "Generalized predictive control (GPC)-ready for use in drive applications?." pp. 1839-1844 vol. 4.

[6] S. Kouro, P. Cortes, R. Vargas et al., "Model Predictive Control—A Simple and Powerful Method to Control Power Converters," *Industrial Electronics, IEEE Transactions on*, vol. 56, no. 6, pp. 1826-1838, 2009.

[7] S. Vazquez, J. I. Leon, L. G. Franquelo et al., "Model Predictive Control: A Review of Its Applications in Power Electronics," *IEEE Industrial Electronics Magazine*, vol. 8, no. 1, pp. 16-31, 2014.

The 2018 International Power Electronics Conference

Constant Switching-Frequency Predictive-Current-Control Method with a Dichotomy Solution for the Grid-tied Inverters

ZhiXun Ma[1,2], Xin Zhang[1*], Jingjing Huang[1,3], Zhao Bin[1], Lyu Jing[4]

1. School of Electrical and Electronic Engineering, Nanyang Technological University, Singapore
2. National Maglev Transportation Engineering Research Center, Tongji University, Shanghai
3. Electrical Engineering, Xi'an University of technology, China
4. Department of Electrical Engineering, Shanghai Jiao Tong University, Shanghai, China

E-mail: jackzhang@ntu.edu.sg*

Abstract-**Model predictive current control with constant switching frequency for a three-phase grid-connected inverter is proposed in this work. Based on a voltage vector dichotomy solution approach, the optimal voltage vector is dynamically calculated in the whole space vector circle plane in a very short time. Therefore, the proposed current control has not only fast dynamic behavior but also good steady state performance. Furthermore, there are no complex coordinate transformation and proportional integral (PI) parameters tuning in the control process. The test results verify the correctness and effectiveness of the proposed current control.**

I. INTRODUCTION

Distributed energy systems has received more and more study due to the fast development of renewable energy generator systems, such as wind turbine, solar cells, biomass generator, etc. Grid-connected inverters which are the interfaces between the renewable energy sources and grid are the core of the distributed energy systems. The performance of the grid-connected inverter directly affects the output power quality of the distributed energy systems. Therefore, the current control of the grid-connected inverters is very important and has received many investigations.

For the voltage source three-phase grid-connected inverter, the conventional closed-loop control method is direct current control which usually contains the inner current feedback loop and coordinate transformation [1]. The direct current control approaches can be further classified by grid-voltage control and virtual-flux control, such as direct power control, voltage oriented control, virtual flux oriented control [2], etc. Besides, proportional and integral (PI) controller are always the most popular implementation tool in the above control methods. However, parameter tuning of PI controller is complicated and the performance mainly depends on the experience of parameter tuning [3]. Recent years, for the fast development of computation ability of embedded systems, predictive control becomes attractive in control of grid-connected inverters, thanks to its relative undependability with the system parameters accuracy [4-6].

Predictive control is the algorithm which predict the future action of a plant to choose the certain optimal control behavior based on some optimization method.

Deadbeat control can be seen as one of the earliest predictive control method [7]. Usually, in deadbeat control, the error can be controlled to zero in the following control interval with an optimization method. As well known, deadbeat control has excellent dynamic control performance. However, it has a high sensitivity to noise and modeling errors. Model predictive control (MPC) does not have the sensitive problem and hence is recently reported as a good candidate for control of grid-connected inverters.

It is easy to contain constrains and nonlinearities of the system in the control algorithm. This is the main advantage of MPC. For the application in power electronics systems, MPC can be concluded in two categories: continuous MPC and finite set MPC (FS-MPC) [8-11]. Generalized predictive control (GPC) can be seen as the typical continuous MPC, which is based on linear models and just extended to include constraints. However, the accuracy of GPC would be degraded of the control for nonlinear systems such as power electronics systems. FS-MPC only considers a finite set of inverter switching states. In every optimization process, these finite set switching states are used to calculate the cost function one by one and the one which minimizes the cost-function is set to the control output. It can be seen that, with a prediction horizon of one step, FS-MPC is very easy to be implemented and has very good dynamic performance for control of power converters. However, comparatively large current ripple is presented in steady state since only one active voltage vector is applied in the whole sampling interval. To overcome this major drawback, duty cycle optimization has been introduced in [12], where one zero voltage vector will be added in the sampling instant to reduce the current ripple. The duration of the nonzero vector is analytically derived by some complicated principles. In [13], in order to realize constant switching frequency FS-MPC is jointed with discrete SVM technique. In each sampling interval, the effective state voltage vectors are employed together with virtual state voltage vectors. However, the performance mainly depends on the number of virtual vectors. Besides, the obtained vector is normally not the optimal one in the vector plane. A cost function optimization algorithm is proposed in[14], which is the foundation of this work.

This paper proposes a constant switching frequency predictive current control (CSF-PCC) method for grid-connected inverters. The proposed CSF-PCC utilizes fast-proper dichotomy based solution algorithm to dynamically select & calculate the optimal voltage vector in the voltage space vector plane in each sampling interval. In addition, the proposed CSF-PCC method is actually a kind of constant frequency FS-MPC method, which can combine the basic FS-MPC with the PWM modulator together. This makes the design of output filter easier for grid-connected inverters compared with conventional FS-MPC controlled inverters. As a result, the proposed method for control of the inverter keeps the dynamic characteristics of FS-MPC and presents very good steady state performance at the same time.

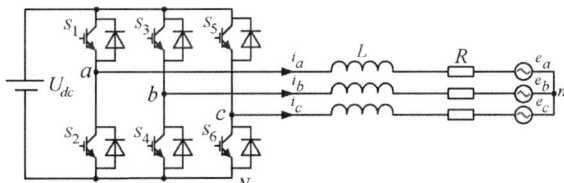

Fig. 1. The topology of a three-phase grid-connected inverter.

II. MODELING OF THE GRID-CONNECTED INVERTER

The topology of a 3ϕ grid-connected inverter is presented at Fig. 1. The inverter is connected with the three-phase supply voltages (e_a, e_b, e_c) through the filter inductances L and resistances R. Therefore, the current dynamic equations can be described as

$$\begin{pmatrix} \frac{di_a}{dt} \\ \frac{di_b}{dt} \\ \frac{di_c}{dt} \end{pmatrix} = -\frac{R}{L}\begin{pmatrix} 1 & 0 & 0 \\ 0 & 1 & 0 \\ 0 & 0 & 1 \end{pmatrix}\begin{pmatrix} i_a \\ i_b \\ i_c \end{pmatrix} + \frac{1}{L}\begin{pmatrix} u_{aN}-e_a-u_{nN} \\ u_{bN}-e_b-u_{nN} \\ u_{cN}-e_c-u_{nN} \end{pmatrix} \quad (1)$$

where i_a, i_b, i_c are the output currents of the grid-connected inverter, u_{aN}, u_{bN}, u_{cN} are the output voltages of the inverter, u_{nN} is the voltage between the neutral point of the grid voltage and the negative point of dc-link, e_a, e_b, e_c are the three-phase grid voltages, respectively. Suppose that the three-phase voltage is balance, that is $e_a+e_b+e_c=0$, the voltage equation in the stationary αβ frame is presented as follows

$$L\frac{di_s}{dt} = u_s - Ri_s - e_s \quad (2)$$

Where u_s is the inverter generated voltage, i_s is the inverter generated current, e_s is the supply line voltage.

The relationship between current vector and the three phase currents can be written as the following equation

$$i_s = \frac{2}{3}(i_a + ai_b + a^2i_c) \quad (3)$$

where $a = e^{j(2\pi/3)}$. The voltage vector is defined in a similar way

$$u_s = \frac{2}{3}(u_{aN} + au_{bN} + a^2u_{cN}) \quad (4)$$

With a certain dc-link voltage, the voltage vector u_s is combined with different switching states of the inverter, and is elaborated as

$$u_s = SU_{dc} \quad (5)$$

where U_{dc} is the dc-link voltage, S is the inverter switching state which can be described as

$$S = \frac{2}{3}(S_a + aS_b + a^2S_c) \quad (6)$$

where S_a, S_b, S_c are the switching states of each inverter leg, and can be defined as 0 if S_x is OFF, or 1 if S_x is ON ($x = a, b, c$), Fig. 1 presents the structure of the inverter.

The predicted current is calculated using the following discrete-time equation

$$i_s(k + 1) = \left(1 - \frac{RT_s}{L}\right)i_s(k) + \frac{T_s}{L}[u_s(k) - e_s(k)] \quad (7)$$

which is obtained from the discretization of equation (2) with a sampling time of T_s. The approximation of the current derivative is described as the difference of the sampling interval

$$\frac{di_s}{dt} \approx \frac{i_s(k+1)-i_s(k)}{T_s}. \quad (8)$$

III. PROPOSED CONSTANT SWITCHING FREQUENCY PREDICTIVE CURRENT CONTROL (CSF-PCC)

The control scheme of CSF-PCC for grid-connected inverters is presented in Fig. 2. Firstly, the three-phase grid voltages and currents are measured and transformed to Cartesian coordinate by Clarke transformation. The predicted currents are obtained using the model predictive equations. Then, compared with the reference current, using dichotomy cost function solution algorithm, the optimal voltage vector is selected and sent to the PWM modulator. At last, the inverter receives the switching state and the currents are regulated as expected.

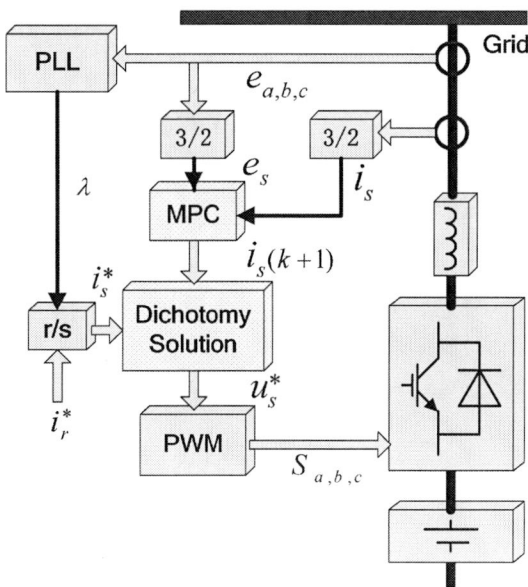

Fig. 2. Block diagram of the proposed CSF-PCC for grid-connected inverter.

Considering current control, the cost function can be defined as the error between the predicted current and the reference current, and it is expressed as

$$G = |i_\alpha^* - i_\alpha(k + 1)| + |i_\beta^* - i_\beta(k + 1)| + g_I \quad (9)$$

2693

$$g_l = \begin{cases} 0, & if \ i_s < I_{s.max} \\ \infty, & else \end{cases}$$

where g_l is added for the over current protection.

The original idea for the optimization algorithm of the cost function is from the dichotomy based computing algorithm[14]. As presented in Fig. 3, all the voltage vectors can be obtained from the inverter are located in a circular voltage vector plane with a radius of U_m which is the maximum available amplitude of voltage. Therefore, the control task is to find an optimal voltage vector from the circuit voltage vector plane in every control interval. In each control interval, with $U_m/2$ amplitude and $\pi/4$ phase different with each other, eight voltage vectors are selected as the vector candidates at the initialization. At the first step of the iteration, according to the optimization of cost-function, one voltage vector can be chosen as the optimal voltage vector from the 8 voltage vectors. Then, according to the former chosen optimal voltage vector, 18 new voltage vectors are generated as the candidates for the optimization of the cost-function. These new voltage vectors are generated as the rule of subtracting/adding their phase & amplitude angle with half of previous step optimal voltage vector. Hence, the search voltage vector plane reduces to half of the previous step. After 14 steps iteration, the optimal voltage vector is found out and sent to the PWM modulator for control of the inverter.

Fig. 4 presents the proposed dichotomy based cost function optimization algorithm calculation flowchart for predictive current control. The input of the optimization algorithm are the currents and grid voltages sample in the stationary frame, current protection value and current control reference. N defines the iteration step for the whole algorithm loop. In this case, it is set to 14, which means the search resolution of voltage vectors is 14 bit. Under the top search loop, there are two parallel inner loops in this algorithm. The calculation resolution of the amplitude (ΔU) and phase angle ($\Delta \theta$) of each voltage vector in the former top loop is defined in these two loops, respectively. In this work, n is the calculation resolution of the phase of the voltage vector in every iteration. It is set to 3. In this work, m is set to 2. The convergence of this algorithm is very fast. In each sampling interval, the optimal voltage space vector can be chosen in a short time. Therefore, this algorithm keeps the fast-dynamic ability and avoid overshoot at the same time.

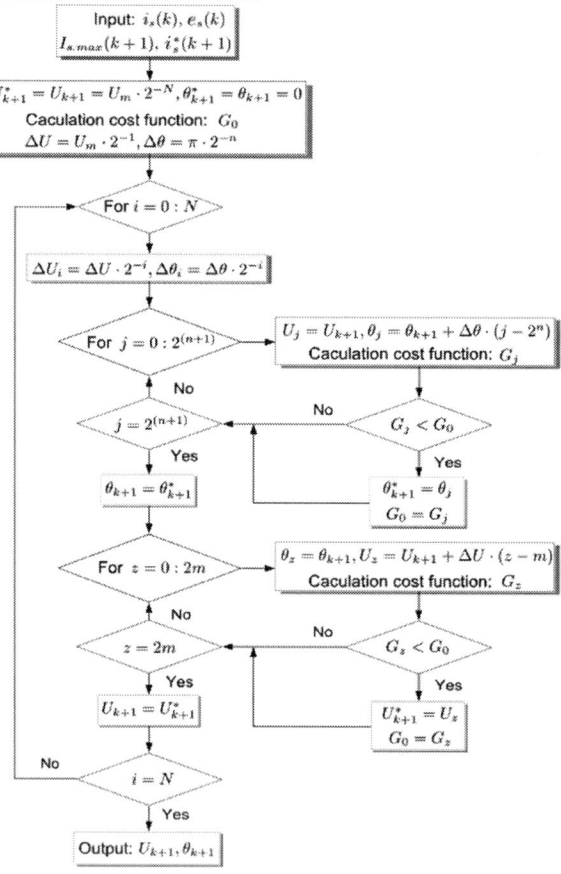

Fig. 4. Flowchart of the dichotomy based solution for constant switching frequency predictive current control.

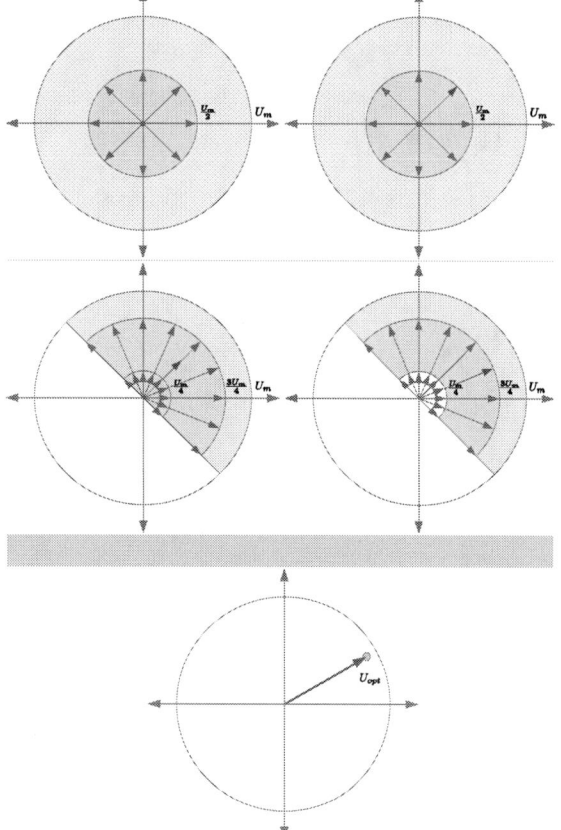

Fig. 3. Optimization algorithm of the cost function.

IV. VALIDATION

The performance of the proposed dichotomy based predictive current control strategy was tested by a 2-kW inverter, and the sampling time for the current control was set to $T_s = 50 \ \mu s$. The PWM switching frequency was 10 kHz. The parameters of the grid-connected inverter system used for MATLAB-based simulation are shown in Table 1.

TABLE I
MAIN PARAMETER VALUES OF THE TESTED GRID CONNECTED INVERTER

Parameters	Values
Nominal power P_N	2 kW
DC-link voltage U_{dc}	300 V
Grid phase voltage peak e_{peak}	155 V (RMS 110V)
Filter inductance L	2 mH
Grid fundamental frequency	50 Hz
DC-link capacitance C	4700 μF
Line resistance R	0.01 Ω

The steady state performance together with dynamic response of the proposed constant switching frequency predictive current control is evaluated by an active current step test. As demonstrated in Fig. 5, the active and reactive current values are set to 20 A and 0 A at the beginning, respectively. At the time of 0.05 S, the active current is given a step to 50 A, and the reactive current is still zero. Fig. 6 shows the detail dynamic process around the time of 0.05 S. From the simulation, we can see that the proposed current control keeps very good steady state performance with a 0.3 ms dynamic process. The grid side voltage and current performance with unity power factor is presented in Fig. 7.

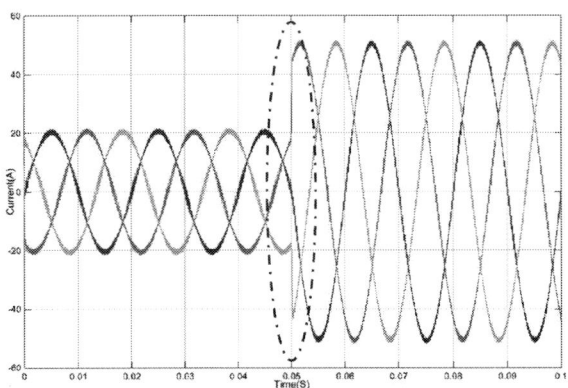

Fig. 5. Steady state and dynamic performance of the constant switching frequency predictive current control.

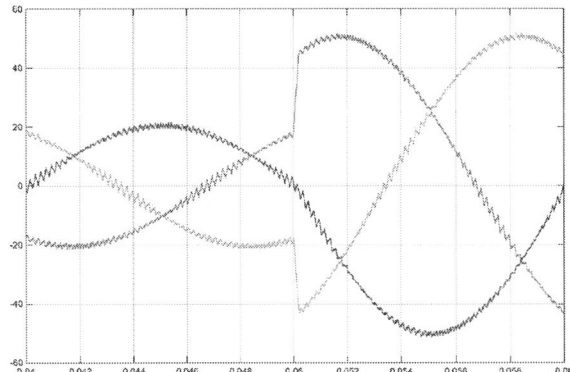

Fig. 6. Dynamic process of the active current step.

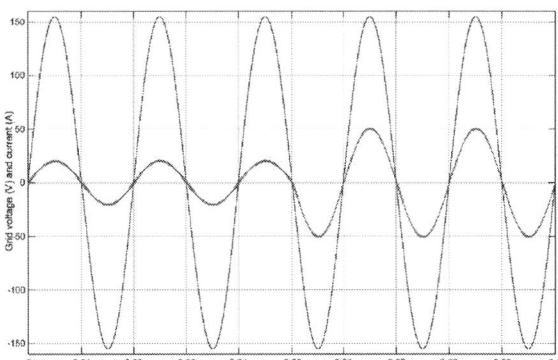

Fig. 7. Grid side voltage and current performance with unity power factor.

V. CONCLUSION

The high computational efficiency and effective dichotomy approach based voltage vector solution algorithm makes MPC have the ability of output constant switching frequency. The proposed control scheme of the grid-connected inverter presents not only a fast-dynamic response but also attractive steady state performance. The dichotomy based solution algorithm is very proper for digital computer system implementation. The proposed control strategy for grid-connected inverters is intuitive and can generate unity power factor based sinusoidal currents. Furthermore, there is not any type PI current controllers in the proposed control method. So it is very easy to be tuned. As a consequence, the proposed MPC scheme for grid-connected inverters is very promising for industrial applications.

REFERENCES

[1] M. P. Kazmierkowski, and L. Malesani, "Current control techniques for three-phase voltage-source PWM converters: a survey," *IEEE Transactions on Industrial Electronics*, vol. 45, no. 5, pp. 691-703, 1998.

[2] M. Malinowski, M. P. Kazmierkowski, and A. M. Trzynadlowski, "A comparative study of control techniques for PWM rectifiers in AC adjustable speed drives," *Power Electronics, IEEE Transactions on*, vol. 18, no. 6, pp. 1390-1396, 2003.

[3] N. B. Lai;, and K. H. Kim, "Robust Control Scheme for Three-phase Grid-connected Inverters with LCL-filter under Unbalanced and Distorted Grid Conditions," *IEEE Transactions on Energy Conversion*, pp. 1-1, 2017.

[4] J. Rodriguez, M. P. Kazmierkowski, J. R. Espinoza *et al.*, "State of the Art of Finite Control Set Model Predictive Control in Power Electronics," *IEEE Transactions on Industrial Informatics*, vol. 9, no. 2, pp. 1003-1016, 2013.

[5] Y. Zhang, and C. Qu, "Model Predictive Direct Power Control of PWM Rectifiers Under Unbalanced Network Conditions," *IEEE Transactions on Industrial Electronics*, vol. 62, no. 7, pp. 4011-4022, 2015.

[6] J. C. Moreno, J. M. E. Huerta, R. G. Gil *et al.*, "A Robust Predictive Current Control for Three-Phase Grid-Connected Inverters," *IEEE Transactions on Industrial Electronics*, vol. 56, no. 6, pp. 1993-2004, 2009.

[7] O. Kukrer, "Deadbeat control of a three-phase inverter with an output LC filter," *IEEE Transactions on Power Electronics*, vol. 11, no. 1, pp. 16-23, 1996.

[8] S. Kouro, P. Cortes, R. Vargas *et al.*, "Model Predictive Control -- A Simple and Powerful Method to Control Power Converters,"

IEEE Transactions on Industrial Electronics, vol. 56, no. 6, pp. 1826-1838, 2009.

[9] J. A. Rohten, J. R. Espinoza, J. A. Muñoz *et al.,* "Model Predictive Control for Power Converters in a Distorted Three-Phase Power Supply," *IEEE Transactions on Industrial Electronics,* vol. 63, no. 9, pp. 5838-5848, 2016.

[10] T. Geyer, "Generalized Model Predictive Direct Torque Control: Long prediction horizons and minimization of switching losses." pp. 6799-6804.

[11] R. Kennel, A. Linder, and M. Linke, "Generalized predictive control (GPC)-ready for use in drive applications?." pp. 1839-1844 vol. 4.

[12] Y. Zhang, W. Xie, Z. Li *et al.,* "Model Predictive Direct Power Control of a PWM Rectifier With Duty Cycle Optimization," *IEEE Transactions on Power Electronics,* vol. 28, no. 11, pp. 5343-5351, 2013.

[13] S. Vazquez, J. I. Leon, L. G. Franquelo *et al.,* "Model Predictive Control with constant switching frequency using a Discrete Space Vector Modulation with virtual state vectors." pp. 1-6.

[14] Z. Ma, S. Saeidi, and R. Kennel, "FPGA Implementation of Model Predictive Control With Constant Switching Frequency for PMSM Drives," *IEEE Transactions on Industrial Informatics,* vol. 10, no. 4, pp. 2055-2063, 2014.

The 2018 International Power Electronics Conference

Observer-based active damping for grid-connected converters with LCL filter

Y. Zhang*, M. G. L. Roes, M. A. M. Hendrix and J. L. Duarte

Department of Electrical Engineering
Eindhoven University of Technology
P.O. Box 513, 5600MB Eindhoven, The Netherlands
*E-mail: ya.zhang@tue.nl

Abstract—**In this paper an add-on control method is proposed to damp oscillations in the LCL filter of a grid-connected converter. An observer is used to estimate the capacitor current, and successively, a conventional active damping technique is applied. Since the capacitor current is estimated from already available measurements, the proposed active damping method does not need an extra current sensor for the capacitor current. Model analysis and simulation validate the effectiveness of the proposed control method.**

Keywords—*observer, damping, converter, LCL filter, control*

I. INTRODUCTION

Pulse width modulation (PWM) converters are widely applied to interface sustainable energy sources, the public grid and local loads [1]–[4]. Because of their good trade-off between attenuating performance and the filter size, LCL filters are widely employed to attenuate PWM switching frequency harmonics [5], [6]. However, the LCL filter also introduces unwanted oscillations at its resonance frequency. Passive damping [7] can reduce these oscillations efficiently but is undesired since it comes at the cost of reduced power efficiency. Active damping [8], [9] is more advantageous because no actual power losses are generated.

Most active damping techniques need the measurements of the capacitor current or voltage, which adds to the system's total hardware cost. In this paper an observer-based active damping technique is presented. It estimates the capacitor current from the LCL filter model and the already existing measurements of the converter output current and the output voltage. Conventional capacitor current based active damping techniques can then be applied to damp the filter's resonance.

II. OBSERVER-BASED ACTIVE DAMPING

A simplified model for a single-phase grid connected converter with LCL filter is shown in Fig. 1, where the proposed add-on active damping technique is indicated in the dashed box. The converter output current and the voltage at the point of common connection (pcc) are used for the capacitor current estimation. Additionally, a high-pass filter (HPF) is applied. This section introduces the algorithm of the proposed damping method.

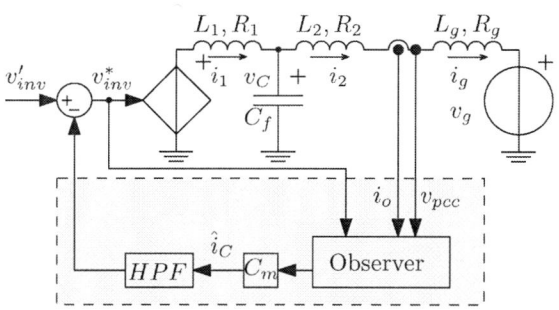

Fig. 1. Proposed observer-based active damping technique applied to a grid-connected converter with LCL filter

A. Modelling

The LCL filter dynamics in Fig. 1 are described by three differential equations, given by

$$L_1 \frac{\mathrm{d}}{\mathrm{d}t} i_1(t) = -R_1 i_1(t) - v_C(t) + v_{inv}^*(t)$$
$$C_f \frac{\mathrm{d}}{\mathrm{d}t} v_C = i_1(t) - i_2(t)$$
$$L_2 \frac{\mathrm{d}}{\mathrm{d}t} i_2(t) = v_C(t) - R_2 i_2(t) - v_{pcc}(t), \quad (1)$$

where i_1, v_c and i_2 are the current through L_1, capacitor voltage and the output current, respectively. A state-space model [10] for the plant in (1) is

$$\dot{x}(t) = Ax(t) + Bu(t) + E\omega(t)$$
$$y(t) = Cx(t), \quad (2)$$

with

$$A = \begin{bmatrix} \frac{-R_1}{L_1} & \frac{-1}{L_1} & 0 \\ \frac{1}{C_f} & 0 & \frac{-1}{C_f} \\ 0 & \frac{1}{L_2} & \frac{-R_2}{L_2} \end{bmatrix}, B = \begin{bmatrix} \frac{1}{L_1} \\ 0 \\ 0 \end{bmatrix}, E = \begin{bmatrix} 0 \\ 0 \\ \frac{-1}{L_2} \end{bmatrix},$$
$$C = \begin{bmatrix} 0 & 0 & 1 \end{bmatrix}, \quad (3)$$

and the state variables, input, disturbance, and output respectively represent

$$x(t) = \begin{bmatrix} i_1(t) & v_C(t) & i_2(t) \end{bmatrix}^T,$$
$$u(t) = v_{inv}^*(t), \ \omega(t) = v_{pcc}(t), \ y(t) = i_o(t). \quad (4)$$

2697

The *zero order hold* method is applied to obtain the discrete-time model for the plant in (2). Accordingly, we have

$$x[k+1] = A_d x[k] + B_d u[k] + E_d \omega[k]$$
$$y[k] = C_d x[k], \tag{5}$$

where $x[k]$, $u[k]$, $\omega[k]$ and $y[k]$ represent the sampled data state variables, input, disturbance, and output in (2). A_d, B_d, E_d and C_d are the resulting matrices from discretization of A, B, E and C.

B. Observer: estimation of capacitor current

The observer in this paper uses the plant model for state estimation [11]. It is described by

$$\hat{x}[k+1] = A_d \hat{x}[k] + B_d u[k] + E_d \omega[k]$$
$$+ L_{ob}(y[k] - C_d \hat{x}[k]), \tag{6}$$

where $\hat{x}[k]$ is the estimate of $x[k]$ and L_{ob} is the observation gain vector/matrix. Since in Fig. 1 $i_C = i_1 - i_2$ holds according to *Kirchhoff's current law*, the estimate of the capacitor current is obtained from

$$\hat{i}_C[k] = C_m \hat{x}[k], \tag{7}$$

with $C_m = \begin{bmatrix} 1 & 0 & -1 \end{bmatrix}$.

C. High-pass filter for active damping

Placing a resistor in series with the capacitor effectively yields passive damping; the damping performance could be theoretically maintained by using a high-pass filter feeding back the capacitor current to the input voltage [12]. However, to resolve delays due to PWM, data sampling and computation, as addressed in [13], a modification is made to the high-pass filter. The filter employed in this paper is

$$K_{hpf}(s) = K_v K(s), \tag{8}$$

where K_v is a gain for tuning and $K(s)$ is from [12], described by

$$K(s) = \frac{R_v C_f L_2 s^2}{C_f L_2 s^2 + C_f R_v s + 1}. \tag{9}$$

The parameters C_f and L_2 are the capacitance and inductance indicated in Fig. 1, and R_v is a *virtual resistor* [12] used to achieve the desired passive damping performance.

III. VALIDATION OF THE CONVERTER MODELLING

Since the estimation of the capacitor current is based on the derived model of the grid-connected converter with LCL filter in (5), it is of vital importance to have an effective model. The outputs of the derived model are compared to the measurements, following the schematics shown in Fig. 2, where the derived plant model in (5) is implemented as indicated in the red dashed box. The circuit and the model are compared under two conditions: weak grid condition and stiff grid condition. In both scenarios, a 10% third harmonic (with respect to the fundamental) is added to the grid voltage at 0.107s to

trigger transient behaviour. The grid voltage waveform is shown in Fig. 3. The parameters of the LCL filter in these tests are listed in Table I and the sampling rate is 20 kHz.

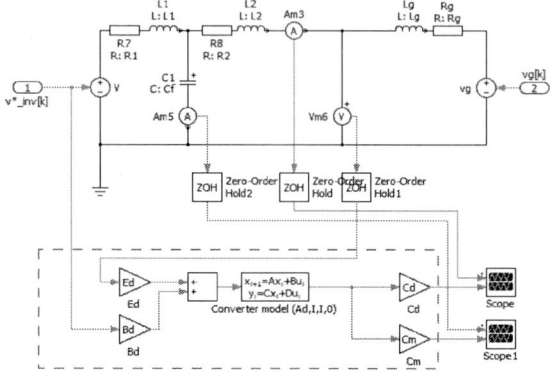

Fig. 2. Schematic in PLECS used to check the derived state space model.

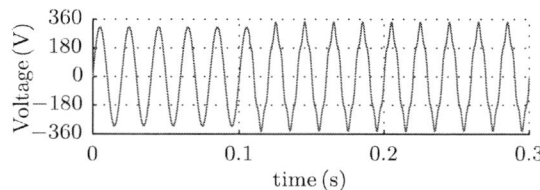

Fig. 3. Waveform of the grid voltage v_g applied to trigger transient behaviour.

A. Weak grid condition

A weak grid condition is created by setting $L_g = 10.44\,\text{mH}$. The capacitor current waveforms from the circuit and from the state space model are shown in Fig. 4. It can be seen that the oscillation of the LCL filter clearly presents in the response of the derived model under the weak grid condition.

B. Stiff grid condition

A stiff grid condition is created by setting $L_g = 0.1\,\text{mH}$. The capacitor current waveforms from the circuit and from the state space model in (5) are shown in Fig. 5.

The difference between the current waveforms from the model (the plant model in (5)) and the circuit (in Fig. 2) can be explained by the delays involved by using zero-order-hold. In an extreme case when the sampling rate is infinite, the derived model should be equivalent to the circuit. In the derived model, the information of the PCC voltage is delayed due to the zero-order-hold block applied (Fig. 2).

2698

The 2018 International Power Electronics Conference

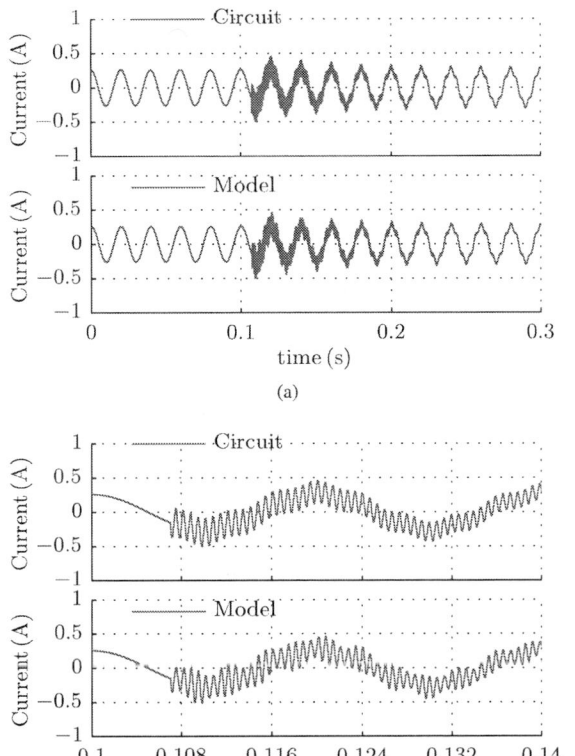

Fig. 4. Capacitor current waveforms from the circuit and from the state space model in (5) under weak grid condition: (a) zoomed-out view ; (b) zoomed-in view.

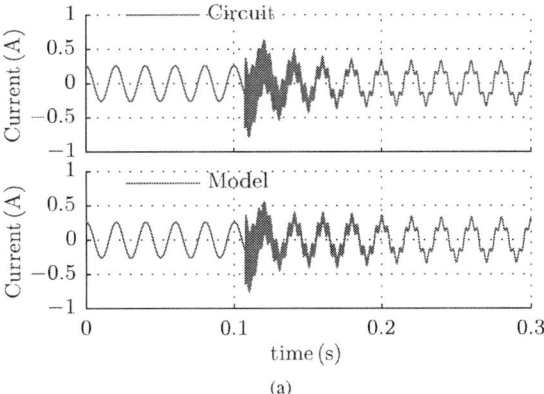

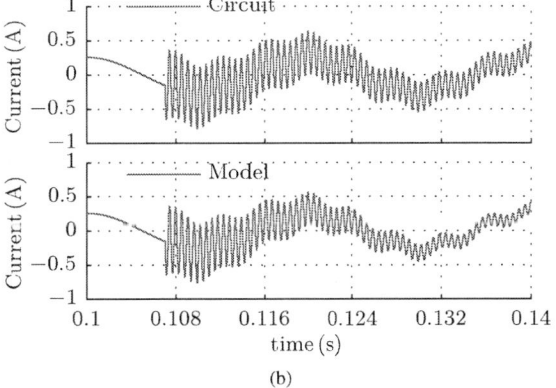

Fig. 5. Capacitor current waveforms from the circuit and from the state space model under stiff grid condition: (a) zoomed-out view ; (b) zoomed-in view.

IV. ANALYSIS OF THE ACTIVE DAMPING TECHNIQUE

In theory the high-pass filter feedback in (8) yields stable resonance damping even without tuning ($K_v = 1$). However, instability may occur due to the delays introduced by the digital implementation [14]. In light of this, stability analysis is performed in this section.

A. Open-loop damping gain

Synthesizing the plant model in (5) and the observer in (6), the state space model from the voltage v_{inv}^* and the pcc voltage v_{pcc} to the estimated capacitor current is

$$\begin{bmatrix} x[k+1] \\ e[k+1] \end{bmatrix} = \underbrace{\begin{bmatrix} A_d & O \\ O & A_d - L_{ob}C_d \end{bmatrix}}_{\bar{A}_d} \begin{bmatrix} x[k] \\ e[k] \end{bmatrix}$$
$$+ \underbrace{\begin{bmatrix} B_d \\ O \end{bmatrix}}_{\bar{B}_d} u[k] + \underbrace{\begin{bmatrix} E_d \\ O \end{bmatrix}}_{\bar{E}_d} \omega[k]$$

$$\hat{i}_C[k] = \underbrace{\begin{bmatrix} C_m & -C_m \end{bmatrix}}_{\bar{C}_d} \begin{bmatrix} x[k] \\ e[k] \end{bmatrix}, \qquad (10)$$

where $e[k]$ is the state estimation error, denoted as $e[k] = x[k] - \hat{x}[k]$. It can be seen from (10) that the estimation

error is uncontrollable; however, it is stabilizable since the eigenvalue of $A_d - L_{ob}C_d$ can be relocated with proper design of the observation gain vector/matrix L_{ob}. The open loop transfer function from the inverter output voltage to the estimated capacitor current is

$$G(z) = \frac{\hat{i}_C(z)}{v_{inv}^*(z)} = \bar{C}_d(zI - \bar{A}_d)\bar{B}_d. \qquad (11)$$

It is suggested that during stability analysis $1.5T_s$ delay should be accounted for PWM, data sampling and computation [14], where T_s is the sampling period. Since fractional-order delay adds to the complexity of stability analysis in the z-domain, this section approximates the worst case delay by 2 sampling periods. Therefore, the open loop gain of the damping path is

$$T(z) = K_{hpf}(z)G(z)z^{-2}, \qquad (12)$$

where $K_{hpf}(z)$ is the discrete-time implementation of $K_{hpf}(s)$ in (8). Combination of (8) and (12) yields

$$T(z) = K_v K(z)G(z)z^{-2}. \qquad (13)$$

It is worth mentioning that two parameters can be tuned in the design of the observer-based active damping: K_v and R_v. The discussion of these two parameters follows in the next sections.

B. Nyquist plot of $T(z)$

Table I lists the parameters of the considered LCL filter. The observer gain vector/matrix L_{ob} is designed in such a way that it can track the plant state variables in $0.2T_g$, where T_g is the grid period. The roots of Bessel Polynomials are adopted as locations of the eigenvalues of $A_d - L_{ob}C_d$. The sampling frequency is chosen as $f_s = 20\,\text{kHz}$. Accordingly, the Nyquist plots of $T(z)$ with different values of R_v are shown in Fig. 6; the tuning gain is set as $K_v = 0.25$ for stability.

As can be seen from Fig. 6, increasing R_v decreases the gain margin; nevertheless, the phase margin increases. Moreover, setting the tuning gain $K_v = 1$ can make the system unstable, referring to the plot with $R_v = 100$ in Fig. 6 (where $K_v = 0.25$ is used).

TABLE I. LCL FILTER PARAMETERS

Description	Symbol	Value
Converter side inductor	L_1, R_1	$5.22\,\text{mH}, 0.2\,\Omega$
Capacitor	C_f	$2.82\,\mu\text{F}$
Grid side inductor	L_2, R_2	$5.22\,\text{mH}, 0.2\,\Omega$

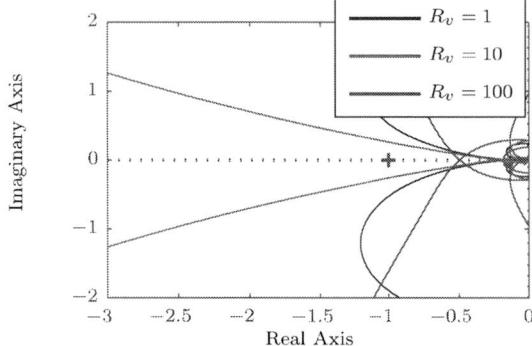

Fig. 6. Nyquist plot of $T(z)$, with different values of R_v, and $K_v = 0.25$.

C. Closed-loop damped plant model

The closed-loop transfer function from v'_{inv} to the converter output current i_o (with reference to Fig. 1) is found to be

$$\bar{P} = \frac{i_o(z)}{v'_{inv}(z)} = \frac{P(z)}{1 + T(z)}, \tag{14}$$

where $P(z)$ is the undamped plant transfer function, described by $P(z) = C_d(zI - A_d)B_d$ according to the undamped plant model in (5).

D. Bode plot of $\bar{P}(z)$

Fig. 7 shows the Bode plots of the damped plant with different values of R_v, with the tuning factor $K_v = 0.25$. It can be seen that increasing the value of R_v yields more damping of the LCL filter resonance.

Fig. 8 shows the Bode plots of the damped plant with different values of K_v, with $R_v = 100$. It can be seen that increasing the value of K_v gives more damping to the LCL filter resonance as well; however, it may make the plant unstable, as depicted in Fig. 6.

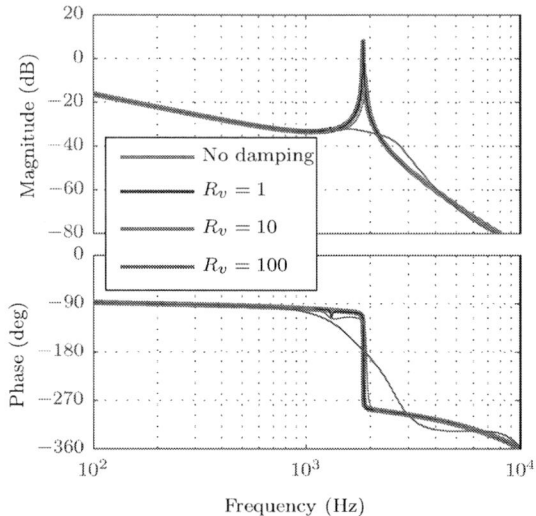

Fig. 7. Bode plot of $\bar{P}(z)$ to show the benefit of active damping, with different values of R_v, and $K_v = 0.25$.

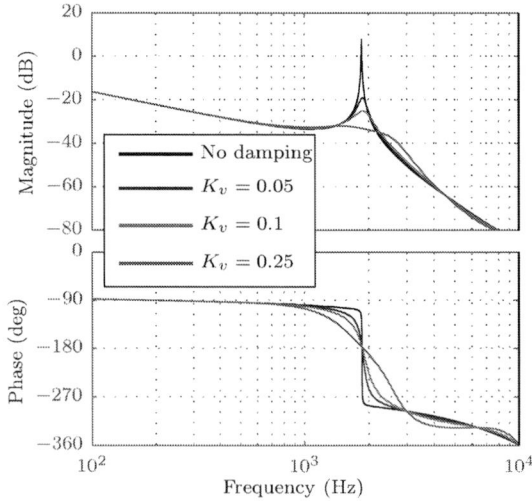

Fig. 8. Bode plot of $\bar{P}(z)$ to show the benefit of active damping, with different values of K_v, and $R_v = 100$.

For a good trade-off between stability margins and LCL filter resonance attenuation effect, $R_v = 100$ and $K_v = 0.25$ are chosen for the active damping design, as summarized in Table II.

V. PR CURRENT CONTROLLER

In this section, a proportional-resonance (PR) current controller is considered to regulate the converter output current. The transfer function of the PR control in the s-domain is

$$G_{pr}(s) = K_p + K_r \frac{s}{s^2 + \omega_o^2}, \tag{15}$$

where K_p is the proportional gain, K_r the resonance gain, and ω_0 is the grid angular frequency. The parameters for the PR control are listed in Table II. The PR is discretized using *forward Euler* and *Backward Euler* methods as in [15] to avoid algebraic loops. The resulting PR control in discrete form is denoted as $G_{pr}(z)$. Ignoring the dynamics caused by the PCC voltage change (with reference to Fig. 1), the closed loop transfer function of the PR current controlled converter is

$$G_{cl}(z) = \frac{\bar{P}(z)G_{pr}(z)z^{-2}}{1 + \bar{P}(z)G_{pr}(z)z^{-2}}, \tag{16}$$

where z^{-2} is introduced to account for delays ($2T_s$). Fig. 9 shows the Nyquist plots of $\bar{P}(z)G_{pr}(z)z^{-2}$ with and without active damping. Consistent with the observation in Fig. 7 and Fig. 8, applying the proposed active damping technique helps to stabilize the current control feedback system.

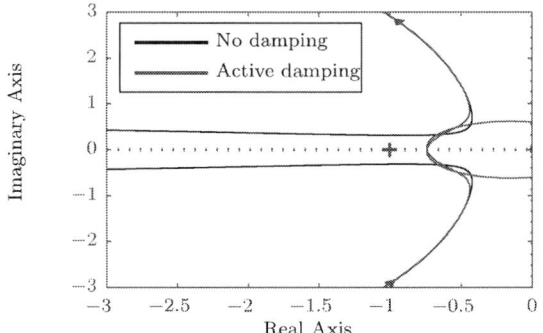

Fig. 9. Nyquist plot of $\bar{P}(z)G_{pr}(z)z^{-2}$ with and without active damping

TABLE II. PARAMETERS OF THE CONTROLLER

Description	Symbol	Value
PR filter	K_p	30
	K_r	6000
	ω_0	$2\pi f_g$
Active damping	K_v	0.25
	R_v	100
Sampling	f_s	20 kHz

VI. SIMULATION RESULTS

Two sets of simulation tests were performed in *Matlab/Simulink/Plecs* to investigate the aforementioned stability discussion. A full bridge converter is used with a 400 V DC input source. A unipolar pulse-width-modulation scheme is applied to the full bridge converter. The switching frequency is $f_{sw} = 20\,\text{kHz}$ and the

desired peak output current is set as 4 A. The LCL filter parameters in simulation are shown in Table I.

A. Weak grid condition

A weak grid condition is created by setting $L_g = 10.44\,\text{mH}$, with reference to the circuit in Fig. 1. Figure 10, Figure 11 and Figure 12 show the simulated transient waveforms of the converter output current i_o under the weak grid condition. In Fig. 10 the active damping loop is disabled at 0.1 s by setting K_v to zero. In Fig. 11 at 0.1 s the parameter K_v is set from 0.25 to 1, and in Fig. 12 at 0.1 s the parameter K_v is set from 0.25 to 2.

It can be seen that the system becomes unstable when the observer-based active damping is disabled in Fig. 10, which is in agreement with the analysis of Fig. 9. Moreover, in accordance with the results in Fig. 6, in Fig. 11 and Fig. 12 it is shown that a large K_v can also make the system unstable.

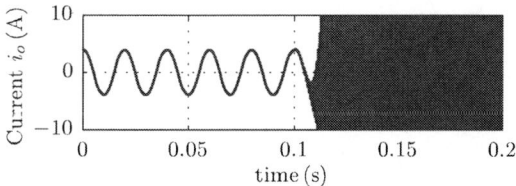

Fig. 10. Simulated transient response of i_o at the weak grid condition. The active damping is disabled at 0.1 s by setting K_v from 0.25 to zero.

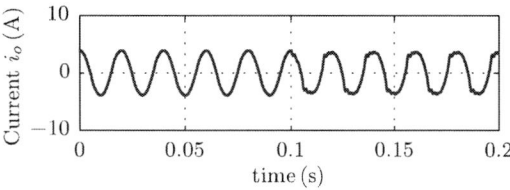

Fig. 11. Simulated transient response of i_o at the weak grid condition. At 0.1 s the parameter K_v is set from 0.25 to 1.

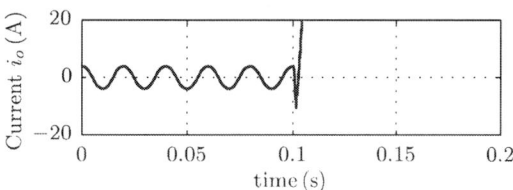

Fig. 12. Simulated transient response of i_o at the weak grid condition. At 0.1 s the parameter K_v is set from 0.25 to 2.

B. Stiff grid condition

A stiff grid condition is created by setting $L_g = 0.1\,\text{mH}$, with reference to the circuit in Fig. 1. Figure 13, Figure 14 and Figure 15 show the simulated transient waveforms of the converter output current i_o under the stiff grid condition. In Fig. 10 the active damping loop is disabled at 0.1 s by setting K_v to zero. In Fig. 14 at 0.1 s

The 2018 International Power Electronics Conference

the parameter K_v is set from 0.25 to 1, and in Fig. 15 at 0.1 s the parameter K_v is set from 0.25 to 2.

Consistent with under the weak grid condition, the system becomes unstable when the active damping is disabled. This is as expected since no grid parameter is used for the observer design. Moreover, an un-decaying harmonic oscillation present in the converter output current waveform when K_v increases to 1 (in Fig. 14) and in a worse case the system becomes unstable (in Fig. 15), which is also consistent with under the weak grid condition.

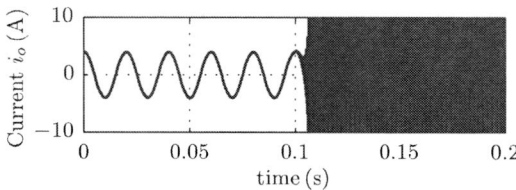

Fig. 13. Simulated transient response of i_o at the stiff grid condition. The active damping is disabled at 0.1 s by setting K_v from 0.25 to zero.

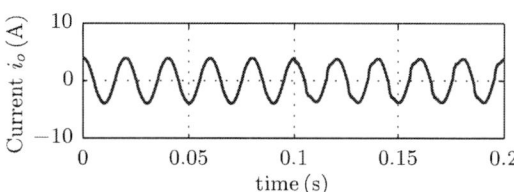

Fig. 14. Simulated transient response of i_o at the stiff grid condition. At 0.1 s the parameter K_v is set from 0.25 to 1.

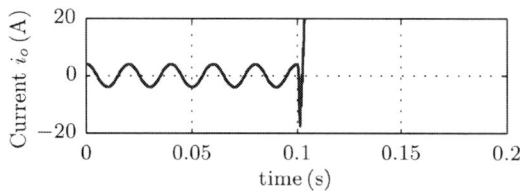

Fig. 15. Simulated transient response of i_o at the stiff grid condition. At 0.1 s the parameter K_v is set from 0.25 to 2.

VII. CONCLUSION

In this paper an add-on control scheme is proposed to damp the resonance of the LCL filter. It estimates the capacitor current from the measurements of the grid-side converter output current, the local voltage, and the LCL filter parameters. Accordingly, a conventional capacitor-current-based active damping technique can be applied. The advantage of this scheme is that there is no need to measure the capacitor current, making it an add-on feature for grid-connected converters since the measurement of the output current and local voltage already exist. A model is derived to define the stability region of the proposed active damping technique.

In order to validate the damping benefit provided by the proposed control technique, a proportional-resonance

current controller is added to the grid-connected converter. The overall system is shown to be stable with the damping technique and unstable without. On top of that, simulation results show that the active damping technique is applicable for both weak and stiff grid operation.

REFERENCES

[1] F. Wang, J. L. Duarte, M. A. M. Hendrix, and P. F. Ribeiro, "Modeling and analysis of grid harmonic distortion impact of aggregated DG inverters," *IEEE Transactions on Power Electronics*, vol. 26, no. 3, pp. 786–797, Mar. 2011.

[2] Y. Yang, K. Zhou, H. Wang, F. Blaabjerg, D. Wang, and B. Zhang, "Frequency adaptive selective harmonic control for grid-connected inverters," *IEEE Transactions on Power Electronics*, vol. 30, no. 7, pp. 3912–3924, Jul. 2015.

[3] M. Castilla, J. Miret, J. Matas, L. G. d. Vicuna, and J. M. Guerrero, "Linear current control scheme with series resonant harmonic compensator for single-phase grid-connected photovoltaic inverters," *IEEE Transactions on Industrial Electronics*, vol. 55, no. 7, pp. 2724–2733, Jul. 2008.

[4] P. M. d. Almeida, J. L. Duarte, P. F. Ribeiro, and P. G. Barbosa, "Repetitive controller for improving grid-connected photovoltaic systems," *IET Power Electronics*, vol. 7, no. 6, pp. 1466–1474, Jun. 2014.

[5] A. Reznik, M. G. Simes, A. Al-Durra, and S. M. Muyeen, "Filter design and performance analysis for grid-interconnected systems," *IEEE Transactions on Industry Applications*, vol. 50, no. 2, pp. 1225–1232, Mar. 2014.

[6] J. Scoltock, T. Geyer, and U. Madawala, "Model predictive direct current control for a grid-connected converter: LCL-filter versus L-filter," in *IEEE International Conference on Industrial Technology (ICIT)*, Feb. 2013, pp. 576–581.

[7] K. A. E. W. Hamza, H. Linda, and L. Cherif, "LCL filter design with passive damping for photovoltaic grid connected systems," in *6th International Renewable Energy Congress (IREC)*, Mar. 2015, pp. 1–4.

[8] P. A. Dahono, "A control method to damp oscillation in the input LC filter," in *IEEE 33rd Annual IEEE Power Electronics Specialists Conference. Proceedings*, vol. 4, 2002, pp. 1630–1635.

[9] J. Dannehl, F. W. Fuchs, S. Hansen, and P. B. Thogersen, "Investigation of active damping approaches for PI-based current control of grid-connected pulse width modulation converters with LCL filters," *IEEE Transactions on Industry Applications*, vol. 46, no. 4, pp. 1509–1517, Jul. 2010.

[10] R. J. Vaccaro, "Tracking Systems," in *Digital Control: A State-Space Approach*. Mcgraw-Hill College, Jan. 1995, pp. 342–348.

[11] R. J. Vaccaro, "Observers," in *Digital Control: A State-Space Approach*. Mcgraw-Hill College, Jan. 1995, pp. 265–272.

[12] H.-c. Chen and P.-t. Cheng, "An active damping technique for multiple grid-connected converters," in *IEEE 3rd International Future Energy Electronics Conference and ECCE Asia (IFEEC - ECCE Asia)*, Jun. 2017, pp. 561–566.

[13] J. Wang, J. D. Yan, L. Jiang, and J. Zou, "Delay-dependent stability of single-loop controlled grid-connected inverters with LCL filters," *IEEE Transactions on Power Electronics*, vol. 31, no. 1, pp. 743–757, Jan. 2016.

[14] W. Yao, Y. Yang, X. Zhang, F. Blaabjerg, and P. C. Loh, "Design and analysis of robust active damping for LCL filters using digital notch filters," *IEEE Transactions on Power Electronics*, vol. 32, no. 3, pp. 2360–2375, Mar. 2017.

[15] R. Teodorescu, F. Blaabjerg, U. Borup, and M. Liserre, "A new control structure for grid-connected LCL PV inverters with zero steady-state error and selective harmonic compensation," in *Nineteenth Annual IEEE Applied Power Electronics Conference and Exposition (APEC)*, vol. 1, 2004, pp. 580–586 Vol.1.

Conduction Loss Analysis and Optimization Design of Full Bridge LLC Resonant Converter

Yugang Yang, Lifei Zhang and Tianshu Ma
Liaoning Technical University, Huludao, China
yangyugang21@126.com

Track number:0672

Abstract-The operating frequency of LLC resonant converter is usually lower than the resonant frequency to achieve zero current switching of secondary rectifier (ZCS), especially in the heavy load condition, the operating frequency is far less than the resonant frequency. Considering the influence of dead time, the resonant tank current formula derived from the resonant frequency point is different from the actual operating state. To solve this problem, this paper presents an improved current formula of resonant tank, and puts forward a design method of the resonant parameters. Finally, a prototype of 48V input and 1kW/ 400V output is built, and the correctness of the theoretical analysis is verified by simulation and experiment.

I. INTRODUCTION

To address the energy crisis, countries are more investing in electric vehicles, but the output voltage range of the power battery pack is wide. The vehicle DC/DC converter has a high requirement for wide input voltage range [1]. Due to the high frequency, high efficiency and high power density, full bridge LLC resonant converter is widely used in many applications, especially in the field of wide input voltage range. Regulating the frequency in a full load range to realize zero voltage switching (ZVS) of H-bridge MOSFETS and zero current switching (ZCS) of diode bridge rectifier greatly reduces the switching loss [2]. But its traditional loss model is established at the resonant frequency point, which does not match the running state of the circuit topology. So there is a certain error in the loss calculation.

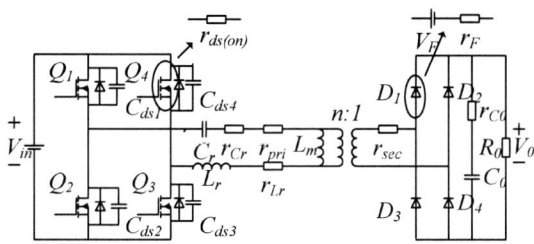

Fig.1 Conduction loss model of LLC resonant converter

Fig.1 shows the conduction loss model of the full bridge LLC resonant converter. In practical engineering applications, devices are not ideal and will bring all kinds of loss. It is necessary to describe the formula of the

resonant tank current and diode bridge rectifier current accurately, because the power loss of LLC resonant converter [3]-[4] is mainly related to both. [5]-[6] give a precise formula of the diode bridge rectifier current. However, the traditional formula [7] is still used in the resonant tank current. The optimization of resonant tank parameter aimed at reducing resonant tank current to reduce the conduction loss and improve the transform efficiency.

This paper points out the traditional formula of resonant tank current is described with low accuracy and puts forward a improved formula. It has a high-accuracy in full load range. According to the guidance of the new formula, a design method of resonant tank parameters is proposed. Finally, a prototype of 48V input and 1kW/400V output is built, and the correctness of the theoretical analysis is verified by simulation and experiment.

II. THE TRADITIONAL FORMULAS OF CURRENT

In order to analyze the conduction loss, the key waveform of each device should be obtained at first. According to the waveform to describe the formula, Fig. 2 shows the main waveform of full bridge LLC resonant converter. The traditional loss calculation formula is received at resonant frequency point, but LLC resonant converter is not only working in resonant frequency point. Considering the influence of dead time, the traditional calculation formulas will cause error. The greater the dead time, the greater the error is.

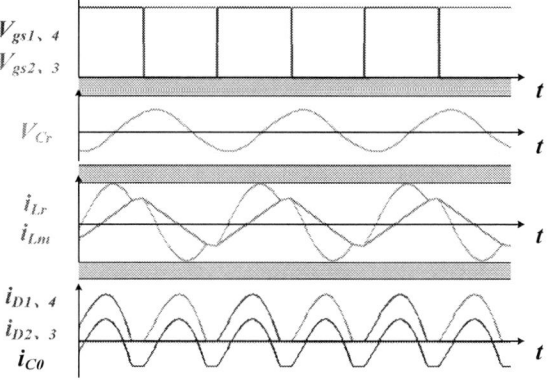

Fig.2 Key Waveform of LLC resonant converter

Project Supported by National Natural Science Foundation of China-Shanxi Coal Based Low Carbon Joint Fund (U1510128)

A. The current of diode bridge rectifier

When the operating frequency deviates from the resonant frequency, the current formula of diode bridge rectifier is as follows:

$$i_{D_{14}}(t)=\begin{cases} n\times I_{pri}\times \sin(\omega_r t), 0\le t<\dfrac{T_r}{2}\\ 0, \dfrac{T_r}{2}\le t<T_s \end{cases} \quad (1)$$

$$i_{D_{23}}(t)=\begin{cases} 0, 0\le t<\dfrac{T_s}{2}\\ n\times I_{pri}\times \sin(\omega_r t), \dfrac{T_s}{2}\le t<\dfrac{T_s+T_r}{2}\\ 0, \dfrac{T_s+T_r}{2}\le t<T_s \end{cases} \quad (2)$$

Where I_{pri} is the max current of the transformer (primary side), ω_r is working angular frequency, T_s is working period, T_r is resonant period, n is turns ratio.

$$\frac{2}{T_s}\times \int_0^{\frac{T_s}{2}} n\times I_{pri}\sin(\omega_s t)\times dt=I_0,\ I_{pri}=\frac{\pi I_0}{2n\left(\dfrac{T_r}{T_s}\right)} \quad (3)$$

Where I_0 is load current.

B. The current of resonant tank

The magnetizing inductance current is an ideal triangle wave in traditional calculation method, the inductance current is expressed as

$$i_{L_m}(t)=\begin{cases} \dfrac{nV_0}{L_m}(t-\dfrac{T_s}{4}), 0\le t<\dfrac{T_s}{2}\\ \dfrac{nV_0}{L_m}\dfrac{T_s}{4}-\dfrac{nV_0}{L_m}(t-\dfrac{T_s}{2}), \dfrac{T_s}{2}\le t<T_s \end{cases} \quad (4)$$

Then the RMS value of resonant tank current is obtained

$$I_{L_r rms}=\frac{1}{8}\frac{V_0}{nR_{eq}}\sqrt{(\frac{\sqrt{2}n^2 R_{eq}}{f_s L_m})^2+\frac{\pi^2}{8}} \quad (5)$$

Where V_0 is rated load voltage, R_{eq} is equivalent resistance.

It is known from (5) that the magnetizing inductance L_m decreases, resulting in I_{Lrrms} increasing and efficiency η decreasing. The RMS value of the resonant tank current affects the optimal design of the parameters, then a more accurate formula of resonant tank current is put forward.

The working period is not equal to the resonant period, plus the dead time, therefore, the error is larger. There are also three irrational assumptions:

(a) The magnetizing inductance current is not an ideal triangular wave in the full load range

(b) After the half resonant period, the working period is larger than the resonant period. During this time, the resonant tank current can be approximated to a straight line with a fixed slope

(c)The impact of dead time was not considered.

In view of the above, the current flowing through the resonant tank can be improved.

When $i_{Lr}(t)=i_{Lm}(t)$, at t_0, there are $i_{Lr}(t_0)=i_{Lm}(t_0)=i_{Lm}(t_0)$; t_1 is the start of the dead time and t_2 is the end,there are $i_{Lr}(t_1)=i_{Lm}(t_1)=i_{Lm}(t_1)$ and $t_d=t_2-t_1$ as shown in Fig.3.

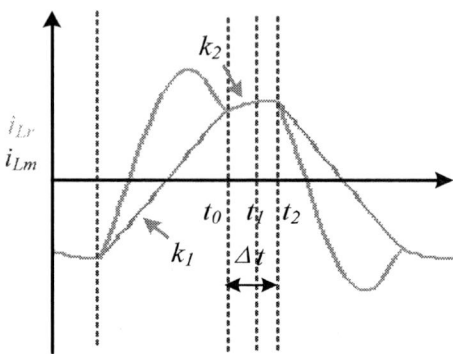

Fig.3 Current waveform of resonant tank when $T_s\ne T_r$

When $T_s\ne T_r$, there is a Δt, the bigger the Δt, the greater the error.

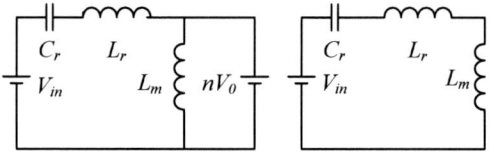

a. LC resonant mode   b. LLC resonant mode

Fig.4 The equivalent schematics of LC resonant mode and LLC resonant mode

Fig.4 shows the equivalent schematics of LC resonant mode and LLC resonant mode. In LC resonant mode and LLC resonant mode, the rising slope of magnetizing inductance current can be shown as follows:

$$\begin{cases} \dfrac{i_L(t_0)+i_L(t_1)}{T_r/2}=\dfrac{nV_0}{L_m}\\ \dfrac{i_L(t_1)-i_L(t_0)}{(T_s-T_r-2t_d)/2}=\dfrac{V_{in}}{L_r+L_m} \end{cases} \quad (6)$$

We can get from (6):

$$\begin{cases} i_L(t_0)=\dfrac{nV_0 T_r}{4L_m}-\dfrac{V_{in}(T_s-T_r-2t_d)}{4(L_m+L_r)}\\ i_L(t_1)=\dfrac{nV_0 T_r}{4L_m}+\dfrac{V_{in}(T_s-T_r-2t_d)}{4(L_m+L_r)} \end{cases} \quad (7)$$

The dead time is short and the magnetizing inductance is several times higher than the resonant inductance, it is considered that the current flowing is a constant value during the dead time through both of the magnetizing inductance and the resonant inductance. So there is $i_{td}=i_L(t_1)=i_L(t_2)$, combining (1) (2) (6) and (7), the improved formula is as following:

2704

$$
i_{L_r} = \begin{cases}
k_1 t - i_L(t_1) + \dfrac{1}{n} \times i_{D_1}(t), 0 \le t < \dfrac{T_r}{2} \\[2mm]
k_2(t - \dfrac{T_r}{2}) + i_L(t_0), \dfrac{T_r}{2} \le t < \dfrac{T_s}{2} - t_d \\[2mm]
i_{t_d}, \dfrac{T_s}{2} - t_d \le t < \dfrac{T_s}{2} \\[2mm]
-k_1(t - \dfrac{T_s}{2}) + i_L(t_1) - \dfrac{1}{n} \times i_{D_2}(t), \dfrac{T_s}{2} \le t < \dfrac{T_s + T_r}{2} \\[2mm]
-k_2(t - \dfrac{T_s + T_r}{2}) - i_L(t_0), \dfrac{T_s + T_r}{2} \le t < T_s - t_d \\[2mm]
i_{t_d}, T_s - t_d \le t < T_s
\end{cases}
$$

(8)

According to the formula (8), in addition to the parameters of resonant tank, the resonant tank current is not only related to the resonant frequency, switching frequency, but also the dead time. And the RMS value of resonant inductance current is obtained:

$$
I_{L,rms} = \sqrt{ \dfrac{1}{T_s} \left\{ \begin{array}{l}
T_r \left[\begin{array}{l} \dfrac{I_{pri}^2}{2} + (\dfrac{k_1^2 T_r^2}{12} - \dfrac{k_1 i_L(t_1) T_r}{2} + i_L^2(t_1)) + \\[2mm] \dfrac{I_{pri}(k_1 T_r - 4 i_L(t_1))}{\pi} \end{array} \right] + \\[6mm]
(T_s - T_r - 2t_d) \left[\begin{array}{l} \dfrac{k_2^2}{3}(\dfrac{T_s - T_r}{2} - t_d)^2 \\[2mm] + k_2 i_L(t_0)(\dfrac{T_s - T_r}{2} - t_d) \\[2mm] + i_L^2(t_0) \end{array} \right] + 2 t_d i_{t_d}^2
\end{array} \right\} }
$$

(9)

According to the formula (9), the resonant tank parameters can be optimized so that the RMS value of the resonant tank current is small, so the conduction loss of the primary side devices, such as the switch tube, can be reduced.

III. Optimal Design of Reaonant Tank Parameter

The parameters of the resonant tank have important influence on the performance of the converter. There are two key parameters of the resonant tank: the quality factor Q and the inductance ratio k, and the other parameters of the resonant tank can be determined by Q and k.

$$
\begin{cases}
Q = \dfrac{\sqrt{L_r / c_r}}{R_{eq}} \\[3mm]
k = \dfrac{L_m}{L_r} \\[3mm]
R_{eq} = \dfrac{8 n^2 R_0}{\pi^2}
\end{cases}
$$

(10)

A. DC voltage gain characteristics

The maximum voltage gain is given according to the input and output voltage design criteria, it is a fixed and known quantity. The design criteria of this experiment is

$M_{max} = 1.2$, Q is bound by K, they satisfy the following equation

$$
Q = \dfrac{0.95}{k M_{max}} \sqrt{k + \dfrac{M_{max}^2}{M_{max}^2 - 1}}
$$

(11)

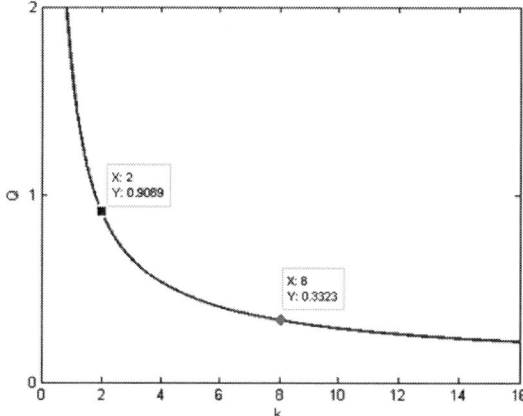

Fig.5 Relationship graph of k, Q

The variation curve of Q obtained from MATLAB simulation is shown in Fig.5. It can be seen from Fig.5: When $k < 2$, Q is increasing rapidly; When $k > 8$, $\dfrac{dQ}{dk}$ is almost constant. In other words, after $k > 8$, the influence of k on Q gradually decreases. The parameters that affect the efficiency of the converter are reduced from k, f and Q to k,f. Then, the value of k can be determined by analyzing the relationship between k and the RMS value of the resonant tank current.

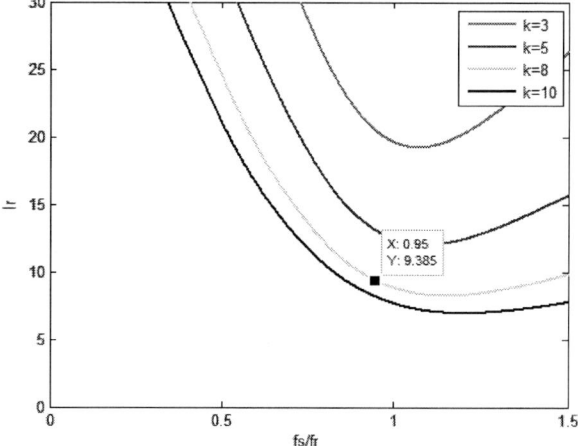

Fig.6 25% full load, the RMS value of resonant current

When the LLC resonant converter working in light load, efficiency is generally low, in order to verify the formula of RMS value of resonant tank current analyzed in the second section of this paper, we use light load for example. Derived from (9) and MATLAB simulation of 25% full load ($I_o = 0.6A$), the RMS value of resonant current at different k is shown in Fig.6. The figure shows: resonant inductance is fixed, as k increases, the RMS value of resonant tank current decreases gradually, and when k increases to a certain extent, I_{Lrrms} decreases more and more slowly, continue to increase the k value is not only giving little help to the improvement of efficiency, but also increasing the range of the operating frequency, which is not conducive to the work of the converter.

B. Parameter calculation of the LLC resonant converter

After determining the key parameters of the resonant tank, other parameters are obtained by following formula combining the design specifications of the LLC resonant converter:

$$\begin{cases} n = \dfrac{V_{in}}{V_0 + 2V_f} \\ C_r = \dfrac{1}{2\pi f_r Q R_{eq}} \\ L_r = \dfrac{Q R_{eq}}{2\pi f_r} \\ L_m = k L_r \end{cases} \quad (12)$$

IV. SIMULATION AND EXPERIMENT

The second section analyzes the key waveform of the LLC resonant converter. A more accurate calculation formula of resonant tank current is obtained. On this basis, the third section gives a new idea of the design of converters parameters. An experimental prototype of 48V input, 1kW/400V output is set up, and relevant data of the design specifications are as follows:

TABLE I
DESIGN SPECIFICATIONS OF LLC RESONANT CONVERTER

Symbol	Meaning	Value
V_{in}/V	Input voltage	48
V_0/V	Output voltage	400
P_0/W	Output power	1000
f_r / kHz	Resonant frequency	100
n	Turns ratio	3:25
R_{eq} / Ω	Equivalent resistance	1.87

According to the theoretical analysis combined with the prototype design specifications, the specific parameters are given as follows:

TABLE II
SPECIFIC PARAMETERS OF LLC RESONANT CONVERTER

Symbols	Meaning	Values	
$C_r / \mu F$	Resonant capacitance	2	2
$L_r / \mu H$	Resonant inductance	1.23	1.23
$L_m / \mu H$	Magnetic inductance	9.72	12.35
K	Inductance ratio	8	10

Fig.7 is a Saber simulation of the converter working at 25% full load. The simulated resonant tank current RMS value is 9.663A. The resonant tank current RMS values calculated with equations (5) and (9) are 8.7582A and 9.385A, the errors are 9.36% and 2.88% respectively. The RMS current value of the resonant tank calculated in this paper is more accurate than that of the original ones. The experimental waveform is shown in Fig.8.

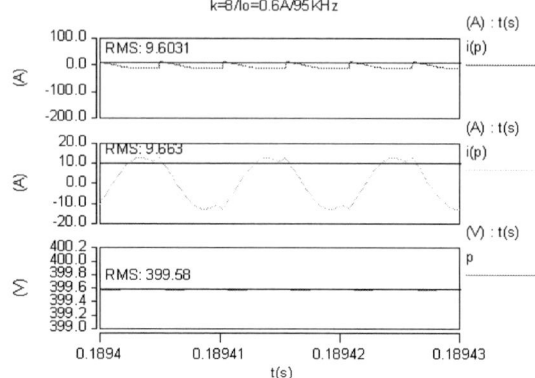

Fig.7 Simulation result

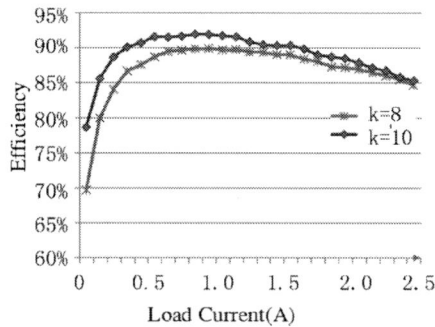

Fig.8 Experimental waveform

The efficiency curve is shown in Fig. 9. The value of Q has little effect on the efficiency of the converter, but the value of k can be improved to increase the efficiency of the converter.

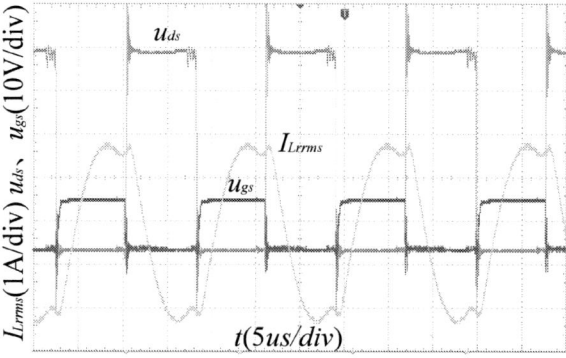

Fig.9 Efficiency curve under rated input voltage at different K

V. CONCLUSION

In this paper, an improved resonant tank current formula is proposed, which is closer to the actual working state of the converter, and can be more accurate in calculating the conduction losses of the converter. On this basis, a design method of the resonant tank parameters is presented. Finally, the accuracy of the improved resonant tank current formula and the feasibility and correctness of the parameter design method are verified by simulation and a 1000W test prototype.

REFERENCES

[1] P. A. Cassani and S. S. Williamson, "Design, Testing, and Validation of a Simplified Control Scheme for a Novel Plug-In Hybrid Electric Vehicle Battery Cell Equalizer," in IEEE Transactions on Industrial Electronics, vol. 57, no. 12, pp. 3956-3962, Dec. 2010.

[2] R. L. Steigerwald, "A comparison of half-bridge resonant converter topologies," IEEE Trans. on Power Electronics, vol. 3, no. 2, pp. 174-182 Apr. 1988.

[3] M. Noah et al., "A novel three-phase LLC resonant converter with integrated magnetics for lower turn-off losses and higher power density," 2017 IEEE Applied Power Electronics Conference and Exposition (APEC), Tampa, FL, 2017, pp. 322-329.

[4] H. H. Choi, K. S. Chung and G. Li, "Analysis on the loss of hybrid transformer winding for multi-output high frequency (300W) LLC resonance converters," 2015 IEEE International Telecommunications Energy Conference (INTELEC), Osaka, 2015, pp. 1-4.

[5] C. H. Yang, T. J. Liang, K. H. Chen, J. S. Li and J. S. Lee, "Loss analysis of half-bridge LLC resonant converter," 2013 1st International Future Energy Electronics Conference (IFEEC), Tainan, 2013, pp. 155-160.

[6] M. Jami, R. Beiranvand, M. Mohamadian and M. Ghasemi, "Optimization the LLC resonant converter for achieving maximum efficiency at a predetermined load value," The 6th Power Electronics, Drive Systems & Technologies Conference (PEDSTC2015), Tehran, 2015, pp. 149-155.

[7] M. J. Zhao, Y. Dai, H. W. Zhang, "Design of the parameters in LLC resonant converter," Journal of Magnetic Materials and Devices, vol. 42, no. 2, pp. 53-57, Dec. 2010.

The 2018 International Power Electronics Conference

Full-Bridge T-type Isolated DC/DC Converter with Wide Input Voltage Range

Dong Liu[1*], Yanbo Wang[1], Fujin Deng[2], Zhe Chen[1]

[1]Department of Energy Technology, Aalborg University, 9220, Aalborg, Denmark
[2]School of Electrical Engineering, Southeast University, 210096, Nanjing, China
*E-mail: dli@et.aau.dk

Abstract— The advent of the silicon carbide (SiC) power device with high voltage stress would simplify the power converter's circuit structure for high voltage applications since two-level topologies would be possible to instead three-level (TL) based topologies. This paper proposes a full-bridge (FB) T-type isolated DC/DC converter composed of four main power switches with high voltage stress (SiC MOSFET) and four auxiliary power switches with low voltage stress (Si MOSFET). Therefore, comparing with the conventional diode-clamped FB TL isolated DC/DC converter, the proposed converter has fewer circuit components and simpler circuit structure. What is more, a corresponding control strategy is proposed, which can not only realize zero-voltage switching (ZVS) but also achieve wide input voltage range. Finally, simulation and experimental results are both presented for verification.

Keywords— *DC/DC converter, Full-bridge (FB), T-type, Wide input voltage range.*

I. INTRODUCTION

Although AC distribution system is most widely used distribution system nowadays [1], [2], DC distribution system is a promising solution for the future power distribution system [3], [4] because the increasing applications of EV infrastructure, and renewable energy. The DC/DC converters are responsible for controlling power flow and converting the voltage level in the DC distribution system, so the research about the efficient and reliable DC/DC converters becomes a hot topic [5] - [7]. The three-level (TL) DC/DC converters become attractive because they can withstand the high input voltage and thus reduce the transmission loss. Many studies have been done on the TL isolated DC/DC converters in topics of extending soft switching range [8], [9], balancing voltages on the input capacitors [10], [11], reducing circulating currents [12], [13], minimizing and balancing currents on the input capacitors [14], and balancing currents among power switches [15].

T-type converter is another type of TL converter, which has been widely applied for inverters [16]. References [17] - [19] discussed about the T-type isolated DC/DC converters recently. Comparing with the conventional diode-clamped TL DC/DC converter, T-type converter has fewer circuit components and simpler circuit structure. The T-type DC/DC converters discussed in [17] - [19] belong to the half-bridge (HB) structure, thus they would be unsuitable for high power applications since the power switches' current stress on

in HB structure is twice of that in FB structure. In addition, a major drawback of the T-type converter is that the main power switches' voltage stress is full input voltage. Fortunately, the advent of SiC power device would result in that the T-type DC/DC converter with SiC power device is possible for the high voltage applications because SiC power device's drain-source breakdown voltage is much higher than Si power device's.

In this paper, a FB T-type isolated DC/DC converter with SiC device is proposed for high-power and high-voltage applications. There are four main power switches with high voltage stress (SiC MOSFET) and four auxiliary power switches with low voltage stress (Si MOSFET) in the proposed converter. What is more, a control strategy composed of two modes is proposed, which not only can realize zero-voltage switching (ZVS) for main and auxiliary power switches but also can fulfill wide input voltage range. Finally, both simulation and experimental results are demonstrated for verification.

II. CIRCUIT STRUCTURE

Fig. 1 presents the proposed converter's circuit structure. In Fig. 1, C_1, C_2 are two input capacitors; V_1, V_2 are two voltages split by C_1, C_2 from input voltage V_{in}; S_1 - S_4 are four main power switches; S_5 - S_8 are four auxiliary power switches; D_1 - D_8 are body diodes of S_1 - S_8; C_{s1} - C_{s8} are parasitic capacitors of S_1 - S_8; T_r is an isolated transformer; L_r is the inductance of series inductor plus leakage inductance of T_r; D_{r1} - D_{r4} are four rectifier diodes; L_o is an output filter inductor; and C_o is an output filter capacitor.

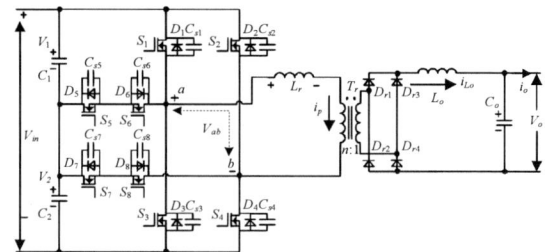

Fig. 1. Proposed converter's circuit structure.

In Fig.1, the input voltage is V_{in}; the voltage between point a and b is V_{ab}; the primary current of T_r is i_p; the current on L_o is i_{Lo}; the output voltage is V_o; the output current is i_o; the turns ratio of T_r is n. Table I shows the comparison results about the primary component number

2708

between the conventional diode-clamped FB TL isolated DC/DC converter and proposed converter.

TABLE I.
COMPARISON RESULTS ABOUT PRIMARY COMPONENT NUMBER

Component number	Diode-clamped FB TL DC/DC converter	Proposed FB T-type DC/DC converter
Power switch	8	8
Clamping diode	4	0
Flying capacitor	2	0
Input capacitor	2	2

III. OPERATION PRINCIPLE

A ZVS control strategy with two working modes is proposed. Figs. 2(a) and 2(b) show main waveforms of mode I and II respectively, in which d_{rv1} - d_{rv8} are driving signals for power switches S_1 - S_8; T_s is one switching period; d_1, d_2 are duty ratios in T_s and are used for operation mode I and II respectively; d_{loss_I} and d_{loss_II} are duty ratio losses in T_s under the operation mode I and II respectively.

Mode I and II are applied for low and high input voltage respectively, which can thus achieve the wide input voltage range. As presented in Fig. 2, V_o is controlled by adjusting d_1, d_2 under mode I and mode II respectively.

Figs. 3 and 4 present operation circuits for explaining the working operations under mode I and II more clearly.

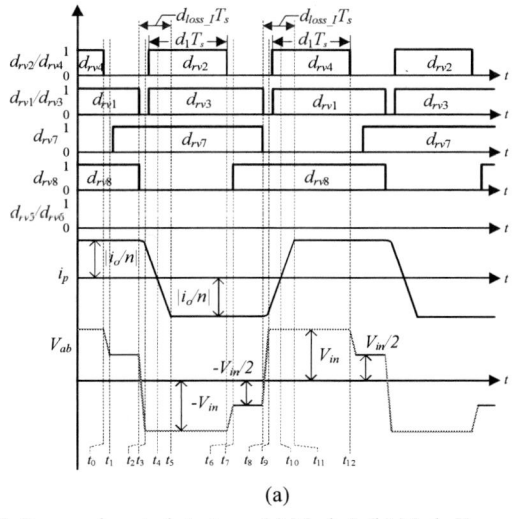

(a)

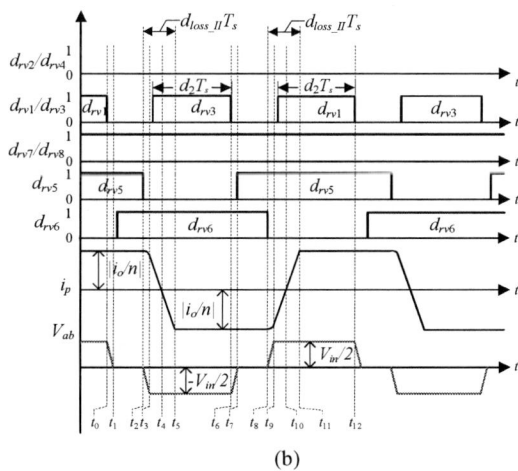

(b)

Fig. 2. Proposed control strategy. (a) Mode I. (b) Mode II.

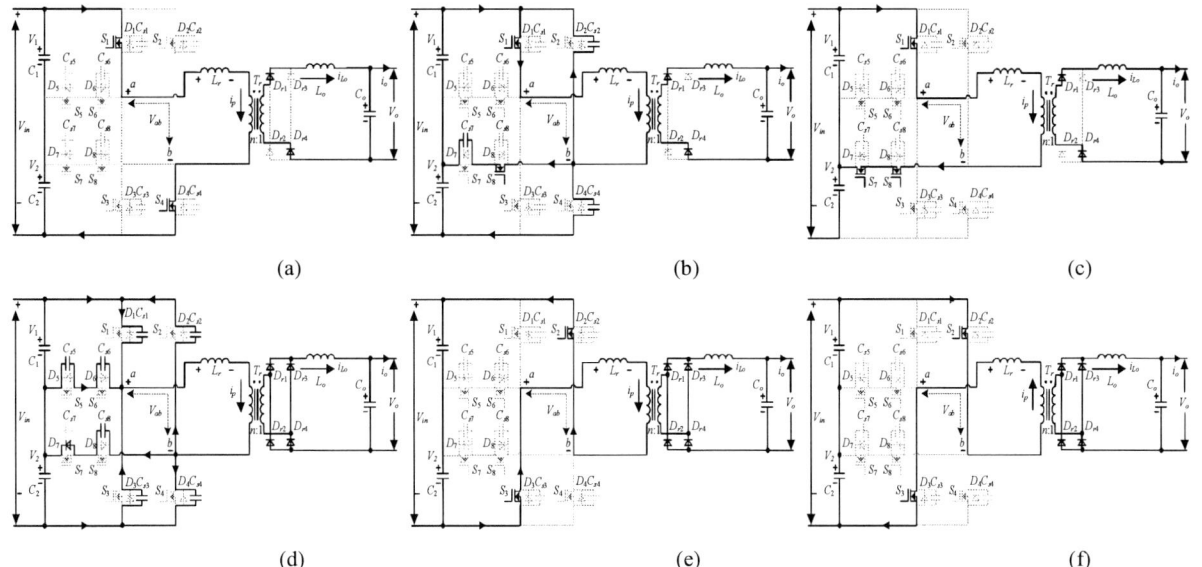

(a)   (b)   (c)

(d)   (e)   (f)

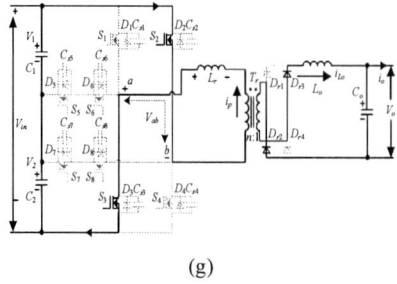

(g)

Fig. 3. Operation circuits under mode I. (a) [before t_0] (b) [t_0-t_1]. (c) [t_1-t_2]. (d) [t_2-t_3]. (e) [t_3-t_4]. (f) [t_4-t_5]. (g) [t_5-t_6].

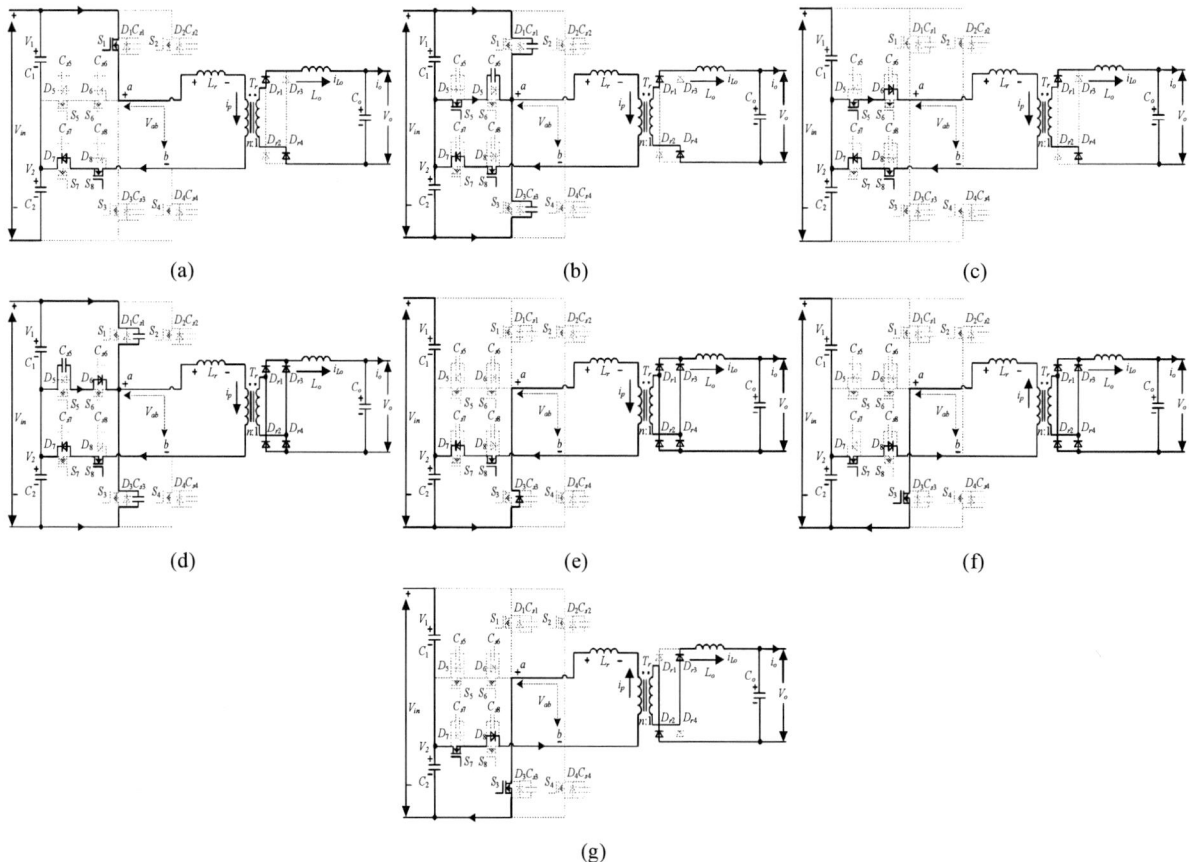

(a)　　　　　　　　　　(b)　　　　　　　　　　(c)

(d)　　　　　　　　　　(e)　　　　　　　　　　(f)

(g)

Fig. 4. Operation circuits under mode II. (a) [before t_0] (b) [t_0-t_1]. (c) [t_1-t_2]. (d) [t_2-t_3]. (e) [t_3-t_4]. (f) [t_4-t_5]. (g) [t_5-t_6].

IV. ANALYSIS OF CHARACTERISTIC AND PERFORMANCE

A. Power Switches' Voltage Stresses

In steady working operations, the voltage stress on main power switches S_1 - S_4 are the input voltage (V_{in}), and the voltage stress on auxiliary power switches S_5 - S_8 are half of the input voltage ($V_{in}/2$).

B. Output Gain

Under mode I and II, the duty ratio losses named d_{loss_I} and d_{loss_II} can be calculated by (1) and (2) respectively.

$$d_{loss\_I} = \frac{2 \cdot L_r \cdot i_o}{n \cdot V_{in} \cdot T_s} \quad (1)$$

$$d_{loss\_II} = \frac{4 \cdot L_r \cdot i_o}{n \cdot V_{in} \cdot T_s} \quad (2)$$

Under mode I, the average output voltage V_{o_I} can be obtained by (3).

$$V_{o\_I} = \frac{V_{in}}{n} \cdot (0.5 + d_1 - 2 \cdot d_{loss\_I}) = \frac{V_{in}}{n} \cdot (0.5 + d_1 - \frac{4 \cdot L_r \cdot i_o}{n \cdot V_{in} \cdot T_s}) \quad (3)$$

Under mode II, the average output voltage V_{o_II} can be obtained by (4).

$$V_{o\_II} = \frac{V_{in}}{n} \cdot (d_2 - d_{loss\_II}) = \frac{V_{in}}{n} \cdot (d_2 - \frac{4 \cdot L_r \cdot i_o}{n \cdot V_{in} \cdot T_s}) \quad (4)$$

The average output voltage $V_{o\_two\_level}$ in the basic FB two-level isolated DC/DC converter utilizing phase-shift control [20] can be calculated by (5).

$$V_{o\_two\_level} = \frac{V_{in}}{n} \cdot (2 \cdot d_{two\_level} - \frac{4 \cdot L_r \cdot i_o}{n \cdot V_{in} \cdot T_s}) \qquad (5)$$

in which d_{two_level} is the overlap time between the leading and lagging power switch divided by one switching period.

Based on equations (3) - (5) and assuming that the basic FB two-level isolated DC/DC converter and proposed converter have the same circuit parameters (n=25:8 L_r=47.7uH, f_s=50kHz) and secondary circuits, theoretical relations between duty ratio and input voltage in proposed converter and FB two-level isolated DC/DC converter are presented in Fig. 5 under working conditions that V_o is 50 V and output power named P_o is 1 kW.

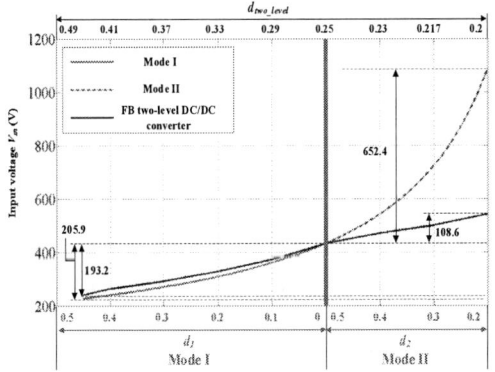

Fig. 5. Theoretical relation curves between the input voltage and duty ratio (V_o = 50 V, P_o = 1 kW, f_s = 50 KHz).

Note: Bottom X axis marked by red color represents the duty ratios in proposed converter; and top X axis represents the duty ratio in FB two-level isolated DC/DC converter.

V. SIMULATION AND EXPERIMENTAL VERIFICATION

A. Simulation Verification

For verification, a simulation model is built in PLECS. The built simulation model's circuit parameters are presented in Table II.

Fig. 6 presents simulations results about V_{in}, V_{ab}, V_o, i_{Lo}, i_p, and i_o under mode I and II.

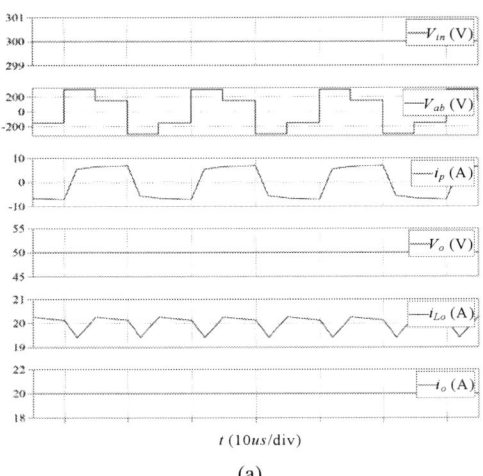

t (10us/div)

(a)

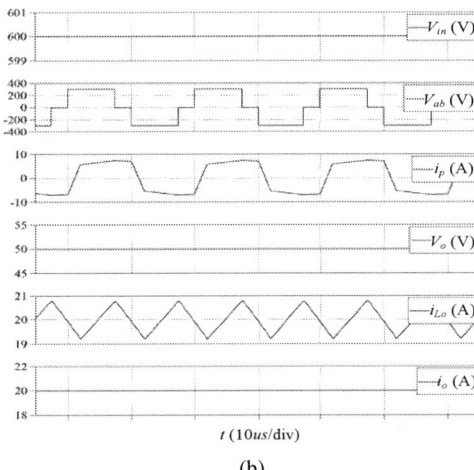

t (10us/div)

(b)

Fig. 6. Simulation results (V_o = 50 V, P_o = 1 kW). (a) V_{in} = 300 V (Mode I). (b) V_{in} = 600 V (Mode II).

Based on simulation results in Fig. 6, it can be obtained: 1) Mode I is applied when V_{in} is low (300 V) as presented in Fig. 6(a); 2) Mode II is applied when V_{in} is high (600 V) as presented in Fig. 6(b); and 3) the ripple current (i_{Lo}) on L_o under mode I is smaller than that under mode II.

B. Experimental Verification

A 1 kW laboratory prototype is established for verification. The established proposed converter's circuit parameters are presented in Table II.

Fig. 7 presents the experimental results about V_{in}, V_{ab}, V_o, and i_p.

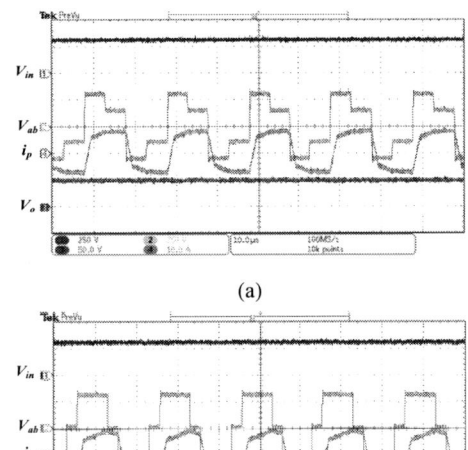

(a)

(b)

Fig. 7. Experimental results about V_{in}, V_{ab}, V_o, i_p (V_o = 50 V, P_o = 1 kW). (a) V_{in} = 300 V (Mode I). (b) V_{in} = 600 V (Mode II).

In Fig. 7, it can be seen: 1) Mode I is applied for the low voltage (V_{in} = 300 V) as presented in Fig. 7 (a); and 2) Mode II is applied for the high voltage (V_{in} = 600 V) as presented in Fig. 7 (a).

Figs. 8 and 9 present the power switches' drain-source voltage. In Figs. 8 and 9, V_{DS_S1} - V_{DS_S8} are drain-source voltages on S_1 - S_8. Based on experimental results in Figs. 8 and 9, it can be obtained: 1) main power switches' voltage stresses are about input voltage; and 2) auxiliary power switches' voltage stresses are about half of input voltage.

Fig. 10 presents ZVS performances about S_2, S_3, S_7 under mode I when V_o is 50 V, V_{in} is 300 V, P_o is 1 kW. Fig. 11 shows ZVS performances of S_3 and S_6 under mode II when V_o is 50 V, V_{in} is 600 V, P_o is 1 kW.

Based on experimental results in Figs. 10 - 11, the followings can be obtained: 1) under mode I, main power switches S_2, S_3 and auxiliary power switch S_7 realize ZVS when P_o is 1 kW respectively; and 2) under mode II, the main power switch S_3 and auxiliary power switch S_6 realize ZVS when P_o is 1 kW. The other main power switches' and auxiliary power switches' ZVS performances under mode I and II are similar to that of 1) and 2).

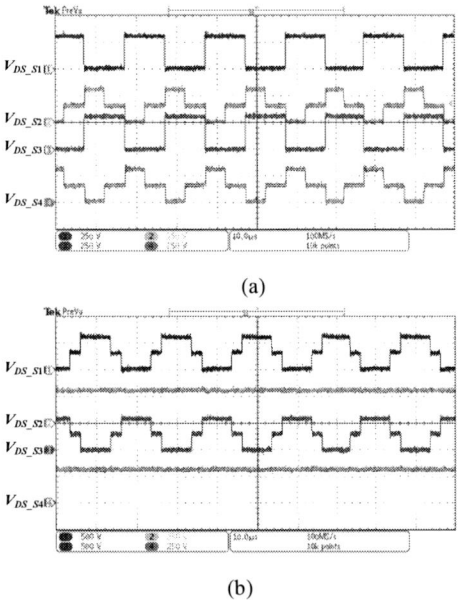

(a)

(b)

Fig. 8. Experimental results about V_{DS_S1} - V_{DS_S4} (V_o = 50 V, P_o = 1 kW). (a) V_{in} = 300 V (Mode I). (b) V_{in} = 600 V (Mode II).

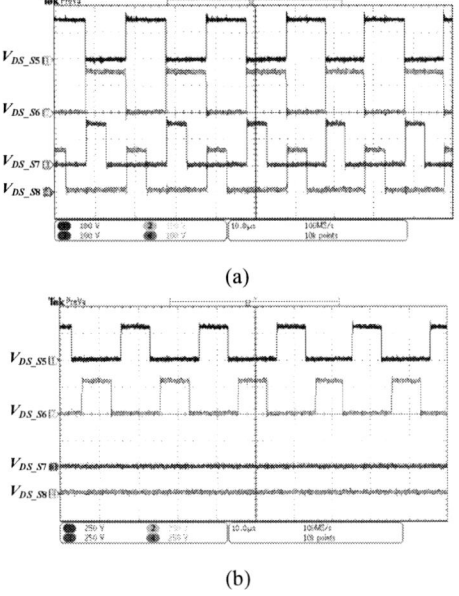

(a)

(b)

Fig. 9. Experimental results about V_{DS_S5} - V_{DS_S8} (V_o = 50 V, P_o = 1 kW). (a) V_{in} = 300 V (Mode I). (b) V_{in} = 600 V (Mode II).

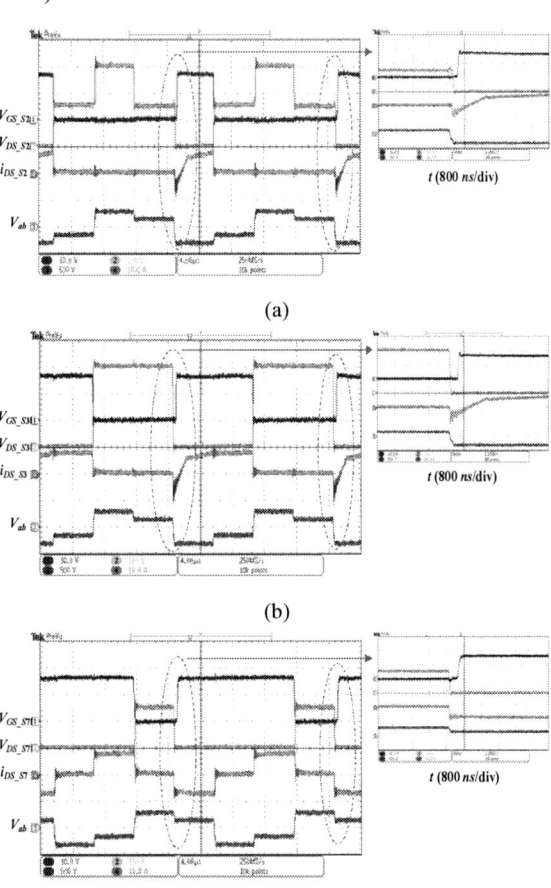

(a)

(b)

(c)

Fig. 10. ZVS performances (V_{in} = 300 V, V_o = 50 V, P_o = 1 kW Mode I). (a) S_2. (b) S_3. (c) S_7.

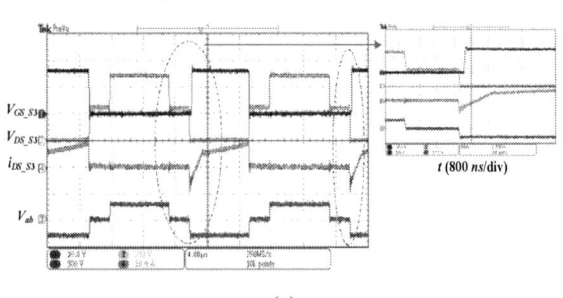

(a)

2712

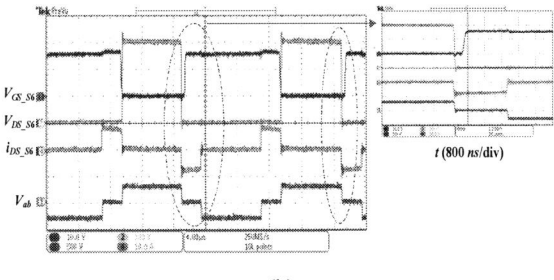

(b)

Fig. 11. ZVS performances (V_{in} = 600 V, V_o = 50 V, P_o = 1 kW Mode II). (a) S_3. (b) S_6.

VI. CONCLUSION

A FB T-type isolated DC/DC converter and corresponding control strategy is proposed in this paper. The proposed converter compromises four main power switches with the voltage stress of input voltage (SiC MOSFET) and four auxiliary power switches with the voltage stress of half of input voltage (Si MOSFET). Therefore, it has fewer circuit components and more compact circuit structure when comparing with the diode-clamped FB TL isolated DC/DC converter. What is more, the proposed control strategy not only can realize ZVS but also can fulfill wide input voltage range. Finally, both simulation and experimental results verify the proposed converter with its corresponding control strategy.

APPENDIX

TABLE II
PARAMETERS OF SIMULATION MODEL AND EXPERIMENTAL PROTOTYPE

Main Power Switches S_1 - S_4	C3M0065090D
Auxiliary Power Switches S_5 - S_8	SPW47N60C3
Rectifier Diodes D_{r1} - D_{r4}	MBR40250TG
Turns Ratio of Transformer T_r (n : 1)	25 : 8
Inductance of Series Inductor plus Leakage Inductance L_r (uH)	47.7
Input Capacitors C_1 and C_2 (uF)	470
Output Filter Capacitor C_o (uF)	470
Output Filter Inductor L_o (uH)	140
Switching Frequency (kHz)	50

REFERENCES

[1] R. Xu, Y. Yu, R. F. Yang, and G. L. Wang, "A novel control method for transformerless H-Bridge cascaded STATCOM with star configuration," *IEEE Trans. on Power Electronics*, vol. 30, no. 3, pp. 1189-1202, Mar. 2015.

[2] Z. Xin, X. Wang, P. C. Loh, and F. Blaabjerg, "Grid-current feedback control for LCL-filtered grid converters with enhanced stability," *IEEE Trans. on Power Electronics*, vol. 32, no. 4, pp. 3216-3228, Apr. 2017.

[3] M. Baran and N. R Mahajan, "DC distribution for industrial systems opportunities and challenges," *IEEE Trans. on Industry Applications*, vol. 39, no. 6, pp. 1596-1601, Nov./Dec. 2003.

[4] F. Chen, R. Burgos, D. Boroyevich and X. Zhang, "Low-frequency common-mode voltage control for systems interconnected with power converters," *IEEE Trans. on Industrial Electronics*, vol. 64, no. 1, pp. 873-882, Jan. 2017.

[5] B. Zhao, Q. Song, W. Liu, and Y. Sun, "Dead-time effect of the high frequency isolated bidirectional full-bridge dc-dc converter: comprehensive theoretical analysis and experimental verification," *IEEE Trans. on Power Electronics*, vol. 29, no. 4, pp. 1667-1680, Apr. 2014.

[6] D. Liu, F. Deng, and Z. Chen, "Five-level active-neutral-point-clamped DC/DC converter for medium voltage DC grids," *IEEE Trans. on Power Electronics*, vol. 32, no. 5, pp. 3402-3412, May. 2017.

[7] X. Guo, D. Sha, Y. Xu, and X. Liao, "Hybrid-bridge-based DAB converter with voltage match control for wide voltage Conversion Gain Application," *IEEE Trans. on Power Electronics*, vol. 33, no. 2, pp. 1378-1388, Feb. 2018.

[8] Y. Shi, and X. Yang, "Wide range soft switching PWM three-level DC-DC converters suitable for industrial applications," *IEEE Trans. on Power Electronics*, vol. 29, no. 2, pp. 603-616, Feb. 2014.

[9] Y. Jang, and M. Jovanovic, "A new three-level soft-switched converter," *IEEE Trans. on Power Electronics*, vol. 20, no. 1, pp. 75-81, Jan. 2005.

[10] B. Lin and S. Zhang, "Analysis and implementation of a three-level hybrid DC-DC converter with the balanced capacitor voltages," *IET Power Electronics*, vol. 9, no. 3, pp. 457-465, Mar. 2016.

[11] X. Yu, K. Jin, and Z. Liu, "Capacitor voltage control strategy for half-bridge three-level DC/DC converter," *IEEE Trans. on Power Electronics*, vol. 29, no. 4, pp. 1557-1561, Apr. 2014.Z.

[12] W. Li, S. Zong, F. Liu, H. Yang, X. He, and B. Wu, "Secondary-side phase-shift-controlled ZVS DC/DC converter with wide voltage gain for high input voltage applications," *IEEE Trans. on Power Electronics*, vol. 28, no. 11, pp. 5128-5139, Nov. 2013.

[13] Guo, D. Sha, and X. Liao, "Hybrid three-level and half-bridge DC-DC converter with reduced circulating loss and output filter inductance," *IEEE Trans. on Power Electronics*, vol. 30, no. 12, pp. 6628-6638, Dec. 2015.

[14] D. Liu, F. Deng, Z. Gong, and Z. Chen, "Input-parallel output-parallel (IPOP) three-level (TL) DC/DC converters with interleaving control strategy for minimizing and balancing capacitor ripple currents," *IEEE Journal of Emerging and Selected Topics in Power Electronics*, vol. 5, no. 3, pp. 1122-1132, Sep. 2017.

[15] D. Liu, F. Deng, Q. Zhang, and Z. Chen, "Periodically swapping modulation (PSM) strategy for three-level (TL) DC/DC Converter with Balanced Switch Currents," *IEEE Trans. on Industrial Electronics*, vol. 65, no. 1, pp. 412-423, Jan. 2018.

[16] M. Schweizer and J. W. Kolar, "Design and implementation of a highly efficient three-level T-type converter for low-voltage applications", *IEEE Trans. on Power Electronics*, vol. 28, no. 2, pp. 899-907, Feb. 2013.

[17] D. Liu, F. Deng, Y. Wang, and Z. Chen, "Improved control strategy for T-type isolated DC/DC converters," *Journal of Power Electronics*, vol. 17, no. 4, pp. 874-883, Jul. 2017.

[18] D. G. Bandeira and I. Barbi, "A T-type isolated zero voltage switching DC-DC converter with capacitive output," *IEEE Trans. on Power Electronics*, vol. 32, no. 6, pp. 4210–4218, June 2017.

[19] D. G. Bandeira, S. A. Mussa, and I. Barbi, "A ZVS-PWM T-type isolated DC-DC converter," in *Proc. 1th Annu. Southern Power Electron. Conf.*, Nov/Dec., 2015.

[20] J. A. Sabate, V. Vlatkovic, R. B. Ridley, F. C. Lee, and B. H. Cho, "Design considerations for high-voltage high-power full-bridge zero-voltage-switched PWM converter," in *Proc. IEEE APEC*, 1990, pp. 275-284.

Research on High Efficiency LLC DC-DC Converter Based on SiC MosFet

Pengcheng Han[1*], Xiaoqiong He[1, 2], Haijun Ren[1], Zhiqing Zhao[1], Xu Peng[1]

1 School of Electrical Engineering, Southwest Jiaotong University, Chengdu, China

2 National Rail Transit Electrification and Automation Engineering Technique Research Center, Chengdu, China

*Email: birdhpc@163.com

Abstract- **In this paper, an efficient H-bridge LLC resonant DC-DC converter based on SiC MosFet is studied. The mathematical model of topological structure is established by using First Harmonic Approximation, (FHA). The steady - state characteristic of H - bridge LLC resonant DCDC converter is analyzed. For improving the efficiency of the converter and realizing the stability of the output voltage under different load conditions, the synchronous rectification and phase shift control strategy are designed. The experiment of H-bridge LLC resonant DCDC converter based on SiC MosFet is designed and ZVS and ZCS without control are realized. And then the efficiency of synchronous rectification and phase shift control is verified while the load changed. Finally, the system efficiency curve is fitted through different loading conditions, and the highest efficiency of the system over 98% has achieved.**

Keywords: **LLC converter; SiC Mosfet; Synchronous Rectifier; Phase shift control**

I. INTRODUCTION

As the frequency growing, the loss of switch exists in the PWM converter. The appearance of resonant converter decrease the loss of switch so that the working frequency of the power semiconductor is permitted to increase to the relatively high level. Based the topology of the resonant converter, the family of it including SRC,PRC, LCC,SPRC and LLC are widely researched [1-3]. Among them, SRC is faced with the problem of efficiency, as the gain of the output voltage remain nearly unchanged when under loading. The PRC is also confronted the efficiency challenge as the resonant current of the PRC and output voltage are independence [4-5]. LLC converter has a relatively wide input range, besides, it could achieve the soft switch in all conditions. Thus, LLC converter is popular because of high efficiency, high reliability and low loss.

Wide bandgap (WBG) power semiconductor device technology, such as SiC and GaN is new technology[6-7]. They are a kind of wide bandgap compound semiconductor material which have greater saturated electron drift velocity, higher critical breakdown electric field and higher thermal conductivity. Own to the high efficiency and low loss, the SiC MOSFET is helpful to rise the efficiency and power density of the converter.

II. CONFIGERATION

The H-bridge primary side and voltage-doubling circuit secondary side are illustrated in Fig.1. In the secondary side, the diode voltage of the rectifier is equal to the output voltage. Capacitor C_1 and C_2 also share the output voltage equally, thus no capacitor filter is needed in the output side. And also, the capacitor of LLC converter is convenient to choose as its

voltage are shared. Therefore, voltage-doubling circuit is suitable for LLC converter when the output voltage requirement strictly.

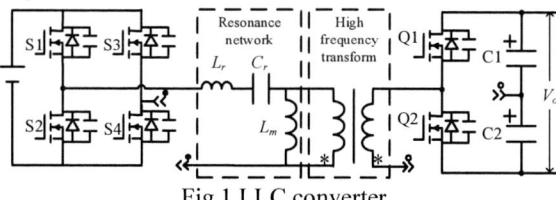

Fig 1 LLC converter

LLC converter is made up of three resonant components, where C_r is resonant capacitor, L_r is resonant inductance and Lm is magnetizing inductance. Based the resonant components, there are two resonant frequency, including f_r and f_m, which is shown in equation (1).

$$\begin{cases} f_r = \dfrac{1}{2\pi\sqrt{L_r C_r}} \\[4mm] f_m = \dfrac{1}{2\pi\sqrt{(L_r + L_m)C_r}} \end{cases} \quad (1)$$

The *ft* and fm divided the working frequency in three part, which is, $f < fm$, $fm < f < fr$ 和 $f > fr$. To realize the soft switch of the converter, the working frequency is restricted in the range of $fm < f < fr$.

III. MATHEMATICAL MODEL

The first step of modeling the resonant DC-DC converter is assuming that:

1) The mid-frequency transformer, switching device (IGBT or MOSFET), rectify diode, inductor and capacitor are all ideal devices.

2) Parasitic capacitor is not included in the resonant tank when working frequency is low.

3) Output voltage ripple is very small, so it can be treated as a constant.

4) Ignore the switching frequency harmonic.

To simplify the dc gain expression, the definition is made: Resonant frequency,

$$f_r = \frac{1}{2\pi\sqrt{L_r C_r}} \quad (2)$$

Characteristic impedance,

$$Z_0 = \sqrt{\frac{L_r}{C_r}} = 2\pi f_r L_r = \frac{1}{2\pi f_r C_r} \quad (3)$$

Quality factor,

$$Q = \frac{Z_0}{R_{eq}} \quad (4)$$

Inductor coefficient,

$$\lambda = \frac{L_r}{L_m} \quad (5)$$

Normalized frequency,

$$f_n = \frac{f_{sw}}{f_r} \quad (6)$$

It can be simplified as

$$
\begin{aligned}
M &= \frac{V_{out}}{V_{dc}} \\
&= \frac{1}{2n} \left| \frac{1}{1 + \lambda(1 - 1/f_n^2) + j(f_n - 1/f_n)Q} \right| \quad (7) \\
&= \frac{1}{2n} \frac{1}{\sqrt{(1 + \lambda - \lambda/f_n^2)^2 + (f_n - 1/f_n)^2 Q^2}}
\end{aligned}
$$

Ignore the transformer ratio n(when n=1), it can be known that there are three Variables, inductor coefficient λ, normalized frequency f_n and Quality factor Q. In Fig 2 and 3, x axis is normalized frequency f_n, y axis is dc voltage gain, dc voltage gain changes when Quality factor Q and inductor coefficient λ change respectively.

In Fig.2, DC voltage gain curve changes gentle gradually with quality Q increasing. In the same switching frequency, dc gain variation is decreasing during the process of light load to heavy load condition. With increasing of normalized frequency f_n, increase first then decrease characteristic of every curve shows that there is a turning point of changing of DC voltage gain, and the frequency of this turning point is increasing with quality factor increasing.

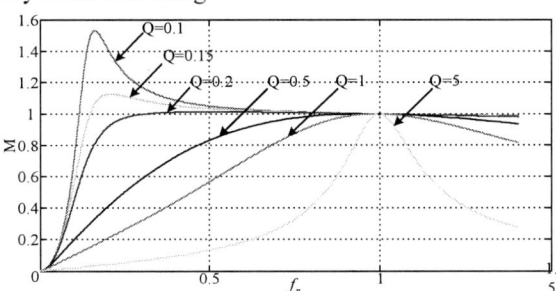

Fig 2 LLC resonant DC-DC converter gain characteristic （n = 1,λ = 0.02）

Analyze the condition when normalized frequency f_n equals to 1. In this condition, f_n =1, that is $f_{sw} = f_r$, f_r is the resonant frequency of L_r and C_r, so the input impedance of resonant tank is L_m paralleled with R_{eq}, the resonant tank is inductive. When f_n >1, that is $f_{sw} > f_r$, so the input impedance of resonant tank Z_{in}>0, the resonant tank is still inductive. When f_n <1, that is $f_{sw} < f_r$, so the characteristic of input impedance Z_{in} is according to the relationship of frequency and load. The physical meaning of the frequency of the turning point is when input impedance is resistive, that is the switching frequency equals to turning frequency, input power in not transited in the energy storage device, so the output power is more, the dc voltage gain reaches the maximum.

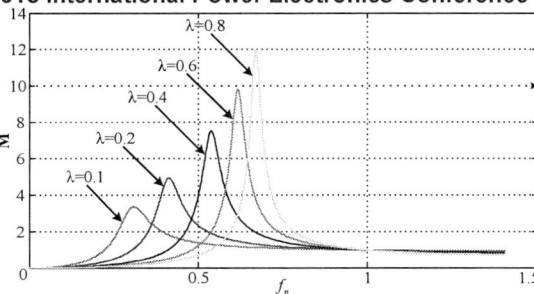

Fig 3 LLC resonant DC-DC converter gain characteristic （n = 1, Q = 0.1)

With the increasing of the inductor coefficient λ, dc voltage gain curve becomes steep. The increase of λ means the increase of L_r or the decrease of L_m, however, any calculation concerns L_r needs to be corrected when L_m changes, including the resonant frequency f_r, so the changes of L_m is discussed. Therefore, the practical request should be considered when choose the inductor coefficient λ. Assume the f_n =1, that is $f_{sw} = f_r$, a stable dc voltage gain can be obtained, and the ZVS of primary side switching devices, if the input voltage changes a lot, range of switching frequency will increase.

IV. SYNCHRONOUS RECTIFICATION AND PHASE SHIFT CONTROL

The secondary Synchronous rectifier circuit is illustrated as Fig.4. The MOSFET is turned on when anti-paralleled diode current is increasing, then the current flow through MOSFET; when the current below a turn off threshold, the MOSFET is turned off.

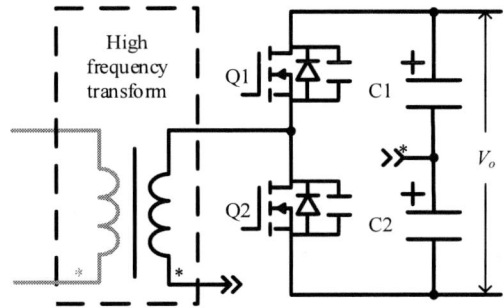

Fig 4 Synchronous rectifier circuit

This process decreases the power loss in the rectifier diode by making the current flow through the MOSFET rather than the diode.

For simplifying the control method, the f is changeless according to the main circuit parameters. The phase shift control is been used in the H-bridge LLC converter. The phase shift control can adjust the duty cycle of the switching device drive signal in the converter to adjust the output voltage. The ZVS can be achieved without adding other auxiliary devices.

On the basic of the normal control which the driving signal of the upper and under switch is opposite, the phase shift control can reduce the input power energy in order to stabilize the output voltage. By adding the status when S1 S3 and S2 S4 turn on simultaneously, the resonant network constitutes a short circuit state so that the input voltage and the resonant network are isolated. The load energy can be reduced.

V. EXPERIMENT

A. The Experiment of H-bridge LLC Without Control

The H-bridge LLC resonant circuit with operating frequency range $f_m < f < f_r$ are shown below. As shown in Fig.5, the converter completes the condition that the primary ZVS and the secondary ZCS. The DC output voltage of the H-bridge LLC converter is twice as large as the input voltage.

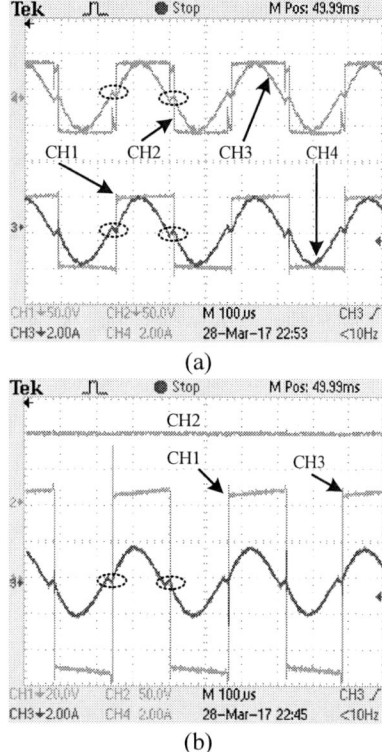

(a)

(b)

Fig 5 H-Bridge LLC DCDC waveform without control
(Fig(a), CH1: Transformer primary voltage; CH2: Transformer secondary voltage; CH3: Transformer primary current; CH4: Transformer secondary current; Fig(b), CH1: Transformer primary voltage; CH2: Output voltage; CH3: Transformer primary current; CH4: Transformer secondary current)

As is shown in Fig.6, the temperature of each device in the experiment was further measured using an optical temperature gauge. The SiC MosFet converter is about 21°C.

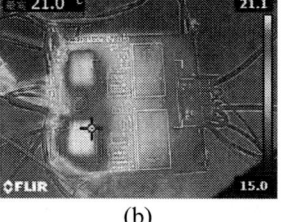

(a)   (b)

Fig 6 H-bridge LLC converter primary side main circuit and corresponding heating condition

The equivalent impedance of the SiC MosFet anti-parallel diode is large. The device temperature is 37.5°C when the secondary side is diode rectifier.

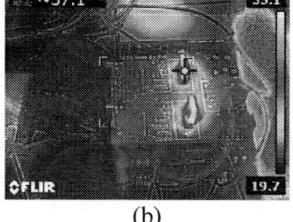

(a)   (b)

Fig 7 LLC resonant converter secondary circuit and unregulated rectifier under the secondary SiC MOSFET

B. The Experiment of Synchronous Rectification

As shown in Fig.8 is the Synchronous rectification waveform. Analyzing the working state in time interval of t_0-t_4. No drive signal is given at t_0. At t_1, the driving signal of the Q_1 is high voltage and the Q1 SiC MosFet is on. The current through the Q1 is 0 and the driving signal of the Q_1 is low voltage at t_2. At t_3, the secondary side Q2 current has just been established, whereas no drive signal is given. Until t_4 the Q2 drive signal is high.

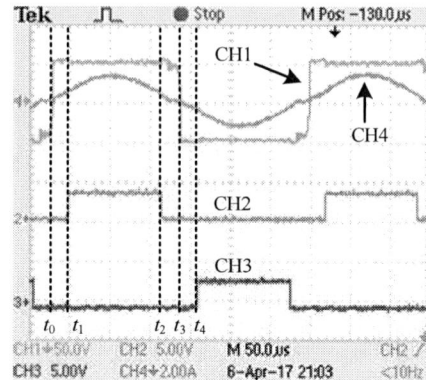

Fig 8 Synchronous rectification waveform
（CH1: Transformer secondary voltage; CH2: Upper SiC MosFet Driving Signal; CH3: The following SiC MosFet driving signal; CH4: Transformer secondary current ）

As shown in Fig.9 is the heat comparison of synchronous rectification. The upper SiC MosFet adopted synchronous rectification is 21.9°C as shown in Fig.9(a). The under SiC MosFet adopts diode rectification is 31.1°C as shown in Fig.9(b). The Synchronous rectification has higher efficiency.

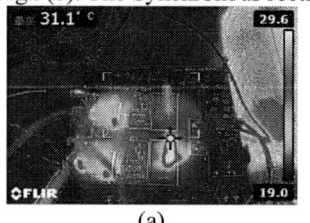
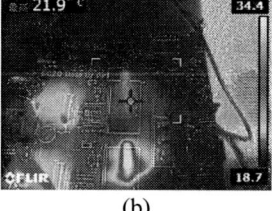

(a)   (b)

Fig 9 Heat comparison of synchronous rectification

C. The Experiment of Phase Shift Control

As shown in Fig.10(a), the feature waveform remains apparent when the phase shift angle is small. And the primary voltage is ZVS and the secondary current is ZCS. As shown in Fig.10(b), the feature waveform remains apparent when the phase shift angle is increased. The primary voltage is ZVS, whereas the secondary side current has a little fluctuations.

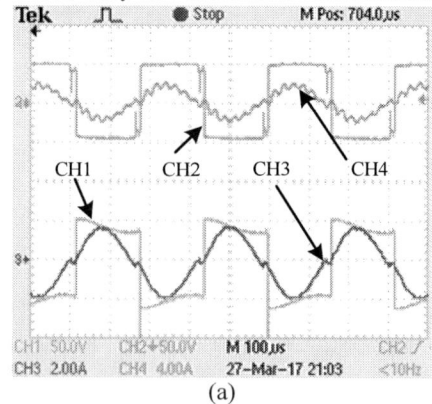

(a)

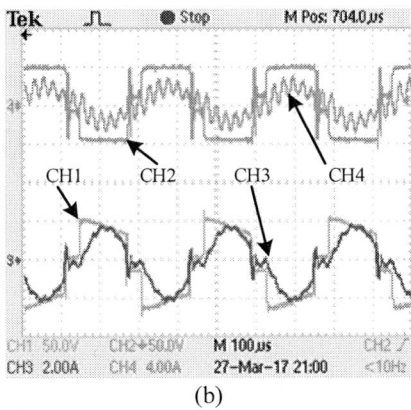

(b)

Fig 10 The experiment waveform of phase shift control
(Fig 10(a) and 10(b) CH1: Transformer primary voltage; CH2: Transformer secondary voltage; CH3: Transformer primary current; CH4: Transformer secondary current;)

As shown in Fig11, when the phase shift angle continues to increase. The resonance time of L_r and C_r current is getting shorter. As the result, lesser the input power to the resonator power supply, smaller the voltage output.

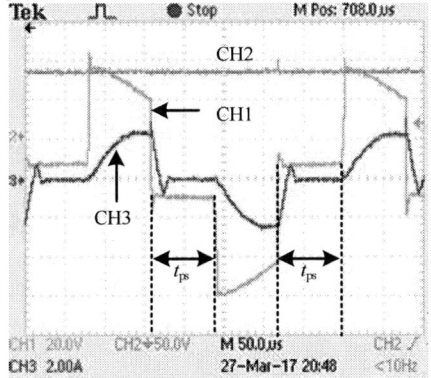

Fig 11 The experiment waveform of increasing shift angle
(CH1: Resonant voltage; CH2: Output voltage; CH3: Transformer primary current)

As shown in Fig.12 is the waveform when the H-bridge LLC converter from no load to 100Ω. The fixed frequency control has been used in the system. The output voltage is stable and reducing the duty cycle of the switch to reduce the gain of voltage when no load. Under load conditions, the output voltage would be reduced. Ensuring the voltage keep constant, it is necessary to increase the duty cycle of the switch to increase the voltage gain.

As shown in the waveform, the LLC characteristic after load changing is still clear and the ZVS is retained. The effect of the phase shift control method is also verified.

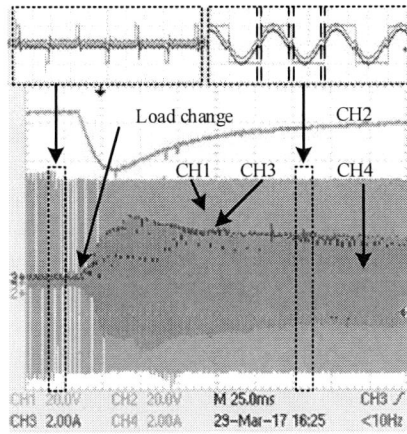

Fig 12 Load changed experiment

The 2018 International Power Electronics Conference

(CH1: Resonant voltage; CH2: Output voltage; CH3: Transformer primary current; CH4: Transformer secondary current)

Four groups of load experiments were selected as efficiency measures. And the efficiency curve is shown in the Fig.13.

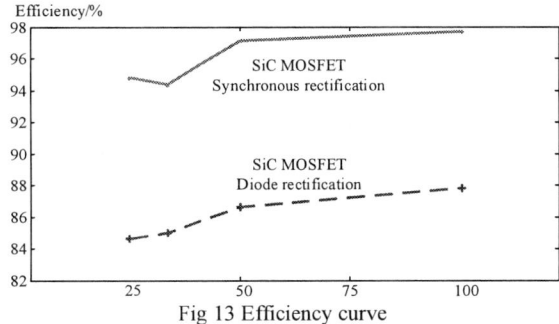

Fig 13 Efficiency curve

VI. CONCLUSION

This paper studied an efficient H-bridge LLC resonant DC-DC converter based on SiC MosFet. By using First Harmonic Approximation, (FHA), the mathematical model of topological structure is established. The steady - state characteristic of H - bridge LLC resonant DCDC converter is analyzed. Draw the following conclusions.

(1) ZVS and ZCS without control are realized by the experiment of H-bridge LLC resonant DCDC converter based on SiC MosFet.

(2) The efficiency of the converter can be improved with the synchronous rectification. The stability of the output voltage can be realized under different load conditions.

(3) The system efficiency curve is fitted through different loading conditions, and the highest efficiency of the system over 98% has achieved.

VII. ACKNOWLEDGMENT

This work is supported by the National Natural Science Foundation of China (Grant Nos.51477144)

REFERENCES

[1] Mohammadi M, Shafiei N, Ordonez M, "LLC synchronous rectification using Coordinate Modulation,"[C] *Applied Power Electronics Conference and Exposition, IEEE*, pp.848-853, 2016.

[2] Wu X, Hua G, Zhang J, et al, "A New Current-Driven Synchronous Rectifier for Series–Parallel Resonant (LLC) DC–DC Converter, "[J]. *IEEE Transactions on Industrial Electronics*, vol.58, no.1, pp.289-297, 2011.

[3] Funaki T, Matsushita M, Sasagawa M, et al, "A Study on SiC Devices in Synchronous Rectification of DC-DC Converter ,"[C] *IEEE Applied Power Electronics Conference and Exposition*, IEEE, pp.339-344, 2007.

[4] Hong L, Ma H, Wang J, et al, "An efficient algorithm strategy for synchronous rectification used in LLC resonant converters,"[C] *Industrial Electronics Society, IECON 2016, Conference of the IEEE*, IEEE, pp.2452-2456, 2016.

[5] Zhang J, Liao J, Wang J, et al, "A Current-Driving Synchronous Rectifier for an LLC Resonant Converter With Voltage-Doubler Rectifier Structure,"[J]. *IEEE Transactions on Power Electronics*, vol.27, no.4, pp.1894-1904, 2012.

[6] Feng W, Lee F C, Mattavelli P, et al, "A Universal Adaptive Driving Scheme for Synchronous Rectification in LLC Resonant Converters,"[J]. *IEEE Transactions on Power Electronics*, vol.27, no.8, pp. 3775-3781, 2012.

[7] Song X, Huang A Q, Sen S, et al, "15kV/40A FREEDM Super-Cascode: A Cost Effective SiC High Voltage and High Frequency Power Switch,"[J]. *IEEE Transactions on Industry Applications*, 2017.

The 2018 International Power Electronics Conference

an Improved Dual Phase Shift Control Strategy for Dual Active Bridge DC-DC Converter with Soft Switching

Miao Hong, GAO Xuanjie, Zeng Chengbi[*] and Duan Shujiang

(College of Electrical Engineering and Information Technology, Sichuan University, Cheng du 610065, China)

*E-mail: 857606631@qq.com

Abstract— **In order to reduce the backflow power of the dual active bridge DC-DC(DAB) converter, the relationship between backflow power versus transmission power, voltage conversion ratio is derived by mathematical analysis of dual active bridge DC-DC bridge and an improved dual phase shift(DPS) control strategy for dual active bridge DC-DC converter with soft switching is proposed. The control strategy is segmented based on the capacity of transmission power, and the optimal shift angle in every range is obtained. The simulation is carried on under MATLAB/Simulink and results show that, the proposed control strategy can reduce backflow power greatly compared to traditional single phase shift(SPS). It also can achieve zero backflow power in a certain range of transmission power.**

Keywords— **DAB dual phase shift soft-switching backflow power improved control.**

I. INTRODUCTION

The dual active bridge DC-DC (DAB) converter (Fig 1) is functionally equivalent to two unidirectional DC-DC transformations. Compared with the traditional unidirectional DC-DC converter, it has the following advantages: it can achieve the bidirectional flow of power, save the number of devices, reduce the volume of the system, reduce the cost and improve the system efficiency. Therefore, DAB converter has been used more and more widely in DC motor drive, uninterruptible power supply, electric vehicle, new energy generation technology, etc [1-3].

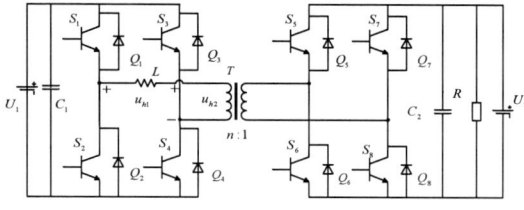

Fig.1 DAB DC-DC converter

At present, the main method to control the power flow of the DAB converter is phase shift control [4-10]. The traditional single phase shift control is simple and easy to

implement. However, it is not easy for DAB to achieve bidirectional power flow. Moreover, when the amplitude of input and output voltage does not match, it is easy to produce backflow power and increase current stress. Due to the presence of backflow power, the loss of the transformer in the DAB is increased and the efficiency of the converter is reduced [11-12]. In order to solve these problems, many scholars have done a lot of research. Document [13] establishes a small signal model of the DAB converter and uses PI closed-loop controller to regulate the power flow between two H bridges in DAB. By these method, the purpose of reducing backflow power is to be achieved. In Paper [14], a method of PWM plus phase shift control is proposed. The method uses two degrees of freedom angle to control the power transmission with reducing the backflow power and increasing the efficiency of the whole power range. A dual phase-shifting control strategy is proposed and analyzed in document [15-16]. The control mode simultaneously regulates two phase shift angles, namely the phase shift angle between the full bridge and the phase shift angle in the two full bridge. As the control variable is added, the DAB transmission power range is increased. In addition, the author established a dual phase shift power optimization control strategy with the target of minimum backflow power and finally achieved the purpose of reducing backflow power. In paper [17], an optimization strategy based on dual phase shift and traditional phase shift control is proposed to ensure the minimum value of the leakage current and realize zero voltage switch (ZVS). In [18], the author compares the traditional phase shift, dual phase shift and model-based phase shift control. Experimental results show that the model-based phase shift control has the best dynamic performance and the dual phase shift control can eliminate the backflow power under the condition of light load. Paper [19] establishes the average model of DAB and small signal model. The PI closed-loop is used to adjust the phase shift angle of the two sides of the transformer to reduce the backflow power. The [20] establishes a mathematical model of DAB to analyze current stress and a comparative analysis of the

2718

performance of traditional single phase and dual phase shift control is made. The author puts forward an optimization strategy of current stress which can minimize the current stress improve system efficiency and the ability of the system to transmission power. The performance is particularly effective in high voltage conversion ratio and in light load operating conditions. Document [21] compares and analyzes the system backflow power, inductor current stress and current RMS characteristics and the realization range of soft switching under various characteristics and obtains the minimum shift phase angle selection by using look-up table method, but this method is not generally applicable. Document [22-25] proposes a three phase-shift control strategy. The control method has three control variables, namely dual phase shift angles of the whole bridge the phase shift of the primary side and the phase shift of the secondary side. The control strategy is more complex because the increase of control variables.

In order to reduce the backflow power of the DAB converter, this paper proposes an improved dual phase shift control strategy for the DAB converter with the soft switching conditions. The main structure of this paper is as follows: In section I is Introduction. In section II, the working principle of the DAB with dual phase shift control is introduced; In section III, the mathematical model of the DAB is established and an improved control strategy of the DAB is proposed to reduce the backflow power; In section IV, the proposed control strategy is simulated and analyzed under MATLAB/Simulink; In section V is the conclusion.

II. DUAL PHASE SHIFT CONTROL PRINCIPLE

The power flow direction of DAB can be controlled according to the power demand. To simplify analysis, the power flow direction in Fig 1 is assumed from U_1 to U_2. The waveform of U_{h1}, U_{h2} and i_L are shown in Fig 2, which U_{h1} is the output voltage of the DAB on the U_1 side. U_{h2} is the voltage which the output voltage is folded to the U_1 side. In the primary side, switches S_1, S_4 or S_2, S_3 have phase shift angle φ. Defining the half cycle internal shift $D_1 = \varphi/\pi$. Between the primary side and the second side, S_1 and S_5 have phase shift angle ϕ. Defining the external shift $D_2 = \phi/\pi$. The following analysis D_1, D_2 satisfy the condition: $0 \le D_1 \le D_2 \le 1$.

From Fig 2, we know that during $t_1 - t_1'$ and $t_4 - t_4'$ the U_1 side voltage U_{h1} is opposite to the inductor current i_L. During $t_1 - t_1'$ and $t_4 - t_4'$ the energy stored in the inductor is recirculation to the U_1 side, which is called the backflow power. The backflow power during $t_1 - t_1'$ is

defined as the left backflow power and $t_4 - t_4'$ is defined as the right backflow power.

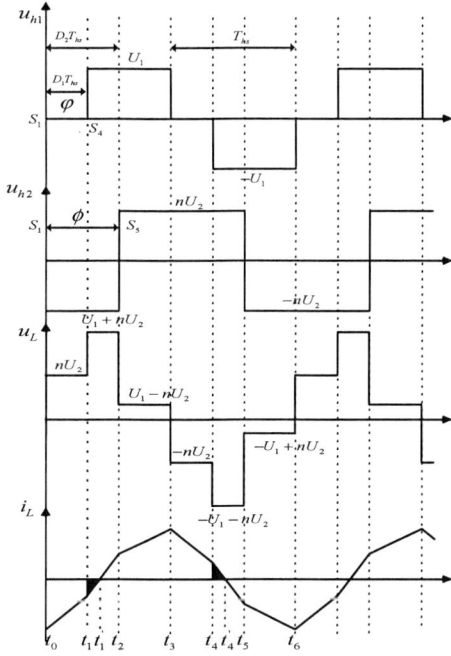

Fig.2 Waveforms of dual phase shift control

A. Analysis of backflow power of the DAB converter

Defining voltage regulation ratio $k = U_1/nU_2$ and switching frequency $f = 1/2T_{hs}$, which T_{hs} is half of the switching cycle, the following mathematical expressions are obtained:

$$i_L(t_0) = -\frac{nU_2}{4fL}\left[k(1-D_1)+2D_2-1\right] \quad (1)$$

$$i_L(t_1) = -\frac{nU_2}{4fL}\left[k(1-D_1)+2D_2-2D_1-1\right] \quad (2)$$

$$i_L(t_2) = \frac{nU_2}{4fL}\left[k(2D_2-D_1-1)+1\right] \quad (3)$$

As $P = \frac{1}{T}\int_0^T u_{h1}i_L(t)dt$, the transmission power under dual phase shift control can be expressed as:

$$P_D = \frac{nU_1U_2}{4fL}\left(-D_1^2+2D_1D_2-D_1-2D_2^2+2D_2\right) \quad (4)$$

Based on the definition of backflow power in the first section, the left backflow power and right backflow power under dual phase shift can be given by as follows, respectively:

$$P_{Dcir\_l} = \frac{nU_1U_2}{16fL(k+1)}\left[k(1-D_1)+2D_2-2D_1-1\right]^2 \quad (5)$$

$$P_{Dir\_r} = \frac{nU_1U_2}{16fLk(k+1)}\left[kD_1-2kD_2+k-1\right]^2 \quad (6)$$

B. Analysis of soft switching range under dual phase shift control

According to equation (2), when the inductance

current $i_L(t_1) \leq 0$ (when $i_L(t_1) = 0$ is the soft switching critical condition), soft switching constraint for the left H bridge can be deduced:

$$D_2 \geq \frac{(k+1)D_1 - k + 1}{2} \qquad (7)$$

Then, the switch S_1 and S_4 of the left H bridge can realize the zero voltage conduction and the soft switch off. Similarly, switch S_2 and S_3 can also achieve zero voltage conduction and soft switch off.

Fig 3 shows the working range of the soft switch of the left H bridge at different voltage conversion ratios obtained by the equation (7). The working range of the soft switch is above the straight line. It can be seen it in Fig 3, the voltage conversion ratio is larger, the working range of the soft switch of the left H bridge is larger.

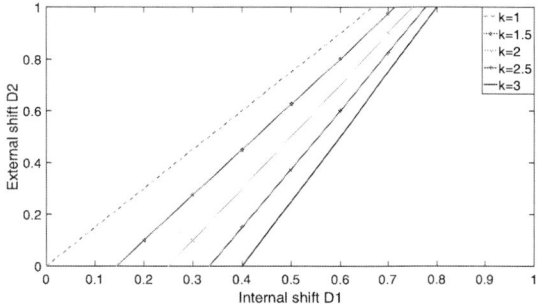

Fig.3 Soft switching range of left H bridge under different voltage conversion ratio k

Similarly, the soft switch constraint conditions of the right H bridge are obtained as follows:

$$D_2 \geq \frac{kD_1 + k - 1}{2k} \qquad (8)$$

Under the condition, the right H bridge switch $S_5 \sim S_8$ will satisfy the zero voltage conduction and the soft switch off condition.

Fig 4 shows the working range of the soft switch of the right H bridge at different voltage conversion ratios obtained by the equation (8). The working range of the soft switch is above the straight line. It can be seen from the Fig 4 that the voltage conversion ratio is lager and the working range of the soft switch of the right H bridge is smaller.

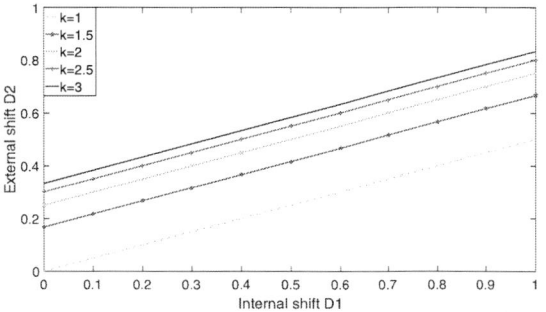

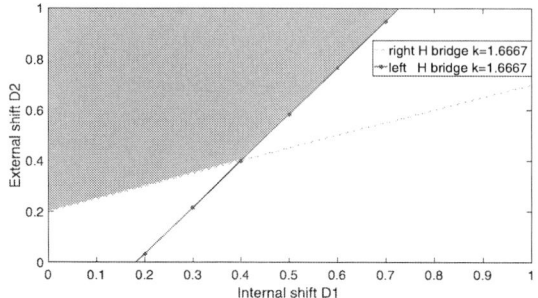

Fig.4 Soft switching range of right H bridge under different voltage conversion ratio k

In this paper, the voltage conversion ratio $k = 1.6667$ is taken as an example. The common range of the left and right H bridge with soft switching working is seen in the shadow area in Fig 5.

Fig.5 Common range of the left and right H bridge with soft switching working when voltage ratio $k = 1.6667$

III. THE MATHEMATICAL MODEL OF THE DAB AND THE IMPROVED PHASE SHIFT CONTROL STRATEGY

A. The mathematical model of the DAB

In order to simplify the analysis, the transmission power is normalized and the maximum transmission power P_N under the traditional single phase shift control is used as the reference value:

$$P_N = \frac{nU_1U_2}{8fL} \qquad (9)$$

According to the equation (4), (5), (6) and (9), the per-unit value of the transmission power and the left and right backflow power under the dual phase shift control can be obtained.

$$P_D' = \frac{P_D}{P_N} = 2\left(-D_1^2 + 2D_1D_2 - D_1 - 2D_2^2 + 2D_2\right) \qquad (10)$$

$$P_{Dir\_l}' = \frac{P_{Dir}}{P_N} = \frac{\left[k(1-D_1) + 2D_2 - 2D_1 - 1\right]^2}{2(k+1)} \qquad (11)$$

$$P_{Dir\_r} = \frac{\left[kD_1 - 2kD_2 + k - 1\right]^2}{2k(k+1)} \qquad (12)$$

The backflow power of the left H bridge may flow into the power side, so it is not expected that there is a lager backflow power on the left side. In the following analysis, the main discussion is to minimize the left backflow power.

Substituting equation (7) to (10), equation (13) can be obtained:

$$D_1 = \frac{(k^2 + k) - \sqrt{2(1-p)(1+k) - k^2 p}}{k^2 + 2k + 2} \qquad (13)$$

Equation (13) shows that when the transmission power $p \leq \dfrac{2k+2}{k^2 + 2k + 2}$, the backflow power at this point is

2720

zero by theoretical. If transmitting power $p > \dfrac{2k+2}{k^2+2k+2}$, the equation (13) has no solution and the minimum backflow power is at the optimum point of (D_1, D_2), which can be determined by Lagrange multiplication. Equation (14) is the basic form of Lagrange multiplication:

$$L(x,y,\lambda) = f(x,y) + \lambda g(x,y) \qquad (14)$$

In this paper, the backflow power is used as the objective function and the transmission power is the conditional function of equality constraint. Equation (14) can be written as:

$$L(D_1, D_2, \lambda) = P'_{Dcir}(D_1, D_2) + \lambda(P_0 - P) \qquad (15)$$

Partial derivatives for D_1, D_2 and λ :

$$\frac{\partial L}{\partial D_1} = \frac{-(k+2)}{k+1}\left[k(1-D_1) + 2D_2 - 2D_1 - 1\right]$$
$$+ \lambda(4D_2 - 4D_1 - 2) \qquad (16)$$
$$= 0$$

$$\frac{\partial L}{\partial D_2} = \frac{2}{k+1}\left[k(1-D_1) + 2D_2 - 2D_1 - 1\right]$$
$$+ \lambda(4D_1 - 8D_2 - 4) \qquad (17)$$
$$= 0$$

$$\frac{\partial L}{\partial \lambda} = 2\left(-D_1^2 + 2D_1D_2 - D_1 - 2D_2^2 + 2D_2\right) - P = 0 \qquad (18)$$

From the above equations:

$$\begin{cases} D_1 = \dfrac{(k+1)\sqrt{(k^2+2k+2)(1-P)}}{k^2+2k+2} \\[4mm] D_2 = \dfrac{k\sqrt{(k^2+2k+2)(1-P)}}{2(k^2+2k+2)} + \dfrac{1}{2} \end{cases} \qquad (19)$$

Therefore the minimum backflow power can be obtained:

$$P_{Dcir} = \frac{\left(k - \sqrt{(k^2+2k+2)(1-P)}\right)^2}{2(k+1)} \qquad (20)$$

B. Improved phase shift control strategy of the DAB

Fig 6 is the closed loop control block diagram of the system. The output power of the converter is determined by the output voltage U_2 and output current I_2 and then the relationship between the output power P and the voltage ratio k determines the magnitude of internal shift angle D_1. Under the condition of given D_1, the D_2 is adjusted by PI regulator to ensure the output voltage is constant.

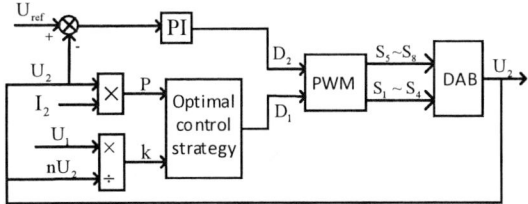

Fig.6 Control block diagram

When the D_1 in Fig 6 is set to zero, it is the traditional single phase shift control. Fig 7 gives the comparison of the backflow power of the DAB converter with dual phase shift control strategy with soft switching and single phase shift control under different transmission power and voltage ratio. As shown in Fig 7, with the increase of the transmission power P or the voltage variation k, the backflow power under the single phase shift control and the dual phase shift control is increasing. But under the proposed dual phase shift control strategy, when the transmission power $P \leq \dfrac{2k+2}{k^2+2k+2}$, the backflow power is zero in theoretically. When the transmission power $P = 1$, P_{cir_SPS} is equal to P_{cir_DPS}, and in other cases, P_{cir_DPS} is less than P_{cir_SPS}. Fig 8 is the control algorithm of optimized strategy. Table I is the D_1, D_2 value corresponding to the minimum backflow power at different transmission power ranges.

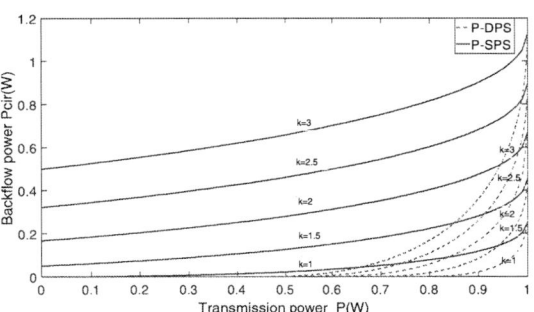

Fig.7 2-D Curve of P_{cir_SPS} and P_{cir_DPS}

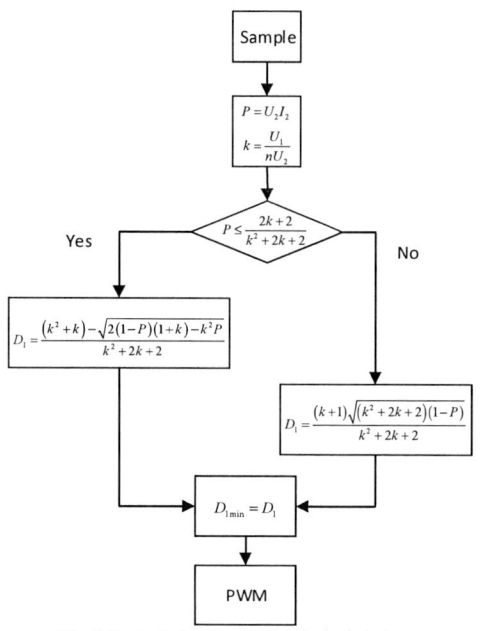

Fig.8 Control algorithm of optimized strategy

TABLE I
MINIMUM BACKFLOW POWER AND D_1 D_2 VALUES AT DIFFERENT RANGES OF P

P	D_1, D_2	P_{cir}
$0 \leq P \leq \dfrac{2k+2}{k^2+2k+2}$	$\begin{cases} D_1 = \dfrac{(k^2+k)-\sqrt{2(1-P)(1+k)-k^2P}}{k^2+2k+2} \\ D_2 = \dfrac{(k+1)D_1-k+1}{2} \end{cases}$	$P_{cir} = 0$
$\dfrac{2k+2}{k^2+2k+2} < P \leq 1$	$\begin{cases} D_1 = \dfrac{(k+1)\sqrt{(k^2+2k+2)(1-P)}}{k^2+2k+2} \\ D_2 = \dfrac{k\sqrt{(k^2+2k+2)(1-P)}}{2(k^2+2k+2)} + \dfrac{1}{2} \end{cases}$	$P_{cir} = \dfrac{\left(k-\sqrt{(k^2+2k+2)(1-P)}\right)^2}{2k+2}$

IV. SIMULATION BASED ON MATLAB

The DAB converter with proposed control strategy is simulated based on MATLAB/Simulink. The simulation parameters are listed as table II.

TABLE II
SIMULATION PARAMETERS

parameters	value
U_1	50V
U_2	30V
L	7.5uH
C_1	2200uF
C_2	2200uF
n	1
k	1.6667
f	50kHz

A. Parameter design of inductance and capacitance

By equation (9), we can get the relationship between the value of the inductor L and the transmission power:

$$L = \frac{nU_1U_2}{8fP} \quad (21)$$

In this paper, $P = 500W$ is taken as an example. We can get the value of inductor L from equation (21): $L = 7.5uH$.

In order to limit the dynamic change of the DC voltage in the load disturbance, the capacitance should be larger and meet the interference performance index ΔU_{max} of the load disturbance and the capacitance should be satisfied:

$$C_{min} = \frac{I_2T_{hs}}{\Delta U_{max}} \quad (22)$$

The output DC current I_2 is obtained by the transmission power P and the output DC voltage U_2 and $I_2 = 16.67A$. ΔU_{max} is set to 5% of the output DC voltage. Therefore $C_{min} = 222uF$ according to equation (22).In the actual circuit, the stray parameters of the components make it necessary to choose the larger capacitance to reach the ripple requirement. So the final selection of the capacitance value is $C = 2200uF$.

B. Theoretical analysis

Because of $k = U_1/nU_2 = 1.6667$, the critical value of the per-unit value of the transmission power is 0.6575 , and the rated power is P=500W , therefore, the critical value of the transmission power is $P_b = 328.75W$ .When $R = 3.6\Omega$, the transmission power $P = 250W$ is less than the critical transmission power $P_b = 328.75W$,so the theoretical value of backflow power is zero. When $R = 2.25\Omega$, the transmission power $P = 400W$ is greater than the critical transmission power $P_b = 328.75W$, The backflow power doesn't equal to zero and the backflow power should be less than the single phase shift control.

C. Simulation result

Fig 9 shows the waveforms of U_{h1} , U_{h2} and i_L of the DAB adopting single phase shift control and the $R = 2.25\Omega$. Fig 10 and 11 show the waveform of U_{h1} , U_{h2} and i_L of the DAB adopting proposed dual phase shift control. Fig 12 is the instantaneous output power waveform corresponding to Fig 9. It can be seen from the figure that the backflow power is approximately $P_{cir} = 390W$ at single phase shift control. Fig 13 is the instantaneous output power waveform corresponding to Fig 10 and it can be seen from the figure that the backflow power is zero. Fig 14 is the instantaneous output power waveform corresponding to Fig 11, we can see that the backflow power is

2722

approximately $P_{cir}=150W$, compared with the single phase shift in Fig 12, it reduces the backflow power about $240W$. From the above analysis, the simulation results are consistent with the theoretical analysis.

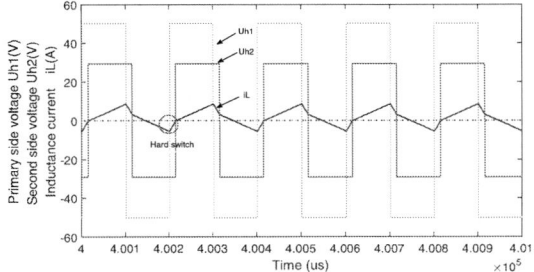

Fig.9 Single phase shift control at $R=2.25\Omega$

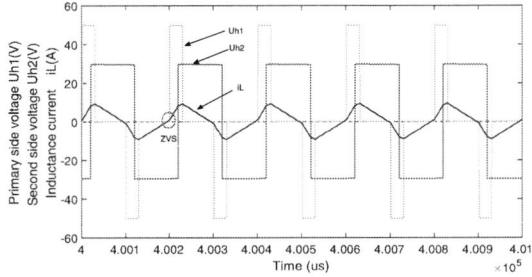

Fig.10 Dual phase shift control at $R=3.6\Omega$

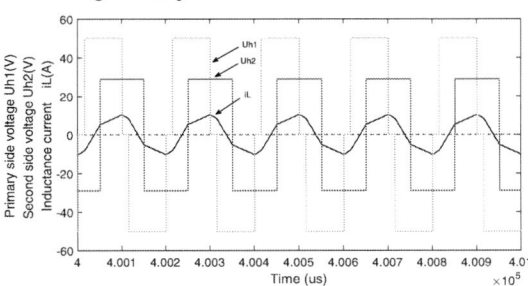

Fig.11 Dual phase shift control at $R=2.25\Omega$

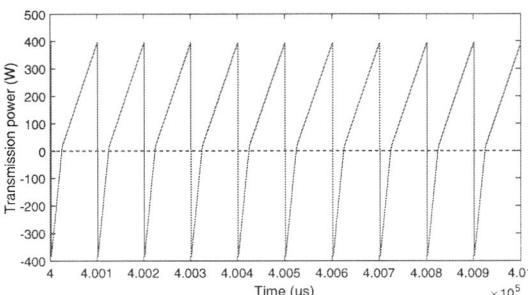

Fig.12 Transient waveform of output power at $R=2.25\Omega$ (SPS)

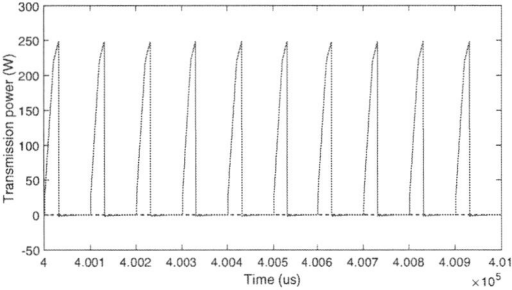

Fig.13 Transient waveform of output power at $R=3.6\Omega$ (DPS)

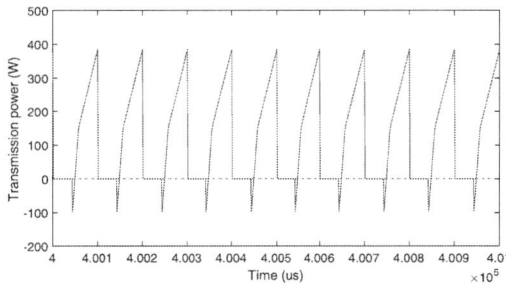

Fig.14 Transient waveform of output power at $R=2.25\Omega$ (DPS)

V. CONCLUSIONS

Based on the mathematical analysis of transmission power and backflow power of DAB converter, a mathematical model of backflow power under soft switching condition is established based on the condition of soft switching. The different expressions of the backflow power at different transmission power are obtained. In this paper, an improved dual phase shifted control strategy for dual active bridge DC-DC converter satisfying soft switching condition is proposed. This strategy not only makes converter work in soft switching condition, but also greatly reduces backflow power. The simulation results are carried out under MATLAB/Simulink, and the simulation results verify the correctness and effectiveness of the proposed control strategy.

ACKNOWLEDGMENT

Project is Supported by Sichuan Science and Technology (2016GZ0391) of China.

REFERENCES

[1] Zhao B, Yu Q, Sun W. Extended-phase-shift control of isolated bidirectional DC–DC converter for power distribution in microgrid[J]. IEEE Transactions on Power Electronics, 2012, 27(11): 4667-4680.

[2] Zhao Biao, Yu Qingguang, Wang Liwen. Novel Grid-connected UPS System With the Electricity Feedback Function and Its Distributed Logic Control Strategy[J]. Proceedings of the CSEE, 2011, 31(31): 85-93(in Chinese).

[3] Camara M B, Gualous H, Gustin F, et al. DC/DC converter design for supercapacitor and battery power management in hybrid vehicle applications—Polynomial control strategy[J]. IEEE Transactions on Industrial Electronics, 2010, 57(2): 587-597.

[4] Rodriguez, Alberto, et al. "Different purpose design strategies and techniques to improve the performance of a Dual Active Bridge with phase-shift control." Control and Modeling for Power

Electronics (COMPEL), 2014 IEEE 15th Workshop on. IEEE, 2014.

[5] Yazdani, Farzad, and Mohammadreza Zolghadri. "Design of dual active bridge isolated bi-directional DC converter based on current stress optimization." Power Electronics, Drive Systems & Technologies Conference (PEDSTC), 2017 8th. IEEE, 2017.

[6] Zhao, Biao, et al. "Current-stress-optimized switching strategy of isolated bidirectional DC–DC converter with dual-phase-shift control." IEEE Transactions on Industrial Electronics 60.10 (2013): 4458-4467.

[7] Kim, Myoungho, et al. "A dual-phase-shift control strategy for dual-active-bridge DC-DC converter in wide voltage range." Power Electronics and ECCE Asia (ICPE & ECCE), 2011 IEEE 8th International Conference on. IEEE, 2011.

[8] Facchinello, Gabriel Grunitzki, et al. "AC-AC hybrid dual active bridge converter for solid state transformer." Power Electronics for Distributed Generation Systems (PEDG), 2016 IEEE 7th International Symposium on. IEEE, 2016.

[9] Shi, Haochen, Huiqing Wen, and Jie Chen. "Reactive power reduction method based on harmonics analysis for dual active bridge converters with 3-level modulated phase-shift control." Power Electronics, Drives and Energy Systems (PEDES), 2016 IEEE International Conference on. IEEE, 2016.

[10] Shi, Haochen, et al. "Minimum-Reactive-Power Scheme of Dual Active Bridge DC-DC Converter With 3-Level Modulated Phase-Shift Control." IEEE Transactions on Industry Applications (2017).

[11] Bai H, Mi C. Eliminate reactive power and increase system efficiency of isolated bidirectional dual-active-bridge DC–DC converters using novel dual-phase-shift control[J]. IEEE Transactions on Power Electronics, 2008, 23(6): 2905-2914.

[12] Mi C, Bai H, Wang C, et al. Operation, design and control of dual H-bridge-based isolated bidirectional DC–DC converter[J]. IET Power Electronics, 2008, 1(4): 507-517.

[13] Gonzalez-Agudelo D, Escobar-Mejía A, Ramirez-Murrillo H. Dynamic model of a dual active bridge suitable for solid state transformers[C]. Power Electronics (CIEP), 2016 13th International Conference on. IEEE, 2016: 350-355.

[14] Wang, Dongzhi, Weige Zhang, and Jingxin Li. "PWM plus phase shift control strategy for dual-active-bridge DC-DC converter in electric vehicle charging/discharging system." Transportation Electrification Asia-Pacific (ITEC Asia-Pacific), 2014 IEEE Conference and Expo. IEEE, 2014.

[15] Zhao Biao, Yu Qingguang, Sun Weixin. Bi-directional Fullbridge DC-DC converters with dual-phase-shifting control and its backflow power characteristic analysis[J]. Proceedings of the CSEE, 2012,32(12):43-50(in Chinese).

[16] Zhang Xun, Wang Guangzhu, Shang Xiujuan, Wang Ting. "Daul phase shift control of backflow power optimization for bidirectional full bridge DC-DC converter" [J]. Proceedings of CSEE,2016, (04): 1090-1097(in Chinese).

[17] Wu Junjuan, Meng Deyue, Shen Yanfeng, Shen Hong, Sun Xiaofeng. Optimal control strategy of dual active bridge DC-DC converter based on the combination of dual phase shift control and traditional phase shift control [J]. Transactions of the Electrotechnical Society, 2016, (19): 97-105(in Chinese).

[18] Bai, Hua, Ziling Nie, and Chris Chunting Mi. "Experimental comparison of traditional phase-shift, dual-phase-shift, and model-based control of isolated bidirectional DC–DC converters." IEEE Transactions on Power Electronics25.6 (2010): 1444-1449.

[19] Feng B, Wang Y, Man J. A novel dual-phase-shift control strategy for dual-active-bridge DC-DC converter[C]. Industrial Electronics Society, IECON 2014-40th Annual Conference of the IEEE. IEEE, 2014: 4140-4145.

[20] Zhao B, Song Q, Liu W, et al. Current-stress-optimized switching strategy of isolated bidirectional DC–DC converter with dual-phase-shift control[J]. IEEE Transactions on Industrial Electronics, 2013, 60(10): 4458-4467.

[21] Kim M, Rosekeit M, Sul S K, et al. A dual-phase-shift control strategy for dual-active-bridge DC-DC converter in wide voltage range[C]. Power Electronics and ECCE Asia (ICPE & ECCE), 2011 IEEE 8th International Conference on. IEEE, 2011: 364-371.

[22] Song Wensheng, Hou Nie, Wu Mingyi, Feng Xiaoyun. The minimum peak current and its virtual power control method of bidirectional full bridge isolated DC/DC converter [J]. Proceedings of the CSEE, 2016,36 (18) (in Chinese).

[23] Hou Nie, Song Wensheng, Wu Mingyi.Normalization of phase shift control and minimum backflow power control of full-bridge isolated DC/DC converter[J]. Proceedings of the CSEE,2016,36(2):499-506(in Chinese).

[24] Gu Hongjie, Jiang Daozhuo, Yin Rui, Huang Liang, Wang Yufen. The three phase shift of bidirectional full bridge DC-DC power characteristics analysis [J]. Chinese power based on 2016,49 (07): 122-127(in Chinese).

[25] Wen, H., and W. Xiao. "Bidirectional dual-active-bridge DC-DC converter with triple-phase-shift control." Applied Power Electronics Conference and Exposition (APEC), 2013 Twenty-Eighth Annual IEEE. IEEE, 2013.

Development of an SiC High-Frequency PWM Inverter Using a Thick Multilayer PCB to Minimize Stray Inductance

Kohsuke Ishikawa[1*], Satoshi Ogasawara[1], Masatsugu Takemoto[1] and Koji Orikawa[1]

1 Graduate School of Information Science and Technology, Hokkaido University, Sapporo, Japan

*E-mail: k.ishikawa@ist.hokudai.ac.jp

Abstract— **Inverters using SiC or GaN power devices can realize high frequency and high efficiency operation. To achieve high efficiency, the switching characteristics of these power devices are important, because stray inductances in inverter main circuit have strong influence on the switching characteristics. To reduce the switching loss and surge voltage, minimization of stray inductance in the main circuit is required for high-frequency PWM inverter. This paper describes design guidelines for high-frequency inverters that realize low inductance. The PCB design guideline on the thick multilayer PCB is derived from the inductance calculation using 3D-FEA. It is shown experimentally, that the stray inductance of designed PCB can be reduced to the same level as the inductance inside the power devices. Experimental results verify that a prototype can achieve high speed switching and can suppress surge voltage. A load test is demonstrated to evaluate main circuit efficiency in half-bridge inverter at 100 kHz.**

Keywords— *SiC-MOSFET, Thick Multilayer PCB, Stray Inductance, PCB design.*

I. INTRODUCTION

In recent years, semiconductor power converters have been widely used due to problems such as depletion of fossil fuels and global warming. Power electronics is a key technology for energy conservation. Variable-speed AC-motor drive system using inverters are one of large application of power electronics. This system has high efficiency and easy maintenance properties, and it has been using in various fields, e.g., industrial applications and home appliances.

Si devices have been used in conventional power converters. However, performance of Si devices are reaching the theoretical limit for the sake of researches and improvement of semiconductor manufacturing technologies in many years. Compared with Si devices, next-generation power semiconductor devices using SiC or GaN have high speed, low loss and high-temperature operation characteristics. Thus, power converters using SiC or GaN devices can realize high frequency and high efficiency. To increase the switching frequency has advantages for power density and miniaturization of the converters.

In the next generation semiconductor devices, the baliga figure of merit which is the performance evaluation index of the power semiconductors is exceedingly higher than Si [1]. In particular, SiC device has a very high thermal conductivity, which is advantageous from the viewpoint of cooling. In addition, it is suitable for manufacturing high breakdown voltage devices, so it is expected to be applied to high voltage and large power applications [2]-[4].

The switching characteristics of power device are important to consider inverter efficiency. Parasitic inductances in inverter main circuit greatly affect the switching characteristics. The impacts of switching characteristics have been discussing in many researches [5]-[8]. The inductance in inverter makes switching speed slow, and generates an induced voltage due to current change di/dt. To increase stray inductance leads to increasing switching losses and surge voltage. Since switching losses are proportional to switching frequency, reduction of the stray inductance in high-frequency PWM inverter is extremely important to achieve high efficiency.

This paper describes development of an SiC high-frequency PWM inverter using a thick multilayer PCB to reduce the stray inductance. The authors have already shown that the stray inductance can be reduced to the same level as the inductance inside the device package by using a thick multilayer PCB. The design guideline for main circuit of high-frequency PWM inverter is discussed to reduce the stray inductance in a thick multilayer PCB. The inductance of a simplified PCB patterns is calculated using 3D-FEM, and the PCB design guideline to reduce the stray inductance in a thick multilayer PCB is derived. To evaluate stray inductance of the PCB designed based on the design guidelines, frequency characteristics of the corresponding impedance is measured. It is confirmed that the inverter main circuit can realize low inductance as 24 nH. From the experimental results, the prototype can achieve high speed switching operations and can suppress surge voltage due to reduction of stray inductance in the PCB. To verify improving the inverter efficiency, load test is demonstrated. As a result, the efficiency reaches 97.6% under the condition of chopper test with half-bridge inverter operation at 100 kHz.

II. Design Guideline for Main Circuit Pattern

A. Parasitic inductances on main circuit

Fig. 1 shows parasitic inductances in an equivalent circuit of a half-bridge inverter. Then, the inductances between power supply and inverter input L_{i1}, L_{i2} are existing in input power cables. The inductances L_{p1}, L_{p2}, L_{p3}, L_{p4} mean the stray inductances in the PCB patterns. The MOSFETs include drain inductances L_{d1}, L_{d2}, source inductances L_{s1}, L_{s2} and gate inductances L_{g1}, L_{g2}. Especially, drain and source inductances are strongly affected to switching characteristics, because these exist in a power loop through DC-link and power devices.

The stray inductance is classified into an inductance existing in the package of the power semiconductor device and an inductance existing in the inverter main circuit. The inductances in MOSFET depend on packaging of the device [9],[10]. Thus, the stray inductances included in PCB pattern and power cables can be reduced by inverter implementation techniques. Generally, DC-link capacitor and snubber capacitor likes C_{dc} shown in Fig. 1 are placed near the power device to reduce the effects of inductance in power cables. This paper discusses the pattern design of thick multilayer PCB to reduce the stray inductance.

B. Design guideline for main circuit pattern

As mentioned above, the stray inductance of the main circuit board greatly affects the switching characteristics in the inverter, it is important to design pattern of the main circuit board [11]-[14]. However, strict optimization is difficult because the pattern shape changes according to some restriction. Therefore, shape of a simplified PCB pattern was considered. The design guideline of the PCB pattern is derived by calculating stray inductance using 3D-FEM.

Fig. 2 shows the simplified analysis model. The pattern inductances are calculated using 3D-FEA software JMAG, by shorting the upper and lower conductors on the receiving side, as the inductance seen from transmitting side. The analysis is performed at 1MHz, and skin effect and proximity effect are considered. The four parameters of pattern length l, pattern width w, pattern thickness t and distance between layers d are changed individually.

Fig. 3(a) shows the inductance by changing the conductor length l. The pattern inductance increases in proportion to the conductor length, it is shown that shortening the pattern length is particularly important. Fig. 3(b) shows the inductance by changing the conductor width w. The pattern inductance decreases in inverse proportion to the conductor width, that is to say, expanding the conductor width is effective for decreasing the inductance. Fig. 3(c) shows the inductance by changing the conductor thickness t. The pattern inductance is constant, and the thickness does not affect to the inductance, but it affects to the dc current density. The currents concentrating on opposing surface by proximity effect may cause it. Thus, it is preferred to select the pattern thickness from current density on minimum pattern width in low-frequency. Fig. 3(d) shows the inductance by changing the distance between conductors d. The pattern

inductance increases in proportion to the distance. Making the interlayer distance as thin as possible within the range permitted by the insulation is more effective.

Based on the analysis results, the inverter PCB pattern is designed according to the following rules:
- The pattern length is shorter.
- The pattern width is larger.
- The distance between the layers is as thin as possible within satisfying the withstand voltage.
- The pattern thickness is determined from minimum pattern width and maximum current density.

The main circuit board have been designed with a thick copper multi-layer board which has 4 layers and 300 μm of pattern thickness. Since the rated maximum current of the DC-link is 17.7 A, if the allowable current density is 3 A/mm², the pattern width has to be wider than 19.6 mm. So, the minimum pattern width is designed to be 20 mm. For minimizing the inductance of the loop formed by the DC link and the switching device, the DC link positive electrode layer and the negative electrode layer are adjacent to each other to make the distance between the layers small. Additionally, the pattern is designed so that the pattern length is as short as possible, and the pattern width is larger.

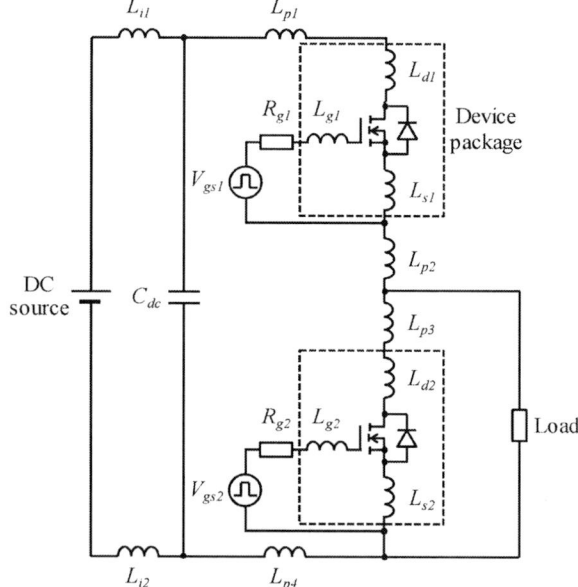

Fig. 1. Equivalent circuit of half-bridge inverter includes parasitic inductances.

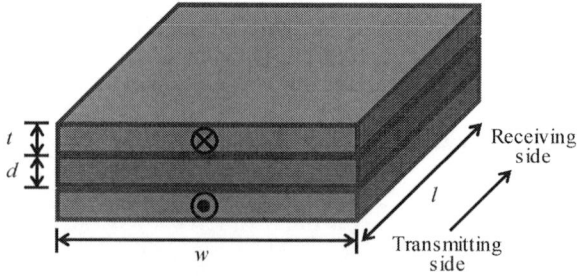

Fig. 2. The simplified analysis model with modified parameters.

III. INVERTER DESIRN TO MINIMIZE STRAY INDUCTANCE

A. Integration of inverter PCB

The inverter components are roughly classified into a main circuit, a drive circuit for driving a power device, a sensor circuit for detecting inverter voltage and current, and a control circuit for controlling gate signals. These circuits are designed separately and connected by long cables via connectors. However, the long cables may cause deterioration of switching characteristics. So, it is preferred to place the gate drive circuit close to the power devices. To miniaturize the inverter and to prevent malfunction of the control circuit affected by electromagnetic noise from the main circuit, it is desirable to integrate the gate circuit, the sensor circuit, and the control circuit on only one control board.

We have designed a prototype inverter using a thick multilayer PCB and an integrated control PCB. Fig. 4 shows the configuration of the inverter PCBs. The stray inductance between the device and gate driving circuits is minimized by placing the leads of the power device through both the main circuit board and the control board and setting the gate drive circuit on the control board around the leads.

On the other hand, to put the control board on top of the main circuit board increases the electrostatic coupling between these boards. Consequently, a voltage fluctuation in the main circuit may affect the control circuit and cause malfunction in the control board. As a countermeasure, a shield layer is provided in the uppermost layer of the main circuit board, and the propagation of noise from the main circuit board to the control board is blocked. To reduce the stray inductance, a laminate bus bar is widely used. Basically, applying the thick copper multilayer board to the main circuit is the same principle as the laminated bus bar. The feature of this main circuit using a thick multilayer PCB is characterized by the additional shield layer.

B. Reduction of board area by modulation

In the integrated structure of control board, many components have to be mounted only on one side of the control board. Therefore, the area of the integrated control board increases, and the main circuit board also becomes large. In addition, the control board has some dead space above the board due to some tall components such as current sensors and film capacitors.

To use this dead space effectively, we adopted a structure, in which each function is modularized to a sub-board and each sub-board is placed vertically on the control board. It is possible to minimize the board area by selecting height of each sub-board to height of tallest component among the parts mounted on the control board. Finally, the modularized structure can provide reduction of the 35% board size and the 22% inverter volume, compared with a control board on which all components are arranged on one side. As a result, the switching characteristics of the inverter can be improved by shortening the pattern length, because board area of the

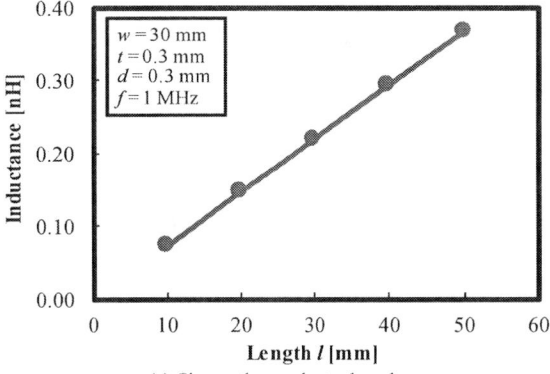

(a) Change the conductor length.

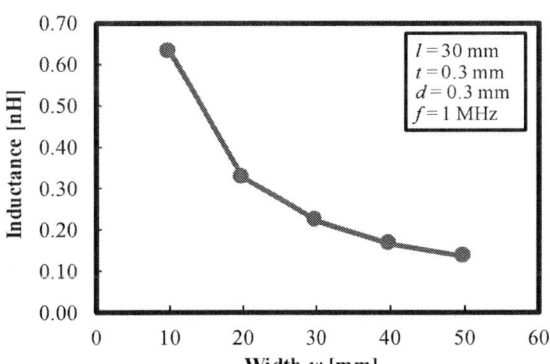

(b) Change the conductor width.

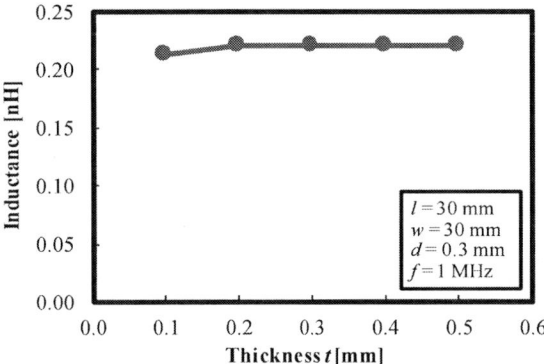

(c) Change the conductor thickness.

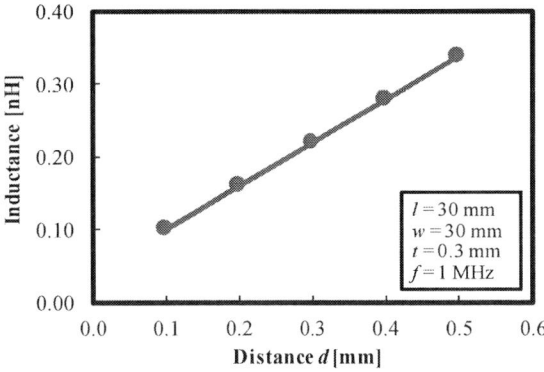

(d) Change the distance between conductors.

Fig. 3. Analysis results of inductance on thick copper boards.

main circuit is also reduced.

The appearance of the prototype which designed according to derived design guideline is shown in Fig. 5. TABLE I shows specification of prototype. The main circuit board is designed to be longer than the control board, so that the electrolytic capacitor, the current sensor, and the terminals are directly attached to the extended part.

IV. IMPEDANCE CHARACTERISTICS OF INVERTER PCB

A. Measurement method of impedance characteristic

Fig. 6 shows the appearance of the impedance measurement system. The impedance characteristic has been measured to evaluate the stray inductance of the designed inverter PCB using an impedance analyzer (E4990A) in frequency range from 10 kHz to 120 MHz. Then, the terminals between drain and source of high side and low side devices are shorted to measure the inductance in the loop which is made by DC-link and power devices. For decrease in measurement errors, the DC-link of the main circuit are connected to an impedance analyzer by copper tapes as shown in Fig. 6. Snubber circuits are ignored to evaluate only the inductance of PCB.

B. Impedance characteristics and equivalent circuit

Fig. 7 shows measured impedance characteristic of the inverter PCB (red-line) and simulated impedance characteristic (blue-line) by circuit simulator. Fig. 8 shows an equivalent circuit calculated from the measured impedance characteristic.

The PCB can be modeled stray inductances of 11 nH and 13 nH, capacitance between the PCB layers of 2000 pF and parasitic dumping resistances of 85 mΩ and 83 mΩ. For the region from 10 kHz to 10 MHz, the PCB acts as an inductance of 11 + 13 = 24 nH. There is an anti-resonance at 32 MHz due to parallel resonance of 2000 pF and 13 nH, and a resonance at 46 MHz is generated by series resonance of 11 nH and 2000 pF. In the higher frequency range, since the impedance of the parallel capacitance becomes very small, the impedance characteristic is obtained as the inductance of 11 nH.

Since the resistance value is determined to match the measured and simulated characteristics at the resonance and anti-resonance frequency, these closely resemble each other in the frequency range of several MHz or more. However, deviation occurs in the frequency region lower than that. The reason is that the actual resistance increases with the frequency by the skin effect, but it is handled as a constant value in the equivalent circuit. From this, it is understood that the resistance at the resonance and anti-resonance frequency is several ten times that in the low frequency region.

According to some literature, TO-247 package which used in the prototype has 18 nH of stray inductance [15]. In section V, an impedance characteristic of the using device is measured. The total stray inductance of designed PCB is 24 nH, which is sum of 11 nH and 13 nH. As a result, the stray inductance of designed PCB can be reduced to the same level as the inductance inside the device.

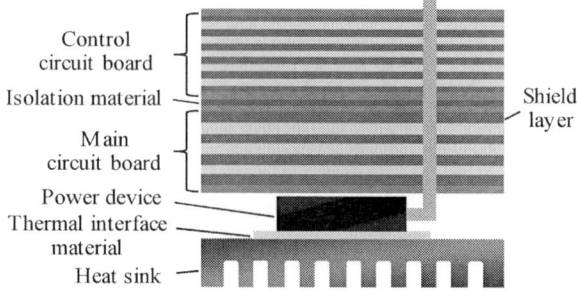

Fig. 4. Configuration of inverter PCBs.

Fig. 5. Appearance of the prototype.

TABLE I
SPECIFICATION OF PROTOTYPE

Parameter	Value
Phase number	Three-phase
Rated output capacity	10 kVA
Rated output voltage	400 Vrms
Rated output current	14.5 Arms
Size	200 mm × 130 mm × 180 mm
Device	SCH2080KE

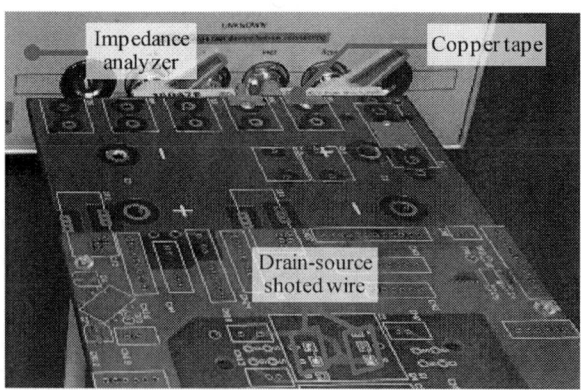

Fig. 6. Appearance of the impedance measurement system.

2728

V. EXPERIMENTAL RESULT

A. Experimental circuit

Fig. 9 shows experimental circuit, and Table II shows experimental condition. Chopper test is adopted to ignore the effects of deadtime. The switching characteristics are measured by oscilloscope (HDO8058) at 2.5 GS/s of sampling rate. For measuring accurately, the probe ground lead should as short as possible [16]. The passive probes PP018 (500 MHz) and HVP120 (400 MHz) are connected via dedicated adaptor that used short wire to measuring terminal close to devices.

By using power meter (PW6001) with external current sensor (PW9100), the inverter efficiency is measured. By changing load resistances, the inverter output power is adjusted. The measurements are performed 10 times, and the efficiency is regarded as the average value.

B. Switching characteristics

The switching waveforms when turn-on and turn-off transient are shown in Fig. 10. TABLE III shows switching characteristics of prototype. Since, the rise time and fall time are 47.3 ns and 40.9 ns, the voltage slew rate dv/dt of drain-source voltage is as fast as about 10 kV/μs. The maximum surge voltage is 21.4 V as 4% of the DC-link voltage in spite of fast switching speed. The PCB's low inductance leads to high switching speed and decrease in surge voltage.

In turn-off transient, with sufficiently large snubber capacitance C_{snub} as compared to device output capacitance C_{oss}, the ringing frequency f_r can be estimated as

$$f_r = \frac{1}{2\pi\sqrt{L_{loop}C_{oss}}} \tag{1}$$

where, L_{loop} is total inductance of loop path that flows high frequency current. Fig. 11 shows frequency analysis result of drain-source voltage v_{DS} during turn-off transient. The frequency f_r is decided as 41.8 MHz from Fig. 12. A value of the C_{oss} is selected as 190 pF referred from datasheet at 564V of the voltage v_{DS} [17]. Based on the equation (1), The inductance L_{loop} can be estimated as 76 nH.

The inverter has snubber capacitors in parallel with the legs, inverter input inductance can ignore, the path consists of the loop inductance is shown in Fig. 11. The loop inductance follows equation (2).

$$L_{loop} = L_p + 2 \cdot (L_d + L_s) + L_{snub} \tag{2}$$

where, L_p is the inductance in PCB patterns, each of L_d and L_s are MOSFET's drain and source inductance, and L_{sunb} is an inductance in a snubber capacitor. These parameters are determined by impedance analysis as shown in Table IV.

The loop inductance estimated from equation (1) and (2) are each of 76 nH and 68 nH, which are approximately equal. Thus, a resonance which device output capacitance and the inductance in loop that consists snubber capacitor caused the ringing during turn-off transient.

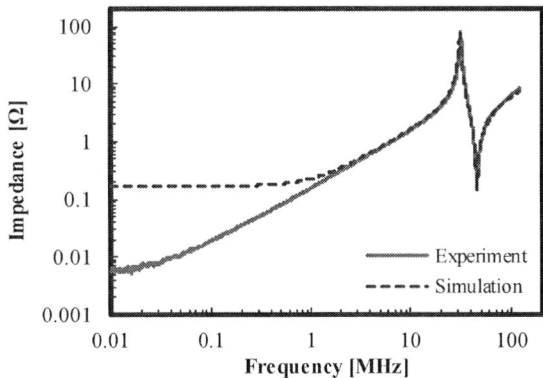

Fig. 7. Impedance characteristic of the inverter PCBs.

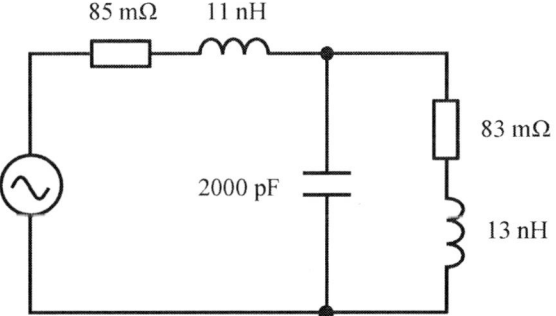

Fig. 8. Equivalent circuit of the inverter PCBs.

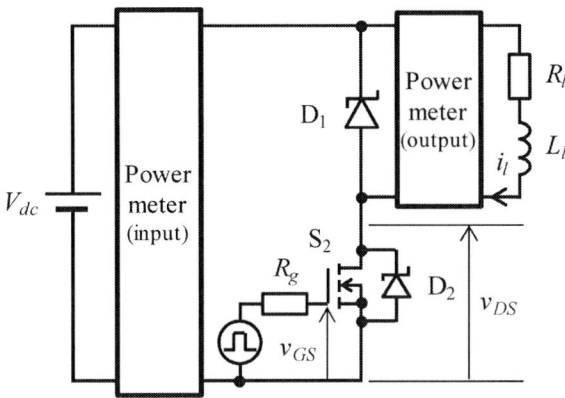

Fig. 9. Experimental circuit.

TABLE II
EXPERIMENTAL CONDITION

Parameter	Symbol	Value
DC-link voltage	V_{dc}	564 V
Load resistance	R_l	24 - 120 Ω
Load inductance	L_l	15 mH
Gate-Source voltage	V_{GS}	+18.2 V / -3.3 V
Internal Gate resistance	$R_{g(int)}$	6.7 Ω
External Gate resistance	$R_{g(ext)}$	3.3 Ω
Switching frequency	f_{sw}	100 kHz

C. Efficiency characteristic

Fig. 13 shows the relationship of output power and the inverter main circuit efficiency at a switching frequency of 100 kHz, by changing load resistance. The inverter achieved 97.6 % of maximum efficiency at 3.3 kW of output power that equal to 80% of the rated output capacity. In this measurement, losses of protection circuit and discharging resistance of DC-link capacitors are including.

The total inverter loss P_{loss} is divided into conduction losses and switching losses, can be obtained as

$$P_{loss} = P_{c(MOS)} + P_{c(SBD)} + P_{sw} \qquad (3)$$

where, $P_{c(MOS)}$ is conduction losses of MOSFET, $P_{c(SBD)}$ is conduction losses of schottky barrier diode and P_{sw} is switching losses of device. The conduction losses of MOSFET is joule loss of on-resistance, can be derived as

$$P_{c(MOS)} = \frac{1}{T} \int_0^T R_{DS(on)}(i_d) \cdot i_d^2 \, dt \qquad (4)$$

where, T is output period, $R_{DS(on)}$ is on-resistance between MOFET's Drain and Source, and i_d is drain current. The conduction losses of schottky barrier diode is caused by forward-voltage drop, is given by

$$P_{c(SBD)} = \frac{1}{T} \int_0^T v_F(i_F) \cdot i_F \, dt \qquad (5)$$

where, v_F is forward-voltage of schottky barrier diode and i_F is current that flows Schottky barrier diode. When a device S_2 is turn-on, load current flows S_2. Similarly, when S2 is turn-off, load current goes round a path that through a diode D_1. According to the equations (3)-(5), the switching losses can be determined.

Fig. 12 also shows breakdown of the losses. In the maximum efficiency point, the conduction losses and the switching losses are each of 13 W and 66 W. In other words, ratio of the conduction losses and the switching losses is 1:5. Theoretically, inverter efficiency reaches maximum when the conduction losses and the switching losses are equal. Thus, the efficiency may be improved in more high output power operation point.

VI. CONCLUSION

In this paper, to reduce the stray inductance of the main circuit, the inverter using the thick copper multilayer board was described. The inductance on simplified patterns is calculated by 3D-FEM, and the design guideline to reduce the stray inductance on thick multilayer PCB is derived.

The impedance characteristics show that the stray inductance of the designed PCB using the design guideline and the modularized structure can reduce the loop inductance to 24 nH, which is the same level as the terminal inductance of the device. The switching characteristics show that the prototype can achieve high speed switching operations and can suppress surge voltage within 4% of DC-link voltage. Additionally, the ringing at

TABLE III
SWITCHING CHARACTERISTICS OF THE PROTOTYPE

Parameter	Value
Rise time	47.3 ns
Fall time	40.9 ns
Maximum turn-off surge	21.4 V
Maximum turn-on surge	-4.6 V
Ringing frequency	41.8 MHz

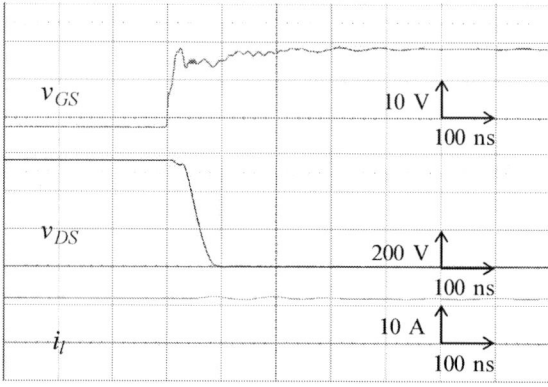

(a) Turn-on transient.

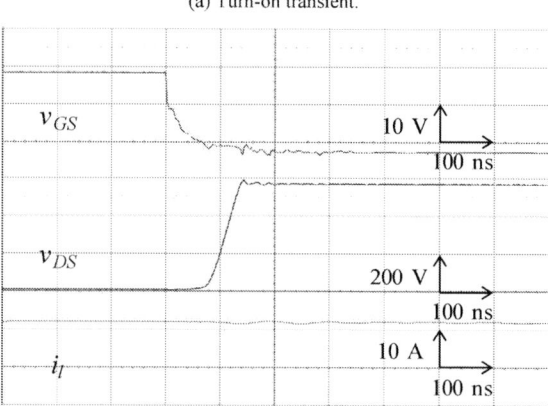

(b) Turn-off transient.

Fig. 10. Switching waveforms when turn-on and turn-off transient.

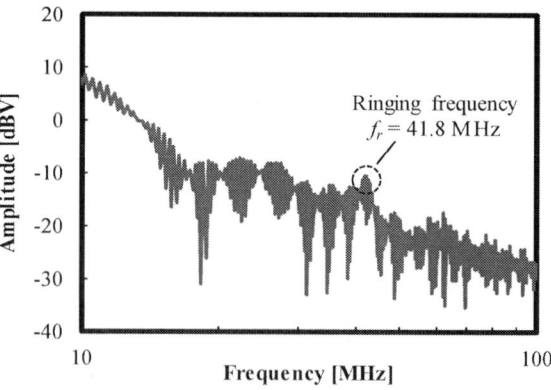

Fig. 11. Frequency analysis result of the waveform of v_{DS} in turn-off transient.

turn-off transient was evaluated quantitatively. As a result, the efficiency of the prototype that operates half-bridge inverter on chopper test method can obtain 97.6% at 100 kHz. Therefore, to reduce the stray inductance can contribute to improve the inverter efficiency at high frequency.

ACKNOWLEDGMENT

This work was supported by Council for Science, Technology and Innovation (CSTI), Cross-ministerial Strategic Innovation Promotion Program (SIP), "Next-generation power electronics" (funding agency: NEDO)

REFERENCES

[1] B. J. Baliga, "Power semiconductor device figure of merit for high-frequency applications," *IEEE Electron Device Lett*, vol. 10, no. 10, pp. 455-457, 1989.

[2] K. Shinai, R. S. Scott, and B. J. Baliga, "Optimum semiconductors for high-power electronics," *IEEE Trans. Electron Devices*, vol. 36, no. 9, pp. 1811-1823, 1989.

[3] C. E. Weitzel et al., "Silicon carbide high-power devices," *IEEE Trans. Electron Devices*, vol. 43, no. 10, pp. 1732-1741, 1996.

[4] F. Hilpert, K. Brinkfeldt and S. Arenz, "Modular integration of a 1200 V SiC inverter in a commercial vehicle wheel-hub drivetrain," *in Proc. EDPC*, pp.1-8, 2014.

[5] J. Wang, H.S.H.Chung, and R. T. H. Li, "Characterization and Experimental Assessment of the Effects of Parasitic Elements on the MOSFET Switching Performance," *IEEE Trans. Power Electron.*, vol. 28, no. 1, pp. 573-590, 2013.

[6] Z. Chen, D. Boroyevich, and R. Burgos, "Experimental parametric study of the parasitic inductance influence on MOSFET switching characteristics," *in Proc. IPEC*, pp. 164-169, 2010.

[7] I. Josifovic, J. P. Gerber, and J.A. Ferreira, "Improving SiC JFET Switching Behavior Under Influence of Circuit Parasitics," *IEEE Trans. Power Electron.*, vol. 27, no. 8, pp. 3843-3854, 2012.

[8] Y. Xiao, H. Shah, T. P. Chow, and R. J. Gutmann, "Analytical modeling and experimental evaluation of interconnect parasitic inductance on MOSFET switching characteristics," *in Proc. APEC*, vol. 1, pp. 516-521, 2004.

[9] L. Zhang, S. Guo, X. Li, Y. Lei, W. Yu, and A. Q. Huang, "Integrated SiC MOSFET module with ultra low parasitic inductance for noise free ultra high speed switching," *in Proc. WiPDA*, pp. 224-229, 2017.

[10] K. Aikawa, T. Shiida, R. Matsumoto, K. Umetani, and E. Hiraki, "Measurement of the common source inductance of typical switching device packages," in Proc. IFEEC, pp. 1172-1177, 2017.

[11] J. Fabre, P. Ladoux, and M. Piton, "Characterization and Implementation of Dual-SiC MOSFET Modules for Future Use in Traction Converters," *IEEE Trans. Power Electron.*, vol. 30, no. 8, pp. 4079-4090, 2015.

[12] K. Wada, M. Ando, and A. Hino, "Design of Dc-side wiring structure for high-speed switching operation using SiC power devices," *in Proc. APEC*, pp. 584-590, 2013.

[13] D. Reusch, and J. Strydom, "Understanding the Effect of PCB Layout on Circuit Performance in a High-Frequency Gallium-Nitride-Based Point of Load Converter," *IEEE Trans. Power Electron.*, vol. 29, no. 4, pp. 2008-2015, 2014.

[14] M. C. Caponet, F. Profumo, R. W. D. Doncker, and A. Tenconi, "Low stray inductance bus bar design and construction for good EMC performance in power electronic circuits," *IEEE Trans. Power Electron.*, vol. 17, no. 2, pp. 225-231, 2002.

[15] G. J. Krausse, "DE-series fast power MOSFET", Direct Energy Inc. Technical Note, 2002.

[16] M. Young, "Methodology for switching characterization evaluation of wide band-gap devices in a phase-leg configuration," *in Proc. APEC*, pp. 2534-2541, 2014.

[17] Rohm Semiconductor, "N-channel SiC power MOSFET co-packaged with SiC-SBD", SCH2080KE datasheet, 2012 [Revised June 2017].

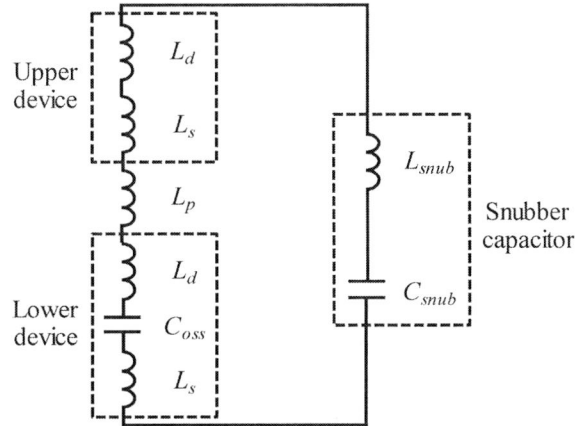

Fig. 12. The loop path that flows high frequency current.

TABLE IV
MEASURED INDUCTANCE OF FIG.12

Parameter	Symbol	Value
Inverter PCB	L_p	14 nH
MOSFET (drain lead)	L_d	6 nH
MOSFET (source lead)	L_s	12 nH
Snubber capacitor	L_{snub}	18 nH
Total	L_{loop}	68 nH

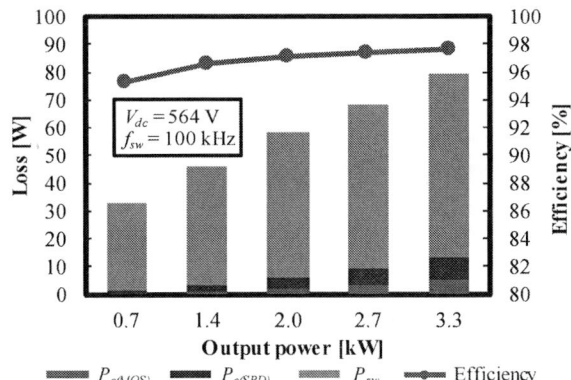

Fig. 13. Output power vs. efficiency characteristic at 100 kHz.

The 2018 International Power Electronics Conference

Fast Switching Planar Power Module With SiC MOSFETs and Ultra-low Parasitic Inductance

Arash Edvin Risseh[1]*, Hans-Peter Nee[1], Konstantin Kostov[2]

1 School of Electrical Engineering, KTH Royal Institute of Technology, Stockholm, Sweden
2 The Mads Clausen Institute, SDU Electrical Engineering, Sonderborg, Denmark
*E-mail: risseh@kth.se

Abstract—Parasitic inductances caused by the package of semiconductor devices in power converters, are limiting the switching speed and giving rise to higher switching losses than necessary. In this study a half-bridge planar power module with Silicon Carbide (SiC) MOSFET bare dies was designed and manufactured for ultra-low parasitic inductance. The circuit structure was simulated and the parasitic inductances were extracted from ANSYS-Q3D. The values were then fed into LT-Spice to simulate the electrical behavior of the half-bridge. The experimental and simulation results were compared to each other and were used to adjust and easily extend the simulation model with additional MOSFETs for higher current capability. It was shown that the proposed planar module, with four parallel SiC MOSFETs at each position, is able to switch 600V and 400A during 40 and 17ns with E_{ON} and E_{OFF} equal to 3.1 and 1.3mJ, respectively. Moreover, unlike the commercial modules, this design allows double-sided cooling to extract the generated heat from the device, resulting in lower operating temperature.

Keywords—*SiC MOSFET, planar power module, bare die, ultra-low parasitic inductance, fast switching, PCB, double sided cooling.*

I. INTRODUCTION

Efforts have been put into development of fast switching speed in power electronics [1]–[4]. Among others, this may be done by adding different components to the circuit, proposing turn-on/off strategies [5], [6], suggesting gate-driver circuits [7], [8], or by using new device- and package-structures [9], [10]. The fundamental issue regarding switching time in power semiconductor devices is the parasitic inductance of the circuit board and the package of the device itself. The parasitic inductance increases the rise- and fall times, which in turn increase the switching losses and also limit the switching frequency. Furthermore, the inductance together with the capacitance in the circuit cause overshoots and oscillations which also give rise to additional losses and thereby higher junction temperature and accelerated component aging. The impact of the parasitic inductance in power converters as well as new packages and in power modules, with low parasitic inductance, has been studied and presented in [11]–[16].

Silicon Carbide (SiC) metal-oxide semiconductor field-effect transistors (MOSFETs) have been developed rapidly during the last decade. SiC MOSFETs have magnificent characteristics such as low input capacitance, very low ON-state resistance and capability of operation at high temperature and frequency. However, the commercially available SiC MOSFET modules are packaged using technology developed for silicon insulated gate bipolar

transistors (IGBTs) featuring only single-side cooling and high parasitic inductance. In order to extensively take advantage of SiC MOSFETs, new double-side cooled packages with ultra-low inductance are required. In order to eliminate the parasitic inductances, the *regular* package of the device may completely be excluded.

In this investigation, a planar power module with four SiC MOSFET bare dies, directly connected to two printed circuit boards (PCBs), was designed and manufactured. The power module was formed as a half-bridge where two chips were placed in parallel at each position, and the upper and lower switches were connected to two separate gate drivers. Wolfspeed's CPM2-1200-0025B (1.2 kV & 98 A & 25 mΩ) was used as SiC MOSFETs in the module.When the functionality of the module was verified the module was exposed to a double-pulse test (DPT). Later on, the measurement results were used to refine and obtain an accurate model of the structure. The new simulation model was then used to *expand* the module with additional bare dies to increase the current capacity. The manufactured module as well as the expanded model has extremely low parasitic inductances which enables fast switching speed and low losses. A similar concept was reported in an earlier study where a module was set up as a DC/DC converter with low voltage and current(15 V & 1 A) [17]. In Section II the structure and the test procedure of the module as well as the experimental results are presented. The parasitic components from the structure extracted from ANSYS Q3D were simulated in LT-Spice and the results are presented in Section III followed by a presentation of the proposed planar module, extended with four chips in Section IV. Section V presents the results showing the waveforms and switching times of the proposed extended module. Finally, in Section VI conclusions are drawn.

II. CONCEPT AND MEASUREMENTS

The package of a MOSFET and interconnections to its substrate create internal and external parasitic elements such as inductances and capacitances. Figure 1 shows a model of a power MOSFET where internal parasitic elements due to the package of the device and external parasitic elements due to the outside circuits, are shown. In order to reduce the parasitic inductances, the chip (bare die) of the MOSFET may directly be attached to a substrate where other components such as gate-drive circuits operate. By excluding the wire-bonds and leads of the package, part of the total parasitic elements will decrease.

The 2018 International Power Electronics Conference

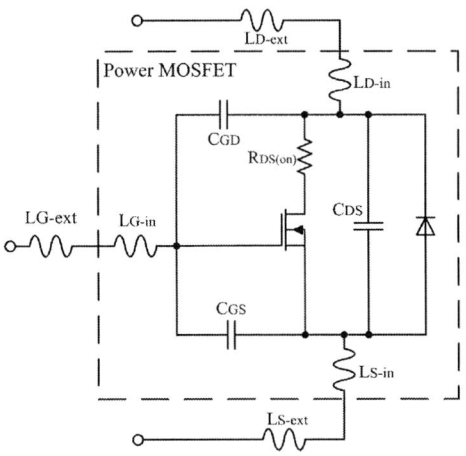

Fig. 1. A model of a power MOSFET including parasitic elements.

A half-bridge module as seen in Fig. 2 was considered. It may be designed with Wolfspeed's CPM2-1200-0025B bare dies. The structure, the size of the bare die pads, and the type of the substrate (FR4) were drawn in a simulation software (ANSYS Q3D) to determine the parasitic elements of the module. Part of the simulated structure, drawn in ANSYS Q3D and the size of the bare die can be seen in Fig. 3. The dies were placed and sandwiched between two PCBs to create connections between the two sides of the dies, as shown in Fig. 4. PCB 1 is the main substrate where the gate-drive circuits and other necessary components are placed. The connections to the Drain pads are also placed on PCB 1. The purpose of PCB 2 is to create connections to the Source and Gate pads and thereby connect them to PCB 1 via connectors, see Figs. 3 and 5 . The entire manufacturing procedure and more details about the structure is described in [18]. The parasitic elements could be determined and extracted from ANSYS Q3D. The values are listed in Tables I and II which confirm that the structure exhibits extremely low inductances. According to the simulations in ANSYS Q3D, the parasitic inductance (L_{stray}) of the module is approximately 0.659 nH which is approximately 96 % lower than most commercial 1.2 kV half-bridge modules.

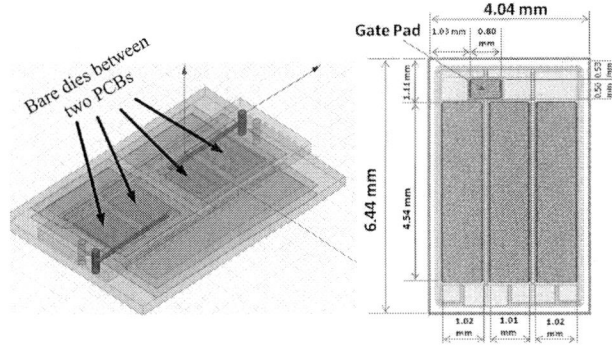

Fig. 3. The layout and the size of the bare die pads (right) and the structure of the planar module with four bare dies from ANSYS Q3D (left).

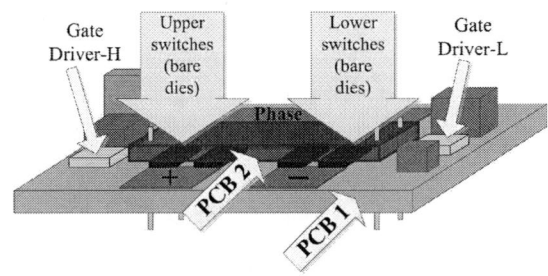

Fig. 4. Structure of the half-bridge planar power module where two bare dies are connected in parallel at each position and can be sandwiched between two sunbstrates.

TABLE I. SIMULATION RESULTS FROM ANSYS Q3D: PARASITIC *inductances* OF THE MODULE.

Parasitic Inductance [nH]	FET 1	FET 2	FET 3	FET 4
GATE	7.20	3.47	2.37	5.96
DRAIN	0.0862	0.0876	0.347	0.346
SOURCE	0.632	0.638	0.249	0.254

TABLE II. SIMULATION RESULTS FROM ANSYS Q3D: PARASITIC *resistances* OF THE MODULE.

Parasitic Resistance [$\mu\Omega$]	FET 1	FET 2	FET 3	FET 4
GATE	8534	4323	3131	7344
DRAIN	58.51	59.14	132.3	132.6
SOURCE	216.5	217.1	122.4	124.2

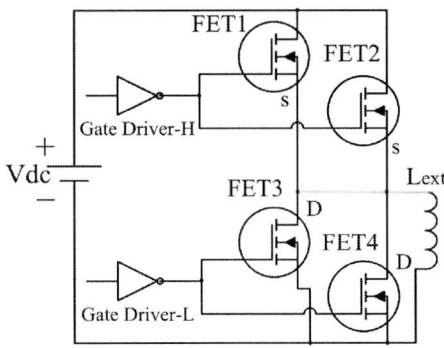

Fig. 2. Schematic diagram of the half-bridge configuration with two SiC MOSFETs in parallel at each position. An external inductor (L_{ext}) was employed to perform double double pulse test.

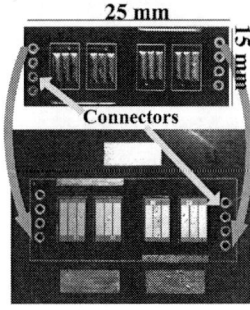

Fig. 5. The figure shows part of PCB 1 (lower) and PCB 2 (upper). SiC MOSFET bare dies are attached to PCB 1. They were then sandwiched and soldered to PCB 1 and PCB 2, forming a planar module. The connector pins were used to transfer the Gate and Source signals.

2733

The low parasitic inductances, as seen in Table I, enable fast switching speed. Since, FR4 has a relatively large thermal resistance it would not be able to extract any significant amount of heat from the chips. In order to generate the lowest possible heat power a regular DPT-setup was configured, see Fig. 6. The low-side MOSFETs act as freewheeling path where the current passes through their channels. The first applied pulse was 6 μs and the second one was 2.5 μs long. Two identical isolated gate drivers were employed to drive the MOSFETs with an external Rg=10 Ω. The gate voltages varied between -3 and +15 V. An inductor (30 μH) was connected in parallel to the low-side switches and the applied voltage was increased stepwise and at 400 V a current of 75 A was flowing through the switches. The components used in the planar module are listed in Table III and measurement results are presented in Fig. 7 and Fig. 8. As seen in these figures, there is one overshoot in V_{DS} reaching just above 500 V and no significant oscillations during the turn-off instant can be observed. The rise time of V_{DS} was measured to approximately 23 ns. During the turn-on, no oscillation can be observed and the turn-on time was measured to approximately 60 ns.

TABLE III.   THE MODULE COMPONENT LIST.

Component	Part number	Function/Remarks
Regular PCB	-	2 mm thick FR4 & 35μm copper
Bare die	CPM2-1200-0025B	Switches
DSP	TMDSCNCD28335	Signal generator
Gate driver and isolator	ADuM3221	-
DC-DC converter	Traco TMR 6-2413 and 2410	Supply gate-drive circuits
Capacitor	GRM188C81E475KE11D	Capacitor bank
Capacitor	107TTA350M	DC-link
Inductor	Core ETD59	The air gap and windings were changed during the experiment to adjust the current

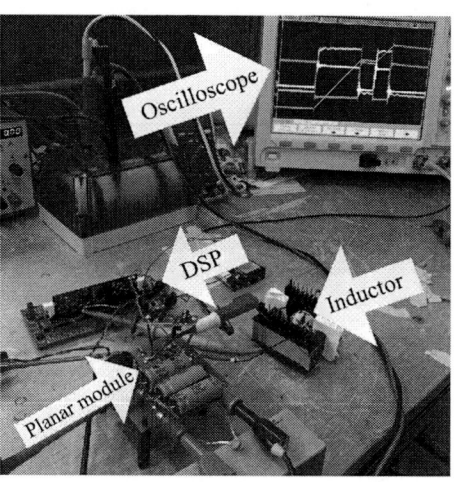

Fig. 6. The figure shows the planar module and the test setup. The gate signals were generated by a digital signal processor (DSP) and were sent directly to two separate digital isolators, acting as gate drivers. The gate signals, V_{DS} and inductor current were measured and studied on the oscilloscope.

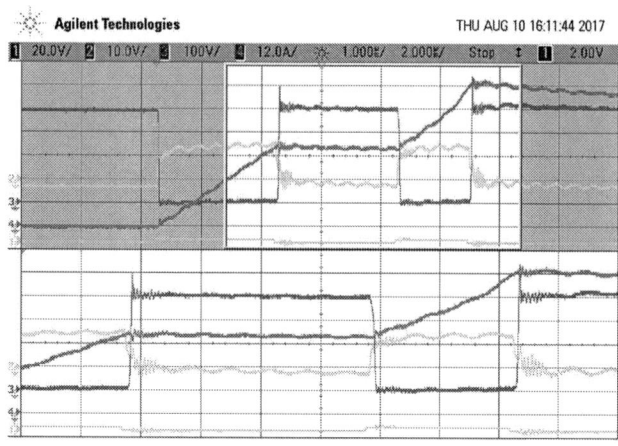

Fig. 7. Measurement results from the DPT. The green pulse shows the gate signal and the lilac shows the V_{DS} of the DUT. The red signal is the inductor current and the yellow one shows the control signal from the controller (3.3 V). Time base for the lower graph is 1 μs/div.

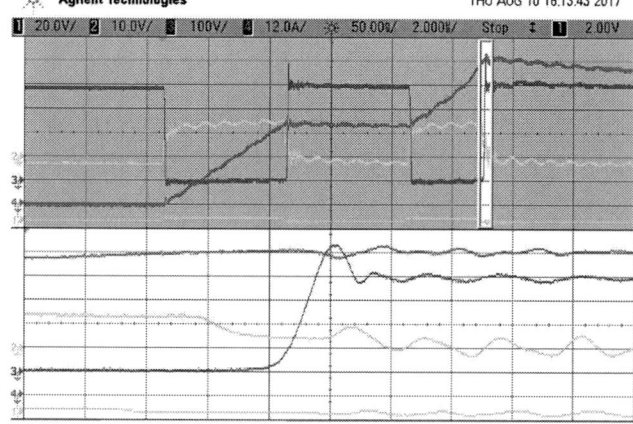

Fig. 8. Measurement results from the DPT. The green pulse shows the gate signal and the lilac shows the V_{DS} of the DUT. The red signal is the inductor current and the yellow one shows the control signal from the controller (3.3 V). The graph is zoomed into the switching instant when the upper switches turn off. Time base for the lower graph is 50 ns/div.

In order to measure the signals, an Agilent Technologies (MSO7104A) oscilloscope was used. Voltages were measured by isolated differential probes and the inductor current was measured by a Rogowski Current Transducer (CWT). Since the switch current flows through the copper plate (35 μm x 10 mm) on a 2 mm PCB, using Rogowski probe would not give an accurate current measurement. Therefore, only the inductor current was measured during the test, see Fig. 7 and Fig. 8.

III.   SIMULATIONS

Results from the experiment were collected and studied in order to create an accurate model of the structure in the simulation software LT-Spice. The parasitic components obtained from ANSYS Q3D and the model of the switches, provided by Wolfspeed, were integrated into LT-Spice.

The 2018 International Power Electronics Conference

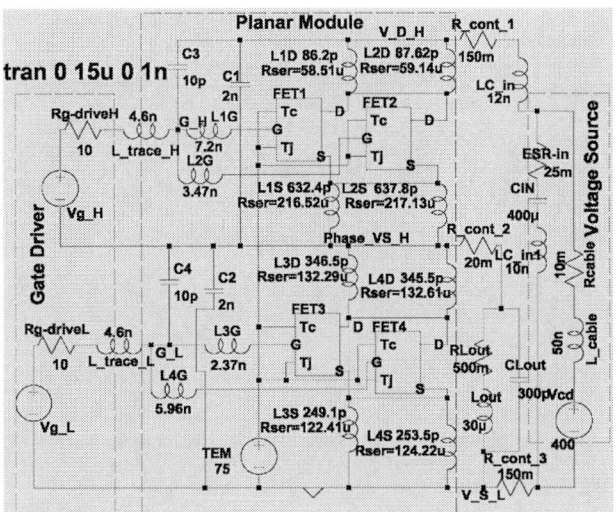

Fig. 9. Schematic diagram of the model of the planar module with two SiC MOSFETs at each position and the parasitic elements from ANSYS Q3D. The model was used for simulations in LT-Spice. In order to resemble the behavior of the real circuit, additional parasitic components were added into this model.

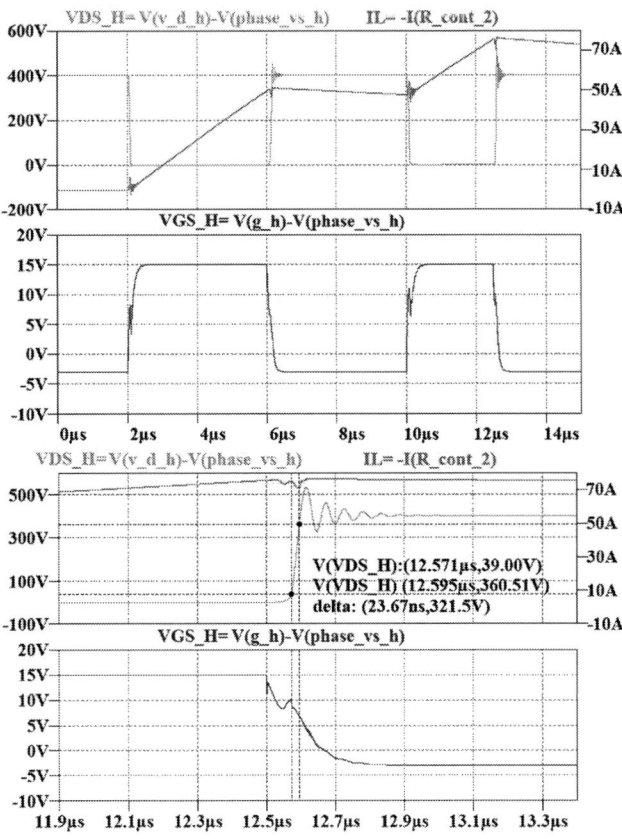

Fig. 10. The figure shows the simulation results from LT-Spice. The inductor current is shown in red, V_{DS} of the DUT in turquoise and VGS in blue. The simulation resembles the measurement results.

The developed model of the manufactured planar power module in LT-Spice can be seen in Fig. 9. Additional parasitic elements, for instance capacitances which were excluded in ANSYS Q3D, were added into this model in order to match the simulation and experimental results. The capacitances C1-C4 are produced because of the two parallel PCBs close to each other. The values of these capacitances are adjusted to resemble the measurement results. Also other parasitic elements like *L_trace_H*, *L_trace_L* and *R_cont_1* etc. were estimated and added into the model. Simulation results based on the developed model can be found in Fig. 10. When adjusting the parasitic components, the focus has been on the fall- and rise times, the amplitude of the overshoot and the frequency of the oscillations of V_{DS} as well as the inductor current level and waveforms.

IV. PROPOSED AND EXPANDED MODEL OF THE PLANAR POWER MODULE

The obtained model in Fig. 9 represents the manufactured planar power module in Figs. 5 and 6, verified by experiments and simulations. Since the total inductance of a copper trace decreases by increased width, a module with additional number of bare dies connected in parallel may be designed. By adding additional bare dies, the width of traces will increase and as result, a lower Drain-to-Source parasitic inductance as well as higher current capability will be obtained. Therefore, the model in Fig. 9 was expanded with two additional bare dies (totally four bare dies in parallel at each position). The new module with four bare dies would manage 400 A. A sketch of the proposed expanded model can be seen in Fig. 11. In real applications, the substrates of the proposed module must have a low thermal resistance to conduct the generated heat to the heat sink. For this purpose ceramic substrates and heat sinks on both sides (double-sided cooling) may be used. In this structure, the same gate signal goes to the up-side (or low-side) MOSFETs through one and the same trace. Therefore, in the model it is assumed that the inductance of the gate trace increases by the previous value plus 1.2 times (worst case) the initial value, which is 3.47 and 2.37 nH for the upper and lower positions, respectively.

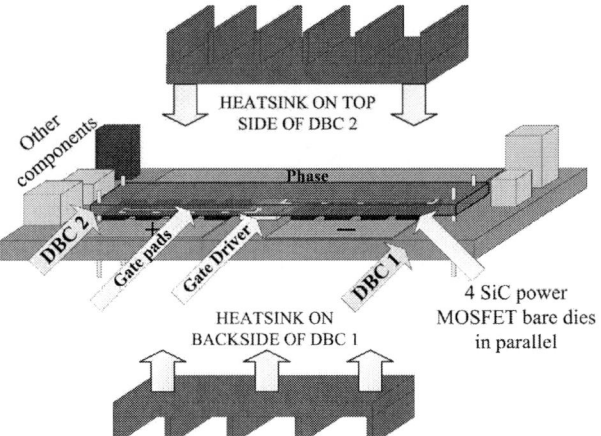

Fig. 11. The proposed structure of the power planar module with four parallel SiC MOSFETs at each position. Unlike regular power modules, the proposed planar module can be cooled from both sides.

2735

The schematic diagram of the proposed module including the parasitic elemets used in LT-Spice is presented in Fig. 12. In this model the values of the parasitic capacitances, caused by the traces, were increased since the area of traces is increased twofold compared to the previous design. Other parasitic elements were kept fixed as in the design with two MOSFETs. The model of the proposed extended module was then simulated in LT-Spice and the results are presented in Fig. 13 and Fig. 14.

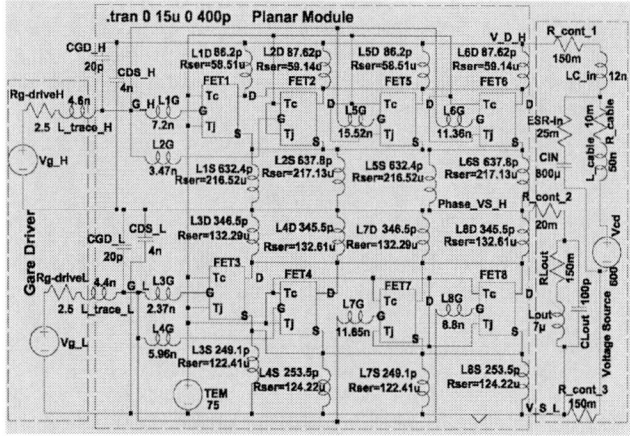

Fig. 12. Schematic diagram of the expanded model of the planar module in Fig. 9 with four SiC MOSFETs bare dies (CPM2-1200-0025B) at each position. This model was used for simulation in LT-spice in order to predict the behavior of the expanded model at $V_{DS}=600$ V and $I_D=400$ A.

V. RESULTS

The simulation was based on the DPT with the same pulse length as used in the experiment. The value of the inductance was adjusted to 7 μH to increase the switch current to approximately 400 A (100 A/switch). Figure 13 shows the gate signals of the upper and lower switches, V_{DS}, and the current as well as the power losses of the upper switches during the entire simulation time (15 μs). Figure 14 shows the same signals zoomed into the switching instants. As seen in these figures the proposed module shows extremely good properties with fast switching times (17.2 and 40.7 ns), and very low oscillations in voltage with an overshoot to approx. 850 V, which is 250 V over the input voltage.

In order to be able to compare the results with commercial modules an external resistance Rg=2.5 Ω was used in the simulation. However, by adding a larger gate resistance and decreasing the switching speed, the overshoot and oscillations in V_{DS} will reduce. Turn-on (E_{ON}) and turn-off (E_{OFF}) energies were, according to IEC 60747-8-4, calculated to approximately 3.1 and 1.3 mJ respectively. Most commercial half-bridge modules with 1.2 kV blocking voltage and 400 A current capability have switching energy losses of approximately 11-13 mJ. In other words, the proposed power module has approximately 63 % lower switching losses at 600 V & 400 A than for instance CAS300M12BM2 from Wolfspeed.

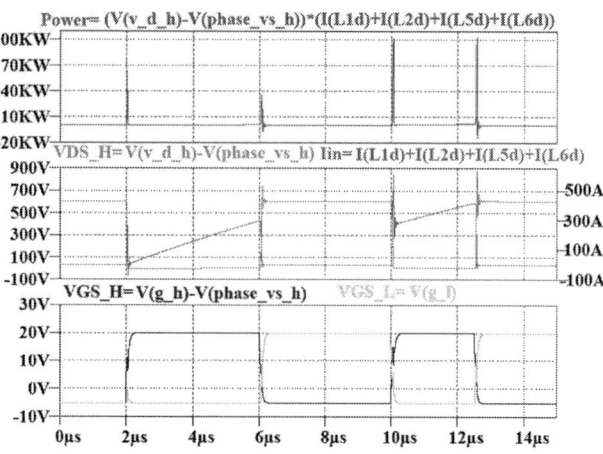

Fig. 13. Simulation result of the proposed planar power module. The total switch current is shown in gray, V_{DS} in turquoise, VGS of the upper switches in blue and VGS of the lower switches in green. The instantaneous power is shown in red.

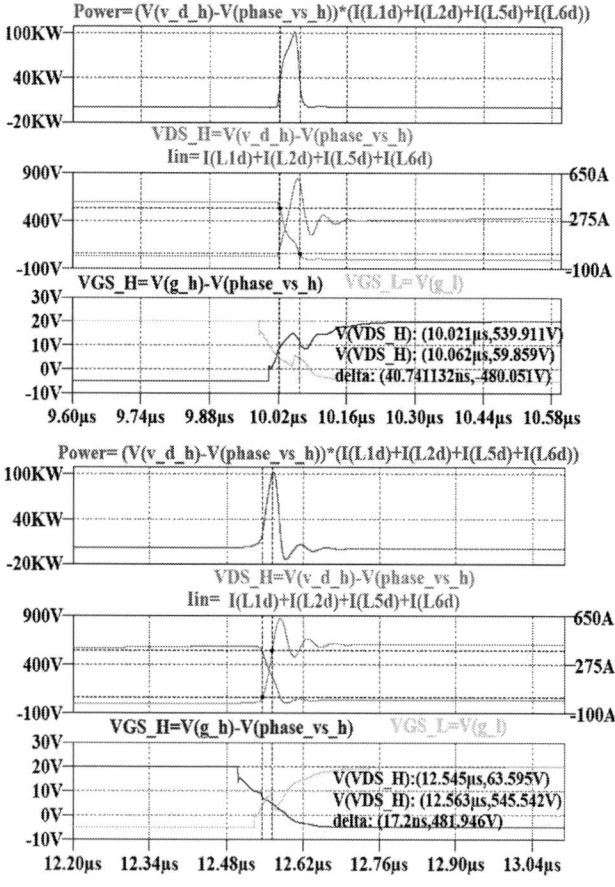

Fig. 14. Simulation result of the proposed planar power module. The total switch current is shown in gray, V_{DS} in turquoise, VGS of the upper switches in blue and VGS of the lower switches in green. The instantaneous power is shown in red. The turn-on and turn-off times are 40.7 ns and 17.2 ns, respectively.

The result from this study may be summarized in Table IV where a selection of parameters of the proposed module was compared to a commercial module from Wolfspeed. It is assumed that 1 mm thickness ceramic is used as substrate material for the proposed module in this comparison. The lower switching losses and the possibility of double-sided cooling in the proposed module will result in lower operating temperature, extended life time of the module and increased power density.

TABLE IV. THE TABLE LISTS AND COMPARES SOME IMPORTANT PARAMETERS BETWEEN THE PROPOSED MODULE AND A COMMERCIAL HALF-BRIDGE MODULE FROM WOLFSPEED.

Module/Parameters	CAS300M12BM2	Proposed planar module (DBC)
Dimensions [mm]	106 x 62 x 30	≈ 40 x 40 x 3
Volume [cm^3]	197	≈ 4.8
Weight [g]	300	≈ 20
V_{DS} [kV]	1.2	1.2
I_D @25°C [A]	423	400
$R_{ds(on)}$ [mΩ]	4.2	6.25
Stray inductance [mH]	15	0.33
Rise time [ns]	68	40
Fall time [ns]	43	17
Double-sided cooling	No	Yes

VI. CONCLUSION

A half-bridge planar power module with two SiC MOSFET bare dies, with very low parasitic inductances was designed, built and tested. The module was exposed to a double-pulse test and at 400 V & 75 A, the rise and fall times were measured to 23 and 60 ns, respectively. Later on, an accurate simulation model of the module was developed and it was adjusted to resemble the experimental results. The model was then expanded with two additional MOSFETs at each positions for higher current capacity. Simulation results showed that the proposed module has approximately 63 % lower switching energy losses than most commercial modules. From the simulation results, E_{ON} and E_{OFF} were measured to 3.1 and 1.3 mJ, respectively. An overshoot and some minor oscillations were observed in V_{DS} and I_D in the double-pulse simulations at 600 V and 400 A. A possible reason for the oscillations is the increased parasitic *capacitances* in this structure compared to the structure with only two bare dies at each position. The Gate-Drain parasitic capacitance in this structure may have a similar effect as the Miller capacitance. It should be noted that in real applications the distance between substrate 1 and 2 has to be filled in order to handle the electric fields and improve the mechanical and thermal properties. However, the filler material will probably increase the parasitic capacitances, which especially may cause issues if it appears between Gate and Drain of the MOSFETs. A solution to this problem may be to employ a multilayer substrate where the gate trace is placed inside substrate 2 in order to increase the distance between the Gate and Drain traces.

REFERENCES

[1] S. Hazra, A. De, L. Cheng, J. Palmour, M. Schupbach, B. A. Hull, S. Allen, and S. Bhattacharya, "High switching performance of 1700-V, 50-A SiC power MOSFET over Si IGBT/BiMOSFET for advanced power conversion applications," *IEEE Trans. Power Electron.*, vol. 31, no. 7, pp. 4742–4754, Jul. 2016.

[2] Q. J. Zhang, G. Wang, C. Jonas, C. Capell, S. Pickle, P. Butler, D. Lichtenwalner, E. V. Brunt, S. Ryu, J. Richmond, B. Hull, J. Casady, S. Allen, and J. Palmour, "Next generation planar 1700 V, 20 mohm; 4H-SiC DMOSFETs with low specific on-resistance and high switching speed," in *European Conference on Silicon Carbide Related Materials (ECSCRM)*, Sep. 2016, pp. 1–1.

[3] J. Colmenares, D. Peftitsis, H. P. Nee, and J. Rabkowski, "Switching performance of parallel-connected power modules with SiC MOSFETs," in *International Power Electronics Conference (IPEC-Hiroshima - ECCE ASIA)*, May 2014, pp. 3712–3717.

[4] H. Sato, F. Kato, H. Nakagawa, H. Yamaguchi, S. Rejeki, and F. Lang, "Development of SiC power module for high-speed switching operation," in *IEEE Electrical Design of Advanced Packaging Systems Symposium (EDAPS)*, Dec. 2013, pp. 13–16.

[5] E. Velander, L. Kruse, and H. P. Nee, "Optimal switching of SiC lateral MOSFETs," in *17th European Conference on Power Electronics and Applications (EPE'15 ECCE-Europe)*, Sep. 2015, pp. 1–10.

[6] E. Velander, L. Kruse, T. Wiik, A. Wiberg, J. Colmenares, and H. P. Nee, "An IGBT turn-on concept offering low losses under motor drive dv/dt constraints based on diode current adaption," *IEEE Trans. Power Electron.*, vol. PP, no. 99, pp. 1–1, Feb. 2017.

[7] P. Nayak and K. Hatua, "Parasitic inductance and capacitance-assisted active gate driving technique to minimize switching loss of SiC MOSFET," *IEEE Trans. Ind. Electron.*, vol. 64, no. 10, pp. 8288–8298, Oct. 2017.

[8] J. Colmenares, D. Peftitsis, J. Rabkowski, D. P. Sadik, and H. P. Nee, "Dual-function gate driver for a power module with SiC Junction Field-Effect transistors," *IEEE Trans. Power Electron.*, vol. 29, no. 5, pp. 2367–2379, May 2014.

[9] N. Rouger, J. Widiez, L. Benaissa, B. Imbert, P. Gondcharton, B. Letowski, and J. Crebier, "3D packaging for vertical power devices," in *CIPS 2014; 8th International Conference on Integrated Power Electronics Systems*, Feb. 2014, pp. 1–6.

[10] Z. Liang, F. Wang, and L. Tolbert, "Development of packaging technologies for advanced SiC power modules," in *IEEE Workshop on Wide Bandgap Power Devices and Applications*, Oct. 2014, pp. 42–47.

[11] Z. Zhang, J. Fu, Y. F. Liu, and P. C. Sen, "Switching loss analysis considering parasitic loop inductance with current source drivers for buck converters," *IEEE Trans. Power Electron.*, vol. 26, no. 7, pp. 1815–1819, Jul. 2011.

[12] T. Meade, D. O'Sullivan, R. Foley, C. Achimescu, M. Egan, and P. McCloskey, "Parasitic inductance effect on switching losses for a high frequency dc-dc converter," in *2008 Twenty-Third Annual IEEE Applied Power Electronics Conference and Exposition*, Feb. 2008, pp. 3–9.

[13] T. Yamamoto, K. Hasegawa, M. Ishida, and K. Takao, "Switching simulation of SiC high-power module with low parasitic inductance," in *International Power Electronics Conference (IPEC-Hiroshima - ECCE ASIA)*, May 2014, pp. 3707–3711.

[14] F. Yang, Z. Liang, Z. J. Wang, and F. Wang, "Design of a low parasitic inductance SiC power module with double-sided cooling," in *IEEE Applied Power Electronics Conference and Exposition (APEC)*, Mar. 2017, pp. 3057–3062.

[15] Z. Liang, "Integrated double sided cooling packaging of planar SiC power modules," in *IEEE Energy Conversion Congress and Exposition (ECCE)*, Sep. 2015, pp. 4907–4912.

[16] K. Takao and S. Kyogoku, "Ultra low inductance power module for fast switching sic power devices," in *IEEE 27th International Symposium on Power Semiconductor Devices IC's (ISPSD)*, May 2015, pp. 313–316.

[17] A. E. Risseh, H. Nee, and K. Kostov, "Electrical performance of directly attached SiC power MOSFET bare dies in a half-bridge configuration," in *IEEE 3rd International Future Energy Electronics Conference and ECCE Asia (IFEEC 2017 - ECCE Asia)*, Jun. 2017, pp. 417–421.

[18] A. E. Risseh, H. P. Nee, and K. Kostov, "Realization of a planar power circuit with silicon carbide MOSFETs on printed circuit board," in *24th International Symposium on Power Electronics, Electrical Drives, Automation and Motion - (SPEEDAM 2018)*, Jun. 2018.

The 2018 International Power Electronics Conference

Experimental Evaluation of Inverter System Consisting of 4-parallel GaN Devices Unit

Yoshiya Ohnuma[1*], Satoshi Miyawaki[1], Fumiya Hattori[2] and Masayoshi Yamamoto[2]
1 Nagaoka Power Electronics, Nagaoka, Japan
2 Electrical Engineering, Nagoya University, Nagoya, Japan
* E-mail: ohnuma@npe.co.jp

Abstract— **In this paper, an inverter system consisting of a converter unit with 4 GaN FETs in parallel is proposed. Most GaN FET packages are used for chip type packages. It is difficult for chip type packages to mount a heatsink since a heatsink requires the complicated configuration at the surface. To resolve this problem, a plastic heatsink is installed in the proposed system. A prototype inverter system with maximum efficiency of over 98% is fabricated, and their system losses are analyzed. As a result, it can be confirmed that most part of losses are no-load loss.**

Keywords—GaN FET, heatsink, no-load loss

I. INTRODUCTION

In recent years, wide band-gap (WBG) semiconductors such as SiC and GaN are commercialized, and miniaturization and high efficiency of power converters are realized[1]-[11]. In order to maximize their performances, WBG switching devices are often used with small chip-shaped packages. These packages can reduce influence of parasitic inductance and improve density of arrangement. On the other hand, there are difficulties in mounting and cooling accompanying high density. For fully demonstrating the performance of the devices, it is necessary to solve these problems[1].

Most of commercially available GaN devices have characteristics that the on-resistance rapidly deteriorates as the junction temperature rises, so that it can be used at the relatively low temperature, and then a heat sink with the small thermal resistance must be connected to the device. However, since the package of the device itself is small, the heat radiation surface is small, and there is a high possibility that the heat sink interferes with other parts because the mounting density is high. For mounting the heat sink on the substrate, it is necessary to sandwich the insulating material. Moreover, the switching speed of the WBG device is faster, and it thus tends to cause false turn-on. Therefore, we have to take care of the pattern wiring of the board and the arrangement of the components.

On the other hand, a structure in which a plurality of switching devices are connected and driven in parallel is attracting attention[2][3]. By parallelizing the devices, there are merits such as improvement in efficiency by reducing

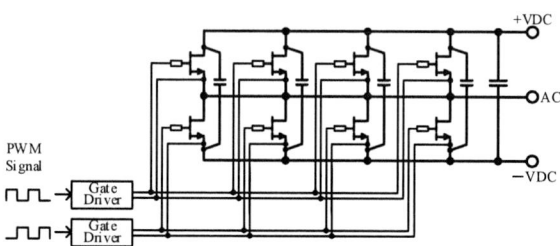

Fig. 1. Circuit diagram.

TABLE 1.
CIRCUIT COMPONENTS OF UNIT BOARD.

Circuit component	Part number (Manufacturer)	Specification
Switching device	GS66508T (GaN Systems)	650V,30A, 50mΩ
Snubber capacitor	GRM55D7U2J473JW31L (Murata Electronics)	630V, 0.22uF
Gate driver (Isolator)	SI8271GB-IS (Silicon Labs)	2500Vrms (isolation)
Isolated power supply	PES1-S12-S9-M (CUI Inc.)	12V-9V,1W 3000V(isolation)

the on-resistance and in the heat dissipation performance by dispersing the heat generation parts. It is thus possible to realize the converter with the larger capacity. However, the problems such as current concentration due to the inconsistent switching timing and the current variation due to the device performance variation may occur.

The authors connect 4 chip-shaped GaN devices in parallel, and fabricate a unit board corresponding to 1 leg of the converter. For example, it is possible to construct a buck-chopper circuit with a single unit or a 3-phase inverter with 3 boards. In this paper, we describe specification, structures and heat radiation designs of 4 parallel GaN unit board, and report on characteristics evaluation by thermal analysis and loss analysis with a buck-chopper circuit and an inverter.

II. UNIT SPECIFICATION AND STRUCTURE

A circuit configuration of a 4-parallel GaN unit board is shown in Fig.1, and main components used for this unit board are described in Table.1. A main circuit, a current sensor, and driver circuits are mounted on the PCB. GaN

2738

The 2018 International Power Electronics Conference

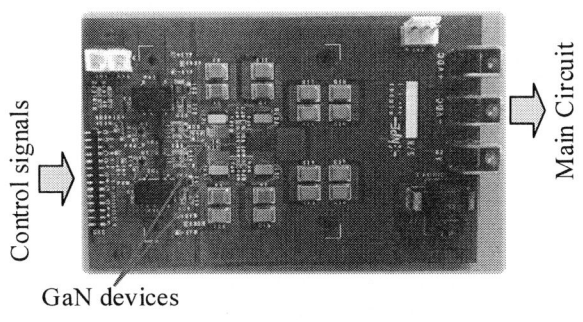

GaN devices

(a) Top view.

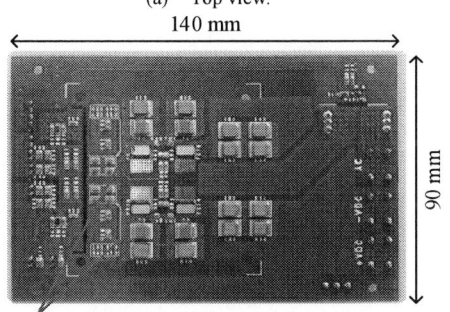

Snubber capacitors

(b) Bottom view.

Fig. 2. PCB design and layout.

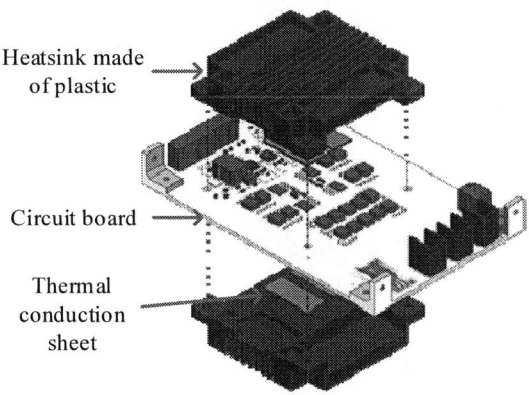

Heatsink made of plastic

Circuit board

Thermal conduction sheet

Fig. 3. Structural drawing.

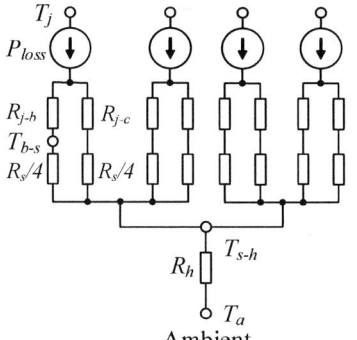

Fig. 4. Equivalent circuit of thermal resistance for one side.

Systems' GS 66508T is used as a switching device. The 4 devices are respectively connected in parallel at the high

and low side and driven by the 1 gate driver. By connecting the terminals of +VDC, -VDC, and AC, it is possible to configure a buck-chopper, boost-chopper circuit, or an inverter easily.

GaN devices show high switching speed, but the gate voltage threshold is generally low. In order to suppress the false turn-on caused by the GaN's characteristics and avoid the adverse influence to the control device, the sufficient insulation is required between the main circuit and the control circuit. In the proposed unit, their countermeasures are taken by using the gate driver with the high withstand voltage and the insulated power supply and inserting the common mode filter.

Photographs of the front side and the back side of the prototype substrate is shown Fig.2. The left side is the control signal input, and the right side is the main circuit connection. The GaN devices mounted at the center of the substrate are arranged point symmetrically at the same position on the front and back. The high side devices are mounted on the front side, and the low side devices are mounted on the back side[4]. By adopting such an arrangement, the following merits can be obtained even when the 4 devices are arranged in parallel.

(1) PN wiring can be the laminate structure, and the loop of the snubber capacitor, high side device, and low side device can be minimized.

(2) Gate wiring from driver to device becomes equal length and the shortest

(3) Output wiring from each device becomes equal length

III. THERMAL DESIGN

A mounting diagram of a board and heat sinks are shown in Fig.3. The heat sink is connected on the top of the GaN devices which are arranged on the front and back. The prototype's heat sink is not an ordinary aluminum material, but an insulated plastic heat sink with the improved thermal conductivity. In order to dissipate the heat generated by the chip-shaped package via the heat sink, it is necessary to uniformly apply pressure to the substrate and the devices. Although the method of fixing the heat sink to the hole of the substrate with the screws or the like is generally used, the flexibility of the copper foil pattern of the PCB is greatly impaired. Moreover, if there are components with different heights are mounted near the devices, it is necessary to adjust the height with the heat spreader in order to avoid interference. In addition, the connection part of the substrate, the components and the heat sink should be cover by the insulation sheet. The plastic heat sink can improve the thermal conductivity from 0.2 W/mK of ordinary plastic to 4.9 W/mK while maintaining the insulation performance by mixing the special filler into the plastic. Note that the flexible heat conduction sheet to fill the gap is inserted into the contact part of the GaN device and the heat sink. As a result, the pressure of the heat sink can be uniformly applied to the devices.

A simple thermal resistance equivalent circuit of the

heat sink is shown in Fig. 4. For this time, the thermal resistance is experimentally calculated by using the prototype board. In the experiment, the switch keeps on for a certain period of time, and the losses are generated by flowing the constant current to the switching devices. The temperature change is measured with the thermocouple, and the thermal resistance is then calculated from the difference of the temperature.

Calculation results of thermal resistances are shown in Table 2. The thermal resistance of the heat sink is 8.7 K/W in the natural air cooling and 4.8 K/W in the forced air cooling respectively. A performance comparison between the plastic heat sink and the aluminum heat sink is shown Table 3. In the comparison of the thermal resistance, the thermal resistance of the plastic is approximately 2 to 3 times larger, which is inferior to that of the aluminum. On the other hand, the weight becomes from 1/2 to 1/3 times lighter than that of aluminum. The plastic heat sinks have larger thermal resistance than the aluminum, but the weight savings are expected. If the high efficiency can be achieved by using the plurality of the GaN devices, it is possible to use the plastic heat sink and to realize the lightweight system.

IV. LOSS ANALYSISY OF BUCK-CHOPPER CIRCUIT

In order to evaluate losses generated from the unit, a buck-chopper circuit is fabricated by using a single unit. The loss analysis is performed, based on the mathematical expression, and their loss are clarified. As a result, it is confirmed that the influence of the no-load loss is large.

A. No-Load loss without LC filter

Experimental and calculated results in a single unit without LC filter are shown in Fig.5 The input voltage is 300 V, and the switching frequency is 10 kHz. The experiments are carried out from 20V to 300V in 20 V step, and the dead time is set to 20 nsec and 100 nsec. Note that only the losses occurring in the GaN device are on target for the hard switching.

Principle of generation for no-load loss without LC filter is shown in Fig.6. When the high side switch turns on after the dead time period, the steep current flows by charging the parasitic capacitance of the device and the PCB of the low side. Since this current has the larger peak than the output current, it becomes a cause of increasing the switching loss of the high side switch. At this time, the energy charged in the parasitic capacitor of the high-side switch is consumed and becomes a loss in the device. When the parasitic inductance of the PCB is low on the main circuit, the no-load loss $P_{no\text{-}load}$ generated from the 4 device and PCB is expressed by the following equation (1).

$$P_{\text{no-load}} = \left(8C_{\text{oss}} + C_{\text{ph}} + C_{\text{pl}}\right)V_{\text{ds}}^2 f_{\text{sw}}$$
$$= 16E_{\text{oss}}f_{\text{sw}} + \left(C_{\text{ph}} + C_{\text{pl}}\right)V_{\text{ds}}^2 f_{\text{sw}} \cdots (1)$$

Table 2. Thermal resistances and values.

Item	Mean	Value [K/W]
$R_{j\text{-}c}$	Thermal resistance of junction to case	0.5
$R_{j\text{-}b}$	Thermal resistance of junction to board	5.0
R_s	Thermal resistance of thermal conduction sheet	0.83
R_h	Thermal resistance of heatsink	8.7 4.8

Table 3. Comparison between a plastic type heatsink and an aluminum type heatsink.

	Plastic type (Development article)	Aluminum type
Volume	70×70×15 mm	Same level of volume
Thermal resistance	8.7 K/W (natural air-c) 4.8 K/W (forced air-c)	2.5 ~ 3.5 K/W (natural air-c) *
Weight	41g	90g~120g *

* Result of our survey

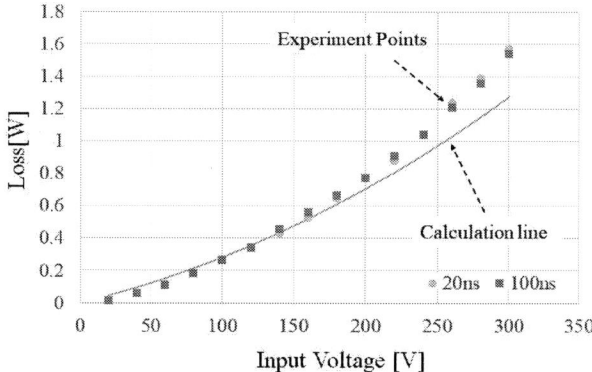

Fig. 5. No-load loss without LC filter.

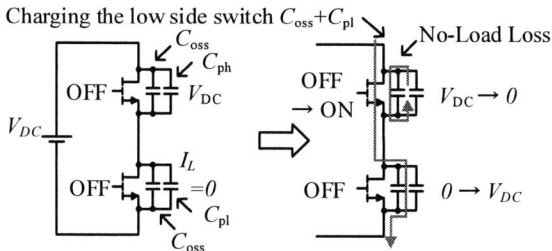

Fig. 6. Principle diagram of no-load loss.

where C_{oss} is the output capacitor, C_{ph} and C_{pl} are the high and low side parasitic capacitor of the inner layer respectively, V_{ds} is the drain-source voltage, f_{sw} is the switching frequency, and E_{oss} is the output capacitor stored energy. The parasitic capacitors C_{ph} and C_{pl} of the inner layer are directly measured by the impedance analyzer, agilent 4249A as shown in Table 4. As plotted in Fig. 5, the measurement result is generally in accord with that of the calculation. There is no difference on the losses which dead times are 20 nsec and 100 nsec. From this, the short circuit between the high and low side, which is caused by the dead time is not occurred, and it is considered that the output capacitor are the cause of losses, and the half of the loss is depend on

2740

the parasitic capacitor of the inner layer of the PCB. Therefore, we have to pay attention to the PCB pattern.

B. No-Load loss with LC filter

Experimental and calculated results in a single unit with LC filter, a buck-chopper circuit are shown in Fig.7, and loss details are described in Fig. 8. The input voltage is 300 V, the output voltage is 150 V, the switching frequency is changed to 10 kHz, 20 kHz, and 30 kHz. The experiments are carried out up to 2 kW each. Note that only the losses occurring in the GaN device are described in Fig.8. From this no-load loss, the loss occurred in the state with no connection to the output of the unit is measured and added to the calculated value as the loss independent of the output current. From the results, it can be confirmed that the no-load loss accounts for most of the loss. Further, the loss increases with respect to the output current is greatly affected by the conduction loss. Since the switching speed of the GaN device is faster, the switching loss does not increase so much. The no-load loss increases in proportion to the switching frequency due to its characteristics. The Approximate estimation is possible from the parasitic capacitor of the device, but the loss also changes depending on the current direction of the load and the magnitude of the current. Therefore, it is difficult to separate from the switching loss and to estimate the loss accurately. Therefore, in this analysis, the loss measured when operating without load is treated as the no-load loss by separating from the switching loss.

V. EXPERIMENTAL RESULTS OF 3-PHASE INVERTER

Experiments are carried out by constructing a 3-phase inverter with 3 units. The test condition is 3.8 kW (4.8 kVA) with an input voltage of 300 V and an output voltage of 200 V. In this case, The RL load is used, and the switching frequency is varied to 10 kHz, 20 kHz, and 30 kHz. The dead time is set to 20 nsec.

Waveforms of input voltage, input current, output voltage, and output current are shown in Fig.9. As can be seen from the experimental results, the sine wave without the distortion is output. Despite the fact that the dead time error compensation is not included, THD 1% or less is achieved due to the small dead time.

Efficiency and loss are shown in Fig.10 and the results of the temperature test of the device is shown in Fig.11. Efficiency of 98% or higher has been achieved in most areas. The case temperature at that time is 52 °C at room temperature 22 °C. From this fact, the validity of the heat radiation design is confirmed. Therefore, the plastic heat sink is effective for the GaN FET.

VI. CONCLUSIONS

In this paper, characteristics of a proposed converter uni, which 4 GaN devices connected in parallel, were evaluated by thermal analysis and loss analysis when a buck-chopper or an inverter were fabricated.

Firstly, we introduced the plastic heat sink and measured the thermal resistance of it. The thermal

Table 4. Parameters for loss calculation.

Item	Mean	Value
E_{oss}	Stored energy of output capacitance	7 μJ*
C_{ph}	parasitic capacitance of Internallayer around high side device	263 pF
C_{pl}	parasitic capacitance of Internallayer around low side device	300 pF

* V_{ds}=400V

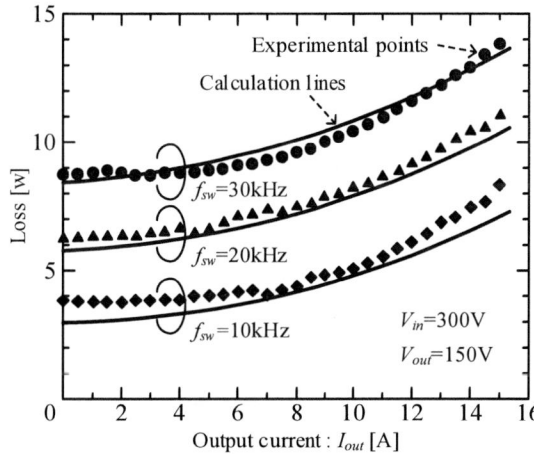

Fig. 7. Comparison between the experimental and calculation results by buck-chopper circuit using single unit.

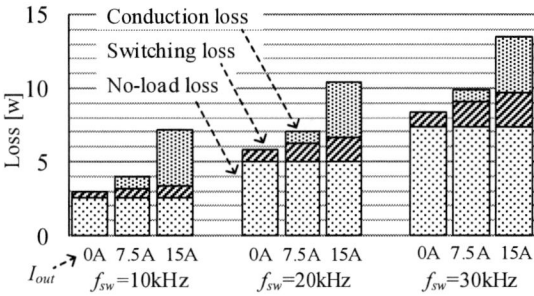

Fig. 8. Loss breakdown comparison of switching frequency by buck-chopper curcuit.

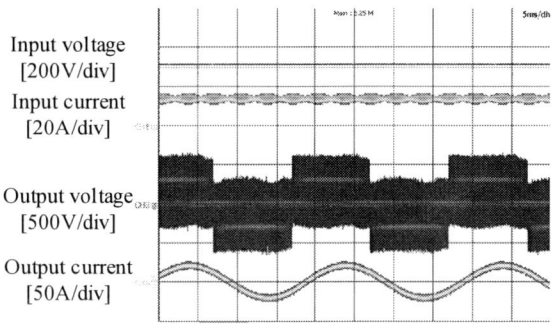

Fig. 9. Experimental waveform of 3-phase inverter.

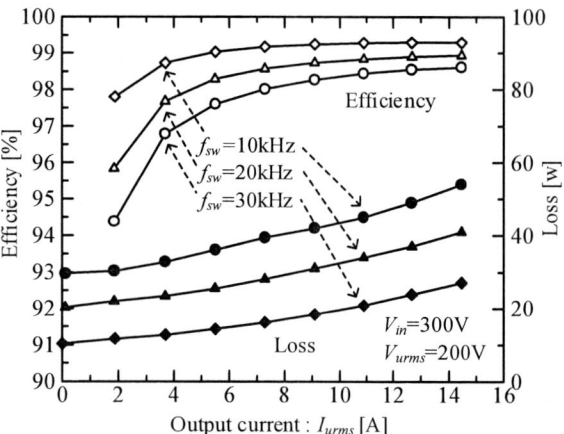

Fig. 10. Efficiency and power loss of 3-phase inverter.

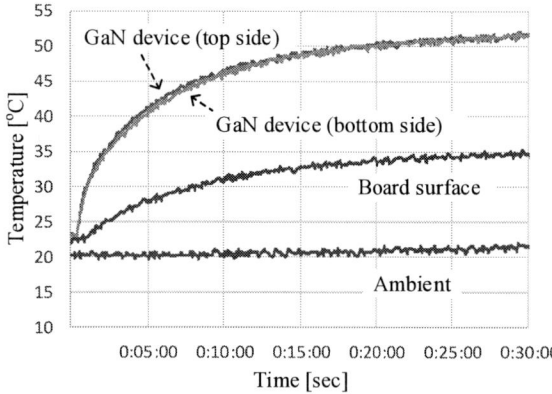

Fig. 11. GaN device temperature of 3-phase inverter.

resistance of the heat sink was 8.7 K/W in the natural air cooling and 4.8 K/W in the forced air cooling respectively. Therefore, by installing this heat sink,resistance of the heat sink was 8.7 K/W in the natural air cooling and 4.8 K/W in the forced air cooling respectively. Therefore, by installing this heat sink, converters with GaN FET can be lighter.

Secondly, we analyzed and experimented the no-load loss without the LC filter, and the half loss is depend on the parasitic capacitor of the inner layer. The multi-layer PCB pattern is thus important for the loss reduction even in kHz switching.

Thirdly, the experiment of the buck-chopper circuit was carried out. The loss was depends on the no-load loss most. Hence, we have to pay attention for this loss to improve the converter efficiency.

Finally, we fabricated a 3-phase inverter the maximum efficiency of 98% or more was achieved, and the usefulness of the prototype with plastic heat sink was confirmed.

REFERENCES

[1] Weijing Du, Xiucheng Huang, Fred C. Lee, Qiang Li and Wenli Zhang : "Avoiding divergent oscillation of cascode GaN device

under high current turn-off condition", IEEE APEC 2016, pp.1002-1009 (2016)

[2] Juncheng Lu, Hua Bai, Alan Brown, Matt McAmmond, Di Chen and Julian Styles : "Design consideration of gate driver circuits and PCB parasitic parameters of paralleled E-mode GaN HEMTs in zero-voltage-switching applications", IEEE APEC 2016, pp.529-535 (2016)

[3] He Li, Xuan Zhang, Lucheng Wen, John Alex Brothers, Chengcheng Yao, Ke Zhu, Jin Wang, Liming Liu, Jing Xu and Joonas Puukko : "Evaluation of high voltage cascode GaN HEMTs in parallel operation", IEEE APEC 2016, pp.990-995 (2016)

[4] Nidhi Haryani, Xuning Zhang, Rolando Burgos and Dushan Boroyevich : "Static and dynamic characterization of GaN HEMT with low inductance vertical phase leg design for high frequency high power applications", IEEE APEC 2016, pp.1024-1031 (2016)

[5] Meneghini, Matteo, Meneghesso, Gaudenzio and Zanoni, Enrico: "Power GaN Devices Materials, Applications and Reliability", Springer, 1st ed, (2016)

[6] Lucas Lu, Guanliang Liu, Kevin Bai: "Critical transient processes of enhancement-mode GaN HEMTs in high-efficiency and high-reliability applications", CES Transactions on Electrical Machines and Systems 2017, Vol. 1, Issue 3, pp. 283 - 291(2017)

[7] Andrew J. Sellers, Cheikh Tine, Roshan L. Kini, Michael R. Hontz, Raghav Khanna, Andrew N. Lemmon, Ali Shahabi;" Effects of parasitic inductance on performance of 600-V GaN devices" Christopher New 2017 IEEE Electric Ship Technologies Symposium (ESTS), pp. 50-55(2017)

[8] Abhijit Kulkarni, Ankit Gupta, Sudip Kumar Mazumder: " Resolving Practical Design Issues in a Single-Phase Grid-Connected GaN-FET Based Differential-Mode Inverter" IEEE Transactions on Power Electronics 2017, Issue 99, pp. 1-1(2017)

[9] Harry C. P. Dymond, Jianjing Wang, Dawei Liu, Jeremy J. O. Dalton, Neville McNeill, Dinesh Pamunuwa, Simon J. Hollis, Bernard H. Stark: "A 6.7-GHz Active Gate Driver for GaN FETs to Combat Overshoot, Ringing, and EMI" IEEE Transactions on Power Electronics 2018, Vol. 33, Issue 1, pp. 581 – 594(2018)

[10] Emre Gurpinar, Alberto Castellazzi: "Trade-off Study of Heat Sink and Output Filter Volume in a GaN HEMT Based Single Phase Inverter" IEEE Transactions on Power Electronics 2017, Issue: 99, pp 1 – 1(2017)

[11] Akinori Hariya, Tomoya Koga, Ken Matsuura, Hiroshige Yanagi, Satoshi Tomioka, Yoichi Ishizuka, Tamotsu Ninomiya: "Circuit Design Techniques for Reducing the Effects of Magnetic Flux on GaN-HEMTs in 5-MHz 100-W High Power-Density LLC Resonant DC–DC Converters" IEEE Transactions on Power Electronics 2017, Vol. 32, Issue 8, pp. 5953 – 5963(2017)

The 2018 International Power Electronics Conference

Impact of the Thermal-Interface-Material Thickness on IGBT Module Reliability in the Modular Multilevel Converter

Yi Zhang*, Huai Wang, Zhongxu Wang, Yongheng Yang, Frede Blaabjerg

Department of Energy Technology, Aalborg University, Aalborg, Denmark

*E-mail: yiz@et.aau.dk

Abstract—The reliability of the Modular Multilevel Converter (MMC) is of great interest in industrial applications, where the dominant failure mechanism of IGBT modules in an MMC system is temperature-related. In this regard, thermal modeling is critical to map the power losses to thermal profiles and then to predict the lifetime. Even though the Thermal Interface Materials (TIMs) in IGBT modules have a considerable influence on the thermal resistance, the thickness of TIMs is often spuriously considered as a constant according to the thermal conductivity or information in the datasheet. This may lead to misleading results in the lifetime prediction. Hence, this paper investigates the impact of the TIM thickness on the estimated lifetime of IGBT modules in an MMC system for offshore wind power applications, including the starting assembly thickness and the Bond-Line Thickness (BLT) of TIMs. In a 30-MW MMC case study, the lifetime of the IGBT modules is discussed with respect to two values of starting thickness and variable BLT from 20 μm to 60 μm. Experiments are also carried out on a scaled-down system to validate the impact of the TIM thicknesses on the reliability prediction.

I. INTRODUCTION

The Modular Multilevel Converter (MMC) is a promising multilevel topology for both High-Voltage (HV) and Medium-Voltage (MV) applications. It encompasses HV Direct Current (HVDC) transmissions, MV drives, MV DC systems for all-electric ships, power conditioners for railway traction feeders, various power quality applications, etc.[1], [2], [3], [4], [5]. However, due to the harsh environments and high-reliability requirements of the MMC, lifetime prediction is significant to support its design and operational management. Especially, for hundreds of the Insulated Gate Bipolar Transistor (IGBT) modules in the MMC, they are regarded as one of the most vulnerable components in an industry survey [6].

Typically, the dominant failure mechanism of the IGBT modules in MMC systems is related to the junction temperature. Hence, the thermal modeling of IGBT devices is very important in order to map the power losses to the thermal profiles. Two kinds of thermal networks are commonly adopted to model the thermal behavior: the physical-structure-based Cauer model and the experiment-based Foster model [8]. The latter is preferable in industrial due to its simplicity. In a thermal model, the junction temperature of the power devices increases when te power losses pass through the thermal impedance. The thermal modeling typically involves in the models of the power module itself, the Thermal Interface

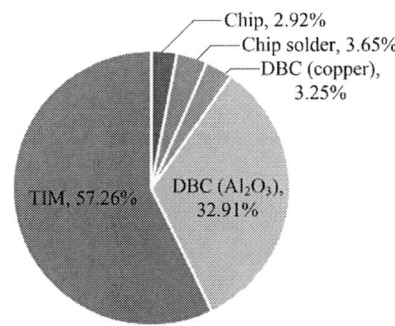

Fig. 1. An overview of thermal resistance share in a specific IGBT module [7].

Material (TIM), and the cooling system. As shown in Fig. 1, the contribution of the TIM to the overall thermal resistance from the junction to the heatsink is over 50 % [7]. However, the thermal resistance of the TIM is often considered as a constant according to its thermal conductivity. In [9], a quantitative analysis of the thermal model of the power devices based on a wind power converter was performed, but a constant thermal resistance of the TIM was assumed without any further discussion. In [10], a new thermal model was proposed to establish a correct transfer function for the filtering effects of power losses. However, the TIM thickness (and thus, the thermal resistance) was also considered as constant. Then, the lifetime prediction of IGBT modules in the MMC system was discussed in [11] with one-year mission profiles, where the thermal resistance of the TIM was neglected for simplification. As the contribution of the TIM to the overall thermal resistance is significant, over-simplified modeling of TIMs in thermal models lead to unrealistic thermal results and therefore lack of confidence in the lifetime prediction. Thus, this paper emphasizes the impact of the TIM for IGBT modules in MMC systems.

More specifically, this paper investigates the impact of the TIM thickness on the predicted lifetime of IGBT modules in MMC systems, where a specific TIM is varied with the starting assembly thickness and the Bond-Line Thickness (BLT). To qualify the impact of the TIM thickness, the lifetime of the IGBT modules is compared under a specific TIM with two

2743

The 2018 International Power Electronics Conference

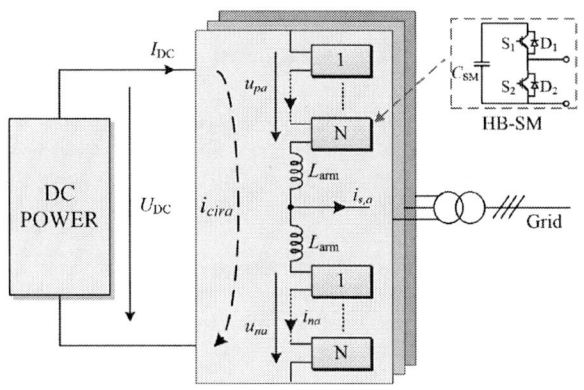

Fig. 2. Circuit configuration of a three-phase MMC system, where each Half-Bridge Sub-Module (HB-SM) is composed of four power devices.

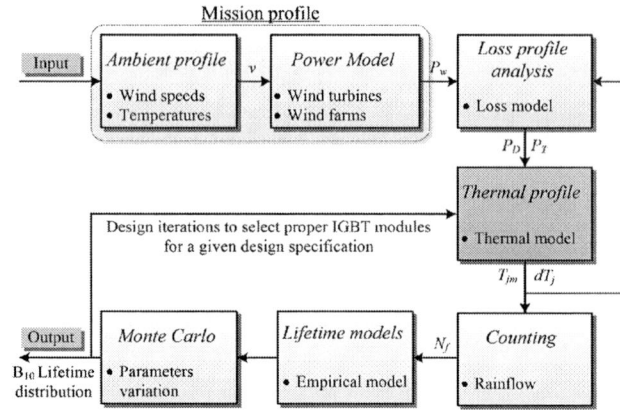

Fig. 3. Mission profile-based lifetime estimation method for IGBT modules in the MMC system.

different values of the starting-thickness and the BLT (20 μm - 60 μm) on a case study of a 30-MW MMC system. Moreover, experiments are carried out on a scaled-down MMC system. The variation of the maximum junction temperature under different values of the thickness of the TIM validates its impact.

The outline of this paper is as follows. Section II briefly introduces the circuit configuration of a 30-MW MMC system and the mission profile-based lifetime prediction method. Section III presents the quantitative lifetime impact on the 30-MW MMC case considering two values of the TIM. Finally, experimental validations are carried on a specially designed scaled-down system. The starting thickness is changed and applied to the IGBT module. The corresponding junction temperature profiles validate the impacts of the TIM.

II. CONFIGURATION OF AN MMC AND MISSION PROFILE BASED LIFETIME PREDICTION METHOD

Fig. 2 shows a three-phase MMC system, where each phase of the MMC consists of two arms and two inductors L_{arm}. In each arm, N identical half-bridge Sub-Modules (SMs) are connected, and each SM consists of four power semiconductor devices (denoted as S_1, D_1, S_2, and D_2) and a capacitor bank C_{SM}. In this topology, each SM just shares an identical portion of the high dc-bus voltage, which enables the MMC to bear several hundred kilo-voltages. Moreover, due to the multilevel operation to achieve the low THD output even with a low switching frequency, this topology is promising in HV and MV applications.

Mission profile-based lifetime prediction method is a widely accepted method to predict the lifetime of power electronic converters, which is shown in Fig. 3. In this method, the mission profiles are translated into loss profiles, thermal profiles, and finally to obtain the lifetime consumption of components or the entire system, where the thermal model is critical to map the power losses into thermal profiles. Even though the TIM has a large contribution to the thermal model, the thermal resistance of the TIM is usually assumed as a constant according to the thermal conductivity in many cases, which

may lead to large uncertainties. As the uncertainties in thermal models lead to unpredictable temperature profiles and further to a wrong predicted lifetime result, the lifetime prediction of the power devices should consider the impact of TIM.

III. CASE STUDY ON A 30-MW MMC

A case study of a 30-MW MMC is discussed in this section, which is utilized to connect the offshore wind farm. The system specifications are listed in Table I. The IGBT module from ABB 5SNA 1200E450350 is chosen as the power device. For the wind farm, ten 3-MW wind turbines V90 [12] are selected. For simplicity, the case study neglects the turbulence and the weak effect of the wind farm. Then, based on the TIM with different thickness values, the translation from the mission profile into lifetime consumption is investigated.

A. Mission Profiles to Loss Profiles

When an MMC is utilized to collect wind energy from an offshore wind farm, the dominant thermal-stress inducer for power devices is the power fluctuation due to the wind profile. Thus, as shown in Fig. 4, an annual wind speed with one-hour average data is selected as the mission profile, in order to evaluate the reliability of the power device in the MMC system.

TABLE I
SPECIFICATIONS OF THE STUDIED MMC SYSTEM

Parameters	Values
System rated active power	P = 30 MW
Rated dc-link voltage	U_{dc} = 31.8 kV
Rated ac grid voltage	U_{ac} = 14 kV
Number of sub-module per arm	N = 12
Arm inductor	L_{arm} = 4 mH
Arm resistor	R_{arm} = 0.0628 Ω
Sub-module capacitor	C_{SM} = 0.8 mF
Switching frequency	f_{sw} = 1 kHz
Fundamental frequency	f_0 = 50 Hz
Modulation index	m = 0.9
Power factor	PF = 1

The 2018 International Power Electronics Conference

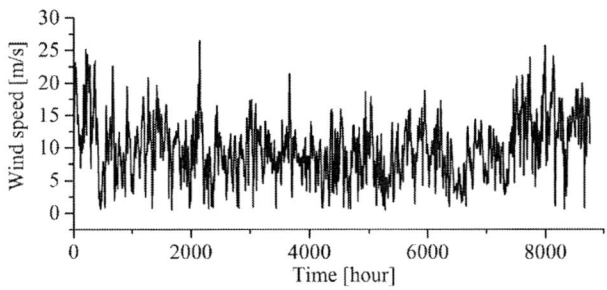

Fig. 4. One year mission profile of the wind speed from a wind farm.

Fig. 5. Annual produced power by each wind turbine.

Then, the wind profile is fed into the manufacturer provided wind turbine model [12] to obtain the annual power production profile. As shown in Fig. 5, the power production fluctuates significantly during a year, which contributes to accumulate fatigue and thus the wear-out of the devices.

As discussed in [13], the power dissipation of a power device includes conduction losses and switching losses. The average conduction loss $P_{\text{cond\_ave}}$ of a power device is

$$P_{\text{cond\_ave}} = f_0 \cdot \int_0^{1/f_0} p_{\text{cond\_inst}}(t)\, dt \tag{1}$$

where the instantaneous conduction loss is

$$p_{\text{cond\_inst}}(t) = u_{\text{cond}}(i_x(t), T_j) \cdot i_x(t) \cdot M(m, t) \tag{2}$$

in which i_x is the conducting current through the power device, the duty ratio $M(m, t)$ is a function of the modulation index m, and the conduction voltage $u_{\text{cond}}(i_x(t), T_j)$ is related to the junction temperature and the conduction current.

Similarly, the average switching loss $P_{\text{sw\_ave}}$ is

$$P_{\text{sw\_ave}} = f_0 \cdot \int_0^{1/f_0} p_{\text{sw\_inst}}(t)\, dt \tag{3}$$

where the instantaneous switching loss is,

$$p_{\text{sw\_inst}}(t) = f_{\text{sw}} \cdot E_{\text{sw}}(i_x(t), T_j) \cdot \left(\frac{U_{\text{SM}}}{U_{\text{ref}}}\right)^{K_v} \tag{4}$$

with f_{sw} being the equivalent switching frequency, U_{SM} being the average capacitor voltage of an SM, K_v representing the voltage coefficient, U_{ref} denoting the reference blocking-voltage in the data-sheet, and E_{sw} showing the switching energy loss.

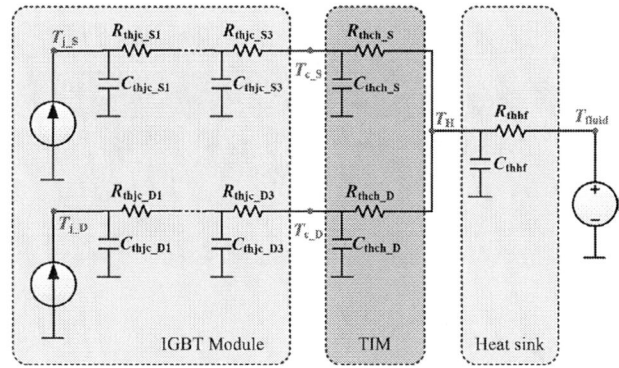

Fig. 6. Thermal model for IGBT modules with TIMs and heatsinks.

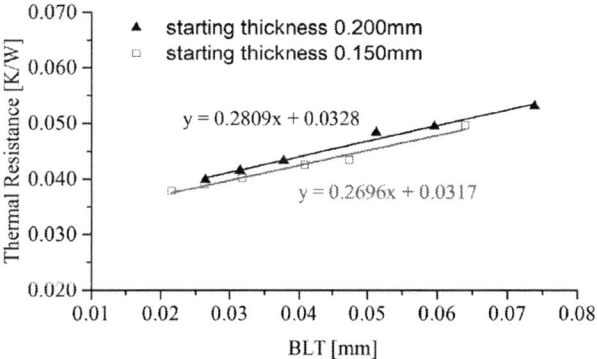

Fig. 7. Thermal resistance of the TIM is varied with the starting assembly thickness and the BLT, data from [14].

B. Thermal Models and the Thermal Profiles

Thermal models are the bridge between the electrical parameters and thermal parameters, which maps the power losses to the junction temperatures of devices, as shown in Fig. 6. As it can be observed, the TIM is considered and it is utilized to fill in the air-gap between the baseplate/case of the IGBT module and the heatsink.

Thus, the TIM has a substantial contribution to the junction-to-ambient thermal resistance, but many cases assume the thermal impedance of the TIM as a constant according to the thermal conductivity. In this paper, a specific TIM (the thermal conductivity is constant) is selected as the study case, where the data comes from [14]. According to the trend of the data shown in Fig. 7, the thermal resistance of the TIM is varied with the starting assembly thickness and BLT. It is indicated that the thicker the BLT is, the larger the TIM thermal resistance is. A similar observation goes to the case of the starting assembly thickness. In other words, the thinner starting thickness results in a smaller thermal resistance.

As a consequence, the trends of the TIM should be substituted into the thermal model. The translation from loss profiles into thermal profiles are the same as [15]. Then, with two values of the starting assembly thickness and various TIM

2745

The 2018 International Power Electronics Conference

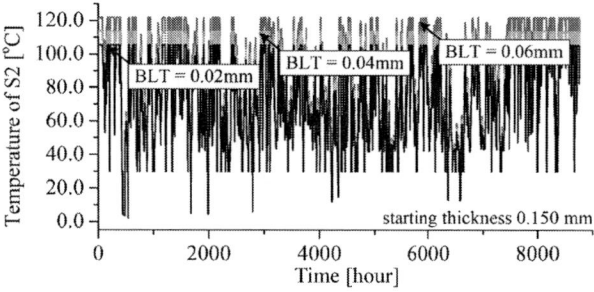

Fig. 8. Thermal profiles of the device S_2 under various BLT values of the TIM (the starting assembly thickness is 0.150mm).

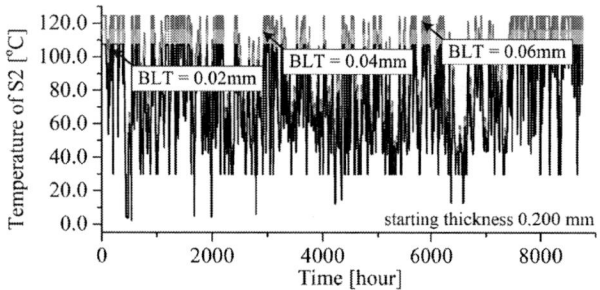

Fig. 9. Thermal profiles of the device S_2 under various BLT values of the TIM (the starting assembly thickness is 0.200mm).

BLT values, the thermal profiles of S_2 are obtained, as shown in Figs. 8 and 9. When the starting assembly thickness is 0.150 mm, the maximum junction temperature of S_2 is about 105 °C (BLT = 0.02 mm) to 121 °C (BLT = 0.06 mm). Moreover, under BLT = 0.06 mm, the maximum junction temperature is increased from 121 °C to roughly 125 °C, when the starting thickness increases from 0.150 mm to 0.200 mm. It reveals that both the BLT and the starting assembly thickness of the TIM have significant impacts on the junction temperatures of the power devices, although the thermal conductivity is the same in the case.

C. Lifetime Consumption Considering the TIM Thickness

Since the thermal loading profiles are random, it is difficult to directly map thermal performance to lifetime. Typically, the thermal stresses are converted into regular thermal cycles using a counting algorithm. In this paper, the rain-flow counting algorithm [16] is adopted. After the cycling counting, the lifetime of the power device is estimated according to its corresponding lifetime model as

$$N_{\text{life}} = A(\Delta T - \Delta T_0)^{-n} \cdot \exp\left(\frac{E_a}{k_B T_m}\right) \qquad (5)$$

where A and n are the device-dependent coefficients, ΔT_0 means the elastic strain, T_m is the mean temperature, and E_a and k_B denote activation energy and the Boltzmann constant, respectively. Noted that the model used in this paper gives the best lifetime prediction in the specific application

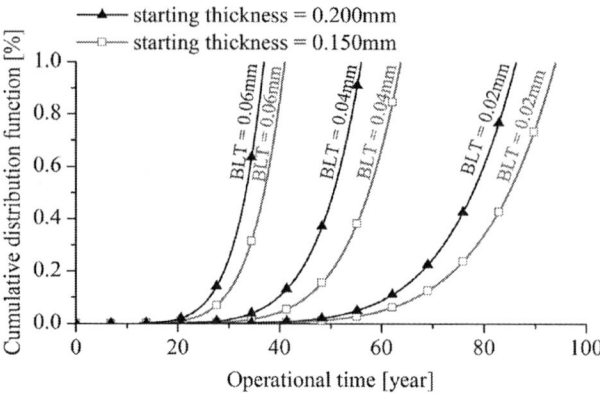

Fig. 10. End-of-life cumulative distribution function of the power device S_2 in an SM of the MMC due to the fatigue.

condition, which is concluded in [17] by comparing with the experimental aging results.

Then, the total lifetime consumption of one year CL_{1year} accumulates according to the Palmgren-Miner's rule. Accordingly, the lifetime of the power device LF is obtained as

$$CL_{\text{1year}} = \sum \frac{100}{N_{\text{life\_1}}} + \frac{100}{N_{\text{life\_2}}} + \cdots + \frac{100}{N_{\text{life\_n}}} (\%) \qquad (6)$$

$$LF = \frac{1}{CL_{\text{1year}}} \qquad (7)$$

The end-of-life distribution of the device S_2 in an SM of the MMC is shown in Fig. 10. When the starting thickness is 0.150 mm, the B_1 lifetime (1% devices is in failure) has a reduction trend with the increase in the BLT. Then, B_1 lifetime has a similar tendency when the starting thickness is 0.200 mm. However, a shorter lifetime is always expected when the starting thickness is 0.200 mm in contrast to 0.150 mm. For example, when the BLT = 0.06 mm, the B_1 lifetime of the starting thickness being 0.200 mm is roughly 30 years, while the B_1 lifetime is 35 years with the starting thickness being 0.150 mm. Therefore, both the starting thickness and the BLT affect the lifetime of the power device. Even if both thickness changes are small, the impact on the lifetime is obvious. Noted that only the starting thickness is related to the cost, while the BLT is solely related to assembling methods. However, the lifetime varies significantly under the same starting thickness. It implies that the lifetime of the power device can be improved without additional costs, which only requires a better understanding of the TIM. Hence, the designing and assembling of the power electronic converters should consider the TIM impact.

IV. EXPERIMENTAL VALIDATIONS

In order to validate the impact of the thickness of the TIM, an SM of the MMC is built up to test the junction temperature under various TIM thickness, which is shown in Fig. 11. An IGBT module from Infineon F4-50R12KS4

2746

The 2018 International Power Electronics Conference

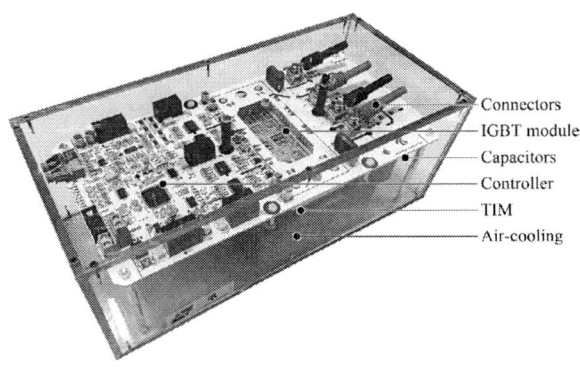

Fig. 11. Experimental setup of an SM for thermal behaviors evaluation in an MMC system.

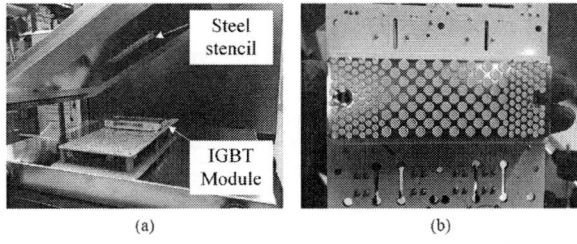

(a)   (b)

Fig. 12. Screen printing equipment to apply TIMs, where the starting thickness of the TIM is controlled by the thickness of the stencil: (a) screen printing equipment and (b) applied TIM on the IGBT module.

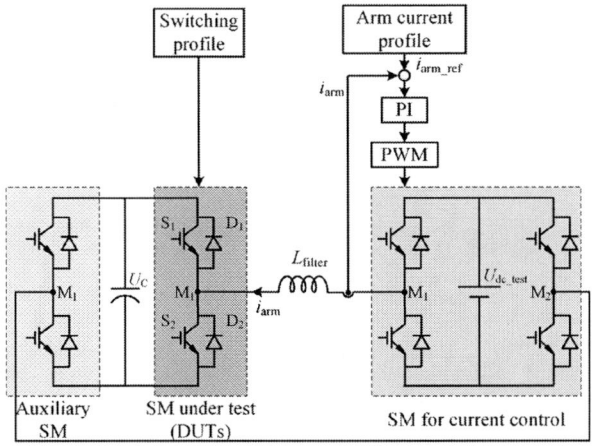

Fig. 13. Experimental platform to evaluate the performance of the TIM's thicknesses in the MMC (PI – Proportional Integral control; PWM – Pulse Width Modulation; U_{dc_test} is the dc voltage of power supply; L_{filter} is the filter inductor.

TABLE II
SPECIFICATIONS OF THE EXPERIMENTAL SETUP

Parameters	Values/features
Thermal-interface-material parameters	
Name	HTSP50T
Color	White
Base	Silicone oil
Conductive components	Powered mental oxides
Thermal conductivity	3.0 W/m·K
Viscosity	Paste
Electrical parameters	
Voltage of dc power supply	U_{dc_test} = 30 V
Capacitor voltage	U_C = 300 V
Filter inductor	L_{filter} = 2 mH
Switching frequency	f_{sw} = 2 kHz
Fundamental frequency	f_0 = 1 Hz
Arm current reference	$i_{arm} = 7.14 + 17.85sin(2\pi f_0 t)$
Mechanical parameters	
Steel stencil thicknesses	80, 100, 120, 150, 200 μm
Screw torque	M_T = 2 N·m

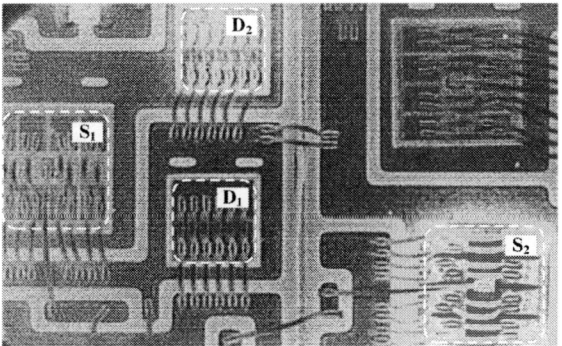

Fig. 14. Thermal distribution of the four power devices in an SM of the MMC measured by an infrared camera.

adopted to develop different starting thickness (80, 100, 120, 150, 200 μm). Then, according to the instructions of [18], the applied TIM is shown in Fig. 12(b).

Furthermore, experiments are carried out referring to the set-up shown in Fig. 13. The configuration consists of three parts, where a full-bridge circuit is utilized to emulate the arm current reference of the MMC, the SM under test (DUTs) is fed into a switching profile, and the auxiliary SM is used to manage the capacitor voltage. In order to avoid the thermal coupling, the auxiliary SM is separated from the SM DUTs. The junction temperatures of power devices are measured by FLIR X8400sc. It should be pointed out that the fundamental frequency is set to 1 Hz in the experiments in order to increase the junction temperature swings. Other experimental parameters are summarized in Table II.

The measured junction temperature distribution is shown in Fig. 14. According to the results, the junction temperature profiles of the four power devices in an SM of the MMC can be obtained, which are shown in Fig. 15. In the thermal profiles of the device S_1 (see Fig. 15(a)), the maximum junction temperature is around 31 °C under 80 μm TIM. When the starting thickness increases to 100 μm, the peak junction

(1200 V/50 A) is selected as the power device in the experimental platform. Moreover, a TIM (HTSP50T, 3.0 W/m·K) with various thickness values is assembled between the power device and the heatsink. The starting thickness of the TIM is controlled by the thickness of the steel stencil, which is shown in Fig. 12(a). In the experiment, five different steel stencils are

2747

The 2018 International Power Electronics Conference

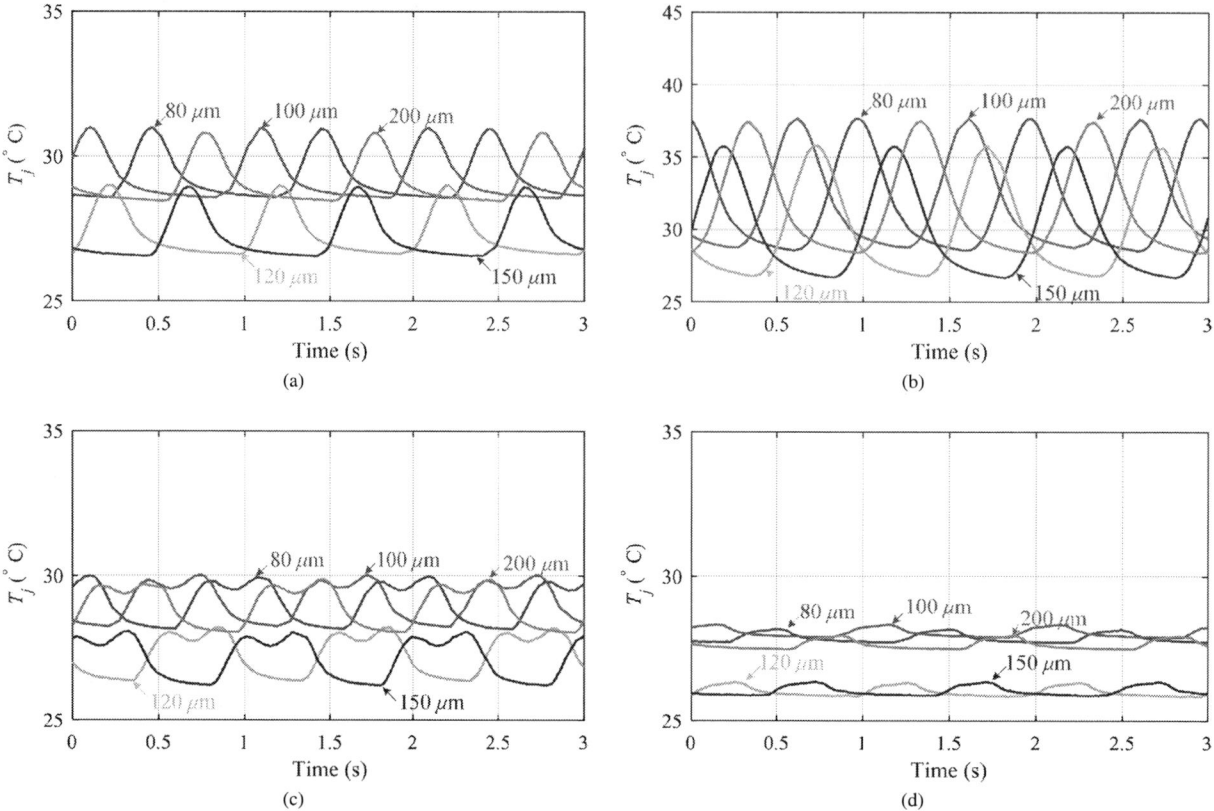

Fig. 15. Junction temperature profiles of the power devices in an SM of the MMC system, where the starting thickness of the TIM changes from 80 μm to 200 μm and the screw torque is fixed to 2 N·m: (a) S_1, (b) S_2, (c)D_1 and (d)D_2.

temperature still keeps at 31 °C approximately. However, with the starting thickness of the TIM increasing to 120 μm and 150 μm, the maximum junction temperature is decreased to 28 °C. Furthermore, the maximum junction temperature goes back to about 31 °C. This indicates that the starting thickness of the TIM will affect the junction temperature. Thus, it leads to a non-neglected influence on the lifetime of the power device. However, the effect of the TIM is not linear. An optimal thickness or an optimal range is expected. Either very thin or thick thickness negatively affects the lifetime of the power devices.

Furthermore, the junction temperature profiles of the devices S_2, D_1 and D_2 have a similar trend, which are shown in Figs. 15(b), (c) and (d). With the increase of the starting thickness of the TIM, the maximum or average junction temperatures decrease first and then increase again. The only difference is that the device S_2 has the largest temperature swings, which comes from the inherently thermal unbalance in the SM of the MMC.

Therefore, according to the above experiments, the impact of the starting thickness of the TIM can be observed. However, the impact is nonlinear, where an optimal thickness design can be obtained as a further study.

V. CONCLUSION

A quantitative analysis of the impact of the TIM thickness on the predicted lifetime of the IGBT module was presented in this paper. In the case study of a 30-MW MMC system for offshore wind applications, a specific TIM was compared based on two different starting assembly thickness values and three BLT types. According to the quantitative results, it can be concluded that both the starting thickness and the BLT of the TIM will affect the lifetime of the power devices. The predicted lifetime can shift as large as 50% with different thickness values of the TIMs. It should be pointed out that only the starting thickness is related to costs, while the BLT is solely related to assembling method. The lifetime shifting significantly with BLT values implies that the reliability of power devices can be improved without additional costs. As a consequence, the TIM thickness should be taken into account carefully in the design and assembly phases. Experimental tests also validated the impact of the TIM thickness. However, the influence is nonlinear. Optimizations may be applied.

REFERENCES

[1] M. A. Perez, S. Bernet, J. Rodriguez, S. Kouro, and R. Lizana, "Circuit topologies, modelling, control schemes and applications of modular

multilevel converters," *IEEE Trans. Power Electron.*, vol. 30, no. 1, pp. 4–17, Jan. 2015.

[2] J. J. Jung, H. J. Lee, and S. K. Sul, "Control strategy for improved dynamic performance of variable-speed drives with modular multilevel converter," *IEEE J. Emerg. Sel. Topics Power Electron.*, vol. 3, no. 2, pp. 371–380, Jun. 2015.

[3] Y. Chen, Z. Li, S. Zhao, X. Wei, and Y. Kang, "Design and implementation of a modular multilevel converter with hierarchical redundancy ability for electric ship mvdc system," *IEEE J. Emerg. Sel. Topics Power Electron.*, vol. 5, no. 1, pp. 189–202, Mar. 2017.

[4] F. Ma, Q. Xu, Z. He, C. Tu, Z. Shuai, A. Luo, and Y. Li, "A railway traction power conditioner using modular multilevel converter and its control strategy for high-speed railway system," *IEEE Trans. Transp. Electrific.*, vol. 2, no. 1, pp. 96–109, Mar. 2016.

[5] H. Akagi, "Classification, terminology, and application of the modular multilevel cascade converter (MMCC)," *IEEE Trans. Power Electron.*, vol. 26, no. 11, pp. 3119–3130, Nov. 2011.

[6] S. Yang, A. Bryant, P. Mawby, D. Xiang, L. Ran, and P. Tavner, "An industry-based survey of reliability in power electronic converters," *IEEE Trans. Ind. Appl.*, vol. 47, no. 3, pp. 1441–1451, May/Jun. 2011.

[7] A. Wintrich, N. Ulrich, T. Werner, and T. Reimann, *Application Manual Power Semiconductors*. Nuremberg, Germany: Semikron Int.GmbH, 2015.

[8] M. B. R. Schnell and S. Geissmann, *Thermal design and temperature ratings of IGBT modules*. ASEA Brown Boveri, Zurich, Switzerland: ABB Application Note, 5SYA 2093-00, 2011.

[9] D. Zhou, F. Blaabjerg, T. Franke, M. Tnnes, and M. Lau, "Comparison of wind power converter reliability with low-speed and medium-speed permanent-magnet synchronous generators," *IEEE Trans. Ind. Electron.*, vol. 62, no. 10, pp. 6575–6584, Oct. 2015.

[10] K. Ma, N. He, M. Liserre, and F. Blaabjerg, "Frequency-domain thermal modeling and characterization of power semiconductor devices," *IEEE Trans. Power Electron.*, vol. 31, no. 10, pp. 7183–7193, Oct. 2016.

[11] H. Liu, K. Ma, Z. Qin, P. C. Loh, and F. Blaabjerg, "Lifetime estimation of MMC for offshore wind power HVDC application," *IEEE J. Emerg. Sel. Topics Power Electron.*, vol. 4, no. 2, pp. 504–511, Jun. 2016.

[12] "Vestas V90 3.0 MW data.pdf." [Online]. Available: https://www.vestas.com/

[13] K. Ma, M. Liserre, and F. Blaabjerg, "Operating and loading conditions of a three-level neutral-point-clamped wind power converter under various grid faults," *IEEE Trans. Ind. Appli.*, vol. 50, no. 1, pp. 520–529, Jan./Feb. 2014.

[14] M. Schulz, S. T. Allen, D. Phan, and P. Wilhelm, "The Crucial Influence of Thermal Interface Material in Power Electronic Design," 2013. [Online]. Available: http://www.imapsfrance.org/puissance2013/4-Henkel.pdf

[15] Y. Zhang, H. Wang, Z. Wang, Y. Yang, and F. Blaabjerg, "The impact of mission profile models on the predicted lifetime of IGBT modules in the modular multilevel converter," in *Proc. IECON 43rd Annual Conf. IEEE Ind. Electron. Society*, 2017.

[16] A. Niesłony, "Determination of fragments of multiaxial service loading strongly influencing the fatigue of machine components," *Mechanical Systems and Signal Processing*, vol. 23, no. 8, pp. 2712–2721, 2009.

[17] Y. Zhang, H. Wang, Z. Wang, and F. Blaabjerg, "Impact of lifetime model selections on the reliability prediction of IGBT modules in modular multilevel converters," in *Proc. the 2017 IEEE Energy Conversion Congress and Exposition (ECCE)*, 2017.

[18] Baginski, *Application of screen print templates to paste thermal grease within IGBT modules*. Warstein: Infineon Application Note, AN-2006-02, 2005.

Nanoscale investigation of the power MOSFET by the AFM/KFM/SCFM

Mizuki Nakajima[1], Yuuki Uchida[1], Nobuo Satoh[1]*, Hidekazu Yamamoto[2]
Chiba Institute of Technology, Narashino, Chiba 275-0016, Japan
*E-mail: satoh.nobuo@it-chiba.ac.jp

Abstract—**Power semiconductor devices progress towards high withstand voltages using wide-band-gap semiconductor materials, and parallel integration enabled by microfabrication techniques. We achieved nanoscale observation of power semiconductor device using a scanning probe microscope based on the combination of atomic force microscopy, Kelvin probe force microscopy, and scanning capacitance force microscopy, which provided high spatial resolution and sensitivity. The nanoscale observations were performed through stability control using the frequency-modulation (FM) detection method under a vacuum pressure environment, with and without bias voltage applied to the power semiconductor device.**

I. INTRODUCTION

In recent years, the development of power semiconductor devices has been prepared. The miniaturization and high efficiency operation make their use as switching and / or rectifying elements in power conversion circuits suitable. Two main approaches are employed to enhance the performance of power semiconductor devices, as discussed below.

The first approach relies on the findings that a high withstand voltage of active device can be obtained using a wide-gap semiconductor. The energy band gap of the silicon (Si) [1], [2] is 1.09 eV [3]. A semiconductor material with an energy band gap larger than twice that of Si, is referred to as a wide-band-gap semiconductor material. Silicon carbide (SiC; 2.8 eV)[3] gallium nitride (GaN; 3.3 eV) [4], [5] and diamond (C; 5.2 eV) [3] are considered for wide-gap semiconductor materials.

The second approach is to employ and optimize the semiconductor microfabrication technology. The miniaturization enabled by the microfabrication technology has led to the development of semiconductor integrated circuits (ICs), which are the basic elements of computers. With the progress of photolithography, ultrafine interconnections and regions have been achieved in semiconductor materials. In addition, the microfabrication technology has enabled to obtain a large-current capacity through connecting devices in parallel within a single semiconductor material.

By employing novel semiconductor materials and fine processing technology, various power semiconductor device structures have been developed and implemented in practical applications. The performance improvement of the above fabrication techniques led to complex internal structures of the employed semiconductor device. Therefore, it becomes more challenging to analyze and evaluate the operation and failure of these devices. With the progress in power semiconductor

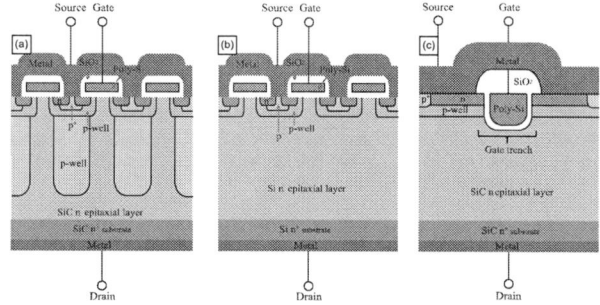

Fig. 1. Schematic structure of various power semiconductor devices. (a) Si-Super Junction MOSFET (planar type), (b) SiC-DMOSFET, and (c) SiC-TMOSFET, respectively.

devices, a continuous advancement of evaluation techniques is required in order to provide approaches to easily reveal origins of device failures and propose design improvements.

Nanoscale observation of devices operating with bias voltage applied is useful for the evaluation. The semiconductor materials have been investigated using various types of scanning probe microscopy (SPM) [6], [7]. It has been reported that a Si-diode with a p-n junction by SPM system [8], [9]. Moreover, a failure analysis of the large-scale integration (LSI) circuits [10], [11] have been performed and the operation of flash memory circuits has been investigated [12]. In addition, the evaluation of an SiC diode by using the Kelvin probe force microscopy (KFM) and scanning capacitance force microscopy (SCFM) was previously reported [13], [14]. Recently, we reported the investigation of a commercial SiC-Schottky barrier diode (SiC-SBD) under an applied bias voltage using the atomic force microscopy (AFM) [15] combined with KFM and SCFM [16].

In this study, in order to achieve stable nanoscale observations under vacuum pressure environment, the regulation of the probe-tip-sample distance with frequency modulation (FM) method was installed in the conventional system. We cut the power semiconductor device of the commercially available TO-247 package, polished the sample surface. We could expose the surface for the nanoscale observations. We could observe three images of a topographic, a surface potential, and a differential capacity, there were simultaneously and stably acquired within the same area. In addition, under the bias voltage was applied to gate electrode in three-terminal power semiconductor device.

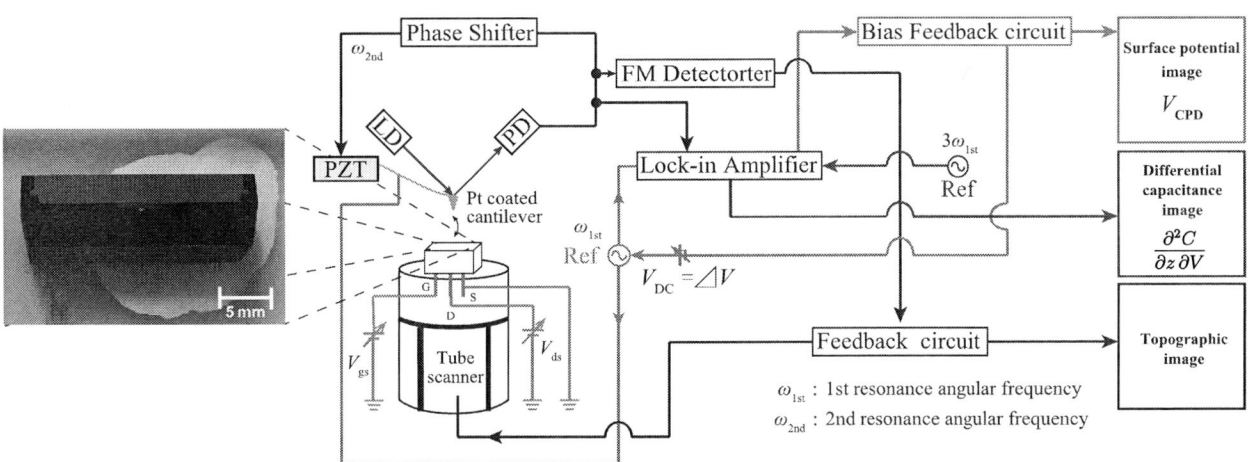

Fig. 2. A cross-sectional photograph of the power device and schematic diagram of the AFM/KFM/SCFM system. The modulated signal $v(t) = V_{dc} + V_{ac}\cos(\frac{1}{3}\omega_{1st}t)$ [V] is applied between the probe-tip and the sample for KFM/SCFM. The signal corresponding to ω_{1st} is then detected, using the lock-in amplifier, from the electrostatic force, and yields a differential capacitance image.

II. EXPERIMENTAL PROCEDURE

A. AFM/KFM/SCFM system

A schematic of the AFM/KFM/SCFM system is shown in Fig. 1. We combined with detective functions both the surface potential and differential capacitance into a commercial AFM (Hitachi High-Technologies, AFM5300E). The micrograph of sample exposing power semiconductor devices for the nanoscale observation is included and shown there.

All results presented were obtained under a vacuum environment, so as to target a high force sensitivity using frequency modulation atomic force microscopy (FM-AFM), which can stably control the distance between the probe-tip and the sample surface [17], [18]. The FM detector contains of a phase-locked loop with a voltage-controlled crystal oscillator, which has a high thermal stability [19]. The simultaneous application of FM-AFM and an electric-properties evaluation method in the same area requires a large measuring time (1.5 hour per scan in this study). The FM detector significantly contributes to the control of the probe-tip during the slow scanning period. Using the feedback controller in the FM-AFM, a surface topographic image can be acquired by scanning the sample's surface, while maintaining a constant value of the frequency shift (Δf). In these experiments, the cantilever oscillated at the secondary resonance frequency (approximately 384 kHz), which is approximately six times larger than the primary resonance frequency of the cantilever of 65.7 kHz. As the secondary resonance mode of the cantilever features stable vibration with a smaller amplitude than that of the primary resonance mode, we expected that a precise distance control between the probe's tip and sample's surface can be achieved. The frequency shift Δf was in the range of approximately -20 to -40 Hz in all experiments. In the system, three images were obtained under in vacuum of 1×10^{-2} Pa at room temperature. The mechanical Q-factor of the cantilever under these conditions increases to over 2000.

KFM allows the measurement of the surface potential on the nanoscale [18], [20], [21], [22], [23], [24]. From the electrostatic force dependence on the applied AC and DC bias voltages, the fundamental wave component of the force F_ω is given by

$$F_\omega = \frac{dC(z)}{dz}(\Delta\Phi - V_{dc})V_{ac}\cos(\omega t). \tag{1}$$

Here, $\Delta\Phi$ is the difference in the work function between the probe-tip and the sample surface (it is the contact-potential difference), and the V_{dc} is the applied DC bias. We noted that $\omega_1 = 2\pi f_1$, $\omega_2 = 2\pi f_2$. The signal derived from F_ω is detected by phase detection using a lock-in amplifier (NF LI5660). This signal is subsequently fed into a custom-built bias feedback circuit set for a proportional-integral (PI) control. Therefore, the KFM method can map the surface potential based on Eq. (1), when the relation ($V_{dc} = \Delta\Phi = V_{CPD}$) (to cancel $\Delta\Phi$ exactly) is applied via the PC analog/digital converter. In the KFM measurements, the reference signal frequency was set to the one-third of the primary resonance frequency (it is approximately 21.9 kHz) of the cantilever, while its amplitude was set to 8 V_{p-p}.

The SCFM [25], [26], [27] exploits the differential capacitance in the metal-oxide-semiconductor (MOS) structure [28], [29], which contains three layers: the metal-coated probe-tip, the oxide layer on the sample surface, and the semiconductor substrate. The induced force component at the angular frequency of 3ω, derived from the carrier density in the sample is:

$$F_{3\omega} = \frac{1}{8}\left.\frac{\partial^2 C(z,V)}{\partial z \partial V}\right|_{V_{dc}} V_{ac}^3 \cos(3\omega t). \tag{2}$$

When ω is too high to react with the minority carriers, the 3ω component amplitude corresponds to the majority-carrier density (the lower the carrier density, the greater the amplitude), and its phase corresponds to the semiconductor type (p^-

or n^- type). Therefore, the carrier-density distribution can be mapped, as the 3ω component is detected from the electrostatic force by the lock-in amplifier. To increase the sensitivity, we followed a previous study, where 3ω corresponded to the second-resonance angular frequency of the AFM cantilever [26]. In the SCFM measurements, the reference signal for the lock-in amplifier was set at the primary resonance frequency (it is approximately 65.7 kHz), three times larger than 21.9 kHz.

Our AFM/KFM/SCFM system is suitable for the simultaneous acquisition of different images within a given region. In the KFM and SCFM measurements, modulated signals were applied to the cantilever. It is noted that a modulation signal is applied to the sample in these case of general measurement. Therefore, it can be allowed dependency observations without/with an applied bias voltage to the three-terminal power device, independently. Stable differential-capacitance images can then be obtained by using the KFM to nullify the high potential difference that arises as an artifact upon the application of the bias between the probe-tip and the sample surface. In addition, both the KFM and SCFM are based on the electrostatic forces, i.e., on the fundamental wave component and the third harmonic component, respectively. All AFM/KFM/SCFM images in this study were simultaneously obtained within the same region.

B. Voltage-current characteristics of the SiC-DMOSFET

The sample employed in this study was the Silicon Carbide Double Diffused Metal Oxide Semiconductor Field Effect Transistor (SiC-DMOSFET) operated at high frequency with the applied high bias voltage, as utilized for power electronics. The SiC-DMOSFET was fabricated with the TO-247 package, which is typically used for power devices. The $V-I$ characteristics of the SiC-DMOSFET, measured by using a curve tracer (IWATSU CS-3100), is shown in Fig. 3. The set breakdown voltage of the device used, it was 1200 V from the data-sheet as the rated value. The breakdown voltage of the device of this package was found to be taken 1650 V from the actual $V-I$ characteristic.

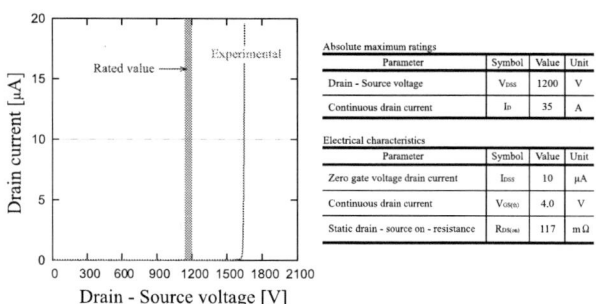

Fig. 3. The voltage - current ($V-I$) characteristics of the SiC DMOSFET used for nanoscale observations. Typical values listed in the data sheet in absolute maximum ratings and electrical characteristics.

III. RESULTS AND DISCUSSION

A. Internal structure of SiC-DMOSFET

Figures 4 show AFM / KFM / SCFM images of 25 $\times$ 25 μm^2 region and a schematic diagram of the internal structure of the SiC-DMOSFET. The topographic image, surface potential image, and differential capacitance image could be observed by the FM-AFM/KFM/SCFM system in simultaneously and same region. Especially, the differential capacitance image by the SCFM was clearly indicated the difference in carrier concentration between the epitaxial layer and substrate in the SiC-DMOSFET. The observations showed that the epitaxial layer had a thickness of 12 μm. As the rated withstand voltage was approximately 1200 V, a breakdown voltage larger than 100 V per 1 μm could be expected.[30]

The positions of the gate-electrode and the oxide film were also clearly observed. In addition, the results enable to measure the contact area between the source-electrode and the wiring metal, and the thickness of the insulating interlayer film. As the position of the channel formation was also suggested, the electric path through which the current is flowed in the SiC-DMOSFET could also be predicted on the schematic diagram of the internal structure.

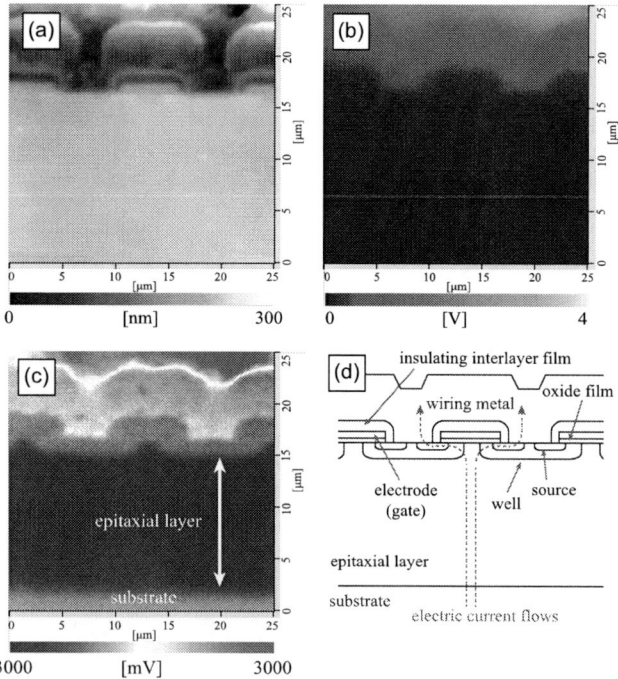

Fig. 4. AFM / KFM / SCFM images of 25 μm square area and internal schematic structure diagram. (a) Topographic image by the AFM, (b) Surface potential image by the KFM, (c) Differential capacitance image by the SCFM, and (d) Schematic view of the internal structure.

2752

B. Observation with applied voltage between electrodes

The gate electrode, the drain electrode, and the source electrode, which are all the electrodes, are connected correspondingly to independent external power supplies. As shown in Fig 5, we note that the applied bias conditions for nanoscale observation, (a) is $V_{gs} = 0$ V and $V_{ds} = 0$ V, (b) is $V_{gs} = 5$ V and $V_{ds} = 0$ V, (c) is $V_{gs} = 5$ V and $V_{ds} = 1$ V.

Firstly, in Fig. 5(a), the signals that correspond to the material and structure of the gate electrode were detected. Secondly, we believe that the formation of a channel by bias voltage (V_{gs}) to the between gate-and-source electrodes could be mapped in Fig. 5(b). Finally, bias voltages were applied between the each electrode, the gate-and-source and the drain-and-source, respectively. It seems that the differential capacitance component corresponding to the movement of the carrier were detected by channel formation as shown in Fig. 5(c). Experimentally, when the $V_{gs} = 5$ V and $V_{ds} = 1$ V were applied to the sample, the display of the power supply showed a current value of 60 mA.

An interesting point is that the capacitance change of the epitaxial layer was measured according to the applied voltage between the drain and source electrodes. As a future task, we clarify the device operation mode by simulation analysis, which based on the visualization of the state change according to each applied bias voltage by FM-AFM/KFM/SCFM.

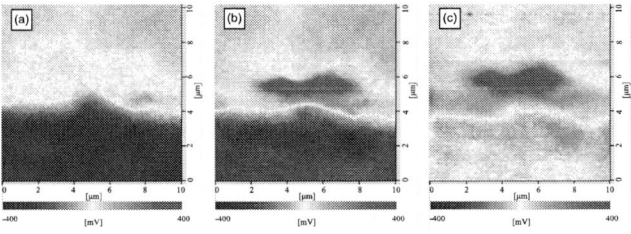

Fig. 5. The differential capacitance images by the SCFM system (scanning range was 10 μm). (a) $V_{gs} = 0$ V and $V_{ds} = 0$ V, (b) $V_{gs} = 5$ V and $V_{ds} = 0$ V, (c) $V_{gs} = 5$ V and $V_{ds} = 1$ V.

IV. CONCLUSIONS

We observed of the internal structure of a SiC-DMOSFET by the scanning probe microscope (SPM) built on combined with AFM/KFM/SCFM that achieved high spatial-resolution and high sensitivity. The nanoscale observations were performed through stable control by the FM detection method under vacuum environment, and without/with bias-voltage applied to the three-terminal power semiconductor device. The internal structure of the SiC-DMOSFET was clarified from the nanoscale observation results by the AFM/KFM/SCFM system. We applied each bias voltage to the gate-source and drain-source electrodes in the SiC-DMOSFET, and we mapped the state of channel formation were shown by the differential capacitance images.

ACKNOWLEDGMENT

This study was partially supported by the MEXT-Supported Program for Strategic Research Foundation at Private Universities 2013-2017 (s131104), and the Cross-Ministerial Strategic Innovation Promotion Program for Next-Generation Power Electronics.

APPENDIX A: SI-PN PATTERN SAMPLE

We identified both the carrier type and density by the SCFM method using the sample with Si - pn pattern [18] for semiconductor evaluation. Fig. 6 shows the observation results of Si - pn pattern. By comparing the observed differential capacitance images, achievement of high resolution was clarified. Fig. 6 (a) shows, in previous study, the differential capacitance image when was observed by using the secondary resonance of the cantilever (the topographic image was obtained with primary resonance, simultaneously). On the other hand, Fig. 6(b) shows, in this study, the differential capacitance image when was observed by using the primary resonance of the cantilever (the topographic image was obtained with secondary resonance, simultaneously). The difference in these resolutions depended on the setting of the reference signal corresponding to the two resonance frequency modes possessed by the cantilever.

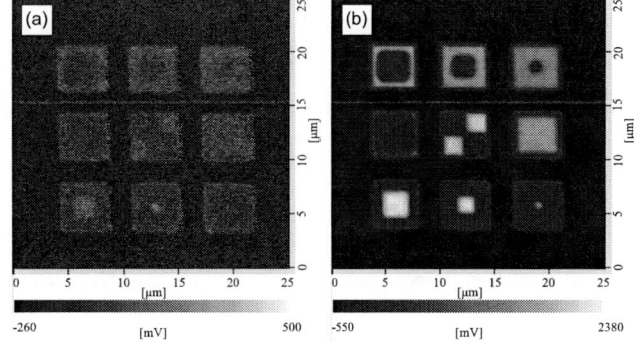

Fig. 6. The comparison of the Si - pn pattern sample due to the differential capacitance images. (a) In previous study, the differential capacitance image when was observed by using the secondary resonance of the cantilever. (b) In this study, the differential capacitance image when was observed by using the primary resonance of the cantilever.

We describe the concrete experimental conditions. The frequency response characteristics of the cantilever used are shown in the Fig. 7. The primary resonance frequency as the $f_1 = 65.7$ kHz and the secondary resonance frequency as the $f_2 = 384$ kHz were acquired, respectively. The differential capacitance values had detected the electrostatic force caused by the voltage modulation.

Figure 6 (a) shows the same setting condition as the previous report [25], by applying the voltage modulation signal having an primary frequency (f_1) which is one-third of the secondary resonance frequency (f_2) between the probe and the sample. On the other hand, Fig. 6(b) shows, the differential capacitance

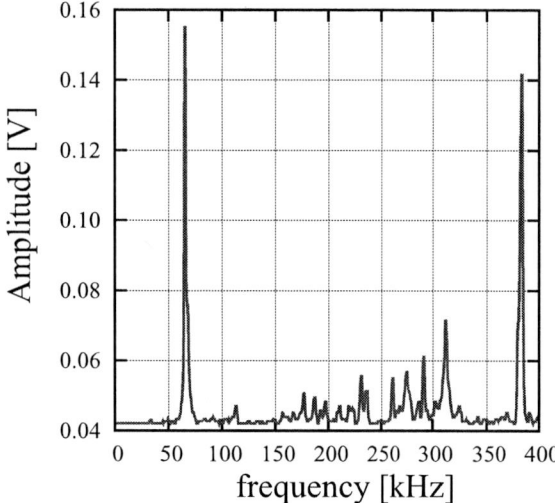

Fig. 7. Frequency response characteristics of the cantilever.

The Taylor expansion of $\partial C(z,V)/\partial z$, neglecting higher-order terms as Vdc is a fixed number, yields

$$
\begin{aligned}
\frac{\partial C(z,V)}{\partial z} &= \left.\frac{\partial C(z,V)}{\partial z}\right|_{V_{\mathrm{dc}}} \\
&+ \left.\frac{\partial^2 C(z,V)}{\partial z \partial V}\right|_{V_{\mathrm{dc}}} \left\{ [V_{\mathrm{dc}} + V_{\mathrm{ac}}\cos(\omega t)] - V_{\mathrm{dc}} \right\} \\
&+ \cdots \\
&\simeq \left.\frac{\partial C(z,V)}{\partial z}\right|_{V_{\mathrm{dc}}} + \left.\frac{\partial^2 C(z,V)}{\partial z \partial V}\right|_{V_{\mathrm{dc}}} V_{\mathrm{ac}}\cos(\omega t).
\end{aligned}
$$

$$(5)$$

REFERENCES

[1] H. Yamamoto, Y. Murakami, H. Nishimura, S. Takahashi, N. Tokuda, K. Shimizu, A. Furukawa, S. Yamakawa, and M. Imaizumi, Denshi Joho Tsushin Gakkai Ronbunshi C *J92-C*, 159 (2009) [in Japanese].
[2] H. Yamamoto and T. Hashizume, *Phys. Status Solidi C 8*, 662 (2011).
[3] N. B. Hannay: "SEMICONDUCTORS" (REINHOLD: New York), (1959).
[4] W. Saito, Y. Takada, M. Kuraguchi, K. Tsuda, I. Omura, T. Ogura, and H. Ohashi, *IEEE Trans. Electron Devices 50*, 2528 (2003).
[5] E. G. Bylander: "MATERIALS FOR SEMICONDUCTOR FUNCTIONS" (HAYDEN: New York), (1971).
[6] S. Morita, R. Wiesendanger, E. Meyer, T. Yatsui and M. Naruse: "NONCONTACT ATOMIC FORCE MICROSCOPY" (SPRINGER: Heidelberg), (2002).
[7] C. F. Quate, Jpn. *J. Appl. Phys. 42*, 4777 (2003).
[8] G. H. Buh, H. J. Chung, J. H. Yi, I. T. Yoon, and Y. Kuk, *J. Appl. Phys. 90*, 443, 2001.
[9] G. H. Buh, H. J. Chung, C. K. Kim, J. H. Yi, I. T. Yoon, and Y. Kuk, *Appl. Phys. Lett. 77*, 106, 2000.
[10] L. Zhang, H. Tanimoto, K. Adachi, and A. Nishiyama, *IEEE Electron Device Lett. 29*, 799, 2008.
[11] L. Zhang, K. Ohuchi, K. Adachi, K. Ishimaru, M. Takayanagi, and A. Nishiyama, *Appl. Phys. Lett. 90*, 192103, 2007.
[12] K. Honda, S. Hashimoto, and Y. Cho, *Nanotechnology 17*, S185, 2006.
[13] H. Bartolf, U. Gysin, T. Glatzel, H. R. Rossmann, T. A. Jung, S. A.Reshanov, A. Schöner, and E. Meyer, *Microelectron. Eng. 148*, 1, 2015.
[14] H. R. Rossmann, U. Gysin, A. Bubendorf, T. Glatzel, S. A. Reshanov, A. Zhang, A. Schoner, T. A. Jung, E. Meyer, and H. Bartolf, *Mater. Sci. Forum 858*, 497, 2016.
[15] G. Binnig, and C. F. Quate, Phys. *Rev. Lett. 56*, 930 (1986).
[16] T. Uruma, N. Satoh, and H. Yamamoto, *Jpn. J. Appl. Phys. 56*, 08LB05, 2017.
[17] T. R. Albrecht, P. Grutter, D. Horne, and D. Rugar, *J. Appl. Phys. 69*, 668, 1991.
[18] H. Sugimura, Y. Ishida, K. Hayashi, O. Takai, and N. Nakagiri, *Appl. Phys. Lett. 80*, 1459, 2002.
[19] K. Kobayashi, H. Yamada, H. Itoh, T. Horiuchi, and K. Matsushige, *Rev. Sci. Instrum. 72*, 4383, 2001.
[20] M. Nonnenmacher, M. P. O'Boyle, and H. K. Wickramasinghe, *Appl. Phys.Lett. 58*, 2921, 1991.
[21] M. Ito, K. Kobayashi, Y. Miyato, K. Matsushige, and H. Yamada, *Appl. Phys. Lett. 102*, 013115, 2013.
[22] Th. Glatzel, S. Sadewasser, and M. Ch. Lux-Steiner, *Appl. Surf. Sci. 210*, 84, 2003.
[23] T. Fukuma, K. Kobayashi, H. Yamada, and K. Matsushige, *Rev. Sci. Instrum. 75*, 4589, 2004.
[24] M. Takihara, T. Igarashi, T. Ujihara, and T. Takahashi, *Jpn. J. Appl. Phys. 46*, 5548, 2007.
[25] K. Kobayashi, H. Yamada, and K. Matsushige, *Appl. Phys. Lett. 81*, 2629, 2002.
[26] K. Kimura, K. Kobayashi, H. Yamada, and K. Matsushige, *Appl. Surf. Sci. 210*, 93, 2003.
[27] K. Kimura, K. Kobayashi, K. Matsushige, K. Usuda, and H. Yamada, *Appl. Phys. Lett. 90*, 083101, 2007.

value is acquired based on the primary resonance frequency. We believe that two effects worked to achieve high resolution. Firstly, when the lock-in amplifier was used, high sensitivity could be realized by phase detection with a lower reference frequency. Secondly, when topographic image was obtained, cantilever osciallation was smaller amplitude. There was contributed to stable control for tip-sample distance regulation.

In the future, we discuss the variation of the SCFM images simultanously KFM observation [31] with applied the DC-bias based on the electrostatic capacitance (C) - voltage (V) measurement, and the contrast reversal [32], [33] of the carrier concentration.

APPENDIX B: PRINCIPLE OF THE SCFM METHOD

A bias voltage (V) applied between the probe tip and sample produces an electrostatic force F similar to that in the KFM measurement, where

$$
\begin{aligned}
F &= -\frac{1}{2}\frac{dC(z)}{dz}[v(t)]^2 \\
&= -\frac{1}{4}\frac{dC(z)}{dz}[2V_{\mathrm{dc}}^2 \\
&\quad + V_{\mathrm{ac}}^2 + 4V_{\mathrm{dc}}V_{\mathrm{ac}}\cos(\omega t) + V_{\mathrm{ac}}^2\cos(2\omega t).
\end{aligned}
$$

$$(3)$$

Here, z is distance and, when C is a function of both z and V, we have

$$
\begin{aligned}
\frac{dC(z)}{dz} &= \frac{\partial C(z,V)}{\partial z} \\
V &= v(t) = V_{\mathrm{dc}} + V_{\mathrm{ac}}\cos(\omega t)
\end{aligned}
$$

$$(4)$$

[28] K. Kimura, K. Kobayashi, H. Yamada, K. Matsushige, and K. Usuda, J. Vac. Sci. Technol. B 24, 1371 (2006).

[29] K. Kimura, K. Kobayashi, H. Yamada, K. Matsushige, and K. Usuda, J. Vac. Sci. Technol. B 23, 1454 (2005).

[30] H. Yamamoto: "WiDE-BANDGAP SEMICONDUCTOR POWER DE-VICES" (CORONA: Tokyo), (2015) [in Japanese].

[31] U. Zerweck, C. Loppacher, T. Otto, S. Grafström, and L. M. Eng, *Phys. Rev. B 71*, 125424, 2005.

[32] R. Stephenson, A. Verhulst, P. De Wolf, M. Caymax, and W. Vandervorst, *Appl. Phys. Lett. 73*, 2597, 1998.

[33] H. Edwards and V. Ukraintsev, *J. Appl. Phys. 87*, 1485, 2000.

Simulation Analysis of Optimum Gate Driving Conditions of IGBTs

Satoshi Sugahara*, Masaki Kawakami and Kousuke Kamakura
Faculty of Engineering, Fukuyama University, Fukuyama, Japan
*E-mail: sugahara@fuee.fukuyama-u.ac.jp

Abstract— The rapid development of wide bandgap semi-conductor devices leads power electronics devices such as inverters to higher power applications. However, EMI caused by the switching operations of the power semi-conductor devices increases. In this paper, the simulation results of the optimum gate driving conditions of IGBTs forming inverters to reduce the EMI noise are discussed.

Keywords— *Electromagnetic interference (EMI), Gate drive, Insulated-gate bipolar transistors (IGBTs), Switching losses*

I. INTRODUCTION

Power electronics devices such as inverters are widely used in various power devices such as industries, railways, solar and wind power generation, hybrid cars, and home appliances. They are important functions to reduce the size and power consumption of these power devices. Figure 1 shows a circuit configuration of a general three-phase inverter. Each phase consists of two parallel pairs of insulated-gate bipolar transistor (IGBT) and freewheeling diode (FWD) in the high side and the low side.

In recent years, development of wide bandgap semiconductor devices such as SiC which is expected to have 10 times the voltage and current capacity of the conventional Si semiconductor switch is progressing. Further, it is studied to extend the operating range of power electronics devices using SiC semiconductor switches to high power, high temperature, wide band region such as 10 kV, several 10 kA, 150°C, several MHz. Therefore, it is expected that power electronics devices will be applied to electric power systems, electric railways, etc., which have been impossible or difficult to apply power electronics devices. This will contribute to miniaturization, high performance and low cost of high power devices.

However, the increase in power and the increase in frequency of the power electronics devices deteriorate the electromagnetic interference (EMI) caused by the switching operation of the power semiconductor switches. Generally, in order to reduce the EMI, resistors are inserted in series with the gate of power semiconductor switches such as IGBTs or MOSFETs to slow the gate drive. Unfortunately, this method has a problem of an increase in switching loss due to an increase in switching time. In response to this problem, methods of nonlinearly

controlling the gate voltage of power semiconductor switches to suppress an increase in switching loss have been studied [1]-[6]. [6] reports a method to achieve both reduction of EMI and suppression of loss increase by adjusting the gate voltage waveform of the IGBTs constituting the inverter in order to control the time derivative values of the collector-emitter voltage at turn-off and the collector current at turn-on.

The purpose of this research is to develop optimum gate driving methods and their circuit systems for reducing the EMI due to the switching operation of the power semiconductor switches and minimizing the increases in the switching losses.

This paper describes the results of verifying the relationship between the EMI reduction effects and switching losses using a circuit simulator when the slopes of the rise and the fall of the gate voltage of the IGBT constituting the inverter are gently varied. In addition, the effect of reducing the switching losses when the rising and the falling waveforms of the gate voltage are nonlinearized and the optimum gate voltage waveform

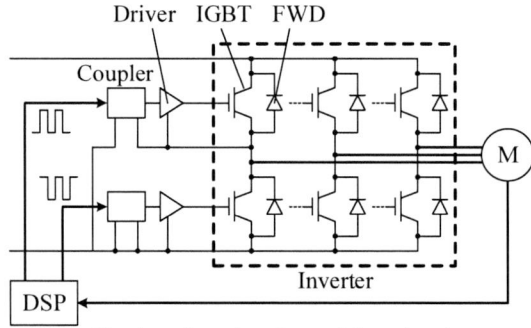

Fig. 1. Circuit configuration of general three-phase inverter.

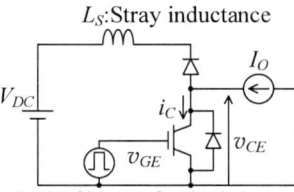

Fig. 2. Equivalent circuit of inverter for simulating switching operation of IGBT.

for minimizing the increase in switching losses are discussed.

II. Switching Noise of IGBT

Figure 2 illustrates an equivalent circuit for analyzing the switching operation of the IGBTs constituting the inverter. This circuit assumes the case where the low side IGBT switches. It consists of low-side IGBT, high- and low-side FWDs. V_{DC} is the DC input voltage, I_O is the output current flowing to the inductive load, v_{GE}, v_{CE} and i_C are the gate-emitter voltage, the collector-emitter voltage and the collector current, respectively. L_S is the stray inductance of the supply line. Table 1 lists simulation conditions. Since this analysis aims to investigate the qualitative relationship between EMI noises and losses associated with the switching operation of the IGBT, it is executed with $V_{DC} = 100$ V and $I_O = 30$ A, which are lower electric power conditions than actual use. L_S is estimated to be 200 nH assuming the supply line of approximately 20 cm in length. PSpice is used for the circuit simulator, IGBT and FWD of standard provided simulation models are applied. Their electric characteristics are shown in Figs. 3 and 4.

Figure 5 shows the simulation waveforms of the gate-emitter voltage v_{GE}, the collector-emitter voltage v_{CE}, and the collector current i_C when the IGBT is hard switched. (a) and (b) show the waveforms at turn-on and turn-off, respectively. It is confirmed that over-current occurs in i_C at turn-on due to superposition of the reverse recovery current of the high-side FWD [6]. The reverse recovery charge of the FWD calculated from (a) and the ones estimated from measured values described in the specification sheet are approximately 7 μC and 6 μC, respectively. Therefore, it is considered that the simulation waveforms in (a) are sufficient to analyze the improvement of the collector over-current.

At turn-off, over-voltage occurs in v_{CE} because of superimposition of the induced electromotive force of L_S. Since the gradient of i_C at turn-off is approximately 375 A/μs in (b), the induced electromotive force of $L_S = 200$ nH is theoretically 75 V. Simulated collector-emitter over-voltage in (b) is approximately 77V. Therefore, it is considered that the simulation waveforms in (b) are sufficient to analyze the improvement of the over-voltage.

III. Relationship Between Switching Noises and Switching Losses

This section describes the results of analyzing the relationship between noise reduction effect and loss increase by varying the slopes of v_{GE} at turn-on and turn-off that simulates conventional noise preventions with the gate resistor.

Figure 6 displays simulation waveforms of v_{GE}, v_{CE}, i_C and power consumption $p_{SW} = v_{CE} \cdot i_C$ when the gradient of v_{GE} with respect to time t is varied. (a) and (b) show the waveforms at turn-on and turn-off, respectively. Figure 7 plots the relationship between the slope dv_{GE}/dt and the noises and losses due to the switching operation.

(a) draws the collector over-current i_{COV} and the

TABLE I
SIMULATION CONDITIONS

Input DC voltage V_{DC} [V]	100
Output current I_O [V]	30
Stray inductance L_S [nH]	200
Gate-emitter voltage amplitude V_{GE} [Ω]	30
IGBT model	Z6MBI30L-060 (Fuji)
FWD model	MR876 (Motorola)

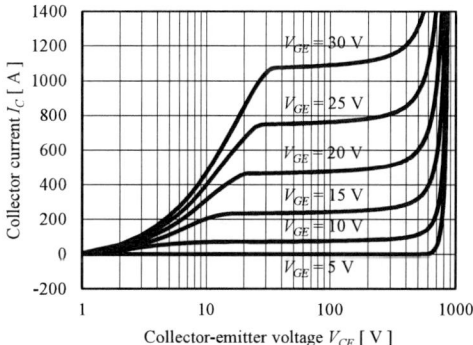

(a) Collector current versus collector-emitter voltage.

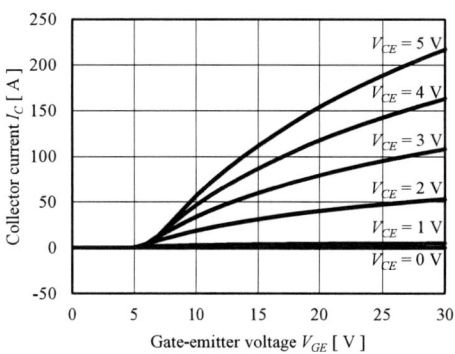

(b) Collector current versus gate-emitter voltage.

Fig. 3. Simulated electric characteristics of IGBT simulation model.

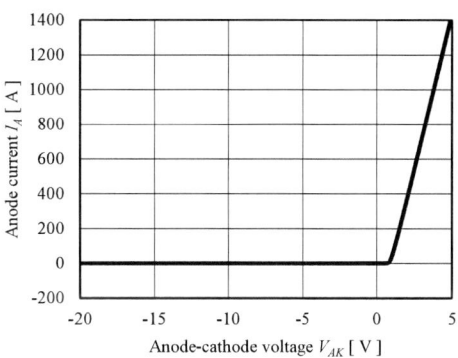

Fig. 4. Simulated electric characteristics of FWD simulation model.

2757

The 2018 International Power Electronics Conference

switching loss E_{ON} at turn-on, and (b) draws the collector-emitter over-voltage v_{CEOV} and the switching loss E_{OFF} at turn-off.

These figures detail that i_{COV} and v_{CEOV} decrease with decreasing dv_{GE}/dt in the region of $dv_{GE}/dt < 30$ V/µs and in the region of $dv_{GE}/dt < 10$ V/µs, respectively. In contrast, E_{ON} and E_{OFF} increase with decreasing dv_{GE}/dt. In order to reduce i_{COV} and v_{CEOV} by decreasing dv_{GE}/dt, it is necessary that dv_{GE}/dt is set below 30 V/µs and below 10 V/µs, respectively.

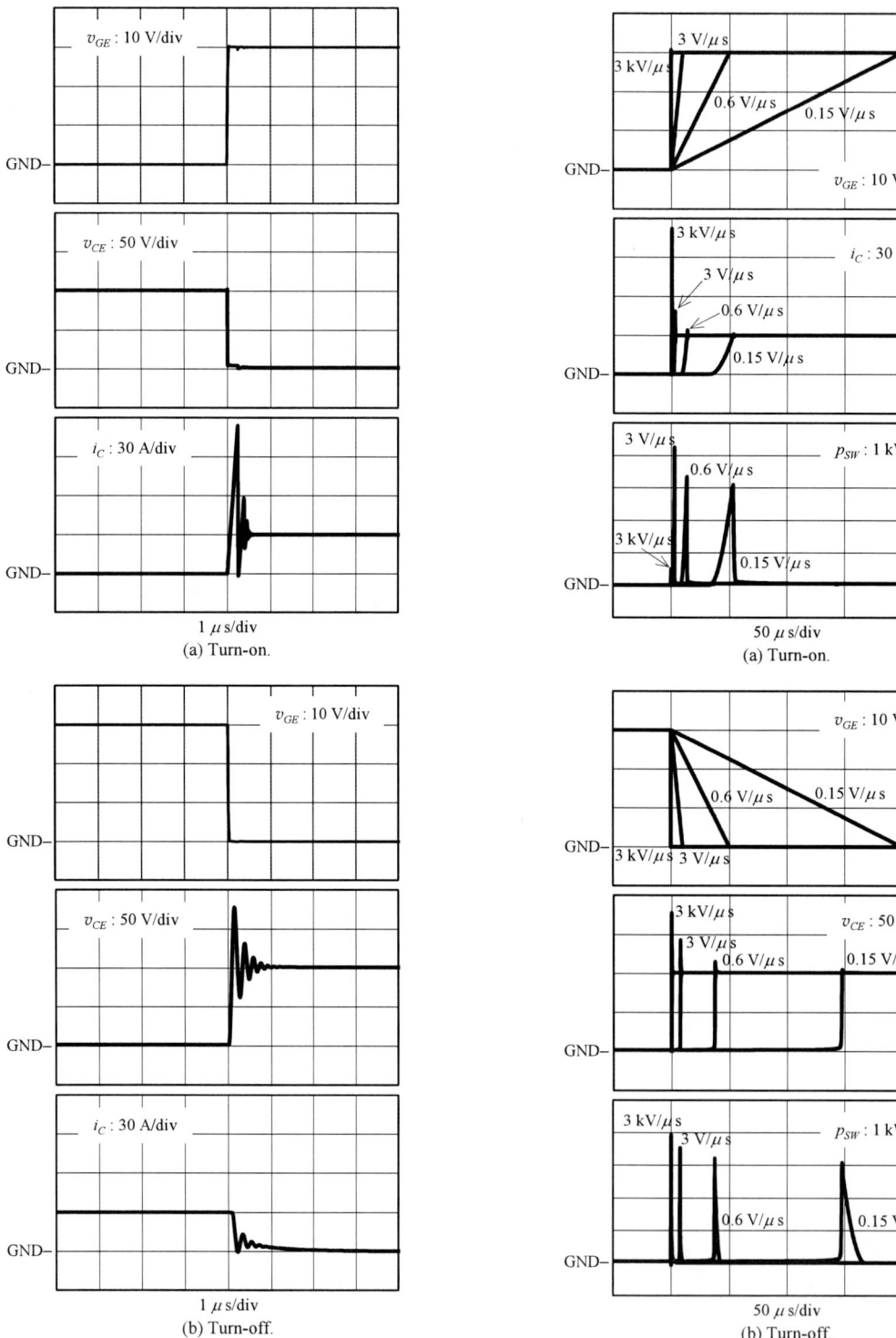

Fig. 5. Simulated waveforms of IGBT at hard switching.

Fig. 6. Simulated waveforms of IGBT when dv_{GE}/dt is varied.

IV. REDUCTION OF SWITCHING LOSSES BY NONLINEAR GATE DRIVE

In this section, the effects of suppressing an increase in the switching loss when the waveform of v_{GE} at the rise and fall timings are nonlinear as shown in Fig. 8 are investigated. v_{GE} is set slopes in the ranges of rising voltages v_{GER1} to v_{GER2} and falling voltages v_{GEF2} to v_{GEF1} near the threshold voltage V_T of the IGBT.

A. Switching Waveforms with Nonlinear Gate Draive

Figures 9 and 10 demonstrate simulated waveforms of v_{GE}, v_{CE}, i_C and p_{SW} when the waveform of v_{GE} at the rise and fall timings are nonlinearized, respectively. The gradient of the v_{GE} slope is 0.15 V/μs. Figure 9(a) and (b) show the cases where v_{GER1} and v_{GER2} in Fig. 8 are set, respectively. Figure 10(a) and (b) show the cases where

v_{GEF1} and v_{GEF2} in Fig. 8 are set, respectively. When v_{GER1} increases or v_{GER2} decreases and they approach the threshold voltage, the collector over-current increases and the turn-on switching loss decreases in Fig. 9. Similarly,

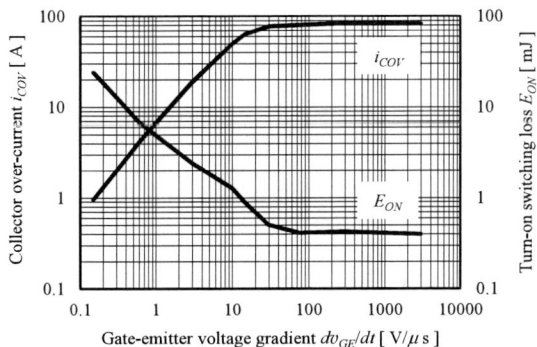

(a) Collector over-current & switching loss at turn-on.

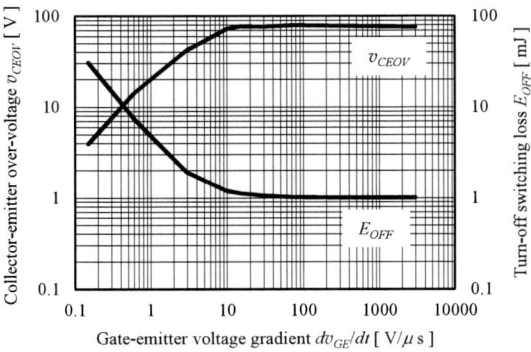

(b) Collector-emitter over-voltage & switching loss at turn-off.
Fig. 7. Switching noise and loss versus dv_{GE}/dt.

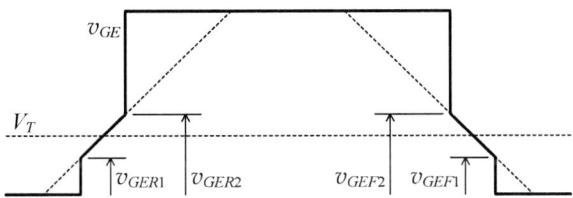

Fig. 8. Ideal gate-emitter voltage waveform to reduce switching losses.

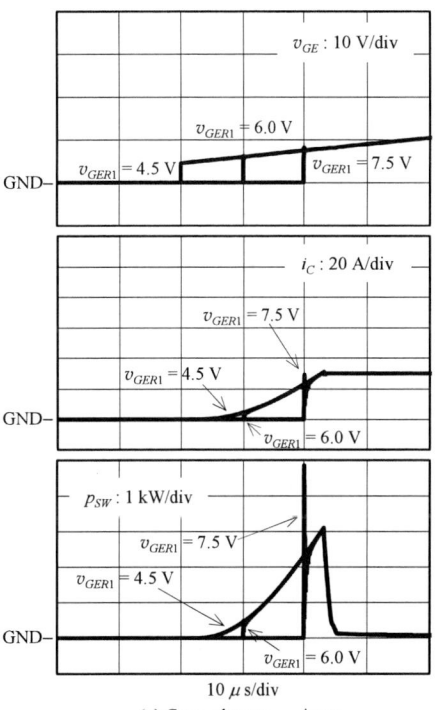

(a) Case where v_{GER1} is set.

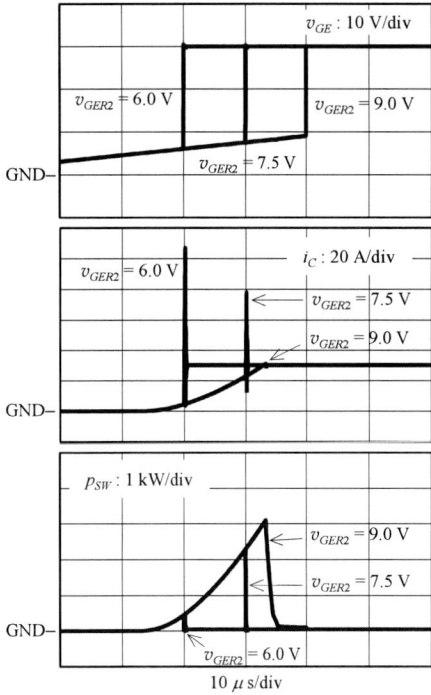

(b) Case where v_{GER2} is set.
Fig. 9. Simulated turn-on waveforms when gate voltage steps are set.

2759

in Fig. 10, when v_{GEF1} increases or v_{GEF2} decreases and they approach the threshold voltage, the collector-emitter over-voltage increases and the turn-off switching loss decreases.

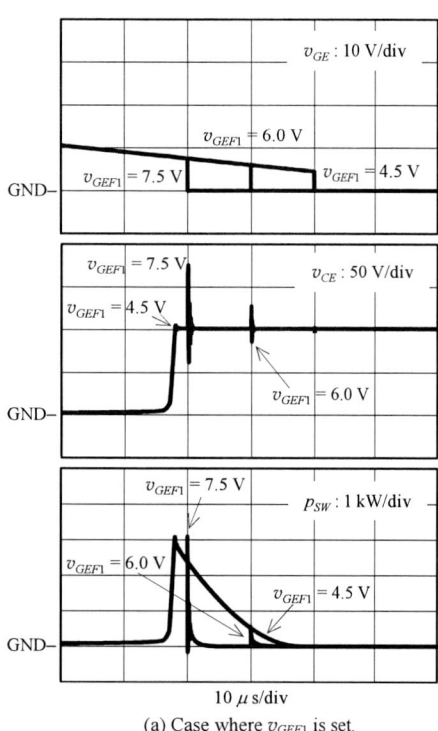

(a) Case where v_{GEF1} is set.

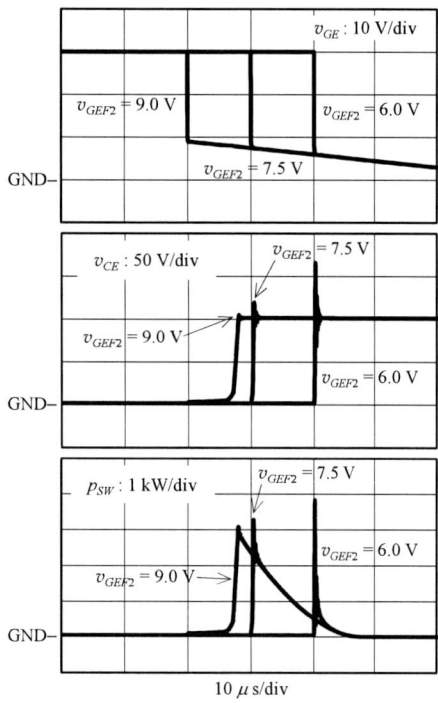

(b) Case where v_{GEF2} is set.

Fig. 10. Simulated turn-off waveforms when gate voltage steps are set.

B. Optimum Gate-Emitter Voltage Steps

Figures 11 and 12 show i_{COV} and E_{ON} at $dv_{GE}/dt = 0.15$ V/μs, 0.6 V/μs, 3 V/μs and 10 V/μs when v_{GER1} and v_{GER2} are varied, respectively. In the range of v_{GER1} below approximately 7 V, i_{COV} is remarkably reduced, but E_{ON} increases with decreasing v_{GER1}. Similarly, in the range of v_{GER2} above approximately 8 to 11 V, i_{COV} is significantly reduced, but E_{ON} increases with v_{GER2}. Here, the values of v_{GER1} and v_{GER2} that minimize E_{ON} while maintaining i_{COV} reduction effect, are defined as the optimum gate-emitter voltage steps $v_{GER1opt}$ and $v_{GER2opt}$, respectively.

Figures 13 and 14 show v_{CEOV} and E_{OFF} at $dv_{GE}/dt = 0.15$ V/μs, 0.6 V/μs, 3 V/μs and 10 V/μs when v_{GEF1} and v_{GEF2} are varied, respectively. In the range of v_{GEF1} below approximately 5 V, v_{CEOV} is remarkably reduced, but E_{OFF} increases with decreasing v_{GEF1}. Similarly, in the range of v_{GEF2} above approximately 8 V, v_{CEOV} is significantly reduced, but E_{OFF} increases with v_{GEF2}. Here, the values of v_{GEF1} and v_{GEF2} that minimize E_{OFF} while maintaining v_{CEOV} reduction effect, are defined as the optimum gate-emitter voltage steps $v_{GEF1opt}$ and $v_{GEF2opt}$, respectively.

Figure 15 shows simulation results of the switching loss reduction effect at turn-on when the rising waveform of v_{GE} is nonlinear. (a) plots the optimum gate-emitter voltage steps $v_{GER1opt}$ and $v_{GER2opt}$. (b) plots i_{COV} when $v_{GER1opt}$ and $v_{GER2opt}$ are set, and a comparison of E_{ON} when $v_{GER1opt}$ and $v_{GER2opt}$ are set and when they are not set. From this figure, it can be seen that nonlinearization of v_{GE} has the effect of reducing the increase of E_{ON}. E_{ON} is reduced to 1/4 at $dv_{GE}/dt = 0.15$ V/μs.

Figure 16 shows simulation results of the switching loss reduction effect at turn-off when the falling waveform of v_{GE} is nonlinear. (a) plots the optimum gate-emitter voltage steps $v_{GEF1opt}$ and $v_{GEF2opt}$. (b) plots v_{CEOV} when $v_{GEF1opt}$ and $v_{GEF2opt}$ are set, and a comparison of E_{OFF} when $v_{GEF1opt}$ and $v_{GEF2opt}$ are set and when they are not set. From this figure, it can be seen that nonlinearization of v_{GE} has the effect of reducing the increase of E_{OFF}. E_{OFF} is reduced to 2/3 at $dv_{GE}/dt = 0.15$ V/μs.

C. Reduction of Switching Losses

Figure 17 analyzes details of the turn-on switching losses reduced by the optimization of the gate-emitter voltage step v_{GER1} and v_{GER2}. (a) estimates the loss components. E_{ONHS} is a hard switching loss without collector over-current reduction. E_{ON12} is an incremental loss when both $v_{GER1opt}$ and $v_{GER2opt}$ are set. E_{ON1} and E_{ON2} are losses reduced by setting $v_{GER1opt}$ and $v_{GER2opt}$, respectively. In (b), the ratio of turn-on loss components is calculated, e_{ON1}, e_{ON2}, e_{ON12} and e_{ONHS} are normalized losses of E_{ON1}, E_{ON2}, E_{ON12} and E_{ONHS}, respectively. It can be seen that the loss reductions E_{ON1} and E_{ON2} increase with decreasing dv_{GE}/dt. At $dv_{GE}/dt = 0.15$ V/μs, e_{ON1} and e_{ON2} expressing the loss reduction effect reached

The 2018 International Power Electronics Conference

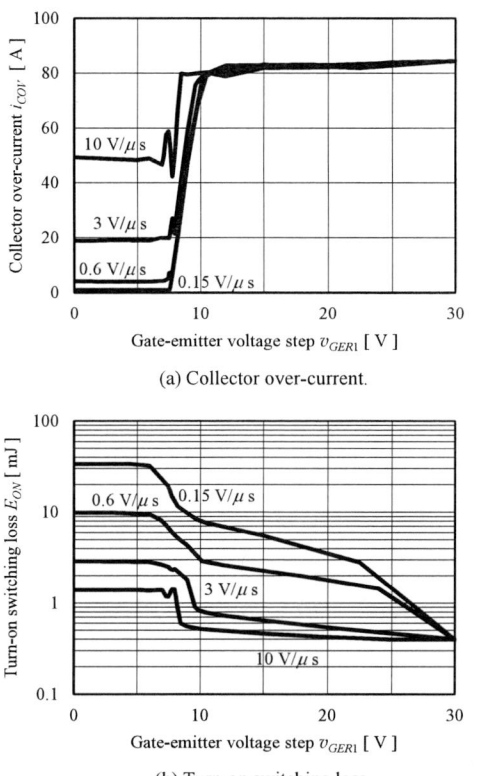

(a) Collector over-current.

(b) Turn-on switching loss.

Fig. 11. Collector over-current and turn-on loss versus v_{GER1}.

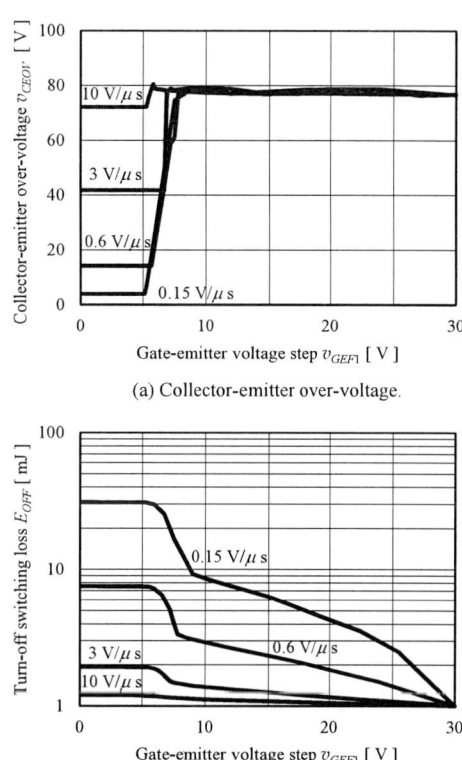

(a) Collector-emitter over-voltage.

(b) Turn-off switching loss.

Fig. 13. Collector-emitter over-voltage and turn-off loss versus v_{GEF1}.

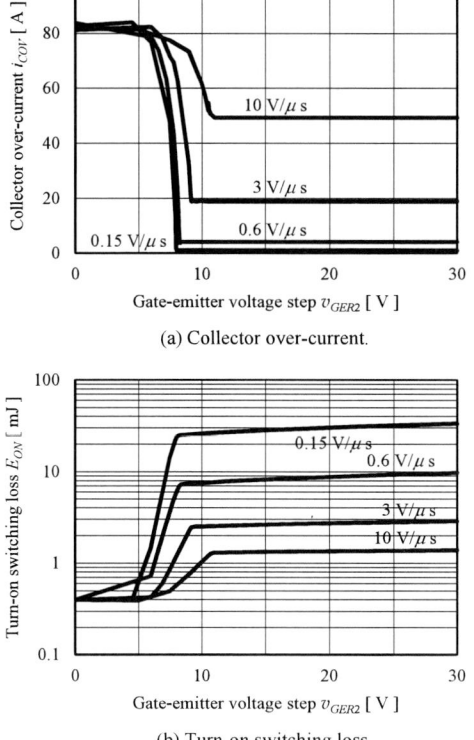

(a) Collector over-current.

(b) Turn-on switching loss.

Fig. 12. Collector over-current and turn-on loss versus v_{GER2}.

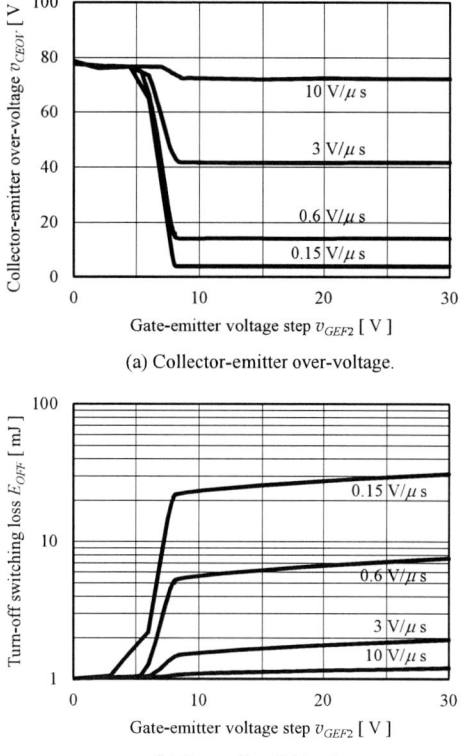

(a) Collector-emitter over-voltage.

(b) Turn-off switching loss.

Fig. 14. Collector-emitter over-voltage and turn-off loss versus v_{GEF2}.

2761

The 2018 International Power Electronics Conference

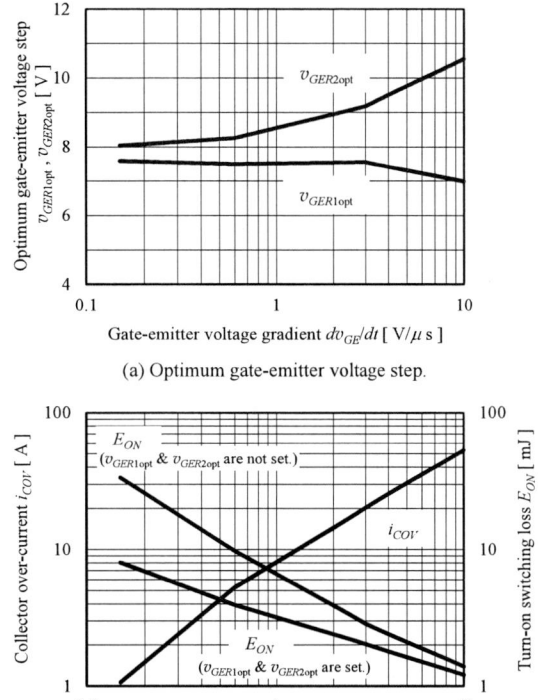

(a) Optimum gate-emitter voltage step.

(b) Collector over-current & switching losses.

Fig. 15. Switching loss reduction of nonlinear gate driving at turn-on.

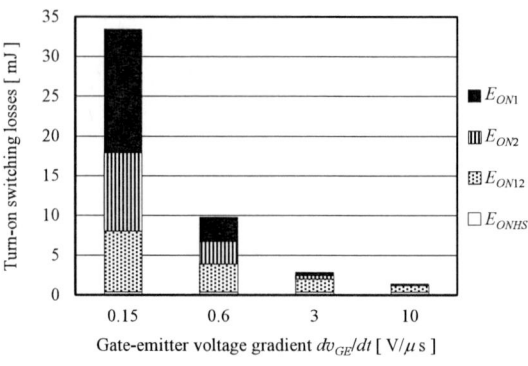

(a) Turn-on switching loss estimation.

(b) Normalized turn-on switching loss estimation.

Fig. 17. Analysis of turn-on switching loss reduction effect.

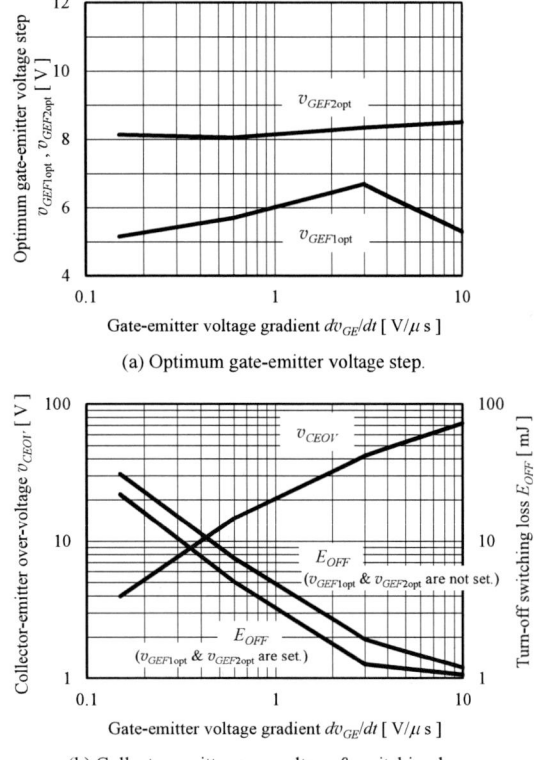

(a) Optimum gate-emitter voltage step.

(b) Collector-emitter over voltage & switching losss.

Fig. 16. Switching loss reduction of nonlinear gate driving at turn-off.

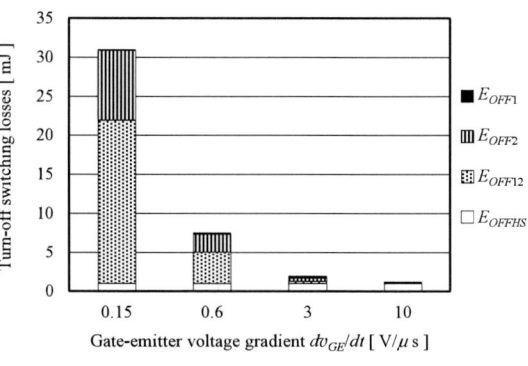

(a) Turn-off switching loss estimation.

(b) Normalized turn-off switching loss estimation.

Fig. 18. Analysis of turn-off switching loss reduction effect.

2762

approximately 46% and 30%, respectively. Therefore, setting both $v_{GER1opt}$ and $v_{GER2opt}$ is effective for reducing turn-on losses.

Figure 18 shows details of the turn-off switching losses reduced by the optimization of the gate-emitter voltage step v_{GEF1} and v_{GEF2}. (a) estimates the loss components. E_{OFFHS} is a hard switching loss without collector-emitter over-voltage reduction. E_{OFF12} is an incremental loss when both $v_{GEF1opt}$ and $v_{GEF2opt}$ are set. E_{OFF1} and E_{OFF2} are losses reduced by setting $v_{GEF1opt}$ and $v_{GEF2opt}$, respectively. (b) calculates the ratio of turn-off loss components, e_{OFF1}, e_{OFF2}, e_{OFF12} and e_{OFFHS} are normalized losses of E_{OFF1}, E_{OFF2}, E_{OFF12} and E_{OFFHS}, respectively. It can be seen from (a) that the loss reduction E_{OFF2} increases with decreasing dv_{GE}/dt, but E_{OFF1} is kept at a very low value. In (b), at $dv_{GE}/dt = 0.15$ V/μs, e_{OFF2} expressing the loss reduction effect is approximately 29%, but e_{OFF1} is only approximately 0.3%. Therefore, it is found that setting $v_{GEF2opt}$ is effective for reducing turn-off losses, but the effect of $v_{GEF1opt}$ can not be expected.

V. CONCLUSIONS

For the IGBTs constituting the inverter, the relationship between the slope of the rise- / fall-timing of the gate voltage and the influence of the decrease in EMIs and the increase in switching losses was simulated. The effect of reducing the switching losses when the rising / falling waveforms of the gate voltage were nonlinearized, and optimum gate voltage waveforms to minimize the increase in switching losses were verified. There are effective ranges of the gate voltage slope for the EMI reduction effect to appear. Switching losses increased contrary to reduction of EMI noises. Switching losses were reduced by nonlinearizing the rising / falling waveforms of the gate voltage. When the rising / falling slopes of the gate voltage were 0.15 V/μs, the turn-on loss and the turn-off loss were reduced to 1/4 and 2/3, respectively, by optimizing the nonlinear gate waveform. Guidelines for optimization of nonlinear gate waveforms for effective loss reduction at turn-on and turn-off were clarified.

In further research, control methods and circuit schemes for the nonlinear gate drive would be studied.

ACKNOWLEDGMENT

The authors would like to thank Fuji Electric Co., Ltd. for constructive discussion and support.

REFERENCES

[1] V. John, B. Suh, and T. Lipo, "High-performance active gate drive for high-power IGBT's," *IEEE Trans. Ind. Appl.*, vol. 35, no. 5, pp. 1108-1117, Sep./Oct. 1999.

[2] S. Park and T. Jahns, "Flexible dv/dt and di/dt control method for insulated gate power switches," in *Rec. IEEE Ind. Appl. Soc. Annu. Meeting*, Oct. 2001, pp. 1038-1045.

[3] P. Palmer and H. Rajamani, "Active Voltage control of IGBTs for high power applications," *IEEE Trans. Power Electron.*, vol. 19, no. 4, pp. 894-901, July 2004.

[4] L. Chen and F. Peng, "Closed-loop gate drive for high power IGBTs," in *Proc. IEEE Appl. Power Electron. Conf. Expo. (APEC)*, Feb. 2009, pp. 1331-1337.

[5] Z. Wang, X. Shi, L. Tolbert, and B. Blalock, "Switching performance improvement of IGBT modules using an active gate driver," in *Proc. IEEE Appl. Power Electron. Conf. Expo. (APEC)*, Mar. 2013.

[6] Y. Lobsiger and J. Kolar, "Closed-loop di/dt and dv/dt IGBT gate driver," *IEEE Trans. Power Electron.*, vol. 30, no. 6, pp. 3402-3417, June 2015.

The 2018 International Power Electronics Conference

Improvement of the I^2t capability for xEV active short circuit protection by combination of RC-IGBT and Leadframe technologies

Keiichi Higuchi, Hayato Nakano, Akihiro Osawa, Akio Kitamura, Shunji Takenoiri,
Daisuke Inoue, Souichi Yoshida, Hiromichi Gohara and Masahito Otsuki
Fuji Electric Co., Ltd, 4-18-1, Tsukama, Matsumoto, Nagano, Japan
*E-mail: higuchi-keiichi@fujielectric.com

Abstract— This paper describes the investigation results of I^2t capability for IGBT module used for xEV powertrain application. I^2t experiments and simulation evaluations are done by RC-IGBT or the conventional FWD with bondwires structure. As a result, RC-IGBT(wire) has 80% higher I^2t capability compared with FWD in the case of comparable diode active area. In addition, RC-IGBT(leadframe) and RC-IGBT(bondwires) are also evaluated by experimental and simulational approaches. As a result, RC-IGBT(leadframe) has additional 33% increase in I^2t compared to RC-IGBT(wire). Over all, I^2t of RC-IGBT (leadframe) is 2.4 times higher than that of FWD (bondwires).

Keywords— I^2t, Leadframe, RC-IGBT

I. INTRODUCTION

xEV market is expected as one of the most prospective market in future[1]. In the terms of components of xEV system, power module is a key device in order to manage important factors as motor control, energy saving and reliability. Currently, high voltage motor is a trend in order to realize benefits as lighter cables, high power and better efficiency for xEV[2]. On the other hands, high voltage of motor should be occasionally higher than that of battery, therefore active short circuit must be concerned.

Regarding this emerged issue, I^2t capability is a key characteristic which is an index to survive power module in active short circuit. In general, large FWD (free wheel diode) chip size can improve I^2t capability however this is caused large power module and inverter system. This doesn't meet market trend as more compact inverter. Therefore, it is important that power module achieves better I^2t capability without any sacrifice of electric characteristics. However, it is difficult to breakthrough with the conventional FWD because I^2t capability is determined by thermal management. Diode active area of FWD is designed by from the view of electric characteristics however RC-IGBT should have different mechanism for I^2t capability because RC-IGBT can realize the same electric characteristics of FWD with lower diode active area.

Regarding this background, this paper describes that new knowledge that RC-IGBT and leadframe combined structure are able to achieve high I^2t capability.

II. EXPLANATION FOR I^2t PHENOMENA

Fig.1 shows an example of electrical system diagram for xEV application. In this system, in the case of voltage of motor is higher than voltage of battery, reverse load current from motor should flow in IGBT module. This phenomena is explained as I^2t. I^2t is the value of joule energy that can be allowed within the range which device can not destroy. The overcurrent is defined by a line frequency sine half wave and one cycle. I^2t is defined with typical waveform shown in Fig.2. When I^2t current flows in IGBT module, diode which is connected in parallel with IGBT is suffered from I^2t current. Therefore, I^2t capability of diode should be important in order to achieve enough capability for I^2t current.

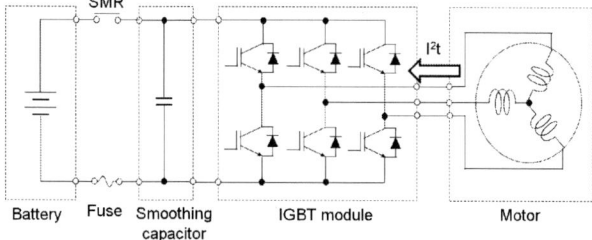

Fig.1 Electrical system diagram for xEV

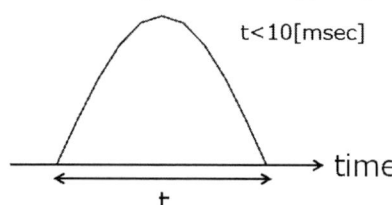

Fig.2 Example of I^2t waveform

III. IGBT MODULE STRUCTURE

As we reported[3][4], RC-IGBT has been achieved significant improvement for electrical and chip design which is ended up the impact for smaller power module. Difference of conventional IGBT and FWD chip set and RC-IGBT is shown in Fig.3. RC-IGBT has FWD region on its surface. I^2t current flows in this region via mechanical connection as bondwires or leadframe. RC-IGBT has lower diode active area however forward voltage drop should be similar with FWD because of current expansion effect inside device[5].

Fig.4 shows module structure comparison between conventional IGBT and FWD set and RC-IGBT. A picture of Fuji power module is shown in Fig.4. It has RC-IGBT and leadframe structure in order to achieve higher I^2t capability by enhancing heat expansion compared with conventional structure.

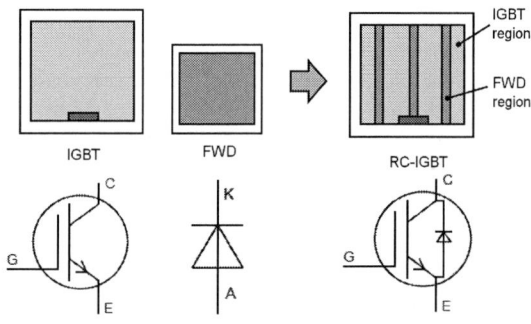

Fig.3 Defference of convenional IGBT and FWD set and RC-IGBT

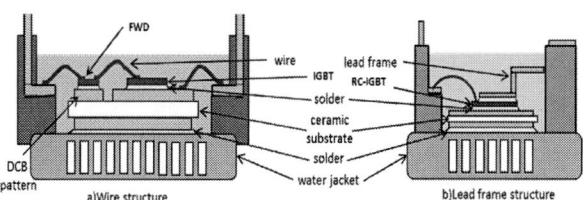

Fig. 4 cross sectional view of conventional IGBT and FWD chip set and RC-IGBT structures

Fig. 5 Fuji power module 1200A/750V 6in1

IV. I^2T CAPABILITY COMPARISON FOR RC-IGBT AND FWD BY BONDWIRES

As the advantage of RC-IGBT for I^2t capability is referred, RC-IGBT and FWD with bondwires are evaluated. The difference between RC-IGBT and FWD is summarized in Table 1. Internal wiring structure is the same with bondwires. The ratio of contact area on chip surface is 1.2 vs, 1.0. This is because RC-IGBT should be larger than FWD in order to meet similar characteristics. Then the number of bondwires is increased. On the other hand, the ratio of diode active area is 0.5 vs, 1.0. This means diode characteristics of RC-IGBT can achieve similar characteristics with a half of active area. This is because current expansion should occur inside diode even though diode active area is less. As a result, the ratio of I^2t capability per diode active area is 1.8 vs, 1.0. RC-IGBT can achieve 80% higher I^2t capability compared with the conventional FWD in the case of I^2t capability divided by diode active area.

Fig.6 shows experimental waveforms for I^2t test. Lower current should have ideal sine curve however waveforms should be disturbed when I^2t current is increased. After non ideal I^2t waveforms are observed, device is destruction.

TABLE 1

Properties	RC-IGBT	IGBT / FWD
Internal wiring structure	bondwires	bondwires
Contact area on chip surface (ratio)	1.2	1.0
Diode active area (ratio)	0.5	1.0
Forward voltage drop (ratio)	1.0	1.0
I^2t / Diode active area (ratio)	1.8	1.0

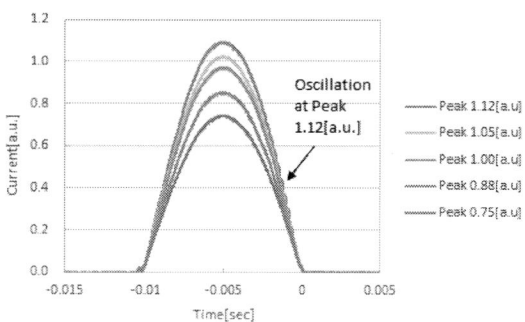

Fig.6 I^2t test waveforms (up to destruction)

Fig.7 shows I^2t capability of multiple size of the conventional FWD. I^2t capability shows linear dependency for diode active area. As a result, RC-IGBT potentially has 2.0 times stronger I^2t capability compared with FWD which has the same diode active area.

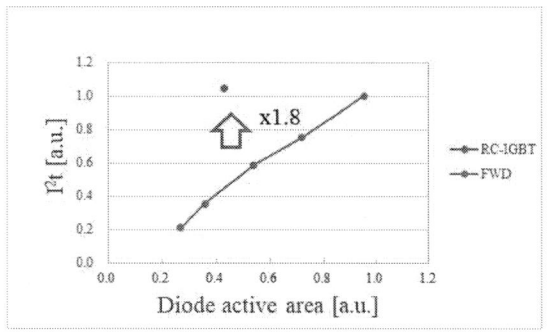

Fig.7 Diode active area dependency of I²t capability

In order to investigate the I²t capability advantage of RC-IGBT, finite element method (FEM) is done. Each diode active area is assigned designated temperature which is generated by I²t load. Fig.8 and Fig.9 show 3D model for FEM, the number of bondwires and contact area of bondwires are the same. The diode active area of RC-IGBT is a half of the conventional FWD.

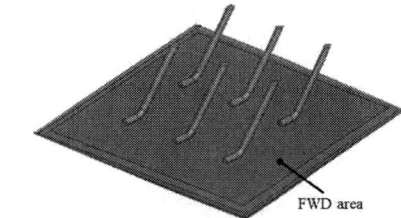

Fig. 8 FEM model of FWD with bondwires

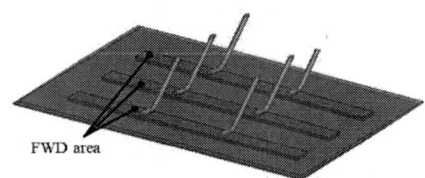

Fig. 9 FEM model of RC-IGBT with bondwires

Fig.10 shows FWD Temperature distribution and Fig.11 shows RC-IGBT Temperature distribution respectively. In order to clarify temperature distribution unnecessary components are not visualized. Temperature distribution of FWD is almost uniform with higher temperature but temperature distribution of RC-IGBT has higher temperature area. This is why RC-IGBT has IGBT area which is not generate any heat. This means IGBT area can cool diode area's heat generation which is ended up with higher I²t capability[6].

Fig.10 FWD temperature distribution by FEM

Fig.11 RC-IGBT temperature distribution by FEM

V. I²T COMPARISON OF BONDWIRES AND LEADFRAME

In previous section, the advantage of I²t capability is explained. The essential mechanism is thermal management during I²t load. Therefore, it is important to enhance cooling effect for RC-IGBT by wiring structure. Table 2 shows device characteristics for experimental result. Each RC-IGBT has leadframe and bondwires, RC-IGBT has leadframe structure has 13.1 times contact area on chip surface compared with RC-IGBT has bondwires. The same design of RC-IGBT is used, diode active area is the same. As a result of I²t test, RC-IGBT has leadframe should have 1.3 times higher I²t capability compared with RC-IGBT has bondwires.

TABLE 2

Properties	Leadframe	bondwires
Device	RC-IGBT	RC-IGBT
Contact area on chip surface (ratio)	13.1	1.0
Diode active area (ratio)	1.0	1.0
I²t / Diode active area (ratio)	1.3	1.0

Fig.12 shows I²t of diode active area dependency with RC-IGBT has leadframe and bondwires respectively. It is clear that leadframe structure has 5.3 times higher I²t capability in the case of compared in the same diode active area.

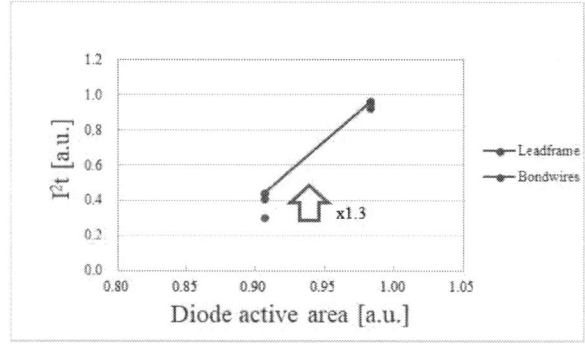

Fig.12 Diode active area dependency of I²t capability

It is important to understand that the advantage of leadframe structure with RC-IGBT, FEM is done. Fig.13 shows 3D model for FEM. In this figure, 3D model doesn't show unnecessary components for better visualization. Leadframe has 13.1 times contact area compared with previous RC-IGBT has bondwires. Diode area is assigned designated power dissipation for heat generation. This FEM aims to clarify heat spread effect by different wiring structures. For 3D model of RC-IGBT has bondwires is used previous 3D model shown in Fig.8.

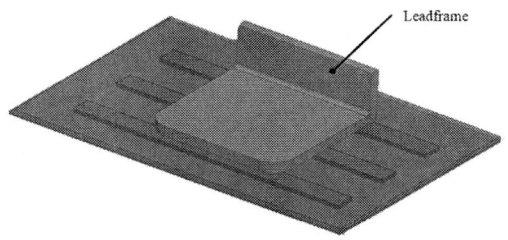

Fig.13 FEM model of RC-IGBT with Leadframe

Fig.14 shows temperature distribution result for RC-IGBT has leadframe and Fig.15 shows temperature distribution result for RC-IGBT has bondwires. The legend and color bar are united in both results. As a result of FEM, RC-IGBT has leadframe can achieve lower temperature compared with RC-IGBT has bondwires. In addition, temperature distribution of RC-IGBT has bonwires shows more concentrated on contact area of bondwires compared with leadframe. This is because leadframe has cooler effect when power dissipation is occurred, then maximum temperature of RC-IGBT has leadframe is decreased approximately 15deg.C compared with RC-IGBT has bondwires. As a result, I^2t capability is obviously improved by leadframe structure because of lower temperature is generated when I^2t current flows in diode active area.

VI. CONCLUSIONS

I^2t capability which is one of the key characteristics of power module is improved by the combination of RC-IGBT and leadframe technologies. As a result of investigation, I^2t capability should be discussed by I^2t capability per diode area. As a result RC-IGBT can improve 80% higher I^2t capability compared with FWD because of wider chip area for heat spread. In addition, leadframe can improves I^2t capability 33% because it has 13.1 times contact area on chip surface compared with bondwires. As a result the combination of RC-IGBT and leadframe can be achieved 2.4 times I^2t capability compared with conventional combination as FWD and bondwires.

REFERENCES

[1] M. Allington et al., "Overview of the Electric Vehicle market and the potential of charge points for demand response", ICF Consulting Services, March 2016.

[2] S. Maduhusoodhanan et al., "Stability Analysis of the High Voltage DC Link between the FEC and DC-DC stage of a Transformer-less Intelligent Power Substation", Proceedings of ECCE 2014, Pittsburgh, PA, USA, pp. 3702-3709.

[3] A.Osawa et al., "The highest power density IGBT module in the world for xEV power train", Proceedings of PCIM Europe 2017, Nuremberg, Germany, pp. 1761-1766.

[4] Kawabata, J. et al. "The New High Power Density 7th Generation IGBT Module for Compact Power Conversion Systems", Proceeding of PCIM Europe 2015.

[5] M.Takahashi, et al. "Extended Power Rating of 1200 V IGBT Module with 7 G RC-IGBT Chip Technologies", Proceeding of PCIM Europe 2016.

[6] M.Otsuki et al., "Advanced thin wafer IGBTs with new thermal management solution", Proceedings of ISPSD 2003, Cambridge, England, pp. 144-147.

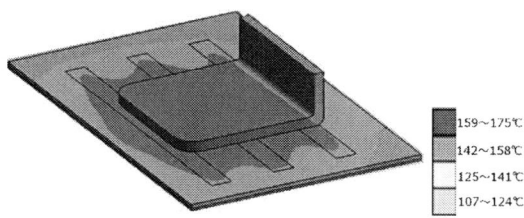

Fig.14 Temperature distribution of RC-IGBT has leadframe by FEM

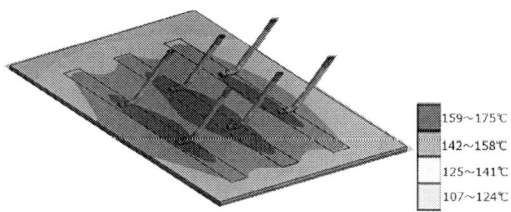

Fig.15 Temperature distribution of RC-IGBT has bondwires by FEM

Investigation of Switching Behavior of an IGBT under Soft Turn-off in Application for Dual-Active Bridge Converters

Eri Ogawa[1*], Yuichi Onozawa[1] and Rik W. De Doncker[2]

[1] Electronic Devices Business Group, Fuji Electric Co., Ltd. , Matsumoto, Japan
[2] Flexible Electrical Networks Research CAMPUS
Institute PGS at E.ON ERC of RWTH Aachen University, Aachen, Germany
*E-mail: ogawa-eri@fujielectric.com

Abstract— **To investigate the behavior of insulated gate bipolar transistors (IGBTs) applied in dual-active bridge (DAB) converters, for first approach, the switching behavior of a 3300 V IGBTs with a snubber capacitor in parallel is analyzed using the software TCAD. Simulations confirmed that a long-tail current appears due to low dV/dt of the collector-emitter voltage. This causes increased turn-off losses even under zero voltage soft-switching (ZVS) conditions. To resolve this issue, reduced tail current structures are proposed and analyzed.**

Keywords— DC-DC Converter, Dual-Active Bridge , Soft - switching, Insulated gate bipolar transistor.

I. INTRODUCTION

The DAB dc-dc converter, depicted in Fig. 1, is one of candidates for electrical energy conversion in direct-current (dc) distribution grids [1-3]. High performance, high efficiency along with soft-switching capability and galvanic isolation are notable benefits of this converter topology.

IGBTs are applied in DAB converters because of their attractive characteristics such as low on-state voltage drop and high switching speed [3, 4]. Recently, IGBTs with high blocking voltage capability up to 3300 V, 4500V and 6500 V are available in the market. However, these IGBTs still appear to have high switching losses. Experiments show that turn-off energy losses are dominant in the overall device energy losses when high blocking voltage IGBTs are used in DAB converters [4]. To reduce turn-off energy losses, a capacitor as turn-off snubber can be used in the DAB. The process of a soft turn-off is shown in Fig. 2. Initially, the switch S_1 conducts the current. When S_1 is turned off, the load current commutates to the snubber capacitor. The upper snubber capacitor is charged while the low-side capacitor is discharged. Hence, V_{top} gradually increases. Due to the low dV/dt across the semiconductor, the turn-off losses are substantially lower than in a hard-switched turn-off. This technique is well known in this converter. However, device manufacturers produce IGBTs, which are optimized and produced for hard-switching inverters.

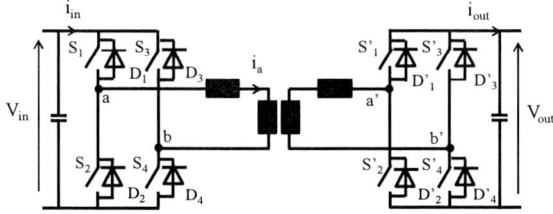

Fig. 1. Single-phase dual-active bridge converter (DAB1).

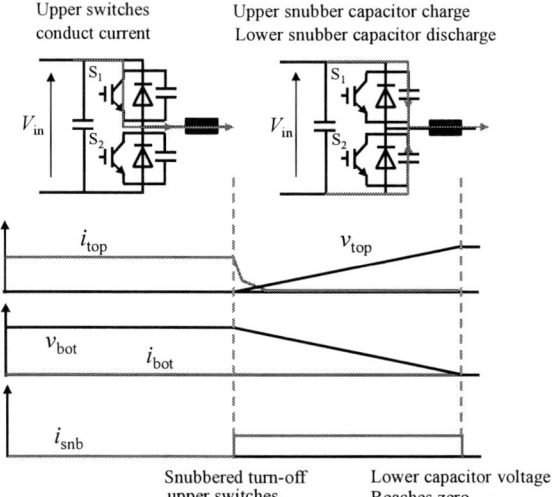

Fig. 2. Process of soft turn-off.

Hence, bipolar devices such as IGBTs show longer tail currents when the snubber capacitor is connected to bipolar devices [5,6]. Thus, it is necessary to choose optimal devices for ZVS operation.

The purpose of this paper is to investigate in detail the switching behavior of a 3300V-IGBT connected in parallel to a snubber capacitor using a device simulator. This study is an approach to understand and optimize the operation of IGBTs applied to the DAB converter.

II. SIMULATION CIRCUIT AND DEVICES

In this research, Technology CAD (TCAD) from Synopsys was used as device simulator. Figure 3 shows double-pulse test circuit used to simulated soft turn-off behavior of an IGBT. Mixed 2-D device and circuit simulations were carried out to study and understand the internal plasma dynamics of the device during switching conditions.

Figure 4 shows the cross section of the field-stop (FS) IGBT used for 2-D simulations as a conventional structure in this study. The device has a rating of 3300V-50A. The IGBT is fabricated using a bulk silicon material as drift region with the collector formed at the backside using boron implantation. Trench gate and p-type floating layers are formed at the surface area. First, the turn-off switching behavior of the IGBT with snubber capacitor and without snubber capacitor were analyzed.

In a second step, to investigate reduced tail current structures, two additional types of the IGBTs were investigated as shown in Table 1. The first one has a 0.88 times thinner drift region compared to conventional one. The other one has a 1.73 times higher resistivity in the drift region. The collector boron concentration was optimized to obtain a similar collector-emitter saturation voltage compared to the conventional structure.

In this paper, the turn-off time is defined from 90 % of gate-emitter voltage to 0.2 % of collector current.

Figure 5 shows simulated hard-switching turn-off waveforms of these three types of IGBTs. The collector current of the two additional types of the IGBTs show a reduced tail current turn-off time as expected. However it should be noted that the initial tail current of the thin thickness and high resistivity IGBTs are larger than that of the conventional one. The reason for this behavior is described below. In the thin thickness IGBT, at the same collector-emitter voltage, the depletion layer boundary is closer to the collector p-layer compared to that of the conventional one. As a result, more carriers are swept out by the electric field in the depletion layer. In the high resistivity IGBT, the wider depletion layer also drives a larger number of carriers. Therefore the collector-emitter current becomes large in these cases.

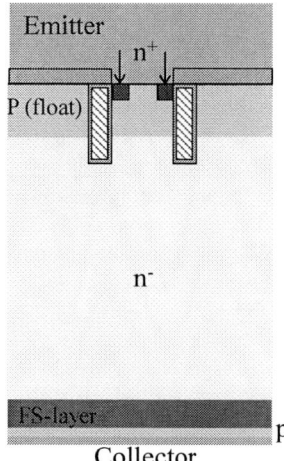

Fig. 4. Cross section of the cell structure of a FS-IGBT, showing the structure, the field-stop (FS) layer and the n⁻ drift region.

Table 1. Different drift region types of IGBTs analyzed in this study

	Drift region type		
	1. Conventional	2. Thin thickness	3. High resistivity
Si thickness	1	0.88	1
Wafer resistivity	1	1	1.73
Collector Boron concentration	1	1	0.9
Turn-off loss	43.15 mJ	43.40 mJ	43.02 mJ
Collector-Emitter saturation voltage	4.96 V	3.78 V	4.99 V

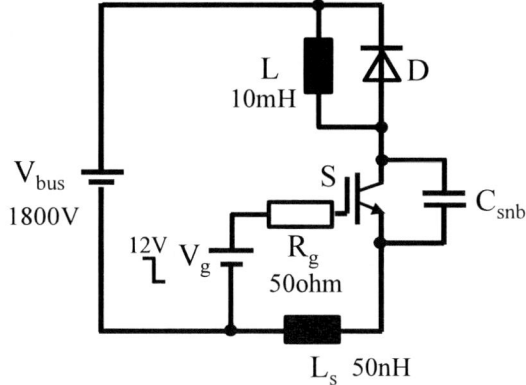

Fig. 3. Simplified diagram of the double-pulse test circuit to characterize soft-switching turn-off performance.

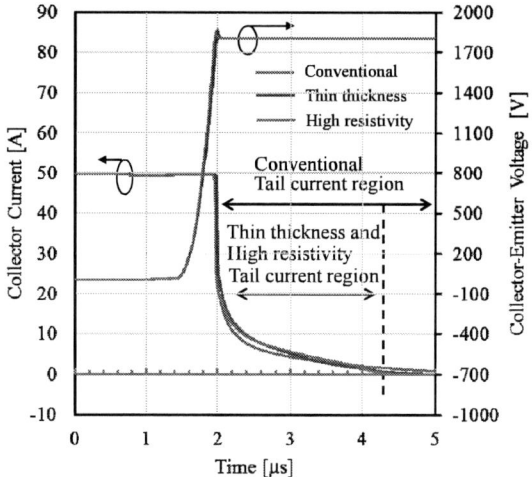

Fig. 5. Simulated hard-switched turn-off waveforms of each type of IGBT

III. TURN-OFF CHARACTERISTICS OF THE CONVENTIONAL STRUCTURE

The first approach is to understand switching behavior of IGBTs with snubber capacitor and without snubber capacitor. Figure 6 shows the turn-off waveforms of the conventional IGBT without a snubber capacitor (hard switching in blue) and connected with a snubber capacitor (soft switching in red) at T_j=125 °C. Switching behaviors and plasma dynamics in the device were analyzed at each time instance.

In both cases the IGBTs are in the same on-state at t=0.0 μs. At t=0.8 μs, the inductor current is constant. The gate-emitter voltage is slightly dropping since the gate-emitter capacitance is discharged. In case of hard-switched turn-off, the collector current is kept constant. However, in case of the IGBT under soft-switched turn-off, the collector current starts to drop since the load current immediately starts charging the snubber capacitor. Hence, the collector current does not remain constant. The gate-emitter voltage is constant from t=1.0 to 1.4 μs. This time period is the so-called "Miller plateau". During this period, the collector current under hard turn-off is still constant. On the other hand, the collector current of the IGBT under soft turn-off continues to decrease as the (inductive) load current, which is kept constant by the inductor, can flow in the snubber capacitor. The amount and rate at which the collector current decreases depends on the capacity of the snubber capacitor. The collector-emitter voltage starts to build-up at t=1.5 μs. The depletion layer starts to expand and the stored carriers are swept out. In case of the IGBT under hard turn-off, the depletion layer spreads gradually from the floating P layer to collector layer as shown in Fig. 7 (a). On the other hand, the surface area is completely depleted under soft turn-off because the collector current is already reduced and there is no electron injection as shown in Fig. 7 (b). Figure 8 (a) shows the cross section of the simulated IGBT cell under hard turn-off at t= 2.0 μs. The electron current still flows and the depletion layer continue to expand. In case of soft-turn off, the stored carriers are swept out very slowly due to the significantly lower dV/dt of the collector-emitter voltage, which depends on capacity of the snubber capacitor.

The depletion layer expansion is much slower in soft turn-off as illustrated in Fig. 8 (b). Therefore, carriers from the bottom of the depletion layer drifting to the collector layer cause a longer tail current. As shown in Fig. 9 (a) the surface area of the IGBT under hard turn-off at t=5.0 μs is completely depleted In contrast, the depletion layer spreads gradually under soft turn-off as in Fig. 9 (b). Figure 10 shows the depletion layer distribution under hard turn-off and soft turn-off at t=8.0 μs. Their widths are 344.1μm and 138.2μm respectively. Clearly, the tail current duration becomes larger under soft-turn-off conditions.

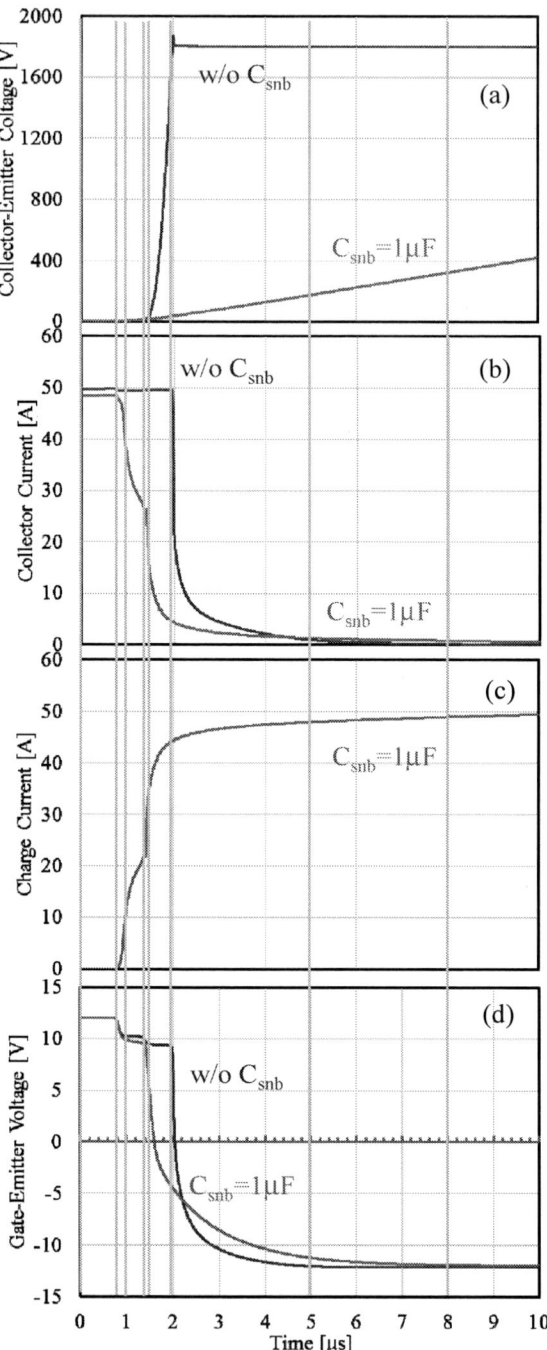

Fig. 6. IGBT turn-off waveforms without a snubber capacitor (hard switching) and connected with a snubber capacitor (soft switching) at T_j=125 °C.
(a) Collector-emitter voltage, (b) collector current, (c) charge current of snubber capacitor, (d) gate-emitter voltage. Dotted lines indicate time instances, from left to right t=0.0, 0.8, 1.0, 1.4, 1.5, 2.0, 5.0 and 8.0 μs.

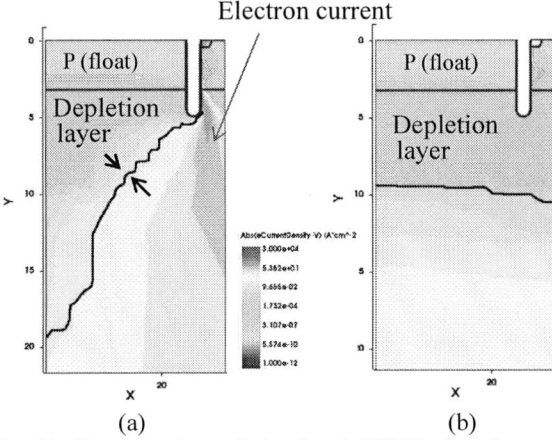

(a)   (b)

Fig. 7. Cross section of simulated IGBTs for electron current distribution at t=1.0 µs (a) under hard turn-off and (b) soft turn-off.

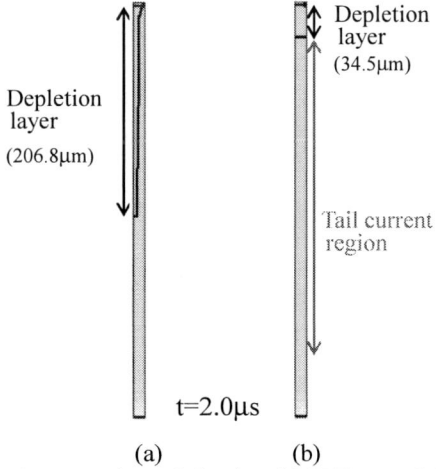

(a)   (b)

Fig. 8. Cross section of simulated IGBTs at t=2.0 µs for (a) under hard turn-off and (b) under soft turn-off.

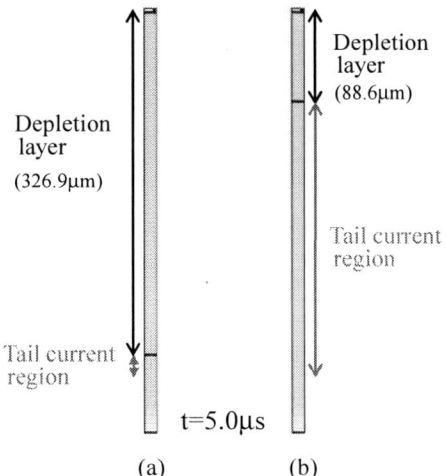

(a)   (b)

Fig. 9. Cross section of simulated IGBTs at t=5.0 µs for (a) under hard turn-off and (b) under soft turn-off.

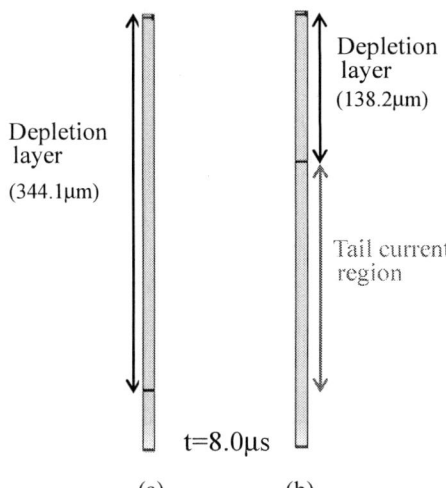

(a)   (b)

Fig. 10. Cross section of simulated IGBTs at t=8.0 µs for (a) under hard turn-off and (b) under soft turn-off.

IV. TURN-OFF CHARACTERISTICS OF THE SHORT TAIL-CURRENT STRUCTURES

To reduce the enlarged tail current, two modified "short tail-current" IGBTs, i.e. one having a thinner thickness and another having a higher resistivity drift region, were analyzed. Table 2 shows the dependency of the IGBTs turn-off losses on snubber capacitance. It can be noted that in all cases the turn-off losses are reduced when the snubber capacitance increases. However, in comparison to the conventional IGBT, the loss reductions obtained with the modified IGBTs are getting smaller at higher capacitance values. Fig. 11 shows turn-off waveforms of conventional and thin thickness structures with a 0.1 µF snubber capacitor. Although the initial tail current is slightly larger than that of the conventional IGBT, the tail current after t=5.4 µs is lower. As a result, lower turn-off losses were obtained in the thinner thickness of IGBT. On the other hand, there is no major difference between the waveforms when a 1 µF snubber capacitance is used, as shown in Fig. 12. The differences between the tail currents become smaller with larger snubber capacitor values, because the collector current is already sufficiently reduced due to the higher snubber capacitor charging current.

Table 2. Dependency of the IGBT turn-off losses on snubber capacitance.

	Drift region type		
	1. Conventional	2. Thin thickness	3. High resistivity
Snubber Capacitance	Turn-off energy losses [mJ]		
0 µF	43.15	43.40	43.02
0.1 µF	20.52	18.78	19.14
0.4 µF	8.92	8.02	8.15
1.0 µF	3.42	3.29	3.32

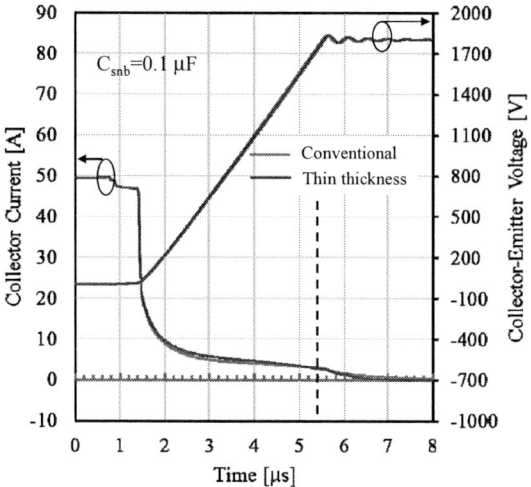

Fig. 11. Turn-off waveforms of conventional and thin thickness IGBTs with 0.1 μF snubber at T$_j$=125℃.

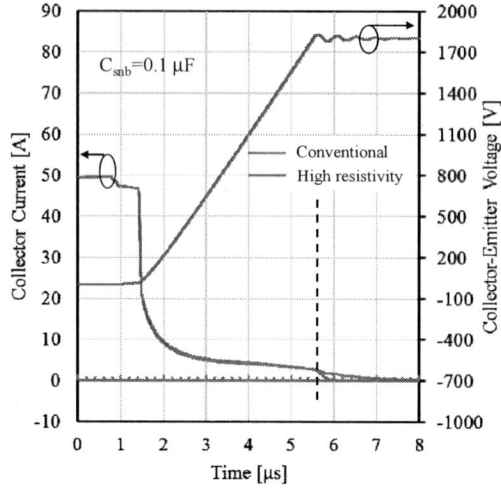

Fig. 13. Turn-off waveforms of conventional and high resistivity IGBTs with 0.1 μF snubber at T$_j$=125℃.

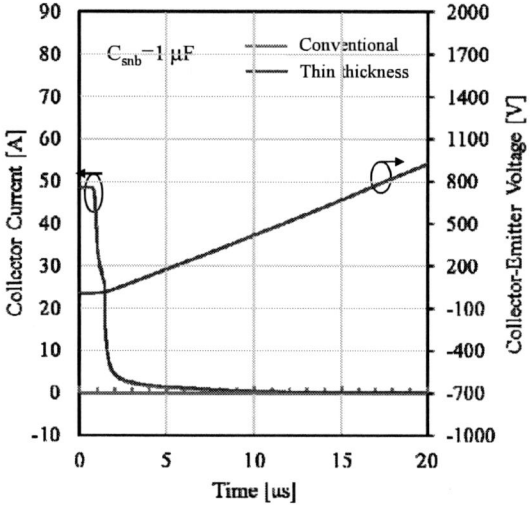

Fig. 12. Turn-off waveforms of conventional and thin thickness IGBTs with 1 μF snubber at T$_j$=125℃.

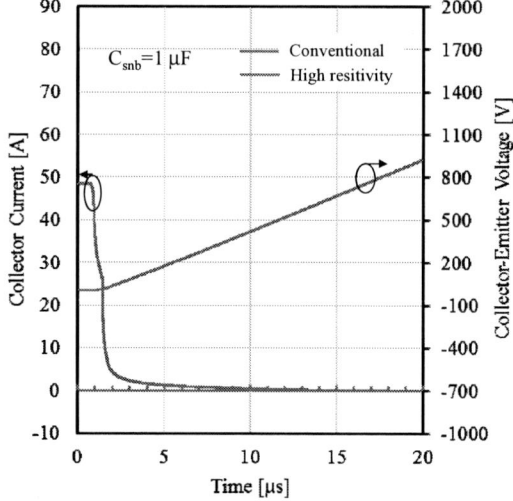

Fig. 14. Turn-off waveforms of conventional and high resistivity IGBTs with 1 μF snubber at T$_j$=125℃.

Similarly, Fig. 13 shows turn-off waveforms of conventional and high resistivity drift region IGBTs with a 0.1 μF snubber capacitor. Similar to the thin thickness IGBT, the initial tail current of the high resistivity IGBT is slightly higher than that of the conventional IGBT. In contrast, the tail current of the high resistivity IGBT after t=5.6 μs is cut off. As a consequence, this IGBT achieves lower turn-off losses. For the same reasons as the thin IGBT, the turn-off losses between the IGBTs are almost the same when using a 1 μF snubber capacitance as shown in Fig. 14.

Figure 15 shows the relationship between snubber capacitance and turn-off losses. Up to 0.4 μF the losses of the modified IGBTs, relative to the conventional IGBT, are approximately 8-10% lower. However, this number diminishes down to 3% at 1.0 μF capacitance.

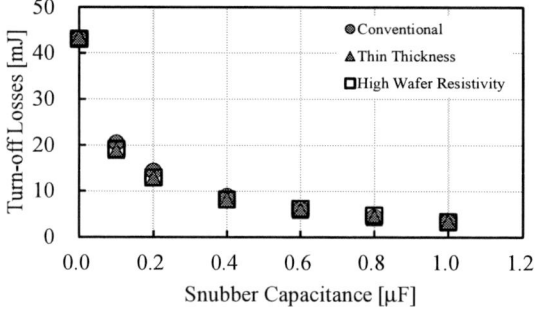

Fig. 15. Dependence of snubber capacitance for each IGBT's turn-off losses.

V. CONCLUSION

The soft turn-off behavior of a 3300 V, 50 A FS-IGBT is analyzed by using TCAD. During the turn-off process, the collector current decreases immediately due to the fact that the load current can flow through the snubber capacitor before the collector-emitter voltage rises. As a result, the turn-off loss in zero-voltage switching circuits can drastically be reduced. However, compared to hard switching, the lower dV/dt of the collector-emitter voltage causes a longer tail current to flow. To reduce the tail current, two modified IGBT designs were investigated; one with a thinner thickness and one with a higher resistivity drift layer.. It is confirmed that both structures are beneficial to reduce the turn-off losses under soft turn-off (snubbered) operation. However, the initial tail current of these structures are slightly increased due to an increase of swept out carriers. Therefore, it is recommended that careful optimization of the carrier distribution.is investigated to further reduce turn-off losses in soft-switching converters

REFERENCES

[1] R. W. De Doncker, D. M. Divan and M. H. Kheraluwala, "A three-phase soft-switched high-power-density DC/DC converter for high-power applications," in *IEEE Transactions on Industry Applications*, vol. 27, no. 1, pp. 63-73, Jan/Feb 1991.

[2] M. Stieneker and R. W. De Doncker, "Dual-active bridge dc-dc converter systems for medium-voltage DC distribution grids," *2015 IEEE 13th Brazilian Power Electronics Conference and 1st Southern Power Electronics Conference (COBEP/SPEC)*, Fortaleza, 2015, pp. 1-6

[3] T. Jimichi, M. Kaymak and R. W. De Doncker, "Comparison of single-phase and three-phase dual-active bridge DC-DC converters with various semiconductor devices for offshore wind turbines," *2017 IEEE 3rd International Future Energy Electronics Conference and ECCE Asia (IFEEC 2017 - ECCE Asia)*, Kaohsiung, 2017, pp. 591-596.

[4] R. T. Naayagi, "Selection of Power Semiconductor Devices for the DAB DC-DC Converter for Aerospace Applications," IEEE PEDS 2015, pp. 499-502, 2015.

[5] R. U. Lenke, "A Contribution to the Design of Isolated DC-DC Converters for Utility Application," Dissertation, 2012, RWTh Aachen University,

[6] R. Lenke, H. van Hoek, S. Taraborrelli, R. W. De Doncker, J. San-Sebastian and I. Etxeberria-Otadui, "Turn-off behavior of 4.5 kV asymmetric IGCTs under zero voltage switching conditions," *Proceedings of the 2011 14th European Conference on Power Electronics and Applications*, Birmingham, 2011, pp. 1-10.

The 2018 International Power Electronics Conference

600 V High Voltage Gate Driver IC (HVIC) with 1.0 MHz High Frequency Operation for LLC Current Resonant Power Supply

Masaharu Yamaji, Masashi Akahane, Takahide Tanaka, Akihiro Jonishi,
Hidetomo Ohashi, Masahiro Sasaki and Hitoshi Sumida
Fuji Electric Co. Ltd., Matsumoto, Nagano, Japan
*E-mail: yamaji-masaharu-m@fujielectric.com

Abstract— **For LLC current resonant power supply a 600 V-class high voltage gate driver IC (HVIC) with a high operating frequency of 1.0MHz has been developed for the first time. A difficult point of this development was to suppress increases in chip temperature and propagation delay time of the HVIC. For a solution to above point we have modified the level shift circuit configuration in the HVIC and the device structure for forming the level shift circuit in the HVIC chip. This paper reports our established techniques in detail.**

Keywords— LLC, HVIC, Level shifter, HVNMOS

I. INTRODUCTION

LLC current resonant power supplies are characterized by soft switching, resonance control with a duty ratio of 50 % and the leakage inductance of a transformer structure. For relatively large capacity power supplies, such as ones for large screen TVs and server devices, the LLC current resonant power supplies are commonly used to meet the requirements for high efficiency, reduced size and lower noise. We have commercialized a control IC for LLC current resonant power supplies, which is called as an LLC-IC. Our LLC-IC can configure compact and thin power supplies ranging from 100 W class to relatively large capacity of 500 W class and offer high efficiency and low noise [1, 2].

Our LLC-IC consists of a 600 V-class high voltage gate driver IC (i.e. HVIC) for switching devices on the high side and low side of a half-bridge circuit in the LLC current resonant converter [3]. Recently, the LLC-IC with a high operating frequency over 400 kHz have been required to minimize the transformer size and the electrolytic output capacitor in the LLC current resonant converter. To achieve the high operating frequency of the LLC-IC, it is necessary to develop an HVIC which can operate at the frequency over 400 kHz.

The high-frequency operation brings about a rise in chip temperature of the HVIC. The HVIC integrates a level shift circuit, namely the level shifter [3], whose

operation causes the chip temperature rise of the HVIC. In order to suppress the temperature rise, it is necessary to reduce the operating current of the level shifter. However, this leads to an increase in the propagation delay time of the HVIC, which will be described later in this paper. Thus, one important issue in developing the HVIC with the high operating frequency is how to suppress the increases in the chip temperature and propagation delay time of the HVIC.

For the HVIC with the high operating frequency, we have established techniques to suppress increases in both the chip temperature and propagation delay time. By applying the source follower configuration to the level shift circuit and modifying the device structure for forming the level shift circuit in the HVIC chip, we have successfully developed the 600V-class HVIC with 1.0 MHz high operating frequency. This paper will show our established techniques in detail.

II. LLC CURRENT RESONANT CONVERTER

In general, an LLC current resonance power supply consists of three parts: an EMI filter, a PFC converter and an LLC converter [1]. Figure 1 shows the circuit diagram

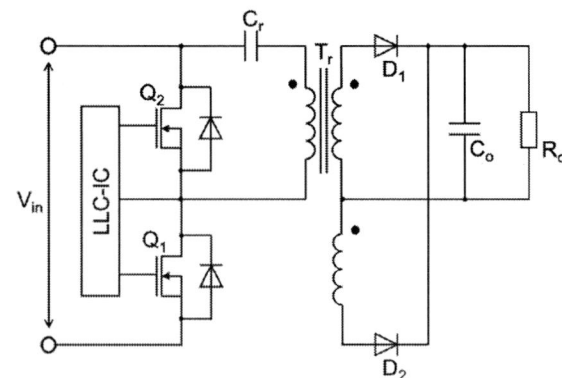

Fig. 1 Circuit diagram of an LLC current resonant converter.

The 2018 International Power Electronics Conference

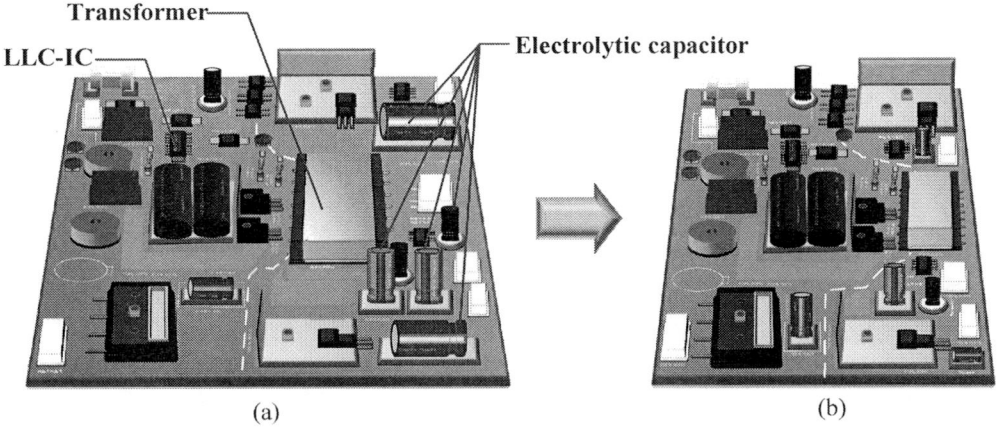

Fig. 2 LCD display circuit board integrating the LLC-IC with (a) 140 kHz operation and (b) 1.0 MHz operation.

of an LLC current resonant converter [2]. This circuit is composed of a half-bridge circuit that connects two MOSFETs (Q1 and Q2) in series, a capacitor for resonance (Cr), a transformer (Tr), output rectifier diodes (D1 and D2) and an electrolytic output capacitor (Co). The LLC-IC is used in the LLC converter and switches Q1 and Q2 for controlling the resonance current. The input voltage (Vin) is generated by the PFC converter which converts the high voltage from the AC voltage over 200 V to DC 390 V, so that the LLC-IC has the withstanding voltage over 600 V.

To minimize the transformer size and the electrolytic output capacitor in the LLC current resonant converter, the LLC-IC with the high operating frequency is requested. The size effect of the LLC current resonant power supply by using the LLC-IC with high operating frequency is estimated in the case of an application of the LCD display circuit board. Figures 2 (a) and (b) show schematic views of the LCD display circuit boards integrating the LLC-IC with the operating frequency of 140 kHz and 1.0 MHz, respectively. The size of the circuit board integrating the LLC-IC with the operating frequency of 1.0 MHz is about 70 % of that of the circuit board integrating the LLC-IC with the operating frequency of 140 kHz. This can be done by minimizing the transformer size and the electrolytic output capacitor. Thus, higher frequency operation of the LLC-IC has an effect on the size shrink of the LLC current resonant power supply unit.

III. HVIC

Figure 3 shows a circuit block diagram of the HVIC chip [3]. The major rise in the chip temperature occurs at the level shifter. The function of level shifter is to

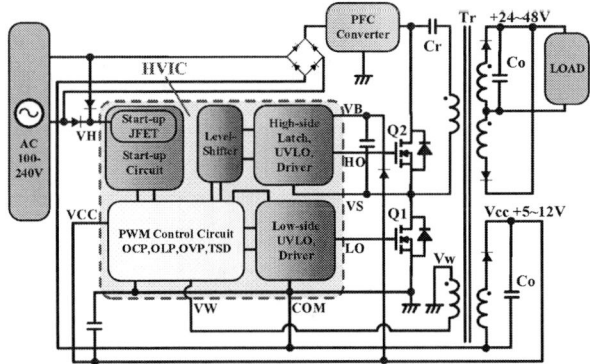

Fig. 3 Circuit block diagram of the HVIC in an LLC-IC.

transfer the signal based on the GND level to the high side circuit driven by the signal based on the source voltage of the high-side device (VS) of the half-bridge circuit. This function can be done by a high voltage n-channel MOSFET (HVNMOS) in the level shifter [4, 5], which switches at the operating frequency of the LLC-IC under the high voltage applied to VB.

The easiest method to suppress the rise in the chip temperature at the level shifter is to reduce the operating current of the level shifter (I_L). However, this method leads to an increase in the propagation delay time of the level shifter as noted above. Therefore, it is necessary to consider the impact on the propagation delay time when the I_L is reduced.

2775

Figure 4 shows the proportion of the propagation delay time occurred in the HVIC, which were obtained from a circuit simulation. The propagation delay time occurs at the level shifter occupies 33.5 % of whole propagation delay time of the HVIC. Therefore, developing methods to suppress increases in both operation temperature and propagation delay time of the level shifter are needed for the HVIC with the high operating frequency.

In order to obtain the key parameters dominating the propagation delay time of the level shifter, the circuit operation of the level shifter is analyzed by using the simple circuit model shown in Fig. 5. From this analysis,

Eq. (1) is derived when the level shift resistor (r) is sufficiently larger than that of the on-resistance of the HVNMOS (R).

$$t_d = -RC_d ln\left(\frac{VB+VS}{2VB}\right) \propto C_d/I_L \qquad (1)$$

We have found that the propagation delay time of the level shifter (td) depends on I_L and the device capacitance (Cd) of the HVNMOS as indicated Eq. (1). Thus, the

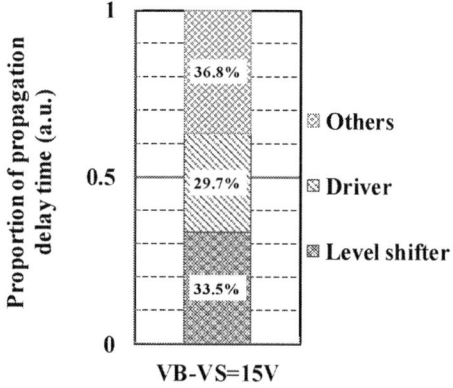

Fig. 4 Details of the propagation delay time of the HVIC.

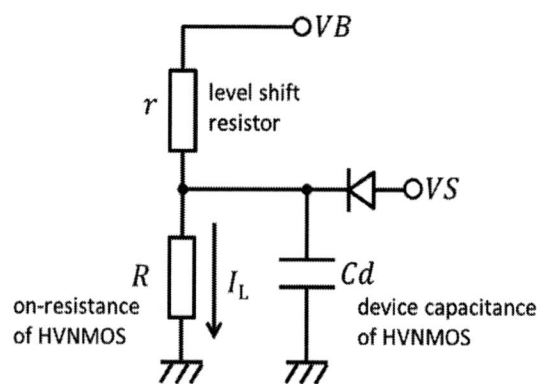

Fig. 5 Circuit model of the HVIC.

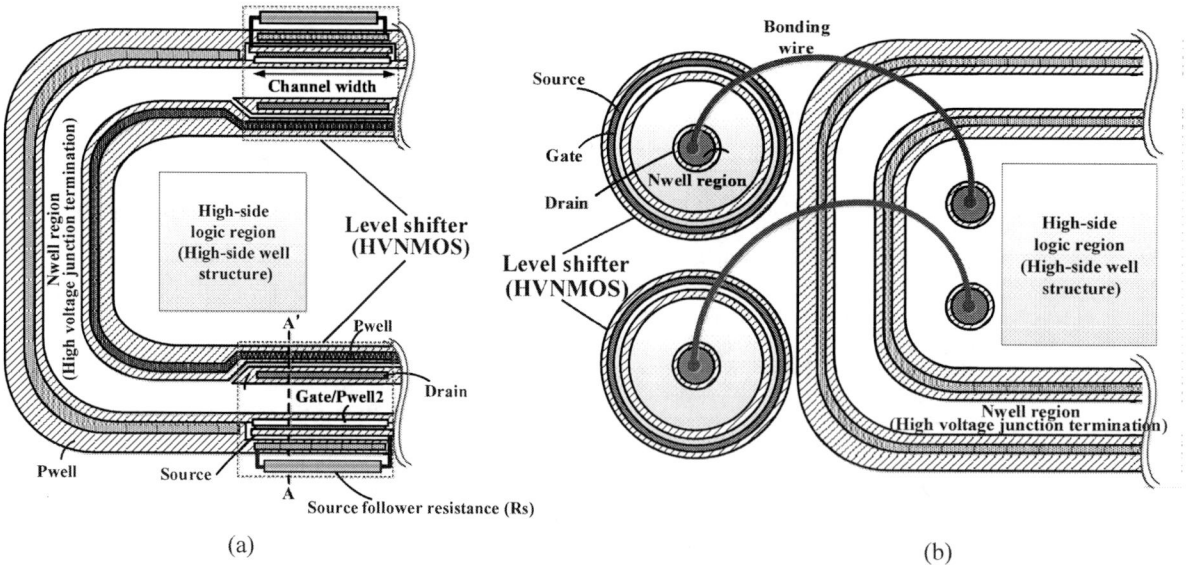

(a)

(b)

Fig. 6 Floor plans of (a) our new developed and (b) conventional level shifters in the HVICs.

reduction of both I_L and Cd is useful to suppress the increase in the chip temperature and propagation delay time of the level shifter.

IV. RESULTS

A. Level shifter

A 600 V-class HVIC was fabricated on a silicon substrate. In order to reduce both the I_L and Cd of the HVNMOS as discussed in section III, the fabricated HVIC has the self-shielding HVJT area, where the level shifter was integrated [6, 7].

Figures 6 (a) and (b) show floor plans of our new developed and conventional level shifters in the HVICs, respectively. In the conventional level shifter, there are two HVNMOSs separated from the high side circuit region and the signals between the HVNMOSs and the circuits in the high side logic region are transferred through the bonding wires [8]. Thus, the Cd of the HVNMOS in the conventional level shifter includes parasitic capacitances between a substrate and the bonding wire as well as junction capacitances formed in the device, and then becomes large. On the other hand, the Cd of the HVNMOS in our developed level shifter includes only junction capacitances and can be suppressed to be low by making the junction area small.

Figure 7 plots experimental results of the Cd/I_L, which is a key parameter for the propagation delay time as shown in Eq. (1), obtained from our developed and the conventional level shifters. In Fig. 7 a horizontal axis is the channel width of the HVNMOS as indicated in Fig. 6 (a), which dominates the I_L. The Cd/I_L of our developed level shifter is smaller than that of the conventional level shifter.

The source follower, in which the source follower resistance (Rs) has a negative temperature coefficient, has been applied to our developed level shifter for reducing the temperature dependence of the I_L. In this paper our developed level shifter is called as the source follower type level shifter.

Figure 8 shows experimental results of the I_L obtained from the level shifters with Rs of 0 ohm and 200 ohm under varying ambient temperature. It is found that the temperature dependence of the I_L in the level shifter with the Rs of 200 ohm is lower than that of the level shifter with the Rs of 0 ohm.

Figures 9 shows the cross section of our developed source follower type level shifter in the fabricated HVIC, which corresponds to the region along the A - A' line in Fig. 6 (a). The Rs formed by the polysilicon resistor is integrated in the vicinity of the HVJT area. Optimized values of the I_L in the level shifter and the Cd of the HVNMOS for the HVIC with the high operating frequency of 1.0 MHz have been obtained by carefully designing the channel width of the HVNMOS in the level shifter. Moreover, parasitic resistances formed by the

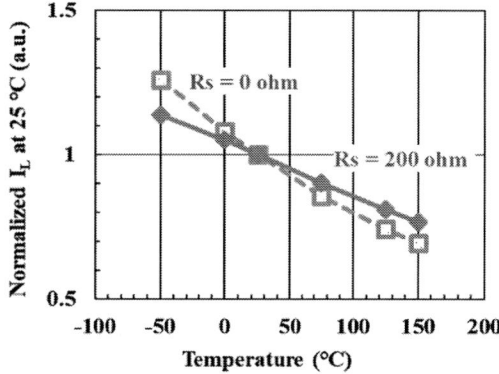

Fig. 8 Experimental results of the I_L obtained from the level shifters with Rs of 0 ohm and 200 ohm.

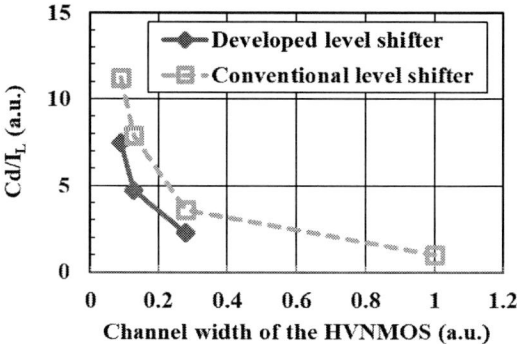

Fig. 7 Experimental results of the Cd/I_L obtained from our developed and the conventional level shifters.

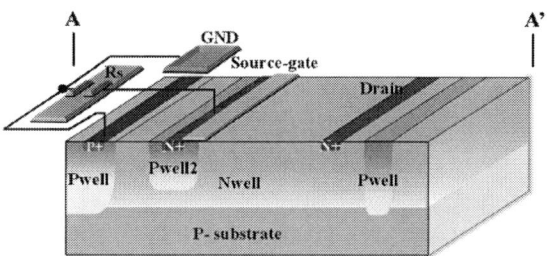

Fig. 9 Cross section of the level shifter in the fabricated HVIC.

Pwell near the source-gate region are prevented from adding to the Rs component by modifying the device structure of the HVNMOS.

B. HVIC

Figure 10 shows the dependence of a chip temperature rise on the operating frequency of the HVIC one minute after the operation started. The results of the HVIC with the conventional level shifter are also shown in Fig. 10. Under the 1.0 MHz operation, the chip temperature rise is 20 °C, while that of the HVIC with the conventional level shifter is about 60 °C under the 400 kHz operation. The chip temperature rise in the HVIC with the source follower type level shifter is drastically suppressed.

The switching waveforms of input and output signals of the HVIC with the source follower type level shifter are shown in Fig. 11. Both propagation delay time at the turn-on and turn-off stages are about 45 ns. The target propagation delay time of the HVIC for satisfying with 1.0MHz operation is below 50 ns [9]. The new HVIC has achieved the target.

Figure 12 shows the output characteristics of the fabricated HVIC under the VS of 600 V. It is observed that the HVIC operated with the switching time of 1.0 μs,

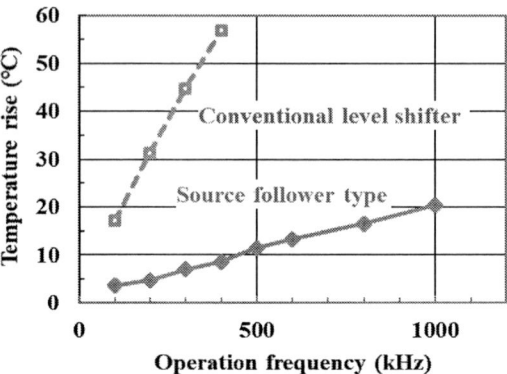

Fig. 10 Dependence of a chip temperature rise on the operating frequency of the HVICs one minute after the operation started.

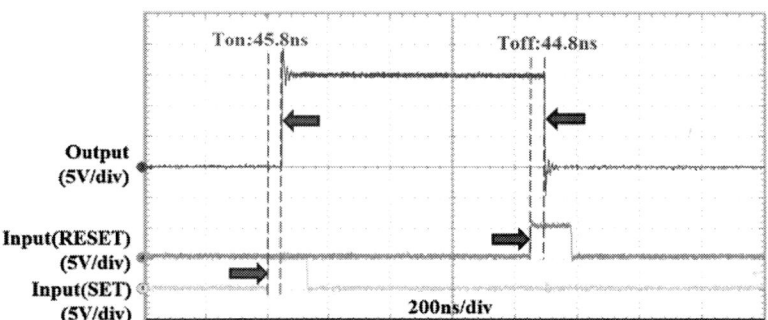

Fig. 11 Switching waveforms of input and output signals of the HVIC with the source follower type level shifter.

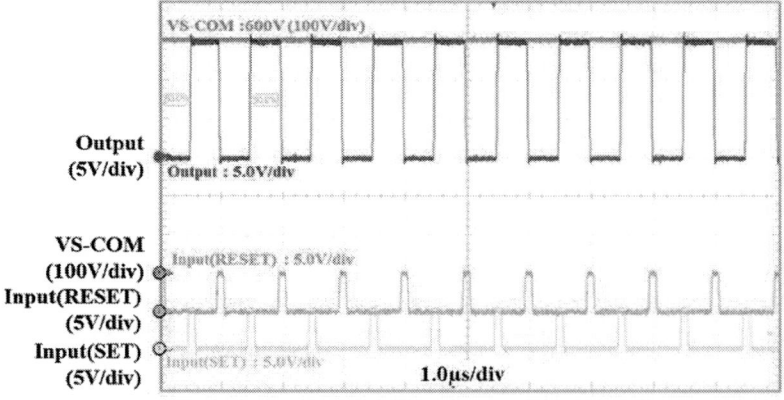

Fig. 12 Output characteristics of the fabricated HVIC under the VS of 600V.

that is, the operation frequency is 1.0 MHz. A 600 V-class HVIC with the high operating frequency of 1.0 MHz has been successfully realized with our new level shifter.

V. CONCLUSIONS

A 600 V-class HVIC with high operating frequency of 1.0 MHz has been developed for LLC current resonant power supplies. Circuit and device techniques to suppress increases in the chip temperature and propagation delay time of the HVIC have been established. Our new HVIC will contribute to making the LLC current resonant power supply unit compact.

REFERENCES

[1] J. Chen, M. Yamadaya, and H. Shiroyama, "2nd generation LLC current resonant control IC," Fuji Electric Review, vol. 59, no. 4, pp. 245-250, 2013.

[2] K. Kawamura, T. Yamamoto, and K. Hojo, "Circuit technology of LLC current resonant power supply," Fuji Electric Review, vol. 60, no. 4, pp. 245-250, 2014.

[3] M. Yamaji, J. Chen, M. Yamadaya, K. Sonobe, A. Jonishi, N. Hiasa, and H. Sumida, "A new 600 V-class power management IC realizing a system downsizing for current resonant type converters," Proceeding of the 2012 PCIM Asia, pp. 207-212, 2012.

[4] M. Akahane, A. Jonishi, M. Yamaji, H. Kanno, T. Tanaka, H. Nishio, and H. Sumida, "A new level up shifter for HVICs with high noise tolerance," Proceeding of the 2014 IPEC, pp. 2302-2309, 2014.

[5] T. Tanaka, M. Yamaji, A. Jonishi, H. Ohashi and H. Sumida, "A new downsized HVIC with high ESD tolerance," Proceeding of the 2017 ISPSD, pp. 175-179, 2017.

[6] T. Fujihira, Y. Yano, S. Obinata, N. Kumagai and K. Sakurai, "Proposal of new interconnection technique for very high-voltage IC's," Japanese Journal of Applied Physics, vol. 35, no. 11 , pp. 5655-5663, 1996.

[7] M. Yamaji, A. Jonishi, T. Tanaka, H. Sumida and Y. Hashimoto, "A 600 V high voltage IC technique with a new self-shielding structure for high noise tolerance and die shrink," IEEE Transaction on Electron Devices, vol. 62, no. 5, pp. 1524-1529, 2015.

[8] M. Yamaji, M. Akahane, and A. Jonishi, "800 V class HVIC technology," Fuji Electric Review, vol. 57, no. 3, pp. 96-102, 2011.

[9] S. Pendharkar, and C. Chey, "Advanced driver and control IC requirements for GaN and SiC Power Devices," ECS Transactions, 50 (3), pp. 189-198, 2012.

The 2018 International Power Electronics Conference

An Integrated Voltage and Current Balancing Strategy of Series-Parallel Connected IGBTs

Xiaotong Du, Fang Zhuo, Haotian Sun, Hao Yi and Yanlin Zhu
School of Electrical Engineering
Xi'an Jiaotong University
Xi'an, China
doxato@stu.xjtu.edu.cn

Abstract- **IGBTs are often connected in series and parallel to meet the requirements of high power applications. Unavoidable differences in circuit parasitic parameters and device parameters lead to voltage and current imbalance even to the destruction of devices. In this paper, an integrated voltage and current balancing strategy of IGBT is proposed, it can suppress turn-off overvoltage meanwhile balance the turn-on current of IGBT. IGBT voltage is clamped by RCD snubber circuit, and current of IGBT is controlled to follow a predefined I_c. Both current and voltage time delay are compensated by FPGA. The strategy integrates the digital global control into the analog control circuit, thus possess the flexibility of the digital circuit and the high reliability and the fast adjustment speed of the analog circuit. Research aims at seeking IGBT series and parallel modules possessing high reliability. Details about the strategy and its superiority are discussed. Experiment results with four series-parallel devices under double pulse test validate the effectiveness of the proposed method.**

Keywords— IGBTs, series-parallel connected, voltage and current balancing

I. INTRODUCTION

IGBT becomes the ideal device in many fields, especially for high power applications such as high-power DC breakers [1-2]. However, in most of case, single IGBT cannot meet the requirement of these high power applications. It is necessary to extend the capacities of IGBT and thus get large power [3-4]. Arranging IGBTS in series and parallel to reach the high power ratings is a common and less costly way. For example, in the switching branches of a solid state DC circuit breaker, in order to satisfy the high shutdown voltage capability and the large current cut-off capability, it is usually necessary to connect the IGBTs in series and in parallel simultaneously [5].

However, high-power IGBT is expensive and very sensitive to turn-off overvoltage, meanwhile the current imbalance between devices can lead to high temperature and high loss to the device, so the voltage sharing problem of series connected IGBTs and current sharing problem of parallel connected IGBT need to be Simultaneously considered [6-8]. So far, several

---

This work was supported by the National Research Program of China (973 Program) under project 2015CB251001 and 2015CB251004

strategies have been proposed to minimize the voltage imbalance [9-12]. Such as passive snubber strategy, active gate control strategy and voltage clamping strategy. Meanwhile there are impedance equalization strategy, derating strategy and gate control circuit as IGBT current sharing strategy [9,13-14]. However, no strategy could realize the current balance and voltage balance at the same time.

This paper proposed an integrated voltage and current balancing strategy of series-parallel connected IGBTs which combines the analog control circuit and an FPGA digital control chip to balance voltage and current sharing at the same time for multiple IGBTs. It effectively improves the system reliability and has high practical value. The strategy is easy to be modular, standardized thus can be extended to a greater number of series and parallel IGBTs.

II. DESCRIPTION OF INTEGRATED VOLTAGE AND CURRENT BALANCING STRATEGY

The block diagram of the integrated voltage and current balancing circuit based on a single IGBT is shown in Fig. 1, each IGBT gate is connected with a drive circuit to control the IGBT turning on or off. The control strategy consists of three control parts: current sense and feedback loop (analog circuit), voltage snubber, and global control loop (digital circuit).

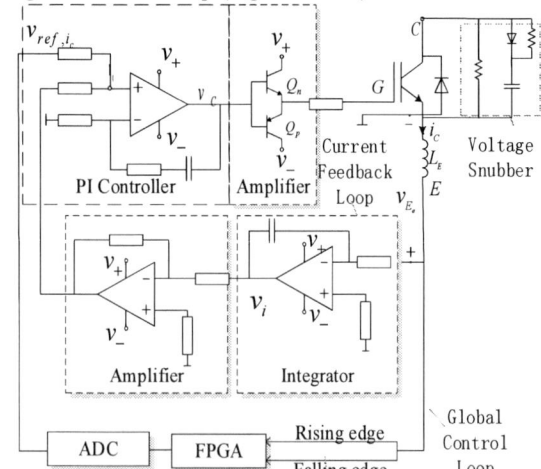

Fig. 1. Block diagram of the balancing strategy

The current sense and feedback loop controls the emitter current of IGBT by adjusting the IGBT gate current. The loop collects the voltage of internal inductance inside the IGBT module, which is proportional to the current change rate. Then it is integrated to obtain the emitter current which is the negative feedback signal. The error of the feedback value and the reference value is adjusted by the PI controller. Finally, through the analog amplifier circuit, the signal is amplified and is sent to the gate as the IGBT gate control signal. The stability of the current feedback loop is proved in [15].

Voltage snubber circuit consists of the snubber capacitor C, resistor R, diode D. Its basic function principle is introduced in many papers such as [16]. With a small Drive signal delay gap, RCD passive snubber circuit could achieve voltage balancing effect.

The global control loop is connected to the current feedback circuit of each IGBT via a high-speed digital-to-analog converter. The global controller is a field programmable gate array (FPGA). FPGA outputs reference signal according to the target current for current feedback and compensate delay difference for both voltage and current. Specifically, when the target current is constant, the reference current is constant. When the target current changes, the reference signal imitate the changing target current signal thus realize balance in static current sharing.

III. THE SUPERIORITY OF THE STRATEGY

A. Current Sense and Feedback Loop

The active current control loop in this strategy has the following superiorities [14-17]:

1. Power loss is small. the loss of the process are mostly generated on the load and IGBT, the current control loop doesn't add any components to the power loop so the power consumption is very small.

2. The capacity waste is small. compared with the derating current sharing method, the method doesn't need significant derating

3.The circuit is simple and convenient. Compared with other active gate control method, the circuit doesn't require current detection equipment or manual operation. Thus reducing the complexity of the circuit.

B. Voltage Snubber Loop

RCD voltage snubber circuit has the following superiorities[18-20]：

1. Circuit structure is simple and easy to design.

2. Even if the voltage is balanced, due to the parasitic parameters on the line, there could be voltage overshoot during turn-off process. The snubber circuit to a certain extent, inhibit this overshoot.

C. Global Control Loop

Taking controlling two IGBTs as an example, Fig. 2 shows the global control block diagram.

Global control is not only shown in the simultaneous control of multiple IGBTs, but also in the simultaneous control of turning-on state current and turning-off state voltage.

1. FPGA controls IGBT emitter current by outputting reference signal to the current feedback loop.

2. FPGA compensate the time delay for both current rising and voltage rising. FPGA receives the rising and falling edges of the voltage across the IGBT inductor, which signifies the time at which the emitter current begins to rise and the time at which the collector-emitter voltage ends to rise. The global controller FPGA performs the comparison operation on the basis of the received data and stores the result, when the next turn-on and turn-off operation comes, the FPGA adjusts the delay of the gate signal sent to each IGBT according to the compensation result, thereby achieving the purpose of optimizing turn-on current balancing and the turn-off voltage balancing.

In order to achieve the same voltage peak during the turn-off process of the IGBT, the voltage imbalance caused by the difference in the gate signal delay and the difference in the snubber capacitance must be fully considered. FPGA identifys the falling edge of IGBT emitter current, which according to the characteristics of IGBT is the end of the voltage rise. Therefore, by directly controlling the end time of voltage rise, the voltage peak is controlled, which is superior to compensating the voltage start rising moment.

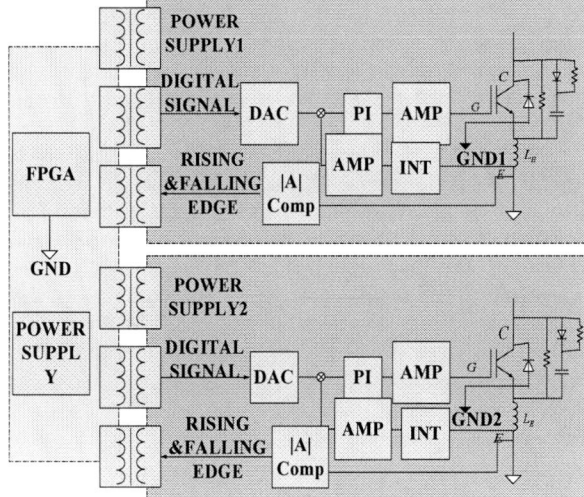

Fig. 2 Global control schematic of integrated balancing strategy

IV. ANALYSIS OF THE OPERATING PROCEDURE

A. Analysis of the turning-on stage

The integrated voltage and current balancing circuit is like the following Fig. 3.

2781

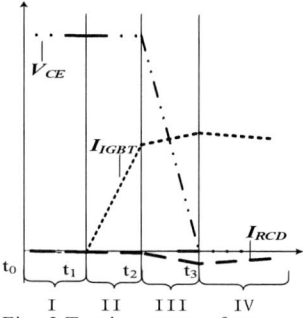

Fig. 3 Turning-on waveform

For each IGBT added into the circuit, its turning-on procedure could be divided into 4 stages.

Stage I is a delayed stage at which neither the voltage nor the current balancing circuit has effects. At moment t_0, the FPGA turns the reference signal from negative to positive, the gate was fed by a positive voltage that is charging the input capacitor and the gate resistance. In this stage, neither the current nor the voltage on the IGBT changes and the IGBT works in the active zone.

During stage II, IGBT current rise. At moment t_1, the gate voltage achieves the threshold, the committer current starts to increase. In this stage, IGBT is in the linear zone and can be considered as a voltage controlled current source.

Stage III shows the drop of the IGBT voltage during which the current balancing circuit is effective. At t_2 moment, IGBT current attend its peak and IGBT enters the saturation zone, the RCD snubber branch gets short circuited little by little, the capacitor C start to charge IGBT. Then a reverse current appears in the RCD branch, the current through the IGBT increase accordingly, and the voltage of the IGBT drops under the control of the RCD circuit.

Stage IV is the complete conduction stage, at which the IGBT is in the saturation region and the current feedback loop controls the IGBT emitter current. The snubber capacitor C, the snubber resistor R and the IGBT constitute the discharge circuit. As the charge on the capacitor C gradually decreases, the current on the discharge circuit decreases, the current of the RCD branch is reduced and the current flowing into the IGBT is gradually reduced as well

B. Analysis of the turning-off stage

Fig. 4 is the waveform of integrated voltage and current balancing circuit operating in the turn-off stage, we can see that the voltage rising stage across the IGBT can be divided into four parts.

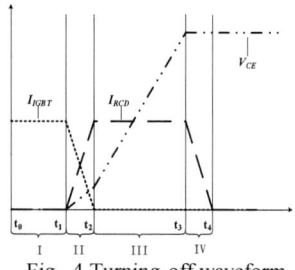

Fig. 4 Turning-off waveform

Stage I is the turn-off delay stage, the current-feedback circuit and the voltage balancing circuit are not functioned. At t_0 moment, the reference signal for feedback loop becomes zero, the gate receives the shutdown signal, drive current of IGBT gate began to decline. The negative drive power, the IGBT input capacitor and the gate resistance composite a first-order discharge circuit, IGBT is always in the saturation zone, as a result the IGBT voltage and current do not change at both ends.

Stage II is the IGBT current dropping stage, the current balancing circuit is active at this stage. IGBT leaves saturation zone, at this stage the IGBT can be equivalent to a current source controlled by the gate voltage, current feedback loop controls the emitter current by controlling the IGBT drive voltage. As the IGBT gate voltage Vge drops, the current flowing through the IGBT is transferred to the RCD branch, and the voltage Vce across the IGBT increases as the current I_{RCD} through the snubber capacitor C changes.

Stage III is the IGBT voltage rise process, the voltage balancing circuit is active at this stage. At this moment, IGBT has been completely turned off, the voltage is no longer affected by the device and drive circuit parameters, but decided by the external circuit. IGBT current I_{IGBT} becomes zero. The current all flows through the RCD branch, I_{RCD} in the continuing role, the capacitor C voltage at both ends continue to rise, therefore, the voltage across the IGBT continues to rise.

Stage IV is RCD current falling stage. At t_3 moment, IGBT voltage reaches the peak and stops rising, the voltage of capacitor C also rises, the current flows through the RCD branch gradually reduced to zero.

V. EXPERIMENTAL VERIFICATION

A. Parameter designing

In order to verify the feasibility and effect of the integrated voltage and current balancing strategy in the actual system, this work decides the low power level double pulse test.

There are two possibilities of connecting schemes as shown in Fig. 5: 1.Two series connected IGBTs form one branch, and then two branches connect in parallel. This scheme needs four voltage snubber circuits. 2. Two IGBTs connect in parallel first and then connect in series with another two IGBTs. This scheme need four current feedback loops. The second connecting scheme is chosen, because the second scheme could economize two voltage buffer circuit and voltage buffer circuit consume more energy than the current balancing loop.

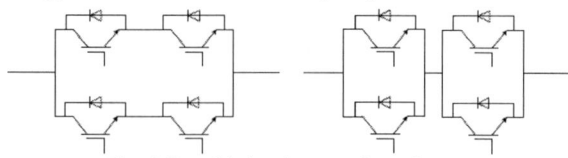

Fig. 5 Two kinds of connecting schemes

The experimental platform is shown in Fig. 5, including the DC power supply V_{DC}, the hollow core inductance L, the anti-parallel diode D, and four IGBTs connected in parallel and series (each using only one IGBT in the module). Using FPGA to adjust the reference signal flexibly and adjust the reference signal transmission time automatically according to the acquisition rise and fall time difference. Parallel IGBTs share the same set of RCD voltage snubber circuit. Each IGBT's gate signal is adjusted by current feedback loop and FPGA. In the experiment, DC power supply voltage sets to 40V, the maximum inductor current reaches 10A.

Article [15-16] outlines the design criteria for the RCD snubber and for the current feedback loop. Experiment uses ITECH's 80V / 40A / 1500W DC power supply, Infineon's 1.2kV / 450A IGBT module FF450R12ME3, 1mH hollow core inductance, anti-parallel diode in Infineon IGBT module F4-75R12MS4 as anti-parallel diode of inductance, uses Altera's Cyclone IV FPGA, uses a bandwidth of 20MHz Pearson110 current sensor to measure the current.

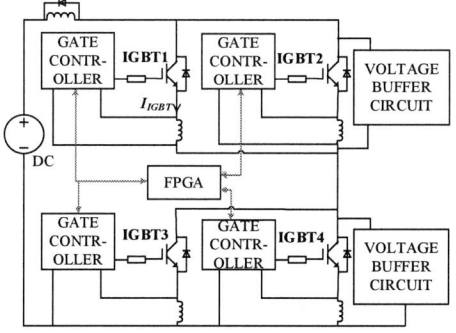

Fig. 6 Schematic diagram of experiment circuit

B. Experiment results discussion

Fig. 7 shows the double pulse test waveforms of the four IGBTs' voltage Vce and the emitter current Ic, under no voltage nor current balancing control. In figure(a), the IGBT collector-emitter voltage waveform can be seen, due to the inevitable existence of delay difference and the differences in circuit parameters, among 4 IGBTs only 2 IGBTs (Parallel connected IGBT3 and IGBT4) bear all the power supply voltage in the off state (2~3.5e-4s), the other two IGBTs in the off stage withstand a small voltage. The turn-off peak voltage reaches a maximum of 70V. From figure (b), the IGBT emitter current waveform can be seen. Four IGBTs show unbalancing current sharing situation, there IGBT2 in a short time (during transition period) bear the full load current, and the steady-state current difference reaches 1A (20%). After the introduction of integrated voltage and current balancing control circuit, shown in Fig. 8, their turn-off voltage and turn-on current have a very good balancing effect, the maximum voltage peak is only 25 V, 4 IGBTs turn-on current are roughly the same, except a little difference in the transition moment.

Although the experimental results only show the situation under which four series-parallel connected IGBTs are tested, it should be noted that in the high power application such as circuit breaker will usually take a few (3 ~ 10) IGBTs as a basic switch module. The aim of this paper is to solve the problem of IGBT voltage

and current sharing problem in a switch module. With the increase of the number of IGBTs in series and parallel, the control method and control circuit structure proposed in this work are still applicable, only need to add corresponding control modules.

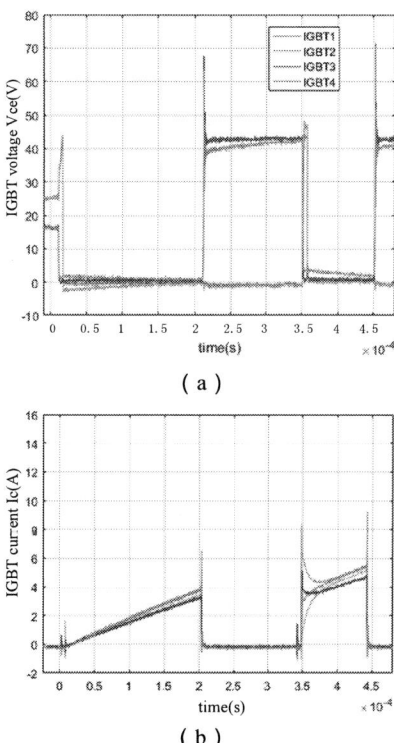

(a)

(b)

Fig. 7 IGBT voltage and current without balancing strategy

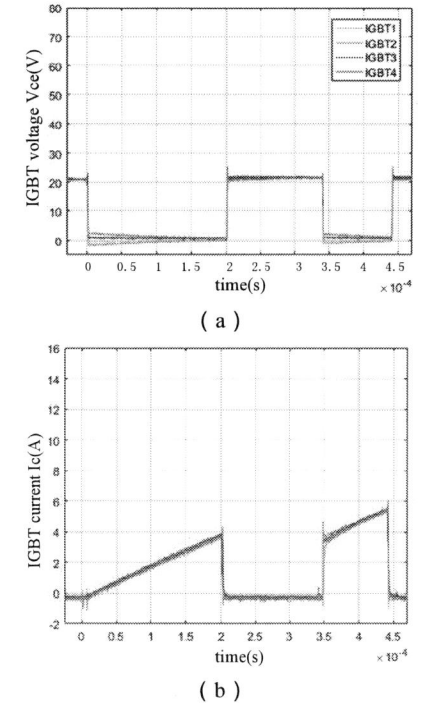

(a)

(b)

Fig. 8 IGBT voltage and current under balancing strategy

VI. CONCLUSION

An integrated balancing strategy of IGBT is proposed and verified by experiment in this paper, which can prevent overvoltage and overcurrent, lessen the imbalance of voltage and current that may occur simultaneously in the serial-parallel IGBT thus protects the device and effectively improves the reliability of the series-parallel connected IGBT module. The proposed integrated control strategy uses a single FPGA controlling multiple IGBTs' current and voltage simultaneously, improves system's integration and reduces their costs. The strategy is easy to be modular and applied to a greater number of series and parallel IGBTs.

ACKNOWLEDGMENT

This work was supported by the National Key Basic Research Program of China (973 Program) under project 2015CB251001 and 2015CB251004.

REFERENCES

[1] B. J. Baliga, Modern Power Devices. NewYork, NY, USA:Wiley, 1987.

[2] Boudreaux R R, Nelms M R. "A comparison of MOSFETs, IGBTs, and MCTs for solid state circuit breakers" *Applied Power Electronics Conference and Exposition*, 227-233, 1996.

[3] F. Chimento,W. Hermansson, and T. Jonsson, "Robustness evaluation of high-voltage press-pack IGBT modules in enhanced short-circuit test," IEEE Trans. Ind. Appl., vol. 48, no. 3, pp. 1046–1053, May/Jun. 2012.

[4] N.Mohan, T. Undeland, andW. Robbins, Power Electronics: Converters, Applications and Design. Hoboken, NJ:Wiley, 1989.

[5] Lei Feng, Ruifeng Gou, Fang Zhuo, Xiaoping Yang and Fan Zhang, "Development of a 10kV solid-state DC circuit breaker based on press-pack IGBT for VSC-HVDC system," 2016 IEEE 8th International Power Electronics and Motion Control Conference (IPEMC-ECCE Asia), Hefei, 2016, pp. 2371-2377.

[6] P. Hofer-Noser and N. Karrer, "Monitoring of paralleled IGBT/diode modules," *IEEE Trans. Power Electron.*, vol. 4, no. 3, pp. 438–444,

[7] U. Schlapbach, "Dynamic paralleling problems in IGBT module construction and application," in *Proc. 6th Int. CIPS,*Mar. pp. 1–7, 2010.

[8] C Das, G Narayanan, A Tiwari, et al. "Experimental study on IGBT voltage and current stresses during switching transitions". Innovative Smart Grid Technologies - Asia. IEEE, 1-6, 2013.

[9] N. Y. A. Shammas, R. Withanage and D. Chamund, "Review of series and parallel connection of IGBTs," IEEE Proceedings - Circuits, Devices and Systems, vol. 153, no. 1, pp. 34-39, Feb. 2006.

[10] R. Withanage and N. Shammas, "Series Connection of Insulated Gate Bipolar Transistors (IGBTs)," IEEE Transactions on Power Electronics, vol. 27, no. 4, pp. 2204-2212, April 2012.

[11] K. Sasagawa, Y. Abe and K. Matsuse, "Voltage-balancing method for IGBTs connected in series," IEEE Transactions on Industry Applications, vol. 40, no. 4, pp. 1025-1030, July-Aug. 2004.

[12] T. C. Lim, B. W. Williams, S. J. Finney and P. R. Palmer, "Series-Connected IGBTs Using Active Voltage Control Technique," *IEEE Transactions on Power Electronics*, vol. 28, no. 8, pp. 4083-4103, Aug. 2013. doi: 10.1109/TPEL.2012.2227812

[13] R Alvarez, S Bernet. "Sinusoidal Current Operation of Delay-Time Compensation for Parallel-Connected IGBTs". *IEEE Trans. Ind. Appl*, vol.50,no.5,pp.3485-3493., 2012

[14] Y. Lobsiger and J. W. Kolar, "Closed-Loop di/dt and dv/dt IGBT Gate Driver," IEEE Transactions on Power Electronics, vol. 30, no. 6, pp. 3402-3417, June 2015.

[15] Y. Chen, F. Zhuo, F. Zhang, W. Pan and Y. Yang, "A novel method for current balancing between parallel-connected IGBTs," 2016 18th European Conference on Power Electronics and Applications (EPE'16 ECCE Europe), Karlsruhe, 2016, pp. 1-8.

[16] Chen, J.-F., Lin, J.-N., and Ai, T.-H.: 'The techniques of the serial and paralleled IGBTs'. Proc. IEEE IECON 22nd Int. Conf., August 1996, Vol. 2, pp. 999–1004

[17] Z. Wang, X. Shi, L. M. Tolbert, F. (. Wang and B. J. Blalock, "A di/dt Feedback-Based Active Gate Driver for Smart Switching and Fast Overcurrent Protection of IGBT Modules," in *IEEE Transactions on Power Electronics*, vol. 29, no. 7, pp. 3720-3732, July 2014.

[18] D. Ning, X. Tong, M. Shen and W. Xia, "The Experiments of Voltage Balancing Methods in IGBTs Series Connection," 2010 Asia-Pacific Power and Energy Engineering Conference, Chengdu, 2010, pp. 1-4.

[19] T. C. Lim, B. W. Williams and S. J. Finney, "Active Snubber Energy Recovery Circuit for Series-Connected IGBTs," in IEEE Transactions on Power Electronics, vol. 26, no. 7, pp. 1879-1889, July 2011.

[20] Fan Zhang, Xu Yang, Yu Ren, Chen Li and Ruifeng Gou, "Voltage balancing optimization of series-connected IGBTs in solid-state breaker by using driving signal adjustment technique," 2015 IEEE 2nd International Future Energy Electronics Conference (IFEEC), Taipei, 2015, pp. 1-5.

Thermal Design and Analysis of a Cable Charger Used for Portable Electronics

Mofan Tian, Xu Yang, Naizeng Wang, Yang Chen, Laili Wang
School of Electrical Engineering, Xi'an Jiaotong University, Xi'an, China
*E-mail: tianmofan@outlook.com

Abstract—As the charging current of mobile electronics increases, the heat dissipation problem is risen for their small portable chargers. By integrating the inductor and heatsink into the cable, a flexible cable-integration charger could be fabricated. In this paper, a thermal resistance model of the cable charger is proposed. Based on this model, the dominant factors related with the thermal performance are extracted, and how they affect the chip junction temperature and the cable surface temperature is further analyzed to ensure its reliability and safety in use. Besides, the method to evaluate the thermal resistance value is obtained, which can be used for the synergetic optimization of the cable charger for better the thermal performance and flexibility performance. Finally, a 3.3/5V prototype is designed, fabricated and tested. The experimental results verified that good thermal performance of the cable charger.

Keywords—Flexible converter, Magnetic integration, Thermal design, Portable electronics.

I. INTRODUCTION

The increasing requirements of mobile electronics for smarter and longer standby time make the battery capability larger and larger. To decrease the charging time, the charging current becomes higher and higher, which brings a heat dissipation problem for the portable charger. To make matters worse, the bulky heatsink cannot be used on the small and light mobile electronics, thus, solving the heat dissipation problem is critical to improve the quick-charge technology. Nowadays, the researches concerned with the charger mainly focus on the circuit, and some high efficiency and high-power circuits have been designed and commercialized, such as a charge pump, multi-phase buck and so on [1][5]. However, an efficient thermal management method of quick-charge charger is lacking, which limits the charging current approaches to higher level.

Thermal design of the charger is directly related to the personal safety in use, so a series of requirements need to be satisfied before it goes into market. For example, the surface temperature of the charger which could be touched by user must below the 39 in case of scald. The operating temperature of the components in the charger must be below a certain threshold to ensure reliability. The common methods to help heat dissipation at natural convection condition can be classified into two categories: 1) using the material with high heat conductivity coefficient [6]. 2) Increasing the contact area with air. The second conception is adopted in the cable-integration charger. The surface of the cable is used as

TABLE I: THERMAL CONDUCTIVITY OF THE MATERIAL

Material	Thermal conductivity (W/(K·m))
Silicon	130
Solder	62
Copper	400
FR4	0.35
Plastic	0.2
Epoxy	0.6
Ferrite sheet	0.133
Aluminum	238

the heatsink to enlarge the heat transfer area. The heat produced by the components could be transferred to the surface and dissipated. Besides, the inductor is transferred from the PCB to the cable, which can decrease the thermal resistance of the core loss and winding loss to dissipate.

In this paper, the thermal resistance model of the cable charger is established. Based on this model, the dominant thermal resistances related with the thermal performance of the charger are extracted. Besides, an elaborate thermal design is performed on the cable-type charger to ensure its surface temperature and operating temperature both below the limit. The method to evaluate the thermal resistance value is also obtained, which can be used for the synergetic optimization of the cable charger for better the thermal performance and flexibility performance.

The remainder of this paper is organized as follow: the structure of the cable charger is presented in section II; a thermal resistance model is established and verified by the finite-element simulation in section III; the good performance of the cable charger is validated by a prototype in section V; finally, the conclusion is drawn and the further work is presented.

II. STRUCTURE OF THE CABLE CHARGER

The structure of the cable-integration charger is showed in Fig.6. A thin PCB [200μm FR4 thickness and 35μm Copper] is used to hold the switching device and capacitors. The cable has a coaxial structure and rectangular cross-section shape. The outer layer made of aluminum foil is used as the heatsink. Note that another thin ferrite sheet layer is embedded in the cable to achieve the inductor.

2785

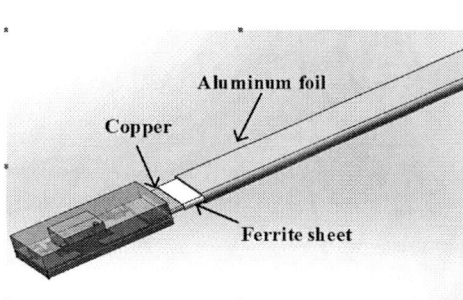

Fig. 1: Sketch of the cable charger. The plastic isolation layer is neglected.

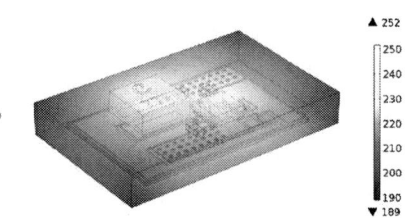

Fig. 2: Structures and finite simulation results of a traditional on-board charger.

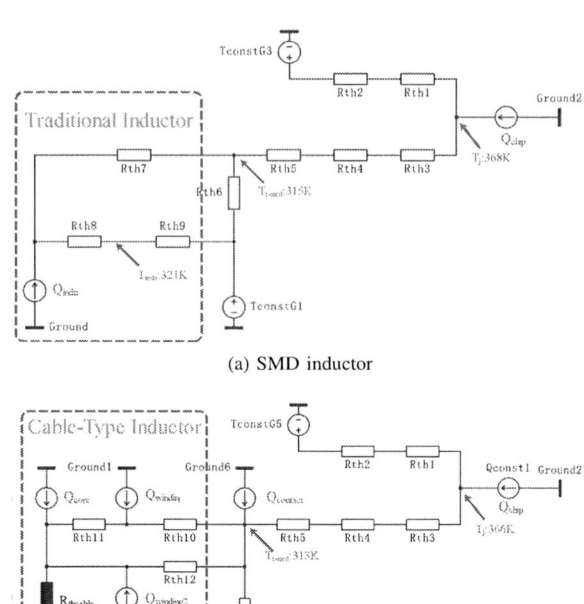

(a) SMD inductor

(b) Cable-type inductor

Fig. 3: Thermal model of the converter with different inductor.

The 3D-model of PCB is established in the finite-element software to simulate the temperature distribution in it, and the cable is excluded in the simulation. A heat dissipation of 0.88W, located at bottom of the chip on die is considered in this model. The boundary conditions on all surface are equivalent to a heat transfer coefficient of 8W/(m²K), which corresponds to natural convection condition. A large number of via holes go through the PCB. The heat conductivities for the different materials are listed in Table I. The simulated temperature distributions in the PCB is shown in Fig.2. It can be seen that the results beyond the allowable junction temperature proving that the charger cannot work without the efficiency heat dissipation method.

When the cable is added, the heat produced by the components could be transferred to the cable surface then dissipated. Owing to its large surface area, the heat could be dissipated easily. It is obvious that the wider the cable, the lower the temperature.

III. THERMAL MODEL AND OPTIMIZATION

Fig.3a is the thermal model of the converter with the conventional SMD inductor and the Fig.3b is the model of the converter with the cable-type inductor., and the corresponding definition of these thermal resistances are list in Table II. The thermal resistances are calculated by finite-element simulation using the method shown in [7]. It should be noted that this method works well only provided the temperature distributes uniformly in one component. Consequently, the copper on the top layer of the PCB is divided into two parts. One part is directly contact with the thermal pad of the switch, which is the main heat source. The others are assumed to be of the same temperature. Meanwhile, the large number of thermal vias make the FR4 board, the copper on the top layer and bottom layer to be regard as isothermal, so the thermal resistance between them is neglected. It can be seen that the model is accurate in the comparison of Fig.3.

As shown in Fig.3(b), the inductor loss is divided into three parts. Q_{core} is the core loss, and the $Q_{winding1}$ and $Q_{winding2}$ represent the winding loss caused by the inner copper and outer aluminum foil separately. The simulated results show that the cable owns a significant advantage on the thermal performance so that the junction temperature and cable temperature are blow the limit. As marked in Fig.3(b) by the solid blue cubic, the thermal resistance from the aluminum foil to the plastic case and the thermal resistance from the plastic case to the ambient are closely related to the temperature limits mentioned above. Fig.4 shows the available region of the two thermal resistances under the temperature limit. The two thermal resistances are dependent on the geometric parameters of

TABLE II: THERMAL CONDUCTIVITY OF THE MATERIAL

Symbol	Definition	Value	Symbol	Definition	Value
R_{th1}	IC plastic thermal resistance, from the IC die to its plastic case	343	R_{th11}	Ferrite thermal resistance, from the inner copper to the ferrite sheet	9.1
R_{th2}	Stuff thermal resistance, from the IC case to the ambient	428	R_{th12}	Aluminum resistance, from the PCB board to the outer aluminum foil	4
R_{th3}	Solder thermal resistance, from the IC die to the solder pad	2	$R_{thcable}$	Packing isolation thermal resistance, from the aluminum foil to the packing isolation	20
R_{th4}	Joints thermal resistance of the IC, from the solder to the top layer of copper on PCB	28.8	R_{thca}	Cable resistance, from the packing isolation to the ambient	30
R_{th5}	Joints thermal resistance of the PCB, from the copper to the PCB board	36	Q_{chip}	The heat source represents the MIC2876 loss	0.762
R_{th6}	PCB thermal resistance, from the PCB to the ambient	20	Q_{indu}	The heat source represents the inductor loss	0.278
R_{th7}	Inductor substrate thermal resistance, from the inductor pad to the PCB board	86	Q_{core}	The heat source represents the core loss of the inductor	0.064
R_{th8}	Inductor thermal resistance, from the inner to surface of the inductor	5	$Q_{winding1}$	The heat source represents the inner copper winding loss	0.106
R_{th9}	Inductor case thermal resistance, from the inductor surface to the ambient	80	$Q_{winding2}$	The heat source represents the outer aluminum foil loss	0.066
R_{th10}	Cable inner thermal resistance, from the PCB board to the inner copper of the cable	20	$Q_{contact}$	The heat represents the contact loss	0.041

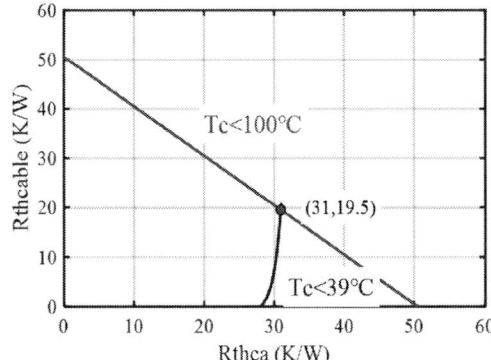

Fig. 4: Thermal resistance boundary under the temperature limit.

the cable.

IV. EXPERIMENT VERIFICATION

As shown in Fig.5(a), a 3.3V/5V prototype is fabricated with the switch chip of MIC2876 from Microchip. Because the process is complicated, the plastic case of the cable charger is not completed in time. In order to verified the thermal performance of the cable charger, an evaluation board of MIC2876 (Fig.5(b)) is also tested, and the results shown in Fig.6. It can be seen that although the PCB board of the cable charger is only one twentieth of the evaluation board, the whole temperature of it is still lower than evaluation board, which can be mainly

(a) Cable charger.

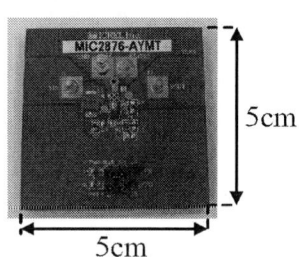

(b) Evaluation board of MIC2876.

Fig. 5: Sketch of the cable charger. The plastic isolation layer is neglected.

ascribed to the large surface area of the cable.

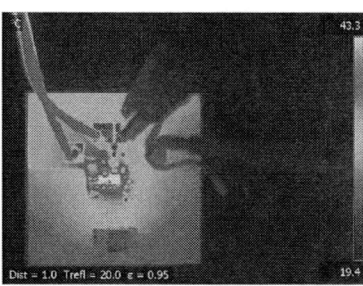

(a) Evaluation board of MIC2876.

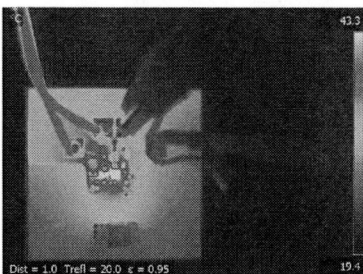

(b) The proposed cable charger.

Fig. 6: Sketch of the cable charger. The plastic isolation layer is neglected.

V. CONCLUSION

In this paper, a thermal resistance model of the cable charger is proposed. Based on this model, the dominant factor related with the thermal performance are extracted, and how they affect the chip junction temperature and the cable surface temperature is analyzed furtherly to ensure its reliability and safety in use. Besides, the method to evaluate the thermal resistance value is obtained, which can be used for the synergetic optimization of the cable charger for better the thermal performance and flexibility performance. A 3.3/5V cable charger is tested and compared with an evaluation board, which validated its good thermal performance.

In further work, the relationship between the thermal resistance and the cable geometric parameters will be studied. Based on that, the electrical performance, flexibility performance and thermal performance will be considered together, so that a high efficiency, high reliability bendable cable charger can be achieved. Meanwhile, the complete prototype will be fabricated to validate it performance.

REFERENCES

[1] G. Gabian, B. Blalock, and D. Costinett, 5V-to-4V integrated buck converter for battery charging applications with an on-chip decoupling capacitor, in 2017 IEEE Applied Power Electronics Conference and Exposition (APEC), 2017, pp. 178-183.

[2] T. C. Huang, R. H. Peng, T. W. Tsai, K. H. Chen, and C. L. Wey, Fast Charging and High Efficiency Switching-Based Charger With Continuous Built-In Resistance Detection and Automatic Energy Deliver Control for Portable Electronics, IEEE J. Solid-State Circuits, vol. 49, no. 7, pp. 1580-1594, Jul. 2014.

[3] Y. S. Hwang, S. C. Wang, F. C. Yang, and J. J. Chen, New Compact CMOS Li-Ion Battery Charger Using Charge-Pump Technique for Portable Applications, IEEE Trans. Circuits Syst. Regul. Pap., vol. 54, no. 4, pp. 705-712, Apr. 2007.

[4] M. G. Jeong, S. H. Kim, and C. Yoo, Switching Battery Charger Integrated Circuit for Mobile Devices in a 130-nm BCDMOS Process, IEEE Trans. Power Electron., vol. 31, no. 11, pp. 7943-7952, Nov. 2016.

[5] M. A. Saket, N. Shafiei, M. Ordonez, M. Craciun, and C. Botting, Low parasitics planar transformer for LLC resonant battery chargers, in 2016 IEEE Applied Power Electronics Conference and Exposition (APEC), 2016, pp. 854-858.

[6] C. Yu, C. Buttay, and . Labour, Thermal Management and Electromagnetic Analysis for GaN Devices Packaging on DBC Substrate, IEEE Trans. Power Electron., vol. 32, no. 2, pp. 906-910, Feb. 2017.

[7] L. Wang, D. Malcolm, W. Liu, and Y. F. Liu, Thermal analysis of a magnetic packaged power module, in 2016 IEEE Applied Power Electronics Conference and Exposition (APEC), 2016, pp. 2095-2101.

Parasitic Inductance Design Considerations to Suppress Gate Voltage Oscillation of Fast Switching Power Semiconductor Devices

Yusuke Sugihara[1], Kimihiro Nanamori[1], Masayoshi Yamamoto[2]and Yasuki Kanazawa[2*]

1 Interdisciplinary Faculty of Science and Engineering, Shimane University, Matsue, Japan
2 Institute of Materials and Systems for Sustainability, Nagoya University, Nagoya, Japan
*E-mail: kanazawa.yasuki@g.mbox.nagoya-u.ac.jp

Abstract— **Fast switching power semiconductor devices such as SiC MOSFET and GaN FET have attracting increasing attention because of superior switching capability which contributes to high efficiency and high power density of power converters. However, fast switching capability generates large switching noise. Particularly, parasitic oscillation observed in the gate voltage is a severe problem due to the gate-source breakdown. This paper hypothesized that the gate resonator selectively takes in parasitic oscillation caused by a power circuit because of its frequency characteristic and analyzed the gate voltage oscillation. As a result, the phenomenon is characterized by an oscillation susceptibility which indicates susceptibility of the gate voltage to a drain voltage oscillation. Moreover, the oscillation susceptibility predicts the parasitic inductance dependency of the gate voltage oscillation. Therefore, the parasitic inductance should be designed to suppress this oscillation considering the oscillation susceptibility. Results obtained by simulation and experiment indicated appropriateness of analysis results.**

Keywords— *Fast Switching, parasitic inductance, parasitic oscillation, SiC MOSFET.*

I. INTRODUCTION

Fast switching power semiconductor devices such as SiC MOSFET and GaN FET have attracting increasing attention because of high breakdown voltage, low on-resistance, superior temperature characteristics and fast switching capability. Their outstanding characteristics achieve high efficiency and high power density of power converters [1]-[4]. However, fast switching capability generates large switching noise, which causes malfunctions of power converters, e.g. a false turn-on and an oscillatory false triggering [5]-[9]. The former causes deterioration of power conversion efficiency and the latter leads to the breakdown of power devices due to an enormous switching loss. Previous research has been revealed that parasitic elements cause these phenomena and proposed that parasitic inductances should be designed to avoid them [7]-[9].

Particularly, the common source inductance of the wiring path shared by a power circuit and a gating circuit is critical problem in fast switching circuits [10], [11]. Therefore, [10] strongly recommends the use of power semiconductor devices with Kelvin source connection to avoid parasitic oscillation, i.e. the common source inductance should be eliminated as much as possible to achieve a stable switching operation.

Although Kelvin source connection eliminates the common source inductance, a cause of false triggering still remains. It is the capacitance between the gate and drain terminal. The gate-drain capacitance capacitively couples a power circuit onto a gating circuit, inducing a gate voltage oscillation due to high dv/dt switching as pointed out in [12]. In order to address this issue, [12] presented design guidelines of parasitic elements for gate-loop stability.

The gate voltage oscillation during the turn-on period is severe problem due to the gate-source breakdown as reported in this paper. In order to suppress the phenomenon, a large gate resistance is generally mounted. However it simultaneously deteriorates switching performance of fast switching power semiconductor devices [12]. Therefore, a design guideline for suppressing the gate voltage oscillation is required without a large gate resistance.

As mentioned before, [7] and [8] have been revealed that designing parasitic inductance prevents from false triggering. Additionally, although the common source inductance and the gate-drain capacitance cause false triggering of power converter, [9] achieves the avoidance of false triggering by optimizing the balance between them, i.e. parasitic inductance design is indispensable to achieve both fast switching and stable switching behavior. Therefore, this paper elucidates the relationship between the gate voltage oscillation and parasitic inductance and discusses parasitic inductance design considerations to suppress this oscillation based on analysis results.

This paper is organized as follows. Section II reports the gate voltage oscillation and hypothesizes that the phenomenon is caused by a parasitic oscillation generated in a power circuit. Section III analyzes an equivalent circuit model yielded by the hypothesis. In Section IV, analysis results are verified by simulation using LTspice and experiment. Section V discusses parasitic inductance design considerations to suppress the gate voltage oscillation. Section VI concludes the paper.

The 2018 International Power Electronics Conference

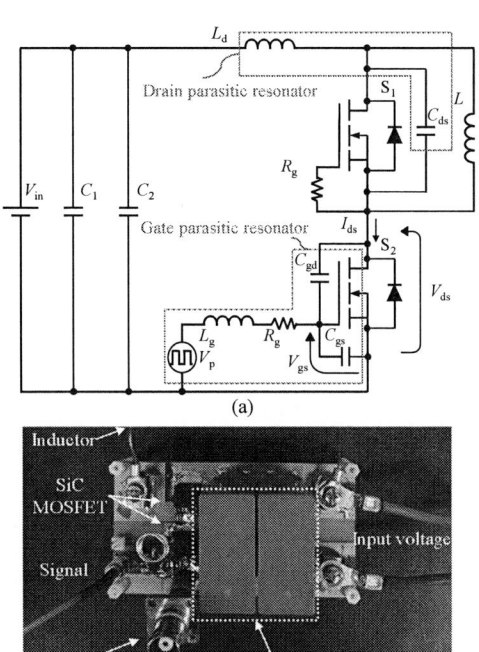

(a)

(b)

Fig. 1. Experimental circuit.
(a) Schematic of experimental circuit.
(b) Photograph of experimental circuit.

TABLE I
CIRCUIT SPECIFICATIONS

Symbol	Meaning	Value/Model number
V_{in}	Input voltage	400V
C_1	Smoothing capacitor	30µF
C_2	Decoupling capacitor	3µF
L	Inductor	34µH
V_p	PWM signal	0/15V
R_g	Gate resistance	2Ω
L_g	Gate inductance	15nH
L_d	Drain inductance	17nH
S_1, S_2	SiC-MOSFET	C3M0065090J

II. GATE VOLTAGE OSCILLATION

This section introduces an example of the gate voltage oscillation. Fig.1 shows the schematic and the photograph of the experimental circuit. This circuit includes SiC MOSFETs with Kelvin source connection. Table I shows circuit parameters. Fig.2 shows experimental waveforms. The gate voltage oscillation was observed as shown in Fig. 2(b). Although the gate voltage rating of SiC MOSFET used in the experiment is 19V, the peak gate voltage reached 24V. Therefore, this phenomenon is the severe problem because of the gate-source breakdown.

The frequency of gate voltage oscillation is 141MHz. Similarly, the drain voltage and current oscillate at the frequency of 138MHz. Hence, the frequency of gate voltage oscillation is almost equal to the resonance frequency of 141MHz which consists of the drain inductance (17nH) of wiring path and the drain-source capacitance (75pF) of S1. Therefore, this paper hypothesizes that the gate voltage oscillation is caused by

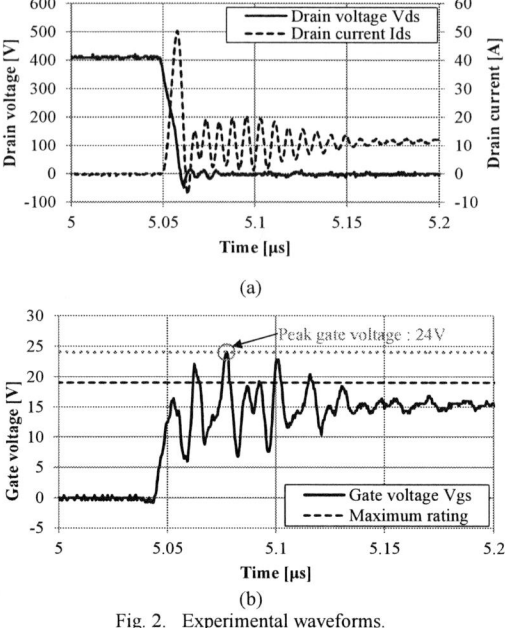

(a)

(b)

Fig. 2. Experimental waveforms.
(a) Drain voltage and current. (b) Gate voltage.

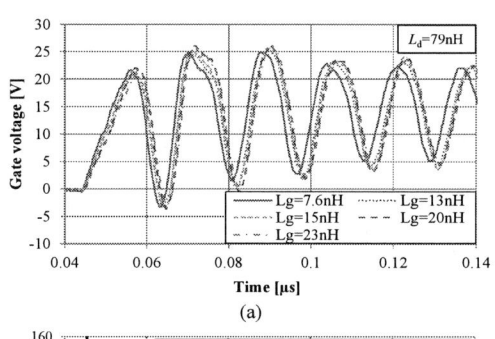

(a)

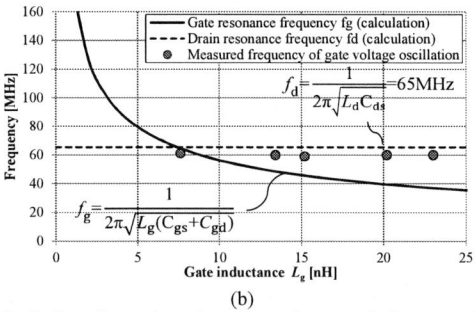

(b)

Fig. 3. Experimental results under various gate inductances L_g.
(a) Experimental waveforms of gate voltage oscillation
(b) Measured frequency of gate voltage oscillation

parasitic oscillation in a drain voltage or current, i.e. the gate resonator selectively takes in parasitic oscillation caused by a power circuit because it has a frequency characteristic.

Next, experiments are carried out to verify the hypothesis. Fig. 3 and 4 show experimental results under various values of the gate inductance L_g and the drain inductance L_d, respectively. Fig. 3(a) indicates that the gate inductance L_g mainly affects the peak value of gate

2790

voltage oscillation. Whereas the gate inductance increases, the frequency of gate voltage oscillation is unchanging as shown in Fig. 3(b), i.e. the measured frequency of gate voltage oscillation is almost constant at the drain resonance frequency f_d instead of the gate resonance frequency f_g. On the other hand, the drain inductance L_d has remarkable influence on the gate voltage oscillation as shown in Fig. 4(a). Furthermore, Fig. 4(b) shows that the frequency of gate voltage oscillation decreases as the drain inductance L_d increases, indicating that the gate voltage oscillation is caused by parasitic oscillation generated in the power circuit.

Because of the presence of the gate parasitic resonator, the gate voltage seems to oscillate at the gate resonance frequency f_g. However, it actually oscillates at the drain resonance frequency f_d. Above facts indicate that the parasitic gate resonator evidently takes in parasitic oscillation caused by the power circuit, which is consistent with the hypothesis. Therefore, the following sections analyze the gate voltage oscillation according to the hypothesis.

III. ANALYSIS OF GATE VOLTAGE OSCILLATION

This section presents an analysis of the gate voltage oscillation. Based on the aforementioned hypothesis, an equivalent circuit model shown in Fig. 5(a) is constructed to elucidate the gate voltage oscillation. The switching device in the equivalent circuit model consists of the gate-source capacitance C_{gs}, the drain-source capacitance C_{ds}, the gate-drain capacitance C_{gd} and the on-resistance R_{on}. The gate resistance R_g and the gate inductance L_g in the gate loop are divided by the semiconductor package, corresponding to the internal gate resistance R_{g_in}, the external gate resistance R_{g_ex}, the internal gate inductance L_{g_in} and the external gate inductance L_{g_ex} in the equivalent circuit model. Furthermore, the common source inductance can be ignored because of Kelvin source connection. Although the equivalent circuit model excludes the drain inductance L_d, L_d corresponds to the frequency of drain-source voltage v_{ds}.

In order to obtain a response of gate voltage to the drain-source voltage v_{ds}, the equivalent circuit model shown in Fig. 5(a) is transformed into the analysis model shown in Fig. 5(b). The impedance of parasitic gate resonator shown in Fig. 5 (b) can be expressed as

$$Z_g = \frac{1}{j\omega C_{gd}} + \frac{R_g + j\omega L_g}{(R_g + j\omega L_g)j\omega C_{gs} + 1}$$
(1)

where R_g and L_g are defined as

$$R_g = R_{g\_ex} + R_{g\_in} \qquad (2)$$
$$L_g = L_{g\_ex} + L_{g\_in}. \qquad (3)$$

The internal gate voltage v_{gs} can be expressed as

$$v_{gs} = \frac{R_g + j\omega L_g}{(R_g + j\omega L_g)j\omega C_{gs} + 1}i_{gd} = \frac{R_g + j\omega L_g}{(R_g + j\omega L_g)j\omega C_{gs} + 1} \cdot \frac{v_{ds}}{Z_g}.$$
(4)

By substituting (1) into (4), a response of gate voltage to the drain-source voltage v_{ds} can be expressed as

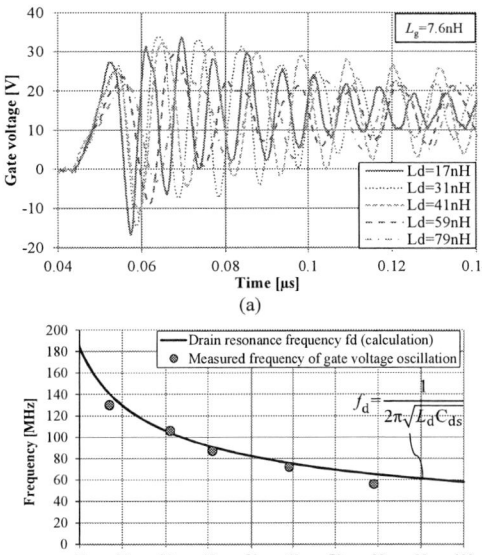

(a)

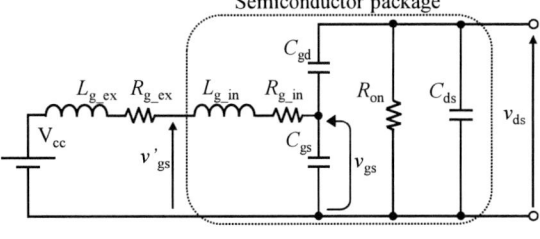

(b)

Fig. 4. Experimental results under various drain inductances L_d.
(a) Experimental waveforms of gate voltage oscillation.
(b) Measured frequency of gate voltage oscillation.

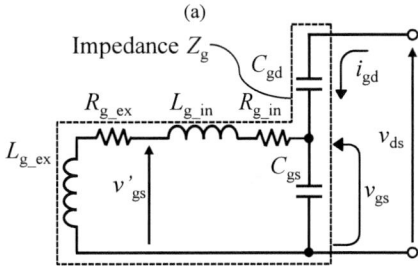

(a)

(b)

Fig. 5. Simple analytical model.
(a) Equivalent circuit model.
(b) Modified equivalent circuit model.

$$v_{gs} = |K_{os}(\omega)|v_{ds}. \qquad (5)$$

A coefficient $|K_{os}(\omega)|$ which indicates susceptibility of the internal gate voltage to a drain voltage oscillation is defined as

$$
K_{os}(\omega) = \frac{C_{gd}R_g^2 - \left[L_g A(\omega) - C_{gs}R_g^2\right]\left[\omega^2 C_{iss}C_{gs}R_g^2 + A(\omega)B(\omega)\right]}{\left[A(\omega)^2 + \left(\omega C_{gs}R_g\right)^2\right]\left[B(\omega)^2 + \left(\omega C_{iss}R_g\right)^2\right]}\omega^2 C_{gd} \\
+ j\omega C_{gd}R_g \frac{\omega^2\left[C_{gs}^2 R_g^2 + C_{gd}L_g A(\omega)\right] + A(\omega)B(\omega)}{\left[A(\omega)^2 + \left(\omega C_{gs}R_g\right)^2\right]\left[B(\omega)^2 + \left(\omega C_{iss}R_g\right)^2\right]}
$$
(6)

where C_{iss}, $A(\omega)$ and $B(\omega)$ are defined as

$$C_{iss} = C_{gs} + C_{gd} \qquad (7)$$
$$A(\omega) = 1 - \omega^2 C_{gs} L_g \qquad (8)$$
$$B(\omega) = 1 - \omega^2 C_{iss} L_g. \qquad (9)$$

Hereafter, the coefficient $|K_{os}(\omega)|$ is referred as to an oscillation susceptibility.

In experiment, a internal gate voltage v_{gs} cannot be observed because the measurement terminal of v_{gs} is hidden inside the semiconductor package. Therefore, an observable gate voltage practically needs to be derived. Similarly, the external gate voltage v'_{gs} can be expressed as

$$v'_{gs} = |K_c(\omega)| v_{gs} \qquad (10)$$

where a correction coefficient $K_c(\omega)$ is defined as

$$K_c(\omega) = \frac{R_g R_{e\_ex} + \omega^2 L_g L_{g\_ex}}{R_g^2 + \omega^2 L_g^2} + j\omega \frac{L_{g\_ex} R_{g\_in} - L_{g\_in} R_{g\_ex}}{R_g^2 + \omega^2 L_g^2}.$$
$$(11)$$

The internal gate voltage is transformed into the external gate voltage by this correction coefficient, i.e. a response of external gate voltage to the drain-source voltage v_{ds} can be expressed as

$$v'_{gs} = |K_{cos}(\omega)| v_{ds} = |K_c(\omega) K_{os}(\omega)| v_{ds}. \qquad (12)$$

Hereafter, the coefficient $|K_{cos}(\omega)|$ which indicates susceptibility of the external gate voltage to a drain voltage oscillation is referred as to an corrected oscillation susceptibility.

Next, simulation using PSIM is carried out to verify consistency of analytical equations. In the simulation, a sinusoidal voltage source simulates a drain-source voltage. Table II shows circuit parameters. Fig. 6 shows the frequency characteristics of each gate voltage calculated by (5) and (12). Simultaneously, simulation results are shown in the same figure. As can be seen in Fig. 6, simulation results agree well with the results acquired by the analytical model. Furthermore, the internal gate voltage v_{gs} is distributed between internal and external gate impedances, i.e. v_{gs} is more susceptible to a drain voltage oscillation than an observable gate voltage v'_{gs}. Therefore, the oscillation susceptibility $|K_{os}|$ of internal gate voltage v_{gs} should be designed to suppress the gate voltage oscillation because v_{gs} directly affects switching behavior.

Analysis results indicate that the gate voltage oscillation has frequency characteristic. The frequency which the gate voltages take a peak value is the same as the gate resonance frequency f_g consists of the input capacitance C_{iss} and the gate inductance L_g, i.e. coincidence of drain and gate resonance frequencies increases the peak value of gate voltage oscillation. Therefore, the drain inductance should be designed to avoid the peak region of $|K_{os}|$, i.e. parasitic inductance should be designed so that parasitic resonance frequencies are separated from each other. This design guideline is the same as the guideline to avoid an oscillatory false triggering [7]-[9]. Therefore, designing

TABLE II
CIRCUIT PARAMETERS

Symbol	Meaning	Value
v_{ds}	Drain-source voltage	30V
C_{gd}	Gate-drain capacitance	150pF
C_{gs}	Gate-source capacitance	650pF
R_{g_ex}, R_{g_in}	Gate resistance	1Ω
L_{g_ex}, L_{g_in}	Gate inductance	5nH

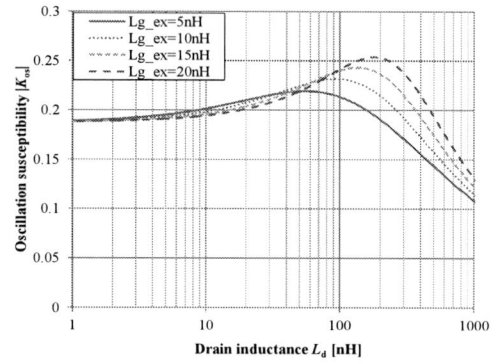

Fig. 6. Frequency characteristic of gate voltage.

TABLE III
ANALYSIS CONDITIONS

Symbol	Meaning	Value
C_{gd}	Gate-drain capacitance	150pF
C_{gs}	Gate-source capacitance	650pF
R_{g_ex}	External gate resistance	2Ω
R_{g_in}	Internal gate resistance	4.7Ω
L_{g_in}	Internal gate inductance	5nH

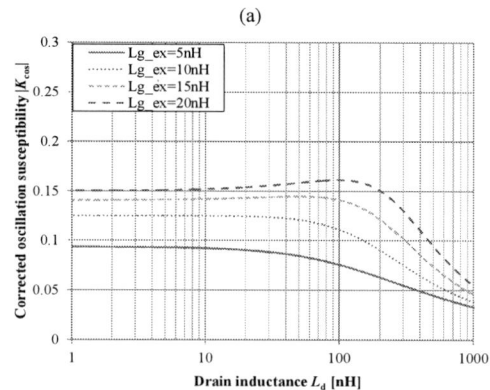

Fig. 7. Drain inductance dependencies of oscillation susceptibilities.
(a) Oscillation susceptibility
(b) Corrected oscillation susceptibility

2792

resonance frequencies is essential to achieve stable switching operation.

IV. MODEL VERIFICATION

In this section, simulation using LTspice and experiments are performed to verify the analysis results.

Table III shows analysis conditions. Parasitic capacitances and internal gate resistance can be extracted from the datasheet [13]. Although parasitic capacitance of switching device has drain-source voltage dependency, C_{gd} and C_{gs} are set as capacitances when the drain-source voltage is 0V. The internal gate inductance is provisionally set as 5nH because of its inaccessible. The drain resonance frequency f_d can be transformed into the drain inductance L_d according to the following equation

$$L_d = \frac{1}{\omega^2 C_{ds}} = \frac{1}{(2\pi f_d)^2 C_{ds}} \qquad (13)$$

where the drain-source capacitance C_{ds} of S_1 is set as capacitance when the drain-source voltage is 400V. Therefore, above equation elucidates the relationship between the drain inductance and the oscillation susceptibility.

As a result, the parasitic inductance dependency of oscillation susceptibility under the condition shown in Table III is calculated by (6) and (11). Fig. 7 shows the drain inductance dependency of oscillation susceptibility under the various external gate inductances L_{g_ex}, and Fig. 8 shows the external gate inductance dependency of oscillation susceptibility under the various drain inductances L_d. As can be seen in Fig. 7 and 8, due to the asymmetry between internal and external impedances, the tendency of the oscillation susceptibility $|K_{os}|$ is different from that of the corrected oscillation susceptibility $|K_{cos}|$. Therefore, analysis results indicate that a response to parasitic oscillation generated in a power circuit is different between internal and external gate voltages.

According to Fig. 7, although the oscillation susceptibility $|K_{os}|$ has the increase tendency, the corrected oscillation susceptibility $|K_{cos}|$ monotonically decreases as the drain inductance L_d increases. Therefore, the result shown in Fig.7 predicts that a measured peak voltage value of v'_{gs} has a decrease tendency. On the other hand, as can be seen in Fig. 8, $|K_{cos}|$ increases as the external gate inductance L_{g_ex} increases, i.e. this result predicts that the peak value of external gate voltage v'_{gs} observed in simulation and experiment monotonically increases.

Unlike analysis results shown in Fig. 6, analysis results shown in Fig. 7 and 8 are not intense because the quality factor of the gate resonator decreases due to the relatively large internal gate resistance of SiC MOSFETs. On the other hand, since GaN FETs have small parasitic capacitances and internal gate resistance, it can be expected that parasitic inductance design to suppress the gate voltage oscillation is more severe.

A. Verification Using LTspice

In this subsection, simulation using the spice model (C3M0065090J) is performed to verify the analysis results under the condition shown in Table IV. Fig. 9

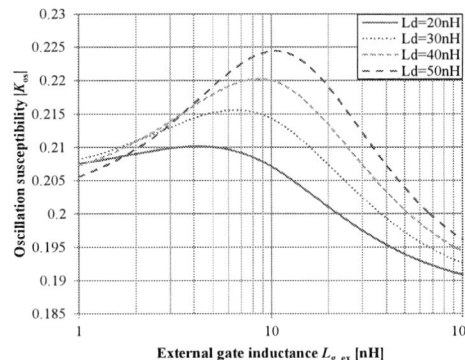

(a)

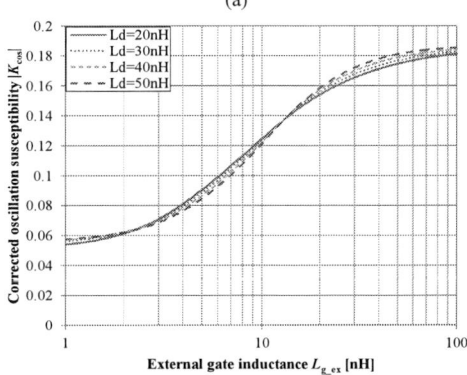

(b)

Fig. 8. External gate inductance dependency of oscillation susceptibility.
(a) Oscillation susceptibility.
(b) Corrected oscillation susceptibility.

TABLE IV
SIMULATION CONDITION

Symbol	Meaning	Value/Model number
V_{in}	Input voltage	400V
I_{ds}	Drain current	10A
V_p	PWM signal	0/15V
R	Gate resistance	2Ω
S_1, S_2	SiC-MOSFET	C3M0065090J

shows parasitic inductance dependencies of peak gate voltage value simulated by LTspice.

Fig. 9(a) shows peak gate voltage values under the various drain inductances L_d. As mentioned before, the corrected oscillation susceptibility $|K_{cos}|$ shown Fig. 7(b) monotonically decreases as the drain inductance L_d increases. In fact, the simulated peak gate voltage value also decreases because the internal gate voltage v_{gs} cannot be observed in the simulation. Similarly, Fig. 9(b) shows the external gate inductance dependency of peak gate voltage value. As predicted by analysis results, the simulated peak gate voltage value increases as the external gate inductance L_{g_ex} increases. Therefore, the tendency of simulated peak gate voltage value is consistent with prediction based on analysis results, indicating the validity of analytical model.

B. Experimental Verification

This subsection experimentally verifies the analysis results. The specification of experimental circuit is shown in Section II. In order to obtain drain and gate inductance dependencies, the variation of parasitic inductance was implemented by elongating each wiring path. As a result, experimental conditions of parasitic inductances are the gate inductance ranged from 7.6 to 23nH and the drain inductance ranged from 17nH to 79nH. In order to verify the external gate inductance dependency of peak gate voltage, experiments were carried out with the drain inductance fixed at 79nH. On the other hand, the external gate inductance L_{g_ex} is fixed at 7.6nH to experimentally investigate the drain inductance dependency of peak gate voltage. Fig. 10 shows parasitic inductance dependencies of measured peak gate voltage value and the oscillation susceptibilities calculated with parameters corresponding to experimental conditions.

Fig. 10(a) shows measured peak voltage values under the various drain inductances. As predicted by the corrected oscillation susceptibility $|K_{cos}|$, the measured peak voltage value decreases as the drain inductance increases. Fig. 10(b) shows measured peak gate voltage values under the various external gate inductances. Similarly, the measured peak voltage value increases as the external gate inductance increases. Therefore, the tendency of measured peak voltage value is also consistent with analytical prediction, indicating the validity of analysis results. Consequently, the next section discusses parasitic inductance design considerations based on analysis results.

V. PARASITIC INDUCTANCE DESIGN CONSIDERATIONS

A. Drain Inductance L_d

This subsection discusses drain inductance design consideration to suppress the gate voltage oscillation. According to analysis results shown in Fig. 7 and 8, the drain inductance should be designed as small as possible even if an apparent peak gate voltage is suppressed by a large drain inductance because the oscillation susceptibility $|K_{os}|$ has the increase tendency, and also because a large drain inductance intensifies the external gate inductance dependency of oscillation susceptibility. Therefore, the drain inductance should be designed considering the oscillation susceptibility $|K_{os}|$ of internal gate voltage v_{gs}.

B. Gate Inductance L_g

This subsection discusses gate inductance design consideration to suppress the gate voltage oscillation. According to analysis results shown in Fig. 8, although the corrected oscillation susceptibility $|K_{cos}|$ increase as the external gate inductance L_{g_ex} increases, the oscillation susceptibility of internal gate voltage decreases in accordance with increase of the external gate inductance L_{g_ex}. Therefore, the external gate inductance need not necessarily be minimized as long as it is designed to avoid the peak region of oscillation susceptibility.

As pointed out in [12], gate inductance should be designed as small as possible to improve the gate-loop stability. Indeed, the peak gate voltage deteriorates in accordance with increase of the external gate inductance L_{g_ex} in simulation and experiment as shown in this paper.

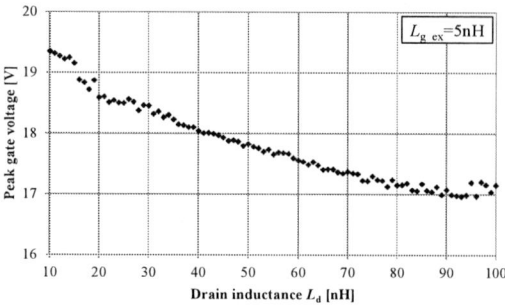

(a)

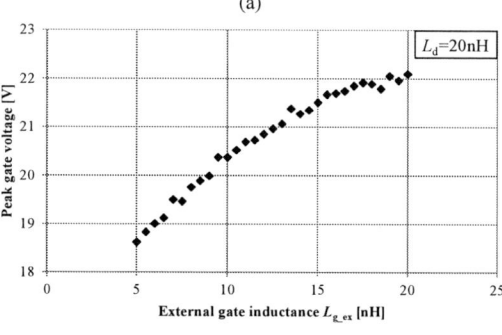

(b)

Fig. 9. Peak gate voltage values simulated by LTspice.
(a) Drain inductance dependency.
(b) External gate inductance dependency.

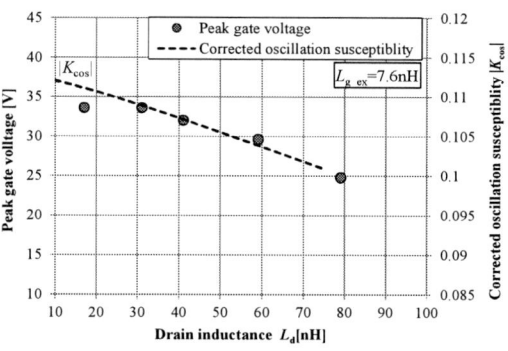

(a)

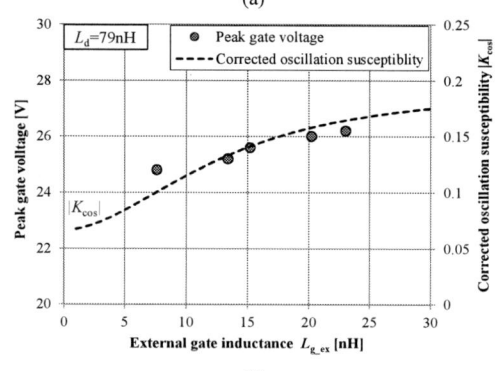

(b)

Fig. 10. Measured peak gate voltage values.
(a) Drain inductance dependency.
(b) External gate inductance dependency.

2794

However, this paper revealed that behavior of internal gate voltage is different from that of observable gate voltage. Therefore, the external gate inductance L_{g_ex} should be designed considering the oscillation susceptibility $|K_{os}|$ of internal gate voltage v_{gs} even if the apparent peak gate voltage increases.

VI. CONCLUSION

Fast switching power semiconductor devices such as SiC MOSFET and GaN FET with attracting features achieve high efficiency and high power density of power converters. However, their fast switching capability generates large switching noise. Particularly, the gate voltage oscillation is severe problem due to the gate-source breakdown. In order to address this issue, this paper hypothesized that the gate resonator selectively takes in parasitic oscillation caused by a power circuit. Based on the hypothesis, this paper analyzed the gate voltage oscillation and revealed that parasitic inductance should be designed to suppress this oscillation considering the oscillation susceptibility which indicates susceptibility of the gate voltage to a drain voltage oscillation. Results obtained by simulation and experiment indicated appropriateness of analysis results.

REFERENCES

[1] C. DiMarino, W. Zhang, N. Haryani, Q. Wang, R. Burgos, D. Boroyevich, "A high-density, high-efficiency 1.2 kV SiC MOSFET module and gate drive circuit," in IEEE 2016 4th Workshop on Wide Bandgap Power Devices and Applications (WiPDA), 2016, pp. 47–52.

[2] C. DiMarino, Z. Chen, M. Danilovic, D. Boroyevich, R. Burgos and P. Mattavelli, "High-temperature characterization and comparison of 1.2 kV SiC power MOSFETs," IEEE Energy Conversion Congress and Exposition (ECCE), pp. 3235–3242, Sept. 2013.

[3] M. A. Khan, G. Simin, S. G. Pytel A. Monti, E. Santi, and J. L. Hudgin, "New developments in gallium nitride and the impact on power electronics," in Proc. IEEE Power Electron. Specialist Conf. (PESC2005), 2005, pp.15-26.

[4] B. Wang, N. Tipirneni, M. Riva, A. Monti, G. Simin, and E. Santi, "An efficient high frequency drive circuit for GaN power HFETs," IEEE Trans. Ind. Appl., Vol. 45, No.2, pp.843-853, 2009.

[5] R. Xie, H. Wang, G. Tang, X. Yang and K. Chen, "An Analysis Model for False Turn-On Evaluation of High-Voltage Enhancement-Mode GaN Transistor in Bridge-Leg Configuration," IEEE Transaction on Power Electronics, vol. 32, issue 8, pp. 6416–6433, Aug. 2017.

[6] H. Ishibashi, H. Umegami, W. Martinez and M. Yamamoto, "An analysis of false turn-on mechanism on high-frequency power devices," IEEE Energy Conversion Congress and Exposition (ECCE), pp. 2247–2253, Sept. 2015.

[7] K. Umetani, K, Yagyu and E. Hiraki, "A design guideline of parasitic inductance for preventing oscillatory false triggering of fast switching GaN-FETs," IEEJ Trans. Elect. Electron Eng. Vol 11, no. S2, pp. S84-S90, Dec. 2016.

[8] Y. Sugihara, Y. Hayashi, K. Aikawa, K. Nanamori, S. Ishiwaki, K. Umetani, E. Hiraki and M. Yamamoto, "Analytical investigation on design instruction to avoid oscillatory false triggering of fast switching SiCMOSFETs," IEEE Energy Conversion Congress and Exposition (ECCE), pp. 5113–5118, Oct. 2017.

[9] R. Matsumoto, K. Umetani and E. Hiraki, "Optimization of the Balance between the Gate-Drain Capacitance and the Common Source Inductance for Preventing the Oscillatory False Triggering of Fast Switching GaN-FETs," IEEE Energy Conversion Congress and Exposition (ECCE), pp. 405–412, Oct. 2017.

[10] W. Zhang, Z. Zhang, F. Wang, D. Costinett, L. Tolbert, B. Blalock, "Common source inductance introduced self-turn-on in MOSFET turn-off transient," IEEE Applied Power Electronics Conference and Exposition (APEC), pp. 837-842, March, 2017.

[11] Z. Wang, J. Zhang, X. Wu, and K. Sheng, "Analysis of stray inductance's influence on SiC MOSFET switching performance," IEEE Energy Conversion Congress and Exposition (ECCE), pp. 2838–2843, Sep. 2014.

[12] X. Wang, Z. Zhao, Y. Zhu, K. Chen and L.Yuan, "A Comprehensive study on the gate-loop stability of the SiC-MOSFET," IEEE Energy Conversion Congress and Exposition (ECCE), pp. 3012–3018, Oct. 2017.

[13] C3M0065090J datasheet.

The Examination of Increasing Operation Speed of Consequent Pole Type Axial Gap Motor for Higher Output Power Density

Toru Ogawa[1*, 2], Tomohira Takahashi[3], Masatsugu Takemoto[2], Satoshi Ogasawara[2],
Hideaki Arita[3] and Akihiro Daikoku[1]
1 Advanced Technology R&D Center, Mitsubishi Electric Corp., Amagasaki, Japan
2 Graduate School of Information Science & Technology, Hokkaido Univ., Sapporo, Japan
3 Himeji Works, Mitsubishi Electric Corp., Himeji, Japan
*E-mail: Ogawa.Toru@dr.MitsubishiElectric.co.jp

Abstract— This paper presents an examination of increasing operation speed of consequent pole type axial gap motor to achieve higher output density. Our research group has been developing consequent pole type axial gap motor with field windings for traction motor of electric vehicles. From the point of fuel economy and layout, smaller and lighter traction motor is required. It is profitable to increase operation speed for reducing the size of motor. It is necessary to suppress line to line voltage to achieve high speed operation, and we examined pole-slot combination. Moreover, as rotor's outer diameter of axial gap motor is larger than that of radial gap motor, the strength of rotor should be considered. We studied the adoption of non-magnetic high tensile strength steel for rotor supporting component. Motor design and the result of spin burst test are presented and we confirmed the operation of target maximum speed and output power density can be possible.

Keywords— *Consequent-Pole, Ferrite PM, Axial Gap Motor, Variable Flux Motor*

I.   INTRODUCTION

The regulation against internal combustion engine vehicles is getting stricter. Electric vehicles (EVs) are getting attention as one of solutions, and a lot of car manufactures and component manufactures are developing traction motor [1-3]. There are a lot of properties required for traction motors for EVs. Variable flux capability and flat shape are two of them. The flat shape is a request from the point of a layout. The other is necessary to achieve a high torque at a low speed, a high efficiency under a low load and a wide constant output with one motor. Generally, a high magnetic flux is necessary for a high torque. On the other hand, if magnetic flux in motor is large, it is difficult to reduce iron loss at a low load. Moreover, a high induced voltage under a high speed operation make it difficult to have a constant output over a wide speed range. Consequently, there is a tradeoff relation between a high torque and a high efficiency or a wide constant output characteristics. To achieve variable flux capability, a lot of types of motors have been developed [4-7]. Our team proposed

and has been developing a consequent pole type ferrite PM axial gap motor with field windings for achieving variable flux capability and flat shape simultaneously [8-11]. There is also requirement of higher output density. A lot of research groups are investigating higher speed motor to increase output density [12, 13]. With regard to axial gap type motor which we are proposing, it is profitable to increase outer diameter of rotor and decrease motor length, therefore the radius and centrifugal force of axial gap motor is larger than those of radial gap type motor in general. So it is necessary to take care of large centrifugal force to increase the operation speed of axial gap type motor. Also it is necessary to suppress line to line voltage to increase operation speed. In this paper, the design to increase operation speed is examined and it is presented that new designed motor can achieve target output by 3D FEA analysis. To make the operation under the maximum speed possible, the material and structure of rotor is also presented. We confirmed the capability of the maximum speed operation with burst test of rotor. We assembled proto sample and test result will be presented.

II.   THE PROPOSED MOTOR

Figure 1 shows the structure of the proposed motor. The motor consists of internal rotor and external stators.

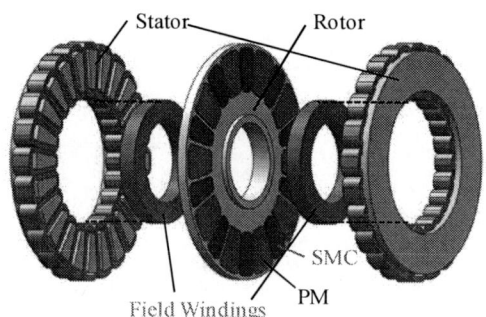

Fig. 1. Outline of the proposed motor

The rotor is composed of permanent magnets (PMs) and rotor cores (SMCs). The SMCs are made of magnetic material and form the consequent pole. All the PMs are magnetized in the same direction and the direction is axial direction. PMs and SMCs are arranged in circumferential direction alternately on the supporting component which is made of non-magnetic material. There is not back yoke on the rotor and magnetic flux go through only axially at the rotor. Two stators are placed on both side of the rotor. Concentrated windings are wound on each tooth. Concentrated winding is chosen for the sake of downsizing. Field windings are placed at the inner area of stator cores to utilize the dead space. DC current is applied to the both field winding to generate magnetic flux. The direction of DC current is the same. Figure 2 shows the cross section of the proposed motor. The arrows indicate the path of magnetic flux generated by DC field winding current (field current). Magnetic flux by field current go around the magnetic path which is formed by motor shaft, motor case, stator core, rotor, stator core and motor case on the other side.

Next, the principle, how the magnetic flux in the motor is altered, is presented. Figure 3 shows how magnetic flux by field current go through in the motor. When field current is not applied (Fig. 3(a)), magnetic flux in motor comes from only PMs. Magnetic flux from PMs goes around the magnetic circuit formed by tooth and back yoke of the stator core, rotor core and tooth and back yoke of the other stator. Consequently magnetic pole appear on SMC and the direction is opposite to that of PM. When field current is applied and the direction is to generate magnetic flux opposite to magnetic flux from PMs (Fig. 3(b)), magnetic flux is strengthened at the SMC. The fundamental component of magnetic flux density at air gap area becomes larger. The direction of this field current is regarded as strengthening direction. On the other hand, when the direction is to generate magnetic flux same direction as that from PM (Fig. 3(c)), magnetic flux is weakened at the SMC. The fundamental component of magnetic flux density becomes smaller. The direction is called weakening direction. As explained, the proposed motor has variable magnetic flux capability.

III. THE EXAMINATION FOR HIGHER OUTPUT DENSITY

Table 1 shows the comparison of target of motor and the condition of design between 1st type motor and new design. The target output density is 5.3 kW/L, which is increased by 1.5 times compared to 1st type motor. Higher output density is achieved by increasing the operation speed, from 5,000 r/min to 12,000 r/min. Armature current density and field current density are same or less than 1st type motor. 24 slots and 20 poles are chosen for 1st type motor. It is difficult to suppress line to line voltage with 20 poles when the maximum speed is increased to 12,000 r/min. To suppress induced line to line voltage, it is necessary to reduce pole number. Our group revealed that slot-pole combination of 12 slots and 10 poles is suitable for the motor which has

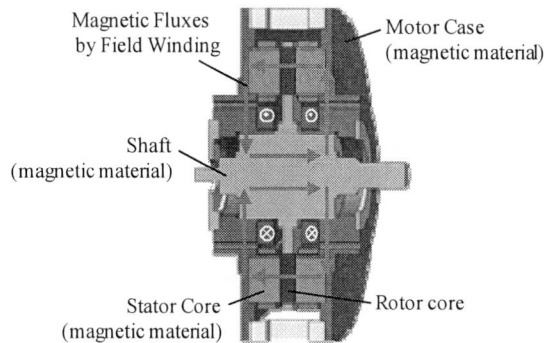

Fig. 2. Magnetic flux by field winding current

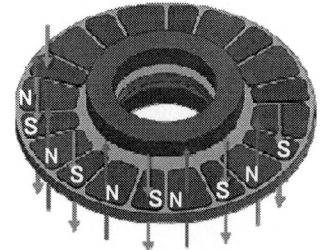

(a) Without field winding current

(b) With strengthening field current

(c) With weakening field current

Fig.3. Control of field flux by field current

consequent pole type rotor [9, 10]. Because winding factors of even harmonic orders are zero, the influence which comes from asymmetric distribution of air gap magnetic flux density can be canceled. As it is difficult to miniaturize the motor size with 10 poles, we investigated changing pole number from 20 to 14. Table 2 shows winding factors of 14 poles motors. The winding factors of 12 slots 10 poles motor are also described. Though the winding factor of fundamental order of 15 slots and 14 poles motor is the largest, winding factors of even orders

Table 1. Target and condition of studied motors

	1st type	new design
Maximum Armature current	74.8 Arms	74.8 Arms
Maximum armature current density	11.9 Arms/mm²	11.9 Arms/mm²
Maximum field current	5.5 A	4.3 A
Maximum field current density	6.95 A/mm²	5.5 A/mm²
Base speed	1,600 r/min	3,360 r/min
Maximum speed	5,000 r/min	12,000 r/min
output power density	3.5 kW/L	5.3 kW/L

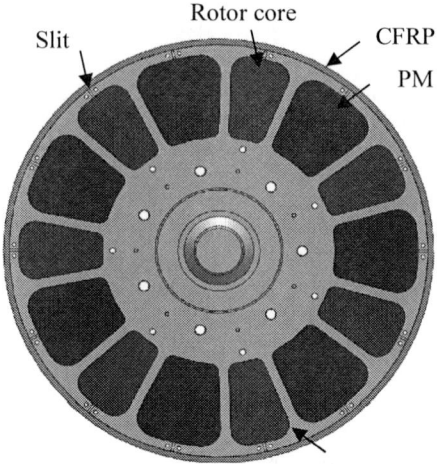

Fig. 4. The top view of the rotor

are not zero, and harmonic component of even order of rotor's magnetomotive force cannot be canceled. Winding factors of even orders of 12 slots and 18 slots motors are zero, so 12 slots or 18 slots are suitable for 14 poles motor. When magnetic flux which flows at each slot is compared, magnetic flux of 12 slots motor is larger. If the thickness of back yoke is optimized for 18 slots motor, magnetic flux density at back yoke of 12 slots motor is large and magnetic flux leaks to cover, which is made of magnetic material. Cover is not made of laminated material, and very large iron loss occurs under high speed operation. Consequently, we chose 18 slots motor, which is suitable to miniaturize the motor.

Next, the outer diameter of rotor is considered. Figure 4 shows the rotor structure of the proposed motor. PMs and rotor cores are glued to rotor supporting component which is made of non-magnetic material. There are slits at the outer area of the rotor supporting component to decrease the eddy current loss, which is generated in the rotor supporting component, by avoiding closing the circuit around the PMs and rotor cores. The rotor supporting component is reinforced by CFRP whose thickness is 3 mm. SUS 304 is used for rotor supporting component of 1st type motor. Mises stress of 1st type motor is 109.2 MPa under the maximum speed, 5,000 r/min. The yield stress of SUS304 is 206 MPa, and can be used. Mises stress is proportional to the square of rotation speed. Mises stress would be larger than 600 MPa under 12,000 r/min. It is obviously impossible to drive the

Table 2. Specification of studied motors

pole number	10	14		
slot number	12	12	15	18
fundamental	0.933	0.933	0.951	0.902
2nd order	0	0	0.100	0
4th order	0	0	0.100	0
5th order	0.067	0.067	0.173	0.038
7th order	0.067	0.067	0.111	0.136
8th order	0	0	0.165	0
10th order	0	0	0.100	0
11th order	0.933	0.171	0.044	0.136
13th order	0.933	0.210	0.021	0.038

motor up to 12,000 r/min without countermeasure. In general, Mises stress is proportional to the outer diameter of the rotor. To decrease Mises stress, the outer diameter is reduced from 250 mm to 210 mm. Moreover, we considered the material of rotor supporting component. There is special material from the Japan Steel Works, non-magnetic high-tensile steel which 18% of Manganese and 18% of Chromium is added (18%Mn18%Cr steel) [14]. The material is developed for retaining ring of electric generator which is required for high yield strength. The yield stress of the material is 1,260 MPa. Figure 5 shows the distribution of displacement of the rotor under the maximum operation speed. The displacement is emphasized by 30 times. The Mises stress under the maximum operation speed is 586.1 MPa, which is smaller than the yield stress of the material.

Figure 6 shows the cross section view of the rotor. The rotor supporting component consists of three plates and they are combined by using rivets. The part is added at the inner area of rotor to prevent the PMs and SMCs from lifting under high speed operation.

Table 3 shows the comparison of motor specifications between 1st type motor and new design. We confirmed motor characteristics using 3D FEA. Figure 7(a) shows waveforms of no load induced line to line voltage, Figure 7(b) shows harmonic components. Harmonic components

Fig. 5. Displacement of the rotor under the maximum rotational speed (new design)

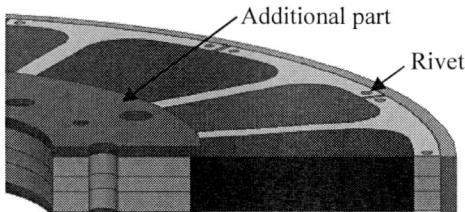

Fig. 6. Cross section view of the rotor

Table 3. Specification of studied motors

	1st type	new design
Pole number	20	14
Slot number	24	18
Outer diameter (coil end included)	250 mm	210 mm
Total motor axial length	76.4 mm	77.8 mm
Air gap	1.0 mm	1.0 mm
Maximum armature current density	11.9 Arms/mm²	11.9 Arms/mm²
Maximum field current density	6.95 A/mm²	5.5 A/mm²
Thickness of PM	13 mm	11 mm

of even orders are small. Total harmonic distortions (THD) of waveforms under without field current condition, the maximum strengthening field condition and the maximum weakening field condition are 0.7, 1.6 and 4.2 respectively and very small. Figure 8 shows no

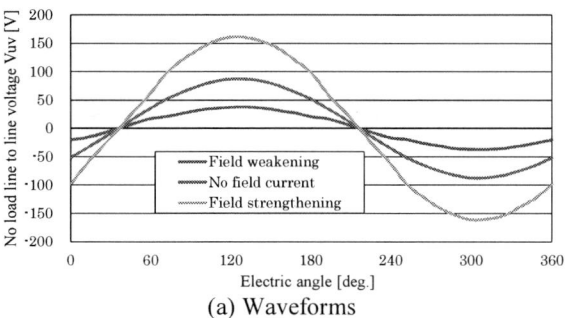

(a) Waveforms

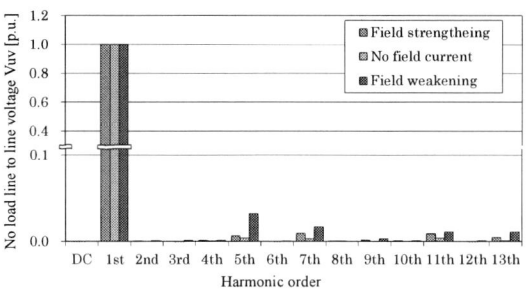

(b) Harmonic components

Fig. 7. No load induced line to line voltage

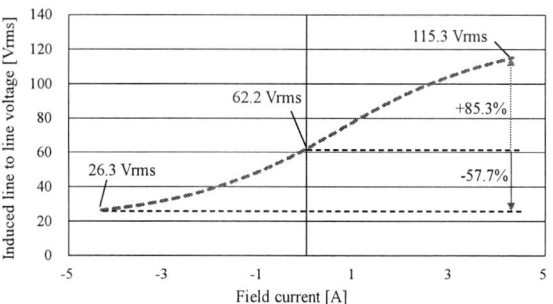

Fig. 8. No load line to line voltage vs. field current

load induced line to line voltage vs. field winding current. No load induced line to line voltage is increased 85.3% under the maximum field strengthening current than that without field winding current, and is decreased 57.7% under the maximum field weakening current. Fig. 9 shows the output torque calculated with changing field current from the maximum field weakening current to the maximum field strengthening current. Armature current is set to the maximum value, 74.8 Arms. When field current is not applied, average torque is 22.6 Nm. By applying the maximum strengthening current, +4.3 A, torque is increased by 81.9% compared to the torque without field current. Torque ripple is 11.9% under the maximum field strengthening condition. On the other hand, torque with the maximum field weakening current is reduced by 65.2%. These results indicate magnetic flux can be controlled widely with changing field current as expected.

Figure 10 shows motor torque speed characteristic. Torque at the base speed, 3,360 r/min, is 41.7 Nm and output power is 14.7 kW. Output power density is 5.4 kW/L. Consequently, new designed motor can achieve the target output power density. Torque at the maximum speed, 12,000 r/min, is 11.9 Nm and output power is 14.9 kW. It is confirmed that motor can be driven at constant output power over wide rotation speed, from 3,360 r/min to 12,000 r/min. It was confirmed that the proposed motor can achieve higher output density by increasing operation speed with using non-magnetic high tensile stainless steel for the rotor supporting component and

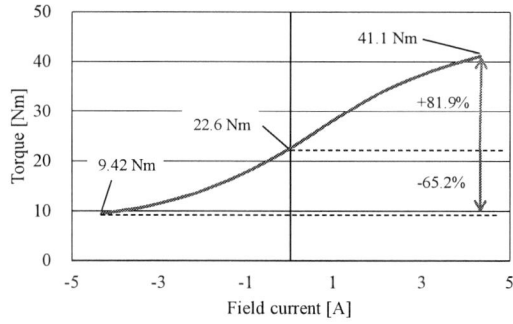

Fig. 9. Field current-torque characteristic

2799

The 2018 International Power Electronics Conference

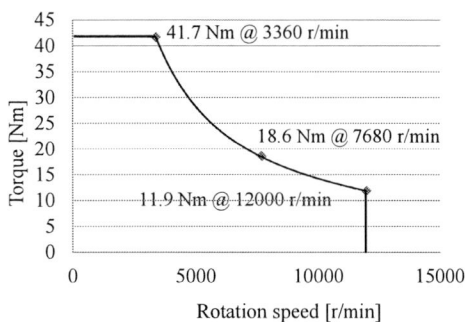

Fig. 10. Speed-torque characteristic

CFRP for reinforcement.

IV. EXPERIMENTAL RESULT

A. High speed burst test

To confirm the strength of the rotor, we performed high speed burst test. We used parts made of SUS304 in place of actual PMs and SMCs, as specific gravities are almost same. Figure 11 shows overview of high speed burst test. Runout of shaft is measured with induced displacement sensor at near the fixing point. Figure 12 shows the runout of shaft vs. rotational speed. Test was performed without CFRP. The rotational speed where the runout inclined rapidly indicates the rotation speed when

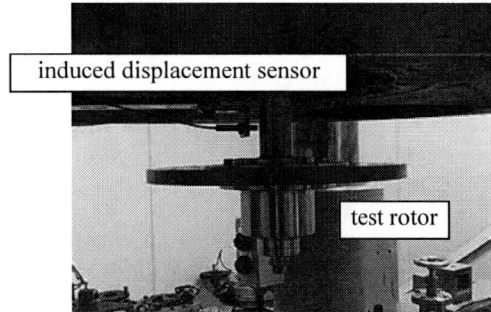

Fig. 11. Overview of high speed burst test

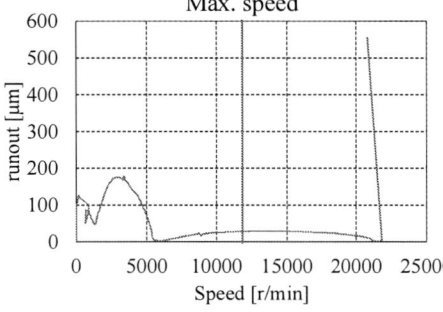

Fig. 12. Rotational speed and runout of the shaft

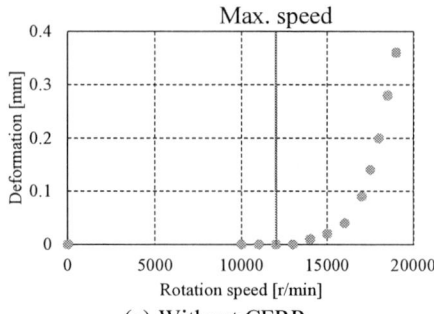

(a) Without CFRP

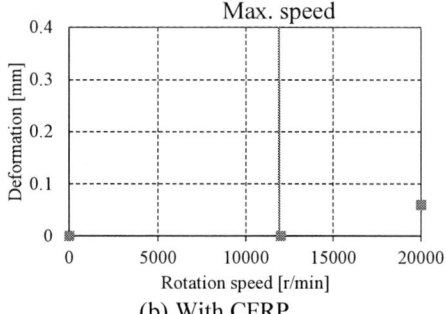

(b) With CFRP

Fig. 13. Deformation of the rotor with respect to rotation speed

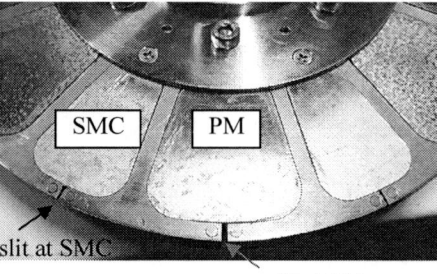

Fig. 14. Deformation of rotor supporting component around slits area

the rotor was burst. The runout inclined rapidly when the rotational speed is around 21,800 r/min.

Figure 13 shows the deformation of rotor supporting component with respect to rotation speed. The outer diameter is measured after increasing rotation speed. Figure 13(a) shows the result measured without CFRP. Even without CFRP, no irreversible deformation occurred at the maximum operation speed, 12,000 r/min and irreversible deformation occurred after 14,000 r/min. Figure 13(b) shows the deformation measured with CFRP. By adding CFRP, the deformation of the rotor can be suppressed by one forth compared to that without CFRP. Figure 14 shows test rotor after rotating 19,000 r/min. tested without CFRP. The mass of PMs is larger than that of SMCs, centrifugal force on PMs is larger than that on SMCs. Slit at PM becomes wider and slit at SMC narrower.

B. Characteristic test

Figure 15 shows photos of proto type machine. It was assembled based on the specification of table 2. Characteristic test is under preparation and test result will be reported.

V. CONCLUSIONS

In this paper, the examination of increasing operation speed for higher output density was presented. The special non-magnetic high tensile stainless steel is applied to rotor supporting component to achieve the maximum operation speed of 12,000 r/min. It was confirmed that the proposed motor can be operated at 12,000 r/min by burst test of the rotor. We designed the motor with 3D-FEA and also confirmed that the proposed motor achieved the maximum speed of 12,000 r/min and more than 1.4 times larger output power density, 5.4 kW/L. The experiment results of proto type machine will be reported.

Fig. 15. Photo of proto type machine

REFERENCES

[1] F. Momen, K. Rahman, and Y. Son, P. Savagian, "Electrical Propulsion System design of Chevrolet Bolt battery electric vehicle," *ECCE 2016*, pp. 1-8, 2016.

[2] T. Kato, N. Limsuwan, C.-Y. Yu, K. Akatsu and R. D. Lorenz, "Rare Earth Reduction Using a Novel Variable Magnetomotive Force Flux-Intensified IPM Machine," *IEEE Trans. on IAS*, vol. 50, no. 3, pp. 1748-1756, 2014.

[3] M. M. Swamy, T. Kume, and A. Maemura, "Extended High-Speed Operation via Electronic Winding-Change Method for AC Motors," *IEEE Trans. on Industry Applications*, vol. 42, no. 3, pp. 742-752, 2006.

[4] I. Ozawa, T. Kosaka, N. Matsui, "Less Rare-Earth Magnet-High Power Density Hybrid Excitation Motor Designed for Hybrid Electric Vehicle Drives," Proc. European Conf. on Power Electronics and Appl., Barcelona, No. 772, 2009.

[5] H. Hijikata, K. Akatsu, Y. Miyama, H. Arita and A. Daikoku, "Matrix Motor with Individual Winding Current Control Capability for Variable Parameters and Iron Loss Suppression," in Proc. Of XXIst International Conference on Electrical Machine (ICEM) 2014, pp.546-552, Sep. 2014.

[6] J. A. Tapia, F. Leonardi, T. Lipo, "Consequent-Pole Permanent-Magnet Machine With Extended Field-Weakening Capability", *IEEE Trans. on Industry Applications*, vol. 39, no. 6, pp. 1704-1709, 2003.

[7] F. G. Capponi, G. D. Donato, G. Borocci, F. Caricchi, "Axial-Flux Hybrid-Excitation Synchronous Machine: Analysis, Design, and Experimental Evaluation", *IEEE Trans. on Industry Applications*, vol. 50, issue 5, pp. 3173-3184, 2014.

[8] T. Takahashi, M. Takemoto, O. Satoshi, T. Ogawa, H. Arita and A. Daikoku, "Operation Characteristics of a Consequent Pole PM type Axial-Gap Motor with DC Field Winding," The Papers of Joint Technical Meeting on "Motor Drive", "Rotating Machinery", "Vehicle Technology" IEE Japan, MD-15-088, RM-15-069, VT-15-016, pp. 81-86, 2015 (in Japanese).

[9] T. Ogawa, H. Arita, M. Nakano, A. Daikoku, T. Takahashi, M. Takemoto and S. Ogasawara, "The Examination of Pole Number Combination of Consequent Pole Type Ferrite PM Axial Gap Motor with Field Winding," The Papers of Technical Meeting on "Rotating Machinery" IEE Japan, RM-15-163, pp. 35-40, Oct. 2015 (in Japanese).

[10] T. Ogawa, T. Takahashi, M. Takemoto, H. Arita, A. Daikoku, and S. Ogasawara, "The Consequent-Pole Type Ferrite Magnet Axial Gap Motor with Field Winding for traction motor used in EV", EVTeC & APE 2016, 6 pages, May, 2016

[11] T. Ogawa, T. Takahashi, M. Takemoto, S. Ogasawara, and A. Daikoku, "The Examination of Pole Geometry of Consequent Pole Type Ferrite PM Axial Gap Motor with Field Winding", IEMDC 2017, May, 2017

[12] T. Yashiro, S. Sano, K. Takizawa, and T. Mizutani, "Development of New Motor for Compact-Class Hybrid Vehicle", EVTeC & APE 2016, 6 pages, May, 2016

[13] K. Ueta, and K. Akatsu, "Study of high-speed SRM with Amorphous steel sheet for EV", ICEMS 2016, Nov. 2016

[14] K. Orita, Y. Ikeda, T. Iwadate, and J. Ishizaka, "Development and Production of 18Mn-18Cr Non-magnetic Retaining Ring with High Yield Strength", ISIJ International, Vol. 30, No. 8, pp. 587-593, 1990

The 2018 International Power Electronics Conference

Basic Study of PMASynRM
with Bonded Magnets for Traction Applications

Marika Kobayashi, Shigeo Morimoto, Masayuki Sanada, and Yukinori Inoue
Osaka Prefecture University, Sakai, Osaka, Japan
E-mail: mb104032@edu.osakafu-u.ac.jp

Abstract— Permanent magnet (PM) synchronous motors for automotive applications commonly use sintered rare-earth magnets in order to produce high power density and high efficiency. However, such magnets have the disadvantages of high cost and instability of the raw materials, which should be reduced. As one possible solution to this problem, this paper examined PM-assisted synchronous reluctance motors (PMASynRMs) with Dy-free bonded magnets. By using the high design flexibility of bonded magnets, the maximum torque and power of the proposed PMASynRM were higher than those of the conventional interior PM synchronous motor with sintered rare-earth magnets.

Keywords— Automotive electric motor, Bonded rare-earth magnet, Permanent magnet (PM) synchronous reluctance motor (PMASynRM).

I. INTRODUCTION

Permanent magnet (PM) synchronous motors (PMSMs) are currently used in various applications, and further performance improvement is required in order to address environmental problems [1]-[3]. In particular, electric motors for automotive applications, such as hybrid vehicles (HVs) and electric vehicles (EVs), require high torque density, high power density, and high efficiency over a wide speed range. In order to achieve such performance, sintered rare-earth magnets (i.e., NdFeB magnets), which have high maximum energy, are commonly used in PMSMs. However, there are several disadvantages associated with the use of such magnets. First, the cost of sintered rare-earth magnets is extremely high because they contain heavy rare-earth materials, such as Dysprosium (Dy), so as to increase the coercivity. Second, there are concerns about the stability of the supply of Dy. Therefore, it is highly desirable to reduce the amount of Dy [4]-[6]. As a possible solution to this problem, the adoption of bonded rare-earth magnets in PMSMs for several applications has been examined in recent years [7]-[11]. Bonded magnets do not contain Dy. As such, bonded magnets are relatively inexpensive in comparison to sintered magnets. Moreover, they have high design flexibility, which allows motors to yield higher reluctance torque, although the magnetic force of bonded magnets are much lower than that of sintered magnets.

This paper proposed a PM-assisted synchronous reluctance motor (PMASynRM) with bonded magnets for automotive applications, where the structure was optimally designed based on the 2-D finite-element analysis taking into account the mechanical strength. The purpose of this study is to develop a PMASynRM with bonded magnets having the performance equivalent to the conventional interior PM synchronous motor (IPMSM) with sintered magnets under the same motor size. This paper investigates the power and efficiency characteristics of the proposed PMASynRM with bonded magnets and discusses their applicability to traction motors.

II. SPECIFICATIONS AND DESIGN OF THE PROPOSED PMASynRM WITH BONDED MAGNETS

A. Targets

The target motor of this paper is an IPMSM with sintered magnets (NdFeB magnets) installed in fourth-generation Prius (Toyota 2015). The maximum torque and power of this target IPMSM are 163 Nm and 53 kW, respectively [12]. In this basic study, the target values were set to be approximately 10% higher than those of the target IPMSM, in consideration of the difference between the analysis value and the actual value. Thus, the target maximum torque and power are 179 Nm and 58 kW, respectively.

B. First Model

Fig. 1 shows the structure of the PMASynRM with bonded magnets, which is the first analysis model in this paper, and Table I shows the common motor specifications. The stator diameter and stack length are the same as those of the target IPMSM. The remanence of bonded magnets is approximately half of that of the sintered magnets contained in the target IPMSM. All of the analysis models in this paper have the same stator structure as the target IPMSM, which has eight poles and distributed windings with 48 slots.

2802

The 2018 International Power Electronics Conference

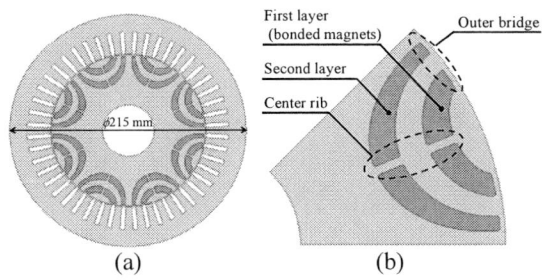

(a)   (b)

Fig. 1. First analysis model. (a) Cross section. (b) Rotor structure (single pole).

TABLE I
COMMON SPECIFICATIONS OF THE PROPOSED
PMASYNRM

Item (Unit)		Value
Number of pole/slot		8/48
Stator diameter (mm)		215
Shaft diameter (mm)		47.3
Air gap length (mm)		0.70
Stack length (mm)		59.5
Number of turns per slot		64
Winding resistance (Ω)※		0.070
Maximum phase current (A)		180
Maximum terminal voltage (V)		468
Bonded rare earth magnets※	Remanence (T)	0.59
	Coercivity (kA/m)	380

※Under condition that temperature is 100°C

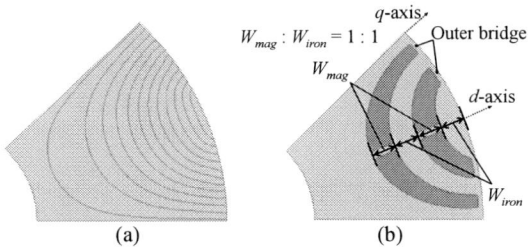

(a)   (b)

Fig. 2. Design concepts of the first rotor structure. (a) Flux lines produced by armature windings. (b) Shape of the bonded magnets.

The rotor structure is designed based on a motor proposed in Reference [11]. The double-layer reverse-arc-shaped magnets were designed based on the characteristics of bonded magnets in order to yield higher reluctance torque. They were designed along magnetic flux lines produced by armature windings on a homogeneous iron rotor, as shown in Fig. 2(a). The length ratio of the iron layer to the magnet layer is 1:1, as shown in Fig. 2(b).

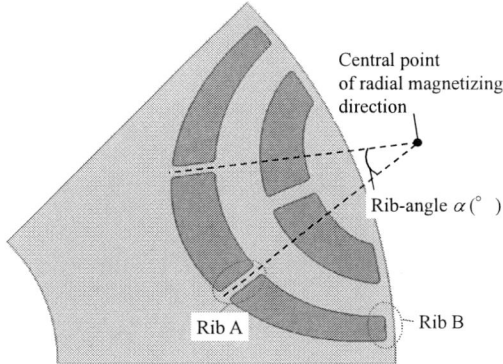

Fig. 3. Two ribs per pole installed in the second magnet layer of the first model.

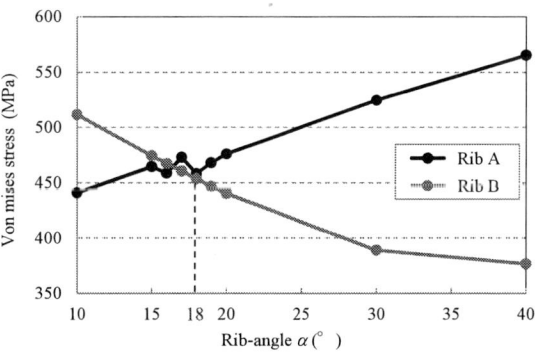

Fig. 4. Von Mises stresses plotted with respect to the rib-angle of Ribs A and B.

C. Rotor Design of the Proposed PMASynRM Considering Mechanical Strength

In this section, the mechanical strength of the rotor was examined using the strength analysis program. The first model had a problem with respect to the mechanical strength because the stress was concentrated on the outer bridge and the center rib of the second-layer magnet. In order to disperse the stress, the installation of two ribs per pole in the second-layer magnet was examined. As shown in Fig. 3, the rib-angle α was defined as the opening angle from the central point of the radial magnetizing direction.

Fig. 4 shows the maximum von Mises stresses of Ribs A and B on the second-layer magnet (see the areas indicated by red ovals in Fig. 3) plotted with respect to α. The von Mises stresses of Ribs A and B were approximately the same, where α is 15 to 20°. Hence, the proposed PMASynRM was obtained by installing two ribs per pole in the second-layer magnet, where α was set to 18°, and changing the width of the outer bridges and the center ribs of the first model. Fig. 5 shows the distributions of von Mises stresses in the first model and the proposed model at the maximum speed of 17,000 min⁻¹. The maximum value of the von Mises stress of the

2803

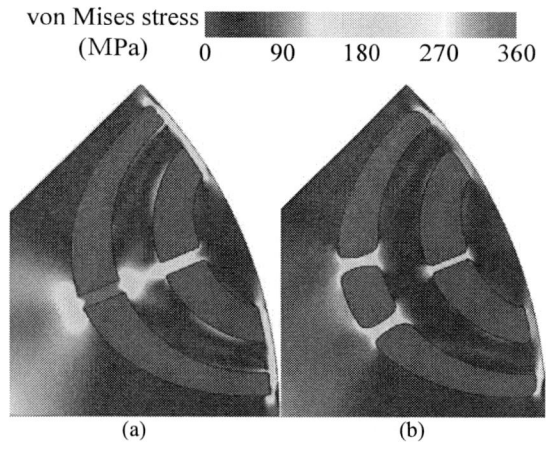

(a)   (b)

Fig. 5. Distribution of von Mises stresses at 17,000 min^{-1}.
(a) First model. (b) Proposed model ($\alpha = 18°$).

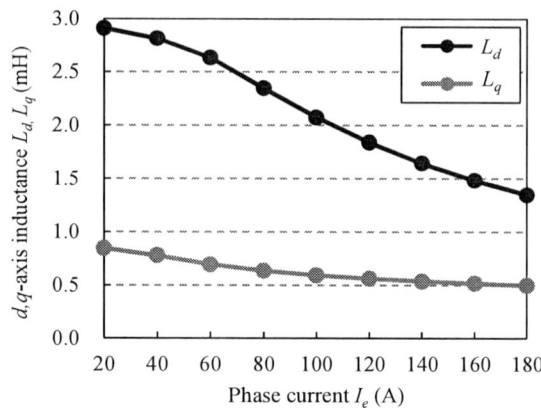

Fig. 6. Inductances plotted with respect to phase current ($\beta = 55°$).

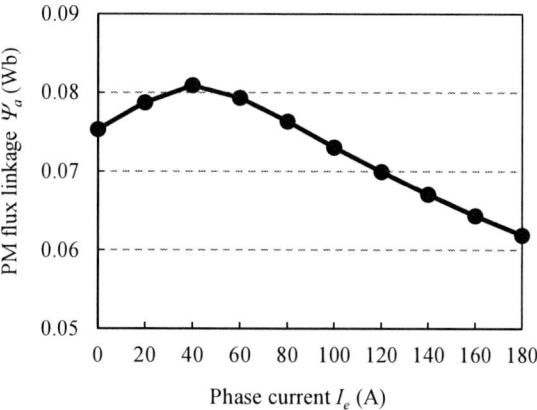

Fig. 7. Permanent magnet flux linkage plotted with respect to phase current ($\beta = 0°$)

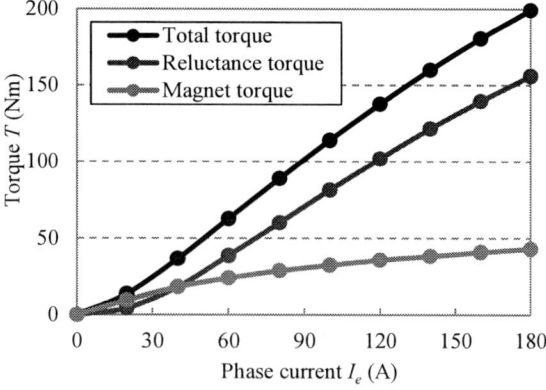

Fig. 8. Torque plotted with respect to phase current under maximum torque per ampere (MTPA) control.

proposed model decreases to 45.2% of that of the first model. In addition, the maximum torque of the proposed model became 98.3% of the first model at the maximum phase current of 180 A. The proposed model was proven to withstand the centrifugal force at the maximum speed of 17,000 min^{-1}.

III. PERFORMANCE EVALUATION OF THE PROPOSED PMASYnRM

A. Motor Parameters

Fig. 6 shows the inductances plotted with respect to the phase current I_e, where L_d and L_q are the d- and q-axis inductances, respectively, and the current phase angle β, which represents leading angle of current vector from the q-axis, was set to 55°. Both inductances tend to decrease as I_e increases, and the rate of decrease of L_q is high because of the higher influence of magnetic saturation on the wider q-axis magnetic path.

Fig. 7 shows the PM flux linkage Ψ_a plotted with respect to the phase current, where Ψ_a was calculated for $\beta = 0°$. The permanent magnet flux leaks through the outer bridges and causes the reduction in Ψ_a. However, the increase of magnetic flux produced by armature windings causes magnetic saturation in the outer bridges, and thus Ψ_a increases as I_e increases in the region of $I_e <$ 40 A. On the other hand, Ψ_a decreases monotonically above 40 A due to magnetic saturation on the entire iron core.

These results revealed that the rotor structure of the proposed PMASynRM easily causes magnetic saturation, which is similar to IPMSMs with sintered magnets.

2804

TABLE II
TORQUE CHARACTERISTICS UNDER MTPA CONTROL

Item (Unit)	Value		
I_e (A)	60	120	180
β (°)	43	52	56
T_m (Nm)	24.1	35.8	43.2
T_r (Nm)	38.7	101.8	155.9
T (Nm)	62.8	137.6	199.0
L_d (mH)	0.67	0.55	0.50
L_q (mH)	2.46	1.77	1.37
$L_q - L_d$ (mH)	1.80	1.21	0.87

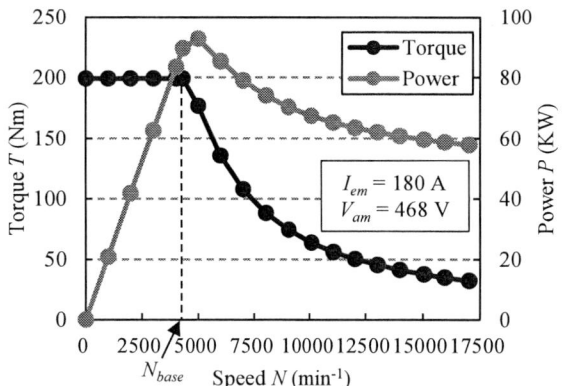

Fig. 9. Torque and power plotted with respect to speed.

TABLE III
TORQUE AND POWER VERSUS SPEED

Speed N (min^{-1})	Torque T (Nm)	Power P (kW)
4298 (N_{base})	199	89.6
5000	176.7	92.7 (P_{max})
17000	32.5	57.8

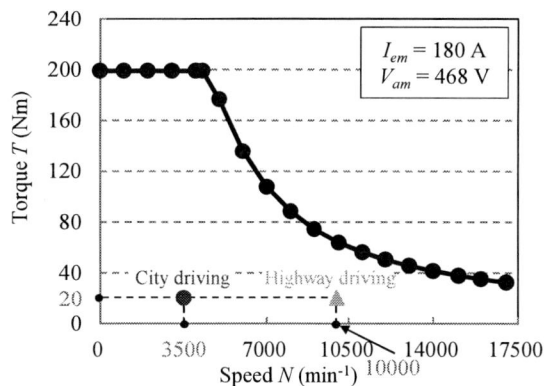

Fig. 10. City and highway driving evaluation points.

B. Torque Characteristics under MTPA Control

Fig. 8 shows the torque plotted with respect to the phase current under maximum torque per ampere (MTPA) control, and Table II shows the torque characteristics under MTPA control, where I_e was set to 60 A, 120 A, and 180 A. The reluctance torque, T_r, increases approximately monotonically against the theory that T_r is proportional to square of I_e. In addition, the magnet torque T_m is approximately constant in the entire I_e region. Such deviation from theory is due to magnetic saturation on the iron core, as mentioned in Section III A. The maximum torque of 199 Nm is achieved at the maximum phase current of 180 A, and the ratio of the reluctance torque to total torque is approximately 78.3%. This proves that the proposed PMASynRM with bonded magnets uses primarily reluctance torque, especially in the region of high I_e.

C. Torque and Output Power versus Speed Characteristics

Fig. 9 shows torque and power plotted with respect to speed, and Table III shows the torque and power at several speeds, where the maximum phase current I_{em} is 180 A, and the maximum terminal voltage V_{am} is 468 V. In Table III, N_{base} is the base speed, and P_{max} is the maximum power.

In the speed region below the base speed N_{base} (under MTPA control), the maximum torque of 199 Nm is higher than the target value by approximately 11.2%. In addition, at a speed of 5,000 min^{-1}, the maximum power P_{max} of 92.7 kW is obtained, which is higher than the target value by approximately 50.9%.

The maximum torque per voltage (MTPV) control is applied above the base speed, and the power decreases as the speed increases. The power at maximum speed decreases by 37.6% compared to P_{max}.

D. Loss and Efficiency Characteristics at Evaluation Points

There are several driving conditions in the actual operation of vehicles. As such, conditions for city driving (3,500 min^{-1}, 20 Nm) and highway driving (10,000 min^{-1}, 20 Nm) were defined as the evaluation points, as shown

in Fig. 10. Fig. 11 shows the loss and efficiency characteristics at both evaluation points under maximum efficiency control, and Table IV shows the current and voltage. In Fig. 10 and Table IV, V_a represents the terminal voltage, and maximum voltage V_{am} was set to 468 V.

Fig. 11 shows that the efficiencies at both evaluation points are higher than 97%, and the proposed PMASynRM achieves high-efficiency performance. At the highway driving evaluation point, the iron loss is higher due to the increase of speed. However, the use of bonded magnets, which have weak magnetic force, leads to low magnetic flux density on the entire iron core, resulting in low iron loss. Accordingly, the proposed PMASynRM produces the highest efficiency of 98.4% at the highway driving evaluation point.

2805

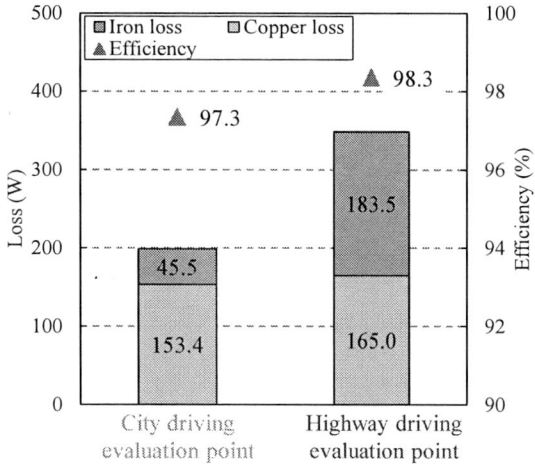

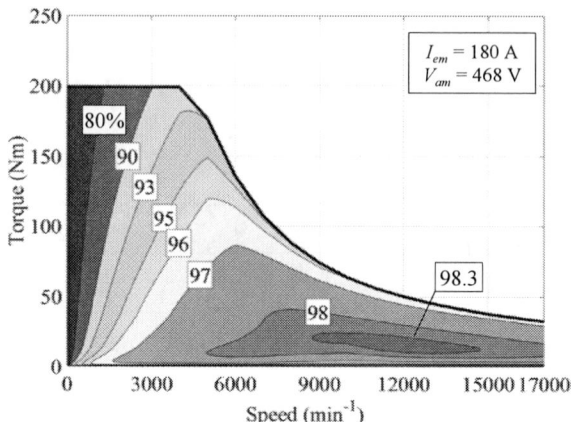

Fig. 12. Efficiency map.

Fig. 11. Loss and efficiency characteristics at the evaluation points under maximum efficiency control.

TABLE IV
CURRENT AND VOLTAGE AT EVALUATION POINTS

Item (Unit)	City driving	Highway driving
I_e (A)	27.0	28.0
β (°)	43.6	48.6
V_a (V)	167.9	445.7

E. Efficiency Map

An efficiency map of the proposed PMASynRM is shown in Fig. 12, where the maximum phase current I_{em} is 180 A, and the maximum terminal voltage V_{am} is 468 V. The efficiency tends to increase as the speed increases, and the efficiency of greater than 90% was obtained across a wide operating range. Moreover, a region of 98.3% efficiency was achieved in the high-speed region of 9,000 to 14,000 min^{-1}. These results revealed that the proposed PMASynRM could produce the high-efficiency driving performance, especially in the high-speed region.

IV. CONCLUSIONS

This paper proposed a PMASynRM with bonded magnets for automotive applications and investigated their performance using 2-D FEM. By installing two ribs per pole in the second-layer magnets, the proposed model was designed to withstand the centrifugal force at a maximum speed of 17,000 min^{-1}. The proposed PMASynRM generates high reluctance torque owing to the adoption of double-layer reversed-arc-shaped magnets. Hence, the maximum torque of the proposed PMASynRM was 11.2% higher than the target value, which was set to be 10% greater than that of the IPMSM with sintered magnets installed in the 2015 Toyota Prius.

Moreover, the maximum power becomes 50.9% larger than the target value.

In addition, the high efficiency of more than 97% was obtained at the two evaluation points considered herein and was highest at the highway driving evaluation point. The use of bonded magnets, which have weak magnetic force, leads to higher efficiency at high speeds because of the low magnetic flux density on the entire iron core, resulting in low iron loss. Moreover, across a wide operating range, the proposed PMASynRM produced a high efficiency greater than 90%.

In conclusion, this study revealed the possibility of the adoption of a PMASynRM with bonded magnets for automotive applications. In the future, we intend to examine the influence of the irreversible demagnetization on the driving performance of the proposed PMASynRM.

REFERENCES

[1] S. Morimoto, Y. Asano, T. Kosaka, and Y. Enomoto, "Recent Technical Trends in IPMSM," *The 2014 International Power Electronics Conference*, pp. 1997-2003, 2014.

[2] X. Liu, H. Chen, J. Zhao, and A. Belahcen, "Research on the Performances and Parameters of Interior PMSM Used for Electric Vehicles," *IEEE Trans. Ind. Electron.*, Vol. 63, No. 6, pp. 3533-3545, 2016.

[3] K. Yamazaki, M. Kumagai, T. Ikemi, and S. Ohki, "A Novel Rotor Design of Interior Permanent-Magnet Synchronous Motors to Cope with Both Maximum Torque and Iron-Loss Reduction," *IEEE Trans. Ind. Appl.*, Vol. 49, No. 6, pp. 2478-2486, 2013.

[4] S. Morimoto, S. Ooi, Y. Inoue, and M. Sanada, "Experimental Evaluation on a Rare-Earth-Free PMASynRM with Ferrite Magnets for Automotive applications," *IEEE Trans. Ind. Electron.*, Vol. 61, No. 10, pp. 5749-5756, 2014.

[5] Critical Raw Materials for the EU 2017, European Commission, Brussels, Belgium.

[6] M. Barcaro, and N. Bianchi, "Interior PM Machines Using Ferrite to Replace Rare-Earth Surface PM Machines," *IEEE Trans. Ind. Appl.*, Vol. 50, No. 2, pp. 979-985, 2014.

[7] R. Tsunata, M. Takemoto, S. Ogasawara, A. Watanabe, T. Ueno, K. Yamada "Development and Evaluation of an Axial Gap Motor Using Neodymium Bonded Magnet," *IEEE Trans. Ind. Appl.*, Vol. 54, No. 1, pp. 254-262, 2018.

[8] L. Ferraris, F. Franchini, E. Poskoviv, and A. Tenconi, "Impact of a Bonded-Magnet Adoption on a Specific Fractional Motor Power and Efficiency," *IEEE Trans. Ind. Appl.*, Vol. 50, No. 5, pp. 3249-3257, 2014.

[9] A. K. Jha, L. Gabuio, A. Kedous-Lebouc, J.-P. Yonnet, and J.-M. Dubus, "Design and Comparison of Outer Rotor Bonded Magnets Halbach Motor With Different Topologies," *IEEE ELMA Conf.*, Sofia, Bulgaria, pp. 6-10, June 1-3, 2017.

[10] M. Kimiabeigi, R. S. Sheridan, J. D. Widmer, A. Walton, M. Farr, B. Scholes, and I. R. Harris, "Production and Application of HPMS Recycled Bonded Permanent Magnets for a Traction Motor Application," *IEEE Trans. Ind. Electron.*, Vol. 65, No. 5, pp. 3795-3804, 2018.

[11] Y. Hamada, S. Morimoto, M. Sanada, Y. Inoue, "A Study of Automotive Application IPMSM with Bonded Rare Earth Magnets for Higher Torque and Demagnetization-Resistance," *The 2017 IEE-Japan Industry Applications Society Conference*, No. 5-005, pp. 9-10, 2017 (in Japanese).

[12] M. Taniguchi, T. Takahisa, K. Takizawa, S. Baba, M. Tsuchida, T. Mizutani, H. Endo, and H. Kimura, "Development of New Hybrid Transaxle for Compact-Class Vehicles," *SAE Technical Paper* 2016-01-1163, 2016.

Study on Rotor Structure Suitable for Improving Power Density and Efficiency in IPMSMs for Automotive Applications

R. Imoto[*], M. Sanada, S. Morimoto and Y. Inoue
Department of Electrical and Information Systems, Osaka Prefecture University, Sakai, Japan
*E-mail: mb104012@edu.osakafu-u.ac.jp

Abstract— Currently, the power density of interior permanent magnet synchronous motors (IPMSMs) for automobiles is improved by downsizing and high-speed rotation. It is necessary to examine the rotor structure and motor size suitable for higher-speed rotation in the future. This study examined the rotor structure of a single-layer V-shaped arrangement of permanent magnets (1V) and a two-layer arrangement (2D). We investigated the influence of iron and copper losses on downsized motors having these rotor structures. As a result, we found that a 2D model with stator diameter reduced by 80 % had the highest efficiency. Furthermore, in consideration of the mechanical strength of the downsized model, the design's influence on driving characteristics was investigated. In addition, we examined the driving characteristics of a model with stronger magnets installed in this model. The 2D model was proved to be suitable for downsizing.

Keywords— *interior permanent magnet synchronous motor (IPMSM), automotive application, rotor structure, improvement of power density and efficiency*

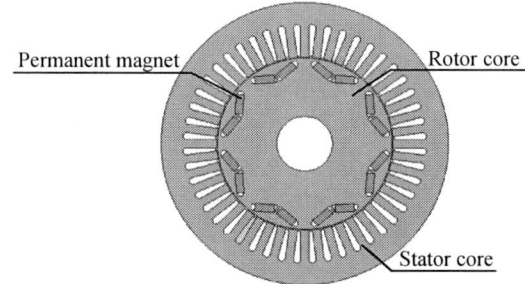

Fig. 1. Cross-section of analysis models.

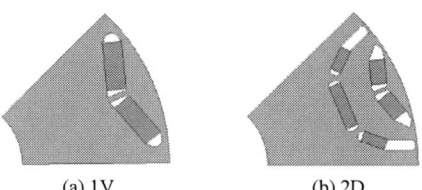

(a) 1V   (b) 2D
Fig. 2. Rotor structure models (single-pole).

I. INTRODUCTION

In recent years, research on hybrid electric vehicles (HEVs) and electric vehicles (EVs) has been actively pursued to mitigate global climate change. The size of traction motors for automotive applications affects the installation space and weight. Therefore, these motors are required to be compact, lightweight, and high efficiency and to produce the same power when they are downsized. To satisfy these requirements, interior permanent magnet synchronous motors (IPMSMs) are widely used for vehicle applications [1], [2]. The design of the rotor structure influences the driving characteristics, and various rotor structures have been proposed to reduce iron and copper losses [3]-[5].

IPMSMs for HEVs are downsized by increasing rotation speed [6]. Further improvement of power density is required in the future. However, the iron loss of IPMSMs increases at high-speed rotation [7], [8]. In this

This result was obtained from the Future Pioneering Program "Development of magnetic material technology for high-efficiency motors," commissioned by the New Energy and Industrial Technology Development Organization (NEDO).

study, we sought a motor structure suitable for small size and high-speed rotation. First, we examined the influence of losses in two driving evaluation points for two types of downsized rotor structures. Next, based on stress analysis of the rotor structure at maximum speed, we increased the width of the ribs, the structures at the ends of the flux barriers. However, the magnetic flux leakage increased, so the power decreased. Therefore, the power was restored by employing a magnet having better characteristics than those of the conventional magnets used in this type of motor. This paper discusses the characteristics of the models based on the results of two-dimensional finite element analysis.

II. INFLUENCE OF DIFFERENCES IN ROTOR STRUCTURE AND MOTOR SIZE ON CHARACTERISTICS

A. Analysis models

Fig. 1 shows a cross-section of the analysis model. The two rotor structures and motor specifications are shown in Fig. 2 and Table I. Both motors have 8-pole rotors and distributed windings in the stator with 48 slots. Model 1V shown in Fig. 2(a) has a single layer of permanent magnets (PMs) arranged in a V shape, which is conventional for IPMSMs used in HEVs [6]. Model 2D shown in Fig. 2(b) has a double layer of PMs and was designed to increase the ratio of reluctance torque to magnet torque [4].

The downsized model is called *x-y* (*x* for rotor structure and *y* for stator diameter reduction ratio). The stack length was adjusted so that each volume became 70 % of the model 1V size used in conventional IPMSMs for HEVs. The rib widths of all rotors were the same as those in the full-size model (100 %). Under driving conditions, the motor temperature was 180 °C, and the magnet temperature inside the motor was 140 °C.

B. Maximum torque characteristics

Fig. 3 shows the torque characteristics at the phase current $I_e = 134$ A (current density of 17.6 A/mm^2) under maximum torque per ampere (MTPA) control. Table II gives the motor parameters, where β is the phase of the current; L_d and L_q are the *d*- and *q*-axis inductance, respectively; Ψ_a is the flux linkage by the PM; and Ψ_{dmin} is the minimum *d*-axis flux linkage. The magnet torque T_m and reluctance torque T_r were calculated from the results of finite element analysis by the following procedure:

1) Ψ_a was calculated as

$$\Psi_a = \frac{T_0}{\sqrt{3}P_n I_e} \qquad (1)$$

where P_n is the number of pole pairs and T_0 is the torque at $\beta = 0$.

2) T_m and T_r were calculated as

$$T_m = \sqrt{3}P_n \Psi_a I_e \cos\beta \qquad (2)$$

$$T_r = T - T_m \qquad (3)$$

As shown in Fig. 2 and Table II, the maximum torque in each model was smaller for models with smaller rotor diameters, so the maximum torques of 1V-80 and 2D-80 were the smallest. This is because Ψ_a and $L_q - L_d$ are smaller in 1V-80 and 2D-80 than in 1V and 2D, so the total torque decreases. The cause is thought to be an increase in magnetic resistance by size reduction, and a decrease in the magnet surface area and number of armature windings. The reluctance torques of 2D, 2D-100, 2D-90, and 2D-80 were larger than those of 1V, 1V-100, 1V-90, and 1V-80, respectively.

Ψ_{dmin} is defined by the following equation:

$$\Psi_{dmin} = \Psi_a - L_d I_{am} \qquad (4)$$

where I_{am} is the maximum armature current. Ψ_{dmin} is an important parameter for determining the maximum operating speed. If $\Psi_{dmin} > 0$, the motor cannot drive above a constant speed under flux weakening control. If

$\Psi_{dmin} < 0$, the motor operation speed approaches infinity by switching to maximum torque per voltage (MTPV) control. However, as Ψ_{dmin} decreases, the maximum power decreases and the constant power speed range (CPSR) narrows. Therefore, if $\Psi_{dmin} = 0$, the motor can attain the highest power and widest CPSR [9]. As a result, 1V-80 and 2D-80 retained high power because their Ψ_{dmin} was closer to 0 than in the other models.

TABLE I
MOTOR MODEL SPECIFICATIONS

Item (Unit)	Value			
Model name	x	x-100	x-90	x-80
Stator outer diameter (mm)	264		238	210
Rotor outer diameter (mm)	160.4		144.2	128.0
Rotor inner diameter (mm)	51			
Air gap length (mm)	0.75			
Motor stack length (mm)	50	35	43	54
Winding resistance[*1] (mΩ)	129	116	96	77
Armature winding (turns/phase)	88		72	56
PM material[*2]	NMX-S34GH			

[*1] The temperature condition is 180 °C.
[*2] The temperature condition is 140 °C.

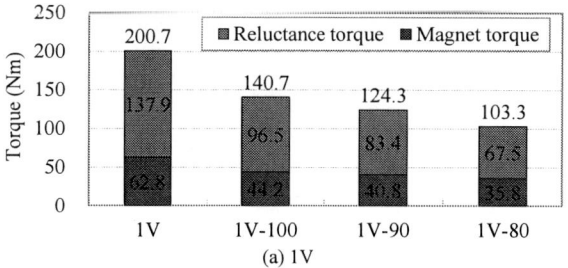

(a) 1V

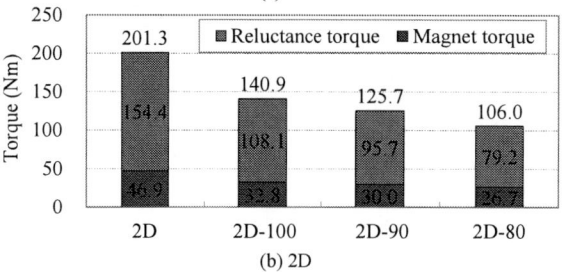

(b) 2D

Fig. 3. Maximum torque characteristics of 1V and 2D ($I_e = 134$ A).

TABLE II
1V AND 2D MOTOR PARAMETERS ($I_e = 134$ A).

(a) 1V

Item (Unit)	1V	1V-100	1V-90	1V-80
β (°)	51	51	49	47
L_d (mH)	0.977	0.684	0.582	0.463
L_q (mH)	2.285	1.600	1.364	1.091
$L_q - L_d$ (mH)	1.309	0.916	0.782	0.628
Ψ_a (Wb)	0.107	0.076	0.067	0.057
Ψ_{dmin} (Wb)	-0.171	-0.120	-0.097	-0.072

(b) 2D

Item (Unit)	2D	2D-100	2D-90	2D-80
β (°)	54	54	53	51
L_d (mH)	0.911	0.638	0.541	0.424
L_q (mH)	2.418	1.693	1.465	1.176
$L_q - L_d$ (mH)	1.507	1.055	0.924	0.752
Ψ_a (Wb)	0.086	0.060	0.054	0.046
Ψ_{dmin} (Wb)	-0.177	-0.124	-0.101	-0.073

C. Speed-versus-torque and power curves

Fig. 4 shows the speed-versus-torque and power curves at the maximum phase current I_{em} = 134 A and the maximum terminal voltage V_{am} = 507 V (the DC bus voltage V_{DC} = 650 V). The speed-versus-torque and power curves were shifted to the high-speed region by the torque reduction. 1V and 2D models of the same size had similar power characteristics. As mentioned in the previous section, the maximum power was higher for the 1V-80 and 2D-80 models with stator diameter reduction of 80 %.

D. Efficiency maps

Because IPMSMs for automotive applications must operate under various conditions, they need to be evaluated using efficiency maps, as shown in Fig. 5. The speed and torque in Fig. 5 are normalized to the maximum speed and maximum torque, respectively.

The efficiency and losses were calculated using the following equations:

$$W_c = 3R_a I_e^2 \tag{5}$$

$$W_h = \sum_{e=1}^{n_e} \left\{ \sum_{k=1}^{n} a(B_k) \times f_k \right\} \times V_e \tag{6}$$

$$W_e = \sum_{e=1}^{n_e} \left\{ \sum_{k=1}^{n} b(B_k, f_k) \times f_k^2 \right\} \times V_e \tag{7}$$

$$W_i = W_h + W_e \tag{8}$$

$$W = W_c + W_i \tag{9}$$

$$\eta = \frac{\omega T - W_i}{\omega T + W_c} \times 100 \tag{10}$$

where W_c (W) is the copper loss, W_h (W) is the hysteresis loss, W_e (W) is the eddy current loss, W_i (W) is the iron loss, W (W) is the total loss, R_a (Ω) is the winding resistance, n_e is the number of elements used in the finite element analysis, e is the index of the elements, n is the maximum harmonic order, $a(B_k)$ and $b(B_k, f_k)$ are coefficients based on loss properties, f_k (Hz) is the k-th order harmonic frequency, V_e (m³) is the volume of the e-th element, η (%) is the efficiency, and ω (rad/s) is the mechanical angular velocity.

As shown in Fig. 5, the efficiency at low speed and high torque depended on the copper loss, so efficiency was highest in the model with the smallest winding resistance. That is, the model with the stator diameter reduction of 80 % had the highest efficiency. In addition, the area of efficiency exceeding 97 % for model 2D was wider than that of 1V. The efficiency at high speed and low torque mainly depended on iron loss. Therefore, iron loss was smaller in 2D than in 1V. As a result, 2D-80 had the widest area of efficiency exceeding 97 %.

E. Comparison of characteristics at driving evaluation point

Table III shows various motor characteristics during both city and highway simulated driving by an HEV.

The characteristics of model 1V are compared with those of other models. Model 1V had the largest Ψ_a. From Table III (A), the phase current I_e required for 1V to obtain the same power was small, and the copper loss was small at city-driving evaluation point. However,

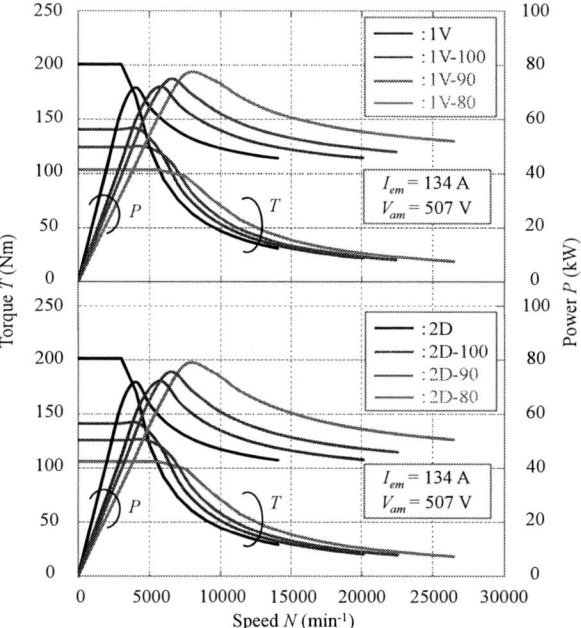

Fig. 4. Speed versus torque and power curves for 1V and 2D.

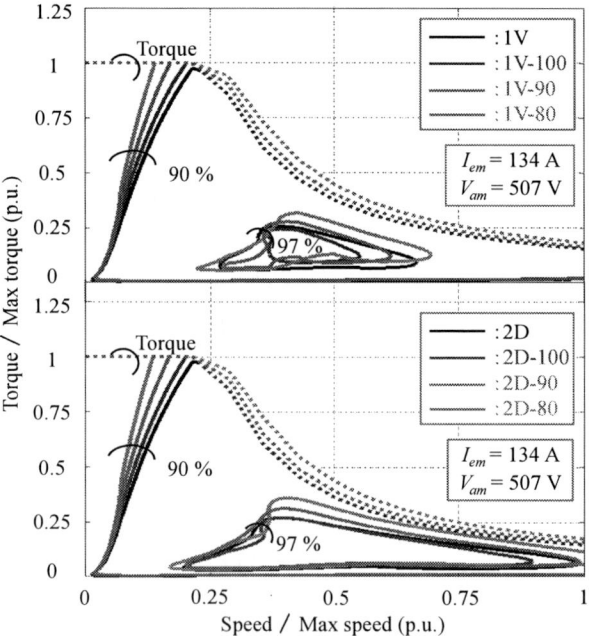

Fig. 5. Efficiency maps for 1V and 2D models.

model 2D had a lower iron loss than did model 1V. Hence, total loss was slightly smaller for 2D than for 1V.

In model 2D, with Ψ_a lower than that in 1V, the phase current I_e required for flux weakening control was small, and the copper loss was small at highway-driving evaluation point. In addition, model 2D had a lower iron loss than did 1V. Therefore, model 2D had 26% less total loss than did 1V.

Comparing model 2D with each downsized model, the copper loss decreased and iron loss increased at both driving evaluation points. This is because the winding resistance in the downsized model was small, and it had to rotate at higher speed to produce the same power. As a result, model 2D-80, with the smallest increase in iron loss relative to the decrease in copper loss, had the highest efficiency.

III. INFLUENCE OF INCREASED RIB WIDTH AND EMPLOYMENT OF STRONGER MAGNETS

A. Analysis models

It is necessary to evaluate the mechanical strength of a rotor during high-speed rotation. To reduce the maximum von Mises stress, we increased the rotor rib width of model 2D-80 to form model 2D-80-2. Then, we investigated the influence of the wider ribs on the driving characteristics. Fig. 6 compares the rotor ribs of 2D-80 and 2D-80-2 and shows that the outer and the inner rib widths of 2D-80-2 were twice as large as those in 2D-80.

The wider ribs were expected to decrease the maximum torque due to increased magnetic flux leakage. Therefore, the maximum torque of 2D-80-2 was increased by employing a stronger magnet, and we called this model 2D-80-2N. Specifications of model 2D-80-2N are shown in Table IV. The stronger magnet was assumed to have a higher remanence calculated from 95 % volume fraction and texture of the main phase with high magnetization shown in references [10] and [11], respectively. The coercivity of the strong magnet was assumed to be 25 % of the anisotropic magnetic field.

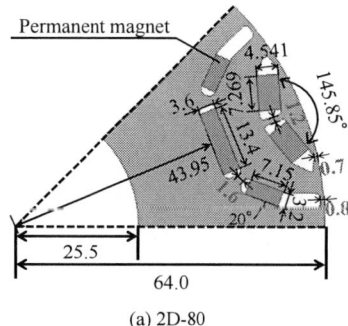

(a) 2D-80

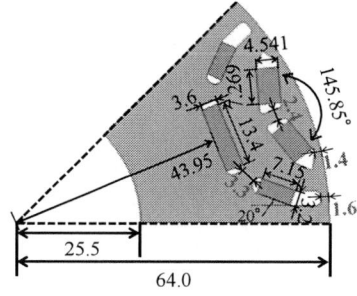

(b) 2D-80-2

Fig. 6. Dimensions of rotor with increased rib width (units: mm).

TABLE III
COMPARISON OF DRIVING CHARACTERISTICS OF ANALYSIS MODELS

(A) CITY-DRIVING EVALUATION POINT ($T = 0.10$ p.u., $N = 0.25$ p.u.)

Item (Unit)	1V	2D	2D-100	2D-90	2D-80
Torque T (Nm)	20	20	14	12.5	10.6
Speed N (min^{-1})	3500	3500	5000	5605	6604
Phase current I_e (A)	18.5	19.4	19.5	20.9	22.9
Phase of current β (°)	33.4	38.9	40.3	40.0	39.5
PM flux linkage Ψ_a (Wb)	0.126	0.110	0.077	0.066	0.053
Terminal voltage V_a (V)	250	240	236	218	201
Copper loss W_c (W)	131.94	145.09	132.21	126.33	121.77
Iron loss W_i (W)	98.76	77.50	89.07	86.42	86.05
Total loss W (W)	230.70	222.58	221.28	212.75	207.81
Efficiency η (%)	96.91	97.02	97.04	97.13	97.21

(B) HIGHWAY-DRIVING EVALUATION POINT ($T = 0.10$ p.u., $N = 0.79$ p.u.)

Item (Unit)	1V	2D	2D-100	2D-90	2D-80
Torque T (Nm)	20	20	14	12.5	10.6
Speed N (min^{-1})	11000	11000	15714	17616	20755
Phase current I_e (A)	26.9	26.6	26.6	26.5	26.5
Phase of current β (°)	66.1	66.1	66.1	61.9	55.9
PM flux linkage Ψ_a (Wb)	0.130	0.110	0.077	0.066	0.053
Terminal voltage V_a (V)	506	505	505	506	506
Copper loss W_c (W)	278.95	272.76	246.02	203.09	163.06
Iron loss W_i (W)	552.56	341.63	418.47	405.70	417.20
Total loss W (W)	831.51	614.40	664.49	608.79	580.27
Efficiency η (%)	96.45	97.36	97.14	97.38	97.50

TABLE IV
SPECIFICATIONS OF 2D-80 MODEL VARIANTS

Item (Unit)	Value	
Model name	2D-80, 2D-80-2	2D-80-2N
Stator outer diameter (mm)	210	
Rotor outer diameter (mm)	128	
Rotor inner diameter (mm)	51	
Air gap length (mm)	0.75	
Motor stack length (mm)	54	
Winding resistance[*1] (mΩ)	77	
Armature winding (turns/phase)	56	
PM material[*2]	NMX-S34GH	Strong magnet (NdFe$_{12}$N$_x$)
PM remanence (T)	1.04	1.39
PM coercivity[*3] (kA/m)	784	1052

[*1] The temperature condition is 180 °C.
[*2] The temperature condition is 140 °C.
[*3] Out of consideration of the demagnetization influence.

B. Reduction of mechanical stress

Fig. 7 shows the von Mises stress distribution caused by centrifugal force on each rotor at the maximum speed (N = 25,472 min⁻¹). The mechanical stress concentrated on the outer and inner ribs of each design. The von Mises stress in the inner rib of the inner magnet in 2D-80 was particularly large. The stress comparison revealed that the maximum von Mises stress could be reduced by 25% by doubling the width of all ribs. However, the maximum von Mises stress in 2D-80-2 was still 1.3 times greater than the yield stress of the electrical steel sheet used. To further reduce the mechanical stress, the radius of curvature of the flux barrier where the stress was concentrated must be increased. Therefore, it is necessary to employ a flux barrier shape that reduces the mechanical stress or use a high-tension electric steel sheet. We plan to examine these options in the future.

C. Maximum torque characteristics

Fig. 8 and Table V show maximum torque characteristics and motor parameters under the same conditions as Fig. 3 and Table II, respectively. The results show that the maximum torque of 2D-80-2 was approximately 84 % of that of 2D-80. This is because $L_q - L_d$ was smaller in 2D-80-2 than in 2D-80, so the reluctance torque decreased. In addition, the flux leakage by the PM increased. Therefore, Ψ_a decreased approximately 33 %, and magnet torque also decreased. From the results, as Ψ_{dmin} decreases, the maximum power decreases and the CPSR narrows in model 2D-80-2.

In contrast, the maximum torque of 2D-80-2N was approximately 96 % of that of 2D-80. This is because the reluctance torque decreased similarly to 2D-80-2, but the strong magnet has a PM remanence B_r approximately 34 % larger than that of the NMX-S34GH, and the magnet torque increased by approximately 12 %. The Ψ_{dmin} value for 2D-80 and 2D-80-2 is almost the same. Therefore, the maximum power and CPSR of 2D-80 and 2D-80-2N have similar characteristics.

D. Speed-versus-torque and power curves

Fig. 9 shows the speed-versus-torque and power curves under the same conditions as Fig. 4. Comparison of 2D-80 with 2D-80-2 shows that the power of 2D-80-2 dropped over the entire speed range. This is because the maximum torque and Ψ_{dmin} decreased, as described in the previous section. In contrast, 2D-80 and 2D-80-2N had similar power characteristics over the entire speed range. This showed that the reduced power could be recovered by employing the stronger PMs.

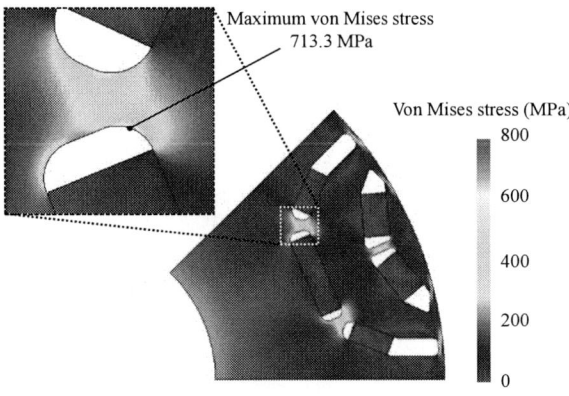

(a) 2D-80

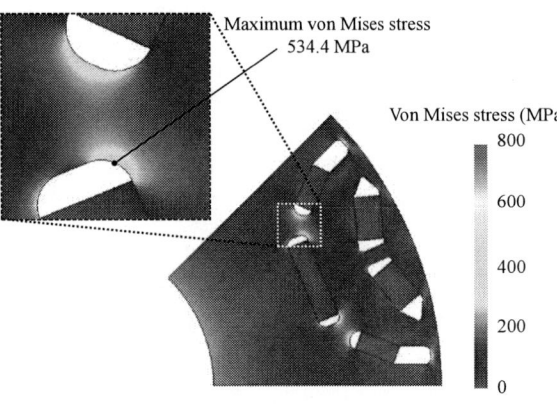

(b) 2D-80-2

Fig. 7. Result of centrifugal force analysis for models 2D-80 and 2D-80-2 (N = 25,472 min⁻¹).

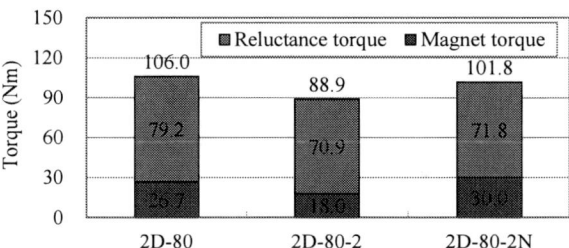

Fig. 8. Maximum torques of 2D-80 models (I_e = 134 A).

TABLE V
MOTOR PARAMETERS OF 2D-80 MODELS (I_e = 134 A).

Item (Unit)	2D-80	2D-80-2	2D-80-2N
β (°)	51	52	50
L_d (mH)	0.424	0.474	0.447
L_q (mH)	1.176	1.153	1.124
$L_q - L_d$ (mH)	0.752	0.679	0.677
Ψ_a (Wb)	0.046	0.031	0.050
Ψ_{dmin} (Wb)	-0.073	-0.096	-0.071

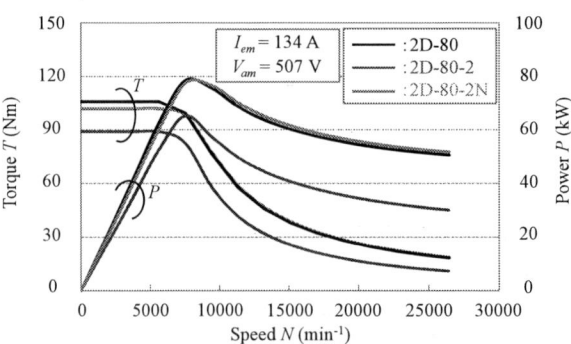

Fig. 9. Speed-versus-torque and power curves for 2D-80 models.

E. Efficiency maps and comparison of driving characteristics

Fig. 10 shows the efficiency maps for the three 2D-80 models. The areas exceeding 90% and 97.1% efficiency for 2D-80-2 are narrower than those for 2D-80. In contrast, the figure shows that the areas exceeding 90% and 97.1% efficiency for 2D-80 and 2D-80-2N were almost equivalent.

Table VI shows various characteristics for both city and highway driving. Comparison of 2D-80 with 2D-80-2 shows that the I_e required for 2D-80-2 to produce the same power was small and the copper loss increased under both driving evaluation points. This is because model 2D-80-2 had the smallest Ψ_a due to large flux leakage. In contrast, 2D-80 and 2D-80-2N experienced almost the same total loss in each driving evaluation point. This is because 2D-80 and 2D-80-2N had approximately the same Ψ_a, and the copper loss was almost the same. From these results, we conclude that model 2D-80-2N, with the wider rib widths and stronger PMs, is a successfully downsized version of model 2D.

IV. CONCLUSIONS

In this paper, we investigated rotor structures suitable for downsizing an IPMSM for automotive applications.

In models 1V-80 and 2D-80, which had stator diameters reduced by 80% with respect to the base designs, the winding resistance was small and the copper loss during simulated driving decreased. However, the smaller motors operated at higher speed to produce the same power as the base design and the iron loss increased. However, model 2D had a lower iron loss than model 1V at high rotation speed.

Model 2D-80-2, with increased rib widths, could reduce the maximum von Mises stress. However, the magnetic flux leakage increased, which decreased the power characteristics of 2D-80-2. In addition, 2D-80-2 had the smallest Ψ_a, increasing the current required to produce the same power, thus increasing the copper loss. However, the power and loss characteristics of 2D-80-2N with stronger PMs were almost equivalent to those of 2D-80. Thus, the design of model 2D-80-2N is suitable for improvement of power density and efficiency.

REFERENCES

[1] J. D. Santiago, H. Bernhoff, B. Ekergard, S. Eriksson, S. Ferhatovic, and R. Waters, "Electrical Motor Drivelines in Commercial All-Electric Vehicles: A Review," *IEEE Trans. on Vehicular Technology*, vol. 61, no. 2, pp. 475-484, 2012.

[2] X. Liu, H. Chen, J. Zhao and A. Belahcen, "Research on the Performances and Parameters of Interior PMSM Used for Electric Vehicles," *IEEE Trans. Ind. Electron.*, vol. 63, no. 6, pp. 3533-3545, 2016.

[3] Y. Katsumi, K. Masaki, I. Takeshi, and O. Shunji, "A Novel Rotor Design of Interior Permanent-Magnet Synchronous Motors to Cope with Both Maximum Torque and Iron-Loss Reduction," *IEEE Trans. on Industry Applications*, vol. 49, no. 6, pp. 2478-2486, 2013.

[4] S. Yoshioka, S. Morimoto, M. Sanada, and Y. Inoue, "Influence of Magnet Arrangement on the Performance of IPMSMs for

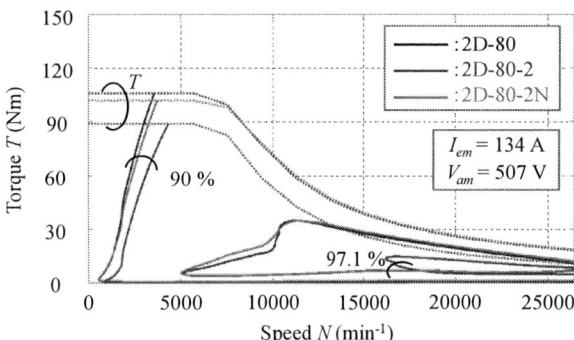

Fig. 10. Efficiency maps for 2D-80 models.

TABLE VI
COMPARISON OF DRIVING CHARACTERISTICS OF 2D-80 MODELS

(A) CITY-DRIVING EVALUATION POINT (T = 0.10 p.u., N = 0.25 p.u.)

Item (Unit)	1V	2D-80	2D-80-2	2D-80-2N
Torque T (Nm)	20	10.6	10.6	10.6
Speed N (min^{-1})	3500	6604	6604	6604
Phase current I_e (A)	18.5	22.9	28.9	22.7
Phase of current β (°)	33.4	39.5	44.3	37.3
PM flux linkage Ψ_a (Wb)	0.126	0.053	0.029	0.054
Terminal voltage V_a (V)	250	201	199	201
Copper loss W_c (W)	131.94	121.77	193.94	119.65
Iron loss W_i (W)	98.76	86.05	84.58	86.61
Total loss W (W)	230.70	207.81	278.51	206.26
Efficiency η (%)	96.91	97.21	96.28	97.23

(B) HIGHWAY-DRIVING EVALUATION POINT (T = 0.10 p.u., N = 0.79 p.u.)

Item (Unit)	1V	2D-80	2D-80-2	2D-80-2N
Torque T (Nm)	20	10.6	10.6	10.6
Speed N (min^{-1})	11000	20755	20755	20755
Phase current I_e (A)	26.9	26.5	32.4	26.7
Phase of current β (°)	66.1	55.9	58.0	56.2
PM flux linkage Ψ_a (Wb)	0.130	0.053	0.030	0.055
Terminal voltage V_a (V)	506	506	506	502
Copper loss W_c (W)	278.95	163.06	243.75	165.53
Iron loss W_i (W)	552.56	417.20	401.49	407.11
Total loss W (W)	831.51	580.27	645.25	572.64
Efficiency η (%)	96.45	97.50	97.23	97.53

Automotive Applications," *in Conf. on IEEE Energy Conversion Congress and Exposition (ECCE-2014)*, pp. 4507-4512, 2014.

[5] Z. Q. Zhu, Fellow, W. Q. Chu, and Y. Guan, "Quantitative Comparison of Electromagnetic Performance of Electrical Machines for HEVs/EVs," *CES Trans. on Electrical Machines and Systems*, vol. 1, no. 1, pp. 37-47, 2017.

[6] M. Olszewski, "Evaluation of the 2010 Toyota Prius Hybrid Electric Drive System," *Oak Ridge Nat. Lab.*, U. S. Dept. Energy, 2011.

[7] M. Arata, Y. Kurihara, D. Misu, and M. Matsubara, "EV and HEV Motor Development in TOSHIBA," *IEEJ Journal of Industry Applications*, vol. 4, no. 3, pp. 152-157, 2014.

[8] J. H. Seo, D. K. Woo, T. K. Chung, and H. K. Jung, "A Study on Loss Characteristics of IPMSM for FCEV Considering the Rotating Field," *IEEE Trans. on Magnetics*, vol. 46, no. 8, pp. 3213-3216, 2010.

[9] S. Morimoto, "Trend of Permanent Magnet Synchronous Machines," *IEEJ Trans.*, 2, pp. 101-108, 2007.

[10] Y. Hirayama, Y. K. Takahashi, S. Hirosawa and K. Hono, "NdFe12Nx Hard-Magnetic Compound with High Magnetization and Anisotropy Field," *Materialia*, vol. 95, pp.70-72, 2015.

[11] Y. Shimizu, S. Morimoto, M. Sanada, and Y. Inoue, "Influence of Permanent Magnet Properties and Arrangement on Performance of IPMSMs for Automotive Applications," *IEEJ Journal of Industry Applications*, vol. 6, pp.401-408, 2017.

The 2018 International Power Electronics Conference

Examination of the Demagnetization Suppression Effect of Placing Flux Barriers in an IPMSM Using Rare-earth Bonded Magnets

Takashi Umeda[1], Masayuki Sanada[1], Shigeo Morimoto[1] and Yukinori Inoue[1]
1 Graduate School of Engineering, Osaka Prefecture University, Sakai-City, Japan
* Email : mb104015@edu.osakafu-u.ac.jp

Abstract—**IPMSM using rare-earth sintered permanent magnets generally have high performance. However, rare-earth sintered magnets are costly, and the stability of the supply of this resource is a concern. In this study, we examine an IPMSM using rare-earth bonded magnets to solve the issue. The use of rare-earth bonded magnets increases the risk of irreversible demagnetization. Therefore, it is necessary to examine the method of demagnetization suppression. In this paper, we design a basic model using rare-earth bonded magnets and evaluate irreversible demagnetization. In addition, we evaluate irreversible demagnetization in the anti-demagnetization models with flux barriers placed at the ends of the magnets. We then analyze the motor characteristics after demagnetization in these models. Finally, based on the obtained results, we examine differences in the demagnetization suppression effect due to the flux barrier shape.**

Keywords—IPMSM, Rare-earth bonded magnet, Demagnetization, and Flux barrier.

I. INTRODUCTION

The interior permanent magnet synchronous motor (IPMSM) is an efficient and high-power motor that is used in a wide range of applications, including electric vehicles, industrial machines, and home appliances. At present, rare-earth sintered magnets (Nd-Fe-B) containing Dysprosium (Dy) are generally used in IPMSMs to enhance coercivity. However, in Japan, Dy must be imported and is very expensive, and so reducing the amount of Dy used in IPMSMs is necessary. As such, an IPMSM using a rare-earth bonded magnet is proposed [1]-[4]. Rare-earth bonded magnets are inexpensive because of Dy free and are highly shapeable. On the other hand, the possibility of irreversible demagnetization is increased because the coercive force is lower than in rare-earth sintered magnets [5].

Therefore, in this study, we investigate the rotor structure suppressing the effect of irreversible demagnetization in an IPMSM using rare-earth bonded magnets [6]-[8].

In this paper, demagnetization analysis is performed on the basic model and anti-demagnetization models having various flux barrier shapes. Then, we analyze

changes in motor characteristics after demagnetization for each model. Based on the results, we examine differences in the demagnetization suppression effect due to the flux barrier shape.

II. BASIC MODEL

Fig. 1 shows the structure of the basic model considered in the present study, and Table I lists its specifications. The model is designed to be used as a motor for the compressor of an air conditioner. The stator has a concentrated winding. Permanent magnet materials used in rare-earth bonded magnets have less favorable magnetic characteristics than rare-earth sintered magnets. Therefore, each magnet is constructed to have a reverse-arc shape so that its surface area is large. This model is referred to herein as Type-RA6. The motor characteristics of Type-RA6 are analyzed using a two-dimensional finite element method and are listed in Table II.

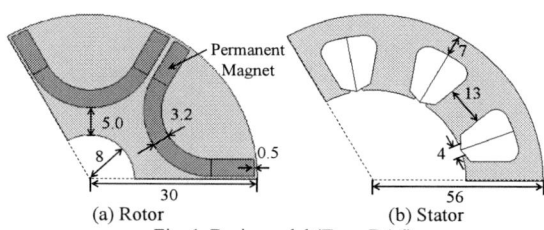

(a) Rotor   (b) Stator
Fig. 1. Basic model (Type-RA6).

TABLE I
SPECIFICATIONS OF THE BASIC MODEL

Item (Unit)	Value
Stator external diameter (mm)	112
Rotor external diameter (mm)	60
Shaft diameter (mm)	16
Stack length (mm)	40
Air gap length (mm)	0.5
PM thickness (mm)	3.2
Armature winding (turn/phase)	264
Space factor (%)	50
Type of rare earth PM	bonded
PM residual magnetization B_r (T)	0.66
Rated current (A)	3.52

2814

TABLE II
CHARACTERISTICS OF THE BASIC MODEL

Item (Unit)	value
Maximum torque (Nm)	3.658
Magnet torque (Nm)	3.357
Reluctance torque (Nm)	0.301
d axis inductance (mH)	17.83
q axis inductance (mH)	28.64
Armature flux linkage by PM (Wb)	0.190
Induced voltage (V)	16.863

A. Demagnetization evaluation

In a permanent magnet, the operating point on the B-H curve depends on the demagnetization field. In the case of irreversible demagnetization of a permanent magnet, when a demagnetization field is removed, the magnetic force is degraded. Therefore, the B-H curve after demagnetization changes from the original. As shown in Fig. 2, when the operating point decreases due to a demagnetizing field and falls below the knickpoint, it returns through the recoil line. The remanence decreases from B_r to B_r'. Therefore, the magnetic flux density of the magnet is analyzed by a two-dimensional finite element method. The demagnetization rate is determined based on the obtained result.

In this study, demagnetization analysis is performed when the flowing current is 100% to 300% of the rated current (current phase $\beta = 90°$). As examples of the demagnetization analysis results, Fig. 3 shows the results when the demagnetization currents are 1.0 p.u. and 3.0 p.u. From these results, irreversible demagnetization occurs at the ends of the magnets when the demagnetization current is 1.0 p.u. This is a significant problem when actually using a motor. Then, as the demagnetization current increases, the region of demagnetization also expands, and, at 3.0 p.u., the magnetization is completely lost at the end of the magnet.

B. Characteristics analysis after demagnetization

Based on the results described in Section IIA, the motor characteristics after demagnetization are analyzed. Then, the deterioration of the motor characteristics due to irreversible demagnetization is evaluated based on the falling rates of the maximum torque and the no-load induced voltage. Table III and Fig. 4 show the analysis results for the motor characteristics after demagnetization, where the result for the no-load induced voltage is obtained at a speed of 500 min⁻¹. The results indicate that the motor performance is decreased by demagnetization when the demagnetization current is 1.0 p.u. The no-load induced voltage decreases approximately 0.3%, and the maximum torque decreases approximately 0.2%. Moreover, the difference between the falling rates for the no-load induced voltage and the maximum torque increases as the demagnetization current increases. The no-load induced voltage decreases depending on the deterioration of the magnet. The maximum torque is expressed as the sum of the magnet torque and the reluctance torque. Although the magnet torque is affected by the deterioration of the magnet, the reluctance torque is determined by the salient pole ratio and there is almost no change at this time.

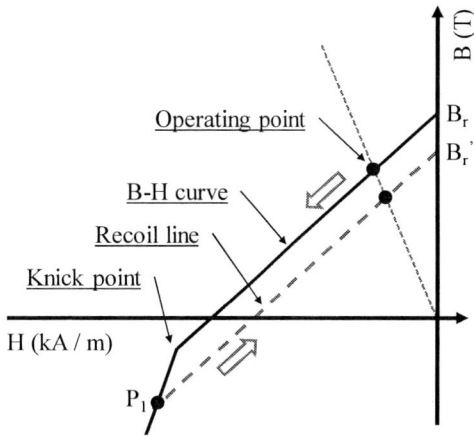

Fig. 2. Demagnetization of a permanent magnet.

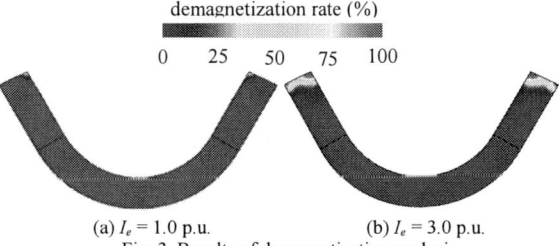

(a) I_e = 1.0 p.u.   (b) I_e = 3.0 p.u.
Fig. 3. Results of demagnetization analysis.

TABLE III
CHARACTERISTICS OF AFTER DEMAGNETIZATION

Item (Unit)	Demagnetization current (p.u.)					
	1.0	1.4	1.8	2.2	2.6	3.0
Maximum torque (Nm)	3.650	3.643	3.632	3.607	3.559	3.461
Magnet torque (Nm)	3.352	3.344	3.334	3.310	3.262	3.165
Reluctance torque (Nm)	0.298	0.298	0.298	0.298	0.297	0.295
Armature flux linkage by PM (Wb)	0.190	0.189	0.189	0.187	0.185	0.179
d axis inductance (mH)	18.02	18.05	18.08	18.16	18.32	18.66
q axis inductance (mH)	28.72	28.75	28.78	28.84	28.97	29.25
Induced voltage (V)	16.81	16.75	16.68	16.52	16.24	15.68

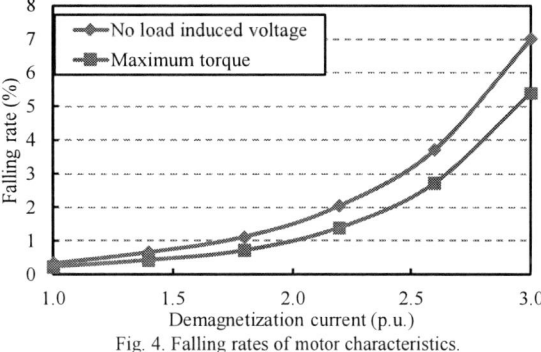

Fig. 4. Falling rates of motor characteristics.

III. ANTI-DEMAGNETIZATION MODELS

Methods of demagnetization suppression in permanent magnets, such as the placement of flux barriers, increasing the magnet thickness, and changing the embedding depth of the magnet, have been examined. In this study, in order to examine the effect of flux barrier shape on the demagnetization suppression effect, we analyze the demagnetization rate and the motor characteristics of after demagnetization.

A. Consideration of flux barrier shape

Fig. 5 shows the magnetic flux when only the stator winding is energized in the basic model. In this case, the magnet part was analyzed as air. The results indicate that the magnetic flux is concentrated at the ends of the magnet. Therefore, irreversible demagnetization occurs from the ends of the magnet. As a countermeasure, we consider two methods. The first is the placement of flux barriers to block the flow of magnetic flux toward the magnet ends, and the second method is the replacement of the magnet ends with flux barriers.

Fig. 6 shows the flux barrier shape of the anti-demagnetization models used in each countermeasure. Other dimensions, such as the rotor external diameter, are the same as in the basic model. Moreover, the stator structure and the winding scheme are the same as in the basic model. In the Type-RAI6, the magnet thickness is increased to 3.4 mm so that the magnet quantity will be the same as that in the basic model. The groove depth x_1 of the rotor surface of the Type-RAg6-x_1 is varied as 0.5 mm, 1.0 mm, and 1.5 mm.

Fig. 7 shows the maximum torque characteristics and the no-load induced voltage for the anti-demagnetization models. The no-load induced voltage is the result when the speed is 500 min^{-1}. The results indicate that the maximum torque of the anti-demagnetization models is lower than that of the Type-RA6. The Type-RAg6 has grooves on the rotor surface, so that leakage flux from the rib decreases and the magnet torque increases. However, since the q-axis magnetic path was narrowed, the reluctance torque decreases. Therefore, the maximum torque decreases slightly compared with that of the Type-RA6. In the Type-RAI6, the flux barrier replaced the magnet ends, so the magnet surface area decreased and the magnet torque decreased. The no-load induced voltage of the Type-RAg6 is higher than the Type-RA6. The reason for this is the same as the reason the magnet torque increases. Since the surface area of the magnet of the Type-RAI6 is decreased, the no-load induced voltage also decreases.

Figs. 8 through 11 show the demagnetization analysis results for the Type-RAg6-x_1 and the Type-RAI6. As a result of preventing the demagnetizing field flux, the Type-RAg6-x_1 shows decreased demagnetization at the ends of the magnet. However, a region with a demagnetization rate of 100% still exists at the magnet ends. Moreover, when the depth of the groove becomes great, the region in which irreversible demagnetization occurs spreads toward the surface of the magnet.

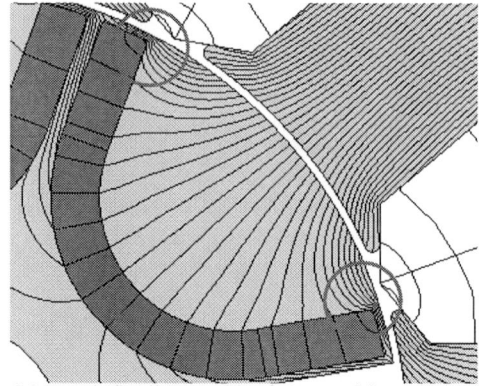

Fig. 5. Magnetic flux lines resulting from energizing only the stator winding (Type-RA6).

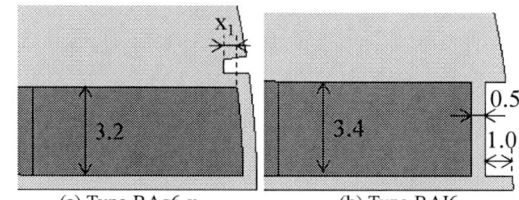

(a) Type-RAg6-x_1 (b) Type-RAI6
Fig. 6. Flux barrier shape of anti-demagnetization models.

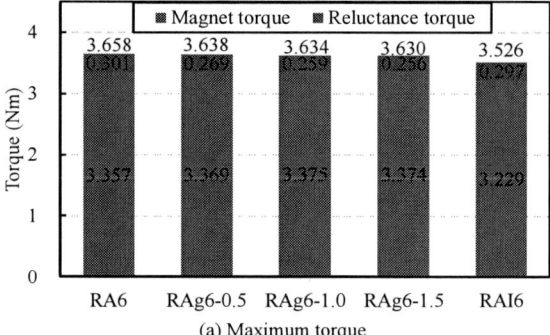

(a) Maximum torque

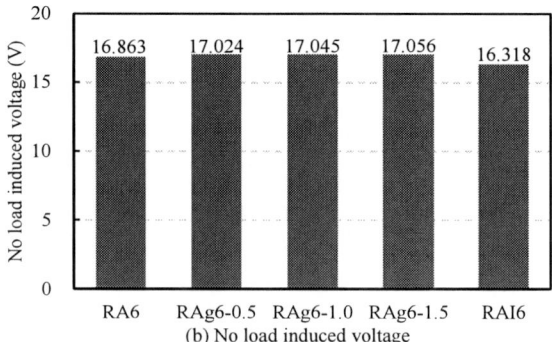

(b) No load induced voltage
Fig. 7. Motor characteristics (Type-RA6, RAg6, RAI6).

The region of the Type-RA6 with a demagnetization rate of 10% extends approximately 2.1 mm along the surface of the magnet. The region of the Type-RAg6-x_1 with a demagnetization rate of 10% extends approximately 2.2 mm when $x_1 = 0.5$ mm, approximately 2.5 mm when $x_1 = 1.0$ mm, and approximately 2.6 mm when $x_1 = 1.5$ mm. The Type-RAI6 significantly improves irreversible

The 2018 International Power Electronics Conference

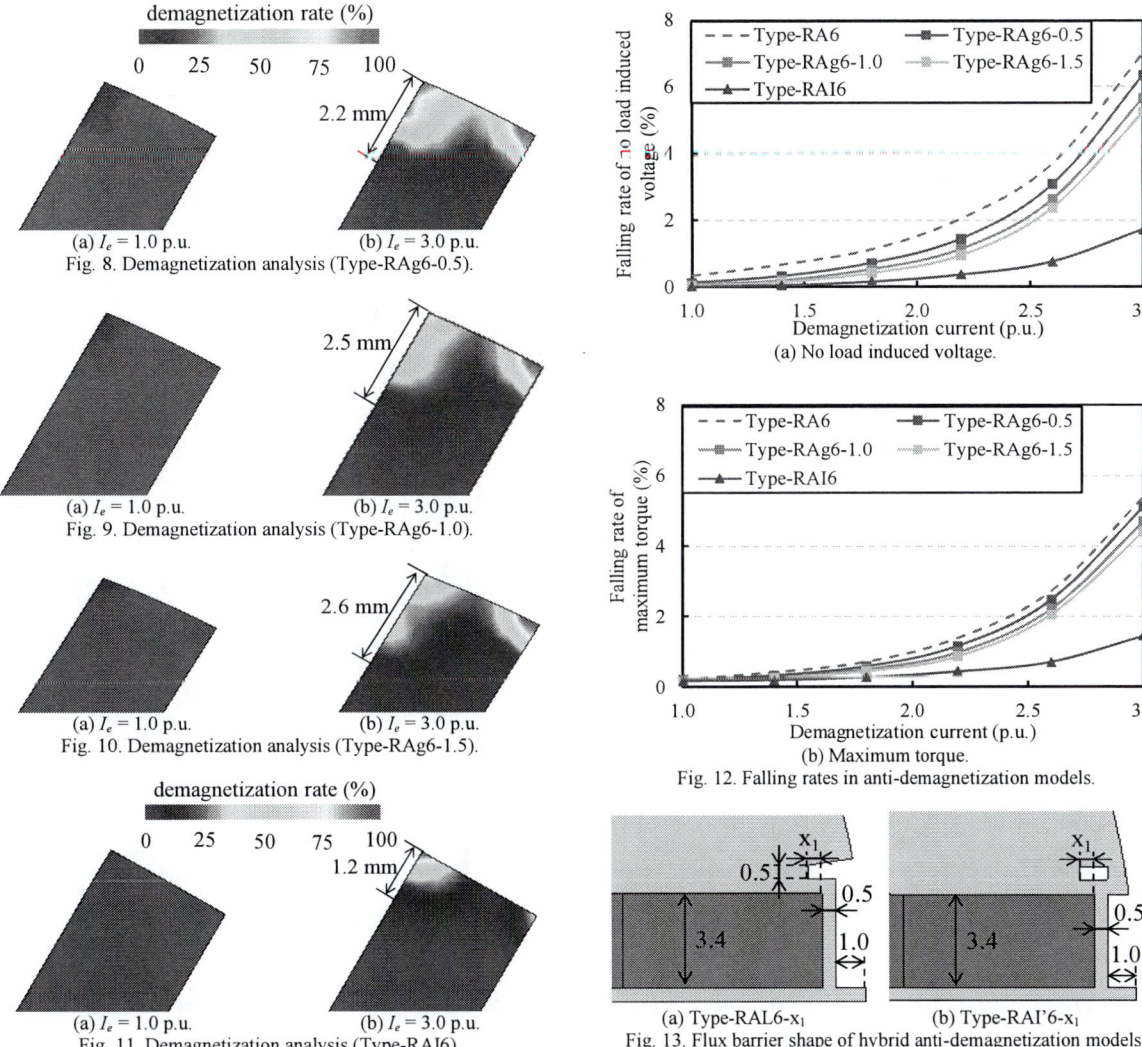

demagnetization rate (%)

0   25   50   75   100

2.2 mm

(a) $I_e = 1.0$ p.u.   (b) $I_e = 3.0$ p.u.
Fig. 8. Demagnetization analysis (Type-RAg6-0.5).

2.5 mm

(a) $I_e = 1.0$ p.u.   (b) $I_e = 3.0$ p.u.
Fig. 9. Demagnetization analysis (Type-RAg6-1.0).

2.6 mm

(a) $I_e = 1.0$ p.u.   (b) $I_e = 3.0$ p.u.
Fig. 10. Demagnetization analysis (Type-RAg6-1.5).

demagnetization rate (%)

0   25   50   75   100

1.2 mm

(a) $I_e = 1.0$ p.u.   (b) $I_e = 3.0$ p.u.
Fig. 11. Demagnetization analysis (Type-RAI6).

(a) No load induced voltage.

(b) Maximum torque.
Fig. 12. Falling rates in anti-demagnetization models.

(a) Type-RAL6-x_1 (b) Type-RAI'6-x_1
Fig. 13. Flux barrier shape of hybrid anti-demagnetization models.

demagnetization as compared with the Type-RA6 and the Type-RAg6-x_1. This is due to replacing the flux barrier and increasing the thickness of the magnet. Moreover, unlike Type-RAg6-x_1, the demagnetization region on the surface of the magnet does not expand. The region with a demagnetization rate of 10% is approximately 1.2 mm from the magnet end.

Next, Fig. 12 shows the falling rates of the maximum torque and the no-load induced voltage of each model. In the Type-RA6, when the demagnetization current is 1.0 p.u., the no-load induced voltage decreases approximately 0.3%, and the maximum torque decreases approximately 0.2%. In anti-demagnetization models, the no-load induced voltage decreases only slightly when the demagnetization current is 1.0 p.u. However, the maximum torque still decreases by 0.15 through 0.2%. The demagnetization suppression effect of the Type-RAI6 is confirmed to become large as the demagnetizing current increases. In the Type-RAg6, the falling rates of the motor characteristics can be suppressed as compared with the Type-RA6, but are still inferior to that of the Type-RAI6.

B. Combination of anti-demagnetization models

We next combine the two flux barrier shapes examined in Section IIIA and examine the effects on the motor characteristics and demagnetization suppression effect. Fig. 13 shows the flux barrier shape of the hybrid anti-demagnetization models. The model in which the two flux barriers are combined is the Type-RAL6-x1, and the model in which the two flux barriers are separated is the Type-RAI'6-x_1. The value of x_1 is varied as 0.5 mm, 1.0 mm, and 1.5 mm. The stator structure is the same as in the basic model. The magnet thickness is increased to 3.4 mm so that the magnet quantity will be the same as that in the basic model.

Fig. 14 shows the maximum torque characteristics and the no-load induced voltage of the hybrid type anti-demagnetization models. The no-load induced voltage is the result when the speed is 500 min^{-1}. Since the two flux barrier shapes are combined, the maximum torque characteristics and the no-load induced voltage are lower than those of the Type-RA6, Type-RAg6-x_1, and Type-RAI6. Differences in motor characteristics between

2817

The 2018 International Power Electronics Conference

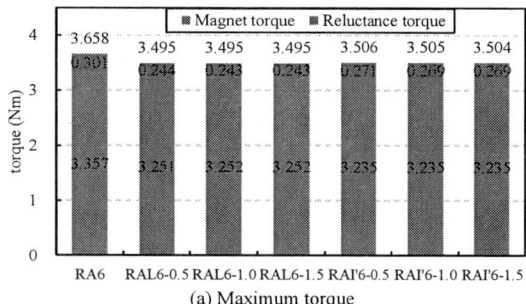

(a) Maximum torque

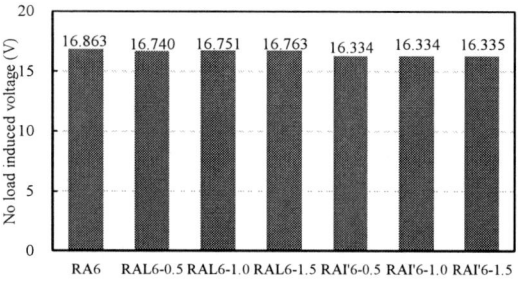

(b) No load induced voltage

Fig. 14. Motor characteristics (Type-RAL6-x_1 and Type-RAI'6-x_1).

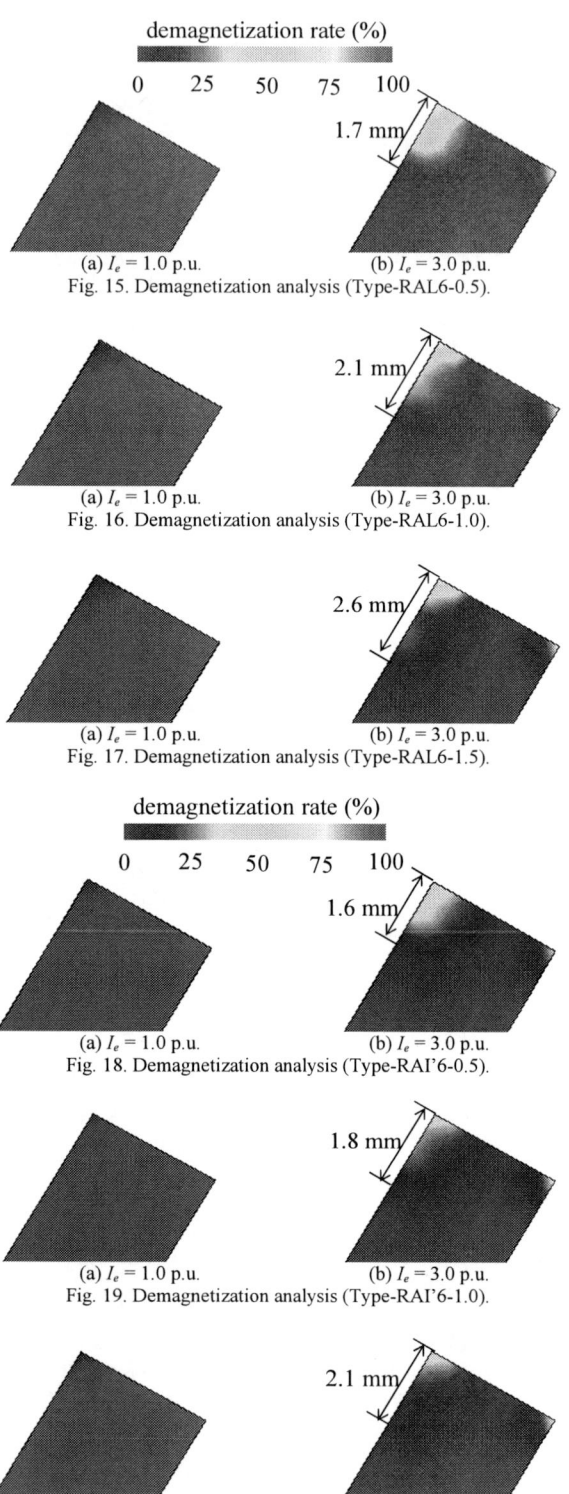

models due to changing the depth of grooves are slight. The maximum torque of the Type-RAI'6-x_1 is slightly larger than that of the Type-RAL6-x_1, although the no-load induced voltage is small.

Next, the results of demagnetization analysis obtained using these analysis models are shown in Figs. 15 through 20. Unlike the previous models, there is no region with a demagnetization rate of 100%. In both the Type-RAL6-x_1 and the Type-RAI'6-x_1, the region of low demagnetization rate expands in the magnet surface direction, as in the case of the Type-RAg6-x_1. In the Type-RAL6-x_1, the region with a demagnetization rate of 10% expands approximately 1.7 mm when $x_1 = 0.5$ mm, approximately 2.1 mm when $x_1 = 1.0$ mm, and approximately 2.6 mm when $x_1 = 1.5$ mm. In the Type-RAI'6-x_1, the region with a demagnetization rate of 10% expands approximately 1.6 mm when $x_1 = 0.5$ mm, approximately 1.8 mm when $x_1 = 1.0$ mm, and approximately 2.1 mm when $x_1 = 1.5$ mm. In both models, the region of irreversible demagnetization can be suppressed the most when $x_1 = 1.5$ mm. Compared with the Type-RAI6, the region of high demagnetization rate is small when the demagnetization current is 3.0 p.u. This is considered to be an effect of the flux barrier, which blocks the flow of the demagnetizing field flux as in the Type-RAg6-x_1. The regions of irreversible demagnetization are large in the Type-RAL6-x_1 and the Type-RAI'6-x_1. Moreover, the Type-RAI'6-x_1 has a greater irreversible demagnetization suppression effect than the Type-RAL6-x_1. The reason for this is thought to be that the flux escape path is made by separating the flux barrier into two parts.

Based on the demagnetization analysis results, the motor characteristics after demagnetization are analyzed. Fig. 21 shows the falling rates of the no-load induced voltage and the maximum torque after demagnetization.

The results indicate that when the demagnetization current is 3.0 p.u. the falling rate of motor performance of the Type-RAI6 is smaller than that of the Type-RAL6-x_1 and the Type-RAI'6-x_1. The reason for this is thought to be that irreversible demagnetization spreads to the

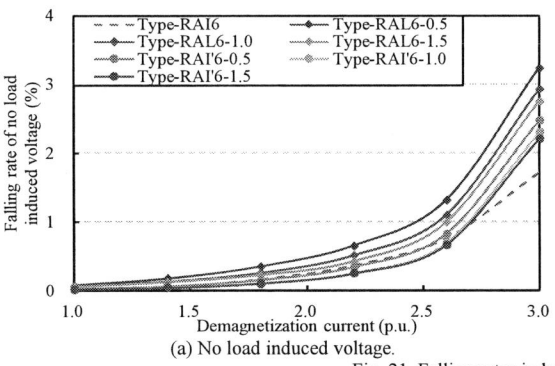 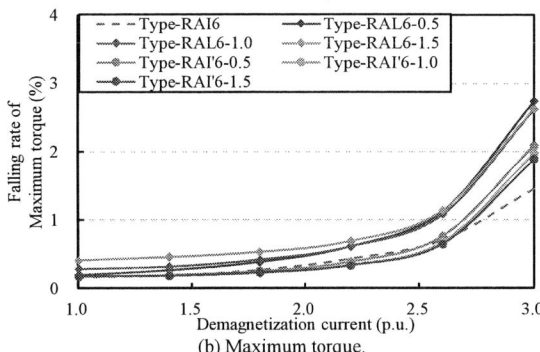

(a) No load induced voltage.   (b) Maximum torque.

Fig. 21. Falling rates in hybrid anti-demagnetization models.

magnet surface as a result of the groove placement. Previous studies have confirmed that if the region of irreversible demagnetization expands toward the surface direction of a magnet, the influence on the motor performance becomes large [10]. On the other hand, the falling rates of the Type-RAI'6-1.0 and the Type-RAI'6-1.5 are smaller than that of the Type-RAI6 until the demagnetization current reaches 2.6 p.u. Therefore, reducing the region of high demagnetization rate by placing the flux barrier so as to block the flow of the demagnetizing field flux is effective, but it is necessary to decide the placement location considering the influence on the motor performance.

In all of the anti-demagnetization models, it was impossible to prevent the decrease of the motor performance when the demagnetization current is 1.0 p.u. In order to prevent performance deterioration when the demagnetization current is 1.0 p.u., it is necessary to consider not only the flux barrier shape but also the magnet thickness and shape.

IV. CONCLUSION

In this paper, in order to realize a rotor structure that suppresses irreversible demagnetization in an IPMSM using rare-earth bonded magnets, the flux barrier shape at the end of the magnet was examined. The motor performance before demagnetization of the anti-demagnetization models was found to decrease compared with the basic model. In the Type-RAg6-x1, although it was possible to suppress the influence of demagnetization as compared with the basic model, the suppression effect was not significant. The influence of demagnetization was found to become 1/3 or less in the Type-RAI6 as compared with the basic model. In the hybrid anti-demagnetization models, the Type-RAI'6-1.5 had the greatest demagnetization suppression effect when the demagnetization current was small. However, when the demagnetization current reached 3.0 p.u., the demagnetization suppression effect was smaller than that of the Type-RAI6. Moreover, in all of the anti-demagnetization models, it was impossible to completely eliminate the decrease in motor performance at the rated current.

In this study, we made the magnet quantity constant and equal to that of the basic model. In the future, in order to make the motor performance of all models equal to that of the basic model, we will consider increasing the volume of the magnets. In addition, we will examine the improvement of the demagnetization suppression effect by increasing the magnet amount. Therefore, we will consider not only the flux barrier shape but also the magnet thickness and shape.

REFERENCES

[1] Ren Tsunata, Masatsugu Takemoto, Satoshi Ogasawara, Asako Watanabe, Tomoyuki Ueno, and Koji Yamada, "Development and Evaluation of an Axial Gap Motor Using Neodymium Bonded Magnet," IEEE Transactions on Industry Applications, Vol. 54, No. 1, pp. 254 – 262, 2018.

[2] Y. Hayashi, H. Mitarai, and Y. Honkura, "Development of a DC Brush Motor with 50% Weight and Volume Reduction Using an Nd-Fe-B Anisotropic Bonded Magnet," IEEE Transactions on Magnetics, Vol. 39, No. 5, pp. 2893 – 2895, 2003.

[3] Daiki Tanaka, Masayuki Sanada, Shigeo Morimoto, and Yukinori Inoue "Comparison of IPMSMs Using Bonded and Sintered Rare Earth Magnets with Different Magnet Arrangements," in Proc. ICEMS2016, LS2D1, 1-6(CD-ROM).

[4] Y. Yoshikawa, T. Ogawa, Y. Okada, S. Tsutsumi, H. Murakami, and S. Morimoto "Some Considerations on the Optimum Design of IPMSM Using a Bonded Rare-Earth Magnet," IEEE Trans. on Industry Applications, Vol. 136, No. 12, pp. 997 – 1004, 2016.

[5] Min-Ro Park, Hae-Joong Kim, Yun-Yong Choi, Jung-Pyo Hong, and Jeong-Jong Lee, "Characteristics of IPMSM According to Rotor Design Considering Nonlinearity of Permanent Magnet," IEEE Transactions on Magnetics, Vol. 52, No. 3, 8101904, 2016.

[6] Dong-Kyun Woo and Byung Hwan Jeong, "Irreversible Demagnetization of Permanent Magnet in a Surface-Mounted Permanent Magnet Motor With Overhang Structure," IEEE Transactions on Magnetics, Vol. 52, No. 4, 8102606, 2016.

[7] Hyung-Kyu Kim and Jin Hur, "Dynamic Characteristic Analysis of Irreversible Demagnetization in SPM- and IPM-Type BLDC Motors," IEEE Transactions on Industry Applications, Vol. 53, No. 2, pp.982 – 990, 2017.

[8] Tanveer Yazdan, Wenliang Zhao, Thomas A. Lipo, and Byung-Il Kwon, "A Novel Technique for Two-Phase BLDC Motor to Avoid Demagnetization," IEEE Transactions on Magnetics, Vol. 52, No. 7, 8106704, 2016.

[9] Dong-Kyun Woo, Dong-Kuk Lim, Han-Kyeol Yeo, Jong-Suk Ro, and Hyun-Kyo Jung, "A 2-D Finite-Element Analysis for a Permanent Magnet Synchronous Motor Taking an Overhang Effect Into Consideration," IEEE Transactions on Magnetics, Vol. 49, No. 8, pp. 4894 – 4899, 2013.

[10] Toshiyuki Endo, Masayuki Sanada, Shigeo Morimoto, and Yukinori Inoue, "Influence of Demagnetized Area Shape in Permanent Magnet on the Performance of IPMSM," IEE-Japan Industry Applications Society Conference, Vol. 3, No. 55, pp. 303 – 306, 2016.

The 2018 International Power Electronics Conference

A Novel Pole-changing Method
with a Multiple Three-phase Inverter

Yuki Hidaka[1]*, Taiga Komatsu[1] and Hideaki Arita[2]

1 Advanced Technology R&D Center, Mitsubishi Electric Corporation, Amagasaki, Japan

2 Himeji Works, Mitsubishi Electric Corporation, Himeji, Japan

*E-mail: Hidaka.Yuki@ea.MitsubishiElectric.co.jp

Abstract— This paper presents a new pole-changing method with a multiple three-phase inverter. Since the number of controllable current phases is much larger than that of conventional methods, a higher winding factor can be obtained in both high and low pole driving modes. To evaluate the magnetic characteristics of the present method, magnetomotive force analysis is conducted based on Fourier series expansion and finite element method (FEM). Moreover, mutual inductance is compared between high and low pole driving using FEM. To validate the effectiveness, an efficiency map is evaluated by using a prototype induction motor.

Keywords—Pole-changing method, induction motor, magnetomotive force, multiple three-phase inverter.

I. INTRODUCTION

Because regulation for reduction of CO_2 emissions hae been strengthened, electrification of vehicles is expected to accelerate [1]. Motors for electric vehicles (EVs) and hybrid EVs (HEVs) are used over a wide speed range [2-3]. In low-speed conditions, such as starting and climbing, a high-torque characteristic is required. On the other hand, available armature currents are limited due to the heat capacities of tips and coils. For this reason, magnetic design must be conducted to raise the motor impedance to improve the inner product between line voltage and current. In high-speed conditions, such as during highway driving and in regeneration mode, a high-output characteristic must be satisfied. Since motors are driven on limited bus-line voltage, low motor impedance is desirable to raise available current. As noted above, required characteristics in low and high speed conditions have a trade-off relationship in principle.

Since the pole-changing techniques can vary operational frequency, many researches have been reported about their effectiveness [4-7]. To apply the pole-changing techniques for motors, the motor driven area can be expanded. The pole-changing method can be mainly divided into two ideas. One is a mechanical switching method in which mechanical switches are prepared for changing the coil winding structure between low and high pole driven modes [4-5]. Although wide speed operation can be realized, torque shock occurs in

the pole-changing period due to on/off operation of the switch. Since the vehicle moves using only motor output in EVs, the above torque shock is directly transmitted to the driver. The other method is using a multiple inverter with more controllable current phases than an ordinal three-phase inverter [6-7]. To change the pole, the input current phase of each stator slot is optimized by current control. In this idea, the current control method of an inverter in the pole-changing period differs from that of the former idea. Torque shock can be regulated by managing the motor drive control.

In this work, we present a novel pole-changing method in which torque shock is regulated by using multiple inverters. In the present method, the controllable current phase is larger than that of the conventional multiple inverter method. In addition, the neutral point of the motor can be set for each three-phase inverter. Thanks to this, fluctuation of line-voltage in armature coils can be prevented to facilitate drive control. In this paper, we denote methodology and structure of the conventional and present pole-changing methods. From several numerical results, it can be found that harmonic mangetomotive force is reduced by using the present method. To validate the effectiveness, an efficiency map is obtained using a prototype induction motor.

II. POLE-CHANGING METHOD

A. Previous work

In the conventional method [7], two three-phase inverters are applied for motor driving to increase the number of controllable current phases as in Fig. 1(a). Each coil connected to the output terminal of an inverter is set as in Fig. 1(b). Fig. 1(b) shows the minimum unit for realizing the pole-changing method.

In the low-speed condition, corresponding to high-pole mode, current phases are controlled as follow.

$$\begin{aligned}
\boldsymbol{I}_1(t) &= [I_{A1}, I_{B1}, I_{C1}] \\
&= \sqrt{2} I_m \left[\cos(\omega t), \cos\left(\omega t - \frac{4}{3}\pi\right), \cos\left(\omega t - \frac{2}{3}\pi\right) \right] \\
\boldsymbol{I}_2(t) &= [I_{A2}, I_{B2}, I_{C2}] \\
&= \sqrt{2} I_m \left[\cos\left(\omega t - \frac{2}{3}\pi\right), \cos(\omega t), \cos\left(\omega t - \frac{4}{3}\pi\right) \right]
\end{aligned} \tag{1}$$

In this situation, magnetomotive force is generated satisfying four-pole driving. In a high speed condition, corresponding to low-pole mode, current phases are controlled as follows:

$$I_1(t)=\sqrt{2}I_m\left[\cos(\omega t),\cos\left(\omega t-\frac{2}{3}\pi\right),\cos\left(\omega t-\frac{4}{3}\pi\right)\right]$$

$$I_2(t)=\sqrt{2}I_m\left[\cos\left(\omega t-\frac{1}{3}\pi\right),\cos(\omega t-\pi),\cos\left(\omega t-\frac{5}{3}\pi\right)\right]$$

(2)

As mensioned above, the driving mode can be changed to a pole ratio of 2:1, as in the conventional method.

On the other hand, in the conventional method, the number of controllable current phases per pole is equal to the concentrated winding structure in which relation of pole and slots is 2:3 in low-speed conditions. So, the harmonic winding factor is increased compared to the ordinal three-phase motor.

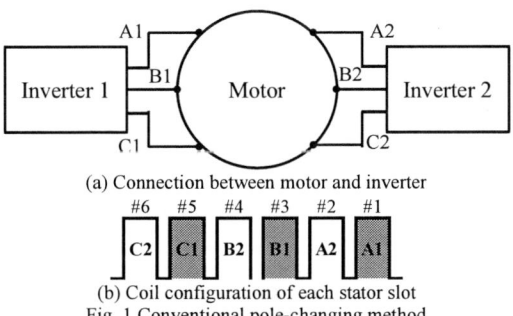

(a) Connection between motor and inverter

(b) Coil configuration of each stator slot
Fig. 1 Conventional pole-changing method

B. The present method

In order to solve the above problems, the number of controllable current phases per pole is increased at both low and high speed conditions in the present method. To raise the controllable current phase, a quadruple three-phase inverter is used in the present method as shown in Fig. 2(a). Thanks to this, harmonic magnetomotive force can be reduced. In the present method, each coil connected to the output terminal of an inverter is set as Fig. 2(b). Fig. 2(b) shows the minimum unit for realizing the pole-changing method.

In low-speed conditions, corresponding to high-pole mode, each coil current is controlled as follows:

$$I_1(t)=\sqrt{2}I_m\left[\cos(\omega t),\cos\left(\omega t-\frac{4}{3}\pi\right),\cos\left(\omega t-\frac{2}{3}\pi\right)\right]$$

$$I_2(t)=\sqrt{2}I_m\left[\cos\left(\omega t-\frac{1}{3}\pi\right),\cos\left(\omega t-\frac{5}{3}\pi\right),\cos(\omega t-\pi)\right]$$

$$I_3(t)=\sqrt{2}I_m\left[\cos\left(\omega t-\frac{2}{3}\pi\right),\cos(\omega t),\cos\left(\omega t-\frac{4}{3}\pi\right)\right]$$

$$I_4(t)=\sqrt{2}I_m\left[\cos(\omega t-\pi),\cos\left(\omega t-\frac{1}{3}\pi\right),\cos\left(\omega t-\frac{5}{3}\pi\right)\right]$$

(3)

From Eq. (3), the magnetomotive force is made for realizing four pole motor driving. In high-speed conditions, corresponding to low-pole mode, each current phase is controlled by an inverter as follows:

$$I_1(t)=\sqrt{2}I_m\left[\cos(\omega t),\cos\left(\omega t-\frac{4}{6}\pi\right),\cos\left(\omega t-\frac{8}{6}\pi\right)\right]$$

$$I_2(t)=\sqrt{2}I_m\left[\cos\left(\omega t-\frac{1}{6}\pi\right),\cos\left(\omega t-\frac{5}{6}\pi\right),\cos\left(\omega t-\frac{9}{6}\pi\right)\right]$$

$$I_3(t)=\sqrt{2}I_m\left[\cos\left(\omega t-\frac{2}{6}\pi\right),\cos(\omega t-\pi),\cos\left(\omega t-\frac{10}{6}\pi\right)\right]$$

$$I_4(t)=\sqrt{2}I_m\left[\cos\left(\omega t-\frac{3}{6}\pi\right),\cos\left(\omega t-\frac{7}{6}\pi\right),\cos\left(\omega t-\frac{11}{6}\pi\right)\right]$$

(4)

As dimensioned above, the pole-changing ratio is 2:1 in the present method. Moreover the number of controllable current phases per pole is equal to (high pole) or higher than (low pole) that of the ordinal three-phase motor. By using the present method, the harmonic winding factor can be reduced from the conventional pole-changing method.

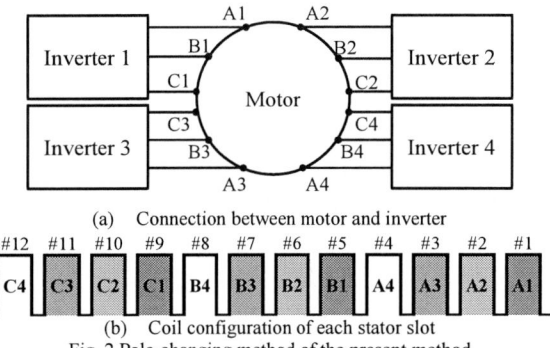

(a)   Connection between motor and inverter

(b)   Coil configuration of each stator slot
Fig. 2 Pole-changing method of the present method

C. Comparison of magnetomotive force

In the present method, A1-C1 coils are turned as in Fig. 3. A2-C2, A3-C3, and A4-C4 coils are wound the same as A1-C1 coils. From Fig. 3, the magnetomotive force of a single coil can be expressed as shown in Fig. 4. From Fig. 4, the Fourier coefficients are given by

$$a_k=\frac{1}{\pi}\int_{-\pi}^{\pi}f(\xi)\cos(v\xi)d\xi$$
$$=\frac{z\cdot i(t)}{\pi}\left\{\int_{\pi/4}^{\pi}-\frac{1}{4}\cos(k\xi)d\xi+\int_{-\pi/4}^{\pi/4}\frac{3}{4}\cos(k\xi)d\xi+\int_{-\pi}^{-\pi/4}-\frac{1}{4}\cos(k\xi)d\xi\right\}$$
$$=\frac{z\cdot i(t)}{k\pi}\sin\left(\frac{k\pi}{4}\right)$$

$$a_0=\frac{z\cdot i(t)}{k\pi}\sin\left(\frac{0*\pi}{4}\right)=0$$

$$b_k=\frac{1}{\pi}\int_{-\pi}^{\pi}f(\xi)\sin(k\xi)d\xi$$
$$=\frac{z\cdot i(t)}{\pi}\left\{\int_{\pi/4}^{\pi}-\frac{1}{4}\sin(k\xi)d\theta+\int_{-\pi/4}^{\pi/4}\frac{3}{4}\sin(k\xi)d\xi+\int_{-\pi}^{-\pi/4}-\frac{1}{4}\sin(k\xi)d\xi\right\}$$
$$=0$$

(5)

In Eq. (5), $sin(k\pi/4)$ in a_k is equal to the short-pitch winding factor. By using Eq. (3), (5), total magnetomotive force at high pole mode is obtained by

$$f(\theta,t)=\frac{6zI_m}{\pi}\left\{\sum_{v=2,14,26...}\frac{1}{v}\sin\left(\frac{v\pi}{4}\right)\cos(\omega t-\theta)+\sum_{v=10,22,34...}\frac{1}{v}\sin\left(\frac{v\pi}{4}\right)\cos(\omega t+\theta)\right\}$$

(6)

In Eq. (6), θ means mechanical angle. For this reason, fundamental frequency corresponds to $v=2$. Moreover the first and second term on the right hand of Eq. (6) are positive and negative phase respectively.

Similarly, the total magnetomotive force at low pole mode can be calculated by using Eq. (4), (5) as follows:

$$f(\theta,t)=\frac{6zI_m}{\pi}\left\{\sum_{v=1,13,25...}\frac{1}{v}\sin\left(\frac{v\pi}{4}\right)\cos(\omega t-\theta)+\sum_{v=11,23,35...}\frac{1}{v}\sin\left(\frac{v\pi}{4}\right)\cos(\omega t+\theta)\right\} \quad (7)$$

In Eq. (7), $sin(v\pi/4)$ is equal to the short pitch winding factor. Moreover fundamental frequency corresponds to $v=1$.

The magnetomotive force of the conventional method can be calculated in the same way. At high-pole mode, the magnetomotive force is obtained by

$$f(\theta,t)=\frac{6zI_m}{\pi}\left\{\sum_{v=2,8,14...}\frac{1}{v}\sin\left(\frac{v\pi}{3}\right)\cos(\omega t-\theta)+\sum_{v=4,10,16...}\frac{1}{v}\sin\left(\frac{v\pi}{3}\right)\cos(\omega t+\theta)\right\} \quad (8)$$

In Eq. (8), $sin(v\pi/3)$ is equal to the short pitch winding factor. Moreover, fundamental frequency corresponds to $v=2$. The first and second terms on the right hand of Eq. (8) are positive and negative phases. The magnetomotive force at low-pole mode can be expressed by

$$f(\theta,t)=\frac{6zI_m}{\pi}\left\{\sum_{v=1,7,13...}\frac{1}{v}\sin\left(\frac{v\pi}{3}\right)\cos(\omega t-\theta)+\sum_{v=5,11,17...}\frac{1}{v}\sin\left(\frac{v\pi}{3}\right)\cos(\omega t+\theta)\right\} \quad (9)$$

In Eq. (9), $sin(v\pi/3)$ is equal to the short pitch winding factor. Moreover, the fundamental frequency corresponds to $v=1$. From Eq. (6)-(9), a comparison of the harmonic winding factor between the conventional and present method are summarized in Table II. The results show that the harmonic winding factor can be reduced by using the present pole-changing method.

On the other hand, the fundamental winding factor of the present method is lower than that of the conventional method at low-pole mode. From this difference, the coil copper loss of the present method is larger than that of the conventional method. This is because the coil pitch shown in Fig. 3 is selected for maximizing the fundamental winding factor at high pole mode to obtain high torque characteristic. In high rotational speed conditions, corresponding to low-pole mode, iron loss is larger than that of the coil copper loss. In contrast, the coil copper loss is larger than that of the iron loss in low-speed condition. For this reason, it is effective to select coil pitch to improve the magnetic characteristics of high-pole mode.

In the above calculation, rotor/stator core shapes and magnetic saturation is not considered in evaluating magnetomotive force. To maximize the advantage of the present pole-changing method, slot and permeance harmonics must be cleared. In the next section, affection of core shapes and magnetic saturation is confirmed by using the finite element method.

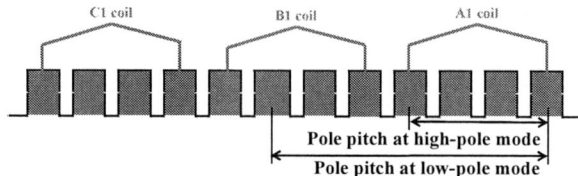

Fig. 3 Stator coil structure of the present method

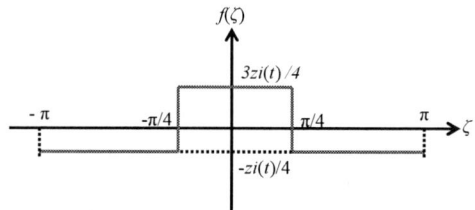

Fig. 4 Magnetomotive force of a single coil

TABLE II
COMPARISON OF HARMONIC WINDING FACTOR BASED ON ELECTRIC DEGREE

Harmonic order		Conventional method		Present method	
		High pole	Low pole	High pole	Low pole
	1	0.866	0.866	1.0	0.707
	2	0.433	-	-	-
	4	0.216	-	-	-
	5	0.173	0.173	0.200	-
	7	0.124	0.124	0.143	-
	8	0.108	-	-	-
	10	0.087	-	-	-
	11	0.079	0.079	0.091	0.064
	13	0.067	0.067	0.077	0.054

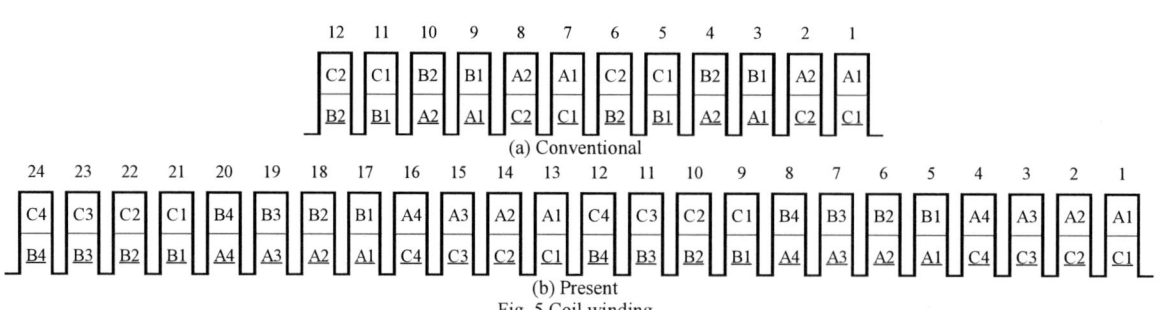

(a) Conventional

(b) Present
Fig. 5 Coil winding

2822

III. NUMERICAL RESULTS

A. Magnetomotive force

In the former section, magnetomotive force is calculated by using Fourier series expansion. In this work, magnetic saturation and rotor/stator core shapes are taken into account for evaluating the magnetomotive force of the present and conventional pole-changing methods. In Fig. 5, coil connections of the test induction motor for the conventional and present methods are shown. Moreover, motor specifications are summarized in Table III. The pole numbers of high and low pole mode are set as 8 and 4 respectively. To arrange the flux level between the conventional and present methods, total armature ampere turn is set to be equal. 30JNE is selected for rotor and stator cores. To remove the affection of secondary current, the material property of the rotor bar is set as air.

The radial direction of air gap magnetic density is compared in Fig. 6. From Fig. 6, intended characteristics can be obtained in both the conventional and present methods. The harmonic order of magnetomotive force based on electrical degree is compared in Fig. 7. From Fig. 7, it can be seen that the calculation results between fourier series expansion and FEM is the same. High order magnetomotive force of the present method is lower than that of the conventional method. From these results, superiority of the present method can be shown, even if core shape is considered. In the next section, the affection of magnetic saturation is inspected using a prototype motor.

TABLE III
MOTOR SPECIFICATIONS

		Conventional	Present
	Pole number	8 / 4	←
	Core material	30JNE	←
Stator	Outer radius (mm)	138	←
	Inner radius (mm)	95.6	←
	Core length (mm)	70.0	←
	Number of slots	12	24
	Coil turn	2 para 4 turn	4 para 8 turn
	Coil throw	#1-3	#1-4
Rotor	Outer radius (mm)	95.0	←
	Inner radius (mm)	17.4	←
	Core length (mm)	70.0	←
	Number of slots	16	32

B. Affection of magnetic saturation

To clear the affection of magnetic saturation, mutual inductance between rotor and stator is evaluated using FEM. The prototype induction motor to which the present pole-changing method is applied is shown in Fig. 8. The pole number of high and low pole mode are set as 8 and 4, respectively. Core length, outer/inner radius and coil turn are the same as Table III. Core back width and coil throw are set for improving high-pole characteristics.

In Fig. 9, mutual inductance is plotted for each phase current value. In Fig. 9, vertical axis of low and high pole mode are normalized by maximum value at each pole mode. From this result, the affection of magnetic saturation differs between low and high pole mode. Magnetic contours at 48Arms phase current conditions are shown in Fig. 10. Magnetic density of the core back in low pole mode is twice as large as that of high-pole mode. In the prototype motor, the magnetic circuit is designed for improving torque density. Since core design is fitted to high-pole driving, magnetic saturation of the core back became noticeable in low-pole mode.

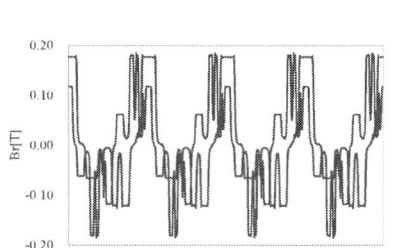

(a) High-pole mode

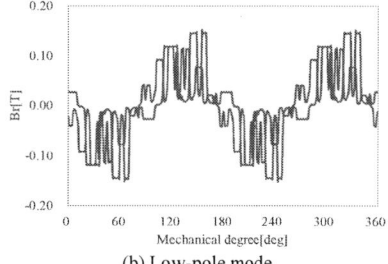

(b) Low-pole mode

Fig. 6 Radial direction of air gap magnetic density

(a) High-pole mode

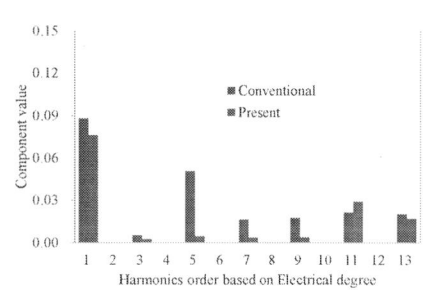

(b) Low-pole mode

Fig. 7 Radial direction of air gap magnetic density in no-load condition

The 2018 International Power Electronics Conference

In the present pole-changing method, only the current phase can be changed in the pole-changing period. There is no difference in mechanical structure and coil pitch between low and high-pole driving. To maximize the performance of the pole-changing motor, loss characteristics must be considered in magnetic design. For an induction motor, motor loss is dominated by copper and iron loss in low and high speed modes, respectively. From this view point, coil pitch is selected for maximizing fundamental winding coefficient to improve torque per current during high-pole driving. Although the winding coefficient of low-pole driving is dropped, operational frequency can be reduced by changing the number of poles. Thanks to this, motor efficiency can be improved over a wide speed range.

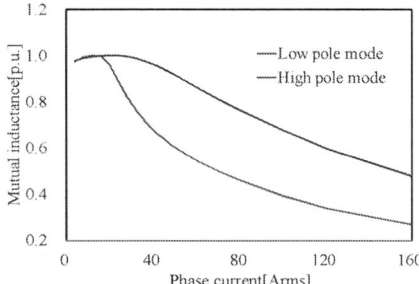

Fig. 9 Contour of magnetic density at no-load condition

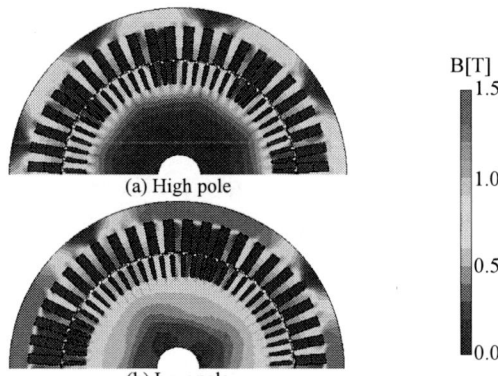

(a) High pole

(b) Low pole

Fig. 10 Magnetic contour at no-load condition

IV. EXPERIMENTAL RESULTS

A. Motor bench

Fig. 11(a) shows an experimental system. Torque is measured using a torque meter. In the load testing, rotational speed is set by the load motor. PE-Inverter and PE-Expert made by Myway Company were selected as the inverters and drive controller. Limit current and voltage were set to 75Arms per phase and 48V-DC. Motor bench is as shown in Fig. 11(b).

B. Efficiency map

Fig. 12 shows the efficiency map both in high and low pole modes. From Fig. 12, it can be seen that high torque and output characteristics can be obtained in high and low pole modes by using the present pole-changing method. Especially, output power of 4000-6000r/min in low-pole mode is twice as large as that of high-pole mode. From this result, it can be realized that low motor impedance can be obtained by changing the operational frequency. Moreover, maximum torque of the high-pole mode is 1.5 times as large as that of low-pole mode. This is because motor impedance became higher by driving in high-pole mode.

The mechanical output power of the low-pole mode is higher than that of the high-pole mode at over 2000r/min. From this result, high-pole driving is used only in low-speed condition. Considering this, torque shock motor control must be focused on the low-frequency driving area. Since the carrier frequency of the inverter depends on the semiconductor device, it can be seen that tip costs can be suppressed for realizing torque shock control. Moreover, heat capacity of the tip can be dispersed by using multiple inverters. For this reason, tip size can be reduced by using multiple inverters.

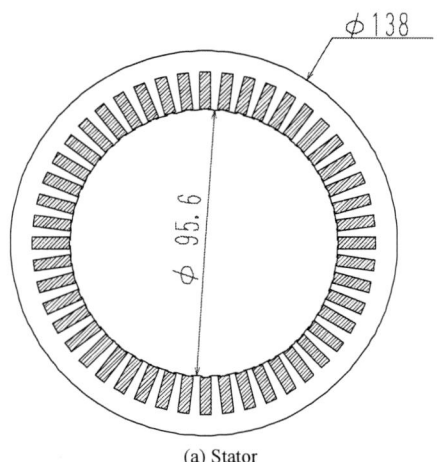

(a) Stator

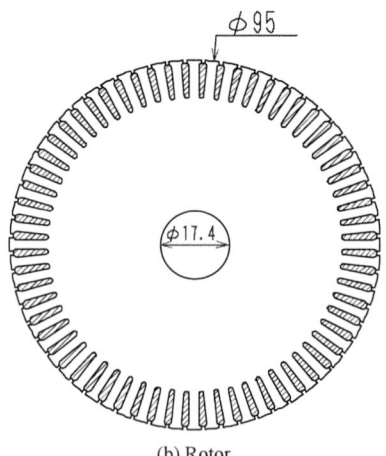

(b) Rotor

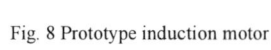

Fig. 8 Prototype induction motor

2824

V. CONCLUSIONS

In this paper, a novel pole-changing method is suggested. In the present method, a quadruple three-phase inverter is used for improving the characteristics of the magnetomotive force. Since there are more controllable current phases per pole than in the conventional pole-changing method, the harmonic magnetomotive force can be reduced. To validate the effectiveness, magnetomotive force is calculated using Fourier series expansion. In order to consider the affection of permeance and slot harmonics, FEM analysis is conducted using a test and prototype induction motor. Moreover, an efficiency map of the prototype induction motor is obtained for validating the superiority. From above results, it can be seen that maximum torque and output power can be improved by using the present pole-changing method.

In the future work, optimization of coil and slot combination is conducted to acquire well-balanced magnetic characteristics at low and high pole modes. Moreover, the drive control method for reducing torque shock in the pole-changing period is developed and tested.

REFERENCES

[1] Responsible Resource Management for a Sustainable World: Findings from the International Resource Panel, June 2012, United Nations Environmental Program.

[2] T. Kato, M. Minowa, H. Hijikata, K. Akatsu and R. D. Lorenz: "Design methodology for variable leakage flux IPM for automobile traction drives", *IEEE Trans. on IAS.*, vol. 51, no. 5, pp. 3811-3821, 2015.

[3] N. Denis, M. R. Dubois, J. P. Trovao and A. Desrochers: "Power split strategy optimization of a plug-in parallel hybrid electric vehicle", *IEEE Trans. on Vehicular technology*, vol. 67, no. 1, pp. 315-326, 2018.

[4] M. Mori, T. Mizuno, T. Ashikaga and I. Matsuda: "A control method of an inverter-fed six-phase pole change induction motor for electric vehicles", *IEEJ Transactions on Industry Applications*, Vol. 117, No.6, pp.688-695 , 1997.

[5] L. M. Melcescu, M. V. Cistelecan, O. Craiu and H. B. Cosan: "A new 4/6 pole-changing double layer winding for three phase electrical machines", *proceedings of XIX international conference on electrical machines*, 2010.

[6] M. V. Cistelecan, L. M. Melcescu, H. B. Cosan and M. Popescu: "Induction Motors with Changeable Pole Windings in the Ratio 1:4", *in Proc. Int. Aegean Conf. Electr. Mach. Power Electron. Electromotion Joint Conf.*, pp.781-786, 2011.

[7] M. Okayasu and K. Sakai: "Novel integrated motor design that supports phase and pole changes using multiphase or single-phase inverters", in 18th European conference on power electronics and application, pp.1-9, 2016.

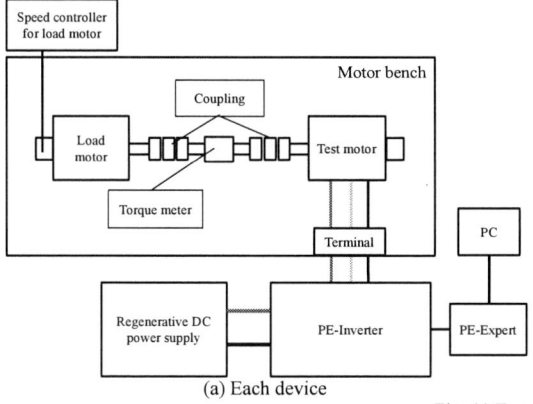

(a) Each device

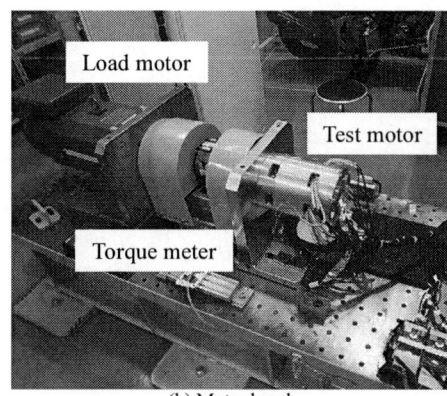

(b) Motor bench

Fig. 11 Experimental system

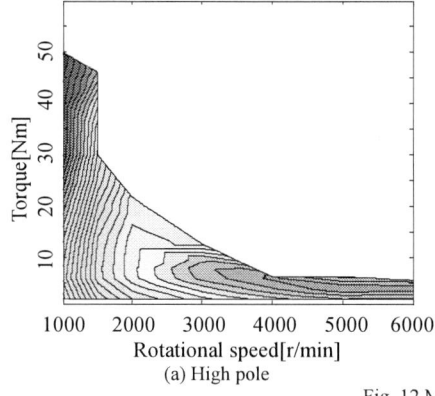

(a) High pole

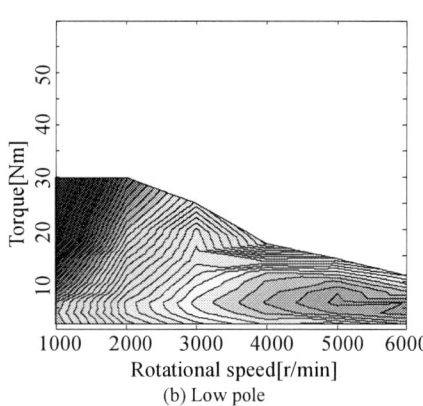

(b) Low pole

Fig. 12 Motor efficiency map

Starting Characteristics of an Ultra-Lightweight Motor Using Magnetic Resonance Coupling

Kenta Takishima[1] and Kazuto Sakai[2*]

1 Graduate School of Science and Engineering, Course of Electricity, Electronics and Communications,
Toyo University, Kawagoe-shi, Saitama, Japan
2 Department of Electrical, Electronic and Communication Engineering, Toyo University, Kawagoe-shi, Saitama, Japan
*E-mail: k_sakai@toyo.jp

Abstract— We propose a novel machine, which can convert electrical energy between the stator and the rotor using magnetic resonance coupling (MRC) to produce an ultra-lightweight motor. MRC was employed for wireless power transfer in the proposed motor, by enabling energy conversion between the stator and the rotor without magnetic cores, resulting in a reduction in the weight of the motor. In this study, we propose a motor design using MRC and describe its operating principles, starting characteristics, and resonance conditions. We analyzed the proposed MRC motor using magnetic analysis to clarify its essential characteristics and verify the resonance conditions calculated from the equivalent circuit. Our results confirmed that the proposed MRC motor without magnetic cores can convert electrical energy between the stator and the rotor, thereby generating torque. Also, we verified that the resonance condition derived from the equivalent circuit is a valid technique to achieve a resonant state.

Keywords— *Aircraft, Electric vehicles, Induction motors, Magnetic resonance.*

I. INTRODUCTION

Electric aircraft have been attracting attention as a potential energy-saving technology in recent studies. However, these aircraft require an ultra-lightweight motor for their practical realization. Conventional motors comprise windings and heavy magnetic cores to generate a high magnetic flux density and convert electromagnetic energy in the air gap effectively.

To reduce the weight of motors, we propose a novel technology that combines electromagnetic resonance coupling theory with induction motor technology. Kurs et al. proposed the idea of using electromagnetic resonance coupling which can transfer electromagnetic energy between two coils placed apart due to the electromagnetic resonance action [1].

In this paper, we describe the principle of a magnetic resonance coupling (MRC) motor without magnetic cores and discuss its starting characteristics. Furthermore, we derive the resonance conditions from an equivalent MRC motor circuit.

We analyze the key characteristics using an equivalent circuit and electromagnetic field analysis. By deriving the resonance conditions from an equivalent circuit, we found that three resonant frequencies exist based on the combination of one coil and a resonant capacitor. The three resonant frequencies can be classified into two categories: a series resonant frequency and a parallel resonant frequency. To understand both the series resonant frequency and the parallel resonant frequency, we performed magnetic field analysis with two power supply patterns with a constant current source and a constant voltage source. The results of our analysis determined the characteristics estimated by the equivalent circuit, and confirmed that the proposed MRC motor without magnetic cores can transfer electrical energy between the stator and the rotor, and produce torque.

II. PRINCIPLES AND BEHAVIOR OF THE MRC MOTOR

The fundamental circuit in an MRC motor comprises primary and secondary multiphase windings, which generate a magnetic field and a surrounding electrical field [2, 3]. The rotating field is generated by the multiphase current of the stator couples with the same induced polarity field as the current of the rotor winding. The electrical rotating field on the rotor, originating from the difference between the rotating field of the stator and mechanical rotating speed, synchronously couples with the rotating field of the stator as shown in Fig. 1.

The MRC motor operates based on fundamental principles that are similar to those of induction motors. Induction motors transfer energy by electromagnetic induction between the stator and the rotor through an air gap. In contrast, MRC motors do not have adequate magnetic paths since they do not use iron cores. Therefore, in this motor design, MRC is used to transfer energy efficiently without electromagnetic induction. MRC can be used to transfer enormous amounts of energy through large air gaps between two detached coils that are connected via a resonant capacitor. Thus, in the proposed MRC motor, two resonant capacitors are connected to the multiphase winding of the stator and to the rotor coil. The resonant frequencies corresponding to these capacitors are the frequency of power supply for the stator and the slip frequency for the rotor, respectively, as shown in Fig. 1. The resonant capacitor values of the stator and the rotor are determined by the inductance and the resonant frequency of each coil. The stator and the rotor are made of a nonmagnetic material and the proposed MRC motor solely comprises of coils and resonant capacitors, making the motor ultra-lightweight.

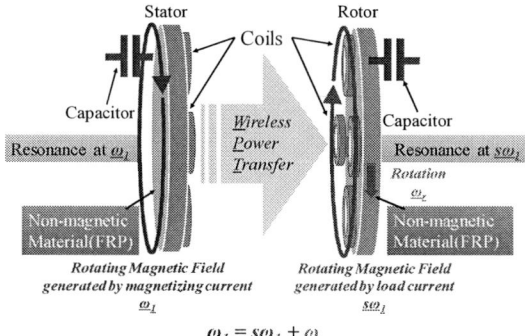

Fig. 1. Conceptual scheme of a magnetic resonance coupling motor.

III. AN EQUIVALENT CIRCUIT AND THE RESONANCE CONDITIONS OF THE PROPOSED MRC MOTOR

An equivalent circuit can be used to qualitatively consider the proposed MRC motor. The equivalent circuit of the MRC motor combines an equivalent MRC circuit and an induction motor. The equivalent MRC motor circuit comprises a winding resistance, a mutual inductance coupled with the coils and the leakage inductance. It also includes capacitors to induce the resonance. Fig. 2 shows an equivalent circuit (not considering the resistance of the windings when the proposed MRC motor starts at a slip of s = 1).

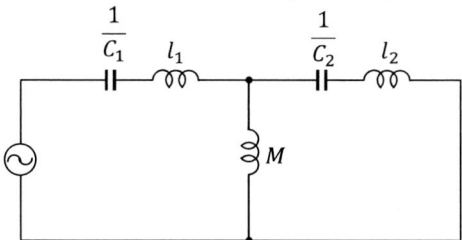

Fig. 2. Equivalent circuit of the proposed MRC motor.

Here, (1) shows Z, which is the impedance viewed from the primary side. The indexes of Z refer to the leakage inductances of the primary and secondary side, the mutual inductance, and the resonance capacitances of the primary and secondary side.

$$Z = Z_{C1} + Z_{l1} + \frac{Z_M \left(Z_{l2} + Z_{C2} \right)}{Z_M + Z_{l2} + Z_{C2}} \quad (1)$$

where l_1, l_2, and M are the primary leakage inductance, secondary leakage inductance, and mutual inductance, and C_1 and C_2 are the capacitances of the primary and secondary resonant capacitors, respectively.

The parallel resonance condition states that the denominator of the impedance in (1) is zero and it is shown in (4). In other words, an essential condition is that the reactance component of the impedance in (1) is maximized.

$$Z_{L2} + Z_{C2} = 0 \quad (2)$$

$$\omega L_2 + \frac{1}{\omega C_2} = 0 \quad (3)$$

$$\omega_o = \frac{1}{\sqrt{L_2 C_2}} \quad (4)$$

The series resonance conditions that state the numerator of the impedance in (1) is zero are shown in (5)–(8). In other words, these are the conditions wherein the reactance component of the impedance in (1) is minimized.

$$\left(Z_{l1} + Z_{C1} \right)\left(Z_M + Z_{l2} + Z_{C2} \right) + Z_M \left(Z_{l2} + Z_{C2} \right) = 0 \quad (5)$$

$$Z_{C1}\left(Z_{L2} + Z_{C2} \right) + Z_{L1}\left(Z_{L2} + Z_{C2} \right) - Z_M{}^2 = 0 \quad (6)$$

$$\left(\omega L_1 - \frac{1}{\omega C_1} \right)\left(\omega L_2 - \frac{1}{\omega C_2} \right) = \left(\omega M \right)^2 \quad (7)$$

$$\left\{ \left(\sqrt{L_1 L_2} + M \right)\sqrt{C_1 C_2}\, \omega^2 - 1 \right\}\left\{ \left(\sqrt{L_1 L_2} - M \right)\sqrt{C_1 C_2}\, \omega^2 - 1 \right\}$$
$$+ 2\sqrt{L_1 L_2 C_1 C_2} - \left(L_1 C_1 + L_2 C_2 \right) = 0 \quad (8)$$

Then, (9) shows the arithmetic and geometric mean.

$$2\sqrt{L_1 L_2 C_1 C_2} = \left(L_1 C_1 + L_2 C_2 \right)$$
$$\left(when \quad L_1 C_1 = L_2 C_2 \right) \quad (9)$$

From the above equations, the two series resonance frequencies are shown in (10) and (11) are obtained.

$$\omega_m = \frac{1}{\sqrt{\left(\sqrt{L_1 + L_2} + M \right)\left(\sqrt{C_1 C_2} \right)}} \quad (10)$$

$$\omega_e = \frac{1}{\sqrt{\left(\sqrt{L_1 + L_2} - M \right)\left(\sqrt{C_1 C_2} \right)}} \quad (11)$$

In the case of a constant current power source, the input voltage V_1 is obtained by the product of the input impedance Z_1 and the input current I_1 as shown in Fig. 3(a). Therefore, when the input impedance reaches a maximum due to the parallel resonant frequency that is given by (4), the input voltage reaches its maximum, and when the input impedance is the minimum due to the series resonant frequencies that are given by (10) and (11), the input voltage becomes the minimum. In contrast, in the case of a constant voltage power source, the input current I_1 is obtained by the quotient of the input impedance Z_1 and the input voltage V_1 as shown in Fig. 3(b). Thus, when the input impedance reaches a maximum such as when it has the parallel resonant frequency, the input current becomes the minimum, and when the input impedance is at the minimum such as when it has the series resonant frequency, the input current becomes the maximum.

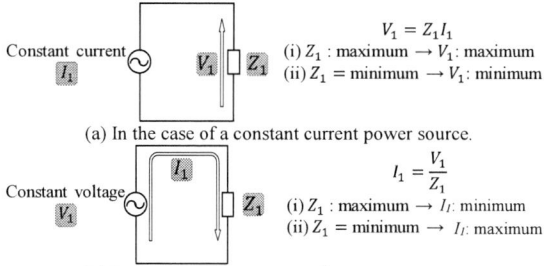

(a) In the case of a constant current power source.

(b) In the case of a constant voltage power source.

Fig. 3. Relationships between input current, input voltage, and input impedance.

Fig. 4 shows the frequency characteristics and resonance frequency points calculated from (4) for a parallel

2827

configuration and (10) and (11) for a series configuration, respectively. As shown in Fig. 4(a), the primary voltage, rotor current, and torque increase dramatically and reach a maximum at parallel resonant frequency, ω_o. On the other hand, as shown in Fig. 4(b), the stator current, rotor current, and torque increase dramatically and reach a maximum at ω_m and ω_e of the series condition. Also, the stator current, rotor current, and torque decrease and reach a minimum at ω_o of the parallel resonance condition.

Thus, when the aim is to find the parallel and series resonant frequencies, it is effective to investigate the characteristics by analyzing a system with a constant current source and a constant voltage source, respectively.

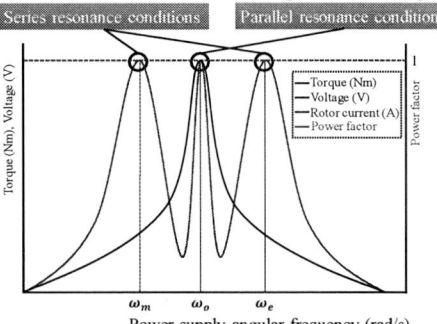

(a) In the case of a constant current power source.

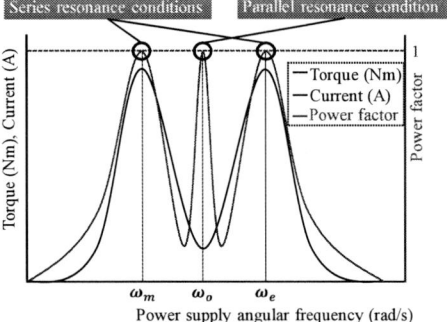

(a) In the case of a constant voltage power source.

Fig. 4. Resonance condition of the MRC motor derived from the equivalent circuit.

IV. VERIFICATION MODEL OF A MRC MOTOR

Fig. 5 shows a basic configuration of the proposed axial-flux-type MRC motor model. Fig. 6 shows the circuits of the MRC motor. Table I shows the specifications of the MRC motor model.

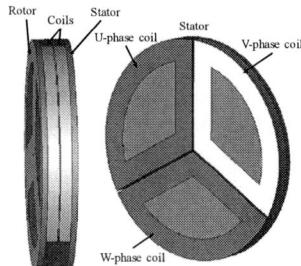

Fig. 5. Analysis model of the proposed MRC motor.

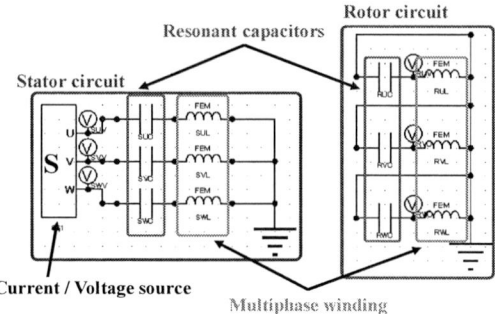

Fig. 6. MRC motor analytical circuit.

TABLE I
SPECIFICATIONS FOR THE MRC MOTOR

Pole/Phase		2/3
Input current (A$_{rms}$)		4
Input voltage (V$_{rms}$)		70
Mechanical gap length (mm)		1
Conductor sectional area(mm^2)	Stator	0.196
	Rotor	0.196
Outer diameter of stator (mm)	Stator	300
	Rotor	300
Series turn of coil (turn/phase)	Stator	250
	Rotor	250
Winding resistance/phase (Ω/phase)	Stator	2.16
	Rotor	2.16
Winding inductance/phase (mH/phase)	Stator	16.1
	Rotor	16.1

V. MOTOR PERFORMANCE

Magnetic field analysis was performed to obtain the frequency characteristics of the torque for the MRC motor model using the JMAG finite element magnetic field analysis software. Capacitor values in the circuit were determined by the inductance of the MRC motor obtained using FEM analysis and the resonance frequency of each winding. The capacitors are shown in Table II. C_1 and C_2 are the capacitor values of the stator and the rotor, respectively.

TABLE II
CALCULATED CAPACITOR VALUE

Resonance frequency	C_1 (μF)	C_2 (μF)
1 kHz	1.59	1.59
2 kHz	0.397	0.397
3 kHz	0.176	0.176
4 kHz	0.0992	0.0992
5 kHz	0.0635	0.0635

A. Starting characteristics in the case of a constant current power source

Fig. 7 shows the starting power factor characteristics of the primary circuit, primary voltage, rotor current, and torque, respectively. As shown in Fig. 7(a), there are three resonant frequencies with which the power factor is 1.0 with each resonant capacitor. Then, in Fig. 7(b–d), it can be seen that a frequency exists, at which the characteristic values varied sharply and reached peaks. These characteristics are similar to the estimated characteristics from the equivalent circuit as shown in Fig. 4(a).

The 2018 International Power Electronics Conference

It can be seen that Fig. 7(b) and Fig. 7(c) exhibit overvoltage and overcurrent conditions. In reality, however, the plan would be to actually drive the motor at a slip frequency of ≤100 Hz, so that the primary voltage would be the square of the slip frequency, and the rotor current and torque would be proportional to the slip frequency. Therefore, when calculating the slip frequency of 100Hz, the primary voltage is 300 V, the rotor current is 40 A, and the torque is 1.8 Nm, approximately.

B. Starting characteristics in the case of a constant voltage power source

In a similar configuration to the constant current power source, Fig. 8 shows the starting power factor characteristics of the primary circuit, stator current, rotor current, and torque, respectively. As shown in Fig. 8(a), there are three resonant frequencies at which the power factor is 1.0 for each resonance capacitor. However, in Fig. 8(b–d), two frequencies were found, at which the characteristics value varied sharply and became peaks. This indicates that a large current flow is associated with two of the three resonance frequencies.

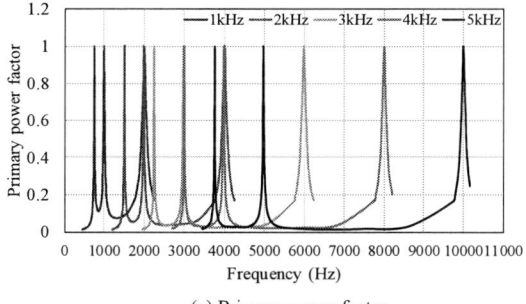

(a) Primary power factor

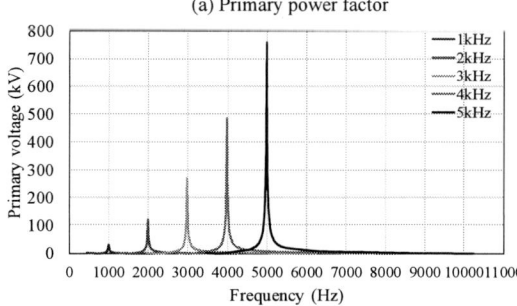

(b) Primary voltage

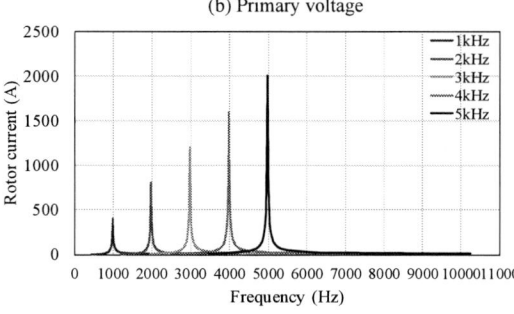

(c) Rotor current

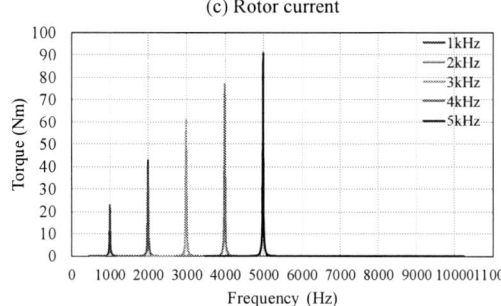

(d) Torque

Fig. 7. Starting characteristics in the case of a constant current power source.

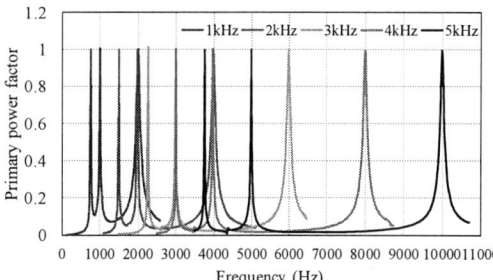

(a) Primary power factor

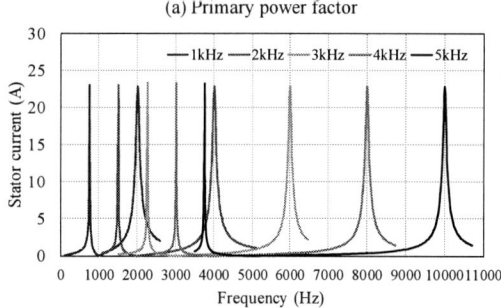

(b) Stator current

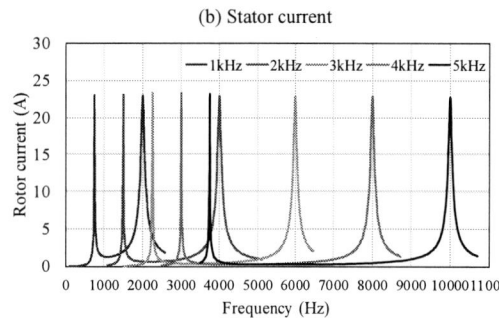

(c) Rotor current

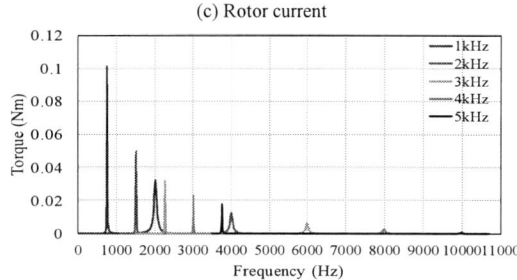

(d) Torque

Fig. 8. Starting characteristics in the case of a constant voltage power source.

These characteristics are similar to the estimated characteristics from the equivalent circuit as shown in Fig. 4(b).

Table III lists the resonance frequencies obtained from (4), (10), and (11) for the range 1–5 kHz. Additionally, as shown in Fig. 8 and Table III, it can be confirmed that the resonances occur at the same resonance frequencies as the calculated resonance frequencies. f_m , f_o , and f_e are the three resonance frequencies calculated from ω_m , ω_o , and ω_e .

TABLE III
RESONANT FREQUENCIES

		Resonance frequency (Hz)
1 kHz	f_m	752
	f_o	995
	f_e	1997
2 kHz	f_m	1503
	f_o	1990
	f_e	3994
3 kHz	f_m	2255
	f_o	2985
	f_e	5991
4 kHz	f_m	3007
	f_o	3979
	f_e	7988
5 kHz	f_m	3758
	f_o	4974
	f_e	9985

VI. SHIFT OF STARTING CHARACTERISTICS IN THE CASE OF A CONSTANT VOLTAGE POWER SOURCE

The three resonance frequencies were calculated from (4), (10), and (11). Then, we discussed whether it is possible to shift the three resonance frequencies in parallel by changing the value of the capacitors using the resonance condition equations: (4), (10), and (11). Table IV shows the calculated capacitor with a shifting resonance frequency point. It is indicated that the resonance frequency point shifts as shown in Fig. 9 and Table V.

TABLE IV
CALCULATED CAPACITORS VALUE FOR SHIFT

Resonance frequency	C_1 (µF)	C_2 (µF)
1 kHz	6.33	6.33
2 kHz	1.58	1.58
3 kHz	0.704	0.704
4 kHz	0.396	0.396
5 kHz	0.253	0.253

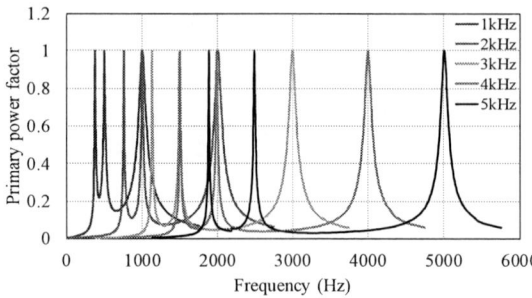

(a) Primary power factor

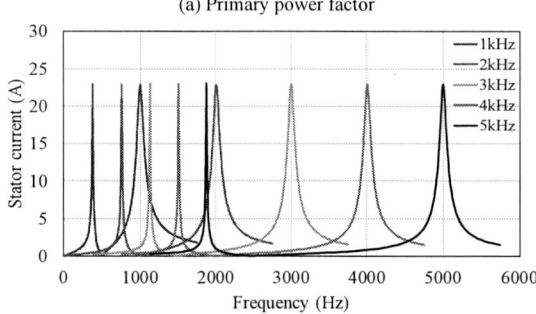

(b) Primary voltage

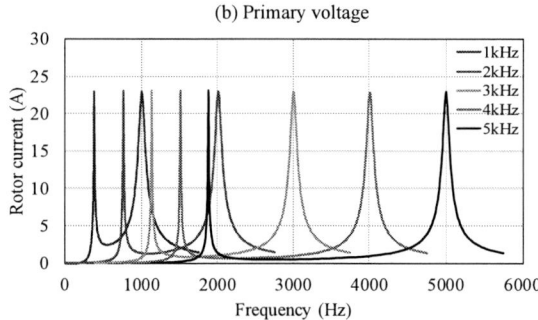

(c) Rotor current

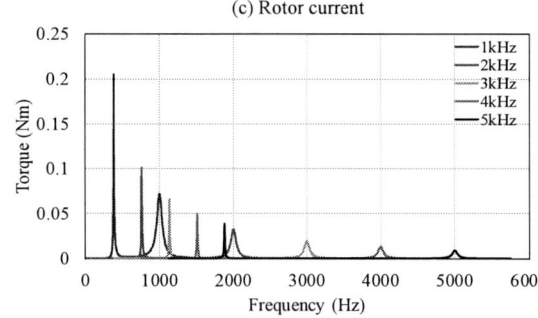

(d) Torque

Fig. 9. Shifted starting characteristics in the case of a constant voltage power source.

TABLE V
SHIFTED RESONANT FREQUENCIES

		Resonance frequency (Hz)
1 kHz	f_m	376
	f_o	498
	f_e	1000
2 kHz	f_m	753
	f_o	996
	f_e	2000
3 kHz	f_m	1129
	f_o	1494
	f_e	3000
4 kHz	f_m	1505
	f_o	1993
	f_e	4000
5 kHz	f_m	1882
	f_o	2491
	f_e	5000

VII. CONCLUSIONS AND FUTURE WORK

We proposed a novel motor without iron cores using MRC. Our analytical results confirmed that the MRC motor is capable of converting electrical energy between the stator and the rotor and producing a torque. Furthermore, the required resonance conditions can be derived from the equivalent circuit and the analysis results verified that the MRC motor reached the resonant state under resonance conditions. The MRC motor has two series resonance frequencies and one parallel resonance frequency for a configuration combining one coil and a resonant capacitor. To investigate the two types of resonant frequencies, magnetic field analysis was performed in the case of the constant current source and in that of the constant voltage source. The results of magnetic field analysis agreed with the characteristics estimated from the equivalent circuit. Furthermore, by changing the resonant capacitance values using the resonance condition equation, it is possible to shift the resonance point.

In the future, we plan to analyze and verify the operating characteristics of the MRC motor with practical experiments.

ACKNOWLEDGMENT

This work was supported by JSPS KAKENHI Grant Number JP17K06313.

REFERENCES

[1] A. Kurs, A. Karalis, R. Moffatt, J. D. Joannopoulos, P. Fisher, and M. Soljačić, "Wireless power transfer via strongly coupled magnetic resonances," Science, vol. 317, no. 5834, pp. 83-86, 2007.

[2] K. Sakai and Y. Sugasawa, "Ultralightweight motor design using electromagnetic resonance coupling," in *Proc. 2016 IEEE ECCE.*, EC-0201

[3] K. Sakai, K. Takishima, and K. Nihei, "Principle and characteristics of an Ultralightweight electromagnetic resonance coupling machine with a cage rotor," in *Proc. 2017 IEEE ECCE.*, EC-0193.

The 2018 International Power Electronics Conference

Design and Basic Characteristics Analysis of Toroidal Winding Axial Gap Induction Motor

Ryosuke Sakai[1]*, Yukihiro Yoshida[2] and Katsubumi Tajima[1]

1 Department of Cooperative Major in Life Cycle Design Engineering, Akita Univ., *1-1, Tegata Gakuen-machi, Akita 010-5802, Japan*
2 Department of Electrical and Electronic Engineering, Akita Univ., *1-1, Tegata Gakuen-machi, Akita 010-5802, Japan*
*E-mail: m8017902@s.akita-u.ac.jp

Abstract— A proposed axial gap induction motor (AGIM) with an improved space factor using toroidal winding exhibited double the output torque of a radial gap induction motor (RGIM) of the same size. The structure of the AGIM's toroidal winding, in consideration of actual machine production, was then studied, and the results of the torque characteristics analyses were compared with those of previous analyses. The actual machinery was made using a powder core, and its maximum torque was 1.6 times that of a finite element analysis RGIM when set at the same torque. The high efficiency confirmed that the AGIM had improved.

Keywords— *induction motor, toroidal winding, axial gap motor, finite element analysis.*

I. INTRODUCTION

Recently, interest in environmental problems represented by global warming is increasing yearly. Therefore, the demand for electric vehicles (EVs) and hybrid electric vehicles (HEVs) has increased because they produce less carbon dioxide emissions than conventional vehicles. Permanent magnet synchronous motors (PMSMs) are widely used for EVs and HEVs because of their high torque and high efficiency. However, owing to the rising price of rare earth metals used in PMSMs, the development of high performance rare earth-free motors is required [1], [2]. As an alternative, induction motors (IMs) have several advantages because of their simple structure with no permanent magnet; they are able to rotate at high speed, have a robust structure, and are not affected by soaring rare earth metal prices. However, it is difficult to generate high torque with IMs because of their magnet-free structure; the excitation current is applied to the primary winding, and there is a risk of heat generation and magnetic saturation. Improvement of the torque density and efficiency of IMs is required [3].

Recently, several effective techniques for increasing torque densities have been reported; it is easier to increase the gap area of axial gap motors, which improves the torque, compared to conventional motors [4]. Toroidal winding, compared to distributed winding, can improve the space factor and reduce coil end length,

which increases torque density [5]. Therefore, it was expected that the torque density could be improved by applying these techniques to IMs.

This study initially proposed an axial gap induction motor (AGIM) with improved output torque characteristic. First, the proposed design had the same dimensions as that of a RGIM with a conventional structure. Second, simulated results of the AGIM's basic properties were compared to those of the RGIM. The proposed AGIM had an output torque density that was 1.60 times that of the conventional RGIM and an efficiency improvement of 1.8 points at the operating point of the maximum torque of the RGIM was obtained.

II. DESIGN METHOD

A. Specifications of RGIM

Fig. 1 shows a conventional RGIM for comparison; the dimensions refer to the actual machine. The motor specifications are shown in Table 1. The motor diameter, the iron core thickness, coil end thickness, and the gap width were 100, 30, 28, 0.35 mm, respectively. The distributed winding formed four poles.

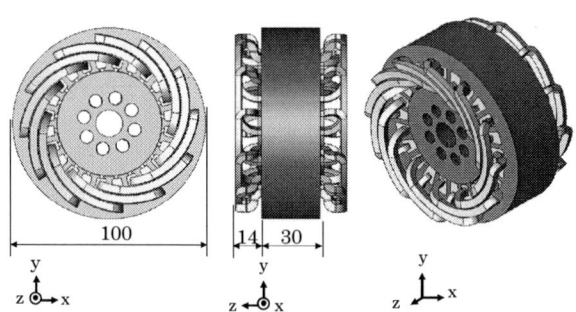

Fig. 1. RGIM to be compared.

TABLE I
SPECIFICATION OF RGIM

Stator diameter	100 mm
Core thickness	30 mm
Motor thickness	58 mm
Air gap length	0.35 mm
Number of poles	4
Number of slots (stator/rotor)	24/34
Winding method	Distributed winding
Number of windings	40
Wire diameter	0.5 mm
Winding space factor	11%
Secondary conductor cross-sectional area	9.62 mm^2
Stator core, Rotor core	Laminated steel sheet
Coil	Cu
Rotor Conductor	Al
Shaft	Air

B. Design of AGIM

Fig. 2 shows the coil arrangement of the proposed AGIM's toroidal winding. The coils were wound concentrically around the slot and the yoke; therefore, the toroidal winding produced an equivalent magnetomotive force to the distributed winding without interfering with other coils. Compared to the RGIM, the winding space factor was increased from 11% to 40%. The proposed AGIM had a single stator and a double rotor, which increased the airgap area to generate a larger torque compared to the conventional RGIM.

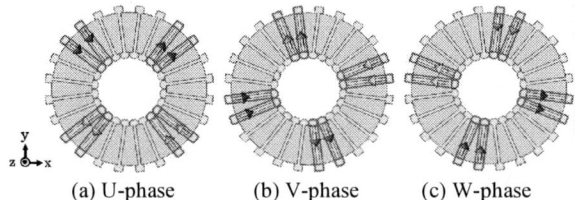

(a) U-phase (b) V-phase (c) W-phase
Fig. 2. Toroidal winding coil arrangement of the proposed axial gap induction motor.

Fig. 3 shows an overview of one side of the rotors. The cross-sectional area of the rotor conductor bar, the inner and outer diameter, and the slot opening width of the rotor are equivalent to those of RGIM rotor. Another rotor of similar shape is prepared to construct a double rotor structure.

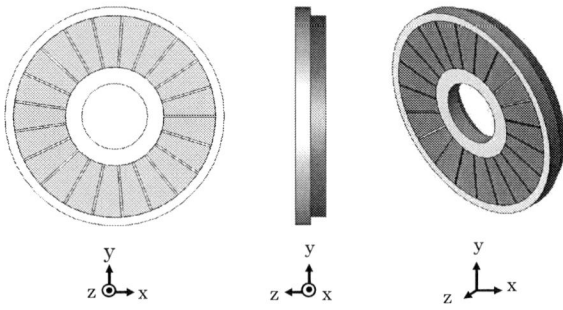

Fig. 3. Overview of rotor.

However, it was expected that the proposed AGIM would be difficult to manufacture using only electromagnetic steel plates because the magnetic flux flows in a three-dimensional (3D) manner and of the shape of the flange at the tip of the teeth. Therefore, a powder core was used to make a stator structure. Fig. 4 shows the BH characteristics of the laminated steel sheet and powder core. The powder core had lower relative magnetic permeability and saturation magnetic flux density than those of the laminated steel sheet used for the RGIM. Fig. 5 shows the designed model and Table II lists the designed motor's materials.

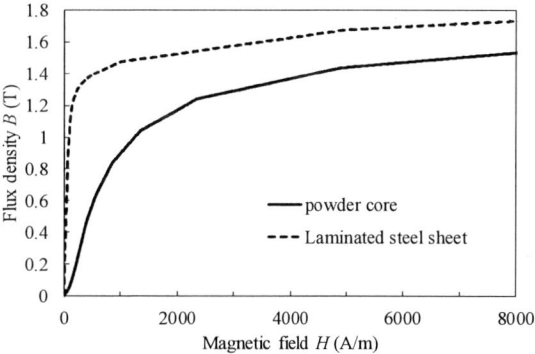

Fig. 4. BH characteristics of laminated steel sheet and powder core.

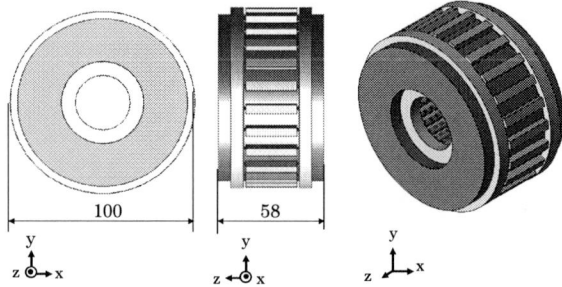

Fig. 5. Overview of AGIM.

TABLE II
MATERIAL OF DESIGNED MOTOR

Stator Core	Powder core
Rotor Core	Powder core
Coil	Cu
Rotor Conductor	Al
Shaft	Air

III. COMPARISON OF TORQUE CHARACTERISTIC

The torque characteristics of the proposed motor and the RGIM were compared. In this study, the magnetic field distribution in the slip state was analyzed by frequency response analysis using JMAG-Designer, Version 16.0, software for three-dimensional finite element analysis (3D-FEA). The frequency range was determined at 50 Hz maximum from the operating frequency of the actual machine. A current amplitude of

The 2018 International Power Electronics Conference

4 A was applied to the motor. Table III shows the analysis conditions of the FEA for the AGIM and RGIM. Fig. 5 shows the torque–slip characteristics of the two motors, and Table IV shows the comparison of the maximum torque and torque density. The maximum torque of the RGIM was 0.617 N·m, while it was 0.988 N·m for the AGIM. The torque density of the RGIM was 1.353 N·m/L, and 2.167 N·m/L for the AGIM. The proposed AGIM provided 1.6 times the maximum torque and torque density of the RGIM, even though the *BH* characteristics of the powder core were lower than that of the laminated steel sheet.

TABLE III
ANALYSIS CONDITIONS

Simulation mode	Frequency response analysis
Frequency range	~0.5–50 Hz
Step	31
Current amplitude	4 A

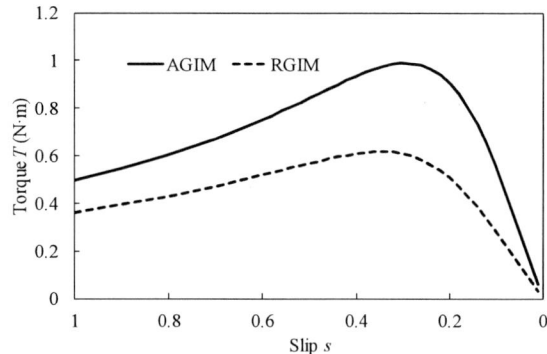

Fig. 5. Torque – slip characteristic.

TABLE IV
COMPARISON OF MAX TORQUE AND TORQUE DENSITY

	RGIM	AGIM
Max torque (N·m)	0.617	0.988
Volume (L)	0.456	0.456
Torque density (N·m/L)	1.353	2.167
Maximum torque ratio	1	1.601
Torque density ratio	1	1.601

IV. COMPARISON OF EFFICIENTRY

As a simple comparison, the current where the maximum torques of the two motors became equal was analyzed. The AGIM input current was reduced to equalize the torque. To analyze the efficiency, it was necessary to analyze the transient response of the FEA because iron loss was calculated by FFT. However, since the induction motor is current induced to the rotor conductor, a very long analysis time was required. Therefore, to shorten the analysis time, the analysis target was replaced with a two-dimensional (2D) model.

The method for replacing the AGIM with a 2D model is illustrated in Fig. 6. A cylinder, with a diameter bisecting the gap area of the stator, was developed into a 2D shape, and the difference between the inner diameter and the outer diameter was given as a thickness. The rotor conductor was expressed by connecting the conductor bar with the end-ring resistance of the external circuit. Fig. 7 shows the 2D analytical model.

The analysis conditions are shown in Table V. The rotation speed was set from the rotation speeds at which the AGIM and RGIM exhibited maximum torque, respectively. The winding resistance was the measured value from the actual machine. The winding resistance of the AGIM was measured by winding it in a stator shape with a 3D printer. The conductor resistance of the rotor was calculated from the shape.

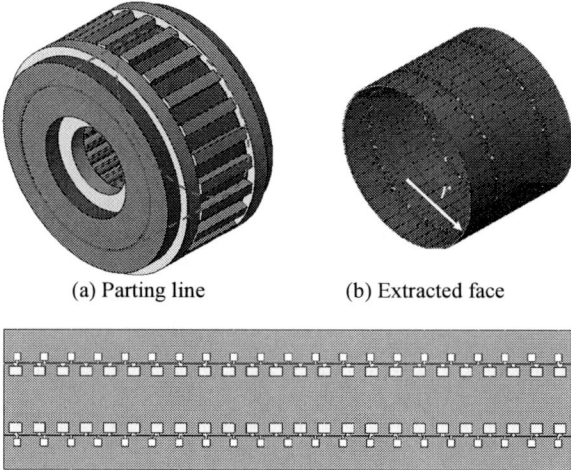

(a) Parting line   (b) Extracted face

(c) Expanded face

Fig. 6. A method of replacing AGIM with a two-dimensional model.

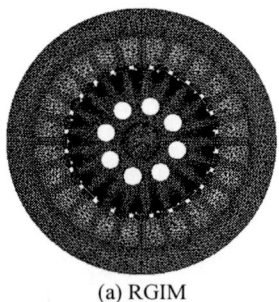

(a) RGIM

(b) AGIM

Fig. 7. Analysis model.

TABLE V
ANALYSIS CONDITIONS

Simulation mode	2D transient analysis
Simulation time (s)	~0–0.40
Time step (ms)	0.15625
Current amplitude (AGIM)	3.2 A
Current amplitude (RGIM)	4 A
Frequency (Hz)	50 Hz
Slip	0.34
Winding resistance (AGIM)	2.097 Ω
Winding resistance (RGIM)	2.838 Ω

Fig. 8. shows the torque of the AGIM and RGIM from the transient response analysis. The torque of the AGIM, T, was calculated as follows:

$$T = F \times r \tag{1}$$

where, F is the electromagnetic force obtained by analysis and r is the motor radius shown in Fig. 6. Equation (1) was used because the AGIM replaced by the 2D model was equal to the linear motor; the result was outputted as electromagnetic force F. The average torque was calculated from the results obtained between 0.35 and 0.4 s, and it was confirmed that the torques of both motors were consistent.

Table VI shows the primary copper loss, secondary copper loss, iron loss, and efficiency of each motor. The efficiency η was calculated as follows:

$$\eta = \frac{\omega T}{\omega T + W_{cp} + W_{cs} + W_i} \tag{2}$$

where, co is the motor angular velocity, W_{cp} is the primary copper loss, W_{cs} is the secondary copper loss, and W_i is the iron loss. The efficiency was improved because the primary copper loss of the AGIM was greatly reduced compared to the RGIM. Secondary copper loss and iron loss increased for the RGIM. Since the AGIM has a double rotor structure, the secondary copper loss and the iron loss of the rotor increased. The powder core had greater iron loss compared to the electromagnetic steel plate at the analyzed frequency. There is a possibility that the magnitude of the iron loss may be reversed when used at high rotation speeds. The efficiency obtained from the result of the loss calculation above was 37.3% for the RGIM and 39.1% for the AGIM; thus, an improvement of 1.8 points was confirmed. The value of this efficiency was calculated at the maximum torque point of the RGIM and not its maximum efficiency. Efficiency could be further improved by undertaking an optimum design study. Because this value was obtained from a 2D analysis, it is not accurate, and loss and efficiency changed owing to slip; therefore, further verification is required.

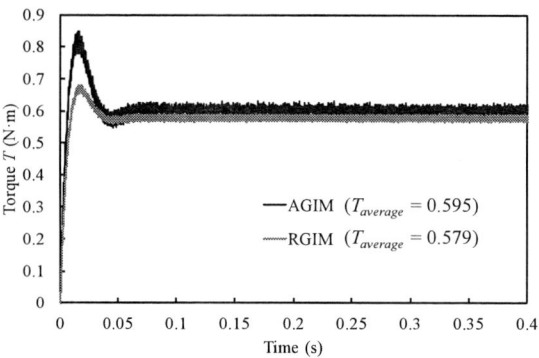

Fig. 8. Torque of the AGIM and RGIM using transient response analysis.

TABLE VI
COMPARISON OF MOTOR LOSS AND EFFICIENCY

	RGIM	AGIM
Primary copper loss (W)	61.7	42.1
Secondary copper loss (W)	31.4	50.7
Iron loss (W)	1.98	3.62
Efficiency (%)	37.3	39.1

V. CONCLUSION

This paper presented an induction motor with an axial gap and toroidal winding structure. The complex shape of the AGIM was manufactured using a powder core. Testing of the proposed AGIM indicated that the maximum torque and the torque density were up to 1.6 times greater than a conventional RGIM. From these results, an advantage, with respect to high torque density, has been demonstrated by comparing the torque density with a conventional RGIM. Future study will include actual machine production and efficiency improvements.

REFERENCES

[1] H. Arihara, K. Akatsu: "A Basic Property of Axial Type Switched Reluctance Motor", ICEMS 2010, pp. 1681 - 1686 (2010).

[2] M. Obata, S. Morimoto, M. Sanada, Y. Inoue: "Characteristic of PMASynRM with Ferrite Magnets for EV/HEV Applications", ICEMS 2012, DS3G2-7, (2012).

[3] T. Nishiyama, K. Endo, A. Matsuda: In-wheel Motor Genri to Sekkei (in Japanese), p.128, (Kagaku Joho Syuppan, Ibaraki, 2004).

[4] Aydin, S. Huang, T. A. Lipo: "Axial flux permanent magnet disc machines: A review", WEMPEC Research Report 2004-10, (2004).

[5] Y. Iwai, Y. Yoshida, K. Tajima: "Consideration of Efficiency Improvement of Ferrite Magnet Motor with Toroidal Winding", *The papers of Technical Meeting on Magnetics IEEJ*, MAG 15-117 (2015).

The 2018 International Power Electronics Conference

Magnet Arrangement suitable for Large Air Gap Length in Linear PM Vernier Motor

Tatsuya Ninomiya[1], Abdulaziz Gasim[1] and Shoji Shimomura[1*]

1 Department of Electrical Engineering, Shibaura Institute of Technology, Tokyo, Japan

*E-mail: simomura@shibaura-it.ac.jp

Abstract— This paper makes basic examination on four models of linear PM vernier motors in order to apply them to a steel-wheel linear motor train. Their models are categorized into two types with PM arrangement. One is the motor that has the PM arrangement based on the original design of the conventional PM vernier motor, and the other is the motor with a spoke type arrangement of PMs, which we have previously proposed. Typical vernier motors have a short air gap length, but the one that is applied to the steel-wheel linear motor train requires a larger air gap length. The examinations find a result that in the motor with the conventional PM arrangement increase of the PM volume hardly contribute to increase of the thrust, but that in the motor with the PM arrangement of a spoke type has a good effect on increasing the thrust.

Keywords— *steel-wheel linear motor train, vernier motor, permanent magnet arrangement.*

I. INTRODUCTION

In Japan, Central Japan Railway co. is planning to construct Linear Chuo Shinkansen, which will be able to run between Tokyo and Nagoya (approximately 300 km) in 40 minutes. That system introduces a magnetic levitation style, and the traction system uses a linear synchronous motor. The car has a superconducting coil as DC exciter, and propulsion coils (armature winding) and levitation coils will be lain along the very long railway. Such a system will make the construction costs extremely high compared with the conventional Shinkansen, but that will be in the relation of trade-off to obtain a maximum speed over 500 km/h.

As for other linear motor train, some subway lines in Japan have introduced a steel-wheel linear motor train, and the introducing it to a new subway line is also planned. In such a subway line, use of a linear motor instead of a rotary motor enabled to miniaturize the whole size of the train car, because the floor height has been lowered by introducing the linear motor. The miniaturization of the train car has made the required diameter of tunnel smaller compared with that of the conventional subway line. Since their steel-wheel linear motor train uses a linear induction motor, two rails and reaction plates between the two rails are only put along the railway. Consequently, use of the steel-wheel linear induction motor train would contribute greatly to the reduction of the construction costs of the subway line.

However, induction motors have efficiency less than that of permanent magnet (PM) motors of synchronous type, as known well. For that reason, a linear PM motor will be expected to be used for the steel-wheel linear motor train in the near future.

If introducing a linear PM motor of moving-armature type having a typical topology to a subway line, a very large amount of PM will be needed to pave them along the railway. That is a fatal disadvantage. We, therefore, have taken interest in the topology of not the surface PM vernier motor[1-3] but the dual excited PM vernier motor (DPMVM)[4-7], which we have studied applying for a traction motor of EV or HEV. The DPMVM is a kind of PM motors of synchronous type and has PMs on two sides of the stator and rotor in rotary machines. This motor in which the PMs have been removed from the rotor side also generates a torque due to an interaction between the armature current and the magnetic flux from the stator side PMs that is modified by permeance variation due to the small rotor teeth. The interaction is called magnetic gear effect. Adopting the novel topology will allow the linear PM motor train to use a small amount of PM and will reduce the construct costs.

We have already study the moving-armature type linear PM motor with such a novel topology to apply for an automatic transferring machine and have reported the verification results on the performance of a prototype of that linear motor[8], which will be called moving-armature type armature-PM-excited linear PM vernier motor (MA-APM-LVM) hereafter. Moreover, we have also proposed applying a spoke arrangement of PM to the MA-APM-LVM to obtain a larger thrust[9]. Although some motors with similar topology have been proposed, the motor types with a spoke type PM on the rotor don't match with our purpose[10-12], and we think that the linear motor of the flux switched type [13-16] will not be able to obtain enough thrust. However, in the MA-APM-LVM, there are some technical challenges to overcome. One of them is to develop the MA-APM-LVM that has a larger air gap length and generates a sufficiently large thrust. For that purpose, this paper examines the relationship between the air gap length and the magnitude of the generated thrust in the MA-APM-LVM based on the results from FEA, and also discusses the influence of the amount of PM or the PM arrangement on the generated thrust.

The 2018 International Power Electronics Conference

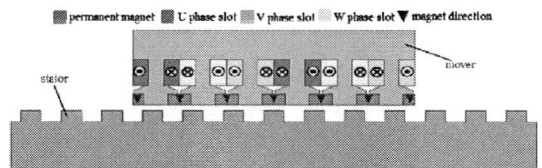

Fig.1 Basic configuration of MA-APM-LVM with the configuration A and also model A-1.

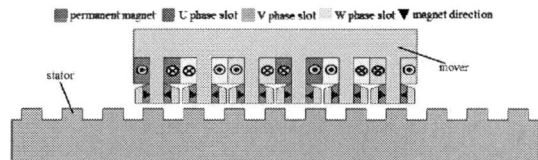

Fig.2 Basic configuration of MA-APM-LVM with the configuration B and also model B-1.

II. BASIC CONFIGURATION OF STUDY MODELS AND FORCE EQUATION

A. Basic Configuratons of Study models

Figures 1 and 2 show two basic configurations of the MA-APM-LVMs that will be examined in this paper, which have the same topology as the linear moto that we have previously proposed[8][9], i.e. the moving-armature linear motor having PMs only on the mover. In the two motors, each mover has the armature winding and some PMs that are set at the slot opening on the mover side. The PM arrangement of the motor shown in Fig. 1 is based on the original design of the conventional rotary type PMVM[4], and that of the motor shown in Fig. 2 adopts the spoke type PM arrangement that we have proposed in [9] to increase the yielded thrust. Each stator is made of typical magnetic steel sheets and only has small teeth as shown as seen from Figs. 1 and 2. The configuration having both the topology only with the mover PMs and the conventional PM arrangement will be called configuration A or simply conf. A, and the configuration having both the topology only with the mover PMs and the spoke type PM arrangement will be called configuration B or simply conf. B.

B. Thrust Equation

As shown in the previous paper[8], the theoretical equation expressing the thrust yielded by using the machine configurations in Figs.1 and 2 is given as

$$f = \frac{\sqrt{3}Nl_a}{\sqrt{2}} \frac{Z_2}{P_p} k_{w(1)} B_1 i_q \quad [\text{N}] \tag{1}$$

where N is the number of series conductors per phase, and l_a is stack length of the move and the stator, and Z_2 is the number of small slots on stator, and P_p is the number of pole pairs, and $k_{w(1)}$ is winding factor for fundamental component, and B_1 is fundamental component of air gap flux density due to PMs on the mover (i.e., under no-load condition), and i_q is q-axis current on d-p reference

frame. As seen from (1), The yielded thrust of the MA-APM-LVM is only proportional to i_q, which has no relation with i_d in the same as a typical surface PM motor and the conventional PMVM[].

III. EXAMINATION TO ENLARGE AIR GAP LENGTH

A. Motors with Configuration A

In general surface PM motor, it is known that increasing the PM volume has an effect on raising the torque or the thrust. Hence, the effect will be examined in two motors: the model A-1 and A-2 shown in Figs.1 and 3, respectively, in the beginning.

The PM of the model A-2 has twice the volume of PM of the model A-1. Owning to the increase of the PM volume, the mover core height of the model A-2 is also increased: 50.0 mm in the model A-1 and 57.5 mm in the model A-2. The other specifications are the same as those of the model A-1 and are listed in Table I.

Figure 4 shows the relationships between air gap length and thrust in the model A-1 and A-2, which were obtained from FEA. As seen from Fig.4, the thrust decreases according to an enlargement of the air gap length, and the thrust will become almost zero around the air gap length of 10 mm. In addition, although the PM of the model A-2 has double volume compared to that of the model A-1, there is only slight difference between two characteristic curves expressing the variation of the thrust.

The air gap flux density waveforms of the model A-1 and A-2 are drawn in Fig. 5. The waveforms have been computed under no-load condition, and Fig. 6 is the results of harmonic analysis for the waveforms. The fundamental component, B_1, contributes to the generation of the thrust as described in (1). Between two waveforms there is almost no difference and the values of B_1 are almost the same: 0.13 T in the model A-1 and 0.136 T in the model A-2 at the air gap length of 2 mm, for example. At the same air gap length as the example of B_1, the thrusts are 151.6 N in the model A-1 and 159.0 N in the model A-2. They are approximately proportional to B_1,

TABLE I
COMMON SPECIFICATION OF THE STUDY MODELS IN FIGS. 1 AND 2

Mover & Armature	Core length	186 mm
	Stack length of core, l_a	110 mm
	Back yoke height	19.2 mm
	Number of armature slots	6
	Number of pole pairs, P_p	1
	Winding configuration	Concentrated winding
	Number of series conductors per phase, N	166
	Rated current	5.1 A (rms)
Stator	Core height	39 mm
	Stack length of core, l_a	110 mm (same as mover core)
	Number of small slots, Z_2	7 (per mover length)
Permanent magnet	Coercive force H_c	1000 kA/m
	Residual magnetic flux density, B_r	1.3 T

2837

The 2018 International Power Electronics Conference

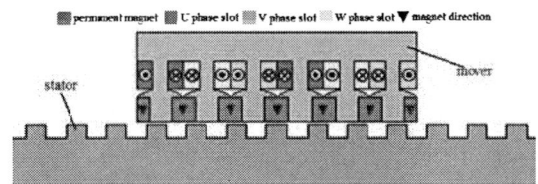

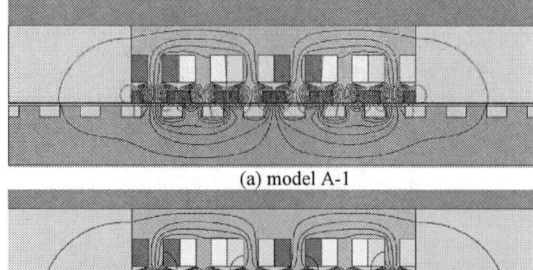

(a) model A-1

(b) model A-2

Fig. 7 Flux lines under condition at rated current and air gap length of 2 mm in model A-1 and A-2.

Fig.3 MA-APM-LVM with the configuration A and double volume PM of model A-1, i.e. model A-2.

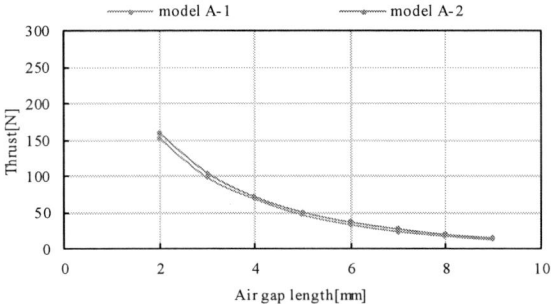

Fig. 4 Thrust versus air gap length at rated current in model A-1 and A-2.

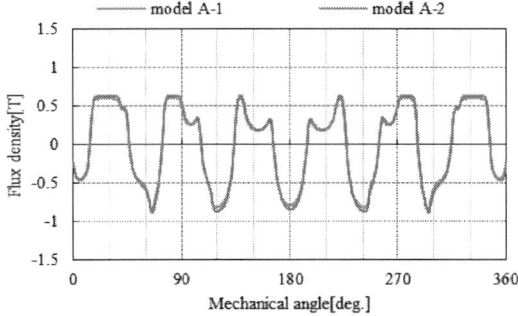

Fig. 5 Air gap flux density waveforms at no-load in model A-1 and A-2.

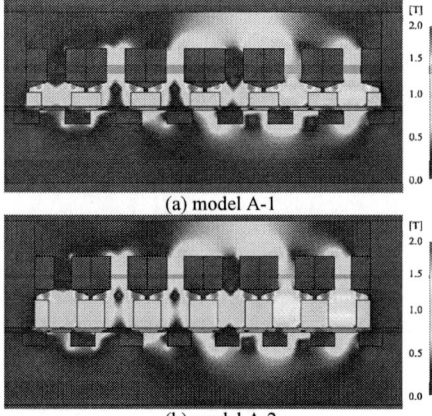

(a) model A-1

(b) model A-2

Fig. 8 Magnetic flux density distribution at rated current and air gap length of 2 mm in model A-1 and A-2.

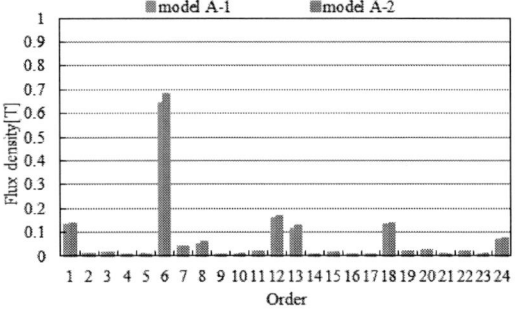

Fig. 6 Harmonic contents of air gap flux density waveforms at no-load in model A-1 and A-2.

thus there isn't big difference between them.

Figure 7 (a) and (b) draw the magnetic flux lines under condition at the rated current and the air gap length of 2 mm in the model A-1 and A-2. As can be seen from the figures, many flux lines surround each PM piece. The flux lines express leakage flux and doesn't contribute tohe generation of the thrust. From the comparison of Figs. 7 (a) and (b), it is seen that the leakage flux of the model A-2 is considerably more than that of the model A-1. Hence, in the motor with the configuration A, increasing the volume of PM only increase the leakage

flux and doesn't lead to an increase of the thrust. In addition, it is seen from Fig. 8, expressing flux density distribution, that the configuration A causes a remarkable magnetic saturation of the mover teeth.

What is obvious from the above examination is that increasing the volume of PM hardly have an effect on raising the thrust in the motor with the configuration A.

B. Comparison between Configuration A and B

This section will make an examination concerning the model B-1, which has the spoke type arrangement of PM as illustrated in Fig. 2. The PM of the model B-1 is the same volume as the model A-1 (Fig.1) with the configuration A. The other specifications except for the PM arrangement are kept in the same.

The air gap density waveform under no-load condition in the model B-1 is drawn in Fig.9 with that in the model A-1 so that they can be compared. The two waveforms are different in amplitude and phase, because of the different arrangement of PM, but their forms are very similar. That means that the two different arrangements of PM have the same effect on yielding thrust.

The thrust characteristics of the model A-1 and B-1 are compared in Fig. 10, which shows the variation of thrust versus air gap length. The two characteristic curves

2838

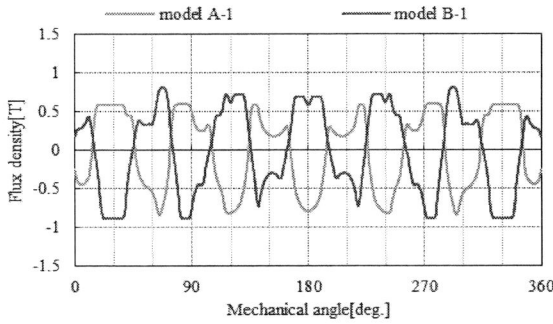

Fig. 9 Air gap flux density waveforms at no-load in model A-1 and B-1.

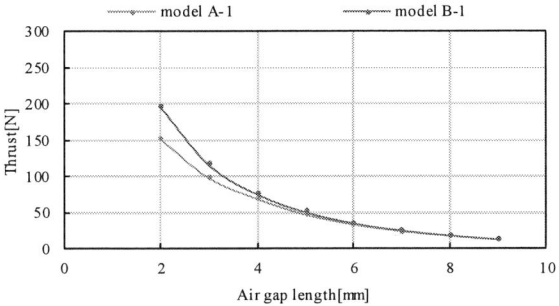

Fig. 10 Thrust versus air gap length at rated current in model A-1 and B-1.

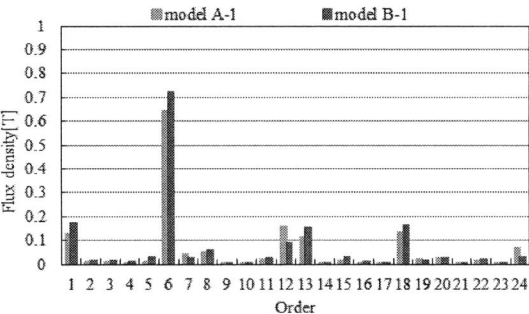

Fig. 11 Harmonic contents of air gap flux density waveforms at no-load in model A-1 and B-1.

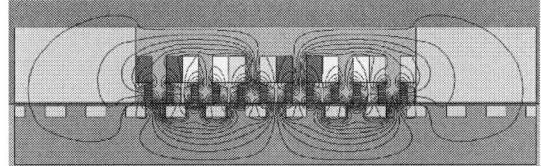

Fig.12 Flux lines at rated current and air gap length of 2 mm in model B-1.

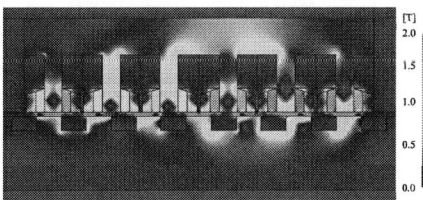

Fig.13 Magnetic flux density distribution at rated current and air gap length of 2 mm in mode B-1

asymptotically approach zero according to increase of the air gap length but there is a big difference between the two characteristic curves in a range around small air gap length. For example, the thrust at the air gap length of 2 mm in the model B-1 is about 30 % greater than that in the model A-1. The increasing rate is nearly equal to that of B_1 shown in Fig. 11, which is the harmonic analysis results of the air gap density waveforms. Their concrete numerical values are summarized as follows: 152 N and 0.13 T in the model A-1, which have been shown also in the previous section, and 196 N and 0.17 T in the model B-1.

Figures 12 and 13 are the diagrams of the flux lines and the magnetic flux density distribution in the model B-1, respectively. The LVM that this paper deals with creates a two-poles distribution of magnetic flux by an effect that the permeance variation due to the small stator teeth modifies the flux from the PM on the mover. One can see such an effect in Figs.7 and 12. Especially, the distribution of the model B-1 with the configuration B appears clearly as can be seen from Fig. 12. In addition, it is seen from the comparison of Fig.8 and Fig.13 that the magnetic saturation of the core in the model B-1 is softened compared with the model A-1 or A-2.

From above examination, it is obvious that the configuration B is superior to the configuration A in obtaining a larger thrust.

C. Improvement of the Motor with Configuration B

This section will consider increasing the thrust by increase of the PM volume in the moto with the configuration B

Figure. 14 is the model B-2 examined, in which the PM is two times the volume in the model B-1, and the mover core height is extended by 62.24 mm to put the PM of the double volume in each slot opening. The other specifications are the same as the model B-1, which are shown in Table I as common specifications.

Figures 15 and 16 respectively shows the flux lines of the model B-2 and the air gap flux density waveforms of the model B-1 and B2, which were computed under no-load condition. The waveform of the model B-1 in Fig 16 is the same as that in Fig.9. In Fig. 15, a magnetic field distribution of two-poles appears clearly. Concerning the air gap flux density waveforms, it is seen that the waveform amplitude of the model B-2 is greater than that of the model B-1. Consequently, the fundamental component, B_1, of the air gap flux density waveform of the model B-2 becomes larger than B_1 of the model B-1, as seen from Fig.17 which is the results of harmonic analysis of the waveforms shown in Fig. 16. The concrete numerical values of B_1 are 0.17 T (the same as above-mentioned value) in the model B-1 and 0.23 T in the model B-2.

The relationship between the air gap length and the thrust yielded is shown in Fig. 18. A change trend of the thrust versus the air gap length in the model B-2 is similar to those of the other models, that is, the thrust decreases according to the increase of the air gap length. However, attention should be paid to a range of relatively small gap length. At the air gap length of 2 mm, the

The 2018 International Power Electronics Conference

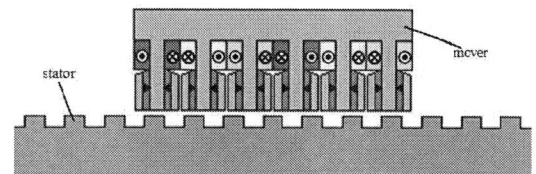

Fig.14 MA-APM-LVM with the configuration B and double volume PM of model B-1, i.e. model B-2.

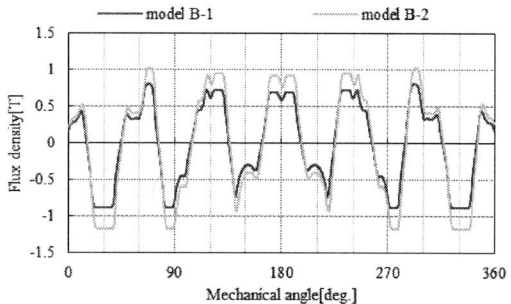

Fig. 15 Flux lines at rated current and air gap length of 2 mm in model B-2.

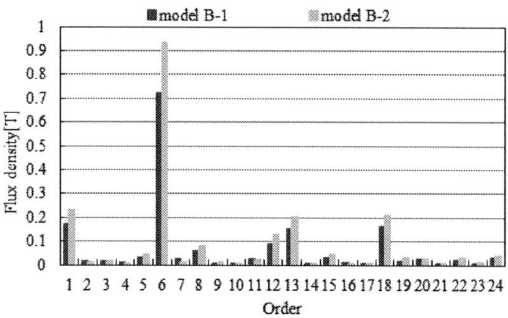

Fig.16 Air gap flux density waveforms at no-load in model B-1 and B-2.

Fig.17 Harmonic contents of air gap flux density waveforms at no-load in model B-1 and B-2.

thrust of the model B-2 is about 34 % larger than that of the model B-1 and reaches about 263 N. Concerning the overall range of air gap length, the increasing rate at each air gap length is listed in Table II. Their increasing rates are about 30 % or greater than 30 %.

Figure 19 shows the flux density distribution on the core under the condition of the air gap length of 2 mm and of the rated current, in the model B-2. Owing the increased volume of PM, the magnetic saturation degree of the core is somewhat high compared to the model B-1, but it seems that the somewhat high magnetic saturation doesn't have a bad influence on the thrust characteristic.

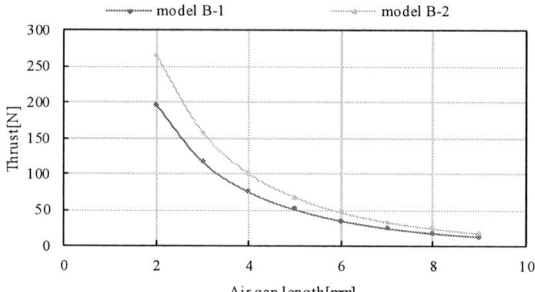

Fig.18 Thrust versus air gap length at rated current in model B-1 and B-2.

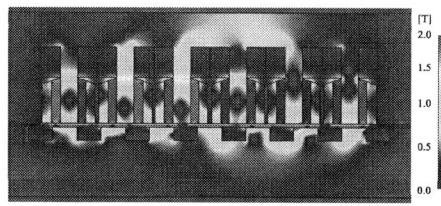

Fig. 19 Magnetic flux density distribution at rated current and air gap length of 2mm in mode B-2

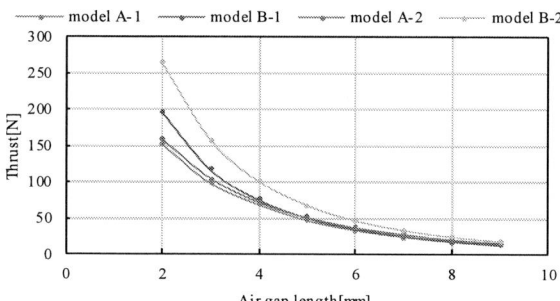

Fig. 20 Thrust versus air gap length at rated current in model A-1, A-2, B-1 and B-2.

TABLE II
INCREASE RATIO OF MODEL B-2 TO MODEL B-1 IN THRUST AT EACH AIR GAP LENGTH

Air gap length (mm)	2	3	4	5	6	7	8	9
Thrust of model B-1 (N)	196	116	75.1	50.6	35.3	25.2	18.3	13.3
Thrust of model B-2 (N)	263	156	100	66.9	46.3	32.9	23.7	17.4
Increasing rate (%)	34.2	34.5	33.2	32.2	31.2	30.6	29.5	30.8

D. Summary of the Exmeriments

The examination results in the four models are summarized in Fig. 20. It is seen that there is no effect that the thrust is increased by increasing the PM volume of the motor with the configuration A, but in the motor with the configuration B the increase of the PM volume leads to a remarkable increase of the thrust.

2840

I. CONCLUSIONS

This paper has examined the relationship between the thrust and the air gap length in two types of moving-armature type armature-PM-excited linear permanent magnet vernier motor (MA-APM-LPMVM). One was a motor with the PM arrangement which was based on the original design of a conventional PMVM, namely, the motor with the configuration A. The other was a motor with a novel spoke type PM arrangement that we previously proposed, namely, the motor with the configuration B.

As a result, it became obvious from the examinations that increasing the PM volume didn't lead to an increase of the thrust in the motor with the configuration A, but an increase of the PM volume in the motor with the configuration B had an extremely good effect on increasing the thrust. The results obtained in this paper will bring a lower cost of the construction and a higher efficiency to the steel-wheel linear motor train system.

In the next step, we will examine a MA-APM-LPMVM of a practical size.

REFERENCES

[1] Toba, A.; Lipo, T.A., "Generic torque-maximizing design methodology of surface permanent-magnet vernier machine,"*Industry Applications, IEEE Transactions on*, vol.36, no.6, pp.1539-1546, Nov/Dec 2000

[2] D. Jang and J. Chang, "Effects of Flux Modulation Poles on the Radial Magnetic Forces in Surface-Mounted Permanent-Magnet Vernier Machines," in *IEEE Transactions on Magnetics*, vol. 53, no. 6, pp. 1-4, June 2017.

[3] S. Hyoseok, N. Niguchi and K. Hirata, "Characteristic Analysis of Surface Permanent-Magnet Vernier Motor According to Pole Ratio and Winding Pole Number," in *IEEE Transactions on Magnetics*, vol. 53, no. 11, pp. 1-4, Nov. 2017.

[4] Ishizaki, A.; Tanaka, T.; Takasaki, K.; Nishikata, S., "Theory and optimum design of PM Vernier motor, "*Electrical Machines and Drives, 1995. Seventh International Conference on (Conf. Publ. No. 412)*, pp.208-212, 11-13 Sep 1995

[5] R. Ishikawa, K. Sato, S. Shimomura and R. Nishimura, "Design of In-Wheel Permanent Magnet Vernier Machine to reduce the armature current density," *2013 International Conference on Electrical Machines and Systems (ICEMS)*, Busan, 2013, pp. 459-464.

[6] Y. Tasaki, R. Hosoya, Y. Kashitani and S. Shimomura, "Design of the vernier machine with permanent magnets on both stator and rotor side," *Proceedings of The 7th International Power Electronics and Motion Control Conference*, Harbin, China, 2012, pp. 302-309.

[7] H. Wang *et al.*, "A Novel Consequent-Pole Hybrid Excited Vernier Machine," in *IEEE Transactions on Magnetics*, vol. 53, no. 11, pp. 1-4, Nov. 2017.

[8] S.Shimomura, M.Takano, "Linear Vernier Machine with Permanent Magnets only on Armature Side," 2013, Applied Mechanics and Materials, 416-417, 233

[9] T. Imada and S. Shimomura, "Magnet arrangement of linear PM vernier machine," *2014 17th International Conference on Electrical Machines and Systems (ICEMS)*, Hangzhou, 2014, pp. 3642-3647.

[10] F. Zhao, T. A. Lipo and B. I. Kwon, "A Novel Dual-Stator Axial-Flux Spoke-Type Permanent Magnet Vernier Machine for Direct-Drive Applications," in *IEEE Transactions on Magnetics*, vol. 50, no. 11, pp. 1-4, Nov. 2014.

[11] B. Kim and T. A. Lipo, "Analysis of a PM vernier motor with spoke structure," *2014 IEEE Energy Conversion Congress and Exposition (ECCE)*, Pittsburgh, PA, 2014, pp. 2358-2365.

[12] D. Li, R. Qu, W. Xu, J. Li and T. A. Lipo, "Design Procedure of Dual-Stator Spoke-Array Vernier Permanent-Magnet Machines," in *IEEE Transactions on Industry Applications*, vol. 51, no. 4, pp. 2972-2983, July-Aug. 2015.

[13] B. Zhang, M. Cheng, R. Cao, Y. Du and G. Zhang, "Analysis of Linear Flux-Switching Permanent Magnet Motor Using Response Surface Methodology," in *IEEE Transactions on Magnetics*, vol. 50, no. 11, pp. 1-4, Nov. 2014.

[14] B. Zhang, M. Cheng, M. Zhang, W. Wang and Y. Jiang, "Comparison of modular linear flux-switching permanent magnet motors with different mover and stator pole pitch," *2017 20th International Conference on Electrical Machines and Systems (ICEMS)*, Sydney, NSW, 2017, pp. 1-5.

[15] R. Cao, M. Cheng, C. Mi, W. Hua, X. Wang and W. Zhao, "Modeling of a Complementary and Modular Linear Flux-Switching Permanent Magnet Motor for Urban Rail Transit Applications," in *IEEE Transactions on Energy Conversion*, vol. 27, no. 2, pp. 489-497, June 2012.

[16] R. Cao, M. Cheng and B. Zhang, "Speed Control of Complementary and Modular Linear Flux-Switching Permanent-Magnet Motor," in *IEEE Transactions on Industrial Electronics*, vol. 62, no. 7, pp. 4056-4064, July 2015.

Micro Electromagnetic Vibration Energy Harvester with Mechanical Spring and Iron Frame for Low Frequency Operation

Yecheng Shen*, Kaiyuan Lu and Yongming Xia
Department of Energy Technology, Aalborg University, Aalborg, Denmark
*E-mail: yec@et.aau.dk

Abstract-A new micro electromagnetic energy harvester with iron frame based on mechanical spring vibration is proposed in this paper for increasing the output voltage. The dynamic model of this new structure and an analytical method for maximizing the output voltage and power from low-frequency vibration sources are derived. Finite Element Analysis (FEA) is carried out for validation. This new structure achieves normalized power density of 1651.3 $\mu Wcm^{-3}g^{-2}$ and the output power at 10.1 Hz can reach 446.5 μW with 0.52 g acceleration amplitude. It demonstrates superior performances compared to previously reported low frequency energy harvesters.

Keywords-micro electromagnetic energy harvester; low-frequency; high normalized power density

I. INTRODUCTION

With the increasing demand on wireless sensors and small-scale electronic devices in the past decade, the self-sustaining power with the mechanism of converting ambient mechanical vibration into electrical energy has achieved extensive concerns. Most recent vibration energy harvesting approaches have been attempted to obtain electricity through transduction mechanism of piezoelectric, magnetostrictive, electrostatic and electromagnetic from the ambient vibration energy sources. Current research has a strong focus on piezoelectric and electromagnetic based energy harvesters. Because of good energy density and easy miniaturized fabrication features, the feasibility of using piezoelectric energy harvester for harvesting vibration energy has been well recognized. However, compared to electromagnetic energy harvesters, piezoelectric devices have larger internal impedances so they have difficulties in providing sufficient power to the load. Often, the optimum resonance frequency of piezoelectric energy harvesters is high. To harvest energy in low-frequency vibration environment such as the vibration energy from human body motion, highly compliant mechanical spring and magnetic sprung structures may make the electromagnetic energy harvester a more suitable candidate.

Based on the Faraday's law of induction, many electromagnetic energy harvesting structures including magnetic levitation and spring or cantilever oscillating with base excitation were investigated [1]-[4]. Nevertheless, the output voltages are inclined to be too low to drive most circuits while working in low-frequency vibration environment. In order to improve the output voltage, the technique of frequency-up conversion has been introduced as those reported in [5], [6]. But, most of those methods are based on adding piezoelectric bimorph cantilever leading to the increased device volume, which is not preferred due to the small dimension requirement for micro electronic device. For enhancing the output voltage, reducing magnetic reluctance by introducing high permeability material in the flux path and then increasing the flux linkage in the coil is another popular method, which has been widely adopted in linear motor designs. Due to the limited volume in micro energy harvesters, the high permeability, low iron loss electrical machine laminations are often difficult to be fabricated and assembled to the desired shape, e.g. a cylinder has difficulties to be seamlessly formed by axially laminated steels. In contrast, the electromagnetic energy harvester with a spring oscillating mechanism could be designed to avoid the iron loss completely in a solid iron core.

In this paper, a mechanical spring based electromagnetic vibration energy harvester that consists of an iron frame with two magnets of the same polarity facing each other in the center is presented to harvest energy from low-frequency vibration sources. The main characteristic of the proposed energy harvester is that it exhibits higher power density and voltage per turn than other electromagnetic energy harvesters; in the meantime, there is no iron loss produced in operation. Electromechanical model of this device is established and the finite element analysis is carried out for performance evaluation. The obtained power density and voltage at the resonant frequency are compared with other low-frequency energy harvesters reported previously in the literatures.

II. THE PROPOSED NEW ENERGY HARVESTER

The proposed micro electromagnetic energy harvester is shown in Fig. 1. The coil and the holder are connected through the slots in the cylindrical iron frame and are attached to the vibration source; inside the iron frame, two magnets and two proof masses form the internal translator. Initiated by the vibration energy source and contributed by the mechanical spring linking the coil and the internal translator, the repetitive linear oscillating motion of the coil along the axial direction will eventually drive the translator to move relatively to the coil with different oscillation displacement amplitude and

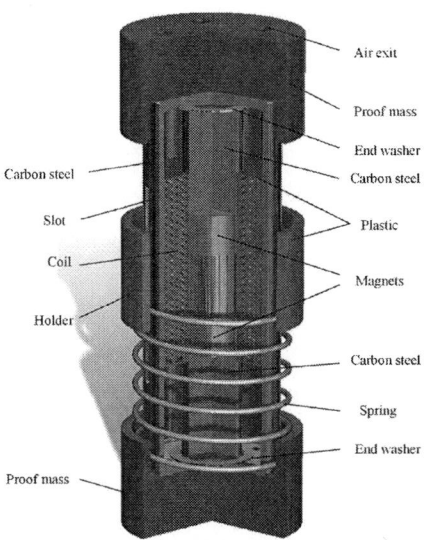

Fig. 1. Cut view of the proposed micro energy harvester.

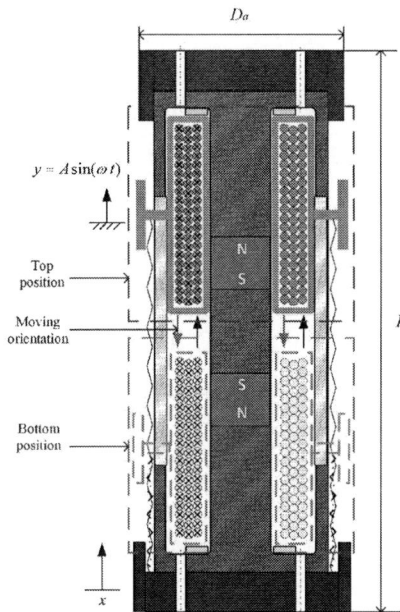

Fig. 2. Schematic of the proposed electromagnetic energy harvester.

phase angle. A schematic drawing of the proposed micro electromagnetic energy harvester is shown in Fig. 2. As a case study, this new structure is designed to have an outer diameter of 9 mm and a height of 25.2 mm. The magnets have a thickness of 2 mm and a diameter of 3 mm. When the system is excited, the translator is forced to move along the axial direction of the device due to the linear force produced by the mechanical spring; the changing relative position between the coil and the internal translator will lead the coil to cut the magnetic flux, generating induced voltage in the coil. Based on the Faraday's law of electromagnetic induction [7], the voltage generated in the coil can be derived as

$$e = -n\frac{d\lambda}{dt} = -n\frac{d\lambda}{dz}\cdot\frac{dz}{dt} \qquad (1)$$

where n is the number of turns of the windings, λ is the flux variation per turn and z is relative displacement of the coil to the iron frame expressed by $z = x - y$. (Labels are indicated in Fig. 2.)

In (1), the generated voltage is related to the number of turns of the coil, the variation of the flux linkage as well as the relative velocity of the moving part. In this device, two magnets with the same polarity facing each other are placed in the center of the translator. The iron frame is used to enhance the flux generated by the magnets. The permanent magnet flux linking the coil can be significantly improved compared to that produced by the structure of e.g. a magnetic levitation harvester where the magnet flux travels in the air only [8]. The generated magnetic flux pattern and achievable flux density values are shown in Fig. 3. The two extreme positions of the coil are as indicated in Fig. 3a and 3b respectively. It may be observed that the permanent magnet flux links the coil in opposite directions at these two extreme positions. When the coil is moving upwards and downwards, the flux density in the iron part remains unchanged due to negligible armature reaction effects in this design; solid

iron material can be used to form the iron frame with no iron loss to be introduced.

TABLE I
DEVICE PARAMETERS

Parameters	Value
Outer diameter of frame (D_a)	9 mm
Height of frame (L_a)	25.2 mm
Diameter of coil	0.2 mm
Turns of coils - n	240
Coil resistance - R_{coil}	1.85 Ω
Magnetization	876 kA/m
Magnet type	NdFeB-N40
Mass of moving part - m	8 g
Stiffness - k	32 N/m
Spring viscous damping - c_v	0.1 Ns/m

III. FINITE ELEMENT ANALYSIS

In order to investigate the magnetic field generated in this device, a 2D axis symmetrical model is built in Comsol (Fig. 3). The dimension details and corresponding parameters of the spring, the proof mass etc. are given in Table I.

To investigate the flux variation in the coil, the magnetic flux linking the coil needs to be properly calculated. The length of the coil is denoted as l and the number of coil layers is N_l; each coil layer has a number of turns of N_t. The rate change of the flux linking the coil could be approximated as follows.

First, the flux travelling through one turn of the coil located at a distance d from the device center is obtained by integrating the magnetic flux density over the turn area as may be expressed by [9]

$$\varphi_t = \frac{\mu_0 \tau}{2}\left[\frac{1}{\sqrt{(r^2+d^2)}} - \frac{d^2}{(r^2+d^2)^{3/2}}\right] \qquad (2)$$

where τ is the magnetic moment, μ_0 is the air permeability, and r is the coil radius.

Differentiating (2) with respect to d yields

$$\frac{d\varphi_t}{d_d} = \frac{\mu_0 \tau}{2}\left[\frac{3d^3}{\left(r^2+d^2\right)^{5/2}} - \frac{3d}{\left(r^2+d^2\right)^{3/2}}\right] \quad (3)$$

Since each individual turn of the layer is located at different distances from the center during the translator oscillation, the total rate change of the flux linking the whole coil will be the sum of the rate change of the flux linking individual turn multiplied by the number of layers as

$$\frac{d\varphi}{d_d} = N_l \frac{\mu_0 \tau}{2}\sum_{n=-N_t/2}^{n=N_t/2} d_n\left[\frac{3d_n^2}{\left(r^2+d_n^2\right)^{5/2}} - \frac{3}{\left(r^2+d_n^2\right)^{3/2}}\right] \quad (4)$$

$$d_n = d + n\left(\frac{l}{n_t-1}\right) \quad (5)$$

The flux linkage could then be obtained by integrating by (4), and its final obtained profile is shown in Fig. 4.

It is convenient to assume the linear oscillation motion profile is ideally sinusoidal, as e.g. $z=4.5\times\sin(20\pi t)$, which means the oscillation is within a range of -4.5 mm to 4.5 mm at a frequency of 10 Hz. A cubic polynomial is used to fit the data of flux linkage variation shown in Fig. 4, which can be expressed as $\varphi= -3304\times z^3+0.4056\times z$. Using this flux linkage and (1), the open circuit voltage e_o may then be obtained and is plotted in Fig. 5. The instant power can be calculated by the following equation. The result is shown in Fig. 6.

$$P_o = \frac{e_o^2}{2\left(R_{in}+R_{load}\right)} \quad (6)$$

where R_{in} is the internal resistance if regarding the energy harvester as a voltage source, which is equal to the coil resistance; R_{load} is the external load resistance of 1.85 Ω, which will maximize the output power on the load.

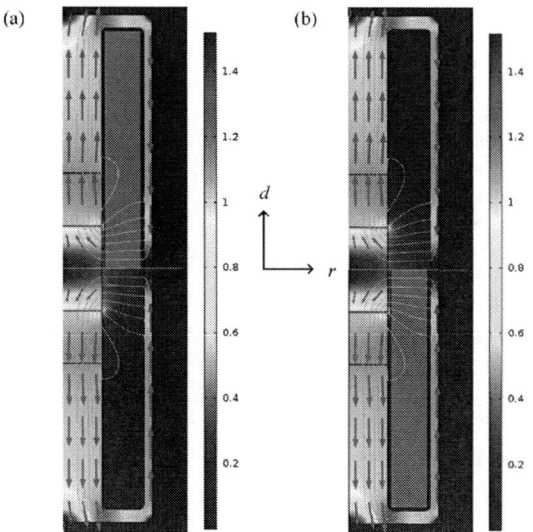

Fig. 3. The flux density distribution in this device, (a) coil at top position and (b) coil at bottom position.

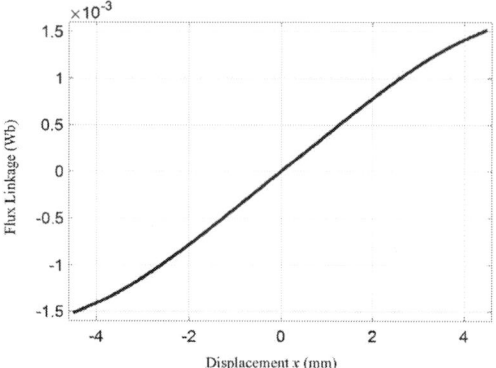

Fig. 4. Flux linkage in coils

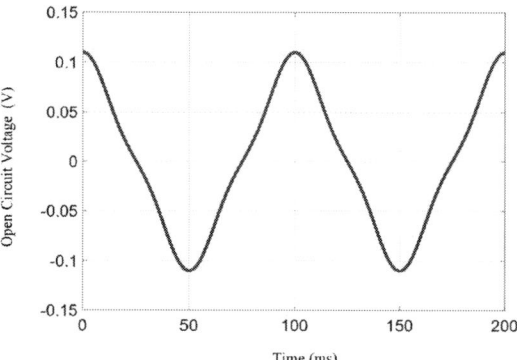

Fig. 5. Open circuit voltage

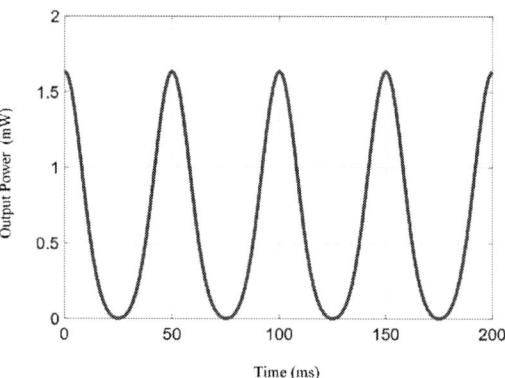

Fig. 6. The instantaneous output power at the optimal resistive load

IV. ELECTROMECHANICAL MODEL AND FREQUENCY RESPONSE ANALYSIS

For the system presented in Fig. 2, assuming the mechanical spring force is linear with an equivalent stiffness of k, when the energy harvester is excited by a sinusoidal signal from the base, the schematic diagram of the equivalent linear vibration system may be shown in Fig. 7 [7], where m is the mass of the translator and c is the total damping factor.

When this equivalent mechanical spring energy harvester is excited by external vibration, the dynamic model of the motion is given by [10]

$$m\frac{d^2z}{dt^2}+c\frac{dz}{dt}+kz=m\frac{d^2y}{dt^2} \tag{7}$$

where z is the relative displacement, y is the absolute displacement, m is the mass of the moving part including the coil and the proof mass. The total damping factor c to (to be used in (7)) consists of the spring viscous damping coefficient c_v and electrical damping ratio c_e. In the low frequency vibration environment, it may be assumed that there is almost no loss occurring in the energy conversion [11]. So the electrical damping coefficient c_e can be expressed as

$$c_e=\frac{(d\varphi/dz)^2}{R} \tag{8}$$

where φ is the total flux linkage of the coil and R is the total resistance including the coil resistance and the load resistance.

By dividing m on the both sides of (7), the equation can be shown as

$$\frac{d^2z}{dt^2}+2\xi\omega_n\frac{dz}{dt}+\omega_n^2z=\frac{d^2y}{dt^2} \tag{9}$$

where ω_n is the nature resonance frequency, $2\xi\omega_n=c/m$, $\omega_n^2=k/m$. Supposing there is a phase angle difference θ in the base signal that drives the responding output in form of $z=Z\sin(\omega t+\theta)$. By substituting it into (9), and multiplying it by $\sin(\omega t)$ and $\cos(\omega t)$, the following equations can be obtained

$$-\omega^2Z+\omega_n^2Z=-\omega^2A\sin(\theta) \tag{10}$$

$$-2\xi\omega_n\omega Z=\omega^2A\cos(\theta) \tag{11}$$

where A is the amplitude of the excitation signal. The amplitude of relative displacement Z can then be expressed as

$$Z=\frac{A\omega^2m}{\sqrt{(k-\omega^2m)^2+c^2\omega^2}} \tag{12}$$

The resonant frequency is defined as the frequency at which the responding oscillation achieves its maximum displacement [10], so it can be derived as

$$\omega_r=\omega_n\sqrt{1-2\xi^2} \tag{13}$$

V. PERFORMANCE SIMULATION AND COMPARISON

According to (12), the amplitude of relative displacement Z has the response as a function of the input frequency as shown in Fig. 8. This figure shows the maximum displacements at three acceleration levels of 0.3 g, 0.5 g and 0.7 g, respectively. It can be observed that with the increasing of the acceleration, the maximum amplitude of the relative displacement exhibits a clear increasing trend, which means it can produce more power at the same frequency. The resonance frequency found for this energy harvester is 9.86 Hz and the resonance frequency at three different acceleration levels are the same. This suggests that the resonant frequency of the mechanical spring structure is not affected by the acceleration.

Many parameters may affect the output power, including: excitation amplitude, frequency, damping

coefficient, and external load resistance. In this paper, assuming the damping coefficient is constant, thus the excitation amplitude and frequency are two main factors determining open circuit voltage.

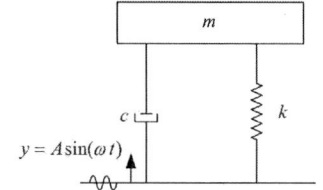

Fig. 7. The equivalent linear vibration system

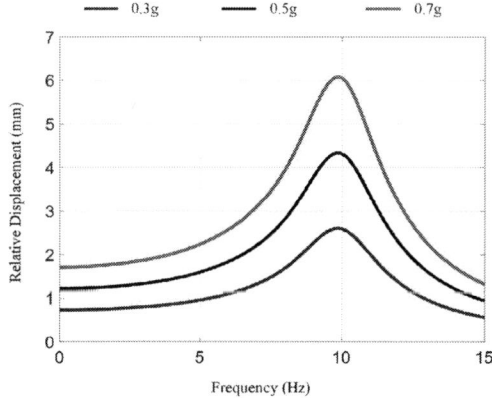

Fig. 8. Frequency response of the relative displacement

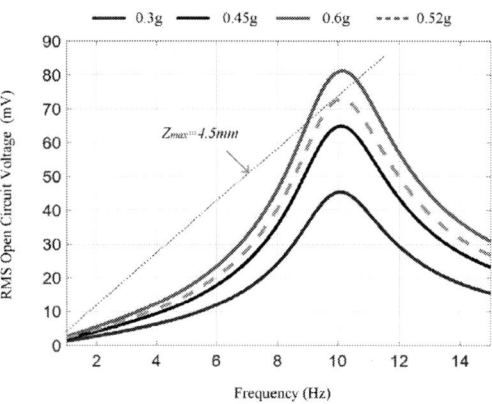

Fig. 9. Open circuit voltage as the frequency at three acceleration level

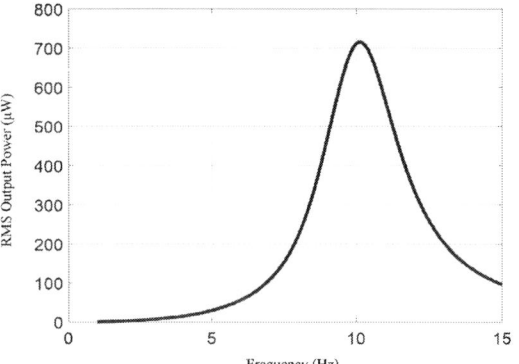

Fig. 10. RMS output power with frequency at 0.52 g excitation

TABLE II
COMPARISON OF ENERGY HARVESTER RESULTS FROM LITERATURE

RMS power (µW)	f (Hz)	Accel. (g)	Vol. (cm³)	PD (µW cm⁻³)	NPD (µW /cm³ g²)	Ref.
716	10.1	0.52	1.6	446.5	1651.3	This work
113.3	3.33	1.26	0.76	149	93.9	[13]
11890 (max)	5.17	2.06	6.47	333	78.47	[14]
2920	6	0.5	30.42	96	460	[15]
12.7	4	2	1.1	11.5	2.875	[16]
14.55	8	0.039	12.5	1.17	770	[17]

In Fig. 9, the open circuit voltages at three acceleration amplitudes of 0.3 g, 0.45 g and 0.6 g are illustrated. It can be seen that the open circuit voltages have the peak values at 10.1 Hz rather than at the resonant frequency for all of the three acceleration levels. This is because there is no noticeable decrease of the oscillation displacement with the increase of the frequency near 10 Hz. However, if the maximum displacement is limited, the region below the dashed line Z_{max}=4.5 mm in Fig. 9 has the achievable voltages. So, it can be figured out that the maximum voltage at 10.1 Hz is achieved at the oscillation amplitude of 4.5 mm. According to (11), the corresponding acceleration amplitude of the input signal is 0.52 g, which is plotted as the green dashed line. Fig. 10 shows the output power as a function of frequency at 0.52 g acceleration amplitude, in the situation that the external load resistance equals the internal resistance. It indicates that the maximum output power is 716 µW at 10.1 Hz.

One of the advantages of the proposed new energy harvester is that it can produce high output voltage. However, there is no obvious criterion of evaluating the output voltage level for different energy harvesters in the past, because the output voltage could be affected by the number of turns, the volume as well as many other factors. Normally, the criterion in vibration energy harvester evaluation for low frequency applications is the ability to match variable amplitude vibrations producing large output power within a small volume. A comparison of the results from the energy harvester presented in this paper with other electromagnetic energy harvesters is given in Table II. The table lists the power density (PD), which is defined as the output power divided by the volume of the energy harvesters. The best evaluation of this work can be done by comparing the output power at the same size and input vibration. However, the data published in previous works varies quite a lot. Moreover, the normalized power density (NPD) is introduced, which is the power density divided by the square of the acceleration due to the linear relationship between them [12]. Normalized power density makes a fairer comparison between different energy harvesters than power density. As shown in the Table II, the electromagnetic energy harvester proposed in this paper has the highest power density and normalized power densities.

Although the energy harvester in this paper has exhibited a high normalized power density, the further optimization is required to improve the flux and decrease the operating frequency, in order to achieve the best performance.

VI. CONCLUSION

In this work, a new micro electromagnetic energy harvester based on mechanical spring is presented. The mechanical model and frequency response analysis have been investigated with respect to harvest energy from low-frequency vibration sources. This proposed energy harvester achieves the output power of 716 µW at 10.1 Hz with 0.52 g acceleration amplitude. The normalized power density of 1651.3 µWcm⁻³g⁻² is higher than those offered by other micro electromagnetic energy harvester.

REFERENCES

[1] W. Deng and Y. Wang, "Systematic parameter study of a nonlinear electromagnetic energy harvester with matched magnetic orientation: Numerical simulation and experimental investigation," *Mechanical Systems and Signal Processing*, vol. 85, pp. 591-600, 2017.

[2] A.R.M. Foisal, B.C. Lee, and G.S. Chung, "Fabrication and performance optimization of an AA size electromagnetic energy harvester using magnetic spring," *IEEE Sensors proceedings*, pp. 1125-1128, 2011.

[3] W. Wang, J. Cao, N. Zhang, J. Lin and W.H. Liao, "Magnetic-spring based energy harvesting from human motions: Design, modeling and experiments," *Energy Conversion and Management*, vol. 132, pp. 189-197, 2017.

[4] B. C. Lee and G. S. Chung, "Design and fabrication of low frequency driven energy harvester using electromagnetic conversion," *Transactions on Electrical and Electronic Materials*, vol. 14, no. 3, pp. 143-147, 2013.

[5] H. Fu and E.M. Yeatman, "A methodology for low-speed broadband rotational energy harvesting using piezoelectric transduction and frequency up-conversion," *Energy*, vol. 125, pp. 152-161, 2017.

[6] F. Cottone, R. Frizzell, S. Goyal, G. Kelly and J. Punch, "Enhanced vibrational energy harvester based on velocity amplification," *Journal of Intelligent Material Systems and Structures*, vol. 25, no. 4, pp. 443-451, 2013.

[7] N. G. Stephen, "On energy harvesting from ambient vibration," *Journal of sound and vibration*, vol. 293, no. 1, pp. 409-425, 2006.

[8] P. Zeng, and A. Khaligh, "A permanent-magnet linear motion driven kinetic energy harvester," *IEEE Transactions on Industrial Electronics*, vol. 60, no 12, pp. 5737-5746, 2013.

[9] T. L. Chow, *Introduction to electromagnetic theory: a modern perspective*. Jones & Bartlett Learning, 2006, pp. 146-150.

[10] B. P. Mann and N. D. Sims, "Energy harvesting from the nonlinear oscillations of magnetic levitation," *Journal of Sound and Vibration*, vol. 319, no. 1-2, pp. 515-530, 2009.

[11] S. Roundy, E.S. Leland, J. Baker, E. Carleton, E. Reilly, E. Lai, B. Otis, J.M. Rabaey, P.K. Wright and V. Sundararajan, "Improving power output for vibration-based energy scavengers," *IEEE Pervasive computing*, vol. 4, no. 1, pp. 28-36, 2005.

[12] S.D. Moss, O.R. Payne, G.A. Hart and C. Ung, "Scaling and power density metrics of electromagnetic vibration energy harvesting devices," *Smart Materials and Structures*, vol. 24, no. 2, p. 023001, 2015.

[13] A. Haroun, I. Yamada and S. Warisawa, "Micro Electromagnetic Vibration Energy Harvester Based on Combined Free/Impact Motion for Low Frequency-Large Amplitude Operation,"

in ASME 2017 International Mechanical Engineering Congress and Exposition, pp.V010T13A023-V010T13A023. American Society of Mechanical Engineers. Nov. 2017.

[14] M. A. Halim, H. Cho and J. Y. Park, "Design and experiment of a human-limb driven, frequency up-converted electromagnetic energy harvester," *Energy Conversion and Management*, vol. 106, pp.393-404, 2015.

[15] M. Salauddin and J. Y. Park, "Design and experiment of human hand motion driven electromagnetic energy harvester using dual Halbach magnet array," *Smart Materials and Structures*, vol. 26, no. 3, p.035011, 2017.

[16] Y. Wang, Q. Zhang, L. Zhao and E. S. Kim, "Non-Resonant Electromagnetic Broad-Band Vibration-Energy Harvester Based on Self-Assembled Ferrofluid Liquid Bearing," *Journal of Microelectromechanical Systems*, vol. 26, no. 4, pp. 809-819, 2017.

[17] C. R. Saha, T. O'donnell, N. Wang and P. McCloskey, "Electromagnetic generator for harvesting energy from human motion," *Sensors and Actuators A: Physical*, vol. 147, no. 1, pp. 248-253, 2008.

AUTHOR INDEX

Aapro, Aapo 3156
Abdollahi, Hessamaldin 1719
Abe, Kazuyuki 1567
Abe, Kensho 767
Abe, Kodai 1741, 3890
Abe, Seiya 2360, 2370
Abe, Takashi 2176
Abrishamifar, Adib 2854
Abuogo, James 1125
Acharya, Anirudh Budnar 2630
Acharya, Sayan 3564
Adachi, Masakazu 2237
Afsharian, Jahangir 1537, 3797
Agarwal, Vivek 3471
Agelidis, Vassilios G. 3215
Agostinelli, Matteo 3140
Ahmad, Hamzeh J. 3273
Aiso, Kohei 3186
Akagi, Hirofumi 2352
Akahane, Masashi 2774
Akama, Yousuke 1741
Akao, Naoki 1217
Akatsu, Kan 711, 3186
Alatise, O 1149
Alenius, Henrik 1704, 4205
Ali, Muhammad 528
Ali, Murad 2317
Allmeling, Jost 422, 2199
Almér, Stefan 555
Alsofyani, Ibrahim Mohd 466
Alvarez, S. 4009
Amano, Koki 94
Amei, Kenji 3182
Amin, Mohammad 759
Amrhein, Wolfgang 3640
An, Ronghui 957, 1524, 3251, 3692, 3924
An, Zheng 4001
Andenna, M. 3596
Andersen, A. E. Michael 1351
Andersen, Michael A. E. 607, 4066
Ando, Akinobu 517
Ando, H. 3665
Ando, Masato 1919
Ando, Takashi 3658
Ang, Simon S. 153
Antivachis, Michael 181
Antonini, Giulio 3588
Antonopoulos, Antonios 2335
Anurag, Anup 3564
Anyapo, Chan 3332
Aoyagi, Kazuki 2237
Aoyama, Masahiro 718, 753
Arai, Takuro 1997
Araumi, Ryunosuke 1877, 3658
Arimatsu, Kenji 1370

Arita, Hideaki 2796, 2820
Arrua, Silvia 1719
Asada, Kazunori 3658
Asama, Junichi 4016
Ashizaki, Yusuke 3450
Ashourloo, Mojtaba 2380
Aso, Shinji 3086
Aware, Mohan 1730
Ayano, Hideki 1080
Azad, A N M Wasekul 2416
Azegami, Kazuya 3723
Azuma, S. 3665
Baba, Teppei 2283
Babasaki, Tadatoshi 207
Bach, Hoang Linh 2410
Baek, Jae-Il 108, 2365, 3100, 3533, 3538
Baek, Miran 1141
Bahat-Treidel, Eldad 3607
Bai, Baodong 2638
Baik, Jeong Min 3063
Bak, Yeongsu 1736, 4104
Bakran, Mark-M. 2476
Bandyopadhyay, Soumya 1426
Barrena, Jon Andoni 759
Barrera-Cardenas, Rene 3431
Bauer, Pavol 1426, 2630
Bauer, Walter 3640
Bayer, Christoph Friedrich 2410
Bellini, M. 4009
Berg, Matias 963, 4205
Bergveld, H.J. 267
Bertoldi, F. 488
Besselmann, Thomas 555
Bezha, Minella 3170
Bhattacharya, Subhashish 3564, 3993
Bhowate, Apekshit 1730
Bhumkittipich, Krischonme 2430
Biela, J. 1896
Biela, Jürgen 1103, 1509, 2301, 3734
Bilal, Ahmad 2193
Bilsalam, A. 1622
Bin, Zhao 2692
Bixel, Paul 238
Blaabjerg, Frede 439, 746, 1183,
    1246, 1711, 1788, 2512, 2604, 2743, 3123,
    3164, 3357
Blanes, José M. 1435
Böcker, Jan 3607
Bojoi, R. 732
Bonyadi, R. 1149
Boroyevich, Dushan 790, 3705, 3749, 3985
Bortis, D. 4080
Bortis, Dominik 181
Boynov, K.O. 161
Braun, Michael 2848, 3074

AUTHOR INDEX

Büdel, Johannes 3034
Bui, M.X. 4174
Bunlaksananusorn, Chanin........................2490
Burgos, Rolando.................... 790, 3705, 3749, 3985
Cai, Kejun..3965
Cai, Panpan 3495
Cai, Xu 1004, 1491, 2245, 4162, 4220
Canales, F. 4009
Cao, Hu...1816, 3484
Cao, Pengpeng...2973
Cao, Qi..100
Cao, Wu..3002, 3010, 3015
Cardenas, Rene Alexander Barrera 1111
Carvalho, Kelly C. M. 3785
Castellazzi, Alberto 130, 2932
Ceballos, Salvador 3117
Celik, Mustafa 1680
Cha, Honnyong.................... 927, 1046, 2619, 3134
Chae, Beomseok 1977
Chailloux, Thibaut 2153
Chang, Chen-Wei..1617
Chang, Chien-Hsuan.....................................2860
Chang, Liuchen 815, 1472, 1793, 2505
Chang, Yung-Ruei 639, 883
Chanmontree, P. 1622
Chao, Yi-Hao..1145
Charalambous, Apollo..................................1634
Charoensuksirikul, Supanut 2113
Chattopadhyay, Ritwik................................3564
Chazal, Hervé 2158
Chen, Ang-Tung 2102
Chen, Bo 1397
Chen, C. 142
Chen, Ching-Chen 1617
Chen, Ching-Jan 2086
Chen, Chuantong 1598
Chen, Dezhi 2638
Chen, Guan-Jung 1341
Chen, Guo..370
Chen, Hao 3112
Chen, I-Lin 2107
Chen, Jiangnan 1157, 1167
Chen, Jiann-Fuh 2653
Chen, Jie 1015, 1177
Chen, Kai-Hui 3081
Chen, Ke...1391
Chen, Kun-Feng 1341
Chen, Min..878
Chen, Minwu 2547
Chen, Nan 2335
Chen, Pingping 1118
Chen, Shen-Li 1145
Chen, Song 2153
Chen, Tang-Jung 1617
Chen, Tao..1872

Chen, Wan-Jung..3544
Chen, Wenjia 4213
Chen, Wenjie 1062, 2854, 3329
Chen, Wu 1504, 2496
Chen, Xiliang 3329
Chen, Xin 1015, 1177
Chen, Xingxing 1051, 3129, 3439
Chen, Yang 2785
Chen, Yangyang 560
Chen, Yaow-Ming 639, 883
Chen, Yenan..1118
Chen, Yen-Wen 2576
Chen, Yufeng 3383
Chen, Yu-Jen 275
Chen, Zhe 1758, 2708
Chen, Zhi..2997
Chen, Zhigang 3040
Cheng, Ching-Hsiang 2086
Cheng, Chun-An 2860
Cheng, Hung-Liang 2860
Cheng, Nie 2625
Cheng, Po-Tai 503, 1038, 2462, 3549
Cheng, Ran 3877
Cheng, Xiangpeng 2435, 3934
Chengbi, Zeng...2718
Chi, Yongning 1491
Chiba, Akira 3627
Chien, Lin-Hao 2102
Chiu, Huang-Jen 2092, 3151
Chiu, Hui-Lung 123
Chiu, Yi-Hao 1145
Cho, Geum-Bae 2145
Cho, In-Ho 3323
Cho, Shin-Young 1530
Cho, Young Joon 137
Cho, Younghoon 1403
Choe, Chanyang 1598
Choi, Byungcho 1465
Choi, Hyun-Jun 383
Choi, Jae Hyuk 1336
Choi, Jaeho 803
Choi, Joon-Ho 982, 1799
Choi, Seung-Hyun 4049
Choi, Sewan 256
Choi, Sung-Jin 1409
Choi, Youn-Ok..2145
Chou, Shih-Feng 1711
Chou, T.-C. 1912
Choudhury, Abhijit 3401
Chuai, Guoming 3025
Chung, Daewoong...1141
Chung, Henry S. H. 917
Chunkag, V. 1622
Collins, Caspar 1931
Cortes, Camilo 2193

AUTHOR INDEX

Corvasce, C. 3596
Cucala, Asuncion P. 2534
Cui, Shenghui 2250, 2484
Cui, Xiang 1125
Cvetkovic, Igor 790, 3985
Czyz, Piotr 396
D'arco, Salvatore 782, 2003
Da Silva, C. 267
Dahidah, Mohamed S A 3215
Dai, T. 1149
Dai, Wenjing 1015
Daikoku, Akihiro 2796
Danqing, Liu 1376
Dao, Ngoc Dat 1212
Dauphin, Benjamin 3644
Davari, Pooya 746
Davletzhanova, Z 1149
De Doncker, Rik W. 375, 388, 598,
     1073, 2250, 2484, 2768, 3729, 3979
Decker, Simon 2848, 3074
Delaforge, Timothé 2158, 3820
Deng, Fujin 1758, 2708
Deng, Jinxin 2992
Deshpande, Prathamesh Pravin.....................4186
Dieckerhoff, Sibylle 3607
Dimarino, Christina 3985
Din, Zakiud..2262
Ding, Yong 815
Dinh, Nguyen Duy 363
Diouf, Fatou 2078
Dirksen, Daniel 2410
Divan, Deepak 4001
Doki, Shinji 1032, 1223, 1228, 1295, 1747, 2224
Dong, Hanjing 987
Dong, Mi 1771
Dong, Qinghua 459
Dong, Xiaofeng 4168
Dong, Zhen 459
Dong, Zheng 3768
Driesen, J. 488
Du, Chao 2204
Du, Xiaotong 1167, 2780
Du, Xizhou 1491
Du, Yan 1472, 2877
Du, Zhijiang 84
Duarte, J. L. 946, 1067, 2697
Duarte, Jorge L. 1447, 3840
Dugal, F. 3596
Dujic, Drazen..........................422, 1484, 1498, 2170
Duong, Truong-Duy 982
Duque, C. A. 1067
Eberle, Wilson 927
Ekman, Jonas 3588
Elbaset, Adel A. 3945
Endegnanew, Atsede G. 2003

Endo, Hiroaki 4151
Endres, Tobias Maximilian 2410
Engelmann, Georges 3979
Enomoto, Bruno Yukio 3785
Eto, Haruhi 2097
Faiz, Muhammad Talib 528
Fajri, Poria 3223
Fan, Dongchen.....................3002, 3010, 3015
Fan, Shengwen 977, 3040
Fan, Weiyan 1386, 1421
Fang, Jingyang 337, 3910
Fang, Ran 4213
Fangfang, Luo 1282
Farkas, Gabor 137
Fayyaz, Asad 130
Felderer, Niklaus 2199
Feng, Chao 2058
Feng, Wei 3678
Ferdowsi, Mehdi 3223
Fernandez, Gabriel...3209
Fernandez-Cardador, Antonio.........................2534
Fischer, F. 3596
Foo, Gilbert 1724
Formentini, A. 4034
Freijedo, Fracisco D. 1498
Friedrichs, Peter 3584
Fuchs, Simon..2301
Fujii, H. 1253
Fujii, Kansuke 3711
Fujii, Keisuke 1189
Fujii, Toshiyuki 2540, 3578
Fujimoto, Hiroshi 77, 663
Fujimoto, Kazuki 2047
Fujimoto, Yasutaka 571, 681
Fujimura, Akira 1080
Fujita, Atsushi 296
Fujita, Goro 363
Fujita, Hideaki.....................626, 1854, 3813, 3940
Fujiwara, Hajime 1381
Fujiwara, Kazuya 3773
Fukuda, Hiroto 2938
Fukuda, Kenji 2558
Fukui, Tomoya 860
Fukuoka, T. 1240
Fukushima, Kentarou 2176
Fukushima, Takafumi.......................................3478
Funabiki, Shigeyuki 2449
Funaki, Tsuyoshi 309, 2181, 3092
Funato, Hirohito 94, 2036
Funato, Hiroki 2073
Furukawa, Keita 3349
Furukawa, Kimihisa..3572
Furukawa, Yudai 4193
Furusho, Yasuaki 3711
Gan, Yiliang..1391

AUTHOR INDEX

Ganisetti, V. K. 2907
Gao, Feng 2016, 3383, 3965
Gao, Xiaonan 1661
Gao, Zhuo 3455
Garrigós, A. 1435
Gasim, Abdulaziz 2836
Gehlot, Deepak 3471
Geng, Hua 542
Geng, Yiwen 619
Gerada, C. 4034
Gheonjian, Anna 2078
Gietler, Harald 3140
Gohara, Hiromichi 2764
Gondo, Ryota 3490
Gong, Bing 3797
Gong, Chunying 1015, 1177
Gong, Z. 267
Gorodnichev, Anton 375
Goto, Akihisa 2449
Goto, Hiroki 3192
Goto, Kazuya 1315
Goto, Yasuyuki 809
Gou, Yating 1157, 1167
Grimm, Ferdinand 2895
Grossner, Ulrike 3588
Gruber, Wolfgang 3632, 3640, 4028
Gu, Lei 632
Gu, Qing 2963
Guajardo, Cristian Andres Garces 1854
Guan, Bo 1032
Guan, Yajuan 2668, 3678
Guan, Yueshi 614, 3780
Guangzhu, Wang 1376
Guerrero, Josep M. 1498, 2668, 3112
Guerrero, M. Josep 3678
Gui, Yonghao 2668
Guidi, Giuseppe 782, 2003
Guillod, Thomas 396
Gunji, Daisuke 663
Guo, Leilei 904
Guo, Yanjie 3338
Guozhao, Duan 2625
Gupta, K. 267
Gurpinar, Emre 130
Gutiérrez, R. 1435
Ha, Jung-Ik 565, 2500
Ha, Sang-Hyun 3466
Haga, Hitoshi 1370, 3890
Hagiwara, Makoto 3273
Hahashi, Yuji 4059
Haider, M. 4080
Halamicek, Michael 831
Halick, Mohamed 416
Hamabe, Yasumasa 1276
Hamada, Shizunori 227

Hamaguchi, Takumi 3507
Hamasaki, S. 1240
Hamasaki, Shin-Ichi 1217, 1276, 2938, 3237
Hameyer, Kay 740
Han, Byung-Moon 466
Han, Jung-Kyu 3107, 3533, 4049, 4054
Han, Pengcheng 1027, 2714
Han, Yang 3112
Hanajiri, Kensuke 663
Hanamoto, Tsuyoshi 1315, 1698
Hancioglu, Oguz Kaan 1680
Handa, Hiroyuki 3762
Handa, Yuuichi 4059
Hane, Yoshiki 2426
Hang, Lijun 1391, 2866
Hanju, Cha 1985
Hao, Liu 3484
Hao, Xiang 1478
Harnefors, Lennart 3684
Hartmann, S. 3596
Haruna, Junnosuke 94, 2036
Hasegawa, Kazunori 1938
Hasegawa, M. 3665
Hasegawa, Ryuta 2011
Hashempour, Mohammad M. 4198
Hashimoto, Kazuki 3757
Hasler, Jean-Philippe 3684
Hata, Katsuhiro 663
Hata, Ryotaro 2149
Hatakeyama, Tomoyuki 1991
Hataya, Morimasa 410
Hatipoglu, E. 3805
Hatsumi, Takuya 94
Hatta, Yoshiyuki 675
Hattori, Fumiya 2738
Hattori, Keisuke 3286
Haung-Jen, Chiu 645
Hayashi, Nobuo 866
Hayashi, Yuji 356
He, Wangpin 560
He, Xiaokun 1504, 2496
He, Xiaoqiong 1027, 2714
He, Yigang 2317
He, Yingjie 3439
Hendrix, M. A. M. 946, 2697
Heo, Jongwon 726
Hidaka, Yuki 2820
Higuchi, Keiichi 2764
Higuchi, Masato 3952
Higuchi, Shinichi 2216
Hikaru, Naruse 3418
Hikihara, T. 3665
Hikihara, Takashi 3654, 3757
Hiller, Marc 3074
Hillers, A. 1896

AUTHOR INDEX

Hillers, André 2301
Hilt, Oliver 3607
Hinz, Arne 598
Hirahara, Hideaki 1960
Hiraki, Eiji.............................410, 1602, 1610
Hirao, Takashi...........................2082, 2137
Hirase, Yuko......................................767
Hirayama, Katsutoshi..........................4193
Hirayama, Tadashi 3406
Hirokawa, Masahiko..................1543, 4133
Hirokawa, Takayuki....................296, 410
Hiromoto, Masayuki............................3644
Hirose, Keiichi...........................593, 822
Hirose, Naoki....................................3791
Hiroshi, Tadano 3431
Hiroshige, Shinichi.............................3369
Hirota, Takashi 3952
Hoang, Tuan V....................................1752
Hoda, Isao 2073
Hofmann, Viktor 2476
Hofmann, Wilfried 3243
Hojo, Masahlde 3369
Holenstein, Thomas.............................3619
Holmes, D. G. 3670
Hong, Miao 2718
Hongpeng, Liu...................1442, 2969
Honjo, Satoshi 2066
Hori, Motohito..................................3396
Hori, Yoichi..............................77, 663
Horie, Shunsuke..................................809
Horikoshi, Takahiro 1997
Hoshi, Nobukazu...............971, 2660, 3855
Hou, Chung-Chuan...............................1617
Hou, Lijun 2901
Houran, Mohamad Abou 1062, 2854
Hsieh, Guan-Chyun 123
Hsieh, Hung-I....................................123
Hsieh, Yao-Ching.........................3151, 3544
Hsu, Chi-Hsuan..................................2653
Hu, Jiewen 3985
Hu, Jingxin 1073, 2250, 2484
Hu, Sheng...3052
Hu, Song 370
Hu, Xihong..............................614, 3780
Hu, Xing 2262
Huang, Bing-Siang 2092
Huang, Bo-Jia....................................3528
Huang, Chien-Chun...............................3151
Huang, Huazhen...................................1125
Huang, Jingjin..........................2980, 4157
Huang, Jingjing 1004, 2688, 2692
Huang, Jun-Xian.........................1626, 3081
Huang, Lang......................................1478
Huang, Pin Yu....................................2165
Huang, Ta-Wei....................................1626

Huang, Wen-Mei 2576
Huang, Xianjin 1131, 2051
Huang, Xiaoliang 84
Huang, Xuehao 3455
Huang, You-Chun 275
Huemer, Mario 3140
Hui, S. Y. Ron 889, 2552
Hung, Chun-Yao 2576
Hung, Shun-Kang 1575
Huo, Chongcan 987
Huo, Junya 1206, 1234
Hussein, Abdallah................... 130, 2932
Huynh, Dang Minh 3086
Hwang, Duck-Hwan 1403
Hwang, Seon-Ik 3323
Hwu, K.I. 851
Hyakutake, Y. 1253
Hyodo, Takashi 2589
Hyunsung, An 1985
Iannuzzi, Diego 2527
Ibuchi, Takaaki 309
Ichinose, H. 1240
Ide, Yuji 3896
Iijima, Ryuji 313, 1111
Iioka, Daisuke 2278
Ikari, Yuki 148
Ikeda, Hidehiro................................. 1315
Ikeda, Yoshinari 3396
Ilves, Kalle 2335
Imai, Kazu 3363
Imai, Makoto 296, 410
Imamori, Satoshi 699
Imaoka, Jun 1087, 1095, 1554, 3773
Imoto, R. 2808
Imtiaz, Abu Saleh 2416
Imura, Takehiro 77, 663
Inaba, Tsuyoshi 4114
Inomata, Kentaro 3952
Inoue, Daisuke 2764
Inoue, Kaoru.............1264, 2186, 4151
Inoue, Kent 348
Inoue, Masamichi 1228
Inoue, Takatoshi 1276
Inoue, Y. 704, 2808
Inoue, Yukinori 1189, 1289, 1329, 2802, 2814, 3197
Irino, Yusuke 244
Ise, Toshifumi 775, 2393, 3762, 3902
Ishibashi, Mikiya 1370
Ishibashi, Naoyuki 1543
Ishibashi, Taku 2292
Ishigaki, Shingo 227
Ishiguro, Takahiro 1997, 2011, 3304
Ishihara, Masataka 1610
Ishii, Y. 1834
Ishii, Yuki 1196

AUTHOR INDEX

Ishikawa, Hiroki 2176, 3412
Ishikawa, Kohsuke.....................................2725
Ishikura, Yuki 1087, 1095, 3717
Isobe, Eisuke...2042
Isobe, Takanori 313, 1111, 3375, 3431
Isozaki, Keisaku 1364
Itaya, Yohei 3450
Ito, Kazuhiko 2540
Ito, Yasuaki 1586, 2324
Ito, Yoichi 3086
Ito, Youichi 439
Itoh, Gimpei 1289
Itoh, Jun-Ichi...................69, 348, 534, 896, 1567,
      2229, 2237, 2519, 2596, 3349, 3797
Iwabuchi, Akio 439
Iwai, Akinobu 2066
Iwaji, Yoshitaka 1301
Iwasaki, Makoto 1666
Iwasaki, Tetsuya 3490
Iwata, Hiroki 3896
Iyasu, Seiji 4059
Iyoda, Isao 2914
Jacobs, Keijo 3292
Jaffar, Hanis Afiqah Binti.............................2956
Jain, Prashant..3471
Janah, Mounia...681
Jang, Duekjin 2619
Jang, Yu-Jin 1655, 3466
Jang, Yun 1736
Jangs, Yujin 1562
Jarutus, Neerakorn 2121
Jehle, Andreas 1509
Jennings, M 1149
Jeong, Seog Y..2564
Jeong, Si-Hoon 289
Jeong, Yeonho 838, 2365, 2376
Jhang, Ying-Yi 3884
Jhou, Yu-Lin 1145
Ji, Guyuan 2921
Jia, Haiyang..998
Jia, Pengyu 977, 3040
Jia, Xu 3025
Jiacheng, Wang 2986
Jiajie, Zang 2986
Jiajie, Zhou 1442
Jian, Jun-Min 2653
Jiang, Jinhai 84
Jiang, Shuai 987
Jiang, Siyue 4168
Jiang, Yanfeng 2058
Jiang, Yongbin 3863
Jianhua, Wang 1282
Jianming, Xu 528
Jianqiao, Zhou 2986
Jianwen, Zhang 2986

Jiaxing, Liu 1376
Jikumaru, Takehiro 177
Jimichi, T. 1834
Jimichi, Takushi 3729
Jin, Nan 904
Jin, Zheming 2668
Jing, Lei 878
Jing, Lyu 2692
Jing, Yang 3383
Jingyu, Song 1282
Jing-Yuan, Lin 645
Jinjun, Liu 4181
Jinshui, Zhang 4181
Jisaki, Jun 3182
Joebges, Philipp 375
Jongudomkarn, Jonggrist 3902
Jonishi, Akihiro 2774
Joryo, Satoshi 1202
Joseph, Anto 1358
Jumayev, S. 161
Jung, Hanul 688
Jung, Hyun-Sam 911
Jung, Jae-Jung 3557
Jung, Jee-Hoon 289, 383
Jung, Jun-Hyung 3323
Jung, Si-Hoon 383
Jungmayr, Gerald 3640
Junior, Lourenço Matakas 3785
Jynu-Jhe, Jhang 645
Kada, Haruya 3890
Kadota, Mitsuhiro 3572
Kai, Masahiko 1803
Kaicheng, Ding 4181
Kaipia, T. 2948
Kaishakuji, Hikaru 2360
Kakigano, Hiroaki 583, 2956
Kamaeguchi, Koki 410
Kamakura, Kousuke 2756
Kamejima, Takayoshi 3286
Kamiya, Naoki 1673
Kamiyama, Naosumi 1955
Kamoshida, Naoki 1111
Kampeerawar, Warayut 3257
Kanai, Naoyuki 3396
Kanaya, Kazuhisa 2011
Kanazawa, Yasuki 2789
Kanchan, R. S. 488
Kandula, Prasad 4001
Kaneko, Satoshi 3396
Kanetani, Kaisei 207
Kang, Dong-Hun 3030
Kang, Feel-Soon 2376
Kang, Kyoung-Suk 922
Kang, Tahyun 1977
Kang, Yong 2997

AUTHOR INDEX

Kanno, Junya 3299
Kano, Fumihisa 2036
Kanoda, Akihiko 3572
Kanzian, Marc...3140
Kapisch, E. B. 1067
Karami, Bagher...2854
Karppanen, J. 2948
Kasai, Yuji 2036
Kashihara, Tatsuki 1741
Katayama, Tatsuji 1346
Kato, Hideaki 1580, 1586, 2324
Kato, Hirokazu..3478
Kato, Koji 439, 1370, 3086
Kato, Toshiji...............................1264, 2186, 4151
Katoh, Kaoru...233
Katoh, Shinji 2176
Katsuki, Akihiko 1543
Katsura, Seiichiro 669
Katsura, Shogo 767
Katsushi, Terazono 3431
Kawabata, Naoki 2887
Kawabata, Shuma 3406
Kawagoe, Natsuki 3490
Kawaguchi, Hironori 517
Kawaguchi, Jun'ichiro 1828
Kawaguchi, Yuki...3572
Kawakami, Masaki 2756
Kawakami, Noriko 1346
Kawamura, Atsuo 318, 1649, 1687, 3916
Kawamura, Itsuo 3396
Kawamura, Kazuki..1567
Kawanishi, Kota 169
Kawashita, Jun 2042
Kayashima, Kazuya.......................................1315
Kaymak, Murat..3729
Kazmi, Syed Muhammad Raza 4168
Ke, Junji 1125
Kennel, Ralph...............................1661, 2895, 3965
Kezuka, Nobutaka 227
Khan, Ashraf Ali 927
Khan, Faisal..446, 2416
Khan, Muhammad Mansoor...........................528
Khan, Usman Ali 927
Khomfoi, Surin...1460
Khubchandani, Vasudha................................845
Kiatsookkanatorn, Paiboon..........................2581
Kida, Masahiro.....................................1586, 2324
Kido, Tatsuya 329
Kikuchi, Ryosuke 1877
Kikuchi, Takaaki 2292
Kikuchi, Takeshi 3578
Kikuma, Toshiaki..3299
Kim, Byeongwoo 256
Kim, Chong-Eun 108, 3538
Kim, Dong-Kwan 1655, 3466, 3538

Kim, Gun-Woo 838
Kim, Hansang 1465
Kim, Heung-Geun 927, 1046, 2619, 3134
Kim, Hideaki...207
Kim, Hyeon-Sik 521
Kim, In-Dong 3229
Kim, Jae-Kuk..3100
Kim, Jang-Mok 3323
Kim, Jin-Hak 1530
Kim, Jin-Young 3229
Kim, Jong-Woo..............................3107, 4049
Kim, Kangsan 256
Kim, Katherine A. 2092, 3063
Kim, Keon Young 4104
Kim, Keon-Woo............108, 1562, 1655, 2365, 2376
Kim, Ki-Mok...2365
Kim, Myong Hwan 2500
Kim, Sanghun 2619
Kim, Sunju 3833
Kim, Yeonjung..1465
Kimura, Hideki 2036
Kimura, Mamoru 1991, 1997
Kimura, Noriyuki............1202, 1259, 2558, 2887, 2914
Kinoshita, Masahiro 3929
Kishimoto, Toshihiko 261
Kishita, Ken 1301
Kitagawa, Wataru 1847, 3507
Kitamura, Akio..2764
Kitamura, Toshinori 2660
Kiyoshi, Ohishi 1673
Kiyota, Kyohei 3182
Klammer, Bianca 3632
Ko, Chien-Tzu 2107
Kobayashi, Hiroyasu 2527, 3490
Kobayashi, Koji 1741
Kobayashi, Marika 2802
Kodaka, Wataru 2589
Kogai, Naoki 1364
Koizumi, Hirotaka 4114
Kolar, J. W. 3805, 4080
Kolar, Johann W. 181, 396, 3619
Kolb, Johannes 2848
Komaru, Yuma 1329
Komatsu, Hiroyoshi 1346
Komatsu, Taiga 2820
Komatsu, Wilson 3785
Komeda, Shohei 3813
Kometani, Haruyuki 711
Kondo, Keiichiro 726, 2047, 2527, 3490
Kondo, Shota 1295
Kondo, Takeshi 4114
Kong, Wei 3460
Kongjeen, Yuttana 2430
Konishi, Akihiro 1602
Konno, Junya..1692

AUTHOR INDEX

Konstantinou, Georgios 3117
Kopta, A. 3596
Kosaka, Takashi 3418
Koseki, K. 1162
Koseki, Takafumi 2042, 2309, 3257
Koshikizawa, Hiroyuki 1567
Kostov, Konstantin 2732
Kouketsu, Masaju 227
Kouno, Yusuke 2176
Kovacevic-Badstübner, Ivana 3588
Kowatari, Hiroki 2660
Koyama, Yushi 2011
Krismer, F. 3805
Krismer, Florian 396
Kubo, Hajime 483
Kubota, Hisao 1196
Kucka, Jakub 1904
Kumada, Keishirou 3396
Kumagai, Shuta 1264
Kumar, Ashish 3993
Kumar, Rajesh 2456
Kumar, S. Gautam 3471
Kumsuwan, Yuttana 2113, 2121
Kunomura, Ken 1803
Kuo, Chun-Ting 1145
Kuraishi, Daigo 3896
Kuraku, Nagendra Vara Prasad 2317
Kuring, Carsten 3607
Kurisaka, Masakatsu 4151
Kurita, Naoyuki 1991
Kurita, Nobuyuki 3640
Kurokawa, Fujio 826, 2097, 2283, 4193
Kurosawa, Nobuhito 1810
Kurumatani, Hiroki 669
Kusaka, Keisuke...............69, 348, 2237, 3349
Kusumah, Ferdi Perdana 3870
Kuwata, Gen...177
Kwon, Min-Jun 114
Kyyrä, Jorma 2193, 3870
Lai, Jih-Sheng 3107, 4049
Lai, Jui-Hung 3081
Lan, Yuanliang 1167
Lana, A. 2948
Le, Hanh-Phuc 213
Le, Hoai Nam 2519
Lee, Byoung-Hee 838
Lee, Byung-Kwon 3030
Lee, Chan 688
Lee, Choongin 565
Lee, Dong-Choon 478, 1212
Lee, Hong-Hee 1752
Lee, Hyong Gun 1336
Lee, Il-Oun 1530
Lee, Jae-Bum 3100
Lee, Jia-You 657, 2107

Lee, Joon-Hee 3557
Lee, Junbae 1141
Lee, June-Hee 466
Lee, Jung-Yong 1403
Lee, Jun-Young 3030
Lee, Jusuk 1336
Lee, Kyo-Beum 466, 1736, 4104, 4109
Lee, Kyoung-Won 2145
Lee, Kyung-Hwan 2500
Lee, Min-Su...108
Lee, Minsub 1141
Lee, Nayoung 1562
Lee, Song-Kai.......................................2102
Lee, T. L. 4198
Lee, Tzung-Lin 2576
Lee, Woo-Cheol 114
Lee, Woo-Seok 1530
Lee, Young-Dal 3466, 3538
Lehn, Peter W. 3203
Lei, Qin 2400, 3742
Leng, Darith 1764
Leubner, Martin 3243
Li, Bodong 878
Li, Chi 790, 3705
Li, Dongsheng 1301
Li, Fei 2611
Li, Fujian 2944
Li, Guanglei 1455
Li, Haijin 2270
Li, Haisi 3040
Li, Haoyu 2901
Li, Hong 2058
Li, Hongchang 337, 3910
Li, Jhih-Sian 3081
Li, Jia 2073
Li, Jianfeng 130
Li, Kaiyuan 1517, 1592
Li, Lei 1172
Li, Li 1771
Li, Ming 2973
Li, Mingshen 2668, 3678
Li, Pengcheng 3698
Li, Shufan 3338
Li, Sinan 889, 2552
Li, T.-Y. 1912
Li, Xiaodong 370
Li, Xiaolu Lucia 3768
Li, Xiaoqiang 3910
Li, Xingshuo 453
Li, Xinying 2646
Li, Yan 2245
Li, Yang 795, 1478
Li, Yangman 2901
Li, Yi-Chan 639, 883
Li, Yongdong 1010, 2386

AUTHOR INDEX

Li, Yong-Jyun 275
Li, Yunwei 3958
Li, Yunwei Ryan 1537
Li, Yuze 2997
Li, Zhenjie 84
Li, Zhenwei 998
Li, Zhiqing 100
Liang, Daniel 1943
Liang, Junrui 4122
Liang, Ning 1157
Liang, Wencai 1131
Liao, Chenglin 3338
Liao, Chih-Yi 657
Liao, Hsuan 2653
Liao, Jian-Tang 4233
Liao, Mengyan 1386, 1421
Liaw, C. M. 2907
Lim, Cheon-Yong 1655, 2376, 3533
Lim, Dae-Sik 1212
Lim, Kyungbae 803
Lim, Young-Cheol 982, 1799
Lln, Chang-Hua 1341, 1777
Lin, Cheng-Hung 2092
Lin, Fei 1131, 1816, 2051, 2058, 3484, 3495
Lin, Jin 3460
Lin, Jing-Yuan 3151
Lin, K.-E. 1912
Lin, Min 4133
Lin, Xiang 3460
Lin, Xiaolan 1027
Lin, Xuerui 1537
Lin, Yu-Hsiu 1575
Lin, Yu-Lin 1145
Lisha, Chen 3958
Liske, Andreas 2848
Liu, Baojin 1051, 2944, 3924
Liu, Bi 1872
Liu, Bo 542, 878
Liu, Chao 2245
Liu, Chunhui 3742
Liu, Cuicui 1157, 1167
Liu, Dong 1758, 2708
Liu, Fang 2611, 2992
Liu, Furong 3052
Liu, He 3215
Liu, Hwa-Dong 1341, 1777
Liu, Jia 775, 3902
Liu, Jiaxin 2016
Liu, Jinjun 957, 1051, 1524, 2435,
     2646, 2681, 3129, 3176, 3251, 3439, 3692,
     3924, 3934
Liu, Junwen 3863
Liu, Kangli 3010, 3015
Liu, Nianzhou 1010
Liu, Ning 2877

Liu, Pang-Jung 2102
Liu, Ruofei 2547
Liu, Shu 3052
Liu, Siqi 1491
Liu, Tao 1478
Liu, Teng 2681, 3176, 3934
Liu, Wei 3164
Liu, Wenzhao 3678
Liu, Xiaosheng 934
Liu, Xicai 1661, 3965
Liu, Xinbo 3455
Liu, Yifu 2400, 3742
Liu, Yu-Chen 2092
Liu, Yuping 1816
Liu, Zeng 957, 1524, 2435, 2681,
     3176, 3251, 3692, 3749, 3924
Liu, Zhiyuan 3495
Liu, Zipeng 2681, 3176
Lo, Jen-Hao 1145
Lomonova, E.A. 161
Lopez-Lopez, Alvaro J. 2534
Lotfi, Nima 3223
Lovison, Giorgio 77
Lu, David H. 2404
Lu, David Hongfei 3390
Lu, Kaiyuan 1183, 1246, 2842
Lu, M. Z. 2907
Lu, Shengli 3145
Lu, Shuai 3698
Lu, Y. 267
Luhtala, Roni 547, 2470, 3156
Lunglmayr, Michael 3140
Luo, Min 422, 2199
Luo, Rui 3129, 3439
Luo, Y. 267
Luong, Hoan-Tien 2145
Lyu, Jing 1004, 4162, 4220
Ma, Baohui 2882
Ma, Jie 1118
Ma, Ke 3877
Ma, Shaokang 542
Ma, Tianshu 2703
Ma, Yue 3717
Ma, Zhixun 917, 2688, 2692, 4157, 4162
Mabuchi, Yuichi 3572
Machavolu, Sawanth Krishna 753
Machida, Yuuki 2449
Maharjan, Laxman 1840
Makishima, Shingo 2047
Mannen, Tomoyuki 1414, 1866
Mantooth, H. Alan 153
Mao, Meiqin 815, 1472, 1793, 2505
Mariéthoz, Sébastien 2158, 3820
Marinescu, Radu-Florin 1822
Marroquí, D. 1435

AUTHOR INDEX

Martinez, Wilmar 2193
Maruta, Hidenori 826
Maruyama, Kouji.........................3396
März, Martin 2410
Masuda, Eisuke 309
Masuda, Mitsuru...........................88
Masuko, Toshitake.......................3723
Matsubayashi, Tatsushi..................207
Matsuda, Akihiro 2329
Matsuda, Tomohiro 1972
Matsudate, Koki..........................2022
Matsui, Nobumasa 826, 2283
Matsui, Nobuyuki........................3418
Matsui, Teruhisa.........................1803
Matsui, Yoshihiro........................1080
Matsui, Yuto 1847, 3791
Matsuki, Yosuke..........................2224
Matsumori, Hiroaki......................3357
Matsumoto, Satoshi..................2360, 2370
Matsumoto, Takashi......................2404
Matsumoto, Toshiaki................2011, 3304
Matsumoto, Yasuaki......................517
Matsumoto, Yohei........................233
Matsumura, Toshiro......................809
Matsuo, Keisuke..........................169
Matsuse, Kouki...........................169
Mattsson, A..............................2948
Mawby, P.................................1149
Mcgrath, B. P............................3670
Meng, Xin 957, 1549, 3251
Menzi, David 181
Mertens, Axel 1904
Messo, Tuomas 547, 963, 1704, 2470, 3156, 4205
Michihira, Masakazu..................992, 3058
Michikoshi, Hisato 2558
Milovanovic, Stefan......................1484
Min, Geon-Hong..........................2500
Minami, Masataka 992, 3058
Mino, Kazuaki...........................3717
Mira, Maria C............................1351
Mishima, Tomokazu...................329, 872
Misra, Mitradatta........................3884
Mitsantisuk, Chowarit....................3332
Miura, Yushi 775, 2393, 3762
Miwa, Yoshihiro..........................404
Miyajima, Hiroki.........................1803
Miyama, Yoshihiro 711
Miyawaki, Satoshi........................2738
Miyazaki, Toshimasa......................1673
Mizumoto, Yuki...........................1810
Mizuno, Takayuki.........................169
Mizuno, Yuji.............................2283
Mizushima, Takuya 1543
Mocevic, Slavko..........................3985
Mochidate, Sae...........................1972

Mogorovic, Marko........................ 2170
Moiannou, Tom 831
Mok, Hyung Soo 1336
Molinas, Marta 759
Moo, Chin-Sien 275, 3544
Moon, Gun-Woo...................... 108, 838, 1562, 1655, 2365, 2376, 3100, 3466, 3533, 3538, 4049, 4054
Mori, Kazuhisa 233
Morimoto, Hiroaki 2540, 3265
Morimoto, S. 704, 2808
Morimoto, Shigeo ...1189, 1289, 1329, 2802, 2814, 3197
Morimoto, Shinya........................ 2210
Morishima, Naoki 2540, 3450
Moriyama, Hiroyuki 1580, 1586, 2324
Morizane, Toshimitsu 1202, 1259, 2558, 2887, 2914
Mortimer, Benedict J. 598
Motegi, Shin-Ichi 992, 3058
Motohashi, Yuto 753
Motoyama, Hiromasa 356
Mouawad, Bassem 130
Mukaiyama, Naoki 2558
Müller-Hellmann, Adolf 598
Muni, Bishnu Prasad..................... 3471
Murakami, Toshiyuki.................... 575
Nabetani, Yoichi 2404
Nada, Kaho 3578
Nagai, Sakahisa 1687
Nagai, Satoshi 534
Nagao, S. 142
Nagaoka, Naoto 3170
Nagaoka, Shingo 118, 4139
Nagasaka, Kuniaki...................... 1692
Nagashima, Takumi 3490
Nagira, Yoshiki 4016
Naina, Sagar........................... 3046
Nakabayashi, Shigeaki 1692
Nakabayashi, Shigeyuki 517
Nakagawa, Hidehiko 767
Nakahara, Kengo........................ 3237
Nakahara, Mizuki 3572
Nakai, Masanobu 3182
Nakajima, Mizuki 2750
Nakajima, Tatsuhito 1997, 3299
Nakamura, Fuminori 2329
Nakamura, Hideyo 1137
Nakamura, Kenji 2426
Nakamura, Kimikazu 4059
Nakamura, M........................... 201
Nakamura, Masashi 471
Nakamura, Ritaka 495
Nakano, Hayato 2764
Nakano, Shigeki 2370
Nakao, Hiroshi......................... 196
Nakao, Kazushige 148, 2914

AUTHOR INDEX

Nakao, Yuta 588
Nakashima, Yoshiyasu 196
Nakatsu, Kinya 2082
Nakazawa, Haruo 2404
Nakazawa, Y. 1253
Nakazawa, Yuji 244
Namba, Akihiro 2082
Nanamori, Kimihiro 2789
Naradhipa, Adhistira M. 3833
Narita, Takayoshi 1580, 1586, 2324
Narushima, Hiroki 693
Nashida, Norihiro 1137
Nasr, Miad 2380
Natori, Kenji 588, 1860
Nawaz, Muhammad 2335
Nazib, A. A. 3670
Nee, Hans-Peter 2732, 3292, 3684
Neubert, Markus 3979
Ngamroo, Issarachai 2287
Ngo, Tung 1724
Nguyen, Bang Le-Huy 1046, 3134
Nguyen, Hong-Quan 3426
Nguyen, Minh-Khai 982, 1799, 2145
Nguyen, Tien-The 1046, 3134
Nho, Eui-Cheol 922
Nicolae, Ileana-Diana 1822
Nicolae, Petre-Marian 1822
Nie, Jintong 2963
Niki, Toru 856
Nimura, Takumi 1295
Ninomiya, Tatsuya 2836
Nishikata, Shoji 4227
Nishimura, Yoshitaka 1137
Nishino, Taisei 1364
Nishiyama, Shigeki 2149
Nishizawa, Koroku 2229
Nishizawa, Shin-Ichi 1938
Niu, Haonan 3025
Niyomsatian, K. 4096
Noah, Mostafa 1087, 1095
Noda, Taku 2176
Noda, Yujiro 324
Noguchi, Toshihiko 718, 753
Noh, Seungjun 1598
Nomura, Naofumi 2216
Nomura, Shinichi 2022
Nonogaki, Midori 2292
Noro, Osamu 767
Norrga, Staffan 3292
Norum, Lars 2630
Noto, Yasuyuki 3711
Notohara, Yasuo 1301
Nuchnoi, S. 4096
Nugroho, Dannisworo S. 3855
Nussbaumer, Thomas 3619

Nuutinen, P. 2948
Obara, Hidemine 1649
Oda, Yoshiho 1586, 2324
Ogasawara, Satoshi 2589, 2725, 2796, 3315
Ogawa, Eri 2768
Ogawa, Kazuki 1580
Ogawa, Takuro 866
Ogawa, Tomoyuki 1828, 3265
Ogawa, Toru 2796
Ogino, Hiroshi 517
Oh, Sehoon 688
Ohashi, Hidetomo 2774
Ohdera, Fumiya 1322
Ohguchi, Hideki 699
Ohishi, Kiyoshi 1741, 3332, 3890, 3896
Ohji, Takahisa 3182
Ohnishi, Haruna 3273
Ohno, Takanobu 971
Ohno, Tatsuki 1649
Ohnuma, Naoto 233
Ohnuma, Takumi 1223
Ohnuma, Yoshiya 2738
Ohta, Kazuki 1223
Ohta, Takahiro 517
Ohtake, Asuka 3286
Ohyama, K. 1253
Ohyama, Kazuhiro 2921
Ohyama, Kazunobu 244
Oi, Kazunobu 1890
Oishi, Kazuki 3644
Oiwa, Takaaki 157, 4042
Oka, Toshiomi 2370
Okamoto, Kenkichiro 1095
Okazaki, Yuhei 2335
Okazawa, Toshio 2066
Oki, Yusuke 1828
Okitsu, Takashi 169
Okuda, Takafumi 3654, 3757
Okuno, Kengo 1586, 2324
Okuyama, Ryota 3450
Omori, Hideki 1202, 1259, 2558, 2887
Omori, Shuto 471
Omura, Ichiro 1938
Onishi, Hiroyuki 4139
Onishi, Masami 2082
Ono, Y. 4080
Onozawa, Yuichi 2768
Ooshima, Masahide 3613
Orikawa, Koji 2589, 2725, 3315
Ortiz-Gonzalez, J. 1149
Osawa, Akihiro 2764
Oshima, Takuya 4088
Osman, Ilham 3971
Ota, Ryosuke 3855
Ouaida, Rémy 2153

AUTHOR INDEX

Ouchi, Takayuki 250
Ouyang, Shaodi 1051, 3129
Ouyang, Ziwei 4066
Owaki, Daiki 809
Paiboon, Supakorn 1642
Pairindra, Worapong 1460
Pan, Pengpeng 1504
Pan, Xuewei 1172
Panda, Sanjib Kumar 4186
Pang, Hui 2343
Papadopoulos, C 3596
Papini, L 4034
Paramalingam, Jan 2329
Parashar, Sanket 3993
Park, Hwa-Pyeong 289
Park, Jin-Hyuk 4104
Park, Jun H. 2564
Park, Kwon-Sik 922
Park, Moo-Hyun 1562, 3100, 3533
Park, Mu-Hyun 838
Park, Sang Uk 1336
Park, Sanghyeon 282
Partanen, J. 2948
Pasterczyk, Robert 2158
Patel, Prashant 3046
Patel, Utsav 3046
Pathmanathan, M. 488
Patwa, Premal 3046
Pauli, Florian 740
Pecharroman, Ramon R. 2534
Pei, Xuejun 2997
Peltoniemi, P. 2948
Peng, Jinjie 939
Peng, Xu 1027, 2714, 3020
Pengxiang, Zeng 4181
Pham, N. Ha 1414
Pidaparthy, Syam Kumar 1465
Pinomaa, P. 2948
Polmai, Sompob 1764, 2490
Pou, Josep 3117
Prabowo, Yos 3564
Prasanth, Sundararajan 416
Prodic, Aleksandar 831
Promyoo, Adisak 2871
Pueschel, Tilo 190
Pyrhonen, J. 161
Qi, Wenlong 889
Qian, Cheng 1472
Qian, Qinsong 3145
Qiao, Liang 3329
Qin, Zian 1925
Qiu, Maohang 878
Qiu, Zhifeng 939
Rabkowski, Jacek 2129
Radman, Karlo 3632

Radwan, Hamdy 3945
Rahimo, M. 3596, 4009
Rahman, Ahmad Arif Bin Abd 2956
Rahman, Faz 3971
Rahmati, Abdolreza 2854
Ramirez-Elizondo, Laura 1426
Ramos, Niño Christopher 3092
Ran, L. 1149
Ran, Li 1931
Rao, Eswar 3471
Rathore, Akshay Kumar 342, 2456
Reinikka, Tommi 1704
Remus, Nico 3243
Ren, Haijun 2714
Ren, Yu 3329
Rencz, Marta 137
Rengarajan, Satish 3564
Riar, Baljit 4074, 4145
Rietmann, Stefan 2301
Rim, Chun T. 2564
Risseh, Arash Edvin 2732
Rivas-Davila, Juan 282, 632, 3848
Robert, Mickaël 2158
Rodriguez-Diaz, Enrique 1498
Roes, M. G. L. 946, 2697
Roinila, Tomi 547, 1704, 1719, 2470, 3156, 4205
Romano, Daniele 3588
Roy, Sourov 446, 2416
Ruan, Liheng 3010, 3015
Rubino, S. 732
Ruf, Andreas 740
Rygg, Atle 759
Sadakata, Hideki 410
Sagawa, Kouhei 2036
Saha, Tarak 4074, 4145
Saito, Tatsuhito 1828, 3265
Saito, Yota 1782
Saitoh, Hiroumi 2278
Sakabe, Tomoki 3058
Sakai, Kazuto 2826
Sakai, Ryosuke 2832
Sakai, Yoshikazu 4114
Sakawaki, Atsushi 244
Sakimoto, Kenichi 767
Sakiyama, Taiki 2186
Sakoda, Kenichi 860
Sakr, Nadim 2078
Sakuma, Kensuke 3522
Sakuraba, Tomokazu 2153
Sakurai, Seiya 3412
Samanta, Suvendu 342
Samermurn, S. 4096
Samizadeh, Mehdi 1062, 2854
Sanada, M. 704, 2808
Sanada, Masayuki ..1189, 1289, 1329, 2802, 2814, 3197

AUTHOR INDEX

Sangwongwanich, Ariya 2512
Sangwongwanich, S...................................4096
Sangwongwanich, Somboon 1642, 2581
Sannomiya, Kenta 1259
Sano, Kenichiro 3299
Sano, Toshiki 3896
Santi, Enrico 1719
Sasaki, Masahiro 2774
Sasaki, Masato...3344
Sasongko, Firman 416
Sathik, Mohamed 416
Sato, Fumihiro...250
Sato, Keisuke 3265
Sato, Kenji 3478
Sato, Mitsuru...118
Sato, Motoki..663
Sato, Takashi 3644
Sato, Yasuhiro 2042
Sato, Yukihiko 588, 1860, 1972, 3514, 3522
Satoh, Nobuo 2750
Sayed, Mahmoud A....................................3945
Schanen, Jean-Luc 2158
Schletz, Andreas.......................................2410
Schülting, Philipp.....................................388
Schweiker, Daniel......................................2848
Schweizer, Mario 555
Schwendemann, Rüdiger...............................3074
See, Kye Yak 2296
Sekiba, Yoichi..2176
Sekimoto, Morimitsu 866
Sekisue, Takayuki 2176
Sekiya, Hiroo 3650, 4127
Semwal, R. R. 1358
Senanayake, Thilak....................................313
Seng, Tan Chuan 416
Seo, Byuong-Jun.......................................922
Seo, Gab-Su..213
Sera, Dezso...2512
Setiadi, Hadi 626
Settels, Sjef J...3840
Severson, Eric L. 4020
Sewergin, Alexander 3979
Sha, Yilin...3329
Shabib, G. 3945
Shamseh, Mohammad Bani 3916
Shan, Zhenyu...977
Shang, Gao..1282
Shao, Chi..2866
Shao, Riming 1793
Sharma, Avinash 2456
Sharma, Sohit 1730
Shen, Yanfeng 1788, 1925
Shen, Yatao..815
Shen, Yecheng 2842
Shen, Zhan.......................................1788, 1925

Sheng, Caiwang 1167
Shi, Gang..4220
Shi, Haixu...4168
Shi, Xiangyue 939
Shi, Yong..2877
Shibata, Naoya 3929
Shigeeda, Hidenori 2540
Shigematsu, Koichi 2176
Shigeuchi, Koji 3514, 3522
Shijo, Takuya 324
Shimada, Takae 250
Shimakage, Toyonari 2292
Shimamoto, Keita 2210
Shimao, Tohihiro 439
Shimaoka, Masahiro 1747
Shimizu, Toshihisa 302, 404, 2137, 2165, 3309, 3357
Shimizu, Toshimasa 1803
Shimomura, Shoji......................................2836
Shimono, Tomoyuki 675
Shimosato, Noboru 261, 3514, 3522
Shimoyama, A. 142
Shin, Sungyong 3418
Shinohara, Atsushi 1308, 1322
Shinohara, Hiroshi 1840
Shinshi, Tadahiko......................................4016
Shintani, Michihiro 3644
Shirai, Ryo..3309
Shirata, Kento 1137
Shiyuan, Yin 2625
Shoyama, Masahito 1095, 1554, 3773
Shujiang, Duan 2718
Shunsuke, Ohasi 3363
Shuto, Masao 699
Si, Yunpeng 2400, 3742
Sihvo, Jussi...2470
Sih-Yi, Lee..645
Silber, Siegfried 4028
Silventoinen, P. 2948
Simanjorang, Rejeki.............................416, 2296
Singh, Amit Kumar 4186
Singh, Vijay Kumar 1698
Son, Yung-Deug 3323
Song, Hongyu 3825
Song, Injong 803
Song, Kai...84
Song, Seung-Min 3229
Song, Shuguang 1051, 3129, 3924
Song, Wensheng..1872
Song, Yang..3698
Song, Yipeng 746
Song, Yubo..3877
Soong, Boon-Hee 1517, 1592
Soong, Theodore 3203
Soontorntaweesub, Kittichot 1764
Spiliotis, K. 488

AUTHOR INDEX

Stieneker, Marco 598, 2484
Stock, Alexander 3034
Stojadinovic, Miloš.........................1103
Su, Huiling 795
Su, Jianhui 2877
Su, Yu-Chen 1038, 3549
Sudo, K. 1240
Suetake, A. 142
Suetsugu, Tadashi 4193
Sueuchi, Yuki 1955
Sugahara, Satoshi...........................2756
Sugahara, T. 142
Suganuma, K. 142
Suganuma, Katsuaki..........................1598
Sugihara, Yusuke 2789
Sugimoto, Hiroya 3627
Sugimoto, Kazushige.........................767
Sugiyama, Takashi 3578
Suh, Yongsug 1977
Sul, Seung-Ki 521, 911, 3557
Sumida, Hitoshi 2774
Sun, Bainan 607
Sun, Chuan 370
Sun, Haotian 2780
Sun, Jianning 2963
Sun, Kai 3460, 4168
Sun, Lejia 2882
Sun, Peng 1125
Sun, Shumin 1455
Sun, Weifeng 3145
Sun, Xiangdong 2204
Sun, Yongping 560
Sun, Yuchong 3650, 4127
Sung, Kyungmin..............................1364
Suntio, Teuvo................................963
Supanyapong, S..............................1622
Surakitbovorn, Kawin 632, 3848
Surinkaew, Tossaporn 2287
Suul, Jon Are 782, 2003
Suwa, Hiroshi 1997
Suwankawin, S...............................4096
Suwankawin, Surapong 2871
Suzuki, Akio 1840
Suzuki, Dai..................................157
Suzuki, Hiromitsu 495
Suzuki, Kazuma 1847, 3501, 3507
Suzuki, Kenichiro...........................511
Suzuki, Toshiki.....................1586, 2324
Suzuki, Yuhei 3390
Suzumori, Hirofumi 2066
Tabata, Yoichiro 329
Tada, Makoto 1580
Tadano, Hiroshi 313, 1111, 3375
Tadano, Yugo 483, 1890
Taguchi, Masashi 826

Taguchi, Yoshiaki...........................3280
Taiyuan, Yin 2625
Tajima, Katsubumi 2832
Tajyuta, Toshihisa 1840
Takahashi, Akihiko 3896
Takahashi, Akiko 2449
Takahashi, Arata 1270
Takahashi, Isseki 575
Takahashi, Masaki 3186
Takahashi, R. 3665
Takahashi, Shotaro 3315
Takahashi, Tomohira 2796
Takahashi, Toshimichi........................227
Takahashi, Yuki 3375
Takakura, Shotaro 1270
Takami, Hiroshi 471
Takamura, Kenya 1381
Takano, Sho 3390
Takasho, Kenta 1890
Takatori, Koji 4139
Takayanagi, Ryohei..........................3396
Takeda, Kodai 2309
Takemoto, Masatsugu 2589, 2725, 2796, 3315
Takenaka, Hiroshi 3304
Takeno, K. 201
Takenoiri, Shunji...........................2764
Takeshita, Takaharu 356, 1847, 3501, 3507, 3791, 3945, 4088
Takeuchi, Norikazu 2292
Takeuchi, Yoko 1828, 3265
Takiguchi, Masashi 3723
Takimoto, Kazuyasu 3304
Takishima, Kenta 2826
Takubo, Hiromu 3390
Takuma, Shunsuke 2596
Takuno, Tsuguhiro 3578
Tamate, Michio..............................3315
Tan, Nguyen Anh 478
Tan, Siew-Chong 889
Tanaka, Akira 1960
Tanaka, Takaaki 2604
Tanaka, Takahide 2774
Tanaka, Toshihiko...................324, 1381
Tanaka, Tsuguhiro 3929
Tanaka, Y. 1162
Tanemo, Masamichi 2022
Tang, Cheng-Yu 639
Tang, Houjun 528
Tang, Ye 3705
Tang, Yi 337, 428, 434, 3910
Taniguchi, Katsumi 3396
Taniguchi, Katsunori 1202
Taniguchi, Tomoisa 866
Tatsumi, Kazuto 1202
Tatsuta, Fujio..............................4227

AUTHOR INDEX

Tatte, Yogesh 1730
Tausif, Ali 3833
Tcai, Anatolii 4109
Techama, Pantarote 2490
Teerakawanich, Nithiphat 3332
Teigelkötter, Johannes 3034
Tenconi, A. 732
Teraoka, Kenji 3086
Tey, Kuan-Chung 511
Thai, Van X. 2564
Thummala, Prasanth 4066
Tian, Mofan 998, 2785
Tian, Wei 1661
Tian, Xiaoyu 1771
Tian, Yanjun 1397
Tibola, Gabriel 1447
Tikka, V. 2948
Toba, Akio 1840
Toi, Takato 2229
Tokumaru, Syohei 2938
Tokusaki, Hiroyuki 2589
Tominaga, Isamu 1692
Tomita, Mutuwo 1295
Tong, Anping 1391, 2866
Tran, Hai N. 3833
Tran, Tan-Tai 1799, 2145
Trescases, O. 267
Trescases, Olivier 2380
Tripathi, Ravi Nath 1698
Troppenz, Maria 3607
Trung, Tran Vu 1666
Tsai, Chang-Lin 3151
Tsai, Meng-Jiang 2462
Tsai, Men-Shen 1575
Tsai, Terng-Wei 639, 883
Tsai, Tsung-Lin 3151
Tsai, Yue-Ting 4198
Tse, Chi K. 3768
Tseng, King Jet 1517, 1592
Tseng, Wei-Jing 1626
Tsuchiya, Taichiro 2329
Tsuji, Hitoshi 3717
Tsuji, M. 1240
Tsuji, Mineo 1217, 1276, 2938
Tsukakoshi, Masahiko 238
Tsumura, Akihiko 3490
Tsuno, Masahito 2558
Tsuruta, Ryoji 495
Tsuruta, Yukinori 318
Tsutsumi, Hirohiko 3723
Tu, Yiming 2435, 2681, 3176, 3439, 3934
Tuji, Mineo 3237
Tumerdem, Ugur 1680
Tumurbaatar, Anudari 1972
Uchida, Junichi 1955

Uchida, Yuuki 2750
Uchino, Yuki 324
Uda, Ryosuke 3578
Udagawa, Ikuto 517, 1692
Ueda, Tetsuzo 3762
Uehara, H. 1253
Uematsu, Takeshi 118, 4139
Uemura, Takamasa 860
Ueno, Tsutomu 4151
Uesugi, Yuma 3412
Ueta, Hiroaki 1883
Umeda, Takashi 2814
Umetani, Kazuhiro 410, 1602, 1610
Unamuno, Eneko 759
Uno, Masatoshi 1782, 2030
Unterrieder, Christoph 3140
Ura, A. 704
Urabe, Shinichi 1782
Urata, Kazuki 302
Ute, Ryo 3773
Valente, G. 4034
Van De Ven, B.A.C. 267
Van Duivenbode, Jeroen 3840
Van Lam, Phi 571
Vasquez, C. Juan 3678
Vasquez, Juan C. 1498
Vass-Varnai, Andras 137
Veerachary, M. 845
Vemulapati, U. 3596, 4009
Vobecky, J. 3596
Vukadinovic, Nenad 831
Vyacheslav, Shkodyrev 1966
Wachi, Tsuneshisa 1997
Wada, Haruhisa 3286
Wada, Keiji 1414, 1866, 1919, 2137, 4059
Wakimoto, Hiroki 2404
Wang, Beibei 795
Wang, Bo 459
Wang, Can 1172
Wang, Chao 2386, 2901
Wang, Congling 3112
Wang, Dong 1183, 1246
Wang, Feng 1157, 1167, 2882
Wang, Fusheng 2611, 2992
Wang, Gaolin 1206, 1234
Wang, Guoxin 1206
Wang, Hanyu 2997
Wang, Hao 2270
Wang, Haoyu 100, 3825
Wang, Hechao 1183, 1246
Wang, Hongjie 4074, 4145
Wang, Huai 1021, 1788, 1925, 2604, 2743, 3123
Wang, Huiying 1234
Wang, Jianing 2611
Wang, Jizhe 826, 2097

AUTHOR INDEX

Wang, Jun...............................3749, 3985
Wang, Kui..............................1010, 2386
Wang, Laili...........................2785, 3863
Wang, Liang..................................3958
Wang, Lifang..................................3338
Wang, Liwei...................................927
Wang, Meng...................................2992
Wang, Naizeng.........................998, 2785
Wang, Panrui..................................3383
Wang, Po-Wei..................................1617
Wang, Qiusheng................................2421
Wang, Shike..........................1524, 3692
Wang, Shinn-Shyong...........................2086
Wang, Shitao..................................2866
Wang, Shunyu..................................3002
Wang, Wei.............................614, 3780
Wang, Wenjie.........................1391, 2866
Wang, Xiaolei.................................1455
Wang, Xiaoqing.................................878
Wang, Xiaoyang.................................453
Wang, Xiongfei.............. 1711, 2673, 3164, 3357, 3684
Wang, Yanbo..........................1758, 2708
Wang, Yangyang................................2505
Wang, Yi.....................1027, 1397, 3495
Wang, Yijie...............614, 934, 3780, 3825
Wang, Youyun..................................2204
Wang, Yu-Chi...................................657
Wang, Yue............................1455, 3863
Wang, Yuncheng................................1177
Wang, Zhongxu.........................2743, 3123
Watanabe, Hiroki..............................896
Watanabe, Shoichiro..........................2042
Wei, Baoze...................................3678
Wei, Feng............................1517, 1592
Wei, Jianzhao.................................2630
Wei, Juan.....................................1131
Wei, Shilei...................................1397
Wei, Wang............................1442, 2969
Wei, Xiaoguang................................2343
Wei, Xiuqin..........................3650, 4127
Wei, Zhang....................................2969
Wellawatta, Thusitha Randima 1409
Wen, Huiqing..................................453
Wen, Po-Hsiang................................3544
Wenbing, Li...................................1282
Wickramasinghe, Harith R. 3117
Wijaya, Febry Pandu...........................3490
Wikström, T. 3596
Winter, Christian.............................388
Wolf, Mihaela.................................3607
Wolski, Kornel................................2129
Wu, Bin.......................................3797
Wu, Heng......................................2673
Wu, Hongfei...................................4168
Wu, Min.......................................3863

Wu, Pei-Lin...................................1145
Wu, Ping-Heng........................503, 3549
Wu, T.-F......................................1912
Wu, Tsai-Fu...................................3884
Wu, Tsung-Hsi.................................3544
Wu, Xiaojie....................................619
Wu, Xiaojun..........................3010, 3015
Wu, Ya'nan....................................2496
Wu, Zhiqian...................................1549
Würfl, Joachim................................3607
Wyss, Jonas...................................3734
Xia, Meng.....................................3484
Xia, Yongming.................................2842
Xiao, Chanjuan................................1131
Xiao, Dan.....................................3971
Xiao, Guochun........................1549, 2944
Xiao, Jianfang................................4157
Xiao, Xi......................................1966
Xiaoxi, Liu...................................2969
Xie, Jingwen..................................3069
Xie, Shaofeng.................................2547
Xie, Xiaogao...................................987
Xie, Zhen............................2611, 2992
Xiong, Wei....................................939
Xu, Binci.....................................2270
Xu, Cai.......................................2986
Xu, Dehong..................1118, 2270, 2569
Xu, Dewei David......................1537, 3797
Xu, Dianguo.....................459, 560, 614, 934,
       1206, 1234, 3780, 3825, 4213
Xu, Guangzhao.................................998
Xu, Huadian...................................2877
Xu, Jin.......................261, 3514, 3522
Xu, Peng......................................1478
Xu, Sheng.....................................3002
Xu, Shuang....................................1793
Xu, Yin-Chi...................................3884
Xu, Yue.......................................3985
Xuan, Yang....................................1478
Xuanjie, Gao..................................2718
Xue, Danhong..................................2435
Yabuuchi, Tatsushi............................233
Yada, Tomoharu................................1381
Yamada, Hiroaki......................324, 1381
Yamada, Koji..................................169
Yamaguchi, Daiki..............................3940
Yamaguchi, Koji...............................1972
Yamaji, Masaharu..............................2774
Yamamoto, Aoto................................2558
Yamamoto, Hidekazu............................2750
Yamamoto, Kichiro....................1308, 1322
Yamamoto, Masaya....................1782, 2030
Yamamoto, Masayoshi.....1087, 1095, 2738, 2789, 3344
Yamamoto, Ryo.................................4016
Yamamoto, Shu.......................1949, 1960

AUTHOR INDEX

Yamamoto, Yuuto 3197
Yamanaka, Daisuke 2329
Yamanaka, Kenji 3369
Yamashita, Hiroki 1196
Yamashita, Yoshinori 3490
Yamazaki, Katsumi...........................693, 699
Yamazaki, Masahiro 207
Yan, Qingzeng 619
Yan, Y.T. 851
Yan, Zhang 4181
Yanagisawa, Yuta 3762
Yang, Chang-Jun 3884
Yang, Cheng-Jhen 639, 883
Yang, Daoshu 1549
Yang, Dongsheng 3357
Yang, Geng 542
Yang, Hong-Tzer 4233
Yang, Hui-Chen 2296
Yang, Mei...3958
Yang, Ming 560
Yang, Peng 1966
Yang, Ping 3112
Yang, Renxin 4220
Yang, Sheng-Ming 651, 3426
Yang, Shunfeng 428
Yang, Shuying 2611
Yang, Xu 998, 1062, 1478, 2785, 2854, 3329
Yang, Ying 2973
Yang, Yongheng............... 439, 1021, 1788, 2512, 2743
Yang, Yugang 2703
Yang, Zebin 1157, 1167
Yang, Zhichang 2058
Yang, Zhihua 3797
Yang, Zhiqing..1073
Yang, Zhongping............. 1131, 1816, 2058, 3484, 3495
Yano, Junya...3723
Yao-Ching, Hsieh 645
Yaoqin, Jia 2441
Yasuda, Takumi 992
Yasuda, Yusuke 2082
Yaxin, Peng 416
Ye, Han 1504, 2496
Yeh, Shun-Hao 4233
Yelaverthi, Dorai Babu...........................4066
Yen, Chih-Ying......................................1145
Yenchamchalit, Kulsomsup.....................2430
Yi, Hao 2780, 2882
Yijie, Hou 2441
Yin, Shiyuan 1455
Yin, Taiyuan 1455
Yin, Zhijian 1021
Yin, Zhonggang 2204
Yingchun, Xu...2441
Yokokura, Yuki.............. 1673, 1741, 3890, 3896
Yokoyama, T. 3665

Yokoyama, Tomoki 1270, 1877, 1883, 2914, 3363, 3658
Yonezawa, Y. 3603
Yonezawa, Yu 196
Yoon, Bo-Kyung 3063
Yoshida, Souichi 2764
Yoshida, Yukihiro 2832
Yoshihara, Hidemasa 219
Yoshihara, Tohru 1997
Yoshikawa, Gaku 3280
Yoshimi, Daisuke 3952
Yoshimura, Eiji 767
Yoshino, Takuma 3363
Yoshino, Teruo 1692, 3916
Yoshioka, Yusuke 4151
Yoshizawa, Daisuke 238
You, Jiang 1386, 1421
You, Zih-Cing 651
Yu, Yong 459
Yuan, Huawei 889
Yuan, Liqiang 2963
Yuan, Xibo 619, 1634
Yuan, Yiqin 977, 3040
Yue, Wang...2625
Yui, Haiyan 699
Yukita, Kazuto.......................................809
Zaijun, Wu 1282
Zaitsu, Toshiyuki 118, 4139
Zaman, Mohammad Shawkat 2380
Zanchetta, P. 4034
Zane, Regan 4066, 4074, 4145
Zdanowski, Mariusz..............................2129
Zeng, Pengxiang 2646
Zhang, Chen 4220
Zhang, Feili 1315
Zhang, Guoqiang...........................1206, 1234
Zhang, H. 142
Zhang, Hailong 3863
Zhang, Hao 1131, 1598
Zhang, Hongyang 3684
Zhang, Jianwen 1004
Zhang, Jianzhong 2262
Zhang, Le 3145
Zhang, Lei 3383
Zhang, Lifei 2703
Zhang, Meng 1966
Zhang, Qianfan......................................3025
Zhang, Runze 1816
Zhang, Shichong 2638
Zhang, Shu 614, 934, 3780
Zhang, Shuai 2944
Zhang, Tengfei 2980
Zhang, Wang 2625
Zhang, Xiaofang 2547

AUTHOR INDEX

Zhang, Xin 917, 953, 1004, 2688, 2692, 2980, 4157, 4162
Zhang, Xinan 1724
Zhang, Xing 2973, 2992
Zhang, Xueguang 4213
Zhang, Y. 946, 2697
Zhang, Yan 2646
Zhang, Yang 1177
Zhang, Yanping 2204
Zhang, Yaqian 2262
Zhang, Yi 2743, 3123
Zhang, Zhe 607, 1351, 3460
Zhang, Zhenbin 1661, 2895, 3965
Zhang, Zhigang 1157, 1167
Zhao, Chongyan 904
Zhao, Fangzhou 1549, 2944
Zhao, Fei 1172
Zhao, Jianfeng 3002, 3010, 3015
Zhao, Juan 2051
Zhao, Shengnan 795
Zhao, Tianshu 3020
Zhao, Tianyang 1172
Zhao, Yuanliang 3698
Zhao, Zhengming 2963
Zhao, Zhibin 1125
Zhao, Zhiqing 2714
Zheng, Deyou 2611
Zheng, Xuemei 2901
Zheng, Zedong 1010, 2386
Zhong, Wenxing 1118, 2569
Zhou, Dao 1758
Zhou, Dehong 428, 434
Zhou, Fulin 3257
Zhou, Jiuyang 2462
Zhou, Lei 2505
Zhou, Sheng-Zhi 370
Zhou, Victor 1943
Zhou, Yan 934
Zhou, Yimin 2547
Zhu, Cailing 3052
Zhu, Chunbo 84
Zhu, Helin 1336
Zhu, Junjie 3145
Zhu, Lianghong 1206, 1234
Zhu, Qingwei 3338
Zhu, Yanlin 2780
Zhu, Ye 2270
Zhujian, Ou 1376
Zhuo, Fang....................... 1157, 1167, 2780, 2882
Zhuyong, Li 2986
Zischler, Sigrid 2410
Zou, Yaohan 3455

IEEE
445 Hoes Lane
Piscataway, NJ 08854-4141

ISBN 978-1-5386-4190-3